Case Studies

Essays

Earth Watch

Health Watch

A Closer Look At

Scientific Inquiry

Links to Everyday Life

NINTH EDITION

BIOLOGY

LIFE ON EARTH WITH PHYSIOLOGY

Teresa Audesirk
UNIVERSITY OF COLORADO DENVER

Gerald Audesirk
UNIVERSITY OF COLORADO DENVER

Bruce E. Byers
UNIVERSITY OF MASSACHUSETTS

Benjamin Cummings
Boston Columbus Indianapolis New York San Francisco Upper Saddle River
Amsterdam Cape Town Dubai London Madrid Milan Munich Paris Montréal Toronto
Delhi Mexico City São Paulo Sydney Hong Kong Seoul Singapore Taipei Tokyo

Editor-in-Chief: Beth Wilbur
Senior Acquisitions Editor: Star MacKenzie
Executive Director of Development: Deborah Gale
Development Editor: Mary Ann Murray
Assistant Editor: Erin Mann
Editorial Assistant: Frances Sink
Executive Managing Editor: Erin Gregg
Managing Editor: Michael Early
Production Supervisor: Camille Herrera
Production Service and Compositor: S4Carlisle Publishing Services
Production Editor: Mary Tindle
Copyeditor: Lorretta Palagi
Interior and Cover Designer: Gary Hespenheide
Illustrators: Imagineering Media Services, Inc., and Steve McEntee of McEntee Art & Design, Inc.
Illustrator Project Manager: Winnie Luong
Art Style Development: Blake Kim
Photo Researcher: Yvonne Gerin
Manufacturing Buyer: Michael Penne
Cover and Text Printer and Binder: Courier Kendallville
Supplements Production Supervisor: Jane Brundage
Supplements Editor: Nina Lewallen Hufford
Senior Media Producer: Jonathan Ballard
Media Production Supervisor: James Bruce
Executive Marketing Manager: Lauren Harp
Cover Photo Credit: Daniel J. Cox/Corbis

ABOUT THE COVER

Since they were reintroduced into Yellowstone National Park in the mid-1990s, wolves have become one of the Park's principal attractions. Many thousands of visitors come to the Park every summer, hoping to see a gray wolf, or at least to hear one howl.

But the wolves of Yellowstone are much more than just another tourist attraction—they are an integral part of the web of life in the Park, and magnificent examples of the wonders of all living things. How can a wolf's feet keep from freezing, and continue to function, in the frigid snow of a Yellowstone winter (Chapter 5)? How do wolves harvest energy from the bodies of their prey and power their own metabolism (Chapters 6 and 8)? Why do wolves howl (Chapter 25)? Why are there many thousands of elk in the Park, millions of trees, and untold billions of grasses, but only a couple of hundred wolves (Chapter 28)? How do the wolves help aspen trees to reproduce and thrive, for the first time in decades (Chapter 30)? Finally, in the ancient wisdom of the Inuit people of northern Canada, "The wolf and the caribou are as one; the caribou feeds the wolf, but it is the wolf that keeps the caribou strong." The same can be said of wolves and elk in Yellowstone. What fundamental principles of biology are expressed in these few phrases (Chapters 14, 15, and 27)?

Biology is so much more than just another class to take. It is the continuing pursuit of understanding the grandeur of life on Earth. Enjoy the quest!

Library of Congress Cataloging-in-Publication Data
Biology: Life on Earth with Physiology/Teresa Audesirk, Gerald Audesirk, Bruce E. Byers. — 9th ed. p. cm.
Includes bibliographical references and index.
ISBN 978-0-321-59846-2
1. Biology. I. Audesirk, Gerald. II. Byers, Bruce E. III. Title.
QH308.2.A93 2010
570—dc22

2009043914

Credits and acknowledgments borrowed from other sources and reproduced, with permission, in this textbook appear on page 947.

ISBN 10: 0-321-59846-6; ISBN 13: 978-0-321-59846-2 (Student edition)
ISBN 10: 0-321-60527-6; ISBN 13: 978-0-321-60527-6 (Professional copy)
ISBN 10: 0-321-61540-9; ISBN 13: 978-0-321-61540-4 (a la Carte)
4 5 6 7 8 9 10—CKR—13 12 11

Benjamin Cummings
is an imprint of

www.pearsonhighered.com

About the Authors

TERRY AND GERRY AUDESIRK

grew up in New Jersey, where they met as undergraduates. After marrying in 1970, they moved to California, where Terry earned her doctorate in marine ecology at the University of Southern California and Gerry earned his doctorate in neurobiology at the California Institute of Technology. As postdoctoral students at the University of Washington's marine laboratories, they worked together on the neural bases of behavior, using a marine mollusk as a model system.

They are now emeritus professors of biology at the University of Colorado Denver, where they taught introductory biology and neurobiology from 1982 through 2006. In their research, funded primarily by the National Institutes of Health, they investigated the mechanisms by which neurons are harmed by low levels of environmental pollutants and protected by estrogen.

Terry and Gerry share a deep appreciation of nature and of the outdoors. They enjoy hiking in the Rockies, walking near their home in Steamboat Springs, and attempting to garden at 7,000 feet in the presence of hungry deer and elk. They are long-time members of many conservation organizations. Their daughter, Heather, provides another welcome focus to their lives.

BRUCE E. BYERS

is a midwesterner transplanted to the hills of western Massachusetts, where he is a professor in the biology department at the University of Massachusetts, Amherst. He's been a member of the faculty at UMass (where he also completed his doctoral degree) since 1993. Bruce teaches introductory biology courses for both nonmajors and majors; he also teaches courses in ornithology and animal behavior.

A lifelong fascination with birds ultimately led Bruce to scientific exploration of avian biology. His current research focuses on the behavioral ecology of birds, especially on the function and evolution of the vocal signals that birds use to communicate. The pursuit of vocalizations often takes Bruce outdoors, where he can be found before dawn, tape recorder in hand, awaiting the first songs of a new day.

With love to Jack and Lori and in loving memory of Eve and Joe

T. A. & G. A.

To Bob and Ruth, with gratitude

B. E. B.

Brief Contents

Detailed Contents

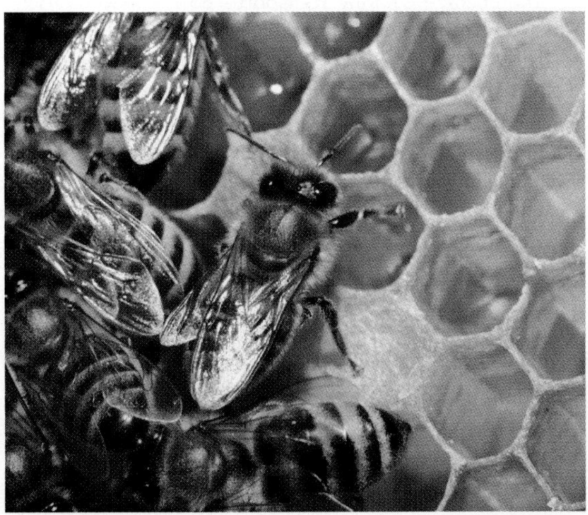

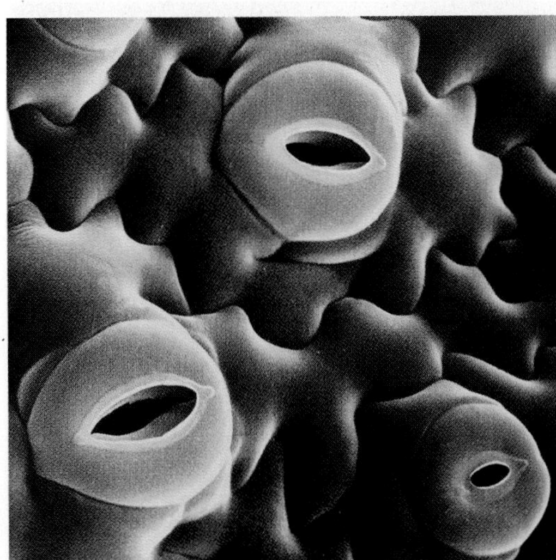

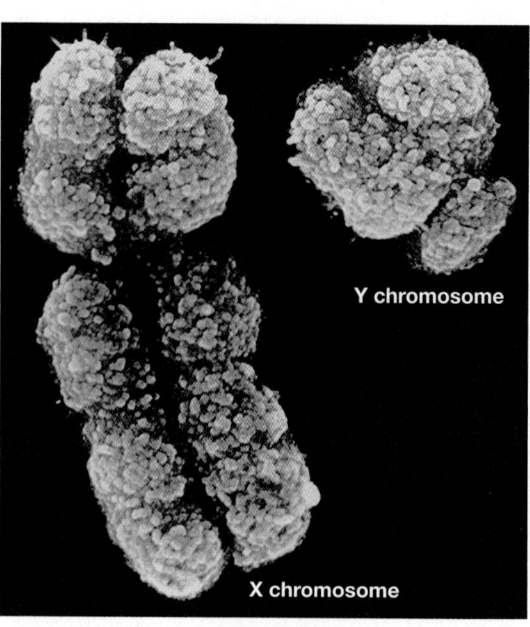
Y chromosome
X chromosome

Unit 3 Evolution and Diversity of Life 265

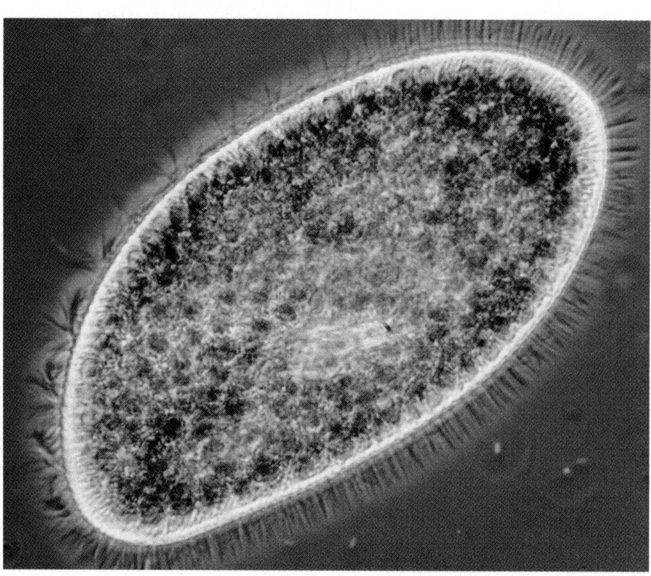

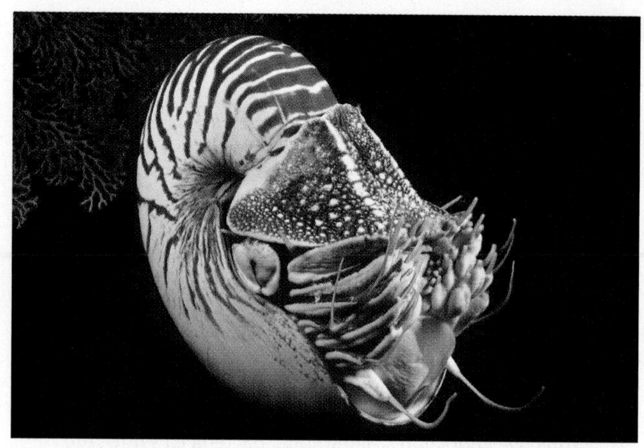

Unit 5 Animal Anatomy and Physiology 603

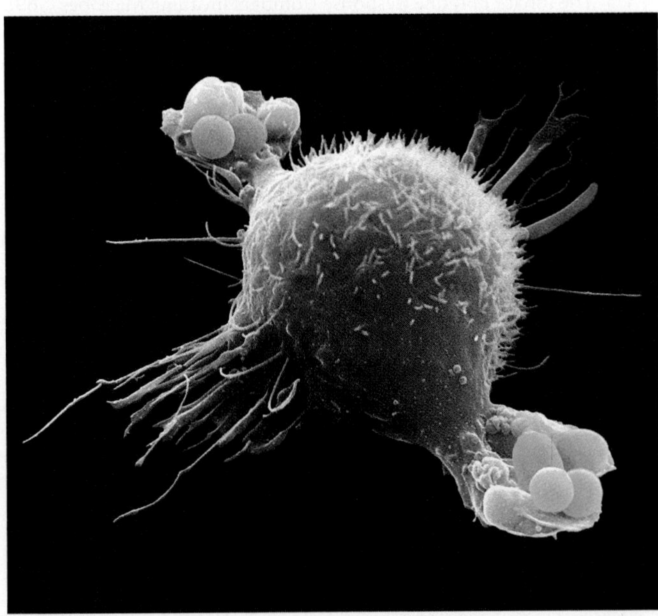

34 Nutrition and Digestion 655

35 The Urinary System 676

36 Defenses Against Disease 691

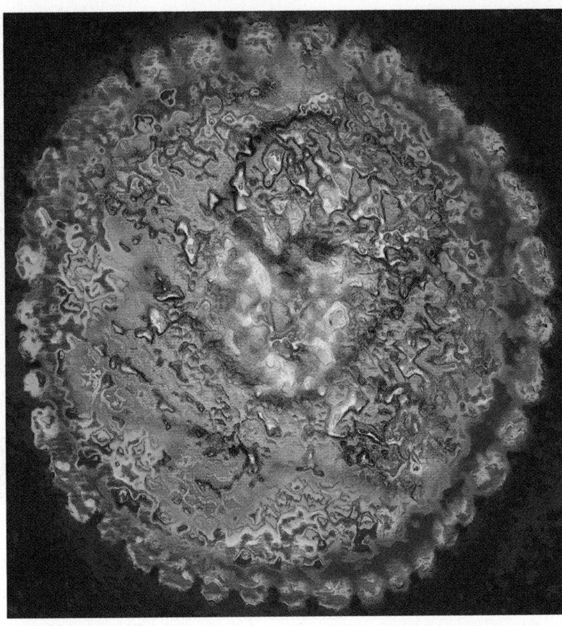

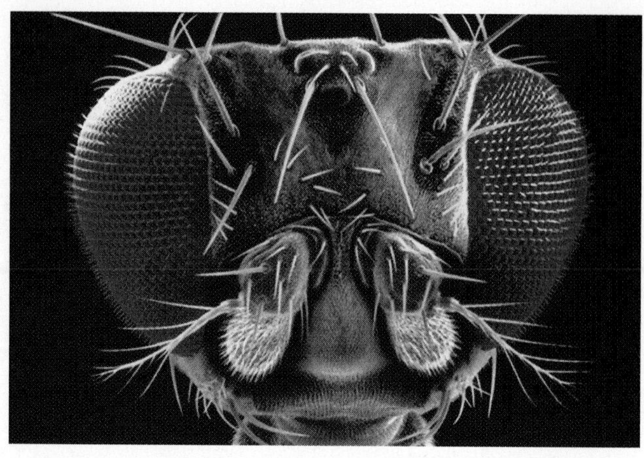

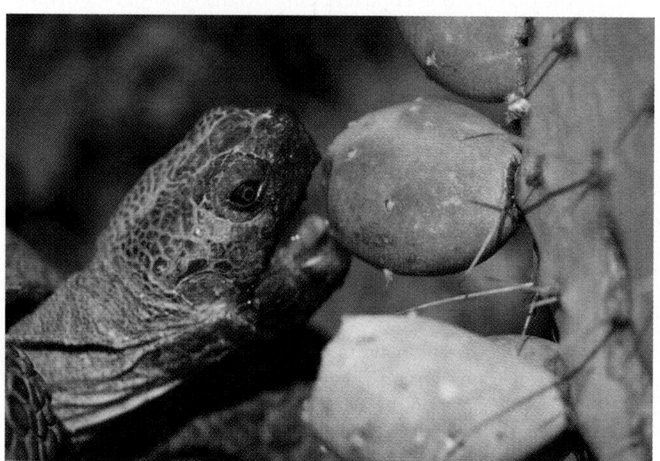

Preface

Global warming, biofuels, bioengineered crops in agriculture, stem cells in medicine, potential flu pandemics, the plight of polar bears and pandas—our students are bombarded with buzzwords. The Internet connects them almost instantly to a wealth of reliable information, but also to floods of falsehoods and hype. The concerns that are sweeping our increasingly interconnected world are very real, urgent, and interrelated. Never have scientifically literate societies been more important to humanity's future. As educators, we feel humbled before this massive challenge. As authors, we feel hopeful that with this Ninth Edition of *Biology: Life on Earth,* we can help our introductory biology students along the path to understanding.

Scientific literacy requires a foundation of factual knowledge that provides a solid and accurate cognitive framework into which new information can be integrated. More importantly, however, it endows people with the mental tools they'll need to apply that knowledge to ever-changing situations. A scientifically literate individual recognizes the interrelatedness of concepts and the need to integrate information from many areas before reaching a conclusion. Scientifically literate citizens can readily grasp and evaluate new information from the popular press, allowing them to make informed choices in the political arena as well as in their personal lives.

In this new edition, we strive to:

- Help instructors present biological information in a way that will foster scientific literacy among their students.
- Help inspire students with a sense of wonder about the natural world and their role in it, which, in turn, will foster a lifetime of inquiry and discovery.
- Help students learn to think about the nature of science, and to recognize the importance of what they are learning to their own lives and to their rapidly changing world.

BIOLOGY: LIFE ON EARTH, NINTH EDITION

. . . Is Organized Clearly and Uniformly

Throughout each chapter, users will find navigational aids to help them explore the topics covered.

- Each chapter opens with an "At a Glance" section, which presents the chapter's major sections and essays. Users can readily see what the chapter covers, and how the topics are organized and related to each other. Instructors can easily assign—and students can easily locate—key topics within the chapter.
- Major sections are introduced as questions, to stimulate thinking about the material that will follow, while minor subheadings are summary statements reflecting their more specific content. An important goal of this organization is to present biology as a hierarchy of closely interrelated concepts rather than as a compendium of independent topics.
- A "Summary of Key Concepts" section pulls together important concepts using the chapter's major headings, allowing instructors and students to review the chapter efficiently.

. . . Engages and Motivates Students

Scientific literacy cannot be imposed on students—they must actively participate in acquiring the necessary information and skills. To be inspired to accomplish this, they must first recognize that biology is about their own lives. A major goal for biology educators is to expand students' perception of what is important in their lives. For example, we want to help students appreciate their own bodies as evolutionary marvels, while gaining a basic understanding of how they function.

Further, we hope that our newly biologically literate students will see the entire outside world through different eyes, viewing ponds, fields, and forests as vibrant and interconnected ecosystems brimming with diverse life-forms rather than as mundane features of their everyday surroundings. If we have done our job, our users will also gain the interest, insight, and information they need to look at how humanity has intervened in the natural world. If they ask the question: "Is this activity sustainable?" and then use their knowledge and critical thinking skills to seek some answers, we can be optimistic about the future.

In support of these goals, the Ninth Edition continues to offer these revised and updated features:

- **Case Studies** Each chapter opens with a case study based on major news events, personal interest stories, or particularly fascinating biological topics. The case study is continued in small, boxed features within the text, and is revisited at the end of each chapter, exploring the topic further in light of what the chapter has covered.
- **BioEthics Questions** Many topics explored in the text have ethical implications for human society, such as human cloning, the cultivation of genetically modified foods, and human impacts on other species. BioEthics questions are identified with an icon that alerts students and instructors to the possibility for further discussion and investigation.
- **Boxed Essays** All of our essays, which explore topics relevant both to the chapter and to the goals of this edition, have been thoroughly revised and updated. Five types of essays enliven this text: "Links to Everyday Life" essays illuminate subjects familiar and interesting to students within the context of the chapter topics; "Earth Watch" essays explore pressing environmental issues; "Health Watch" essays cover important and intriguing

medical topics; "Scientific Inquiry" essays explain how scientific knowledge is acquired; and finally, essays titled "A Closer Look At . . ." allow students to explore selected topics in greater depth.
- **Critical Thinking Questions** At the end of each "Case Study Revisted" section, in selected figure captions, and in the "Applying the Concepts" section at the end of each chapter, our critical thinking questions stimulate students to think about science rather than to simply memorize facts.
- **End-of-Chapter Questions** The questions that conclude each chapter allow students to review the material in different formats. Fill-in-the-blank, essay, and critical thinking questions help students to study and test what they've learned. All of the answers to these questions have been revised by the authors and are included in the Study Area of the MasteringBiology Web site. Answers to figure caption and fill-in-the-blank questions appear at the end of the text.
- **Key Terms and a Complete Glossary** Boldfaced key terms are defined clearly within the text as they are introduced. These terms are also listed at the end of each chapter, providing students with a quick reference to the key vocabulary for the chapter. This edition features an outstanding glossary, thoroughly revised by the authors. All key terms and many additional important terms are carefully defined.

. . . Is a Comprehensive Learning Package

The Ninth Edition of *Biology: Life on Earth* is not merely a substantially improved textbook. Thanks to the unprecedented support of our new publishers, Pearson Benjamin Cummings, it has become a superb and complete learning package, providing new and innovative teaching aids for instructors and new learning aids for students, as described in the sections that follow.

WHAT'S NEW IN THIS EDITION?

The Ninth Edition of *Biology: Life on Earth* has undergone substantial revision to help it convey the wonder and relevance of biology even more effectively than in the previous editions.

We have added entirely new features, including:

- **"Have You Ever Wondered?" Questions** This new feature explores high-interest questions that students commonly ask their instructors. One question appears in each chapter, along with a clear, scientific explanation that relates to the chapter's topic. Questions such as "Have you ever wondered . . . why it hurts so much to do a belly flop; why cyanide is so deadly; why dogs lick their wounds; and if the food you eat has been genetically modified?" pique student interest, and demonstrate to them that the subjects being covered in the text are directly related to their everyday experiences.
- **"Case Study Continued" Sections** As in previous editions, "Case Study" and "Case Study Revisited" sections open and conclude each chapter. In the Ninth Edition, we've added brief "Case Study Continued" sections that relate the case study to material covered at relevant places in the chapter.
- **Fill-in-the-Blank Questions** The "Thinking Through the Concepts" review material now begins with a set of fill-in-the-blank questions, which allow students to better test their recall of key terms and concepts, and cover a wider range of chapter topics than the multiple-choice questions that they replace. But, never fear—an extensive collection of multiple-choice questions for each chapter is readily available on the MasteringBiology Web site.

The text and art have been thoroughly revised to include the latest important developments in biology, as well as to make the text more clear and consistent than ever:

- **New Case Studies** There are six entirely new case studies, covering topics such as "Are Viruses Alive?" (Chapter 1) and "The Migration of the Monarchs" (Chapter 30). All of the case studies retained from the Eighth Edition have been revised for relevance, interest, and currency.
- **Feature Essays** All of our boxed essays—including Health Watch, Earth Watch, Links to Everyday Life, Scientific Inquiry, and A Closer Look At . . . —have been revised and updated.
- **Extensively Revised Text** Each chapter has been reviewed and revised, and has undergone extensive developmental editing for clarity, consistency, and optimal organization.
- **Reinvigorated Art Program** All of the text's art has been extensively reviewed by an expert developmental editor working closely with the authors to ensure that the art clearly explains the concepts and exactly reflects the wording used in the text. Nearly 70% of our art has been revised, and many pieces of complex art have been entirely and beautifully redrawn for clarity and consistency in presentation.
- **A Wealth of New Photos** Throughout the text, returning adopters will find a that we have painstakingly researched eye-catching and informative photographs to illuminate the text, replacing well over 300 of the photos from the last edition.

Chapter-by-Chapter Summary of Important Changes

Unit 1: The Life of a Cell

- **Chapter 1: An Introduction to Life on Earth** features a new case study that explores the question "Are Viruses Alive?," setting the stage for this chapter's discussion of the properties of living things.
- **Chapter 2: Atoms, Molecules, and Life** also begins with a new case study, "Crushed by Ice," which relates the properties of water to the demise of the *Endurance*, the ill-fated ship that carried explorer Ernest Shackleton to the Antarctic in 1914. This chapter also presents a "Links to Everyday Life" essay about the disastrous final flight of

the *Hindenburg*, dramatically illustrating how the differing atomic structures of helium and hydrogen influence their chemical properties.

- **Chapter 4: Cell Structure and Function** features beautifully redrawn art of plant and animal cells, as well as a new table that highlights the structure and function of cytoskeletal components.
- In **Chapter 5: Cell Membrane Structure and Function**, new art of both membrane structure and diffusion through membranes helps clarify these critical concepts.
- **Chapter 6: Energy Flow in the Life of a Cell** is revised to better explain how enzyme function is regulated.
- In **Chapter 7: Capturing Solar Energy: Photosynthesis**, the light reactions are depicted with greater precision and clarity. We now provide coverage of CAM photosynthesis, including art comparing the C_4 and CAM photosynthetic pathways. Our "Earth Watch" essay on biofuels stimulates thinking about how the energy that plants capture from sunlight can be used for fuel, and whether current methods for producing biofuels are sustainable.
- **Chapter 8: Harvesting Energy: Glycolysis and Cellular Respiration** has been reorganized, and summary art has been redrawn to make this challenging subject even more accessible.

Unit 2: Inheritance

- The chapters in Unit 2 have been reorganized for better conceptual flow. This unit now begins with cellular reproduction, followed by Mendelian genetics, DNA, gene regulation, and biotechnology.
- **Chapter 9: The Continuity of Life: Cellular Reproduction** has been reorganized to reflect its opening role in the unit, and features an engaging new case study on animal cloning.
- **Chapter 10: Patterns of Inheritance** includes a new "Health Watch" essay on muscular dystrophy and its inheritance.
- **Chapter 12: Gene Expression and Regulation** now opens with a case study about cystic fibrosis. The ingenious experiments demonstrating that most genes code for proteins are now described in a "Scientific Inquiry" essay.
- **Chapter 13: Biotechnology** has been extensively updated to reflect new advances in this explosively expanding field.

Unit 3: Evolution and Diversity of Life

- The major topics of **Chapter 14: Principles of Evolution** have been reorganized to develop this critical topic in the most logical way. This chapter also features an "Earth Watch" essay that describes evolutionary changes that can be directly related to human modification of the biosphere.
- **Chapter 17: The History of Life** includes enhanced art and text coverage of the ribozyme.
- Systematics is undergoing extensive revision based on new genetic information. **Chapter 18: Systematics: Seeking Order Amidst Diversity** helps guide readers through these changes with new art and text describing the importance of cladistics.
- In **Chapter 19: The Diversity of Prokaryotes and Viruses**, users will find both the text and art describing viral replication revised to further clarify this fascinating topic.
- **Chapter 20: The Diversity of Protists** features a clear and careful re-rendering of the malarial life cycle, and coverage of an important taxonomic group, the Rhizarians.
- **Chapter 21: The Diversity of Plants** includes beautifully redrawn plant life cycles throughout the chapter, clarified with numbered steps for each stage of the plant's life.
- Fungal phyla are organized in a comparative table in **Chapter 22: The Diversity of Fungi**, which also now highlights the mycorrhizal fungi (Glomeromycetes), essential components of many of Earth's forests. Every fungal life cycle has been redrawn for better consistency and clarity.
- In **Chapter 24: Animal Diversity II: Vertebrates**, the clades of chordates are described, and a table allows for easy comparison of craniate groups.

Unit 4: Behavior and Ecology

- **Chapter 25: Animal Behavior** provides a new "Links to Everyday Life" essay that features rats trained to detect buried land mines, educating students both about how land mines plague some war-torn countries and about the amazing sensory capabilities of animals.
- **Chapter 26: Population Growth and Regulation** includes a new "Health Watch" essay that describes boom-and-bust population cycles in hazardous algal blooms.
- **Chapter 27: Community Interactions** features beautifully redrawn stages of succession, and a new "Scientific Inquiry" essay that describes the fascinating research leading to the discovery of a parasite that makes some tropical ants resemble berries.
- In **Chapter 28: How Do Ecosystems Work?**, all of the biogeochemical cycles have been redrawn by an outstanding nature artist, with painstaking attention to consistency and clarity.
- **Chapter 29: Earth's Diverse Ecosystems** opens with a new case study highlighting sustainable agriculture in the Tropics. The art for this exceptionally well-illustrated chapter is almost entirely redrawn, and features an abundance of new photos that showcase Earth's diverse ecosystems.
- **Chapter 30: Conserving Earth's Biodiversity** opens with a new case study describing the magnificent migration of monarch butterflies and the challenges they face. This chapter features updated coverage of the loss of species as humanity's ecological footprint grows, and expanded coverage of the increasing threat of global warming to biodiversity.

Unit 5: Animal Anatomy and Physiology

- **Chapter 31: Homeostasis and the Organization of the Animal Body** begins with a new case study about professional football player Korey Stringer, who died of heat stroke at the age of 27, illustrating the importance of homeostasis and the potentially fatal consequences that can result when homeostasis fails. The illustration program in this chapter has been extensively revised, including striking new artwork and new micrographs of tissue types.
- **Chapter 32: Circulation** features a new and inspiring case study about Donald Arthur, a former "couch potato" who became a marathoner after receiving a heart transplant. A new rendering of blood clotting clearly illustrates the steps involved in the process.
- **Chapter 33: Respiration** offers clearer illustrations and explanations of countercurrent exchange, the dynamics of breathing in birds, and the mechanisms of gas transport in blood.
- **Chapter 34: Nutrition and Digestion** features new art of the digestive systems of earthworms and ruminants.
- **Chapter 35: The Urinary System** opens with the story of Anthony DeGiulio, who decided to make a difference by donating a kidney to a stranger, starting a kidney donor chain that resulted in four lives being saved. Both the text and art for the essay "A Closer Look at The Nephron and Urine Formation" have been revised and updated, providing a clearer and more thorough explanation of how human nephrons regulate the composition of urine.
- **Chapter 36: Defenses Against Disease** begins with a new case study on flesh-eating bacteria and what makes these microbes so destructive and virulent. The discussion about the biology of influenza, including swine flu (H1N1) and avian flu (H5N1), has been updated. Coverage of the innate and adaptive immune responses has been expanded, as has coverage of the generation of antibody diversity. Our revision also features new art and text on the lymphatic system.
- **Chapter 37: Chemical Control of the Animal Body: The Endocrine System** begins with an extensively updated case study on anabolic steroids, including their abuse by well-known athletes. The chapter now includes a Health Watch essay describing research into the possibility of using stem cells to cure diabetes.
- **Chapter 38: The Nervous System** is now a stand-alone chapter (the senses are described in Chapter 39), making this chapter shorter and more accessible to students. We have extensively revised illustrations of the human brain, the structure and function of synapses, and the role of ion channels in the resting potential, action potential, and synaptic transmission.
- **Chapter 39: The Senses** is now a separate chapter, beginning with a fascinating case study on how cochlear implants endow people who were once profoundly deaf with the ability to hear. The chapter includes new art and descriptions of sensory transduction and the vestibular system. Coverage of the perception of pain has been expanded and updated and is the basis for a brief essay describing why chili peppers taste hot.
- **Chapter 40: Action and Support: The Muscles and Skeleton** begins with a new case study on Olympic athletes and the role of muscle fiber types in athletic ability. We have added new art and descriptions of the sliding filament mechanism of muscle contraction and also of the role of muscle fiber action potentials and calcium ions in the control of contraction.
- **Chapter 41: Animal Reproduction** begins with a new case study about the increasingly crucial efforts to breed endangered species. The table of contraceptive techniques has been updated, and the mechanisms of erectile dysfunction drugs, such as Viagra™, are explained.
- **Chapter 42: Animal Development** features a new Scientific Inquiry essay tracing the discovery of fundamental mechanisms of animal development, using amphibians as model systems. We have expanded our coverage of homeobox genes, including new art mapping these genes in the fruit fly. The Health Watch essay on stem cells has been updated and includes a discussion of the possibility of using induced pluripotent stem cells in therapeutic applications.

Unit 6: Plant Anatomy and Physiology

- **Chapter 43: Plant Anatomy and Nutrient Transport** is enhanced with extensively revised art of plant anatomy, primary and secondary growth, root uptake of water and minerals, and the mechanisms of opening and closing stomata. The supporting text descriptions have been revised to complement the new art.
- **Chapter 44: Plant Reproduction and Development** includes new and redrawn art of the flowering plant life cycle, the development of male and female gametophytes, pollination and fertilization, and seed development. The chapter also includes a lighthearted look at the botanical confusion in the produce department of grocery stores, where many fruits are sold as vegetables.
- **Chapter 45: Plant Responses to the Environment** features several figures that have been redrawn for clarity in presentation.

ACKNOWLEDGMENTS

When Pearson Education reorganized, *Biology: Life on Earth* had the great good fortune to be acquired by Benjamin Cummings. Our new Editor-in-Chief, Beth Wilbur, accepted the surprise bundle placed on her doorstep with warmth and grace and nurtured it to maturity. Both Beth and Senior Acquisitions Editor Star MacKenzie did a great job of overseeing the process, listening, smoothing out rough spots, and, along with Deborah Gale, Executive Director of Development, helping to coordinate this complex and multifaceted endeavor.

Our Developmental Editor, Mary Ann Murray, carefully examined every word and every piece of art with an awe-inspiring attention to detail. Working tirelessly with the authors (who sometimes got tired), she has brought the text and art to an unprecedented standard of clarity and consistency. Lorretta Palagi, our Copyeditor, put the finishing touches on the text, combining a painstaking attention to detail with tactful suggestions. Many new photos enliven the text, thanks to our Photo Researcher, Yvonne Gerin, who skillfully and cheerfully provided us with an array of selections, and again, to Mary Ann Murray, who examined each photo and helped us to select those that best convey the wonder of life on Earth.

The extensively revised art program, also overseen by Mary Ann, was implemented by two excellent art houses—Imagineering Media Services, under the careful direction of Project Manager Winnie Luong, and McEntee Art and Design. Artist Steve McEntee deserves special praise for his exceptional rendering of natural landscapes, plants, and animals. Thanks to Blakely Kim for his help in developing new art styles for the Ninth Edition. We owe our beautifully redesigned text and cover to Gary Hespenheide. The authors are grateful to Gary and our editors for allowing us to participate actively in the design process and in the selection of the cover photo, and are also grateful for the efforts of Lauren Harp, Executive Marketing Manager, to get this book into your hands.

The production of this text would not have been possible without the considerable efforts of Camille Herrera, Production Supervisor, in coordination with Erin Gregg, Executive Managing Editor, and Michael Early, Managing Editor. Camille and Mary Tindle, Project Editor, S4Carlisle Publishing Services, brought the art, photos, and manuscript together into a seamless whole, graciously handling last-minute changes. In his role as Manufacturing Buyer, Michael Penne's expertise has served us well.

The supplements have evolved into an endeavor fully as important as the text itself. Nina Lewallen Hufford, ably assisted by Erin Mann and Frances Sink, skillfully coordinated the enormous effort of producing a truly outstanding package that complements and supports the text. Jane Brundage oversaw the production of the Study Card that accompanies the text, while James Bruce took the lead on the Instructor's Resource CD-ROM. Finally, thanks to Senior Media Producer Jonathan Ballard for his efforts in developing the outstanding MasteringBiology Web site that accompanies this text.

We are delighted to be under the wings of the Pearson Benjamin Cummings team. This Ninth Edition of *Biology: Life on Earth* reflects their exceptional skill and dedication.

With thanks,

TERRY AUDESIRK, GERRY AUDESIRK,
AND BRUCE BYERS

NINTH EDITION REVIEWERS

Mike Aaron, *Shelton State Community College*
Ana Arnizaut-Vilella, *Mississippi University for Women*
Dan Aruscavage, *State University of New York, Potsdam*
Isaac Barjis, *New York City College of Technology*
Mike Barton, *Centre College*
Robert Benard, *American International College*
Heather Bennett, *Illinois College*
Karen E. Bledsoe, *Western Oregon University*
David Brown, *Marietta College*
Diep Burbridge, *Long Beach City College*
Judy A. Chappell, *Luzerne County Community College*
Art Conway, *Randolph-Macon College*
Clifton Cooper, *Linn-Benton Community College*
Brian E. Corner, *Augsburg College*
Peter Cumbie, *Winthrop University*
Joe Demasi, *Massachusetts College*
Christy Donmoyer, *Winthrop University*
Doug Florian, *Trident Technical College*
Cynthia Galloway, *Texas A&M University–Kingsville*
Janet Gaston, *Troy University*
Charles Good, *Ohio State University, Lima*
Joan-Beth Gow, *Anna Maria College*
Mary Ruth Griffin, *University of Charleston*
Wendy Grillo, *North Carolina Central University*
Robert Hatherill, *Del Mar College*
Kathleen Hecht, *Nassau Community College*
Wiley Henderson, *Alabama A&M University*
Kristy Y. Johnson, *The Citadel*
Ross Johnson, *Chicago State University*
Ragupathy Kannan, *University of Arkansas, Fort Smith*
Patrick Larkin, *Sante Fe College*
Mary Lipscomb, *Virginia Polytechnic and State University*
Richard W. Lo Pinto, *Fairleigh Dickinson University*
Jonathan Lochamy, *Georgia Perimeter College*
Paul Lonquich, *California State University Northridge*
Bernard Majdi, *Waycross College*
Cindy Malone, *California State University Northridge*
Barry Markillie, *Cape Fear Community College*
Daniel Matusiak, *St. Charles Community College*
Jeanne Minnerath, *St. Mary's University of Minnesota*
Russell Nemecek, *Columbia College, Hancock*
Murad Odeh, *South Texas College*
Luis J. Pelicot, *City University of New York, Hostos*
Larry Pilgrim, *Tyler Junior College*
Therese Poole, *Georgia State University*
Cameron Russell, *Tidewater Community College*
Tim Sellers, *Keuka College*
Rick L. Simonson, *University of Nebraska, Kearney*
Howard Singer, *New Jersey City University*
Anthony Stancampiano, *Oklahoma City Community College*
Jose G. Tello, *Long Island University*
Richard C. Tsou, *Gordon College*
Jim Van Brunt, *Rogue Community College*
Jerry G. Walls, *Louisiana State University, Alexandria*
Holly Walters, *Cape Fear Community College*
Winfred Watkins, *McLennan Community College*
Richard Whittington, *Pellissippi State Technical Community College*

Roger K. Wiebusch, *Columbia College*
Taek H. You, *Campbell University*
Martin Zahn, *Thomas Nelson Community College*
Michelle Zurawski, *Moraine Valley Community College*

PREVIOUS EDITION REVIEWERS

W. Sylvester Allred, *Northern Arizona University*
Judith Keller Amand, *Delaware County Community College*
William Anderson, *Abraham Baldwin Agriculture College*
Steve Arch, *Reed College*
George C. Argyros, *Northeastern University*
Kerri Lynn Armstrong, *Community College of Philadelphia*
G. D. Aumann, *University of Houston*
Vernon Avila, *San Diego State University*
J. Wesley Bahorik, *Kutztown University of Pennsylvania*
Peter S. Baletsa, *Northwestern University*
John Barone, *Columbus State University*
Bill Barstow, *University of Georgia–Athens*
Michael C. Bell, *Richland College*
Colleen Belk, *University of Minnesota, Duluth*
Gerald Bergtrom, *University of Wisconsin*
Arlene Billock, *University of Southwestern Louisiana*
Brenda C. Blackwelder, *Central Piedmont Community College*
Melissa Blamires, *Salt Lake Community College*
Raymond Bower, *University of Arkansas*
Robert Boyd, *Auburn University*
Michael Boyle, *Seattle Central Community College*
Marilyn Brady, *Centennial College of Applied Arts and Technology*
Virginia Buckner, *Johnson County Community College*
Arthur L. Buikema, Jr., *Virginia Polytechnic Institute*
J. Gregory Burg, *University of Kansas*
William F. Burke, *University of Hawaii*
Robert Burkholter, *Louisiana State University*
Matthew R. Burnham, *Jones County Junior College*
Kathleen Burt-Utley, *University of New Orleans*
Linda Butler, *University of Texas–Austin*
W. Barkley Butler, *Indiana University of Pennsylvania*
Jerry Button, *Portland Community College*
Bruce E. Byers, *University of Massachusetts–Amherst*
Sara Chambers, *Long Island University*
Nora L. Chee, *Chaminade University*
Joseph P. Chinnici, *Virginia Commonwealth University*
Dan Chiras, *University of Colorado–Denver*
Nicole A. Cintas, *Northern Virginia Community College*
Bob Coburn, *Middlesex Community College*
Joseph Coelho, *Culver Stockton College*
Martin Cohen, *University of Hartford*
Jay L. Comeaux, *Louisiana State University*
Walter J. Conley, *State University of New York at Potsdam*
Mary U. Connell, *Appalachian State University*
Jerry Cook, *Sam Houston State University*
Sharon A. Coolican, *Cayuga Community College*
Joyce Corban, *Wright State University*
Ethel Cornforth, *San Jacinto College–South*
David J. Cotter, *Georgia College*
Lee Couch, *Albuquerque Technical Vocational Institute*
Donald C. Cox, *Miami University of Ohio*
Patricia B. Cox, *University of Tennessee*
Peter Crowcroft, *University of Texas–Austin*
Carol Crowder, *North Harris Montgomery College*
Mitchell B. Cruzan, *Portland State University*
Donald E. Culwell, *University of Central Arkansas*
Robert A. Cunningham, *Erie Community College, North*
Karen Dalton, *Community College of Baltimore County–Catonsville Campus*
Lydia Daniels, *University of Pittsburgh*
David H. Davis, *Asheville-Buncombe Technical Community College*
Jerry Davis, *University of Wisconsin, LaCrosse*
Douglas M. Deardon, *University of Minnesota*
Lewis Deaton, *University of Southwestern Louisiana*
Lewis Deaton, *University of Louisiana–Lafayette*
Fred Delcomyn, *University of Illinois–Urbana*
David M. Demers, *University of Hartford*
Lorren Denney, *Southwest Missouri State University*
Katherine J. Denniston, *Towson State University*
Charles F. Denny, *University of South Carolina–Sumter*
Jean DeSaix, *University of North Carolina–Chapel Hill*
Ed DeWalt, *Louisiana State University*
Daniel F. Doak, *University of California–Santa Cruz*
Matthew M. Douglas, *University of Kansas*
Ronald J. Downey, *Ohio University*
Ernest Dubrul, *University of Toledo*
Michael Dufresne, *University of Windsor*
Susan A. Dunford, *University of Cincinnati*
Mary Durant, *North Harris College*
Ronald Edwards, *University of Florida*
Rosemarie Elizondo, *Reedley College*
George Ellmore, *Tufts University*
Joanne T. Ellzey, *University of Texas–El Paso*
Wayne Elmore, *Marshall University*
Thomas Emmel, *University of Florida*
Carl Estrella, *Merced College*
Nancy Eyster-Smith, *Bentley College*
Gerald Farr, *Southwest Texas State University*
Rita Farrar, *Louisiana State University*
Marianne Feaver, *North Carolina State University*
Susannah Feldman, *Towson University*
Linnea Fletcher, *Austin Community College–Northridge*
Charles V. Foltz, *Rhode Island College*
Dennis Forsythe, *The Citadel*
Douglas Fratianne, *Ohio State University*
Scott Freeman, *University of Washington*
Donald P. French, *Oklahoma State University*
Harvey Friedman, *University of Missouri–St. Louis*
Don Fritsch, *Virginia Commonwealth University*

Teresa Lane Fulcher, *Pellissippi State Technical Community College*
Michael Gaines, *University of Kansas*
Irja Galvan, *Western Oregon University*
Gail E. Gasparich, *Towson University*
Farooka Gauhari, *University of Nebraska–Omaha*
John Geiser, *Western Michigan University*
George W. Gilchrist, *University of Washington*
David Glenn-Lewin, *Iowa State University*
Elmer Gless, *Montana College of Mineral Sciences*
Charles W. Good, *Ohio State University–Lima*
Margaret Green, *Broward Community College*
David Grise, *Southwest Texas State University*
Martha Groom, *University of Washington*
Lonnie J. Guralnick, *Western Oregon University*
Martin E. Hahn, *William Paterson College*
Madeline Hall, *Cleveland State University*
Georgia Ann Hammond, *Radford University*
Blanche C. Haning, *North Carolina State University*
Richard Hanke, *Rose State College*
Helen B. Hanten, *University of Minnesota*
John P. Harley, *Eastern Kentucky University*
William Hayes, *Delta State University*
Stephen Hedman, *University of Minnesota*
Jean Helgeson, *Collins County Community College*
Alexander Henderson, *Millersville University*
Timothy L. Henry, *University of Texas–Arlington*
James Hewlett, *Finger Lakes Community College*
Alison G. Hoffman, *University of Tennessee–Chattanooga*
Kelly Hogan, *University of North Carolina–Chapel Hill*
Leland N. Holland, *Paso-Hernando Community College*
Laura Mays Hoopes, *Occidental College*
Dale R. Horeth, *Tidewater Community College*
Michael D. Hudgins, *Alabama State University*
David Huffman, *Southwest Texas State University*
Joel Humphrey, *Cayuga Community College*
Donald A. Ingold, *East Texas State University*
Jon W. Jacklet, *State University of New York–Albany*
Rebecca M. Jessen, *Bowling Green State University*
James Johnson, *Central Washington University*
J. Kelly Johnson, *University of Kansas*
Florence Juillerat, *Indiana University–Purdue University at Indianapolis*
Thomas W. Jurik, *Iowa State University*
Arnold Karpoff, *University of Louisville*
L. Kavaljian, *California State University*
Joe Keen, *Patrick Henry Community College*
Jeff Kenton, *Iowa State University*
Hendrick J. Ketellapper, *University of California, Davis*
Jeffrey Kiggins, *Blue Ridge Community College*
Aaron Krochmal, *University of Houston–Downtown*
Harry Kurtz, *Sam Houston State University*
Kate Lajtha, *Oregon State University*
Tom Langen, *Clarkson University*
Stephen Lebsack, *Linn-Benton Community College*
Patricia Lee-Robinson, *Chaminade University of Honolulu*
David E. Lemke, *Texas State University*
William H. Leonard, *Clemson University*
Edward Levri, *Indiana University of Pennsylvania*
Graeme Lindbeck, *University of Central Florida*
Jerri K. Lindsey, *Tarrant County Junior College–Northeast*
Jason L. Locklin, *Temple College*
John Logue, *University of South Carolina–Sumter*
William Lowen, *Suffolk Community College*
Ann S. Lumsden, *Florida State University*
Steele R. Lunt, *University of Nebraska–Omaha*
Daniel D. Magoulick, *The University of Central Arkansas*
Cindy Malone, *California State University–Northridge*
Paul Mangum, *Midland College*
Richard Manning, *Southwest Texas State University*
Mark Manteuffel, *St. Louis Community College*
Ken Marr, *Green River Community College*
Kathleen A. Marrs, *Indiana University–Purdue University Indianapolis*
Michael Martin, *University of Michigan*
Linda Martin-Morris, *University of Washington*
Kenneth A. Mason, *University of Kansas*
Margaret May, *Virginia Commonwealth University*
D. J. McWhinnie, *De Paul University*
Gary L. Meeker, *California State University, Sacramento*
Thoyd Melton, *North Carolina State University*
Joseph R. Mendelson III, *Utah State University*
Karen E. Messley, *Rockvalley College*
Timothy Metz, *Campbell University*
Steven Mezik, *Herkimer County Community College*
Glendon R. Miller, *Wichita State University*
Hugh Miller, *East Tennessee State University*
Neil Miller, *Memphis State University*
Christine Minor, *Clemson University*
Jeanne Mitchell, *Truman State University*
Lee Mitchell, *Mt. Hood Community College*
Jack E. Mobley, *University of Central Arkansas*
John W. Moon, *Harding University*
Nicole Moore, *Austin Peay University*
Richard Mortenson, *Albion College*
Gisele Muller-Parker, *Western Washington University*
James Mulrooney, *Central Connecticut State University*
Kathleen Murray, *University of Maine*
Robert Neill, *University of Texas*
Harry Nickla, *Creighton University*
Daniel Nickrent, *Southern Illinois University*
Jane Noble-Harvey, *University of Delaware*
David J. O'Neill, *Community College of Baltimore County–Dundalk Campus*
James T. Oris, *Miami University of Ohio*
Marcy Osgood, *University of Michigan*
C. O. Patterson, *Texas A&M University*
Fred Peabody, *University of South Dakota*
Charlotte Pedersen, *Southern Utah University*
Harry Peery, *Tompkins-Cortland Community College*

Rhoda E. Perozzi, *Virginia Commonwealth University*
Gary B. Peterson, *South Dakota State University*
Bill Pfitsch, *Hamilton College*
Ronald Pfohl, *Miami University of Ohio*
Robert Kyle Pope, *Indiana University South Bend*
Bernard Possident, *Skidmore College*
Ina Pour-el, *DMACC–Boone Campus*
Elsa C. Price, *Wallace State Community College*
Marvin Price, *Cedar Valley College*
Kelli Prior, *Finger Lakes Community College*
Jennifer J. Quinlan, *Drexel University*
James A. Raines, *North Harris College*
Paul Ramp, *Pellissippi State Technical College*
Robert N. Reed, *Southern Utah University*
Wenda Ribeiro, *Thomas Nelson Community College*
Elizabeth Rich, *Drexel University*
Mark Richter, *University of Kansas*
Robert Robbins, *Michigan State University*
Jennifer Roberts, *Lewis University*
Frank Romano, *Jacksonville State University*
Chris Romero, *Front Range Community College*
Paul Rosenbloom, *Southwest Texas State University*
Amanda Rosenzweig, *Delgado Community College*
K. Ross, *University of Delaware*
Mary Lou Rottman, *University of Colorado–Denver*
Albert Ruesink, *Indiana University*
Connie Russell, *Angelo State University*
Marla Ruth, *Jones County Junior College*
Christopher F. Sacchi, *Kutztown University*
Eduardo Salazar, *Temple College*
Doug Schelhaas, *University of Mary*
Brian Schmaefsky, *Kingwood College*
Alan Schoenherr, *Fullerton College*
Brian W. Schwartz, *Columbus State University*
Edna Seaman, *University of Massachusetts, Boston*
Patricia Shields, *George Mason University*
Marilyn Shopper, *Johnson County Community College*
Anu Singh-Cundy, *Western Washington University*
Linda Simpson, *University of North Carolina–Charlotte*
Steven Skarda, *Linn-Benton Community College*
Russel V. Skavaril, *Ohio State University*
John Smarelli, *Loyola University*
Mark Smith, *Chaffey College*
Dale Smoak, *Piedmont Technical College*
Jay Snaric, *St. Louis Community College*
Phillip J. Snider, *University of Houston*
Shari Snitovsky, *Skyline College*
Gary Sojka, *Bucknell University*
John Sollinger, *Southern Oregon University*
Sally Sommers Smith, *Boston University*
Jim Sorenson, *Radford University*
Mary Spratt, *University of Missouri, Kansas City*
Bruce Stallsmith, *University of Alabama–Huntsville*
Benjamin Stark, *Illinois Institute of Technology*
William Stark, *Saint Louis University*
Barbara Stebbins-Boaz, *Willamette University*
Kathleen M. Steinert, *Bellevue Community College*
Barbara Stotler, *Southern Illinois University*
Nathaniel J. Stricker, *Ohio State University*
Martha Sugermeyer, *Tidewater Community College*
Gerald Summers, *University of Missouri–Columbia*
Marshall Sundberg, *Louisiana State University*
Bill Surver, *Clemson University*
Eldon Sutton, *University of Texas–Austin*
Peter Svensson, *West Valley College*
Dan Tallman, *Northern State University*
David Thorndill, *Essex Community College*
William Thwaites, *San Diego State University*
Professor Tobiessen, *Union College*
Richard Tolman, *Brigham Young University*
Sylvia Torti, *University of Utah*
Dennis Trelka, *Washington and Jefferson College*
Sharon Tucker, *University of Delaware*
Gail Turner, *Virginia Commonwealth University*
Glyn Turnipseed, *Arkansas Technical University*
Lloyd W. Turtinen, *University of Wisconsin, Eau Claire*
Robert Tyser, *University of Wisconsin, La Crosse*
Robin W. Tyser, *University of Wisconsin, LaCrosse*
Kristin Uthus, *Virginia Commonwealth University*
Rani Vajravelu, *University of Central Florida*
F. Daniel Vogt, *State University of New York–Plattsburgh*
Nancy Wade, *Old Dominion University*
Susan M. Wadkowski, *Lakeland Community College*
Jyoti R. Wagle, *Houston Community College–Central*
Lisa Weasel, *Portland State University*
Michael Weis, *University of Windsor*
DeLoris Wenzel, *University of Georgia*
Jerry Wermuth, *Purdue University–Calumet*
Diana Wheat, *Linn-Benton Community College*
Jacob Wiebers, *Purdue University*
Carolyn Wilczynski, *Binghamton University*
Lawrence R. Williams, *University of Houston*
P. Kelly Williams, *University of Dayton*
Roberta Williams, *University of Nevada–Las Vegas*
Emily Willingham, *University of Texas–Austin*
Sandra Winicur, *Indiana University–South Bend*
Bill Wischusen, *Louisiana State University*
Michelle Withers, *Louisiana State University*
Chris Wolfe, *North Virginia Community College*
Stacy Wolfe, *Art Institutes International*
Colleen Wong, *Wilbur Wright College*
Wade Worthen, *Furman University*
Robin Wright, *University of Washington*
Taek You, *Campbell University*
Brenda L. Young, *Daemen College*
Cal Young, *Fullerton College*
Tim Young, *Mercer University*
Martin Zahn, *Thomas Nelson Community College*
Izanne Zorin, *Northern Virginia Community College–Alexandria*

Explore the Fascinating World of Biology: Life on Earth

The **Ninth Edition** of ***Biology: Life on Earth*** engages students by asking and answering questions about the fascinating world of biology.

NEW! ***Have you ever wondered?*** feature poses high-interest questions that students commonly ask themselves or instructors. Each question is addressed with a clear, scientifically-accurate response.

Have you ever wondered...?

why it hurts so much to do a belly flop?
(Chapter 2, page 29)

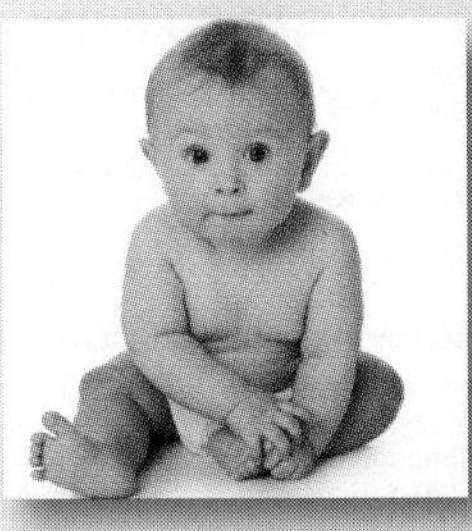

how many cells are in the human body?
(Chapter 4, page 62)

why bruises turn color?
(Chapter 12, page 233)

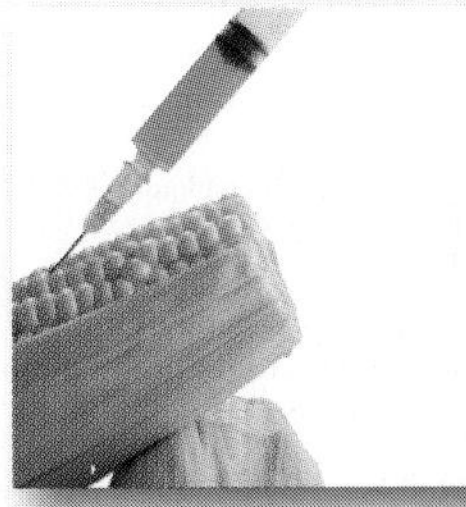

if the food you eat has been genetically modified?
(Chapter 13, page 249)

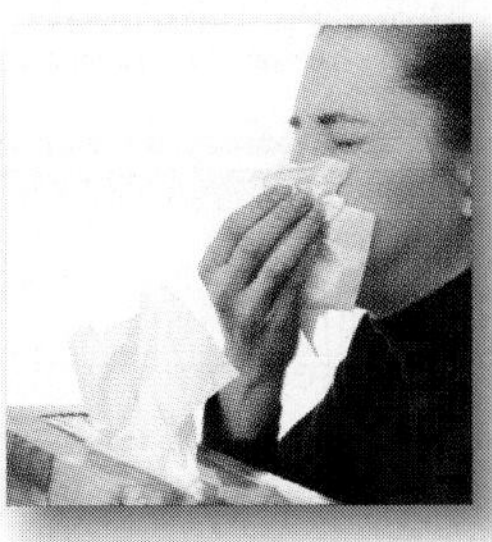

why antibiotics can't cure a cold?
(Chapter 15, page 297)

Have your students ever wondered
how is biology relevant to my everyday life?

Case Studies appear throughout each chapter, tying together important themes.

Page 20

Atoms, Molecules, and Life

Chapter 2

▲ The *Endurance* trapped in ice

Case Study

Crushed by Ice

IT TOOK JUST 5 MINUTES for the ship *Endurance* to disappear into the ocean depths, but the captain did not go down with his ship, nor was his crew in danger of drowning. They watched—miserable bu[t] where they h[a]

Captained Sir Ernest Sha[ckleton] of scientists a[nd] from Plymout[h] goal: to trave[l] South Pole. Fi[…] the Weddell S[ea] unseasonably around it, lite[rally] slowly crushi[ng] nights onboa[rd] contorted—g[…] splintering. W[…] open the dec[k] Shackleton o[…] dogsleds to h[…] they set up camp.

On November 21, 1915, the *Endurance* shifted, its stern rising skyward as it slid benea[th] the ice. Shackleton and crew watched in a kind [of] despairing awe from a safe distance. They wer[e] to endure three more months of ceaseless cold on their frozen camp, waiting for the ice to release its grip and the sea to become passabl[e] for the smaller boats they had rescued from th[e] doomed ship.

The *Endurance* was trapped and ultimately sunk as it was crushed by expanding ice floating on the Antarctic sea. Most liquids, including water, contract and become denser as they cool. But as water freezes, it undergoes an unusual transformation that produces a solid that is less dense and, therefore, floats on the remaining liquid. How does this happen? Why was the *Endurance* crushed as ice froze around it? How do the unique properties of water influence life on Earth?

Human-interest case studies open each chapter, engaging students in real world biology from the start. The Ninth Edition includes several new case studies, including the blood doping scandal at the 2008 Tour de France (Chapter 8) and concerns about threatened Monarch butterfly populations (Chapter 30).

Page 30

Case Study continued

Crushed by Ice

The seawater in which the *Endurance* sank provides an excellent illustration of the ability of water to dissolve other molecules. Seawater contains more than 70 different elements, most in trace quantities. The most abundant dissolved elements are sodium and chlorine (NaCl; comprising about 3.2% of seawater by weight) and smaller amounts of sulfur, calcium, magnesium, pota[ssium …] water also contains dissolved ga[s…] carbon dioxide. Oxygen in seaw[ater …] respire, and carbon dioxide allow[…] makes energy available to aquati[c …]

Page 33

Case Study continued

Crushed by Ice

Shackleton and his crew did not lack drinking water during the months they spent waiting; they could simply melt chunks of ice. Although the ice had solidified from salty ocean water, Na^+ and Cl^- ions (and the other substances dissolved in ocean water) do not fit into the lattice of water molecules that make up the crystalline structure of ice (see Fig. 2-14). As ice forms, the remaining water becomes increasingly salty and requires increasingly low temperatures to freeze.

NEW! *Case Study Continued* sections are interspersed throughout each chapter, relating the case study to the chapter topic at hand.

Page 33

Case Study revisited

Crushed by Ice

Although they failed to traverse Antarctica by dogsled, the heroic and ultimately successful efforts of Shackleton and his crew to return home alive are a remarkable true-life adventure. By studying Figure 2-14 you can see why the *Endurance* was crushed—the orderly arrangement of water molecules in ice causes them to occupy more space than they did as a liquid; thus, water expands as it forms ice. You have witnessed this firsthand if you've ever left a full water bottle in the freezer and later found it cracked open by the expanding ice. The expansion of ice also explains why it floats and why the ice formed a safe campground for the crew of the *Endurance*.

Many animals rely on ocean ice, from penguins and walruses that breed on the ice sheets, to polar bears that use the ice as a staging ground to hunt seals. The seals themselves are protected from their polar bear predators by the ice cover as they hunt for fish beneath it, becoming vulnerable when they surface to breathe through holes that they maintain in the ice.

Consider This

Many of water's unique properties are the result of its polar covalent bonds, which allow water molecules to form hydrogen bonds with each other. What if water molecules were nonpolar? How would this have affected the voyage of the *Endurance*?

A ***Case Study Revisited*** section ends each chapter, summing up the case study in the context of what the student has learned from the chapter. The *Consider This* question within the *Revisited* section allows students to further reflect on themes brought up in the Case Study, encouraging deeper critical thinking.

Have your students ever asked you

will this be on the test?

Questions throughout the text provide opportunities for students to quiz themselves in preparation for tests.

Every major section heading is written as a question. As students read the material, the answers to the question are revealed. This format helps to stimulate analytical thinking and models the importance of asking questions in scientific exploration.

> 9.3 HOW IS THE DNA IN EUKARYOTIC CHROMOSOMES ORGANIZED?
>
> Eukaryotic chromosomes are separated from the cytopl in a membrane-bound nucleus. Further, eukaryotic cell ways have multiple chromosomes the smallest numbe

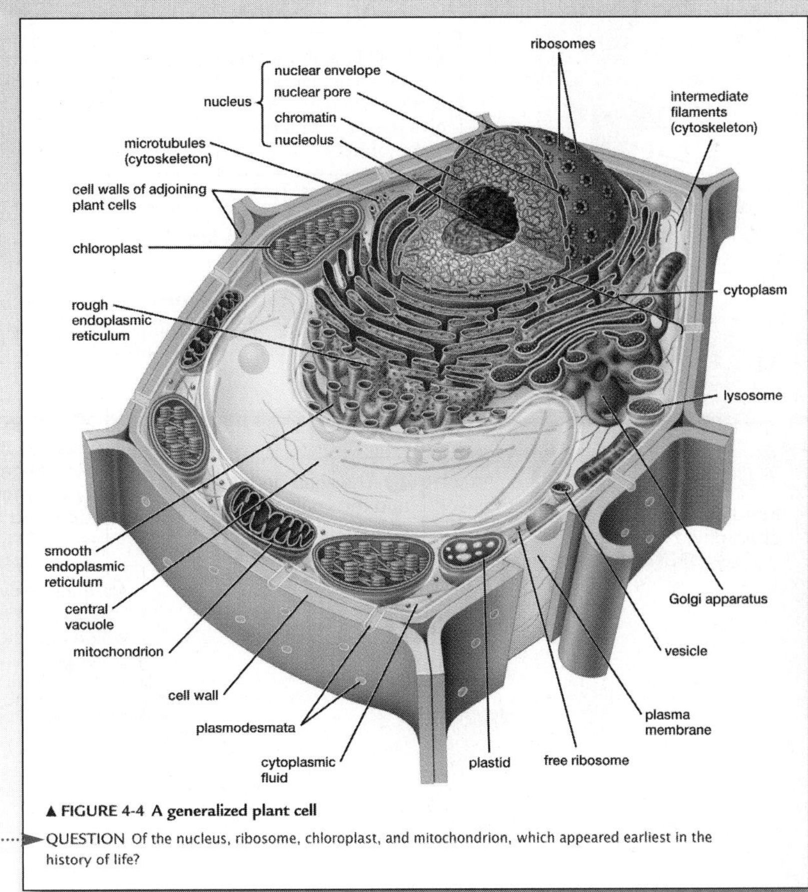

▲ FIGURE 4-4 A generalized plant cell

QUESTION Of the nucleus, ribosome, chloroplast, and mitochondrion, which appeared earliest in the history of life?

Figure caption questions ask the student to think critically about the art to ensure they understand the concept being illustrated. Suggested answers appear at the end of the text.

> **Applying the Concepts**
>
> 1. BioEthics Insects are the largest group of animals on Earth. Insect diversity is greatest in the Tropics, where habitat destruction and species extinction are occurring at an alarming rate. What biological, economic, and ethical arguments can you advance to persuade people and governments to preserve this biological diversity?
> 2. Discuss at least three ways in which the ability to fly has contributed to the success and diversity of insects.
> 3. Discuss and defend the attributes you would use to define biological success among animals. Are humans a biological success by these standards? Why?

***Applying The Concepts* questions** are provided at the end of each chapter, challenging students to synthesize information beyond the chapter material and inviting them to apply this knowledge to new situations. Suggested answers appear on the MasteringBiology® Website.

Expanded *BioEthics* questions appear within relevant essay boxes and figure captions, and are included in the end of chapter *Applying the Concepts* section. These questions present bioethical issues and prompt students to carefully consider various points of view.

Have your students ever wondered

how to visualize biological concepts?

The significantly revised and invigorated **art program** not only helps draw students into the material, but also teaches concepts more effectively by bringing biological concepts to life.

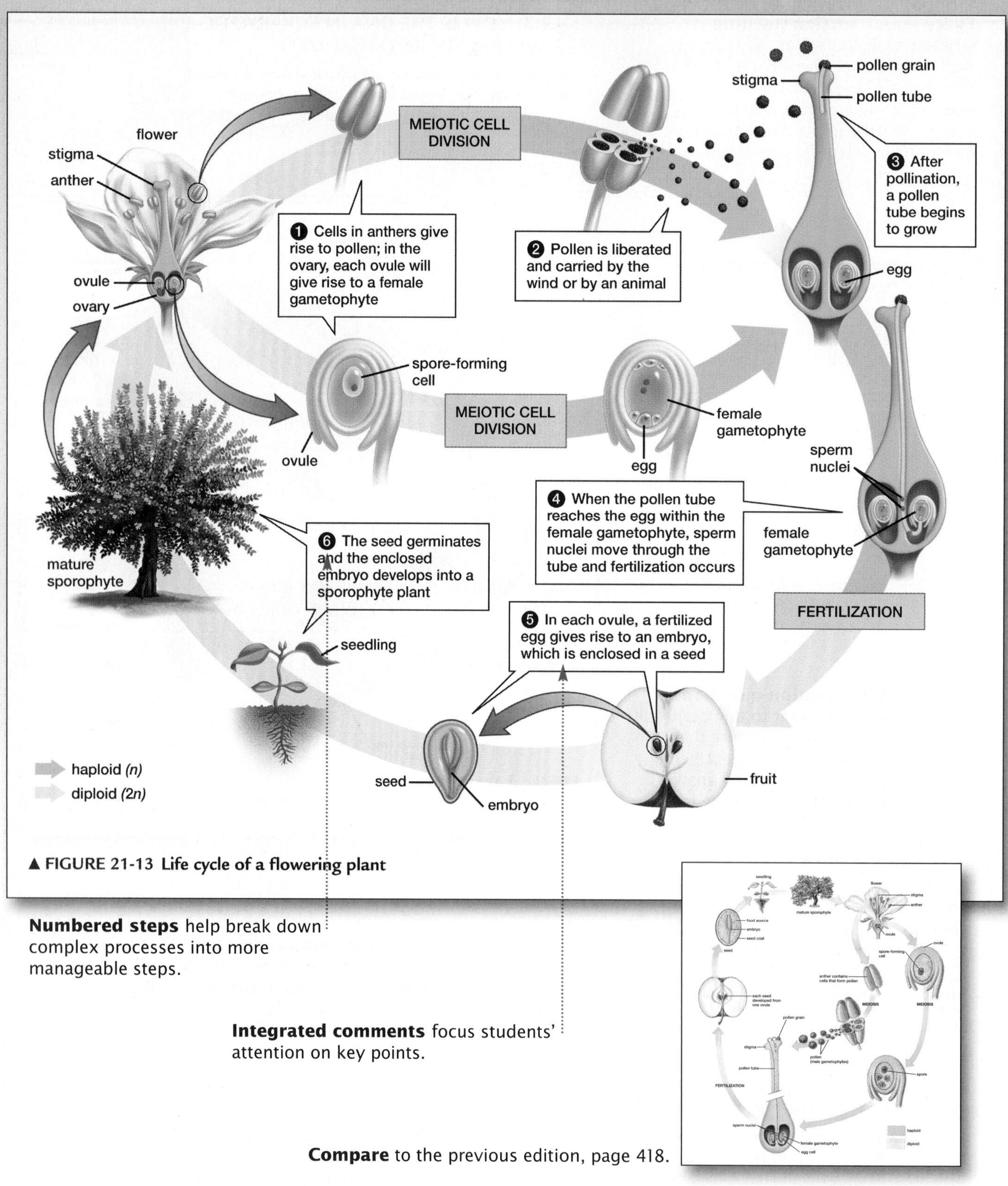

▲ **FIGURE 21-13 Life cycle of a flowering plant**

Numbered steps help break down complex processes into more manageable steps.

Integrated comments focus students' attention on key points.

Compare to the previous edition, page 418.

NEW! Eye-catching photos throughout the text provide an enhanced view into the world of biology and convey the wonder of life on Earth.

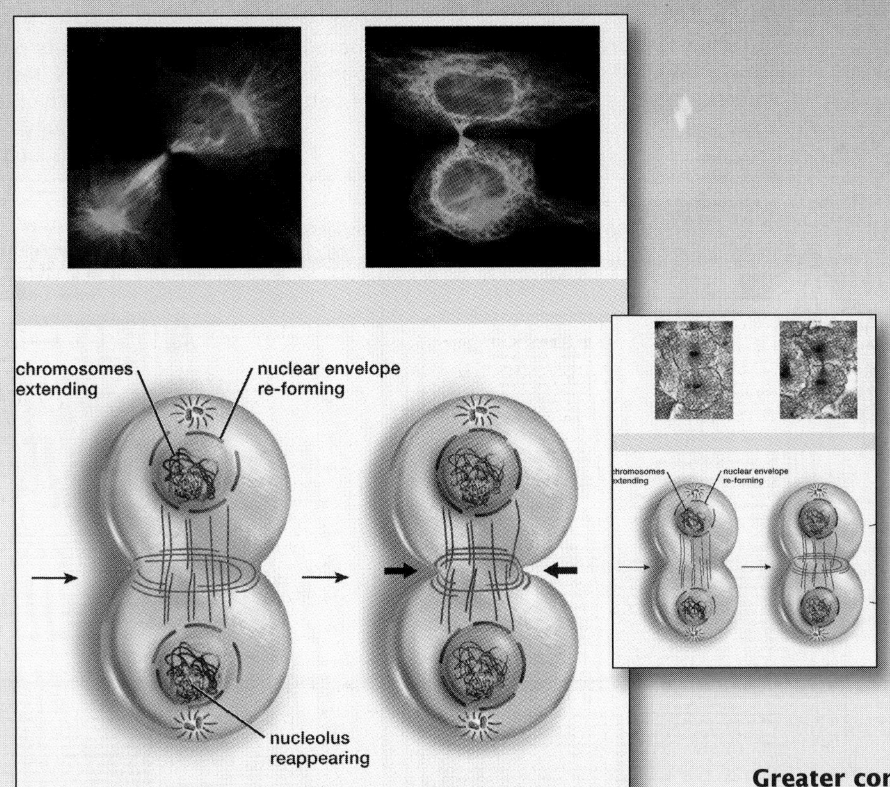

Compare to the previous edition, page 199.

Greater consistency in color and illustration format makes it easier for students to intuitively make connections among figures.

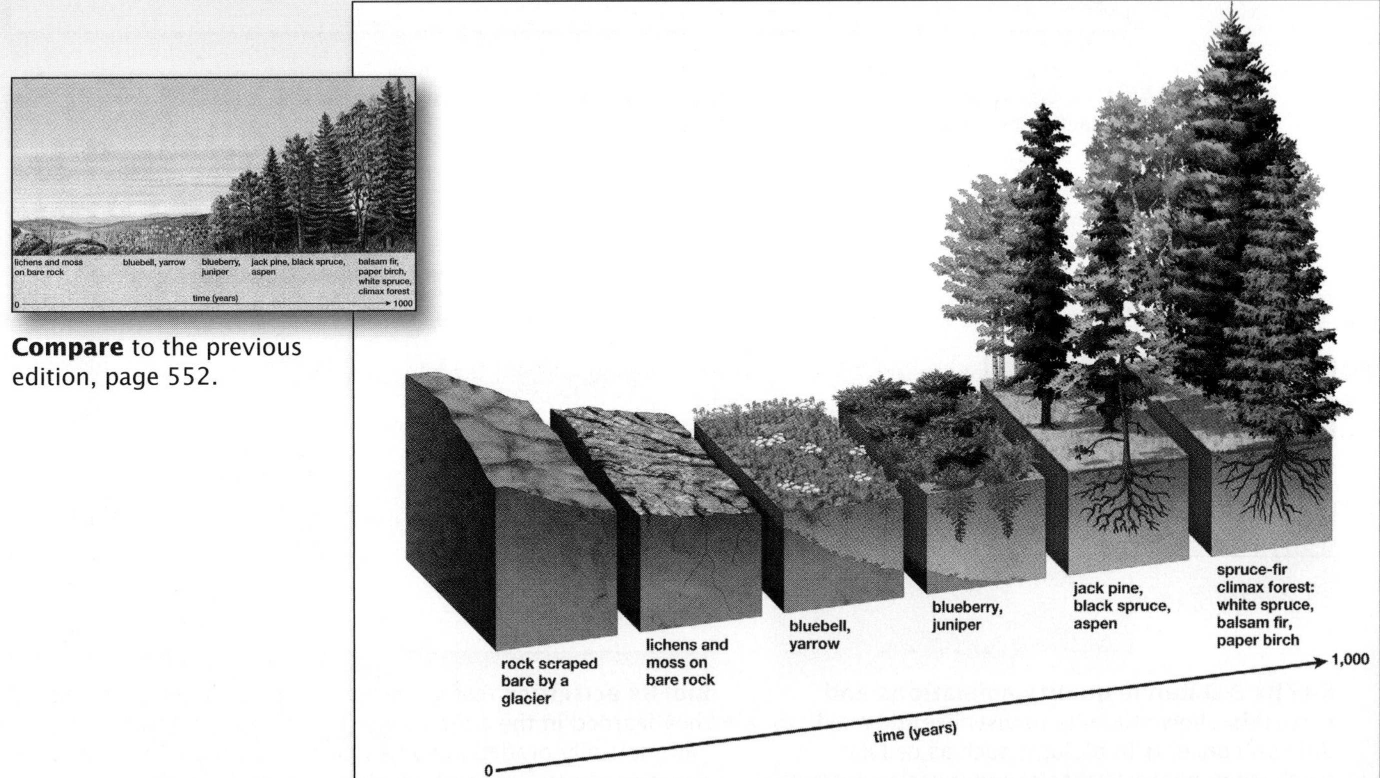

Compare to the previous edition, page 552.

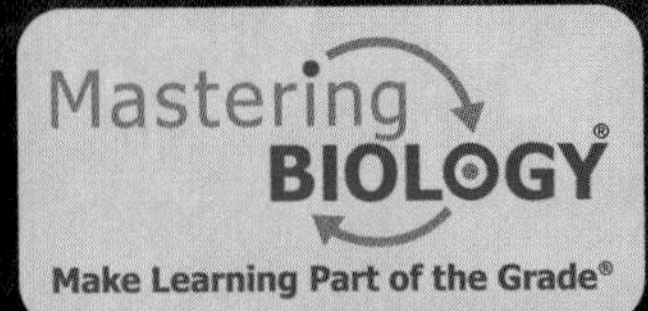

Have you ever wanted to help students make the most of their study time?

NEW! **MasteringBiology** offers access to Pearson eText, one of the many resources available in the self-study area that students can access 24/7. Other resources include practice quizzes, and personalized study plans. Go to www.masteringbiology.com to get started.

Highlight function lets students highlight what they want to remember.

Annotation feature allows students to take notes.

Google™-based search function lets students search the eText.

Zoom lets students zoom in and out for better viewing.

Hyperlinks connect to quizzes, tests, activities, BioFlix 3-D movie-quality animations, and more.

Pearson eText

Available in the Study Area, Pearson eText gives students access to the text whenever and wherever they can access the Internet. The eText pages look exactly like the printed text, and include powerful interactive and customization functions.

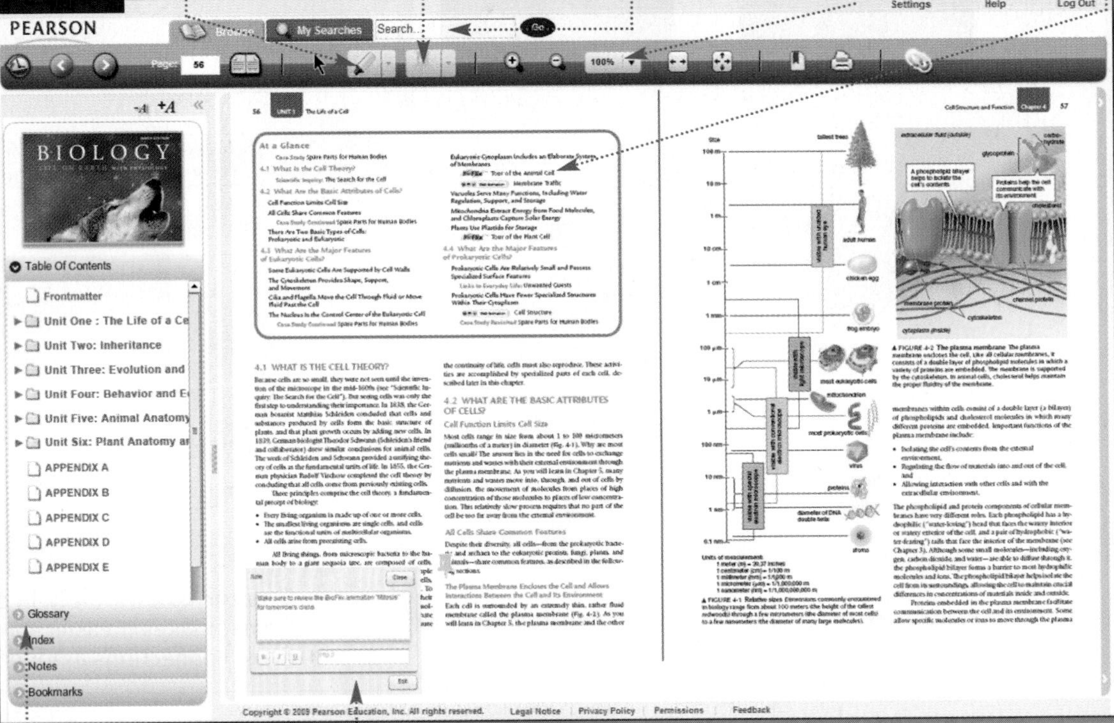

Interactive glossary provides pop-up definitions and terms.

Instructor notes allow you to share notes and highlights with the class.

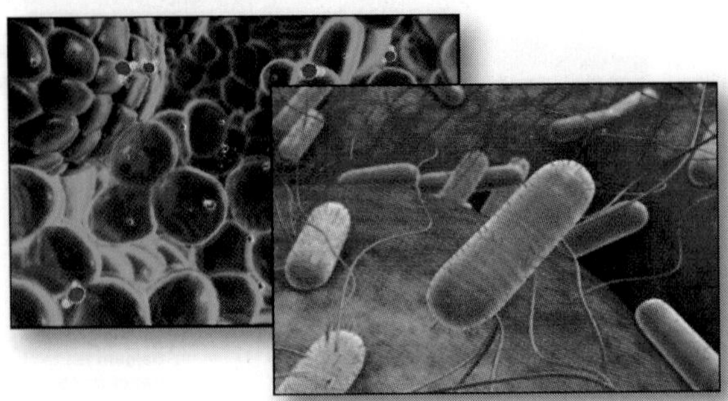

BioFlix 3-D movie-quality animations and tutorials allow students to visualize the most difficult concepts in biology, such as cellular respiration, photosynthesis, and muscle contraction.

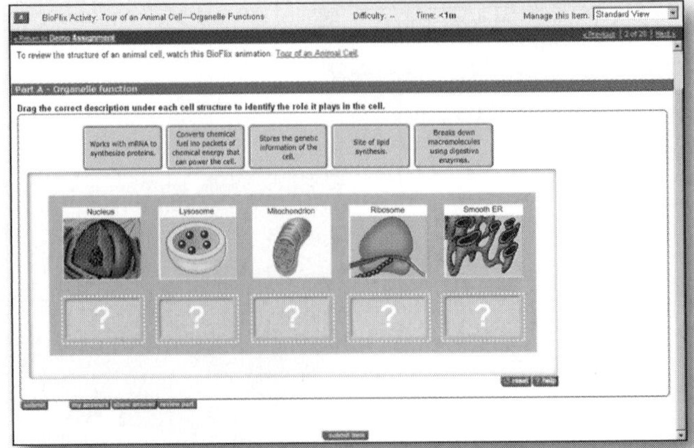

BioFlix activities test students understanding of concepts they learned in the animations. These highly interactive, automatically graded activities include a wide variety of question types—none of which are multiple choice.

Have you ever wondered

how to increase participation and accountability in your non-majors biology course?

MasteringBiology® helps you customize assessment to perfectly match your teaching style and course goals.

MasteringBiology is an online assessment and tutorial system designed to help instructors teach more efficiently, and is pedagogically proven to help students learn. With MasteringBiology, you can...

- **Break free from only offering multiple-choice assessment** but still get automatically graded assignments.
- **Personalize your course** by editing pre-loaded assignments or uploading your own.
- **Take the guesswork out of lectures, quizzes, and tests** by utilizing current data on your students' performance.
- **Assign auto-graded activities** in MasteringBiology that appeal to diverse learning styles: BioFlix™ Activities, *New York Times* Articles, *Discovery Channel* Videos, coaching activities, GraphIt! Activities, Pre-Lecture Quizzes, Post-Lecture Quizzes, and Test Bank Questions.

Customize content with questions and answers that can be easily edited.

Color bar data uses green to indicate correct answers. Red shows the percentage of students who requested an answer. Orange indicates the average number of wrong answers per student.

Pre-lecture quizzes

Help students to arrive at lecture prepared and keep them on track all term.

At-a-glance statistics reveal data for your class as well as national results.

Wrong answer summary gives unique insight into your students' misunderstandings and allows for just-in-time teaching adjustments.

Part A

Which type of organism obtains energy by metabolizing molecules produced by other organisms?

ANSWER:

- autotrophs
- heterotrophs
- decomposers
- B and C
- A, B, and C

Answer Stats:	Students	% Correct	% Unfinished	% Req'd Solution	Wrong/student	Hints/student
Overall	297	98.7%	0.3%	1%	0.8	0
MBIODEMOGRADES	5	80%	0%	20%	1	0

Wrong Answers for MBIODEMOGRADES		
% Wrong	Answer	Response
60%	decomposers	
20%	heterotrophs	
20%	autotrophs	

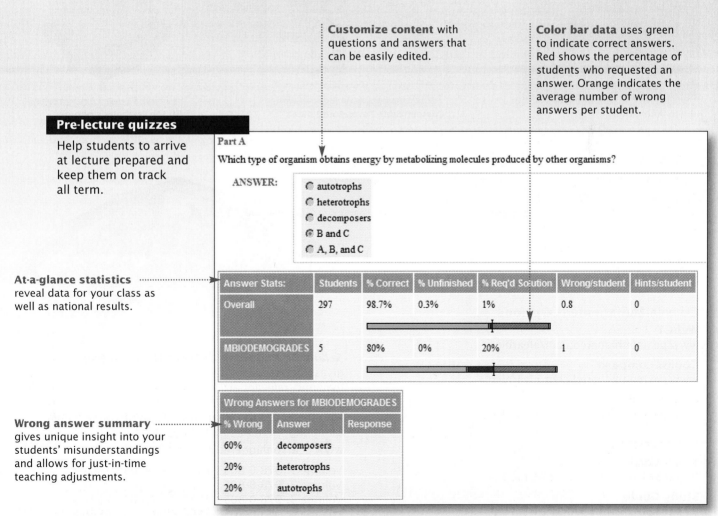

Pre-lecture Quiz Student Performance Data

Have you ever wondered
what additional tools are available for your biology course?

Instructor Resource DVD

978-0-321-61193-2 ▪ 0-321-61193-4

Instructor resources are available on DVD and online.

Assets for each chapter include:

- All textbook figures, photos, and tables in JPEG and PowerPoint® versions. Figures are optimized for projection in labeled, unlabeled, and editable label versions.
- PowerPoint Lecture Outline Presentations for each chapter, which contain extensive lecture notes and figures from the text.
- 250+ Instructor Animations, including dynamic BioFlix.™
- Assessment tools including the Test Bank in TestGen® software and Microsoft® Word format, and PowerPoint CRS Clicker Questions.
- The Instructor Guide in Microsoft Word format.
- *Discovery Channel* Videos.

Discovery Channel Videos

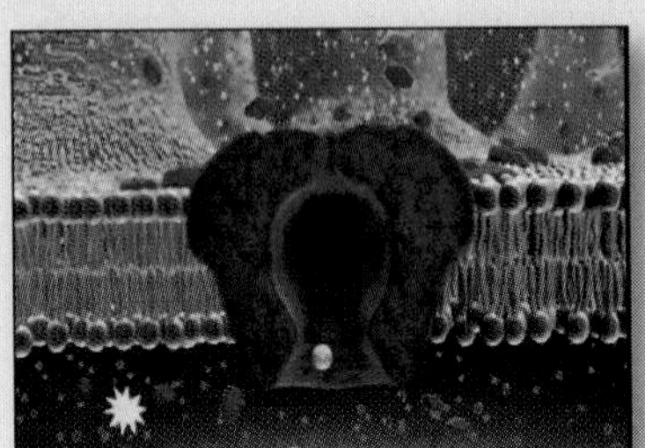
BioFlix Animations

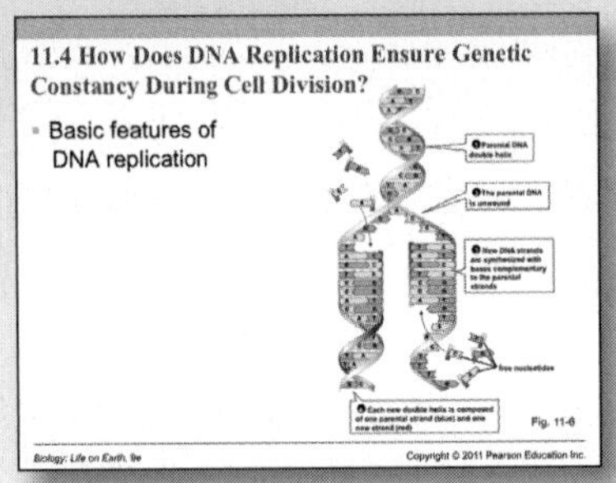

Customizable PowerPoint Slides

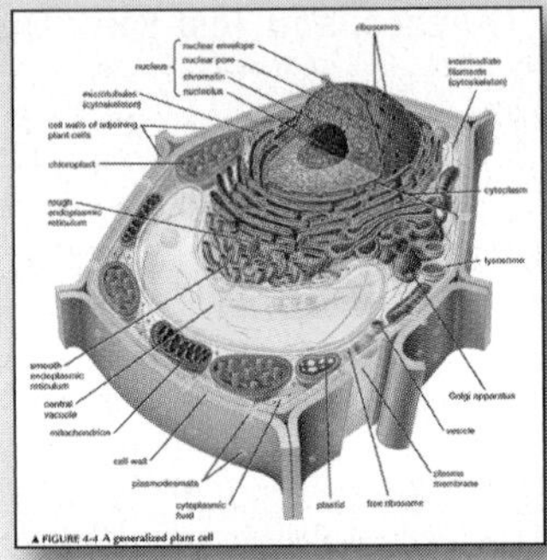
All the textbook's photos, figures, and tables

Additional Resources

For Instructors

Transparency Acetates
978-0-321-61192-5 ▪ 0-321-61192-6

TestGen®
978-0-321-61194-9 ▪ 0-321-61194-2

MasteringBiology®
www.masteringbiology.com

Course Management Options

WebCT
www.pearsonhighered.com/elearning

CourseCompass™
www.pearsonhighered.com/elearning

Blackboard
www.pearsonhighered.com/elearning

For Students

Study Card
978-0-321-61173-4 ▪ 0-321-61173-X

Study Guide
978-0-321-61179-6 ▪ 0-321-61179-9

MasteringBiology
www.masteringbiology.com

Scientific American Current Issues in Biology gives your students the best of both worlds—accessible, dynamic, relevant articles from *Scientific American* magazine that present key issues in biology, paired with the authority, reliability, and clarity of Benjamin Cummings' non-majors biology texts.

www.pearsonhighered.com/currentissues

Volume 1: 978-0-8053-7507-7 ▪ 0-8053-7507-4
Volume 2: 978-0-8053-7108-6 ▪ 0-8053-7108-7
Volume 3: 978-0-8053-7527-5 ▪ 0-8053-7527-9
Volume 4: 978-0-8053-3566-8 ▪ 0-8053-3566-8
Volume 5: 978-0-321-54187-1 ▪ 0-321-54187-1
Volume 6: 978-0-321-59849-3 ▪ 0-321-59849-0

An Introduction to Life on Earth

Chapter 1

Case Study

Are Viruses Alive?

IN 1996 IN GABON, AFRICA, a hunting party discovered a recently dead chimpanzee in the bush and butchered it for food. Within a month, the village was devastated. Thirty-seven residents, including the hunters and many of their family members, had fallen ill. Before the scourge had ended, more than half of the victims had died.

The villagers had fallen victim to one of Earth's most sinister pathogens—the Ebola virus (see the inset photo, below). The very mention of Ebola hemorrhagic fever strikes fear in anyone familiar with its symptoms, which begin with fever, headache, joint and muscle aches, and stomach pains, progressing to severe vomiting, bloody diarrhea, and organ failure. Internal hemorrhaging can leave victims bleeding from every orifice. Death usually occurs within 7 to 16 days after the onset of symptoms, and there is no cure. Ebola fever is so contagious and deadly (with up to a 90% fatality rate) that heroic precautions must be taken to protect caregivers from contact with the body fluids of their patients (see the photo to the right).

Ebola fever is one of many infectious diseases caused by viruses. The existence of viruses, entities far smaller than most bacteria, was first surmised in the late nineteenth century. The name "virus" is derived from the Latin word for "poison" because of the disease-causing effects of viruses. Some viruses, such as smallpox and polio, have been largely eradicated, whereas others, such as the common cold and influenza, remain with us but are usually only temporarily debilitating. The viruses of greatest concern to public health officials are the human immunodeficiency virus (HIV), which causes AIDS and kills about 2 million people annually, and the avian influenza virus (H5N1; "bird flu"). Although bird flu's total death toll was about 250 people as of early 2009, its mortality rate is higher than 60%. Health officials are concerned that bird flu may mutate into a far more infectious form, causing a worldwide epidemic.

Measured by the misery they cause, viruses are an unmitigated success. Viruses infect every known form of life on Earth and are the most abundant biological entity on the planet. Viruses can spread from one organism to another, replicate rapidly, and mutate. This tendency to mutate helps them to evade our attempts to develop vaccines against them. In spite of this, most scientists will tell you that viruses are not alive. Why? What is life, anyway?

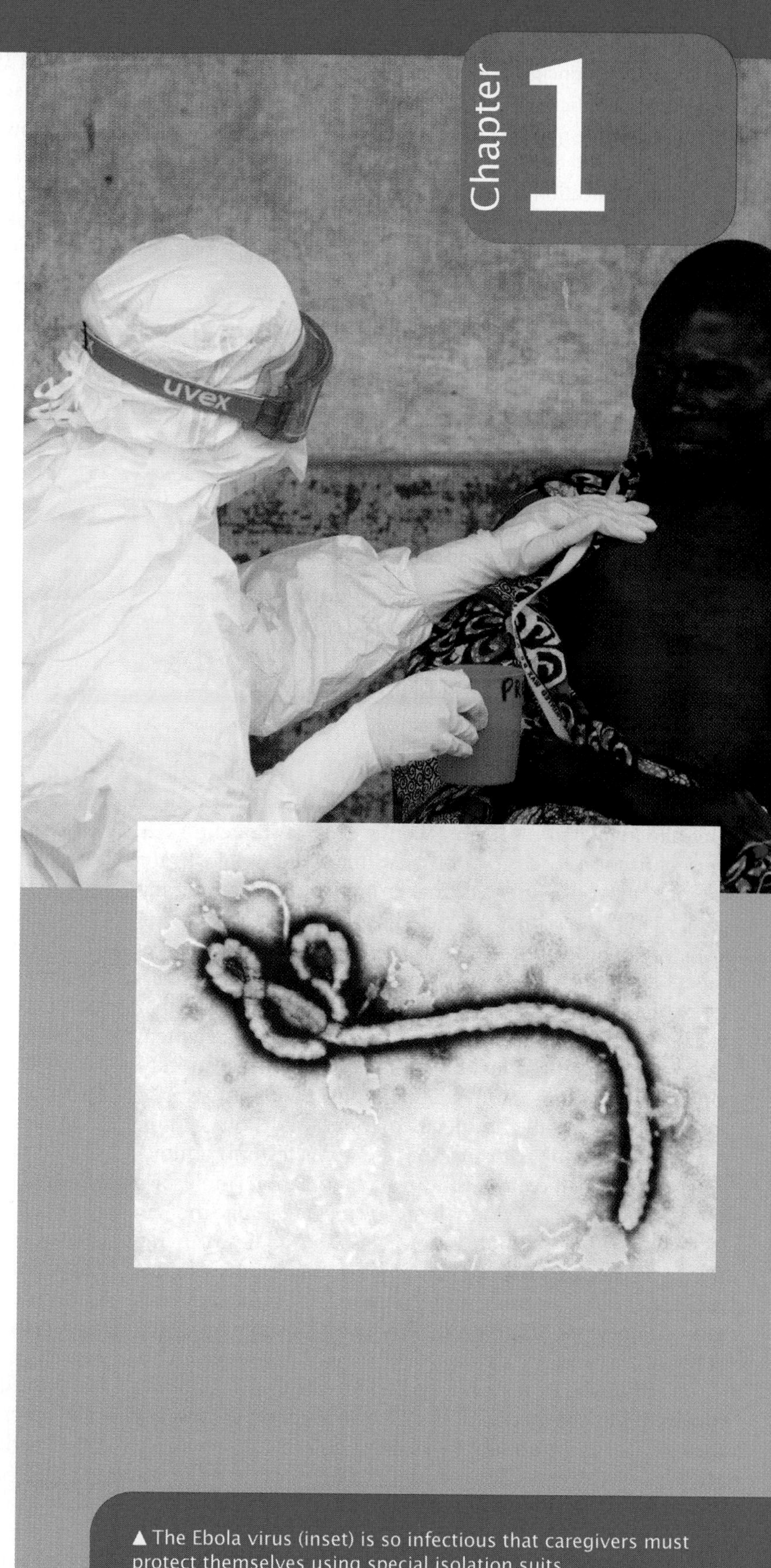

▲ The Ebola virus (inset) is so infectious that caregivers must protect themselves using special isolation suits.

At a Glance

1.1 HOW DO SCIENTISTS STUDY LIFE?

Life Can Be Studied at Different Levels

Just as mud can be formed into bricks that provide the building blocks of a wall, which in turn provide the basis of a structure, the living and nonliving worlds exist at different levels of organization. Each level provides the building blocks for the one above it, and each higher level incorporates many components of each level beneath it (**Fig. 1-1**).

All matter on Earth consists of atoms of substances called **elements,** each of them unique. An **atom** is the smallest particle of an element that retains all the properties of that element. For example, a diamond is a form of the element carbon. The smallest possible unit of a diamond is an individual carbon atom. Atoms may combine in specific ways to form assemblies called **molecules;** for example, one carbon atom can combine with two oxygen atoms to form a molecule of carbon dioxide. Although many simple molecules form spontaneously, only living things manufacture extremely large and complex molecules. The bodies of living things, or **organisms,** are composed primarily of complex molecules called **organic molecules,** meaning that they contain a framework of carbon to which at least some hydrogen is bound.

Although atoms and molecules form the building blocks of life, the quality of life itself emerges on the level of the cell. Just as an atom is the smallest unit of an element, so the **cell** is the smallest unit of life (**Fig. 1-2**). Although many forms of life consist of single cells, in multicellular organisms, cells of similar type combine to form structures known as **tissues;** for example, muscle cells working together form muscle tissue. Different tissues, in turn, may combine to form **organs** (the heart, for example). A group of organs united by a common overall function are called **organ systems** (for example, the heart is part of the circulatory system). Multicellular organisms often have several organ systems.

Levels of organization extend beyond the individual. Within a given area, a group of organisms of the same type (the same species) constitutes a **population.** All organisms capable of breeding with one another are collectively called a **species.** A collection of populations of different species that interact with one another makes up a **community** (see Fig. 1-1). A community and the nonliving environment that surrounds it constitute an **ecosystem.** Finally, the entire surface of Earth and the living things that dwell in and on it is called the **biosphere.**

Biologists work at many different levels, depending on the question they are investigating. For example, to find out how antelope digest their food, a biologist might study the organs of the antelope digestive system or, at a smaller level, the cells that line the digestive tract. Delving deeper, a researcher might investigate the biological molecules secreted into the digestive tract that break down the animal's food. On the other hand, to find out whether habitat destruction is reducing the number of antelope, biologists would investigate

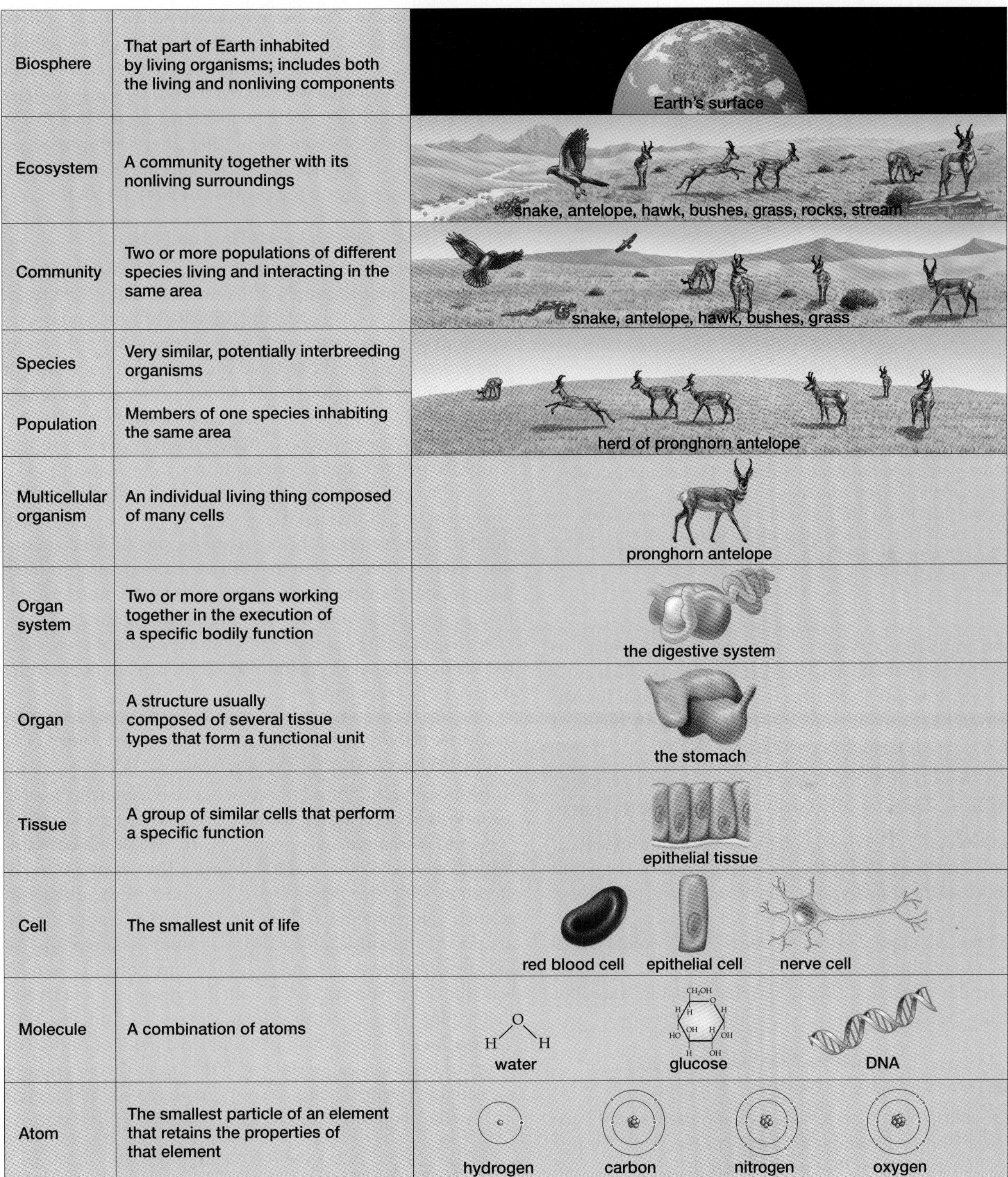

Biosphere
That part of Earth inhabited by living organisms; includes both the living and nonliving components
Earth's surface
Ecosystem
A community together with its nonliving surroundings
snake, antelope, hawk, bushes, grass, rocks, stream
Community
Two or more populations of different species living and interacting in the same area
snake, antelope, hawk, bushes, grass
Species
Very similar, potentially interbreeding organisms
Population
Members of one species inhabiting the same area
herd of pronghorn antelope
Multicellular organism
An individual living thing composed of many cells
pronghorn antelope
Organ system
Two or more organs working together in the execution of a specific bodily function
the digestive system
Organ
A structure usually composed of several tissue types that form a functional unit
the stomach
Tissue
A group of similar cells that perform a specific function
epithelial tissue
Cell
The smallest unit of life
red blood cell
epithelial cell
nerve cell
Molecule
A combination of atoms
water
glucose
DNA
Atom
The smallest particle of an element that retains the properties of that element
hydrogen
carbon
nitrogen
oxygen

▲ **FIGURE 1-1 Levels of organization of matter** Interactions among the components of each level and the levels below it allow the development of the next higher level of organization.

QUESTION What current environmental change is likely to affect the entire biosphere?

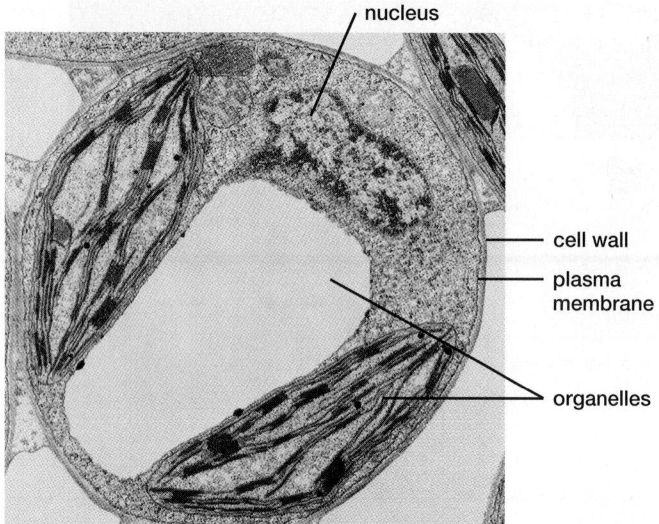

▲ **FIGURE 1-2 The cell is the smallest unit of life** This artificially colored micrograph of a plant cell (a eukaryotic cell) shows the supporting cell wall that surrounds plant (but not animal) cells. Just inside the cell wall, the plasma membrane (found in all cells) has control over which substances enter and leave. The cell also contains several types of specialized organelles, including the nucleus.

antelope populations as well as the interacting populations of other species that make up the community to which antelope belong. It is safe to say that, because biological investigations occur at so many different levels, biology is the most diverse science you will ever encounter.

Scientific Principles Underlie All Scientific Inquiry

Harvard biologist E. O. Wilson defines science as "the organized systematic enterprise that gathers knowledge about the world and condenses the knowledge into testable laws and principles." Scientific inquiry is based on a small set of assumptions. Although these assumptions can never be proven absolutely, they have been so thoroughly tested and validated that we might call them scientific principles. These principles are natural causality, uniformity in space and time, and common perception.

Natural Causality Is the Principle That All Events Can Be Traced to Natural Causes

Over the course of human history, two approaches have been taken to the study of life and other natural phenomena. The first assumes that some events happen through the intervention of supernatural forces beyond our understanding. Ancient Greeks believed that the god Zeus hurled lightning bolts from the sky. Throughout the middle ages, many people were convinced that life arose spontaneously from nonliving matter; for example, that maggots came from rotting meat. Epileptic seizures were once thought to be the result of a visitation from the gods.

Science, in contrast, adheres to the second approach, or the principle of **natural causality,** which states that all events can be traced to natural causes that are potentially within our ability to comprehend. Today, we realize that lightning is an electrical discharge, that maggots are the larval form of flies, and that epilepsy is a brain disorder. The principle of natural causality has an important corollary: The evidence we gather has not been deliberately distorted to fool us. This corollary may seem obvious, yet some people have argued that fossils are not evidence of evolution; rather, they were placed on Earth by God to test our faith. The enormous accomplishments of science rely on the premise of natural causality.

Natural Laws Apply to Every Time and Place

A second fundamental principle of science is that natural laws—laws derived from the study of nature—are uniform in space and time. The laws of gravity, the behavior of light, and the interactions of atoms, for example, are the same today as they were a billion years ago, and they hold true everywhere on Earth, or in our universe. The principle of uniformity in space and time is especially vital to biology, because many important biological events, such as the evolution of today's diversity of living things, happened before humans were around to observe them.

Some people believe that each of the different types of organisms was individually created at one time in the past by the direct intervention of God, a philosophy called creationism. Scientists freely admit that this idea cannot be absolutely disproved, but creationism is contrary to both natural causality and uniformity in time. The overwhelming success of science in explaining natural events through natural causes has led scientists to reject creationism as an explanation for the diversity of life on Earth.

Scientific Inquiry Is Based on the Assumption That People Perceive Natural Events in Similar Ways

A third basic assumption of science is that, generally, people are able to perceive and measure events, and that these perceptions and measurements provide us with reliable, objective information about the natural world. The assumption of common, objective perception is, to some extent, unique to science. Value systems, such as those involved in appreciating art, poetry, and music, do not assume common perception. We may perceive the colors and shapes in a painting in a similar way (the objective aspect of art), but may disagree about its aesthetic value (the subjective, humanistic aspect of art; **Fig. 1-3**). Values differ among individuals, often as a result of cultural or religious beliefs. Because value systems are subjective and not measurable, science cannot answer certain types of philosophical or ethical questions, such as the morality of abortion.

The Scientific Method Is the Basis for Scientific Inquiry

Given these assumptions, how do biologists study the workings of life? Scientific inquiry is a rigorous method for making observations of specific phenomena and searching for the order that underlies those phenomena. Biology and other sciences commonly use the **scientific method,** which consists of six interrelated elements: observation, question, hypothesis, prediction, experiment, and conclusion (**Fig. 1-4**). All scientific inquiry begins with an **observation** of a specific phenome-

▲ **FIGURE 1-3 Value systems differ** Science assumes that people will agree about the colors and shapes in this painting, which are objective aspects of perspection. But subjective questions such as "What does this mean?" or "Is this beautiful?" or "Do I want this on my wall?" will be answered in different ways by different observers.

QUESTION This unique piece of contemporary art by artist and geneticist Hunter Cole is called "Endosymbiosis." The painting features mitochondria and chloroplasts, organelles that are likely to have evolved over millions of years from bacteria living within other cells (see pp. 325–326). Does this objective information change your subjective perception of the art?

non. The observation, in turn, leads to a **question:** "How did this happen?" Then, in a flash of insight—or more often after long, hard thought—a hypothesis is formulated. A **hypothesis** is a supposition, based on previous observations, that is offered as an answer to the question and a natural explanation for the observed phenomenon. To be useful, the hypothesis must lead to a **prediction,** typically expressed in "If . . . then" language. The prediction is tested by carefully controlled manipulations called **experiments.** These experiments produce results that, when analyzed, either support or refute the hypothesis, allowing the scientist to reach a **conclusion** about the validity of the hypothesis. A single experiment is never an adequate basis for a conclusion; the results must be repeatable not only by the original researcher but also by others.

Simple experiments, at least those in the life sciences, test the assertion that a single factor—a **variable**—is the cause of a specific observation. To be scientifically valid, the experiment must rule out other possible variables as the cause of the observation. For this reason, biologists design **controls** into their experiments. Control situations, in which all the variables not being tested remain constant, are then compared with the experimental situation, in which only the variable being tested is changed. In the early 1600s, Francesco Redi used the scientific method to test the hypothesis that flies do not arise spontaneously from rotting meat, and this method is still used today, as illustrated by Malte Andersson's experiment to test the hypothesis that female widowbirds prefer to mate with males with long tails (see "Scientific Inquiry: Controlled Experiments, Then and Now" on pp. 6–7).

You probably use some variation of the scientific method to solve everyday problems (see Fig. 1-4). For example, late for an important date, you rush to your car, turn the ignition key, and make the *observation* that it won't start. Your *question* "Why won't the car start?" immediately leads to a *hypothesis:* The battery is dead. Your hypothesis leads to the *prediction:* A new battery will allow you to start your car. Quickly, you design an *experiment:* You replace your battery with the battery from your roommate's new car and try to start your car again. Your car starts immediately, supporting your hypothesis. But wait, you didn't control for other variables, such as a loose battery cable. To test this, you might replace your old battery, tighten the cables securely, and then attempt to restart the car. If your car repeatedly refuses to start with the old battery and tightened cables but then starts immediately when you put in a new battery, you have isolated a single variable—the battery. Although you are very late for your date, you can now safely draw the conclusion that your old battery is dead.

The scientific method is powerful, but it is important to recognize its limitations. In particular, scientists can seldom be sure that they have controlled for *all* possible variables or made all the possible observations supporting a given hypothesis.

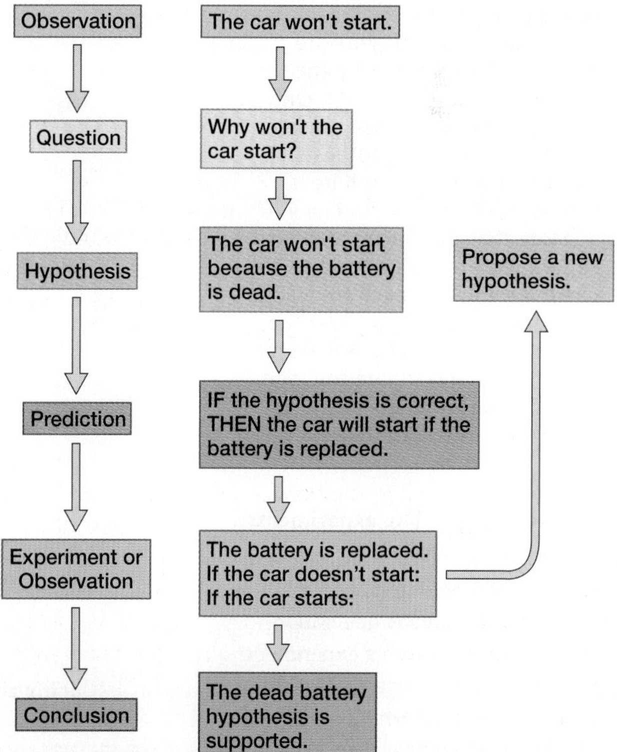

▲ **FIGURE 1-4 The scientific method** The steps of the scientific method are on the left, with a "real-life" example illustrated on the right.

Scientific Inquiry

Controlled Experiments, Then and Now

A classic experiment by the Italian physician Francesco Redi (1621–1697) beautifully demonstrates the scientific method and helps to illustrate the principle of natural causality, on which modern science is based. Redi investigated why maggots (which are the larval form of flies) appear on spoiled meat. In Redi's time, the appearance of maggots on meat was considered to be evidence of spontaneous generation, the production of living things from nonliving matter.

Redi *observed* that flies swarm around fresh meat and that maggots appear on meat left out for a few days. He formed a testable *hypothesis:* The flies produce the maggots. In his *experiment,* Redi wanted to test just one variable—the access of flies to the meat. Therefore, he took two clean jars and filled them with similar pieces of meat. He left one jar open (the control jar) and covered the other with gauze to keep out flies (the experimental jar). He did his best to keep all the other variables the same (for example, the type of jar, the type of meat, and the temperature). After a few days, he observed maggots on the meat in the open jar, but saw none on the meat in the covered jar. Redi *concluded* that his hypothesis was correct and that maggots are produced by flies, not by the nonliving meat (**Fig. E1-1**). Only through controlled experiments could the age-old hypothesis of spontaneous generation be laid to rest.

More than 300 years after Redi's experiment, today's scientists still use the same approach to design their experiments. Consider the experiment that Malte Andersson designed to investigate the long tails of male widowbirds. Andersson *observed* that male, but not female, widowbirds have extravagantly long tails, which they display while flying across African grasslands. This observation led Andersson to ask the *question:* Why do the males, and only the males, have such long tails? His *hypothesis* was that males have long tails because females prefer to mate with long-tailed males, which therefore have more offspring than shorter-tailed males. From this hypothesis, Andersson *predicted* that if his hypothesis were true, then more females would build

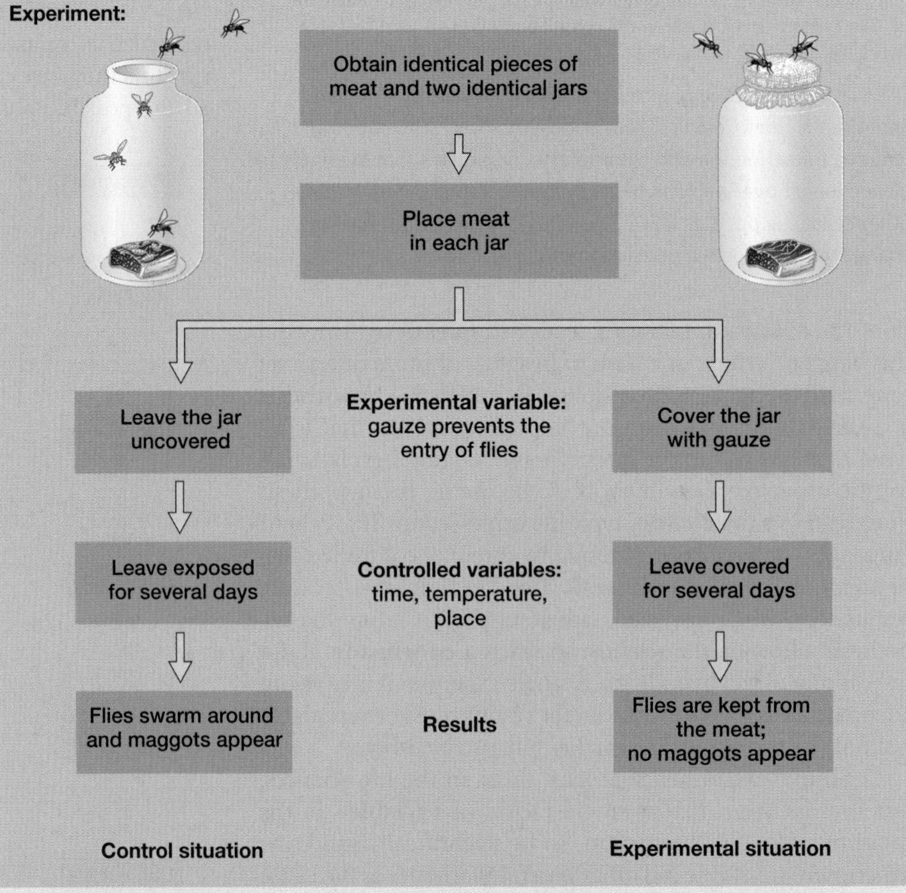
Observation: Flies swarm around meat left in the open; maggots appear on the meat.
Question: Where do maggots on the meat come from?
Hypothesis: Flies produce the maggots.
Prediction: IF the hypothesis is correct, THEN keeping the flies away from the meat will prevent the appearance of maggots.
Experiment:
Obtain identical pieces of meat and two identical jars
Place meat in each jar
Leave the jar uncovered
Experimental variable: gauze prevents the entry of flies
Cover the jar with gauze
Leave exposed for several days
Controlled variables: time, temperature, place
Leave covered for several days
Flies swarm around and maggots appear
Results
Flies are kept from the meat; no maggots appear
Control situation
Experimental situation
Conclusion: The experiment supports the hypothesis that flies are the source of maggots and that spontaneous generation of maggots does not occur.

▶ **FIGURE E1-1 The experiments of Francesco Redi**

QUESTION Redi's experiment falsified spontaneous generation of maggots, but did his experiment convincingly demonstrate that flies produce maggots? What kind of follow-up experiment would be needed to more conclusively determine the source of maggots?

nests on the territories of males with artificially lengthened tails than on the territories of males with artificially shortened tails. He then captured some males, trimmed their tails to about half their original length, and released them (*experimental* group 1). Another group of males had the tail feathers that had been removed from the first group glued on as tail extensions (*experimental* group 2). Finally, Andersson had two *control* groups. In one, the tail was cut and then glued back in place (to control for the effects of capturing the birds and manipulating their feathers). In the other, the birds were simply captured and released. The experimenter was doing his best to make sure that tail length was the only variable that was changed. After a few days, Andersson counted the number of nests that females had built on each male's territory. He found that males with lengthened tails had the most nests on their territories, males with shortened tails had the fewest, and control males (with normal-length tails) had an intermediate number (**Fig. E1-2**). Andersson *concluded* that his hypothesis was correct, and that female widowbirds prefer to mate with males that have long tails.

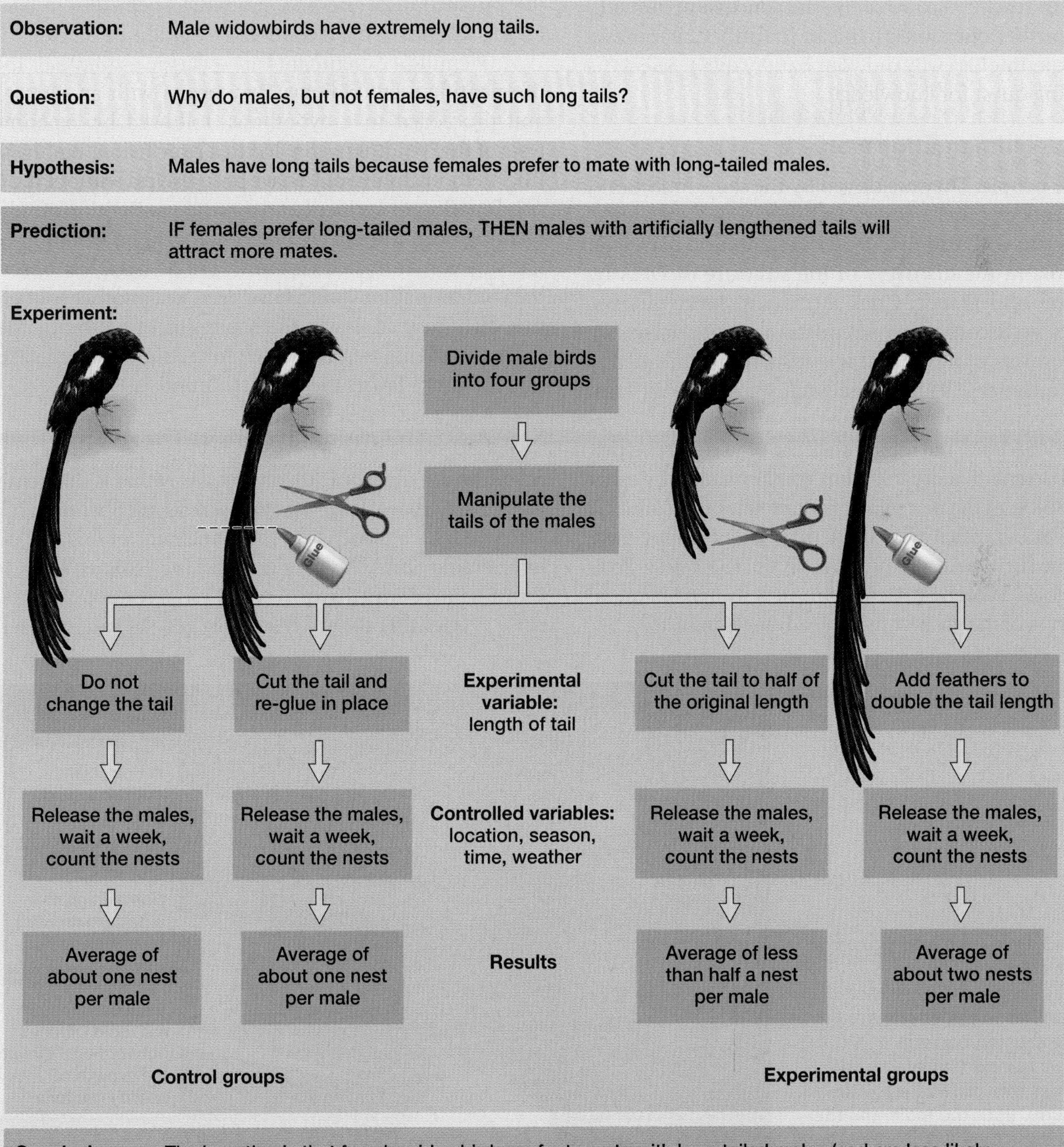
Observation: Male widowbirds have extremely long tails.
Question: Why do males, but not females, have such long tails?
Hypothesis: Males have long tails because females prefer to mate with long-tailed males.
Prediction: IF females prefer long-tailed males, THEN males with artificially lengthened tails will attract more mates.
Experiment:
Divide male birds into four groups
Manipulate the tails of the males
Glue
Do not change the tail
Cut the tail and re-glue in place
Experimental variable: length of tail
Cut the tail to half of the original length
Add feathers to double the tail length
Release the males, wait a week, count the nests
Release the males, wait a week, count the nests
Controlled variables: location, season, time, weather
Release the males, wait a week, count the nests
Release the males, wait a week, count the nests
Average of about one nest per male
Average of about one nest per male
Results
Average of less than half a nest per male
Average of about two nests per male
Control groups
Experimental groups
Conclusion: The hypothesis that female widowbirds prefer to mate with long-tailed males (and are less likely to mate with short-tailed males) is supported.

▲ **FIGURE E1-2 The experiments of Malte Andersson**

Therefore, scientific conclusions must always remain subject to revision if new experiments or observations demand it.

Communication Is Crucial to Science

No matter how well designed an experiment is, it is useless if it is not communicated thoroughly and accurately. Redi's experimental design and conclusions survive today only because he carefully recorded his methods and observations. If experiments are not communicated to other scientists in enough detail, they cannot be repeated to verify the conclusions. Without verification, scientific findings cannot be safely used as the basis for new hypotheses and further experiments.

A fascinating aspect of scientific inquiry is that whenever a scientist reaches a conclusion, the conclusion immediately raises further questions that lead to further hypotheses and more experiments (why did your battery die?). Science is a never-ending quest for knowledge.

Science Is a Human Endeavor

Scientists are people. They are driven by the same ambitions, pride, and fears as other people, and they sometimes make mistakes. As you will read in Chapter 11, ambition played an important role in the discovery of the structure of DNA by James Watson and Francis Crick. Accidents, lucky guesses, controversies with competing scientists, and, of course, the intellectual powers of individual scientists contribute greatly to scientific advances. To illustrate what we might call "real science," let's consider an actual case.

To study bacteria, microbiologists use pure cultures—that is, plates of bacteria that are free from contamination by other bacteria or molds. At the first sign of contamination, a culture is usually thrown out, often with mutterings about sloppy technique. On one such occasion, however, in the late 1920s, Scottish bacteriologist Alexander Fleming turned a ruined bacterial culture into one of the greatest medical advances in history.

One of Fleming's bacterial cultures became contaminated with a mold (a type of fungus) called *Penicillium*. Before throwing out the culture dish, Fleming made the observation that no bacteria were growing near the mold (**Fig. 1-5**). Why not? Fleming formulated the hypothesis that *Penicillium* releases a substance that kills bacteria. To test this hypothesis, Fleming performed an experiment. He grew *Penicillium* in a liquid nutrient broth. He then filtered out the mold and placed some of the remaining liquid in a plate with an uncontaminated bacterial culture. Sure enough, something in the liquid killed the bacteria, supporting his hypothesis and leading to the conclusion that *Penicillium* secretes something that kills bacteria. Further research into these mold extracts resulted in the production of the first antibiotic—penicillin, a bacteria-killing substance that has since saved millions of lives.

Fleming's experiments are a classic example of the use of scientific methodology. They began with an observation and proceeded to a hypothesis, followed by experimental tests of the hypothesis that led to a conclusion. But the scientific method alone would have been useless without the lucky combination of accident and a brilliant scientific mind. Had Fleming been a "perfect" microbiologist, he would not have had any contaminated cultures. Had he been less observant, the contamination would have been just another spoiled culture dish. Instead, it was the beginning of antibiotic therapy for bacterial diseases. As French microbiologist Louis Pasteur said, "Chance favors the prepared mind."

Scientific Theories Have Been Thoroughly Tested

Scientists use the word "theory" in a way that is different from its everyday usage. If Dr. Watson were to ask Sherlock Holmes, "Do you have a theory as to the perpetrator of this foul deed?" in scientific terms, he would be asking Holmes for a hypothesis—an "educated guess" based on observable evidence, or clues. A **scientific theory** is far more general and more reliable

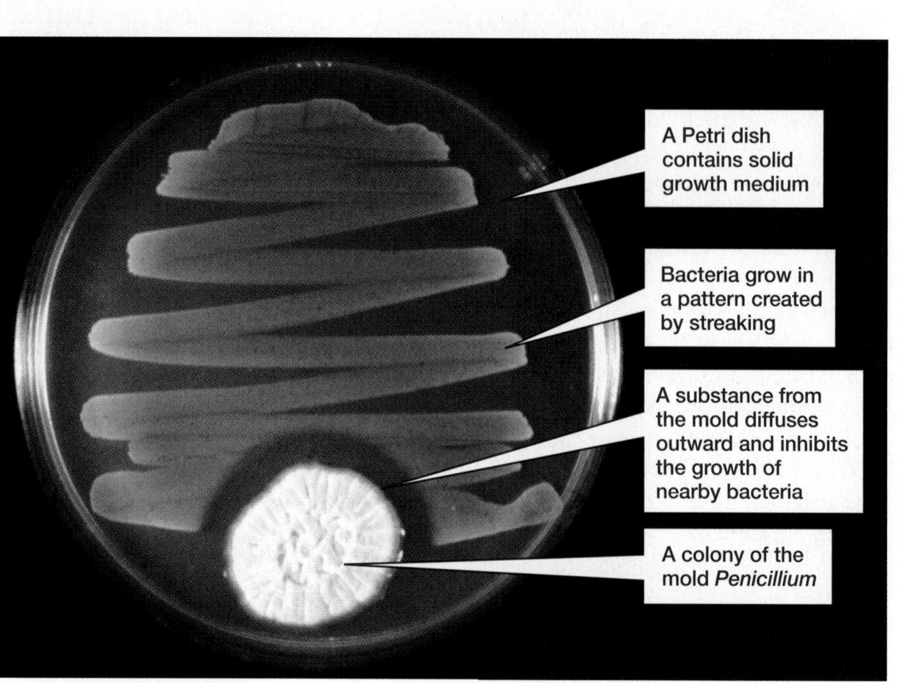
A Petri dish contains solid growth medium
Bacteria grow in a pattern created by streaking
A substance from the mold diffuses outward and inhibits the growth of nearby bacteria
A colony of the mold *Penicillium*

▶ **FIGURE 1-5 Penicillin kills bacteria** A fuzzy white colony of the mold *Penicillium* has secreted a substance (penicillin) that has inhibited the growth of the disease-causing bacteria *Staphylococcus aureus* (smeared back and forth across this plate of jelly-like growth medium). Both the mold and the bacteria are visible only when they grow at high densities.

QUESTION Why might some molds produce substances that are toxic to bacteria?

than a hypothesis. Far from being an educated guess, a scientific theory is a general explanation of important natural phenomena, developed through extensive and reproducible observations. In common English, it is more like a principle or a natural law. For example, scientific theories such as the atomic theory (that all matter is composed of atoms) and the theory of gravitation (that objects exert attraction for one another) are fundamental to the science of physics. Likewise, the cell theory (that all living things are composed of cells) and the theory of evolution (discussed in Section 1.2) are fundamental to the study of biology. Scientists describe fundamental principles as "theories" rather than "facts" because a basic premise of scientific inquiry is that it must be performed with an open mind. If compelling evidence arises, a scientific theory will be modified.

A modern example of the need to keep an open mind in the light of new scientific evidence is the discovery of prions, which are infectious proteins (see the case study for Chapter 3). Before the early 1980s, all known infectious disease agents possessed genetic material—either DNA or the related molecule, RNA. When neurologist Stanley Prusiner from the University of California at San Francisco published evidence in 1982 that scrapie (an infectious disease that causes the brains of sheep to degenerate) is actually caused and transmitted by a protein with no associated genetic material, his results were met with widespread disbelief. Prions have since been found to cause "mad cow disease" (also called BSE for "bovine spongiform encephalitis"), which has killed not only cattle but also about 200 people who ate infected beef. Prior to the discovery of prions, the concept of an infectious protein was unknown to science. But by being willing to modify accepted beliefs to accommodate new data, scientists maintained the integrity of the scientific process while expanding our understanding of how diseases can occur. For his pioneering work, Stanley Prusiner was awarded the Nobel Prize in Physiology or Medicine in 1997.

Science Is Based on Reasoning

Scientific theories arise through **inductive reasoning,** the process of creating a generalization as a result of making many observations that support it, and none that contradict it. Simplistically, the theory that Earth exerts gravitational forces on objects began from repeated observations of objects falling down toward Earth and from a complete lack of observations of objects "falling up." The cell theory arises from the observation that all organisms that possess the attributes of life are composed of one or more cells, and that nothing that is not composed of cells shares all these attributes.

Once a scientific theory has been formulated, it then can be used to support deductive reasoning. In science, **deductive reasoning** is the process of generating hypotheses about how a specific experiment or observation will turn out, based on a well-supported generalization such as a scientific theory. For example, based on the cell theory, if scientists discover a new organism that exhibits all the attributes of life, scientists can confidently deduce (hypothesize) that it will be composed of cells. Of course, the new organism must be carefully scrutinized under the microscope to determine its cellular structure; if compelling new evidence arises, a scientific theory can be modified.

Scientific Theories Are Formulated in Ways That Can Potentially Be Disproved

A major difference between a scientific theory and a belief based on faith is that a scientific theory can be disproved or falsified, whereas a faith-based assertion cannot. The potential to be falsified is why scientists continue to refer to basic precepts of science as "theories." To explain falsifiability, let's look at elves. The scientific approach to elves is that they do not exist because no objective evidence supports their existence. People with faith in elves might describe them as creatures so secretive that they can never be captured or otherwise detected, or they might claim that elves manifest themselves only to those who believe in them. The scientific theory that elves do not exist could be falsified if someone captured an elf or provided other repeatable, objective evidence of their existence. In contrast, faith-based assertions—for example, that elves exist or that each creature on Earth was separately created—are formulated in ways that can never be disproved or "falsified." For this reason, they are articles of faith rather than science.

1.2 EVOLUTION: THE UNIFYING THEORY OF BIOLOGY

In the words of biologist Theodosius Dobzhansky, "Nothing in biology makes sense, except in the light of evolution." Why don't snakes have legs? Why are there dinosaur fossils but no living dinosaurs? Why are monkeys so like us, not only in appearance but also in the structure of their genes and proteins? The answers to these questions, and thousands more, lie in the processes of evolution, which we will examine in detail in Unit 3.

Evolution, the process by which modern organisms descended, with modifications, from pre-existing forms of life, not only explains the origin of diverse forms of life, but accounts for the remarkable similarities among different organisms as well. Ever since the theory of evolution was formulated in the mid-1800s by two English naturalists, Charles Darwin and Alfred Russel Wallace, it has been supported by fossils, geological studies, radioactive dating of rocks, genetics, molecular biology, biochemistry, and breeding experiments. People who refer to evolution as "just a theory" profoundly misunderstand the concept of a scientific theory.

Three Natural Processes Underlie Evolution

The **scientific theory of evolution** states that modern organisms descended, with modification, from preexisting lifeforms. The most important force in evolution is **natural selection,** the process by which organisms with specific traits that help them cope with the rigors of their environment reproduce more successfully than do others that lack these traits. The changes that occur during evolution are a result of natural selection acting on the inherited variation that occurs among individuals in a population, causing changes in the population over successive generations. The variation upon which natural selection acts is a result of small differences in the genetic makeup of the individuals within the population.

Evolution arises as a consequence of three natural occurrences: genetic variation among members of a population owing to differences in their DNA, inheritance of those variations by the offspring of parents who carry the variation, and natural selection, the enhanced reproduction of organisms with variations that help them cope with their environment.

Genetic Variability Among Organisms Is Inherited

Look around at your classmates and notice how different they are, or go to an animal shelter and observe the differences among the dogs in size, shape, and coat color. Although some of this variation (particularly among your classmates) is due to differences in environment and lifestyle, much of it is influenced by genes. For example, most of us could pump iron for the rest of our lives and never develop a body like that of "Mr. Universe."

But what are genes? The hereditary information of all known forms of life is contained within a type of molecule called **deoxyribonucleic acid,** or **DNA** (**Fig. 1-6**). An organism's DNA, which is contained in **chromosomes** in each cell, is the cell's genetic "blueprint" or molecular "instruction manual," a guide to the construction and the operation of its body. Genes are segments of DNA; each gene directs the formation of one of the crucial molecular components of the organism's body. When an organism reproduces, it passes a copy of its chromosomes containing DNA to its offspring.

The accuracy of the DNA copying process is astonishingly high; in people only about 25 mistakes, called **mutations,** occur for every billion subunits of the DNA molecule that are copied. Mutations can also result from damage to DNA, for example, by ultraviolet light, radioactive particles, or toxic chemicals such as those in cigarette smoke. These occasional errors can alter the information in genes or alter the collections of genes within chromosomes. Some mutations are harmful. For example, mutations in skin cells caused by too much ultraviolet light or in lung cells exposed to poisons in cigarette smoke can cause cancer.

On rare occasions, a mutation will occur when a sperm or egg cell is being formed, allowing the change to be passed to the organism's offspring. As a result, each cell in the new individual's body will carry this hereditary mutation. Some of these mutations will prevent the new organism from developing, while other changes in genetic material cause diseases such as Down syndrome. Many mutations have no observable effect or may change the organism in a way that is harmless. On rare occasions, however, a mutation occurs that helps an organism survive and reproduce. Mutations that occurred hundreds of thousands of years ago have been passed from parent to offspring through countless generations, accounting for individual differences in height, body proportions, facial features, and the color of skin, hair, and eyes.

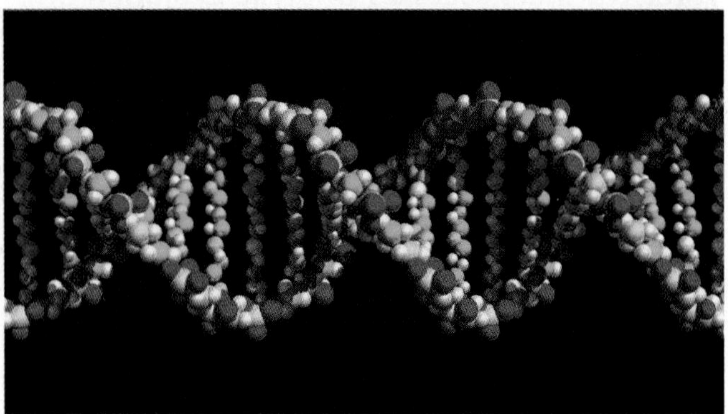

▲ **FIGURE 1-6 DNA** A model of DNA, the molecule of heredity. As James Watson, its codiscoverer, put it: "A structure this pretty just had to exist."

Natural Selection Tends to Preserve Genes That Help an Organism Survive and Reproduce

Organisms that best meet the challenges of their environment will usually leave the most offspring. These offspring will have inherited the genes that made their parents successful. Thus, natural selection preserves genes that help organisms flourish in their environment. For example, we can hypothesize that modern beavers have long front teeth because of an ancient mutation passed to offspring of a single pair of beavers that caused their offspring's teeth to grow larger than usual. Beavers with this mutation could chew down trees more efficiently, build bigger dams and lodges, and eat more bark than could "normal" beavers that lacked the mutation. Because these big-toothed beavers obtained more food and better shelter, they were able to raise more offspring who inherited their parents' genes for larger front teeth. Over time, less-successful, smaller-toothed beavers became increasingly scarce; after many generations, all beavers had large front teeth.

Structures, physiological processes, or behaviors that aid in survival and reproduction in a particular environment are called **adaptations.** Most of the features that we admire so much in our fellow life-forms, such as the long limbs of deer, the wings of eagles, and the mighty trunks of redwood trees, are adaptations that help them escape predators, capture prey, reach the sunlight, or accomplish other feats that ensure their survival and reproduction. These features were molded by millions of years of natural selection acting on random mutations.

Over time, the interplay of environment, genetic variation, and natural selection inevitably results in evolution: a change in the genetic makeup of a species. Evolution has been documented innumerable times both in laboratory settings and in the wild. For example, antibiotics act as agents of natural selection on bacterial populations, causing the evolution of antibiotic-resistant forms. Lawn mowers have caused changes in the genetic makeup of populations of dandelions, favoring those that produce flowers on very short stems. Scientists have documented the spontaneous emergence of entirely new species of plants due to mutations that alter their chromosome number, preventing them from interbreeding with their parent species.

What helps an organism survive today may become a liability tomorrow. If environments change—for example, as global warming occurs—the genetic makeup that best adapts organisms to their environments will also change over time. When random new mutations increase the fitness of an organism in the altered environment, these mutations will spread throughout the population. Populations within a species that live in different environments will be subjected to different types of natural selection. If the differences are great enough and continue for long enough, they may eventually cause the

▲ **FIGURE 1-7 A fossil *Triceratops*** This *Triceratops* died in what is now Montana about 70 million years ago. No one is certain what caused the extinction of the dinosaurs, but we do know that they were unable to evolve new adaptations rapidly enough to keep up with changes in their environment.

populations to become sufficiently different from one another to prevent interbreeding—a new species will have evolved.

If, however, favorable mutations do not occur, a changing environment may doom a species to extinction. Dinosaurs (**Fig. 1-7**) are extinct not because they were failures—after all, they flourished for 100 million years—but because they did not evolve rapidly enough to adapt to changing conditions. In recent decades, human activities have drastically accelerated the rate of environmental change. Adaptive mutations are relatively rare. Thus, many species are unable to adapt to rapid environmental change, and the rate of extinction has increased dramatically as a result. This concept is explored further in "Earth Watch: Why Preserve Biodiversity?"

1.3 WHAT ARE THE CHARACTERISTICS OF LIVING THINGS?

The word "biology" comes from the Greek root "bio" meaning "life" and "logy" meaning "the study of" (you will find many more word roots in Appendix I). But exactly what is life? If you look up "life" in a dictionary, you will find definitions such as "the quality that distinguishes a vital and functioning being from a dead body," but you won't find out what that "quality" is. Life emerges as a result of incredibly complex, ordered interactions among nonliving molecules. How did life originate? Although scientists have several hypotheses as to how life on Earth first arose (see Chapter 17), these are, of course, untestable. Life is an intangible quality that defies simple definition, even among biologists. However, most will agree that living things share certain characteristics that, taken together, are not shared by nonliving objects. Characteristics of life include the following:

- Living things are composed of cells that have a complex, organized structure.
- Living things actively maintain their complex structure and their internal environment, a process called homeostasis.
- Living things respond to stimuli from their environment.
- Living things acquire and use materials and energy from their environment and convert them into different forms.
- Living things grow.
- Living things reproduce themselves, using the molecular blueprint of DNA.
- Living things, as a whole, have the capacity to evolve.

Let's explore these characteristics in more detail.

Living Things Are Complex, Organized, and Composed of Cells

In Chapter 4 you will learn how researchers in the early 1800s, observing life with primitive microscopes, devised the **cell theory,** which states that the cell is the basic unit of life. Even a single cell has an elaborate internal structure (see Fig. 1-2). All cells contain **genes,** units of heredity that provide the information required to control the life of the cell. Cells also contain structures called **organelles** that are specialized to carry out specific functions such as moving the cell, obtaining energy, or synthesizing large molecules. Cells are always surrounded by a thin **plasma membrane** that encloses the **cytoplasm** (organelles and the fluid surrounding them) and separates the cell from the outside world. Some organisms, mostly invisible to the naked eye, consist of just one cell. Your body—and the bodies of organisms that are most familiar to us—is composed of many cells that are specialized and elaborately organized to perform specific functions. The water flea beautifully illustrates the complexity found in a multicellular form of life that is smaller than the letter "o" in this text (**Fig. 1-8**).

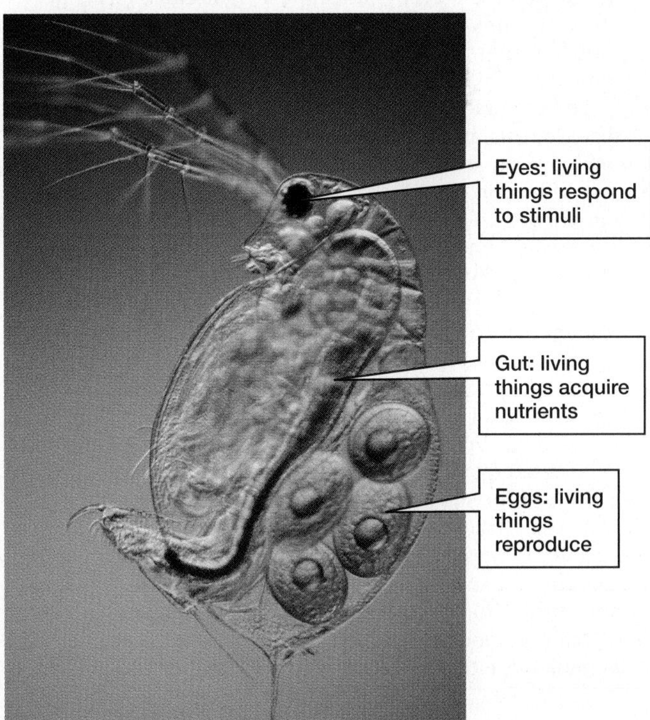
Eyes: living things respond to stimuli
Gut: living things acquire nutrients
Eggs: living things reproduce

▲ **FIGURE 1-8 Life is both complex and organized** The water flea, *Daphnia longispina*, is only 1 millimeter ($\frac{1}{1,000}$ meter) in length, yet it has legs, a mouth, a digestive tract, reproductive organs, light-sensing eyes, and a simple brain.

Earth Watch

Why Preserve Biodiversity?

"The loss of species is the folly our descendants are least likely to forgive us."

—E. O. Wilson,
Professor, Harvard University

What is biodiversity, and why should we be concerned with preserving it? **Biodiversity** refers to the total number of species within a given region. Over the 3.5-billion-year history of life on Earth, evolution has produced an estimated 8 to 10 million unique and irreplaceable species. Of these, scientists have named only about 1.4 million, and only a tiny fraction of this number have been studied. Evolution has not, however, merely churned out millions of independent species. Over thousands of years, organisms in a given area have been molded by forces of natural selection exerted by other living species as well as by the nonliving environment in which they live. The outcome is the community; a highly complex web of interdependent organisms whose interactions sustain one another. By participating in the natural cycling of water, oxygen, and other nutrients, and by producing rich soil and purifying wastes, these communities contribute to the sustenance of human life as well.

The tropics are home to the vast majority of all the species on Earth, perhaps 7 or 8 million of them, living in complex communities. The rapid destruction of habitats in the tropics—from rain forests to coral reefs—as a result of human activities is driving large numbers of species to extinction (**Fig. E1-3**). Most of these species have never been named, and others never even discovered. Aside from ethical concerns over eradicating irreplaceable forms of life, as we drive unknown organisms to extinction, we lose potential sources of medicine, food, and raw materials for industry.

For example, a wild relative of corn that is not only very disease resistant but also perennial (that is, lasts more than one growing season) was found growing only on a 25-acre plot of land in Mexico that was scheduled to be cut and burned within a week of the discovery. The genes of this plant might one day enhance the disease resistance of corn or create a perennial corn plant. The rosy periwinkle, a flowering plant found in the tropical forest of the island of Madagascar (off the eastern coast of Africa; see Fig. E1-3 inset) produces two substances that are now synthesized and marketed for the treatment of childhood leukemia and Hodgkin's disease, a cancer of the lymphatic organs. The periwinkle's wild habitat has been extensively deforested. Only about 3% of the world's flowering plants have been examined for substances that might fight cancer or other diseases. Closer to home, loggers of the Pacific Northwest frequently cut and burned the Pacific yew tree as a "nuisance species" until the active ingredient for the anticancer drug Taxol® was discovered in its bark.

Many ecologists are also concerned that as species are eliminated, either locally or through total extinction, the communities of which they were a part will change, becoming less stable and more vulnerable to damage by diseases or adverse environmental conditions. Some experimental evidence supports this viewpoint, but the interactions within communities are so complex that these hypotheses are difficult to test. Clearly, some species have a much larger role than others in preserving the stability of a given ecosystem. Which species are most crucial in each ecosystem? No one knows. Human activities have increased the natural rate of extinction by a factor of at least 100 and possibly by as much as 1,000 times the "pre-human" rate. By reducing biodiversity to support increasing numbers of people and wasteful standards of living, we have ignorantly embarked on an uncontrolled global experiment, using planet Earth as our laboratory. Stanford ecologists Paul and Anne Ehrlich compare the loss of biodiversity to the removal of rivets from the wing of an airplane. The rivet-removers continue to assume that there are far more rivets than needed, until one day, when the airplane takes off, they are proven tragically wrong. As human activities drive species to extinction with little knowledge of the role each plays in the complex web of life, we run the risk of removing "one rivet too many."

▲ **FIGURE E1-3 Biodiversity threatened** Destruction of tropical rain forests threatens Earth's greatest storehouse of biological diversity. (Inset) The rosy periwinkle's habitat has been dramatically reduced by logging in Madagascar.

Living Things Maintain Relatively Constant Internal Conditions Through Homeostasis

Complex, organized structures are not easy to maintain. Whether we consider the molecules of your body or the books and papers on your desk, organization tends to disintegrate into chaos unless energy is used to sustain it (we explore this concept further in Chapter 6). To stay alive and function effectively, organisms must keep the conditions within their bodies fairly constant; in other words, they must maintain **homeostasis** (derived from Greek words meaning "to stay the same"). For example, organisms must precisely regulate the amount of water and salts within their cells. Their bodies must also be maintained at appropriate temperatures for biological functions to occur. Among warm-blooded animals, vital organs such as the brain and heart are kept at a warm, constant temperature despite wide fluctuations in outside temperature. Homeostasis is maintained by a variety of mechanisms. In the case of temperature regulation, these include sweating during hot weather and exercise, dousing oneself with cool water (**Fig. 1-9**), metabolizing more food in cold weather, basking in the sun, or even adjusting a thermostat.

Of course, not everything stays the same throughout an organism's life. Major changes, such as growth and reproduction, occur; but these are not failures of homeostasis. Rather, they are specific, genetically programmed parts of the organism's life cycle.

Living Things Respond to Stimuli

To stay alive, reproduce, and maintain homeostasis, organisms must perceive and respond to stimuli in their internal and external environments. Animals have evolved elaborate sensory organs and muscular systems that allow them to detect and respond to light, sound, touch, chemicals, and many other stimuli from their surroundings. Internal stimuli are perceived by receptors for stretch, temperature, pain, and various chemicals. For example, when you feel hungry, you perceive contractions of your empty stomach and low levels of sugars and fats in your blood. You then respond to external stimuli by choosing appropriate objects to eat, such as a sandwich rather than a plate. Yet animals, with their elaborate nervous systems and motile bodies, are not the only organisms that perceive and respond to stimuli. The plants on your windowsill grow toward light, and even the bacteria in your intestines can move toward favorable conditions and away from harmful substances.

▲ **FIGURE 1-9 Living things maintain homeostasis** Evaporative cooling by water, both from sweat and from a bottle, helps Lance Armstrong (seven-time winner of the Tour de France bicycle race) maintain temperature homeostasis.

Living Things Acquire and Use Materials and Energy

Organisms need materials and energy to maintain their high level of complexity and organization, to grow, to maintain homeostasis, and to reproduce. Organisms acquire the materials they need, called **nutrients,** from air, water, or soil, or from other living things. Nutrients include minerals, oxygen, water, and all the other chemical building blocks that make up biological molecules. These nutrients are obtained from the environment, where they are continuously exchanged and recycled among living things and their nonliving surroundings (**Fig. 1-10**).

To sustain life, organisms must obtain **energy**—the ability to do work, such as carrying out chemical reactions, growing leaves in the spring, or contracting a muscle. Ultimately, the energy that sustains nearly all life comes from sunlight. Plants and some single-celled organisms capture the energy of sunlight directly and store it in energy-rich molecules, such as sugars, using a process called **photosynthesis.** Only photosynthetic organisms can capture the energy of sunlight. Organisms that cannot photosynthesize, such as animals and fungi, acquire their energy prepackaged in molecules of the bodies of other organisms. Thus, energy flows in a one-way path from the sun through nearly all forms of life. The energy is eventually released as heat, which cannot be used to power life (see Fig. 1-10).

Living Things Grow

At some time in its life cycle, every organism becomes larger—that is, it grows. Although this characteristic is obvious in most animals and plants, even single-celled bacteria grow to about double their original size before they divide. In all cases, growth involves the conversion of materials acquired from the environment into the specific molecules of the organism's body.

Living Things Reproduce Themselves

Organisms reproduce, giving rise to offspring of the same type and creating the continuity of life. The processes by which reproduction occurs differ widely among different

▶ **FIGURE 1-10 The flow of energy and the recycling of nutrients** Energy is acquired from sunlight (yellow arrow) by photosynthetic organisms, is transferred through organisms that consume other forms of life (red arrows), and is lost as heat (orange arrows) in a one-way flow. In contrast, nutrients (purple arrows) are recycled among organisms and the nonliving environment.

forms of life, but the result is the same—the perpetuation of the parents' genes.

Living Things, Collectively, Have the Capacity to Evolve

Populations of organisms evolve in response to changing environments. Although the genetic makeup of a single organism remains essentially the same over its lifetime, the genetic makeup of a population will change over time as a result of natural selection.

Case Study continued

Are Viruses Alive?

Viruses possess only two of the characteristics of life described here—they reproduce and they evolve.

1.4 HOW DO SCIENTISTS CATEGORIZE THE DIVERSITY OF LIFE?

Although all living things share the general characteristics discussed earlier, evolution has produced an amazing variety of life-forms. Scientists generally categorize organisms into three major groups, or **domains:** Bacteria, Archaea, and Eukarya (**Fig. 1-11**). This classification reflects fundamental differences among the cell types that compose them. Members of both the Bacteria and the Archaea usually consist of single, simple cells. Systematists (scientists who classify organisms based on their evolutionary relatedness) have not been able to agree about how to distinguish kingdoms within these two domains, in part because the cells of different organisms within each domain are so simple and have so much in common. In contrast, members of the Eukarya have bodies composed of one or more highly complex cells. This domain includes three major subdivisions or **kingdoms**—the Fungi, Plantae, and Animalia—as well as a diverse collection of mostly single-celled organisms collectively known as "protists" (see Fig. 1-11).

Categories within kingdoms are phylum, class, order, family, genus, and species. These groupings form a hierarchy in which each category includes all of those below it. Within the final category, the species, all members are so similar that they can interbreed. Biologists use a **binomial system** (from Latin, meaning "two names") for naming species. Why? The animal in Figure 1-8, for example, has the common name *water flea*, but because there are many types of water fleas, people who study them need to be more precise. Each type of organism is assigned a scientific name with two parts: its genus and its species. The genus name is always capitalized, and the species name is not; both are italicized. So *Daphnia longispina*, the water flea in Figure 1-8, is in the genus *Daphnia* (which includes many other water fleas) and the species *longispina* (which refers to a long spine that characterizes this particular type of water flea). People are classified as *Homo sapiens*; we are the only members of this genus and species.

Have you ever wondered 

Why Scientists Study Obscure Things—Like Water Fleas?

The more we learn about life, the more we realize how interdependent various forms of life are. *Daphnia* eat microscopic algae, and they in turn are eaten by juvenile fish, allowing the fish to survive, grow, and provide food for other fish, birds, mammals, and people. In addition, by consuming suspended microscopic algae, water fleas help keep lake water relatively clear; this allows sunlight to penetrate the water, allowing submerged plants to thrive. These plants provide food and habitat for a variety of aquatic insects that support adult fish populations. *Daphnia* are also easy to raise in the laboratory and sensitive to toxic chemicals, so scientists sometimes test water quality by determining whether *Daphnia* can thrive in it. The bottom line? You name it, and scientists have good reasons for studying it.

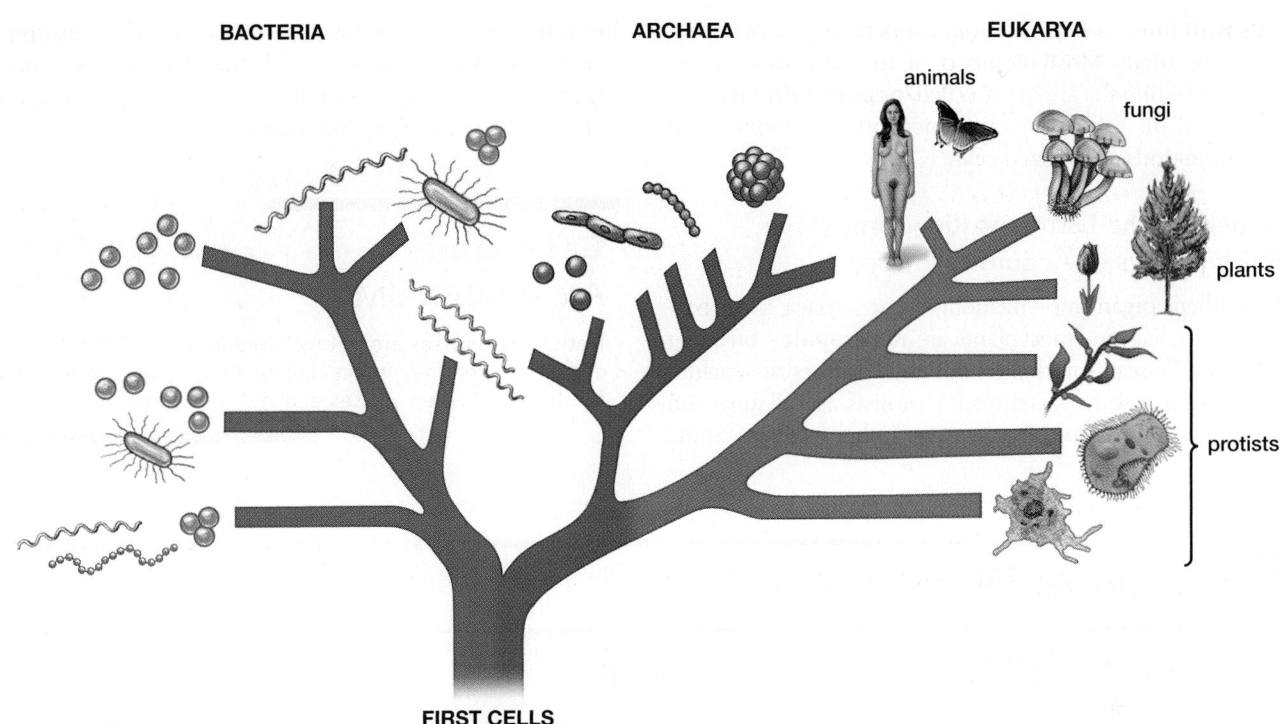

▲ **FIGURE 1-11 The domains and kingdoms of life**

This binomial system of naming organisms allows scientists worldwide to communicate very precisely about any given organism. In the following paragraphs, we provide a brief introduction to the domains and kingdoms of life. You will learn far more about life's incredible diversity and how it evolved in Unit 3.

There are exceptions to any simple set of criteria used to characterize the domains and kingdoms, but three characteristics are particularly useful: cell type, the number of cells in each organism, and how it acquires energy (**Table 1-1**).

The Domains Bacteria and Archaea Consist of Prokaryotic Cells; the Domain Eukarya Is Composed of Eukaryotic Cells

There are two fundamentally different types of cells: **prokaryotic** and **eukaryotic.** "Karyotic" refers to the **nucleus** of a cell, a membrane-enclosed sac containing the cell's genetic material (see Fig. 1-2). "Pro" means "before" in Greek; prokaryotic cells almost certainly evolved before eukaryotic cells (and, as we will see in Chapter 17, eukaryotic cells almost certainly evolved from prokaryotic cells). Prokaryotic cells do not have a nucleus; their genetic material resides in their cytoplasm. They are usually small—only 1 or 2 micrometers in diameter—and lack membrane-bound organelles. The domains Bacteria and Archaea consist of prokaryotic cells; the cells of Eukarya, however, as the name implies, are eukaryotic. "Eu" means "true" in Greek, and eukaryotic cells possess a "true" membrane-enclosed nucleus. Eukaryotic cells are generally larger than prokaryotic cells and contain a variety of other organelles.

Bacteria and Archaea Are Mostly Unicellular; Members of the Kingdoms Fungi, Plantae, and Animalia Are Nearly All Multicellular

Most members of the domains Bacteria and Archaea are single celled, or **unicellular,** although a few live in strands or mats

Table 1-1 Some Characteristics Used in Classification of Organisms

Domain	Kingdom	Cell Type	Cell Number	Energy Acquisition
Bacteria	(Under discussion)	Prokaryotic	Unicellular	Autotrophic or heterotrophic
Archaea	(Under discussion)	Prokaryotic	Unicellular	Autotrophic or heterotrophic
Eukarya	Fungi	Eukaryotic	Multicellular	Heterotrophic
	Plantae	Eukaryotic	Multicellular	Autotrophic
	Animalia	Eukaryotic	Multicellular	Heterotrophic
	"Protists"*	Eukaryotic	Uni- and multicellular	Autotrophic or heterotrophic

*The "protists" are a diverse collection of organisms that includes several kingdoms under discussion.

of cells with little communication, cooperation, or organization among them. Most members of the kingdoms Fungi, Plantae, and Animalia are many celled, or **multicellular;** their lives depend on intimate communication and cooperation among numerous specialized cells.

Members of the Different Kingdoms Have Different Ways of Acquiring Energy

Photosynthetic organisms—including some archaea, some bacteria, some protists, and most plants, are **autotrophic**—meaning "self-feeding." Organisms that cannot photosynthesize—including some archaea, some bacteria, some protists, and all fungi and animals—are **heterotrophic,** meaning "other feeding." Some heterotrophs, such as archaea, bacteria, and fungi, absorb individual food molecules from outside their bodies; others, including most animals, ingest chunks of food and break them down to molecules in their digestive tracts.

Case Study continued
Are Viruses Alive?

Notice that viruses are not included in our classification system of life. They infect cells of every known form of life, but how and when viruses evolved is unknown.

Links to *Everyday Life*

Knowledge of Biology Illuminates Life

Some people regard science as a "dehumanizing" activity, thinking that too deep an understanding of the world robs us of vision and awe. Nothing could be further from the truth, as we repeatedly discover in our own lives. For example, lupine flowers have two lower petals that form a tube enclosing both the male and the female reproductive parts (**Fig. E1-4**). In young lupine flowers, the weight of a bee on these petals forces pollen (carrying sperm) out of the tube and onto the bee's abdomen (see Fig. E1-4, inset). In older flowers ready to be fertilized, the female reproductive part protrudes through the lower petals. When a pollen-dusted bee visits, it usually leaves behind a few grains of pollen that contain the lupine's sperm.

Do these insights into lupine flowers detract from our appreciation of them? Far from it. We look on lupines with greater delight, because we understand some of the interplay of form and function, bee and flower, that shaped the evolution of the flower. Several years ago, as we crouched beside a wild lupine in Olympic National Park in Washington State, an elderly man stopped to ask what we were looking at so intently. He listened with interest as we explained the structure to him; he then went off to another patch of lupines to watch the bees foraging. He too felt the increased sense of wonder that comes with understanding.

Throughout this text, we try to convey to you that sense of understanding and wonder, and we hope that you, too, will experience it. We also emphasize that biology is not a completed work but an exploration that has really just begun. We cannot urge you strongly enough, even if you are not contemplating a career in biology, to join in the journey of biological discoveries throughout your life. Don't think of biology as just another course to take, just another set of facts to memorize. Biology is a pathway to a new understanding of yourself and of the life on Earth around you.

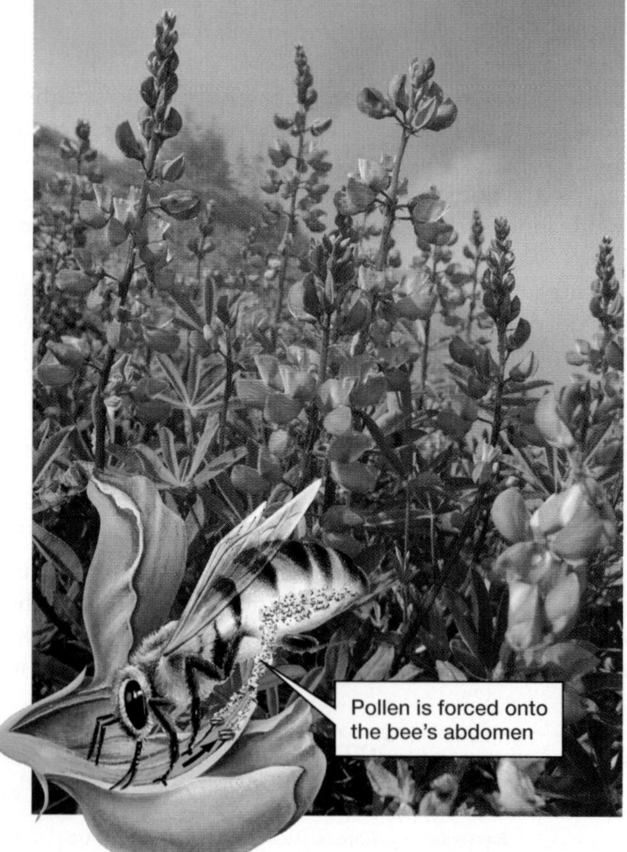
Pollen is forced onto the bee's abdomen

▲ **FIGURE E1-4 Wild lupines** Thousands of people visit Hurricane Ridge in Washington State's Olympic National Park each summer to gaze in awe at Mt. Olympus, but few bother to investigate the small wonders at their feet. (Inset) A lupine flower deposits pollen on a foraging bee's abdomen.

Case Study revisited
Are Viruses Alive?

Researchers got their first blurry glimpse of viruses in the mid-1930s, through the newly invented electron microscope. They discovered that the simplest viruses contain genetic material (DNA or RNA) surrounded by a coat made of proteins. This knowledge provided the first indication that viruses are not composed of cells. Many viruses, including Ebola and herpes, are somewhat more complex, with an outer envelope derived partly from the membrane of the cells they infect (**Fig. 1-12**). Protein spikes protrude from the viral surface and help them attach to and infect the cells of their victims. But this additional complexity still doesn't approach that of an actual cell.

Viruses seem to be inert packages of complex biological chemicals—that is, until they infect a living cell. After penetrating the host cell, the virus hijacks its metabolic processes, using the cell's energy, organelles, and raw materials to produce new viruses. The newly formed viruses then emerge from the host cell, often rupturing it and sometimes acquiring an envelope made of the host's plasma membrane. Errors in copying viral genetic material, or accidental recombination of viral genetic material with genetic material from different viruses or from the host cell itself, produce mutations that allow viruses to evolve. Through evolution, viruses sometimes become more infectious or more deadly, or they may gain the ability to infect new hosts.

The uncomplicated structure of viruses, coupled with amazing advances in biotechnology, has now allowed researchers to synthesize viruses in the laboratory using materials available from chemical supply companies. The first virus to be synthesized was the small, relatively simple poliovirus. This feat was accomplished by Eckard Wimmer and coworkers at the University of New York at Stony Brook in 2002. Did these researchers create life in the laboratory? A few scientists would say "yes"—defining life by its ability to reproduce and to evolve. The more stringent definition of life provided in this chapter explains why most scientists agree that viruses aren't exactly a "life-form": They are not composed of cells, they do not obtain or use their own energy or nutrients, they do nothing to maintain homeostasis, they do not respond to stimuli, nor do they grow. As virologist Luis Villarreal of the University of California at Irvine states, "Viruses are parasites that skirt the boundaries between life and inert matter."

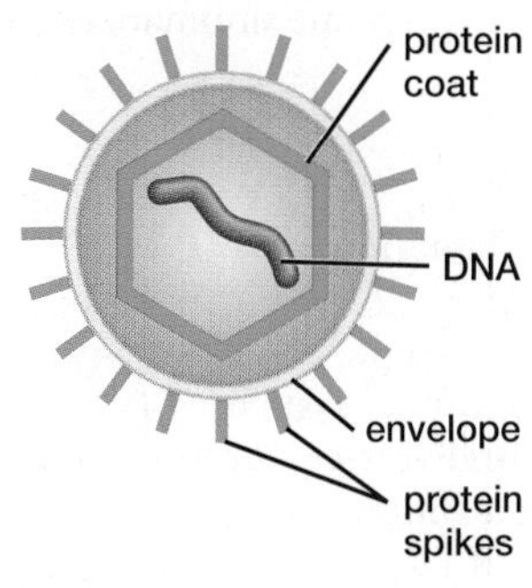

▲ **FIGURE 1-12 A herpes virus**

BioEthics Consider This

Wimmer and coworkers' announcement that they had synthesized the poliovirus created considerable controversy both within and outside of the scientific community. People raised concerns that this research jeopardized the eradication of polio, which has nearly been eliminated by widespread vaccination. They feared that deadly and highly contagious viruses such as Ebola could now be synthesized by bioterrorists. The researchers responded that they were merely applying already established knowledge and techniques to demonstrate the principle that viruses are basically chemical entities that can be synthesized in the laboratory. Do you think scientists should synthesize viruses? What are the implications of forbidding such research?

CHAPTER REVIEW

Summary of Key Concepts

1.1 How Do Scientists Study Life?
Scientists identify a hierarchy of levels of organizations, as illustrated in Figure 1-1. Biology is based on the scientific principles of natural causality, uniformity in space and time, and common perception. Knowledge in biology is acquired through application of the scientific method, which starts with an observation leading to a question, which leads to a hypothesis. The hypothesis leads to a prediction that is tested by controlled experiments. The experimental results, which must be repeatable, either support or refute the hypothesis, leading to a conclusion about the validity of the hypothesis. A scientific theory is a general explanation of natural phenomena, developed through extensive and reproducible experiments and observations.

1.2 Evolution: The Unifying Theory of Biology
Evolution is the scientific theory that modern organisms descended, with modification, from preexisting life-forms. Evolution occurs as a consequence of genetic differences—arising from mutations—among members of a population, inheritance of those variations by offspring, and natural selection of the variations that best adapt an organism to its environment.

1.3 What Are the Characteristics of Living Things?
Organisms possess the following characteristics: Their structure is complex and organized; they maintain homeostasis; they respond to stimuli; they acquire energy and materials from the environment; they grow; they reproduce; and they have the capacity to evolve. Photosynthetic organisms capture and store the energy of sunlight in energy-rich molecules. Most obtain all of their nutrients from their nonliving environment. Organisms that cannot photosynthesize, and therefore eat other organisms, obtain all of their energy and most of their nutrients from the bodies of the organisms that they eat.

1.4 How Do Scientists Categorize the Diversity of Life?
Organisms can be grouped into three major categories, called domains: Archaea, Bacteria, and Eukarya. Within the Eukarya are three kingdoms, Fungi, Plantae, and Animalia, and unicellular eukaryotes known collectively as "protists." Features used to classify organisms include the cell type (eukaryotic or prokaryotic), cell number (unicellular or multicellular), and energy acquisition (autotrophic or heterotrophic). The genetic material of eukaryotic cells is enclosed within a membrane-bound nucleus. Prokaryotic cells do not have a nucleus. Food obtained by heterotrophic organisms is either ingested in chunks or absorbed molecule by

molecule from the environment. The features of the domains and kingdoms are summarized in Table 1-1.

Key Terms

Thinking Through the Concepts

Fill-in-the-Blank

1. The smallest particle of an element that retains all the properties of that element is a(n) __________. The smallest unit of life is the __________. Most molecules found in cells have frameworks of __________ atoms, and are called __________. Cells of a specific type within multicellular organisms combine to form __________.
2. A(n) __________ is a general explanation of natural phenomena supported by extensive, reproducible tests and observations. In contrast, a(n) __________ is a proposed explanation for observed events. To answer specific questions about life, biologists use a general process called the __________.
3. An important scientific theory that explains why organisms are at once so similar and also so diverse is the theory of __________. This theory explains life's diversity as having originated primarily through the process of __________.
4. The molecule that guides the construction and operation of an organism's body is called (complete term) __________, abbreviated as __________. This large molecule contains discrete segments with specific instructions; these segments are called __________.
5. Fill in the following about living things: Living things maintain a relatively stable internal environment by the process of __________; living things respond to __________; living things acquire and use __________ and __________ from the environment. Living things are composed of cells whose structure is both __________ and __________.

Review Questions

1. List the hierarchy of organization of life from an atom to a multicellular organism, briefly explaining each level.
2. What is the difference between a scientific theory and a hypothesis? Explain how scientists use each operation. Why do scientists refer to basic principles as "theories," not "facts"?
3. Explain the differences between inductive and deductive reasoning, and provide an example, real or hypothetical, of each.
4. Describe the scientific method. In what ways do you use the scientific method in everyday life?
5. What are the differences between a salt crystal and a tree? Which is living? How do you know?
6. Define and explain the terms *natural selection, evolution, mutation, creationism,* and *population.*
7. What is evolution? Briefly describe how evolution occurs.
8. Define *homeostasis.* Why must organisms continuously acquire energy and materials from the external environment to maintain homeostasis?

Applying the Concepts

1. Review the properties of life, and then discuss whether humans are unique.
2. Review Alexander Fleming's experiment that led to the discovery of penicillin. Devise an appropriate control for the experiment in which he applied filtered medium from a *Penicillium* culture to plates of bacteria.
3. Science is based on principles, including uniformity in space and time and common perception. Assume that humans encounter intelligent beings from a planet in another galaxy where they evolved under very different conditions. Discuss the two principles just mentioned, and explain how they would affect the nature of scientific observations on the different planets as well as communications about these observations.
4. Identify two different types of organisms that you have seen interacting, for example, a caterpillar on a plant such as a milkweed, or a beetle in a flower. Now, form a single, simple hypothesis about this interaction. Use the scientific method and your imagination to design an experiment that tests this hypothesis. Be sure to identify variables and control for them.
5. Explain an instance in which your understanding of a phenomenon enhances your appreciation of it.

MB *Go to www.masteringbiology.com for practice quizzes, activities, eText, videos, current events, and more.*

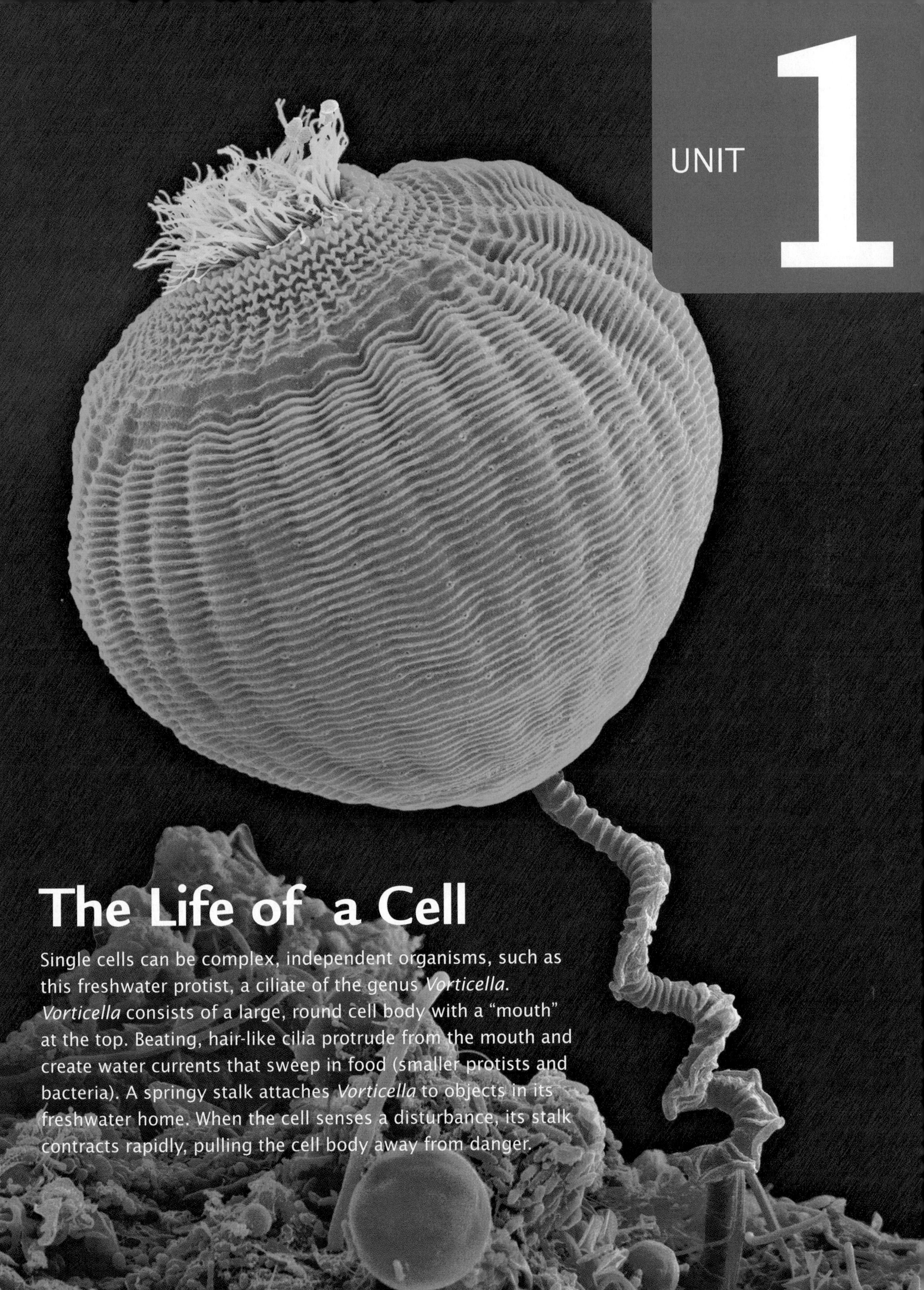

UNIT 1

The Life of a Cell

Single cells can be complex, independent organisms, such as this freshwater protist, a ciliate of the genus *Vorticella*. *Vorticella* consists of a large, round cell body with a "mouth" at the top. Beating, hair-like cilia protrude from the mouth and create water currents that sweep in food (smaller protists and bacteria). A springy stalk attaches *Vorticella* to objects in its freshwater home. When the cell senses a disturbance, its stalk contracts rapidly, pulling the cell body away from danger.

Chapter 2

Atoms, Molecules, and Life

▲The *Endurance* trapped in ice.

Case Study

Crushed by Ice

IT TOOK JUST 5 MINUTES for the ship *Endurance* to disappear into the ocean depths, but the captain did not go down with his ship, nor was his crew in danger of drowning. They watched—miserable but safe—from the surrounding ice, where they had been camped for 3 weeks.

Captained by the Irish explorer and adventurer Sir Ernest Shackleton, the *Endurance* and her crew of scientists and seasoned sailors had departed from Plymouth, England, in August 1914. Their goal: to traverse Antarctica by dogsled via the South Pole. Five months later, as the ship crossed the Weddell Sea toward its landing site, unseasonably cold weather froze the waters around it, literally trapping the ship in the ice, and slowly crushing it. The crew spent sleepless nights onboard as the wood around their bunks contorted—groaning, cracking, and sometimes splintering. When the pressure threatened to split open the decks beneath the sailors' feet, Shackleton ordered the crew to evacuate, using dogsleds to haul their supplies to the ice, where they set up camp.

On November 21, 1915, the *Endurance* shifted, its stern rising skyward as it slid beneath the ice. Shackleton and crew watched in a kind of despairing awe from a safe distance. They were to endure three more months of ceaseless cold on their frozen camp, waiting for the ice to release its grip and the sea to become passable for the smaller boats they had rescued from the doomed ship.

The *Endurance* was trapped and ultimately sunk as it was crushed by expanding ice floating on the Antarctic sea. Most liquids, including water, contract and become denser as they cool. But as water freezes, it undergoes an unusual transformation that produces a solid that is less dense and, therefore, floats on the remaining liquid. How does this happen? Why was the *Endurance* crushed as ice froze around it? How do the unique properties of water influence life on Earth?

At a Glance

2.1 WHAT ARE ATOMS?

If you cut a diamond (a form of carbon) into pieces, each piece would still be carbon. If you could make finer and finer divisions of these pieces, you would eventually produce a pile of carbon atoms, each so small that 100 million of them placed in a row would span only about 0.4 inch (1 centimeter).

Atoms, the Basic Structural Units of Elements, Are Composed of Still Smaller Particles

Carbon is an example of an **element**—a substance that can neither be separated into simpler substances, nor converted into other substances by ordinary chemical means. (You won't be able to modify elements in your college chemistry lab!) Elements, both alone and in combination, form all matter. **Atoms** are the fundamental structural units of elements, and each atom retains all the properties of its element. Atoms, in turn, are composed of a central atomic nucleus with electrons orbiting around it (**Fig. 2-1**). The **atomic nucleus** (often called simply the "nucleus") contains two types of subatomic particles: positively charged **protons** and uncharged **neutrons.** The orbiting **electrons** are far lighter, negatively charged particles. Atoms generally have an equal number of electrons and protons, making them electrically neutral.

Ninety-two types of atoms occur naturally, each forming the structural unit of a different element. The number of protons in the nucleus—called the **atomic number**—is the defining characteristic of each element. For example, every hydrogen atom has one proton in its nucleus; every carbon atom has six protons; every oxygen atom has eight; these atoms therefore have atomic numbers of 1, 6, and 8, respectively. Balancing the positive charge of the protons is an equal number of electrons. Each element has unique chemical and physical properties based on the number and configuration of its subatomic particles. For example, some elements, such as oxygen and hydrogen, are gases at room temperature; others, such as lead, are extremely dense solids. Some react readily with other atoms; others remain inert. Most elements are present in only small quantities in the biosphere, and relatively few are essential to life on Earth. The periodic table (Appendix II) organizes the elements according to their atomic number (rows) and their chemical properties (columns). **Table 2-1** lists the most common elements in living things. Four elements—oxygen, carbon, hydrogen, and nitrogen—make up about 96% of the weight of the human body, and a similar percentage of the bodies of most other organisms as well. For this reason, these elements are sometimes called the "building blocks" of life on Earth.

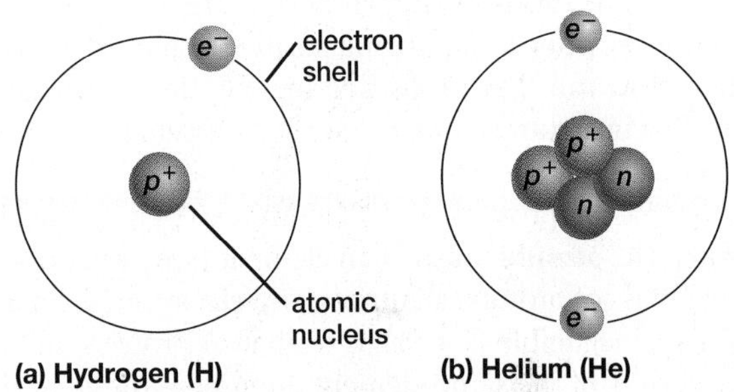

▲ **FIGURE 2-1 Atomic models** Structural representations of the two smallest atoms, **(a)** hydrogen and **(b)** helium. In these simplified models, the electrons (pale blue) are represented as miniature planets, circling in specific orbits around a nucleus that contains protons (gold) and neutrons (dark blue).

Table 2-1 Common Elements in Living Organisms

Element	Atomic Number[1]	Atomic Mass[2]	% by Weight in the Human Body[3]
Oxygen (O)	8	16	65
Carbon (C)	6	12	18.5
Hydrogen (H)	1	1	9.5
Nitrogen (N)	7	14	3.0
Calcium (Ca)	20	40	1.5
Phosphorus (P)	15	31	1.0
Potassium (K)	19	39	0.35
Sulfur (S)	16	32	0.25
Sodium (Na)	11	23	0.15
Chlorine (Cl)	17	35	0.15
Magnesium (Mg)	12	24	0.05
Iron (Fe)	26	56	Trace
Fluorine (F)	9	19	Trace
Zinc (Zn)	30	65	Trace

[1]Atomic number: number of protons in the atomic nucleus.

[2]Atomic mass: total mass of protons, neutrons, and electrons (negligible); numbers are rounded.

[3]Approximate percentage of this element, by weight, in the human body.

All atoms except hydrogen have more than one proton in their nuclei, but protons, being positively charged, repel one another. How can they be packed together in a nucleus? The answer lies in neutrons, which act as "peacemakers" among the protons, allowing them to exist side by side. Neutrons have the same mass as protons, and all atoms except hydrogen have neutrons. The **atomic mass** of an element (see Appendix II) is the total mass of protons, neutrons, and electrons (the mass of electrons is negligible compared to that of protons and neutrons). Atoms of the same element often have different numbers of neutrons; when this occurs, the atoms are called **isotopes** of each other. The atomic mass of such elements is an average number that takes into account the relative abundance of that element's isotopes. Most isotopes are stable, but some are **radioactive;** that is, they spontaneously break apart, forming different atoms and releasing energy in the process. Radioactive isotopes are extremely useful tools for studying biological processes (see "Scientific Inquiry: Radioactivity in Research").

Electrons Travel Within Specific Regions Called Electron Shells That Correspond to Different Energy Levels

As you may know from experimenting with magnets, like poles repel while opposite poles attract one another. In a similar way, electrons, which are negatively charged, repel one another but are attracted to the positively charged protons of the nucleus. An atom with many protons in its nucleus requires many electrons to balance these protons. These electrons orbit the atomic nucleus within restricted three-dimensional spaces called **electron shells.** For simplicity, we have drawn these shells as increasingly large rings around the nucleus (**Fig. 2-2**). The farther away an electron shell is from the nucleus, the higher the energy of the electrons that occupy it.

The electrons in an atom fill the shell closest to the nucleus first, and then begin to occupy higher-level shells. The electron shell closest to the atomic nucleus can only contain two electrons, and these electrons carry the least possible energy. This first shell is the only shell in hydrogen and helium atoms (see Fig. 2-1). The second shell, corresponding to a higher energy level, can hold up to eight electrons. Thus, a carbon atom with six electrons has two electrons in the first shell and four electrons in its second shell (see Fig. 2-2). Notice that the electrons prefer to orbit singly in each of four regions as they fill the outer shells; as additional electrons are added to balance the positive charges of larger nuclei, the electrons begin to pair up.

Nuclei and electron shells play complementary roles in atoms. Nuclei (assuming they are not radioactive) provide

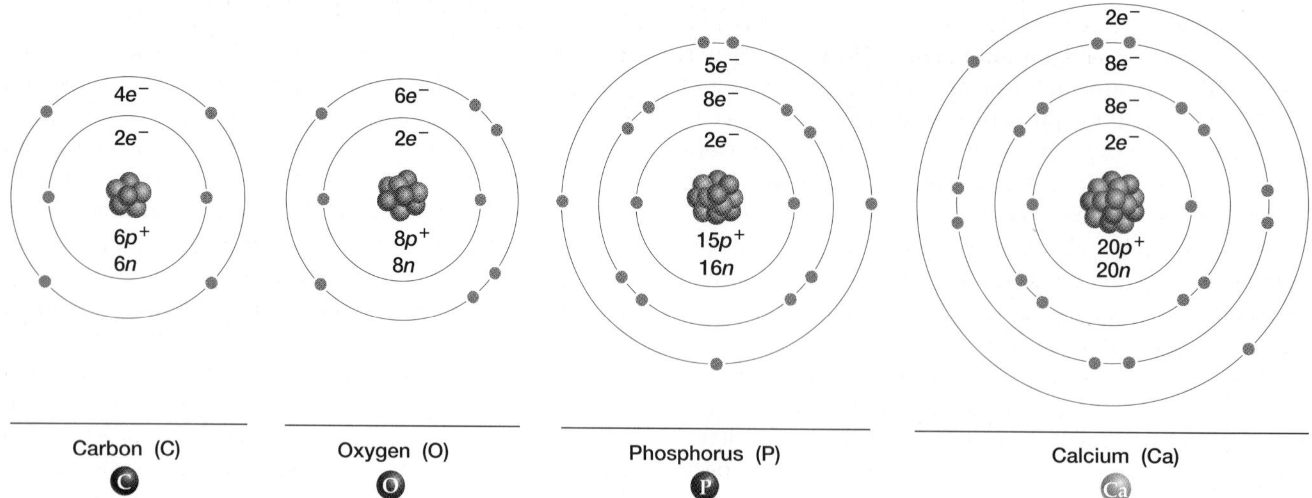

▲ **FIGURE 2-2 Electron shells in atoms** Most biologically important atoms have at least two shells of electrons. The first shell, closest to the nucleus, can hold two electrons; the next shell contains a maximum of eight electrons. More-distant shells can hold larger numbers of electrons.

QUESTION Why do atoms that tend to react with other atoms have outer shells that are not full?

Scientific Inquiry

Radioactivity in Research

How do biologists know that DNA is the genetic material of cells? How do paleontologists measure the ages of fossils? How do botanists know how plants transport sugars made in their leaves during photosynthesis to other parts of the plant? These discoveries, and many more, were made possible only through the use of radioactive isotopes. During radioactive decay, the process by which a radioactive isotope spontaneously breaks apart, an isotope emits particles that can be detected with sensitive electronic devices. A particularly fascinating and medically important use of radioactive isotopes is in positron emission tomography, or PET, scans (**Fig. E2-1**). In one common application of PET scans, a subject is given glucose sugar with a radioactive isotope of fluorine attached. As the isotope decays, it emits two bursts of energy that travel in opposite directions. Detectors in a ring around the subject's head capture the emissions, recording the slightly different arrival times of the two energy bursts from each decaying particle.

A powerful computer then calculates the location within the brain where the decay occurred and generates a color-coded map of the frequency of decays within a given "slice" of the brain. The more active a brain region is, the more glucose it uses as an energy source, and the more radioactivity is concentrated there. For example, tumor cells grow and divide rapidly, using large amounts of glucose, so they show up in PET scans as "hot spots" (**Fig. E2-1c**). Normal brain regions activated by a specific mental task (for example, a math problem) will also have higher glucose demands that can be detected by PET scans. Thus, physicians can use PET to locate brain disorders, and researchers can use it to study what parts of the brain are activated by different mental processes.

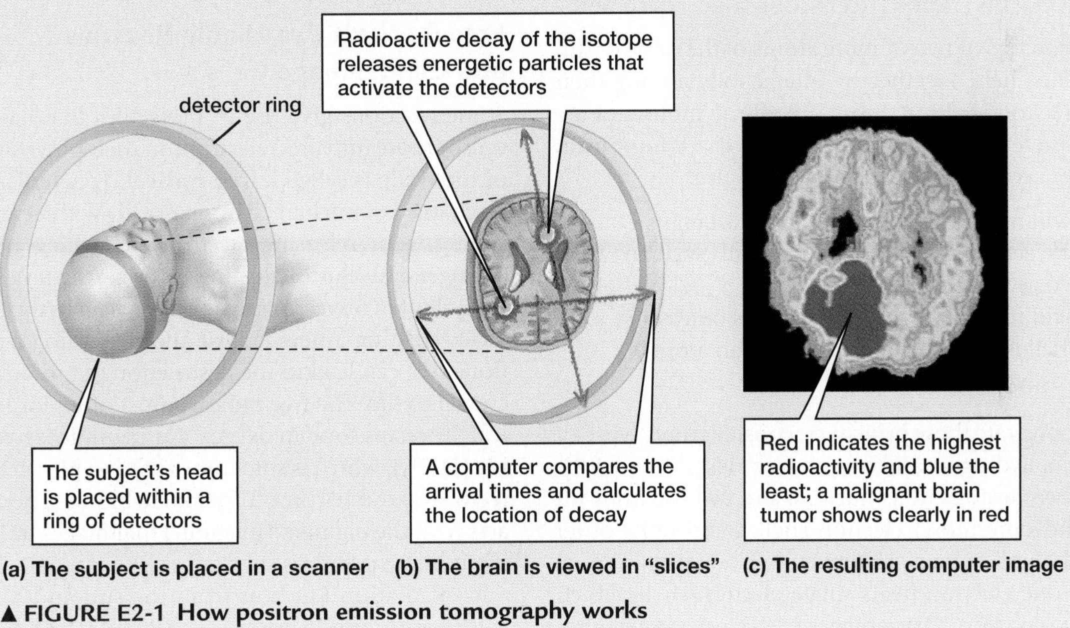
Radioactive decay of the isotope releases energetic particles that activate the detectors
detector ring
The subject's head is placed within a ring of detectors
A computer compares the arrival times and calculates the location of decay
Red indicates the highest radioactivity and blue the least; a malignant brain tumor shows clearly in red
(a) The subject is placed in a scanner
(b) The brain is viewed in "slices"
(c) The resulting computer image

▲ **FIGURE E2-1 How positron emission tomography works**

stability, whereas the electron shells allow attractive interactions, called **chemical bonds,** with other atoms. Nuclei resist disturbance by outside forces; ordinary sources of energy—such as heat, electricity, and light—hardly affect them at all. Because its nucleus is stable, a carbon atom remains carbon whether it is part of a diamond, carbon dioxide, or sugar. Electron shells, however, are dynamic, capable of capturing and releasing energy and forming chemical bonds.

Life Depends on the Ability of Electrons to Capture and Release Energy

Because electron shells correspond to energy levels, when an atom is excited by energy, such as heat or light, this energy causes an electron to jump from a lower-energy electron shell to a higher-energy shell. Soon afterward, the electron spontaneously falls back into its original electron shell, releasing the energy, often as light (**Fig. 2-3**).

We make use of this property of electrons every day. When we switch on an incandescent light bulb, electricity flowing through the filament in the bulb heats it, and the heat energy bumps electrons in the metal filament into higher-energy electron shells. As the electrons drop back down into their original shells, they emit their captured energy as light. Life itself depends on the ability of electrons to capture and release energy, as you will learn in Chapters 7 and 8, when we discuss photosynthesis and cellular respiration.

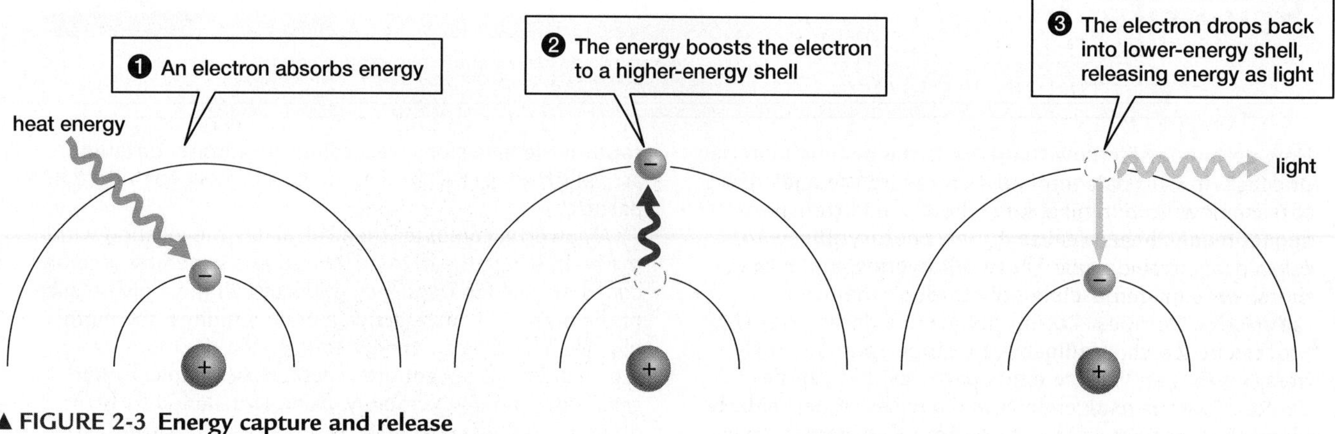

❶ An electron absorbs energy
❷ The energy boosts the electron to a higher-energy shell
❸ The electron drops back into lower-energy shell, releasing energy as light
heat energy
light

▲ **FIGURE 2-3 Energy capture and release**

2.2 HOW DO ATOMS INTERACT TO FORM MOLECULES?

Atoms Interact with Other Atoms When There Are Vacancies in Their Outermost Electron Shells

A **molecule** consists of two or more atoms of the same or different elements, held together by interactions among their outermost electron shells. A substance whose molecules are formed of different types of atoms is called a **compound.** Atoms behave according to two basic principles:

- An atom will not react with other atoms when its outermost electron shell is completely full. Such an atom is described as being inert.
- An atom will react with other atoms if its outermost electron shell is only partially full. Such an atom is described as being reactive.

To demonstrate these principles, consider three types of atoms: helium, hydrogen, and oxygen (see Figs. 2-1 and 2-2). Helium has two protons in its nucleus, and two electrons fill its single (and outermost) electron shell. Because its outer shell is full, helium is inert. Hydrogen has one proton in its nucleus and one electron in its single electron shell, which can hold two electrons. Oxygen has six electrons in its outer shell, which can hold eight. Hydrogen and oxygen atoms, both with partially empty outer shells, are reactive. One way for these atoms to gain stability is by reacting with each other. The single electrons from each of two hydrogen atoms can fill the outer shell of an oxygen atom, forming water (H_2O; see Fig. 2-6b later in the chapter). In fact, hydrogen reacts so vigorously with oxygen that the space shuttle and other rockets use liquid hydrogen as fuel to power liftoff. (See "Links to Everyday Life: Hmm . . . What Gas Should I Use to Inflate My Blimp—Or Fill My Car?")

An atom with an outermost electron shell that is partially full can gain stability by losing electrons (emptying the shell completely), gaining electrons (filling the shell), or sharing electrons with another atom (allowing both atoms to behave as though they had full outer shells). In many instances, reactions that fill or empty outer electron shells produce stable chemical bonds (described later). However, certain reactions that occur routinely in the body generate molecules called free radicals that are highly unstable and reactive.

Free Radicals Are Highly Reactive and Can Damage Cells

Some reactions give rise to molecules that have atoms with one or more unpaired electrons in their outer shells. This type of molecule is called a **free radical.** Free radicals, which can be charged or uncharged, react readily with nearby molecules, capturing or releasing electrons to achieve a more stable arrangement. But in stealing or donating an electron, a free radical often leaves an unpaired electron in the molecule it attacks, creating a new radical and beginning a series of reactions that can lead to the destruction of biological molecules crucial to life. The free radicals most often formed by biological processes contain oxygen. An example is hydrogen peroxide (H_2O_2), which is sometimes used to bleach hair or whiten teeth. You can become a "peroxide blonde" because H_2O_2 reacts with the pigment (melanin) that makes hair dark, changing it into a colorless substance.

Radiation (such as from the sun and X-rays), chemicals in automobile exhaust and cigarette smoke, and industrial metals such as mercury and lead can enter our bodies and generate free radicals. Most free radicals, however, are generated by reactions that are essential to life because they supply cells with energy, using oxygen in the process. Cell death caused by free radicals contributes to a variety of human ailments, including heart disease and nervous system disorders such as Alzheimer's disease. By damaging genetic material, free radicals can also cause some forms of cancer. The gradual deterioration of the body that accompanies aging is believed by many scientists to result, at least in part, from free-radical damage accumulating over a lifetime of exposure (**Fig. 2-4**). Fortunately molecules called **antioxidants** react with free radicals and render them harmless. Our bodies synthesize several antioxidants, and others can be obtained from a healthy diet. Vitamins E and C are antioxidants, as are a variety of other substances found in fruits and vegetables. To learn about another

Links to *Everyday Life*

Hmm . . . What Gas Should I Use to Inflate My Blimp—Or Fill My Car?

On May 6, 1937, the luxury passenger blimp *Hindenburg* was descending to land in Lakehurst, New Jersey, having departed from Germany three days earlier. As its landing lines were dropped, the crowd of journalists, photographers, and a radio announcer gathered to chronicle its landing watched in horror as flames suddenly appeared. Within 40 seconds, the craft was fully consumed by fire (**Fig. E2-2**), and thirty six people were killed.

The *Hindenburg* was filled with hydrogen gas (H_2). Because of their half empty outer electron shells, the hydrogen atoms in H_2 combine vigorously with oxygen in the atmosphere; in other words, hydrogen burns. The highly combustible nature of hydrogen makes it a dangerous gas to use in blimps, but gives it the potential to replace gasoline as a fuel for vehicles. Burning hydrogen produces only water as a reaction product. Gasoline, however, consists of chains of carbons bonded to hydrogens, and the burning of gasoline not only combines oxygen and hydrogen atoms to produce water, it also combines oxygen with carbon atoms. This generates carbon dioxide gas, now recognized as a major contributor to global warming.

Should cars be fueled with hydrogen? If fossil fuels are used in the energy-intensive stages of generating, compressing, transporting, storing, and delivering hydrogen to cars, this could completely negate the reduction in carbon dioxide emissions that should be an advantage of burning hydrogen. But if solar, wind, or other renewable energy sources could be used to produce and deliver hydrogen, it might become a sustainable and environmentally-friendly replacement for gasoline.

▲ **FIGURE E2-2 The hydrogen-filled *Hindenburg* goes up in flames**

QUESTION Filling the *Hindenburg* with hydrogen rather than helium gave it some extra lift—explain why.

Incidentally, you need not worry that the blimps floating above stadiums at sporting events will burst into flames, because modern blimps are filled with helium. With its full outer electron shell, helium is nonreactive and cannot burn.

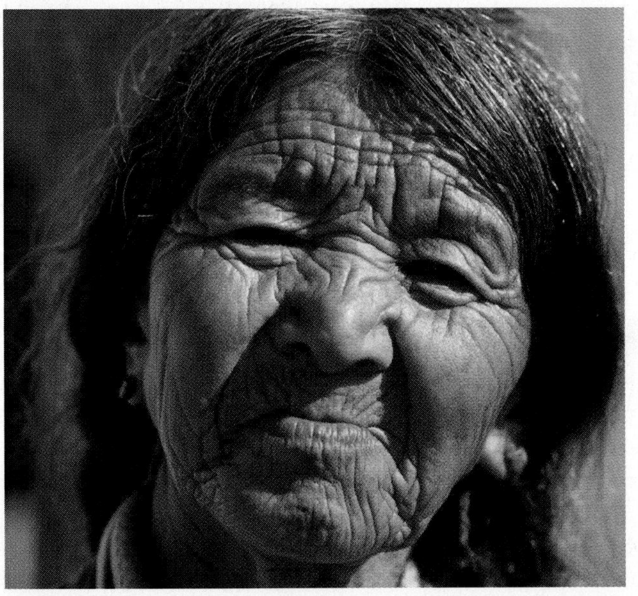

▲ **FIGURE 2-4 Free-radical damage** Aging is partly a result of accumulating free-radical damage to the biological molecules that compose our bodies. For example, sunlight can generate free radicals in skin, damaging molecules that give skin its elasticity and contributing to wrinkles as we age.

QUESTION How do free radicals damage biological molecules?

source of antioxidants, see "Health Watch: Might Chocolate Actually Be Good for You?"on p. 26.

Chemical Bonds Hold Atoms Together in Molecules

Chemical bonds, the attractive forces that hold atoms together in molecules, arise from interactions between adjacent atoms as they gain, lose, or share electrons. The atoms of each element have chemical bonding properties that arise from the configuration of electrons in their outer shells. A **chemical reaction** is a process by which new chemical bonds are formed or existing bonds are broken, converting one substance into another. There are three major types of chemical bonds: ionic bonds, covalent bonds, and hydrogen bonds (**Table 2-2**).

Ionic Bonds Form Among Charged Atoms Called Ions

Atoms that have an almost empty outermost electron shell, as well as atoms that have an almost full outermost shell, can become stable by losing electrons (emptying their outermost shells) or by gaining electrons (filling their outermost shells). The formation of table salt (sodium chloride

Health Watch

Might Chocolate Actually Be Good for You?

Many fruits and vegetables contain vitamins C and E as well as other antioxidants. But did you know that chocolate (**Fig. E2-3**) also contains antioxidants, and so might actually be a type of health food? Although it is extremely difficult to do controlled experiments on the effects of antioxidants in the human diet, there is evidence that diets high in antioxidants may be beneficial. For example, the low incidence of heart disease among the French has been attributed in part to antioxidants in wine, which many French people consume regularly. They also eat considerably more fruits and vegetables than Americans do.

Now, amazingly, researchers have given us an excuse to eat chocolate and feel good about it! Cocoa powder (the dark, bitter powder made from the seeds inside cacao pods; see Fig. E2-3 inset) contains high concentrations of flavenoids, which are powerful antioxidants, chemically related to those found in wine. No studies have yet been done to determine whether eating large quantities of chocolate actually reduces the risk of cancer or heart disease, but there will certainly be no shortage of volunteers for this research. Be aware that the most sinfully delicious chocolates are also high in fat and sugar, and becoming overweight by indulging could counteract any positive effects of the pure cocoa powder. However, slim "chocoholics" have reason to relax and enjoy!

▲ **FIGURE E2-3 Chocolate** Cocoa powder is derived from the cacao beans found inside cacao pods (inset), which grow on trees in tropical regions of the Americas.

or NaCl) demonstrates this principle. Sodium (Na) has only one electron in its outermost electron shell, and chlorine (Cl) has seven electrons in its outer shell, making it one electron short of a full shell (**Fig. 2-5a**).

Sodium can become stable by losing the electron from its outer shell to chlorine. This reaction also fills the outer shell of chlorine. Atoms that have lost or gained electrons, altering the balance between protons and electrons, are charged, and are called **ions.** To form sodium chloride, sodium loses an electron, becoming a positively charged sodium ion (Na^+), while chlorine picks up the electron and becomes a negatively charged chloride ion (Cl^-).

The sodium and chloride in NaCl are held together by **ionic bonds:** the electrical attraction between positively and negatively charged ions (**Fig. 2-5b**). The ionic bonds between sodium and chloride ions result in crystals composed of a

Table 2-2 Common Types of Bonds in Biological Molecules

Type	Type of Interaction	Example
Ionic bond	An electron is transferred, creating positive and negative ions that attract one another	Occurs between the sodium (Na^+) and chloride (Cl^-) ions of table salt (NaCl)
Covalent bond	Electrons are shared	
Nonpolar	Equal sharing	Occurs between the two oxygen atoms in oxygen gas (O_2)
Polar	Unequal sharing	Occurs between the hydrogen and oxygen atoms of a water molecule (H_2O)
Hydrogen bond	A slightly positive hydrogen in a polar molecule attracts the slightly negative pole of a nearby polar molecule	Occurs between water molecules; slightly positive charges on hydrogen atoms attract slightly negative charges on oxygen atoms of adjacent molecules

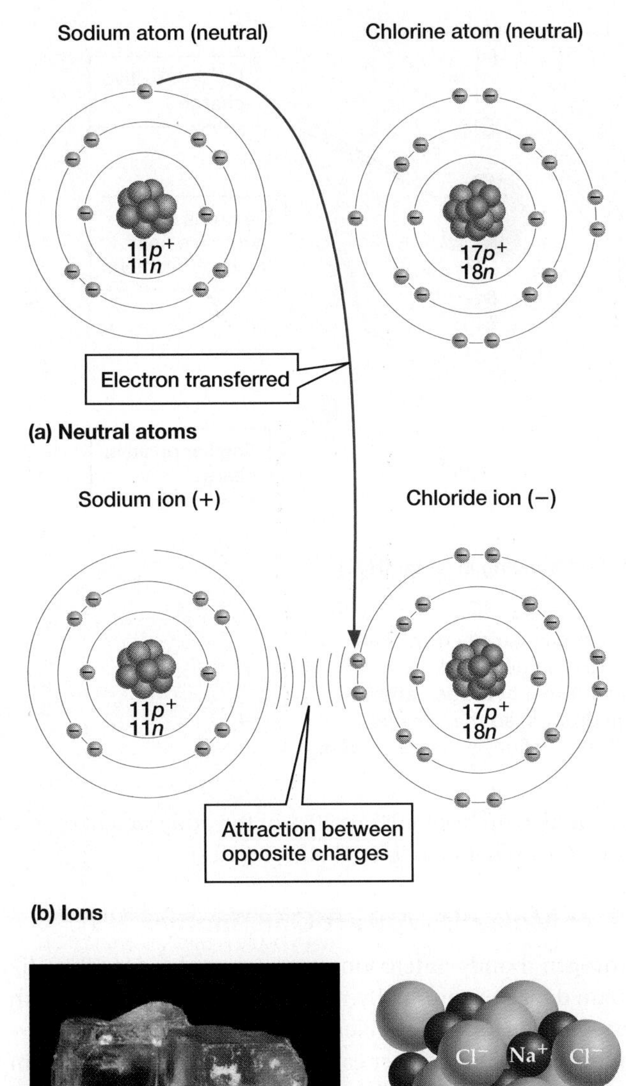

(c) An ionic compound: NaCl

▲ **FIGURE 2-5 The formation of ions and ionic bonds**
(a) Sodium has only one electron in its outer electron shell; chlorine has seven. **(b)** Sodium can become stable by losing an electron, and chlorine can become stable by gaining an electron. Sodium then becomes a positively charged ion and chlorine a negatively charged ion. **(c)** Because oppositely charged particles attract one another, the resulting sodium and chloride ions nestle closely together in a crystal of salt, NaCl. The arrangement of ions in salt causes it to form cubic crystals.

repeating, orderly array of the two ions (**Fig. 2-5c**). As we will see later, water will often break ionic bonds.

Covalent Bonds Form Between Uncharged Atoms That Share Electrons

An atom with a partially full outermost electron shell can become stable by sharing electrons with another atom, forming a **covalent bond** (**Fig. 2-6**).

Most Biological Molecules Utilize Covalent Bonding

Because biological molecules must function in a watery environment where ionic bonds rapidly dissociate (break apart), the atoms in most biological molecules, such as those found in proteins, sugars, and fats, are joined by covalent bonds. The bonds formed by the most common atoms in biological molecules are illustrated in **Table 2-3**.

How Electrons Are Shared Determines Whether a Covalent Bond Is Nonpolar or Polar

The electrons in covalent bonds may be shared equally between the two atoms, or they may spend more time near one of the two atoms. If the electrons are shared roughly equally, the bond is called a **nonpolar covalent bond.** Consider the hydrogen atom, which has one proton in its nucleus and one electron in its electron shell (which can hold two electrons). A hydrogen atom can become reasonably stable if it shares its single electron with another hydrogen atom, forming a molecule of hydrogen gas (H_2) in which each atom behaves almost as if it had two electrons in its electron shell (**Fig. 2-6a**). Because the two nuclei in H_2 are identical, the shared electrons spend equal time near each nucleus. Therefore, not only is the molecule as a whole electrically neutral or uncharged, but each end, or pole, of the molecule is also uncharged. Other examples of nonpolar molecules include oxygen gas (O_2), nitrogen gas (N_2), carbon dioxide (CO_2), and biological molecules such as fats (see pp. 43–44).

In contrast, some pairs of atoms form covalent bonds in which the shared electrons spend more time near one of the atoms. This atom thus acquires a slight negative charge, and the other atom takes on a slight positive charge. This situation produces a **polar covalent bond** (**Fig. 2-6b**). Although the molecule as a whole is electrically neutral, it has charged poles. In water (H_2O), for example, an electron is shared between each hydrogen atom and the central oxygen atom. However, the shared electrons spend far more time near the oxygen atom than near either of the hydrogen atoms. By attracting electrons, the oxygen pole of a water molecule becomes slightly negative, leaving each hydrogen atom slightly positive (see Fig. 2-6b). Water is an example of a polar molecule.

Hydrogen Bonds Are Attractive Forces Between Polar Molecules

As you might guess, opposite charges on polar molecules attract one another. A **hydrogen bond** is the attraction between the slightly positive hydrogen in a polar molecule and the slightly negative pole of a nearby polar molecule.

Polar molecules result when hydrogen atoms form bonds with nitrogen, oxygen, or fluorine. Although fluorine is rare in biological molecules, O—H and N—H bonds are common. Molecules such as water, sugars, proteins, and DNA are polar and can form hydrogen bonds with other polar molecules or with different polar regions of the same large molecule. Water

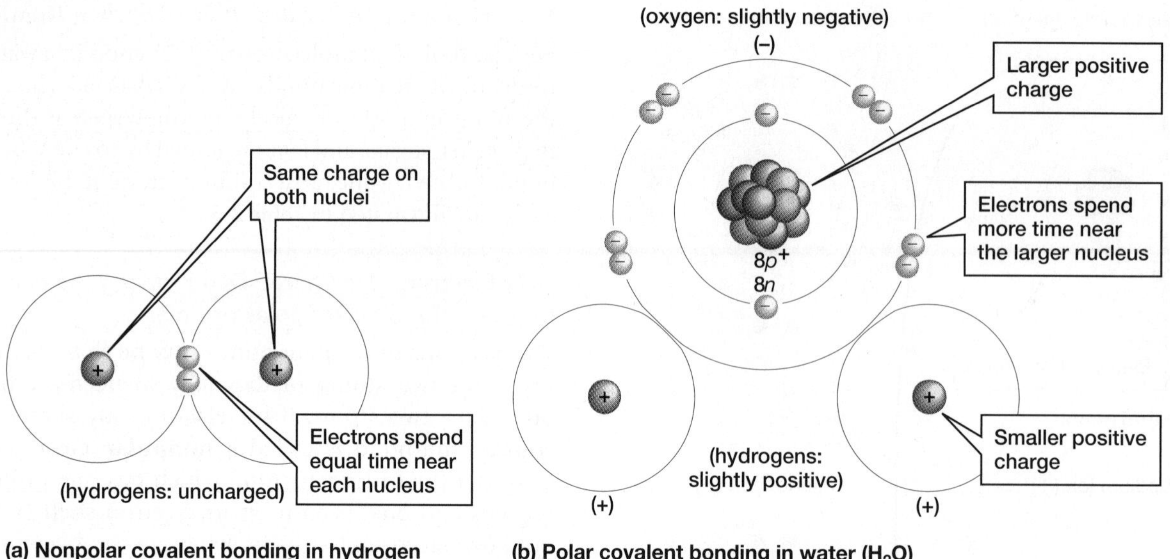

(a) Nonpolar covalent bonding in hydrogen gas (H_2)

(b) Polar covalent bonding in water (H_2O)

▲ **FIGURE 2-6 Covalent bonds involve shared electrons (a)** In hydrogen gas, an electron from each hydrogen atom is shared, forming a single nonpolar covalent bond. **(b)** Oxygen lacks two electrons to fill its outer shell, so oxygen can form polar covalent bonds with two hydrogen atoms, creating water. Oxygen exerts a greater pull on the electrons than does hydrogen, so the "oxygen end" of the molecule has a slight negative charge and the "hydrogen end" has a slight positive charge.

molecules form hydrogen bonds with one another between their positive hydrogen poles and negative oxygen poles (**Fig. 2-7**). As we will see shortly, hydrogen bonds give water several unusual properties that are essential to life on Earth.

2.3 WHY IS WATER SO IMPORTANT TO LIFE?

As naturalist Loren Eiseley eloquently stated, "If there is magic on this planet, it is contained in water." Water is extraordinarily abundant on Earth, has unusual properties, and is so essential to life that it merits special consideration. Life is likely to have arisen in the waters of the primeval Earth, and water makes up from 60% to 90% of the body weight of most organisms. What is so special about water?

Water Molecules Attract One Another

Hydrogen bonds interconnect water molecules. But, like square dancers continually moving from one partner to the next—joining and then releasing hands as they go—hydrogen bonds in liquid water easily break and re-form, allowing the water to flow. **Cohesion** among water molecules, caused by hydrogen bonding, produces **surface tension,** the tendency for the water surface to resist being broken. Surface tension can support fallen leaves, some spiders and water insects, and even a running basilisk lizard (**Fig. 2-8a**).

Table 2-3 Bonding Patterns of Atoms Commonly Found in Biological Molecules

Atom	Capacity of Outer Electron Shell	Electrons in Outer Shell	Number of Covalent Bonds Usually Formed	Common Bonding Patterns
Hydrogen	2	1	1	—H
Carbon	8	4	4	C with four single bonds; C with two single and one double bond; =C=; —C≡
Nitrogen	8	5	3	N with three single bonds; —N=; N≡
Oxygen	8	6	2	—O—; O=
Phosphorus	8	5	5	P with three single bonds and one double bond
Sulfur	8	6	2	—S—

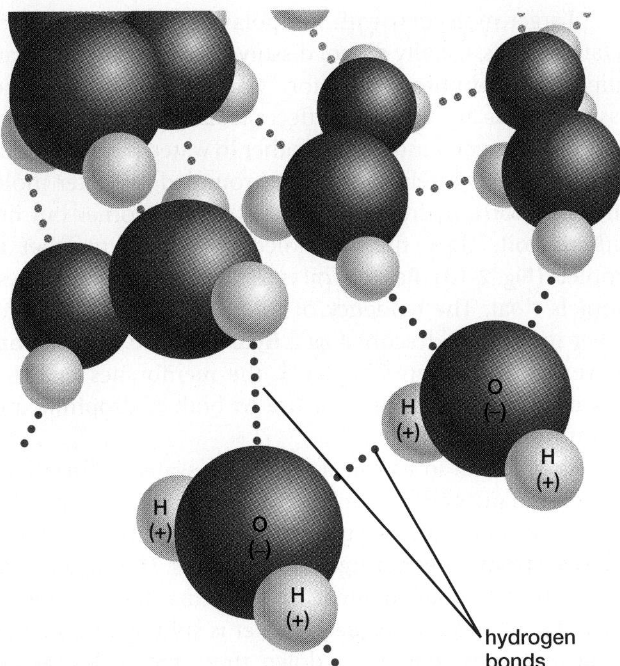

▲ **FIGURE 2-7 Hydrogen bonds in water** The partial charges on different parts of water molecules produce weak attractive forces called hydrogen bonds (dotted lines) between the oxygen and hydrogen atoms in adjacent water molecules. As water flows, these bonds are constantly breaking and re-forming.

> **Have you ever wondered ?**
>
> *Why It Hurts So Much to Do a Belly Flop?*
>
> The slap of a belly flop provides firsthand experience of the power of cohesion among water molecules. Because of the hydrogen bonds that interconnect its molecules, the surface of water resists being broken. When you suddenly force water molecules apart over a large area using your body, the result can be a bit painful.

The cohesion of water plays a crucial role in the life of land plants. How does water absorbed through a plant's roots ever reach the leaves, especially if the plant is a 300-foot-tall redwood tree (**Fig. 2-8b**)? Water molecules in plants fill tiny tubes that connect the roots, stem, and leaves. Water molecules evaporating from the leaves pull water up these tubes, much like a chain being dragged up from the top. The system works because the hydrogen bonds that link water molecules are stronger than the weight of the water in the tube and so the water chain doesn't break. Without the cohesion of water, there would be no land plants as we know them, and terrestrial life would have evolved quite differently.

Water also exhibits **adhesion,** a tendency to stick to surfaces having slight charges that attract the polar water molecules. If you touch water with the end of a very narrow glass tube, adhesion will draw the water a short distance up into the tube. In a similar way, adhesion also helps water in plants move up the thin tubes from roots to leaves.

Water Interacts with Many Other Molecules

A **solvent** is a substance that **dissolves** (completely surrounds and disperses) other substances. A solvent containing one or more dissolved substances is called a **solution.** Molecules involved in biochemical reactions are often in solution; this allows them to move around freely and meet one another to react. Water is an excellent solvent; its polar nature attracts it to other polar molecules and to ions, causing it to surround them. Water dissolves many molecules that are important to life, including proteins, salts, sugars, oxygen, and carbon dioxide.

To illustrate water's role as a solvent, let's look at how table salt dissolves. A crystal of table salt is held together by the electrical attraction between positively charged sodium ions and negatively charged chloride ions (see Fig. 2-5c). If a salt crystal is dropped into water, the positively charged hydrogen poles of water molecules will congregate around the negatively charged chloride ions, and the negatively charged oxygen poles of water molecules will attract the positively charged sodium ions. As water molecules enclose the sodium

(a) Cohesion causes surface tension

(b) Cohesion helps water to reach treetops

◀ **FIGURE 2-8 Cohesion among water molecules (a)** A basilisk lizard uses surface tension as it races across the water's surface to escape a predator. **(b)** Cohesion caused by hydrogen bonding allows water evaporating from the leaves of plants to pull water up from the roots.

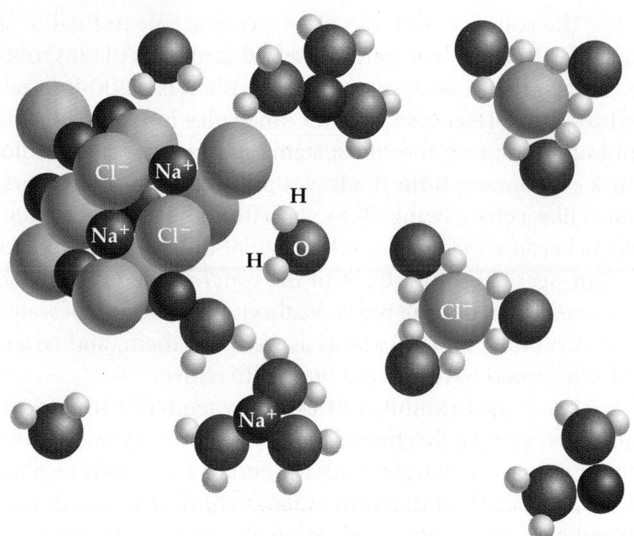

▲ **FIGURE 2-9 Water as a solvent** When a salt crystal is dropped into water, the water surrounds the sodium and chloride ions with oppositely charged poles of its molecules. The ions disperse as the surrounding water molecules isolate them from one another, and the crystal gradually dissolves.

and chloride ions, shielding them from interacting with each other, the ions separate from the crystal and drift away in the water—thus, the salt dissolves (**Fig. 2-9**).

Water also dissolves other polar molecules because its positive and negative poles are attracted to their oppositely charged poles. Because of their electrical attraction for water molecules, ions and polar molecules are described as **hydrophilic** (from the Greek "hydro," meaning "water," and "philic" meaning "loving"). Many biological molecules are hydrophilic and dissolve readily in water.

Gases such as oxygen and carbon dioxide are essential for biochemical reactions but are nonpolar—so how do they dissolve? These molecules are so small that they fit into the spaces among water molecules without disrupting the hydrogen bonds linking the water molecules. Fish swimming below the ice on a frozen lake rely on oxygen that dissolved before the ice formed, and the CO_2 they release dissolves into the water.

Case Study continued
Crushed by Ice

The seawater in which the *Endurance* sank provides an excellent illustration of the ability of water to dissolve other molecules. Seawater contains more than 70 different elements, most in trace quantities. The most abundant dissolved elements are sodium and chlorine (NaCl; comprising about 3.2% of seawater by weight) and smaller amounts of sulfur, calcium, magnesium, potassium, and nitrogen. Ocean water also contains dissolved gases such as oxygen and carbon dioxide. Oxygen in seawater allows aquatic animals to respire, and carbon dioxide allows photosynthesis, which makes energy available to aquatic organisms.

Larger molecules with nonpolar covalent bonds, such as fats and oils, usually do not dissolve in water and hence are called **hydrophobic** (Greek for "water-fearing"). Nevertheless, water has an important effect on such molecules. When oil molecules encounter one another in water, their nonpolar surfaces nestle closely together, surrounded by water molecules that form hydrogen bonds with one another but not with the oil. Thus, the oil molecules remain together in droplets (**Fig. 2-10**). Because oil is less dense than water, these droplets float. The tendency of oil molecules to clump together in water is described as a **hydrophobic interaction.** As we will discuss in Chapter 5, the membranes of living cells owe much of their structure to both hydrophilic and hydrophobic interactions.

In addition to its role as a solvent, water is directly involved in many of the chemical reactions that occur in living cells. For example, the oxygen that plants release into the air is derived from water during photosynthesis. The chemical reactions in cells that manufacture proteins, fats, or nucleic acids liberate water; conversely, water is split apart and used in the reactions that break down these molecules, as described in later chapters.

Water-Based Solutions Can Be Acidic, Basic, or Neutral

Although water is generally regarded as a stable compound, it has a slight tendency to spontaneously form ions that, in turn, can reunite to form water. At any given time, a small

▲ **FIGURE 2-10 Oil and water don't mix** Yellow oil has just been poured into this bottle of water and is rising to the surface. Oil floats because it is less dense than water, and it forms droplets in water because it is a hydrophobic, nonpolar molecule that is not attracted to water's polar molecules.

fraction of water molecules have split into hydroxide ions (OH^-) and hydrogen ions (H^+) (**Fig. 2-11**).

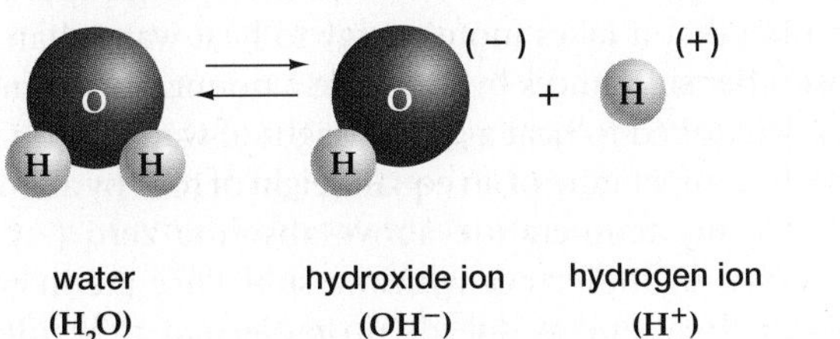

▲ **FIGURE 2-11 Some water is ionized**

A hydroxide ion is negatively charged because it has gained an electron from the hydrogen atom. By losing an electron, the hydrogen atom is converted into a positively charged hydrogen ion. Pure water contains equal concentrations of hydrogen ions and hydroxide ions.

In many solutions, however, the concentrations of H^+ and OH^- are not the same. If the concentration of H^+ exceeds the concentration of OH^-, the solution is **acidic.** An **acid** is a substance that releases hydrogen ions when it dissolves in water. When hydrochloric acid (HCl), for example, is added to pure water, almost all of the HCl molecules separate into H^+ and Cl^-. Therefore, the concentration of H^+ greatly exceeds the concentration of OH^-, and the resulting solution is acidic. Many acidic substances, such as lemon juice and vinegar, have a sour taste because the sour receptors on your tongue are specialized to respond to an excess of H^+. Bacteria common in your mouth can form a coating on your teeth and break down sugar from trapped food, producing lactic acid. The excess H^+ released by the acid can cause tooth enamel to dissolve and erode away, creating a cavity. Orange juice and many soft drinks are quite acidic, but generally don't remain in your mouth long enough to damage teeth.

If the concentration of OH^- is greater than H^+, the solution is **basic.** A **base** is a substance that combines with hydrogen ions, reducing their number. If, for instance, sodium hydroxide (NaOH) is added to water, the NaOH molecules separate into Na^+ and OH^-. Some OH^- ions combine with H^+ to produce H_2O, reducing the number of H^+ ions and creating a basic solution.

The degree of acidity is expressed on the **pH scale** (**Fig. 2-12**), in which neutrality (equal numbers of H^+ and OH^-) is indicated by the number 7. Pure water, with equal concentrations of H^+ and OH^-, has a pH of 7. Acids have a pH below 7; bases have a pH above 7. Each unit on the pH scale represents a tenfold change in the concentration of H^+. Thus, a soft drink with a pH of 3 has a concentration of H^+ that is 10,000 times that of water, which has a pH of 7.

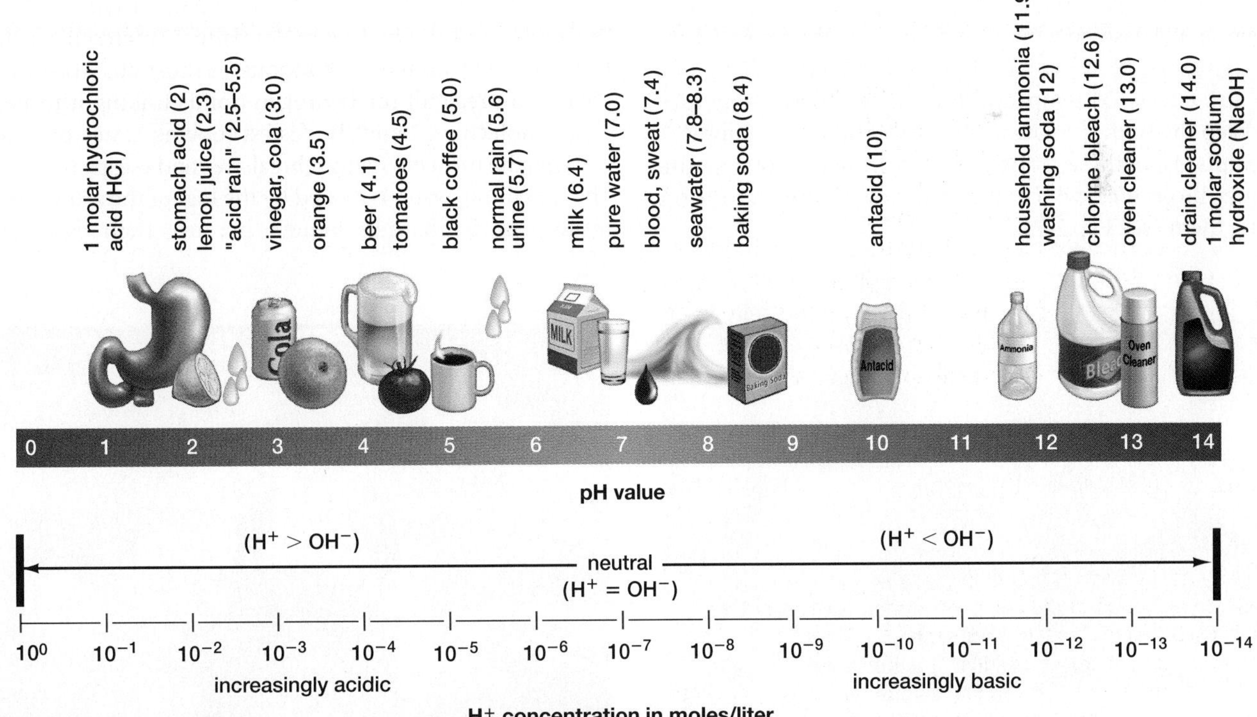

▲ **FIGURE 2-12 The pH scale** The pH scale reflects the concentration of hydrogen ions in a solution. The pH (upper scale) is the negative log of the H^+ concentration (lower scale). Each unit on the scale represents a tenfold change. Lemon juice, for example, is about 10 times more acidic than orange juice, and the most severe acid rains in the northeastern United States are almost 1,000 times more acidic than normal rainfall. Except for the inside of your stomach, nearly all the fluids in your body are finely adjusted to a pH of about 7.4.

A Buffer Helps Maintain a Solution at a Relatively Constant pH

In most mammals, including humans, both the cell interior (cytoplasm) and the fluids that bathe the cell are nearly neutral (pH about 7.3 to 7.4). Small increases or decreases in pH may cause drastic changes in both the structure and function of biological molecules, leading to the death of cells or entire organisms. Nevertheless, living cells seethe with chemical reactions that take up or give off H^+. How, then, does the pH remain constant overall? The answer lies in the many buffers found in living organisms. A **buffer** is a compound that tends to maintain a solution at a constant pH by accepting or releasing H^+ in response to small changes in H^+ concentration. If the H^+ concentration rises, buffers combine with them; if the H^+ concentration falls, buffers release H^+. Thus, the original concentration of H^+ is maintained. Common buffers in living organisms include phosphate ($H_2PO_4^-$ and HPO_4^{2-}) and bicarbonate (HCO_3^-), both of which can accept or release H^+ depending on the circumstances. Blood pH is precisely regulated by bicarbonate ions.

Water Moderates the Effects of Temperature Changes

Your body and the bodies of other organisms can survive only within a limited temperature range. As we will see in Chapter 6, high temperatures may damage enzymes that promote the chemical reactions essential to life. Low temperatures are also dangerous because enzyme action slows as the temperature drops. Fortunately, water has important properties that moderate the effects of temperature changes. These properties help keep the bodies of organisms within tolerable temperature limits. All of these properties result from the polar nature of water molecules, which allows hydrogen bonds to form among them (see Fig. 2-7).

It Takes a Lot of Energy to Heat Water

The energy required to heat 1 gram of a substance by 1 °C is called its **specific heat.** Water has a very high specific heat, meaning that it takes more energy to heat water than to heat most other substances by the same amount. For example, the energy required to heat a given weight of water by 1 °C would raise the temperature of an equal weight of rock by about 50 °C.

At any temperature above absolute zero (−273 °C), the atoms or molecules within a substance are in constant motion. Heat energy speeds up molecular movement. But increasing the rate of movement of water molecules requires that the hydrogen bonds between them be broken at a higher rate. A considerable amount of heat energy must be expended to break these bonds, and this energy is then unavailable to raise the temperature of the water. The high specific heat of water allows organisms, whose bodies are mostly water, to inhabit hot, sunny environments without overheating. The sunbathers in **Figure 2-13a** are taking advantage of this property of water.

It Takes a Lot of Energy to Cause Water to Evaporate

Sunbathers take advantage of another property of water—its high heat of vaporization. The **heat of vaporization** is the amount of heat needed to cause a substance to evaporate (to change from a liquid to a vapor). It takes a great deal of heat energy (539 calories per gram) to convert liquid water to water vapor; in fact, water's heat of vaporization is one of the highest known. This is also due to the polar nature of water molecules and the hydrogen bonds that interconnect them. For a water molecule to evaporate, it must absorb sufficient energy to break all the hydrogen bonds linking it to nearby water molecules. Only the fastest-moving water molecules, carrying the most energy, can break free and escape into the air. The remaining water is cooled by the loss of these high-energy molecules. Sunbathers benefit because their evaporating

(a) Water's high specific heat protects sunbathers

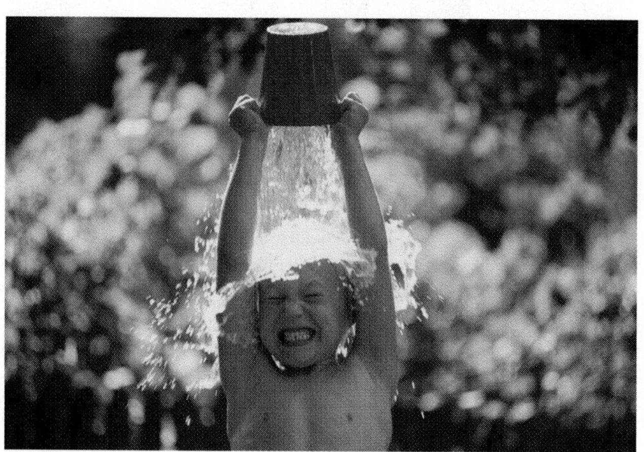

(b) Water's high heat of vaporization cools the body

▲ **FIGURE 2-13 Water's high specific heat and high heat of vaporization affect human behavior (a)** Because our bodies are mostly water, sunbathers can absorb a great deal of heat without sending their body temperatures skyrocketing, a result of water's high specific heat. **(b)** Water's high heat of vaporization (evaporative cooling) and specific heat together make water a very effective coolant on a hot day.

Case Study continued
Crushed by Ice

Shackleton and his crew did not lack drinking water during the months they spent waiting; they could simply melt chunks of ice. Although the ice had solidified from salty ocean water, Na^+ and Cl^- ions (and the other substances dissolved in ocean water) do not fit into the lattice of water molecules that make up the crystalline structure of ice (see Fig. 2-14). As ice forms, the remaining water becomes increasingly salty and requires increasingly low temperatures to freeze.

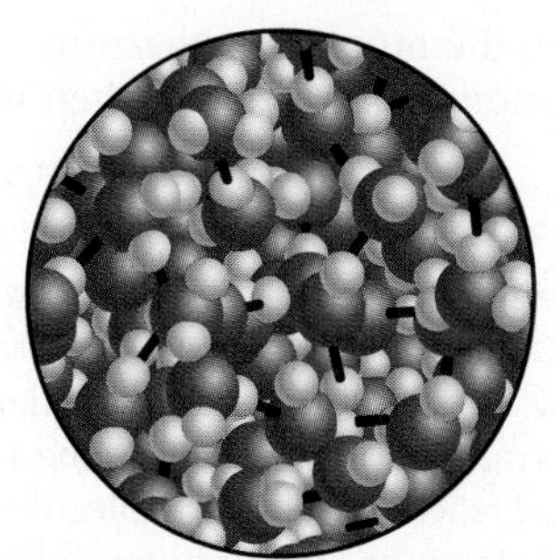

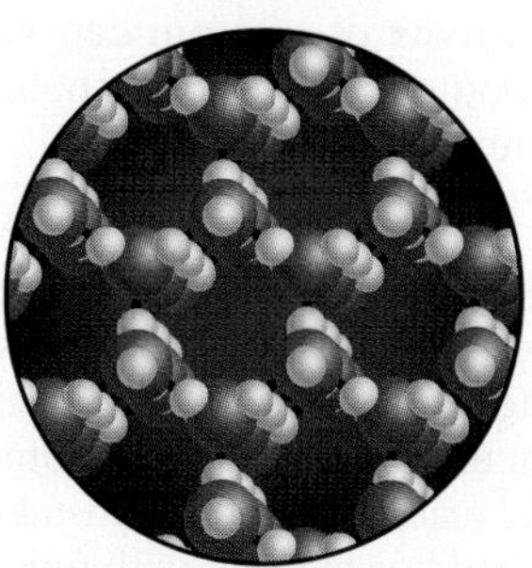

▲ FIGURE 2-14 **Liquid water (left) and ice (right)**

perspiration carries away a lot of heat without much loss of water. Children playing in water on a hot summer day are cooled as heat energy from their bodies breaks the hydrogen bonds of water, first as the water is warmed and then as it evaporates from their skin (**Fig. 2-13b**).

Water Forms an Unusual Solid: Ice

Even solid water is unusual. Most liquids become denser when they solidify, but ice is less dense than liquid water. The molecules of liquid water are continually moving past one another, breaking and re-forming hydrogen bonds as they shift position. When water freezes, however, each molecule forms stable hydrogen bonds with four other water molecules, creating an open, hexagonal (six-sided) arrangement. This keeps the water molecules further apart than their average distance in liquid water, so ice is less dense than liquid water (**Fig. 2-14**). The beautiful six-sided symmetry of snowflakes is an outgrowth of the hexagonal array of water molecules in ice crystals.

When a pond or lake starts to freeze in winter, the ice floats on top, forming an insulating layer that delays the freezing of the rest of the water. This insulation allows fish and other aquatic animals to survive in the liquid water below. If ice were to sink, many ponds and lakes around the world would freeze solid from the bottom up during the winter, killing plants, fish, and other underwater organisms. The ocean floor at higher latitudes would be covered with extremely thick layers of ice that would never melt.

Case Study revisited
Crushed by Ice

Although they failed to traverse Antarctica by dogsled, the heroic and ultimately successful efforts of Shackleton and his crew to return home alive are a remarkable true-life adventure. By studying Figure 2-14 you can see why the *Endurance* was crushed—the orderly arrangement of water molecules in ice causes them to occupy more space than they did as a liquid; thus, water expands as it forms ice. You have witnessed this firsthand if you've ever left a full water bottle in the freezer and later found it cracked open by the expanding ice. The expansion of ice also explains why it floats and why the ice formed a safe campground for the crew of the *Endurance*.

Many animals rely on ocean ice, from penguins and walruses that breed on the ice sheets, to polar bears that use the ice as a staging ground to hunt seals. The seals themselves are protected from their polar bear predators by the ice cover as they hunt for fish beneath it, becoming vulnerable when they surface to breathe through holes that they maintain in the ice.

Consider This

Many of water's unique properties are the result of its polar covalent bonds, which allow water molecules to form hydrogen bonds with each other. What if water molecules were nonpolar? How would this have affected the voyage of the *Endurance*?

CHAPTER REVIEW

Summary of Key Concepts

2.1 What Are Atoms?

An element is a substance that can neither be broken down nor converted to different substances by ordinary chemical means. The smallest possible particle of an element is the atom, which is itself composed of a central nucleus, containing protons and neutrons, and electrons outside the nucleus. All atoms of a given element have the same number of protons, different from that in every other element. Neutrons cluster with protons in most atomic nuclei, contributing to the atomic weight of the nucleus but not affecting its charge. Electrons orbit the nucleus within specific regions called electron shells. Shells at increasing distances from the nucleus contain electrons with increasing amounts of energy. Each shell can contain a fixed maximum number of electrons. The chemical reactivity of an atom depends on the number of electrons in its outermost electron shell; an atom is most stable when its outermost shell is full.

2.2 How Do Atoms Interact to Form Molecules?

Atoms gain stability by filling or emptying their outermost electron shell by acquiring, losing, or sharing electrons during chemical reactions. This results in stable attractions called chemical bonds among atoms, forming molecules. Some chemical reactions produce unstable and reactive atoms or molecules, called free radicals, with unpaired electrons in their outer shells. These may react with important biological molecules, such as DNA, destroying them. Antioxidants help protect against free-radical damage.

Chemical bonds may be ionic, covalent, or hydrogen bonds. Atoms that have lost or gained electrons are negatively or positively charged particles called ions. Ionic bonds are electrical attractions between charged ions, holding them together in crystals. When two atoms share electrons, covalent bonds form. In a nonpolar covalent bond, the two atoms share electrons equally. In a polar covalent bond, one atom attracts the electron more strongly than the other atom does, giving the molecule slightly positive and negative poles. Polar covalent bonds allow hydrogen bonding, the attraction between the slightly positive hydrogen on one molecule and the slightly negative regions of other polar molecules.

2.3 Why Is Water So Important to Life?

Water has many unique properties that allowed life as we know it to evolve. Water is polar and dissolves polar substances and ions. Water forces nonpolar substances, such as fat, to form clumps. Water also participates directly in some chemical reactions. Water molecules cohere to each other using hydrogen bonds, and adhere to many surfaces. Because of its high specific heat and heat of vaporization, water helps maintain a fairly stable temperature despite wide temperature fluctuations in the environment. Water is almost unique in being less dense in its frozen than in its liquid state.

Key Terms

acid *31*
acidic *31*
adhesion *29*
antioxidant *24*
atom *21*
atomic mass *22*
atomic nucleus *21*
atomic number *21*
base *31*
basic *31*
buffer *32*
chemical bond *23*
chemical reaction *25*
cohesion *28*
compound *24*
covalent bond *27*
dissolve *29*
electron *21*
electron shell *22*
element *21*
free radical *24*
heat of vaporization *32*
hydrogen bond *27*
hydrophilic *30*
hydrophobic *30*
hydrophobic interaction *30*
ion *26*
ionic bond *26*
isotope *22*
molecule *24*
neutron *21*
nonpolar covalent bond *27*
pH scale *31*
polar covalent bond *27*
proton *21*
radioactive *22*
solution *29*
solvent *29*
specific heat *32*
surface tension *28*

Thinking Through the Concepts

Fill-in-the-Blank

1. An atom consists of an atomic nucleus composed of positively charged __________ and uncharged __________. The number of positively charged particles in the nucleus determines the __________ of the element. Orbiting around the nucleus are __________ that occupy confined spaces called __________.
2. An atom that has lost or gained one or more electrons is called a(n) __________. If an atom loses an electron it takes on a(n) __________ charge. If it gains an electron it takes on a(n) __________ charge. Atoms with opposite charges attract one another, forming __________ bonds.
3. Atoms of the same element that differ in the number of neutrons in their nuclei are called __________ of one another. Some of these atoms spontaneously break apart, releasing __________ and forming different atoms in the process. Atoms that behave this way are described as being __________.
4. An atom with an outermost electron shell that is either completely full or completely empty is described as __________. Atoms with partially full outer electron shells are __________. Covalent bonds are formed when atoms __________ electrons, filling their outer shells.
5. Highly reactive atoms or molecules with unpaired electrons in their outer shells are called __________. These can damage biological molecules, including the hereditary molecule called __________. Some of this damage can be prevented by __________.
6. Water is described as __________ because each water molecule has partial negative and positive charges. This property allows water molecules to form __________ bonds with one another. The bonds between water molecules give water a high __________ that produces surface tension.

Review Questions

1. What are the six most abundant elements found in the human body by percent weight?
2. Distinguish between atoms and molecules; between elements and compounds; and among protons, neutrons, and electrons.
3. Compare and contrast covalent bonds and ionic bonds.
4. Why can water absorb a great amount of heat with little increase in its temperature?
5. Describe how water dissolves a salt. How does this phenomenon compare with the effect of water on a hydrophobic substance such as corn oil?
6. Define *pH*, *acid*, *base*, and *buffer*. How do buffers reduce changes in pH when hydrogen ions or hydroxide ions are added to a solution? Why is this phenomenon important in organisms?

Applying the Concepts

1. Fats and oils do not dissolve in water; polar and ionic molecules dissolve easily in water. Detergents and soaps

help clean by dispersing fats and oils in water so that they can be rinsed away. From your knowledge of the structure of water and the hydrophobic nature of fats, what general chemical structures (for example, polar or nonpolar parts) must a soap or detergent have, and why?

2. If the density of ice were greater than that of liquid water, how would this affect aquatic life? Describe some impacts this would have on terrestrial organisms.
3. How does sweating help you regulate your body temperature? Why do you feel hotter and more uncomfortable on a hot, humid day than on a hot, dry day?
4. Free radicals are formed when organisms use oxygen to metabolize sugar to make high-energy molecules. Ross Hardison, of Pennsylvania State University, stated: "Keeping oxygen under control while using it in energy production has been one of the great compromises struck in the evolution of life on Earth." What did he mean by this?
5. Some people have proposed towing huge icebergs to regions where people are short of drinking water. How can ice that formed from salty ocean water be good to drink?
6. Earth's atmosphere is about 20% oxygen (O_2), 79% nitrogen (N_2), and 0.38% carbon dioxide (CO_2). Explain why blimps will float in air if they are filled with either hydrogen (H_2) or helium (He). Which gas is preferable for filling blimps, and why?

MB *Go to www.masteringbiology.com for practice quizzes, activities, eText, videos, current events, and more.*

Biological Molecules

Chapter 3

▲ Although this cow is healthy, mad cow disease leaves animals drooling and often unable to stand.

Case Study

Puzzling Proteins

"YOU KNOW, LISA, I think something is wrong with me," the vibrant, 22-year-old scholarship winner told her sister. It was 2001 when Charlene began to lose her memory and experience mood swings. During the next three years, her symptoms worsened. Charlene's hands shook, she was subject to uncontrollable episodes of biting and striking people, and became unable to walk or swallow. In June 2004, Charlene became the first U.S. resident to die of the human form of mad cow disease, which she had almost certainly contracted more than 10 years earlier while living in England.

It was not until the mid-1990s that health officials recognized that mad cow disease, or bovine spongiform encephalitis (BSE), could be spread to people who ate meat from infected cattle. Although millions of people may have eaten this tainted beef, only about 200 people worldwide have died from the human version of BSE, or variant Creutzfeldt-Jakob disease (vCJD). For those infected, however, the disease is always fatal, riddling brains of both humans with vCJD and cows with BSE with microscopic holes, giving the brain a spongy appearance.

Where did mad cow disease come from? For centuries, it was known that sheep could suffer from a spongiform encephalitis called scrapie. Scientists believe that a mutated form of scrapie became capable of infecting cattle, perhaps as early as the early 1980s. At that time, cattle feed often contained parts from sheep, some of whom may have harbored scrapie. Since 1986, when BSE was first identified in British cattle, millions of cattle there have been slaughtered and their bodies burned to prevent its spread.

Why is mad cow disease particularly fascinating to scientists? In the early 1980s, Dr. Stanley Prusiner, a researcher at the University of California—San Francisco, startled the scientific world by providing evidence that a protein, which lacks genetic material (DNA or RNA), was the cause of scrapie, and that this protein could transmit the disease to experimental animals in the laboratory. He dubbed the infectious proteins "prions" (pronounced PREE-ons), short for "proteinaceous infectious particles."

What are proteins? How do they differ from DNA and RNA? How can a protein with no hereditary material infect another organism, increase in number, and produce a fatal disease? Read on to find out.

At a Glance

3.1 WHY IS CARBON SO IMPORTANT IN BIOLOGICAL MOLECULES?

You have probably seen fruits and vegetables grown without synthetic fertilizers or pesticides labeled as "organic." But in chemistry, **organic** describes molecules that have a carbon skeleton bonded to some hydrogen atoms. The term is derived from the ability of living organisms to synthesize and use this general type of molecule. **Inorganic** molecules, which include carbon dioxide (which lacks hydrogen atoms) and all molecules without carbon (such as water and salt), are far less diverse and generally far simpler than organic molecules.

Life is characterized by an amazing variety of molecules interacting in ways that are dazzlingly complex. Molecular interactions are governed by the structures of molecules and the chemical properties that arise from these structures. Life is a dynamic condition; as molecules within cells interact with one another, their structures and chemical properties change. Collectively, these precisely orchestrated changes give cells the ability to acquire and utilize nutrients, to eliminate wastes, to move and grow, and to reproduce.

The versatile carbon atom is the key to the tremendous variety of organic molecules that make life on Earth possible. The carbon atom has four electrons in its outermost shell, which can accommodate eight electrons. Therefore, a carbon atom can become stable by bonding with up to four other atoms, or by forming double or even triple bonds. As a result, organic molecules can assume complex shapes, including branched chains, rings, sheets, and helices.

Attached to the carbon backbone of organic molecules are **functional groups,** groups of atoms that help determine the characteristics and chemical reactivity of the molecules. Functional groups are less stable than the carbon backbone and more likely to participate in chemical reactions. A large number of different functional groups may be attached to organic molecules; the six most common functional groups found in biological molecules are shown in **Table 3-1**.

Although the molecules of life are incredibly diverse, most contain the same basic set of functional groups. Further, most large organic molecules are synthesized from repeating subunits, as described in the following section.

3.2 HOW ARE ORGANIC MOLECULES SYNTHESIZED?

Although a complex molecule could be made by combining atom after atom following an extremely detailed blueprint, life takes a modular approach, preassembling smaller molecules and hooking them together. Just as trains are made by joining a series of train cars, small organic molecules (for example, sugars or amino acids) are joined to form longer molecules (for example, starches or proteins)—like cars in a train. The individual subunits are called **monomers** (from Greek words meaning "one part"); chains of monomers are called **polymers** ("many parts").

Table 3-1 Important Functional Groups in Biological Molecules

Group	Structure	Properties	Found In
Hydroxyl	—O—H	Polar; involved in dehydration and hydrolysis reactions	Sugars, starches, nucleic acids, alcohols, some acids, and steroids
Carbonyl	—C=O	Polar; tends to make parts of molecules hydrophilic (water-soluble)	Sugars, some hormones, some vitamins
Carboxyl	—C(=O)O⁻	Acidic; the negatively charged oxygen may bond H^+, forming carboxylic acid (—COOH); involved in peptide bonds	Amino acids, fatty acids
Amino	—N(H)H	Basic; may bond an additional H^+, becoming positively charged; involved in peptide bonds	Amino acids, nucleic acids
Sulfhydryl	—S—H	Forms disulfide bonds in proteins	Some amino acids; many proteins
Phosphate	—O—P(=O)(O⁻)—O⁻	Acidic; links nucleotides in nucleic acids; energy-carrier group in ATP (this ionized form occurs in cellular environments)	Nucleic acids, phospholipids

Biological Polymers Are Formed by Removing Water and Split Apart by Adding Water

The subunits of large biological molecules are usually joined by a chemical reaction called **dehydration synthesis,** literally meaning "to form by removing water." In dehydration synthesis, a hydrogen ion (H^+) is removed from one subunit and a hydroxyl ion (OH^-) is removed from a second subunit, leaving openings in the outer electron shells of atoms in the two subunits. These openings are filled when the subunits share electrons, creating a covalent bond that links them. The hydrogen ion and the hydroxyl ion combine to form a molecule of water (H_2O), as shown in **Figure 3-1** using arbitrary subunits.

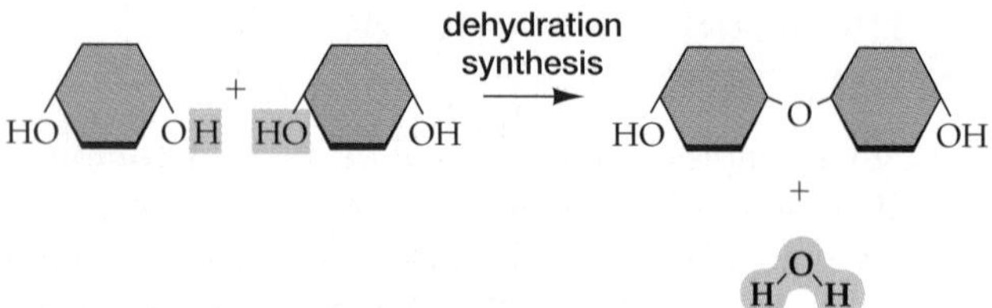

▲ **FIGURE 3-1 Dehydration synthesis**

The reverse reaction, **hydrolysis** ("to break apart with water"), splits the molecule back into its original subunits, with water donating a hydrogen ion to one subunit and a hydroxyl ion to the other (**Fig. 3-2**).

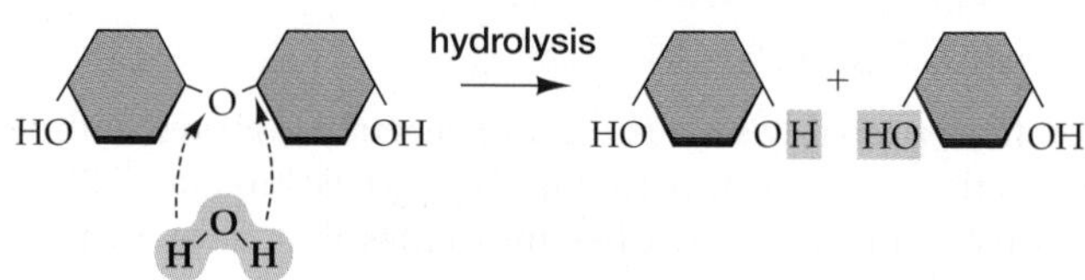

▲ **FIGURE 3-2 Hydrolysis**

Digestive enzymes use hydrolysis to break down food. For example, the starch in a cracker is composed of a series of glucose (simple sugar) molecules (see Fig. 3-8). Enzymes in our saliva and small intestines promote hydrolysis of the starch into individual sugar molecules that can be absorbed into the body.

In spite of the tremendous diversity of biological molecules, most of them fall into one of four general categories: carbohydrates, lipids, proteins, and nucleotides/nucleic acids (**Table 3-2**).

3.3 WHAT ARE CARBOHYDRATES?

Carbohydrate molecules are composed of carbon, hydrogen, and oxygen in the approximate ratio of 1:2:1. This formula explains the origin of the word "carbohydrate," which literally means "carbon plus water." All carbohydrates are either small, water-soluble **sugars** or polymers of sugar, such as starch. If a carbohydrate consists of just one sugar molecule, it is called a **monosaccharide** (Greek for "single

Table 3-2 Principal Types of Biological Molecules

Type and Structure of Molecule	Principal Subtypes and Structures	Example	Function
Carbohydrate: Most contain carbon, oxygen, and hydrogen, in the approximate formula $(CH_2O)_n$*	*Monosaccharide:* Simple sugar, often with the formula $C_6H_{12}O_6$	Glucose Fructose	Important energy source for cells; subunit of polysaccharides
	Disaccharide: Two monosaccharides bonded together	Sucrose	Energy-storage molecule in fruits and honey Principal sugar transported throughout the bodies of land plants
	Polysaccharide: Chain of monosaccharides (usually glucose)	Starch	Energy storage in plants
		Glycogen	Energy storage in animals
		Cellulose	Structural material in plants
Lipid: Contains a high proportion of carbon and hydrogen; most lipids are nonpolar and insoluble in water	*Triglyceride:* Three fatty acids bonded to glycerol	Oil, fat	Energy storage in animals and some plants
	Wax: Variable numbers of fatty acids bonded to a long-chain alcohol	Waxes in plant cuticle	Waterproof covering on leaves and stems of land plants
	Phospholipid: Polar phosphate group and two fatty acids bonded to glycerol	Phosphatidylcholine	Component of cell membranes
	Steroid: Four fused rings of carbon atoms with functional groups attached	Cholesterol	Component of the membranes of eukaryotic cells; precursor for other steroids such as testosterone and bile salts
Protein: Consists of one or more chains of amino acids; may have up to four levels of structure that determine its function	*Peptide:* Short chain of amino acids	Oxytocin	Hormone consisting of nine amino acids; functions include stimulating uterine contractions during childbirth
	Polypeptide: Long chain of amino acids; also called "protein"	Hemoglobin	Globular protein composed of four subunit peptides; transport of oxygen in vertebrate blood
Nucleotide/nucleic acid: *Nucleotide:* Consists of a sugar, a base, and a phosphate group	*Nucleotide:* Composed of a five-carbon sugar (ribose or deoxyribose), a nitrogen-containing base, and a phosphate group	Adenosine triphosphate (ATP)	Principal short-term energy carrier molecule in cells
		Cyclic adenosine monophosphate (cAMP)	Intracellular messenger
Nucleic acid: A polymer made of nucleotides	*Nucleic acid:* A polymer of nucleotide subunits joined by covalent bonds between their phosphate and sugar groups	Deoxyribonucleic acid (DNA)	Genetic material of all cells
		Ribonucleic acid (RNA)	In cells, essential for protein synthesis using the genetic sequence copied from DNA; genetic material of some viruses

**n* signifies the number of carbons in the molecule's backbone.

sugar"). When two monosaccharides are linked, they form a **disaccharide** ("two sugars"), and a polymer of many monosaccharides is called a **polysaccharide** ("many sugars"). Sugars and starches are often used to store energy in cells, while other carbohydrates strengthen cell walls in plants, fungi, and bacteria, or form a supportive armor over the bodies of insects, crabs, and their relatives.

If you've stirred sugar into coffee or tea, you know that it dissolves in water. This is because the hydroxyl functional groups of sugars are polar and form hydrogen bonds with polar water molecules (**Fig. 3-3**).

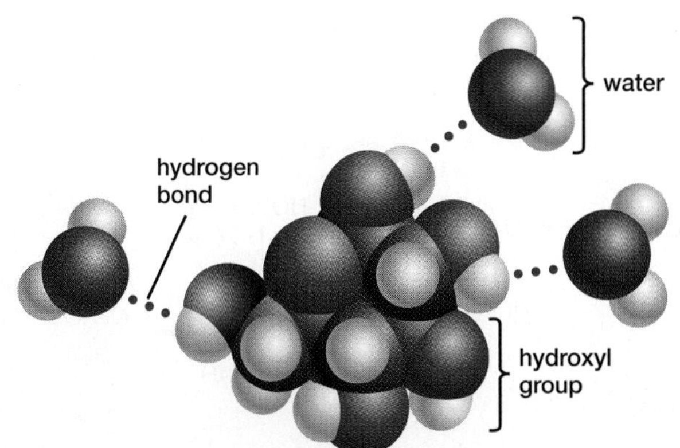

▲ **FIGURE 3-3 Sugar dissolving in water** Glucose dissolves as the polar hydroxyl groups of each sugar molecule form hydrogen bonds with nearby water molecules.

There Are Several Monosaccharides with Slightly Different Structures

Monosaccharides have a backbone of three to seven carbon atoms. Most of these carbon atoms have both a hydrogen (—H) and a hydroxyl group (—OH) attached to them; therefore,

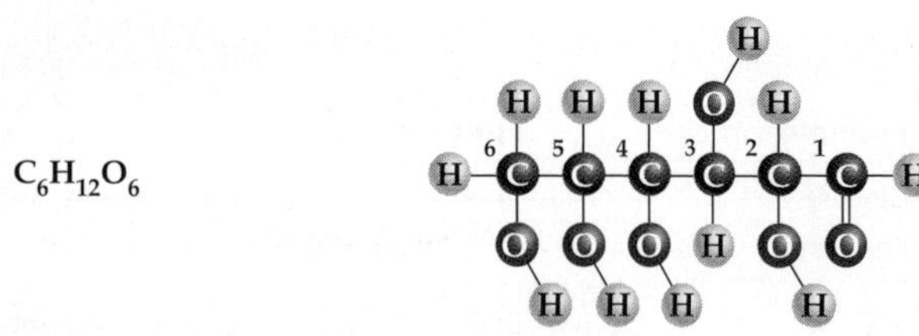

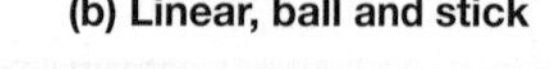

(a) Chemical formula

(b) Linear, ball and stick

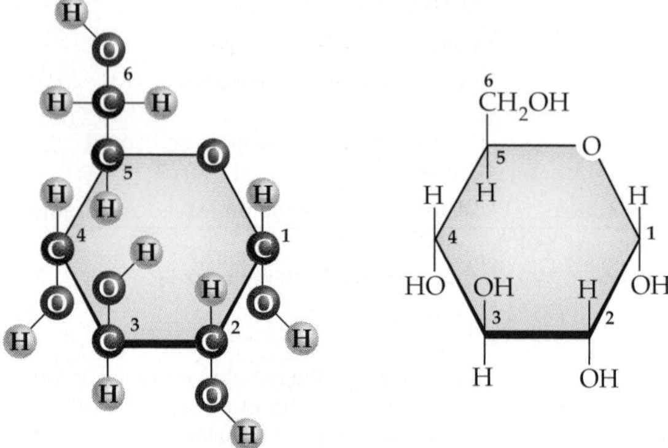

(c) Ring, ball and stick

(d) Ring, simplified

▲ **FIGURE 3-4 Glucose structure depictions** Chemists draw the same molecule in different ways. **(a)** The chemical formula for glucose; **(b)** the linear form, which occurs when glucose is a solid crystal; **(c, d)** two depictions of the ring form of glucose, which forms when glucose dissolves in water. In (d), each unlabeled joint is a carbon atom; the carbons are numbered for easy reference. Figure 3-3 shows a space-filling depiction of the ring structure of glucose.

carbohydrates generally have the approximate chemical formula $(CH_2O)_n$, where n is the number of carbons in the backbone. When dissolved in water, such as in the cytoplasmic fluid of a cell, the carbon backbone of a sugar usually forms a ring. **Figure 3-4** and Figure 3-3 show various ways of depicting the chemical structure of the common sugar **glucose.** When you see simplified structures, keep in mind that every unlabeled "joint" in a ring or chain is actually a carbon atom.

Glucose is the most common monosaccharide in living organisms and is a subunit of many polysaccharides. Glucose has six carbons, so its chemical formula is $C_6H_{12}O_6$. Many organisms synthesize other monosaccharides that have the same chemical formula as glucose but slightly different structures. These include fructose ("fruit sugar" found in fruits, corn syrup, and honey) and galactose (part of lactose, or "milk sugar") (**Fig. 3-5**).

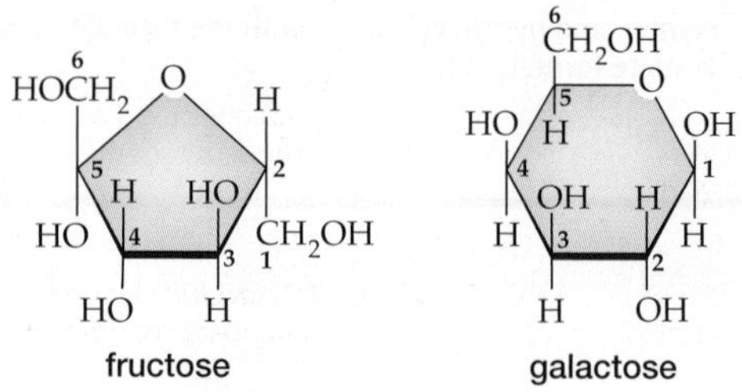

fructose

galactose

▲ **FIGURE 3-5 Monosaccharides**

Other common monosaccharides, such as ribose and deoxyribose (found in the nucleic acids of RNA and DNA, respectively), have five carbons. Notice in **Figure 3-6** that deoxyribose ("deoxy" means "remove oxygen") has one fewer oxygen atom than ribose because one of the hydroxyl functional groups in ribose is replaced by a hydrogen atom.

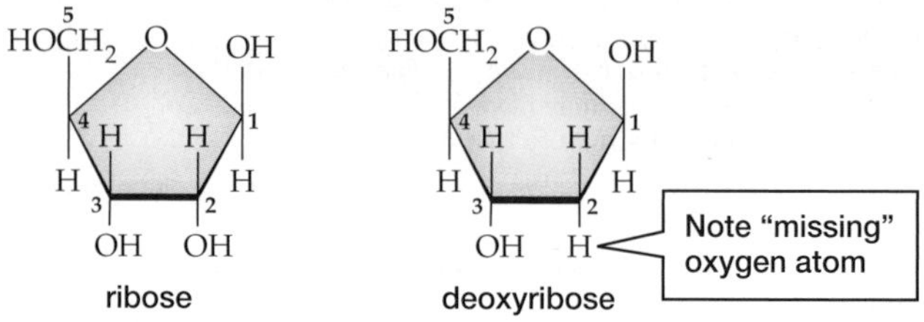

ribose

deoxyribose

Note "missing" oxygen atom

▲ **FIGURE 3-6 Five-carbon monosaccharides**

Disaccharides Consist of Two Single Sugars Linked by Dehydration Synthesis

Monosaccharides can be linked by dehydration synthesis to form disaccharides or polysaccharides (**Fig. 3-7**). Disaccharides are often used for short-term energy storage, especially in plants. When energy is required, the disaccharides are broken apart into their monosaccharide subunits by hydrolysis (see Fig. 3-2), then broken down further to release energy stored in their chemical bonds. Many foods contain disaccharides. Perhaps you had toast and coffee with cream and sugar at breakfast. You stirred **sucrose** (glucose plus fructose, used as

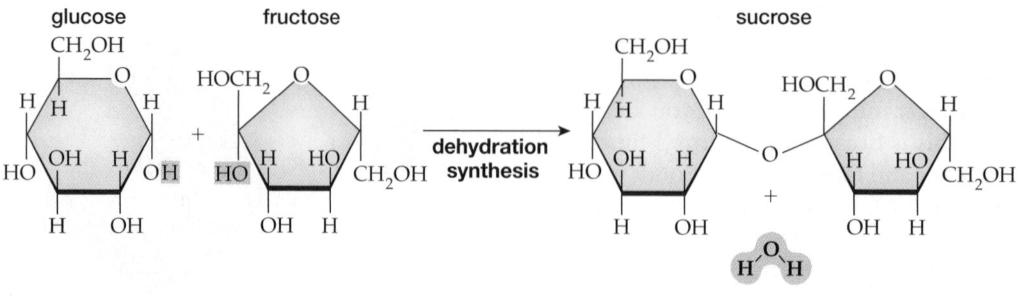

glucose + fructose → dehydration synthesis → sucrose + H_2O

▲ **FIGURE 3-7 Synthesis of a disaccharide** Sucrose is synthesized by a dehydration reaction in which a hydrogen is removed from glucose and a hydroxyl group is removed from fructose, forming a water molecule and leaving the two monosaccharide rings joined by single bonds to the remaining oxygen atom.

QUESTION Describe hydrolysis of this molecule.

an energy storage molecule in sugarcane and sugar beets) into your coffee, then added cream containing **lactose** (glucose plus galactose). The disaccharide **maltose** (glucose plus glucose) is rare in nature, but it is formed when enzymes in your digestive tract hydrolyze starch, such as that in your toast. Other digestive enzymes then hydrolyze each maltose into two glucose molecules that cells can absorb and break down to liberate energy.

If you are on a diet, you may have used an artificial sugar substitute in your coffee. These interesting molecules are described in "Links to Everyday Life: Fake Foods."

Polysaccharides Are Chains of Single Sugars

Try chewing a cracker for a long time. Does it taste sweeter the longer you chew? It should, because over time, enzymes in saliva cause hydrolysis of the **starch** (a polysaccharide) in crackers into its component glucose sugar molecules (monosaccharides), which taste sweet. Plants often use starch (**Fig. 3-8**) as an energy-storage molecule, and animals commonly store **glycogen,** a similar polysaccharide. Both consist of polymers of glucose molecules. Starch is commonly formed in roots and seeds; a cracker contains starch from wheat seeds. Starch consists of branched chains of up to half a million glucose subunits. Glycogen, stored in the liver and muscles of animals (including ourselves), is also a chain of glucose subunits, but is more highly branched than starch.

Many organisms use polysaccharides as structural materials. One of the most important structural polysaccharides is **cellulose,** which makes up most of the cell walls of plants, the fluffy white bolls of a cotton plant, and about half the bulk of a tree trunk (**Fig. 3-9**). Ecologists estimate that about 1 trillion tons of cellulose are synthesized by organisms each year, making it the most abundant organic molecule on Earth.

Cellulose, like starch, is a polymer of glucose, but in cellulose, unlike starch, every other glucose is "upside down" (compare Fig. 3-8c with Fig. 3-9d). Although most animals can easily digest starch, no vertebrates have enzymes that can break down cellulose. As you will learn in Chapter 6, enzymes must precisely fit the molecules they break down, and the bonding pattern in cellulose prevents any vertebrate enzyme from attacking it. A few animals, such as cows, harbor cellulose-digesting microbes in their digestive tracts and can benefit from the glucose subunits they release. In humans, cellulose fibers pass intact through the digestive system, providing no nutrients, but providing "roughage" that helps prevent constipation.

Links to *Everyday Life*

Fake Foods

In societies blessed with an overabundance of food, obesity is a serious health problem. One goal of food scientists is to modify biological molecules to make them noncaloric; sugar (at 4 Cal/gram) and fat (at 9 Cal/gram) are prime candidates. Several artificial sweeteners such as aspartame (Nutrasweet™) and sucralose (Splenda™) add a sweet taste to foods, while providing few or no calories. The artificial oil called olestra is completely indigestible, ensuring that potato chips made with olestra have no fat calories and far fewer total calories than normal chips (**Fig. E3-1**). How are these "nonbiological molecules" made?

Aspartame is a combination of two amino acids: aspartic acid and phenylalanine (see Fig. 3-18b). For unknown reasons, aspartame is far more effective than sugar in stimulating the sweet taste receptors on our tongues. Sucralose is a modified sucrose molecule in which three hydroxyl groups are replaced with chlorine atoms. Sucralose activates our sweet taste receptors 600 times as effectively as sucrose, but our enzymes cannot digest it, so it provides no calories.

To understand olestra, look at Figure 3-13b and you will see that oils contain a glycerol backbone whose three carbon atoms are each bonded to a fatty acid. In Olestra, the glycerol molecule is replaced by a sucrose molecule with six to eight fatty acids attached to its carbon atoms. The fatty acid chains—extending outward from the rings of the sucrose molecule like the arms of an octopus—prevent digestive enzymes from breaking down the molecule into absorbable fragments. Although it is not digested, olestra adds the same appealing feel and flavor to foods that oil does.

▲ **FIGURE E3-1 Artificial foods** The sucralose in Splenda™ and the olestra in fat-free potato chips are synthetic, indigestible versions of sugar and oil, respectively, that are intended to help people control weight.

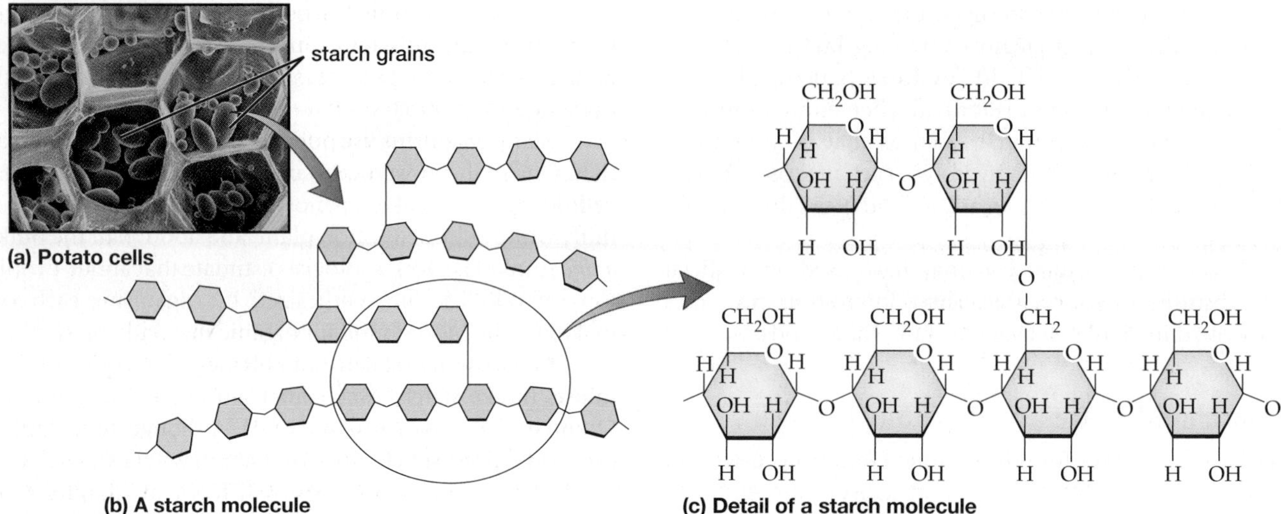

▲ FIGURE 3-8 Starch is an energy-storage plant polysaccharide made of glucose subunits **(a)** Starch grains inside individual potato cells. **(b)** A small section of a single starch molecule, which occurs as branched chains of up to half a million glucose subunits. **(c)** The precise structure of the circled portion of the starch molecule in (b). Notice the linkage between the individual glucose subunits for comparison with cellulose (Fig. 3-9).

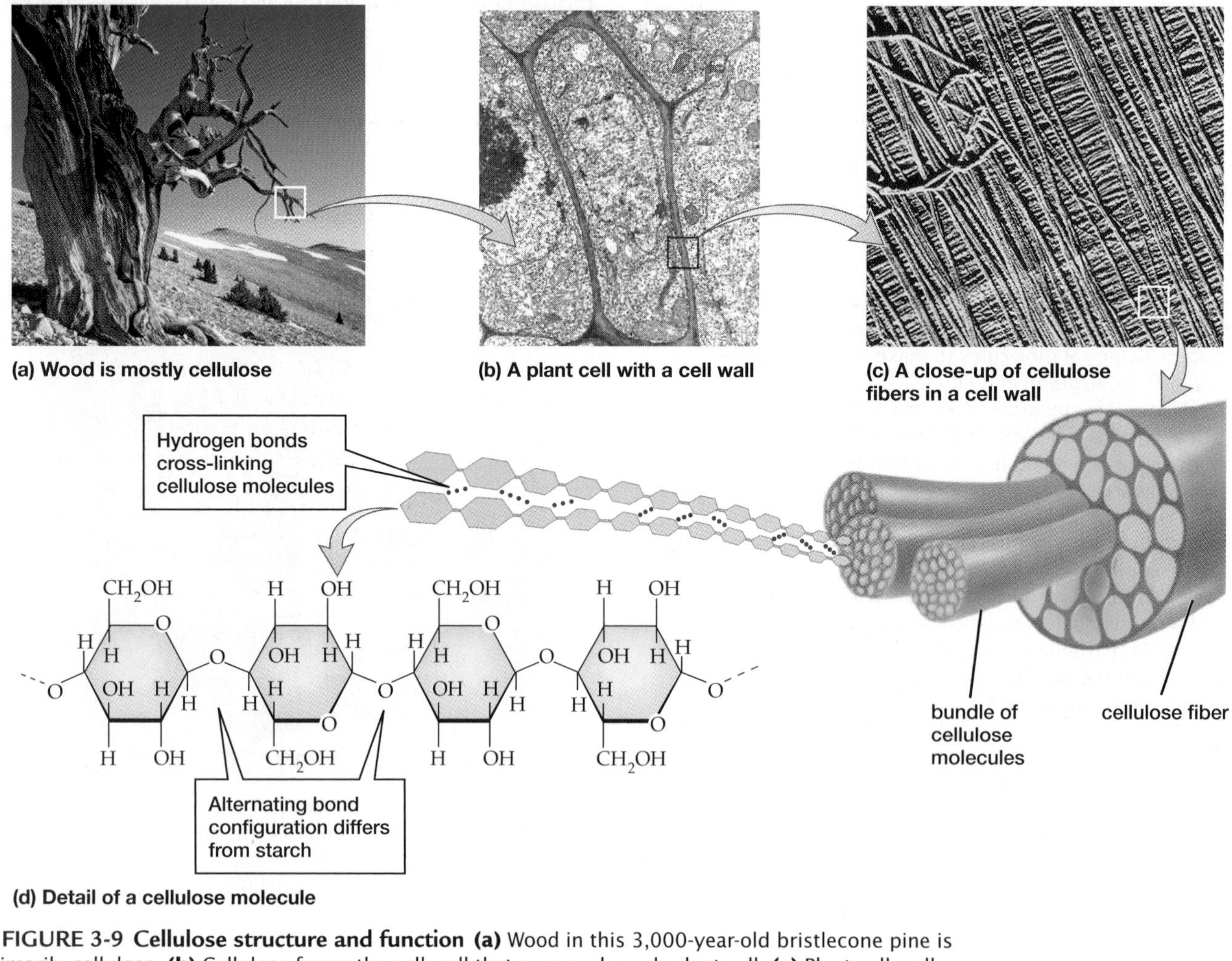

(a) Wood is mostly cellulose

(b) A plant cell with a cell wall

(c) A close-up of cellulose fibers in a cell wall

Hydrogen bonds cross-linking cellulose molecules

bundle of cellulose molecules

cellulose fiber

Alternating bond configuration differs from starch

(d) Detail of a cellulose molecule

▲ FIGURE 3-9 Cellulose structure and function **(a)** Wood in this 3,000-year-old bristlecone pine is primarily cellulose. **(b)** Cellulose forms the cell wall that surrounds each plant cell. **(c)** Plant cell walls often consist of cellulose fibers in layers that run at angles to each other and resist tearing in both directions. **(d)** Cellulose is composed of glucose subunits. Compare this structure with Figure 3-8c and notice that every other glucose molecule in cellulose is "upside down."

Have you ever wondered?

How Termites Can Digest Wood?

The answer to this is that, although they can eat wood, termites can't digest it by themselves. Like cows, termites harbor diverse populations of cellulose-digesting bacteria and protists in their guts. In this mutually beneficial relationship, the termites provide a home, and the microbes earn their keep by breaking down cellulose into its glucose subunits, which it shares with the termite. Different types of termites harbor unique assemblages of cellulose-digesting microbes. Some protists that live in termites are so dependent on cellulose that they will die if the termite is fed starch.

The supportive outer coverings (exoskeletons) of insects, crabs, and spiders are made of **chitin,** a polysaccharide in which the glucose subunits bear a nitrogen-containing functional group (**Fig. 3-10**). Chitin also stiffens the cell walls of many fungi, including mushrooms.

Carbohydrates may also form parts of larger molecules; for example, the plasma membrane that surrounds each cell is studded with proteins to which carbohydrates are attached. Nucleic acids, discussed later, also contain sugar molecules.

3.4 WHAT ARE LIPIDS?

Lipids are a diverse group of molecules that contain regions composed almost entirely of hydrogen and carbon, with nonpolar carbon–carbon and carbon–hydrogen bonds. These nonpolar regions make lipids hydrophobic and insoluble in water. Some lipids are used to store energy, some form waterproof coverings on plant or animal bodies, some serve as the primary component of cellular membranes, and still others are hormones.

Lipids are classified into three major groups: (1) oils, fats, and waxes, which are similar in structure and contain only carbon, hydrogen, and oxygen; (2) phospholipids, which are structurally similar to oils, but which also contain phosphorus and nitrogen; and (3) the "fused-ring" family of steroids that are composed of carbon, hydrogen, and oxygen.

Oils, Fats, and Waxes Are Lipids Containing Only Carbon, Hydrogen, and Oxygen

Oils, fats, and waxes are related in three ways. First, they contain only carbon, hydrogen, and oxygen. Second, they contain one or more **fatty acid** subunits, which are long chains of carbon and hydrogen with a carboxylic acid group (—COOH) at one end. Finally, most do not have ring structures.

Fats and **oils** are used primarily as energy-storage molecules; they contain more than twice as many calories per gram as do carbohydrates and proteins (9 Cal/gram for fats; 4 Cal/gram for carbohydrates and proteins). The bear in **Figure 3-11a** has accumulated enough fat to last him through his winter hibernation. To avoid accumulating excess fat, some people turn to foods manufactured with fat substitutes such as olestra, as described in "Links to Everyday Life: Fake Foods."

Fats and oils are formed by dehydration synthesis linking three fatty acid subunits to one molecule of **glycerol,** a three-carbon molecule (**Fig. 3-12**). This structure gives fats and oils their chemical name: **triglycerides.**

Fats (such as butter and lard) are produced primarily by animals, whereas oils (such as corn, canola, and soybean oil) are found primarily in the seeds of plants. The difference between a fat, which is a solid at room temperature, and an oil, which is liquid, lies in the structure of their fatty acid subunits. In fats, the carbons of fatty acids are joined by single bonds, with hydrogen atoms at all the other bonding sites. Such fatty acids are described as **saturated,** because they contain as many hydrogen atoms as possible. Lacking double bonds between carbons, these saturated fatty acid chains are

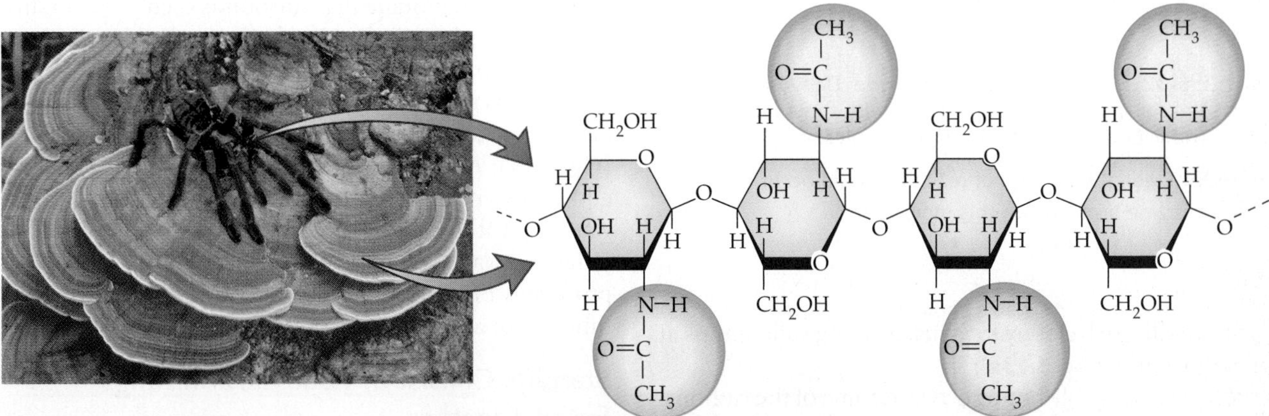

▲ **FIGURE 3-10 Chitin: a unique polysaccharide** Chitin has the same bonding configuration of glucose molecules as cellulose does. In chitin, however, the glucose subunits have a nitrogen-containing functional group (green) instead of a hydroxyl group. Tough, flexible chitin supports the otherwise soft bodies of arthropods (insects, spiders, and their relatives) and most fungi, such as this shelf fungus.

(a) Fat

(b) Wax

▲ **FIGURE 3-11 Lipids in nature (a)** A fat grizzly bear ready to hibernate. If this bear stored the same amount of energy in carbohydrates instead of fat, he probably would be unable to walk. **(b)** Wax is a highly saturated lipid that remains very firm at normal outdoor temperatures. Its rigidity allows it to be used to form the strong but thin-walled hexagons of this honeycomb.

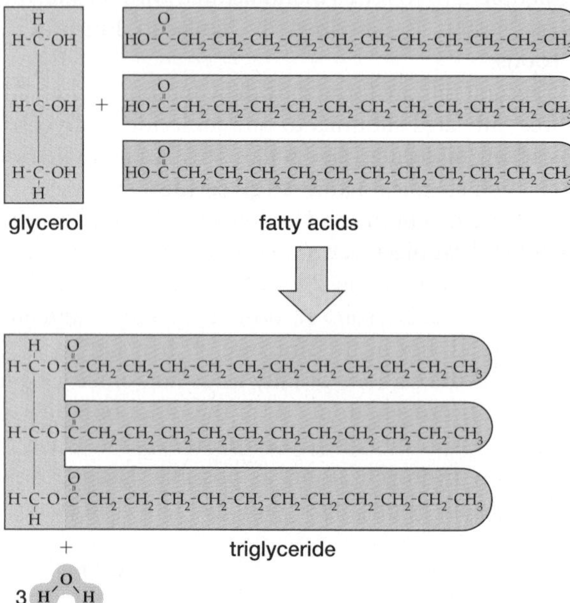

▲ **FIGURE 3-12 Synthesis of a triglyceride** Dehydration synthesis links a single glycerol molecule with three fatty acids to form a triglyceride and three water molecules.

QUESTION Predict what kind of reaction would occur if a fat-digesting enzyme broke down this molecule.

straight and can pack closely together, forming solid lumps at room temperature (**Fig. 3-13a**).

If there are double bonds between some of the carbons, and consequently fewer hydrogens, the fatty acid is **unsaturated.** The double bonds in the unsaturated fatty acids of oils produce kinks in the fatty acid chains (**Fig. 3-13b**). Oils are liquid at room temperature because these kinks prevent oil molecules from packing closely together. The commercial process of hydrogenation—which breaks some of the double bonds and adds hydrogens to the carbons—can convert liquid oils to solids, but with some health consequences (see Health Watch: "Cholesterol, Trans Fats, and Your Heart" on p. 46).

Although **waxes** are chemically similar to fats, humans and most other animals do not have the appropriate enzymes to break them down. Waxes are highly saturated and therefore solid at normal outdoor temperatures. Waxes form a waterproof coating over the leaves and stems of land plants. Animals synthesize waxes as waterproofing for mammalian fur and insect exoskeletons. Honey bees use waxes to build elaborate honeycomb structures where they store honey and lay their eggs (**Fig. 3-11b**).

Phospholipids Have Water-Soluble "Heads" and Water-Insoluble "Tails"

The plasma membrane that surrounds each cell contains several types of **phospholipids.** A phospholipid is similar to an oil, except that one of the three fatty acids that form the "fatty acid tails" is replaced by a phosphate group attached to a variable polar functional group that typically contains nitrogen (**Fig. 3-14**). A phospholipid has two dissimilar ends. At one end are two nonpolar fatty acid "tails," which are insoluble in water, and at the other is a phosphate–nitrogen "head" that is polar and water soluble. As you will learn in Chapter 5, these properties of phospholipids are crucial to the structure and function of all the membranes of cells.

Steroids Consist of Four Carbon Rings Fused Together

In contrast to fats and oils, which lack rings, all **steroids** are composed of four rings of carbon atoms, fused together, with various functional groups protruding from them (**Fig. 3-15**).

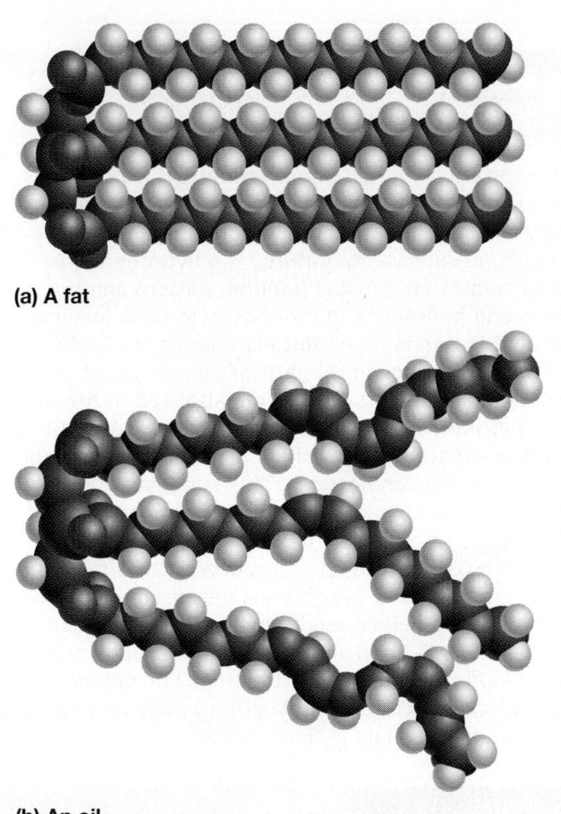

▲ **FIGURE 3-13 A fat and an oil** **(a)** Fats have straight chains of carbon atoms in their fatty acid tails. **(b)** The fatty acid tails of oils have double bonds between some of their carbon atoms, creating kinks in the chains. Oils are liquid at room temperature because their kinky tails keep the molecules farther apart.

One type of steroid is cholesterol. Cholesterol (**Fig. 3-15a**) is a vital component of the membranes of animal cells. It makes up about 2% of the human brain, where it is an important component of the material used to insulate nerve cells. Cholesterol is used by cells to synthesize other steroids, including the female sex hormone estrogen (**Fig. 3-15b**), the male sex hormone testosterone (**Fig. 3-15c**), and bile that assists in fat digestion. Too much of the wrong form of cholesterol can contribute to heart disease, as explained in "Health Watch: Cholesterol, Trans Fats, and Your Heart."

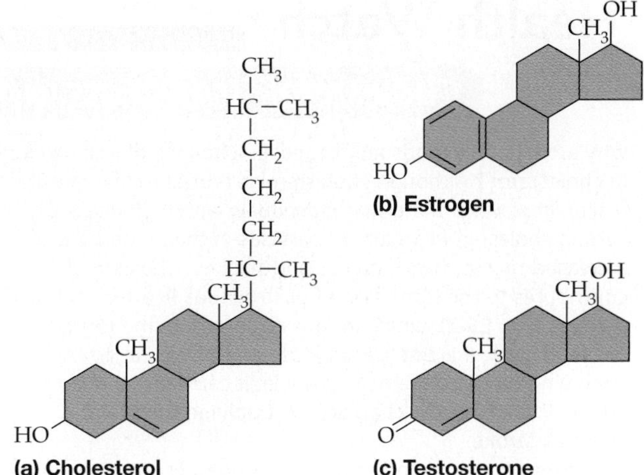

▲ **FIGURE 3-15 Steroids** All steroids have a similar, nonpolar molecular structure with four fused carbon rings. Differences in steroid function result from differences in functional groups attached to the rings. **(a)** Cholesterol, the molecule from which other steroids are synthesized; **(b)** estradiol (estrogen), the female sex hormone; **(c)** the male sex hormone testosterone. Note the similarities in structure between estrogen and testosterone.

QUESTION Why are steroid hormones able to penetrate plasma membranes and nuclear membranes of cells to exert their effects?

3.5 WHAT ARE PROTEINS?

Proteins are molecules composed of one or more chains of amino acids. Proteins perform many functions; this diversity of function is made possible by the diversity of protein structures (**Table 3-3**). Most cells contain hundreds of different **enzymes,** proteins that promote chemical reactions. Other types of proteins form structures on or outside the body.

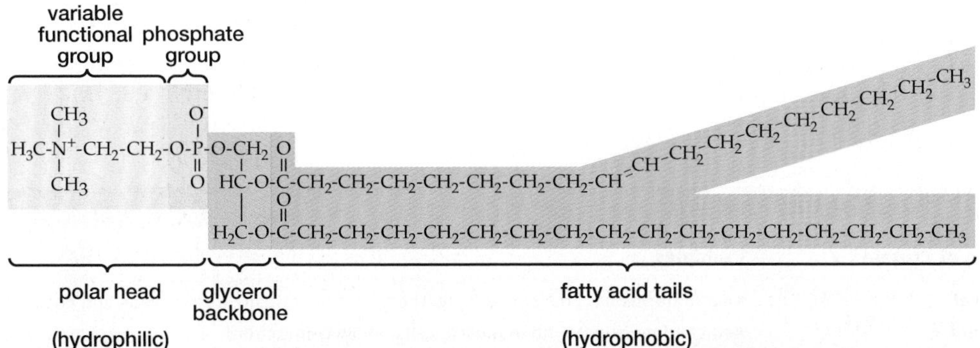

▲ **FIGURE 3-14 Phospholipids** Phospholipids have two hydrophobic fatty acid tails attached to the glycerol backbone. The third position on glycerol is occupied by a polar "head" consisting of a phosphate group to which a second (usually nitrogen-containing) variable functional group is attached. The phosphate group bears a negative charge, and the nitrogen-containing group bears a positive charge, making the heads hydrophilic.

Health Watch

Cholesterol, Trans Fats, and Your Heart

Why are so many foods advertised as "cholesterol free" or "low in cholesterol"? Although cholesterol is crucial to life, medical researchers have found that individuals with high levels of certain cholesterol-containing particles in their blood are at increased risk for heart attacks and strokes. Cholesterol contributes to the formation of obstructions in arteries, called plaques (**Fig. E3-2**), which in turn can promote the formation of blood clots. If a clot breaks loose and blocks an artery supplying blood to the heart muscle, it can cause a heart attack. If the clot blocks an artery supplying the brain, it causes a stroke.

Cholesterol is found in animal-derived foods including egg yolks, sausages, bacon, whole milk, and butter. Perhaps you have heard of "good" and "bad" cholesterol. Because cholesterol molecules are nonpolar, they do not dissolve in blood but are transported in packets surrounded by polar carrier molecules consisting of protein and phospholipids. These carriers are called lipoproteins (lipids plus proteins).

Lipoproteins that consist of more protein and less lipid are described as "high-density lipoproteins" ("HDL"), because protein is denser than lipid. The "good cholesterol" is carried by HDL. These particles are taken up by the liver, where the cholesterol is metabolized (used in bile synthesis, for example). The bad "low-density lipoprotein" ("LDL") cholesterol packets have less protein and more cholesterol. This form circulates to cells throughout the body and can be deposited in artery walls.

Animals, including humans, can synthesize all the cholesterol their bodies require. Most of the cholesterol in human blood is synthesized by the body. Because of genetic differences, however, some people's bodies manufacture more cholesterol, and some (but not all) people respond to a high-cholesterol diet by manufacturing less. Lifestyle also contributes to blood cholesterol content; exercise tends to increase HDL cholesterol, whereas obesity and smoking increase LDL levels. A high ratio of HDL to LDL is correlated with a reduced risk of heart disease. A cholesterol screening test will distinguish between these two forms.

Perhaps you've heard about **trans fats** as dietary villains. These are produced when oils are artificially hardened by combining them with hydrogen to make them solid at room temperature. The hydrogenation process creates an unusual bonding pattern among carbons and hydrogens in the fatty acid tails, leaving some double bonds intact but eliminating the kinks that these bonds produce in the tails of oils. Research has revealed that trans fats are not metabolized in the same way as naturally-occurring fats, and (for reasons still under investigation) can both increase LDL cholesterol and decrease HDL cholesterol. This suggests that they may place consumers at a higher risk of heart disease.

Until recently, artificially produced trans fats were abundant in commercial food products such as margarine, cookies, crackers, and french fries, because they extended the shelf life and helped retain the flavor of packaged products. The U.S. Food and Drug Administration now requires food labels to include the trans fat content, and most food manufacturers and fast-food chains are reducing or eliminating trans fats in their products.

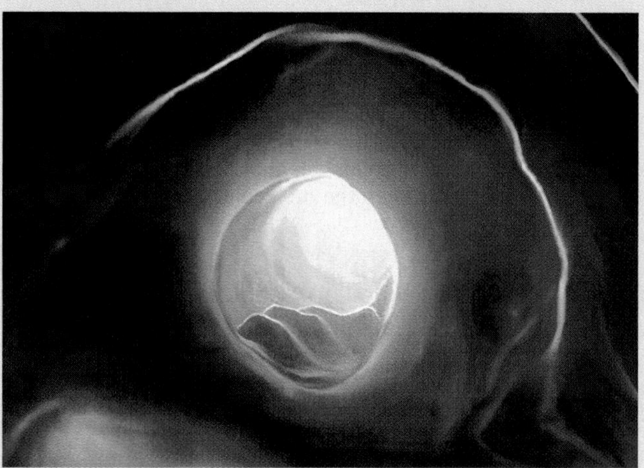

▲ **FIGURE E3-2 Plaque** A plaque deposit (rippled structure) partially blocks a carotid artery.

Table 3-3 Functions of Proteins

Function of Protein	Examples
Structural	Keratin (forms hair, nails, scales, feathers, and horns); silk (forms webs and cocoons)
Movement	Actin and myosin (found in muscle cells; allow contraction)
Defense	Antibodies (found in the bloodstream; fight disease organisms, some neutralize venoms); venoms (found in venomous animals; deter predators and disable prey)
Storage	Albumin (in egg white; provides nutrition for an embryo)
Signaling	Insulin (secreted by the pancreas; promotes glucose uptake into cells)
Catalyzing reactions	Amylase (found in saliva and the small intestine; digests carbohydrates)

These include keratin, the principal protein of hair, horns, nails, scales, and feathers (**Fig. 3-16**). Silk proteins are secreted outside the body by silk moths and spiders to make cocoons and webs. Still other proteins provide a source of amino acids for developing animals, such as albumin protein in egg white and casein protein in milk. The protein hemoglobin transports oxygen in the blood. Contractile proteins, such as actin and myosin in muscle, allow animal bodies to move. Some hormones, including insulin and growth hormone, are small proteins, as are antibodies (which help fight disease and infection) and many toxins (such as rattlesnake venom) produced by animals.

(a) Hair

(b) Horn

(c) Silk

▲ **FIGURE 3-16 Structural proteins** Common structural proteins include keratin, which is the predominant protein found in **(a)** hair, **(b)** horn, and **(c)** the silk of a spider web.

Proteins Are Formed from Chains of Amino Acids

Proteins are polymers consisting of chains of **amino acids** joined by peptide bonds. All amino acids have the same fundamental structure, consisting of a central carbon bonded to three different functional groups: a nitrogen-containing amino group ($—NH_2$), a carboxylic acid group ($—COOH$), and an "R" group that varies among different amino acids (**Fig. 3-17**).

Twenty amino acids are commonly found in the proteins of organisms, and the R group gives each amino acid distinctive properties (**Fig. 3-18**). Some amino acids are hydrophilic and water soluble because their R groups are polar. Others are hydrophobic, with nonpolar R groups that are insoluble in water. The sulfhydryl R group of one amino acid, cysteine (**Fig. 3-18c**), can form covalent **disulfide bonds** with the sulfur in another cysteine molecule. Although peptide bonds (described later) link amino acids to form the chains that make up a protein, disulfide bonds can link different chains of amino acids to one another or interconnect different parts of the same amino acid chain, causing the protein to bend or fold.

Amino Acids Are Joined to Form Chains by Dehydration Synthesis

Like polysaccharides and lipids, proteins are formed by dehydration synthesis. The nitrogen in the amino group ($—NH_2$) of one amino acid is joined to the carbon in the carboxylic acid group ($—COOH$) of another amino acid by a single covalent bond, and water is liberated (**Fig. 3-19**). This bond is called a **peptide bond,** and the resulting chain of two amino acids is called a **peptide.** More amino acids are added, one by one, until the protein chain is complete. Amino acid chains in living cells vary in length from three to thousands. The terms "protein" and "polypeptide" are often reserved for long chains of more than 50 amino acids, whereas the term "peptide" usually refers to shorter chains.

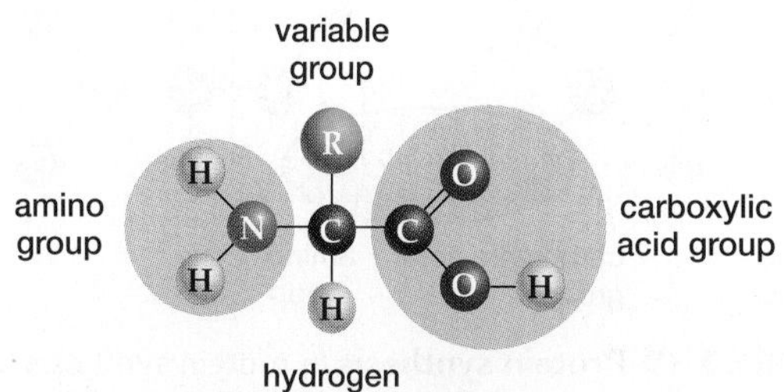

▲ **FIGURE 3-17 Amino acid structure**

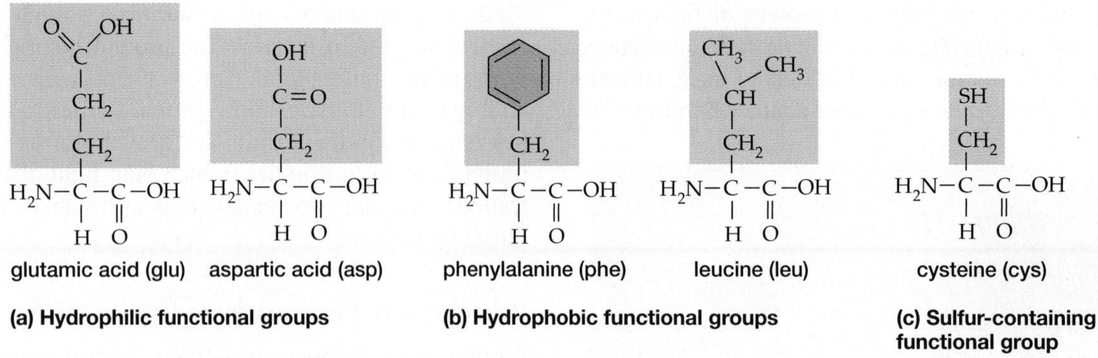

▲ **FIGURE 3-18 Amino acid diversity** The diversity of amino acids is caused by the variable R group (colored blue), which may be **(a)** hydrophilic or **(b)** hydrophobic. The R group of cysteine **(c)** has a sulfur atom that can form covalent bonds with the sulfur in other cysteines, creating disulfide bonds that can bend a protein or link adjacent polypeptide chains.

A Protein Can Have Up to Four Levels of Structure

Proteins come in many shapes, and exhibit up to four levels of protein structure, with each structural level built on the level below it. A molecule of hemoglobin, the oxygen-carrying protein in red blood cells, illustrates all four structural levels (**Fig. 3-20**). A protein's **primary structure** refers to the sequence of amino acids that make up the protein (**Fig. 3-20a**). This sequence is specified by genes within molecules of DNA. Different types of proteins have different sequences of amino acids.

The amino acid sequence in turn causes each polypeptide chain to assume one of two types of simple, repeating **secondary structures.** Secondary structures are maintained by hydrogen bonds between the polar portions of amino acids. Many proteins, such as keratin in hair and the subunits of the hemoglobin molecule, acquire a coiled, spring-like secondary structure called a **helix** (**Fig. 3-20b**). Hydrogen bonds that form between the oxygens of the carbonyl functional groups (with slightly negative charges) and the hydrogens of the amino functional groups (with slightly positive charges) hold the turns of the coils together. Other proteins, such as silk, contain polypeptide chains that repeatedly fold back upon themselves, with hydrogen bonds holding adjacent segments of the polypeptide together in a **pleated sheet** arrangement (**Fig. 3-21**).

Case Study continued
Puzzling Proteins

Normal, noninfectious prion proteins have a secondary structure that is primarily helical. They perform a variety of beneficial functions in the animal body, including in the brain. Infectious prion proteins, however, fold into pleated sheets, and force normal prion proteins to fold into pleated sheets as well. The sheets are so stable that they cannot be denatured by high temperatures nor by most chemical agents. More importantly, they are unaffected by the enzymes that break down normal prion protein. Because they are so stable, infectious prions accumulate harmfully in the brain.

In addition to their secondary structures, proteins assume **tertiary structures,** consisting of folds determined by interactions of the amino acid functional groups with one another and with their surroundings (**Fig. 3-20c**). Tertiary structure is determined both by the protein's secondary structure and its environment—for example, whether the protein is dissolved in the watery cytoplasmic fluid within a cell or in the lipids of cellular membranes, or both. A protein in water folds in a way that exposes its hydrophilic (polar, "water-loving") amino acids to the water and causes its hydrophobic amino acids to cluster together in the center of the molecule. In

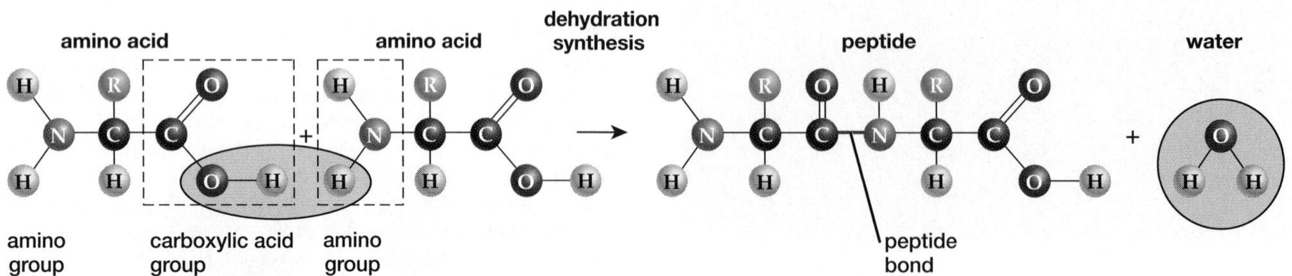

▲ **FIGURE 3-19 Protein synthesis** In protein synthesis, a dehydration reaction joins the carbon of the carboxylic acid group of one amino acid to the nitrogen of the amino group of a second amino acid, releasing water. The resulting covalent bond between amino acids is a peptide bond.

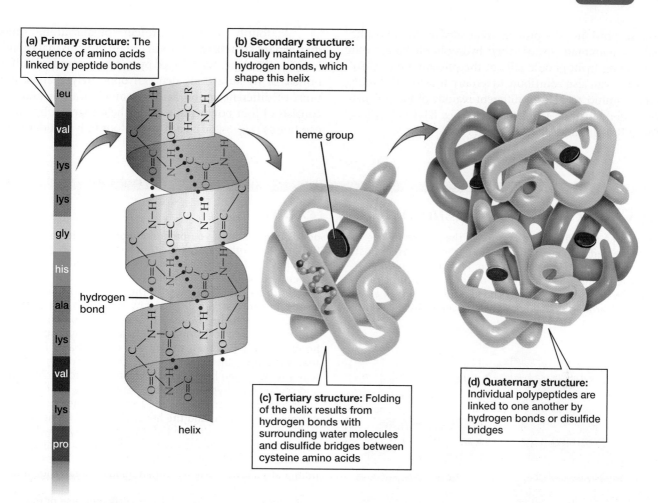

(a) **Primary structure:** The sequence of amino acids linked by peptide bonds

leu
val
lys
lys
gly
his
ala
lys
val
lys
pro

(b) **Secondary structure:** Usually maintained by hydrogen bonds, which shape this helix

hydrogen bond

helix

heme group

(c) **Tertiary structure:** Folding of the helix results from hydrogen bonds with surrounding water molecules and disulfide bridges between cysteine amino acids

(d) **Quaternary structure:** Individual polypeptides are linked to one another by hydrogen bonds or disulfide bridges

▲ **FIGURE 3-20 The four levels of protein structure** Levels of protein structure are illustrated here with hemoglobin, the oxygen-carrying protein in red blood cells (red disks represent the iron-containing heme group that binds oxygen). Levels of protein structure are generally determined by the amino acid sequence of the protein, interactions among the R groups of the amino acids, and interactions among the R groups and their surroundings.

QUESTION Why do most proteins, when heated, lose their ability to function?

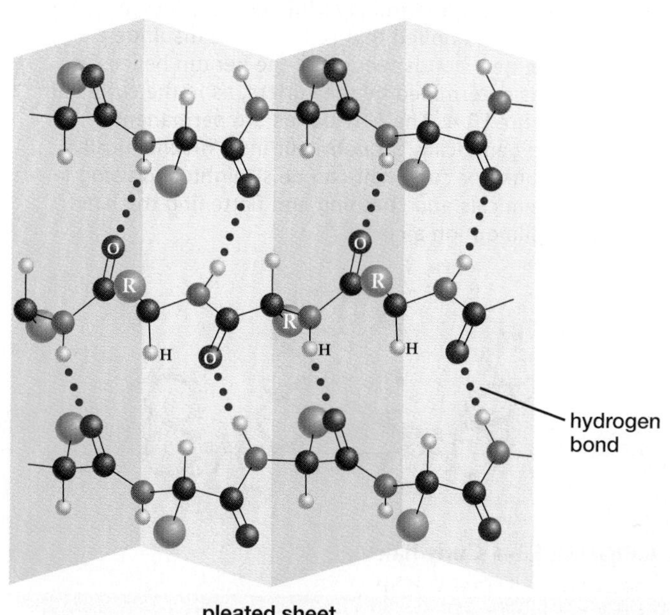

◀ **FIGURE 3-21 The pleated sheet is an example of protein secondary structure** In a pleated sheet, a single polypeptide chain is folded back upon itself repeatedly (loops of peptide connecting the ends of the chains in the sheet are not shown). Adjacent segments of the folded polypeptide are linked by hydrogen bonds (dotted lines), creating a sheet-like configuration. The R groups (green) project alternately above and below the sheet. Despite its accordion-pleated appearance, each peptide chain is in a fully extended state and cannot easily be stretched farther. Silk strands are stretchy because segments of pleated sheets are interspersed with regions of spirally structured proteins, which are highly elastic.

contrast, portions of a protein embedded in the phospholipid cell membrane expose their hydrophobic parts to the surrounding hydrophobic tails of the phospholipid. Disulfide bonds can also contribute to tertiary structure by linking cysteine amino acids from different regions of the polypeptide, as in the hair protein keratin (see Fig. E3-3 in "A Closer Look at Proteins and Hair Texture").

A fourth level of protein organization, called **quaternary structure,** occurs in certain proteins that contain individual polypeptides linked by hydrogen bonds, disulfide bonds, or by attractions between oppositely charged portions of different amino acids. Hemoglobin, for example, consists of four polypeptide chains held together by hydrogen bonds (**Fig. 3-20d**). Each of the four polypeptides holds

A Closer Look At *Proteins and Hair Texture*

Pull out a strand of hair, and notice the root or follicle that was embedded in the scalp. Hair is composed mostly of a helical protein called keratin. Living cells in the hair follicle produce new keratin at the rate of 10 turns of the protein helix every second. The keratin proteins in a hair entwine around one another, held together by disulfide bonds (**Fig. E3-3**). If you pull gently on the hair, you will find it to be both strong and stretchy. When hair stretches, hydrogen bonds that create the helical structure of keratin are broken, allowing the strand of protein to be extended. Most of the covalent disulfide bonds, in contrast, are distorted by stretching, but do not break. When tension is released, these disulfide bonds return the hair to its normal length, and the hydrogen bonds re-form. Wet the hair and you will probably find that it is limp, a bit longer, and easier to stretch. In wet hair, the hydrogen bonds of the keratin helices are broken and replaced by hydrogen bonds between the amino acids and the water molecules surrounding them, so the protein is denatured and the helices collapse. If you roll wet hair around a curler and allow it to dry, hydrogen bonds will re-form in slightly different places, holding the hair in a curl. However, the slightest moisture (even humid air) allows these hydrogen bonds to rearrange into their natural configuration.

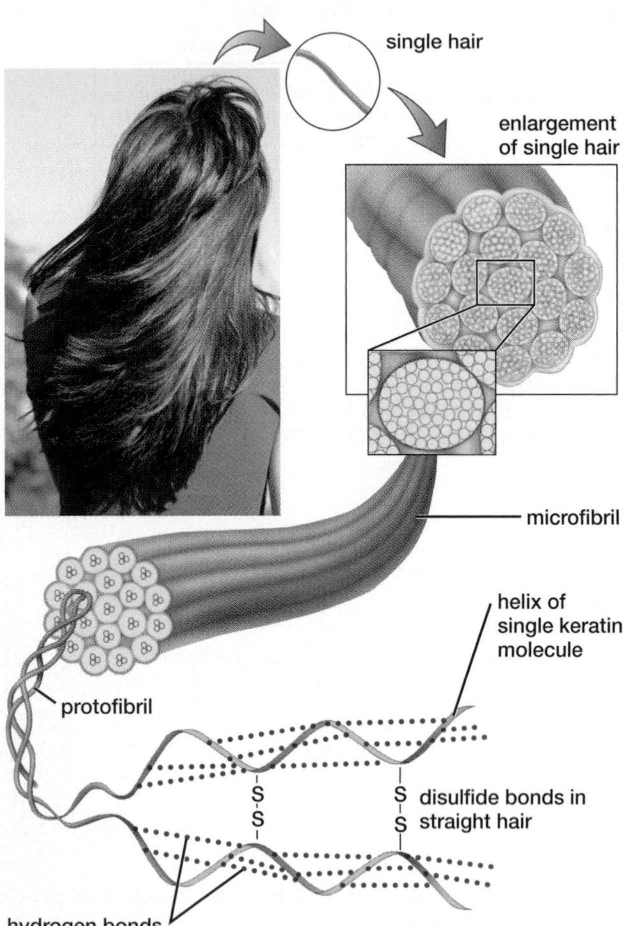

▲ **FIGURE E3-3 The structure of hair** At the microscopic level, a single hair is organized into bundles of protofibrils within larger bundles called microfibrils. Each protofibril consists of keratin molecules held in a helical shape by hydrogen bonds, with different keratin strands cross-linked by disulfide bonds. These bonds give hair both elasticity and strength. Straight hair is depicted here.

If your hair is naturally kinky or curly (because of the sequence of amino acids specified by your genes), the disulfide bonds within and between the individual keratin helices form at locations that bend the keratin molecules, producing curls or kinks (**Fig. E3-4**).

In straight hair, the disulfide bonds occur in places that do not distort the keratin (as shown in Fig. E3-3). When straight hair is given a "permanent wave," two lotions are applied. The first breaks disulfide bonds, denaturing the protein. After the hair is rolled tightly onto curlers, a second solution is applied that re-forms the disulfide bonds. These new bonds reconnect the keratin helices at new positions determined by the curlers, as in the curly hair shown in Figure E3-4. The new bonds are permanent, transforming genetically straight hair into "biochemically curly" hair. Kinky or curly hair can be straightened using the same chemicals and combing and flattening the hair instead of rolling it on a curler.

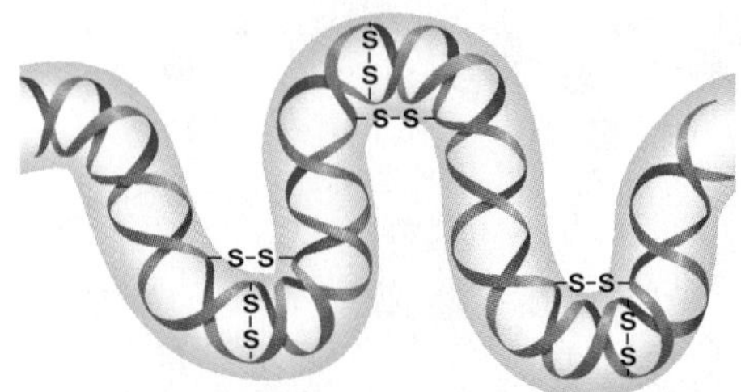

▲ **FIGURE E3-4 Curly hair**

an iron-containing organic molecule called a heme group (red disks in Figs. 3-20c,d), which can bind one molecule (two atoms) of oxygen.

The Functions of Proteins Are Related to Their Three-Dimensional Structures

Within a protein, the exact type, position, and number of amino acids bearing specific R groups determine both the structure of the protein and its biological function. In hemoglobin, for example, amino acids bearing specific R groups must be present in precisely the right places to hold the iron-containing heme group that binds oxygen. In contrast, the polar amino acids on the outside of a hemoglobin molecule serve mostly to keep it dissolved in the cytoplasmic fluid of a red blood cell. Therefore, as long as they are hydrophilic, changes in these amino acids will not alter the protein's function. A mutation that replaces a hydrophilic with a hydrophobic amino acid, however, can have catastrophic effects on the hemoglobin molecule, causing the painful and sometimes life-threatening disorder called sickle-cell anemia (see pp. 190–191).

The importance of higher level protein structure becomes obvious when a protein is **denatured,** meaning that its secondary, tertiary, or quaternary structures are altered while leaving the primary structure intact. A denatured protein has different properties and will no longer perform its function. In a fried egg, for example, the heat of the frying pan causes so much movement of the atoms in the albumin protein that hydrogen bonds are ripped apart. Due to the loss of its secondary and tertiary structure, the egg white changes from clear to white and its texture changes from liquid to solid. When hair is given a permanent wave, the disulfide bonds of keratin are altered, denaturing the natural protein as described in "A Closer Look at Proteins and Hair Texture." Bacteria and viruses are often destroyed by denaturing their proteins. This can be accomplished using heat, ultraviolet rays, or solutions that are salty or acidic. For example, canned foods are sterilized by heating them, water is sometimes sterilized using ultraviolet light, and dill pickles are preserved from bacterial attack in their salty, acidic brine.

3.6 WHAT ARE NUCLEOTIDES AND NUCLEIC ACIDS?

A **nucleotide** is a molecule with a three-part structure: a five-carbon sugar, a phosphate functional group, and a nitrogen-containing **base** that differs among nucleotides. Nucleotides fall into two general classes: deoxyribose nucleotides and ribose nucleotides, depending on which type of sugar they contain (see Fig. 3-6). Bases all have carbon and nitrogen atoms linked in rings, with functional groups attached to some of the carbon atoms. The bases in deoxyribose nucleotides can be adenine, guanine, cytosine, or thymine. Adenine and guanine have double rings, whereas cytosine and thymine have single rings (see Fig. 3-24 and Fig. 11-3). A nucleotide with the adenine base is illustrated in **Figure 3-22**.

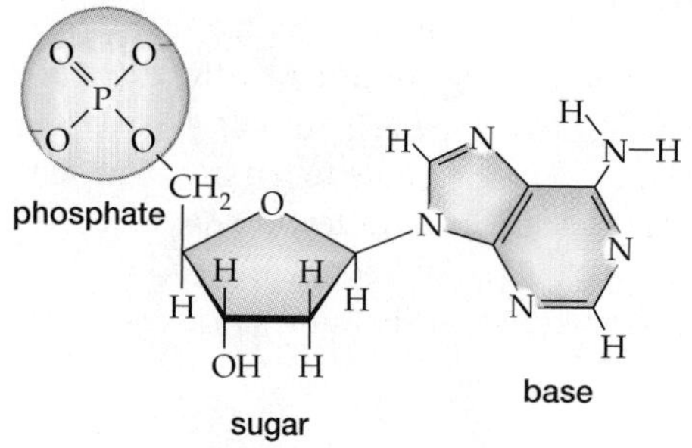

▲ **FIGURE 3-22 Deoxyribose nucleotide**

Nucleotides may function as individual energy-carrier molecules or intracellular messengers, or as subunits in polymers called nucleic acids (see below).

Nucleotides Act as Energy Carriers and Intracellular Messengers

Adenosine triphosphate (ATP) is a ribose nucleotide with three phosphate functional groups (**Fig. 3-23**). This unstable molecule carries energy from place to place within a cell, storing the energy in bonds between its phosphate groups. Energy is released to drive energy-demanding reactions (to synthesize a protein, for example) when the bond linking the last phosphate to ATP is broken.

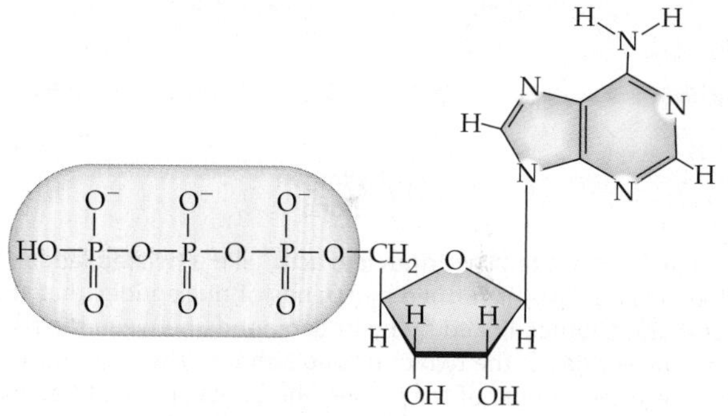

▲ **FIGURE 3-23 The energy-carrier molecule adenosine triphosphate (ATP)**

Other nucleotides (NAD^+ and FAD) are known as "electron carriers" that transport energy in the form of high-energy electrons. You will learn more about these nucleotides in Chapters 6, 7, and 8. The nucleotide cyclic adenosine monophosphate (cAMP) acts as an intracellular messenger molecule, often initiating a series of reactions inside cells.

DNA and RNA, the Molecules of Heredity, Are Nucleic Acids

Single nucleotides (monomers) may be strung together in long chains, forming polymers called **nucleic acids.** In nucleic acids, an oxygen in the phosphate functional group of one nucleotide is covalently bonded to the sugar of the next. The polymer of deoxyribose nucleotides, called **deoxyribonucleic acid (DNA),** can contain millions of nucleotides (**Fig. 3-24**). DNA is found in the chromosomes of all cells. Its sequence of nucleotides, like the dots and

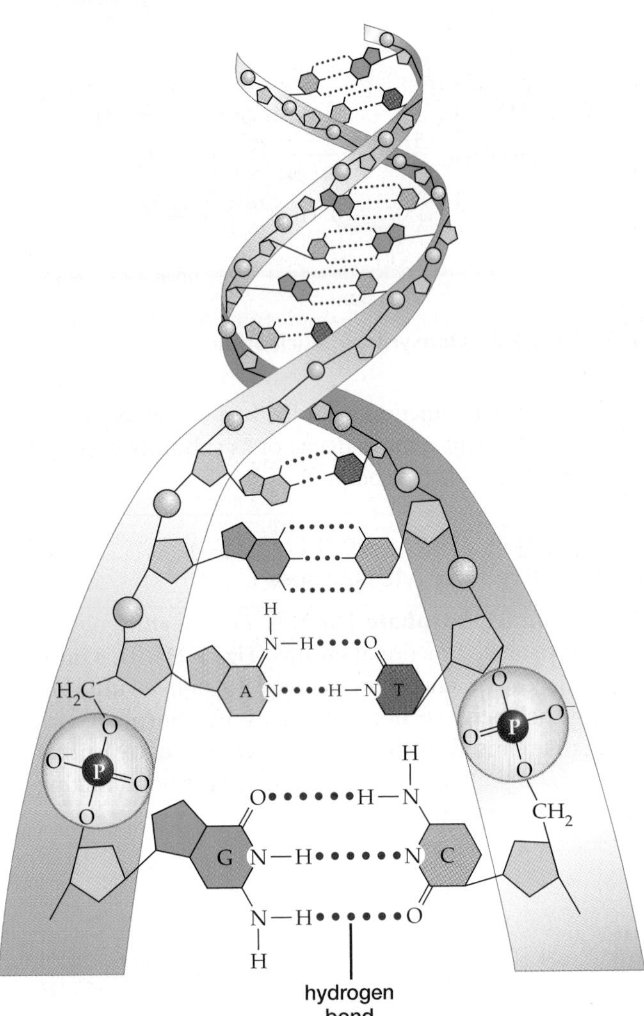

▲ **FIGURE 3-24 Deoxyribonucleic acid** Like a twisted ladder, the double helix of DNA is formed by strands of nucleotides that spiral around one another, linked by hydrogen bonds between the bases of the nucleotides in the two opposing strands. These connected bases form the "rungs" of the ladder. The bases are identified by their initials: A: adenine; C: cytosine; T: thymine; G: guanine.

dashes of a biological Morse code, spells out the genetic information needed to construct the proteins of each organism. A DNA molecule consists of two strands of nucleotides entwined in the form of a double helix and linked by hydrogen bonds (see Fig. 3-24). Single-stranded chains of ribose nucleotides, called **ribonucleic acid (RNA),** are copied from the DNA and direct the synthesis of proteins (see Chapters 11 and 12).

Case Study continued
Puzzling Proteins

All cells use DNA as a blueprint for producing more cells, and different viruses use either DNA or RNA. Before the discovery of prions, however, no infectious agent had ever been discovered that completely lacked genetic material composed of nucleic acids. Scientists were extremely skeptical of Prusiner's hypothesis that proteins could effectively "reproduce themselves" until repeated studies found no trace of genetic material associated with prions.

Case Study revisited
Puzzling Proteins

Prusiner and his associates identified a misfolded version of a protein found throughout the animal kingdom as the culprit in scrapie, BSE, and vCJD. But how does the infectious prion "reproduce" itself? Prusiner and other researchers have strong experimental evidence supporting a radical hypothesis: The misfolded prion interacts with normal prion proteins and causes them to change configuration, misfolding into the pleated sheet secondary structure of the infectious form. These new misfolded "converts" then go on to transform more normal prion proteins in an ever-expanding chain reaction. As in Charlene's case, it can be years after infection before enough of the proteins have been transformed to cause disease symptoms. Researchers are investigating how the abnormal folding occurs, how it can induce misfolding of normal prion proteins, and how misfolded prions cause disease.

Stanley Prusiner's efforts led to the recognition of a totally new disease process. A major strength of the scientific method is that hypotheses can be tested experimentally. If repeated experiments support the hypothesis, then even well-established scientific principles—for example, that all infectious agents must have genetic material—must be reexamined and sometimes discarded. Although a few scientists still insist that proteins can't be infectious, Prusiner's efforts have been so convincing to the scientific community that he was awarded a Nobel Prize in 1997.

Consider This

A disorder called "chronic wasting disease" has been reported among both wild and captive elk and deer populations in several western states in the United States. Like scrapie and BSE, wasting disease is a fatal brain disorder caused by prions. No human cases have been confirmed. If you were a hunter in an affected region, would you eat deer or elk meat? Explain why or why not.

CHAPTER REVIEW

Summary of Key Concepts

3.1 Why Is Carbon So Important in Biological Molecules?

Organic molecules are so diverse because the carbon atom is able to form many types of bonds. This ability, in turn, allows organic molecules (molecules with a backbone of carbon) to form complex shapes, including branched chains, helices, sheets, and rings. The presence of functional groups produces further diversity among biological molecules (see Table 3-1).

3.2 How Are Organic Molecules Synthesized?

Most large biological molecules are polymers synthesized by linking together many smaller monomer subunits. The subunits are connected by covalent bonds through dehydration synthesis; the chains may be broken apart by hydrolysis reactions. The most important organic molecules fall into one of four classes: carbohydrates, lipids, proteins, and nucleotides/nucleic acids (see Table 3-2).

3.3 What Are Carbohydrates?

Carbohydrates include sugars, starches, chitin, and cellulose. Sugars (monosaccharides and disaccharides) are used for temporary storage of energy and for the construction of other molecules. Starches and glycogen are polysaccharides that provide longer-term energy storage in plants and animals, respectively. Cellulose forms the cell walls of plants, and chitin strengthens the exoskeletons of many invertebrates and the cell walls of fungi. Other polysaccharides form the cell walls of bacteria.

3.4 What Are Lipids?

Lipids are nonpolar, water-insoluble molecules of diverse chemical structure, including oils, fats, waxes, phospholipids, and steroids. Lipids are used for energy storage (oils and fats), as waterproofing for the outside of many plants and animals (waxes), as the principal component of cellular membranes (phospholipids and cholesterol), and as hormones (steroids).

3.5 What Are Proteins?

Proteins are chains of amino acids that possess primary, secondary, tertiary, and sometimes quaternary structure. Both the structure and the function of a protein are determined by the sequence of amino acids in the chain as well as by how these amino acids interact with their surroundings and with each other. Proteins can be enzymes (which promote and guide chemical reactions), structural molecules (e.g., hair, horn), hormones (e.g., insulin), or transport molecules (e.g., hemoglobin).

3.6 What Are Nucleotides and Nucleic Acids?

A nucleotide is composed of a phosphate group, a five-carbon sugar (ribose or deoxyribose), and a nitrogen-containing base. Molecules formed from single nucleotides include intracellular messengers, such as cyclic AMP, and energy-carrier molecules, such as ATP. The nucleic acid molecules deoxyribonucleic acid (DNA) and ribonucleic acid (RNA), which carry the hereditary blueprint for cells and viruses (including the instructions for constructing proteins), are chains of nucleotides.

Key Terms

adenosine triphosphate (ATP) *51*
amino acid *47*
base *51*
carbohydrate *38*
cellulose *41*
chitin *43*
dehydration synthesis *38*
denatured *51*
deoxyribonucleic acid (DNA) *51*
disaccharide *39*
disulfide bond *47*
enzyme *45*
fat *43*
fatty acid *43*
functional group *37*
glucose *40*
glycerol *43*
glycogen *41*
helix *48*
hydrolysis *38*
inorganic *37*
lactose *41*
lipid *43*
maltose *41*
monomer *37*
monosaccharide *38*
nucleic acid *51*
nucleotide *51*
oil *43*
organic *37*
peptide *47*
peptide bond *47*
phospholipid *44*
pleated sheet *48*
polymer *37*
polysaccharide *39*
primary structure *48*
protein *45*
quaternary structure *50*
ribonucleic acid (RNA) *52*
saturated *43*
secondary structure *48*
starch *41*
steroid *44*
sucrose *40*
sugar *38*
tertiary structure *48*
trans fat *46*
triglyceride *43*
unsaturated *44*
wax *44*

Thinking Through the Concepts

Fill-in-the-Blank

1. In organic molecules made of chains of subunits, each subunit is called a(n) __________, and the chains are called __________. Carbohydrates consisting of long chains of sugars are called __________. These sugar chains can be broken down by __________reactions. Three types of carbohydrates consisting of long glucose chains are __________, __________, and __________. Three examples of disaccharides are __________, __________, and __________.

2. Fill in the following with the appropriate type of lipid: Unsaturated, liquid at room temperature: __________; bees use to make honeycombs: __________; stores energy in animals: __________; sex hormones are synthesized from these: __________; the LDL form of this contributes to heart disease: __________; a major component of cell membranes that has polar heads: __________.

3. Proteins are synthesized by a reaction called __________ synthesis, which releases __________. Subunits of proteins are called __________. The sequence of protein subunits is called the __________ structure of the protein. Two regular configurations of secondary protein structure are __________ and __________. Which of these secondary structures is characteristic of infectious prion proteins? __________ When a protein's secondary or higher-order structure is destroyed, the protein is said to be __________.

4. Fill in the following with the most specific term describing the bonds involved: Creates kinks in the carbon chains of fatty acids in oils: __________; maintain the helical structure of many proteins: __________; links polypeptide chains and can cause proteins to bend: __________; join the two strands of the double helix of DNA: __________; link amino acids to form the primary structure of proteins: __________.
5. A nucleotide consists of three parts: __________, __________, and __________. A nucleotide that acts as an energy carrier is __________. The four bases found in deoxyribose nucleotides are __________, __________, __________, and __________. Two important nucleic acids are __________ and __________. The functional group that joins nucleotides in nucleic acids is __________.
6. Infectious prions are unique as infectious agents because they contain no __________. Some people have developed a prion disease called __________. These victims almost certainly ate meat from cows that had a disease called __________. Cows may originally have contracted the disease from sheep who had the disease called __________.

Review Questions

1. What does the term "organic" mean to a chemist?
2. List the four principal types of biological molecules and give an example of each.
3. What roles do nucleotides play in living organisms?
4. One way to convert corn oil to margarine (a solid at room temperature) is to add hydrogen atoms, decreasing the number of double bonds in the molecules of oil. What is this process called? Why does it work?
5. Describe and compare dehydration synthesis and hydrolysis. Give an example of a substance formed by each chemical reaction, and describe the specific reaction in each instance.
6. Distinguish among the following: monosaccharide, disaccharide, and polysaccharide. Give two examples of each, and list their functions.
7. Describe the synthesis of a protein from amino acids. Then describe primary, secondary, tertiary, and quaternary structures of a protein.
8. Where in nature do we find cellulose? Chitin? In what way(s) are these two polymers similar? Different?
9. Which kinds of bonds between keratin molecules are altered when hair is (a) wet and allowed to dry on curlers and (b) given a permanent wave?

Applying the Concepts

1. In Chapter 2, you learned that hydrophobic molecules tend to cluster when immersed in water. In this chapter, you read that a phospholipid has a hydrophilic head and hydrophobic tails. What do you think would be the configuration of phospholipids that are immersed in water? What if they were immersed in oil?
2. Fat contains twice as many calories per unit weight as carbohydrate does, so fat is an efficient way for animals, who must move about, to store energy. Compare the way fat and carbohydrates interact with water, and explain why this interaction also gives fat an advantage for weight-efficient energy storage.
3. Some people think that consuming fat and sugar substitutes is a good way to reduce weight, whereas others object to these dietary "shortcuts." Explain and justify one of these two positions.
4. Saliva from infected deer can transmit chronic wasting disease, caused by misfolded prion proteins. The infectious prions can persist in soil. How might the disease pass from an infected animal to others in the herd?

MB *Go to www.masteringbiology.com for practice quizzes, activities, eText, videos, current events, and more.*

Cell Structure and Function

Chapter 4

Case Study

Spare Parts for Human Bodies

"I DON'T THINK I've ever screamed so loud in my life." Jennifer looks back on the terrible day when boiling oil from a deep-fat fryer spilled onto her 10-month-old baby, burning over 70% of his body. "The 911 operator said to get his clothes off, but they'd melted onto him. I pulled his socks off and the skin came right with them." A few decades ago, this child's burns would have been fatal. Now, the only evidence of the burn on his chest is slightly crinkled skin. Zachary Jenkins was saved by the bioengineering marvel of artificial skin.

Skin consists of several specialized cell types with complex interactions. The outer cells of skin are masters of multiplication, allowing minor burns to heal without a trace. However, if the deeper (dermal) layers of the skin are completely destroyed, healing occurs very slowly, from the edges of the burn. Deep burns are often treated by grafting skin, including some dermis taken from other sites on the body; but for extensive burns, the lack of healthy skin makes this approach impossible. Although cells from a piece of the patient's healthy skin can be grown in culture and placed over the affected area, growing enough cells to cover extensive burns can take weeks. Until recently, the only other option was to use skin from human cadavers or pigs. At best, these tissues serve as temporary "biological bandages" because the victim's body eventually rejects them. Extensive and disfiguring scars are a common legacy.

The availability of bioengineered skin has radically changed the prognosis for burn victims. The child in this chapter's opening photograph was treated with a bioengineered artificial skin (TransCyte™) that uses skin cells obtained from the donated foreskins of infants who were circumcised at birth. This tissue, which would otherwise be discarded, provides a source of rapidly dividing cells. Cultured in the laboratory on nylon mesh, the cells secrete growth factors and protein fibers that support skin. Although the foreskin cells die, the supporting substances stimulate regeneration of the victim's own skin and new blood vessels. A newer form of bioengineered artificial skin, called Orcel™, retains living foreskin cells on a biodegradable spongy protein matrix.

Bioengineered artificial skin demonstrates our increasing power to manipulate cells, the fundamental units of life. If scientists can shape cells into artificial, but living, skin, might they also be able to sculpt cells into working bones, livers, kidneys, and bladders?

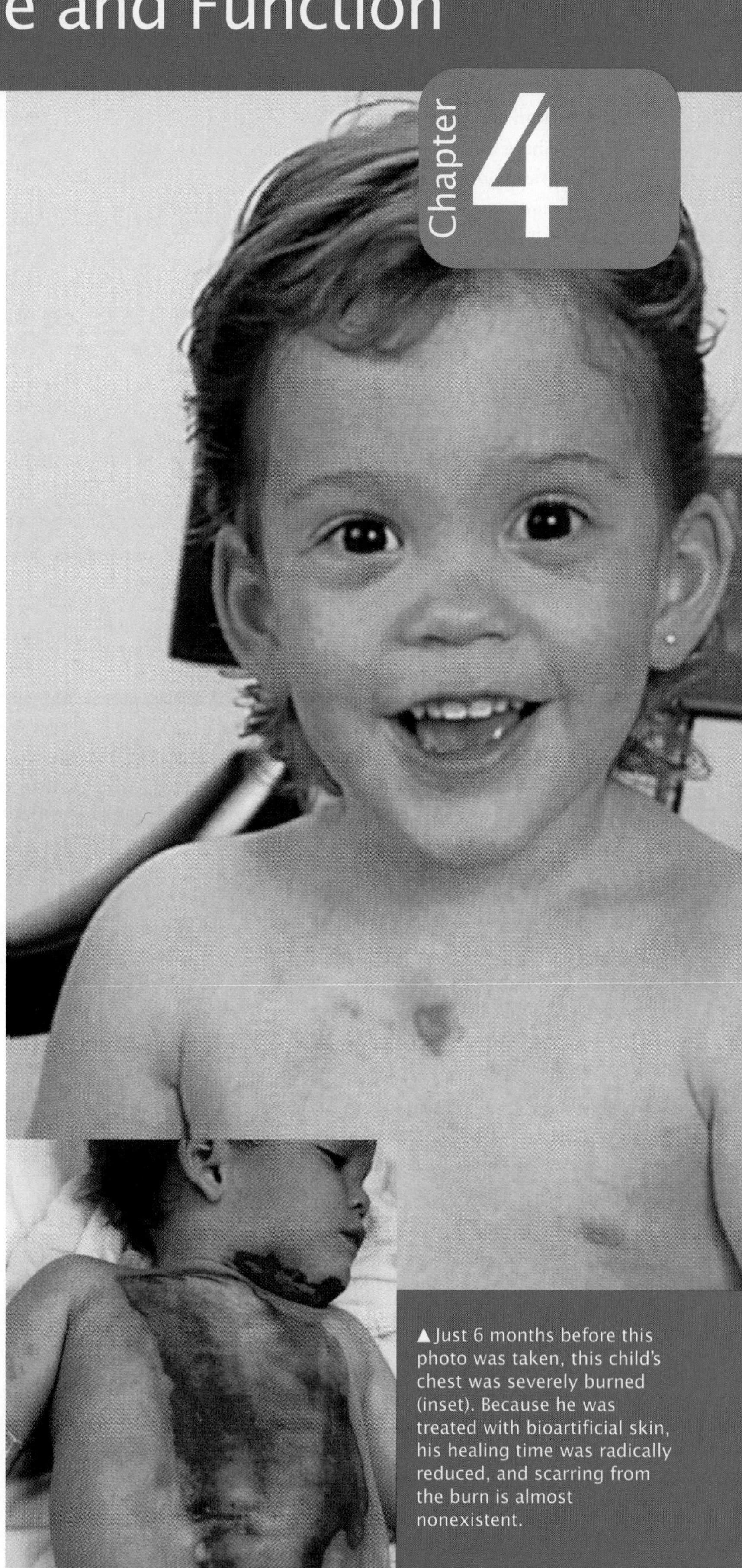

▲Just 6 months before this photo was taken, this child's chest was severely burned (inset). Because he was treated with bioartificial skin, his healing time was radically reduced, and scarring from the burn is almost nonexistent.

At a Glance

4.1 WHAT IS THE CELL THEORY?

Because cells are so small, they were not seen until the invention of the microscope in the mid-1600s (see "Scientific Inquiry: The Search for the Cell" on pp. 58–59). But seeing cells was only the first step to understanding their importance. In 1838, the German botanist Matthias Schleiden concluded that cells and substances produced by cells form the basic structure of plants, and that plant growth occurs by adding new cells. In 1839, German biologist Theodor Schwann (Schleiden's friend and collaborator) drew similar conclusions for animal cells. The work of Schleiden and Schwann provided a unifying theory of cells as the fundamental units of life. In 1855, the German physician Rudolf Virchow completed the **cell theory** by concluding that all cells come from previously existing cells.

Three principles comprise the cell theory, a fundamental precept of biology:

- Every living organism is made up of one or more cells.
- The smallest living organisms are single cells, and cells are the functional units of multicellular organisms.
- All cells arise from preexisting cells.

All living things, from microscopic bacteria to the human body to a giant sequoia tree, are composed of cells. Whereas each bacterium consists of a single, relatively simple cell, the human body consists of trillions of complex cells, specialized to perform an enormous variety of functions. To survive, all cells must obtain energy and nutrients from their environment, synthesize a variety of proteins and other molecules necessary for their growth and repair, and eliminate wastes. Most cells need to interact with other cells. To ensure the continuity of life, cells must also reproduce. These activities are accomplished by specialized parts of each cell, described later in this chapter.

4.2 WHAT ARE THE BASIC ATTRIBUTES OF CELLS?

All types of cells are derived from common ancestors, and must perform similar functions. As a result, they have many similarities in size and structure.

Cell Function Limits Cell Size

Most cells range in size from about 1 to 100 micrometers (millionths of a meter) in diameter (**Fig. 4-1**). Why are most cells small? The answer lies in the need for cells to exchange nutrients and wastes with their external environment through the plasma membrane. As you will learn in Chapter 5, many nutrients and wastes move into, through, and out of cells by diffusion, the movement of molecules from places of high concentration of those molecules to places of low concentration. This relatively slow process requires that no part of the cell be too far away from the external environment.

All Cells Share Common Features

Despite their diversity, all cells—from the prokaryotic bacteria and archaea to the eukaryotic protists, fungi, plants, and animals—share common features, as described in the following sections.

The Plasma Membrane Encloses the Cell and Allows Interactions Between the Cell and Its Environment

Each cell is surrounded by an extremely thin, rather fluid membrane called the **plasma membrane** (**Fig. 4-2**). As you will learn in Chapter 5, the plasma membrane and the other

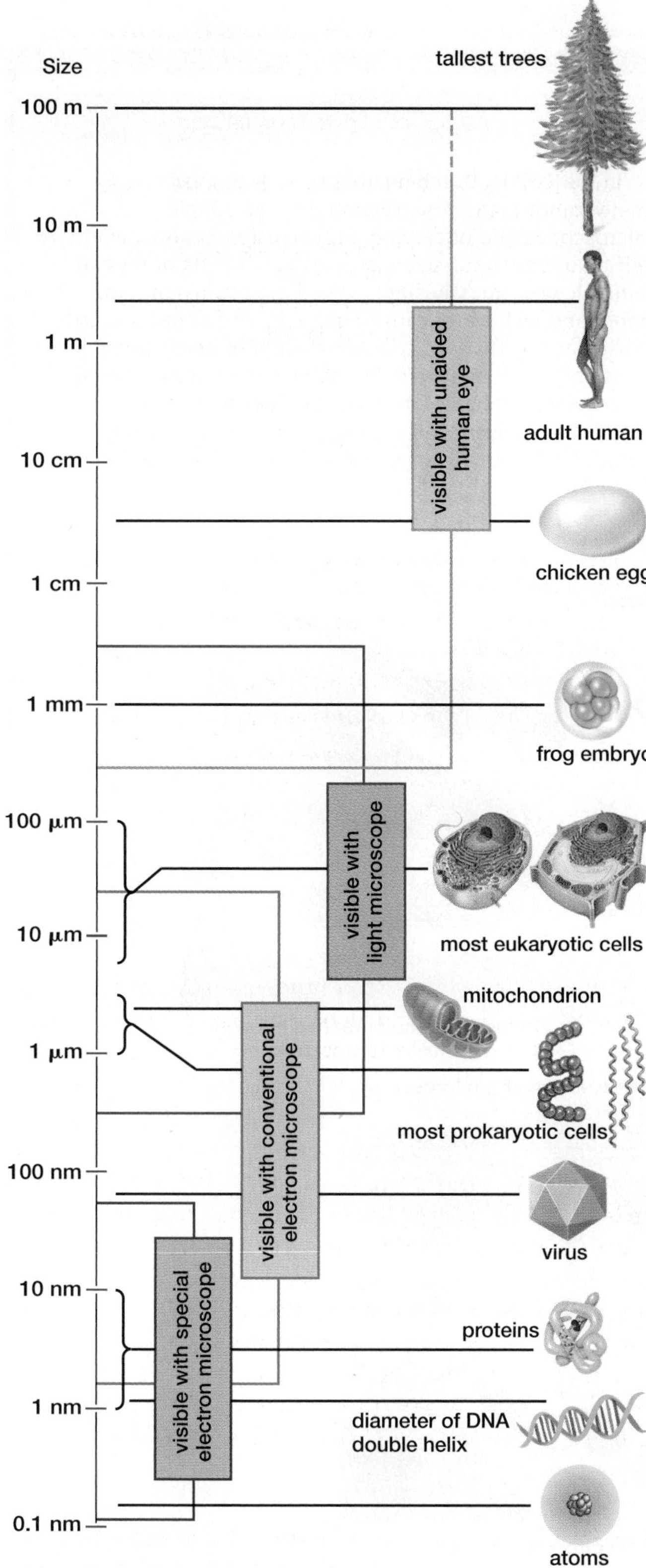

▲ FIGURE 4-1 **Relative sizes** Dimensions commonly encountered in biology range from about 100 meters (the height of the tallest redwood trees) through a few micrometers (the diameter of most cells) to a few nanometers (the diameter of many large molecules).

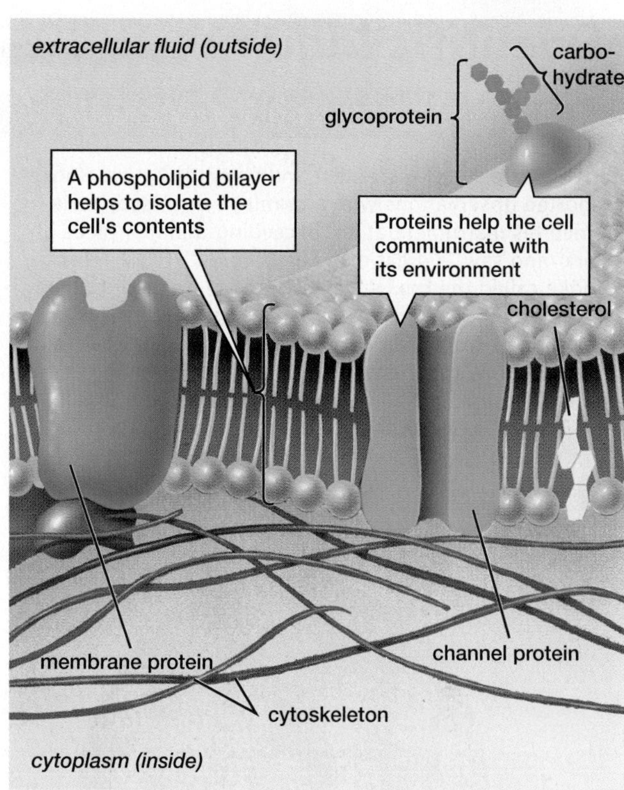

▲ FIGURE 4-2 **The plasma membrane** The plasma membrane encloses the cell. Like all cellular membranes, it consists of a double layer of phospholipid molecules in which a variety of proteins is embedded. The membrane is supported by the cytoskeleton. In animal cells, cholesterol helps maintain the proper fluidity of the membrane.

membranes within cells consist of a double layer (a bilayer) of phospholipids and cholesterol molecules in which many different proteins are embedded. Important functions of the plasma membrane include:

- Isolating the cell's contents from the external environment.
- Regulating the flow of materials into and out of the cell.
- Allowing interaction with other cells and with the extracellular environment.

The phospholipid and protein components of cellular membranes have very different roles. Each phospholipid has a hydrophilic ("water-loving") head that faces the watery interior or watery exterior of the cell, and a pair of hydrophobic ("water-fearing") tails that face the interior of the membrane (see pp. 44–45). Although some small molecules—including oxygen, carbon dioxide, and water—are able to diffuse through it, the phospholipid bilayer forms a barrier to most hydrophilic molecules and ions. The phospholipid bilayer helps isolate the cell from its surroundings, allowing the cell to maintain crucial differences in concentrations of materials inside and outside.

Proteins embedded in the plasma membrane facilitate communication between the cell and its environment. Some allow specific molecules or ions to move through the plasma

Scientific Inquiry

The Search for the Cell

In 1665, the English scientist and inventor Robert Hooke reported observations with a primitive microscope. He aimed his instrument at an "exceeding thin . . . piece of Cork" and saw "a great many little Boxes" (**Fig. E4-1a**). Hooke called the boxes "cells," because he thought they resembled the tiny rooms, or cells, occupied by monks. Cork comes from the dry outer bark of the cork oak, and we now know that he was looking at the nonliving cell walls that surround all plant cells. Hooke wrote that in the living oak and other plants, "These cells [are] fill'd with juices."

In the 1670s, Dutch microscopist Anton van Leeuwenhoek was constructing his own simple microscopes and observing a previously unknown world. A self-taught amateur scientist, his descriptions of myriad "animalcules" (mostly single-celled organisms) in rain, pond, and well water caused quite an uproar because in those days, water was consumed without being treated. Eventually, van Leeuwenhoek made careful observations of an enormous range of microscopic specimens, including blood cells, sperm, and the eggs of small insects such as aphids and fleas. His discoveries struck a blow to the idea

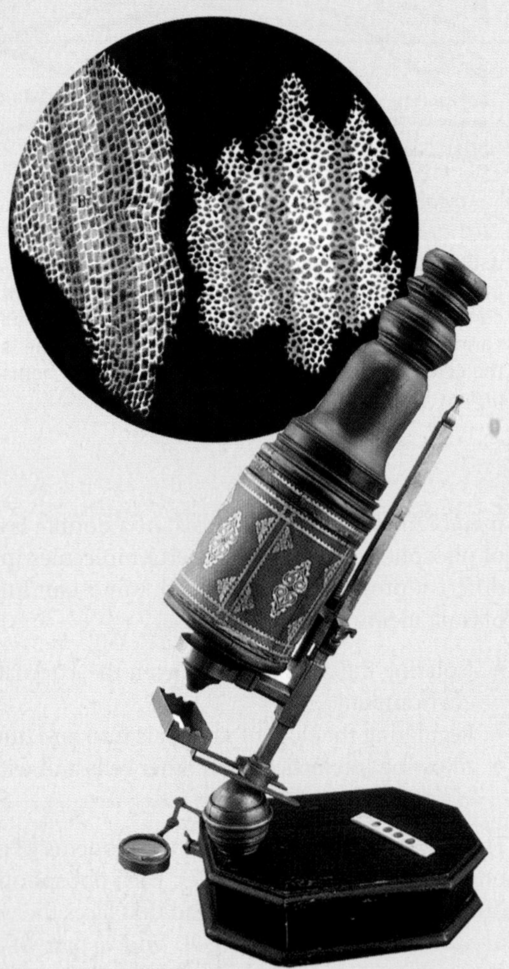

(a) 17th-century microscope and cork cells

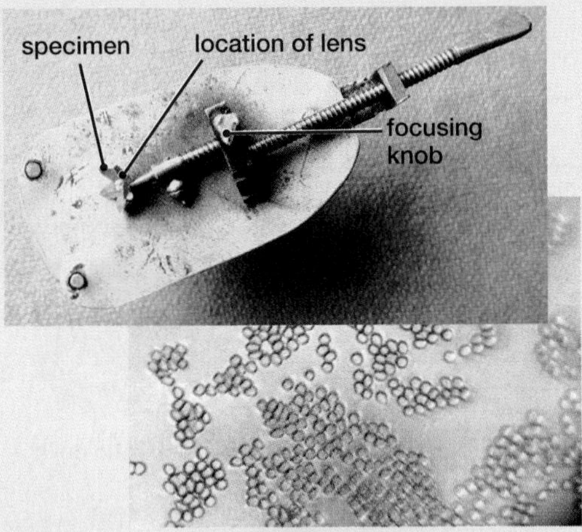

blood cells photographed through Leeuwenhoek's microscope

(b) Leeuwenhoek's microscope

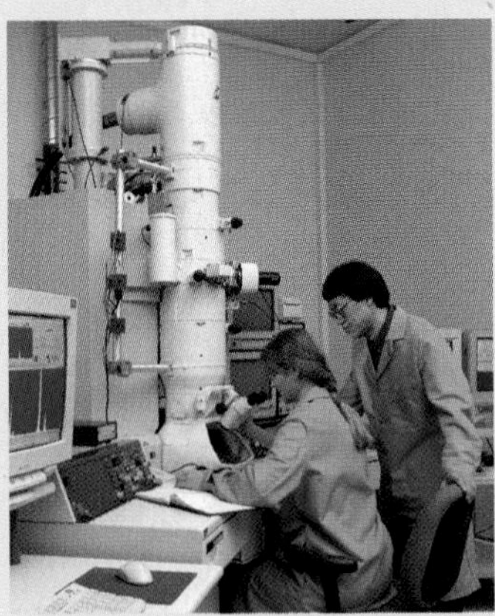

(c) Electron microscope

▲ **FIGURE E4-1 Microscopes yesterday and today (a)** Robert Hooke's drawings of cork cells, as he viewed them with an early light microscope similar to the one shown here. Only the cell walls remain. **(b)** One of Leeuwenhoek's microscopes, and a photograph of blood cells taken through a Leeuwenhoek microscope. The specimen is viewed through a tiny hole just underneath the lens. **(c)** This electron microscope performs as a transmission electron microscope (TEM) and a scanning electron microscope (SEM).

of spontaneous generation; at that time, fleas were believed to emerge spontaneously from sand or dust, and weevils from grain. Although they appear much more primitive than Hooke's microscopes, Leeuwenhoek's microscopes provided much clearer images and higher magnification (**Fig. E4-1b**).

Ever since the pioneering efforts of Robert Hooke and Anton van Leeuwenhoek, biologists, physicists, and engineers have collaborated in the development of a variety of advanced microscopes to view the cell and its components.

Light microscopes use lenses, usually made of glass or quartz, to focus light rays that either pass through or bounce off a specimen, thereby magnifying its image. Light microscopes provide a wide range of images, depending on how the specimen is illuminated and whether it has been stained (**Fig. E4-2a**).

The resolving power of light microscopes—that is, the smallest structure that can be seen—is about 1 micrometer (a millionth of a meter).

Electron microscopes (**Fig. E4-1c**) use beams of electrons instead of light, which are focused by magnetic fields rather than by lenses. Some types of electron microscopes can resolve structures as small as a few nanometers (billionths of a meter). Transmission electron microscopes pass electrons through a thin specimen and can reveal the details of interior cell structure, including organelles and plasma membranes (**Fig. E4-2c**). Scanning electron microscopes bounce electrons off specimens that have been coated with metals and provide three-dimensional images. These can be used to view the surface details of structures that range in size from entire organisms down to cells and even organelles (**Figs. E4-2b,d**).

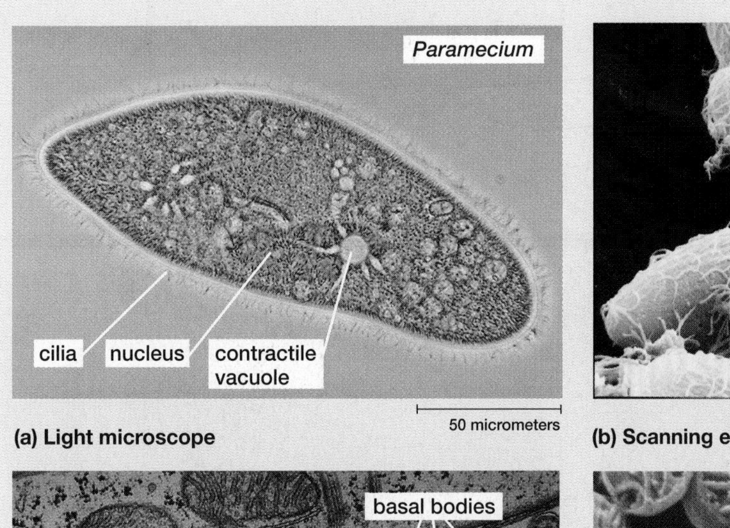

(a) Light microscope

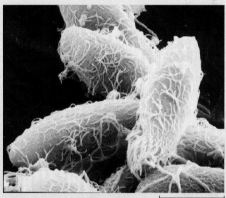

(b) Scanning electron microscope

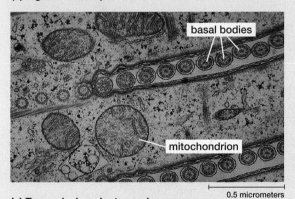

(c) Transmission electron microscope

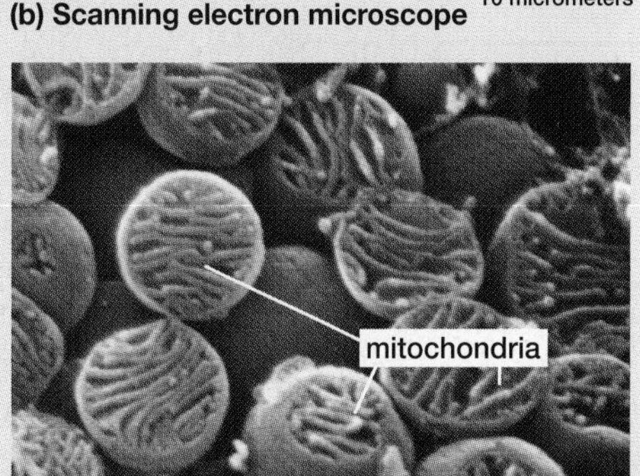

(d) Scanning electron microscope

▲ **FIGURE E4-2 A comparison of microscope images (a)** A living *Paramecium* (a single-celled freshwater organism) photographed through a light microscope. **(b)** A false-color scanning electron micrograph (SEM) of *Paramecium*. **(c)** A transmission electron micrograph (TEM) showing the basal bodies at the bases of the cilia that cover *Paramecium*. Mitochondria are also visible. **(d)** An SEM at much higher magnification showing mitochondria (many of which are sliced open) within the cytoplasm. You will see both TEMs and SEMs throughout this text. Any colors you see in electron micrographs have been added artificially to enhance the structures.

membrane, while others promote chemical reactions inside the cell. Some membrane proteins join cells to one another and others receive and respond to signals from molecules (such as hormones) in the fluid surrounding the cell. Glycoproteins embedded in the plasma membrane extend carbohydrate branches outward from the cell (see Fig. 4-2). Some of these glycoproteins—called major histocompatibility complex (MHC) proteins—are responsible for identifying the cell as part of a unique individual. In Chapter 5, we discuss the plasma membrane in detail.

All Cells Contain Cytoplasm

The **cytoplasm** consists of all the fluid and structures that lie inside the plasma membrane, but outside of the nucleus (**Figs. 4-3** and **4-4**). The fluid portion of the cytoplasm in both prokaryotic and eukaryotic cells, called the **cytoplasmic fluid,** contains water, salts, and an assortment of organic molecules, including proteins, lipids, carbohydrates, sugars, amino acids, and nucleotides (described in Chapter 3). Most of the cell's metabolic activities—the biochemical reactions that support life—occur in the cell cytoplasm. Protein synthesis is one example. Cells must synthesize a variety of proteins, such as those of the cytoskeleton (described later), proteins in the plasma membrane (described in Chapter 5), and all the enzymes that promote biochemical reactions (described in Chapter 6).

Case Study continued
Spare Parts for Human Bodies

Why is skin from cadavers and pigs rejected by burn victims, while synthetic skin substitutes are not? Skin from a different person or animal contains MHC glycoproteins that differ from those of the burn victim. This causes the victim's immune system to treat the skin as foreign, attack it, and eventually reject it. Fortunately, the response is sufficiently gradual that the skin can still help the burn heal before it is rejected. TransCyte™ has no remaining cells to stimulate an immune response, and the developers of Orcel™ hypothesize that the MHC proteins on the original foreskin cells are lost during the extensive culturing process.

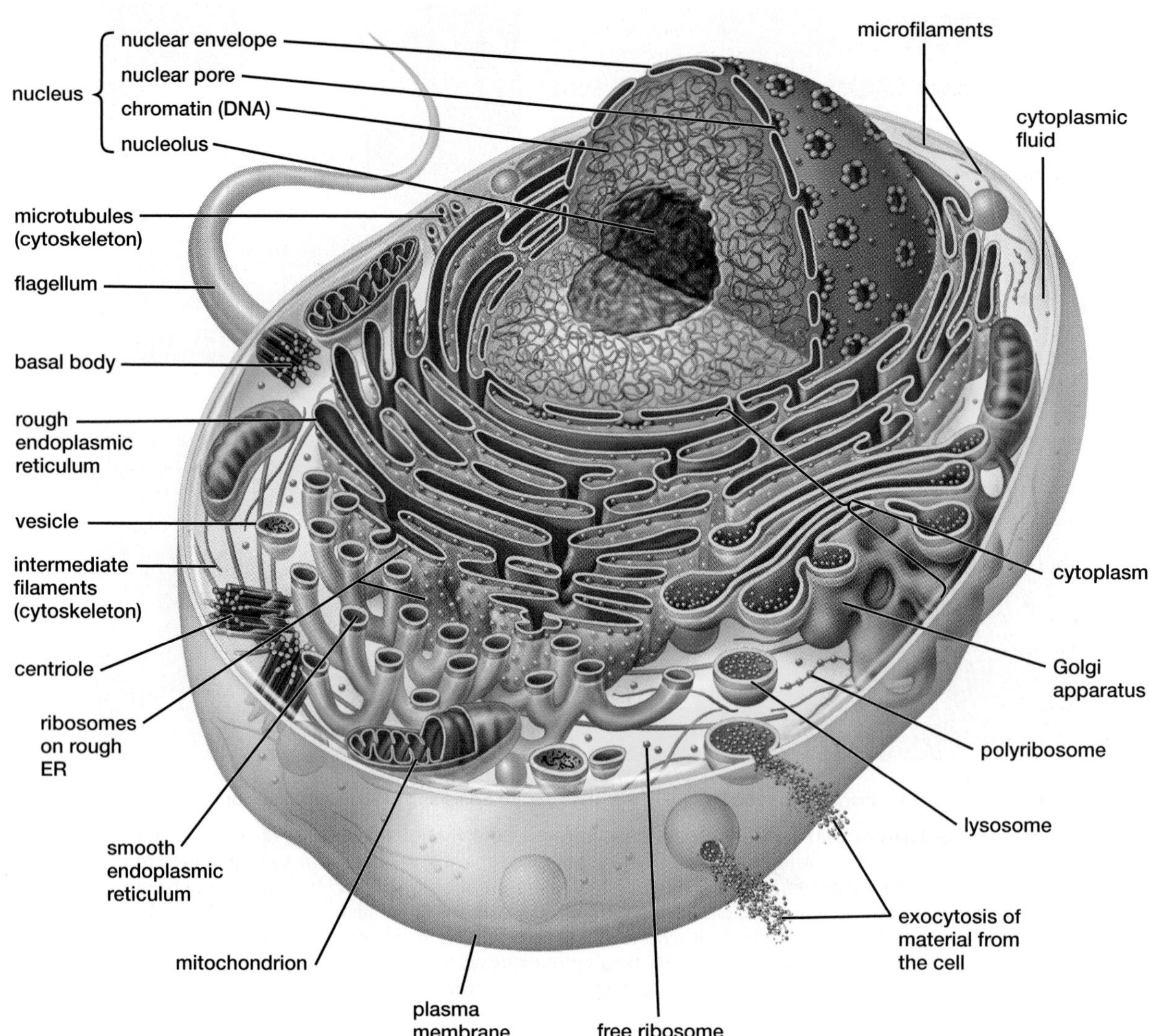

▲ **FIGURE 4-3 A generalized animal cell**

All Cells Use DNA as a Hereditary Blueprint and RNA to Copy the Blueprint and Guide Construction

Each cell contains genetic material, an inherited blueprint that stores the instructions for making the other parts of the cell and for producing new cells. The genetic material in cells is **deoxyribonucleic acid (DNA).** This fascinating molecule, described in detail in Chapter 11, contains genes consisting of precise sequences of nucleotides. During cell division, the original or "parent cells" pass exact copies of their DNA to their newly formed offspring, or "daughter cells." **Ribonucleic acid (RNA)** is chemically related to DNA and comes in different forms that copy the blueprint of genes on DNA and help construct proteins based on this blueprint. All cells contain both DNA and RNA.

All Cells Obtain Raw Materials and Energy from Their Environment

The major elements in biological molecules, including carbon, hydrogen, oxygen, nitrogen, sulfur, and phosphorus, as well as trace amounts of many minerals, ultimately come from the environment, which includes air, water, rocks, and other organisms. To maintain their incredible complexity, cells must continuously acquire and expend energy. As we explain in Chapters 7 and 8, essentially all of the energy powering life on Earth originates in sunlight. Only cells that are capable of photosynthesis can harness this energy. The energy stored by photosynthetic cells fuels the metabolic activities of nearly all other forms of life. Thus, all cells obtain the materials to synthesize the molecules of life and the energy to power this synthesis from their living and nonliving environment.

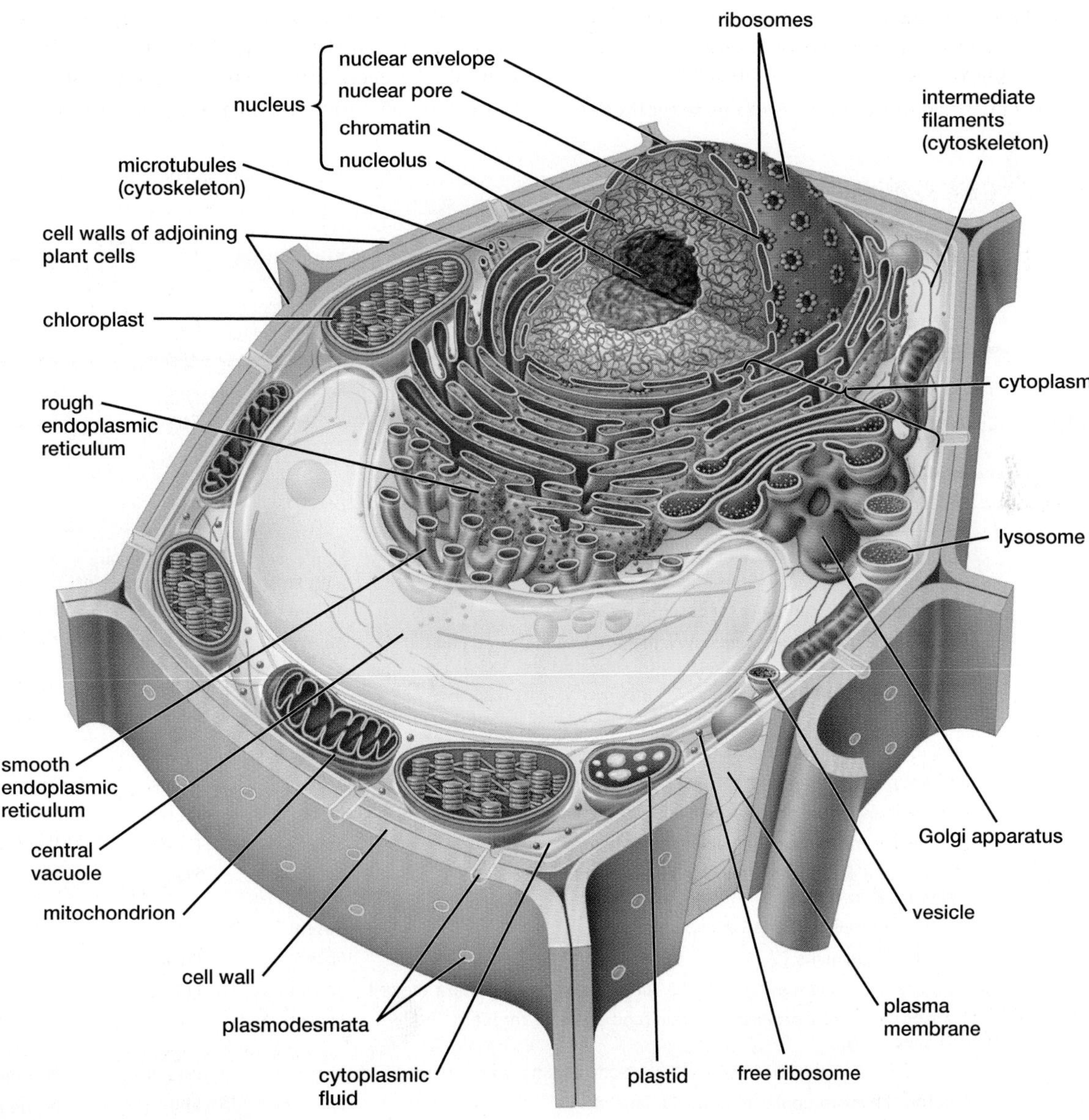

▲ **FIGURE 4-4 A generalized plant cell**

QUESTION Of the nucleus, ribosome, chloroplast, and mitochondrion, which appeared earliest in the history of life?

There Are Two Basic Types of Cells: Prokaryotic and Eukaryotic

All forms of life are composed of only two fundamentally different types of cells. **Prokaryotic** cells ("before the nucleus"; see Fig. 4-19) form the bodies of **bacteria** and **archaea,** the simplest forms of life. **Eukaryotic** cells ("true nucleus"; see Figs. 4-3 and 4-4) are far more complex and comprise the bodies of animals, plants, fungi, and protists.

Have you ever wondered 

How Many Cells Are in the Human Body?

Well, scientists have wondered that too, and they can't agree. Estimates range from 10 trillion to 100 trillion. Interestingly, there seems to be a consensus that there are far more bacterial cells in the body than there are human cells—somewhere between 10 and 20 times as many. Does that make us primarily prokaryotic? Hardly. Most of the bacteria are guests that reside in our digestive tracts.

As their names imply, one striking difference between prokaryotic and eukaryotic cells is that the genetic material of eukaryotic cells is contained within a membrane-enclosed nucleus. In contrast, the genetic material of prokaryotic cells is not enclosed within a membrane. Other membrane-enclosed structures, called organelles, contribute to the far greater structural complexity of eukaryotic cells. **Table 4-1** summarizes the features of prokaryotic and eukaryotic cells, which we discuss in the sections that follow.

4.3 WHAT ARE THE MAJOR FEATURES OF EUKARYOTIC CELLS?

Eukaryotic cells comprise the bodies of animals, plants, protists, and fungi; as you might imagine, these cells are extremely diverse. Within the body of any multicellular organism there exists a huge variety of eukaryotic cells specialized to perform different functions. In contrast, the single-celled bodies of protists and some fungi must be sufficiently complex to perform all the necessary activities needed to sustain life, grow, and reproduce independently. Here we emphasize plant and

Table 4-1 **Functions and Distribution of Cell Structures**

Structure	Function	Prokaryotes	Eukaryotes: Plants	Eukaryotes: Animals
Cell Surface				
Cell wall	Protects and supports the cell	Present	Present	Absent
Cilia	Move the cell through fluid or move fluid past the cell surface	Absent	Absent	Present
Flagella	Move the cell through fluid	Present[1]	Present[2]	Present
Plasma membrane	Isolates the cell contents from the environment; regulates movement of materials into and out of the cell; allows communication with other cells	Present	Present	Present
Organization of Genetic Material				
Genetic material	Encodes the information needed to construct the cell and to control cellular activity	DNA	DNA	DNA
Chromosomes	Contain and control the use of DNA	Single, circular, no proteins	Many, linear, with proteins	Many, linear, with proteins
Nucleus	Membrane-bound container for chromosomes	Absent	Present	Present
Nuclear envelope	Encloses the nucleus; regulates movement of materials into and out of the nucleus	Absent	Present	Present
Nucleolus	Synthesizes ribosomes	Absent	Present	Present
Cytoplasmic Structures				
Mitochondria	Produce energy by aerobic metabolism	Absent	Present	Present
Chloroplasts	Perform photosynthesis	Absent	Present	Absent
Ribosomes	Provide the sites for protein synthesis	Present	Present	Present
Endoplasmic reticulum	Synthesizes membrane components, proteins, and lipids	Absent	Present	Present
Golgi apparatus	Modifies and packages proteins and lipids; synthesizes some carbohydrates	Absent	Present	Present
Lysosomes	Contain intracellular digestive enzymes	Absent	Present	Present
Plastids	Store food, pigments	Absent	Present	Absent
Central vacuole	Contains water and wastes; provides turgor pressure to support the cell	Absent	Present	Absent
Other vesicles and vacuoles	Transport secretory products; contain food obtained through phagocytosis	Absent	Present	Present
Cytoskeleton	Gives shape and support to the cell; positions and moves cell parts	Absent	Present	Present
Centrioles	Produce the microtubules of cilia and flagella	Absent	Absent (in most)	Present

[1]Some prokaryotes have structures called flagella, which lack microtubules and move in a fundamentally different way than do eukaryotic flagella.

[2]A few types of plants have flagellated sperm.

animal cells; the specialized structures of protists and fungi are covered in more detail in Chapters 20 and 22, respectively.

Eukaryotic cells differ from prokaryotic cells in many ways. For example, they are usually larger than prokaryotic cells—typically more than 10 micrometers in diameter. The cytoplasm of eukaryotic cells includes a variety of **organelles**—membrane-enclosed structures, such as the nucleus and mitochondria, that perform specific functions within the cell. The **cytoskeleton,** a network of protein fibers, gives shape and organization to the cytoplasm of eukaryotic cells. Many of the organelles are attached to the cytoskeleton.

Figures 4-3 and 4-4 illustrate the structures that are found in animal and plant cells, respectively (although not all individual cells possess all the features shown). Although they have many structures in common, some structures are unique to one type or the other. For example, plant cells are all surrounded by a cell wall, and they contain chloroplasts, plastids, and a central vacuole, none of which are found in animal cells. Animal cells contain centrioles, which are absent in plant cells. Many animal cells also bear cilia, which are rarely found in plant cells. You may want to refer to Figures 4-3 and 4-4, and Table 4-1, as we describe cell structure in greater detail.

Some Eukaryotic Cells Are Supported by Cell Walls

The outer surfaces of plants, fungi, and some protists are covered with nonliving, relatively stiff coatings called **cell walls** that support and protect the delicate plasma membrane. Single-celled protists that live in the ocean may have cell walls made of cellulose, protein, or glassy silica (see Chapter 20). Plant cell walls are composed of cellulose and other polysaccharides, whereas fungal cell walls are usually made of chitin (see p. 43). Prokaryotic cells also have cell walls, but made of different polysaccharides.

Cell walls are produced by the cells they surround. Plant cells secrete cellulose through their plasma membranes, forming the cell wall (see Fig. 4-4). The walls of adjacent cells are joined by the "middle lamella," a layer made primarily of the polysaccharide pectin. Pectin has a gelatinous texture and is often added to fruit jellies to make them gel. Cell walls support and protect otherwise fragile cells. For example, cell walls allow plants and mushrooms to resist the forces of gravity and wind and to stand erect on land. Cell walls are usually porous, allowing oxygen, carbon dioxide, and water carrying dissolved molecules to flow easily through them. The structure governing the interactions between a cell and its external environment is the plasma membrane, located just beneath the cell wall (when a cell wall is present). The plasma membrane is covered in detail in Chapter 5.

The Cytoskeleton Provides Shape, Support, and Movement

Organelles and other structures within eukaryotic cells do not drift about the cytoplasm haphazardly; most are attached to the network of protein fibers that make up the cytoskeleton (Fig. 4-5). Even individual enzymes, which are often a part of a complex series of reactions, may be fastened in sequence to the cytoskeleton, so that molecules can be passed from one enzyme to the next in the correct order for a particular chemical transformation. The cytoskeleton is composed of three types of protein fibers: thin **microfilaments,** medium-sized **intermediate filaments,** and thick **microtubules** (Table 4-2).

The cytoskeleton performs the following important functions:

- **Cell shape** In cells without cell walls, the cytoskeleton, particularly networks of intermediate filaments, determines the shape of the cell.
- **Cell movement** Cell movement occurs as microfilaments or microtubules assemble, disassemble, or slide past one another. Examples of moving cells include single-celled protists propelled by cilia, swimming sperm, and contracting muscle cells.

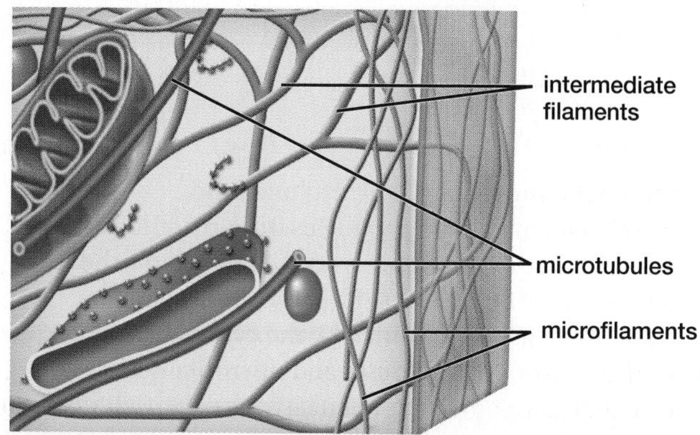

(a) Cytoskeleton

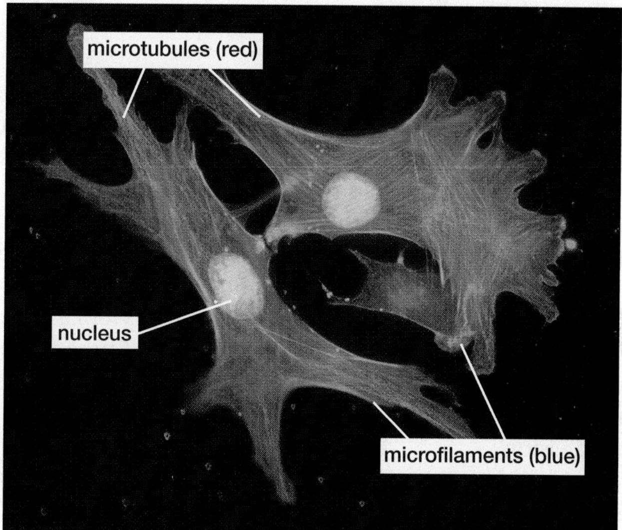

(b) Light micrograph showing the cytoskeleton

▲ FIGURE 4-5 **The cytoskeleton (a)** Eukaryotic cells are given shape and organization by the cytoskeleton, which consists of three types of proteins: microtubules, intermediate filaments, and microfilaments. **(b)** In this light micrograph, cells treated with fluorescent stains reveal microtubules and microfilaments, as well as the nucleus.

Table 4-2 Components of the Cytoskeleton

	Structure	Protein Type and Structure	Functions
Microfilaments	Twisted double strands of protein subunits; about 7 nm in diameter	Actin subunits	Involved in muscle contraction; allow for changes in cell shape; facilitate cytoplasmic division in animal cells
Intermediate filaments	Helical subunits twisted around one another and bundled into clusters of four, which may be further twisted together	Proteins vary with function and cell type subunits	Provide supporting framework within the cell; support the plasma membrane; anchor some organelles within the cytoplasm; attach some cells together
Microtubules	Tubes consisting of spirals of two-part protein subunits; about 25 nm in diameter	Tubulin subunit	Allow movement of chromosomes during cell division; form centrioles and basal bodies; are a major component of cilia and flagella

- **Organelle movement** Microtubules and microfilaments move organelles from place to place within a cell. For example, vesicles budded off the endoplasmic reticulum (ER) and Golgi apparatus (described later) are guided by the cytoskeleton.
- **Cell division** Microtubules and microfilaments are essential to cell division in eukaryotic cells. Cell division is covered in detail in Chapter 9.

Cilia and Flagella Move the Cell Through Fluid or Move Fluid Past the Cell

Both **cilia** (singular, cilium; Latin for "eyelash") and **flagella** (singular, flagellum; "whip") are slender extensions of the plasma membrane, supported internally by microtubules of the cytoskeleton. Each cilium and flagellum contains a ring of nine pairs of microtubules, with another pair in the center (**Fig. 4-6**). Cilia and flagella arise from a **basal body,** which anchors them to the plasma membrane. Basal bodies are derived from **centrioles** (found in pairs in the cytoplasm; see Fig. 4-3). Both consist of barrel-shaped rings of nine microtubule triplets. Centrioles are absent in most plants. They are present in animal cells, where they appear to participate in cell division.

How do cilia and flagella move? In cilia and flagella, tiny "arms" of protein attach neighboring pairs of microtubules (see Fig. 4-6). These arms can flex, using energy released from adenosine triphosphate (ATP) to power their movement. Flexing of the arms slides one pair of microtubules past the adjacent pairs, causing the cilia or flagellum to bend. Cilia and flagella often move almost continuously; the energy to power this motion is supplied by mitochondria, usually found in abundance near the basal bodies.

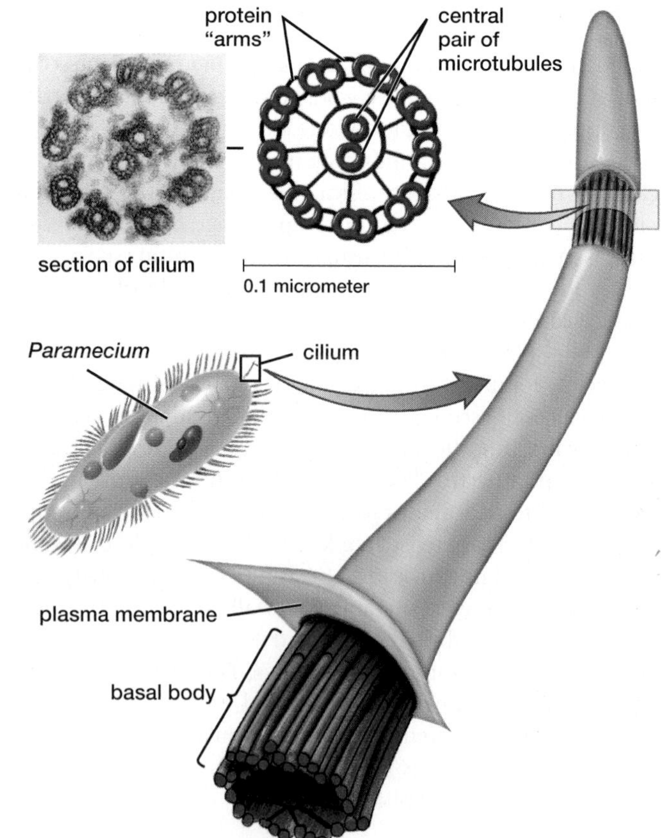

▲ **FIGURE 4-6 Cilia and flagella** Both cilia and flagella contain an outer ring of nine fused pairs of microtubules surrounding a central unfused pair. The outer pairs have "arms" made of protein that interact with adjacent pairs to provide the force for bending. Cilia and flagella arise from basal bodies, composed of triplets of microtubules, which are located just beneath the plasma membrane.

Cilia and flagella differ in length and number. In general, cilia are shorter and more numerous than flagella. The force generated by cilia can be compared to that created by oars on the sides of a rowboat (**Fig. 4-7a,** left). Flagella are longer than cilia, and cells with flagella usually have only one or two. The force generated by a flagellum can be compared to that created by the engine on a motorboat (**Fig. 4-7b,** left).

Some unicellular organisms, such as *Paramecium* (see Figs. E4-2a,b), use cilia to swim through water; others use flagella. In multicellular animals, cilia usually move fluids and suspended particles past a surface. Ciliated cells line such diverse structures as the gills of oysters (where they move water rich in food and oxygen over the gills), the oviducts of female mammals (where they move an egg through fluid from the ovary to the uterus), and the respiratory tracts of most land vertebrates (where they clear mucus that carries debris and microorganisms out of the trachea and lungs; see Fig. 4-7a, right). Nearly all animal sperm and a few types of plant sperm cells rely on flagella for movement (see Fig. 4-7b, right).

The Nucleus Is the Control Center of the Eukaryotic Cell

A cell's DNA stores all the information needed to construct the cell and direct the countless chemical reactions necessary for life and reproduction. The genetic information in DNA is used selectively by the cell, depending on its stage of development, its environment, and the function of the cell in a multicellular body. In eukaryotic cells, DNA is housed within the nucleus.

The **nucleus** is an organelle (usually the largest in the cell) consisting of three major parts: the nuclear envelope, chromatin, and the nucleolus, shown in **Figure 4-8** and described in the following sections.

The Nuclear Envelope Allows Selective Exchange of Materials

The nucleus is isolated from the rest of the cell by a **nuclear envelope** that consists of a double membrane. The membrane is perforated with tiny protein-lined nuclear pores. Water, ions, and small molecules can pass freely through the pores, but the passage of large molecules—particularly proteins, pieces of ribosomes, and RNA—is regulated by specialized gatekeeper proteins called the **nuclear pore complex** that line each nuclear pore. Ribosomes stud the outer nuclear membrane, which is continuous with membranes of the rough ER, described later (see also Figs. 4-3 and 4-4).

Chromatin Contains DNA, Which Codes for the Synthesis of Proteins

Because the nucleus is darkly colored by stains used in light microscopy, early microscopists, not knowing its function, named the nuclear material **chromatin,** meaning "colored substance." Biologists have since learned that chromatin consists of DNA associated with proteins. Eukaryotic DNA and its associated proteins form long strands called **chromosomes** ("colored bodies"). When cells divide, each chromosome coils upon itself, becoming thicker and shorter. The resulting condensed chromosomes are easily visible, even with a light microscope (**Fig. 4-9**).

The genes on DNA provide a blueprint or a "molecular code" for a huge variety of proteins. Some of these proteins form structural components of the cell. Others regulate the movement of materials through cell membranes, and still others are enzymes that promote chemical reactions within the cell that are responsible for growth and repair, nutrient and energy acquisition and use, and reproduction.

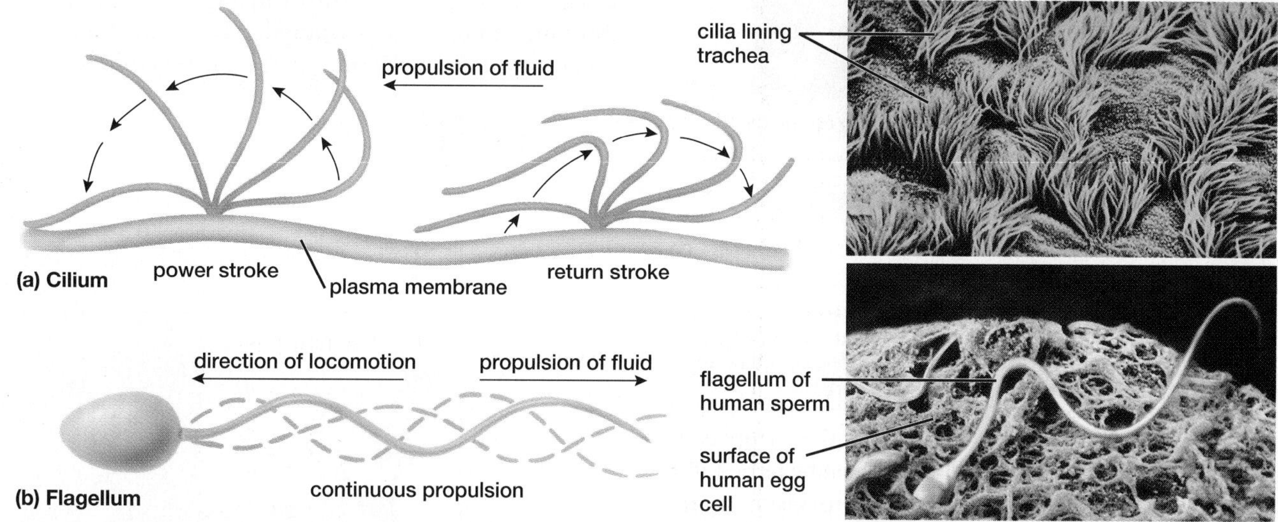

▲ **FIGURE 4-7 How cilia and flagella move (a)** (left) Cilia usually "row," providing a force parallel to the plasma membrane. Their movement resembles the arms of a swimmer doing the breast stroke. (right) SEM of cilia lining the trachea (which conducts air to the lungs); these cilia sweep out mucus and trapped particles. **(b)** (left) Flagella move in a wavelike motion, providing continuous propulsion perpendicular to the plasma membrane. In this way, a flagellum attached to a sperm can move the sperm straight ahead. (right) SEM of a human sperm cell on the surface of a human egg cell.

QUESTION What problems would arise if the trachea were lined with flagella?

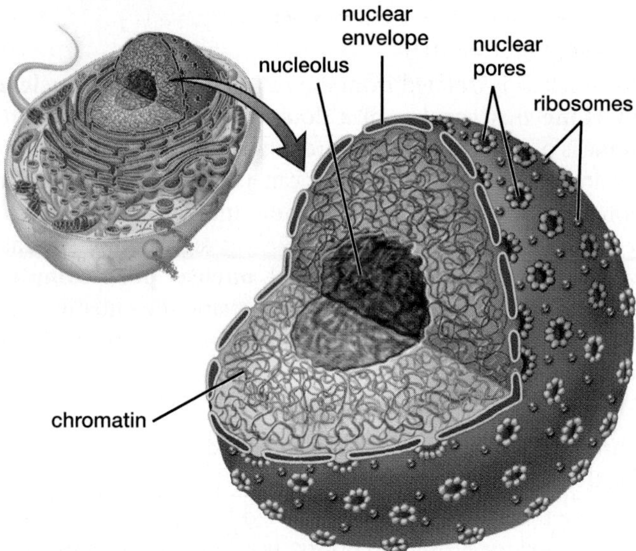

(a) The nucleus

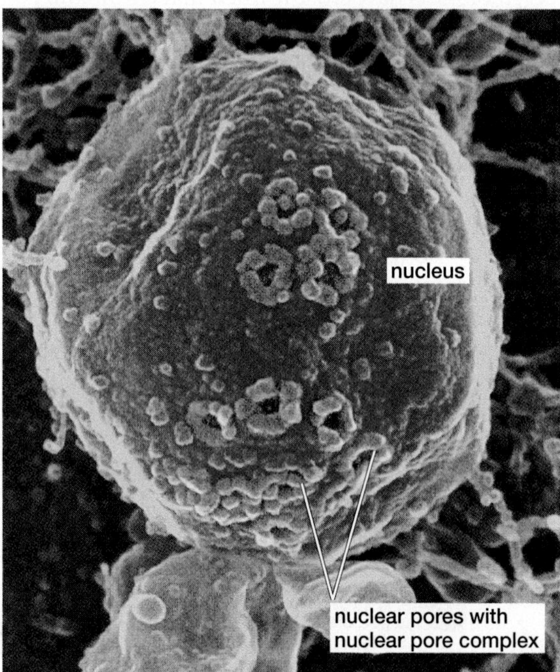

(b) Nucleus of a yeast cell

▲ FIGURE 4-8 **The nucleus (a)** The nucleus is bounded by a double outer membrane perforated by pores. Inside are chromatin and a nucleolus. **(b)** SEM of the nucleus of a yeast cell. The "gatekeeper proteins" of the nuclear pore complex are colored pink. These proteins line the nuclear pores.

Because proteins are synthesized in the cytoplasm, copies of the protein blueprints on DNA must be ferried out through the nuclear membrane into the cytoplasm. To accomplish this, genetic information is copied from DNA into molecules of messenger RNA (mRNA), which travel through the pores of the nuclear envelope into the cytoplasm. This information, coded by the sequence of nucleotides in mRNA, is then used to direct the synthesis of cellular proteins, a process that occurs on ribosomes (**Fig. 4-10**). We take a closer look at these processes in Chapter 12.

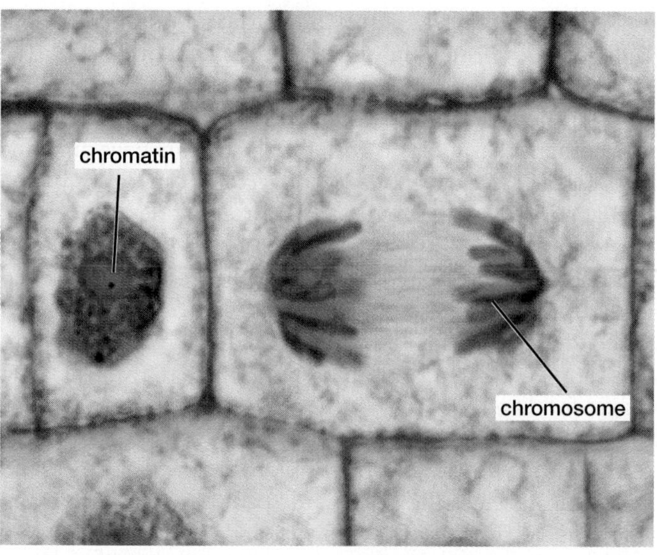

▲ FIGURE 4-9 **Chromosomes** Chromosomes, seen here in a light micrograph of a dividing cell (on the right) in an onion root tip, contain the same material (DNA and proteins) as the chromatin seen in adjacent nondividing cells.

QUESTION Why does the chromatin form discrete structures (chromosomes) in dividing cells?

The Nucleolus Is the Site of Ribosome Assembly

Eukaryotic nuclei have at least one **nucleolus** (plural, nucleoli, meaning "little nuclei"; see Fig. 4-8a). The nucleolus is the site of ribosome synthesis. It consists of ribosomal RNA, proteins, ribosomes in various stages of synthesis, and DNA that carries the genes coding for ribosomal RNA.

A **ribosome** is a small particle composed of ribosomal RNA and proteins that serves as a kind of "workbench" for the synthesis of proteins within the cell cytoplasm. Just as a workbench can be used to construct many different objects, any ribosome can be used to synthesize any of the thousands of proteins made by a cell. In electron micrographs, ribosomes

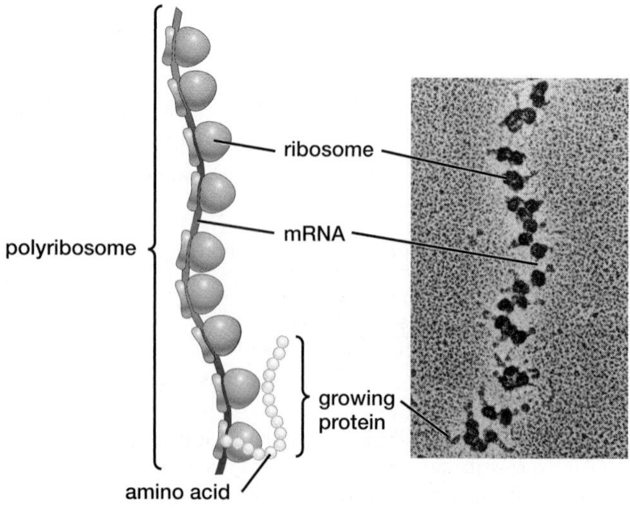

▲ FIGURE 4-10 **Ribosomes** Ribosomes in this TEM (right) are strung along a messenger RNA molecule, forming a polyribosome. The ribosomes are synthesizing a protein, indicated by the chain of amino acids.

appear as dark granules, either singly, studding the membranes of the nuclear envelope and endoplasmic reticulum (**Fig. 4-11**), or as polyribosomes ("many ribosomes") clustered along strands of mRNA (see Fig. 4-10).

Case Study continued
Spare Parts for Human Bodies

To replace the skin cells killed by a burn requires a tremendous proliferation of new cells, generated by those surrounding the burned area. The process of cell division parcels out each of the DNA-containing chromosomes of the nucleus to the daughter cells. As each cell prepares to divide, it grows in size, forming more of all the new organelles and other cellular components required by its offspring. This growth relies on DNA in the nucleus, which provides a blueprint for proteins, including the enzymes that promote the synthesis of other biological molecules. Within the nucleolus of the nucleus, DNA directs the synthesis of ribosomes, which will serve as cytoplasmic workbenches for protein construction. Through its crucial role in synthesizing new cells, the nucleus helps burn victims heal.

Eukaryotic Cytoplasm Includes an Elaborate System of Membranes

All eukaryotic cells have an elaborate system of membranes that enclose the cell and create internal compartments within the cytoplasm. Imagine a large factory with many rooms and some separate buildings. Each room houses specialized machinery, and some rooms are interconnected to allow a complex product to be manufactured in stages. Some products must be moved between buildings before they are finished. The factory must import raw materials, but it makes and repairs its own machinery and exports some of its products.

In a comparable way, specialized regions within the cytoplasm, demarcated by membranes, separate a variety of biochemical reactions from one another and process different types of molecules in specific ways. The fluid property of membranes allows them to fuse with one another, so that these internal compartments can interconnect, exchange pieces of membrane among themselves, and transfer their contents to different compartments for various types of processing. Membrane sacs called **vesicles** ferry membrane and specialized contents among the separate regions of the membrane system. Vesicles may also fuse with the plasma membrane, exporting their contents outside the cell. How do the vesicles know where to go within the complex membrane system? Researchers have discovered that specific proteins embedded in membranes serve as "mailing labels" that specify the address to which the sac and its contents are being sent.

The cell's membrane system includes the plasma membrane, nuclear membrane, endoplasmic reticulum, Golgi apparatus, lysosomes, vesicles, and vacuoles, which we explore further in the sections that follow.

The Endoplasmic Reticulum Forms Membrane-Enclosed Channels Within the Cytoplasm

The **endoplasmic reticulum (ER)** is a series of interconnected membranes that form a labyrinth of flattened sacs and channels within the cytoplasm ("reticulum" means "network" and "endoplasmic" means "within the cytoplasm"; see Fig. 4-11). The ER membrane typically makes up at least 50% of the total membrane of the cell. Among its many functions is serving as the site for the synthesis of cell membrane proteins and phospholipids. This is important because ER membrane is constantly being budded off and transported to the Golgi apparatus, lysosomes, and the plasma membrane. Eukaryotic cells have two forms of ER: rough and smooth. Parts of the rough ER are continuous with the nuclear membrane, and some smooth ER is continuous with rough ER (see Fig. 4-3).

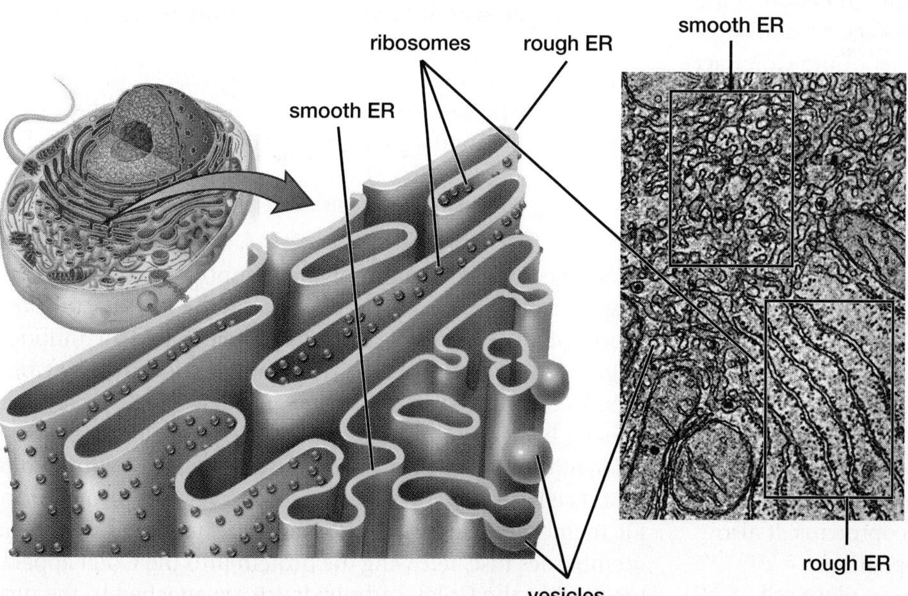

(a) Endoplasmic reticulum may be rough or smooth

(b) Smooth and rough ER

◀ **FIGURE 4-11 Endoplasmic reticulum (a)** In some cells, rough and smooth ER are thought to be linked, as depicted in the drawing. In others, smooth ER may be separate. Ribosomes (orange dots) stud the cytoplasmic face of the rough ER membrane. **(b)** A TEM of smooth and rough ER with vesicles.

Smooth Endoplasmic Reticulum Smooth ER lacks ribosomes and is specialized for different activities in different cells. In some cells, smooth ER manufactures large quantities of lipids such as steroid hormones made from cholesterol. For example, sex hormones are produced by smooth ER in mammalian reproductive organs. Smooth ER is also abundant in liver cells, where it contains enzymes that detoxify harmful drugs such as alcohol and metabolic by-products such as ammonia. Other enzymes in the smooth ER of the liver break down glycogen (a carbohydrate stored in the liver) into glucose molecules that provide energy. Smooth ER stores calcium in all cells, but in skeletal muscles it is enlarged and specialized to store large amounts of calcium that are required for muscle contraction.

Rough Endoplasmic Reticulum The ribosomes on rough ER are sites of protein synthesis. For example, the various proteins embedded in cellular membranes are manufactured here. Ribosomes on rough ER are also sites for manufacturing proteins such as digestive enzymes and protein hormones (for example, insulin) that some cells export. As these proteins are synthesized, they are inserted through the ER membrane into the interior compartment. Proteins synthesized either for secretion outside the cell or to be used elsewhere within the cell move through the ER channels. Here they are chemically modified and folded into their proper three-dimensional structures. Eventually the proteins accumulate in pockets of membrane that bud off as vesicles that carry their protein cargo to the Golgi apparatus.

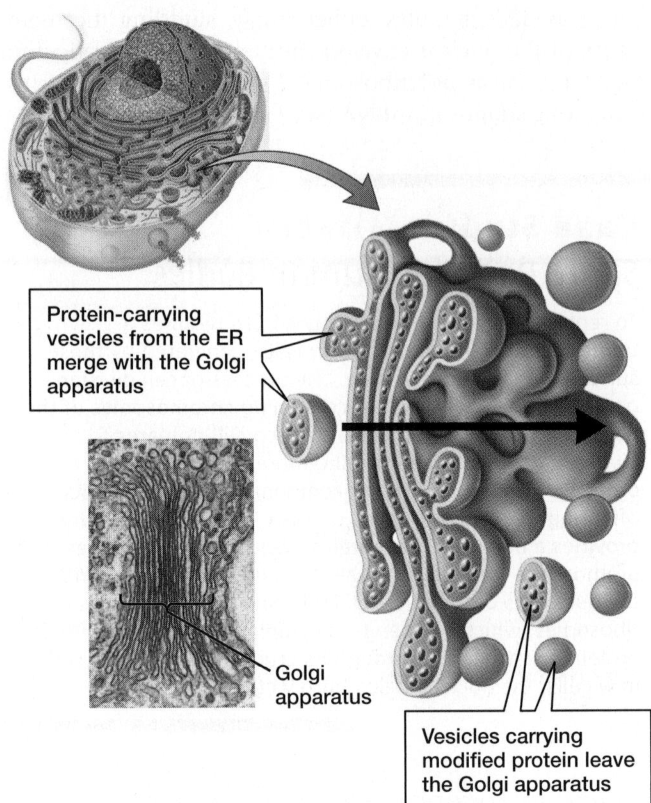
Protein-carrying vesicles from the ER merge with the Golgi apparatus
Golgi apparatus
Vesicles carrying modified protein leave the Golgi apparatus

▲ FIGURE 4-12 **The Golgi apparatus** The Golgi apparatus is a stack of flattened membranous sacs. Vesicles transport both cell membrane and the enclosed material from the ER to the Golgi. The arrow shows the direction of movement of materials through the Golgi as they are modified and sorted. Vesicles bud from the face of the Golgi opposite the ER.

The Golgi Apparatus Sorts, Chemically Alters, and Packages Important Molecules

Named for the Italian physician and cell biologist Camillo Golgi, who discovered it in the late 1800s, the **Golgi apparatus** (or **Golgi**) is a specialized set of membranes derived from the endoplasmic reticulum. It resembles a stack of flattened and interconnected sacs (**Fig. 4-12**). The main function of the Golgi apparatus is to modify, sort, and package proteins produced by the rough ER. The compartments of the Golgi act like the finishing rooms of a factory, where the final touches are added to products to be packaged and exported. Vesicles from the rough ER fuse with one face of the Golgi apparatus, adding their membranes to the Golgi and emptying their contents into the Golgi sacs. Within the Golgi compartments, some of the proteins synthesized in the rough ER are modified further; for example, carbohydrates may be added to form glycoproteins. Some large proteins are cleaved into smaller fragments. Finally, vesicles bud off from the opposite face of the Golgi, carrying away finished products for use in the cell or export out of the cell.

The Golgi apparatus performs the following functions:

- Modifies some molecules; an important role is to add carbohydrates to proteins to make glycoproteins. It also breaks up some proteins into smaller peptides.
- Synthesizes some polysaccharides used in plant cell walls, such as cellulose and pectin.
- Separates various proteins and lipids received from the ER according to their destinations. For example, the Golgi apparatus separates the digestive enzymes that are bound for lysosomes from the cholesterol used in new membrane synthesis, and from the protein hormones that the cell will secrete.
- Packages the finished molecules into vesicles that are then transported to other parts of the cell or to the plasma membrane for export.

Secreted Proteins Are Modified as They Move Through the Cell

To understand how some of the components of the membrane system work together, let's look at the manufacture and export of an extremely important protein called an antibody (**Fig. 4-13**). Antibodies are glycoproteins produced by white blood cells that bind to foreign invaders (such as disease-causing bacteria) and help destroy them. Antibody proteins are synthesized on ribosomes of the rough ER within white blood cells, and they are packaged into vesicles formed from ER membrane. These vesicles travel to the Golgi, where the membranes fuse, releasing the protein into the Golgi apparatus. Within the Golgi, carbohydrates are attached to the protein, which is then repackaged into vesicles formed from

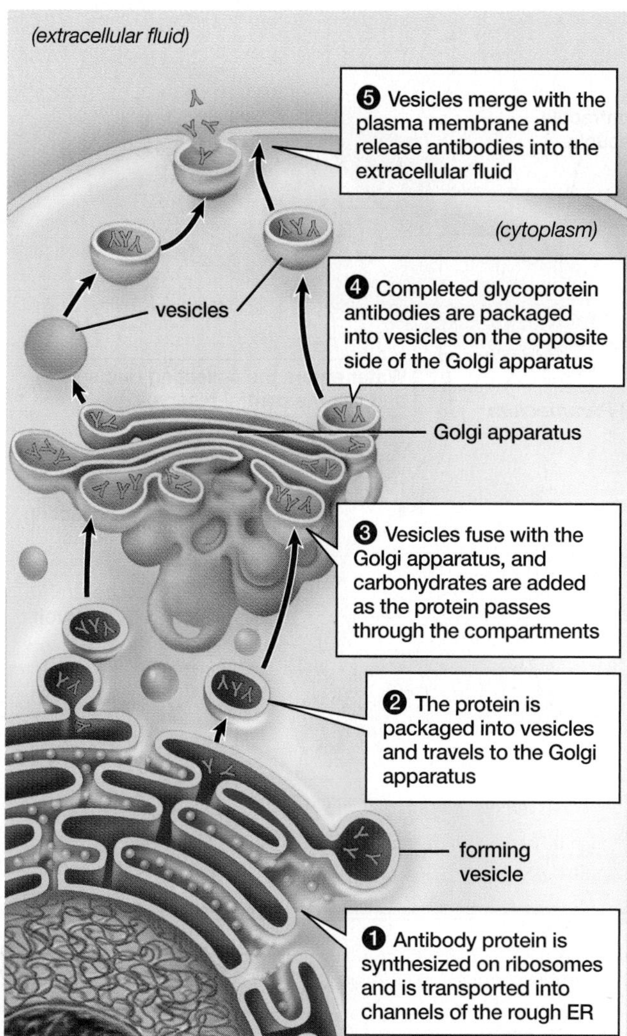

▲ FIGURE 4-13 **A protein is manufactured and exported** Here, we illustrate the formation of an antibody as an example of this process.

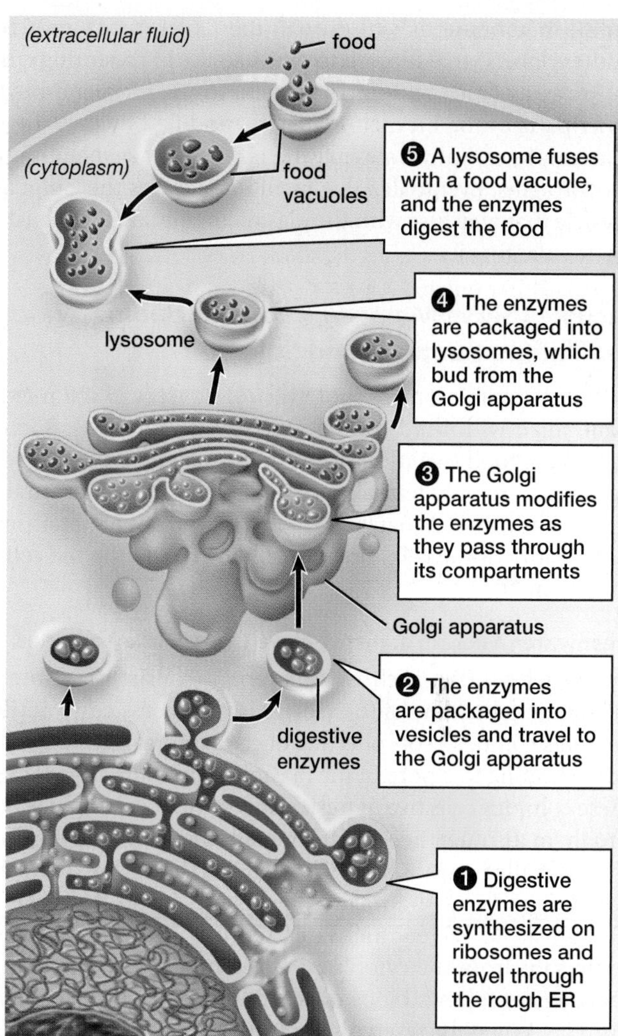

▲ FIGURE 4-14 **Formation and function of lysosomes and food vacuoles**

QUESTION Why is it advantageous for all membranes associated with the cell to have a fundamentally similar composition?

membrane acquired from the Golgi. The vesicle containing the completed antibody travels to the plasma membrane and fuses with it, releasing the antibody outside the cell, where it will make its way into the bloodstream to help defend the body against infection.

Lysosomes Serve as the Cell's Digestive System

Some of the proteins manufactured in the ER and sent to the Golgi apparatus are intracellular digestive enzymes that can break proteins, fats, and carbohydrates into their component subunits. In the Golgi, these enzymes are packaged in membrane-enclosed vesicles called **lysosomes** (**Fig. 4-14**). One major function of lysosomes is to digest food particles, which range from individual proteins to complete microorganisms.

As you will learn in Chapter 5, many cells "eat" by phagocytosis—that is, by engulfing particles just outside the cell using extensions of the plasma membrane. The plasma membrane with its enclosed food then pinches off inside the cytoplasm and forms a **food vacuole.** Lysosomes recognize these food vacuoles and merge with them. The contents of the two vesicles mix, and lysosomal enzymes digest the food into small molecules such as amino acids, monosaccharides, and fatty acids that can be used within the cell. Lysosomes also digest worn out or defective organelles. These are enclosed in vesicles of ER membrane, which fuse with lysosomes, exposing the organelle to digestive enzymes that break it down into its component molecules. These molecules are released into the cytoplasm where they can be reused in metabolic processes.

Membrane Is Exchanged Throughout the Cell

Throughout the cell, membrane is continually exchanged among the nuclear envelope, rough and smooth ER, Golgi apparatus, lysosomes, food vacuoles, and the plasma membrane. By reviewing Figures 4-13 and 4-14, you can get an idea of how the membranes interconnect. For example, the ER synthesizes the phospholipids and proteins that make up the plasma membrane and buds off some of this membrane in vesicles, which then fuse with the Golgi membranes. Some of the ER membrane that fuses with the Golgi carries protein "address labels" that direct it to return to the ER, restoring crucial proteins (such as some enzymes) to the ER membrane. Other parts of the ER membrane are modified by the Golgi; for example, carbohydrates may be added to make membrane glycoproteins. Eventually, this membrane leaves the Golgi as a vesicle that may fuse with the plasma membrane, replenishing or enlarging it.

Vacuoles Serve Many Functions, Including Water Regulation, Support, and Storage

Most cells contain one or more **vacuoles**—sacs of cell membrane filled with fluid containing various molecules. Some, such as the food vacuoles that form during phagocytosis (see Fig. 4-14), are only temporary. However, many cells contain permanent vacuoles that have important roles in maintaining the integrity of the cell, most notably by regulating the cell's water content.

Freshwater Microorganisms Have Contractile Vacuoles

Freshwater protists such as *Paramecium* consist of a single eukaryotic cell. Many of these organisms possess **contractile vacuoles** composed of collecting ducts, a central reservoir, and a tube leading to a pore in the plasma membrane (**Fig. 4-15**). These complex cells live in fresh water, which constantly leaks into them through their plasma membranes (we describe this process, called osmosis, in Chapter 5). The influx of water would soon burst the fragile organism if it did not have a mechanism to excrete the water. Cellular energy is used to pump salts from the cytoplasm of the protist into collecting ducts. Water follows by osmosis and drains into the central reservoir. When the reservoir of the contractile vacuole is full, it contracts, squirting the water out through a pore in the plasma membrane.

Plant Cells Have Central Vacuoles

Three-quarters or more of the volume of many plant cells is occupied by a large **central vacuole** (see Fig. 4-4). The central vacuole has several functions. Filled mostly with water, the central vacuole is involved in the cell's water balance. It also provides a "dump site" for hazardous wastes, which plant cells often cannot excrete. Some plant cells store poisonous substances, such as sulfuric acid, in their vacuoles. These poisons deter animals from munching on the otherwise tasty leaves.

Vacuoles may also store sugars and amino acids not immediately needed by the cell. Blue or purple pigments stored in central vacuoles are responsible for the colors of many flowers. As you will learn in Chapter 5, dissolved substances attract water into the vacuole. The water pressure (called turgor pressure) within the vacuole pushes the fluid portion of the cytoplasm up against the cell wall with considerable force. Cell walls are usually somewhat flexible, so both the overall shape and the rigidity of the cell depend on turgor pressure within the cell. Turgor pressure thus provides support for the non-woody parts of plants. (Look ahead to Fig. 5-10 to see what happens when you forget to water your houseplants.)

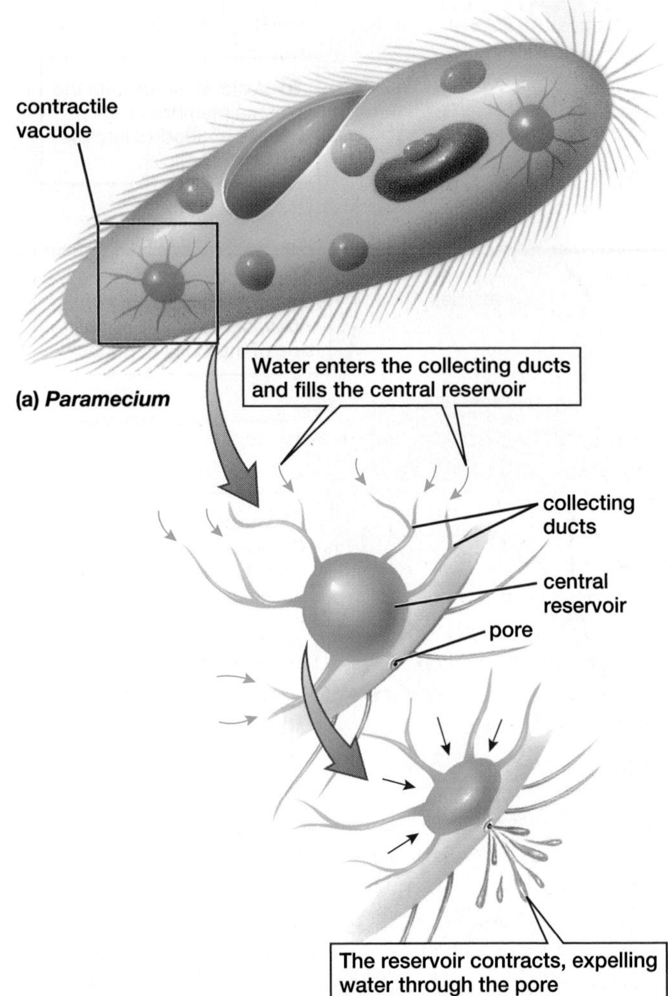
contractile vacuole
(a) *Paramecium*
Water enters the collecting ducts and fills the central reservoir
collecting ducts
central reservoir
pore
The reservoir contracts, expelling water through the pore
(b) Contractile vacuole

▲ **FIGURE 4-15 Contractile vacuoles** Many freshwater protists contain contractile vacuoles. **(a)** The single-celled protist *Paramecium* lives in freshwater ponds and lakes. **(b)** An enlargement of the contractile vacuole showing its structure and activity as it collects and expels water.

Mitochondria Extract Energy from Food Molecules, and Chloroplasts Capture Solar Energy

Every cell requires a continuous supply of energy to manufacture complex molecules and structures, to acquire nutrients from the environment and excrete waste materials, to move, and to reproduce. All eukaryotic cells have mitochondria that capture energy stored in sugar by producing high-energy ATP molecules. The cells of plants (and some protists) also have chloroplasts, which can capture energy directly from sunlight and store it in sugar molecules.

Most biologists accept the hypothesis that both mitochondria and chloroplasts evolved from prokaryotic bacteria that took up residence long ago within the cytoplasm of other prokaryotic cells, by a process called endosymbiosis (literally, "living together inside"). The **endosymbiont hypothesis** of mitochondrial and chloroplast evolution is discussed in more detail in Chapter 17. Mitochondria and chloroplasts are similar to each other and to prokaryotic cells in several ways. Both are about the size of some prokaryotic cells (1 to 5 micrometers in diameter). Both are surrounded by a double membrane; the outer membrane may have come from the original host cell and the inner membrane from the guest cell. Both have assemblies of enzymes that synthesize ATP, as would have been needed by an independent cell. Finally, both possess DNA and ribosomes that more closely resemble prokaryotic than eukaryotic DNA and ribosomes.

Mitochondria Use Energy Stored in Food Molecules to Produce ATP

All eukaryotic cells contain **mitochondria** (singular, mitochondrion), which are sometimes called the "powerhouses of the cell" because they extract energy from food molecules and store it in the high-energy bonds of ATP. As you will see in Chapter 8, different amounts of energy can be released from a food molecule, depending on how it is broken down. The breakdown of food molecules begins with enzymes in the cytoplasmic fluid and does not use oxygen. This **anaerobic** ("without oxygen") breakdown does not convert much food energy into ATP energy. Mitochondria enable a eukaryotic cell to use oxygen to break down high-energy molecules even further. These **aerobic** ("with oxygen") reactions generate energy much more effectively; about 16 times as much ATP is generated by aerobic metabolism in the mitochondria than by anaerobic metabolism in the cytoplasmic fluid. Not surprisingly, mitochondria are found in large numbers in metabolically active cells, such as muscle, and they are less abundant in cells that are less active, such as those of cartilage.

Mitochondria possess a pair of membranes (**Fig. 4-16**). The outer membrane is smooth, whereas the inner membrane forms deep folds called cristae (singular, crista, meaning "crest"). The mitochondrial membranes enclose two fluid-filled spaces: the intermembrane compartment between the inner and outer membranes and the matrix, or inner compartment, within the inner membrane. Some of the reactions that break down high-energy molecules occur in the fluid of the matrix inside the inner membrane; the rest are conducted by a series of enzymes attached to the membranes of the cristae within the intermembrane compartment. The role of mitochondria in energy production is described in detail in Chapter 8.

Chloroplasts Are the Sites of Photosynthesis

Photosynthesis, which captures sunlight and provides the energy to power life on Earth, occurs in the chloroplasts of eukaryotic cells (and on membranes of prokaryotic cells, described later). **Chloroplasts** (**Fig. 4-17**) are specialized organelles surrounded by a double membrane. The inner membrane of the chloroplast encloses a fluid called the stroma. Within the stroma are interconnected stacks of hollow, membranous sacs. The individual sacs are called thylakoids, and a stack of sacs is a granum (plural, grana).

The thylakoid membranes contain the green pigment molecule **chlorophyll** (which gives plants their green color) as well as other pigment molecules. During photosynthesis, chlorophyll captures the energy of sunlight and transfers it to other molecules in the thylakoid membranes. These molecules transfer the energy to ATP and other energy carriers. The energy carriers diffuse into the stroma, where

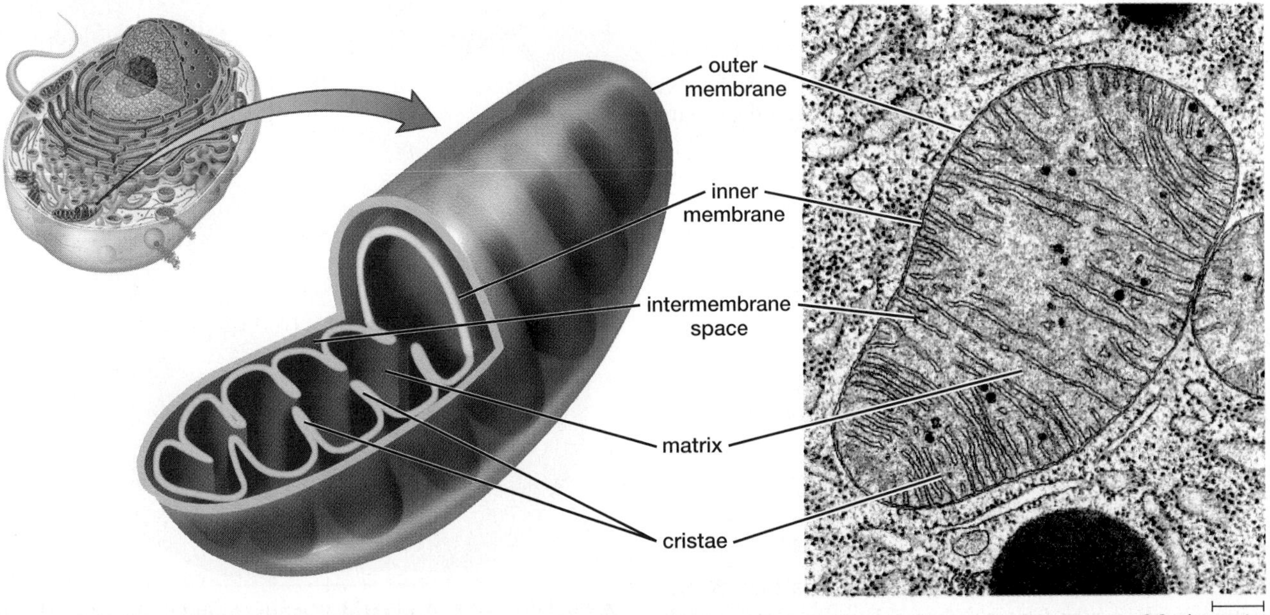

▲ **FIGURE 4-16 A mitochondrion** Mitochondria contain two membranes enclosing two fluid compartments: the intermembrane compartment between the outer and inner membranes, and the matrix within the inner membrane. The outer membrane is smooth, but the inner membrane forms deep folds called cristae. These structures are visible in the TEM on the right.

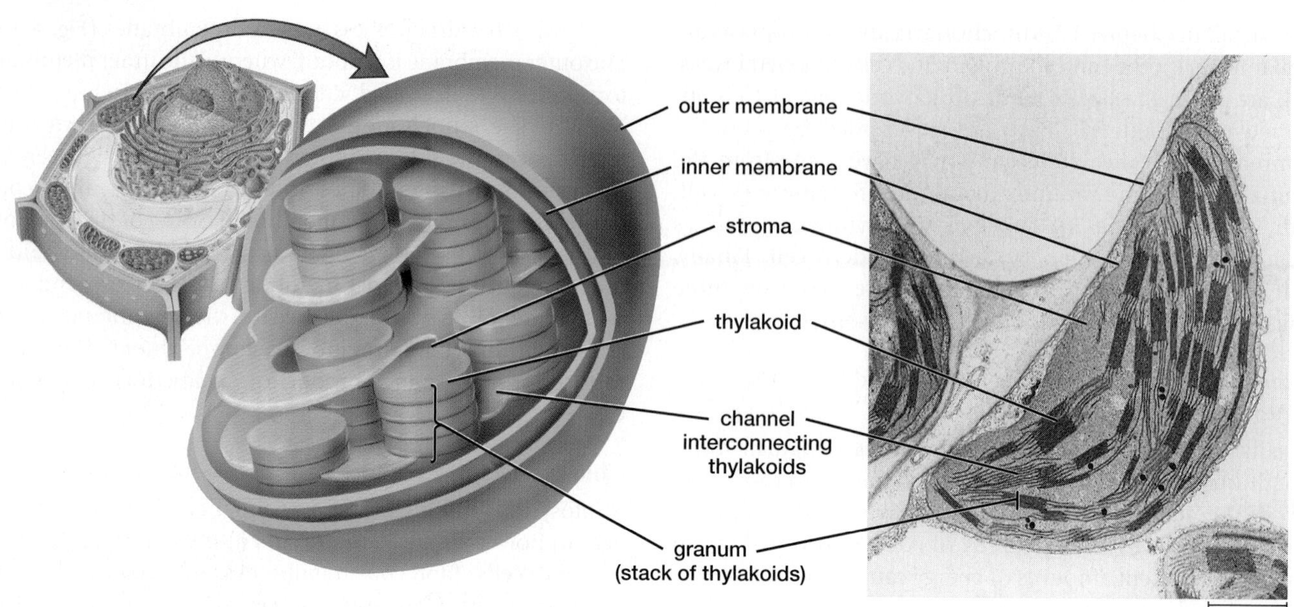

▲ FIGURE 4-17 **A chloroplast** Chloroplasts are surrounded by a double membrane. The fluid stroma is enclosed by the inner membrane; within the stroma are stacks of thylakoid sacs called grana. Chlorophyll is embedded in the membranes of the thylakoids.

their energy is used to drive the synthesis of sugar from carbon dioxide and water.

Plants Use Plastids for Storage

Chloroplasts are highly specialized **plastids,** organelles found only in plants and photosynthetic protists (Fig. 4-18). Plants and photosynthetic protists use nonchloroplast types of plastids as storage containers for various molecules, including pigments that give ripe fruits their yellow, orange, or red colors. In plants that continue growing from one year to the next, plastids store photosynthetic products from the summer for use during the following winter and spring. Most plants convert the sugars made during photosynthesis into starch, which is also stored in plastids. Potatoes, for example, are composed almost entirely of cells stuffed with starch-filled plastids (see Fig. 4-18).

BioFlix™ Tour of an Animal Cell

BioFlix™ Tour of a Plant Cell

4.4 WHAT ARE THE MAJOR FEATURES OF PROKARYOTIC CELLS?

Prokaryotic Cells Are Relatively Small and Possess Specialized Surface Features

Most prokaryotic cells are small, generally less than 5 micrometers in diameter (most eukaryotic cells are from 10 to 100 micrometers in diameter) with a simple internal structure compared to eukaryotic cells (Fig. 4-19; compare this to Figs. 4-3 and 4-4). Nearly all prokaryotic cells are

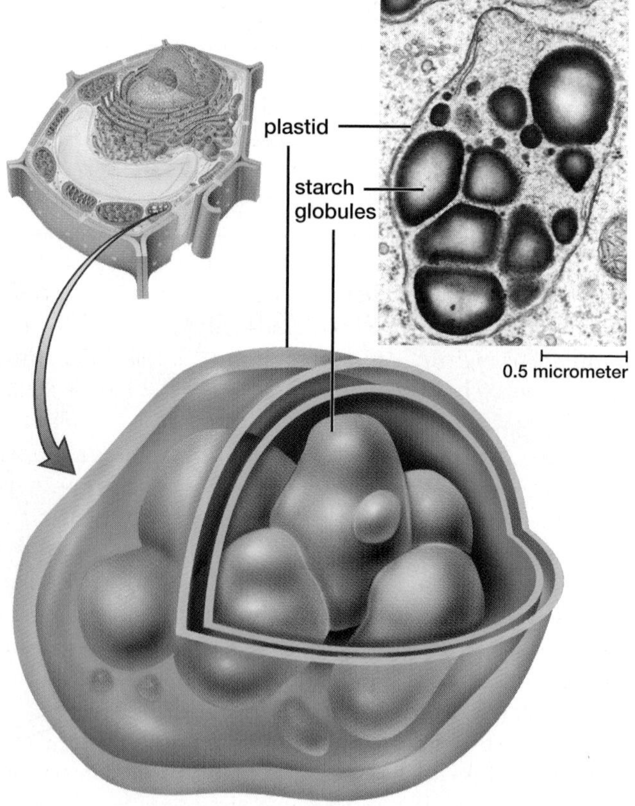

▲ FIGURE 4-18 **A plastid** Plastids, found in the cells of plants and photosynthetic protists, are surrounded by a double outer membrane. Chloroplasts are the most familiar; other types store various materials, such as the starch filling these plastids in potato cells, seen on the right in a TEM.

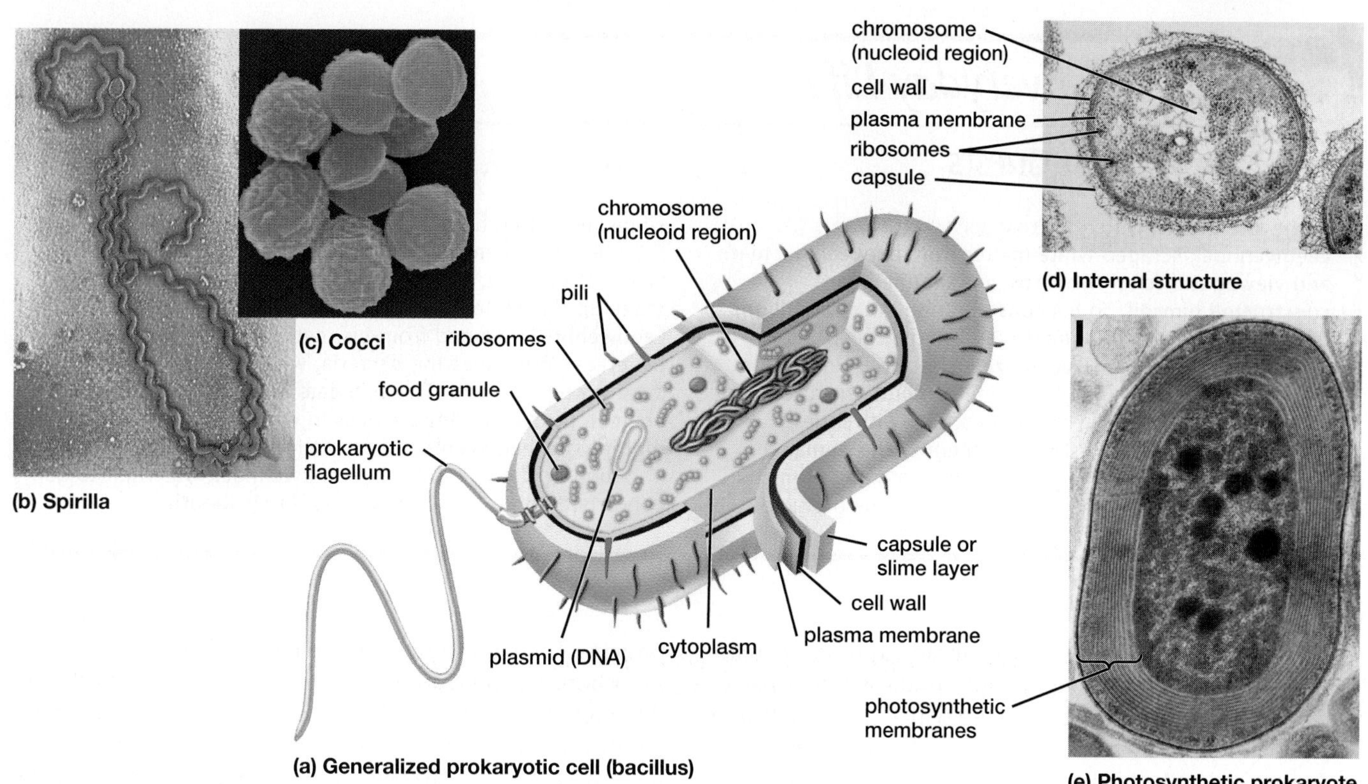

▲ FIGURE 4-19 **Prokaryotic cells are simpler than eukaryotic cells** Prokaryotes assume a variety of shapes, including **(a)** rod-shaped bacilli, **(b)** spiral-shaped spirilla, and **(c)** spherical cocci. Internal structures are revealed in the TEMs in **(d)** and **(e)**. Some photosynthetic bacteria have internal membranes where photosynthesis occurs, as shown in **(e)**.

surrounded by a stiff cell wall, which protects the cell and confers its characteristic shape. Most prokaryotes take the form of rod-shaped bacilli (**Fig. 4-19a**), spiral-shaped spirilla (**Fig. 4-19b**), or spherical cocci (**Fig. 4-19c**). Several types of antibiotics, including penicillin, fight bacterial infections by interfering with cell wall synthesis, causing the bacteria to rupture. Although no prokaryotic cells possess cilia, some bacteria and archaea are propelled by flagella whose structure differs from eukaryotic flagella.

Bacteria that infect other organisms—such as those that cause tooth decay, diarrhea, pneumonia, or urinary tract infections—possess surface features that help them adhere to specific host tissues, such as the surface of a tooth or the lining of the small intestine, lungs, or bladder. These surface features include capsules and slime layers, which are polysaccharide coatings that some bacteria secrete outside their cell walls. When Anton van Leeuwenhoek observed material scraped from his teeth under his simple microscope, he observed many bacteria that adhere using slime layers (see "Links to Everyday Life: Unwanted Guests" on p. 74). Capsules and slime layers also help some prokaryotic cells avoid drying out. Pili (literally, "hairs") are surface proteins jutting outward from the prokaryotic cell wall (see Fig. 4-19a). Some types of bacteria form sex pili, hollow protein tubes used to exchange genetic material (DNA) between bacterial cells. The features of prokaryotic cells are discussed and illustrated in more detail in Chapter 19.

Prokaryotic Cells Have Fewer Specialized Structures Within Their Cytoplasm

The cytoplasm of most prokaryotic cells is rather homogeneous in appearance when compared to eukaryotic cells. In the central region of the cell is an area called the **nucleoid** (see Fig. 4-19a). Within the nucleoid, prokaryotic cells generally have a single, circular chromosome consisting of a long coiled strand of DNA that carries essential genetic information. The nucleoid is not separated from the cytoplasm by a membrane, and so is not a true nucleus as seen in eukaryotic cells. Most prokaryotic cells also contain small rings of DNA called plasmids, which are located outside the nucleoid. Plasmids usually carry genes that give the cell special properties; for example, some disease-causing bacteria possess plasmids that allow them to inactivate antibiotics, making them much more difficult to kill.

Prokaryotic cells lack nuclei and the other membrane-enclosed organelles (such as chloroplasts, mitochondria, ER, Golgi, and other components of the membrane system) that eukaryotic cells possess. Nonetheless, some prokaryotic cells do use membranes to organize the enzymes responsible for a series of biochemical reactions. Such enzymes are situated in

Links to *Everyday Life*

Unwanted Guests

In the late 1600s, Dutch microscopist Anton van Leeuwenhoek scraped white matter from between his teeth and viewed it through a microscope that he had constructed himself. To his consternation, he saw millions of cells that he called "animalcules," single-celled organisms that we now recognize as bacteria. Annoyed at the presence of these life-forms in his mouth, he attempted to kill them with vinegar and hot coffee—with little success. The warm, moist environment of the human mouth, particularly the crevices of the teeth and gums, is an ideal habitat for a variety of bacteria. Some produce slime layers that help them and others adhere to the tooth. Each divides repeatedly to form a colony of offspring. Thick layers of bacteria, slime, and glycoproteins from saliva make up the white substance, called plaque, that Leeuwenhoek scraped from his teeth. Sugar in foods and beverages nourishes the bacteria, which break down the sugar into lactic acid. The acid eats away at the tooth enamel, producing tiny crevices in which the bacteria multiply further, eventually producing a cavity. So, although he didn't know why, Leeuwenhoek was right to be concerned about the "animalcules" in his mouth!

a particular sequence along the membrane to promote reactions in the proper order. For example, photosynthetic bacteria possess internal membranes in which light-capturing proteins and enzymes that catalyze the synthesis of high-energy molecules are embedded in a specific order (**Fig. 4-19e**). In prokaryotic cells, reactions that harvest energy from the breakdown of sugars are catalyzed by enzymes that may be either anchored along the inner plasma membrane or floating free in the cytoplasm.

Bacterial cytoplasm contains ribosomes (see Fig. 4-19a). Although their function is similar to that of eukaryotic ribosomes, bacterial ribosomes are smaller and contain different proteins. These ribosomes resemble those found in eukaryotic mitochondria and chloroplasts, providing support for the endosymbiont hypothesis described earlier. Prokaryotic cytoplasm also may contain food granules that store energy-rich molecules, such as glycogen, but these are not enclosed by membranes.

You might wish to go back and look at Table 4-1 at this point to review the differences between prokaryotic and eukaryotic cells. As you compare these cell types, think about the endosymbiont hypothesis. You'll learn more about this concept in Chapter 17.

Case Study revisited

Spare Parts for Human Bodies

Bioengineering of tissues and organs such as skin requires the coordinated efforts of biochemists, biomedical engineers, cell biologists, and physicians.

In laboratories across the world, teams of scientists are working to grow not only skin but also bone, cartilage, heart valves, bladders, and breast tissue on plastic scaffolds. To heal broken bones, for example, teams of researchers are working to develop degradable plastics, incorporating protein growth factors into the material. These growth factors would encourage nearby bone cells and tiny blood vessels to invade the plastic as it breaks down, eventually replacing it with the patient's own bone.

Researchers continue to refine tissue-culturing techniques and to develop better scaffolding materials with the goal of duplicating entire organs. Bioartificial bladders have recently been created using muscle and bladder-lining cells taken from patients with improperly functioning bladders (**Fig. 4-20**). The cells are seeded onto a bladder-shaped scaffold composed of collagen, are incubated for about 7 weeks, and are then transplanted into a patient. The bladders, designed by Dr. Anthony Atala, a surgeon at Wake Forest University, have performed successfully in the seven patients who received them. A biotechnology company has completed animal studies and hopes to begin further tests of these bladders in humans in 2009, with the goal of bringing the bioengineered "Neo-Bladder" to the medical market.

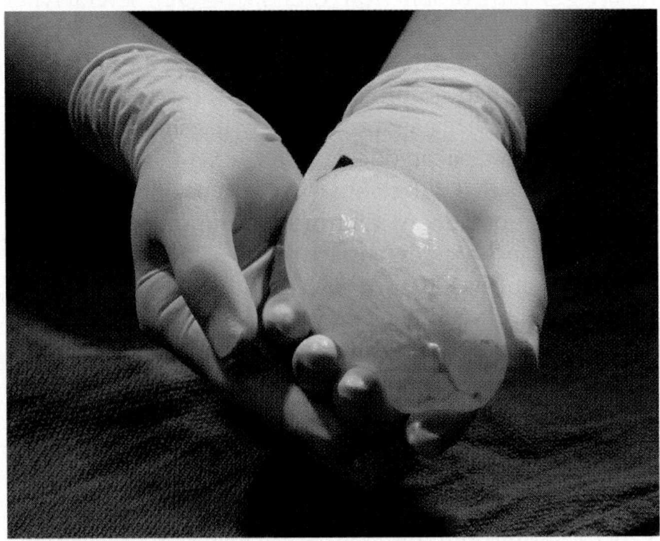

▲ FIGURE 4-20 **A laboratory-grown bladder**

An exciting new development in tissue engineering is "bioprinting." Using a specially designed instrument (a distant relative of the familiar ink-jet printer), Gabor Forgacs, a biophysicist at the University of Missouri–Columbia, hopes to engineer organs without using scaffolds. His "printer" ejects spheres less than 0.5 millimeter in diameter, which contain from 10,000 to 40,000 cells, onto a gel within which the cells can migrate and interact with one another. The bioprinting technique uses the inherent ability of cells to communicate with one another using physical and chemical signals, and to organize themselves into functional units. In an early test, chicken heart muscle cells were printed onto a dish. After about 90 hours, the heart muscle cells had migrated together and were communicating with one another, contracting in unison, as does a beating heart.

Forgacs and his team are currently working on printing out blood vessels using spheres composed of the three types of cells that comprise a blood vessel. The spheres surround a central core of collagen that will eventually be removed to allow blood to flow through. Over time, the three cell types spontaneously organize themselves around the collagen core just as they do in a natural vessel. Bioprinting offers the possibility of one day building a complete organ from a patient's own cells, and the near future holds the almost certain promise of bioengineered blood vessels.

BioEthics Consider This

Forgacs foresees one of the first uses of his cell-printing technique to be the formation of tissues from human cells on which to test new drugs. What do you think are the advantages of this approach compared to testing drugs on laboratory animals?

CHAPTER REVIEW

Summary of Key Concepts

4.1 What Is the Cell Theory?

The principles of the cell theory are as follows: Every living organism is made up of one or more cells, the smallest living organisms are single cells, cells are the functional units of multicellular organisms, and all cells arise from preexisting cells.

4.2 What Are the Basic Attributes of Cells?

Cells are small because they must exchange materials with their surroundings by diffusion, a slow process that requires that the interior of the cell never be too far from the plasma membrane. All cells are surrounded by a plasma membrane that regulates the interchange of materials between the cell and its environment. All cells use DNA as a genetic blueprint and RNA to direct protein synthesis based on this blueprint. All cells obtain the materials to generate the molecules of life, and the energy to power this synthesis, from their living and nonliving environment. There are two fundamentally different types of cells: prokaryotic and eukaryotic. Prokaryotic cells are small and lack membrane-enclosed organelles. Eukaryotic cells have a variety of organelles, including a nucleus.

4.3 What Are the Major Features of Eukaryotic Cells?

Cells of plants, fungi, and some protists are supported by porous cell walls outside the plasma membrane. All eukaryotic cells have an internal cytoskeleton of protein filaments. The cytoskeleton organizes and shapes the cell, and moves and anchors organelles. Some eukaryotic cells have cilia or flagella, extensions of the plasma membrane that contain microtubules in a characteristic pattern. These structures move fluids past the cell or move the cell through a fluid environment.

Genetic material (DNA) is contained within the nucleus, which is bounded by the double membrane of the nuclear envelope. Pores in the nuclear envelope regulate the movement of molecules between nucleus and cytoplasm. The genetic material is organized into strands called chromosomes, which consist of DNA and proteins. The nucleolus consists of ribosomal RNA and ribosomal proteins, as well as the genes that code for ribosome synthesis. Ribosomes are composed of ribosomal RNA and protein and are the sites of protein synthesis.

The membrane system of a cell includes the plasma membrane, endoplasmic reticulum (ER), Golgi apparatus, vacuoles, and vesicles. Endoplasmic reticulum forms a series of interconnected membranous compartments. The ER is a major site of membrane synthesis within the cell. Smooth ER, which lacks ribosomes, manufactures lipids such as steroid hormones, detoxifies drugs and metabolic wastes, breaks glycogen into glucose, and stores calcium. Rough ER, which bears ribosomes, manufactures and modifies proteins. The Golgi apparatus is a series of membranous sacs derived from the ER. The Golgi processes and modifies materials synthesized in the rough ER. Substances modified in the Golgi are packaged into vesicles for transport elsewhere in the cell. Lysosomes are vesicles that contain digestive enzymes, which digest food particles and defective organelles.

All eukaryotic cells contain mitochondria—organelles that use oxygen to complete the metabolism of food molecules, capturing much of their energy as ATP. Cells of plants and some protists contain plastids. Storage plastids store pigments or starch. Chloroplasts are specialized plastids that capture solar energy during photosynthesis, allowing plant cells to manufacture sugar from water and carbon dioxide. Both mitochondria and chloroplasts probably originated from bacteria.

Many eukaryotic cells contain sacs, called vacuoles, that are bounded by a single membrane and function to store food or wastes, excrete water, or support the cell. Some protists have contractile vacuoles, which collect and expel water. Plants use central vacuoles to support the cell as well as to store wastes and toxic materials.

BioFlix™ Tour of an Animal Cell

BioFlix™ Tour of a Plant Cell

4.4 What Are the Major Features of Prokaryotic Cells?

Prokaryotic cells are generally much smaller than eukaryotic cells and have a much simpler internal structure. Most are surrounded by a relatively stiff cell wall. The cytoplasm of prokaryotic cells lacks membrane-enclosed organelles (although some photosynthetic bacteria have extensive internal membranes). A single, circular strand of DNA is found in the nucleoid. Table 4-1 compares prokaryotic cells to the eukaryotic cells of plants and animals.

Study Note

Figures 4-3, 4-4, and 4-19 illustrate the overall structure of animal, plant, and prokaryotic cells, respectively. Table 4-1 lists the principal organelles, their functions, and their occurrence in animals, plants, and prokaryotes.

Key Terms

aerobic *71*
anaerobic *71*
archaea *62*
bacteria *62*
basal body *64*
cell theory *56*
cell wall *63*
central vacuole *70*
centriole *64*
chlorophyll *71*
chloroplast *71*
chromatin *65*
chromosome *65*
cilium (plural, **cilia**) *64*
contractile vacuole *70*
cytoplasm *60*
cytoplasmic fluid *60*
cytoskeleton *63*
deoxyribonucleic acid (DNA) *61*
endoplasmic reticulum (ER) *67*
endosymbiont hypothesis *70*
eukaryotic *62*
flagellum (plural, **flagella**) *64*
food vacuole *69*
Golgi apparatus *68*
intermediate filament *63*
lysosome *69*
microfilament *63*
microtubule *63*
mitochondrion (plural, **mitochondria**) *71*
nuclear envelope *65*
nuclear pore complex *65*
nucleoid *73*
nucleolus (plural, **nucleoli**) *66*
nucleus *65*
organelle *63*
plasma membrane *56*
plastid *72*
prokaryotic *62*
ribonucleic acid (RNA) *61*
ribosome *66*
vacuole *70*
vesicle *67*

Thinking Through the Concepts

Fill-in-the-Blank

1. The plasma membrane is composed of what two major types of molecules? __________ and __________ Which type of molecule is responsible for each of the following functions? Isolation from the surroundings: __________; interactions with other cells: __________; movement of hydrophilic molecules through the membrane: __________.
2. The __________ is composed of a network of protein fibers. The three types of protein fibers are __________, __________, and __________. Which of these supports cilia? __________
3. After each description, fill in the appropriate term: "Workbenches" of the cell: __________; comes in rough and smooth forms: __________; site of ribosome production: __________; a stack of flattened membranous sacs: __________; outermost layer of plant cells: __________; ferries blueprints for protein production from the nucleus to the cytoplasm: __________.
4. Antibody proteins are synthesized on ribosomes associated with the __________. The antibody proteins are packaged into membranous sacs called __________ and are then transported to the __________. Here, what type of molecule is added to the protein? __________ After the antibody is completed, it is packaged into vesicles that fuse with the __________ membrane.
5. After each description, fill in the appropriate structure: "Powerhouses" of the cell: __________; capture solar energy: __________; outermost structure of plant cells: __________; region of prokaryotic cell containing DNA: __________; propel fluid past cells: __________; consists of cytoplasmic fluid and the organelles within it: __________.
6. Two organelles that are believed to have evolved from prokaryotic cells are __________ and __________. Evidence for this hypothesis is that both have __________ membranes, both have groups of enzymes that synthesize __________, both have their own __________, and their __________ is similar to that of prokaryotic cells.

Review Questions

1. What are the three parts of the cell theory?
2. Which organelles are common to both plant and animal cells, and which are unique to each?
3. Define *stroma* and *matrix*.
4. Describe the nucleus and the function of each of its components, including the nuclear envelope, chromatin, chromosomes, DNA, and the nucleolus.
5. What are the functions of mitochondria and chloroplasts? Why do scientists believe that these organelles arose from prokaryotic cells?
6. What is the function of ribosomes? Where in the cell are they found? Are they limited to eukaryotic cells?
7. Describe the structure and function of the endoplasmic reticulum and the Golgi apparatus and how they work together.
8. How are lysosomes formed? What is their function?
9. Diagram the structure of cilia and flagella. Describe how each one beats and what their beating accomplishes.

Applying the Concepts

1. If samples of muscle tissue were taken from the legs of a world-class marathon runner and a sedentary individual, which would you expect to have a higher density of mitochondria? Why?
2. One of the functions of the cytoskeleton in animal cells is to give shape to the cell. Plant cells have a fairly rigid cell wall surrounding the plasma membrane. Does this mean that a cytoskeleton is unnecessary for a plant cell?
3. Most cells are very small. What physical and metabolic constraints limit cell size? What problems would an enormous cell encounter? What adaptations might help a very large cell survive?

MB ***Go to www.masteringbiology.com for practice quizzes, activities, eText, videos, current events, and more.***

Cell Membrane Structure and Function

Chapter 5

Case Study

Vicious Venoms

WELL BEFORE SCHOOL LET OUT, 13-year-old Justin Schwartz was anticipating his three-week stay at a camp near Yosemite National Park. On July 21, 2002, after a four-and-a-half mile hike, Justin rested on some sunny rocks, hands hanging loosely at his sides. Suddenly, he felt a piercing pain in his left palm. A 5-foot rattlesnake—probably feeling threatened by Justin's dangling arm—had struck without warning.

His campmates stared in alarm as the snake slithered into the undergrowth, but Justin focused on his hand, where the pain was becoming agonizing and the palm was beginning to swell. He suddenly felt weak and dizzy. As counselors and campmates spent the next 4 hours carrying him down the trail, pain and discoloration spread up Justin's arm, and his hand felt as if it were going to burst. A helicopter whisked him to a hospital, where he fell unconscious. A day later, he regained consciousness at the University of California Davis Medical Center. There, Justin spent over a month undergoing 10 surgeries. These relieved the enormous pressure from swelling in his arm, removed dead muscle tissue, and began the long process of repairing the extensive damage to his hand and arm.

Diane Kiehl's ordeal began as she dressed for an informal Memorial Day celebration with her family on May 27, 2006, pulling on blue jeans that she had tossed on the bathroom floor the previous night. Feeling a sting on her right thigh, she ripped off the jeans and watched in dismay as a long-legged brown recluse spider crawled out. Living in an old house in the Kansas countryside, Diane had grown accustomed to spiders as frequent—although uninvited—guests. The two small puncture wounds seemed merely a minor annoyance until the next day, when an extensive, itchy rash appeared at the site. By the third day, intermittent pain pierced like a knife through her thigh. A physician gave her painkillers, steroids to reduce the swelling, and antibiotics to combat bacteria introduced by the spider's mouthparts. The next 10 days were a nightmare of pain from the growing sore, now covered with oozing blisters and underlain with clotting blood. It took 4 months for the lesion to heal. Even after a year, Diane sometimes felt pain in the large scar that remained.

How do rattlesnake and brown recluse spider venoms cause leaky blood vessels, disintegrating skin and tissue, and sometimes life-threatening symptoms throughout the body? What do venoms have to do with cell membranes?

▲ A rattlesnake prepares to strike. (Inset) A brown recluse spider.

5.1 HOW IS THE STRUCTURE OF A MEMBRANE RELATED TO ITS FUNCTION?

Cell Membranes Isolate the Cell Contents While Allowing Communication with the Environment

As you now know, all cells, as well as organelles within eukaryotic cells, are surrounded by membranes. Cell membranes perform several crucial functions:

- They selectively isolate the cell's contents from the external environment, allowing concentration gradients of dissolved substances to be produced across the membrane.
- They regulate the exchange of essential substances between the cell and the extracellular fluid, or between membrane-enclosed organelles and the surrounding cytoplasm.
- They allow communication between cells.
- They create attachments within and between cells.
- They regulate many biochemical reactions.

These are formidable tasks for a structure so thin that 10,000 membranes stacked atop one another would scarcely equal the thickness of this page. The key to membrane function lies in its structure. Membranes are not simply uniform sheets; they are complex, heterogeneous structures whose different parts perform very distinct functions. Membranes vary from one tissue type to another, and they change dynamically in response to their surroundings.

All the membranes of a cell have a similar basic structure: proteins floating in a double layer of phospholipids (see pp. 44–45). Phospholipids are responsible for the isolating function of membranes, whereas proteins are responsible for selectively exchanging substances and communicating with the environment, controlling biochemical reactions associated with the cell membrane, and forming attachments.

Membranes Are "Fluid Mosaics" in Which Proteins Move Within Layers of Lipids

Prior to the 1970s, although cell biologists knew that cell membranes consist primarily of proteins and lipids, how these molecules create the structure and function of membranes remained unknown. In 1972, cell researchers S. J. Singer and G. L. Nicolson developed the **fluid mosaic model** of cell membranes, which is now known to be accurate. According to this model, each membrane consists of a mosaic, or "patchwork," of different proteins that constantly shift and flow within a viscous (thick, sticky) fluid formed by a double layer of phospholipids (**Fig. 5-1**). Although the components of the plasma membrane remain relatively constant, the overall distribution of proteins and various types of phospholipids can change over time. Let's look more closely at the structure of membranes.

The Phospholipid Bilayer Is the Fluid Portion of the Membrane

As you learned in Chapter 3, a phospholipid consists of two very different parts: a "head" that is polar and hydrophilic (attracted to water) and a pair of fatty acid "tails" that are nonpolar and hydrophobic (not attracted to water). Membranes contain many different phospholipids of the general type shown in **Figure 5-2.** Notice that in this particular phospholipid, a double bond (which makes the lipid unsaturated) creates a kink in the fatty acid tail.

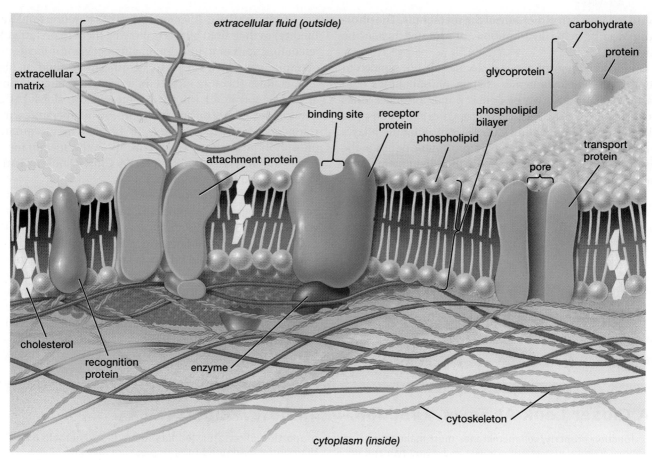

▲ **FIGURE 5-1 The plasma membrane** The plasma membrane is a bilayer of phospholipids interspersed with cholesterol molecules that form a fluid matrix in which various proteins (blue) are embedded. Many proteins have carbohydrates attached to them, forming glycoproteins. Recognition, attachment, receptor, and transport proteins are illustrated.

All cells are surrounded by water. Single-celled organisms may live in fresh water or in the ocean, water saturates the cell walls of plants, and animal cells are bathed in a weakly salty extracellular fluid that filters out of the blood. The fluid portion of the cytoplasm is mostly water. Thus, plasma membranes separate the watery cytoplasmic fluid from its watery external environment, and internal membranes surround watery compartments within the cell. Surrounded by water, phospholipids spontaneously arrange themselves into a double layer called a **phospholipid bilayer** (Fig. 5-3). Hydrogen bonds form between water and the hydrophilic phospholipid heads, causing the heads to orient outward

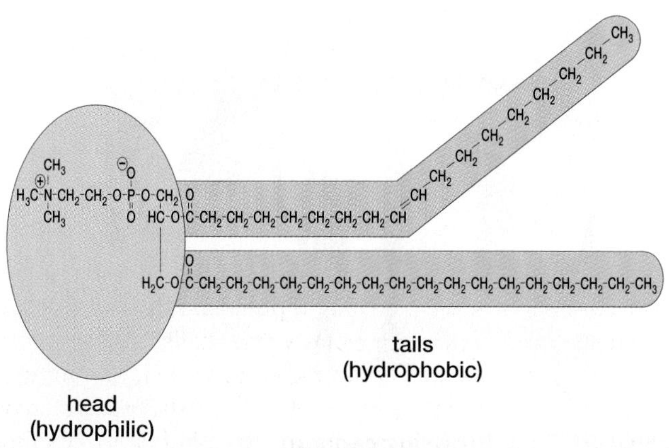

▲ **FIGURE 5-2 A phospholipid** Notice that a double bond in one of the fatty acid tails causes the tail to bend.

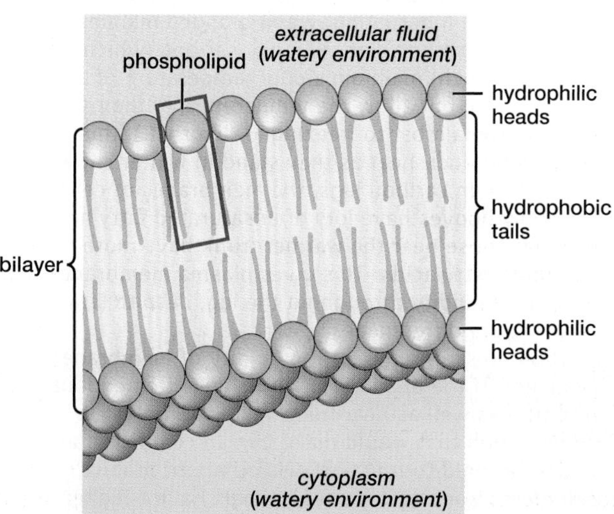

▲ **FIGURE 5-3 The phospholipid bilayer of the cell membrane**

toward the water on either side of the membrane. The phospholipid tails, being hydrophobic, are located inside the bilayer.

Individual phospholipid molecules are not bonded to one another, and membranes contain phospholipids with unsaturated fatty acids whose double bonds introduce kinks into their tails (see Fig. 5-2). These properties allow phospholipids to move about within each layer, making the bilayer reasonably fluid. Membranes with more double bonds in the tails of their phospholipids are more fluid than membranes whose phospholipids have fewer double bonds (**Fig. 5-4**).

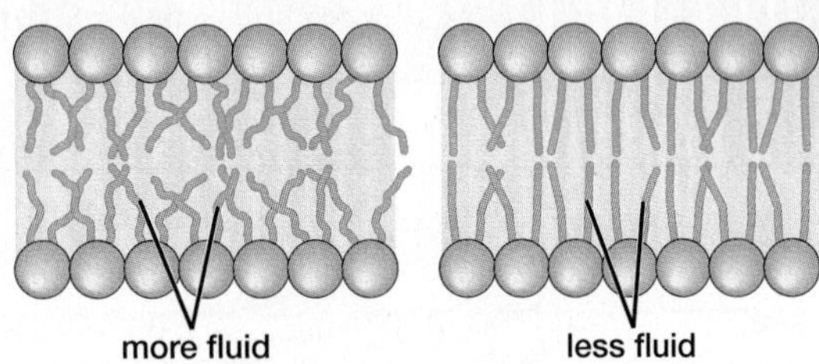

▲ **FIGURE 5-4 Kinks increase fluidity**

Cells may have different degrees of saturation (and thus fluidity) in their lipid bilayers, allowing them to perform different functions or to function in different environments. For example, membranes tend to become more fluid at high temperatures (because molecules move faster) and less fluid at low temperatures (molecules move more slowly). Cell membranes of organisms living in low-temperature environments are therefore likely to be rich in unsaturated phospholipids, whose kinky tails allow the membrane to retain the necessary fluidity (see "A Closer Look At Form, Function, and Phospholipids").

Most biological molecules, including salts, amino acids, and sugars, are hydrophilic; that is, they are polar and water soluble. Such substances cannot easily pass through the nonpolar, hydrophobic fatty acid tails of the phospholipid bilayer. The phospholipid bilayer is largely responsible for the first of the five membrane functions listed earlier: selectively isolating the cell's contents from the external environment.

The isolation provided by the plasma membrane is not complete, however. As we will describe later, very small molecules—such as water, oxygen, and carbon dioxide—as

A Closer Look At *Form, Function, and Phospholipids*

To function properly, cell membranes must maintain an optimal fluidity. Just as butter melts in a frying pan and oil solidifies in a freezer, the fluidity of cell membranes is sensitive to changing temperatures. Many organisms—including protists and bacteria, plants, "cold-blooded" animals (such as frogs, fish, and snakes), and hibernating mammals—experience large fluctuations in body temperature. Membrane fluidity is influenced by the relative amounts of saturated and unsaturated fatty acid tails in membrane phospholipids. Cells can modify their membrane phospholipid composition to maintain the proper fluidity at different temperatures. At higher temperatures, phospholipids with more saturated fatty acids are inserted into cell membranes, while at lower temperatures, unsaturated fatty acids are added.

A few non-hibernating, warm-blooded mammals have similar adaptations. For example, caribou (which live in the far north) maintain a core temperature of about 100°F (38°C), but allow the temperature of their lower legs to drop almost to freezing (32°F or 0°C), thus conserving body heat as they stand in winter snow (**Fig. E5-1**). In caribou legs, the membranes of cells near the chilly hooves have lots of unsaturated fatty acids, whereas those near the warmer trunk have more saturated fatty acids. This gives plasma membranes the necessary fluidity throughout the leg, despite large temperature differences.

High pressure reduces the fluidity of membranes. Denizens of the deep ocean experience tremendous pressure as well as near-freezing temperatures, a combination that would make human cell membranes much too rigid to function. As you might predict, cell membranes of deep-sea animals have a higher proportion of unsaturated fats (more like vegetable oil than butter). Adapted to high pressure and low temperatures, deep-sea animals rarely survive when brought to the surface. At surface pressure, researchers hypothesize that their cell membranes become too fluid and leaky, preventing their cells from maintaining normal gradients, which leads to death.

▲ **FIGURE E5-1 Browsing caribou**

Case Study continued
Vicious Venoms

Some of the most devastating effects of certain snake and spider venoms occur because they contain enzymes, called phospholipases, that break down phospholipids in cell membranes, causing the cells to die.

well as larger, uncharged, lipid-soluble molecules, can pass through the phospholipid bilayer.

In most animal cells, the phospholipid bilayer of membranes also contains cholesterol (see Fig. 5-1). Some cellular membranes have relatively few cholesterol molecules; others have as many cholesterol molecules as they do phospholipids. Cholesterol affects membrane structure and function in several ways: It stabilizes the lipid bilayer, making it less fluid at higher temperatures, less solid at lower temperatures, and also less permeable to water-soluble substances such as ions or monosaccharides.

The flexible, fluid nature of the bilayer is very important for membrane function. As you breathe, move your eyes, and turn the pages of this book, cells in your body change shape. If plasma membranes were stiff instead of flexible, cells would break open and die. Further, as you learned in Chapter 4, membranes within eukaryotic cells are in constant motion. Membrane-enclosed compartments ferry substances into the cell, carry materials within the cell, and expel them to the outside, merging membranes in the process. This flow and merger of membranes is made possible by the fluid nature of the phospholipid bilayer.

A Variety of Proteins Form a Mosaic Within the Membrane

Thousands of different membrane proteins are embedded within or attached to the surfaces of the phospholipid bilayer of cell membranes. Many of the proteins in plasma membranes have carbohydrate groups attached to the portion that is exposed on the outer membrane surface (see Fig. 5-1). These are called **glycoproteins** ("glyco" is from the Greek, meaning "sweet," and refers to the carbohydrate portion with its sugar subunits). Most plasma membrane proteins are embedded, at least partially, in the phospholipid bilayer, but some adhere to the bilayer surface. Membrane proteins may be grouped into five major categories based on their function: receptor proteins, recognition proteins, enzymatic proteins, attachment proteins, and transport proteins.

Most cells bear dozens of types of **receptor proteins** (some of which are glycoproteins) spanning their plasma membranes. To perform their functions, cells must respond to messages sent by other cells. These messages are often molecules (such as hormones) carried in the bloodstream. After diffusing into the extracellular fluid, these message molecules bind to specific sites on receptor proteins, which convey the message to the interior of the cell (**Fig. 5-5**). When the appropriate molecule binds to the receptor, the receptor is "activated"—often by changing shape—and this, in turn, causes a response inside the cell.

The response can take many forms. The shape of a protein inside the cell may be modified, converting it from an inactive form to an active form. This change, in turn, may trigger a sequence of chemical reactions within the cell that alters cellular activities. For example, when the hormone epinephrine (adrenaline) binds to a specific membrane receptor on muscle cells, it stimulates the muscle cells to break down glycogen into glucose, providing more energy for muscle contraction. When other receptor proteins bind messenger molecules, they may cause ion channels to open, or they may start sequences of reactions that stimulate cells to divide or to secrete hormones. Communication among nerve cells also depends on receptors, and receptor proteins allow cells of the immune system to recognize and attack foreign invaders that can cause disease.

Have you ever wondered?
What to Do If You Are Bitten by a Poisonous Snake?

If there are a variety of poisonous snakes in your area, notice the color pattern and size of the snake that bit you, to help identify it as you dial 911. Dr. David Erk, a physician at the Wyoming Medical Center who has a particular interest in snakebites, provides the following advice: First, get medical help as quickly as possible. To slow the spread of venom and the swelling it causes, keep the bitten limb immobilized, ideally just below the level of the heart. Remove constricting jewelry and clothing from around the bite. Don't cut the bite, and don't apply suction, a tourniquet, heat, or ice. Don't drink coffee or alcohol or take any medications. Do try to stay as calm as possible.

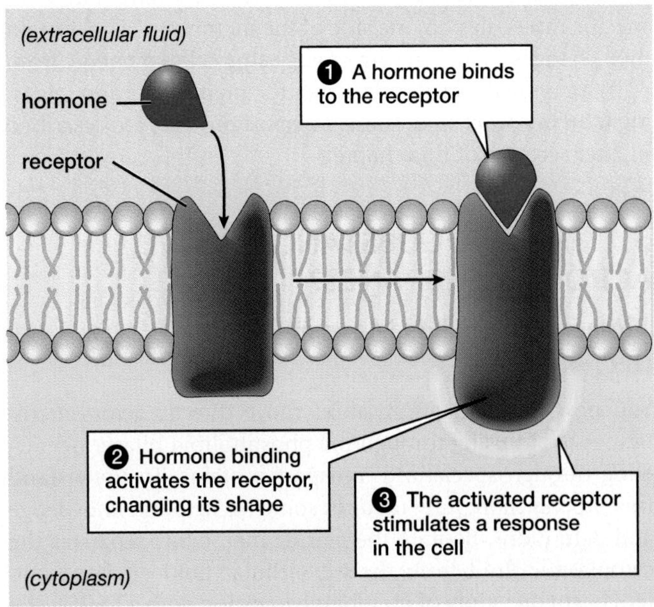

▲ **FIGURE 5-5 Receptor protein activation**

Recognition proteins are glycoproteins that serve as identification tags (see Fig. 5-1). The cells of each individual bear unique glycoproteins that identify them as "self." Immune cells ignore self and attack invading cells, such as bacteria, that have different recognition proteins on their membranes. Recognition glycoproteins on the surfaces of red blood cells bear different sugar groups and determine whether the blood is type A, B, AB, or O (see Table 10-1 on p. 187). Transfusions as well as transplanted organs must have their most important recognition glycoproteins matched to those of the recipient to minimize attack by the immune system.

Enzymes are proteins that promote chemical reactions that synthesize or break apart biological molecules, as described in Chapter 6. Although many enzymes are located within the cytoplasm, some span cell membranes and others are attached to membrane surfaces (see Fig. 5-1). Plasma membrane enzymes include those that are responsible for synthesizing the proteins and carbohydrates of the extracellular matrix (a web of protein and glycoprotein fibers that fills spaces between animal cells).

A diverse group of **attachment proteins** anchors cell membranes in various ways. Some span the plasma membrane, linking the cytoskeleton inside the cell with the extracellular matrix outside (see Fig. 5-1), thus anchoring the cell in place within a tissue. Some attachment proteins help maintain cell shape by linking the plasma membrane to the underlying cytoskeleton, and some help cells adhere to and move along surfaces. Other types of attachment proteins form connections between adjacent cells, as described later in this chapter (see Figs. 5-17 and 5-18).

Transport proteins regulate the movement of hydrophilic molecules through the plasma membrane. Some transport proteins, called **channel proteins,** form channels whose central pores allow water molecules or specific ions to pass through the membrane along their concentration gradients (see Fig. 5-1). Other transport proteins, called **carrier proteins,** have binding sites that can temporarily attach to specific molecules on one side of the membrane. The protein then changes shape (in some cases using cellular energy from ATP), moving the molecule across the membrane, and releasing it on the other side. These transport proteins are described in later sections of this chapter.

5.2 HOW DO SUBSTANCES MOVE ACROSS MEMBRANES?

Molecules in Fluids Move in Response to Gradients

You now know that substances move directly across membranes by diffusing through the phospholipid bilayer or traveling through specialized transport proteins. To understand this process more fully requires some background knowledge and definitions. Because the plasma membrane separates the cytoplasmic fluid from the extracellular fluid environment, let's begin our study of membrane transport with a look at the characteristics of fluids, starting with a few definitions:

- A **fluid** is any substance whose molecules can flow past one another; as a result, fluids have no defined shape. Fluids include gases, liquids, and also cell membranes, whose molecules can flow past one another.
- A **solute** is a substance that can be dissolved (dispersed into individual atoms, molecules, or ions) in a **solvent,** which is a fluid (usually a liquid) capable of dissolving the solute. Water, in which all biological processes occur, dissolves so many different solutes that it is sometimes called the "universal solvent."
- The **concentration** of a substance defines the amount of solute in a given amount of solvent. The concentration of sugar solution, for example, is a measurement of the number of molecules of that sugar contained in a given volume of solution.
- A **gradient** is a physical difference in properties such as temperature, pressure, electrical charge, or concentration of a particular substance in a fluid between two adjoining regions of space. It requires energy to create gradients. Over time, gradients tend to break down unless energy is supplied to maintain them or unless an effective barrier separates them. For example, gradients in temperature cause a flow of energy from a higher-temperature region to a lower-temperature region. Electrical gradients can drive the movement of ions. Gradients of concentration or pressure cause molecules or ions to move from one region to the other in a manner that tends to equalize the difference. Cells use energy and cell membrane proteins to generate **concentration gradients** of various molecules and ions dissolved in their cytoplasmic fluid.

It is important to be aware that, at temperatures above absolute zero (−459.4°F, or −273°C), atoms, molecules, and ions are in constant random motion. As the temperature rises, their rate of motion increases, and temperatures that support life cause these particles to move rapidly indeed. As a result of this motion, molecules and ions in solution are continuously bombarding one another and the structures surrounding them. Over time, random movements produce a net movement from regions of high concentration to regions of low concentration, a process called **diffusion.** In a nonliving system, if nothing opposes this movement (opposing factors include electrical charge, pressure differences, or physical barriers), then the random movement of molecules will continue until the substance is evenly dispersed throughout the fluid.

To envision how the random movement of molecules or ions within a fluid eventually breaks down concentration gradients, consider a sugar cube dissolving in hot coffee, or perfume molecules moving from an open bottle into the air. In each case you have a concentration gradient. If you leave the perfume jar open or forget your coffee, eventually you will have an empty perfume bottle and a fragrant room, or cold, but uniformly sweet, coffee. In an analogy to gravity, we describe molecules moving from regions of high concentration to regions of low concentration as moving "down" their concentration gradients.

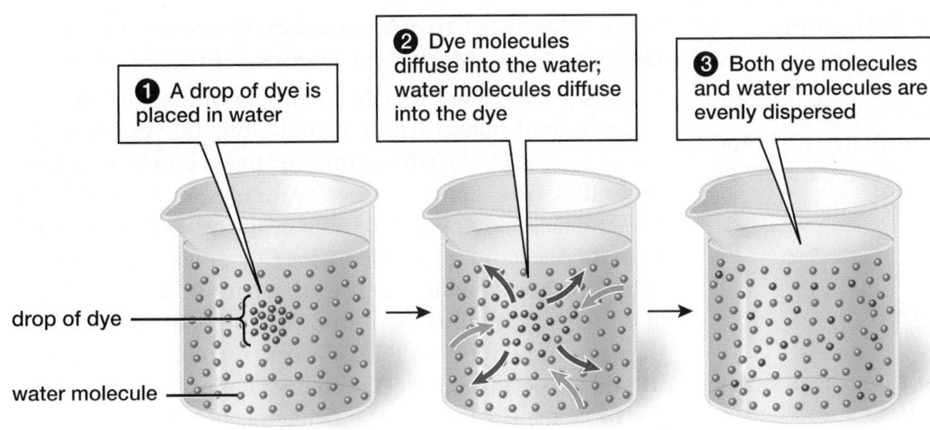

1 A drop of dye is placed in water
2 Dye molecules diffuse into the water; water molecules diffuse into the dye
3 Both dye molecules and water molecules are evenly dispersed
drop of dye
water molecule

◀ **FIGURE 5-6 Diffusion of a dye in water**

To watch diffusion in action, place a drop of food coloring in a glass of water (**Fig. 5-6**). Random motion propels dye molecules both into and out of the dye droplet, but there is a net transfer of dye into the water, and of water into the dye. The net movement of dye will continue until it is uniformly dispersed in the water. If you compare the diffusion of a dye in hot and cold water, you will see that heat increases the diffusion rate; this is because heat causes molecules to move faster.

Summing Up The Principles of Diffusion

- Diffusion is the net movement of molecules down a gradient from high to low concentration.
- The greater the concentration gradient, the faster the rate of diffusion.
- The higher the temperature, the faster the rate of diffusion.
- If no other processes intervene, diffusion will continue until the concentrations become equal throughout; that is, until the concentration gradient is eliminated.

Movement Through Membranes Occurs by Passive Transport and Energy-Requiring Transport

Gradients of ions and molecules across cell membranes are crucial for life; a cell without them is dead. Proteins in cell membranes expend energy to create and maintain these concentration gradients, because many of the biochemical processes of life rely on them. For example, neurons depend on the flow of specific ions down their concentration gradients to produce the electrical signals that underlie sensation and movement. Plasma membranes are described as **selectively permeable** because they selectively allow only certain ions or molecules to pass through (permeate). The selective permeability of the plasma membrane creates a barrier that helps maintain the gradients that characterize all cells.

In its role as gatekeeper of the cell, the cell's plasma membrane permits substances to move through it in two different ways: passive transport and energy-requiring transport (**Table 5-1**). **Passive transport** involves diffusion of substances across cell membranes down their concentration gradients, whereas **energy-requiring transport** requires that the cell expend energy to move substances across membranes or into or out of the cell.

Table 5-1 Transport Across Membranes

Passive Transport	Diffusion of substances across a membrane down a gradient of concentration, pressure, or electrical charge; does not require cellular energy.
Simple diffusion	Diffusion of water, dissolved gases, or lipid-soluble molecules through the phospholipid bilayer of a membrane.
Facilitated diffusion	Diffusion of water, ions, or water-soluble molecules through a membrane via a channel or carrier protein.
Osmosis	Diffusion of water across a selectively permeable membrane from a region of higher free water concentration to a region of lower free water concentration.
Energy-Requiring Transport	Movement of substances into or out of a cell using cellular energy, usually supplied by ATP.
Active transport	Movement of individual small molecules or ions against their concentration gradients through membrane-spanning proteins.
Endocytosis	Movement of particles or large molecules into a cell; occurs as the plasma membrane engulfs the substance in a membranous sac that pinches off, entering the cytoplasm.
Exocytosis	Movement of particles or large molecules out of a cell; occurs as the plasma membrane encloses the material in a membranous sac that moves to the cell surface and fuses with the plasma membrane, allowing its contents to diffuse out.

Passive Transport Includes Simple Diffusion, Facilitated Diffusion, and Osmosis

Diffusion can occur within a fluid or across a membrane that is permeable to the substance and that separates two fluid compartments. Many molecules cross plasma membranes by diffusion, driven by concentration differences between the cytoplasm and the extracellular fluid.

Some Molecules Move Across Membranes by Simple Diffusion

Very small molecules with no net charge, such as water, oxygen, and carbon dioxide—as well as lipid-soluble molecules that include alcohol, vitamins A, D, and E, and steroid hormones—diffuse directly across the phospholipid bilayer down their concentration gradients. This process is called **simple diffusion** (Fig. 5-7a). A larger concentration gradient, higher temperature, smaller molecular size, and greater solubility in lipids all increase the rate of simple diffusion.

How can water—a polar molecule—diffuse directly through the hydrophobic (literally "water-fearing") phospholipid bilayer? Because water molecules are so small and so abundant in both the cytoplasm and the extracellular fluid, some water molecules stray into the thicket of phospholipid tails, where their random movements bring them through to the far side of the membrane. Simple diffusion of water through the phospholipid bilayer is relatively slow, but in many types of cells, water molecules traverse membranes far more rapidly by facilitated diffusion through transport proteins, as described in the following section.

Some Molecules Cross Membranes by Facilitated Diffusion Using Membrane Transport Proteins

Many substances are unable to simply diffuse through the phospholipid bilayer. These include ions (for example, K^+, Na^+, Cl^-, and Ca^{2+}), which form hydrogen bonds with water molecules. The attached water molecules tend to hold the ions in place, either inside or outside the cell. Molecules such as monosaccharides (simple sugars) are polar, and also form hydrogen bonds with water; thus, their attraction to water and their size inhibits them from entering the bilayer. These ions and molecules can diffuse across membranes only with the aid of specific transport proteins: channel proteins or carrier proteins. The process of moving through cell membranes down a concentration gradient with the aid of transport proteins is called **facilitated diffusion.**

Channel proteins form pores (channels) in the lipid bilayer through which certain ions or water can flow down their concentration gradients (Fig. 5-7b). Channel proteins have a specific interior diameter related to the diameter of the ion whose movement they facilitate, and electrical charges that attract the ion to the amino acids that line the pore. For example, the Na^+ channel is lined with negative charges to attract Na^+. Because cells must maintain gradients of many ions across their membranes, many ion channels include protein "gates" that are opened or closed, depending on the needs of the cell.

Many cells have specialized water channel proteins called **aquaporins** (literally, "water pores"; Fig. 5-7c). The small size of these channels, coupled with positively charged amino acids (which attract the negative pole of water molecules) inside the channels, makes them selective for water molecules. The movement of water through a membrane, either by simple diffusion or facilitated diffusion through

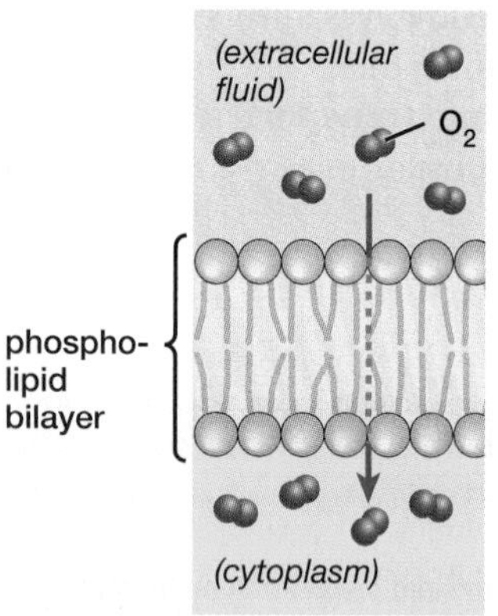

(a) Simple diffusion through the phospholipid bilayer

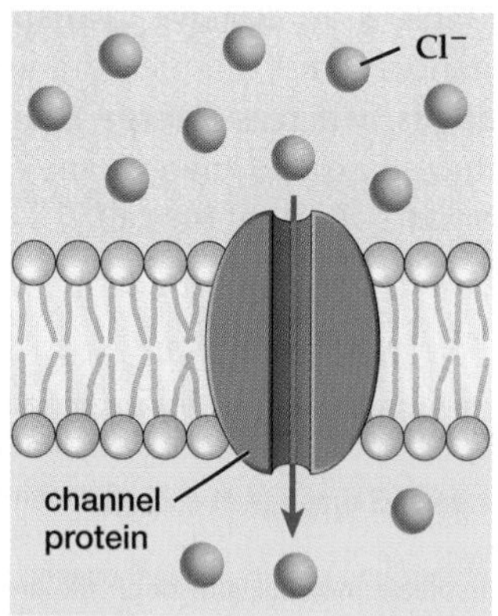

(b) Facilitated diffusion through channel proteins

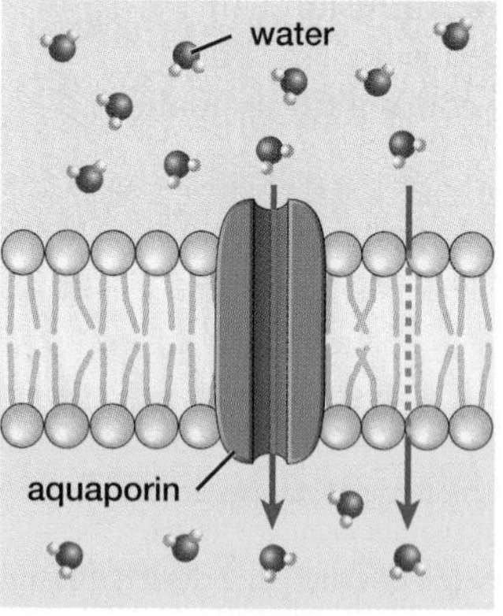

(c) Osmosis through aquaporins or the phospholipid bilayer

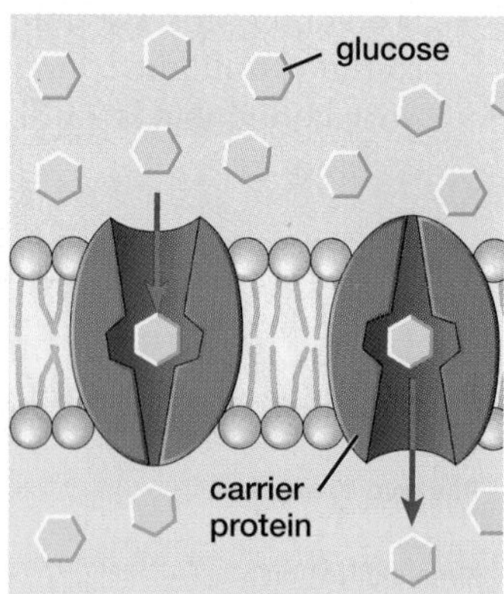

(d) Facilitated diffusion through carrier proteins

▲ FIGURE 5-7 **Types of diffusion through the plasma membrane** **(a)** Molecules that are very small, uncharged, or lipid-soluble move directly through the phospholipid bilayer by simple diffusion. Here, oxygen molecules diffuse from the extracellular fluid into the cell down their concentration gradient (red arrow). **(b)** Facilitated diffusion through channel proteins allows ions to cross membranes. Here, chloride ions diffuse down their concentration gradient into the cell through chloride channels. **(c)** Osmosis is the diffusion of water. Water molecules can diffuse directly through the phospholipid bilayer by simple diffusion, or they can pass far more rapidly by facilitated diffusion through water channels called aquaporins. **(d)** Carrier proteins have binding sites for specific molecules (such as the glucose shown here). Binding of the transported molecule causes the carrier to change shape, shuttling the molecule across the membrane down its concentration gradient.

aquaporins, has a special name: osmosis. To learn more about water channels, see "Scientific Inquiry: The Discovery of Aquaporins" on p. 86.

Carrier proteins have active sites that bind specific molecules from the cytoplasmic fluid or the extracellular fluid, such as particular sugars or small proteins. The binding triggers a change in the shape of the carrier that allows the molecules to pass through the protein and across the membrane. These carrier proteins do not use cellular energy, and they can transfer molecules only down their concentration gradients (**Fig. 5-7d**).

Osmosis Is the Diffusion of Water Across Selectively Permeable Membranes

Osmosis is the movement of water across a selectively permeable membrane in response to gradients of concentration, pressure, or temperature. Here, we will focus on osmosis from a region of higher water concentration to a region of lower water concentration. Osmosis may occur directly through the phospholipid bilayer, or (more rapidly) through aquaporin channels composed of membrane-spanning proteins.

What do we mean when we describe a solution as having a "high water concentration" or a "low water concentration"? The answer is simple: Pure water has the highest possible water concentration. Any substance that dissolves in water (any solute) displaces some of the water molecules in a given volume, and it also forms hydrogen bonds with many more of the water molecules, preventing them from moving across a water-permeable membrane. Therefore, the higher the concentration of dissolved substances, the lower the concentration of water that is available to move across a membrane. Consequently, there will be a net movement of water molecules from the solution with with more free water molecules (containing less solute) into the solution with fewer free water molecules (containing more solute). For example, water will move by osmosis from a solution with less dissolved sugar into a solution with more dissolved sugar. Because there are more free water molecules in the less-concentrated sugar solution, more water molecules will collide with—and thus move through—the water-permeable membrane on that side. The concentration of solute in water determines the solution's "osmotic strength"; the higher the solute concentration, the higher the osmotic strength.

Scientists use the word "tonicity" to compare the concentrations of substances dissolved in water across a membrane that is selectively permeable to water. Solutions with equal concentrations of solute (and thus equal concentrations of water) are described as being **isotonic** to one another (the prefix "iso" means "same"). When isotonic solutions are separated by a water-permeable membrane, there is no net movement of water between them (**Fig. 5-8a**).When a membrane selectively permeable to water separates solutions with different concentrations of a solute, the solution that contains a greater concentration of solute is described as being **hypertonic** (the prefix "hyper" means "greater than") to the less concentrated solution (**Fig. 5-8b**). The more dilute solution is described as **hypotonic** ("hypo" means "below"; **Fig. 5-8c**). Water tends to move from hypotonic solutions into hypertonic solutions.

> **Summing Up** **The Principles of Osmosis**
>
> - Osmosis is the movement of water across a selectively water-permeable membrane by simple diffusion or by facilitated diffusion through aquaporins.
> - Water moves across a selectively water-permeable membrane down its concentration gradient from the side with a higher concentration of free water molecules to the side with a lower concentration of free water molecules.
> - Dissolved substances reduce the concentration of free water molecules in a solution.
> - If two solutions are separated by a membrane that is selectively permeable to water, the solution with a higher concentration of solute is hypertonic and has a higher osmotic strength, and the solution with the lower concentration of solute is hypotonic and has a lower osmotic strength.

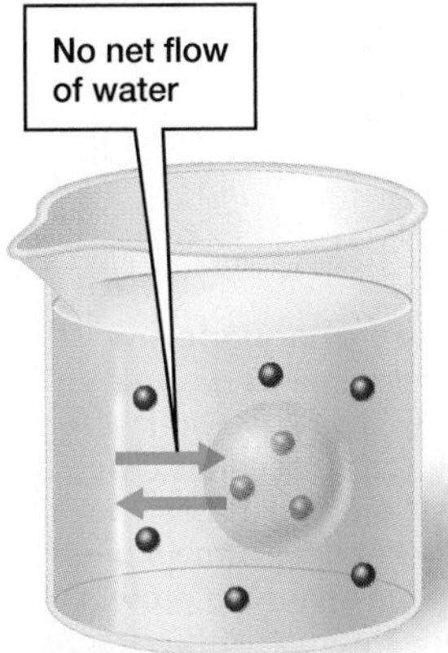
No net flow of water
(a) A balloon in an isotonic solution

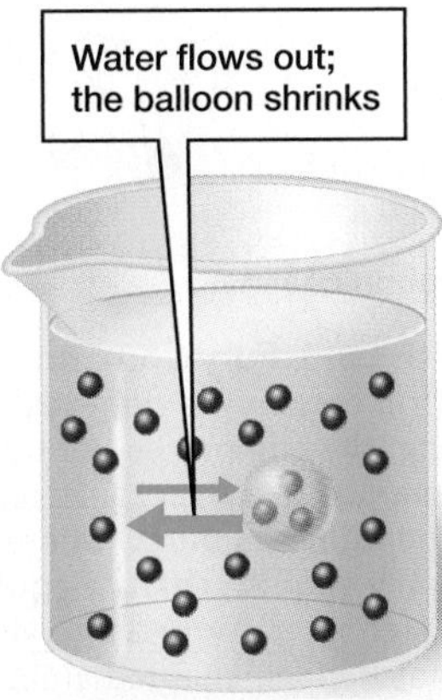
Water flows out; the balloon shrinks
(b) A balloon in a hypertonic solution

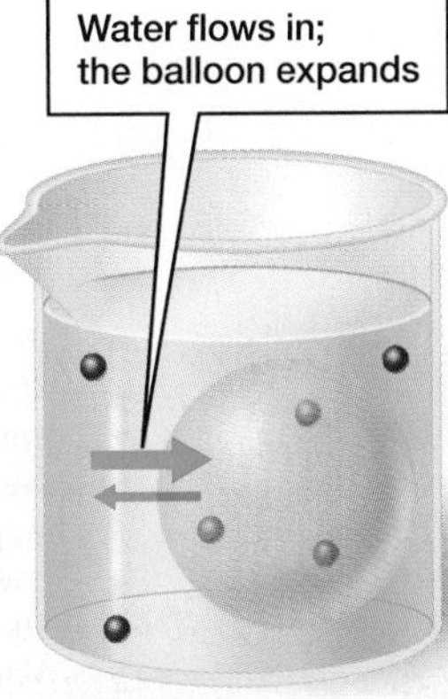
Water flows in; the balloon expands
(c) A balloon in a hypotonic solution

◀ FIGURE 5-8 **The effect of solute concentration on osmosis** We start with three balloons made of membrane that is selectively permeable to water (not to sugar; red balls). We place equal volumes of the same concentration of sugar water in each balloon. We then immerse each in a beaker filled with a different concentration of sugar water—**(a)** isotonic, **(b)** hypertonic, or **(c)** hypotonic—to that in the balloon. The figure shows the results after an hour. The blue arrows indicate relative water movement into and out of the balloon.

Scientific Inquiry

The Discovery of Aquaporins

The French microbiologist Louis Pasteur's observation that "Chance favors the prepared mind" is as true today as it was when he first expressed it in the 1800s. Scientists have long realized that osmosis directly through the phospholipid bilayer is much too slow to account for water movement across certain cell membranes, including those of kidney tubules (which reabsorb tremendous quantities of water that the kidney filters from the blood) and those of red blood cells (see Fig. 5-9). But attempts to identify selective transport proteins for water repeatedly failed, partly because water is abundant on both sides of the plasma membrane, and partly because water can also move directly through the phospholipid bilayer.

Then, as often happens in science, chance and prepared minds met. In the mid-1980s, Dr. Peter Agre (**Fig. E5-2**), then at the Johns Hopkins School of Medicine in Maryland, was attempting to determine the structure of a glycoprotein on red blood cells. The protein he isolated was contaminated, however, with large quantities of another protein. Instead of discarding the unknown protein, he and his coworkers collaborated to determine its structure. They found that it was similar to previously identified membrane channel proteins of unknown function. Agre and colleagues investigated the protein by forcing frog eggs (which are only slightly permeable to water) to insert the protein into their plasma membranes. Whereas eggs without the mystery protein swelled only slightly when placed in a hypotonic solution, those with the protein swelled rapidly and burst in the same solution (**Fig. E5-3a**). Further studies showed that no other ions or molecules could move through this channel, which was then dubbed "aquaporin." In 2000, Agre and other research teams reported the three-dimensional structure of aquaporin and described how specific polar amino acids in its interior attract water and allow billions of water molecules to move through the channel in single file every second, while repelling other ions or molecules (**Fig. E5-3b**).

▲ **FIGURE E5-2 Peter Agre**

Many types of aquaporins have now been identified, and aquaporins have been found in all forms of life that have been investigated. For example, the plasma membrane of the central vacuole of plant cells is rich in aquaporins, which allow it to fill rapidly when water is available (see Fig. 5-10). In 2003, Agre shared the Nobel Prize in Chemistry for his discovery—the result of chance, careful observation, persistence, and perhaps a bit of what Agre himself describes modestly as "sheer blind luck."

(a) Frog eggs

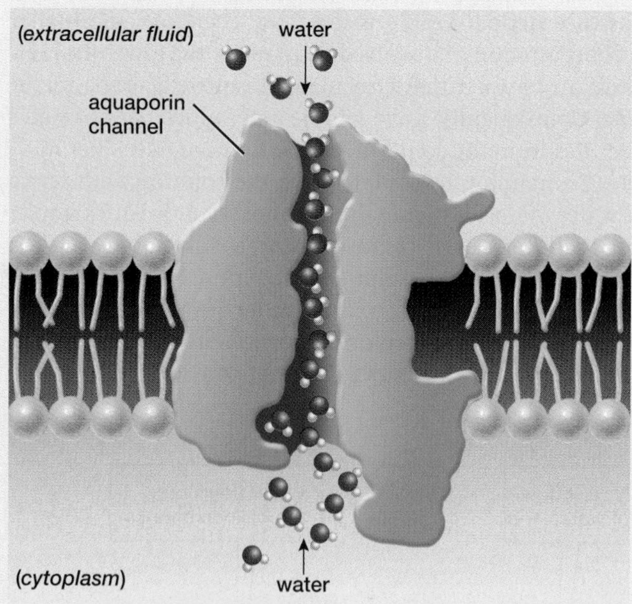

(b) Aquaporin channel

▲ **FIGURE E5-3 Function and structure of aquaporins** **(a)** The frog egg on the right has aquaporins in its plasma membrane; the egg on the left does not. They have both been immersed in a hypotonic solution for 30 seconds. The egg on the right has burst, whereas that on the left has swollen only slightly. **(b)** The aquaporin channel (shown here in cross-section) is formed by a membrane-spanning protein. Within the channel, charged amino acids interact with water molecules, promoting their movement in either direction while repelling other substances.

Osmosis Across the Plasma Membrane Plays an Important Role in the Lives of Cells

The extracellular fluid of animals is usually isotonic to the cytoplasmic fluid of their cells, so there is no net tendency for water to enter or leave the cells. Although the concentrations of specific solutes are rarely the same inside and outside the cells, the total concentration of all dissolved particles is equal; therefore, the water concentration is equal inside and outside of cells.

If you immerse red blood cells in salt solutions of varying solute concentrations, you can observe the effects of water movement across their cell membranes. In an isotonic salt solution, the sizes of the cells remain constant (**Fig. 5-9a**). If the salt solution is hypertonic to the cytoplasmic fluid of the red blood cells, water will leave the cells by osmosis, and the cells will shrivel (**Fig. 5-9b**). Conversely, if the salt solution is very dilute and hypotonic to the red blood cell cytoplasmic fluid, water will enter the cells, causing them to swell (**Fig. 5-9c**). If red blood cells are placed in pure water, they will continue to swell until they burst.

Osmosis across plasma membranes is crucial to many biological processes, including water uptake by the roots of plants, absorption of dietary water from the intestine, and the reabsorption of water in kidneys. Organisms that live in fresh water must expend energy to counteract osmosis. Protists such as *Paramecium* have contractile vacuoles that eliminate the water that continuously leaks into their cytoplasmic fluid, which is hypertonic to the fresh water of the ponds in which they live. Cellular energy is used to pump salts from the cytoplasmic fluid into the contractile vacuole. This causes water to follow by osmosis, filling the vacuole, which then contracts, squirting the water out through a pore in the plasma membrane (see Fig. 4-15).

Nearly every living plant cell is supported by water that enters through osmosis. As you learned in Chapter 4, most plant cells have a large, membrane-enclosed central vacuole. The membrane that encloses this vacuole is rich in aquaporins. Dissolved substances stored in the vacuole make its contents hypertonic to the surrounding cytoplasmic fluid, which in turn is usually hypertonic to the extracellular fluid that bathes the cells. Water therefore flows into the cytoplasmic fluid and then into the vacuole by osmosis. The water pressure within the vacuole, called **turgor pressure,** pushes the cytoplasm up against the cell wall with considerable force (**Fig. 5-10a**).

Cell walls are often flexible, so many plant cells depend on turgor pressure for support. If you forget to water a houseplant, the central vacuole and cytoplasmic fluid of each cell loses water and the plasma membrane shrinks away from its cell wall as the vacuole collapses. Just as a balloon goes limp when its air leaks out, the plant droops as its cells lose turgor pressure (**Fig. 5-10b**). Now you know why grocery stores are always spraying their leafy produce: to keep it looking perky and fresh with full central vacuoles.

Energy-Requiring Transport Includes Active Transport, Endocytosis, and Exocytosis

Without concentration gradients across its membranes, a cell is dead. By building gradients and then allowing them to run down under specific circumstances, cells regulate

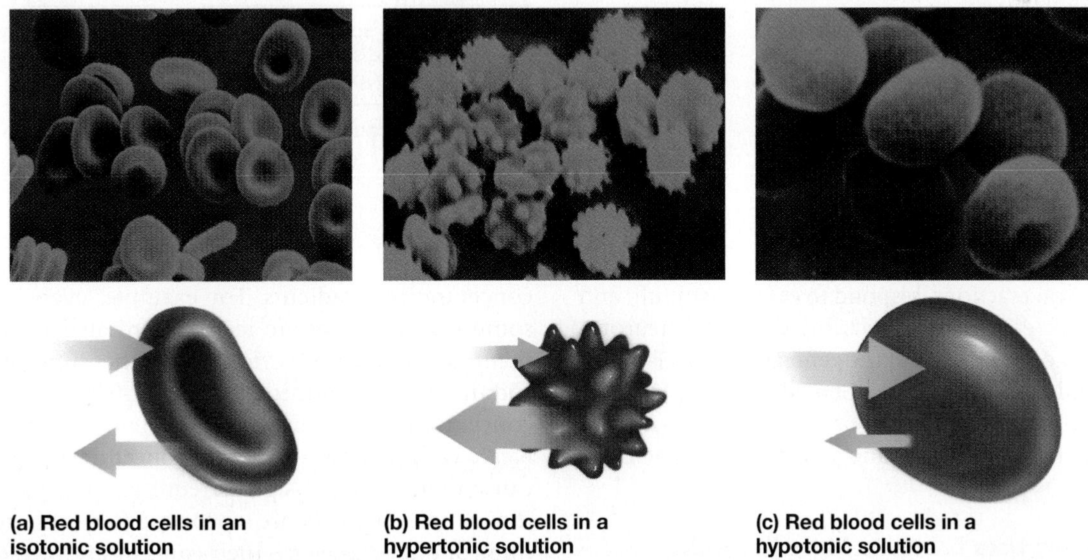

(a) Red blood cells in an isotonic solution
(b) Red blood cells in a hypertonic solution
(c) Red blood cells in a hypotonic solution

▲ FIGURE 5-9 **The effects of osmosis on red blood cells (a)** The cells are immersed in an isotonic solution and retain their normal dimpled shape. **(b)** The cells shrink when placed in a hypertonic solution, as more water moves out than flows in. **(c)** The cells expand when placed in a hypotonic solution.

QUESTION A student pours some distilled water into a sample of blood. Returning later, she looks at the blood under a microscope and sees no blood cells at all. What happened?

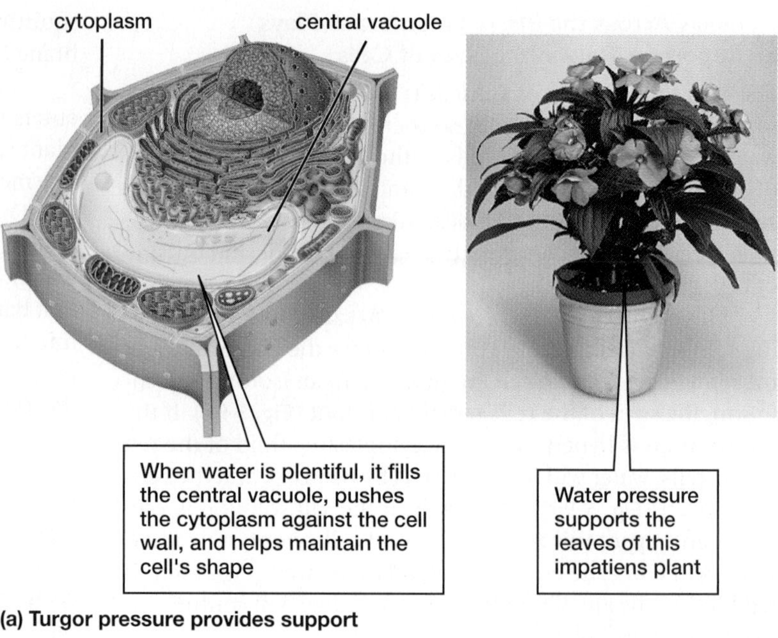
cytoplasm
central vacuole
When water is plentiful, it fills the central vacuole, pushes the cytoplasm against the cell wall, and helps maintain the cell's shape
Water pressure supports the leaves of this impatiens plant

(a) Turgor pressure provides support

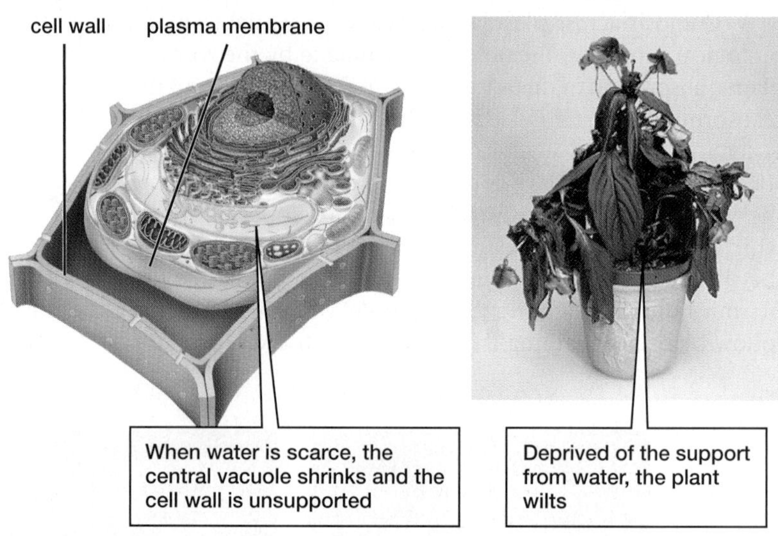
cell wall
plasma membrane
When water is scarce, the central vacuole shrinks and the cell wall is unsupported
Deprived of the support from water, the plant wilts

(b) Loss of turgor pressure causes the plant to wilt

▶ **FIGURE 5-10 Turgor pressure in plant cells** Aquaporins allow water to move rapidly in and out of the central vacuoles of plant cells. The cell and plant in part **(a)** are supported by water turgor pressure, whereas those in part **(b)** have lost pressure due to dehydration.

QUESTION If a plant cell is placed in water containing no solutes, will the cell eventually burst? What about an animal cell? Explain.

their biochemical reactions, respond to external stimuli, and obtain chemical energy. The electrical signals of neurons, the contraction of muscles, and the generation of ATP in mitochondria and chloroplasts (see Chapters 7 and 8) all rely on concentration gradients of ions. But gradients do not form spontaneously—they require active transport across a membrane.

Active Transport Uses Energy to Move Molecules Against Their Concentration Gradients

During **active transport,** membrane proteins use cellular energy to move molecules or ions across the plasma membrane against their concentration gradients (**Fig. 5-11**). All cells need to move some materials "uphill" against concentration gradients. For example, every cell requires some nutrients that are less concentrated in the environment than in the cell's cytoplasm. Other substances, such as sodium and calcium ions, are maintained at much lower concentrations inside the cell than in the extracellular fluid. Nerve cells maintain large ion concentration gradients because their electrical signals require rapid, passive flow of ions when channels are opened. After these ions diffuse into (or out of) the cell, their concentration gradients must be restored by active transport.

Active-transport proteins span the width of the membrane and have two active sites. One active site—which may face the inside or outside of the plasma membrane, depending on the transport protein—binds a particular molecule or

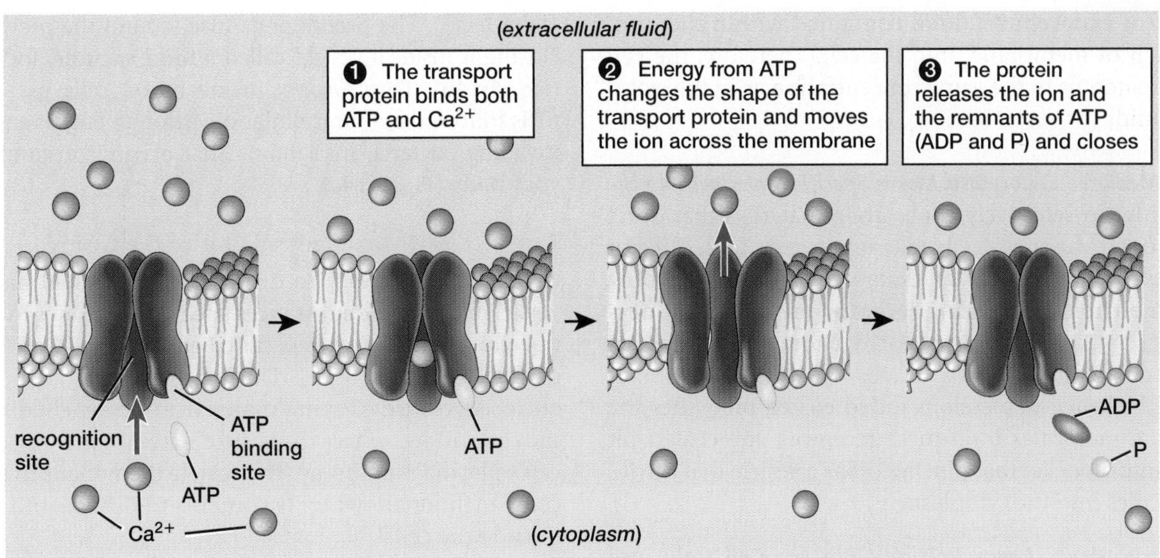

▲ **FIGURE 5-11 Active transport** Active transport uses cellular energy to move molecules across the plasma membrane against a concentration gradient. A transport protein (blue) has an ATP binding site and a recognition site for the molecules to be transported; in this case, calcium ions (Ca^{2+}). Notice that when ATP donates its energy, it loses its third phosphate group and becomes ADP + P.

ion, such as a calcium ion. As shown in **Figure 5-11** ❶, the second site, always on the inside of the membrane, binds an energy-carrier molecule, usually adenosine triphosphate (ATP; see p. 51). The ATP donates energy to the protein, causing the protein to change shape and move the calcium ion across the membrane (**Fig. 5-11** ❷). The energy for active transport comes from a high-energy bond that links the last of the three phosphate groups in ATP. Upon releasing its stored energy, ATP becomes ADP (adenosine diphosphate) plus a free phosphate (**Fig. 5-11** ❸). Active-transport proteins are often called pumps—in an analogy to water pumps—because they use energy to move ions or molecules uphill against a concentration gradient.

Cells Engulf Particles or Fluids by Endocytosis

A cell may need to acquire materials from its extracellular environment that are too large to move directly through the membrane. These materials are engulfed by the plasma membrane and are transported within the cell inside vesicles. This energy-requiring process is called **endocytosis** (Greek for "into the cell"). Here, we describe three forms of endocytosis based on the size and type of material acquired and the method of acquisition: pinocytosis, receptor-mediated endocytosis, and phagocytosis.

Pinocytosis Moves Liquids into the Cell In **pinocytosis** ("cell drinking"), a very small patch of plasma membrane dimples inward as it surrounds extracellular fluid and buds off into the cytoplasm as a tiny vesicle (**Fig. 5-12**). Pinocytosis moves

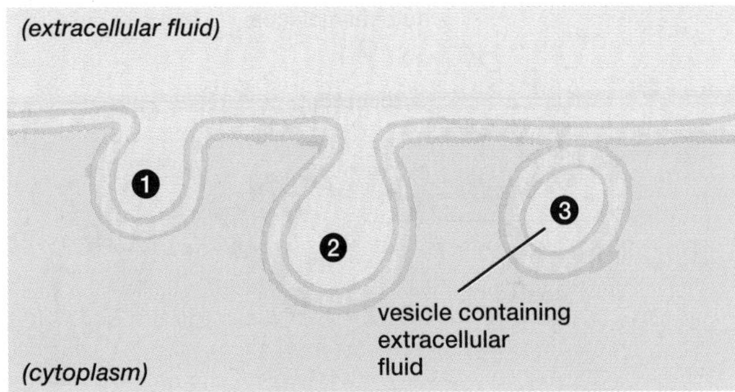

❶ A dimple forms in the plasma membrane, which ❷ deepens and surrounds the extracellular fluid. ❸ The membrane encloses the extracellular fluid, forming a vesicle.

(a) Pinocytosis

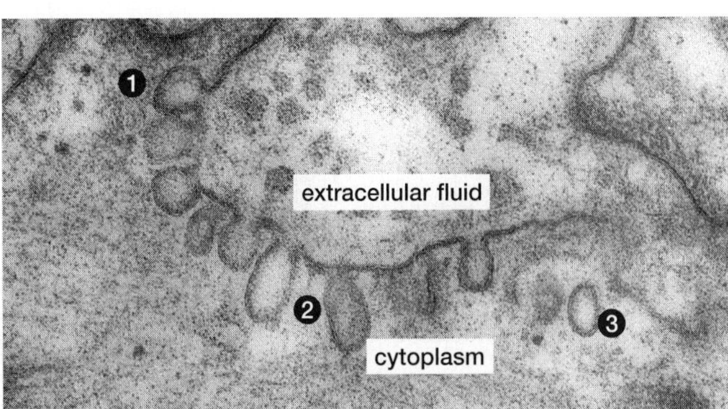

(b) TEM of pinocytosis

▲ **FIGURE 5-12 Pinocytosis** The circled numbers correspond to both **(a)** the diagram and **(b)** the transmission electron micrograph.

a droplet of extracellular fluid, contained within the dimpling patch of membrane, into the cell. Therefore, the cell acquires materials in the same concentration as in the extracellular fluid.

Receptor-Mediated Endocytosis Moves Specific Molecules into the Cell In order to selectively concentrate materials that don't move through channels, cells take up specific molecules or complexes of molecules (for example, packets containing protein and cholesterol) using **receptor-mediated endocytosis** (Fig. 5-13). Receptor-mediated endocytosis relies on specialized receptor proteins located on the plasma membrane in thickened depressions called coated pits. After the appropriate molecules bind these receptors, the coated pit deepens into a pocket that pinches off as a vesicle that carries the molecules into the cytoplasm.

Phagocytosis Moves Large Particles into the Cell Cells use **phagocytosis** (which means "cell eating") to pick up large particles, including whole microorganisms (Fig. 5-14a). When the freshwater protist *Amoeba*, for example, senses a tasty *Paramecium*, the *Amoeba* extends parts of its surface membrane. These membrane extensions are called pseudopods (Latin for "false feet"). The pseudopods fuse around the prey, thus enclosing it inside a vesicle, called a **food vacuole,** for digestion (Fig. 5-14b). Like *Amoeba*, white blood cells use phagocytosis followed by intracellular digestion to engulf and destroy invading bacteria, in a drama that occurs constantly within your body (Fig. 5-14c).

Exocytosis Moves Material Out of the Cell

Cells also use energy to dispose of undigested particles of waste, or to secrete substances such as hormones into the extracellular fluid, a process called **exocytosis** (Greek for "out of the cell"; Fig. 5-15). During exocytosis, a membrane-enclosed vesicle carrying material to be expelled moves to the cell surface, where the vesicle's membrane fuses with the cell's plasma membrane. The vesicle then opens to the extracellular fluid, allowing its contents to diffuse into the fluid outside the cell.

Exchange of Materials Across Membranes Influences Cell Size and Shape

As you learned in Chapter 4, most cells are too small to be seen with the naked eye; they range from about 1 to

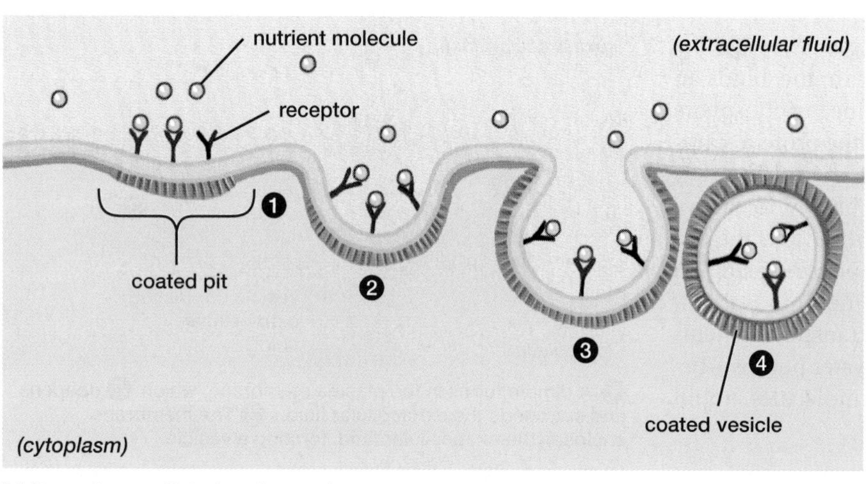
nutrient molecule
(extracellular fluid)
receptor
coated pit
coated vesicle
(cytoplasm)
1 Receptor proteins for specific molecules or complexes of molecules are localized at coated pit sites.
2 The receptors bind the molecules and the membrane dimples inward.
3 The coated pit region of the membrane encloses the receptor-bound molecules.
4 A vesicle ("coated vesicle") containing the bound molecules is released into the cytoplasm.

(a) Receptor-mediated endocytosis

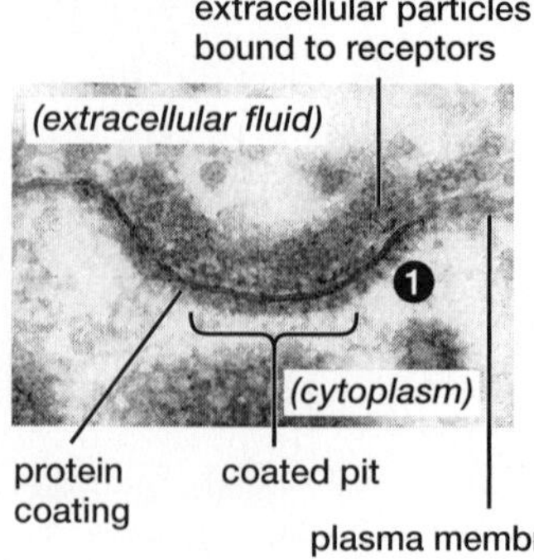
extracellular particles bound to receptors
(extracellular fluid)
(cytoplasm)
protein coating
coated pit
plasma membrane

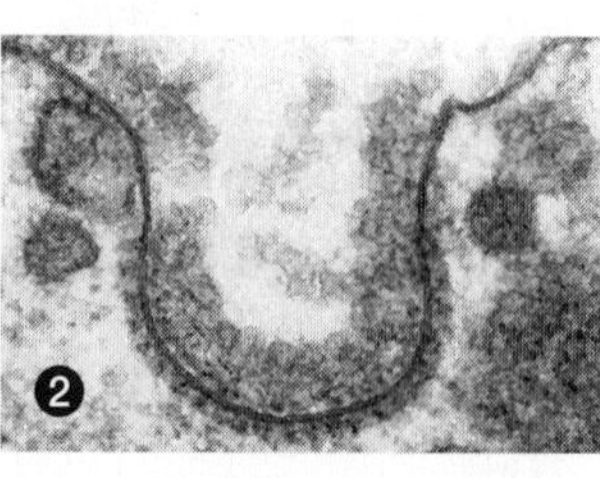

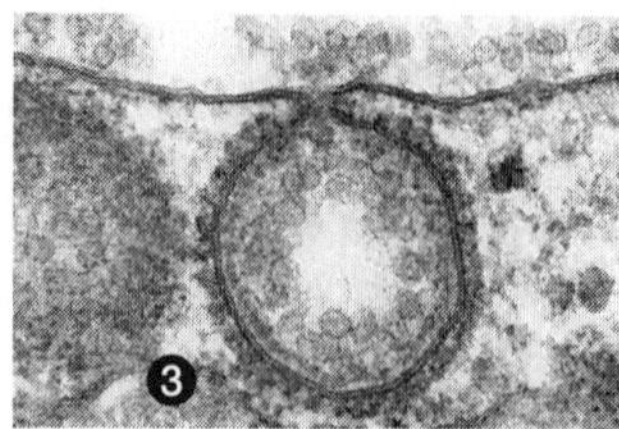

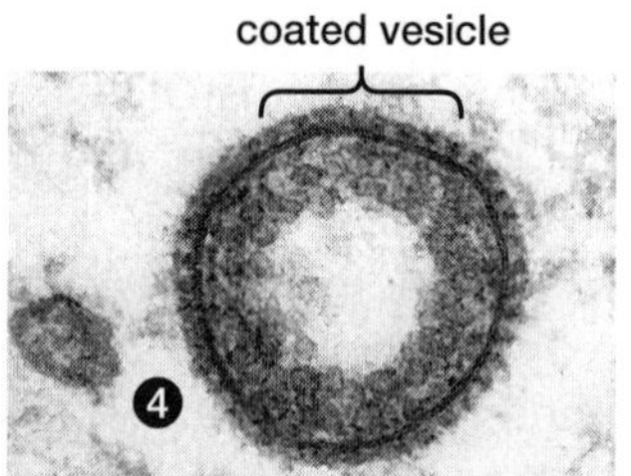
coated vesicle
0.1 micrometer

(b) TEM of receptor-mediated endocytosis

▲ FIGURE 5-13 **Receptor-mediated endocytosis** The circled numbers correspond to both **(a)** the diagram and **(b)** the transmission electron micrograph.

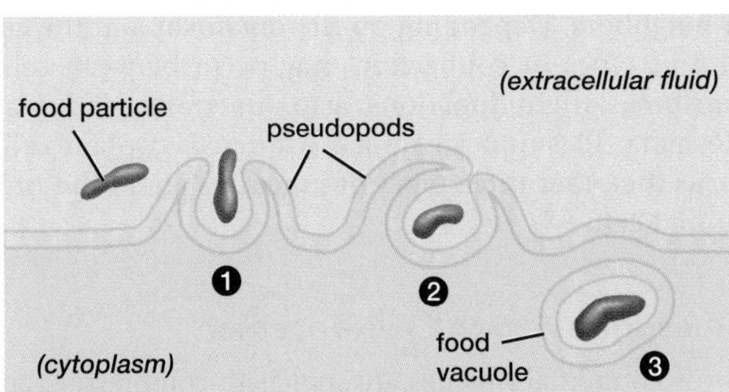

1 The plasma membrane extends pseudopods toward an extracellular particle (for example, food). 2 The ends of the pseudopods fuse, encircling the particle. 3 A vesicle called a food vacuole is formed containing the engulfed particle.

(a) Phagocytosis

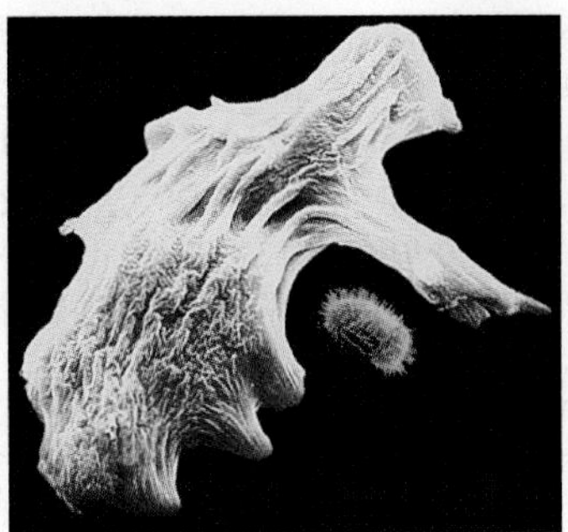

(b) An *Amoeba* engulfs a *Paramecium*

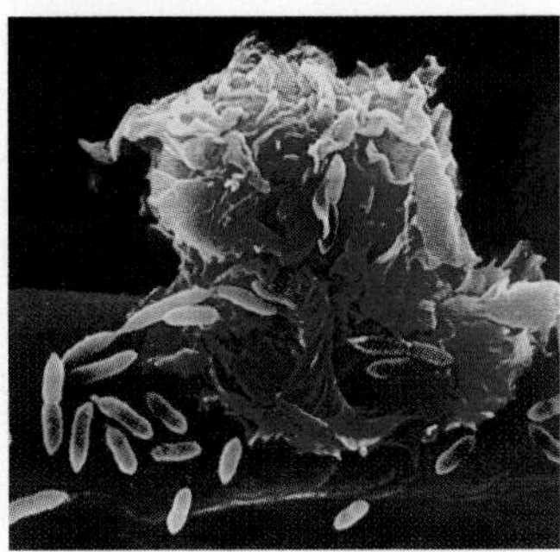

(c) A white blood cell ingests bacteria

▲ FIGURE 5-14 **Phagocytosis (a)** The mechanism of phagocytosis. The photographs of **(b)** an *Amoeba* and **(c)** a white blood cell were taken with a scanning electron microscope, which provides a three-dimensional image. The color is computer generated to help distinguish the cells.

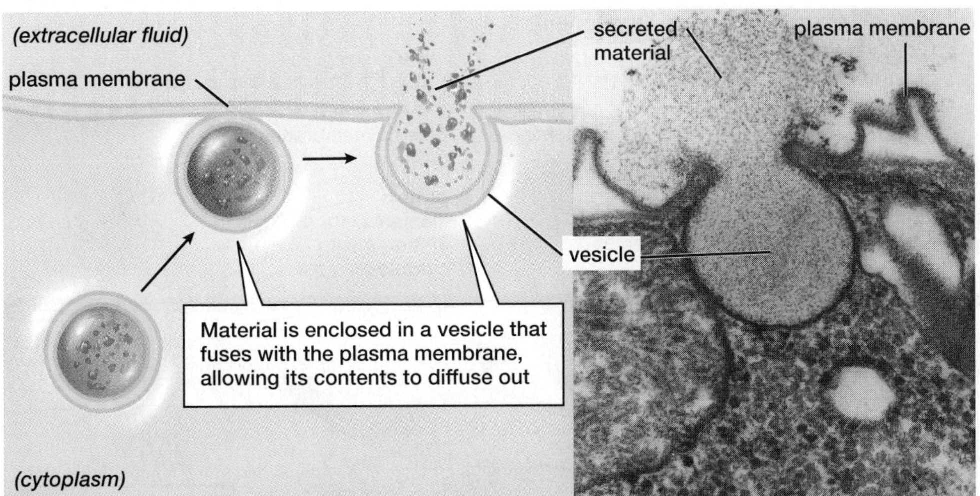

◀ FIGURE 5-15 **Exocytosis** Exocytosis is functionally the reverse of endocytosis.

QUESTION How does exocytosis differ from diffusion of materials out of a cell?

100 micrometers (millionths of a meter) in diameter (see Fig. 4-1). Why? As a roughly spherical cell becomes larger, its innermost regions become farther away from the plasma membrane, which is responsible for acquiring all the cell's nutrients and eliminating its waste products. Much of the exchange occurs by the slow process of diffusion. In a hypothetical giant cell 8.5 inches (20 centimeters) in diameter, oxygen molecules would take more than 200 days to diffuse to the center of the cell, by which time the cell would be long dead for lack of oxygen. In addition, as a sphere enlarges, its volume increases more rapidly than does its surface area. So a large, roughly spherical cell (which would require more nutrients and produce more wastes) would have a relatively smaller area of membrane to accomplish this exchange than would a small spherical cell (**Fig. 5-16**).

In a very large, roughly spherical cell, the surface area of the plasma membrane would be too small, and the diffusion distances within the cell too large, to supply the

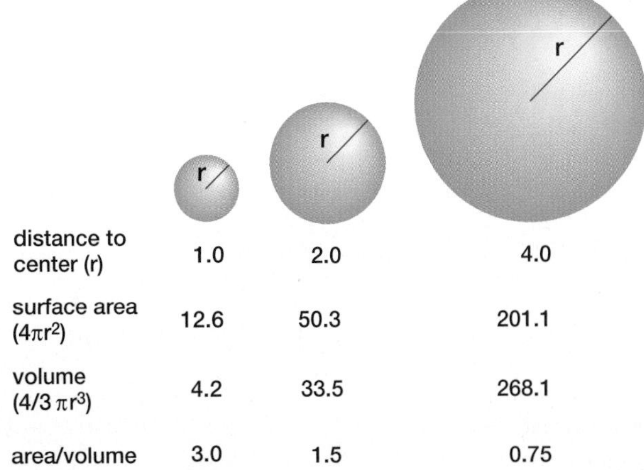

distance to center (r)	1.0	2.0	4.0
surface area ($4\pi r^2$)	12.6	50.3	201.1
volume ($4/3\ \pi r^3$)	4.2	33.5	268.1
area/volume	3.0	1.5	0.75

▲ FIGURE 5-16 **Surface area and volume relationships** As the size of a sphere increases, its volume increases much more than its surface area. Thus, a large spherical cell will have a relatively small surface area through which to obtain nutrients and eliminate wastes. As a result, roughly spherical cells are extremely small.

cell's metabolic needs. This constraint limits the size of most cells. However, some cells, such as nerve and muscle cells, can become very large because their elongated shape increases their membrane surface area, keeping the ratio of surface area to volume relatively high. In another example, cells lining the small intestine have plasma membranes that project out in fringe-like folds called **microvilli** (see Fig. 5-17a, middle). These microvilli create an enormous surface area of membrane to absorb nutrients from digested food.

BioFlix™ Membrane Transport

5.3 HOW DO SPECIALIZED JUNCTIONS ALLOW CELLS TO CONNECT AND COMMUNICATE?

In multicellular organisms, specialized structures on plasma membranes hold together clusters of cells and provide avenues through which cells communicate with their neighbors. Depending on the organism and the cell type, four types of connection may occur between cells: desmosomes, tight junctions, gap junctions, and plasmodesmata. Plasmodesmata are restricted to plant cells, whereas the other three types of junctions are found only in animal cells.

Desmosomes Attach Cells Together

Many of an animal's tissues are stretched, compressed, and bent as the animal moves. Cells in the skin, intestine, urinary bladder, and other organs must adhere firmly to one another to avoid tearing under the stresses of movement. Such animal tissues have junctions called **desmosomes,** which link adjacent cells (**Fig. 5-17a**). In a desmosome, the membranes of adjacent cells are tacked together by a complex of different attachment proteins. Intermediate filament proteins of the cytoskeleton are linked to the insides of the desmosomes and extend into the interior of each cell, further strengthening the attachment.

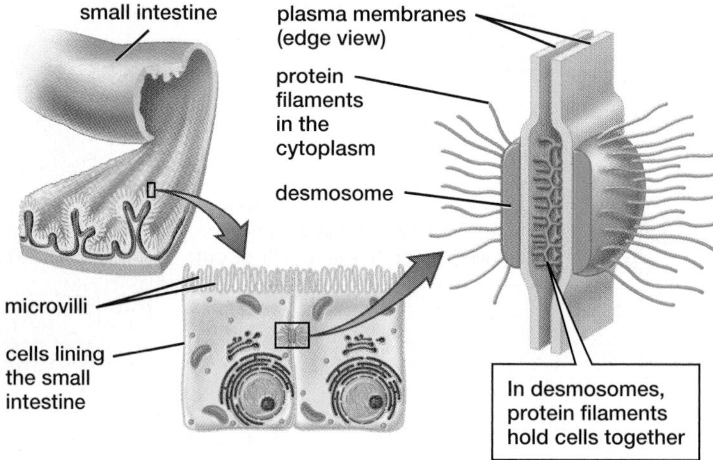

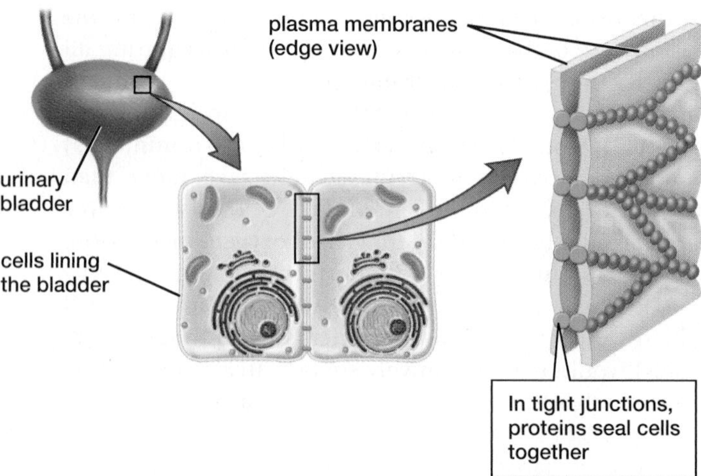

▶ **FIGURE 5-17 Cell attachment structures (a)** Cells lining the small intestine are attached by desmosomes. Protein filaments bound to the inside surface of each desmosome extend into the cytoplasm and attach to other filaments inside the cell, strengthening the connection between cells. **(b)** Tight junctions prevent leakage between cells, such as those of the urinary bladder.

Tight Junctions Make Cell Attachments Leakproof

The animal body contains many tubes and sacs that must hold their contents without leaking; for example, leaky skin or a leaky urinary bladder would spell disaster for the rest of the body. When cells must create a leakproof barrier, the spaces between them are blocked with special attachment proteins that are embedded in the membranes of each of adjacent linked cells. These proteins form **tight junctions** by adhering to one another, joining cells along defined pathways, almost as if the two membranes had been stitched together (**Fig. 5-17b**).

Gap Junctions and Plasmodesmata Allow Direct Communication Between Cells

Multicellular organisms must coordinate the actions of their component cells. In animals, a large proportion of the cells of the body communicate through protein channels formed by attachment proteins that connect the insides of adjacent cells. These cell-to-cell channels are called **gap junctions** (**Fig. 5-18a**). Hormones, nutrients, ions, and electrical signals can pass through the channels formed by gap junctions.

Virtually all of the living cells of plants are connected to one another by **plasmodesmata.** Plasmodesmata are holes in the walls of adjacent plant cells. These openings, lined with plasma membrane and filled with cytoplasmic fluid, link the insides of adjacent cells (**Fig. 5-18b**). Many plant cells have thousands of plasmodesmata, allowing water, nutrients, and hormones to pass quite freely from one cell to another.

Throughout this text, we will return many times to the concepts of membrane structure and transport described in this chapter. Understanding the diversity of membrane lipids and proteins is the key to understanding not just the isolated cell but also entire organs, which could not perform as they do without the specialized membrane properties of their component cells.

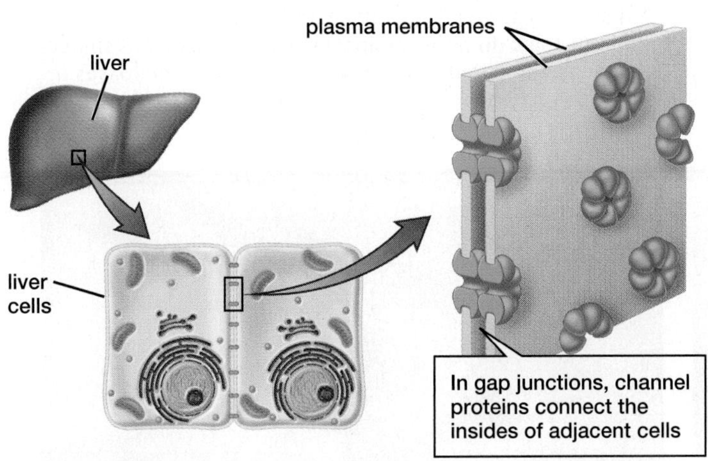
plasma membranes
liver
liver cells
In gap junctions, channel proteins connect the insides of adjacent cells
(a) Gap junctions

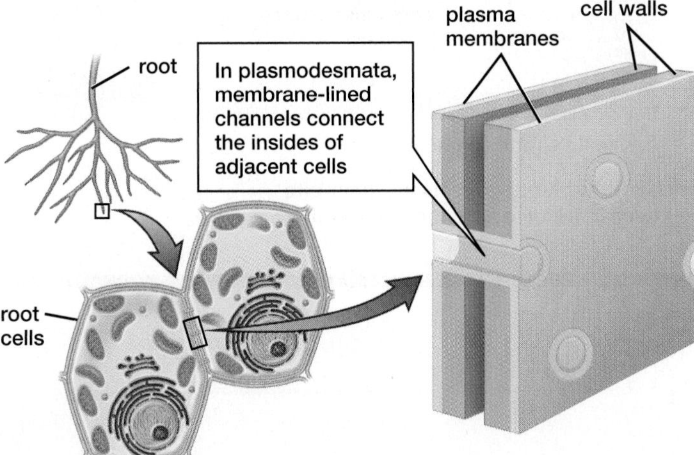
plasma membranes
cell walls
root
In plasmodesmata, membrane-lined channels connect the insides of adjacent cells
root cells
(b) Plasmodesmata

◀ **FIGURE 5-18 Cell communication structures (a)** Gap junctions, such as those between cells of the liver, contain cell-to-cell channels that interconnect the cytoplasm of adjacent cells. **(b)** Plant cells are interconnected by plasmodesmata, which link the cytoplasm of adjacent cells.

Case Study revisited
Vicious Venoms

Both rattlesnake and brown recluse spider venoms are complex mixtures of poisonous proteins. In each case, the proteins responsible for the worst symptoms are enzymes. As you will learn in Chapter 6, enzymes cause the breakdown of biological molecules while remaining unchanged themselves.

Enzymes are often named after the molecules they break apart, with the suffix "–ase" added to identify the protein as an enzyme. Several of the toxic enzymes in snake and spider venoms are phospholipases; the name tells us that they break down phospholipids. You now know that within cell membranes, the fluid bilayer portion—which allows the membrane to maintain the gradients that are crucial to life—consists of phospholipids.

Although the phospholipases and the other toxic proteins that form the "witches brew" of spider and rattlesnake venoms differ among species, in both cases, the venom attacks cell membranes, causing cells to rupture and die. Cell death destroys the tissue around a rattlesnake or brown recluse spider bite (**Fig. 5-19**). Phospholipases also attack the membranes of capillary cells, rupturing these tiny blood vessels and releasing blood under the skin around the wound. They also attack the membranes of red blood cells (which carry oxygen throughout the body), and so these venoms can cause anemia (an inadequate number of red blood cells), although this is far more common in rattlesnake than in brown recluse bites. In extreme cases, capillary damage can lead to bleeding—not just at the wound site, but internally as well. Rattlesnake phospholipases also attack muscle cells; this caused extensive damage in Justin's forearm.

Justin required large quantities of antivenin, which contains specialized proteins that bind and neutralize the toxins in snake venom. Unfortunately, no antivenin is available for brown recluse bites, and treatment generally consists of preventing infection, controlling pain and swelling, and waiting patiently—sometimes for months—for the wound to heal.

Although both snake and spider bites can have serious consequences, very few of the spiders and snakes found in the Americas are dangerous to people. The best defense is to learn which of these live in your area and where they prefer to hang out. If your activities bring you to such places, wear protective clothing—and always look before you reach! Knowledge can help us to comfortably coexist with spiders and snakes, to avoid their bites, and to keep our cell membranes intact.

Consider This

Phospholipases and other digestive enzymes are found in animal digestive tracts (including those of humans, snakes, and spiders) as well as in snake and spider venom. What different roles do phospholipases play in snake and spider venom as compared to the role of these phospholipases in digestive enzymes?

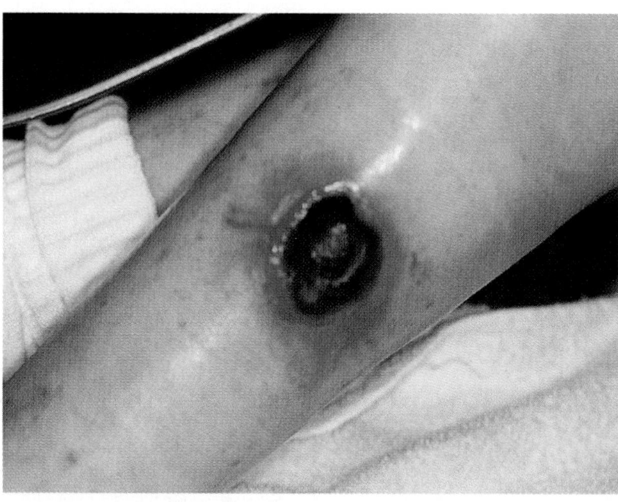

(a) Brown recluse spider bite

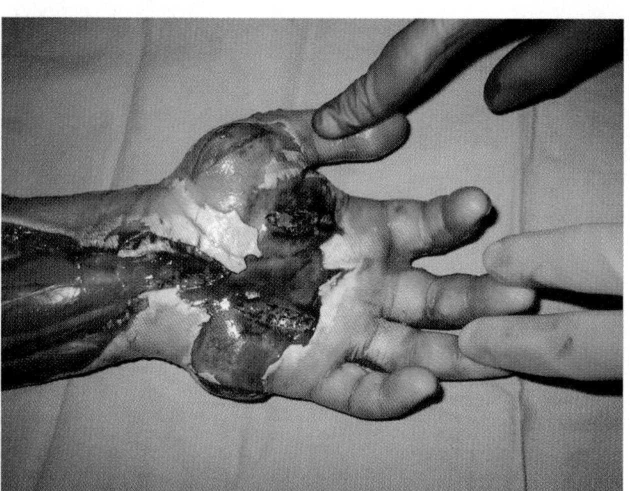

(b) Justin's rattlesnake bite

▲ FIGURE 5-19 **Phospholipases in venoms can destroy cells** Both **(a)** a brown recluse spider bite and **(b)** Justin's hand 36 hours after the rattlesnake bite show extensive tissue destruction caused by phospholipases.

CHAPTER REVIEW

Summary of Key Concepts

5.1 How Is the Structure of a Membrane Related to Its Function?

The plasma membrane consists of a bilayer of phospholipids in which a variety of proteins are embedded. It isolates the cytoplasm from the external environment, regulates the flow of materials into and out of the cell, allows communication between cells, allows attachments within and between cells, and regulates many biochemical reactions. There are five major types of membrane protein: (1) receptor proteins, which bind molecules and trigger changes within the cell; (2) recognition proteins, which label the cell; (3) enzymes, which promote chemical reactions; (4) attachment proteins, which anchor the plasma membrane to the cytoskeleton and extracellular matrix or bind cells to one another; and (5) transport proteins, which regulate the movement of most water-soluble substances through the membrane.

5.2 How Do Substances Move Across Membranes?

Diffusion is the movement of particles from regions of higher concentration to regions of lower concentration. In simple diffusion, water, dissolved gases, and lipid-soluble molecules diffuse through the phospholipid bilayer. During facilitated diffusion, carrier proteins or channel proteins allow water and water-soluble molecules to cross the membrane down their concentration gradients, without expending cellular energy.

Osmosis is the diffusion of water across a selectively permeable membrane down its concentration gradient through the phospholipid bilayer or through aquaporins.

Energy-requiring transport includes active transport, in which carrier proteins use cellular energy (ATP) to drive the movement of molecules across the plasma membrane against concentration gradients. Extracellular fluid, large molecules, and food particles may be acquired by endocytosis, which includes pinocytosis, receptor-mediated endocytosis, and phagocytosis. The secretion of substances, such as hormones, and the excretion of particulate cellular wastes are accomplished by exocytosis.

BioFlix™ Membrane Transport

5.3 How Do Specialized Junctions Allow Cells to Connect and Communicate?

Animal cell junctions include (1) desmosomes, which attach adjacent cells and prevent tissues from tearing apart during ordinary movements; (2) tight junctions, which leak-proof the spaces between adjacent cells; and (3) gap junctions that connect the cytoplasm of adjacent cells. Plasmodesmata interconnect the cytoplasm of adjacent plant cells.

Key Terms

active transport *88*
aquaporin *84*
attachment protein *82*
carrier protein *82*
channel protein *82*
concentration *82*
concentration gradient *82*
desmosome *92*
diffusion *82*
endocytosis *89*
energy-requiring transport *83*
enzyme *82*
exocytosis *90*
facilitated diffusion *84*
fluid *82*
fluid mosaic model *78*
food vacuole *90*
gap junction *93*
glycoprotein *81*
gradient *82*
hypertonic *85*
hypotonic *85*
isotonic *85*
microvilli *92*
osmosis *85*
passive transport *83*
phagocytosis *90*
phospholipid bilayer *79*
pinocytosis *89*
plasmodesmata *93*
receptor-mediated endocytosis *90*
receptor protein *81*
recognition protein *82*
selectively permeable *83*
simple diffusion *84*
solute *82*
solvent *82*
tight junction *93*
transport protein *82*
turgor pressure *87*

Thinking Through the Concepts

Fill-in-the-Blank

1. Membranes consist of a bilayer of __________ in which are suspended proteins of five major categories based on function: __________, __________, __________, __________, and __________ proteins.
2. A membrane that is permeable to some substances but not to others is described as being __________. The movement of a substance through a membrane down its concentration gradient is called __________. When applied to water, this process is called __________. In many cells, water travels through channels called __________. The process by which substances are moved through the membrane against their concentration gradient is called __________.
3. Membranes that are more fluid have phospholipids with more __________ in their fatty acid tails. Two environmental factors that decrease membrane fluidity are high __________ and low __________.
4. Facilitated diffusion involves either __________ proteins or __________ proteins. Diffusion directly through the phospholipid bilayer is called __________ diffusion, and molecules that take this route must be soluble in __________ or be very small and have no net electrical charge.
5. Energy-requiring transport includes the following three processes: __________, __________, and __________. Which of these uses ATP directly? __________
6. After each molecule, place the term that most specifically describes the process by which it moves through a plasma membrane: Carbon dioxide: __________; ethyl alcohol: __________; a sodium ion: __________; glucose: __________.

7. The general process by which fluids or particles are transported into cells is called __________. Does this process require energy? __________ The specific term for engulfing fluid is __________, and the term for engulfing particles is __________. The engulfed substances are taken into the cell in membrane-lined sacs called __________.

Review Questions

1. Describe and diagram the structure of a plasma membrane. What are the two principal types of molecules in plasma membranes, and what is the general function of each?
2. What are the five categories of proteins commonly found in plasma membranes, and what is the function of each one?
3. Define *diffusion,* and compare that process to osmosis. How do these two processes help plant leaves remain firm?
4. Define *hypotonic, hypertonic,* and *isotonic.* What would be the fate of an animal cell immersed in each of these three types of solution?
5. Describe the following types of transport processes in cells: simple diffusion, facilitated diffusion, active transport, pinocytosis, receptor-mediated endocytosis, phagocytosis, and exocytosis.
6. Name the protein that allows facilitated diffusion of water. What experiment demonstrated the function of this protein?
7. Imagine a container of glucose solution, divided into two compartments (A and B) by a membrane that is permeable to water and glucose but not to sucrose. If some sucrose is added to compartment A, how will the contents of compartment B change? Explain.
8. Name four types of cell-to-cell junctions, and describe the function of each. Which function in plants, and which in animals?

Applying the Concepts

1. Different cells have somewhat different plasma membranes. The plasma membrane of a *Paramecium,* for example, is only about 1% as permeable to water as the plasma membrane of a human red blood cell. Hypothesize about why this is the case. Is *Paramecium* likely to have aquaporins in its plasma membrane? Explain your answer.
2. Predict and sketch the configuration of phospholipids that are placed in vegetable oil. Explain your prediction.
3. The fluid portion of blood, in which red blood cells are suspended, is called plasma. Is the plasma likely to be isotonic, hypertonic, or hypotonic to the red blood cells? Explain.
4. Some cells in the nervous system wrap themselves around parts of neurons, serving as insulation for the electrical signals that flow inside neurons. These signals along the inside of neurons are carried by ions. Given the general roles of protein and lipids in the cell membrane, which component would you predict to be more abundant in these cells? Explain.

MB® *Go to www.masteringbiology.com for practice quizzes, activities, eText, videos, current events, and more.*

Energy Flow in the Life of a Cell

Chapter 6

Case Study

Energy Unleashed

Picture a huge crowd that includes people powering along in wheelchairs, a woman battling cancer, a supporter of wildlife dressed as a rhino, a 91-year-old man proceeding at a slow shuffle, a fireman wearing full firefighting gear to honor his fallen comrades, a man with only one leg using crutches, and blind people guided by the sighted. All participate in a 26-mile journey; a personal odyssey for each individual and a collective testimony to human hope, persistence, and endurance.

The 20,000-plus runners in the New York Marathon collectively expend more than 50 million Calories, and travel roughly 520,000 miles. Once finished, they douse their overheated bodies with water and refuel on high-energy drinks and snacks. Finally, cars, buses, and airplanes—burning vast quantities of fuel and releasing enormous amounts of heat—carry the runners back to their homes throughout the world.

What exactly is energy? Do our bodies use it according to the same principles that govern energy use in the engines of cars and airplanes? Why do our bodies generate heat, and why do we give off more heat when exercising than when watching TV?

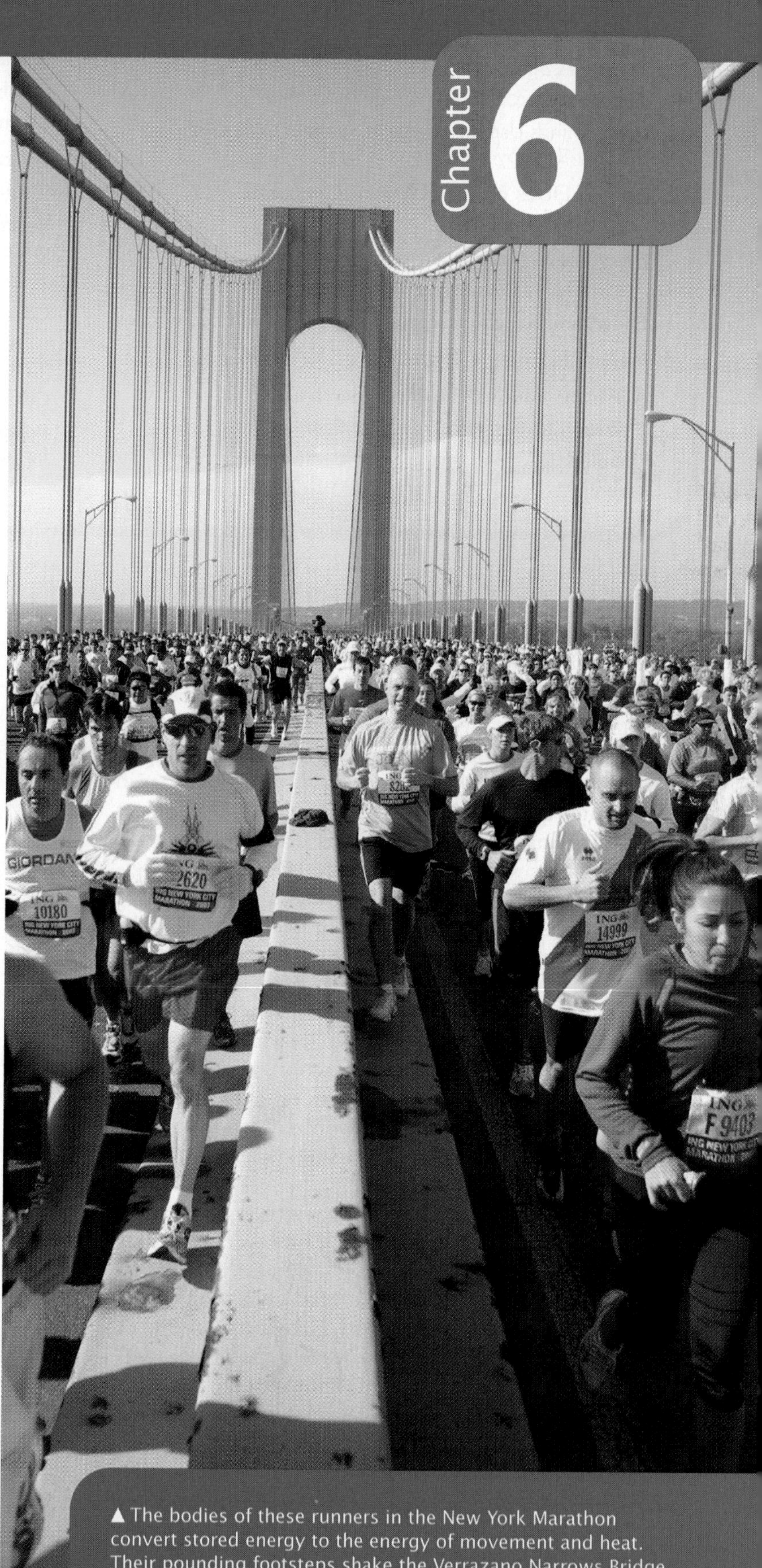

▲ The bodies of these runners in the New York Marathon convert stored energy to the energy of movement and heat. Their pounding footsteps shake the Verrazano Narrows Bridge.

At a Glance

6.1 WHAT IS ENERGY?

Energy is the capacity to do work. **Work,** in turn, is the transfer of energy to an object, causing the object to move. The objects upon which energy acts are not always easy to see or even to measure. It is obvious that the marathoners in our case study are doing work by moving things: Their chests heave, their arms pump, and their legs stride, pushing their bodies relentlessly forward for 26 miles. **Chemical energy,** which is energy that is contained in molecules and released by chemical reactions, powers this muscular work. Molecules that provide chemical energy—including sugar, glycogen, and fat—are stored in the runners' bodies. Cells utilize specialized molecules such as ATP to accept, briefly store, and transfer energy from one chemical reaction to the next. Muscle contraction results from interactions among specialized proteins that are driven by chemical energy released from molecules of ATP. The synchronized contractions of muscle cells move runners' bodies, using chemical energy to do work.

There are two fundamental types of energy: potential energy and kinetic energy, each of which takes several forms. **Potential energy,** or stored energy, includes chemical energy maintained in the bonds that bind atoms together in molecules, electrical energy housed in a battery, and positional energy stored in a penguin poised to plunge (**Fig. 6-1**). **Kinetic energy** is the energy of movement. It includes light (movement of photons), heat (movement of molecules), electricity (movement of electrically charged particles), and any movement of larger objects—the plummeting penguin (see Fig. 6-1), your eyes flicking across this page, and marathon runners as they strive to complete a grueling race. Under the right conditions, kinetic energy can be transformed into potential energy, and vice versa. For example, a penguin converts kinetic energy of movement into potential energy of position when it climbs back up from the water onto the ice. During photosynthesis (see Chapter 7), the kinetic energy of light is captured and transformed into the potential energy of chemical bonds. To understand energy flow and change, we need to

▲ **FIGURE 6-1 From potential to kinetic energy** Perched atop an ice floe, the body of the penguin has potential energy. As it dives, the potential energy is converted to the kinetic energy of motion of the penguin's body. Finally, some of this kinetic energy transferred to the water causes the water to splash and ripple.

know more about the properties and behavior of energy, described by the laws of thermodynamics.

The Laws of Thermodynamics Describe the Basic Properties of Energy

The **laws of thermodynamics** describe the quantity (the total amount) and the quality (the usefulness) of energy. The **first law of thermodynamics** states that energy can neither be created nor destroyed by ordinary processes (nuclear reactions, in which matter is converted into energy, are the exception). Energy can, however, change form—for example, from light energy to heat and chemical energy. If you had a **closed system,** where neither energy nor matter could enter or leave, and if you could measure energy in all its forms both before and after a particular process occurred, you would find the total energy before and after the process to be unchanged. Therefore, the first law of thermodynamics is often called the **law of conservation of energy.**

To illustrate the first law, consider your car. Before you turn the ignition key, the energy in the car is all potential energy, stored in the chemical bonds of its fuel. As you drive, only about 25% of this potential energy is converted into the kinetic energy of motion. According to the first law of thermodynamics, however, energy is neither created nor destroyed. So where is the "missing" energy? The burning gas not only moves the car but also heats up the engine, the exhaust system, and the air around the car. The friction of tires on the pavement slightly heats the road. So, as the first law dictates, no energy is lost; the total amount of energy remains the same, although it has changed in form.

The **second law of thermodynamics** states that when energy is converted from one form to another, the amount of useful energy decreases. Put another way, the second law states that all reactions or physical changes cause energy to be converted from more-useful into less-useful forms. For example, the 75% of the energy stored in gasoline that did not move the car was converted into heat (**Fig. 6-2**). Heat is a less-usable form of energy because it merely increases the random movement of molecules in the car, the air, and the road.

Similarly, the heat energy that runners liberate to the air when food "burns" in their bodies cannot be harnessed to allow them to run farther or faster. Thus, the second law tells us that no energy conversion process, including those that occur

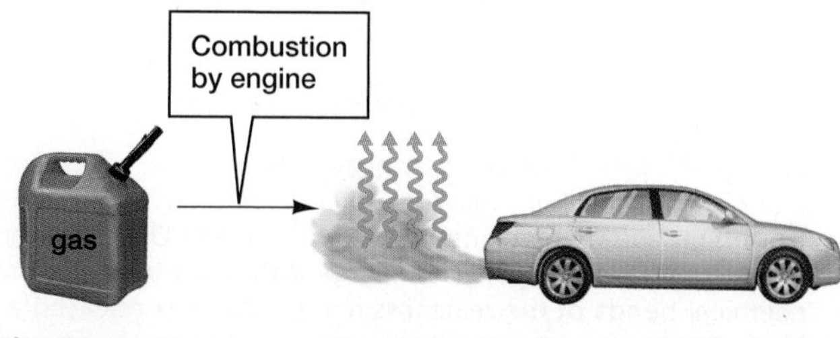

▲ **FIGURE 6-2 Energy conversions result in a loss of useful energy**

Case Study continued
Energy Unleashed

Like a car's engine, the marathoner's body is only about 25% efficient in converting chemical energy into movement; much of the other 75% is lost as heat. A highly trained marathoner completing the race in 2 hours generates enough heat to raise his or her body temperature by about 1.8°F (1°C) every 3 minutes. If unable to dissipate the heat, the runner would collapse from overheating within 10 minutes. Fortunately—thanks to the high heat of vaporization of water (see pp. 32–33)—most of this excess body heat (up to 98% in hot, dry weather) is dissipated by the evaporative cooling of sweat.

in the body, is 100% efficient in using energy to achieve a specific outcome.

The second law of thermodynamics also tells us something about the organization of matter. Useful energy tends to be stored in highly ordered matter, and whenever energy is used within a closed system, there is an overall increase in randomness and disorder of matter. We all experience this in our homes. Without energy-demanding cleaning and organizing efforts, dirty dishes accumulate; books, newspapers, and clothes collect in confusion on the floor; and the bed remains rumpled.

In the case of chemical energy, the eight carbon atoms in a single molecule of gasoline have a much more orderly arrangement than do the carbon atoms of the eight separate, randomly moving molecules of carbon dioxide and the nine molecules of water that are formed when the gasoline burns. The same is true for the glycogen molecules stored in a runner's muscles, which are converted from highly organized chains of sugar molecules into simpler water and carbon dioxide as they are used by the muscles. This tendency toward loss of complexity, orderliness, and useful energy—and the concurrent increase in randomness, disorder, and less-useful energy—is called **entropy.** To counteract entropy, energy must be infused into the system from an outside source.

When the eminent Yale scientist Evelyn Hutchinson stated, "Disorder spreads through the universe, and life alone battles against it," he made an eloquent reference to entropy and the second law of thermodynamics. Fortunately, Earth is not a closed system, for life as we know it depends on a constant infusion of energy from a source that is 93 million miles away—the sun.

Living Things Use the Energy of Sunlight to Create the Low-Entropy Conditions of Life

If you think about the second law of thermodynamics, you may wonder how life can exist at all. If chemical reactions, including those inside living cells, cause the amount of unusable energy to increase, and if matter tends toward increasing randomness and disorder, how can organisms accumulate the usable energy and precisely ordered molecules that characterize life? The answer is that nuclear reactions in the sun generate kinetic energy in the form of sunlight, a process that also produces vast increases in entropy within the sun in the form of heat. In fact, the temperature of the sun's core is estimated to be 27 million degrees F (16 million degrees C).

Living things use a continuous input of solar energy to synthesize complex molecules and maintain orderly structures, to "battle against disorder." The highly organized, low-entropy systems that characterize life do not violate the second law of thermodynamics because they are achieved through a continuous influx of usable light energy from the sun. The solar reactions that provide light energy here on Earth cause a far greater loss of useful energy in the sun, which will eventually burn out (but fortunately, not for billions of years). Because the solar energy that powers life on Earth occurs with an enormous net increase in solar entropy, life does not violate the second law of thermodynamics.

6.2 HOW DOES ENERGY FLOW IN CHEMICAL REACTIONS?

A **chemical reaction** is a process that forms or breaks the chemical bonds that hold atoms together. Chemical reactions convert one set of chemical substances, the **reactants,** into another set, the **products.** All chemical reactions either release energy or require an overall (net) input of energy. A reaction is **exergonic** (Greek for "energy out") if it releases energy; that is, if the starting reactants contain more energy than the end products. All exergonic reactions give off some of their energy as heat (**Fig. 6-3**).

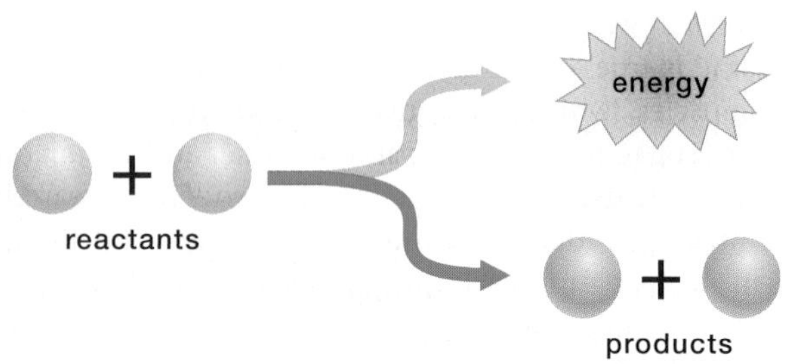

▲ **FIGURE 6-3 An exergonic reaction**

A reaction is **endergonic** ("energy in") if it requires a net input of energy; that is, if the products contain more energy than the reactants. Endergonic reactions require a net influx of energy from an outside source (**Fig. 6-4**).

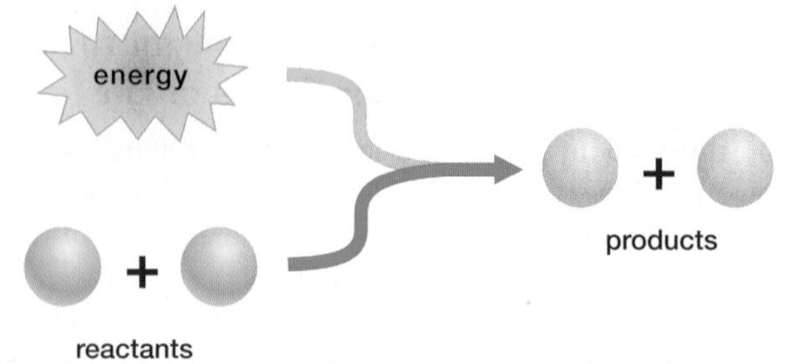

▲ **FIGURE 6-4 An endergonic reaction**

In the following sections, we examine the exergonic process of burning sugar and the endergonic process of photosynthesis.

Exergonic Reactions Release Energy

In an exergonic reaction, the reactants contain more energy than the products. Sugar can be burned, as any cook can tell you. As it is burned, sugar (glucose, for example) undergoes the same overall reaction as sugar in the body of a runner (or nearly any other form of life): sugar ($C_6H_{12}O_6$) is combined with oxygen (O_2) to produce carbon dioxide (CO_2) and water (H_2O), releasing energy (**Fig. 6-5**).

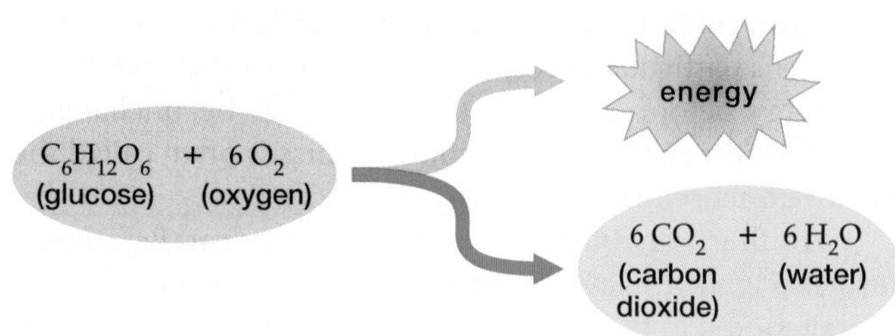

▲ **FIGURE 6-5 Reactants and end products of burning glucose**

Because molecules of sugar contain much more energy than do molecules of carbon dioxide and water, the reaction releases energy. Once ignited, sugar will continue to burn. It may be helpful to think of exergonic reactions as running "downhill," from high energy to low energy, as shown in **Figure 6-6.**

Heat, generated as a by-product of every biochemical transformation, is used by people and other "warm-blooded" animals to maintain a high body temperature. An elevated

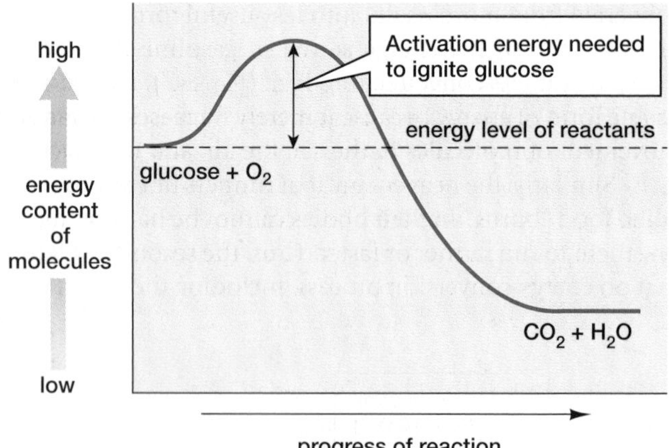

▲ **FIGURE 6-6 Activation energy in exergonic reactions** An exergonic ("downhill") reaction, such as burning sugar, proceeds from high-energy reactants (here, glucose and O_2) to low-energy products (CO_2 and H_2O). The energy difference between the chemical bonds of the reactants and products is released as heat. To start the reaction, however, an initial input of energy—the activation energy—is required.

QUESTION In addition to heat and sunlight, what are some other possible sources of activation energy?

Case Study continued
Energy Unleashed

Marathoners rely on glycogen, stored in their muscles and liver, for the energy to power their run. Glycogen consists of chains of glucose molecules, which are first cleaved from the chain and then broken down into carbon dioxide and water. This exergonic reaction generates heat and the ATP required for muscle contraction. The carbon dioxide is exhaled as the runners breathe rapidly to supply their muscles with adequate oxygen. The water produced (and a lot more that the runners drink during the race) is lost as sweat. As the sweat evaporates, it carries away the heat generated by the chemical reactions in their muscles, preventing the runners' bodies from overheating.

body temperature speeds up biochemical reactions, allowing animals to move faster and respond more quickly to stimuli than if their body temperatures were lower.

All Chemical Reactions Require Activation Energy to Begin

Although burning sugar releases energy, sugar doesn't burst into flames by itself. This observation leads to an important concept: All chemical reactions, even those that can continue spontaneously, require energy to get started. Think of a rock sitting at the top of a hill. It will remain there indefinitely unless something gives it a push to start it rolling down. In chemical reactions, the energy "push" is called the **activation energy** (see Fig. 6-6). Chemical reactions require activation energy to get started because shells of negatively charged electrons surround all atoms and molecules. For two molecules to react with each other, their electron shells must be forced together, despite their mutual electrical repulsion. Forcing the electron shells together requires activation energy.

Activation energy can be provided by the kinetic energy of moving molecules. At any temperature higher than absolute zero (−460°F, or −273°C), atoms and molecules are in constant motion. Reactive molecules moving with sufficient speed collide hard enough to force their electron shells to mingle and react. Because molecules move faster as the temperature increases, most chemical reactions occur more readily at high temperatures. For example, heat provided by a match can set sugar on fire. The combination of sugar with oxygen then releases enough of its own heat to sustain the reaction, allowing it to continue spontaneously. How is adequate activation energy generated in the body to allow it to "burn" sugar? Keep this question in mind—you'll find the answer later in this chapter.

Endergonic Reactions Require a Net Input of Energy

In contrast to what happens when sugar is burned, many reactions in living systems result in products that contain more energy than the reactants do. Sugar, produced by photosynthetic organisms such as plants, contains far more energy than does the carbon dioxide and water from which it was formed. The protein in a muscle cell contains more energy than the individual amino acids that were joined together to synthesize it. In other words, synthesizing complex biological molecules requires an input of energy; these reactions are endergonic. As we will see in Chapter 7, photosynthesis in green plants uses solar energy to produce sugar from water and carbon dioxide (**Fig. 6-7**). The oxygen produced by this reaction is used when glucose is broken down by cells to release its stored energy.

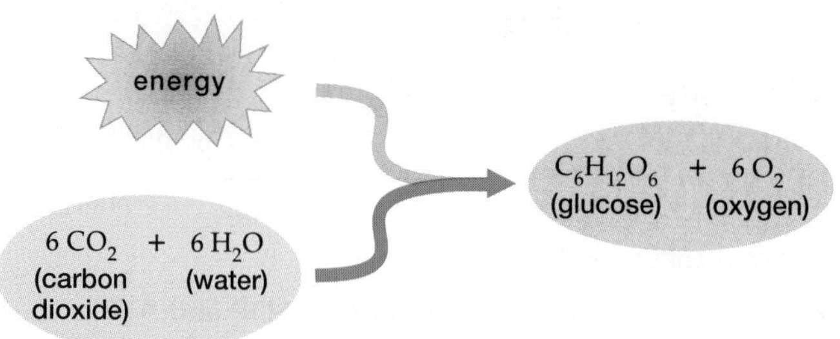

▲ **FIGURE 6-7 Photosynthesis**

Endergonic reactions are not spontaneous; we might call them "uphill" reactions, because the reactants contain less energy than the products do. Going from low energy to high energy is like pushing a rock up to the top of the hill. Where do we and other animals get the energy to fuel endergonic reactions such as synthesizing muscle protein and other complex biological molecules?

6.3 HOW IS ENERGY TRANSPORTED WITHIN CELLS?

Most organisms are powered by the breakdown of sugar (usually glucose). By combining glucose with oxygen and releasing carbon dioxide and water, cells acquire chemical energy from the glucose molecule. This energy is used to do cellular work, such as constructing complex biological molecules and powering muscle contraction. But glucose cannot be used directly to fuel these endergonic processes. Instead, the energy released by glucose breakdown is first transferred to an energy-carrier molecule. **Energy-carrier molecules** are high-energy, unstable molecules that are synthesized at the site of an exergonic reaction, capturing some of the released energy. These energy carriers work something like rechargeable batteries; they pick up an energy charge at an exergonic reaction, move to another location within the cell, and then release the energy to drive an endergonic reaction. Because energy-carrier molecules are unstable, they are used only to capture and transfer energy within cells. They cannot ferry energy from cell to cell, nor are they used for long-term energy storage.

ATP Is the Principal Energy Carrier in Cells

Many exergonic reactions in cells produce **adenosine triphosphate,** or **ATP,** the most common energy-carrier molecule in the body. Because it provides energy to drive such a wide variety of endergonic reactions, ATP is sometimes called

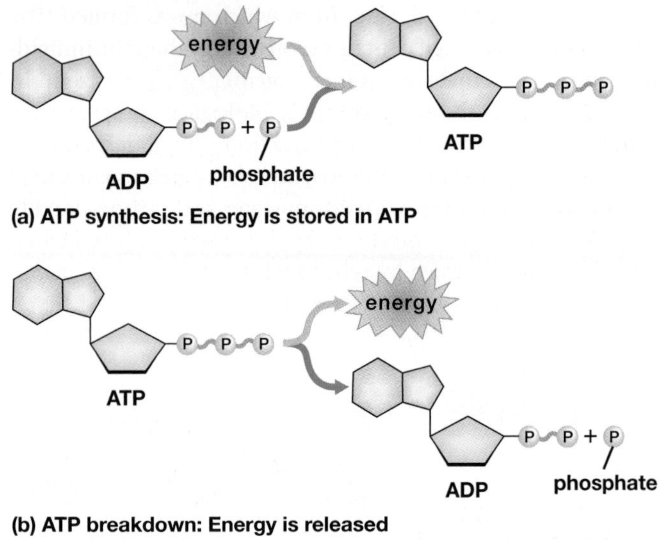

energy
ADP phosphate
ATP

(a) ATP synthesis: Energy is stored in ATP

energy
ATP
ADP phosphate

(b) ATP breakdown: Energy is released

▲ **FIGURE 6-8 The interconversion of ADP and ATP**
(a) Energy is captured when a phosphate group (P) is added to adenosine diphosphate (ADP) to make adenosine triphosphate (ATP). **(b)** Energy to accomplish cellular work is released when ATP is broken down into ADP and P.

the "energy currency" of cells. As you learned in Chapter 3, ATP is a nucleotide composed of the nitrogen-containing base adenine, the sugar ribose, and three phosphate groups (see Fig. 3-23). Energy released in cells during glucose breakdown or other exergonic reactions is used to combine the relatively low-energy molecules **adenosine diphosphate (ADP)** and phosphate (HPO_4^{2-}, also called P) into the much less stable, high-energy molecule ATP (**Fig. 6-8a**). The formation of ATP is endergonic—it requires an input of energy which is captured in this newly formed high energy molecule.

ATP stores energy within its chemical bonds and diffuses throughout the cell, carrying the energy to sites where energy-requiring reactions occur. Its energy is liberated as the ATP is broken down, regenerating ADP and P (**Fig. 6-8b**). The life span of an ATP molecule in a living cell is very short. If the molecules of ATP that you use just sitting at your desk all day could be captured (instead of recycled), they would weigh around 90 pounds (about 40 kg). A marathon runner may recycle the equivalent of a pound of ATP every minute, so without rapid reconversions, it would be a very brief run. As you can see, ATP is *not* a long-term energy-storage molecule. Relatively stable molecules such as glycogen and fat can store energy for hours, days, or—in the case of fat—years.

Electron Carriers Also Transport Energy Within Cells

ATP is not the only energy-carrier molecule within cells. In some exergonic reactions, including both glucose breakdown and the light-capturing stage of photosynthesis, some energy is transferred to electrons. These energetic electrons, along with hydrogen ions (H^+; present in the cytoplasmic fluid) are captured by special energy-carrier molecules called **electron carriers.** Common electron carriers include nicotinamide adenine dinucleotide (NADH) and its relative, flavin adenine dinucleotide ($FADH_2$). The loaded electron carriers then donate their high-energy electrons to other molecules, which are often involved in pathways that generate ATP. You will learn more about electron carriers and their role in cellular metabolism in Chapters 7 and 8.

Coupled Reactions Link Exergonic with Endergonic Reactions

In a **coupled reaction,** an exergonic reaction provides the energy needed to drive an endergonic reaction (**Fig. 6-9**). During photosynthesis, for example, plants use sunlight (from exergonic reactions in the sun's core) to drive the endergonic synthesis of high-energy glucose molecules from lower energy reactants (carbon dioxide and water). Nearly all organisms use the energy released by exergonic reactions (such as the breakdown of glucose into carbon dioxide and water) to drive endergonic reactions (such as the synthesis of proteins from amino acids). Because energy is lost as heat every time it is transformed, in coupled reactions the energy released by exergonic reactions must exceed the energy needed to drive the endergonic reactions.

The exergonic and endergonic portions of coupled reactions often occur in different places within a cell, so there must be some way to transfer the energy from the exergonic reactions that release energy to the endergonic reactions that require it. In coupled reactions, energy is transferred from place to place by energy-carrier molecules such as ATP. In its role as an intermediary in coupled reactions, ATP is constantly being synthesized to capture the energy released during exergonic reactions and then broken down to power endergonic reactions, as shown in Figure 6-9.

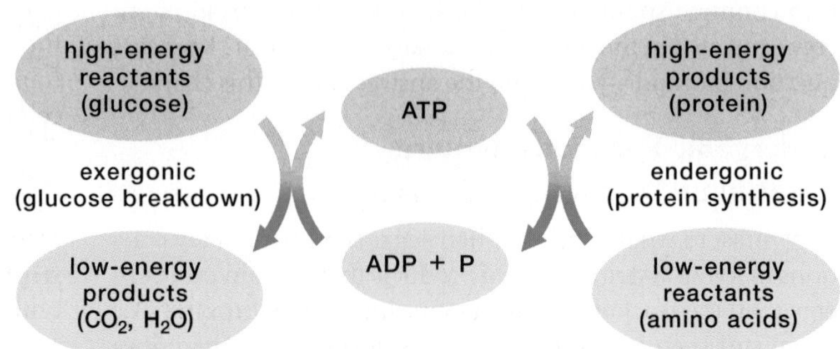

► **FIGURE 6-9 Coupled reactions within living cells**
Exergonic reactions (such as glucose breakdown) drive the endergonic reaction that synthesizes ATP from ADP and P. The ATP molecule carries its chemical energy to a part of the cell where energy is needed to drive an endergonic reaction (such as protein synthesis). The ADP and P are rejoined to form ATP via endergonic reactions.

6.4 HOW DO ENZYMES PROMOTE BIOCHEMICAL REACTIONS?

You now know that cells couple reactions together, powering energy-requiring endergonic reactions with the energy released by exergonic reactions. You have also learned that cells synthesize energy-carrier molecules that capture energy from exergonic reactions and transport it to endergonic reactions. But how do cells promote and control the multitude of biochemical reactions required to maintain the precise chemical balance necessary for life?

At Body Temperatures, Spontaneous Reactions Proceed Too Slowly to Sustain Life

In general, the speed at which a reaction occurs is determined by its activation energy; that is, how much energy is required to start the reaction (see Fig. 6-6). Some reactions, such as table salt dissolving in water (see Fig. 2-9), have low activation energies and occur readily at human body temperature (approximately 98.6°F, or 37°C). In contrast, you could maintain sugar at body temperature for decades and it would appear unchanged.

The reaction of sugar with oxygen to yield carbon dioxide and water is exergonic, but it has a high activation energy. The heat of a match flame increases the rate of movement of sugar and nearby oxygen molecules, causing them to collide with sufficient force to overcome their activation energy and react; the sugar burns. The energy released by this exergonic reaction speeds up the movement of more sugar and oxygen molecules, allowing the sugar to continue to burn on its own. At the temperatures found in living organisms, however, sugar and many other energy-containing molecules would almost never break down and give up their energy. Promoting reactions with high activation energies is the job of catalysts.

Catalysts Reduce Activation Energy

Catalysts are molecules that speed up the rate of a reaction without themselves being used up or permanently altered. A catalyst speeds up a reaction by reducing the reaction's activation energy (**Fig. 6-10**). For example, consider the catalytic converters on automobile exhaust systems. Flaws in the gasoline combustion process generate poisonous carbon monoxide (CO). Carbon monoxide reacts spontaneously but slowly with oxygen in the air to form carbon dioxide:

$$2\,CO + O_2 \rightarrow 2\,CO_2 + \text{heat energy}$$

In heavy traffic, the spontaneous reaction of CO with O_2 can't keep pace with the CO emitted, and unhealthy levels of CO accumulate. Enter the catalytic converter. Catalysts such as platinum in the converter provide a specialized surface upon which O_2 and CO combine more readily, hastening the conversion of CO to CO_2 and reducing air pollution.

All catalysts share three important properties:

- Catalysts speed up reactions by lowering the activation energy required for the reaction to begin.

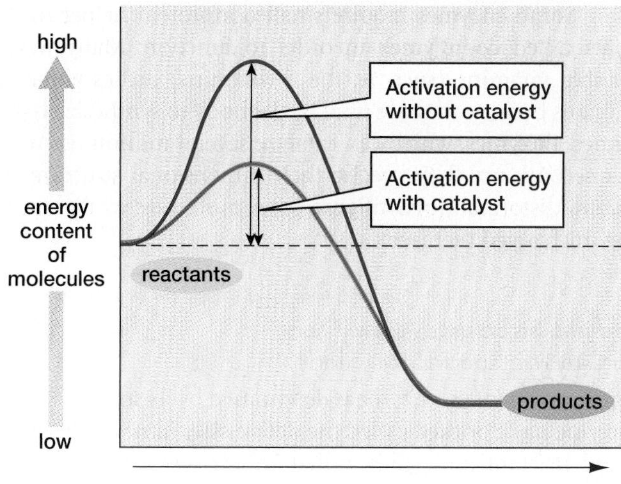

▲ **FIGURE 6-10 Catalysts such as enzymes lower activation energy** A high activation energy (red curve) means that reactant molecules must collide very forcefully in order to react. Catalysts lower the activation energy of a reaction (blue curve), so a much higher proportion of molecules moves fast enough to react when the molecules collide. Therefore, the reaction proceeds much more rapidly.

QUESTION Can a catalyst make an endergonic reaction occur spontaneously?

- Catalysts can speed up both exergonic and endergonic reactions, but they cannot make an endergonic reaction occur spontaneously. An endergonic reaction will still need a net input of energy, with or without an assist from a catalyst.
- Catalysts are not consumed or permanently changed by the reactions they promote.

Enzymes Are Biological Catalysts

Inorganic catalysts, such as the platinum in a catalytic converter, speed up a great number of chemical reactions. Indiscriminately speeding up dozens of reactions would not be useful, and indeed would almost certainly be deadly, to living organisms. Instead, cells employ highly specific biological catalysts called **enzymes,** which are composed primarily of proteins. A given enzyme catalyzes only a few types of chemical reactions, at most. The majority of enzymes catalyze a single reaction involving specific molecules, while leaving even very similar molecules unchanged.

Both exergonic and endergonic reactions are catalyzed by enzymes. The synthesis of ATP from ADP and P, for example, is catalyzed by the enzyme ATP synthase. This enzyme is responsible for capturing some of the energy released during the series of reactions that break down glucose and then storing it in ATP. When energy is required to drive endergonic reactions, ATP is broken down by ATPase. (Notice that some enzymes are named by adding the suffix "ase" to a description of what the enzyme does [ATP synthase] and some others by adding "ase" to the molecule upon which the enzyme acts [ATPase].)

Some enzymes require small nonprotein helper molecules called **coenzymes** in order to function. Many water-soluble vitamins (such as the B vitamins) are essential to humans because they are used by the body to synthesize coenzymes. Enzymes, which can catalyze several million reactions per second, use their precise three-dimensional structures to orient, distort, and reconfigure other molecules while emerging unchanged themselves.

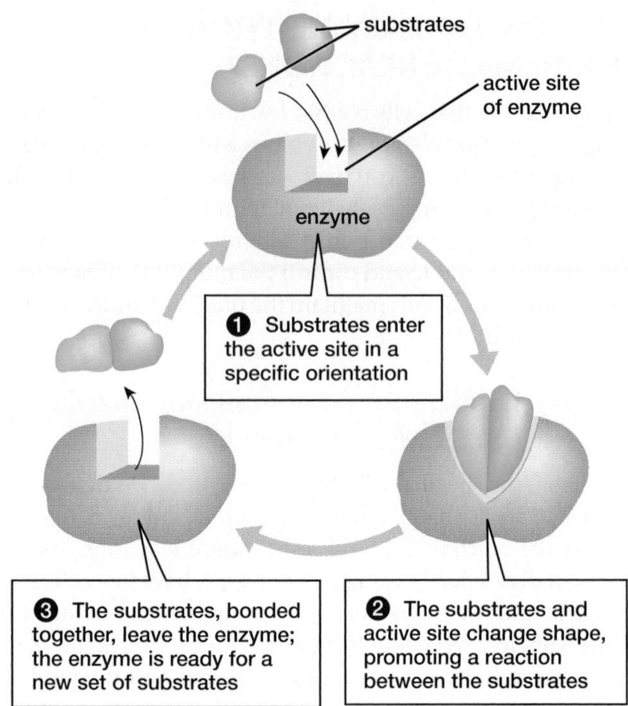

▲ **FIGURE 6-11 The cycle of enzyme–substrate interactions**

Enzyme Structures Allow Them to Catalyze Specific Reactions

The function of an enzyme is determined by its structure. Each enzyme has a pocket, called the **active site,** into which one or more reactant molecules, called **substrates,** can enter. You may recall from Chapter 3 that proteins have complex three-dimensional shapes (see Fig. 3-20). Their primary structure is determined by the precise order in which amino acids are linked. The chain of amino acids then folds around itself in a configuration (often a helix or a pleated sheet) called secondary structure. The protein then acquires the additional twists and bends of tertiary structure. Some enzymes also contain peptide subunits joined into a quaternary structure. The order of amino acids, and the precise way in which the amino acid chains of a protein are twisted and folded, creates the distinctive shape of the active site and a specific distribution of electrical charges within the site.

Because the enzyme and its substrate fit together precisely, only certain molecules can enter the active site. Take the enzyme amylase, for example. Amylase breaks down starch molecules by hydrolysis, but leaves cellulose molecules intact, even though both consist of chains of glucose molecules. A different bonding pattern between the glucose molecules in cellulose prevents them from fitting into the enzyme's active site. (If you chew a soda cracker long enough, you'll notice a sweet flavor caused by the release of sugar molecules from the starch in the cracker by amylase in your saliva.) The stomach enzyme pepsin is selective for proteins, attacking them at many sites along their amino acid chains. Certain other protein-digesting enzymes (trypsin, for example) will break bonds only between specific amino acids. Digestive systems manufacture several different enzymes that work together to completely break down dietary protein into its individual amino acids.

How does an enzyme catalyze a reaction? First, both the shape and the charge of the active site allow substrates to enter the enzyme only in specific orientations (**Fig. 6-11** ❶). Second, when substrates enter the active site, both the substrate and active site change shape (**Fig. 6-11** ❷). Certain amino acids within the enzyme's active site may temporarily bond with atoms of the substrates, or electrical interactions between the amino acids in the active site and substrates may distort chemical bonds within the substrates. The combination of substrate selectivity, substrate orientation, temporary chemical bonds, and the distortion of existing bonds promotes the specific chemical reaction catalyzed by a particular enzyme. When the reaction between the substrates is finished, the product(s) no longer fit properly into the active site and drift away (**Fig. 6-11** ❸). The enzyme reverts to its original configuration, and it is ready to accept another set of substrates.

Enzymes, Like All Catalysts, Lower Activation Energy

How do enzymes speed up the rate of chemical reactions? The breakdown or synthesis of a molecule within a cell usually occurs in many small, discrete steps, each catalyzed by a different enzyme (see Fig. 6-12). Each of these enzymes lowers the activation energy for its particular reaction, allowing the reaction to occur readily at body temperature. Think of a rock climber ascending a steep cliff by finding a series of hand- and footholds that allow her to scale it one small step at a time. In a similar manner, a series of reaction steps, each requiring a small amount of activation energy and each catalyzed by an enzyme that lowers activation energy, allows the overall reaction (in this case, oxidizing sugar) to surmount its high overall activation energy "cliff" and to proceed at body temperature.

Enzymes Control the Rate of Energy Release and Capture Some Energy in ATP

Ignite sugar and it will go up in flames as it combines rapidly with oxygen, releasing carbon dioxide and water. The same overall reaction occurs in our cells, but they are in no danger of catching on fire. Thanks to a series of chemical transformations, each catalyzed by a different enzyme, the energy stored in sugar is released gradually. Some is lost as heat, keeping our bodies warm, while some is harnessed to power endergonic reactions that lead to ATP synthesis.

What Makes a Firefly Glow?

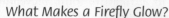

The almost magical glow of fireflies comes from specialized cells in their abdomens that are bioluminescent, meaning that they produce "biological light." These cells are rich in ATP and the fluorescent chemical luciferin (Latin, meaning "light-bringer"). Luciferin and ATP serve as substrates for the enzyme luciferase. In the presence of oxygen, luciferase catalyzes a reaction that modifies luciferin and uses the energy from ATP to boost electrons briefly into a higher-energy electron shell. As they fall back into their original shell, the electrons emit their excess energy as light. The process produces so little heat that these beetles are often described as emitting "cold light."

6.5 HOW DO CELLS REGULATE THEIR METABOLIC REACTIONS?

Cells are incredibly complex chemical factories. The **metabolism** of a cell is the sum of all its myriad chemical reactions. Many of these reactions are linked in sequences called **metabolic pathways** (Fig. 6-12). In a metabolic pathway, an initial reactant molecule is modified by an enzyme, forming a slightly different intermediate molecule, which is modified by another enzyme to form a second intermediate, and so on, until an end product is produced. Within metabolic pathways, molecules are both synthesized and broken down, so that the end product may be a waste molecule (such as carbon dioxide formed when glucose is broken down). Alternatively, the end product may be a molecule (such as an amino acid; see Fig. 6-14) synthesized by the metabolic pathway. Photosynthesis (Chapter 7) is a metabolic pathway that results in the synthesis of high-energy molecules, including glucose. Another metabolic pathway, glycolysis, begins the breakdown of glucose (Chapter 8). Different metabolic pathways often use some of the same molecules; as a result, all the thousands of metabolic pathways within a cell are directly or indirectly interconnected (see Fig. 6-12).

Reaction Rates Tend to Increase as Substrate or Enzyme Levels Increase

In a test tube containing an enzyme and its substrate (under constant, ideal conditions), the rate of the reaction will depend on how many substrate molecules encounter the active sites of enzyme molecules in a given time period. As you might guess, if you added either more enzyme or more substrate molecules, the reaction rate would increase. For a given amount of enzyme, as substrate levels increase, the reaction rate will increase until the active sites of all the enzyme molecules are being continuously occupied by new substrate molecules. To increase the rate further, more enzymes would be needed.

But consider the needs of a cell. When glucose molecules flood in after a meal, it would not be desirable to metabolize them all and form far more unstable ATP molecules than the cell can use. To be effective, the metabolic reactions within cells must be precisely regulated; they must occur at the proper time and proceed at the proper rate. This fine-tuning of metabolic reactions is accomplished in several ways.

Case Study continued
Energy Unleashed

If a runner preparing for a marathon eats a huge dish of pasta, a few hours later the starch-digesting enzymes in his or her small intestine will encounter an enormous quantity of carbohydrates. Because of their abundance, these substrate molecules will encounter the active sites of amylase enzymes far more often than before the meal, and the rate at which amylase enzymes break down carbohydrate will increase dramatically.

Cells Regulate Enzyme Synthesis

Cells exert tight control over all the types of proteins they produce. Genes that code for specific proteins are turned on or off depending on the need for that protein, a process described in detail in Chapter 12. Enzyme proteins govern all of the cell's metabolic activities, and these activities must be able to change to meet the cell's changing needs. Some enzymes, therefore, are synthesized in larger quantities when more of their substrate is available. Larger amounts of the enzyme then speed up the rate at which a substance is metabolized, because more substrate molecules are bound to enzyme molecules at

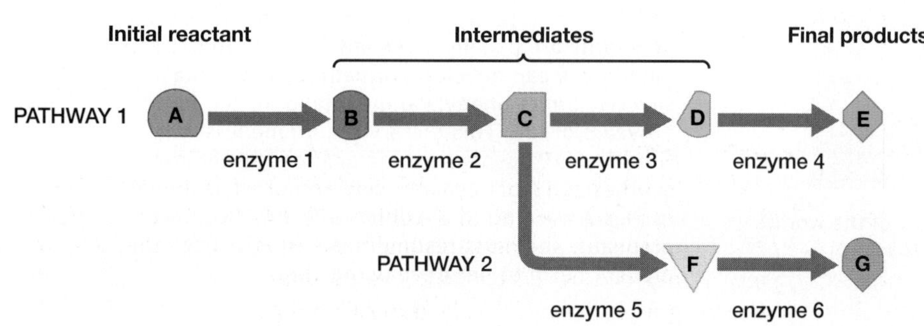

◀ FIGURE 6-12 **Simplified metabolic pathways** The initial reactant molecule (A) undergoes a series of reactions, each catalyzed by a specific enzyme. The product of each reaction serves as the reactant for the next reaction in the pathway. Metabolic pathways are commonly interconnected such that the product of a step in one pathway may serve as a reactant for the next reaction or for a reaction in another pathway.

any given time. For example, the livers of people who regularly drink large quantities of alcohol produce more alcohol dehydrogenase, an enzyme that breaks down alcohol. Unfortunately, this and other liver enzymes convert alcohol into toxic substances, so that alcoholics frequently suffer liver damage.

Mutations (accidental changes) in genes can alter enzyme production, as described in "Health Watch: Lacking an Enzyme Can Lead to Lactose Intolerance or Phenylketonuria."

Cells Regulate Enzyme Activity

Enzymes are complex proteins, and many have properties that cause them to be regulated by their chemical surroundings. This allows enzymes to become active when and where they are needed.

Some Enzymes Are Synthesized in Inactive Forms

Some enzymes are synthesized in an inactive form that is activated under the conditions found where the enzyme is needed. Examples are the protein-digesting enzymes pepsin and trypsin, mentioned earlier. Cells synthesize and release each of these enzymes in an inactive form, preventing the enzyme from digesting and killing the cell that manufactures it. In the stomach, acid modifies the shape of pepsin, exposing its active site and allowing it to begin breaking down proteins that have been eaten. Trypsin, which helps complete protein digestion, works best in the more basic (higher pH) conditions of the small intestine (see Fig. 6-15a). The small intestine secretes an enzyme that acts on trypsin, altering its configuration and allowing it to function.

Enzyme Activity May Be Inhibited Competitively or Noncompetitively

It is not in the cell's best interest to have its enzymes churning out products all the time. Many enzymes need to be inhibited so that the cell does not use up all of its substrate or become overwhelmed by too much product. You'll recall that

Health Watch

Lacking an Enzyme Can Lead to Lactose Intolerance or Phenylketonuria

Is it difficult for you to imagine life without milk, ice cream, or even pizza (**Fig. E6-1**)? Although some people consider these foods to be staples of the U.S. diet, they are not enjoyed by most of the world's population. Why? About 75% of people worldwide, including 25% of people in the United States, lose the ability to digest lactose, or milk sugar, in early childhood. Roughly 75% of African Americans, Hispanics, and Native Americans, as well as 90% of Asian Americans, experience **lactose intolerance.**

▲ **FIGURE E6-1 Risky behavior?** For most of the world's adults, drinking a glass of milk invites unpleasant consequences.

From an evolutionary perspective, this makes perfect sense. The lactose enzyme, called lactase, is found in the small intestines of all healthy young children. After being weaned in early childhood, our early ancestors no longer consumed milk—the main source of lactose. Because it takes energy to synthesize enzymes, losing the ability to synthesize an unnecessary enzyme is adaptive. However, some populations experienced mutations that conferred the ability to continue to digest lactose into adulthood, such as populations in northern Europe, who raised cattle for milk and made dairy products a regular part of their diet. Their descendants continue to enjoy this healthy food.

Compared to other consequences of enzyme deficiency, the inability to tolerate milk is a relatively minor inconvenience. Because enzymes are crucial to all aspects of cellular life, mutations that render them nonfunctional may cause life-threatening disorders or may prevent an embryo from developing at all. One serious result of a defective enzyme is **phenylketonuria (PKU),** a disorder that occurs most frequently in Caucasians (affecting about 1 in 10,000). This mutation cripples the enzyme that oxidizes the common amino acid phenylalanine, causing a buildup of phenylalanine in the blood. Untreated, PKU causes severe mental retardation. Today, in the United States, newborns are usually screened for PKU, and affected individuals can develop normally on a diet that severely limits phenylalanine intake for the first 10 years of life. This dietary requirement is why NutraSweet™ (aspartame; an artificial sweetener synthesized from aspartic acid and phenylalanine) carries a warning. If a woman with PKU becomes pregnant, she must resume this very restricted diet to prevent harm to her developing child.

for an enzyme to catalyze a reaction, its substrate must bind to the enzyme's active site (**Fig. 6-13a**). There are two general ways in which enzymes can be inhibited.

In **competitive inhibition,** a substance that is not the enzyme's normal substrate binds to the active site of the enzyme, competing with the substrate for the active site (**Fig. 6-13b**). For example, along the metabolic pathway that breaks down glucose, one enzyme in the sequence of reactions is competitively inhibited by the product molecule formed two steps further along. This product diffuses in and out of the active site, so the binding is reversible. This means that the normal substrate and the inhibitor can each displace the other if its concentration is high enough, which helps control the rate of glucose breakdown.

In **noncompetitive inhibition,** a molecule binds to a noncompetitive inhibitor site on the enzyme that is distinct from the active site. As a result, the enzyme active site is distorted, making it less able to catalyze the reaction (**Fig. 6-13c**). Many noncompetitive inhibitor molecules are poisons, as described later.

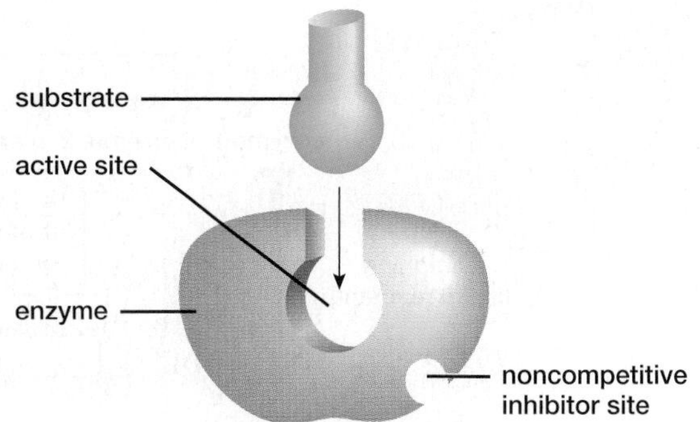
substrate
active site
enzyme
noncompetitive inhibitor site
(a) A substrate binding to an enzyme

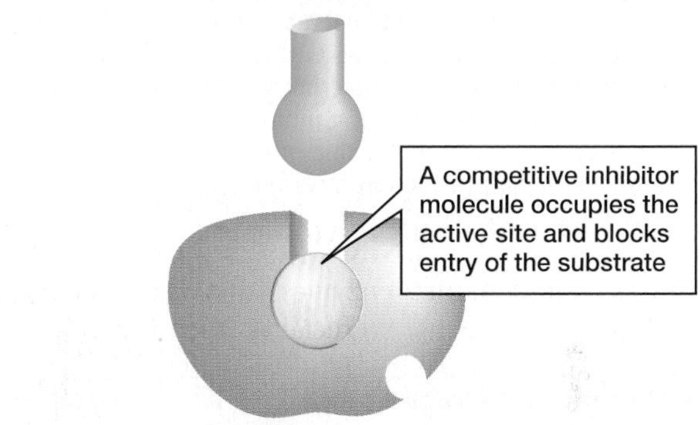
A competitive inhibitor molecule occupies the active site and blocks entry of the substrate
(b) Competitive inhibition

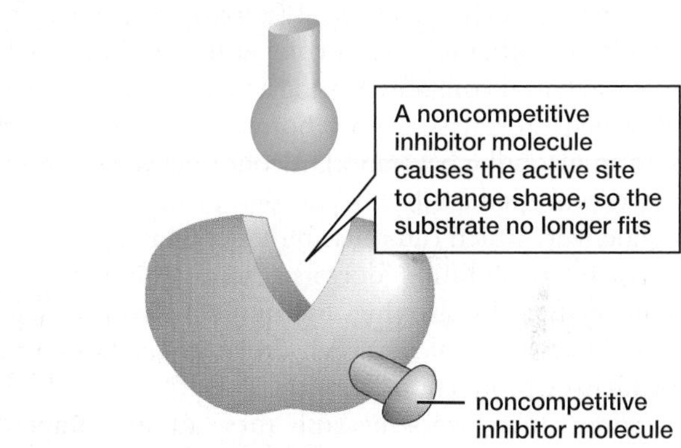
A noncompetitive inhibitor molecule causes the active site to change shape, so the substrate no longer fits
noncompetitive inhibitor molecule
(c) Noncompetitive inhibition

▲ **FIGURE 6-13 Competitive and noncompetitive enzyme inhibition (a)** The normal substrate fits easily into the enzyme's active site when the enzyme is not being inhibited. **(b)** In competitive inhibition, a competitive inhibitor molecule that resembles the substrate blocks the active site. **(c)** In noncompetitive inhibition, a molecule binds to a different site on the enzyme, distorting the active site so it no longer fits its substrate.

Some Enzymes Are Controlled by Allosteric Regulation

Some enzymes, called allosteric enzymes, switch readily and spontaneously between two different configurations ("allosteric" literally means "other shape")—one configuration is active, while the other is inactive. Such enzymes are controlled by **allosteric regulation,** which occurs when molecules described as "allosteric activators" or "allosteric inhibitors" bind reversibly to the regulatory sites of the enzyme—sites distinct from the enzyme's active site. Reversible binding of the activator and inhibitor molecules means that they bind temporarily, so that the number of enzymes being activated (or inhibited) is proportional to the numbers of activator (or inhibitor) molecules that are present at any given time.

Allosteric *activators* stabilize the enzyme in its active form; when there are large numbers of allosteric activators present, the enzyme activity will be high. Allosteric *inhibitors* bind to a different regulatory site that stabilizes the enzyme in its inactive form. Thus, the overall level of allosteric enzyme activity is regulated by the relative quantities of activators and inhibitors available. An example of an allosteric activator molecule is ADP, from which ATP is synthesized. An abundance of ADP in a cell means that a lot of ATP has been used up, and more ATP is needed. ADP activates allosteric enzymes in metabolic pathways that produce ATP, stabilizing them in their active state.

An important form of allosteric regulation is feedback inhibition. **Feedback inhibition** (**Fig. 6-14**) causes a metabolic pathway to stop producing its product when the product concentration reaches an optimal level, much as a thermostat turns off a heater when a room becomes warm enough. In feedback inhibition, the activity of an enzyme near the beginning of a metabolic pathway is inhibited by the end product of that pathway. The end product acts as an allosteric inhibitor molecule.

In the metabolic pathway illustrated in Figure 6-14, a series of reactions, each catalyzed by a different enzyme, converts one amino acid to another. As levels of the end product amino acid increase, it increasingly encounters and binds to the allosteric regulatory site on an enzyme early in the pathway, inhibiting it. Thus, when enough of the end product is present, the pathway is slowed or halted. Another allosteric inhibitor is ATP, which inhibits enzymes in metabolic pathways that lead to ATP synthesis. When a cell has all the ATP it needs, there is enough ATP around to block ATP production. As ATP is used up, the pathways that produce it become active again.

Poisons, Drugs, and Environmental Conditions Influence Enzyme Activity

Poisons and drugs that act on enzymes usually inhibit them, either competitively or noncompetitively. Environmental

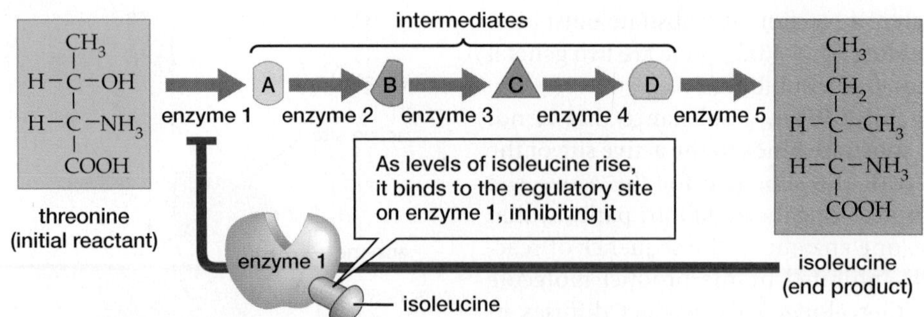

▲ **FIGURE 6-14 Allosteric regulation of an enzyme by feedback inhibition** Here, we illustrate a metabolic pathway that converts one amino acid into another by way of a series of intermediate molecules (colored shapes), each acted upon by a different enzyme (arrows). In this example, the first enzyme in the pathway that converts threonine into isoleucine is inhibited by high concentrations of isoleucine, which acts as an allosteric inhibitor molecule. If a cell lacks isoleucine, the reactions proceed. As isoleucine builds up, it inhibits enzyme 1, blocking the pathway. When concentrations of isoleucine drop and fewer isoleucine molecules are available to inhibit the enzyme, it becomes active again, and isoleucine production resumes.

conditions can denature enzymes, distorting the three-dimensional structure that is crucial for their function.

Some Poisons and Drugs Compete with the Substrate for the Active Site of the Enzyme

Some poisons, such as methanol (a toxic alcohol used as a solvent and antifreeze), act as competitive inhibitors of enzymes. Methanol competes for the active site of the enzyme alcohol dehydrogenase, whose normal substrate is ethanol (found in alcoholic beverages). Alcohol dehydrogenase can break down methanol, but in the process it produces formaldehyde, which can cause blindness. Taking advantage of competitive inhibition, doctors administer ethanol to victims of methanol poisoning. By competing with methanol for the active site of alcohol dehydrogenase, ethanol blocks formaldehyde production.

Some drugs work because they act as competitive inhibitors of enzymes. For example, ibuprofen (Advil™) acts as a competitive inhibitor of an enzyme that catalyzes the synthesis of molecules that contribute to swelling, pain, and fever. Some anticancer drugs are competitive inhibitors of enzymes. Rapidly dividing cancer cells are constantly synthesizing new strands of DNA. Some anticancer drugs resemble the subunits that comprise DNA. These drugs compete with the normal subunits, tricking enzymes into building defective DNA, which in turn prevents the cancer cells from proliferating. Unfortunately, these drugs also interfere with the growth of other rapidly dividing cells, including those in hair follicles and lining the digestive tract. This explains why hair loss and nausea are side effects of some types of cancer chemotherapy drugs.

Some Poisons and Drugs Bind Permanently to the Enzyme

Some poisons and drugs bind irreversibly to enzymes. These irreversible inhibitors may permanently block the enzyme's active site, or they may attach to another part of the enzyme, noncompetitively changing the enzyme's shape or charge so that it can no longer bind its substrates properly.

For example, some nerve gases and insecticides permanently block the active site of acetylcholinesterase, an enzyme that breaks down acetylcholine (a substance that nerve cells release to activate muscles). This allows acetylcholine to build up and overstimulate muscles, causing paralysis. Death ensues because victims are unable to breathe. Penicillin kills bacteria because it is an irreversible competitive inhibitor of an enzyme that helps bacteria produce their cell walls, without which they will burst. Penicillin is harmless to animal cells, which lack cell walls.

Other poisons—including arsenic, mercury, and lead—are noncompetitive inhibitors that bind permanently to other parts of various enzymes, inactivating them.

The Activity of Enzymes Is Influenced by the Environment

The complex three-dimensional structures of enzymes are sensitive to environmental conditions. You may recall from Chapter 3 that much of the three-dimensional structure of proteins is produced by hydrogen bonds between partially charged amino acids. These bonds only occur within a narrow range of chemical and physical conditions, including the proper pH, temperature, and salt concentration. Thus, most enzymes have a very narrow range of conditions in which they function optimally. When conditions fall outside this range, the enzyme becomes **denatured,** meaning that it loses the exact three-dimensional structure required for it to function properly.

In humans, cellular enzymes generally work best at a pH around 7.4, the level maintained in and around our cells (**Fig. 6-15a**). For these enzymes, an acid pH alters the charges on amino acids by adding hydrogen ions to them, which in turn will change the enzyme's shape and compromise its ability to function. Stomach acid kills many bacteria by denaturing their enzymes. Enzymes that operate in the human digestive tract, however, may function outside of the pH range maintained within cells. The protein-digesting enzyme pepsin, for example, requires the acidic conditions of the stomach (pH around 2). In contrast, the protein-digesting enzyme trypsin, found in the small intestine where alkaline conditions prevail, works best at a pH close to 8 (see Fig. 6-15a).

In addition to pH, temperature affects the rate of enzyme-catalyzed reactions, which are slowed by lower temperatures and accelerated by moderately higher temperatures. This is because the rate of movement of molecules determines how likely they are to encounter the active site of an enzyme (**Fig. 6-15b**). Cooling the body can drastically slow human metabolic reactions. In one actual case, a young boy who fell through the ice on a lake was rescued and survived unharmed after 20 minutes under water. Although at normal body temperature the brain dies after about 4 minutes without oxygen, the boy's body temperature and metabolic rate were lowered by the icy water, drastically reducing his need for oxygen.

In contrast, when temperatures rise too high, the hydrogen bonds that regulate protein shape may be broken apart by the excessive molecular motion, denaturing the protein. Think of the protein in egg white, and how its appearance and texture are completely altered by cooking. Far lower temperatures than those required to fry an egg can still be too hot to allow enzymes to function properly. Excessive heat may be fatal; every summer, many children in the United States die from heat stroke when left unattended in overheated cars.

Bacteria and fungi, which exist in nearly all our food, can cause it to spoil. Food remains fresh in the refrigerator or freezer because cooling slows the enzyme-catalyzed reactions upon which these microorganisms rely to grow and reproduce. Before the advent of refrigeration, meat was commonly preserved by using concentrated salt solutions (think of bacon or salt pork), which kill most bacteria. Salts dissociate into ions, which form bonds with amino acids in enzyme proteins. Too much (or too little) salt interferes with the three-dimensional structure of enzymes, destroying their activity. Dill pickles are very well preserved in a vinegar-salt solution, which combines both highly salty and acidic conditions. Organisms that live in saline environments, as you might predict, have enzymes whose configuration depends on a relatively high concentration of salt ions.

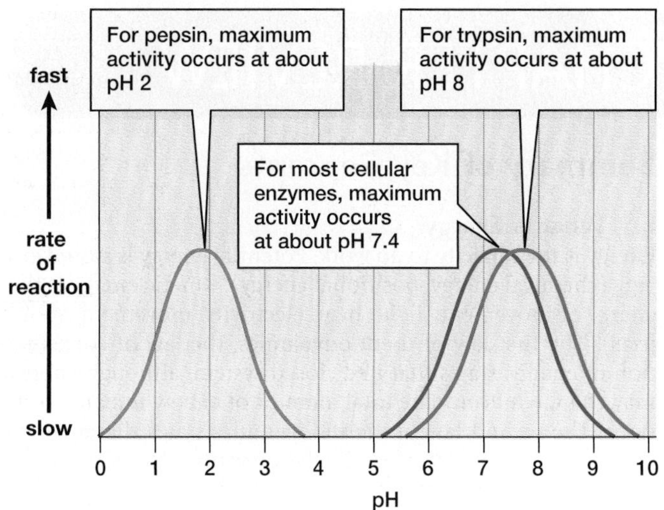

(a) Effect of pH on enzyme activity

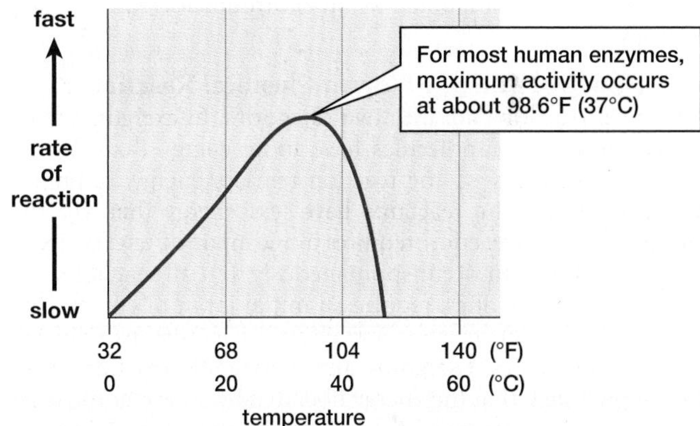

(b) Effect of temperature on enzyme activity

▲ **FIGURE 6-15 Human enzymes function best within narrow ranges of pH and temperature (a)** The digestive enzyme pepsin, released into the stomach, works best at an acidic pH. Trypsin, released into the small intestine, works best at a basic pH. Most enzymes from cells, however, work best at the precisely regulated pH found in the extracellular fluid and cytoplasm (about 7.4). **(b)** The maximum activity of most human enzymes occurs at human body temperature.

Case Study revisited
Energy Unleashed

Needless to say, during the course of a 26-mile race, a marathoner burns a great deal of glucose to provide enough ATP to power muscles through roughly 34,000 running steps. People store glucose molecules linked together in long branched chains of glycogen, primarily in the muscles and liver. Adults typically store about 3.5 ounces (100 grams) of glycogen in the liver, and another 9.9 ounces (280 grams) in muscles. Trained distance athletes can increase liver glycogen storage capacity by more than 50%, and can more than double the glycogen storage capacity of their muscles.

This is important because during a marathon, a runner depletes essentially all of his or her body's stored glycogen. At this point, often about 90 minutes into the race, the runner may experience extreme muscle fatigue, loss of motivation, and occasionally even hallucinations, because both muscles and brain are starving for energy. Runners describe this sensation as "hitting the wall" or "bonking." To store the greatest possible amount of glycogen, endurance athletes practice "carbo-loading," consuming large quantities of carbohydrates (starches and sugars) during the three days preceding the race. By packing their livers and muscles with glycogen before the race, and by consuming energy drinks during the race, some runners manage to cross the finish line before they hit the wall.

Consider This

When a runner's body temperature begins to rise, the body activates several mechanisms, including sweating and circulating more blood to the skin. Compare this response to overheating with feedback inhibition in enzymes.

CHAPTER REVIEW

Summary of Key Concepts

6.1 What Is Energy?

Energy is the capacity to do work. Potential energy is stored energy (chemical energy, positional energy). Kinetic energy is the energy of movement (light, heat, electricity, movement of objects). The first law of thermodynamics, the law of conservation of energy, states that in a closed system, although energy may change in form, the total amount of energy remains constant. The second law of thermodynamics states that any use of energy causes a decrease in the quantity of useful energy and an increase in entropy (disorder and less-useful energy). The highly organized, low-entropy systems that characterize life do not violate the second law of thermodynamics, because they are achieved through a continuous influx of usable energy from the sun, accompanied by an enormous increase in solar entropy.

6.2 How Does Energy Flow in Chemical Reactions?

Chemical reactions fall into two categories. In exergonic reactions, the reactant molecules have more energy than do the product molecules, so the reaction releases energy. In endergonic reactions, the reactants have less energy than do the products, so the reaction requires a net input of energy. Exergonic reactions can occur spontaneously, but all reactions, including exergonic ones, require an initial input of energy (the activation energy) to overcome electrical repulsions between reactant molecules. Exergonic and endergonic reactions may be coupled such that the energy liberated by an exergonic reaction drives the endergonic reaction. Organisms couple exergonic reactions, such as capturing light or breaking down sugar, with endergonic reactions, such as synthesizing organic molecules.

6.3 How Is Energy Transported Within Cells?

Energy released by chemical reactions within a cell is captured and transported within the cell by unstable energy-carrier molecules, such as ATP and the electron carriers NADH and $FADH_2$. These molecules are the major means by which cells couple exergonic and endergonic reactions occurring at different places in the cell.

6.4 How Do Enzymes Promote Biochemical Reactions?

Cells control their metabolic reactions by regulating the synthesis and the use of enzyme proteins, which are biological catalysts that help overcome activation energy. High activation energies slow many reactions, even exergonic ones, to an imperceptible rate under normal environmental conditions. Catalysts lower the activation energy and thereby speed up chemical reactions without being permanently changed themselves. Organisms synthesize enzyme catalysts that promote one or a few specific reactions. The reactants temporarily bind to the active site of the enzyme, making it easier to form the new chemical bonds of the products and, thus, reducing activation energy. Enzymes also allow the breakdown of energy-rich molecules such as glucose in a series of small steps so that energy is released gradually and can be captured in ATP for use in endergonic reactions.

6.5 How Do Cells Regulate Their Metabolic Reactions?

Enzyme action is regulated in several ways; these include altering the rate of enzyme synthesis, activating previously inactive enzymes, competitive and noncompetitive inhibition, and allosteric regulation, which includes feedback inhibition. Many poisons act as enzyme inhibitors. These include methanol, some nerve gases, and insecticides. Ibuprofen and cancer chemotherapy drugs also act as competitive enzyme inhibitors. Environmental conditions—including pH, salt concentration, and temperature—can promote or inhibit enzyme function by altering the enzyme's three-dimensional structure.

Key Terms

activation energy *101*
active site *104*
adenosine diphosphate (ADP) *102*
adenosine triphosphate (ATP) *101*
allosteric regulation *107*
catalyst *103*
chemical energy *98*
chemical reaction *100*
closed system *99*
coenzyme *104*
competitive inhibition *107*
coupled reaction *102*
denature *108*
electron carrier *102*
endergonic *100*
energy *98*
energy-carrier molecule *101*
entropy *99*
enzyme *103*
exergonic *100*
feedback inhibition *107*
first law of thermodynamics *99*
kinetic energy *98*
lactose intolerance *106*
law of conservation of energy *99*
laws of thermodynamics *99*
metabolic pathway *105*
metabolism *105*
noncompetitive inhibition *107*
phenylketonuria (PKU) *106*
potential energy *98*
product *100*
reactant *100*
second law of thermodynamics *99*
substrate *104*
work *98*

Thinking Through the Concepts

Fill-in-the-Blank

1. According to the first law of thermodynamics, energy can neither be __________ nor __________. Energy occurs in two major forms: __________, the energy of movement, and __________, or stored energy.
2. According to the second law of thermodynamics, when energy changes forms it tends to be converted from __________ useful to __________ useful forms. This leads to the conclusion that matter spontaneously tends to become less __________. This tendency is called __________.
3. The energy needed to start any chemical reaction is called __________ energy. This energy is required to force the __________ of the reactants together. This energy is usually supplied by __________.
4. Once started, some reactions release energy and are called __________ reactions. Others require a net input of energy and are called __________ reactions. Which type will continue spontaneously once it starts? __________ Which type allows the formation of complex biological molecules from simpler molecules (for example, proteins from amino acids)? __________ When the energy released from one reaction provides the energy required by another, the two reactions are described as __________ reactions.

5. The abbreviation "ATP" stands for __________. This molecule is the principal __________ molecule in living cells. The molecule is synthesized by cells from __________ and __________. This synthesis requires __________, which is then stored in the molecule of ATP.
6. Enzymes are (type of biological molecule) __________. Enzymes promote reactions in cells by lowering the __________. Each enzyme possesses a specialized region called a(n) __________ into which the reactant molecules fit. Each of these specialized regions has a distinctive __________ and a distinctive distribution of __________ that make it quite specific for its substrate molecules.

Review Questions

1. Explain why organisms do not violate the second law of thermodynamics. What is the ultimate energy source for most forms of life on Earth?
2. Define *metabolism*, and explain how reactions can be coupled to one another.
3. What is activation energy? How do catalysts affect activation energy? How do catalysts change the rate of reactions?
4. Describe some exergonic and endergonic reactions that occur regularly in plants and animals.
5. Describe the structure and function of enzymes. How is enzyme activity regulated?

Applying the Concepts

1. One of your nerdiest friends walks in while you are vacuuming. Trying to impress her, you casually mention that you are infusing energy into your room to create a lower-entropy state, with the energy coming from electricity. She comments that, ultimately, you are taking advantage of the increase in entropy of the sun to clean your room. What is she talking about? (Look for clues in Chapter 7.)
2. As you learned in Chapter 3, the subunits of virtually all organic molecules are joined by condensation reactions and can be broken apart by hydrolysis reactions. Why, then, does your digestive system produce separate enzymes to digest proteins, fats, and carbohydrates?
3. Suppose someone tried to refute evolution using the following argument: "According to evolutionary theory, organisms have increased in complexity through time. However, the increase in complexity contradicts the second law of thermodynamics. Therefore, evolution is impossible." Is this a valid statement?
4. When a brown bear eats a salmon, does the bear acquire all the energy contained in the body of the fish? Why or why not? What implications do you think this answer would have for the relative abundance (by weight) of predators and their prey?

MB® *Go to www.masteringbiology.com for practice quizzes, activities, eText, videos, current events, and more.*

Capturing Solar Energy: Photosynthesis

Chapter 7

▲ A giant meteorite may have ended the reign of *Tyrannosaurus* and the other dinosaurs.

Case Study

Did the Dinosaurs Die from Lack of Sunlight?

ABOUT 65 MILLION YEARS AGO, the Cretaceous period came to a violent end, and life on Earth suffered a catastrophic blow. Within a short time, most of the species on the planet became extinct. This devastating mass extinction eliminated more than 70% of the species then existing, including the dinosaurs. *Triceratops*, *Tyrannosaurus*, and all the other dinosaur species disappeared forever. Land and sea were nearly emptied of life, and it would be many millions of years before new species arose to take the places of those that had disappeared.

Most scientists believe that this devastation began when a gigantic meteorite, roughly six miles in diameter, entered the atmosphere and crashed into Earth. As the meteorite plowed into the ocean at the tip of the Yucatán Peninsula in southeastern Mexico, it dug a crater a mile deep and 120 miles long. Any organism in the immediate area was, of course, killed by the blast wave from the impact. This kind of direct destruction, however, must have been limited to a relatively small area. How, then, did the meteorite impact eliminate thousands of species all over the globe? In all likelihood, the most serious damage was done not by the meteorite itself, but by the lingering effects of its sudden arrival. In particular, the meteorite's most damaging long-term effect was disruption of the most important chemical reaction on Earth: photosynthesis.

What exactly does photosynthesis do? What makes it so important that interrupting it may have brought down the mighty dinosaurs? To find out, read on.

At a Glance

7.1 WHAT IS PHOTOSYNTHESIS?

All cells require energy to live, but as you learned in Chapter 6, the first law of thermodynamics tells us that energy cannot be created. Therefore, life relies on energy from outside sources. For nearly all forms of life on Earth, energy is derived from sunlight, either directly or indirectly. The only organisms capable of directly trapping this abundant energy source are those that perform **photosynthesis,** the process by which solar energy is trapped and stored as chemical energy in the bonds of organic molecules such as sugar. The evolution of photosynthesis made life as we know it possible. This amazing process provides not only fuel for life, but the oxygen required to "burn" this fuel, as we will describe in Chapter 8. Photosynthesis occurs in plants, photosynthetic protists, and certain types of bacteria. Here, we will concentrate on the most familiar of these: land plants.

Leaves and Chloroplasts Are Adaptations for Photosynthesis

Photosynthesis in plants takes place within **chloroplasts,** most of which are contained within leaf cells. The leaves of most land plants are only a few cells thick, a structure that is elegantly adapted to the demands of photosynthesis (**Fig. 7-1**). The flattened shape of leaves exposes a large surface area to the sun, and their thinness ensures that sunlight can penetrate to reach the light-trapping chloroplasts inside. Both the upper and lower surfaces of a leaf consist of a layer of transparent cells, the **epidermis.** The outer surface of both epidermal layers is covered by the **cuticle,** a transparent, waxy, waterproof covering that reduces the evaporation of water from the leaf.

A leaf obtains the CO_2 necessary for photosynthesis from the air, through adjustable pores in the epidermis called **stomata** (singular, **stoma;** Greek for "mouth"; **Fig. 7-2** on page 115). Inside the leaf are layers of cells collectively called **mesophyll** (which means "middle of the leaf"). The mesophyll cells contain most of a leaf's chloroplasts and, consequently, photosynthesis occurs principally in these cells. Vascular bundles, which form veins in the leaf, (**Fig. 7-1b**), supply water and minerals to the mesophyll cells and carry the sugars they produce to other parts of the plant. Cells that surround these bundles are called **bundle sheath cells,** and in most plants, these lack chloroplasts.

A single mesophyll cell often has 40 to 50 chloroplasts (**Fig. 7-1c**), which are so small (about 5 micrometers in diameter) that 2,500 of them lined up would span an average thumbnail. As described in Chapter 4, chloroplasts are organelles that consist of a double outer membrane enclosing a semifluid medium, the **stroma.** Embedded in the stroma are disk-shaped, interconnected membranous sacs called **thylakoids** (**Fig. 7-1d**). Each of these sacs encloses a fluid-filled region called the thylakoid space. The chemical reactions of photosynthesis that depend on light (light reactions) occur in and adjacent to the membranes of the thylakoids. The reactions of the Calvin cycle that capture carbon from CO_2 and produce sugar occur in the surrounding stroma.

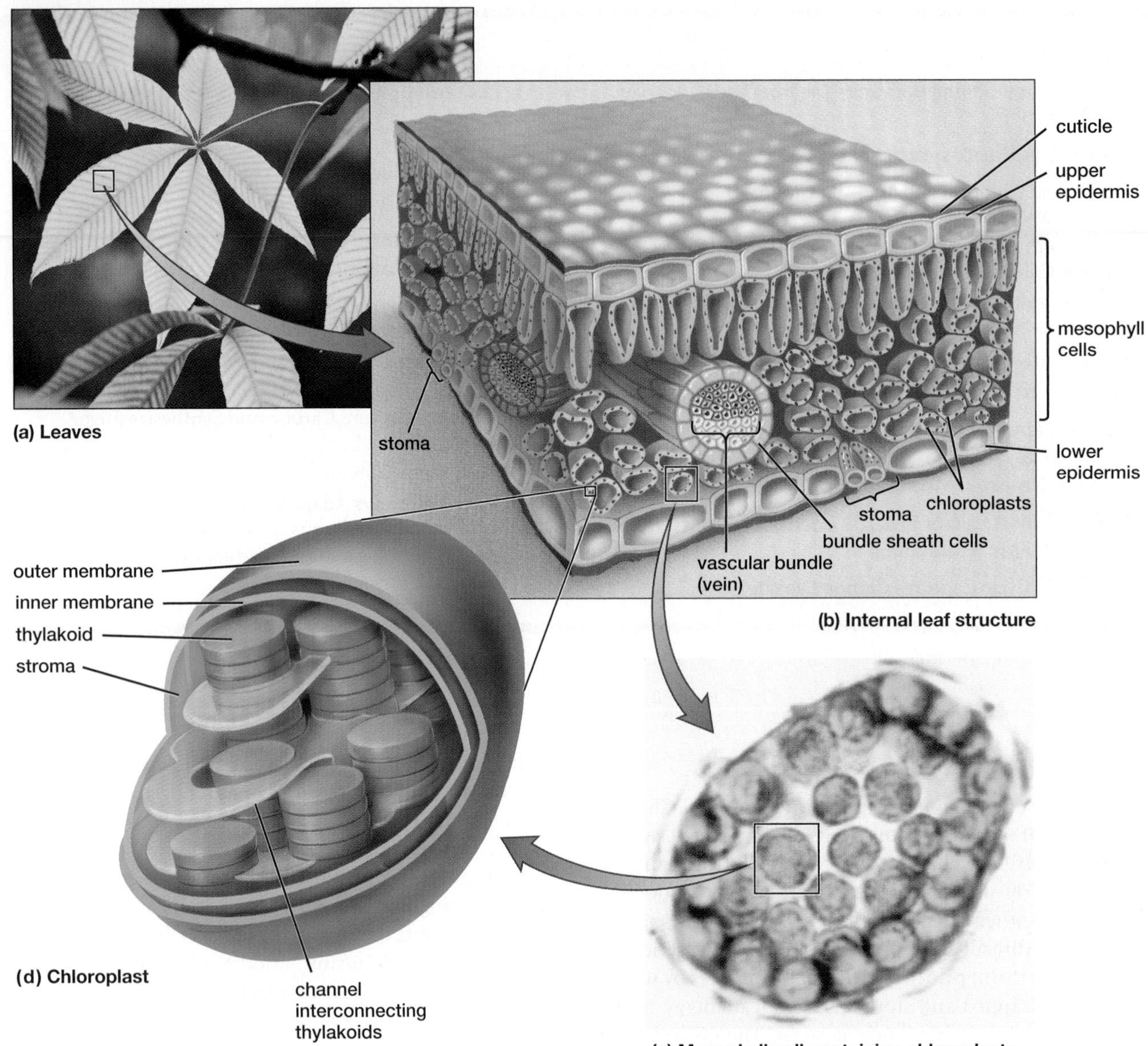

▲ **FIGURE 7-1 An overview of photosynthetic structures (a)** In land plants, photosynthesis occurs primarily in the leaves. **(b)** A section of a leaf, showing mesophyll cells, where chloroplasts are concentrated, and the waterproof cuticle that coats the leaf on both surfaces. **(c)** A light micrograph of a single mesophyll cell, packed with chloroplasts. **(d)** A single chloroplast, showing the stroma and thylakoids where photosynthesis occurs.

Photosynthesis Consists of the Light Reactions and the Calvin Cycle

Starting with the simple molecules of carbon dioxide (CO_2) and water (H_2O), photosynthesis converts the energy of sunlight into chemical energy stored in the bonds of glucose ($C_6H_{12}O_6$) and releases oxygen (O_2) as a by-product (**Fig. 7-3**). The simplest overall chemical reaction for photosynthesis is:

$$6\ CO_2 + 6\ H_2O + \text{light energy} \rightarrow C_6H_{12}O_6 + 6\ O_2$$

This simple equation obscures the fact that photosynthesis actually involves dozens of individual reactions catalyzed by dozens of enzymes. These reactions occur in two distinct stages: the light reactions and the Calvin cycle. Each stage occurs within different regions of the chloroplast, but the two processes are connected through an important link: energy-carrier molecules.

In the **light reactions,** chlorophyll and other molecules embedded in the membranes of the chloroplast thylakoids capture sunlight energy and convert some of it into chemical energy stored in the energy-carrier molecules ATP (adenosine triphosphate) and NADPH (nicotinamide adenine dinucleotide phosphate). Water is split apart, and oxygen gas is released as a by-product. In the reactions of the **Calvin cycle,** enzymes in the fluid stroma located outside the thylakoids use CO_2 from the atmosphere and chemical energy from the energy-carrier

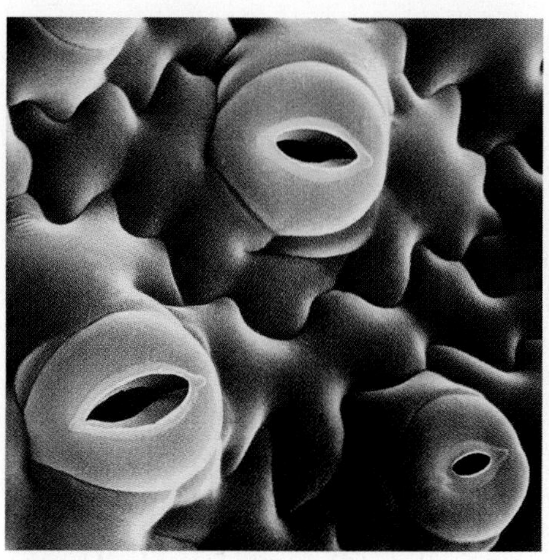

(a) Stomata open

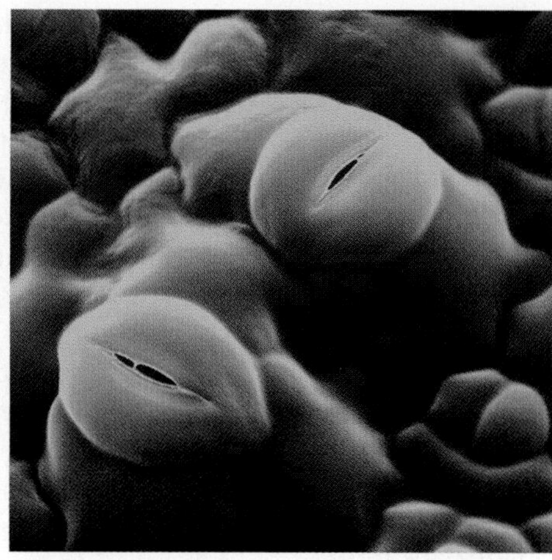

(b) Stomata closed

◀ **FIGURE 7-2 Stomata (a)** Open stomata allow CO_2 to diffuse in and oxygen to diffuse out. **(b)** Closed stomata reduce water loss by evaporation, but prevent CO_2 from entering and oxygen from leaving.

molecules to drive the synthesis of a three-carbon sugar that will be used to make glucose. Figure 7-3 shows the relationship between the light reactions and the Calvin cycle, illustrating the interdependence of the two processes and placing each in its appropriate location within the chloroplast. Simply put, the "photo" part of photosynthesis refers to the capture of light energy by the light reactions in the thylakoid membranes. These reactions use light energy to "charge up" the energy-carrier molecules ADP (adenosine diphosphate) and $NADP^+$ (the energy-depleted form of NADPH) to form ATP and NADPH. The "synthesis" part of photosynthesis refers to the Calvin cycle, which captures carbon and uses it to synthesize sugar, powered by energy provided by ATP and NADPH. The depleted carriers ADP and $NADP^+$ are then recharged by the light reactions into ATP and NADPH, which will fuel the synthesis of more sugar molecules.

Now that you have an overview of photosynthesis, let's look at each stage in greater detail.

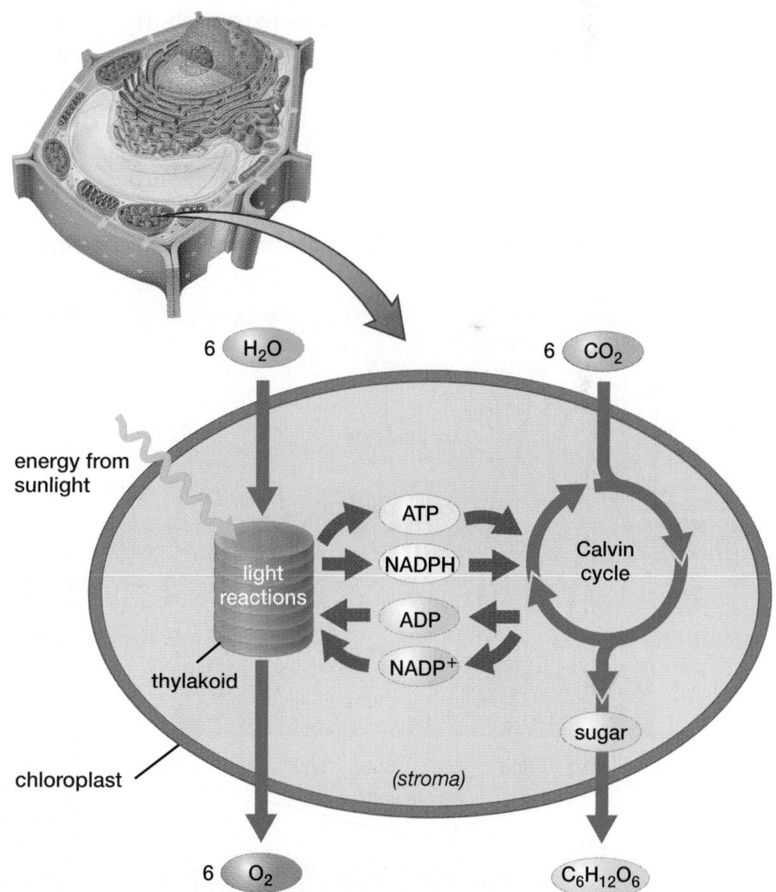

▲ **FIGURE 7-3 An overview of the relationship between the light reactions and the Calvin cycle** The simple molecules that provide the raw ingredients for photosynthesis (H_2O and CO_2) enter at different stages of the process and are used in different parts of the chloroplast. You can also see here that the O_2 liberated by photosynthesis is derived from H_2O, while the carbon used in the synthesis of sugar is obtained from CO_2. In later figures in this chapter, you will see smaller versions of this figure, indicating where the specific processes are occurring.

Case Study continued

Did the Dinosaurs Die from Lack of Sunlight?

More than 2 billion years ago, some bacterial (prokaryotic) cells, through chance mutations in their genetic makeup, acquired the ability to harness the energy of sunlight. Exploiting this abundant energy source, early photosynthetic cells filled the seas. As their numbers increased, oxygen began to accumulate in the atmosphere. Later, plants evolved, made the transition to land, and eventually grew in luxuriant profusion. By the time of the dinosaurs, plants had become abundant and provided enough food to support plant-eating giants, such as the 85-foot-long, 35-ton *Apatosaurus*, on which *Tyrannosaurus* may have preyed.

7.2 LIGHT REACTIONS: HOW IS LIGHT ENERGY CONVERTED TO CHEMICAL ENERGY?

The light reactions capture the energy of sunlight, storing it as chemical energy in two different energy-carrier molecules: ATP and NADPH. The molecules that make these reactions possible, including light-capturing pigments and enzymes, are anchored in a precise array within the membranes of the thylakoid. As you read this section, notice how the thylakoid membranes and their enclosed spaces support these crucial reactions.

Light Is Captured by Pigments in Chloroplasts

The sun emits energy in a broad spectrum of electromagnetic radiation. The **electromagnetic spectrum** ranges from short-wavelength gamma rays, through ultraviolet, visible, and infrared light, to very long-wavelength radio waves (**Fig. 7-4**). Light and other electromagnetic waves are composed of individual packets of energy called **photons.** The energy of a photon corresponds to its wavelength: short-wavelength photons, such as gamma and X-rays, are very energetic, whereas longer-wavelength photons such as radio waves carry lower energies. Visible light consists of wavelengths with energies that are strong enough to alter biological **pigment molecules** (light-absorbing molecules) such as chlorophyll, but weak enough not to break the bonds of crucial molecules such as DNA. It is no coincidence that these wavelengths, with just the right amount of energy, not only power photosynthesis, but also stimulate the pigments in our eyes, allowing us to see.

One of three events occurs when a specific wavelength of light strikes an object such as a leaf; the light may be absorbed (captured), reflected (bounced back), or transmitted (passed through). Light that is absorbed can heat up the object or drive biological processes, such as photosynthesis. Light that is reflected or transmitted is not captured, and thus can reach the eyes of an observer, allowing her to perceive this light as the color of the object .

Chloroplasts contain various pigment molecules that absorb different wavelengths of light. **Chlorophyll *a*,** the key light-capturing pigment molecule in chloroplasts, strongly absorbs violet, blue, and red light, but reflects green, thus giving green leaves their color (see Fig. 7-4). Chloroplasts also contain other molecules, collectively called **accessory pigments,** which absorb additional wavelengths of light energy and transfer them to chlorophyll *a*. Chlorophyll *b* is a slightly different form of green chlorophyll that serves as an accessory pigment, absorbing blue and red-orange wavelengths of light that are missed by chlorophyll *a*, and reflecting yellow-green light. **Carotenoids** are accessory pigments found in all chloroplasts. They absorb blue and green light and appear mostly yellow or orange because they reflect these wavelengths to our eyes (see Fig. 7-4).

Although carotenoids are present in leaves, their color is usually masked by the more abundant green chlorophyll. In the autumn, as leaves begin to die, chlorophyll breaks down before carotenoids do, revealing the bright yellow and orange carotenoid pigments as fall colors. The chlorophyll in the aspen leaves in **Figure 7-5** has broken down and faded, revealing yellow carotenoids. (Red and purple fall leaf colors are not involved in photosynthesis.)

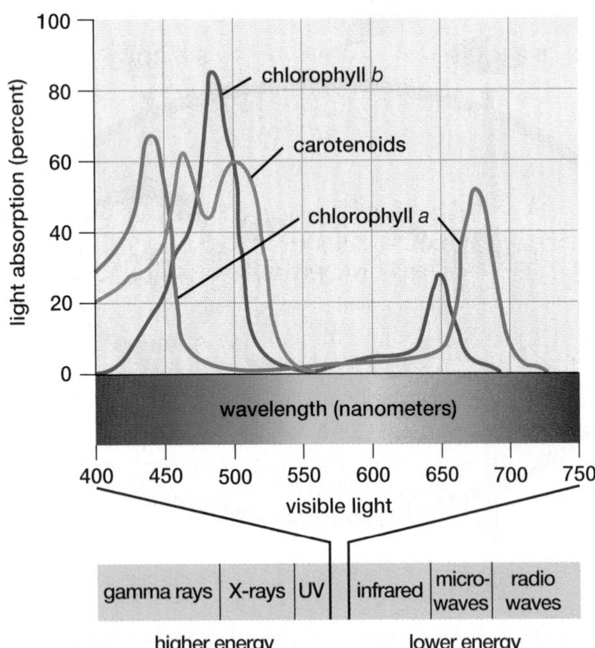

▲ **FIGURE 7-4 Light and chloroplast pigments** Visible light, a small part of the electromagnetic spectrum, consists of wavelengths that correspond to the colors of the rainbow. Chlorophyll *a* and *b* (green and blue curves, respectively) strongly absorb violet, blue, and red light. Carotenoids (orange curve) absorb blue and green wavelengths.

▲ **FIGURE 7-5 Loss of chlorophyll reveals yellow carotenoid pigments** As winter approaches, chlorophyll in these aspen leaves breaks down, revealing yellow carotenoid pigments.

Have you ever wondered 

What Color Plants Might Be on Other Planets?

Biologist Nancy Kiang and her colleagues at NASA have developed hypotheses about alien plant colors. M-type stars, the most abundant type in our galaxy, emit light that is redder and dimmer than that of our sun. Assuming photosynthetic organisms happened to evolve on an Earth-like planet circling such a star, to glean enough energy, they would very possibly require pigments that would absorb light throughout the entire visible spectrum. Reflecting almost none of the light back to our eyes, these alien photosynthesizers would appear black.

You may have heard of the carotenoid beta-carotene. This pigment helps capture light in chloroplasts, and it produces the orange color of vegetables such as carrots. Beta-carotene is converted by animals into vitamin A, which forms the basis of the light-capturing pigment in the eyes of animals, including people. In a beautiful symmetry, the carotenoids that capture light energy in plants are converted into substances that capture light in animals as well.

The Light Reactions Occur in Association with the Thylakoid Membranes

Light energy is captured and converted into chemical energy by the light reactions that occur in and on the thylakoid membranes. These membranes contain many **photosystems,** each consisting of a cluster of chlorophyll and accessory pigment molecules surrounded by various proteins. Two photosystems—photosystem II (PS II) and photosystem I (PS I)—work together during the light reactions. (These photosystems are named in the order in which they were discovered; however, the order in which they proceed within the light reactions is from photosystem II to photosystem I.)

Each type of photosystem has a different electron transport chain located adjacent to it. These **electron transport chains (ETC)** each consist of a series of electron carrier molecules embedded in the thylakoid membrane. In this discussion, we will refer to these as ETC II and ETC I, according to the photosystem with which they are associated. Thus, within the thylakoid membrane, the overall path of electrons is as follows: PS II $\rightarrow$ ETC II $\rightarrow$ PS I $\rightarrow$ ETC I $\rightarrow$ $NADP^+$ (see Fig. 7-7).

You can think of the light reactions as a sort of pinball game. Energy (light) is transferred to a ball (an electron) by spring-driven pistons (chlorophyll molecules). The ball is propelled upward (enters a higher energy level). As the ball bounces back downhill, the energy it releases can be used to turn a wheel (generate ATP) or ring a bell (generate NADPH). With this overall scheme in mind, let's look more closely at the actual sequence of events in the light reactions.

Photosystem II Uses Light Energy to Create a Hydrogen Ion Gradient and Split Water

The light reactions begin when photons of light are absorbed by pigment molecules clustered in photosystem II (**Fig. 7-6** ❶). The energy hops from one pigment molecule to the next until it is funneled into the reaction center (**Fig. 7-6** ❷). The **reaction center** within each photosystem consists of a pair of specialized chlorophyll *a* molecules and a **primary electron acceptor** molecule embedded in a complex of proteins. When it reaches the reaction center, the energy from light boosts an electron from one of the reaction center chlorophylls to the primary electron acceptor, which captures the energized electron (**Fig. 7-6** ❸).

The reaction center of photosystem II must be continuously supplied with electrons to replace the ones that jump out of it when they are energized by light. These replacement electrons come from water (see photosystem II in Fig. 7-6). An enzyme associated with PS II splits water molecules, liberating electrons that will replace those lost by the reaction center chlorophyll molecules. This reaction also releases hydrogen ions that contribute to the H^+ gradient that drives ATP synthesis (see Fig. 7-7 ❷). For every two molecules of water that are split, one molecule of oxygen gas (O_2) is produced.

Once the primary electron acceptor captures the electron, it passes it on to the first molecule of ETC II. The electron then travels from one electron carrier molecule to the next, losing energy as it goes (**Fig. 7-6** ❹). Some of this energy is harnessed to pump hydrogen ions (H^+) across the thylakoid membrane and into the thylakoid space, where they will be used to generate ATP (to be discussed shortly; **Fig. 7-6** ❺ and see Fig. 7-7). Finally, the energy-depleted electron leaves ETC II and enters the reaction center of photosystem I, where it replaces the electron ejected when light strikes photosystem I.

Photosystem I Generates NADPH

Meanwhile, light has also been striking the pigment molecules of PS I (**Fig. 7-6** ❻). As in PS II, light energy is captured by these pigment molecules and is funneled to a chlorophyll *a* molecule in the reaction center. This ejects an energized electron that is picked up by the primary electron acceptor of PS I (**Fig. 7-6** ❼). (The energized electron is immediately replaced by an energy-depleted electron from ETC II.) From the primary electron acceptor of PS I, the energized electron is passed along ETC I (**Fig. 7-6** ❽) until it reaches $NADP^+$. The energy-carrier molecule NADPH is formed when each $NADP^+$ molecule (dissolved in the fluid stroma) picks up two energetic electrons, along with one hydrogen ion (**Fig. 7-6** ❾).

The Hydrogen Ion Gradient Generates ATP by Chemiosmosis

Figure 7-7 shows the movement of electrons through the thylakoid membrane. It also shows how the energy of these electrons is used to create the H^+ gradient that drives ATP synthesis, in a process called **chemiosmosis.** As the energized electron travels along ETC II, some of the energy it liberates is used to pump hydrogen ions into the thylakoid space

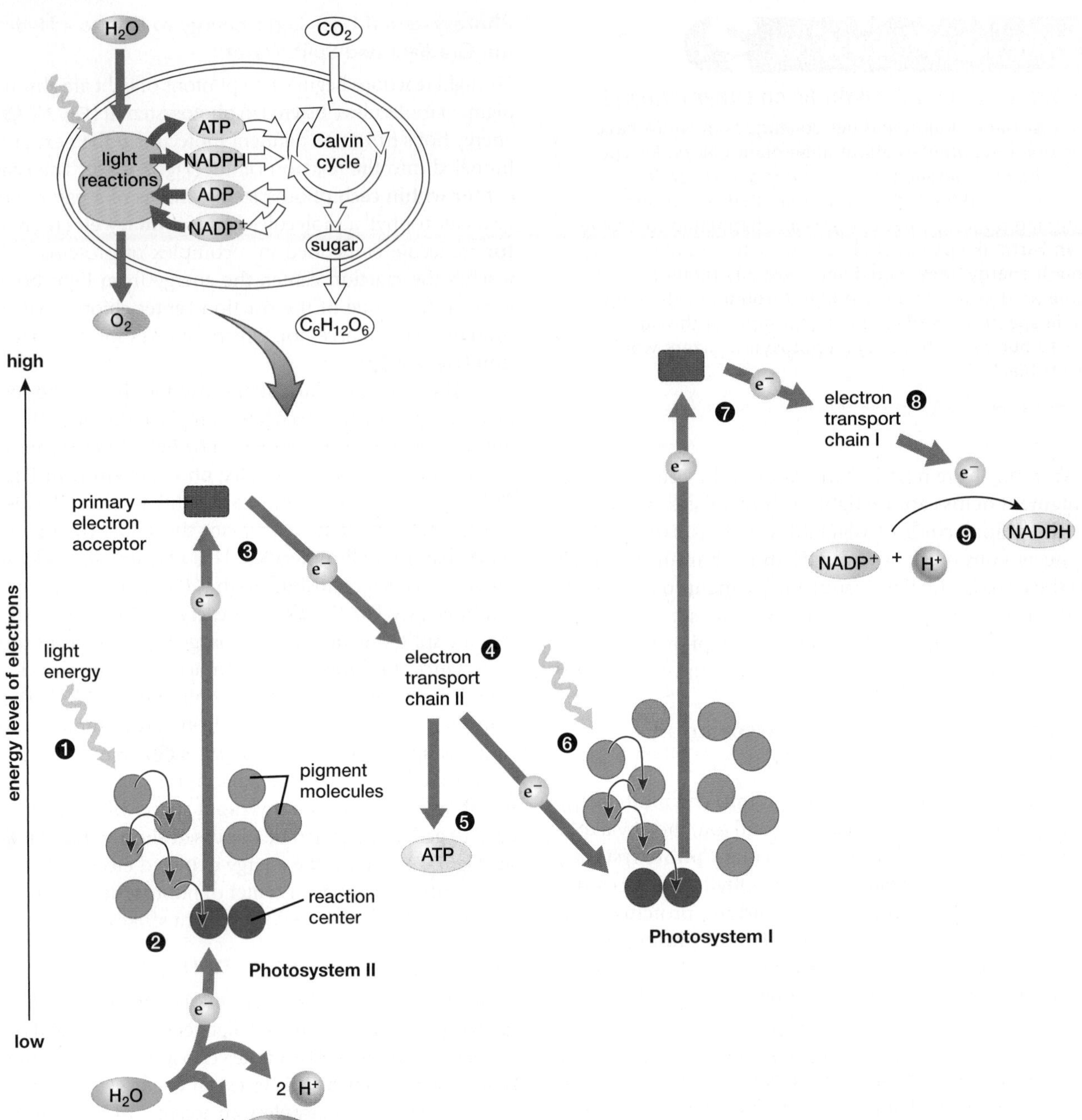

▲ **FIGURE 7-6 Energy transfer and the light reactions of photosynthesis** Light reactions involve a series of molecules and reactions that occur within or adjacent to the thylakoid membrane. ❶ Light is absorbed by photosystem II, and the energy is passed to an electron in one of the reaction center chlorophyll *a* molecules. ❷ The energized electron is ejected from the chlorophyll *a* molecule. ❸ The electron is captured by a primary electron acceptor, also in the reaction center. ❹ The electron is passed through the electron transport chain (ETC II). ❺ As the ETC II transfers the electron along, some of the released energy is used to pump H^+ into the thylakoid space. In a separate process, the resulting H^+ gradient is used to generate ATP. The energy-depleted electron then enters the reaction center of photosystem I, replacing an ejected electron. ❻ Light strikes photosystem I, and the energy is passed to the electron in the reaction center chlorophyll molecules. ❼ The energized electron is ejected from the reaction center and is captured by the primary electron acceptor. ❽ The electron moves down the ETC I. ❾ NADPH is formed when $NADP^+$ in the stroma picks up two energized electrons, along with one H^+.

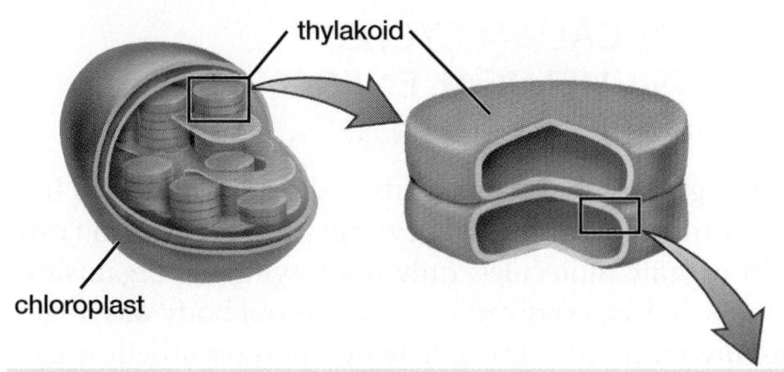

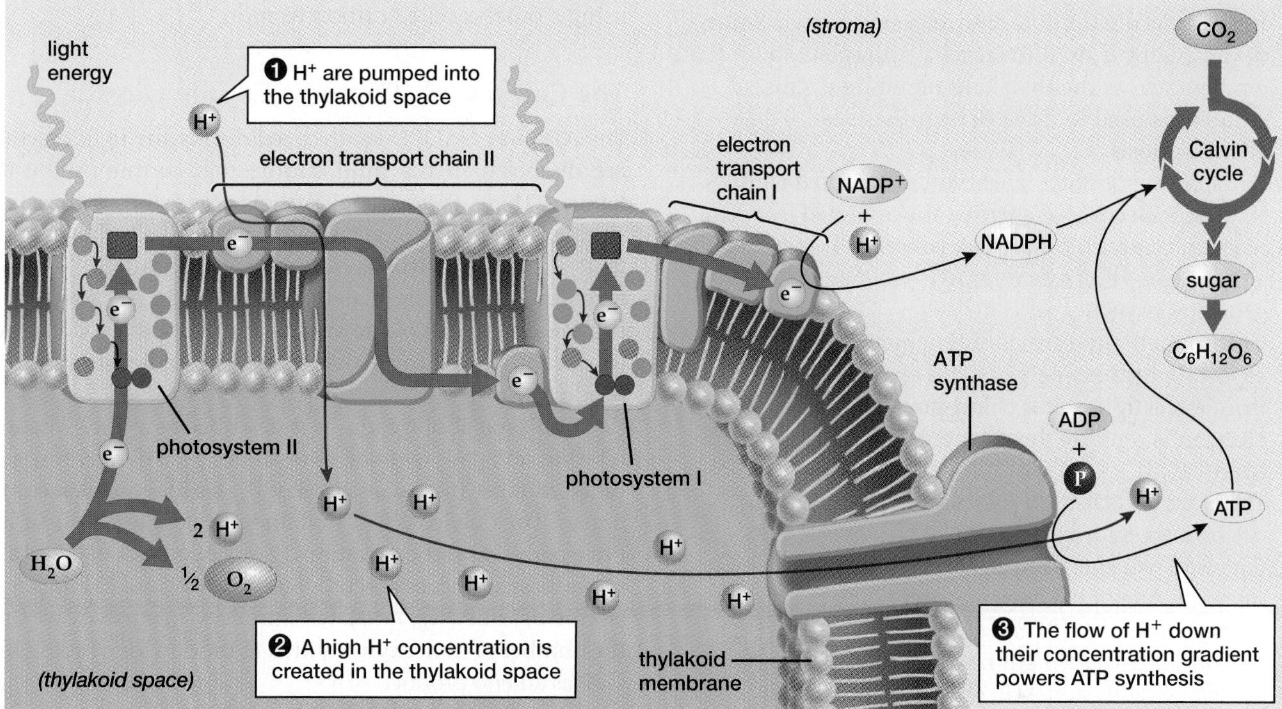

▲ **FIGURE 7-7 Events of the light reactions occur in and near the thylakoid membrane** Here, you can trace the pathway of an energized electron as it travels through the molecules associated with the thylakoid membrane. ❶ Energy released as the electron passes through ETC II is harnessed to pump H^+ across the thylakoid membrane and into the thylakoid space. ❷ A concentration gradient of H^+ is generated as electrons are passed along the ETC II, and their energy is used to pump H^+ across the membrane into the thylakoid space. ❸ During chemiosmosis, H^+ flow down their concentration gradient through an ATP synthase channel, which uses energy from the gradient to generate ATP. About one ATP molecule is synthesized for every three hydrogen ions that pass through the channel.

(**Fig. 7-7** ❶), creating a high concentration of H^+ inside the space (**Fig. 7-7** ❷), and a low concentration in the surrounding stroma. During chemiosmosis, H^+ flow back down their concentration gradient through a special channel that spans the thylakoid membrane. This channel, called **ATP synthase,** generates ATP from ADP and phosphate dissolved in the stroma as the H^+ flow through the channel (**Fig. 7-7** ❸).

How is a gradient of H^+ used to synthesize ATP? Compare the H^+ gradient to water stored behind a dam at a hydroelectric plant (**Fig. 7-8**). The water flows out over turbines, rotating them. The turbines convert the energy of moving water into electrical energy. Hydrogen ions in the thylakoid space (like water stored behind a dam) can move down their gradient into the stroma only through ATP synthase channels. Like turbines generating electricity, ATP synthase captures the energy liberated by the flow of H^+ and uses it to drive ATP synthesis from ADP plus phosphate.

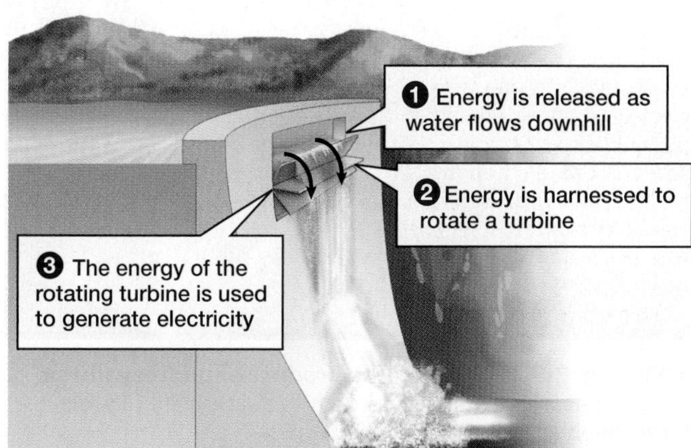

▲ **FIGURE 7-8 Energy stored in a water "gradient" can be used to generate electricity**

Summing Up **Light Reactions**

- Chlorophyll and carotenoid pigments of PS II absorb light that energizes and ejects an electron from a reaction center chlorophyll molecule. The energized electron is caught by the primary electron acceptor molecule.
- The electron is passed from the primary electron acceptor to the adjacent ETC II, where it moves from molecule to molecule, releasing energy. Some of the energy is used to create a hydrogen ion gradient across the thylakoid membrane. This gradient is used to drive ATP synthesis by chemiosmosis.
- The splitting of water by enzymes associated with PS II releases electrons (replacing those ejected from the reaction center chlorophylls), supplies H^+ that enhance the H^+ gradient for ATP production, and generates oxygen.
- Chlorophyll and carotenoid pigments in PS I absorb light that energizes and ejects an electron from a reaction center chlorophyll molecule. The electron is replaced by an energy-depleted electron from ETC II.
- The energized electron passes from the primary electron acceptor into the ETC I.
- For every two energized electrons that exit the ETC I, one molecule of the energy carrier NADPH is formed, from $NADP^+$ and H^+.
- The products of the light reactions are the energy carriers NADPH and ATP; O_2 is released as a by-product (**Fig. 7-9**).

▲ **FIGURE 7-9 Oxygen is a by-product of photosynthesis** Oxygen released by the leaves of this aquatic plant (*Elodea*) forms bubbles in the surrounding water.

QUESTION Would you expect bubbles of oxygen to form in the dark? Explain.

7.3 THE CALVIN CYCLE: HOW IS CHEMICAL ENERGY STORED IN SUGAR MOLECULES?

Although we inhale some carbon dioxide with every breath, and drink it in carbonated beverages, our bodies can't capture it in organic molecules; only photosynthetic organisms can do this. In fact, every carbon atom in your body was originally captured from the atmosphere by a photosynthetic organism using a process called carbon fixation.

The Calvin Cycle Captures Carbon Dioxide

The ATP and NADPH synthesized during the light reactions are dissolved in the fluid stroma that surrounds the thylakoids. There, these energy carriers power the synthesis of a simple, three-carbon sugar (glyceraldehyde-3-phosphate or G3P) from CO_2. This is accomplished through a series of reactions, called the Calvin cycle, which is catalyzed by enzymes dissolved in the stroma. For every three CO_2 molecules captured by the Calvin cycle, one molecule of G3P is produced.

This metabolic pathway is described as a "cycle" because it begins and ends with the same molecule, a five-carbon sugar called ribulose bisphosphate (RuBP). RuBP is recycled continuously. The Calvin cycle is best understood if we divide it into three parts: (1) carbon fixation, (2) the synthesis of G3P (used to synthesize glucose and other molecules), and (3) the regeneration of RuBP (**Fig. 7-10**). For simplicity, in our cycle we show the number of molecules required to produce one molecule of G3P as an end product.

Carbon Fixation During **carbon fixation,** carbon from CO_2 is incorporated or "fixed" into a larger organic molecule. The Calvin cycle uses the enzyme **rubisco** to combine each CO_2 molecule with an RuBP molecule. This produces an unstable six-carbon molecule that immediately splits in half, forming two molecules of PGA (phosphoglyceric acid; a three-carbon molecule). Because carbon fixation generates this three-carbon PGA molecule, the Calvin cycle is often referred to as the **C_3 pathway** (**Fig. 7-10**).

Case Study continued

Did the Dinosaurs Die from Lack of Sunlight?

Any event that reduces light availability—such as the dust, smoke, and ash that would accompany a meteorite collision with Earth—decreases the ability of plants to synthesize glucose, starch, and other high-energy molecules that they use as food. Because the food that plants synthesized for themselves also fed the dinosaurs, a loss of the ability to photosynthesize would spell disaster for both.

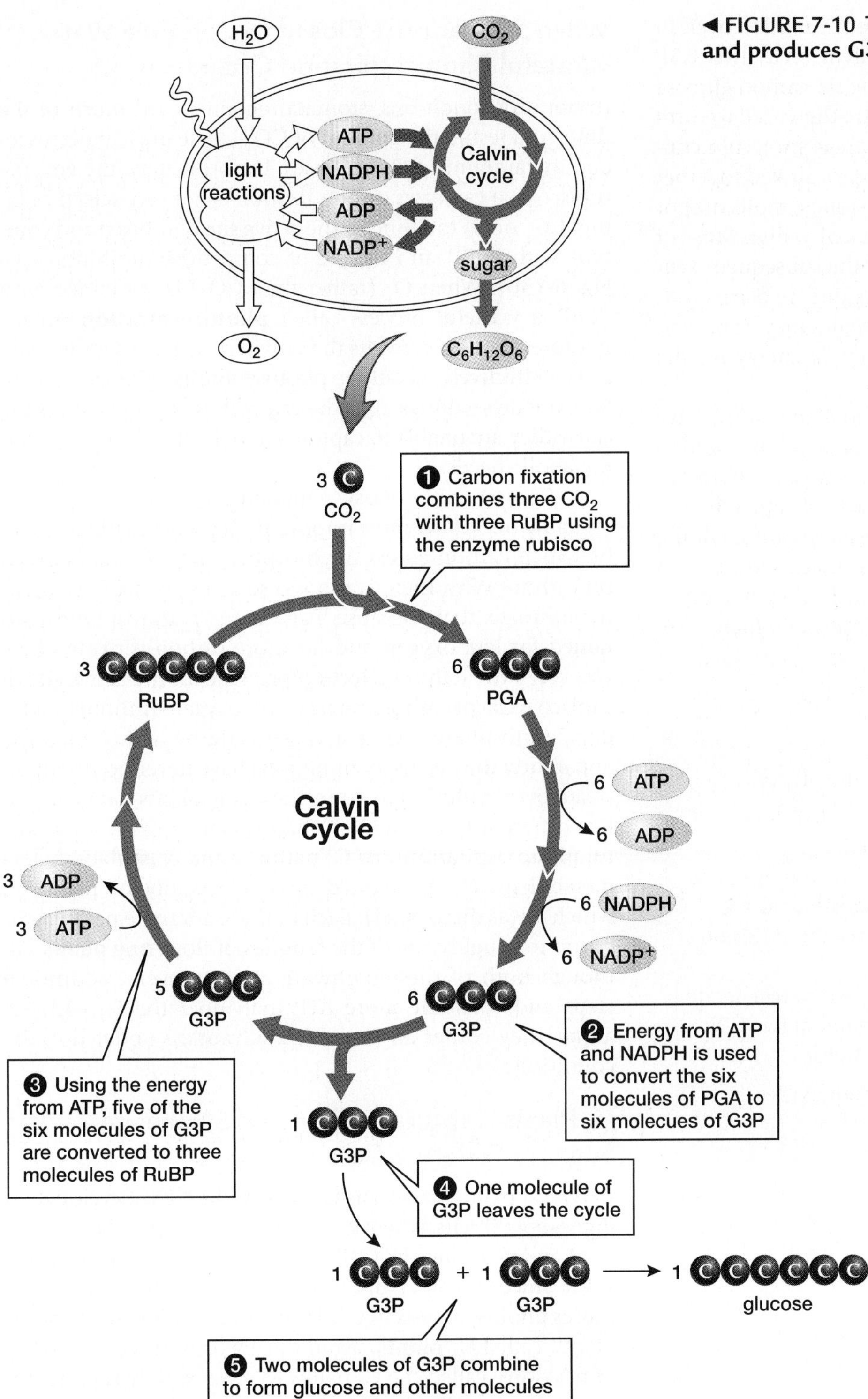

◀ **FIGURE 7-10 The Calvin cycle fixes carbon from CO_2 and produces G3P**

Synthesis of G3P In a series of enzyme-catalyzed reactions, energy donated by ATP and NADPH (generated by the light reactions) is used to convert six PGA molecules into six molecules of the three-carbon sugar G3P (**Fig. 7-10** ❷).

Regeneration of RuBP Through a series of enzyme-catalyzed reactions requiring ATP from the light reactions, five of the six G3P molecules are used to regenerate the RuBP necessary to repeat the cycle (**Fig. 7-10** ❸). The remaining G3P molecule, which is the end product of photosynthesis, exits the cycle (**Fig. 7-10** ❹).

Carbon Fixed During the Calvin Cycle Is Used to Synthesize Sugar

Using simple "carbon accounting," if you start and end one turn of the cycle with three molecules of RuBP and capture three molecules of CO_2, one molecule of G3P is left over. In reactions that occur outside of the Calvin cycle, two G3P molecules can be combined to form one six-carbon glucose molecule (**Fig. 7-10** ❺). Most of these are then used to form sucrose (table sugar; a disaccharide storage molecule consisting of a glucose linked to a fructose), or linked together in long chains to form starch (another storage molecule) or cellulose (a major component of plant cell walls). Most of the synthesis of glucose from G3P and the subsequent synthesis of more complex molecules from glucose occurs outside the chloroplast. Later, glucose molecules may be broken down during cellular respiration to provide energy for the plant cells.

In recent years, countries throughout the world have been attempting to reduce their reliance on fossil fuels and to obtain more of their energy from the products of photosynthesis by converting plant material into oil and ethanol. These "biofuels" have the potential advantage of not adding more carbon dioxide (a "greenhouse gas" that contributes to global warming) to the atmosphere, but do they live up to their promise? We explore this in "Earth Watch: Biofuels—Are Their Benefits Bogus?" on p. 124.

Summing Up **The Calvin Cycle**

The Calvin cycle can be divided into three stages:

- **Carbon fixation:** Three RuBP capture three CO_2, forming six PGA.
- **G3P synthesis:** A series of reactions, driven by energy from ATP and NADPH (from the light reactions), produces six G3P, one of which leaves the cycle and is available to form glucose.
- **RuBP regeneration:** ATP energy is used to regenerate three RuBP molecules from the remaining five G3P molecules, allowing the cycle to continue.

Two G3P molecules produced by the Calvin cycle combine to form glucose.

BioFlix™ Photosynthesis

7.4 WHY DO SOME PLANTS USE ALTERNATE PATHWAYS FOR CARBON FIXATION?

Plant leaf structure represents a compromise between obtaining adequate light and CO_2 and minimizing water loss through evaporation. Most leaves have a large surface area for intercepting light, a waterproof cuticle to reduce evaporation, and adjustable stomata. When there is adequate water, the stomata open, letting in CO_2. If the plant is in danger of drying out, the stomata close (see Fig. 7-2). Closing the stomata reduces evaporation, but prevents CO_2 intake and also limits the ability of the leaf to release O_2, a by-product of photosynthesis.

When Stomata Are Closed to Conserve Water, Wasteful Photorespiration Occurs

In hot, dry conditions, stomata remain closed much of the time. As a result, the amount of CO_2 inside the leaf decreases and the amount of O_2 increases. Unfortunately, the enzyme rubisco that catalyzes carbon fixation is not very selective. Either CO_2 or O_2 can bind to the active site of rubisco and combine with RuBP, an example of competitive inhibition (see Fig. 6-13b). When O_2 (rather than CO_2) is combined with RuBP, a wasteful process called **photorespiration** occurs. Photorespiration prevents the Calvin cycle from synthesizing sugar, effectively derailing photosynthesis. Plants, particularly fragile seedlings, may die under these circumstances because they are unable to capture enough energy to meet their metabolic needs.

Rubisco is the most abundant enzyme on Earth, and arguably one of the most important. It catalyzes the reaction by which carbon enters the biosphere, and all life is based on carbon. Why, then, is rubisco so nonselective? Scientists hypothesize that, because Earth's early atmosphere contained far less oxygen and far more carbon dioxide, there was very little natural selection pressure for the active site of rubisco to favor carbon dioxide over oxygen. Although in today's atmosphere such a change would be highly adaptive, apparently the necessary mutations have never occurred. Instead, over evolutionary time, flowering plants have evolved two different but related mechanisms to circumvent wasteful photorespiration: the C_4 pathway and crassulacean acid metabolism (CAM; named after the family of plants in which it was discovered). Each of these alternate pathways is found in roughly 5% of the families of flowering plants. Although both of these pathways involve several additional steps and consume more ATP than does the C_3 pathway alone, they confer an important advantage under hot, dry conditions.

C_4 Plants Capture Carbon and Synthesize Sugar in Different Cells

In typical plants, known as **C_3 plants,** the Calvin cycle occurs in mesophyll cells, where it both fixes carbon (via the C_3 pathway) and generates the G3P molecules used to synthesize the sugar glucose. Under dry, hot conditions, photorespiration causes photosynthesis in C_3 plants to slow dramatically. Some plants, called **C_4 plants,** avoid this by using a separate series of reactions, called the C_4 pathway, to selectively capture carbon in mesophyll cells. This fixed carbon is then shuttled into bundle sheath cells (see Fig. 7-1) where it enters the Calvin cycle with little competition from oxygen (**Fig. 7-11a**).

How is carbon captured and shuttled? C_3 and C_4 leaves have both structural and biochemical differences. Unlike C_3 plants (whose bundle sheath cells lack chloroplasts), C_4 plants have chloroplasts in both their bundle sheath cells and their mesophyll cells. Another difference is that the mesophyll chloroplasts of C_4 plants lack the enzymes used in the Calvin cycle; only the bundle sheath chloroplasts of C_4 plants contain Calvin cycle enzymes.

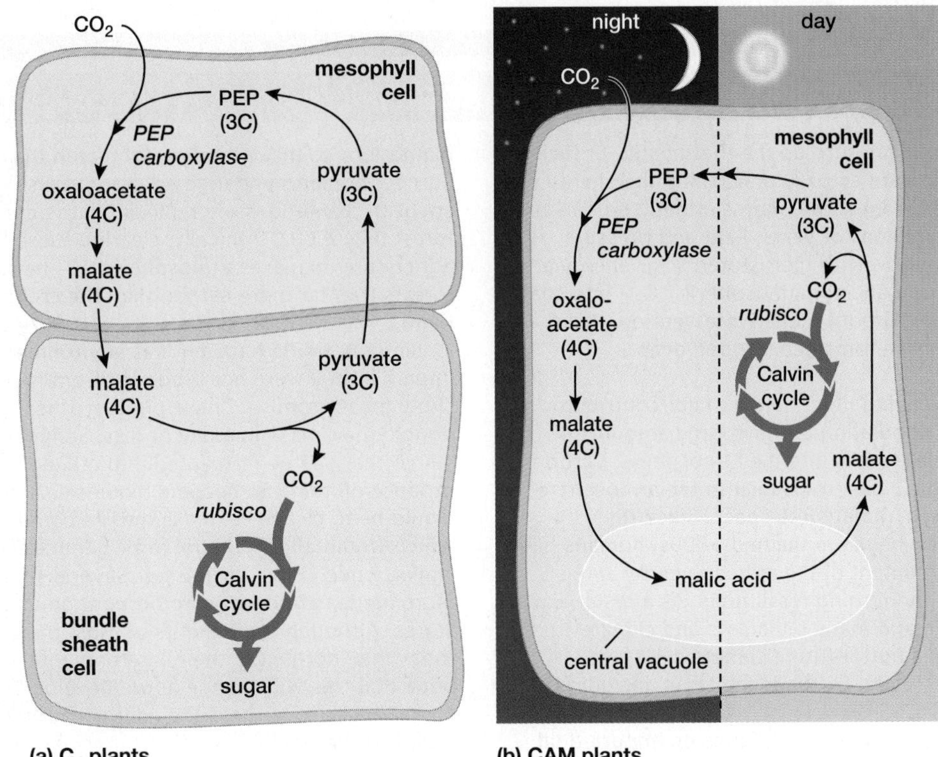

▲ **FIGURE 7-11 The C_4 pathway and the CAM pathway** Some of the reactions illustrated here occur in the cytoplasm, whereas others, including all those of the Calvin cycle, occur in chloroplasts. If you compare parts **(a)** and **(b)**, you will see that both use the C_4 pathway, in which the three-carbon (3C) PEP molecule is combined (by the selective enzyme PEP carboxylase) with a one-carbon CO_2 molecule to become the four-carbon (4C) oxaloacetate. The oxaloacetate molecule is rearranged to generate the four-carbon malate molecule that will release CO_2 to be used in the Calvin cycle. After releasing its carbon as CO_2, the malate is transformed into a three-carbon pyruvate, which is eventually converted back to the three-carbon PEP. By creating a high level of CO_2 in **(a)** the bundle sheath cells of C_4 plants and **(b)** the mesophyll cells of CAM plants (during the day, when stomata are closed), each of these specialized photosynthetic pathways minimizes photorespiration.

QUESTION Why do C_3 plants have an advantage over C_4 plants under conditions that are *not* hot and dry?

In the **C_4 pathway,** mesophyll cells of C_4 plants fix carbon using an enzyme called PEP carboxylase that (unlike rubisco) is highly selective for CO_2 over O_2. PEP carboxylase catalyzes a reaction between CO_2 and a three-carbon molecule called phosphoenolpyruvate (PEP). This produces the four-carbon molecule oxaloacetate, from which the C_4 pathway gets its name. Oxaloacetate is rapidly converted into another four-carbon molecule, malate, which diffuses from the mesophyll cells into bundle sheath cells. The malate effectively acts as a shuttle for CO_2.

In the bundle sheath cells, malate is broken down, forming the three-carbon molecule pyruvate and releasing CO_2. This creates a high CO_2 concentration in the bundle sheath cells (up to 10 times higher than atmospheric CO_2). The high level of CO_2 allows rubisco in the C_3 pathway of the Calvin cycle to fix carbon, and thus generate sugar, with little competition from oxygen. The pyruvate is then actively transported back to the mesophyll cells. Here, more ATP energy is used to convert pyruvate back into PEP, allowing the cycle to continue.

CAM Plants Capture Carbon and Synthesize Sugar at Different Times

Like C_4 plants, CAM plants use the C_4 pathway to selectively capture CO_2 by combining it with PEP to form oxaloacetate. They also convert the oxaloacetate to malate, which breaks down, releasing CO_2 for use in the Calvin cycle to synthesize sugar. But in contrast to C_4 plants, CAM plants do not use different cell types to capture carbon and to synthesize sugar. Instead, they perform both of these activities in the same mesophyll cells, but at different times; carbon fixation occurs at night, and sugar synthesis occurs during the day (**Fig. 7-11b**).

At night, when temperatures are cooler and humidity is higher, the stomata of CAM plants open, allowing CO_2 to diffuse in, where it is captured in mesophyll cells using the C_4 pathway. The resulting oxaloacetate is converted to malate, which is then shuttled into the central vacuole where it is stored as malic acid until daytime. During the day, when stomata are closed to conserve water, the malic acid leaves the vacuole and re-enters the cytoplasm as

Earth Watch

Biofuels—Are Their Benefits Bogus?

When you drive your car, turn up the thermostat, or flick on your desk lamp, you are actually unleashing the energy of prehistoric sunlight trapped by photosynthetic organisms. Over hundreds of millions of years, heat and pressure converted their bodies—with their stored solar energy and carbon captured from ancient atmospheric CO_2—into coal, oil, and natural gas. Without human intervention, these fossil fuels would have remained trapped deep underground.

Global warming is big news, and a major contributor is a growing human population burning large amounts of fossil fuels that release CO_2 into the atmosphere. Carbon dioxide, a greenhouse gas, traps heat in the atmosphere that would otherwise radiate into space. Since the industrial revolution began in the mid-1800s, humans have increased the CO_2 content of the atmosphere by almost 37%, much of this by burning fossil fuels. As a result, Earth is growing warmer, and many biologists and climate scientists fear that a hotter future climate will place extraordinary stresses on Earth's inhabitants, including humans (see pp. 545–549).

To reduce CO_2 emissions and reliance on imported oil, governments throughout the world are promoting the use of biofuels. Bioethanol is produced by fermenting sugar- or starch-rich plants, such as sugar cane or corn, to produce alcohol (fermentation is described on pp. 136–139). Biodiesel fuel is made primarily from oil derived from plants such as soybeans, canola, or palms. Because the carbon stored in them was removed from the modern atmosphere by photosynthesis, burning biofuels simply restores CO_2 that was recently present in the atmosphere. Is this a solution to global warming?

The environmental and social benefits of burning food crops in our gas tanks are hotly debated. One concern is that this added demand will drive up food prices. In Nebraska, for example, ethanol refineries now consume about one-third of the state's corn production. But growing and converting this crop into ethanol uses up large quantities of fossil fuel, making it, at best, only slightly better than burning gasoline. The demand for corn as fuel has caused farmers to convert wheat and soybean fields to corn fields. In Indonesia, plans for enormous new palm plantations to make biodiesel threaten the rain forests of Borneo, home to endangered orangutans. In Brazil, soybean plantations are replacing large expanses of rain forest (**Fig. E7-1**). Ironically, clearing these forests for agriculture increases atmospheric CO_2 because rain forests trap far more carbon than the crops that replace them.

Biofuels would have far less environmental and social impact if they were not produced from food crops. Algae show great promise. These photosynthesizers can double their mass in a day or less. Some algae produce starch that can be fermented into ethanol; others produce oil that can become biodiesel. Another solution would be to cleave cellulose into its component sugars, which would allow ethanol to be generated from corn stalks, wood chips, or grasses. Several small cellulose biorefineries are now being brought online in the United States. Although the benefits of most biofuels produced today may not justify their environmental costs, there is hope that this will change as we develop better technologies to harness the energy captured by photosynthesis.

▲ FIGURE E7-1 A Brazilian rain forest has been felled to make way for soybean fields

malate. The malate is broken down, forming pyruvate and releasing CO_2, which enters the Calvin cycle to produce sugar. The pyruvate is then regenerated into PEP using ATP energy.

Alternate Pathways Adapt Plants to Different Environmental Conditions

Because they are so much more efficient at fixing carbon, you might think that C_4 plants and CAM plants should have taken over the world by now. However, plants using these pathways consume more energy than do C_3 plants. They only have the advantage when light energy is abundant but water is not. In environments where water is abundant or light levels are low, plants using the C_3 carbon fixation pathway dominate. These differing adaptations explain why your spring lawn of lush Kentucky bluegrass (a C_3 plant) may be taken over by spiky crabgrass (a C_4 plant) during a long, hot, dry summer. Plants using C_4 photosynthesis include corn, sugarcane, sorghum, some grasses, and some thistles. CAM plants include cacti, most succulents (such as the jade plant), and pineapples.

Case Study revisited

Did the Dinosaurs Die from Lack of Sunlight?

Paleontologists (scientists who study fossils) have cataloged the extinction of approximately 70% of the species existing at that time by the disappearance of their fossils at the end of the Cretaceous period. In sites from around the globe, researchers have found a thin layer of clay deposited around 65 million years ago; the clay has about 30 times the typical levels of a rare element called iridium, which is found in high concentrations in some meteorites. The clay also contains soot, such as would have been deposited in the aftermath of massive fires.

Did a meteorite wipe out the dinosaurs? Many scientists believe it did. Certainly, the evidence of an enormous meteorite impact, dated to 65 million years ago, is clear on the Yucatán Peninsula. But other scientists believe that more gradual climate changes, possibly triggered by intense volcanic activity, produced conditions that would no longer support the enormous reptiles. Volcanoes also spew out soot and ash, and iridium is found in higher levels in Earth's molten mantle than on its surface, so furious volcanic activity could also explain the iridium layer.

Either scenario would significantly reduce the amount of sunlight and immediately impact the rate of photosynthesis. Large herbivores (plant eaters), such as *Triceratops*, which would have needed to consume hundreds of pounds of vegetation daily, would suffer if plant growth slowed significantly. Predators such as *Tyrannosaurus*, which fed on herbivores, would also be deprived of food. In the Cretaceous as now, interrupting this vital flow of energy would be catastrophic.

Consider This

Design an experiment to test the effects of light-blocking soot (such as may have filled Earth's atmosphere after a huge meteor strike) on photosynthesis. What might you measure to determine the relative amounts of photosynthesis that occurred under normal versus sooty conditions?

CHAPTER REVIEW

Summary of Key Concepts

7.1 What Is Photosynthesis?

Photosynthesis captures the energy of sunlight and uses it to convert the inorganic molecules of carbon dioxide and water into a high-energy sugar molecule, releasing oxygen as a by-product. In plants, photosynthesis takes place in the chloroplasts, using two major reaction sequences: the light reactions and the Calvin cycle.

7.2 Light Reactions: How Is Light Energy Converted to Chemical Energy?

The light reactions occur in the thylakoids of chloroplasts. Light excites electrons in chlorophyll molecules located in photosystems II and I. Energetic electrons jump to a primary electron acceptor, then move into adjacent electron transport chains. Energy lost as the electrons move through ETC II is used to pump hydrogen ions into the thylakoid space, creating an H^+ gradient across the thylakoid membrane. Hydrogen ions move down this concentration gradient through ATP synthase channels in the membrane, driving ATP synthesis. For every two energized electrons that pass through ETC I, one m olecule of the energy carrier NADPH is formed from $NADP^+$ and H^+. Electrons lost from photosystem II are replaced by electrons liberated by splitting water, which also generates H^+ and O_2.

7.3 The Calvin Cycle: How Is Chemical Energy Stored in Sugar Molecules?

The Calvin cycle, which occurs in the stroma of chloroplasts, uses energy from ATP and NADPH generated during the light reactions to drive the synthesis of G3P; two molecules of G3P combine to form glucose. The Calvin cycle has three major parts: (1) **Carbon fixation**: Carbon dioxide combines with ribulose bisphosphate (RuBP) to form phosphoglyceric acid (PGA). (2) **Synthesis of G3P**: PGA is converted to glyceraldehyde-3-phosphate (G3P), using energy from ATP and NADPH. (3) **Regeneration of RuBP**: Five molecules of G3P are used to regenerate three molecules of RuBP, using ATP energy. One molecule of G3P exits the cycle; G3P may be used to synthesize glucose and other molecules.

BioFlix™ Photosynthesis

7.4 Why Do Some Plants Use Alternate Pathways for Carbon Fixation?

The enzyme rubisco, which catalyzes the reaction between RuBP and CO_2, can also catalyze a reaction, called photorespiration, between RuBP and O_2. Under hot, dry conditions, stomata remain closed, CO_2 concentrations fall, and O_2 concentrations rise, causing wasteful photorespiration, which prevents carbon fixation. C_4 and CAM plants minimize photorespiration. In the mesophyll cells of C_4 plants, CO_2 combines with phosphoenolpyruvic acid (PEP) to form a four-carbon oxaloacetate molecule, which is converted to malate and travels into adjacent bundle sheath cells. Here, the malate is broken down, releasing CO_2. This creates a high concentration of CO_2 in the bundle sheath cells. The CO_2 is then fixed using the Calvin cycle. Mesophyll cells of CAM plants open their stomata at night, and generate malate using the CO_2 that diffuses into the leaf. The mesophyll cells store the malate as malic acid in their central vacuoles. During the day, when stomata are closed, mesophyll cells break down the malate, releasing CO_2 for the Calvin cycle.

Key Terms

accessory pigment *116*
ATP synthase *119*
bundle sheath cells *113*
C_3 pathway *120*
C_3 plant *122*
C_4 pathway *123*
C_4 plant *122*
Calvin cycle *114*
carbon fixation *120*
carotenoids *116*

chemiosmosis *117*
chlorophyll *a* *116*
chloroplast *113*
cuticle *113*
electromagnetic spectrum *116*
electron transport chain (ETC) *117*
epidermis *113*
light reactions *114*
mesophyll *113*
photon *116*
photorespiration *122*
photosynthesis *113*
photosystem *117*
pigment molecule *116*
primary electron acceptor *117*
reaction center *117*
rubisco *120*
stoma (plural, **stomata**) *113*
stroma *113*
thylakoid *113*

Thinking Through the Concepts

Fill-in-the-Blank

1. Plant leaves contain pores called __________ that allow the plant to release __________ and take in __________. In hot dry weather, these pores are closed to prevent __________. Photosynthesis occurs in organelles called __________ that are concentrated within the __________ cells of most plant leaves. In plants that use the C_4 pathway, these organelles are also found in __________ cells.
2. Chlorophyll captures wavelengths of light that correspond to which three colors? __________, __________, and __________ What color does chlorophyll reflect? __________ Accessory pigments that reflect yellow and orange are called __________.
3. During the first stage of photosynthesis, light is captured by pigment molecules clustered in __________ and funneled into the reaction center, a region containing special __________ molecules and a(n) __________ molecule. An electron energized by light passes through a series of molecules called the __________. Energy lost during transfers is used to create a gradient of __________. The process that uses this gradient to generate ATP is called __________.
4. The oxygen produced as a by-product of photosynthesis is derived from __________, and the carbons used to make glucose are derived from __________. The process of capturing carbon is called __________. In most plants, the enzyme that catalyzes carbon capture is __________, which is quite nonselective, binding __________ as well as CO_2.
5. Fill in the blanks with "C_4 plants," "CAM plants," or "both": Live in environments that are hot and dry: __________; capture carbon using the C_4 pathway: __________; carbon capture and the Calvin cycle occur in separate cell types: __________; carbon capture and the Calvin cycle occur in the same mesophyll cells: __________.
6. Light reactions generate the energy-carrier molecules __________ and __________, which are then used in the __________ cycle. Carbon fixation combines carbon dioxide with the five-carbon sugar __________. Two molecules of __________ can be combined to produce the six-carbon sugar __________.

Review Questions

1. Write the overall equation for photosynthesis. Does the overall equation differ between C_3 and C_4 plants?
2. Draw a simplified diagram of a chloroplast, and label it. Explain specifically how chloroplast structure is related to its function.
3. Summarize the light reactions and the Calvin cycle. In what part of the chloroplast does each occur?
4. What are the differences in carbon fixation among C_3, C_4, and CAM plants? Under what conditions does each mechanism of carbon fixation work most effectively?
5. Trace the flow of energy in chloroplasts from sunlight to ATP.

Applying the Concepts

1. Suppose an experiment is performed in which plant I is supplied with normal carbon dioxide but with water that contains radioactive oxygen atoms. Plant II is supplied with normal water but with carbon dioxide that contains radioactive oxygen atoms. Each plant is allowed to perform photosynthesis, and the oxygen gas and sugars produced are tested for radioactivity. Which plant would you expect to produce radioactive sugars, and which plant would you expect to produce radioactive oxygen gas? Why?
2. You continuously monitor the photosynthetic oxygen production from the leaf of a plant illuminated by white light. Explain how oxygen production will change—and why—if you place filters in front of the light source that transmit (a) red, (b) blue, and (c) green light onto the leaf.
3. Based on your knowledge of pH gleaned from Chapter 2, if you were to measure the pH in the space surrounded by the thylakoid membrane in an actively photosynthesizing plant, would you expect it to be acidic, basic, or neutral? Explain your answer.
4. You are called before the Ways and Means Committee of the House of Representatives to explain why the U.S. Department of Agriculture should continue to fund photosynthesis research. How would you justify the expense of using genetic engineering to modify the enzyme rubisco, making it selective for CO_2 and preventing it from reacting with O_2? Describe potential applied benefits of this research.

MB *Go to www.masteringbiology.com for practice quizzes, activities, eText, videos, current events, and more.*

Harvesting Energy: Glycolysis and Cellular Respiration

Chapter 8

Case Study

When Athletes Boost Their Blood Counts: Do Cheaters Prosper?

JUST AN HOUR before the start of the 12th stage of the 2008 Tour de France, the Saunier-Duval team was in high spirits. Their star cyclist, Riccardo Riccò, had led the pack in two stages, including the grueling Stage 9, which crested two mountains over a 139-mile (224-km) course. But their optimism turned to consternation as they watched their top rider escorted away by police. Minutes later, the entire team withdrew from the race. Riccò's offense? Blood doping.

Blood doping packs more oxygen-carrying red blood cells into the bloodstream with the goal of enhancing athletic performance. Riccò accomplished this by injecting the drug CERA (continuous erythropoietin receptor activator), a synthetic version of erythropoietin. The natural hormone erythropoietin (EPO) stimulates bone marrow to produce red blood cells. A healthy body produces just enough EPO to ensure that red blood cells are replaced as they age and die. If a person moves to a higher altitude where each lungful of air provides less oxygen, the body compensates by increasing EPO and producing more red blood cells. An injection of synthetic EPO, however, can stimulate the production of a huge number of extra red blood cells, increasing the oxygen-carrying capacity of the blood beyond normal levels, and enhancing performance in sports that require endurance, such as cycling.

Why is endurance improved by extra oxygen molecules in the bloodstream? Think about this question as we examine the role of oxygen in supplying energy to muscle cells.

▲ Riccardo Riccò celebrates victory in Stage 6 of the 2008 Tour de France. He later admitted to taking a new drug that increased the oxygen-carrying capacity of his blood.

At a Glance

8.1 HOW DO CELLS OBTAIN ENERGY?

Cells require a continuous supply of energy to power the multitude of metabolic reactions that are essential just to stay alive. To drive a reaction, however, energy must be in a usable form. Most cellular energy is stored in the chemical bonds of energy-carrier molecules, especially **adenosine triphosphate (ATP).** In this chapter, we describe the cellular reactions that transfer energy from energy-storage molecules, particularly glucose, to energy-carrier molecules, such as ATP. Cells break down glucose in two stages: glycolysis, which liberates a small quantity of ATP, followed by cellular respiration, which produces far more ATP.

As you read this chapter, think back to the principles of energy use in cells explained in Chapter 6. As you may recall from our discussion of the second law of thermodynamics, every time a spontaneous reaction occurs, the amount of useful energy is decreased, and heat is generated. Cells are relatively efficient at capturing chemical energy during glucose breakdown, storing about 40% of the energy released by glucose in ATP molecules and releasing the rest as heat. If cells were as inefficient as most gasoline engines (25% or less), animals would need to eat far more voraciously to remain active.

You may also recall from Chapter 6 that enzymes are needed to promote biochemical reactions. Although for simplicity we do not show enzymes in the pathways illustrated here, each stage in the conversion of one molecule to another requires a specific enzyme to help overcome activation energy and allow the body to regulate the rate of the reaction.

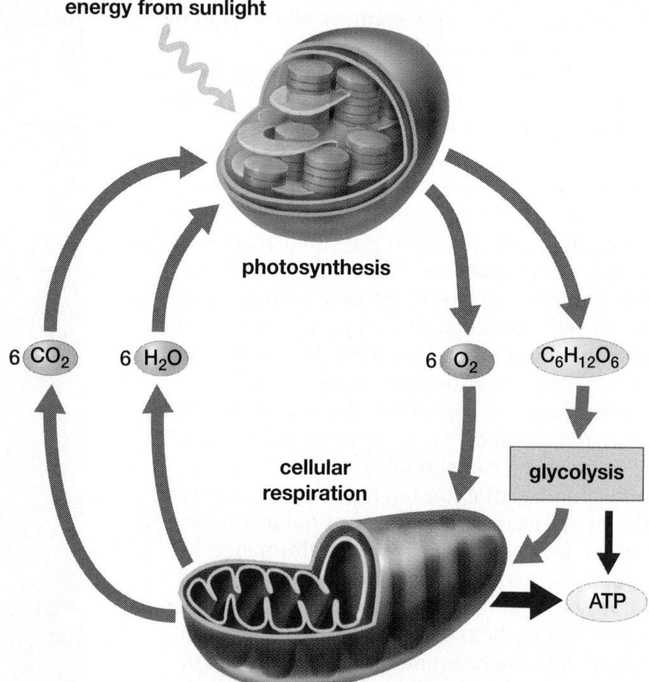

▲ **FIGURE 8-1 Photosynthesis captures the energy released by glycolysis and cellular respiration** Energy from sunlight is captured in glucose by photosynthesis, which occurs in the chloroplasts of green plants. Then glycolysis (which occurs in the cytoplasmic fluid) and cellular respiration (which occurs in the mitochondria) liberate the chemical energy stored in glucose. As indicated by the thickness of the red arrows, cellular respiration yields far more energy than does glycolysis. The products of photosynthesis are used in cellular respiration, which in turn releases molecules used in photosynthesis.

Photosynthesis Is the Ultimate Source of Cellular Energy

Figure 8-1 illustrates the interrelationship between photosynthesis and the breakdown of glucose by glycolysis and cellular respiration. Most organisms, whether or not they engage in photosynthesis, rely on this process to provide both high energy molecules and the oxygen used to break them down. As you learned in Chapter 7, photosynthesis, which occurs in chloroplasts, captures the energy of sunlight and uses it to form glucose ($C_6H_{12}O_6$) from the simple molecules CO_2 and H_2O. Oxygen is liberated as a by-product. In mitochondria, cells break down glucose to supply energy, which is captured in the energy-carrier molecule ATP. In forming ATP during cellular respiration, cells use oxygen and liberate both water and carbon dioxide—raw materials for photosynthesis.

The chemical equations for glucose formation by photosynthesis and the breakdown of glucose are almost perfectly symmetrical:

Photosynthesis:

$$6\ CO_2 + 6\ H_2O + \text{light energy} \rightarrow C_6H_{12}O_6 + 6\ O_2$$

Complete Glucose Breakdown:

$$C_6H_{12}O_6 + 6\ O_2 \rightarrow 6\ CO_2 + 6\ H_2O + \text{chemical energy (ATP)} + \text{heat}$$

Glucose Is a Key Energy-Storage Molecule

Most cells can metabolize a variety of organic molecules to produce ATP. In this chapter, we focus on the breakdown of glucose, which all cells can use as an energy source. Humans and many other animals store energy in molecules such as glycogen (a carbohydrate consisting of long chains of glucose molecules) and fat (see pp. 43–44). When cells use molecules such as glycogen, starch, or fat to produce ATP, they first convert the molecules to glucose or to other compounds that enter the metabolic pathway that is used to break down glucose (see "Health Watch: Why Can You Get Fat by Eating Sugar?" on p. 137).

An Overview of Glucose Breakdown

Glucose breakdown proceeds in stages, as summarized below and illustrated in **Figure 8-2:**

- The first stage is glycolysis (from the Greek, "glyco," meaning "sweet," and "lysis," meaning "to split apart"). **Glycolysis** begins the breakdown of glucose (a six-carbon sugar), producing two molecules of pyruvate (a three-carbon molecule). Some of the energy in glucose is used to generate two ATP molecules. Glycolysis does not require oxygen and proceeds in the same way under both **aerobic** (with oxygen) and **anaerobic** (without oxygen) conditions. The reactions of glycolysis occur in the cytoplasmic fluid and are described in section 8.2.
- If oxygen is available, the second stage of glucose breakdown is cellular respiration. During **cellular respiration,** the two pyruvate molecules produced by glycolysis are broken down into six carbon dioxide

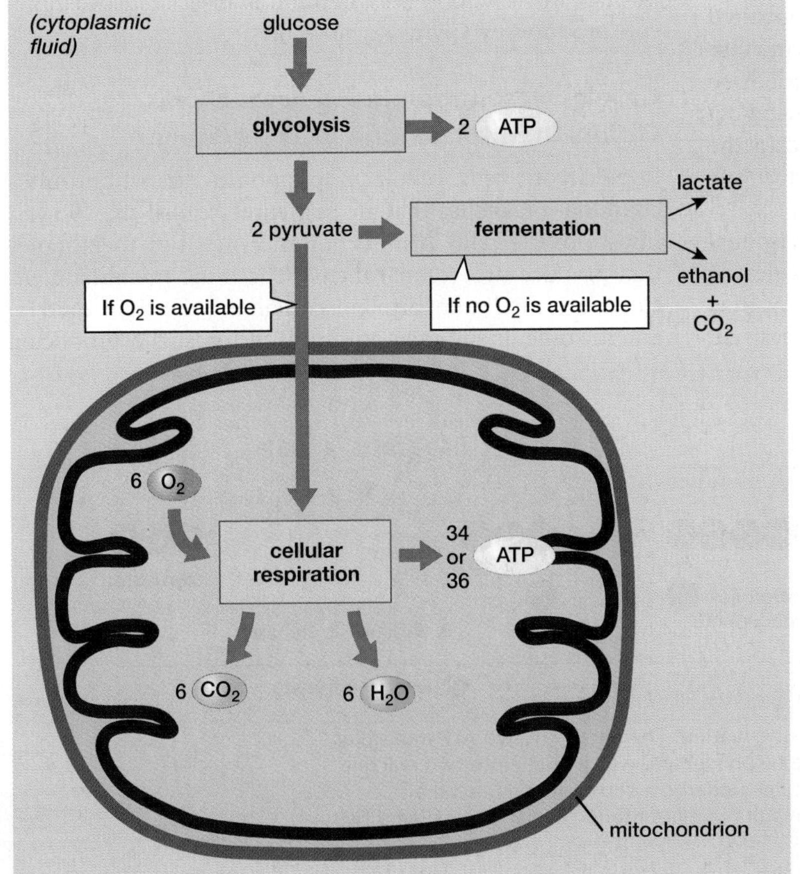

◀ **FIGURE 8-2 A summary of glucose breakdown**

molecules and six water molecules. Oxygen is used during the final stage of cellular respiration, which produces an additional 34 or 36 ATP molecules for every two pyruvate molecules that enter. In eukaryotic cells, the reactions of cellular respiration occur in **mitochondria,** organelles specialized for the aerobic breakdown of pyruvate. Cellular respiration is described in section 8.3.

- If oxygen is not available, the second stage of glucose breakdown is fermentation, which does not generate any additional chemical energy. During **fermentation,** pyruvate does not enter the mitochondria; it instead remains in the cytoplasmic fluid and is converted either into lactate, or into ethanol and CO_2. Fermentation is described in section 8.4.

8.2 WHAT HAPPENS DURING GLYCOLYSIS?

Glycolysis has two parts: glucose activation and energy harvesting, each with several steps (**Fig. 8-3**).

Before glucose can be broken down, it must be activated. Activating glucose uses up the energy from two molecules of ATP. When ATP energy is used, **adenosine diphosphate (ADP)** is formed (ADP is the energy-depleted form of ATP, with one fewer phosphate group). Glucose activation converts a stable glucose molecule into an "activated" molecule of fructose bisphosphate (**Fig. 8-3** ❶). Fructose is a sugar molecule similar to glucose; "bisphosphate" ("bis" means "two") refers to the two phosphate groups acquired from the ATP molecules. Although the formation of fructose bisphosphate costs the cell two ATP molecules, this initial investment of energy is necessary to produce greater energy returns later. Because energy from ATP is now stored in the bonds linking the phosphate groups to the sugar, fructose bisphosphate is a highly unstable molecule.

Next, in the energy-harvesting steps, fructose bisphosphate is split apart into two, three-carbon molecules of glyceraldehyde-3-phosphate, or G3P (**Fig. 8-3** ❷; you may recognize G3P from the Calvin cycle of photosynthesis in Chapter 7). Each G3P molecule, which retains one phosphate with its high-energy bond, then undergoes a series of reactions that convert the G3P to pyruvate. During these reactions, two ATPs are generated for each G3P, for a total of four ATPs per glucose molecule. Because two ATPs were used up to activate the glucose molecule in the first place, there is a net gain of only two ATPs per glucose molecule. Along the pathway from G3P to pyruvate, two high-energy electrons and a hydrogen ion (H^+) are added to the "empty" electron-carrier **nicotinamide adenine dinucleotide (NAD^+)** to produce the high-energy electron-carrier molecule **NADH.** For more on glycolysis, see "A Closer Look at Glycolysis."

Summing Up Glycolysis

- Each molecule of glucose is broken down into two molecules of pyruvate.
- These reactions produce a net of two molecules of ATP and two molecules of NADH.

8.3 WHAT HAPPENS DURING CELLULAR RESPIRATION?

In most organisms, if oxygen is present, glycolysis is followed by the second stage of glucose breakdown, cellular respiration. In this series of reactions, the pyruvate produced by glycolysis is broken down, extracting a great deal more energy and also releasing carbon dioxide and water. As you read through the reactions, keep in mind that each glucose molecule produces two pyruvate molecules.

Cellular Respiration in Eukaryotic Cells Occurs in Mitochondria in Three Stages

In eukaryotic cells, cellular respiration occurs within mitochondria, organelles that are sometimes called the "powerhouses of the cell." A mitochondrion has two membranes that produce two compartments. The inner membrane encloses a central compartment containing the fluid **matrix,** and the outer membrane surrounds the organelle, producing

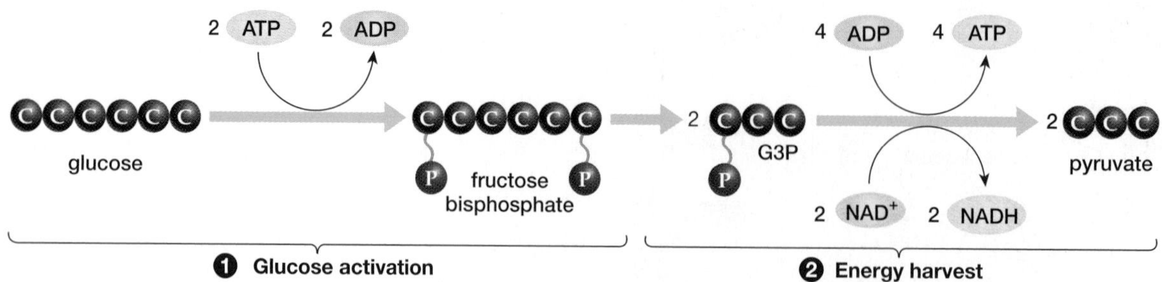

▲ **FIGURE 8-3 The essentials of glycolysis** ❶ Glucose activation: The energy of two ATP molecules is used to convert glucose to the highly reactive fructose bisphosphate, which splits into two reactive molecules of G3P. ❷ Energy harvest: The two G3P molecules undergo a series of reactions that generate four ATP and two NADH molecules. Glycolysis results in a net production of two ATP and two NADH molecules for each glucose molecule.

A Closer Look At *Glycolysis*

Glycolysis is a series of enzyme-catalyzed reactions that break down a single molecule of glucose into two molecules of pyruvate. To help you follow the reactions in **Figure E8-1,** we show only the "carbon skeletons" of glucose and the molecules produced during glycolysis. Each blue arrow represents a reaction catalyzed by at least one enzyme.

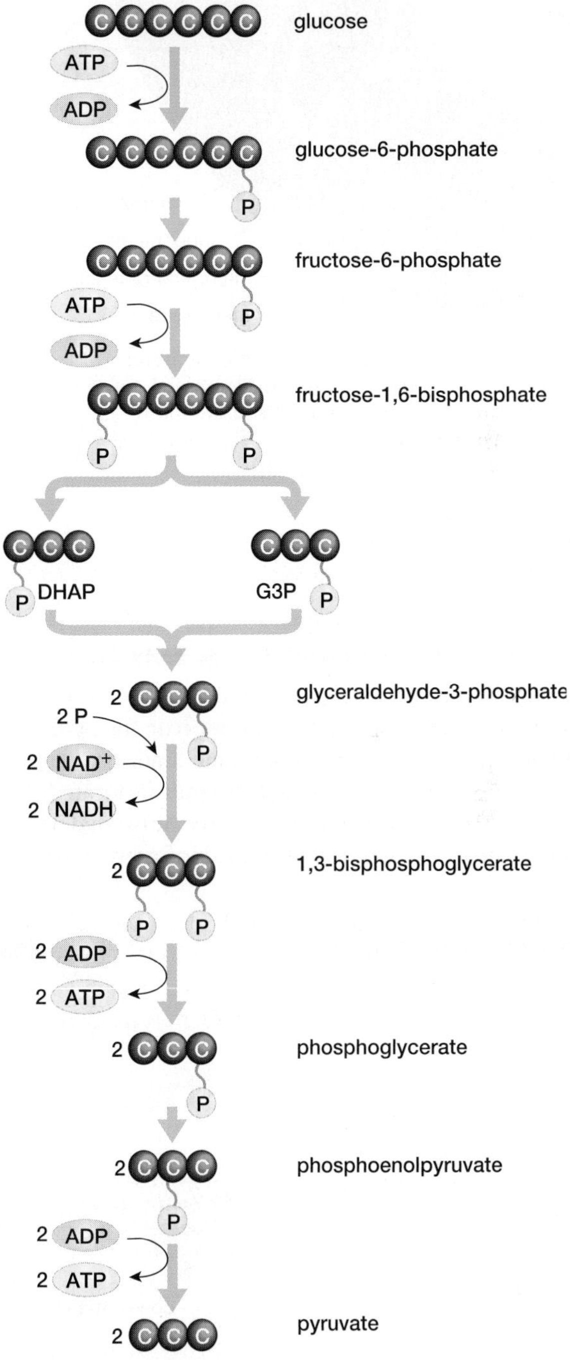

❶ A glucose molecule is energized by the addition of a high-energy phosphate from ATP.

❷ The molecule is slightly rearranged, forming fructose-6-phosphate.

❸ A second phosphate is added from another ATP.

❹ The resulting molecule, fructose-1,6-bisphosphate, is split into two three-carbon molecules, one DHAP (dihydroxyacetone phosphate) and one G3P. Each has one phosphate attached.

❺ DHAP rearranges into G3P. From now on, there are two molecules of G3P going through the same reactions.

❻ Each G3P undergoes two almost-simultaneous reactions. Two electrons and a hydrogen ion are donated to NAD^+ to make the energized carrier NADH, and an inorganic phosphate (P) is attached to the carbon skeleton with a high-energy bond. The resulting molecules of 1,3-bisphosphoglycerate have two high-energy phosphates.

❼ One phosphate from each bisphosphoglycerate is transferred to ADP to form ATP, for a net of two ATPs. This transfer compensates for the initial two ATPs used in glucose activation.

❽ After another rearrangement, the second phosphate from each phosphoenolpyruvate is transferred to ADP to form ATP, leaving pyruvate as the final product of glycolysis. There is a net profit of two ATPs from each glucose molecule.

glucose
ATP → ADP
glucose-6-phosphate
fructose-6-phosphate
ATP → ADP
fructose-1,6-bisphosphate
DHAP
G3P
glyceraldehyde-3-phosphate
2 P
2 NAD^+ → 2 NADH
1,3-bisphosphoglycerate
2 ADP → 2 ATP
phosphoglycerate
phosphoenolpyruvate
2 ADP → 2 ATP
pyruvate

▲ FIGURE E8-1 **Glycolysis**

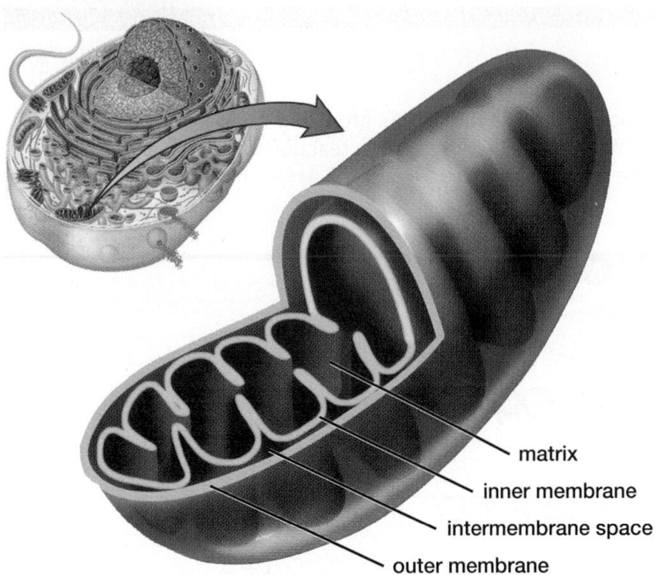

▲ **FIGURE 8-4 A mitochondrion** The double membrane of this organelle creates two spaces. The intermembrane space lies between the outer and inner membranes, and the matrix is enclosed by the inner membrane.

an **intermembrane space** between the inner and outer membranes (**Fig. 8-4**).

First, Pyruvate Is Broken Down in the Mitochondrial Matrix, Releasing Energy and CO_2

Pyruvate, the end product of glycolysis, is synthesized in the cytoplasmic fluid. Before cellular respiration can occur, the pyruvate must be transported from the cytoplasmic fluid into the mitochondrial matrix, where the necessary enzymes are located.

The reactions of the mitochondrial matrix, illustrated in **Figure 8-5**, occur in two stages: the formation of acetyl CoA and the Krebs cycle. Acetyl CoA consists of a two-carbon acetyl group attached to a molecule called coenzyme A (CoA). To generate acetyl CoA, pyruvate is split, forming an acetyl group and releasing CO_2. The acetyl group reacts with CoA, forming acetyl CoA (**Fig. 8-5** ❶). During this reaction, two high-energy electrons and a hydrogen ion are transferred to NAD^+, forming NADH.

The next set of mitochondrial matrix reactions forms a cyclic pathway known as the **Krebs cycle** (**Fig. 8-5** ❷). (The cycle is named after its discoverer Hans Krebs, a biochemist who won a Nobel Prize for this work in 1953.) The Krebs cycle is also called the **citric acid cycle** because citrate (the dissolved, ionized form of citric acid) is the first molecule produced in the cycle.

The Krebs cycle begins by combining acetyl CoA with a four-carbon molecule to form the six-carbon citrate; coenzyme A is released. Like an enzyme, CoA is not permanently altered during these reactions and is reused many times. As the Krebs cycle proceeds, enzymes within the mitochondrial matrix break down the acetyl group, releasing two CO_2 molecules (whose carbons come from the acetyl group) and regenerating the four-carbon molecule for use in future cycles.

During the Krebs cycle, chemical energy released by breaking down each acetyl group (formed from pyruvate generated during glycolysis) is captured in energy-carrier molecules. Each acetyl group produces one ATP, three NADH, and one **$FADH_2$. FAD** is **flavin adenine dinucleotide,** a high-energy electron carrier similar to NAD^+. During the Krebs cycle, FAD picks up two energetic electrons along with two H^+, forming $FADH_2$. Remember that for each glucose molecule, two pyruvate molecules are formed, so the energy generated per glucose molecule is twice the energy generated for each pyruvate. These reactions are shown in more detail in "A Closer Look at the Mitochondrial Matrix Reactions."

▶ **FIGURE 8-5 Reactions in the mitochondrial matrix** ❶ Pyruvate reacts with CoA, forming acetyl CoA and releasing CO_2. During this reaction, two high-energy electrons and a hydrogen ion are added to NAD^+, forming NADH. ❷ When acetyl CoA enters the Krebs cycle, CoA is released for reuse. The Krebs cycle produces one ATP, three NADH, one $FADH_2$, and two CO_2 from each acetyl CoA.

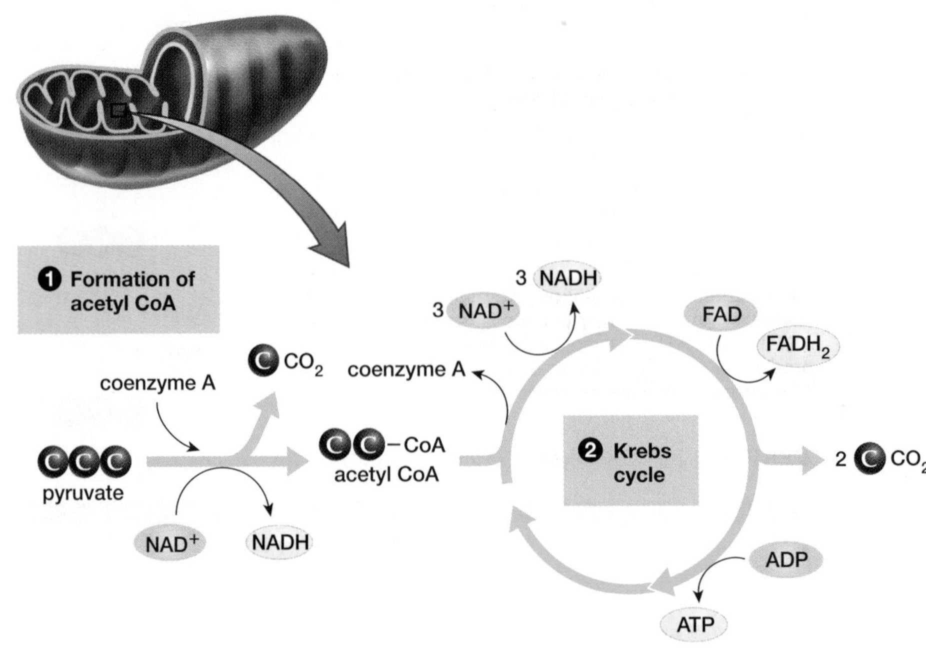

A Closer Look At *The Mitochondrial Matrix Reactions*

Mitochondrial matrix reactions occur in two stages: the formation of acetyl coenzyme A and the Krebs cycle (**Fig. E8-2**). Here we show only the carbon skeletons of the molecules. The blue arrows represent enzyme-catalyzed reactions. Recall that glycolysis produces two pyruvates from each glucose molecule, so each set of matrix reactions occurs twice during the breakdown of a single glucose molecule.

First Stage: Formation of Acetyl Coenzyme A

Pyruvate is split to form CO_2 and an acetyl group. The acetyl group attaches to CoA to form acetyl CoA. Simultaneously, NAD^+ receives two energized electrons and a hydrogen ion to make NADH. Acetyl CoA enters the Krebs cycle.

Second Stage: The Krebs Cycle

1. Acetyl CoA donates its acetyl group to the four-carbon molecule oxaloacetate, forming citrate. CoA is released. Water donates hydrogen to the CoA molecule and oxygen to citrate.
2. Citrate is rearranged to form isocitrate.
3. Isocitrate forms α-ketoglutarate by releasing CO_2. Two energetic electrons and a H^+ are captured by NAD^+ to form NADH.
4. Alpha-ketoglutarate forms succinate by releasing CO_2. Two energetic electrons and a H^+ are captured by NAD^+ to form NADH, and additional energy is captured in a molecule of ATP. At this point, all three carbons of the original pyruvate have been released as CO_2.
5. Succinate is converted to fumarate. Two energetic electrons and two H^+ are captured by FAD, forming $FADH_2$.
6. Fumarate is converted to malate, which contains two additional hydrogens and one additional oxygen, derived from water.
7. Malate is converted to oxaloacetate. Two energetic electrons and a H^+ are captured by NAD^+ to form NADH.

For each acetyl CoA, the Krebs cycle produces two CO_2, one ATP, three NADH, and one $FADH_2$. The formation of each acetyl CoA prior to the Krebs cycle also generates one CO_2 and one NADH. So for each pyruvate molecule supplied by glycolysis, the mitochondrial matrix reactions produce three CO_2, one ATP, four NADH, and one $FADH_2$. Because each glucose molecule produces two pyruvates, the number of high-energy products and CO_2 per glucose molecule will be double that from a single pyruvate. The high-energy electron-carrier molecules NADH and $FADH_2$ will deliver their high-energy electrons to the electron transport chain, where their energy will be used to synthesize ATP by chemiosmosis.

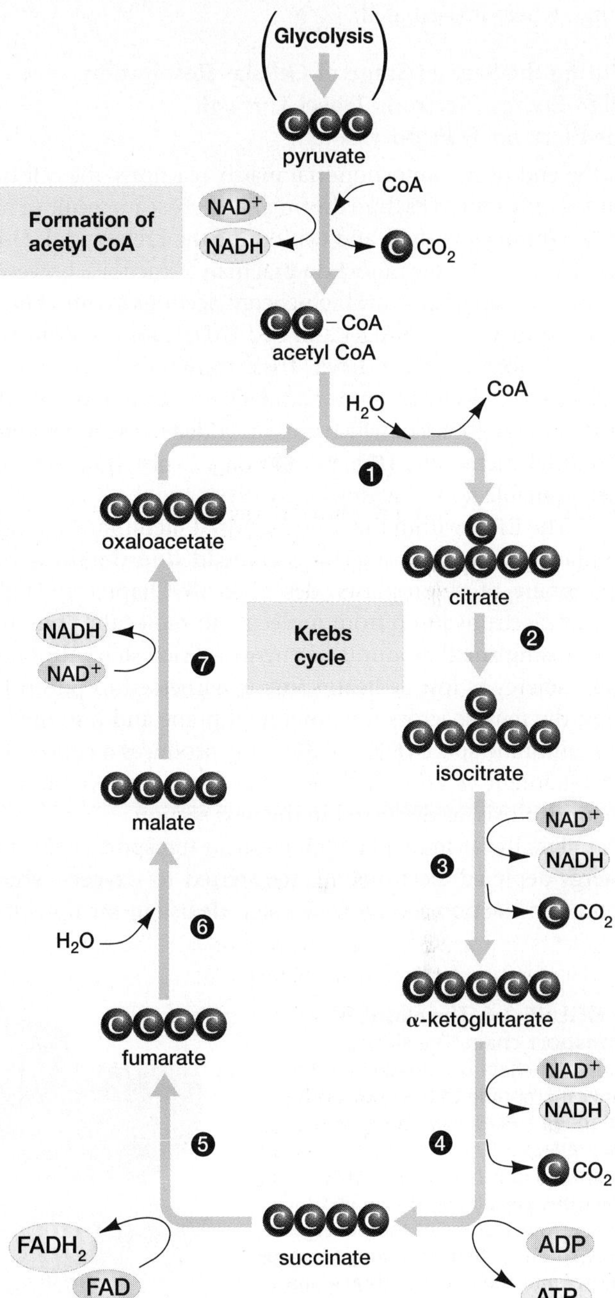

▲ **FIGURE E8-2 The mitochondrial matrix reactions**

During the mitochondrial matrix reactions, CO_2 is generated as a waste product. In your body, the CO_2 diffuses out of your cells and into your blood, which carries it to your lungs, where it is exhaled.

During the Second Stage of Cellular Respiration, High-Energy Electrons Travel Through the Electron Transport Chain

At the end of the mitochondrial matrix reactions, the cell has gained only four ATPs from the original glucose molecule (a net of two during glycolysis and two during the Krebs cycle). During glycolysis and the mitochondrial matrix reactions, however, the cell has captured many high-energy electrons in carrier molecules: a total of 10 NADH and two $FADH_2$ for every glucose molecule that is broken down. These carriers each release two high-energy electrons into an **electron transport chain (ETC),** many copies of which are embedded in the inner mitochondrial membrane (**Fig. 8-6** ❶). The depleted carriers are then available for recharging by glycolysis and the Krebs cycle.

The ETCs within the mitochondrial membrane are very similar in function to the ETCs embedded in the thylakoid membrane of chloroplasts, described in Chapter 7. High-energy electrons jump from molecule to molecule along the ETC, losing small amounts of energy at each step. Although some energy is lost as heat, some is harnessed to pump H^+ from the matrix across the inner membrane and into the intermembrane space (**Fig. 8-6** ❷). This produces a concentration gradient of H^+, which is used to generate ATP during chemiosmosis, as discussed in the next section.

Finally, at the end of the electron transport chain, the energy-depleted electrons are transferred to oxygen, which acts as an electron acceptor. This step clears out the transport chain, leaving it ready to accept more electrons. The energy-depleted electrons, oxygen, and hydrogen ions combine to form water. One water molecule is produced for every two electrons that traverse the ETC (**Fig. 8-6** ❸).

Without oxygen, which we get from the air we breathe, electrons would be unable to move through the ETC, and H^+ would not be pumped across the inner membrane. The H^+ gradient would soon dissipate, and ATP synthesis by chemiosmosis would stop. Because our cells are so metabolically active, they cannot survive without a steady supply of oxygen to allow ATP generation to continue.

Case Study continued
When Athletes Boost Their Blood Counts: Do Cheaters Prosper?

When people and other animals exercise vigorously, they are unable to get enough air into their lungs, enough oxygen into their blood, and enough blood circulating to their muscles to allow cellular respiration to meet all their energy needs. As oxygen demand exceeds oxygen supply, muscles must rely on glycolysis (yielding far less ATP) for short periods of intense exercise. This is why athletes, desperate for a competitive edge, may turn to illegal blood doping to increase the ability of their blood to carry oxygen.

The Third Stage of Cellular Respiration Generates ATP by Chemiosmosis

Why pump hydrogen ions across a membrane? As you may recall from Chapter 7, **chemiosmosis** is the process by which

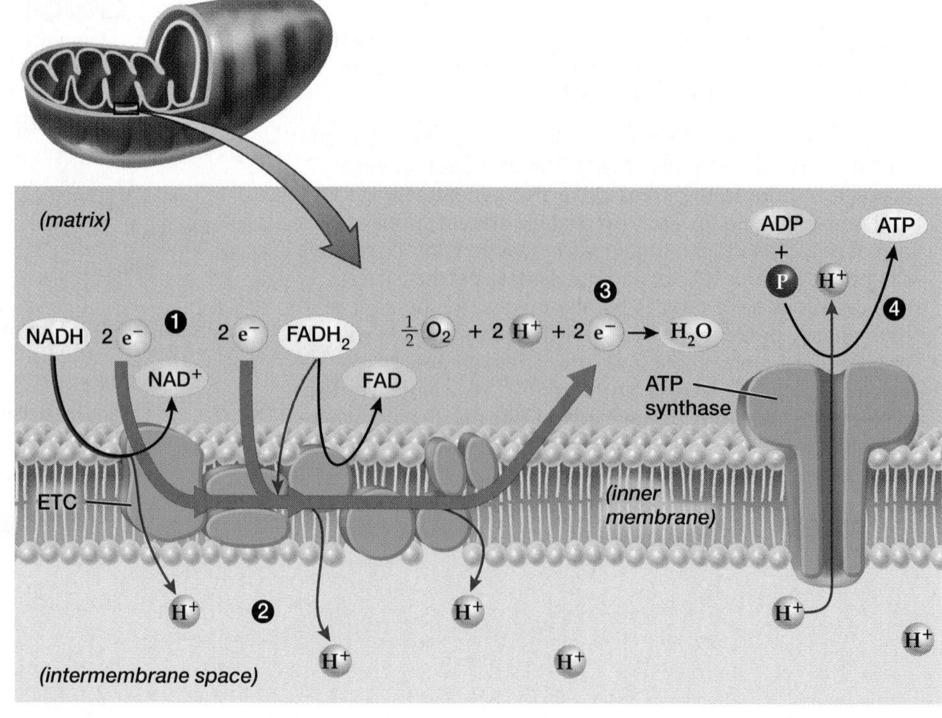

▶ **FIGURE 8-6 The electron transport chain** The electron transport chain is embedded in the inner membrane of the mitochondrion. ❶ NADH and $FADH_2$ donate their energetic electrons and hydrogen ions to the ETC. As the electrons pass along the chain (thick gray arrows), some of their energy is used to pump hydrogen ions from the matrix into the intermembrane space (thin red arrows). ❷ This creates a hydrogen ion gradient that is used to drive ATP synthesis. ❸ At the end of the electron transport chain, the energy-depleted electrons combine with hydrogen ions and oxygen and in the matrix to form water. ❹ The hydrogen ions flow down their concentration gradient from the intermembrane space and into the matrix through ATP synthase channels, generating ATP from ADP and phosphate.

QUESTION How would the rate of ATP production be affected by the absence of oxygen?

energy is first used to generate a gradient of H^+, and then captured in the bonds of ATP as H^+ flows down its gradient. As the ETC pumps H^+ across the inner membrane, it produces a high concentration of H^+ in the intermembrane space and a low concentration in the matrix (see Fig. 8-6 ❷). According to the second law of thermodynamics, energy must be expended to produce this nonuniform distribution of H^+, sort of like charging a battery. This energy is released when hydrogen ions are allowed to flow down their concentration gradient, a process somewhat comparable to using the energy stored in a battery to light a bulb.

As in the thylakoid membranes of chloroplasts, the inner membranes of mitochondria are only permeable to H^+ at ATP synthase channels. As hydrogen ions flow from the intermembrane space into the matrix through these ATP-synthesizing enzymes, the flow of ions generates ATP from ADP and phosphate dissolved in the matrix, providing the energy to synthesize 32 or 34 molecules of ATP for each molecule of glucose (**Fig. 8-6** ❹).

The ATP synthesized in the mitochondrial matrix during chemiosmosis enters the surrounding cytoplasmic fluid. These ATP molecules provide most of the energy needed by the cell. Simultaneously, ADP moves from the cytoplasmic fluid into the mitochondrial matrix, replenishing the supply of ADP and allowing more ATP synthesis.

Summing Up **Cellular Respiration**

- In the mitochondrial matrix, each pyruvate molecule is converted into acetyl CoA, producing one NADH per pyruvate molecule and releasing one CO_2.
- As each acetyl CoA passes through the Krebs cycle, its energy is captured in one ATP, three NADH, and one $FADH_2$. Its carbons are released in two CO_2 molecules.
- By the end of the matrix reactions, the two pyruvate molecules produced from each glucose molecule during glycolysis have been completely broken down, yielding a total of two ATP and 10 high-energy electron carriers: eight NADH and two $FADH_2$. Carbon atoms have been released in six molecules of CO_2.
- The NADH and $FADH_2$ release their energetic electrons into the ETC embedded in the inner mitochondrial membrane.
- As they pass through the ETC, energy from the high-energy electrons is harnessed to pump H^+ into the intermembrane space.
- As they leave the ETC, the energy-depleted electrons combine with hydrogen ions and oxygen to form water.
- During chemiosmosis, hydrogen ions in the intermembrane space flow down their concentration gradient through ATP synthase channels, which synthesize ATP.

Have you ever wondered 

Why Cyanide Is So Deadly?

Cyanide is a common weapon in old murder mysteries, where its hapless victims fall dead almost instantly. Cyanide exerts its deadly effects by reacting with the final protein in the ETC, blocking oxygen from accepting electrons from this protein. If the energy-depleted electrons that emerge from the ETC are not carried away by oxygen, no new high-energy electrons can enter the chain, no hydrogen can be pumped across the membrane, and ATP production by chemiosmosis comes to a screeching halt. We are so dependent on the large numbers of ATP molecules produced by cellular respiration that blocking it with cyanide can kill a person within a few minutes.

A Summary of Glucose Breakdown in Eukaryotic Cells

Figure 8-7 summarizes the breakdown of one glucose molecule in a eukaryotic cell with oxygen present, showing the energy produced during each stage. Glycolysis occurs in the cytoplasmic fluid, producing two, three-carbon pyruvate molecules and releasing a small fraction of the chemical energy stored in glucose. Some of this energy is used to generate two ATP molecules, and some is captured in two NADH, which (if oxygen is present) will feed their electrons into the ETC during cellular respiration, generating more ATP.

During cellular respiration, the two pyruvate molecules enter a mitochondrion. First, each reacts with coenzyme A (CoA). This captures high-energy electrons in two NADH, produces two molecules of acetyl CoA, and liberates two molecules of carbon dioxide. The acetyl CoA molecules enter the Krebs cycle. The Krebs cycle releases four molecules of carbon dioxide, produces two ATP, and captures high-energy electrons in six NADH and two $FADH_2$. These electrons are passed into the ETC, where their energy is used during chemiosmosis to generate a gradient of H^+, yielding a net of 32 or 34 ATP. As energy-depleted electrons exit the ETC, they are picked up by the hydrogen ions released from the high-energy electron carriers NADH and $FADH_2$, and combine with oxygen to form water. The total energy captured from the breakdown of a single glucose molecule, starting with glycolysis and continuing through cellular respiration, is 36 or 38 ATP. The ATP enters the cytoplasmic fluid for use in cellular metabolic activities.

Why do we provide two different numbers for ATP? Glycolysis, which occurs in the cytoplasmic fluid, produces two NADH molecules, which must be actively transported into a mitochondrion to the ETC. In most body cells (including those of the brain and skeletal muscle) this transport uses up one ATP molecule per NADH (two ATPs per glucose molecule), so these cells generate a total of 36 ATPs during the breakdown of one glucose molecule. The heart and liver cells of mammals, however, use a more efficient transport mechanism that allows them to generate 38 ATPs per glucose molecule.

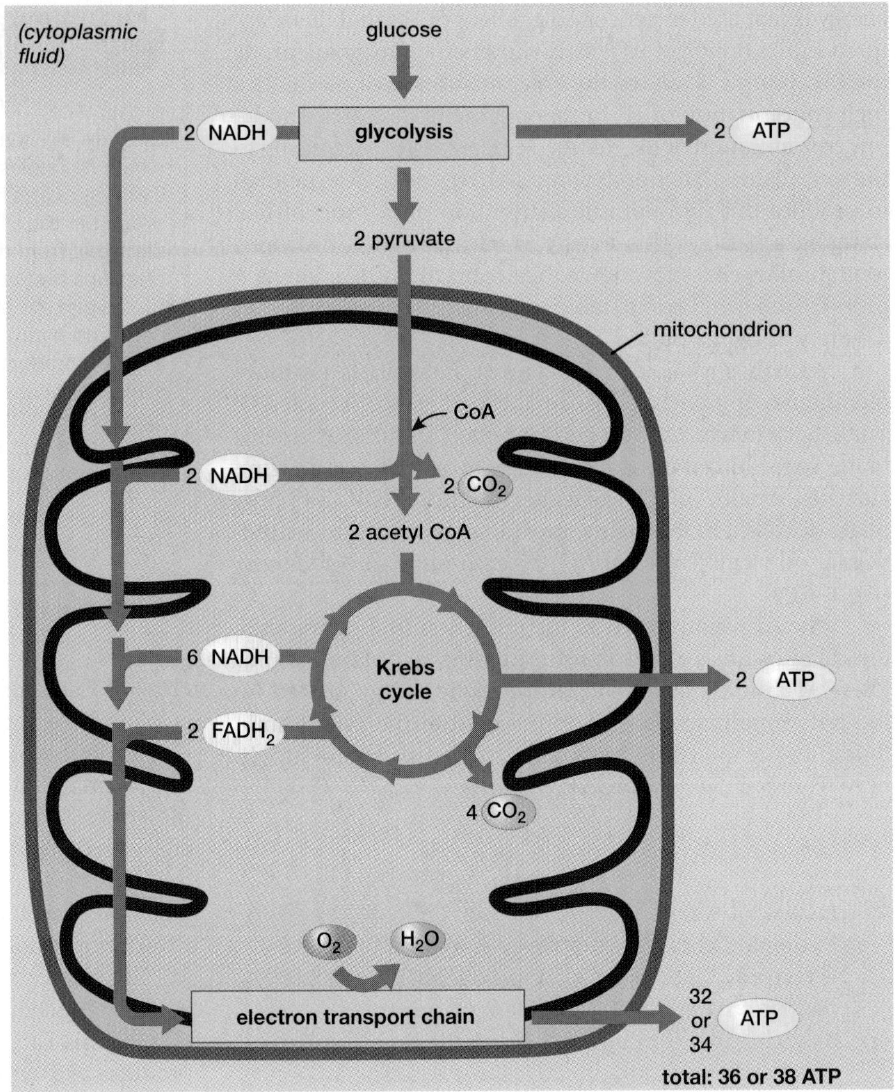

▶ **FIGURE 8-7 The energy sources and ATP harvest from glycolysis and cellular respiration** Here we follow the energy captured from one glucose molecule, which generates two molecules of pyruvate (during glycolysis) that then enter cellular respiration, which encompasses all of the reactions shown inside the mitochondrion. Notice that most of the ATP derived from glucose breakdown comes as a result of high-energy electrons donated by NADH and $FADH_2$ to the electron transport chain, which allows chemiosmosis to occur.

The breakdown of glucose is a multistep process with many intermediate molecules. By regulating enzymes (see pp. 105–107), cells can promote reactions that divert these intermediate molecules into other pathways. For example, if the cell has adequate supplies of ATP, it can use extra glucose and other sugars to generate fat, as described in "Health Watch: Why Can You Get Fat by Eating Sugar?"

BioFlix™ Cellular Respiration

8.4 WHAT HAPPENS DURING FERMENTATION?

Virtually every organism on Earth uses glycolysis, providing evidence that it is one of the most ancient of all biochemical pathways. Scientists hypothesize that the earliest forms of life appeared under anaerobic conditions (before the evolution of oxygen-liberating photosynthesis). These pioneering life-forms relied on glycolysis for energy production, breaking down the organic molecules formed under the conditions that existed on Earth before life appeared (see pp. 318–320). Many microorganisms still thrive in places where oxygen is rare or absent, such as in the stomach and intestines of animals (including humans), deep in soil, or in bogs and marshes. Some microorganisms lack the enzymes for cellular respiration; these are unable to use oxygen even under aerobic conditions. A few types of bacteria, such as those that cause tetanus, are actually poisoned by oxygen.

Why Is Fermentation Necessary?

Following glycolysis, under anaerobic conditions, fermentation occurs. Fermentation does not produce more ATP, so why is it necessary? Fermentation is required to convert the NADH produced during glycolysis back to NAD^+, which needs to be continuously recycled to allow glycolysis to continue.

Under aerobic conditions, most organisms use cellular respiration and regenerate NAD^+ by donating the energetic electrons from NADH to the ETC. But under anaerobic conditions, with no oxygen to allow the ETC to function, the cell must regenerate the NAD^+ using fermentation. During fermentation, the electrons from NADH (along with some

Health Watch

Why Can You Get Fat by Eating Sugar?

People fortunate enough to have access to abundant food often wonder why everything they eat seems to turn to fat. Why don't we just excrete excess high-energy molecules that we don't need? Why do we continue to feel hungry even if we are already carrying a great deal of excess weight?

From an evolutionary perspective, overeating during times of easy food availability is a highly adaptive behavior. During times of famine, which were common during our early history, heavier people are more likely to survive, while leaner individuals succumb to starvation. It is only recently (from an evolutionary vantage point) that many people have had continuous access to high-calorie food. Under these conditions, the drive to eat and the adaptation of storing excess food as fat leads to obesity, a growing health problem in the United States, Mexico, and many other countries throughout the world.

Why don't we simply store energy in glucose or ATP? As you may recall from Chapter 3, fats store twice as much energy per unit weight as do proteins or carbohydrates such as glucose. Because using fat as a storage molecule actually reduces the weight we must carry, this adaptation was important to our prehistoric ancestors who needed to move quickly to catch prey or to avoid becoming prey themselves. We don't use ATP for long-term energy storage because the amount of energy stored in its bonds makes ATP too unstable.

Gaining fat by eating sugar or other carbohydrates is common among animals. For example, a ruby-throated hummingbird weighs 0.035 to 0.07 of an ounce (2 to 3 grams; about as much as a penny). During late summer, it feeds voraciously on the sugary nectar of flowers (**Fig. E8-3**), storing enough fat to almost double its weight. The energy in this fat allows it to make its long migration from the eastern United States, across the Gulf of Mexico and into Mexico or Central America for the winter. If the hummingbird stored sugar instead of fat, it might be too heavy to fly.

Because fat storage is so important for survival, cells have metabolic pathways that can chemically transform a variety of foods into fat if the foods are consumed in excessive amounts. In addition to breaking down glucose, the biochemical pathways described in this chapter participate in producing fat as well. In Chapter 3 we described the structure of a fat: three fatty acids attached to a glycerol backbone (see Fig 3-12). When you eat a candy bar loaded with sucrose (table sugar), for example, this disaccharide is first broken down into its component simple sugars: glucose and fructose. Both enter glycolysis (the fructose enters a bit further along than glucose), and both are converted into G3P. If cells have plenty of ATP, some of this G3P is diverted and used to make the glycerol backbone of fat. As the sugars continue to be broken down, acetyl CoA is formed (see Fig. 8-5). Excess acetyl CoA molecules are used as raw materials to synthesize the fatty acids that will be linked to the glycerol to form a fat molecule. Starches, such as those in potatoes or bread, are actually long chains of glucose molecules, so you can now see how eating large amounts of starch, as well as sugar, can make you fat.

▲ FIGURE E8-3 A ruby-throated hummingbird feeds on nectar

hydrogen ions) are donated to pyruvate, changing it chemically. If the supply of NAD^+ were to be exhausted—which would happen quickly without fermentation—glycolysis would stop, energy production would cease, and the organism would rapidly die.

Depending on which metabolic pathway they have evolved, organisms use one of two types of fermentation to regenerate NAD^+: lactic acid fermentation, which produces lactic acid from pyruvate, or alcoholic fermentation, which generates alcohol and CO_2 from pyruvate. Because fermentation does not break down glucose completely and does not use the energy of NADH to produce more ATP, organisms that rely on fermentation must consume far more sugar to generate the same amount of ATP than do organisms using cellular respiration.

Some Cells Ferment Pyruvate to Form Lactate

Fermentation of pyruvate to lactate is called **lactic acid fermentation** (in the cytoplasmic fluid, lactic acid is dissolved and becomes ionized, forming lactate). Lactic acid fermentation happens in muscles that are working hard enough to use up all their available oxygen.

When deprived of adequate oxygen, muscles do not immediately stop working. After all, animals exercise most vigorously when fighting, fleeing, or pursuing prey. During these activities, the ability to continue just a little bit longer can make the difference between life and death. When muscles are sufficiently low on oxygen, glycolysis supplies its meager two ATPs per glucose molecule, which may provide the energy needed for a final, brief burst of speed. To regenerate NAD^+, muscle cells ferment pyruvate molecules to lactate, using electrons from NADH and hydrogen ions (**Fig. 8-8**).

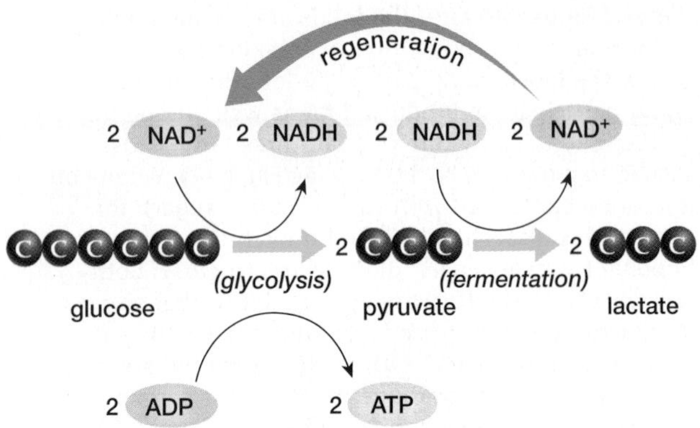

▲ **FIGURE 8-8 Glycolysis followed by lactic acid fermentation**

For example, if you're breathing hard after sprinting to arrive at class on time, your muscles have been relying on glycolysis for part of their energy, and your lungs are working to restore the oxygen needed for cellular respiration. As oxygen is replenished, lactate is converted back into pyruvate both in muscle cells, where it will be used for cellular respiration, and in the liver, where the pyruvate is converted back into glucose. This glucose may then be released into the bloodstream to be used as an energy source by cells throughout the body.

A variety of microorganisms use lactic acid fermentation, including the bacteria that convert milk into yogurt, sour cream, and cheese. Because acids taste sour, the lactic acid contributes to the distinctive tastes of these foods. The acid also denatures milk protein, altering its three-dimensional structure. This causes the milk to thicken, so that sour cream and yogurt are semisolid in texture.

Some Cells Ferment Pyruvate to Form Alcohol and Carbon Dioxide

Many microorganisms, such as yeast (single-celled fungi), engage in **alcoholic fermentation** under anaerobic conditions. As in lactic acid fermentation, the NAD^+ must be regenerated to allow glycolysis to continue. During alcoholic fermentation, H^+ and electrons from NADH are used to convert pyruvate into ethanol and CO_2 (rather than lactate). This releases NAD^+, which is then available to accept more high-energy electrons during glycolysis (**Fig. 8-9**).

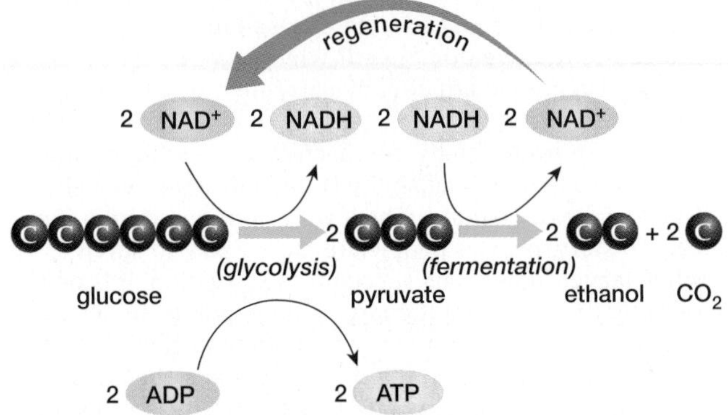

▲ **FIGURE 8-9 Glycolysis followed by alcoholic fermentation**

In sparkling wines, fermentation continues in the bottle, trapping CO_2 and producing the tiny bubbles for which champagne is noted (**Fig. 8-10**). See "Links to Everyday Life: A Jug of Wine, a Loaf of Bread, and a Nice Bowl of Sauerkraut" for more about how people put fermentation to use.

▲ **FIGURE 8-10 Fermentation in action** Sparkling wines rely on alcoholic fermentation for their alcohol; the CO_2 it releases can cause this explosive decompression.

QUESTION Some species of bacteria use aerobic respiration and other species use anaerobic (fermenting) respiration. In an oxygen-rich environment, would either type be at a competitive advantage? What about in an oxygen-poor environment?

Case Study continued
When Athletes Boost Their Blood Counts: Do Cheaters Prosper?

Why is the average speed of the 5,000-meter run in the Olympics slower than that of the 100-meter dash? During the dash, runners' leg muscles use more ATP than cellular respiration can supply. But anaerobic fermentation can only provide ATP for a short dash. Longer runs must be aerobic, and thus slower, to prevent lactic acid buildup from causing extreme fatigue, muscle pain, and cramps.

Links to *Everyday Life*

A Jug of Wine, a Loaf of Bread, and a Nice Bowl of Sauerkraut

Life would be a little less interesting without fermentation. Persian poet Omar Khayyam (1048–1122) described his vision of paradise on Earth as "A Jug of Wine, a Loaf of Bread—and Thou Beside Me." People have long exploited yeast's ability to ferment the sugars in fruit to form alcohol; historical evidence suggests that wine and beer were commercially produced at least 5,000 years ago. Yeasts (single-celled fungi) are opportunists; they will engage in efficient cellular respiration if oxygen is available, but switch to alcoholic fermentation if they run out of oxygen. Thus, wine must be fermented in special containers that prevent air from entering (so cellular respiration doesn't occur) but allow carbon dioxide to escape (so the containers don't explode). Sparkling (fizzy) wines and champagne are made by adding more yeast and sugar just before the wine is bottled, so that final fermentation occurs in the sealed bottle, trapping carbon dioxide (see Fig. 8-10).

Fermentation also gives bread its airy texture. Bread contains yeast, flour, and water. Dry yeast cells are awakened from their dormant state by water, and the yeast cells multiply rapidly while metabolizing the sugars present in flour. The yeasts release CO_2 during cellular respiration while oxygen is present, and they also release CO_2 during alcoholic fermentation after they have used up the available oxygen. The CO_2 forms tiny pockets of gas in the bread dough. Kneading distributes the multiplying yeast cells evenly throughout the bread and makes the dough stretchy and resilient so it traps the gas, resulting in an evenly porous texture that is solidified by baking. Although both wine and bread are produced by alcoholic fermentation, bacteria that utilize lactic acid fermentation are responsible for other culinary staples.

For thousands of years, people have relied on microorganisms that produce lactic acid to convert milk into sour cream, yogurt, and a wide variety of cheeses (**Fig. E8-4**). In addition, lactate fermentation by salt-loving bacteria converts the sugars in cucumbers and cabbage to lactic acid. The result: dill pickles and sauerkraut, excellent partners for other fermented foods.

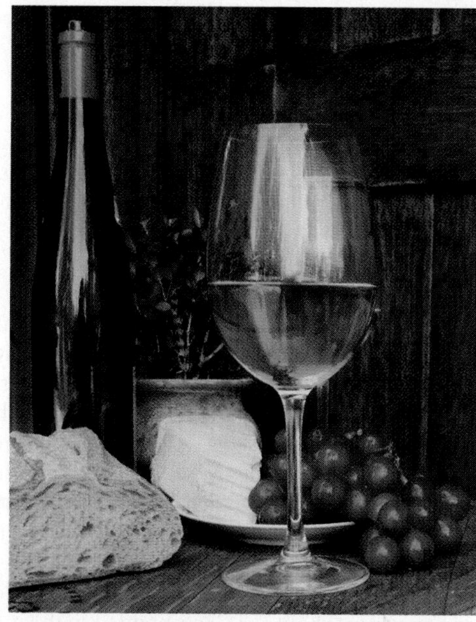

▲ FIGURE E8-4 **Without fermentation, there would be no wine, bread, or cheese**

Case Study revisited

When Athletes Boost Their Blood Counts: Do Cheaters Prosper?

Although runners can do a 100-meter dash without adequate oxygen, distance runners and other long-distance athletes, such as cross-country skiers and cyclists, must pace themselves. They must rely on aerobic cellular respiration for most of the race, saving the anaerobic sprint for the finish. Training for distance events focuses on increasing the capacity of the athletes' respiratory and circulatory systems to deliver enough oxygen to their muscles. For this reason, it is the distance athletes who most often resort to blood doping, to increase the oxygen-carrying capacity of their blood. This delivers more oxygen to their muscles so that cellular respiration can generate the maximum amount of ATP from glucose.

The EPO-mimicking drug CERA—that Riccò now admits taking—helped keep his muscles supplied with ATP by stimulating overproduction of oxygen-carrying red blood cells. In the particularly demanding mountain stages of the Tour de France, which Riccò won, his "clean" competitors were at a disadvantage as their leg muscles became painfully laden with lactate from fermentation sooner than did Riccò's.

Because EPO is produced naturally in the human body, its abuse is hard to detect. CERA, intended for use by people with anemia (who have too few red blood cells), was new on the market at the time of the 2008 Tour de France, and Riccò may have assumed it would be undetectable. But CERA's manufacturer, the pharmaceutical firm Hoffman-La Roche, had provided samples of the drug to the World Anti-Doping Agency before it was marketed, allowing researchers to develop urine tests to identify users.

Consider This

BioEthics Some athletes move to high-altitude locations to train for races run at lower altitudes, because the low oxygen levels present at high altitudes stimulate increased production of red blood cells. Is this cheating? Explain your reasoning. Advances in gene therapy may one day make it possible to modify athletes' cells so that they have extra copies of the gene that produces EPO. Would this be cheating?

CHAPTER REVIEW

Summary of Key Concepts

8.1 How Do Cells Obtain Energy?

Cells produce chemical energy by breaking down glucose into lower-energy compounds and capturing some of the released energy as ATP. During glycolysis, glucose is broken down in the cytoplasmic fluid, forming pyruvate and generating a small quantity of ATP and the high-energy electron carrier NADH. If oxygen is available, pyruvate is broken down through cellular respiration in the mitochondria, generating much more ATP than does glycolysis.

8.2 What Happens During Glycolysis?

During glycolysis, a molecule of glucose is activated by adding high-energy phosphates from two ATP molecules to form fructose bisphosphate. Then, in a series of reactions, the fructose bisphosphate is broken down into two molecules of pyruvate. This produces a net yield of two ATP and two NADH.

8.3 What Happens During Cellular Respiration?

If oxygen is available, cellular respiration can occur. Pyruvate is transported into the matrix of the mitochondria. Here, it reacts with coenzyme A to form acetyl CoA plus CO_2. The acetyl CoA enters the Krebs cycle and releases the remaining two carbons as CO_2. One ATP, three NADH, and one $FADH_2$ are also formed for each acetyl group that goes through the cycle. During the reactions in the mitochondrial matrix, each molecule of glucose that originally entered glycolysis produces a total of two ATP, eight NADH, and two $FADH_2$ (see Fig. 8-7).

The NADH and $FADH_2$ deliver their high-energy electrons to the electron transport chain (ETC) embedded in the inner mitochondrial membrane. The energy of the electrons is used to pump hydrogen ions across the inner membrane from the matrix to the intermembrane space. At the end of the ETC, the depleted electrons combine with oxygen and hydrogen ions, forming water. This is the oxygen-requiring step of cellular respiration. During chemiosmosis, the hydrogen ion gradient created by the ETC is used to produce ATP, as the hydrogen ions diffuse back across the inner membrane through channels that contain ATP synthase. Electron transport and chemiosmosis yield 32 or 34 additional ATP, for a net yield of 36 or 38 ATP per glucose molecule during combined glycolysis and cellular respiration.

BioFlix™ Cellular Respiration

8.4 What Happens During Fermentation?

Glycolysis uses up NAD^+ to produce NADH as glucose is broken down into pyruvate. For these reactions to continue, NAD must be continuously recycled. Under anaerobic conditions, NADH cannot release its high-energy electrons to the ETC, because there is no oxygen to accept them at the end of the chain. So, NAD is regenerated from NADH by fermenting pyruvate to ethanol and CO_2, or to lactate, depending on the organism. Yeast and certain other microorganisms generate ethanol and CO_2. Humans and other animals generate lactate.

Key Terms

adenosine diphosphate (ADP) *130*
adenosine triphosphate (ATP) *128*
aerobic *129*
alcoholic fermentation *138*
anaerobic *129*
cellular respiration *129*
chemiosmosis *134*
citric acid cycle *132*
electron transport chain (ETC) *134*
fermentation *130*
flavin adenine dinucleotide (FAD or $FADH_2$) *132*
glycolysis *129*
intermembrane space *132*
Krebs cycle *132*
lactic acid fermentation *137*
matrix *130*
mitochondrion *130*
nicotinamide adenine dinucleotide (NAD^+ or NADH) *130*

Thinking Through the Concepts

Fill-in-the-Blank

1. The complete breakdown of glucose in the presence of oxygen occurs in two stages: __________ and __________. The first of these stages occurs in the __________ of the cell, and the second stage occurs in organelles called __________. Which stage generates the most ATP? __________
2. Conditions in which oxygen is absent are described as __________. In the absence of oxygen, some microorganisms break down glucose using __________, which generates only __________ molecules of ATP. This process is followed by __________, in which no more ATP is produced, but the electron-carrier molecule __________ is regenerated so it can be used in further glucose breakdown.
3. Yeast in bread dough and alcoholic beverages use a type of fermentation that generates __________ and __________. Muscles pushed to their limit use __________ fermentation. Which form of fermentation is used by microorganisms that produce yogurt, sour cream, and sauerkraut? __________
4. The hormone __________ causes production of extra __________ cells, which increase the ability of the blood to carry __________. This gives athletes a competitive edge because their muscle cells can engage in __________ for energy production for a longer time during vigorous exercise.
5. During cellular respiration, the electron transport chain pumps H^+ out of the mitochondrial __________ into the __________, producing a large __________ of H^+. The ATP produced by cellular respiration is generated by a process called __________. During this process, H^+ travels through membrane channels linked to __________.
6. The cyclic portion of cellular respiration is called the __________ cycle, or the __________ cycle. The molecule that enters this cycle is __________. How many ATP molecules are generated by the cycle per molecule of glucose? __________ What two types of high-energy electron-carrier molecules are generated during the cycle? __________ and __________

Review Questions

1. Starting with glucose ($C_6H_{12}O_6$), write the overall reactions for aerobic respiration.
2. Draw and label a mitochondrion, and explain how its structure relates to its function.
3. What role do the following play in the breakdown of glucose: glycolysis, mitochondrial matrix, inner membrane of mitochondria, fermentation, and NAD^+?
4. Outline the two major stages of glycolysis. How many ATP molecules (overall) are generated per glucose molecule during glycolysis? Where in the cell does glycolysis occur?
5. Under what conditions does fermentation occur? What are its possible products? What is its function?
6. What molecule is the end product of glycolysis? How are the carbons of this molecule used during the Krebs cycle? In what form is most of the energy from the Krebs cycle captured?
7. Describe the mitochondrial electron transport chain and the process of chemiosmosis.
8. Why is oxygen necessary for cellular respiration to occur?
9. Compare the structure of chloroplasts (described in Chapter 7) to that of mitochondria, and describe how the similarities in structure relate to similarities in function. Also describe any differences in structure and function between chloroplasts and mitochondria.

Applying the Concepts

1. Some years ago a freight train overturned, spilling a load of grain. Because the grain was unusable, it was buried in the railroad embankment. Although there is no shortage of other food, the local bear population has become a nuisance by continually uncovering the grain. Yeasts are common in the soil. What do you think has happened to the grain to make the bears keep digging it up, and how is their behavior related to human cultural evolution?
2. Why can't a victim of cyanide poisoning survive by using anaerobic respiration?
3. Some species of bacteria that live at the surface of sediment on the bottom of lakes are capable of using either glycolysis plus fermentation or cellular respiration to generate ATP. There is very little circulation of water in lakes during the summer. Predict and explain what will happen to the bottommost water of lakes as the summer progresses, and describe how this situation will affect energy production by bacteria.
4. Dumping large amounts of raw sewage into rivers or lakes typically leads to massive fish kills, although sewage itself is not toxic to fish. Similar fish kills also occur in shallow lakes that become covered with ice during the winter. What kills the fish? How might you reduce fish mortality after raw sewage is accidentally released into a small pond?
5. Different types of cells respire at different rates. Explain why this is useful. How could you predict the relative respiratory rates of different tissues by examining cells through a microscope?
6. Imagine a hypothetical situation in which a starving cell reaches the stage where every bit of its ATP has been depleted and converted to ADP plus phosphate. If at this point you place the cell in a solution containing glucose, will it recover and survive? Explain your answer based on what you know about glucose breakdown.

MB® *Go to www.masteringbiology.com for practice quizzes, activities, eText, videos, current events, and more.*

UNIT 2

Inheritance

Inheritance provides for both similarity and difference. All dogs share many similarities because their genes are nearly identical. The enormous variety of sizes, fur length and color, and bodily proportions result from tiny differences in their genes.

The Continuity of Life: Cellular Reproduction

Chapter 9

▲ Snuppy (pronounced Snoopy; center), the first cloned dog, poses for a family portrait with his genetic donor, a male Afghan hound (left), and his surrogate mother, a Labrador retriever (right).

Case Study

Send in the Clones

IN FEBRUARY 2008, the Korean biotech company RNL Bio offered to clone your pet dog for a mere $150,000. But some people may have preferred to take their chances in the June online auction hosted by BioArts of Mill Valley, California, where the bidding for "Best Friends Again" dog cloning began at only $100,000 (the winning bids were $170,000, $155,000, and $140,000).

Cloning is the production of one or more individual organisms (**clones**) that are genetically identical to a preexisting individual. Frogs were the first animals to be cloned, back in the 1950s. Cloning pets got its start in 1997 when Joan Hawthorne and John Sperling, the founder of the University of Phoenix, decided that Hawthorne's aging dog, Missy, was special—so special that Sperling invested millions of dollars to clone her. Although the original "Missyplicity Project" at the wittily named Genetic Savings and Clone (GSC) cloned a few cats, GSC failed to clone Missy. In 2006, GSC shut down.

In 2005, however, researchers at the South Korean National University produced Snuppy—a cloned Afghan hound (see the photo to the left). About 2 years later, the Korean team successfully produced three clones of Missy. In 2008, BioArts, the successor company to GSC, announced the "Best Friends Again" auction.

Not all cloned dogs will be pets. For example, Korean teams have produced seven clones of South Korea's top drug-sniffing dog. No one knows if either the ability or the temperament to be superb sniffer dogs is genetically determined, but government officials believe that the expense of cloning will be more than worth it if the clones are as good as their "genetic parent."

Valuable livestock are another favorite subject for cloning. For example, in 2006 the Texas firm ViaGen cloned the 10-time world champion barrel-racing horse, Scamper. Why couldn't Scamper's owner, Charmayne James, just breed Scamper the old-fashioned way? By the time you've finished this chapter, you'll know why many famous horses don't sire offspring that are as athletic as they are. In any case, Scamper is a gelding—a castrated male—so he can't breed.

Why is a clone nearly identical to its genetic parent, while offspring produced by sexual reproduction are often so different from their biological parents and from each other? Keep that question in mind as we explore two types of cellular reproduction—mitotic and meiotic cell division—that provide for both constancy and variability in eukaryotic organisms.

At a Glance

9.1 WHY DO CELLS DIVIDE?

"All cells come from cells." This insight, first stated by the German physician Rudolf Virchow in the mid-1800s, captures the critical importance of cellular reproduction for all living organisms. Cells reproduce by **cell division,** in which a parent cell normally gives rise to two **daughter cells.** In typical cell division, each daughter cell receives a complete set of hereditary information, usually identical to the hereditary information of the parent cell, and about half the cytoplasm.

Cell Division Transmits Hereditary Information to Each Daughter Cell

The hereditary information of all living cells is **deoxyribonucleic acid (DNA),** contained in one or more **chromosomes.** A single molecule of DNA consists of a long chain composed of smaller subunits called **nucleotides** (Fig. 9-1a; see also pp. 51 and 203). Each nucleotide consists of a phosphate, a sugar (deoxyribose), and one of four bases—adenine (A), thymine (T), guanine (G), or cytosine (C). The DNA in a chromosome consists of two long strands of nucleotides wound around each other, as a ladder would look if it were twisted into a corkscrew shape. This structure is called a double helix (Fig. 9-1b). The units of inheritance, called **genes,** are segments of DNA ranging from a few hundred to many thousands of nucleotides long. Like the letters of an alphabet, in a language with very long sentences, the specific sequences of nucleotides in genes spell out the instructions for making the proteins of a cell. We will see in Chapters 11 and 12 how DNA encodes genetic information and how a cell regulates which genes it uses at any given time.

For a cell to survive, it must have a complete set of genetic instructions. Therefore, when a cell divides, it cannot simply split its set of genes in half and give each daughter cell half a set. Rather, the cell must first replicate its DNA to make two identical copies, much like making a photocopy of an instruction manual. Each daughter cell then receives a complete "DNA manual" containing all the genes.

Cell Division Is Required for Growth and Development

The familiar form of cell division in eukaryotic cells, in which each daughter cell is genetically identical to the parent cell, is called mitotic cell division (see sections 9.4 and 9.5). Since your conception as a single fertilized egg, mitotic cell division has produced all the cells in your body, and continues every day in many organs. After cell division, the daughter cells may grow and divide again, or they may **differentiate,** becoming specialized for specific functions, such as contraction (muscle cells), fighting infections (white blood cells), or producing digestive enzymes (cells of the pancreas, stomach, and intestine). This repeating pattern of divide, grow, and (possibly) differentiate, then divide again, is called the **cell cycle** (see sections 9.2 and 9.4).

Most multicellular organisms have three categories of cells, based on their abilities to divide and differentiate:

- **Stem cells** Most of the daughter cells formed by the first few cell divisions of a fertilized egg, and a few cells in adults, including cells in the heart, skin, intestines, brain, and bone marrow, are **stem cells.** Stem cells have two important characteristics: self-renewal and the ability to differentiate into a variety of cell types. Stem cells self-renew because they retain the ability to divide, perhaps for the entire life of the organism. Usually, when a stem cell divides, one of its daughters remains a stem cell, maintaining the population of stem cells. The other daughter cell often undergoes several rounds of cell division, but the resulting cells eventually differentiate into specialized cell types. Some stem cells in early embryos can produce any of the specialized cell types of the entire body.
- **Other cells capable of dividing** Many cells of the bodies of embryos, juveniles, and adults can also divide, but typically differentiate into only one or two different cell types. Dividing cells in your liver, for example, can only become more liver cells.

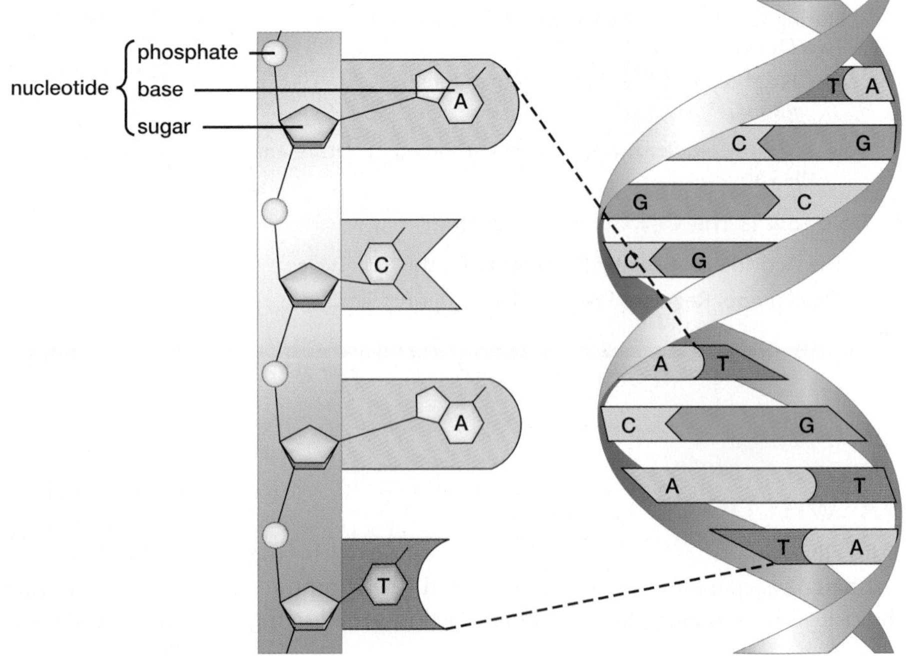

▶ FIGURE 9-1 The structure of DNA
(a) A nucleotide consists of a phosphate, a sugar, and one of four bases—adenine (A), thymine (T), guanine (G), or cytosine (C). A single strand of DNA consists of a long chain of nucleotides held together by bonds between the phosphate of one nucleotide and the sugar of the next. **(b)** Two DNA strands twist around one another to form a double helix.

- **Permanently differentiated cells** Some cells differentiate and never divide again. For example, most of the cells in your heart and brain cannot divide.

Cell Division Is Required for Sexual and Asexual Reproduction

Sexual reproduction in eukaryotic organisms occurs when offspring are produced by the fusion of **gametes** (sperm and eggs) from two adults. Cells in the adult's reproductive system undergo a specialized type of cell division called meiotic cell division, which we will describe in section 9.8, to produce daughter cells with exactly half of the genetic information of their parent cells (and of the "ordinary" body cells of the rest of the adult organism). In animals, these cells become sperm or eggs. When a sperm fertilizes an egg, the resulting offspring once again contain the full complement of hereditary information typical of that species.

Reproduction in which offspring are formed from a single parent, without having a sperm fertilize an egg, is called **asexual reproduction.** Asexual reproduction produces offspring that are genetically identical to the parent. Bacteria (**Fig. 9-2a**) and single-celled eukaryotic organisms, such as *Paramecium* commonly found in ponds (**Fig. 9-2b**), reproduce asexually by cell division, in which two new cells arise from each preexisting cell. Some multicellular organisms can also reproduce by asexual reproduction. A *Hydra* reproduces by growing a small replica of itself, called a bud, on its body (**Fig. 9-2c**). Eventually, the bud is able to live independently and separates from its parent. Many plants and fungi reproduce both asexually and sexually. The beautiful aspen groves of Colorado, Utah, and New Mexico develop asexually from

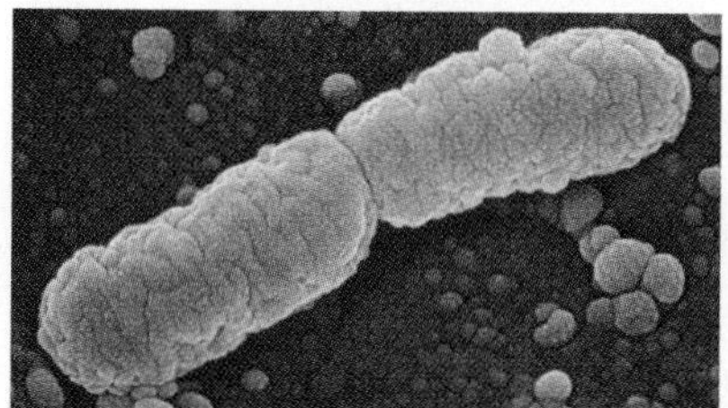

(a) Dividing bacteria

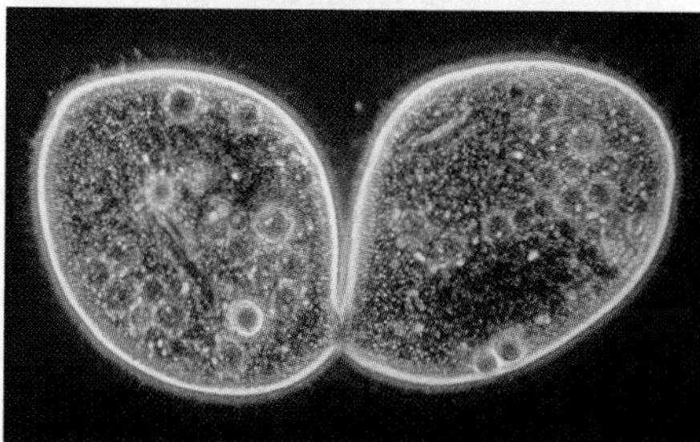

(b) Cell division in *Paramecium*

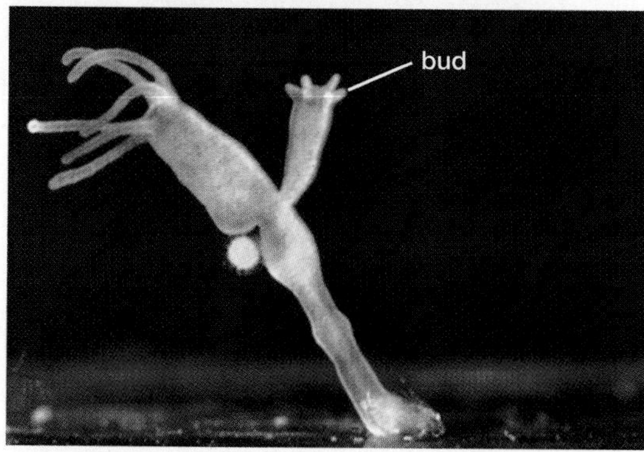
bud

(c) *Hydra* reproduces asexually by budding

The trees in this grove have already lost their leaves
The trees in this grove have begun to change color
The trees in this grove are still green

(d) A grove of aspens often consists of genetically identical trees produced by asexual reproduction

▲ FIGURE 9-2 **Cell division enables asexual reproduction** **(a)** Bacteria reproduce asexually by dividing in two. **(b)** In unicellular eukaryotic microorganisms, such as the protist *Paramecium*, cell division produces two new, independent organisms. **(c)** *Hydra*, a freshwater relative of the sea anemone, grows a miniature replica of itself (a bud) on its side. When fully developed, the bud breaks off and assumes independent life. **(d)** Trees in an aspen grove are often genetically identical. Each tree grows up from the roots of a single ancestral tree. This photo shows three separate groves near Aspen, Colorado. In fall, the timing of the changing color of their leaves shows the genetic identity within a grove and the genetic difference between groves.

shoots growing up from the root system of a single parent tree (**Fig. 9-2d**). Although a grove looks like a population of separate trees, it is often a single individual whose multiple trunks are interconnected by a common root system. Aspen can also reproduce by seeds, which result from sexual reproduction.

Both prokaryotic and eukaryotic cells have cell cycles that include growth, metabolic activity, DNA replication, and cell division. However, prokaryotic and eukaryotic cells have major structural and functional differences, including in the organization of their DNA—the structure, size, number, and location of their chromosomes. Therefore, we will first describe the organization of the prokaryotic chromosome and the prokaryotic cell cycle. We will then examine the eukaryotic chromosome and the essential processes of the eukaryotic cell cycle. Finally, we will describe the two types of cell division in eukaryotes, mitotic cell division and meiotic cell division.

9.2 WHAT OCCURS DURING THE PROKARYOTIC CELL CYCLE?

The DNA of a prokaryotic cell is contained in a single, circular chromosome about a millimeter or two in circumference. Unlike eukaryotic chromosomes, prokaryotic chromosomes are not contained in a membrane-bound nucleus (see pp. 65–67).

The prokaryotic cell cycle consists of a relatively long period of growth—during which the cell also replicates its DNA—followed by a type of cell division called **binary fission,** which means "splitting in two" (**Fig. 9-3a**). The prokaryotic chromosome is usually attached at one point to the plasma membrane of the cell (**Fig. 9-3b ❶**). During the growth phase of the prokaryotic cell cycle, the DNA is replicated, producing two identical chromosomes that become attached to the plasma membrane at nearby, but separate, sites (**Fig. 9-3b ❷**). As the cell grows, new plasma membrane is added between the attachment sites of the chromosomes, pushing them apart (**Fig. 9-3b ❸**). When the cell has approximately doubled in size, the plasma membrane around the middle of the cell grows inward between the two attachment sites (**Fig. 9-3b ❹**). Fusion of the plasma membrane along the equator of the cell completes binary fission, producing two daughter cells, each containing one of the chromosomes (**Fig. 9-3b ❺**). Because DNA replication produces two identical DNA molecules, the two daughter cells are genetically identical to one another and to the parent cell.

Under ideal conditions, binary fission in prokaryotes occurs rapidly. For example, the common intestinal bacterium *Escherichia coli* (usually called simply *E. coli*) can grow, replicate its DNA, and divide in about 20 minutes. Luckily, the environment in our intestines isn't ideal for bacterial growth; otherwise, the bacteria would soon outweigh the rest of our bodies!

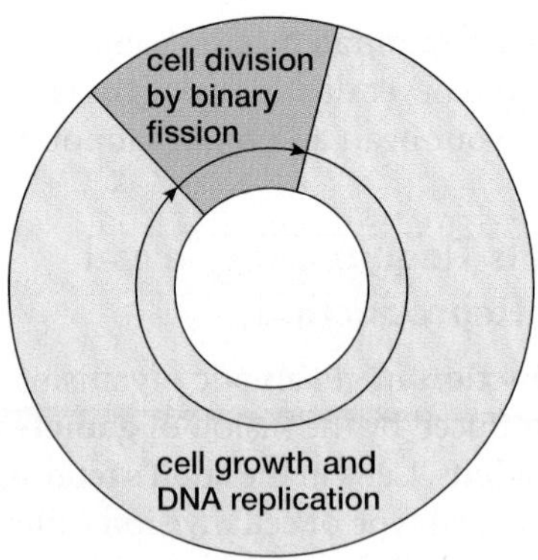

(a) The prokaryotic cell cycle

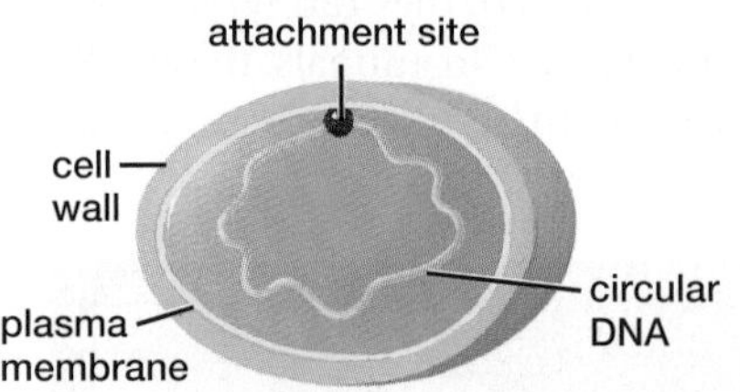

❶ The circular DNA double helix is attached to the plasma membrane at one point.

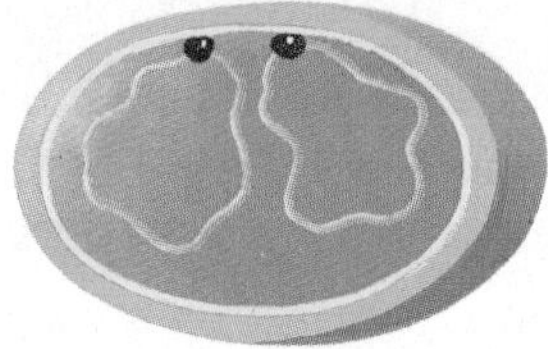

❷ The DNA replicates and the two DNA double helices attach to the plasma membrane at nearby points.

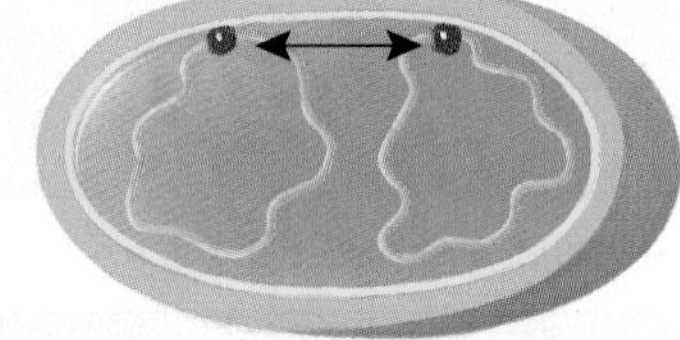

❸ New plasma membrane is added between the attachment points, pushing them farther apart.

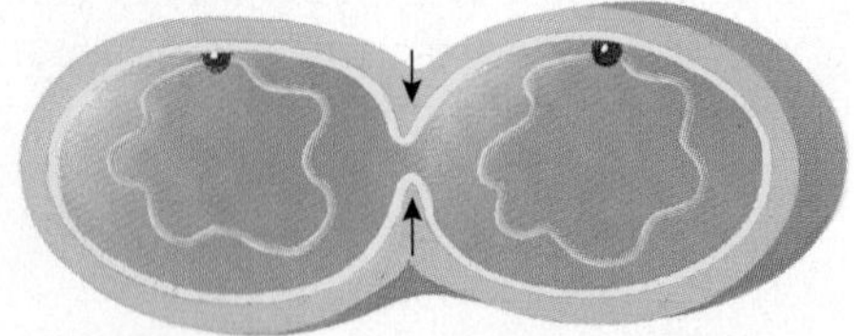

❹ The plasma membrane grows inward at the middle of the cell.

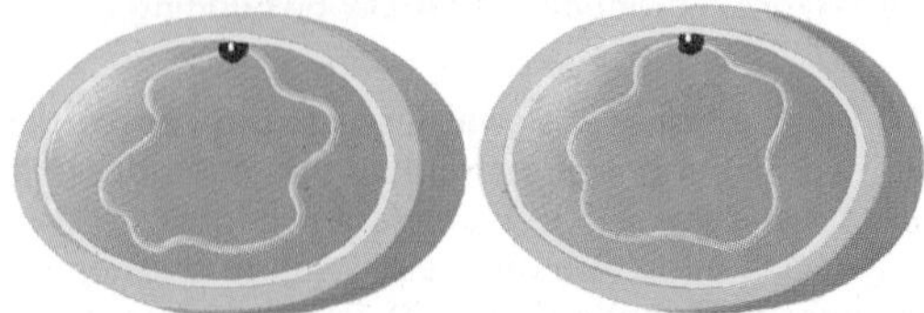

❺ The parent cell divides into two daughter cells.

(b) Binary fission

▶ **FIGURE 9-3 The prokaryotic cell cycle (a)** The prokaryotic cell cycle consists of growth and DNA replication, followed by binary fission. **(b)** Binary fission in prokaryotic cells.

9.3 HOW IS THE DNA IN EUKARYOTIC CHROMOSOMES ORGANIZED?

Eukaryotic chromosomes are separated from the cytoplasm in a membrane-bound nucleus. Further, eukaryotic cells always have multiple chromosomes—the smallest number, 2, is found in the cells of females of a species of ant, but most animals have dozens and some ferns have more than 1,200! Finally, eukaryotic chromosomes usually contain more DNA than do prokaryotic chromosomes. Human chromosomes, for example, are about 10 to 80 times longer than the typical prokaryotic chromosome, and contain 10 to 50 times more DNA. The complex events of eukaryotic cell division are largely an evolutionary solution to the problem of sorting out a large number of long chromosomes. Therefore, we will begin by taking a closer look at the structure of the eukaryotic chromosome.

The Eukaryotic Chromosome Consists of a Linear DNA Double Helix Bound to Proteins

Each human chromosome contains a single DNA double helix, about 50 million to 250 million nucleotides long. If their DNA were completely relaxed and extended, human chromosomes would be about 0.6 to 3.0 inches long (15 to 75 millimeters). Laid end to end, the total DNA in a single human cell would be about 6 feet long (1.8 meters).

Fitting this huge amount of DNA into a nucleus only a few micrometers in diameter is no trivial task. For most of a cell's life, the DNA in each chromosome is wound around proteins called histones. These DNA/histone beads further coil up and attach to other proteins, reducing the length of the DNA about 1,000-fold (**Fig. 9-4**). When a cell needs to read some of its genetic information, it temporarily frees specific regions of DNA from these proteins, but the DNA is soon repacked. However, even this enormous degree of compaction still leaves the chromosomes much too long to be sorted out and moved into the daughter nuclei during cell division. Just as thread is easier to organize when it is wound onto spools, sorting and transporting chromosomes is easier when they are condensed and shortened. During cell division, still more proteins fold up the DNA of each chromosome into compact structures that are about 10 times shorter than they are during the rest of the cell cycle, to about 4 micrometers in length (less than 2 ten-thousands of an inch; see Fig. 9-4).

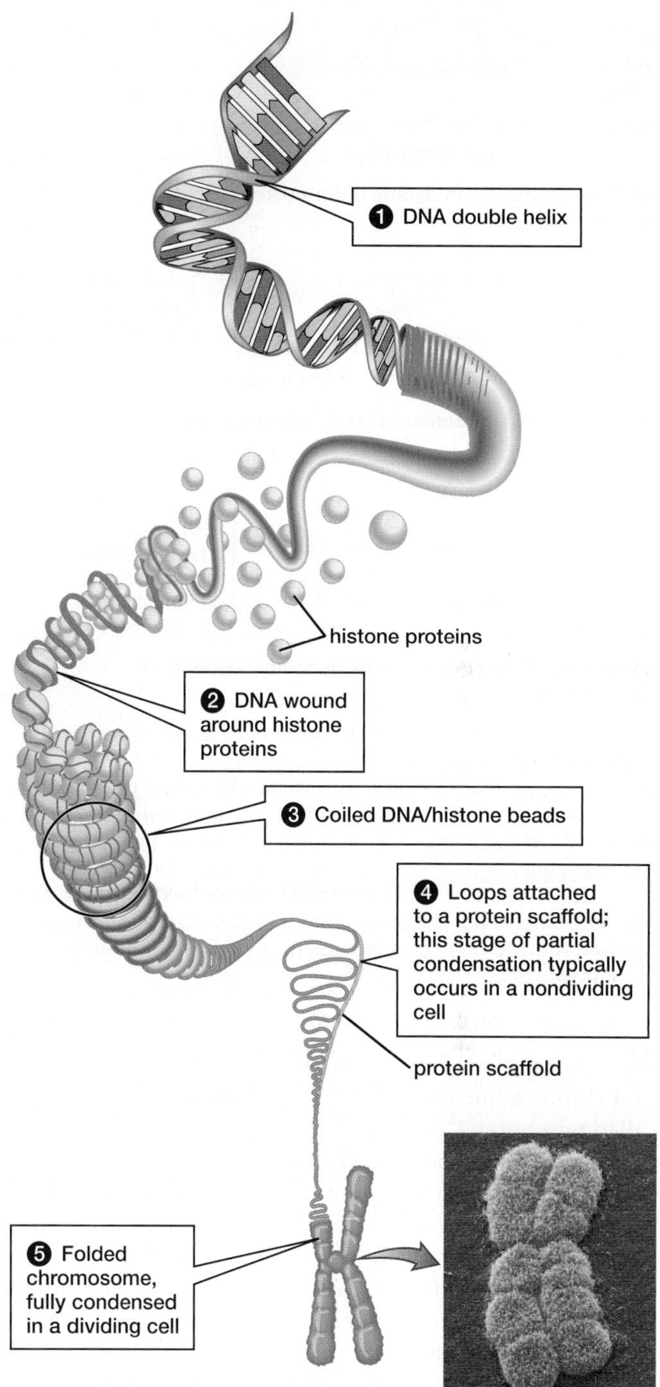

▲ **FIGURE 9-4 Chromosome structure** ❶ A eukaryotic chromosome contains a single DNA double helix. ❷ The DNA is wound around proteins called histones, reducing the length by about a factor of 6. ❸ Other proteins coil up the DNA/histone beads, much like a Slinky toy, reducing the length by another factor of 6 or 7. ❹ These coils are attached in loops to protein "scaffolding" to complete the chromosome as it occurs during most of the life of a cell. The wrapping, coiling, and looping make the chromosome roughly 1,000 times shorter than the DNA molecule it contains. ❺ During cell division, still other proteins fold up the chromosome, yielding about another 10-fold condensation. (Inset) The fuzzy edges visible in the electron micrograph are loops of folded chromosome.

Genes Are Segments of the DNA of a Chromosome

Genes are sequences of DNA from hundreds to thousands of nucleotides long. Each gene occupies a specific place, or **locus** (plural, **loci**), on a chromosome (**Fig. 9-5a**). Chromosomes vary in the number of genes they contain. The largest human chromosome, chromosome 1, contains more than 3,000 genes, whereas one of the smallest human chromosomes, chromosome 22, contains only about 600 genes.

In addition to genes, every chromosome has specialized regions that are crucial to its structure and function: two

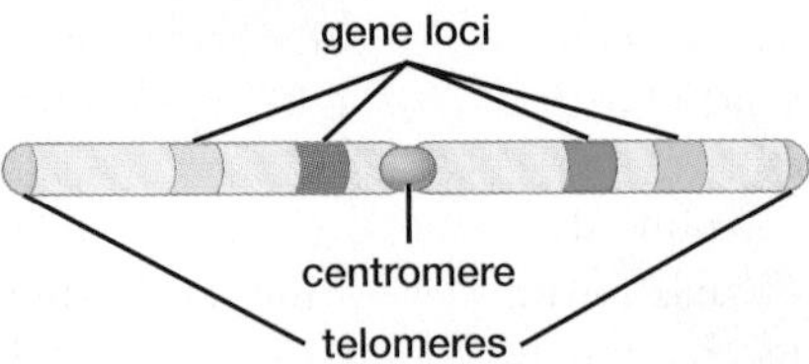

(a) A eukaryotic chromosome before DNA replication

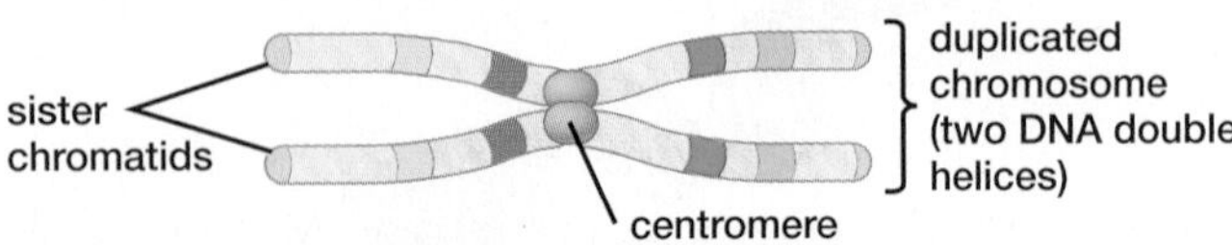

(b) A eukaryotic chromosome after DNA replication

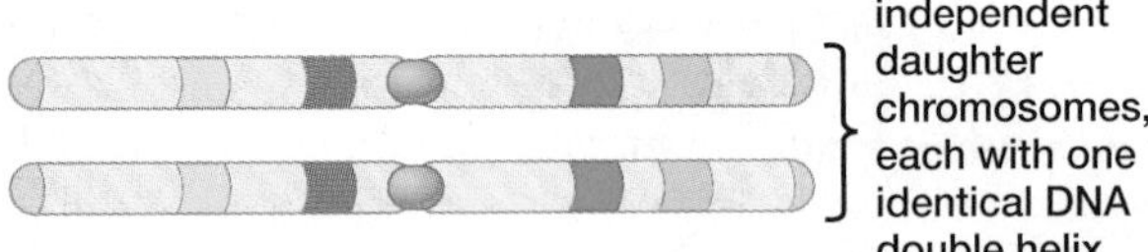

(c) Separated sister chromatids become independent chromosomes

▲ **FIGURE 9-5 The principal features of a eukaryotic chromosome during cell division** **(a)** Before DNA replication, each chromosome consists of a single DNA double helix. Genes are segments of the DNA, usually hundreds to thousands of nucleotides in length. The ends of the chromosome are protected by telomeres. **(b)** The two sister chromatids of a duplicated chromosome are held together by the centromere. **(c)** The sister chromatids separate during cell division to become two independent, genetically identical, chromosomes.

telomeres and one centromere (see Fig. 9-5a). The two ends of a chromosome consist of repeated nucleotide sequences called **telomeres** ("end body" in Greek), which are essential for chromosome stability. Without telomeres, the ends of chromosomes might be removed by DNA repair enzymes, or the ends of two or more chromosomes might become connected, forming long, unwieldy structures that probably could not be distributed properly to the daughter cells during cell division. The second specialized region of the chromosome is the **centromere.** As we will see, the centromere has two principal functions: (1) It temporarily holds two daughter DNA double helices together after DNA replication, and (2) it is the attachment site for microtubules that move the chromosomes during cell division.

Duplicated Chromosomes Separate During Cell Division

Prior to cell division, the DNA within each chromosome is replicated, by a mechanism that we will describe in Chapter 11. At the end of DNA replication, a **duplicated chromosome** consists of two identical DNA double helices, now called sister **chromatids,** which are attached to each other at the centromere (**Fig. 9-5b**).

During mitotic cell division, the two sister chromatids separate, each becoming an independent chromosome that is delivered to one of the two daughter cells (**Fig. 9-5c**).

Eukaryotic Chromosomes Usually Occur in Pairs with Similar Genetic Information

The chromosomes of each eukaryotic species have characteristic shapes, sizes, and staining patterns. When we view an entire set of stained chromosomes from a single cell—its **karyotype**—we see that the nonreproductive cells of many organisms, including humans, contain pairs of chromosomes (**Fig. 9-6**). With one exception that we will discuss shortly, both members of each pair are the same length and have the same staining pattern. This similarity in size, shape, and staining occurs because each chromosome in a pair carries the same genes arranged in the same order. Chromosomes that contain the same genes are called **homologous chromosomes,** or **homologues,** from Greek words that mean "to say the same thing." Cells with pairs of homologous chromosomes are called **diploid,** meaning "double."

Homologous Chromosomes Are Usually Not Identical

Despite their name, homologues usually don't say exactly the "same thing." Why not? A cell might make a mistake when it

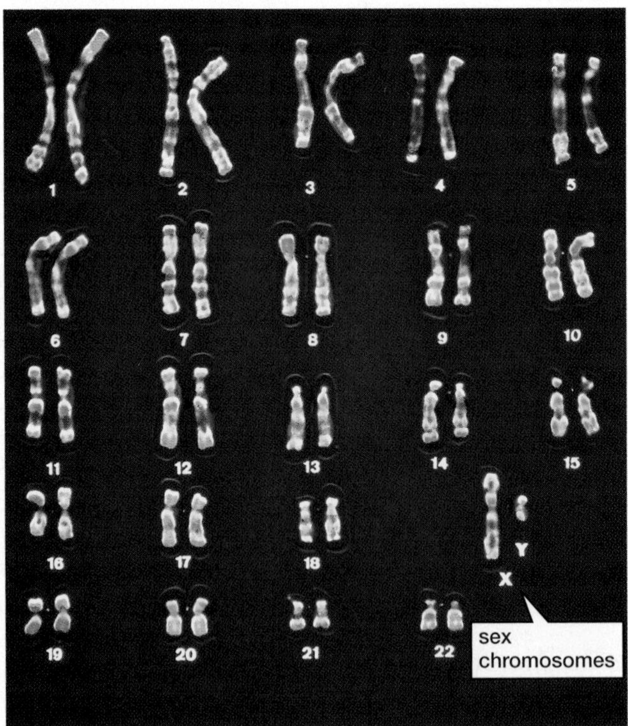

▲ **FIGURE 9-6 The karyotype of a human male** Staining and photographing the entire set of duplicated chromosomes within a single cell produces a karyotype. Pictures of the individual chromosomes are cut out and arranged in descending order of size. The chromosome pairs (homologues) are similar in both size and staining pattern and have similar genetic material. Chromosomes 1 through 22 are the autosomes; the X and Y chromosomes are the sex chromosomes. Note that the Y chromosome is much smaller than the X chromosome. If this were a female karyotype, it would have two X chromosomes.

copies the DNA of one homologue but not the other (see Chapter 11). Or, for example, a ray of ultraviolet light from the sun might zap the DNA of one homologue, thus changing it. These changes in the sequence of nucleotides in DNA are called **mutations,** and will make one homologue be a little different, genetically, than its pair. A given mutation might have happened yesterday, or perhaps 10,000 years ago and have been inherited ever since. If we think of the DNA as an instruction manual for building a cell or an organism, then mutations are like misspelled words in the manual. Although some misspellings might not matter much, others might have serious consequences. For example, single-letter misspellings in crucial genes cause genetic diseases such as sickle-cell anemia and cystic fibrosis. Occasionally, however, a mutation may actually improve the DNA manual and spread throughout a population, as the organisms carrying the mutation survive and reproduce better than other members of its species (see Unit 3).

As shown in Figure 9-6, a typical human cell has 23 pairs of chromosomes, for a total of 46. There are two copies of chromosome 1, two copies of chromosome 2, and so on, up through chromosome 22. These chromosomes—which have similar appearance, similar (but probably not identical) DNA sequences, and are paired in diploid cells of both sexes—are called **autosomes.** The cell also has two **sex chromosomes:** either two X chromosomes (in females) or an X and a Y chromosome (in males). The X and Y chromosomes are quite different in size (see Fig. 9-6) and in genetic composition. Thus, sex chromosomes are an exception to the rule that homologous chromosomes contain the same genes. However, in a male, the X and Y chromosomes behave as a pair during meiotic cell division.

Not All Cells Have Paired Chromosomes

Most cells in our bodies are diploid. However, during sexual reproduction, cells in the ovaries or testes undergo meiotic cell division (see section 9.8) to produce gametes (sperm or eggs). Gametes contain only one member of each pair of autosomes and one of the two sex chromosomes. Cells that contain only one of each type of chromosome are called **haploid** (meaning "half"). In humans, a haploid cell contains one each of the 22 autosomes, plus either an X or Y sex chromosome, for a total of 23 chromosomes. (Think of a haploid cell as one that contains half the diploid number of chromosomes, or one of each type of chromosome.)

When a sperm fertilizes an egg, fusion of the two haploid cells produces a diploid cell with two copies of each type of chromosome. Because one member of each pair of homologues was inherited from the mother (in her egg), these are usually called the maternal chromosomes. The chromosomes inherited from the father (in his sperm) are called paternal chromosomes.

In biological shorthand, the number of different types of chromosomes in a species is called the haploid number and is designated n. For humans, $n = 23$ because we have 23 different types of chromosomes (autosomes 1 to 22 plus one sex chromosome). Diploid cells contain $2n$ chromosomes. Thus, the body cells of humans have 46 (2×23) chromosomes. Every species has a specific number of chromosomes in its cells, from just a handful (e.g., 6 in mosquitoes) to hundreds (in shrimp and some plants).

Not all organisms are diploid. The bread mold *Neurospora,* for example, has haploid cells for most of its life cycle. Some plants, on the other hand, have more than two copies of each type of chromosome, with $4n$, $6n$, or even more chromosomes per cell.

9.4 WHAT OCCURS DURING THE EUKARYOTIC CELL CYCLE?

Newly formed cells usually acquire nutrients from their environment, synthesize additional cellular components, and grow larger. After a variable amount of time—depending on the organism, the type of cell, and the nutrients available—the cell may divide. Each daughter cell may then enter another cell cycle, producing additional cells. Many cells, however, divide only if they receive signals, such as growth hormones, that cause them to enter another cell cycle. Other cells may differentiate and never divide again.

The Eukaryotic Cell Cycle Consists of Interphase and Cell Division

The eukaryotic cell cycle is divided into two major phases: interphase and cell division (**Fig. 9-7**). During **interphase,** the cell acquires nutrients from its environment, grows, and

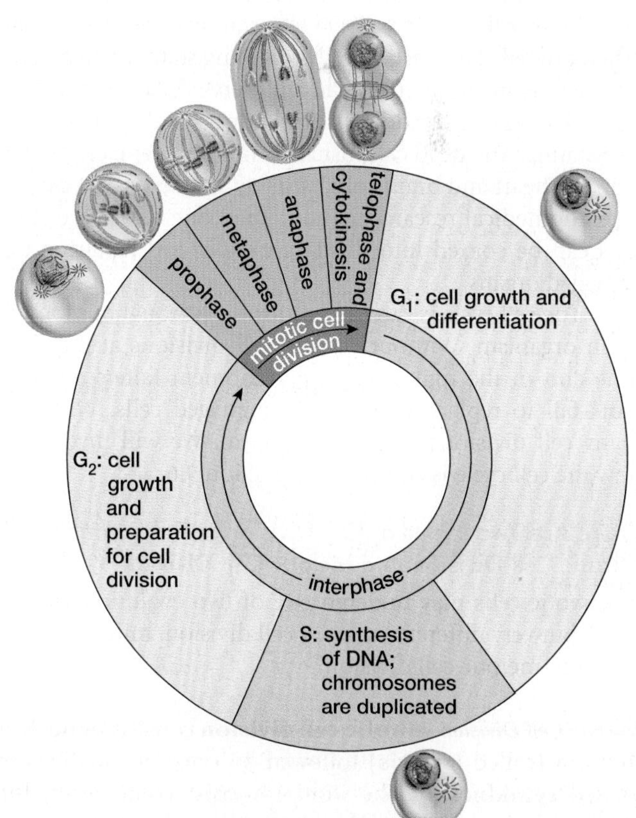

▲ **FIGURE 9-7 The eukaryotic cell cycle** The eukaryotic cell cycle consists of interphase and mitotic cell division.

duplicates its chromosomes. With the exception of meiotic cell division, cell division parcels one copy of each chromosome and usually about half the cytoplasm (including mitochondria, ribosomes, and other organelles) into each of the two daughter cells.

During Interphase, a Cell Grows in Size, Replicates Its DNA, and Often Differentiates

Most eukaryotic cells spend the majority of their time in interphase. For example, some cells in human skin, which divide about once a day, spend roughly 22 hours in interphase. Interphase itself contains three subphases: G_1 (the first *g*ap in DNA synthesis and the first *g*rowth phase), S (DNA synthesis), and G_2 (the second *g*ap in DNA synthesis and the second *g*rowth phase).

Immediately after it is formed by cell division, a newly formed daughter cell enters the G_1 portion of interphase. During G_1, a cell carries out one or more of three activities. First, it almost always grows in size. Second, it often specializes, or differentiates, to perform a specific function. For example, most nerve cells grow long strands, called axons, that allow them to connect with other cells, while liver cells produce bile, proteins such as clotting factors, and enzymes that detoxify many poisonous materials. Third, the cell is sensitive to internal and external signals that help it "decide" whether to divide. If that decision is positive, the cell enters the S phase, when DNA synthesis (replication) occurs. The cell then proceeds to the G_2 phase, during which it may grow some more and then synthesize the proteins needed for cell division.

Many cells, such as liver cells, can be recalled from the differentiated state back into the dividing state, whereas others, such as most heart muscle and nerve cells, never divide again. That is one reason why heart attacks and strokes are so devastating: The dead cells usually cannot be replaced. However, the heart and brain contain a few stem cells that *can* divide. Biomedical researchers hope that, one day, these stem cells can be coaxed into dividing rapidly and repairing the damaged organs.

The cell cycle is carefully controlled throughout the life of an organism. Without enough cell divisions at the right time and in the right organs, development falters or body parts fail to replace worn-out or damaged cells. With too many cell divisions, cancers may form. We will investigate how the cell cycle is controlled in section 9.6.

There Are Two Types of Cell Division in Eukaryotic Cells: Mitotic Cell Division and Meiotic Cell Division

Eukaryotic cells may undergo one of two evolutionarily related, but very different, types of cell division: mitotic cell division or meiotic cell division.

Mitotic Cell Division **Mitotic cell division** consists of nuclear division (called **mitosis**) followed by cytoplasmic division (called **cytokinesis**). The word "mitosis" comes from the Greek for "thread"; during mitosis, chromosomes condense and appear as thin, thread-like structures when viewed through a light microscope. Cytokinesis (from the Greek words for "cell movement") is the process by which the cytoplasm is divided between the two daughter cells. As we will see in section 9.5, mitosis gives each daughter nucleus one copy of the parent cell's replicated chromosomes, and cytokinesis usually places one of these nuclei into each daughter cell. Hence, mitotic cell division typically produces two daughter cells that are genetically identical to each other and to the parent cell, and usually contain about equal amounts of cytoplasm.

Mitotic cell division takes place in all types of eukaryotic organisms. It is the mechanism of asexual reproduction in eukaryotic cells, including unicellular organisms such as yeast, *Amoeba*, and *Paramecium*, and multicellular organisms such as *Hydra* and aspens. Mitotic cell division followed by differentiation of the daughter cells allows a fertilized egg to grow into an adult with perhaps trillions of specialized cells. Mitotic cell division also allows an organism to maintain its tissues, many of which require frequent replacement; to repair damaged parts of the body, for example, after a wound; and sometimes even to regenerate body parts. Mitotic cell division is also the mechanism whereby stem cells reproduce.

Meiotic Cell Division **Meiotic cell division** is a prerequisite for sexual reproduction in all eukaryotic organisms. In animals, meiotic cell division occurs only in ovaries and testes. The process of meiotic cell division involves a specialized nuclear division called **meiosis** and two rounds of cytokinesis to produce four daughter cells that can become gametes (eggs or sperm). Gametes carry half of the genetic material of the parent. As we shall see in section 9.8, the cells produced by meiotic cell division are not genetically identical to each other *or* to the original cell. During sexual reproduction, fusion of two gametes, one from each parent, reconstitutes a full complement of genetic material, forming a genetically unique offspring that is similar to both parents, but identical to neither.

9.5 HOW DOES MITOTIC CELL DIVISION PRODUCE GENETICALLY IDENTICAL DAUGHTER CELLS?

Mitotic cell division consists of mitosis (nuclear division) and cytokinesis (cytoplasmic division) (**Fig. 9-8**). After interphase (**Fig. 9-8a**), when the cell's chromosomes have been duplicated and all other necessary preparations for division have been made, mitotic cell division can occur. For convenience, biologists divide mitosis into four phases, based on the appearance and behavior of the chromosomes: prophase, metaphase, anaphase, and telophase. As with most biological processes, however, these phases are not really discrete events. Rather, they form a continuum, each phase merging into the next.

Cytokinesis usually occurs during telophase. However, mitosis sometimes occurs without cytokinesis, producing cells with multiple nuclei. This is fairly common in fungi and in certain stages of development in plants and flies.

During Prophase, the Chromosomes Condense, the Spindle Microtubules Form, and the Chromosomes Are Captured by the Spindle Microtubules

The first phase of mitosis is called **prophase** (meaning "the stage before" in Greek). During prophase, three major events occur: (1) The duplicated chromosomes condense, (2) the spindle microtubules form, and (3) the chromosomes are captured by the spindle microtubules (**Figs. 9-8b,c**).

Recall that chromosome duplication occurs during the S phase of interphase. Therefore, when mitosis begins, each chromosome already consists of two sister chromatids attached to one another at the centromere. During prophase, the duplicated chromosomes coil up and condense. In addition, the nucleolus, a structure within the nucleus where ribosomes assemble, disappears.

After the duplicated chromosomes condense, the **spindle microtubules** begin to assemble. In all eukaryotic cells, the proper movement of chromosomes during mitosis depends on these spindle microtubules. In animal cells, the spindle microtubules originate from a region in which a pair of microtubule-containing structures called **centrioles** is located. During interphase, a new pair of centrioles forms near the previously existing pair. During prophase, the centriole pairs migrate to opposite sides of the nucleus (see Fig. 9-8b). When the cell divides, each daughter cell will receive a pair of centrioles. Each centriole pair serves as a central point from which the spindle microtubules radiate, both inward toward the nucleus and outward toward the plasma membrane. These points are called spindle poles (see Fig. 9-8c). Although the cells of plants, fungi, many algae, and certain mutant fruit flies do not contain centrioles, they nevertheless form functional spindles during mitotic cell division, showing that centrioles are not required for spindle formation.

As the spindle microtubules form into a complete basket around the nucleus, the nuclear envelope disintegrates, releasing the duplicated chromosomes. Each sister chromatid has a protein-containing structure at its centromere called a **kinetochore.** In each duplicated chromosome, the kinetochores of the two sister chromatids are arranged back-to-back, facing away from one another. The kinetochore of one sister chromatid binds to the ends of spindle microtubules leading to one pole of the cell, while the kinetochore of the other sister chromatid binds to spindle microtubules leading to the opposite pole of the cell (see Fig. 9-8c). The microtubules that bind to kinetochores are called kinetochore microtubules, to distinguish them from microtubules that do not bind to a kinetochore (see below). When the sister chromatids separate later in mitosis, the newly independent chromosomes will move along the kinetochore microtubules to opposite poles.

Other spindle microtubules, called polar microtubules, do not attach to chromosomes; rather, they have free ends that overlap along the cell's equator. As we will see, the polar microtubules will push the two spindle poles apart later in mitosis.

During Metaphase, the Chromosomes Line Up Along the Equator of the Cell

At the end of prophase, the two kinetochores of each duplicated chromosome are connected to spindle microtubules leading to opposite poles of the cell. As a result, each duplicated chromosome is connected to both spindle poles. During **metaphase** (the "middle stage"), the two kinetochores on a duplicated chromosome engage in a "tug of war." During this process, the microtubules lengthen or shorten, until each chromosome lines up along the equator of the cell, with one kinetochore facing each pole (**Fig. 9-8d**).

During Anaphase, Sister Chromatids Separate and Are Pulled to Opposite Poles of the Cell

At the beginning of **anaphase** (**Fig. 9-8e**), the sister chromatids separate, becoming independent daughter chromosomes. This separation allows motor proteins in each kinetochore to pull the chromosome poleward, while simultaneously nibbling off the end of the attached microtubule, thereby shortening it (a mechanism appropriately called "Pac-Man" movement). One of the two daughter chromosomes derived from each original parental chromosome moves to each pole of the cell. Because the daughter chromosomes are identical copies of the parental chromosomes, each cluster of chromosomes that forms at opposite poles of the cell contains one copy of every chromosome that was in the parent cell.

At about the same time, the polar microtubules emanating from each pole grab onto one another where they overlap at the equator. These polar microtubules then simultaneously lengthen and push on one another, which forces the poles of the cell apart, so that the cell assumes an oval shape (see Fig. 9-8e).

During Telophase, Nuclear Envelopes Form Around Both Groups of Chromosomes

When the chromosomes reach the poles, **telophase** (the "end stage") begins (**Fig. 9-8f**). The spindle microtubules disintegrate and a nuclear envelope forms around each group of chromosomes. The chromosomes revert to their extended state, and nucleoli begin to form. In most cells, cytokinesis occurs during telophase, isolating each daughter nucleus in its own daughter cell (**Fig. 9-8g**).

During Cytokinesis, the Cytoplasm Is Divided Between Two Daughter Cells

In animal cells, microfilaments attached to the plasma membrane form a ring around the equator of the cell (see Fig. 9-8f). During cytokinesis, the ring contracts and constricts the cell's equator, much like the drawstring on a pair of sweatpants tightens the waist when pulled (see Fig. 9-8g). Eventually the "waist" constricts completely, dividing the cytoplasm into two new daughter cells (**Fig. 9-8h**).

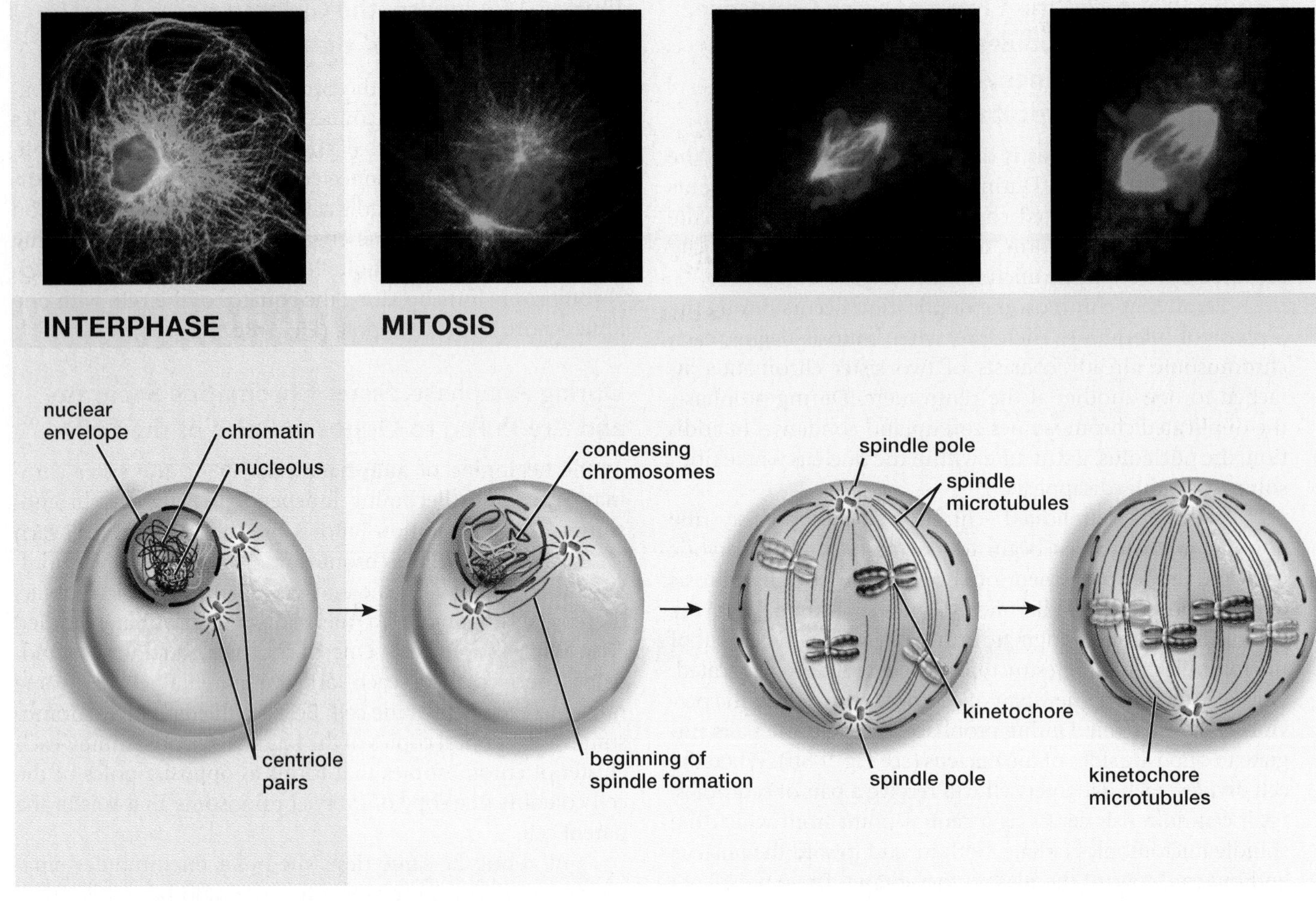

(a) Late Interphase Duplicated chromosomes are in the relaxed uncondensed state; duplicated centrioles remain clustered.

(b) Early Prophase Chromosomes condense and shorten; spindle microtubules begin to form between separating centriole pairs.

(c) Late Prophase The nucleolus disappears; the nuclear envelope breaks down; some spindle microtubules attach to the kinetochore (blue) of each sister chromatid.

(d) Metaphase Kinetochore microtubules line up the chromosomes at the cell's equator.

▲ **FIGURE 9-8 Mitotic cell division in an animal cell**

QUESTION What would the consequences be if one set of sister chromatids failed to separate at anaphase?

Cytokinesis in plant cells is quite different, perhaps because their stiff cell walls make it impossible to divide one cell into two by pinching at the waist. Instead, carbohydrate-filled vesicles, which bud off the Golgi apparatus, line up along the cell's equator between the two nuclei (**Fig. 9-9**). The vesicles fuse, producing a structure called the **cell plate,** which is shaped like a flattened sac, surrounded by membrane and filled with sticky carbohydrates. When enough vesicles have fused, the edges of the cell plate merge with the plasma membrane around the circumference of the cell. The two sides of the cell plate membrane form new plasma membranes between the two daughter cells. The carbohydrates formerly contained in the vesicles remain between the plasma membranes as part of the cell wall.

Following cytokinesis, eukaryotic cells enter G_1 of interphase, thus completing the cell cycle.

Case Study continued

Send in the Clones

Mitotic cell division is essential for cloning, because mitosis produces daughter nuclei that are genetically identical to the parent nucleus. Therefore, nuclei taken from almost any of Missy's or Scamper's cells will produce clones that are genetically identical to their respective dog or horse "nuclear donor." Cloning is explored more thoroughly in "Scientific Inquiry: Carbon Copies—Cloning in Nature and the Lab" on pp. 156–157.

BioFlix™ Mitosis

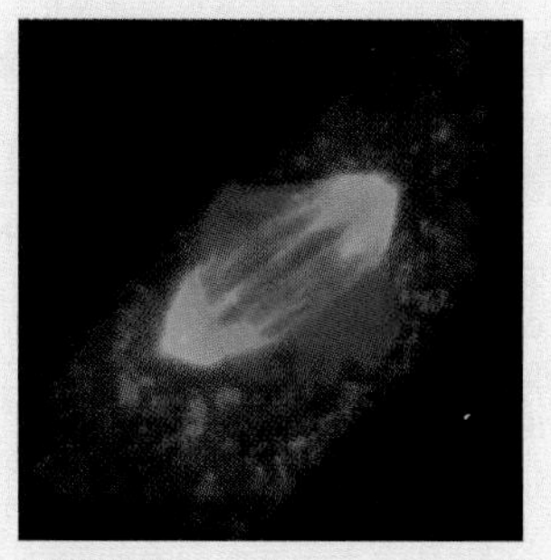
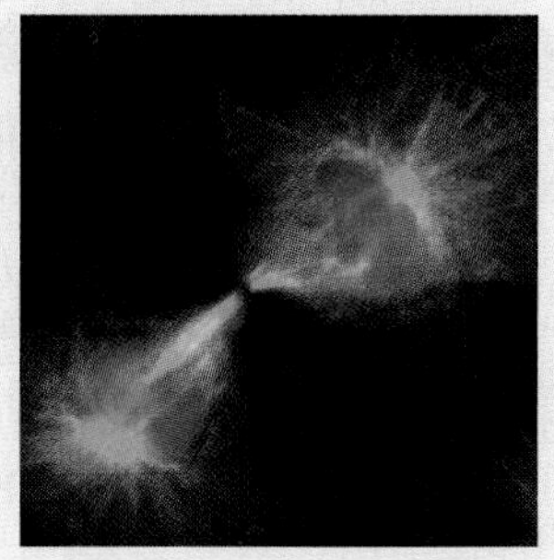
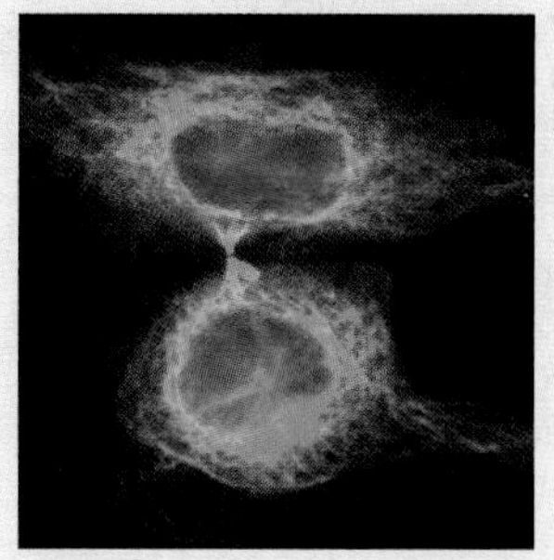
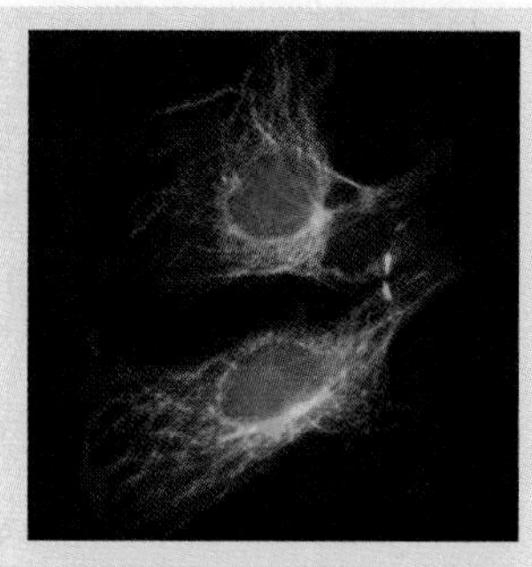

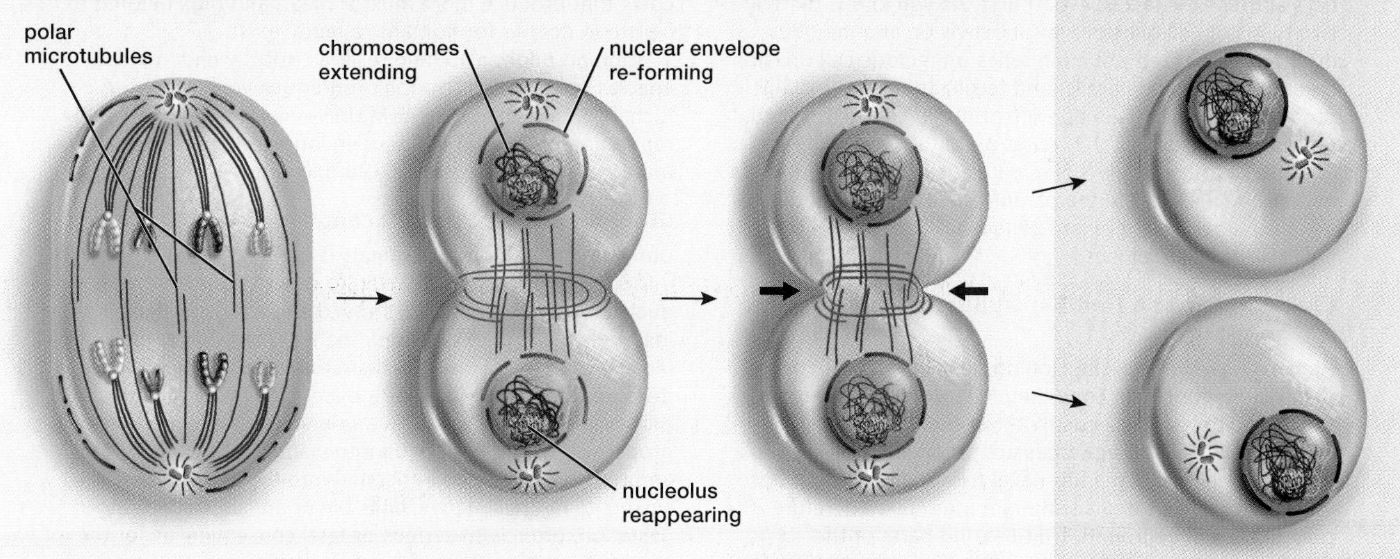

(e) Anaphase
Sister chromatids separate and move to opposite poles of the cell; polar microtubules push the poles apart.

(f) Telophase
One set of chromosomes reaches each pole and begins to decondense; nuclear envelopes start to form; nucleoli begin to reappear; spindle microtubules begin to disappear; microfilaments form rings around the equator.

(g) Cytokinesis
The ring of microfilaments contracts, dividing the cell in two; each daughter cell receives one nucleus and about half of the cytoplasm.

(h) Interphase of daughter cells
Spindles disappear, intact nuclear envelopes form, and the chromosomes extend completely.

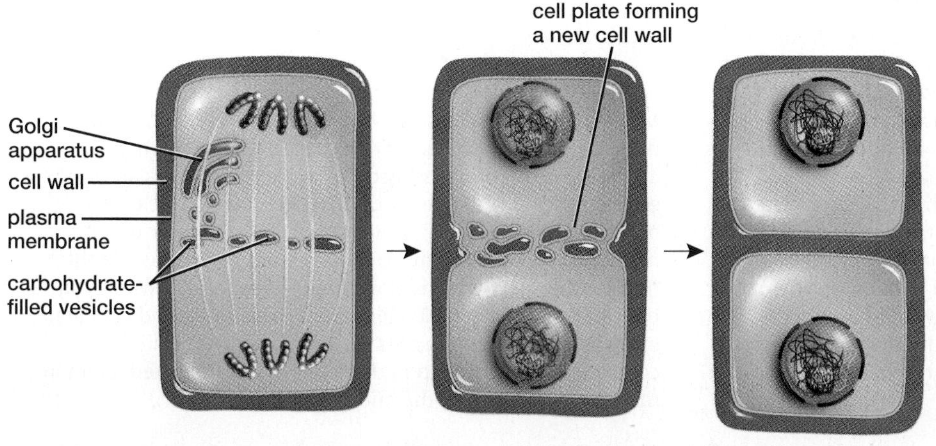

❶ Carbohydrate-filled vesicles bud off the Golgi apparatus and move to the equator of the cell.

❷ The vesicles fuse to form a new cell wall (red) and plasma membrane (yellow) between the daughter cells.

❸ Complete separation of the daughter cells.

◀ **FIGURE 9-9 Cytokinesis in a plant cell**

Scientific Inquiry

Carbon Copies—Cloning in Nature and the Lab

The word "cloning" usually brings to mind images of Dolly the sheep or even *Star Wars: Attack of the Clones*, but nature has been quietly cloning for hundreds of millions of years. How are clones produced, either in nature or in the lab? Why is cloning such a controversial topic? And why is cloning included in a chapter on cell division?

Cloning in Nature: The Role of Mitotic Cell Division

Let's address the last question first. As you know, there are two types of cell division: mitotic division and meiotic division. Sexual reproduction relies on meiotic cell division, the production of gametes, and fertilization, and usually produces genetically unique offspring. In contrast, asexual reproduction (see Fig. 9-2) relies on mitotic cell division. Because mitotic cell division creates daughter cells that are genetically identical to the parent cell, offspring produced by asexual reproduction are genetically identical to their parents—they are clones.

Cloning Plants: A Familiar Application in Agriculture

Humans have been in the cloning business a lot longer than you might think. For example, consider navel oranges, which don't produce seeds. Without seeds, how do they reproduce? Navel orange trees are propagated by cutting a piece of stem from an adult navel tree and grafting it onto the top of the root of a seedling orange tree. Therefore, the cells of the aboveground, fruit-bearing parts of the resulting tree are clones of the original navel orange stem. All navel oranges originated from a single mutant bud of an orange tree discovered in Brazil in the early 1800s, and propagated asexually ever since. Two navel orange trees were brought from Brazil to Riverside, California, in the 1870s—one of them is still there! All American navel orange trees are clones of these two trees.

Cloning Adult Mammals

In the 1950s, John Gurdon and his colleagues destroyed the nuclei of frog eggs, and then inserted new nuclei, taken from embryonic frog cells, into them. Some of the resulting cells developed into complete frogs. By the 1990s, several labs had cloned mammals using nuclei from embryos, but it wasn't until 1996 that Dr. Ian Wilmut of the Roslin Institute in Edinburgh, Scotland, cloned the first adult mammal, the famous Dolly (**Fig. E9-1**).

In agriculture, it is important to clone adults, because only in adults can we see the traits that we wish to propagate (such as milk production in cows, or speed and strength in horses). Any valuable traits of the adult that are genetically determined will also be expressed in all of its clones. Cloning of embryos would usually not be useful, because the embryonic cells would have been produced by sexual reproduction in the first place, and normally no one could tell if the embryo had any especially desirable traits.

For some medical applications, too, cloning adults is essential. Suppose that a pharmaceutical company genetically engineered (see Chapter 13) a cow that secreted a valuable molecule, such as an antibiotic, in its milk. These techniques are extremely expensive and somewhat hit or miss, so the company might successfully produce only one profitable cow. This cow could then be cloned, creating a whole herd of antibiotic-producing cows. Cloned cows that produce more milk or meat, and pigs tailored to be organ donors for humans, already exist.

Cloning might also help rescue critically endangered species, many of which don't reproduce well in zoos. As Richard Adams of Texas A&M University put it, "You could repopulate the world [with an endangered species] in a matter of a couple of years. Cloning is not a trivial pursuit."

Cloning: An Imperfect Technology

Unfortunately, cloning mammals is inefficient and beset with difficulties. An egg is subjected to severe trauma when its nucleus is sucked out or destroyed and a new nucleus is inserted (see Fig. E9-1). Often, the egg may simply die. Molecules in the cytoplasm that are needed to control development may be lost or moved to the wrong places, so that even if the egg survives and divides, it may not develop properly. If the eggs develop into viable embryos, the embryos must then be implanted into the uterus of a surrogate mother. Many clones die or are aborted during gestation, often with serious or fatal consequences for the surrogate mother. Even if the clone survives gestation and birth, it may have defects, commonly a deformed head, lungs, or heart. Given the high failure rate—it took 277 tries to produce Dolly, and more than 1,000 embryos implanted into 123 female dogs to produce Snuppy—cloning mammals is an expensive proposition.

To make things even more problematic, some "successful" clones may have hidden defects. Dolly, for example, developed arthritis when she was 5½ and was euthanized with a serious lung disease when she was 6½, so her problems occurred at a relatively young age (the typical life span of a sheep is 11 to 16 years), although no one knows if these health problems occurred because she was a clone.

The Future of Cloning

Modern cloning technology has now successfully cloned dogs, cows, cats, sheep, horses, and a variety of other animals. As the process becomes more routine, it also brings ethical questions. While hardly anyone objects to cloning navel oranges, and few would refuse antibiotics or other medicinal products from cloned livestock, some people think that cloning pets is a frivolous luxury—especially when you consider that almost 10 million unwanted dogs and cats are euthanized in the United States every year. We will briefly consider human cloning in the "Case Study Revisited: Send in the Clones" section at the end of this chapter.

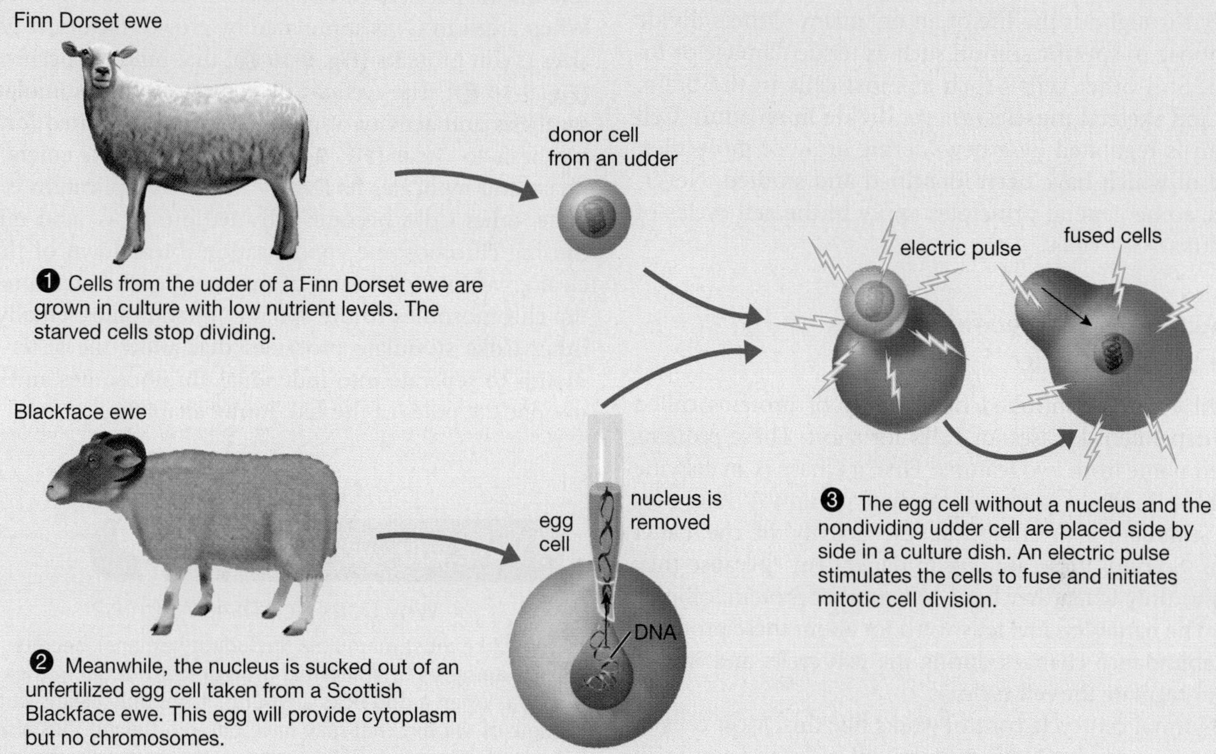

Finn Dorset ewe
donor cell from an udder
electric pulse
fused cells
❶ Cells from the udder of a Finn Dorset ewe are grown in culture with low nutrient levels. The starved cells stop dividing.
Blackface ewe
nucleus is removed
egg cell
DNA
❸ The egg cell without a nucleus and the nondividing udder cell are placed side by side in a culture dish. An electric pulse stimulates the cells to fuse and initiates mitotic cell division.
❷ Meanwhile, the nucleus is sucked out of an unfertilized egg cell taken from a Scottish Blackface ewe. This egg will provide cytoplasm but no chromosomes.

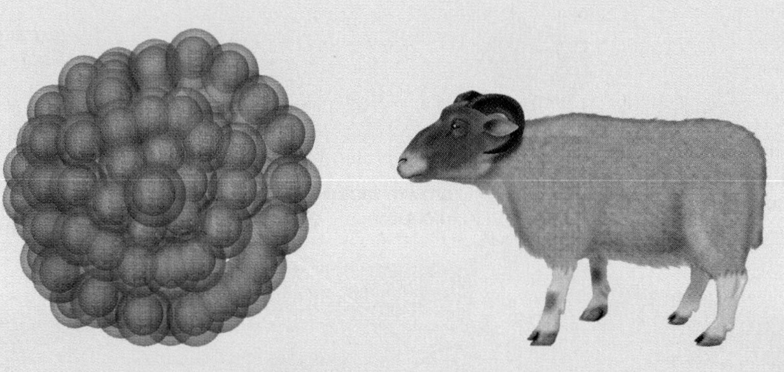

❹ The cell divides, forming an embryo that consists of a ball of cells.
❺ The ball of cells is implanted into the uterus of another Blackface ewe.

❻ The Blackface ewe gives birth to Dolly, a female Finn Dorset lamb, a genetic twin of the Finn Dorset ewe.

▲ **FIGURE E9-1 The making of Dolly**

9.6 HOW IS THE CELL CYCLE CONTROLLED?

Some cells, such as those of the stomach lining, divide frequently throughout the life of an organism. Others divide in response to specific stimuli such as tissue damage or infection. Still other cells—such as most cells in the brain, heart, and skeletal muscles—never divide in an adult. Cell division is regulated by a bewildering array of molecules, not all of which have been identified and studied. Nevertheless, some general principles apply to the cell cycles of most eukaryotic cells.

The Activities of Specific Enzymes Drive the Cell Cycle

The cell cycle is controlled by a family of proteins called cyclin-dependent kinases, or Cdks for short. These proteins get their name from two features: First, a kinase is an enzyme that phosphorylates (adds a phosphate group to) other proteins, stimulating or inhibiting the activity of the target protein. Second, these are "cyclin dependent" because they are active only when they bind still another protein, called a cyclin. The name "cyclin" tells you a lot about these proteins: Their abundance changes during the cell cycle, and in fact helps to regulate the cell cycle.

Normal cell-cycle control works like this. Most cells in your body are in the G_1 phase of the cell cycle, and a cell will usually divide only if it receives signals from hormone-like molecules called growth factors. For example, if you cut your skin, platelets (cell fragments in the blood that are involved in blood clotting) accumulate at the wound site and release growth factors, including platelet-derived growth factor and epidermal growth factor. These growth factors bind to receptors on the surfaces of cells deep in the skin (**Fig. 9-10 ❶**). When a cell in G_1 is stimulated by growth factors, it synthesizes cyclin proteins (**Fig. 9-10 ❷**) that bind to specific Cdks (**Fig. 9-10 ❸**). The cyclin–Cdk complexes then stimulate the synthesis and activity of proteins that are required for DNA synthesis to occur (**Fig. 9-10 ❹**). The cell thus enters the S phase and replicates its DNA. After DNA replication is complete, other Cdks become activated during G_2 and mitosis, causing chromosome condensation, breakdown of the nuclear envelope, formation of the spindle, and attachment of the chromosomes to the spindle microtubules. Finally, still other Cdks stimulate processes that allow the sister chromatids to separate into individual chromosomes and move to opposite poles of the cell during anaphase.

> **Have you ever wondered?**
>
> *Why Dogs Lick Their Wounds?*
>
> Dogs, like most mammals (including humans), secrete large amounts of epidermal growth factor (EGF) in their saliva. When a dog licks a wound, it not only cleans out some of the dirt that may have entered the cut, but also leaves EGF behind. EGF speeds up the synthesis of cyclins, thereby stimulating the division of cells that regenerate the skin, helping to heal the wound more rapidly.

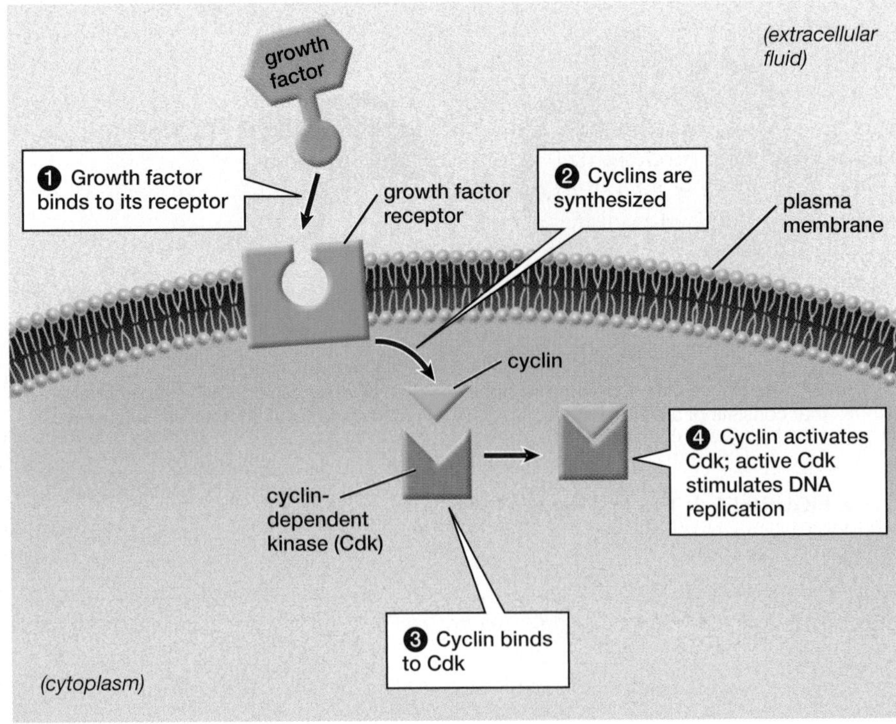

▶ **FIGURE 9-10 Growth factors stimulate cell division** Progress through the cell cycle is under the overall control of cyclin and cyclin-dependent kinases (Cdks). In most cases, growth factors stimulate synthesis of cyclin proteins, which activate Cdks, starting a cascade of events that lead to DNA replication and cell division.

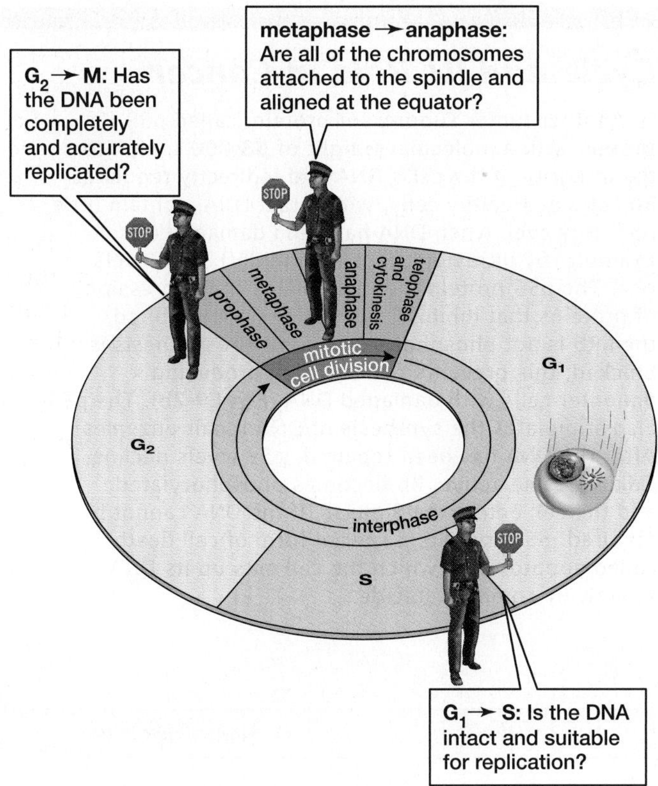
metaphase → anaphase: Are all of the chromosomes attached to the spindle and aligned at the equator?
G_2 → M: Has the DNA been completely and accurately replicated?
STOP
prophase
metaphase
anaphase
telophase and cytokinesis
mitotic cell division
G_1
G_2
interphase
S
G_1 → S: Is the DNA intact and suitable for replication?

▲ **FIGURE 9-11 Control of the cell cycle** Three major checkpoints regulate a cell's transitions from one phase of the cell cycle to the next: (1) G_1 to S, (2) G_2 to mitosis (M), and (3) metaphase to anaphase.

Checkpoints Regulate Progress Through the Cell Cycle

When unregulated, cell division can be dangerous. If a cell contains mutations in its DNA or if daughter cells receive too many or too few chromosomes, the daughter cells may die, or they may become cancerous. To prevent this, the eukaryotic cell cycle has three major **checkpoints** (Fig. 9-11). At each checkpoint, protein complexes in the cell determine whether the cell has successfully completed a specific phase of the cycle:

- **G_1 to S:** Is the cell's DNA intact and suitable for replication?
- **G_2 to mitosis:** Has the DNA been completely and accurately replicated?
- **Metaphase to anaphase:** Are all the chromosomes attached to the spindle and aligned properly at the equator of the cell?

The checkpoint proteins usually regulate the production of cyclins or the activity of cyclin-dependent kinases, or both, thereby regulating progression from one phase of the cell cycle to the next. Defective checkpoint control is a primary cause of cancer, as we explore in "A Closer Look at Control of the Cell Cycle and Its Role in Cancer" on pp. 160–161.

9.7 WHY DO SO MANY ORGANISMS REPRODUCE SEXUALLY?

There are a number of very successful organisms that routinely reproduce asexually. For example, the molds *Penicillium* (which synthesizes penicillin) and *Aspergillus niger* (used in the commercial manufacture of vitamin C) reproduce by "mitospores"—clouds of tiny cells produced by mitosis—and in fact have never been observed to reproduce sexually. Many of the grasses and weeds in your lawn can reproduce by sprouting whole new plants from their stems or roots. Some, like Kentucky bluegrass and dandelions, even bear flowers that can produce seeds without being fertilized! Clearly, asexual reproduction must work pretty well.

Why, then, do nearly all eukaryotic organisms, including bluegrass and dandelions, also reproduce sexually? As we have seen, mitosis can only produce clones of genetically identical offspring. In contrast, sexual reproduction shuffles genes to produce genetically unique offspring. The nearly universal presence of sexual reproduction provides evidence for the tremendous evolutionary advantage that DNA exchange among individuals confers on a species.

Case Study continued

Send in the Clones

Consider a dandelion producing seeds without using sexual reproduction. That dandelion, and all of its offspring, will be genetically identical to one another (except for mutations), probably genetically a little different from all other dandelions, and incapable of sharing genes with any other dandelion. You may have a few such "microspecies" clones in your own backyard!

Mutations in DNA Are the Ultimate Source of Genetic Variability

As we will see in Chapter 11, the fidelity of DNA replication minimizes errors as the DNA is duplicated, but changes in DNA do occur, producing mutations. Although most mutations are either neutral or harmful, they are also the raw material for evolution. Mutations in gametes may be passed to offspring and become a part of the genetic makeup of the species. Such mutations form **alleles,** alternate forms of a given gene that may produce differences in structure or function—such as black, brown, or blond hair in humans, or different mating calls in frogs. As we saw earlier, most eukaryotic organisms are diploid, containing pairs of homologous chromosomes. Homologous chromosomes have the same genes, but each homologue may have the same alleles of some genes and different alleles of other

A Closer Look At *Control of the Cell Cycle and Its Role in Cancer*

As we described in section 9.6, the cell cycle is tightly controlled. Both during embryonic development, and during maintenance and repair of the adult body, progression through the cell cycle is regulated primarily by two interacting processes: (1) production of, and responses to, growth factors that generally speed up the cell cycle; and (2) intracellular checkpoints that stop the cell cycle if problems such as mutations in DNA or misalignment of the chromosomes have occurred. Most cancers develop because one or both of these processes goes awry.

Many different molecules control the cell cycle; here, we will focus on the proteins produced from two important types of genes, called oncogenes and tumor suppressor genes. We will first examine their roles in normal cell cycle control, and then we will explore how defective oncogenes and tumor suppressor genes lead to cancer. Because of the importance of the G_1 to S checkpoint in cancer, we will discuss only their effects at this checkpoint.

Normal Control of the G_1 to S Checkpoint

Proto-oncogenes

Any gene whose protein product tends to promote mitotic cell division is called a proto-oncogene. The genes for growth factors, growth factor receptors, and some of the cyclins and Cdks are proto-oncogenes. In most cases, progress through the cell cycle begins when a growth-stimulating protein, such as epidermal growth factor (EGF), binds to a receptor on the surface of a cell (see Fig. 9-10). This stimulates the synthesis of cyclins, which bind to Cdks and activate them. Thus, these proto-oncogenes are essential to the normal control of the cell cycle. We will shortly examine what happens when a proto-oncogene mutates into an oncogene.

Tumor suppressor genes

The protein products of tumor suppressor genes prevent uncontrolled cell division and the production of daughter cells with mutated DNA, both of which are common in tumors. Let's examine the activity of the proteins produced by two important tumor suppressor genes, *Rb* and *p53*. (We will follow the usual convention of italicizing the gene [*Rb*] and using normal fonts for the protein [Rb].)

Recall that Cdks regulate the activity of other proteins by adding a phosphate group to them. One such protein is Rb. Normally, Rb inhibits transcription of several genes whose protein products are required for DNA synthesis. Phosphorylation of Rb by Cdks relieves this inhibition in the G_1 phase of the cell cycle, allowing the cell to proceed to the S phase and replicate its DNA (**Fig. E9-2a**). This chain of events, from growth factor stimulation to phosphorylation of Rb, ensures that the cell cycle starts up only when the body needs it to.

Another tumor suppressor protein, called p53 ("a **p**rotein with a molecular weight of **53**,000"), monitors the integrity of the cell's DNA and indirectly regulates Rb activity. Healthy cells, with intact DNA, contain little p53. However, when DNA has been damaged (for example, by ultraviolet rays in sunlight), p53 levels rise. The p53 protein then stimulates the expression of proteins that inhibit Cdks. If Cdks are inhibited, then Rb is not phosphorylated and DNA synthesis is blocked; this prevents the cell from producing daughter cells with damaged DNA (**Fig. E9-2b**). The p53 also stimulates the synthesis of DNA repair enzymes. After the DNA has been repaired, p53 levels decline, Cdks become active, Rb becomes phosphorylated, and the cell enters the S phase. If the DNA cannot be repaired, p53 triggers a special form of cell death called apoptosis, in which the cell cuts up its DNA and effectively commits suicide.

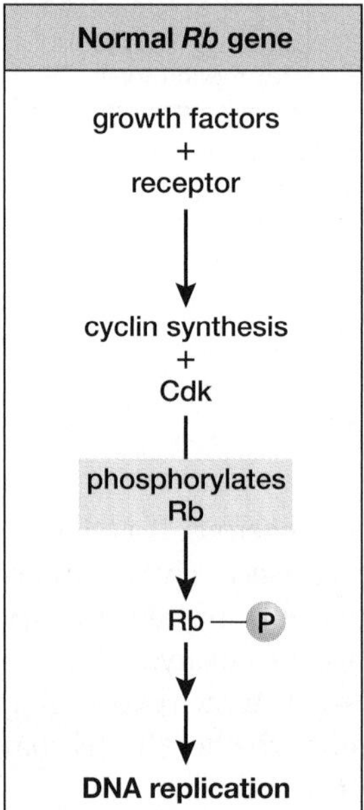

(a) Normal operation of the G_1 to S checkpoint

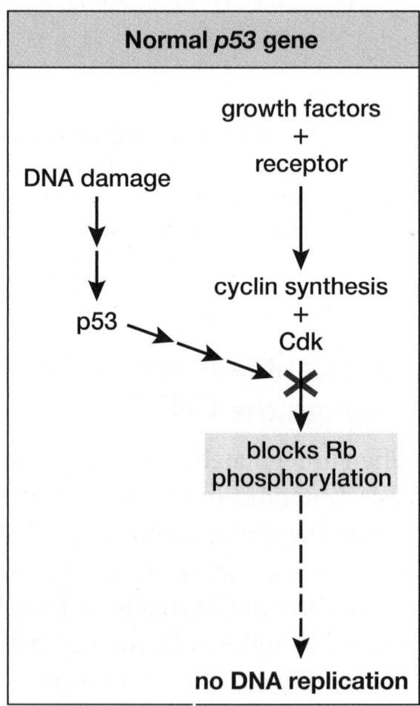

(b) DNA damage prevents progression through the G_1 to S checkpoint

▲ **FIGURE E9-2 Controlling the transition from G_1 to S** **(a)** The Rb protein inhibits DNA synthesis. Toward the end of the G_1 phase, cyclin levels rise. These activate Cdk, which then adds a phosphate group to the Rb protein. Phosphorylated Rb no longer inhibits DNA synthesis, so the cell enters the S phase. **(b)** Damaged DNA stimulates increased levels of the p53 protein, which triggers a cascade of events that inhibit Cdk, thereby preventing entry into the S phase until the DNA has been repaired.

Defective Control of the G_1 to S Checkpoint Can Produce Cancerous Cells

Oncogenes

A mutation may convert a harmless, indeed essential, proto-oncogene into a cancer-causing oncogene ("a gene that causes cancer"). For example, mutated receptors for growth factors may be "turned on" all the time, regardless of the presence or absence of a growth factor (**Fig. E9-3a**). Mutations in cyclin genes may cause cyclins to be continuously synthesized at a high rate, regardless of growth factor activity (see Fig. E9-3a). In either case, a cell may skip right through the G_1 to S checkpoint and divide much more frequently than it should. Rapid, uncontrolled cell division, of course, is one of the hallmarks of cancerous cells.

Inactive tumor suppressor genes

Many carcinogens mutate the *Rb* or *p53* genes, so that the proteins encoded by these genes can't do their jobs (**Fig. E9-3b**). Mutated Rb mimics phosphorylated Rb, which permits uncontrolled DNA synthesis. Mutated p53 is inactive, so Cdks remain active, phosphorylating Rb and allowing damaged DNA to be replicated (see Fig. E9-3b). With either mutation, replication proceeds, whether or not the DNA has been damaged. Not surprisingly, about half of all cancers—including tumors of the breast, lung, brain, pancreas, bladder, stomach, and colon—have mutations in the *p53* gene. Many others, including tumors of the eye (retinoblastoma, hence the name *Rb*), lung, breast, and bladder, have a mutated *Rb* gene.

From Mutated Cell to Cancer

Do overactive oncogenes or inactive tumor suppressor genes, and the mutated DNA that usually accompanies them, doom a person to cancer? Not necessarily. Many mutations cause the surface of a cell to "look foreign" to the cells of the immune system, which then kills the mutated cell. Occasionally, however, a renegade cell survives and reproduces. Because mitotic cell division usually faithfully transmits genetic information from cell to cell, the descendants of the original cancerous cell will themselves be cancerous.

Why does medical science, which has conquered smallpox, measles, and a host of other diseases, have such a difficult time curing cancer? Both normal and cancerous cells use the same machinery for cell division, so treatments that slow down the multiplication of cancer cells also inhibit the maintenance of essential body parts, such as the stomach, intestine, and blood cells. Truly effective and selective treatments for cancer must target cell division only in cancerous cells. Although great strides have been made in the fight to cure cancer, much remains to be done.

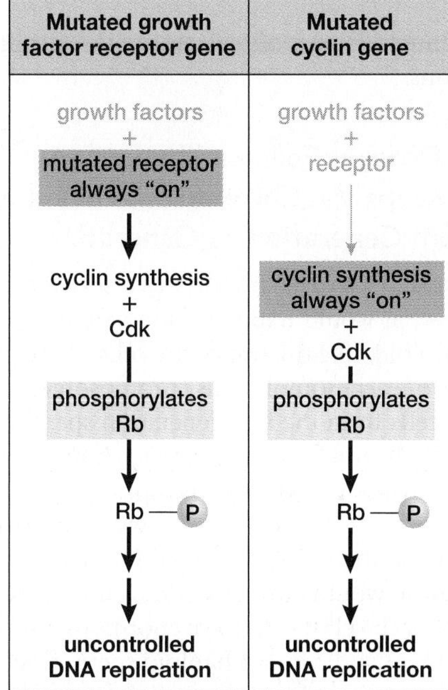

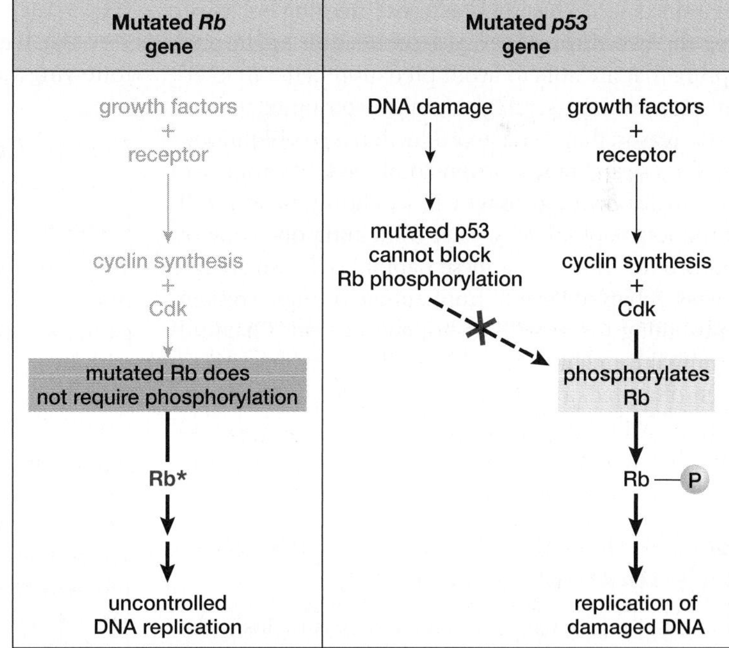

▲ **FIGURE E9-3 Actions of oncogenes and tumor suppressor genes (a)** Oncogenes permit DNA replication and cell division without stimulation by growth factors. **(b)** Mutated *Rb* tumor suppressor genes (left) permit DNA replication without stimulation by growth factors. Mutated *p53* tumor suppressor genes (right) permit replication of damaged DNA.

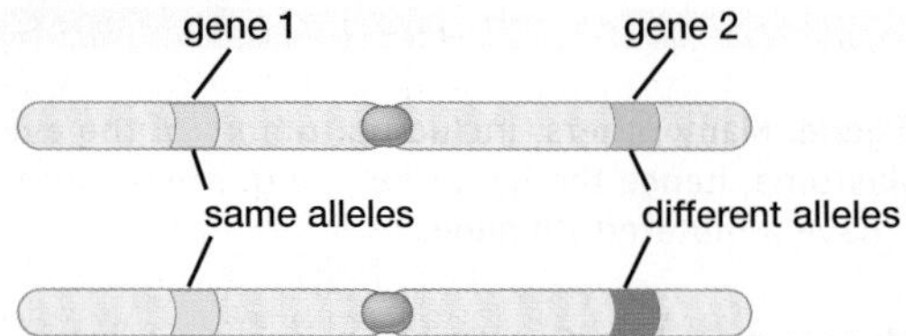

▲ **FIGURE 9-12 Homologous chromosomes may have the same or different alleles of individual genes** The two homologues have the same genes at the same locations (loci). The homologues may have the same allele of a gene at certain loci (left) and different alleles of a gene at other loci (right).

genes (**Fig. 9-12**). We'll explore the consequences of having paired genes—and more than one allele of each gene—in Chapter 10.

Sexual Reproduction May Combine Different Parental Alleles in a Single Offspring

Let's consider a hypothetical evolutionary problem: Camouflage coloration can help an animal avoid predation only if it stays still when it sees a predator. Camouflaged animals that constantly jump around and brightly colored animals that "freeze" when a predator appears will probably both be eaten. How might a single animal combine camouflage coloration and freezing behavior? Let's suppose that a ground-nesting bird has better-than-average camouflage color, while another bird of the same species has more effective freezing behavior. Combining the two through sexual reproduction might produce offspring that are able to avoid predation better than either parent. Combining useful, genetically determined traits is probably one reason that sexual reproduction is so ubiquitous.

How does sexual reproduction combine traits from two parents in a single offspring? As we will see shortly, meiotic cell division produces haploid cells, each containing one copy of each chromosome. In animals, these haploid cells usually become gametes. A haploid sperm from animal A might contain alleles contributing to camouflage coloration, and a haploid egg from animal B might contain alleles that favor freezing at the first sign of a predator. Fusion of these gametes may produce an animal with camouflage coloration that also becomes motionless when a predator approaches.

9.8 HOW DOES MEIOTIC CELL DIVISION PRODUCE HAPLOID CELLS?

The key to sexual reproduction in eukaryotes is meiosis, the production of haploid nuclei with unpaired chromosomes from diploid parent nuclei with paired chromosomes.

Meiosis Separates Homologous Chromosomes, Producing Haploid Daughter Nuclei

In meiotic cell division (meiosis followed by cytokinesis), each daughter cell receives one member of each pair of homologous chromosomes. Therefore, meiosis (from a Greek word meaning "to diminish") reduces the number of chromosomes in a diploid cell by half. For example, each diploid cell in your body contains 23 *pairs* of chromosomes; meiotic cell division produces sperm or eggs with 23 chromosomes, one from each pair.

Meiosis evolved from mitosis, so many of the structures and events of meiosis are similar or identical to those of mitosis. However, meiosis differs from mitosis in a major way: During meiosis, the cell undergoes *one* round of DNA replication followed by *two* nuclear divisions. One round of DNA replication produces two chromatids in each duplicated chromosome. Because diploid cells have pairs of homologous chromosomes—with two chromatids per homologue—a single round of DNA replication creates four chromatids for each type of chromosome (**Fig. 9-13a**).

The first division of meiosis (called meiosis I) separates the pairs of homologues and sends one of each pair into each of two daughter nuclei, producing two haploid nuclei. Each homologue, however, still consists of two chromatids (**Fig. 9-13b**).

A second division (called meiosis II) separates the chromatids and parcels one chromatid into each of two more daughter nuclei. Therefore, at the end of meiosis, there are four haploid daughter nuclei, each with one copy of each homologous chromosome. Because each nucleus is usually contained in a different cell, meiotic cell division normally produces four haploid cells from a single diploid parent cell (**Fig. 9-13c**).

We'll explore the stages of meiosis in more detail in the following sections.

Meiotic Cell Division Followed by Fusion of Gametes Keeps the Chromosome Number Constant from Generation to Generation

Why is meiotic cell division so important to sexual reproduction? Consider what would happen if the parent organisms' gametes were diploid, like the rest of the cells of the parents, with two copies of each homologous chromosome. Fertilization of diploid cells from the first-generation sperm and egg would result in a cell with four copies of each homologue, giving the offspring twice as many chromosomes as its parents. These offspring would produce gametes with four copies of each homologue, so their offspring would have eight copies. The next generation would have 16 copies, and so on. After a few generations, each cell would have enormous amounts of DNA. On the other hand, when a haploid sperm fuses with a haploid egg, the resulting offspring are diploid, just like their parents (**Fig. 9-14**).

Meiosis I Separates Homologous Chromosomes into Two Haploid Daughter Nuclei

The phases of meiosis have the same names as the roughly equivalent phases in mitosis, followed by I or II to distinguish the two nuclear divisions that occur in meiosis

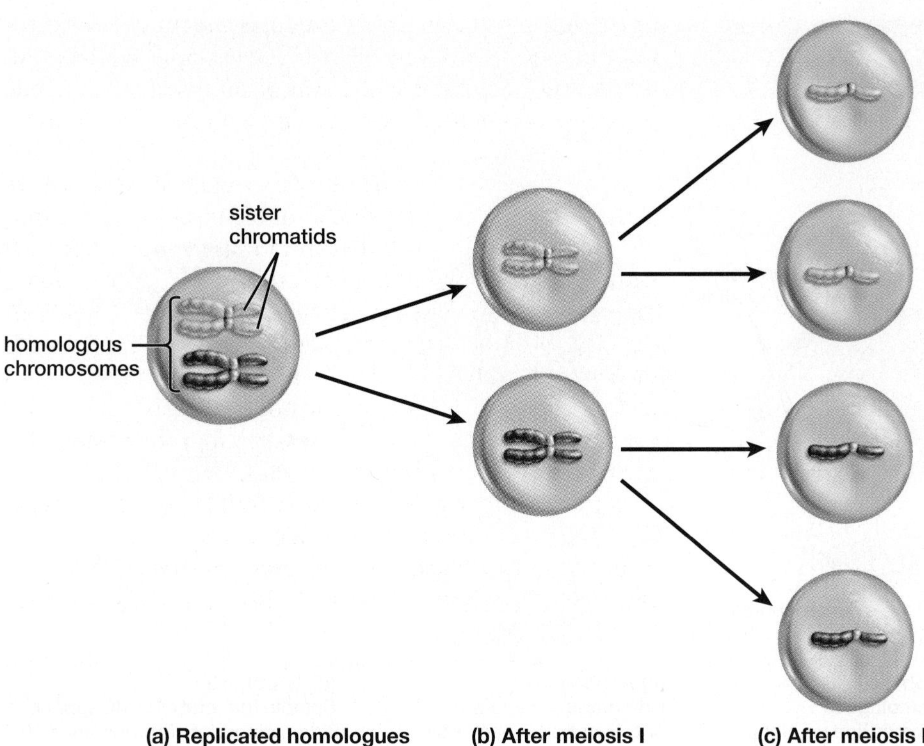

◀ **FIGURE 9-13 Meiosis halves the number of chromosomes (a)** Both members of a pair of homologous chromosomes are replicated prior to meiosis. **(b)** During meiosis I, each daughter cell receives one member of each pair of homologues. **(c)** During meiosis II, sister chromatids separate into independent chromosomes, and each daughter cell receives one of these chromosomes.

(**Fig. 9-15**). In the descriptions that follow, we assume that cytokinesis accompanies the nuclear divisions. As in mitosis, the chromosomes are duplicated during interphase prior to meiosis, and the sister chromatids of each chromosome are attached to one another at the centromere when meiosis begins.

During Prophase I, Homologous Chromosomes Pair Up and Exchange DNA

During mitosis, homologous chromosomes move completely independently of each other. In contrast, during prophase I of meiosis, homologous chromosomes line up side by side and exchange segments of DNA (**Fig. 9-15a** and **Fig. 9-16**). We'll call one homologue the "maternal homologue" and the other the "paternal homologue," because one was originally inherited from the organism's mother and the other from the organism's father.

During prophase I, proteins bind the maternal and paternal homologues together so that they align precisely along their entire length. Enzymes then cut through the DNA of the paired homologues and graft the cut ends together, often joining part of one of the chromatids of the maternal homologue to part of one of the chromatids of the paternal homologue, and vice versa. The binding proteins and enzymes then depart, leaving crosses, or **chiasmata** (singular, **chiasma**), where the maternal and paternal chromosomes have exchanged parts (see Fig. 9-16). In human cells, each pair of homologues usually forms two or three chiasmata during prophase I. The mutual exchange of DNA between maternal and paternal chromosomes at chiasmata is called **crossing over.** If the chromosomes have different alleles, then crossing over creates genetic **recombination,** the formation of new combinations of alleles on a chromosome.

In addition to being the sites of crossing over, the arms of the homologues remain temporarily entangled at the chiasmata. This keeps the two homologues together until they are pulled apart during anaphase I.

As in mitosis, the spindle microtubules begin to assemble outside the nucleus during prophase I. Near the end of prophase I, the nuclear envelope breaks down and spindle microtubules invade the nuclear region, capturing the chromosomes by attaching to their kinetochores.

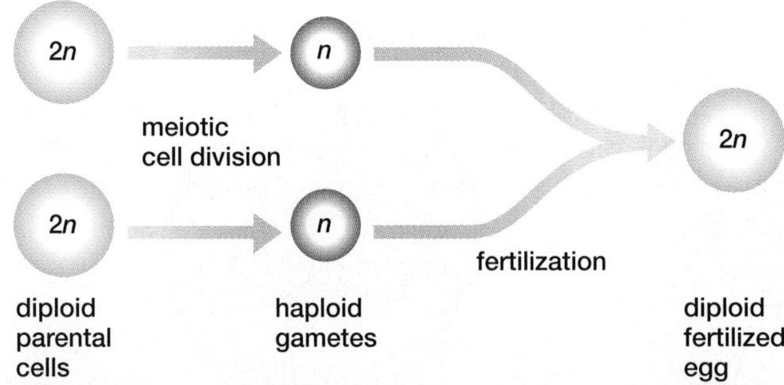

▲ **FIGURE 9-14 Meiotic cell division is essential for sexual reproduction** In sexual reproduction, specialized reproductive cells undergo meiosis to produce haploid cells. In animals, these cells become gametes (sperm or eggs). When an egg is fertilized by a sperm, the resulting fertilized egg, or zygote, is once again diploid.

MEIOSIS I

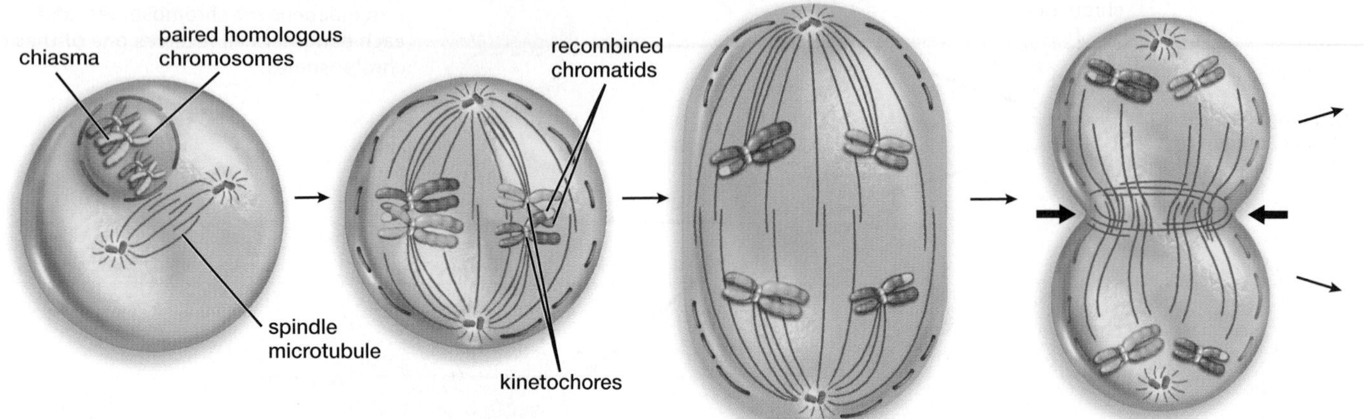

(a) Prophase I
Duplicated chromosomes condense. Homologous chromosomes pair up and chiasmata occur as chromatids of homologues exchange parts by crossing over. The nuclear envelope disintegrates, and spindle microtubules form.

(b) Metaphase I
Paired homologous chromosomes line up along the equator of the cell. One homologue of each pair faces each pole of the cell and attaches to the spindle microtubules via the kinetochore (blue).

(c) Anaphase I
Homologues separate, one member of each pair going to each pole of the cell. Sister chromatids do not separate.

(d) Telophase I
Spindle microtubules disappear. Two clusters of chromosomes have formed, each containing one member of each pair of homologues. The daughter nuclei are therefore haploid. Cytokinesis commonly occurs at this stage. There is little or no interphase between meiosis I and meiosis II.

▲ **FIGURE 9-15** The homologous chromosomes of a diploid cell are separated, producing four haploid daughter cells. Two pairs of homologous chromosomes are shown. Yellow chromosomes are from one parent and violet chromosomes are from the other parent.

QUESTION What would be the consequences for the resulting gametes if one pair of homologues failed to separate at anaphase I?

During Metaphase I, Paired Homologous Chromosomes Line Up at the Equator of the Cell

During metaphase I, interactions between the kinetochores and the spindle microtubules move the paired homologues to the equator of the cell (**Fig. 9-15b**). Unlike in mitosis, where *individual* duplicated chromosomes line up along the equator, in meiosis, *homologous pairs* of duplicated chromosomes, held together by chiasmata, line up along the equator during metaphase I. Which member of a pair of homologous chromosomes faces which pole of the cell is random—the

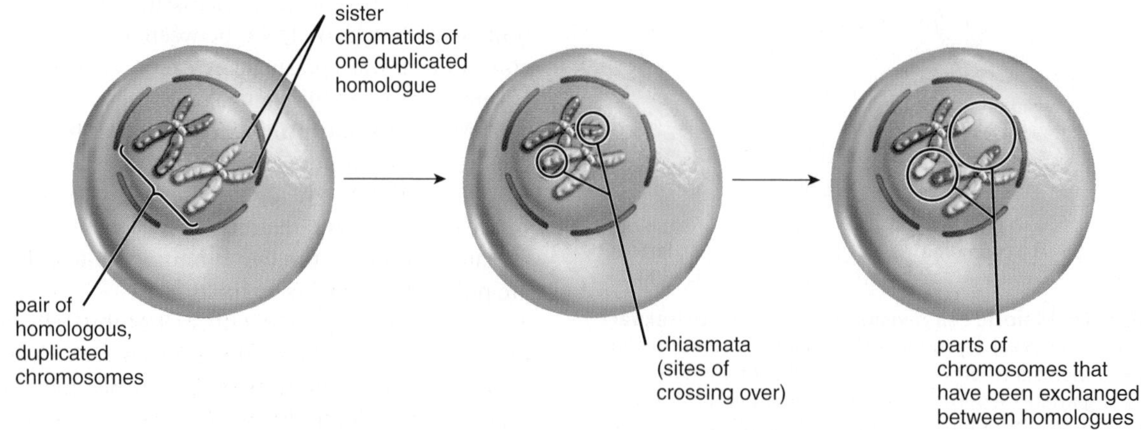

▲ **FIGURE 9-16 Crossing over** Nonsister chromatids of different members of a homologous pair of chromosomes exchange DNA at chiasmata.

MEIOSIS II

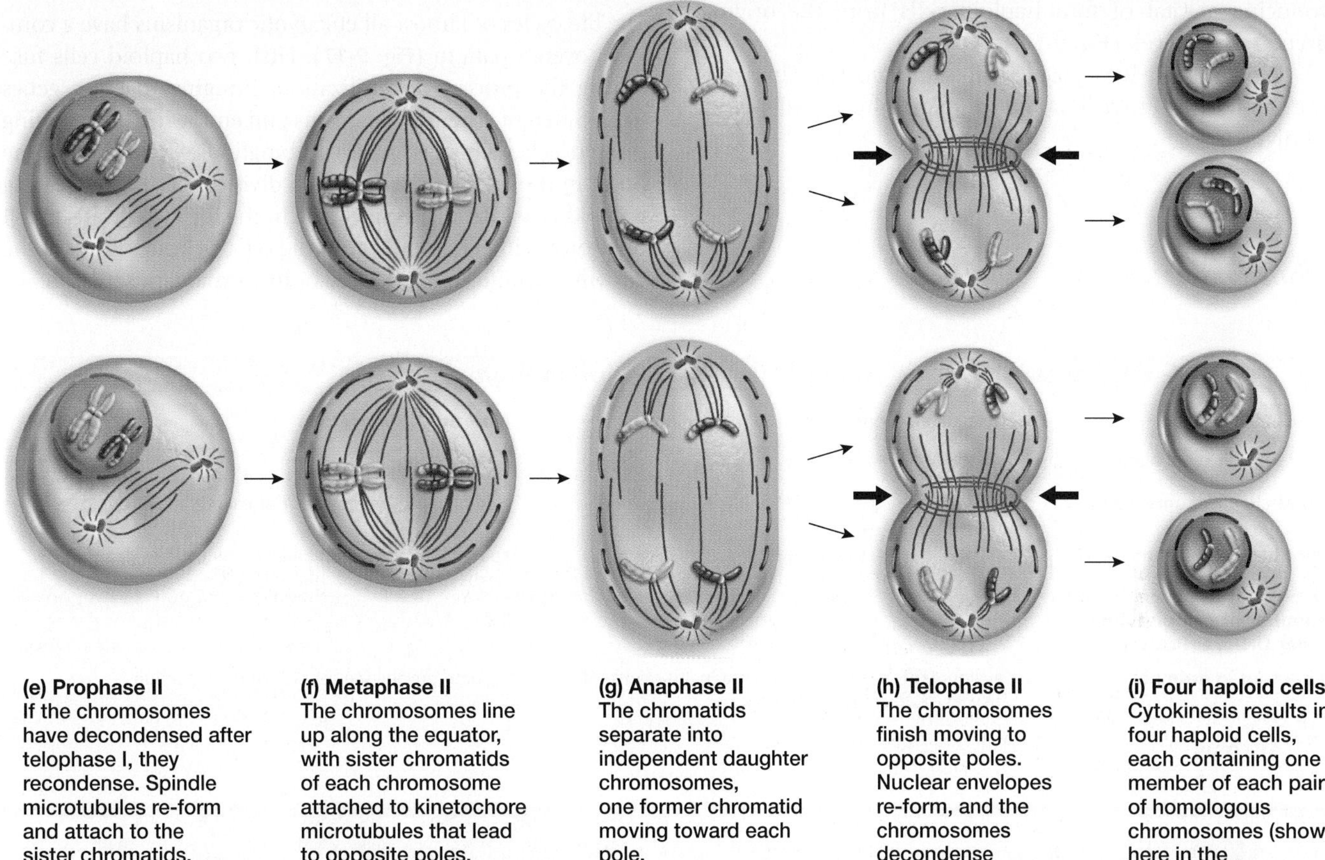

(e) Prophase II If the chromosomes have decondensed after telophase I, they recondense. Spindle microtubules re-form and attach to the sister chromatids.

(f) Metaphase II The chromosomes line up along the equator, with sister chromatids of each chromosome attached to kinetochore microtubules that lead to opposite poles.

(g) Anaphase II The chromatids separate into independent daughter chromosomes, one former chromatid moving toward each pole.

(h) Telophase II The chromosomes finish moving to opposite poles. Nuclear envelopes re-form, and the chromosomes decondense again (not shown here).

(i) Four haploid cells Cytokinesis results in four haploid cells, each containing one member of each pair of homologous chromosomes (shown here in the condensed state).

maternal homologue may face "north" for some pairs and "south" for other pairs. This randomness (also called independent assortment), together with genetic recombination caused by crossing over, is responsible for the genetic diversity of the haploid cells produced by meiosis.

During Anaphase I, Homologous Chromosomes Separate

Anaphase in meiosis I differs considerably from anaphase in mitosis. In anaphase of mitosis, the *sister chromatids separate* and move to opposite poles. In contrast, in anaphase I of meiosis, the sister chromatids of each duplicated homologue remain attached to each other and move to the same pole, but the *homologues separate* as the chiasmata untangle, and move to opposite poles (**Fig. 9-15c**). One duplicated chromosome of each homologous pair (still consisting of two sister chromatids) moves to each pole of the dividing cell.

During Telophase I, Two Haploid Clusters of Duplicated Chromosomes Form

At the end of anaphase I, the cluster of chromosomes at each pole contains one member of each pair of homologous chromosomes. Therefore, each cluster contains the haploid number of chromosomes. In telophase I, the spindle microtubules disappear. Cytokinesis commonly occurs during telophase I (**Fig. 9-15d**). In many, but not all, organisms, nuclear envelopes re-form. Telophase I is usually followed immediately by meiosis II, with little or no intervening interphase. The chromosomes do not replicate between meiosis I and meiosis II, but they may temporarily decondense.

Meiosis II Separates Sister Chromatids into Four Daughter Nuclei

During meiosis II, the sister chromatids of each duplicated chromosome separate in a process that is virtually identical to mitosis, though it takes place in haploid cells. During prophase II, the spindle microtubules re-form (**Fig. 9-15e**). If the chromosomes decondensed at the end of meiosis I, they recondense. As in mitosis, the kinetochores of the sister chromatids of each duplicated chromosome attach to spindle microtubules extending to opposite poles of the cell. During metaphase II, the duplicated chromosomes line up at the cell's equator (**Fig. 9-15f**). During anaphase II, the sister chromatids separate and are towed to opposite poles (**Fig. 9-15g**). Telophase II and cytokinesis conclude meiosis II as nuclear envelopes re-form, the chromosomes decondense into their extended state, and the cytoplasm divides (**Fig. 9-15h**). Commonly, both daughter cells produced in meiosis I undergo meiosis II, producing a total of four haploid cells from the original parental diploid cell (**Fig. 9-15i**).

Now that we have covered all of the processes in detail, examine **Table 9-1** to review and compare mitotic and meiotic cell division.

BioFlix™ Meiosis

9.9 WHEN DO MITOTIC AND MEIOTIC CELL DIVISION OCCUR IN THE LIFE CYCLES OF EUKARYOTES?

The life cycles of almost all eukaryotic organisms have a common overall pattern (**Fig. 9-17**). First, two haploid cells fuse during the process of fertilization, bringing together genes from different parental organisms and endowing the resulting diploid cell with new gene combinations. Second, at some point in the life cycle, meiotic cell division occurs, re-creating haploid cells. Third, at some time in the life cycle, mitotic cell division of either haploid or diploid cells, or both, results in the growth of multicellular bodies, or in asexual reproduction.

Table 9-1 A Comparison of Mitotic and Meiotic Cell Division in Animal Cells

Feature	Mitotic Cell Division	Meiotic Cell Division
Cells in which it occurs	Body cells	Gamete-producing cells
Final chromosome number	Diploid—2*n*; two copies of each type of chromosome (homologous pairs)	Haploid—1*n*; one member of each homologous pair
Number of daughter cells	Two, identical to the parent cell and to each other	Four, containing recombined chromosomes due to crossing over
Number of cell divisions per DNA replication	One	Two
Function in animals	Development, growth, repair, and maintenance of tissues; asexual reproduction	Gamete production for sexual reproduction

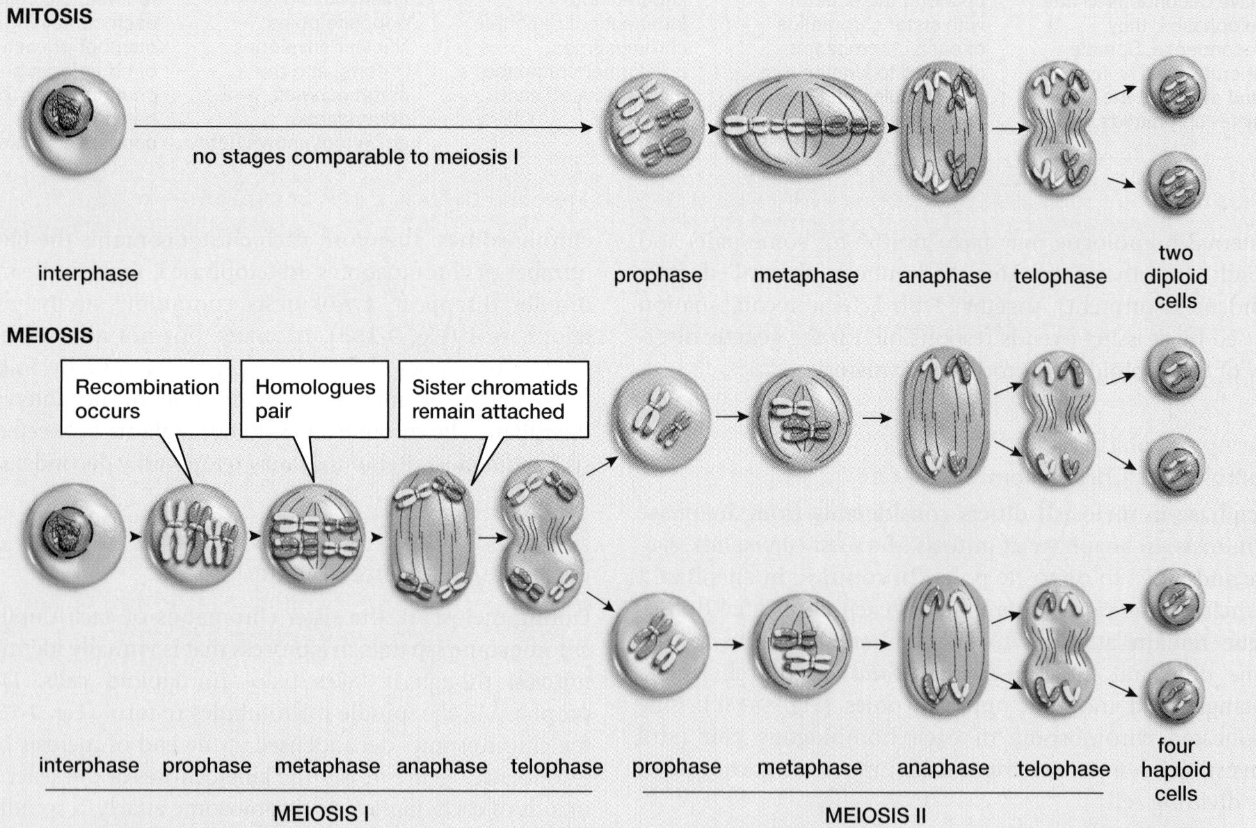

In these diagrams, comparable phases are aligned. In both mitosis and meiosis, chromosomes are replicated during interphase. Meiosis I, with the pairing of homologous chromosomes, formation of chiasmata, exchange of chromosome parts, and separation of homologues to form haploid daughter nuclei, has no counterpart in mitosis. Meiosis II, however, is virtually identical to mitosis in a haploid cell.

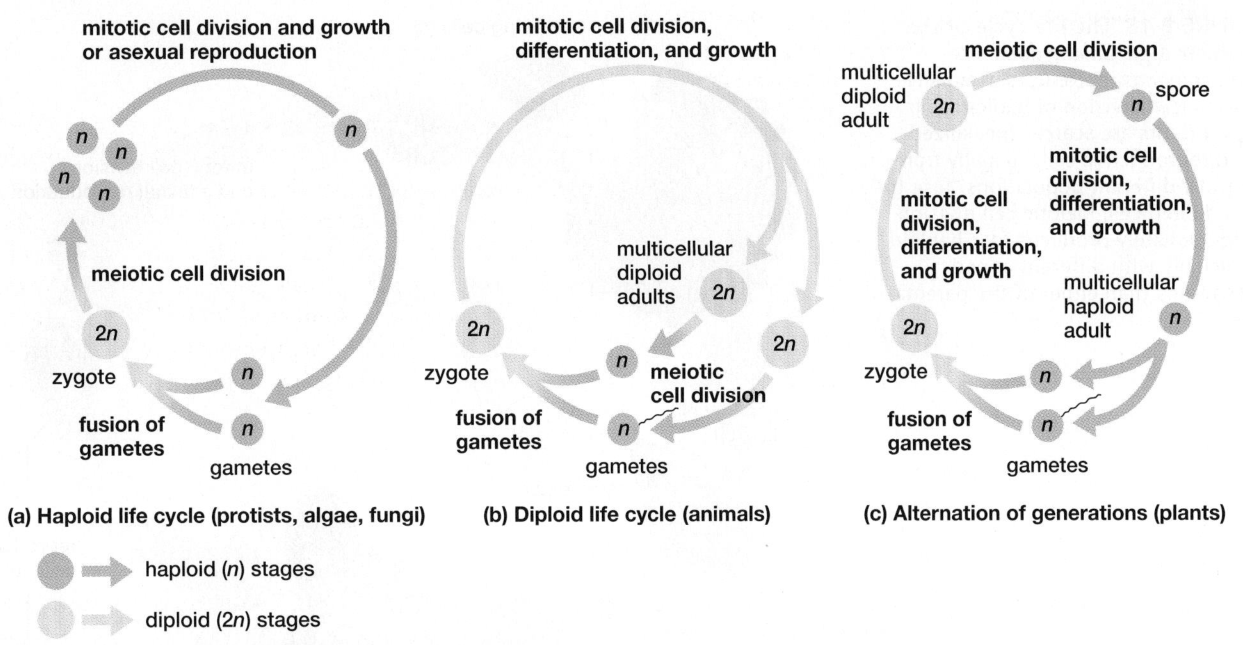

▲ **FIGURE 9-17 The three major types of eukaryotic life cycles** The lengths of the arrows correspond roughly to the proportion of the life cycle spent in each stage. **(a)** Almost all of a haploid life cycle is spent in the haploid state. Meiosis usually occurs soon after fertilization. **(b)** Almost all of a diploid life cycle is spent in the diploid state. Meiosis usually occurs just before fertilization. **(c)** In alternation of generations life cycles, significant portions of the cycle are spent in both haploid and diploid states. Meiosis occurs at the transition between diploid and haploid stages.

The seemingly vast differences between the life cycles of, say, ferns and humans are caused by variations in three aspects: (1) the interval between meiotic cell division and the fusion of haploid cells, (2) at what points in the life cycle mitotic and meiotic cell division occur, and (3) the relative proportions of the life cycle spent in the diploid and haploid states. These aspects of life cycles are interrelated, and we can conveniently label life cycles according to the relative dominance of diploid and haploid stages.

In Haploid Life Cycles, the Majority of the Cycle Consists of Haploid Cells

Some eukaryotes, such as many fungi and unicellular algae, spend most of their life cycles in the haploid state, with single copies of each type of chromosome (**Fig. 9-17a** and **Fig. 9-18**). Asexual reproduction by mitotic cell division produces a population of identical, haploid cells. Under certain environmental conditions, specialized reproductive haploid cells are produced. Two of these reproductive cells fuse, forming a diploid zygote. The zygote immediately undergoes meiosis, producing haploid cells again. In organisms with haploid life cycles, mitotic cell division never occurs in diploid cells.

In Diploid Life Cycles, the Majority of the Cycle Consists of Diploid Cells

Most animals have life cycles that are just the reverse of the haploid cycle. Virtually the entire animal life cycle is spent in the diploid state (**Fig. 9-17b** and **Fig. 9-19**). Haploid gametes (sperm in males, eggs in females) are formed by meiotic cell division. These fuse to form a diploid fertilized egg, called a zygote. Growth and development of the zygote to the adult organism result from mitotic cell division and differentiation of diploid cells.

In Alternation of Generations Life Cycles, There Are Both Diploid and Haploid Multicellular Stages

The life cycle of plants is called alternation of generations, because it alternates between multicellular diploid and multicellular haploid forms. In the typical pattern (**Fig. 9-17c** and **Fig. 9-20**), a multicellular diploid adult stage (the "diploid generation") gives rise to haploid cells, called spores, through meiotic cell division. These spores then undergo mitotic cell division and differentiation of the daughter cells to produce a multicellular haploid adult stage (the "haploid generation"). At some point, certain haploid cells differentiate into haploid gametes. Two gametes then fuse to form a diploid zygote. The zygote grows by mitotic cell division into another multicellular diploid adult stage.

In some plants, such as ferns, both the haploid and diploid stages are free-living, independent plants. Flowering plants, however, have reduced haploid stages, represented only by the pollen grain and a small cluster of cells in the ovary of the flower.

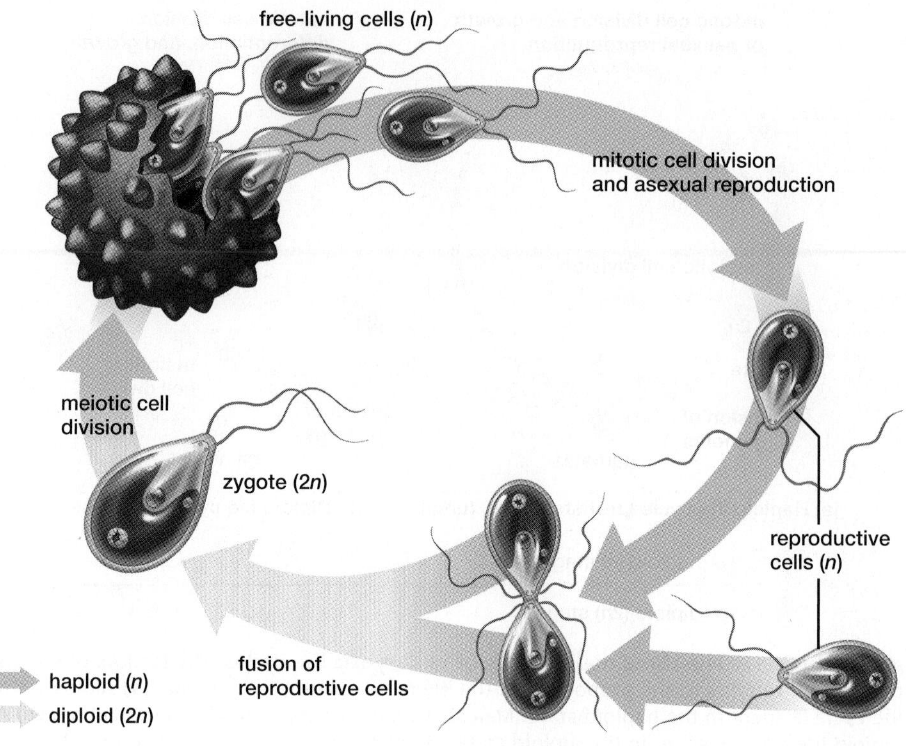

▶ **FIGURE 9-18 The life cycle of the unicellular alga *Chlamydomonas*** *Chlamydomonas* reproduces asexually by mitotic cell division of haploid cells. When nutrients are scarce, specialized haploid reproductive cells (usually from genetically different populations) fuse to form a diploid cell. Meiotic cell division then immediately produces four haploid cells, usually with different genetic compositions than either of the parental strains.

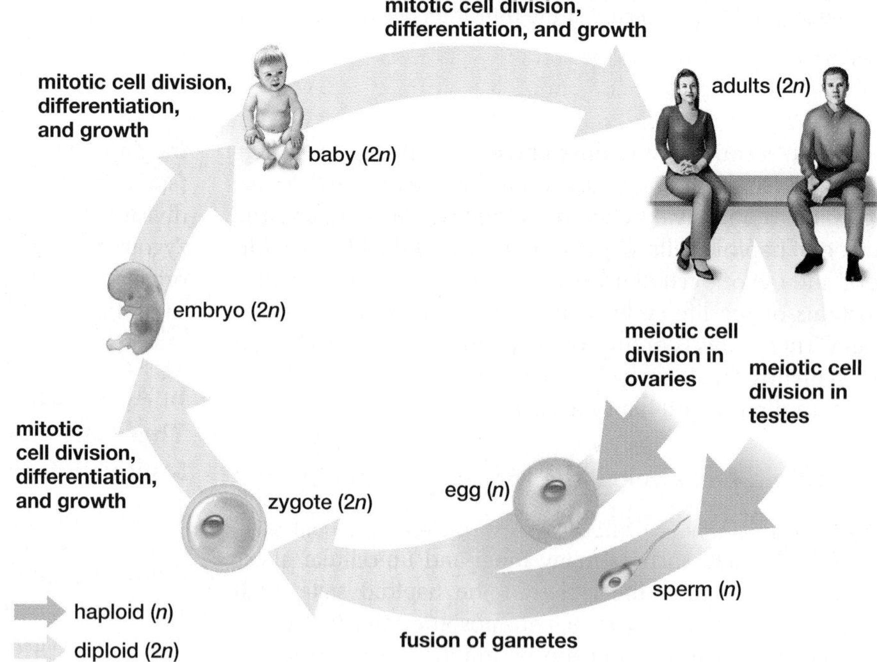

▶ **FIGURE 9-19 The human life cycle** Through meiotic cell division, the two sexes produce gametes—sperm in males and eggs in females—that fuse to form a diploid zygote. Mitotic cell division and differentiation of the daughter cells produce an embryo, child, and ultimately a sexually mature adult. The haploid stages last only a few hours to a few days; the diploid stages may survive for a century.

9.10 HOW DO MEIOSIS AND SEXUAL REPRODUCTION PRODUCE GENETIC VARIABILITY?

Genetic variability among organisms is essential for survival and reproduction in a changing environment and, therefore, for evolution. Mutations occurring randomly over millions of years are the original sources of genetic variability within the populations of organisms that exist today. However, mutations are rare events. Therefore, the genetic variability that occurs from one generation to the next results almost entirely from meiosis and sexual reproduction.

Shuffling of Homologues Creates Novel Combinations of Chromosomes

How does meiosis produce genetic diversity? One mechanism is the random distribution of maternal and paternal homologues to the daughter cells during meiosis I. Remember, at metaphase I the paired homologues line up at the cell's equator. In each pair of homologues, the maternal chromosome faces one pole and the paternal chromosome faces the opposite pole, but which homologue faces which pole is random, and is not affected by the distribution of the homologues of other chromosome pairs.

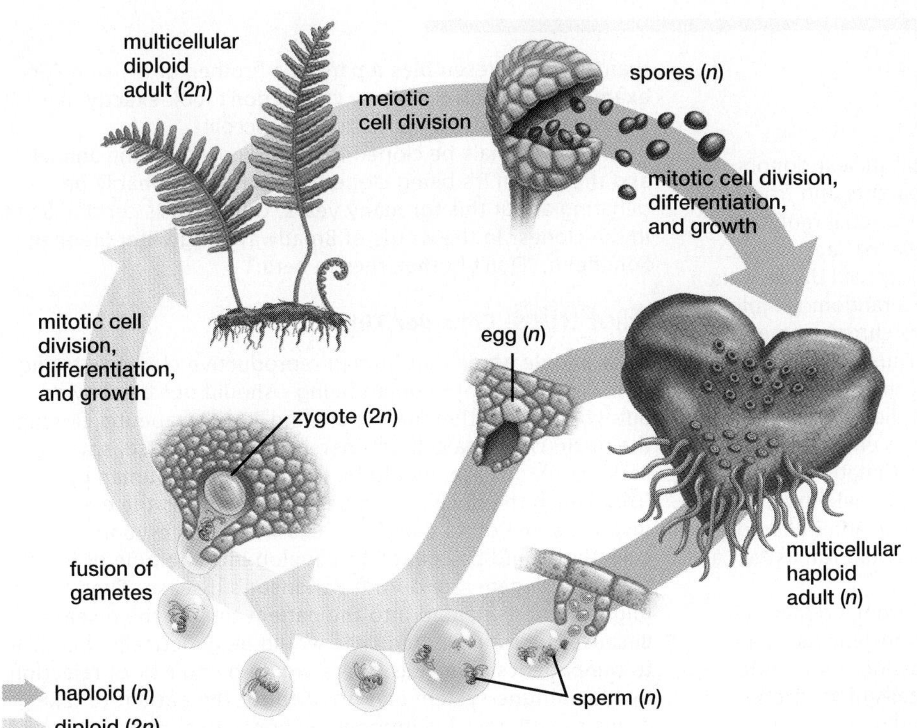

◀ **FIGURE 9-20 Alternation of generations in plants** In plants, such as this fern, specialized cells in the multicellular diploid adult stage undergo meiotic cell division to produce haploid spores. The spores undergo mitotic cell division and differentiation of the daughter cells to produce a multicellular haploid adult stage. Sometime later, perhaps many weeks later, some of these haploid cells differentiate into sperm and eggs. These fuse to form a diploid zygote. Mitotic cell division and differentiation once again give rise to a multicellular diploid adult stage.

Let's consider meiosis in mosquitoes, which have three pairs of homologous chromosomes ($n = 3$, $2n = 6$). At metaphase I, the chromosomes can align in four configurations (**Fig. 9-21a**). Therefore, anaphase I can produce eight possible sets of chromosomes ($2^3 = 8$; **Fig. 9-21b**).

When each of these chromosome clusters undergoes meiosis II, it produces two gametes. Therefore, a single mosquito, with three pairs of homologous chromosomes, can produce gametes with eight different chromosome sets. A single human, with 23 pairs of homologous chromosomes, can theoretically produce gametes with more than 8 million (2^{23}) different combinations of maternal and paternal chromosomes.

(a) The four possible chromosome arrangements at metaphase of meiosis I

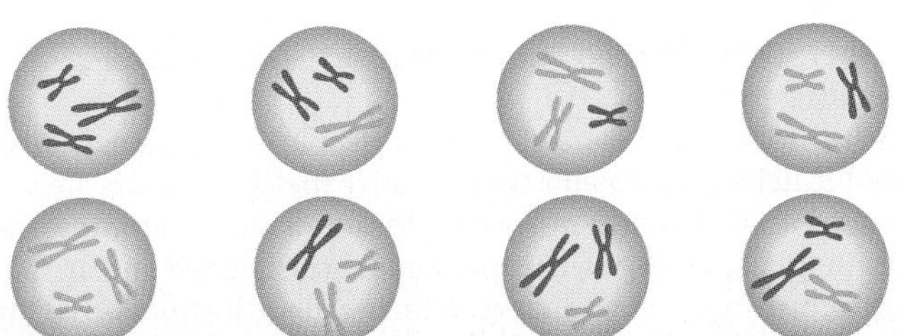

(b) The eight possible sets of chromosomes after meiosis I

▲ **FIGURE 9-21 Random separation of homologous pairs of chromosomes produces genetic variability** For clarity, the chromosomes are depicted as large, medium, and small. The paternal chromosomes are colored yellow and the maternal chromosomes are colored violet.

Crossing Over Creates Chromosomes with Novel Combinations of Genes

In addition to the genetic variation resulting from the random assortment of parental chromosomes, crossing over during meiosis produces chromosomes with combinations of alleles that differ from those of either parent. In fact, some of these new combinations may never have existed before, because homologous chromosomes cross over in new and different places at each meiotic division. In humans, although 1 in 8 million gametes could have the same combination of maternal and paternal chromosomes, in reality, none of those chromosomes will be purely maternal or purely paternal. Even though a man produces about 100 million sperm each day, he may never produce two that carry exactly the same combinations of alleles. In all probability, every sperm and every egg is genetically unique.

Fusion of Gametes Adds Further Genetic Variability to the Offspring

At fertilization, two gametes, each containing a unique combination of alleles, fuse to form a diploid offspring. Even if we ignore crossing over, every human can theoretically produce about 8 million different gametes based solely on the random separation of the homologues. Therefore, fusion of gametes from just two people could produce 8 million × 8 million, or 64 trillion, genetically different children, which is far more people than have ever existed on Earth! Put another way, the chances that your parents could produce another child that is genetically the same as you are about $1/8{,}000{,}000 \times 1/8{,}000{,}000$, or about 1 in 64 trillion! When we factor in the almost endless variability produced by crossing over, we can confidently say that (except for identical twins) there never has been, and never will be, anyone just like you.

Case Study revisited
Send in the Clones

Why are clones genetically identical to their nuclear donors, while children are so different from their parents and from one another? As you've learned, barring the occasional mutation, mitotic cell division produces daughter cells that are genetically identical to the parent cell. Meiotic cell division, in contrast, produces haploid cells that have a random sampling of one member of each pair of homologous chromosomes. Crossing over further shuffles the genes. Thus, when a sperm fertilizes an egg, the offspring receive half their genes from the mother and half from the father. Two children of the same parents will never be genetically the same as either parent or as one another (except for identical twins). Genetic variability produced by sexual reproduction is the reason why many famous racehorses—including Seabiscuit, War Admiral, and Secretariat—never sired any offspring who could run as fast as they could.

During cloning, the nucleus is removed from a donor cell and implanted into an egg from which the nucleus has been removed or destroyed. The resulting cell divides by mitotic cell division to form a clone, so all of the cells of the clone have the same nuclear genes as the donor did.

However, everything about an organism—structure, metabolism, personality, learning ability, and other aspects of behavior—are influenced by both genes and environment. The uterus in which the cloned fetus develops, the rearrangement of cells as the embryo progresses from fertilized egg to complex newborn, and the conditions in which the organism lives (nutrition, family conditions, friends, world events, etc.) are important in ways that we have barely begun to understand. Further, in mammals all of a cell's mitochondria are inherited from the mother, in the cytoplasm of her egg cell. Mitochondria possess their own small supply of DNA (see pp. 70–71 and pp. 325–326), and mitochondrial DNA affects how well a mitochondrion functions. Therefore, a clone will usually have different mitochondrial genes than its nuclear donor. Nevertheless, clones are likely to resemble their nuclear donors, both in appearance and behavior, far more than any child resembles a parent, a brother, or a sister. For example, although the Missy clones don't look exactly like Missy, all three share her taste for broccoli!

Should animals be cloned? Does it depend on the animal and the reason it's being cloned? People will probably be debating about this for many years. One thing is certain: Send in the clones? In the words of Broadway songwriter Stephen Sondheim, "Don't bother, they're here."

BioEthics Consider This

Most people agree that human reproductive cloning—trying to make a duplicate human being—should not be allowed. But what about "therapeutic cloning"? In therapeutic cloning, the nucleus from a cell of a person with a disease, say Parkinson's disease, would be inserted into a human egg from which the nucleus has been removed. As the egg divides, some of its daughter cells will be embryonic stem cells that could be coaxed to develop into the type of brain cells that degenerate during Parkinson's disease. These cells might be transplanted into the patient to cure the disease. Because the transplanted cells would be genetically identical to the patient's own cells, there would be no risk of rejection by his immune system and no need for the patient to take drugs to suppress his immune response (which would make him more susceptible to diseases). Some people are fervent supporters of therapeutic cloning for medical treatments, whereas others are equally fervently opposed, on the grounds that a human embryo would be created, and destroyed, for each patient's treatment.

In late 2007, researchers may have discovered a way around this dilemma. They reported inserting just the right mix of genes into adult cells, causing them to become "induced pluripotent stem cells"; that is, stem cells that could be forced to differentiate into any type of cell of the entire body. But what if induced pluripotent stem cells turn out not to be as versatile as embryonic stem cells? Or what if pluripotent stem cells are found to be dangerous, for example, if the procedure might make them cancerous? Should therapeutic cloning of human embryos then be pursued as an option?

CHAPTER REVIEW

Summary of Key Concepts

9.1 Why Do Cells Divide?

Growth of multicellular eukaryotic organisms and replacement of cells that die during an organism's life occur through cell division and differentiation of the daughter cells. Asexual reproduction also occurs through cell division. A specialized type of cell division, called meiotic cell division, is required for sexual reproduction in eukaryotes.

9.2 What Occurs During the Prokaryotic Cell Cycle?

The prokaryotic cell cycle consists of growth, replication of the DNA of its single, circular chromosome, and division by binary fission. The two resulting daughter cells are genetically identical to the parent cell.

9.3 How Is the DNA in Eukaryotic Chromosomes Organized?

Each chromosome in a eukaryotic cell consists of a single DNA double helix and proteins that organize the DNA. Genes are segments of DNA found at specific locations on a chromosome. During cell growth, the chromosomes are extended and accessible for use by enzymes that read their genetic instructions. During cell division, the chromosomes are duplicated and condense into short, thick structures. Eukaryotic cells typically contain pairs of chromosomes, called homologues, that appear virtually identical because they carry the same genes with similar, although usually not identical, nucleotide sequences. Cells with pairs of homologous chromosomes are diploid. Cells with only one member of each chromosome pair are haploid.

9.4 What Occurs During the Eukaryotic Cell Cycle?

The eukaryotic cell cycle consists of interphase and cell division. During interphase, the cell grows and duplicates its chromosomes. Interphase is divided into G_1 (growth phase 1), S (DNA synthesis), and G_2 (growth phase 2). During G_1, many cells differentiate to perform a specific function. Some differentiated cells can reenter the dividing state; other cells stay differentiated for the life of the organism and never divide again. Most eukaryotic cells divide by mitotic cell division, which produces two daughter cells that are genetically identical to their parent cell. Specialized reproductive cells undergo meiotic cell division, each producing four haploid daughter cells.

9.5 How Does Mitotic Cell Division Produce Genetically Identical Daughter Cells?

The chromosomes are duplicated during interphase, prior to mitosis. A replicated chromosome consists of two identical sister chromatids that remain attached to one another at the centromere during the early stages of mitosis. Mitosis (nuclear division) consists of four phases, usually accompanied by cytokinesis (cytoplasmic division) during the last phase (see Fig. 9-8):

- **Prophase** The chromosomes condense and their kinetochores attach to spindle microtubules, called kinetochore microtubules, that form at this time.
- **Metaphase** Kinetochore microtubules move the chromosomes to the equator of the cell.
- **Anaphase** The two chromatids of each duplicated chromosome separate. The kinetochore microtubules move them to opposite poles of the cell. Meanwhile, polar microtubules force the cell to elongate.
- **Telophase** The chromosomes decondense, and nuclear envelopes re-form around each new daughter nucleus.
- **Cytokinesis** Cytokinesis normally occurs at the end of telophase and divides the cytoplasm into approximately equal halves, each containing a nucleus. In animal cells, a ring of microfilaments pinches the plasma membrane in along the equator. In plant cells, new plasma membrane forms along the equator by the fusion of vesicles produced by the Golgi apparatus.

BioFlix™ Mitosis

9.6 How Is the Cell Cycle Controlled?

Complex interactions among many proteins, particularly cyclins and cyclin-dependent protein kinases, drive the cell cycle. There are three major checkpoints at which progress through the cell cycle is regulated: between G_1 and S, between G_2 and mitosis, and between metaphase and anaphase. These checkpoints ensure that the DNA is intact and replicated accurately and that the chromosomes are properly arranged for mitosis before the cell divides.

9.7 Why Do So Many Organisms Reproduce Sexually?

Genetic differences among organisms originate as mutations, which, when preserved within a species, produce alternate forms of genes, called alleles. Alleles in different individuals of a species may be combined in offspring through sexual reproduction, creating variation among the offspring and potentially improving their likelihood of surviving and reproducing in their turn.

9.8 How Does Meiotic Cell Division Produce Haploid Cells?

Meiosis separates homologous chromosomes and produces haploid cells with only one homologue from each pair. During interphase before meiosis, chromosomes are duplicated. The cell then undergoes two specialized cell divisions—meiosis I and meiosis II—to produce four haploid daughter cells (see Fig. 9-15).

Meiosis I

During prophase I, homologous duplicated chromosomes, each consisting of two chromatids, pair up and exchange parts by crossing over. During metaphase I, homologues move together as pairs to the cell's equator, one member of each pair facing opposite poles of the cell. Homologous chromosomes separate during anaphase I, and two nuclei form during telophase I. Cytokinesis also usually occurs during telophase I. Each daughter nucleus receives only one member of each pair of homologues and, therefore, is haploid. The sister chromatids remain attached to each other throughout meiosis I.

Meiosis II

Meiosis II usually occurs in both daughter nuclei and resembles mitosis in a haploid cell. The duplicated chromosomes move to the cell's equator during metaphase II. The two chromatids of each chromosome separate and move to opposite poles of the cell during anaphase II. This second division produces four haploid nuclei. Cytokinesis normally occurs during or shortly after telophase II, producing four haploid cells.

BioFlix™ Meiosis

9.9 When Do Mitotic and Meiotic Cell Division Occur in the Life Cycles of Eukaryotes?

Most eukaryotic life cycles have three parts: (1) Sexual reproduction combines haploid gametes to form a diploid cell. (2) At some point in the life cycle, diploid cells undergo meiotic cell division to produce haploid cells. (3) At some point in the life cycle, mitosis of either a haploid cell, or a diploid cell, or both, results in the growth of multicellular bodies. When these stages occur, and what proportion of the life cycle is occupied by each stage, varies greatly among different species.

9.10 How Do Meiosis and Sexual Reproduction Produce Genetic Variability?

The random shuffling of homologous maternal and paternal chromosomes creates new chromosome combinations. Crossing over creates chromosomes with allele combinations that may never have occurred before on single chromosomes. Because of the separation of homologues and crossing over, a parent probably never produces any two gametes that are completely identical. The fusion of two such genetically unique gametes adds further genetic variability to the offspring.

Key Terms

allele *159*
anaphase *153*
asexual reproduction *147*
autosome *151*
binary fission *148*
cell cycle *146*
cell division *145*
cell plate *154*
centriole *153*
centromere *150*
checkpoint *159*
chiasma (plural, **chiasmata**) *163*
chromatid *150*
chromosome *146*
clone *144*
cloning *144*
crossing over *163*
cytokinesis *152*
daughter cell *145*
deoxyribonucleic acid (DNA) *146*
differentiate *146*
diploid *150*
duplicated chromosome *150*
gamete *147*
gene *146*
haploid *151*
homologous chromosome *150*
homologue *150*
interphase *151*
karyotype *150*
kinetochore *153*
locus (plural, **loci**) *149*
meiosis *152*
meiotic cell division *152*
metaphase *153*
mitosis *152*
mitotic cell division *152*
mutation *151*
nucleotide *146*
prophase *153*
recombination *163*
sex chromosome *151*
sexual reproduction *147*
spindle microtubule *153*
stem cell *146*
telomere *150*
telophase *153*

Thinking Through the Concepts

Fill-in-the-Blank

1. Prokaryotic cells divide by a process called ___________.
2. Growth and development of eukaryotic organisms occur through ___________ cell division and ___________ of the resulting daughter cells.
3. Most plants and animals have pairs of chromosomes that have similar appearance and genetic composition. These pairs are called ___________. Pairs of chromosomes that are the same in both males and females are called ___________. Pairs that are different in males and females are called ___________.
4. The four phases of mitosis are ___________, ___________, ___________, and ___________. Division of the cytoplasm into two cells, called ___________, usually occurs during which phase? ___________
5. Chromosomes attach to spindle microtubules at structures called ___________. Some spindle microtubules, called ___________ microtubules, do not bind to chromosomes, but have free ends that overlap along the equator of the cell. These microtubules push the poles of the cell apart.
6. Meiotic cell division produces ___________haploid daughter cells from each diploid parental cell. The first haploid nuclei are produced at the end of ___________. In animals, the haploid daughter cells produced by meiotic cell division are called ___________.
7. During___________ of meiosis I, homologous chromosomes intertwine, forming structures called ___________. These structures are the sites of what event? ___________
8. Three processes that promote genetic variability of offspring during sexual reproduction are ___________, ___________, and ___________.

Review Questions

1. Diagram and describe the eukaryotic cell cycle. Name the various phases, and briefly describe the events that occur during each.
2. Define mitosis and cytokinesis. What changes in cell structure would result if cytokinesis does not occur after mitosis?
3. Diagram the stages of mitosis. How does mitosis ensure that each daughter nucleus receives a full set of chromosomes?
4. Define the following terms: *homologous chromosome* (*homologue*), *centromere, kinetochore, chromatid, diploid,* and *haploid.*
5. Describe and compare the process of cytokinesis in animal cells and in plant cells.
6. How is the cell cycle controlled? Why must the progression of cells through the cell cycle be regulated?
7. Diagram the events of meiosis. At which stage do homologous chromosomes separate?
8. Describe crossing over. At which stage of meiosis does it occur? Name two functions of chiasmata.
9. In what ways are mitosis and meiosis similar? In what ways are they different?
10. Describe or diagram the three main types of eukaryotic life cycles. When do meiotic cell division and mitotic cell division occur in each?
11. Describe how meiosis provides for genetic variability. If an animal had a haploid number of two (no sex chromosomes), how many genetically different gametes could it produce? (Assume no crossing over.) If it had a haploid number of five?

Applying the Concepts

1. Most nerve cells in the adult human central nervous system, as well as heart muscle cells, do not divide. In contrast, cells lining the inside of the small intestine divide frequently. Discuss this difference in terms of why damage to the nervous system and heart muscle cells (such as damage caused by a stroke or heart attack) is so dangerous. What do you think might happen to tissues such as the intestinal lining if some disorder blocked mitotic cell division in all cells of the body?
2. Cancer cells divide out of control. Side effects of chemotherapy and radiation therapy that fight cancers include loss of hair and of the intestinal lining, producing severe nausea. Note that cells in hair follicles and intestinal lining divide frequently. What can you infer about the mechanisms of these treatments? What would you look for in an improved cancer therapy?
3. Some animal species can reproduce either asexually or sexually, depending on the state of the environment. Asexual reproduction tends to occur in stable, favorable environments; sexual reproduction is more common in unstable or unfavorable circumstances. Discuss the advantages and disadvantages of sexual and asexual reproduction.

MB® *Go to www.masteringbiology.com for practice quizzes, activities, eText, videos, current events, and more.*

Patterns of Inheritance

Chapter 10

Case Study

Sudden Death on the Court

FLO HYMAN, over 6 feet tall, graceful and athletic, was one of the best woman volleyball players of all time. Star of the 1984 silver medal American Olympic volleyball team, Hyman later joined a professional Japanese team. In 1986, taken out of a game for a short breather, she died at the age of 31, sitting quietly on the bench. How could this happen to someone so young and fit?

Flo Hyman had a genetic disorder called Marfan syndrome, which affects about 1 in 5,000 to 10,000 people, possibly including Abraham Lincoln, the classical composer and pianist Sergei Rachmaninoff, and the Egyptian pharaoh Akhenaten. People with Marfan syndrome are typically tall and slender, with unusually long limbs and large hands and feet. For some people with Marfan syndrome, these characteristics lead to fame and fortune. Unfortunately, Marfan syndrome can also be deadly.

Hyman died from a ruptured aorta, the massive artery that carries blood from the heart to most of the body. Why did Hyman's aorta break? What does a weak aorta have in common with tallness and large hands? Marfan syndrome is caused by a mutation in the gene that encodes a protein called fibrillin, an essential component of connective tissue. Many parts of your body contain connective tissue, including the tendons that attach muscles to bones, ligaments (such as the fibers that hold the lens in place in your eye), and the tough walls of arteries. Fibrillin forms long fibers that give strength to connective tissue. Normal fibrillin also "traps" certain types of growth factors, keeping them from excessively stimulating cell division (see p. 158) in, for example, bone-forming cells. Defective fibrillin cannot trap these growth factors, with the result that the arms, legs, hands, and feet of people with Marfan syndrome tend to become unusually long. The combination of defective fibrillin and high concentrations of growth factors also weakens bone, cartilage, and artery walls.

As we described in p. 158, diploid organisms, including people, generally have two copies of each gene, one on each homologous chromosome. One defective copy of the fibrillin gene is enough to cause Marfan syndrome. What does this tell us about the inheritance of Marfan syndrome? Are all inherited diseases caused by a single defective copy of a gene? To find out, we must go back in time to a monastery in Moravia and visit the garden of Gregor Mendel.

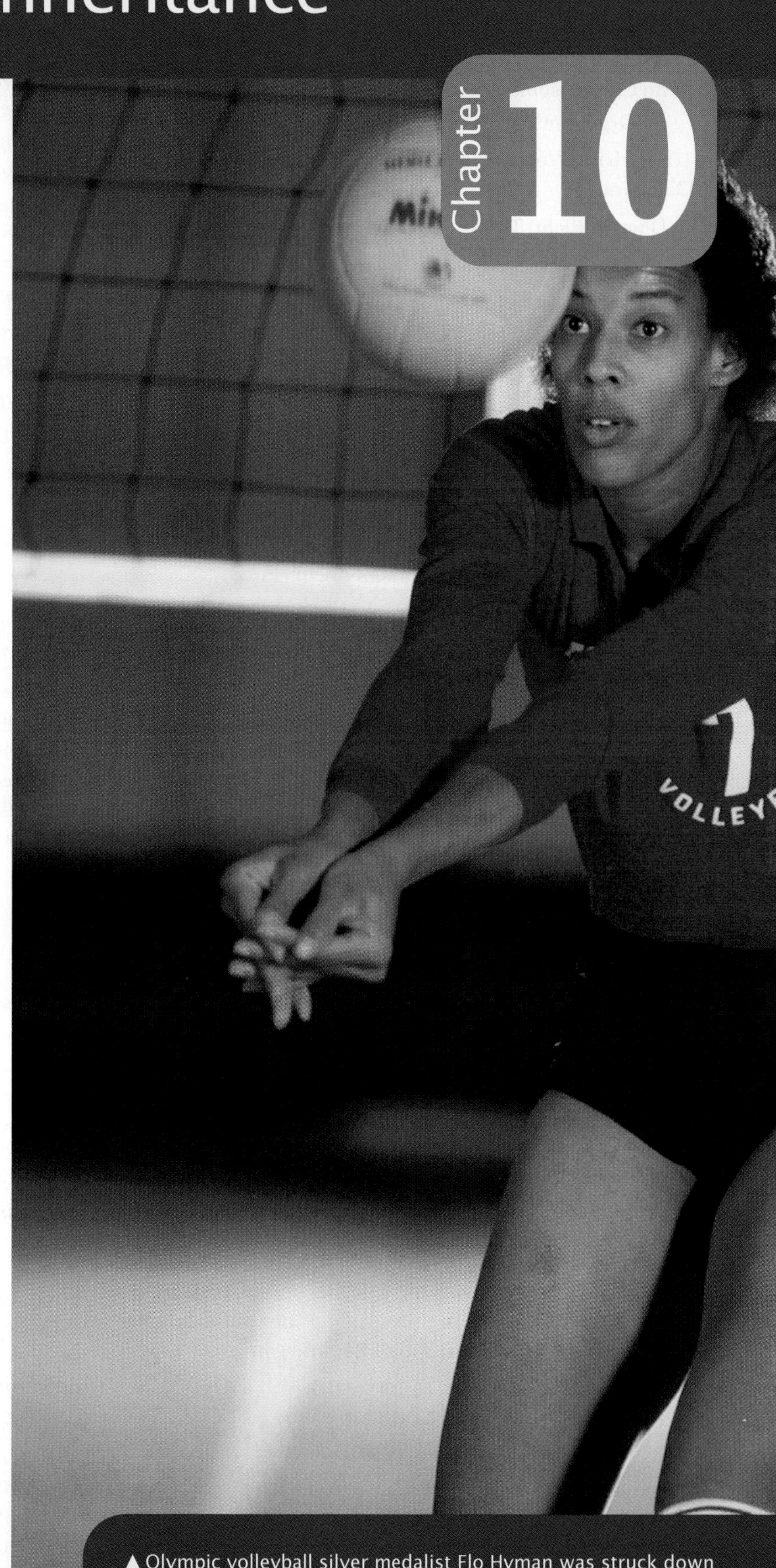

▲ Olympic volleyball silver medalist Flo Hyman was struck down by Marfan syndrome at the height of her career.

At a Glance

10.1 WHAT IS THE PHYSICAL BASIS OF INHERITANCE?

Inheritance is the process by which the characteristics of individuals are passed to their offspring. We will begin our exploration of inheritance with a brief overview of the structures—genes and chromosomes—that form its physical basis. In this chapter, we will confine our discussion to diploid organisms, including most plants and animals, that reproduce sexually by the fusion of haploid gametes.

Genes Are Sequences of Nucleotides at Specific Locations on Chromosomes

A chromosome consists of a single double helix of DNA, packaged with a variety of proteins. Segments of DNA ranging from a few hundred to many thousands of nucleotides in length are the units of inheritance—the **genes**—that encode the information needed to produce proteins, cells, and entire organisms. Therefore, genes are parts of chromosomes. A gene's physical location on a chromosome is called its **locus** (plural, **loci**). The chromosomes of diploid organisms occur in pairs called homologues. Both members of a pair of homologues carry the same genes, located at the same loci (**Fig. 10-1**). However, the nucleotide sequences of a given gene may differ in different members of a species, or even on the two homologues of a single individual. These different versions of a gene at a given locus are called **alleles** (see Fig. 10-1).

Mutations Are the Source of Alleles

Think of genes as very long sentences, written in an alphabet of nucleotides instead of letters. Alleles arise as mutations

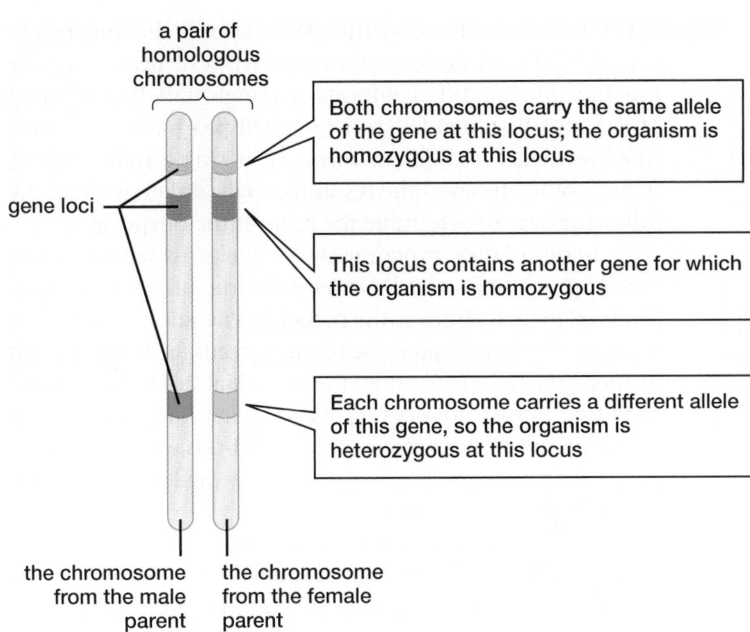
a pair of homologous chromosomes
Both chromosomes carry the same allele of the gene at this locus; the organism is homozygous at this locus
gene loci
This locus contains another gene for which the organism is homozygous
Each chromosome carries a different allele of this gene, so the organism is heterozygous at this locus
the chromosome from the male parent
the chromosome from the female parent

◀ **FIGURE 10-1 The relationships among genes, alleles, and chromosomes** Each homologous chromosome carries the same set of genes. Each gene is located at the same position, or locus, on its chromosome. Differences in nucleotide sequences at the same gene locus produce different alleles of the gene. Diploid organisms have two alleles of each gene, one on each homologue. The alleles on the two homologues may be the same or different.

that slightly change the spelling of these nucleotide sentences. If a mutation occurs in the cells that become sperm or eggs, it can be passed on from parent to offspring.

Most of the alleles in an organism's DNA first appeared as mutations in the reproductive cells of the organism's ancestors, perhaps hundreds or even millions of years ago, and have been inherited, generation after generation, ever since. A few alleles, which we will call "new mutations," may have occurred in the reproductive cells of the organism's own parents.

An Organism's Two Alleles May Be the Same or Different

Because a diploid organism has pairs of homologous chromosomes, and both members of a pair contain the same gene loci, the organism has two copies of each gene. If both homologues have the same allele at a given gene locus, the organism is said to be **homozygous** at that locus. ("Homozygous" comes from Greek words meaning "same pair.") For example, the chromosomes shown in Figure 10-1 are homozygous at two loci. If two homologous chromosomes have different alleles at a locus, the organism is **heterozygous** ("different pair") at that locus. The chromosomes in Figure 10-1 are heterozygous at one locus. Organisms that are heterozygous at a specific locus are often called **hybrids.**

10.2 HOW WERE THE PRINCIPLES OF INHERITANCE DISCOVERED?

The common patterns of inheritance and many essential facts about genes, alleles, and the distribution of alleles in gametes and zygotes during sexual reproduction were discovered by an Austrian monk, Gregor Mendel (**Fig. 10-2**), in the mid-1800s. This was long before DNA, chromosomes, or meiosis had been discovered. Because his experiments are succinct, elegant examples of science in action, let's follow Mendel's paths of discovery.

Doing It Right: The Secrets of Mendel's Success

There are three key steps to any successful experiment in biology: choosing the right organism with which to work, designing and performing the experiment correctly, and analyzing the data properly. Mendel was the first geneticist to complete all three steps.

Mendel chose the edible pea as the subject for his experiments on inheritance (**Fig. 10-3**). Stamens, the male reproductive structures of a flower, produce pollen. Each pollen grain contains sperm. Pollination allows sperm to fertilize the female gamete, or egg, located within the ovary at the base of

▲ **FIGURE 10-2 Gregor Mendel**

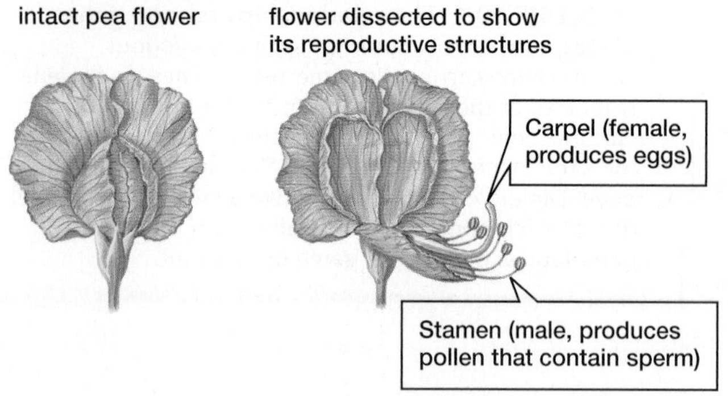

intact pea flower

flower dissected to show its reproductive structures

Carpel (female, produces eggs)

Stamen (male, produces pollen that contain sperm)

▲ **FIGURE 10-3 Flowers of the edible pea** In the intact pea flower (left), the lower petals form a container enclosing the reproductive structures—the stamens (male) and carpel (female). Pollen normally cannot enter the flower from outside, so peas usually self-pollinate and, hence, self-fertilize. If the flower is opened (right), it can be cross-pollinated by hand.

the flower's female reproductive structure, called the carpel. In pea flowers, the petals enclose all of the reproductive structures, preventing another flower's pollen from entering (see Fig. 10-3). Therefore, the egg cells in a pea flower must be fertilized by sperm from the pollen of the same flower. When an organism's sperm fertilize its own eggs, the process is called **self-fertilization.**

Mendel, however, often wanted to mate two different pea plants to see what types of offspring they would produce. To do this, he opened one pea flower and removed its stamens, preventing self-fertilization. Then, he dusted the sticky tip of the carpel with pollen from the flower of another plant. When sperm from one organism fertilize eggs from a different organism, the process is called **cross-fertilization.**

Mendel's experimental design was simple, but brilliant. Earlier researchers had generally tried to study inheritance by simultaneously considering all of the features of entire organisms, including features that differed only slightly between organisms. Not surprisingly, the investigators were often confused rather than enlightened. Mendel, however, chose to study individual characteristics (called traits) that had unmistakably different forms, such as white versus purple flowers. He also started out by studying only one trait at a time.

Mendel followed the inheritance of these traits for several generations, counting the numbers of offspring with each type of trait. By analyzing these numbers, the basic patterns of inheritance became clear. Today, quantifying experimental results and applying statistical analysis are essential tools in virtually every field of biology, but in Mendel's time, numerical analysis was an innovation.

10.3 HOW ARE SINGLE TRAITS INHERITED?

To study inheritance, a researcher needs to begin with parental organisms with easily identified traits that are inherited consistently from generation to generation. Organisms are called **true-breeding** when they possess a trait, such as purple flowers, that is always inherited unchanged by all of its offspring that are produced by self-fertilization. Even in the mid-1800s, commercial seed dealers sold many types of peas that were true-breeding for different forms of a single trait. In his first set of experiments, Mendel cross-fertilized pea plants that were true-breeding for different forms of a single trait, such as flower color. He saved the resulting seeds and grew them the following year to determine the traits of the offspring.

In one of these experiments, Mendel cross-fertilized true-breeding, white-flowered plants with true-breeding, purple-flowered plants. This was the parental generation, denoted by the letter P. When he grew the resulting seeds, he found that all the first-generation offspring (the "first filial," or F_1 generation) produced purple flowers (**Fig. 10-4**). What had happened to the white color? The flowers of the F_1 hybrids were just as purple as their parents. The white color seemed to have disappeared in the F_1 generation.

Mendel then allowed the F_1 flowers to self-fertilize, collected the seeds, and planted them the next spring. In the second generation (F_2), Mendel counted 705 plants with purple flowers and 224 plants with white flowers. These numbers are approximately three-fourths purple flowers and one-fourth white flowers, or a ratio of about 3 purple to 1 white (**Fig. 10-5**). This result showed that the capacity to produce white flowers had not disappeared in the F_1 plants, but had only been "hidden."

Mendel allowed the F_2 plants to self-fertilize and produce yet a third (F_3) generation. He found that all the white-flowered F_2 plants produced white-flowered offspring; that is, they were true-breeding. For as many generations as he had time and patience to raise, white-flowered parents always gave rise only to white-flowered offspring. In contrast, when purple-flowered F_2 plants self-fertilized, their offspring were of two types. About one-third were true-breeding for purple, but the other two-thirds were hybrids that produced both purple- and white-flowered offspring, again in the ratio of 3 purple to 1 white. Therefore, the F_2 generation included one-quarter true-breeding purple plants, one-half hybrid purple, and one-quarter true-breeding white.

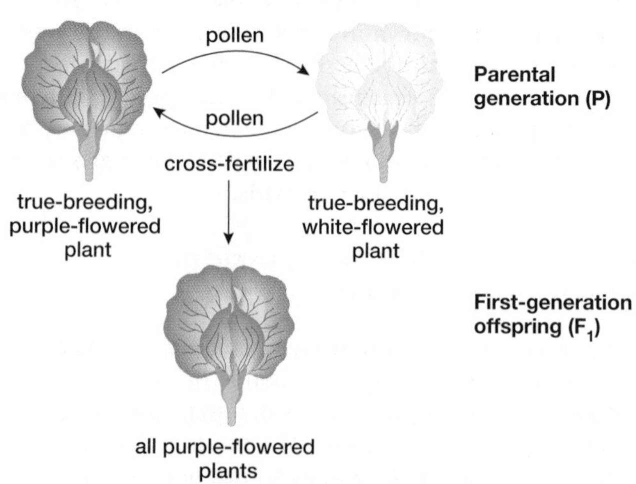

▲ **FIGURE 10-4 Cross of pea plants true-breeding for white or purple flowers** All of the offspring bear purple flowers.

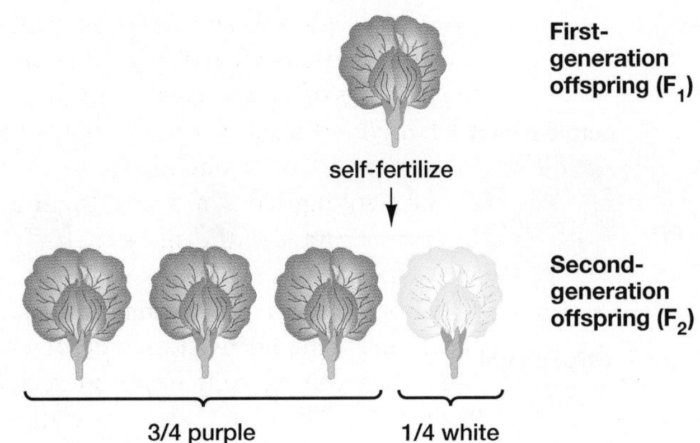

▲ **FIGURE 10-5 Self-fertilization of F_1 pea plants with purple flowers** Three-quarters of the offspring bear purple flowers and one-quarter bear white flowers.

The Inheritance of Dominant and Recessive Alleles on Homologous Chromosomes Can Explain the Results of Mendel's Crosses

Mendel's results, supplemented by modern knowledge of genes and homologous chromosomes, allow us to develop a five-part hypothesis to explain the inheritance of single traits:

- Each trait is determined by pairs of discrete physical units called genes. Each organism has two alleles for each gene, one on each homologous chromosome. True-breeding, white-flowered peas have different alleles of the flower-color gene than true-breeding, purple-flowered peas do.
- When two different alleles are present in an organism, one—the **dominant** allele—may mask the expression of the other—the **recessive** allele. The recessive allele, however, is still present. In the edible pea, the allele for purple flowers is dominant, and the allele for white flowers is recessive.
- The pairs of alleles on homologous chromosomes separate, or segregate, from each other during meiosis. This is known as Mendel's **law of segregation.** As a result, each gamete receives only one allele of each pair. (Unlike in animals, in plants, the gametes do not form immediately following meiosis [see Figs. 9-17c and 9-20]; however, later in its life cycle, the plant eventually does produce gametes containing one allele of each pair.) When a sperm fertilizes an egg, the resulting offspring receives one allele from the father (in his sperm) and one from the mother (in her egg).
- Chance determines which allele is included in a given gamete. Because homologous chromosomes separate at random during meiosis, the distribution of alleles to the gametes is also random.
- True-breeding organisms have two copies of the same allele for a given gene, and are therefore homozygous for that gene. All of the gametes from a homozygous individual receive the same allele for that gene (**Fig. 10-6a**). Hybrid organisms have two different alleles for a given gene and are, therefore, heterozygous for that gene. Half of a heterozygote's gametes will contain one allele for that gene and half will contain the other allele (**Fig. 10-6b**).

Let's see how this hypothesis explains the results of Mendel's experiments with flower color (**Fig. 10-7**). Using letters to represent the different alleles, we will assign the uppercase letter *P* to the dominant allele for purple flower color and the lowercase letter *p* to the recessive allele for white flower color. A homozygous purple-flowered plant has two alleles for purple flower color (*PP*), whereas a homozygous white-flowered plant has two alleles for white flower color (*pp*). All the sperm and eggs produced by a *PP* plant carry the *P* allele; all the sperm and eggs of a *pp* plant carry the *p* allele (**Fig. 10-7a**).

The F_1 offspring were produced when *P* sperm fertilized *p* eggs or when *p* sperm fertilized *P* eggs. In both cases, the F_1 offspring were *Pp*. Because *P* is dominant over *p*, all of the offspring were purple (**Fig. 10-7b**).

For the F_2 generation, Mendel allowed the heterozygous F_1 plants to self-fertilize. Each gamete produced by a heterozygous *Pp* plant has an equal chance of receiving either the *P* allele or the *p* allele. That is, a heterozygous plant produces equal numbers of *P* and *p* sperm and equal numbers of *P* and *p* eggs. When a *Pp* plant self-fertilizes, each type of sperm has an equal chance of fertilizing each type of egg (**Fig. 10-7c**). Therefore, the F_2 generation contained three types of offspring: *PP*, *Pp*, and *pp*. The three types occurred in the approximate proportions of one-quarter *PP* (homozygous purple), one-half *Pp* (heterozygous purple), and one-quarter *pp* (homozygous white).

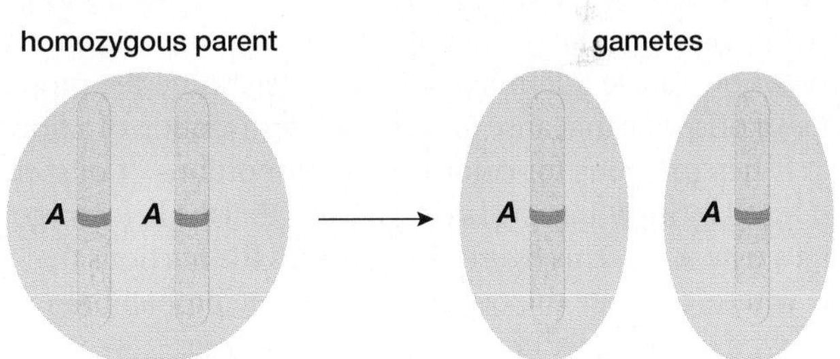

(a) Gametes produced by a homozygous parent

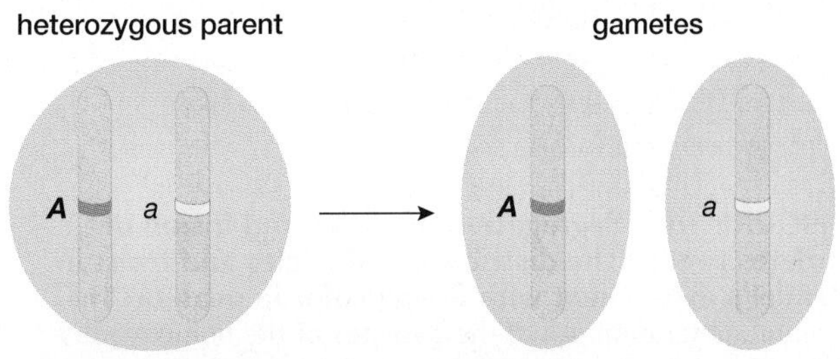

(b) Gametes produced by a heterozygous parent

▲ **FIGURE 10-6 The distribution of alleles in gametes** **(a)** All of the gametes produced by homozygous organisms contain the same allele. **(b)** Half of the gametes produced by heterozygous organisms contain one allele, and half of the gametes contain the other allele.

Have you ever wondered

Why Dogs Vary So Much in Size?

All dogs evolved from wolves. Although all wolves are about the same size, dogs vary in size more than any other mammal—from huge Great Danes and Irish wolfhounds to minuscule Chihuahuas and toy poodles. In 2007, researchers found that toy breeds are usually homozygous for a "small" allele that encodes insulin-like growth factor (IGF), a protein that helps to regulate body size in many mammals. Large breeds usually carry at least one "large" allele. The "small" allele probably does not exist in wolves.

Two organisms that look alike may actually carry different combinations of alleles. The actual combination of alleles carried by an organism (for example, *PP* or *Pp*) is its **genotype.** The organism's traits, including its outward appearance, behavior, digestive enzymes, blood type, or any other observable or measurable feature, make up its **phenotype.** As we have seen, plants with either the *PP* or the *Pp* genotype have the phenotype of purple flowers. Therefore, the F_2 generation of Mendel's peas consisted of three genotypes (one-quarter *PP*, one-half *Pp*, and one-quarter *pp*) but only two phenotypes (three-quarters purple and one-quarter white).

Simple "Genetic Bookkeeping" Can Predict Genotypes and Phenotypes of Offspring

The **Punnett square method,** named after R. C. Punnett, a famous geneticist of the early 1900s, is a convenient way to predict the genotypes and phenotypes of offspring. **Figure 10-8** shows how to use a Punnett square to determine the proportions of offspring that arise from the self-fertilization of a flower that is heterozygous for color (or the proportions of offspring produced by breeding any two organisms that are heterozygous for a single trait). This figure also provides the fractions that allow you to calculate the outcomes using the probabilities that each type of sperm will fertilize each type of egg.

As you use these "genetic bookkeeping" techniques, keep in mind that in a real experiment, the offspring will occur only in *approximately* the predicted proportions, because

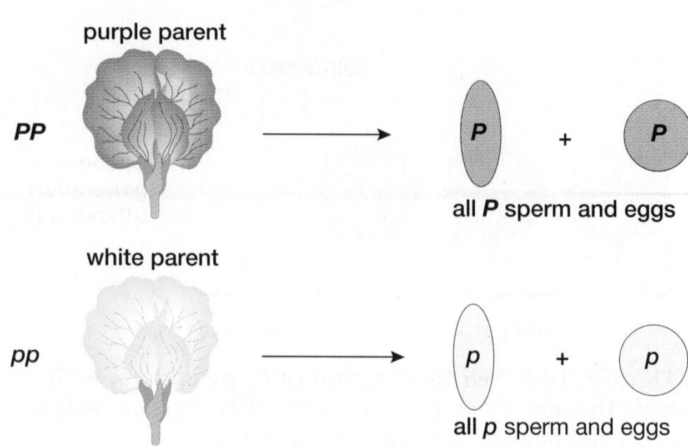

(a) Gametes produced by homozygous parents

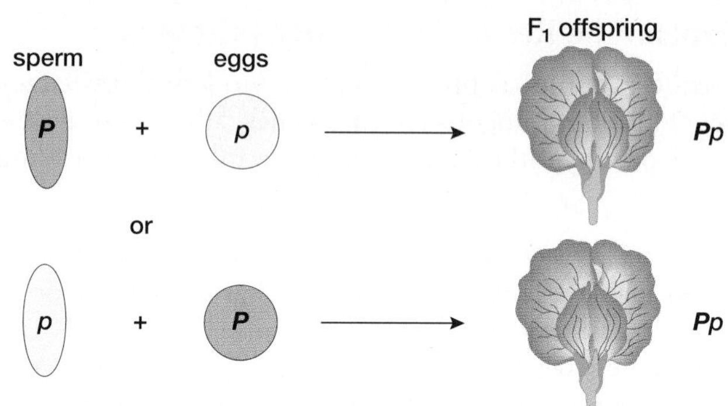

(b) Fusion of gametes produces F_1 offspring

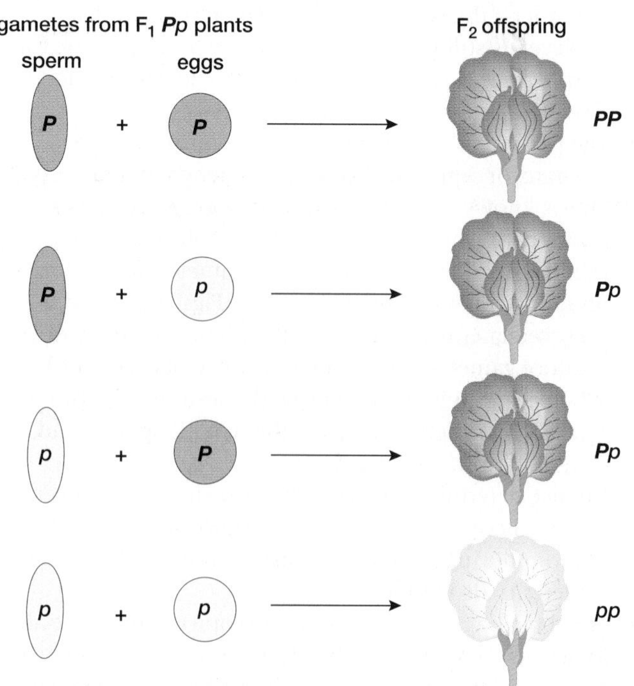

(c) Fusion of gametes from the F_1 generation produces F_2 offspring

▶ **FIGURE 10-7 Segregation of alleles and fusion of gametes predict the distribution of alleles and traits in Mendel's experiment with flower color in peas (a)** The parental generation. All of the gametes of the homozygous parents contain the same allele: only *P* alleles in the gametes of *PP* parents and only *p* alleles in the gametes of *pp* parents. **(b)** The F_1 generation: Fusion of gametes containing the *P* allele with gametes containing the *p* allele produces only *Pp* offspring. **(c)** The F_2 generation: Half of the gametes of heterozygous *Pp* parents contain the *P* allele and half contain the *p* allele. Fusion of these gametes produces *PP*, *Pp*, and *pp* offspring.

sperm and eggs with different alleles encounter one another randomly. Let's consider an example. Each time a baby is conceived, it has a 50:50 chance of becoming a boy or a girl. However, many families with two children do not have one girl and one boy. The 50:50 ratio of girls to boys occurs only if we average the genders of the children in many families.

Mendel's Hypothesis Can Be Used to Predict the Outcome of New Types of Single-Trait Crosses

You have probably recognized that Mendel used the scientific method: making an observation and using it to formulate a hypothesis. But does Mendel's hypothesis accurately predict the results of further experiments? Based on the hypothesis that the heterozygous F_1 plants had one allele for purple flowers and one for white (that is, they had the *Pp* genotype), Mendel predicted the outcome of cross-fertilizing *Pp* plants with homozygous recessive white plants (*pp*): There should be equal numbers of *Pp* (purple) and *pp* (white) offspring. This is indeed what he found.

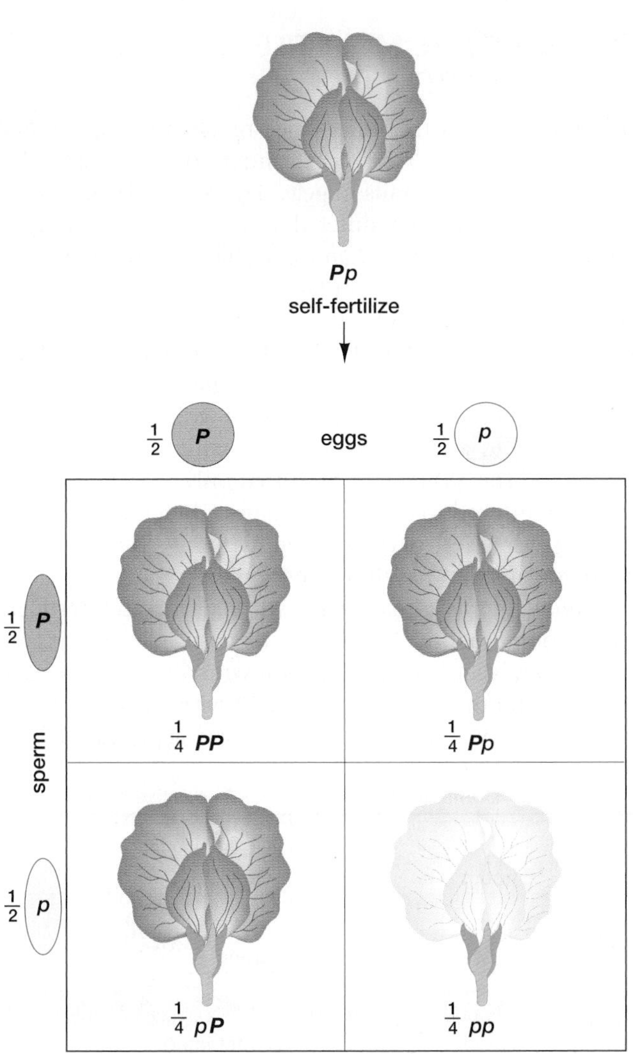

(a) Punnett square of a single-trait cross

sperm		eggs		offspring genotypes	genotypic ratio (1:2:1)	phenotypic ratio (3:1)
$\frac{1}{2}$ *P*	×	$\frac{1}{2}$ *P*	=	$\frac{1}{4}$ *PP*	$\frac{1}{4}$ *PP*	$\frac{3}{4}$ purple
$\frac{1}{2}$ *P*	×	$\frac{1}{2}$ *p*	=	$\frac{1}{4}$ *Pp*	$\frac{1}{2}$ *Pp*	
$\frac{1}{2}$ *p*	×	$\frac{1}{2}$ *P*	=	$\frac{1}{4}$ *pP*		
$\frac{1}{2}$ *p*	×	$\frac{1}{2}$ *p*	=	$\frac{1}{4}$ *pp*	$\frac{1}{4}$ *pp*	$\frac{1}{4}$ white

(b) Using probabilities to determine the offspring of a single-trait cross

◀ **FIGURE 10-8 Determining the outcome of a single-trait cross**
(a) The Punnett square allows you to predict both genotypes and phenotypes of specific crosses; here we use it for a cross between pea plants that are heterozygous for a single trait—flower color.

(1) Assign letters to the different alleles; use uppercase for dominant alleles and lowercase for recessive alleles.
(2) Determine all the types of genetically different gametes that can be produced by the male and female parents.
(3) Draw the Punnett square, with the columns labeled with possible genotypes of the eggs and the rows labeled with the possible genotypes of the sperm. (We have included the fractions of these genotypes with each label.)
(4) Fill in the genotype of the offspring in each box by combining the genotype of the sperm in its row with the genotype of the egg in its column. (Multiply the fraction of sperm of each type in the row headers by the fraction of eggs of each type in the column headers.)
(5) Count the number of offspring with each genotype. Note that *Pp* is the same as *pP*.
(6) Convert the number of offspring of each genotype to a fraction of the total number of offspring. In this example, out of four fertilizations, only one is predicted to produce the *pp* genotype, so one-quarter of the total number of offspring produced by this cross is predicted to be white. To determine phenotypic fractions, add the fractions of genotypes that would produce a given phenotype. For example, purple flowers are produced by $\frac{1}{4}$*PP* + $\frac{1}{4}$*Pp* + $\frac{1}{4}$*pP*, for a total of three-quarters of the offspring.

(b) Probabilities may also be used to predict the outcome of a single-trait cross. Determine the fractions of eggs and sperm of each genotype, and multiply these fractions together to calculate the fraction of offspring of each genotype. When two genotypes produce the same phenotype (e.g., *Pp* and *pP*), add the fractions of each genotype to determine the phenotypic fraction.

QUESTION If you crossed a heterozygous *Pp* plant with a homozygous recessive *pp* plant, what would be the expected ratio of offspring? How does this differ from the offspring of a *PP* × *pp* cross? Try working this out before you read further in the text.

Case Study continued
Sudden Death on the Court

When a person with Marfan syndrome has children with a person without the syndrome, those children have a 50% chance of inheriting the condition. Do you think that Marfan is inherited as a dominant or a recessive allele? Why? Check your reasoning in "Case Study Revisited: Sudden Death on the Court" at the end of the chapter.

This type of experiment also has practical uses. Cross-fertilization of an organism with a dominant phenotype (in this case, a purple flower) but an unknown genotype with a homozygous recessive organism (a white flower) tests whether the organism with the dominant phenotype is homozygous or heterozygous; logically enough, this is called a **test cross** (**Fig. 10-9**). When crossed with a homozygous recessive (*pp*), a homozygous dominant (*PP*) produces all phenotypically dominant offspring, whereas a heterozygous dominant (*Pp*) yields offspring with both dominant and recessive phenotypes in a 1:1 ratio.

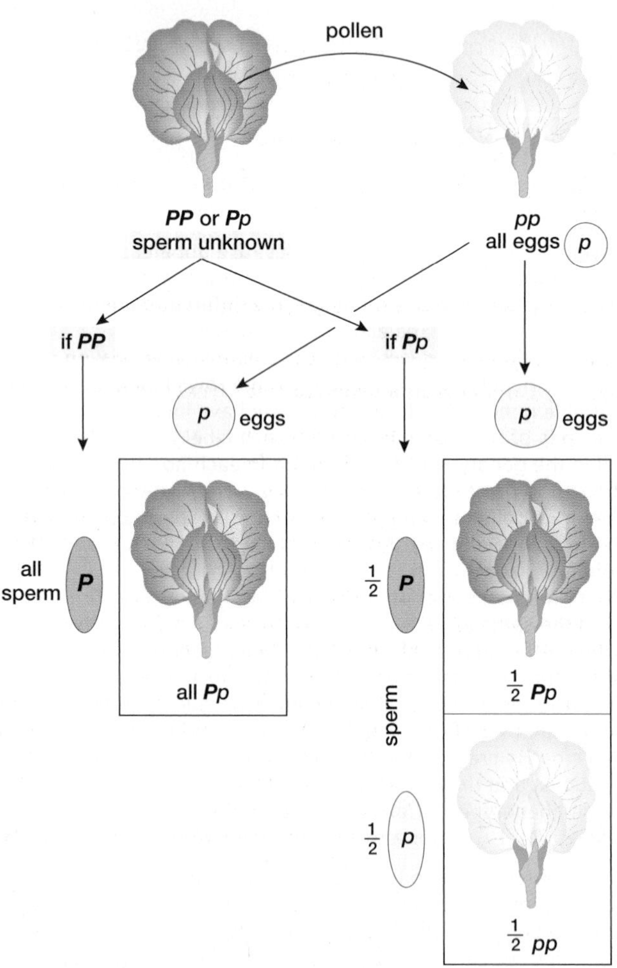

▲ **FIGURE 10-9 Punnett square of a test cross** An organism with a dominant phenotype may be either homozygous or heterozygous. Crossing such an organism with a homozygous recessive organism can determine whether the dominant organism was homozygous (left) or heterozygous (right).

10.4 HOW ARE MULTIPLE TRAITS INHERITED?

Having determined the inheritance modes for single traits, Mendel then turned to the more complex question of the inheritance of multiple traits in peas (**Fig. 10-10**). He began by crossbreeding plants that differed in two traits—for example, seed color (yellow or green) and seed shape (smooth or wrinkled). From other crosses of plants with these traits, Mendel already knew that the smooth allele of the seed shape gene (*S*) is dominant to the wrinkled allele (*s*), and that the yellow allele of the seed color gene (*Y*) is dominant to the green allele (*y*). He crossed a true-breeding plant with smooth, yellow seeds (*SSYY*) to a true-breeding plant with wrinkled, green seeds (*ssyy*). The *SSYY* plant produced only *SY* gametes, and the *ssyy* plant produced only *sy* gametes. Therefore, all the F_1 offspring were heterozygotes: genotypically *SsYy* with the phenotype of smooth, yellow seeds.

Allowing these heterozygous F_1 plants to self-fertilize, Mendel found that the F_2 generation consisted of 315 plants with smooth, yellow seeds; 101 with wrinkled, yellow seeds; 108 with smooth, green seeds; and 32 with wrinkled, green seeds—a ratio of about 9:3:3:1. The offspring produced from

Trait	Dominant form	Recessive form
Seed shape	smooth	wrinkled
Seed color	yellow	green
Pod shape	inflated	constricted
Pod color	green	yellow
Flower color	purple	white
Flower location	at leaf junctions	at tips of branches
Plant size	tall (about 6 feet)	dwarf (about 8 to 16 inches)

▲ **FIGURE 10-10 Traits of pea plants studied by Gregor Mendel**

other crosses of plants that were heterozygous for two traits also had phenotypic ratios of about 9:3:3:1.

Mendel Hypothesized That Traits Are Inherited Independently

Mendel realized that these results could be explained if the genes for seed color and seed shape were inherited independently of each other and did not influence each other during gamete formation. If so, then for each trait, three-quarters of the offspring should show the dominant phenotype and one-quarter should show the recessive phenotype. This result is just what Mendel observed. He found 423 plants with smooth seeds (of either color) and 133 with wrinkled seeds (a ratio of about 3:1); in this same group of plants, 416 produced yellow seeds (of either shape) and 140 produced green seeds (also about 3:1). **Figure 10-11** shows how a Punnett square or probability calculation can be used to determine the outcome of a cross between organisms that are heterozygous for two traits.

The independent inheritance of two or more traits is called the **law of independent assortment.** Multiple traits are inherited independently if the alleles of one gene are distributed to gametes independently of the alleles for other genes. Independent assortment will occur when the traits being studied are controlled by genes on different pairs of homologous chromosomes. Why? From Chapter 9, recall the movement of chromosomes during meiosis. When paired homologous chromosomes line up during metaphase I, which homologue faces which pole of the cell is random, and the orientation of one homologous pair does not influence other pairs. Therefore, when the homologues separate during anaphase I, which allele of a gene on homologous pair 1 moves "north" does not affect which allele of a gene on homologous pair 2 moves "north"; that is, the alleles of genes on different chromosomes are distributed, or assorted, independently (**Fig. 10-12**).

In an Unprepared World, Genius May Go Unrecognized

In 1865, Gregor Mendel presented the results of his experiments on inheritance to the Brünn Society for the Study of Natural Science, and they were published the following year. His paper did not mark the beginning of genetics. In fact, it didn't make any impression at all on the study of biology during his lifetime. Mendel's experiments, which eventually spawned one of the most important scientific theories in all of biology, simply vanished from the scene. Apparently, very few biologists read his paper, and those who did failed to recognize its significance.

It was not until 1900 that three biologists—Carl Correns, Hugo de Vries, and Erich Tschermak—working independently of one another and knowing nothing of Mendel's work, rediscovered the principles of inheritance. No doubt to their intense disappointment, when they searched the scientific literature

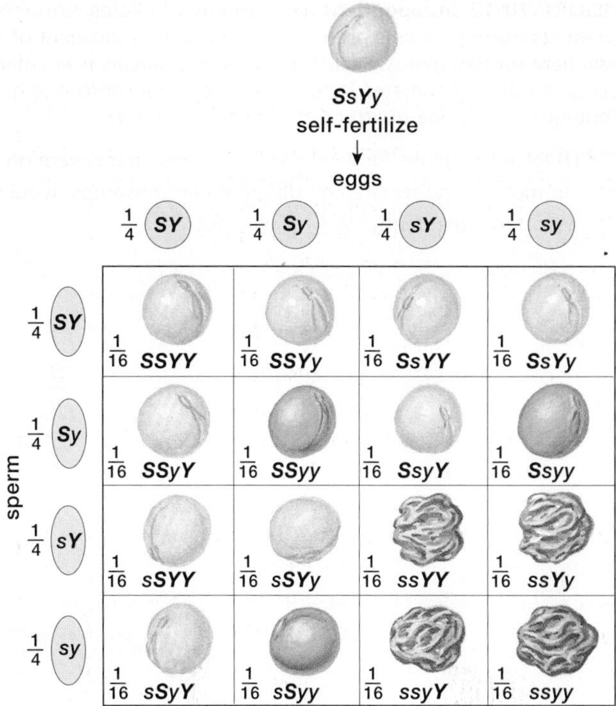

(a) Punnett square of a two-trait cross

seed shape		seed color		phenotypic ratio (9:3:3:1)
$\frac{3}{4}$ smooth	×	$\frac{3}{4}$ yellow	=	$\frac{9}{16}$ smooth yellow
$\frac{3}{4}$ smooth	×	$\frac{1}{4}$ green	=	$\frac{3}{16}$ smooth green
$\frac{1}{4}$ wrinkled	×	$\frac{3}{4}$ yellow	=	$\frac{3}{16}$ wrinkled yellow
$\frac{1}{4}$ wrinkled	×	$\frac{1}{4}$ green	=	$\frac{1}{16}$ wrinkled green

(b) Using probabilities to determine the offspring of a two-trait cross

▲ **FIGURE 10-11 Predicting genotypes and phenotypes for a cross between parents that are heterozygous for two traits** In pea seeds, yellow color *(Y)* is dominant to green *(y)*, and smooth shape *(S)* is dominant to wrinkled *(s)*. **(a)** Punnett square analysis. In this cross, an individual heterozygous for both traits self-fertilizes. In a cross involving two independent genes, the gamete types consist of all of the possible combinations of alleles of the two genes—*S* with *Y*, *S* with *y*, *s* with *Y*, and *s* with *y*. Place these gamete combinations as the labels for the rows and columns in the Punnett square, and then calculate the offspring as explained in Figure 10-8. Note that the Punnett square predicts both the frequencies of combinations of traits ($\frac{9}{16}$ smooth yellow, $\frac{3}{16}$ smooth green, $\frac{3}{16}$ wrinkled yellow, and $\frac{1}{16}$ wrinkled green) and the frequencies of individual traits ($\frac{3}{4}$ yellow, $\frac{1}{4}$ green, $\frac{3}{4}$ smooth, and $\frac{1}{4}$ wrinkled). **(b)** Probability theory states that the probability of two independent events is the product (multiplication) of their individual probabilities. Seed *shape* is independent of seed *color*. Therefore, multiplying the independent probabilities of the genotypes or phenotypes for each trait produces the predicted frequencies for the combined genotypes or phenotypes of the offspring. These ratios are identical to those generated by the Punnett square.

QUESTION Can the genotype of a plant bearing smooth yellow seeds be revealed by a test cross with a plant bearing wrinkled green seeds?

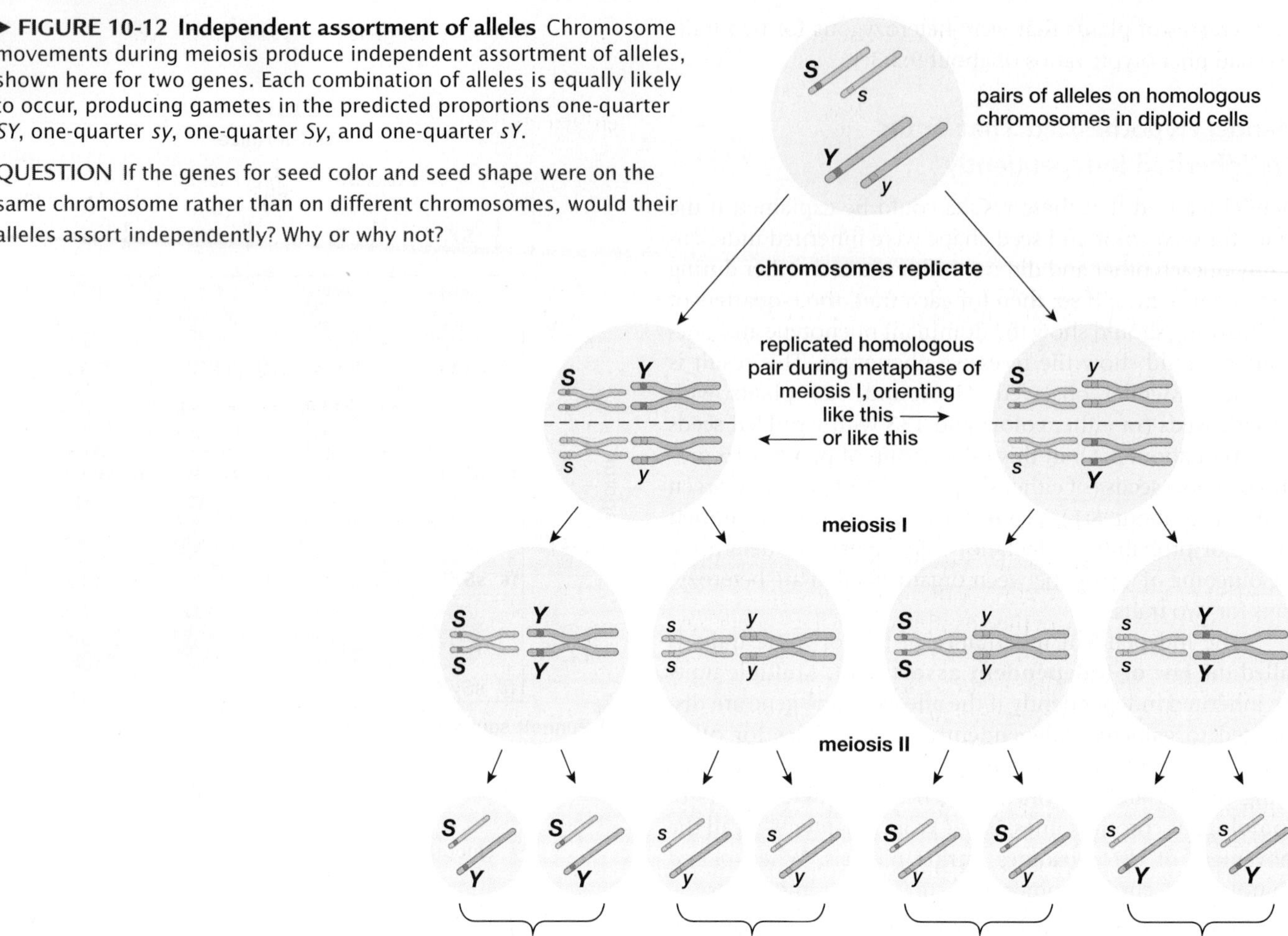

▶ **FIGURE 10-12 Independent assortment of alleles** Chromosome movements during meiosis produce independent assortment of alleles, shown here for two genes. Each combination of alleles is equally likely to occur, producing gametes in the predicted proportions one-quarter *SY*, one-quarter *sy*, one-quarter *Sy*, and one-quarter *sY*.

QUESTION If the genes for seed color and seed shape were on the same chromosome rather than on different chromosomes, would their alleles assort independently? Why or why not?

before publishing their results, they found that Mendel had scooped them more than 30 years earlier. To their credit, they graciously acknowledged the important work of the Austrian monk, who had died in 1884.

10.5 HOW ARE GENES LOCATED ON THE SAME CHROMOSOME INHERITED?

Gregor Mendel knew nothing about the physical nature of genes or chromosomes. Much later, when scientists discovered that chromosomes are the vehicles of inheritance, it became obvious that there are many more traits (and therefore many more genes) than there are chromosomes. We now know that genes are parts of chromosomes, and that each chromosome contains many genes, up to several thousand in a really large chromosome. These facts have important implications for inheritance.

Genes on the Same Chromosome Tend to Be Inherited Together

We now know that chromosomes, not individual genes, assort independently during meiosis I. Therefore, genes located on *different chromosomes* assort independently into gametes. In contrast, genes on the *same chromosome* tend to be inherited together, a phenomenon called gene **linkage.** One of the first pairs of linked genes to be discovered was found in the sweet pea, a different species from Mendel's edible pea. In sweet peas, the gene for flower color (purple vs. red) and the gene for pollen grain shape (round vs. long) are carried on the same chromosome. Thus, the alleles for these genes normally assort together into gametes during meiosis and are inherited together.

Consider a heterozygous sweet pea plant with purple flowers and long pollen (**Fig. 10-13**). Note that the dominant purple allele of the flower-color gene and the dominant long allele of the pollen-shape gene are located on one homologous chromosome (see Fig. 10-13, top). The recessive red allele of the flower-color gene and the recessive round allele of the pollen-shape gene are located on the other homologue (see Fig. 10-13, bottom). Therefore, the gametes produced by this plant are likely to have either purple and long alleles or red and round alleles. This pattern of inheritance does not conform to the law of independent assortment, because the alleles for flower color and pollen shape do not segregate independently of one another, but tend to stay together during meiosis.

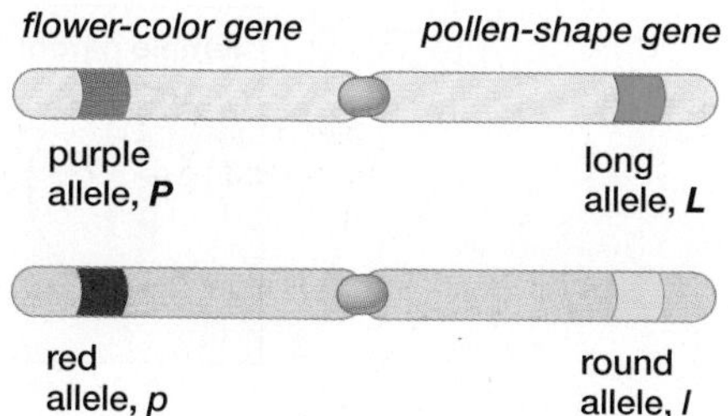

▲ **FIGURE 10-13 Linked genes on homologous chromosomes in the sweet pea** The genes for flower color and pollen shape are on the same chromosome, so they tend to be inherited together.

Crossing Over Creates New Combinations of Linked Alleles

Although they *tend* to be inherited together, genes on the same chromosome do not *always* stay together. If you cross-fertilized two sweet peas with the chromosomes shown in Figure 10-13, you might expect that all of the offspring would have either purple flowers with long pollen grains, or red flowers with round pollen grains. (Try working this out with a Punnett square.) However, in reality you would usually find a few plants in which the genes for flower color and pollen shape have been inherited as if they were not linked. That is, some of the offspring will have purple flowers and round pollen, and some will have red flowers and long pollen. How can this be?

As you learned in Chapter 9, during prophase I of meiosis, homologous chromosomes sometimes exchange parts, a process called crossing over (see Fig. 9-16). In most chromosomes, at least one exchange between each homologous pair occurs during each meiotic cell division. The exchange of corresponding segments of DNA during crossing over produces new allele combinations on both homologous chromosomes. Then, when homologues separate at anaphase I, the chromosomes that the haploid daughter cells receive will have different sets of alleles than the chromosomes of the parent cell did.

Crossing over during meiosis explains **genetic recombination:** the appearance of new combinations of alleles that were previously linked. Let's revisit the sweet pea during meiosis. During prophase I, the replicated, homologous chromosomes pair up (**Fig. 10-14a**). Each homologue will have one or more regions where crossing over will occur. Imagine that crossing over exchanges the alleles for flower color between nonsister chromatids of the two homologues (**Fig. 10-14b**). At anaphase I, the separated homologues will each now have one chromatid bearing a piece of DNA from a chromatid of the other homologue (**Fig. 10-14c**). During meiosis II, four types of chromosomes will be distributed to the four daughter cells: two unchanged chromosomes and two recombined chromosomes (**Fig. 10-14d**).

Therefore, some gametes will be produced with each of four chromosome configurations: *PL* and *pl* (the original parental types), and *Pl* and *pL* (the recombined chromosomes). If a sperm with a *Pl* chromosome fertilizes an egg with a *pl* chromosome, the offspring plant will have purple flowers (*Pp*) and round pollen (*ll*). If a sperm with a *pL*

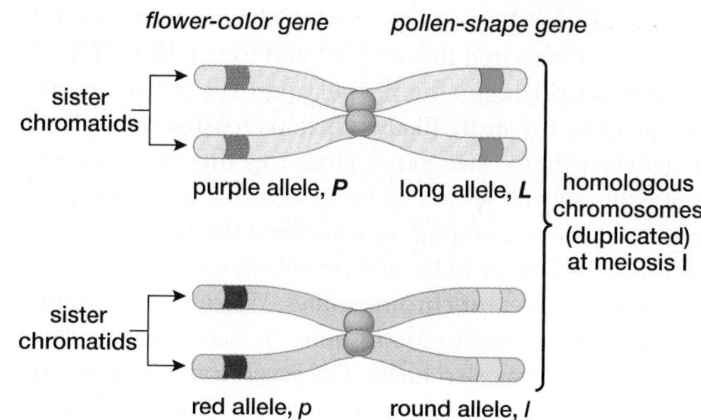

(a) Replicated chromosomes in prophase of meiosis I

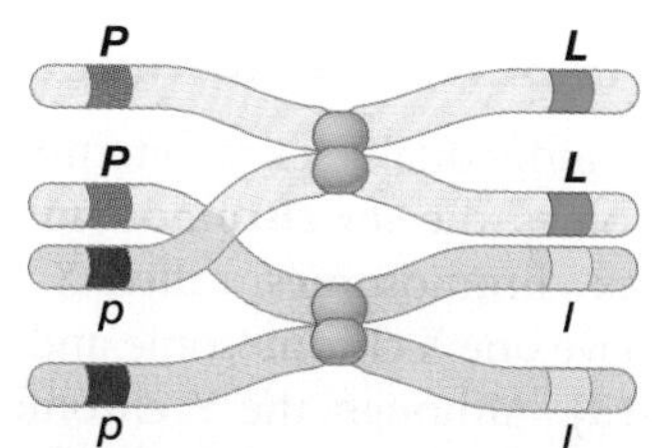

(b) Crossing over during prophase I

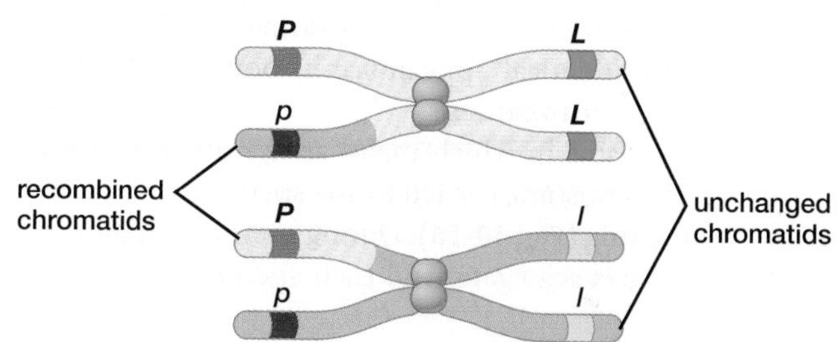

(c) Homologous chromosomes separate at anaphase I

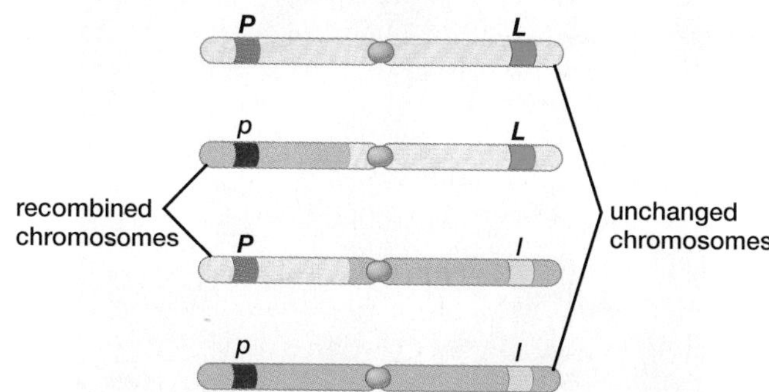

(d) Unchanged and recombined chromosomes after meiosis II

▲ **FIGURE 10-14 Crossing over recombines alleles on homologous chromosomes (a)** During prophase of meiosis I, replicated homologous chromosomes pair up. **(b)** Nonsister chromatids of the two homologues exchange parts by crossing over. **(c)** When the homologous chromosomes separate during anaphase of meiosis I, one chromatid of each of the homologues now contains a piece of DNA from a chromatid of the other homologue. **(d)** After meiosis II, two of the haploid daughter cells receive unchanged chromosomes, and two receive recombined chromosomes. The recombined chromosomes contain allele arrangements that did not occur in the original parental chromosomes.

chromosome fertilizes an egg with a *pl* chromosome, then the offspring will have red flowers (*pp*) and long pollen (*Ll*).

Not surprisingly, the farther apart the genes are on a chromosome, the more likely it is that crossing over will occur between them. Two genes close together on a chromosome will rarely be separated by a crossover, but if two genes are very far apart, crossing over between the genes occurs so often that they seem to be independently assorted, just as if they were on different chromosomes. When Gregor Mendel discovered independent assortment, he was not only clever and careful, he was also lucky. The seven traits that he studied were controlled by genes on only four different chromosomes. He observed independent assortment because the genes that were on the same chromosome were far apart.

10.6 HOW IS SEX DETERMINED?

In mammals, an individual's sex is determined by a special pair of chromosomes, the **sex chromosomes.** Females have two identical sex chromosomes, called **X chromosomes,** whereas males have one X chromosome and one **Y chromosome** (**Fig. 10-15**). Although the Y chromosome is much smaller than the X chromosome, a small part of both sex chromosomes is homologous. As a result, the X and Y chromosomes pair up during prophase of meiosis I and separate during anaphase I. The other chromosomes, which occur in pairs that have identical appearance in both males and females, are called **autosomes.**

For organisms in which males are XY and females are XX, the sex chromosome carried by the sperm determines the sex of the offspring (**Fig. 10-16**). During sperm formation, the sex chromosomes segregate, and each sperm receives either the X or the Y chromosome (plus one member of each pair of autosomes). The sex chromosomes also segregate during egg formation, but because females have two X chromosomes, every egg receives one X chromosome (along with one member of each pair of autosomes). Thus, a male offspring is produced if an egg is fertilized by a Y-bearing sperm, and a female offspring is produced if an egg is fertilized by an X-bearing sperm.

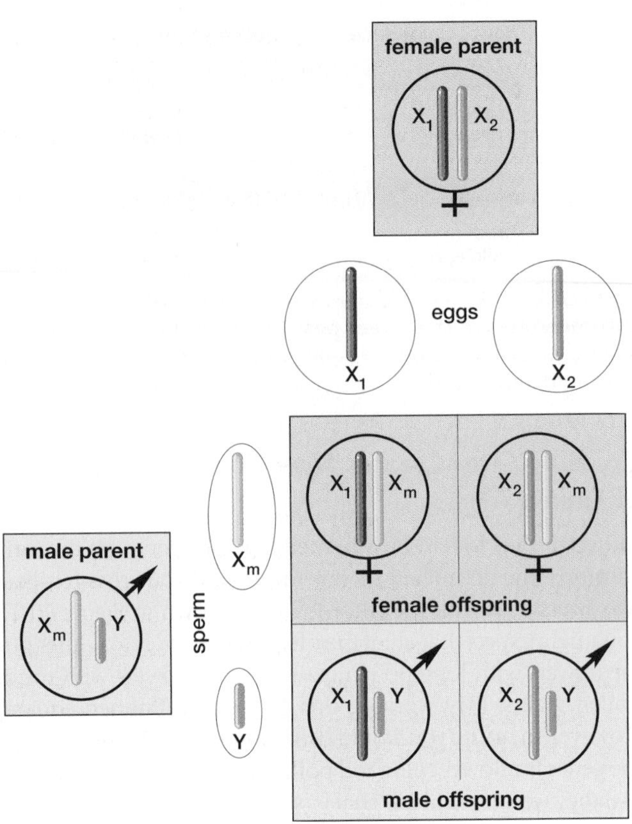

▲ **FIGURE 10-16 Sex determination in mammals** Male offspring receive their Y chromosome from their father; female offspring receive the father's X chromosome (labeled X_m). Both male and female offspring receive an X chromosome (either X_1 or X_2) from their mother.

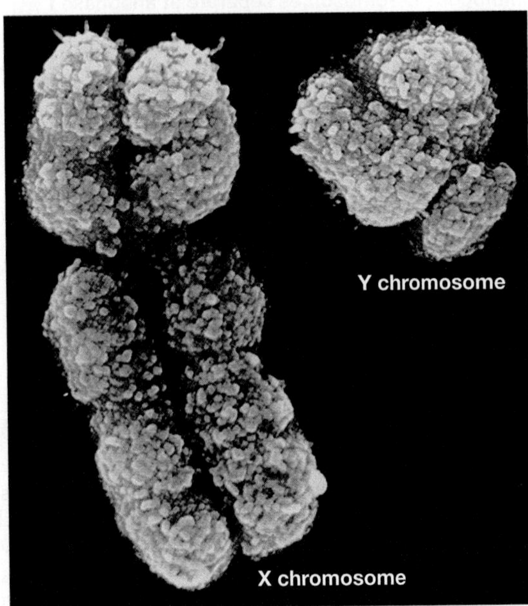

▲ **FIGURE 10-15 Photomicrograph of human sex chromosomes** The Y chromosome (right), which carries relatively few genes, is much smaller than the X chromosome (left).
Image courtesy of Indigo® Instruments: http://www.indigo.com.

10.7 HOW ARE SEX-LINKED GENES INHERITED?

Sex-Linked Genes Are Found Only on the X or Only on the Y Chromosome

Genes that are on one sex chromosome but not on the other are said to be **sex-linked.** In many animal species, the Y chromosome carries only a few genes. In humans, the Y chromosome contains a few dozen genes (probably less than 80), many of which play a role in male reproduction. The most well-known Y-linked gene is *SRY,* the sex-determining region on the *Y* chromosome, discovered in 1990. During embryonic life, the action of this gene sets in motion the entire male developmental pathway. Under normal conditions, *SRY* causes the male gender to be 100% linked to the Y chromosome.

In contrast to the small Y chromosome, the human X chromosome contains more than 1,000 genes, few of which have a specific role in female reproduction. Most of the genes on the X chromosome have no counterpart on the Y chromosome, including traits that are important in both sexes, such as color vision, blood clotting, and certain structural proteins in muscles.

How does linkage of genes on the X chromosome affect inheritance? Because they have two X chromosomes, females can be either homozygous or heterozygous for genes on the X chromosome, and dominant versus recessive relationships among alleles will be expressed. Males, in contrast, fully express all the alleles they have on their single X chromosome, whether those alleles are otherwise dominant or recessive.

Let's look at a familiar example: red-green color blindness (**Fig. 10-17**). Color blindness is caused by recessive alleles of either of two genes located on the X chromosome. The normal, dominant alleles of these genes (we will call them *C*) encode proteins that allow one set of color-vision cells in the eye, called cones, to be most sensitive to red light and another set to be most sensitive to green light (**Fig. 10-17a**). There are several defective recessive alleles of these genes (we will call them *c*). In the most extreme cases, one of the genes will actually be missing from an X chromosome, or the defective alleles will encode proteins that make both sets of cones equally sensitive to both red and green light. Therefore, the affected person cannot distinguish red from green (**Fig. 10-17b**).

How is color blindness inherited? A man can have the genotype *C*Y or *c*Y, meaning that he has a color-vision allele *C* or *c* on his X chromosome and no corresponding gene on his Y chromosome. He will therefore have normal color vision if his X chromosome bears the *C* allele, or be color-blind if it bears the *c* allele. A woman may be *CC*, *Cc*, or *cc*. Women with *CC* or *Cc* genotypes will have normal color vision; only women with *cc* genotypes will be color-blind. Roughly 7% of men have defective color vision. Among women, about 93% are homozygous normal *CC*, 7% are heterozygous normal *Cc*, and less than 0.5% are color-blind *cc*.

A color-blind man (*c*Y) will pass his defective allele only to his daughters, because only his daughters inherit his X chromosome. Usually, however, the daughters will have normal color vision, because they will also inherit a normal *C* allele from their mother, who is very likely homozygous normal *CC*.

A heterozygous woman (*Cc*), although she has normal color vision, will pass her defective allele to half of her sons (**Fig. 10-17c**). These sons will be color-blind (*c*Y). The other half of her sons will inherit her functional allele and have normal color vision (*C*Y).

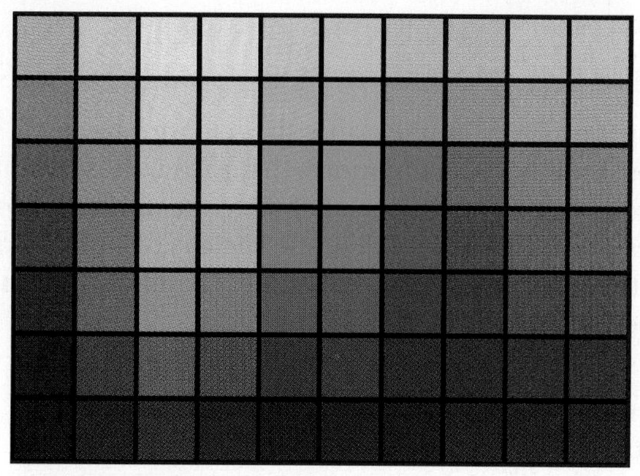

(a) Normal color vision

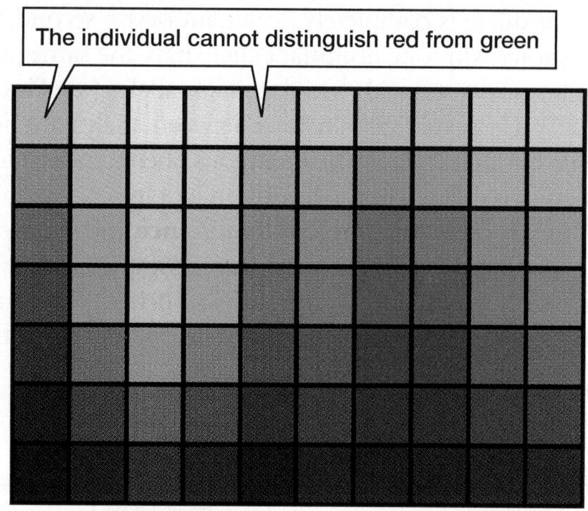

(b) Red-green color blindness

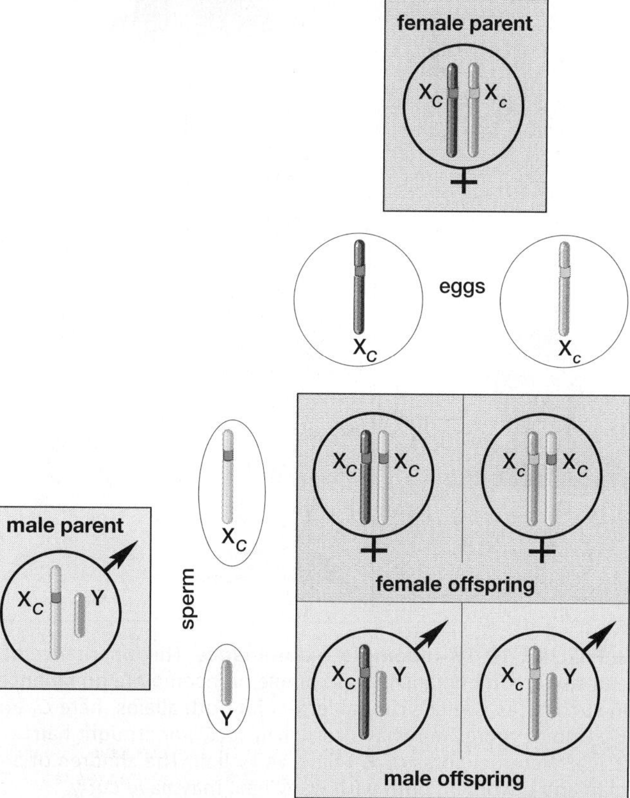

(c) Expected children of a man with normal color vision (*C*Y), and a heterozygous woman (*Cc*)

▶ **FIGURE 10-17 Sex-linked inheritance of color blindness**
Color grids help people compare **(a)** normal color vision to **(b)** red-green color deficiency. Usually, an affected person isn't really color "blind"—he sees most of the colors that normal people see, but not as well. **(c)** A Punnett square showing the inheritance of color blindness from a heterozygous woman (*Cc*) to her sons.

10.8 DO THE MENDELIAN RULES OF INHERITANCE APPLY TO ALL TRAITS?

In our discussion of inheritance thus far, we have made some major simplifying assumptions: that each trait is completely controlled by a single gene, that there are only two possible alleles of each gene, and that one allele is completely dominant to the other, recessive, allele. Most traits, however, are influenced in more varied and subtle ways.

Incomplete Dominance: The Phenotype of Heterozygotes Is Intermediate Between the Phenotypes of the Homozygotes

When one allele is completely dominant over a second allele, heterozygotes with one dominant allele have the same phenotype as homozygotes with two dominant alleles (see Figs. 10-8 and 10-9). However, relationships between alleles are not always this simple. When the heterozygous phenotype is intermediate between the two homozygous phenotypes, the pattern of inheritance is called **incomplete dominance.** In humans, hair texture is influenced by a gene with two incompletely dominant alleles, which we will call C_1 and C_2 (**Fig. 10-18**). A person with two copies of the C_1 allele has curly hair; two copies of the C_2 allele produce straight hair. Heterozygotes, with the C_1C_2 genotype, have wavy hair. If two wavy-haired people marry, they might have children with any of the three hair types, with probabilities of one-quarter curly (C_1C_1), one-half wavy (C_1C_2), and one-quarter straight (C_2C_2).

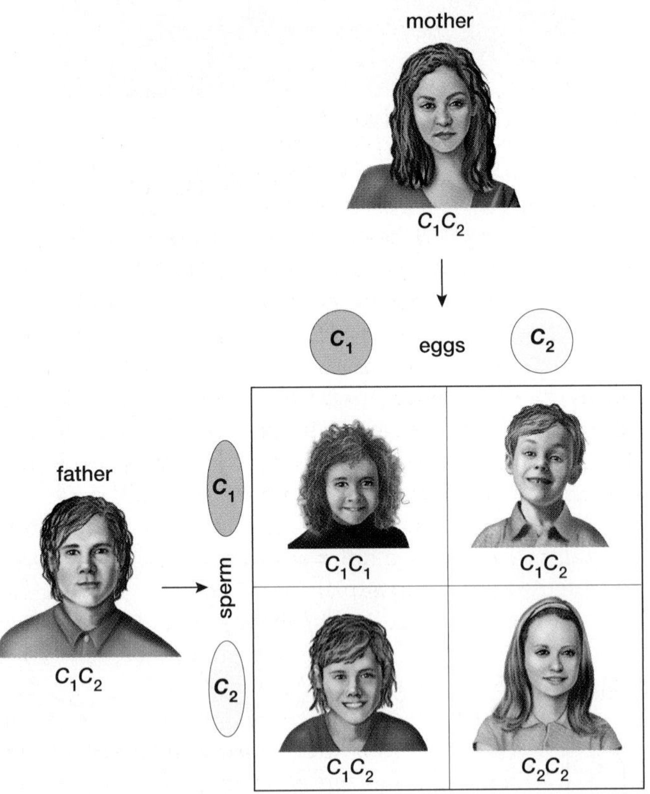

▲ **FIGURE 10-18 Incomplete dominance** The inheritance of hair texture in humans is an example of incomplete dominance. In such cases, we use capital letters for both alleles, here C_1 and C_2. Homozygotes may have curly hair (C_1C_1) or straight hair (C_2C_2). Heterozygotes (C_1C_2) have wavy hair. The children of a man and a woman, both with wavy hair, may have curly, straight, or wavy hair, in the approximate ratio of $\frac{1}{4}$ curly: $\frac{1}{2}$ wavy: $\frac{1}{4}$ straight.

A Single Gene May Have Multiple Alleles

Alleles arise through mutation, and the same gene in different individuals may have different mutations, each producing a new allele. Therefore, although an *individual* can have at most two different alleles, a *species* may have **multiple alleles** of many of its genes. There are multiple alleles for many human genetic disorders, including Marfan syndrome, Duchenne muscular dystrophy (see the "Health Watch: Muscular Dystrophy" on pp. 194–195), and cystic fibrosis (see the Case Study for Chapter 12).

The human blood types are an example of multiple alleles of a single gene, with an added twist to the pattern of inheritance. The blood types A, B, AB, and O arise as a result of three different alleles of a single gene on chromosome 9 (for simplicity, we will designate the alleles *A*, *B*, and *o*). This gene codes for an enzyme that adds sugar molecules to the ends of glycoproteins that protrude from the surfaces of red blood cells. Alleles *A* and *B* code for enzymes that add different sugars to the glycoproteins (we'll call the resulting molecules glycoproteins A and B, respectively). Allele *o* codes for a nonfunctional enzyme that doesn't add any sugar molecule.

A person may have one of six genotypes: *AA, BB, AB, Ao, Bo,* or *oo* (**Table 10-1**). Alleles *A* and *B* are dominant to *o*. Therefore, people with genotypes *AA* or *Ao* have only type A glycoproteins and have type A blood. Those with genotypes *BB* or *Bo* synthesize only type B glycoproteins and have type B blood. Homozygous recessive *oo* individuals lack both types of glycoproteins and have type O blood. In people with type AB blood, both enzymes are present, so the plasma membranes of their red blood cells have both A and B glycoproteins. When heterozygotes express phenotypes of *both* of the homozygotes (in this case, both A and B glycoproteins), the pattern of inheritance is called **codominance,** and the alleles are said to be codominant to one another.

People make antibodies to the type of glycoprotein(s) that they lack. These antibodies are proteins in blood plasma that bind to foreign glycoproteins by recognizing different sugar molecules. The antibodies cause red blood cells that bear foreign glycoproteins to clump together and rupture. The resulting clumps and fragments can clog small blood vessels and damage vital organs such as the brain, heart, lungs, or kidneys. This means that blood type must be determined and matched carefully before a blood transfusion is made.

Type O blood, lacking any sugars, is not attacked by antibodies in A, B, or AB blood, so it can be transfused safely to all other blood types. (The antibodies present in transfused blood become too diluted to cause problems.) People with type O blood are called "universal donors." But O blood carries antibodies to both A and B glycoproteins, so type O individuals can receive transfusions of only type O blood. Table 10-1 summarizes blood types and transfusion characteristics.

Table 10-1 Human Blood Group Characteristics

Blood Type	Genotype	Red Blood Cells	Has Plasma Antibodies to:	Can Receive Blood from:	Can Donate Blood to:	Frequency in U. S.
A	*AA* or *Ao*	A glycoprotein	B glycoprotein	A or O (no blood with B glycoprotein)	A or AB	40%
B	*BB* or *Bo*	B glycoprotein	A glycoprotein	B or O (no blood with A glycoprotein)	B or AB	10%
AB	*AB*	Both A and B glycoproteins	Neither A nor B glycoprotein	AB, A, B, O (universal recipient)	AB	4%
O	*oo*	Neither A nor B glycoprotein	Both A and B glycoproteins	O (no blood with A or B glycoprotein)	O, AB, A, B (universal donor)	46%

Many Traits Are Influenced by Several Genes

If you look around your class, you are likely to see people of varied heights, skin colors, and body builds. Traits such as these are not governed by single genes but are influenced by interactions among two or more genes, as well as by interactions with the environment. Many traits, such as human height, weight, eye color, and skin color, may have several phenotypes or even seemingly continuous variation that cannot be split up into convenient, easily defined categories. This is an example of **polygenic inheritance,** a form of inheritance in which the interaction of two or more genes contributes to a single phenotype.

As you might imagine, the more genes that contribute to a single trait, the greater the number of phenotypes and the finer the distinctions among them. When three or more pairs of genes contribute to a trait, differences between phenotypes are fairly small. For example, although no one fully understands the inheritance of human skin color, it is probably controlled by at least three different genes, each with pairs of incompletely dominant alleles (**Fig. 10-19a**). If the environment also contributes significantly to the trait (for instance, exposure to sunlight alters skin color), there can be virtually continuous variation in phenotype (**Fig. 10-19b**).

Single Genes Typically Have Multiple Effects on Phenotype

We have just seen that a single phenotype may result from the interaction of several genes. The reverse is also true: Single genes commonly have multiple phenotypic effects, a phenomenon called **pleiotropy.** A good example is the *SRY* gene on the Y chromosome. The *SRY* gene codes for a protein that activates other genes; those genes in turn code for proteins that switch on male development in an embryo. Under the influence of the genes activated by the SRY protein, sex organs develop into testes. The testes, in turn, secrete sex hormones that stimulate the development of both internal and external male reproductive structures, such as the prostate gland and penis.

The Environment Influences the Expression of Genes

An organism is not just the sum of its genes. In addition to its genotype, the environment in which an organism lives profoundly affects its phenotype. A striking example of environmental effects on gene action occurs in Siamese cats. All Siamese cats are born with pale fur, but within the first few weeks, the ears, nose, paws, and tail turn dark (**Fig. 10-20**). A Siamese cat actually has the genotype for dark fur all over its body. However, the enzyme that produces the dark pigment is inactive at temperatures above about 93°F (34°C). Unborn kittens are warm all over, inside their mother's uterus, so newborn Siamese kittens have pale fur on their entire bodies. After they are born, the ears, nose, paws, and tail become cooler than the rest of the body, and dark pigment is produced there.

Most environmental influences are more complicated and subtle than this. Complex environmental influences are particularly common in human characteristics. The polygenic trait of skin color is modified by the environmental effects of sun exposure (see Fig. 10-19b). Height, another polygenic trait, is influenced by nutrition.

The interactions between complex genetic systems and varied environmental conditions can create a continuum of phenotypes that defies precise analysis into genetic and environmental components. The human generation time is long, and the number of offspring per couple is small. Add to these factors the countless subtle ways in which people respond to their environments, and you can see why it may be exceedingly difficult to determine the exact genetic basis of complex human traits such as intelligence, or abilities in music, art, or athletics.

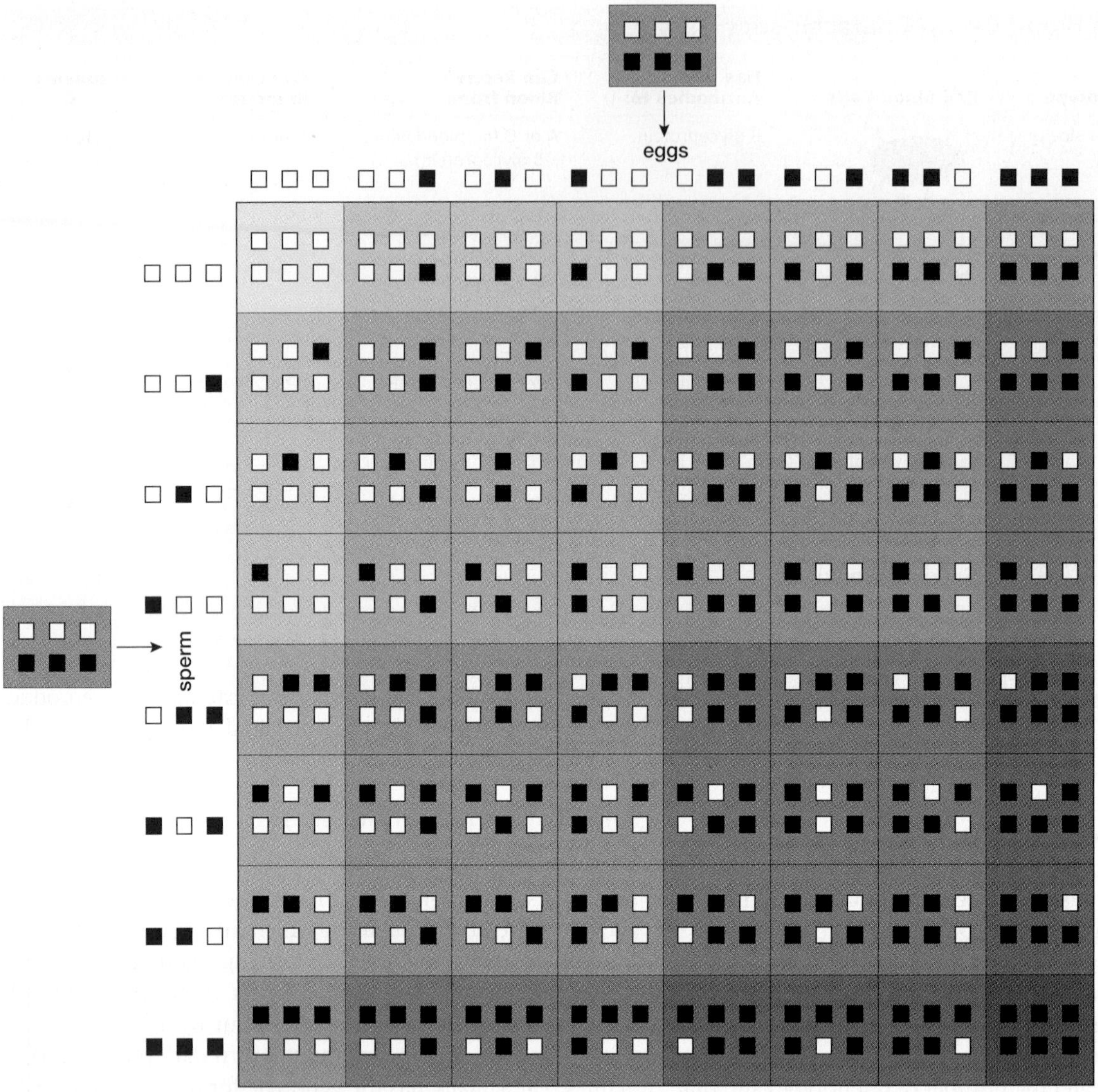

eggs

sperm

(a) Punnett square of the inheritance of skin color

(b) Humans show a wide range of skin tones, from almost white to very dark brown

▲ **FIGURE 10-19 Polygenic inheritance of skin color in humans (a)** At least three separate genes, each with two incompletely dominant alleles, determine human skin color (the inheritance is actually more complex than this). The black boxes denote "dark skin" alleles and the white boxes denote "pale skin" alleles. The backgrounds of each cell of the Punnett square indicate the depth of skin color expected from each genotype. **(b)** The combination of complex polygenic inheritance and environmental effects (especially exposure to sunlight) produces almost infinite gradation of human skin colors.

▲ **FIGURE 10-20 Environmental influence on phenotype** The distribution of dark fur in the Siamese cat is an interaction between genotype and environment, producing a particular phenotype. Newborn Siamese kittens have pale fur everywhere on their bodies. In an adult Siamese, the allele for dark fur is expressed only in the cooler areas (nose, ears, paws, and tail).

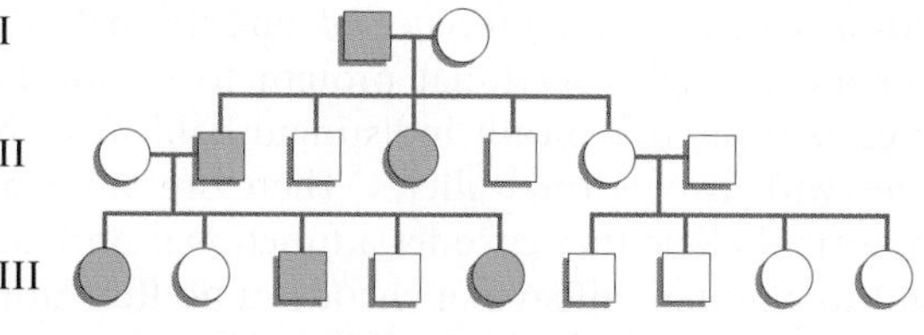

(a) A pedigree for a dominant trait

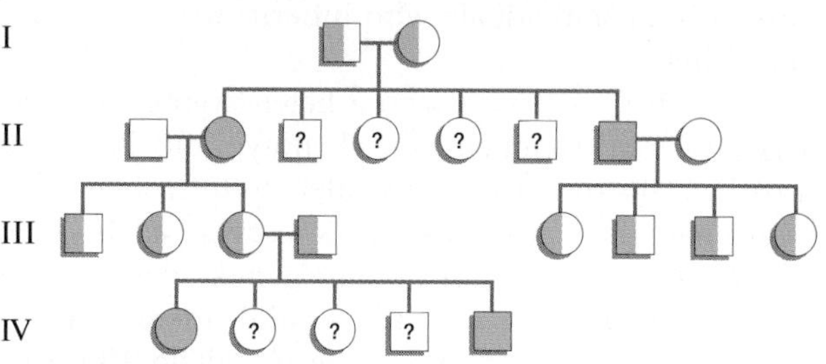

(b) A pedigree for a recessive trait

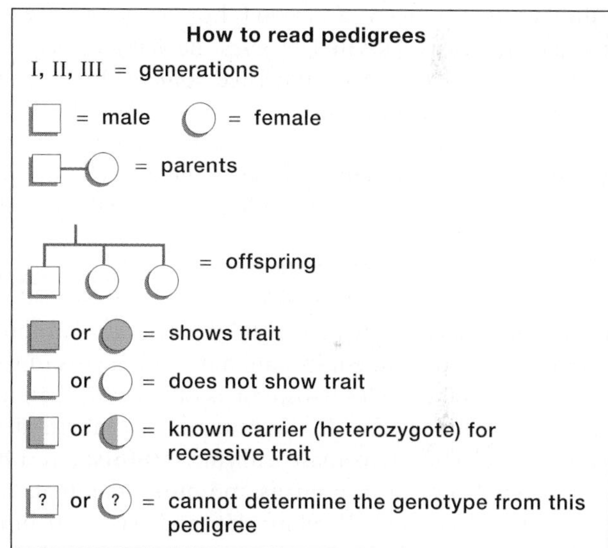

▲ **FIGURE 10-21 Family pedigrees (a)** A pedigree for a dominant trait. Note that any offspring showing a dominant trait must have at least one parent with the trait (see Figs. 10-8, 10-9, and 10-11). **(b)** A pedigree for a recessive trait. Any individual showing a recessive trait must be homozygous recessive. If that person's parents did not show the trait, then both of the parents must be heterozygotes (carriers). Note that the genotype cannot be determined for some offspring, who may be either carriers or homozygous dominants.

Case Study continued

Sudden Death on the Court

In Marfan syndrome, a single defective fibrillin allele causes increased height, long limbs, large hands and feet, weak walls in the aorta, and often dislocated lenses in one or both eyes—a striking example of pleiotropy.

10.9 HOW ARE HUMAN GENETIC DISORDERS INVESTIGATED?

Many human diseases are influenced by genetics to a greater or lesser degree. Because experimental crosses with humans are out of the question, human geneticists search medical, historical, and family records to study past crosses. Records extending across several generations can be arranged in the form of family **pedigrees,** diagrams that show the genetic relationships among a set of related individuals (**Fig. 10-21**). Careful analysis of pedigrees can reveal whether a particular trait is inherited in a dominant, recessive, or sex-linked pattern. Since the mid-1960s, analysis of human pedigrees, combined with molecular genetic technology, has produced great strides in understanding human genetic diseases. For instance, geneticists now know the genes responsible for dozens of inherited diseases, including sickle-cell anemia, hemophilia, muscular dystrophy, Marfan syndrome, and cystic fibrosis. Research in molecular genetics has increased our ability to predict genetic diseases and perhaps even to cure them, a topic we explore further in Chapter 13.

10.10 HOW ARE HUMAN DISORDERS CAUSED BY SINGLE GENES INHERITED?

Many common human traits, such as freckles, long eyelashes, and a widow's peak hairline, are inherited in a simple Mendelian fashion; that is, each trait appears to be controlled by a single gene with a dominant and a recessive allele. Here, we will concentrate on a few examples of medically important genetic disorders and the ways in which they are transmitted from one generation to the next.

Some Human Genetic Disorders Are Caused by Recessive Alleles

The human body depends on the actions of thousands of enzymes and other proteins. A mutation in an allele of the gene coding for one of these proteins can impair or destroy its function. However, the presence of one normal allele may generate enough functional protein to enable heterozygotes to be phenotypically indistinguishable from homozygotes with two normal alleles. Therefore, for many genes, a normal allele that encodes a functional protein is dominant to a mutant allele that encodes a nonfunctional protein. Put another way, a mutant allele of these genes is recessive to a normal allele. Thus, an abnormal phenotype occurs only in individuals who inherit two copies of the mutant allele.

A **carrier** is a person who is heterozygous for a recessive genetic trait: That person is phenotypically healthy but can pass on his or her recessive allele to offspring. Geneticists estimate that each of us carries recessive alleles of 5 to 15 genes that would cause serious genetic defects in homozygotes. Every time we have a child, there is a 50:50 chance that we will pass on the defective allele. This is usually harmless, because an unrelated man and woman are unlikely to possess a defective allele of the same gene, so they are unlikely to produce a child who is homozygous recessive for a genetic disease. Related couples, however (especially first cousins or closer), have inherited some of their genes from recent common ancestors, and so are more likely to carry a defective allele of the same gene. If they are heterozygous for the same defective recessive allele, such couples have a 1 in 4 chance of having a child affected by the genetic disorder (see Fig. 10-21).

Albinism Results from a Defect in Melanin Production

An enzyme called tyrosinase is needed to produce melanin—the dark pigment in skin, hair, and the iris of the eye. The gene that encodes tyrosinase is called *TYR*. If an individual is homozygous for a mutant *TYR* allele that encodes a defective tyrosinase enzyme, **albinism** results (**Fig. 10-22**). Albinism in humans and other mammals results in almost completely white skin and hair, and pink eyes (without melanin in the iris, you can see the color of the blood vessels in the retina).

Sickle-Cell Anemia Is Caused by a Defective Allele for Hemoglobin Synthesis

Sickle-cell anemia results from a mutation in the hemoglobin gene. The hemoglobin protein, which gives red blood cells their color, transports oxygen in the blood. In sickle-cell anemia, a change in a single nucleotide results in an incorrect amino acid at a crucial position in the hemoglobin molecule, altering its properties (see section 12.4 in Chapter 12). When oxygen concentrations in the blood are low (such as in exercising muscles), sickle-cell hemoglobin molecules in each red blood cell stick together. The resulting clumps force the red blood cell out of its normal disk shape (**Fig. 10-23a**) into a longer, sickle shape (**Fig. 10-23b**). Sickled cells are more fragile than normal red blood cells, making them likely to break; they also tend to clump, clogging capillaries. Tissues downstream of the clog do not receive enough oxygen and cannot have their wastes removed. Paralyzing strokes can result if blocks occur in blood vessels in the brain. The condition also causes anemia, because so many red blood cells are destroyed before their usual life span is completed.

People homozygous for the sickle-cell allele synthesize only defective hemoglobin. Therefore, many of their red blood cells become sickled. However, although heterozygotes have about half normal and half abnormal hemoglobin, they usually have few sickled cells and are not seriously affected—in fact, some world-class athletes are heterozygous for the sickle-cell allele. Because only people who are homozygous for the sickle-cell allele usually show symptoms, sickle-cell anemia is considered to be a recessive disorder.

About 20% to 40% of sub-Saharan Africans and 8% of African Americans are heterozygous for sickle-cell anemia, but the allele is very rare in Caucasians. Why? The prevalence of the sickle-cell allele in people with African origins occurs because heterozygotes have some resistance to the parasite that causes malaria, which is common in Africa and other places with warm climates, but not in colder regions such as most of Europe (see p. 300).

If two heterozygous carriers have children, every conception will have a 1 in 4 chance of producing a child who is homozygous for the sickle-cell allele and consequently will have sickle-cell anemia. Modern DNA techniques can distinguish the normal hemoglobin allele from the sickle-cell

(a) Human

(b) Wallaby

▶ **FIGURE 10-22 Albinism (a)** These sisters show that albinism and normal pigmentation can occur in the same family. **(b)** The albino wallaby in the foreground is healthy and safe in a zoo, but its bright white fur would make it very conspicuous to predators in the wild.

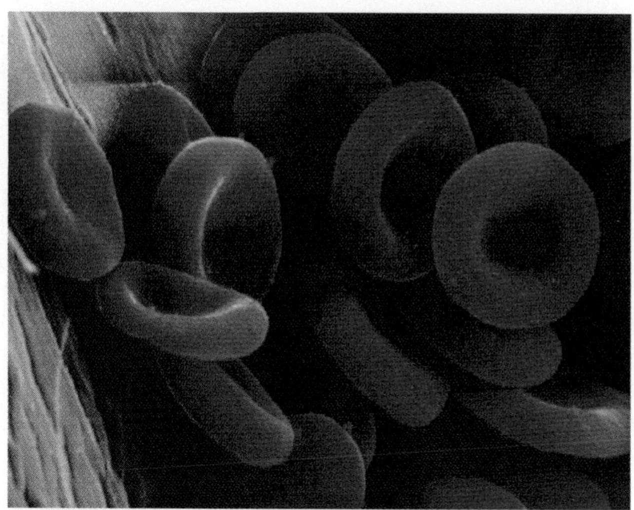

(a) Normal red blood cells

(b) Sickled red blood cells

▲ **FIGURE 10-23 Sickle-cell anemia (a)** Normal red blood cells are disk shaped with indented centers. **(b)** When blood oxygen is low, the red blood cells in a person with sickle-cell anemia take on a slender, curved shape resembling a sickle. In this shape they are fragile and tend to clump together, clogging capillaries.

allele, and analysis of fetal cells allows medical geneticists to diagnose sickle-cell anemia in fetuses. These methods are described in Chapter 13.

Some Human Genetic Disorders Are Caused by Dominant Alleles

Many serious genetic disorders, such as Huntington disease, are caused by dominant alleles. For dominant diseases to be passed on to offspring, at least one parent must have the disease; thus, at least some people with dominant diseases must remain healthy enough to grow up and have children. Alternatively, the dominant allele may result from a new mutation in an unaffected person's eggs or sperm. In this situation, neither parent has the disease.

How can a mutant allele be dominant to the normal allele? Some dominant alleles encode an abnormal protein that interferes with the function of the normal one. For example, some proteins must link together into long chains in order to perform their function in the cell. The abnormal protein may enter into a chain but prevent addition of new protein links. These shortened chain fragments may be unable to properly perform a needed function. Other dominant alleles may encode proteins that carry out new, toxic reactions. Finally, dominant alleles may encode a protein that is overactive, performing its function at inappropriate times and places.

Huntington Disease Is Caused by a Defective Protein That Kills Cells in Specific Brain Regions

Huntington disease is a dominant disorder that causes a slow, progressive deterioration of parts of the brain, resulting in a loss of coordination, flailing movements, personality disturbances, and eventual death. The symptoms of Huntington disease typically do not appear until 30 to 50 years of age. Therefore, many people pass the allele to their children before they suffer the first symptoms. Geneticists isolated the Huntington gene in 1993, and a few years later, identified the gene's product, a protein they named "huntingtin." The function of normal huntingtin remains unknown. Mutant huntingtin seems both to interfere with the action of normal huntingtin and to form large aggregates in nerve cells that ultimately kill the cells.

Some Human Genetic Disorders Are Sex-Linked

As we described earlier, the X chromosome contains many genes that have no counterpart on the Y chromosome. Because males have only one X chromosome, they have only one allele for each of these genes. This single allele will be expressed with no possibility of its activity being "hidden" by expression of another allele.

A son receives his X chromosome from his mother and passes it only to his daughters. Thus, sex-linked disorders caused by a recessive allele have a unique pattern of inheritance. Such disorders appear far more frequently in males and typically skip generations: An affected male passes the trait to a phenotypically normal, carrier daughter, who in turn bears some affected sons. The most familiar genetic defects due to recessive alleles of X-chromosome genes are red-green color blindness (see Fig. 10-17), muscular dystrophy, and **hemophilia** (**Fig. 10-24**).

Hemophilia is caused by a recessive allele on the X chromosome that results in a deficiency in one of the proteins needed for blood clotting. People with hemophilia bruise easily and may bleed extensively from minor injuries. Hemophiliacs often have anemia due to blood loss. Nevertheless, even before modern treatment with clotting factors, some hemophiliac males survived to pass on their defective allele to their daughters, who in turn could pass it to their sons. Muscular dystrophy, a fatal degeneration of the muscles in

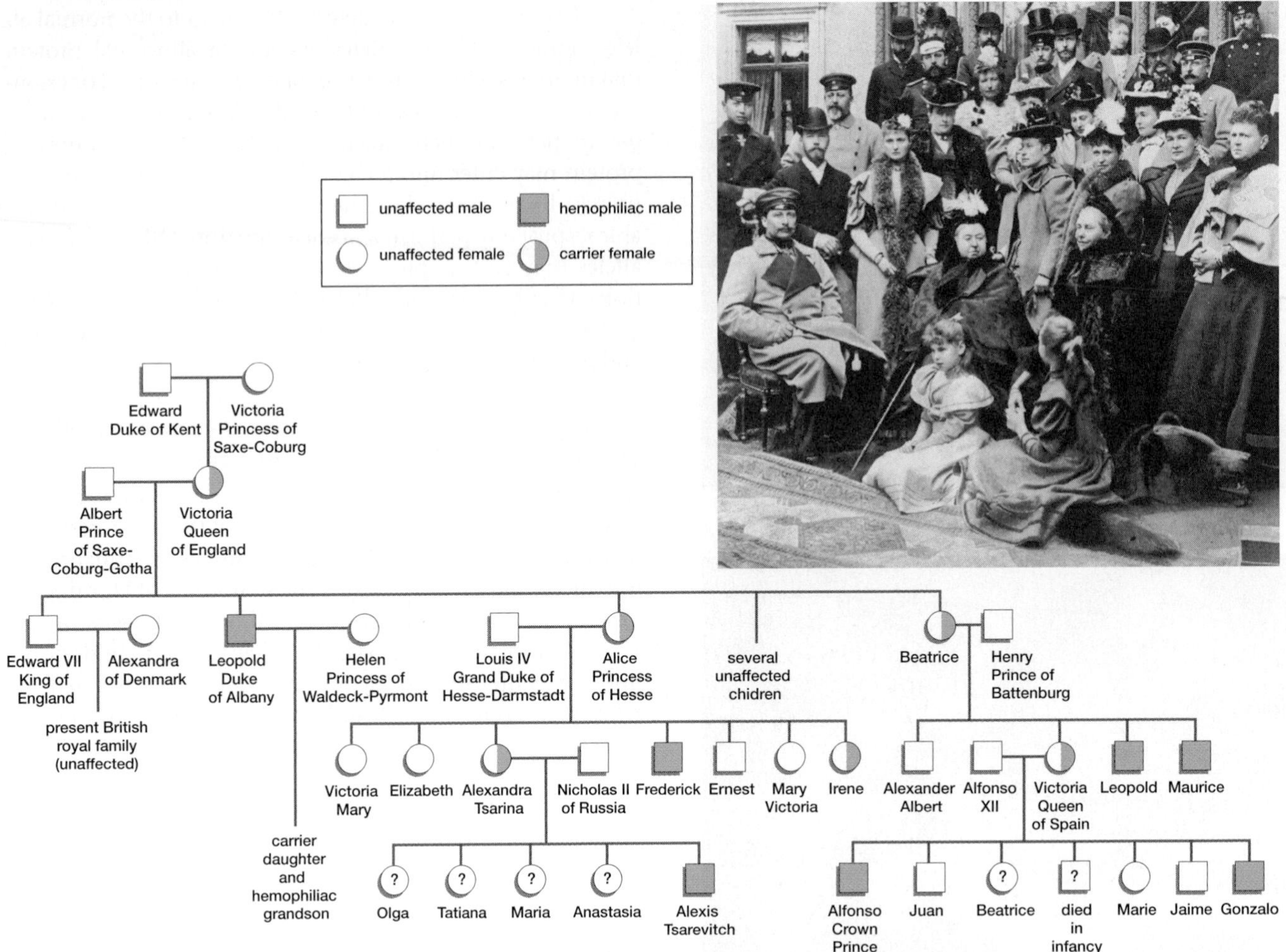

▲ **FIGURE 10-24 Hemophilia among the royal families of Europe** A famous genetic pedigree involves the transmission of sex-linked hemophilia from Queen Victoria of England (seated center front, with cane, in 1885) to her offspring and eventually to virtually every royal house in Europe, because of the extensive intermarriage of her children to the royalty of other European nations. Because Victoria's ancestors were free of hemophilia, the hemophilia allele must have arisen as a mutation either in Victoria herself or in one of her parents (or as a result of marital infidelity).

QUESTION Why is it not possible that a mutation in Victoria's husband, Albert, was the original source of hemophilia in this family pedigree?

young boys, is another recessive sex-linked disorder that we describe in "Health Watch: Muscular Dystrophy" on pp. 194–195.

10.11 HOW DO ERRORS IN CHROMOSOME NUMBER AFFECT HUMANS?

In Chapter 9, we examined the intricate mechanisms of meiotic cell division, which ensure that each sperm and egg receives only one chromosome from each homologous pair. Not surprisingly, this elaborate dance of the chromosomes occasionally misses a step, resulting in gametes that have too many or too few chromosomes (**Fig. 10-25**). Such errors in meiosis, called **nondisjunction,** can affect the number of either sex chromosomes or autosomes. Most embryos that arise from the fusion of gametes with abnormal chromosome numbers spontaneously abort, accounting for 20% to 50% of all miscarriages, but some embryos with abnormal numbers of chromosomes survive to birth or beyond.

Some Genetic Disorders Are Caused by Abnormal Numbers of Sex Chromosomes

Because the X and Y chromosomes pair up during meiosis, sperm usually carry either an X or a Y chromosome. Nondisjunction of sex chromosomes in males produces sperm with either no sex chromosome (often called "O" sperm), or two sex chromosomes (the sperm may be XX, YY, or XY, depending on whether the nondisjunction occurred in meiosis I or II). Nondisjunction of the sex chromosomes in females produces O or XX eggs instead of eggs with one X chromosome. When normal gametes fuse with these defective sperm or eggs, the zygotes have normal numbers of autosomes but abnormal numbers of sex chromosomes (**Table 10-2**). The most common abnormalities are XO, XXX, XXY, and XYY. (Genes on the X chromosome

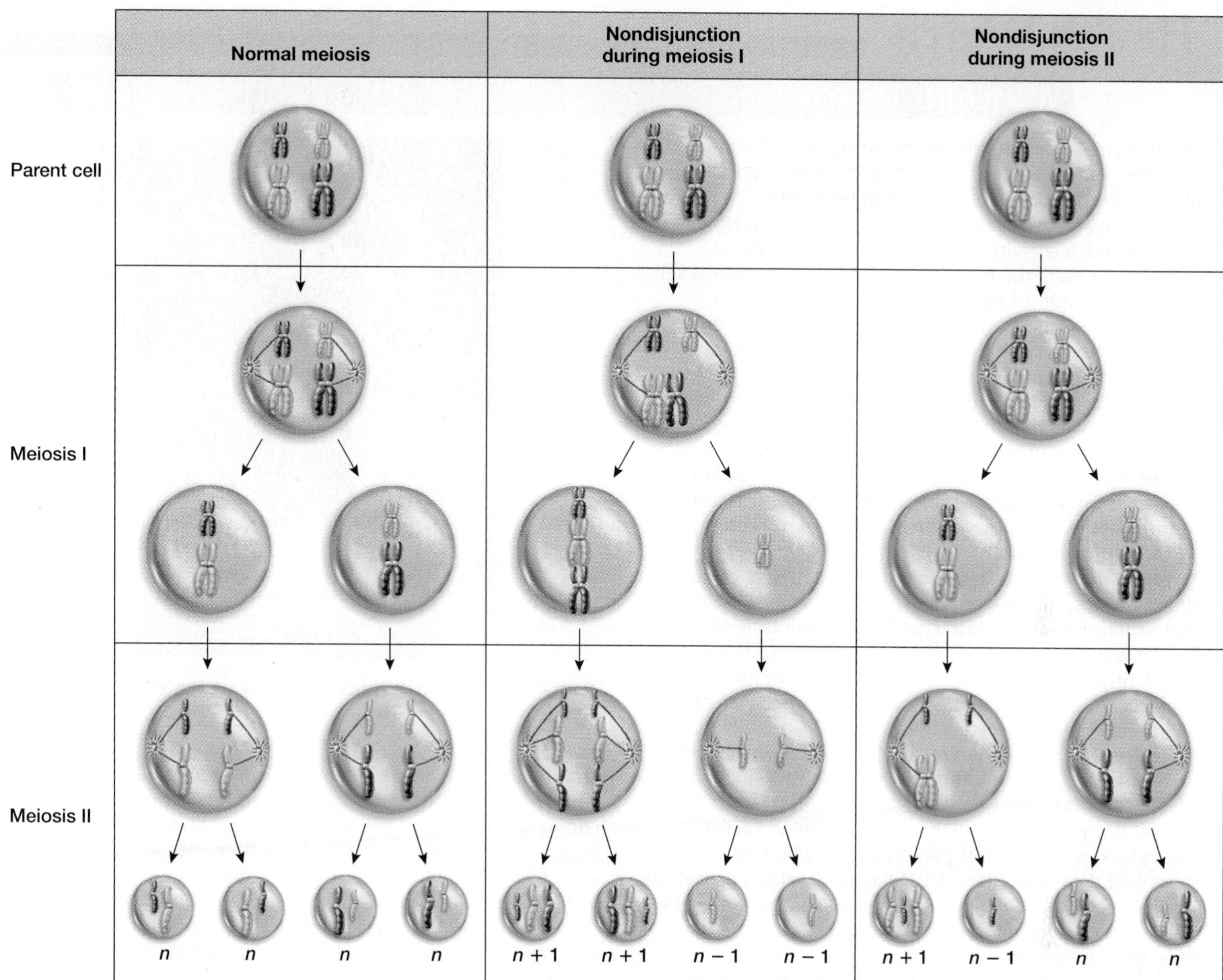

▲ **FIGURE 10-25 Nondisjunction during meiosis** Nondisjunction may occur either during meiosis I or meiosis II, resulting in gametes with too many ($n + 1$) or too few ($n - 1$) chromosomes.

Table 10-2 Effects of Nondisjunction of the Sex Chromosomes During Meiosis

Nondisjunction in Father			
Sex Chromosomes of Defective Sperm	**Sex Chromosomes of Normal Egg**	**Sex Chromosomes of Offspring**	**Phenotype**
O (none)	X	XO	Female—Turner syndrome
XX	X	XXX	Female—Trisomy X
XY	X	XXY	Male—Klinefelter syndrome
YY	X	XYY	Male—Jacob syndrome

Nondisjunction in Mother			
Sex Chromosomes of Normal Sperm	**Sex Chromosomes of Defective Egg**	**Sex Chromosomes of Offspring**	**Phenotype**
X	O (none)	XO	Female—Turner syndrome
Y	O (none)	YO	Dies as embryo
X	XX	XXX	Female—Trisomy X
Y	XX	XXY	Male—Klinefelter syndrome

Health Watch

Muscular Dystrophy

When Olympic weightlifter Matthias Steiner of Germany won the gold medal in 2008 with a "clean and jerk" lift of almost 569 pounds (258 kilograms), he exerted tremendous forces on his own body (**Fig. E10-1**). How could his muscles withstand these stresses? For that matter, why don't the muscles in your body rip apart when you climb a flight of stairs?

Muscle cells are firmly tied together by a very long protein called dystrophin. The almost 3,700 amino acids of dystrophin form a supple yet strong rod that connects the cytoskeleton inside a muscle cell to proteins in its plasma membrane, which in turn attach to proteins that form a fibrous support surrounding each muscle. Thus, when a muscle contracts, the muscle cells remain intact because the forces are distributed fairly evenly throughout each cell in the muscle and to the extracellular support proteins.

Unfortunately, about 1 in 3,500 boys synthesizes seriously defective dystrophin proteins. As these boys use their muscles, the lack of functioning dystrophin means that ordinary muscle contraction tears the muscle cells. The cells die and are replaced by fat and connective tissue (**Fig. E10-2**). By the age of 7 or 8, the boys can no longer walk. Death usually occurs in the early 20s because of heart and respiratory difficulties.

These boys suffer from **muscular dystrophy,** which literally means "degeneration of the muscles." The most severe form is called Duchenne muscular dystrophy; a less severe, but eventually still fatal, form is called Becker muscular dystrophy, after the physicians who first described these disorders. Muscular dystrophy is caused by a defective allele of the dystrophin gene (usually called the *DMD* gene because of its involvement in ***D****uchenne* ***M****uscular* ***D****ystrophy*).

Girls almost never have muscular dystrophy. Why not? Because the dystrophin gene is on the X chromosome, and muscular dystrophy alleles are recessive. Therefore, a boy will suffer muscular dystrophy if he has a defective

▲ **FIGURE E10-1 Matthias Steiner wins the gold with a lift of 569 pounds**

dystrophin allele on his single X chromosome, but a girl, with two X chromosomes, would need two defective copies to suffer the disorder. This virtually never happens, because a girl would have to receive a defective dystrophin allele from both her mother, on one of her X chromosomes, and from her father, on his X chromosome. With their early disability and death, boys with muscular dystrophy almost never reproduce.

This scenario makes genetic sense, but it may seem to defy the concept of evolution by natural selection. How can a lethal allele be so common? Shouldn't natural selection have almost completely eliminated the defective

are essential to survival, so any embryo without at least one X chromosome spontaneously aborts very early in development.)

Turner Syndrome (XO)

About 1 in every 3,000 phenotypically female babies has only one X chromosome, a condition known as **Turner syndrome.** At puberty, hormone deficiencies prevent XO females from menstruating or developing secondary sexual characteristics, such as enlarged breasts. Treatment with estrogen promotes physical development. However, because most women with Turner syndrome lack mature eggs, hormone treatment does not make it possible for them to bear children. Additional characteristics of women with Turner syndrome include short stature, folds of skin around the neck, and increased risk of cardiovascular disease, kidney defects, and hearing loss. Because women with Turner syndrome have only one X chromosome, they display X-linked recessive disorders, such as hemophilia and color blindness, much more frequently than do XX women.

Trisomy X (XXX)

About 1 in every 1,000 women has three X chromosomes, a condition known as **trisomy X,** or triple X. Most such women have no detectable defects, except for a tendency to be tall and a higher incidence of learning disabilities. Unlike women with Turner syndrome, most trisomy X women are fertile and, interestingly enough, almost always bear normal XX and XY children. Some unknown mechanism must operate during meiosis to prevent an extra X chromosome from being included in their eggs.

Klinefelter Syndrome (XXY)

About 1 male in every 1,000 is born with two X chromosomes and one Y chromosome. Most of these men go through life never realizing that they have an extra X chromosome. However, at puberty, some show mixed secondary sexual characteristics, including partial breast development, broadening of the hips, and small testes. These symptoms are known as **Klinefelter syndrome.** XXY men are often infertile because of low sperm count but are not impotent. They are usually diagnosed when

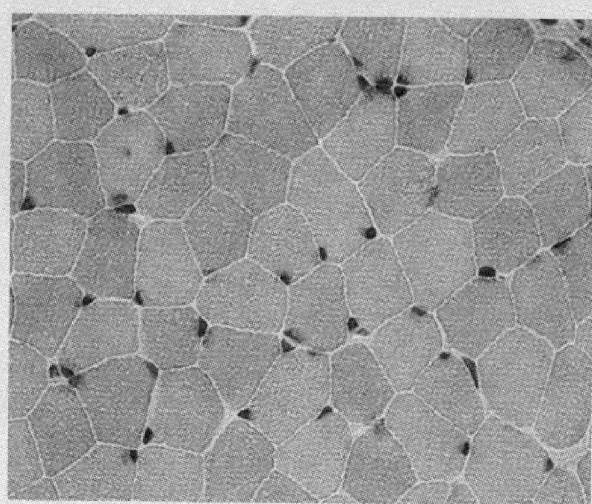

(a) Normal muscle

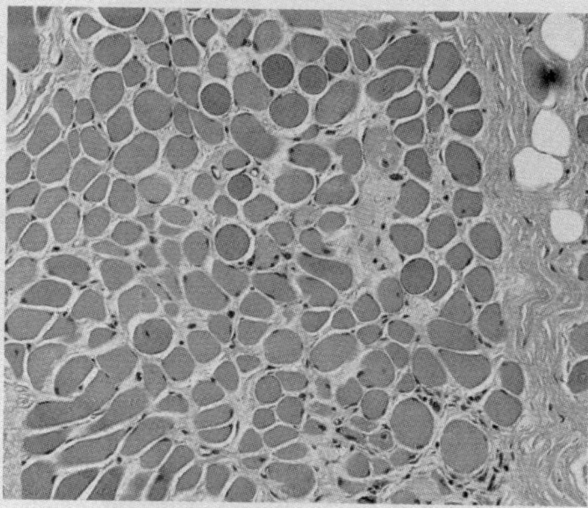

(b) Degenerating muscle in muscular dystrophy

▲ **FIGURE E10-2 Defective dystrophin proteins cause muscle degeneration (a)** A normal muscle is packed with specialized muscle cells, with very little space between the cells. **(b)** In muscular dystrophy, the individual cells are scattered in the midst of fat, white blood cells, and connective tissue.

dystrophin alleles? Actually, natural selection *does* rapidly eliminate the defective alleles. However, the dystrophin gene is enormous—about 2.2 *million* nucleotides long, compared to about 27 *thousand* nucleotides for the average human gene. In fact, the dystrophin gene occupies about 2% of the entire X chromosome.

Why does this matter? Remember, alleles arise as mutations in DNA. The longer the gene, the greater the chances for a mistake during DNA replication. It is a tribute to the astounding accuracy of DNA copying that we don't all suffer from muscular dystrophy, and you shouldn't be surprised to learn that the mutation rate for the dystrophin gene is a hundred times greater than average. Therefore, about one-third of the boys with muscular dystrophy receive a new mutation that occurred in an X chromosome of a reproductive cell of their mother, and two-thirds inherit a preexisting allele in one of their mother's X chromosomes. The new mutations counterbalance natural selection, resulting in the steady incidence of about 1 in 3,500 boys.

Is muscular dystrophy hopeless, then? Right now, there are no cures, although several treatments are available that slow the progress of the disorder, prolong life, and make the affected boys more comfortable. An exciting possibility based on stem cells, however, has been discovered in dogs with a genetic disorder very similar to Duchenne muscular dystrophy. As we described in Chapter 9, stem cells can differentiate into many different types of mature cells. Giulio Cossu and colleagues at the San Raffaele Scientific Institute in Milan, Italy, found that injecting these dogs with stem cells taken from the blood vessels of healthy dogs allows them to synthesize normal dystrophin and retain muscle function long after the dogs would normally be unable to walk. The Muscular Dystrophy Association is funding research in similar human stem cells, with the goal of eventually treating human patients.

an XXY man and his partner seek medical help because they are unable to have children.

Jacob Syndrome (XYY)

Jacob syndrome (XYY) occurs in about 1 male in every 1,000. You might expect that an extra Y chromosome, which has few active genes, would not make very much difference, and this seems to be true in most cases. However, XYY males often have high levels of testosterone, develop severe acne, and are tall (about two-thirds of XYY males are over 6 feet tall, compared with the average male height of 5 feet 9 inches).

Some Genetic Disorders Are Caused by Abnormal Numbers of Autosomes

Nondisjunction of the autosomes may also occur, producing eggs or sperm missing an autosome or with two copies of an autosome. Fusion with a normal gamete (bearing one copy of each autosome) leads to an embryo with either one or three copies of the affected autosome. Embryos that have only one copy of any of the autosomes abort so early in development that the woman never knows she was pregnant. Embryos with three copies of an autosome (trisomy) also usually spontaneously abort. However, a small fraction of embryos with three copies of chromosomes 13, 18, or 21 survive to birth. In the case of trisomy 21, the child may live into adulthood.

Trisomy 21 (Down Syndrome)

In about 1 of every 900 births, the child inherits an extra copy of chromosome 21, a condition called **trisomy 21,** or **Down syndrome.** Children with Down syndrome have several distinctive physical characteristics, including weak muscle tone, a small mouth held partially open because it cannot accommodate the tongue, and distinctively shaped eyelids (**Fig. 10-26**). Much more serious defects include low resistance to infectious diseases, heart malformations, and varying degrees of mental retardation, often severe.

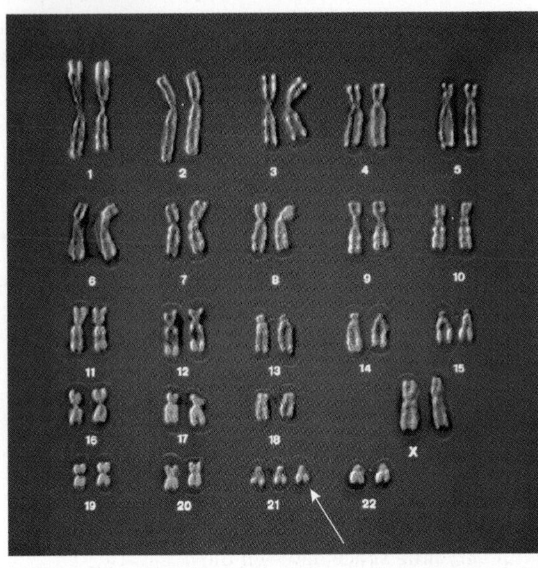

(a) Karyotype showing three copies of chromosome 21

(b) Girl with Down syndrome

▲ **FIGURE 10-26 Trisomy 21, or Down syndrome (a)** This karyotype of a Down syndrome child reveals three copies of chromosome 21 (arrow). **(b)** The younger of these two sisters shows the facial features common in people with Down syndrome.

The frequency of nondisjunction increases with the age of the parents, especially the mother; more than 3% of children born to women over 45 years of age have Down syndrome. Nondisjunction in sperm accounts for about 10% of the cases of Down syndrome, and there is only a small increase in defective sperm with increasing age of the father. Since the 1970s, it has become more common for couples to delay having children, increasing the probability of trisomy 21. Trisomy can be diagnosed before birth by examining the chromosomes of fetal cells and, with less certainty, by biochemical tests and ultrasound examination of the fetus (see "Health Watch: Prenatal Genetic Screening" in pp. 258–259).

Case Study revisited
Sudden Death on the Court

Medical examinations revealed that Flo Hyman's father and sister have Marfan syndrome, but her mother and brother do not. Does this prove that Hyman inherited the defective allele from her father? As you learned in this chapter, diploid organisms, including people, generally have two alleles of each gene, one on each homologous chromosome. One defective fibrillin allele is enough to cause Marfan syndrome. What can we conclude from these data?

First, if even one defective fibrillin allele produces Marfan syndrome, then Hyman's mother must carry two normal alleles, because she does not have Marfan syndrome. Second, because Hyman's father has Marfan syndrome, it is very likely that Hyman inherited a defective fibrillin allele from him. The fact that her sister also has Marfan syndrome makes this virtually certain. Third, is Marfan syndrome inherited as a dominant or a recessive condition? Once again, if only one defective allele is enough to cause Marfan, then this allele must be dominant and the normal allele must be recessive. Finally, if Hyman had borne children, could they have inherited Marfan syndrome from her? For a dominant disorder, any children who inherited her defective allele would develop Marfan syndrome. Therefore, on average, half of her children would have had Marfan syndrome. (Try working this out with a Punnett square.)

At the beginning of this chapter, we suggested that Lincoln, Rachmaninoff, and Akhenaten also had Marfan syndrome. You may wonder how anyone could be sure if they actually had the disorder. Well, you're right—nobody really knows. The "diagnosis" is based on photographs and descriptions. For example, Rachmaninoff's hands could span 13 white keys on a piano (try it for yourself!); most modern pianists cannot play some of Rachmaninoff's music as it was originally written, because the chords are too large. As for Lincoln, one visitor to the White House is said to have remarked, "Mr. President, what long legs you have!" To which Lincoln reportedly replied, "Just long enough to reach the ground, Madam." Although other genetic conditions can contribute to tall stature, long limbs and large hands, the fact that these men were otherwise healthy, successful, and fertile points to Marfan syndrome.

BioEthics Consider This

At this time, Marfan syndrome cannot be detected in an embryo, although researchers are working on it. Some other genetic disorders, such as cystic fibrosis and sickle-cell anemia, can easily be detected, in adults, children, and embryos. In these recessive diseases, if two heterozygotes have children together, each of their children has a 25% chance of having the disorder. Although there is no cure, there will probably be better treatments within a few years. If you and your spouse were both heterozygotes, would you seek prenatal diagnosis of an embryo? What would you do if your embryo were destined to be born with Marfan syndrome?

CHAPTER REVIEW

Summary of Key Concepts

10.1 What Is the Physical Basis of Inheritance?

The units of inheritance are genes, which are segments of DNA found at specific locations (loci) on chromosomes. Genes may exist in two or more slightly different, alternative forms, called alleles. When both homologous chromosomes carry the same allele at a given locus, the organism is homozygous for that particular gene. When the two homologous chromosomes have different alleles at a given locus, the organism is heterozygous for that gene.

10.2 How Were the Principles of Inheritance Discovered?

Gregor Mendel deduced many principles of inheritance in the mid-1800s, before the discovery of DNA, genes, chromosomes, or meiosis. He did this by choosing an appropriate experimental subject, designing his experiments carefully, following progeny for several generations, and analyzing his data statistically.

10.3 How Are Single Traits Inherited?

A trait is an observable or measurable feature of the organism's phenotype, such as hair texture or blood type. Traits are inherited in particular patterns that depend on the alleles that parents pass on to their offspring. Each parent provides its offspring with one allele of every gene, so that the offspring inherits a pair of alleles for every gene. The combination of alleles in the offspring determines whether it displays a particular phenotype. Dominant alleles mask the expression of recessive alleles. The masking of recessive alleles can result in organisms with the same phenotype but different genotypes. Organisms with two dominant alleles (homozygous dominant) have the same phenotype as do organisms with one dominant and one recessive allele (heterozygous). Because each allele segregates randomly during meiosis, we can predict the relative proportions of offspring with a particular trait, using Punnett squares or probability.

10.4 How Are Multiple Traits Inherited?

If the genes for two traits are located on separate chromosomes, their alleles assort independently of one another into the egg or sperm; that is, the distribution of alleles of one gene into the gametes does not affect the distribution of the alleles of the other gene. Thus, crossing two organisms that are heterozygous at two loci on separate chromosomes produces offspring with nine different genotypes. If the alleles are typical dominant and recessive alleles, these progeny will display only four different phenotypes (see Fig. 10-11).

10.5 How Are Genes Located on the Same Chromosome Inherited?

Genes on the same chromosome (encoded on the same DNA double helix) are linked to one another and therefore tend to be inherited together. However, crossing over will result in some recombination of alleles on each chromosome. Crossing over will occur more often the farther apart on a chromosome the genes lie.

10.6 How Is Sex Determined?

In many animals, sex is determined by sex chromosomes, often designated X and Y. In mammals, females have two X chromosomes, whereas males have one X and one Y chromosome. The rest of the chromosomes, identical in the two sexes, are called autosomes. Males include either an X or a Y chromosome in their sperm, whereas females always include an X chromosome in their eggs. Therefore, gender is determined by the sex chromosome in the sperm that fertilizes an egg.

10.7 How Are Sex-Linked Genes Inherited?

Sex-linked genes are genes found on the X or Y chromosome. In mammals, the Y chromosome has many fewer genes than the X chromosome, so most sex-linked genes are found on the X chromosome. Because males have only one copy of X chromosome genes, recessive traits on the X chromosome are more likely to be phenotypically expressed in males.

10.8 Do the Mendelian Rules of Inheritance Apply to All Traits?

Not all inheritance follows the simple dominant-recessive pattern:

- In incomplete dominance, heterozygotes have a phenotype that is intermediate between the two homozygous phenotypes.
- If we examine the genes of many members of a given species, we find that many genes have more than two alleles; that is, there are multiple alleles of the gene.
- Codominance is a relationship among the alleles of a single gene in which two alleles independently contribute to the observed phenotype.
- Many traits are determined by several different genes that all contribute to the phenotype, a phenomenon called polygenic inheritance.
- Pleiotropy occurs when a single gene has multiple effects on an organism's phenotype.
- The environment influences the phenotypic expression of all traits.

10.9 How Are Human Genetic Disorders Investigated?

The genetics of humans is similar to the genetics of other animals, but is more difficult to study because experimental crosses are not feasible. Analysis of family pedigrees and, more recently, molecular genetic techniques are used to determine the mode of inheritance of human traits.

10.10 How Are Human Disorders Caused by Single Genes Inherited?

Some genetic disorders are inherited as recessive traits; therefore, only homozygous recessive persons show symptoms of the disease. Heterozygotes are called carriers; they carry the recessive allele but do not express the trait. Some other diseases are inherited as simple dominant traits. In such cases, only one copy of the dominant allele is needed to cause disease symptoms. Some human genetic disorders are sex-linked.

10.11 How Do Errors in Chromosome Number Affect Humans?

Errors in meiosis can result in gametes with abnormal numbers of sex chromosomes or autosomes. Many people with abnormal numbers of sex chromosomes have distinguishing physical characteristics. Abnormal numbers of autosomes typically lead to spontaneous abortion early in pregnancy. In rare instances, the fetus may survive to birth, but mental and physical deficiencies always occur, as is the case with Down syndrome (trisomy 21). The likelihood of abnormal numbers of chromosomes increases with increasing age of the mother and, to a lesser extent, the father.

Key Terms

albinism *190*
allele *174*
autosome *184*
carrier *190*
codominance *186*
cross-fertilization *176*
dominant *177*
Down syndrome *195*
gene *174*
genetic recombination *183*
genotype *178*
hemophilia *191*
heterozygous *175*
homozygous *175*
Huntington disease *191*
hybrid *175*
incomplete dominance *186*
inheritance *174*
Jacob syndrome *195*
Klinefelter syndrome *194*
law of independent assortment *181*
law of segregation *177*
linkage *182*
locus (plural, loci) *174*
multiple alleles *186*
muscular dystrophy *194*
nondisjunction *192*
pedigree *189*
phenotype *178*
pleiotropy *187*
polygenic inheritance *187*
Punnett square method *178*
recessive *177*
self-fertilization *176*
sex chromosome *184*
sex-linked *184*
sickle-cell anemia *190*
test cross *180*
trisomy 21 *195*
trisomy X *194*
true-breeding *176*
Turner syndrome *194*
X chromosome *184*
Y chromosome *184*

Thinking Through the Concepts

Fill-in-the-Blank

1. The physical position of a gene on a chromosome is called its __________. Alternative forms of a gene are called __________. These alternative forms of genes arise as __________, which are changes in the nucleotide sequence of a gene.
2. An organism is described as *Rr*: red. *Rr* is the organism's __________, while red color is its __________. This organism would be (homozygous/heterozygous) for this color gene.
3. The inheritance of multiple traits depends on the locations of the genes that control the traits. If the genes are on different chromosomes, then the traits are inherited (as a group/independently). If the genes are located close together on a single chromosome, then the traits tend to be inherited (as a group/independently). Genes on the same chromosome are said to be __________.
4. Many organisms, including mammals, have both autosomes and sex chromosomes. In mammals, males have __________ sex chromosomes and females have __________ chromosomes. The sex of offspring depends on which chromosome is present in the (sperm/egg).
5. Genes that are present on one sex chromosome but not the other are called __________.
6. If the phenotype of heterozygotes is intermediate between the phenotypes of the two homozygotes, this pattern of inheritance is called __________. If heterozygotes express phenotypes of both homozygotes (not intermediate, but showing both traits), this is called __________. In __________, many genes, usually with similar effects on phenotype, control the inheritance of a trait.

Review Questions

1. Define the following terms: *gene, allele, dominant, recessive, true-breeding, homozygous, heterozygous, cross-fertilization,* and *self-fertilization.*
2. Explain why genes located on the same chromosome are said to be linked. Why do alleles of linked genes sometimes separate during meiosis?
3. Define *polygenic inheritance.* Why does polygenic inheritance sometimes allow parents to produce offspring that are notably different in skin color than either parent?
4. What is sex linkage? In mammals, which sex would be most likely to show recessive sex-linked traits?
5. What is the difference between a phenotype and a genotype? Does knowledge of an organism's phenotype always allow you to determine the genotype? What type of experiment would you perform to determine the genotype of a phenotypically dominant individual?
6. In the pedigree of part (a) of Figure 10-21, do you think that the individuals showing the trait are homozygous or heterozygous? How can you tell from the pedigree?
7. Define *nondisjunction,* and describe the common syndromes caused by nondisjunction of sex chromosomes and autosomes.

Applying the Concepts

1. Sometimes the term *gene* is used rather casually. Compare the terms *allele* and *gene.*
2. **BioEthics** Mendel's numbers seemed almost too perfect to be real; some believe he may have cheated a bit on his data. Perhaps he continued to collect data until the numbers matched his predicted ratios, then stopped. Recently, there has been much publicity over violations of scientific ethics, including researchers plagiarizing others' work, using other scientists' methods to develop lucrative patents, or just plain fabricating data. How important is this issue for society? What are the boundaries of ethical scientific behavior? How should the scientific community or society "police" scientists? What punishments would be appropriate for violations of scientific ethics?
3. Although American society has been described as a "melting pot," people often engage in "assortative mating," in which they marry others of similar height, socioeconomic status, race, and IQ. Discuss the consequences to society of assortative mating among humans. Would society be better off if people mated more randomly? Explain.

Genetics Problems

1. In certain cattle, hair color can be red (homozygous R_1R_1), white (homozygous R_2R_2), or roan (a mixture of red and white hairs, heterozygous R_1R_2).
   a. When a red bull is mated to a white cow, what genotypes and phenotypes of offspring could be obtained?
   b. If one of the offspring bulls in part (a) were mated to a white cow, what genotypes and phenotypes of offspring could be produced? In what proportion?
2. The palomino horse is golden in color. Unfortunately for horse fanciers, palominos do not breed true. In a series of matings between palominos, the following offspring were obtained:

   65 palominos
   32 cream-colored
   34 chestnut (reddish brown)

   What is the probable mode of inheritance of palomino coloration?
3. In the edible pea, tall (*T*) is dominant to short (*t*), and green pods (*G*) are dominant to yellow pods (*g*). List the types of gametes and offspring that would be produced in the following crosses:
   a. *TtGg* × *TtGg*
   b. *TtGg* × *TTGG*
   c. *TtGg* × *Ttgg*
4. In tomatoes, round fruit (*R*) is dominant to long fruit (*r*), and smooth skin (*S*) is dominant to fuzzy skin (*s*). A true-breeding round, smooth tomato (*RRSS*) was crossbred with a true-breeding long, fuzzy tomato (*rrss*). All the F_1 offspring were round and smooth (*RrSs*). When these F_1 plants were bred, the following F_2 generation was obtained:

   Round, smooth: 43
   Long, fuzzy: 13

   Are the genes for skin texture and fruit shape likely to be on the same chromosome or on different chromosomes? Explain your answer.
5. In the tomatoes of Problem 4, an F_1 offspring (*RrSs*) was mated with a homozygous recessive (*rrss*). The following offspring were obtained:

   Round, smooth: 583 Long, fuzzy: 602
   Round, fuzzy: 21 Long, smooth: 16

   What is the most likely explanation for this distribution of phenotypes?
6. In humans, hair color is controlled by two interacting genes. The same pigment, melanin, is present in both brown-haired and blond-haired people, but brown hair has much more of it. Brown hair (*B*) is dominant to blond (*b*). Whether any melanin can be synthesized depends on another gene. The dominant form of this second gene (*M*) allows melanin synthesis; the recessive form (*m*) prevents melanin synthesis. Homozygous recessives (*mm*) are albino. What will be the expected proportions of phenotypes in the children of the following parents?
   a. *BBMM* × *BbMm*
   b. *BbMm* × *BbMm*
   c. *BbMm* × *bbmm*
7. In humans, one of the genes determining color vision is located on the X chromosome. The dominant form (*C*) produces normal color vision; red-green color blindness (*c*) is recessive. If a man with normal color vision marries a color-blind woman, what is the probability of them having a color-blind son? A color-blind daughter?
8. In the couple described in Problem 7, the woman gives birth to a color-blind but otherwise normal daughter. The husband sues for a divorce on the grounds of adultery. Will his case stand up in court? Explain your answer.

MB *Go to www.masteringbiology.com for practice quizzes, activities, eText, videos, current events, and more.*

DNA: The Molecule of Heredity

chapter 11

▲ Ordinary bull or incredible hulk? A tiny change in DNA makes all the difference.

Case Study

Muscles, Mutations, and Myostatin

NO, THE BULL in the top photo hasn't been pumping iron—he's a Belgian Blue, and they always have bulging muscles. What makes a Belgian Blue look like a bodybuilder, compared to an ordinary bull, such as the Hereford in the bottom photo?

When any mammal develops, its cells divide many times, enlarge, and become specialized for a specific function. The size, shape, and cell types in any organ are precisely regulated during development, so that you don't wind up with a head the size of a basketball, or have hair growing on your liver. Muscle development is no exception. When you were very young, cells destined to form your muscles multiplied, fused together to form long, relatively thick cells with multiple nuclei, and synthesized the specialized proteins that cause muscles to contract and thereby move your skeleton. A protein called myostatin, found in all mammals, puts the brakes on this process. The word "myostatin" literally means "to make muscles stay the same," and that is exactly what myostatin does. As muscles develop, myostatin slows down—and eventually stops—the multiplication of these pre-muscle cells. Myostatin also regulates the ultimate size of muscle cells and, therefore, their strength. A bodybuilder can bulk up by lifting weights, which *enlarges* the muscle cells, but doesn't usually add many *more* cells.

Belgian Blues have more, and larger, muscle cells than ordinary cattle do. Why? You may already have guessed—they don't produce normal myostatin. And why not? As you will learn in this chapter, proteins are synthesized from the genetic directions contained in **deoxyribonucleic acid (DNA).** The DNA of a Belgian Blue is very slightly different from the DNA of normal cattle—it has a change, or mutation, in the DNA of its myostatin gene. As a result, it produces defective myostatin. Belgian Blue pre-muscle cells multiply more than normal, and the cells become extra large as they differentiate, thereby producing remarkably buff cattle.

How does DNA contain the instructions for traits such as muscle size, flower color, or gender? How are these instructions usually passed unchanged from generation to generation? And why do the instructions sometimes change? The answers to these questions lie in the structure and function of DNA.

At a Glance

11.1 HOW DID SCIENTISTS DISCOVER THAT GENES ARE MADE OF DNA?

By the late 1800s, scientists had learned that genetic information exists in discrete units that they called genes. However, they didn't know what a gene was. Scientists merely knew that genes determine many of the heritable differences among individuals within a species. For example, genes for flower color determine whether roses are red, pink, yellow, or white. By the early 1900s, studies of dividing cells provided strong evidence that genes are parts of chromosomes (see pp. 149–150, 174–175). Soon, biochemists found that eukaryotic chromosomes are composed only of protein and DNA. One of these substances must carry the cell's hereditary blueprint, but which one?

Transformed Bacteria Revealed the Link Between Genes and DNA

In the late 1920s, a British researcher named Frederick Griffith was trying to make a vaccine to prevent bacterial pneumonia, a major cause of death at that time. Making a vaccine against some types of infectious bacteria is very difficult (for example, modern vaccines against anthrax are neither completely safe nor completely effective), but this was not known back in the 1920s. Some antibacterial vaccines consist of a weakened strain of the bacteria, which can't cause illness. Injecting this weakened but living strain into an animal may stimulate immunity against the disease-causing strains. Other vaccines use disease-causing (virulent) bacteria that have been killed by exposure to heat or chemicals.

Griffith was trying to make a vaccine using two strains of the *Streptococcus pneumoniae* bacterium. One strain, R, did not cause pneumonia when injected into mice (**Fig. 11-1a**). The other strain, S, was deadly when injected, causing pneumonia and killing the mice in a day or two (**Fig. 11-1b**). As expected, when the S-strain was killed and injected into mice, it did not cause disease (**Fig. 11-1c**). Unfortunately, neither the live R-strain nor the killed S-strain provided immunity against live S-strain bacteria.

Griffith also tried mixing living R-strain bacteria together with heat-killed S-strain bacteria and injecting the mixture into mice (**Fig. 11-1d**). Because neither of these bacterial strains causes pneumonia on its own, he expected the mice to remain healthy. To his surprise, the mice sickened and died. When he autopsied the mice, he recovered *living* S-strain bacteria from them. Griffith hypothesized that some substance in the heat-killed S-strain changed the living, harmless R-strain bacteria into the deadly S-strain, a process he called transformation. The transformed S-strain cells then multiplied and caused pneumonia.

Griffith never discovered an effective pneumonia vaccine, so in that sense his experiments were a failure (in fact, an effective and safe vaccine against most forms of *Streptococcus pneumoniae* was not developed until the late 1970s). However, Griffith's experiments marked a turning point in our understanding of genetics, because other researchers suspected that the substance that causes transformation might be the long-sought molecule of heredity.

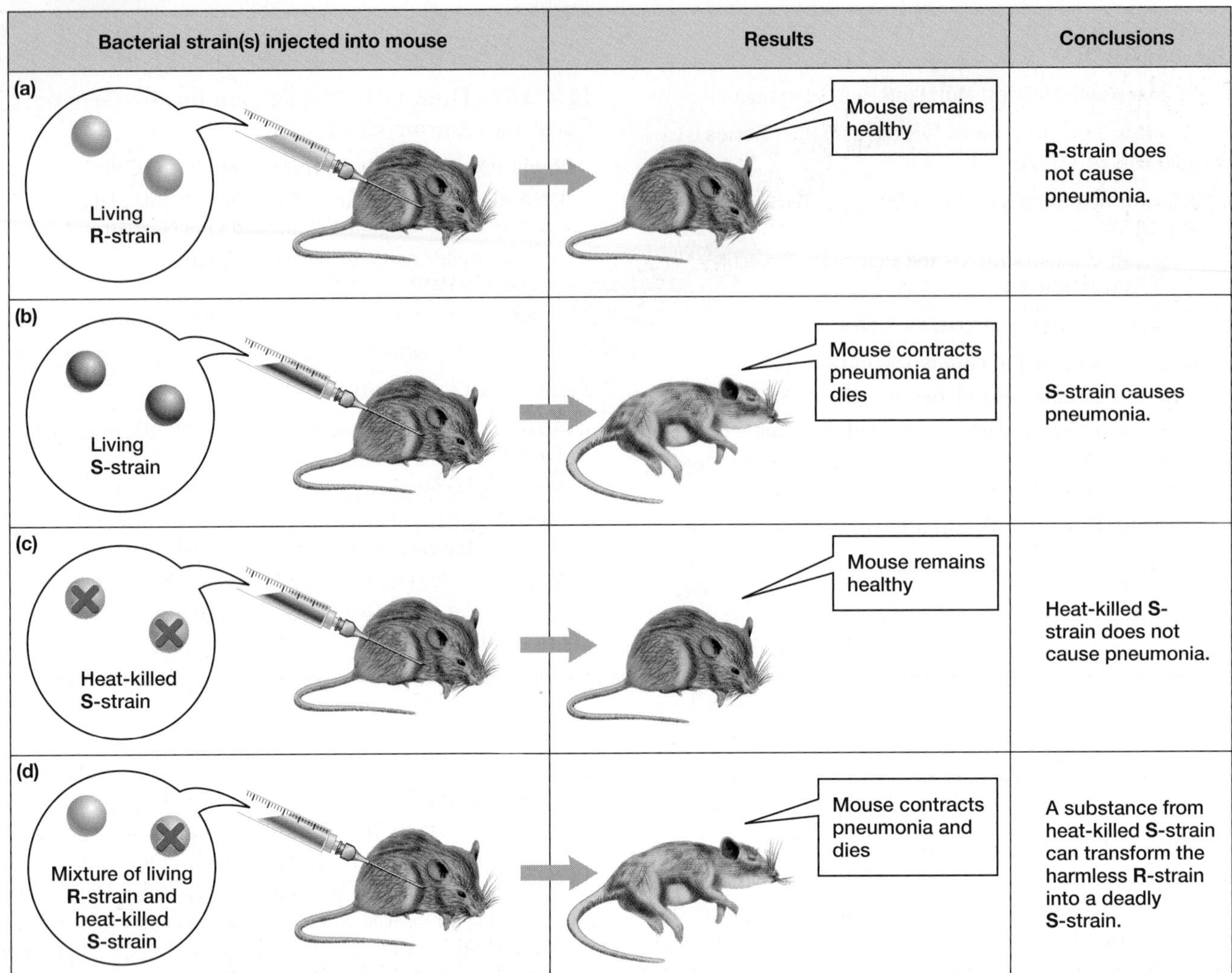

Bacterial strain(s) injected into mouse
Results
Conclusions
(a)
Living R-strain
Mouse remains healthy
R-strain does not cause pneumonia.
(b)
Living S-strain
Mouse contracts pneumonia and dies
S-strain causes pneumonia.
(c)
Heat-killed S-strain
Mouse remains healthy
Heat-killed S-strain does not cause pneumonia.
(d)
Mixture of living R-strain and heat-killed S-strain
Mouse contracts pneumonia and dies
A substance from heat-killed S-strain can transform the harmless R-strain into a deadly S-strain.

▲ **FIGURE 11-1 Transformation in bacteria** Griffith's discovery that bacteria can be transformed from harmless to deadly laid the groundwork for the discovery that genes are composed of DNA.

The Transforming Molecule Is DNA

In 1933, J. L. Alloway discovered that the mice played no role in transformation, which occurred just as well when live R-strain bacteria were mixed with dead S-strain bacteria in culture dishes. A decade later, Oswald Avery, Colin MacLeod, and Maclyn McCarty discovered that the transforming molecule is DNA. Avery, MacLeod, and McCarty isolated DNA from S-strain bacteria, mixed it with live R-strain bacteria, and produced live S-strain bacteria. To show that transformation was caused by DNA, and not by traces of protein contaminating the DNA, they treated some samples with protein-destroying enzymes. These enzymes did not prevent transformation. However, treating samples with DNA-destroying enzymes did prevent transformation.

This discovery helps us to interpret the results of Griffith's experiments. Heating S-strain cells killed them but did not completely destroy their DNA. When killed S-strain bacteria were mixed with living R-strain bacteria, fragments of DNA from the dead S-strain cells entered into some of the R-strain cells and became incorporated into the chromosome of the R-strain bacteria (**Fig. 11-2**). If these fragments of DNA contained the genes needed to cause disease, an R-strain cell would be transformed into an S-strain cell. Thus, Avery, MacLeod, and McCarty concluded that genes are made of DNA.

DNA, Not Protein, Is the Molecule of Heredity

Nevertheless, not everyone was persuaded. Some still thought that genes are made of protein, and that the transforming DNA molecules from S-strain bacteria caused a mutation in the genes of R-strain bacteria. Others hypothesized that DNA might be the hereditary molecule of bacteria, but not of other organisms. However, evidence continued to accumulate that DNA is the genetic material in many, or perhaps all, organisms. For example, before dividing, a eukaryotic cell duplicates its chromosomes (see pp. 151–152) and exactly doubles its DNA content—just what would be expected if genes are made of DNA. Finally, virtually all of the remaining skeptics were convinced by a superb set of experiments by Alfred Hershey

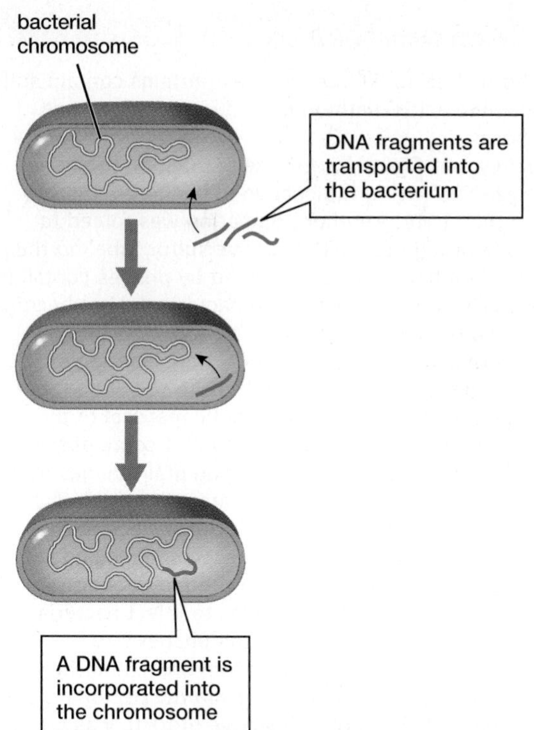
bacterial chromosome
DNA fragments are transported into the bacterium
A DNA fragment is incorporated into the chromosome

▲ **FIGURE 11-2 Molecular mechanism of transformation**
Most bacteria have a single circular chromosome made of DNA. Transformation may occur when a living bacterium takes up pieces of DNA from its environment and incorporates those fragments into its chromosome.

phosphate
sugar
base = adenine

phosphate
sugar
base = guanine

phosphate
sugar
base = thymine

phosphate
sugar
base = cytosine

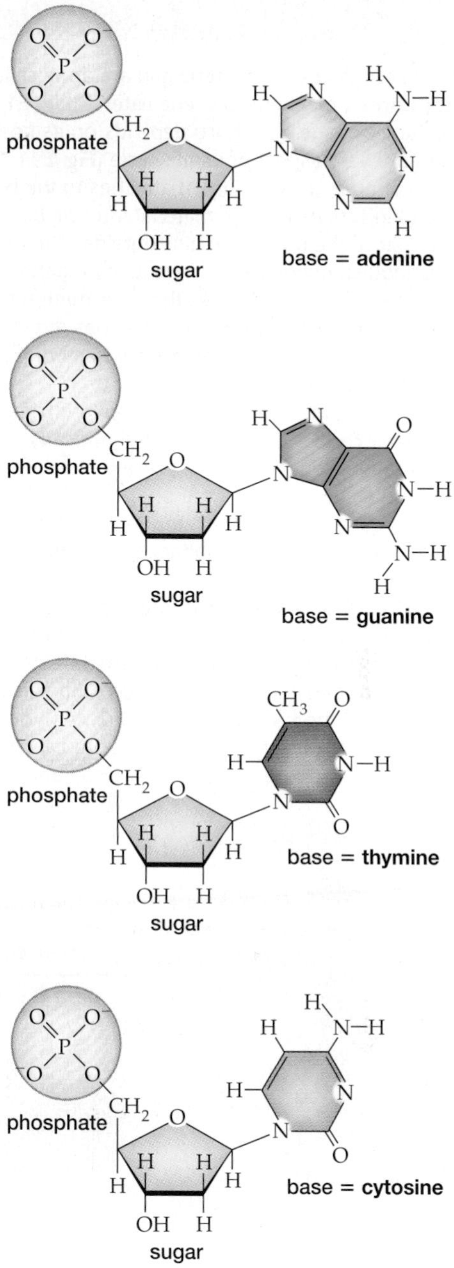

▲ **FIGURE 11-3 DNA nucleotides**

and Martha Chase, conclusively showing that DNA is the hereditary molecule of certain viruses (see "Scientific Inquiry: DNA Is the Hereditary Molecule of Bacteriophages" on pp. 204–205).

11.2 WHAT IS THE STRUCTURE OF DNA?

Knowing that genes are made of DNA does not answer critical questions about inheritance: How does DNA encode genetic information? How is DNA replicated so that a cell can pass its hereditary information to its daughter cells? The secrets of DNA function and, therefore, of heredity itself, are found in the three-dimensional structure of the DNA molecule.

DNA Is Composed of Four Nucleotides

As you learned in Chapter 3, DNA consists of four small subunits called **nucleotides.** Each nucleotide in DNA has three parts: a phosphate group, a sugar called deoxyribose, and one of four nitrogen-containing **bases: adenine (A), guanine (G), thymine (T),** or **cytosine (C)** (Fig. 11-3).

In the 1940s, when biochemist Erwin Chargaff of Columbia University analyzed the amounts of the four bases in DNA from organisms as diverse as bacteria, sea urchins, fish, and humans, he found a curious consistency. The DNA of any given species contains equal amounts of adenine and thymine, as well as equal amounts of guanine and cytosine. This consistency, often called "Chargaff's rule," certainly seemed significant, but it would be almost another decade before anyone figured out what it meant about DNA structure.

DNA Is a Double Helix of Two Nucleotide Strands

Determining the structure of any biological molecule is no simple task, even for scientists today. Nevertheless, in the late 1940s, several scientists began to investigate the structure of DNA. British scientists Maurice Wilkins and Rosalind Franklin used X-ray diffraction to study the DNA molecule. They bombarded crystals of purified DNA with X-rays and

Scientific Inquiry

DNA Is the Hereditary Molecule of Bacteriophages

Certain viruses infect only bacteria and are aptly called **bacteriophages,** meaning "bacteria eaters" (**Fig. E11-1**). A bacteriophage ("phage" for short) depends on its host bacterium for every aspect of its life cycle (**Fig. E11-1b**). When a phage encounters a bacterium, it attaches to the bacterial cell wall and injects its genetic material into the bacterium. The outer coat of the phage remains outside. The bacterium cannot distinguish phage genes from its own genes, so it "reads" the phage genes and uses that information to produce more phages. Finally, one of the phage genes directs the synthesis of an enzyme that ruptures the bacterium, freeing the newly manufactured phages.

Even though many bacteriophages have intricate structures (see **Fig. E11-1a**), they are chemically very simple, containing only DNA and protein. Therefore, one of these two molecules must be the phage genetic material. In the early 1950s, Alfred Hershey and Martha Chase used the chemical simplicity of bacteriophages to deduce that their genetic material is DNA.

Hershey and Chase knew that infected bacteria should contain phage genetic material, so if they could "label" phage DNA and protein, and separate the infected bacteria from the phage coats left outside, they could see which molecule entered the bacteria (**Fig. E11-2**). As you learned in Chapter 3, DNA and protein both contain carbon, oxygen, hydrogen, and nitrogen. DNA also contains phosphorus but not sulfur, whereas proteins contain sulfur (in the amino acids methionine and cysteine) but not phosphorus.

Hershey and Chase forced one population of phages to synthesize DNA using radioactive phosphorus, thereby labeling their DNA. Another population was forced to synthesize protein using radioactive sulfur, labeling their protein. When bacteria were infected by phages containing radioactively labeled protein, the bacteria did not become radioactive. However, when bacteria were infected by phages containing radioactive DNA, the bacteria did become radioactive. Hershey and Chase concluded that DNA, and not protein, was the genetic material of phages.

Hershey and Chase also reasoned that some of the labeled genetic material from the "parental" phages might be incorporated into the genetic material of the "offspring" phages. (You will learn more about this in section 11.4.) In a second set of experiments, the researchers again labeled DNA in one phage population and protein in another phage population, and allowed the phages to infect bacteria. After enough time had passed so that the phages had reproduced, the bacteria were broken open, and the offspring phages were separated from the bacterial debris. Radioactive DNA, but not radioactive protein, was found in the offspring phages. This second experiment confirmed the results of the first: DNA is the hereditary molecule.

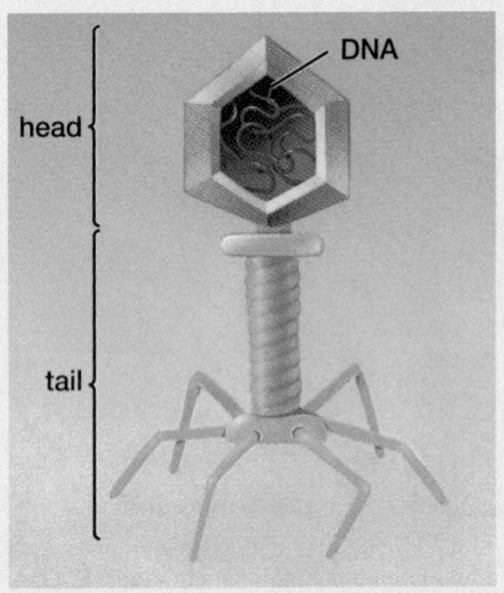

(a) Structure of a bacteriophage

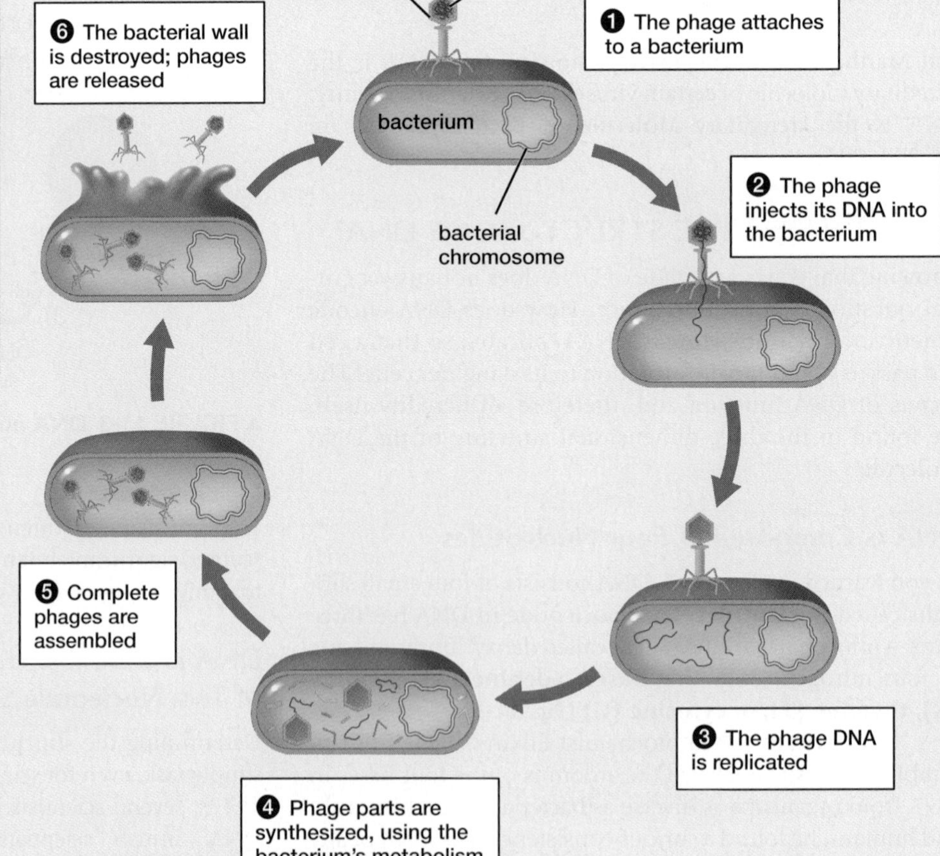

(b) Bacteriophage life cycle

▲ FIGURE E11-1 **Bacteriophages** **(a)** Many bacteriophages have complex structures, including a head containing genetic material, tail fibers that attach to the surface of a bacterium, and an elaborate apparatus for injecting their genetic material into the bacterium. **(b)** The life cycle of a bacteriophage. The bacteriophage uses the metabolism of the bacterium to produce more phages.

Observations:
1. Bacteriophage viruses consist of only DNA and protein.
2. Bacteriophages inject their genetic material into bacteria, forcing the bacteria to synthesize more phages.
3. The outer coat of bacteriophages stays outside of the bacteria.
4. DNA contains phosphorus but not sulfur.
    a. DNA can be "labeled" with radioactive phosphorus.
5. Protein contains sulfur but not phosphorus.
    a. Protein can be "labeled" with radioactive sulfur.

Question: Is DNA or protein the genetic material of bacteriophages?

Hypothesis: DNA is the genetic material.

Prediction:
1. If bacteria are infected with bacteriophages containing radioactively labeled DNA, the bacteria will be radioactive.
2. If bacteria are infected with bacteriophages containing radioactively labeled protein, the bacteria will not be radioactive.

Experiment:

Radioactive phosphorus (^{32}P)

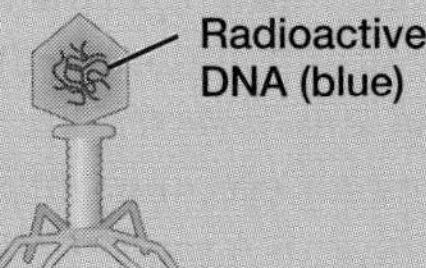

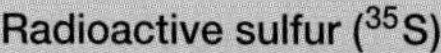

Radioactive sulfur (^{35}S)

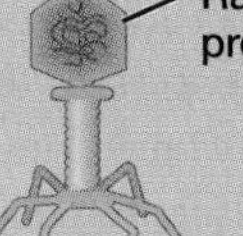

❶ Label the phages with ^{32}P or ^{35}S.

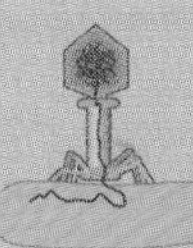

❷ Infect the bacteria with the labeled phages; the phages inject their genetic material into the bacteria.

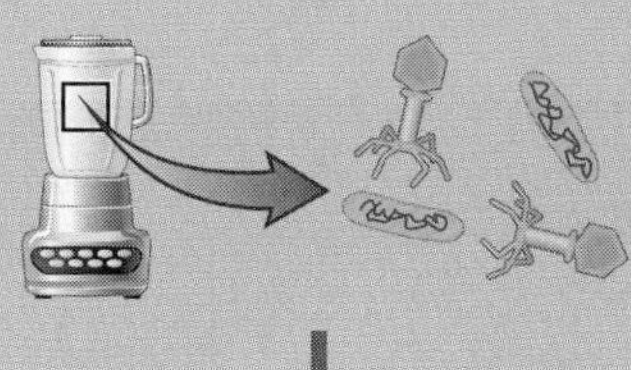

❸ Whirl in a blender to break off the phage coats from the bacteria.

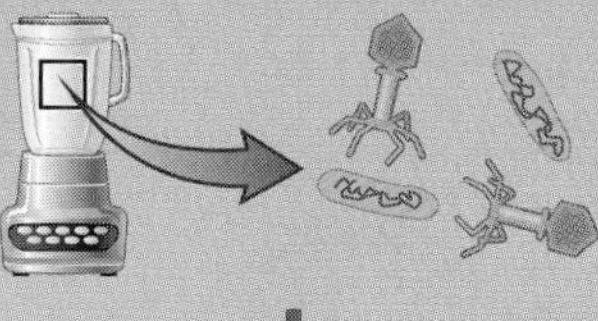

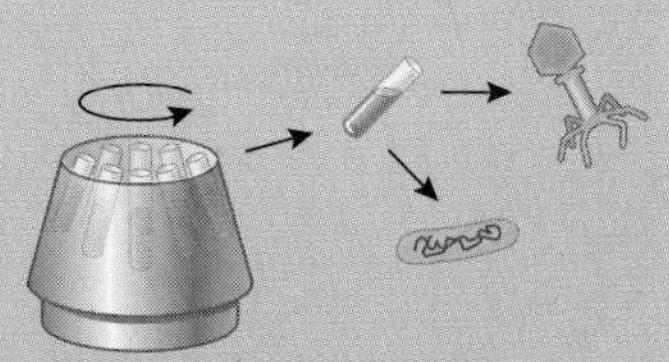

❹ Centrifuge to separate the phage coats (low density: stay in the liquid) from the bacteria (high density: sink to the bottom as a "pellet")

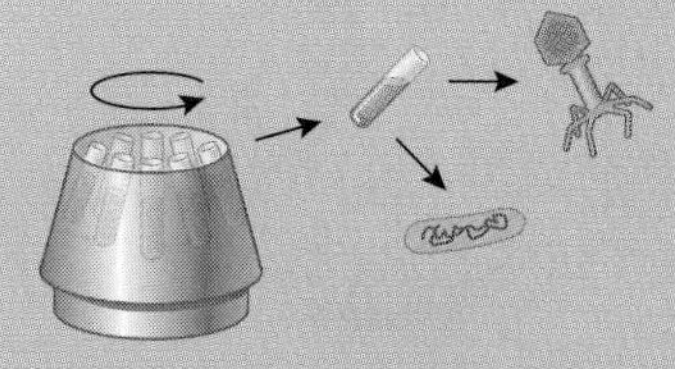

Results: Bacteria are radioactive; phage coats are not.

❺ Measure the radioactivity of the phage coats and bacteria.

Results: Phage coats are radioactive; bacteria are not.

Conclusion: Infected bacteria are labeled with radioactive phosphorus but not with radioactive sulfur, supporting the hypothesis that the genetic material of bacteriophages is DNA, not protein.

▲ **FIGURE E11-2 The Hershey-Chase experiment** By radioactively labeling either the DNA or the protein of bacteriophages, Hershey and Chase tested whether the genetic material of phages is DNA (left side of the experiment) or protein (right side).

recorded how the X-rays bounced off the DNA molecules (**Fig. 11-4a**). As you can see, the resulting "diffraction" pattern does not provide a direct picture of DNA structure. However, experts like Wilkins and Franklin (**Figs. 11-4b,c**) could extract a lot of information about DNA from the pattern. First, a molecule of DNA is long and thin, with a uniform diameter of 2 nanometers (2 billionths of a meter). Second, DNA is helical; that is, it is twisted like a corkscrew or a spiral staircase. Third, the DNA molecule consists of repeating subunits.

The chemical and X-ray diffraction data did not provide enough information for researchers to work out the structure of DNA; some good guesses were also needed. Combining Wilkins and Franklin's data with a knowledge of how complex organic molecules bond together and an intuition that "important biological objects come in pairs," James Watson and Francis Crick proposed a model for the structure of DNA (see "Scientific Inquiry: The Discovery of the Double Helix" on p. 208). They suggested that the DNA molecule consists of two separate DNA polymers of linked nucleotides, called **strands** (**Fig. 11-5**). Within each DNA strand, the phosphate group of one nucleotide bonds to the sugar of the next nucleotide in the same strand. This bonding pattern produces a "backbone" of alternating, covalently bonded sugars and phosphates. The nucleotide bases protrude from this **sugar-phosphate backbone.**

All of the nucleotides within a single DNA strand are oriented in the same direction. Therefore, the two ends of a DNA strand differ; one end has a "free" or unbonded sugar, and the other end has a "free" or unbonded phosphate (see **Fig. 11-5a**). Picture a long line of cars stopped on a crowded one-way street at night; the cars' headlights (free phosphates) always point forward and their taillights (free sugars) always point backward. If the cars are jammed tightly together, a pedestrian standing in front of the line of cars will see only the headlights on the first car; a pedestrian at the back of the line will see only the taillights of the last car.

Hydrogen Bonds Between Complementary Bases Hold Two DNA Strands Together in a Double Helix

Watson and Crick proposed that two DNA strands are held together by hydrogen bonds that form between the bases protruding from the sugar-phosphate backbones of each strand (see Fig. 11-5a). These bonds give DNA a ladder-like structure, with the sugar-phosphate backbones on the outside (forming the uprights of the ladder) and the nucleotide bases on the inside (forming the rungs of the ladder). However, the DNA strands are not straight. Instead, they are twisted about each other to form a **double helix,** resembling a ladder twisted lengthwise into the shape of a circular staircase (see **Fig. 11-5b**). Further, the two strands in a DNA double helix are oriented in opposite directions, or are antiparallel. In the diagram of Figure 11-5a, note that the left-hand DNA strand has a free phosphate group at the top and a free sugar on the bottom, while these are reversed in the right-hand DNA strand. Again imagine an evening traffic jam, this time on a crowded two-lane highway. A traffic helicopter pilot overhead would see only the headlights on cars in one lane, and only the taillights of cars in the other lane.

Take a closer look at the pairs of bases that form each rung of the double helix ladder. Adenine forms hydrogen bonds only with thymine, and guanine forms hydrogen bonds only with cytosine (see Figs. 11-5a,b). These A–T and G–C pairs are called **complementary base pairs.** All of the bases of the two strands of a DNA double helix are complementary to each other. For example, if one strand reads A-T-T-C-C-A-G-G-C-T, then the other strand must read T-A-A-G-G-T-C-C-G-A.

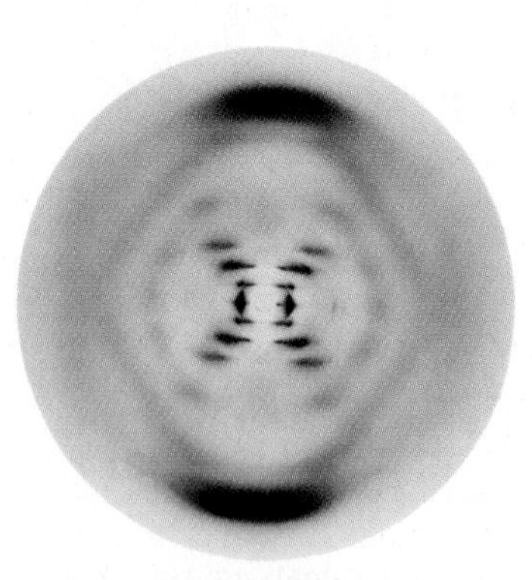

(a) Diffraction pattern of DNA

(b) Maurice Wilkins

(c) Rosalind Franklin

▲ **FIGURE 11-4 X-ray diffraction studies of DNA (a)** The X formed of dark spots is characteristic of helical molecules such as DNA. Measurements of various aspects of the pattern indicate the dimensions of the DNA helix; for example, the distance between the dark spots corresponds to the distance between turns of the helix. **(b)** Maurice Wilkins and **(c)** Rosalind Franklin discovered many of the features of DNA by carefully examining such X-ray diffraction patterns. Wilkins shared the Nobel Prize in Physiology or Medicine with Watson and Crick in 1962. Franklin died in 1958; because Nobel Prizes are not awarded posthumously, her contributions often do not receive the recognition they deserve.

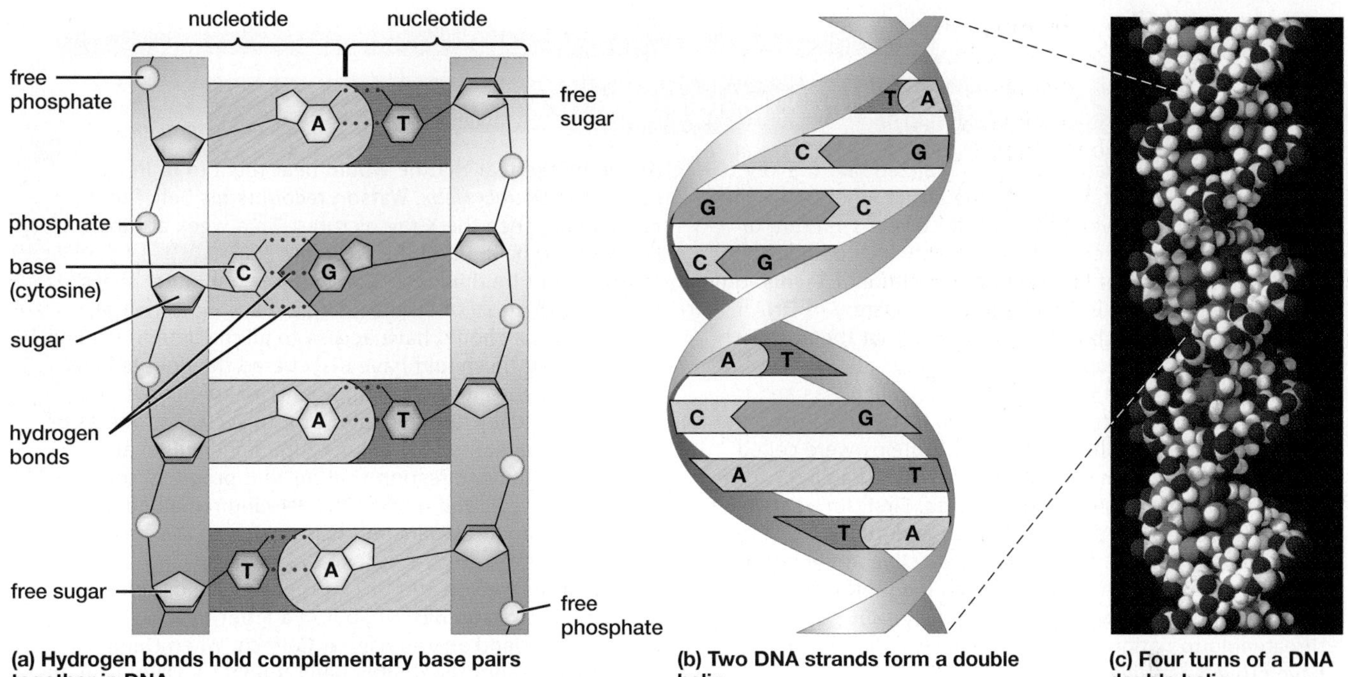

▲ **FIGURE 11-5 The Watson-Crick model of DNA structure (a)** Hydrogen bonding between complementary base pairs holds the two strands of DNA together. Three hydrogen bonds hold guanine to cytosine, and two hydrogen bonds hold adenine to thymine. Note that each strand has a free phosphate on one end and a free sugar on the opposite end. Further, the two strands run in opposite directions. **(b)** Strands of DNA wind about each other in a double helix, like a twisted ladder, with the sugar-phosphate backbone forming the uprights and the complementary base pairs forming the rungs. **(c)** A space-filling model of DNA structure.

QUESTION Which do you think would be more difficult to break apart: an A–T base pair or a C–G base pair?

Complementary base pairs explain "Chargaff's rule"—that the DNA of a given species contains equal amounts of adenine and thymine, as well as equal amounts of cytosine and guanine. Because an A in one DNA strand always pairs with a T in the other strand, the amount of A always equals the amount of T. Similarly, because a G in one strand always pairs with a C in the other DNA strand, the amount of G always equals the amount of C.

Finally, look at the sizes of the bases: Because adenine and guanine consist of two fused rings, they are large, whereas thymine and cytosine, composed of only a single ring, are small. Because the double helix has only A–T and G–C pairs, all the rungs of the DNA ladder are the same width. Therefore, the double helix has a constant diameter, just as the X-ray diffraction pattern predicted.

The structure of DNA was solved. On March 7, 1953, at the Eagle Pub in Cambridge, England, Francis Crick proclaimed to the lunchtime crowd, "We have discovered the secret of life." This claim was not far from the truth. Although further data would be needed to confirm the details, within just a few years, their DNA model revolutionized biology, including genetics, evolution, and medicine. As we will see in later chapters, the revolution continues today.

11.3 HOW DOES DNA ENCODE INFORMATION?

Look again at the structure of DNA shown in Figure 11-5. Can you see why many scientists had trouble believing that DNA could be the carrier of genetic information? Consider the many characteristics of just one organism. How can the color of a bird's feathers, the size and shape of its beak, its ability to make a nest, its song, and its ability to migrate all be determined by a molecule with just four simple subunits?

The answer is that it's not the *number* of different subunits but their *sequence* that's important. Within a DNA strand, the four types of bases can be arranged in any order, and each unique sequence of bases represents a unique set of genetic instructions. An analogy might help: You don't need a lot of unique letters to make up a language. English has 26 letters, but Hawaiian has only 12, and the binary language of computers uses only two "letters" (0 and 1, or "off" and "on"). Nevertheless, all three languages can spell out thousands of different words. A stretch of DNA that is just 10 nucleotides long can have more than a million possible sequences of the four bases. Because an organism has millions (in bacteria) to billions (in plants or animals) of nucleotides, DNA molecules can encode a staggering amount of information.

Scientific Inquiry

The Discovery of the Double Helix

In the early 1950s, many biologists realized that the key to understanding inheritance lay in the structure of DNA. They also knew that whoever deduced the correct structure of DNA would receive recognition, probably including a Nobel Prize. Linus Pauling of the California Institute of Technology was the person most likely to solve the mystery of DNA structure. Pauling probably knew more about the chemistry of large organic molecules than any person alive. Like Rosalind Franklin and Maurice Wilkins, Pauling was an expert in X-ray diffraction techniques. In 1950, he used these techniques to show that many proteins were coiled into single-stranded helices (see Fig. 3-20b). Pauling, however, had two important handicaps. First, for years he had concentrated on protein research and, therefore, had little data about DNA. Second, he was active in the peace movement. At that time, some government officials considered such activity to be potentially subversive and threatening to national security. This latter handicap may have proved decisive.

The second most likely competitors were Wilkins and Franklin, the British scientists who had set out to determine the structure of DNA by using X-ray diffraction patterns. In fact, they were the only scientists who had good data about the general shape of the DNA molecule. Unfortunately for them, their methodical approach was slow.

The door was open for the eventual discoverers of the double helix—James Watson and Francis Crick, two scientists with neither Pauling's tremendous understanding of chemical bonds nor Franklin and Wilkins' expertise in X-ray analysis. Watson and Crick did no experiments in the ordinary sense of the word. Instead, they spent their time thinking about DNA, trying to construct a molecular model that made sense and fit the data. Because they were working in England and because Wilkins shared Franklin's data with them (perhaps against her wishes), Watson and Crick were familiar with all the X-ray information relating to DNA.

The X-ray data were just what Pauling lacked. Because of his presumed subversive tendencies, the U.S. State Department refused to issue Pauling a passport to leave the United States, so he could neither attend meetings at which Wilkins presented the X-ray data nor visit England to talk with Franklin and Wilkins directly. Watson and Crick knew that Pauling was working on DNA structure and were driven by the fear that he would beat them to it. In his book, *The Double Helix,* Watson recounts his belief that, had Pauling seen the X-ray pictures, "in a week at most, Linus would have [had] the structure."

You might be thinking, "But wait just a minute! That's not fair. If the goal of science is to advance knowledge, then everyone should have access to all the data. If Pauling was the best, he should have discovered the double helix first." Perhaps so. But after all, scientists are people, too. Although virtually all scientists want to see the progress of science and benefits for humanity, each individual also wants to be the one responsible for that progress and to receive the credit and glory. Linus Pauling remained in the dark about the X-ray data and was beaten to the correct structure.

Soon after Watson and Crick proposed the double helix (**Fig. E11-3**), Watson described it in a letter to Max Delbruck, a friend and adviser at Caltech. When Delbruck told Pauling about the double helix model for DNA, Pauling graciously congratulated Watson and Crick on their brilliant solution. The race was over.

▲ **FIGURE E11-3 The discovery of DNA** James Watson and Francis Crick with a model of DNA structure.

Of course, to make sense, the letters of a language must be in the correct order. Similarly, a gene must have the right bases in the right sequence. Just as "friend" and "fiend" mean different things, and "fliend" doesn't mean anything, different sequences of bases in DNA may encode very different pieces of information, or no information at all.

In Chapter 12, we will describe how the information in DNA is used to produce the structures of living cells. In the remainder of this chapter, we will examine how DNA is replicated during cell division to ensure accurate copying of this genetic information.

Case Study continued
Muscles, Mutations, and Myostatin

All "normal" mammals have a DNA sequence that encodes a functional myostatin protein, which limits their muscle growth. Belgian Blue cattle have a mutation that changes a "friendly" gene to a nonsensical "fliendly" one that no longer codes for a functional protein, so they have excessive muscle development.

11.4 HOW DOES DNA REPLICATION ENSURE GENETIC CONSTANCY DURING CELL DIVISION?

Replication of DNA Is a Critical Event in the Cell Cycle

In the 1850s, Austrian pathologist Rudolf Virchow realized that "all cells come from [preexisting] cells." All of the trillions of cells of your body are the offspring of other cells, going all the way back to when you were a fertilized egg. Moreover, almost every cell of your body contains identical genetic information—the same genetic information present in that fertilized egg. When cells reproduce by mitotic cell division, each daughter cell receives a nearly perfect copy of the parent cell's genetic information. Consequently, before cell division, the parent cell must synthesize two exact copies of its DNA. A process called **DNA replication** produces these two identical DNA double helices.

DNA Replication Produces Two DNA Double Helices, Each with One Original Strand and One New Strand

How does a cell accurately copy its DNA? In their paper describing DNA structure, Watson and Crick included one of the greatest understatements in all of science: "It has not escaped our notice that the specific [base] pairing we have postulated immediately suggests a possible copying mechanism for the genetic material." In fact, base pairing is the foundation of DNA replication. Remember, the rules for base pairing are that an adenine on one strand must pair with a thymine on the other strand, and a cytosine must pair with a guanine. If one strand reads A–T–G, for example, then the other strand must read T–A–C. Therefore, the base sequence of each strand contains all the information needed to replicate the other strand.

Conceptually, DNA replication is quite simple (**Fig. 11-6**). The essential ingredients are (1) the parental DNA strands, (2) **free nucleotides** that were previously synthesized in the cytoplasm and imported into the nucleus, and (3) a variety of enzymes that unwind the parental DNA double helix and synthesize new DNA strands.

First, enzymes called **DNA helicases** (meaning "enzymes that break apart the double helix") pull apart the parental DNA double helix, so that the bases of the two DNA strands no longer form base pairs with one another. Now DNA strands complementary to the two parental strands must be synthesized. Other enzymes, called **DNA polymerases** ("enzymes that synthesize a DNA polymer"), move along each separated parental DNA strand, matching bases on the strand with complementary free nucleotides. For example, DNA polymerase pairs an exposed adenine in the parental strand with a thymine. DNA polymerase also connects these free nucleotides with one another to form two new DNA strands, each complementary to one of the parental DNA strands. Thus, if a parental DNA strand reads T–A–G, DNA polymerase will synthesize a new DNA strand with the complementary sequence A–T–C. For more information on how DNA is replicated, refer to "A Closer Look at DNA Structure and Replication" on pp. 210–212.

When replication is complete, one parental DNA strand and its newly synthesized, complementary daughter

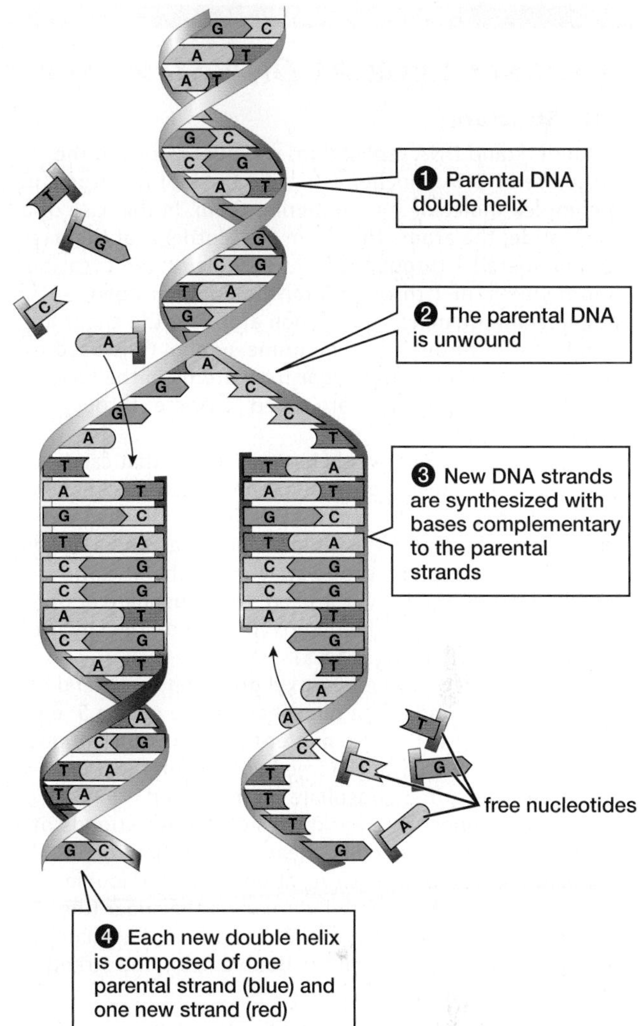
❶ Parental DNA double helix
❷ The parental DNA is unwound
❸ New DNA strands are synthesized with bases complementary to the parental strands
free nucleotides
❹ Each new double helix is composed of one parental strand (blue) and one new strand (red)

▲ **FIGURE 11-6 Basic features of DNA replication** During replication, the two strands of the parental DNA double helix separate. Free nucleotides that are complementary to those in each strand are joined to make new daughter strands. Each parental strand and its new daughter strand then form a new double helix.

Case Study continued

Muscles, Mutations, and Myostatin

Through a complex mechanism involving many other molecules, myostatin prevents pre-muscle cells from replicating their DNA. Therefore, the cells stop dividing, and the number of mature muscle cells is limited. The mutated myostatin of Belgian Blue cattle fails to inhibit DNA replication, so the pre-muscle cells continue to divide, producing more muscle cells.

A Closer Look At *DNA Structure and Replication*

DNA Structure

To understand DNA replication, we must return to the structure of DNA. Biochemists keep track of the atoms in a complex molecule by numbering them. In the case of a nucleotide, the atoms that form the "corners" of the base are numbered 1 through 6 for the single-ringed cytosine and thymine, or 1 through 9 for the double-ringed adenine and guanine. The carbon atoms of the sugar are numbered 1′ through 5′. The prime symbol (′) is used to distinguish atoms in the sugar from atoms in the base. The carbons in the sugar are called "1-prime" through "5-prime" (**Fig. E11-4**).

The sugar of a nucleotide has two "ends" that can be involved in synthesizing the sugar-phosphate backbone of a DNA strand: a 3′ end, which has a free —OH (hydroxyl) group attached to the 3′ carbon of the sugar, and a 5′ end, which has a phosphate group attached to the 5′ carbon. When a DNA strand is synthesized, the phosphate of one nucleotide bonds with the hydroxyl group on the sugar of the next nucleotide (**Fig. E11-5**).

This still leaves a free hydroxyl group on the 3′ end of one nucleotide, and a free phosphate group on the 5′ end of the other nucleotide. No matter how many nucleotides are joined, there is always a free hydroxyl on the 3′ end of the strand and a free phosphate on the 5′ end.

The sugar-phosphate backbones of the two strands of a double helix are antiparallel, meaning that they run in opposite directions. Therefore, at one end of a double helix, one strand has a free hydroxyl on the sugar (the 3′ end), whereas the other strand has a free phosphate (the 5′ end). At the other end of the double helix, the strand ends are reversed (**Fig. E11-6**).

DNA Replication

DNA replication involves three major actions (**Fig. E11-7**). First, the DNA double helix must be opened up so that the base sequence can be read. Then, new DNA strands with base sequences complementary to the two original strands must be synthesized. In eukaryotic cells, these new DNA strands are synthesized in fairly short pieces. Therefore, the third step in DNA replication is to stitch the pieces together to form a continuous new strand of DNA. Each step is carried out by a distinct set of enzymes.

DNA helicase separates the parental DNA strands

Acting in concert with several other enzymes, DNA helicase breaks the hydrogen bonds between complementary base pairs that hold the two parental DNA strands together. This separates and unwinds the parental double helix, forming a replication "bubble" (**Fig. E11-7** ❶ and ❷). Each replication bubble contains a replication "fork" at each end, where the two parental DNA strands are just beginning to be unwound. Within the replication bubble, the bases of the parental DNA strands are no longer paired with one another.

DNA polymerase synthesizes new DNA strands

Replication bubbles are essential because they allow a second enzyme, DNA polymerase, to gain access to the bases of each DNA strand (**Fig. E11-7** ❸). At each replication fork, a complex of DNA polymerase and other proteins binds to each parental strand. Therefore, there will be two DNA polymerase complexes, one on each parental strand. DNA polymerase recognizes an unpaired base in the parental strand and matches it up with a complementary base in a free nucleotide. Then, DNA polymerase catalyzes the formation of new covalent bonds, linking the phosphate of the incoming free nucleotide (the 5′ end) to the sugar of the most recently added nucleotide (the 3′ end) of the growing daughter strand. In this way, DNA polymerase synthesizes the sugar-phosphate backbone of the daughter strand.

Why make replication bubbles, rather than simply starting at one end of a double helix and letting one DNA polymerase molecule copy the DNA in one continuous

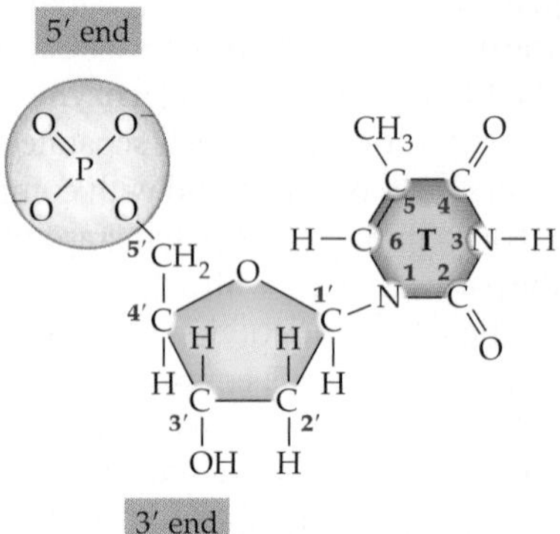

▲ **FIGURE E11-4 Numbering of carbon atoms in a nucleotide**

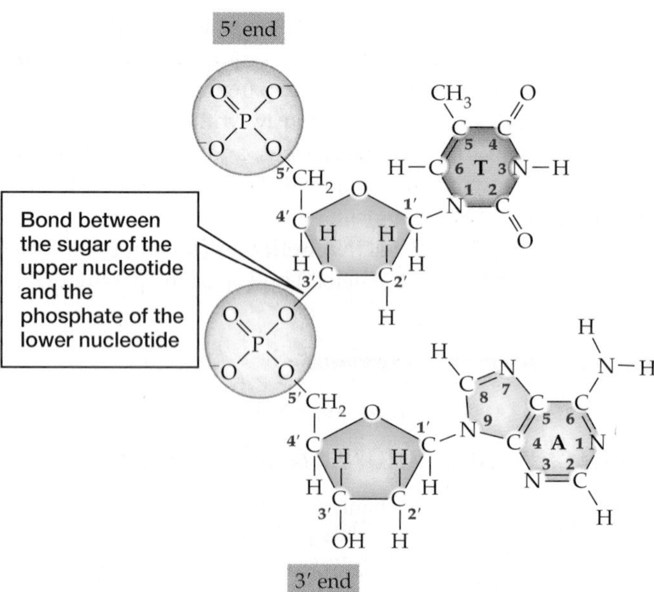

▲ **FIGURE E11-5 Numbering of carbon atoms in a dinucleotide**

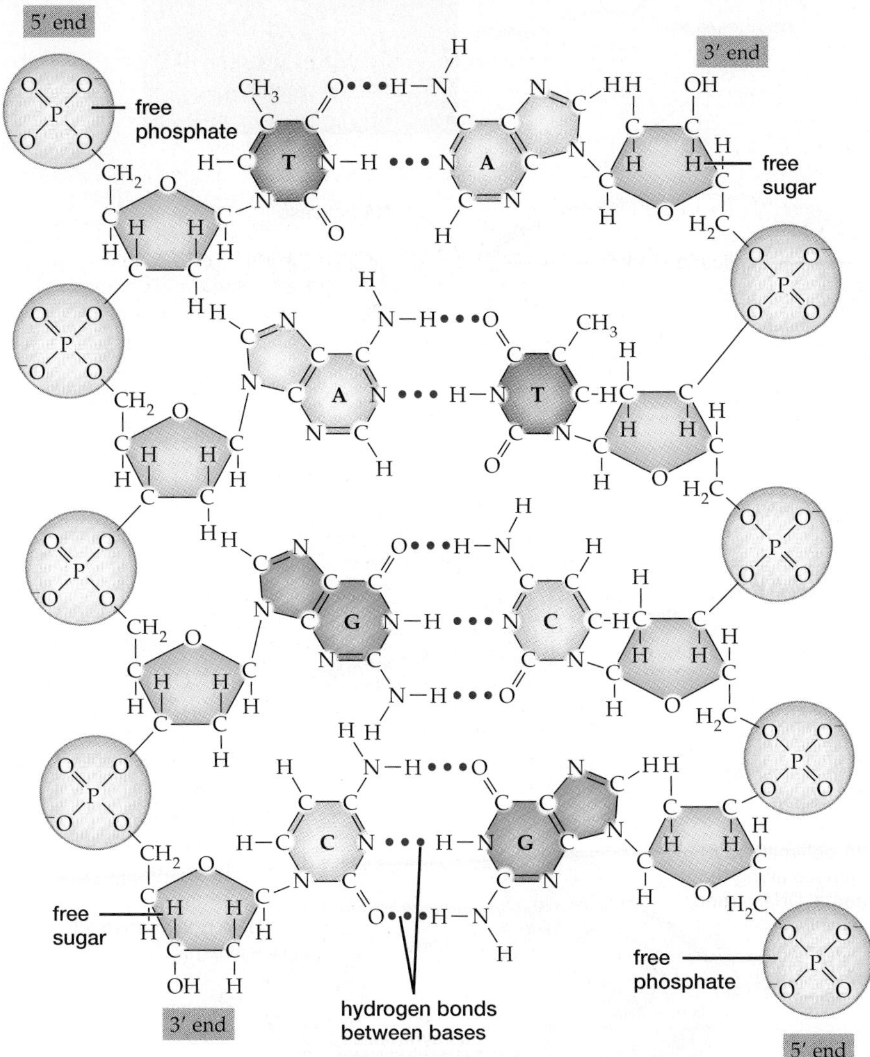

▲ **FIGURE E11-6 The two strands of a DNA double helix are antiparallel**

piece all the way to the other end? Well, eukaryotic chromosomes are *very* long: Human chromosomes range from about 23 million bases in the relatively tiny Y chromosome, to about 246 million bases in chromosome 1. Eukaryotic DNA is copied at a rate of about 50 nucleotides per second, so it would take about 5 to 57 days to copy human chromosomes in one continuous piece. To replicate an entire chromosome in a reasonable time, many DNA helicase enzymes open up many replication bubbles, allowing many DNA polymerase enzymes to copy the parental strands in fairly small pieces. The bubbles grow as DNA replication progresses, and they merge when they contact one another.

DNA polymerase always moves away from the 3′ end of a parental DNA strand (this is the end with the free hydroxyl group of the sugar) toward the 5′ end (with a free phosphate group). New nucleotides are always added to the 3′ end of the daughter strand. In other words, DNA polymerase moves 3′ → 5′ on a parental strand, and simultaneously moves 5′ → 3′ on the daughter strand. Finally, because the two strands of the parental DNA double helix are oriented in opposite directions, the DNA polymerase molecules move in opposite directions on the two parental strands (see Fig. E11-7 ❸).

DNA helicase and DNA polymerase work together (**Fig. E11-7 ❹**). A DNA helicase "lands" on the double helix and moves along, unwinding the double helix and separating the strands. Because the two DNA strands run in opposite directions, as a DNA helicase enzyme moves toward the 5′ end of one parental strand, it is simultaneously moving toward the 3′ end of the other parental strand. Now visualize two DNA polymerases landing on the two separated strands of DNA. One DNA polymerase (call it polymerase #1) can follow behind the helicase toward the 5′ end of the parental strand and can synthesize a continuous daughter DNA strand, until it runs into another replication bubble. This continuous daughter DNA strand is called the leading strand. On the other parental strand, however, DNA polymerase #2 moves *away from* the helicase: in ❸ of Figure E11-7, note that the helicase moves to the left, whereas DNA polymerase #2 moves to the right. Therefore, DNA synthesis on this strand will be discontinuous: DNA polymerase #2 will synthesize a short new DNA strand, called the lagging strand, but meanwhile, the helicase continues to move to the left, unwinding more of the double helix (see Fig. E11-7 ❹ and ❺). Additional DNA polymerases (#3, #4, and so on) must land on this strand and synthesize more short lagging strands.

DNA ligase joins segments of DNA

Multiple DNA polymerases synthesize pieces of DNA of varying lengths. Each chromosome may form hundreds of replication bubbles. Within each bubble, there will be one leading strand, tens to hundreds of thousands of nucleotide pairs long, and dozens to thousands of lagging strands, each about 100 to 200 nucleotide pairs long. Therefore, a cell might synthesize millions of pieces of DNA while replicating a single chromosome. How are all of these pieces sewn together? This is the job of the third major enzyme, **DNA ligase** ("an enzyme that ties DNA together"; **Fig. E11-7 ❺**). Many DNA ligase enzymes stitch the fragments of DNA together until each daughter strand consists of one long, continuous DNA polymer.

(continued)

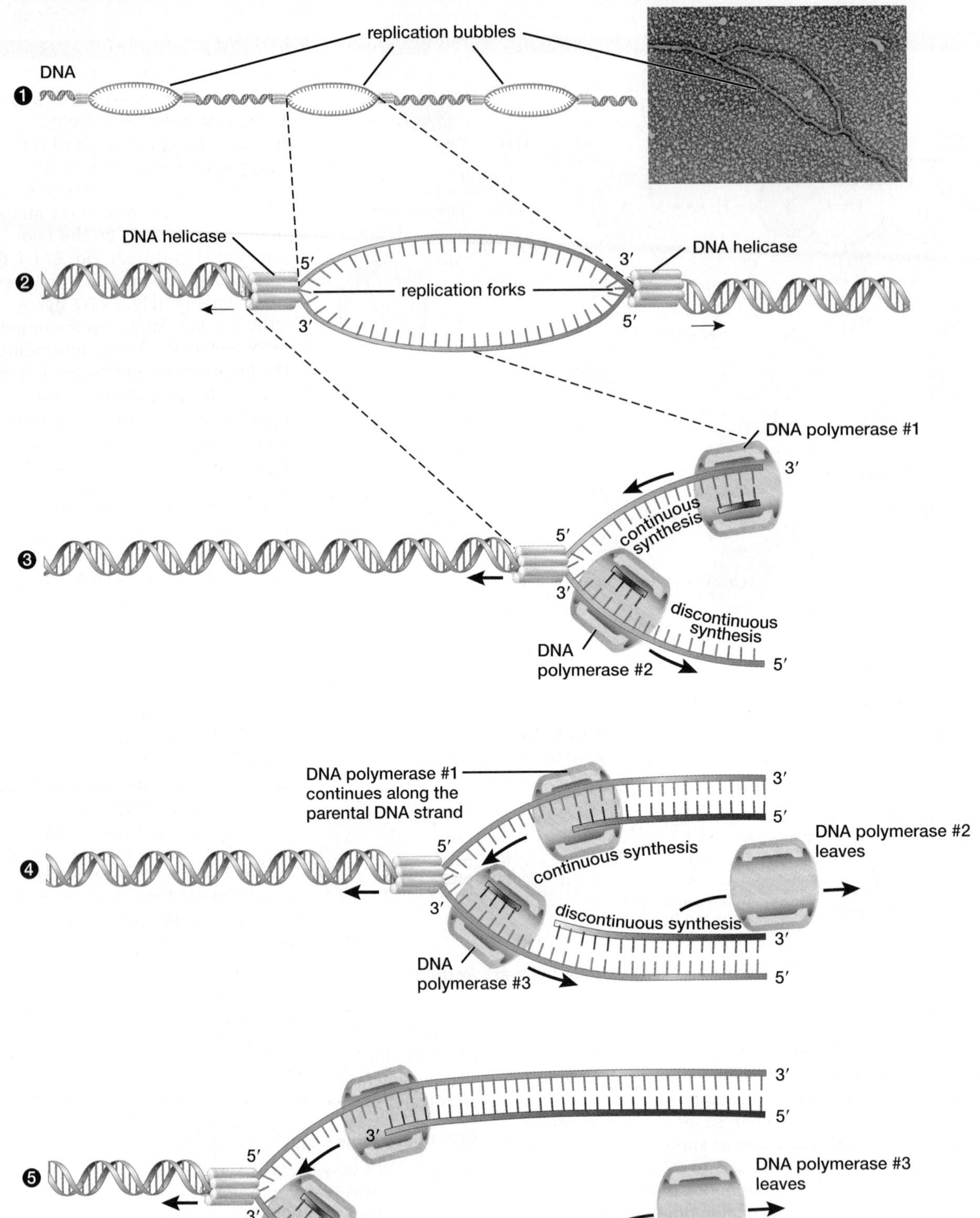

▲ **FIGURE E11-7 DNA replication** ❶ DNA helicase enzymes separate the parental strands of a chromosome to form replication bubbles. ❷ Each replication bubble consists of two replication forks, with unwound DNA strands between the forks. ❸ DNA polymerase enzymes synthesize new pieces of DNA. ❹ DNA helicase moves along the parental DNA double helix, unwinding it and enlarging the replication bubble. DNA polymerases in the replication bubble synthesize daughter DNA strands. ❺ DNA ligase joins the small DNA segments into a single daughter strand.

QUESTION During synthesis, why doesn't DNA polymerase move away from the replication fork on both strands?

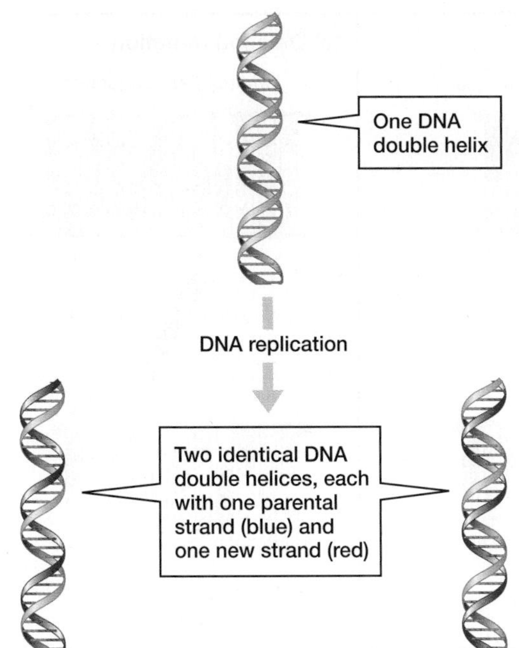

▲ **FIGURE 11-7 Semiconservative replication of DNA**

DNA strand wind together into one double helix. At the same time, the other parental strand and its daughter strand wind together into a second double helix. In forming a new double helix, the process of DNA replication conserves one parental DNA strand and produces one newly synthesized strand. Hence, the process is called **semiconservative replication** (Fig. 11-7).

If no mistakes have been made, the base sequences of both new DNA double helices are identical to the base sequence of the parental DNA double helix and, of course, to each other.

BioFlix™ DNA Replication

11.5 HOW DO MUTATIONS OCCUR?

Nothing is perfect, including the DNA in your cells. Changes in the sequence of bases in DNA, often resulting in a defective gene, are called **mutations.** In most cells, mutations are minimized by highly accurate DNA replication, "proofreading" of the newly synthesized DNA, and the repair of any changes in DNA that may occur even when the DNA is not being replicated.

Accurate Replication and Proofreading Produce Almost Error-Free DNA

The specificity of hydrogen bonding between complementary base pairs makes DNA replication highly accurate. DNA polymerase incorporates incorrect bases about once in every 1,000 to 100,000 base pairs. However, completed DNA strands contain only about one mistake in every 100 million to 1 billion base pairs (in humans, usually less than one per chromosome per replication). This phenomenally low error rate is ensured by a variety of DNA repair enzymes that proofread each daughter strand during and after its synthesis. For example, some forms of DNA polymerase recognize a base pairing mistake as it is made. This type of DNA polymerase will pause, fix the mistake, and then continue synthesizing more DNA.

Mistakes Do Happen

Despite this amazing accuracy, neither humans nor any other organisms have error-free DNA. In addition to rare mistakes made during normal DNA replication, a variety of environmental conditions can damage DNA. For example, certain chemicals (such as components of cigarette smoke) and some types of radiation (such as ultraviolet rays in sunlight and X-rays) increase the frequency of base pairing errors during replication, or even induce changes in DNA composition between replications. Most of these changes in DNA sequence are fixed by repair enzymes in the cell. However, some inevitably remain.

Mutations Range from Changes in Single Nucleotide Pairs to Movements of Large Pieces of Chromosomes

During replication, a pair of bases is occasionally mismatched. Usually, repair enzymes recognize the mismatch, cut out the incorrect nucleotide, and replace it with a nucleotide that bears a complementary base. Sometimes, however, the enzymes replace the original nucleotide instead of the mismatched one. The resulting base pair is complementary, but incorrect. These **nucleotide substitutions** are also called **point mutations,** because individual nucleotides in the DNA sequence are changed (Fig. 11-8a). An **insertion mutation** occurs when one or more nucleotide pairs are inserted into the DNA double helix (Fig. 11-8b). A **deletion mutation** occurs when one or more nucleotide pairs are removed from the double helix (Fig. 11-8c).

Pieces of chromosomes ranging in size from a single nucleotide pair to massive pieces of DNA are occasionally rearranged. An **inversion** occurs when a piece of DNA is cut out of a chromosome, turned around, and reinserted into the gap (Fig. 11-8d). Finally, a **translocation** results when a chunk of DNA, often very large, is removed from one chromosome and attached to another one (Fig. 11-8e).

Mutations May Have Varying Effects on Function

Mutations are often harmful, much as randomly changing words in the middle of Shakespeare's *Hamlet* would probably interrupt the flow of the play. If they are really damaging, a cell or an organism inheriting such a mutation may quickly die. Some mutations, however, have no effect or, in very rare instances, are even beneficial, as you will learn in Chapter 12. Mutations that are beneficial, at least in certain environments, may be favored by natural selection, and are the basis for the evolution of life on Earth (see Unit 3).

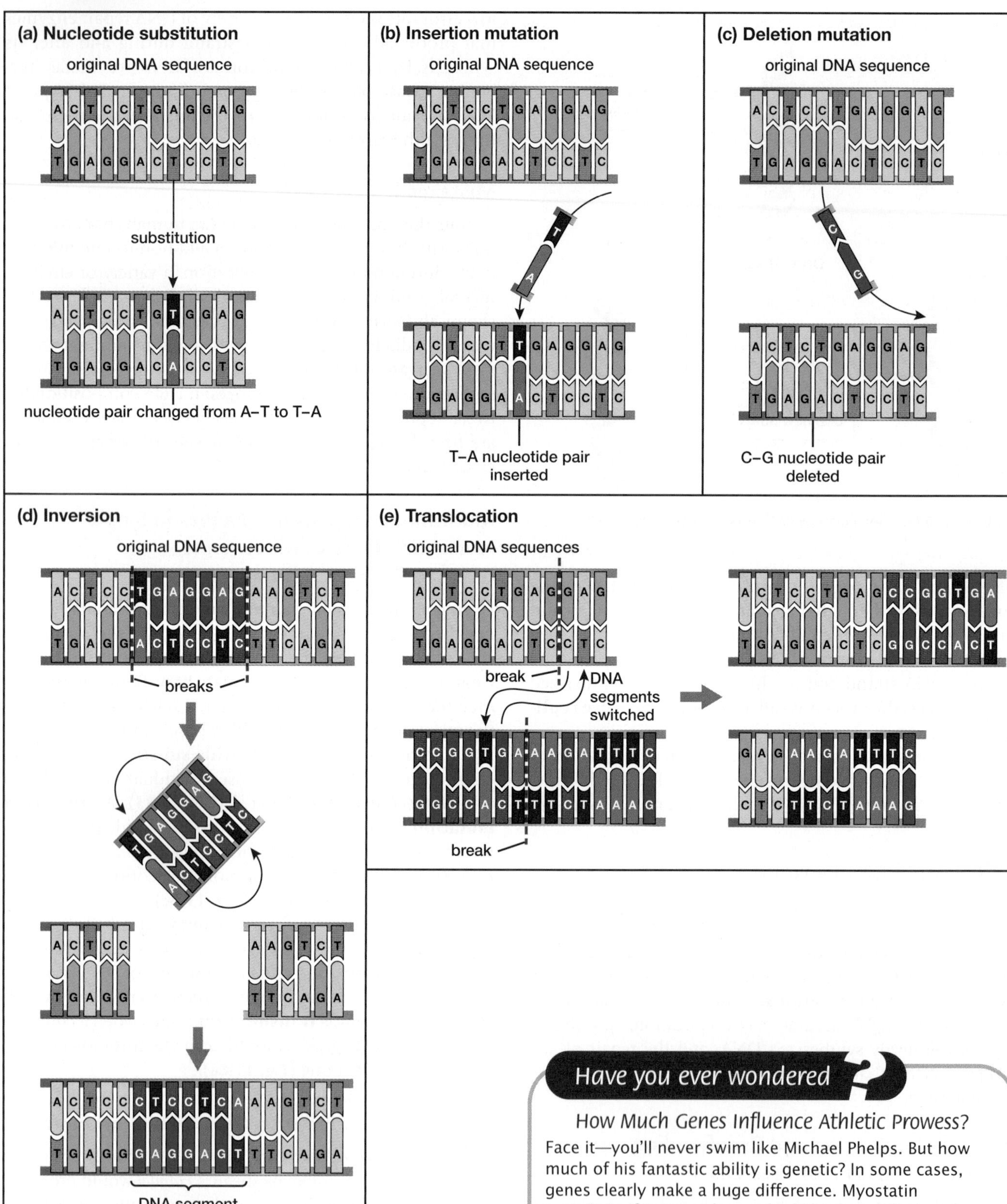

▲ **FIGURE 11-8 Mutations (a)** Nucleotide substitution. **(b)** Insertion mutation. **(c)** Deletion mutation. **(d)** Inversion mutation. **(e)** Translocation of pieces of DNA between two different chromosomes. In parts (a) through (d), the original DNA bases are in pale colors with black letters; mutations are in deep colors with white letters.

Have you ever wondered?

How Much Genes Influence Athletic Prowess?

Face it—you'll never swim like Michael Phelps. But how much of his fantastic ability is genetic? In some cases, genes clearly make a huge difference. Myostatin mutations, for example, can boost speed and strength, as we explain in the Case Study Revisited section. The effects of most individual genes, however, are small. Over 240 genes contribute to athletic performance. Do super-athletes have all the best "athletic alleles"? How many athletic alleles do Phelps, and you, have? Right now, no one can tell.

Case Study revisited

Muscles, Mutations, and Myostatin

Belgian Blue cattle have a deletion mutation in their myostatin gene. The result is that their cells stop synthesizing the myostatin protein about halfway through. (In Chapter 12, we'll explain why some mutations cause short, or truncated, proteins to be synthesized.) Several other breeds of "double muscled" cattle have the same mutation, but some have totally different mutations. Other animals, including some breeds of dogs (**Fig. 11-9**), also have myostatin mutations. They are generally different than the mutations in any of the breeds of cattle, but produce similar phenotypic effects.

▲ **FIGURE 11-9 Myostatin mutation in a dog** Like Belgian Blue cattle, "bully" whippets have defective myostatin, resulting in enormous muscles.

What they all have in common is that their myostatin proteins are nonfunctional. This is an important feature of the language of DNA: The nucleotide words must be spelled just right, or at least really close, for the resulting proteins to function. In contrast, any one of an enormous number of mistakes will render the proteins useless.

Humans have myostatin, too; not surprisingly, mutations can occur in the human myostatin gene. As you probably know, a child inherits two copies of most genes, one from each parent. A few years ago, a child was born in Germany who inherited a point mutation in his myostatin gene from both parents. This particular point mutation results in short, inactive, myostatin proteins. At 7 months, the boy already had well-developed calf, thigh, and buttock muscles. At 4 years, he could hold a 7-pound (3.2-kilogram) dumbbell in each hand, with his arms fully extended horizontally out to his sides (try it—it's not that easy for many adults).

Consider This

Mutations may be neutral, harmful, or beneficial. Into which category do myostatin mutations fall? Belgian Blue cattle are born so muscular and, consequently, so large, that they usually must be delivered by cesarean section. Myostatin-deficient mice have small, brittle tendons, so their oversize muscles are weakly attached to their bones. On the other hand, heterozygous whippets, with one normal and one defective myostatin allele, can run faster than ordinary whippets. The German boy's mother and some other relatives, presumably heterozygous for the myostatin mutation, were top-level athletes or unusually strong, or both. Is it possible that myostatin mutations are harmful in homozygotes but beneficial in heterozygotes? How do you think that natural selection might operate in this situation?

CHAPTER REVIEW

Summary of Key Concepts

11.1 How Did Scientists Discover That Genes Are Made of DNA?

By the turn of the century, scientists knew that genes must be made of either protein or DNA. Studies by Griffith showed that genes can be transferred from one bacterial strain into another. This transfer could transform the bacterial strain from harmless to deadly. Avery, MacLeod, and McCarty showed that DNA was the molecule that could transform bacteria. Hershey and Chase found that DNA is also the hereditary material of bacteriophage viruses. Thus, genes must be made of DNA.

11.2 What Is the Structure of DNA?

DNA consists of nucleotides that are linked into long strands. Each nucleotide consists of a phosphate group, the five-carbon sugar deoxyribose, and a nitrogen-containing base. Four types of bases occur in DNA: adenine, guanine, thymine, and cytosine. The sugar of one nucleotide is linked to the phosphate of the next nucleotide, forming a sugar-phosphate backbone for each strand. The bases stick out from this backbone. Two nucleotide strands wind together to form a DNA double helix, which resembles a twisted ladder. The sugar-phosphate backbones form the sides of the ladder. The bases of each strand pair up in the middle of the helix, held together by hydrogen bonds, forming the rungs of the ladder. Only complementary base pairs can bond together in the helix: Adenine bonds with thymine, and guanine bonds with cytosine.

11.3 How Does DNA Encode Information?

Genetic information is encoded as the sequence of bases in a DNA molecule. Just as a language can form thousands of words and complex sentences from a small number of letters by varying the sequence and number of letters in each word and sentence, so too can DNA encode large amounts of information with varying sequences and numbers of bases in different genes. Because DNA molecules are usually millions to billions of nucleotides long, DNA can encode huge amounts of information in its base sequence.

11.4 How Does DNA Replication Ensure Genetic Constancy During Cell Division?

When cells reproduce, they must replicate their DNA so that each daughter cell receives all the original genetic information. During DNA replication, enzymes unwind the two parental DNA strands. Then DNA polymerase enzymes bind to each parental DNA strand. Free nucleotides form hydrogen bonds with complementary bases on the parental strands, and DNA polymerase links the free nucleotides to form new DNA strands. Therefore, the sequence of nucleotides in each newly formed strand is complementary to the sequence of a parental strand. Replication is semiconservative because, when DNA replication is complete, both new DNA double helices consist of one parental DNA strand and one newly synthesized, complementary daughter strand. The two new DNA double helices are, therefore, duplicates of the parental DNA double helix.

BioFlix™ DNA Replication

11.5 How Do Mutations Occur?

Mutations are changes in the base sequence in DNA. DNA polymerase and other repair enzymes "proofread" the DNA, minimizing the number of mistakes during replication, but mistakes do occur. Other base changes occur as a result of radiation and damage from certain chemicals. Mutations include substitutions, insertions, deletions, inversions, and translocations. Most mutations are harmful or neutral, but a few are beneficial and may be favored by natural selection.

Key Terms

adenine (A) *203*
bacteriophage *204*
base *203*
complementary base pair *206*
cytosine (C) *203*
deletion mutation *213*
deoxyribonucleic acid (DNA) *200*
DNA helicase *209*
DNA ligase *211*
DNA polymerase *209*
DNA replication *209*
double helix *206*
free nucleotide *209*
guanine (G) *203*
insertion mutation *213*
inversion *213*
mutation *213*
nucleotide substitution *213*
nucleotide *203*
point mutation *213*
semiconservative replication *213*
strand *206*
sugar-phosphate backbone *206*
thymine (T) *203*
translocation *213*

Thinking Through the Concepts

Fill-in-the-Blank

1. DNA consists of subunits called __________. Each subunit consists of three parts: __________, __________, and __________.
2. The subunits of DNA are assembled by linking the __________ of one nucleotide to the __________ of the next. As it is found in chromosomes, two DNA polymers are wound together into a structure called a(n) __________.
3. The "base-pairing rule" in DNA is that adenine pairs with __________, and guanine pairs with __________. Bases that can form pairs in DNA are called __________.
4. When DNA is replicated, two new DNA double helices are formed, each consisting of one parental strand and one new, daughter strand. For this reason, DNA replication is called __________.
5. The DNA double helix is unwound by an enzyme called __________. Daughter DNA strands are synthesized by the enzyme __________. In eukaryotic cells, the daughter DNA strands are synthesized in pieces; these pieces are joined by the enzyme __________.
6. Sometimes, mistakes are made during DNA replication. If uncorrected, these mistakes are called __________. When a single nucleotide is changed, this is called a(n) __________, or __________.

Review Questions

1. Draw the general structure of a nucleotide. Which parts are identical in all nucleotides, and which can vary?
2. Name the four types of nitrogen-containing bases found in DNA.
3. Which bases are complementary to one another? How are they held together in the double helix of DNA?
4. Describe the structure of DNA. Where are the bases, sugars, and phosphates in the structure?
5. Describe the process of DNA replication.
6. How do mutations occur? Describe the principal types of mutations.

Applying the Concepts

1. As you learned in "Scientific Inquiry: The Discovery of the Double Helix," scientists in different laboratories often compete with one another to make new discoveries. Do you think this competition helps promote scientific discoveries? Sometimes researchers in different laboratories collaborate with one another. What advantages does collaboration offer over competition? What factors might provide barriers to collaboration and lead to competition?
2. Genetic information is encoded in the sequence of nucleotides in DNA. Let's suppose that the nucleotide sequence on one strand of a double helix encodes the information needed to synthesize a hemoglobin molecule. Do you think that the sequence of nucleotides on the other strand of the double helix also encodes useful information? Why? (An analogy might help. Suppose that English were a "complementary language," with letters at opposite ends of the alphabet complementary to one another; that is, A is complementary to Z, B to Y, C to X, and so on. Would a sentence composed of letters complementary to "To be or not to be?" make sense?) Finally, why do you think DNA is double-stranded?
3. **BioEthics** Today, scientific advances are being made at an astounding rate, and nowhere is this more evident than in our understanding of the biology of heredity. Using DNA as a starting point, do you believe that there are limits to the knowledge that people should acquire? Defend your answer.

Go to www.masteringbiology.com for practice quizzes, activities, eText, videos, current events, and more.

Gene Expression and Regulation

chapter 12

Case Study

Cystic Fibrosis

IF ALL YOU knew was her music, you'd think Alice Martineau had it made—a young, pretty singer-songwriter signed to a recording contract with Sony Music Entertainment. But genetics dealt Alice Martineau a double whammy—two copies of a defective recessive allele that encodes a crucially important protein called CFTR. Martineau—like about 30,000 Americans, 3,000 Canadians, and 20,000 Europeans—had cystic fibrosis. Before modern medical care, most people with cystic fibrosis died by age 4 or 5; even now, the average life span is only 35 to 40 years.

CFTR is a channel protein that is permeable to chloride and is found in many parts of the body, including the sweat glands, lungs, and intestines. Let's look at its role in perspiration. When sweat is first secreted by glands deep in the skin, it contains a lot of salt (sodium chloride), about as much as in blood. Most of the salt, however, is reclaimed as the sweat moves through tubes leading from the secreting cells to the surface of the skin. Although researchers do not fully understand the precise mechanism, CFTR is required for reabsorption of both chloride and sodium. Mutations in the *CFTR* gene, however, produce defective CFTR proteins that prevent reabsorption of chloride and sodium, so salt stays in the sweat.

Salty sweat usually isn't very harmful. Unfortunately, the cells lining the airways in the lungs have the same CFTR proteins. Normally, the airways are covered with a thin film of watery mucus, which traps bacteria and debris. "Natural antibiotic" proteins in the fluid kill many bacteria. The bacteria are then swept out of the lungs by cilia on the cells lining the airways. CFTR is essential for fluid secretion in the airway. Defective CFTR proteins cause the mucus to be "dehydrated" and so thick that the cilia can't move it out of the lungs. Therefore, the airways are partially clogged and bacteria multiply, causing chronic lung infections. People with cystic fibrosis cough frequently, trying to clear their airways. Alice Martineau credited coughing with strengthening her vocal cords and helping to produce her deep, strong voice.

In this chapter, we will examine the processes by which the instructions in genes are translated into proteins. When a gene mutates, how does that affect the structure and function of the encoded protein, such as CFTR? Why might different mutations in the same gene have different consequences?

▲ Although she was diagnosed early in life with cystic fibrosis, Alice Martineau hoped that ". . . people will realise when they hear the music, I am a singer-songwriter who just happens to be ill."

At a Glance

12.1 HOW IS THE INFORMATION IN DNA USED IN A CELL?

Information, by itself, doesn't do anything. For example, a blueprint may describe the structure of a house in great detail, but unless that information is translated into action by construction workers, no house will be built. Likewise, although the base sequence of DNA, the "molecular blueprint" of every cell, contains an incredible amount of information, DNA cannot carry out any action on its own. So, how does DNA determine whether you have black, blond, or red hair, or have normal lung function or cystic fibrosis?

Most Genes Contain the Information for the Synthesis of a Single Protein

Long before they found out that genes are made of DNA, biologists tried to determine how genes affect the phenotype of individual cells and entire organisms. Starting with studies of the inheritance of human metabolic disorders in the early 1900s and culminating in a brilliant series of experiments with common bread molds in the 1940s, biologists discovered that most genes contain the information needed to direct the synthesis of a single protein (see "Scientific Inquiry: One Gene, One Protein" on pp. 220–221). Proteins, in turn, are the "molecular workers" of the cell, forming many of its cellular structures and the enzymes that catalyze its chemical reactions. Therefore, there must be a flow of information from DNA to protein.

DNA Provides Instructions for Protein Synthesis via RNA Intermediaries

The DNA of a eukaryotic cell is housed in the nucleus, but protein synthesis occurs on ribosomes in the cytoplasm (see see pp. 66–67). Therefore, DNA cannot directly guide protein synthesis. There must be an intermediary, a molecule that carries information from DNA in the nucleus to the ribosomes in the cytoplasm. This molecule is **ribonucleic acid,** or **RNA.**

RNA is similar to DNA but differs structurally in three respects: (1) RNA is usually single-stranded; (2) RNA has the sugar ribose instead of deoxyribose in its backbone; and (3) RNA has the base uracil instead of the base thymine (**Table 12-1**).

DNA codes for the synthesis of many types of RNA, three of which play specific roles in protein synthesis: **messenger RNA (mRNA), ribosomal RNA (rRNA),** and **transfer RNA (tRNA)** (**Fig. 12-1**). There are several other types of RNA, including RNA used as the genetic material in some viruses, such as HIV; enzymatic RNA, called ribozymes, that catalyze various reactions, including cutting apart other molecules of RNA; and two types of RNA that we will briefly discuss later in this chapter: Xist RNA, which

Table 12-1 **A Comparison of DNA and RNA**

	DNA	RNA	
Strands	2	1	
Sugar	Deoxyribose	Ribose	
Types of bases	adenine (A), thymine (T) cytosine (C), guanine (G)	adenine (A), uracil (U) cytosine (C), guanine (G)	
Base pairs	DNA–DNA	RNA–DNA	RNA–RNA
	A–T	A–T	A–U
	T–A	U–A	U–A
	C–G	C–G	C–G
	G–C	G–C	G–C
Function	Contains genes; the sequence of bases in most genes determines the amino acid sequence of a protein	Messenger RNA (mRNA): carries the code for a protein-coding gene from DNA to ribosomes Ribosomal RNA (rRNA): combines with proteins to form ribosomes, the structures that link amino acids to form a protein Transfer RNA (tRNA): carries amino acids to the ribosomes	

prevents most of the genetic information in one of the X chromosomes of female mammals from being used (see section 12.5), and microRNA, which plays a role in regulating development and fighting disease (see "Scientific Inquiry: RNA—It's Not Just a Messenger Anymore" on p. 234). Here we will focus on the functions of mRNA, rRNA, and tRNA in protein synthesis.

Messenger RNA Carries the Code for Protein Synthesis from DNA to Ribosomes

Messenger RNA carries the code for the amino acid sequence of a protein from DNA to the ribosomes, which synthesize the protein specified by the mRNA base sequence (**Fig. 12-1a**). In eukaryotic cells, the DNA remains safely stored in the nucleus, like a valuable document in a library,

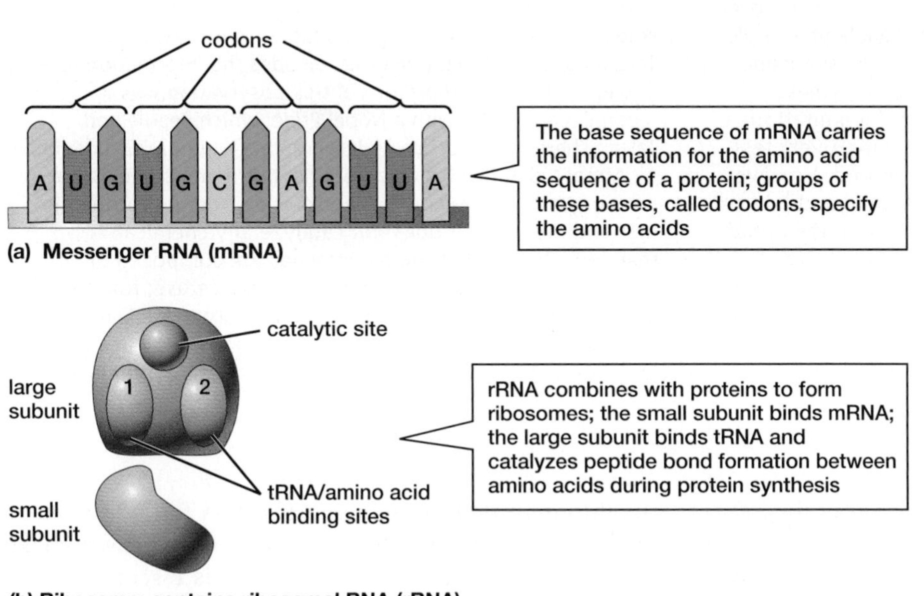
codons
A U G U G C G A G U U A
The base sequence of mRNA carries the information for the amino acid sequence of a protein; groups of these bases, called codons, specify the amino acids
(a) Messenger RNA (mRNA)
catalytic site
large subunit
1
2
rRNA combines with proteins to form ribosomes; the small subunit binds mRNA; the large subunit binds tRNA and catalyzes peptide bond formation between amino acids during protein synthesis
tRNA/amino acid binding sites
small subunit
(b) Ribosome: contains ribosomal RNA (rRNA)

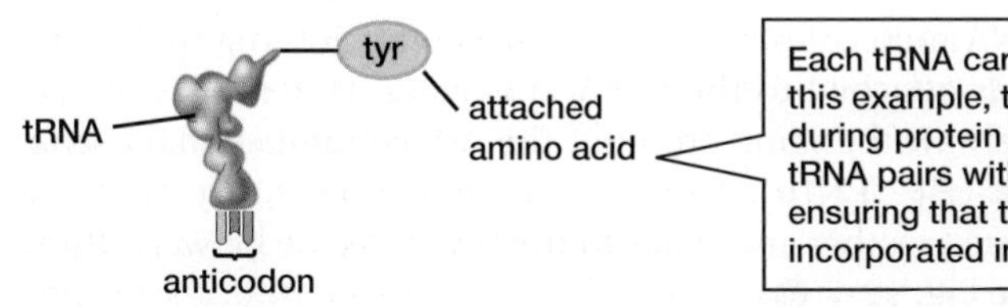
tyr
tRNA
attached amino acid
Each tRNA carries a specific amino acid (in this example, tyrosine [tyr]) to a ribosome during protein synthesis; the anticodon of tRNA pairs with a codon of mRNA, ensuring that the correct amino acid is incorporated into the protein
anticodon
(c) Transfer RNA (tRNA)

◀ FIGURE 12-1 **Cells synthesize three major types of RNA that are required for protein synthesis**

Scientific Inquiry

One Gene, One Protein

In Chapter 10, we saw that genes can determine, or at least influence, traits as varied as hair texture, flower color, and inherited diseases such as sickle-cell anemia. But how? Just as Gregor Mendel discovered the principles of inheritance using edible peas as an understandable "model system," later biologists sought to learn how genes work using model systems with easily measurable, unambiguous phenotypes. These model systems were the metabolic pathways by which cells synthesize complex molecules (see pp. 105–106).

Many metabolic pathways synthesize molecules in a series of linked steps, with each step catalyzed by a specific protein enzyme. Within a metabolic pathway, the product produced by one enzyme becomes the substrate of the next enzyme, like a molecular assembly line (see Fig. 6-12). How do genes encode the information needed to produce these pathways?

The first hint came from children who are born with defective metabolic pathways. For example, defects in the metabolism of two amino acids, phenylalanine and tyrosine, can cause albinism (no pigmentation in skin or hair; see Fig. 10-22) or several diseases with symptoms as varied as urine that turns brown when exposed to the air (alkaptonuria) or phenylalanine accumulation in the brain, leading to mental retardation (phenylketonuria). In the early 1900s, the English physician Archibald Garrod studied the inheritance of these inborn errors of metabolism. He hypothesized that (1) each inborn error of metabolism is caused by a defective version of a specific enzyme; (2) each defective enzyme is caused by a defective allele of a single gene; and (3) therefore, at least some genes must encode the information needed for the synthesis of enzymes.

Given the technology of his time and the limitations of human genetic studies, Garrod couldn't definitively test his hypotheses, and they were largely ignored. However, in the early 1940s, geneticists George Beadle and Edward Tatum used the metabolic pathways of a common bread mold, *Neurospora crassa*, to show that Garrod was correct.

Although we usually encounter *Neurospora* growing on stale bread, it can survive on a much simpler diet. All it needs is an energy source such as sugar, a few minerals, and vitamin B_6. Therefore, *Neurospora* manufactures the enzymes needed to make virtually all of its own organic molecules, including amino acids. (In contrast, we humans cannot synthesize many vitamins or 9 of the 20 common amino acids; we must obtain these from our food.) Beadle and Tatum used *Neurospora* to test the hypothesis that many of an organism's genes encode the information needed to synthesize enzymes.

If this hypothesis was correct, a mutation in a specific gene might disrupt the synthesis of a specific enzyme, which would disable one of the mold's metabolic pathways. Thus, a mutant mold couldn't synthesize some of the organic molecules, such as amino acids, that it needs to survive. The mutant *Neurospora* could grow on a simple medium of sugar, minerals, and vitamin B_6 only if the missing organic molecules were provided.

Beadle and Tatum induced mutations by exposing *Neurospora* to X-rays, and then studied the inheritance of the metabolic pathway that synthesizes the amino acid arginine (**Fig. E12-1**). In normal molds, arginine is synthesized from citrulline, which in turn is synthesized from ornithine (**Fig. E12-1a**). Mutant A could grow only if supplemented with arginine, but not if supplemented with citrulline or ornithine (**Fig. E12-1b**). Therefore, this strain had a defect in the enzyme that converts citrulline to arginine. Mutant B could grow if supplemented with either arginine or citrulline, but not with ornithine (see Fig. E12-1b). This mutant had a defect in the enzyme that converts ornithine to citrulline. Because a mutation in a single gene affected only a single enzyme within a single metabolic pathway, Beadle and Tatum concluded that one gene encodes the information for one enzyme. The importance of this observation was recognized in 1958 with a Nobel Prize, which Beadle and Tatum shared with Joshua Lederberg, one of Tatum's students.

Almost all enzymes are proteins, but many proteins are not enzymes. For example, keratin is a structural protein in hair and nails, but it does not catalyze any chemical reactions. In addition, many enzymes are composed of more than one protein subunit. DNA polymerase, for instance, consists of more than a dozen proteins. Thus, the "one gene, one enzyme" relationship proposed by Beadle and Tatum was later clarified to "one gene, one protein."

while mRNA, like a molecular photocopy, carries the information to the cytoplasm to be used in protein synthesis. As we will see shortly, groups of three bases in mRNA, called codons, specify which amino acids will be incorporated into a protein.

Ribosomal RNA and Proteins Form Ribosomes

Ribosomes, the structures that carry out translation, are composed of rRNA and many different proteins. Each ribosome consists of two subunits—one small and one large. The small subunit has binding sites for mRNA, a "start" (methionine) tRNA (see the next section for a description of tRNA), and several other proteins that collectively make up the "preinitiation complex," which is essential for assembling the ribosome and starting protein synthesis (see Fig. 12-7 ➊). The large subunit has binding sites for two tRNA molecules and a catalytic site for joining the amino acids attached to the tRNA molecules. Unless they are actively synthesizing proteins, the two subunits remain separate (**Fig. 12-1b**). During protein synthesis, the subunits come together and clasp an mRNA molecule between them (see Fig. 12-7 ➌).

(As you know from Chapter 3, a protein is a chain of amino acids joined by peptide bonds. Depending on their length, proteins may be called peptides [short chains] or polypeptides [long chains]. In this text, we usually call any chain of amino acids, regardless of length, a peptide or protein.) There are exceptions to the "one gene, one protein" rule, including several in which the final product of a gene isn't protein, but ribonucleic acid (RNA). Nevertheless, most genes do encode the amino acid sequence of a protein.

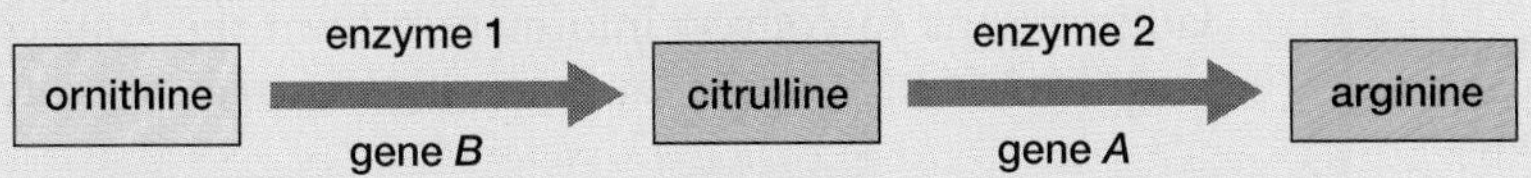

(a) The metabolic pathway for synthesizing the amino acid arginine

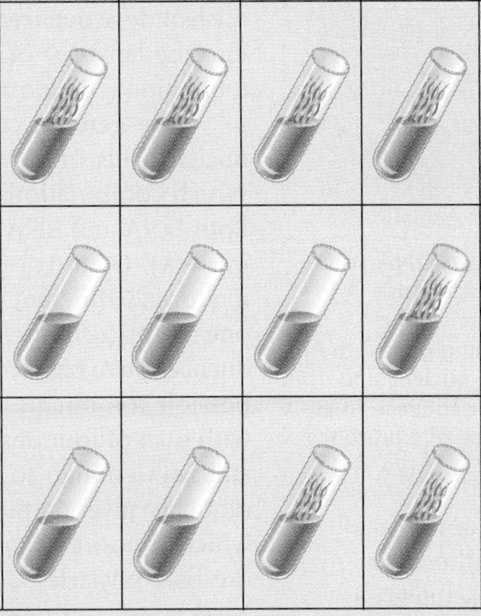

Genotype of *Neurospora*		Supplements Added to Medium				Conclusions
		none	ornithine	citrulline	arginine	
Normal *Neurospora*						Normal *Neurospora* can synthesize arginine, citrulline, and ornithine.
Mutants with single gene defect	A					Mutant A grows only if arginine is added. It cannot synthesize arginine because it has a defect in enzyme 2; gene *A* is needed for synthesis of arginine.
	B					Mutant B grows if either arginine or citrulline is added. It cannot synthesize arginine because it has a defect in enzyme 1. Gene *B* is needed for synthesis of citrulline.

(b) Growth of normal and mutant *Neurospora* on a simple medium with different supplements

▲ **FIGURE E12-1 Beadle and Tatum's experiments with *Neurospora* mutants (a)** Each step in the metabolic pathway for arginine synthesis is catalyzed by a different enzyme. **(b)** By analyzing which supplements allowed mutant molds to grow on simple nutrient medium, Beadle and Tatum concluded that a single gene codes for the synthesis of a single enzyme.

QUESTION What result would you expect for a mutant that lacks an enzyme needed to produce ornithine?

Transfer RNA Carries Amino Acids to the Ribosomes

Transfer RNA delivers the appropriate amino acids to the ribosome for incorporation into a protein. Every cell synthesizes at least one (and sometimes several) tRNAs for each amino acid. Twenty enzymes in the cytoplasm, one for each amino acid, recognize the tRNA molecules and use the energy of ATP to attach the correct amino acid to one end (**Fig. 12-1c**). These "loaded" tRNA molecules deliver their amino acids to a ribosome. A group of three bases, called an anticodon, protrudes from each tRNA. Complementary base pairing between codons of mRNA and anticodons of tRNA directs the correct amino acids to be used to synthesize a protein (see section 12.3).

Overview: Genetic Information Is Transcribed into RNA and Then Translated into Protein

Information in DNA is used to direct the synthesis of proteins in two steps, called transcription and translation (**Fig. 12-2** and **Table 12-2**):

1. In **transcription** (**Fig. 12-2a**), the information contained in the DNA of a specific gene is copied into messenger RNA (mRNA), transfer RNA (tRNA), or ribosomal RNA (rRNA). Thus, a gene is a segment of DNA that can be copied, or transcribed, into RNA. In eukaryotic cells, transcription occurs in the nucleus.

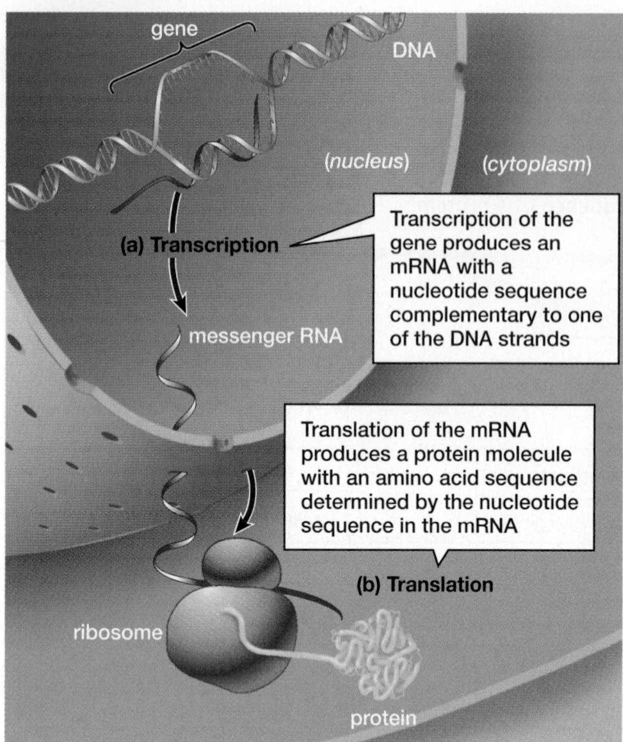
gene
DNA
(nucleus)
(cytoplasm)
(a) Transcription
Transcription of the gene produces an mRNA with a nucleotide sequence complementary to one of the DNA strands
messenger RNA
Translation of the mRNA produces a protein molecule with an amino acid sequence determined by the nucleotide sequence in the mRNA
(b) Translation
ribosome
protein

▲ **FIGURE 12-2 Genetic information flows from DNA to RNA to protein (a)** During transcription, the nucleotide sequence in a gene specifies the nucleotide sequence of a complementary RNA molecule. For protein-encoding genes, the product is an mRNA molecule that exits from the nucleus and enters the cytoplasm. **(b)** During translation, the sequence in an mRNA molecule specifies the amino acid sequence of a protein.

2. The base sequence of mRNA encodes the amino acid sequence of a protein. During protein synthesis, or **translation (Fig. 12-2b)**, this mRNA base sequence is decoded. Ribosomal RNA combines with dozens of proteins to form a ribosome. Transfer RNA molecules bring amino acids to the ribosome. Messenger RNA binds to the ribosome, where base pairing between mRNA and tRNA converts the base sequence of mRNA into the amino acid sequence of the protein. In eukaryotic cells, ribosomes are found in the cytoplasm and, thus, translation occurs there as well.

It's easy to confuse the terms transcription and translation. Comparing their common English meanings with their biological meanings may help you to remember the difference. In English, to *transcribe* means to make a written copy of something, almost always in the same language. In an American courtroom, for example, verbal testimony is transcribed into a written copy, and both the testimony of the witnesses and the transcriptions are in English. In biology, transcription is the process of copying information from DNA to RNA using the common "language" of nucleotides. In contrast, the common English meaning of *translation* is to convert words from one language to a different language. In biology, translation means to convert information from the "nucleotide language" of RNA to the "amino acid language" of proteins.

The Genetic Code Uses Three Bases to Specify an Amino Acid

We will investigate both transcription and translation in more detail in sections 12.2 and 12.3. First, however, let's see how geneticists broke the language barrier: How does the language of nucleotide sequences in DNA and messenger RNA translate into the language of amino acid sequences in proteins? This translation relies on a "dictionary" called the genetic code.

The **genetic code** translates the sequence of bases in nucleic acids into the sequence of amino acids in proteins. Which combinations of bases code for which amino acids? Both DNA and RNA contain four different bases: A, T (or U in RNA), G, and C (see Table 12-1). However, proteins are made of 20 different amino acids, so one base cannot code for one amino acid: there are not enough different bases. If a sequence of two bases codes for an amino acid, there would be 16 possible combinations (each of four possible first bases paired with each of four possible second bases, or $4 \times 4 = 16$). This still isn't enough to code for 20 amino acids. A three-base sequence gives 64 possible combinations ($4 \times 4 \times 4 = 64$), which is more than enough. Based on this mathematical exercise, physicist George Gamow hypothesized that three bases specify a single amino acid. In 1961, Francis Crick and three coworkers demonstrated that this hypothesis is correct.

For any language to be understood, its users must know what the words mean, where words start and stop, and where sentences begin and end. To decipher the "words" of the genetic code, Marshall Nirenberg and Heinrich Matthaei ground up bacteria and isolated the components needed to synthesize proteins. To this mixture, they added artificial mRNA, allowing them to control which "words" were to be translated. They could then see which amino acids were incorporated into proteins. For example, an mRNA strand composed entirely of uracil (UUUUUUUU . . .) directed the mixture to synthesize a protein composed solely of the amino acid phenylalanine.

Table 12-2 Transcription and Translation

Process	Information for the Process	Product	Major Enzyme or Structure Involved in the Process	Type of Base Pairing Required
Transcription (synthesis of RNA)	A segment of one DNA strand	One RNA molecule (e.g., mRNA, tRNA, or rRNA)	RNA polymerase	RNA with DNA: RNA bases pair with DNA bases as an RNA molecule is synthesized
Translation (synthesis of a protein)	mRNA	One protein molecule	Ribosome (also requires tRNA)	mRNA with tRNA: a codon in mRNA forms base pairs with an anticodon in tRNA

Therefore, the triplet UUU must specify phenylalanine. Because the genetic code was deciphered by using these artificial mRNAs, it is usually written in terms of the base triplets in mRNA (rather than in DNA) that code for each amino acid (**Table 12-3**). These mRNA triplets are called **codons.**

What about punctuation? How does a cell recognize where codons start and stop, and where the code for an entire protein starts and stops? Translation begins with the codon AUG, appropriately known as the **start codon.** Because AUG also codes for the amino acid methionine, all proteins originally begin with methionine (though it may be removed after the protein is synthesized). Three codons—UAG, UAA, and UGA—are **stop codons.** When the ribosome encounters a stop codon, it releases both the newly synthesized protein and the mRNA. Because all codons consist of three bases, and the beginning and end of a protein are specified, then punctuation ("spaces") between codons is unnecessary. Why? Consider what would happen if English used only three-letter words: a sentence such as THEDOGSAWTHECAT would be perfectly understandable, even without spaces between the words.

Because the genetic code has three stop codons, 61 nucleotide triplets remain to specify only 20 amino acids. Therefore, most amino acids are specified by several different codons. For example, six codons—UUA, UUG, CUU, CUC, CUA, and CUG—all specify leucine (see Table 12-3). However, each codon specifies one, and only one, amino acid; UUA always specifies leucine, never isoleucine, glycine, or any other amino acid.

How are codons used to direct protein synthesis? Decoding the codons of mRNA is the job of tRNA and ribosomes. Remember that tRNA transports amino acids to the ribosome, and that there are unique tRNA molecules that carry each different type of amino acid. Each of these unique tRNAs has three exposed bases, called an **anticodon,** which are complementary to the bases of a codon in mRNA. For example, the mRNA codon GUU forms base pairs with the anticodon CAA of a tRNA that has the amino acid valine attached to its end. As we will see in section 12.3, a ribosome can then incorporate valine into a growing protein chain.

12.2 HOW IS THE INFORMATION IN A GENE TRANSCRIBED INTO RNA?

Transcription (**Fig. 12-3**) consists of three steps: (1) initiation, (2) elongation, and (3) termination. These three steps correspond to the three major parts of most genes in both eukaryotes and prokaryotes: (1) a promoter region at the beginning of the gene, where transcription is started, or initiated; (2) the "body" of the gene where elongation of the RNA strand occurs; and (3) a termination signal at the end of the gene, where RNA synthesis ceases, or terminates.

Transcription Begins When RNA Polymerase Binds to the Promoter of a Gene

The enzyme **RNA polymerase** synthesizes RNA. To initiate transcription, RNA polymerase must first locate the beginning of a gene. Near the beginning of every gene is an untranscribed sequence of DNA called the **promoter.** In eukaryotic cells, a promoter consists of two main parts: (1) a short sequence of bases, often TATAAA, that binds RNA polymerase, and (2) one or more other sequences, often called transcription factor binding sites or response elements. When specific cellular proteins, called transcription factors, attach to one of these binding sites, they enhance or suppress binding of RNA polymerase to the promoter and, consequently, they enhance or suppress transcription of the gene. We will return to the important topic of gene regulation in section 12.5.

Table 12-3 The Genetic Code (Codons of mRNA)

First Base	Second Base: U		C		A		G		Third Base
U	UUU	Phenylalanine (Phe)	UCU	Serine (Ser)	UAU	Tyrosine (Tyr)	UGU	Cysteine (Cys)	U
U	UUC	Phenylalanine	UCC	Serine	UAC	Tyrosine	UGC	Cysteine	C
U	UUA	Leucine (Leu)	UCA	Serine	UAA	Stop	UGA	Stop	A
U	UUG	Leucine	UCG	Serine	UAG	Stop	UGG	Tryptophan (Trp)	G
C	CUU	Leucine	CCU	Proline (Pro)	CAU	Histidine (His)	CGU	Arginine (Arg)	U
C	CUC	Leucine	CCC	Proline	CAC	Histidine	CGC	Arginine	C
C	CUA	Leucine	CCA	Proline	CAA	Glutamine (Gln)	CGA	Arginine	A
C	CUG	Leucine	CCG	Proline	CAG	Glutamine	CGG	Arginine	G
A	AUU	Isoleucine (Ile)	ACU	Threonine (Thr)	AAU	Asparagine (Asp)	AGU	Serine (Ser)	U
A	AUC	Isoleucine	ACC	Threonine	AAC	Asparagine	AGC	Serine	C
A	AUA	Isoleucine	ACA	Threonine	AAA	Lysine (Lys)	AGA	Arginine (Arg)	A
A	AUG	Methionine (Met) Start	ACG	Threonine	AAG	Lysine	AGG	Arginine	G
G	GUU	Valine (Val)	GCU	Alanine (Ala)	GAU	Aspartic acid (Asp)	GGU	Glycine (Gly)	U
G	GUC	Valine	GCC	Alanine	GAC	Aspartic acid	GGC	Glycine	C
G	GUA	Valine	GCA	Alanine	GAA	Glutamic acid (Glu)	GGA	Glycine	A
G	GUG	Valine	GCG	Alanine	GAG	Glutamic acid	GGG	Glycine	G

▶ FIGURE 12-3 Transcription is the synthesis of RNA from instructions in DNA A gene is a segment of a chromosome's DNA. One of the DNA strands will serve as the template for the synthesis of an RNA molecule with bases complementary to the bases in the DNA strand.

QUESTION If the other DNA strand of this molecule were the template strand, in which direction would the RNA polymerase travel?

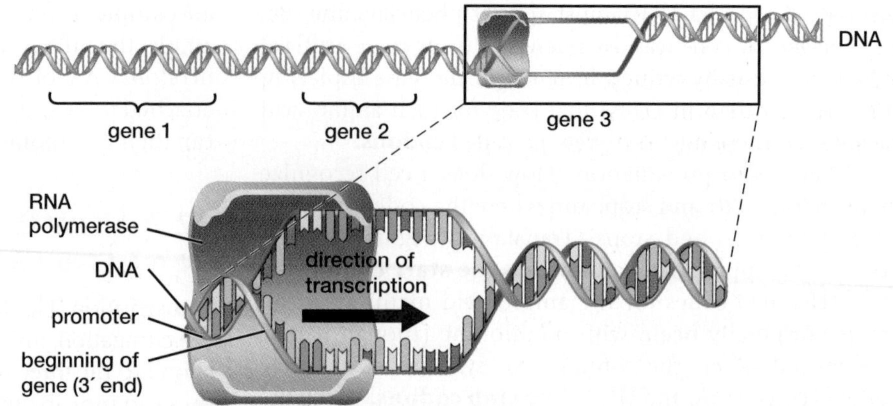

1 Initiation: RNA polymerase binds to the promoter region of DNA near the beginning of a gene, separating the double helix near the promoter.

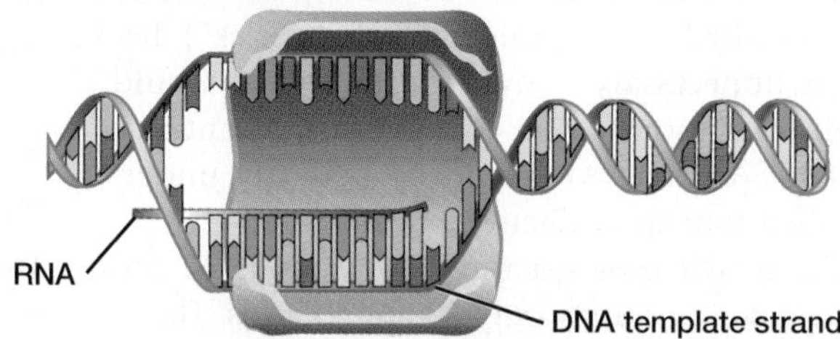

2 Elongation: RNA polymerase travels along the DNA template strand (blue), unwinding the DNA double helix and synthesizing RNA by catalyzing the addition of ribose nucleotides into an RNA molecule (red). The nucleotides in the RNA are complementary to the template strand of the DNA.

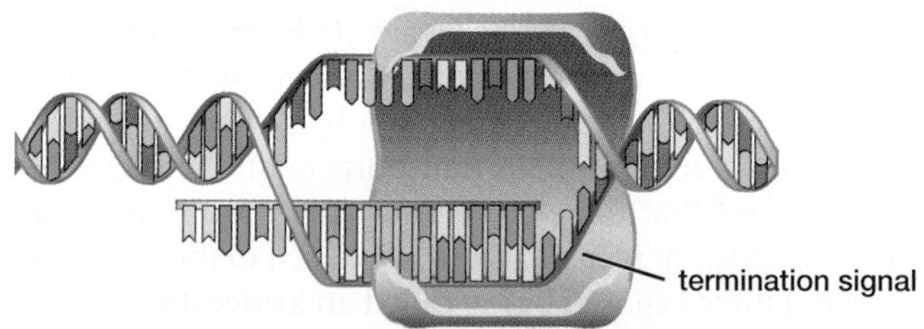

3 Termination: At the end of the gene, RNA polymerase encounters a DNA sequence called a termination signal. RNA polymerase detaches from the DNA and releases the RNA molecule.

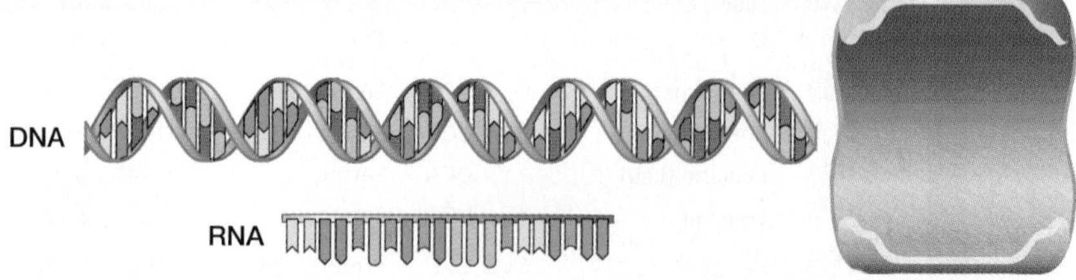

4 Conclusion of transcription: After termination, the DNA completely rewinds into a double helix. The RNA molecule is free to move from the nucleus to the cytoplasm for translation, and RNA polymerase may move to another gene and begin transcription once again.

When RNA polymerase binds to the promoter region of a gene, the DNA double helix at the beginning of the gene unwinds and transcription begins (**Fig. 12-3 1**).

Elongation Generates a Growing Strand of RNA

RNA polymerase then travels down one of the DNA strands, called the **template strand,** synthesizing a single strand of RNA with bases complementary to those in the DNA (**Fig. 12-3 2**). Like DNA polymerase (see p. 211), RNA polymerase always travels along the DNA template strand starting at the 3′ end of a gene and moving toward the 5′ end. Base pairing between RNA and DNA is the same as between two strands of DNA, except that uracil in RNA pairs with adenine in DNA (see Table 12-1).

After about 10 nucleotides have been added to the growing RNA chain, the first nucleotides in the RNA molecule separate from the DNA template strand. This separation allows the two DNA strands to rewind into a double helix (**Fig. 12-3** ❸). As transcription continues to elongate the RNA molecule, one end of the RNA drifts away from the DNA, while RNA polymerase keeps the other end attached to the template strand of the DNA (**Fig. 12-3** ❸, and **Fig. 12-4**).

Transcription Stops When RNA Polymerase Reaches the Termination Signal

RNA polymerase continues along the template strand of the gene until it reaches a sequence of DNA bases known as the termination signal. At this point, RNA polymerase releases the completed RNA molecule and detaches from the DNA (**Fig. 12-3** ❸ and ❹). The RNA polymerase is then free to bind to the promoter region of another gene and to synthesize another RNA molecule.

12.3 HOW IS THE BASE SEQUENCE OF MESSENGER RNA TRANSLATED INTO PROTEIN?

Messenger RNA Synthesis Differs Between Prokaryotes and Eukaryotes

The first step in synthesizing a protein is to produce a messenger RNA molecule with the base sequence specified by the gene that encodes the amino acid sequence of the protein. Prokaryotic and eukaryotic cells differ considerably in how they produce a functional mRNA molecule from the instructions in their DNA.

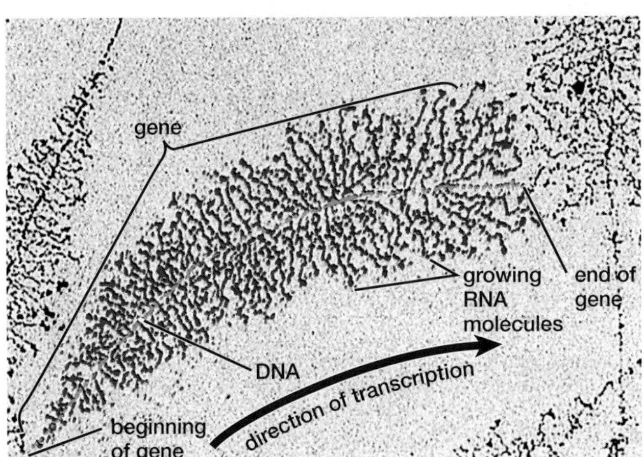

▲ **FIGURE 12-4 RNA transcription in action** This colorized electron micrograph shows the progress of RNA transcription in the egg of an African clawed toad. In each treelike structure, the central "trunk" is DNA and the "branches" are RNA molecules. A series of RNA polymerase molecules (too small to be seen in this micrograph) are traveling down the DNA, synthesizing RNA as they go. The beginning of the gene is on the left. The short RNA molecules on the left have just begun to be synthesized; the long RNA molecules on the right are almost finished.

QUESTION Why do you think so many mRNA molecules are being synthesized from the same gene?

Messenger RNA Synthesis in Prokaryotes

Prokaryotic genes are typically compact: All the nucleotides of a gene code for the amino acids in a protein. What's more, most or all of the genes for a complete metabolic pathway sit end to end on the chromosome (**Fig. 12-5a**). Therefore, prokaryotic cells commonly transcribe a single, very long mRNA from a series of adjacent genes. Because prokaryotic cells do not have a nuclear membrane separating their DNA from the cytoplasm (see Fig. 4-19), transcription and translation are usually not separated, either in space or in time. In most cases, as an mRNA molecule begins to separate from the DNA during transcription, ribosomes immediately begin translating the mRNA into protein (**Fig. 12-5b**).

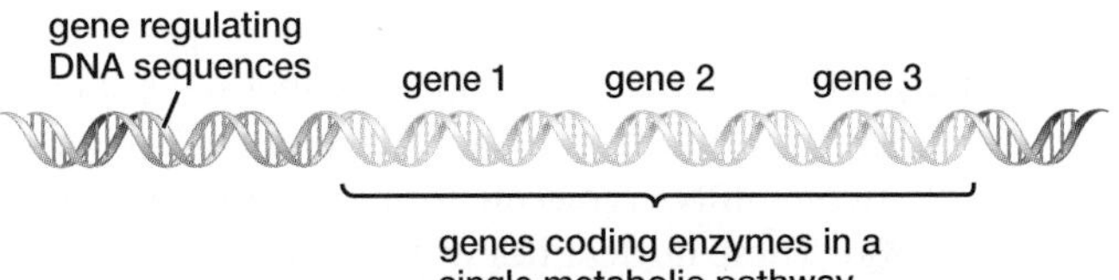

(a) Gene organization on a prokaryotic chromosome

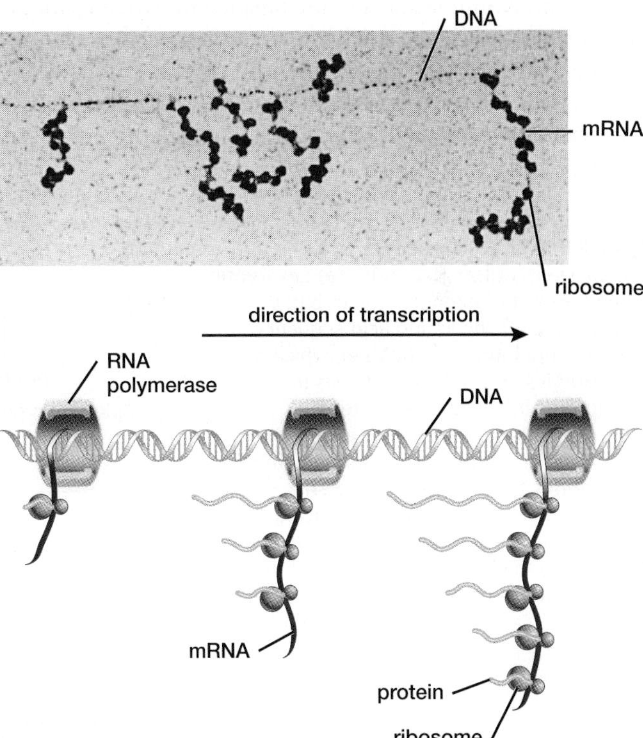

(b) Simultaneous transcription and translation in prokaryotes

▲ **FIGURE 12-5 Messenger RNA synthesis in prokaryotic cells (a)** In prokaryotes, many or all of the genes for a complete metabolic pathway lie side by side on the chromosome. **(b)** Transcription and translation are simultaneous in prokaryotes. In the electron micrograph, RNA polymerase (not visible at this magnification) travels from left to right on a strand of DNA. As it synthesizes a messenger RNA molecule, ribosomes bind to the mRNA and immediately begin synthesizing a protein (not visible). The diagram below the micrograph shows all of the key molecules involved.

Messenger RNA Synthesis in Eukaryotes

In contrast, the DNA of eukaryotic cells is contained in the nucleus, whereas the ribosomes reside in the cytoplasm. Further, the genes that encode the proteins needed for a metabolic pathway in eukaryotes are not clustered together as they are in prokaryotes, but may be dispersed among several chromosomes. Finally, each eukaryotic gene typically consists of two or more segments of DNA with nucleotide sequences that encode for a protein, interrupted by other nucleotide sequences that are not translated into protein. The coding segments are called **exons,** because they are *ex*pressed in protein, and the noncoding segments are called **introns,** because they are "*intr*agenic," meaning "within a gene" (**Fig. 12-6a**). Most eukaryotic genes have introns; the dystrophin gene (see pp. 194–195) contains almost 80!

Transcription of a eukaryotic gene produces a very long RNA strand, often called a precursor mRNA or pre-mRNA, which starts before the first exon and ends after the last exon (**Fig. 12-6b ❶**). More nucleotides are added at the beginning and end of this pre-mRNA molecule, forming a "cap" and "tail" (**Fig. 12-6b ❷**). These nucleotides will help to move the finished mRNA through the nuclear envelope to the cytoplasm, to bind the mRNA to a ribosome, and to prevent cellular enzymes from breaking down the mRNA molecule. To convert this pre-mRNA molecule into the finished mRNA, enzymes in the nucleus cut the pre-mRNA apart at the junctions between introns and exons, splice together the protein-coding exons, and discard the introns (**Fig. 12-6b ❸**). The finished mRNA molecules then leave the nucleus and enter the cytoplasm through pores in the nuclear envelope (**Fig. 12-6b ❹**). In the cytoplasm, the mRNA binds to ribosomes, which synthesize a protein specified by the mRNA base sequence.

Possible Functions of Intron-Exon Gene Structure

Why do eukaryotic genes contain introns and exons? This gene structure appears to serve at least two functions. The first function is to allow a cell to produce multiple proteins from a single gene by splicing exons together in different ways. For example, a gene called *CT/CGRP* is transcribed in both the thyroid and the brain. In the thyroid, one splicing arrangement results in the synthesis of the hormone calcitonin, which helps regulate calcium concentrations in the blood. In the brain, a different splicing arrangement results in the synthesis of a protein used as a messenger for communication between nerve cells. Alternative splicing may occur in the RNA transcribed from more than half of all eukaryotic genes. Therefore, in eukaryotes, the "one gene, one protein" rule must be reworded as "one gene, one *or more* proteins."

The second function of interrupted genes is more speculative, but is supported by some good experimental evidence: Fragmented genes may provide a quick and efficient way for eukaryotes to evolve new proteins with new functions. Chromosomes sometimes break apart, and their parts may reattach to different chromosomes. If the breaks occur within the noncoding introns of genes, exons may be moved intact from one chromosome to another. Most such errors would be harmful. But sometimes, the accidental exchange of exons among genes produces new eukaryotic genes that will enhance the survival and reproduction of the organism that carries them.

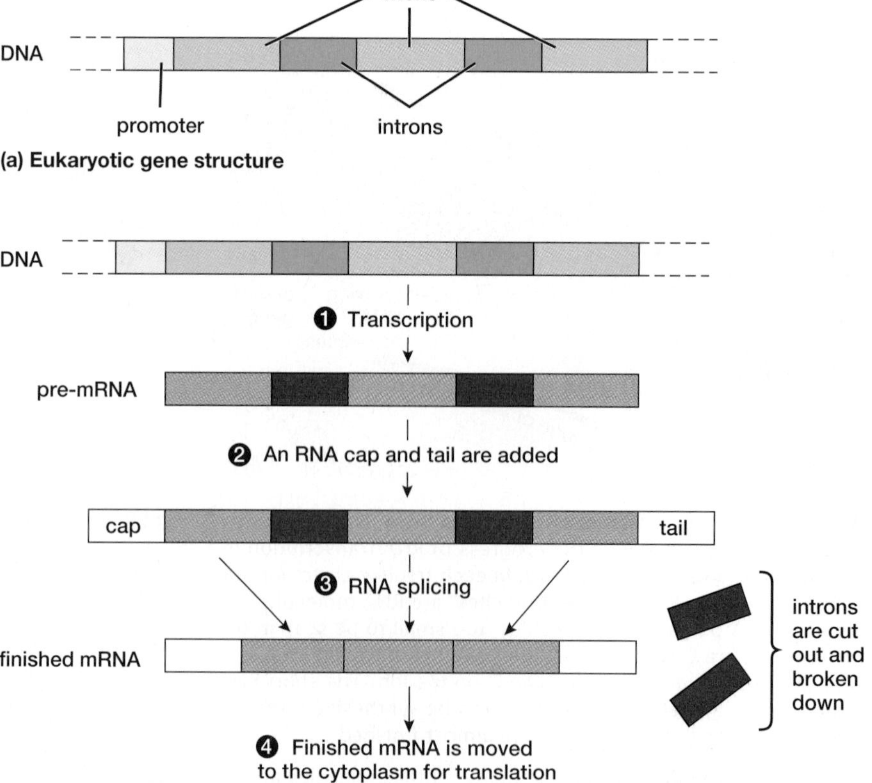

▶ **FIGURE 12-6 Messenger RNA synthesis in eukaryotic cells (a)** Eukaryotic genes consist of exons (medium blue), which code for the amino acid sequence of a protein, and introns (dark blue), which do not. **(b)** Messenger RNA synthesis in eukaryotes involves several steps: ❶ transcribing the gene into a long pre-mRNA molecule; ❷ adding additional RNA nucleotides to form the cap and tail; ❸ cutting out the introns and splicing the exons together into the finished mRNA; and ❹ moving the finished mRNA out of the nucleus into the cytoplasm for translation.

Health Watch

Genetics, Evolution, and Medicine

All life on Earth is related through evolution, sometimes closely (dogs and foxes), sometimes distantly (bacteria and people). Although often still similar, the genes of distantly related organisms may differ by many bases. Medicine takes advantage of these differences to develop antibiotics to treat bacterial infections.

Streptomycin and neomycin, commonly prescribed antibiotics, bind to a specific sequence of RNA in the small subunits of the ribosomes of certain bacteria, thereby inhibiting protein synthesis. Without adequate protein synthesis, the bacteria die. Patients infected by these bacteria don't die, however, because the small subunits of human eukaryotic ribosomes have a different base sequence than the bacteria's prokaryotic ribosomes do.

You have probably heard of antibiotic resistance, in which bacteria that are frequently exposed to antibiotics evolve defenses against those antibiotics. Bacteria evolve resistance against neomycin and related antibiotics rather rapidly. How? Well, if eukaryotic ribosomes are insensitive to neomycin, then eukaryotic ribosomes must function perfectly well with a different RNA sequence than prokaryotic ribosomes have. Bacteria that are resistant to neomycin and its relatives have a mutation that changes just a single base in their ribosomal RNA from adenine to guanine, which is precisely the base found at the comparable position in eukaryotic ribosomal RNA.

As this example illustrates, genetics, mutations, the mechanisms of protein synthesis, and evolution are important not only to biologists, but to physicians, too. In fact, a whole discipline called evolutionary medicine has arisen that uses the evolutionary relationships between people and microbes to help fight disease.

During Translation, mRNA, tRNA, and Ribosomes Cooperate to Synthesize Proteins

We will describe translation only in eukaryotic cells (**Fig. 12-7**), but the differences between eukaryotes and prokaryotes are crucial to the action of many antibiotics commonly used to treat bacterial infections (see "Health Watch: Genetics, Evolution, and Medicine").

Like transcription, translation has three steps: (1) initiation, (2) elongation of the protein chain, and (3) termination.

Initiation: Translation Begins When tRNA and mRNA Bind to a Ribosome

A preinitiation complex—composed of a small ribosomal subunit, a methionine (start) tRNA, and several other proteins (**Fig. 12-7 ❶**)—binds to the beginning of an mRNA molecule. The preinitiation complex scans the mRNA until it finds a start (AUG) codon, which forms base pairs with the UAC anticodon of the methionine tRNA (**Fig. 12-7 ❷**). A large ribosomal subunit then attaches to the small subunit, sandwiching the mRNA between the two subunits and holding the methionine tRNA in its first tRNA binding site (**Fig. 12-7 ❸**). The ribosome is now fully assembled and ready to begin translation.

Elongation: Amino Acids Are Added One at a Time to the Growing Protein Chain

A ribosome holds two mRNA codons in alignment with the two tRNA binding sites of the large subunit. A second tRNA, with an anticodon complementary to the second codon of the mRNA, moves into the second tRNA binding site on the large subunit (**Fig. 12-7 ❹**). The catalytic site of the large subunit breaks the bond holding the first amino acid (methionine) to its tRNA and forms a peptide bond between this amino acid and the amino acid attached to the second tRNA (**Fig. 12-7 ❺**). Interestingly, ribosomal RNA, and not one of the proteins of the large subunit, catalyzes the formation of the peptide bond. Therefore, this enzymatic RNA is often called a "ribozyme."

After the peptide bond is formed, the first tRNA is "empty," and the second tRNA carries a two-amino-acid chain. The ribosome then releases the empty tRNA and shifts to the next codon on the mRNA molecule (**Fig. 12-7 ❻**). The tRNA holding the elongating chain of amino acids also shifts, moving from the second to the first binding site of the ribosome. A new tRNA, with an anticodon complementary to the third codon of the mRNA, binds to the empty second site (**Fig. 12-7 ❼**). The catalytic site on the large subunit now links the third amino acid onto the growing protein chain (**Fig. 12-7 ❽**). The empty tRNA leaves the ribosome, the ribosome shifts to the next codon on the mRNA, and the process repeats, one codon at a time.

Termination: A Stop Codon Signals the End of Translation

A stop codon in the mRNA molecule signals the ribosome to terminate protein synthesis. Stop codons do not bind to tRNA. Instead, proteins called "release factors" bind to the ribosome

Case Study continued

Cystic Fibrosis

Recall from Chapter 4 that proteins destined to be inserted into the plasma membrane are synthesized by ribosomes on rough endoplasmic reticulum (ER) and move into the ER for processing. The most common defective allele causing cystic fibrosis produces a misshapen CFTR protein that is broken down within the ER. Four other common mutant alleles code for a stop codon in the middle of the protein, so translation terminates partway through. These alleles cause a complete lack of the CFTR protein and, consequently, very severe cystic fibrosis.

Initiation:

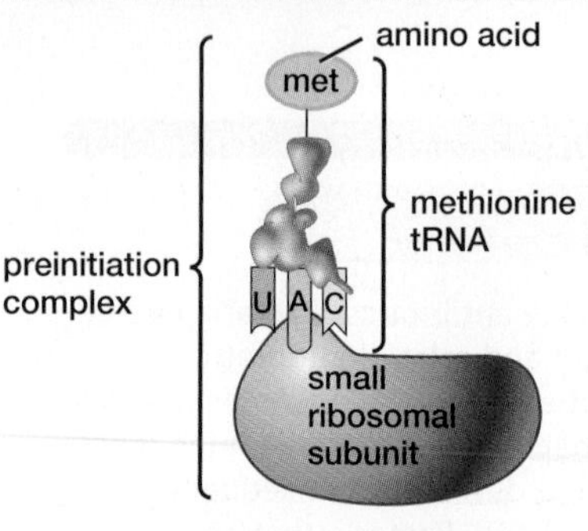

❶ A tRNA with an attached methionine amino acid binds to a small ribosomal subunit, forming a preinitiation complex.

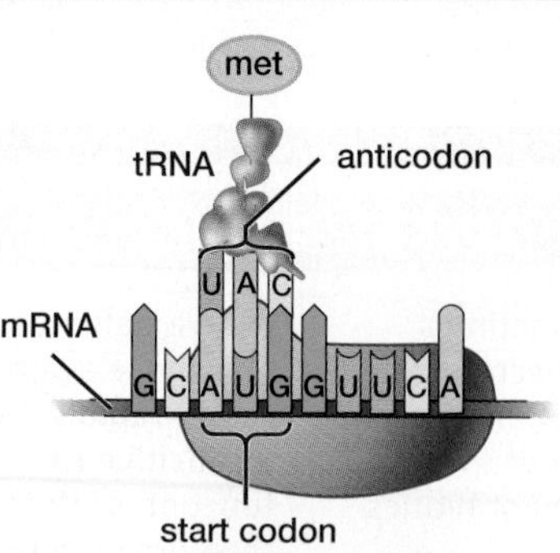

❷ The preinitiation complex binds to an mRNA molecule. The methionine (met) tRNA anticodon (UAC) base-pairs with the start codon (AUG) of the mRNA.

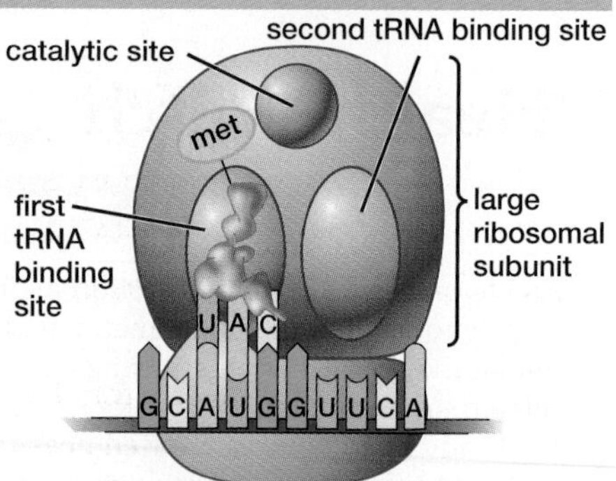

❸ The large ribosomal subunit binds to the small subunit. The methionine tRNA binds to the first tRNA site on the large subunit.

Elongation:

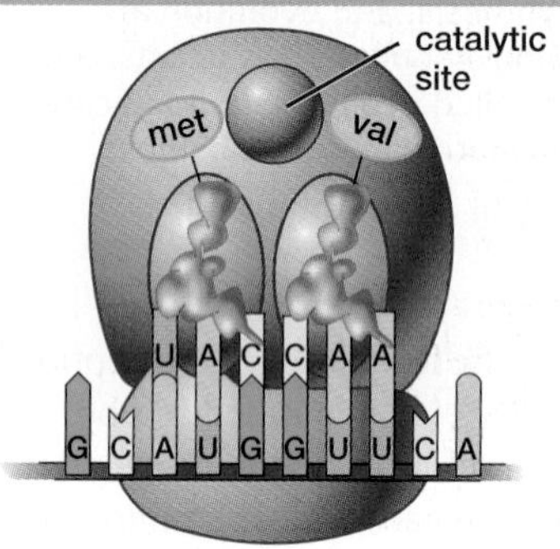

❹ The second codon of mRNA (GUU) base-pairs with the anticodon (CAA) of a second tRNA carrying the amino acid valine (val). This tRNA binds to the second tRNA site on the large subunit.

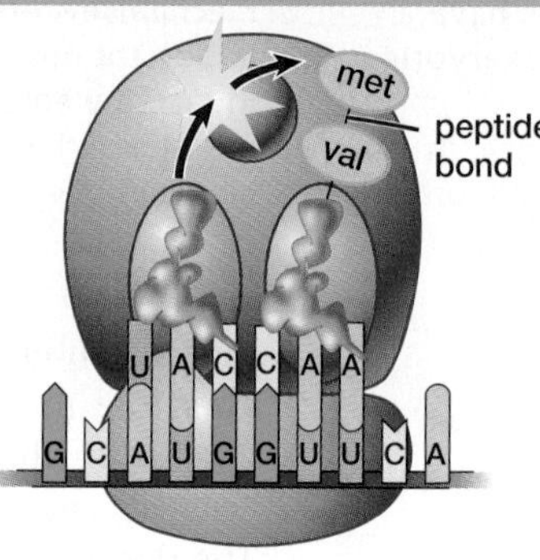

❺ The catalytic site on the large subunit catalyzes the formation of a peptide bond linking the amino acids methionine and valine. The two amino acids are now attached to the tRNA in the second binding site.

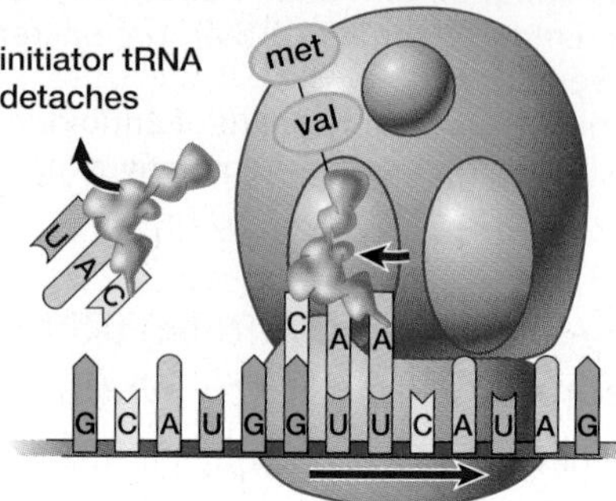

❻ The "empty" tRNA is released and the ribosome moves down the mRNA, one codon to the right. The tRNA that is attached to the two amino acids is now in the first tRNA binding site and the second tRNA binding site is empty.

Termination:

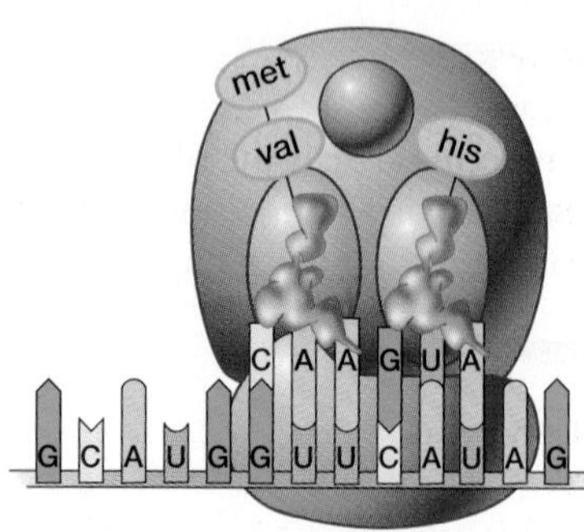

❼ The third codon of mRNA (CAU) base-pairs with the anticodon (GUA) of a tRNA carrying the amino acid histidine (his). This tRNA enters the second tRNA binding site on the large subunit.

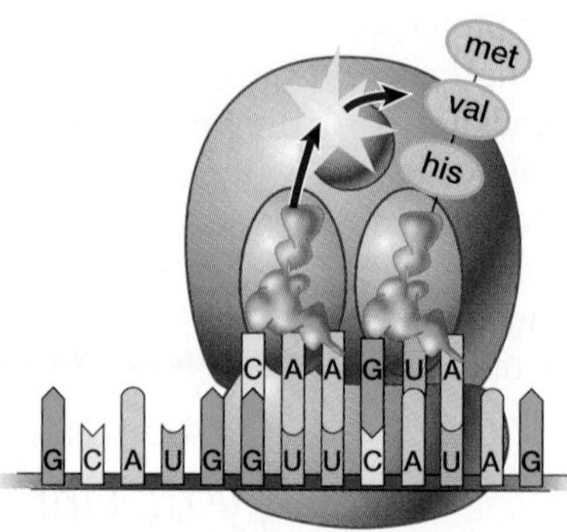

❽ The catalytic site forms a peptide bond between valine and histidine, leaving the peptide attached to the tRNA in the second binding site. The tRNA in the first site leaves, and the ribosome moves one codon over on the mRNA.

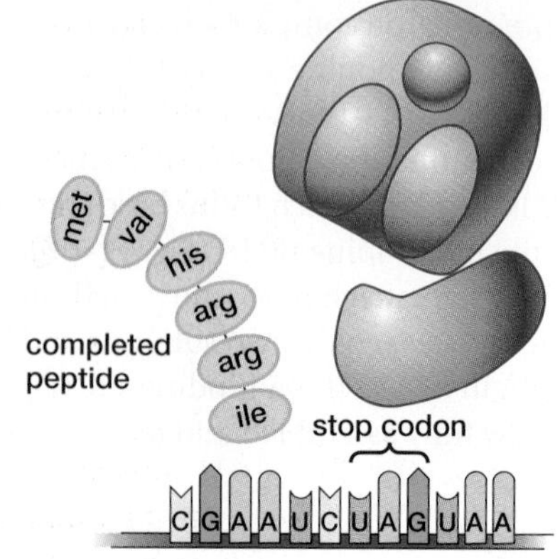

❾ This process repeats until a stop codon is reached; the mRNA and the completed peptide are released from the ribosome, and the subunits separate.

▲ **FIGURE 12-7 Translation is the process of protein synthesis** Translation decodes the base sequence of an mRNA into the amino acid sequence of a protein.

QUESTION Examine step ❾. If mutations changed all of the guanine molecules visible in the mRNA sequence shown here to uracil, how would translated peptide differ from the one shown?

when it encounters a stop codon, forcing the ribosome to release the finished protein chain and the mRNA (**Fig. 12-7** ⑨). The ribosome disassembles into its large and small subunits, which can then be used to translate another mRNA.

BioFlix™ Protein Synthesis

Summing Up **Decoding the Sequence of Bases in DNA into the Sequence of Amino Acids in Protein Requires Transcription and Translation**

Let's summarize how a eukaryotic cell decodes the genetic information stored in its DNA to synthesize a protein (**Fig. 12-8**):

a. With some exceptions, such as the genes for tRNA and rRNA, each gene codes for the amino acid sequence of a protein. The DNA of the gene consists of the template strand, which is transcribed into mRNA, and its complementary strand, which is not transcribed.
b. Transcription of a protein-coding gene produces an mRNA molecule that is complementary to the template strand of the DNA of a gene. Starting from the first AUG, each codon within the mRNA is a sequence of three bases that specifies an amino acid or a "stop."
c. Enzymes in the cytoplasm attach the appropriate amino acid to each tRNA based on the tRNA's anticodon.
d. The mRNA moves out of the nucleus to a ribosome in the cytoplasm. Transfer RNAs carry their attached amino acids to the ribosome. There, the bases in tRNA anticodons bind to the complementary bases in mRNA codons, so the amino acids attached to the tRNAs line up in the sequence specified by the codons. The ribosome joins the amino acids with peptide bonds to form a protein. When a stop codon is reached, the finished protein is released from the ribosome.

This decoding chain, from DNA bases to mRNA codons to tRNA anticodons to amino acids, results in the synthesis of a protein with an amino acid sequence determined by the base sequence of a gene.

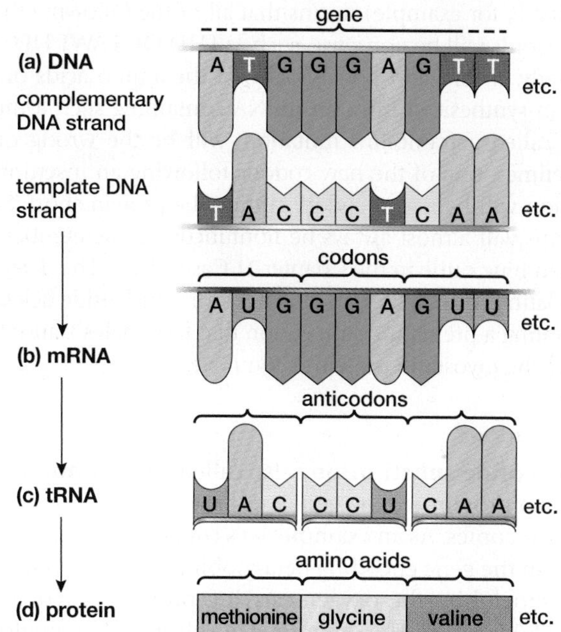

▲ **FIGURE 12-8 Complementary base pairing is critical to the process of decoding genetic information (a)** The DNA of a gene contains two strands; only the template strand is used by RNA polymerase to synthesize an RNA molecule. **(b)** Bases in the template strand of DNA are transcribed into a complementary mRNA. Codons are sequences of three bases that specify an amino acid or a stop during protein synthesis. **(c)** Unless it is a stop codon, each mRNA codon forms base pairs with the anticodon of a tRNA molecule that carries a specific amino acid. **(d)** The amino acids borne by the tRNAs are joined to form a protein.

12.4 HOW DO MUTATIONS AFFECT PROTEIN FUNCTION?

As we saw in Chapter 11, mistakes during DNA replication, ultraviolet rays in sunlight, chemicals in cigarette smoke, and a host of other environmental factors may change the sequence of bases in DNA. These changes are called **mutations.** The consequences for an organism's structure and function depend on how the mutation affects the functioning of the protein encoded by the mutated gene.

Mutations May Have a Variety of Effects on Protein Structure and Function

Most mutations may be categorized as substitutions, deletions, insertions, inversions, or translocations (see pp. 213–214).

Inversions and Translocations

Inversions and **translocations** occur when pieces of DNA—sometimes most or even all of a chromosome—are broken apart and reattached, sometimes within a single chromosome, sometimes to a different chromosome. These mutations may be relatively benign if entire genes, including their promoters, are merely moved from one place to another. However, if a gene is split in two, it will no longer code for a complete, functional protein. For example, almost half the cases of severe hemophilia are caused by an inversion in the gene that encodes a protein required for blood clotting.

Deletions and Insertions

The effects of **deletion mutations** and **insertion mutations** usually depend on how many nucleotides are removed or added, respectively. Why? Think back to the genetic code: Three nucleotides encode a single amino acid. Therefore, adding or deleting three nucleotides will add or delete a single amino acid to the encoded protein. In many cases, this doesn't alter the function of the protein very much. In contrast, deletions and insertions of one or two nucleotides, or any deletion or insertion that isn't a multiple of three nucleotides, can have particularly catastrophic effects, because all of the codons that follow the deletion or insertion will be altered.

Recall our English sentence with all three-letter words: THEDOGSAWTHECAT. Deleting or inserting a letter (deleting the first E, for example) means that all of the following three-letter words will be nonsense, such as THD OGS AWT HEC AT. Similarly, most—and possibly all—of the amino acids of the protein synthesized from an mRNA containing such a mutation, called a frameshift mutation, will be the wrong ones. Sometimes, one of the new codons following an insertion or deletion will be a stop codon, cutting the protein short. Such proteins will almost always be nonfunctional. Remember the Belgian Blue cattle in the Chapter 11 Case Study? The defective myostatin gene of a Belgian Blue has an 11-nucleotide deletion, generating a premature stop codon that terminates translation before the myostatin protein is complete.

Substitutions

A **nucleotide substitution** (also called a **point mutation**) within a protein-coding gene can produce one of four possible outcomes. As an example, let's consider mutations that occur in the gene encoding beta-globin, one of the subunits of hemoglobin, the oxygen-carrying protein in red blood cells (**Table 12-4**). The other type of subunit in hemoglobin is called alpha-globin; a normal hemoglobin molecule consists of two alpha and two beta subunits. In all but the last example, we will consider the results of mutations that occur in the sixth codon of the beta-globin gene (CTC in DNA, GAG in mRNA), which specifies glutamic acid—a charged, hydrophilic, water-soluble amino acid.

- **The protein may be unchanged.** Recall that most amino acids can be encoded by several different codons. If a mutation changes the beta-globin DNA base sequence from CTC to CTT, this sequence still codes for glutamic acid. Therefore, the protein synthesized from the mutated gene remains the same.
- **The new protein may be functionally equivalent to the original one.** Many proteins have regions whose exact amino acid sequence is relatively unimportant. In beta-globin, the amino acids on the outside of the protein must be hydrophilic to keep the protein dissolved in the cytoplasm of red blood cells. Exactly *which* hydrophilic amino acids are on the outside doesn't matter much. For example, a family in the Japanese town of Machida was found to contain a mutation from CTC to GTC, replacing glutamic acid (hydrophilic) with glutamine (also hydrophilic). Hemoglobin containing this mutant beta-globin protein—known as Hemoglobin Machida—functions well. Mutations such as the ones in Hemoglobin Machida and in the previous example are called **neutral mutations** because they do not detectably change the function of the encoded protein.
- **Protein function may be changed by an altered amino acid sequence.** A mutation from CTC to CAC replaces glutamic acid (hydrophilic) with valine (hydrophobic). This substitution is the genetic defect that causes sickle-cell anemia (see see pp. 190–191). The valines on the outside of the hemoglobin molecules cause them to clump together, distorting the shape of the red blood cells. These changes can cause serious illness.
- **Protein function may be destroyed by a premature stop codon.** A particularly catastrophic mutation occasionally occurs in the 17th codon of the beta-globin gene (TTC in DNA, AAG in mRNA). This codon specifies the amino acid lysine. A mutation from TTC to ATC (UAG in mRNA) results in a stop codon, halting translation of beta-globin mRNA before the protein is completed. People who inherit this mutant gene from both their mother and their father do not synthesize any functional beta-globin protein; they manufacture hemoglobin consisting entirely of alpha-globin subunits. This "pure alpha" hemoglobin does not bind oxygen very well. This condition, called beta-thalassemia, can be fatal unless treated with regular blood transfusions throughout life.

Mutations Provide the Raw Material for Evolution

Mutations in gametes (sperm or eggs) can be passed on to future generations. In humans, mutation rates range from about 1 in every 100,000 gametes to 1 in 1,000,000 gametes. For reference, a man releases about 300 to 400 million sperm per ejaculation; each ejaculate contains about 600 sperm with new mutations. Although most mutations are neutral or potentially harmful, mutations are essential for evolution, because these random changes in DNA sequence are the ultimate source of

Table 12-4 Effects of Mutations in the Hemoglobin Gene

	DNA (Template Strand)	mRNA	Amino Acid	Properties of Amino Acid	Functional Effect on Protein	Disease
Original codon 6	CTC	GAG	Glutamic acid	Hydrophilic	Normal protein function	None
Mutation 1	CT**T**	GA**A**	Glutamic acid	Hydrophilic	Neutral; normal protein function	None
Mutation 2	**G**TC	**C**AG	Glutamine	Hydrophilic	Neutral; normal protein function	None
Mutation 3	C**A**C	G**U**G	Valine	Hydrophobic	Loses water solubility; compromises protein function	Sickle-cell anemia
Original codon 17	TTC	AAG	Lysine	Hydrophilic	Normal protein function	None
Mutation 4	**A**TC	**U**AG	Stop codon	Ends translation after amino acid 16	Synthesizes only part of the protein; eliminates protein function	Beta-thalassemia

Case Study continued
Cystic Fibrosis

Why hasn't natural selection eliminated mutated *CFTR* alleles? Perhaps because the mutated alleles provide protection against cholera and typhoid. The normal CFTR protein is activated by cholera toxin, causing excessive chloride secretion from intestinal cells. Water follows by osmosis, and so cholera victims have debilitating, and often fatal, diarrhea. Defective CFTR proteins cannot be activated by cholera toxin. The CFTR protein is also the site by which typhoid bacteria enter cells, but typhoid bacteria cannot enter via mutated CFTR proteins. Such protection cannot make up for the devastating effects of cystic fibrosis. However, heterozygotes, with one normal and one mutated *CFTR* allele, have nearly normal CFTR function, but might be less severely affected by cholera and typhoid. This "heterozygote advantage" may explain the high frequency of mutated *CFTR* alleles (about 4% of people of European descent carry mutated *CFTR* alleles), as it does for the sickle-cell anemia allele (see p. 190).

all genetic variation. New base sequences undergo natural selection as organisms compete to survive and reproduce. Occasionally, a mutation proves beneficial in an organism's interactions with its environment. Through reproduction over time, the mutant base sequence may spread throughout the population, as organisms possessing it outcompete and outreproduce rivals bearing the original base sequence. This process is described in detail in Unit 3.

12.5 HOW ARE GENES REGULATED?

The complete human genome contains 20,000 to 25,000 genes. Each of these genes is present in most of your body cells, but any individual cell expresses (transcribes and, if the gene product is a protein, translates) only a small fraction of them. Some genes are expressed in all cells, because they encode proteins or RNA molecules that are essential for the life of any cell. For example, all cells need to synthesize proteins, so they all transcribe tRNA genes, rRNA genes, and genes for ribosomal proteins. Other genes are expressed exclusively in certain types of cells, at certain times in an organism's life, or under specific environmental conditions. For example, even though every cell in your body contains the gene for casein, the major protein in milk, that gene is expressed only in mature women, only in certain cells in the breast, and only when a woman is breast-feeding.

Regulation of gene expression may occur at the level of transcription (which genes are used to make mRNA in a given cell), translation (how much protein is made from a particular type of mRNA), or protein activity (how long the protein lasts in a cell and how rapidly protein enzymes catalyze specific reactions). Although these general principles apply to both prokaryotic and eukaryotic organisms, there are some differences as well, as we describe below.

Gene Regulation in Prokaryotes

Prokaryotic DNA is often organized in packages called **operons,** in which the genes for related functions lie close to one another (**Fig. 12-9a**). An operon consists of four parts: (1) a **regulatory gene,** which controls the timing

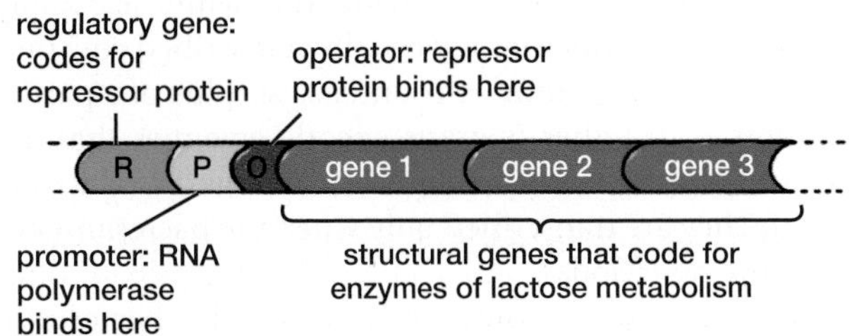

(a) Structure of the lactose operon

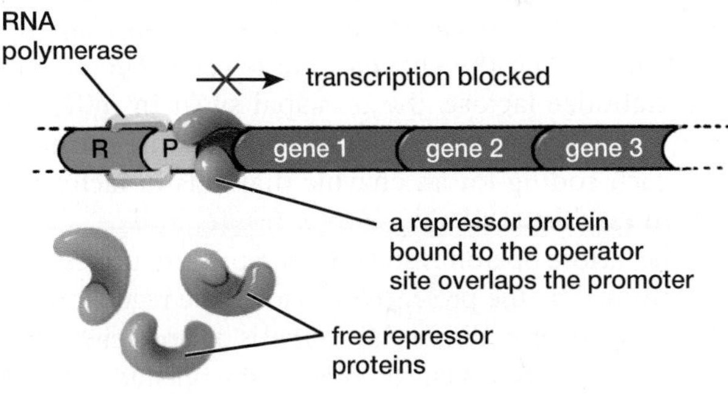

(b) Lactose absent

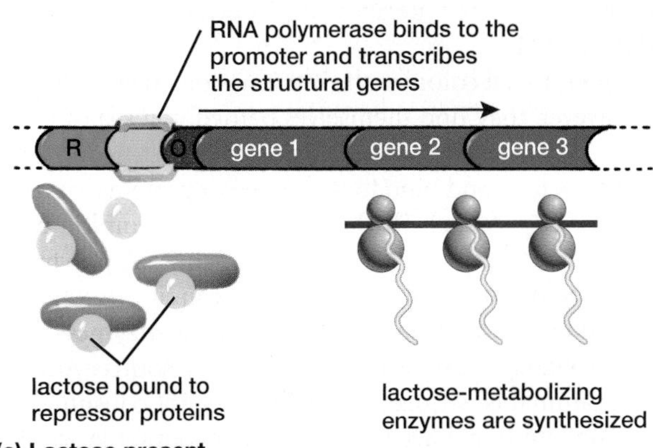

(c) Lactose present

▲ **FIGURE 12-9 Regulation of the lactose operon (a)** The lactose operon consists of a regulatory gene, a promoter, an operator, and three structural genes that code for enzymes necessary for lactose metabolism. **(b)** In the absence of lactose, repressor proteins bind to the operator of the lactose operon. RNA polymerase can bind to the promoter but cannot move past the repressor protein to transcribe the structural genes. **(c)** When lactose is present, it binds to the repressor proteins, making the repressor proteins unable to bind to the operator. RNA polymerase binds to the promoter, moves past the unoccupied operator, and transcribes the structural genes.

or rate of transcription of other genes; (2) a promoter, which RNA polymerase recognizes as the place to start transcribing; (3) an **operator,** which governs access of RNA polymerase to the promoter or to the (4) **structural genes,** which actually encode the related enzymes or other proteins. Whole operons are regulated as units, so that functionally related proteins are synthesized simultaneously when the need arises.

Prokaryotic operons may be regulated in a variety of ways, depending on the functions that they control. Some operons encode enzymes that are needed by the cell just about all the time, such as the enzymes that synthesize many amino acids. These operons are usually transcribed continuously, unless the bacterium encounters a surplus of a particular amino acid. Other operons encode enzymes that are needed only occasionally, for instance, to digest a relatively rare food. They are transcribed only when the bacterium encounters the rare food.

Consider the common intestinal bacterium, *Escherichia coli (E. coli)*. This bacterium must live on whatever types of nutrients its host eats, and it can synthesize many different enzymes to metabolize a wide variety of foods. The genes that code for these enzymes are transcribed only when the enzymes are needed. The enzymes that metabolize lactose, the principal sugar in milk, are a case in point. The **lactose operon** contains three structural genes, each coding for an enzyme that aids in lactose metabolism (see Fig. 12-9a).

The lactose operon is shut off, or repressed, unless specifically activated by the presence of lactose. The regulatory gene of the lactose operon directs the synthesis of a protein, called a **repressor protein,** which can bind to the operator site. RNA polymerase, although still able to bind to the promoter, cannot get past the repressor protein to transcribe the structural genes. Consequently, lactose-metabolizing enzymes are not synthesized (**Fig. 12-9b**).

When *E. coli* colonize the intestines of a newborn mammal, however, they find themselves bathed in a sea of lactose whenever their host nurses from its mother. Lactose molecules enter the bacteria and bind to the repressor proteins, changing their shape (**Fig. 12-9c**). The lactose-repressor complex cannot attach to the operator site. Therefore, when RNA polymerase binds to the promoter of the lactose operon, it can transcribe the structural genes for lactose-metabolizing enzymes, allowing the bacteria to use lactose as an energy source. After the young mammal is weaned, it usually does not consume milk again. The intestinal bacteria no longer encounter lactose, the repressor proteins are free to bind to the operator, and the genes for lactose metabolism are shut down.

Gene Regulation in Eukaryotes

Eukaryotic gene regulation is similar to prokaryotic regulation in some respects. In both, not all genes are transcribed and translated all the time. Further, controlling the rate of transcription is an important mechanism of gene regulation in both. However, the compartmentalization of the DNA in a membrane-bound nucleus, the variety of cell types in multicellular eukaryotes, a very different organization of the genome, and complex processing of RNA transcripts all distinguish gene regulation in eukaryotes from regulation in prokaryotes.

Expression of genetic information by a eukaryotic cell is a multistep process, beginning with transcription of DNA and commonly ending with a protein that performs a particular function. Regulation of gene expression can occur at any of these steps, which are illustrated in **Figure 12-10**:

1. **Cells can control the frequency at which an individual gene is transcribed.** The rate at which cells transcribe specific genes depends on the demand for the protein or RNA product that they encode. Gene transcription differs between organisms, between cell types in a given organism, and within a given cell at different stages in the organism's life or when it is stimulated by different environmental conditions (see the following section).
2. **The same gene may be used to produce different mRNAs and protein products.** As we described in section 12.3, the same gene may be used to produce several different protein products, depending on how the pre-mRNA is spliced to form the finished mRNA that will be translated on the ribosomes.
3. **Cells may control the stability and translation of messenger RNAs.** Some mRNAs are long lasting and are translated into protein many times. Others are translated only a few times before they are degraded. Recently, molecular biologists have discovered that small "regulatory RNA" molecules may block translation of some mRNAs, or may even target some mRNAs for destruction (see "Scientific Inquiry: RNA—It's Not Just a Messenger Anymore" on p. 234).
4. **Proteins may require modification before they can carry out their functions.** Many proteins must be modified before they become active. For instance, the protein-digesting enzymes produced by cells in your stomach wall and pancreas are initially synthesized in an inactive form, which prevents the enzymes from digesting the cells that produce them. After these inactive forms are secreted into the digestive tract, portions of the enzymes are snipped out to unveil the enzyme's active site, allowing the enzymes to digest the proteins in food (their substrate) into smaller proteins or even into individual amino acids (the resulting products). Other modifications, such as adding or removing phosphate groups, can temporarily activate or inactivate a protein, allowing second-to-second control of the protein's activity. Similar regulation of protein structure and function occurs in prokaryotic cells.
5. **Cells can control the rate at which proteins are degraded.** By preventing or promoting a protein's degradation, a cell can rapidly adjust the amount of a particular protein it contains. Protein degradation may also be regulated in prokaryotic cells.

◀ **FIGURE 12-10 An overview of information flow in a eukaryotic cell**

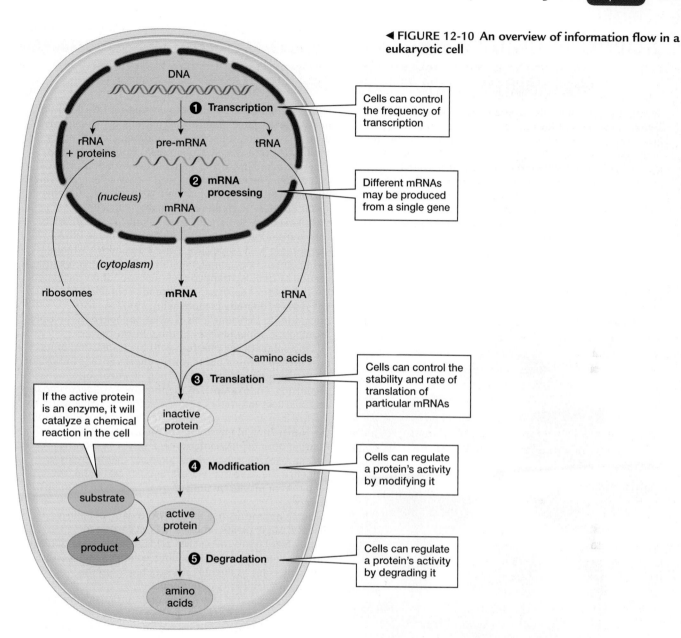

Eukaryotic Cells May Regulate the Transcription of Individual Genes, Regions of Chromosomes, or Entire Chromosomes

In eukaryotic cells, transcriptional regulation can operate on at least three levels: the individual gene, regions of chromosomes, or entire chromosomes.

Regulatory Proteins That Bind to a Gene's Promoter Alter Its Rate of Transcription

The promoter regions of virtually all genes contain several different transcription factor binding sites, or response elements. Therefore, whether these genes are transcribed depends on which transcription factors are synthesized by the cell and whether those transcription factors are active. For example, when cells are exposed to free radicals (see pp. 24–25), a transcription factor binds to antioxidant

Have you ever wondered 

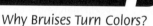

Why Bruises Turn Colors?

Bruises typically progress from purple to green to yellow. This sequence is visual evidence of the control of gene expression. If you bang your shin on a chair, blood vessels break and release red blood cells, which burst and spill their hemoglobin. Hemoglobin and its iron-containing heme group are dark bluish-purple in the deoxygenated state, so fresh bruises are purple. Heme is toxic to the liver, kidneys, brain, and blood vessels. However, heme stimulates transcription of the heme oxygenase gene. Heme oxygenase converts heme to biliverdin, which is green. A second enzyme, which is always present because its gene is always expressed, converts biliverdin to bilirubin, which is yellow. Eventually, bilirubin moves to the liver, where it is excreted into the bile. You can follow the detoxification of heme by watching your bruise change color.

Scientific Inquiry

RNA—It's Not Just a Messenger Anymore

In recent years, molecular biologists have discovered an entirely new class of RNA molecules, called "regulatory RNA." Various types of regulatory RNA have many different functions. Here, we will focus on only one function, called RNA interference. RNA interference was discovered, accidentally, in petunias. Trying to make petunias with more vibrant flower colors, researchers inserted extra genes for flower pigment production into petunias. Much to their surprise, many of the petunias produced variegated flowers (**Fig. E12-2**). Why?

As you know, messenger RNA is transcribed from DNA and translated into protein, which then performs cellular functions such as catalyzing biochemical reactions. How much protein is synthesized depends both on how much mRNA is made and on how rapidly and for how long the mRNA is translated. Enter RNA interference. Organisms as diverse as plants, roundworms, and humans transcribe small RNA strands, called microRNA, from hundreds of different microRNA genes. MicroRNAs are not translated into protein. Instead, after processing by cellular enzymes, microRNAs give rise to very short RNA strands, usually about 20 to 25 nucleotides in length, which are complementary to parts of specific mRNAs and interfere with translation of the mRNA (hence, the name "RNA interference"). In some cases, these small RNA strands base-pair with the complementary mRNA, forming a little section of double-stranded RNA that cannot be translated. In other cases, the short RNA strands combine with enzymes to degrade complementary mRNA, also preventing translation. This is what happened in the cells forming the white patches of the petunia flowers.

▲ **FIGURE E12-2 RNA interference in petunias** The white patches on this petunia flower consist of cells in which RNA interference blocked production of the enzymes that produce purple pigment. (The variegated petunias sold in your local greenhouse are made by a different process.)

Why should a cell interfere with the translation of its own mRNA? It turns out that RNA interference is important in the development of many, if not all, eukaryotic organisms. For example, in the roundworm *Caenorhabditis elegans*, RNA interference is crucial for a juvenile worm to mature into an adult. (Andrew Fire and Craig Mello shared the Nobel Prize in Physiology or Medicine in 2006 for figuring out how RNA interference works in roundworms.) Recent research in mammals suggests that microRNAs influence the development of the heart and brain, the secretion of insulin by the pancreas, and even learning and memory.

Some organisms use microRNA to defend against disease. Many plants produce microRNA that is complementary to the nucleic acids (usually RNA) of plant viruses. When the microRNA finds complementary viral RNA molecules, it directs enzymes to cut up the viral RNA, thereby preventing viral reproduction.

RNA interference also shows great promise in medicine. Clinical trials are under way testing the safety and efficacy of RNA interference in diseases as diverse as macular (retinal) degeneration, hepatitis B, and a respiratory virus that is the most common cause of pneumonia in infants. Basic research in animals suggests that RNA interference may also be used to treat prostate cancer and defects in cholesterol metabolism. Some experts expect RNA interference to reach patients by 2010.

response elements in the promoters of several genes. As a result, the cell produces enzymes that break down free radicals to harmless substances.

Many transcription factors require activation before they can affect gene transcription. One of the best-known examples is the role that the sex hormone, estrogen, plays in controlling egg production in birds. The gene for albumin, the major protein in egg white, is not transcribed in the winter when birds are not breeding and estrogen levels are low. During the breeding season, the ovaries in female birds release estrogen, which enters cells in the oviduct and binds to a transcription factor (usually called an estrogen receptor). The estrogen–receptor complex then attaches to an estrogen response element in the promoter of the albumin gene, making it easier for RNA polymerase to bind to the promoter and initiate transcription of mRNA. The mRNA is then translated into large amounts of albumin. Similar activation of gene transcription by steroid hormones occurs in other animals, including humans. The importance of hormonal regulation of transcription during development is illustrated by genetic defects in which receptors for sex hormones are nonfunctional (see "Health Watch: Sex, Aging, and Mutations" on pp. 236–237).

Some Regions of Chromosomes Are Condensed and Not Normally Transcribed

Certain parts of eukaryotic chromosomes are in a highly condensed, compact state in which most of the DNA seems to be inaccessible to RNA polymerase. Some of these regions are structural parts of chromosomes that don't contain genes. Other tightly condensed regions contain functional genes that are not currently being transcribed. When the product of a gene is needed, the portion of the chromosome containing that gene becomes "decondensed"—loosened so that the DNA is accessible to RNA polymerase and transcription can occur.

Large Parts of Chromosomes May Be Inactivated, Preventing Transcription

In some cases, most of a chromosome may be condensed, making it inaccessible to RNA polymerase. An example occurs in the sex chromosomes of female mammals. Male mammals usually have an X and a Y chromosome (XY), and females usually have two X chromosomes (XX). As a consequence, females have the capacity to synthesize mRNA from genes on their two X chromosomes, whereas males, with only one X chromosome, may produce only half as much. In 1961, the geneticist Mary Lyon hypothesized that one of the two X chromosomes in females was inactivated in some way, so that its genes were not expressed. We now know that about 85% of the genes on an inactivated X chromosome are not transcribed. Early in development (about the 16th day in humans), one X chromosome in each of a female's cells begins to produce large amounts of a regulatory RNA molecule called Xist, which coats most of the chromosome, condenses it into a tight mass, and prevents transcription. The condensed X chromosome, called a **Barr body** after its discoverer, Murray Barr, appears as a discrete spot in the nuclei of the cells of female mammals (**Fig. 12-11**).

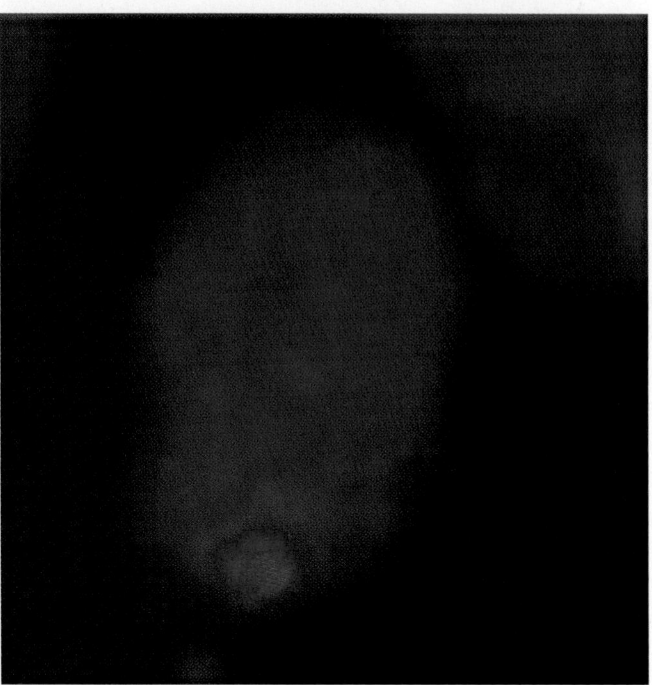

▲ **FIGURE 12-11 A Barr body** The red spot at the bottom of the nucleus is an inactivated X chromosome called a Barr body. In this fluorescence micrograph, the Barr body is stained with a dye that binds to the Xist RNA coating the inactivated X chromosome.

Usually, large clusters of cells (with each cluster descended from a single cell of the very early embryo) have the same X chromosome inactivated. As a result, the bodies of female mammals (including women) are composed of patches of cells in which one of the X chromosomes is fully active, and patches of cells in which the other X chromosome is active. The results of this phenomenon are easily observed in calico cats (**Fig. 12-12**). The X chromosome of a cat contains a gene encoding an enzyme that produces fur pigment. There are two common alleles of this gene: One produces orange fur and the other produces black fur. If one X chromosome in a female cat has the orange allele and the other X chromosome has the black allele, the cat will have patches of orange and black fur. These patches represent areas of skin that developed from cells in the early embryo in which different X chromosomes were inactivated. Calico coloring is almost exclusively found in female cats. Because male cats normally have only one X chromosome, which is active in all of their cells, male cats usually have black fur or orange fur, but not both.

▲ **FIGURE 12-12 Inactivation of the X chromosome regulates gene expression** This female calico kitten carries a gene for orange fur on one X chromosome and a gene for black fur on her other X chromosome. Inactivation of different X chromosomes produces the black and orange patches. The white color is due to an entirely different gene that prevents pigment formation altogether.

Health Watch

Sex, Aging, and Mutations

Sometime in her early to mid-teens, a girl usually goes through puberty: Her breasts swell, her hips widen, and she begins to menstruate. In rare instances, however, a girl may develop all of the outward signs of womanhood, but does not menstruate. Eventually, she reports this symptom to her physician, who may take a blood sample to do a chromosome test. In some cases, the chromosome test gives what might seem to be an impossible result: the girl's sex chromosomes are XY, a combination that would usually give rise to a boy. The reason she has not begun to menstruate is that she lacks ovaries and a uterus, but instead has testes inside her abdominal cavity. She has about the same concentrations of androgens (male sex hormones, such as testosterone) circulating in her blood as would be found in a boy her age. In fact, androgens have been present since early in her development. However, her cells cannot respond to them—a condition called androgen insensitivity (**Fig. E12-3**).

The affected gene codes for a protein known as an androgen receptor. In normal males, androgens bind to the receptor molecules. The hormone–receptor combination stimulates the transcription of genes that help to produce many male features, including the formation of a penis and the descent of the testes into sacs outside the body cavity. There are more than 200 mutant alleles of the androgen receptor gene. The most serious are mutations that create a premature stop codon. As you know, these mutations are likely to have catastrophic effects on protein structure and function. Because the androgen receptor gene is on the X chromosome, a person who is genetically male (XY) inherits a single allele for the androgen receptor. If this allele is seriously defective, the person will not synthesize functional androgen receptor proteins. The person's cells will be unable to respond to testosterone, and male characteristics will not develop. Thus, a mutation that changes the nucleotide sequence of a single gene, causing a single type of defective protein to be produced, can cause a person who is genetically male to look like and perceive herself to be a woman (see **Fig. E12-3**).

A second type of mutation provides clues to why people age. Why will your hair whiten, skin wrinkle,

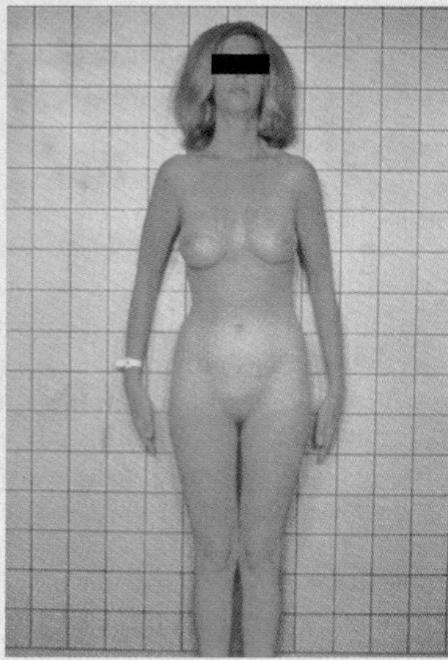

▲ **FIGURE E12-3 Androgen insensitivity leads to female features** This individual has an X and a Y chromosome. She has testes that produce testosterone, but a mutation in her androgen receptor genes makes her cells unable to respond to testosterone, resulting in her female appearance.

Case Study revisited
Cystic Fibrosis

Researchers have identified more than 1,200 defective alleles of the *CFTR* gene, all of which are recessive to the functional *CFTR* allele. People who are heterozygous, with one normal *CFTR* allele and one copy of any of the defective alleles, produce enough functional *CFTR* protein for adequate chloride transport. Therefore, they are phenotypically normal—that is, they produce watery secretions in their lungs and do not develop cystic fibrosis. Someone with two defective alleles will have proteins that don't work properly and will develop the disease.

Genetic diseases such as cystic fibrosis can't be "cured" in the way that an infection can be cured by killing offending bacteria or viruses. Typically, genetic diseases are treated by replacing the lost function, such as giving insulin to diabetics, or by relieving the symptoms. In cystic fibrosis, the most common treatments relieve some of the symptoms. These include antibiotics, medicines that open the airways, and physical therapy to drain the lungs. In the next chapter, we will see that biotechnology offers some hope of replacing the lost function by delivering functioning *CFTR* genes to the cells lining the airways.

What ultimately happens to a person with cystic fibrosis depends on how defective the mutant alleles are. Canadian triathlete Lisa Bentley, for example, has a relatively mild case of cystic fibrosis (**Fig. 12-13**). However, during a 9-hour triathlon, she produces copious amounts of very salty sweat. It is a constant challenge for Bentley to keep her body supplied with salt during the race. Nevertheless, Bentley has won 11 triathlons, including the Canadian Ironman Triathlon in 2007 and the Australian Ironman Triathlon five straight years, from 2002 through 2006. Bentley carefully controls her diet, especially her salt intake. Vigorous exercise helps to clear out her lungs. She also scrupulously avoids situations where she might be exposed to contagious diseases. Her cystic fibrosis hasn't kept her from becoming one of the finest athletes in the world. Alice Martineau, on the other hand, had extremely defective alleles. By the age of 30, Martineau needed a triple transplant—heart, lung, and liver. She died before a suitable donor could be found.

joints ache, and eyes cloud as you become elderly? A small number of people inherit Werner syndrome, which causes a type of premature aging (**Fig. E12-4**). Most people with Werner syndrome have inherited two defective recessive alleles for a gene that codes for a protein involved in DNA replication and repair. If cells cannot replicate DNA accurately and repair errors in DNA, mutations will accumulate rapidly throughout the body. This is what happens in people with Werner syndrome, who often die of aging-related conditions while still in their 40s or 50s.

The fact that an overall increase in mutations caused by defective replication and repair enzymes produces symptoms of old age provides support for one hypothesis to explain many of the symptoms of normal aging: During a reasonably long, 80-year life span, mutations gradually accumulate because of environmentally induced DNA damage and mistakes in DNA replication. Eventually, these mutations interfere with nearly every aspect of body functioning and contribute to death from "old age." In Werner syndrome, this process occurs much more rapidly. Disorders such as androgen insensitivity and Werner syndrome provide insights into the impact of mutations, the function of specific genes and their protein products, the ways hormones regulate gene transcription, and even the causes of aging.

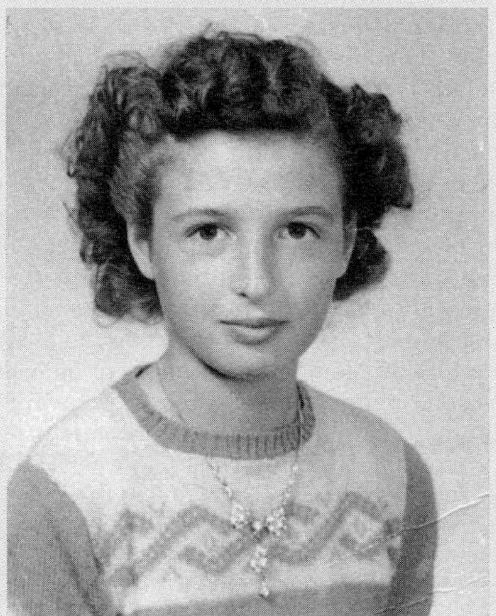

(a) Werner's syndrome patient at age 13

(b) At age 56

◀ FIGURE E12-4 **Werner syndrome** This condition is the result of a mutation that interferes with proper DNA replication and repair, increasing the incidence of mutations throughout the body.

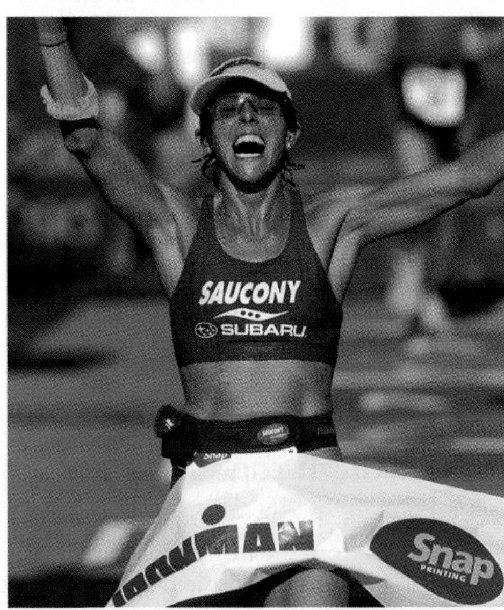

▲ FIGURE 12-13 **Lisa Bentley, sometimes called the Iron Queen, wins another triathlon**

BioEthics Consider This

The vast majority of biomedical research is funded by government agencies, such as the National Institutes of Health (NIH). The NIH funds research on genetic disorders such as cystic fibrosis and muscular dystrophy; infectious diseases such as AIDS and tuberculosis; and a host of other conditions, including heart disease and cancer. The NIH also funds research on diseases, such as malaria, that are uncommon in the United States but claim hundreds of thousands of lives each year in impoverished countries. How do you think that NIH funds should be spent—in proportion to disease severity and incidence in the United States, which might mean that almost all NIH funds should be spent on heart disease and cancer; or according to the prospective life span of the victims, so that diseases of the young, such as cystic fibrosis, receive much more? Does the United States have a responsibility to help people in other countries suffering from diseases that are rare in America?

CHAPTER REVIEW

Summary of Key Concepts

12.1 How Is the Information in DNA Used in a Cell?

Genes are segments of DNA that can be transcribed into RNA and, for most genes, translated into protein. Transcription produces the three types of RNA needed for translation: messenger RNA (mRNA), transfer RNA (tRNA), and ribosomal RNA (rRNA). Messenger RNA carries the genetic information of a gene from the nucleus to the cytoplasm, where ribosomes use the information to synthesize a protein. Ribosomes contain rRNA and proteins organized into large and small subunits. There are many different tRNAs. Each tRNA binds a specific amino acid and carries it to a ribosome for incorporation into a protein. The genetic code consists of codons, sequences of three bases in mRNA that specify either an amino acid in the protein chain or the end of protein synthesis (stop codons).

12.2 How Is the Information in a Gene Transcribed into RNA?

Within an individual cell, only certain genes are transcribed. When the cell requires the product of a gene, RNA polymerase binds to the promoter region of the gene and synthesizes a single strand of RNA. This RNA is complementary to the template strand in the gene's DNA double helix. Cellular proteins called transcription factors may bind to DNA near the promoter and enhance or suppress transcription of a given gene.

12.3 How Is the Base Sequence of Messenger RNA Translated into Protein?

In prokaryotic cells, all of the nucleotides of a protein-coding gene code for amino acids; therefore, the RNA transcribed from the gene is the mRNA that will be translated on a ribosome. In eukaryotic cells, protein-coding genes consist of two parts: exons, which code for amino acids in a protein, and introns, which do not. The entire gene, including both introns and exons, is transcribed into a pre-mRNA molecule. Therefore, the introns of the pre-mRNA must be cut out and the exons spliced together to produce a finished mRNA.

In eukaryotes, mRNA carries the genetic information from the nucleus to the cytoplasm, where ribosomes can use this information to synthesize a protein. Ribosomes contain rRNA and proteins organized into large and small subunits. These subunits come together at the first AUG (start) codon of the mRNA molecule to form the complete protein-synthesizing machine. Transfer RNAs deliver the appropriate amino acids to the ribosome for incorporation into the growing protein. Which tRNA binds and, consequently, which amino acid is delivered depend on base pairing between the anticodon of the tRNA and the codon of the mRNA. Two tRNAs, each carrying an amino acid, bind simultaneously to the ribosome; the large subunit catalyzes the formation of peptide bonds between the amino acids. As each new amino acid is attached, one tRNA detaches, and the ribosome moves over one codon, binding to another tRNA that carries the next amino acid specified by mRNA. Addition of amino acids to the growing protein continues until a stop codon is reached, causing the ribosome to disassemble and to release both the mRNA and the newly formed protein.

BioFlix ™ Protein Synthesis

12.4 How Do Mutations Affect Protein Function?

A mutation is a change in the nucleotide sequence of a gene. Mutations can be caused by mistakes in base pairing during replication, by chemical agents, and by environmental factors such as radiation. Common types of mutations include inversions, translocations, deletions, insertions, and substitutions (point mutations). Mutations may be neutral or harmful, but in rare cases a mutation will promote better adaptation to the environment and thus will be favored by natural selection.

12.5 How Are Genes Regulated?

The expression of a gene requires that it be transcribed and translated, and the resulting protein must perform some action within the cell. Which genes are expressed in a cell at any given time is regulated by the function of the cell, the developmental stage of the organism, and the environment. Control of gene regulation can occur at many steps. The amount of mRNA synthesized from a particular gene can be regulated by increasing or decreasing the rate of its transcription, as well as by changing the stability of the mRNA itself. Rates of translation of mRNAs can also be regulated. Regulation of transcription and translation affects how many protein molecules are produced from a particular gene. Even after they are synthesized, many proteins must be modified before they can function. Proteins also vary in how rapidly they are degraded in a cell. In addition to regulation of individual genes, cells can regulate transcription of groups of genes. For example, entire chromosomes or parts of chromosomes may be condensed and inaccessible to RNA polymerase, whereas other portions may be expanded, allowing transcription to occur.

Key Terms

anticodon *223*
Barr body *235*
codon *223*
deletion mutation *229*
exon *226*
genetic code *222*
insertion mutation *229*
intron *226*
inversion *229*
lactose operon *232*
messenger RNA (mRNA) *218*
mutation *229*
neutral mutation *230*
nucleotide substitution *230*
operator *232*
operon *231*
point mutation *230*
promoter *223*
regulatory gene *231*
repressor protein *232*
ribonucleic acid (RNA) *218*
ribosomal RNA (rRNA) *218*
ribosome *220*
RNA polymerase *223*
start codon *223*
stop codon *223*
structural gene *232*
template strand *224*
transcription *221*
transfer RNA (tRNA) *218*
translation *222*
translocation *229*

Thinking Through the Concepts

Fill-in-the-Blank

1. Synthesis of RNA from the instructions in DNA is called __________. Synthesis of a protein from the instructions in messenger RNA is called __________. Which structure in the cell is the site of protein synthesis? __________

2. The three types of RNA that are essential for protein synthesis are__________, __________, and__________. Another type of RNA, which can interfere with translation, is called __________.
3. The genetic code uses __________ (how many?) bases to code for a single amino acid. This short sequence of bases in messenger RNA is called a(n) __________. The complementary sequence of bases in transfer RNA is called a(n) __________.
4. The enzyme __________ synthesizes RNA from the instructions in DNA. DNA has two strands, but for any given gene, usually only one strand, called the __________ strand, is transcribed. To begin transcribing a gene, this enzyme binds to a specific sequence of DNA bases located at the beginning of the gene. This DNA sequence is called the __________. Transcription ends when the enzyme encounters a DNA sequence at the end of the gene called the __________.
5. Protein synthesis begins when messenger RNA binds to a ribosome. Translation begins with the __________ codon of messenger RNA and continues until a(n) __________ codon is reached. Individual amino acids are brought to the ribosome by __________ RNA. These amino acids are linked into protein by __________ bonds.
6. There are several different types of mutations in DNA. If one nucleotide is substituted for another, this is called a(n) __________ mutation. __________ mutations occur if nucleotides are added in the middle of a gene. __________ mutations occur if nucleotides are removed from the middle of a gene.

Review Questions

1. How does RNA differ from DNA?
2. What are the three types of RNA that are essential to protein synthesis? What is the function of each?
3. Define the following terms: *genetic code*, *codon*, and *anticodon*. What is the relationship among the bases in DNA, the codons of mRNA, and the anticodons of tRNA?
4. How is mRNA formed from a eukaryotic gene?
5. Diagram and describe protein synthesis.
6. Explain how complementary base pairing is involved in both transcription and translation.
7. Describe some mechanisms of gene regulation.
8. Define *mutation*. Are most mutations likely to be beneficial or harmful? Explain your answer.

Applying the Concepts

1. **BioEthics** As you have learned in this chapter, many factors influence gene expression, including hormones. The use of anabolic steroids and growth hormones among athletes has created controversy in recent years. Hormones certainly affect gene expression, but, in the broadest sense, so do vitamins and foods. What do you think are appropriate guidelines for the use of hormones? Should athletes take steroids or growth hormones? Should children at risk of being unusually short be given growth hormones? Should parents be allowed to request growth hormones for a child of normal height in the hope of producing a future basketball player?
2. About 40 years ago, some researchers reported that they could transfer learning from one animal (a flatworm) to another by feeding trained animals to untrained animals. Further, they claimed that RNA was the active molecule of learning. Given your knowledge of the roles of RNA and protein in cells, do you think that a *specific* memory (for example, remembering the base sequences of codons of the genetic code) could be encoded by a *specific* molecule of RNA and that this RNA molecule could transfer that memory to another person? In other words, in the future, could you learn biology by popping an RNA pill? If so, how would this work? If not, can you propose a reasonable hypothesis for the results with flatworms? How would you test your hypothesis?
3. Androgen insensitivity is inherited as a simple recessive trait, because one copy of the normal androgen receptor allele produces sufficient amounts of androgen receptors. Given this information and your knowledge of the chromosomal basis of inheritance, can androgen insensitivity be inherited, or must it arise as a new mutation each time it occurs? If it can be inherited, would inheritance be through the mother or through the father? Why?

MB *Go to www.masteringbiology.com for practice quizzes, activities, eText, videos, current events, and more.*

Biotechnology

Chapter 13

▲ DNA profiling proved that Dennis Maher, shown here with attorney Aliza Kaplan, was innocent of the crimes for which he spent 19 years in prison.

Case Study

Guilty or Innocent?

"IT'S OK TO CRY," Innocence Project attorney Aliza Kaplan told Dennis Maher, on his way to court for his release from prison in 2003. Maher seemed calm as District Attorney Martha Coakley asked the judge to dismiss all of the charges for which Maher had been imprisoned for 19 years, 2 months, and 29 days. The judge ordered Maher's immediate release. Although outwardly composed in court, Maher and his family hugged and wept in the hall outside. "We're a bunch of crybabies," said his father, Donat.

Nineteen years earlier, Maher had been convicted of two counts of rape and one of attempted rape. As it turned out, his only crimes were to live in the vicinity of the rapes, to wear a red sweatshirt, and to look like the real assailant. All three victims picked Maher out of lineups. How can three people all identify the wrong man? It was dark, the assaults were swift, and, obviously, the women were tremendously stressed. In fact, contrary to popular belief, eyewitness testimony is really very unreliable. Various studies have found error rates for eyewitness identification from about 35% to 80%, depending on the conditions used in the experiments.

You have probably already guessed what led to Maher's eventual exoneration—DNA evidence. In 1993, while watching the *Phil Donahue Show* in prison, Maher heard about the Innocence Project, founded in 1992 by Barry Scheck and Peter Neufeld of the Benjamin Cardozo School of Law at Yeshiva University. He wrote to Scheck asking for help. Scheck agreed, but the Innocence Project hit a brick wall—no biological evidence was available for any of the cases.

Finally, 7 years later, an Innocence Project law student found the semen-stained underwear of one of the rape victims, lost in a box in a courthouse storage room. A few months later, a semen specimen from the second rape turned up as well. DNA profiling proved that Maher was not the assailant in either case.

In this chapter, we'll investigate the techniques of biotechnology that now pervade much of modern life. How do crime scene investigators decide that two DNA samples match? How can biotechnology diagnose inherited disorders? Should biotechnology be used to change the genetic makeup of crops, livestock, and even people?

At a Glance

13.1 WHAT IS BIOTECHNOLOGY?

Biotechnology is the use, and especially the alteration, of organisms, cells, or biological molecules to produce food, drugs, or other goods. Therefore, some aspects of biotechnology are ancient. For example, people have used yeast cells to produce bread, beer, and wine for the past 10,000 years. Selective breeding of plants and animals has an equally long history: 8,000- to 10,000-year-old fragments of squash, found in a dry cave in Mexico, have larger seeds and thicker rinds than wild squash, suggesting selective breeding for higher nutritional content. Prehistoric art and animal remains indicate that dogs, sheep, goats, pigs, and cattle were domesticated and selectively bred at least 10,000 years ago. Selective breeding causes domestic plants and animals to differ genetically from their wild relatives; for instance, the short legs and floppy ears of beagles are genetically determined, and differ enormously from the comparable characteristics of wolves, the ancestors of all dogs.

Even today, selective breeding remains an important tool. Modern biotechnology, however, also frequently uses **genetic engineering,** a term that refers to more direct methods for modifying genetic material. Genetically engineered cells or organisms have had genes deleted, added, or changed. Genetic engineering can be used to learn more about how cells and genes work; to develop better treatments for diseases; to produce valuable biological molecules, including hormones and vaccines; and to improve plants and animals for agriculture.

A key tool in genetic engineering is **recombinant DNA,** which is DNA that has been altered to contain genes or portions of genes from different organisms. Large amounts of recombinant DNA can be produced in bacteria, viruses, or yeast, and then transferred into other species. Plants and animals that express DNA that has been modified or derived from other species are called **transgenic,** or **genetically modified organisms (GMOs).**

Modern biotechnology includes many methods of manipulating DNA, whether or not the DNA is subsequently put into a cell or an organism. For example, determining the nucleotide sequence of specific pieces of DNA is crucial to forensic science, to the diagnosis of inherited disorders, and to studies of evolutionary relationships among organisms.

In this chapter, we will provide an overview of modern biotechnology. Our emphasis will be on applications of biotechnology and their impacts on society, but we will also briefly describe some of the important methods used in those applications. We will organize our discussion around five major themes: (1) recombinant DNA mechanisms found in nature; (2) biotechnology in criminal forensics, principally DNA matching; (3) biotechnology in agriculture, specifically the production of transgenic plants and animals; (4) the Human Genome Project and its applications; and (5) biotechnology in medicine, focusing on the diagnosis and treatment of inherited disorders.

13.2 HOW DOES DNA RECOMBINE IN NATURE?

Most people think that a species' genetic makeup is constant, except for the occasional mutation, but genetic reality is far more fluid. Many natural processes can transfer DNA from one organism to another, sometimes even to organisms of different species.

Sexual Reproduction Recombines DNA

As we saw in Chapter 9, homologous chromosomes exchange DNA by crossing over during meiosis I. As a result, each chromosome in a gamete contains a mixture of alleles from the two parental chromosomes. Thus, every egg and sperm contain recombinant DNA, derived from the two parents. When a sperm fertilizes an egg, the resulting offspring also contains recombinant DNA.

Transformation May Combine DNA from Different Bacterial Species

Bacteria can undergo several types of recombination that allow gene transfer between species (**Fig. 13-1**). In **transformation,** bacteria pick up pieces of DNA from the environment (**Fig. 13-1b**). The DNA may be part of the chromosome from another bacterium, even from another species. Transformation may also occur when bacteria pick up tiny circular DNA molecules called **plasmids** (**Fig. 13-1c**). Many types of bacteria contain plasmids, ranging in length from about 1,000 to 100,000 nucleotides. For comparison, the *E. coli* chromosome is around 4,600,000 nucleotides long. A single bacterium may contain dozens or even hundreds of copies of a plasmid. When the bacterium dies, it releases these plasmids into the environment, where they may be picked up by other bacteria of the same or different species. In addition, living bacteria can often pass plasmids directly to other living bacteria. Transfer of plasmids from bacteria to yeast may also occur, moving genes from a prokaryotic cell to a eukaryotic cell.

What use are plasmids? A bacterium's chromosome contains all the genes the cell normally needs for basic survival. However, genes carried by plasmids may permit the bacteria to thrive in novel environments. Some plasmids contain genes that allow bacteria to metabolize unusual energy sources, such as petroleum. Other plasmids may carry genes that enable bacteria to grow in the presence of antibiotics, such as penicillin. In environments where antibiotic use is high, particularly in

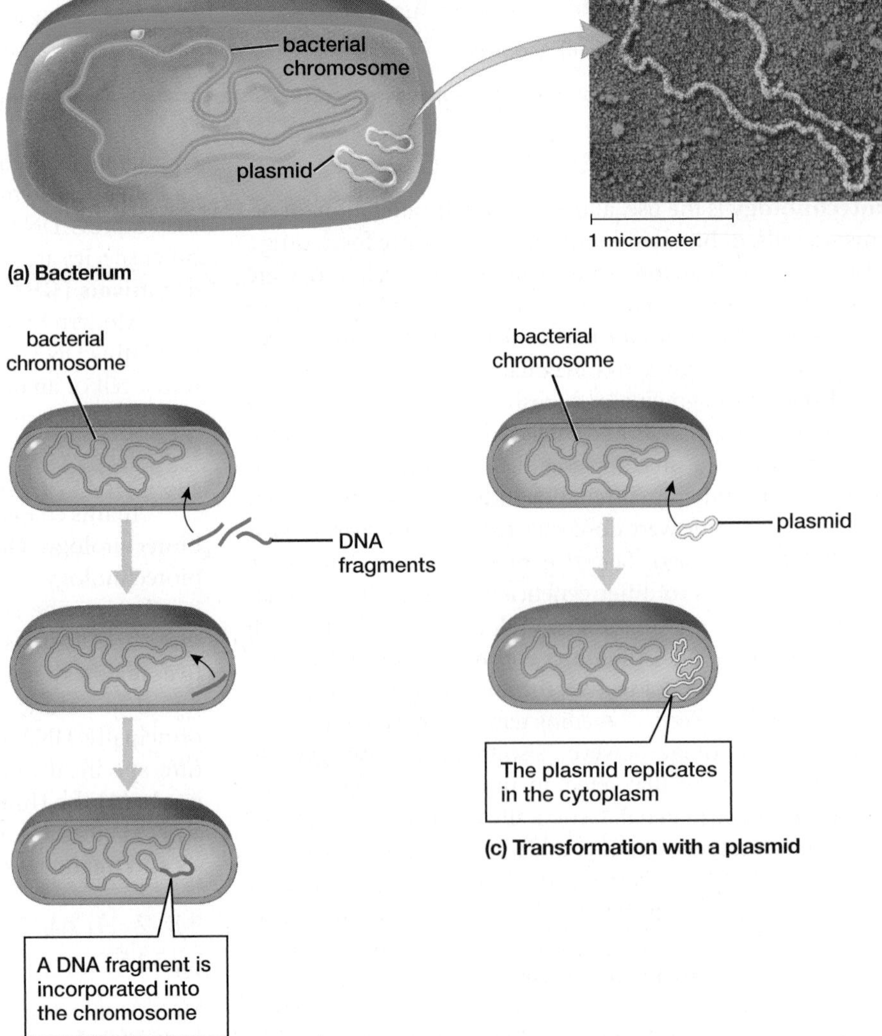

▶ **FIGURE 13-1 Transformation in bacteria (a)** In addition to their large circular chromosome, bacteria commonly possess small rings of DNA called plasmids, which often carry additional useful genes. Bacterial transformation occurs when living bacteria take up **(b)** fragments of chromosomes or **(c)** plasmids.

hospitals, bacteria carrying antibiotic-resistance plasmids can quickly spread among patients and health care workers, making antibiotic-resistant infections a serious problem.

Viruses May Transfer DNA Between Species

Viruses, which are often little more than genetic material encased in a protein coat, can reproduce only inside cells. A virus attaches to specific molecules on the surface of a suitable host cell (**Fig. 13-2** ❶). The rabies virus, for example, attaches to the receptors on muscle cells that the nervous system uses to stimulate contraction of the muscle. Both HIV (the AIDS virus) and cold viruses attach to molecules that are involved in the immune response to an infection. Usually the virus then enters the cytoplasm of the host cell (**Fig. 13-2** ❷), where it releases its genetic material (**Fig. 13-2** ❸). Unable to distinguish which genetic information is its own, and which is from the virus, the host cell replicates the viral genetic material (DNA or sometimes RNA) and synthesizes viral proteins (**Fig. 13-2** ❹). The replicated genes and viral proteins assemble inside the cell (**Fig. 13-2** ❺), forming new viruses that are released and may infect new cells (**Fig. 13-2** ❻).

Some viruses can transfer genes from one organism to another. In these instances, the viral DNA is inserted into one of the host cell's chromosomes (see Fig. 13-2 ❸). The viral DNA may remain there for days, months, or even years. Every time the cell divides, it replicates the viral DNA along with its own DNA. When new viruses are finally produced, some of the host cell's genes may be attached to the viral DNA. If such recombinant viruses infect another cell and insert their DNA into the new host cell's chromosomes, pieces of the previous host cell's DNA will also be inserted.

Most viruses infect and replicate only in the cells of specific bacterial, animal, or plant species. Therefore, most of the time, viruses move host DNA between different individuals of a single, or fairly closely related, species. However, some viruses may infect species only distantly related to one another; for example, influenza infects birds, pigs, and humans. Gene transfer between viruses that infect multiple species can produce extremely lethal recombined viruses. This happened in

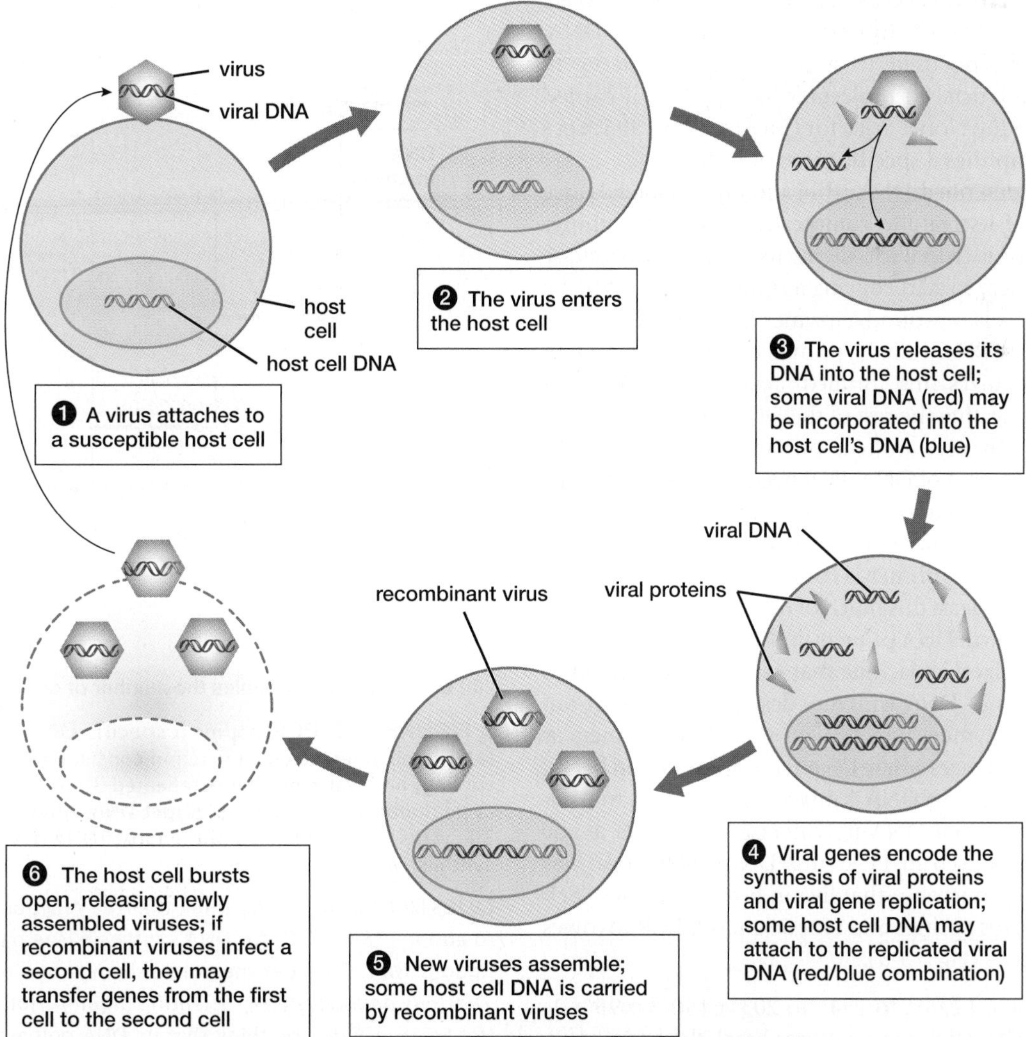

▲ FIGURE 13-2 **The life cycle of a typical virus** In some cases, viral infections may transfer DNA from one host cell to another.

1957 and again in 1968, when recombination between bird flu viruses and human flu viruses caused global epidemics that killed hundreds of thousands of people.

13.3 HOW IS BIOTECHNOLOGY USED IN FORENSIC SCIENCE?

Applications of DNA biotechnology vary, depending on the goals of those who use it. Forensic scientists need to identify victims and criminals; biotechnology firms need to identify specific genes and insert them into organisms such as bacteria, cattle, or crop plants; and biomedical firms and physicians need to detect defective alleles and, ideally, devise ways to fix them or to insert normally functioning alleles into patients. We will begin by describing a few common methods of manipulating DNA, using their application to forensic DNA analysis as a specific example. Later, we will investigate how biotechnology is used in agriculture and medicine.

The Polymerase Chain Reaction Amplifies DNA

Developed by Kary Mullis of the Cetus Corporation in 1986, the **polymerase chain reaction (PCR)** can be used to make billions, even trillions, of copies of selected pieces of DNA. Further, PCR can be used to amplify selected pieces of DNA. PCR is so crucial to molecular biology that it earned Mullis a share in the Nobel Prize for Chemistry in 1993. Let's see how PCR amplifies a specific piece of DNA.

When we described DNA replication in Chapter 11, we omitted some of its real-life complexity. One of the things we did not discuss is crucial to PCR: By itself, DNA polymerase doesn't know where to start copying a strand of DNA. When a DNA double helix is unwound, enzymes first put a little piece of complementary RNA, called a primer, on each strand. DNA polymerase recognizes this "primed" region of DNA as the place to start replicating the rest of the DNA strand.

PCR generally uses artificial primers made of DNA. To copy a specific segment of DNA, PCR needs two DNA primers, which are complementary to the two ends of the DNA segment. One primer is complementary to one strand of the double helix and the other primer is complementary to the other strand. Therefore, at least this much of the DNA sequence of interest must be known. DNA primers for PCR are manufactured in a DNA synthesizer, a machine that can be programmed to make short pieces of DNA with any desired sequence of nucleotides. DNA polymerase recognizes these DNA primers as the nucleotide sequences where DNA replication should begin.

In a small test tube, DNA is mixed with primers, free nucleotides, and a special DNA polymerase that works at the high temperatures used in PCR (this DNA polymerase was first isolated from microbes that live in hot springs; see "Scientific Inquiry: Hot Springs and Hot Science"). PCR involves the following steps (**Fig. 13-3**):

1. The test tube is heated to 194° to 203°F (90° to 95°C) (**Fig. 13-3a** ❶). High temperatures break the hydrogen bonds between complementary bases, separating the DNA into single strands.

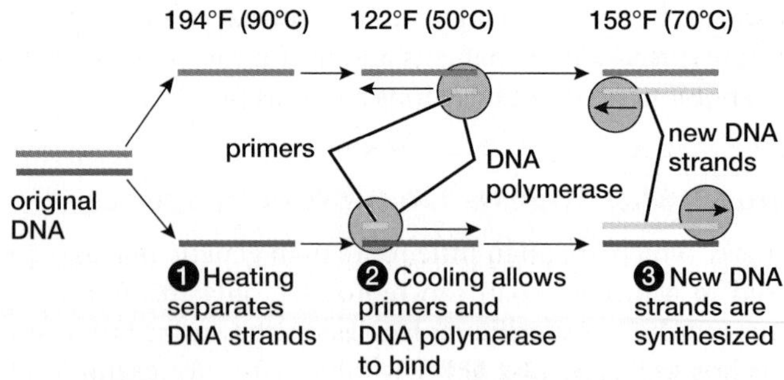

(a) One PCR cycle

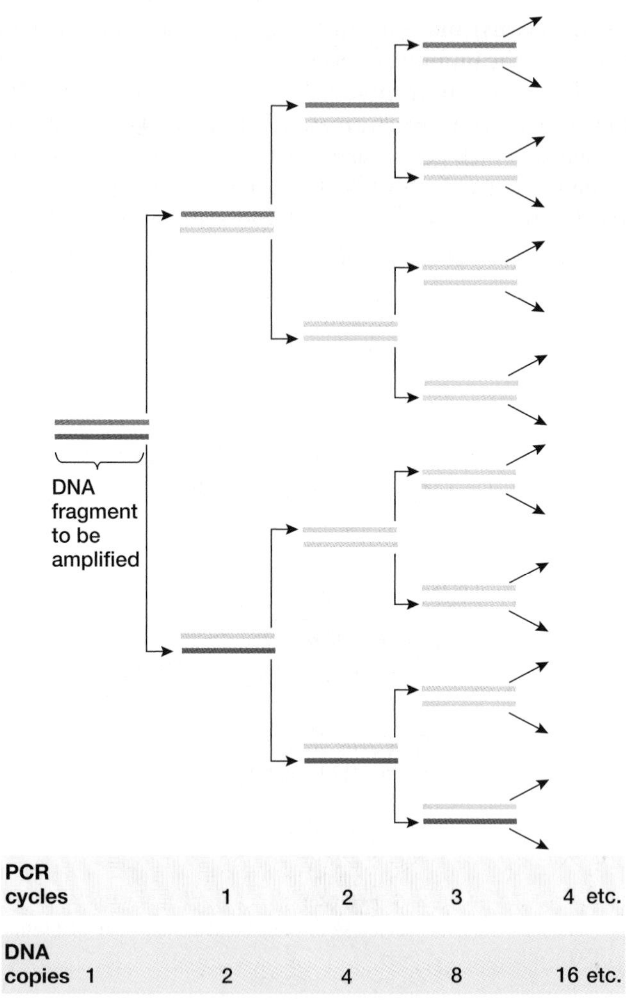

PCR cycles		1	2	3	4 etc.
DNA copies	1	2	4	8	16 etc.

(b) Each PCR cycle doubles the number of copies of the DNA

▲ **FIGURE 13-3 PCR copies a specific DNA sequence**
(a) The polymerase chain reaction consists of a cycle of heating, cooling, and warming that is repeated 30 to 40 times. **(b)** Each cycle doubles the amount of target DNA. After a little more than 30 cycles, a billion copies of the target DNA have been synthesized.

QUESTION Why do you think that the reaction is warmed up to about 158°F (70°C) for DNA synthesis [see part (a) of the figure]? *Hint:* Consider the normal living conditions of *Thermus aquaticus* (see "Scientific Inquiry: Hot Springs and Hot Science"); do you think that its DNA polymerase would work most rapidly at about 122°F (50°C) or 158°F (70°C)?

Scientific Inquiry

Hot Springs and Hot Science

At a hot spring, such as those found in Yellowstone National Park, water literally boils out of the ground, gradually cooling as it flows to the nearest stream (**Fig. E13-1**). You might think that such springs, scalding hot and often containing poisonous metals and sulfur compounds, must be lifeless. However, closer examination often reveals a diversity of microorganisms, each adapted to a different temperature zone in the spring. Back in 1966, in a Yellowstone hot spring, Thomas Brock of the University of Wisconsin discovered *Thermus aquaticus*, a bacterium that lives in water as hot as 176°F (80°C).

When Kary Mullis first developed the polymerase chain reaction, he encountered a major technical difficulty. The DNA solution must be heated almost to boiling to separate the double helix into single strands, then cooled so DNA polymerase can synthesize new DNA, and this process must be repeated again and again. "Ordinary" DNA polymerase, like most proteins, is ruined, or denatured, by high temperatures. Therefore, new DNA polymerase had to be added after every heat cycle, which was expensive and labor intensive. Enter *Thermus aquaticus*. Like other organisms, it replicates its DNA when it reproduces. But because it lives in hot springs, it has a particularly heat-resistant DNA polymerase. When DNA polymerase from *T. aquaticus* is used in PCR, it needs to be added to the DNA solution only once, at the start of the reaction.

▲ **FIGURE E13-1 Thomas Brock surveys Mushroom Spring** The colors in hot springs arise from minerals dissolved in the water and from various types of microbes that live at different temperatures.

2. The temperature is lowered to about 122°F (50°C) (**Fig. 13-3a ❷**), which allows the two primers to form complementary base pairs with the original DNA strands.
3. The temperature is raised to 158° to 162°F (70° to 72°C) (**Fig. 13-3a ❸**). DNA polymerase uses the free nucleotides to make copies of the DNA segment bounded by the primers.
4. This cycle is repeated, usually 30 to 40 times, until the reactants have been used up.

PCR synthesizes DNA in a geometric progression ($1 \rightarrow 2 \rightarrow 4 \rightarrow 8$, etc.), so 20 PCR cycles make about a million copies, and a little over 30 cycles make a billion copies (**Fig. 13-3b**). Each cycle takes only a few minutes, so PCR can produce billions of copies of a DNA segment in a single afternoon, starting, if necessary, from a single molecule of DNA. The DNA is then available for forensics, cloning, making transgenic organisms, or many other purposes in biotechnology.

Differences in Short Tandem Repeats Can Identify Individuals by Their DNA

In many criminal investigations, PCR is used to amplify the DNA so that there is enough to compare the DNA left at a crime scene with a suspect's DNA. How is this comparison done? After years of painstaking work, forensics experts have found that small, repeating segments of DNA, called **short tandem repeats (STRs),** can be used to identify people with astonishing accuracy. It may be helpful to think of STRs as very short, stuttering genes (**Fig. 13-4**). Each STR is *short* (consisting of 2 to 5 nucleotides), *repeat*ed (up to 50 times), and *tandem* (having all the repetitions right alongside one another). As with any gene, different people may have different alleles of the STRs. In the case of an STR, each allele simply has a different number of repeats of the same few nucleotides. In the United States, the Department of Justice established a standard set of 13 different STRs, each with repeats four nucleotides long, to identify individuals by DNA samples. Worldwide, different people may have as few as five to as many 38 repeats of a given STR. As we will see, STR analysis is fairly simple, making it ideal for forensic use.

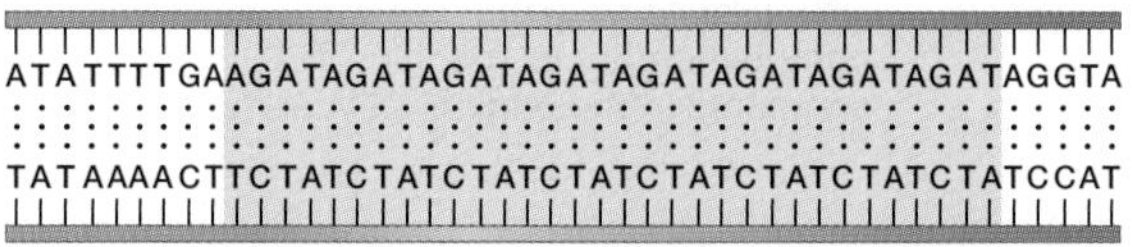

Eight side-by-side (tandem) repeats of the same four-nucleotide sequence

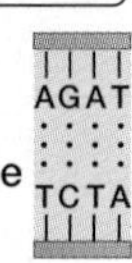

▲ **FIGURE 13-4 Short tandem repeats are common in noncoding regions of DNA** This STR contains the sequence AGAT, repeated from 7 to 15 times in different individuals.

Case Study continued
Guilty or Innocent?

When a law student finally located semen samples in the Dennis Maher case, the Innocence Project team needed to find out if the samples collected from the rape victims came from Maher. Fortunately, DNA doesn't degrade very fast, so even old DNA samples, such as those in the Maher case, usually have STRs that are mostly intact. First, lab technicians amplified the DNA with PCR so that they had enough material to analyze. Then, they determined whether the DNA from the semen samples matched Maher's DNA.

Forensics labs use PCR primers that amplify only the STRs and the DNA immediately surrounding them. Because STR alleles vary in how many repeats they contain, they vary in size: An STR allele with more repeats is larger than one with fewer repeats. Therefore, a forensic lab needs to identify each STR in a DNA sample and then must determine its size to find out which alleles occur in the sample.

Modern forensics labs use sophisticated and expensive machines to determine how many repeats each STR in their samples contain. Most of these machines, however, are based on two methods that are used in molecular biology labs around the world: first, separating DNA segments by size, and second, labeling specific DNA segments of interest.

Gel Electrophoresis Separates DNA Segments

A mixture of DNA pieces is separated by a technique called **gel electrophoresis** (**Fig. 13-5**). First, a laboratory technician loads the DNA into shallow grooves, or wells, in a slab of agarose, a carbohydrate purified from certain types of seaweed (**Fig. 13-5 ①**). Agarose forms a gel, which is simply a

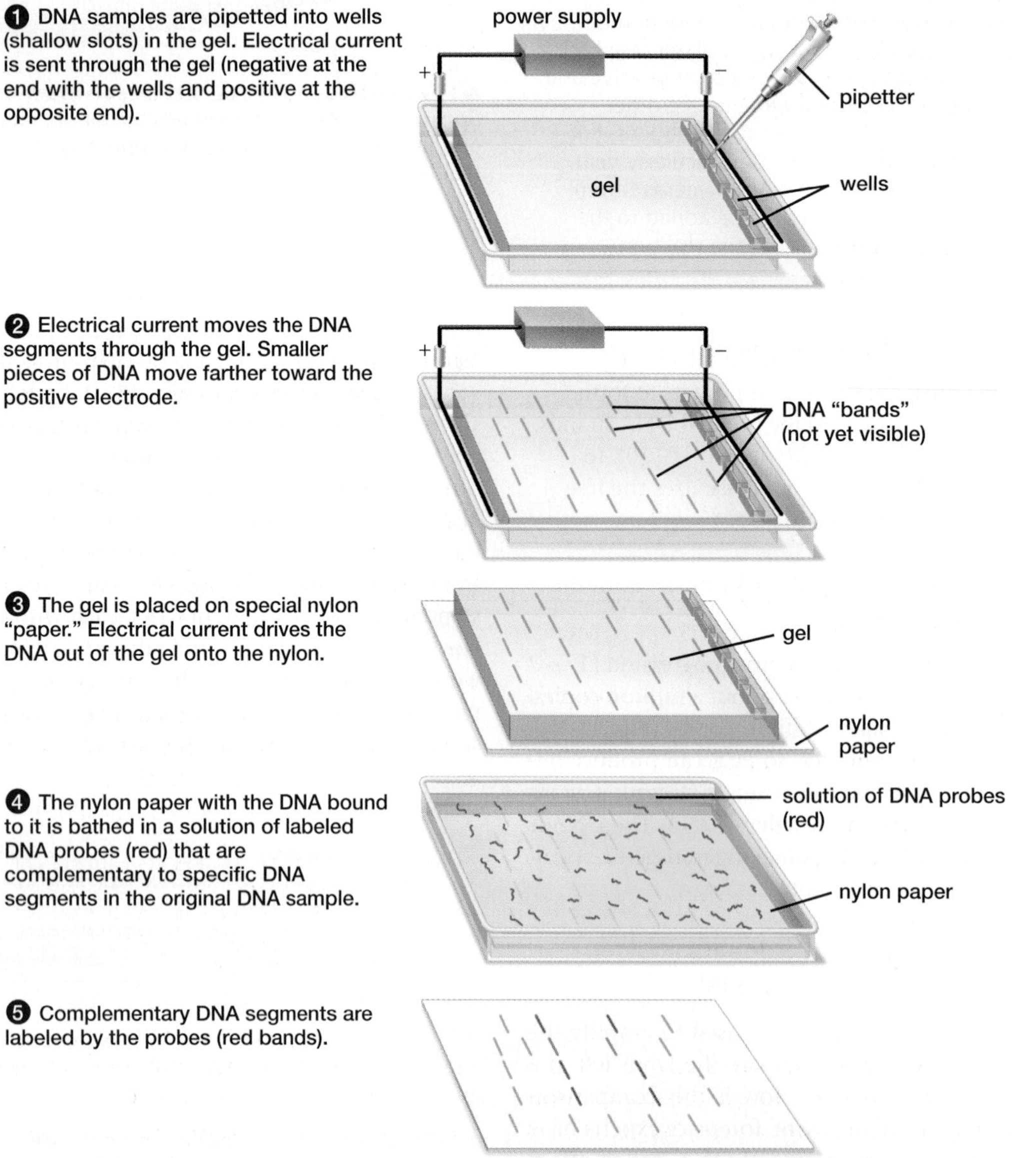

① DNA samples are pipetted into wells (shallow slots) in the gel. Electrical current is sent through the gel (negative at the end with the wells and positive at the opposite end).
power supply
+
−
pipetter
gel
wells
② Electrical current moves the DNA segments through the gel. Smaller pieces of DNA move farther toward the positive electrode.
+
−
DNA "bands" (not yet visible)
③ The gel is placed on special nylon "paper." Electrical current drives the DNA out of the gel onto the nylon.
gel
nylon paper
④ The nylon paper with the DNA bound to it is bathed in a solution of labeled DNA probes (red) that are complementary to specific DNA segments in the original DNA sample.
solution of DNA probes (red)
nylon paper
⑤ Complementary DNA segments are labeled by the probes (red bands).

▲ **FIGURE 13-5 Gel electrophoresis and labeling with DNA probes separates and identifies segments of DNA.**

meshwork of fibers with holes of various sizes between the fibers. The gel is put into a chamber with electrodes connected to each end. One electrode is made positive and the other negative; therefore, current will flow between the electrodes through the gel. How does this process separate pieces of DNA? Remember, the phosphate groups in the backbones of DNA are negatively charged. When electrical current flows through the gel, the negatively charged DNA fragments move toward the positively charged electrode. Because smaller fragments slip through the holes in the gel more easily than larger fragments do, they move more rapidly toward the positively charged electrode. Eventually the DNA fragments are separated by size, forming distinct bands on the gel (**Fig. 13-5 ❷**).

DNA Probes Are Used to Label Specific Nucleotide Sequences

Unfortunately, the DNA bands are invisible. There are several dyes that stain DNA, but these are often not very useful in either forensics or medicine. Why not? Because there may be many DNA fragments of approximately the same size; for example, five or six different STRs with the same numbers of repeats might be mixed together in the same band. How can a technician identify a *specific* STR? Well, how does *nature* identify sequences of DNA? Right—by base pairing!

When the gel is finished running, the technician treats it with chemicals that break apart the double helices into single DNA strands. These DNA strands are transferred out of the gel onto a piece of paper made of nylon (**Fig. 13-5 ❸**). Because the DNA samples are now single-stranded, pieces of synthetic DNA, called **DNA probes,** can base-pair with specific DNA fragments in the sample. DNA probes are short pieces of single-stranded DNA that are complementary to the nucleotide sequence of a given STR (or any other DNA of interest in the gel). The DNA probes are labeled, either by radioactivity or by attaching colored molecules to them. Therefore, a given DNA probe will label certain DNA sequences, but not others (**Fig. 13-6**).

To visualize a specific STR, the paper is bathed in a solution containing a DNA probe that will base-pair with, and therefore bind to, only that particular STR (**Fig. 13-5 ❹**). Any extra DNA probe is then washed off the paper. The result: The DNA probe shows where that specific STR ran in the gel (**Fig. 13-5 ❺**). (Labeling the DNA fragments with radioactive or colored DNA probes is standard procedure in most research applications. In modern forensic applications, however, the STRs are usually directly labeled with colored molecules during the PCR reaction. Therefore, the STRs are immediately visible in the gel and do not have to be stained with DNA probes.)

Unrelated People Almost Never Have Identical DNA Profiles

DNA samples run on STR gels produce a pattern, called a **DNA profile** (**Fig. 13-7**). The positions of the bands on the gel are determined by the numbers of repeats of each STR allele. If the same STRs are analyzed, then a particular DNA sample will produce the same profile every time.

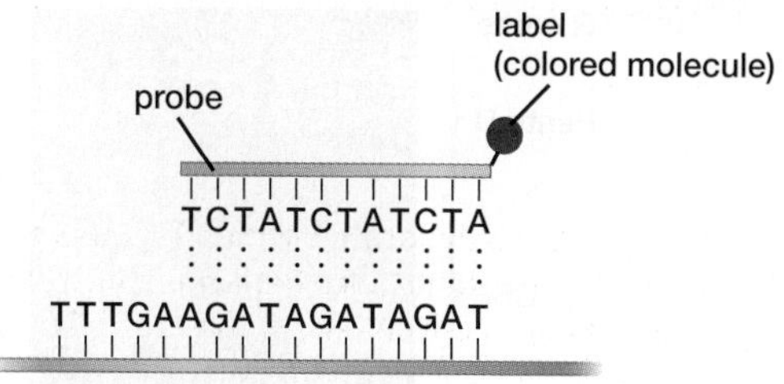

STR #1: The probe base-pairs and binds to the DNA

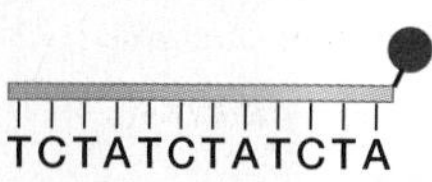

STR #2: The probe cannot base-pair with the DNA, so it does not bind

▲ **FIGURE 13-6 DNA probes base-pair with complementary DNA sequences** A short, single-stranded piece of DNA is labeled with a colored molecule (red ball). This labeled DNA will base-pair with a target strand of DNA with a complementary base sequence (top), but not with a noncomplementary strand (bottom).

What does a DNA profile tell us? As with any gene, every person has two alleles of each STR, one on each homologous chromosome. The two alleles of a given STR might have the same number of repeats (the person would be homozygous for that STR gene) or a different number of repeats (the person would be heterozygous). For example, in the enlargement of the D16 STR on the right side of Figure 13-7, the first person's DNA has a single band at 12 repeats (this person is homozygous for the D16 STR), but the second person's DNA has two bands—at 13 and 12 repeats (this person is heterozygous for the D16 STR). If you look closely at all of the DNA samples in Figure 13-7, you will see that, although the DNA from some people had the same repeats for one of the STRs (e.g., the fourth and fifth samples for D16), no one's DNA had the same repeats for all four STRs.

Remember, in the United States, a standard set of 13 STRs is used for DNA identification, A perfect DNA profile match would mean that both alleles of all 13 STRs would be identical between the crime scene DNA and the suspect's DNA. How definitive is an STR match? A perfect match of both alleles for all 13 STRs means that there is far less than one chance in a trillion that the two DNA samples matched purely by random chance. For complicated statistical reasons, there are probably a few unrelated people in the world who have the same DNA profile, but the odds of anyone who could possibly be a likely suspect in a criminal case being misidentified is extremely small. Finally, a *mismatch* in DNA profiles is absolute proof that two samples did not come from the same source. In the Maher case, when DNA profiling showed that the STR alleles in the semen did not match Maher's, that ruled him out as the perpetrator.

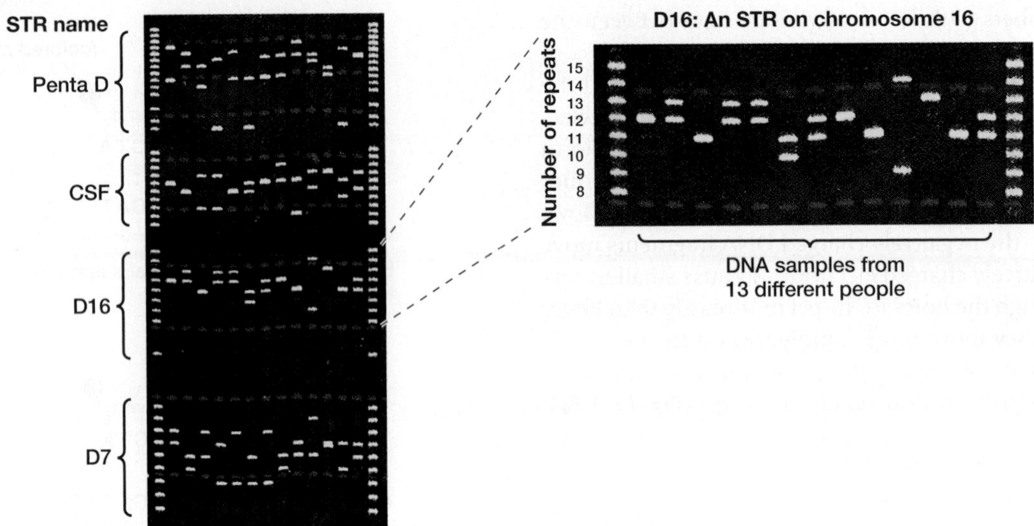

▲ **FIGURE 13-7 DNA profiling** The lengths of short tandem repeats of DNA form characteristic patterns on a gel. This gel displays four different STRs (Penta D, CSF, etc.). The columns of evenly spaced yellow bands on the far left and far right sides of the gel show the number of repeats in the individual STRs. DNA samples from 13 different people were run between these standards, resulting in one or two yellow bands in each vertical lane. The position of each band corresponds to the number of repeats in that STR allele (more repeats means more nucleotides, which in turn means that the allele is larger). (Photo courtesy of Dr. Margaret Kline, National Institute of Standards and Technology.)

QUESTION For any single person, a given STR always has either one or two bands. Why? Further, single bands are always about twice as bright as each band of a pair. For example, in the D16 STR on the right, the single bands of the first and third DNA samples are twice as bright as the pairs of bands of the second, fourth, and fifth samples. Why?

In the United States, anyone convicted of certain crimes (assault, burglary, attempted murder, etc.) must give a blood sample. Using the standard array of STRs, technicians determine the criminal's DNA profile, and code the results as the number of repeats of each STR. The profile is stored in computer files at a state agency, at the FBI, or both. (On *CSI* and other TV crime shows, when the actors refer to "CODIS," that acronym stands for "Combined DNA Index System," a DNA profile database kept on FBI computers.) Because all forensic labs use the same 13 STRs, computers can easily determine if DNA left behind at another crime scene, even years before or years later (after the criminal may have been released from prison), matches one of the profiles stored in the CODIS database. If the STRs match, then the odds are overwhelming that the crime scene DNA was left by the person with the matching profile. If there aren't any matches, the crime scene DNA profile will remain on file. Sometimes, years later, a new DNA profile will match an archived crime-scene profile, and a "cold case" will be solved (see "Case Study Revisited" at the end of this chapter).

Case Study continued
Guilty or Innocent?

If you are a regular viewer of *CSI*, you may have heard an investigator say that a parent and child have "seven alleles in common." However, each person has two alleles of each of the 13 STR loci used in the United States—one allele at each locus inherited from the mother, the other allele inherited from the father. Further, even unrelated people usually have a few of the same STRs (see Fig. 13-7). So how many alleles should a father share with his child? The answer, of course, is at least 13, not merely 7.

13.4 HOW IS BIOTECHNOLOGY USED IN AGRICULTURE?

The main goals of agriculture are to grow as much food as possible, as cheaply as possible, with minimal loss from pests such as insects and weeds. Many commercial farmers and seed suppliers have turned to biotechnology to achieve these goals. Many people, however, feel that the risks of genetically modified food to human health or the environment are not worth the benefits. We will explore this controversy later in section 13.7.

Many Crops Are Genetically Modified

Currently, almost all of the genetically modified organisms used in agriculture are plants. According to the U.S. Department of Agriculture (USDA), in 2008, about 80% of the corn, 86% of the cotton, and 92% of the soybeans grown in the United States were transgenic; that is, they contained genes from other species (**Table 13-1**). Globally, more than 280 million acres of land were planted with transgenic crops in 2007, a 40% increase in just 3 years.

Table 13-1 **Genetically Engineered Crops with USDA Approval**

Genetically Engineered Trait	Potential Advantage	Examples of Bioengineered Crops with USDA Approval
Resistance to herbicide	Application of herbicide kills weeds but not crop plants, producing higher crop yields	Beet, canola, corn, cotton, flax, potato, rice, soybean, tomato
Resistance to pests	Crop plants suffer less damage from insects, producing higher crop yields	Corn, cotton, potato, rice, soybean
Resistance to disease	Plants are less prone to infection by viruses, bacteria, or fungi, producing higher crop yields	Papaya, potato, squash
Sterile	Transgenic plants cannot cross with wild varieties, making them safer for the environment and more economically productive for the seed companies that produce them	Chicory, corn
Altered oil content	Oils can be made healthier for human consumption or can be made similar to more expensive oils (such as palm or coconut)	Canola, soybean
Altered ripening	Fruits can be more easily shipped with less damage, producing higher returns for the farmer	Tomato

Crops are most commonly modified to improve their resistance to insects, herbicides, or both. Many herbicides kill plants by inhibiting an enzyme that is used by plants, fungi, and some bacteria—but not animals—to synthesize amino acids such as tyrosine, tryptophan, and phenylalanine. Without these amino acids, the plants cannot synthesize proteins, and they die. Many herbicide-resistant transgenic crops have been given a bacterial gene encoding an enzyme that functions even in the presence of these herbicides, so the plants continue to synthesize normal amounts of amino acids and proteins. Herbicide-resistant crops allow farmers to kill weeds without harming their crops. Less competition from weeds means more water, nutrients, and light for the crops, and hence, larger harvests.

The insect resistance of many crops has been enhanced by giving them a gene, called *Bt*, from the bacterium *Bacillus thuringiensis*. The protein encoded by the *Bt* gene damages the digestive tract of insects, but not mammals. Transgenic *Bt* crops often suffer far less damage from insects (**Fig. 13-8**) than do regular crops, so farmers can apply less pesticide to their fields.

How would a seed company make a transgenic plant? Let's examine the process, using insect-resistant *Bt* plants as an example.

▲ **FIGURE 13-8** ***Bt*** **plants resist insect attack** Transgenic cotton plants expressing the *Bt* gene (right) resist attack by bollworms, which eat cotton seeds. The transgenic plants therefore produce far more cotton than nontransgenic plants (left).

Have you ever wondered 

If the Food You Eat Has Been Genetically Modified?

In addition to obvious things—tortilla chips, soy sauce, cooking oil, and margarine—corn and soy products are found in an amazing variety of foods. For example, corn syrup is an ingredient in foods as diverse as soda, ketchup, and bran flakes; corn starch is found in many baked goods. According to the Grocery Manufacturers of America, about 70% of the foods in your local supermarket contain products made from genetically modified (GM) crops. Many countries, including the European Union, require labeling of GM foods, but the U.S. Food and Drug Administration does not, so you can't tell if a food contains GM ingredients just by looking at the label. Therefore, unless you have been extremely motivated and careful, you have probably eaten GM foods.

The Desired Gene Is Cloned

Cloning a gene usually involves two tasks: (1) obtaining the gene and (2) inserting it into a plasmid so that huge numbers of copies of the gene can be made.

Two common methods are used to obtain a gene. For a long time, the only practical method was to isolate the gene from the organism that makes it. Now, biotechnologists can often synthesize the gene—or a modified version of it—in the lab, using DNA synthesizers.

Once the gene has been obtained, why insert it into a plasmid? Plasmids are replicated when the bacteria multiply. Therefore, once the desired gene has been inserted into a plasmid, producing huge numbers of copies of the gene is as simple as raising lots of bacteria. Inserting the gene into a plasmid also allows it to be easily separated from the bacteria, achieving partial purification of the gene, free of the DNA of the bacterial chromosome. Finally, plasmids may be taken up by other bacteria (this is important when making transgenic *Bt* plants) or injected into animal eggs.

One common method for making transgenic plants employs the bacterium *Agrobacterium tumefaciens*, which infects many plant species. *A. tumefaciens* often contains a specialized plasmid called the Ti (tumor-inducing) plasmid. (As the name implies, Ti plasmids normally cause plant tumors. However, biotechnologists can disable Ti plasmids by taking out the tumor-causing genes, so the plasmids become harmless.) When *A. tumefaciens* infects a plant cell, the Ti plasmid inserts itself into one of the plant cell's chromosomes. Thereafter, any time that plant cell divides, it replicates the Ti plasmid DNA as well, and all of its daughter cells inherit the Ti DNA. Most other plasmids don't insert themselves into their host cell's chromosomes, so the plasmid's genes aren't always passed down faithfully to the daughter cells when the host cell divides.

To insert a cloned gene into a plant cell, biotechnologists first must insert the gene into a Ti plasmid, then transform *A. tumefaciens* bacteria with the new recombinant plasmid, infect plant cells with the transformed bacteria, and finally make entire plants from these individual recombinant plant cells, as we describe below.

Restriction Enzymes Cut DNA at Specific Nucleotide Sequences

Genes are inserted into plasmids through the action of **restriction enzymes** isolated from bacteria. Each restriction enzyme cuts DNA at a specific nucleotide sequence. Many restriction enzymes cut "straight across" a double helix of DNA. Others make a "staggered" cut, snipping the DNA in a different location on each of the two strands, so that single-stranded sections hang off the ends of the DNA. These single-stranded regions are commonly called "sticky ends," because they can base-pair with, and thus stick to, other single-stranded pieces of DNA with complementary bases (**Fig. 13-9**).

Cutting Two Pieces of DNA with the Same Restriction Enzyme Allows the Pieces to Be Joined Together

To insert the *Bt* gene into a Ti plasmid, the same restriction enzyme is used to cut the DNA on either side of the *Bt* gene and to split open the circle of Ti plasmid DNA (**Fig. 13-10 ❶**). As a result, the ends of the *Bt* gene and the opened-up Ti plasmid both have complementary nucleotides in their sticky ends, and can base-pair with each other. When the cut *Bt* genes and Ti plasmids are mixed together, some of the *Bt* genes will be temporarily inserted between the ends of the Ti plasmids, held together by their complementary sticky ends. Adding DNA ligase (see pp. 211–212) permanently bonds the *Bt* gene into the Ti plasmid, forming a recombinant Ti plasmid (**Fig. 13-10 ❷**).

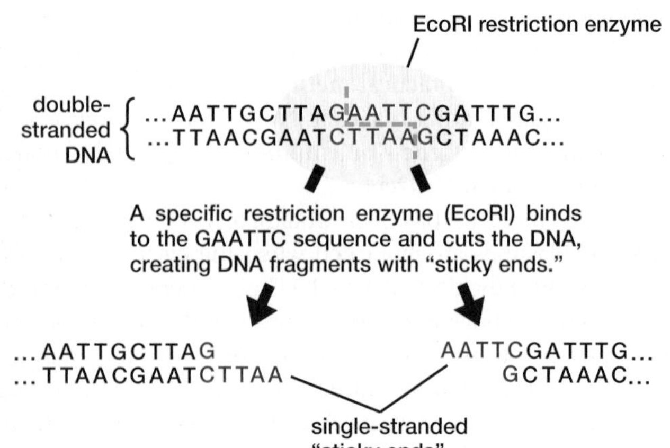

▲ **FIGURE 13-9 Restriction enzymes cut DNA at specific nucleotide sequences**

Plasmid-Transformed Bacteria Insert the *Bt* Gene into a Plant

To make insect-resistant plants, *Bt* genes are inserted into disabled Ti plasmids. *A. tumefaciens* bacteria are then transformed with the recombinant Ti plasmids (**Fig. 13-10 ❸**). When the transformed bacteria enter a plant cell, the Ti plasmids insert their DNA, including the *Bt* gene, into the plant cell's chromosomes, so that the plant cells now permanently carry the *Bt* gene (**Fig. 13-10 ❹**). Hormones are used to stimulate the transgenic plant cells to divide and differentiate into entire plants. These plants are bred to one another, or to other plants, to create commercially valuable plants that resist insect attack.

Another way of making transgenic plants is to use a "gene gun." Tiny metallic particles are coated with the desired DNA, often as part of a recombinant plasmid, and shot into a lump of plant cells growing in a Petri dish. This process is literally "hit or miss." Many of the cells are destroyed, many others never receive the DNA pellets, and only some of those that do receive the DNA successfully insert the DNA into a chromosome. However, because gene gun technology is quite straightforward and even a single successful cell can be grown into a transgenic plant, gene guns are increasingly used to produce genetically modified plants.

Genetically Modified Plants May Be Used to Produce Medicines

Similar techniques can be used to insert medically useful genes into plants, producing medicines down on the "pharm." For example, a plant can be engineered to produce harmless proteins that are normally found in disease-causing bacteria or viruses. If these proteins resisted digestion in the stomach and small intestine, simply eating such plants could act as a vaccination against the disease organisms. Several years ago, such "edible vaccines" were touted as a great way to provide vaccinations—no need to produce purified vaccines, no refrigeration needed, and, of course, no needles. Recently, however, many biomedical researchers have warned that edible vaccine plants are not really a good idea, because there is no simple way to control the dose: Too little and the user doesn't develop decent immunity; too much, and the vaccine proteins might be harmful. Nevertheless, producing vaccine proteins in plants is still worthwhile; pharmaceutical companies just have to extract and purify the proteins before use. Plant-produced vaccines against hepatitis B, tooth decay, avian flu, and certain types of diarrhea are now in clinical trials.

Molecular biologists can also engineer plants to produce human antibodies to combat various diseases. When a microbe invades your body, it takes several days for your immune system to respond and produce enough antibodies to

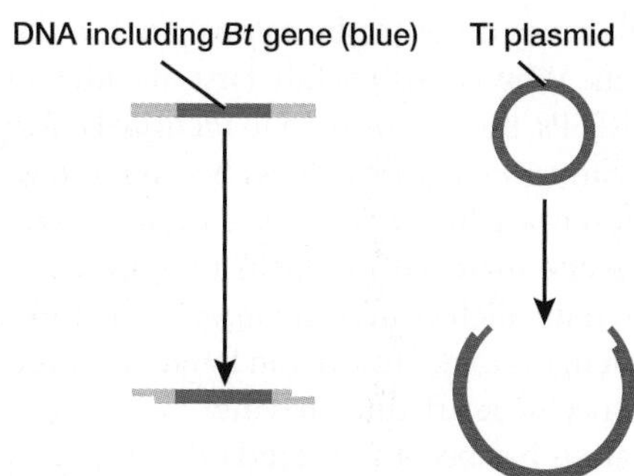

❶ The DNA containing the *Bt* gene and the Ti plasmid are cut with the same restriction enzyme.

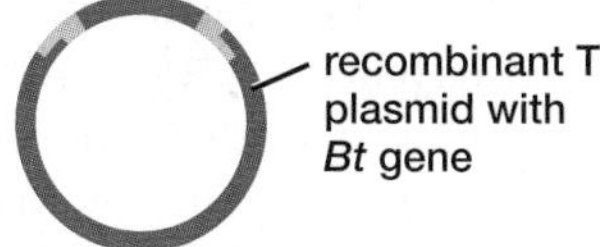

❷ *Bt* genes and Ti plasmids, both with the same complementary sticky ends, are mixed together; DNA ligase bonds the *Bt* genes into the plasmids.

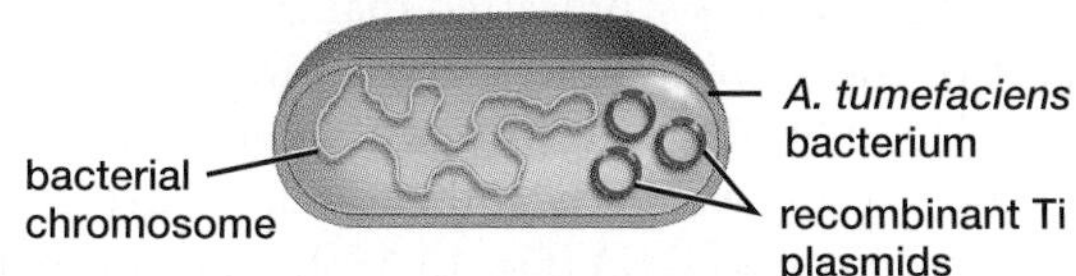

❸ Bacteria are transformed with the recombinant plasmids.

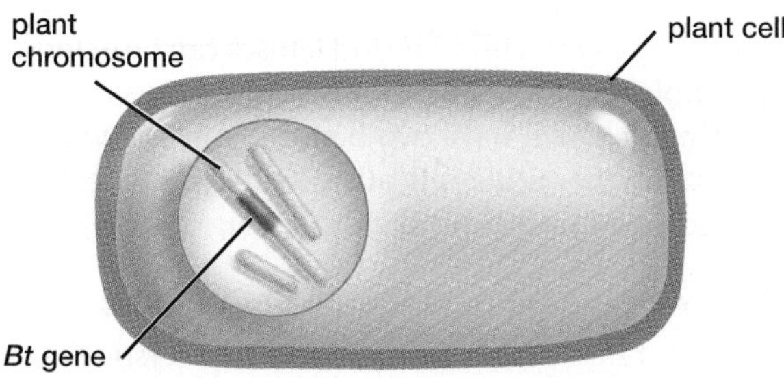

❹ Transgenic bacteria enter the plant cells, and *Bt* genes are inserted into the chromosomes of the plant cells.

▲ **FIGURE 13-10 Using plasmids to insert a bacterial gene into a plant chromosome**

overcome the infection. Meanwhile, you feel terrible, and might even die if the disease is serious enough. A direct injection of large quantities of the right antibodies might be able to cure the disease very quickly. Although none have yet entered medical practice, plant-derived antibodies against bacteria that cause tooth decay and non-Hodgkin's lymphoma (a cancer of the lymphatic system) are in clinical trials. Ideally, such "plantibodies" can be produced very cheaply, making such therapies available to rich and poor alike.

Genetically Modified Animals May Be Useful in Agriculture and Medicine

Producing transgenic animals usually involves injecting the desired DNA, often incorporated into a disabled virus that cannot produce disease, into a fertilized egg. The egg is allowed to divide a few times in culture before being implanted into a surrogate mother. If the offspring are healthy and express the foreign gene, they are then bred to one another produce homozygous transgenic organisms. So far, it has proven difficult to produce commercially valuable transgenic livestock, but several companies in countries around the world are working on it.

One example is Nexia Biotech, which has engineered a herd of goats to carry the genes for spider silk, and to secrete silk protein into their milk. The resulting BioSteel® is five times stronger than steel and twice as strong as Kevlar®, the fiber usually used in bulletproof vests. Nexia has encountered difficulties in spinning BioSteel® into commercially usable silk, but has successfully manufactured "nanofibers" for medical and microelectronic applications.

Biotechnologists are also developing animals that will produce medicines, such as human antibodies or other essential proteins. For example, there are sheep whose milk contains a protein, alpha-1-antitrypsin, that may prove valuable in treating cystic fibrosis. Other livestock have been engineered so that their milk contains erythropoietin (a hormone that stimulates red blood cell synthesis), clotting factors (for treatment of hemophilia), or clot-busting proteins (to treat heart attacks caused by blood clots in the coronary arteries).

13.5 HOW IS BIOTECHNOLOGY USED TO LEARN ABOUT THE HUMAN GENOME?

Genes influence virtually all the traits of human beings, including gender, size, hair color, intelligence, and susceptibility to disease organisms and toxic substances in the environment. To begin to understand how our genes influence our lives, the Human Genome Project was launched in 1990, with the goal of determining the nucleotide sequence of all the DNA in our entire set of genes, called the human genome.

By 2003, this joint project of molecular biologists in several countries sequenced the human genome with an accuracy of about 99.99%. The human genome contains between 20,000 and 25,000 genes, comprising approximately 2% of the DNA. Some of the other 98% consists of promoters and regions that regulate how often individual genes are transcribed, but it's not really known what most of our DNA does.

Why did scientists want to sequence the human genome? First, many genes were discovered whose functions are unknown. Using the genetic code to translate the DNA sequences of novel genes, biologists can predict the amino acid sequences of the proteins they encode. Comparing these proteins to familiar proteins whose functions are already known will enable us to find out what many of these genes do.

Second, knowing the nucleotide sequences of human genes will have an enormous impact on medical practice. In 1990, fewer than 100 genes known to be associated with human diseases had been discovered. By 2006, this number had jumped to more than 1,800.

Third, there is no single "human genome" (or else all of us would be genetically identical). Most of the DNA of everyone on the planet is the same, but we each also carry our own unique set of alleles. Some of those alleles can cause or predispose people to develop various medical conditions, including sickle-cell anemia, cystic fibrosis, breast cancer, alcoholism, schizophrenia, heart disease, Alzheimer's disease, and many others. Ongoing human genome research involves sequencing the DNA of many different people, looking for small differences in allele DNA sequences that might contribute to diseases, susceptibility to toxic pollutants or chemicals in tobacco smoke, or responses to drugs. Researchers hope to use this information to help diagnose genetic disorders or predispositions, and to devise treatments or even cures in the future. Knowing how different people respond to drug treatments—for example, different people's bodies break down drugs at tremendously different rates—will someday allow custom-tailored treatments for diseases.

Fourth, the Human Genome Project, along with companion projects that have sequenced the genomes of organisms as diverse as bacteria, mice, and chimpanzees, helps us to appreciate our place in the evolution of life on Earth. For example, the DNA of humans and chimps differs by only about 1%. Comparing the similarities and differences may help biologists to understand what genetic differences help to make us human, and why we are susceptible to certain diseases that chimps are not.

13.6 HOW IS BIOTECHNOLOGY USED FOR MEDICAL DIAGNOSIS AND TREATMENT?

For more than a decade, biotechnology has been routinely used to diagnose some inherited disorders, even in fetuses (see "Health Watch: Prenatal Genetic Screening" on pp. 258–259). More recently, medical researchers have begun using biotechnology in an attempt to cure, or at least treat, genetic diseases.

DNA Technology Can Be Used to Diagnose Inherited Disorders

A person inherits a genetic disease because he or she inherits one or more defective alleles, which differ from normal, functional alleles because they have different nucleotide sequences. Two methods are commonly used to find out if a person carries a normal allele or a malfunctioning allele.

Restriction Enzymes May Cut Different Alleles of a Gene at Different Locations

Recall that restriction enzymes cut DNA only at specific nucleotide sequences. Because chromosomes are so large, any given restriction enzyme usually cuts the DNA of a chromosome in many places. What if two homologous chromosomes have different alleles of several genes, and some alleles have nucleotide sequences that *can* be cut by a restriction enzyme, whereas others have nucleotide sequences that *cannot* be cut by the same enzyme? The result will be a mixture of DNA segments of various lengths. These are called **restriction fragment length polymorphisms (RFLPs;** pronounced "riff-lips"). This rather daunting phrase simply means that *restriction* enzymes have cut DNA into *fragments* that vary in *length*, and that homologous chromosomes (from the same person or from different people) may differ (or be *polymorphic*) in the lengths of the fragments. Why is this useful? First, if different people have different RFLPs, this can be used to identify DNA samples. In fact, in the early 1990s, before STRs became the gold standard in DNA forensics, RFLPs were used to determine if DNA from a crime scene matched the DNA of a suspect. Second, medically important alleles can sometimes be identified by differences in the lengths of the restriction fragments produced by cutting with a specific restriction enzyme.

RFLP analysis has become a standard technique for diagnosing sickle-cell anemia, even in an embryo. Sickle-cell anemia is caused by a point mutation in which thymine replaces adenine near the beginning of the globin gene (see Table 12-4). A restriction enzyme called MstII cuts DNA just "ahead" of the globin gene locus (cut #1 in **Figs. 13-11a,b**). It also makes a cut in about the middle of both the normal and sickle-cell alleles (cut #3 in Figs. 13-11a,b). Crucially, MstII also makes a cut near the beginning of the normal globin allele (cut #2 in Fig. 13-11a), but, because of the point mutation, cannot make this cut in the sickle-cell allele (Fig. 13-11b). How can this difference be used to diagnose sickle-cell anemia?

A DNA probe is synthesized that is complementary to the part of the globin allele spanning the second cut site. When sickle-cell DNA is cut with MstII and run on a gel, this probe labels a single band near the top of the gel, consisting of very large pieces of DNA (**Fig. 13-11c**). When normal DNA is cut with MstII, the probe labels two bands, one with small pieces of DNA (near the bottom of the gel) and one with pieces that are not quite as large as the sickle-cell pieces. The genotypes of parents, children, and fetuses can be determined by this simple test. Someone who is homozygous for the normal globin allele will have two bands. Someone who is homozygous for the sickle-cell allele will have one band. A heterozygote will have three bands (see Fig. 13-11c).

Different Alleles Bind to Different DNA Probes

The Chapter 12 case study introduced you to cystic fibrosis, a disease caused by a defect in a protein, called CFTR, that normally helps to move chloride ions across the plasma membrane. There are more than 1,200 different *CFTR* alleles, all at the same locus, each encoding a different, defective CFTR protein. People with either one or two normal alleles synthesize enough functioning chloride transport proteins that they do not develop cystic fibrosis. People with two defective alleles (they may be the same or different alleles) do not synthesize fully functional transport proteins and develop cystic fibrosis.

How can anyone hope to diagnose a disorder with 1,200 different alleles? Fortunately, 32 alleles account for about 90% of the cases of cystic fibrosis—the rest of the alleles are extremely rare. Although 32 alleles is still a lot, each of these defective alleles has a different nucleotide sequence. As you know, a DNA strand will form perfect base pairs only with a perfectly complementary strand. Several companies now produce cystic fibrosis "arrays," which are pieces of specialized filter paper to which single-stranded DNA probes are bound (**Fig. 13-12**). Each probe is complementary to one strand of a different *CFTR* allele (**Fig. 13-12a**). A person's DNA is tested for cystic fibrosis by cutting

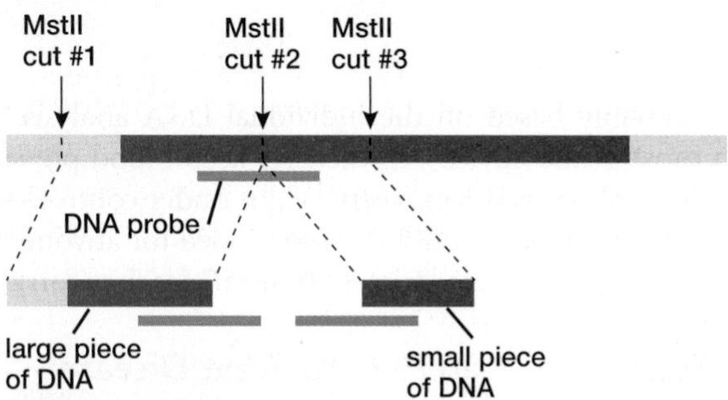

(a) MstII cuts the normal globin allele into two pieces that can be labeled by a probe

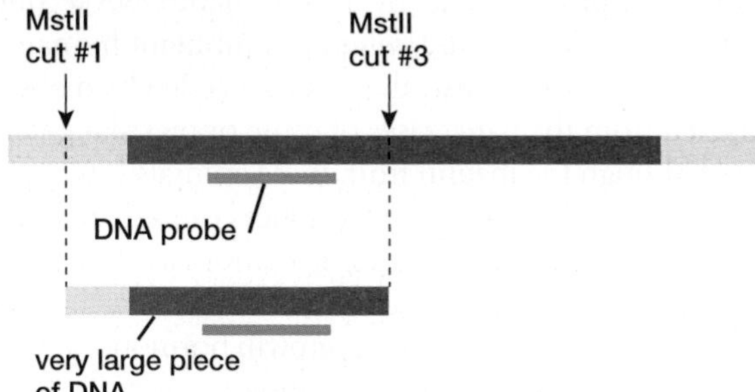

(b) MstII cuts the sickle-cell allele into one very large piece that can be labeled by the same probe

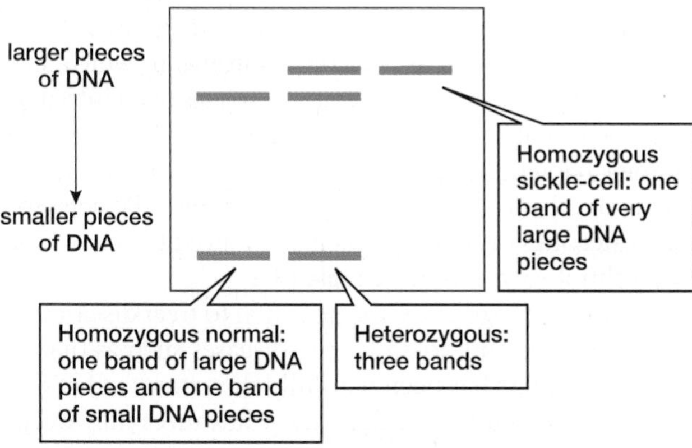

(c) Analysis of globin alleles by gel electrophoresis

▲ **FIGURE 13-11 Diagnosing sickle-cell anemia with restriction enzymes** The globin gene locus is shown in red; adjacent DNA on the chromosome is shown in yellow. **(a)** The restriction enzyme MstII cuts the normal globin allele somewhat "ahead" of the globin locus (cut site #1), near the beginning of the allele (cut site #2), and about in the middle of the allele (cut site #3). A DNA probe (blue) is synthesized that is complementary to the DNA on both sides of cut site #2. This probe can therefore label one large and one small piece of DNA from the normal allele. **(b)** MstII cuts the sickle-cell allele only at cut sites #1 and #3. The DNA probe will label only one, very large piece of DNA from the sickle-cell allele. **(c)** DNA is extracted from cells and cut with MstII. The cut DNA is separated on the gel and labeled with the DNA probe. The large single pieces of DNA from sickle-cell alleles run close to the beginning (top) of the gel, while the smaller pieces from normal alleles run farther into the gel (lower down). The pattern of DNA bands shows the genotype of the person from whom the DNA sample was obtained.

QUESTION Why do heterozygotes have three bands? Would these bands have the same intensity (brightness) as the bands from homozygotes?

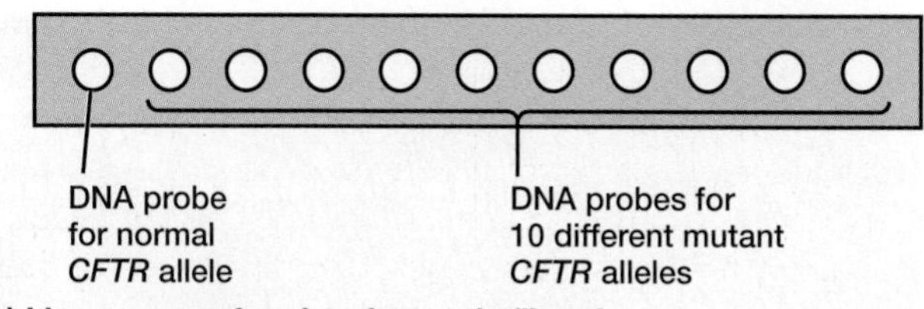

(a) Linear array of probes for cystic fibrosis

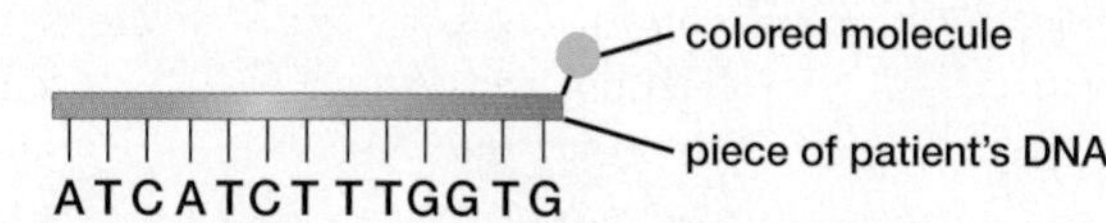

(b) *CFTR* allele labeled with a colored molecule

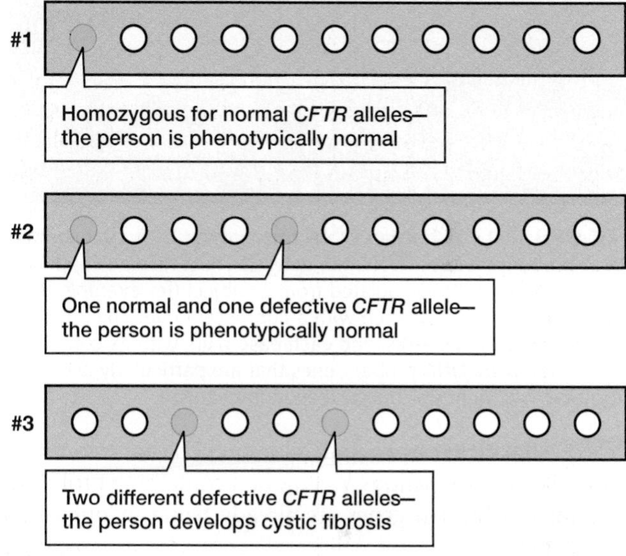

(c) Linear arrays with labeled DNA samples from three different people

▲ **FIGURE 13-12 A cystic fibrosis diagnostic array** **(a)** A "linear array" for cystic fibrosis consists of special paper to which DNA probes complementary to the normal *CFTR* allele (far left spot) and several of the most common defective *CFTR* alleles (the other 10 spots) are attached. **(b)** DNA from a patient is cut into small pieces, separated into single strands, and the *CFTR* alleles are labeled with colored molecules. **(c)** The array is bathed in a solution of the patient's labeled DNA. Different spots on the array are labeled, depending on which *CFTR* alleles the patient possesses.

it into small pieces, separating the pieces into single strands, and labeling the strands with a colored molecule (**Fig. 13-12b**). The array is then bathed in the resulting solution of labeled DNA fragments. Only a perfect complementary strand of the person's DNA will bind to any given probe on the array, thereby showing which *CFTR* alleles the person possesses (**Fig. 13-12c**).

Although not yet practical for routine medical use, an expanded version of this type of DNA analysis may one day offer customized medical care. Different people have different alleles of hundreds of genes; these may cause them to be more or less susceptible to many diseases, or to respond more or less well to various treatments. Someday, physicians might be able to use an array containing hundreds, even thousands, of probes for disease-related alleles, to determine which susceptibility

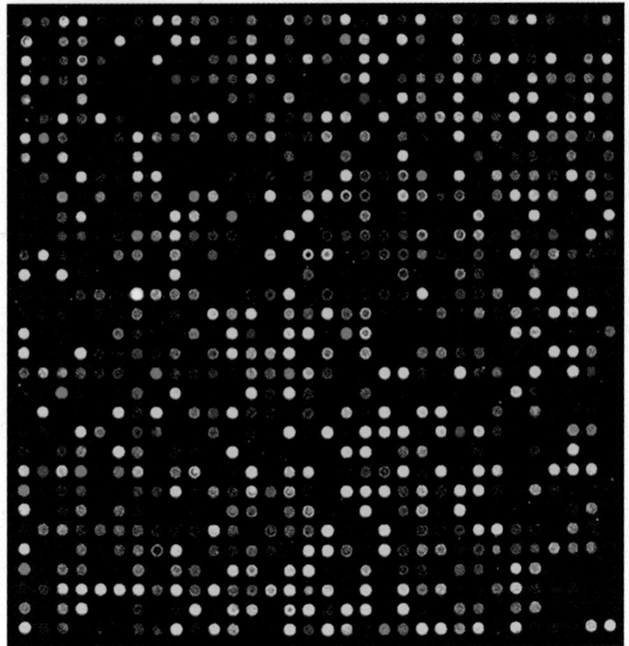

▲ **FIGURE 13-13 A human DNA microarray** Each spot contains a DNA probe for a specific human gene. In most research applications, messenger RNA is isolated from a subject (for example, from a human cancer), and labeled with a fluorescent dye. The mRNA is then poured onto the array, and each base-pairs with its complementary template DNA probe. Genes that are particularly active in the cancer will "light up" the corresponding DNA.

alleles each patient carries, and tailor medical care accordingly. Sound like science fiction? Well, arrays containing probes for thousands of human genes are already being manufactured (**Fig. 13-13**).

Several companies produce small arrays tailored to investigate gene activity in specific diseases, such as breast cancer. Some hospitals use them to provide patients with treatments that are most likely to succeed with their specific cancer.

Finally, as of 2008, at least three companies—deCODE, 23andMe, and Navigenics—offered individualized DNA screening to the public, at prices ranging from about $1,000 to $2,500. These screening tests look for alleles that might predispose an individual to develop heart disease, breast cancer, arthritis, or other diseases. The companies even offer health advice, presumably based on the individual DNA analysis, although most of the advice—to exercise, keep blood pressure and cholesterol levels down, keep weight under control, and refrain from smoking—would be a good idea for anyone, regardless of whatever disease susceptibility alleles they carry.

DNA Technology Can Help to Treat Disease

DNA technology can be used to treat disease as well as diagnose it. Thanks to recombinant DNA technology, several medically important proteins are now routinely made in bacteria. The first human protein made by recombinant DNA technology was insulin. Prior to 1982, when recombinant human insulin was first licensed for use, the insulin needed by diabetics was extracted from the pancreases of cattle or pigs slaughtered for meat. Although the insulin from these animals is very similar to human insulin, the slight differences caused an allergic reaction in about 5% of diabetics. Recombinant human insulin does not cause allergic reactions.

Other human proteins, such as growth hormone and clotting factors, can also be produced in transgenic bacteria. Before recombinant DNA technology, some of these proteins were obtained from either human blood or human cadavers; these sources are expensive and sometimes dangerous. As you know, blood can be contaminated by HIV, the virus that causes AIDS. Cadavers may also contain several hard-to-diagnose infectious diseases, such as Creutzfeldt-Jakob syndrome, in which an abnormal protein can be passed from the tissues of an infected cadaver to a patient and cause irreversible, fatal brain degeneration (see the Chapter 3 Case Study). Engineered proteins grown in bacteria or other cultured cells circumvent these dangers. Some of the categories of human proteins produced by recombinant DNA technology are listed in **Table 13-2**.

Biotechnology offers the potential to treat diseases such as cystic fibrosis and possibly cure diseases such as diabetes, although progress has been painfully slow thus far. Let's look at two specific examples of how these advances may treat, or even cure, life-threatening illnesses.

Using Biotechnology to Treat Cystic Fibrosis

Cystic fibrosis causes devastating effects in the lungs, where the lack of chloride transport causes the usually thin, watery

Table 13-2 Examples of Medical Products Produced by Recombinant DNA Methods

Type of Product	Purpose	Product	Genetic Engineering
Human hormones	Used in the treatment of diabetes and growth deficiency	Humulin™ (human insulin)	Human gene inserted into bacteria
Human cytokines (regulate immune system function)	Used in bone marrow transplants and to treat cancers and viral infections, including hepatitis and genital warts	Leukine™ (granulocyte-macrophage colony stimulating factor)	Human gene inserted into yeast
Antibodies (immune system proteins)	Used to fight infections, cancers, diabetes, organ rejection, and multiple sclerosis	Herceptin™ (antibodies to a protein expressed in some breast cancer cells)	Recombinant antibody genes inserted into cultured hamster cells
Viral proteins	Used to generate vaccines against viral diseases and for diagnosing viral infections	Energiz-B™ (hepatitis B vaccine)	Viral gene inserted into yeast
Enzymes	Used in the treatment of heart attacks, cystic fibrosis, and other diseases, and in the production of cheeses and detergents	Activase™ (tissue plasminogen activator)	Human gene inserted into cultured hamster cells

mucus lining the airways to become thick, clogging the airways (see the Case Study in Chapter 12). Several research groups are developing methods to deliver the normal *CFTR* allele to the cells of the lungs, get them to synthesize functioning CFTR chloride transport proteins, and insert the CFTR proteins into their plasma membranes.

To treat cystic fibrosis, the researchers first disable a suitable virus, so that the treatment itself doesn't cause disease. Cold viruses are often used, because they normally infect cells of the respiratory tract. The DNA of the functional *CFTR* allele is inserted into the DNA of the virus. The recombinant viruses are suspended in a solution and sprayed into the patient's nose or dripped directly into the lungs through a nasal tube. The viruses enter cells of the lungs and release the normal *CFTR* allele into the cells. The cells then manufacture functional CFTR proteins, insert them into their plasma membranes, and transport chloride into the fluid lining the airways.

Clinical trials of these treatments have been reasonably successful, but for only a few weeks. The patients' immune systems probably attack the viruses and eliminate them—and the helpful genes they carry. Because lung cells are continually replaced over time, a single dose "wears off" as the modified cells die. Researchers are trying both to increase the expression of the normal *CFTR* alleles in the viruses and to extend the effective lifetime of a single treatment.

Using Biotechnology to Cure Severe Combined Immune Deficiency

Like the cells of the lung, the vast majority of cells in the human body eventually die and are replaced by new cells. In many cases, the new cells come from special populations of cells called stem cells; when they divide, stem cells give rise to daughter cells that can differentiate into several different types of mature cells (see p. 146). In the brain, for example, stem cells give rise to several types of nerve cells and multiple types of non-nervous, support cells as well. It's possible that some stem cells, under the right conditions in the laboratory, might be able to give rise to *any* cell type of the entire body! For now, we will look at a more limited function of stem cells: producing or replacing just one or two types of cells.

All the cells of the immune system (mostly white blood cells) originate in the bone marrow. Some go on to produce antibodies, some kill cells that have been infected by viruses, and some regulate the actions of the other cells. As mature cells die, they are replaced by new cells that arise from division of stem cells in the bone marrow.

Severe combined immune deficiency (SCID) is a rare disorder in which a child fails to develop an immune system. About 1 in 80,000 children is born with some form of SCID. Infections that would be trivial in a normal child become life threatening. In some cases, if the child has an unaffected relative with a similar genetic makeup, a bone marrow transplant from the healthy relative can give the child functioning stem cells, so that he or she can develop a working immune system. Most SCID victims, however, die before their first birthday.

Most forms of SCID are single-gene, recessive defects. In some cases, children are homozygous recessive for a defective allele that normally codes for an enzyme called adenosine deaminase. In 1990, the first test of human gene therapy was performed on such a SCID patient, 4-year-old Ashanti DeSilva. Some of her white blood cells were removed, genetically altered with a virus containing a functional version of her defective allele, and then returned to her bloodstream. Now, Ashanti is a healthy adult, with a reasonably functional immune system. However, as the altered white blood cells die, they must be replaced with new ones; therefore, Ashanti needs repeated treatments. She is also given regular injections of a form of adenosine deaminase, which makes it difficult to assess the exact benefits of her gene therapy. Nevertheless, Ashanti receives a much smaller dose of adenosine deaminase than is typically given to an adult, so the gene therapy, although not perfect, is certainly making a major difference.

In 2005, Italian researchers appeared to have completely cured Ashanti's type of SCID in six children. Instead of inserting a normal copy of the adenosine deaminase gene into mature white blood cells, the Italian team inserted the gene into stem cells. Because the "cured" stem cells should continue to multiply and churn out new white blood cells, these children will probably have functioning immune systems for the rest of their lives. (In 1990, when Ashanti received her pioneering treatments, stem cell research was in its infancy. At that time, it would not have been possible to isolate her stem cells and correct their adenosine deaminase genes.)

13.7 WHAT ARE THE MAJOR ETHICAL ISSUES OF MODERN BIOTECHNOLOGY?

Modern biotechnology offers the promise—some would say the threat—of greatly changing our lives, and the lives of many other organisms on Earth. As Spider-Man noted, "With great power comes great responsibility." Is humanity capable of handling the responsibility of biotechnology? Here we will explore two important debates about biotechnology: the use of genetically modified organisms in agriculture and prospects for genetically modifying human beings.

Should Genetically Modified Organisms Be Permitted in Agriculture?

The aims of "traditional" and "modern" agricultural biotechnology are the same: to modify the genetic makeup of living organisms to make them more useful. However, there are three principal differences. First, traditional biotechnology is usually slow; many generations of selective breeding are necessary before significantly useful new traits appear in plants or animals. Genetic engineering, in contrast, can potentially introduce massive genetic changes in a single generation. Second, traditional biotechnology almost always recombines genetic material from the same, or at least very closely related, species, whereas genetic engineering can recombine DNA from very different species in one organism. Finally, traditional biotechnologists had no way to manipulate the DNA sequence of genes themselves. Genetic engineering, however, can produce new genes never before seen on Earth.

The best transgenic crops have clear advantages for farmers. Herbicide-resistant crops allow farmers to rid their fields of weeds, which reduce harvests by 10% or more, through the use of powerful, nonselective herbicides at virtually any stage of crop growth. Insect-resistant crops decrease the need to apply synthetic pesticides, saving the cost of the pesticides themselves, as well as tractor fuel and labor. Therefore, transgenic crops may produce larger harvests at less cost. These savings may be passed along to the consumer. Transgenic crops also have the potential to be more nutritious than "standard" crops (see "Health Watch: Golden Rice").

However, many people strenuously object to transgenic crops or livestock. For example, in November 2005, citizens in Switzerland voted to ban the cultivation of transgenic crops (although food made from transgenic crops, grown elsewhere, can be imported and sold). The principal concerns are that GMOs may be harmful to human health or dangerous to the environment.

Are Foods from GMOs Dangerous to Eat?

In most cases, there is no reason to think that GMOs are dangerous to eat. For example, tests have shown that the protein encoded by the *Bt* gene is not toxic to mammals, and should not prove a danger to human health. If growth-enhanced livestock are ever marketed, they will simply have more meat, composed of exactly the same proteins that exist in nontransgenic animals, so they shouldn't be dangerous either.

On the other hand, some people might be allergic to genetically modified plants. In the 1990s, a gene from Brazil nuts was inserted into soybeans in an attempt to improve the balance of amino acids in soybean protein. It was soon discovered that people allergic to Brazil nuts would probably also be allergic to the transgenic soybeans. These transgenic soybean plants never made it to the farm. The U.S. Food and Drug Administration now monitors all new transgenic crop plants for allergenic potential.

In 2003, the U.S. Society of Toxicology studied the risks of genetically modified plants and concluded that the current transgenic plants pose no significant dangers to human health. The society also recognized that past safety does not guarantee future safety, and recommended continued testing and evaluation of all new genetically modified plants. Similar conclusions were drawn by the U.S. National Academy of Sciences in 2004, which found that "the process of genetic engineering has not been shown to be inherently dangerous but . . . any technique, including genetic engineering, carries the potential to result in unintended changes in the composition of the food."

Are GMOs Hazardous to the Environment?

The environmental effects of GMOs are much more debatable. One clear positive effect of *Bt* crops is that farmers apply less pesticide to their fields. This should translate into less pollution of the environment, and less harm to the farmers as well. For example, in 2002 and 2003, Chinese farmers planting *Bt* rice reduced pesticide use by 80% compared to farmers planting conventional rice. Further, they suffered no instances of pesticide poisoning, compared to about 5% of farmers planting conventional rice. A 10-year study in Arizona showed that *Bt* cotton allowed farmers to use less pesticide while obtaining the same yields of cotton.

On the other hand, *Bt* or herbicide-resistance genes might spread outside a farmer's fields. Because these genes are incorporated into the genome of the transgenic crop, the genes will be in its pollen, too. A farmer cannot control where pollen from a transgenic crop will go. In 2006, researchers at the U.S. Environmental Protection Agency identified "escaped" herbicide-resistant grasses more than 2 miles away from a test plot in Oregon. Based on genetic analyses, some of the herbicide-resistance genes escaped in pollen (most grasses are wind-pollinated) and some escaped in seeds (most grasses have very lightweight seeds).

Does this matter? Many crops, including corn, sunflowers, and, in Eastern Europe and the Middle East, wheat, barley and oats, have wild relatives living nearby. Suppose these wild plants interbred with transgenic crops and became resistant to herbicides or pests. Would they become significant weed problems? Would they displace other plants in the wild, because they would be less likely to be eaten by insects? Even if transgenic crops have no close relatives in the wild, bacteria and viruses can carry genes from one plant to another, even between unrelated species. Could viruses spread unwanted genes into wild plant populations? No one really knows the answers to these questions.

In 2002, a committee of the U.S. National Academy of Sciences studied the potential impact of transgenic crops on the environment. The committee pointed out that crops modified by both traditional breeding methods and recombinant DNA technologies have the potential to cause major changes in the environment. In addition, the committee found that the United States does not have an adequate system for monitoring changes in ecosystems that might be caused by transgenic crops. It recommended more thorough screening of transgenic plants before they are used commercially, and sustained ecological monitoring of both the agricultural and natural environments after commercialization.

What about transgenic animals? Unlike pollen, most domesticated animals, such as cattle or sheep, are relatively immobile. Further, most have few wild relatives with which they might exchange genes, so the dangers to natural ecosystems appear minimal. However, some transgenic animals, especially fish, have the potential to pose more significant threats, because they can disperse rapidly and are nearly impossible to recapture. If they were more aggressive, grew faster, or matured faster than wild fish, they might rapidly replace native populations. One possible way out of this dilemma, pioneered by a company called Aqua Bounty, is to raise only sterile transgenic fish, so that any escapees would die without reproducing and thus have minimal impact on natural ecosystems.

Should the Genome of Humans Be Changed by Biotechnology?

Many of the ethical implications of human applications of biotechnology are fundamentally the same as those connected with other medical procedures. For example, long before biotechnology enabled prenatal testing for cystic fibrosis or sickle-cell anemia, trisomy 21 (Down syndrome) could be diagnosed in embryos by simply counting the chromosomes in cells taken from amniotic fluid (see "Health Watch: Prenatal Genetic Screening" on pp. 258–259). Whether parents should use such

Health Watch

Golden Rice

Rice is the principal food for about two-thirds of the people on Earth (**Fig. E13-2**). Rice provides carbohydrates and some protein, but is a poor source of many vitamins, including vitamin A. Unless people eat enough fruits and vegetables, they often lack sufficient vitamin A, and may suffer from poor vision, immune system defects, and damage to the respiratory, digestive, and urinary tracts. According to the World Health Organization, more than 100 million children suffer from vitamin A deficiency; as a result, each year 250,000 to 500,000 children become blind, principally in Asia, Africa, and Latin America. Vitamin A deficiency typically strikes the poor, because rice may be all they can afford to eat. In 1999, biotechnology provided a possible remedy: rice genetically engineered to contain beta-carotene, a pigment that makes daffodils bright yellow and that the human body easily converts into vitamin A.

Creating a rice strain with high levels of beta-carotene wasn't simple. However, funding from the Rockefeller Institute, the European Community Biotech Program, and the Swiss Federal Office for Education and Science enabled molecular biologists Ingo Potrykus and Peter Beyer to tackle the task. They inserted three genes into the rice genome, two from daffodils and one from a bacterium. As a result, "Golden Rice" grains synthesize beta-carotene (**Fig. E13-3**, upper right).

The trouble was, the original Golden Rice didn't make very much beta-carotene, so people would have had to eat enormous amounts to get enough vitamin A. The Golden Rice community didn't give up. It turns out that daffodils aren't the best source for genes that direct beta-carotene synthesis. Golden Rice 2, with genes from corn, produces more than 20 times more beta-carotene than the original Golden Rice (compare the rice in the upper right and the left-hand sections of Fig. E13-3). About 3 cups of cooked Golden Rice 2 should provide enough beta-carotene to equal the full recommended daily amount of vitamin A. Golden Rice 2 was given, free, to the Humanitarian Rice Board for experiments and planting in Southeast Asia.

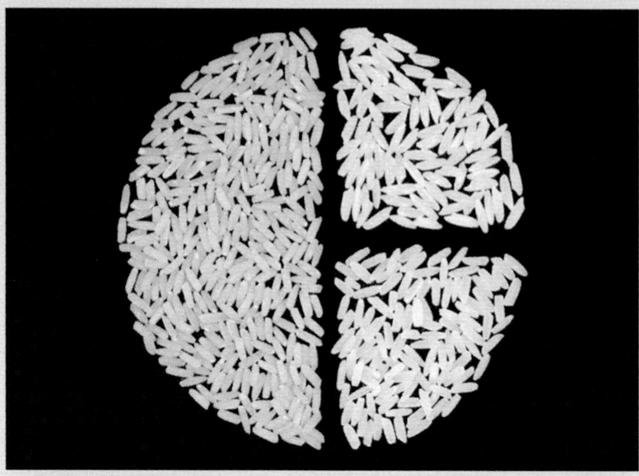

▲ **FIGURE E13-3 Golden Rice** Conventional milled rice is white or very pale tan (lower right). The original Golden Rice (upper right) was pale golden-yellow because of its increased beta-carotene content. Second-generation Golden Rice 2 (left) is much deeper yellow, because it contains about 20 times more beta-carotene than did the original Golden Rice.

▲ **FIGURE E13-2 A field of dreams?** For hundreds of millions of people, rice provides the major source of calories, but not enough vitamins and minerals. Can biotechnology improve the quality of rice and, hence, the quality of life for these people?

However, Golden Rice faces other hurdles. First, many people strongly resist large-scale planting of Golden Rice (or many other transgenic crops). Second, the strain of rice used to make Golden Rice is not a strain that is acceptable to people in southeast Asia. Getting these genes into the popular Asian strains required years of traditional genetic crosses. By 2007, the International Rice Research Institute succeeded in incorporating the carotene-synthesizing genes of Golden Rice into Asian rice strains, and the first field trials of these Golden Rice varieties began in the Philippines in April 2008.

Is Golden Rice the best way, or the only way, to solve the problems of malnutrition in poor people? Perhaps not. For one thing, many poor people's diets are deficient in many nutrients, not just vitamin A. To help solve that problem, the Bill and Melinda Gates Foundation is funding research to increase the levels of vitamin E, iron, and zinc in rice. Further, not all poor people eat mostly rice. In parts of Africa, sweet potatoes are the main source of starches. Eating orange, instead of white, sweet potatoes, has dramatically increased vitamin A intake for many of these people. Finally, in many parts of the world, governments and humanitarian organizations have started vitamin A supplementation programs. In some parts of Africa and Asia, as many as 80% of the children receive large doses of vitamin A a few times when they are very young. Some day, the combination of these efforts may result in a world in which no children suffer blindness from the lack of a simple nutrient in their diets.

Health Watch

Prenatal Genetic Screening

Prenatal diagnosis of a variety of genetic disorders, including cystic fibrosis, sickle-cell anemia, and Down syndrome, requires samples of fetal cells or chemicals produced by the fetus. Presently, three techniques are commonly used to obtain samples for prenatal diagnosis: amniocentesis, chorionic villus sampling, and maternal blood collection.

Amniocentesis

The human fetus, like all animal embryos, develops in a watery environment. A waterproof membrane called the amnion surrounds the fetus and holds the fluid. As the fetus develops, it releases various chemicals (often in its urine) and sheds some of its cells into the amniotic fluid. When a fetus is 15 weeks or older, amniotic fluid can be collected safely by a procedure called **amniocentesis.**

First, the physician determines the position of the fetus by ultrasound scanning. You probably already know how bats find moths at night: They shriek extremely high-frequency sound (far above the limits of human hearing) and listen for echoes bouncing off the moth's body. To locate a fetus with ultrasound, high-frequency sound is broadcast into a pregnant woman's abdomen, and sophisticated instruments convert the resulting echoes into a real-time image of the fetus (**Fig. E13-4**). Using the ultrasound image as a guide, the physician carefully inserts a sterilized needle through the abdominal wall, the uterus, and the amnion (being sure to avoid the fetus and placenta), and withdraws 10 to 20 milliliters of amniotic fluid (**Fig. E13-5**). Amniocentesis carries a slight risk of miscarriage, about 0.5% or less.

Chorionic Villus Sampling

The chorion is a membrane that is produced by the fetus and becomes part of the placenta. The chorion produces many small projections, called villi. In **chorionic villus sampling (CVS),** a physician inserts a small tube into the uterus through the mother's vagina and suctions off a few villi for analysis (see Fig. E13-5). The loss of a few villi does not harm the fetus. CVS has two major advantages over amniocentesis. First, it can be done much earlier in pregnancy—as early as the 8th week, but usually between the 10th and 12th weeks. This is especially important if the woman is contemplating a therapeutic abortion if her fetus has a major defect. Second, the sample contains far more fetal cells than can be obtained by amniocentesis. However, CVS appears to have a slightly greater risk of causing miscarriages than amniocentesis does. Also, because the chorion is outside of the amniotic sac, CVS does not obtain a sample of the amniotic fluid. Finally, in some cases chorionic cells have chromosomal abnormalities that are in fact not present in the fetus, which complicates karyotyping. For these reasons, CVS is less commonly performed than amniocentesis.

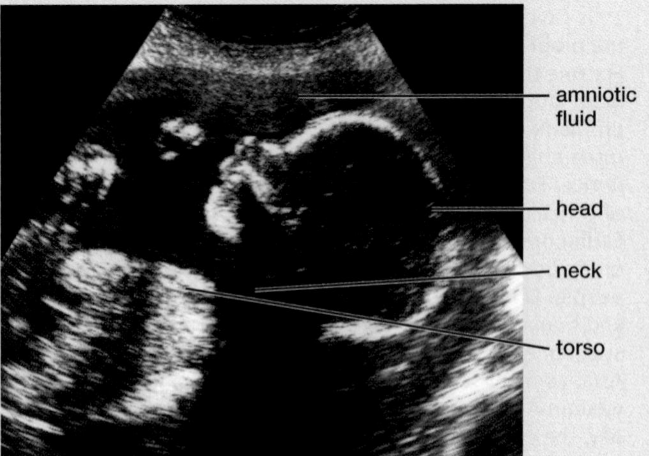

▲ FIGURE E13-4 **A human fetus imaged with ultrasound**

Maternal Blood Collection

A tiny number of fetal cells cross the placenta and enter the mother's bloodstream as early as the 6th week of pregnancy. Separating fetal cells (perhaps as few as one per milliliter of blood) from the huge numbers of maternal cells is challenging, but it can be done. Surprisingly, there is also fetal DNA floating free in the mother's blood. In addition, a variety of proteins and other chemicals produced by the fetus may enter the mother's bloodstream.

Analyzing the Samples

Both amniotic fluid and maternal blood are briefly centrifuged to separate the cells from the fluids (amniotic fluid or plasma, respectively). Biochemical analysis may be performed to measure the concentrations of various proteins, hormones, or enzymes in the fluids. For example, if the amniotic fluid contains high concentrations of an embryonic protein called alpha-fetoprotein, this indicates that the fetus may have certain nervous system disorders, such as spina bifida, in which the spinal cord is incomplete, or anencephaly, in which major portions of the brain fail to develop. Specific combinations of alpha-fetoprotein, estrogen, and other chemicals in maternal serum (called a triple screen, quad screen, or penta screen, depending on whether three, four, or five chemicals are analyzed) indicate the likelihood of Down syndrome, spina bifida, or certain other disorders. However, these biochemical assays do not provide definitive diagnoses, so other tests, such as karyotyping or DNA analysis of fetal cells, or highly detailed ultrasound examination of the fetus, are employed to ascertain whether or not the fetus actually has one of these defects.

Fetal cells, or at least fetal DNA, are required for karyotyping or DNA analysis. Prenatal diagnosis using fetal cells or fetal DNA in maternal blood was mostly an experimental procedure as of 2008, although several companies now offer paternity testing based on fetal cells and DNA in maternal blood. Amniotic fluid contains small numbers of fetal cells, so to obtain enough cells for karyotyping or DNA analysis, the usual procedure is to grow the cells in culture for a week or two. The large number of fetal cells obtained by CVS means that karyotyping and DNA analyses can usually be performed

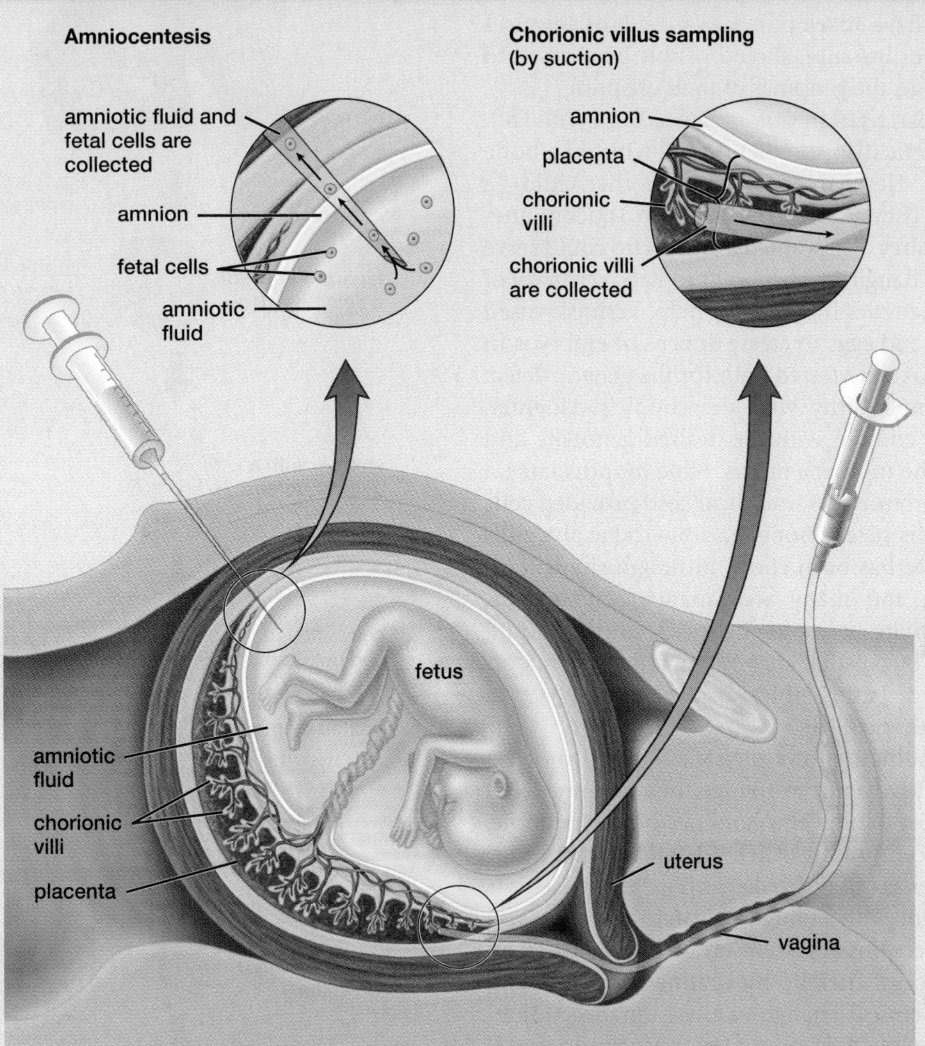

▲ **FIGURE E13-5 Prenatal sampling techniques** The two most common ways of obtaining samples for prenatal diagnosis are amniocentesis and chorionic villus sampling. (In reality, CVS is usually performed when the fetus is much younger than the one depicted in this illustration.)

without culturing the cells first (although some labs do culture the cells for a few days). Karyotyping the fetal cells can show if there are too many or too few copies of the chromosomes, and if any chromosomes show structural abnormalities. Down syndrome, for example, results from the presence of three copies of chromosome 21 (see pp. 195–196).

Various biotechnology techniques can be used to analyze fetal DNA for many defective alleles; for example, RFLP analysis for sickle-cell anemia (see Fig. 13-11) or DNA array analysis for cystic fibrosis (see Fig. 13-12). Although not yet commonly available, some labs use PCR to amplify fetal DNA before testing for Down syndrome or a few other genetic disorders. If widely adopted, PCR amplification would eliminate the need to culture fetal cells obtained by amniocentesis, sometimes for as long as 2 weeks, to obtain a large enough supply of cells and DNA.

You may think that prenatal diagnosis is useful mainly when the parents are contemplating an abortion if the fetus has a serious genetic disorder. However, prenatal diagnosis can also be used to provide better care for an affected infant. For example, if the fetus is homozygous for the sickle-cell allele, some therapeutic measures can be taken beginning at birth. In particular, regular doses of penicillin greatly reduce bacterial infections that otherwise kill about 15% of homozygous children. Further, knowing that a child has the disorder ensures correct diagnosis and rapid treatment during a "sickling crisis," when malformed red blood cells clump and block blood flow. Fetuses with spina bifida are often delivered by planned cesarean section, both to reduce damage to the spinal cord during birth and so that a neurosurgical team can be standing by in case immediate surgery on the infant is necessary. Comparable medical or behavioral interventions can be planned for several other disorders as well.

information as a basis for abortion or to prepare to care for the affected child generates enormous controversy. Other ethical concerns, however, have arisen purely as a result of advances in biotechnology. For instance, should people be allowed to select, or even change, the genomes of their offspring?

On July 4, 1994, a girl in Colorado was born with Fanconi anemia, a genetic disorder that is fatal without a bone marrow transplant. Her parents wanted another child—a very special child. They wanted one without Fanconi anemia, of course, but they also wanted a child who could serve as a donor for their daughter. They went to Yury Verlinsky of the Reproductive Genetics Institute for help. Verlinsky used the parents' sperm and eggs to create dozens of embryos in culture. The embryos were tested both for the genetic defect and for tissue compatibility with the couple's daughter. Verlinsky chose an embryo with the desired genotype and implanted it into the mother's uterus. Nine months later, a son was born. Blood from his umbilical cord provided cells to transplant into his sister's bone marrow. Today, the girl's bone marrow failure has been cured, although she will always have anemia and many accompanying symptoms. Were these appropriate uses of genetic screening? Should dozens of embryos be created, knowing that the vast majority will be discarded? Is this ethical if it is the only way to save the life of another child?

Today's technology allows physicians only to select among existing embryos, not to change their genomes. But technologies do exist to alter the genomes of, for example, bone marrow stem cells to cure SCID. What if biotechnology could change the genes of fertilized eggs (**Fig. 13-14**)? This is not possible yet, but with enough research, the time will come. Suppose it were possible to insert functional *CFTR* alleles into human eggs, thereby preventing cystic fibrosis. Would this be an ethical change to the human genome? What about making bigger and stronger football players? If and when the technology is developed to cure diseases, it will be difficult to prevent it from being used for nonmedical purposes. Who will determine which uses are appropriate and which are trivial vanity?

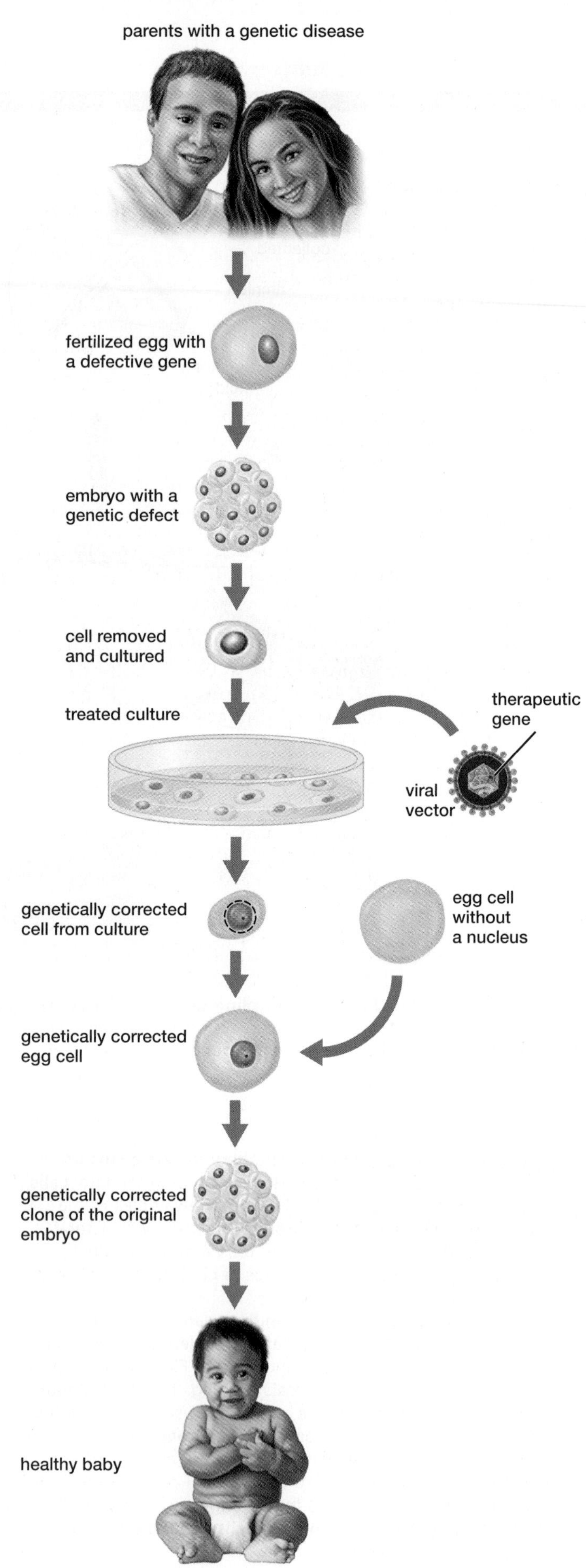

▶ **FIGURE 13-14 Using biotechnology to correct genetic defects in human embryos** In this hypothetical example, human embryos would be derived from eggs fertilized in culture dishes using sperm and eggs from a man and woman, one or both of whom have a genetic disorder. When an embryo containing a defective gene grows into a small cluster of cells, a single cell would be removed from the embryo and the defective gene replaced using an appropriate vector, usually a disabled virus. The repaired nucleus could then be implanted into another egg (taken from the same woman) whose nucleus had been removed. The repaired egg cell would then be implanted in the woman's uterus for normal development.

Case Study revisited
Guilty or Innocent?

DNA profiling is often considered the gold standard of evidence, because of its extremely low error rate. Thanks to the power of DNA profiling, Dennis Maher is now a free man.

Many people exonerated after years in prison have a hard time adjusting to life "outside." Maher, however, is a real success story. Soon after his release, he got a job servicing trucks for Waste Management, a trash pickup and recycling company. He met his wife, Melissa, through an online dating service, using the name DNADennis. They have two children, one named Aliza after his Innocence Project attorney. He's even a movie star of sorts, having been featured in *After Innocence*, an award-winning film following the lives of eight exonerated men. Does Maher harbor any resentment against the system that cost him 19 years of his life? Sure. As he puts it, "I lost what I can never get back." But he also says, "I don't have time to be angry. If I'm an angry person, I won't have the things I have in my life."

If you've ever watched the *CSI* or *Cold Case* TV shows, you know that DNA profiling also helps police all over the country to solve cases. Most police departments in large cities now have a "cold case unit" that takes a fresh look at stale evidence, sometimes leading to convictions many years after a perpetrator seemed to have gotten away with a crime. In Pasadena, California, for example, a 72-year-old woman was murdered in her home in 1987, before DNA profiling had been invented. Although police thought they knew the assailant, they didn't have enough evidence to prove it, so the man was freed. In 2007, however, Pasadena set up an Unsolved Homicide Case Unit. Their first success was to use DNA profiling to identify, arrest, and convict the murderer, now serving 15 years to life in prison.

Consider This

Who are the "heroes" in these stories? There are the obvious ones, of course—Barry Scheck and Peter Neufeld, the founders of the Innocence Project (**Fig. 13-15**), attorney Aliza Kaplan, the law students who spent many hours slogging through poorly labeled boxes searching for evidence, the members of the Pasadena Police Department, and, of course, Dennis Maher himself. But what about Thomas Brock, who discovered *Thermus aquaticus* and its unusual lifestyle in Yellowstone hot springs? Or molecular biologist Kary Mullis, who discovered PCR? Or the hundreds of biologists, chemists, and mathematicians who developed procedures for gel electrophoresis, DNA labeling, and statistical analysis of sample matching?

Scientists often say that science is worthwhile for its own sake, and that it is difficult or impossible to predict which discoveries will lead to the greatest benefits for humanity. Nonscientists, when asked to pay the costs of scientific projects, are sometimes skeptical of such claims. How do you think that public support of science should be allocated? Forty-five years ago, would you have voted to give Thomas Brock public funds to see what types of organisms lived in hot springs?

▲ **FIGURE 13-15 Peter Neufeld and Barry Scheck, cofounders of the Innocence Project**

CHAPTER REVIEW

Summary of Key Concepts

13.1 What Is Biotechnology?

Biotechnology is the use, and especially the alteration, of organisms, cells, or biological molecules to produce food, drugs, or other goods. Modern biotechnology generates altered genetic material by genetic engineering. Genetic engineering frequently involves the production of recombinant DNA by combining DNA from different organisms. When DNA is transferred from one organism to another, the recipients are called transgenic or genetically modified organisms (GMOs). Major applications of modern biotechnology include increasing our understanding of gene function, treating disease, improving agriculture, and solving crimes.

13.2 How Does DNA Recombine in Nature?

DNA recombination occurs naturally through processes such as sexual reproduction; bacterial transformation, in which bacteria acquire DNA from plasmids or other bacteria; and viral infection, in which viruses incorporate fragments of DNA from their hosts and transfer the fragments to members of the same or other species.

13.3 How Is Biotechnology Used in Forensic Science?

Specific regions of very small quantities of DNA, such as might be obtained at a crime scene, can be amplified by the polymerase chain reaction (PCR). The most common regions used in forensics are short tandem repeats (STRs). The STRs are separated by gel electrophoresis and made visible with DNA probes. The pattern of STRs, called a DNA profile, can be used to match DNA found at a crime scene with DNA from suspects with extremely high accuracy.

13.4 How Is Biotechnology Used in Agriculture?

Many crop plants have been modified by the addition of genes that promote herbicide resistance or pest resistance. The most common procedure uses restriction enzymes to insert the gene into a Ti plasmid from the bacterium *Agrobacterium tumefaciens*. The genetically modified Ti plasmid is then used to transform the bacteria, which are allowed to infect plant cells. The Ti plasmid inserts the new gene into one of the plant chromosomes. Using cell cultures, entire plants are produced from the transgenic cells and eventually grown commercially. Plants may also be modified to produce human proteins, vaccines, or antibodies. Transgenic animals may be produced as well, with properties such as faster growth, increased production of valuable products such as milk, or the ability to produce human proteins, vaccines, or antibodies.

13.5 How Is Biotechnology Used to Learn About the Human Genome?

Techniques of biotechnology were used to discover the complete nucleotide sequence of the human genome. This knowledge will be used to learn the identities and functions of new genes, to discover medically important genes, to explore genetic variability among individuals, and to better understand the evolutionary relationships between humans and other organisms.

13.6 How Is Biotechnology Used for Medical Diagnosis and Treatment?

Biotechnology may be used to diagnose genetic disorders such as sickle-cell anemia or cystic fibrosis. For example, in the diagnosis of sickle-cell anemia, restriction enzymes cut normal and defective globin alleles in different locations. The resulting DNA fragments of different lengths may then be separated and identified by gel electrophoresis. In the diagnosis of cystic fibrosis, DNA probes complementary to various cystic fibrosis alleles are placed on a DNA array. Base-pairing of a patient's DNA to specific probes on the array identifies which alleles are present in the patient.

Inherited diseases are caused by defective alleles of crucial genes. Genetic engineering may be used to insert functional alleles of these genes into normal cells, stem cells, or even into eggs to correct the genetic disorder.

13.7 What Are the Major Ethical Issues of Modern Biotechnology?

The use of genetically modified organisms in agriculture is controversial for two major reasons: consumer safety concerns and the potentially harmful effects on the environment. In general, GMOs contain proteins that are harmless to mammals, are readily digested, or are already found in other foods. The transfer of potentially allergenic proteins to normally nonallergenic foods can be avoided by thorough testing. Environmental effects of GMOs are more difficult to predict. It is possible that foreign genes, such as those for pest resistance or herbicide resistance, might be transferred to wild plants, with resulting damage to agriculture and/or disruption of ecosystems. If they escape, highly mobile transgenic animals might displace their wild relatives.

Genetically selecting or modifying human embryos is highly controversial. As technologies improve, society may be faced with decisions about the extent to which parents should be allowed to correct or enhance the genomes of their children.

Key Terms

amniocentesis *258*
biotechnology *241*
chorionic villus sampling (CVS) *258*
DNA probe *247*
DNA profile *247*
gel electrophoresis *246*
genetic engineering *241*
genetically modified organism (GMO) *241*
plasmid *242*
polymerase chain reaction (PCR) *244*
recombinant DNA *241*
restriction enzyme *250*
restriction fragment length polymorphism (RFLP) *252*
short tandem repeat (STR) *245*
transformation *242*
transgenic *241*

Thinking Through the Concepts

Fill-in-the-Blank

1. __________ are organisms that contain DNA that has been modified, usually using recombinant DNA technology, or that was derived from other species.
2. __________ is the process whereby bacteria pick up DNA, from the same or other species, from their environment. This DNA may be part of a chromosome, or it may be tiny circles of DNA called __________.
3. The __________ is a technique for multiplying DNA in the laboratory.
4. Matching DNA samples in forensics uses a specific set of small "genes" called __________. The alleles of these genes in different people vary in the __________ of the allele. The pattern of these alleles that a given person possesses is called his or her __________.
5. Pieces of DNA can be separated according to size by a process known as __________. The identity of a specific sample of DNA is usually determined by binding a synthetic piece of DNA called a(n) __________, which binds to the sample DNA by __________.

Review Questions

1. Describe three natural forms of genetic recombination, and discuss the similarities and differences between recombinant DNA technology and these natural forms of genetic recombination.
2. What is a plasmid? How are plasmids involved in bacterial transformation?
3. What is a restriction enzyme? How can restriction enzymes be used to splice a piece of human DNA into a plasmid?
4. Describe the polymerase chain reaction.
5. What is a short tandem repeat? How are short tandem repeats used in forensics?
6. How does gel electrophoresis separate pieces of DNA?
7. How are DNA probes used to identify specific nucleotide sequences of DNA? How are they used in the diagnosis of genetic disorders?
8. Describe several uses of genetic engineering in agriculture.
9. Describe several uses of genetic engineering in human medicine.
10. Describe amniocentesis and chorionic villus sampling, including the advantages and disadvantages of each. What are their medical uses?

Applying the Concepts

1. **BioEthics** In a 2004 Web survey conducted by the Canadian Museum of Nature, 84% of the people polled said they would eat a genetically modified banana that contained a vaccine against an infectious disease, whereas only 47% said they would eat a GM banana with extra vitamin C produced by the action of a rat gene. Do you think that this difference in acceptance of GMOs is scientifically valid? Why or why not?
2. **BioEthics** As you may know, many insects have evolved resistance to common pesticides. Do you think that insects might evolve resistance to *Bt* crops? If this is a risk, do you think that *Bt* crops should be planted anyway? Why or why not?
3. **BioEthics** Discuss the ethical issues that surround the release of genetically modified organisms (plants, animals, or bacteria) into the environment. What could go wrong? What precautions might prevent the problems you listed from occurring? What benefits do you think would justify the risks?
4. **BioEthics** If you were contemplating having a child, would you want both yourself and your spouse tested for the cystic fibrosis gene? If both of you were carriers, how would you deal with this decision?

MB *Go to www.masteringbiology.com for practice quizzes, activities, eText, videos, current events, and more.*

UNIT 3

Evolution and Diversity of Life

All of Earth's species, including this strikingly colored chameleon, are linked by descent from a common ancestor.

Principles of Evolution

Chapter 14

▲ This massive, earthbound ostrich has wings, a legacy of its evolutionary heritage.

Case Study

What Good Are Wisdom Teeth?

HAVE YOU HAD YOUR WISDOM TEETH REMOVED YET? If not, it's probably only a matter of time. Almost all of us will visit an oral surgeon to have our wisdom teeth extracted. There's just not enough room in our jaws for these rearmost molars, and removing them is the best way to prevent the pain, infections, and gum disease that can accompany the emergence of wisdom teeth. Removal is harmless because we don't really need wisdom teeth—they're pretty much useless.

If you've already suffered through a wisdom tooth extraction, you may have found yourself wondering why we even *have* these extra molars. Biologists hypothesize that we have them because our apelike ancestors had them and we inherited them, even though we don't need them. The presence in a living species of structures that have no current essential function, but that *are* useful in other living species, demonstrates shared ancestry among these species.

Some excellent evidence of the connection between useless traits and evolutionary ancestry is provided by flightless birds. Consider the ostrich, a bird that can grow to 8 feet tall and weigh 300 pounds (see the photo at left). These massive creatures cannot fly. Nonetheless, they have wings, just as sparrows and ducks do. Why do ostriches have wings? Because the common ancestor of sparrows, ducks, and ostriches had wings, as do all of its descendants, even those that cannot fly. The bodies of today's organisms may contain now-useless hand-me-downs from their ancestors.

At a Glance

14.1 HOW DID EVOLUTIONARY THOUGHT DEVELOP?

When you began studying biology, you may not have seen a connection between your wisdom teeth and an ostrich's wings. But the connection is there, provided by the concept that unites all of biology: **evolution,** or change over time in the characteristics of populations.

Modern biology is based on our understanding that life has evolved, but early scientists did not recognize this fundamental principle. The main ideas of evolutionary biology became widely accepted only after the publication of Charles Darwin's work in the late nineteenth century. Nonetheless, the intellectual foundation on which these ideas rest developed gradually over the centuries before Darwin's time. (You may wish to refer to the timeline in **Fig. 14-1** as you read the following historical account.)

Early Biological Thought Did Not Include the Concept of Evolution

Pre-Darwinian science, heavily influenced by theology, held that all organisms were created simultaneously by God and that each distinct life-form remained fixed and unchanging from the moment of its creation. This explanation of how life's diversity arose was elegantly expressed by the ancient Greek philosophers, especially Plato and Aristotle. Plato (427–347 B.C.) proposed that each object on Earth was merely a temporary reflection of its divinely inspired "ideal form." Plato's student Aristotle (384–322 B.C.) categorized all organisms into a linear hierarchy that he called the "Ladder of Nature" (**Fig. 14-2**).

These ideas formed the basis of the view that the form of each type of organism is permanently fixed. This view reigned unchallenged for nearly 2,000 years. By the eighteenth century,

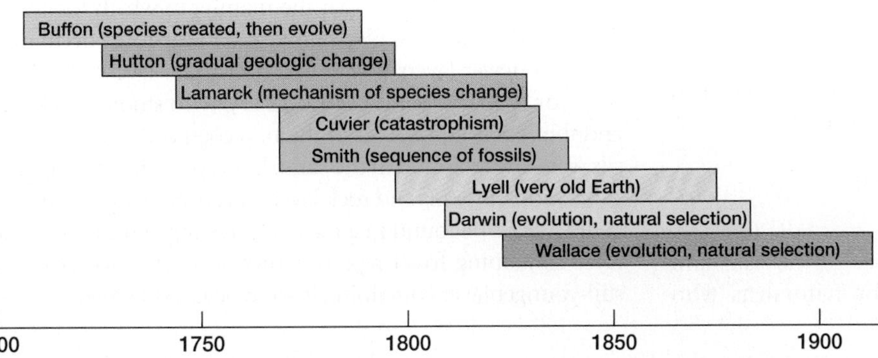

◀ **FIGURE 14-1 A timeline of the roots of evolutionary thought** Each bar represents the life span of a scientist who played a key role in the development of modern evolutionary biology.

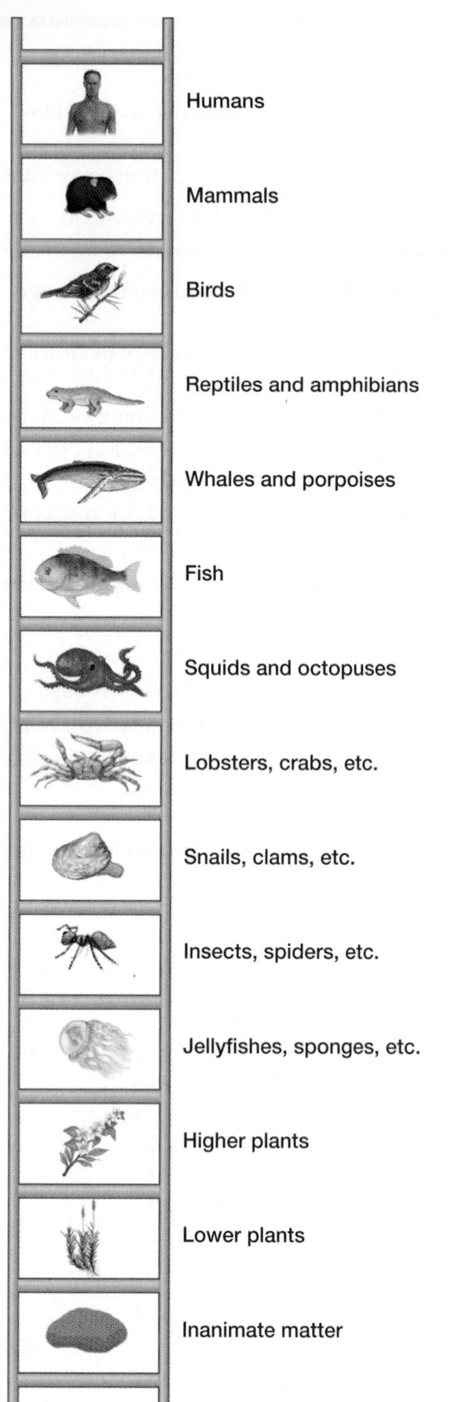

▲ **FIGURE 14-2 Aristotle's Ladder of Nature** In Aristotle's view, fixed, unchanging species can be arranged in order of increasing closeness to perfection, with inferior types at the bottom and superior types above.

however, several lines of newly emerging evidence began to erode the dominance of this static view of creation.

Exploration of New Lands Revealed a Staggering Diversity of Life

The Europeans who explored and colonized Africa, Asia, and the Americas were often accompanied by naturalists who observed and collected the plants and animals of these previously unknown (to Europeans) lands. By the 1700s, the accumulated observations and collections of the naturalists had begun to reveal the true scope of life's variety. The number of species, or different types of organisms, was much greater than anyone had suspected.

Stimulated by the new evidence of life's incredible diversity, some eighteenth-century naturalists began to take note of some fascinating patterns. They noticed, for example, that each area had its own distinctive set of species. In addition, the naturalists saw that some of the species in a given location closely resembled one another, yet differed in some characteristics. To some scientists of the day, the differences between the species of different geographical areas and the existence of clusters of similar species within areas seemed inconsistent with the idea that species were fixed and unchanging.

A Few Scientists Speculated That Life Had Evolved

A few eighteenth-century scientists went so far as to speculate that species had, in fact, changed over time. For example, the French naturalist Georges Louis LeClerc (1707–1788), known by the title Comte de Buffon, suggested that perhaps the original creation provided a relatively small number of founding species and that some modern species had been "conceived by Nature and produced by Time"—that is, they had changed over time through natural processes.

Fossil Discoveries Showed That Life Has Changed over Time

As Buffon and his contemporaries pondered the implications of new biological discoveries, developments in geology cast further doubt on the idea of permanently fixed species. Especially important was the discovery, during excavations for roads, mines, and canals, of rock fragments that resembled parts of living organisms. People had known of such objects since the fifteenth century, but most thought they were ordinary rocks that wind, water, or people had worked into lifelike forms. As more and more organism-shaped rocks were discovered, however, it became obvious that they were **fossils,** the preserved remains or traces of organisms that had died long ago (**Fig. 14-3**). Many fossils are bones, wood, shells, or their impressions in mud that have been petrified, or converted to stone. Fossils also include other kinds of preserved traces, such as tracks, burrows, pollen grains, eggs, and feces.

By the beginning of the nineteenth century, some pioneering investigators realized that the manner in which fossils were distributed in rock was also significant. Many rocks occur in layers, with newer layers positioned over older layers. The British surveyor William Smith (1769–1839), who studied rock layers and the fossils embedded in them, recognized that certain fossils were always found in the same layers of rock. Further, the organization of fossils and rock layers was consistent: Fossil type A could always be found in a rock layer resting beneath a younger layer containing fossil type B, which in turn rested beneath a still-younger layer containing fossil type C, and so on.

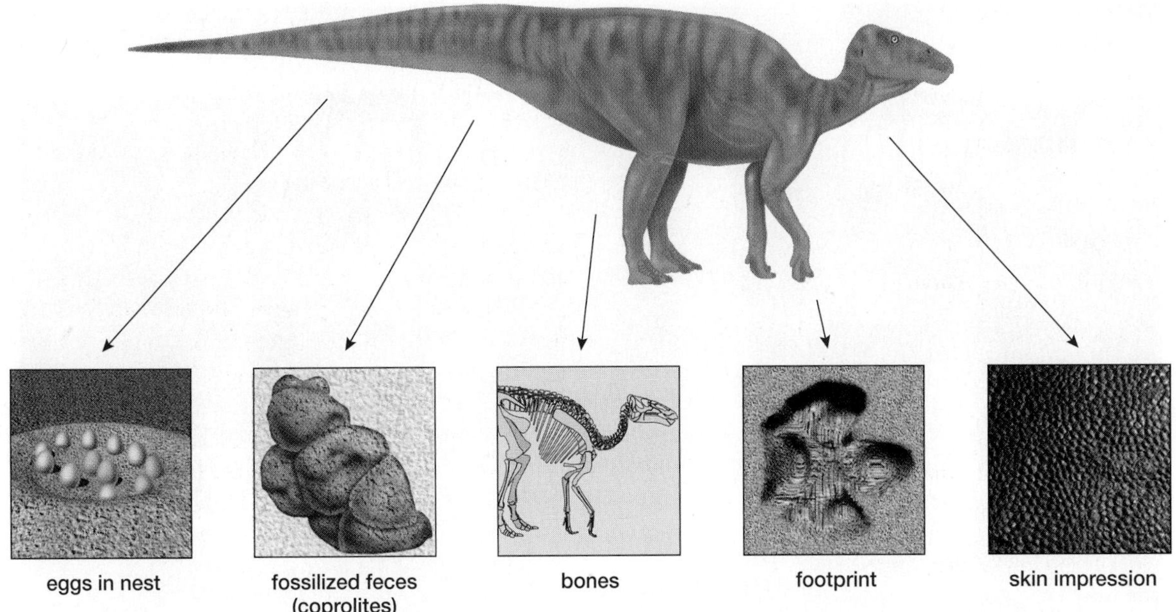

▲ **FIGURE 14-3 Types of fossils** Any preserved part or trace of an organism is a fossil.

Scientists of the period also discovered that fossil remains showed a remarkable progression. Most fossils found in the oldest layers were very different from modern organisms, and the resemblance to modern organisms gradually increased in progressively younger rocks. Many of the fossils were from plant or animal species that had gone *extinct*; that is, no members of the species still lived on Earth (**Fig. 14-4**).

Putting all of these facts together, some scientists came to an inescapable conclusion: Different types of organisms had lived at different times in the past.

Some Scientists Devised Nonevolutionary Explanations for Fossils

Despite the growing fossil evidence, many scientists of the period did not accept the proposition that species changed and new ones had arisen over time. To account for extinct species while preserving the notion of a single creation by God, Georges Cuvier (1769–1832) advanced the idea of *catastrophism*. Cuvier, a French paleontologist, hypothesized that a vast supply of species was created initially. Successive catastrophes (such as the Great Flood described in the Bible) produced layers of rock and destroyed many species, fossilizing some of their remains in the process. The organisms of the modern world, he speculated, are the species that survived the catastrophes.

Geology Provided Evidence That Earth Is Exceedingly Old

Cuvier's hypothesis of a world shaped by successive catastrophes was challenged by the work of the geologist Charles Lyell (1797–1875). Lyell, building on the earlier thinking of James Hutton (1726–1797), considered the forces of wind, water, and volcanoes and concluded that there was no need to invoke catastrophes to explain the findings of geology. Don't flooding rivers lay down layers of sediment? Don't lava flows produce layers of basalt? Shouldn't we conclude, then, that layers of rock are evidence of ordinary natural processes, occurring repeatedly over long periods of time? This concept, that Earth's present landscape was produced by past action of the same gradual geological processes that we observe today, is called *uniformitarianism*. Acceptance of uniformitarianism by scientists of the time had a profound impact, because the idea implies that Earth is very old.

Before the 1830 publication of Lyell's evidence in support of uniformitarianism, few scientists suspected that Earth could be more than a few thousand years old. Counting generations in the Old Testament, for example, yields a maximum age of 4,000 to 6,000 years. An Earth this young poses problems for the idea that life has evolved. For example, ancient writers such as Aristotle described wolves, deer, lions, and other organisms that were identical to those present in Europe more than 2,000 years later. If organisms had changed so little over that time, how could whole new species possibly have arisen if Earth was created only a couple of thousand years before Aristotle's time?

But if, as Lyell suggested, rock layers thousands of feet thick were produced by slow, natural processes, then Earth must be old indeed, many millions of years old. Lyell, in fact, concluded that Earth was eternal. Modern geologists estimate that Earth is about 4.5 billion years old (see "Scientific Inquiry: How Do We Know How Old a Fossil Is?" on p. 324).

Lyell (and his intellectual predecessor Hutton) showed that there was enough time for evolution to occur. But what was the mechanism? What process could cause evolution?

Some Pre-Darwin Biologists Proposed Mechanisms for Evolution

One of the first scientists to propose a mechanism for evolution was the French biologist Jean Baptiste Lamarck (1744–1829). Lamarck was impressed by the sequences of organisms in the rock layers. He observed that older fossils tend

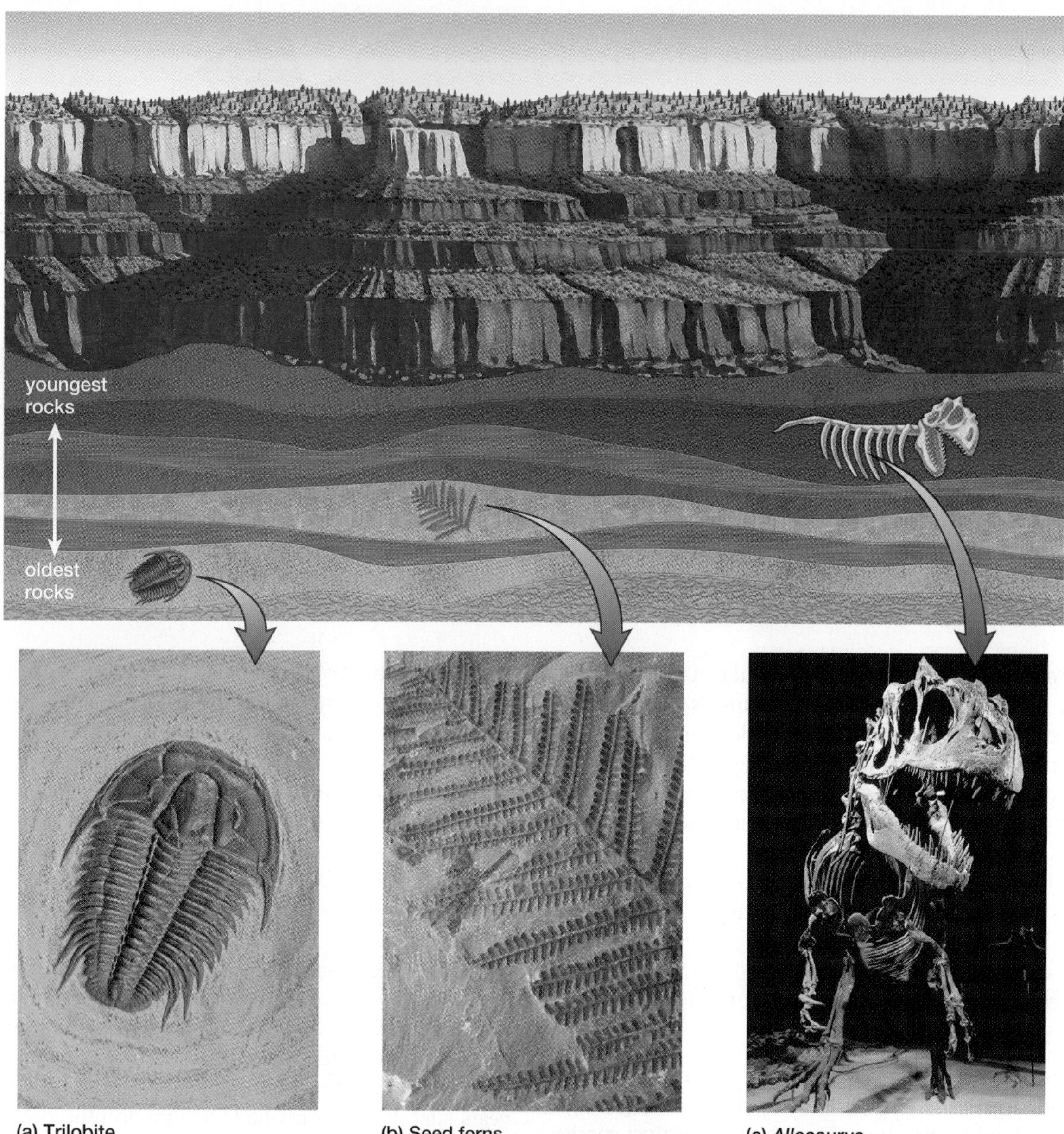

(a) Trilobite (b) Seed ferns (c) *Allosaurus*

▲ **FIGURE 14-4 Fossils of extinct organisms** Fossils provide strong support for the idea that today's organisms were not created all at once but arose over time by the process of evolution. If all species had been created simultaneously, we would not expect **(a)** the earliest trilobites to be found in older rock layers than **(b)** the earliest seed ferns, which in turn would not be expected in older layers than **(c)** dinosaurs, such as *Allosaurus.* Trilobites first appeared about 520 million years ago, seed ferns about 380 million years ago, and dinosaurs about 230 million years ago.

to be simple, whereas younger fossils tend to be more complex and more like existing organisms. In 1801, Lamarck hypothesized that organisms evolved through the inheritance of acquired characteristics, a process in which the bodies of living organisms are modified through the use or disuse of parts, and these modifications are inherited by offspring. Why would bodies be modified? Lamarck proposed that all organisms possess an innate drive for perfection. For example, if ancestral giraffes tried to improve their lot by stretching upward to feed on leaves that grow high up in trees, their necks became slightly longer as a result. Their offspring would inherit these longer necks and then stretch even farther to reach still higher leaves. Eventually, this process would produce modern giraffes with very long necks indeed.

Today, we understand how inheritance works and can see that Lamarck's proposed evolutionary process could not work as he described it. Acquired characteristics are not inherited. The fact that a prospective father pumps iron doesn't mean that his child will look like a champion body-builder. Remember, though, that in Lamarck's time the principles of inheritance had not yet been discovered. (Gregor Mendel was born a few years before Lamarck's death, and his work with inheritance in pea plants was not universally recognized until 1900—see pp. 181–182.) In any case, Lamarck's insight that inheritance plays an important role in evolution had an important influence on the later biologists who discovered the key mechanism of evolution.

Darwin and Wallace Proposed a Mechanism of Evolution

By the mid-1800s, a growing number of biologists had concluded that present-day species had evolved from earlier ones. But how? In 1858, Charles Darwin (1809–1882) and Alfred Russel Wallace (1823–1913), working separately, provided convincing evidence that evolution was driven by a simple yet powerful process.

Although their social and educational backgrounds were very different, Darwin and Wallace were quite similar in some respects. Both had traveled extensively in the tropics and had studied the plants and animals living there. Both found that some species differed in only a few features (**Fig. 14-5**). Both were familiar with the fossils that had been discovered, which showed a trend toward increasing complexity through time. Finally, both were aware of the studies of Hutton and Lyell, who had proposed that Earth is extremely ancient. These facts suggested to both Darwin and Wallace that species change over time. Both men sought a mechanism that might cause such evolutionary change.

Of the two, Darwin was the first to propose a mechanism for evolution, which he described in a paper he wrote in 1842. But he did not submit the paper for publication, perhaps because he was fearful of the controversy that publication would cause. Some historians wonder if Darwin would ever have publicized his ideas had he not received, 16 years later, a draft of a paper by Wallace that outlined ideas remarkably similar to Darwin's own. Darwin realized that he could delay no longer.

In separate but similar papers that were presented to the Linnaean Society in London in 1858, Darwin and Wallace each described the same mechanism for evolution. Initially, their papers had little impact. The secretary of the society, in fact, wrote in his annual report that nothing very interesting happened that year. Fortunately, the next year, Darwin published his monumental book, *On the Origin of Species by Means of Natural Selection*, which attracted a great deal of attention to the new ideas about how species evolve. (To learn more about Darwin's life, see "Scientific Inquiry: Charles Darwin—Nature Was His Laboratory" on p. 272.)

14.2 HOW DOES NATURAL SELECTION WORK?

Darwin and Wallace proposed that life's huge variety of excellent designs arose by a process of descent with modification, in which individuals in each generation differ slightly from the members of the preceding generation. Over long stretches of time, these small differences accumulate to produce major transformations.

Darwin and Wallace's Theory Rests on Four Postulates

The chain of logic that led Darwin and Wallace to their proposed process of evolution turns out to be surprisingly simple and

(a) Large ground finch, beak suited to large seeds

(b) Small ground finch, beak suited to small seeds

(c) Warbler finch, beak suited to insects

(d) Vegetarian tree finch, beak suited to leaves

◄ **FIGURE 14-5 Darwin's finches, residents of the Galapágos Islands**
Darwin studied a group of closely related species of finches on the Galapágos Islands. Each species specializes in eating a different type of food and has a beak of characteristic size and shape, because natural selection has favored the individuals best suited to exploit each local food source efficiently.

Scientific Inquiry

Charles Darwin—Nature Was His Laboratory

In 1831, when Charles Darwin was 22 years old (**Fig. E14-1**), he secured a position as "gentleman companion" to Captain Robert Fitzroy of the HMS *Beagle*. The Beagle soon embarked on a 5-year surveying expedition along the coastline of South America and then around the world.

Darwin's voyage on the *Beagle* sowed the seeds for his theory of evolution. In addition to his duties as companion to the captain, Darwin served as the expedition's official naturalist, whose task was to observe and collect geological and biological specimens. The *Beagle* sailed to South America and made many stops along its coast. There Darwin observed the plants and animals of the tropics and was stunned by the greater diversity of species compared with that of Europe.

Although he had boarded the *Beagle* convinced of the permanence of species, Darwin's experiences soon led him to doubt it. He discovered a snake with rudimentary hind limbs, calling it "the passage by which Nature joins the lizards to the snakes" (**Fig. E14-2**). Darwin also noticed that penguins used their wings to paddle through the water rather than fly through the air, and he observed a snake that had no rattles but vibrated its tail like a rattlesnake. If a creator had individually created each animal in its present form, to suit its present environment, what could be the purpose behind these makeshift arrangements?

Perhaps the most significant stopover of the voyage was the month spent on the Galapágos Islands off the northwestern coast of South America. There, Darwin found huge tortoises. Different islands were home to distinctively different types of tortoises. Darwin also found several types of finches and, as with the tortoises, different islands had slightly different finches. Could the differences in these organisms have arisen after they became isolated from one another on separate islands? The diversity of tortoises and finches haunted him for years afterward.

▲ **FIGURE E14-1 A painting of Charles Darwin as a young man**

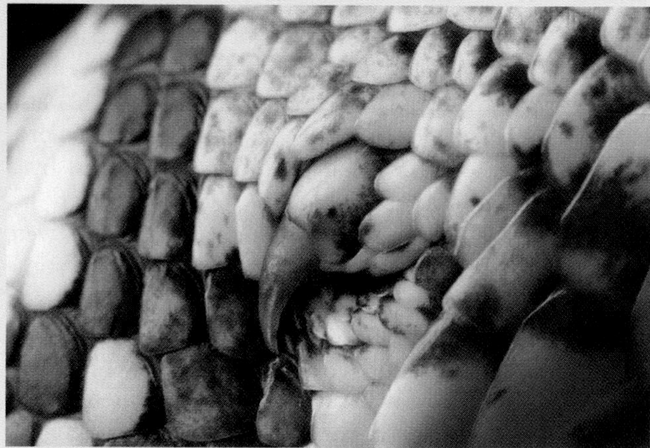

▲ FIGURE E14-2 **The vestigial remnants of hind legs in a snake** Some snakes have small "spurs" where their distant ancestors had rear legs. In some species, these vestigial structures even retain claws.

In 1836, Darwin returned to England and became established as one of the foremost naturalists of his time. But the problem of how isolated populations come to differ from each other gnawed constantly at his mind.

From his experience as a naturalist, Darwin realized that members of a species typically compete with one another for survival. He also recognized that the competition's winners and losers were determined not by chance, but by the characteristics and abilities of the individual competitors. As his colleague Alfred Wallace put it, "Those which, year by year, survived this terrible destruction must be, on the whole, those which have some little superiority enabling them to escape each special form of death to which the great majority succumbed." Darwin understood that if these "little superiorities" were inherited by their possessors' offspring, species would evolve by natural selection. He wrote that "under these circumstances favorable variations would tend to be preserved, and unfavorable ones to be destroyed" and that "the result . . . would be the formation of new species. Here, then, I had at last got a theory by which to work."

When Darwin finally published *On the Origin of Species* in 1859, his evidence had become truly overwhelming. Although its full implications would not be realized for decades, Darwin's theory of evolution by natural selection has become a unifying concept for virtually all of biology.

straightforward. It is based on four postulates about **populations,** or all the individuals of one species in a particular area.

Postulate 1 Individual members of a population differ from one another in many respects.
Postulate 2 At least some of the differences among members of a population are due to characteristics that may be passed from parent to offspring.
Postulate 3 In each generation, some individuals in a population survive and reproduce successfully but others do not.
Postulate 4 The fate of individuals is not determined entirely by chance or luck. Instead, an individual's likelihood of survival and reproduction depends on its characteristics. Individuals with advantageous traits survive longest and leave the most offspring, a process known as **natural selection.**

Darwin and Wallace understood that if all four postulates were true, populations would inevitably change over time. If members of a population have different traits, and if the individuals that are best suited to their environment leave more offspring, and if those individuals pass their favorable traits to the next generation, then the favorable traits will be more common in subsequent generations. The characteristics of the population will change slightly with each generation. This process is evolution by natural selection.

Are the four postulates true? Darwin thought so, and devoted much of *On the Origin of Species* to describing supporting evidence. Let's briefly examine each postulate, in some cases with the advantage of knowledge that was not available to Darwin and Wallace.

Postulate 1: Individuals in a Population Vary

The accuracy of postulate 1 is apparent to anyone who has glanced around a crowded room. People differ in size, eye color, skin color, and many other physical features. Similar variability is present in populations of other organisms, although it may be less obvious to the casual observer (**Fig. 14-6**). We now know that the variations in natural populations arise purely by chance, as a result of random mutations in DNA (see pp. 213–214, 229–231). Thus, the differences among individuals extend to the molecular level. The reason that DNA tests can match blood from a crime scene to a suspect is that each person's exact DNA sequence is unique.

Postulate 2: Traits Are Passed from Parent to Offspring

The principles of genetics had not yet been discovered when Darwin published *On the Origin of Species*. Therefore, although observation of people, pets, and farm animals seemed to show that offspring generally resemble their parents, Darwin and Wallace did not have scientific evidence in support of postulate 2. Mendel's later work, however, demonstrated conclusively that particular traits can be passed to offspring. Since Mendel's time, genetics researchers have produced a detailed picture of how inheritance works.

▲ **FIGURE 14-6 Variation in a population of snails** Although these snails are all members of the same population, no two are exactly alike.

QUESTION Is sexual reproduction required to generate the variability in structures and behaviors that is necessary for natural selection?

Postulate 3: Some Individuals Fail to Survive and Reproduce

Darwin's formulation of postulate 3 was heavily influenced by Thomas Malthus's *Essay on the Principle of Population* (1798), which described the perils of unchecked growth of human populations. Darwin was keenly aware that organisms can produce far more offspring than are required merely to replace the parents. He calculated, for example, that a single pair of elephants would multiply to a population of 19 million in 750 years if each descendant had six offspring.

But we aren't overrun with elephants. The number of elephants, like the number of individuals in most natural populations, tends to remain relatively constant. Therefore, more organisms must be born than survive long enough to reproduce. In each generation, many individuals must die young. Even among those that survive, many must fail to reproduce, produce few offspring, or produce less-vigorous offspring that, in turn, fail to survive and reproduce. As you might expect, whenever biologists have measured reproduction in a population, they have found that some individuals have more offspring than others.

Postulate 4: Survival and Reproduction Are Not Determined by Chance

If unequal reproduction is the norm in populations, what determines which individuals leave the most offspring? A large amount of scientific evidence has shown that reproductive success depends on an individual's characteristics. For example, scientists found that larger male elephant seals in a California population had more offspring than smaller males (because females preferred to mate with large males). In a Colorado population, snapdragon plants with white flowers had more offspring than plants with yellow flowers (because pollinators found white flowers more attractive). In a hospital patient's body, antibiotic-resistant tuberculosis bacteria reproduced more rapidly than antibiotic-sensitive ones (because the patient was being treated with antibiotics). These results, and hundreds of other similar ones, show that in the competition to survive and reproduce, winners are determined not by chance but by the traits they possess.

Natural Selection Modifies Populations over Time

Observation and experiment suggest that the four postulates of Darwin and Wallace are sound. Logic suggests that the resulting consequence ought to be change over time in the characteristics of populations. In *On the Origin of Species,* Darwin proposed the following example: "Let us take the case of a wolf, which preys on various animals, securing [them] by . . . fleetness. . . . The swiftest and slimmest wolves would have the best chance of surviving, and so be preserved or selected. . . . Now if any slight innate change of habit or structure benefited an individual wolf, it would have the best chance of surviving and of leaving offspring. Some of its young would probably inherit the same habits or structure, and by the repetition of this process, a new variety might be formed." The same logic applies to the wolf's prey; the fastest or most alert would be most likely to avoid predation and would pass these traits to its offspring.

Notice that natural selection acts on individuals within a population. Selection's influence on the fates of individuals eventually has consequences for the population as a whole. Over generations, the population changes as the percentage of individuals inheriting favorable traits increases. An individual cannot evolve, but a population can.

Although it is easier to understand how natural selection would cause changes within a species, under the right circumstances, the process might produce entirely *new* species. We will discuss the circumstances that might give rise to new species in Chapter 16.

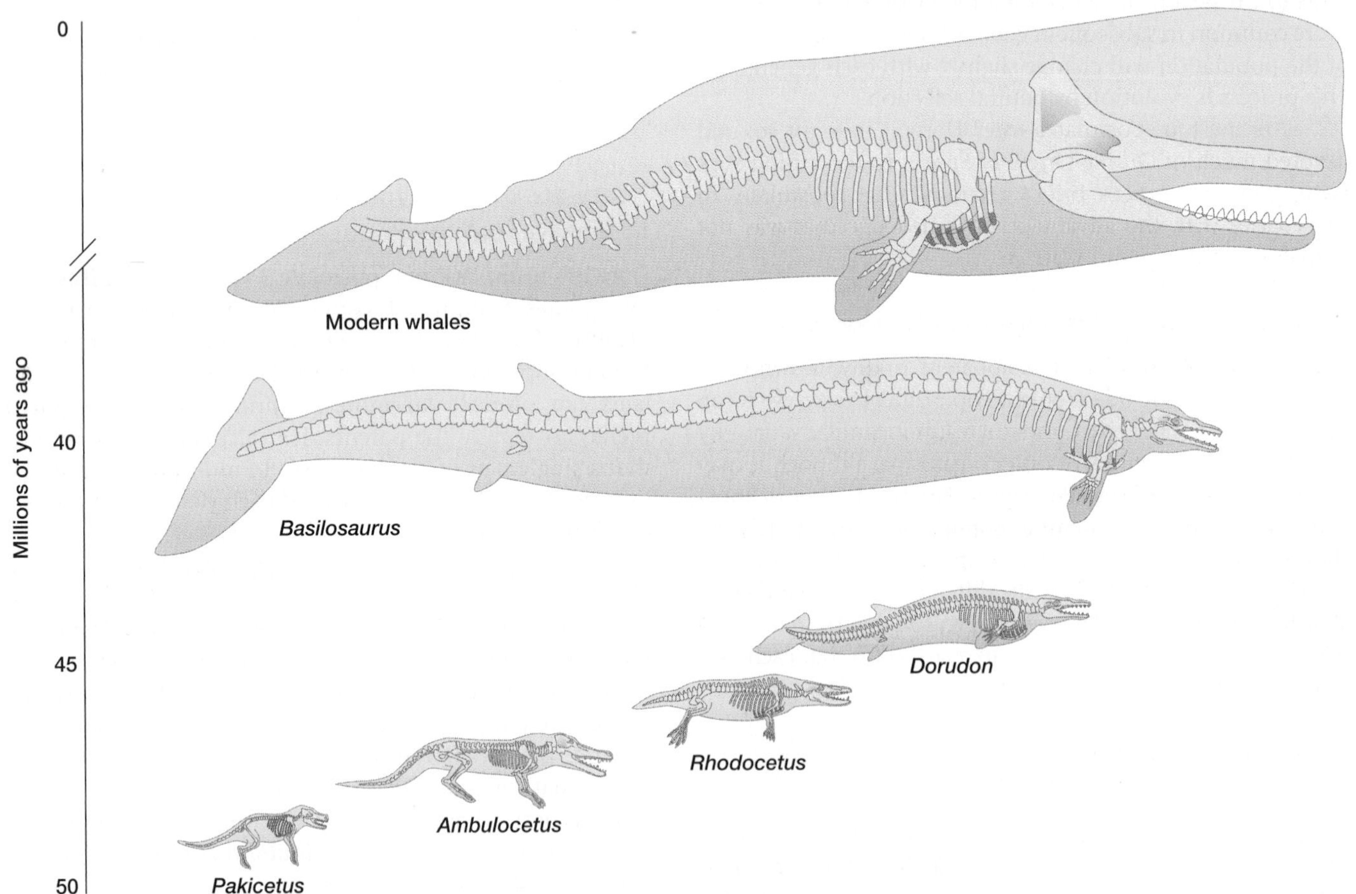

▲ **FIGURE 14-7 The evolution of the whale** During the past 50 million years, whales have evolved from four-legged land dwellers, to semi-aquatic paddlers, to fully aquatic swimmers with shrunken hind legs, to today's sleek ocean dwellers.

QUESTION The fossil history of some kinds of modern organisms, such as sharks and crocodiles, shows that their structure and appearance have changed very little over hundreds of millions of years. Is this lack of change evidence that such organisms have not evolved over that time?

14.3 HOW DO WE KNOW THAT EVOLUTION HAS OCCURRED?

Today, evolution is an accepted scientific theory. You may recall from Chapter 1 that a scientific theory is a general explanation of important natural phenomena, developed through extensive, reproducible observations. An overwhelming body of evidence supports the conclusion that evolution has occurred. The key lines of evidence come from fossils, comparative anatomy (the study of how body structures differ among species), embryology, biochemistry, and genetics.

Fossils Provide Evidence of Evolutionary Change over Time

If it is true that many fossils are the remains of species ancestral to modern species, we might expect to find progressive series of fossils that start with an ancient organism, progress through several intermediate stages, and culminate in a modern species. Such series have indeed been found. For example, fossils of the ancestors of modern whales illustrate stages in the evolution of an aquatic species from land-dwelling ancestors (**Fig. 14-7**). Series of fossil giraffes, elephants, horses, and mollusks also show the evolution of body structures over time. These fossil series suggest that new species evolved from, and replaced, previous species.

Comparative Anatomy Gives Evidence of Descent with Modification

Fossils provide snapshots of the past that allow biologists to trace evolutionary changes, but careful examination of today's organisms can also uncover evidence of evolution. Comparing the bodies of organisms of different species can reveal similarities that can be explained only by shared ancestry and differences that could result only from evolutionary change during descent from a common ancestor. In this way, the study of comparative anatomy has supplied strong evidence that different species are linked by a common evolutionary heritage.

Homologous Structures Provide Evidence of Common Ancestry

A body structure may be modified by evolution to serve different functions in different species. The forelimbs of birds and mammals, for example, are variously used for flying, swimming, running over several types of terrain, and grasping objects, such as branches and tools. Despite this enormous diversity of function, the internal anatomy of all bird and mammal forelimbs is remarkably similar (**Fig. 14-8**). It seems inconceivable that the same bone arrangements would be used to serve such diverse functions if each animal had been created separately. Such similarity is exactly what we would

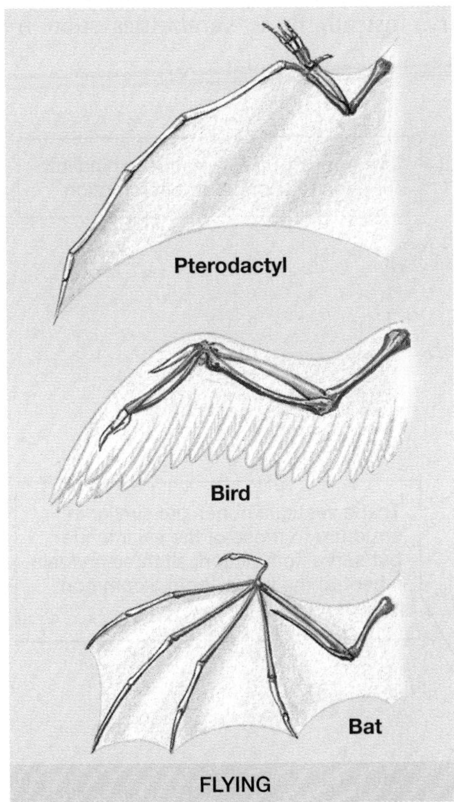

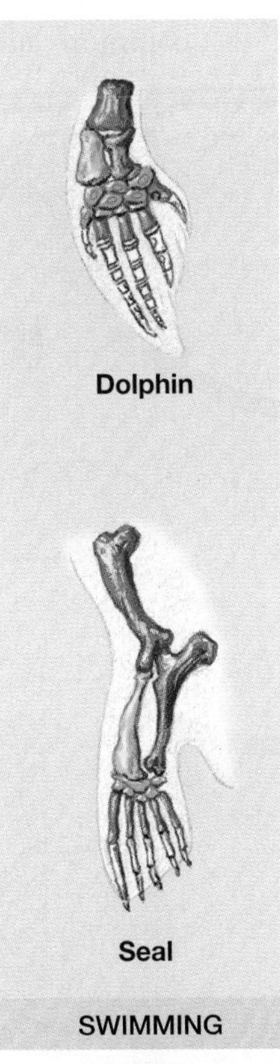

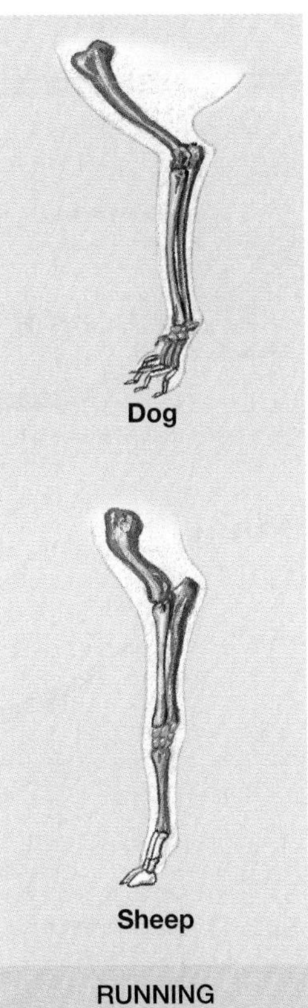

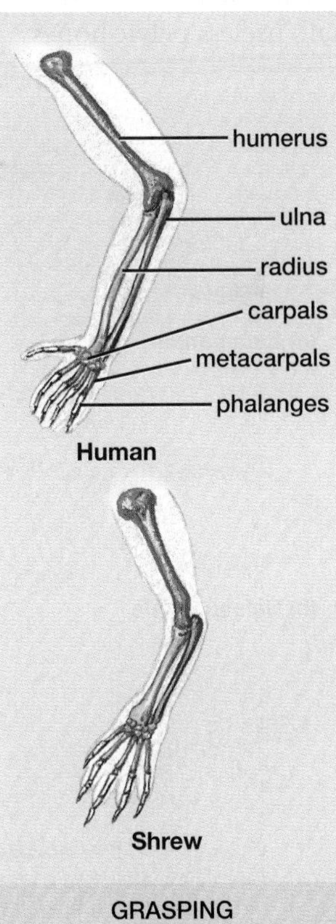

▲ FIGURE 14-8 **Homologous structures** Despite wide differences in function, the forelimbs of all of these animals contain the same set of bones, inherited from a common ancestor. The different colors of the bones highlight the correspondences among the various species.

QUESTION Compile a list of human vestigial structures. For each structure, name the corresponding homologous structure in a nonhuman species.

expect, however, if bird and mammal forelimbs were derived from a common ancestor. Through natural selection, the ancestral forelimb has undergone different modifications in different kinds of animals. The resulting internally similar structures are called **homologous structures,** meaning that they have the same evolutionary origin despite any differences in current function or appearance.

Functionless Structures Are Inherited from Ancestors

Evolution by natural selection also helps explain the curious circumstance of **vestigial structures** that serve no apparent purpose. Examples include such things as molar teeth in vampire bats (which live on a diet of blood and, therefore, don't chew their food) and pelvic bones in whales and certain snakes (**Fig. 14-9**). Both of these vestigial structures are clearly homologous to structures that are found in—and used by—other vertebrates (animals with a backbone). Their continued existence in animals that have no use for them is best explained as a sort of "evolutionary baggage." For example, the ancestral mammals from which whales evolved had four legs and a well-developed set of pelvic bones (see Fig. 14-7). Whales do not have hind legs, yet they have small pelvic and leg bones embedded in their sides. During whale evolution, losing the hind legs provided an advantage, better streamlining the body for movement through water. The result is the modern whale with small, useless pelvic bones.

Have You Ever Wondered

Why Backaches Are So Common?

Between 70% and 85% of people will experience lower back pain at some point in life and, for many people, the condition is chronic. This state of affairs is an unfortunately painful consequence of the evolutionary process. We walk upright on two legs, but our distant ancestors walked on all fours. Thus, natural selection formed our vertically oriented spine by remodeling one whose normal orientation was parallel to the ground. Our spinal anatomy evolved some modifications in response to its new posture but, as is often the case with evolution, the changes involved some trade-offs. The arrangements of bone and muscle that permit our smooth, bipedal gait also generate vertical compression of the spine, and the resulting pressure can, and frequently does, cause painful damage to muscle and nerve tissues.

Some Anatomical Similarities Result from Evolution in Similar Environments

The study of comparative anatomy has demonstrated the shared ancestry of life by identifying a host of homologous structures that different species have inherited from common ancestors, but comparative anatomists have also identified many anatomical similarities that do not stem from common ancestry. Instead, these similarities stem from

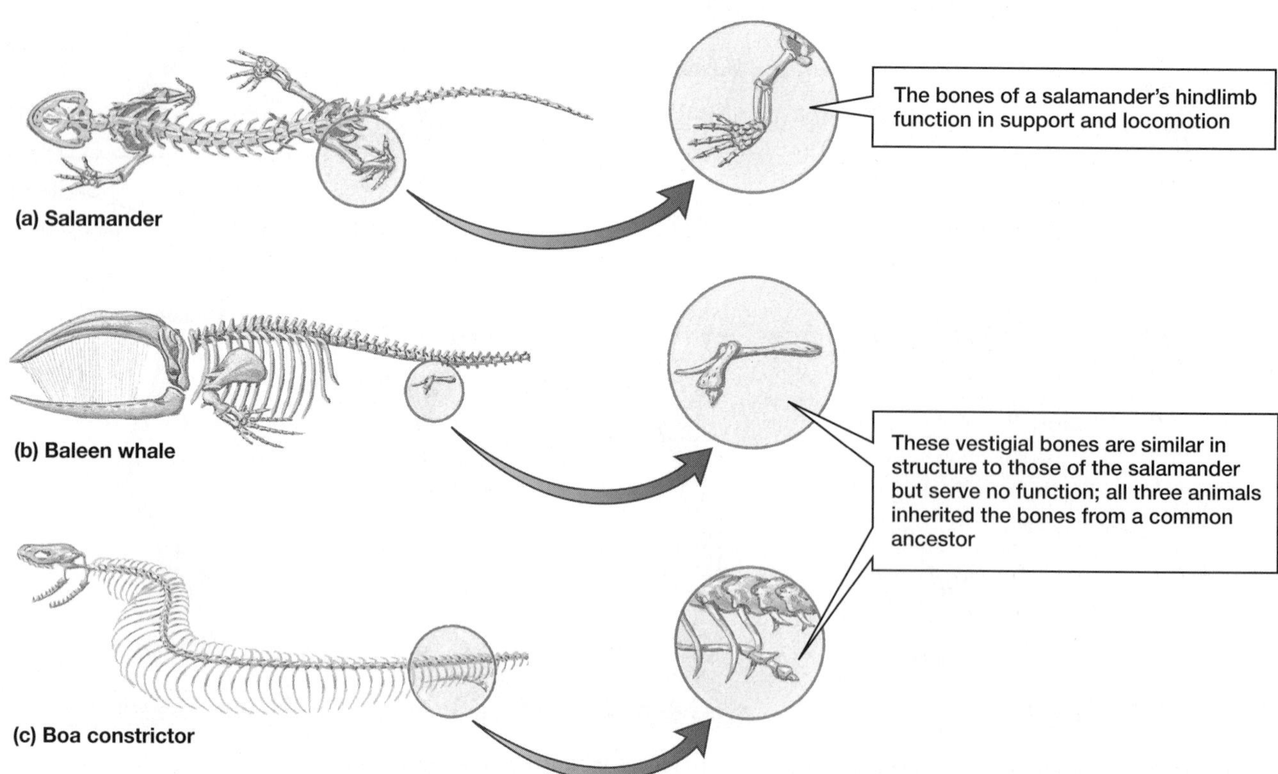

▲ FIGURE 14-9 **Vestigial structures** Many organisms have vestigial structures that serve no apparent function. The **(a)** salamander, **(b)** baleen whale, and **(c)** boa constrictor all inherited hind limb bones from a common ancestor. These bones remain functional in the salamander but are vestigial in the whale and snake.

convergent evolution, in which natural selection causes non-homologous structures that serve similar functions to resemble one another. For example, both birds and insects have wings, but this similarity did not arise from evolutionary modification of a structure that both birds and insects inherited from a common ancestor. Instead, the similarity arose from modification of two different, non-homologous structures that eventually gave rise to superficially similar structures. Because natural selection favored flight in both birds and insects, the two groups evolved superficially similar structures—wings—that are useful for flight. Such outwardly similar but non-homologous structures are called **analogous structures** (Fig. 14-10). Analogous structures are typically very different in internal anatomy, because the parts are not derived from common ancestral structures.

(a) Damselfly

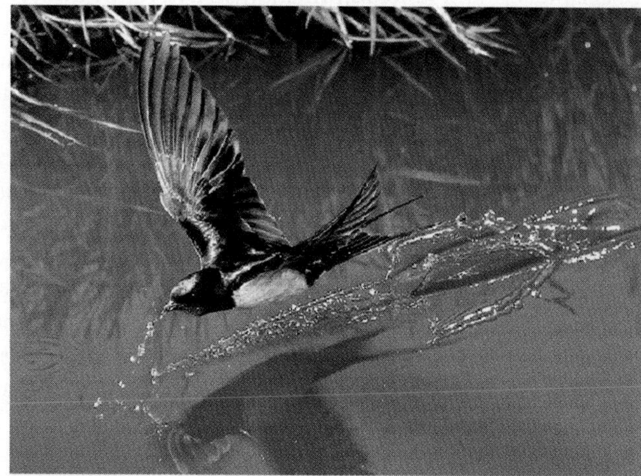

(b) Swallow

▲ **FIGURE 14-10 Analogous structures** Convergent evolution can produce outwardly similar structures that differ anatomically, such as the wings of **(a)** insects and **(b)** birds.

QUESTION Are a peacock's tail and a dog's tail homologous structures or analogous structures?

Embryological Similarity Suggests Common Ancestry

In the early 1800s, German embryologist Karl von Baer noted that all vertebrate embryos (developing organisms in the period from fertilization to birth or hatching) look quite similar to one another early in their development (**Fig. 14-11**). In their early embryonic stages, fish, turtles, chickens, mice, and humans all develop tails and gill slits (also called gill grooves). But among this group of animals, only fish retain gills as adults, and only fish, turtles, and mice retain substantial tails.

Why do vertebrates that are so different have similar developmental stages? The only plausible explanation is that ancestral vertebrates possessed genes that directed the development of gills and tails. All of their descendants still have those genes. In fish, these genes are active throughout development, resulting in adults with fully developed tails and gills. In humans and chickens, these genes are active only during early developmental stages, and the structures are lost or inconspicuous in adults.

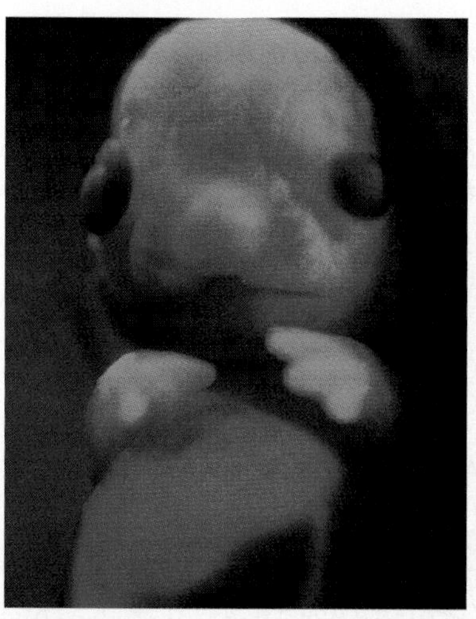

(a) Lemur

(b) Pig

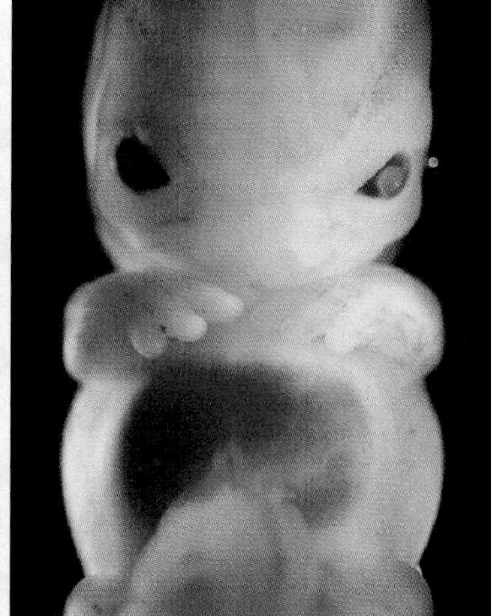

(c) Human

▲ **FIGURE 14-11 Embryological stages reveal evolutionary relationships** Early embryonic stages of a **(a)** lemur, **(b)** pig, and **(c)** human, showing strikingly similar anatomical features.

Modern Biochemical and Genetic Analyses Reveal Relatedness Among Diverse Organisms

Biologists have been aware of anatomical and embryological similarities among organisms for centuries, but it took the emergence of modern technology to reveal similarity at the molecular level. Biochemical similarities among organisms provide perhaps the most striking evidence of their evolutionary relatedness. Just as relatedness is revealed by homologous anatomical structures, it is also revealed by homologous molecules.

Today's scientists have access to a powerful tool for revealing molecular homologies: DNA sequencing. It is now possible to quickly determine the sequence of nucleotides in a DNA molecule and to compare the DNA of different organisms. For example, consider the gene that encodes the protein cytochrome *c* (see Chapters 11 and 12 for information on DNA and how it encodes proteins). Cytochrome *c* is present in all plants and animals (and many single-celled organisms) and performs the same function in all of them. The sequence of nucleotides in the gene for cytochrome *c* is similar in these diverse species (**Fig. 14-12**). The widespread presence of the same complex protein, encoded by the same gene and performing the same function, is evidence that the common ancestor of plants and animals had cytochrome *c* in its cells. At the same time, though, the sequence of the cytochrome *c* gene differs slightly in different species, showing that variations arose during the independent evolution of Earth's multitude of plant and animal species.

Some biochemical similarities are so fundamental that they extend to all living cells. For example:

- All cells have DNA as the carrier of genetic information.
- All cells use RNA, ribosomes, and approximately the same genetic code to translate that genetic information into proteins.
- All cells use roughly the same set of 20 amino acids to build proteins.
- All cells use ATP as a cellular energy carrier.

The most plausible explanation for such widespread sharing of complex and specific biochemical traits is that the traits are homologies. That is, they arose only once, in the common ancestor of all living things, from which all of today's organisms inherited them.

Case Study continued

What Good Are Wisdom Teeth?

Just as anatomical homology can lead to vestigial structures such as human wisdom teeth and the wings of flightless birds, genetic homology can lead to vestigial DNA sequences. For example, many mammal species produce an enzyme, L-gulonolactone oxidase, that catalyzes the last step in the production of vitamin C. The species that produce the enzyme are able to do so because they all inherited the gene that encodes it from a common ancestor. Humans, however, do not produce L-gulonolactone oxidase, so we can't produce vitamin C ourselves and must consume it in our diets. But even though we don't produce the enzyme, our cells do contain a stretch of DNA with a sequence very similar to that of the enzyme-producing gene present in rats and most other mammals. The human version, though, does not encode the enzyme (or any protein). We inherited this stretch of DNA from an ancestor that we share with other mammal species, but in us, the sequence has undergone a change that rendered it nonfunctional. It remains as a vestigial trait, evidence of our shared ancestry.

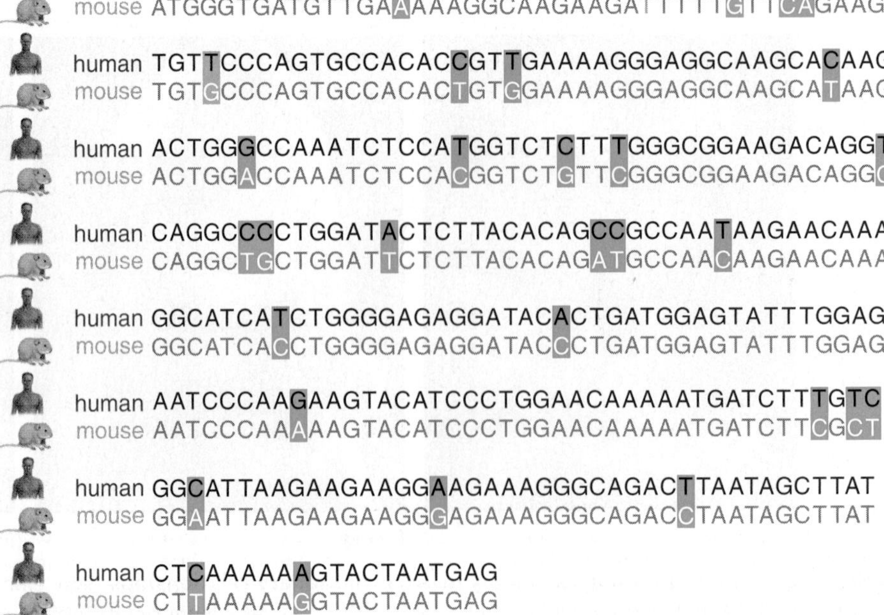

▶ **FIGURE 14-12 Molecular similarity shows evolutionary relationships** The DNA sequences of the genes that code for cytochrome *c* in a human and a mouse. Of the 315 nucleotides in the gene, only 30 (shaded blue) differ between the two species.

14.4 WHAT IS THE EVIDENCE THAT POPULATIONS EVOLVE BY NATURAL SELECTION?

We have seen that evidence of evolution comes from many sources. But what is the evidence that evolution occurs by the process of natural selection?

Controlled Breeding Modifies Organisms

One line of evidence supporting evolution by natural selection is **artificial selection,** the breeding of domestic plants and animals to produce specific desirable features. The various dog breeds provide a striking example of artificial selection (**Fig. 14-13**). Dogs descended from wolves, and even today, the two will readily crossbreed. With few exceptions, however, modern dogs do not resemble wolves. Some breeds are so different from one another that they would be considered separate species if they were found in the wild. Humans produced these radically different dogs in a few thousand years by doing nothing more than repeatedly selecting individuals with desirable traits for breeding. Therefore, it is quite plausible that natural selection could, by an analogous process acting over hundreds of millions of years, produce the spectrum of living organisms. Darwin was so impressed by the connection between artificial selection and natural selection that he devoted a chapter of *Origin of Species* to the topic.

(a) Gray wolf

(b) Diverse dogs

▲ FIGURE 14-13 **Dog diversity illustrates artificial selection** A comparison of **(a)** the ancestral dog (the gray wolf, *Canis lupus*) and **(b)** various breeds of dog. Artificial selection by humans has caused great divergence in the size and shape of dogs in only a few thousand years.

Evolution by Natural Selection Occurs Today

The logic of natural selection gives us no reason to believe that evolutionary change is limited to the past. After all, inherited variation and competition for access to resources are certainly not limited to the past. If Darwin and Wallace were correct that those conditions lead inevitably to evolution by natural selection, then scientific observers and experimenters ought to be able to detect evolutionary change as it occurs. And they have. Next, we will consider some examples that give us a glimpse of natural selection at work.

Brighter Coloration Can Evolve When Fewer Predators Are Present

On the island of Trinidad, guppies live in streams that are also inhabited by several species of larger, predatory fish that frequently dine on guppies (**Fig. 14-14**). In upstream portions of these streams, however, the water is too shallow for the predators, and guppies are free of danger from predators. When scientists compared male guppies in an upstream area with ones in a downstream area, they found that the upstream guppies were much more brightly colored than the downstream

▲ FIGURE 14-14 **Guppies evolve to become more colorful in predator-free environments** Male guppies (top) are more brightly colored than females (bottom). Some male guppies are more colorful than others. In some environments, brighter males are selected; in other environments, duller males are selected.

guppies. The scientists knew that the source of the upstream population was guppies that had found their way up into the shallower waters many generations earlier.

The explanation for the difference in coloration between the two populations stems from the sexual preferences of female guppies. The females prefer to mate with the most brightly colored males, so the brightest males have a large advantage when it comes to reproduction. In predator-free areas, male guppies with the bright colors that females prefer have more offspring than duller males. Bright color, however, makes guppies more conspicuous to predators and, therefore, more likely to be eaten. Thus, where predators are common, they act as agents of natural selection by eliminating the bright-colored males before they can reproduce. In these areas, the duller males have the advantage and produce more offspring. The color difference between the upstream and downstream guppy populations is a direct result of natural selection.

Natural Selection Can Lead to Pesticide Resistance

Natural selection is also evident in numerous instances of insect pests evolving resistance to the pesticides with which we try to control them. For example, a few decades ago, Florida homeowners were dismayed to realize that roaches were ignoring a formerly effective poison bait called Combat®. Researchers discovered that the bait had acted as an agent of natural selection. Roaches that liked it were consistently killed; those that survived inherited a rare mutation that caused them to dislike glucose, a type of sugar found in the corn syrup used as bait in Combat®. By the time researchers identified the problem, the formerly rare mutation had become common in Florida's urban roach population. (For additional examples of how humans influence evolution, see "Earth Watch: People Promote High-Speed Evolution.")

Unfortunately, the evolution of pesticide resistance in insects is a common example of natural selection in action. Such resistance has been documented in more than 500 species of crop-damaging insects, and virtually every pesticide has fostered the evolution of resistance in at least one insect species. We pay a heavy price for this evolutionary phenomenon. The additional pesticides that farmers apply in their attempts to control resistant insects cost almost $2 billion each year in the United States alone and add millions of tons of poisons to Earth's soil and water.

Experiments Can Demonstrate Natural Selection

In addition to observing natural selection in the wild, scientists have also devised numerous experiments that confirm the action of natural selection. For example, one group of evolutionary biologists released small groups of *Anolis sagrei* lizards onto 14 small Bahamian islands that were previously uninhabited by lizards (**Fig. 14-15**). The original lizards came from a population on Staniel Cay, an island with tall vegetation, including plenty of trees. In contrast, the islands to which the small colonial groups were introduced had few or no trees and were covered mainly with small shrubs and other low-growing plants.

The biologists returned to those islands 14 years after releasing the colonists and found that the original small groups of lizards had given rise to thriving populations of hundreds of individuals. On all 14 of the experimental islands, lizards had legs that were shorter and thinner than lizards from the original source population on Staniel Cay. In just over a decade, it appeared, the lizard populations had changed in response to new environments.

▲ FIGURE 14-15 Anole leg size evolves in response to a changed environment

Why had the new lizard populations evolved shorter, thinner legs? Long legs allow greater speed, but shorter legs allow for more agility and maneuverability on narrow surfaces. So, natural selection favors legs that are as long and thick as possible while still allowing sufficient maneuverability. When the lizards were moved from an environment with thick-branched trees to an environment with only thin-branched bushes, the individuals with formerly favorable long legs were at a disadvantage. In the new environment, more agile, shorter-legged individuals were better able to escape predators and survive to produce a greater number of offspring. Thus, members of subsequent generations had shorter legs on average.

Selection Acts on Random Variation to Favor the Phenotypes That Work Best in Particular Environments

Two important points underlie the evolutionary changes just described:

- **The variations on which natural selection works are produced by chance mutations.** The bright coloration in Trinidadian guppies, distaste for glucose in Florida cockroaches, and shorter legs in Bahamian lizards were not *produced* by the female mating preferences, poisoned corn syrup, or thinner branches. The mutations that produced each of these beneficial traits arose spontaneously.

Earth Watch

People Promote High-Speed Evolution

You probably don't think of yourself as a major engine of evolution. Nonetheless, as you go about the routines of your daily life, you are contributing to what is perhaps today's most significant cause of rapid evolutionary change. Human activity has changed Earth's environments tremendously, and when environments change, populations adapt or perish. The biological logic of natural selection, spelled out so clearly by Darwin, tells us that environmental change leads inevitably to evolutionary change. Thus, by changing the environment, humans have become a major agent of natural selection.

Unfortunately, many of the evolutionary changes we have caused have turned out to be bad news for us. Our liberal use of pesticides has selected for resistant pests that frustrate efforts to protect our food supply. By overmedicating ourselves with antibiotics and other drugs, we have selected for resistant "supergerms" and diseases that are ever more difficult to treat. Heavy fishing in the world's oceans has favored smaller fish that can slip through nets more easily, thereby selecting for slow-growing individuals that remain small even as mature adults. As a result, fish of many commercially important species are now so small that our ability to extract food from the sea is compromised.

Our use of pesticides, antibiotics, and fishing technology has caused evolutionary changes that threaten our health and welfare, but the scope of these changes may be dwarfed by those that will arise from human-caused modification of Earth's climate. Human activities, especially activities that use energy derived from fossil fuels, modify the climate by contributing to global warming. In coming years, species' evolution will be influenced by environmental changes associated with a warming climate, such as reduced ice and snow, longer growing seasons, and shifts in the life cycles of other species that provide food or shelter.

There is growing evidence that global warming is already causing evolutionary change. Warming-related evolution has been found in a number of plant and animal populations. For example, researchers have discovered that, in northern populations of a mosquito species, the insects' genetically programmed response to changing day length has shifted during the past four decades. (Mosquitoes use day length as a cue to tell them what time of year it is.) As a result, shorter days are now required to stimulate mosquito larvae to enter their overwintering pupal stage, and the transition takes place much later in the autumn. The delay allows the larvae to take advantage of the longer feeding and growing season produced by global warming.

In Canada, red squirrels in a colony closely monitored by scientists now produce litters 18 days earlier, on average, than they did 10 years ago (**Fig. E14-3**). The change is tied to a warming climate, because spring now arrives earlier and spruce trees produce earlier crops of seeds, the squirrels' only food. Squirrels that breed earlier gain a competitive edge by better exploiting the warmer weather and more abundant food. And, because the time at which a squirrel gives birth is influenced by the animal's genetic makeup, early-breeding squirrels pass to their offspring the genes that confer this advantage. As a result, the genetic makeup of the squirrel population is changing, and early-breeding squirrels are becoming more common than later-breeding ones.

The available evidence suggests that global climate change will have an enormous evolutionary impact, potentially affecting the evolution of almost every species. How will these evolutionary changes affect us and the ecosystems on which we depend? This question is not readily answerable, because the path of evolution is not predictable. We can hope, however, that careful monitoring of evolving species and increased understanding of evolutionary processes will help us take appropriate steps to safeguard our health and well-being as Earth warms.

▲ **FIGURE E14-3 Red squirrels have evolved in response to global warming**

- **Natural selection favors organisms that are best adapted to a particular environment.** Natural selection is not a process for producing ever-greater degrees of perfection. Natural selection does not select for the "best" in any absolute sense, but only for what is best in the context of a particular environment, which varies from place to place and which may change over time. A trait that is advantageous under one set of conditions may become disadvantageous if conditions change. For example, in the presence of poisoned corn syrup, a distaste for glucose yields an advantage to a cockroach, but under natural conditions avoiding glucose would cause the insect to bypass good sources of food.

Case Study revisited
What Good Are Wisdom Teeth?

Wisdom teeth are but one of many human anatomical structures that appear to no longer serve an important function (Fig. 14-16). Darwin himself noted many of these "useless, or nearly useless" traits in the very first chapter of *Origin* and declared them to be prime evidence that humans had evolved from earlier species.

Body hair is another vestigial human trait. It seems to be an evolutionary relic of the fur that kept our distant ancestors warm (and that still warms our closest evolutionary relatives, the great apes). Not only do we retain useless body hair, we also still have arrector pili, the muscle fibers that allow other mammals to puff up their fur for better insulation. In humans, these vestigial structures just give us goose bumps.

Though humans don't have and don't need a tail, we nonetheless have a tailbone. The tailbone consists of a few tiny vertebrae fused into a small structure at the base of the backbone, where a tail would be if we had one. People born without a tailbone or who have theirs surgically removed suffer no ill effects.

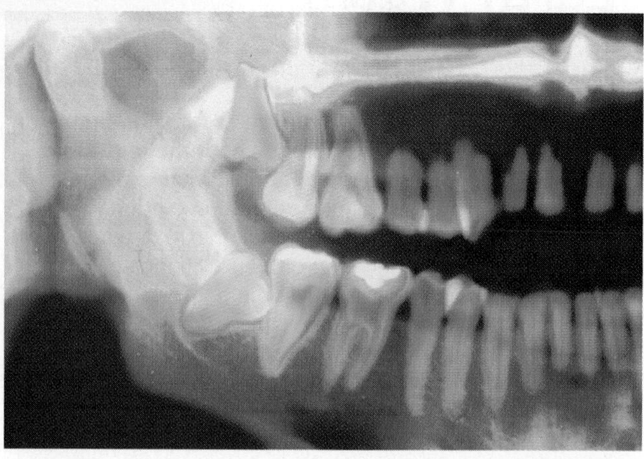

▲ FIGURE 14-16 **Wisdom teeth** Squeezed into a jaw that is too short to contain them, wisdom teeth often become impacted—unable to erupt through the surface of the gum. The leftmost upper and lower teeth in this X-ray image are impacted wisdom teeth.

Consider This

Advocates of creationism argue that there are no vestigial organs because if a structure can do *anything*, it cannot be considered functionless, even if its removal has no effect. Thus, according to this view, wisdom teeth are not evidence of evolution, because they *can* be used to chew if not removed. Do you find this argument persuasive?

CHAPTER REVIEW

Summary of Key Concepts

14.1 How Did Evolutionary Thought Develop?

Historically, the most common explanation for the origin of species was the divine creation of each species in its present form, and species were believed to remain unchanged after their creation. This view was challenged by evidence from fossils, geology, and biological exploration of the tropics. Since the middle of the nineteenth century, scientists have realized that species originate and evolve by the operation of natural processes that change the genetic makeup of populations.

14.2 How Does Natural Selection Work?

Charles Darwin and Alfred Russel Wallace independently proposed the theory of evolution by natural selection. Their theory expresses the logical consequences of four postulates about populations. If (1) populations are variable, (2) the variable traits can be inherited, (3) there is differential (unequal) reproduction, and (4) differences in reproductive success depend on the traits of individuals, then the characteristics of successful individuals will be "naturally selected" and become more common over time.

14.3 How Do We Know That Evolution Has Occurred?

Many lines of evidence indicate that evolution has occurred, including the following:

- Fossils of ancient species tend to be simpler in form than modern species. Sequences of fossils have been discovered that show a graded series of changes in form. Both of these observations would be expected if modern species evolved from older species.
- Species thought to be related through evolution from a common ancestor possess many similar anatomical structures. An example is the forelimbs of amphibians, reptiles, birds, and mammals.
- Stages in early embryological development are quite similar among very different types of vertebrates.
- Similarities in such biochemical traits as the use of DNA as the carrier of genetic information support the notion of descent of related species through evolution from common ancestors.

14.4 What Is the Evidence That Populations Evolve by Natural Selection?

Similarly, many lines of evidence indicate that natural selection is the chief mechanism driving changes in the characteristics of species over time, including the following:

- Inheritable traits have been changed rapidly in populations of domestic animals and plants by selectively breeding organisms with desired features (artificial selection).

The immense variations in species produced in a few thousand years of artificial selection by humans makes it almost inevitable that much larger changes would be wrought by hundreds of millions of years of natural selection.

- Evolution can be observed today. Both natural and human activities drastically change the environment over short periods of time. Inherited characteristics of species have been observed to change significantly in response to such environmental changes.

Key Terms

analogous structure *277*
artificial selection *279*
convergent evolution *277*
evolution *267*
fossil *268*
homologous structure *276*
natural selection *273*
population *273*
vestigial structure *276*

Thinking Through the Concepts

Fill-in-the-Blank

1. The flipper of a seal is homologous with the __________ of a bird, and both of these are homologous with the __________ of a human. The wing of a bird and the wing of a butterfly are described as __________ structures that arose as a result of __________ evolution. Remnants of structures in animals that have no use for them, such as the small hind leg bones of whales, are described as __________ structures.
2. The finding that all organisms share the same genetic code provides evidence that all descended from a(n) __________. Further evidence is provided by the fact that all cells use roughly the same set of __________ to build proteins, and all cells use the molecule __________ as an energy carrier.
3. Georges Cuvier espoused a concept called __________ to explain layers of rock with embedded fossils. Charles Lyell, building on the work of James Hutton, proposed an alternative explanation called __________, which states that layers of rock and many other geological features can be explained by gradual processes that occurred in the past just as they do in the present. This concept provided important support for evolution because it required that Earth be extremely __________.
4. The process by which inherited characteristics of populations change over time is called __________. Variability among individuals is the result of chance changes called __________ that occur in the hereditary molecule __________.
5. The process by which individuals with traits that provide an advantage in their natural habitats are more successful at reproducing is called __________. People who breed animals or plants can produce large changes in their characteristics in a relatively short time, a process called __________.
6. Darwin's postulate 2 states: __________. The work of __________ provided the first experimental evidence for this postulate.

Review Questions

1. Selection acts on individuals, but only populations evolve. Explain why this statement is true.
2. Distinguish between catastrophism and uniformitarianism. How did these hypotheses contribute to the development of evolutionary theory?
3. Describe Lamarck's theory of inheritance of acquired characteristics. Why is it invalid?
4. What is natural selection? Describe how natural selection might have caused differential reproduction among the ancestors of a fast-swimming predatory fish, such as the barracuda.
5. Describe how evolution occurs through the interactions among the reproductive potential of a species, the normally constant size of natural populations, variation among individuals of a species, natural selection, and inheritance.
6. What is convergent evolution? Give an example.
7. How do biochemistry and molecular genetics contribute to the evidence that evolution occurred?

Applying the Concepts

1. In discussions of untapped human potential, it is commonly said that the average person uses only 10% of his or her brain. Is this conclusion likely to be correct? Explain your answer in terms of natural selection.
2. Both the theory of evolution by natural selection and the theory of special creation (which states that all species were simultaneously created by God) have had an impact on evolutionary thought. Discuss why one is considered to be a scientific theory and the other is not.
3. Does evolution through natural selection produce "better" organisms in an absolute sense? Are we climbing the "Ladder of Nature"? Defend your answer.
4. In what sense are humans currently acting as agents of selection on other species? Name some organisms that are *favored* by the environmental changes humans cause.
5. Darwin and Wallace's discovery of natural selection is one of the great revolutions in scientific thought. Some scientific revolutions spill over and affect the development of philosophy and religion. Is this true of evolution? Does (or should) the idea of evolution by natural selection affect the way humans view their place in the world?

MB® *Go to www.masteringbiology.com for practice quizzes, activities, eText, videos, current events, and more.*

How Populations Evolve

Chapter 15

▲ *Staphylococcus aureus*, a common source of human infections, is among the many bacterial species that have evolved resistance to antibiotics.

Case Study

Evolution of a Menace

ON A FEBRUARY DAY IN 2008, a 20-year-old student arrived at the health center at Western Washington University. He had been bothered by a lingering cough for a couple of weeks and, when his symptoms worsened to include a fever and vomiting, he sought medical attention. The health center staff quickly determined that the student had pneumonia, and began treatment. His condition, however, deteriorated, and he was transferred to the local hospital. A few days later, he died.

Why couldn't doctors save a previously healthy young man from a normally curable disease? Because the victim's pneumonia was caused by methicillin-resistant *Staphylococcus aureus* (MRSA) bacteria. *Staphylococcus aureus*, sometimes referred to as "staph," is a common bacterium that can infect the skin, blood, or respiratory system. Many staph infections can be successfully treated with antibiotics, but MRSA bacteria are antibiotic resistant and cannot be killed by many of the most commonly used antibiotics. Until about 10 years ago, MRSA infections occurred almost exclusively in hospitals. Today, however, resistant staph is widespread, and more than 10% of MRSA infections occur in non-hospital settings.

In the United States, MRSA infections kill about 19,000 people each year. And, unfortunately, *Staphylococcus* is by no means the only disease-causing bacterium that is becoming less susceptible to antibiotic drugs. For example, antibiotic resistance has also appeared in the bacteria that cause tuberculosis, a disease that kills almost 2 million people each year. In an increasing number of tuberculosis cases, the disease does not respond to any of the drugs commonly used to treat it. Drug resistance is also common in the bacteria that cause food poisoning, blood poisoning, dysentery, pneumonia, gonorrhea, meningitis, and urinary tract infections. We are experiencing a global onslaught of resistant "supergerms," and are facing the specter of diseases that cannot be cured.

Many physicians and scientists believe that the most effective way to combat the rise of resistant diseases is to reduce the use of antibiotics. Why might such a strategy be effective? Because the upsurge of antibiotic resistance is a consequence of evolutionary change in populations of bacteria, and the agent of this change is natural selection applied by antibiotic drugs. To understand how this crisis arose and to devise a strategy to resolve it, we must have a clear understanding of the mechanisms by which populations evolve.

At a Glance

15.1 HOW ARE POPULATIONS, GENES, AND EVOLUTION RELATED?

If you live in an area with a seasonal climate and you own a dog or cat, you have probably noticed that your pet's fur gets thicker and heavier as winter approaches. Has the animal evolved? No. The changes that we see in an individual organism over the course of its lifetime are not evolutionary changes. Instead, evolutionary changes occur from generation to generation, causing descendants to be different from their ancestors.

Furthermore, we can't detect evolutionary change across generations by looking at a single set of parents and offspring. For example, if you observed that a 6-foot-tall man had an adult son who stood 5 feet tall, could you conclude that humans were evolving to become shorter? Obviously not. Rather, if you wanted to learn about evolutionary change in human height, you would begin by measuring many humans of many generations to see if the average height is changing over time. Evolution is a property not of individuals but of populations. (A **population** is a group that includes all the members of a species living in a given area.)

The recognition that evolution is a population-level phenomenon was one of Darwin's key insights. But populations are composed of individuals, and the actions and fates of individuals determine which characteristics will be passed to descendant populations. In this fashion, inheritance provides the link between the lives of individual organisms and the evolution of populations. We will therefore begin our discussion of the processes of evolution by reviewing some principles of genetics as they apply to individuals. We will then extend those principles to the genetics of populations.

Genes and the Environment Interact to Determine Traits

Each cell of every organism contains genetic information encoded in the DNA of its chromosomes. Recall from Chapter 9 that a *gene* is a segment of DNA located at a particular place on a chromosome. The sequence of nucleotides in a gene encodes the sequence of amino acids in a protein, usually an enzyme that catalyzes a particular reaction in the cell. At a given gene's location, different members of a species may have slightly different nucleotide sequences, called *alleles*. Different alleles generate different forms of the same enzyme. In this way, various alleles of the gene that influences eye color in humans, for example, help produce eyes that are brown, or blue, or green, and so on.

In any population of organisms, there are usually two or more alleles of each gene. An individual of a diploid species whose alleles of a particular gene are both the same is *homozygous* for that gene, and an individual with different alleles for that gene is *heterozygous*. The specific alleles borne on an organism's chromosomes (its *genotype*) interact with the environment to influence the development of its physical and behavioral traits (its *phenotype*).

Let's illustrate these principles with an example. A black hamster's coat is colored black because a chemical reaction in its hair follicles produces a black pigment. When we say that a hamster has the allele for a black coat, we mean that a particular stretch of DNA on one of its chromosomes contains a sequence of nucleotides that codes for the enzyme that catalyzes the pigment-producing reaction. A hamster with the allele for a brown coat has a different sequence of nucleotides at the corresponding chromosomal position. That different sequence codes for an enzyme that cannot produce black pigment. If a hamster is homozygous for the black allele or is heterozygous (one black allele and one brown allele), its fur contains the pigment and is black. But if a hamster is homozygous for the brown allele, its hair follicles produce no black pigment and its coat is brown (**Fig. 15-1**). Because the hamster's coat is black even when only one copy of the black allele is present, the black allele is considered *dominant* and the brown allele *recessive*.

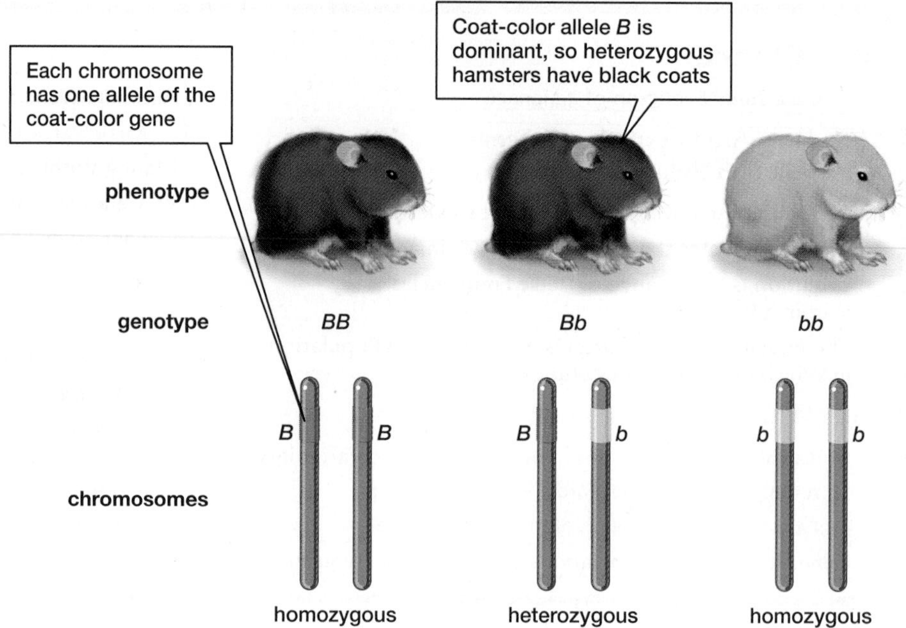

▶ **FIGURE 15-1 Alleles, genotype, and phenotype in individuals** An individual's particular combination of alleles is its genotype. The word "genotype" can refer to the alleles of a single gene (as shown here), to a set of genes, or to all of an organism's genes. An individual's phenotype is determined by its genotype and environment. Phenotype can refer to a single trait, a set of traits, or all of an organism's traits.

The Gene Pool Is the Sum of the Genes in a Population

In studying evolution, looking at the process from the point of view of a gene has proven to be an enormously effective tool. In particular, evolutionary biologists have made excellent use of the tools of a branch of genetics called population genetics, which deals with the frequency, distribution, and inheritance of alleles in populations. To take advantage of this powerful aid to understanding evolution, you will need to learn a few of the basic concepts of population genetics.

Population genetics defines the **gene pool** as the sum of all the genes in a population. In other words, the gene pool consists of all the alleles of all the genes in all the individuals of a population. Each particular gene can also be considered to have its own gene pool, which consists of all the alleles of that specific gene in a population (**Fig. 15-2**). If we added up all the copies of each allele of that gene in all the individuals in a population, we could determine the relative proportion of each allele, a number called the **allele frequency.** For example, the population of 25 hamsters portrayed in Figure 15-2 contains 50 alleles of the gene that controls coat color (because hamsters are diploid and each hamster thus has two copies of each gene). Twenty of those 50 alleles are of the type that codes for black coats, so the frequency of that allele in the population is 0.40 (or 40%), because 20/50 = 0.40.

Evolution Is the Change of Allele Frequencies Within a Population

A casual observer might define evolution on the basis of changes in the outward appearance or behaviors of the members of a population. A population geneticist, however, looks at a population and sees a gene pool that just happens to be divided into the packages that we call individual organisms. So many of the outward changes that we observe in the individuals that make up the population can also be viewed as the visible expression of underlying changes to the gene pool. A population geneticist, therefore, defines evolution as the changes in allele frequencies that occur in a gene pool over time. Evolution is change in the genetic makeup of populations over generations.

The Equilibrium Population Is a Hypothetical Population in Which Evolution Does Not Occur

It is easier to understand what causes populations to evolve if the characteristics of a population that would *not* evolve are considered first. In 1908, English mathematician Godfrey H. Hardy and German physician Wilhelm Weinberg independently developed a simple mathematical model now known as the **Hardy–Weinberg principle** (for more information about the model, see "A Closer Look at the Hardy–Weinberg Principle" on p. 288). This model showed that, under certain conditions, allele frequencies and genotype frequencies in a population will remain constant no matter how many generations pass. In other words, this population will not evolve. Population geneticists use the term **equilibrium population** for this hypothetical nonevolving population in which allele frequencies do not change, as long as the following conditions are met:

- There must be no mutation.
- There must be no **gene flow.** That is, there must be no movement of alleles into or out of the population (as would be caused, for example, by the movement of organisms into or out of the population).
- The population must be very large.
- All mating must be random, with no tendency for certain genotypes to mate with specific other genotypes.
- There must be no natural selection. That is, all genotypes must reproduce with equal success.

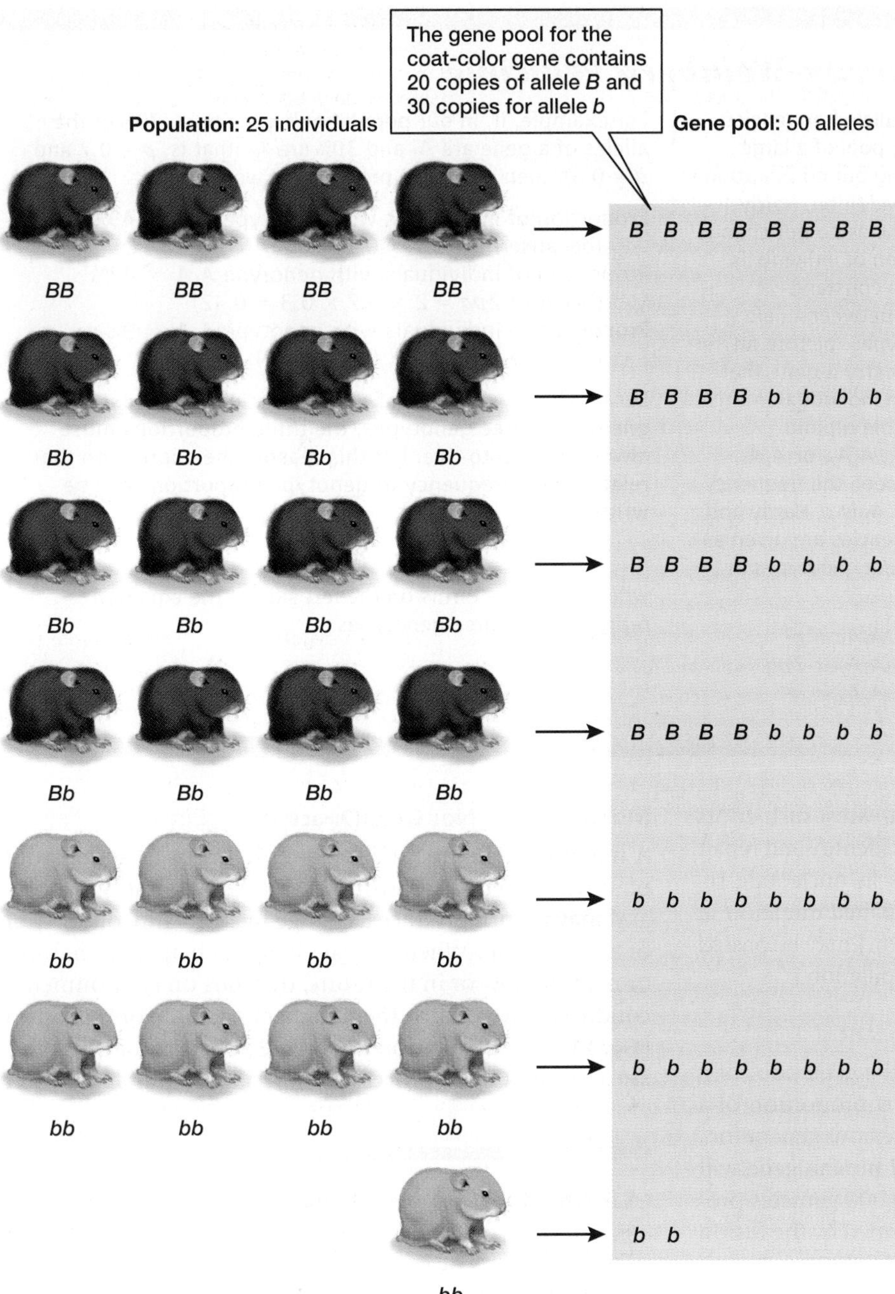

◀ **FIGURE 15-2 A gene pool** In diploid organisms, each individual in a population contributes two alleles of each gene to the gene pool.

Under these conditions, allele frequencies in a population will remain the same indefinitely. If one or more of these conditions is violated, then allele frequencies may change: The population will evolve.

As you might expect, few if any natural populations are truly in equilibrium. What, then, is the importance of the Hardy–Weinberg principle? The Hardy–Weinberg conditions are useful starting points for studying the mechanisms of evolution. In the following sections, we will examine some of the conditions, show that natural populations often fail to meet them, and illustrate the consequences of such failures. In this way, we can better understand both the inevitability of evolution and the processes that drive evolutionary change.

15.2 WHAT CAUSES EVOLUTION?

Population genetics theory predicts that the Hardy–Weinberg equilibrium can be disturbed by deviations from any of its five conditions. Therefore, we might predict five major causes of evolutionary change: mutation, gene flow, small population size, nonrandom mating, and natural selection.

Mutations Are the Original Source of Genetic Variability

A population remains in evolutionary equilibrium only if there are no **mutations** (changes in DNA sequence). Most mutations occur during cell division, when a cell must make a copy of its DNA. Sometimes, errors occur during the copying process and

A Closer Look At *The Hardy–Weinberg Principle*

The Hardy–Weinberg principle states that allele frequencies will remain constant over time in the gene pool of a large population in which there is random mating but no mutation, no gene flow, and no natural selection. In addition, Hardy and Weinberg showed that if allele frequencies do not change in an equilibrium population, the proportion of individuals with a particular genotype will also remain constant.

To better understand the relationship between allele frequencies and the occurrence of genotypes, picture an equilibrium population whose members carry a gene that has two alleles, A_1 and A_2. Note that each individual in this population must carry one of three possible diploid genotypes (combinations of alleles): A_1A_1, A_1A_2, or A_2A_2.

Suppose that in our population's gene pool, the frequency of allele A_1 is p and the frequency of allele A_2 is q. Hardy and Weinberg demonstrated that if allele frequencies are given as p and q, then the proportions of the different genotypes in the population can be calculated as:

Proportion of individuals with genotype $A_1A_1 = p^2$
Proportion of individuals with genotype $A_1A_2 = 2pq$
Proportion of individuals with genotype $A_2A_2 = q^2$

For example, if, in our population's gene pool, 70% of the alleles of a gene are A_1 and 30% are A_2 (that is, $p = 0.7$ and $q = 0.3$), then genotype proportions would be:

Proportion of individuals with genotype A_1A_1 = 49%
(because $p^2 = 0.7 \times 0.7 = 0.49$)
Proportion of individuals with genotype A_1A_2 = 42%
(because $2pq = 2 \times 0.7 \times 0.3 = 0.42$)
Proportion of individuals with genotype A_2A_2 = 9%
(because $q^2 = 0.3 \times 0.3 = 0.09$)

Because every member of the population must possess one of the three genotypes, the three proportions must always add up to one. For this reason, the expression that relates allele frequency to genotype proportions can be written as:

$$p^2 + 2pq + q^2 = 1$$

where the three terms on the left side of the equation represent the three genotypes.

the copied DNA does not match the original. Most such errors are quickly corrected by cellular systems that identify and repair DNA copying mistakes, but some changes in nucleotide sequence slip past the repair systems. An unrepaired mutation in a cell that gives rise to gametes (eggs or sperm) may be passed to offspring and enter the gene pool of a population.

Inherited Mutations Are Rare But Important

How significant is mutation in changing the gene pool of a population? For any given gene, only a tiny proportion of a population inherits a mutation from the previous generation. For example, a mutant version of a typical human gene will appear in only about one out of every 100,000 gametes produced and, because new individuals are formed by the fusion of two gametes, in only about one of every 50,000 newborns. Therefore, mutation by itself generally causes only very small changes in the frequency of any particular allele.

Despite the rarity of inherited mutations of any particular gene, the cumulative effect of mutations is essential to evolution. Most organisms have a large number of different genes, so even if the rate of mutation is low for any one gene, the sheer number of possibilities means that each new generation of a population is likely to include some mutations. For example, geneticists estimate that humans have between 20,000 and 25,000 different genes, so each person carries between 40,000 and 50,000 alleles. Thus, even if each allele has, on average, only a one in 100,000 chance of mutation, most newborn individuals will probably carry one or two mutations overall. These mutations are new alleles—new variations on which other evolutionary processes can work. As such, they are the foundation of evolutionary change. Without mutations there would be no evolution.

Mutations Are Not Goal Directed

A mutation does not arise as a result of, or in anticipation of, environmental necessities. A mutation simply happens and may in turn produce a change in a structure or function of an organism. Whether that change is helpful or harmful or neutral, now or in the future, depends on environmental conditions over which the organism has little or no control (**Fig. 15-3**). The mutation provides a potential for evolutionary change. Other processes, especially natural selection,

Case Study continued
Evolution of a Menace

If mutations are rare and random, why do mutant alleles that confer antibiotic resistance arise so commonly in bacterial populations? The seemingly inevitable emergence of resistance results in large measure from the huge size of bacteria populations and from their very short generation times. A single drop of human saliva contains about 150 million bacteria, and an entire human body contains hundreds of trillions. With so many bacteria present, even an extremely rare resistance mutation that occurs in only a tiny percentage of the population will be present in some individuals. Also, because many mutations occur during cell division (when DNA replicates and copying errors can occur), rapid reproduction creates many opportunities for mutations to arise. Bacteria reproduce very rapidly, as often as every 15 minutes in some species. This rapid reproduction, taking place in huge populations, results in a high likelihood that mutations leading to antibiotic resistance will be present in bacterial populations.

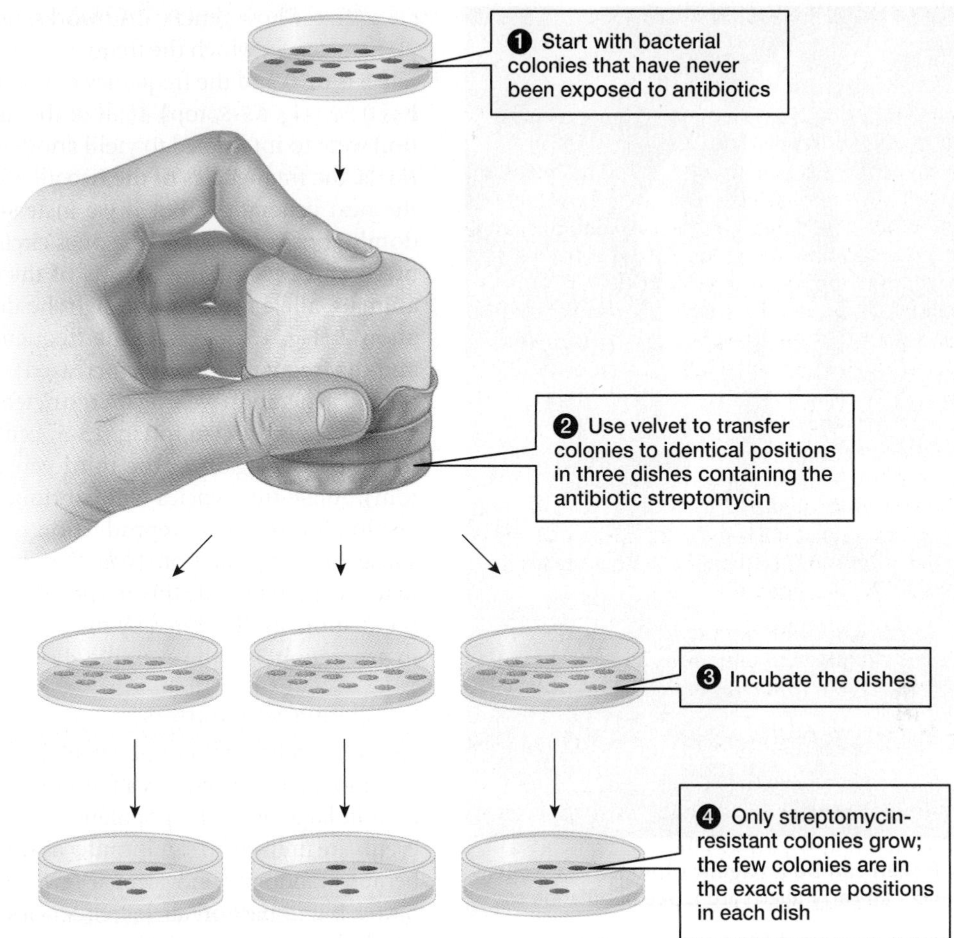

1 Start with bacterial colonies that have never been exposed to antibiotics
2 Use velvet to transfer colonies to identical positions in three dishes containing the antibiotic streptomycin
3 Incubate the dishes
4 Only streptomycin-resistant colonies grow; the few colonies are in the exact same positions in each dish

▲ **FIGURE 15-3 Mutations occur spontaneously** This experiment demonstrates that mutations occur spontaneously and not in response to environmental conditions. When bacterial colonies that have never been exposed to antibiotics are exposed to the antibiotic streptomycin, only a few colonies grow. The observation that these surviving colonies grow in the exact same positions in all dishes shows that the mutations for resistance to streptomycin were present in the original dish before exposure to streptomycin.

QUESTION If it were true that mutations *do* occur in response to the presence of antibiotic, how would the result of this experiment have differed from the actual result?

may act to spread the mutation through the population or to eliminate it from the population.

Gene Flow Between Populations Changes Allele Frequencies

The movement of alleles between populations, known as gene flow, changes how alleles are distributed among populations. When individuals move from one population to another and interbreed at the new location, alleles are transferred from one gene pool to another. In baboons, for example, individuals routinely move to new populations. Baboons live in social groupings called troops. Within each troop, all the females mate with a handful of dominant males. Juvenile males usually leave the troop. If they are lucky, they join and perhaps even become dominant in another troop. Thus, the male offspring of one troop carry alleles to the gene pools of other troops.

Although movement of individuals is a common cause of gene flow, alleles can move between populations even if organisms do not. Flowering plants, for example, cannot move, but their seeds and pollen can (**Fig. 15-4**). Pollen, which contains sperm cells, may be carried long distances by wind or by animal pollinators. If the pollen ultimately reaches the flowers of a different population of its species, it may fertilize eggs and add its collection of alleles to the local gene pool. Similarly, seeds may be borne by wind or water or animals to distant locations where they can germinate to become part of a population far from their place of origin.

The main evolutionary effect of gene flow is to increase the genetic similarity of different populations of a species. To see why, picture two glasses, one containing fresh water and the other salty seawater. If a few spoonfuls are transferred from the seawater glass to the freshwater glass, the liquid in the freshwater glass becomes saltier. In the same way, movement of alleles from one population to another tends to

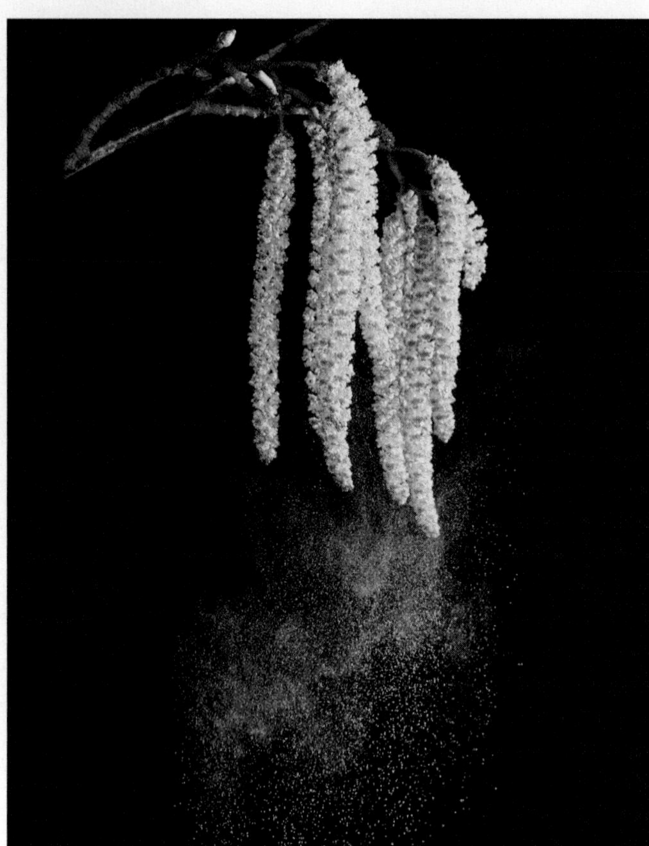

▲ **FIGURE 15-4 Pollen can be an agent of gene flow** Pollen, drifting on the wind, can carry alleles from one population to another.

change the gene pool of the destination population so that it is more similar to the source population.

If alleles move continually back and forth between different populations, the gene pools of the different populations will, in effect, mix. This mixing prevents the development of large differences in the genetic compositions of the populations. But if gene flow between populations of a species is blocked, the resulting genetic differences may grow so large that one of the populations becomes a new species, a process we will discuss in Chapter 16.

Allele Frequencies May Drift in Small Populations

Allele frequencies in populations can be changed by chance events other than mutations. For example, if bad luck prevents some members of a population from reproducing, their alleles will ultimately be removed from the gene pool, altering its makeup. What kinds of bad-luck events can randomly prevent some individuals from reproducing? Seeds can fall into a pond and never sprout; flowers can be destroyed by a hailstorm; organisms can be killed by a fire or by a volcanic eruption. Any event that arbitrarily cuts lives short or otherwise allows only a random subset of a population to reproduce can cause random changes in allele frequencies. The process by which chance events change allele frequencies is called **genetic drift.**

To see how genetic drift works, imagine a population of 20 hamsters in which the frequency of the black coat-color allele *B* is 0.50 and the frequency of the brown coat-color allele *b* is 0.50 (**Fig. 15-5,** top). If all of the hamsters in the population were to interbreed to yield another population of 20 animals, the frequencies of the two alleles would not change in the next generation. But if we instead allow only two, randomly chosen hamsters (the ones circled in Fig. 15-5, top) to breed and become the parents of the next generation of 20 animals, allele frequencies might be quite different in generation 2 (**Fig. 15-5,** center; the frequency of *B* has decreased and the frequency of *b* has increased). And if breeding in the second generation were again restricted to two randomly chosen hamsters (circled in Fig. 15-5, center), allele frequencies might change again in the third generation (**Fig. 15-5,** bottom). Allele frequencies will continue to change in random fashion for as long as reproduction is restricted to a random subset of the population. Note that the changes caused by genetic drift can include the disappearance of an allele from the population, as illustrated by the disappearance of the *B* allele in generation 3 of the example shown in Figure 15-5.

Population Size Matters

Genetic drift occurs to some extent in all populations, but it occurs more rapidly and has a bigger effect in small populations than in large ones. If a population is sufficiently large, chance events are unlikely to significantly alter its genetic composition, because random removal of a few individuals' alleles won't have a big impact on allele frequencies in the population as a whole. In a small population, however, a particular allele may be carried by only a few organisms. Chance events could eliminate most or all examples of such an allele from the population.

To see how population size affects genetic drift, let's return to generation 1 of our imaginary hamster population (see Fig. 15-5, top). Three-quarters of the hamsters are black and one-quarter are brown; the frequencies of alleles *B* and *b* are each 50%. Now imagine two additional populations with same coat color and allele frequencies. However, imagine that one population has only eight individuals in it, whereas the other has 20,000.

Now let's picture reproduction in our two populations. Let's select, at random, a quarter of the individuals in each population and allow them to reproduce, such that each pair of hamsters produces eight offspring and then dies. In the large population, 5,000 hamsters reproduce, yielding a new generation of 20,000. What are the chances that all 20,000 members of the new generation will be brown? Just about nil; that would happen only if all 5,000 randomly chosen breeders were brown (*bb*) hamsters. In fact, it would be extremely unlikely for even 15,000 offspring hamsters to be brown, or for 12,000 to be brown. The most likely outcome is that about a quarter of the breeders will be brown and three-quarters black, which would yield a new generation that is also 25% brown and 75% black (with allele frequencies of $B = 50\%$ and $b = 50\%$), just as in the original population. In the large population, then, we would not expect a major change in allele frequencies from generation to generation.

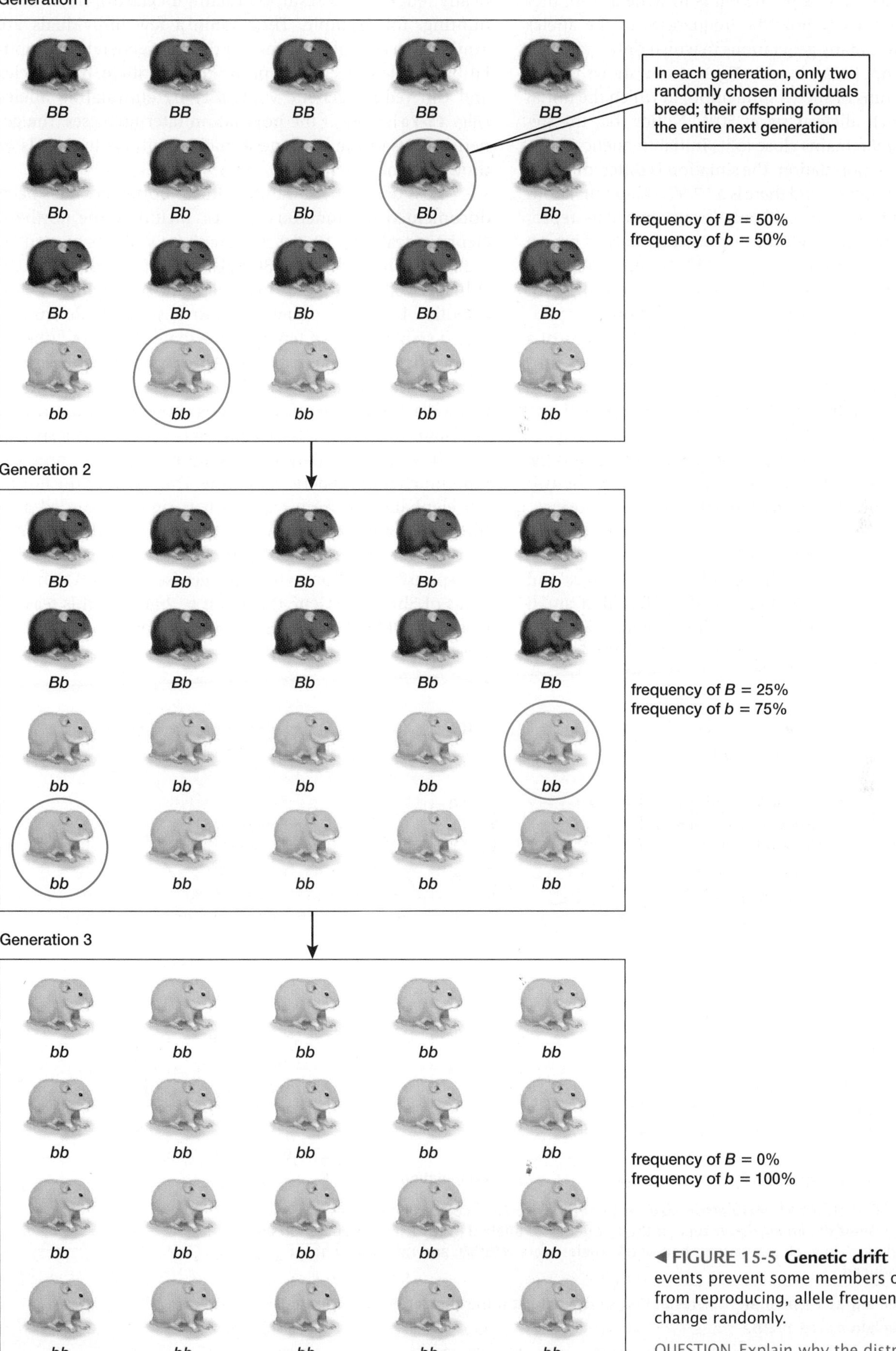

◀ **FIGURE 15-5 Genetic drift** If chance events prevent some members of a population from reproducing, allele frequencies can change randomly.

QUESTION Explain why the distribution of genotypes in generation 2 is as shown.

One way to test this prediction is to write a computer program that simulates how the frequencies of the alleles could change over many generations in which only a random subset of the population breeds. **Figure 15-6a** shows the results from four runs of such a simulation, in which the initial frequency of each allele is set at 50%. Notice that the frequency of allele *B* remains close to its initial frequency.

In the small population, the situation is different. Only two hamsters reproduce, and there is a 12.5% chance that both reproducers will be brown. If only brown hamsters reproduce, then the next generation will consist entirely of brown hamsters—an uncommon but not terribly unlikely outcome. It is thus possible, within a single generation, for the allele for black coat color to disappear from the population.

Figure 15-6b shows the fate of allele *B* in four runs of a simulation of our small population. In one of the four runs (red line), allele *B* reaches a frequency of 100% in the second generation, meaning that all the hamsters in the second and following generations are black. In another run, the frequency of *B* drifts to zero in the third generation (blue line), and the population subsequently is all brown. Thus, one of the two hamster phenotypes disappeared in half of the simulations.

A Population Bottleneck Can Cause Genetic Drift

Two causes of genetic drift, the population bottleneck and the founder effect, further illustrate the effect that small population size may have on the allele frequencies of a species. In a **population bottleneck,** a population is drastically reduced as a result of a natural catastrophe or overhunting, for example. Then, only a few individuals are available to contribute genes to the next generation. Population bottlenecks can rapidly change allele frequencies and can reduce genetic variability by eliminating alleles (**Fig. 15-7a**). Even if the population later increases, the genetic effects of the bottleneck may remain for hundreds or thousands of generations.

Loss of genetic variability due to bottlenecks has been documented in numerous species, including the northern elephant seal (**Fig. 15-7b**). The elephant seal was hunted almost to extinction in the 1800s; by the 1890s, only about 20 still survived. Dominant male elephant seals typically monopolize breeding; therefore, with a single male mating with a stable group of females, one male may have fathered all the offspring at this extreme bottleneck point. Since then, elephant seals have increased in number to about 30,000 individuals, but biochemical analysis shows that all northern elephant seals are genetically almost identical. Other species of seals, whose populations have always remained large, exhibit much more genetic variability. The rescue of the northern elephant seal from extinction is rightly regarded as a triumph of conservation. With very little genetic variation, however, the elephant seal has much less potential to evolve in response to environmental changes (see "Earth Watch: The Perils of Shrinking Gene Pools" on p. 294). For this reason, regardless of how many elephant seals there are, the species must be considered to be threatened with extinction.

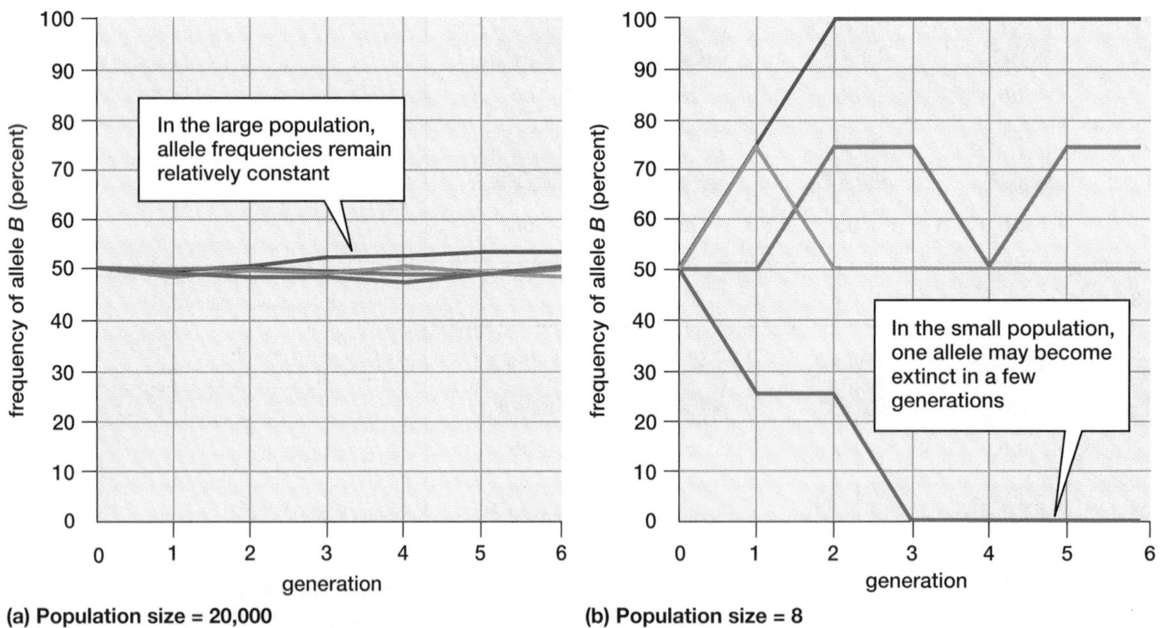

▲ **FIGURE 15-6 The effect of population size on genetic drift** Each colored line represents one computer simulation of the change over time in the frequency of allele *B* in **(a)** a large population and **(b)** a small population. In each population, half of the alleles were *B* (50%), and randomly chosen individuals reproduced.

EXERCISE Sketch a graph that shows the result you would predict if the simulation were run four times with a population size of 20.

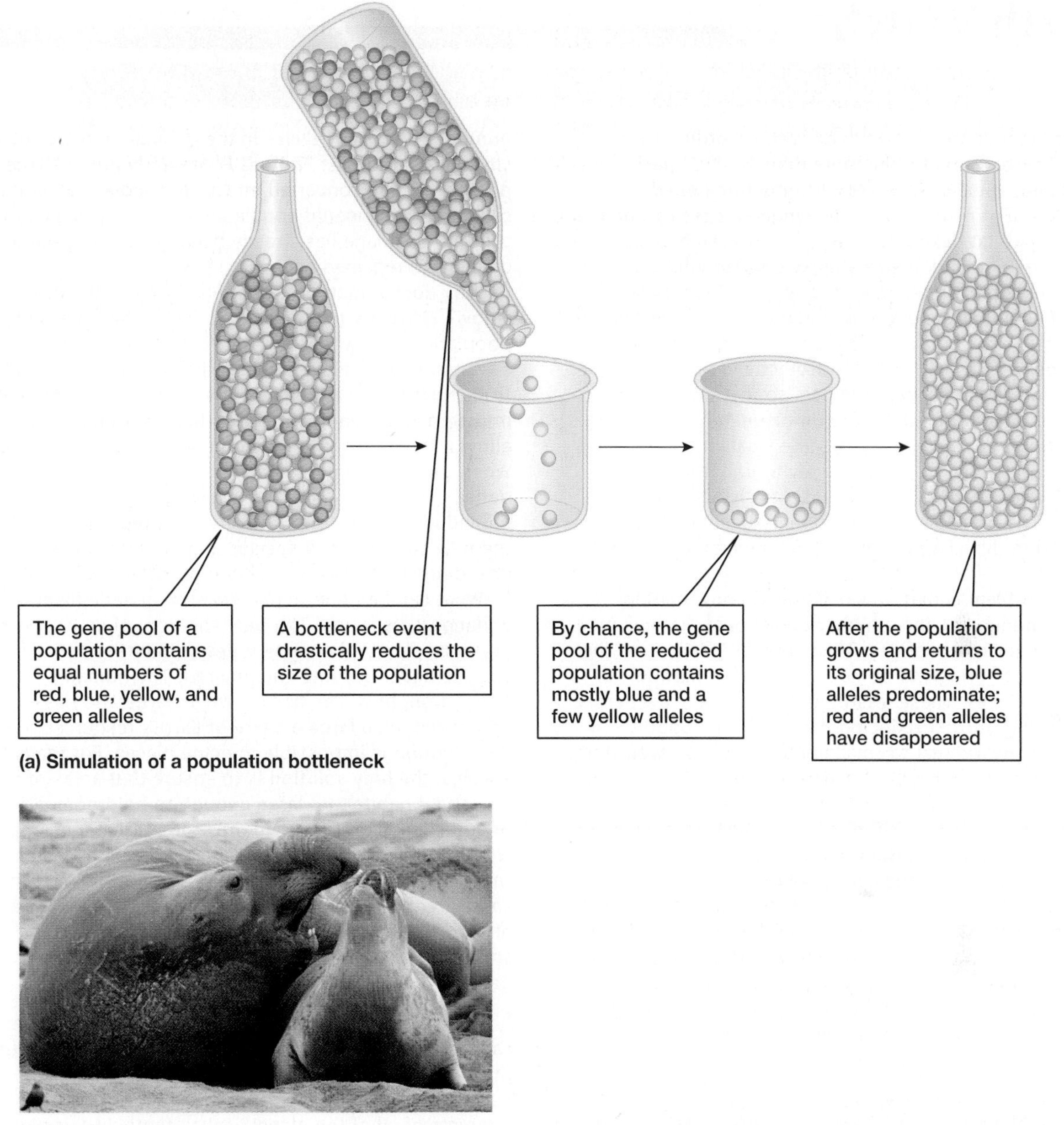

The gene pool of a population contains equal numbers of red, blue, yellow, and green alleles
A bottleneck event drastically reduces the size of the population
By chance, the gene pool of the reduced population contains mostly blue and a few yellow alleles
After the population grows and returns to its original size, blue alleles predominate; red and green alleles have disappeared
(a) Simulation of a population bottleneck
(b) Elephant seals

▲ **FIGURE 15-7 Population bottlenecks reduce variation (a)** A population bottleneck may drastically reduce genotypic and phenotypic variation because the few organisms that survive may all carry similar sets of alleles. **(b)** The northern elephant seal passed through a population bottleneck in the recent past. As a result, the population's genetic diversity is extremely low.

QUESTION If a population grows large again after a bottleneck, genetic diversity will eventually increase. Why?

Isolated Founding Populations May Produce Bottlenecks

The **founder effect** occurs when isolated colonies are founded by a small number of organisms. A flock of birds, for instance, that becomes lost during migration or is blown off course by a storm may settle on an isolated island. The small founder group may, by chance, have allele frequencies that are very different from the frequencies of the parent population. If it does, the gene pool of the future population in the new location will be quite unlike that of the larger population from which it sprang. For example, a set of genetic defects known as Ellis–van Creveld syndrome is far more common among the Amish inhabitants of Lancaster County, Pennsylvania, than among the general population. Today's Lancaster County Amish are descended from only 200 or so eighteenth-century immigrants, and one couple among these immigrants is known to have carried the Ellis–van Creveld allele

Earth Watch

The Perils of Shrinking Gene Pools

Many of Earth's species are in danger. According to the World Conservation Union, more than 16,000 species of plants and animals alone are currently threatened with extinction. For most of these endangered species, the main threat is habitat destruction. When a species' habitat shrinks, its population size almost invariably follows suit.

Many people, organizations, and governments are concerned about the plight of endangered species and are working to protect them and their habitats. The hope is that these efforts will not only protect endangered species, but will also restore their numbers so that they are no longer in danger of extinction. Unfortunately, however, a population that has already become small enough to warrant endangered status is likely to undergo evolutionary changes that increase its chances of going extinct. The principles of evolutionary genetics that we've explored in this chapter can help us understand these changes.

One problem is that, in small populations, mating choices are limited and a high proportion of matings may be between close relatives. This inbreeding increases the odds that offspring will be homozygous for harmful recessive alleles. These less-fit individuals may die before reproducing, further reducing the size of the population.

The greatest threat to small populations, however, stems from their inevitable loss of genetic diversity (**Fig. E15-1**). From our discussion of population bottlenecks, it is apparent that, when populations shrink to very small sizes, many of the alleles that were present in the original population will not be represented in the gene pool of the remnant population. Furthermore, we have seen that genetic drift in small populations will cause many of the surviving alleles to subsequently disappear permanently from the population (see Fig. 15-6b). Because genetic drift is a random process, many of the lost alleles will be advantageous ones that were previously favored by natural selection. Inevitably, the number of different alleles in the population grows ever smaller. As ecologist Thomas Foose aptly put it, "Gene pools are being converted into gene puddles." Even if the size of an endangered population eventually begins to grow, the damage has already been done; lost genetic diversity is regained only very slowly.

▲ **FIGURE E15-1 Only a few hundred Sumatran rhinoceroses remain**

Why does it matter if a population's genetic diversity is low? There are two main risks. First, the fitness of the population as a whole is reduced by the loss of advantageous alleles that underlie adaptive traits. A less-fit population is unlikely to thrive. Second, a genetically impoverished population lacks the variation that will allow it to adapt when environmental conditions change. When the environment changes, as it inevitably will, a genetically uniform species is less likely to contain individuals well suited to survive and reproduce under the new conditions. A species unable to adapt to changing conditions is at very high risk of extinction.

What can be done to preserve the genetic diversity of endangered species? The best solution, of course, is to preserve plenty of diverse types of habitat so that species never become endangered. The human population, however, has grown so large and has appropriated so large a share of Earth's resources that this solution is impossible in many places. For many species, the only solution is to ensure that areas of preserved habitat are large enough to hold populations of sufficient size to contain most of a threatened species' total genetic diversity. If, however, circumstances dictate that preserved areas are small, it is important that the small areas be linked by corridors of the appropriate habitat, so that gene flow among populations in the small preserves can increase the spread of new and beneficial alleles.

BioEthics Does it matter that human activities are causing species to go extinct? Some bioethicists argue that, because humans have the power to extinguish species, we have an ethical obligation to protect the interests of all of the planet's inhabitants. In this view, it is unethical to allow any species to go extinct. For those who believe in the sanctity of other species, the biodiversity crisis poses profound ethical dilemmas. In many cases, the habitat destruction that endangers other species also helps make space for the farmland, housing, and workplaces needed by our growing human population. How can we reconcile the conflict between valid human needs and the needs of endangered species? Furthermore, it is becoming clear that, even with the best of intentions, we cannot save all of the species currently threatened with extinction. The resources available to preserve and manage protected habitats are limited, and we must make choices that will allow some species to survive while others perish. If all species are precious, how can we make such terrible choices? Who should decide which species will live and which will die, and what criteria should be used?

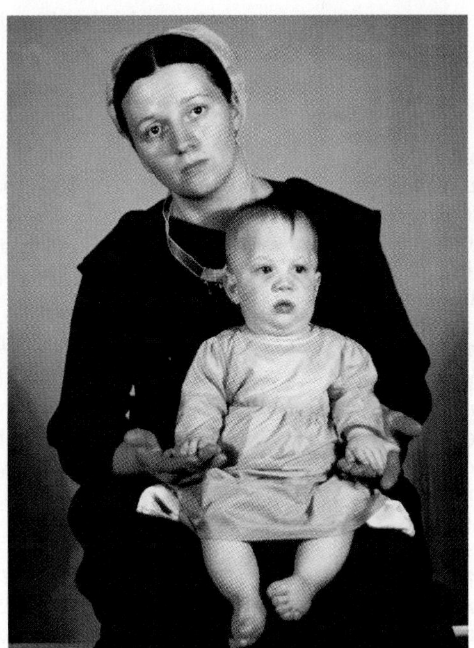

(a) A child with Ellis–van Creveld syndrome

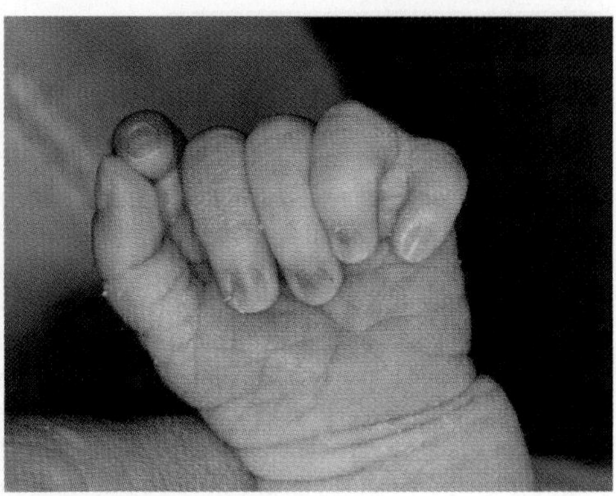

(b) A six-fingered hand

▶ **FIGURE 15-8 A human example of the founder effect** **(a)** An Amish woman with her child, who suffers from a set of genetic defects known as Ellis–van Creveld syndrome. Symptoms of the syndrome include short arms and legs, **(b)** extra fingers, and, in some cases, heart defects. The founder effect accounts for the prevalence of Ellis–van Creveld syndrome among the Amish residents of Lancaster County, Pennsylvania.

(**Fig. 15-8**). In such a small founder population, this single occurrence meant that the allele was carried by a comparatively high proportion of the Amish founder population (1 or 2 carriers out of 200, versus perhaps 1 in 1,000 in the general population). This high initial allele frequency, combined with subsequent genetic drift, has led to extraordinarily high levels of Ellis–van Creveld syndrome among this Amish group.

Mating Within a Population Is Almost Never Random

Nonrandom mating by itself will not alter allele frequencies in a population. Nonetheless, it can have large effects on the distribution of different genotypes, and thus on the distribution of phenotypes, in the population. Certain genotypes may become more common, which can affect the outcome of natural selection.

The effects of nonrandom mating can play a significant role in evolution, because organisms seldom mate strictly randomly. For example, many organisms have limited mobility and tend to remain near their place of birth, hatching, or germination. In such species, most of the offspring of a given parent live in the same area and thus, when they reproduce, there is a good chance that they will be related to their reproductive partners. Sexual reproduction among relatives is called *inbreeding*.

Because relatives are genetically similar, inbreeding tends to increase the number of individuals that inherit the same alleles from both parents and are therefore homozygous for many genes. This increase in homozygotes can have harmful effects, such as increased occurrence of genetic diseases or defects. Many gene pools include harmful recessive alleles that persist in the population because their negative effects are masked in heterozygous carriers (which have only a single copy of the harmful allele). Inbreeding, however, increases the odds of producing homozygous offspring with two copies of the harmful allele.

In animals, nonrandom mating can also arise if individuals have preferences or biases that influence their choice of mates. The snow goose is a case in point. Individuals of this species come in two "color phases"; some snow geese are white, while others are blue-gray (**Fig. 15-9**). Although both white and blue-gray geese belong to the same species, mate choice is not random with respect to color. The birds exhibit a strong tendency to mate with a partner of the same color. This preference for mates that are similar is known as *assortative mating*.

All Genotypes Are Not Equally Beneficial

In a hypothetical equilibrium population, individuals of all genotypes survive and reproduce equally well; no genotype has any advantage over the others. This condition, however, is probably met only rarely, if ever, in real populations. Even though some alleles are neutral, in the sense that organisms possessing any of several alleles are equally likely to survive and reproduce, not all alleles are neutral in all environments. Any time an allele

▲ **FIGURE 15-9 Nonrandom mating among snow geese** Snow geese, which have either white plumage or blue-gray plumage, are most likely to mate with other birds of the same color.

Case Study continued
Evolution of a Menace

Antibiotic resistance evolves by natural selection. To see how, imagine a hospital patient with an infected wound. A doctor decides to treat the infection with an intravenous drip of penicillin. As the antibiotic courses through the patient's blood vessels, millions of bacteria die before they can reproduce. A few bacteria, however, carry a rare allele that codes for a variant enzyme that destroys any penicillin that comes into contact with the bacterial cell. The bacteria carrying this rare allele are able to survive and reproduce, and their offspring inherit the penicillin-destroying allele. After a few generations, the frequency of the penicillin-destroying allele has soared to nearly 100%, and the frequency of the normal allele has declined to near zero. As a result of natural selection imposed by the antibiotic's killing power, the population of bacteria within the patient's body has evolved. The gene pool of the population has changed, and natural selection, in the form of bacterial destruction by penicillin, has caused the change.

provides, in Alfred Russel Wallace's words, "some little superiority," natural selection favors the individuals who possess it. That is, those individuals have higher reproductive success.

Antibiotic Resistance Illustrates Key Points about Evolution

The example of antibiotic resistance highlights some important features of natural selection and evolution.

Natural selection does not cause genetic changes in individuals. Alleles for antibiotic resistance arise spontaneously in some bacteria, long before the bacteria encounter an antibiotic. Antibiotics do not cause resistance to appear; their presence merely favors the survival of bacteria with antibiotic-destroying alleles over that of bacteria without such alleles.

Natural selection acts on individuals, but it is populations that are changed by evolution. The agent of natural selection, antibiotics, acts on individual bacteria. As a result, some individuals reproduce and some do not. However, it is the population as a whole that evolves as its allele frequencies change.

Evolution is change in the allele frequencies of a population, owing to unequal success at reproduction among organisms bearing different alleles. In evolutionary terminology, the **fitness** of an organism is measured by its reproductive success. In our example, the antibiotic-resistant bacteria had greater fitness than the nonresistant bacteria did, because the resistant bacteria produced greater numbers of viable (able to survive) offspring.

Evolution is not progressive; it does not make organisms "better." The traits favored by natural selection change as the environment changes. Resistant bacteria are favored only when antibiotics are present. At a later time, when the environment no longer contains antibiotics, resistant bacteria may be at a disadvantage relative to other bacteria. Similarly, the long necks and legs of male giraffes are helpful when the animals battle to establish dominance, but are a hindrance to drinking (**Fig. 15-10**). The length of male

(a) A contest for dominance

(b) Drinking at a water hole

▲ **FIGURE 15-10 A compromise between opposing environmental pressures (a)** A male giraffe with a long neck and long legs is at a definite advantage in combat to establish dominance. **(b)** But a giraffe's long appendages force it to assume an extremely awkward and vulnerable position when drinking. Thus, drinking and male–male contests generate opposing evolutionary pressures.

Table 15-1 Causes of Evolution

Process	Consequence
Mutation	Creates new alleles; increases variability
Gene flow	Increases similarity of different populations
Genetic drift	Causes random change of allele frequencies; can eliminate alleles
Nonrandom mating	Changes genotype frequencies but not allele frequencies
Natural and sexual selection	Increases frequency of favored alleles; produces adaptations

Have you ever wondered

Why Antibiotics Can't Cure a Cold?

Antibiotics kill or inhibit the reproduction of bacteria, usually by blocking a metabolic pathway that the bacteria use to replicate DNA, synthesize proteins, or build a cell wall. Colds and flu, however, are caused by viruses. Viruses do not have metabolism (as you will learn in Chapter 19) and are not affected by antibiotics. So, taking antibiotics when you have a cold or the flu won't help you get better.

giraffe necks and legs represents an evolutionary compromise between the advantage of being able to win contests with other males and the disadvantage of vulnerability while drinking water.

Table 15-1 summarizes the different causes of evolution.

15.3 HOW DOES NATURAL SELECTION WORK?

Natural selection is not the only evolutionary force. As we have seen, mutation provides variability in heritable traits, and the chance effects of genetic drift may change allele frequencies. Further, evolutionary biologists are now beginning to appreciate the power of random catastrophes in shaping the history of life on Earth; massively destructive events may exterminate thriving and failing species alike. Nevertheless, it is natural selection that shapes the evolution of populations as they adapt to their changing environment. For this reason, we will examine natural selection in more detail.

Natural Selection Stems from Unequal Reproduction

The British economist Herbert Spencer, writing in 1864, coined the phrase "survival of the fittest" to summarize the process that Darwin had named **natural selection.** Natural selection, however, favors traits that increase their possessors' survival only to the extent that improved survival leads to improved reproduction. A trait that improves survival may, for example, increase the likelihood that an individual survives long enough to reproduce, or might increase an organism's life span and, therefore, its number of opportunities to reproduce. But ultimately, it is reproductive success that determines the future of an individual's alleles, and the prevalence in the next generation of the traits associated with those alleles. Thus, the main driver of natural selection is differences in reproduction: Individuals bearing certain alleles leave more offspring (who inherit those alleles) than do other individuals with different alleles.

Natural Selection Acts on Phenotypes

Although we have defined evolution as changes in the genetic composition of a population, it is important to recognize that natural selection does not act directly on the genotypes of individual organisms. Rather, natural selection acts on phenotypes, the structures and behaviors displayed by the members of a population. This selection of phenotypes, however, inevitably affects the genotypes present in a population, because phenotypes and genotypes are closely tied. For example, we know that a pea plant's height is strongly influenced by the plant's alleles of certain genes. If a population of pea plants were to encounter environmental conditions that favored taller plants, then taller plants would leave more offspring. These offspring would carry the alleles that contributed to their parents' height. Thus, if natural selection favors a particular phenotype, it will necessarily also favor the underlying genotype.

BioFlix™ Mechanisms of Evolution

Some Phenotypes Reproduce More Successfully Than Others

As we have seen, natural selection simply means that some phenotypes reproduce more successfully than others do. This simple process is such a powerful agent of change because only the fittest phenotypes pass traits to subsequent generations. But what makes a phenotype fit? Successful phenotypes are those that have the best adaptations to their particular environment. **Adaptations** are characteristics that help an individual survive and reproduce.

An Environment Has Nonliving and Living Components

Individual organisms must cope with an environment that includes not only physical factors but also the other organisms with which the individual interacts. The nonliving (*abiotic*) component of the environment includes such factors as climate, availability of water, and minerals in the soil. The abiotic environment plays a large role in determining the traits that help an organism to survive and reproduce. However, adaptations also arise because of interactions with other organisms—the living (*biotic*) component of the environment. As Darwin wrote, "The structure of every organic being is related . . . to that of all other organic beings, with which it comes into competition for food or residence, or from which it has to escape, or on which it preys." A simple example illustrates this concept.

Consider a buffalo grass plant growing in a small patch of soil in the eastern Wyoming plains. Its roots must be able to take up enough water and minerals for growth and reproduction, and to that extent, it must be adapted to its abiotic environment. But even in the dry prairies of Wyoming, this requirement is relatively trivial, provided that the plant is alone and protected in its square yard of soil. In reality, however, many other plants—other buffalo grass plants as well as other grasses, sagebrush bushes, and annual wildflowers—also sprout in that same patch of soil. If our buffalo grass is to survive, it must compete with the other plants for resources. Its long, deep roots and efficient methods of mineral uptake have evolved not so much because the plains are dry as because the buffalo grass must share the dry prairies with other plants. Further, buffalo grass must also coexist with animals that wish to eat it, such as the cattle that graze the prairie (and the bison that grazed it in the past). As a result, buffalo grass is extremely tough. Silica compounds reinforce its leaves, an adaptation that discourages grazing. Over time, tougher, hard-to-eat plants survived better and reproduced more than did less-tough plants—another adaptation to the biotic environment.

Competition Acts As an Agent of Selection

As the buffalo grass example shows, one of the major agents of natural selection in the biotic environment is **competition** with other organisms for scarce resources. Competition for resources is most intense among members of the same species. As Darwin wrote in *On the Origin of Species*, "The struggle almost invariably will be most severe between the individuals of the same species, for they frequent the same districts, require the same food, and are exposed to the same dangers." In other words, no competing organism has such similar requirements for survival as does another member of the same species. Different species may also compete for the same resources, although generally to a lesser extent than do individuals within a species.

Both Predator and Prey Act As Agents of Selection

When two species interact extensively, each exerts strong selection on the other. When one evolves a new feature or modifies an old one, the other typically evolves new adaptations in response. This constant, mutual feedback between two species is called **coevolution.** Perhaps the most familiar form of coevolution is found in predator–prey relationships.

Predation includes any situation in which one organism eats another. In some instances, coevolution between predators (those who do the eating) and prey (those who are eaten) is a sort of "biological arms race," with each side evolving new adaptations in response to "escalations" by the other. Darwin used the example of wolves and deer: Wolf predation selects against slow or careless deer, thus leaving faster, more-alert deer to reproduce and pass on these traits. The resulting alert, swift deer select in turn against slow, clumsy wolves, because such predators cannot acquire enough food.

Sexual Selection Favors Traits That Help an Organism Mate

In many animal species, males have conspicuous features such as bright colors, long feathers or fins, or elaborate antlers. Males may also exhibit complex courtship behaviors or sing loud, complex songs. Although these extravagant features typically play a role in mating, they also seem to be at odds with efficient survival and reproduction. Exaggerated ornaments and displays may help males gain access to females, but they also make the males more conspicuous and thus vulnerable to predators. Darwin was intrigued by this apparent contradiction. He coined the term **sexual selection** to describe the special kind of selection that acts on traits that help an animal acquire a mate.

Darwin recognized that sexual selection could be driven either by sexual contests among males or by female preference for particular male phenotypes. Male–male competition for access to females can favor the evolution of features that provide an advantage in fights or ritual displays of aggression (**Fig. 15-11**). Female mate choice provides a second source of sexual selection. In animal species in which females actively choose their mates from among males, females often seem to prefer males with the most elaborate ornaments or most extravagant displays (**Fig. 15-12**). Why?

One hypothesis is that male structures, colors, and displays that do not enhance survival might instead provide a female with an outward sign of a male's condition. Only a vigorous, energetic male can survive when burdened with conspicuous coloration or a large tail that might make him more vulnerable to predators. Conversely, males that are sick or under parasitic attack are dull and frumpy compared with healthy males. A female that chooses the brightest, most ornamented male is also choosing the healthiest, most vigorous male. By doing so, she gains fitness if, for example, the most vigorous male

▲ **FIGURE 15-11 Competition between males favors the evolution, through sexual selection, of structures for ritual combat** Two male bighorn sheep spar during the fall mating season. In many species, the losers of such contests are unlikely to mate, while winners enjoy tremendous reproductive success. QUESTION If we studied a population of bighorn sheep and were able to identify the father and mother of each lamb born, would you predict that the difference in number of offspring between the most reproductively successful adult and the least successful adult would be greater for males or for females?

▲ **FIGURE 15-12 The peacock's showy tail has evolved through sexual selection** The ancestors of today's peahens were apparently picky when deciding on a male with which to mate, favoring males with longer and more colorful tails.

provides superior parental care to offspring or if he carries alleles for disease resistance that will be inherited by offspring and help ensure their survival. Females thus gain a reproductive advantage by choosing the most highly ornamented males, and the traits (including the exaggerated ornament) of these flashy males will be passed to subsequent generations.

Selection Can Influence Populations in Three Ways

Natural selection and sexual selection can lead to various patterns of evolutionary change. Evolutionary biologists group these patterns into three categories (**Fig. 15-13**):

- **Directional selection** favors individuals with an extreme value of a trait and selects against both average individuals and individuals at the opposite extreme. For example, directional selection might favor small size and select against both average and large individuals in a population.

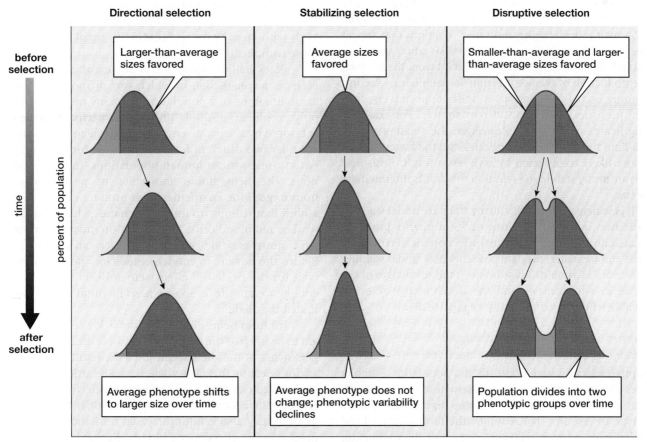

▲ **FIGURE 15-13 Three ways that selection affects a population over time** A graphical illustration of three ways natural and/or sexual selection, acting on a normal distribution of phenotypes, can affect a population over time. In all graphs, the blue areas represent individuals that are selected against—that is, the individuals that do not reproduce as successfully as do the individuals in the purple range.

QUESTION When selection is directional, is there any limit to how extreme the trait under selection will become? Why or why not?

- **Stabilizing selection** favors individuals with the average value of a trait (for example, intermediate body size) and selects against individuals with extreme values.
- **Disruptive selection** favors individuals at both extremes of a trait (for example, both large and small body sizes) and selects against individuals with intermediate values.

Directional Selection Shifts Character Traits in a Specific Direction

If environmental conditions change in a consistent way, a species may respond by evolving in a consistent direction. For example, if the climate becomes colder, mammal species may evolve thicker fur. The evolution of antibiotic resistance in bacteria is an example of directional selection: When antibiotics are present in a bacterial species' environment, individuals with greater resistance reproduce more prolifically than do individuals with less resistance.

Stabilizing Selection Acts Against Individuals Who Deviate Too Far from the Average

Directional selection can't go on forever. What happens once a species is well adapted to a particular environment? If the environment is unchanging, most new variations that appear will be harmful. Under these conditions, we expect species to be subject to stabilizing selection, which favors the survival and reproduction of average individuals. Stabilizing selection commonly occurs when a trait is under opposing environmental pressures from two different sources. For example, among lizards of the genus *Aristelliger*, the smallest lizards have a hard time defending territories, but the largest lizards are more likely to be eaten by owls. As a result, *Aristelliger* lizards are under stabilizing selection that favors intermediate body size.

It is widely assumed that many traits are under stabilizing selection. Although the long necks of giraffes probably originated under directional sexual selection for advantage in combat among males, they are probably now under stabilizing selection, as a compromise between the advantages of being able to win contests and the disadvantage of being vulnerable while drinking water (see Fig. 15-10).

Disruptive Selection Adapts Individuals Within a Population to Different Habitats

Disruptive selection may occur when a population inhabits an area with more than one type of useful resource. In this situation, the most adaptive characteristics may be different for each type of resource. For example, the food source of the black-bellied seedcracker (**Fig. 15-14**), a small, seed-eating bird found in the forests of Africa, includes both hard seeds and soft seeds. Cracking hard seeds requires a large, stout beak, but a smaller, pointier beak is a more efficient tool for processing soft seeds. Consequently, black-bellied seedcrackers have beaks in one of two sizes. A bird may have a large beak or small beak, but very few birds have a medium-sized beak; individuals with intermediate-sized beaks have a lower survival rate than individuals with either large or small beaks. Disruptive selection in black-bellied seedcrackers thus favors birds with large beaks and birds with small beaks, but not those with medium-sized beaks.

▲ **FIGURE 15-14 Black-bellied seedcrackers** As a result of disruptive selection, each black-bellied seedcracker has either a large (left) or small (right) beak.

Black-bellied seedcrackers represent an example of *balanced polymorphism*, in which two or more phenotypes are maintained in a population. In many cases of balanced polymorphism, multiple phenotypes persist because each is favored by a separate environmental factor. For example, consider two different forms of hemoglobin that are present in some African human populations. In these populations, the hemoglobin molecules of people who are homozygous for a particular allele produce defective hemoglobin that clumps up into long chains, which distort and weaken red blood cells. This distortion causes a serious illness known as sickle-cell anemia, which can kill its victims. Before the advent of modern medicine, people homozygous for the sickle-cell allele were unlikely to survive long enough to reproduce. So why hasn't natural selection eliminated the allele?

Far from being eliminated, the sickle-cell allele is present in nearly half the population in some areas of Africa. The persistence of the allele seems to be the result of counterbalancing selection that favors heterozygous carriers of the allele. Heterozygotes, who have one allele for defective hemoglobin and one allele for normal hemoglobin, suffer from mild anemia but they also exhibit increased resistance to malaria, a deadly disease affecting red blood cells that is widespread in equatorial Africa. In areas of Africa with high risk of malaria infection, heterozygotes must have survived and reproduced more successfully than either type of homozygote. As a result, both the normal hemoglobin allele and the sickle-cell allele have been preserved.

Case Study revisited
Evolution of a Menace

The evolution of antibiotic resistance in populations of bacteria—such as the MRSA bacteria that cause so many deadly infections—is a direct consequence of natural selection by antibiotic drugs. When a population of disease-causing bacteria begins to grow in a human body, physicians try to halt population growth by introducing an antibiotic drug to the bacteria's environment. Although many bacteria are killed, some surviving bacteria have genomes with a mutant allele that confers resistance. Bacteria carrying the "resistance allele" produce a disproportionately large share of offspring, which inherit the allele. Soon, resistant bacteria predominate within the population.

By introducing massive quantities of antibiotics into the bacteria's environment, humans have accelerated the pace of the evolution of antibiotic resistance. Each year, U.S. physicians write more than 100 million prescriptions for antibiotics; the Centers for Disease Control and Prevention estimates that about half of these prescriptions are unnecessary.

Although medical use and misuse of antibiotics is the most important source of natural selection for antibiotic resistance, antibiotics also pervade the environment outside our bodies. Our food supply, especially meat, contains a portion of the 20 million pounds of antibiotics that are fed to farm animals each year. In addition, Earth's soils and water are laced with antibiotics that enter the environment through human and animal wastes, and from the antibacterial soaps and cleansers that are now routinely used in many households and workplaces. As a result of this massive alteration of the environment, resistant bacteria are now found not only in hospitals and the bodies of sick people but also in our food, water, and soil. Susceptible bacteria are under constant attack, and resistant strains have little competition. In our fight against disease, we have rashly overlooked some basic principles of evolutionary biology and are now paying a heavy price.

Consider This

Because natural selection acts only on existing variation among phenotypes, antibiotic resistance could not evolve if bacteria in natural populations did not already carry alleles that help them resist attack by antibiotic chemicals. Why are such alleles present (albeit at low levels) in bacterial populations? Conversely, if resistance alleles are beneficial, why are they rare in natural populations of bacteria?

CHAPTER REVIEW

Summary of Key Concepts

15.1 How Are Populations, Genes, and Evolution Related?

Evolution is change in the frequencies of alleles in a population's gene pool. Allele frequencies in a population will remain constant over generations only if the following conditions are met: (1) There is no mutation, (2) there is no gene flow, (3) the population is very large, (4) all mating is random, and (5) all genotypes reproduce equally well (that is, there is no natural selection). These conditions are rarely, if ever, met in nature. Understanding what happens when they are not met helps reveal the mechanisms of evolution.

15.2 What Causes Evolution?

Evolutionary change is caused by mutation, gene flow, small population size, nonrandom mating, and natural selection.

- Mutations are random, undirected changes in DNA composition. Although most mutations are neutral or harmful to the organism, some prove advantageous in certain environments. Mutations are rare and do not change allele frequencies very much, but they provide the raw material for evolution.
- Gene flow is the movement of alleles between different populations of a species. Gene flow tends to reduce differences in the genetic composition of different populations.
- In any population, chance events kill or prevent reproduction by some of the individuals. If the population is small, chance events may affect the survival and reproduction of a disproportionate number of individuals who bear a particular allele, thereby greatly changing the allele frequency in the population; this is genetic drift.
- Nonrandom mating, such as assortative mating and inbreeding, can change the distribution of genotypes in a population, in particular by increasing the proportion of homozygotes.
- The survival and reproduction of organisms are influenced by their phenotypes. Because phenotype depends at least partly on genotype, natural selection tends to favor the reproduction of certain alleles at the expense of others.

15.3 How Does Natural Selection Work?

Natural selection is driven by differences in reproductive success among different genotypes. Natural selection proceeds from the interactions of organisms with both the biotic and abiotic parts of their environments. When two or more species exert mutual environmental pressures on each other for long periods of time, both of them evolve in response. Such coevolution can result from any type of relationship between organisms, including competition and predation. Phenotypes that help organisms mate can evolve by sexual selection.

BioFlix™ Mechanisms of Evolution

Key Terms

Thinking Through the Concepts

Fill-in-the-Blank

1. The __________ provides a simple mathematical model for a nonevolving population, also called a(n) __________ population, in which __________ frequencies do not change over time. Are such populations likely to be found in nature? __________
2. Different versions of the same gene are called __________. These versions arise as a result of changes in the sequence of __________ that form the gene. These changes are caused by __________. An individual with two identical copies of a given gene is described as being __________ for that gene, while an individual with two different versions of that gene is described as __________.
3. An organism's __________ refers to the specific alleles found within its chromosomes, while the traits that these alleles produce are called its __________. Which of these does natural selection act on? __________
4. A random form of evolution is called __________. This form of evolution will only occur in populations that are __________. Two important causes of this form of evolution are the __________ and __________. Which of these would apply to a population started by a breeding pair that was stranded on an island? __________
5. Competition is most intense between members of __________. Predators and their prey act as agents of __________ on one another, resulting in a form of evolution called __________. This results in the evolution of characteristics called __________ that help both predators and their prey survive and reproduce.
6. The evolutionary fitness of an organism is measured by its success at __________. The fitness of an organism can change if its __________ changes.

Review Questions

1. What is a gene pool? How would you determine the allele frequencies in a gene pool?
2. Define *equilibrium population*. Outline the conditions that must be met for a population to stay in genetic equilibrium.
3. How does population size affect the likelihood of changes in allele frequencies by chance alone? Can significant changes in allele frequencies (that is, evolution) occur as a result of genetic drift?
4. If you measured the allele frequencies of a gene and found large differences from those predicted by the Hardy–Weinberg principle, would that prove that natural selection is occurring in the population you are studying? Review the conditions that lead to an equilibrium population, and explain your answer.
5. People like to say that "you can't prove a negative." Study the experiment in Figure 15-3 again, and comment on what it demonstrates.
6. Describe the three ways in which natural selection can affect a population over time. Which way(s) is (are) most likely to occur in stable environments, and which way(s) might occur in rapidly changing environments?
7. What is sexual selection? How is sexual selection similar to and different from other forms of natural selection?

Applying the Concepts

1. In North America, the average height of adult humans has been increasing steadily for decades. Is directional selection occurring? What data would justify your answer?
2. Malaria is rare in North America. In populations of African Americans, what would you predict is happening to the frequency of the hemoglobin allele that leads to sickling in red blood cells? How would you go about determining whether your prediction is true?
3. By the 1940s, the whooping crane population had been reduced to fewer than 50 individuals. Thanks to conservation measures, its numbers are now increasing. What special evolutionary problems do whooping cranes face now that they have passed through a population bottleneck?
4. In many countries, conservationists are trying to design national park systems so that "islands" of natural area (the big parks) are connected by thin "corridors" of undisturbed habitat. The idea is that this arrangement will allow animals and plants to migrate between refuges. Why would such migration be important?
5. A preview question for Chapter 16: A *species* is all the populations of organisms that potentially interbreed with one another but that are reproductively isolated from (cannot interbreed with) other populations. Using the five conditions of the Hardy–Weinberg principle as a starting point, what factors do you think would be important in the splitting of a single ancestral species into two modern species?

MB® *Go to www.masteringbiology.com for practice quizzes, activities, eText, videos, current events, and more.*

The Origin of Species

Chapter 16

Case Study

Lost World

THE STEEP, rain-drenched slopes of Vietnam's Annamite Mountains are remote and forbidding, cloaked in tropical mists that lend an air of mystery and concealment to the forested peaks. As it turns out, this remote refuge conceals a most astonishing biological surprise: the saola, a hoofed, horned mammal that was unknown to science until the early 1990s. The discovery of a new species of large mammal at this late date was a complete shock. After centuries of human exploration and exploitation of every corner of the world's forests, deserts, and savannas, scientists were certain that no large mammal species could have escaped detection. As long ago as 1812, French naturalist Georges Cuvier wrote that "there is little hope of discovering new species of large quadrupeds." And yet, the saola—3 feet high at the shoulder, weighing up to 200 pounds and sporting 20-inch black horns—remained hidden in Annamite Mountain forests, outside the realm of scientific knowledge.

It is surprising that the saola stayed hidden from scientists for so long because people tend to notice large animals. But as the discovery of the saola showed, even relatively conspicuous organisms may remain unknown if they live in a sufficiently remote area. The Annamite Mountains, isolated by inhospitable terrain and the wars fought in Vietnam during the last century, concealed the saola and other undiscovered plant, mammal, and reptile species. Similarly, the remote Foja Mountains of New Guinea hid a new bird species and dozens of new species of butterflies, frogs, and flowering plants, which were not discovered until a 2006 scientific expedition to the region. A recent survey of unexplored underwater habitats in Indonesia revealed more than 50 new species of fish, coral, and shrimp. A curious biologist might wonder why these distinctive species are concentrated in particular geographic areas. But before we can fully consider that question, we will need to explore the evolutionary process by which new species arise.

▲The saola, unknown to science until 1992, is one of a number of previously undiscovered species recently found in the mountains of Vietnam.

At a Glance

16.1 WHAT IS A SPECIES?

Although Darwin brilliantly explained how evolution shapes complex, amazingly well-designed organisms, his ideas did not fully explain life's diversity. In particular, the process of natural selection cannot by itself explain how living things came to be divided into groups, with each group distinctly different from all other groups. When we look at big cats, we don't see a continuous array of different tiger phenotypes that gradually grades into a lion phenotype. We see lions and tigers as separate, distinct types with no overlap. Each distinct type is known as a species. (Would you like to name a species? See "Links to Everyday Life: Biological Vanity Plates.")

Biologists Need a Clear Definition of Species

Before we can study the origin of species, we must first clarify our definition of the term. Throughout most of human history, "species" was a poorly defined concept. In pre-Darwinian Europe, the word "species" simply referred to one of the "kinds" produced by the biblical creation. In this view, humans could not possibly know the criteria of the creator, but could only attempt to distinguish among species on the basis of visible differences in structure. In fact, "species" is Latin for "appearance."

On a coarse scale, it is easy to use quick visual comparisons to distinguish species. For example, warblers are clearly different from eagles, which are obviously different from ducks. But it is far more difficult to distinguish among different species of warblers, eagles, or ducks. How do scientists make these finer distinctions?

Each Species Evolves Independently

Today, biologists define a **species** as a group of populations that evolves independently. Each species follows a separate evolutionary path because alleles do not move between the gene pools of different species. This definition, however, does not clearly state the standard by which such evolutionary independence is judged. The most widely used standard defines species as "groups of actually or potentially interbreeding natural populations, which are reproductively isolated from other such groups." This definition, known as the *biological species concept*, is based on the observation that **reproductive isolation** (inability to successfully breed outside the group) ensures evolutionary independence.

The biological species concept has at least two major limitations. First, because the definition is based on patterns of sexual reproduction, it does not help us determine species boundaries among asexually reproducing organisms. Second, it is not always practical or even possible to directly observe whether members of two different groups interbreed. Thus, a biologist who wishes to determine if a group of organisms is a separate species must often make the determination without knowing for sure if group members breed with organisms outside the group.

Despite the limitations of the biological species concept, most biologists accept it for identifying species of sexually reproducing organisms. Nonetheless, scientists who study bacteria and other organisms that mainly reproduce asexually must use alternative definitions. Even some biologists who study sexually reproducing organisms prefer species definitions that do not depend on a property (reproductive isolation) that can be difficult to measure. Several such alternatives to the biological species concept have been proposed; one that has gained many adherents is described in Chapter 18 (p. 351).

Appearance Can Be Misleading

Biologists have found that some organisms with very similar appearances belong to different species. For example, the cordilleran flycatcher and the Pacific-slope flycatcher are so similar that even experienced birdwatchers cannot tell them apart (**Fig. 16-1**). Until recently, these birds were considered to be a single species. However, research revealed that the two kinds of birds do not interbreed and are in fact two different species.

Links to *Everyday Life*

Biological Vanity Plates

Looking for a special gift for a friend or loved one? Why not name a species after him or her? Or, for that matter, name one after yourself! Thanks to the BIOPAT project (www.biopat.de), anyone with $3,500 can be immortalized in the Latin name of a newly discovered plant or animal.

Typically, the scientist who discovers and describes a new species is entitled to choose its Latin name. Scientists usually choose a name that describes a trait of the species or perhaps the location where it was found. Sometimes, however, more whimsical choices are made. For example, a recently discovered beetle was named *Agra schwarzeneggeri*, in honor of the California governor and former film star, a frog was named *Hyla stingi* after the British rock star, and a spider was named *Aptostichus stephencolberti* to recognize the satiric news commentator. *Agathidium bushi* and *Agathidium cheneyi* are beetles named for the former U.S. president and vice president.

If you donate money to the BIOPAT project, the name of a new species will be entirely up to you. In return for a contribution that supports efforts to discover and conserve endangered species, the people at BIOPAT will offer you a selection of newly discovered but unnamed species. You can choose your species and pick a name, which is then given an appropriate Latin ending and published in a scientific journal. Your chosen name becomes the official, recognized scientific name of the new species.

In perhaps the most extraordinary example of purchased species naming rights, a newly discovered monkey was named after an online casino (**Fig. E16-1**). In return for a contribution of $650,000, the new species was named *Callicebus aureipalatii*; the species name is Latin for golden palace. The payment will be used for management of the Madidi National Park in Bolivia, where the new species was discovered.

▲ **FIGURE E16-1 The Golden Palace monkey is named after a casino**

Superficial similarity can sometimes hide multiple species. Researchers recently discovered that the species known until now as the two-barred flasher butterfly is actually a group of at least 10 different species. The caterpillars of the different species do differ in appearance, but the adult butterflies are all so similar that their species identities went undetected for more than two centuries after the butterfly was first described and named.

Conversely, differences in appearance do not always mean that two populations belong to different species. For example, bird field guides published in the 1970s list the myrtle warbler and Audubon's warbler as distinct species (**Fig. 16-2**).

(a) Cordilleran flycatcher

(b) Pacific-slope flycatcher

▲ **FIGURE 16-1 Members of different species may be similar in appearance (a)** The cordilleran flycatcher and **(b)** Pacific-slope flycatcher are different species.

(a) Myrtle warbler

(b) Audubon's warbler

▲ **FIGURE 16-2 Members of a species may differ in appearance (a)** The myrtle warbler and **(b)** Audubon's warbler are members of the same species.

These birds differ in geographic range and in the color of their throat feathers. More recently, scientists determined that these birds are local varieties of the same species. The reason: Where their ranges overlap, these warblers interbreed freely.

16.2 HOW IS REPRODUCTIVE ISOLATION BETWEEN SPECIES MAINTAINED?

What prevents different species from interbreeding? The traits that prevent interbreeding and maintain reproductive isolation are called **isolating mechanisms.** Isolating mechanisms give a clear benefit to individuals. An individual that mates with a member of another species will probably produce no offspring (or unfit or sterile offspring), thereby wasting its reproductive effort and failing to contribute to future generations. Thus, natural selection favors traits that prevent mating across species boundaries. Mechanisms that prevent mating between species are called **premating isolating mechanisms.**

When premating isolating mechanisms fail or have not yet evolved, members of different species may mate. If, however, all resulting hybrid offspring die during development, then the two species are still reproductively isolated from one another. Even if hybrid offspring are able to survive, if these hybrids are less fit than their parents or are themselves infertile, the two species may still remain separate, with little or no gene flow between them. Mechanisms that prevent the formation of vigorous, fertile hybrids between species are called **postmating isolating mechanisms.**

Premating Isolating Mechanisms Prevent Mating Between Species

Reproductive isolation can be maintained by a variety of mechanisms, but those that prevent mating attempts are especially effective. We next describe the most important types of such premating isolating mechanisms.

Members of Different Species May Be Prevented from Meeting

Members of different species cannot mate if they never get near one another. *Geographic isolation* prevents interbreeding between populations that do not come into contact because they live in different, physically separated places (**Fig. 16-3**). However, we cannot determine if geographically

► **FIGURE 16-3 Geographic isolation** To determine if these two squirrels are members of different species, we must know if they are "actually or potentially interbreeding." Unfortunately, it is hard to tell, because **(a)** the Kaibab squirrel lives only on the north rim of the Grand Canyon and **(b)** the Abert squirrel lives only on the south rim. The two populations are geographically isolated but still quite similar. Have they diverged enough since their separation to become reproductively isolated? Because they remain geographically isolated, we cannot say for sure.

(a) Kaibab squirrel

(b) Abert squirrel

separated populations are actually distinct species. Should the physical barrier separating the two populations disappear (a new channel might connect two previously isolated lakes, for example), the reunited populations might interbreed freely and not be separate species after all. If they cannot interbreed, then other mechanisms, such as those considered below, must have developed during their isolation. Geographic isolation, therefore, is usually considered to be a mechanism that allows new species to form rather than a mechanism that maintains reproductive isolation between species.

Different Species May Occupy Different Habitats

Two populations that use different resources may spend time in different habitats within the same general area and thus exhibit *ecological isolation*. White-crowned sparrows and white-throated sparrows, for example, have extensively overlapping geographic ranges. The white-throated sparrow, however, frequents dense thickets, whereas the white-crowned sparrow inhabits fields and meadows, seldom penetrating far into dense growth. The two species may coexist within a few hundred yards of one another and yet seldom meet during the breeding season. A more dramatic example is provided by the more than 300 species of fig wasp (**Fig. 16-4**). In most cases, fig wasps of a given species breed in (and pollinate) the fruits of one particular species of fig, and each fig species hosts only one or two species of pollinating wasp. Thus, fig wasps of different species only rarely encounter one another during breeding, and pollen from one fig species is not ordinarily carried to flowers of a different species.

Although ecological isolation may slow down interbreeding, it seems unlikely that it could prevent gene flow entirely. Other mechanisms normally also contribute to reproductive isolation.

▲ **FIGURE 16-4 Ecological isolation** This female fig wasp is carrying fertilized eggs from a mating that took place within a fig. She will find another fig of the same species, enter it through a pore, lay eggs, and die. Her offspring will hatch, develop, and mate within the fig. Because each species of fig wasp reproduces only in its own particular fig species, each wasp species is reproductively isolated.

Different Species May Breed at Different Times

Even if two species occupy similar habitats, they cannot mate if they have different breeding seasons, a phenomenon called *temporal (time-related) isolation*. For example, the spring field cricket and the fall field cricket both occur in many areas of North America but, as their names suggest, the former species breeds in spring and the latter in autumn. As a result, the two species do not interbreed.

In plants, the reproductive structures of different species may mature at different times. For example, Bishop pines and Monterey pines grow together near Monterey on the California coast (**Fig. 16-5**), but the two species release

(a) Bishop pine

(b) Monterey pine

▲ **FIGURE 16-5 Temporal isolation (a)** Bishop pines and **(b)** Monterey pines coexist in nature. In the laboratory they produce fertile hybrids. In the wild, however, they do not interbreed, because they release pollen at different times of the year.

their sperm-containing pollen (and have eggs ready to receive the pollen) at different times: The Monterey pine releases pollen in early spring, the Bishop pine in summer. For this reason, the two species never interbreed under natural conditions.

Different Species May Have Different Courtship Signals

Among animals, elaborate courtship colors and behaviors not only serve as recognition and evaluation signals between male and female, but also prevent mating with members of other species. Signals and behaviors that differ from species to species create *behavioral isolation*. The distinctive plumage and vocalizations of a male bird, for example, may attract females of his own species, whereas females of other species treat these displays with indifference. For example, the extravagant plumes and arresting pose of a courting male greater bird of paradise are conspicuous indicators of his species and there is little chance that females of another species will be mistakenly attracted (**Fig. 16-6**). Among frogs, males are often impressively indiscriminate, jumping on every female in sight, regardless of the species, when the spirit moves them. Females, however, approach only male frogs that utter the call appropriate to their species. If they do find themselves in an unwanted embrace, they utter the "release call," which causes the male to let go. As a result, few hybrids are produced.

Differing Sexual Organs May Foil Mating Attempts

In rare instances, a male and a female of different species attempt to mate. Their attempt is likely to fail. Among animal species with internal fertilization (in which the sperm is deposited inside the female's reproductive tract), the male's and female's sexual organs simply may not fit together. Incompatible body shapes may also make copulation between species impossible. For example, snails of species whose shells have left-handed spirals may be unable to successfully copulate with snails whose shells have right-handed spirals (**Fig. 16-7**). Among plants, differences in flower size or structure may prevent pollen transfer between species because the differing flowers may attract different pollinators. Isolating mechanisms of this type are called *mechanical incompatibilities*.

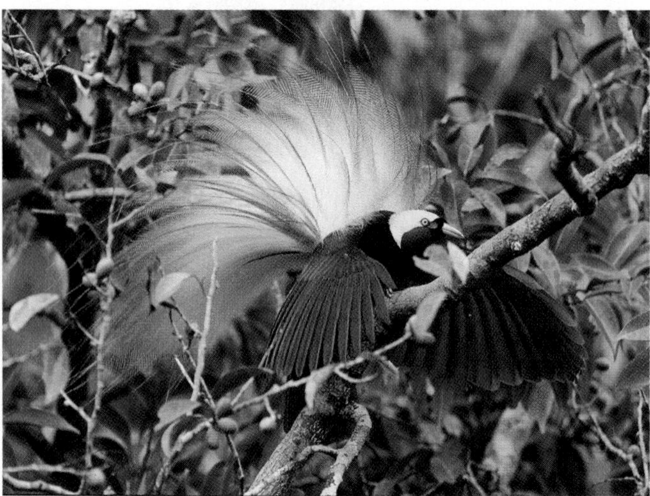

▲ **FIGURE 16-6 Behavioral isolation** The mate-attraction display of a male greater bird of paradise includes distinctive posture, movements, plumage, and vocalizations that do not resemble those of other bird-of-paradise species.

Postmating Isolating Mechanisms Limit Hybrid Offspring

Premating isolation sometimes fails. When it does, members of different species may mate, and the sperm of one species may reach the egg of another species. Such matings, however, often fail to produce vigorous, fertile hybrid offspring, owing to postmating isolating mechanisms.

One Species' Sperm May Fail to Fertilize Another Species' Eggs

Even if a male inseminates a female of a different species, his sperm may not be able to fertilize her eggs, an isolating mechanism called *gametic incompatibility*. For example, in animals with internal fertilization, fluids in the female reproductive tract may weaken or kill sperm of other species.

Gametic incompatibility may be an especially important isolating mechanism in species, such as marine invertebrate animals and wind-pollinated plants, that reproduce by scattering gametes in the water or in the air. For example, sea

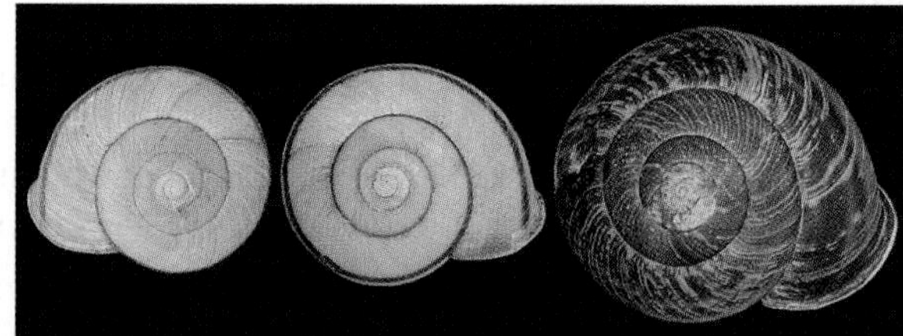

(a) Shells of three snail species

(b) Matching coils

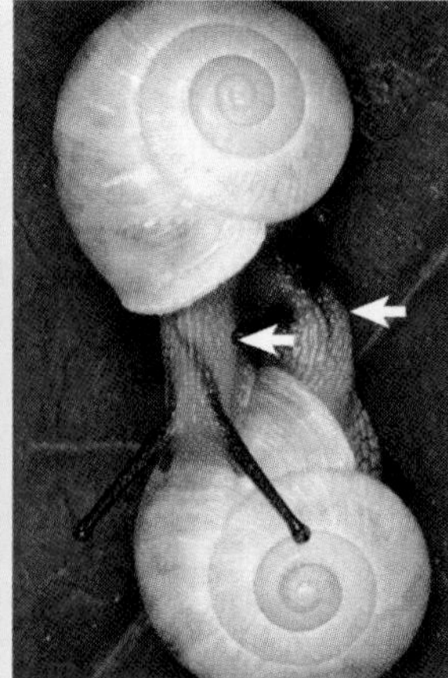

(c) Non-matching coils

▲ **FIGURE 16-7 Mechanical incompatibility (a)** The shells of different snail species may coil in different directions. Among the three closely related species shown, two have shells that coil left and one has a shell that coils right. **(b)** Two snails with matching coils can mate, but **(c)** snails of different species with mismatched coiling cannot mate because the mismatch keeps their genitals (arrows) apart.

urchin sperm cells contain a protein that allows them to bind to eggs. The structure of the protein differs among species so that sperm of one sea urchin species cannot bind to the eggs of another species. In abalones (a type of mollusk), eggs are surrounded by a membrane that can be penetrated only by sperm containing a particular enzyme. Each abalone species has a distinctive version of the enzyme, so hybrids are rare, even though several species of abalones coexist in the same waters and spawn during the same period. Among plants, similar chemical incompatibility may prevent the germination of pollen from one species that lands on the stigma (pollen-catching structure) of the flower of another species.

Hybrid Offspring May Fail to Survive or Reproduce

If cross-species fertilization does occur, the resulting hybrid may be unable to survive, a situation called *hybrid inviability*. The genetic instructions directing development of the two species may be so different that hybrids abort early in development. For example, captive leopard frogs can be induced to mate with wood frogs, and the matings generally yield fertilized eggs. The resulting embryos, however, inevitably fail to survive more than a few days.

In other animal species, a hybrid might survive but display behaviors that are mixtures of the two parental types. In attempting to do some things the way species A does them and other things the way species B does them, the hybrid may be hopelessly uncoordinated and therefore unable to reproduce. Hybrids between certain species of lovebirds, for example, have great difficulty learning to carry nest materials during flight and probably could not reproduce in the wild.

Hybrid Offspring May Be Infertile

Most animal hybrids, such as the mule (a cross between a horse and a donkey) and the liger (a zoo-based cross between a male lion and a female tiger), are sterile (**Fig. 16-8**). *Hybrid infertility* prevents hybrids from passing on their genetic material to offspring, which blocks gene flow between the two parent populations. A common reason for hybrid infertility is the failure of chromosomes to pair properly during meiosis, so that eggs and sperm fail to develop. (To learn how hybrid inviability and infertility can threaten endangered species, see "Earth Watch: Hybridization and Extinction" on p. 310.)

▲ **FIGURE 16-8 Hybrid infertility** This liger, the hybrid offspring of a lion and a tiger, is sterile. The gene pools of its parent species remain separate.

Table 16-1 summarizes the different types of isolating mechanisms.

16.3 HOW DO NEW SPECIES FORM?

Despite his exhaustive exploration of the process of natural selection, Charles Darwin did not propose a complete mechanism of **speciation,** the process by which new species form. One scientist who did play a large role in describing the process of speciation was Ernst Mayr of Harvard University, an ornithologist (expert on birds) and a pivotal figure in the history of evolutionary biology. Mayr developed the biological species concept discussed above. He was also among the first to recognize that speciation depends on two factors acting on a pair of populations: isolation and genetic divergence.

- **Isolation of populations:** If individuals move freely between two populations, interbreeding and the resulting gene flow will cause changes in one population to soon become widespread in the other as well. Thus, two populations cannot grow increasingly different unless something happens to block interbreeding between them. Speciation depends on isolation.

Table 16-1 Mechanisms of Reproductive Isolation

Premating isolating mechanisms: factors that prevent organisms of two species from mating
- **Geographic isolation:** The species cannot interbreed because a physical barrier separates them.
- **Ecological isolation:** The species do not interbreed even if they are within the same area because they occupy different habitats.
- **Temporal isolation:** The species cannot interbreed because they breed at different times.
- **Behavioral isolation:** The species do not interbreed because they have different courtship and mating rituals.
- **Mechanical incompatibility:** The species cannot interbreed because their reproductive structures are incompatible.

Postmating isolating mechanisms: factors that prevent organisms of two species from producing vigorous, fertile offspring after mating
- **Gametic incompatibility:** Sperm from one species cannot fertilize eggs of another species.
- **Hybrid inviability:** Hybrid offspring fail to survive to maturity.
- **Hybrid infertility:** Hybrid offspring are sterile or have low fertility.

Earth Watch

Hybridization and Extinction

The main cause of extinction is environmental change, especially habitat destruction. Some species with small populations, however, are also threatened by a less obvious danger: hybridization. Although premating isolating mechanisms ensure that, for the most part, members of one species cannot interbreed with members of a different species, matings between members of different species are nonetheless possible. Between-species matings and the resulting hybrid offspring are especially common in birds and plants.

How can hybrid mating be dangerous to endangered species? Recall that postmating isolating mechanisms ensure that, in most cases, hybrid offspring will survive poorly and may even be sterile. Now, picture what happens when contact between two species produces hybrids, and one of the species has a much smaller population than the other. If the hybrid offspring fail to survive and reproduce, the numbers of each species will decline, but the decline will have a proportionally larger impact on the small population. When the more abundant species moves into the range of the rare species, the impact on the rare species can be severe. Even if the hybrid offspring survive well, high numbers of hybrids can overwhelm the rare species, essentially absorbing the rare species into the abundant species.

Damage from hybridization is most likely to occur when formerly isolated small populations come into contact with larger populations of a closely related species. For example, the plant *Clarkia lingulata* is extremely rare, known to exist only in two sites in the Sierra Nevada mountains of California. Unfortunately, it readily hybridizes with its abundant relative *Clarkia biloba* to produce sterile hybrid offspring. Because several populations of *C. biloba* grow near the *C. lingulata* populations, extinction by hybridization is a real possibility for this rare species.

▲ **FIGURE E16-2 Ethiopian wolves** Fewer than 500 Ethiopian wolves remain. Among the threats to their continued existence is hybridization with wild dogs.

Human activities often cause contact between an endangered species and a more abundant species with which it can hybridize. For example, the rare Hawaiian duck, found only on the Hawaiian Islands, hybridizes freely with mallard ducks, a non-native species introduced to Hawaii to provide hunters with a new game species. Similarly, the endangered Ethiopian wolf (**Fig. E16-2**) is threatened by interbreeding with feral domestic dogs, and the endangered European wildcat is at risk from hybridization with domestic cats. In these cases and others, a species first declined in number due to habitat destruction and then became vulnerable to further damage by hybridization with a more numerous species that was present as a result of human activities.

- **Genetic divergence of populations:** It is not sufficient for two populations simply to be isolated. They will become separate species only if, during the period of isolation, they evolve sufficiently large genetic differences. The differences must be large enough that, if the isolated populations are reunited, they can no longer interbreed and produce vigorous, fertile offspring. That is, speciation is complete only if divergence results in evolution of an isolating mechanism. Such differences can arise by chance (genetic drift), especially if at least one of the isolated populations is small (see pp. 290–292). Large genetic differences can also arise through natural selection, if the isolated populations experience different environmental conditions.

Speciation always requires isolation followed by divergence, but these steps can take place in several different ways. Evolutionary biologists group the different pathways to speciation into two broad categories: **allopatric speciation,** in which two populations are geographically separated from one another, and **sympatric speciation,** in which two populations share the same geographic area.

Geographic Separation of a Population Can Lead to Allopatric Speciation

New species can arise by allopatric speciation when an impassible barrier physically separates different parts of a population.

Organisms May Colonize Isolated Habitats

A small population can become isolated if it moves to a new location (**Fig. 16-9**). For example, some members of a population of land-dwelling organisms might colonize an oceanic island. The colonists might be birds, flying insects, fungal spores, or wind-borne seeds blown by a storm. More earthbound organisms could reach the island on a drifting "raft" of vegetation torn from the mainland coast. Whatever the means, such colonization must occur regularly, given the presence of living things on even the remotest islands.

▲ **FIGURE 16-9 Allopatric isolation and divergence** In allopatric speciation, some event causes a population to be divided by an impassable geographic barrier. One way the division can occur is by colonization of an isolated island. The two now-separated populations may diverge genetically. If the genetic differences between the two populations become large enough to prevent interbreeding, then the two populations constitute separate species.

EXERCISE Make a list of events or processes that could cause geographic subdivision of a population. Are the items on your list sufficient to account for formation of the millions of species that have inhabited Earth?

Isolation by colonization need not be limited to islands. For example, different coral reefs may be separated by miles of open ocean, so any reef-dwelling sponges, fishes, or algae that were carried by ocean currents to a distant reef would be effectively isolated from their original populations. Any bounded habitat, such as a lake, a mountaintop, or a parasite's host can isolate arriving colonists.

Geological and Climate Changes May Divide Populations

Isolation can also result from landscape changes that divide a population. For example, rising sea levels might transform a coastal hilltop into an island, isolating the residents. New rock from a volcanic eruption can divide a previously continuous sea or lake, splitting populations. A river that changes course can also divide populations, as can a newly formed mountain range. Climate shifts, such as those that happened in past ice ages, can change the distribution of vegetation and strand portions of populations in isolated patches of suitable habitat. You can probably imagine many other scenarios that could lead to the geographic subdivision of a population.

Over the history of Earth, many populations have been divided by continental drift. Earth's continents float on molten rock and slowly move about the surface of the planet. On a number of occasions during Earth's long history, continental landmasses have broken into pieces that subsequently moved apart (see Fig. 17-12). Each of these breakups must have split a multitude of populations.

Natural Selection and Genetic Drift May Cause Isolated Populations to Diverge

If two populations become geographically isolated for any reason, there will be no gene flow between them. If the environments of the locations differ, then natural selection may favor different traits in the different locations, and the populations may accumulate genetic differences. Alternatively, genetic differences may arise if one or more of the separated populations is small enough that substantial genetic drift occurs, which may be especially likely in the aftermath of a founder event (in which a few individuals become isolated from the main body of the species). In either case, genetic differences between the separated populations may eventually become large enough to make interbreeding impossible. At that point, the two populations will have become separate species. Most evolutionary biologists believe that geographic isolation followed by allopatric speciation has been the most common source of new species, especially among animals.

Genetic Isolation Without Geographic Separation Can Lead to Sympatric Speciation

Genetic isolation—limited gene flow—is required for speciation, but populations can become genetically isolated without geographic separation. Thus, new species can arise by sympatric speciation.

Ecological Isolation Can Reduce Gene Flow

If a geographic area contains two distinct types of habitats (each with distinct food sources, places to raise young, and so on), different members of a single species may begin to specialize in one habitat or the other. If conditions are right, natural selection in the two different habitats may lead to the evolution of different traits in the two groups. Eventually, these differences may become large enough to prevent members of the

Case Study continued
Lost World

It is not unexpected that the forests of New Guinea's Foja Mountains would be home to a variety of distinctive species. New Guinea is, after all, an island. It is likely that, in the past, populations colonized the island and became genetically isolated from mainland populations, thereby initiating the process of speciation. Similarly, the distinctive species recently discovered on Indonesian coral reefs arose because reefs form in shallow seas, and are typically surrounded by deep, open-ocean waters that many reef-dwelling organisms cannot easily cross. But what about the saola and the other unique species of the Ammanite Mountains? How might populations inhabiting mainland forests in Vietnam have become isolated from other populations? Try to think of a possible answer to this question. We will return to this topic in the "Case Study Revisited" section at the end of this chapter.

two groups from interbreeding, and the formerly single species will have split into two species. Such a split seems to be taking place right before biologists' eyes, so to speak, in the case of the fruit fly *Rhagoletis pomonella*.

Rhagoletis is a parasite of the American hawthorn tree. This fly lays its eggs in the hawthorn's fruit; when the maggots hatch, they eat the fruit. About 150 years ago, entomologists (scientists who study insects) noticed that *Rhagoletis* had begun to infest apple trees, which were introduced into North America from Europe. Today, it appears that *Rhagoletis* is splitting into two species—one that breeds on apples, and one that breeds on hawthorns (**Fig. 16-10**). The two groups have evolved substantial genetic differences, some of which—such as those that affect the timing of emergence of the adult flies—are important for survival on a particular host plant.

The two kinds of flies will become two species only if they maintain reproductive separation. Apple trees and hawthorns typically grow in the same areas, and flies, after all, can fly. So why don't apple flies and hawthorn flies interbreed and cancel out any genetic differences between them? First, female flies usually lay their eggs in the same type of fruit in which they developed. Males also tend to prefer the same type of fruit in which they developed. Therefore, apple-liking males will encounter and mate with apple-liking females. Second, apples mature 2 or 3 weeks later than do hawthorn fruits, and the two types of flies emerge with timing appropriate for their chosen host fruit. Thus, the two varieties of flies have very little chance of meeting. Although some interbreeding between the two types of flies occurs, it seems they are well on their way to speciation. Will they make it? Entomologist Guy Bush suggests, "Check back with me in a few thousand years."

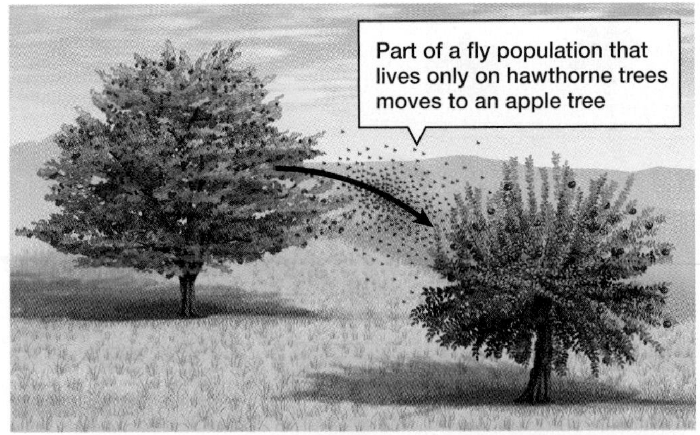

The flies living on the apple tree do not encounter the flies living on the hawthorne tree, so the populations diverge

▲ **FIGURE 16-10 Sympatric isolation and divergence** In sympatric speciation, some event blocks gene flow between two parts of a population that remains in a single geographic area. One way in which this genetic isolation can occur is if a portion of a population begins to use a previously unexploited resource, as when some members of an insect population shift to a new host plant species (as has occurred in the fruit fly species *Rhagoletis pomonella*). The two now-isolated populations may diverge genetically. If the genetic differences between the two populations become large enough to prevent interbreeding, then the two populations constitute separate species.

QUESTION How might future scientists test whether the current *R. pomonella* has become two species?

Mutations Can Lead to Genetic Isolation

In some instances, new species can arise nearly instantaneously as a result of mutations that change the number of chromosomes in an organism's cells. The acquisition of multiple copies of each chromosome is known as **polyploidy** and has been a frequent cause of sympatric speciation. In general, polyploid individuals cannot mate successfully with normal diploid individuals. Thus, a polyploid mutant is genetically isolated from its parent species. If, however, it could somehow reproduce and leave offspring, its descendents could form a new, reproductively isolated species.

Polyploid plants are more likely than polyploid animals to be able to reproduce; speciation by polyploidy is therefore more common in plants than in animals. Many plants can either self-fertilize or reproduce asexually, or both. Most animals, however, cannot self-fertilize or reproduce asexually. Therefore, a polyploid plant is much more likely than a polyploid animal to become the founding member of a new, polyploid species.

Under Some Conditions, Many New Species May Arise

In the same way that the history of your family can be represented by a family tree, the history of life can be represented by an *evolutionary tree*. The trunk at the base of the evolutionary tree of life represents Earth's earliest organisms, and each of the topmost branches represents one of today's living species. Each fork in the branches represents a speciation event, when one species split into two. Hypotheses and discoveries about the evolutionary relationships among species are often communicated by depictions of a portion of life's evolutionary tree (**Fig. 16-11a**).

In some cases, many new species have arisen in a relatively short time (**Fig. 16-11b**). This process, called **adaptive radiation,** can occur when populations of one species invade a variety of new habitats and evolve in response to the differing environmental pressures in those habitats.

Adaptive radiation has occurred many times and in many groups of organisms, typically when species encounter a wide variety of unoccupied habitats. For example, episodes of adaptive radiation took place when some wayward finches colonized the Galápagos Islands, when a population of cichlid fish reached isolated Lake Malawi in Africa, and when an ancestral silversword plant species arrived at the Hawaiian Islands (**Fig. 16-12**). These events gave rise to adaptive radiations of 13 species of Darwin's finches in the Galápagos, more than 300 species of cichlids in Lake Malawi, and 30 species of silversword plants in Hawaii. In these examples, the invading species faced no competitors except other members of their own species, and all the available habitats and food sources were rapidly exploited by new species that evolved from the original invaders.

How Many Species Inhabit the Planet?

One way to determine the number of species on Earth might be to simply count them. You could comb the scientific literature to find all the species that scientists have discovered and named, and then tally up the total number. If you did that, you'd end up with a count of roughly 1.5 million species. But you still wouldn't know how many species are on Earth.

Why doesn't counting work? Because most of the planet's species remain undiscovered. Relatively few scientists are engaged in the search for new species, and many undiscovered species are small and inconspicuous, or live in poorly explored habitats such as the floor of the ocean or the topmost branches of tropical rain forests. So, no one knows the actual number of species on Earth. But biologists agree that the number must be much higher than the number of named species. Estimates range from 2 million to 100 million or more.

16.4 WHAT CAUSES EXTINCTION?

Every living organism must eventually die, and the same is true of species. Just like individuals, species are "born" (through the process of speciation), persist for some period of time, and then perish. The ultimate fate of any species is **extinction,** the death of the last of its members. In fact, at least 99.9% of all the species that have ever existed are now extinct. The natural course of evolution, as revealed by fossils, is continual turnover of species as new ones arise and old ones become extinct.

The immediate cause of extinction is probably always environmental change, in either the nonliving or the living parts of the environment. Environmental changes that can lead to extinction include habitat destruction and increased competition among species. In the face of such changes,

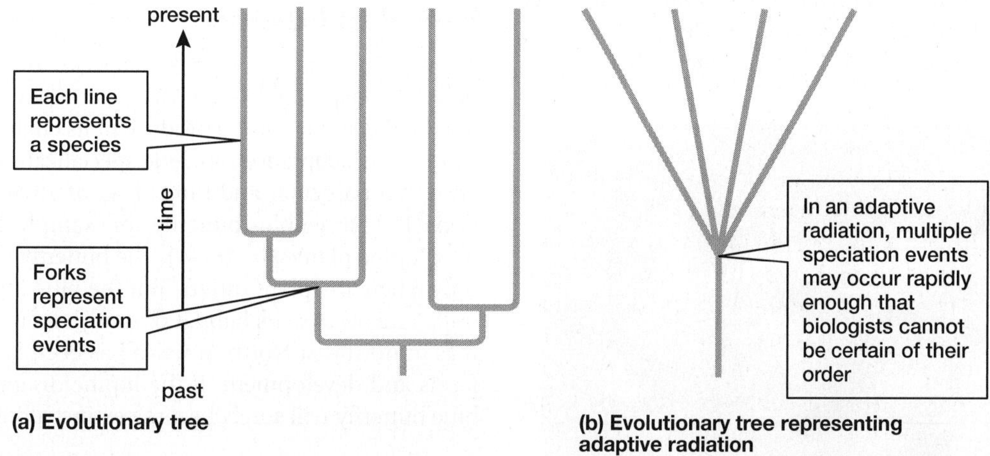

▲ **FIGURE 16-11 Interpreting evolutionary trees** Evolutionary history is often represented by **(a)** an evolutionary tree, a graph in which the vertical axis plots time. In **(b)**, an evolutionary tree representing an adaptive radiation, many lines may branch from a single point. This pattern reflects biologists' uncertainty about the order in which the multiple speciation events of the radiation took place. With more research, it may be possible to replace the "starburst" pattern with a more informative tree.

▲ **FIGURE 16-12 Adaptive radiation** About 30 species of silversword plants inhabit the Hawaiian Islands. These species are found nowhere else, and all of them descended from a single ancestral population within a few million years. This adaptive radiation has led to a collection of closely related species of diverse form and appearance, with an array of adaptations for exploiting the many different habitats in Hawaii, from warm, moist rain forests to cool, barren volcanic mountaintops.

species with small geographic ranges or highly specialized adaptations are especially susceptible to extinction.

Localized Distribution Makes Species Vulnerable

Species vary widely in their range of distribution and, hence, in their vulnerability to extinction. Some species, such as herring gulls, white-tailed deer, and humans, inhabit entire continents or even the whole Earth; others, such as the Devil's Hole pupfish (**Fig. 16-13**), have extremely limited ranges. Obviously, if a species inhabits only a very small area, any disturbance of that area could easily result in extinction. If Devil's Hole dries up due to a drought or well drilling nearby, its pupfish will immediately vanish. Conversely, wide-ranging species will not succumb to local environmental catastrophes.

▲ **FIGURE 16-13 Very localized distribution can endanger a species** The Devil's Hole pupfish is found in only one spring-fed water hole in the Nevada desert. This and other isolated small populations are at high risk of extinction.

Overspecialization Increases the Risk of Extinction

Another factor that may make a species vulnerable to extinction is overspecialization. Each species evolves adaptations that help it survive and reproduce in its environment. In some cases, these adaptations include specializations that favor survival in a particular and limited set of environmental conditions. The Karner blue butterfly, for example, feeds only on the blue lupine plant (**Fig. 16-14**). The butterfly is therefore found only where the plant thrives. But the blue lupine has become quite rare because its habitat of sandy, open woods and clearings in northeast North America has been largely replaced by farms and development. If the lupine disappears, the Karner blue butterfly will surely become extinct along with it.

Interactions with Other Species May Drive a Species to Extinction

As described in Chapter 15, interactions such as competition and predation serve as agents of natural selection (see p. 298).

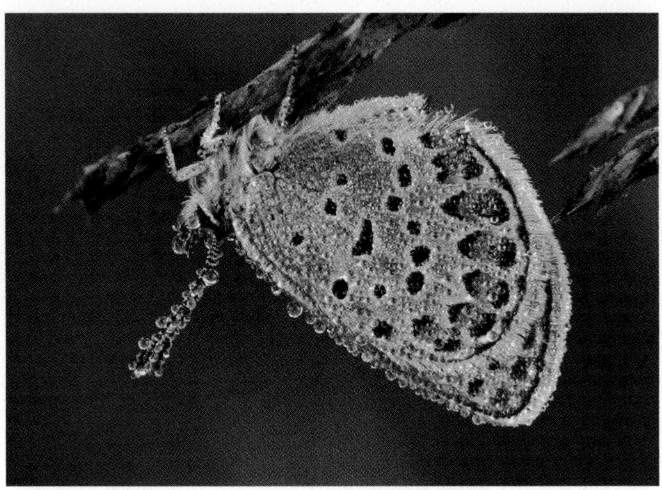

▲ **FIGURE 16-14 Extreme specialization places species at risk** The Karner blue butterfly feeds exclusively on the blue lupine, found in dry forests and clearings in the northeastern United States. Such behavioral specialization renders the butterfly extremely vulnerable to any environmental change that may exterminate its single host plant species.

QUESTION If specialization puts a species at risk for extinction, how could this hazardous trait have evolved?

In some cases, these same interactions can lead to extinction rather than to adaptation.

Organisms compete for limited resources in all environments. If a species' competitors evolve superior adaptations and the species doesn't evolve fast enough to keep up, it may become extinct. A particularly striking example of extinction through competition occurred in South America, beginning about 2.5 million years ago. At that time, the isthmus of Panama rose above sea level and formed a land bridge between North America and South America. After the previously separated continents were connected, the mammal species that had evolved in isolation on each continent were able to mix. Many species did indeed expand their ranges, as North American mammals moved southward and South American mammals moved northward. As they moved, each species encountered resident species that occupied the same kinds of habitats and exploited the same kinds of resources. The ultimate result of the ensuing competition was that the North American species diversified and underwent an adaptive radiation that displaced the vast majority of the South American species, many of which went extinct. Clearly, evolution had bestowed on the North American species some (as yet unknown) set of adaptations that enabled their descendants to exploit resources more efficiently and effectively than their South American counterparts could.

Habitat Change and Destruction Are the Leading Causes of Extinction

Habitat change, both contemporary and prehistoric, is the single greatest cause of extinctions. Present-day habitat destruction due to human activities is proceeding at a rapid pace. Many biologists believe that we are presently in the midst of the fastest-paced and most widespread episode of species extinction in the history of life. Loss of tropical forests is especially devastating to species diversity. As many as half the species presently on Earth may be lost during the next 50 years as the tropical forests that contain them are cut for timber or to clear land for cattle and crops. In Chapter 17, we will discuss extinctions due to prehistoric habitat change.

Case Study revisited
Lost World

One possible explanation for the distinctive collection of species found in the Annamite Mountains of Vietnam lies in the geological history of the region. During the ice ages that have occurred repeatedly during the past million years or so, the area covered by tropical forests must have shrunk dramatically. Organisms that depended on the forests for survival would have been restricted to any remaining "islands" of forest, isolated from their fellows in other, distant patches of forest. What is now the Annamite Mountain region may well have been an isolated forest during periods of glacial advance. As we learned in this chapter, this kind of isolation can set the stage for allopatric speciation and may have created the conditions that gave rise to the saola and other unique denizens of Vietnamese forests.

Ironically, we have discovered the lost world of Vietnamese animals at a moment when that world is in danger of disappearing. Economic development in Vietnam has brought logging and mining to ever more remote regions of the country, and Annamite Mountain forests are being cleared at an unprecedented rate. The increasing local human population means that animals are hunted heavily; most of our knowledge of the saola comes from carcasses found in local markets. All of the newly discovered mammals of Vietnam are quite rare, seen only infrequently even by local hunters. Fortunately, the Vietnamese government has established a number of national parks and nature preserves in key areas. Only time will tell if these measures are sufficient to ensure the survival of the mysterious mammals of the Annamites.

Consider This

The All Species Foundation is a nonprofit organization that promotes the goal of finding and naming all of Earth's undiscovered species within the next 25 years. According to the foundation, this task "deserves to be one of the great scientific goals of the new century." The foundation estimates the cost of the job at between $700 and $2,000 per species, with perhaps millions of undiscovered species remaining to be found. Do you think the search for undiscovered species should continue? What value or benefit to humans does the search for new species provide?

CHAPTER REVIEW

Summary of Key Concepts

16.1 What Is a Species?

According to the biological species concept, a species consists of all the populations of organisms that are potentially capable of interbreeding under natural conditions and that are reproductively isolated from other populations.

16.2 How Is Reproductive Isolation Between Species Maintained?

Reproductive isolation between species may be maintained by one or more of several mechanisms, collectively known as premating isolating mechanisms and postmating isolating mechanisms. Premating isolating mechanisms include geographic isolation, ecological isolation, temporal isolation, behavioral isolation, and mechanical incompatibility. Postmating isolating mechanisms include gametic incompatibility, hybrid inviability, and hybrid infertility.

16.3 How Do New Species Form?

Speciation, the formation of new species, takes place when gene flow between two populations is reduced or eliminated and the populations diverge genetically. Most commonly, speciation is allopatric—gene flow is restricted by geographic isolation. However, speciation can also be sympatric—gene flow is restricted by ecological isolation or by mutations that cause polyploidy. Whether genetic isolation initially arises allopatrically or sympatrically, speciation is completed by subsequent genetic divergence of the separated populations through genetic drift or natural selection.

16.4 What Causes Extinction?

Factors that cause extinctions include competition among species and habitat destruction. Localized distribution and overspecialization increase a species' vulnerability to extinction.

Key Terms

adaptive radiation 313
allopatric speciation 310
extinction 313
isolating mechanism 306
polyploidy 312
postmating isolating mechanism 306
premating isolating mechanism 306
reproductive isolation 304
speciation 309
species 304
sympatric speciation 310

Thinking Through the Concepts

Fill-in-the-Blank

1. A species is a group of __________ that evolves __________. The biological species concept identifies species on the basis of their __________. The biological species concept cannot be applied to species that reproduce __________.
2. Fill in the following with the appropriate isolating mechanism: Occurs when members of two populations have different courtship behaviors: __________; occurs when hybrid offspring fail to survive to reproduce: __________; occurs when members of two populations have different breeding seasons: __________; occurs when sperm from one species fails to fertilize the eggs of another species: __________; occurs when the sexual organs of two species are incompatible: __________.
3. Formation of a new species occurs when two populations of an existing species first become __________ and then __________. The process in which geographic separation of parts of a population leads to the formation of new species is called __________. Isolated populations may diverge through the action of __________ or __________.
4. The process by which many new species arise in a relatively short period of time is known as __________. This process often occurs when a species arrives in a previously unoccupied __________.
5. A species may be at higher risk of extinction if its geographic range includes a(n) __________area, or if its food or habitat requirements are __________. The leading direct cause of extinction is __________.

Review Questions

1. Define the following terms: *species, speciation, allopatric speciation*, and *sympatric speciation*. Explain how allopatric and sympatric speciation might work, and give a hypothetical example of each.
2. Many of the oak tree species in central and eastern North America hybridize (interbreed). Are they "true species"?
3. Review the material on the possibility of sympatric speciation in *Rhagoletis* flies. What types of genotypic, phenotypic, or behavioral data would convince you that the two forms have become separate species?
4. A drug called colchicine prevents cell division after the chromosomes have doubled at the start of meiosis. Describe how you would use colchicine to produce a new polyploid plant species.
5. What are the two major types of reproductive isolating mechanisms? Give examples of each type, and describe how they work.

Applying the Concepts

1. Why do you suppose there are so many *endemic* species—that is, species found nowhere else—on islands? Why have the overwhelming majority of recent extinctions occurred on islands?
2. A biologist you've met claims that the fact that humans are pushing other species into small, isolated populations is good for biodiversity because these are the conditions that lead to new speciation events. Comment.
3. Southern Wisconsin is home to several populations of gray squirrels (*Sciurus carolinensis*) with black fur. Design a study to determine if they are actually a separate species.
4. It is difficult to gather data on speciation events in the past or to perform interesting experiments about the process of speciation. Does this difficulty make the study of speciation "unscientific"? Should we abandon the study of speciation?

MB® *Go to www.masteringbiology.com for practice quizzes, activities, eText, videos, current events, and more.*

The History of Life

Chapter 17

Case Study

Little People, Big Story

MUCH OF OUR KNOWLEDGE of the evolutionary history of life comes from the work of paleontologists, the scientists who study fossils. Not long ago, a small group of paleontologists, digging beneath the floor of a cave on the Indonesian island of Flores, discovered what they at first believed to be the fossil skeleton of a human child. Closer examination of the skeleton, however, revealed that it instead belonged to a fully grown adult, but one that was no more than 3 feet tall. The researchers gave this extraordinary creature the nickname "Hobbit."

Unlike today's small humans, such as pygmies or pituitary dwarves, Hobbit had a very small brain, smaller even than the brain of a typical chimpanzee. Furthermore, the shapes and arrangements of the bones in Hobbit's wrist, shoulder, and other parts of its skeleton were unlike those of anatomically modern humans. On the basis of these findings, many researchers have concluded that Hobbit was not simply a small *Homo sapiens*, but was instead a different, related species. The new species was dubbed *Homo floresiensis* by its discoverers.

Some paleontologists are skeptical of the conclusion that the specimen represents a new human species. The skeptics contend that Hobbit might simply be a small-bodied *H. sapiens* who was victimized by a skeleton-deforming disease. Hobbit proponents, however, maintain that the available evidence supports the claim that *H. floresiensis* is a distinct species. The scientific disagreement about Hobbit's status may not be fully resolved until additional specimens are found.

If Hobbit is indeed *H. floresiensis*, we shared Earth with close relatives until much more recently than was previously thought. Hobbit's bones are about 18,000 years old. Until they were unearthed, scientists believed that, by well before 18,000 years ago, we were the only surviving member of the human family tree. Hobbit's discovery, however, raises the possibility that, not too long ago, in the forests of Flores, representatives of our species encountered members of another, tiny human species.

Although the tale of *H. floresiensis* is especially significant in our human-centered view of the world, it is but one thread among the millions that together make up the story of life's evolution. We therefore turn our attention from our hobbit-like cousin to a brief tour of some of the highlights of life's history.

▲ The skull of *Homo floresiensis*, a recently discovered diminutive human relative, is dwarfed by the skull of a modern *Homo sapiens*.

17.1 HOW DID LIFE BEGIN?

Pre-Darwinian thought held that all species were simultaneously created by God a few thousand years ago. Further, until the nineteenth century most people thought that new members of species sprang up all the time, through **spontaneous generation** from both nonliving matter and other, unrelated forms of life. In 1609, a French botanist wrote, "There is a tree . . . frequently observed in Scotland. From this tree leaves are falling; upon one side they strike the water and slowly turn into fishes, upon the other they strike the land and turn into birds." Medieval writings abound with similar observations. Microorganisms were thought to arise spontaneously from broth, maggots from meat, and mice from mixtures of sweaty shirts and wheat.

Experiments Refuted Spontaneous Generation

In 1668, the Italian physician Francesco Redi disproved the maggots-from-meat hypothesis simply by keeping flies (whose eggs hatch into maggots) away from uncontaminated meat (see "Scientific Inquiry: Controlled Experiments, Then and Now," on pp. 6–7). In the mid-1800s, Louis Pasteur in France and John Tyndall in England disproved the broth-to-microorganism idea by showing that microorganisms did not appear in sterile broth unless the broth was first exposed to existing microorganisms in the surrounding environment (**Fig. 17-1**). Although Pasteur and Tyndall's work effectively demolished the notion of spontaneous generation, it did not address the question of how life on Earth originated in the first place. Or, as the biochemist Stanley Miller put it, "Pasteur never proved it didn't happen once, he only showed that it doesn't happen all the time."

The First Living Things Arose from Nonliving Ones

Modern scientific ideas about the origin of life began to emerge in the 1920s, when Alexander Oparin in Russia and John B. S. Haldane in England noted that today's oxygen-rich atmosphere would not have permitted the spontaneous formation of the complex organic molecules necessary for life. Oxygen reacts readily with other molecules, disrupting chemical bonds. Thus, an oxygen-rich environment tends to keep molecules simple.

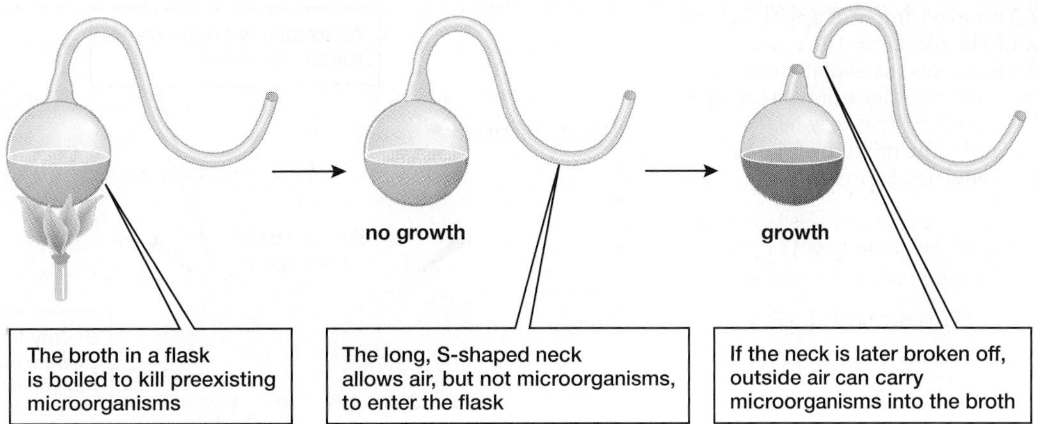
no growth
growth
The broth in a flask is boiled to kill preexisting microorganisms
The long, S-shaped neck allows air, but not microorganisms, to enter the flask
If the neck is later broken off, outside air can carry microorganisms into the broth

▲ **FIGURE 17-1 Spontaneous generation refuted** Louis Pasteur's experiment disproving the spontaneous generation of microorganisms in broth.

Oparin and Haldane speculated that the atmosphere of the young Earth must have contained very little oxygen and that, under such atmospheric conditions, complex organic molecules could have arisen through ordinary chemical reactions. Some kinds of molecules could persist in the lifeless environment of early Earth better than others and would therefore become more common over time. This chemical version of the "survival of the fittest" is called *prebiotic* (meaning "before life") evolution. In the scenario envisioned by Oparin and Haldane, prebiotic chemical evolution gave rise to progressively more complex molecules and eventually to living organisms.

Organic Molecules Can Form Spontaneously Under Prebiotic Conditions

Inspired by the ideas of Oparin and Haldane, Stanley Miller and Harold Urey set out in 1953 to simulate prebiotic evolution in the laboratory. They knew that, on the basis of the chemical composition of the rocks that formed early in Earth's history, geochemists had concluded that the early atmosphere probably contained virtually no oxygen gas, but did contain other substances, including methane (CH_4), ammonia (NH_3), hydrogen (H_2), and water vapor (H_2O). Miller and Urey simulated the oxygen-free atmosphere of early Earth by mixing these components in a flask. An electrical discharge mimicked the intense energy of early Earth's lightning storms. In this experimental microcosm, the researchers found that simple organic molecules appeared after just a few days (**Fig. 17-2**). The experiment showed that small molecules likely present in the early atmosphere can combine to form larger organic molecules if electrical energy is present (recall from Chapter 6 that reactions that synthesize biological molecules from smaller ones are endergonic—they consume energy). Similar experiments by Miller and others have produced amino acids, short proteins, nucleotides, adenosine triphosphate (ATP), and other molecules characteristic of living things.

In recent years, new evidence has convinced most geochemists that the actual composition of Earth's early atmosphere probably differed from the mixture of gases used in the pioneering Miller-Urey experiment. This improved understanding of the early atmosphere, however, has not undermined the basic finding of the Miller-Urey experiment. Additional experiments with more realistic (but still oxygen-free) simulated atmospheres have also yielded organic molecules. In addition, these experiments have shown that electricity is not the only suitable energy source. Other energy sources that were available on early Earth, such as heat or ultraviolet (UV) light, have also been shown to drive the formation of organic molecules in experimental simulations of prebiotic conditions. Thus, even though we may never know exactly what the earliest atmosphere was like, we can be confident that organic molecules formed on early Earth.

Additional organic molecules probably arrived from space when meteorites and comets crashed into Earth's surface. Analysis of present-day meteorites recovered from impact craters on Earth has revealed that some meteorites contain relatively high concentrations of amino acids and other simple organic molecules. Laboratory experiments suggest that these molecules could have formed in interstellar space before plummeting to Earth. When small molecules known to be present in space were placed under space-like conditions of very low temperature and pressure and bombarded with UV light, larger organic molecules were produced.

Organic Molecules Can Accumulate Under Prebiotic Conditions

Prebiotic synthesis was neither very efficient nor very fast. Nonetheless, in a few hundred million years, large quantities of organic molecules accumulated in the early Earth's oceans. Today, most organic molecules have a short life because they are either digested by living organisms or they react with atmospheric oxygen. Early Earth, however, lacked both life and free oxygen, so molecules would not have been exposed to these threats.

Still, the prebiotic molecules must have been threatened by the sun's high-energy UV radiation, because early Earth lacked an ozone layer. The ozone layer is a region high in today's atmosphere that is enriched with ozone molecules, which form when incoming solar energy splits some

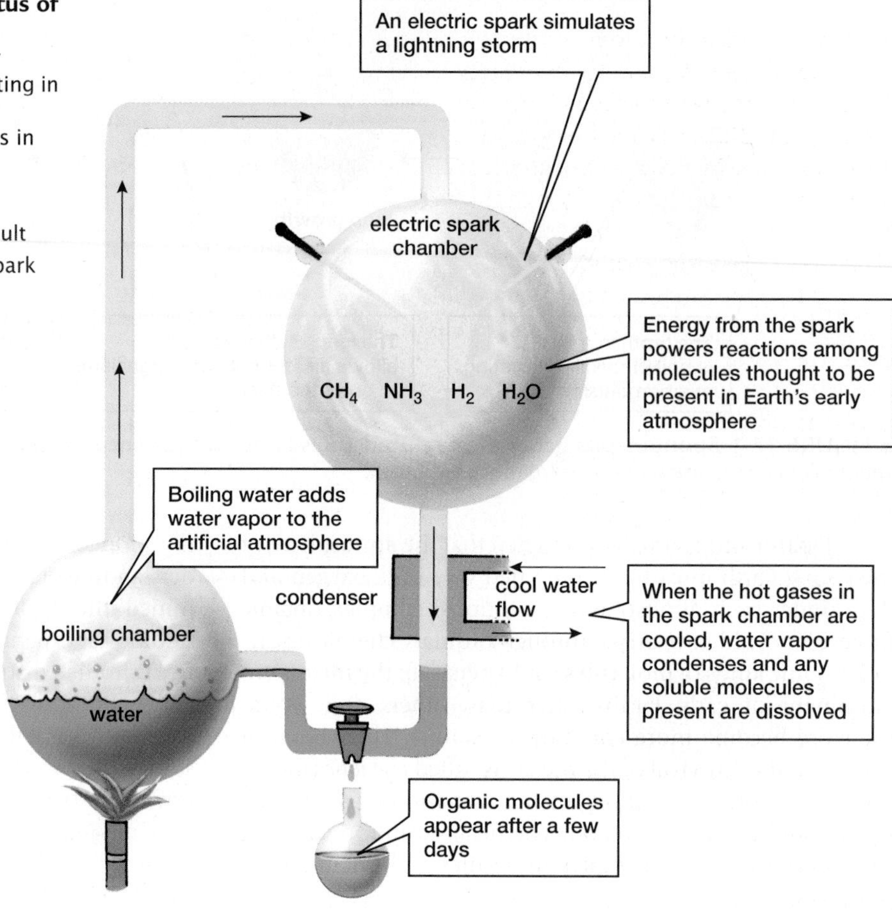

An electric spark simulates a lightning storm
electric spark chamber
Energy from the spark powers reactions among molecules thought to be present in Earth's early atmosphere
CH_4 NH_3 H_2 H_2O
Boiling water adds water vapor to the artificial atmosphere
condenser
cool water flow
When the hot gases in the spark chamber are cooled, water vapor condenses and any soluble molecules present are dissolved
boiling chamber
water
Organic molecules appear after a few days

▶ **FIGURE 17-2 The experimental apparatus of Stanley Miller and Harold Urey** Life's very earliest stages left no fossils, so evolutionary historians have pursued a strategy of re-creating in the laboratory the conditions that may have prevailed on early Earth. The mixture of gases in the spark chamber simulates Earth's early atmosphere.

QUESTION How would the experiment's result change if oxygen (O_2) were included in the spark chamber?

O_2 molecules in the outer atmosphere into individual oxygen atoms (O) that then react with O_2 to form O_3 (ozone). The resulting high-altitude layer of ozone molecules absorbs some of the sun's UV light before it reaches Earth's surface. Early Earth, however, had no ozone layer, because there was little or no oxygen gas in the atmosphere and therefore no ozone formation.

Before the ozone layer formed, UV bombardment must have been fierce. UV radiation, as we have seen, can provide energy for the formation of organic molecules but it can also break them apart. Some places, however, such as those beneath rock ledges or at the bottoms of even fairly shallow seas, would have been protected from UV radiation. In these locations, organic molecules may have accumulated.

Clay May Have Catalyzed the Formation of Larger Organic Molecules

In the next stage of prebiotic evolution, simple molecules combined to form larger molecules. The chemical reactions that formed the larger molecules required that the reacting molecules be packed closely together. Scientists have proposed several processes by which the required high concentrations might have been achieved on early Earth. One possibility is that small molecules accumulated on the surfaces of clay particles, which may have a small electrical charge that attracts dissolved molecules with the opposite charge. Clustered on a clay particle, small molecules would have been sufficiently close together to allow chemical reactions between them. Researchers have demonstrated the plausibility of this scenario with experiments in which adding clay to solutions of dissolved small organic molecules catalyzed the formation of larger, more complex molecules. Such molecules might have formed on clay at the bottom of early Earth's oceans or lakes and gone on to become the building blocks of the first living organisms.

RNA May Have Been the First Self-Reproducing Molecule

Although all modern organisms use DNA to encode and store genetic information, it is unlikely that DNA was the earliest informational molecule. DNA can reproduce itself only with the help of large, complex protein enzymes, but the instructions for building these enzymes are encoded in DNA itself. For this reason, the origin of DNA's role as life's information storage molecule poses a "chicken and egg" puzzle: DNA requires proteins, but those proteins require DNA. It is thus difficult to construct a plausible scenario for the origin of self-replicating DNA from prebiotic molecules. It is therefore likely that the current DNA-based system of information storage evolved from an earlier system.

RNA Can Act as a Catalyst

A prime candidate for the first self-replicating informational molecule is RNA. In the 1980s, Thomas Cech and Sidney Altman, working with the single-celled organism *Tetrahymena*, discovered a cellular reaction that was catalyzed not by a protein, but by a small RNA molecule. Because this special RNA molecule performed a function previously thought to be performed only by protein enzymes, Cech and Altman decided to give their catalytic RNA molecule the name **ribozyme** (Fig. 17-3).

In the years since their discovery, researchers have found dozens of naturally occurring ribozymes that catalyze a variety of different reactions, including cutting other RNA molecules and splicing together different RNA fragments. Ribozymes have also been found in the protein-manufacturing machinery of cells, where they help catalyze the attachment of amino acid molecules to growing proteins. In addition, researchers have been able to synthesize various ribozymes in the laboratory, including some that can catalyze the replication of small RNA molecules.

Earth May Once Have Been an RNA World

The discovery that RNA molecules can act as catalysts for diverse reactions, including RNA replication, provides support for the hypothesis that life arose in an "RNA world." According to this view, the current era of DNA-based life was preceded by one in which RNA served as both the information-carrying genetic molecule and the catalyst for its own replication. This RNA world may have emerged after hundreds of millions of years of prebiotic chemical synthesis, during which RNA nucleotides would have been among the molecules synthesized. After reaching a sufficiently high concentration, perhaps on clay particles, the nucleotides probably bonded together to form short RNA chains.

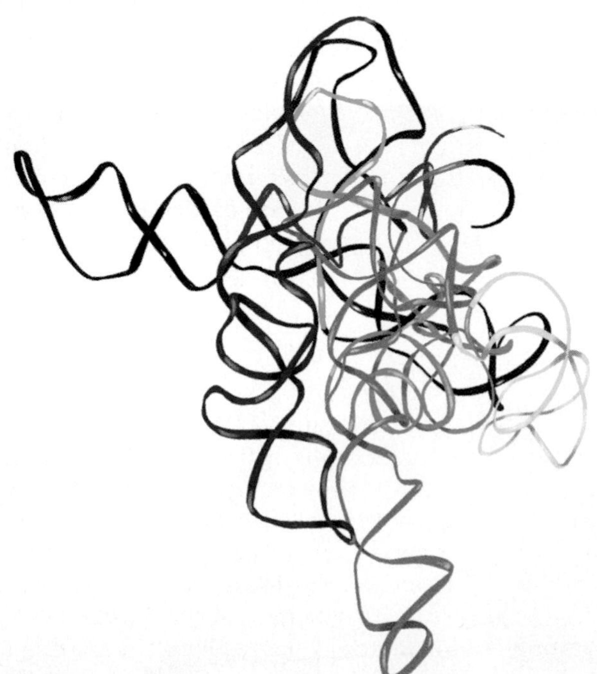

▲ **FIGURE 17-3 A ribozyme** This RNA molecule, isolated from the single-celled organism *Tetrahymena*, acts like an enzyme, catalyzing metabolic reactions.

Let's suppose that, purely by chance, one of these RNA chains was a ribozyme that could catalyze the production of copies of itself. This first self-reproducing ribozyme probably wasn't very good at its job and produced copies with lots of errors. These mistakes were the first mutations. Like modern mutations, most undoubtedly ruined the catalytic abilities of the "daughter molecules," but a few may have been improvements. Such improvements set the stage for the evolution of RNA molecules, as variant ribozymes with increased speed and accuracy of replication reproduced, copying themselves and displacing less efficient molecules. Molecular evolution in the RNA world proceeded until, by some still unknown chain of events, RNA gradually receded into its present role as an intermediary between DNA and protein enzymes.

Membrane-Like Vesicles May Have Enclosed Ribozymes

Self-replicating molecules alone do not constitute life; these molecules must be contained within some kind of enclosing membrane. The precursors of the earliest biological membranes may have been simple structures that formed spontaneously from purely physical, mechanical processes. For example, chemists have shown that if water containing proteins and lipids is agitated to simulate waves beating against ancient shores, the proteins and lipids combine to form hollow structures called *vesicles*. These hollow balls resemble living cells in several respects. They have a well-defined outer boundary that separates their internal contents from the external solution. If the composition of the vesicle is right, a "membrane" forms that is remarkably similar in appearance to a real cell membrane. Under certain conditions, vesicles can absorb material from the external solution, grow, and even divide.

If a vesicle happened to surround the right ribozymes, it would form something resembling a living cell. We could call it a **protocell,** structurally similar to a cell but not a living thing. In the protocell, ribozymes and any other enclosed molecules would have been protected from free-roaming ribozymes in the primordial soup. Nucleotides and other small molecules might have diffused across the membrane and been used to synthesize new ribozymes and other complex molecules. After sufficient growth, the vesicle may have divided, with a few copies of the ribozymes becoming incorporated into each daughter vesicle. If this process occurred, the evolution of the first cells would be nearly complete.

Was there a particular moment when a nonliving protocell gave rise to something alive? Probably not. Like most evolutionary transitions, the change from protocell to living cell was a continuous process, with no sharp boundary between one state and the next.

But Did All This Happen?

The above scenario, although plausible and consistent with many research findings, is by no means certain. One of the most striking aspects of origin-of-life research is a great diversity of assumptions, experiments, and contradictory hypotheses. Researchers disagree as to whether life arose in quiet pools, in the sea, in moist films on the surfaces of crystals, or in hot deep-sea vents. A few researchers even argue that life arrived on Earth from space. Can we draw any firm conclusions from the research conducted so far? No, but we can make a few observations.

First, the experiments of Miller and others show that amino acids, nucleotides, and other organic molecules, along with simple membrane-like structures, are likely to have formed in abundance on early Earth. Second, chemical evolution had long periods of time and huge areas of the Earth available to it. Given sufficient time and a sufficiently large pool of reactant molecules, even extremely rare events can occur many times. Given the vast expanses of time and space available, each small step on the path from primordial soup to living cell had ample opportunity to take place.

Most biologists accept that the origin of life was probably an inevitable consequence of the working of natural laws. We should emphasize, however, that this proposition cannot be definitively tested. The origin of life left no record, and researchers exploring this mystery can proceed only by developing a hypothetical scenario and then conducting laboratory investigations to determine if the scenario's steps are chemically and biologically possible and plausible.

17.2 WHAT WERE THE EARLIEST ORGANISMS LIKE?

When Earth first formed about 4.5 billion years ago, it was quite hot (**Fig. 17-4**). A multitude of meteorites smashed into the forming planet, and the kinetic energy of these extraterrestrial rocks was converted into heat on impact. Still more heat was released by the decay of radioactive atoms. The rock composing Earth melted, and heavier elements such as iron and nickel sank to the center of the planet, where they remain molten even today. It must have taken hundreds of millions of years for Earth to cool enough to allow water to exist as a liquid. Nonetheless, it appears that life arose in fairly short order once liquid water was available.

The oldest fossil organisms found so far are in rocks that are about 3.5 billion years old. (Their age was determined using radiometric dating techniques; see "Scientific Inquiry: How Do We Know How Old a Fossil Is?" on p. 324.) Chemical traces in older rocks have led some paleontologists to believe that life is even older, perhaps as old as 3.9 billion years.

The period in which life began is known as the Precambrian era. This interval was designated by geologists and paleontologists, who have devised a hierarchical naming system of eras, periods, and epochs to delineate the immense span of geological time (**Table 17-1**).

The First Organisms Were Anaerobic Prokaryotes

The first cells to arise in Earth's oceans were **prokaryotes,** cells whose genetic material was not contained within a nucleus. These cells probably obtained nutrients and energy by absorbing organic molecules from their environment. There was no oxygen gas in the atmosphere, so the cells must have metabolized the organic molecules anaerobically. You may recall from Chapter 8 that anaerobic metabolism yields only small amounts of energy.

Thus, the earliest cells were primitive anaerobic bacteria. As these bacteria multiplied, they must have eventually used up the organic molecules produced by prebiotic chemical reactions. Simpler molecules, such as carbon dioxide and

▲ **FIGURE 17-4 Early Earth** Life began on a planet characterized by abundant volcanic activity, frequent electrical storms, repeated meteorite strikes, and an atmosphere that lacked oxygen gas.

Table 17-1 The History of Life on Earth

Era	Period	Epoch	Millions of years ago	Major events
Cenozoic	Quaternary	Recent	0.01–present	Evolution of genus *Homo*
		Pleistocene	1.8–0.01	
	Tertiary	Pliocene	5–1.8	Widespread flourishing of birds, mammals, insects, and flowering plants
		Miocene	23–5	
		Oligocene	38–23	
		Eocene	54–38	
		Paleocene	65–54	
Mesozoic	Cretaceous		146–65	Flowering plants appear and become dominant Mass extinction of marine and terrestrial life, including dinosaurs
	Jurassic		208–146	Dominance of dinosaurs and conifers First birds
	Triassic		245–208	First mammals and dinosaurs Forests of gymnosperms and tree ferns
Paleozoic	Permian		286–245	Massive marine extinctions, including trilobites Flourishing of reptiles and the decline of amphibians
	Carboniferous		360–286	Forests of tree ferns and club mosses Dominance of amphibians and insects First reptiles and conifers
	Devonian		410–360	Fishes and trilobites flourish First amphibians, insects, seeds, and pollen
	Silurian		440–410	Many fishes, trilobites, and mollusks First vascular plants
	Ordovician		505–440	Dominance of arthropods and mollusks in the ocean Invasion of land by plants and arthropods First fungi
	Cambrian		544–505	Marine algae flourish Origin of most marine invertebrate phyla First fishes
Precambrian			About 1,000	First animals (soft-bodied marine invertebrates)
			1,200	First multicellular organisms
			2,000	First eukaryotes
			2,200	Accumulation of free oxygen in the atmosphere
			3,500	Origin of photosynthesis (in cyanobacteria)
			3,900–3,500	First living cells (prokaryotes)
			4,000–3,900	Appearance of the first rocks on Earth
			4,600	Origin of the solar system and Earth

Scientific Inquiry

How Do We Know How Old a Fossil Is?

Early geologists could date rock layers and their accompanying fossils only in a *relative* way: Fossils found in deeper layers of rock were generally older than those found in shallower layers. With the discovery of radioactivity, it became possible to determine *absolute* dates, within certain limits of uncertainty. The nuclei of radioactive elements spontaneously break down, or decay, into other elements. For example, carbon-14 (usually written ^{14}C) decays by emitting an electron to become nitrogen-14 (^{14}N). Each radioactive element decays at a rate that is independent of temperature, pressure, or the chemical compound of which the element is a part. The time it takes for half of a radioactive element's nuclei to decay at this characteristic rate is called its *half-life*. The half-life of ^{14}C, for example, is 5,730 years.

How are radioactive elements used in determining the age of rocks? If we know the rate of decay and measure the proportion of decayed nuclei to undecayed nuclei, we can estimate how much time has passed since these radioactive elements became trapped in the rock. This process is called *radiometric dating*.

A particularly straightforward radiometric dating technique measures the decay of potassium-40 (^{40}K), which has a half-life of about 1.25 billion years, into argon-40 (^{40}Ar). Potassium-40 is commonly found in volcanic rocks such as granite and basalt, and the argon-40 it decays into is a gas. Suppose that a volcano erupts with a massive lava flow, covering the countryside. All the ^{40}Ar, being a gas, will bubble out of the molten lava, so when the lava first cools and solidifies into rock, it will not contain any ^{40}Ar (**Fig. E17-1**). Over time, however, any ^{40}K present in the hardened lava will decay into ^{40}Ar, with half of the ^{40}K decaying every 1.25 billion years. This ^{40}Ar gas will be trapped in the rock. A geologist could take a sample of the rock and measure the ratio of ^{40}K to ^{40}Ar to determine the rock's age. For example, if the analysis finds equal amounts of the two elements, the geologist will conclude that the lava hardened 1.25 billion years ago (see Fig. E17-1). With appropriate care, such age estimates are quite reliable. If a fossil is found beneath a lava flow dated at, say, 500 million years, then we know that the fossil is at least that old.

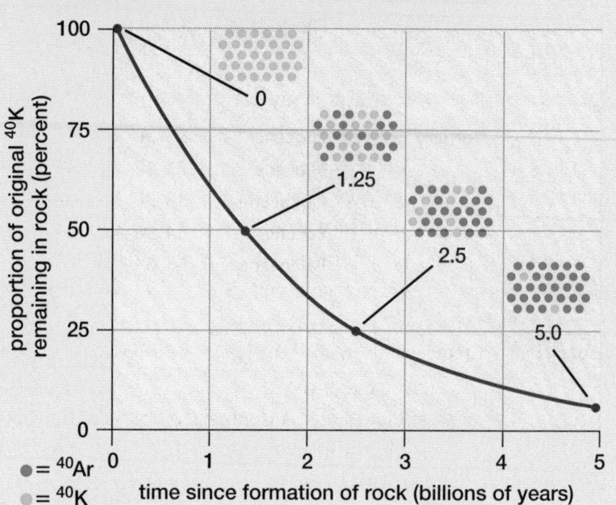

▲ **FIGURE E17-1 The relationship between time and the decay of radioactive ^{40}K to ^{40}Ar**

QUESTION Uranium-235, with a half-life of 713 million years, decays to lead-207. If you analyze a rock and find that it contains uranium-235 and lead-207 in a ratio of 1:1, how old is the rock?

water, would still have been very abundant, as was energy in the form of sunlight. What was lacking, then, was not materials or energy itself, but energetic molecules—molecules in which energy is stored in chemical bonds.

Some Organisms Evolved the Ability to Capture the Sun's Energy

Eventually, some cells evolved the ability to use the energy of sunlight to drive the synthesis of complex, high-energy molecules from simpler molecules; in other words, photosynthesis appeared. Photosynthesis requires a source of hydrogen, and the very earliest photosynthetic bacteria probably used hydrogen sulfide gas dissolved in water for this purpose (much as today's purple photosynthetic bacteria do). Eventually, however, Earth's supply of hydrogen sulfide (which is produced mainly by volcanoes) must have run low. The shortage of hydrogen sulfide set the stage for the evolution of photosynthetic bacteria that were able to use the planet's most abundant source of hydrogen—water (H_2O).

Photosynthesis Increased the Amount of Oxygen in the Atmosphere

Water-based photosynthesis converts water and carbon dioxide to energetic molecules of sugar, releasing oxygen as a by-product. The emergence of this new method for capturing energy introduced significant amounts of free oxygen to the atmosphere for the first time. At first, the newly liberated oxygen was quickly consumed by reactions with other molecules in the atmosphere and in Earth's crust (surface layer). One especially common reactive atom in the crust was iron, and much of the new oxygen combined with iron atoms to form huge deposits of iron oxide (also known as rust). As a result, iron oxide is abundant in rocks formed during this period.

After all the accessible iron had turned to rust, the concentration of oxygen gas in the atmosphere began to increase. Chemical analysis of rocks suggests that significant amounts of oxygen first appeared in the atmosphere about 2.3 billion years ago, produced by bacteria that were probably very similar to modern cyanobacteria. (Because Earth's supply of oxygen molecules is continually recycled, you will undoubtedly breathe in some oxygen molecules today that were expelled 2 billion years ago by one of these early cyanobacteria.)

Aerobic Metabolism Arose in Response to the Oxygen Crisis

Oxygen is potentially very dangerous to living things, because it can react with organic molecules, breaking them down. Many of today's anaerobic bacteria perish when exposed to what is for them a deadly poison—oxygen. The accumulation of oxygen in the atmosphere of early Earth probably exterminated many organisms and fostered the evolution of cellular mechanisms for detoxifying oxygen. This crisis for evolving life also provided the environmental pressure for the next great advance in the Age of Microbes: the ability to use oxygen in metabolism. This ability not only provides a defense against the chemical action of oxygen, but actually channels oxygen's destructive power through aerobic respiration to generate useful energy for the cell (see Chapter 8 for more information on aerobic respiration). Because the amount of energy available to a cell is vastly increased when oxygen is used to metabolize food molecules, aerobic cells had a significant selective advantage.

Some Organisms Acquired Membrane-Enclosed Organelles

Hordes of bacteria would offer a rich food supply to any organism that could eat them. Paleobiologists speculate that, once this potential prey population appeared, predation would have evolved quickly. These early predators were probably prokaryotes that had evolved to become larger than typical bacteria. In addition, they must have lost the rigid cell wall that surrounds most bacterial cells, so that their flexible plasma membrane was in contact with the surrounding environment. Thus, the predatory cells were able to envelop smaller bacteria in an infolded pouch of membrane and in this fashion engulf whole bacteria as prey.

These early predators were probably capable of neither photosynthesis nor aerobic metabolism. Although they could capture large food particles, namely bacteria, they metabolized them inefficiently. By about 1.7 billion years ago, however, one predator probably gave rise to the first eukaryotic cell. Eukaryotic cells differ from prokaryotic cells in that they have an elaborate system of internal membranes, many of which enclose organelles such as a nucleus that contains the cell's genetic material. Organisms composed of one or more eukaryotic cells are known as **eukaryotes.**

The Internal Membranes of Eukaryotes May Have Arisen Through Infolding of the Plasma Membrane

The internal membranes of eukaryotic cells may have originally arisen through inward folding of the cell membrane of a single-celled predator. If, as in most of today's bacteria, the DNA of the eukaryotes' ancestor was attached to the inside of its cell membrane, an infolding of the membrane near the point of DNA attachment may have pinched off and become the precursor of the cell nucleus.

In addition to the nucleus, other key eukaryotic structures include the organelles used for energy metabolism: mitochondria and (in plants and algae) chloroplasts. How did these organelles evolve?

Mitochondria and Chloroplasts May Have Arisen from Engulfed Bacteria

The **endosymbiont hypothesis** proposes that early eukaryotic cells acquired the precursors of mitochondria and chloroplasts by engulfing certain types of bacteria. These cells and the bacteria trapped inside them (*endo* means "within") gradually entered into a *symbiotic* relationship, a close association between different types of organisms over an extended time. How might this have happened?

Let's suppose that an anaerobic predatory cell captured an aerobic bacterium for food, as it often did, but for some reason failed to digest this particular prey (**Fig. 17-5** ❶). The

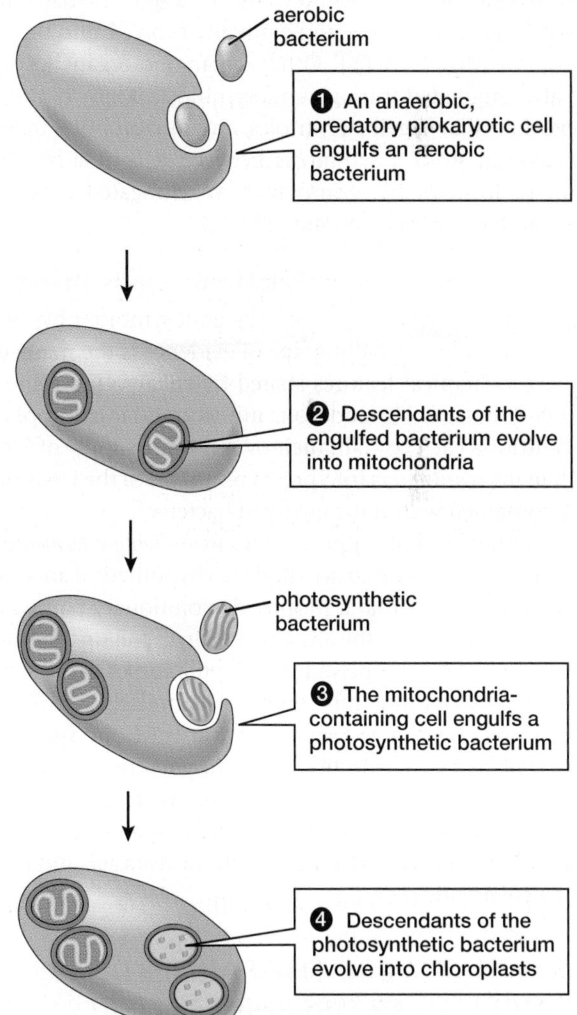

▲ FIGURE 17-5 **The probable origin of mitochondria and chloroplasts in eukaryotic cells**

QUESTION Scientists have identified a living bacterium believed to be descended from the endosymbiont that gave rise to mitochondria. Would you expect the DNA sequence of this modern bacterium to be most similar to the sequence of DNA from a plant chloroplast, an animal cell nucleus, or a plant mitochondrion?

aerobic bacterium remained alive and well. In fact, it was better off than ever, because the cytoplasm of its predator-host was chock-full of half-digested food molecules, the remnants of anaerobic metabolism. The aerobe absorbed these molecules and used oxygen to metabolize them, thereby gaining enormous amounts of energy and reproducing prolifically. So abundant were the aerobes' food resources, and so bountiful their energy production, that the aerobes must have leaked energy, probably as ATP or similar molecules, back into their host's cytoplasm. The anaerobic predatory cell with its symbiotic bacteria could now metabolize food aerobically, gaining a great advantage over other anaerobic cells and leaving a greater number of offspring. Eventually, the endosymbiotic bacterium lost its ability to live independently of its host, and the mitochondrion was born (**Fig. 17-5 ❷**).

One of these successful new cellular partnerships managed a second feat: It captured a photosynthetic bacterium and similarly failed to digest its prey (**Fig. 17-5 ❸**). The bacterium flourished in its new host and gradually evolved into the first chloroplast (**Fig. 17-5 ❹**). Other eukaryotic organelles may have also originated through endosymbiosis. Many biologists believe that cilia, flagella, centrioles, and microtubules may all have evolved from a symbiosis between a spirilla-like bacterium (a form of bacterium with an elongated corkscrew shape) and an early eukaryotic cell.

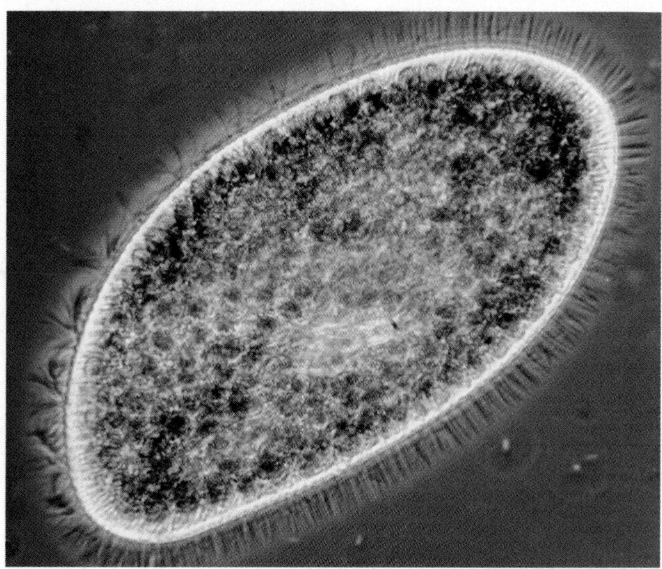

▲ **FIGURE 17-6 Symbiosis within a modern cell** The ancestors of the chloroplasts in today's plant cells may have resembled *Chlorella*, the green, photosynthetic, single-celled algae living symbiotically within the cytoplasm of the *Paramecium* pictured here.

Evidence for the Endosymbiont Hypothesis Is Strong

Several types of evidence support the endosymbiont hypothesis. A particularly compelling line of evidence is the many distinctive biochemical features shared by eukaryotic organelles and living bacteria. In addition, mitochondria, chloroplasts, and centrioles each contain their own minute supply of DNA, which many researchers interpret as remnants of the DNA originally contained within the engulfed bacteria.

Another kind of support comes from *living intermediates*, organisms alive today that are similar to hypothetical ancestors and thus help show that a proposed evolutionary pathway is plausible. For example, the amoeba *Pelomyxa palustris* lacks mitochondria but hosts a permanent population of aerobic bacteria that carry out much the same role. Similarly, a variety of corals, some clams, a few snails, and at least one species of *Paramecium* harbor a collection of photosynthetic algae in their cells (**Fig. 17-6**). These examples of modern cells that host bacterial endosymbionts suggest that similar symbiotic associations could have occurred almost 2 billion years ago and led to the first eukaryotic cells.

17.3 WHAT WERE THE EARLIEST MULTICELLULAR ORGANISMS LIKE?

Once predation had evolved, increased size became an advantage. In the marine environments to which life was restricted, a larger cell could easily engulf a smaller cell and would also be difficult for other predatory cells to ingest. Larger organisms can also generally move faster than smaller ones, making successful predation and escape more likely. But enormous single cells have problems. Oxygen and nutrients going into the cell and waste products going out must diffuse through the plasma membrane. The larger a cell becomes, the less surface membrane is available per unit volume of cytoplasm (see Fig. 5-16).

There are only two ways that an organism larger than a millimeter or so in diameter can survive. First, it can have a low metabolic rate so that it doesn't need much oxygen or produce much carbon dioxide. This strategy seems to work for certain very large single-celled algae. Alternatively, an organism can be multicellular; that is, it can consist of many small cells packaged into a larger, unified body.

Some Algae Became Multicellular

The oldest fossils of multicellular organisms are about 1.2 billion years old and include impressions of the first multicellular algae, which arose from single-celled eukaryotic organisms containing chloroplasts. Multicellularity would have provided at least two advantages for these seaweeds. First, large, many-celled algae would have been difficult for single-celled predators to engulf. Second, specialization of cells would have provided the potential for staying in one place in the brightly lit waters of the shoreline, as rootlike structures burrowed in sand or clutched onto rocks, while leaflike structures floated above in the sunlight. The green, brown, and red algae lining our shores today—some, such as the brown kelp, more than 200 feet long—are the descendants of these early multicellular algae.

Animal Diversity Arose in the Precambrian Era

In addition to fossil algae, billion-year-old rocks have yielded fossil traces of animal tracks and burrows. This evidence of early animal life notwithstanding, fossils of animal bodies first appear in Precambrian rocks laid down between 610 million and 544 million years ago. Some of these ancient invertebrate animals (animals lacking a backbone) are quite different in appearance from any animals that appear in later fossil layers and may represent types of animals that left no descendants. Other fossils in these rock layers, however, appear to be ancestors of today's animals. Ancestral sponges and jellyfish appear in the oldest layers, followed later by ancestors of worms, mollusks, and arthropods.

The full range of modern invertebrate animals, however, does not appear in the fossil record until the Cambrian period, marking the beginning of the Paleozoic era, about 544 million years ago. (The phrase "fossil record" is a shorthand reference to the entire collection of all fossil evidence that has been found to date.) These Cambrian fossils reveal an adaptive radiation (see p. 313) that had already yielded a diverse array of complex body plans. Almost all of the major groups of animals on Earth today were already present in the early Cambrian. The apparently sudden appearance of so many different kinds of animals suggests that these groups actually arose earlier, but their early evolutionary history is not preserved in the fossil record.

Predation Favored the Evolution of Improved Mobility and Senses

The early diversification of animals was probably driven in part by the emergence of predatory lifestyles. For example, coevolution of predator and prey favored animals that were more mobile than their evolutionary predecessors. Mobile predators gained an advantage from an ability to travel over wide areas in search of suitable prey; mobile prey benefited if they were able to make a speedy escape. The evolution of efficient movement was often associated with the evolution of greater sensory capabilities and more complex nervous systems. Senses for detecting touch, chemicals, and light became highly developed, along with nervous systems capable of handling the sensory information and directing appropriate behaviors.

By the Silurian period (440 million to 410 million years ago), life in Earth's seas included an array of anatomically complex animals, including mud-skimming, armored trilobites, ammonites, and the chambered nautilus (**Fig. 17-7**). The nautilus survives today in almost unchanged form in deep Pacific waters.

(a) Silurian scene

(b) Trilobite

(c) Ammonite

(d) *Nautilus*

▲ **FIGURE 17-7 Diversity of ocean life during the Silurian period (a)** Life characteristic of the oceans during the Silurian period, 440 million to 410 million years ago. Among the most common fossils from that time are **(b)** the trilobites and their predators, the nautiloids, and **(c)** the ammonites. **(d)** This living *Nautilus* is very similar in structure to the Silurian nautiloids, showing that a successful body plan may exist virtually unchanged for hundreds of millions of years.

Skeletons Improved Mobility and Protection

In many Paleozoic animal species, mobility was enhanced in part by the origin of hard external body coverings known as **exoskeletons.** Exoskeletons improved mobility by providing hard surfaces to which muscles attached; these attachments made it possible for animals to use their muscles to move appendages used to swim about or move over the sea floor. Exoskeletons also provided support for animals' bodies and protection from predators.

About 530 million years ago, one group of animals—the fishes—developed a new form of body support and muscle attachment: an internal skeleton. These early fishes were inconspicuous members of the ocean community, but by 400 million years ago, fishes were a diverse and prominent group. By and large, the fishes proved to be faster than the invertebrates, with more acute senses and larger brains. Eventually, they became the dominant predators of the open seas.

17.4 HOW DID LIFE INVADE THE LAND?

A compelling subplot in the long tale of life's history is the story of life's invasion of land after more than 3 billion years of a strictly watery existence. In moving to solid ground, organisms had many obstacles to overcome. Life in the sea provides buoyant support against gravity, but on land an organism must bear its weight against the crushing force of gravity. The sea provides ready access to life-sustaining water, but a terrestrial organism must find adequate water. Sea-dwelling plants and animals can reproduce by means of mobile sperm or eggs, or both, which swim or drift to each other through the water. The gametes of land-dwellers, however, must be protected from drying out.

Despite the obstacles to life on land, the vast empty spaces of the Paleozoic landmass represented a tremendous evolutionary opportunity. The potential rewards of terrestrial life were especially great for plants. Water strongly absorbs light, so even in the clearest water, photosynthesis is limited to the upper few hundred meters of depth, and usually much less. Out of the water, the dazzling brightness of the sun permits rapid photosynthesis. Furthermore, terrestrial soils are rich storehouses of nutrients, whereas seawater tends to be low in some nutrients, particularly nitrogen and phosphorus. Finally, the Paleozoic sea swarmed with plant-eating animals, but the land was devoid of animal life. The plants that first colonized the land would have had ample sunlight, untouched nutrient sources, and no predators.

Some Plants Became Adapted to Life on Dry Land

In moist soils at the water's edge, a few small green algae began to grow, taking advantage of the sunlight and nutrients. They didn't have large bodies to support against the force of gravity, and, living right in the film of water on the soil, they could easily obtain water. About 475 million years ago, some of these algae gave rise to the first multicellular land plants. Initially simple, low-growing forms, land plants rapidly evolved solutions to two of the main difficulties of plant life on land: obtaining and conserving water and staying upright despite gravity and winds. Water-resistant coatings on aboveground parts reduced water loss by evaporation, and rootlike structures delved into the soil, mining water and minerals. Specialized cells formed tubes called vascular tissues to conduct water from roots to leaves. Extra-thick walls surrounding certain cells enabled stems to stand erect.

Primitive Land Plants Retained Swimming Sperm and Required Water to Reproduce

Reproduction out of water presented challenges. As do animals, plants produce sperm and eggs, which must be able to meet to produce the next generation. The first land plants had swimming sperm, presumably much like those of today's mosses and ferns. Consequently, the earliest plants were restricted to swamps and marshes, where the sperm and eggs could be released into the water or to areas with abundant rainfall, where the ground would occasionally be covered with water. Later, plants with swimming sperm prospered during periods in which the climate was warm and moist. For example, the Carboniferous period (360 million to 286 million years ago) was characterized by vast forests of giant tree ferns and club mosses (**Fig. 17-8**). The coal we mine today is derived from the fossil remains of those forests.

Seed Plants Encased Sperm in Pollen Grains

Meanwhile, some plants inhabiting drier regions had evolved a means of reproduction that no longer depended on water. The eggs of these plants were retained on the parent plant, and the sperm were encased in drought-resistant pollen grains that traveled on the wind from plant to plant. When the pollen grains landed near an egg, they released sperm cells directly into living tissue, eliminating the need for a surface film of water. The fertilized egg remained on the parent plant, where it developed inside a seed, which provided protection and nutrients for the developing embryo within.

The earliest seed-bearing plants appeared in the late Devonian period (375 million years ago) and produced their seeds along branches, without any specialized structures to hold them. By the middle of the Carboniferous period, however, a new kind of seed-bearing plant had arisen. These plants, called **conifers,** protected their developing seeds inside cones. Conifers, which as wind-pollinated plants did not depend on water for reproduction, flourished and spread during the Permian period (286 to 245 million years ago), when mountains rose, swamps drained, and the climate became much drier. The conifers' good fortune, however, was not shared by the tree ferns and giant club mosses, which, with their swimming sperm, largely went extinct.

Flowering Plants Enticed Animals to Carry Pollen

About 140 million years ago, during the Cretaceous period, the flowering plants appeared, having evolved from a group of conifer-like plants. Many flowering plants are pollinated by insects and other animals, and this mode of pollination seems to have conferred an evolutionary advantage. Flower pollination by animals can be far more efficient than pollination

◀ **FIGURE 17-8 The swamp forest of the Carboniferous period** The treelike plants in this artist's reconstruction are tree ferns and giant club mosses, most species of which are now extinct.

QUESTION Why are today's ferns and club mosses so small in comparison to their giant ancestors?

by wind. Wind-pollinated plants must produce an enormous amount of pollen because the vast majority of pollen grains fail to reach their target. Today, flowering plants dominate the land, except in cold northern regions, where conifers still prevail.

Some Animals Became Adapted to Life on Dry Land

Soon after land plants evolved, providing potential food sources for other organisms, animals emerged from the sea. The first animals to move onto land were **arthropods** (the group that today includes insects, spiders, scorpions, centipedes, and crabs). Why arthropods? The answer seems to be that they already possessed certain structures that, purely by chance, were suited to life on land. Foremost among these structures was an exoskeleton, such as the shell of a lobster or crab. Exoskeletons are both waterproof and strong enough to support a small animal against the force of gravity.

For millions of years, arthropods had the land and its plants to themselves, and for tens of millions of years more, they were the dominant land animals. Dragonflies with a wingspan of 28 inches (70 centimeters) flew among the Carboniferous tree ferns, while millipedes 6.5 feet (2 meters) long munched their way across the swampy forest floor. Eventually, however, the arthropods' splendid isolation came to an end.

Amphibians Evolved from Lobefin Fishes

About 400 million years ago, a group of Silurian fishes called the lobefins appeared, probably in fresh water. **Lobefins** had two important features that would later enable their descendants to colonize land: (1) stout, fleshy fins with which they crawled about on the bottoms of shallow, quiet waters, and (2) an outpouching of the digestive tract that could be filled with air, like a primitive lung. One group of lobefins inhabited very shallow ponds and streams, which shrank during droughts and often became oxygen poor. By taking air into their lungs, these lobefins could obtain oxygen anyway. Some began to use their fins to crawl from pond to pond in search of prey or water, as some modern fish do today (**Fig. 17-9**).

The benefits of feeding on land and moving from pool to pool favored the evolution of a group of animals that could stay out of water for longer periods and that could move about more effectively on land. With improvements in lungs and legs, **amphibians** evolved from lobefins, first appearing in the fossil record about 350 million years ago. To an amphibian, the Carboniferous swamp forests were a kind of paradise: no predators to speak of, abundant prey, and a warm, moist climate. As had the insects and millipedes, some amphibians

▲ **FIGURE 17-9 A fish that walks on land** Some modern fishes, such as this mudskipper, walk on land. As did the ancient lobefin fishes that gave rise to amphibians, mudskippers use their strong pectoral fins to move across dry areas in their swampy habitats.

QUESTION Does the mudskipper's ability to walk on land constitute evidence that lobefin fishes were the ancestors of amphibians?

evolved gigantic size, including salamanders more than 10 feet (3 meters) long.

Despite their success, the early amphibians were not fully adapted to life on land. Their lungs were simple sacs without very much surface area, so they had to obtain some of their oxygen through their skin. Therefore, their skin had to be kept moist, a requirement that restricted them to swampy habitats where they wouldn't dry out. Further, amphibian sperm and eggs could not survive in dry surroundings and had to be deposited in watery environments. So, although amphibians could move about on land, they could not stray too far from the water's edge. Along with the tree ferns and club mosses, amphibians declined when the climate turned dry at the beginning of the Permian period about 286 million years ago.

Reptiles Evolved from Amphibians

As the conifers were evolving on the fringes of the swamp forests, a group of amphibians was also evolving adaptations to drier conditions. These amphibians ultimately gave rise to the **reptiles,** which had three major adaptations to life on land. First, reptiles evolved shelled, waterproof eggs that enclosed a supply of water for the developing embryo. Thus, eggs could be laid on land without the reptiles' having to venture back to the dangerous swamps full of fish and amphibian predators. Second, ancestral reptiles evolved scaly, water-resistant skin that helped prevent the loss of body water to the dry air. Finally, reptiles evolved improved lungs that were able to provide the entire oxygen supply of an active animal. As the climate dried during the Permian period, reptiles became the dominant land vertebrates, relegating amphibians to the swampy backwaters where most remain today.

A few tens of millions of years later, the climate returned to more moist and stable conditions. This period saw the evolution of some very large reptiles, in particular, the dinosaurs (**Fig. 17-10**). The variety of dinosaur forms was enormous—large and small, fleet-footed and ponderous, predators and plant eaters. Dinosaurs were among the most successful animals ever, if we consider persistence as a measure of success. They flourished for more than 100 million years, until about 65 million years ago, when the last dinosaurs went extinct. No one is certain why they died out, but the aftereffects of a gigantic meteorite's impact with Earth seem to have been the final blow (as discussed in section 17.5).

Even during the age of dinosaurs, many reptiles remained quite small. One major difficulty faced by small reptiles is maintaining a high body temperature. A warm body is advantageous for an active animal, because warmer nerves and muscles work more efficiently. But a warm body loses heat to the environment unless the air is also warm. Heat loss is a big problem for small animals, which have a larger surface area per unit of volume than do larger animals. Many species of small reptiles have evolved slow metabolisms and cope with the heat-loss problem by confining activity to times when the air is sufficiently warm. One group of reptiles, however, followed a different evolutionary pathway. Members of this group, the birds, evolved insulation, in the form of feathers. (Birds were formerly placed in their own class, separate from reptiles. For more information on why birds are now understood to be a type of reptile, see "A Closer Look at Phylogenetic Trees" on pp. 348–349.)

In ancestral birds, feathers, which arose through evolutionary modification of scales, helped retain body heat. Consequently, these animals could be active in cool habitats and during the night, when their scaly relatives became sluggish. Later, some ancestral birds evolved longer, stronger feathers on their forelimbs, perhaps under selection for better ability

▶ **FIGURE 17-10 A reconstruction of a Cretaceous forest** By the Cretaceous period, flowering plants dominated terrestrial vegetation. Dinosaurs, such as the predatory pack of 6-foot-long *Velociraptors* shown here, were the preeminent land animals. Although small by dinosaur standards, *Velociraptors* were formidable predators with great running speed, sharp teeth, and deadly, sickle-like claws on their hind feet.

to glide from trees or to jump after insect prey. Ultimately, feathers evolved into structures capable of supporting powered flight. Fully developed, flight-capable feathers are present in 150-million-year-old fossils, so the earlier insulating structures that eventually developed into flight feathers must have been present well before that time.

Reptiles Gave Rise to Mammals

Unlike the egg-laying reptiles, **mammals** evolved live birth and the ability to feed their young with secretions of the mammary (milk-producing) glands. Ancestral mammals also developed hair, which provided insulation. Because the uterus, mammary glands, and hair do not fossilize, we may never know when these structures first appeared, or what their intermediate forms looked like. Recently, however, a team of paleontologists found bits of hair preserved in coprolites, which are fossilized animal feces. These coprolites, found in the Gobi Desert of China, were deposited by an anonymous predator 55 million years ago, so mammals have presumably had hair for at least that many years.

The earliest fossil mammal unearthed thus far is almost 200 million years old. Early mammals, which can be distinguished by distinctive skeletal features, thus coexisted with the dinosaurs. They were mostly small creatures. The largest known mammal from the dinosaur era was about the size of a modern raccoon, but most early mammal species were far smaller than that. When the dinosaurs went extinct, however, mammals colonized the habitats left empty by the extinctions. Mammal species prospered, diversifying into the array of modern forms that we see today.

17.5 WHAT ROLE HAS EXTINCTION PLAYED IN THE HISTORY OF LIFE?

If there is a lesson in the great tale of life's history, it is that nothing lasts forever. The story of life can be read as a long series of evolutionary dynasties, with each new dominant group rising, ruling the land or the seas for a time and, inevitably, falling into decline and extinction. Dinosaurs are the most famous of these fallen dynasties, but the list of extinct groups known only from fossils is impressively long. Despite the inevitability of extinction, however, the overall trend has been for species to arise at a faster rate than they disappear, so the number of species on Earth has tended to increase over time.

Evolutionary History Has Been Marked by Periodic Mass Extinctions

Over much of life's history, the origin and disappearance of species has proceeded in a steady, relentless manner. This slow and steady turnover of species, however, has been interrupted by episodes of **mass extinction** (**Fig. 17-11**). These mass extinctions are characterized by the relatively sudden disappearance of a wide variety of species over a large part of Earth. The worst episode of all, which occurred 245 million years ago at the end of the Permian period, wiped out more than 90% of the world's species, and life came perilously close to disappearing altogether.

Climate Change Contributed to Mass Extinctions

Mass extinctions have had a profound impact on the course of life's history, repeatedly redrawing the picture of life's diversity. What could have caused such dramatic changes in the fortunes of so many species? Many evolutionary biologists believe that changes in climate must have played an important role. When the climate changes, as it has done many times over the course of Earth's history, organisms that are adapted for survival in one climate may be unable to survive in a drastically different climate. In particular, at times when warm climates gave way to drier, colder climates with more variable temperatures, species may have gone extinct after failing to adapt to the harsh new conditions.

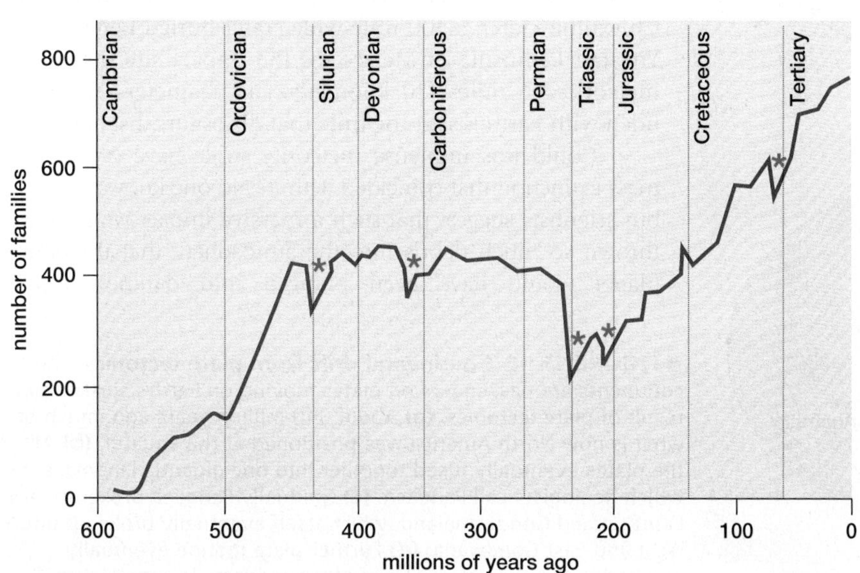

◀ **FIGURE 17-11 Mass extinctions** This graph plots the number of marine-animal groups against time, as reconstructed from the fossil record. Notice the general trend toward an increasing number of groups, punctuated by periods of sometimes rapid extinction. Five of these declines, marked by asterisks, are so steep that they qualify as catastrophic mass extinctions.

QUESTION If extinction is the ultimate fate of all species, how can the total number of species have increased over time?

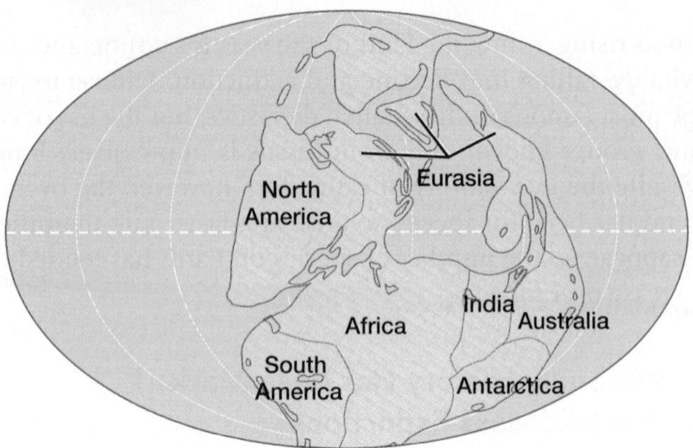

(a) 340 million years ago

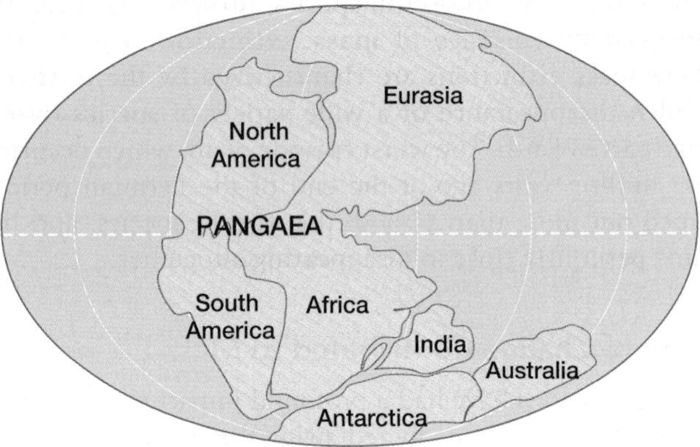

(b) 225 million years ago

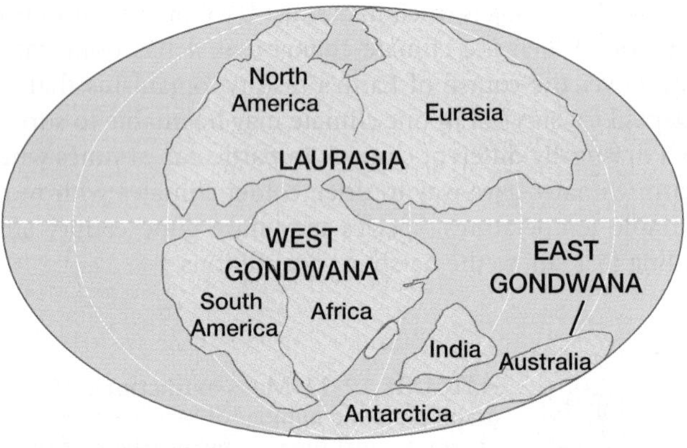

(c) 135 million years ago

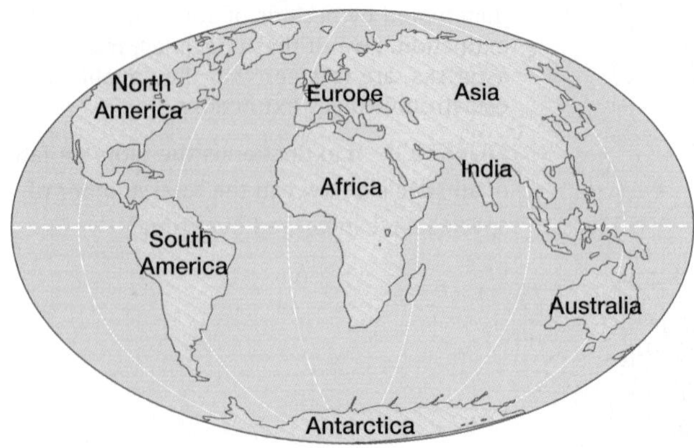

(d) Present

One cause of climate change is the shifting positions of continents. These movements are sometimes called *continental drift.* Continental drift is caused by **plate tectonics,** in which the Earth's surface, including the continents and the seafloor, is divided into plates that rest atop a viscous but fluid layer and move slowly about. As the plates wander, their positions may change in latitude (**Fig. 17-12**). For example, 340 million years ago, much of North America was located at or near the equator, an area characterized by consistently warm and wet tropical weather. But as time has passed, plate tectonics has carried the continent up into temperate and arctic regions. As a result, the once tropical climate was replaced by a regime of seasonal changes, cooler temperatures, and less rainfall. Plate tectonics continues today; the Atlantic Ocean, for example, widens by a few centimeters each year.

Catastrophic Events May Have Caused the Worst Mass Extinctions

Geological data indicate that most mass extinction events coincided with periods of climatic change. To many scientists, however, the rapidity of mass extinctions suggests that the slow process of climate change could not, by itself, be responsible for such large-scale disappearances of species. Perhaps more sudden events also play a role. For example, catastrophic geological events, such as massive volcanic eruptions, could have had devastating effects. Geologists have found evidence of past volcanic eruptions so huge that they make the 1980 Mount St. Helens explosion look like a firecracker by comparison. Even such gigantic eruptions, however, would directly affect only a relatively small portion of Earth's surface.

The search for the causes of mass extinctions took a fascinating turn in the early 1980s when Luis and Walter Alvarez proposed that the extinction event of 65 million years ago, which wiped out the dinosaurs and many other species, was caused by the impact of a huge meteorite. The Alvarezes' idea was met with great skepticism when it was first introduced, but geological research since that time has generated a great deal of evidence that a massive impact did indeed occur 65 million years ago. In fact, researchers have identified the Chicxulub crater, a 100-mile-wide crater buried beneath the Yucatan Peninsula of Mexico, as the impact site of a giant meteorite—6 miles (10 kilometers) in diameter—that collided with Earth just at the time that dinosaurs disappeared.

Could this immense meteorite strike have caused the mass extinction that coincided with it? No one knows for sure, but scientists suggest that such a massive impact would have thrown so much debris into the atmosphere that the entire planet would have been plunged into darkness for a

◀ **FIGURE 17-12 Continental drift from plate tectonics** The continents are passengers on plates moving on Earth's surface as a result of plate tectonics. **(a)** About 340 million years ago, much of what is now North America was positioned at the equator. **(b)** All the plates eventually fused together into one gigantic landmass, which geologists call Pangaea. **(c)** Gradually Pangaea broke up into Laurasia and Gondwanaland, which itself eventually broke up into West and East Gondwana. **(d)** Further plate motion eventually resulted in the current positions of the modern-day continents.

Have you ever wondered

If Extinct Species Can Be Revived by Cloning?

Scientists have cloned a number of animal species, including mice, dogs, cats, horses, and cows. Could the technology of cloning be used to bring back extinct species? In principle, yes, provided that perfectly preserved DNA of the extinct species is available. Such DNA could be transferred to an egg from a closely related, living species, and the egg implanted in a surrogate mother of that species.

For example, researchers have suggested that it might be possible to clone a woolly mammoth, using an elephant surrogate mother and DNA extracted from 20,000-year-old mammoths found frozen beneath the Siberian tundra. Most scientists, however, believe that any DNA recovered from a fossil mammoth would be far too degraded for use in cloning, and synthesizing an entire mammoth genome (its sequence is now almost fully known) is beyond the capabilities of current technology. The odds of success might be greater for another proposed project, which would use DNA from a preserved museum specimen to revive the Tasmanian tiger, an Australian mammal that has been extinct for more than 70 years. If cloning recently extinct species proves to be possible, would you support doing it?

period of years. With little light reaching the planet, temperatures would have dropped precipitously and the photosynthetic capture of energy (on which all life ultimately depends) would have declined drastically. The worldwide "impact winter" would have spelled doom for the dinosaurs and a host of other species.

17.6 HOW DID HUMANS EVOLVE?

Scientists are intensely interested in the origin and evolution of humans. The outline of human evolution that we present in this section represents an interpretation that is widely shared among paleontologists. However, fossil evidence of human evolution is comparatively scarce and therefore open to a variety of interpretations. Thus, some paleontologists would disagree with aspects of the scenario we present.

Humans Inherited Some Early Primate Adaptations for Life in Trees

Humans are members of a mammal group known as **primates,** which also includes lemurs, monkeys, and apes. The oldest primate fossils are 55 million years old, but because primate fossils are relatively rare compared with those of many other animals, the first primates probably arose considerably earlier but left no fossil record. Early primates probably fed on fruits and leaves, and were adapted for life in the trees. Many modern primates retain the tree-dwelling lifestyle of their ancestors (**Fig. 17-13**). The common heritage of humans and other primates is reflected in a set of physical characteristics that was present in the earliest primates and that persists in many modern primates, including humans.

Binocular Vision Provided Early Primates with Accurate Depth Perception

One of the earliest primate adaptations seems to have been large, forward-facing eyes (see Fig. 17-13). Jumping from branch to branch is risky business unless an animal can accurately judge where the next branch is located. Accurate depth

(a) Tarsier

(b) Lemur

(c) Macaque

◀ **FIGURE 17-13 Representative primates** The **(a)** tarsier, **(b)** lemur, and **(c)** lion-tail macaque monkey all have a relatively flat face, with forward-looking eyes providing binocular vision. All also have color vision and grasping hands. These features, retained from the earliest primates, are shared by humans.

perception was made possible by binocular vision, provided by forward-facing eyes with overlapping fields of view. Another key adaptation was color vision. We cannot, of course, tell if a fossil animal had color vision, but since modern primates have excellent color vision, it seems reasonable to assume that earlier primates did as well. Many primates feed on fruit, and color vision helps to detect ripe fruit among a bounty of green leaves.

Early Primates Had Grasping Hands

Early primates had long, grasping fingers that could wrap around and hold onto tree limbs. This adaptation to tree dwelling was the basis for later evolution of human hands that could perform both a *precision grip* (used by modern humans for delicate maneuvers such as manipulating small objects, writing, and sewing) and a *power grip* (used for powerful actions, such as swinging a club or thrusting with a spear).

A Large Brain Facilitated Hand–Eye Coordination and Complex Social Interactions

Primates have brains that are larger, relative to their body size, than the brains of almost all other animals. No one really knows for certain which environmental factors favored the evolution of large brains. It seems reasonable, however, that controlling and coordinating rapid locomotion through trees, the dexterous movements of the hands in manipulating objects, and binocular, color vision would be facilitated by increased brain power. Most primates also have fairly complex social systems, which require relatively high intelligence. If sociality promoted increased survival and reproduction, the benefits to individuals of successful social interaction might have favored the evolution of a larger brain.

The Oldest Hominin Fossils Are from Africa

On the basis of comparisons of DNA from modern chimps, gorillas, and humans, researchers estimate that the **hominin** line (humans and their fossil relatives) diverged from the ape lineage sometime between 5 million and 8 million years ago. The fossil record, however, suggests that the split must have occurred at the early end of that range. Paleontologists working in the African country of Chad in 2002 discovered fossils of a hominin, *Sahelanthropus tchadensis*, that lived more than 6 million years ago (**Fig. 17-14**). *Sahelanthropus* is clearly a hominin because it shares several anatomical features with later members of the group. But because this oldest known member of our family also exhibits other features that are more characteristic of apes, it may represent a point on our family tree that is close to the split between apes and hominins.

In addition to *Sahelanthropus,* two other hominin species—*Ardipithecus ramidus* and *Orrorin tugenensis*—are known from fossils appearing in rocks that are between 4 million and 6 million years old. Our knowledge of these hominins is limited, however, because only a few specimens have been found thus far, and most of these recent discoveries typically include only small portions of skeletons. A more extensive

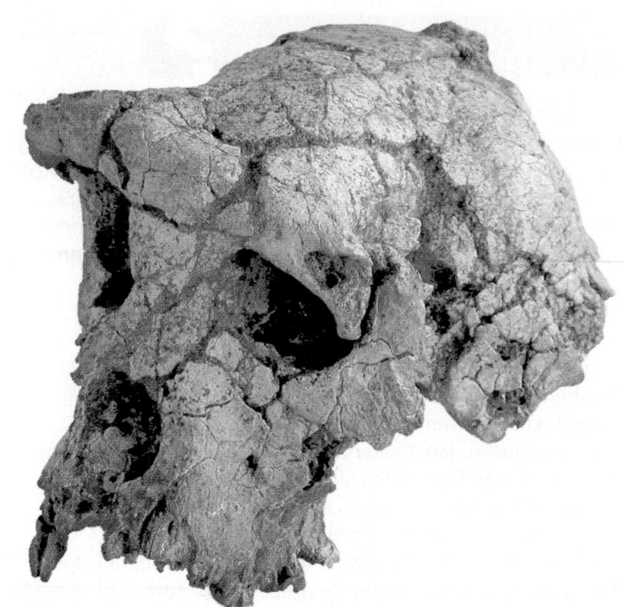

▲ **FIGURE 17-14 The earliest hominin** This nearly complete skull of *Sahelanthropus tchadensis*, which is more than 6 million years old, is the oldest hominin fossil yet found.

record of early hominin evolution does not begin until about 4 million years ago. That date marks the beginning of the fossil record of the genus *Australopithecus* (**Fig. 17-15**), a group of African hominin species with brains larger than those of their prehominin forebears but still much smaller than those of modern humans.

The Earliest Hominins Could Stand and Walk Upright

The earliest australopithecines (as the various species of *Australopithecus* are collectively known) had legs that were shorter, relative to their height, than those of modern humans, but their knee joints allowed them to straighten their legs fully, permitting efficient bipedal (upright, two-legged) locomotion. Footprints almost 4 million years old, discovered in Tanzania by anthropologist Mary Leakey, show that even the earliest australopithecines could, and at least sometimes did, walk upright. Upright posture may have evolved even earlier. The discoverers of *Sahelanthropus* and *Orrorin* argue that the leg and foot bones of these earliest hominins have characteristics that indicate bipedal locomotion, but this conclusion will remain speculative until more complete skeletons of these species are found.

The reasons for the evolution of bipedal locomotion among the early hominins remain poorly understood. Perhaps hominins that could stand upright gained an advantage in gathering or carrying food in their forest habitat. Whatever its cause, the early evolution of upright posture was extremely important in the evolutionary history of hominins, because it freed their hands from use in walking. Later hominins were thus able to carry weapons, manipulate tools, and eventually achieve the cultural revolutions produced by modern *Homo sapiens*.

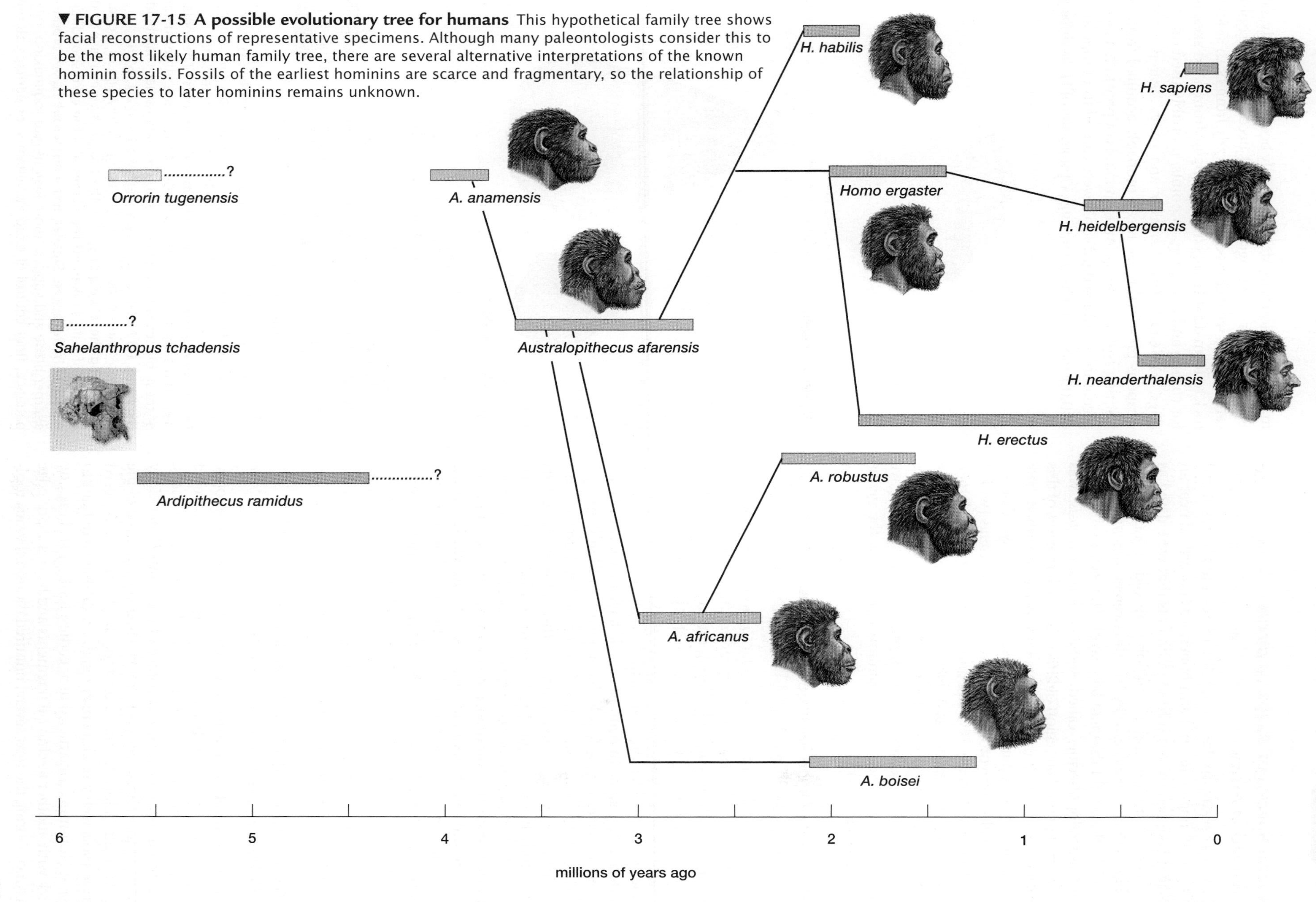

▼ **FIGURE 17-15 A possible evolutionary tree for humans** This hypothetical family tree shows facial reconstructions of representative specimens. Although many paleontologists consider this to be the most likely human family tree, there are several alternative interpretations of the known hominin fossils. Fossils of the earliest hominins are scarce and fragmentary, so the relationship of these species to later hominins remains unknown.

Several Species of *Australopithecus* Emerged in Africa

The oldest australopithecine species, represented by fossilized teeth, skull fragments, and arm bones, was unearthed near an ancient lake bed in Kenya from sediments that were dated as between 3.9 million and 4.1 million years old. It was named *Australopithecus anamensis* by its discoverers (*anam* means "lake" in the local Ethiopian language). The second most ancient australopithecine, called *Australopithecus afarensis*, was discovered in the Afar region of Ethiopia. Fossil remains of this species as old as 3.9 million years have been unearthed. The *A. afarensis* line apparently gave rise to at least two distinct forms: small, omnivorous species such as *A. africanus* (which was similar to *A. afarensis* in size and eating habits), and larger, herbivorous species such as *A. robustus* and *A. boisei*. All of the australopithecine species had gone extinct by 1.2 million years ago. Before disappearing, however, one of these species gave rise to a new branch of the hominin family tree, the genus *Homo* (see Fig. 17-15).

The Genus *Homo* Diverged from the Australopithecines 2.5 Million Years Ago

Hominins that are sufficiently similar to modern humans to be placed in the genus *Homo* first appear in African fossils that are about 2.5 million years old. Among the earliest African *Homo* fossils are *H. habilis* (see Fig. 17-15), a species whose body and brain were larger than those of the australopithecines but that retained the apelike long arms and short legs of their australopithecine ancestors. In contrast, the skeletal anatomy of *H. ergaster*, a species whose fossils first appear 2 million years ago, has limb proportions more like those of modern humans. This species is believed by many paleoanthropologists (scientists who study human origins) to be on the evolutionary branch that led ultimately to our own species, *H. sapiens*. In this view, *H. ergaster* was the common ancestor of two distinct branches of hominins. The first branch led to *H. erectus*, which was the first hominin species to leave Africa. The second branch from *H. ergaster* ultimately led to *H. heidelbergensis*, some of which migrated to Europe and gave rise to the Neanderthals, *H. neanderthalensis*. Meanwhile, back in Africa, another branch split off from the *H. heidelbergensis* lineage. This branch became *H. sapiens*—modern humans.

The Evolution of *Homo* Was Accompanied by Advances in Tool Technology

Hominin evolution is closely tied to the development of tools, a hallmark of hominin behavior. The oldest tools discovered so far were found in 2.5-million-year-old East African rocks, concurrent with the early emergence of the genus *Homo*. Early *Homo*, whose cheek teeth were much smaller than those of the genus's australopithecine ancestors, might first have used stone tools to break and crush tough foods that were hard to chew. Hominins constructed their earliest tools by striking one rock with another to chip off fragments and leave a sharp edge behind. During the next several hundred thousand years, toolmaking techniques in Africa gradually became more advanced. By 1.7 million years ago, tools had become more sophisticated. Flakes were chipped symmetrically from both sides of a rock to form double-edged tools ranging from hand axes, used for cutting and chopping, to points, probably used on spears (**Figs. 17-16a,b**). *Homo ergaster* and other bearers of these weapons presumably ate meat, probably acquired from both hunting and scavenging for the remains of prey killed by other predators. Double-edged tools were carried to Europe at least 600,000 years ago by migrating populations of *H. heidelbergen-*

(a) *Homo habilis*

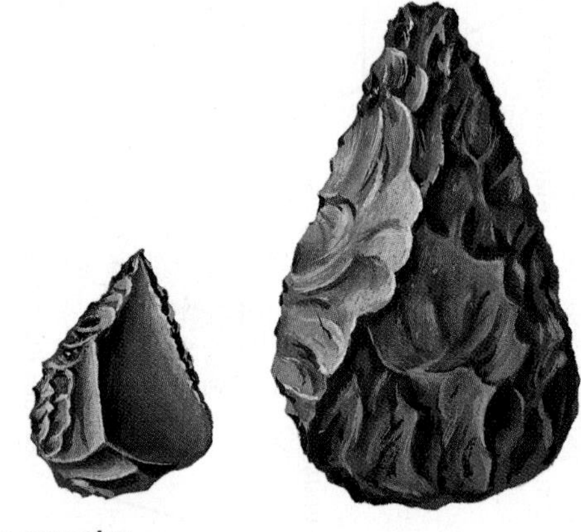

(b) *Homo ergaster*

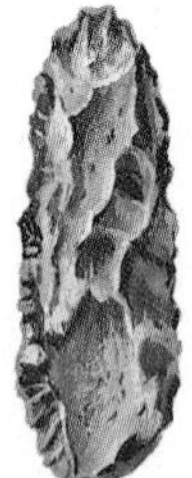

(c) *Homo neanderthalensis*

▲ FIGURE 17-16 **Representative hominin tools (a)** *Homo habilis* produced only fairly crude chopping tools called hand axes, usually unchipped on one end to hold in the hand. **(b)** *Homo ergaster* manufactured much finer tools. The tools were typically sharp all the way around the stone; at least some of these blades were probably tied to spears rather than held in the hand. **(c)** Neanderthal tools were works of art, with extremely sharp edges made by flaking off tiny bits of stone. In comparing these weapons, note the progressive increase in the number of flakes taken off the blades and the corresponding decrease in flake size. Smaller, more numerous flakes produce a sharper blade and suggest more insight into toolmaking, more patience, finer control of hand movements, or perhaps all three.

sis, and the Neanderthal descendants of these emigrants took stone-tool construction to new heights of skill and delicacy (**Fig. 17-16c**).

Neanderthals Had Large Brains and Excellent Tools

Neanderthals first appeared in the European fossil record about 150,000 years ago. By about 70,000 years ago, they had spread throughout Europe and western Asia. By 30,000 years ago, however, the species was extinct.

Contrary to the popular image of a hulking, stoop-shouldered "caveman," Neanderthals were quite similar to modern humans in many ways. Although more heavily muscled, Neanderthals walked fully erect, were dexterous enough to manufacture finely crafted stone tools, and had brains that, on average, were slightly larger than those of modern humans. Many European Neanderthal fossils show heavy brow ridges and a broad, flat skull, but others, particularly from areas around the eastern shores of the Mediterranean Sea, are somewhat more physically similar to *H. sapiens*.

Despite the physical and technological similarities between Neanderthals and *H. sapiens*, there is no solid archaeological evidence that Neanderthals ever developed an advanced culture that included such characteristically human endeavors as art, music, and rituals. Some anthropologists argue that, because their skeletal anatomy shows that they were physically capable of making the sounds required for speech, Neanderthals might have acquired language. This interpretation of Neanderthal anatomy, however, is not unanimously accepted. In general, the available evidence of the Neanderthal way of life is limited and open to different interpretations, and anthropologists are engaged in a sometimes heated debate about how advanced Neanderthal culture became.

Though some anthropologists argue that Neanderthals were simply a variety of *H. sapiens*, most agree that Neanderthals were a separate species. Dramatic evidence in support of this hypothesis has come from researchers who have isolated DNA from various Neanderthal skeletons that are between 20,000 and 38,000 years old. These extractions of ancient DNA have allowed researchers to compare the nucleotide sequences of Neanderthal genes with the sequences of the same genes in both fossil and modern humans. The comparisons have shown that Neanderthal sequences are very different from those of both modern and fossil humans, but that modern and fossil humans share similar sequences. These findings indicate that the evolutionary branch leading to Neanderthals diverged from the ancestral human line about 500,000 years ago, hundreds of thousands of years before the emergence of modern *H. sapiens*. This early divergence supports the conclusion that Neanderthals were a distinct species, *H. neanderthalensis*.

Modern Humans Emerged Less Than 200,000 Years Ago

The fossil record shows that anatomically modern humans appeared in Africa at least 160,000 years ago and possibly as long as 195,000 years ago. The location of these fossils suggests that *Homo sapiens* originated in Africa, but most of our knowledge about our own early history comes from European and Middle Eastern fossils of *H. sapiens*, collectively known as Cro-Magnons (after the district in France in which their remains were first discovered). Cro-Magnons appeared about 90,000 years ago. They had domed heads, smooth brows, and prominent chins (just like us). Their tools were precision instruments similar to the stone tools used until recently in many parts of the world.

Behaviorally, Cro-Magnons seem to have been similar to, but more sophisticated than, Neanderthals. Artifacts from 30,000-year-old Cro-Magnon archaeological sites include elegant bone flutes, graceful carved ivory sculptures, and evidence of elaborate burial ceremonies (**Fig. 17-17**). Perhaps the most remarkable accomplishment of Cro-Magnons is the magnificent art left in caves in places such as Altamira in Spain and Lascaux and Chauvet in France (**Fig. 17-18**). The oldest cave paintings so far found are more than 30,000 years old, and even the oldest ones make use of sophisticated artistic techniques. No one knows exactly why these paintings were made, but they attest to minds as capable as our own.

Cro-Magnons and Neanderthals Lived Side by Side

Cro-Magnons coexisted with Neanderthals in Europe and the Middle East for perhaps as many as 50,000 years before the Neanderthals disappeared. Some researchers believe that Cro-Magnons interbred extensively with Neanderthals, so Neanderthals were essentially absorbed into the human genetic

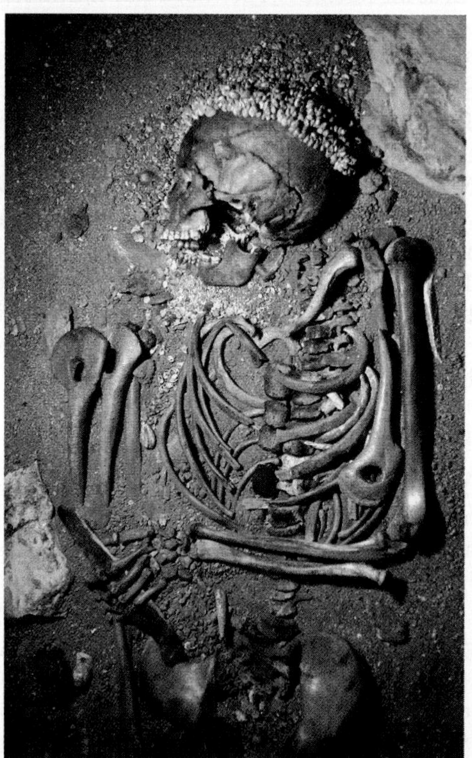

▲ **FIGURE 17-17 Paleolithic burial** This 24,000-year-old grave shows evidence that Cro-Magnon people ritualistically buried their dead. The body was covered with a dye known as red ocher, then buried with a headdress made of snail shells and a flint tool in its hand.

▲ **FIGURE 17-18 The sophistication of Cro-Magnon people** Cave paintings by Cro-Magnons have been remarkably preserved by the relatively constant underground conditions of a cave in Lascaux, France.

mainstream. Other scientists disagree, citing mounting evidence such as the fossil DNA described earlier, and suggest that later-arriving Cro-Magnons simply overran and displaced the less-well-adapted Neanderthals.

Neither hypothesis does a good job of explaining how the two kinds of hominins managed to occupy the same geographical areas for such a long time. The persistence in one area of two similar but distinct groups for tens of thousands of years seems inconsistent with both interbreeding and direct competition. Perhaps the competition between *H. neanderthalensis* and *H. sapiens* was indirect, so that the two species were able to coexist for a time in the same habitat, until the superior ability of *H. sapiens* to exploit the available resources slowly drove Neanderthals to extinction.

Several Waves of Hominins Emigrated from Africa

The human family tree is rooted in Africa, but hominins found their way out of Africa on numerous occasions. For example, *H. erectus* reached tropical Asia almost 2 million years ago and apparently thrived there, eventually spreading across Asia (**Fig. 17-19a**). Similarly, *H. heidelbergensis* made it to Europe at least 780,000 years ago. It is increasingly clear that the genus *Homo* made repeated long-distance emigrations, beginning as soon as sufficiently capable limb anatomy evolved. What is less clear is how all this wandering is related to the origin of modern *H. sapiens*. According to the "African replacement" hypothesis (the basis of the scenario outlined earlier), *H. sapiens* emerged in Africa and dispersed less than 150,000 years ago, spreading into the Near East, Europe, and Asia and replacing all other hominins (see Fig. 17-19a). But some paleoanthropologists believe that populations of *H. sapiens* evolved simultaneously in many regions from the already widespread populations of *H. erectus*. According to this "multiregional origin" hypothesis, continued migrations and interbreeding among *H. erectus* populations in different regions of the world maintained them as a single species as they gradually evolved into *H. sapiens* (**Fig. 17-19b**). Although an increasing number of studies of modern human DNA support the African replacement model of the origin of our

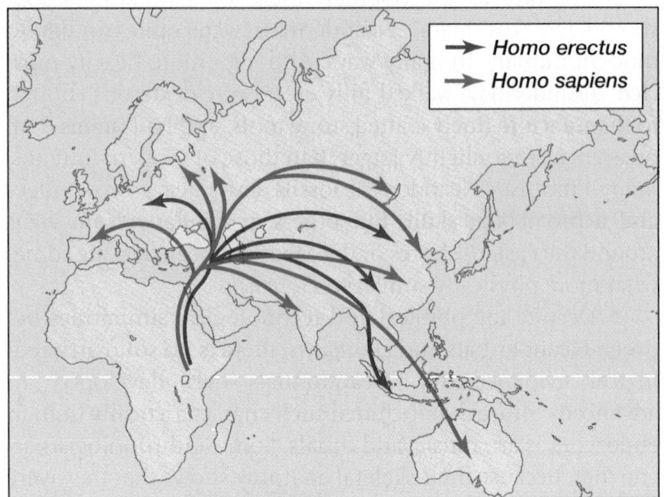

(a) African replacement hypothesis

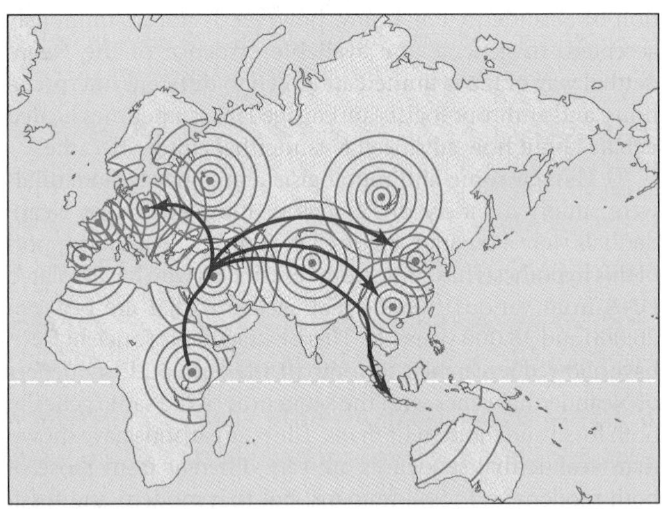

(b) Multiregional hypothesis

▲ **FIGURE 17-19 Competing hypotheses for the evolution of *Homo sapiens* (a)** The "African replacement" hypothesis suggests that *H. sapiens* evolved in Africa, then migrated throughout the Near East, Europe, and Asia, displacing the other hominin species that were present in those regions. **(b)** The "multiregional" hypothesis suggests that populations of *H. sapiens* evolved in many regions simultaneously from the already widespread populations of *H. erectus*.

QUESTION Paleontologists recently discovered fossil hominins with features characteristic of modern humans in 160,000-year-old sediments in Africa. Which hypothesis does this new evidence support?

Case Study continued
Little People, Big Story

What is the ancestry of *H. floresiensis*, the Hobbit of Flores Island in Indonesia? Some clues point toward an intriguing scenario. First, the only other evidence of early hominin habitation on Flores consists of stone tools found at an 840,000-year-old site. The age and relative crudeness of the tools suggest that they were probably left by *H. erectus*, the only hominin known to be present in Asia that long ago. So, *H. floresiensis* may have descended from a population of *H. erectus* that became isolated on Flores. This conclusion is supported by some anatomical similarities between *H. floresiensis* and *H. erectus*. Oddly, *H. floresiensis* is more similar to *H. erectus* specimens from a distant, 1.8-million-year-old site in Central Asia than it is to much younger *H. erectus* specimens found in relatively nearby sites on other Indonesian islands. Perhaps *H. floresiensis* descended from a very early wave of *H. erectus* migrants.

species, both hypotheses are consistent with the fossil record. Therefore, the question remains unsettled.

The Evolutionary Origin of Large Brains May Be Related to Meat Consumption

The main physical features that distinguish us from our closest relatives, the apes, are our upright posture and large, highly developed brains. As described earlier, upright posture arose very early in hominin evolution, and hominins walked upright for several million years before large-brained *Homo* species arose. What circumstances might have caused the evolution of increased brain size? Many explanations have been proposed, but little direct evidence is available; hypotheses about the evolutionary origins of large brains are necessarily speculative.

One proposed explanation for the origin of large brains suggests that they evolved in response to increasingly complex social interactions. In particular, fossil evidence suggests that, beginning about 2 million years ago, hominin social life began to include a new type of activity—the cooperative hunting of large game. The resulting access to significant amounts of meat must have fostered a need to develop methods for distributing this valuable, limited resource among group members. Some anthropologists hypothesize that the individuals best able to manage this social interaction would have been more successful at gaining a large share of meat and using their share advantageously. Perhaps this social management was best accomplished by individuals with larger, more powerful brains, and natural selection therefore favored such individuals. Observations of chimpanzee societies have shown that the distribution of group-hunted meat often involves intricate social interactions in which meat is used to form alliances, repay favors, gain access to sexual partners, placate rivals, and so on. Perhaps the mental skill required to plan, assess, and remember such interactions was the driving force behind the evolution of our large, clever brains.

The Evolutionary Origin of Human Behavior Is Highly Speculative

Even after the evolution of comparatively large brains in species such as *H. erectus*, more than a million years passed before the origin of modern humans and their extremely large brains. And even after the first appearance of modern *H. sapiens*, more than 100,000 years passed before the appearance of any archaeological evidence of the distinctively human characteristics that were made possible by a large brain: language, abstract thought, and advanced culture. The evolutionary origin of these human traits is another unresolved question, in part because direct evidence of the transition to advanced culture may never be found. Early humans capable of language and symbolic thought would not necessarily have created artifacts that indicated these capabilities. We can uncover some clues by studying our ape relatives, which possess less-complex versions of many human behaviors and mental processes. Their behavior might resemble that of ancestral hominins. Nonetheless, the late, seemingly rapid origin of advanced human culture remains a puzzle.

The Cultural Evolution of Humans Now Far Outpaces Biological Evolution

In recent millennia, human evolution has come to be dominated by *cultural evolution*, the evolution of information and behaviors that are transmitted from generation to generation by learning. Our recent evolutionary success, for example, was engendered not so much by new physical adaptations as by a series of cultural and technological revolutions. The first such revolution was the development of tools, which began with the early hominins. Tools increased the efficiency with which food and shelter could be acquired and thus increased the number of individuals that could survive within a given ecosystem. About 10,000 years ago, human culture underwent a second revolution as people discovered how to grow crops and domesticate animals. This agricultural revolution dramatically increased the amount of food that could be extracted from the environment, and the human population surged, increasing from about 5 million at the dawn of agriculture to around 750 million by 1750. The subsequent Industrial Revolution gave rise to the modern economy and its attendant improvements in public health. Longer lives and lower infant mortality led to truly explosive population growth, and today our population is 6.8 billion and still growing.

Human cultural evolution and the accompanying increases in human population have had profound effects on the continuing biological evolution of other life-forms. Our agile hands and minds have transformed much of Earth's terrestrial and aquatic habitats. Humans have become the most powerful agent of natural selection. In the words of the late evolutionary biologist Stephen Jay Gould, "We have become, by the power of a glorious evolutionary accident called intelligence, the stewards of life's continuity on Earth. We did not ask for this role, but we cannot renounce it. We may not be suited for it, but here we are."

Case Study revisited

Little People, Big Story

The discovery of the hobbit-sized *Homo floresiensis* was exciting to many people in part because it suggested that our species may have more close relatives than we previously suspected and that at least some of them lived tantalizingly close to the present. Plus, the idea of a society of tiny humans seems to have an inherent appeal. But the discovery also raises a host of fascinating evolutionary questions.

For example, how can we account for the arrival on Flores of the *H. erectus* population that left tools behind and that may be ancestral to *H. floresiensis*? Unlike some islands, Flores was never connected to the mainland. Archaeologists generally agree that hominins did not construct boats until 60,000 years ago at the earliest. So how did *H. erectus* get to Flores nearly 800,000 years before the invention of boats? Perhaps they drifted across on clumps of floating vegetation.

Another interesting question about *H. floresiensis* is the cause of their small size. Large animal species that are isolated on islands sometimes evolve smaller body size. For example, the now-extinct elephants of Flores were only about 4 feet tall. Biologists suggest that the absence of large predators on most islands eliminates much of the benefit of large size, conferring an advantage on smaller individuals that require less food. Did this kind of dynamic foster the evolution of small stature in *H. floresiensis*? Are the bodies of hominins, which have weapons to defend against predators and tools to aid in acquiring food, subject to the same evolutionary pressures that shape the bodies of other animals?

Consider This

Homo floresiensis was found on an island. If you were searching for evidence of other undiscovered recent hominin species, would you concentrate your search on islands? Why or why not? In which regions of the world would you search?

CHAPTER REVIEW

Summary of Key Concepts

17.1 How Did Life Begin?

Before life arose, lightning, ultraviolet light, and heat formed organic molecules from water and the components of primordial Earth's atmosphere—methane, ammonia, hydrogen, and water vapor. The organic molecules formed probably included nucleic acids, amino acids, short proteins, and lipids. By chance, some molecules of RNA may have had enzymatic properties, catalyzing the assembly of copies of themselves from nucleotides in Earth's waters. These may have been the forerunners of life. Protein-lipid vesicles enclosing these ribozymes may have formed the first protocells.

17.2 What Were the Earliest Organisms Like?

The oldest fossils, about 3.5 billion years old, are of prokaryotic cells that fed by absorbing organic molecules that had been synthesized in the environment. Because there was no free oxygen in the atmosphere, their energy metabolism must have been anaerobic. As the cells multiplied, they depleted the organic molecules that had been formed by prebiotic synthesis. Some cells developed the ability to synthesize their own food molecules by using simple inorganic molecules and the energy of sunlight. These earliest photosynthetic cells were probably ancestors of today's cyanobacteria.

Photosynthesis releases oxygen as a by-product, and by about 2.2 billion years ago significant amounts of free oxygen were accumulating in the atmosphere. Aerobic metabolism, which generates more cellular energy than does anaerobic metabolism, probably arose about this time.

Eukaryotic cells had evolved by about 1.7 billion years ago. The first eukaryotic cells probably arose as symbiotic associations between predatory prokaryotic cells and other bacteria. Mitochondria may have evolved from aerobic bacteria engulfed by predatory cells. Similarly, chloroplasts may have evolved from photosynthetic cyanobacteria.

17.3 What Were the Earliest Multicellular Organisms Like?

Multicellular organisms evolved from eukaryotic cells and first appeared in the seas about 1.2 billion years ago. Multicellularity offers several advantages, including greater size. In plants, increased size offered some protection from predation. Specialization of cells allowed plants to anchor themselves in the nutrient-rich, well-lit waters of the shore. For animals, multicellularity allowed more efficient predation and more effective escape from predators. These in turn provided environmental pressures for faster locomotion, improved senses, and greater intelligence.

17.4 How Did Life Invade the Land?

The first land organisms were probably algae. The first multicellular land plants appeared about 475 million years ago. Life on land required special adaptations for support of the body, reproduction, and the acquisition, distribution, and retention of water, but the land also offered abundant sunlight and protection from aquatic herbivores. Soon after land plants evolved, arthropods invaded the land. Absence of predators and abundant land plants for food probably facilitated the invasion of the land by animals.

The earliest land vertebrates evolved from lobefin fishes, which had leglike fins and a primitive lung. A group of lobefins evolved into the amphibians about 350 million years ago. Reptiles evolved from amphibians, with several further adaptations for life on land: internal fertilization, waterproof eggs that could be laid on land, water-resistant skin, and better lungs. One reptile group, the birds, evolved feathers that provided insulation and facilitated flight. Mammals, whose bodies are insulated by hair, descended from a reptile group.

17.5 What Role Has Extinction Played in the History of Life?

The history of life has been characterized by constant turnover of species as species go extinct and are replaced by new ones. Mass extinctions, in which large numbers of species disappear within a relatively short time, have occurred periodically. Mass extinctions were probably caused by some combination of climate change and catastrophic events, such as volcanic eruptions and meteorite impacts.

17.6 How Did Humans Evolve?

One group of mammals evolved into the tree-dwelling primates. Some primates descended from the trees, and these were the ancestors of apes and humans. The oldest known hominin fossils are between 6 million and 7 million years old and were found in Africa. The australopithecines arose in Africa about 4 million years ago. These hominins walked erect, had larger brains than did their forebears, and fashioned primitive tools. One group of australopithecines gave rise to a line of hominins in the genus *Homo*. *Homo* arose in Africa, but populations of several *Homo* species migrated from Africa and spread to other geographic areas. In the last of these migrations, *Homo sapiens*, characterized by a large brain and advanced tool technology, dispersed from Africa to Asia and Europe.

Key Terms

amphibian *329*
arthropod *329*
conifer *328*
endosymbiont hypothesis *325*
eukaryote *325*
exoskeleton *328*
hominin *334*
lobefin *329*
mammal *331*
mass extinction *331*
plate tectonics *332*
primate *333*
prokaryote *322*
protocell *321*
reptile *330*
ribozyme *321*
spontaneous generation *318*

Thinking Through the Concepts

Fill-in-the-Blank

1. Because there was no oxygen in the earliest atmosphere, the first cells must have derived energy by __________ metabolism of organic molecules. Oxygen was introduced into the atmosphere when some microbes developed the ability to __________ and released oxygen as a by-product. Oxygen was __________ to many of the earliest cells, but some evolved the ability to use oxygen in __________ respiration, which provided far more __________.
2. The molecule __________ became a candidate for the first self-replicating information-carrying molecule when Tom Cech and Sidney Altman discovered that some of these molecules can act as __________, which they called __________.
3. Complex cells that contain a nucleus and other organelles are called __________ cells. A compelling explanation for the origin of these complex cells is the __________ hypothesis. One observation that supports this hypothesis is that mitochondria have their own __________.
4. The sperm of early land plants had to reach the egg by __________, limiting them to __________ environments. An important adaptation to dry land was the evolution of __________, which enclosed sperm in a drought-resistant coat.
5. Early plants that protected their seeds within cones are called __________. These relied on __________ to carry their pollen. This type of plant still dominates in __________ regions. Later, some plants evolved __________, which attracted animals, particularly __________ that carried their pollen. Animal pollination is much more __________ than wind pollination.
6. The first animals to live on land were __________ because their external skeletons, also called __________, supported the animals' weight, while protecting their bodies from __________.
7. Amphibians gave rise to __________, which had three important adaptations to life on dry land: shelled, waterproof __________; scaly, water-resistant __________; and more efficient __________.

Review Questions

1. What is the evidence that life might have originated from nonliving matter on early Earth? What kind of evidence would you like to see before you would accept this hypothesis?
2. Explain the endosymbiont hypothesis for the origin of chloroplasts and mitochondria.
3. Name two advantages of multicellularity for plants and two for animals.
4. What advantages and disadvantages would terrestrial existence have had for the first plants to invade the land? For the first land animals?
5. Outline the major adaptations that emerged during the evolution of vertebrates, from fish to amphibians to reptiles to birds and mammals. Explain how these adaptations increased the fitness of the various groups for life on land.
6. Outline the evolution of humans from early primates. Include in your discussion such features as binocular vision, grasping hands, bipedal locomotion, toolmaking, and brain expansion.

Applying the Concepts

1. What is cultural evolution? Is cultural evolution more or less rapid than biological evolution? Why?
2. Do you think that studying our ancestors can shed light on the behavior of modern humans? Why or why not?
3. A biologist would probably answer the age-old question "What is life?" by saying, "The ability to self-replicate." Do you agree with this definition? If so, why? If not, how would you define life in biological terms?
4. Extinctions have occurred throughout the history of life on Earth. Why should we care if humans are causing a mass extinction event now?
5. The "African replacement" and "multiregional origin" hypotheses of the evolution of *Homo sapiens* make contrasting predictions about the extent and nature of genetic divergence among human races. One predicts that races are old and highly diverged genetically; the other predicts that races are young and little diverged genetically. What data would help you determine which hypothesis is closer to the truth?
6. In biological terms, what do you think was the most significant event in the history of life? Explain your answer.

MB® *Go to www.masteringbiology.com for practice quizzes, activities, eText, videos, current events, and more.*

Systematics: Seeking Order Amidst Diversity

Chapter 18

▲ Biologists studying the evolutionary history of type 1 human immunodeficiency virus (HIV-1) discovered that the virus, which causes AIDS, probably originated in chimpanzees.

Case Study

Origin of a Killer

ONE OF THE WORLD'S MOST FRIGHTENING DISEASES is also one of its most mysterious. Acquired immune deficiency syndrome (AIDS) appeared seemingly out of nowhere, and when it was first recognized in the early 1980s, no one knew what caused it or where it came from. Scientists raced to solve the mystery and, within a few years, had identified the infectious agent that causes AIDS: human immunodeficiency virus (HIV). Once HIV had been identified, researchers turned their attention to the question of its origin.

Finding the source of HIV required an evolutionary approach. To ask, "Where did HIV come from?" is really to ask, "What kind of virus was the ancestor of HIV?" Biologists who examine questions of ancestry are known as *systematists*. Systematists strive to categorize organisms according to their evolutionary history, building classifications that accurately reflect the structure of the tree of life. When a systematist concludes that two species are closely related, it means that the two species share a recent common ancestor from which both species evolved.

The systematists who explored the ancestry of HIV discovered that its closest relatives are found not among other viruses that infect humans, but among those that infect monkeys and apes. In fact, the latest research on HIV's evolutionary history has concluded that the closest relative of HIV-1 (the type of HIV that is most responsible for the worldwide AIDS epidemic) is a virus strain that infects a particular chimp subspecies that inhabits a limited range in West Africa. Therefore, the ancestor of the virus that we now know as HIV-1 did not evolve from a preexisting human virus but must have somehow jumped from West African chimpanzees to humans.

At a Glance

18.1 HOW ARE ORGANISMS NAMED AND CLASSIFIED?

To study and discuss organisms, biologists must name them. The branch of biology that is concerned with naming and classifying organisms is known as **taxonomy.** The basis of modern taxonomy was established by the Swedish naturalist Carl von Linné (1707–1778), who called himself Carolus Linnaeus, a Latinized version of his given name. One of Linnaeus' most durable achievements was the introduction of the two-part scientific name.

Each Species Has a Unique, Two-Part Name

The **scientific name** of an organism designates its genus and species. A **genus** is a group that includes a number of very closely related species; each **species** within a genus includes populations of organisms that can potentially interbreed under natural conditions. For example, the genus *Sialia* (bluebirds) includes three species: the eastern bluebird (*Sialia sialis*), the western bluebird (*Sialia mexicana*), and the mountain bluebird (*Sialia currucoides*) (**Fig. 18-1**). Although the three species are similar, bluebirds normally breed only with members of their own species.

(a) Eastern bluebird

(b) Western bluebird

(c) Mountain bluebird

▲ **FIGURE 18-1 Three species of bluebird** Despite their obvious similarity, these three species of bluebird—from left to right, the eastern bluebird (*Sialia sialis*), the western bluebird (*Sialia mexicana*), and the mountain bluebird (*Sialia currucoides*)—remain distinct because they do not interbreed.

In a scientific name, the genus name appears first, followed by the species name. By convention, scientific names are always underlined or *italicized*. The first letter of the genus name is always capitalized, and the first letter of the species name is always lowercase. The species name is never used alone but is always paired with its genus name.

Each two-part scientific name is unique, so referring to an organism by its scientific name rules out any chance of ambiguity or confusion. For example, the bird *Gavia immer* is commonly known in North America as the common loon, in Great Britain as the northern diver, and by still other names in non-English-speaking countries. But the Latin scientific name *Gavia immer* is recognized by biologists worldwide, overcoming language barriers and allowing precise communication.

Classification Originated as a Hierarchy of Categories

In addition to devising a method for naming species, Linnaeus also developed a method for classifying them. He placed each species into a series of hierarchically arranged categories on the basis of its resemblance to other species. The categories form a nested hierarchy in which each level includes all of the other levels below it.

The Linnaean classification system came to include eight major categories, or *taxonomic ranks*: **domain, kingdom, phylum, class, order, family,** genus, and species. Because the ranks form a nested hierarchy, each domain contains a number of kingdoms; each kingdom contains a number of phyla; each phylum includes a number of classes; each class includes a number of orders; and so on. As we move down the hierarchy, smaller and smaller groups are included.

Modern Classification Emphasizes Patterns of Evolutionary Descent

Prior to the 1859 publication of Darwin's *On the Origin of Species*, classification served mainly to facilitate the study and discussion of organisms, much as a library catalog facilitates our ability to find a book. But after Darwin demonstrated that all organisms are linked by common ancestry, biologists began to recognize that classification ought to reflect and describe the pattern of evolutionary relatedness among organisms. Today, the process of classification focuses almost exclusively on reconstructing **phylogeny,** or evolutionary history. The science of reconstructing phylogeny is known as **systematics.** Systematists communicate their findings about phylogeny by constructing evolutionary trees (see Fig. 16-11).

As systematists have increasingly focused their efforts on building evolutionary trees, they have de-emphasized the Linnaean classification system. Systematists still name groups, which they call **clades,** that include species linked by descent from a common ancestor. Clades, like the taxonomic ranks in the Linnaean classification system, can be arranged in a hierarchy, with smaller clades nested within larger ones (**Fig. 18-2**). Many systematists, however, do not assign taxonomic ranks to the clades they name. These systematists concentrate on using data to construct accurate evolutionary trees, rather than on more subjective evaluations of whether a given clade should be called a kingdom, a phylum, a class, an order, or a family. As a result, use of Linnaean taxonomic ranks is declining. Nonetheless, the Linnaean ranks, with their long tradition of use in classical taxonomy, still appear in much scientific discourse.

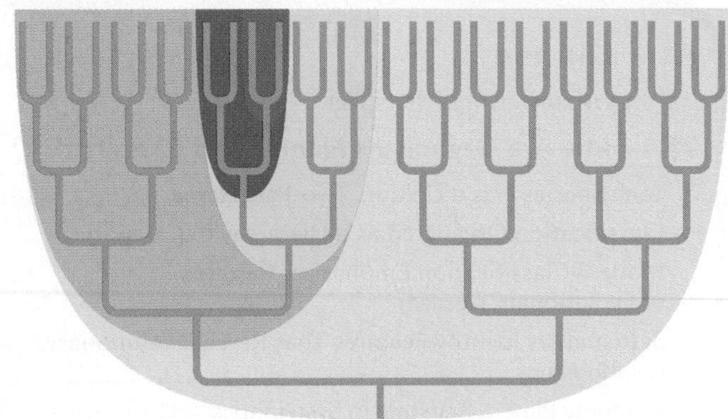

▲ **FIGURE 18-2 Clades form a nested hierarchy** Any group that includes all the descendants of a common ancestor is a clade. Some of the clades represented on this evolutionary tree are outlined in different colors. Note that smaller clades nest within larger clades.

Systematists Identify Features That Reveal Evolutionary Relationships

Systematists seek to reconstruct the tree of life, but they must do so without much direct knowledge of evolutionary history. Because systematists can't see into the past, they must infer the past as best they can, on the basis of similarities among living organisms. Not just any similarity will do, however. Some observed similarities stem from convergent evolution (see pp. 276–277) in organisms that are not closely related, and such similarities are not useful for inferring evolutionary history. Instead, systematists use similarities that exist because two kinds of organisms both inherited a characteristic from a common ancestor. Therefore, the scientists who devise classifications must distinguish informative similarities caused by common ancestry from uninformative similarities that result from convergent evolution. In the search for informative similarities, biologists look at many kinds of characteristics.

Anatomy Plays a Key Role in Systematics

Historically, the most important and useful distinguishing characteristics have been anatomical. Systematists look carefully at similarities in both external body structure (see Fig. 18-1) and internal structures, such as skeletons and muscles. For example, homologous structures such as the finger bones of dolphins, bats, seals, and humans provide evidence of a common ancestor (see Fig. 14-8). To detect relationships between more closely related species, biologists may use microscopes to discern finer details—for example, the number and shape of the "teeth" on the tongue-like radula of a snail, the shape and position of the bristles on a marine worm, or the external structure of pollen grains of a flowering plant (**Fig. 18-3**).

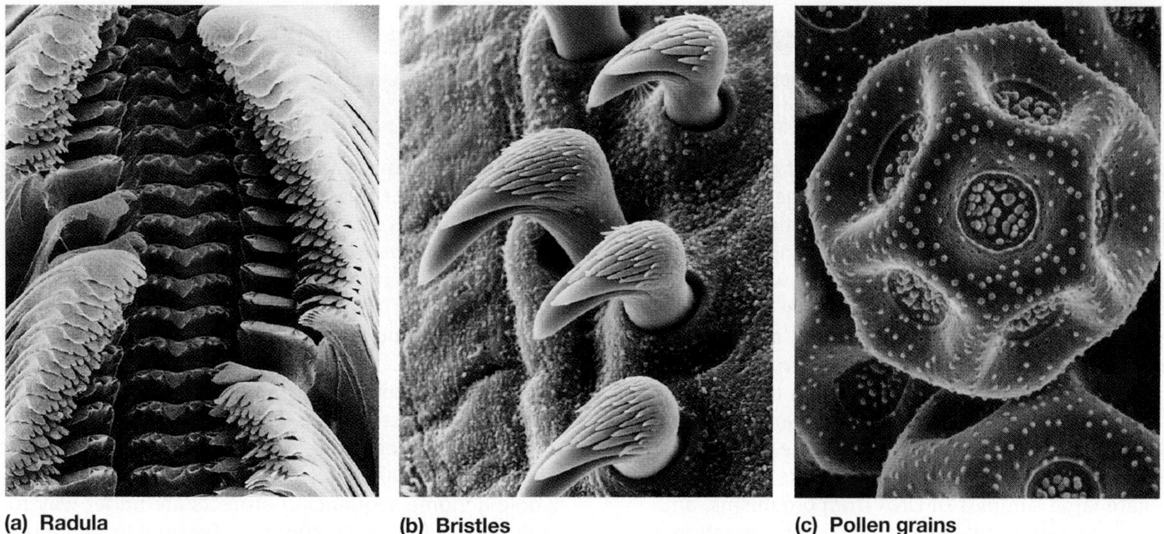

▲ **FIGURE 18-3 Microscopic structures may be used to classify organisms (a)** The "teeth" on a snail's tongue-like radula (a structure used in feeding), **(b)** the bristles on a marine worm, and **(c)** the shape and surface features of pollen grains are characteristics that are potentially useful in classification. Such finely detailed structures can reveal similarities between species that are not apparent in larger and more obvious structures.

Molecular Similarities Are Also Useful for Reconstructing Phylogeny

The anatomical characteristics shared by related organisms are expressions of underlying genetic similarities, so it stands to reason that evolutionary relationships among species must also be reflected in genetic similarities. Of course, direct genetic comparisons were not possible for most of the history of biology. Since the 1980s, however, advances in the techniques of molecular genetics have revolutionized studies of evolutionary relationships.

As a result of these technical advances, today's systematists can use the nucleotide sequences of DNA (that is, organisms' genotypes) to investigate relatedness among different types of organisms. Closely related species have similar DNA sequences. In some cases, similarity of DNA sequences will be reflected in the structure of chromosomes. For example, both the DNA sequences and the chromosomes of chimpanzees and humans are extremely similar, showing that these two species are very closely related (**Fig. 18-4**). Some of the key methods of genetic analysis are explored in "Scientific Inquiry: Molecular Genetics Reveals Evolutionary Relationships" on p. 346. The process by which systematists use genetic and anatomical similarities to

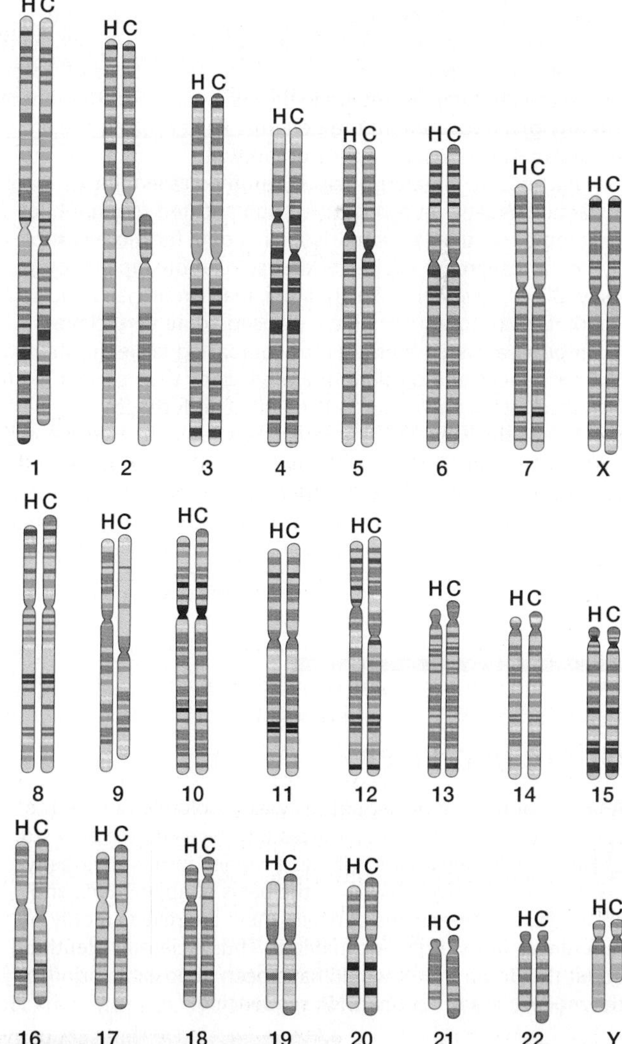

► **FIGURE 18-4 Human and chimp chromosomes are similar** Chromosomes from different species can be compared by means of banding patterns that are revealed by staining. The comparison illustrated here, between human chromosomes (left member of each pair; H) and chimpanzee chromosomes (C), reveals that the two species are genetically very similar. In fact, the entire genomes of both species have been sequenced and are nearly 99% identical. The numbering system shown is that used for human chromosomes; note that human chromosome 2 corresponds to a combination of two chimp chromosomes.

Scientific Inquiry

Molecular Genetics Reveals Evolutionary Relationships

Evolution results from the accumulation of inherited changes in populations. Because DNA is the molecule of heredity, evolutionary changes must be reflected in changes in DNA. Systematists have long known that comparing DNA within a group of species would be a powerful method for inferring evolutionary relationships, but for most of the history of systematics, direct access to genetic information was nothing more than a dream. Today, however, **DNA sequencing**—determining the sequence of nucleotides in segments of DNA—is comparatively cheap, easy, and widely available. The *polymerase chain reaction* (PCR; see pp. 244–245) allows systematists to easily accumulate large samples of DNA from organisms, and automated machinery makes sequence determination a comparatively simple task. Sequencing has rapidly become one of the primary tools for uncovering phylogeny.

The logic underlying molecular systematics is straightforward. It is based on the observation that when a single species divides into two species, the gene pool of each resulting species begins to accumulate mutations. The particular mutations in each species, however, will differ because the species are now evolving independently, with no gene flow between them. As time passes, more and more genetic differences accumulate. So, a systematist who has obtained DNA sequences from representatives of both species can compare the two species' nucleotide sequences at any given location in the genome. Fewer differences indicate more closely related organisms.

Putting the simple principles outlined above into practice usually involves more sophisticated thinking. For example, sequence comparisons become far more complex when a researcher wishes to assess relationships among, say, 20 or 30 species. Fortunately, mathematicians and computer programmers have devised some very clever computer-assisted methods for comparing large numbers of sequences and deriving the phylogeny that is most likely to account for the observed sequence differences.

Molecular systematists must also use care in choosing which segment of DNA to sequence. Different parts of the genome evolve at different rates, and it is crucial to sequence a DNA segment whose rate of change is well matched to the phylogenetic question at hand. In general, slowly evolving genes work best for comparing distantly related organisms, and rapidly changing portions of the genome are best for analyzing closer relationships. It is sometimes difficult to find any single gene that will yield sufficient information to provide an accurate picture of evolutionary change across the genome, so sequences from several different genes are often needed to construct reliable phylogenies, such as the one shown in **Figure E18-1.**

Today, sequence data are piling up with unprecedented rapidity, and systematists have access to sequences from an ever-increasing number of species. The entire genomes of more than 700 species have been sequenced, and whole-genome sequencing projects are under way for an additional 3,000 or so species. The Human Genome Project has been completed, and human DNA sequences are now a matter of public record. The revolution in molecular biology has fostered a great leap forward in our understanding of evolutionary history.

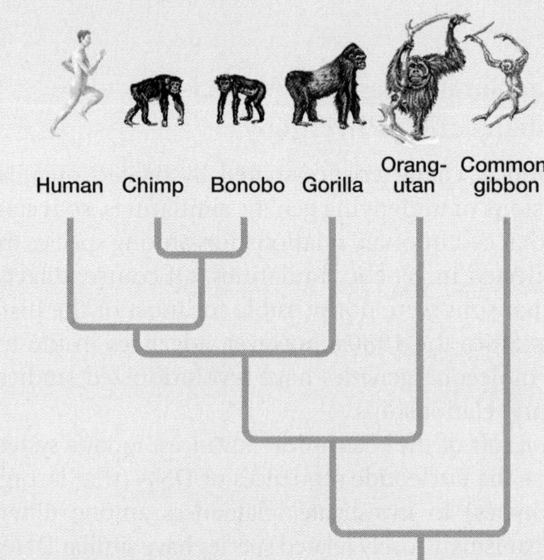

▲ **FIGURE E18-1 Relatedness can be determined by comparing DNA sequences** This evolutionary tree was derived from the nucleotide sequences of several different genes that are common to humans and apes.

Case Study continued

Origin of a Killer

Analysis of nucleotide sequences was absolutely required to construct the phylogeny of viruses that revealed the ancestor of HIV. The closely related viruses included in the phylogeny are all but indistinguishable on the basis of appearance and structure; the differences between them are revealed only by differences in their genetic material. Thus, scientific sleuthing about the origin of HIV would have been impossible prior to the modern era of routine DNA sequencing.

reconstruct evolutionary history are discussed in "A Closer Look at Phylogenetic Trees" on pp. 348–349. Some findings about genetic variation in humans are described in "Links to Everyday Life: Small World."

18.2 WHAT ARE THE DOMAINS OF LIFE?

If we picture the common ancestor of all living things as the trunk at the very base of the tree of life, we might ask: Which clades arose from the earliest branching of the trunk? Each of the earliest branches must have given rise to a huge clade of descendant species. These early-branching clades will be the largest ones that systematists can distinguish.

Before about 1970, systematists concluded from the available evidence that the earliest split in the tree of life divided all species into two groups: Animalia and Plantae. According to this view, bacteria, fungi, and photosynthetic eukaryotes were considered to be plants, and all other organisms were classified as animals. However, as knowledge of life's evolutionary history expanded, it became apparent that this early view had vastly oversimplified evolutionary history.

A Three-Domain System More Accurately Reflects Life's History

As new data accumulated and understanding of phylogeny grew, scientific assessment of life's fundamental categories was gradually revised. A key element of this revision stemmed from the pioneering work of microbiologist Carl Woese, who showed that biologists had overlooked a key event in the early history of life, one that demanded a new and more accurate classification of life.

Woese and other biologists interested in the evolutionary history of microorganisms studied the biochemistry of prokaryotic organisms. The researchers, by studying nucleotide sequences of the RNA that is found in organisms' ribosomes, discovered that prokaryotes actually include two very different kinds of organisms. Woese dubbed these two groups the Bacteria and the Archaea (**Fig. 18-5**).

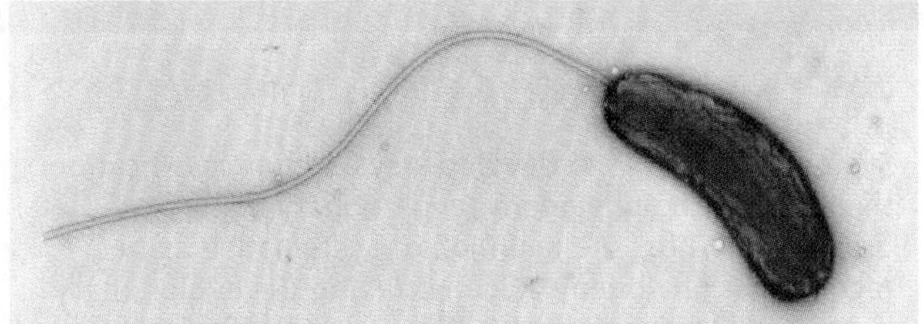

(a) A bacterium

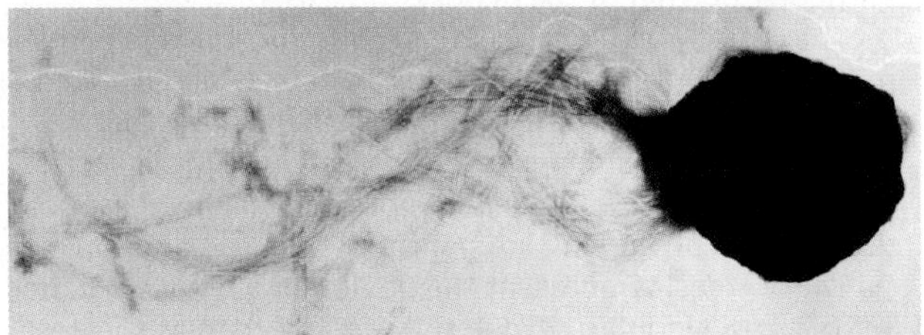

(b) An archaean

▲ **FIGURE 18-5 Two domains of prokaryotic organisms** Although similar in appearance, **(a)** *Vibrio cholerae* and **(b)** *Methanococcus jannaschii* are less closely related than a mushroom and an elephant. *Vibrio* is in the domain Bacteria, and *Methanococcus* is in Archaea.

Despite superficial similarities in their appearance under the microscope, Bacteria and Archaea are radically different. The two groups are no more closely related to one another than either one is to any eukaryote. The tree of life split into three parts very early in the history of life, long before the appearance of plants, animals, and fungi. This early

Links to *Everyday Life*

Small World

In light of humankind's intense curiosity about the origin of our species, it is not surprising that systematists have devoted a great deal of attention to uncovering the evolutionary history of *Homo sapiens*. Though much of this inquiry has focused on revealing the evolutionary connections between modern humans and the species to which we are most closely related, the techniques and methods of systematics have also been used to assess the evolutionary relationships among different populations within our species. Biologists have compared DNA sequences from human populations in many different parts of the world; different investigators have compared different portions of the human genome. As a result, a large amount of data has been gathered, and some interesting findings have emerged.

First, genetic divergence among human populations is very low compared to that of other animal species. For example, the range of genetic differences among all of Earth's humans is only one-tenth the size of the differences among the deer mice of North America (and many other species have even more genetic variability than deer mice). Clearly, all humans are genetically very similar, and the differences among different human populations are tiny.

It is also increasingly apparent that most of the human genetic variability that does exist can be found in African populations. The range of genetic differences found *within* sub-Saharan African populations is greater than the range between African populations and any non-African population. For many genes, all known variants are found in Africa, and no non-African population contains any distinctive variants; rather, non-African populations contain subsets of the African set. This finding strongly suggests that *Homo sapiens* originated in Africa, and that we have not lived anywhere else long enough to evolve very many differences from our African ancestors.

A Closer Look At *Phylogenetic Trees*

Systematists strive to develop a system of classification that reflects the phylogeny (evolutionary history) of organisms. Thus, the main job of systematists is to reconstruct phylogeny. Reconstructing the evolutionary history of all of Earth's organisms is, of course, a huge task, so each systematist typically chooses to work on some particular portion of the history.

The result of a phylogenetic reconstruction is usually represented by a diagram. These diagrams can take a number of different forms, but all of them show the sequence of branching events in which ancestral species split to give rise to descendant species. For this reason, diagrams of phylogeny are generally tree-like.

These trees can present the phylogeny of any specified set of *taxa* (singular, *taxon*). A taxon is a named species, such as *Homo sapiens*, or a named group of species, such as primates or beetles or ferns. Thus, phylogenetic trees can show evolutionary history at different scales. Systematists might reconstruct, for example, a tree of 10 species in a particular genus of clams, or a tree of 25 clades of animals, or a tree of the three domains of life.

After selecting the taxa to include, a systematist is ready to begin building a tree. Most systematists use the *cladistic* approach to reconstruct phylogenetic trees. Under the cladistic approach, relationships among taxa are revealed by the occurrence of similarities known as *synapomorphies*. A synapomorphy is a trait that is similar in two or more taxa because these taxa inherited a "derived" version of the trait that had changed from its original state in a common ancestor. The formation of synapomorphies is illustrated in **Figure E18-2.**

In the imaginary scenario illustrated in Figure E18-2, we can easily identify synapomorphies because we know the ancestral state of the trait and the subsequent changes that took place. In real life, however, systematists would not have direct knowledge of the ancestor, which lived in the distant past and whose identity is unknown. Without this direct knowledge, a systematist observing a similarity between two taxa is faced with a challenge. Is the observed similarity a synapomorphy, or does it have some other cause, such as convergent evolution or common inheritance of the ancestral state? The cladistic approach provides methods for identifying synapomorphies, but the possibility of mistaken interpretation remains. To guard against such errors, systematists use numerous traits to build a tree, thereby minimizing the influence of any single trait.

In the last phase of the tree-building process, the systematist compares different possible trees. For example, three taxa can be arranged in three different branching patterns (**Fig. E18-3**). Each branching pattern represents a different hypothesis about the evolutionary history of taxa A, B, and C. Which hypothesis is most likely to represent the true history of the three taxa? The one in which the taxa on adjacent branches share synapomorphies. For example, imagine that a systematist identified a number of synapomorphies that are shared by taxa A and B but are not found in C, but has found no synapomorphies that link taxa B and C or taxa A and C. In this case, tree 1 in Figure E18-3 is the best-supported hypothesis.

With large numbers of taxa, the number of possible trees grows dramatically. Similarly, a large number of traits also complicates the problem of identifying the tree best supported by the data. Fortunately, however, systematists have developed sophisticated computer programs to help cope with these complications.

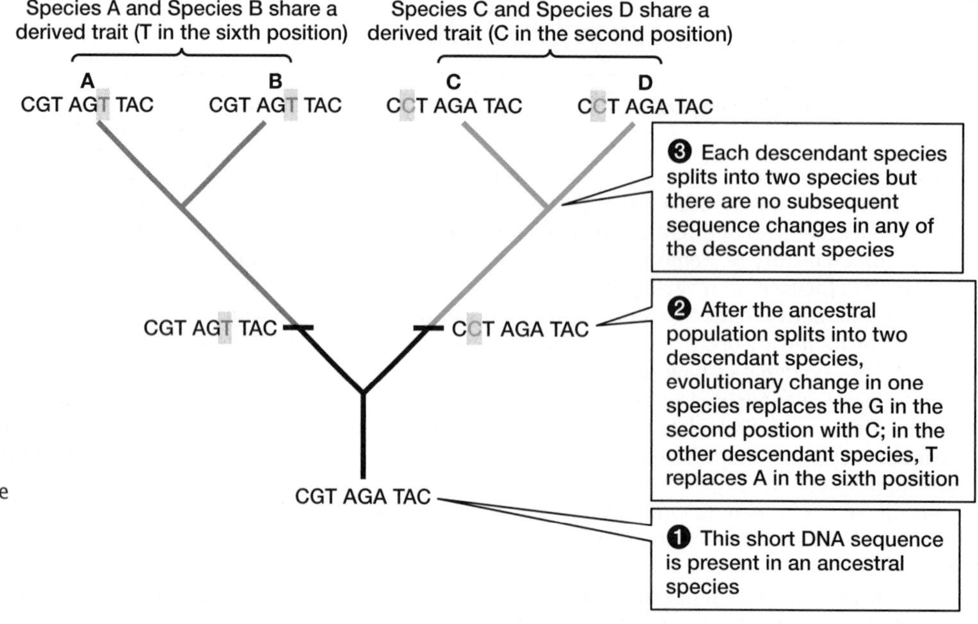
Species A and Species B share a derived trait (T in the sixth position)
Species C and Species D share a derived trait (C in the second position)
A
CGT AGT TAC
B
CGT AGT TAC
C
CCT AGA TAC
D
CCT AGA TAC
❸ Each descendant species splits into two species but there are no subsequent sequence changes in any of the descendant species
CGT AGT TAC
CCT AGA TAC
❷ After the ancestral population splits into two descendant species, evolutionary change in one species replaces the G in the second postion with C; in the other descendant species, T replaces A in the sixth position
CGT AGA TAC
❶ This short DNA sequence is present in an ancestral species

▶ **FIGURE E18-2 Related taxa are linked by shared derived traits (synapomorphies)** A derived trait is one that has been modified from the ancestral version of the trait. When two or more taxa share a derived trait, the shared trait is said to be a synapomorphy. The hypothetical scenario illustrated here shows how synapomorphies arise.

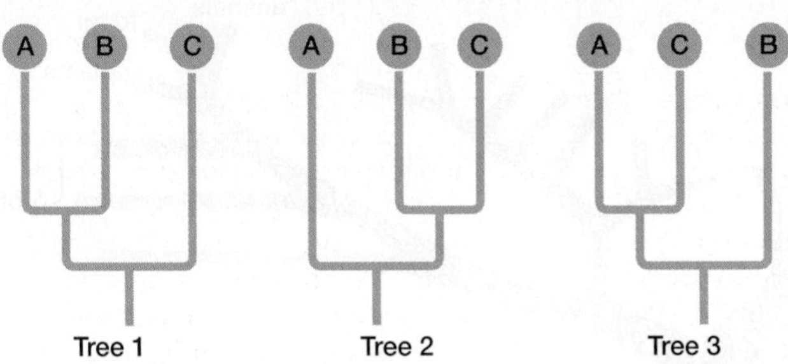

▲ **FIGURE E18-3 The three possible trees for three taxa**

Under the cladistic approach, phylogenetic trees play a key role in classification. Each designated group should contain only organisms that are more closely related to one another than to any organisms outside the group. So, for example, the members of the clade Canidae (which includes dogs, wolves, foxes, and coyotes) are more closely related to each other than to any member of any other clade. Another way to state this principle is to say that each designated group should contain *all* of the living descendants of a common ancestor (**Fig. E18-4a**). In the terminology of cladistic systematics, such groups are said to be *monophyletic*.

Some names, especially names that predate the cladistic approach, designate groups that contain some, but not all, of the descendants of a common ancestor. Such groups are *paraphyletic*. For example, the group historically known as the reptiles—snakes, lizards, turtles, and crocodilians—is paraphyletic. To see why, examine the tree in **Figure E18-4b.** Find the branch that represents the common ancestor of snakes, lizards, turtles, and crocodilians (it is at the base of the tree). Then examine the tree again and make a list of all of the descendants of that common ancestor. Your list, if you performed this mental exercise correctly, includes the birds. That is, birds are part of the monophyletic group that includes all living descendants of the common ancestor that gave rise to snakes, lizards, turtles, and crocodilians. Therefore, the reptiles (Reptilia) constitute a monophyletic clade only if birds are included in the group. If we omit the birds, Reptilia is paraphyletic and, according to cladistic principles, is not a valid group name. Nonetheless, you will probably continue to encounter the word "reptiles" used in its older, technically incorrect sense, if only because so many people are accustomed to using it that way.

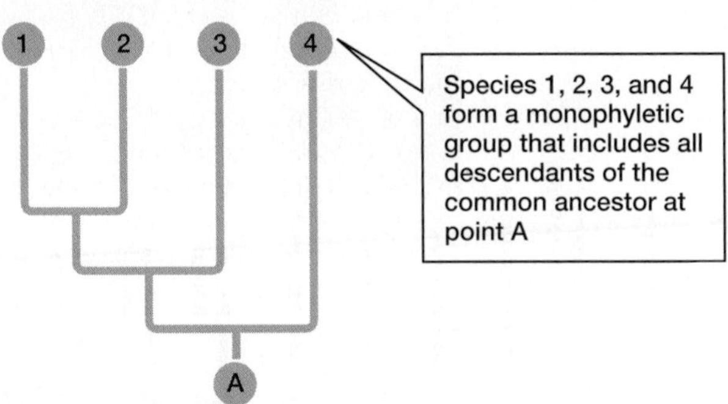

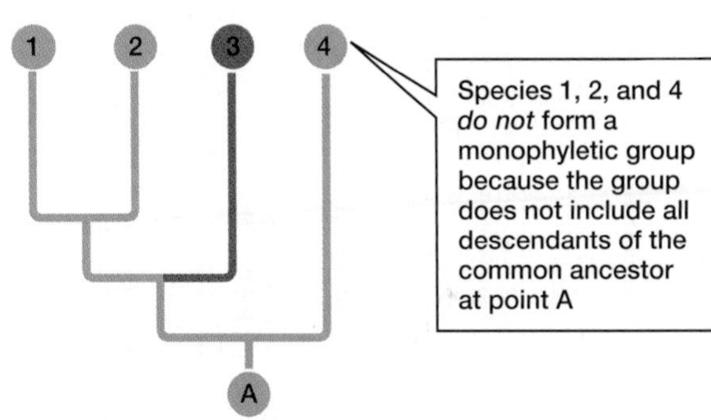

(a) Monophyletic and paraphyletic groups

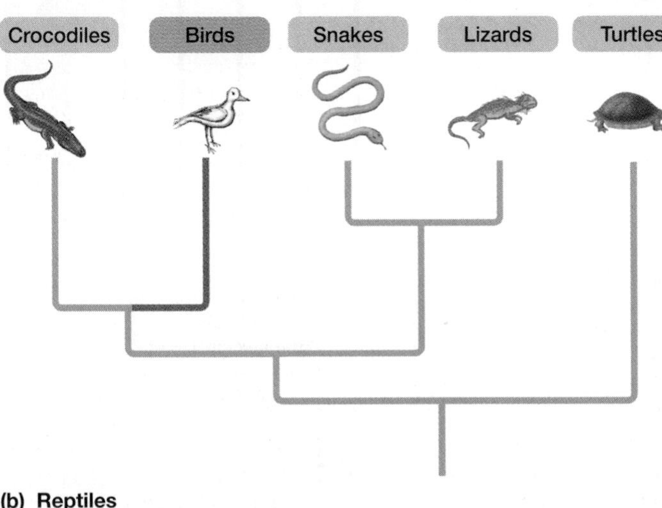

(b) Reptiles

▲ **FIGURE E18-4 Reptiles are a monophyletic group only if birds are included** Only groups that contain all of the descendants of a common ancestor are considered to be monophyletic groups.

EXERCISE Consider the following list of groups: (1) protists, (2) fungi, (3) great apes (chimpanzees, pygmy chimpanzees, gorillas, orangutans, and gibbons), (4) seedless plants (ferns, mosses, and liverworts), (5) prokaryotes (bacteria and archaea), and (6) animals. Using Figures 18-6, 18-7, E18-1, 21-5, 22-4, and 23-1 for reference, identify the monophyletic groups on the list.

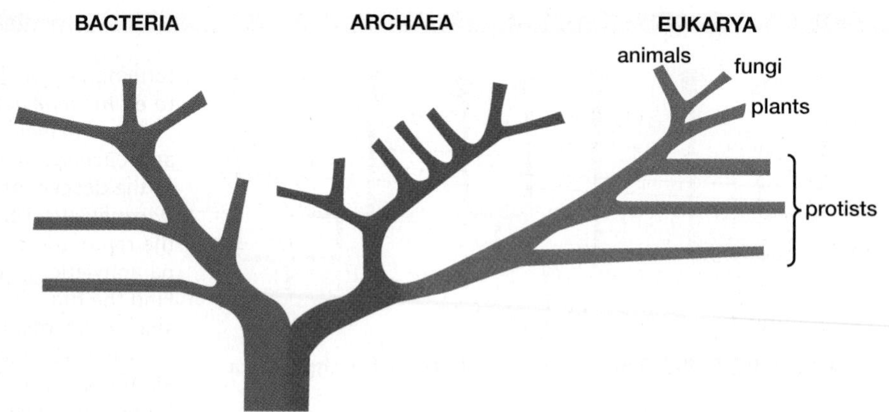

▶ **FIGURE 18-6 The tree of life** The three domains of life represent the three main "trunks" on the tree of life.

split is now reflected in a classification scheme that divides life into three domains: **Bacteria, Archaea,** and **Eukarya** (Fig. 18-6).

This text's descriptions of the diversity of life—which appear in Chapters 19 through 24—begin with an overview of the prokaryotic domains Archaea and Bacteria. Within the domain Eukarya, we provide separate discussions of three clades—plants, fungi, and animals—and of "protists" (a generic term that designates the diverse collection of eukaryote organisms that are not plants, fungi, or animals). **Figure 18-7** shows the evolutionary relationships among some members of the domain Eukarya.

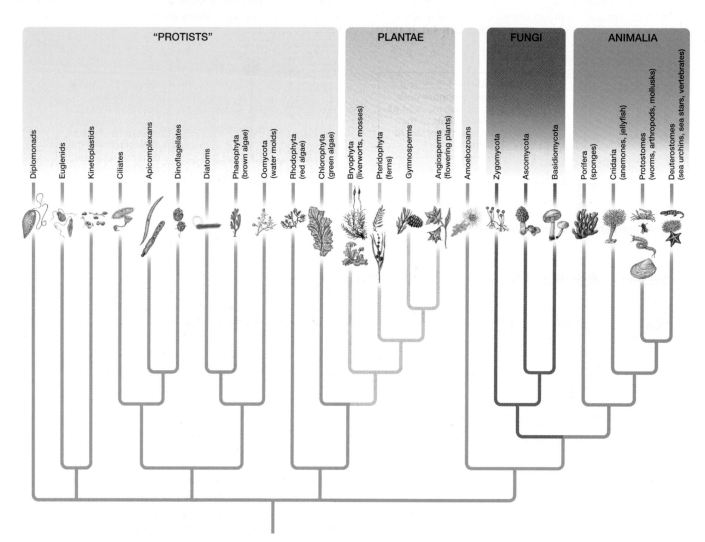

▲ **FIGURE 18-7 An evolutionary tree of eukaryotes** Some of the major evolutionary lineages within the domain Eukarya are shown. The term "protist" refers to the many eukaryotes that are not plants, animals, or fungi.

18.3 WHY DO CLASSIFICATIONS CHANGE?

As the emergence of the three-domain system shows, the hypotheses of evolutionary relationships on which classification is based are subject to revision as new data become available. Even the largest, most inclusive clades—which represent old branchings of the tree of life—must sometimes be rearranged. Such changes at the top levels of classification occur only rarely, but at the other end of the classification hierarchy, among species designations, revisions are more frequent.

Species Designations Change When New Information Is Discovered

As researchers uncover new information, systematists regularly propose changes in species-level classifications. For example, until recently, systematists recognized two species of elephant, the African elephant and the Indian elephant. Now, however, they recognize three elephant species; the former African elephant is now divided into two species, the savanna elephant and the forest elephant. Why the change? Genetic analysis of elephants in Africa revealed that there is little gene flow between forest-dwelling and savanna-dwelling elephants. The two groups of elephants are no more genetically similar than lions and tigers.

The Biological Species Definition Can Be Difficult or Impossible to Apply

In some cases, systematists find themselves unable to say with certainty where one species ends and another begins. As discussed in Chapter 16, asexually reproducing organisms pose a particular challenge to systematists, because the criterion of interbreeding (the basis of the biological species definition that we have used in this text) cannot be used to distinguish among species. The irrelevance of this criterion in studies of asexual organisms leaves plenty of room for investigators to disagree about which asexual populations constitute a species, especially when comparing groups with similar phenotypes. For instance, some systematists recognize 200 species of the British blackberry (a plant that can produce seeds parthenogenetically—that is, without fertilization), but others recognize only 20 species.

The difficulty of applying the biological species definition to asexual organisms is a serious problem for systematists. After all, a significant portion of Earth's organisms reproduces without sex. Most bacteria, archaea, and protists, for example, reproduce asexually most of the time. Some systematists argue that we need a more universally applicable definition of species, one that won't exclude asexual organisms and that doesn't depend on the criterion of reproductive isolation.

The Phylogenetic Species Concept Offers an Alternative Definition

A number of alternative species definitions have been proposed over the history of evolutionary biology, but none has been sufficiently compelling to displace the biological species definition. One alternative definition, however, has been gaining adherents in recent years. The *phylogenetic species concept* defines a species as "the smallest diagnosable group that contains all the descendants of a single common ancestor." In other words, if we draw an evolutionary tree that describes the pattern of ancestry among a collection of organisms, each distinctive branch on the tree constitutes a separate species, regardless of whether the individuals represented by that branch can interbreed with individuals from other branches. As you might suspect, rigorous application of the phylogenetic species concept would vastly increase the number of different species recognized by systematists.

Proponents and opponents of the phylogenetic species concept are currently engaged in a vigorous debate about its merits. Perhaps one day it will replace the biological species concept as the "textbook definition" of species, or perhaps we will continue to rely on the biological species concept. In the meantime, classifications will continue to be debated and revised as systematists learn more and more about evolutionary relationships, particularly with the application of techniques derived from molecular biology. Although the precise evolutionary relationships of many organisms continue to elude us, classification is enormously helpful in ordering our thoughts and investigations into the diversity of life on Earth.

18.4 HOW MANY SPECIES EXIST?

Scientists do not know even within an order of magnitude how many species share our world. Each year, between 7,000 and 10,000 new species are named, most of them insects, many from tropical rain forests. The total number of named species is currently about 1.5 million. However, many scientists believe that 7 million to 10 million species may exist, and estimates range as high as 100 million. This total range of species diversity is known as **biodiversity.** Of all the species that have been identified thus far, about 5% are prokaryotes and protists. An additional 22% are plants and fungi, and the rest are animals. This distribution has little to do with the actual abundance of these organisms and a lot to do with the size of the organisms, how easy they are to classify, how accessible they are, and the number of scientists studying them. Historically, systematists have chiefly focused on large or conspicuous organisms in temperate regions, but biodiversity is greatest among small, inconspicuous organisms in the Tropics. In addition to the overlooked species on land and in shallow waters, an entire "continent" of species lies largely unexplored on the deep-sea floor. From the limited samples available, scientists estimate that hundreds of thousands of unknown species may reside there.

Although about 5,000 species of prokaryotes have been described and named, prokaryotic diversity remains largely unexplored. Consider a study by Norwegian scientists, who analyzed DNA to count the number of different bacteria species present in a small sample of forest soil. To distinguish among species, the researchers arbitrarily defined bacterial DNA as coming from separate species if it differed by at least 30% from that of any other bacterial DNA in the sample. Using this criterion, they reported more than 4,000 species of bacteria in their soil sample and an equal number of species in a sample of shallow marine sediment.

Our ignorance of the full extent of life's diversity adds a new dimension to the tragedy of the destruction of tropical rain forests. Although these forests cover only about 6% of Earth's land area, they are believed to be home to two-thirds of the world's existing species, most of which have never been studied or named. Because these forests are being destroyed so rapidly, Earth is losing many species that we will never even know existed! For example, in 1990, a new species of primate, the black-faced lion tamarin, was discovered in a small patch of dense rain forest on an island just off the east coast of Brazil (**Fig. 18-8**). Had the patch of forest been cut before this squirrel-sized monkey was discovered, its existence would have remained undocumented. At current rates of deforestation, most of the tropical rain forests, with their undescribed wealth of life, will be gone within the next century.

▶ **FIGURE 18-8 The black-faced lion tamarin** Researchers estimate that no more than 400 individuals remain in the wild; captive breeding may be the black-faced lion tamarin's only hope for survival.

Case Study revisited

Origin of a Killer

What evidence has persuaded evolutionary biologists that HIV originated in apes and monkeys? To understand the evolutionary thinking behind this conclusion, examine the evolutionary tree shown in **Figure 18-9.** This tree illustrates the phylogeny of HIV and its close relatives, the simian immunodeficiency viruses (SIVs), as revealed by a comparison of RNA sequences among different viruses.

Notice the positions on the tree of the four human viruses (two strains of HIV-1 and two of HIV-2; a strain is a genetically distinct subgroup of a particular type of virus). The branch leading to strain 1 of HIV-1 is directly adjacent to the branch leading to strain 1 of chimpanzee SIV. The adjacent branches indicate that strain 1 of HIV-1 is more closely related to a chimpanzee virus than to strain 2 of HIV-1. Similarly, strain 1 of HIV-2 is more closely related to pig-tailed macaque SIV than to strain 2 of HIV-2. Both HIV-1 and HIV-2 are more closely related to ape or monkey viruses than to one another.

The only way for the evolutionary history shown in the tree to have emerged is if viruses jumped between host species. If HIV had evolved strictly within human hosts, the human viruses would be each other's closest relatives. Because the human viruses do not cluster together on the phylogenetic tree, we can infer that cross-species infection occurred, probably on multiple occasions. The most likely means of transmission is human consumption of monkeys (HIV-2) and chimpanzees (HIV-1). Recently, a strain of SIV that is especially closely related to HIV-1 was found in members of a population of chimps that inhabit forests in the southeastern corner of the Central African country of Cameroon. It is likely that viruses from this chimp population made the chimp-to-human jump that started the HIV epidemic.

Consider This

Can understanding the evolutionary origin of HIV help researchers devise better ways to treat and control the spread of AIDS? How might such understanding influence strategies for treatment and prevention? More generally, how can evolutionary thinking help advance medical research?

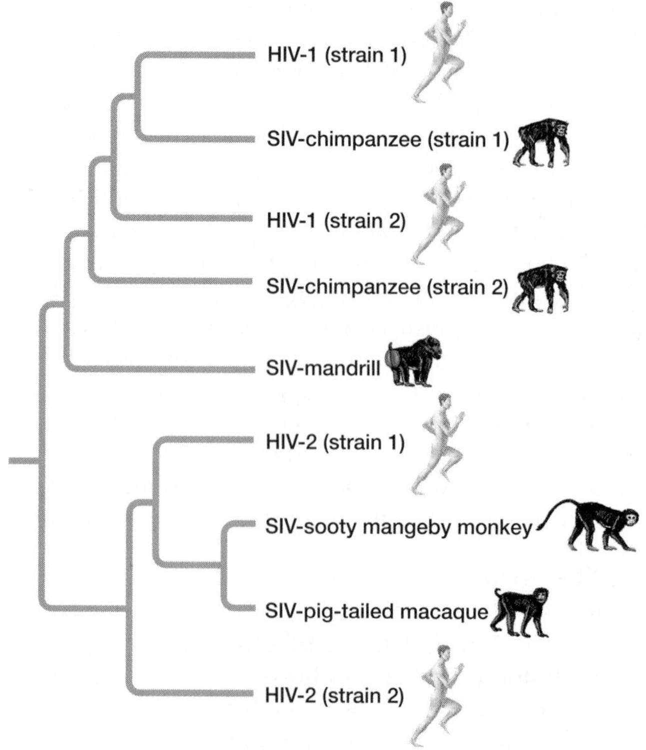

▶ **FIGURE 18-9 Evolutionary analysis helps reveal the origin of HIV** In this phylogeny of some immunodeficiency viruses, the viruses with human hosts do not cluster together. This lack of congruence between the evolutionary histories of the viruses and their host species suggests that the viruses must have jumped between host species. (Note that SIV stands for *simian immunodeficiency virus.*)

CHAPTER REVIEW

Summary of Key Concepts

18.1 How Are Organisms Named and Classified?

The scientific name of an organism is composed of its genus name and species name. Systematists use anatomical and molecular similarities among organisms to reconstruct the evolutionary relationships among species, and depict the results of their reconstructions in tree diagrams. On the basis of these evolutionary trees, systematists name clades (groups that include the species descended from a common ancestor). Clades nest within larger clades to form a nested hierarchy of categories. In Linnaean classification, the different clades in a hierarchy are assigned taxonomic ranks. The eight major ranks, in order of decreasing inclusiveness, are domain, kingdom, phylum, class, order, family, genus, and species.

18.2 What Are the Domains of Life?

The three domains of life, each representing one of three main branches of the tree of life, are Bacteria, Archaea, and Eukarya. Within the domain Eukarya, Fungi, Plantae, and Animalia are monophyletic clades. The assemblage known as "protists" comprises a variety of distinct clades.

18.3 Why Do Classifications Change?

Classifications are subject to revision as new information is discovered. Species boundaries may be hard to define, particularly in the case of asexually reproducing species. However, systematics is essential for precise communication and contributes to our understanding of the evolutionary history of life.

18.4 How Many Species Exist?

Although only about 1.5 million species have been named, estimates of the total number of species range up to 100 million. New species are being identified at the rate of 7,000 to 10,000 annually, mostly in tropical rain forests.

Key Terms

Archaea *350*
Bacteria *350*
biodiversity *351*
clade *344*
class *344*
DNA sequencing *346*
domain *344*
Eukarya *350*
family *344*
genus *343*
kingdom *344*
order *344*
phylogeny *344*
phylum *344*
scientific name *343*
species *343*
systematics *344*
taxonomy *343*

Thinking Through the Concepts

Fill-in-the-Blank

1. The science of naming and classifying organisms is called __________. The related science of reconstructing and depicting evolutionary history is called __________. A group consisting of all organisms descended from a particular common ancestor is a(n) __________.
2. A scientific name consists of a(n) __________ name followed by a(n) __________ name. Both parts of a scientific name are in the __________ language. The first letter of the first word in a scientific name is always __________, and both parts of the name are written in __________ letters.
3. In Linnaean classification, the eight major taxonomic ranks, in descending order of inclusiveness, are __________, __________, __________, __________, __________, __________, __________, and __________. The three domains of life are __________, __________, and __________.
4. Systematists determine the evolutionary relationships among species mainly on the basis of similarities in __________ and __________. The biological species definition is difficult to apply to organisms that __________. An alternative species definition, known as the __________, instead designates species on the basis of __________.
5. The number of named species is about__________, but the actual number of species on Earth is estimated to be between __________ and __________.

Review Questions

1. What contributions did Linnaeus and Darwin make to modern taxonomy?
2. What features would you study to determine whether a dolphin is more closely related to a fish or to a bear?
3. What techniques might you use to determine whether the extinct cave bear is more closely related to a grizzly bear or to a black bear?
4. Only a small fraction of the total number of species on Earth has been scientifically described. Why?
5. In England, "daddy longlegs" refers to a long-legged fly, but the same name refers to a spider-like animal in the United States. How do scientists attempt to avoid such confusion?

Applying the Concepts

1. There are many areas of disagreement about the classification of organisms. For example, there is no consensus about whether the red wolf is a distinct species or about how many kingdoms are within the domain Bacteria. What difference does it make whether biologists consider the red wolf a species, or into which kingdom a bacterial species falls? As Shakespeare put it, "What's in a name?"
2. **BioEthics** The pressures created by human population growth and economic expansion place storehouses of biological diversity such as the Tropics in peril. The seriousness of the situation is clear when we consider that probably only 1 out of every 20 tropical species is known to science at present. What arguments can you make for preserving biological diversity in poor

and developing countries, such as those in many areas of the Tropics? Does such preservation require that these countries sacrifice economic development? Suggest some solutions to the conflict between the growing demand for resources and the importance of conserving biodiversity.

3. During major floods, only the topmost branches of submerged trees may be visible above the water. If you were asked to sketch the branches below the surface of the water solely on the basis of the positions of the exposed tips, you would be attempting a reconstruction somewhat similar to the "family tree" by which taxonomists link various organisms according to their common ancestors (analogous to branching points). What sources of error do both exercises share? What advantages do modern taxonomists have?

4. The Florida panther, found only in the Florida Everglades, is currently classified as an endangered species, protecting it from human activities that could lead to its extinction. It has long been considered a subspecies of cougar (mountain lion), but recent mitochondrial DNA studies have shown that the Florida panther may actually be a hybrid between American and South American cougars. Should the Florida panther be protected by the Endangered Species Act?

MB® *Go to www.masteringbiology.com for practice quizzes, activities, eText, videos, current events, and more.*

The Diversity of Prokaryotes and Viruses

Chapter 19

Case Study

Agents of Death

IN THE FALL OF 2001, a long-standing fear became a terrible reality when residents of the United States were attacked with a biological weapon. The weapon, which killed five people and caused six others to become gravely ill, was simply a culture of bacteria. The bacteria had been placed in envelopes and mailed to the Hart Office Building in Washington, D.C., and to media company offices, where they were unknowingly inhaled by victims who opened the harmless-looking envelopes. The attack, though comparatively small, nonetheless dramatically illustrated the potential destructive power of a larger attack.

The bacterium used in the attack was *Bacillus anthracis*, which causes the disease anthrax. Anthrax bacteria normally infect domestic animals such as goats and sheep, but can also infect humans. The bacterium is a dangerous, often deadly infectious agent with properties that make it especially attractive to developers of biological weapons. Anthrax bacteria are easily isolated from infected animals, are cheap and easy to culture in large quantities, and, once produced, can be dried into a powder that remains viable for years. The powder is easily "weaponized" by packing it into a missile warhead or other delivery device, and a small volume of bacteria can infect a very large number of people. Areas contaminated with anthrax bacteria are very difficult to decontaminate.

Since the attack, it has become apparent that much of our ability to defend against such biological assaults depends on our understanding of the microbes (as single-celled organisms are collectively known) that can be used as biological weapons. Scientific investigation of microbes can provide the knowledge required to detect an attack, destroy dangerous microbes in the environment, and prevent and treat infections. Fortunately, biologists already know quite a bit about microorganisms. In this chapter, we'll explore some of that knowledge.

▲ Workers prepare to decontaminate the Hart Office Building in Washington, D.C., after it was attacked with a biological weapon.

At a Glance

19.1 WHICH ORGANISMS ARE MEMBERS OF THE DOMAINS ARCHAEA AND BACTERIA?

Earth's first organisms were prokaryotes, single-celled organisms that lacked organelles such as the nucleus, chloroplasts, and mitochondria. (See Chapter 4 for a comparison of prokaryotic and eukaryotic cells.) For the first 1.5 billion years or more of life's history, all life was prokaryotic. Even today, prokaryotes are extraordinarily abundant. A drop of seawater contains hundreds of thousands of prokaryotic organisms, and a spoonful of soil contains billions. The average human body is home to trillions of prokaryotes, which live on the skin, in the mouth, and in the stomach and intestines. In terms of abundance, prokaryotes are Earth's predominant form of life.

Bacteria and Archaea Are Fundamentally Different

Two of life's three domains, **Bacteria** and **Archaea,** consist entirely of prokaryotes. Bacteria and archaea are superficially similar in appearance under the microscope, but have striking structural and biochemical differences that reveal the ancient evolutionary separation between the two groups. For example, the cell walls of bacterial cells contain molecules of *peptidoglycan*, a polysaccharide that also incorporates some amino acids, which helps strengthen the cell wall. Peptidoglycan is unique to bacteria, and the cell walls of archaea do not contain it. Bacteria and archaea also differ in the structure and composition of their plasma membranes, ribosomes, and RNA polymerases, as well as in the mechanics of basic processes such as transcription and translation.

Classification of Prokaryotes Within Each Domain Is Difficult

The sharp differences between archaea and bacteria make distinguishing the two domains a straightforward matter, but it is more challenging to identify clades (groups of species linked by descent from a common ancestor) within each domain. The challenge arises because prokaryotes are very small and structurally simple; they do not exhibit the huge array of anatomical differences that can be used to infer the evolutionary history of plants, animals, and other eukaryotes. Consequently, prokaryotes have historically been classified on the basis of such features as shape, means of locomotion, pigments, nutrient requirements, the appearance of colonies (groups of individuals that descended from a single cell), and staining properties. For example, the Gram stain, a staining technique, distinguishes two types of cell-wall construction in bacteria. Depending on the results of the stain, these bacteria are classified as either *gram-positive* or *gram-negative.*

In recent years, our understanding of evolutionary history within the prokaryotic domains has been greatly improved by comparisons of DNA and RNA nucleotide sequences. New sequence data have revealed that some aspects of earlier classifications of bacteria and archaea did not accurately reflect evolutionary history. Systematists have therefore begun to redraw the prokaryote evolutionary tree, but have not yet reached a consensus about the evolutionary relationships within Bacteria and Archaea. The effort to develop a robust classification of prokaryotes remains a work in progress.

Prokaryotes Differ in Size and Shape

Both bacteria and archaea are usually very small, ranging from about 0.2 to 10 micrometers in diameter. (In comparison, the diameters of eukaryotic cells range from about 10 to 100 micrometers.) About 250,000 average-sized bacteria or archaea could congregate on the period at the end of this sentence, though a few species of bacteria are larger. The largest known bacterium (*Thiomargarita namibiensis*) is as much as 700 micrometers in diameter, making it visible to the naked eye.

The cell walls that surround prokaryotic cells give characteristic shapes to different types of bacteria and archaea. The most common shapes are spherical, rodlike, and corkscrew-shaped (**Fig. 19-1**).

(a) Spherical

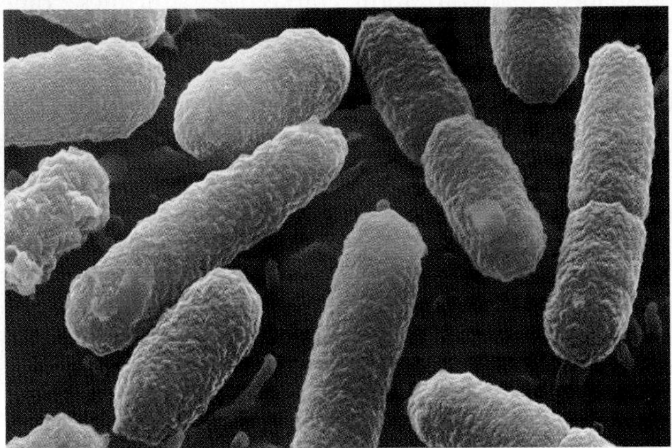

(b) Rod-shaped

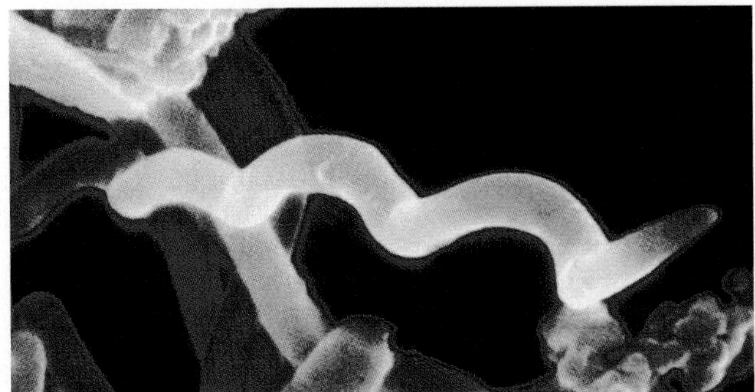

(c) Corkscrew-shaped

▲ FIGURE 19-1 **Three common prokaryote shapes** **(a)** Spherical bacteria of the genus *Micrococcus*, **(b)** rod-shaped bacteria of the genus *Escherichia*, and **(c)** corkscrew-shaped bacteria of the genus *Borrelia*.

19.2 HOW DO PROKARYOTES SURVIVE AND REPRODUCE?

The abundance of prokaryotes is due in large part to adaptations that allow members of the two prokaryotic domains to inhabit and exploit a wide range of environments. In this section, we discuss some of the traits that help prokaryotes survive and thrive.

Some Prokaryotes Are Motile

Many bacteria and archaea adhere to a surface or drift passively in liquid surroundings, but some are motile—they can move about. Many of these motile prokaryotes have **flagella** (singular, flagellum). Prokaryote flagella may appear singly at one end of a cell, in pairs (one at each end of the cell), as a tuft at one end of the cell (**Fig. 19-2a**), or scattered over the entire cell surface. Flagella can rotate rapidly, propelling the organism

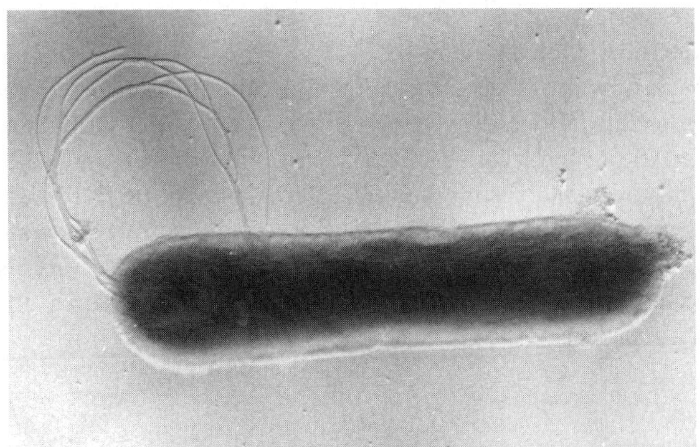

(a) A flagellated archaean

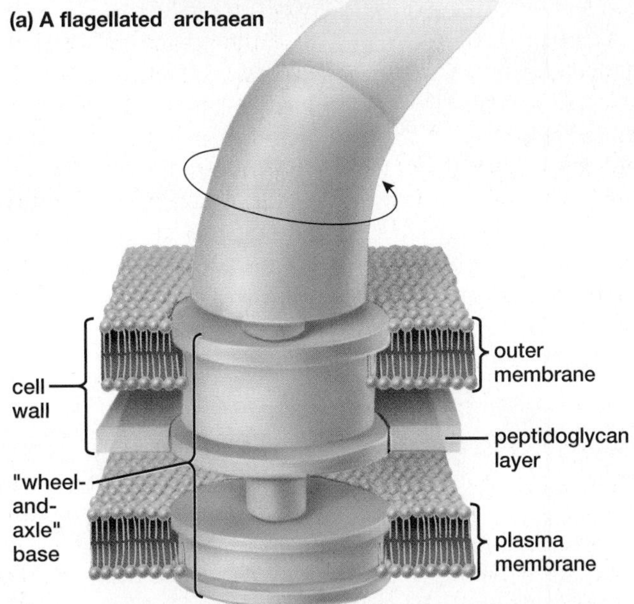

(b) The structure of the bacterial flagellum

▲ FIGURE 19-2 **The prokaryote flagellum** **(a)** A flagellated archaean of the genus *Aquifex* uses its flagella to move toward favorable environments. **(b)** In bacteria, a unique "wheel-and-axle" arrangement anchors the flagellum within the cell wall and plasma membrane, enabling the flagellum to rotate rapidly. This diagram represents the flagellum of a gram-negative bacterium; the flagella of gram-positive bacteria lack the outermost "wheels."

through its liquid environment. The use of flagella to move allows prokaryotes to disperse into new habitats, migrate toward nutrients, and leave unfavorable environments.

The structure of prokaryote flagella is different from the structure of eukaryotic flagella (see pp. 64–65 for a description of the eukaryotic flagellum). In bacterial flagella, a unique wheel-like structure embedded in the bacterial membrane and cell wall allows the flagellum to rotate (**Fig. 19-2b**). Archaeal flagella are thinner than bacterial flagella and are constructed of different proteins. The structure of the archaeal flagellum, however, is not yet as well understood as that of the bacterial flagellum.

Many Bacteria Form Films on Surfaces

The cell walls of some prokaryote species are surrounded by sticky layers of protective slime, composed of polysaccharide or protein, which protects the cells and helps them adhere to surfaces. In many cases, slime-secreting prokaryotes of one or more species aggregate in colonies to form communities known as **biofilms.** One familiar biofilm is dental plaque, which is formed by the bacteria that inhabit the mouth (**Fig. 19-3**). The protection afforded by biofilms helps defend the embedded bacteria against a variety of attacks, including those launched by antibiotics and disinfectants. As a result, biofilms formed by bacteria harmful to humans can be very difficult to eradicate. Many infections of the human body take the form of biofilms, including those responsible for tooth decay, gum disease, and ear infections.

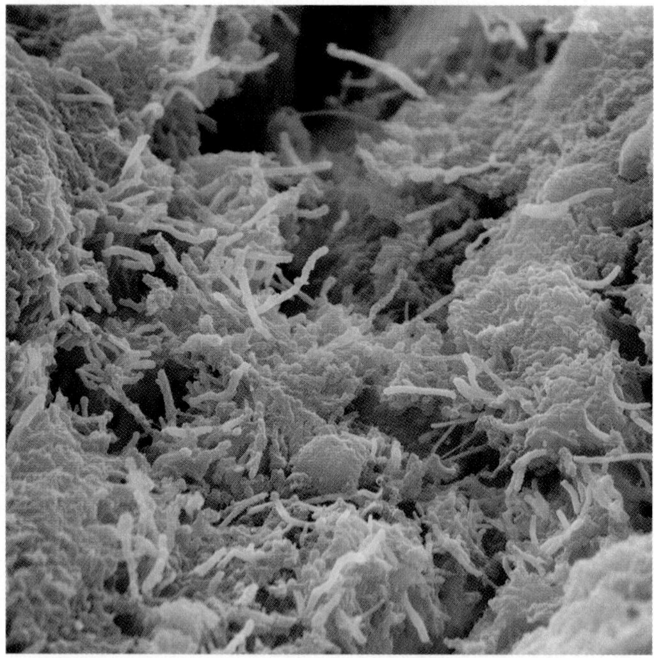

▲ FIGURE 19-3 **The cause of tooth decay** Bacteria in the human mouth form a slimy biofilm that helps them cling to tooth enamel and protects them from threats in the environment. In this micrograph, individual bacteria (colored green or yellow) are visible, embedded in the brown biofilm. The bacteria-laden biofilm can cause tooth decay.

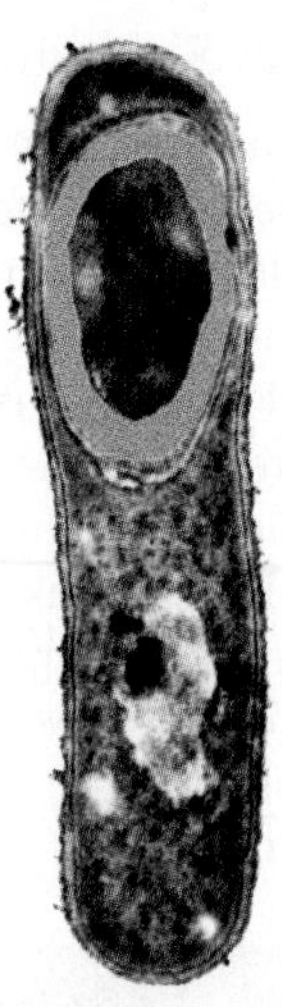

▲ FIGURE 19-4 **Spores protect some bacteria** A resistant endospore has formed inside a bacterium of the genus *Clostridium*, which causes the potentially fatal food poisoning called botulism.

QUESTION What might explain the observation that most species of endospore-forming bacteria live in soil?

Protective Endospores Allow Some Bacteria to Withstand Adverse Conditions

When environmental conditions become inhospitable, many rod-shaped bacteria form protective structures called **endospores.** An endospore, which forms inside a bacterium, contains genetic material and a few enzymes encased within a thick protective coat (**Fig. 19-4**). After an endospore forms, the bacterial cell that contains it breaks open, and the spore is released to the environment. Metabolic activity ceases until the spore encounters favorable conditions, at which time metabolism resumes and the spore develops into an active bacterium.

Endospores are resistant to even extreme environmental conditions. Some can withstand boiling for an hour or more. Endospores are also able to survive for extraordinarily long periods. In the most extraordinary example of such longevity, scientists recently discovered endospores that had been sealed

Case Study continued
Agents of Death

Endospores are one of the main reasons that the bacterial disease anthrax has become an agent of biological terrorism. The bacterium that causes anthrax forms endospores, which provide the means by which terrorists can disperse the bacteria. The spores can be stored indefinitely and can survive harsh conditions that would kill other kinds of bacteria. Anthrax spores can even survive detonation of an explosive device that would disperse them into the atmosphere, where they would remain dormant but viable until inhaled by a potential victim.

inside rock for 250 million years. After being carefully extracted from their rocky "tomb," the spores were incubated in test tubes. Amazingly, live bacteria developed from the ancient spores, which were older than the oldest dinosaur fossils.

Prokaryotes Are Specialized for Specific Habitats

Prokaryotes occupy virtually every habitat, including those where extreme conditions keep out other forms of life. For example, some bacteria thrive in near-boiling environments, such as the hot springs of Yellowstone National Park (**Fig. 19-5**). Many archaea live in even hotter environments, including springs where the water actually boils or in deep-ocean vents, where superheated water is spewed through cracks in Earth's crust at temperatures of up to 230°F (110°C). Prokaryotes can also survive at the extremely high pressures found deep beneath Earth's surface and in very cold environments, such as in Antarctic sea ice.

Even extreme chemical conditions fail to impede invasion by prokaryotes. Thriving colonies of bacteria and archaea live in the Dead Sea, where a salt concentration seven times that of the oceans precludes all other life, and in waters that are as acidic as vinegar or as alkaline as household ammonia. Of course, rich prokaryote communities also reside in a full range of more moderate habitats, including in and on the healthy human body.

No single species of prokaryote, however, is as versatile as these examples may suggest. In fact, most prokaryotes are specialists. One species of archaea that inhabits deep-sea vents, for example, grows optimally at 223°F (106°C) and stops growing altogether at temperatures below 194°F (90°C). Clearly, this species could not survive in a less extreme habitat. Bacteria that live on the human body are also specialized; different species colonize the skin, the mouth, the respiratory tract, the large intestine, and the urogenital tract.

▲ **FIGURE 19-5 Some prokaryotes thrive in extreme conditions** Hot springs harbor bacteria and archaea that are both heat and mineral tolerant. Several species of cyanobacteria paint these hot springs in Yellowstone National Park with vivid colors; each is confined to a specific area determined by temperature range.

QUESTION Some of the enzymes that have important uses in molecular biology procedures are extracted from prokaryotes that live in hot springs. Can you guess why?

Have you ever wondered

What Causes Bad Breath?

Unpleasant breath odors are caused mainly by prokaryotes that live in the mouth. The warm, moist human mouth cavity hosts a diverse microbial community that includes more than 600 prokaryote species. Many of these species acquire energy and nutrients by breaking down mucus, food particles, and dead cells. The by-products of this breakdown can include foul-smelling gases, some of which are also found in feces or decaying bodies.

The highest concentration of bad-breath prokaryotes is found at the base of the tongue. This location may be especially hospitable to microbes owing to the accumulation of mucus that drains down the back of the throat from the nose. Thus, mouthwashes used to control bad breath are most effective if gargled with the tongue thrust forward, so that the mouthwash can reach the base of the tongue.

Prokaryotes Have Diverse Metabolisms

Prokaryotes are able to colonize diverse habitats partly because they have evolved diverse methods of acquiring energy and nutrients from the environment. For example, unlike eukaryotes, many prokaryotes are **anaerobes;** their metabolisms do not require oxygen. Their ability to inhabit oxygen-free environments allows prokaryotes to exploit habitats that are off-limits to eukaryotes. Some anaerobes, such as many of the archaea found in hot springs and the bacterium that causes tetanus, are actually poisoned by oxygen. Others are opportunists, engaging in anaerobic respiration when oxygen is lacking and switching to aerobic respiration (a more efficient process; see pp. 130–136) when oxygen becomes available. Many prokaryotes, of course, are strictly aerobic, and require oxygen at all times.

Whether aerobic or anaerobic, different prokaryote species can extract energy from an amazing array of substances. Prokaryotes subsist not only on the sugars, carbohydrates, fats, and proteins that we usually think of as foods, but also on compounds that are inedible or even poisonous to humans, including petroleum, methane (the main component of natural gas), and solvents such as benzene and toluene. Prokaryotes can even metabolize inorganic molecules, including hydrogen, sulfur, ammonia, iron, and nitrite. The process of metabolizing inorganic molecules sometimes yields by-products that are useful to other organisms. For example, certain bacteria release sulfates or nitrates, crucial plant nutrients, into the soil.

Some species of bacteria, such as the *cyanobacteria* (**Fig. 19-6**), use photosynthesis to capture energy directly from sunlight. Like green plants, photosynthetic bacteria possess chlorophyll. Most species produce oxygen as a by-product of photosynthesis, but some, known as the sulfur bacteria, use hydrogen sulfide (H_2S) instead of water (H_2O)

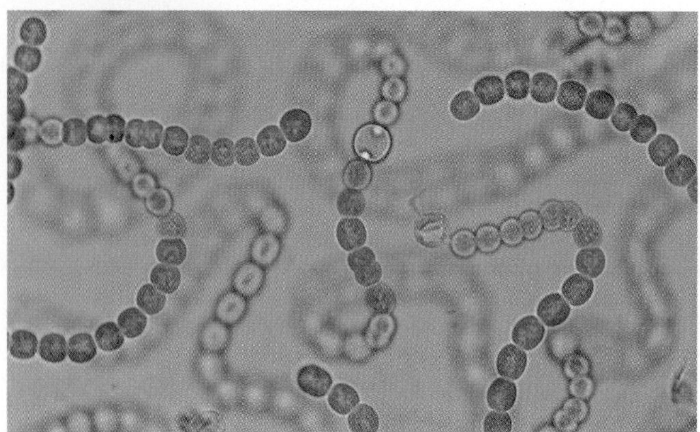

▲ **FIGURE 19-6 Cyanobacteria** Micrograph of cyanobacterial filaments.

in photosynthesis, releasing sulfur instead of oxygen. No photosynthetic archaea are known.

Prokaryotes Reproduce by Binary Fission

Most prokaryotes reproduce asexually by **binary fission,** a form of cell division that is much simpler than mitotic cell division (see pp. 152–155). Binary fission produces genetically identical copies of the original cell (**Fig. 19-7**). Under ideal conditions, some prokaryotic cells can divide about once every 20 minutes, potentially giving rise to sextillions (10^{21}) of offspring in a single day. This rapid reproduction allows prokaryotes to exploit temporary habitats, such as a mud puddle or warm pudding.

Rapid reproduction also allows bacterial populations to evolve quickly. Recall that many mutations, the source of genetic variability, are the result of mistakes in DNA replication during cell division (see pp. 159, 162). Thus, the rapid, repeated cell division of prokaryotes provides ample opportunity for new mutations to arise and also allows mutations that enhance survival to spread quickly.

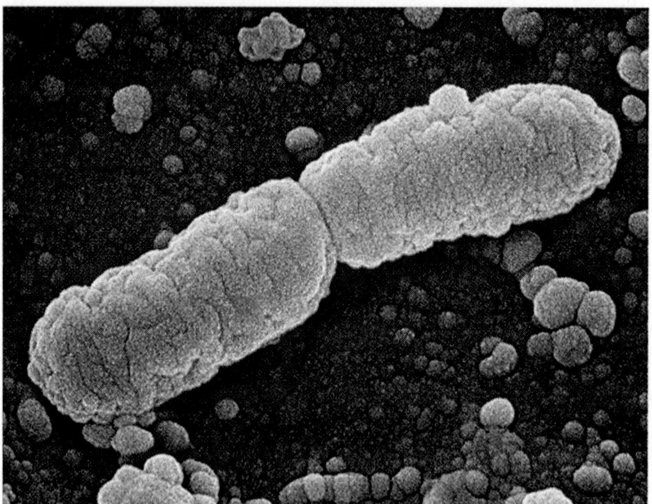

▲ **FIGURE 19-7 Reproduction in prokaryotes** Prokaryotic cells reproduce by binary fission. In this color-enhanced electron micrograph, an *Escherichia coli*, a normal inhabitant of the human intestine, is dividing.

QUESTION What is the main advantage of binary fission, compared to sexual reproduction?

Prokaryotes May Exchange Genetic Material Without Reproducing

Although prokaryote reproduction is generally asexual and does not involve genetic recombination, some bacteria and archaea nonetheless exchange genetic material. In these species, DNA is transferred from a donor to a recipient in a process called **conjugation.** The plasma membranes of two conjugating prokaryotes fuse temporarily to form a cytoplasmic bridge across which DNA travels. In bacteria, donor cells may use specialized extensions called *sex pili* that attach to a recipient cell, drawing it closer to allow conjugation (**Fig. 19-8**). Conjugation produces new genetic combinations that may allow the resulting bacteria to survive under a greater variety of conditions. In some cases, genetic material may be exchanged even between individuals of different species.

Much of the DNA transferred during bacterial conjugation is contained within a structure called a **plasmid,** a small, circular DNA molecule that is separate from the single bacterial chromosome. Plasmids may carry genes for antibiotic resistance or alleles of genes also found on the main bacterial chromosome. Researchers in molecular genetics have made extensive use of bacterial plasmids, as described in Chapter 13.

19.3 HOW DO PROKARYOTES AFFECT HUMANS AND OTHER ORGANISMS?

Although they are largely invisible to us, prokaryotes play a crucial role in life on Earth. Plants and animals (including humans) are utterly dependent on prokaryotes. Prokaryotes help plants and animals obtain vital nutrients and help break down and recycle wastes and dead organisms. We could not survive without

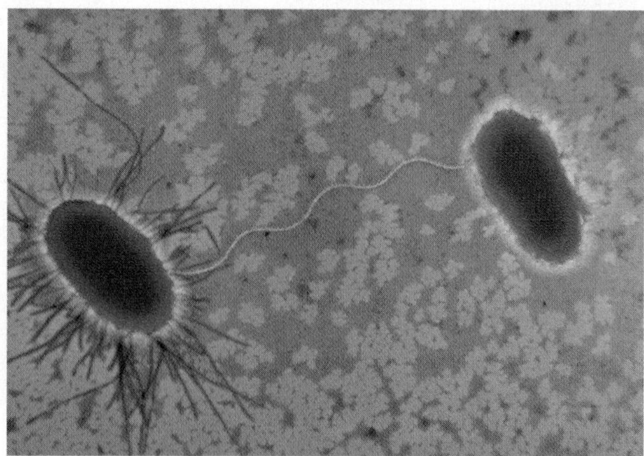

▲ **FIGURE 19-8 Conjugation: prokaryotic "mating"** During conjugation, one prokaryote acts as a donor, transferring DNA to the recipient. In this photo, two *Escherichia coli* are connected by a long sex pilus. The sex pilus will retract, drawing the recipient bacterium (at right) to the donor bacterium. The donor bacterium is bristling with nonsex pili that help it attach to surfaces.

prokaryotes, but their impact on us is not always beneficial. Some of humanity's most deadly diseases stem from microbes.

Prokaryotes Play Important Roles in Animal Nutrition

Many eukaryotic organisms depend on close associations with prokaryotes. For example, most animals that eat leaves—including cattle, rabbits, koalas, and deer—can't themselves digest cellulose, the principal component of plant cell walls. Instead, these animals depend on certain bacteria that have the unusual ability to break down cellulose. Some of these bacteria live in the animals' digestive tracts, where they help liberate nutrients from plant tissue that the animals are unable to break down themselves. Without the bacteria, leaf-eating animals could not survive.

Prokaryotes also have important impacts on human nutrition. Many foods, including cheese, yogurt, and sauerkraut, are produced by the action of bacteria. Bacteria also inhabit your intestines. These bacteria feed on undigested food and synthesize such nutrients as vitamin K and vitamin B_{12}, which the human body absorbs.

Prokaryotes Capture the Nitrogen Needed by Plants

Humans could not live without plants, and plants are entirely dependent on bacteria. In particular, plants are unable to capture nitrogen from that element's most abundant reservoir, the atmosphere. Plants need nitrogen to grow. To acquire it, they depend on **nitrogen-fixing bacteria,** which live both in soil and in specialized nodules, small, rounded lumps on the roots of certain plants (legumes, which include alfalfa, soybeans, lupines, and clover; **Fig. 19-9**). The nitrogen-fixing bacteria capture nitrogen gas (N_2) from air trapped in the soil and combine it with hydrogen to produce ammonium (NH_4^+), a nitrogen-containing nutrient that plants can use directly.

Prokaryotes Are Nature's Recyclers

Prokaryotes play a crucial role in recycling waste. Most prokaryotes obtain energy by breaking down complex organic molecules (molecules that contain carbon and hydrogen). Such prokaryotes find a plentiful source of organic molecules in the waste products and dead bodies of plants and animals. By consuming and thereby decomposing these wastes, prokaryotes prevent wastes from accumulating in the environment. In addition, decomposition by prokaryotes releases the nutrients contained in wastes. Once released, the nutrients become available for reuse by living organisms.

Prokaryotes perform their recycling service wherever organic matter is found. They are important decomposers in lakes and rivers, in the oceans, and in the soil and groundwater of forests, grasslands, deserts, and other terrestrial environments. The recycling of nutrients by prokaryotes and other decomposers provides the basis for continued life on Earth.

Prokaryotes Can Clean Up Pollution

Many of the pollutants that are produced as by-products of human activity are organic compounds. As such, these pollutants can potentially serve as food for archaea and bacteria. Many of them are, in fact, consumed; the range of compounds

(a) Nodules on roots

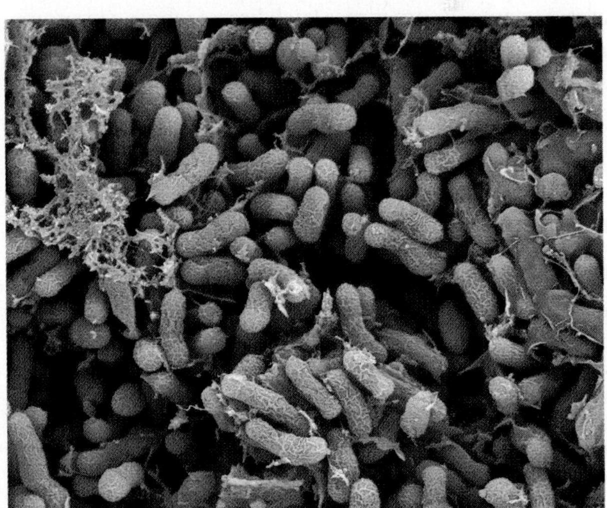

(b) Nitrogen-fixing bacteria within nodules

▲ **FIGURE 19-9 Nitrogen-fixing bacteria in root nodules (a)** Special chambers called nodules on the roots of a legume provide a protected and constant environment for nitrogen-fixing bacteria. **(b)** This scanning electron micrograph shows the nitrogen-fixing bacteria inside cells within the nodules.

QUESTION If all of Earth's nitrogen-fixing prokaryotes were to die suddenly, what would happen to the concentration of nitrogen gas in the atmosphere?

consumed by prokaryotes is staggering. Nearly anything that human beings can synthesize can be broken down by some prokaryote, including detergents, many toxic pesticides, and harmful industrial chemicals such as benzene and toluene.

Even oil can be broken down by prokaryotes. Soon after the tanker *Exxon Valdez* spilled 11 million gallons of crude oil into Prince William Sound, Alaska, in 1989, researchers sprayed oil-soaked beaches with a fertilizer that encouraged the growth of natural populations of oil-eating bacteria. Within 15 days, the oil deposits on these beaches were noticeably reduced in comparison with unsprayed areas.

The practice of manipulating conditions to stimulate breakdown of pollutants by living organisms is known as **bioremediation.** Improved methods of bioremediation could dramatically increase our ability to clean up toxic waste sites and polluted groundwater. A great deal of current research is therefore devoted to identifying prokaryote species that are especially effective for bioremediation and discovering practical methods for manipulating these organisms to improve their effectiveness.

Some Bacteria Pose a Threat to Human Health

Despite the benefits some bacteria provide, the feeding habits of certain bacteria threaten our health and well-being. These **pathogenic** (disease-producing) bacteria synthesize toxic substances that cause disease symptoms. (So far, no pathogenic archaea have been identified.)

Some Anaerobic Bacteria Produce Dangerous Poisons

Some bacteria produce toxins that attack the nervous system. One such toxin is produced by *Clostridium tetani*, the bacterium that causes tetanus, a sometimes fatal disease whose symptoms include painful, uncontrolled contraction of muscles throughout the body. *C. tetani* are anaerobic bacteria that survive as spores until introduced into a favorable, oxygen-free environment. A deep puncture wound may allow tetanus bacteria to penetrate a human body and reach a place where they will be protected from contact with oxygen. As they multiply, the bacteria release their toxin into the body's bloodstream.

Humans Battle Bacterial Diseases Old and New

Bacterial diseases have had a significant impact on human history. Perhaps the most infamous example is bubonic plague, or "Black Death," which killed 100 million people during the mid-fourteenth century. In many parts of the world, one-third or more of the population died. Plague is caused by a highly infectious bacterium that is spread by fleas that feed on infected rats and then move to human hosts. Although bubonic plague has not reemerged as a large-scale epidemic, about 2,000 to 3,000 people worldwide are still diagnosed with the disease each year.

Some bacterial pathogens seem to emerge suddenly. Lyme disease, for example, was unknown until 1975. This disease, named after the town of Old Lyme, Connecticut, where it was first described, is caused by the spiral-shaped bacterium *Borrelia burgdorferi*. The bacterium is carried by deer ticks, which transmit it to the humans they bite. At first, the symptoms resemble flu, with chills, fever, and body aches. If untreated, weeks or months later the victim may experience rashes, bouts of arthritis, and in some cases abnormalities of the heart and nervous system. Both physicians and the general public are becoming more familiar with the disease, so more victims are receiving treatment before serious symptoms develop.

Perhaps the most frustrating pathogens are those that come back to haunt us long after we believed that we had them under control. Tuberculosis, a bacterial disease once almost vanquished in developed countries, is again on the rise in the United States and elsewhere. Two sexually transmitted bacterial diseases, gonorrhea and syphilis, have reached epidemic proportions around the globe. Cholera, a water-transmitted bacterial disease that flourishes when raw sewage contaminates drinking water or fishing areas, is under control in developed countries but remains a major killer in poorer parts of the world.

Common Bacterial Species Can Be Harmful

Some pathogenic bacteria are so widespread and common that we cannot expect to ever be totally free of their damaging effects. For example, different species of the abundant streptococcus bacterium produce several ailments. One streptococcus causes tooth decay. Another causes pneumonia by stimulating an immune response that clogs the lungs with fluid. Yet another streptococcus has gained fame as the "flesh-eating bacterium." About 500 to 1,000 Americans each year develop necrotizing fasciitis (as the "flesh-eating" infection is more properly known), and about 15% of these victims die. The streptococci enter through broken skin and produce toxins that either destroy flesh directly or stimulate an overwhelming and misdirected attack by the immune system against the body's own cells. A limb can be destroyed in hours, and in some cases, only amputation can halt the rapid tissue destruction. In other cases, these rare strep infections sweep through the body, causing death within a matter of days.

Case Study continued

Agents of Death

Botulinum toxin is a bacterial toxin that could become a biological weapon. This paralyzing neurotoxin is produced by the anaerobic bacterium *Clostridium botulinum*. *C. botulinum* occurs naturally in soil, but may also thrive in a sealed container of canned food that has been improperly sterilized. Such food-borne *C. botulinum* is dangerous because botulinum toxin is among the most toxic substances known; a single gram is enough to kill 15 million people. Unfortunately, this toxicity may attract the attention of potential terrorists. Bioterrorism experts warn that botulinum toxin could be intentionally dispersed in food or as an airborne aerosol. Treating the victims of a large-scale outbreak of botulism would pose a huge challenge to our emergency health care capabilities.

One of the most common bacterial inhabitants of the human digestive system, *Escherichia coli*, is also capable of doing harm. Different populations of *E. coli* may differ genetically, and some genetic differences can transform this normally benign species into a pathogen. One particularly notorious strain, known as O157:H7, infects about 70,000 Americans each year, about 60 of whom die from its effects. Most O157:H7 infections result from consumption of contaminated beef. About a third of the cattle in the United States carry O157:H7 in their intestinal tracts, and the bacteria can be transmitted to humans when a slaughterhouse inadvertently grinds some gut contents into hamburger. Produce such as lettuce, spinach, and tomatoes can also become contaminated with O157:H7 if farm fields are exposed to animal feces carried from nearby ranches or feedlots in dust or runoff. Once in a human digestive system, O157:H7 bacteria attach firmly to the wall of the intestine and begin to release a toxin that causes intestinal bleeding and that spreads to and damages other organs as well. The best defense against O157:H7 is to cook all meat thoroughly. (For more tips on protecting yourself from food-borne bacteria, see "Links to Everyday Life: Unwelcome Dinner Guests.")

19.4 WHAT ARE VIRUSES, VIROIDS, AND PRIONS?

The particles known as **viruses** are generally found in close association with living organisms, but most biologists do not consider them to be alive. Viruses lack many of the traits that characterize life. For example, they are not cells—nor are they composed of cells. Further, they cannot, on their own, accomplish the basic tasks that living cells perform. Viruses have no ribosomes on which to make proteins, no cytoplasm, no ability to synthesize organic molecules, and no capacity to extract and use the energy stored in such molecules. They possess no membranes of their own and cannot grow or reproduce on their own. The simplicity of viruses seems to place them outside the realm of living things.

A Virus Consists of a Molecule of DNA or RNA Surrounded by a Protein Coat

Viruses are tiny; most are much smaller than even the smallest prokaryotic cell (**Fig. 19-10**). Virus particles are so small (0.05–0.2 micrometer in diameter) that they can be seen only under the enormous magnification of an electron microscope. Under such magnification, one can see that viruses assume a great variety of shapes (**Fig. 19-11**).

Links to *Everyday Life*

Unwelcome Dinner Guests

Although the prospect of biological weapons is chilling, you are far more likely to encounter harmful microorganisms from a more mundane source: your food. The nutrients that you consume during meals and snacks can also provide sustenance for a wide variety of disease-causing bacteria and protists. Some of these invisible diners may accompany your lunch to your digestive tract and take up residence there, causing unpleasant symptoms. The Centers for Disease Control and Prevention estimates that U.S. residents experience an astonishing 76 million cases of food-borne illness each year, resulting in 325,000 hospitalizations and 5,200 deaths.

The most frequent culprits in food-borne diseases are bacteria. Species in the genera *Escherichia* (especially the O157:H7 variant of *E. coli*), *Salmonella*, *Listeria*, *Streptococcus*, and *Campylobacter* are responsible for an especially large number of illnesses, with *Campylobacter* currently claiming the largest number of victims.

How can you protect yourself from the bacteria and protists that share our food supply? It's easy: Clean, cook, and chill. Cleaning helps prevent the spread of pathogens. Wash your hands before preparing food, and wash all utensils and cutting boards after preparing each item. Thorough cooking is the best way to ensure that any bacteria and protists present in food are killed. Meats, in particular, must be thoroughly cooked; try to avoid eating meat that is still pink inside (**Fig. E19-1**). Fish should be cooked until it is opaque and flakes easily with a fork; cook eggs until both white and yolk are firm. Finally, keep food cold. Pathogens multiply most rapidly at temperatures between 40° and 140°F. So get your groceries home from the store and into the refrigerator or freezer as quickly as possible. Don't leave cooked leftovers unrefrigerated for more than 2 hours. Thaw frozen foods in the refrigerator, not at room temperature. A little bit of attention to food safety can save you from unwelcome guests in your food.

▲ FIGURE E19-1 Rare beef is a haven for dangerous bacteria

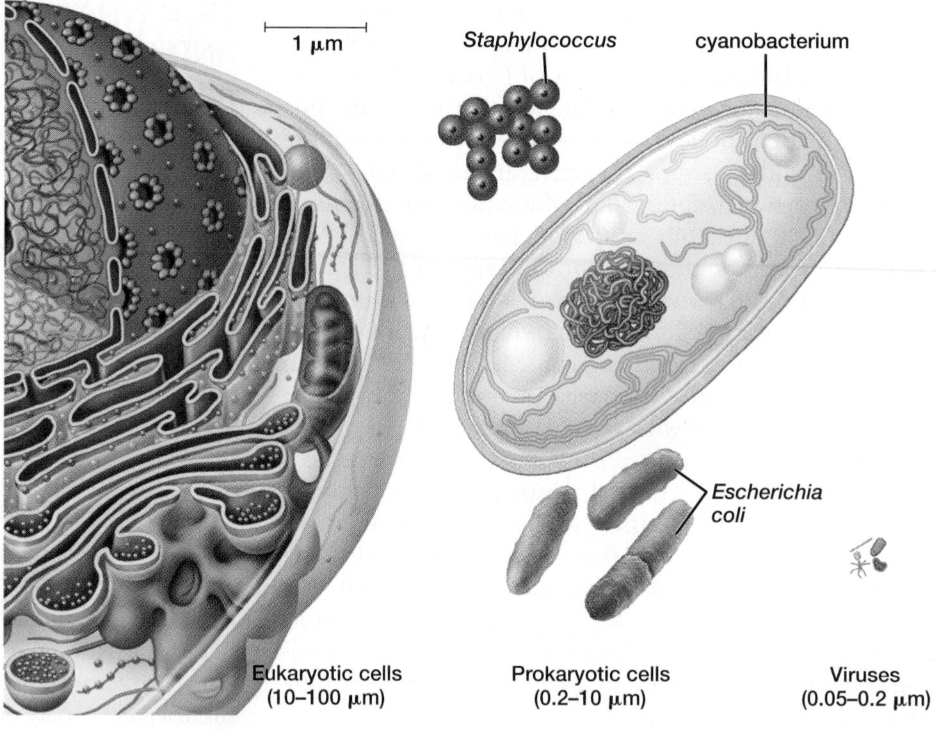

▶ **FIGURE 19-10 The sizes of microorganisms** The relative sizes of eukaryotic cells, prokaryotic cells, and viruses (1 μm = 1/1,000 millimeter).

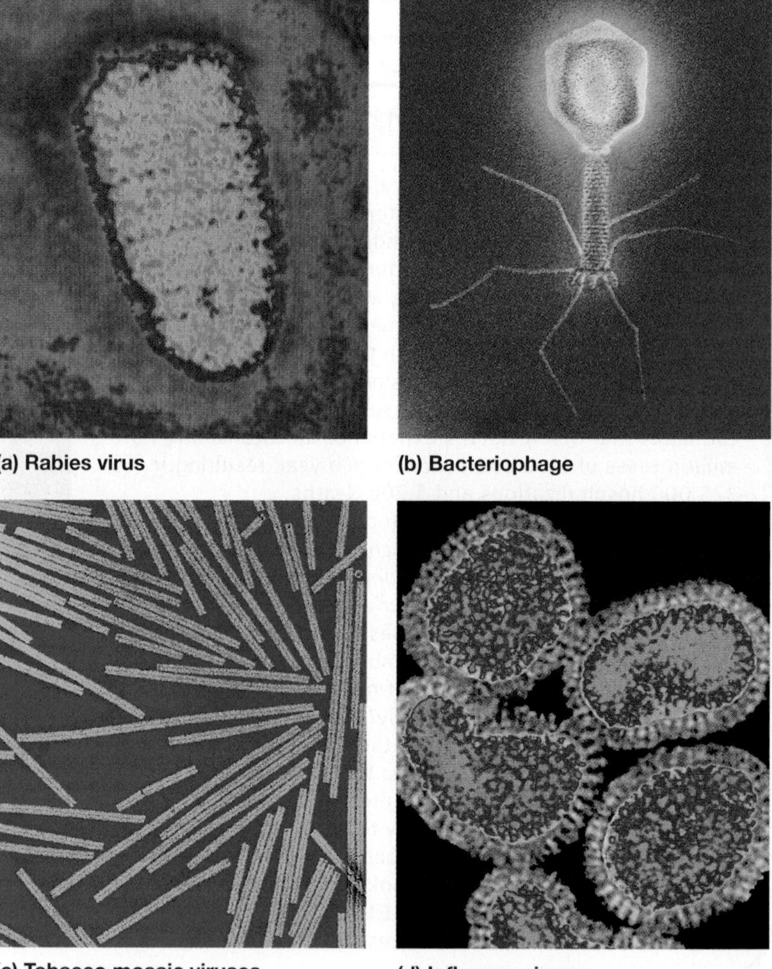

▶ **FIGURE 19-11 Viruses come in a variety of shapes** Viral shape is determined by the nature of the virus's protein coat.

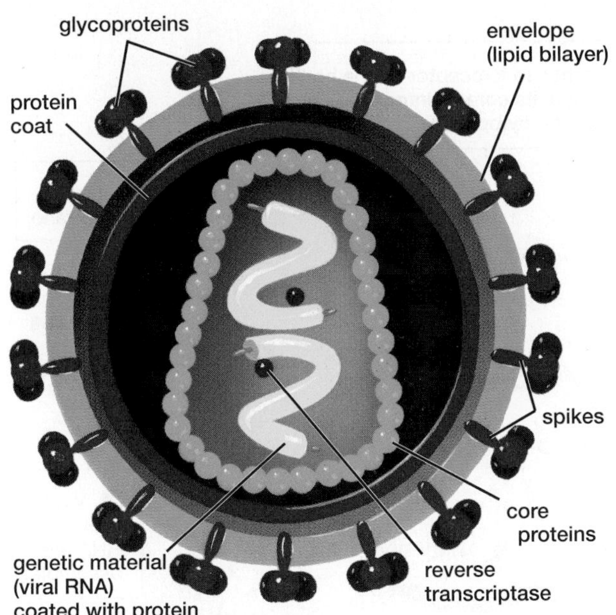

▲ **FIGURE 19-12 Viral structure and replication** A cross-section of HIV, the virus that causes AIDS. Inside, a protein coat surrounds genetic material and molecules of reverse transcriptase, an enzyme that catalyzes the transcription of DNA from the viral RNA template after the virus enters a host cell. This virus is among those that also have an outer envelope that is formed from the host cell's plasma membrane. Spikes made of glycoprotein (protein and carbohydrate) project from the envelope and help the virus attach to its host cell.

QUESTION Why are viruses unable to replicate outside of a host cell?

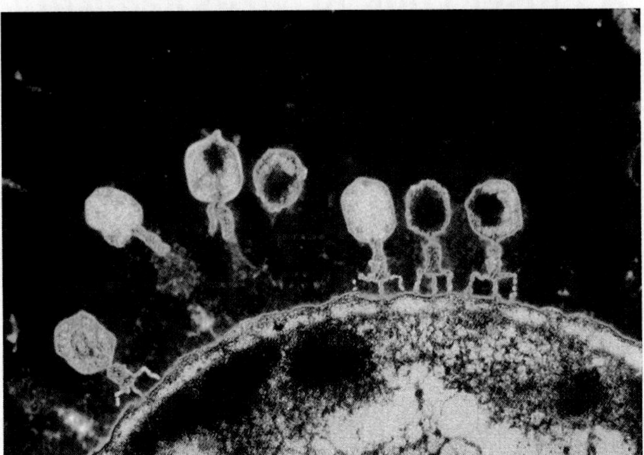

▲ **FIGURE 19-13 Some viruses infect bacteria** In this electron micrograph, bacteriophages are seen attacking a bacterium. They have injected their genetic material inside, leaving their protein coats clinging to the bacterial cell wall.

QUESTION In biotechnology, viruses are often used to transfer genes from the cells of one species to the cells of another. Which properties of viruses make them useful for this purpose?

Viruses consist of two major parts: a molecule of hereditary material and a coat of protein surrounding the molecule. Depending on the type of virus, the hereditary molecule may be either DNA or RNA and may be single-stranded or double-stranded, linear or circular. The protein coat may be surrounded by an envelope formed from the plasma membrane of the host cell (**Fig. 19-12**).

Viruses Are Parasites

Viruses are parasites of living cells. (Parasites live in or on other organisms, harming them in the process.) A virus can reproduce only inside a **host** cell—the cell that a virus or other infectious agent infects. Viral reproduction begins when a virus penetrates a host cell. After the virus enters the host cell, the viral genetic material takes command. The hijacked host cell uses the instructions encoded in the viral genes to produce the components of new viruses. The pieces are rapidly assembled, and an army of new viruses bursts forth to invade and conquer neighboring cells (see "A Closer Look at Virus Replication" on p. 366).

Viruses Are Host Specific

Each type of virus is specialized to attack a specific host cell. As far as we know, no organism is immune to all viruses. Even bacteria fall victim to viral invaders called **bacteriophages** (**Fig. 19-13**). Bacteriophages may soon become important in treating diseases caused by bacteria, because many disease-causing bacteria have become increasingly resistant to antibiotics. Treatments based on bacteriophages could also take advantage of the viruses' specificity, attacking only the targeted bacteria and not the many other harmless or beneficial bacteria in the body.

In multicellular organisms such as plants and animals, different viruses specialize in attacking particular cell types. Viruses responsible for the common cold, for example, attack the membranes of the respiratory tract, and rabies viruses attack nerve cells. One type of herpes virus specializes in the mucous membranes of the mouth and lips, causing cold sores; a second type produces similar sores on or near the genitals. Herpes viruses take up permanent residence in the body, erupting periodically (typically during times of stress) as infectious sores. The devastating disease AIDS (acquired immune deficiency syndrome), which cripples the body's immune system, is caused by a virus that attacks a specific type of white blood cell that controls the body's immune response. Viruses have also been linked to some types of cancer, such as T-cell leukemia (a cancer of the white blood cells), liver cancer, and cervical cancer.

Viral Infections Are Difficult to Treat

Because viruses are closely tied to the cellular machinery of their hosts, the illnesses they cause are difficult to treat. The antibiotics that are often effective against bacterial infections are useless against viruses, and antiviral agents may destroy host cells as well as viruses. Despite the difficulty of attacking viruses as they "hide" within cells, a number of antiviral drugs have been developed. Many of these drugs destroy or block the function of enzymes that the targeted virus requires for replication.

A Closer Look At *Virus Replication*

Viruses multiply, or replicate, using their own genetic material, which—depending on the virus—consists of single-stranded or double-stranded RNA or DNA. This material serves as a blueprint for the viral proteins and genetic material required to make new viruses. Viral enzymes may participate in replication as well, but the overall process depends on the biochemical machinery that the host cell uses to make its own proteins and replicate its own DNA. Viruses cannot replicate outside of living cells.

The process of viral replication varies considerably among the different types of viruses, but most modes of replication are variations of a general sequence of events:

Penetration To replicate, a virus must enter a host cell. Some viruses are engulfed by a host cell (endocytosis) after binding to receptors on the cell's plasma membrane that stimulate endocytosis. Other viruses are coated with an envelope that can fuse with the host's membrane. The viral genetic material is then released into the cytoplasm.

Synthesis Viruses redirect the host cell's protein synthesis machinery to produce many copies of the virus's proteins, and the virus's genetic material is replicated many times. Transcription of the viral genome to messenger RNA uses nucleotides from the host cell, and protein synthesis uses the host cell's ribosomes, transfer RNAs, and amino acids.

Assembly The viral genetic material and enzymes are surrounded by their protein coat.

Release Viruses emerge from the host cell by "budding" from the plasma membrane or by bursting the cell.

Here, two types of viral life cycles are depicted. **Figure E19-2a** shows the human immunodeficiency virus (HIV), the *retrovirus* that causes AIDS. Retroviruses are so named because a key step in their replication uses single-stranded RNA as a template to make double-stranded DNA, a process that reverses the normal DNA-to-RNA pathway. Retroviruses accomplish this reverse transcription by using a viral enzyme called *reverse transcriptase*. **Figure E19-2b** shows the herpes virus, which contains double-stranded DNA that is transcribed into mRNA.

envelope
coat
core
RNA
reverse transcriptase
(cytoplasm)
(nucleus)
DNA
RNA
mRNA
❶ A virus attaches to a receptor on the host's plasma membrane; its core disintegrates, and viral RNA enters the cytoplasm
❷ Viral reverse transcriptase produces DNA, using viral RNA as a template
❸ DNA enters the nucleus and is incorporated into the host chromosomes; it is transcribed into mRNA and more viral RNA, which move to the cytoplasm
❹ Viral proteins are synthesized, using mRNA
❺ Viral proteins and RNA are assembled
❻ Viruses bud from the plasma membrane

(a) HIV virus, a retrovirus, invades a white blood cell

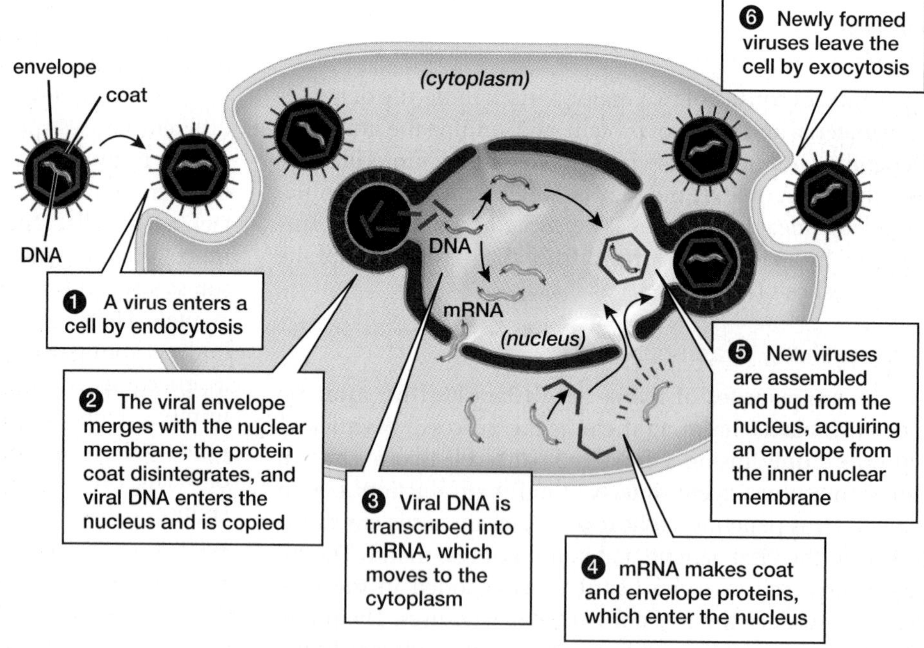
envelope
coat
DNA
(cytoplasm)
DNA
mRNA
(nucleus)
❶ A virus enters a cell by endocytosis
❷ The viral envelope merges with the nuclear membrane; the protein coat disintegrates, and viral DNA enters the nucleus and is copied
❸ Viral DNA is transcribed into mRNA, which moves to the cytoplasm
❹ mRNA makes coat and envelope proteins, which enter the nucleus
❺ New viruses are assembled and bud from the nucleus, acquiring an envelope from the inner nuclear membrane
❻ Newly formed viruses leave the cell by exocytosis

(b) Herpes virus, a double-stranded DNA virus, invades a skin cell

▲ **FIGURE E19-2 How viruses replicate**

Case Study continued
Agents of Death

The difficulty of treating viral infections makes the possibility of virus-based biological weapons troubling. Of particular concern is the smallpox virus. Smallpox has been eradicated as a naturally occurring disease, and the only known cultures of the virus are held at two well-guarded government labs, one in Russia and one in the United States; nevertheless, other samples may exist in unknown locations. Because of this possibility, plans to destroy the remaining stocks of the virus have been indefinitely deferred so that the stored viruses can be used in research to develop a more effective smallpox vaccine.

Unfortunately, the benefits of most antiviral drugs are limited because many viruses quickly evolve resistance to the drugs. Mutation rates can be very high in viruses, in part because many viruses lack mechanisms for correcting errors that occur during replication of genetic material. It is thus common that when a population of viruses is under attack by an antiviral drug, a mutation will arise that confers resistance to the drug. The resistant viruses prosper and replicate in great numbers, eventually spreading to new human hosts. Ultimately, resistant viruses predominate, and a formerly helpful antiviral drug is rendered ineffective.

Some Infectious Agents Are Even Simpler Than Viruses

Viroids are infectious particles that lack a protein coat and consist of nothing more than short, circular strands of RNA. Despite their simplicity, viroids are able to enter the nucleus of a host cell and direct the synthesis of new viroids. About a dozen crop diseases, including cucumber pale fruit disease, avocado sunblotch, and potato spindle tuber disease, are caused by viroids.

Prions are even more puzzling than viroids. In the 1950s, physicians studying the Fore, a primitive tribe in New Guinea, were puzzled to observe numerous cases of a fatal degenerative disease of the nervous system, which the Fore called *kuru*. The symptoms of kuru—loss of coordination, dementia, and ultimately death—were similar to those of the rare but more widespread Creutzfeldt-Jakob disease in humans and of scrapie and bovine spongiform encephalopathy, diseases of domestic livestock (see the Case Study in Chapter 3). Each of these diseases typically results in brain tissue that is spongy—riddled with holes. The researchers in New Guinea eventually determined that kuru was transmitted by ritual cannibalism; members of the Fore tribe honored their dead by consuming their brains. This practice has since stopped, and kuru has virtually disappeared. Clearly, kuru was caused by an infectious agent transmitted by infected brain tissue—but what was that agent?

In 1982, neurologist Stanley Prusiner published evidence that scrapie (and, by extension, kuru, Creutzfeldt-Jakob disease, and a number of other, similar afflictions) is caused by an infectious agent that consists only of protein. This idea seemed preposterous at the time, because most scientists believed that infectious agents must contain genetic material such as DNA or RNA in order to replicate. But Prusiner and his colleagues were able to isolate the infectious agent from scrapie-infected hamsters and demonstrate that it contained no nucleic acids. The researchers called these infectious protein particles *prions* (**Fig. 19-14**).

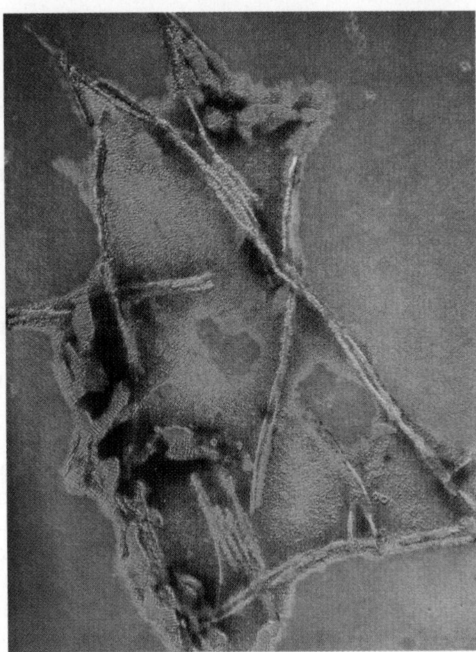

▲ **FIGURE 19-14 Prions: puzzling proteins** A section from the brain of a cow infected with bovine spongiform encephalopathy contains fibrous clusters of prion proteins.

How can a protein replicate itself and be infectious? Not all researchers are convinced that this is possible. However, recent findings have sketched the outline of a possible mechanism for prion replication. It turns out a prion is a protein produced by normal nerve cells. Some copies of this normal protein molecule, for reasons still poorly understood, become folded into the wrong shape and are thus transformed into infectious prions. Once present, prions can apparently induce other, normal copies of the protein molecule to become transformed into prions. Eventually, the concentration of prions in nerve tissue may get high enough to cause cell damage and degeneration. Why would a slight alteration to a normally benign protein turn it into a dangerous cell-killer? No one knows.

Another peculiarity of prion-caused diseases is that they can be inherited as well as transmitted by infection. Recent research has shown that certain small mutations in the gene that codes for "normal" prion protein increase the likelihood that the protein will fold into its abnormal form. If one of these mutations is genetically passed on to offspring, the tendency to develop a prion disease may be inherited.

No One Is Certain How These Infectious Particles Originated

The origin of viruses, viroids, and prions is obscure. Some scientists believe that the huge variety of mechanisms for self-replication among these particles reflects their status as evolutionary remnants of the very early history of life, before evolution settled on the more familiar large, double-stranded DNA molecules. Another possibility is that viruses, viroids, and prions may be the degenerate descendants of parasitic cells. These ancient parasites may have been so successful at exploiting their hosts that they eventually lost the ability to synthesize all of the molecules required for survival and became dependent on the host's biochemical machinery. Whatever the origin of these infectious particles, their success poses a continuing challenge to living things.

Case Study revisited

Agents of Death

Although anthrax is thought to be the most likely biological weapon, many other infectious agents also have the potential to become weapons. These include the viruses that cause smallpox and Ebola hemorrhagic fever, and the bacteria that cause plague. Evidence also exists that some countries are trying to use genetic engineering to "improve" pathogens; for example, by adding antibiotic-resistance genes to plague bacteria so that victims of an attack would be more difficult to treat and more likely to die.

Before 2001, humankind relied on politics, diplomacy, and widespread revulsion at the concept of biological warfare to protect us from its terrifying destructive potential. Now, however, it is painfully clear that humanity includes people willing to unleash biological weapons. Unfortunately, little expertise is required to culture pathogenic bacteria or viruses, and the necessary supplies and equipment are easily acquired. Given the difficulty of preventing biological weapons from falling into the wrong hands, much current research is focused on developing tools to detect attacks and render them harmless.

It is not easy to detect a biological attack, because pathogens are invisible and symptoms may take hours or days to appear. Nonetheless, rapid detection is crucial if an effective response is to be mounted, and a variety of new detection technologies are being rapidly developed. Detectors must be able to distinguish released pathogens from the multitude of harmless microbes that ordinarily inhabit the air, water, and soil. One promising approach relies on sensors that incorporate living human immune cells that have been genetically altered to glow when receptor molecules in their cell membranes bind to a particular pathogen.

Once an attack is detected, the main task is to care for those who have been exposed. Thus, developing fast-acting, easily distributed treatments is a top priority for researchers. For example, biologists have intensively investigated the mechanism by which the toxin released by anthrax bacteria attacks and damages cells. Better understanding of this process has improved investigators' ability to block it, and has yielded several promising ideas for antidotes that could be used in conjunction with antibiotics as a treatment for anthrax exposure.

Consider This

The threat of biological attack has prompted a debate: Should large numbers of people be immunized against potential agents of attack for which vaccinations exist? Mass vaccinations are costly and would inevitably cause some deaths due to occasional adverse reactions. Is the increased protection and peace of mind that would come with vaccinations worth the price?

CHAPTER REVIEW

Summary of Key Concepts

19.1 Which Organisms Are Members of the Domains Archaea and Bacteria?

Members of the domains Archaea and Bacteria—the archaea and bacteria—are unicellular and prokaryotic. Although archaea and bacteria are morphologically similar, they are not closely related and differ in several fundamental features, including cell-wall composition, ribosomal RNA sequence, and membrane lipid structure. A cell wall determines the characteristic shapes of prokaryotes: round, rodlike, or spiral.

19.2 How Do Prokaryotes Survive and Reproduce?

Certain types of bacteria can move about using flagella; others form spores that disperse widely and withstand inhospitable environmental conditions. Bacteria and archaea have colonized nearly every habitat on Earth, including hot, acidic, very salty, and anaerobic environments.

Prokaryotes obtain energy in a variety of ways. Some, including the cyanobacteria, rely on photosynthesis. Others break down inorganic or organic molecules to obtain energy. Many are anaerobic, able to obtain energy from fermentation when oxygen is not available. Prokaryotes reproduce by binary fission and may exchange genetic material by conjugation.

19.3 How Do Prokaryotes Affect Humans and Other Organisms?

Some bacteria are pathogenic, causing disorders such as pneumonia, tetanus, botulism, and the sexually transmitted infections gonorrhea and syphilis. Most prokaryotes, however, are harmless to humans and play important roles in natural ecosystems. Some live in the digestive tracts of animals that eat leaves, where the prokaryotes break down cellulose. Nitrogen-fixing bacteria enrich the soil and aid in plant growth. Many others live off the dead bodies and wastes of other organisms, liberating nutrients for reuse.

19.4 What Are Viruses, Viroids, and Prions?

Viruses are parasites consisting of a protein coat that surrounds genetic material. They are noncellular and unable to move, grow, or reproduce outside a living cell. They invade cells of a specific host and use the host cell's energy, enzymes, and ribosomes to produce more virus particles, which are liberated when the cell ruptures. Many viruses are pathogenic to humans, including those causing colds and flu, herpes, AIDS, and certain forms of cancer.

Viroids are short strands of RNA that can invade a host cell's nucleus and direct the synthesis of new viroids. To date, viroids are known to cause only certain diseases of plants.

Prions have been implicated in diseases of the nervous system, such as kuru, Creutzfeldt-Jakob disease, and scrapie. Prions are unique in that they lack genetic material. They are composed solely of mutated prion protein, which may act as an enzyme, catalyzing the formation of more prions from normal prion protein.

Key Terms

anaerobe *359*
Archaea *356*
Bacteria *356*
bacteriophage *365*
binary fission *360*
biofilm *358*
bioremediation *362*
conjugation *360*
endospore *358*
flagellum *357*
host *365*
nitrogen-fixing bacterium *361*
pathogenic *362*
plasmid *360*
prion *367*
viroid *367*
virus *363*

Thinking Through the Concepts

Fill-in-the-Blank

1. __________ have peptidoglycan in their __________, but __________ do not.
2. The size of prokaryotic cells is __________ than the size of eukaryotic cells. The most common shapes of prokaryotes are __________, __________, and __________.
3. Many prokaryotes use __________ to move about. Some prokaryotes secrete slime that protects them when they aggregate in communities called __________. Other prokaryotes can survive long periods and extreme conditions by producing protective structures called __________.
4. __________ bacteria inhabit environments that lack oxygen. __________ bacteria capture energy from sunlight.
5. Prokaryotes reproduce by __________, and may sometimes exchange genetic material through the process of __________.
6. The plant nutrient ammonium is produced by __________ bacteria in the soil. Prokaryotes that live in the digestive tracts of cows and rabbits break down __________ in the leaves that those mammals eat.
7. Diseases caused by pathogenic bacteria include __________, __________, and __________. Harmful strains of *E. coli* can be transmitted to humans by consumption of __________ or __________.
8. A virus consists of a molecule of __________ or __________ surrounded by a(n) __________ coat. A virus cannot reproduce unless it enters a(n) __________ cell. A virus that infects bacteria is known as a(n) __________.

Review Questions

1. Describe some of the ways in which prokaryotes obtain energy and nutrients.
2. What are nitrogen-fixing bacteria, and what role do they play in ecosystems?
3. What is an endospore? What is its function?
4. What is conjugation? What role do plasmids play in conjugation?
5. Why are prokaryotes especially useful in bioremediation?
6. Describe the structure of a typical virus. How do viruses replicate?
7. Describe some examples of how prokaryotes are helpful to humans and some examples of how they are harmful to humans.
8. How do archaea and bacteria differ? How do prokaryotes and viruses differ?

Applying the Concepts

1. In some developing countries, antibiotics can be purchased without a prescription. Why do you think this is done? What biological consequences would you predict?
2. Before the discovery of prions, many (perhaps most) biologists would have agreed with the statement, "It is a fact that no infectious organism or particle can exist that lacks nucleic acid (such as DNA or RNA)." What lessons do prions teach us about nature, science, and scientific inquiry? You may wish to review Chapter 1 to help answer this question.
3. Argue for and against the statement "Viruses are alive."

MB *Go to www.masteringbiology.com for practice quizzes, activities, eText, videos, current events, and more.*

The Diversity of Protists

Chapter 20

▲ The photosynthetic protist *Caulerpa taxifolia* is an unwanted invader in temperate seas.

Case Study

Green Monster

IN CALIFORNIA, it is a crime to possess, transport, or sell *Caulerpa*. Is *Caulerpa* an illicit drug or some kind of weapon? No—it is merely a small green seaweed. Why, then, do California's lawmakers want to ban it from their state?

The story of *Caulerpa*'s rise to public enemy status begins in the early 1980s at the Wilhelmina Zoo in Stuttgart, Germany. There, the keepers of a saltwater aquarium found that the tropical seaweed *Caulerpa taxifolia* was an attractive companion and background for the tropical fish on display. Even better, years of captive breeding at the zoo had yielded a strain of the seaweed that was well suited to life in an aquarium. The new strain was very hardy and could survive in waters considerably cooler than the tropical waters in which wild *Caulerpa* is found. The aquarium strain was tough but attractive, and the zoo staff was happy to send cuttings to other institutions that wished to use it in aquarium displays.

One institution that received a cutting was the Oceanographic Museum of Monaco, which occupies a stately building that stands directly on the shore of the Mediterranean Sea. In 1984, a visiting marine biologist discovered a small patch of *Caulerpa* growing in the waters just below the museum. Presumably, someone cleaning an aquarium had dumped the water into the sea and thereby inadvertently introduced *Caulerpa* to the Mediterranean.

By 1989, the *Caulerpa* patch had grown to cover a few acres. It grew as a continuous mat that seemed to exclude most of the other organisms that normally inhabit the Mediterranean sea bottom. The local herbivores, such as sea urchins and fish, did not feed on *Caulerpa*.

It soon became apparent that *Caulerpa* spreads rapidly, is not controlled by predation, and displaces native species. By the mid-1990s, biologists were alarmed to find *Caulerpa* all along the Mediterranean coast from Spain to Italy. Today, it grows in extensive beds throughout the Mediterranean and covers an ever-expanding area on the sea floor.

Despite the threat it poses to ecosystems, *Caulerpa* is a fascinating creature. We will return to *Caulerpa* and its biology after an overview of the protists, a group that includes green seaweeds such as *Caulerpa*, along with a host of other organisms.

20.1 WHAT ARE PROTISTS?

Two of life's domains, Bacteria and Archaea, contain only prokaryotes. The third domain, Eukarya, includes all eukaryotic organisms. The most conspicuous Eukarya are plants, fungi, and animals, which we will discuss in Chapters 21 through 24. The remaining eukaryotes constitute a diverse collection of organisms collectively known as **protists** (Table 20-1). Protists do not form a clade—a group consisting of all the descendents of a particular common ancestor—so systematists do not use the term "protist" as a formal group name. Instead, protist is a term of convenience that refers to any eukaryote that is not a plant, animal, or fungus.

Most Protists Are Single Celled

Most protists are single celled and are invisible to us as we go about our daily lives. If we could somehow shrink to their microscopic scale, we might be more impressed with their beautiful forms, their varied lifestyles, their astonishingly diverse modes of reproduction, and the structural and physiological innovations that are possible within the limits of a single cell. In reality, however, the small size of protists makes them challenging to observe. A microscope and a good supply of patience are required to appreciate the majesty of protists.

Although most protists are single celled, some are visible to the naked eye, and a few are genuinely large. Some large protists are aggregations or colonies of single-celled individuals; others are multicellular organisms.

Protists Use Diverse Modes of Nutrition

Three major modes of nutrition are represented among protists. Protists may ingest their food, absorb nutrients from their surroundings, or capture solar energy directly by photosynthesis.

Protists that ingest their food are generally predators. Predatory single-celled protists may have flexible cell membranes that can change shape to surround and engulf food such as bacteria. Protists that feed in this manner typically use finger-like extensions called **pseudopods** to engulf prey (Fig. 20-1).

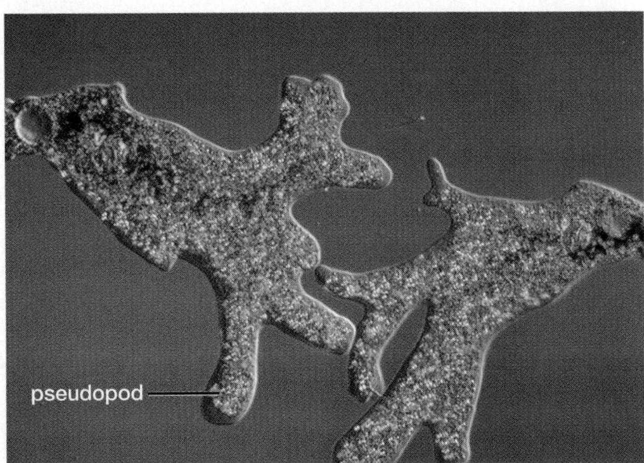

▲ **FIGURE 20-1 Pseudopods** Some single-celled protists, such as these amoebas, can extend projections that are used to engulf food or move the organism about.

Other predatory protists create tiny currents that sweep food particles into mouth-like openings in the cell. Whatever the means by which food is ingested, once it is inside the protist cell, it is typically packaged into a membrane-surrounded *food vacuole* for digestion.

Protists that absorb nutrients directly from the surrounding environment may be free living or may live inside the bodies of other organisms. The free-living types live in soil and other environments that contain dead organic matter, where they act as decomposers. Most absorptive feeders, however, live inside other organisms. In most cases, these protists are parasites whose feeding activity harms the host species.

Photosynthetic protists are abundant in oceans, lakes, and ponds. Most float suspended in the water, but some live in close association with other organisms, such as corals or clams. These associations appear to be mutually beneficial; some of the solar energy captured by the photosynthetic protists is used by the host organism, which provides shelter and protection for the protists.

Protist photosynthesis takes place in chloroplasts. As described in Chapter 17, chloroplasts are the descendants of

Table 20-1 The Major Groups of Protists

Group	Subgroup	Locomotion	Nutrition	Representative Features	Representative Genus
Excavates	Diplomonads	Swim with flagella	Heterotrophic	Lack mitochondria; inhabit soil or water, or may be parasitic	*Giardia* (intestinal parasite of mammals)
	Parabasalids	Swim with flagella	Heterotrophic	Lack mitochondria; parasites or commensal symbionts	*Trichomonas* (causes the sexually transmitted infection trichomoniasis)
Euglenozoans	Euglenids	Swim with one flagellum	Autotrophic; photosynthetic	Have an eyespot; all fresh water	*Euglena* (common pond-dweller)
	Kinetoplastids	Swim with flagella	Heterotrophic	Inhabit soil or water, or may be parasitic	*Trypanosoma* (causes African sleeping sickness)
Stramenopiles (chromists)	Water molds	Swim with flagella (gametes)	Heterotrophic	Filamentous	*Plasmopara* (causes downy mildew)
	Diatoms	Glide along surfaces	Autotrophic; photosynthetic	Have silica shells; most marine	*Navicula* (glides toward light)
	Brown algae	Nonmotile	Autotrophic; photosynthetic	Seaweeds of temperate oceans	*Macrocystis* (forms kelp forests)
Alveolates	Dinoflagellates	Swim with two flagella	Autotrophic; photosynthetic	Many bioluminescent; often have cellulose	*Gonyaulax* (causes red tide)
	Apicomplexans	Nonmotile	Heterotrophic	All parasitic; form infectious spores	*Plasmodium* (causes malaria)
	Ciliates	Swim with cilia	Heterotrophic	Include the most complex single cells	*Paramecium* (fast-moving pond-dweller)
Rhizarians	Foraminiferans	Extend thin pseudopods	Heterotrophic	Have calcium carbonate shells	*Globigerina*
	Radiolarians	Extend thin pseudopods	Heterotrophic	Have silica shells	*Actinomma*
Amoebozoans	Lobose amoebas	Extend thick pseudopods	Heterotrophic	Have no shells	*Amoeba* (common pond-dweller)
	Acellular slime molds	Sluglike mass oozes over surfaces	Heterotrophic	Form multinucleate plasmodium	*Physarum* (forms a large, bright orange mass)
	Cellular slime molds	Amoeboid cells extend pseudopods; sluglike mass crawls over surfaces	Heterotrophic	Form pseudoplasmodium with individual amoeboid cells	*Dictyostelium* (often used in laboratory studies)
Red algae		Nonmotile	Autotrophic; photosynthetic	Some deposit calcium carbonate; mostly marine	*Porphyra* (used to make sushi wrappers)
Green algae		Swim with flagella (some species)	Autotrophic; photosynthetic	Closest relatives of land plants	*Ulva* (sea lettuce)

ancient photosynthetic bacteria that took up residence inside a larger cell in a process known as *endosymbiosis*. In addition to the original instance of endosymbiosis that created the first protist chloroplast, there have been several different later occurrences of *secondary endosymbiosis* in which a nonphotosynthetic protist engulfed a photosynthetic, chloroplast-containing protist. Ultimately, most components of the engulfed species disappeared, leaving only a chloroplast surrounded by four membranes. Two of these membranes are from the original, bacteria-derived chloroplast; one is from the engulfed protist; and one is from the food vacuole that originally contained the engulfed protist. Multiple occurrences of secondary endosymbiosis account for the presence of photosynthetic species in a number of different, unrelated protist groups.

Past classifications of protists grouped species according to their mode of nutrition, but improved understanding of the evolutionary history of protists has revealed that the old categories did not accurately reflect phylogeny. Nonetheless, biologists still use terminology that refers to groups of protists that share particular characteristics but are not necessarily related. For example, photosynthetic protists are collectively known as **algae** (singular, alga), and single-celled, nonphotosynthetic protists are collectively known as **protozoa** (singular, protozoan).

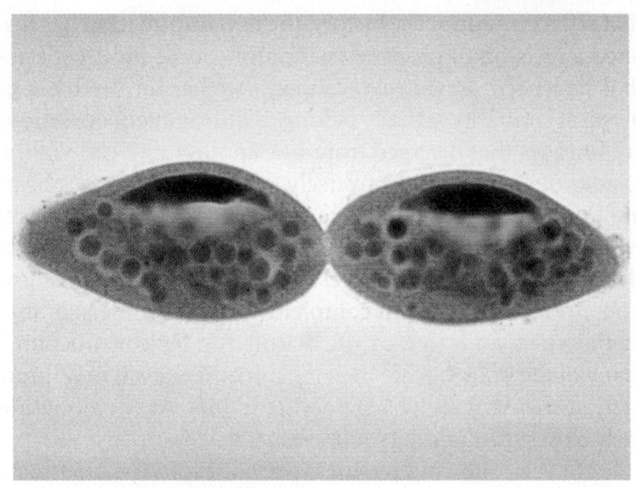

(a) Reproducing by cell division

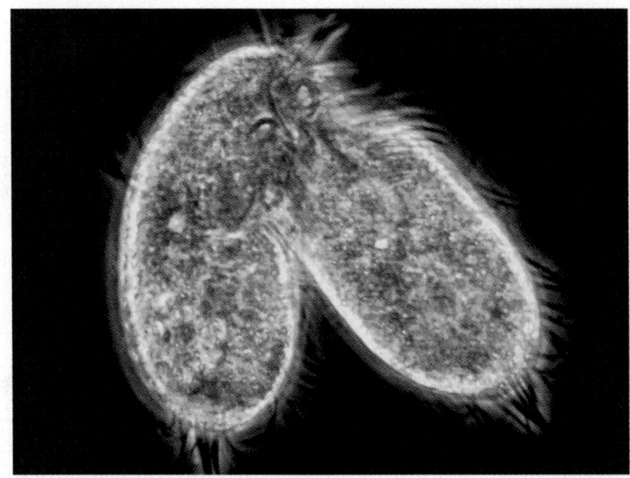

(b) Exchanging genetic material

▲ FIGURE 20-2 **Protistan reproduction and gene exchange**
(a) *Paramecium*, a ciliate, reproduces asexually by cell division. **(b)** Two *Euplotes* ciliates exchange genetic material across a cytoplasmic bridge.

QUESTION What do biologists mean when they say that sex and reproduction are uncoupled in most protists?

Protists Use Diverse Modes of Reproduction

Most protists reproduce asexually; an individual divides by mitotic cell division to yield two individuals that are genetically identical to the parent cell (**Fig. 20-2a**). Many protists, however, are also capable of sexual reproduction, in which two individuals contribute genetic material to an offspring that is genetically different from either parent. Nonreproductive processes that combine the genetic material of different individuals are also common among protists (**Fig. 20-2b**).

In many protist species that are capable of sexual reproduction, most reproduction is nonetheless asexual. Sexual reproduction occurs only infrequently, at a particular time of year or under certain circumstances, such as a crowded environment or a shortage of food. The details of sexual reproduction and the resulting life cycles vary tremendously among different types of protists. However, protist reproduction never includes the formation and development of an embryo, as occurs during reproduction in plants and animals.

Protists Affect Humans and Other Organisms

Protists have important impacts, both positive and negative, on human lives. (Some of the particular protists responsible for these impacts are described later in this chapter.) The primary positive impact actually benefits all living things and stems from the ecological role of photosynthetic marine protists. Just as plants do on land, photosynthetic protists in the oceans capture solar energy and make it available to the other organisms in the ecosystem. Thus, the marine ecosystems on which humans depend for food in turn depend on protists. Further, in the process of using photosynthesis to capture energy, the protists release oxygen gas that helps replenish the atmosphere.

On the negative side of the ledger, many human diseases are caused by parasitic protists. The diseases caused by protists include some of humanity's most prevalent ailments and some of its deadliest afflictions. Protists also cause a number of plant diseases, some of which attack crops that are important to humans. In addition to causing diseases, some marine protists release toxins that can accumulate to harmful levels in coastal areas.

20.2 WHAT ARE THE MAJOR GROUPS OF PROTISTS?

Genetic comparisons are helping systematists gain a better understanding of the evolutionary history of protist groups. Because systematists strive to devise classification systems that reflect evolutionary history, the new information has fostered a revision of protist classification. Some protist species that had been previously grouped together on the basis of physical similarity actually belong to independent evolutionary lineages that diverged from one another very early in the history of eukaryotes. Conversely, some protist groups bearing little physical resemblance to one another have been revealed to share a common ancestor, and have therefore been classified together. The process of revising protist classification, however, is far from complete. Thus, our understanding of the eukaryotic tree of life is still "under construction"; many of the branches are in place, but others await new information that will allow systematists to place them alongside their closest evolutionary relatives.

In the following sections, we will explore a brief sampling of protist diversity.

▲ **FIGURE 20-3 *Giardia:* The curse of campers** A diplomonad (genus *Giardia*) that may infect water—causing gastrointestinal disorders for the people who drink it—is shown here in a human small intestine.

Excavates Lack Mitochondria

Excavates are named for a feeding groove that gives the appearance of having been "excavated" from the surface of the cell. Excavates also lack mitochondria. It is likely that the excavates' ancestors did possess mitochondria, but these organelles were lost early in the evolutionary history of the group. The two largest groups of excavates are the diplomonads and the parabasalids.

Diplomonads Have Two Nuclei

The single cells of **diplomonads** have two nuclei and move about by means of multiple flagella. A parasitic diplomonad, *Giardia*, is an increasing problem in the United States, particularly to hikers who drink from apparently pure mountain streams. *Cysts* (tough structures that enclose the organism during one phase of its life cycle) of this diplomonad are released in the feces of infected humans, dogs, or other animals; a single gram of feces may contain 300 million cysts. Once outside the animal's body, the cysts may enter freshwater streams and even community reservoirs. If a mammal drinks infected water, the cysts develop into the adult form in the small intestine of their mammalian host (**Fig. 20-3**). In humans, infections can cause severe diarrhea, dehydration, nausea, vomiting, and cramps. Fortunately, these infections can be cured with drugs, and deaths from *Giardia* infections are uncommon.

Parabasalids Include Mutualists and Parasites

Parabasalids are anaerobic, flagellated protists named for the presence in their cells of a distinctive structure call the parabasal body. All known parabasalids live inside animals. For example, this group includes several species that inhabit the digestive systems of some species of wood-eating termites. The termites cannot themselves digest wood, but the parabasalids can. Thus, the insect and the protist are in a mutually beneficial relationship. The termite delivers food to the parabasalids in its gut; as the parabasalids digest the food, some of the energy and nutrients released become available for use by the termite.

In other cases, the host animal does not benefit from a parabasalid's presence but is instead harmed. For example, in humans, the parabasalid *Trichomonas vaginalis* causes the sexually transmitted infection trichomoniasis (**Fig. 20-4**). *Trichomonas* inhabits the mucous layers of the urinary and reproductive tracts, and uses its flagella to move along them. When conditions are favorable, a *Trichomonas* population can grow rapidly. Infected females may experience unpleasant symptoms, including vaginal itching and discharge. Infected males do not usually display symptoms, but they can transmit the infection to sexual partners.

Euglenozoans Have Distinctive Mitochondria

In most **euglenozoans,** the folds of the inner membrane of the cell's mitochondria have a distinctive shape that appears, under the microscope, as a stack of disks. Two major groups of euglenozoans are the euglenids and the kinetoplastids.

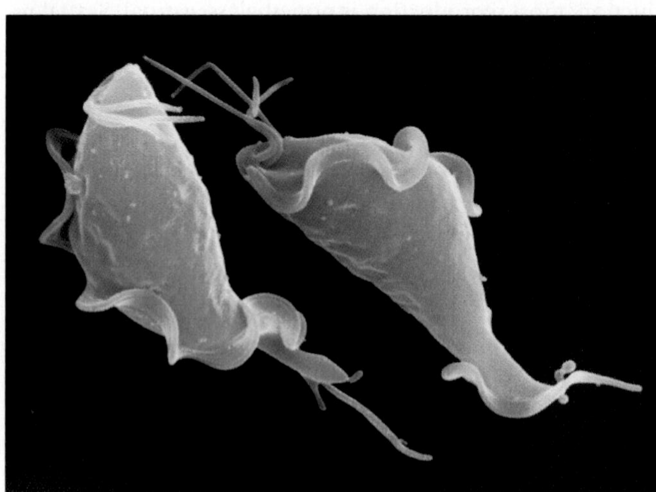

▲ **FIGURE 20-4 Trichomonas causes a sexually transmitted infection** The parabasalid *Trichomonas vaginalis* infects the urinary and reproductive tracts of men and women. Women, however, are far more likely to experience unpleasant symptoms.

Euglenids Lack a Rigid Covering and Swim by Means of Flagella

Euglenids are single-celled protists that live mostly in fresh water and are named after the group's best-known representative, *Euglena*, a complex single cell that moves about by whipping its flagellum through water (**Fig. 20-5**). Many euglenids are photosynthetic, but some species instead absorb or engulf food. Euglenids lack a rigid outer covering, so some can move by wriggling as well as by whipping their flagella. Some euglenids also possess simple light-sensing organelles consisting of a photoreceptor, called an *eyespot*, and an adjacent patch of pigment. The pigment shades the photoreceptor only when light strikes from certain directions, enabling the organism to determine the direction of the light source. Using information from the photoreceptor, the flagellum propels the protist toward light levels appropriate for photosynthesis.

Some Kinetoplastids Cause Human Diseases

The DNA in the mitochondria of **kinetoplastids** is arranged in distinctive structures called kinetoplasts. Most kinetoplastids possess at least one flagellum, which may propel the organism, sense the environment, or ensnare food. Some kinetoplastids are free living, inhabiting soil and water; others live inside other organisms in a relationship that may be mutually beneficial or parasitic. A dangerous parasitic kinetoplastid in the genus *Trypanosoma* is responsible for African sleeping sickness, a potentially fatal disease (**Fig. 20-6**). Like many parasites, this organism has a complex life cycle, part of which is spent in the tsetse fly. While feeding on the blood of a mammal, an infected fly can transmit saliva containing the trypanosome to the mammal. The parasite then develops in the new host (which may be a human) and enters the bloodstream. The trypanosome may then be ingested by another tsetse fly that bites the host, thus beginning a new cycle of infection.

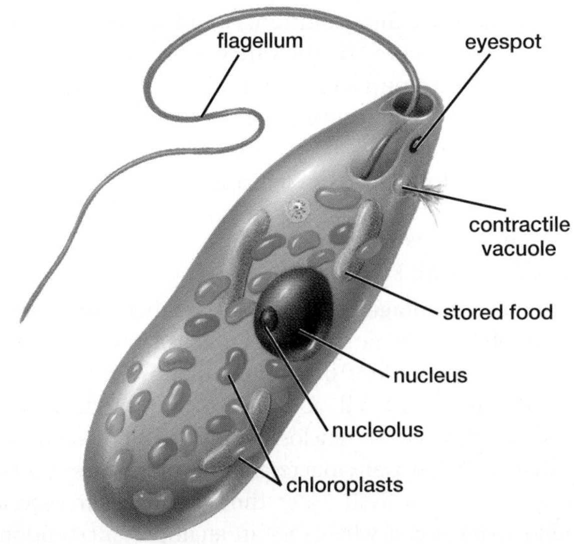

▲ **FIGURE 20-5 *Euglena*, a representative euglenid** *Euglena*'s elaborate single cell is packed with green chloroplasts, which will disappear if the protist is kept in darkness.

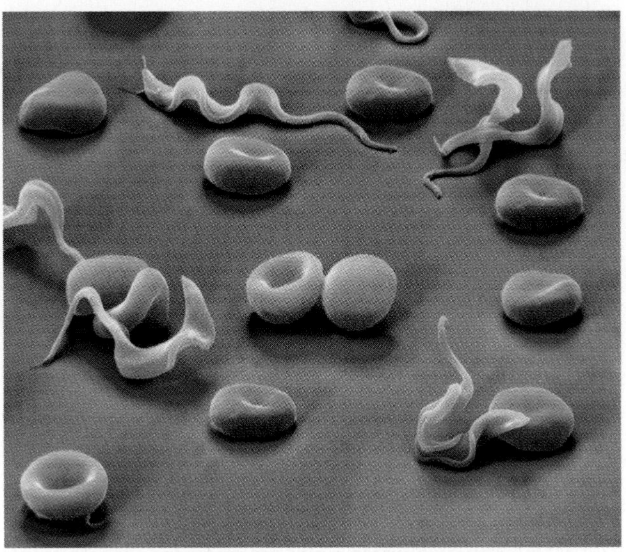

▲ **FIGURE 20-6 A disease-causing kinetoplastid** This photomicrograph shows human blood that is heavily infested with the corkscrew-shaped, parasitic kinetoplastid *Trypanosoma*, which causes African sleeping sickness. Note that the *Trypanosoma* are larger than red blood cells.

Stramenopiles Include Photosynthetic and Nonphotosynthetic Organisms

The **stramenopiles** (also known as *chromists*) form a clade whose shared ancestry was discovered through genetic comparison. All members of the group have fine, hair-like projections on their flagella (though in many stramenopiles, flagella are present only at certain stages of the life cycle). Despite their shared evolutionary history, however, stramenopiles display a wide range of forms. Some are photosynthetic and some are not; most are single celled, but some are multicellular. Three major stramenopile groups are the water molds, the diatoms, and the brown algae.

Water Molds Have Had Important Impacts on Humans

The **water molds,** or *oomycetes*, form a small group of protists, many of which are shaped as long filaments that aggregate to form cottony tufts. These tufts are superficially similar to structures produced by some fungi, but this resemblance is due to convergent evolution (see pp. 276–277), not shared ancestry. Many water molds are decomposers that live in water and damp soil. Some species have profound economic impacts on humans. For example, a water mold causes a disease of grapes, known as *downy mildew* (**Fig. 20-7**). Its inadvertent introduction into France from the United States in the late 1870s nearly destroyed the French wine industry. Another oomycete has destroyed millions of avocado trees in California; still another is responsible for *late blight*, a devastating disease of potatoes. When this protist was accidentally introduced into Ireland in about 1845, it destroyed nearly the entire potato crop, causing a devastating potato famine during which as many as 1 million people in Ireland starved and many more emigrated to the United States.

▲ **FIGURE 20-7 A parasitic water mold** Downy mildew, a plant disease caused by the water mold *Plasmopara*, nearly destroyed the French wine industry in the 1870s.

QUESTION Although water molds are stramenopiles, they look and function very much like fungi. What is the cause of this similarity?

Diatoms Encase Themselves Within Glassy Walls

The **diatoms,** photosynthetic stramenopiles found in both fresh and salt water, produce protective shells of *silica* (glass), some of exceptional beauty (**Fig. 20-8**). These shells consist of top and bottom halves that fit together like a pillbox or petri dish. Accumulations of diatoms' glassy walls over millions of years have produced fossil deposits of "diatomaceous earth" that may be hundreds of meters thick. This slightly abrasive substance is widely used in products such as toothpaste and metal polish.

▲ **FIGURE 20-8 Some representative diatoms** This photomicrograph illustrates the intricate, microscopic beauty and variety of the glassy walls of diatoms.

Diatoms form part of the **phytoplankton,** the single-celled photosynthesizers that float passively in the upper layers of Earth's lakes and oceans. Phytoplankton play an immensely important ecological role. Marine phytoplankton account for nearly 70% of all the photosynthetic activity on Earth, absorbing carbon dioxide, recharging the atmosphere with oxygen, and supporting the complex web of aquatic life.

Brown Algae Dominate in Cool Coastal Waters

Though most photosynthetic protists, such as diatoms, are single celled, some form multicellular aggregations that are commonly known as seaweeds. Although some seaweeds seem to resemble plants, they lack many of the distinctive features of plants. For example, none of the seaweeds have roots or shoots, and none form embryos during reproduction.

The stramenopiles include one group of seaweeds, the brown algae, which are named for the brownish-yellow pigments that (in combination with green chlorophyll) increase the seaweed's light-gathering ability. Almost all brown algae are marine. The group includes the dominant seaweed species that dwell along rocky shores in the temperate (cooler) oceans of the world, including the eastern and western coasts of the United States. Brown algae live in habitats ranging from nearshore, where they cling to rocks that are exposed at low tide, to far offshore. Several species use gas-filled floats to support their bodies (**Fig. 20-9a**). Some of the giant kelp found along the Pacific coast reach heights of 325 feet (100 meters), and may grow more than 6 inches (15 centimeters) in a single day. With their dense growth and towering height, kelp form undersea forests that provide food, shelter, and breeding areas for marine animals (**Fig. 20-9b**).

Alveolates Include Parasites, Predators, and Phytoplankton

The **alveolates** are single-celled organisms that have distinctive, small cavities beneath the surface of their cells. As with the stramenopiles, the evolutionary link among the alveolates was long obscured by the variety of structures and ways of life among the group members, but was revealed by molecular comparisons. Some alveolates are photosynthetic, some are parasitic, and some are predatory. The major alveolate groups are the dinoflagellates, apicomplexans, and ciliates.

Dinoflagellates Swim by Means of Two Whip-Like Flagella

Though most **dinoflagellates** are photosynthetic, there are also some nonphotosynthetic species. Dinoflagellates are named for the two whip-like flagella that propel them (**Fig. 20-10**). One flagellum encircles the cell, and the second projects behind it. Some dinoflagellates are enclosed only by a cell membrane; others have cellulose walls that resemble armor plates. Although some species live in fresh water, dinoflagellates are especially abundant in the ocean, where they are an important component of the phytoplankton and a food source for larger organisms. Many dinoflagellates are bioluminescent, producing a brilliant blue-green light when disturbed by motion in the water.

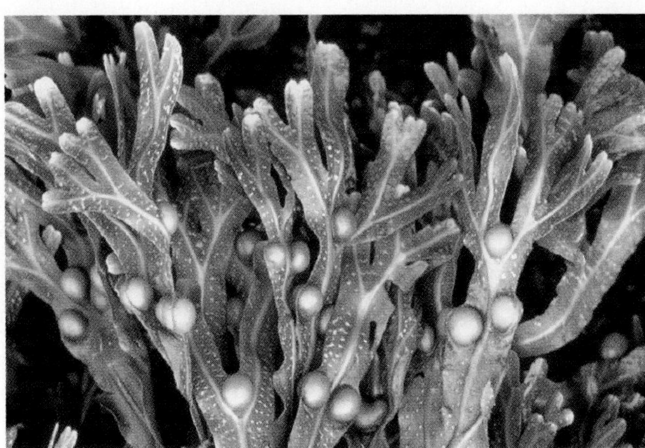

(a) *Fucus*

(b) Kelp forest

▲ **FIGURE 20-9 Brown algae, a multicellular protist**
(a) *Fucus*, a genus found near shores, is shown here exposed at low tide. Notice the gas-filled floats, which provide buoyancy in water. **(b)** The giant kelp *Macrocystis* forms underwater forests off southern California.

Warm water that is rich in nutrients may bring on a dinoflagellate population explosion. Dinoflagellates can become so numerous that the water is dyed red by the color of their bodies, causing a "red tide" (**Fig. 20-11**). During red tides, fish die by the thousands, suffocated by clogged gills or by the oxygen depletion that results from the decay of billions of dinoflagellates. One type of dinoflagellate, *Pfiesteria*, even eats fish directly, first secreting chemicals that dissolve the victim's flesh. But dinoflagellate explosions can benefit oysters, mussels, and clams, which have a feast, filtering millions of the protists from the water and consuming them. In the process, however, the mollusks' bodies accumulate concentrations of a nerve toxin produced by the dinoflagellates. Dolphins, seals, sea otters, and humans who eat affected mollusks may be stricken with potentially lethal paralytic shellfish poisoning.

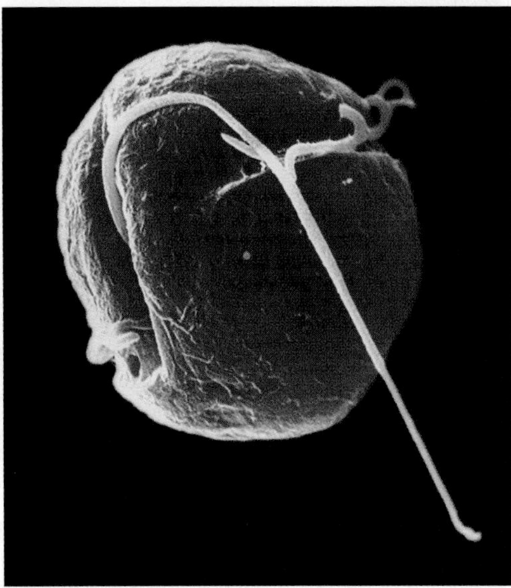

▲ **FIGURE 20-10 A dinoflagellate** This dinoflagellate has two flagella: a longer one that extends from a slot, and a shorter one that lies in a groove that encircles the cell.

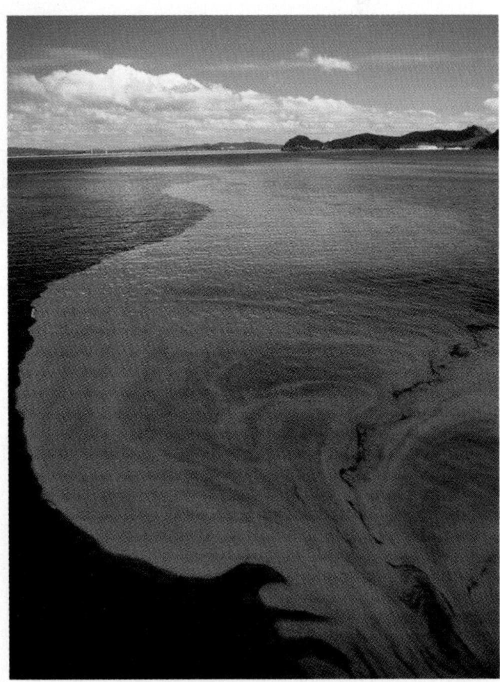

▲ **FIGURE 20-11 A red tide** The explosive reproductive rate of certain dinoflagellates under the right environmental conditions can produce concentrations so great that their microscopic bodies dye the seawater red or brown.

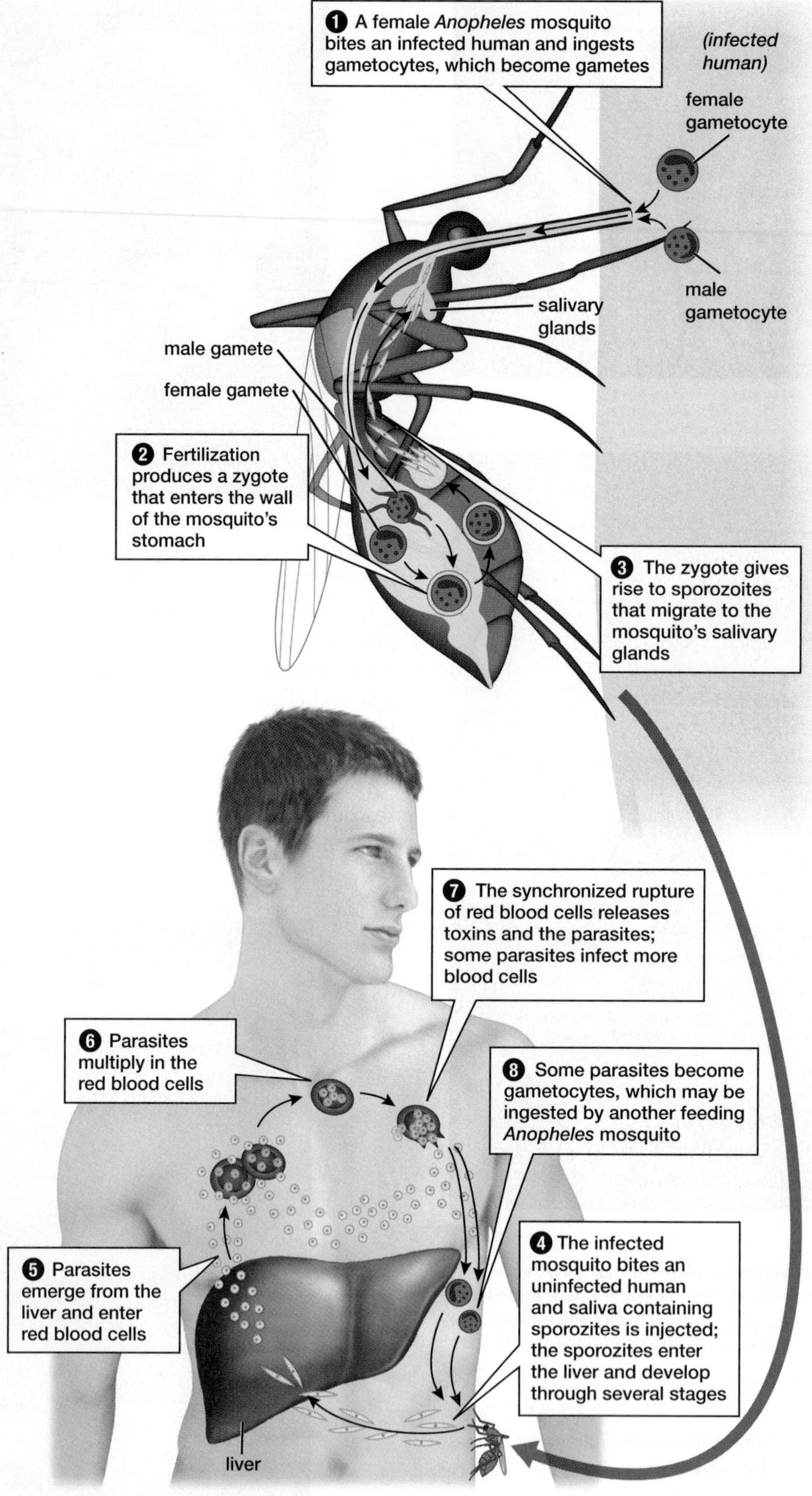

1 A female *Anopheles* mosquito bites an infected human and ingests gametocytes, which become gametes
(infected human)
female gametocyte
male gametocyte
salivary glands
male gamete
female gamete
2 Fertilization produces a zygote that enters the wall of the mosquito's stomach
3 The zygote gives rise to sporozoites that migrate to the mosquito's salivary glands
7 The synchronized rupture of red blood cells releases toxins and the parasites; some parasites infect more blood cells
6 Parasites multiply in the red blood cells
8 Some parasites become gametocytes, which may be ingested by another feeding *Anopheles* mosquito
5 Parasites emerge from the liver and enter red blood cells
4 The infected mosquito bites an uninfected human and saliva containing sporozoites is injected; the sporozoites enter the liver and develop through several stages
liver

▲ FIGURE 20-12 **The life cycle of the malaria parasite**

Case Study continued
Green Monster

Red tides have become increasingly common in recent years. One reason for this increased incidence is that dinoflagellate species that can cause red tides have been inadvertently spread around the world by humans. The dinoflagellates travel mainly in seawater that is pumped into the ballast tanks of cargo ships and then discharged at distant ports. When released into new waters, populations of toxin-producing dinoflagellates may grow explosively, just as the invasive *Caulerpa* seaweed often spreads uncontrollably when introduced to environments free of its normal predators and parasites.

Apicomplexans Are Parasitic and Have No Means of Locomotion

All **apicomplexans** (sometimes known as *sporozoans*) are parasitic, living inside the bodies and sometimes inside the individual cells of their hosts. They form infectious spores—resistant structures transmitted from one host to another through food, water, or the bite of an infected insect. As adults, apicomplexans have no means of locomotion. Many have complex life cycles, a common feature of parasites. A well-known example is the malarial parasite *Plasmodium* (**Fig. 20-12**). Parts of its life cycle are spent in the body of a female *Anopheles* mosquito. The mosquito is not harmed by the presence of *Plasmodium,* and may eventually bite a human and pass the protist to the unfortunate victim. The protist develops in the victim's liver, then enters the blood, where it reproduces rapidly in red blood cells. The release of large quantities of spores through the rupture of the blood cells causes the recurrent fever of malaria. Uninfected mosquitoes may acquire the parasite by feeding on the blood of a malaria victim, spreading the parasite when they bite another person.

Ciliates Are the Most Complex of the Alveolates

Ciliates, which inhabit fresh and salt water, represent the peak of unicellular complexity. They possess many specialized organelles, including **cilia** (singular, cilium), the short hair-like outgrowths for which they are named. The cilia may cover the cell or may be localized. In the well-known freshwater genus *Paramecium,* rows of cilia cover the organism's entire body surface (**Fig. 20-13**). Their coordinated beating propels the cell through the water at a rate of 1 millimeter per second—a protistan speed record. Although only a single cell, *Paramecium* responds to its environment as if it had a well-developed nervous system. Confronted with a noxious chemical or a physical barrier, the cell immediately backs up by reversing the beating of its cilia and then proceeds in a new direction. Some ciliates, such as *Didinium,* are accomplished predators (**Fig. 20-14**).

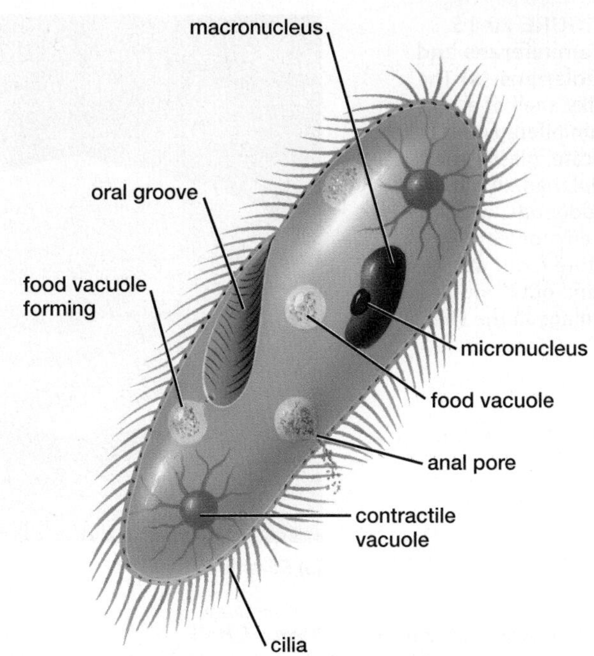

▲ FIGURE 20-13 **The complexity of ciliates** The ciliate *Paramecium* illustrates some important ciliate organelles. The oral groove acts as a mouth, food vacuoles—miniature digestive systems—form at its apex, and waste is expelled by exocytosis through an anal pore. Contractile vacuoles regulate water balance.

▲ FIGURE 20-14 **A microscopic predator** In this scanning electron micrograph, the predatory ciliate *Didinium* attacks a *Paramecium.* Notice that the cilia of *Didinium* are confined to two bands, whereas *Paramecium* has cilia over its entire body. Ultimately, the predator will engulf and consume its prey. This microscopic drama could take place on a pinpoint with room to spare.

Rhizarians Have Thin Pseudopods

Protists in a number of different groups possess flexible plasma membranes that they can extend in any direction to form finger-like projections called pseudopods, which they use in locomotion and for engulfing food. The pseudopods of **rhizarians** are thin and thread-like. In many species in this group, the pseudopods extend through hard shells. Rhizarians include the foraminiferans and the radiolarians.

(a) Foraminifera

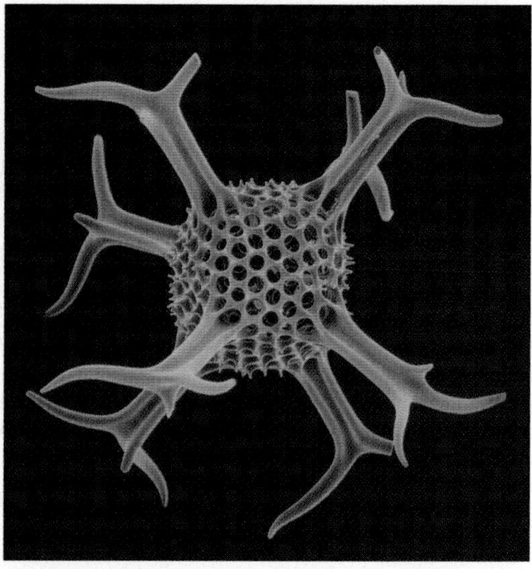

(b) A radiolarian

▶ **FIGURE 20-15 Foraminiferans and radiolarians (a)** The chalky shell of a foraminiferan, and **(b)** the delicate, glassy shell of a radiolarian. In life, thin pseudopods, which sense the environment and capture food, would extend out through the openings in the shells.

Fossil Foraminiferan Shells Form Chalk

The **foraminiferans** are primarily marine protists that produce beautiful shells. Their shells are constructed mostly of calcium carbonate (chalk; Fig. 20-15a). These elaborate shells are pierced by myriad openings through which pseudopods extend. The chalky shells of dead foraminiferans, sinking to the ocean bottom and accumulating over millions of years, have resulted in immense deposits of limestone, such as those that form the famous White Cliffs of Dover, England.

Radiolarians Have Glassy Shells

Like foraminiferans, **radiolarians** have thin pseudopods that extend through hard shells. The shells of radiolarians, however, are made of glass-like silica (Fig. 20-15b). In some areas of the ocean, radiolarian shells raining down over vast stretches of time have accumulated to form thick layers of "radiolarian ooze."

Amoebozoans Inhabit Aquatic and Terrestrial Environments

Amoebozoans move by extending finger-shaped pseudopods, which may also be used for feeding. They generally do not have shells. The major groups of amoebozoans are the amoebas and the slime molds.

Amoebas Have Thick Pseudopods and No Shells

Amoebas, sometimes known as *lobose amoebas* to distinguish them from other protists that have pseudopods, are common in freshwater lakes and ponds (Fig. 20-16). Many amoebas are predators that stalk and engulf prey, but some species are parasites. One parasitic form causes amoebic dysentery, a disease that is prevalent in warm climates. The dysentery-causing amoeba multiplies in the intestinal wall, triggering severe diarrhea.

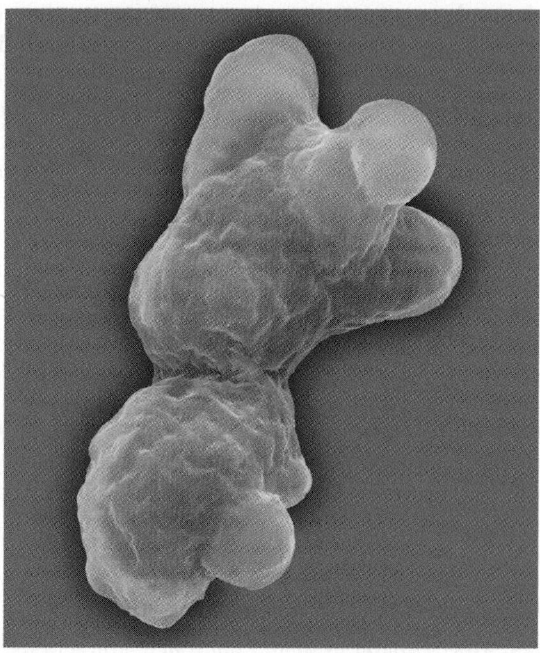

▲ **FIGURE 20-16 An amoeba** Lobose amoebas are active predators that move through water to engulf food with thick, blunt pseudopods.

Slime Molds Are Decomposers That Inhabit the Forest Floor

The physical form of *slime molds* seems to blur the boundary between a colony of different individuals and a single, multicellular individual. The life cycle of the slime mold consists of two phases: a mobile feeding stage and a stationary reproductive stage called a *fruiting body*. There are two main types of slime mold: acellular and cellular.

Acellular Slime Molds Form a Multinucleate Mass of Cytoplasm Called a Plasmodium The **acellular slime molds,** also known as *plasmodial* slime molds, consist of a mass of cytoplasm that may spread thinly over an area of several square yards. Although the mass contains thousands of diploid nuclei, the nuclei are not confined in separate cells surrounded by plasma membranes, as in most multicellular organisms. This structure, called a **plasmodium,** explains why these protists are described as "acellular" (without cells). The plasmodium oozes through decaying

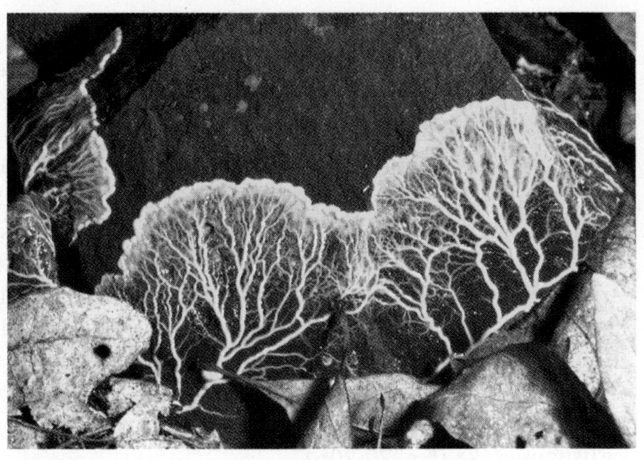

(a) Plasmodium

(b) Fruiting bodies

▲ **FIGURE 20-17 An acelluar slime mold (a)** A plasmodium oozes over a stone on the damp forest floor. **(b)** When food becomes scarce, the mass differentiates into fruiting bodies in which spores are formed.

leaves and rotting logs, engulfing food such as bacteria and particles of organic material. The mass may be bright yellow or orange; a large plasmodium can be rather startling (**Fig. 20-17a**). Dry conditions or starvation stimulate the plasmodium to form a fruiting body, on which haploid spores are produced (**Fig. 20-17b**). The spores are dispersed and germinate under favorable conditions, eventually giving rise to a new plasmodium.

Cellular Slime Molds Live as Independent Cells But Aggregate into a Pseudoplasmodium When Food Is Scarce The **cellular slime molds** live in soil as independent haploid cells that move and feed by producing pseudopods. In the best-studied genus, *Dictyostelium*, individual cells release a chemical signal when food becomes scarce. This signal attracts nearby cells into a dense aggregation that forms a slug-like mass called a **pseudoplasmodium** ("false plasmodium") because, unlike a true plasmodium, it consists of individual cells (**Fig. 20-18**). A pseudoplasmodium can be viewed as a colony of individuals, because the cells that compose it are not all genetically identical. In some ways, however, a pseudoplasmodium is more like a multicellular organism, because its cells differentiate into different cell types, with different types serving different functions. A pseudoplasmodium moves about in slug-like fashion, migrating toward a sunlit spot, where its cells differentiate to

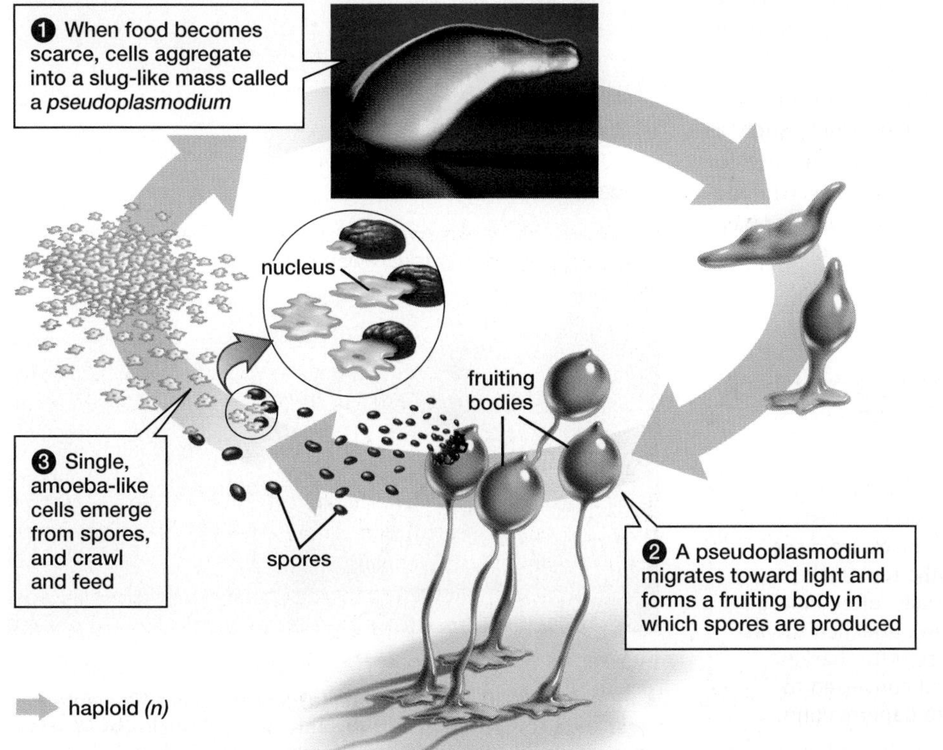
❶ When food becomes scarce, cells aggregate into a slug-like mass called a *pseudoplasmodium*
nucleus
fruiting bodies
❸ Single, amoeba-like cells emerge from spores, and crawl and feed
spores
❷ A pseudoplasmodium migrates toward light and forms a fruiting body in which spores are produced
haploid *(n)*

◀ **FIGURE 20-18 The life cycle of a cellular slime mold**

▲ **FIGURE 20-19 Red algae** Red coralline algae from the Mediterranean Sea. Coralline algae, which deposit calcium carbonate within their bodies, also contribute to coral reefs in tropical waters.

convert the structure to a fruiting body. Haploid spores formed within the fruiting body are dispersed by wind and germinate directly into new single-celled individuals.

Red Algae Live Primarily in Clear Tropical Oceans

The red algae are multicellular, photosynthetic seaweeds (**Fig. 20-19**). These protists range in color from bright red to nearly black, and they derive their color from red pigments that mask their green chlorophyll. Red algae are found almost exclusively in marine environments. They dominate in deep, clear tropical waters, where their red pigments absorb the deeply penetrating blue-green light and transfer this light energy to chlorophyll, where it is used in photosynthesis.

Some species of red algae deposit calcium carbonate (which forms limestone) in their tissues, and contribute to the formation of reefs. Red algae also contain certain gelatinous substances with commercial uses, including carrageenan (used as a thickening agent in products such as paints, cosmetics, and ice cream) and agar (a substrate for growing bacterial colonies in laboratories). However, the major importance of these and other algae lies in their photosynthetic ability; the energy they capture helps support nonphotosynthetic organisms in marine ecosystems.

Have you ever wondered 

What Sushi Wrappers Are Made Of?

If you like sushi, you've probably eaten a sushi roll, in which rice and other items are surrounded by a tasty, blackish-green wrapper. The wrapper is made from the dried bodies of a multicellular protist, the red alga *Porphyra*. *Porphyra* is grown commercially, often in large coastal "farms" where the seaweed grows attached to vast nets suspended from the ocean's surface. After harvest, the seaweeds are shredded, pulped, and converted to dried sheets in a process very similar to papermaking.

Green Algae Live Mostly in Ponds and Lakes

Green algae, a large and diverse group of photosynthetic protists, include both multicellular and unicellular species. Most species live in freshwater ponds and lakes, but some live in the seas. Some green algae, such as *Spirogyra*, form thin filaments from long chains of cells (**Fig. 20-20a**). Other species of green algae form colonies containing clusters of cells that are somewhat interdependent and constitute a structure intermediate between unicellular and multicellular forms. These colonies range from a few cells to a few thousand cells, as in species of *Volvox*. Most green algae are small, but some marine species are large. For example, the green alga *Ulva*, or sea lettuce, is similar in size to the leaves of its namesake (**Fig. 20-20b**).

The green algae are of special interest because, unlike other groups that contain multicellular, photosynthetic protists, green algae are closely related to plants. Plants share a common ancestor with some types of green algae, and many researchers believe that the very earliest plants were similar to today's multicellular green algae.

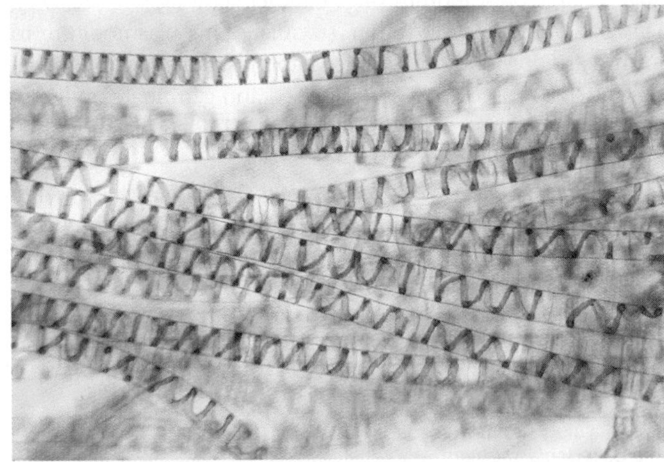

(a) ***Spirogyra***

(b) ***Ulva***

▲ **FIGURE 20-20 Green algae (a)** *Spirogyra* is a filamentous green alga composed of strands only one cell thick. **(b)** *Ulva* is a multicellular green alga that assumes a leaflike shape.

Case Study revisited
Green Monster

Caulerpa taxifolia, the invasive seaweed that threatens to overrun the Mediterranean, is a green alga. This species and other members of its genus have very unusual bodies. Outwardly, they appear plantlike, with rootlike structures that attach to the seafloor and with other structures that look like stems and leaves, rising to a height of several inches. Despite its seeming similarity to a plant, however, a *Caulerpa* body consists of a single, extremely large cell. The entire body is surrounded by a single, continuous cell membrane. The interior consists of cytoplasm that contains numerous cell nuclei but is not subdivided. That a single cell can take such a complex shape is extraordinary.

A potential problem with *Caulerpa*'s single-celled organization might arise when its body is damaged, perhaps by wave action or when a predator takes a bite out of it. When the cell membrane is breached, there is nothing to prevent all of the cytoplasm from leaking out, an event that would be fatal. But *Caulerpa* has evolved a defense against this potential calamity. Shortly after the cell membrane breaks, it is quickly filled with a "wound plug" that closes the gap. After the plug is established, the cell begins to grow and regenerates any lost portion of the body.

This ability to regenerate is a key component of the ability of aquarium-strain *Caulerpa taxifolia* to spread rapidly in new environments. If part of a *Caulerpa* body breaks off and drifts to a new location, it can regenerate a whole new body. The regenerated individual becomes the founder of a new, quickly growing colony—and these quickly growing colonies might appear anywhere in the world. Authorities in many countries worry that the aquarium strain of *Caulerpa* could invade their coastal waters, unwittingly transported by ships from the Mediterranean or released by careless aquarists. In fact, invasive *Caulerpa* is no longer restricted to the Mediterranean. It has been found in two locations on the California coast, and in at least eight bodies of water in Australia. Local authorities in both countries have attempted to control the invading algae, but it is impossible to say if their efforts will be successful. *Caulerpa taxifolia* is a resourceful foe.

Consider This

Is it important to stop the spread of *Caulerpa*? Governments invest substantial resources to combat introduced species and prevent their populations from increasing and dispersing. Why might this be a wise use of funds? Can you think of some arguments against spending time and money for this purpose?

CHAPTER REVIEW

Summary of Key Concepts

20.1 What Are Protists?

"Protist" is a term of convenience that refers to any eukaryote that is not a plant, animal, or fungus. Most protists are single, highly complex eukaryotic cells, but some form colonies, and some, such as seaweeds, are multicellular. Protists exhibit diverse modes of nutrition, reproduction, and locomotion. Photosynthetic protists form much of the phytoplankton, which plays a key ecological role. Some protists cause human diseases; others are crop pests.

20.2 What Are the Major Groups of Protists?

Protist groups include excavates (diplomonads and parabasalids), euglenozoans (euglenids and kinetoplastids), stramenopiles (water molds, diatoms, and brown algae), alveolates (dinoflagellates, apicomplexans, and ciliates), rhizarians (foraminiferans and radiolarians), amoebozoans (amoebas and slime molds), red algae, and green algae (the closest relatives of plants).

Key Terms

acellular slime mold *380*
alga (plural, algae) *373*
alveolate *376*
amoeba *380*
amoebozoan *380*
apicomplexan *379*
cellular slime mold *381*
ciliate *379*
cilium (plural, cilia) *379*
diatom *376*
dinoflagellate *376*
diplomonad *374*
euglenid *375*
euglenozoan *374*
excavate *374*
foraminiferan *380*
kinetoplastid *375*
parabasalid *374*
phytoplankton *376*
plasmodium *380*
protist *371*
protozoan (plural, protozoa) *373*
pseudoplasmodium *381*
pseudopod *371*
radiolarian *380*
rhizarian *379*
stramenopile *375*
water mold *375*

Thinking Through the Concepts

Fill-in-the-Blank

1. Many predatory protists engulf prey with finger-shaped extensions called __________. Protists that absorb nutrients from their surroundings may act as __________ of dead organic matter, or as harmful __________ of larger living organisms.
2. Photosynthetic protists are collectively known as __________; nonphotosynthetic, single-celled protists are collectively known as __________.
3. Protist chloroplasts surrounded by four-layer membranes arose evolutionarily through __________, in which an ancestral nonphotosynthetic protist engulfed but did not digest a(n) __________.
4. The disease-causing parasite *Giardia* is a member of the __________ group; the protist that causes malaria is a member of the __________ group; and the protist that causes sleeping sickness is a member of the __________ group.

5. The plant diseases downy mildew and late blight are caused by protists in the __________ group. Slime molds are members of the __________ group.
6. Protists that make up a large proportion of Earth's phytoplankton include __________ and __________. The protist group most closely related to land plants is __________.

Review Questions

1. List the major differences between prokaryotes and protists.
2. What is secondary endosymbiosis?
3. What is the importance of dinoflagellates in marine ecosystems? What can happen when they reproduce rapidly?
4. What is the major ecological role played by single-celled algae?
5. Which protist group consists entirely of parasitic forms?
6. Which protist groups include seaweeds?
7. Which protist groups include species that use pseudopods?

Applying the Concepts

1. Recent research shows that ocean water off southern California has become 2° to 3°F (1° to 1.5°C) warmer during the past four decades, possibly due to the greenhouse effect. This warming has indirectly led to a depletion of nutrients in the water and thus a decline in photosynthetic protists such as diatoms. What effects is this warming likely to have on life in the oceans?
2. The internal structure of many protists is much more complex than that of cells of multicellular organisms. Does this mean that the protist is engaged in more-complex activities than the multicellular organism is? If not, why are protistan cells more complicated?
3. Why would the lives of multicellular animals be impossible if prokaryotic and protistan organisms did not exist?

MB *Go to www.masteringbiology.com for practice quizzes, activities, eText, videos, current events, and more.*

The Diversity of Plants

Chapter 21

Case Study

Queen of the Parasites

THE FLOWER OF THE STINKING CORPSE LILY makes a strong impression. For one thing, it's huge; a single flower may be 3 feet across. It also has a rather strange appearance, consisting largely of fleshy lobes that are almost fungus-like. But the thing that makes a stinking corpse lily almost impossible to ignore is its aroma, which has been described as "a penetrating smell more repulsive than any buffalo carcass in an advanced stage of decomposition."

Close examination of a stinking corpse lily reveals that it has no visible leaves, roots, or stems. In fact, it is a parasite, and its body is completely embedded in the tissue of its host, a vine of the genus *Tetrastigma*. Without leaves, the stinking corpse lily cannot produce any food of its own, but instead draws all of its nutrition from its host. The parasite becomes visible outside the body of its host only when one of its cabbage-shaped flower buds pushes through the surface of the host's stem and its gigantic, stinking flower opens for a week or so before shriveling and falling off. If a male and a female flower happen to be open simultaneously and close together, the female flower may be fertilized and produce seeds. A seed that is dispersed in the droppings of an animal that has consumed it, and that happens to land on a *Tetrastigma* stem, may germinate and penetrate a new host.

When you think of plants, you might first think of their most obvious feature: green leaves that capture solar energy by photosynthesis. It may seem odd, then, that this chapter about plants begins with a peculiar plant that does not photosynthesize. Oddities such as the stinking corpse lily, however, serve as reminders that evolution does not always follow a predictable pathway, and that even an adaptation as seemingly valuable as the ability to live on sunlight can be discarded.

▲ The huge, foul-smelling flower of the stinking corpse lily is a treat for visitors to Asian rain forests.

At a Glance

21.1 WHAT ARE THE KEY FEATURES OF PLANTS?

Plants are the most conspicuous living things in almost every landscape on Earth. Unless you are in a polar region, a harsh desert, or a densely populated urban area, you live surrounded by plants. The plants that dominate Earth's forests, grasslands, parks, lawns, orchards, and farm fields are such familiar parts of the backdrop to our daily lives that we tend to take them for granted. But if we take some time to look more closely at our green companions, we might gain a greater appreciation for the adaptations that have made them so successful, and for the properties that make them essential to our own survival.

What distinguishes members of the plant kingdom from other organisms? Perhaps the most noticeable feature of plants is their green color. The color comes from the presence of the pigment chlorophyll in many plant tissues. Chlorophyll plays a crucial role in photosynthesis, the process by which plants use energy from sunlight to convert water and carbon dioxide to sugar. Chlorophyll and photosynthesis, however, are not unique to plants; they are also present in many types of protists and prokaryotes. Instead, the feature that distinguishes plants from other photosynthetic organisms is their multicellular embryos.

Plants Have Multicellular, Dependent Embryos

The multicellular embryo of a plant is retained within and receives nutrients from the tissues of the parent plant. That is, a plant embryo is attached to and dependent on its parent as it grows and develops. Such multicellular, dependent embryos are not found among photosynthetic protists; they distinguish plants from algae.

Plants Have Alternating Multicellular Haploid and Diploid Generations

Plant reproduction is characterized by a type of life cycle called **alternation of generations** (**Fig. 21-1**). In organisms with alternation of generations, separate diploid and haploid generations alternate with one another. (Recall that a diploid organism has two sets of chromosomes; a haploid organism, one set.) In the diploid generation, the body consists of diploid cells and is known as the **sporophyte.** (In plants, the multicellular embryos described in the previous section are part of the diploid sporophyte generation.) Certain cells of sporophytes undergo meiosis to produce haploid reproductive cells called *spores*. The haploid spores develop into multicellular, haploid bodies called **gametophytes.**

A gametophyte ultimately produces male and female haploid *gametes* (sperm and eggs) by mitosis. Gametes, like spores, are reproductive cells but, unlike spores, an individual gamete by itself cannot develop into a new individual. Instead, two gametes of opposite sexes must meet and fuse to form a new individual. In plants, gametes produced by gametophytes fuse to form a diploid *zygote* (a fertilized egg), which develops into a diploid embryo. The embryo develops into a mature sporophyte, and the cycle begins again.

21.2 HOW DO PLANTS AFFECT OTHER ORGANISMS?

As plants survive, grow, and reproduce, they alter and influence Earth's landscape and atmosphere in ways that are tremendously beneficial to the rest of the planet's inhabitants, including humans. Humans also reap additional benefits by actively exploiting plants.

Plants Play a Crucial Ecological Role

The complex ecosystems that host terrestrial life could not be maintained without the help of plants. Plants make vital contributions to the food, air, soil, and water that sustain life on land.

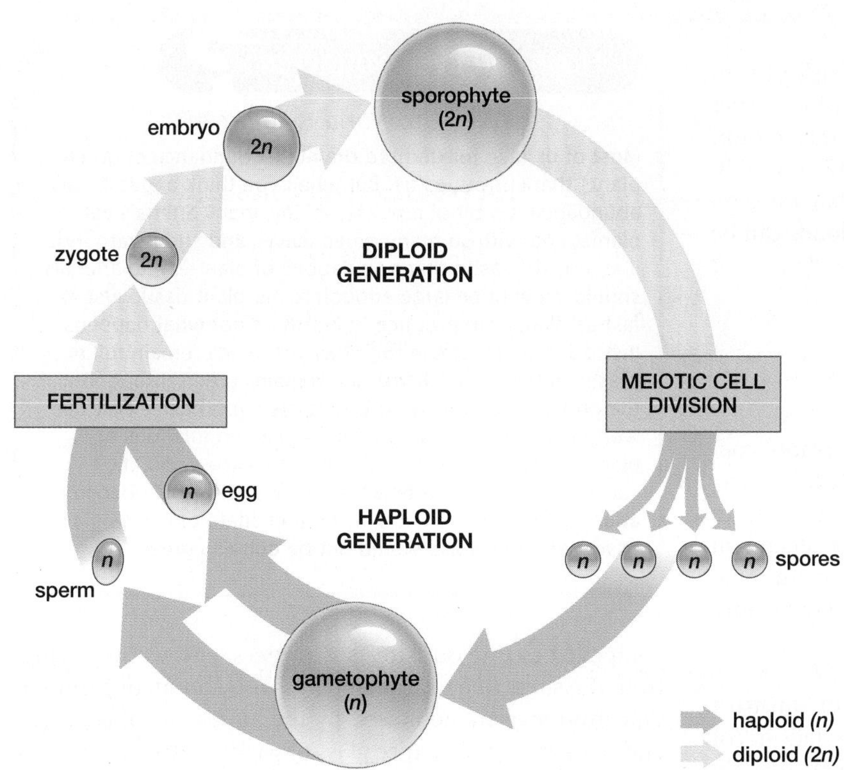

◀ **FIGURE 21-1 Alternation of generations in plants** As shown in this generalized depiction of a plant life cycle, a diploid sporophyte generation produces haploid spores through meiotic cell division. The spores develop into a haploid gametophyte generation that produces haploid gametes by mitotic cell division. The fusion of these gametes results in a diploid zygote that develops into the sporophyte plant.

Plants Capture Energy That Other Organisms Use

Plants provide food, directly or indirectly, for all of the animals, fungi, and nonphotosynthetic microbes on land. Plants use photosynthesis to capture solar energy, and they convert part of the captured energy into leaves, shoots, seeds, and fruits that are eaten by other organisms. Many of these consumers of plant tissue are themselves eaten by still other organisms. Plants are the main providers of energy and nutrients to terrestrial ecosystems, and life on land depends on plants' ability to manufacture food from sunlight.

Plants Help Maintain the Atmosphere

In addition to their role as food suppliers, plants make essential contributions to the atmosphere. For example, plants produce oxygen gas as a by-product of photosynthesis, and by doing so they continually replenish oxygen in the atmosphere. Without plants' contribution, atmospheric oxygen would be rapidly depleted by the oxygen-consuming respiration of Earth's multitude of organisms.

Plants Build Soil

Plants also help create and maintain soil. When a plant dies, its stems, leaves, and roots become food for fungi, prokaryotes, and other decomposers. Decomposition breaks the plant tissue into tiny particles of organic matter that become part of the soil. Organic matter improves the ability of soil to hold water and nutrients, thereby making the soil more fertile and better able to support the growth of living plants. The roots of those living plants help hold the soil together and keep it in place. Soils from which vegetation has been removed are susceptible to erosion by wind and water (**Fig. 21-2**).

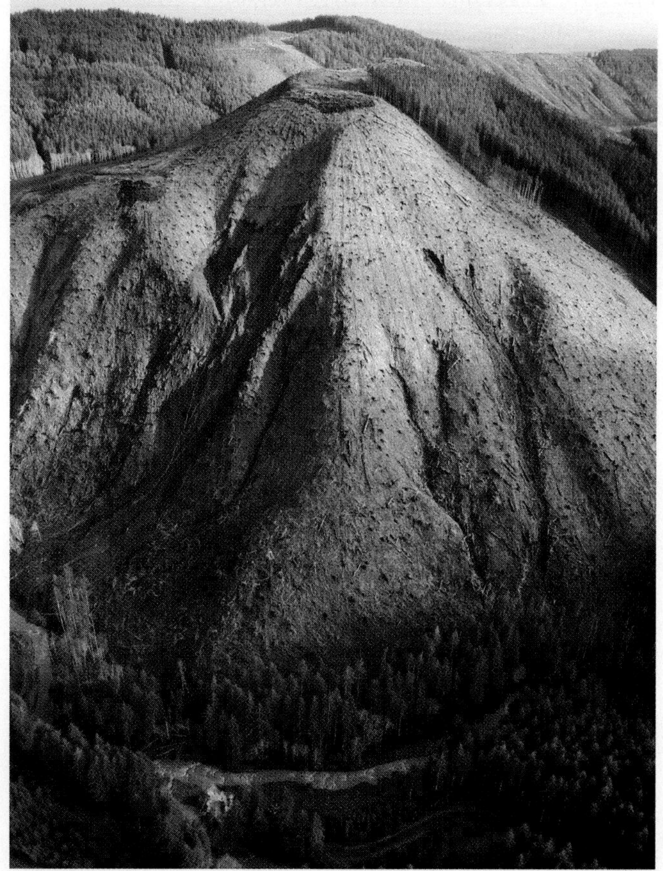

▲ **FIGURE 21-2 Plants protect soil** Damage to natural vegetation, such as the deforestation of this mountainside, leaves the underlying soil vulnerable to erosion.

Plants Help Keep Ecosystems Moist

Plants take up water from the soil, and retain much of it in their tissues. By doing so, plants slow the rate at which water escapes from terrestrial ecosystems, and increase the amount of water available to meet the needs of the ecosystems' inhabitants. By reducing the amount of water runoff, plants also reduce the chances of destructive flooding. Thus, floods can be more frequent in areas in which forests, grasslands, or marshes have been destroyed by human activities.

Plants Provide Humans with Necessities and Luxuries

All inhabitants of terrestrial ecosystems depend on plants' contributions to those ecosystems, but reliance on plants is especially pronounced for humans. It would be difficult to exaggerate the degree to which human populations depend on plants. Neither our explosive population growth nor our rapid technological advance would have been possible without plants.

Plants Provide Shelter, Fuel, and Medicine

Plants are the source of the wood that is used to construct housing for a large portion of Earth's human population. For much of human history, wood was also the main fuel for warming dwellings and for cooking. Wood is still the most important fuel in many parts of the world. Coal, another important fuel, is composed of the remains of ancient plants that have been transformed by geological processes.

Plants have also supplied many of the medicines on which modern health care depends. Important drugs that were originally found in and extracted from plants include aspirin, the heart medication digoxin, the cancer treatments Taxol® and vinblastine, the malaria drug quinine, the painkillers codeine and morphine, and many more.

In addition to harvesting useful material from wild plants, humans have domesticated a host of useful plant species. Through generations of selective breeding, people have modified the seeds, stems, roots, flowers, and fruits of favored plant species in order to provide themselves with food and fiber. It is difficult to imagine life without corn, rice, potatoes, apples, tomatoes, cooking oil, cotton, and the myriad other staples that domestic plants provide.

Plants Provide Pleasure

Despite the obvious contributions of plants to human well-being, our relationship with plants seems to be based on something more profound than their ability to help us meet our material needs. Though we appreciate the practical value of wheat and wood, our most emotionally powerful connections with plants are purely sensual. Many of life's pleasures come to us courtesy of our plant partners. We delight in the beauty and fragrance of flowers, and present them to others as symbols of our most sublime and inexpressible emotions. Quite a few of us spend hours of our leisure time tending gardens and lawns, for no reward other than the pleasure and satisfaction we derive from observing the fruits of our labor. In our homes, we reserve space not only for members of our families, but also for our houseplant companions. We feel compelled to line our streets with trees, and we seek refuge from the stress of daily life in parks with abundant plant life. Our mornings are enhanced by the aroma of coffee or tea, and our evenings by a nice glass of wine. Clearly, plants help fill our desires as well as our needs.

Have you ever wondered 

Why the World Is So Green?

Most of us take for granted the huge abundance of green plants that surrounds us. But when you think about it, this abundance is a bit of a puzzle. Many, many animals eat plants, and with so many green leaves and shoots around, it stands to reason that populations of plant-eating animals should grow to be large enough to eat plant tissue just as fast as plants can produce it. But that's not what happens. Instead, forests remain full of leaves, fields remain full of grass, and most of Earth's land remains green throughout the growing season. Why? Ecologists hypothesize that the world is green because predators keep populations of plant-eating animals in check. If not for the population control provided by predators, plant-eaters might become abundant enough to eat a far greater share of Earth's plant tissue, and the world would not be quite so green.

21.3 WHAT IS THE EVOLUTIONARY ORIGIN OF PLANTS?

The ancestors of plants were photosynthetic protists, perhaps similar to the modern green algae known as stoneworts (**Fig. 21-3**). Stoneworts are plants' closest living relatives. The evolutionary relationship between plants and stoneworts has been revealed by DNA comparisons, and is reflected in other similarities between plants and green algae. For example, green algae and plants use the same type of chlorophyll and accessory pigments in photosynthesis. In addition, both plants and green algae store food as starch and have cell walls made of cellulose. In contrast, the photosynthetic pigments, food-storage molecules,

▲ **FIGURE 21-3** ***Chara*, a stonewort** The green algae known as stoneworts are plants' closest living relatives.

and cell walls of other photosynthetic protists, such as the red algae and the brown algae, differ from those of plants.

The Ancestors of Plants Were Aquatic

Like modern stoneworts, the protists that gave rise to plants presumably lacked true roots, stems, leaves, and complex reproductive structures such as flowers or cones, features that appeared only later in the evolutionary history of plants. In addition, the ancestors of plants were confined to watery habitats.

For the ancestors of plants, life in water had many advantages. For example, in water, a body is bathed in a nutrient-rich solution, is supported by buoyancy, and is not likely to dry out. In addition, life in water facilitates reproduction, because gametes (sex cells) and zygotes (fertilized sex cells) can be carried by water currents or propelled by flagella.

Case Study continued

Queen of the Parasites

The stinking corpse lily, with its huge, 3-foot-wide flowers, apparently evolved from an ancestor with tiny flowers. A recent analysis of DNA sequences revealed that the plant group most closely related to the Rafflesiaceae (the family that includes the stinking corpse lily) is the spurges, a family of plants with mostly tiny flowers. The analysis also showed that the common ancestor of spurges and corpse lilies probably had flowers that were about $\frac{1}{80}$th the size of modern stinking corpse lily flowers. Thus, by the standard of evolutionary timescales, the flower of the stinking corpse lily evolved its massive size very rapidly.

21.4 HOW HAVE PLANTS ADAPTED TO LIFE ON LAND?

Despite the benefits of aquatic environments, early plants invaded habitats on land. Today, most plants live on land. The move to land brought its own advantages, including access to sunlight unimpeded by water that might block its rays, and access to nutrients contained in surface rocks. However, the move to land also imposed some challenges; plants could no longer rely on watery surroundings to provide support, moisture, access to nutrients, and transportation for gametes and zygotes. As a result, life on land has favored the evolution in plants of traits that help meet these environmental challenges: structures that support the body and conserve water, conducting cells that transport water and nutrients to all parts of the plant, and processes that disperse gametes and zygotes by methods that are independent of water.

Plant Bodies Resist Gravity and Drying

Some of the key adaptations to life on land arose early in plant evolution, and they are now common to virtually all land plants. They include:

- Roots or rootlike structures that anchor the plant and absorb water and nutrients from the soil.
- A waxy **cuticle** that covers the surfaces of leaves and stems and that limits the evaporation of water (see Fig. 7-1).
- Pores called **stomata** (singular, stoma) in the leaves and stems that open to allow gas exchange but close when water is scarce, reducing the amount of water lost to evaporation (see Fig. 7-2).

Other key adaptations occurred somewhat later in the transition to terrestrial life, and are now widespread but not universal among plants (most nonvascular plants, a group described later, lack them):

- Conducting cells that transport water and minerals upward from the roots and that move photosynthetic products from the leaves to the rest of the plant body
- The stiffening substance **lignin,** which is a rigid polymer that impregnates the conducting cells and supports the plant body, helping the plant expose maximum surface area to sunlight

Plant Embryos Are Protected, and Some Plants Have Sex Cells That Disperse Without Water

The most widespread groups of plants, collectively known as seed plants, are characterized by especially well-protected and well-provisioned embryos and by waterless dispersal of sex cells. The key adaptations of these plant groups are seeds, pollen, and, in the flowering plants, flowers and fruits.

Early seed plants gained an advantage over their competitors by producing seeds, which provided protection and nourishment for developing embryos and the potential for more effective dispersal. Early seed plants also produced dry, microscopic pollen grains that allowed wind, instead of water, to carry the male gametes. Later came the evolution of flowers, which enticed animal pollinators that delivered pollen more precisely than did wind. Fruits also attracted animal foragers, which consumed the fruit and dispersed its indigestible seeds in their feces.

21.5 WHAT ARE THE MAJOR GROUPS OF PLANTS?

Two major groups of land plants arose from ancient algal ancestors (**Fig. 21-4** and **Table 21-1**). Members of one group, the **nonvascular plants** (also called *bryophytes*), require a moist environment to reproduce and thus straddle the boundary between aquatic and terrestrial life, much like the amphibians of the animal kingdom. The other group, the **vascular plants** (also called *tracheophytes*), has been able to colonize drier habitats.

Nonvascular Plants Lack Conducting Structures

Nonvascular plants retain some characteristics of their algal ancestors. They lack true roots, leaves, and stems. They do possess rootlike anchoring structures called *rhizoids* that bring water and nutrients into the plant body, but nonvascular plants lack well-developed structures for conducting water and nutrients. They must instead rely on slow diffusion or poorly developed conducting tissues to distribute water and other nutrients. As a result, their body size is limited. Size is also limited by the absence of any stiffening agent in their bodies. Without such material, they cannot grow upward very far. Most nonvascular plants are less than 1 inch (2.5 centimeters) tall.

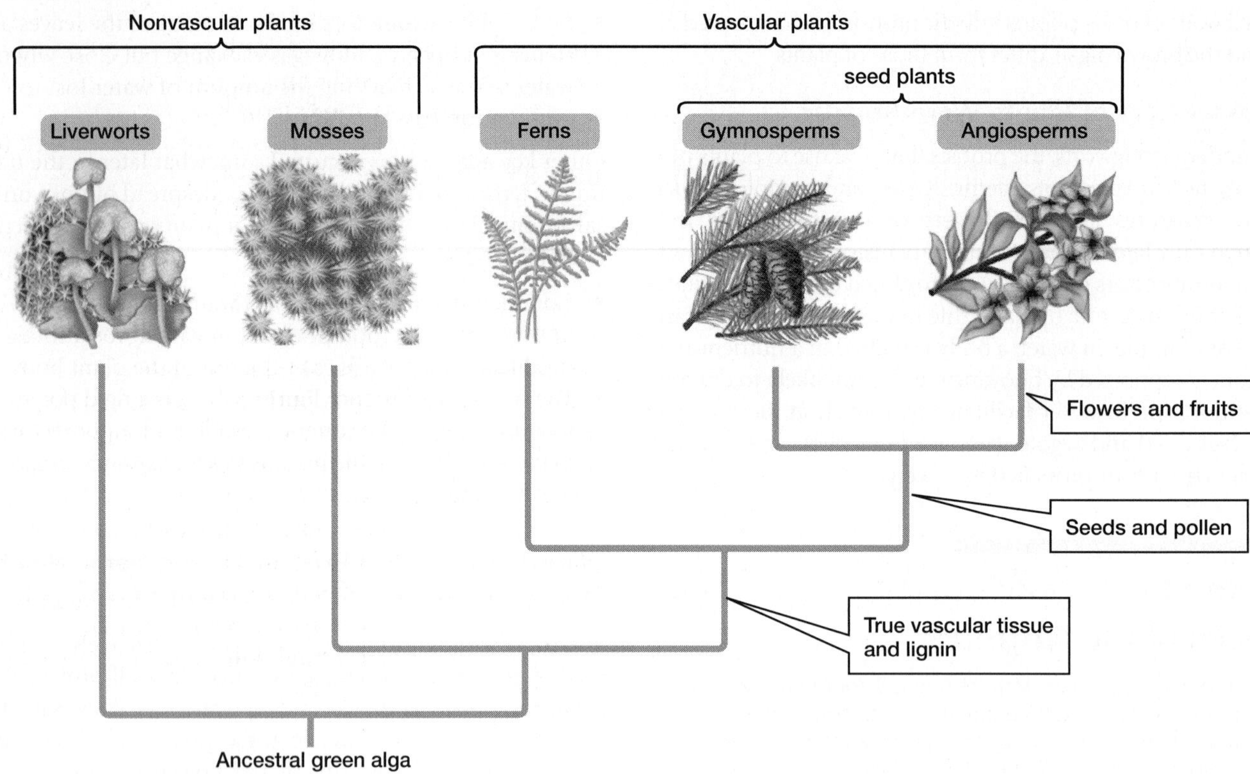

▲ **FIGURE 21-4 Evolutionary tree of some major plant groups**

Nonvascular Plants Include the Hornworts, Liverworts, and Mosses

The nonvascular plants include three groups: hornworts, liverworts, and mosses. Hornworts and liverworts are named for their shapes. Hornwort sporophytes generally have a spiky shape that appears hornlike to some observers (**Fig. 21-5a**). The gametophytes of certain liverwort species have a lobed form reminiscent of the shape of a liver (**Fig. 21-5b**). Hornworts and liverworts are most abundant in areas where moisture is plentiful, such as in moist forests and near the banks of streams and ponds.

Mosses are the most diverse and abundant of the nonvascular plants (**Fig. 21-5c**). Like hornworts and liverworts, mosses are most likely to be found in moist habitats. Some mosses, however, have a waterproof covering that retains

Table 21-1 Features of the Major Plant Groups

Group	Subgroup	Relationship of Sporophyte and Gametophyte	Transfer of Reproductive Cells	Early Embryonic Development	Dispersal	Water and Nutrient Transport Structures
Nonvascular Plants		The gametophyte is dominant—the sporophyte develops from a zygote retained on a gametophyte	Motile sperm swim to a stationary egg retained on a gametophyte	Occurs within the archegonium of a gametophyte	Haploid spores are carried by wind	Absent
Vascular Plants	Ferns	The sporophyte is dominant—it develops from a zygote retained on a gametophyte	Motile sperm swim to a stationary egg retained on a gametophyte	Occurs within the archegonium of a gametophyte	Haploid spores are carried by wind	Present
	Conifers	The sporophyte is dominant—the microscopic gametophyte develops within a sporophyte	Wind-dispersed pollen carries sperm to a stationary egg in a cone	Occurs within a protective seed containing a food supply	Seeds containing a diploid sporophyte embryo are dispersed by wind or animals	Present
	Flowering plants	The sporophyte is dominant—the microscopic gametophyte develops within a sporophyte	Pollen, dispersed by wind or animals, carries sperm to a stationary egg within a flower	Occurs within a protective seed containing a food supply; the seed is encased within fruit	Fruit, carrying seeds, is dispersed by animals, wind, or water	Present

(a) Hornwort

(b) Liverwort

(c) Moss

(d) *Sphagnum* bog

moisture, preventing water loss. Many of these mosses are also able to survive the loss of much of the water in their bodies; they dehydrate and become dormant during dry periods but absorb water and resume growth when moisture returns. Such mosses can survive in deserts, on bare rock, and in far northern and southern latitudes where humidity is low and liquid water is scarce for much of the year.

Mosses of the genus *Sphagnum* are especially widespread, living in moist habitats in northern regions around the world. In many of these wet northern habitats, *Sphagnum* is the most abundant plant, forming extensive mats (**Fig. 21-5d**). Because decomposition is slow in cold climates and because *Sphagnum* contains compounds that inhibit bacteria, dead *Sphagnum* may decay only very slowly. As a result, partially decayed moss tissue can accumulate in deposits that can, over thousands of years, become hundreds of feet thick. These deposits are known as peat. Peat has long been harvested for use as fuel, a practice that continues today in some northern areas. Now, however, peat is more often harvested for use in horticulture. Dried peat can absorb many times its own weight in water, making it useful as a soil conditioner and as a packing material for transporting live plants.

The Reproductive Structures of Nonvascular Plants Are Protected

Nonvascular plants require moisture to reproduce, but they have evolved some traits that facilitate reproduction on land (**Fig. 21-6**). For example, the reproductive structures of nonvascular plants are enclosed, which prevents the gametes from drying out. There are two types of reproductive structures: **archegonia** (singular, archegonium), in which eggs develop, and **antheridia** (singular, antheridium), where sperm are formed (**Fig. 21-6 ❶**). In some nonvascular plant species, both archegonia and antheridia are located on the same plant; in other species, each individual plant is either male or female.

In all nonvascular plants, the sperm must swim to the egg, which emits a chemical attractant, through a film of water (**Fig. 21-6 ❷**). (Nonvascular plants that live in drier areas must time their reproduction to coincide with rains.) The fertilized egg (zygote) is retained in the archegonium, where the embryo grows and matures into a small diploid sporophyte that remains attached to the parent gametophyte plant (**Fig. 21-6 ❸**). At maturity, the sporophyte produces haploid spores by meiosis within a capsule (**Fig. 21-6 ❹**). When the capsule is opened, spores are released and dispersed by the wind (**Fig. 21-6 ❺**). If a spore lands in a suitable environment, it may develop into another haploid gametophyte plant (**Fig. 21-6 ❻**).

◀ FIGURE 21-5 **Nonvascular plants** The plants shown here are less than a half-inch (about 1 centimeter) in height. **(a)** The hornlike sporophytes of hornworts grow upward from the gametophyte body. **(b)** Liverworts grow in moist, shaded areas. The palmlike structures on the female plants shown here hold eggs. Male plants produce sperm that swim through a film of water to reach and fertilize the eggs. **(c)** Moss plants, showing the stalks that carry spore-bearing capsules. **(d)** Mats of *Sphagnum* moss cover moist bogs in northern regions.

QUESTION Why are all nonvascular plants short?

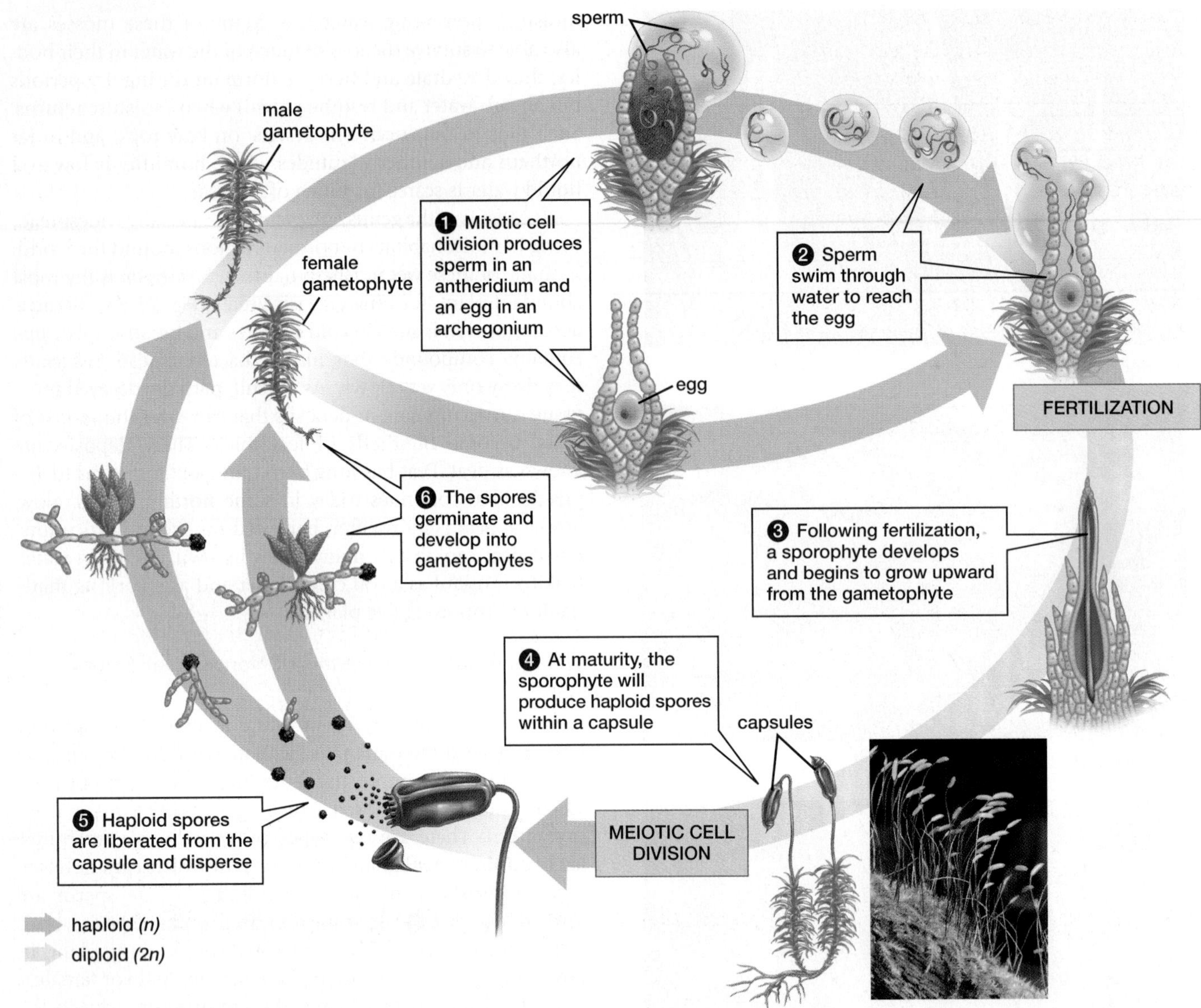
sperm
male gametophyte
female gametophyte
1 Mitotic cell division produces sperm in an antheridium and an egg in an archegonium
egg
2 Sperm swim through water to reach the egg
FERTILIZATION
3 Following fertilization, a sporophyte develops and begins to grow upward from the gametophyte
4 At maturity, the sporophyte will produce haploid spores within a capsule
capsules
MEIOTIC CELL DIVISION
5 Haploid spores are liberated from the capsule and disperse
6 The spores germinate and develop into gametophytes
haploid (n)
diploid (2n)

▲ **FIGURE 21-6 Life cycle of a moss** The photo shows moss plants; the short, leafy green plants are haploid gametophytes; the reddish brown stalks are diploid sporophytes.

Vascular Plants Have Conducting Cells That Also Provide Support

Vascular plants are distinguished by specialized groups of tube-shaped conducting cells. These cells are impregnated with the stiffening substance lignin and serve both supportive and conducting functions. They allow vascular plants to grow taller than nonvascular plants, both because of the extra support provided by lignin and because the conducting cells allow water and nutrients absorbed by the roots to move to the upper portions of the plant. Another difference between vascular plants and nonvascular plants is that in vascular plants, the diploid sporophyte is the larger, more conspicuous generation; in nonvascular plants, the haploid gametophyte is more evident.

The vascular plants can be divided into two groups: the seedless vascular plants and the seed plants.

The Seedless Vascular Plants Include the Club Mosses, Horsetails, and Ferns

Like the nonvascular plants, seedless vascular plants have swimming sperm and require water for reproduction. As their name implies, they do not produce seeds but rather propagate by spores. Present-day seedless vascular plants—the club mosses, horsetails, and ferns—are much smaller than their ancestors, which dominated the landscape in the Carboniferous period (350 million to 290 million years ago; see Fig. 17-8). The seedless vascular plants were once the predominant plants in the landscape, but today, the more versatile seed plants are more prominent.

Club Mosses and Horsetails Are Small and Inconspicuous

The club mosses, which despite their common name are not actually mosses, are now limited to representatives a few inches in height (**Fig. 21-7a**). Their leaves are small and scale-like, resembling the leaflike structures of mosses. Club mosses of the genus *Lycopodium*, commonly known as ground pine, form a beautiful ground cover in some temperate coniferous and deciduous forests.

Modern horsetails belong to a single genus, *Equisetum*, that contains only 15 species, most less than 3 feet tall (**Fig. 21-7b**). The bushy branches of some species lend them the common name horsetails; the leaves are reduced to tiny scales on the branches. They are also called "scouring rushes" because all species of *Equisetum* deposit large amounts of silica (glass) in their outer layer of cells, giving them an abrasive texture. Early European settlers of North America used horsetails to scour pots and floors.

Ferns Are Broad-Leaved and More Diverse

The ferns, with 12,000 species, are the most diverse of the seedless vascular plants (**Fig. 21-7c**). In the tropics, tree ferns still reach heights reminiscent of their ancestors from the Carboniferous period (**Fig. 21-7d**). Ferns are the only seedless vascular plants that have broad leaves.

In ferns, as in the nonvascular plants, reproduction requires moisture (**Fig. 21-8**). Gametes are produced in archegonia and antheridia on the tiny fern gametophyte (**Fig. 21-8 ❶**).

▲ **FIGURE 21-7 Some seedless vascular plants** Seedless vascular plants are found in moist woodland habitats. **(a)** The club mosses (sometimes called ground pines) grow in temperate forests. This specimen is releasing spores. **(b)** The giant horsetail extends long, narrow branches in a series of rosettes at regular intervals along the stem. Its leaves are insignificant scales. At right is a cone-shaped spore-forming structure. **(c)** The leaves of this deer fern are emerging from coiled, immature leaves called fiddleheads. **(d)** Although most fern species are small, some, such as this tree fern, retain the large size that was common among ferns of the Carboniferous period.

QUESTION In each of these photos, is the pictured structure a sporophyte or a gametophyte?

Sperm are released into water and swim to reach an egg in an archegonium (**Fig. 21-8 ❷**). If fertilization occurs, the resulting zygote develops into a sporophyte plant, which grows upward from its parent, the gametophyte (**Fig. 21-8 ❸**). On a mature sporophyte fern plant, which is much larger than the gametophyte, haploid spores are produced in structures called *sporangia* that form on special leaves of the sporophyte (**Fig. 21-8 ❹**). The spores are dispersed by the wind (**Fig. 21-8 ❺**); if a spore lands in a spot with suitable conditions, it germinates and develops into a gametophyte plant (**Fig. 21-8 ❻**).

The windborne spores of ferns make them especially effective at colonizing locations that lack abundant plant life. For example, following the massive volcanic eruption that destroyed most life on the island of Krakatau in 1883, ferns soon blanketed the previously denuded landscape. Similarly, fern abundance increased dramatically following the catastrophic asteroid impact that caused the extinction of dinosaurs and many other species about 65 million years ago. Paleontologists have discovered that fossil fern spore abundance is dramatically high in 65-million-year-old rocks at many locations around the world; these "spore spikes" are interpreted as evidence that massive fires followed the asteroid impact, burning up most vegetation and creating an opening for widespread colonization by ferns.

Although no fern species are widely cultivated as food crops, some ferns are edible. Especially tasty are the tightly coiled "fiddleheads" of some species, formed when the developing sporophyte plant first pokes through the soil surface (see Fig. 21-7c). Boiled or sautéed, edible fiddleheads make an appetizing dish. In times before the advent of refrigerated transportation made fresh vegetables widely available year-round, fiddleheads were often the first spring vegetable available each year to residents of temperate regions.

The Seed Plants Are Aided by Two Important Adaptations: Pollen and Seeds

The seed plants are distinguished from nonvascular plants and seedless vascular plants by their production of pollen and

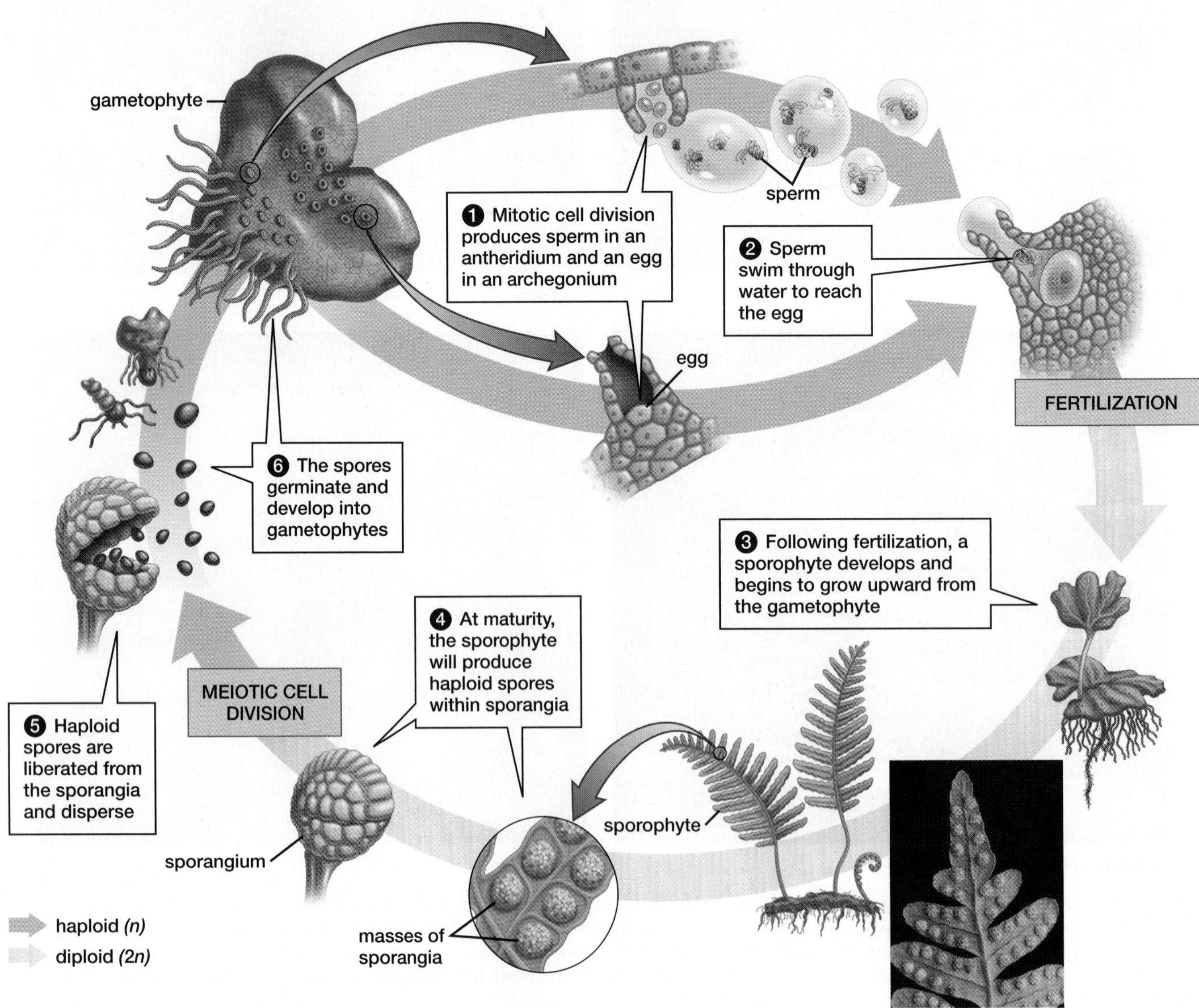
gametophyte
sperm
❶ Mitotic cell division produces sperm in an antheridium and an egg in an archegonium
❷ Sperm swim through water to reach the egg
egg
FERTILIZATION
❻ The spores germinate and develop into gametophytes
❸ Following fertilization, a sporophyte develops and begins to grow upward from the gametophyte
❹ At maturity, the sporophyte will produce haploid spores within sporangia
MEIOTIC CELL DIVISION
❺ Haploid spores are liberated from the sporangia and disperse
sporophyte
sporangium
haploid (n)
diploid (2n)
masses of sporangia

▲ **FIGURE 21-8 Life cycle of a fern** The photo shows clusters of sporangia on the underside of a fern leaf.

seeds. **Pollen** grains are tiny male gametophytes that carry sperm-producing cells. Pollen grains are dispersed by wind or by animal pollinators such as bees. In this way, sperm move through the air to fertilize egg cells. This airborne transport means that the distribution of seed plants is not limited by the need for water through which sperm can swim to the egg.

Analogous to the eggs of birds and reptiles, **seeds** consist of an embryonic sporophyte plant, a supply of food for the embryo, and a protective outer coat (**Fig. 21-9**). The *seed coat* maintains the embryo in a state of suspended animation or dormancy until conditions are proper for growth. The stored food helps sustain the emerging plant until it develops roots and leaves and can make its own food by photosynthesis. Some seeds possess elaborate adaptations that allow them to be dispersed by wind, water, and animals.

In seed plants, gametophytes (which produce the sex cells) are tiny. The female gametophyte is a small group of haploid cells that produces the egg. The male gametophyte is the pollen grain.

Seed plants are grouped into two general types: gymnosperms, which lack flowers, and angiosperms, the flowering plants.

Gymnosperms Are Nonflowering Seed Plants

Gymnosperms evolved earlier than the flowering plants. Early gymnosperms coexisted with the forests of seedless vascular plants that prevailed during the Carboniferous period. During the subsequent Permian period (290 million to 248 million years ago), however, gymnosperms became the predominant plant group and remained so until the rise of the flowering plants more than 100 million years later. Despite their success, most of these early gymnosperms are now extinct. Today, four groups of gymnosperms survive: the ginkgos, the cycads, the gnetophytes, and the conifers.

Only One Ginkgo Species Survives

Ginkgos have a long evolutionary history. They were widespread during the Jurassic period, which began 208 million years ago. Today, however, they are represented by the single species *Ginkgo biloba*, the maidenhair tree. Ginkgo trees are either male or female; female trees bear foul-smelling, fleshy seeds the size of cherries (**Fig. 21-10a**). Ginkgos have been maintained by cultivation, particularly in Asia; if not for this cultivation, they might be extinct today. Because they are more resistant to pollution than are most other trees, ginkgos (usually the male trees) have been extensively planted in U.S. cities. Recently, the leaves of the ginkgo have gained attention as an herbal supplement that purportedly improves memory.

Cycads Are Restricted to Warm Climates

Like ginkgos, cycads were diverse and abundant in the Jurassic period but have since dwindled. Today approximately 160 species survive, most of which dwell in tropical or subtropical climates. Cycads have large, finely divided leaves and bear a superficial resemblance to palms or large ferns (**Fig. 21-10b**). Most cycads are about 3 feet (1 meter) in height, although some species can reach 65 feet (20 meters).

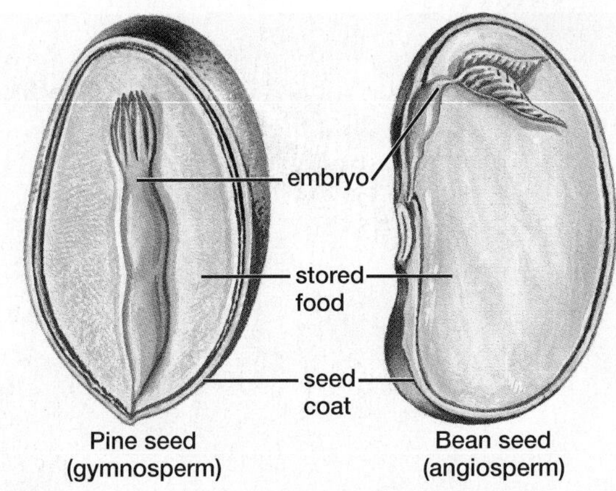

(a) Seeds

(b) Dandelion

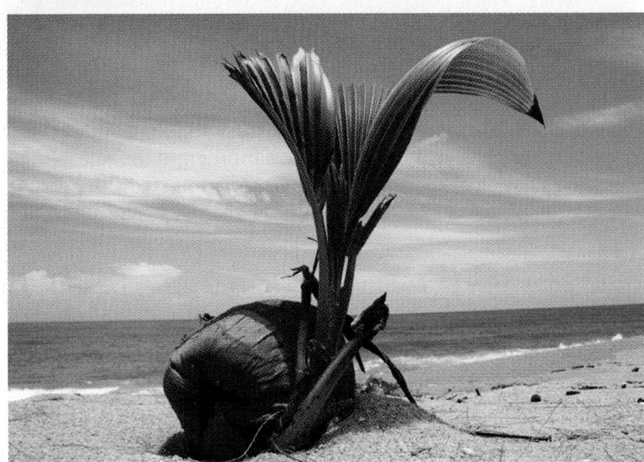

(c) Coconut

▲ **FIGURE 21-9 Seeds (a)** Seeds from a gymnosperm (left) and an angiosperm (right). Both consist of an embryonic plant and stored food confined within a seed coat. Seeds exhibit diverse adaptations for dispersal, including **(b)** the dandelion's tiny, tufted seeds that float in the air and **(c)** the massive, armored seeds (protected inside the fruit) of the coconut palm, which can survive prolonged immersion in seawater as they traverse oceans.

QUESTION Can you think of some adaptations that help protect seeds from destruction by animal consumption?

(a) Gingko

(b) Cycad

(c) Gnetophyte

(d) Conifer

▲ **FIGURE 21-10 Gymnosperms (a)** This ginkgo, or maidenhair tree, is female and bears fleshy seeds the size of large cherries. **(b)** A cycad. Common in the age of dinosaurs, these are now limited to about 160 species. Like ginkgos, cycads have separate sexes. **(c)** The leaves of the gnetophyte *Welwitschia* may be hundreds of years old. **(d)** The needle-shaped leaves of conifers are protected by a waxy surface layer.

Cycads grow slowly and live for a long time; one Australian specimen is estimated to be 5,000 years old.

The tissues of cycads contain potent toxins. Despite the presence of these toxins, people in some parts of the world use cycad seeds, stems, and roots for food. Careful preparation and processing removes the toxins before the plants are consumed. Nonetheless, cycad toxins are the suspected cause of neurological problems that occur with some frequency in populations that use cycads for food. Cycad toxins can also harm grazing livestock.

About half of all cycad species are classified as threatened or endangered. The main threats to cycads are habitat destruction, competition from introduced species, and harvesting for the horticultural trade. A large specimen of a rare cycad can sell for thousands of dollars. Because cycads grow slowly, recovery of endangered populations is uncertain.

Gnetophytes Include the Odd *Welwitschia*

The gnetophytes include about 70 species of shrubs, vines, and small trees. Leaves of gnetophyte species in the genus *Ephedra* contain alkaloid compounds that act in humans as stimulants and appetite suppressants. For this reason, *Ephedra* is widely used as an energy booster and weight-loss aid. However, following reports of sudden deaths of *Ephedra* users and publication of several studies linking *Ephedra* consumption to increased risk of heart problems, the U.S. Food and Drug Administration banned the sale of products containing this gnetophyte.

The gnetophyte *Welwitschia mirabilis* is among the most distinctive of plants (**Fig. 21-10c**). Found only in the extremely dry deserts of southwest Africa, *Welwitschia* has a deep taproot that can extend as far as 100 feet (30 meters) down into the soil. Above the surface, the plant has a fibrous stem. Two (and only two) leaves grow from the stem. The leaves are never shed and remain on the plant for its entire life, which can be very long. The oldest *Welwitschia* are more than 2,000 years old, and a typical life span is about 1,000 years. The strap-like leaves continue to grow for that entire period, spreading over the ground. The older portions of the leaves, whipped by the wind for centuries, may shred or split, giving the plant its characteristic gnarled and well-worn appearance.

Conifers Are Adapted to Cool Climates

Though the other gymnosperm groups are drastically reduced from their former prominence, the **conifers** still dominate large areas of our planet. Conifers, whose 500 species include pines, firs, spruce, hemlocks, and cypresses, are most abundant in the cold latitudes of the far north and at high elevations where conditions are dry. Not only is rainfall limited in these areas, but water in the soil remains frozen and unavailable during the long winters.

Conifers are adapted to these dry, cold conditions in three ways. First, most conifers retain green leaves throughout the year, enabling these plants to continue photosynthesizing and growing slowly during times when most other plants become dormant. For this reason, conifers are often called evergreens. Second, conifer leaves are actually thin needles covered with a thick, waterproof surface that minimizes evaporation (**Fig. 21-10d**). Finally, conifers produce an "antifreeze" in their sap that enables them to continue transporting nutrients in below-freezing temperatures. This substance gives them their fragrant piney scent.

Conifer Seeds Develop in Cones

Reproduction is similar in all conifers, so let's examine the reproductive cycle of a pine tree (**Fig. 21-11**). The tree itself is the diploid sporophyte, and it develops both male and female cones (**Fig. 21-11 ❶**). Male cones are relatively small (typically about ¾ inch long), delicate structures consisting of scales in which pollen (the male gametophyte) develops. Each female cone consists of a series of woody scales arranged in a spiral around a central axis. At the base of each scale are two **ovules** (unfertilized seeds), within which diploid spore-forming cells arise.

Male cones release pollen during the reproductive season and then disintegrate (**Fig. 21-11 ❷**). The amount of

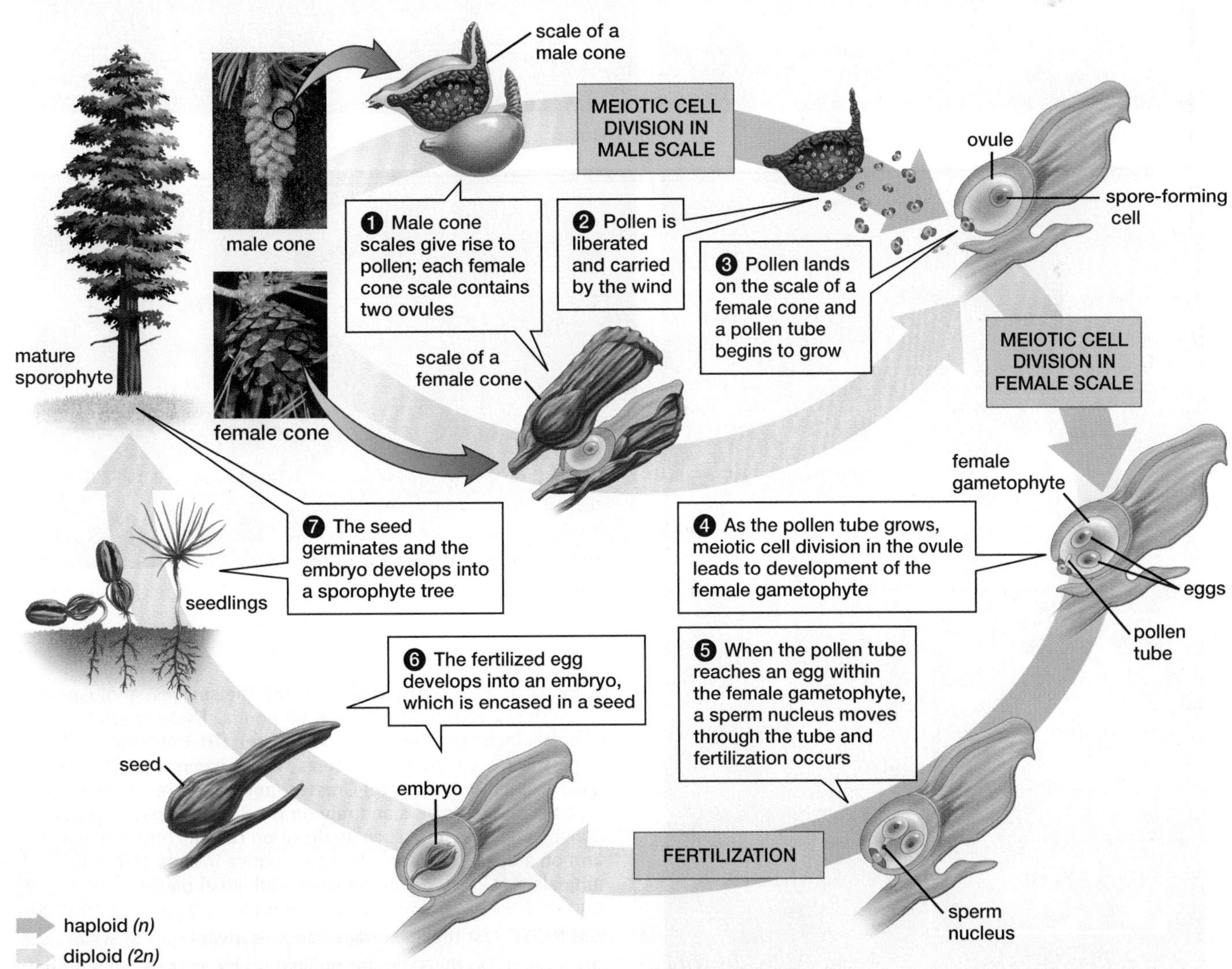
scale of a male cone
MEIOTIC CELL DIVISION IN MALE SCALE
ovule
spore-forming cell
male cone
❶ Male cone scales give rise to pollen; each female cone scale contains two ovules
❷ Pollen is liberated and carried by the wind
❸ Pollen lands on the scale of a female cone and a pollen tube begins to grow
MEIOTIC CELL DIVISION IN FEMALE SCALE
mature sporophyte
scale of a female cone
female cone
female gametophyte
❼ The seed germinates and the embryo develops into a sporophyte tree
❹ As the pollen tube grows, meiotic cell division in the ovule leads to development of the female gametophyte
eggs
seedlings
pollen tube
❻ The fertilized egg develops into an embryo, which is encased in a seed
❺ When the pollen tube reaches an egg within the female gametophyte, a sperm nucleus moves through the tube and fertilization occurs
seed
embryo
FERTILIZATION
sperm nucleus
haploid (n)
diploid (2n)

▲ **FIGURE 21-11 Life cycle of the pine**

pollen released is immense; inevitably, some pollen grains land by chance on female cone scales (**Fig. 21-11** ❸). In the aftermath of such a pollination event, a pollen grain sends out a pollen tube that slowly burrows into an ovule. As the pollen tube grows, the diploid spore-forming cell in the ovule undergoes meiosis to produce haploid spores, one of which gives rise to a haploid female gametophyte, within which egg cells develop (**Fig. 21-11** ❹). After nearly 14 months, the tube finally reaches the egg cell and releases the sperm that fertilize it (**Fig. 21-11** ❺). The resulting zygote becomes enclosed in a seed as it develops into an embryo—a tiny embryonic sporophyte plant (**Fig. 21-11** ❻). The seed is liberated when the cone matures and its scales separate. If it lands in a suitable patch of soil, it may germinate and grow into a sporophyte tree (**Fig. 21-11** ❼).

Angiosperms Are Flowering Seed Plants

Flowering plants, or **angiosperms,** have been Earth's predominant plants for more than 100 million years. The group is incredibly diverse, with more than 230,000 species. Angiosperms range in size from the diminutive duckweed (**Fig. 21-12a**) to the

(a) Duckweed

(c) Grass

(b) Eucalyptus

(d) Butterfly weed

▲ **FIGURE 21-12 Angiosperms (a)** The smallest angiosperm is the duckweed, found floating on ponds. These specimens are about ⅛ inch (3 millimeters) in diameter. **(b)** The largest angiosperms are eucalyptus trees, which can reach 325 feet (100 meters) in height. **(c)** Grasses (and many trees) have inconspicuous flowers and rely on wind for pollination. More conspicuous flowers, such as those on **(d)** this butterfly weed and on a eucalyptus tree (**b**, inset), entice insects and other animals that carry pollen between individual plants.

EXERCISE List the advantages and disadvantages of wind pollination. Do the same for pollination by animals. Why do both types of pollination persist among the angiosperms?

towering eucalyptus tree (**Fig. 21-12b**). From desert cactus to tropical orchids to grasses to parasitic stinking corpse lilies, angiosperms rule over the plant kingdom. Their enormous success is due in part to three major adaptations: flowers, fruits, and broad leaves.

Flowers Attract Pollinators

Flowers, the structures in which both male and female gametes are formed, may have evolved when gymnosperm ancestors formed an association with animals (most likely insects) that carried their pollen from plant to plant. According to this scenario, the relationship between these ancient gymnosperms and their animal pollinators was so beneficial that natural selection favored the evolution of showy flowers that advertised the presence of pollen to insects and other animals (**Figs. 21-12b,d**). The animals benefited by eating some of the protein-rich pollen, whereas the plant benefited from the animals' unwitting transportation of pollen from plant to plant. With this animal assistance, many flowering plants no longer needed to produce prodigious quantities of pollen and send it flying on the fickle winds to ensure fertilization. But there are nonetheless many wind-pollinated angiosperms (**Fig. 21-12c**).

In the angiosperm life cycle (**Fig. 21-13**), flowers develop on the dominant sporophyte plant. In the flower, female gametophytes develop from ovules within a structure called the *ovary*; male gametophytes (pollen) are formed inside a structure called the *anther* (**Fig. 21-13** ❶). During the reproductive season, pollen is released from the anthers and carried away on the wind or by animal pollinators (**Fig. 21-13** ❷). If a pollen grain lands on a *stigma,* a sticky pollen-catching structure of the flower, a pollen tube begins to grow from the pollen grain (**Fig. 21-13** ❸). The tube bores through the stigma and extends toward the female gametophyte, within which egg cells have developed. Fertilization occurs when the pollen tube reaches the egg cells

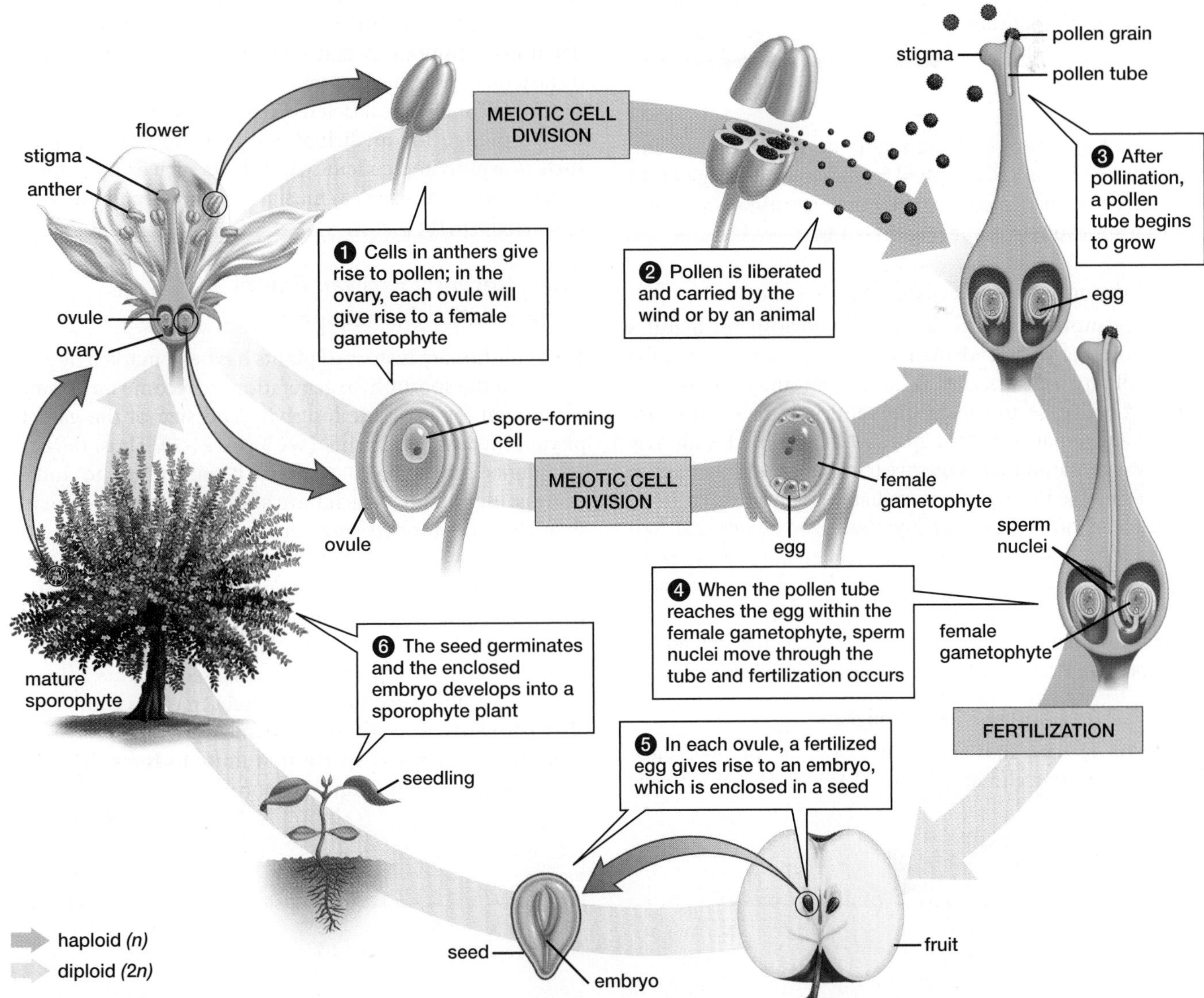

▲ **FIGURE 21-13 Life cycle of a flowering plant**

Case Study continued
Queen of the Parasites

Why does the flower of a stinking corpse lily smell like rotting meat? Though the smell is utterly revolting to humans, it is attractive to blowflies and other insects that normally feed on and lay their eggs in decaying flesh. When such insects visit a male stinking corpse lily, they may carry away pollen that can fertilize a nearby female flower.

In many angiosperm species, flowers contain nectar that provides food for animal pollinators. But no such nectar reward awaits a fly that enters the flower of stinking corpse lily. Instead, the fly attracted by the flower's stench searches in vain for putrefying meat, its movement guided toward the flower's cache of sticky pollen by grooves and hairs inside the flower. Eventually, the fly departs, coated in pollen. In essence, the fly has been tricked by the plant into providing a service for no reward. Thus, the stinking corpse lily is a double parasite: It exploits both the *Tetrastigma* vines that provide its food and the flies that facilitate its reproduction.

(**Fig. 21-13** ❹). The resulting zygote develops into an embryo enclosed in a seed formed from the ovule (**Fig. 21-13** ❺). After it is dispersed, the seed may germinate and give rise to a sporophyte plant (**Fig. 21-13** ❻).

Fruits Encourage Seed Dispersal

The ovary surrounding the seeds of an angiosperm matures into a **fruit,** the second adaptation that has contributed to the success of angiosperms. Just as flowers encourage animals to transport pollen, so, too, many fruits entice animals to disperse seeds. If an animal eats a fruit, many of the enclosed seeds may pass through the animal's digestive tract unharmed, perhaps to fall at a suitable location for germination. Not all fruits, however, depend on edibility for dispersal. Dog owners are well aware, for example, that some fruits (called burs) disperse by clinging to animal fur. Other fruits, such as those of maples, form wings that carry the seed through the air. The variety of dispersal mechanisms made possible by fruits has helped the angiosperms invade nearly all possible terrestrial habitats.

Broad Leaves Capture More Sunlight

The third feature that gives angiosperms an advantage in warmer, wetter climates is broad leaves. When water is plentiful, as it is during the warm growing season of temperate and tropical climates, broad leaves provide an advantage by collecting more sunlight for photosynthesis. In regions with seasonal variation in growing conditions, many trees and shrubs drop their leaves during periods when water is in short supply, because being leafless reduces evaporative water loss. In temperate climates, such periods occur during the fall and winter, at which time most temperate angiosperm trees and shrubs drop their leaves. In the tropics and subtropics, most angiosperms are evergreen, but species that inhabit certain tropical climates where periods of drought are common may drop their leaves to conserve water during the dry season.

The advantages of broad leaves are offset by some evolutionary costs. In particular, broad, tender leaves are much more appealing to herbivores than are the tough, waxy needles of conifers. As a result, angiosperms have developed a range of defenses against mammalian and insect herbivores. These adaptations include physical defenses such as thorns, spines, and resins that toughen the leaves. But the evolutionary struggle for survival has also led to a host of chemical defenses—compounds that make plant tissue poisonous or distasteful to potential predators. Many of the compounds responsible for chemical defense have properties that humans have exploited for medicinal and culinary uses. Medicines such as aspirin and codeine, stimulants such as nicotine and caffeine, and spices such as mustard and pepper are all derived from angiosperm plants.

More Recently Evolved Plants Have Smaller Gametophytes

The evolutionary history of plants has been marked by a tendency for the sporophyte generation to become increasingly prominent, and for the longevity and size of the gametophyte generation to shrink (see Table 21-1). Thus, the earliest plants are believed to have been similar to today's nonvascular plants, which have a sporophyte that is smaller than the gametophyte and remains attached to it. In contrast, plants that originated somewhat later, such as ferns and the other seedless vascular plants, feature a life cycle in which the sporophyte is dominant, and the gametophyte is a much smaller, independent plant. Finally, in the most recently evolved group of plants, the seed plants, gametophytes are microscopic and barely recognizable as an alternate generation. These tiny gametophytes, however, still produce the eggs and sperm that unite to form the zygote that develops into the diploid sporophyte.

Case Study revisited

Queen of the Parasites

The approximately 17 parasitic plant species of the genus *Rafflesia*, which includes the stinking corpse lily, are found in the moist forests of Southeast Asia, a habitat that is disappearing rapidly as forests are cleared for agriculture and development. The geographic range of the stinking corpse lily is limited to the dwindling forests of the Malaysian peninsula and the Indonesian islands of Borneo and Sumatra; the species is rare and endangered. The government of Indonesia has established some parks and reserves that help protect the stinking corpse lily, but—as is often the case in developing countries—a forest that is protected on paper may still be vulnerable in reality.

Perhaps the best hope for the continued survival of the largest *Rafflesia* is the growing realization among the rural residents of Sumatra and Borneo that the spectacular, putrid-smelling flowers of the stinking corpse lily might lure interested tourists to their countries. Under an innovative conservation program that seeks to take advantage of this potential for ecotourism, people who live in the vicinity of the stinking corpse lily can become caretakers of the plants. These assigned caretakers watch over the plants and, in return, may charge a small fee to curious visitors. Local inhabitants have been given an economic incentive to protect this rare parasitic plant.

Consider This

A parasitic lifestyle is unusual among plants, but it is not exactly rare. Fifteen different plant families contain parasitic species, and systematists estimate that parasitism has evolved at least nine different times over the evolutionary history of plants. Given the obvious benefits of photosynthesis, why has parasitism (which is often accompanied by loss of photosynthetic capability) evolved repeatedly in photosynthetic plants?

CHAPTER REVIEW

Summary of Key Concepts

21.1 What Are the Key Features of Plants?

Plants are photosynthetic, multicellular organisms that exhibit alternation of generations, in which a haploid gametophyte generation alternates with a diploid sporophyte generation. Unlike their green algae relatives, plants have multicellular, dependent embryos.

21.2 How Do Plants Affect Other Organisms?

Plants play a key ecological role, capturing energy for use by inhabitants of terrestrial ecosystems, replenishing atmospheric oxygen, creating and stabilizing soils, and slowing the loss of water from ecosystems. Plants are also exploited by humans to provide food, fuel, building materials, medicines, and aesthetic pleasure.

21.3 What Is the Evolutionary Origin of Plants?

Photosynthetic protists, probably aquatic green algae, gave rise to the first plants. Ancestral plants were probably similar to modern multicelluar algae such as stoneworts, which are plants' closest living relatives.

21.4 How Have Plants Adapted to Life on Land?

Early plants invaded terrestrial habitats, and modern plants exhibit a number of key adaptations for terrestrial existence: rootlike structures for anchorage and for absorption of water and nutrients; a waxy cuticle to slow the loss of water through evaporation; stomata that can open, allowing gas exchange, and that can also close, preventing water loss; conducting cells to transport water and nutrients throughout the plant; and a stiffening substance, called lignin, to impregnate the conducting cells and support the plant body.

Plant reproductive structures suitable for life on land include a reduced male gametophyte (pollen) that allows wind to replace water in carrying sperm to eggs; seeds that nourish, protect, and help disperse developing embryos; flowers that attract animals, which carry pollen more precisely and efficiently than wind; and fruits that entice animals to disperse seeds.

21.5 What Are the Major Groups of Plants?

Two major groups of plants, nonvascular plants and vascular plants, arose from their ancient algal ancestors. Nonvascular plants, including the hornworts, liverworts, and mosses, are small, simple land plants that lack conducting cells. Although some have adapted to dry areas, most live in moist habitats. Nonvascular plant reproduction requires water through which the sperm swim to the egg.

In vascular plants, a system of conducting cells—stiffened by lignin—conducts water and nutrients absorbed by the roots into the upper portions of the plant and supports the body as well. Owing to this support system, seedless vascular plants, including the club mosses, horsetails, and ferns, can grow larger than nonvascular plants. As in nonvascular plants, the sperm of seedless vascular plants must swim to the egg for sexual reproduction to occur, and the gametophyte lacks conducting cells.

Vascular plants with seeds have two major additional adaptive features: pollen and seeds. Seed plants are often classified into two categories: gymnosperms and angiosperms. Gymnosperms include ginkgos, cycads, gnetophytes, and the highly successful conifers. These plants were the first fully terrestrial plants to evolve. Their success on dry land is partially due to the evolution of the male gametophyte into the pollen grain. Pollen protects and transports the male gamete, eliminating the need for the sperm to swim to the egg. The seed, a protective resting structure containing an embryo and a supply of food, is a second important adaptation contributing to the success of seed plants.

Angiosperms, the flowering plants, dominate much of the land today. In addition to pollen and seeds, angiosperms also produce flowers and fruits. The flower allows angiosperms to use animals as pollinators. In contrast to wind, animals can in some cases carry pollen farther and with greater accuracy and less waste. Fruits may attract animal consumers, which incidentally disperse the seeds in their feces.

There has been a general evolutionary trend toward a reduction in size of the haploid gametophyte, which is dominant in nonvascular plants but microscopic in seed plants.

Key Terms

alternation of generations *386*
angiosperm *398*
antheridium (plural, **antheridia**) *391*
archegonium (plural, **archegonia**) *391*
conifer *397*
cuticle *389*
flower *399*
fruit *400*
gametophyte *386*
gymnosperm *395*
lignin *389*
nonvascular plant *389*
ovule *397*
pollen *395*
seed *395*
sporophyte *386*
stoma (plural, **stomata**) *389*
vascular plant *389*

Thinking Through the Concepts

Fill-in-the-Blank

1. Scientists hypothesize that the ancestors of plants were __________. There are two major types of plants; those that lack conducting cells are called __________, and those with conducting cells are called __________. All plants produce multicellular __________ and exhibit a complex life cycle called __________.
2. Plant adaptations to life on land include a(n) __________, which reduces evaporation of water, and __________, which open to allow gas exchange but close when __________ is scarce. In addition, the bodies of vascular plants gain increased support from __________ impregnated with the polymer __________; these structures also help __________ and __________ to move within the plant body.
3. Seedless vascular plants must reproduce when conditions are wet because their sperm must __________. Two adaptations that allow seed plants to reproduce more efficiently on dry land are __________ and __________. The seed plants fall into two major categories: the nonflowering __________ and the flowering __________. Flowers were favored by natural selection because they __________. Fruits were favored by natural selection because they __________.
4. Three groups of nonvascular plants are __________, __________, and __________. Three groups of seedless vascular plants are __________, __________, and __________. Today, the most diverse group of plants is __________.

Review Questions

1. What is meant by "alternation of generations"? What two generations are involved? How does each reproduce?
2. Explain the evolutionary changes in plant reproduction that adapted plants to increasingly dry environments.
3. Describe evolutionary trends in the life cycles of plants. Emphasize the relative sizes of the gametophyte and sporophyte.
4. From which algal group did green plants probably arise? Explain the evidence that supports this hypothesis.
5. List the structural adaptations necessary for the invasion of dry land by plants. Which of these adaptations are possessed by nonvascular plants? By ferns? By gymnosperms and angiosperms?
6. The number of species of flowering plants is greater than the number of species in the rest of the plant kingdom. What feature(s) are responsible for the enormous success of angiosperms? Explain why.
7. List the adaptations of gymnosperms that have helped them become the dominant trees in dry, cold climates.
8. What is a pollen grain? What role has it played in helping plants colonize dry land?
9. The majority of all plants are seed plants. What is the advantage of a seed? How do plants that lack seeds meet the needs served by seeds?

Applying the Concepts

1. You are a geneticist working for a firm that specializes in plant biotechnology. Explain what *specific* parts (fruit, seeds, stems, roots, etc.) of the following plants you would try to alter by genetic engineering, what changes you would try to make, and why: (a) corn, (b) tomatoes, (c) wheat, and (d) avocados.
2. Prior to the development of synthetic drugs, more than 80% of all medicines were of plant origin. Even today, indigenous tribes in remote Amazonian rain forests can provide a plant product to treat virtually any ailment. Herbal medicine is also widely and successfully practiced in China. Most of these drugs are unknown to the Western world. But the forests from which much of this plant material is obtained are being converted to agriculture. We are in danger of losing many of these potential drugs before they can be discovered. What steps can you suggest to preserve these natural resources while also allowing nations to direct their own economic development?
3. Only a few hundred of the hundreds of thousands of species in the plant kingdom have been domesticated for human use. One example is the almond. The domestic almond is nutritious and harmless, but its wild precursor can cause cyanide poisoning. The oak makes potentially nutritious seeds (acorns) that contain very bitter-tasting tannins. If we could breed the tannin out of acorns, they might become a delicacy. Why do you suppose we have failed to domesticate oaks?

MB *Go to www.masteringbiology.com for practice quizzes, activities, eText, videos, current events, and more.*

The Diversity of Fungi

Chapter 22

Case Study

Humongous Fungus

WHAT IS THE LARGEST organism on Earth? A reasonable guess might be the world's largest animal, the blue whale, which can be 100 feet long and weigh 300,000 pounds. But the blue whale is dwarfed by the General Sherman tree, a giant sequoia specimen that is 275 feet high and whose weight is estimated at 6,200 *tons*. Even these two behemoths, however, are pip-squeaks compared to the real record-holder, the fungus *Armillaria ostoyae*, also known as the honey mushroom.

The largest known *Armillaria* is a specimen in Oregon that spreads over 2,200 acres (about 3.4 square miles) and probably weighs even more than the General Sherman tree. Despite its huge size, no one has actually seen the monster fungus, because it is largely underground. Its only aboveground parts are brown mushrooms that sprout occasionally from the creature's gigantic body. Just beneath the surface, however, the fungus spreads through the soil by means of long, string-like structures called rhizomorphs. These rhizomorphs extend until they encounter the tree roots on which *Armillaria* subsists.

How can researchers be sure that the Oregon fungus is truly one single individual and not many intertwined individuals? The strongest evidence is genetic. Researchers gathered *Armillaria* tissue samples from throughout the area thought to be inhabited by a single individual and compared DNA extracted from the samples. All were genetically identical, demonstrating that they came from the same individual.

It may seem strange that the world's largest organisms went unnoticed until very recently, but the lives of fungi typically take place outside of our view. Nonetheless, fungi play a fascinating role in human affairs. Read on to find out more about the inconspicuous but often influential fungi.

▲ These honey mushrooms are part of the visible portion of the largest organism on Earth.

22.1 WHAT ARE THE KEY FEATURES OF FUNGI?

When you think of a fungus, you probably picture a mushroom. Most fungi, however, do not produce mushrooms. And even in fungus species that do produce mushrooms, the mushrooms are just temporary reproductive structures that extend from a main body that is typically concealed beneath the soil or inside a piece of decaying wood. So, to fully appreciate the fungi, we must look beyond the conspicuous structures we encounter on the forest floor, at the edges of our lawns, and as a pizza topping. A closer look at the fungi reveals a group of mostly multicellular organisms that play a key role in the web of life and whose way of life differs in fascinating ways from that of plants or animals.

Fungal Bodies Consist of Slender Threads

The body of almost all fungi is a **mycelium** (plural, mycelia; **Fig. 22-1a**), which is an interwoven mass of one-cell-thick, threadlike filaments called **hyphae** (singular, hypha; **Fig. 22-1b,c**). In some species, hyphae consist of single elongated cells with numerous nuclei; in other species, hyphae are subdivided—by partitions called **septa** (singular, septum)—into many cells, each containing one or more nuclei. Pores in the septa allow

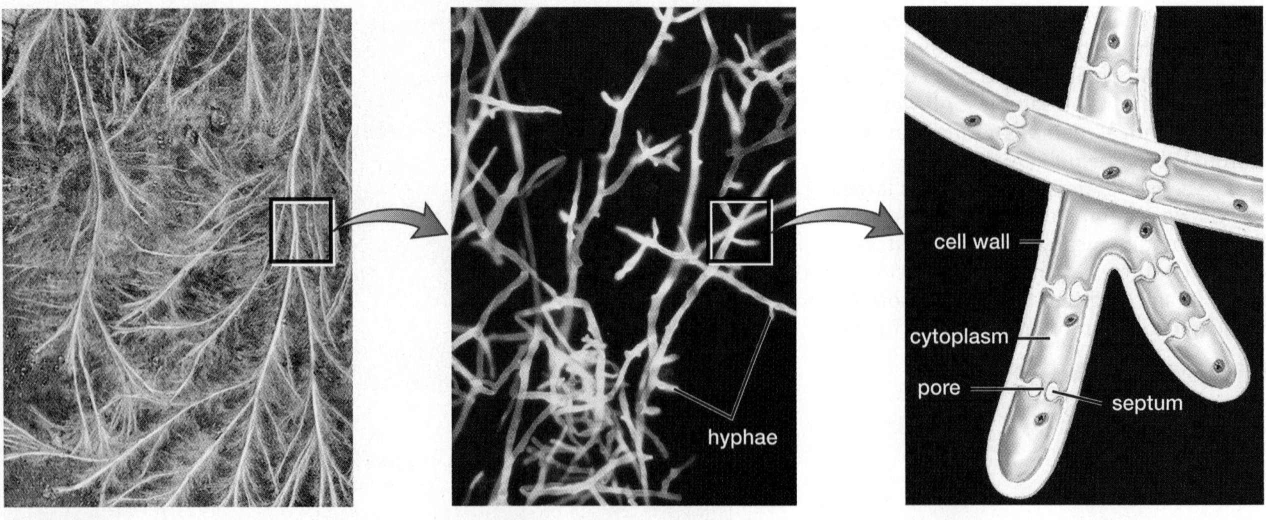

(a) Mycelium (b) Hyphae (c) Hypha cross-section

▲ **FIGURE 22-1 The filamentous body of a fungus** **(a)** A fungal mycelium spreads over decaying vegetation. The mycelium is composed of **(b)** a tangle of microscopic hyphae, only one cell thick, portrayed in cross-section **(c)** to show their internal organization.

QUESTION Which features of a fungus's body structure are adaptations related to its method of acquiring nutrients?

cytoplasm to stream between cells, distributing nutrients. Like plant cells, fungal cells are surrounded by cell walls. Unlike plant cells, however, fungal cell walls are strengthened by *chitin*, the same substance found in the exoskeletons of arthropods.

Fungi cannot move. They compensate for this lack of mobility with hyphae that can grow rapidly in any direction within a suitable environment. In this way, the fungal mycelium can quickly infuse itself into aging bread or cheese, beneath the bark of decaying logs, or into the soil. Periodically, the hyphae grow together and differentiate into reproductive structures that project above the surface beneath which the mycelium grows. These structures, including mushrooms, puffballs, and the powdery molds on spoiled food, represent only a fraction of the complete fungal body, but are typically the only part of the fungus that we can easily see.

Fungi Obtain Their Nutrients from Other Organisms

Like animals, fungi survive by breaking down nutrients stored in the bodies or wastes of other organisms. Some fungi digest the bodies of dead organisms. Others are parasitic, feeding on living organisms and causing disease. Others live in close, mutually beneficial relationships with other organisms that provide food. There are even a few predatory fungi, which attack tiny worms in soil (**Fig. 22-2**).

Unlike animals, fungi do not ingest food. Instead, they secrete enzymes that digest complex molecules outside their bodies, breaking down the molecules into smaller subunits that can be absorbed. Fungal filaments can penetrate deeply into a source of nutrients and, because the filaments are only one cell thick, they present an enormous surface area through which to secrete enzymes and absorb nutrients. This mode of securing nutrition serves fungi well. Almost every biological material can be consumed by at least one fungal species, so nutritional support for fungi is likely to be present in nearly every habitat.

Fungi Propagate by Spores

Fungi develop from **spores**—tiny, lightweight reproductive packages that are extraordinarily mobile, even though most lack a means for self-propulsion. Spores are distributed far and wide as hitchhikers on the outside of animal bodies, as passengers inside the digestive systems of animals that have eaten them, or as airborne drifters, cast aloft by chance or shot into the atmosphere by elaborate reproductive structures (**Fig. 22-3**).

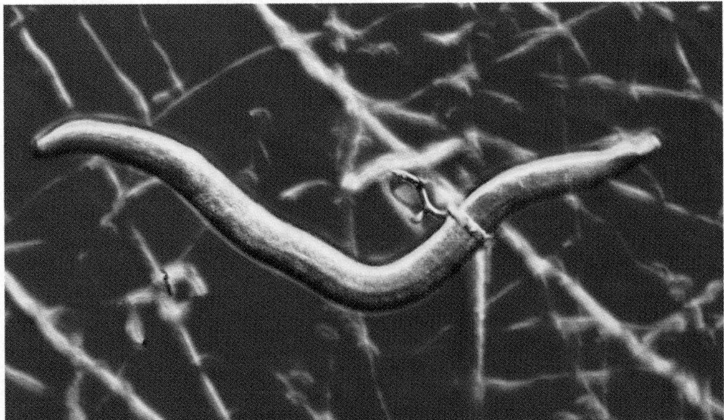

▲ **FIGURE 22-2 Nemesis of nematodes** *Arthrobotrys*, the nematode (roundworm) strangler, traps its prey in a noose-like modified hypha. When a nematode wanders into the noose, its presence stimulates the noose cells to swell with water. In a fraction of a second, the noose constricts, trapping the worm. Fungal hyphae then penetrate and feast on their prey.

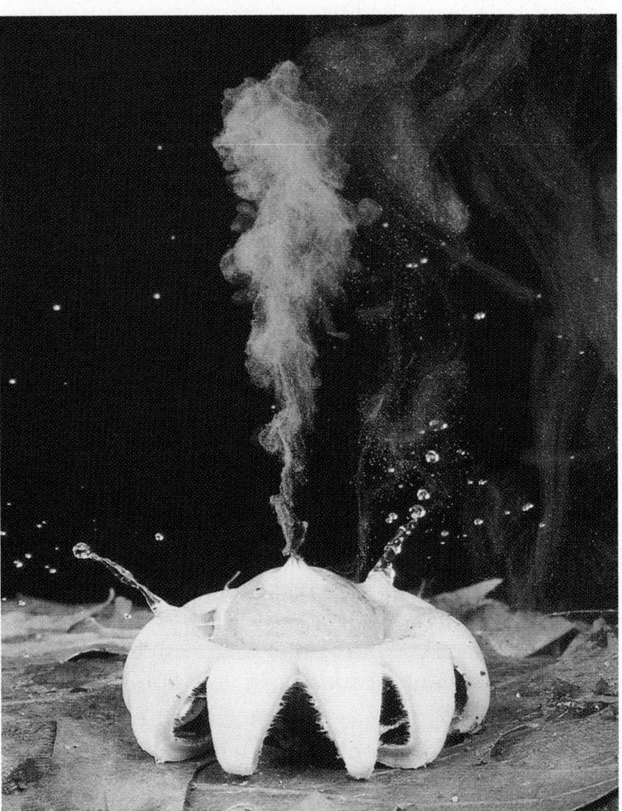

(a) Earthstar

(b) *Pilobolus*

▲ **FIGURE 22-3 Some fungi can eject spores (a)** A ripe earthstar mushroom, struck by a drop of water, releases a cloud of spores that will be dispersed by air currents. **(b)** The delicate, translucent reproductive structures of *Pilobolus*, which inhabits horse manure, literally blow their tops when ripe, dispersing the black, spore-containing caps up to 3 feet away. Spores that adhere to grass remain there until consumed by a grazing herbivore, perhaps a horse. Later (likely some distance away), the horse will deposit a fresh pile of manure containing *Pilobolus* spores that have passed unharmed through its digestive tract.

Spores are often produced in great numbers (a single giant puffball may contain 5 trillion spores; see Fig. 22-9a later in the chapter). The fungal combination of prodigious reproductive capacity and highly mobile spores ensures that fungi are ubiquitous in terrestrial environments and accounts for the inevitable growth of fungi on every uneaten sandwich and container of leftovers.

Most Fungi Can Reproduce Both Asexually and Sexually

In general, fungi are capable of both asexual and sexual reproduction. For the most part, fungi reproduce asexually under stable conditions, with sexual reproduction occurring mainly under conditions of environmental change or stress. Both asexual and sexual reproduction ordinarily involve the production of spores within special fruiting bodies that project above the mycelium.

Asexual Reproduction Produces Haploid Spores by Mitosis

The bodies and spores of fungi are haploid (contain only a single copy of each chromosome). A haploid mycelium produces haploid asexual spores by mitosis. If an asexual spore is deposited in a favorable location, it will begin mitotic divisions and develop into a new mycelium. This simple reproductive cycle results in the rapid production of genetically identical clones of the original mycelium.

Sexual Reproduction Produces Haploid Spores by Meiosis

Diploid structures form only during a brief period of the sexual portion of the fungal life cycle. Sexual reproduction begins when a filament of one mycelium comes into contact with a filament from a second mycelium that is of a different, but compatible, mating type (the different mating types of fungi are analogous to the different sexes of animals, except that there are often more than two mating types). If conditions are suitable, the two hyphae may fuse, so that nuclei from the two different hyphae share a common cell. This merger of hyphae is followed (immediately in some species, after some delay in others) by fusion of the two different haploid nuclei to form a diploid zygote. The zygote then undergoes meiosis to form haploid sexual spores. These spores are dispersed, germinate, and divide by mitosis to form new haploid mycelia. Unlike the cloned offspring produced by asexual spores, these sexually produced fungal bodies are genetically distinct from either parent.

22.2 WHAT ARE THE MAJOR GROUPS OF FUNGI?

Fungi and animals are more closely related to one another than either is to plants. That is, the common ancestor of fungi and animals lived more recently than did the common ancestor of plants, animals, and fungi (see Fig. 18-7). A person eating a salad of lettuce leaves topped by a sliced mushroom is more closely related to the mushroom than the mushroom is to the lettuce.

Fungi are very diverse. Nearly 100,000 species of fungi have been described, but this number represents only a fraction of the true diversity of these organisms. Many new species are discovered and described each year, and mycologists estimate that the number of undiscovered species of fungus is well over a million. Fungus species are classified into five phyla: Chytridiomycota (chytrids), Zygomycota (zygomycetes), Glomeromycota (glomeromycetes), Basidiomycota (basidiomycetes), and Ascomycota (ascomycetes) (**Fig. 22-4** and **Table 22-1**). Recent analysis of DNA sequences has revealed that two of these groups—the chytrids and the zygomycetes—probably do not constitute clades. (A clade is a group consisting of all the

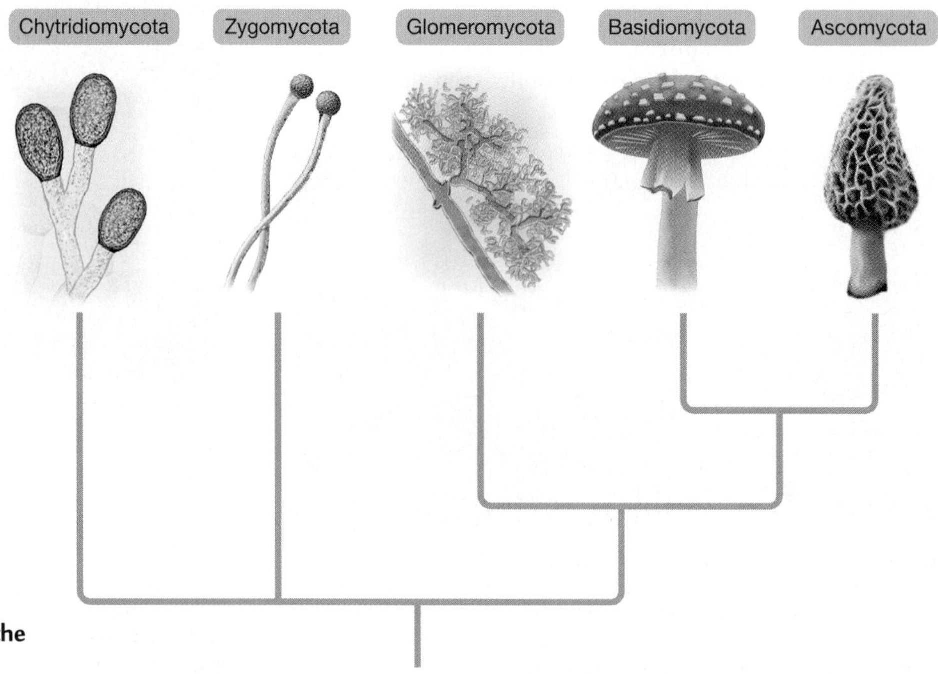

▶ **FIGURE 22-4 Evolutionary tree of the major groups of fungi**

Table 22-1 **The Phyla of Fungi**

Common Name (Phylum)	Reproductive Structures	Cellular Characteristics	Economic and Health Impacts	Representative Genera
Chytrids (Chytridiomycota)	Form haploid or diploid flagellated spores	Septa are absent	Contribute to the decline of frog populations	*Batrachochytrium* (frog pathogen)
Zygomycetes (Zygomycota)	Form diploid sexual zygospores	Septa are absent	Cause soft fruit rot and black bread mold	*Rhizopus* (causes black bread mold); *Pilobolus* (dung fungus)
Glomeromycetes (Glomeromycota)	Form haploid asexual spores, often in clusters	Septa are absent	Form mycorrhizae (mutualistic, symbiotic associations with plant roots)	*Glomus* (widespread mycorrhizal partner)
Basidiomycetes (Basidiomycota)	Sexual reproduction involves formation of haploid basidiospores on club-shaped basidia	Septa are present	Cause smuts and rusts on crops; include some edible mushrooms	*Amanita* (poisonous mushroom); *Polyporus* (shelf fungus)
Ascomycetes (Ascomycota)	Form haploid sexual ascospores in saclike ascus	Septa are present	Cause molds on fruit; can damage textiles; cause Dutch elm disease and chestnut blight; include yeasts and morels	*Saccharomyces* (yeast); *Ophiostoma* (causes Dutch elm disease)

descendents of a particular common ancestor.) Because systematists prefer to give formal names only to clades, classification may soon be revised to place the species currently classified as chytrids or zygomycetes into several new phyla. The revision, however, is still in progress, so our description of fungal diversity will follow the traditional classification.

Chytrids Produce Swimming Spores

Unlike other types of fungi, most **chytrids** live in water. The chytrids (**Fig. 22-5**) are further distinguished from other fungi by their swimming spores, which require water for dispersal (even soil-dwelling chytrids require a film of water for reproduction). A chytrid spore propels itself through the water by means of a single flagellum located on one end of the spore. No other fungus group has flagella.

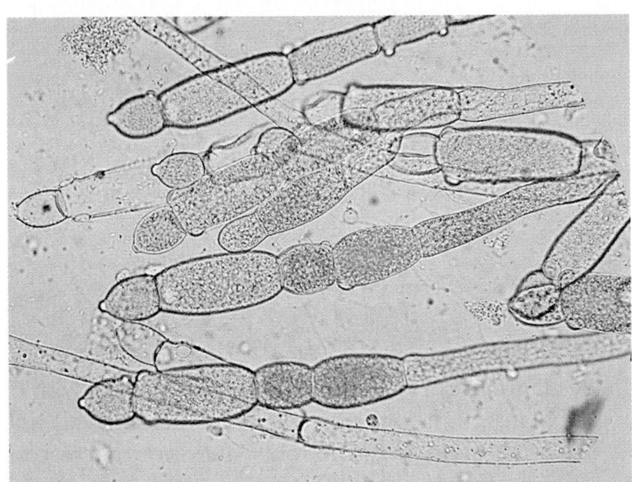

▲ **FIGURE 22-5 Chytrid filaments** These filaments of the chytrid fungus *Allomyces* are in the midst of sexual reproduction. The orange structures visible on many of the filaments will release male gametes; the clear swollen structures will release female gametes. Chytrid gametes are flagellated, and these swimming reproductive structures aid dispersal of members of this mostly aquatic phylum.

The oldest known fossil fungi are chytrids that were found in rocks more than 600 million years old. Ancestral fungi may well have been similar in habit to today's aquatic and marine chytrids, so fungi (like plants and animals) probably originated in a watery environment before colonizing land.

Most chytrid species feed on dead aquatic plants or other debris in watery environments, but some species are parasites of plants or animals. One such parasitic chytrid is believed to be a major cause of the current worldwide die-off of frogs, which threatens many species and has apparently already caused the extinction of several. No one yet understands exactly why this fungal disease emerged as a major cause of death in frogs. One hypothesis is that frog populations under stress from pollution and other environmental challenges might be more susceptible to infection by chytrids. (For more on the decline of frogs, see "Earth Watch: Frogs in Peril" on pp. 456–457.)

Zygomycetes Can Reproduce by Forming Diploid Spores

The **zygomycetes** generally live in soil or on decaying plant or animal material. This group includes species belonging to the genus *Rhizopus*, which cause the familiar annoyances of soft fruit rot and black bread mold. The life cycle of the black bread mold, which reproduces both asexually and sexually, is depicted in **Figure 22-6.** Asexual reproduction in zygote fungi is initiated by the formation of haploid spores in black spore cases called **sporangia** (singular, sporangium; **Fig. 22-6 ❶**). These spores disperse through the air and, if they land on a suitable substrate (such as a piece of bread), germinate to form new haploid hyphae.

If extensions from two hyphae of different mating types (designated "+" and "−") come into contact, sexual reproduction may ensue (**Fig. 22-6 ❷**). The two hyphae fuse to

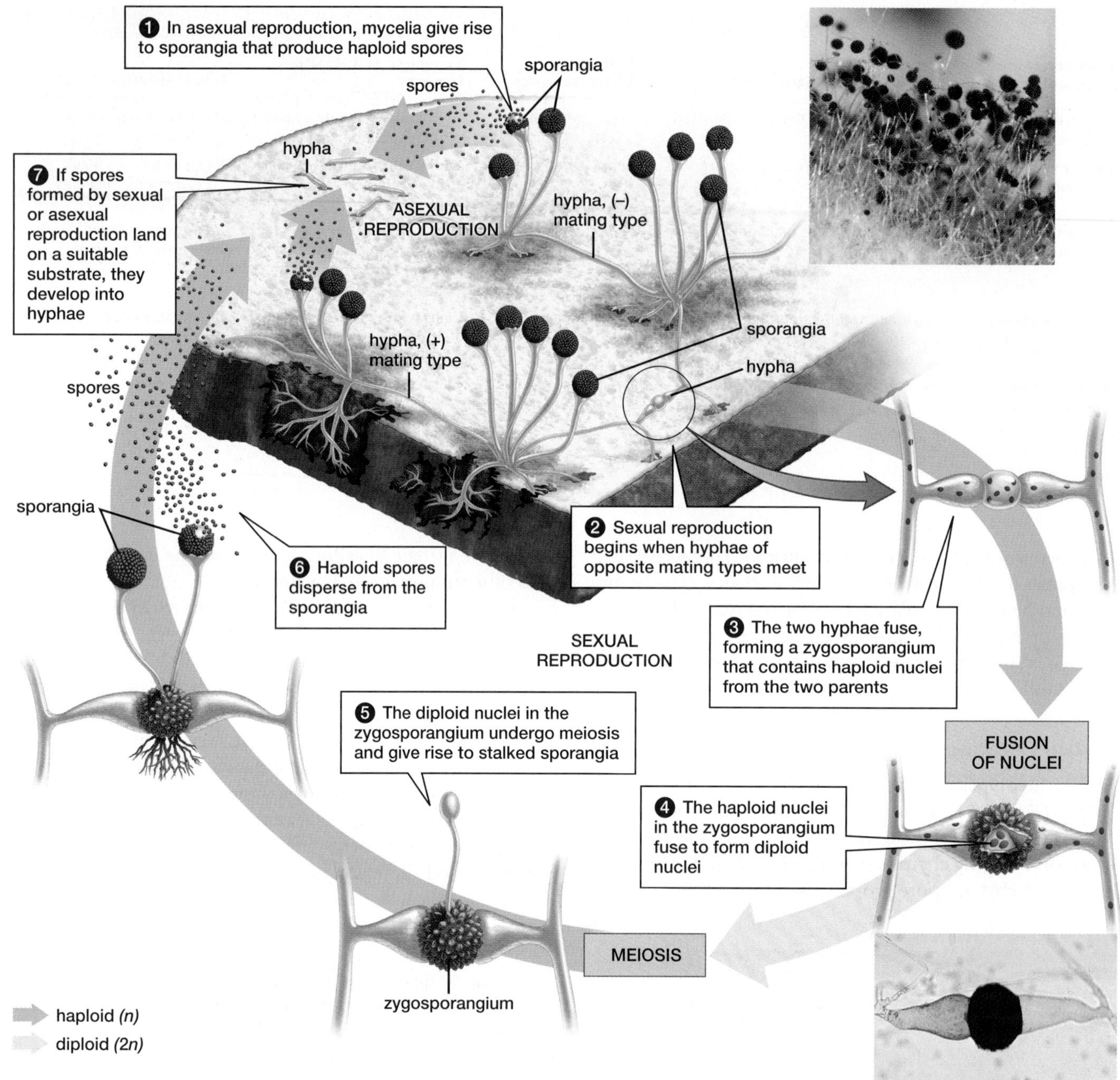

1 In asexual reproduction, mycelia give rise to sporangia that produce haploid spores
sporangia
spores
hypha
7 If spores formed by sexual or asexual reproduction land on a suitable substrate, they develop into hyphae
ASEXUAL REPRODUCTION
hypha, (–) mating type
sporangia
hypha, (+) mating type
hypha
spores
sporangia
2 Sexual reproduction begins when hyphae of opposite mating types meet
6 Haploid spores disperse from the sporangia
SEXUAL REPRODUCTION
3 The two hyphae fuse, forming a zygosporangium that contains haploid nuclei from the two parents
5 The diploid nuclei in the zygosporangium undergo meiosis and give rise to stalked sporangia
FUSION OF NUCLEI
4 The haploid nuclei in the zygosporangium fuse to form diploid nuclei
MEIOSIS
zygosporangium
haploid (n)
diploid (2n)

▲ **FIGURE 22-6 The life cycle of a zygomycete**

form a *zygosporangium* that contains multiple haploid nuclei from the two parents (**Fig. 22-6** ❸). As the zygosporangium develops, it becomes tough and resistant, and can remain dormant for long periods until environmental conditions are favorable for growth. Inside the zygosporangium, the haploid nuclei fuse to produce diploid nuclei (**Fig. 22-6** ❹). When conditions are favorable, the diploid nuclei undergo meiosis and give rise to stalked sporangia (**Fig. 22-6** ❺). The sporangia produce haploid spores that disperse (**Fig. 22-6** ❻), germinate, and develop into new haploid hyphae (**Fig. 22-6** ❼).

Glomeromycetes Associate with Plant Roots

Almost all **glomeromycetes** live in intimate contact with the roots of plants. In fact, the hyphae of glomeromycetes actually penetrate the cells of the roots, and form microscopic branching structures inside the cell (**Fig. 22-7**). This invasion of the plant's cells does not appear to harm the plant, which exhibits no signs of disease. To the contrary, glomeromycetes provide benefits to the plants they inhabit. This type of beneficial association between fungi and plant roots is known as a mycorrhiza, and is described in more detail later in this chapter.

▲ **FIGURE 22-7 Glomeromycete in a plant cell** Glomeromycete hyphae penetrate the cells of plants with which the fungus forms mutually beneficial associations. Inside the host plant's root cells, the fungus develops characteristic branching structures.

Glomeromycete reproduction is not fully understood; sexual reproduction by a member of the phylum is yet to be observed. During asexual reproduction, glomeromycetes produce clusters of spores by mitotic cell division at the tips of hyphae that typically remain outside the host plant cell. When the spores germinate, hyphae grow into the surrounding soil, but the new fungus survives only if its germinating hyphae reach a plant root.

Basidiomycetes Produce Club-Shaped Reproductive Structures

Basidiomycetes are called the **club fungi** because they produce club-shaped reproductive structures. Members of this phylum typically reproduce sexually (**Fig. 22-8**). Hyphae of different mating types fuse (**Fig. 22-8** ❶) to form hyphae in which each cell contains two nuclei, one from each parent (**Fig. 22-8** ❷). These hyphae grow into an

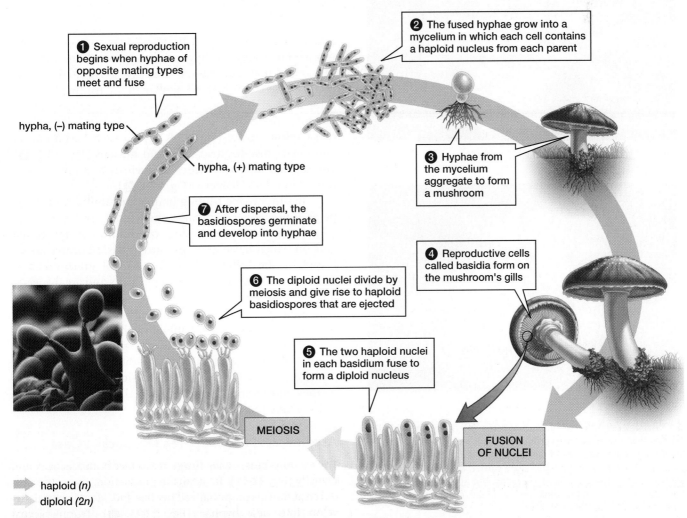

❶ Sexual reproduction begins when hyphae of opposite mating types meet and fuse

hypha, (–) mating type

hypha, (+) mating type

❷ The fused hyphae grow into a mycelium in which each cell contains a haploid nucleus from each parent

❸ Hyphae from the mycelium aggregate to form a mushroom

❹ Reproductive cells called basidia form on the mushroom's gills

❺ The two haploid nuclei in each basidium fuse to form a diploid nucleus

FUSION OF NUCLEI

MEIOSIS

❻ The diploid nuclei divide by meiosis and give rise to haploid basidiospores that are ejected

❼ After dispersal, the basidiospores germinate and develop into hyphae

haploid *(n)*

diploid *(2n)*

▲ **FIGURE 22-8 The life cycle of a typical basidiomycete** The photo shows two basidiospores attached to a basidium.

QUESTION If two spores from the same sporangium each germinate and the resulting hyphae come into contact, can sexual reproduction follow?

(a) Puffball

(b) Shelf fungi

(c) Stinkhorns

▲ **FIGURE 22-9 Diverse basidiomycetes (a)** The giant puffball *Lycoperdon giganteum* may produce up to 5 trillion spores. **(b)** Shelf fungi, some the size of dessert plates, are conspicuous on trees. **(c)** The spores of stinkhorns are carried on the outside of a slimy cap that smells terrible to humans, but appeals to flies. The flies lay their eggs on the stinkhorn, and inadvertently disperse the spores that stick to their bodies.

QUESTION Are the structures shown in these photos haploid or diploid?

▲ **FIGURE 22-10 A mushroom fairy ring** Mushrooms emerge in a fairy ring from an underground fungal mycelium, growing outward from a central point where a single spore germinated, perhaps centuries ago.

underground mycelium that, in response appropriate environmental conditions, gives rise to an aboveground fruiting body that consists of densely aggregated hyphae (**Fig. 22-8** ❸). Some of the hyphae in the fruiting body develop into club-shaped reproductive cells called **basidia** (singular, basidium) that, like their precursor cells, contain two haploid nuclei (**Fig. 22-8** ❹). In each basidium, the two nuclei fuse to yield a diploid nucleus (**Fig. 22-8** ❺). The diploid nucleus divides by meiosis and gives rise to four haploid *basidiospores* (**Fig. 22-8** ❻).

Basidiomycete fruiting bodies are familiar to most of us as mushrooms, puffballs, shelf fungi, and stinkhorns (**Fig. 22-9**). On the undersides of mushrooms are leaflike gills on which basidia are produced. Basidiospores are released by the billions from the gills of mushrooms or through openings in the tops of puffballs and are dispersed by wind and water.

Falling on fertile ground, a mushroom basidiospore may germinate and form haploid hyphae (**Fig. 22-8** ❼). In many cases, these hyphae grow outward from the original spore in a roughly circular pattern as the older hyphae in the center die. The subterranean body periodically sends up numerous mushrooms, which emerge in a ringlike pattern called a fairy ring (**Fig. 22-10**).

Ascomycetes Form Spores in a Saclike Case

The **ascomycetes,** or **sac fungi,** reproduce both asexually and sexually (**Fig. 22-11**). In asexual reproduction, spores are produced at the tips of specialized hyphae and, after dispersal, develop into new hyphae (**Fig. 22-11** ❶). During sexual reproduction, spores are produced by a more complex sequence of events that begins, in a typical ascomycete, when hyphae of different mating types (+ and −) come into contact (**Fig. 22-11** ❷). The two hyphae form special reproductive

Case Study continued

Humongous Fungus

Because the underground bodies of basidiomycetes such as *Armillaria* grow at a relatively steady rate, the age of a mycelium can be estimated by measuring the area over which its aboveground reproductive structures spread. On the basis of such measurements, it is apparent that basidiomycetes can live for hundreds of years. Some are even older than that. For example, the researchers who discovered the gigantic *Armillaria* in Oregon estimate that it took at least 2,400 years to grow to its current size.

structures that become linked by a connecting bridge. Haploid nuclei move across the bridge from the (−) reproductive structure to the (+) one, so that the (+) structure contains the nuclei from both parents (**Fig. 22-11** ❸). The (+) structures that now contain the pooled nuclei develop into hyphae that are incorporated into a fruiting body (**Fig. 22-11** ❹). At the tips of some of these hyphae, a saclike case called an **ascus** (plural, asci) forms (**Fig. 22-11** ❺). At this stage, each ascus contains two haploid nuclei. These nuclei fuse to yield a single diploid nucleus (**Fig. 22-11** ❻), which then divides by meiosis to yield four haploid nuclei (**Fig. 22-11** ❼). These four nuclei divide by mitosis and develop into eight haploid spores known as *ascospores* (**Fig. 22-11** ❽). Eventually, the ascus ruptures, liberating its ascospores. If the spores land in an appropriate location, they may germinate and develop into hyphae (**Fig. 22-11** ❾).

Some ascomycetes live in decaying forest vegetation and form either beautiful cup-shaped reproductive structures (**Fig. 22-12a**) or corrugated, mushroom-like fruiting bodies called *morels* (**Fig. 22-12b**). This phylum also includes many of the colorful molds that attack stored food and destroy fruit and grain crops and other plants, as well as the species that

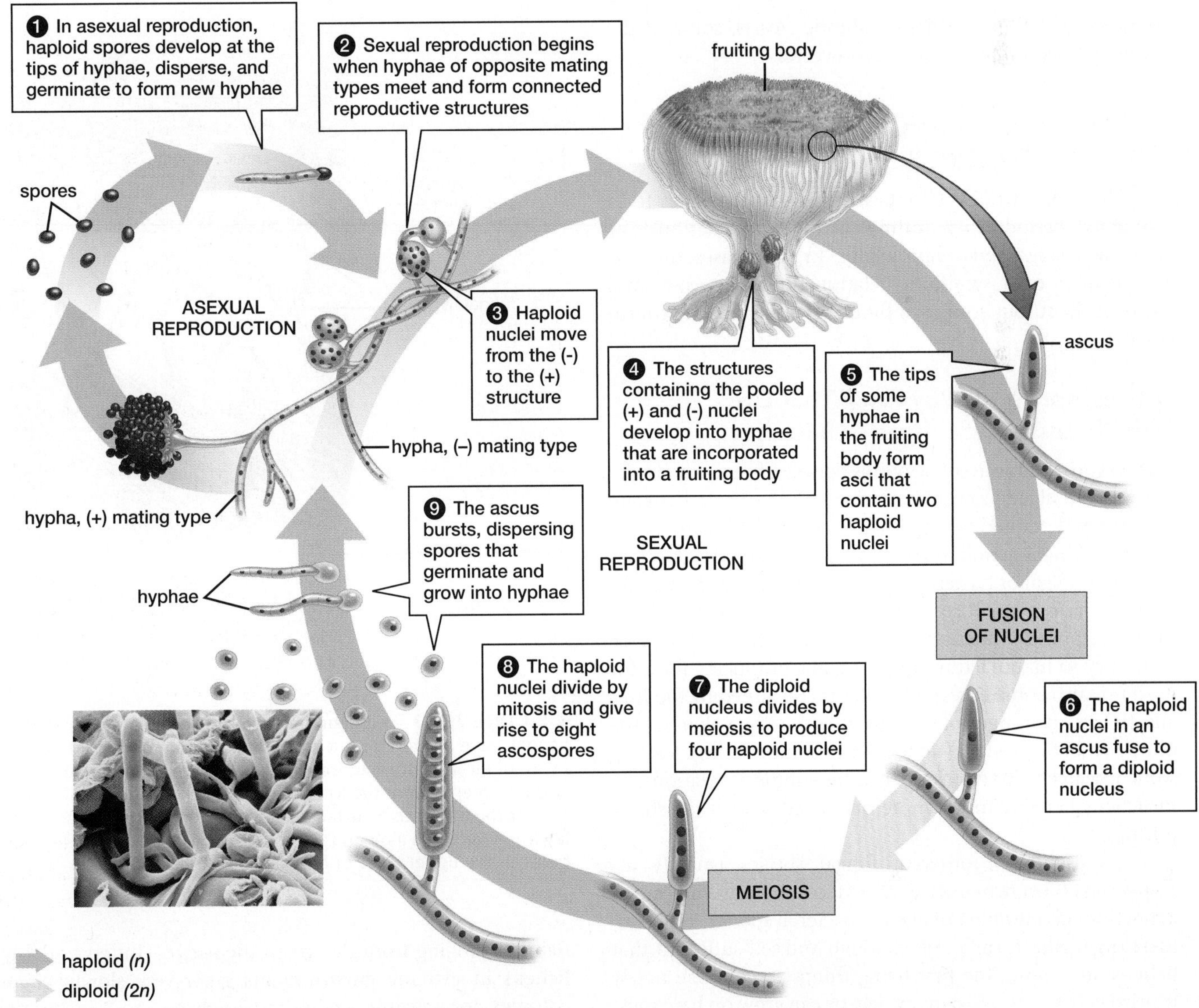
❶ In asexual reproduction, haploid spores develop at the tips of hyphae, disperse, and germinate to form new hyphae
❷ Sexual reproduction begins when hyphae of opposite mating types meet and form connected reproductive structures
fruiting body
spores
ASEXUAL REPRODUCTION
❸ Haploid nuclei move from the (-) to the (+) structure
hypha, (–) mating type
hypha, (+) mating type
❹ The structures containing the pooled (+) and (-) nuclei develop into hyphae that are incorporated into a fruiting body
❺ The tips of some hyphae in the fruiting body form asci that contain two haploid nuclei
ascus
❾ The ascus bursts, dispersing spores that germinate and grow into hyphae
hyphae
SEXUAL REPRODUCTION
FUSION OF NUCLEI
❻ The haploid nuclei in an ascus fuse to form a diploid nucleus
❽ The haploid nuclei divide by mitosis and give rise to eight ascospores
❼ The diploid nucleus divides by meiosis to produce four haploid nuclei
MEIOSIS
haploid (n)
diploid (2n)

▲ **FIGURE 22-11 The life cycle of a typical ascomycete** Some asci rising from hyphae are shown in the photo.

(a) Cup fungus

(b) Morels

▲ FIGURE 22-12 **Diverse ascomycetes (a)** The cup-shaped fruiting body of the scarlet cup fungus. **(b)** The morel, an edible delicacy. (Consult an expert before sampling any wild fungus—some are deadly!)

produces penicillin, the first antibiotic. Yeasts, some of the few unicellular fungi, are also ascomycetes.

22.3 HOW DO FUNGI INTERACT WITH OTHER SPECIES?

Many fungi live in direct contact with another species for a prolonged period. Such intimate, long-term relationships are known as *symbiotic* relationships. In many cases, the fungal member of a symbiotic relationship is parasitic and harms its host. But some symbiotic relationships are mutually beneficial.

Lichens Are Formed by Fungi That Live with Photosynthetic Algae or Bacteria

Lichens are symbiotic associations between fungi and single-celled green algae or cyanobacteria (**Fig. 22-13**). Lichens are sometimes described as fungi that have learned to garden, because the fungal member of the partnership "tends" the photosynthetic algal or bacterial partner by providing shelter and protection from harsh conditions. In this protected environment, the photosynthetic member of the partnership uses solar energy to manufacture simple sugars, producing food for itself but also some excess food that is consumed by the fungus. In fact, the fungus often consumes the lion's share of the photosynthetic product (up to 90% in some species), leading some researchers to conclude that the symbiotic relationship in lichens is really much more one sided than it is usually portrayed.

Thousands of different fungal species (mostly ascomycetes) form lichens (**Fig. 22-14**), combining with one of a much smaller number of algal or bacterial species. Together, these organisms form a unit so tough and self-sufficient that lichens are among the first living things to colonize newly formed volcanic islands; many lichens can grow on bare rock. Brightly colored lichens also invade other inhospitable habitats ranging from deserts to the Arctic. Understandably, lichens in extreme environments grow very slowly; arctic colonies, for example, expand as slowly as 1 to 2 inches per 1,000 years. Despite their slow growth, lichens can persist

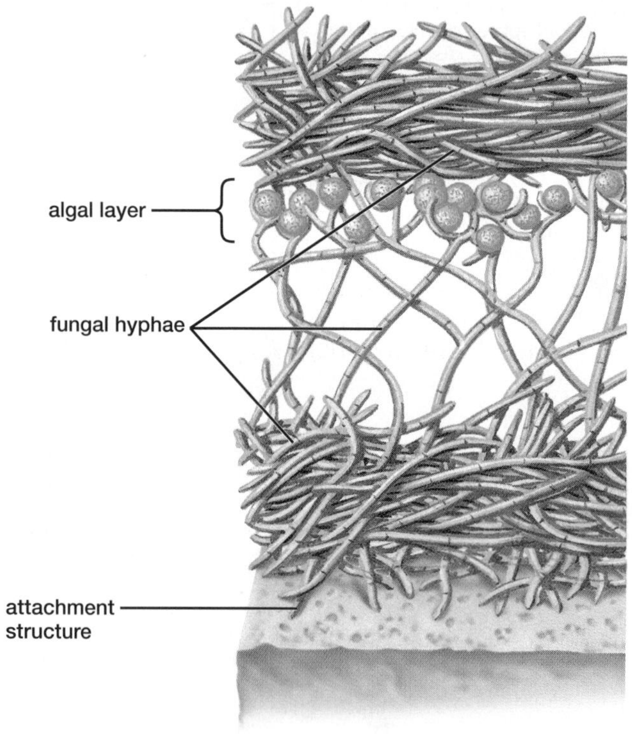

▲ FIGURE 22-13 **The lichen: a symbiotic partnership** Most lichens have a layered structure bounded on the top and bottom by an outer layer formed from fungal hyphae. The fungal hyphae emerge from the lower layer, forming attachments that anchor the lichen to a surface, such as a rock or a tree. An algal layer in which the alga and fungus grow in close association lies beneath the upper layer of hyphae.

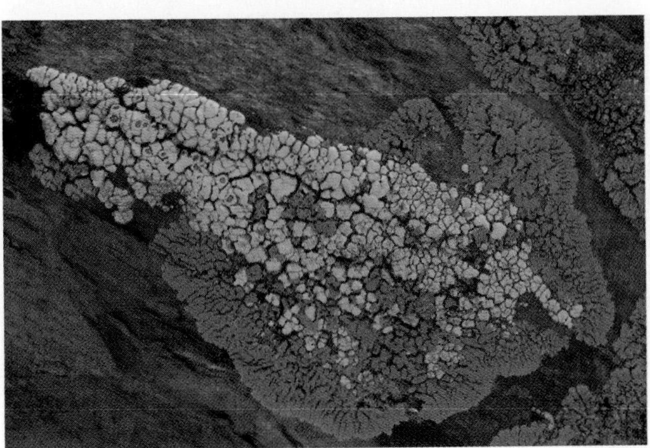

(a) Encrusting lichen

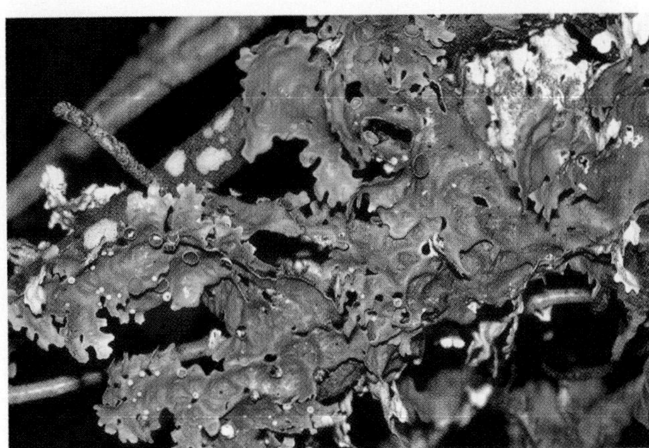

(b) Leafy lichen

▲ **FIGURE 22-14 Diverse lichens (a)** A colorful encrusting lichen, growing on dry rock, illustrates the tough independence of this symbiotic combination of fungus and algae. Pigments produced by the fungal partner are responsible for the bright orange color. **(b)** A leafy lichen grows from a dead tree branch.

for long periods of time; some arctic lichens are more than 4,000 years old.

Mycorrhizae Are Fungi Associated with Plant Roots

Mycorrhizae (singular, mycorrhiza) are important symbiotic associations between fungi and plant roots. More than 5,000 species of mycorrhizal fungi (including representatives of all the major groups of fungi) can grow in intimate association with plant roots, including those of most tree species. The hyphae of mycorrhizal fungi surround the plant root and invade the root cells (**Fig. 22-15**).

Mycorrhizae Help Feed Plants

The association between plants and mycorrhizae benefits both the fungi and their plant partners. The mycorrhizal fungi receive energy-rich sugar molecules that are produced photosynthetically by plants and passed from their roots to the fungi. In return, the fungi digest and absorb minerals and organic nutrients from the soil, passing some of them directly into the root cells. Experiments have shown that phosphorus and nitrogen, key nutrients that are crucial for plant growth, are among the molecules that mycorrhizae transport from soil to roots. Mycorrhizal fungi also absorb water and pass it to the plant—an advantage for plants in dry, sandy soils.

▲ **FIGURE 22-15 Mycorrhizae enhance plant growth** Hyphae of mycorrhizae entwining about the root of an aspen tree. Plants grow significantly better in a symbiotic association with these fungi, which help make nutrients and water available to the roots.

The partnership between mycorrhizae and plants makes a crucial contribution to the health of Earth's plants. Plants that are deprived of mycorrhizal fungi tend to be smaller and less vigorous than are plants with mycorrhizal partners. Thus, the presence of mycorrhizae increases the overall productivity of Earth's plant communities and thereby increases their ability to support the animals and other organisms that depend on them.

Mycorrhizae May Have Helped Plants Invade Land

Some scientists believe that mycorrhizal associations may have been important in the invasion of land by plants more than 400 million years ago. Such a relationship between an aquatic fungus and a green alga (ancestral to terrestrial plants) could have helped the alga acquire the water and mineral nutrients it needed to survive out of water.

The fossil record is consistent with the hypothesis that mycorrhizae played a role in plants' colonization of land. The oldest fossils of terrestrial fungi are about 460 million years old, roughly the same age as the oldest fossils of terrestrial plants. This finding suggests that fungi and plants came ashore at the same time, perhaps together. In addition, plant fossils formed not long after the invasion of land exhibit distinctive root structures similar to those that form today in response to the presence of mycorrhizae. These fossils show that fully developed mycorrhizae were present very early in the evolution of land plants, suggesting that a simpler plant–fungus association might have been present even earlier.

Endophytes Are Fungi That Live Inside Plant Stems and Leaves

The intimate association between fungi and plants is not limited to root mycorrhizae. Fungi have also been found living inside the aboveground tissues of virtually every plant species that has been tested for their presence. Some of these *endophytes* (organisms that live inside other organisms) are parasites that cause plant diseases, but many, perhaps most, are beneficial to the host plant. The best-studied examples of beneficial fungal endophytes are the ascomycete species that live inside the leaf cells of many species of grass. These fungi produce substances that are distasteful or toxic to insects and grazing mammals and thus help protect the grass plants from those predators.

The antipredator protection provided by the fungal endophytes is sufficiently effective that agricultural scientists are working hard to discover a way to grow grasses free of the endophytes. Horses, cows, and other agriculturally important grazers tend to avoid eating grasses that contain the endophytes. When the only available food is endophyte-containing grass, animals that eat it experience poor health and slow growth.

Some Fungi Are Important Decomposers

Some fungi, acting as mycorrhizae and endophytes, play a major role in the growth and preservation of plant tissue. Other fungi, however, play a similarly major role in its destruction, by acting as decomposers. Alone among organisms, fungi can digest both lignin and cellulose, the molecules that make up wood. When a tree or other woody plant dies, only fungi are capable of decomposing its remains.

Fungi are Earth's undertakers, consuming not only dead wood but the dead of all kingdoms. The fungi that are *saprophytes* (feeding on dead organisms) return the dead tissues' component substances to the ecosystems from which they came. The extracellular digestive activities of saprophytic fungi liberate nutrients that can be used by plants. If fungi and bacteria were suddenly to disappear, the consequences would be disastrous. Nutrients would remain locked in the bodies of dead plants and animals, the recycling of nutrients would grind to a halt, soil fertility would rapidly decline, and waste and organic debris would accumulate. In short, ecosystems would collapse.

22.4 HOW DO FUNGI AFFECT HUMANS?

The average person gives little thought to fungi, except perhaps for an occasional, momentary appreciation for the mushrooms on a pizza. Nonetheless, fungi affect our lives in more ways than you might imagine.

Fungi Attack Plants That Are Important to People

Fungi cause the majority of plant diseases, and some of the plants that they infect are important to humans. For example, fungal pathogens have a devastating effect on the world's food supply. Especially damaging are the basidiomycete plant pests descriptively called *rusts* and *smuts*, which cause billions of dollars' worth of damage to grain crops annually (**Fig. 22-16**). Fungal diseases also affect the appearance of our landscape. The American elm and the American chestnut—two tree species that were once prominent in many of America's parks, yards, and forests—were destroyed on a massive scale by the ascomycetes that cause Dutch elm disease and chestnut blight. Today, few people can recall the graceful forms of large elms and chestnuts, which are now almost entirely absent from the landscape.

▲ **FIGURE 22-16 Corn smut** This basidiomycete pathogen destroys millions of dollars' worth of corn each year. Even a pest like corn smut has its admirers, though. In Mexico this fungus is known as *huitlacoche* and is considered to be a great delicacy.

Fungi continue to attack plant tissues long after they have been harvested for human use. To the dismay of homeowners, a host of different fungal species attack wood, causing it to rot. Some ascomycete molds secrete the enzymes cellulase, which attacks cellulose, and protease, which attacks proteins. These fungal enzymes can cause significant damage to cotton and wool textiles, especially in warm, humid climates where molds flourish.

The fungal impact on agriculture and forestry is not entirely negative, however. Fungal parasites that attack insects and other arthropod pests can be an important ally in pest control (**Fig. 22-17a**). Farmers who wish to reduce their dependence on toxic and expensive chemical pesticides are increasingly turning to biological methods of pest control, including the application of "fungal pesticides." Fungal pathogens are currently used to control termites,

Case Study continued

Humongous Fungus

The *Armillaria* fungus species that grew to massive size in Oregon harms trees in the forests it inhabits. As the fungus feeds on roots, it causes "root rot" that weakens or kills trees. This root rot provides aboveground evidence of *Armillaria*'s existence; the giant Oregon specimen was first identified by examining aerial photos to find forested areas with many dead trees.

(a) Pest-killing fungus

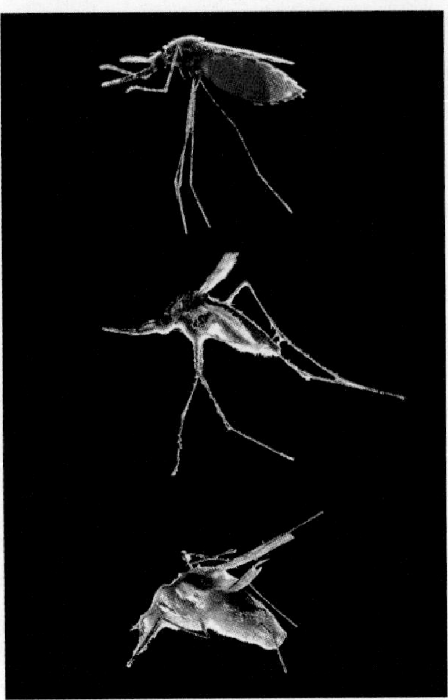

(b) Disease-fighting fungus

▲ **FIGURE 22-17 Helpful fungal parasites (a)** Fungi such as *Cordyceps* are used by farmers to control insect pests. **(b)** A healthy malaria-carrying mosquito (top) infected by *Beauveria* is transformed to a fungus-encrusted corpse in less than 2 weeks.

rice weevils, tent caterpillars, aphids, citrus mites, and other pests. In addition, biologists have discovered that certain fungi attack and kill the mosquito species that transmit malaria (**Fig. 22-17b**). Plans are under way to enlist these fungi in the fight against malaria, one of the world's deadliest diseases.

Fungi Cause Human Diseases

The fungi include parasitic species that attack humans directly. Some of the most familiar fungal diseases are those caused by ascomycetes that attack the skin, resulting in athlete's foot, jock itch, and ringworm. These diseases, though unpleasant, are not life threatening and can usually be treated with antifungal ointments. Prompt treatment can also usually control another common fungal disease, vaginal infections caused by the yeast *Candida albicans* (**Fig. 22-18**). Fungi can also infect the lungs if victims inhale spores of disease-causing fungal species such as those that cause valley fever and histoplasmosis. Like other fungal infections, these diseases can, if promptly diagnosed, be controlled with antifungal drugs. If untreated, however, they can develop into serious, systemic infections. Singer Bob Dylan, for instance, became gravely ill with histoplasmosis when a fungus infected the pericardial membrane surrounding his heart.

Fungi Can Produce Toxins

In addition to their role as agents of infectious disease, some fungi produce toxins that are dangerous to humans. Of particular concern are toxins produced by fungi that grow on grains and other foodstuffs that have been stored in overly moist conditions. For example, molds of the genus *Aspergillus* produce highly toxic, carcinogenic compounds known as aflatoxins. Some foods, such as peanuts, seem especially susceptible to attack by *Aspergillus*. Since aflatoxins were discovered in the 1960s, food growers and processors have developed methods for reducing the growth of *Aspergillus* in stored crops, so aflatoxins have been largely eliminated from the nation's peanut butter supply.

One infamous toxin-producing fungus is the ascomycete *Claviceps purpurea*, which infects rye plants and causes a disease known as ergot. This fungus produces several toxins, which can affect humans if infected rye is ground into flour and consumed. This happened frequently in northern Europe in the Middle Ages, with devastating effects. At that time, ergot poisoning was typically fatal, and victims experienced terrible symptoms before dying. One ergot toxin is a vasoconstrictor, which constricts blood vessels and reduces blood flow. The effect can be so extreme

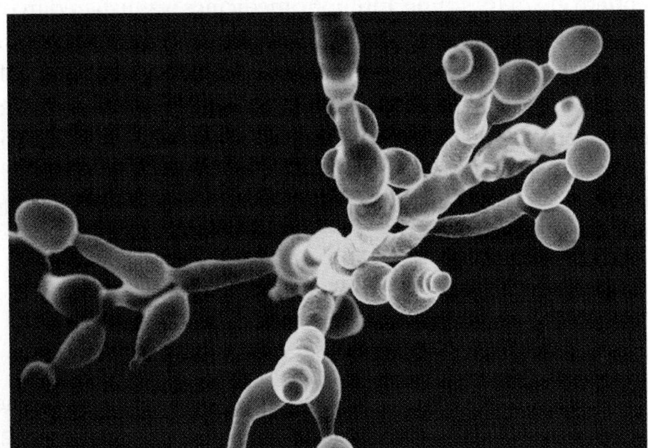

▲ **FIGURE 22-18 The unusual yeast** Most yeasts are single-celled, but some, such as *Candida* (shown here) may form multicellular filaments. *Candida* is a common cause of vaginal infections. Yeasts usually reproduce by budding; budding cells are visible at the tips of some of the filaments in the photo.

▲ FIGURE 22-19 **Penicillium** Penicillium growing on an orange. Reproductive structures, which coat the fruit's surface, are visible, while hyphae, beneath, draw nourishment from inside. The antibiotic penicillin was first isolated from this fungus.

QUESTION Why do some fungi produce antibiotic chemicals?

that gangrene develops and limbs actually shrivel and fall off. Other ergot toxins cause symptoms that include a burning sensation, vomiting, convulsive twitching, and vivid hallucinations. Today, new agricultural techniques have effectively eliminated ergot poisoning, but the hallucinogenic drug LSD, which is derived from a component of the ergot toxins, still remains as a legacy of this disease.

Many Antibiotics Are Derived from Fungi

Fungi have also had positive impacts on human health. The modern era of lifesaving antibiotic medicines was ushered in by the discovery of penicillin, which is produced by an ascomycete mold (**Fig. 22-19**; see Fig. 1-5). Penicillin is still used, along with other fungi-derived antibiotics such as oleandomycin and cephalosporin, to combat bacterial diseases. Other important drugs are also derived from fungi, including cyclosporin, which is used to suppress the immune response after organ transplants so that the body is less likely to reject the transplanted organs.

Fungi Make Important Contributions to Gastronomy

Fungi make important contributions to human nutrition. The most obvious components of this contribution are the fungi that we consume directly: wild and cultivated basidiomycete mushrooms and ascomycetes, such as morels and the rare and prized truffle (**Fig. 22-20**). (Use caution if you gather your own wild fungi; see "Links to Everyday Life: Collect Carefully.")

▲ FIGURE 22-20 **The truffle** Truffles are the underground, spore-containing structures of an ascomycete that forms a mycorrhizal association with the roots of oak trees.

The role of fungi in cuisine also has less-visible manifestations. For example, some of the world's most famous cheeses, including Roquefort, Camembert, Stilton, and Gorgonzola, gain their distinctive flavors from ascomycete molds that grow on them as they ripen. Perhaps the most important and pervasive fungal contributors to our food supply, however, are the single-celled ascomycetes (a few species are basidiomycetes) known as *yeasts*.

Wine and Beer Are Made Using Yeasts

The discovery that yeasts could be harnessed to enliven our culinary experience is surely a key event in human history. Among the many foods and beverages that depend on yeasts for their

Have you ever wondered?

Why Truffles Are So Expensive?

Although many fungi are prized as food, none is as avidly sought as the truffle. The finest Italian truffles may sell for as much as $1,500 per pound, and unusually large specimens can fetch spectacular prices. A 3.3-pound Italian white truffle recently sold at auction for $330,000!

Why such high prices? Truffles develop underground, and it takes some work to find one. In fact, humans can't do it alone and need help from other species. Some animals, especially pigs, are attracted to the aroma of a mature truffle. If a pig follows the smell to a truffle, it will dig the fungus up and devour it. That's why, traditionally, truffle collectors have used muzzled pigs to hunt their quarry.

Today, trained dogs are the most common assistants to truffle hunters. Dogs are necessary even on the farms where much of today's truffle crop is laboriously grown. The difficulty of cultivating and harvesting truffles accounts in part for their high price. And no one has figured out how to cultivate the prized white truffle. The only way to acquire one is to search a suitable forest, accompanied by a trained, truffle-hunting dog or pig.

Links to *Everyday Life*

Collect Carefully

In the early 1980s, doctors at a California hospital noticed a curious trend. Over a period of a few months, the number of patients admitted for treatment of poisoning had risen dramatically, and many of those admitted had died. What caused this sudden outbreak of poisoning? Further investigation revealed that, in almost all cases, the victims were recent immigrants from Laos or Cambodia. Struggling to adjust to their new country, they had been thrilled to find that California's forests contained mushrooms that looked just like the ones they had collected for food back in Asia. Unfortunately, the similarity was superficial; the mushrooms were, in fact, poisonous species. The immigrants' nostalgic pursuit of "comfort food" had tragic consequences.

In general, immigrants from countries where mushroom collecting is common have proved to be especially likely to be poisoned by toxic mushrooms. But they are not the only victims. Each year, a number of small children, inexperienced collectors, and unlucky guests at gourmet dinners make unexpected trips to the hospital after eating poisonous wild mushrooms.

It can be fun and rewarding to collect wild mushrooms, which offer some of the richest and most complex flavors a human can experience. But if you decide to go collecting, be careful, because some of the deadliest poisons known to humankind are found in mushrooms. Especially noted for their poisons are certain species in the genus *Amanita*, which have evocative common names such as death cap and destroying angel (**Fig. E22-1**). These names are apt, because even a single bite of one of these mushrooms can be lethal.

Damage from *Amanita* toxins is most severe in the liver, where the toxins tend to accumulate. Often, a victim of *Amanita* poisoning can be saved only by undergoing a liver transplant. So be sure to protect your health by inviting an expert to join your mushroom-hunting expeditions.

▲ **FIGURE E22-1 The destroying angel** Mushrooms produced by the basidiomycete *Amanita virosa* can be deadly.

production are bread, wine, and beer, which are consumed so widely that it is difficult to imagine a world without them. All derive their special qualities from fermentation by yeasts. Fermentation occurs when yeasts extract energy from sugar and, as by-products of the metabolic process, emit carbon dioxide and ethyl alcohol (see "Links to Everyday Life: A Jug of Wine, a Loaf of Bread, and a Nice Bowl of Sauerkraut" on p. 139).

As yeasts consume the fruit sugars in grape juice, the sugars are converted to alcohol, and wine is the result. Eventually, the increasing concentration of alcohol kills the yeasts, ending fermentation. If the yeasts die before all available grape sugar is consumed, the wine will be sweet; if the sugar is exhausted, the wine will be dry.

Beer is brewed from grain (usually barley), but yeasts cannot effectively consume the carbohydrates that compose grain kernels. For the yeasts to do their work, the barley grains must have sprouted (recall that grains are actually seeds). Germination converts the kernels' carbohydrates to sugar, so the sprouted barley provides an excellent food source for the yeasts. As with wine, fermentation converts sugars to alcohol, but beer brewers capture the carbon dioxide by-product as well, giving the beer its characteristic bubbly carbonation.

Yeasts Make Bread Rise

In bread making, carbon dioxide is the crucial fermentation product. The yeasts added to bread dough do produce alcohol as well as carbon dioxide, but the alcohol evaporates during baking. In contrast, the carbon dioxide is trapped in the dough, where it forms the bubbles that give bread its light, airy texture (and saves us from a life of eating sandwiches made of crackers). So the next time you're enjoying a slice of French bread with Camembert cheese and a nice glass of chardonnay, or a slice of pizza and a cold bottle of your favorite brew, you might want to quietly give thanks to the yeasts. Our diets would certainly be a lot duller without the help we get from fungal assistants.

Case Study revisited

Humongous Fungus

Why do *Armillaria* fungi grow so large? Their size is due in part to their ability to form rhizomorphs, which consist of hyphae bundled together inside a protective rind. The enclosed hyphae carry nutrients to the rhizomorphs, allowing them to extend long distances through nutrient-poor areas to reach new sources of food. The *Armillaria* fungus can thus grow beyond the boundaries of a particular food-rich area.

Another factor that may contribute to the gigantic size of the Oregon *Armillaria* is the climate in which it was found. In this dry region, fungal fruiting bodies form only rarely, so the colossal *Armillaria* rarely produces spores. In the absence of spores that might grow into new individuals, the existing individual faces little competition for resources, and is free to grow and fill an increasingly large area.

The discovery of the Oregon specimen is merely the latest chapter in a long-running, good-natured "fungus war" that began in 1992 with the discovery of the first humongous fungus, a 37-acre *Armillaria gallica* growing in Michigan. Since that initial landmark discovery, research groups in Michigan, Washington, and Oregon have engaged in a friendly competition to find the largest fungus. Will the current record ever be topped? Stay tuned.

Consider This

Because the entire Oregon *Armillaria* grew from a single spore, all of its cells are genetically identical. Its parts, however, are not all physiologically dependent on one another, so it is unlikely that any substances are transported through the entire 3.4-square-mile mycelium. And there is no continuous skin or bark or membrane that covers the entire mycelium and separates it from the environment as a unit. Is the fungus's genetic unity sufficient evidence for it to be considered a single individual, or is greater physiological integration required? Do you think that the claim of "world's largest organism" is valid?

CHAPTER REVIEW

Summary of Key Concepts

22.1 What Are the Key Features of Fungi?

Fungal bodies generally consist of filamentous hyphae, which are either multicellular or multinucleated and form large, intertwined networks called *mycelia*. Fungal nuclei are generally haploid. A cell wall of chitin surrounds fungal cells.

All fungi are heterotrophic, secreting digestive enzymes outside their bodies and absorbing the liberated nutrients.

Fungal reproduction is varied and complex. Asexual reproduction can occur through asexual spore formation. Sexual spores form after compatible haploid nuclei fuse to form a diploid zygote, which undergoes meiosis to form haploid sexual spores. Both asexual and sexual spores produce haploid mycelia through mitosis.

22.2 What Are the Major Groups of Fungi?

The major phyla of fungi and their characteristics are summarized in Table 22-1.

22.3 How Do Fungi Interact with Other Species?

A lichen is a symbiotic association between a fungus and algae or cyanobacteria. This self-sufficient combination can colonize bare rock. Mycorrhizae are associations between fungi and the roots of most vascular plants. The fungus derives photosynthetic nutrients from the plant roots and, in return, carries water and nutrients into the root from the surrounding soil. Endophytes are fungi that grow inside the leaves or stems of plants and that may help protect the plants that harbor them. Saprophytic fungi are extremely important decomposers in ecosystems. Their filamentous bodies penetrate rich soil and decaying organic material, liberating nutrients through extracellular digestion.

22.4 How Do Fungi Affect Humans?

Many plant diseases are caused by parasitic fungi. Some parasitic fungi can help control insect crop pests. Others can cause human diseases, including ringworm, athlete's foot, and common vaginal infections. Some fungi produce toxins that can harm humans. Nonetheless, fungi add variety to the human food supply, and fermentation by fungi helps make wine, beer, and bread.

Key Terms

ascomycete *410*
ascus (plural, asci) *411*
basidiomycete *409*
basidium (plural, basidia) *410*
chytrid *407*
club fungus *409*
glomeromycete *408*
hypha (plural, hyphae) *404*
lichen *412*
mycelium (plural, mycelia) *404*
mycorrhiza (plural, mycorrhizae) *413*
sac fungus *410*
septum (plural, septa) *404*
sporangium (plural, sporangia) *407*
spore *405*
zygomycete *407*

Thinking Through the Concepts

Fill-in-the-Blank

1. The portions of a fungus that are visible to the naked eye are often structures specialized for __________. These structures release tiny __________, which are dispersed to produce new fungi.
2. The fungal body is a(n) __________, and is composed of microscopic threads called __________ that may be subdivided into many cells by __________. The cells walls of fungi are strengthened by __________.
3. In fungi, asexual spores are produced by __________ cell division and have __________ set(s) of chromosomes. Sexual spores are produced by __________ cell division

in a(n) __________ and have __________ set(s) of chromosomes.

4. Fill in the blanks in the following sentences with the common names of fungal phyla. Almost all __________ live in intimate association with plant roots. Flagellated, swimming spores are produced by __________. Mushrooms and puffballs are reproductive structures of __________.
5. __________ are symbiotic associations of fungi and green algae. __________ are symbiotic, mutually beneficial associations of fungi and plant roots. Some fungi are __________ that live inside the aboveground tissues of plants.
6. Fungi are the only decomposers that can digest __________. The fungi that cause breads to rise and wines to ferment are __________. Human ailments caused by fungi include __________ and __________.

Review Questions

1. Describe the structure of the fungal body. How do fungal cells differ from most plant and animal cells?
2. What portion of the fungal body is represented by mushrooms, puffballs, and similar structures? Why are these structures elevated above the ground?
3. What two plant diseases, caused by parasitic fungi, have had an enormous impact on forests in the United States? In which phylum are these fungi found?
4. List some fungi that attack crops. To which phyla do they belong?
5. Describe asexual reproduction in fungi.
6. List the phyla of fungi, describe some key features of each, and give one example of each.
7. Describe how a fairy ring of mushrooms is produced. Why is the diameter related to its age?
8. Describe two symbiotic associations between fungi and organisms from other clades. In each case, explain how each partner in these associations is affected.

Applying the Concepts

1. Dutch elm disease in the United States is caused by an *exotic*—that is, an organism (in this case, a fungus) introduced from another part of the world. What damage has this introduction done? What other fungal pests fall into this category? Why are parasitic fungi particularly likely to be transported out of their natural habitat? What can governments do to limit this importation?
2. The discovery of penicillin revolutionized the treatment of bacterial diseases. However, penicillin is now rarely prescribed. Why is this?
3. The discovery of penicillin was the result of a chance observation by an observant microbiologist, Alexander Fleming. How would you search systematically for new antibiotics produced by fungi? Where would you look for these fungi?
4. Fossil evidence indicates that mycorrhizal associations between fungi and plant roots existed in the late Paleozoic era, when the invasion of land by plants began. This evidence suggests an important link between mycorrhizae and the successful invasion of land by plants. Why might mycorrhizae have been important fungi in the colonization of terrestrial habitats by plants?
5. General biology texts in the 1960s included fungi in the plant kingdom. Why do biologists no longer consider fungi as legitimate members of the plant kingdom?
6. What ecological consequences would occur if humans, using a new and deadly fungicide, destroyed all fungi on Earth?

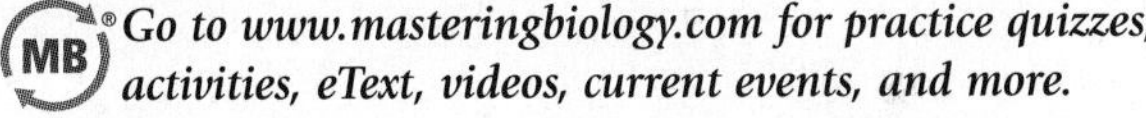

Go to www.masteringbiology.com for practice quizzes, activities, eText, videos, current events, and more.

Animal Diversity I: Invertebrates

Chapter 23

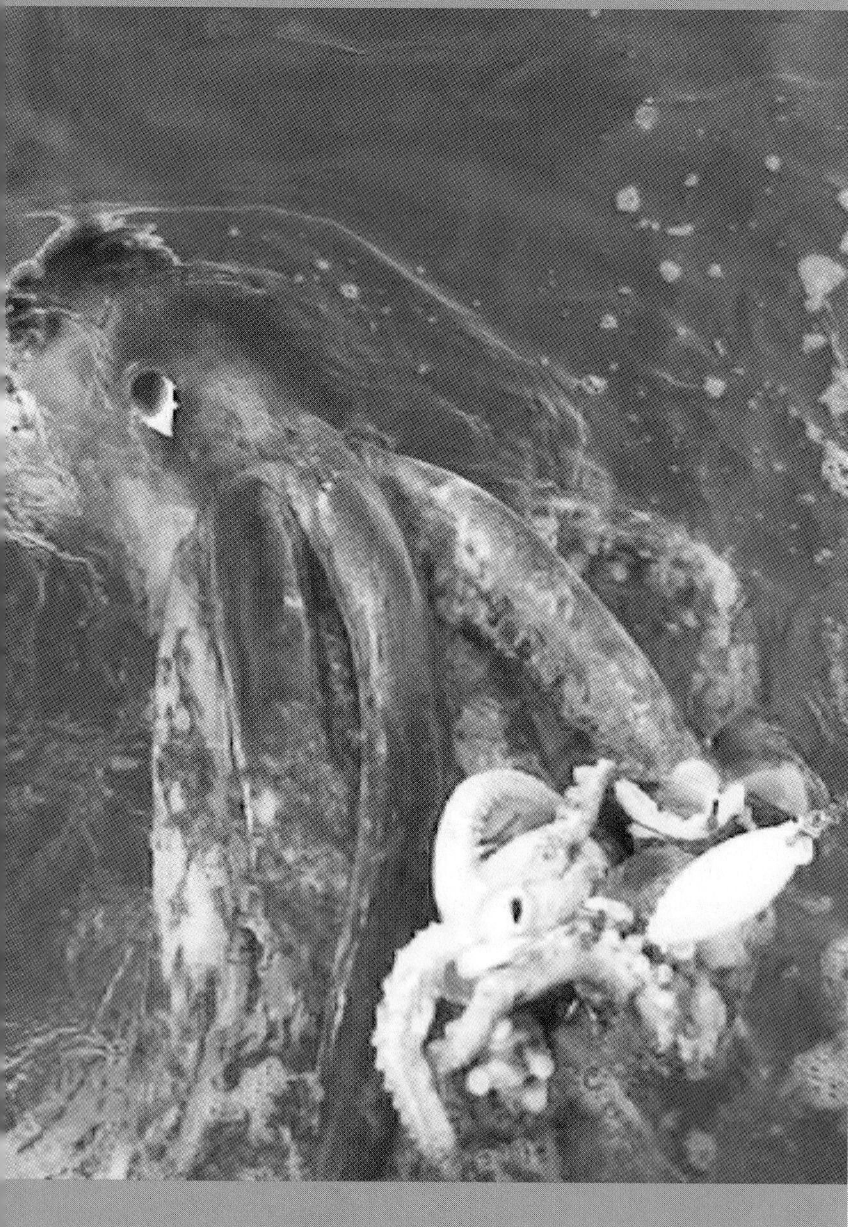

▲ Although the giant squid is Earth's largest invertebrate animal, a single, brief video is our only observation of a living giant squid in its natural habitat.

Case Study

The Search for a Sea Monster

EVERYONE LOVES A GOOD mystery, and a mystery involving a gigantic, fearsome predator is even better. Consider *Architeuthis*, the giant squid. It is one of the world's largest invertebrate animals, reaching lengths of 40 feet (13 meters) or more. Each of its huge eyes can be as large as a human head. The squid's ten tentacles, two of which are longer than the others, are covered with powerful suckers. The suckers contain sharp, clawlike hooks to better grasp prey, which is then pulled toward the squid's mouth, where a heavily muscled beak tears the food apart.

The giant squid is one of the most imposing organisms on Earth, yet we know almost nothing about its habits and lifestyle. What does it eat? Does it swim with its head elevated or pointed downward? How does it mate? Does it live alone or in groups? Even these basic questions about the behavior of the giant squid remain unanswered, because a single brief video recording is our only observation of an adult giant squid alive in its natural habitat.

Our limited scientific knowledge of the giant squid comes almost entirely from specimens that have been found dead or dying, washed up on beaches, caught in fishermen's nets, or contained in the stomachs of sperm whales—whose diet includes the occasional giant squid. More than 200 such specimens have been reported during the past century, and written accounts of squid corpses go back to the sixteenth century. Live squids, however, remain elusive because they inhabit deep ocean waters, beyond the reach of human divers.

Clyde Roper, a biologist at the Smithsonian Institution, has devoted much of his professional life to a quest to view and study live giant squids. We'll explore some of Dr. Roper's undersea search methods at the close of this chapter, following a survey of the extraordinary diversity of invertebrate animal life.

At a Glance

23.1 WHAT ARE THE KEY FEATURES OF ANIMALS?

It is difficult to devise a concise definition of the term "animal." No single feature uniquely defines animals, so the group is defined by a list of characteristics. None of these characteristics is unique to animals, but together they distinguish animals from members of other clades:

- Animals are multicellular.
- Animal cells lack a cell wall.
- Animals obtain their energy by consuming other organisms.
- Animals typically reproduce sexually. Although animal species exhibit a tremendous diversity of reproductive styles, most are capable of sexual reproduction.
- Animals are motile (able to move about) during some stage of their lives. Even the relatively stationary sponges have a free-swimming *larval* stage (a juvenile form).
- Most animals are able to respond rapidly to external stimuli as a result of the activity of nerve cells, muscle tissue, or both.

23.2 WHICH ANATOMICAL FEATURES MARK BRANCH POINTS ON THE ANIMAL EVOLUTIONARY TREE?

By the Cambrian period, which began 544 million years ago, most of the animal phyla that currently populate Earth were already present. Unfortunately, the Precambrian fossil record is very sparse, and does not reveal much about the early evolutionary history of animals. Therefore, animal systematists have looked to features of animal anatomy, embryological development, and DNA sequences for clues about the phylogeny of animals. These investigations have shown that certain features mark major branching points on the animal evolutionary tree and represent milestones in the evolution of the different body plans of modern animals (**Fig. 23-1**). In the following sections, we will explore these evolutionary milestones and their legacies in the bodies of modern animals.

Lack of Tissues Separates Sponges from All Other Animals

One of the earliest major innovations in animal evolution was the appearance of **tissues**—groups of similar cells integrated into a functional unit, such as a muscle. Today, the bodies of almost all animals include tissues; the only animals that have retained the ancestral lack of tissues are the sponges. In sponges, individual cells may have specialized functions, but they act more or less independently and are not organized into true tissues. This unique feature of sponges suggests that the split between sponges and the evolutionary branch leading to all other animal phyla must have occurred very early in the history of animals. An ancient common ancestor without tissues gave rise to both the sponges and the animals that have tissues.

Animals with Tissues Exhibit Either Radial or Bilateral Symmetry

The evolutionary advent of tissues coincided with the first appearance of body symmetry; all animals with true tissues also have symmetrical bodies. An animal is said to be symmetrical if it can be bisected along at least one plane such that the resulting halves are mirror images of one another. Unlike the asymmetrical sponges, any symmetrical animal has an upper, or *dorsal,* surface and a lower, or *ventral,* surface.

The symmetrical, tissue-bearing animals can be divided into two groups, one containing animals that exhibit

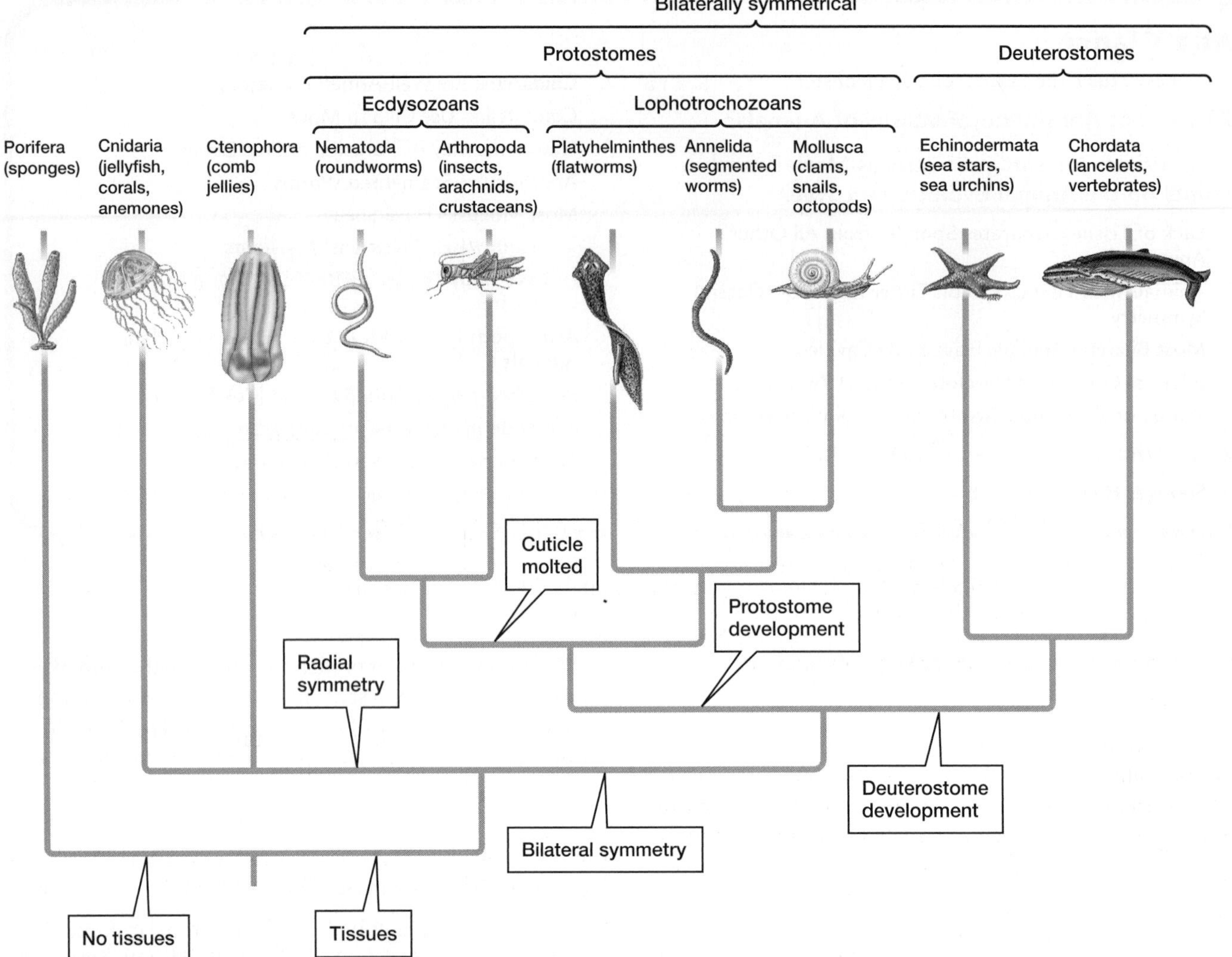

▲ **FIGURE 23-1 An evolutionary tree of some major animal phyla**

radial symmetry (**Fig. 23-2a**) and one with animals that exhibit **bilateral symmetry** (**Fig. 23-2b**). In radial symmetry, any plane through a central axis divides the object into roughly equal halves. In contrast, a bilaterally symmetrical animal can be divided into roughly mirror-image halves only along one particular plane through the central axis.

The difference between radially and bilaterally symmetrical animals reflects another major branching point in the animal evolutionary tree. This split separated the ancestors of the radially symmetrical cnidarians (sea jellies, anemones, and corals) and ctenophores (comb jellies) from the ancestors of the remaining animal phyla, all of which are bilaterally symmetrical.

Radially Symmetrical Animals Have Two Embryonic Tissue Layers; Bilaterally Symmetrical Animals Have Three

The distinction between radial and bilateral symmetry in animals is closely tied to a corresponding difference in the number of tissue layers, called germ layers, that arise during embryonic development. Embryos of animals with radial symmetry have two germ layers: an inner layer of **endoderm** (which gives rise to the tissues that line most hollow organs) and an outer layer of **ectoderm** (which gives rise to nerve tissue and to the tissues that cover the body and line respiratory surfaces and the gut). Embryos of bilaterally symmetrical animals add a third germ layer. Between the endoderm and ectoderm is a layer of **mesoderm** (which forms muscle and, when present, the circulatory and skeletal systems).

The parallel evolution of symmetry type and number of germ layers helps us make sense of the potentially puzzling case of the echinoderms (starfish, sea cucumbers, and sea urchins). Adult echinoderms are radially symmetrical, yet our evolutionary tree places them squarely within the bilaterally symmetrical group. It turns out that echinoderms have three germ layers, as well as several other characteristics (some described later) that unite them with the bilaterally symmetrical animals. So, the immediate ancestors of echinoderms must have been bilaterally symmetrical, and the group subsequently evolved radial symmetry (a case of convergent evolution). To this day larval echinoderms retain bilateral symmetry.

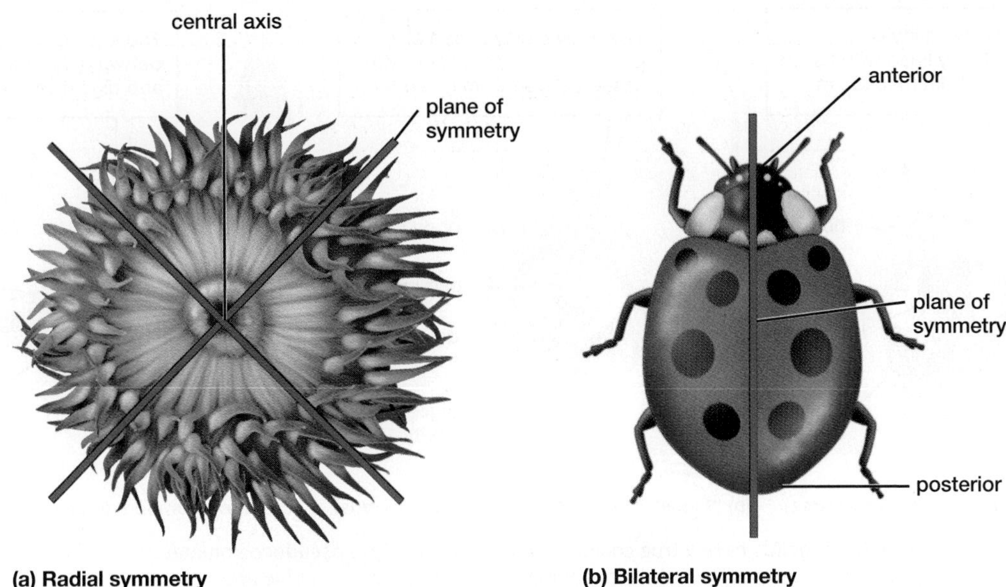

▲ FIGURE 23-2 **Body symmetry and cephalization (a)** Animals with radial symmetry, such as this sea anemone, lack a well-defined head. Any plane that passes through the central axis divides the body into mirror-image halves. **(b)** Animals with bilateral symmetry, such as this beetle, have an anterior head end and a posterior tail end. The body can be split into two mirror-image halves only along a particular plane that runs down the midline.

Bilaterally Symmetrical Animals Have Heads

Radially symmetrical animals tend either to be *sessile* (fixed to one spot, like sea anemones) or to drift around on currents (like sea jellies). Such animals may encounter food or threats from any direction, so a body that essentially faces all directions at once is advantageous. In contrast, most bilaterally symmetrical animals are *motile* (move under their own power). Resources such as food are most likely to be encountered by the part of the animal that is closest to the direction of movement. The evolution of bilateral symmetry was therefore accompanied by **cephalization,** the concentration of sensory organs and a brain in a defined head region. Cephalization produces an *anterior* (head) end, where sensory cells, sensory organs, clusters of nerve cells, and organs for ingesting food are concentrated. The other end of a cephalized animal is designated *posterior* and may feature a tail (see Fig. 23-2b).

Most Bilateral Animals Have Body Cavities

The members of many bilateral animal phyla have a fluid-filled cavity between the digestive tube (or gut, where food is digested and absorbed) and the outer body wall. In an animal with a body cavity, the gut and body wall are separated by a space, creating a "tube-within-a-tube" body plan. Body cavities are absent in radially symmetrical animals, so it is likely that this feature arose sometime after the split between radially and bilaterally symmetrical animals.

A body cavity can serve a variety of functions. In the earthworm it acts as a kind of skeleton, providing support for the body and a framework against which muscles can act. In other animals, internal organs are suspended within the fluid-filled cavity, which serves as a protective buffer between the organs and the outside world.

Body Cavity Structure Varies Among Phyla

The most widespread type of body cavity is a **coelom,** a fluid-filled cavity that is completely lined with a thin layer of tissue that develops from mesoderm (**Fig. 23-3a**). Phyla whose members have a coelom are called *coelomates*. The annelids (segmented worms), arthropods (insects, spiders, crustaceans), mollusks (clams and snails), echinoderms, and chordates (which include humans) are coelomates.

Some animals have a body cavity that is *not* completely surrounded by mesoderm-derived tissue. This type of cavity is known as a **pseudocoelom,** and phyla whose members have one are collectively known as *pseudocoelomates* (**Fig. 23-3b**). The roundworms (nematodes) are the largest pseudocoelomate group.

Some phyla of bilateral animals have no body cavity at all and are known as *acoelomates*. For example, flatworms have no cavity between their gut and body wall; instead, the space is filled with solid tissue (**Fig. 23-3c**).

Bilateral Organisms Develop in One of Two Ways

Among the bilateral animal phyla, embryological development follows a variety of pathways. These diverse developmental pathways, however, can be grouped into two categories, known as **protostome** and **deuterostome** development. In protostome development, the body cavity forms within the space between the body wall and the digestive cavity. In deuterostome development, the body cavity forms as an outgrowth of the digestive cavity. The two types of development

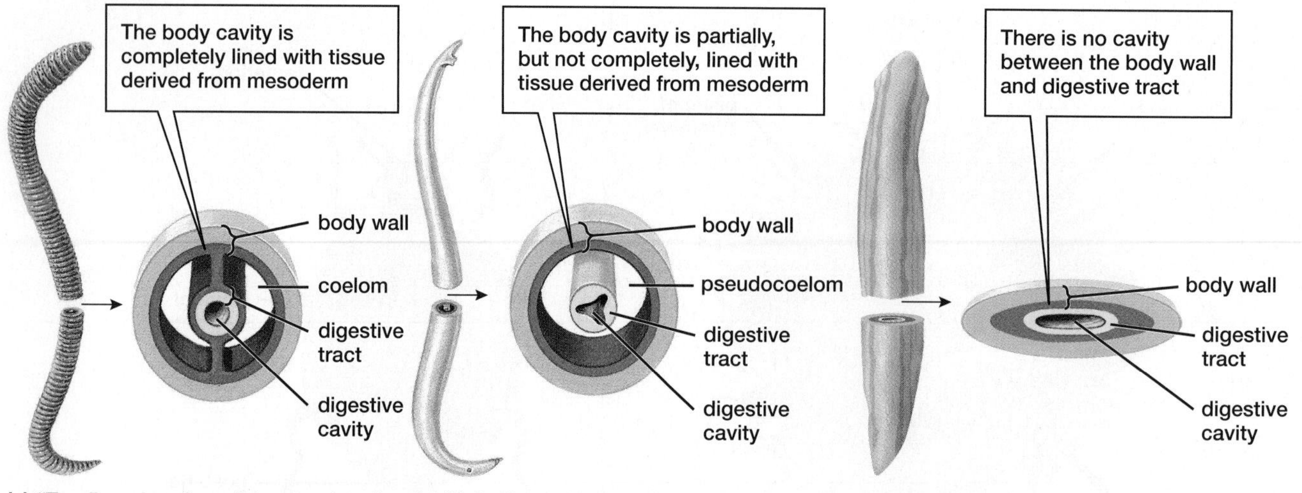
The body cavity is completely lined with tissue derived from mesoderm
body wall
coelom
digestive tract
digestive cavity
The body cavity is partially, but not completely, lined with tissue derived from mesoderm
body wall
pseudocoelom
digestive tract
digestive cavity
There is no cavity between the body wall and digestive tract
body wall
digestive tract
digestive cavity
(a) "True" coelom (annelids, chordates)
(b) "False" or pseudocoelom (roundworms)
(c) No coelom (cnidarians, flatworms)

▲ **FIGURE 23-3 Body cavities (a)** Annelids have a true coelom. **(b)** Roundworms are pseudocoelomates. **(c)** Flatworms have no cavity between the body wall and digestive tract. (Tissues shown in blue are derived from ectoderm, those in red from mesoderm, and those in yellow from endoderm.)

also differ in the pattern of cell division immediately after fertilization and in the method by which the mouth and anus are formed. Protostomes and deuterostomes represent distinct evolutionary branches within the bilateral animals. Annelids, arthropods, and mollusks exhibit protostome development; echinoderms and chordates are deuterostomes.

Protostomes Include Two Distinct Evolutionary Lines

The protostome animal phyla fall into two groups, which correspond to two different lineages that diverged early in the evolutionary history of protostomes. One group, the *ecdysozoans,* includes phyla such as the arthropods and roundworms, whose members have bodies covered by an outer layer that is periodically shed. The other group is known as the *lophotrochozoans* and includes phyla whose members have a special feeding structure called a lophophore, as well as phyla whose members pass through a particular type of developmental stage called a trochophore larva. The flatworms, annelids, and mollusks are examples of lophotrochozoan phyla.

23.3 WHAT ARE THE MAJOR ANIMAL PHYLA?

It's easy to overlook the differences among the multitude of small, boneless animals in the world. Even Carolus Linnaeus, the originator of modern biological classification, recognized only two phyla of animals without backbones (insects and worms). Today, however, biologists recognize about 27 phyla of boneless animals, some of which are summarized in **Table 23-1**.

For convenience, biologists often place animals in one of two major categories: **vertebrates,** those with a backbone (or vertebral column), and **invertebrates,** those lacking a backbone. The vertebrates—fish, amphibians, reptiles, and mammals (see Chapter 24)—are perhaps the most conspicuous animals from a human point of view, but less than 3% of all known animal species are vertebrates. The vast majority of animals are invertebrates.

The earliest animals probably originated from colonies of protists whose members had become specialized to perform distinct roles within the colony. In our survey of the invertebrate animals, we will begin with the sponges, whose body plan most closely resembles the probable ancestral protist colonies.

Sponges Have a Simple Body Plan

Sponges (Porifera) are found in most marine and aquatic environments. Most of Earth's 5,000 or so sponge species live in salt water, where they inhabit ocean waters warm and cold, deep and shallow. In addition, some sponges live in freshwater habitats such as lakes and rivers. Adult sponges live attached to rocks or other underwater surfaces. They do not generally move, though researchers have demonstrated that some species, at least when captive in aquaria, are able to move about (very slowly—a few millimeters per day). Sponges come in a variety of shapes and sizes. Some species have a well-defined shape, but others grow free-form over underwater rocks (**Fig. 23-4**). The largest sponges can grow to more than 3 feet (1 meter) in height.

Sponges may reproduce asexually by **budding,** in which the adult produces miniature versions of itself that drop off and assume an independent existence. Alternatively, sponges may reproduce sexually through the fusion of sperm and eggs. Fertilized eggs develop inside the adult into active larvae that escape through the openings in the sponge body. Water currents disperse the larvae to new areas, where they settle and develop into adult sponges.

(a) Encrusting sponge

(b) Tubular sponge

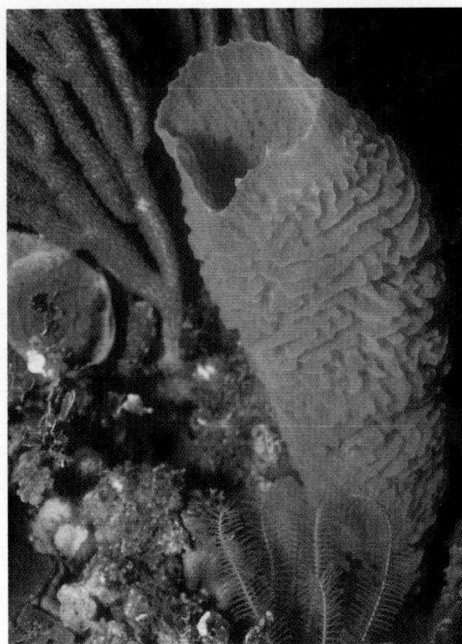

(c) Flared sponge

▲ **FIGURE 23-4 The diversity of sponges** Sponges come in a wide variety of sizes, shapes, and colors. Some, such as **(a)** this encrusting sponge, grow in a free-form pattern over undersea rocks. Others may be **(b)** tubular or **(c)** vase-shaped.

QUESTION Sponges are often described as the most "primitive" of animals. How can such a primitive organism have become so diverse and abundant?

Sponges Lack Tissues

Sponge bodies do not contain true tissues and organs; in some ways, a sponge resembles a colony of single-celled organisms. The colony-like properties of sponges were revealed in an experiment performed by embryologist H. V. Wilson in 1907. Wilson mashed a sponge through a piece of silk, thereby breaking it apart into single cells and cell clusters. He then placed these tiny bits of sponge into seawater and waited for 3 weeks. By the end of the experiment, the cells had reaggregated into a functional sponge, demonstrating that individual sponge cells had been able to survive and function independently.

A sponge's body is perforated by numerous tiny pores, through which water enters, and by fewer, large openings, through which water is expelled. Within the sponge, water travels through canals. As water passes through the sponge, oxygen is extracted, microorganisms are filtered out and taken into individual cells where they are digested, and wastes are released (**Fig. 23-5**).

Sponge Cells Are Specialized for Different Functions

Sponges have three major cell types, each with a specialized role. Flattened *epithelial cells* cover their outer body surfaces.

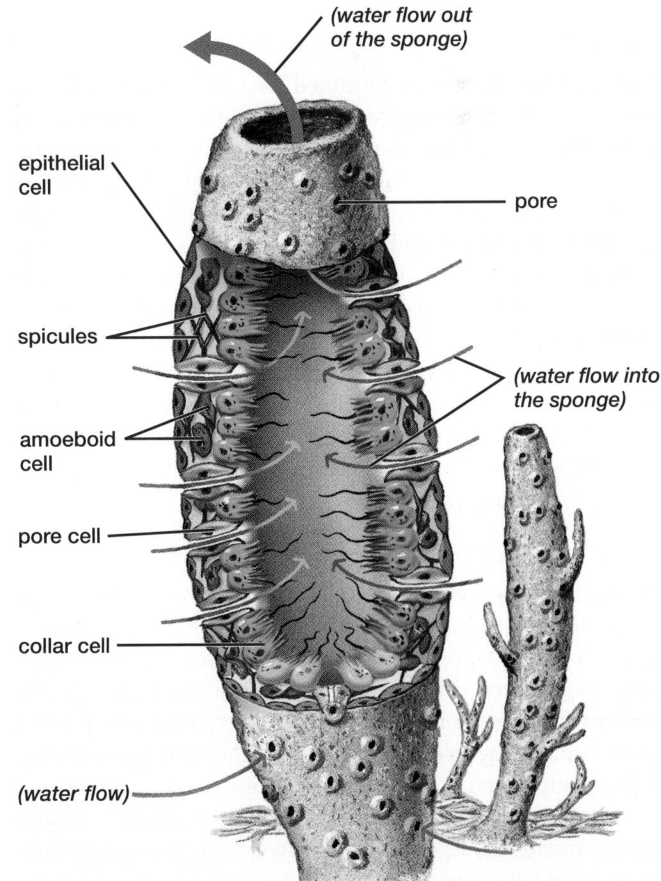

► **FIGURE 23-5 The body plan of sponges** Water enters through numerous tiny pores in the sponge body and exits through a larger opening. Microscopic food particles are filtered from the water.

Table 23-1 **Comparison of the Major Animal Phyla**

Common Name (Phylum)		Sponges (Porifera)	Hydra, Anemones, Sea Jellies (Cnidaria)	Flatworms (Platyhelminthes)
Body Plan	**Level of Organization**	Cellular—lack tissues and organs	Tissue—lack organs	Organ system
	Germ layers	Absent	Two	Three
	Symmetry	Absent	Radial	Bilateral
	Cephalization	Absent	Absent	Present
	Body cavity	Absent	Absent	Absent
	Segmentation	Absent	Absent	Absent
Internal Systems	**Digestive system**	Intracellular	Gastrovascular cavity; some intracellular	Gastrovascular cavity
	Circulatory system	Absent	Absent	Absent
	Respiratory system	Absent	Absent	Absent
	Excretory system (fluid regulation)	Absent	Absent	Canals with ciliated cells
	Nervous system	Absent	Nerve net	Head ganglia with longitudinal nerve cords
	Reproduction	Sexual; asexual (budding)	Sexual; asexual (budding)	Sexual (some hermaphroditic); asexual (body splits)
	Support	Endoskeleton of spicules	Hydrostatic skeleton	Hydrostatic skeleton
	Number of known species	5,000	9,000	20,000

Some epithelial cells are modified into *pore cells*, which surround pores, controlling their size and regulating the entry of water. The pores close when harmful substances are present. *Collar cells* maintain a flow of water through the sponge by beating flagella that extend into the inner cavity. The collars that surround these flagella act as fine sieves, filtering out microorganisms that are then ingested by the cell. Some of the food is passed to the *amoeboid cells*. These cells roam freely between the epithelial and collar cells, digesting and distributing nutrients, producing reproductive cells, and secreting small skeletal projections called *spicules*. Spicules may be composed of calcium carbonate (chalk), silica (glass), or protein and form an internal skeleton that provides support for the sponge's body (see Fig. 23-5).

Some Sponges Contain Chemicals Useful to Humans

Because sponges remain in one spot and have no protective shell, they are vulnerable to predators such as fish, turtles, and sea slugs. Many sponges, however, contain chemicals that are toxic or distasteful to potential predators. Fortunately for people, a number of these chemicals have proved to be valuable medicines. For example, the drug spongistatin, a compound first isolated from a sponge, is an emerging treatment for the fungal infections that frequently sicken AIDS patients. Other medicines derived from sponges include some promising new anticancer drugs. The discovery of these and other medicines has raised hopes that, as more species are screened by researchers, sponges will become a major source of new drugs.

Cnidarians Are Well-Armed Predators

The *cnidarians* (Cnidaria) include sea jellies (also inappropriately known as jellyfish), sea anemones, corals, and hydrozoans. Like sponges, the roughly 9,000 known species of cnidarians are confined to watery habitats, and most are marine. Most species are small, from a few millimeters to a few inches in diameter, but the largest sea jellies can be 8 feet across and have tentacles 150 feet long. All cnidarians are carnivorous predators.

Cnidarians Have Tissues and Two Body Types

The cells of cnidarians are organized into distinct tissues, including contractile tissue that acts like muscle. The nerve cells are organized into tissue called a *nerve net*, which branches through the body and controls the contractile tissue to bring about movement and feeding behavior. However, most cnidarians lack true organs and have no brain.

Cnidarians come in a variety of forms (**Fig. 23-6**), all of which are actually variations on two basic body plans: the *polyp* (**Fig. 23-7a**) and the *medusa* (**Fig. 23-7b**). The generally tubular polyp is adapted to a life spent quietly attached to rocks. The polyp has tentacles that reach upward to grasp and immobilize prey. The medusa floats in the water and is carried by currents, its bell-shaped body trailing tentacles like multiple fishing lines.

Segmented Worms (Annelida)	Snails, Clams, Squid (Mollusca)	Insects, Arachnids, Crustaceans (Arthropoda)	Roundworms (Nematoda)	Sea Stars, Sea Urchins (Echinodermata)
Organ system	Organ system	Organ system	Organ system	Organ system
Three	Three	Three	Three	Three
Bilateral	Bilateral	Bilateral	Bilateral	Bilateral larvae, radial adults
Present	Present	Present	Present	Absent
Coelom	Coelom	Coelom	Pseudocoel	Coelom
Present	Absent	Present	Absent	Absent
Separate mouth and anus	Separate mouth and anus	Separate mouth and anus	Separate mouth and anus	Separate mouth and anus (usually)
Closed	Open	Open	Absent	Absent
Absent	Gills, lungs	Tracheae, gills, or lungs	Absent	Tube feet, skin gills, respiratory tree
Nephridia	Nephridia	Excretory glands resembling nephridia	Excretory gland	Absent
Head ganglia with paired ventral cords; ganglia in each segment	Well-developed brain in some cephalopods; several paired ganglia, most in the head; nerve network in the body wall	Head ganglia with paired ventral nerve cords; ganglia in segments, some fused	Head ganglia with dorsal and ventral nerve cords	Head ganglia absent; nerve ring and radial nerves; nerve network in the skin
Sexual (some hermaphroditic)	Sexual (some hermaphroditic)	Usually sexual	Sexual (some hermaphroditic)	Sexual (some hermaphroditic); asexual by regeneration (rare)
Hydrostatic skeleton	Hydrostatic skeleton	Exoskeleton	Hydrostatic skeleton	Endoskeleton of plates beneath outer skin
9,000	50,000	1,000,000	12,000	6,500

(a) Anemone

(b) Sea jelly

(c) Coral

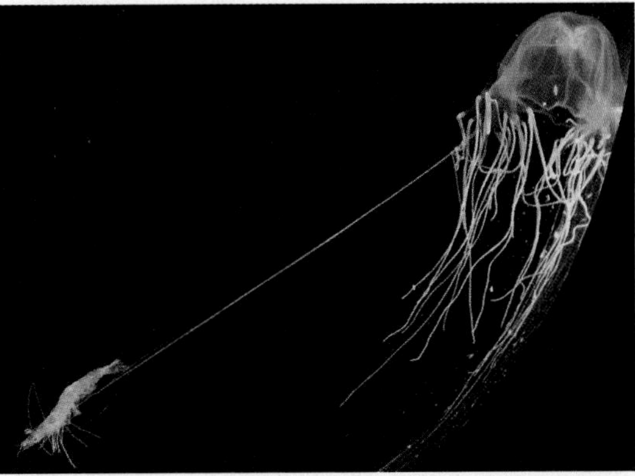

(d) Sea wasp

◀ **FIGURE 23-6 Cnidarian diversity (a)** A red-spotted anemone spreads its tentacles to capture prey. **(b)** A small sea jelly. **(c)** A close-up of coral reveals the extended tentacles of the polyps. **(d)** A sea wasp, a type of sea jelly whose stinging cells contain one of the most toxic of all known venoms. The venom quickly kills prey, such as the the shrimp shown here, that brush against a sea wasp's tentacles.

QUESTION In each of these photos, is the pictured organism a polyp or a medusa?

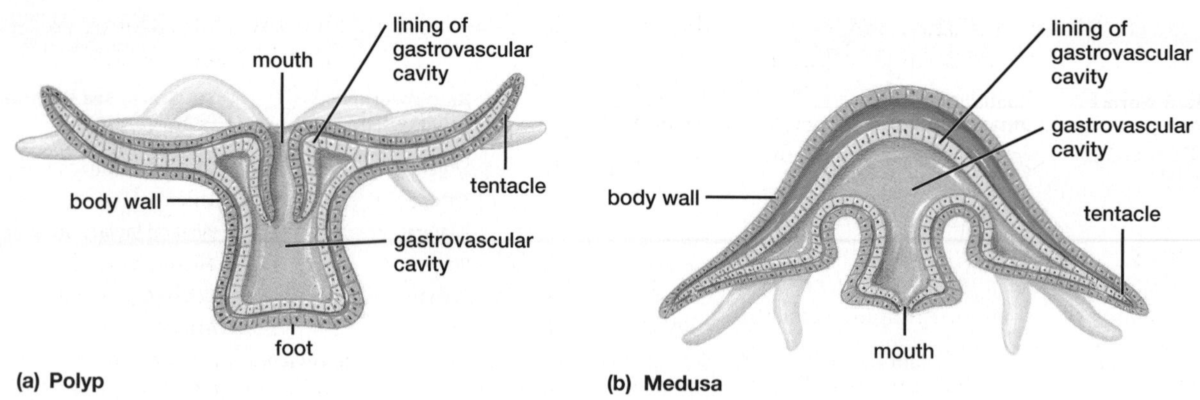

▲ **FIGURE 23-7 Polyp and medusa (a)** The polyp form is seen in sea anemones (see Fig. 23-6a) and the individual polyps within a coral (see Fig. 23-6c). **(b)** The medusa form, seen in the sea jelly (see Fig. 23-6b), resembles an inverted polyp. (Tissues shown in blue are derived from ectoderm, those in yellow from endoderm.)

Many cnidarian life cycles include both polyp and medusa stages, though some species live only as polyps and others only as medusae. Both polyps and medusae develop from just two germ layers—the interior endoderm and the exterior ectoderm; between those layers is a jelly-like substance. Polyps and medusae are radially symmetrical, with body parts arranged in a circle around the mouth and digestive cavity (see Fig. 23-2a). This arrangement of parts is well suited to these sessile or free-floating animals because it enables them to respond to prey or threats from any direction.

Reproduction varies considerably among different types of cnidarians, but one pattern is fairly common in species with both polyp and medusa stages. In such species, polyps typically reproduce by asexual budding that gives rise to new polyps. Under certain conditions, however, budding may instead give rise to medusae. After a medusa grows to maturity, it may release gametes (sperm or eggs) into the water. If a sperm and egg should meet, they may fuse to form a zygote that develops into a free-swimming, ciliated larva. The larva may eventually settle on a hard surface, where it develops into a polyp.

Cnidarians Have Stinging Cells

Cnidarian tentacles are armed with *cnidocytes*, cells containing structures that, when stimulated by contact, explosively inject poisonous or sticky filaments into prey (**Fig. 23-8**). These stinging cells, found only in cnidarians, are used to capture prey. Cnidarians do not hunt actively. Instead, they wait for

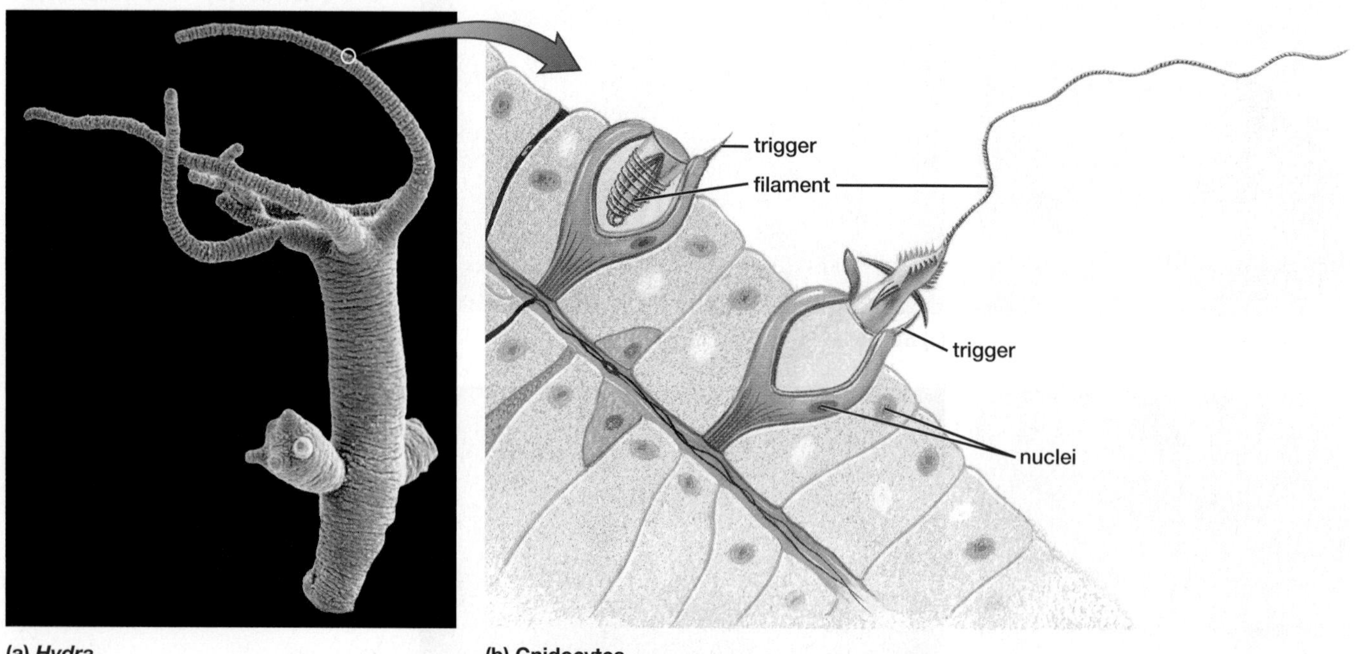

▲ **FIGURE 23-8 Cnidarian weaponry: the cnidocyte (a)** Cnidarian tentacles, such as those of the hydra, contain cnidocyte cells. **(b)** At the slightest touch to its trigger, a structure within a cnidocyte cell violently expels a poisoned filament, which may penetrate prey.

their victims to blunder, by chance, into the grasp of their enveloping tentacles. Stung and firmly grasped, the prey is forced through an expandable mouth into a digestive sac, the *gastrovascular cavity*. Digestive enzymes secreted into this cavity break down some of the food, and further digestion occurs within the cells lining the cavity. Because the gastrovascular cavity has only a single opening, undigested material is expelled through the mouth when digestion is completed. Although this two-way traffic prevents continuous feeding, it is adequate to support the low energy demands of these animals.

The venom of some cnidarians can cause painful stings in humans unfortunate enough to come into contact with them, and the stings of a few sea jelly species can even be life threatening. The most deadly of these species is the sea wasp, *Chironex fleckeri* (see Fig. 23-6d), which is found in the waters off northern Australia and Southeast Asia, and can grow to be about 12 inches (30 centimeters) in diameter. The amount of venom in a single sea wasp could kill up to 60 people, and the victim of a serious sting may die within minutes of being stung.

Many Corals Secrete Hard Skeletons

One group of cnidarians, the corals, is of particular ecological importance (see Fig. 23-6c). In many coral species, polyps form colonies, and each member of the colony secretes a hard skeleton of calcium carbonate. The skeletons persist long after the organisms die, serving as a base to which other individuals may attach themselves. The cycle continues until, after thousands of years, massive coral reefs are formed.

Coral reefs are found in both cold and warm oceans. Cold-water reefs form in deep waters and, though widely distributed, have only recently attracted the attention of researchers and are not yet well studied. The more familiar warm-water coral reefs are restricted to clear, shallow waters in the tropics. Here, coral reefs form undersea habitats that are the basis of an ecosystem of stunning diversity and unparalleled beauty.

Comb Jellies Use Cilia to Move

The roughly 150 species of radially symmetrical *comb jellies* (Ctenophora) are superficially similar in appearance to some cnidarians, but form a distinct evolutionary lineage. Most comb jellies are less than an inch (2.5 cm) in diameter, but a few species can grow to more than 3 feet (1 m) across. Comb jellies move by means of cilia, which are arranged in eight rows known as combs. Although most comb jellies are colorless and transparent or translucent, light scattered by the beating cilia of the combs can appear as eight ever-changing rainbows of color on the comb jelly's body.

All comb jellies are carnivores. Most species float in coastal or oceanic waters and eat tiny invertebrate animals (in some cases including other, smaller comb jellies), which are captured with sticky tentacles. (Comb jellies lack the stinging cells that characterize cnidarians.) Almost all comb jellies are hermaphroditic; each individual releases both sperm and eggs to the surrounding water. Fertilized eggs float freely in the water until larvae emerge and gradually develop into adult comb jellies.

Flatworms May Be Parasitic or Free Living

The *flatworms* (Platyhelminthes) are aptly named; they have a flat, ribbon-like shape. Many of the approximately 20,000 flatworm species are parasites (**Fig. 23-9a**). (**Parasites** are organisms that live in or on the body of another organism, called a *host*, which is harmed as a result of the relationship.) Nonparasitic, free-living flatworms inhabit aquatic, marine, and moist terrestrial habitats. They tend to be small and inconspicuous

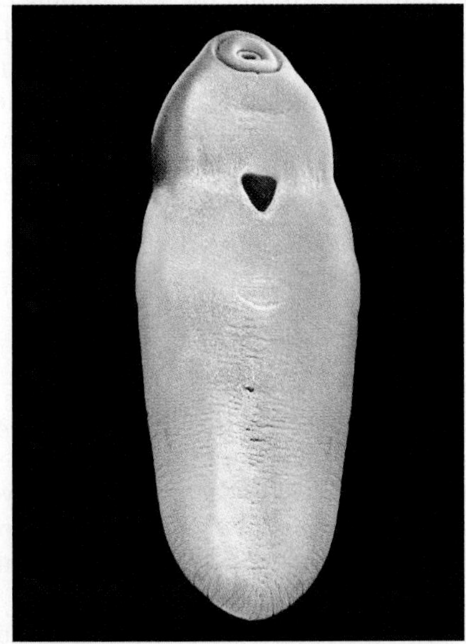

(a) Fluke

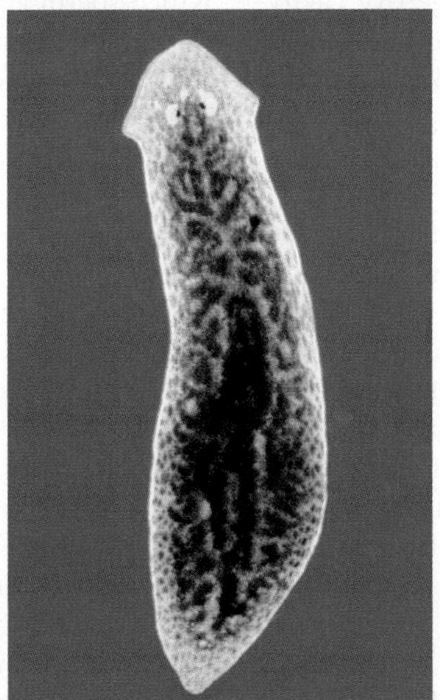

(b) Freshwater flatworm

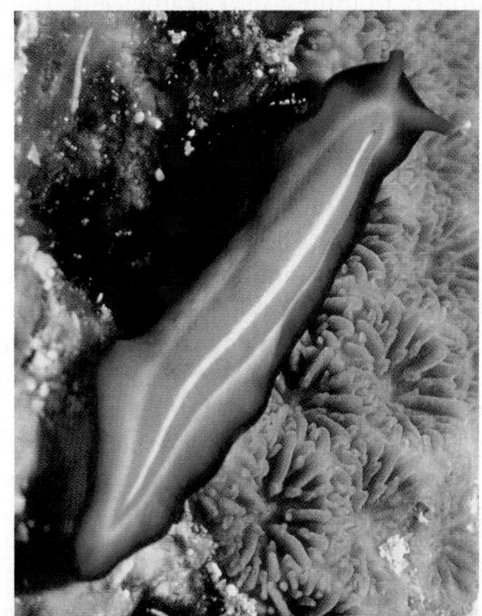

(c) Marine flatworm

▲ **FIGURE 23-9 Flatworm diversity (a)** This fluke is an example of a parasitic flatworm. **(b)** Eyespots are clearly visible in the head of this free-living, freshwater flatworm. **(c)** Many of the flatworms that inhabit tropical coral reefs are brightly colored.

(**Fig. 23-9b**), but some are brightly colored, spectacularly patterned residents of tropical coral reefs (**Fig. 23-9c**).

Flatworms can reproduce both sexually and asexually. Free-living species may reproduce by cinching themselves around the middle until they separate into two halves, each of which regenerates its missing parts. All flatworm species can reproduce sexually; most are **hermaphroditic**—that is, they possess both male and female sexual organs. This trait is a great advantage to parasitic flatworms because it allows a worm to reproduce through self-fertilization, a great advantage to a worm that may be the only individual present in its host.

Flatworms Have Organs But Lack Respiratory and Circulatory Systems

Unlike cnidarians, flatworms have well-developed organs, in which tissues are grouped into functional units. For example, most free-living flatworms have sense organs, including eyespots (see Fig. 23-9b) that detect light and dark, and cells that respond to chemical and tactile stimuli. To process information, a flatworm has clusters of nerve cells called **ganglia** (singular, ganglion) in its head, forming a simple brain. Paired neural structures called **nerve cords** conduct nervous signals to and from the ganglia.

Despite the presence of some organs, flatworms lack respiratory and circulatory systems. In the absence of a respiratory system, gas exchange is accomplished by direct diffusion between body cells and the environment. This mode of respiration is possible because the small size and flat shape of flatworm bodies ensure that no body cell is very far from the surrounding environment. Without a circulatory system, nutrients move directly from the digestive tract to body cells. The digestive cavity has a branching structure that reaches all parts of the body and allows digested nutrients to diffuse into nearby cells. The digestive cavity has only one opening to the environment, so undigested waste must pass out through the same opening that also serves as a mouth.

Flatworms Are Bilaterally Symmetrical

Flatworms are bilaterally symmetrical, rather than radially symmetrical (see Fig. 23-2). This body plan and its accompanying cephalization foster active movement. Cephalized, bilaterally symmetrical animals possess an anterior end, which is the first part of a moving animal to encounter the environment ahead. Consequently, sense organs are concentrated in the anterior portion of the body, thus enhancing the animal's ability to respond appropriately to any stimuli it encounters (for example, eating food items and retreating from obstacles).

Some Flatworms Are Harmful to Humans

Some parasitic flatworms can infect humans. For example, tapeworms can infect people who eat improperly cooked beef, pork, or fish that has been infected by the worms. Tapeworm larvae form encapsulated resting structures, called *cysts*, in the muscles of these animals. The cysts hatch in the human digestive tract, where the young tapeworms attach themselves to the lining of the intestine. There they may grow to a length of more than 20 feet (7 meters), absorbing digested nutrients directly through their outer surface and eventually releasing packets of eggs that are shed in the host's feces. If pigs, cows, or fish eat food contaminated with infected human feces, the eggs hatch in the animal's digestive tract, releasing larvae that burrow into its muscles and form cysts, thereby continuing the infective cycle (**Fig. 23-10**).

Another group of parasitic flatworms is the *flukes* (see Fig. 23-9a). Of these, the most devastating are liver flukes (common in Asia) and blood flukes, such as those of the genus *Schistosoma*, which cause the disease schistosomiasis. Like most parasites, flukes have a complex life cycle that includes an intermediate host (a snail, in the case of *Schistosoma*). Prevalent in Africa and parts of South America, schistosomiasis affects an estimated 200 million people worldwide. Its symptoms include diarrhea, anemia, and possible brain damage.

Annelids Are Segmented Worms

The bodies of *annelids* (Annelida) are divided into a series of similar repeating segments. Externally, these segments appear as ringlike depressions on the surface. Internally, most of the segments contain identical copies of nerves, excretory structures, and muscles. **Segmentation** is advantageous for locomotion, because the body compartments are each controlled by separate muscles and collectively are capable of more complex movement than could be achieved with only one set of muscles to control the whole body.

Sexual reproduction is common among annelids. Some species are hermaphroditic; others have separate sexes. Fertilization may be external or internal. External fertilization, in which sperm and eggs are released into the surrounding environment, is found mainly in species that live in water. In internal fertilization, two individuals copulate and sperm are transferred directly from one to the other. In hermaphroditic species, sperm transfer may be mutual, with each partner both donating and receiving sperm. In addition, some annelids can reproduce asexually, typically by fragmentation in which the body breaks into two pieces, each of which regenerates the missing part.

Annelids Are Coelomates and Have Organ Systems

Unlike flatworms, annelids have a fluid-filled true coelom between the body wall and the digestive tract (see Fig. 23-3a). The incompressible fluid in the coelom of many annelids is confined by the partitions between the segments and serves as a **hydrostatic skeleton,** a supportive framework against which muscles can act. The hydrostatic skeleton allows earthworms to burrow through soil.

Annelids have well-developed organ systems. For example, annelids have a **closed circulatory system** that distributes gases and nutrients throughout the body. In closed circulatory systems (including yours), blood remains confined to the heart and blood vessels. In the earthworm, for

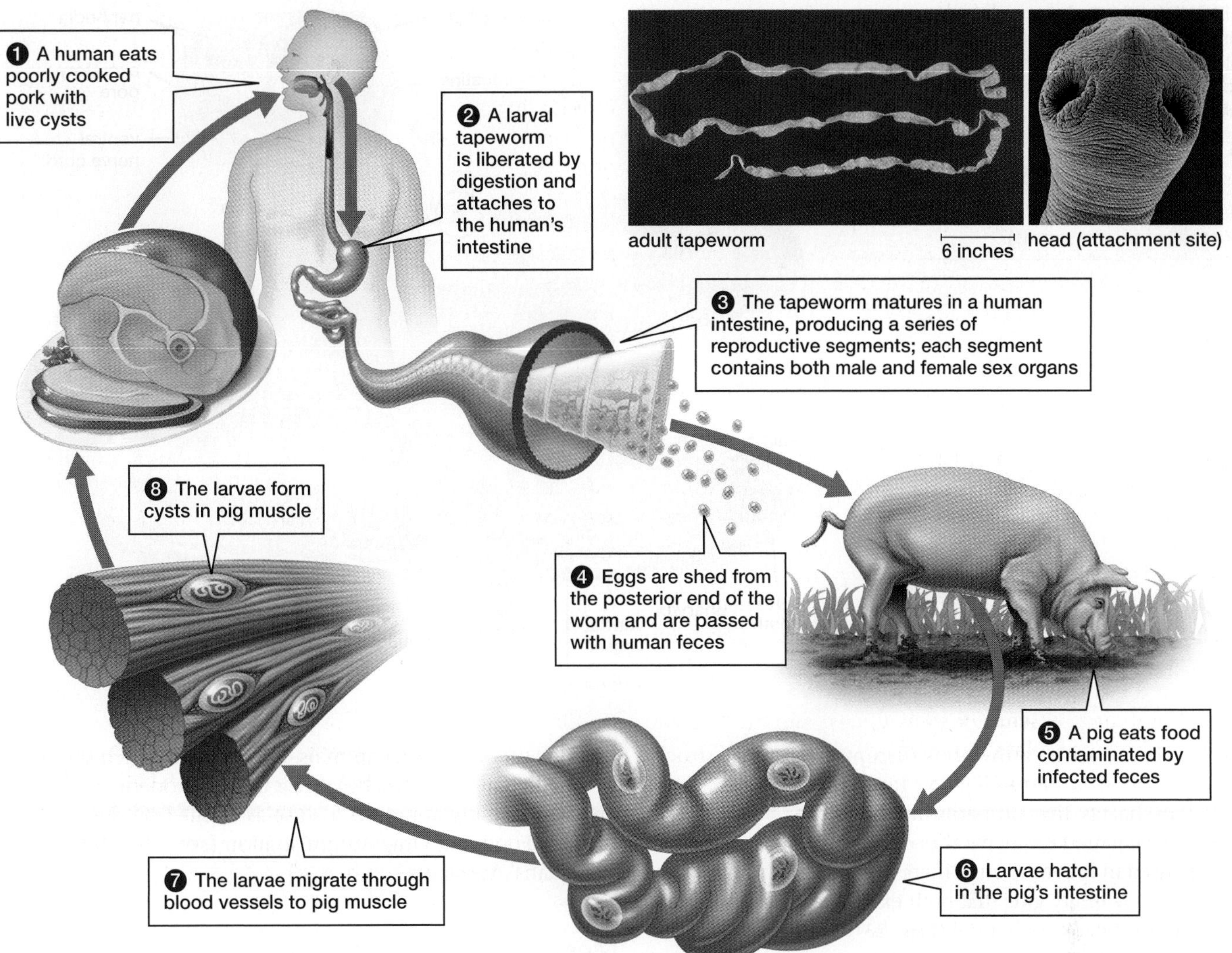
1 A human eats poorly cooked pork with live cysts
2 A larval tapeworm is liberated by digestion and attaches to the human's intestine
adult tapeworm
6 inches
head (attachment site)
3 The tapeworm matures in a human intestine, producing a series of reproductive segments; each segment contains both male and female sex organs
4 Eggs are shed from the posterior end of the worm and are passed with human feces
5 A pig eats food contaminated by infected feces
6 Larvae hatch in the pig's intestine
7 The larvae migrate through blood vessels to pig muscle
8 The larvae form cysts in pig muscle

▲ **FIGURE 23-10 The life cycle of the human pork tapeworm**

QUESTION Why have tapeworms evolved a long, flat shape?

example, blood with oxygen-carrying hemoglobin is pumped through well-developed vessels by five pairs of "hearts" (**Fig. 23-11**). These hearts are actually short segments of specialized blood vessels that contract rhythmically. The blood is filtered and wastes are removed by excretory organs called *nephridia* (singular, nephridium) and excreted to the environment through small pores. Nephridia resemble the individual tubules of the vertebrate kidney. The annelid nervous system consists of a simple brain in the head and a series of repeating paired segmental ganglia joined by a pair of ventral nerve cords that pass along the length of the body. The annelid digestive system includes a tubular gut that runs from the mouth to the anus. This kind of digestive tract, with two openings and a one-way digestive path, is much more efficient than the single-opening digestive systems of cnidarians and flatworms. Digestion in annelids occurs in a series of compartments, each specialized for a different phase of food processing (see Fig. 23-11).

Annelids Include Oligochaetes, Polychaetes, and Leeches

The 9,000 species of annelids fall into three main subgroups: the *oligochaetes*, the *polychaetes*, and the *leeches*. The oligochaetes include the familiar earthworm and its relatives. Charles Darwin, arguably the greatest of all biologists, devoted substantial time to the study of earthworms. He was especially impressed by earthworms' role in improving soil fertility. More than a million earthworms may live in an acre of land, tunneling through soil, consuming and excreting soil particles and organic matter. These actions help ensure that air and water can move easily through the soil and that organic matter is continually mixed into it, creating conditions that are favorable for plant growth. In Darwin's view, the activity of the earthworms has had such a significant impact on agriculture that "it may be doubted whether there are many other animals which have played so important a part in the history of the world." The impact of earthworms, however, can also be negative. In some areas of North America,

▶ **FIGURE 23-11 An annelid, the earthworm** This diagram shows an enlargement of segments, many of which are repeating similar units separated by partitions.

QUESTION What advantage does a digestive system with two openings have relative to digestive systems with only a single opening (like that of the flatworms)?

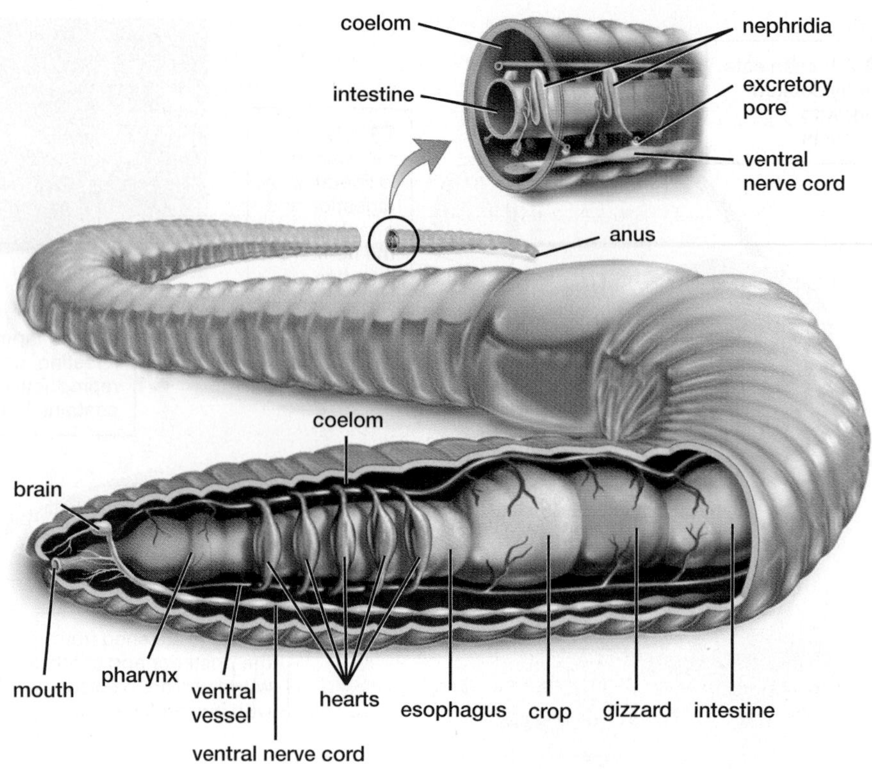

non-native, invasive earthworms disrupt the normal structure of forest soils, harming native forest plants.

Polychaetes live primarily in the ocean. Some polychaetes have paired fleshy paddles on most of their segments that are used in locomotion. Others live in tubes from which they project feathery gills that both exchange gases and filter the water for microscopic food (**Figs. 23-12a,b**).

Leeches (**Fig. 23-12c**) live in freshwater or moist terrestrial habitats and are either carnivorous or parasitic. Carnivorous leeches prey on smaller invertebrates; parasitic leeches suck the blood of larger animals. One parasitic leech species, the medicinal leech, has become a tool of modern medicine and was recently approved for use as a "medical device" by the U.S. Food and Drug Administration (see "Health Watch: Physicians' Assistants" on p. 434).

Most Mollusks Have Shells

If you have ever enjoyed a bowl of clam chowder, a plate of oysters on the half shell, or some sautéed scallops, you are

(a) Polychaete gills

(b) Deep-sea polychaete

(c) Leech

▲ **FIGURE 23-12 Diverse annelids (a)** The brightly colored gills of a polychaete annelid. The rest of the worm's body is hidden inside a tube embedded in the coral that is visible in the background. **(b)** This polychaete lives near deep-sea vents where the water temperature may reach 175°F (80°C). **(c)** This leech, a freshwater annelid, shows numerous segments. The sucker encircles its mouth, allowing it to attach to its prey.

QUESTION Why does pouring salt on a leech harm it?

indebted to *mollusks* (Mollusca). Mollusks are very diverse; in terms of number of known species, the 50,000 mollusks are second (albeit a distant second) only to the arthropods. These diverse species exhibit a range of lifestyles, from sessile forms such as mussels that spend their adult lives in one spot, filtering microorganisms from the water, to active, voracious predators of the ocean depths, such as the giant squid. Most mollusks are protected by hard shells of calcium carbonate. Others, however, lack shells. Some unshelled mollusks pursue prey and escape predators by moving swiftly; others are incapable of rapid movement, but are protected by bodies that incorporate toxic or distasteful chemicals. Mollusks, with the exception of some snails and slugs, are water dwellers.

Most Mollusks Have Open Circulatory Systems

The circulatory systems of most mollusks include a feature not seen in annelids: the **hemocoel,** or blood cavity. Blood empties into the hemocoel, where it bathes the internal organs directly. This arrangement is known as an **open circulatory system.** Mollusks also have a *mantle*, an extension of the body wall that forms a chamber for the gills and, in shelled species, secretes the shell (**Fig. 23-13**). The mollusk nervous system, like that of annelids, consists of ganglia connected by nerves, but many more of the ganglia are concentrated in the brain. Reproduction is sexual; some species have separate sexes, and others are hermaphroditic.

Among the many subgroups of mollusks, we will discuss three in more detail: gastropods, bivalves, and cephalopods.

Gastropods Are One-Footed Crawlers

Snails and slugs—collectively known as *gastropods*—crawl on a muscular *foot*, and many are protected by shells that vary widely in form and color (**Fig. 23-14a**). Not all gastropods are shelled, however. Sea slugs, for example, lack shells, but their brilliant colors warn potential predators that they are poisonous or foul tasting (**Fig. 23-14b**).

Gastropods feed with a *radula*, a flexible ribbon of tissue studded with spines that is used to scrape algae from rocks or to grasp larger plants or prey. Most snails use gills, typically enclosed in a cavity beneath the shell, for respiration. Gases can also diffuse readily through the skin of most gastropods. The few gastropod species that live in terrestrial habitats (including the destructive garden snails and slugs) use a simple lung for breathing.

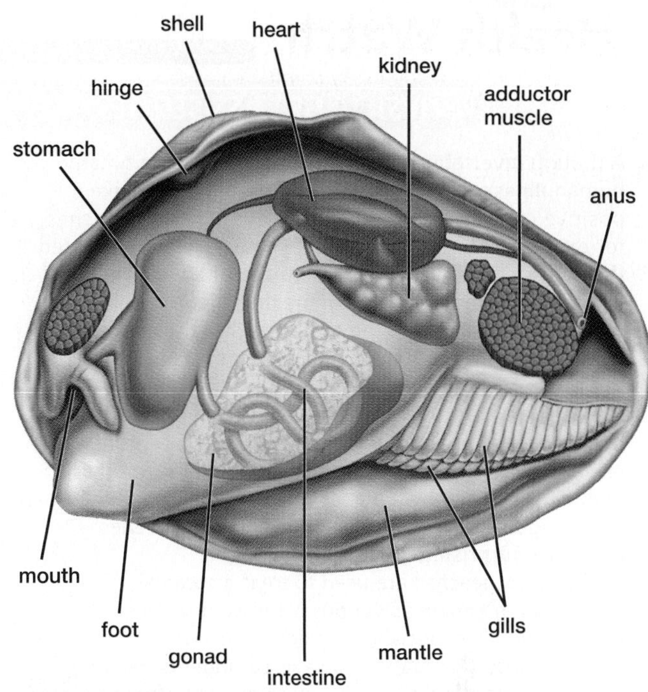

▲ **FIGURE 23-13 A bivalve mollusk** The body plan of a clam, showing the mantle, foot, gills, shell, and other features that are seen in most (but not all) mollusk species. The clam uses its adductor muscles to close its shell.

Bivalves Are Filter Feeders

Included among the *bivalves* are scallops, oysters, mussels, and clams (**Fig. 23-15**). Bivalves possess two shells connected by a flexible hinge. A strong muscle clamps the shells closed

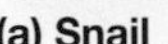
(a) Snail

(b) Sea slugs

▲ **FIGURE 23-14 The diversity of gastropod mollusks (a)** A Florida tree snail displays a brightly striped shell and eyes at the tip of stalks that retract instantly if touched. **(b)** The brilliant colors of many sea slugs warn potential predators that they are distasteful.

Health Watch

Physicians' Assistants

Although invertebrate animals cause or transmit many human illnesses, some have been recruited to make a positive contribution to human health. Consider leeches, for example. For more than 2,000 years, healers enlisted these parasitic annelids for treatment of a wide range of illnesses. For much of human medical history, treatment with leeches was based on the hope that the creatures would suck out the "tainted" blood that was believed to be the primary cause of disease. As the actual causes of disease were discovered, however, medical use of leeches declined. By the beginning of the twentieth century, leeches no longer had a place in the toolkit of modern medicine and had become a symbol of the ignorance of an earlier age. Today, however, medical use of leeches is making a surprising comeback.

Currently, leeches are used to treat a surgical complication known as venous insufficiency. This complication is especially common in reconstructive surgery, such as surgery to reattach a severed finger or repair a disfigured face. In such cases, surgeons are often unable to reconnect all of the veins that would normally carry blood away from tissues. Eventually, new veins will grow, but in the meantime blood may accumulate in the repaired tissue. Unless the excess blood is removed, it will coagulate, causing clots that can deprive the tissue of the oxygen and nutrients it needs to live. Fortunately, leeches can help. Applied to the affected area, the leeches get right to work, making small, painless incisions and sucking blood into their stomachs. To aid them in their blood-removal task, the leeches' saliva contains a mixture of chemicals that dilate blood vessels and prevent blood from clotting. Although the chemical brew in the saliva is an adaptation that helps leeches consume blood more efficiently, it also helps the patient by promoting blood flow in the damaged tissue. In this way, leeches provide a painless, effective treatment for venous insufficiency and have thereby regained their status as medical assistants to humanity.

Another invertebrate animal that has been recruited for medical duty is the blowfly or, more precisely, blowfly larvae, commonly known as maggots (**Fig. E23-1**). Blowfly maggots have proved to be effective at ridding wounds and ulcers of dead and dying tissue. If such tissue is not removed, it can interfere with healing or lead to infection. Traditionally, dead tissue in wounds is removed by a physician wielding a scalpel, but maggots offer an increasingly common alternative treatment. In this treatment, a bandage containing day-old, sterile maggots is applied to the wound. The maggots consume dead or dying tissue, secreting digestive enzymes that do not harm healthy skin or bone. After a few days, the maggots have grown to the size of rice kernels and are removed. The treatment is repeated until the wound is clean.

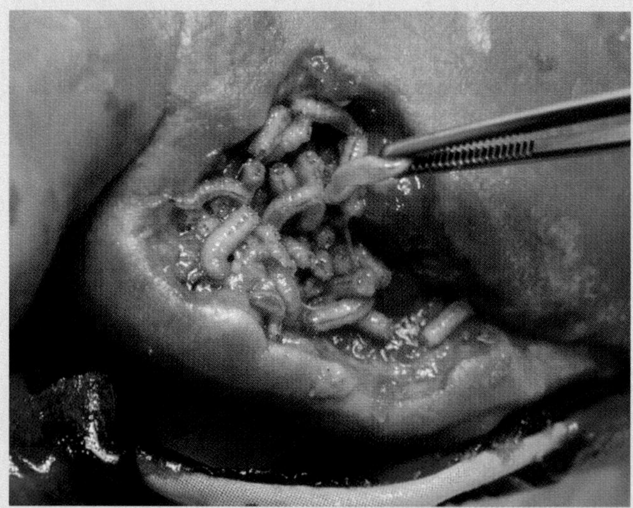

▲ **FIGURE E23-1 Blowfly maggots can clean wounds**

(a) Scallop

(b) Mussels

▲ **FIGURE 23-15 The diversity of bivalve mollusks (a)** This scallop parts its hinged shells to allow the intake of water from which food will be filtered. The blue spots visible along the mantle just inside the upper and lower shells are simple eyes. The upper shell is covered with an encrusting sponge. **(b)** Mussels attach to rocks in dense aggregations exposed at low tide. White barnacles are attached to the mussel shells and surrounding rock.

Case Study continued
The Search for a Sea Monster

Although the giant squid is a very large invertebrate animal, there is another species that is even larger. In 2007, a crew fishing off the coast of New Zealand accidentally caught a colossal squid, *Mesonychoteuthis hamiltoni*, that was 30 feet long and weighed more than 1,000 pounds. The creature was by far the heaviest squid so far measured. When biologists later dissected this specimen, they found that its beak, though quite big, was actually smaller than some colossal squid beaks that had been previously found in the stomachs of sperm whales. So it is likely that colossal squids can grow to be even larger than the massive individuals found thus far, and larger than the largest giant squids.

in response to danger; this muscle is what you are served when you order scallops in a restaurant.

Clams use a muscular foot for burrowing in sand or mud. In mussels, which live attached to rocks, the foot is smaller and helps secrete threads that anchor the animal to the rocks. Scallops lack a foot and move by a sort of jet propulsion achieved by flapping their shells together.

Bivalves are filter feeders, using their gills as both respiratory and feeding structures. Water circulates over the gills, which are covered with a thin layer of mucus that traps microscopic food particles. Food is conveyed to the mouth by the beating of cilia on the gills.

Cephalopods Are Marine Predators

The *cephalopods* include octopuses, nautiluses, cuttlefish, and squids (**Fig. 23-16**). The largest invertebrates, the giant squid and the colossal squid, belong to this group. All cephalopods are predatory carnivores, and all are marine. In these mollusks, the foot has evolved into tentacles with well-developed sensory abilities for detecting prey. Prey are grasped by suction disks on the tentacles and may be immobilized by a paralyzing venom in the saliva before being torn apart by beaklike jaws.

Cephalopods move rapidly by jet propulsion, which is generated by the forceful expulsion of water from the mantle cavity. Octopuses may also travel along the seafloor by using their tentacles like multiple undulating legs. The rapid movements and generally active lifestyles of cephalopods are made possible in part by their closed circulatory systems. Cephalopods are the only mollusks with closed circulation, which transports oxygen and nutrients more efficiently than do open circulatory systems.

Cephalopods have highly developed brains and sensory systems. The cephalopod eye rivals our own in complexity. The cephalopod brain, especially that of the octopus, is (for an invertebrate brain) exceptionally large and complex. It is enclosed in a skull-like case of cartilage and endows the octopus with highly developed capabilities to learn and remember. In the laboratory, octopuses can rapidly learn to associate certain symbols with food and to open a screw-cap jar to obtain food.

(a) Octopus

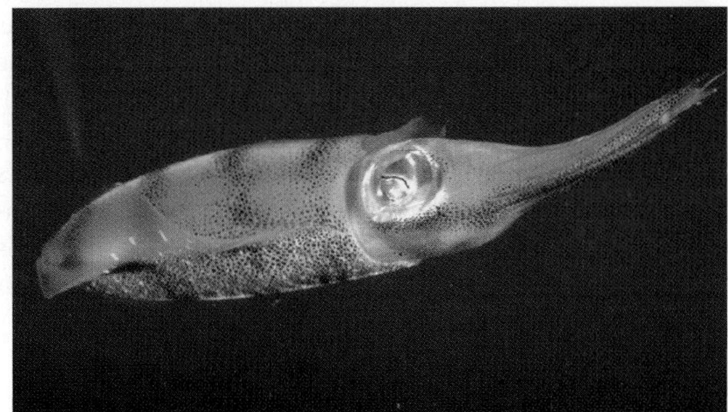

(b) Squid

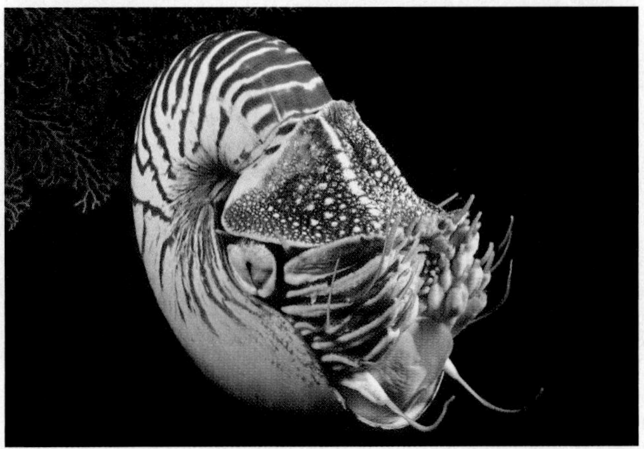

(c) Nautilus

▶ **FIGURE 23-16 The diversity of cephalopod mollusks**
(a) An octopus can crawl rapidly by using its eight suckered tentacles, and it can alter its color and skin texture to blend with its surroundings. In emergencies, this mollusk can jet backward by vigorously contracting its mantle. Octopuses and squid can emit clouds of dark purple ink to confuse pursuing predators. **(b)** A squid can move by contracting its mantle to generate jet propulsion, which pushes the animal backward through the water. **(c)** The chambered nautilus secretes a shell with internal, gas-filled chambers that provide buoyancy. Note the well-developed eyes and the tentacles used to capture prey.

Arthropods Are the Most Diverse and Abundant Animals

In terms of both number of individuals and number of species, no other animal phylum comes close to the *arthropods* (Arthropoda), which include insects, arachnids, myriapods, and crustaceans. About 1 million arthropod species have been discovered, and scientists estimate that millions more remain undescribed.

Arthropods Have Appendages and an External Skeleton

All arthropods have paired, jointed appendages and an **exoskeleton,** an external skeleton that encloses the arthropod body like a suit of armor. Secreted by the *epidermis* (the outer layer of skin), the exoskeleton is composed chiefly of protein and a polysaccharide called *chitin* (see Fig. 3-10). The external skeleton protects against predators and is responsible for arthropods' greatly increased agility relative to their wormlike ancestors. The exoskeleton provides rigid attachment sites for muscles, but also becomes thin and flexible at joints, thereby increasing the range of movement of the appendages. This combination of rigidity and flexibility makes possible the flight of the bumblebee and the intricate, delicate manipulations of the spider as it weaves its web (**Fig. 23-17**). The exoskeleton also contributed enormously to the arthropod invasion of the land by providing a watertight covering for delicate, moist tissues such as those used for gas exchange. (Arthropods were the earliest terrestrial animals; see p. 329.)

Like a suit of armor, however, the arthropod exoskeleton poses some problems. First, because it cannot expand as the animal grows, the exoskeleton periodically must be shed, or **molted,** and replaced with a larger one (**Fig. 23-18**). Molting uses energy and leaves the animal temporarily vulnerable until the new skeleton hardens. The exoskeleton is also heavy, and its weight increases exponentially as the animal grows. It is no coincidence that the largest arthropods are crustaceans (crabs and lobsters), whose watery habitat supports much of their weight.

▲ **FIGURE 23-18 The exoskeleton must be molted periodically** A cicada emerges from its outgrown exoskeleton (left).

Arthropods Have Specialized Segments and Adaptations for Active Lifestyles

Arthropods are segmented, but their segments tend to be few and specialized for different functions such as sensing the environment, feeding, and movement (**Fig. 23-19**). For example, in insects, sensory and feeding structures are concentrated on the front segment, known as the *head,* and digestive structures are largely confined to the *abdomen,* which is the rear segment. Between the head and the abdomen is the *thorax,* the segment to which structures used in locomotion, such as wings and walking legs, are attached.

Efficient gas exchange is required to supply adequate oxygen to the muscles, which allows the rapid flight, swimming, or running displayed by many arthropods. In aquatic arthropods such as crustaceans, gas exchange is accomplished by gills. In terrestrial arthropods, gas exchange is performed either by lungs (in arachnids) or by *tracheae* (singular, trachea), a network of narrow, branching respiratory tubes that

▲ **FIGURE 23-17 The exoskeleton allows precise movements** A net-throwing spider begins to wrap a captured insect in silk. Such dexterous manipulations are made possible by the exoskeleton and jointed appendages characteristic of arthropods.

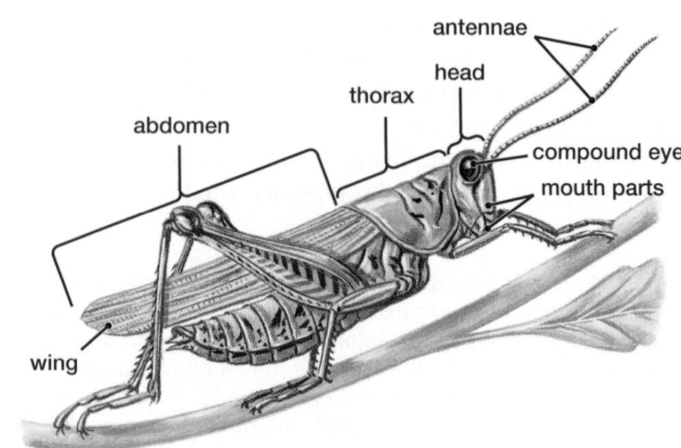

▲ **FIGURE 23-19 Segments are fused and specialized in insects** Insects, such as this grasshopper, show fusion and specialization of body segments into a distinct head, thorax, and abdomen. Segments are visible on the abdomen beneath the wings.

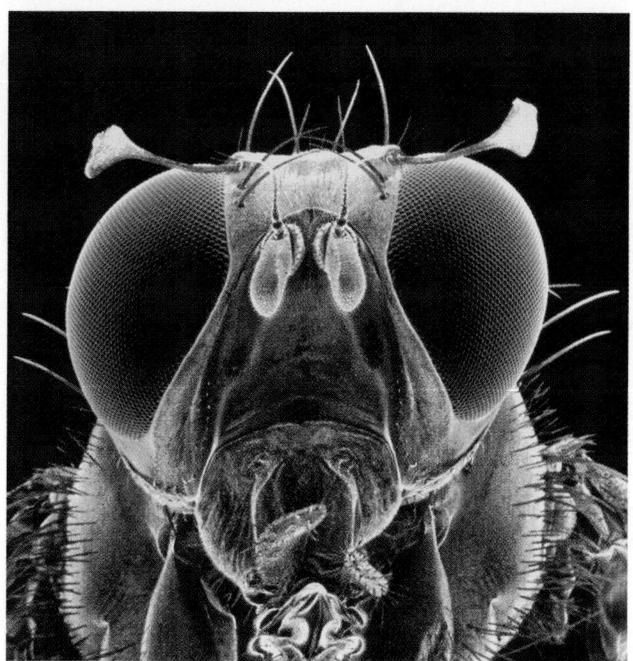

▲ **FIGURE 23-20 Arthropods possess compound eyes** This scanning electron micrograph shows the compound eye of a fruit fly. Compound eyes consist of an array of similar light-gathering and sensing elements whose orientation gives the arthropod a wide field of view. Insects have reasonably good image-forming ability and good color discrimination.

open to the surrounding environment and that penetrate all parts of the body. Most arthropods have open circulatory systems, like those of mollusks, in which blood directly bathes the organs in a hemocoel (see Fig. 32-1a).

Most arthropods possess well-developed sensory and nervous systems. Arthropod sensory systems often include **compound eyes,** which have multiple light detectors (**Fig. 23-20**), and acute chemical and tactile senses. The arthropod nervous system is similar in plan to that of annelids, but more complex. It consists of a brain composed of fused ganglia, and a series of additional ganglia along the length of the body that are linked by a ventral nerve cord. This well-developed nervous system, combined with sophisticated sensory abilities, has permitted the evolution of complex behaviors in many arthropods.

Insects Are the Only Flying Invertebrates

The number of described *insect* species is about 850,000, roughly three times the total number of known species in all other groups of animals combined (**Fig. 23-21**). Insects have a single pair of antennae and three pairs of legs, usually supplemented by two pairs of wings. Insects' capacity for flight distinguishes them from all other invertebrates and has contributed to their enormous success (**Fig. 23-21c**). As anyone who has pursued a fly can testify, flight helps in escaping from predators. It also allows the insect to find widely dispersed

(a) Aphid

(b) Ant

(c) Beetle flying

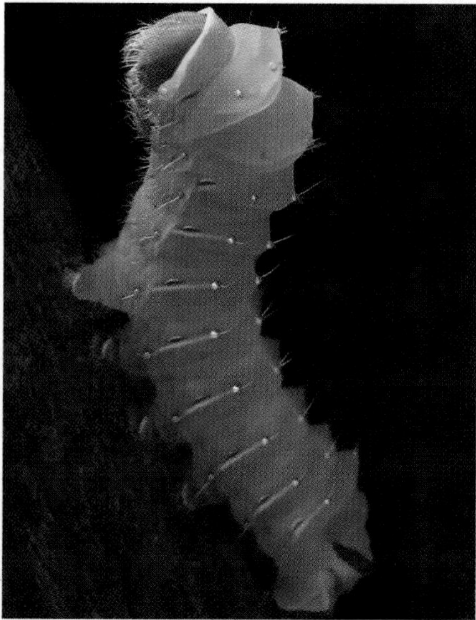

(d) Moth larva

◀ **FIGURE 23-21 The diversity of insects (a)** The rose aphid sucks sugar-rich juice from plants. **(b)** The bullet ant can inflict an extremely painful sting. **(c)** A June beetle displays its two pairs of wings as it comes in for a landing. The outer wings protect the abdomen and the inner wings, which are relatively thin and fragile. **(d)** Caterpillars are larval forms of moths or butterflies. This caterpillar larva of a silkmoth can produce clicking sounds with its mouthparts. The sounds may serve to warn predators that the caterpillar is distasteful.

food. Swarms of locusts can travel 200 miles a day in search of food; researchers tracked one swarm on a journey that totaled almost 3,000 miles. Flight requires rapid and efficient gas exchange, which insects accomplish by means of tracheae.

During their development, insects undergo **metamorphosis,** a radical change from a juvenile body form to an adult body form. In insects with complete metamorphosis, the immature stage, called a **larva** (plural, larvae), is worm shaped (for example, the maggot of a blowfly or the caterpillar of a moth or butterfly; **Fig. 23-21d**). The larva hatches from an egg, grows by eating voraciously and shedding its exoskeleton several times, and then forms a nonfeeding stage called a **pupa** (plural, pupae). Encased in an outer covering, the pupa undergoes a radical change in body form, emerging in its adult, winged form. The adults mate and lay eggs, continuing the cycle. Metamorphosis may include a change in diet as well as in shape, thereby eliminating competition for food between adults and juveniles, and, in some cases, allowing the insect to exploit different foods when they are most available. For instance, a caterpillar that feeds on new green shoots in springtime metamorphoses into a butterfly that drinks nectar from the summer's blooming flowers. Some insects, such as grasshoppers and crickets, undergo a more gradual metamorphosis (called incomplete metamorphosis), hatching as young that bear some resemblance to the adult, then gradually acquiring more adult features as they grow and molt.

Biologists classify the amazing diversity of insects into several dozen groups. We will describe three of the largest ones here.

Butterflies and Moths The butterflies and moths make up what is perhaps the most conspicuous and best-studied group of insects. The brightly colored, often iridescent wing patterns of many butterfly and moth species arise from pigments and light-refracting structures in the scales that cover the wings of all members of this group. (The scales are the powdery substance that rubs off onto your hand when you handle a butterfly or moth.) Butterflies fly mainly during the day, moths at night (though there are exceptions to this general rule, such as the hummingbird-like hawk moths that are often seen feeding by day in flower gardens).

The evolution of butterflies and moths has been closely tied to the evolution of flowering plants. Butterflies and moths at all stages of life feed almost exclusively on flowering plants. Many species of flowering plants depend in turn on butterflies and moths for pollination.

Bees, Ants, and Wasps The bees, ants, and wasps are known to many people by their painful stings. Many species in this group are equipped with a stinger that extends from the abdomen and can be used to inject venom into the victim of a sting. The venom may be extremely toxic but, fortunately for human stinging victims, each insect carries only a tiny amount. Nonetheless, the amount is often sufficient to cause considerable pain. Only females have stingers, which are used by many stinging species to help defend a nest from attack by a potential predator. Defense, however, is not the only use for stingers. Many wasps, for example, act as parasites when reproducing: A wasp lays an egg inside the body of another species, typically a moth or butterfly caterpillar, which becomes food for the wasp larva after it hatches. Before laying its egg, the wasp may sting the caterpillar, paralyzing it.

The social behavior of some ant and bee species is extraordinarily intricate. They may form huge colonies with complex organization in which individuals specialize in particular tasks such as foraging, defense, reproduction, or rearing larvae. The organization and division of labor in these insect societies may require levels of communication and learning comparable to those present in vertebrates. The remarkable tasks accomplished by social insects include the manufacture and storage of food (honey) by honeybees and "farming" by ant species that cultivate fungi in underground chambers or "milk" aphids by inducing them to secrete a nutritious liquid.

Beetles Roughly one-third of all known insect species are beetles. Beetles exhibit a huge variety of shapes, sizes, and lifestyles. All beetles, however, have a hard, protective, exoskeletal structure that covers their wings. Many destructive agricultural pests are beetles, such as the Colorado potato beetle, the grain weevil, and the Japanese beetle. However, others, such as lady beetles, are predators that help control insect pests.

One of this group's most impressive adaptations is found in the bombardier beetle. This species defends itself against ants and other enemies by emitting a toxic spray from a nozzle-like structure at the end of its abdomen. The beetle is able to precisely aim the spray, which emerges with explosive force and at temperatures higher than 200°F (93°C). To avoid harming itself, the beetle produces this hot, toxic brew only when needed, by combining two harmless substances.

Have you ever wondered

Which Insect Has the Most Painful Sting?

If you have ever been stung by a bee, you know that it hurts. But how *much* did it hurt, relative to other possible stings? Fortunately for those interested in the answer to such a question, biologist Justin Schmidt has devised the Sting Pain Index. The index ranges from 0 (painless) to 4 (most painful) and is based mostly on Dr. Schmidt's own experience in getting stung. He has rated 78 species so far. On the scale, a honeybee rates a score of 2, about the same as a bald-faced hornet, but higher than a sweat bee or a fire ant. Paper wasps are more painful, with a score of about 3. More painful still is the sting of a tarantula hawk, a wasp of the southwestern United States that ordinarily uses its sting to attack and paralyze tarantula spiders, which then serve as food for the wasp's larvae. But the most painful sting of all belongs to the bullet ant, a resident of Central and South American rain forests (see Fig. 23-21b). The sting of a bullet ant causes waves of excruciating, throbbing, burning pain that last up to 24 hours.

Most Arachnids Are Predatory Meat Eaters

The *arachnids* include spiders, mites, ticks, and scorpions (**Fig. 23-22**). All arachnids have eight walking legs, and most are carnivorous. Many subsist on a liquid diet of blood or predigested prey. For example, spiders, the most numerous arachnids, first immobilize their prey with a paralyzing venom. They then inject digestive enzymes into the helpless victim (typically an insect) and suck in the resulting soup. Arachnids breathe by using either tracheae, lungs, or both.

In contrast to the compound eyes of insects and crustaceans, arachnids have simple eyes, each with a single lens. Most spiders have eight eyes placed in such a way as to give them a panoramic view of predators and prey. The eyes are sensitive to movement, and in some spider species—especially those that hunt actively and have no webs—the eyes can probably form images. Most spider perception, however, is not through their eyes but through sensory hairs. Some of a spider's hairs are touch sensitive and help the animal perceive prey, mates, and surroundings. Other hairs are sensitive to chemicals and function as organs of smell and taste. Hairs also respond to vibrations in the air, ground, or web, allowing spiders to detect nearby movement by predators, prey, or other spiders.

Among the distinctive features of spiders is their production of protein threads known as silk. Spiders manufacture silk in special glands in their abdomens and use it to perform a variety of functions, such as building webs that capture prey, wrapping up and immobilizing captured prey (see Fig. 23-17), constructing protective shelters for themselves, making cocoons to surround their eggs, and making "draglines" that connect a spider to a web or other surface and support its weight if it drops from its perch. Spider silk is an amazingly light, strong, and elastic fiber. It can be stronger than a steel wire of the same size, yet is as elastic as rubber. Human engineers have long sought to develop a fiber with this combination of strength and elasticity. Despite careful study of the structure of spider silk, no comparable man-made substance has been successfully manufactured. Some researchers have applied the techniques of biotechnology to the problem, inserting spider genes that code for silk proteins into mammal or bacterial cells that can be maintained in the lab. The goal is to produce large quantities of silk.

Myriapods Have Many Legs

The *myriapods* include the centipedes and millipedes, whose most prominent feature is an abundance of legs (**Fig. 23-23**). Most millipede species have between 100 and 300 legs; the species with the largest number of legs can have up to 750. Centipedes are not quite as leggy; a typical species has around 70 legs, but some species have fewer. Both centipedes and millipedes have one pair of antennae. The legs and antennae of centipedes are longer and more delicate than those of millipedes. Myriapods have very simple eyes that detect light and dark but do not form images. In some species, the number of eyes can be high—up to 200—but other species lack eyes altogether. Myriapods respire by means of tracheae.

Myriapods inhabit terrestrial environments exclusively, living mostly in the soil or leaf litter or under logs and rocks.

(a) Spider

(b) Scorpion

(c) Ticks

▲ **FIGURE 23-22 The diversity of arachnids (a)** The tarantula is one of the largest spiders but is relatively harmless. **(b)** Scorpions, found in warm climates such as the deserts of the southwestern United States, paralyze their prey with venom from a stinger at the tip of the abdomen. A few species can harm humans. **(c)** Ticks before (left) and after feeding on blood. The uninflated exoskeleton is flexible and folded, allowing the animal to become grotesquely bloated while feeding.

(a) Centipede

(b) Millipede

▲ **FIGURE 23-23 The diversity of myriapods (a)** Centipedes and **(b)** millipedes are common nocturnal arthropods. Each segment of a centipede's body holds one pair of legs, while each millipede segment has two pairs.

Centipedes are generally carnivorous, capturing prey (mostly other arthropods) with their frontmost legs, which are modified into sharp claws that inject poison into prey. Bites from large centipedes can be painful to humans. In contrast, most millipedes are not predators but instead feed on decaying vegetation and other debris. When attacked, many millipedes defend themselves by secreting a foul-smelling, distasteful liquid.

Most Crustaceans Are Aquatic

The *crustaceans*, including crabs, crayfish, lobsters, shrimp, and barnacles, are the only arthropods whose members live primarily in the water (**Fig. 23-24**). Crustaceans range in size from microscopic species that live in the spaces between grains of sand to the largest of all arthropods, the Japanese spider crab, with legs spanning nearly 12 feet (4 meters). Crustaceans have two pairs of sensory antennae, but the rest of their appendages are highly variable in form and number, depending on the habitat and lifestyle of the species. Most crustaceans have compound eyes similar to those of insects, and nearly all respire by means of gills.

Roundworms Are Abundant and Mostly Tiny

Although you may be blissfully unaware of their presence, *roundworms* (Nematoda) are nearly everywhere. Roundworms,

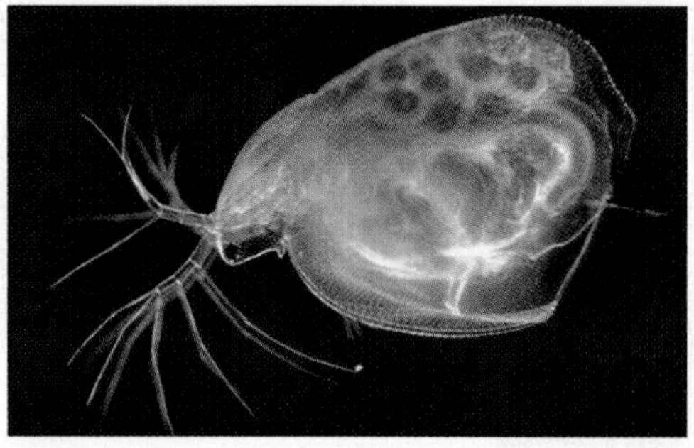

(a) Water flea

(b) Sowbug

(c) Hermit crab

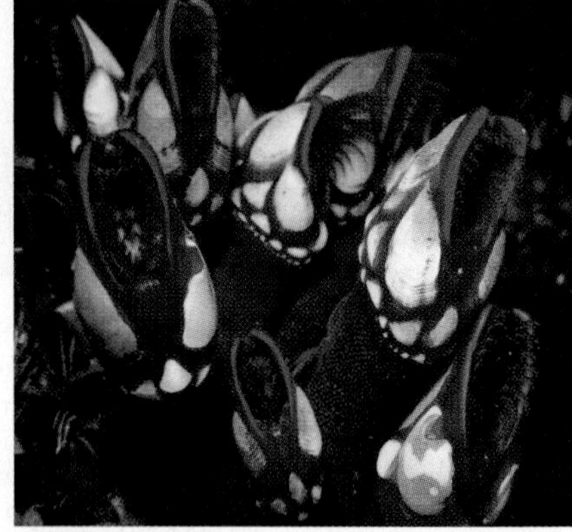

(d) Barnacles

► **FIGURE 23-24 The diversity of crustaceans (a)** The microscopic water flea is common in freshwater ponds. Notice the eggs developing within the body. **(b)** The sowbug, found in dark, moist places such as under rocks, leaves, and decaying logs, is one of the few crustaceans to invade the land successfully. **(c)** The hermit crab protects its soft abdomen by inhabiting an abandoned snail shell. **(d)** The gooseneck barnacle uses a tough, flexible stalk to anchor itself to rocks, boats, or even animals such as whales. Other types of barnacles attach with shells that resemble miniature volcanoes (see Fig. 23-15b). Early naturalists thought barnacles were mollusks until they observed barnacles' jointed legs (seen here extending into the water).

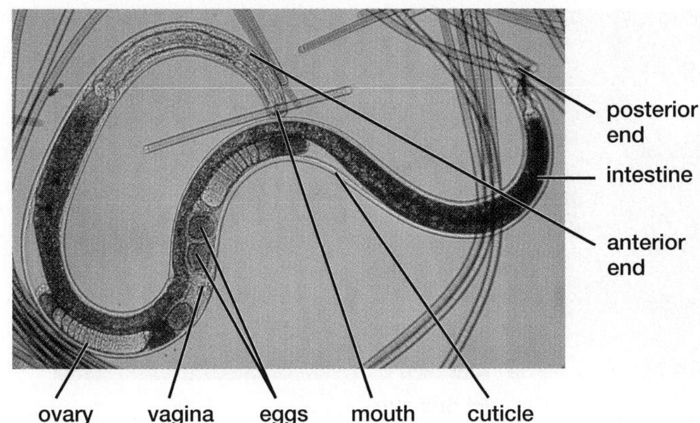

▲ **FIGURE 23-25 A freshwater nematode** Eggs can be seen inside this female freshwater nematode, which feeds on algae.

also called *nematodes*, have colonized nearly every habitat on Earth, and they play an important role in breaking down organic matter. They are extremely numerous; a single rotting apple may contain 100,000 roundworms. Billions thrive in each acre of topsoil. In addition, almost every plant and animal species hosts several parasitic nematode species.

In addition to being abundant and ubiquitous, roundworms are diverse. Although only about 12,000 roundworm species have been named, there may be as many as 500,000. Most, such as the one shown in **Figure 23-25,** are microscopic, but some parasitic forms reach a meter in length.

Roundworms are Pseudocoelomates and Have a Simplified Body Plan

Roundworms have a rather simple body plan, featuring a tubular gut and a fluid-filled pseudocoelom that surrounds the organs and forms a hydrostatic skeleton (see Fig. 23-3b). A tough, flexible, nonliving cuticle encloses and protects the thin, elongated body and is periodically molted. The molting of roundworms reveals that they share a common evolutionary heritage with arthropods and other ecdysozoan phyla. Sensory organs in the roundworm head transmit information to a simple "brain," composed of a nerve ring.

Like flatworms, nematodes lack circulatory and respiratory systems. Because most nematodes are extremely thin and have low energy requirements, diffusion suffices for gas exchange and distribution of nutrients. Most nematodes reproduce sexually, and the sexes are separate; the male (usually smaller than the female) fertilizes the female by placing sperm inside her body.

A Few Roundworm Species Are Harmful to Humans

During your life, you may become host to one of the 50 species of roundworms that infect humans. Most such worms are relatively harmless, but there are important exceptions. For example, hookworm larvae (found in soil in some tropical regions) can bore into human feet, enter the bloodstream, and travel to the intestine, where they cause continuous bleeding. Another dangerous nematode parasite, *Trichinella*, causes the disease trichinosis. *Trichinella* worms can infect people who eat improperly cooked infected pork, which can contain up to 15,000 larval cysts per gram (**Fig. 23-26a**). The cysts hatch in the human digestive tract and invade blood vessels and muscles, causing bleeding and muscle damage.

Parasitic nematodes can also endanger domestic animals. Dogs, for example, are susceptible to heartworm, which is transmitted by mosquitoes (**Fig. 23-26b**). In the southern United States, and increasingly in other parts of the country, heartworm poses a severe threat to the health of unprotected pets.

Echinoderms Have a Calcium Carbonate Skeleton

Echinoderms (Echinodermata) are found only in marine environments, and their common names tend to evoke their saltwater habitats: sand dollars, sea urchins, sea stars (or starfish), sea cucumbers, and sea lilies (**Fig. 23-27**). The name "echinoderm" (Greek, "hedgehog skin") stems from the bumps or spines that extend from the skin of most echinoderms. These spines are especially well developed in sea urchins and much reduced in sea stars and sea cucumbers. Echinoderm bumps and spines are actually extensions of an **endoskeleton** (internal

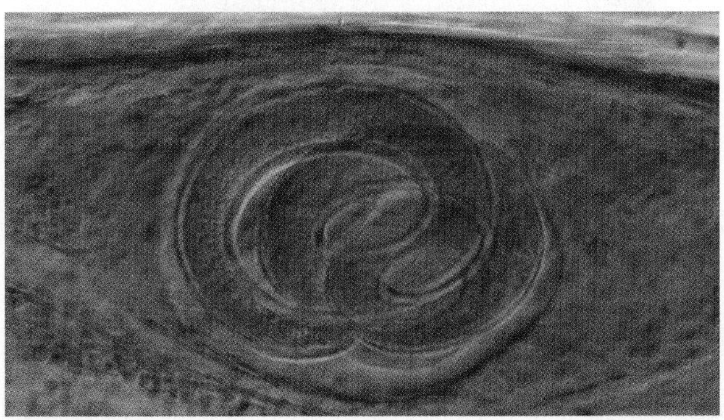

(a) ***Trichinella***

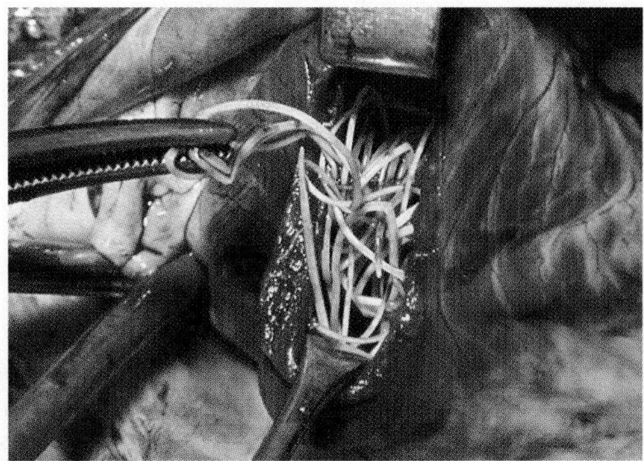

(b) Heartworms

▲ **FIGURE 23-26 Some parasitic nematodes (a)** Encysted larva of the *Trichinella* worm in the muscle tissue of a pig, where it may live for up to 20 years. **(b)** Adult heartworms in the heart of a dog. The juveniles are released into the bloodstream, where they may be ingested by mosquitoes and passed to another dog by the bite of an infected mosquito.

(a) Sea cucumber (b) Sea urchin (c) Sea star

▲ **FIGURE 23-27 The diversity of echinoderms (a)** A sea cucumber feeds on debris in the sand. **(b)** The sea urchin's spines are actually projections of the internal skeleton. **(c)** The sea star typically has five arms.

skeleton) composed of plates of calcium carbonate that lie beneath the outer skin.

Echinoderms Are Bilaterally Symmetrical as Larvae and Radially Symmetrical as Adults

Echinoderms exhibit deuterostome development and are linked by common ancestry with the other deuterostome phyla, including the chordates (described later). Deuterostomes form a group of branches on the larger evolutionary tree of bilaterally symmetrical animals, but in echinoderms bilateral symmetry is expressed only in embryos and free-swimming larvae. An adult echinoderm, in contrast, is radially symmetrical and lacks a head. This absence of cephalization is consistent with the sluggish existence of echinoderms. Most echinoderms move very slowly as they feed on algae or small particles sifted from sand or water. Some echinoderms are slow-motion predators. Sea stars, for example, slowly pursue even slower-moving prey, such as snails or clams.

Echinoderms Have a Water-Vascular System

Echinoderms move on numerous tiny *tube feet,* delicate cylindrical projections that extend from the lower surface of the body and terminate in a suction cup. Tube feet are part of a unique echinoderm feature, the *water-vascular system,* which functions in locomotion, respiration, and food capture (**Fig. 23-28**). Seawater enters through an opening (the *sieve plate*) on the animal's upper surface and is conducted through a circular central canal, from which branch a number of radial canals. These canals conduct water to the tube feet, each of which is controlled by a muscular squeeze bulb known as an *ampulla*. Contraction of the bulb forces water into the tube

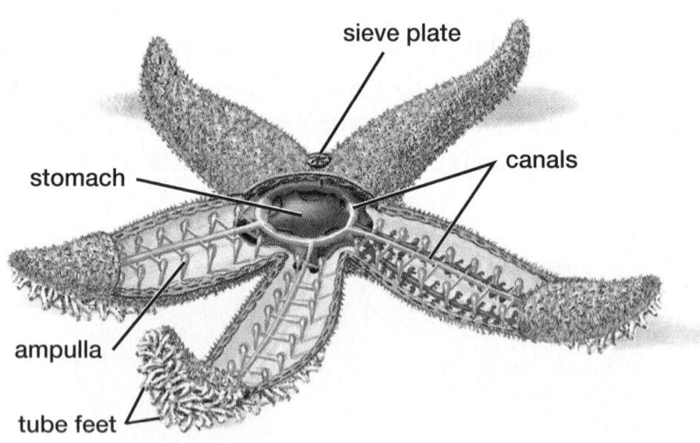

(a) Starfish body plan

(b) Starfish consuming a mussel

▲ **FIGURE 23-28 The water-vascular system of echinoderms (a)** Changing pressure inside the seawater-filled water-vascular system extends or retracts the tube feet. **(b)** The sea star often feeds on mollusks such as this mussel. A feeding sea star attaches numerous tube feet to the mussel's shells, exerting a relentless pull. Then, the sea star turns the delicate tissue of its stomach inside out, extending it through its centrally located ventral mouth. The stomach can fit through an opening in the bivalve shells which measures less than 1 millimeter. Once insinuated between the shells, the stomach tissue secretes digestive enzymes that weaken the mollusk, causing it to open further. Partially digested food is transported to the upper portion of the stomach, where digestion is completed.

foot, causing it to extend. The suction cup may be pressed against the seafloor or a food object, to which it adheres tightly until its internal pressure is released.

Some Echinoderm Organ Systems Are Simplified

Echinoderms have a relatively simple nervous system with no distinct brain. Movements are loosely coordinated by a system consisting of a nerve ring that encircles the esophagus, radial nerves to the rest of the body, and a nerve network through the epidermis. In sea stars, simple receptors for light and chemicals are concentrated on the arm tips, and sensory cells are scattered over the skin. In some brittle star species, light receptors are associated with tiny lenses, smaller than the width of a human hair, that gather light and focus it on receptors. The optical quality of these "microlenses" is excellent, far superior to that of any human-created lens of comparable size.

Echinoderms lack a circulatory system, although movement of the fluid in their well-developed coelom serves this function. Gas exchange occurs through the tube feet and, in some forms, through numerous tiny "skin gills" that project through the epidermis. Most species have separate sexes and reproduce by shedding sperm and eggs into the water, where fertilization occurs.

Many echinoderms are able to regenerate lost body parts, and these regenerative powers are especially potent in sea stars. In fact, a single arm of a sea star is capable of developing into a whole animal, provided that part of the central body is attached to it. Before this ability was widely appreciated, mussel fishermen often tried to rid mussel beds of predatory sea stars by hacking them into pieces and throwing the pieces back. Needless to say, the strategy backfired.

The Chordates Include the Vertebrates

The *chordates* (Chordata) include the vertebrate animals and also include a few groups of invertebrates, such as the sea squirts and the lancelets. We will discuss these invertebrate chordates and their vertebrate relatives in Chapter 24.

Case Study revisited
The Search for a Sea Monster

Clyde Roper's search for the giant squid led him to organize three major expeditions. The first of these searched waters near the Azores Islands in the Atlantic Ocean. Because sperm whales are known to prey upon giant squid, Roper believed that the whales might lead him to the squids. To test this idea, he and his team affixed video cameras to sperm whales, thus allowing the scientists to see what the whales were seeing. These "crittercams" revealed a great deal of new information about sperm whale behavior but, alas, no footage of giant squids.

The next Roper-led expedition took place in the Kaikoura Canyon, an area of very deep water (3,300 feet, or 1,000 meters) off the coast of New Zealand. The scientists chose this spot because deep-sea fishing boats had recently captured several giant squid in the vicinity. Cameras were again deployed on sperm whales; but this time the mobile cameras were supplemented by a stationary, baited camera and an unmanned, remote-controlled submarine. Again, however, a large investment of time, money, and equipment yielded no squid sightings.

A few years later, Roper assembled a team of scientists for a return to Kaikoura Canyon. This time, the group was able to use Deep Rover, a one-person submarine that could carry an observer to depths of 2,200 feet. The scientists used Deep Rover to explore the canyon, following sperm whales in hopes that the huge mammals would lead them to giant squid. Unfortunately, the team again failed to find a squid.

Roper has pursued his search for the giant squid with extraordinary persistence, and he is not alone in seeking a glimpse of the creature. Other research teams have also been on the lookout for giant squid, and it was one of these groups that finally secured the first (and so far only) visual record of living giant squid. Working off the coast of Japan, the researchers placed a video camera on a long, baited fishing line. Long hours of dragging the line through the water at a depth of 3,000 feet were eventually rewarded with images of a giant squid that attacked the bait (**Fig. 23-29**).

Consider This

Steve O'Shea, another scientist interested in the giant squid, captured a few juvenile giant squid in 2002. The tiny animals, only a few millimeters long, survived in captivity for just a few hours, but their identity as giant squid was confirmed by comparing their DNA to that of preserved adult specimens. O'Shea believes that with more research and experience, he could learn to raise the young animals to adulthood. Given that research funds are limited, which approach is better? Would we learn more from viewing wild adult squid in the depths of the ocean, or by capturing tiny juveniles from surface waters and figuring out how to raise them in the lab?

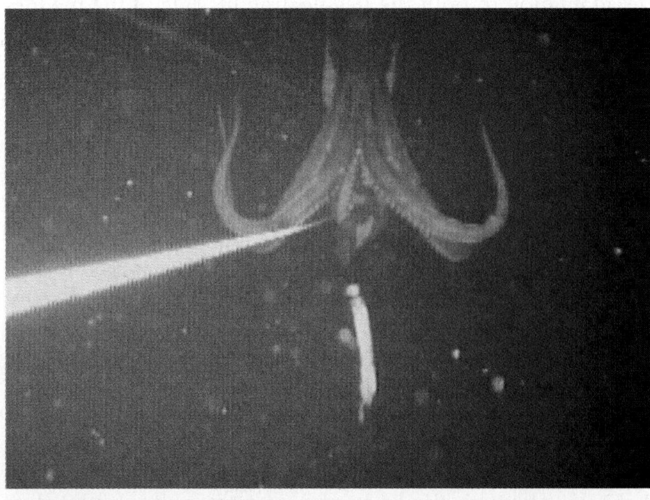

▲ **FIGURE 23-29 A giant squid approaches a baited line**

CHAPTER REVIEW

Summary of Key Concepts

23.1 What Are the Key Features of Animals?

Animals are multicellular, sexually reproducing organisms that acquire energy by consuming other organisms. Most animals can perceive and react rapidly to environmental stimuli and are motile at some stage in their lives. Their cells lack a cell wall.

23.2 Which Anatomical Features Mark Branch Points on the Animal Evolutionary Tree?

The earliest animals had no tissues, a feature retained by modern sponges. All other modern animals have tissues. Animals with tissues can be divided into radially symmetrical and bilaterally symmetrical groups. During embryonic development, radially symmetrical animals have two germ layers; bilaterally symmetrical animals have three. Bilaterally symmetrical animals also tend to have sense organs and clusters of neurons concentrated in the head, a process called cephalization. Bilateral phyla can be divided into two main groups, one of which undergoes protostome development, the other of which undergoes deuterostome development. Protostome phyla can in turn be divided into ecdysozoans and lophotrochozoans. Some phyla of bilaterally symmetrical animals lack body cavities, but most have either pseudocoeloms or true coeloms.

23.3 What Are the Major Animal Phyla?

The bodies of sponges (Porifera) are typically free-form in shape and are sessile. Sponges have relatively few types of cells. Despite the division of labor among the cell types, there is little coordination of activity. Sponges lack the muscles and nerves required for coordinated movement, and digestion occurs exclusively within the individual cells.

The hydrozoans, anemones, and sea jellies (Cnidaria) have tissues. A simple network of nerve cells directs the activity of contractile cells, allowing loosely coordinated movements. Digestion is extracellular, occurring in a central gastrovascular cavity with a single opening. Cnidarians exhibit radial symmetry, an adaptation to both the free-floating lifestyle of the medusa and the sedentary existence of the polyp.

Flatworms (Platyhelminthes) have a distinct head with sensory organs and a simple brain. A system of canals forming a network through the body aids in excretion. They lack a body cavity.

The segmented worms (Annelida) are the most complex of the worms, with a well-developed closed circulatory system and excretory organs that resemble the basic unit of the vertebrate kidney. The segmented worms have a compartmentalized digestive system, like that of vertebrates, which processes food in a sequence. Annelids also have a true coelom, a fluid-filled space between the body wall and the internal organs.

The snails, clams, and squid (Mollusca) lack a skeleton; some forms protect the soft, moist, muscular body with a single shell (many gastropods and a few cephalopods) or a pair of hinged shells (the bivalves). The lack of a waterproof external covering limits this phylum to aquatic and moist terrestrial habitats. Although the body plan of gastropods and bivalves limits the complexity of their behavior, the cephalopod's tentacles are capable of precisely controlled movements. The octopus has the most complex brain and the best-developed learning capacity of any invertebrate.

Arthropods (Arthropoda), the insects, arachnids, millipedes and centipedes, and crustaceans, are the most diverse and abundant animals on Earth. They have invaded nearly every available terrestrial and aquatic habitat. Jointed appendages and well-developed nervous systems make possible complex, finely coordinated behavior. The exoskeleton (which conserves water and provides support) and specialized respiratory structures (which remain moist and protected) enable the insects and arachnids to inhabit dry land. The diversification of insects has been enhanced by their ability to fly. Crustaceans, which include the largest arthropods, are restricted to moist, usually aquatic habitats and respire using gills.

The pseudocoelomate roundworms (Nematoda) possess a separate mouth and anus and a cuticle layer that is molted.

The sea stars, sea urchins, and sea cucumbers (Echinodermata) are an exclusively marine group. Like other complex invertebrates and chordates, echinoderm larvae are bilaterally symmetrical; however, the adults show radial symmetry. This, in addition to a primitive nervous system that lacks a definite brain, adapts them to a relatively sedentary existence. Echinoderm bodies are supported by a nonliving internal skeleton that sends projections through the skin. The water-vascular system, which functions in locomotion, feeding, and respiration, is a unique echinoderm feature.

The chordates (Chordata) include two invertebrate groups, the lancelets and sea squirts, as well as the vertebrates.

Key Terms

Thinking Through the Concepts

Fill-in-the-Blank

1. Bilaterally symmetrical animals have __________ embryonic tissue layers, known as __________, __________, and __________. Radially symmetrical animals have __________ tissue layers; they lack the __________ layer.
2. Animals that have an anterior and posterior end are said to be __________. The anterior end of such animals often contains structures used for __________ and __________. Coelomate animals have a body

__________ that is completely lined with tissue derived from __________.

3. Lophotrochozoans and ecdysozoans (molting animals) are two large clades of animals that exhibit __________ development. The other major type of development found in bilateral animals is known as __________ development, and is present in __________ and __________.

4. Animals that lack a backbone are described as __________; those that possess a backbone are __________. The vast majority of all animals fall into which of these two groups? __________ The only animals that lack tissues are __________, whose bodies resemble a colony of __________. Sea anemones and corals are __________. Earthworms and leeches are __________.

5. Three major groups within the mollusks are the two-shelled clams and scallops called __________; the foot-crawling snails and slugs called __________; and the tentacled squid and octopuses called __________. Members of the largest animal phylum are called __________. Three important groups within this phylum are the six-legged, often flying __________; the eight-legged spiders and mites called __________; and the mostly aquatic __________.

6. Phyla that contain animals with segmented bodies include __________ and __________. In a(n) __________ system, blood is confined to blood vessels. In a(n) __________ system, blood bathes internal organs within a cavity called the __________.

7. For each of the following distinctive structures, name the animal group in which it is found: mantle: __________; cnidocyte: __________; water-vascular system: __________; paired, jointed appendages: __________.

Review Questions

1. List the distinguishing characteristics of each of the phyla discussed in this chapter, and give an example of each.
2. Briefly describe each of the following adaptations, and explain its adaptive significance: bilateral symmetry, cephalization, closed circulatory system, coelom, radial symmetry, segmentation.
3. Describe and compare respiratory systems in the three major arthropod groups.
4. Describe the advantages and disadvantages of the arthropod exoskeleton.
5. In which of the three major mollusk groups is each of the following characteristics found?
   a. two shells
   b. a radula
   c. tentacles
   d. some sessile members
   e. the best-developed brains
   f. numerous eyes
6. Give three functions of the water-vascular system of echinoderms.
7. To what lifestyle is radial symmetry an adaptation? Bilateral symmetry?

Applying the Concepts

1. **BioEthics** Insects are the largest group of animals on Earth. Insect diversity is greatest in the Tropics, where habitat destruction and species extinction are occurring at an alarming rate. What biological, economic, and ethical arguments can you advance to persuade people and governments to preserve this biological diversity?
2. Discuss at least three ways in which the ability to fly has contributed to the success and diversity of insects.
3. Discuss and defend the attributes you would use to define biological success among animals. Are humans a biological success by these standards? Why?

MB *Go to www.masteringbiology.com for practice quizzes, activities, eText, videos, current events, and more.*

Animal Diversity II: Vertebrates

Chapter 24

Case Study

Fish Story

ON DECEMBER 22, 1938, Marjorie Courtney-Latimer received a phone call that would lead to one of the most spectacular discoveries in biological history. The call was from a local fisherman whom Courtney-Latimer, the curator of a small museum in South Africa, had asked to collect some fish specimens for the museum. His boat had returned from its most recent voyage and was waiting at the town dock. Dutifully, Courtney-Latimer went to the boat and began sorting through the fish that were strewn across the deck. Later, she wrote, "I noticed a blue fin sticking up from beneath the pile. I uncovered the specimen, and, behold, there appeared the most beautiful fish I had ever seen." In addition to its beauty, the fish had some odd features, including fins that were stumpy and lobed, unlike the fins of any other living species.

Courtney-Latimer did not recognize the strange fish, but she knew it was unusual. She tried to find a place to refrigerate it, but in her small town she was unable to find a cold storage facility willing to store a fish. In the end, she was able to save only the skin. Undaunted, she made some drawings of the fish and used them to attempt an identification. To her amazement, the creature did not resemble any species known to inhabit the waters off South Africa, but did seem similar to members of a family of fishes known as coelacanths. The only problem with this assessment was that coelacanths were known only from fossils. The earliest coelacanth fossils were found in 400-million-year-old rocks and, as far as anyone knew, the group had been extinct for 80 million years!

Perplexed, Courtney-Latimer sent her drawings to J. L. B. Smith, a fish expert at Rhodes University. Smith was astounded when he saw the sketch, later writing that "a bomb seemed to burst in my brain." Although bitterly disappointed that the specimen's bones and internal organs had been lost, Smith arranged to view the preserved skin. Ultimately, he confirmed the astonishing news that coelacanths still swam in Earth's waters.

▲ Would you be shocked to learn that *Tyrannosaurus* still walked the Earth? The discovery of modern coelacanth fishes was no less surprising.

At a Glance

24.1 WHAT ARE THE KEY FEATURES OF CHORDATES?

Humans are members of the phylum Chordata (**Fig. 24-1**), which includes not only other bony animals like the birds and apes, but also the tunicates (sea squirts) and small fish-like creatures called lancelets. What characteristics do we share with these creatures that seem so different from us?

All Chordates Share Four Distinctive Structures

All chordates have deuterostome development (which is also characteristic of echinoderms; see pp. 423–424) and are further united by four features that all possess at some stage of their lives: a dorsal nerve chord, a notochord, pharyngeal gill slits, and a post-anal tail.

Dorsal Nerve Cord

The **nerve cord** of chordates lies above the digestive tract, running lengthwise along the dorsal (upper) portion of the body. In contrast, the nerve cords of other animals lie in a ventral position, below the digestive tract (see Fig. 23-11). A chordate's nerve cord is hollow—its center is filled with fluid rather than with nerve tissue. During embryonic development in chordates, the nerve cord develops a thickening at its anterior end that becomes a brain.

Notochord

The **notochord** is a stiff but flexible rod that extends along the length of the body, located between the digestive tract and the nerve cord. It provides support for the body and an attachment site for muscles. In many chordates, the notochord is present only during early stages of development and disappears as a skeleton develops.

Pharyngeal Gill Slits

Pharyngeal gill slits are located in the pharynx (the cavity behind the mouth). In some chordates the slits form functional openings for gills (organs for gas exchange in water); in others they appear only as grooves during an early stage of development.

Post-Anal Tail

The **post-anal tail** is a posterior extension of the chordate body that extends past the anus and contains muscle tissue and the rearmost portion of the nerve cord. Other animals lack this kind of tail.

This list of distinctive chordate structures may seem puzzling because, although humans are chordates, at first glance we seem to lack every feature except the nerve cord. But evolutionary relationships are sometimes seen most clearly during early stages of development, and it is during our embryonic phase that we develop, and subsequently lose, our notochord, our gill slits, and our tails (**Fig. 24-2**).

24.2 WHICH CLADES MAKE UP THE CHORDATES?

Chordates include three clades (evolutionary groups that include all of the descendants of a common ancestor): the lancelets, the tunicates, and the craniates.

Lancelets Are Marine Filter-Feeders

The 25 species of lancelets (Cephalochordata) form a group of invertebrate chordates. Lancelets are small (2 inches, or about 5 centimeters, long), fishlike animals that retain all four chordate features as adults (**Fig. 24-3a**). An adult lancelet spends most of its time half-buried in the sandy sea bottom, with only the anterior end of its body exposed. The motion of cilia in the pharynx draws seawater into the lancelet's mouth. As the water passes through the pharyngeal gill slits, a film of mucus filters tiny food particles from the water. The captured food particles are transported to the lancelet's digestive tract.

Tunicates Include Sea Squirts and Salps

The tunicates (Urochordata) form a larger group of marine invertebrate chordates. Tunicates are small, with lengths ranging from a few millimeters to 1 foot (30 centimeters). The group includes immobile, filter-feeding, vase-shaped animals known as sea squirts (**Fig. 24-3b**). Much of a sea squirt's body

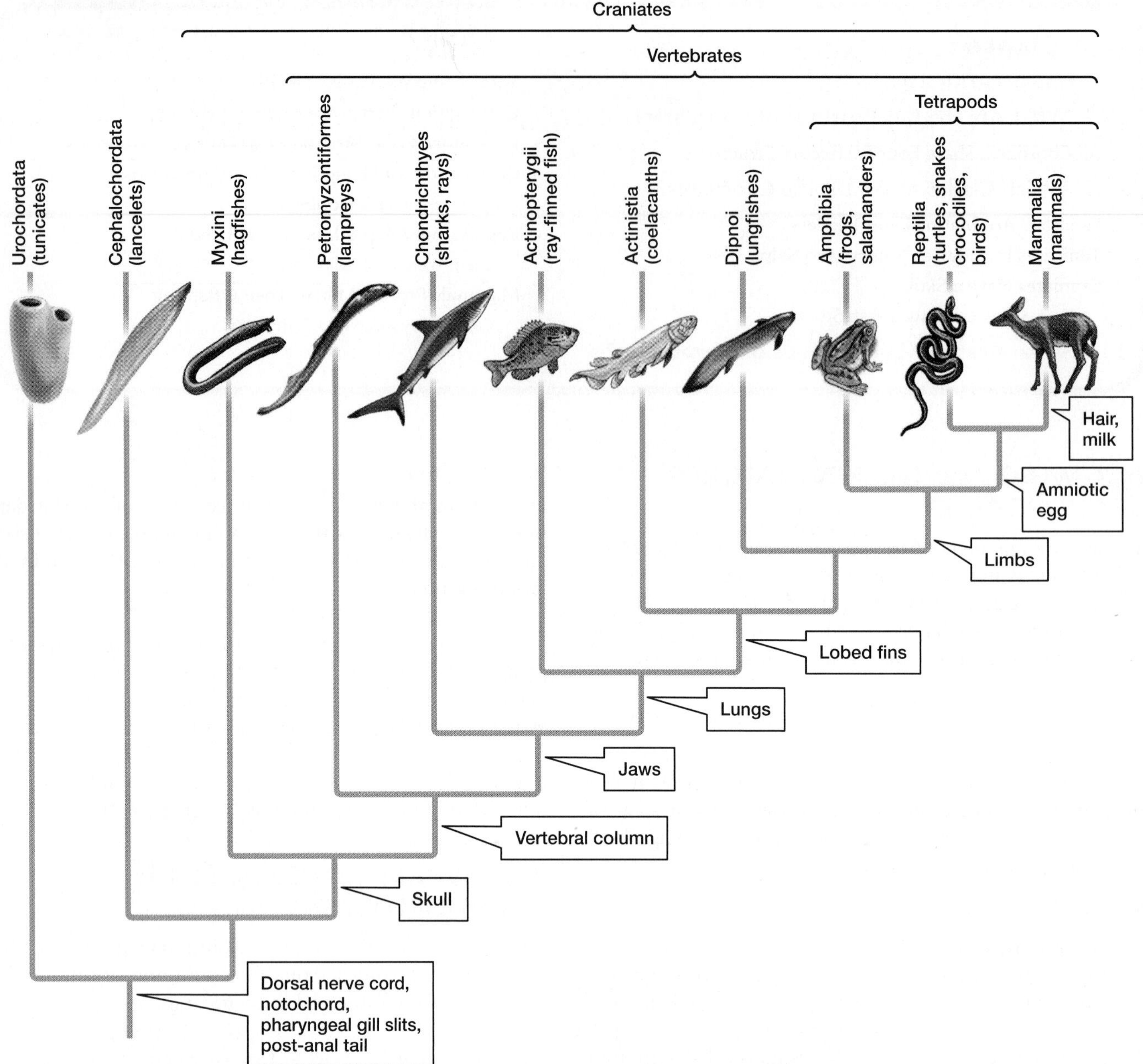

▲ **FIGURE 24-1 An evolutionary tree of the chordates** The talking boxes indicate the points at which key traits first appeared.

is occupied by its pharynx, which is like a basket perforated by gill slits and lined with mucus. Water enters the sea squirt's body through an *incurrent siphon,* passes into the pharyngeal basket at its top, moves through the gill slits, and exits the body through an *excurrent siphon.* Food particles are trapped in the basket's mucous lining.

Adult sea squirts are sessile—they live firmly attached to a surface. Their ability to move is limited to forceful contractions of their saclike bodies, which can send a jet of seawater into the face of anyone who plucks one from its undersea home; hence, the name sea squirt. Although adult sea squirts are immobile, their larvae swim actively and possess the four chordate features (see Fig. 24-3b, left). Some tunicates remain mobile throughout their lives. For example, barrel-shaped tunicates known as salps live in the open ocean and move by contracting an encircling band of muscle, which forces a jet of water out of the back of the animal and propels it forward.

Craniates Have a Skull

The **craniates** include all chordates that have a skull that encloses the brain. The skull may be composed of bone or **cartilage,** a tissue that resembles bone but is less brittle and more flexible. The earliest known craniates, whose fossils were found in 530-million-year-old rocks, resembled lancelets but had brains, skulls, and eyes. However, the mouths of the

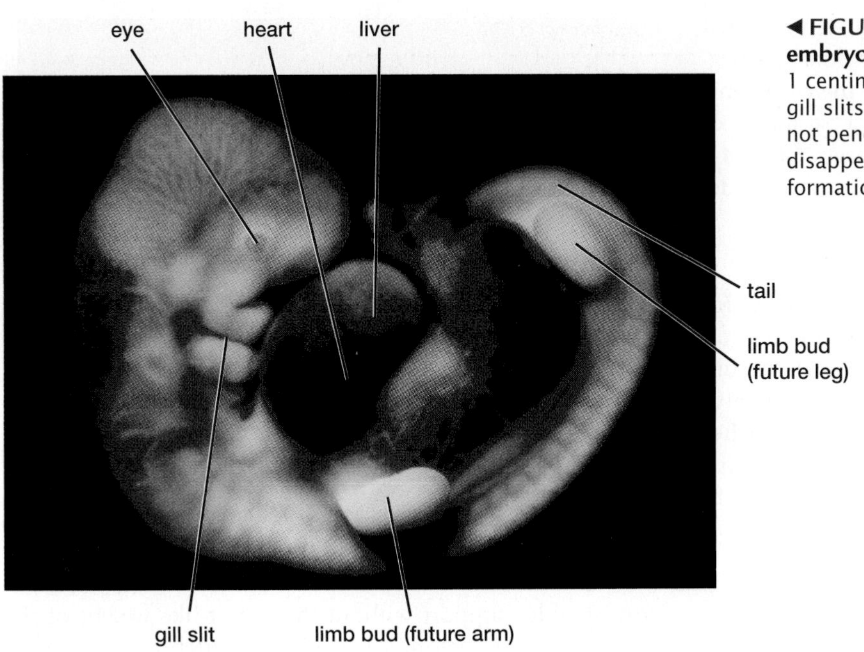

◀ **FIGURE 24-2 Chordate features in the human embryo** This 5-week-old human embryo is about 1 centimeter long and clearly shows a tail and external gill slits (more properly called grooves, since they do not penetrate the body wall). Although the tail will disappear completely, the gill grooves contribute to the formation of the lower jaw.

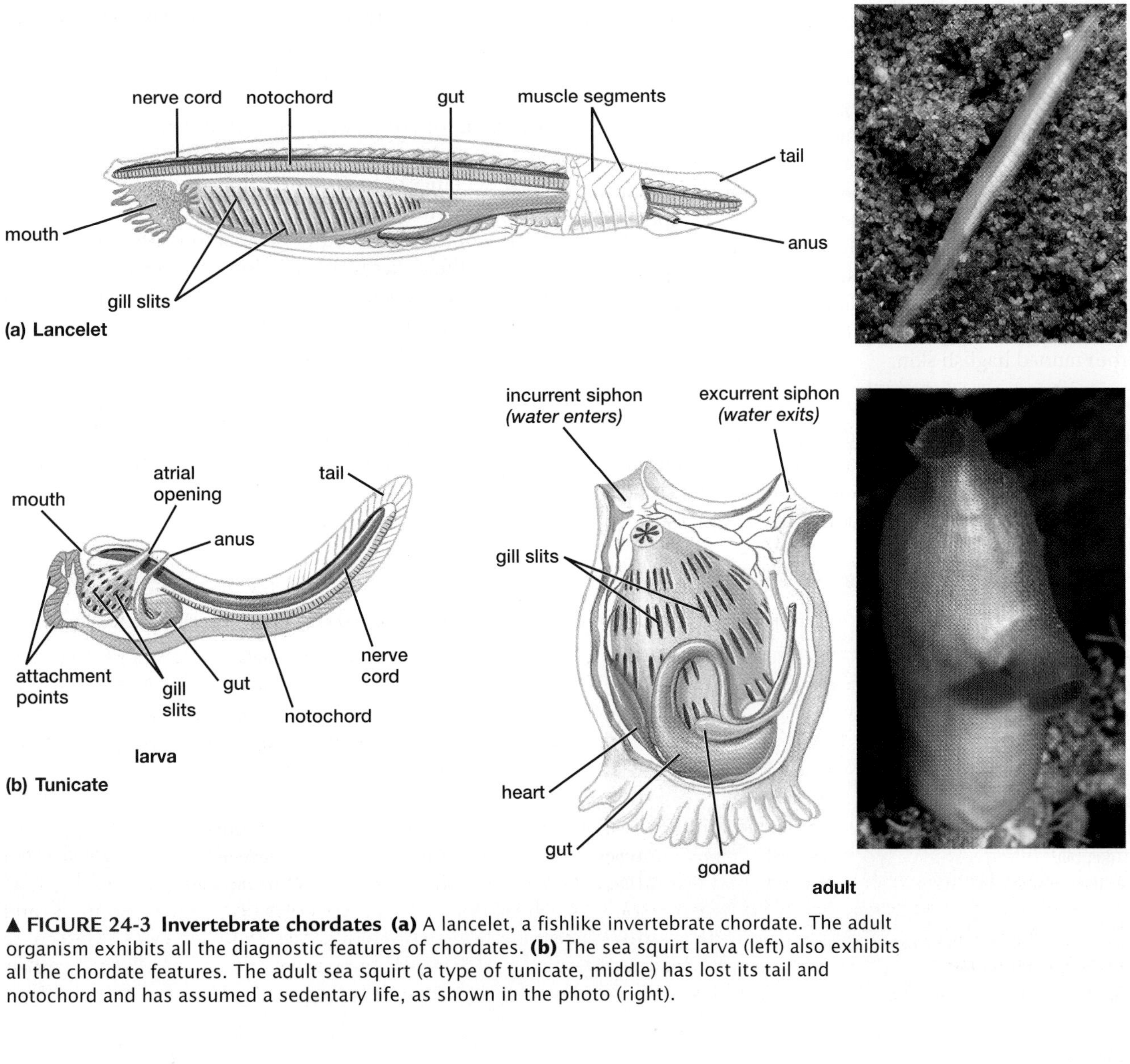

▲ **FIGURE 24-3 Invertebrate chordates (a)** A lancelet, a fishlike invertebrate chordate. The adult organism exhibits all the diagnostic features of chordates. **(b)** The sea squirt larva (left) also exhibits all the chordate features. The adult sea squirt (a type of tunicate, middle) has lost its tail and notochord and has assumed a sedentary life, as shown in the photo (right).

earliest craniates lacked jaws (movable skeletal elements that frame the mouth opening). Today, craniates include the hagfishes and the **vertebrates**—animals in which the embryonic notochord is replaced during development by a backbone, or **vertebral column,** composed of bone or cartilage. **Table 24-1** summarizes some characteristics of the craniate groups described in the rest of this chapter.

Hagfishes Are Slimy Residents of the Ocean Floor

As did ancestral craniates, hagfishes (Myxini) lack jaws. Instead, they use a tonguelike, tooth-bearing structure to grind and tear food. A hagfish body is stiffened by a notochord, but its skeleton is limited to a few small cartilaginous elements, one of which forms a rudimentary skull. Because hagfishes lack skeletal elements that surround the nerve cord to form a vertebral column, most systematists do not consider them to be vertebrates. Instead, hagfishes represent the craniate group that is most closely related to the vertebrates.

The 20 or so species of hagfishes are exclusively marine (**Fig. 24-4**). They respire using gills, have a two-chambered heart, and are ectothermic—that is, they depend on heat from the external environment to regulate their body temperature. (Gills, two-chambered hearts, and ectothermy are also found in all vertebrate fishes.) Hagfishes live near the ocean floor, often burrowing in the mud, and feed primarily on worms. They will, however, eagerly attack dead and dying fish, using their teeth to burrow into a fish's body and consume its soft internal organs.

Hagfishes are regarded with great disgust by some fishermen because they secrete massive quantities of slime as a defense against predators. Despite their well-deserved reputation as "slimeballs of the sea," hagfishes are avidly pursued by many commercial fishermen, because the leather industry in some parts of the world provides a market for hagfish skin. Most leather items that purport to be "eel skin" are in fact made from tanned hagfish skin.

▲ **FIGURE 24-4 Hagfishes** Hagfishes live in communal burrows in mud, feeding on worms.

Vertebrates Have a Backbone

The vertebral column of a vertebrate supports its body, provides attachment sites for muscles, and protects the delicate nerve cord and brain. It is also part of a living internal skeleton that can grow and repair itself. Because this internal skeleton provides support without the armor-like weight of the arthropod exoskeleton, it has allowed vertebrates to achieve great size and mobility.

The early history of vertebrates was characterized by an array of strange, now-extinct jawless fishes, many of which were protected by bony armor plates. About 425 million years ago, jawless fishes gave rise to a group of fish that possessed an important new structure: jaws. Jaws allowed fish to grasp, tear, or crush their food, permitting them to exploit a much wider range of food sources than could jawless fish. Today, most (but not all) vertebrates have jaws.

Vertebrates have other adaptations that have contributed to their successful invasion of most habitats. One such adaptation is paired appendages. These first appeared as fins in fish and served as stabilizers for swimming. Over millions of years, some fins were modified by natural selection into legs that allowed animals to crawl onto dry land, and later into wings that allowed some to take to the air. Another adaptation that has contributed to the success of vertebrates is an increase in the size and complexity of their brains and sensory structures, which allow vertebrates to perceive their environment in detail and to respond to it in a great variety of ways.

Table 24-1 Comparison of Craniate Groups

Group	Fertilization	Respiration	Heart Chambers	Body Temperature Regulation
Hagfishes (Myxini)	External	Gills	Two	Ectothermic
Lampreys (Petromyzontiformes)	External	Gills	Two	Ectothermic
Cartilaginous fishes (Chondrichthyes)	Internal	Gills	Two	Ectothermic
Ray-finned fishes (Actinopterygii)	External	Gills	Two	Ectothermic
Coelacanths (Actinistia)	Internal	Gills	Two	Ectothermic
Lungfishes (Dipnoi)	External	Gills and lungs	Two	Ectothermic
Amphibians (Amphibia)	External or internal[1]	Skin, gills, and lungs	Three	Ectothermic
Reptiles (Reptilia)	Internal	Lungs	Three[2]	Ectothermic[3]
Mammals (Mammalia)	Internal	Lungs	Four	Endothermic

[1]External in most frogs and toads; internal in caecilians and most salamanders.
[2]Except for birds and crocodilians, which have four chambers.
[3]Except for birds, which are endothermic.

Case Study continued
Fish Story

In the years since Courtney-Latimer's discovery that coelacanths are not extinct, scientists have had the opportunity to investigate the creature's anatomy. The body of the living coelacanth has some unusual features. For example, adult coelacanths retain a notochord, the body-stiffening rod that most other vertebrates lose during embryonic development. In addition, a coelacanth's brain is very small relative to its body size. The brain of a 90-pound (40-kilogram) coelacanth weighs only 1 or 2 grams (less than a tenth of an ounce). The tiny brain occupies less than 2% of the space in the cranial cavity; the rest is filled with fat. Although the evolution of large brains may have contributed to the spread and diversification of many vertebrate groups, coelacanths demonstrate that a small brain does not necessarily prevent evolutionary success.

24.3 WHAT ARE THE MAJOR GROUPS OF VERTEBRATES?

Today, vertebrates include lampreys, cartilaginous fishes, ray-finned fishes, coelacanths, lungfishes, amphibians, reptiles, and mammals.

Some Lampreys Parasitize Fish

Like hagfishes, lampreys (Petromyzontiformes) are jawless. A lamprey is recognizable by the large, rounded sucker that surrounds its mouth and by the single nostril on the top of its head. The nerve cord of a lamprey is protected by segments of cartilage, so lampreys are considered to be true vertebrates. They live in both fresh and salt waters, but the marine forms must return to fresh water to spawn.

Some lamprey species are parasitic. A parasitic lamprey uses its tooth-lined mouth to attach itself to a larger fish (**Fig. 24-5**).

▲ **FIGURE 24-5 Lampreys** Some lampreys are parasitic, attaching to fish (such as this carp) with sucker-like mouths lined with rasping teeth (inset).

Using rasping teeth on its tongue, the lamprey excavates a hole in the host's body wall, through which it sucks blood and body fluids. Beginning in the 1920s, lampreys spread into the Great Lakes. There, in the absence of effective predators, they have multiplied prodigiously and greatly reduced commercial fish populations, including the lake trout. Vigorous measures to control the lamprey population have allowed some recovery of the other fish populations of the Great Lakes.

Cartilaginous Fishes Are Marine Predators

The cartilaginous fishes (Chondrichthyes) include 625 marine species, among them the sharks, skates, and rays (**Fig. 24-6**). Unlike hagfishes and lampreys (but like all other vertebrates), cartilaginous fishes have jaws. They are graceful predators whose skeleton is formed entirely of cartilage. Their bodies

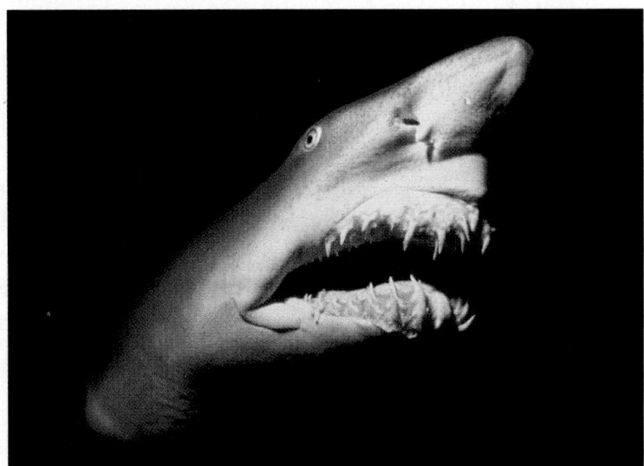

(a) Shark

(b) Ray

▲ **FIGURE 24-6 Cartilaginous fishes (a)** A sand tiger shark displaying several rows of teeth. As the frontmost teeth are lost, they are replaced by the new ones behind them. Both sharks and rays lack a swim bladder and tend to sink toward the bottom when they stop swimming. **(b)** The tropical blue-spotted stingray swims by graceful undulations of lateral extensions of the body.

are protected by leathery skin roughened by tiny scales. Although some must swim to circulate water through their gills, most can pump water across their gills. In contrast to the external fertilization that characterizes reproduction in almost all other fish, cartilaginous fish have internal fertilization, in which a male deposits sperm directly into a female's reproductive tract. Some cartilaginous fishes are very large. A whale shark, for example, can grow to more than 45 feet (14 meters) in length, and a manta ray may be more than 20 feet (6 meters) wide.

Although some sharks feed by filtering plankton (tiny animals and protists) from the water, most are predators of larger prey such as other fishes, marine mammals, sea turtles, crabs, or squid. Many sharks attack their prey with strong jaws that contain several rows of razor-sharp teeth; the back rows move forward as the front teeth are lost to age and use (see Fig. 24-6a).

Most sharks avoid humans, but large sharks of some species can be dangerous to swimmers and divers. However, shark attacks on people are rare. A U.S. resident is 30 times more likely to die from a lightning strike than from a shark attack, and a beachgoer is far more likely to drown than to be bitten by a shark. Nonetheless, shark attacks do occur. During 2008, for example, there were 59 documented attacks in the world, 4 of them fatal.

Skates and rays are mostly bottom dwellers with flattened bodies, wing-shaped fins, and thin tails (see Fig. 24-6b). Most skates and rays eat invertebrates. Some species defend themselves with a spine near their tail that can inflict dangerous wounds, and others produce a powerful electrical shock that can stun their prey.

Ray-Finned Fishes Are the Most Diverse Vertebrates

Just as our size bias makes us overlook the most diverse invertebrate groups, our habitat bias makes us overlook the most diverse vertebrates. The most diverse and abundant vertebrates are not the birds or the predominantly terrestrial mammals. The vertebrate diversity crown belongs instead to the lords of the oceans and fresh water, the ray-finned fishes (Actinopterygii). About 24,000 species have been identified, and scientists estimate that perhaps twice this number exist, with many undiscovered species inhabiting deep waters and remote areas. Ray-finned fishes are found in nearly every watery habitat, both freshwater and marine.

Ray-finned fishes are distinguished by the structure of their fins, which are formed by webs of skin supported by bony spines. In addition, ray-finned fishes have skeletons made of bone, a trait they share with the lobe-finned fishes and limbed vertebrates discussed later in this chapter. The skin of ray-finned fishes is covered with interlocking scales that provide protection while allowing for flexibility. Most ray-finned fishes have a swim bladder, a sort of internal balloon that allows a fish to float effortlessly at any level in the water. The swim bladder evolved from lungs, which were present (along with gills) in the ancestors of modern ray-finned fishes.

The ray-finned fishes include not only a large number of species but also a huge variety of different forms and lifestyles (**Fig. 24-7**). These range from snakelike eels to flattened flounders; from sluggish bottom feeders that probe the seafloor to speedy, streamlined predators that range in open water; from brightly colored reef dwellers to translucent,

▼ **FIGURE 24-7 The diversity of ray-finned fishes** Ray-finned fishes have colonized nearly every aquatic habitat. **(a)** This female deep-sea anglerfish attracts prey with a living lure that projects just above her mouth. The fish is ghostly white; at the 6,000-foot (1,800-meter) depth where anglers live, no light penetrates and thus colors are unnecessary. Male deep-sea anglerfish are extremely small and remain attached to the female as permanent parasites, always available to fertilize her eggs. Two parasitic males can be seen attached to this female. **(b)** This tropical green moray eel lives in rocky crevices. The small fish (a banded cleaner goby) on its lower jaw eats parasites that cling to the moray's skin. **(c)** The tropical sea horse may anchor itself with its prehensile tail (adapted for grasping) while feeding on small crustaceans.

QUESTION With regard to water regulation (maintaining the proper amount of water in the body), how does the challenge faced by a freshwater fish differ from that faced by a saltwater fish?

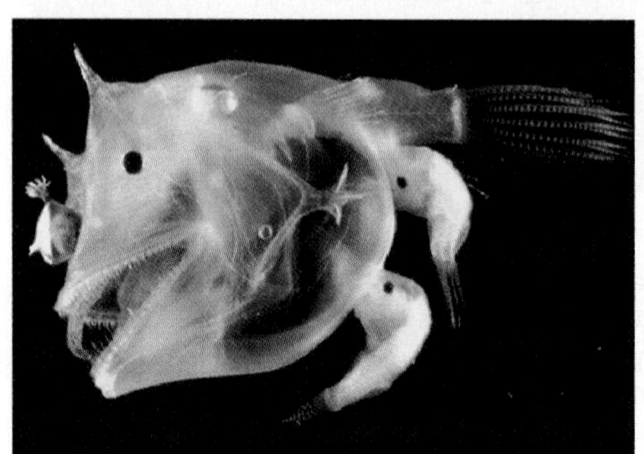

(a) Anglerfish

(b) Moray eel

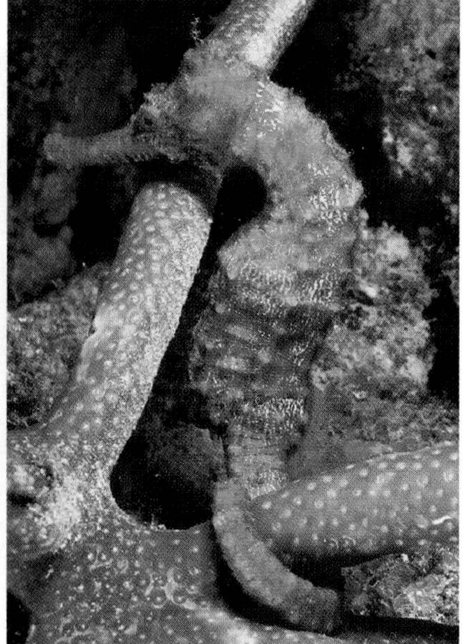

(c) Sea horse

luminescent deep-sea dwellers; from the massive 3,000-pound (1,350-kilogram) mola to the tiny stout infantfish, which weighs in at about 0.00003 ounce (1 milligram).

Ray-finned fishes are an extremely important source of food for humans. Unfortunately, however, our appetite for ray-finned fishes, combined with increasingly effective high-tech methods for finding and catching them, has had a devastating impact on fish populations. Populations of almost all economically important ray-finned fish species have declined drastically. Large, predatory fish such as tuna and cod have been especially hard hit; today's populations of these species contain less than 10% of the number present before commercial fishing began. If overfishing continues, fish stocks are likely to collapse. The solution to this problem—catch fewer fish—is simple in concept but very difficult to implement in practice, due to economic and political factors.

Coelacanths and Lungfishes Have Lobed Fins

Although almost all fish with bony skeletons belong to the ray-finned group, some bony fishes are members of two other groups, the coelacanths (Actinistia) and the lungfishes (Dipnoi). Coelacanths are discussed at length in this chapter's Case Study (see the chapter-opening photo). The six species of lungfishes are found in freshwater habitats in Africa, South America, and Australia (**Fig. 24-8a**). Lungfishes have both gills and lungs. They tend to live in stagnant waters that may be low in oxygen, and their lungs allow them to supplement their supply of oxygen by breathing air directly. Lungfishes of several species are able to survive even if the pools they inhabit dry up completely. These fish burrow into mud and seal themselves in mucus-lined chambers (**Fig. 24-8b**). There, they breathe through their lungs as their metabolic rate declines drastically. When the rains return and the pools refill, the lungfishes leave their burrows and resume their normal underwater way of life.

Lungfishes and coelacanths are sometimes called *lobefins*, because members of both groups have fleshy fins that contain rod-shaped bones surrounded by a thick layer of muscle. This shared trait is indicative of the groups' shared ancestry, though the two lineages have been evolving separately for hundreds of millions of years.

In addition to the coelacanths and lungfishes, several other lineages of lobefins arose early in the evolutionary history of jawed fish. Members of one of these other lineages evolved modified fleshy fins that, in an emergency, could be used as legs, allowing the fish to drag itself from a drying puddle to a deeper pool. This lineage left descendents that survive today. These survivors are the tetrapods: amphibians, reptiles, and mammals. Instead of fins, tetrapods have limbs that can support their weight on land and digits (fingers or toes) on the ends of those limbs.

Amphibians Live a Double Life

The early tetrapods that made the first vertebrate invasion of the land were amphibians. Today, the 6,300 species of amphibians (Amphibia) straddle the boundary between aquatic and terrestrial existence (**Fig. 24-9**). The limbs of amphibians show varying degrees of adaptation to movement on land, from the belly-dragging crawl of salamanders to the long leaps of frogs. A three-chambered heart (in contrast to the two-chambered heart of fishes) circulates blood more efficiently, and lungs replace gills in most adult forms. Amphibian lungs, however, are poorly developed and must be supplemented by the skin, which serves as an additional respiratory organ. This respiratory function requires that the skin remain moist, a constraint that greatly restricts the range of amphibian habitats on land.

Many amphibians are also tied to moist habitats by their breeding behavior, which requires water. For example, as in most fishes, fertilization in frogs and toads is generally external and takes place in water, where the sperm can swim to the eggs. The eggs must remain moist, because they are protected only by a jelly-like coating that leaves them vulnerable to water loss by evaporation. Different amphibian species keep their eggs moist in different ways, but many species simply lay

(a) Lungfish

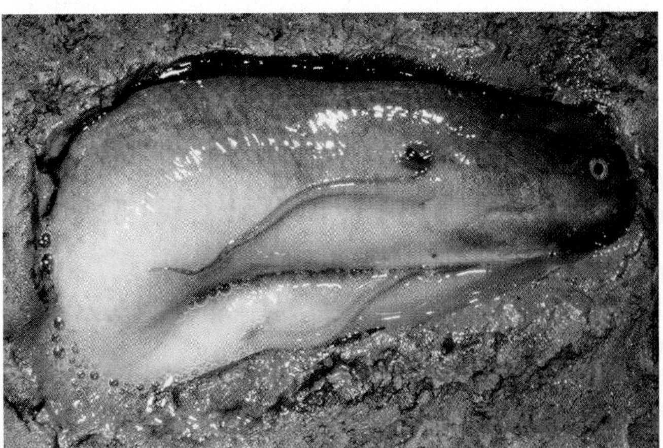

(b) Sealed in a burrow

▲ FIGURE 24-8 **Lungfishes are lobe-finned fish** Among the fishes, **(a)** lungfishes are the group most closely related to land-dwelling vertebrates. **(b)** Lungfishes may wait out dry periods sealed in burrows in the mud.

▲ **FIGURE 24-9 "Amphibian" means "double life"** The double life of amphibians is illustrated by the bullfrog's transition from **(a)** a completely aquatic larval tadpole to **(b)** an adult leading a semiterrestrial life. **(c)** The red salamander is restricted to moist habitats in the eastern United States. **(d)** Caecilians are legless, mostly burrowing amphibians.

QUESTION What advantages might amphibians gain from their "double life"?

their eggs in water. In some amphibian species, fertilized eggs develop into aquatic larvae such as the tadpoles of some frogs and toads. These aquatic larvae undergo a dramatic transformation into semiterrestrial adults, a metamorphosis that gives the amphibians their name, which means "double life." Their double life and their thin, permeable skin have made amphibians particularly vulnerable to pollutants and to environmental degradation, as described in "Earth Watch: Frogs in Peril" on p. 456–457.

Frogs and Toads Are Adapted for Jumping

The frogs and toads, with 5,600 species, are the most diverse group of amphibians. Adult frogs and toads move about by hopping and leaping, and their bodies are well adapted for this mode of locomotion. Their hind legs are quite long relative to their body size (much longer than their forelegs), and they do not have a tail. The names "frog" and "toad" do not describe distinct evolutionary groups, but are instead used informally to distinguish two combinations of characteristics that are common among members of this branch of the amphibians. In general, frogs have smooth, moist skin, live in or near water, and have long hind limbs suitable for leaping; toads have bumpy, drier skin, live on land, and have shorter hind limbs suitable for hopping. Many frogs and toads (and other amphibians) contain toxic substances that make them distasteful to predators. In a few species, such the golden poison dart frog of South America (see Fig. 27-7), the protective chemical is extremely toxic. The toxin from a single golden poison dart frog could kill several adult humans.

Most Salamanders Have Tails

Most salamanders have a lizard-like body: slender, with four legs of roughly equal size and a long tail (**Fig. 24-9c**). Some salamanders, however, have only small, vestigial legs; members of these species may have an eel-like appearance. Most of the roughly 550 species of salamanders live on land, often in moist, protected places, such as beneath rocks or logs on a forest floor.

(a) Snake

(b) Alligator

(c) Tortoise

▲ FIGURE 24-10 **The diversity of reptiles (other than birds)** **(a)** This scarlet king snake has a color pattern very similar to that of the poisonous coral snake, which potential predators avoid. This mimicry helps the harmless king snake elude predation. **(b)** The outward appearance of the American alligator, found in swampy areas of the South, is almost identical to that of 150-million-year-old fossil alligators. **(c)** The tortoises of the Galápagos Islands, Ecuador, may live to be more than 100 years old.

But members of some species are fully aquatic and spend their entire lives in the water. Even members of land-dwelling species generally move to ponds or streams to breed. In almost all salamander species, eggs hatch into aquatic larvae that use external gills to breathe. In some species, the larvae do not metamorphose, but instead retain the larval form even as adults.

Alone among vertebrates, salamanders can regenerate lost limbs. This ability has attracted the attention of researchers interested in regenerative medicine, which seeks treatments that enable human bodies to repair or regenerate damaged tissues and organs. The researchers hope that increased understanding of the mechanisms by which salamanders regenerate limbs will lead to effective treatments for humans.

Caecilians Are Limbless, Burrowing Amphibians

The caecilians form a small (175 species) group of legless amphibians that live in tropical regions. At first glance, a caecilian's appearance is reminiscent of an earthworm, though the larger species, which can be up to 5 feet (1.5 meters) long, might be mistaken for a snake (**Fig. 24-9d**). Most caecilians are burrowing animals that live underground, though a few species are aquatic. Caecilians have eyes, but they are very small and often covered by skin. As a result, caecilian vision is probably limited to detecting light.

Reptiles Are Adapted for Life on Land

The reptiles (Reptilia) include lizards, snakes, alligators, crocodiles, turtles, and birds (**Fig. 24-10**). Reptiles evolved from an amphibian ancestor about 250 million years ago. One group of early reptiles, the dinosaurs, ruled the land for nearly 150 million years.

Reptiles Haves Scales and Shelled Eggs

Some reptiles, particularly desert dwellers such as tortoises and lizards, are completely independent of their aquatic origins. They achieved this independence through a series of adaptations, three of which are especially notable: (1) Reptiles evolved a tough, scaly skin that resists water loss and protects the body. (2) Reptiles evolved internal fertilization, in which the male deposits sperm within the female's body. (3) Reptiles evolved a shelled **amniotic egg,** which can be buried in sand or dirt, far from water. The shell prevents the egg from drying out on land. An internal membrane, the **amnion,** encloses the embryo in the watery environment that all developing animals require (**Fig. 24-11**).

In addition to these features, reptiles have more efficient lungs than do amphibians, and do not use their skin as a respiratory organ. Reptile circulatory systems include three-chambered or (in birds, alligators, and crocodiles) four-chambered hearts that segregate oxygenated and deoxygenated blood more effectively than do amphibian hearts. The reptile skeleton includes features that provide better support and more efficient movement on land than are provided by the amphibian skeleton.

▲ FIGURE 24-11 **The amniotic egg** A crocodile struggles free of its egg. The amniotic egg encapsulates the developing embryo in a fluid-filled membrane (the amnion), ensuring that development occurs in a watery environment, even if the egg is far from water.

Earth Watch

Frogs in Peril

Frogs and toads have lived in Earth's ponds and swamps for nearly 150 million years, somehow surviving the Cretaceous catastrophe that extinguished the dinosaurs and so many other species about 65 million years ago. Their evolutionary longevity, however, doesn't protect them from the environmental changes wrought by human activities. During the past two decades, herpetologists (biologists who study reptiles and amphibians) from around the world have documented an alarming decline in amphibian populations. Thousands of species of frogs, toads, and salamanders are dramatically decreasing in number, and many have gone extinct.

This is a worldwide phenomenon; population crashes have been reported from every part of the globe **(Fig. E24-1)**. Yosemite toads and yellow-legged frogs are disappearing from the mountains of California. Tiger salamanders have been nearly wiped out in the Colorado Rockies. Leopard frogs, formerly abundant throughout North America, are becoming rare in the United States. Logging and development have destroyed the habitats of amphibians from the Pacific Northwest to the tropics, but even amphibians in protected areas are dying. In the Cape Peninsula National Park in South Africa, the only remaining population of Rose's ghost frog has shrunk dramatically and the species is now critically endangered. The gastric brooding frog of Australia fascinated biologists by swallowing its eggs, brooding them in its stomach, and later regurgitating fully formed offspring. This species was abundant and seemed safe in a national park. Suddenly, in 1980, the gastric brooding frog disappeared and hasn't been seen since.

The causes of the worldwide decline in amphibian diversity are not fully understood, but researchers have discovered that frogs and toads in many places are succumbing to infection by a pathogenic fungus. The fungus has been found in the skin of dead and dying frogs in widespread locations, including Australia, Central America, and the western United States. In those places, discovery of the fungus has coincided with massive frog and toad die-offs, and most herpetologists agree that the fungus is causing the deaths.

It seems unlikely, however, that the fungus alone is responsible for the worldwide decline of amphibians. For one thing, die-offs have occurred in many places where the fungus has not been found. In addition, many herpetologists believe that the fungal epidemic would not have arisen if the frogs and toads had not first been weakened by other stresses. So, if the fungus is not doing all of the damage on its own, what are the other possible causes of amphibian decline? All of the most likely causes stem from human modification of the biosphere—the portion of Earth that sustains life.

Habitat destruction, especially the draining of wetlands that are hospitable to amphibian life, is one major cause of the decline. Amphibians are also vulnerable to toxic substances in the environment. For example, researchers found that frogs exposed to trace amounts of atrazine, a widely used herbicide that is found in virtually all fresh water in the United States, suffered severe damage to their reproductive tissues. The unique biology of amphibians makes them especially susceptible to poisons in the environment. Amphibian bodies at all stages of life are protected only by a thin, permeable skin that pollutants can easily penetrate. To make matters worse, the double life of many amphibians exposes their permeable skin to a wide range of aquatic

Lizards and Snakes Share a Common Evolutionary Heritage

Lizards and snakes together form a distinct lineage containing about 6,800 species. The common ancestor of snakes and lizards had limbs, which are retained by most lizards but have been lost in snakes. The limbed ancestry of snakes is revealed by remnants of hind limb bones that are present in some snake species.

Most lizards are small predators that eat insects or other small invertebrates, but a few lizard species are quite large. The Komodo dragon, for example, can reach 10 feet (3 meters) in length and weigh more than 200 pounds (90 kilograms). These giant lizards live in Indonesia and have powerful jaws and inch-long teeth that enable them to prey on large animals including deer, goats, and pigs. The Komodo dragon, however, does not rely on its teeth alone to kill its prey. It also produces a potent venom that flows from a gland in the Komodo dragon's jaw into the wound of a bitten victim. If an animal bitten by a Komodo dragon is not immediately killed, the venom ensures that it will likely die soon after the attack. The lizard simply waits patiently until its wounded, poisoned prey dies.

Most snakes are active, predatory carnivores and have a variety of adaptations that help them acquire food. For example, many snakes have special sense organs that help track prey by sensing small temperature differences between a prey's body and its surroundings. Some snake species immobilize prey with venom that is delivered through hollow teeth. Snakes also have a distinctive jaw joint that allows the jaws to distend so that the snake can swallow prey much larger than its head.

Alligators and Crocodiles Are Adapted for Life in Water

Crocodilians, as the 21 species of alligators and crocodiles are collectively known, are found in coastal and inland waters of the warmer regions of Earth. They are well adapted to an aquatic lifestyle, with eyes and nostrils located high on their heads so that they are able to remain submerged for long periods with only the uppermost portion of the head above the water's surface. Crocodilians have strong jaws and conical teeth that they use to crush and kill the fish, birds, mammals, turtles, and amphibians that they eat.

and terrestrial habitats and to a correspondingly wide range of environmental toxins.

Amphibian eggs can also be damaged by ultraviolet (UV) light, according to research by Andrew Blaustein, an ecologist at Oregon State University. Blaustein demonstrated that the eggs of some species of frogs in the Pacific Northwest are sensitive to damage from UV light and that the most sensitive species are experiencing the most drastic declines. Unfortunately, many parts of Earth are subject to intense UV radiation levels, because atmospheric pollutants have caused a thinning of the protective ozone layer.

Another disturbing trend among frogs and toads is the increasing incidence of grotesquely deformed individuals. Researchers at the Environmental Protection Agency recently demonstrated that developing frogs exposed to current natural levels of UV light grow deformed limbs much more frequently than do frogs protected from UV radiation. Other researchers have shown that deformities are more common in frogs exposed to low concentrations of commonly used pesticides. In addition, a growing body of evidence suggests that some deformities—especially the most common one, extra limbs—are caused by parasitic infections during embryonic development. Many frogs with extra legs are infested with a parasitic flatworm, and researchers have shown that tadpoles experimentally infected with the flatworm developed into deformed adults. Why have the parasites, which have long coexisted with frogs, suddenly begun causing so many deformities? A likely explanation is that exposure to UV radiation, pesticides, and herbicides weakens frogs' immune response. A compromised immune system leaves a developing tadpole more vulnerable to parasitic infection.

Many scientists believe that the troubles of amphibians signal an overall deterioration of Earth's ability to support life. According to this line of reasoning, the highly sensitive amphibians are providing an early warning of environmental degradation that will eventually affect more resistant organisms as well. Equally worrisome is the observation that amphibians are not just sensitive indicators of the health of the biosphere but also crucial components of many ecosystems. They may keep insect populations in check, in turn serving as food for larger carnivores. Their decline will further disrupt the balance of these delicate communities.

▲ **FIGURE E24-1 Amphibians in danger** The corroboree toad is rapidly declining in its native Australia. The thin water-permeable and gas-permeable skin of frogs and toads (and the jelly-like coating around their eggs) make them vulnerable to both air and water pollutants.

Parental care is extensive in crocodilians, which bury their eggs in mud nests. Parents guard the nest until the young hatch and then carry their newly hatched young in their mouths, moving them to safety in the water. Young crocodilians may remain with their mother for several years.

Turtles Have Protective Shells

The 240 species of turtles occupy a variety of habitats, including deserts, streams and ponds, and the ocean. This variety of habitats has fostered a variety of adaptations, but all turtles are protected by a hard, boxlike shell that is fused to the vertebrae, ribs, and collarbone. Turtles have no teeth but have instead evolved a horny beak. The beak is used to eat a variety of foods; some turtles are carnivores, some are herbivores, and some are scavengers. The largest turtle, the leatherback, is an ocean dweller that can grow to 6 feet (2 meters) or more in length and feeds largely on jellyfish. Leatherbacks and other marine turtles must return to land to breed, and often undertake extraordinary long-distance migrations to reach the beaches on which they bury their eggs.

Birds Are Feathered Reptiles

One very distinctive group of reptiles is the birds (**Fig. 24-12**). Although the 9,600 species of birds have traditionally been classified as a group separate from reptiles, biologists have shown that birds are really a subset of the reptiles (see p. 349 for a more complete explanation). The first birds appear in the fossil record roughly 150 million years ago (**Fig. 24-13**) and are distinguished from other reptiles by feathers, which are essentially a highly specialized version of reptilian body scales. Modern birds retain scales on their legs—evidence of the ancestry they share with the rest of the reptiles.

Bird anatomy and physiology are dominated by adaptations that help them fly. In particular, birds are exceptionally light for their size. Lightweight bones reduce the weight of the bird skeleton, and many bones present in other reptiles have been lost in the course of evolution or fused with other bones. Bird reproductive organs shrink considerably during nonbreeding periods, and female birds possess only a single ovary, further minimizing weight. Feathers serve as lightweight extensions of the wing and the tail surfaces that provide the lift and

(a) Hummingbird

(b) Frigate bird

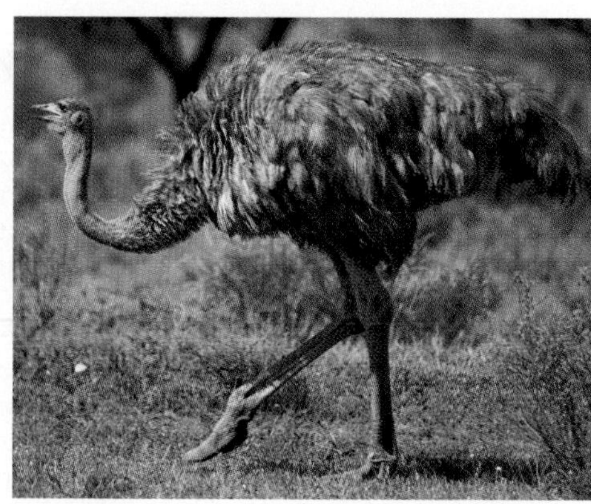
(c) Ostrich

▲ FIGURE 24-12 **The diversity of birds** **(a)** The delicate hummingbird beats its wings about 60 times per second and weighs about 0.15 ounce (4 grams). **(b)** This young frigate bird, a fish-eater from the Galápagos Islands, has nearly outgrown its nest. **(c)** The ostrich, the largest of all birds, weighs more than 300 pounds (135 kilograms); its eggs weigh more than 3 pounds (1,500 grams).

QUESTION Although the ancestor of all birds could fly, many bird species—such as the ostrich—cannot. Why do you suppose flightlessness has evolved repeatedly among birds?

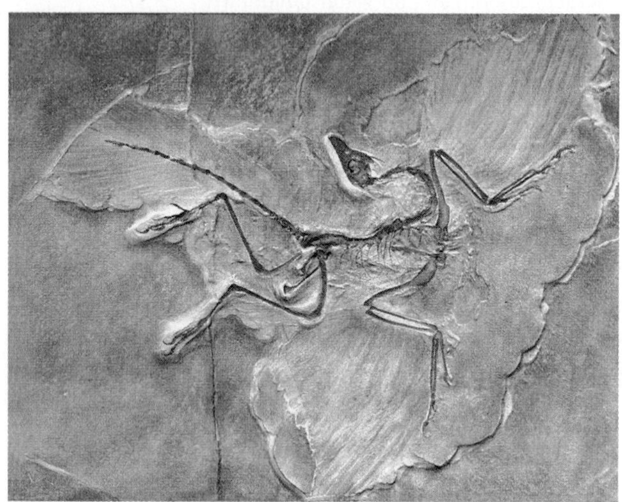

▲ FIGURE 24-13 **Archaeopteryx, the earliest-known bird** *Archaeopteryx* is preserved in 150-million-year-old limestone. Feathers, a feature unique to birds, are clearly visible; but traits characteristic of birds' ancestors are also apparent: unlike a modern bird, *Archaeopteryx* had teeth, a tail, and claws on its forelimbs.

control required for flight; feathers also provide lightweight protection and insulation for the body. The nervous system of birds accommodates the special demands of flight by providing extraordinary coordination and balance, combined with acute eyesight.

Birds are also able to maintain body temperatures high enough to allow their muscles and metabolic processes to operate at peak efficiency, supplying the power to fly regardless of the temperature of the external environment. This physiological ability to maintain an internal temperature that is usually higher than that of the surrounding environment is characteristic of both birds and mammals, which are therefore sometimes described as warm blooded or endothermic. In contrast, the body temperature of ectothermic (cold-blooded) animals—invertebrates, fish, amphibians, and reptiles other than birds—varies with the temperature of their environment, though these animals may exert some control of their body temperature by their behavior (such as basking in the sun or seeking shade).

Warm-blooded animals such as birds have a high metabolic rate, which increases their demand for energy and requires efficient oxygenation of tissues. Therefore, birds must eat frequently and possess circulatory and respiratory adaptations that help meet the need for efficiency. A bird's four-chambered heart prevents mixing of oxygenated and deoxygenated blood. The respiratory system of birds is supplemented by air sacs that provide a continuous supply of oxygenated air to the lungs, even while the bird exhales.

Have you ever wondered 

Which Vertebrates Have Gone into Space?

Humans have walked on the moon and now travel almost routinely into orbit around Earth. But before a person was first launched into space, other animals were sent there, to make sure that a vertebrate body could survive such journeys. In the 1940s and 1950s, rhesus monkeys, mice, and dogs were launched into space as passengers aboard missiles. The first living thing to orbit Earth was a dog named Laika, who was aboard the Soviet satellite *Sputnik 2* during its historic journey in November 1957. In the decades since, the roster of space-faring vertebrates has grown to include dozens of dogs, monkeys, and mice, as well as cats, rats, chimpanzees, and various species of tortoises, frogs, fish, and salamanders. Many of these animal travelers, especially in the early days of space travel, did not survive the experience. Without their sacrifices, however, human space travel would not have become possible.

Mammals Provide Milk to Their Offspring

One branch of the tetrapod evolutionary tree gave rise to a group that evolved hair and diverged to form the mammals (Mammalia). The mammals first appeared approximately 250 million years ago but did not diversify and become prominent on land until after the dinosaurs went extinct roughly 65 million years ago. In most mammals, fur protects and insulates the warm body. Like birds, alligators, and crocodiles, mammals have four-chambered hearts that increase the amount of oxygen delivered to the tissues. Legs that evolved for running rather than crawling make many mammals fast and agile.

Mammals are named for the milk-producing **mammary glands** used by all female mammals to suckle their young. In addition to these unique glands, the mammalian body has sweat, scent, and sebaceous (oil-producing) glands, none of which is found in other vertebrates. The mammalian nervous system has contributed significantly to the success of the mammals by making possible behavioral responses to changing and varied environments. The mammalian brain is more highly developed than in any other vertebrate group, giving mammals unparalleled curiosity and learning ability. Their highly developed brain allows mammals to alter their behavior on the basis of experience and helps them survive in a changing environment. Relatively long periods of parental care after birth allow some mammals to learn extensively under parental guidance. Humans and other primates are exceptional examples. In fact, the large brains of humans have been the major factor leading to human domination of Earth.

The 4,600 species of mammals include three main evolutionary lineages: monotremes, marsupials, and placental mammals.

Monotremes Are Egg-Laying Mammals

Unlike other mammals, **monotremes** lay eggs rather than giving birth to live young. This group includes only three species: the platypus and two species of spiny anteaters, also known as echidnas (**Fig. 24-14**). Monotremes are found only in Australia (the platypus and short-beaked echidna) and New Guinea (the long-beaked echidna).

Echidnas are terrestrial and eat insects or earthworms that they dig out of the ground. Platypuses forage for food in the water, diving below the surface to capture small vertebrate and invertebrate animals. Platypus bodies are well adapted for this aquatic lifestyle, with a streamlined shape, webbed feet, a broad tail, and a fleshy bill that is used to probe for food.

Monotreme eggs have leathery shells and are incubated for 10 to 12 days by the mother. Echidnas have a special pouch for incubating eggs, but platypus eggs are held for incubation between the mother's tail and belly. Newly hatched monotremes are tiny and helpless and feed on milk secreted by the mother. Monotremes, however, lack nipples. Milk from the mammary glands oozes through ducts on the mother's abdomen and soaks the fur around the ducts. The young then suck the milk from the fur.

Marsupial Diversity Reaches Its Peak in Australia

In all mammals except the monotremes, embryos develop in the uterus, a muscular organ in the female reproductive tract. The lining of the uterus combines with membranes derived from the embryo to form the **placenta,** a structure that allows gases, nutrients, and wastes to be exchanged between the circulatory systems of the mother and embryo.

In **marsupials,** embryos develop in the uterus for only a short period. Marsupial young are born at a very immature stage of development. Immediately after birth, a marsupial crawls to a nipple, firmly grasps it, and, nourished by milk, completes its development. In most, but not all, marsupial species, this post-birth development takes place in a protective pouch.

Only one marsupial species, the Virginia opossum, is native to North America. The majority of the 275 species of marsupials are found in Australia, where marsupials such as kangaroos have come to be seen as emblematic of the island

(a) Platypus

(b) Spiny anteater

▲ **FIGURE 24-14 Monotremes (a)** Monotremes, such as this platypus, lay leathery eggs resembling those of reptiles. Platypuses live in burrows that they dig in the banks of rivers, lakes, or streams. **(b)** The short limbs and heavy claws of spiny anteaters (also known as echidnas) help them unearth insects and earthworms to eat. The stiff spines that cover a spiny anteater's body are modified hairs.

(a) Wallaby

(b) Wombat

(c) Tasmanian devil

▲ **FIGURE 24-15 Marsupials (a)** Marsupials, such as the wallaby, give birth to extremely immature young who develop within the mother's protective pouch. **(b)** The wombat is a burrowing marsupial whose pouch opens toward the rear of its body to prevent dirt and debris from entering the pouch during tunnel digging. One of the wombat's predators is **(c)** the Tasmanian devil, the largest carnivorous marsupial.

(a) Whale

(b) Bat

(c) Cheetah

(d) Orangutan

◀ **FIGURE 24-16 The diversity of placental mammals (a)** A humpback whale gives its offspring a boost. **(b)** A bat, the only type of mammal capable of true flight, navigates at night by using a kind of sonar. Large ears help the animal detect echoes as its high-pitched cries bounce off nearby objects. **(c)** Mammals are named after the mammary glands with which females nurse their young, as illustrated by this mother cheetah. **(d)** Orangutans are gentle, intelligent apes that occupy swamp forests in limited areas of the Tropics but are endangered by hunting and habitat destruction.

continent. Kangaroos are the largest and most conspicuous of Australia's marsupials; the largest species, the red kangaroo, may be 7 feet tall (about 2 meters) and can make 30-foot (9-meter) leaps when moving at top speed. Though kangaroos are perhaps the most familiar marsupials, the group encompasses species with a range of sizes, shapes, and lifestyles, including koalas, wombats, and the Tasmanian devil (**Fig. 24-15**).

Placental Mammals Inhabit Land, Air, and Sea

Most mammal species are **placental** mammals, so named because their placentas are far more complex than those of marsupials. Compared to marsupials, placental mammals retain their young in the uterus for a much longer period, so that offspring complete their embryonic development before being born.

The placental mammals have evolved a remarkable diversity of form. The bat, mole, antelope, whale, seal, monkey, and cheetah exemplify the radiation of mammals into nearly all habitats, with bodies finely adapted to their varied lifestyles (**Fig. 24-16**). The largest groups of placental mammals, in terms of number of species, are the rodents and the bats.

Rodents account for almost 40% of all mammal species. Most rodent species are rats or mice, but the group also includes squirrels, hamsters, guinea pigs, porcupines, beavers, woodchucks, chipmunks, and voles. The largest rodent, the capybara, is found in South America and can weigh up to 110 pounds (50 kilograms). Capybara meat is widely consumed in South America, mostly from animals harvested by hunters, but increasingly from animals raised on ranches.

About 20% of mammal species are bats, the only mammals to have evolved wings and powered flight. Bats are nocturnal and spend the daylight hours roosting in caves, rock crevices, trees, or people's houses. Most bat species have evolved adaptations for feeding on a particular kind of food. Some bats eat fruit; others feed on nectar from night-blooming flowers. Most bats are predators, including species that hunt frogs, fish, or even other bats. A few species (vampire bats) subsist entirely on blood that they lap up from incisions they make in the skin of sleeping mammals or birds. Most predatory bats, however, feed on flying insects, which they detect by echolocation. To echolocate, a bat emits short pulses of high-pitched sound (too high for humans to hear). The sounds bounce off objects in the surrounding environment to produce echoes, which the bat hears and uses to identify and locate insect prey.

Case Study revisited
Fish Story

After Marjorie Courtney-Latimer's discovery of the coelacanth, J. L. B. Smith dedicated himself to searching for more coelacanth specimens in the waters off South Africa. He didn't find one until 1952, when fishermen from the Comoro Islands, having seen leaflets that offered a reward for a coelacanth, contacted Smith with the news that they had one in their possession. Smith immediately booked a flight to the Comoros, and reportedly wept for joy upon holding the 88-pound coelacanth awaiting him.

In the years since Smith's trip, about 200 additional coelacanths have been caught by fishermen, mostly in waters around the Comoros but also around nearby Madagascar and off the coasts of Mozambique and South Africa. Scientists thought that the fish's range was restricted to this relatively small area in the western Indian Ocean, and it was therefore something of a shock when a few specimens were discovered in Indonesia, more than 6,000 miles away. DNA tests showed that these Indonesian coelacanths were members of a second species.

Although the coelacanth specimens have revealed a great deal about the creatures' anatomy, their habits and behavior remain comparatively mysterious. Observations from research submarines suggest that coelacanths spend much of their time in caves and beneath rocky overhangs at depths of 300 to 1,200 feet. Radio tracking suggests that they may venture into open water at night, presumably to forage for food. Almost all individuals observed (or caught) have been at least 3 feet long, which suggests that young fish must travel to sites away from the main adult populations to mature, though no such location has been discovered.

The known populations of coelacanths are small, consisting of a few hundred individuals, and appear to be declining. Part of this decline is due to fishing, though coelacanths are mostly caught accidentally by fishermen searching for more commercially desirable species. Conservation efforts in South Africa and the Comoros thus focus largely on introducing fishing methods that will reduce the chances of accidentally snaring a coelacanth.

Consider This

Many accounts of coelacanths refer to them as "living fossils," a term that is also applied to alligators, gingko trees, horseshoe crabs, and other species whose modern appearance closely resembles that of ancient fossils. The implication of the living fossil designation is that these organisms have evolved very little over a very long period. Do you think this implication is accurate? Is it accurate to say that "living fossils" have evolved more slowly or undergone less evolutionary change than other species?

CHAPTER REVIEW

Summary of Key Concepts

24.1 What Are the Key Features of Chordates?

All chordates possess a notochord; a dorsal, hollow nerve cord; pharyngeal gill slits; and a post-anal tail at some stage in their development.

24.2 Which Clades Make Up the Chordates?

The phylum Chordata includes three clades: the lancelets, the tunicates, and the craniates. The lancelets are invertebrate filter-feeders that live partially buried in sandy seafloors. Tunicates are also invertebrate filter-feeders, and include the sessile sea squirts and the motile salps. Craniates include all animals with skulls: the hagfishes and the vertebrates. Hagfishes are jawless, eel-shaped craniates that lack a backbone and, therefore, are not vertebrates.

24.3 What Are the Major Groups of Vertebrates?

Lampreys are jawless vertebrates; the best-known lamprey species are parasites of fish.

Most amphibians have simple lungs for breathing air. Most are confined to relatively damp terrestrial habitats by their need to keep their skin moist, their use of external fertilization, and their aquatic larvae.

Reptiles, with well-developed lungs, dry skin covered with relatively waterproof scales, internal fertilization, and an amniotic egg with its own water supply, are well adapted to the driest terrestrial habitats. One group of reptiles, the birds, has additional adaptations, such as an elevated body temperature, that allow the muscles to respond rapidly regardless of the temperature of the environment. The bird body is molded for flight, with feathers, lightweight bones, efficient circulatory and respiratory systems, and well-developed eyes.

Mammals have insulating hair and (except for monotreme mammals) give birth to live young that are nourished with milk. The mammalian nervous system is the most complex in the animal kingdom, providing mammals with enhanced learning ability that helps them adapt to changing environments.

Key Terms

amnion *455*
amniotic egg *455*
cartilage *448*
craniate *448*
mammary gland *459*
marsupial *459*
monotreme *459*
nerve cord *447*
notochord *447*
pharyngeal gill slit *447*
placenta *459*
placental *461*
post-anal tail *447*
vertebral column *450*
vertebrate *450*

Thinking Through the Concepts

Fill-in-the-Blank

1. In chordates, the nerve cord is __________ and runs along the __________ side of the body. During at least one stage of a chordate's life, it has a tail that extends past its __________ and its body is stiffened by a(n) __________ that runs along its length.
2. Animals that are chordates but not vertebrates include __________, __________, and __________. Craniates are animals that have a(n) __________. Animals that are craniates but not vertebrates include __________.
3. Both gills and lungs are present in adult __________. Sharks and rays have internal skeletons composed of __________. The vertebrate group with the largest number of species is __________. Lampreys have teeth but lack __________.
4. Among tetrapod groups, hair is found in __________; the skin is a respiratory organ in __________; shelled, amniotic eggs are found in __________; aquatic larvae with gills are found in __________.
5. The only mammals that lay eggs are __________. The only vertebrates that regenerate lost limbs are __________. The only mammals with powered flight are __________.

Review Questions

1. Briefly describe each of the following adaptations, and explain the adaptive significance of each: vertebral column, jaws, limbs, amniotic egg, feathers, placenta.
2. List the vertebrate groups that have each of the following:
   a. a skeleton of cartilage
   b. a two-chambered heart
   c. amniotic egg
   d. warm-bloodedness
   e. a four-chambered heart
   f. a placenta
   g. lungs supplemented by air sacs
3. List four distinguishing features of chordates.
4. Describe the ways in which amphibians are adapted to life on land. In what ways are amphibians still restricted to a watery or moist environment?
5. List the adaptations that distinguish reptiles from amphibians and help reptiles adapt to life in dry terrestrial environments.
6. List the adaptations of birds that contribute to their ability to fly.
7. How do mammals differ from birds, and what adaptations do they share?
8. How has the mammalian nervous system contributed to the success of mammals?

Applying the Concepts

1. Are hagfishes vertebrates or invertebrates? On which characteristics did you base your answer? Is it important to be able to place them in one category or the other? Why?
2. Is the decline of amphibian populations of concern to humans? What about the increase in frog deformities? Why is it important to understand the causes of these phenomena?

MB *Go to www.masteringbiology.com for practice quizzes, activities, eText, videos, current events, and more.*

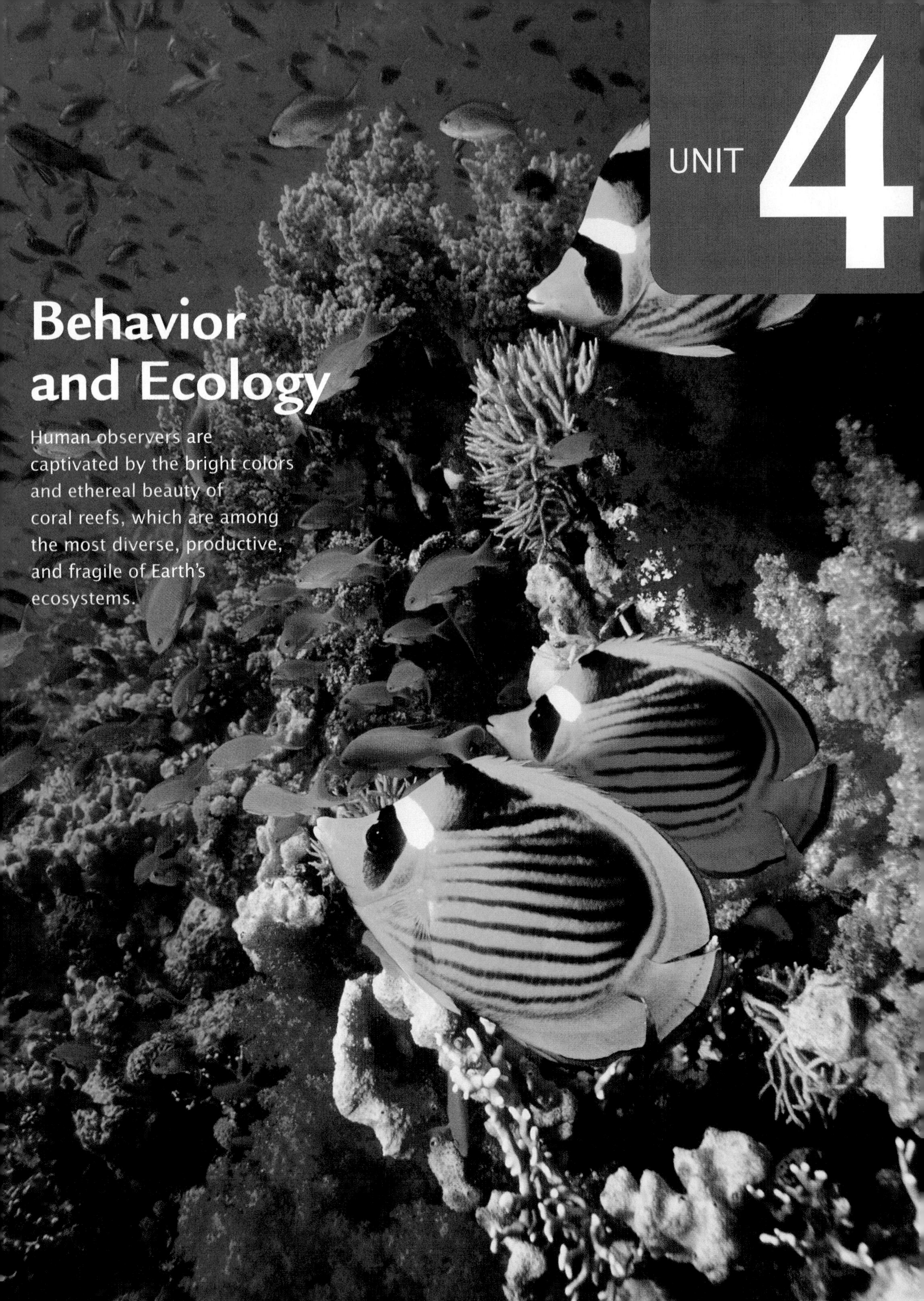

UNIT 4

Behavior and Ecology

Human observers are captivated by the bright colors and ethereal beauty of coral reefs, which are among the most diverse, productive, and fragile of Earth's ecosystems.

Animal Behavior

Chapter 25

▲ Both this male scorpionfly and this male human are exceptionally attractive to females of their species. The secret to their sex appeal may be that both have highly symmetrical bodies.

Case Study

Sex and Symmetry

WHAT MAKES A MAN SEXY? According to a growing body of research, it's his symmetry. Female sexual preference for symmetrical males was first documented in insects. For example, biologist Randy Thornhill found that symmetry accurately predicts the mating success of male Japanese scorpionflies (see the inset photo). In Thornhill's experiments and observations, the most successful males were those whose left and right wings were equal or nearly equal in length. Males with one wing longer than the other were less likely to copulate; the greater the difference between the two wings, the lower the likelihood of success.

Thornhill's work with scorpionflies led him to wonder if the effects of male symmetry also extend to humans. To test the hypothesis that female humans find symmetrical males more attractive, Thornhill and colleagues began by measuring symmetry in some young adult males. Each man's degree of symmetry was assessed by measurements of his ear length and the width of his foot, ankle, hand, wrist, elbow, and ear. From these measurements, the researchers derived an index that summarized the degree to which the size of these features differed between the right and left sides of the body.

The researchers next gathered a panel of heterosexual female observers who were unaware of the nature of the study and showed them photos of the faces of the measured males. As predicted by the researchers' hypothesis, the panel judged the most symmetrical men to be most attractive. Apparently, a man's attractiveness to women is correlated with his body symmetry.

Why might females prefer symmetrical males? Consider this question as you read about animal behavior.

At a Glance

25.1 HOW DO INNATE AND LEARNED BEHAVIORS DIFFER?

Behavior is any observable activity of a living animal. For example, a moth flies toward a bright light, a honeybee flies toward a cup of sugar-water, and a housefly flies toward a piece of rotting meat. Bluebirds sing, wolves howl, and frogs croak. Mountain goats butt heads in ritual combat; chimpanzees groom one another; ants attack a termite that approaches an anthill. Humans dance, play sports, and wage wars. Even the most casual observer sees many examples of animal behavior each day, and a careful observer encounters a virtually limitless number of fascinating behaviors.

Innate Behaviors Can Be Performed Without Prior Experience

Innate behaviors are performed in reasonably complete form the first time an animal of the right age and motivational state encounters a particular stimulus. (The proper motivational state for feeding, for example, would be hunger.) Scientists can demonstrate that a behavior is innate by depriving an animal of the opportunity to learn it. For example, red squirrels, which in the wild bury nuts in the fall for retrieval during the winter, can be raised from birth in a bare cage on a liquid diet, providing them with no experience of nuts, digging, or burying. Nonetheless, such a squirrel will, when presented with nuts for the first time, carry one to the corner of its cage, and then make covering and patting motions with its forefeet, demonstrating that nut burying is an innate behavior.

Some innate behaviors can also be recognized by their occurrence immediately after birth, before any opportunity for learning presents itself. Consider, for example, the common cuckoo, a bird species in which females lay eggs in the nests of other bird species, to be raised by the unwitting adoptive parents. Soon after a cuckoo egg hatches, the cuckoo chick performs the innate behavior of shoving the nest owner's eggs (or baby birds) out of the nest, eliminating its competitors for food (**Fig. 25-1**).

Learned Behaviors Require Experience

Natural selection may favor innate behaviors in many circumstances. For instance, it is clearly to the advantage of a gull chick to peck at its parent's bill as soon as possible after hatching, because pecking stimulates the parent to feed the chick. But in other circumstances, rigidly fixed behavior patterns may be less useful. For example, a male red-winged blackbird presented with a stuffed female blackbird will often attempt to copulate with the stuffed bird, a behavior that obviously will produce no offspring. In many situations, a degree of behavioral flexibility is advantageous.

(a) A cuckoo chick ejects an egg

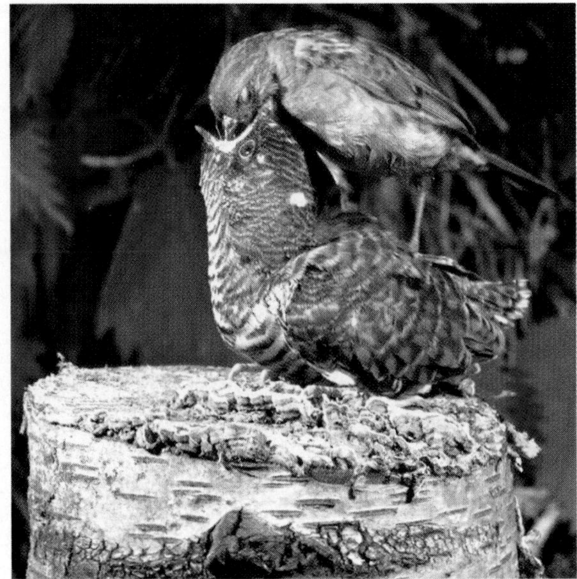

(b) A foster parent feeds a cuckoo

▲ **FIGURE 25-1 Innate behavior (a)** The cuckoo chick, just hours after it hatches and before its eyes have opened, evicts the eggs of its foster parents from the nest. **(b)** The parents, responding to the stimulus of the cuckoo chick's wide-gaping mouth, feed the chick, unaware that it is not related to them.

QUESTION The cuckoo chick benefits from its innate behavior, but the foster parent harms itself with its innate response to the cuckoo chick's begging. Why hasn't natural selection eliminated this disadvantageous innate behavior?

The capacity to make changes in behavior on the basis of experience is called **learning.** This deceptively simple definition encompasses a vast array of phenomena. A toad learns to avoid distasteful insects; a baby shrew learns which adult is its mother; a human learns to speak a language; a sparrow learns to use the stars for navigation. Each of the many examples of animal learning represents the outcome of a unique evolutionary history, so learning is as diverse as animals themselves. Nonetheless, it can be useful to categorize types of learning, as long as we keep in mind that the categories are only rough guides; many examples of learning will not fit neatly into any category.

Habituation Is a Decline in Response to a Repeated Stimulus

A common form of simple learning is **habituation,** defined as a decline in response to a repeated stimulus. The ability to habituate prevents an animal from wasting its energy and attention on irrelevant stimuli. This form of learning is displayed by even the simplest animals. For example, a sea anemone will retract its tentacles when touched but gradually stops retracting if touching is repeated frequently (**Fig. 25-2**).

The ability to habituate is clearly adaptive. If a sea anemone withdrew every time it was brushed by a strand of waving seaweed, the animal would waste a great deal of energy, and its retracted posture would prevent it from snaring food. Humans habituate to many stimuli; city dwellers habituate to nighttime traffic sounds, as country dwellers do to choruses of crickets and tree frogs. Each may initially find the other's habitat unbearably noisy at first, but each eventually habituates.

Conditioning Is a Learned Association Between a Stimulus and a Response

A more complex form of learning is **trial-and-error learning,** in which animals acquire new and appropriate responses to stimuli through experience. Many animals are faced with naturally occurring rewards and punishments and can learn to modify their responses to them. For example, a hungry toad that captures a bee quickly learns to avoid future encounters with bees (**Fig. 25-3**). After only one experience with a stung tongue, a toad ignores bees and even other insects that resemble them.

Trial-and-error learning is an important factor in the behavioral development of many animal species and often occurs during play and exploratory behavior. This type of learning also plays a key role in human behavior—allowing, for example, a child to learn which foods taste good or bad, that a stove can be hot, and not to pull a cat's tail.

Some interesting properties of trial-and-error learning have been revealed by a laboratory technique known as **operant conditioning.** During operant conditioning, an animal learns to perform a behavior (such as pushing a lever or pecking a button) to receive a reward or to avoid punishment. This technique is most closely associated with the American comparative psychologist B. F. Skinner, who designed the "Skinner box," in which an animal is isolated and allowed to train itself. The box might contain a lever that,

After many touches, the anemone habituates and no longer responds
Touched for the first time, the anemone withdraws

◀ **FIGURE 25-2 Habituation in a sea anemone**

when pressed, ejects a food pellet. If the animal accidentally bumps the lever, a food reward appears. After a few such occurrences, the animal learns the connection between pressing the lever and receiving food and begins to press the lever repeatedly.

Operant conditioning has been used to train animals to perform tasks far more complex than pressing a lever, and has revealed that species differ in their propensity to learn associations, more easily learning those that are relevant to their own needs. For example, if a rat is given a distinctively flavored food

❶ A naive toad is presented with a bee.

❷ While trying to eat the bee, the toad is stung painfully on the tongue.

❸ Presented with a harmless robber fly, which resembles a bee, the toad cringes.

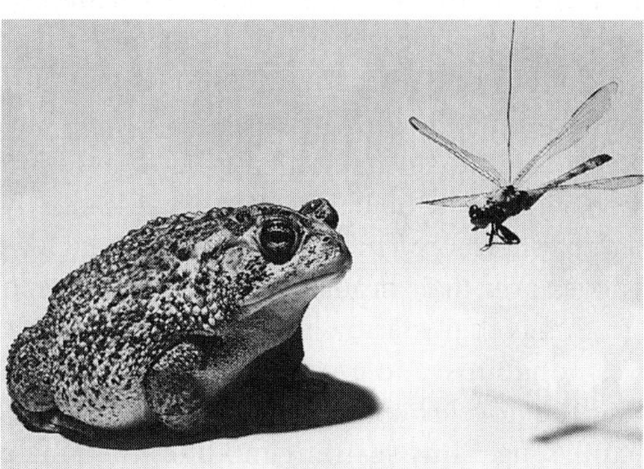

❹ The toad is presented with a dragonfly.

❺ The toad immediately eats the dragonfly, demonstrating that the learned aversion is specific to bees and insects resembling bees.

▲ **FIGURE 25-3 Trial-and-error learning in a toad**

containing a substance that makes the rat sick, the animal learns to avoid eating that food in the future. In contrast, it is very difficult to train a rat to rear up on its hind legs in response to a particular sound or visual cue. The difference can be explained by asking which learning task is more likely to benefit a wild rat. Clearly, avoiding sickening foods is beneficial to animals such as rats that eat a wide variety of foods. However, the animal gains no obvious benefit from learning to stand up in response to a noise. In general, the learning abilities of each species have evolved to support its particular mode of life.

Insight Is Problem Solving Without Trial and Error

In certain situations, animals seem able to solve problems suddenly, without the benefit of prior experience. This kind of sudden problem solving is sometimes called **insight learning,** because it seems at least superficially similar to the process by which humans mentally manipulate concepts to arrive at a solution. We cannot, of course, know for sure if non-human animals experience similar mental states when they solve problems.

In 1917, the animal behaviorist Wolfgang Kohler showed that a hungry chimpanzee, without any training, could stack boxes to reach a banana suspended from the ceiling. This type of problem solving was once believed to be limited to very intelligent types of animals such as primates, but similar abilities may also be present in species that we tend to view as less intelligent. For example, Robert Epstein and colleagues performed an experiment that showed that pigeons may be capable of insight learning. In the experiment, pigeons (whose wings had been clipped to prevent flight) were first trained to perform two unrelated tasks in return for food rewards. The tasks were to push a small box around the cage and to peck at a small plastic banana. Later, the trained birds were presented with a novel situation: a plastic banana that hung above their reach in a cage that also contained a small box. Many of the pigeons pushed the box to a position beneath the plastic banana and climbed atop the box to peck the faux fruit. Apparently, a pigeon trained to execute the necessary physical movements can also solve the suspended banana problem.

There Is No Sharp Distinction Between Innate and Learned Behaviors

Although the terms "innate" and "learned" can help us describe and understand behaviors, these words also have the potential to lull us into an oversimplified view of animal behavior. In practice, no behavior is totally innate or totally learned. Instead, all behaviors are mixtures of the two.

Seemingly Innate Behavior Can Be Modified by Experience

Behaviors that seem to be performed correctly on the first attempt without prior experience can later be modified by experience. For example, a newly hatched gull chick is able to peck at a red spot on its parent's beak (**Fig. 25-4**), an innate behavior that causes the parent to regurgitate food for the chick to eat. Biologist Niko Tinbergen studied this pecking behavior and found that the pecking response of very young chicks was triggered by the long, thin shape and red color of the parent's bill. In fact, when Tinbergen offered newly hatched chicks a thin, red rod with white stripes painted on it, they pecked at it more often than at a real beak. Within a few days, however, the chicks learned enough about the appearance of their parents that they began pecking more frequently at models more closely resembling the parents. After one week, the young gulls recognized their parents' appearance enough to prefer models of their own species to models of a closely related species. Eventually, the young birds learned to beg only from their own parents.

▲ **FIGURE 25-4 Innate behaviors can be modified by experience** A gull chick pecks at the red spot on its mother's bill, causing her to regurgitate food.

Habituation (a decline in response to a repeated stimulus) can also fine-tune an organism's innate responses to environmental stimuli. For example, young birds crouch down when a hawk flies over but ignore harmless birds such as geese. Early scientific observers hypothesized that only the very specific shape of predatory birds provoked crouching. Using an ingenious model (**Fig. 25-5**), Niko Tinbergen and Konrad Lorenz (two of the founding fathers of **ethology**, the study of animal behavior) tested and confirmed this hypothesis. When moved in one direction, the model resembled a goose and chicks ignored it. When its movement was reversed, however, the model resembled a hawk and elicited crouching behavior from the chicks. Further research, however, revealed that newborn chicks instinctively crouch when *any* object moves over their heads. Over time, their response habituates to things that soar by harmlessly and frequently, such as leaves, songbirds, and geese. Predators are much less common, and the novel shape of a hawk continues to elicit instinctive crouching. Thus, learning modifies the innate response, making it more advantageous.

Learning May Be Governed by Innate Constraints

Learning always occurs within boundaries that help increase the chances that only the appropriate behavior is acquired.

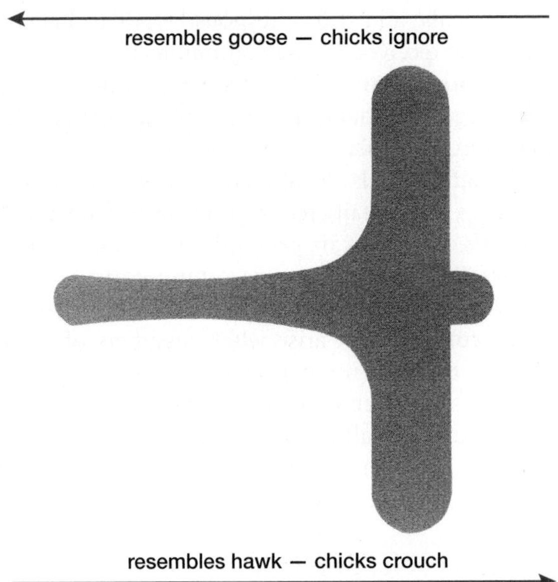

▲ **FIGURE 25-5 Habituation modifies innate responses** The model used by Konrad Lorenz and his student Niko Tinbergen to investigate the response of chicks to the shape of objects flying overhead. The chicks' response depended on which direction the model moved. Moving toward the right, the model resembles a predatory hawk, but when it moves left, it resembles a harmless goose.

For example, even though young robins hear the singing of sparrows, warblers, finches, and other bird species that share nesting areas with robins, the young birds do not imitate the songs of these other species. Instead, young robins learn only the songs of adult robins. The robin's ability to learn songs is limited to those of its own species, and the songs of other species are excluded from the learning process.

The innate constraints on learning are perhaps most strikingly illustrated by **imprinting,** a special form of learning in which an animal's nervous system is rigidly programmed to learn a certain thing only at a certain period of development. This causes a strong association to be formed during a particular stage, called a sensitive period, in the animal's life. During this stage, the animal is primed to learn specific information, which is then incorporated into behaviors that are not easily altered by further experience.

Imprinting is best known in birds such as geese, ducks, and chickens. These birds learn to follow the animal or object that they most frequently encounter during an early sensitive period. In nature, a mother bird is likely to be nearby during the sensitive period, so her offspring imprint on her. In the laboratory, however, these birds may imprint on a toy train or other moving object (**Fig. 25-6**). If given a choice, however, they select an adult of their own species.

All Behavior Arises Out of Interactions Between Genes and Environment

Many early ethologists saw innate behaviors as rigidly controlled by genetic factors and viewed learned behaviors as determined exclusively by an animal's environment. Today, however, ethologists realize that, just as no behavior is wholly innate or wholly learned, no behavior can be caused strictly by genes or strictly by the environment. Instead, all behavior develops out of an interaction between genes and the environment. The relative contributions of heredity and learning vary among animal species and among behaviors within an individual.

The precise nature of the link among genes, environments, and behaviors is not well understood in most cases. The chain of events between the transcription of genes and the performance of a behavior may be so complex that we will never decipher it fully. Nonetheless, a great deal of evidence demonstrates the existence of both genetic and environmental components in the development of behaviors. For example, consider bird migration. It is well known that migratory birds must learn by experience how to navigate with celestial cues; young, developing birds deprived of the opportunity to observe the apparent rotation of the night sky are not, as adults, able to orient correctly at night. However, this kind of learning is not the only factor involved.

Bird Migration Behavior Has an Inherited Component

At the close of the summer, many birds leave their breeding habitats and head for their winter territory, which may be hundreds or even thousands of miles away. Many of these migrating birds are traveling for the first time, because they were hatched only a few months earlier. Amazingly, these naive birds depart at the proper time, head in the proper direction, and locate the proper wintering location, even though they

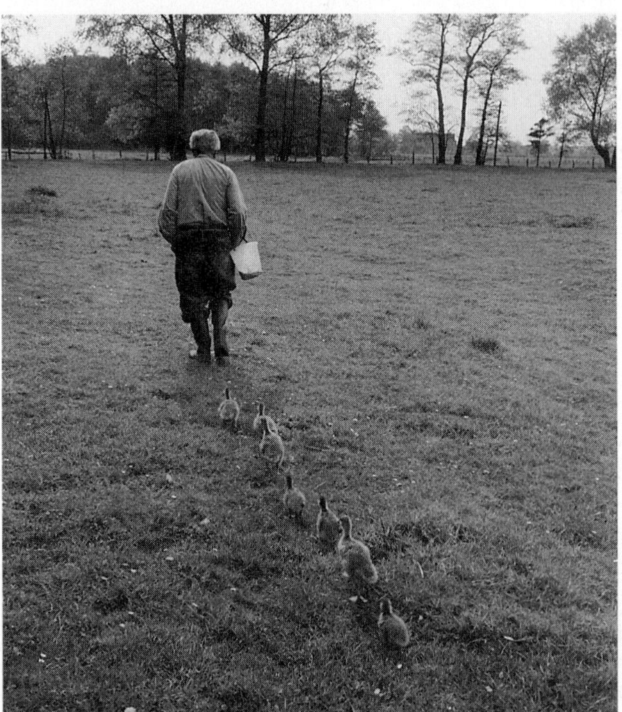

▲ **FIGURE 25-6 Konrad Lorenz and imprinting** Konrad Lorenz, known as the "father of ethology," is followed by goslings that imprinted on him shortly after they hatched. They follow him as they would their mother.

often do not simply follow more experienced birds (which typically depart a few weeks in advance of the first-year birds). Somehow, these young birds execute a very difficult task the first time they try it. Thus, it seems that birds must be born with the ability to migrate; it must be "in their genes." Indeed, birds hatched and raised in isolation indoors still orient in the proper migratory direction when autumn comes, apparently without the need for any learning or experience.

The conclusion that birds must have a genetically controlled ability to migrate in the right direction has been further supported by hybridization experiments with blackcap warblers. This species breeds in Europe and migrates to Africa, but populations from different areas travel by different routes. Blackcaps from western Europe travel in a southwesterly direction to reach Africa, whereas birds from eastern Europe travel to the southeast (**Fig. 25-7**). If birds from the two populations are crossbred in captivity, however, the hybrid offspring exhibit migratory orientation due south, which is intermediate between the orientations of the two parents. This result suggests that parental genes—of which offspring inherit a mixture—influence migratory direction.

25.2 HOW DO ANIMALS COMMUNICATE?

Animals frequently broadcast information. The sounds uttered, movements made, and chemicals emitted by animals can reveal their location, level of aggression, readiness to mate, and so on. If this information evokes a response from other individuals, and if that response tends to benefit the sender and the receiver, then a communication channel can form. **Communication** is defined as the production of a signal by one organism that causes another organism to change its behavior in a way beneficial to both.

Although animals of different species may communicate (picture a cat, its tail erect and bushy, hissing at a dog), most animals communicate primarily with members of their own species. Potential mates may communicate, as may parents and offspring. Communication is also often used to help resolve the conflicts that arise when members of a species compete directly with one another for food, space, and mates.

The ways in which animals communicate are astonishingly diverse and use all of the senses. In the following sections, we will look at communication by visual displays, sound, chemicals, and touch.

Visual Communication Is Most Effective over Short Distances

Animals with well-developed eyes use visual signals to communicate. Visual signals can be *active*, in which a specific movement (such as baring fangs) or posture (such as lowering the head) conveys a message (**Fig. 25-8**). Alternatively, visual signals may be *passive*, in which case the size, shape, or color of the animal conveys important information, commonly about its sex and reproductive state. For example, when female mandrills become sexually receptive, they develop a large, brightly colored swelling on their buttocks (**Fig. 25-9**). Active and passive signals can be combined, as illustrated by the lizard in **Figure 25-10**.

Like all forms of communication, visual signals have both advantages and disadvantages. On the plus side, they are instantaneous, and active signals can be rapidly changed to convey a variety of messages in a short period. Visual communication is quiet and unlikely to alert distant predators,

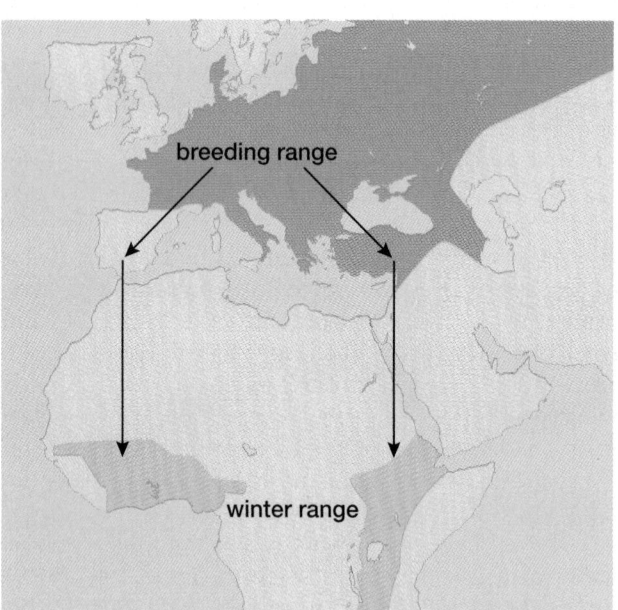

▲ **FIGURE 25-7 Genes influence migratory behavior** Blackcap warblers from western Europe begin their fall migration by flying in a southwesterly direction, but those from eastern Europe fly to the southeast when they begin migrating.

QUESTION If young blackcap warblers from a wild population in western Europe were transported to eastern Europe and reared to adulthood in a normal environment, in which direction would you expect them to orient?

▲ **FIGURE 25-8 Active visual signals** The wolf signals aggression by lowering its head, ruffling the fur on its neck and along its back, facing its opponent with a direct stare, and exposing its fangs. These signals can vary in intensity, communicating different levels of aggression.

▲ **FIGURE 25-9 A passive visual signal** The female mandrill's colorfully swollen buttocks serve as a passive visual signal that she is fertile and ready to mate.

although the signaler does make itself conspicuous to those nearby. On the negative side, visual signals are generally ineffective in dense vegetation or in darkness, and are limited to close-range communication.

Communication by Sound Is Effective over Longer Distances

The use of sound overcomes many of the shortcomings of visual displays. Like visual displays, sound signals reach receivers almost instantaneously. But unlike visual signals, sound can be transmitted through darkness, dense forests, and murky water. Sound signals can also be effective over longer distances than visual signals. For example, the low, rumbling calls of African elephants can be heard by elephants several miles away, and the songs of humpback whales are audible for hundreds of miles. Likewise, the howls of a wolf pack carry for miles on a still night. Even the small kangaroo rat produces a sound (by striking the desert floor with its hind feet) that is audible 150 feet (45 meters) away. The advantages of long-distance transmission, however, are offset by an important disadvantage: Predators and other unwanted receivers can also detect a sound signal from a distance, and can use the signal to find the location of the signaler.

Sound signals are similar to visual displays in that they can be varied to convey rapidly changing messages. An individual can convey different messages by varying the pattern, volume, or pitch of a sound. In a study of vervet monkeys in Kenya in the 1960s, ethologist Thomas Struhsaker found that the monkeys produced different calls in response to threats from each of their major predators: snakes, leopards, and eagles. Later, other researchers reported that the response of listening vervet monkeys to each of these calls is appropriate to the particular predator. For example, the "bark" that warns of a leopard or other four-legged carnivore causes monkeys on the ground to take to trees and those in trees to climb higher. The "rraup" call, which advertises the presence of an eagle or other hunting bird, causes monkeys on the ground to look upward and take cover, whereas monkeys already in trees drop to the shelter of lower, denser branches. The "chutter" call that indicates the presence of a snake causes the monkeys to stand up and search the ground for the predator.

The use of sound is by no means limited to birds and mammals. Male crickets produce species-specific songs that attract female crickets of the same species. The annoying whine of the female mosquito as she prepares to bite alerts nearby males that she may soon have the blood meal necessary for laying eggs. Male water striders vibrate their legs, sending species-specific patterns of vibrations through the water, attracting mates and repelling other males (**Fig. 25-11**). Many species of fish produce croaks, grunts, or other sounds. From these rather simple signals to the complexities of human language, sound is one of the most important forms of communication.

Have you ever wondered 

Why Fireflies Flash?

In many parts of the world, warm nights are enlivened by the myriad flashing lights of fireflies. Humans are entranced and entertained by the pulsing luminescence of these flying beetles, but what does the display do for the fireflies? It's a mating signal. Each of the many species of firefly has its own distinctive pattern of flashes, which helps ensure that a nocturnal firefly searching for a mate can find a partner of its own species. Firefly evolution has produced a method of visual signaling that works in the dark.

▲ **FIGURE 25-10 Active and passive visual signals combined** A South American anole lizard raises his head high in the air (an active visual signal), revealing a colored throat pouch (a passive visual signal) that warns others to keep their distance.

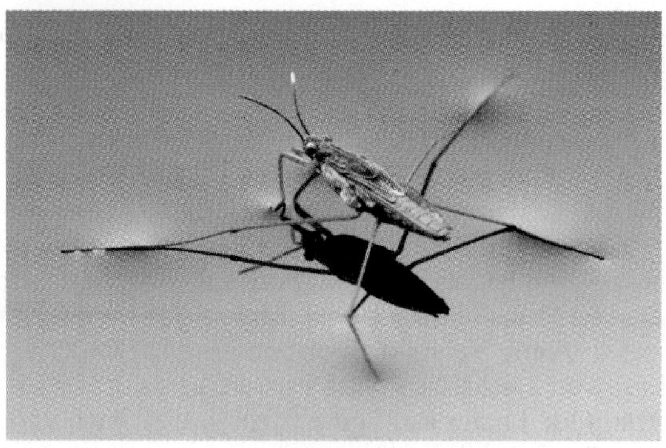

▲ **FIGURE 25-11 Communication by vibration** The light-footed water strider relies on the surface tension of water to support its weight. By vibrating its legs, the water strider sends signals that radiate out over the surface of the water. These vibrations advertise the strider's species and sex to others nearby.

Chemical Messages Persist Longer But Are Hard to Vary

Chemical substances that are produced by individuals and that influence the behavior of other members of the species are called **pheromones**. Pheromones can carry messages over long distances and take little energy to produce. Unlike visual or sound signals that may attract predators, pheromones are typically not detectable by other species. In addition, a pheromone can act as a kind of signpost, persisting over time and conveying a message long after the signaling animal has departed. Wolf packs, hunting over areas of nearly 400 square miles (about 1,000 square kilometers), warn other packs of their presence by marking the boundaries of their travels with urine that contains pheromones. As anyone who has walked a dog can attest, the domesticated dog reveals its wolf ancestry by staking out its neighborhood with urine that carries the chemical message "I live in this area." (For an example of how humans take advantage of animals' ability to detect chemical messages, see "Links to Everyday Life: Mine Finders.")

Chemical communication requires animals to synthesize a different substance for each message. As a result, chemical signaling systems communicate fewer and simpler messages than do sight- or sound-based systems. In addition, pheromone signals cannot easily convey rapidly changing messages. Nonetheless, chemicals effectively convey critical information.

Many pheromones cause an immediate change in the behavior of the animal that detects them. For example, foraging termites that discover food lay a trail of pheromones from the food to the nest, and other termites follow the trail (**Fig. 25-12**). Pheromones can also stimulate physiological changes in the animal that detects them. For example, the queen honeybee produces a pheromone called *queen substance*, which prevents other females in the hive from becoming sexually mature. Similarly, mature males of some mouse species produce urine containing a pheromone that influences female reproductive physiology. The pheromone stimulates newly mature females

Links to *Everyday Life*

Mine Finders

Although people detect their surroundings mainly by sight and sound, animals of many other species have a highly developed sense of smell, with odor-detection abilities that far exceed ours. In some cases, people have taken advantage of animal olfactory abilities to solve human problems. Consider, for example, the problem of unexploded land mines. More than 100 million of these explosive devices remain buried in countries around the world, where they were planted during past wars and forgotten. They pose a major threat to the safety of millions of mostly poor, rural people. Unfortunately, the process of removing mines is slow, expensive, and very dangerous. Animals can help; for example, dogs can smell the explosive in a land mine. Dogs, however, are heavy enough to detonate a mine, so dogs trained to find mines are at risk of getting themselves blown up.

Recently, however, a new, better, animal assistant has been drafted to help find mines. In Mozambique, home to as many as 11 million mines, rats find mines (**Fig. E25-1**). In particular, Gambian giant pouched rats have been trained to sniff out mines and, in return for a banana or peanut reward, scratch the ground vigorously when they find a mine. The rats are very good at detecting mines, and are too light to detonate the ones they find. They also work very quickly; in 1 hour, two rats can search an area that would consume 2 weeks' time for a trained human with a metal detector. Rats have proved to be an unlikely ally in the battle to solve one of humanity's most dangerous self-inflicted problems.

▲ **FIGURE E25-1 A Gambian giant pouched rat at work, detecting land mines**

▲ **FIGURE 25-12 Communication by chemical messages** A trail of pheromones, secreted by termites from their own colony, orients foraging termites toward a source of food.

to become fertile and sexually receptive. It will also cause a female mouse that is newly pregnant by another male to abort her litter and become sexually receptive to the new male.

Humans have harnessed the power of pheromones to combat insect pests. The sex attractant pheromones of some agricultural pests, such as the Japanese beetle and the gypsy moth, have been successfully synthesized. These synthetic pheromones can be used to disrupt mating or to lure these insects into traps. Controlling pests with pheromones has major environmental advantages over conventional pesticides, which kill beneficial as well as harmful insects and foster the evolution of pesticide-resistant insects. In contrast, each pheromone is specific to a single species and does not promote the spread of resistance, because insects resistant to the attraction of their own pheromones do not reproduce successfully.

Communication by Touch Helps Establish Social Bonds

Communication by physical contact often serves to establish and maintain social bonds among group members. This function is especially apparent in humans and other primates, which have many gestures—including kissing, nuzzling, patting, petting, and grooming—that serve important social functions (**Fig. 25-13a**). Touch may even be essential to human well-being. For example, research has shown that when the limbs of premature human infants were stroked and moved for 45 minutes daily, the infants were more active, responsive, and emotionally stable and gained weight more rapidly than did premature infants who received standard hospital treatment.

Communication by touch is not limited to primates, however. For example, in many other mammal species, close physical contact helps cement the bond between parent and offspring. In addition, species in which sexual activity is preceded or accompanied by physical contact can be found throughout the animal kingdom (**Fig. 25-13b**).

25.3 HOW DO ANIMALS COMPETE FOR RESOURCES?

The contest to survive and reproduce stems from the scarcity of resources relative to the reproductive potential of populations. The resulting competition underlies many of the most frequent types of interactions between animals.

Aggressive Behavior Helps Secure Resources

One of the most obvious manifestations of competition for resources such as food, space, or mates is **aggression**, or antagonistic behavior, between members of the same species. Aggressive behavior includes physical combat between rivals. A fight, however, can injure its participants; even the victorious animal might not survive to pass on its genes. As a result, natural selection has favored the evolution of symbolic displays or rituals for resolving conflicts. Aggressive displays allow the competitors to assess each other and determine a winner on the basis of size, strength, and motivation, rather than on the basis of wounds inflicted. Thanks to communication, most aggressive encounters end without physical damage to the participants.

(a) Baboons

(b) Land snails

◄ **FIGURE 25-13 Communication by touch (a)** An adult olive baboon grooms a juvenile. Grooming both reinforces social relationships and removes debris and parasites from the fur. **(b)** Touch is also important in sexual communication. These land snails engage in courtship behavior that will culminate in mating.

(a) A male baboon

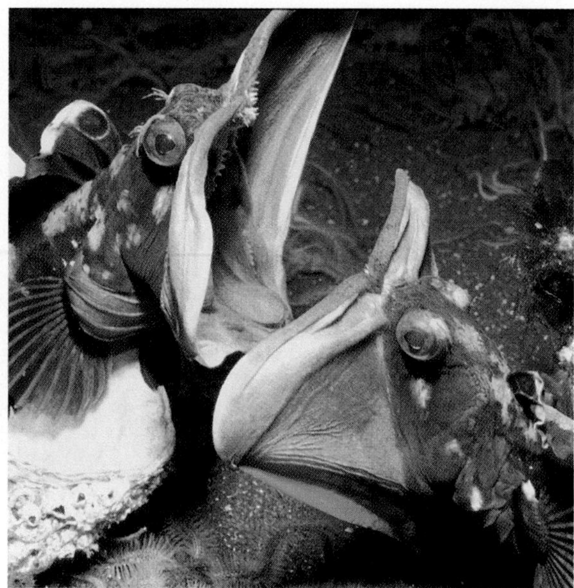

(b) Sarcastic fringeheads

▲ **FIGURE 25-14 Aggressive displays (a)** Threat display of the male baboon. Despite the potentially lethal fangs so prominently displayed, aggressive encounters between baboons rarely cause injury. **(b)** The aggressive display of many male fish, such as these sarcastic fringeheads, includes elevating the fins and flaring the gill covers, thus making the body appear larger.

During aggressive displays, animals may exhibit weapons, such as claws and fangs (**Fig. 25-14a**), and often make themselves appear larger (**Fig. 25-14b**). Competitors often stand upright and erect their fur, feathers, ears, or fins (see Fig. 25-8). These visual displays are typically accompanied by vocal signals such as growls, croaks, roars, or chirps. Fighting tends to be a last resort when displays fail to resolve a dispute.

In addition to aggressive visual and vocal displays, many animal species engage in ritualized combat. Deadly weapons may clash harmlessly (**Fig. 25-15**) or may not be used at all. In many cases, these encounters involve shoving rather than slashing. The ritual thus allows contestants to assess the strength and the motivation of their rivals, and the loser slinks away in a submissive posture that minimizes the size of its body.

▲ **FIGURE 25-15 Displays of strength** Ritualized combat of fiddler crabs. Oversized claws, which could severely injure another animal, grasp harmlessly. Eventually one crab, sensing greater vigor in his opponent, retreats unharmed.

Dominance Hierarchies Help Manage Aggressive Interactions

Aggressive interactions use a lot of energy, can cause injury, and can disrupt other important tasks, such as finding food, watching for predators, or raising young. Thus, there are advantages to resolving conflicts with minimal aggression. In a **dominance hierarchy**, each animal establishes a rank that determines its access to resources. Although aggressive encounters occur frequently while the dominance hierarchy is being established, once each animal learns its place in the hierarchy, disputes are infrequent, and the dominant individuals obtain most access to the resources needed for reproduction, including food, space, and mates. For example, domestic chickens, after some initial squabbling, sort themselves into a reasonably stable "pecking order." Thereafter, all birds in the group defer to the dominant bird, all but the dominant bird give way to the second most dominant, and so on. In wolf packs, one member of each sex is the dominant, or "alpha," individual to whom all

▲ **FIGURE 25-16 A dominance hierarchy** The dominance hierarchy of the male bighorn sheep is signaled by the size of the horns. The backward-curving horns, which have clearly not evolved to inflict injury, are used in ritualized combat.

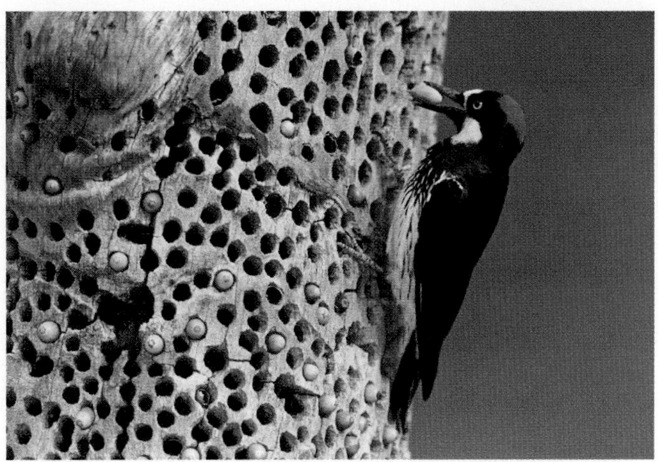

▲ **FIGURE 25-18 A feeding territory** Acorn woodpeckers live in communal groups that excavate acorn-sized holes in dead trees, and stuff the holes with green acorns for dining during the lean winter months. The group defends the trees vigorously against other groups of acorn woodpeckers and against acorn-eating birds of other species, such as jays.

others of that sex are subordinate. Among male bighorn sheep, dominance is reflected in horn size (**Fig. 25-16**).

Perhaps the most thoroughly studied dominance hierarchy is that of chimpanzees. Ethologist Jane Goodall (**Fig. 25-17**) has devoted more than 30 years to meticulously observing chimpanzee behavior in the field at Gombe National Park in Tanzania, describing and documenting the animals' complex social organization. Chimps live in groups, and dominance hierarchies among males are a key aspect of their social life. Many males devote a significant amount of time to maintaining their position in the hierarchy, largely by way of an aggressive *charging display* in which a male rushes forward, throws rocks, leaps up to shake vegetation, and otherwise seeks to intimidate rival males.

Animals May Defend Territories That Contain Resources

In many animal species, competition for resources takes the form of **territoriality**, the defense of an area where important resources are located. The defended area may include places to mate, raise young, feed, or store food. Territorial animals generally restrict most or all of their activities to the defended area and advertise their presence there. Territories may be defended by males, females, a mated pair, or entire social groups (as in the defense of a nest by social insects). However, territorial behavior is most commonly seen in adult males, and territories are usually defended against members of the same species, who compete most directly for the resources being protected.

▲ **FIGURE 25-17 Jane Goodall observing chimps**

Territories are as diverse as the animals defending them. For example, a territory can be a tree where a woodpecker stores acorns (**Fig. 25-18**), a small depression in a lake floor used as a nesting site by a cichlid fish, a hole in the sand that is home to a crab, or an area of forest providing food for a squirrel.

Territoriality Reduces Aggression

Acquiring and defending a territory requires considerable time and energy, yet territoriality is seen in animals as diverse as worms, arthropods, fish, birds, and mammals. The fact that organisms as distantly related as worms and humans independently evolved similar behavior suggests that territoriality provides some important advantages. Although the particular benefits depend on the species and the type of territory it defends, we can make some broad generalizations. First, as with dominance hierarchies, once a territory is established through aggressive interactions, relative peace prevails as boundaries are recognized and respected. One reason for this stability is that an animal is highly motivated to defend its territory and will often defeat even larger, stronger animals that attempt to invade it. Conversely, an animal outside its territory is much less secure and more easily defeated. This principle was demonstrated by Niko Tinbergen in an experiment using stickleback fish (**Fig. 25-19**).

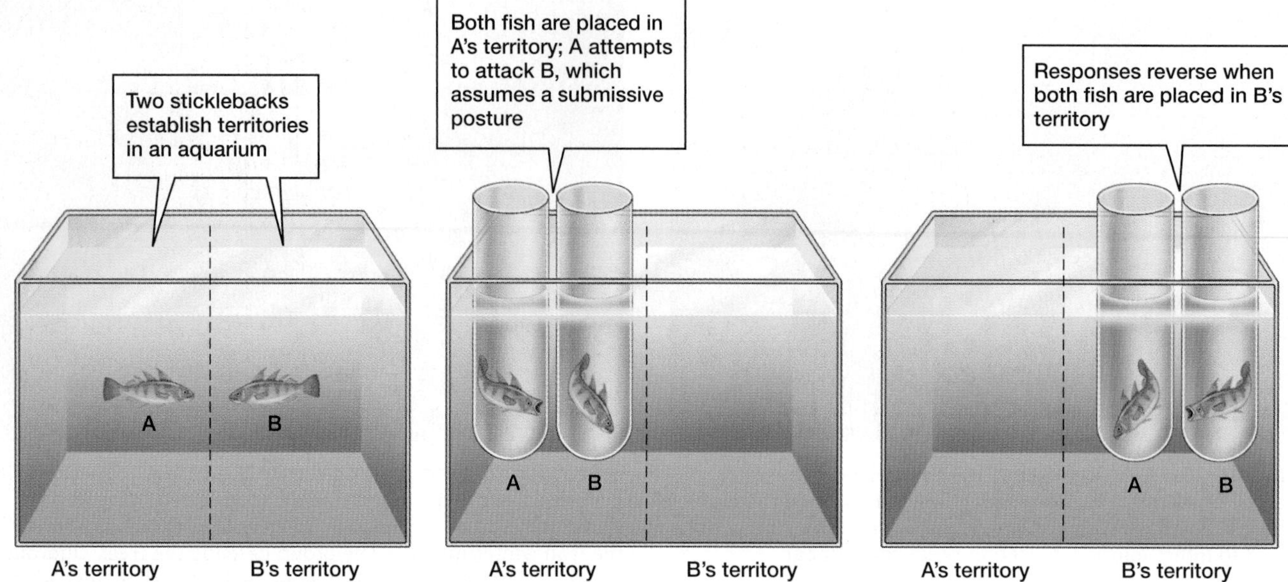

▲ **FIGURE 25-19 Territory ownership and aggression** Niko Tinbergen's experiment demonstrating the effect of territory ownership on aggressive motivation.

Competition for Mates May Be Based on Territories

For males of many species, successful territorial defense has a direct impact on reproductive success. In these species, males defend territories, and females are attracted to high-quality territories, which might have features such as large size, abundant food, and secure nesting areas. Males who successfully defend the best territories have the greatest chance of mating and passing on their genes. For example, experiments have shown that male stickleback fish that defend large territories are more successful in attracting mates than are males that defend small territories. Females that select males with the best territories increase their own reproductive success and pass their genetic traits (typically including their mate-selection preferences) to their offspring.

Animals Advertise Their Occupancy

Territories are advertised through sight, sound, and smell. If a territory is small enough, its owner's mere presence, reinforced by aggressive displays toward intruders, can provide sufficient defense. A mammal that has a territory but cannot always be present may use pheromones to scent-mark the boundaries of its territory. For example, male rabbits use pheromones secreted by anal and chin glands to mark their territories. Hamsters rub the areas around their dens with secretions from glands in their flanks.

Vocal displays are a common form of territorial advertisement. Male sea lions defend a strip of beach by swimming up and down in front of it, calling continuously. Male crickets produce a specific pattern of chirps to warn other males away from their burrows. Birdsong is a striking example of territorial defense. The husky trill of the male seaside sparrow is part of an aggressive display, warning other males to steer clear of his territory (**Fig. 25-20**). In fact, male sparrows that are unable to sing are unable to defend territories. The importance of singing to seaside sparrows' territorial defense was elegantly demonstrated by ornithologist M. Victoria McDonald, who captured territorial males and performed an operation that left them temporarily unable to sing but still able to utter the other, shorter and quieter signals in their vocal repertoires. The songless males were unable to defend territories or attract mates, but regained their lost territories when they recovered their singing ability.

25.4 HOW DO ANIMALS FIND MATES?

In many sexually reproducing animal species, mating involves copulation or other close contact between males and females. Before animals can successfully mate, however, they must identify one another as members of the same

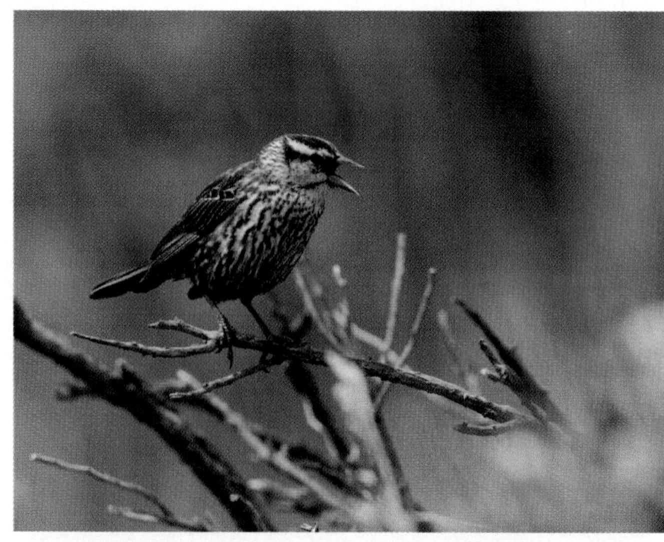

▲ **FIGURE 25-20 Defense of a territory by song** A male seaside sparrow announces ownership of his territory.

(a) A bowerbird bower

(b) A male frigate bird

◀ **FIGURE 25-21 Sexual displays** **(a)** During courtship, a male gardener bowerbird builds a bower out of twigs and decorates it with colorful items that he gathers. **(b)** A male frigate bird inflates his scarlet throat pouch to attract passing females.

QUESTION The male bowerbird provides no protection, food, or other resources to his mate or offspring. Why, then, do females carefully compare the bowers of different males before choosing a mate?

species, as members of the opposite sex, and as being sexually receptive. In many species, finding an appropriate potential partner is only the first step. Often the male must demonstrate his quality before the female will accept him as a mate. The need to fulfill all of these requirements has resulted in the evolution of a diverse and fascinating array of courtship behaviors.

Signals Encode Sex, Species, and Individual Quality

Individuals that waste energy and gametes by mating with members of the wrong sex or wrong species are at a disadvantage in the contest to reproduce. Thus, natural selection favors behaviors by which animals communicate their sex and species to potential mates.

Many Mating Signals Are Acoustic

Animals often use sounds to advertise their sex and species. Consider the raucous nighttime chorus of male tree frogs, each singing a species-specific song. Male grasshoppers and crickets also advertise their sex and species by their calls, as does the female mosquito with her high-pitched whine.

Signals that advertise sex and species may also be used by potential mates in comparisons among rival suitors. For example, the male bellbird uses its deafening song to defend large territories and to attract females from great distances. A female flies from one territory to another, alighting near each male in his tree. The male, beak gaping, leans directly over the flinching female and utters an ear-splitting note. The female apparently endures this noise to compare the songs of the various males, perhaps choosing the loudest as a mate.

Visual Mating Signals Are Also Common

Many species use visual displays for courting. The firefly, for example, flashes a message that identifies its sex and species. Male fence lizards bob their heads in a species-specific rhythm, and females distinguish and prefer the rhythm of their own species. The elaborate construction projects of the male gardener bowerbird and the scarlet throat of the male frigate bird serve as flashy advertisements of sex, species, and male quality (**Fig. 25-21**). Sending these extravagant signals must be risky, because they make it much easier for predators to locate the sender. For males, the added risk is an evolutionary necessity, because females won't mate with males that lack the appropriate signal. Females, in contrast, typically do not need to attract males or assume the risk associated with a conspicuous signal, so in many species females are drab in comparison to males (**Fig. 25-22**).

The intertwined functions of sex recognition and species recognition, advertisement of individual quality, and synchronization of reproductive behavior commonly require a complex series of signals, both active and passive, by both

▲ **FIGURE 25-22 Sex differences in guppies** As in many animal species, the male guppy (left) is brighter and more colorful than the female.

❶ A male, inconspicuously colored, leaves the school of males and females to establish a breeding territory.

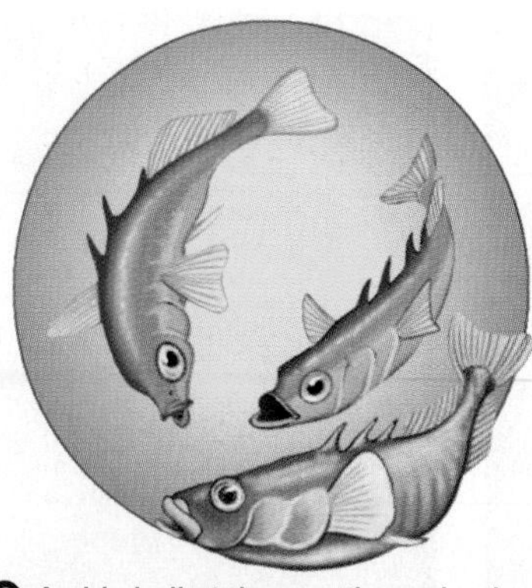

❷ As his belly takes on the red color of the breeding male, he displays aggressively at other red-bellied males, exposing his red underside.

❸ Having established a territory, the male begins nest construction by digging a shallow pit that he will fill with bits of algae cemented together by a sticky secretion from his kidneys.

❹ After he tunnels through the nest to make a hole, his back begins to take on the blue courting color that makes him attractive to females.

❺ An egg-carrying female displays her enlarged belly to him by assuming a head-up posture. Her swollen belly and his courting colors are passive visual displays.

❻ Using a zigzag dance, he leads her to the nest.

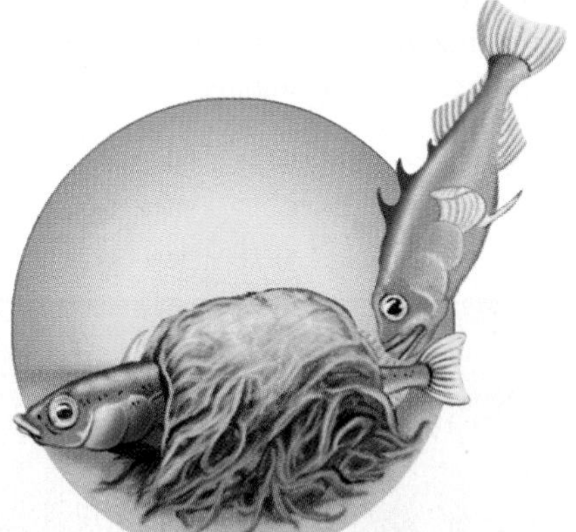

❼ After she enters, he stimulates her to release eggs by prodding at the base of her tail.

❽ He enters the nest as she leaves and deposits sperm, which fertilize the eggs.

▲ **FIGURE 25-23 Courtship of the three-spined stickleback**

Case Study continued
Sex and Symmetry

If females are attracted to symmetrical males, we might expect females to assess the symmetry of male visual signals that function in mate attraction. For example, each male house finch has a patch of bright red feathers on the crown of his head, and researchers have shown that males whose crown patches are brightly colored are more likely to attract a mate than are males with dull patches. But males with patches that are both colorful *and* highly symmetrical have the highest mating success of all.

sexes. Such signals are beautifully illustrated by the complex underwater "ballet" executed by the male and female three-spined stickleback fish (**Fig. 25-23**).

Chemical Signals Can Bring Mates Together

Pheromones can also play an important role in reproductive behavior. A sexually receptive female silk moth, for example, sits quietly and releases a chemical message that can be detected by males up to 3 miles (5 kilometers) away. The exquisitely sensitive and selective receptors on the antennae of the male silk moth respond to just a few molecules of the substance, allowing him to travel upwind along a concentration gradient to find the female (**Fig. 25-24a**).

Water is an excellent medium for dispersing chemical signals, and fish commonly use a combination of pheromones and elaborate courtship movements to ensure the synchronous release of gametes. Mammals, with their highly developed sense of smell, often rely on pheromones released by the female during her fertile periods to attract males (**Fig. 25-24b**).

25.5 WHY DO ANIMALS PLAY?

Many animals play. Pygmy hippopotamuses push one another, shake and toss their heads, splash in the water, and pirouette on their hind legs. Otters delight in elaborate acrobatics. Bottlenose dolphins balance fish on their snouts, throw objects, and carry them in their mouths while swimming. Baby vampire bats chase, wrestle, and slap each other with their wings. Pigface, a giant African softshell turtle who lived at the National Zoo in Washington, D.C., for more than 50 years, would spend hours each day batting a ball around his enclosure. Even octopuses have been seen playing a game: pushing objects away from themselves and into a current, then waiting for the objects to drift back, only to push them back into the current to start the cycle over again.

Animals Play Alone or with Other Animals

Play can be solitary, as when a single animal manipulates an object, such as a cat with a ball of yarn, or the dolphin with its fish, or a macaque monkey making and playing with a snowball. Play can also be social. Often, young of the same species play together, but parents may join them (**Fig. 25-25a**). Social play typically includes chasing, fleeing, wrestling, kicking, and gentle biting (**Fig. 25-25b,c**).

Play seems to lack any clear immediate function, and is abandoned in favor of feeding, courtship, and escaping from danger. Young animals play more frequently than do adults. Play typically borrows movements from other behaviors (attacking, fleeing, stalking, and so on) and uses considerable

(a) Antennae detect pheromones

(b) Noses detect pheromones

▲ **FIGURE 25-24 Pheromone detectors (a)** Male moths find females not by sight but by following airborne pheromones released by females. These odors are sensed by receptors on the male's huge antennae, whose enormous surface area maximizes the chances of detecting the female scent. **(b)** When dogs meet, they typically sniff each other near the base of the tail. Scent glands there broadcast information about the bearer's sex and interest in mating.

QUESTION Female dogs use a pheromone to signal readiness to mate, but female mandrills (see Fig. 25-9) signal mating readiness with a visual signal. What differences would you predict between the two species' methods of searching for food?

(a) Chimpanzees

(b) Brown bears

(c) Arctic foxes

▲ **FIGURE 25-25 Young animals at play**

energy. Also, play is potentially dangerous. Many young humans and other animals are injured, and some are killed, during play. In addition, play can distract an animal from the presence of danger while making it conspicuous to predators. So, why do animals play?

Play Aids Behavioral Development

It is likely that play has survival value and that natural selection has favored those individuals who engage in playful activities. One of the best explanations for the survival value of play is the practice hypothesis. It suggests that play allows young animals to gain experience in behaviors that they will use as adults. By performing these acts repeatedly in play, the animal practices skills that will later be important in hunting, fleeing, or social interactions.

Recent research supports and extends that proposal. Play is most intense early in life when the brain develops and crucial neural connections form. John Byers, a zoologist at the University of Idaho, has observed that species with large brains tend to be more playful than species with small brains. Because larger brains are generally linked to greater learning ability, this relationship supports the idea that adult skills are learned during juvenile play. Watch children roughhousing or playing tag, and you will see how play fosters strength and coordination and develops skills that might have helped our hunting ancestors survive. Quiet play with other children, with dolls, blocks, and other toys, helps children prepare to interact socially, nurture their own children, and deal with the physical world.

25.6 WHAT KINDS OF SOCIETIES DO ANIMALS FORM?

Sociality is a widespread feature of animal life. Most animals interact at least a little with other members of their species. Many spend the bulk of their lives in the company of others, and a few species have developed complex, highly structured societies.

Group Living Has Advantages and Disadvantages

Living in a group has both costs and benefits, and a species will not evolve social behavior unless the benefits of doing so outweigh the costs. Benefits to social animals include:

- Increased abilities to detect, repel, and confuse predators.
- Increased hunting efficiency or increased ability to spot localized food resources.
- Advantages resulting from the potential for division of labor within the group.
- Increased likelihood of finding mates.

On the negative side, social animals may encounter:

- Increased competition within the group for limited resources.
- Increased risk of infection from contagious diseases.

- Increased risk that offspring will be killed by other members of the group.
- Increased risk of being spotted by predators.

Sociality Varies Among Species

The degree to which animals of the same species cooperate varies from one species to the next. Some types of animals, such as the mountain lion, are basically solitary; interactions between adults consist of brief aggressive encounters and mating. Other types of animals cooperate on the basis of changing needs. For example, the coyote is solitary when food is abundant but hunts in packs when food becomes scarce.

Many animals form loose social groups, such as pods of dolphins, schools of fish, flocks of birds, and herds of musk oxen (**Fig. 25-26**). Participation in such groups can provide benefits. For example, the characteristic spacing of fish in schools or the V-pattern of geese in flight provides a hydrodynamic or aerodynamic advantage for each individual in the group, reducing the energy required for swimming or flying. Some biologists hypothesize that herds of antelope or schools of fish confuse predators; their myriad bodies make it difficult for the predator to focus on and pursue a single individual.

A Few Species Form Complex Societies

A small number of species, most of which are insects or mammals, form highly integrated cooperative societies. As you read the following section, you may notice that some cooperative societies are based on behavior that seems to sacrifice the individual for the good of the group. There are many examples: Young, mature Florida scrub jays may remain at their parents' nest and help them raise subsequent broods instead of breeding; worker ants often die in defense of their nest; ground squirrels may sacrifice their own lives to warn the rest of their group about an approaching predator. These behaviors are examples of **altruism**—behavior that decreases the reproductive success of one individual to benefit another.

Forming Groups with Relatives Fosters the Evolution of Altruism

How might altruistic behavior evolve? When individuals perform self-sacrificing deeds, why aren't the alleles that contribute to this behavior eliminated from the gene pool? One possibility is that other members of the group are close relatives of the altruistic individual. Because close relatives share alleles, the altruistic individual may promote the survival of its own alleles through behaviors that maximize the survival of its close relatives. This concept is called **kin selection.** Kin selection helps explain the self-sacrificing behaviors that contribute to the success of cooperative societies. Cooperative behavior is illustrated in the following sections, which describe two examples of complex societies, one in an insect species and one in a mammal species.

Honeybees Live Together in Rigidly Structured Societies

Perhaps the most puzzling of all animal societies are those of the bees, ants, and termites. Scientists have long struggled to explain the evolution of a social structure in which most individuals never breed, but instead labor intensively to feed and protect the offspring of a different individual. Whatever its evolutionary explanation, the intricate organization of a social insect colony is fascinating. In these communities, the individual is a mere cog in an intricate, smoothly running machine and could not survive by itself.

Individual social insects are born into one of several castes within the society. These castes are groups of similar individuals that perform a specific function. For example, honeybees emerge from their larval stage into one of three major preordained roles. One role is that of *queen*. Only one queen is tolerated in a hive at any time. Her functions are to produce eggs (up to 1,000 per day for a lifetime of 5 to 10 years) and regulate the lives of the workers. Male bees, called *drones*, serve merely as mates for the queen. Soon after the queen hatches, drones lured by her sex pheromones swarm around her, and she mates with as many as 15 of them. This relatively

◀ **FIGURE 25-26 Cooperation in loosely organized social groups** A herd of musk oxen functions as a unit when threatened by predators such as wolves. Males form a circle, horns pointed outward, around the females and young.

brief "orgy" supplies her with sperm that will last a lifetime, enough to fertilize more than 3 million eggs. Their sexual chore accomplished, the drones become superfluous and are eventually driven out of the hive or killed.

The hive is run by the third class of bees, sterile female *workers*. A worker's tasks are determined by her age and by conditions in the colony. A newly emerged worker starts life as a "waitress," carrying food such as honey and pollen to the queen, to other workers, and to developing larvae. As she matures, special glands begin to produce wax, and she becomes a builder, constructing perfectly hexagonal cells of wax in which the queen deposits her eggs and the larvae develop. She also takes shifts as a "maid," cleaning the hive and removing the dead, and as a guard, protecting the hive against intruders. Her final role in life is that of a forager, gathering pollen and nectar, food for the hive. She spends nearly half of her 2-month life in this role. At times during her foraging phase, she may act as a scout, seeking new and rich sources of nectar. If she finds one, she returns to the hive and communicates its location to other foragers. She communicates by means of the **waggle dance**, an elegant form of symbolic expression (**Fig. 25-27**).

Pheromones play a major role in regulating the lives of social insects. Honeybee drones are drawn irresistibly to the queen's sex pheromone (*queen substance*), which she releases during her mating flights. Back at the hive, she uses the same substance to maintain her position as the only fertile female. The queen substance is licked off her body and passed among the workers, rendering them sterile. The queen's presence and health are signaled by her continuing production of queen substance; a decrease in production alters the behavior of the workers. Almost immediately they begin building extra-large *royal cells*. The workers feed the larvae that develop in these cells a special glandular secretion known as *royal jelly*. This unique food alters the development of the growing larvae so that, instead of a worker, a new queen emerges from the royal cell. The old queen then leaves the hive, taking a swarm of workers with her to establish residence elsewhere. If more than one new queen emerges, a battle to the death ensues. The victorious queen takes over the hive.

Naked Mole Rats Form a Complex Vertebrate Society

The nervous systems of vertebrates are far more complex than those of insects, and we might therefore expect vertebrate societies to be proportionately more complex. With the exception of human society, however, they are not. Perhaps the most unusual society among non-human mammals is that of the naked mole rat (**Fig. 25-28**). These nearly blind, nearly hairless relatives of guinea pigs live in large underground colonies in southern Africa and have a form of social organization much like that of honeybees. The colony is dominated by the queen, a single reproducing female to whom all other members are subordinate.

The queen is the largest individual in the colony and maintains her status by aggressive behavior, particularly shoving. She prods and shoves lazy workers, stimulating them to become more active. As in honeybee hives, there is a division of labor among the workers, in this case based on size. Small, young rats clean the tunnels, gather food, and dig

▲ **FIGURE 25-27 Bee language: the waggle dance** A forager, returning from a rich source of nectar, performs a waggle dance that communicates the distance and direction of the food source as other foragers crowd around her, touching her with their antennae. The bee moves in a straight line while shaking her abdomen back and forth ("waggling") and buzzing her wings. She repeats this dance over and over in the same location, circling back in alternating directions.

▲ **FIGURE 25-28 A naked mole rat queen rests atop a group of workers**

more tunnels. Tunnelers line up head to tail and pass excavated dirt along the completed tunnel to an opening. Just below the opening, a larger mole rat flings the dirt into the air, adding it to a cone-shaped mound. Biologists observing this behavior from the surface dubbed it "volcanoing." In addition to volcanoing, large mole rats defend the colony against predators and members of other colonies.

If another female begins to become fertile, the queen apparently senses changes in the estrogen levels of the subordinate female's urine. The queen then selectively shoves the would-be breeder, causing stress that prevents her rival from ovulating. Although all adult males are fertile, large males are more likely to mate with the queen than are small ones. When the queen dies, a few of the females gain weight and begin shoving one another. The aggression may escalate until a rival is killed. Ultimately, a single female becomes dominant. Her body lengthens, and she assumes the queenship and begins to breed. Litters averaging 14 pups are produced about four times a year. During the first month, the queen nurses her pups, and the workers feed the queen. Then the workers begin feeding the pups solid food.

25.7 CAN BIOLOGY EXPLAIN HUMAN BEHAVIOR?

The behaviors of humans, like those of all other animals, have an evolutionary history. Thus, the techniques and concepts of ethology can help us understand and explain human behavior. Human ethology, however, will remain a less rigorous science than animal ethology. We cannot treat people as laboratory animals, devising experiments that control and manipulate the factors that influence their attitudes and actions. In addition, some observers argue that human culture has been freed from the constraints of its evolutionary past for so long that we cannot explain our behavior solely in terms of biological evolution. Nevertheless, many scientists have taken an ethological, evolutionary approach to human behavior, and their work has had a major impact on our view of ourselves.

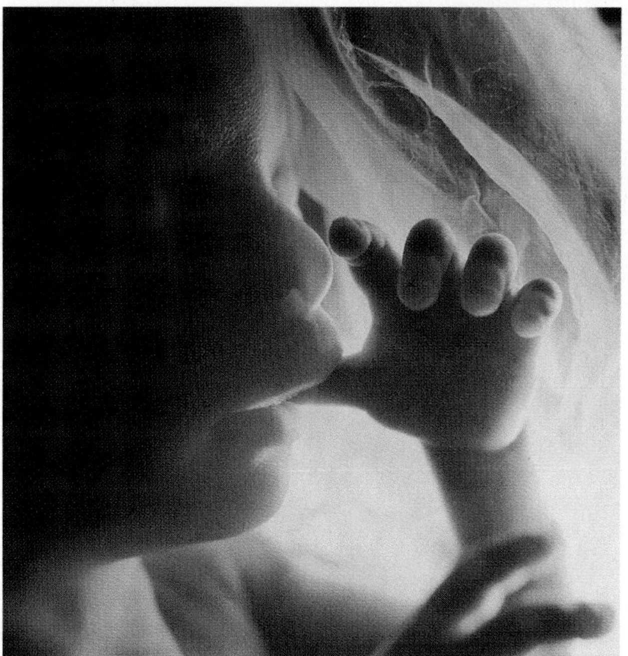

▲ **FIGURE 25-29 A human instinct** Thumb sucking is a difficult habit to discourage in young children, because sucking on appropriately sized objects is an instinctive, food-seeking behavior. This fetus sucks its thumb at about 4 months of development.

The Behavior of Newborn Infants Has a Large Innate Component

Because newborn infants have not had time to learn, we can assume that much of their behavior is innate. The rhythmic movement of an infant's head in search of its mother's breast is an innate behavior that is expressed in the first days after birth. Sucking, which can be observed even in a human fetus, is also innate (**Fig. 25-29**). Other behaviors seen in newborns and even premature infants include grasping with the hands and feet and making walking movements when the body is held upright and supported.

Another example is smiling, which can occur soon after birth. Initially, smiling can be induced by almost any object looming over the newborn. This initial indiscriminate response, however, is soon modified by experience. Infants up to 2 months old will smile in response to a stimulus consisting of two dark, eye-sized spots on a light background, which at that stage of development is a more potent stimulus for smiling than is an accurate representation of a human face. But as the child's development continues, learning and further development of the nervous system interact to limit the response to more correct representations of a face.

Newborns in their first 3 days of life can be conditioned to produce certain rhythms of sucking when their mother's voice is used as reinforcement. In experiments, infants preferred their own mothers' voices to other female voices, as indicated by their responses (**Fig. 25-30**). The infant's ability to learn his or her mother's voice and respond positively to it within days of birth has strong parallels to imprinting and may help initiate bonding with the mother.

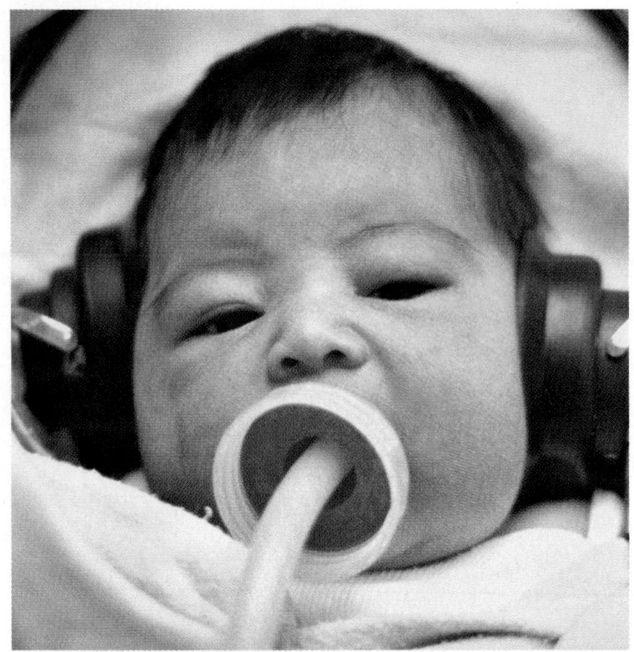

▲ **FIGURE 25-30 Newborns prefer their mother's voice** Using a nipple connected to a computer that plays audio tapes, researcher William Fifer demonstrated that newborns can be conditioned to suck at specific rates in order to listen to their own mothers' voices through headphones. For example, if the infant sucks faster than normal, her mother's voice is played; if she sucks more slowly, another woman's voice is played. Researchers found that infants easily learned and were willing to work hard at this task just to listen to their own mothers' voices, presumably because they had become used to her voice in the womb.

Young Humans Acquire Language Easily

One of the most important insights from studies of animal learning is that animals tend to have an inborn predilection for specific types of learning that are important to their species' mode of life. In humans, one such inborn predilection is for the acquisition of language. Young children are able to acquire language rapidly and nearly effortlessly; they typically acquire a vocabulary of 28,000 words before the age of 8. Research suggests that we are born with a brain that is already primed for this early facility with language. For example, a human fetus begins responding to sounds during the third trimester of pregnancy, and researchers have demonstrated that infants are able to distinguish among consonant sounds by 6 weeks after birth. In one experiment, infants sucked on a pacifier that contained a force transducer to record the sucking rate, and were conditioned to suck at a higher rate in response to playback of adult voices making various consonant sounds. When one sound (such as "ba") was presented repeatedly, the infants became habituated and decreased their sucking rate. But when a new sound (such as "pa") was presented, sucking rate increased, revealing that the infants perceived the new sound as different.

Behaviors Shared by Diverse Cultures May Be Innate

Another way to study the innate bases of human behavior is to compare simple acts performed by people from diverse cultures. This comparative approach, pioneered by ethologist Irenäus Eibl-Eibesfeldt, has revealed several gestures that seem to form a universal, and therefore probably innate, human signaling system. Such gestures include facial expressions for pleasure, rage, and disdain, and greeting movements such as an upraised hand or the "eye flash" (in which the eyes are widely opened and the eyebrows rapidly elevated). The evolution of the neural pathways underlying these gestures presumably depended on the advantages that accrued to both senders and receivers from sharing information about the emotional state and intentions of the sender. A species-wide method of communication was perhaps especially important before the advent of language and later remained useful during encounters between people who shared no common language.

Certain complex social behaviors are widespread among diverse cultures. For example, the incest taboo (avoidance of mating with close relatives) seems to be universal across human cultures (and even across many species of non-human primates). It seems unlikely, however, that a shared belief could be encoded in our genes. Some biologists have suggested that the taboo is instead a cultural expression of an evolved, adaptive behavior. According to this hypothesis, close contact among family members early in life suppresses sexual desire, and this response arose because of the negative consequences of inbreeding (such as a higher incidence of genetic diseases). The hypothesis does not require us to assume an innate social belief, but rather proposes that we inherit a learning program that causes us to undergo a kind of imprinting early in life.

Humans May Respond to Pheromones

Although the main channels of human communication are through the eyes and ears, humans also seem to respond to chemical messages. The possible existence of human pheromones was hinted at in the early 1970s, when biologist Martha McClintock found that the menstrual cycles of roommates and close friends tended to become synchronized. McClintock suggested that the synchrony resulted from some chemical signal between the women, but almost 30 years passed before she and her colleagues uncovered more conclusive evidence that a pheromone was at work.

In 1998, McClintock's research group asked nine female volunteers to wear cotton pads in their armpits for 8 hours each day during their menstrual cycles. The pads were then disinfected with alcohol and swabbed above the upper lips of another set of 20 female subjects (who reported that they could detect no odors other than alcohol on the pads). The subjects were exposed to the pads in this way each day for 2 months, with half the group sniffing secretions from women in the early (preovulation) part of the menstrual cycle, while the other half was exposed to secretions from later

Case Study continued
Sex and Symmetry

Does symmetry have a scent? In one study, researchers measured the body symmetry of 80 men and then issued a clean T-shirt to each one. Each subject wore his shirt to bed for two consecutive nights. A panel of 82 women sniffed the shirts and rated their scents for "pleasantness" and "sexiness." Which shirts had the sexiest, most pleasant scents? The ones worn by the most symmetrical men. The researchers concluded that women can identify symmetrical men by their scent.

in the cycle (postovulation). Women exposed to early-cycle secretions had shorter-than-usual menstrual cycles, and women exposed to late-cycle secretions had delayed menstruation. It appears that women release different pheromones, with different effects on receivers, at different points in the menstrual cycle.

Although McClintock's experiment offers strong evidence for the existence of human pheromones, little else is known about chemical communication in humans. The actual molecules that caused the effects documented by McClintock remain unknown, as does their function. (What benefit would a woman gain by influencing the menstrual cycles of other women?) Receptors for chemical messages have not yet been found in humans, and we don't know if the "menstrual pheromones" are the first known example of an important communication system or merely an isolated case of a vestigial ability. Despite the hopeful advertisements for "sex attraction pheromones" on late-night television, chemical communication in humans is a scientific mystery awaiting a solution.

Studies of Twins Reveal Genetic Components of Behavior

Twins present an opportunity to examine the hypothesis that differences in human behavior are related to genetic differences. If a particular behavior is heavily influenced by genetic factors, we would expect to find that *identical twins* (which arise from a single fertilized egg and have identical genes) are more likely to share the behavior than are *fraternal twins* (which arise from two individual eggs fertilized by different sperm and are no more similar genetically than are other siblings). Data from twin studies, and from other within-family investigations, have tended to confirm the heritability of many human behavioral traits. These studies have documented a significant genetic component for traits such as activity level, alcoholism, sociability, anxiety, intelligence, dominance, and even political attitudes. On the basis of tests designed to measure many aspects of personality, identical twins are about two times more similar in personality than are fraternal twins.

The most fascinating twin findings come from observations of identical twins separated soon after birth, reared in different environments, and reunited for the first time as adults. Identical twins reared apart have been found to be as similar in personality as those reared together, indicating that the differences in their environments had little influence on their personality development. They have been found to share nearly identical taste in jewelry, clothing, humor, food, and names for children and pets. In some cases these separated twins share personal idiosyncrasies such as giggling, nail biting, drinking patterns, hypochondria, and mild phobias.

Biological Investigation of Human Behavior Is Controversial

The field of human behavioral genetics is controversial, especially among nonscientists, because it challenges the long-held belief that environment is the most important determinant of human behavior. As discussed earlier in this chapter, we now recognize that all behavior has some genetic basis and that complex behavior in non-human animals typically combines elements of both innate and learned behaviors. Thus, it seems certain that our own behavior is influenced by both our evolutionary history and our cultural heritage. The debate over the relative importance of heredity and environment in determining human behavior continues and is unlikely ever to be fully resolved. Human ethology is not yet recognized as a rigorous science, and it will always be hampered because we can neither view ourselves with detached objectivity nor experiment with people as if they were laboratory rats. Despite these limitations, there is much to be learned about the interaction of learning and innate tendencies in humans.

Case Study revisited
Sex and Symmetry

In the experiment described at the beginning of this chapter, women found males with the most symmetrical bodies to be the most attractive. But how did the women know which males were most symmetrical? After all, the researchers' measurement of male symmetry was based on small differences in the sizes of body parts that the female judges did not even see during the test.

Perhaps male body symmetry is reflected in facial symmetry, and females prefer symmetrical faces. To test this hypothesis, a group of researchers used computers to alter photos of male faces, either increasing or decreasing their symmetry (**Fig. 25-31**). Then heterosexual female observers rated each face for attractiveness. The observers had a strong preference for more symmetrical faces.

Why would females prefer to mate with symmetrical males? The most likely explanation is that symmetry indicates good physical condition. Disruptions of normal embryological development can cause bodies to be asymmetrical, so a highly symmetrical body indicates healthy, normal development. Females that mate with individuals whose health and vitality are announced by their symmetrical bodies might have offspring that are similarly healthy and vital.

Consider This

Is our perception of human beauty determined by cultural standards, or is it part of our biological makeup, the product of our evolutionary heritage? What evidence would persuade you that beauty is a biological phenomenon or that it is a cultural one?

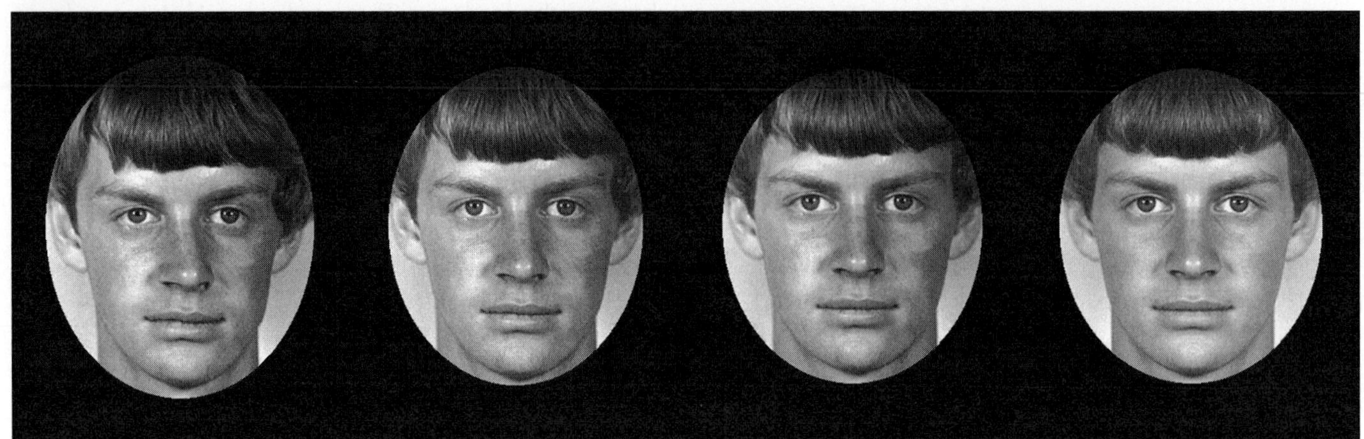

▲ **FIGURE 25-31 Faces of varying symmetry** Researchers used sophisticated software to modify facial symmetry. From left: a face modified to be less symmetrical; the original, unmodified face; a face modified to be more symmetrical; a perfectly symmetrical face.

CHAPTER REVIEW

Summary of Key Concepts

25.1 How Do Innate and Learned Behaviors Differ?

Although all animal behavior is influenced by both genetic and environmental factors, biologists distinguish innate behaviors, whose development is not highly dependent on external factors, and learned behaviors, which require more extensive experiences with environmental stimuli in order to develop. Innate behaviors can be performed properly the first time an animal encounters the appropriate stimulus, whereas learned behavior changes in response to the animal's social and physical environment.

The distinction between innate and learned behavior is often blurred in naturally occurring behaviors. Learning allows animals to modify innate responses so that they occur only with appropriate stimuli. Imprinting, a form of learning with innate constraints, is possible only at a certain time in an animal's development.

25.2 How Do Animals Communicate?

Communication allows animals of the same species to interact effectively in their quest for mates, food, shelter, and other resources. Animals communicate through visual signals, sound, chemicals (pheromones), and touch. Visual communication is quiet and can convey rapidly changing information. Visual signals are active (body movements) or passive (body shape and color). Sound communication can also convey rapidly changing information, and it is effective when vision is impossible. Pheromones can be detected after the sender has departed, conveying simple messages over time. Physical contact reinforces social bonds and is a part of mating.

25.3 How Do Animals Compete for Resources?

Although many competitive interactions are resolved through aggression, serious injuries are rare. Most aggressive encounters are settled by displays that communicate the motivation, size, and strength of the combatants.

Some species establish dominance hierarchies that minimize aggression. On the basis of initial aggressive encounters, each animal acquires a status in which it defers to more dominant individuals and, in turn, dominates subordinates. When resources are limited, dominant animals obtain the largest share and are most likely to reproduce.

Territoriality, a behavior in which animals defend areas where important resources are located, also minimizes aggressive encounters. In general, territorial boundaries are respected, and the best-adapted individuals defend the richest territories and produce the most offspring.

25.4 How Do Animals Find Mates?

Successful reproduction requires that animals recognize the species, sex, and sexual receptivity of potential mates. In many species, animals also assess the quality of potential mates. These requirements have contributed to the evolution of sexual displays that use all forms of communication.

25.5 Why Do Animals Play?

Animals of many species engage in seemingly wasteful (and sometimes dangerous) play behavior. Play behavior in young animals has been favored by natural selection, probably because it provides opportunities to practice and perfect behaviors that will later be crucial for survival and reproduction.

25.6 What Kinds of Societies Do Animals Form?

Social living has both advantages and disadvantages, and species vary in the degree to which their members cooperate. Some species form cooperative societies. The most rigid and highly organized are those of the social insects such as the honeybee, in which the members follow rigidly defined roles throughout life. These roles are maintained through both genetic programming and the influence of certain pheromones. Naked mole rats exhibit the most complex and rigid vertebrate social interactions, resembling those of social insects.

25.7 Can Biology Explain Human Behavior?

Researchers are increasingly investigating whether human behavior is influenced by evolved, genetically inherited factors. This emerging field is controversial. Because we cannot freely experiment on humans, and because learning plays a major role in nearly all human behavior, investigators must rely on studies of newborn infants, comparative cultural studies, and studies of identical and fraternal twins. Evidence is mounting that our genetic heritage plays a role in personality, intelligence, simple universal gestures, our responses to certain stimuli, and our tendency to learn specific things such as language at particular stages of development.

Key Terms

aggression *473*
altruism *481*
behavior *465*
communication *470*
dominance hierarchy *474*
ethology *468*
habituation *466*
imprinting *469*
innate *465*
insight learning *468*
kin selection *481*
learning *466*
operant conditioning *466*
pheromone *472*
territoriality *475*
trial-and-error learning *466*
waggle dance *482*

Thinking Through the Concepts

Fill-in-the-Blank

1. In general, animal behaviors arise from an interaction between the animal's __________ and its __________. Some behaviors are performed correctly the first time an animal encounters the proper __________. Such behaviors are described as __________.
2. Play is almost certainly an adaptive behavior because it uses considerable __________, it can distract the animal from watching for __________, and it may cause __________. The most likely explanation for why animals play is the "__________ hypothesis," which states that play teaches the young animal __________ that will be useful as a(n) __________. This hypothesis is supported by the observation that animals that have larger __________ and are most capable of __________ are more likely to play.
3. One of the simplest forms of learning is __________, defined as a decline in response to a(n) __________, harmless stimulus. A different type of learning in which an animal's nervous system is rigidly programmed to learn a certain behavior during a certain period in its life is called __________. The time frame during which such learning occurs is called the __________.
4. Animals often deal with competition for resources through __________ behavior. Such conflicts are often resolved through __________, which allow the competing animals to assess each other without __________ each other. Animals that resolve conflicts this way often display their __________, and make their bodies appear __________.
5. The defense of an area where important resources are located is called __________. Examples of important resources that may be defended include places to __________, __________, __________, and __________. Such resources are most commonly defended by which sex? __________ Are these spaces usually defended blank against members of the same or of different species? __________
6. After each form of communication, list a major advantage in the first blank, and a major disadvantage in the second blank: Pheromones: __________; __________. Visual displays: __________; __________.

Review Questions

1. Explain why neither the term "innate" nor "learned" adequately describes the behavior of any given organism.
2. Explain why animals play. Include the features of play in your answer.
3. List four senses through which animals communicate, and give one example of each form of communication. For each sense listed, present both advantages and disadvantages of that form of communication.
4. A bird will ignore a squirrel in its territory, but will act aggressively toward a member of its own species. Explain why.
5. Why are most aggressive encounters among members of the same species relatively harmless?
6. Discuss the advantages and disadvantages of group living.
7. In what ways do naked mole rat societies resemble those of the honeybee?

Applying the Concepts

1. Male mosquitoes orient toward the high-pitched whine of the female, and female mosquitoes, the only sex that sucks blood, are attracted to the warmth, humidity, and carbon dioxide exuded by their prey. Using this information, design a mosquito trap or killer that exploits a mosquito's innate behaviors. Then, design one for moths.
2. You raise honeybees but are new at the job. Trying to increase honey production, you introduce several queens into the hive. What is the likely outcome? What different things could you do to increase production?
3. Describe and give an example of a dominance hierarchy. What role does it play in social behavior? Give a human parallel, and describe its role in human society. Are the two roles similar? Why or why not? Repeat this exercise for territorial behavior in humans and in another animal.
4. You are manager of an airport. Planes are being endangered by large numbers of flying birds, which can be sucked into and disable the engines. Without harming the birds, what might you do to discourage them from nesting and flying near the airport and its planes?

MB® *Go to www.masteringbiology.com for practice quizzes, activities, eText, videos, current events, and more.*

Population Growth and Regulation

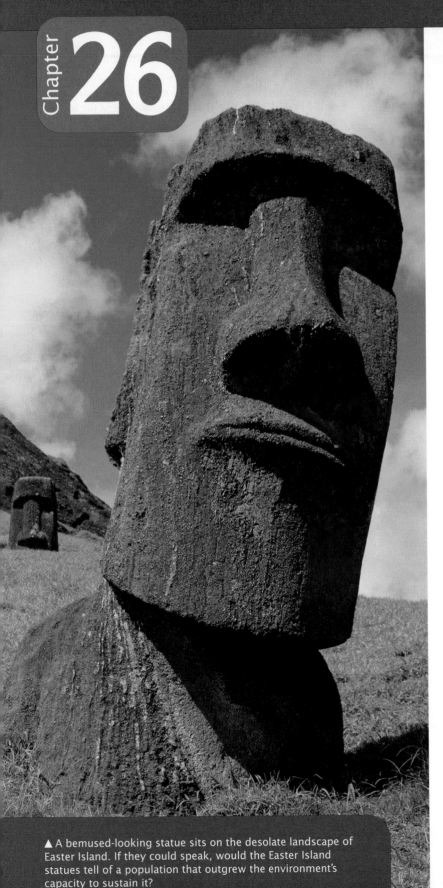

Chapter 26

▲ A bemused-looking statue sits on the desolate landscape of Easter Island. If they could speak, would the Easter Island statues tell of a population that outgrew the environment's capacity to sustain it?

Case Study

The Mystery of Easter Island

WHY DO CIVILIZATIONS VANISH? Among those who have pondered this question were the first Europeans to reach Easter Island. When they arrived in the 1700s, these voyagers were mystified by the enormous stone statues that dominate the island's barren landscape. The island's few inhabitants had no record or memory of the statues' creators, and did not possess the technology to transport and erect such huge and heavy structures. Moving statues that weigh up to 80 tons over the 6 miles from the nearest stone quarry, and then maneuvering them into an upright position, would have required long ropes and strong timbers. Easter Island, however, was devoid of anything that could have furnished the needed wood or fibers for rope. There were almost no trees, and none of island's shrubs grew higher than 10 feet.

An important clue to the mystery of Easter Island was revealed by scientists who studied the pollen grains from layers of ancient sediments. Because the age of each sediment layer can be determined by radioactive carbon dating (see p. 324), and because each plant species can be identified by the unique appearance of its pollen, pollen analysis can show how vegetation has changed over time. The researchers discovered that the island was once covered by a diverse forest, including toromiro trees that make excellent firewood; hauhau trees that could supply fiber for rope; and Easter Island palm trees, with long, straight trunks that would have made good rollers for moving statues.

The island's first settlers probably arrived sometime between the years 800 and 1000 A.D., and by about 1400, almost all of Easter Island's trees were gone. Most researchers agree that the demise of the forest began with the arrival of humans, who cleared land for agriculture and used the trees for firewood and construction materials. Apparently, the culture responsible for the statues disappeared along with the forest. Could there have been a connection between the two disappearances?

At a Glance

26.1 HOW DOES POPULATION SIZE CHANGE?

A **population** consists of all the members of a particular species that live within an ecosystem. On Easter Island, for example, the Easter Island palm trees, the hauhau trees, and the toromiro trees each constituted a different population. Populations do not exist in isolation, however, and each forms an integral part of a larger **community,** defined as a group of interacting populations. Communities, in turn, exist within **ecosystems,** which include all the living and nonliving components of a defined geographical area. An ecosystem can be as small as a pond or as large as an ocean; it can be a field, a forest, or an island. The **biosphere** can be viewed as the enormous ecosystem that encompasses all of Earth's habitable surface. **Ecology** (from the Greek "oikos," meaning "a place to live") is the study of the interrelationships of organisms with each other and with their nonliving environment. We begin this ecology unit with an overview of populations.

Studies of undisturbed ecosystems show that some populations tend to remain relatively stable in size over time, while others fluctuate in a roughly cyclical pattern, and still others vary sporadically in response to complex environmental variables. In contrast to most non-human species, the global human population has grown for centuries. Here, we examine how and why populations grow, and then examine the factors that control this growth.

Population Size Is the Outcome of Opposing Forces

Populations change through births, deaths, or migration. The natural increase of a population is the difference between births and deaths. The net migration of a population is the difference between **immigration** (migration into the population) and **emigration** (migration out of the population). A population grows when the sum of natural increase (births minus deaths) and net migration (immigration minus emigration) is positive. A population shrinks when the sum of natural increase and net migration is negative. A simple equation for the change in population size within a given time span is shown in **Figure 26-1**.

Although populations can be significantly influenced by migration, in most natural populations, birth and death rates are the primary factors that influence population growth, and we focus on these factors in the calculations that follow.

The size of any population results from the interaction between two major opposing factors that determine birth and death rates: biotic potential and environmental resistance. **Biotic potential** is the theoretical maximum rate at which a population could increase, assuming ideal conditions that allow a maximum birth rate and a minimum death rate. **Environmental resistance** refers to the curbs on population growth that are set by the living and nonliving environment. Examples of environmental resistance include interactions among species, such as competition, predation, and parasitism (described in a later section). Environmental resistance also encompasses the always-limited availability of nutrients, energy, and space, as well as natural events such as storms, fires, freezing weather, floods, and droughts.

For long-lived organisms in nature, the interaction between biotic potential and environmental resistance usually results in a balance between the size of a population and the resources available to support it. To understand how populations grow and how their size is regulated, let's examine each of these factors in more detail.

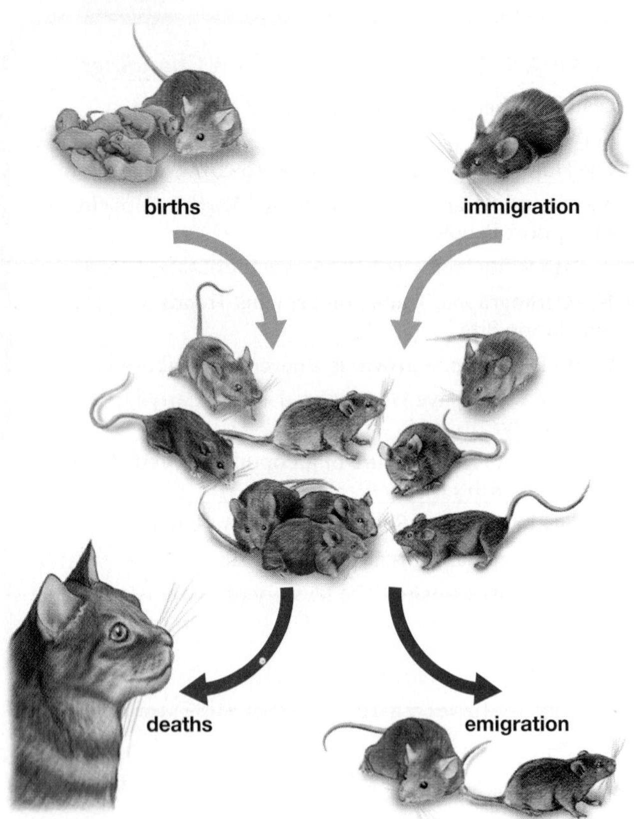

▲ **FIGURE 26-1 Population change** Populations change in size as individuals are added through births or immigration, and are removed by death or emigration.

Biotic Potential Can Produce Exponential Growth

Evolutionarily successful organisms possess traits that make them well adapted to their environment, and they pass these inherited traits on to as many healthy offspring as possible. As a result of natural selection, all organisms have evolved the capacity to reproduce themselves manyfold over their lifetimes. If environmental resistance is reduced, populations can grow extremely rapidly.

Population Growth Is a Function of the Birth Rate, the Death Rate, and Population Size

The change in population size over time is a function of birth rate, death rate, and the number of individuals in the original population. The **birth rate** (b) and the **death rate** (d) are often expressed as the number of births or deaths per individual during a specific unit of time, such as a month or a year.

The **growth rate** (r) of a population is a measure of the change in population size per individual per unit of time. The growth rate is also called the *rate of natural increase*, because it is based solely on births and deaths, and does not take migration into account. This value is determined by subtracting the death rate (d) from the birth rate (b):

$$\underset{\text{(growth rate)}}{r} = \underset{\text{(birth rate)}}{b} - \underset{\text{(death rate)}}{d}$$

If the birth rate exceeds the death rate, the population will grow. If the death rate exceeds the birth rate, the growth rate will be negative and the population will decline. To calculate the annual growth rate of a population of 1,000, in which 150 births (150 is 15% of 1,000, or 0.15) and 50 deaths (50 is 5% of 1,000, or 0.05) occur each year, we can use this simple equation:

$$\underset{\text{(growth rate)}}{r} = \underset{\text{(birth rate)}}{0.15} - \underset{\text{(death rate)}}{0.05} = 0.1 = 10\% \text{ per year}$$

To determine population growth (G), which is the number of individuals added to a population in a given time period, we multiply the growth rate (r) by the original population size (N):

$$\underset{\text{(growth per unit time)}}{G} = \underset{\text{(growth rate)}}{r} \times \underset{\text{(population size)}}{N}$$

In this example, population growth (rN) equals $0.1 \times 1{,}000 = 100$ during the first year. If this growth rate is constant, then the second year, the population size (N) starts at 1,100, and during the second year it will grow by 110 individuals ($0.1 \times 1{,}100$). During the third year, 121 individuals are added, and so on.

If Births Exceed Deaths by a Constant Percentage, Population Growth Produces a J-Curve

This pattern of continuously accelerating increase in population size is called **exponential growth.** Over a given time period, an exponentially growing population grows by a fixed percentage of its size at the beginning of that time period. Thus, an increasing number of individuals are added to the population during each succeeding time period, causing population size to grow at an ever-accelerating pace. This will occur in any population where each individual, over the course of its life span, produces (by a constant number) more than one offspring that survives to reproduce. Although the number of offspring an individual produces each year varies from millions (for an oyster) to one or fewer (for a human), each organism—whether working alone or as part of a sexually reproducing pair—has the potential to replace itself many times over during its reproductive lifetime.

This high biotic potential evolved because it helps ensure that, in a world filled with forces of environmental resistance, some offspring survive to reproduce. Several factors influence biotic potential, including the following:

- The age at which the organism first reproduces.
- The frequency at which reproduction occurs.
- The average number of offspring produced each time the organism reproduces.
- The length of the organism's reproductive life span.
- The death rate of individuals under ideal conditions.

Whether people or bacteria, if the size of a population that is fulfilling (or even approaching) its biotic potential is

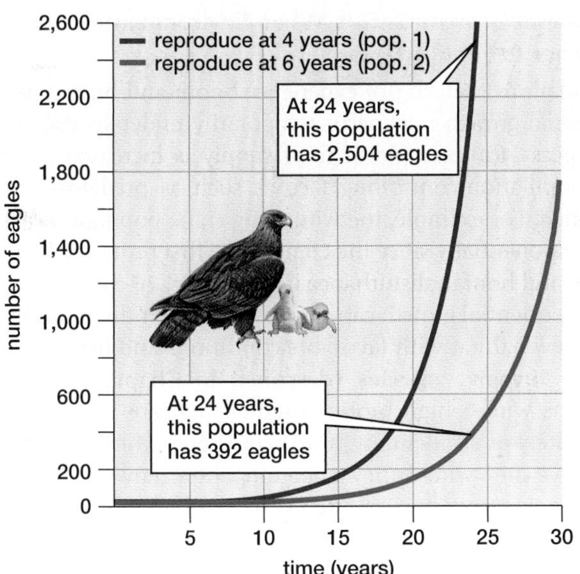

Time (years)	Number of eagles (pop. 1)	Number of eagles (pop. 2)
0	2	2
6	8	4
12	52	18
18	362	86
24	2,504	392
30	17,314	1,764

◀ **FIGURE 26-2 Exponential growth curves are J-shaped** The graph shows the growth of two hypothetical populations of eagles, each starting with a single pair, but differing in the age at which breeding begins. Although both curves show the J-shape of exponential growth, at 24 years, the population that begins reproducing at age 4 is over six times that of the population that begins reproducing at age 6. During the 30-year lifetime of the founding pairs, the population that begins reproducing at age 4 reaches 10 times the size of the population that begins reproducing at age 6 (see the table).

graphed against time, a characteristic shape called a **J-curve** will be produced. To illustrate the J-curve, we use the example of the biotic potential of golden eagles, a relatively long-lived, rather slowly reproducing species. We will assume that golden eagles live for 30 years and that each pair produces two offspring annually after reaching sexual maturity.

Figure 26-2 illustrates the growth of two hypothetical eagle populations (each founded by a single breeding pair) during the founding pair's 30-year life span. The red line shows the J-curve of population growth if reproduction begins at age 4, and the blue line plots population growth assuming that reproduction begins at age 6. In each case, exponential growth occurs and the shapes of the curves are identical. Note that after 24 years, however, the eagle population whose members begin reproducing at 4 years of age is over six times as large as the population whose members begin reproducing at age 6; at 30 years, it would be 10 times as large (see the table in Fig. 26-2). As this example illustrates, delayed reproduction significantly slows population growth. This also applies to humans; if each woman bears three children in her early teens, the population will grow far more rapidly than if each woman bears five children, but has her first after age 30.

Even under ideal conditions, however, deaths are inevitable, and the biotic potential of a population takes minimum death rates into account. **Figure 26-3** compares three

Have you ever wondered?

Why So Many Flies Plague Your Picnic?

Flies swarming around your food can definitely diminish your appetite. Where do they all come from? Female houseflies can lay about 120 eggs at a time. The eggs hatch and mature within 2 weeks, so seven generations can easily occur during the warmer months of the year. If unhindered by environmental resistance, the seventh generation would contain roughly 6 trillion flies—all descended from a single pregnant female! Thank goodness for birds, cold winters, and other forms of environmental resistance.

Does the high biotic potential of house flies provide any insight into the potential growth of the human population? In 1708, Johann Sebastian Bach, the famous German composer, had his first of 20 children, 10 of whom lived to adulthood. If his surviving offspring had each had 10 surviving children and so on to the present time, Bach's descendants today would number roughly 12 billion (making reasonable assumptions about the factors influencing biotic potential), about 80% more than Earth's entire current population.

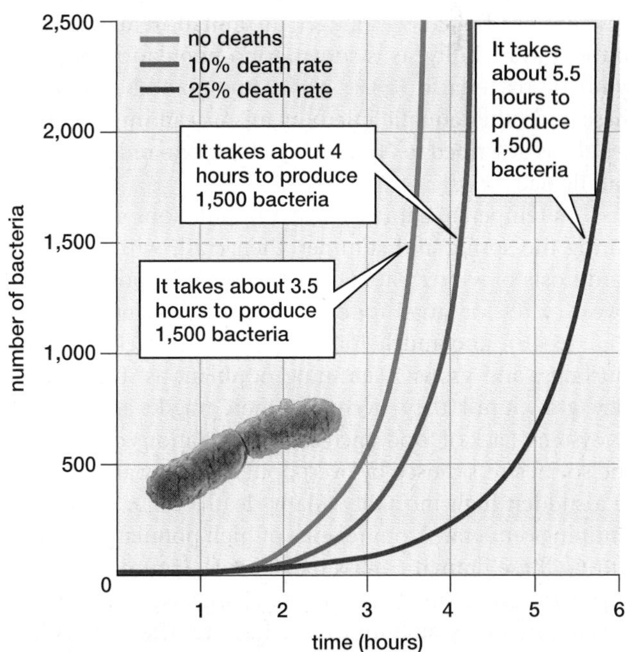

▲ **FIGURE 26-3 The effect of death rates on population growth** This graph assumes that a bacterial population doubles every 20 minutes. Notice that both show the characteristic J-shape of exponential growth curves, although it takes the population with the higher death rate longer to achieve any given size.

hypothetical bacterial populations experiencing different death rates. Notice that the three curves have the same J shape. As long as births exceed deaths, the population eventually becomes enormous, but as the death rate increases, it takes longer to reach any given population size.

26.2 HOW IS POPULATION GROWTH REGULATED?

In 1859, Charles Darwin wrote: "There is no exception to the rule that every organic being naturally increases at so high a rate that, if not destroyed, the Earth would soon be covered by the progeny of a single pair." Clearly, population growth cannot continue indefinitely.

Exponential Growth Occurs Only Under Special Conditions

Natural populations fluctuate in growth rate, and so never exhibit perfect exponential growth. Under special circumstances, however, natural populations can approximate exponential growth, producing J-shaped growth curves.

Exponential Growth Occurs in Populations That Exhibit Boom-and-Bust Cycles

Exponential growth occurs in populations that undergo regular cycles in which rapid population growth is followed by a sudden, massive die-off. These **boom-and-bust cycles** occur in a variety of organisms for complex and varied reasons. Many short-lived, rapidly reproducing species—from photosynthetic microbes to insects—have seasonal population cycles that are linked to changes in rainfall, temperature, or nutrient availability, as is shown for a population of photosynthetic bacteria in **Figure 26-4a**. Boom-and-bust cycles in these and other aquatic microorganisms can impact human health, as described in "Health Watch: Boom-and-Bust Cycles Can Be Bad News."

In temperate climates, insect populations grow rapidly during the spring and summer, then crash with the killing hard frosts of winter. More complex factors produce roughly 4-year cycles for small rodents such as voles and lemmings (**Fig. 26-4b**), and much longer population cycles in hares, muskrats, and grouse. Lemming populations, for example, may grow until they overgraze their fragile arctic tundra ecosystem. Lack of food, increasing populations of predators, and social stress caused by overpopulation may all contribute to a sudden high mortality. Many deaths occur as waves of lemmings emigrate from regions of high population density. During these dramatic mass movements, lemmings are easy targets for predators. Many drown as they encounter bodies of water that they attempt to swim across. The reduced lemming population eventually contributes to a decline in predator numbers, as well as a recovery of the plant community on which the lemmings feed. These responses, in turn, set the stage for the next round of exponential growth in the lemming population (see Fig. 26-4b).

Exponential Growth Occurs When Environmental Resistance Is Reduced

In populations that do not experience boom-and-bust cycles, exponential growth may occur temporarily under special circumstances—for example, if food supply is increased, or if other population-controlling factors, such as predators, are eliminated. For example, the whooping crane population has grown exponentially since the cranes were first protected from hunting and human disturbance in 1940 (**Fig. 26-5**).

Exponential growth can also occur when individuals invade a new habitat with favorable conditions and little competition. **Invasive species** (described in Chapter 27) are organisms with a high biotic potential that are introduced (deliberately or accidentally) into ecosystems where they did not evolve and where they encounter little environmental resistance. Invasive species often show explosive population

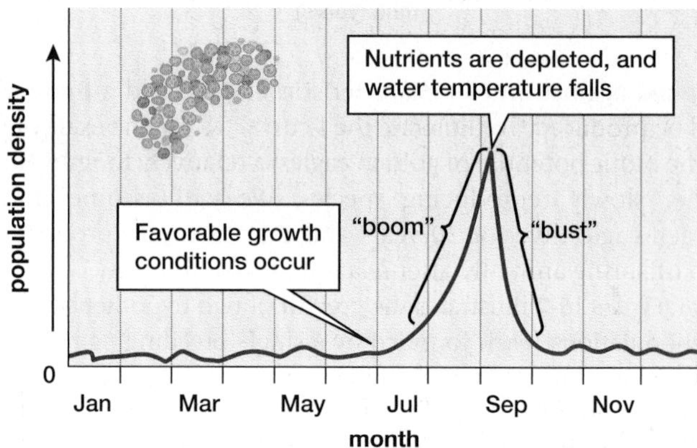

(a) A boom-and-bust cycle in photosynthetic bacteria

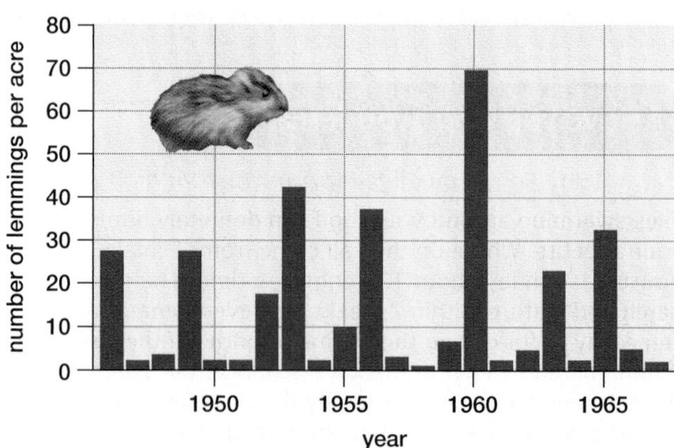

(b) Boom-and-bust cycles in a lemming population in Alaska

▲ FIGURE 26-4 **Boom-and-bust population cycles (a)** The density of a hypothetical population of photosynthetic bacteria in a lake over time during an annual boom-and-bust cycle. These microorganisms persist at a low level throughout the fall, winter, and spring. Early in July, conditions become favorable for growth, and exponential growth occurs through the month of August. In early September, when conditions change, the population plummets. **(b)** This Alaska lemming population follows a roughly 4-year cycle of boom and bust (data from Point Barrow, Alaska).

QUESTION What factors might make the data in the lemming graph somewhat erratic and irregular?

Health Watch

Boom-and-Bust Cycles Can Be Bad News

In August of 2003, in western Lake Erie (one of the Great Lakes spanning the U.S.–Canadian border), populations of the photosynthetic bacteria *Microcystis* exploded. Although they are microscopic—50 in a row would just span the period at the end of this sentence—their bodies formed a thick scum on the lake and washed ashore in rotting, foul-smelling mats.

During the past three decades, late summer population explosions of sometimes toxic, photosynthetic protists and bacteria have occurred with increasing frequency throughout the world. Collectively called *harmful algal blooms (HABs)*, these population booms of toxic microorganisms kill fish and sicken people. They also cause major economic losses to the shellfish industry because clams, mussels, and scallops feed on these organisms and concentrate the poisons in their bodies, posing a hazard to consumers. Some of the most serious poisons are neurotoxins produced by a handful of species of dinoflagellates (a single-celled protist), such as *Karenia brevis* (**Fig. E26-1**), which can reach densities of 20 million per liter of water. This and other protist species can cause red tides that result in massive fish kills (see Fig. 20-11).

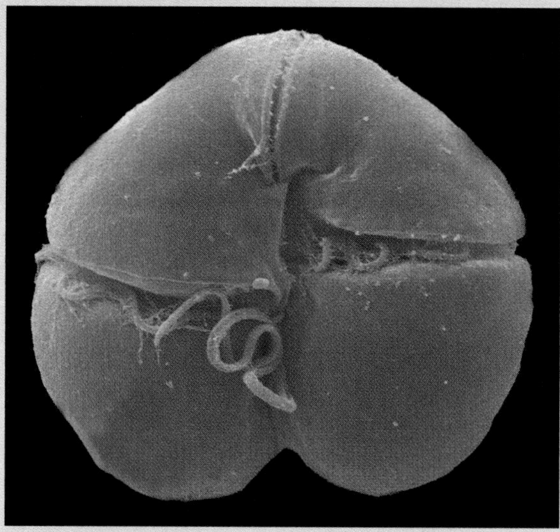

▲ **FIGURE E26-1 The cause of red tides** The dinoflagellate *Karenia brevis* (seen in this artificially-colored scanning electron micrograph) causes red tides in the United States.

Many of the bacteria and protists that produce harmful algal blooms are natural residents of lakes and coastal waters. What causes these populations to "bloom and boom"? The reasons are complex, and vary with the species, but warm water temperatures and adequate nutrients, such as phosphorus and nitrogen, are required. Runoff of these nutrients from human agricultural activities is increasing the frequency and intensity of HABs throughout the world. Global warming may be adding to the problem by fostering more rapid growth of these protists and extending their growing season. The booming populations go "bust" when the enormous populations of cells deplete the local water of nutrients, and falling water temperatures in the autumn and winter further decrease their reproductive rate.

Government agencies are increasing research efforts to improve their ability to monitor HABs, for example, by using satellite images that show changes in water color caused by the high density of microorganisms. They are also funding research into the conditions that stimulate HABs, to allow earlier and more accurate forecasting. Studies of ways to suppress the growth of organisms causing HABs (without harming other forms of life) are also under way.

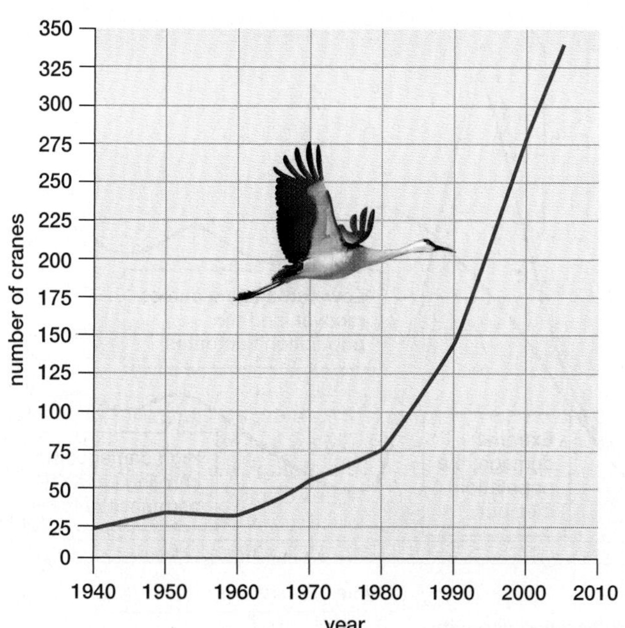

◀ **FIGURE 26-5 Exponential growth of wild whooping cranes** Hunting and habitat destruction had reduced the world's whooping crane population to about 20 individuals before they were protected in 1940. By 2005, their wild population had grown to 340 individuals. Note the J-curve characteristic of exponential growth. Data from the U.S. Fish and Wildlife Service. 2008. *International recovery plan for the whooping crane.*

growth. As described later, the rats that accompanied Polynesian settlers to Easter Island are an example. In another case, people introduced 100 cane toads into Australia in 1935 to control beetles that were destroying sugar cane. Cane toad females lay 7,000 to 35,000 eggs at a time, and in their new environment, the cane toads encountered few predators. Spreading outward from their release point, they now inhabit an area of nearly 400,000 square miles and are migrating rapidly into new habitats, threatening native species by eating or displacing them. Their population, now estimated at well over 200 million, continues to grow exponentially.

As you will learn in the next section, all populations that exhibit exponential growth must eventually either stabilize or drop. If the decline is sudden and dramatic, it is described as a *population crash.*

Environmental Resistance Limits Population Growth

Imagine a sterile culture dish in which nutrients are constantly replenished and wastes removed. If a small number of living skin cells were added, they would attach to the bottom and begin reproducing by mitotic cell division. If you counted the cells daily under a microscope and graphed their numbers, for a while, your graph would resemble the J-curve characteristic of exponential growth. But as the cells began to occupy all the available space on the dish, their growth rate would slow and eventually drop to zero, causing the population size to remain constant.

Logistic Growth Occurs When New Populations Stabilize as a Result of Environmental Resistance

Your graph of skin cell numbers will now resemble the one in **Figure 26-6a**. This growth pattern, described as **logistic population growth,** is characteristic of populations that increase up to the maximum number that their environment can sustain, and then stabilize. The maximum population size that can be sustained by an ecosystem for an extended period of time without damage to the ecosystem is called its **carrying capacity *(K)*.** The curve that results

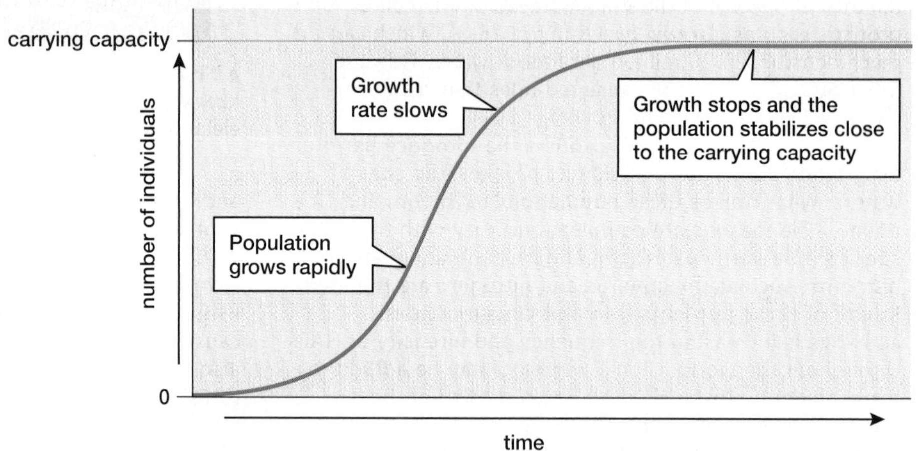

(a) An S-shaped growth curve stabilizes at carrying capacity

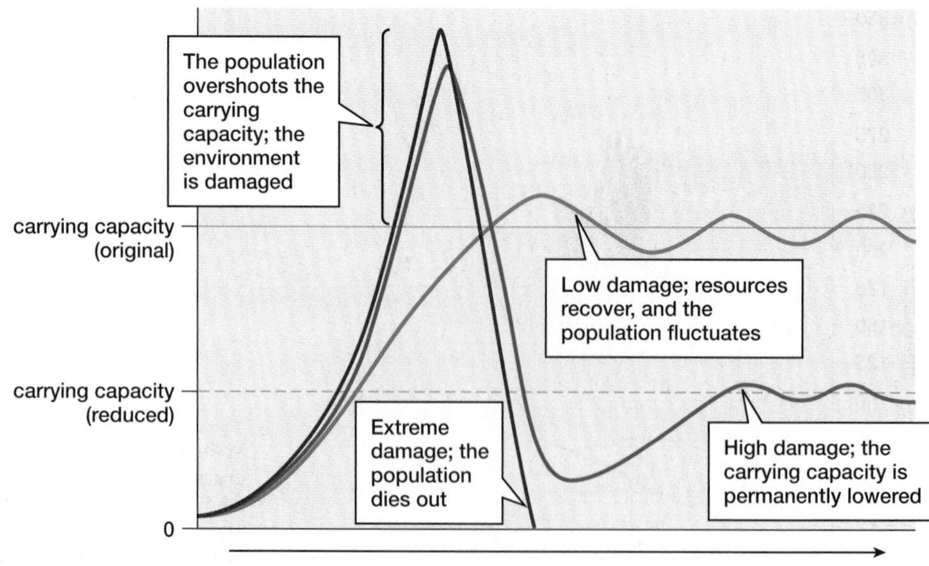

(b) Consequences of exceeding carrying capacity

▶ **FIGURE 26-6 The S-curve of logistic population growth** **(a)** During logistic growth, a population will remain small for a time before its numbers rise dramatically. Eventually, the growth rate slows as the population encounters increasing density-dependent environmental resistance. Population growth finally ceases at or near carrying capacity *(K)*. The result is a curve shaped like a "lazy S." **(b)** Populations can overshoot carrying capacity *(K)*, but only for a limited time. Three possible results are illustrated.

A Closer Look At *Logistic Population Growth*

The mathematical formula that produces a logistic growth curve consists of the formula for exponential growth ($G = rN$) multiplied by a factor that imposes limits on this growth. For natural populations, these limits are imposed by the environment. The formula includes the variable (K) that represents the carrying capacity of the ecosystem. The equation for the S-curve for logistic population growth is:

$$G = rN\frac{(K - N)}{K}$$

The new factor $(K - N)/K$ limits exponential growth ($G = rN$) based on the carrying capacity (K). To Understand this, let's start with the value ($K - N$). When we subtract the current population (N) from the carrying capacity (K), we get the number of individuals that can still be added to the current population before it reaches carrying capacity. Now, if we divide this new number by K, we get the fraction of the carrying capacity that can still be added to the current population before it reaches carrying capacity and stops growing ($G = 0$). When N is very small, $(K - N)/K$ is close to 1, and the equation resembles that for exponential growth. This produces the initial portion of the S-curve, which is essentially a J-curve. As N increases over time, however, $K - N$ will approach zero. The growth rate will slow, and the steeply rising portion of the initial J-curve will begin to level off. When the population size (N) equals the carrying capacity (K), population growth will cease ($G = 0$), as shown on the final portion of the S-curve (see Fig. 26-6a).

when logistic growth is graphed is called an **S-curve,** after its general shape. The formula for logistic growth (which is based on the assumption that the population size cannot overshoot K) is explained in "A Closer Look at Logistic Population Growth."

In nature, an increase in population size (N) above carrying capacity (K) can be sustained for a short time. This is dangerous, however, because a population above carrying capacity is living at the expense of resources that cannot regenerate as fast as they are being depleted. A small overshoot above K is likely to be followed by a decrease in both K and N, until the resources recover and the original K is restored.

If a population far exceeds the carrying capacity of its environment, the consequences are more severe, because in this situation, the excess demands placed on the ecosystem are likely to destroy essential resources (such as Easter Island's forests), which may be unable to recover. This can permanently and severely reduce K, causing the population to decline to a fraction of its former size, or to disappear entirely (**Fig. 26-6b**). For example, when reindeer were introduced onto an island with no large predators, their population increased rapidly, seriously overgrazing the vegetation they relied on for food. As a result, the reindeer population plummeted, as shown in **Figure 26-7**.

Case Study continued

The Mystery of Easter Island

There are no witnesses to the disastrous saga of Easter Island, but if the island's carrying capacity were exceeded by an expanding human population, increased demand for wood may have caused destruction of the island's forests. This would have eliminated habitats for native species, which would also have been increasingly hunted for food, driving many to extinction. Deforestation is likely to change the local climate, making it both hotter and drier. The barren landscape that replaced the lush forests of Easter Island is a dramatic example of what may happen when overpopulation permanently and dramatically reduces the capacity of a region to sustain people and other forms of life. Islands are particularly vulnerable to such drastic events, partly because their populations cannot emigrate. For Earth's expanding human population, however, the entire planet is an island.

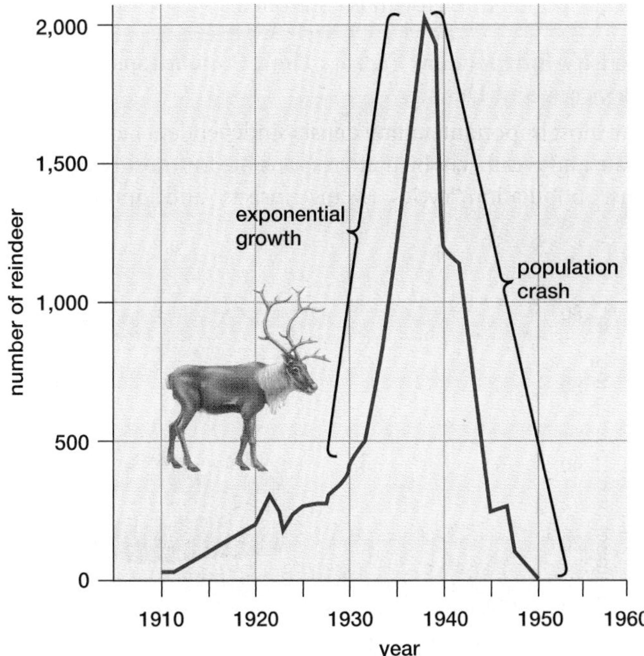

▲ **FIGURE 26-7 The effects of exceeding carrying capacity** In 1911, 25 reindeer were introduced onto St. Paul, an island off the Alaskan coast. Food was plentiful, and there were no reindeer predators on the island. The herd grew exponentially (note the initial J-curve) until it reached 2,046 reindeer in 1938. At this point, the small island was seriously overgrazed, food was scarce, and the population declined dramatically. By 1950, only 8 reindeer remained, far below the island's original carrying capacity. Data from Scheffer, V.B. 1951. *The rise and fall of a reindeer herd.* Science Monthly 73:356–362.

Logistic population growth can occur in nature when a species moves into a new habitat, as ecologist John Connell documented for barnacles as they colonized bare rock along a rocky ocean shoreline (**Fig. 26-8**). Initially, new settlers may find ideal conditions that allow their population to grow almost exponentially. As population density increases, however, individuals begin to compete, particularly for space, energy, and nutrients.

These forms of environmental resistance can reduce the birth rate and increase the death rate (reducing the average life span), as has been demonstrated experimentally in laboratory populations of fruit flies (**Fig. 26-9**). As environmental resistance increases, population growth slows and eventually stops. In nature, conditions are never completely stable, so both carrying capacity and the sizes of different populations within a community will vary somewhat from year to year.

Under natural conditions, where humans have not intervened, environmental resistance usually maintains populations at or below the carrying capacity of their environment. The factors that generate environmental resistance fall into two broad categories: density-independent and density-dependent factors. **Density-independent** factors limit population size regardless of the population density (the number of individuals per unit of area). **Density-dependent** factors, in contrast, increase in effectiveness as the population density increases. Nutrients, energy, and space, the primary determinants of carrying capacity, are all density-dependent regulators of population size. In the following sections, we will look more closely at how population growth is controlled.

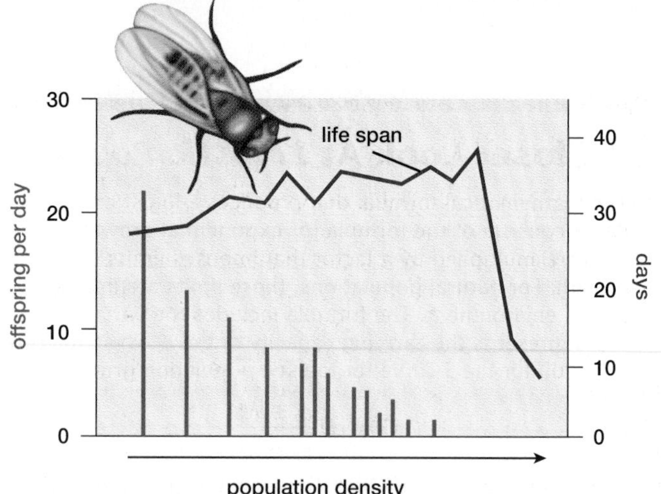

▲ **FIGURE 26-9 Density-dependent environmental resistance** In response to crowding, laboratory populations of fruit flies show a decrease in both reproductive rate and life span. In this graph, population density (horizontal axis; measured by the number of fruit flies occupying a given space in laboratory containers) increases from left to right. Notice that the number of offspring produced per day (blue bars) decreases as population density increases. The life span remains relatively constant until population density reaches a critical level, where death rates then increase, causing life span to drop off dramatically (red line). Data from Pearl, R., Miner, J. R., and Parker, S. L. 1927. *Experimental studies on the duration of life. XI. Density of population and life duration in* Drosophila. American Naturalist 61:289–318.

Density-Independent Factors Limit Populations Regardless of Density

The most important natural density-independent factors are climate and weather, which are responsible for most boom-and-bust population cycles. Many insects and annual plant populations are limited in size by the number of individuals that can be produced before the first hard freeze. Such populations are controlled by the climate because they typically do not reach carrying capacity before winter sets in. Weather can also cause significant variations within natural populations from year to year. Hurricanes, droughts, floods, and fire can have profound effects on local populations, particularly those of small, short-lived species, regardless of their population density.

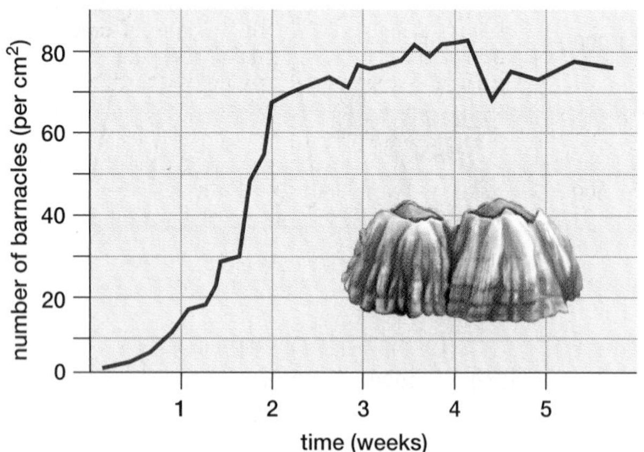

▲ **FIGURE 26-8 A logistic curve in nature** Barnacles are crustaceans whose larvae are carried in ocean currents to rocky seashores where they settle, attach permanently to rock, and grow into the shelled adult form. On a bare rock, the number of settling larvae produce a logistic growth curve as competition for space limits their population density. Data from Connell, J. H. 1961. *Effects of competition, predation by* Thais lapillus *and other factors on natural populations of the barnacle* Balanus balanoides. Ecological Monographs 31:61–104.

Human activities can also limit the growth of natural populations in density-independent ways. Pesticides and pollutants can cause drastic declines in natural populations. Before it was banned in the 1970s, the pesticide DDT drastically reduced populations of predatory birds, including eagles, ospreys, and pelicans; a variety of pollutants continue to adversely affect wildlife, as you will learn in Chapter 28. Although well-regulated hunting can help maintain animal populations in a healthy balance with available resources, overhunting has driven entire animal species to extinction. Examples from the United States are the once-abundant passenger pigeon and the colorful Carolina parakeet. Habitat destruction by humans, a density-independent factor, is the single greatest threat to wildlife worldwide

Density-Dependent Factors Become More Effective as Population Density Increases

Populations of organisms with a life span of more than a year have evolved adaptations that allow them to survive density-independent controls imposed by seasonal changes, such as the onset of winter. Many mammals, for example, develop thick coats and store fat for the winter; some also hibernate.

Migration is another coping mechanism; many birds migrate long distances to find food and a hospitable climate. Most trees and bushes survive the rigors of winter by entering a period of dormancy, dropping their leaves and drastically slowing their metabolic activities.

For such long-lived species in undisturbed habitats, the most important elements of environmental resistance are density dependent. Density-dependent factors exert a negative feedback effect on population size, because they become increasingly effective as the population density increases. Examples of density-dependent factors are various community interactions, introduced in the following sections and covered in more detail in Chapter 27.

Predators Exert Density-Dependent Controls on Populations **Predators** are organisms that eat other organisms, called their **prey.** Often, prey are killed directly and eaten (**Fig. 26-10a**), but not always. When deer browse on the buds of maple trees, or when gypsy moth larvae feed on the leaves of oaks, the trees are harmed but not killed.

(a) Predators often kill weakened prey

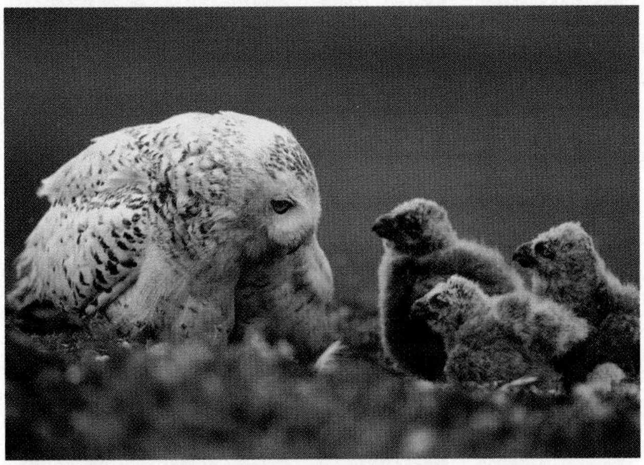

(b) Predator populations increase when prey are abundant

▲ **FIGURE 26-10 Predators help control prey populations** **(a)** A pack of grey wolves has brought down an elk that may have been weakened by age or parasites. **(b)** The snowy owl produces more chicks when prey, such as lemmings, are abundant.

Predation becomes increasingly important as prey populations grow, because most predators eat a variety of prey, depending on what is most abundant and easiest to find. Coyotes might eat more field mice when the mouse population is high, but switch to eating more ground squirrels as the mouse population declines. In this way, predators often exert density-dependent population control over more than one prey population.

Predator populations often grow as their prey becomes more abundant, which make them even more effective as control agents. For predators such as the arctic fox and snowy owl, which rely heavily on lemmings for food, the number of offspring produced is determined by the abundance of prey. Snowy owls (**Fig. 26-10b**) hatch up to 12 chicks when lemmings are abundant, but may not reproduce at all in years when the lemming population has crashed.

In some cases, an increase in predators might cause a dramatic decline in the prey population, which in turn may result in a decline in the predator population. This pattern can result in out-of-phase **population cycles** of predators and prey. In natural ecosystems, both predators and prey are subjected to a variety of other influences, so clear-cut examples of such cycles in nature are rare. However, out-of-phase population cycles of predators and their prey have been demonstrated under controlled laboratory conditions (**Fig. 26-11**).

Predators may contribute to the overall health of prey populations by culling those that are poorly adapted, weakened by age, or unable to find adequate food and shelter. In this way, predation may maintain prey populations near a density that can be sustained by the resources of the ecosystem.

In some instances, predators may maintain populations of their prey at well below carrying capacity. For example, the

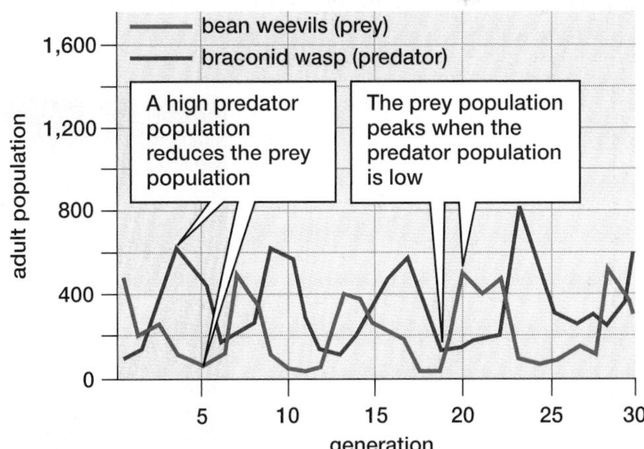

▲ **FIGURE 26-11 Experimental predator–prey cycles** Tiny braconid wasps lay their eggs on bean weevil larvae, which provide food for the newly hatched wasp larvae. A large weevil population ensures a high survival rate for wasp offspring, increasing the predator population. Then, under intense predation, the weevil population plummets, reducing the food available to the next generation of wasps, whose population plummets as a result. Reduced predation then allows the weevil population to increase rapidly, and so on. Data from Utida, S. 1957. *Cyclic fluctuations of population density intrinsic to the host-parasite system.* Ecology 38:442–449.

prickly pear cactus (introduced into Australia from South America in the late 1800s) spread uncontrollably, rendering millions of acres of pasture useless for grazing. In the 1920s, cactus moths (predators of the prickly pear) were imported from Argentina and released to feed on the cacti. Within a few years, the cacti were virtually eliminated. Today, this predatory moth continues to maintain its prey cacti at low population densities, well below carrying capacity.

BioFlix Population Ecology

Parasites Spread More Rapidly Among Dense Populations A **parasite** feeds on a larger organism, its **host,** harming it. Although some parasites kill their hosts, many parasites benefit by having their host remain alive. Parasites include tapeworms that live in the intestines of mammals, and ticks that cling to their skin, sucking blood. Disease-causing organisms are also considered parasites. Most parasites cannot travel long distances, so they spread more readily among hosts in dense populations. For example, plant diseases spread readily through acres of densely planted crops, and childhood diseases spread rapidly through schools and day-care centers. Parasites influence population size by weakening their hosts and making them more susceptible to death from other causes, such as harsh weather or predators. Organisms weakened by parasites are also less likely to reproduce. Parasites, like predators, more often contribute to the death of less-fit individuals, producing a balance in which the host population is regulated but not eliminated.

This balance can be destroyed if parasites or predators are introduced into regions where local prey species have had no opportunity to evolve defenses against them. With little environmental resistance, these organisms reproduce explosively to the detriment of their hosts or prey. The smallpox virus, inadvertently carried by traveling Europeans, caused heavy losses of life among native inhabitants of the continental United States, Hawaii, South America, and Australia. The chestnut blight fungus, introduced from Asia, has almost eliminated wild chestnut trees from U.S. forests. Introduced rats almost certainly contributed to the loss of native birds on Easter Island, and introduced rats and mongooses have exterminated several of Hawaii's native bird populations.

Competition for Resources Helps Control Populations The resources that determine carrying capacity—space, energy, and nutrients—may be inadequate to support all the organisms that need them. **Competition,** defined as the interaction among individuals who attempt to use the same limited resource, limits population size in a density-dependent manner. There are two major forms of competition: **interspecific competition** (competition among individuals of different species; see pp. 513–514) and **intraspecific competition** (competition among individuals of the same species). Because the needs of members of the same species for water and nutrients, shelter, breeding sites, light, and other resources are almost identical, intraspecific competition is an important density-dependent mechanism of population control.

Organisms have evolved several ways to deal with intraspecific competition. Some organisms, including most plants and many insects, engage in **scramble competition,** a kind of free-for-all with resources as the prize. For example, gypsy moth females each lay a mass of up to 1,000 eggs on tree trunks in eastern North America. As the eggs hatch, armies of caterpillars crawl up the tree (**Fig. 26-12**). Huge outbreaks of this invasive species can completely strip large trees of their leaves in a few days. Under these conditions, competition for food may be so great that most of the caterpillars die

(a) Gypsy moths laying eggs

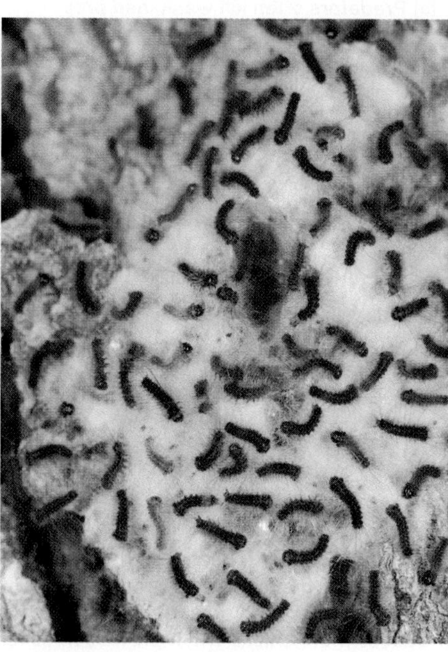
(b) Gypsy moth caterpillars

▶ **FIGURE 26-12 Scramble competition (a)** Gypsy moths gather on tree trunks to lay egg masses. **(b)** Hundreds of caterpillars will hatch from each egg mass.

before they can metamorphose into egg-laying moths. Another example of scramble competition is seen in plants, whose seeds may sprout in dense clusters. As they grow, those that sprouted first begin to shade the smaller ones, and their larger root systems absorb much of the available water from the soil, causing the younger ones to wither and die.

Many animals (and even a few plants) have evolved **contest competition,** in which social or chemical interactions determine access to important resources. Territorial species, including wolves, many fish, rabbits, and songbirds, defend an area that contains important resources such as food or places to raise their young. When the population exceeds the size that can be sustained by available resources, only the best-adapted individuals are able to defend territories that supply adequate food and shelter. Those without territories may not reproduce (reducing the future population), or they may fail to obtain adequate food or shelter and become easy prey for predators.

As population densities increase and competition becomes more intense, some animals react by emigrating. Large numbers leave their homes to colonize new areas, and many, sometimes most, die in the quest. For example, mass movements of lemmings seem to occur in response to overcrowding. Emigrating swarms of locusts periodically plague parts of Africa, consuming all of the vegetation in their path (**Fig. 26-13**).

Density-Independent and Density-Dependent Factors Interact to Regulate Population Size

The size of a population at any given time is the result of complex interactions between density-independent and density-dependent forms of environmental resistance. For example, a stand of pines weakened by drought (a density-independent factor) may more readily fall victim to the pine beetle (density dependent). Likewise, a caribou weakened by hunger (density dependent) and attacked by parasites (density dependent) is more likely to be killed by an exceptionally cold winter (density independent).

Human activities are increasingly imposing density-independent limitations on natural populations. Examples are bulldozing grasslands and their prairie dog towns to build shopping malls, or cutting rain forests for agriculture. These activities also reduce the carrying capacity of the environment, which in turn exerts density-dependent limits on future population sizes.

▲ **FIGURE 26-13 Emigration** In response to overcrowding and lack of food, locusts emigrate in swarms, devouring all vegetation, and even each other, as they go.

QUESTION What benefits does mass emigration give to animals such as locusts or lemmings? Can you see any parallels in human emigration?

26.3 HOW ARE POPULATIONS DISTRIBUTED IN SPACE AND TIME?

Populations of different types of organisms show characteristic spacing of their members, determined by their behavioral characteristics and their environments. In addition, each population exhibits patterns of reproduction and survival that are characteristic of the species.

Populations Exhibit Different Spatial Distributions

The spatial distribution of a population refers to how its individuals are distributed within a given area. Spatial distribution may vary with time, changing with the breeding season, for example. Ecologists recognize three major types of spatial distribution: clumped, uniform, and random (**Fig. 26-14**).

Populations whose members live in groups exhibit a **clumped distribution.** Examples include family or social groupings, such as elephant herds, wolf packs, prides of lions, flocks of birds, and schools of fish (**Fig. 26-14a**). What are the advantages of clumping? Birds in flocks benefit from many eyes to spot food, such as a tree full of fruit. Schooling fish and flocks of birds also may confuse predators with their sheer numbers. Predators, in turn, sometimes hunt in groups, cooperating to bring down larger prey (see Fig. 26-10a). Some species form temporary groups to mate and care for their young. Other plant or animal populations cluster not for social reasons, but because resources are localized. Cottonwood trees, for example, are found clustered along streams and rivers in grasslands.

Organisms with a **uniform distribution** maintain a relatively constant distance between individuals. This is most common among animals that exhibit territorial behaviors that evolved to maintain their access to limited resources. Territorial behavior is more common among animals during their breeding seasons. Seabirds may space their nests evenly along the shore, just out of reach of one another (**Fig. 26-14b**). Among plants, mature desert creosote bushes are often spaced very evenly. Research has shown that this spacing results from competition among their root systems, which occupy a roughly circular area around each plant. The roots efficiently absorb water and other nutrients from the soil, preventing survival of bushes that germinate close by. A uniform distribution

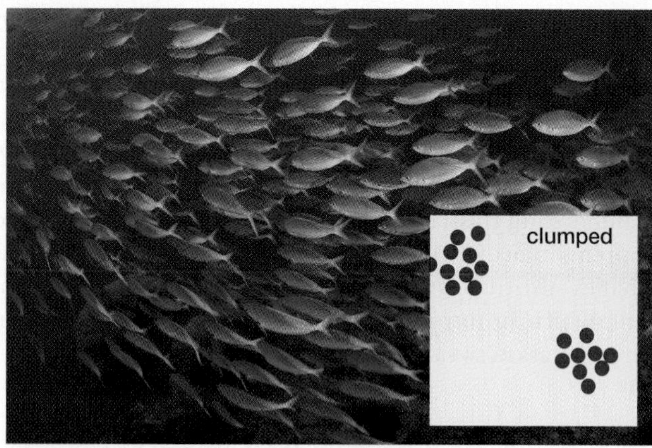

(a) Clumped distribution

(b) Uniform distribution

(c) Random distribution

▲ **FIGURE 26-14 Population distributions (a)** This school of fish may confuse predators with their numbers. **(b)** These gannets occupy evenly spaced nests along the ocean shore. **(c)** Because of uniformly good growing conditions, many rain forest plants grow wherever their seeds fall.

helps these bushes obtain the nutrients and water they need to survive and grow.

Organisms with a **random distribution** are relatively rare. Such individuals do not form social groups. The resources they need are more or less equally available throughout the area they inhabit, and those resources are not scarce enough to require territorial spacing. Trees and other plants in rain forests come close to being randomly distributed (**Fig. 26-14c**). There are probably no vertebrate species that maintain a random distribution throughout the year; most interact socially, at least during the breeding season.

Survivorship in Populations Follows Three Basic Patterns

Animals of different species differ considerably in their chances of dying at any given phase of their life cycle. Some species produce many offspring that are provided with very few resources; most of these offspring die before they can reproduce. Others produce few offspring, which are each given far more resources and often survive to reproduce. To determine the pattern of survivorship, researchers construct **survivorship tables,** which track groups of organisms, born at the same time, throughout their lives, recording how many survive in each succeeding year (or other unit of time; **Fig. 26-15a**). If these numbers are graphed, they reveal the **survivorship curves** characteristic of the species in the particular environment where the data were collected. Three types of survivorship curve—described as late loss, constant loss, and early loss, according to the part of the life cycle during which most deaths occur—are shown in **Figure 26-15b.**

Late-loss populations produce convex survivorship curves. Such populations have relatively low juvenile death rates, and many or most individuals survive to old age. Late-loss survivorship curves are characteristic of humans and other large and long-lived animals such as elephants and mountain sheep. These species produce relatively few offspring, which are protected and nourished by their parents during early life.

Constant-loss populations produce survivorship graphs that are relatively straight lines. In these populations, individuals have an equal chance of dying at any time during their life span. This phenomenon is seen in some birds such as gulls and the American robin, and in laboratory populations of organisms that reproduce asexually, such as hydra and bacteria.

Early-loss populations produce concave survivorship curves. These curves are characteristic of organisms that produce large numbers of offspring, which receive little or no care or resources from the parents after they hatch or germinate. Many of these species engage in scramble competition for resources early in life. The death rate is very high among the young, but those that reach adulthood have a reasonable chance of surviving to old age. Most invertebrates, most plants, and many fish exhibit such early-loss survivorship curves. Even some mammalian populations fall into this category; in some

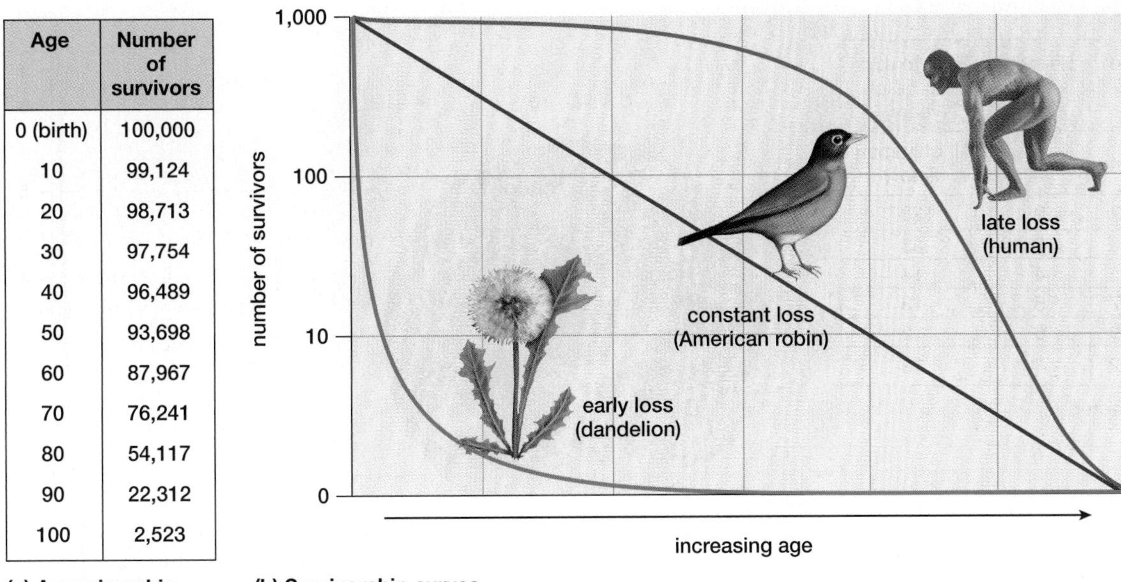

Age	Number of survivors
0 (birth)	100,000
10	99,124
20	98,713
30	97,754
40	96,489
50	93,698
60	87,967
70	76,241
80	54,117
90	22,312
100	2,523

(a) A survivorship table

(b) Survivorship curves

▲ **FIGURE 26-15 Survivorship tables and survivorship curves (a)** A survivorship table for the U.S. population in 2005, showing how many people are expected to remain alive at increasing ages for each 100,000 people born. Plotting these data will produce a curve, similar to the blue curve in part (b). **(b)** Three types of survivorship curve are shown. Because the life spans differ, the percentages of survivors (rather than ages) are used. Data from Centers for Disease Control and Prevention. 2008. *National vital statistics reports.* 56(9).

populations of black-tailed deer, for example, 75% of the population dies within the first 10% of its life span.

26.4 HOW IS THE HUMAN POPULATION CHANGING?

No force on Earth rivals that of humans. We possess enormous brainpower and dexterous hands that can shape the environment to our demands. Our drive to reproduce persists from prehistoric times, when bearing many offspring offered the only hope for the continued survival of our species. Ironically, this drive now threatens us and the biosphere on which we depend.

Demographers Track Changes in the Human Population

Demography is the study of the changing human population. Demographers measure human populations in different countries and world regions, track population changes, and make comparisons between developing and developed countries. They examine birth and death rates by race, sex, education levels, and socioeconomic status both within and between countries. Demographers not only track past and current trends, they attempt to explain these changes, evaluate their impacts, and make predictions for the future. The data gathered by demographers are useful for formulating policies in areas such as public health, housing, education, employment, immigration, and environmental protection.

The Human Population Continues to Grow Rapidly

Compare the graph of human population growth in **Figure 26-16** with the exponential growth curves in Figures 26-2 and 26-3. The time spans are different, but each has the J-curve characteristic of exponential growth. The population initially grew slowly; it took roughly 200,000 years for the human population to reach 1 billion. In the table within Figure 26-16, note the decreasing amount of time required to add billions of people; an estimated 6% of all the humans who have ever lived on Earth are alive today. But also notice that we have been adding billions at a relatively constant rate since the 1970s. This suggests that, although the human population continues to grow rapidly, it may no longer be growing exponentially. Are humans starting to enter the final bend of the S-shaped logistic growth curve (see Fig. 26-6a) that will eventually lead to a stable population? Time will tell. However, Earth's human population (over 6.8 billion in 2009) grows by about 75 million each year; this is over 1.4 million people every week. Why hasn't environmental resistance put an end to our continued population growth?

Like all populations, humans have encountered environmental resistance; but unlike non-human populations, we have responded to environmental resistance by devising ways to overcome it. As a result, the human population has grown for an unprecedented time span. To accommodate our growing numbers, we have altered the face of the globe. Human

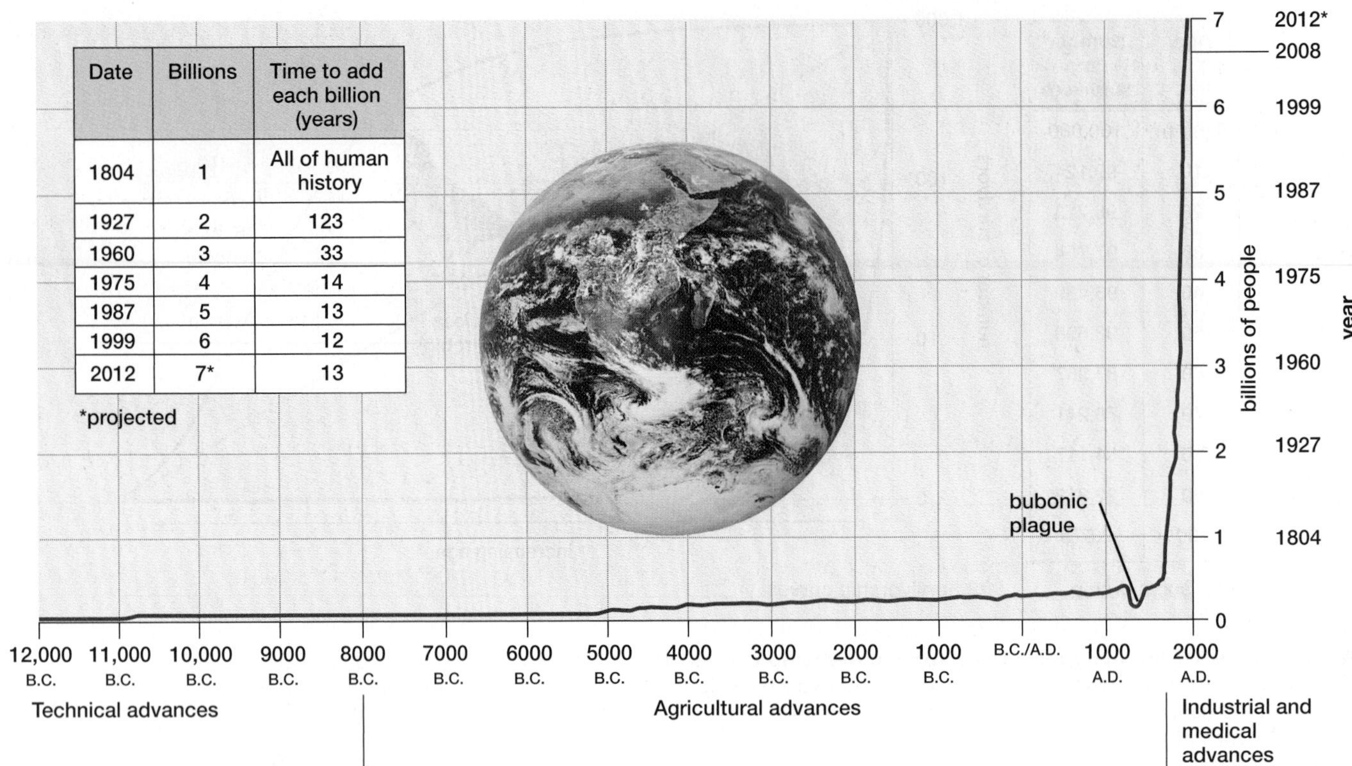

Date	Billions	Time to add each billion (years)
1804	1	All of human history
1927	2	123
1960	3	33
1975	4	14
1987	5	13
1999	6	12
2012	7*	13

*projected

▲ **FIGURE 26-16 Human population growth** The human population from the Stone Age to the present has shown continued exponential growth as various advances overcame environmental resistance. Note the dip in the fourteenth century caused by the bubonic plague. Note also the time intervals over which additional billions were added. (Inset) Earth is an island of life in a sea of emptiness; its space and resources are limited.

QUESTION The human population continues to grow rapidly, but there is evidence that we have already exceeded Earth's carrying capacity at current levels of technology. What do you think this curve will look like when we reach the year 2500? 3000? Explain.

population growth has been spurred by a series of advances, each of which circumvented some type of environmental resistance, increasing Earth's carrying capacity for people. Is there an ultimate limit to these carrying capacity? Have we already reached or perhaps exceeded it? We explore these questions in "Earth Watch: Have We Exceeded Earth's Carrying Capacity?"

A Series of Advances Have Increased Earth's Capacity to Support People

Early humans discovered fire, invented tools and weapons, built shelters, and designed protective clothing, a series of technical advances that increased carrying capacity. Tools and weapons allowed them to hunt more effectively and increased their food supply, while shelter and clothing increased the habitable areas of the globe.

Domesticated crops and animals had supplanted hunting and gathering in many parts of the world by about 8000 B.C. These agricultural advances provided people with a larger, more dependable food supply, further increasing Earth's carrying capacity for humans. An increased food supply resulted in a longer life span and more childbearing years, but a high death rate from disease continued to restrain population growth.

Case Study continued
The Mystery of Easter Island

Although agriculture has increased Earth's overall carrying capacity for humans, agricultural activity can damage the environment. For example, when forests are cleared to make room for fields and pastures, life-sustaining resources provided by the forests are lost. After Polynesians colonized Easter Island, they began clearing its forests to grow crops. Many experts believe that Polynesians deliberately brought rats with them to new islands as a source of food. The rats on Easter Island flourished, killing ground-nesting birds and feeding on seeds of the native palm trees. The loss of seeds would have contributed to the loss of Easter Island's forest ecosystem and the communities of wildlife it supported.

Human population growth continued slowly for thousands of years until major industrial and medical advances permitted a population explosion. These advances began in England in the mid-eighteenth century, spreading throughout

Europe and North America during the nineteenth and into the twentieth century. Medical progress dramatically decreased the death rate by reducing environmental resistance caused by disease. The discovery of bacteria and their role in infection resulted in better control of bacterial diseases through improved sanitation and, later, antibiotics. Vaccines for diseases such as smallpox reduced deaths from viral infections.

The Demographic Transition Explains Trends in Population Size

Today, countries are often described as either developed or developing. People in **developed countries**—including Australia, New Zealand, Japan, and countries in North America and Europe—benefit from a relatively high standard of living, with access to modern technology and medical care, including contraception. Average income is relatively high, education and employment opportunities are available to both sexes, and death rates from infectious diseases are low. But fewer than 20% of the world's people live in developed countries. Most of Earth's people, who live in the **developing countries** of Central and South America, Africa, and much of Asia, lack these advantages.

The rate of population growth in countries that are now developed has changed over time in reasonably predictable stages, producing a pattern called the **demographic transition** (**Fig. 26-17**). Before major industrial and medical advances occurred, these countries were in the *pre-industrial stage*, with relatively small and stable populations whose high birth rates were balanced by high death rates. During the following *transitional stage*, food production increased and health care improved. This caused death rates to fall, while birth rates remained high, leading to an explosive natural rate of population increase. During the *industrial stage*, birth rates fell as more people moved from small farms to cities (where children were less important as a source of labor), contraceptives became more readily available, and opportunities for women to work outside the home increased. Most developed countries are now in the *post-industrial stage* of the demographic transition, and, with the exception of the United States, their populations are relatively stable, with low birth and death rates.

If immigration and emigration rates are balanced, a population will eventually stabilize if parents, on average, have just the number of children required to replace themselves; this is called **replacement-level fertility (RLF).** Replacement level fertility is 2.1 children per woman (rather than exactly 2) because not all children survive to maturity.

World Population Growth Is Unevenly Distributed

In developing countries, such as most countries in Central and South America, Asia (excluding China and Japan), and Africa (excepting those African countries devastated by the AIDS epidemic), medical advances have decreased death rates and increased life span, but birth rates remain relatively high. China is unique. Although it is a developing country, years ago, its government recognized the negative impacts of continued population growth and instituted social reforms

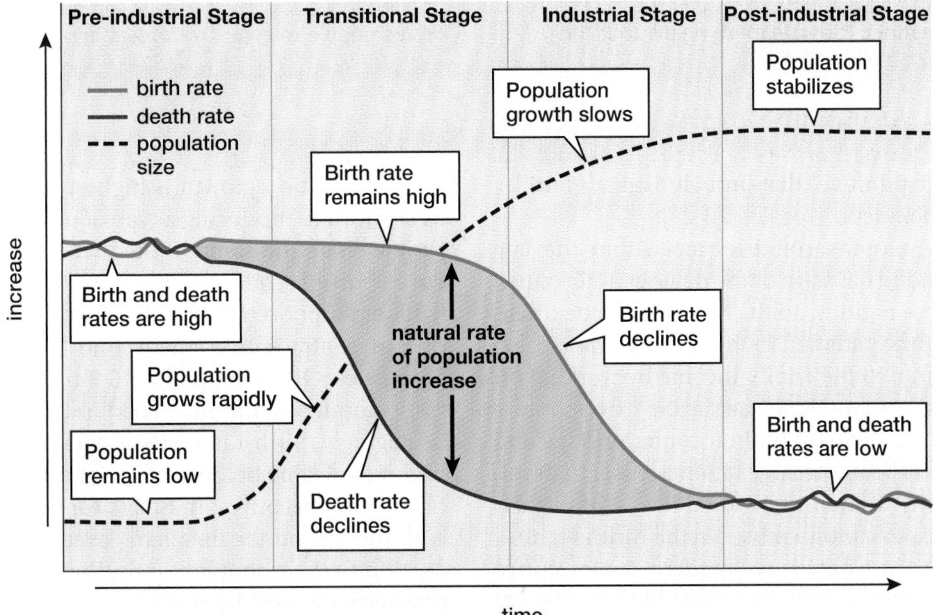

▲ **FIGURE 26-17 The demographic transition** A population that undergoes a demographic transition proceeds through four stages, beginning at a relatively stable and small size with high birth and death rates. Death rates decline first, causing the overall population to increase. Then birth rates decline, causing the population to stabilize at a higher number with relatively low birth and death rates.

Earth Watch

Have We Exceeded Earth's Carrying Capacity?

In Côte d'Ivoire (Ivory Coast), a small country in West Africa, the government is waging a battle to protect some of its rapidly dwindling tropical rain forest from thousands of illegal hunters, farmers, and loggers. Officials burn the homes of the squatters, who immediately return and rebuild. One illegal resident is Sep Djekoule. "I have ten children and we must eat," he explains. "The forest is where I can provide for my family, and everybody has that right." His words illustrate the conflict between population growth and environmental protection, between the "right" to have many children and the ability to provide for them using Earth's finite resources. How many people can Earth support?

Ecologists agree that the concept of a carrying capacity for humans is nebulous. On one hand, we have increased Earth's carrying capacity for humans through agricultural and other advances. On the other hand, our steadily increasing expectations for comfort, mobility, and convenience all reduce carrying capacity, as each individual seeks far more than the minimum resources needed to sustain life. Technology, properly applied, can reduce our impact on other organisms and the environment by improving agricultural efficiency, conserving energy and water, reducing pollutants, and recycling more of what we use. However, no amount of technological innovation will compensate for our biotic potential, which we must restrain if we expect Earth to continue to support us.

An international group of scientists and professionals from many fields are working with the Global Footprint Network to assess humanity's impact on global ecosystems. This enormous project compares human demand for resources (including agricultural products, fish and other wild foods, wood, space, and energy) with the ability of the world's ecosystems to supply these resources in a sustainable manner. "Sustainable" means that the resources could be renewed indefinitely, and the ability of the biosphere to supply them would not be diminished over time.

To evaluate human impact, the researchers estimate the area of biologically productive land needed to supply the resource demands and absorb the wastes of an average person. This area is called an **ecological footprint.** Although the most recent data are for 2005, these experts estimate that during 2008, Earth's 6.7 billion people consumed 140% of the resources that were sustainably available. In other words, to avoid damaging Earth's resources, our current population at its present living standards would require at least 1.4 Earths. Supporting the world population sustainably at the average standard of living in the United States would require 4.7 Earths.

As you have learned in this chapter, a population that exceeds carrying capacity damages its ecosystem and reduces its future ability to sustain that population. As we have used our technological prowess to overcome environmental resistance, humanity's collective ecological footprint has grown large enough to trample Earth's sustainable resource base, reducing Earth's capacity to support us.

For example, each year, overgrazing and deforestation decrease the productivity of land, especially in developing countries. The area of cropland available to support each person has declined by over half in the past 50 years, and the United Nations estimates that over 850 million people currently lack adequate food. A significant portion of the world's agricultural land suffers from erosion, reducing its ability to support both crops and grazing livestock (**Fig. E26-2**). The quest for farmland drives people to clear-cut forests in places where the soil

(many of them quite punitive) that brought China's fertility rate below replacement level (see Fig. 26-21).

Most other developing countries are within the late transitional or the industrial stage of the demographic transition. In many of these nations, adult children provide financial security for aging parents. Young children may also contribute significantly to the family income by working on farms or sometimes in factories. Social factors drive population growth in countries where children confer prestige and where religious beliefs promote large families. Also, in developing countries, many women who would like to limit their family size lack access to contraceptives. In the African nation of Nigeria, only 8% of women use modern contraceptive methods, and the average woman bears six children. Nigeria is suffering from soil erosion, water pollution, and the loss of forests and wildlife. With 45% of its 148 million people under the age of 15, continued population growth is inevitable, for reasons described later.

Population growth is highest in the countries that can least afford it, producing a type of positive feedback. As more people share the same limited resources, poverty increases. Poverty diverts children away from schools and into activities that help support their families. The lack of education and access to contraceptives then contributes to continued high birth rates. Of the more than 6.8 billion people on Earth in 2009, about 5.6 billion resided in developing countries. Encouragingly, birth rates in some developing countries have begun to decline because of social changes and increased access to contraceptives. In Brazil, for example, the fertility rate is about 1.9 and the Brazilian government projects a slow reduction in their population beginning around 2030. But the prospects for world population stabilization in the near future are dim. For the year 2050, the United Nations predicts that the population will be over 9 billion and growing (although much more slowly than at present), with 7.9 billion people living in the developing nations (**Fig. 26-18**).

is poorly suited for agriculture. The demand for wood also causes large areas to be deforested annually, causing the runoff of much-needed fresh water, the erosion of precious topsoil, the pollution of rivers, and an overall reduction in the ability of the land to support future crops or forests. Our appetite for food, wood, and recently, biofuels (crops, such as soybeans, grown to provide fuel), drives the destruction of tens of millions of acres of rain forest annually, and causes the extinction of species on an unprecedented scale (see pp. 587–589).

In many developing countries, including India and China (each home to over 1.1 billion people), clean fresh water is in short supply. In these countries, underground water stores are being depleted to irrigate cropland far faster than they are refilled by water from rain and snow seeping back into the soil. Because irrigated land supplies about 40% of human food crops, water shortages can rapidly lead to food shortages.

The total world catch of wild fish peaked in the late 1980s and has been gradually declining, despite increased investments in equipment, improved technology for finding fish, and increased harvests of smaller and less-desirable fish species. Almost 70% of commercial ocean fish populations are fully used or overfished, and many formerly abundant fish populations (such as cod harvested off New England, Canada, and in the North Sea) have declined dramatically because of overfishing.

Our present population, at its present level of technology, is clearly "overgrazing" the biosphere. As the 5.5 billion people in less-developed countries strive to increase their standard of living, the damage to Earth's ecosystems accelerates. In estimating how many people Earth can—or should—support, it is important to recognize that people desire far more from life than simply surviving. But, because of our numbers, the standard of living in developed countries is already an unattainable luxury for most of Earth's inhabitants.

Inevitably, the human population will stop growing. Either we must voluntarily reduce our birth rate, or various forces of environmental resistance, including disease and starvation, will dramatically increase human death rates. The choice is ours. Hope for the future lies in recognizing the signs of human overgrazing and acting to reduce our population before we irrevocably damage our biosphere.

▲ **FIGURE E26-2 Overgrazing can lead to the loss of productive land** Human activities, including overgrazing, deforestation, and poor agricultural practices, reduce the productivity of the land. (Inset) An expanding human population, coupled with a loss of productive land, can lead to tragedy.

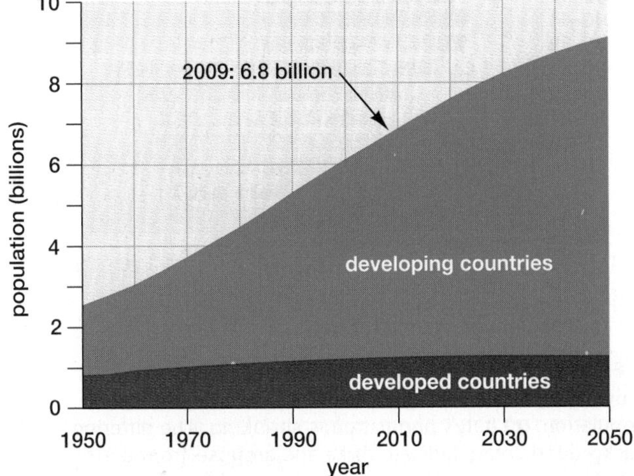

▲ **FIGURE 26-18 Historical and projected world population**

The Current Age Structure of a Population Predicts Its Future Growth

Data gathered by demographers allow the **age structure** of human populations to be graphed. Age structure diagrams show age groups on the vertical axis and the numbers (or percentages) of individuals in each age group on the horizontal axis, with males and females shown on opposite sides. Age-structure diagrams all rise to a peak that reflects the maximum human life span, but the shape of the rest of the diagram reveals whether the population is expanding, stable, or shrinking. If adults in the reproductive age group (15 to 44 years) are having more children (the 0- to 14-year age group) than are needed to replace themselves, the population is above RLF and is expanding; its age structure will be roughly triangular (**Fig. 26-19a**). If the adults of reproductive age have just the number of children needed to replace themselves, the population is at RLF. A population that has been at RLF for many years will have an age-structure diagram with relatively straight sides (**Fig. 26-19b**). In shrinking populations, the reproducing adults have fewer children than are required to replace themselves, causing the age structure diagram to narrow at the base (**Fig. 26-19c**).

Figure 26-20 shows the average age structures for the populations of developed and developing countries for the year 2009, with projections for 2050. Even if rapidly growing countries were to achieve RLF immediately, their population increase would continue for decades, because today's children create a momentum for future growth as they enter their reproductive years and begin their own families—even if they have only two children per couple. This momentum fuels China's population growth of 0.5% annually (adding over 6.6 million people in 2008) even with a fertility rate of 1.6, well below RLF. In a stable human population, fewer than 20% of individuals will fall into the prereproductive (0- to 14-year) age group. In Mexico, this age group makes up 32% of the population, and in many African nations, children comprise well over 40% of the population.

Fertility in Europe Is Below Replacement Level

Figure 26-21 illustrates growth rates for various world regions. In Europe, the average annual change in population is 0%, with an average fertility rate of 1.5—substantially below RLF—as many women delay or forgo having children.

This situation raises concerns about the availability of future workers and taxpayers to support the resulting temporary increase in the percentage of elderly people. Several European countries are offering or considering incentives (such as large tax breaks) for couples to have children at an earlier age, which shortens the generation time and increases the population. Japan's government is also concerned about the country's low fertility rate (1.3) and provides subsidies to encourage larger families. This is occurring even though Japan, which is about the size of the U.S. state of Montana, is home to nearly 128 million people (42% of the entire U.S. population).

Although a reduced population will ultimately offer tremendous benefits for both people and the biosphere that sustains them, current economic structures in countries throughout the world are based on growing populations. The difficult adjustments necessitated as populations decline—or merely stabilize—motivate governments to adopt policies that encourage more childbearing and continued growth.

The U.S. Population Is Growing Rapidly

With a population of over 307 million, and a growth rate of about 1% per year, the United States is the fastest-growing developed country in the world (**Fig. 26-22**). Between 2006 and 2008, the United States added nearly 3 million people annually (over 8,000 each day), with legal and illegal immigration accounting for an estimated 40% of the population increase. The average fertility rate of new immigrants to the United

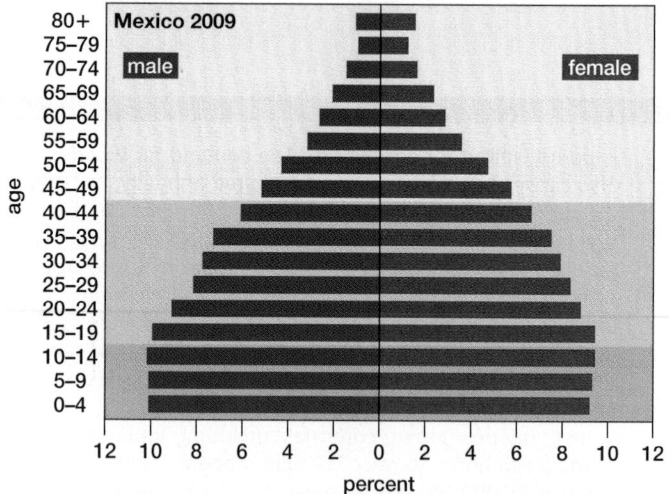

(a) Population pyramid for Mexico

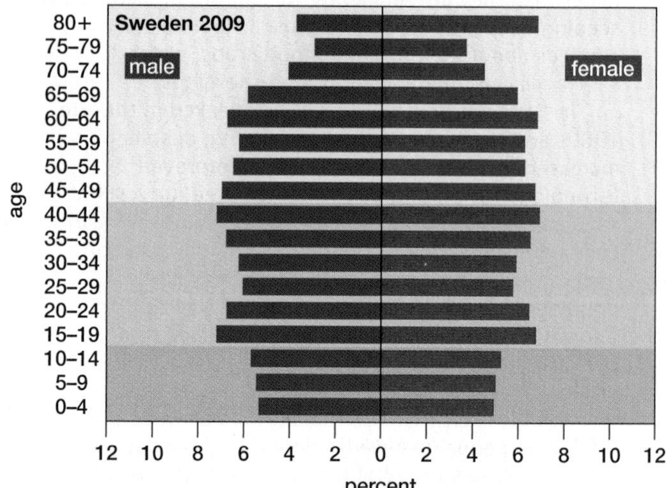

(b) Population pyramid for Sweden

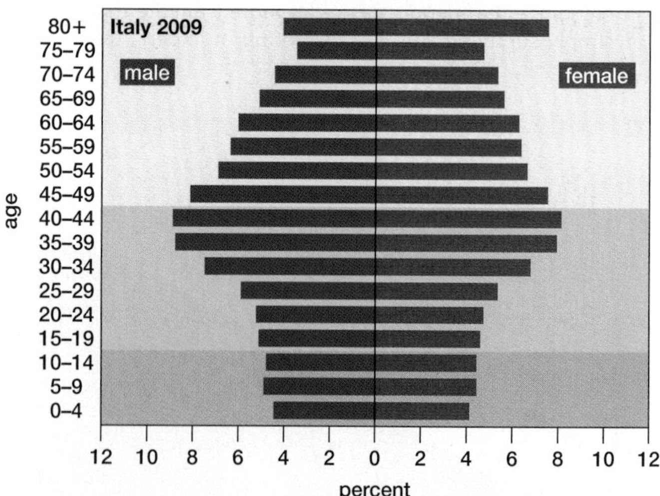

(c) Population pyramid for Italy

▲ **FIGURE 26-19 Age-structure diagrams (a)** Mexico is growing by about 1.6% per year. **(b)** Sweden has a stable population. **(c)** Italy's population is shrinking. The differing background colors indicate three age groups—from bottom to top: prereproductive (0 to 14 years), reproductive (15 to 44 years), and postreproductive (45 to 80+ years). Because the topmost bars include everyone over age 80, they are disproportionately large. These data also show that women, on average, live longer than men. Data from the U.S. Census Bureau.

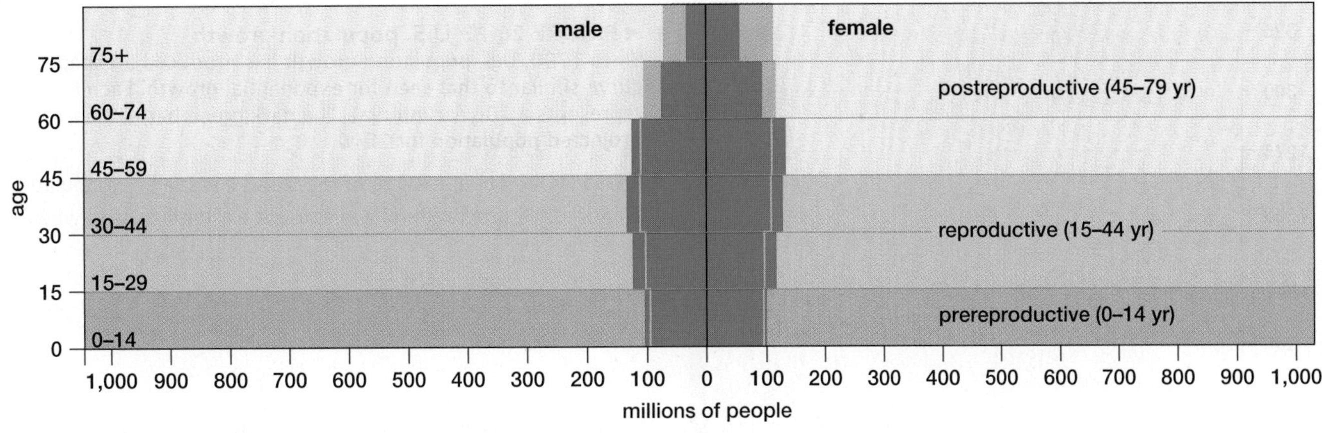

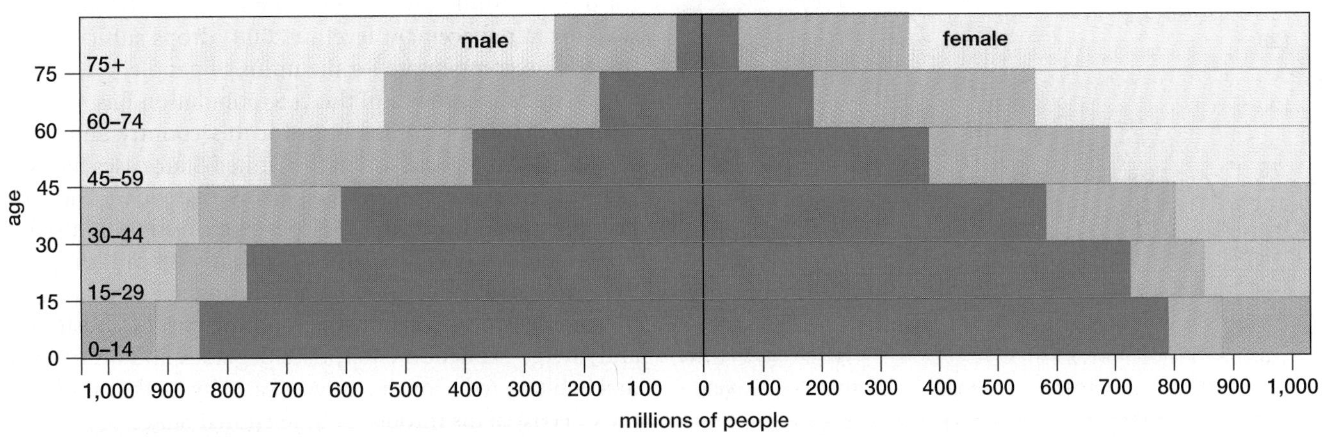

▲ **FIGURE 26-20 Age-structure diagrams of developed and developing countries**
Note that the predicted difference in the number of children compared to parents in developing countries is smaller in 2050, as these populations approach RLF. The large numbers of young people who will be entering their childbearing years will cause continued growth. Data from the U.S. Census Bureau.

QUESTION How does a fertility rate above RLF produce a positive feedback effect (in which a change creates a situation that amplifies itself) on population growth?

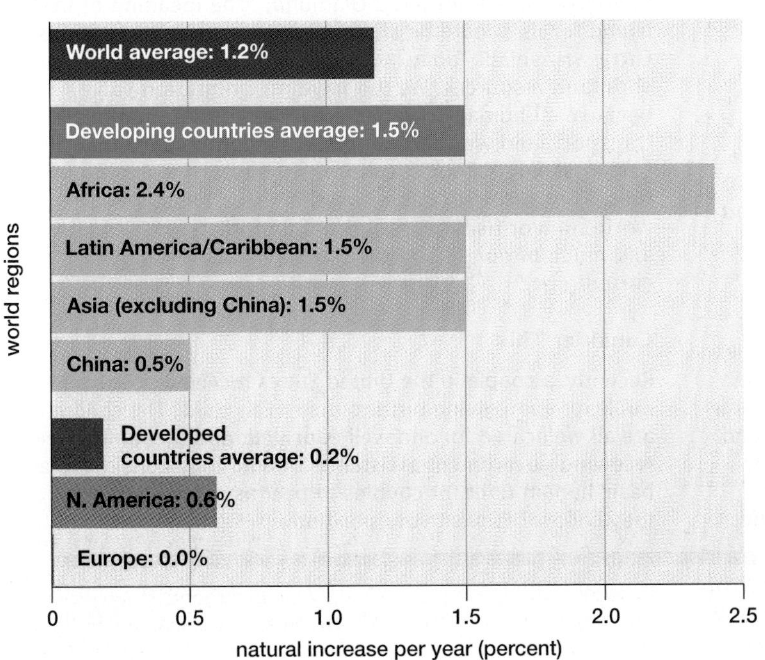

◀ **FIGURE 26-21 Population change by world regions** Growth rates shown are due to natural increase (births − deaths) expressed as the percentage increase per year for various regions of the world (developing regions are indicated in green, developed regions in red). These figures do not include immigration or emigration. Data from the Population Reference Bureau. 2008. *World population data sheet.*

QUESTION Why are there such big population differences between developed and developing countries?

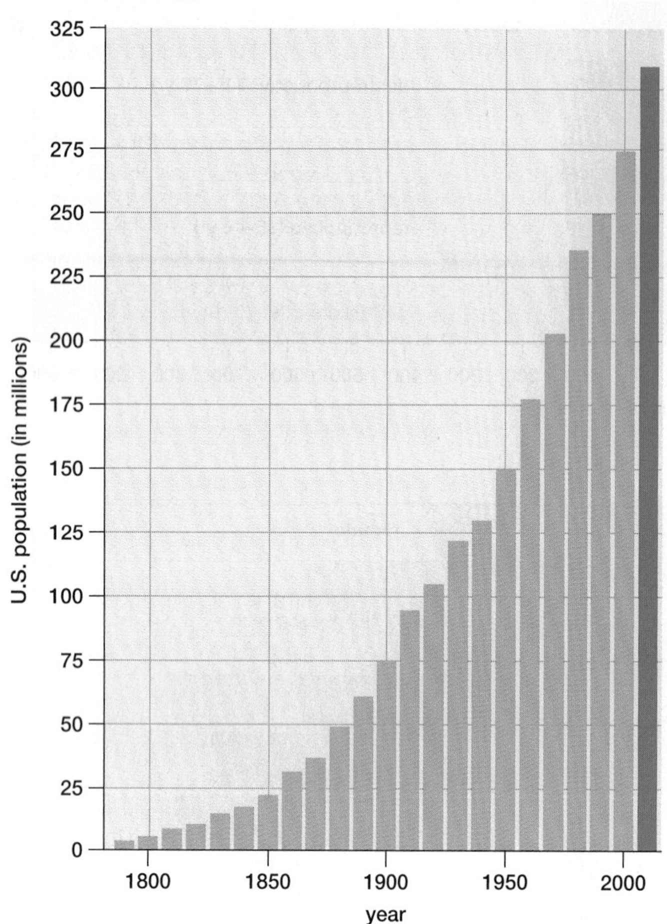

◀ **FIGURE 26-22 U.S. population growth**
Since 1790, U.S. population growth has produced a J-shaped curve similar to that seen for exponential growth. Each bar represents a 10-year interval. The dark green bar is the projected population for 2010.

QUESTION At what stage of the S-curve is the U.S. population? What factors do you think will cause it to stabilize, and when?

States is above RLF, compounding their impact on growth. Immigration will ensure continued U.S. population growth for the indefinite future, unless the U.S. fertility rate (which was just at replacement level in 2008) drops sufficiently below RLF to compensate for the influx of people.

The rapid growth of the U.S. population has major environmental implications both for this country and for the world. The average person in the United States uses five times as much energy as the average person worldwide. The 3 million people added to the U.S. population in 2008 used 2.5 times as much energy as all of the nearly 18 million people that expanded India's population during the same year. The inexorable spread of housing, commercial establishments, and energy-extraction enterprises degrades and destroys natural habitats, reducing the carrying capacity for the non-human life of ecosystems throughout the United States.

Case Study revisited
The Mystery of Easter Island

Radioactive carbon dating of sediments, examination of pollen, and archeological excavations suggest that when Polynesians arrived on Easter Island, they found a forested paradise. Ancient bones provide evidence that 6 species of land-dwelling birds and 25 species of seabirds nested on the island. The now-extinct Easter Island palm tree provided edible nuts and sap that could be made into syrup and wine. Its trunks could be made into large canoes from which the islanders harpooned porpoises, whose bones were abundant in refuse heaps dated between 900 and 1300.

The changing composition of refuse heaps over time attests to the changing diet of the natives as their island ecosystem became degraded and their choices increasingly limited. Palms disappeared around 1400, and porpoise bones disappeared from refuse heaps at around 1500, probably as the last remaining canoes were no longer usable. All of Easter Island's land-dwelling birds were exterminated long ago, and only one species of seabird still nests there. The evidence supports the hypothesis that the population grew until it exceeded the environment's capacity to support it. Like the reindeer on St. Paul Island (see Fig. 26-7), Easter Island's human population damaged the ecosystem on which it depended.

What can we learn from Easter Island? According to biologist and author Jared Diamond, "The meaning of Easter Island for us should be chillingly obvious. Easter Island is Earth writ small. Today, again, a rising population confronts shrinking resources. We too have no emigration valve, because all human societies are linked by international transport, and we can no more escape into space than the Easter Islanders could flee into the ocean. If we continue to follow our present course, we shall have exhausted the world's major fisheries, tropical rain forests, fossil fuels, and much of our soil by the time my sons reach my current age."

Consider This

Recently, a couple in the United States received positive publicity upon giving birth to their 18th child. The children are all well cared for and well educated, and the family is not receiving government assistance. Should it be considered a basic human right for couples to bear as many children as they choose? Explain your position.

CHAPTER REVIEW

Summary of Key Concepts

26.1 How Does Population Size Change?

Individuals join populations through birth or immigration and leave through death or emigration. The ultimate size of a stable population results from interactions among biotic potential (the maximum population growth rate) and environmental resistance (factors that limit population growth).

All organisms have the biotic potential to more than replace themselves over their lifetime, resulting in population growth. Populations in ideal circumstances tend to grow exponentially, with increasing numbers of individuals added during each successive time period. Populations cannot sustain exponential growth for long; they either stabilize or undergo periodic boom-and-bust cycles as a result of environmental resistance.

26.2 How Is Population Growth Regulated?

Environmental resistance restrains population growth by increasing the death rate or decreasing the birth rate. The maximum size at which a population may be sustained indefinitely by an ecosystem is called the carrying capacity (*K*), determined by limited resources such as space, nutrients, and light. Environmental resistance generally maintains populations at or below the carrying capacity. In nature, populations can overshoot *K* temporarily by depleting their resource base. Depending on the amount of damage to critical resources, this leads to (1) the population oscillating around *K*; (2) the population crashing and then stabilizing at a reduced *K*; or (3) the population being eliminated from the area.

Population growth is restrained by density-independent forms of environmental resistance (weather and climate) and density-dependent forms of resistance (competition, predation, and parasitism).

BioFlix Population Ecology

26.3 How Are Populations Distributed in Space and Time?

Populations can be distributed in three general ways: clumped, uniform, and random. Clumped distribution may occur for social reasons or around limited resources. Uniform distribution is often the result of territorial spacing. Random distribution is rare, occurring only when individuals do not interact socially and when resources are abundant and evenly distributed.

Populations show specific survivorship curves that describe the likelihood of survival at any given age. Late-loss (convex) curves are characteristic of long-lived species with few offspring, which receive parental care. Species with constant-loss curves have an equal chance of dying at any age. Early-loss (concave) curves are typical of organisms that produce numerous offspring, most of which die before reaching maturity.

26.4 How Is the Human Population Changing?

The human population has exhibited exponential growth for an unprecedented time, the result of a combination of high birth rates and technological, agricultural, industrial, and medical advances that have overcome several types of environmental resistance and increased Earth's carrying capacity for humans. Age-structure diagrams depict numbers of males and females in various age groups. Expanding populations have pyramidal age structures; stable populations show rather straight-sided age structures; and shrinking populations have age structures that are constricted at the base.

Most of Earth's people live in developing countries with growing populations. Although birth rates have declined considerably in many places, momentum from previous high birth rates ensures continued substantial population growth. The United States is the fastest-growing developed country, owing both to higher birth rates and high immigration rates. Recently, scientists have concluded that the demands of Earth's 6.7 billion people already exceed the sustainably available resources. As the population continues to grow, and as people in less-developed countries strive to increase their standard of living, the damage will accelerate. Unlike other animals, people can make conscious decisions to reverse damaging trends.

Key Terms

age structure *505*
biosphere *489*
biotic potential *489*
birth rate *490*
boom-and-bust cycle *492*
carrying capacity (*K*) *494*
clumped distribution *499*
community *489*
competition *498*
constant-loss population *500*
contest competition *499*
death rate *490*
demographic transition *503*
demography *501*
density-dependent *496*
density-independent *496*
developed country *503*
developing country *503*
early-loss population *500*
ecological footprint *504*
ecology *489*
ecosystem *489*
emigration *489*
environmental resistance *489*
exponential growth *490*
growth rate *490*
host *498*
immigration *489*
interspecific competition *498*
intraspecific competition *498*
invasive species *492*
J-curve *491*
late-loss population *500*
logistic population growth *494*
parasite *498*
population *489*
population cycle *497*
predator *497*
prey *497*
random distribution *500*
replacement-level fertility (RLF) *503*
scramble competition *498*
S-curve *495*
survivorship curve *500*
survivorship table *500*
uniform distribution *499*

Thinking Through the Concepts

Fill-in-the-Blank

1. Two examples of density-independent forms of environmental resistance are __________ and __________. Two important types of density-dependent environmental resistance are __________ and __________.

2. Graphs that plot how the numbers of individuals born at the same time change over time are called __________. The specific type of curve that applies to a dandelion that releases 300 seeds, most of which never germinate, is called __________. The curve for humans is an example of __________.

3. The type of growth that occurs in a population that grows by 0.1% per year is __________. Does this form of growth add the same number of individuals each year? __________ What shape of curve is generated if this type of growth is graphed? __________ Can this type of growth be sustained indefinitely? __________
4. The maximum population size that can be sustained indefinitely without damaging the environment is called the __________. A growth curve in which a population first grows logarithmically and then levels off at (or below) this maximum sustainable size is called a(n) __________ curve, or a(n) __________ curve.
5. The type of spatial distribution likely to occur when resources are localized is __________. The type of spatial distribution that results when pairs of animals defend breeding territories is __________. The least common form of distribution is __________.
6. An expanding population has an age-structure diagram shaped like a(n) __________. If the sides of an age-structure diagram are roughly vertical, the population is __________. The shape of the age-structure diagram for developing countries collectively is __________.
7. A population grows whenever the number of __________ plus __________ exceeds the number of __________ plus __________. The growth rate of a population increases whenever the age at which the organism first reproduces __________, when the frequency of reproduction __________, and when the length of the organism's reproductive life span __________.

Review Questions

1. Define biotic potential and environmental resistance.
2. Draw the growth curve of a population before it encounters significant environmental resistance. What is the name of this type of growth, and what is its distinguishing characteristic?
3. Distinguish between density-independent and density-dependent forms of environmental resistance.
4. What is logistic population growth? What is *K*?
5. Describe three different possible consequences of exceeding carrying capacity. Sketch these scenarios on a graph. Explain your answer.
6. List three density-dependent forms of environmental resistance, and explain why each is density dependent.
7. Distinguish between populations showing concave and convex survivorship curves.
8. Draw the general shape of age-structure diagrams characteristic of (a) expanding, (b) stable, and (c) shrinking populations. Label all the axes. Explain why you can predict near-term future growth by the current age structure of populations.
9. Given that the U.S. birth rate is currently around replacement-level fertility, why is our population growing?
10. Discuss some of the factors that can make the transition from a growing to a stable population economically difficult.

Applying the Concepts

1. **BioEthics** The United States has a long history of accepting large numbers of immigrants. Discuss the pros and cons of allowing high levels of legal immigration. What should be done about illegal immigration? What are the implications of immigration for population stabilization?
2. Explain natural selection in terms of biotic potential and environmental resistance.
3. What factors encourage rapid population growth in developing countries? What will it take to change this growth?
4. Contrast age structures in rapidly growing versus stable human populations. Why would a rapidly growing population continue to grow even if families all immediately started having only two children? For how long would the population increase?
5. Why is the concept of carrying capacity difficult to apply to human populations?
6. **BioEthics** Search the Internet for "ecological footprint" and calculate your ecological footprint. For five of your daily activities, explain how and why each contributes to your footprint.

MB *Go to www.masteringbiology.com for practice quizzes, activities, eText, videos, current events, and more.*

Community Interactions

Chapter 27

Case Study

Mussels Muscle In

ON JANUARY 31, 1989, the residents of Monroe, Michigan, found themselves without water. Monroe's schools, industries, and businesses were closed for two days while workers labored to resolve the problem. How can a town on the shore of Lake Erie lack water? The answer—the intake pipes of its water treatment plant were clogged with hundreds of millions of zebra mussels.

Zebra mussels belong to a large group of "two-shelled," or "bivalve," mollusks that includes clams and scallops. Named for the striped pattern on their shells, zebra mussels can range in size from barely visible to about 2 inches long. Like other mussels, they attach to surfaces using sticky threads.

Where did the mussels come from? About three years earlier, a cargo ship had dumped its ballast water in Lake St. Clair (between Lake Huron and Lake Erie). The water, taken aboard in the Black Sea of southeastern Europe, contained stowaways—millions of microscopic zebra mussel larvae. In the Great Lakes, these invaders found an ideal habitat with plenty of food and no major predators or competitors. Their population exploded, spreading to all of the Great Lakes, the Ohio and Mississippi Rivers, and their tributaries. To make matters worse, a few years after the zebra mussel arrived, the quagga mussel (a close relative of the zebra mussel) was introduced by the same means and began to spread in a similar manner.

Both zebra mussels and quagga mussels present serious ecological problems to the bodies of water where they are introduced. As filter feeders, the mussels remove large quantities of phytoplankton, which are photosynthetic microorganisms that form the basis of aquatic food chains. As phytoplankton become scarce, populations of the microscopic animals (zooplankton) that feed on phytoplankton decline, reducing the food available to small fish. If small fish populations decline, the larger fish that feed on them will also become less abundant, and so on up through the food chain. Thus, as zebra and quagga mussels filter microscopic organisms from the water, they alter the entire community structure of the lake or river.

Think about the zebra and quagga mussels as you read about the community interactions that characterize healthy ecosystems. Why have these unwelcome imports been so enormously successful? What impact, if any, do they have on one another? Can anything control them?

▲ Quagga mussels cover a sandal found at Lake Mead in Nevada, far from the Great Lakes where these mussels were first introduced.

At a Glance

27.1 WHY ARE COMMUNITY INTERACTIONS IMPORTANT?

An ecological **community** consists of all the interacting populations within an ecosystem. Because there are direct or indirect links between all forms of life in a given area, a community can encompass the entire **biotic,** or living, portion of an ecosystem. In Chapter 26, you learned that community interactions such as competition, predation, and parasitism can limit the size of populations. A community's interacting web of life tends to maintain a balance between resources and the numbers of individuals consuming them.

When populations interact and influence each other's ability to survive and reproduce, they act as agents of natural selection on one another. For example, in killing prey that is easiest to catch, predators leave behind individuals with better defenses against predation. These better-adapted individuals produce the most offspring, and over time, their inherited characteristics increase within the prey population. Thus, as community interactions limit population size, they simultaneously shape the bodies and behaviors of the interacting populations. The process by which two interacting species act as agents of natural selection on one another is called **coevolution.**

The most important community interactions are competition, predation, parasitism, and mutualism. If we consider these interactions as involving two different species, they can be classified according to whether each of the species is harmed or helped by the interaction, as shown in **Table 27-1**.

Table 27-1 Interactions Among Species

Type of Interaction	Effect on Species A	Effect on Species B
Competition between A and B	Harms	Harms
Predation by A on B	Benefits	Harms
Parasitism by A on B	Benefits	Harms
Mutualism between A and B	Benefits	Benefits

27.2 WHAT IS THE RELATIONSHIP BETWEEN THE ECOLOGICAL NICHE AND COMPETITION?

The concept of the ecological niche is important to our understanding of how competition within and between species selects for adaptations in body form and behavior. Although the word "niche" may call to mind a small cubbyhole, in ecology, it means much more.

The Ecological Niche Defines the Place and Role of Each Species in Its Ecosystem

Each species occupies a unique **ecological niche** that encompasses all aspects of its way of life. One important aspect of the ecological niche is the organism's physical home, or habitat. The primary habitat of a white-tailed deer in the United States, for example, is the eastern deciduous forest. In addition, an ecological niche includes all the physical environmental conditions necessary for the survival and reproduction of a given species. These can include nesting or denning sites, climate, the type of nutrients the species requires, its optimal temperature range, the amount of water it needs, the pH and salinity of the water or soil it may inhabit, and (for plants) the degree of sun or shade it can tolerate. Finally, the ecological niche also encompasses the entire "role" that a given species performs within an ecosystem, including what it eats (or whether it obtains energy from photosynthesis) and the other species with which it competes. Although different species share many aspects of their niche with others, no two species ever occupy exactly the same ecological niche within the same community, as explained in the following sections.

Competition Occurs Whenever Two Organisms Attempt to Use the Same, Limited Resources

Competition is an interaction that occurs between individuals within a species or between individuals of different species as they attempt to use the same, limited resources, particularly energy, nutrients, or space. The more the ecological niches of two species overlap, the greater the amount of competition between them. **Interspecific competition** refers to competitive interactions between members of different species, such as occurs if they feed on the same things or require similar breeding areas. Interspecific competition is detrimental to all of the species involved because it reduces their access to resources that are in limited supply. The degree of interspecific competition depends on how similar the requirements of the species are.

Adaptations Reduce the Overlap of Ecological Niches Among Coexisting Species

Just as no two organisms can occupy exactly the same physical space at the same time, no two species can inhabit exactly the same ecological niche simultaneously and continuously. This important concept, called the **competitive exclusion principle,** was formulated in 1934 by Russian biologist G. F. Gause. This principle leads to the hypothesis that if a researcher forces two species with the same niche to compete for limited resources, inevitably, one will outcompete the other, and the species that is less well adapted to the experimental conditions will die out.

Gause used two species of the protist *Paramecium* (*P. aurelia* and *P. caudatum*) to demonstrate this principle. Grown separately in laboratory flasks, both species thrived on the same bacteria and fed in the same region of their flasks (**Fig. 27-1a**). But when Gause placed the two species together, one (*P. aurelia*) always eliminated, or "competitively excluded," the other (*P. caudatum*; **Fig. 27-1b**). Gause then repeated the experiment, replacing *P. caudatum* with a different species, *P. bursaria*, which tended to feed in a different part of the flask. In that case, the two species of *Paramecium* were able to coexist indefinitely because they occupied slightly different niches.

Ecologist Robert MacArthur further explored the competitive exclusion principle, under natural conditions, by carefully observing five species of North American warbler. These birds all hunt for insects and nest in a variety of spruce trees. Although the niches of these birds appear to overlap considerably, MacArthur found that each species concentrates its search for food in specific regions within spruce trees, employs different hunting tactics, and nests at a slightly different time. By dividing up the resources provided by the spruce trees they share, the warblers minimize the overlap of their niches and reduce interspecific competition (**Fig. 27-2**).

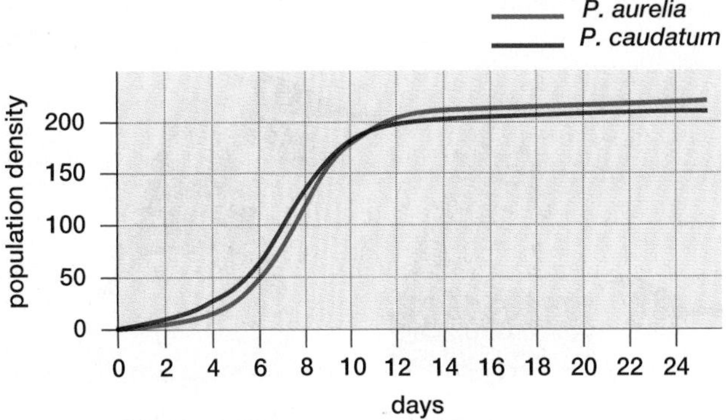

(a) Grown in separate flasks

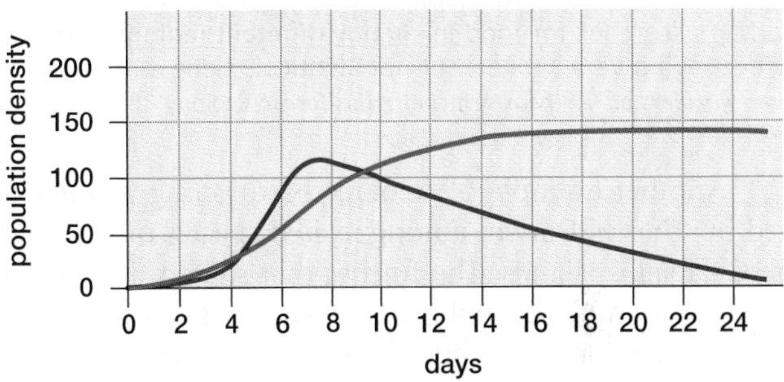

(b) Grown in the same flask

▲ **FIGURE 27-1 Competitive exclusion (a)** Raised separately with a constant food supply, both *Paramecium aurelia* and *P. caudatum* show the S-curve typical of a population that initially grows rapidly and then stabilizes. **(b)** Raised together and forced to occupy the same niche, *P. aurelia* consistently outcompetes *P. caudatum* and causes that population to die off. Data from Gause, G.F. 1934. *The Struggle for Existence.* Baltimore: Williams & Wilkins.

QUESTION Explain how competitive exclusion could contribute to the threat posed by an invasive species.

Had they not evolved in competition with one another, each of the species of warbler that MacArthur observed would probably have evolved to search throughout the entire spruce tree for its spider and insect prey. With more food available, each population would likely have been larger. But when species with similar ecological niches coexist and compete, each species occupies a smaller niche than it would by itself. This phenomenon, called **resource partitioning,** is an evolutionary adaptation that reduces interspecific competition. Resource partitioning is the outcome of the coevolution of species with extensive (but not total) niche overlap. Individuals with fewer competitors can utilize more resources and leave more offspring. As a result, over evolutionary time, competing species evolve physical and behavioral adaptations that reduce their competitive interactions.

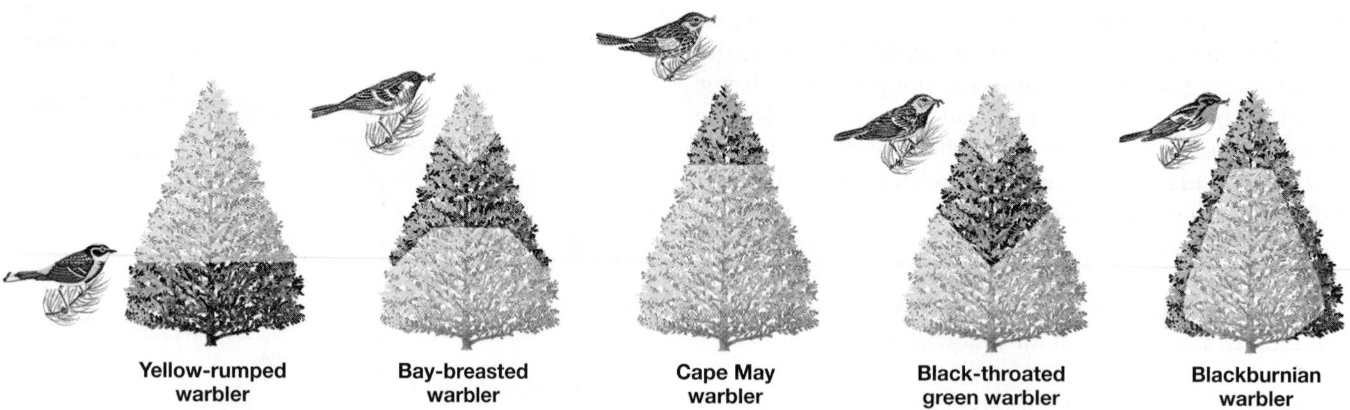

▲ **FIGURE 27-2 Resource partitioning** Each of these five insect-eating species of North American warblers searches for food in slightly different regions within spruce trees. They reduce competition by occupying very similar, but not identical, niches. Data from MacArthur, R.H. 1958. *Population ecology of some warblers of Northeastern coniferous forest.* Ecology 39:599–619.

Another example of resource partitioning was discovered by Charles Darwin among related species of finches of the Galápagos Islands. The finches that shared the same island had evolved different bill sizes and shapes and different feeding behaviors that reduced the competition among them (see p. 271).

Interspecific Competition May Reduce the Population Size and Distribution of Each Species

Although natural selection leads to a reduction of niche overlap between different species, those with similar niches still compete for limited resources, causing both populations to be restricted. A classic study of the effects of interspecific competition was performed by ecologist Joseph Connell, using barnacles (shelled crustaceans that attach permanently to rocks and other surfaces).

Barnacles of the genus *Chthamalus* share rocky ocean shores with barnacles of the genus *Balanus*, and their niches overlap considerably. Both live in the **intertidal zone**, an area of the shore that is alternately covered and exposed by the tides. Connell found that *Chthamalus* dominates the upper shore and *Balanus* dominates the lower. When he scraped off *Balanus*, the *Chthamalus* population increased, spreading downward into the area that its competitor had once inhabited. Where the habitat is appropriate for both genera, *Balanus* conquers because it is larger and grows faster. But *Chthamalus* tolerates drier conditions, giving it a competitive advantage on the upper shore, where only high tides submerge the barnacles. As this example illustrates, interspecific competition can limit both the size and the distribution of competing populations.

Competition Within a Species Is a Major Factor Controlling Population Size

Individuals of the same species have the same requirements for resources and thus occupy the same ecological niche. For this reason, **intraspecific competition**—competition among individuals of the same species—is the most intense form of competition. As explained in Chapter 26, intra-specific competition exerts strong density-dependent environmental resistance, limiting population size. The evolutionary result of intraspecific competition is that individuals who are better equipped to obtain scarce resources are more likely to reproduce successfully, passing their heritable traits to their offspring.

Case Study continued
Mussels Muscle In

The zebra and quagga mussels accidentally imported from Europe have niches that overlap extensively with those of North American freshwater mussels and clams. Where the European mussels have invaded, the indigenous populations have declined. Both native and imported species compete directly for space, but the zebra and quagga mussels reproduce far more rapidly, so their populations have expanded at the expense of the native mollusks.

In fact, the invasive mussels literally cover the natives; more than 10,000 zebra mussels have been found attached to a single native mussel! Zebra and quagga mussels also compete with native bivalves for food, by filtering the same types of microorganisms from the water. This interspecific competition has caused several species of native freshwater clams to be nearly eliminated from Lake St. Clair and western Lake Erie.

27.3 WHAT ARE THE RESULTS OF INTERACTIONS BETWEEN PREDATORS AND THEIR PREY?

Predators eat other organisms. Although we generally think of predators as being **carnivores** (animals that eat other animals), ecologists sometimes include **herbivores** (animals that eat plants) in this general category. Here, we define predation in this more general way, to include the zebra mussel that filters microscopic photosynthetic protists from water,

(a) Pika

(b) Long-eared bat

(c) Northern goshawk

▲ FIGURE 27-3 **Forms of predation (a)** A pika, whose preferred food is grass, is a small relative of the rabbit and lives in the Rocky Mountains. **(b)** A long-eared bat uses a sophisticated echolocation system to hunt moths, which in turn have evolved special sound detectors and behaviors to avoid capture. **(c)** A Northern goshawk feasts on a smaller bird.

QUESTION Describe some examples of coevolution of predators and prey.

the grass-eating pika (**Fig. 27-3a**), the bat homing in on a moth (**Fig. 27-3b**), and the more familiar example of a hawk eating a smaller bird (**Fig. 27-3c**). Predators are generally less abundant than their prey; you will learn why in Chapter 28.

Predator–Prey Interactions Shape Evolutionary Adaptations

To survive, predators must feed and prey must avoid becoming food. Therefore, predator and prey populations exert intense selective pressure on one another, resulting in coevolution. As prey become more difficult to catch, predators must become more adept at hunting. Coevolution has endowed the mountain lion with tearing teeth and claws, and has given the hunted fawn dappled spots that serve as camouflage, as well as the behavior of lying still as it awaits its mother's return from feeding. Coevolution has produced the keen eyesight of the hawk and owl, which is countered by the earthy colors of their mouse and ground squirrel prey. Evolution under predation pressure has also produced the toxins of the poison dart frog, the coral snake, and milkweed (see Figs. 27-7, 27-9b, and 27-12b).

In the following sections, we examine a few of the evolutionary results of predator–prey interactions. In "Earth Watch: Invasive Species Disrupt Community Interactions" on p. 520, we describe what happens when natural checks and balances are circumvented by transporting predatory or competing organisms into ecological communities whose members are not adapted to deal with them.

Some Predators and Prey Have Evolved Counteracting Behaviors

Bat and moth adaptations (see Fig. 27-3b) provide excellent examples of how both body structures and behaviors are molded by coevolution. Most bats are nighttime hunters that emit pulses of sound that are so high pitched that people can't hear them. Using the echoes that occur as their sounds bounce back from nearby objects, bats create a sonar image of their surroundings, which allows them to navigate and detect prey.

Under selection pressure from this unusual prey-locating system, some moths (a favorite prey of bats) have evolved ears that are particularly sensitive to the pitches used by echolocating bats. When they hear a bat, these moths take evasive action, flying erratically or dropping to the ground. The bats, in turn, have evolved the ability to counter this defense by switching the frequency of their sound pulses away from the moth's sensitivity range. Some moths interfere with the bats' echolocation by producing their own high-frequency clicks. In still another coevolutionary adaptation, a bat that is hunting a clicking

Case Study continued

Mussels Muscle In

In 1990, a round goby was discovered in the St. Clair River. Like zebra and quagga mussels, this fish is native to southeastern Europe, and probably arrived at the Great Lakes in the same fashion as the mussels—in the ballast water of ships. Recognizing its natural prey, this 5-inch-long predator began feasting on small zebra and quagga mussels, rapidly expanding its range into all five of the Great Lakes.

Is the introduction of the invasive mussels' natural predator a solution to the mussel problem? Unfortunately, no. The gobies ignore the largest mussels, which produce the most eggs. Further, the gobies are not picky predators. In addition to mussels, they will eat the eggs and young of many fish, including native smallmouth bass, walleye, and perch.

(a) A camouflaged fish

(b) A camouflaged bird

▲ **FIGURE 27-4 Camouflage by blending in (a)** The sand dab is a flat, bottom-dwelling ocean fish with a mottled color that closely resembles the sand on which it rests. **(b)** This nightjar bird on its nest in Central America is barely visible among the surrounding leaf litter.

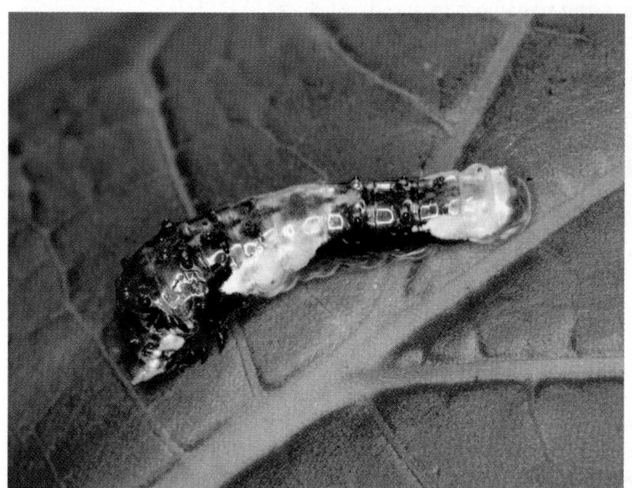

(a) A camouflaged caterpillar

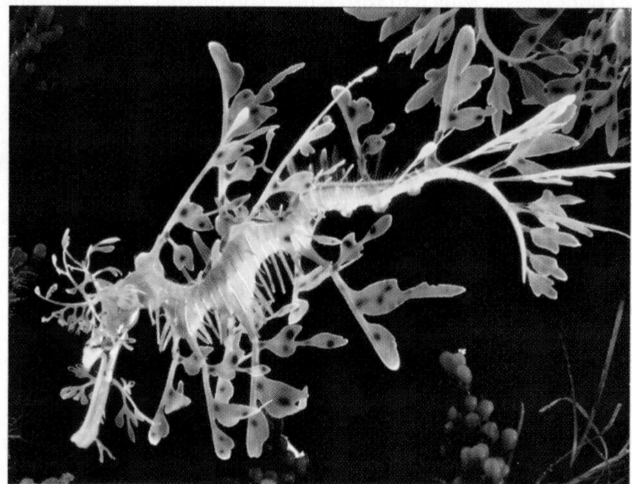

(b) A camouflaged leafy sea dragon

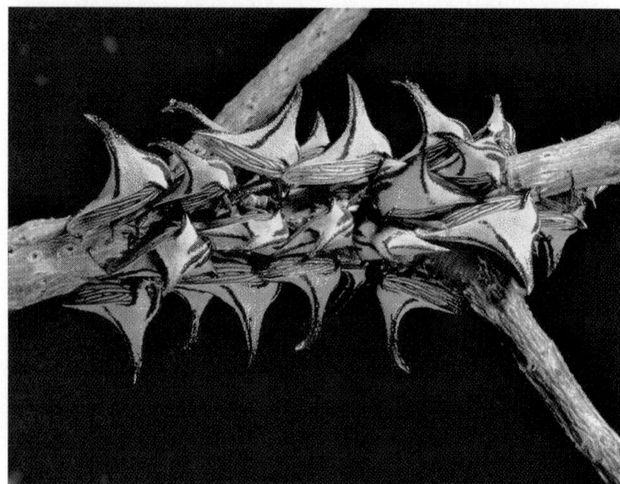

(c) Camouflaged treehoppers

(d) Camouflaged cacti

▲ **FIGURE 27-5 Camouflage by resembling specific objects (a)** A citrus swallowtail butterfly caterpillar, whose color and shape resemble a bird dropping, sits motionless on a leaf. **(b)** The leafy sea dragon (an Australian "seahorse" fish) has evolved extensions of its body that mimic the algae in which it often hides. **(c)** Florida treehopper insects avoid detection by resembling thorns on a branch. **(d)** These cacti of the American Southwest are appropriately called "living rock cacti."

QUESTION In general, how might such camouflage have evolved?

moth may turn off its own sound pulses temporarily (thus avoiding detection) and follow the moth's clicks to capture it.

Camouflage Conceals Both Predators and Their Prey

An old saying goes that the best hiding place may be right out in plain sight. Both predators and prey have evolved colors, patterns, and shapes that resemble their surroundings. Such disguises, called **camouflage,** render plants and animals inconspicuous, even when they are in full view (**Fig. 27-4**).

Some animals closely resemble specific objects such as leaves, twigs, seaweed, thorns, or even bird droppings (**Figs. 27-5a–c**). Camouflaged animals tend to remain motionless; a fleeing bird dropping would be quite conspicuous. Whereas many camouflaged animals resemble parts of plants, a few desert plants have evolved to resemble rocks, which helps them to avoid predation by animals seeking the water they store in their bodies (**Fig. 27-5d**).

Predators that ambush are also aided by camouflage. For example, a spotted cheetah becomes inconspicuous in the grass as it watches for grazing antelope. The frogfish closely resembles the sponges and algae-covered rocks on which it lurks, motionless, awaiting smaller fish to eat (**Fig. 27-6**).

(a) A camouflaged cheetah

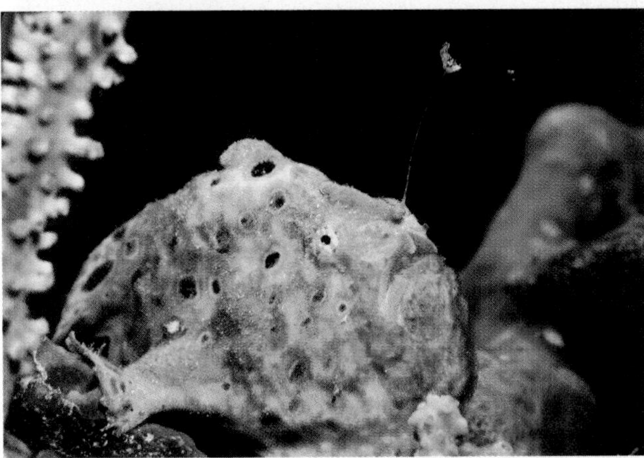

(b) A camouflaged frogfish

▲ **FIGURE 27-6 Camouflage assists predators (a)** As it waits for prey, a cheetah blends into the background of the grass. **(b)** Combining camouflage and aggressive mimicry, a frogfish waits in ambush, its camouflaged body matching the sponge-encrusted rock on which it rests. Above its mouth dangles a lure that closely resembles a small fish. The lure attracts small predators, who will find themselves prey. The frogfish can expand its mouth by a factor of 12 in a few thousands of a second, sucking in its prey.

Bright Colors Often Warn of Danger

Some animals have evolved very differently, exhibiting bright **warning coloration.** These animals may be capable of inflicting a painful sting or may be bad-tasting and poisonous, as is the poison dart frog (**Fig. 27-7**). The bright colors seem to declare "Eat me at your own risk!"

Some Prey Organisms Gain Protection Through Mimicry

Mimicry refers to a situation in which members of one species have evolved to resemble another species. By sharing a similar warning-color pattern, several poisonous species may all benefit. Mimicry among different distasteful species is called *Müllerian mimicry* (after the German zoologist Johann Müller). For example, toxic monarch butterflies have wing patterns strikingly similar to those of equally distasteful

> **Have you ever wondered?**
>
> **Why Some Wild Animals Freeze in the Middle of a Road as Your Car Approaches?**
>
> Animals that are prey have evolved not only camouflaged coloring, but also behaviors that enhance their survival. In response to a predator, they will often remain motionless (making the camouflage more effective) and only bolt at the last minute, when it becomes obvious that the predator has spotted or smelled them. Your car, resembling a very large predator, is likely to evoke both of these instinctive behaviors. Although they work well in natural situations, these responses increase the animal's chances of becoming roadkill.

▲ **FIGURE 27-7 Warning coloration** The South American poison dart frog advertises its poisonous skin with bright and contrasting color patterns.

(a) Monarch (distasteful)

(b) Viceroy (distasteful)

▲ **FIGURE 27-8 Müllerian mimicry** Nearly identical warning coloration protects both **(a)** the distasteful monarch and **(b)** the equally distasteful viceroy butterfly.

(a) Bee (venomous)

Hoverfly (nonvenomous)

(b) Coral snake (venomous)

Scarlet king snake (nonvenomous)

▲ **FIGURE 27-9 Batesian mimicry (a)** A bee, which is capable of stinging (left), is mimicked by the stingless hoverfly (right). **(b)** The warning coloration of the venomous coral snake (left) is mimicked by the harmless scarlet king snake (right).

viceroy butterflies (**Fig. 27-8**). Birds that become ill from consuming one species are more likely to avoid the other as well. A toad that is stung while attempting to eat a bee is likely to avoid not only bees, but other black and yellow striped insects (such as yellow jacket wasps) without ever tasting one. A shared color pattern thus helps all similarly colored species avoid predation.

Once warning coloration evolved, there arose a selective advantage for harmless animals to resemble venomous ones, an adaptation called *Batesian mimicry* (after the English naturalist Henry Bates). Through Batesian mimicry, the harmless hoverfly avoids predation by resembling a bee (**Fig. 27-9a**), and the nonvenomous scarlet king snake is protected by brilliant warning coloration closely resembling that of the highly venomous coral snake (**Fig. 27-9b**).

Certain prey species use another form of mimicry: **startle coloration**. Several insects and even some vertebrates (such as the false-eyed frog) have evolved patterns of color that closely resemble the eyes of a much larger, and possibly dangerous, animal (**Fig. 27-10**). If a predator gets close, the prey suddenly flashes its eyespots, startling the predator and allowing the prey to escape.

A sophisticated variation on the theme of prey mimicking dangerous animals is seen in snowberry flies, which are hunted by territorial jumping spiders (**Fig. 27-11**). When a fly spots an approaching spider, it spreads its wings, moving them back and forth in a jerky dance. Seeing this display, the spider is likely to flee the fly. Why? Researchers have observed that the markings on the fly's wings look like the legs of a jumping spider, and the fly's jerky movements mimic the behavior of a jumping spider driving another spider from its territory. Thus, natural selection has finely tuned both the behavior and the appearance of the fly to avoid predation by jumping spiders.

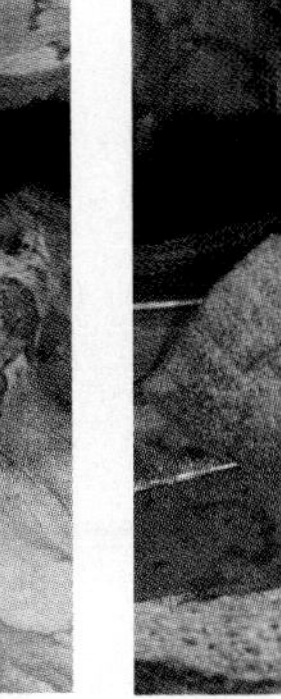

(a) False-eyed frogs

(b) Peacock moth

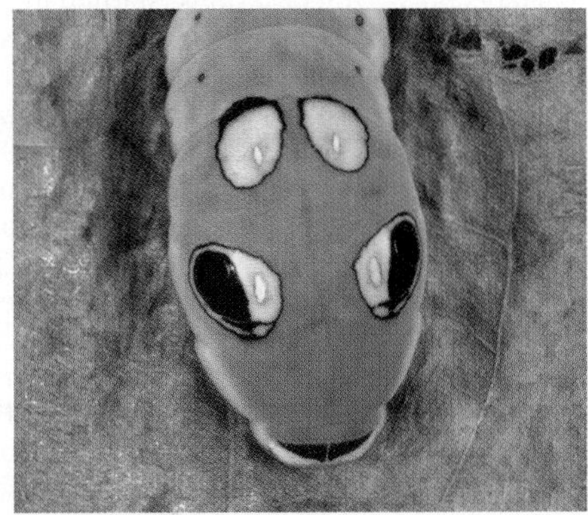

(c) Swallowtail caterpillar

▲ **FIGURE 27-10 Startle coloration (a)** When threatened, the false-eyed frog raises its rump, which resembles the eyes of a large predator. **(b)** The peacock moth from Trinidad is well camouflaged, but should a predator approach, it suddenly opens its wings to reveal spots resembling large eyes. **(c)** Predators of this caterpillar larva of the Eastern tiger swallowtail butterfly are deterred by its resemblance to a snake. The caterpillar's head is the "snake's" nose, and it bears two sets of eyespots.

(a) Jumping spider (predator)

(b) Snowberry fly (prey)

▲ **FIGURE 27-11 A prey mimics its predator (a)** When a jumping spider approaches, **(b)** the snowberry fly spreads its wings, revealing a pattern that resembles spider legs. The fly enhances the effect by performing a jerky, side-to-side dance that resembles the leg-waving display of a jumping spider defending its territory.

Earth Watch

Invasive Species Disrupt Community Interactions

Invasive species are species that are introduced into an ecosystem in which they did not evolve, and that are harmful to human health, the environment, or the economy of a region. Invasive species often spread widely because they find few effective forms of environmental resistance, such as strong competitors, predators, or parasites, in their new environment. Their unchecked population growth may seriously damage the ecosystem as they outcompete or prey on native species.

Not all non-native species become pests; the ones that do typically reproduce rapidly, disperse widely, and thrive under a relatively wide range of environmental conditions. Invasive plants often spread by sprouting from roots as well as seeds, and some aquatic forms generate new plants from fragments. Invasive animals often eat a wide variety of foods. By evading the checks and balances imposed by millennia of coevolution, invasive species are wreaking havoc on natural ecosystems throughout the world. Some examples follow.

English house sparrows were introduced into the United States on several occasions, starting in the 1850s, in the hope that they would control caterpillars feeding on shade trees. In 1890, European starlings were released into Central Park in New York City by a group attempting to introduce all the birds mentioned in the works of Shakespeare. Both bird species have spread throughout the continental United States. Their success has reduced the populations of some native songbirds, such as bluebirds and purple martins, with which they compete for nesting sites. Red fire ants from South America were accidentally introduced into Alabama on shiploads of lumber in the 1930s and have since spread throughout the southern United States. Fire ants kill native ants, birds, and young reptiles. Their mounds can ruin farm fields, and their fiery stings and aggressive temperament can make backyards uninhabitable. Imported cane toads have proven unstoppable as they spread through northeastern Australia, outcompeting native frogs and killing potential predators with their toxic secretions (**Fig. E27-1a**).

Invasive plants also threaten natural communities. In the 1930s and 1940s, the Japanese vine kudzu was planted extensively in the southern United States to control erosion. Today, kudzu is a major pest, killing trees and all other vegetation in its path, and engulfing any stationary object, such as an abandoned house (**Fig. E27-1b**). Both water hyacinth and purple loosestrife were introduced as ornamental plants. Water hyacinth now clogs waterways in the southern United States, slowing boat traffic and displacing natural vegetation. Purple loosestrife aggressively invades wetlands, where it outcompetes native plants and reduces food and habitat for native animals (**Fig. E27-1c**).

Invasive species rank second only to habitat destruction in pushing endangered species toward extinction. Recently, wildlife officials have made cautious attempts to reestablish biological checks and balances by importing predators or parasites (called biocontrols) to attack invasive species. This is fraught with danger, because new imports can have unpredicted and possibly disastrous effects on native wildlife. The cane toad, for example, was introduced into Australia from South America in the 1930s to control introduced beetles that threatened the sugarcane crop.

Despite the risks of imported biocontrols, there are often few realistic alternatives, because poisons kill native and non-native organisms indiscriminately. Biologists now carefully screen proposed biocontrols to make sure they are specific for the intended invasive species, and there have been a number of successful introductions. For example, imported beetles are now among the most important ways of controlling purple loosestrife in North America. Professionals and volunteers work together to raise and release these beetles by the millions, helping to restrain this invasive weed.

(a) Cane toad

(b) Kudzu

(c) Purple loosestrife

▲ FIGURE E27-1 **Invasive species (a)** The cane toad in Australia outcompetes native toads and frogs. **(b)** The Japanese vine kudzu will rapidly cover entire trees and houses. **(c)** Purple loosestrife displaces native plants and reduces food and habitat for native animals in wetlands.

(a) Bombardier beetle

(b) Monarch caterpillar

▲ FIGURE 27-12 **Chemical warfare** **(a)** The bombardier beetle sprays a hot toxic brew when its leg is pinched by a forceps. **(b)** A monarch caterpillar feeds on milkweed that contains a powerful toxin.

QUESTION Why is the caterpillar colored with bright stripes?

Predators May Use Mimicry to Attract Prey

Some predators have evolved **aggressive mimicry**, a "wolf-in-sheep's-clothing" approach, in which they entice their prey to come close by resembling something attractive to the prey. For example, by using a rhythm of flashes that is unique to each species, female fireflies attract males to mate. But in one species, the females sometimes mimic the flashing pattern of a different species, attracting males that they kill and eat. The frogfish (see Fig. 27-6b) is not only camouflaged, but exhibits aggressive mimicry by dangling, just above its mouth, a wriggling lure that resembles a small fish. Fish attracted by the lure are eaten in a split second if they get too close.

Predators and Prey May Engage in Chemical Warfare

Predators and prey use toxins for attack and defense. The venom of spiders and some snakes, such as the coral snake (see Fig. 27-9b), serves both to paralyze prey and to deter predators.

Certain mollusks (including squid, octopus, and some sea slugs) emit clouds of ink when attacked. These colorful chemical "smoke screens" confuse predators and mask the prey's escape. A dramatic example of chemical defense is seen in the bombardier beetle. In response to the bite of an ant, the beetle releases secretions from special glands into an abdominal chamber. There, enzymes catalyze an explosive chemical reaction that shoots a toxic, boiling-hot spray onto the attacker (**Fig. 27-12a**).

Plants have evolved a variety of chemical adaptations that deter their herbivorous predators. Many, such as the milkweed, synthesize toxic and distasteful chemicals. As plants evolved defensive toxins, certain insects evolved increasingly efficient ways to detoxify or even use these substances. The result is that nearly every toxic plant is eaten by at least one type of insect. For example, monarch butterflies lay their eggs on milkweed; when their larvae hatch, they consume this poisonous plant (**Fig. 27-12b**). The caterpillars not only tolerate the milkweed poison, but store it in their tissues as a defense against their own predators. The stored toxin is retained in the metamorphosed monarch butterfly (see Fig. 27-8a). Viceroy butterflies (see Fig. 27-8b) use a similar strategy, storing a bitter compound from willows (eaten by the larvae) in the tissues of the adult.

Grasses embed tough silicon (glassy) substances in their blades that make them difficult to chew, selecting for grazing animals with longer, harder teeth. Over evolutionary time, as grasses evolved tougher blades that discouraged predation, horses evolved longer teeth with thicker enamel coatings that resist wear and abrasion from the tough grasses.

27.4 WHAT IS PARASITISM?

Parasites live in or on their prey, which are called **hosts,** usually harming or weakening them but not immediately killing them. Parasites are generally much smaller and more numerous than their hosts. Familiar parasites include tapeworms, fleas, ticks, and the many types of disease-causing protists, bacteria, and viruses. Many parasites, such as the protist that causes malaria, have complex life cycles involving two or more hosts (see Fig. 20-12). The newly discovered roundworm parasite highlighted in "Scientific Inquiry: A Parasite Makes Ants Berry Appealing to Birds" on p. 522 appears to use both ant and bird hosts during its life cycle. There are very few parasitic vertebrates, but the lamprey (see Fig. 24-5), which attaches itself to a host fish and sucks its blood, is one example.

Scientific Inquiry

A Parasite Makes Ants Berry Appealing to Birds

On a field trip to the rain forest of Panama in May 2005, ecologist Steve Yanoviak and fellow researchers were examining a colony of tree-dwelling black ants (*Cephalotes atratus*) and found a few with strikingly red abdomens (**Fig. E27-2**). When sliced open, the abdomens disgorged hundreds of eggs in which tiny parasitic roundworms were developing. Both the roundworm species and its effect on the ants were new to science, so the team eagerly pursued their chance observation.

To test the hypothesis that only adult ants with red abdomens harbor roundworms, the researchers dissected 300 all-black adults and found none to be infected, while all those with red abdomens were infected. Noticing that the infected abdomens broke off readily, they measured the force required to pluck abdomens from infected ants versus those of normal ants, and discovered that only infected abdomens could be pulled off without dislodging the ant from its twig. Although ants of this species usually bite and emit foul-smelling, distasteful chemicals when handled (as a protection against insect-eating birds), the infected ants did not exhibit these behaviors. Further, their red abdomens strongly resembled, in both color and shape, berries from trees on which the ants foraged (see Fig. E27-2).

Based on these observations, Yanoviak and coworkers hypothesized that the roundworms trick berry-eating birds into dispersing their eggs by causing the ants' abdomens to resemble berries. To investigate this, the team dissected ants at all stages of development, and found that the parasites infect larval ants and develop as the ant matures. Once the roundworms produce eggs that are ready to be released, the infected ants' abdomens become enlarged and red.

To test whether the eggs could survive passing through a bird's digestive tract, and thus be dispersed in its droppings, the researchers fed an infected ant abdomen to a chicken, and later, found hundreds of healthy roundworm eggs in its feces. But how did the larval ants become infected? The researchers hypothesized that ant larvae might eat bird droppings containing roundworm eggs. The team collected food particles that adult ants were carrying to their colony to feed their larvae, and discovered that, indeed, many of these were bits of bird droppings.

▲ **FIGURE E27-2 An ant resembles berries** The abdomen of a parasitized ant closely resembles berries on trees in which the ants live. Birds eating these berries may pluck off the ant's abdomen.

Based on their observations and experiments, the researchers hypothesized that the roundworms spend most of their life cycle inside the abdomens of ants, but then rely on fruit-eating birds to disperse their eggs, a strategy that would help them to survive. If the roundworms only parasitized ants, then as they spread throughout an ant colony they would be likely to kill the entire colony, and themselves with it. Fruit-eating birds, however, visit many trees and leave droppings that are collected by ants from many colonies. So, natural selection would favor roundworms whose larvae caused their host ants' abdomens to swell and mimic ripe fruit, causing birds to disperse them widely.

As always in science, new findings trigger new questions; for example, how does the roundworm infection cause the ant abdomen to turn red? We can be sure that future studies will answer this question and also raise more questions in the process.

Parasites and Their Hosts Act as Agents of Natural Selection on One Another

The variety of infectious bacteria and viruses and the precision of the immune system that counters their attacks are evidence of the powerful forces of coevolution between parasitic microorganisms and their hosts.

Some specific examples of parasitic interactions influencing evolution can be observed today. One is a parasitic protist, common in Africa, that causes a disease called nagana in cattle. Some breeds of African cattle from nagana-infested regions have evolved a partial immunity to it and generally survive infection, whereas different breeds of cattle imported to Africa usually die from nagana if not treated. Likewise, the malaria parasite, which spends part of its life cycle in red blood cells, has exerted strong selective pressure on human populations in malaria-infested regions. A mutation in the hemoglobin gene causes red blood cells to become distorted and resistant to infection by this protist. Although individuals who inherit two copies of this mutation will have sickle-cell anemia, in some regions of Africa, 20% to 40% of the human population carries the sickle-cell gene because of the protection it confers against malaria (see p. 190).

27.5 WHAT IS MUTUALISM?

Mutualism refers to interactions between species in which both benefit. Many mutualistic relationships are symbiotic;

(a) Lichen

(b) Clownfish

▲ FIGURE 27-13 **Mutualism (a)** This brightly colored lichen growing on bare rock is a mutualistic relationship between an alga and a fungus. **(b)** The clownfish snuggles unharmed among the stinging tentacles of the anemone.

that is, they involve a close, long-term physical association between the participating species. If you see colored patches on rocks, they are probably lichens, a mutualistic association between an alga and a fungus (**Fig. 27-13a**). The fungus provides support and protection while obtaining food from the photosynthetic alga, whose bright colors are actually light-trapping pigments. Mutualistic associations also occur in the digestive tracts of cows and termites, where protists and bacteria find food and shelter. The microorganisms break down cellulose, making its component sugar molecules available both to themselves and to the animals that harbor them. In our own intestines, mutualistic bacteria synthesize vitamins, such as vitamin K, which we absorb and use. Plants called legumes benefit by providing chambers in their roots that house nitrogen-fixing bacteria (see Fig. 19-9). These bacteria are among the few organisms that can acquire nitrogen gas from the air and chemically modify it into a form that plants can use as a nutrient.

Many mutualistic relationships are not as intimate and extended as those just discussed. Consider, for example, the relationship between plants and the insects that pollinate them. The insects fertilize the plants by carrying plant sperm (found in pollen grains), and benefit by sipping nectar and sometimes eating pollen. Both the bee and hoverfly shown in Figure 27-9a are important plant pollinators. The clownfish (coated with a layer of protective mucus) takes shelter among the venomous tentacles of certain species of anemones (**Fig. 27-13b**). In this mutualistic association, the anemone provides the clownfish with protection from predators, while the clownfish cleans its anemone, defends it from predators, and may bring it bits of food.

27.6 HOW DO KEYSTONE SPECIES INFLUENCE COMMUNITY STRUCTURE?

In some communities, a particular species, called a **keystone species**, plays a major role in determining community structure—a role that is out of proportion to its abundance in the community. If the keystone species is removed from its community, normal community interactions are significantly altered and the relative abundance of other species changes dramatically. It is difficult to experimentally identify keystone species because the community must be carefully monitored for years with the species present, and for additional years after it is removed. Nonetheless, ecological studies have provided evidence that keystone species are important in a variety of communities.

In 1969, Robert Paine, an ecologist at the University of Washington, removed predatory *Pisaster ochraceous* sea stars (**Fig. 27-14a**) from sections of Washington State's rocky intertidal coast. Native mussels, a favored prey of *Pisaster*, became so abundant that they outcompeted other invertebrates and algae that normally coexist with the mussels in intertidal communities.

Some species are found to be keystone species only after they are removed as a direct or indirect result of human activities. For example, the lobster may be a keystone species off the east coast of Canada. Overfishing of the lobster allowed the population of its prey, sea urchins, to expand enormously. The population explosion of sea urchins nearly eliminated certain types of algae on which the urchins feed, leaving large expanses of bare rock where a diverse community formerly existed.

The sea otter appears to be a keystone species along the coast of western Alaska. Starting around 1990, otter numbers declined drastically, allowing an increase in their sea urchin prey. The sea urchins then overgrazed the kelp forests that provide critical undersea habitat for a variety of marine species. What killed the sea otters? Killer whales, which formerly fed primarily on seals and sea lions, were seen increasingly dining on sea otters as their favored prey disappeared. Scientists hypothesize that seal and sea lion populations have declined, in turn, because of overfishing by humans in the North Pacific, depleting the food supply of these fish-eaters.

In the African savanna, the elephant is a keystone predator. By grazing on small trees and bushes (**Fig. 27-14b**),

(a) Sea stars

(b) African elephant

▲ **FIGURE 27-14 Keystone species (a)** The sea star *Pisaster ochraceous* is a keystone species along the rocky coast of the Pacific Northwest. **(b)** The elephant is a keystone species on the African savanna.

elephants prevent the encroachment of forests and help maintain the grassland community, along with its diverse population of grazing mammals and their predators (described in Chapter 29). In Chapter 30, you will learn about the role of another keystone species, the wolf, described in "Earth Watch: Restoring a Keystone Predator" on p. 594.

27.7 SUCCESSION: HOW DO COMMUNITY INTERACTIONS CAUSE CHANGE OVER TIME?

In a mature terrestrial ecosystem, the populations that make up the community interact with one another and with their nonliving environment in intricate ways. But this tangled web of life did not spring fully formed from bare rock or naked soil; rather, it emerged in stages over a long period, by a process called succession. **Succession** is a structural change in a community and its nonliving environment over time. It is a kind of "community relay" in which assemblages of plants and animals replace one another in a sequence that is somewhat predictable.

Succession is preceded and started by an ecological **disturbance**, an event that disrupts the ecosystem by altering its community, its **abiotic** (nonliving) structure, or both. The precise changes that occur during succession are as diverse as the environments in which succession occurs, but we can recognize certain general stages. Succession starts with a few hardy plants called **pioneers.** The pioneers alter the ecosystem in ways that favor competing plants, which gradually displace them. If allowed to continue, succession progresses to a diverse and relatively stable **climax community.** Alternatively, recurring disturbances maintain many communities in earlier, or **subclimax,** stages of succession. Our discussion of succession will focus on plant communities, which dominate the landscape and provide both food and habitat for animals.

There Are Two Major Forms of Succession: Primary and Secondary

Succession takes two major forms: primary and secondary. During **primary succession,** a community gradually forms where there is no trace of a previous community. The disturbance that sets the stage for primary succession may be a glacier scouring the landscape down to bare rock, or it may be a volcano producing an entirely new island or covering an ecosystem with new rock as lava hardens (**Fig. 27-15a**). This building of a community "from scratch" typically requires thousands or even tens of thousands of years.

During **secondary succession,** a new community develops after an existing ecosystem is disturbed in a way that leaves significant remnants of the previous community behind, such as soil and seeds. For example, beavers, landslides, or people may dam streams, causing marshes, ponds, or lakes to form. A landslide or avalanche may strip a swath of trees from a mountainside. When Mount St. Helens in Washington State erupted in 1980, it left patchy remnants of forest and a thick layer of nutrient-rich ash that encouraged a rapid proliferation of new life (**Fig. 27-15b**). Fire is another common disturbance. Fires produce nutrient-rich ash, and spare some trees and many healthy roots. Some seeds withstand fire and even require it in order to sprout, so fires allow rather rapid regeneration of forests and other communities (**Fig. 27-15c**). In the following sections, we look at specific examples of succession that illustrate the process in more detail.

(a) Mt. Kilauea, Hawaii

(b) Mt. St. Helens, Washington State

(c) Yellowstone National Park, Wyoming

▲ FIGURE 27-15 **Succession in progress (a)** Primary succession. Left: The Hawaiian volcano Mount Kilauea has erupted repeatedly since 1983, sending rivers of lava over the surrounding countryside. Right: A pioneer fern takes root in a crack in hardened lava. **(b)** Secondary succession. Left: On May 18, 1980, the explosion of Mount St. Helens in Washington State devastated the surrounding pine forest ecosystem. Right: Twenty years later, life abounds on the once-barren landscape. Because traces of the former ecosystem remained, this is an example of secondary succession. **(c)** Secondary succession. Left: In the summer of 1988, extensive fires swept through the forests of Yellowstone National Park in Wyoming. Right: Trees and flowering plants are thriving in the sunlight, and wildlife populations are rebounding as secondary succession occurs.

QUESTION People have suppressed fires for decades. What are the implications of fire suppression for forest ecosystems and succession?

Primary Succession Can Begin on Bare Rock

Figure 27-16 illustrates primary succession on Isle Royale, Michigan, an island in northern Lake Superior that was scraped down to bare rock by glaciers that retreated roughly 10,000 years ago. Weathering produces cycles of freezing and thawing that cause rocks to crack, producing fissures, and erodes rock surface layers, producing small particles. Rainwater, which is naturally acidic and an excellent solvent, dissolves some of the minerals from rock particles, making them available to plants.

Weathered rock provides a place for lichens, which are pioneer species, to attach where there are no competitors and plenty of sunlight. Lichens obtain energy through photosynthesis and acquire some of their mineral nutrients by dissolving rock with an acid that they secrete. As the lichens spread over the rock surface, species of mosses that are relatively drought tolerant begin growing in the rock cracks. Fortified by nutrients liberated by the lichens, the mosses form a dense mat that traps dust, tiny rock particles, and bits of organic debris. The mosses will eventually cover and kill many of the lichens that made their growth possible.

As some mosses die each year, their bodies add nutrients to a thin layer of new soil, while the living mosses mat acts like a sponge, absorbing and trapping moisture. Within the moss, seeds of larger plants, such as bluebell and yarrow, germinate. As these plants die, their bodies contribute to the growing layer of soil.

As woody shrubs such as blueberry and juniper take advantage of the newly formed soil, the mosses and remaining lichens may be shaded out and buried by decaying leaves and vegetation. Eventually, trees such as jack pine, black spruce, and aspen take root in the deeper crevices, and the sun-loving shrubs are shaded out. Within the forest, shade-tolerant seedlings of taller or faster-growing trees thrive, including balsam fir, paper birch, and white spruce. In time, they tower over and replace the original trees, which are intolerant of shade. After a thousand years or more, a tall climax forest thrives on what was once bare rock.

An Abandoned Farm Will Undergo Secondary Succession

Figure 27-17 illustrates secondary succession on an abandoned farm in the southeastern United States. The pioneer species, sun-loving, fast-growing annual plants such as crabgrass, ragweed, and Johnson grass, take root in the rich soil. Such species generally produce large numbers of easily dispersed seeds that help them colonize open spaces, but they don't compete well against longer-lived (perennial) species that gradually grow larger and shade out the pioneers.

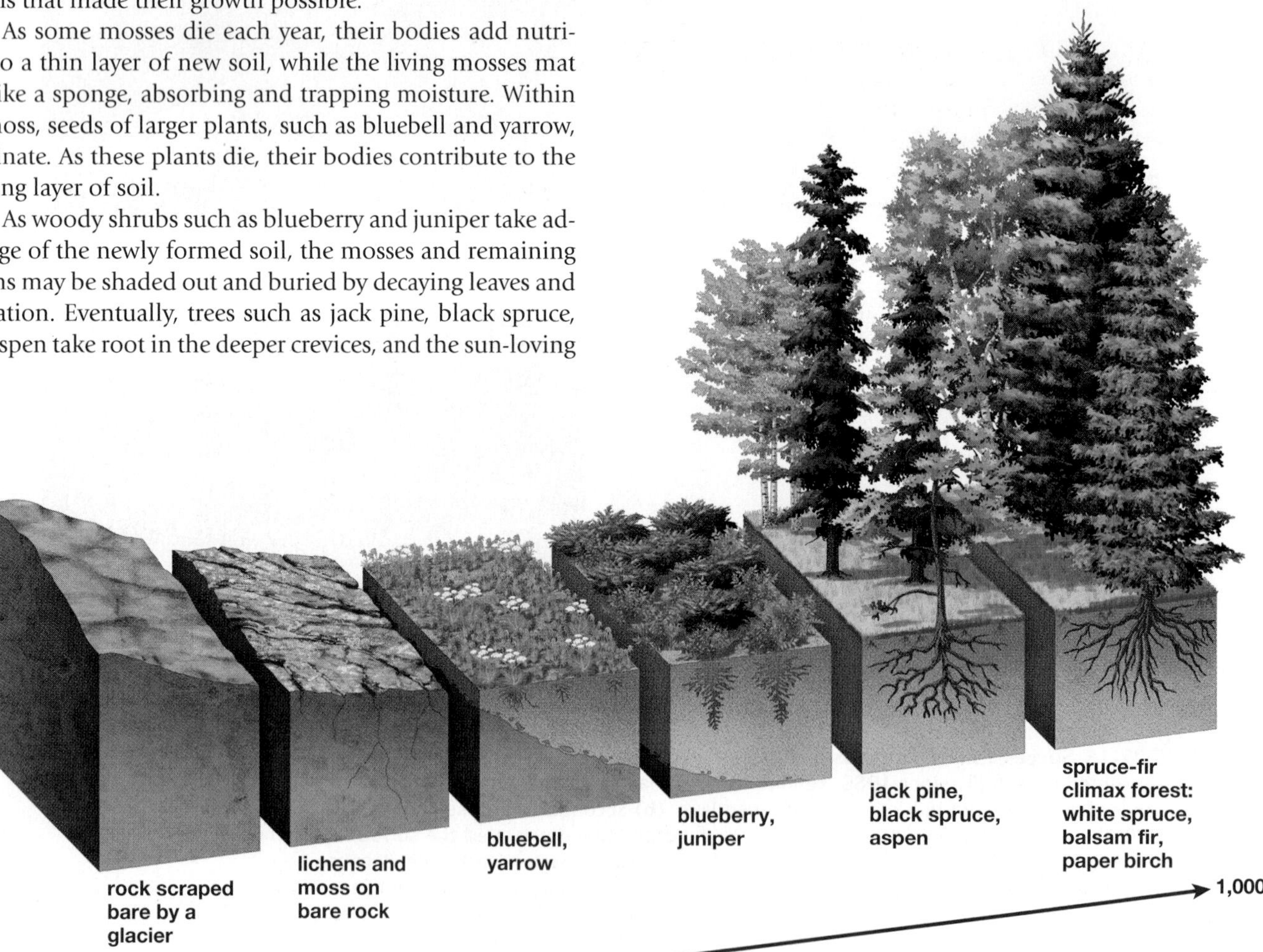

▲ **FIGURE 27-16 Primary succession** Primary succession as it occurs on bare rock exposed as glaciers retreated from Isle Royale in upper Michigan (in the far northern portion of the midwestern United States). Notice that the soil deepens over time, gradually burying the bedrock and allowing trees to take root.

After a few years, perennial plants such as asters, goldenrod, Queen Anne's lace, and perennial grasses move in, followed by woody shrubs such as blackberry and smooth sumac. These plants reproduce rapidly and dominate for many years as pine and cedar trees are becoming established. Then, roughly two decades after a field is abandoned, an evergreen forest dominated by pines takes over and persists for several more decades.

As is common during succession, the new forest alters conditions in ways that favor its competitors. The shade of the pine forest inhibits the growth of its own seedlings, while favoring the growth of hardwood trees, whose seedlings are shade tolerant. Slow-growing hardwoods such as oak and hickory, which take root beneath the pines, begin to replace them after about 70 years, as the pines die of old age. Roughly a century after the field was abandoned, the region is covered by relatively stable climax forest dominated by oak and hickory, which will persist unless the forest is disturbed, for example, by fire or logging.

Succession Also Occurs in Ponds and Lakes

In freshwater ponds or lakes, succession occurs not only through changes within the pond or lake, but also through an influx of nutrients from outside the ecosystem. Sediments and nutrients carried in by runoff from the surrounding land have a particularly large impact on small freshwater lakes, ponds, and bogs, which gradually undergo succession to dry land (**Fig. 27-18**). In forests, meadows may be produced by lakes undergoing succession. As the lake fills in from the edges, grasses colonize the newly formed soil. As the lake shrinks and the area of meadow expands, trees will encroach around the meadow's edges. If you return to a forest lake 20 years after your first visit, it is likely to be smaller as a result of succession.

Succession Culminates in a Climax Community

Succession ends with a relatively stable climax community, which perpetuates itself if it is not disturbed by external forces (such as fire, parasites, introduced species, or human activities). The populations within a climax community have ecological niches that allow them to coexist without replacing one another. In general, climax communities have more species and more types of community interactions than do early stages of succession. The plant species that dominate climax communities generally live longer and tend to be larger than pioneer species; this trend is particularly evident in ecosystems where forest is the climax community.

In your travels, you have undoubtedly noticed that the type of climax community varies dramatically from one area to the next. For example, if you drive through Colorado or Wyoming, you will see a shortgrass prairie climax community on the eastern plains (in those rare areas where it has not been replaced by farms), pine-spruce forests in the mountains, tundra on the mountain summits, and sagebrush-dominated communities in the western valleys. The exact nature of the

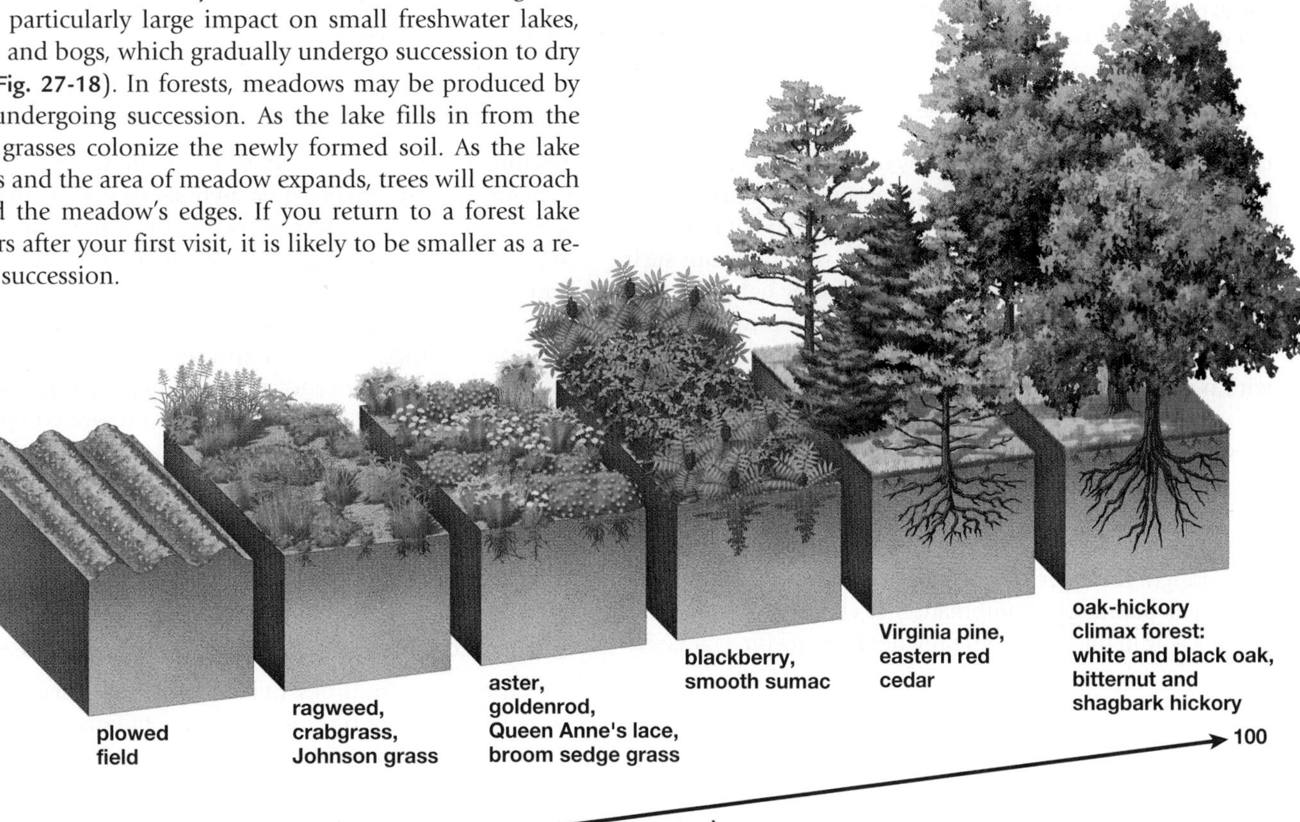

▲ **FIGURE 27-17 Secondary succession** Secondary succession as it might occur on a plowed, abandoned farm field in North Carolina (in the southeastern United States). Notice that a thick layer of soil is present from the beginning, which greatly speeds up the process compared to primary succession.

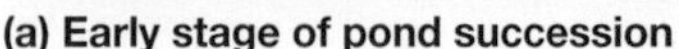

(a) Early stage of pond succession

(b) Late stage of pond succession

▲ **FIGURE 27-18 Succession in a small freshwater pond** In small ponds, succession is speeded by an influx of materials from the surroundings. **(a)** In this small pond, dissolved minerals carried by runoff from the surroundings support aquatic plants, whose seeds or spores were carried in by the winds or by birds and other animals. **(b)** Over time, the decaying bodies of aquatic plants build up soil that provides anchorage for more terrestrial plants. Eventually, the pond is entirely converted to dry land.

climax community is determined by numerous geological and climatic variables, including temperature, rainfall, elevation, latitude, type of rock (which determines the type of nutrients available), and exposure to sun and wind. Natural events such as windstorms, avalanches, and fires started by lightning may destroy sections of climax forest, reinitiating secondary succession and producing a patchwork of various successional stages within an ecosystem.

In many forests throughout the United States, rangers now allow fires set by lightning to run their course, recognizing that this natural process is important for the maintenance of the entire ecosystem. Fires liberate nutrients and kill some (but usually not all) of the trees. As a result, more nutrients and sunlight reach the forest floor, encouraging the growth of subclimax plants. The combination of climax and subclimax regions within the ecosystem provides habitats for a far larger number of species.

Some Ecosystems Are Maintained in a Subclimax Stage

Some ecosystems are maintained in a subclimax stage by frequent disturbances. The tallgrass prairie that once covered northern Missouri and Illinois is a subclimax stage of an ecosystem whose climax community is deciduous forest. The subclimax prairie was maintained by periodic fires, some set by lightning and others deliberately set by Native Americans to increase grazing land for bison. Forest now encroaches, and limited prairie preserves are maintained by carefully managed burning.

Agriculture also depends on carefully selected subclimax communities. Grains are specialized grasses characteristic of the early stages of succession, and farmers spend a great deal of time, energy, and herbicides to prevent competitors, such as other grasses, wildflowers, and woody shrubs, from taking over. The suburban lawn is a painstakingly maintained subclimax ecosystem. Mowing (a disturbance) destroys woody colonizers, and people also use herbicides to selectively kill pioneer species such as crabgrass and dandelions.

Climax Communities Create Earth's Biomes

The climax communities that form during succession are strongly influenced by climate and geography. Extensive areas of characteristic climax plant communities are called **biomes,** and include deserts, grasslands, and a variety of forests. These biomes dominate broad geographical regions with similar climates. The factors that influence climate as well as the biomes that form in the different climates are described in Chapter 29.

Case Study revisited
Mussels Muscle In

Zebra and quagga mussels (**Fig. 27-19**) have all the characteristics that make invasive species successful in their introduced homes. They outcompete native mussels and clams for both food and space, and have few predators to control them. They have phenomenal reproductive potential—each adult female of either species may lay several hundred thousand eggs each year. Their microscopic larvae can be carried for miles in water currents, allowing them to rapidly colonize the length of a river. Because these mussels can survive out of water for days, those that cling to hulls of boats can be easily portaged to other lakes and rivers, where they rapidly establish new colonies.

Although zebra and quagga mussels are both native to southeastern Europe, their geographic ranges usually do not overlap. However, the ecological niches that these closely related species occupy overlap extensively. As a result, the two species have begun to compete with each other in their introduced locations. Ecologists who study these species hypothesize that the quagga may outcompete the zebra mussel because it is better adapted to a range of conditions—it is able to colonize deeper water, and can tolerate both lower temperatures and lower oxygen levels. But because they both damage freshwater ecosystems in the same way, if the quagga outcompetes the zebra mussel, this will have few (if any) practical benefits.

The spread of the zebra and quagga mussels continues. In 2008, both species were reported in reservoirs in Colorado and California, almost certainly transported by recreational boaters. Despite major efforts to identify selective biocontrols, such as selective diseases or predators, there is presently no way to restrain the explosive growth rate of these mussel populations. Some researchers suggest that lack of phytoplankton, their primary food, may eventually slow their proliferation in the Great Lakes. Lake Michigan, for example, has become strikingly clearer since the mussels' invasion, as its phytoplankton population has been depleted. But because phytoplankton directly or indirectly support most of the wildlife within the Great Lakes, food depletion is not an appealing solution to the mussel problem.

▲ **FIGURE 27-19 A quagga mussel (left) and zebra mussel (right)**

Consider This

Invasive mussels can undermine the entire community structure of any lake into which they are introduced. Because mussel larvae settle rapidly on boat hulls, a boat that has been in the water for 24 hours must be washed with a hot-water pressure washer or dried for at least five days before being launched in another body of water. Do you think there is any practical way to restrict the spread of these mussels to still more lakes and reservoirs? What measures would be required to accomplish this?

CHAPTER REVIEW

Summary of Key Concepts

27.1 Why Are Community Interactions Important?

Ecological communities consist of all the interacting populations within an ecosystem. Community interactions influence population size, and the interacting populations within communities act upon one another as agents of natural selection. Thus, community interactions shape the bodies and behaviors of members of the interacting populations.

27.2 What Is the Relationship Between the Ecological Niche and Competition?

The ecological niche defines all aspects of a species' habitat and interactions with its living and nonliving environments. Each species occupies a unique ecological niche. Interspecific competition occurs when the niches of two species within a community overlap. When two species with the same niche are forced (under laboratory conditions) to occupy the same ecological niche, one species always outcompetes the other. Species within natural communities have evolved in ways that avoid excessive niche overlap, with behavioral and physical adaptations that allow resource partitioning. Interspecific competition limits both the population size and the distribution of competing species. Intraspecific competition also limits populations because individuals of the same species occupy the same ecological niche and compete with one another for all of their needs.

27.3 What Are the Results of Interactions Between Predators and Their Prey?

Predators eat other organisms and are generally larger and less abundant than their prey. Predators and prey act as strong agents of selection on one another. Prey animals have evolved a variety of protective colorations that render them either inconspicuous (camouflage) or startling (startle coloration) to their predators. Some prey are poisonous, distasteful, or venomous, and exhibit warning coloration by which they are readily recognized and avoided by predators, while others have evolved to resemble other, more distasteful organisms through mimicry. Both predators and prey have evolved a variety of toxic chemicals for attack and defense. Plants that are preyed on have evolved elaborate defenses, ranging from poisons to overall toughness. These defenses, in turn, have selected for predators that can detoxify poisons and grind down tough tissues.

27.4 What Is Parasitism?

In parasitism, the parasite feeds on a larger, less abundant host, harming it but not killing it immediately. Parasites include disease-causing microorganisms as well as animals such as fleas and tapeworms.

27.5 What Is Mutualism?

Mutualism benefits both interacting species. Some mutualistic interactions are close and extended, such as that between cows and the microorganisms that live in their digestive tracts, which help them to digest cellulose. Other mutualistic interactions are more temporary, such as those that occur between plants and the animals that pollinate them.

27.6 How Do Keystone Species Influence Community Structure?

Keystone species have a greater influence on community structure than can be predicted by their numbers. For example, if the African elephant were driven to extinction, the African grasslands it now inhabits might revert to forests.

27.7 Succession: How Do Community Interactions Cause Change over Time?

Succession is a structural change in a community and its nonliving environment over time. During succession, plants alter the environment in ways that favor their competitors, thus producing a somewhat predictable progression of dominant species. Primary succession, which may take thousands of years, occurs where no remnant of a previous community existed, such as on bare rock. Secondary succession occurs much more rapidly because it builds on the remains of a disrupted community, such as an abandoned field or remnants of a forest after a fire. Uninterrupted succession ends with a climax community, which tends to be self-perpetuating unless acted on by outside forces, such as fire or human activities. Some ecosystems, including tallgrass prairie and farm fields, are maintained in relatively early, subclimax stages of succession by periodic disruptions.

Key Terms

abiotic *524*
aggressive mimicry *521*
biome *528*
biotic *512*
camouflage *517*
carnivore *514*
climax community *524*
coevolution *512*
community *512*
competition *513*
competitive exclusion principle *513*
disturbance *524*
ecological niche *512*
herbivore *514*
host *521*
interspecific competition *513*
intertidal zone *514*
intraspecific competition *514*
invasive species *520*
keystone species *523*
mimicry *517*
mutualism *522*
parasite *521*
pioneer *524*
predator *514*
primary succession *524*
resource partitioning *513*
secondary succession *524*
startle coloration *519*
subclimax *524*
succession *524*
warning coloration *517*

Thinking Through the Concepts

Fill-in-the-Blank

1. Organisms that interact serve as agents of __________ on one another. This results in __________, which is the process by which species evolve adaptations to one another. Four types of community interactions described in this chapter are __________, __________, __________, and __________.
2. Predators may be meat eaters, called __________, or plant eaters, called __________. Both predators and their prey may blend into their surroundings by using __________. Predators are generally __________ and less __________ than their prey.
3. Competition occurs whenever two different populations within a community have overlapping ecological__________. Competition between two species is called __________, and this may lead to evolutionary changes that reduce competition. This phenomenon is called __________.
4. Fill in the types of coloration or mimicry: Used by a prey to signal that it is distasteful: __________; used by a moth with large eyespots on its wings:__________; mimicry of a poisonous animal by a nonpoisonous animal: __________; mimicry used by a predator to attract its prey: __________.
5. Fill in the appropriate type of community interaction: Bacteria, living in the human gut, that synthesize vitamin K: __________; bacteria that cause illness: __________; a deer eating grass: __________; a tick sucking blood: __________; a bee pollinating a flower:__________; kudzu covering trees:__________.
6. A somewhat predictable change in community structure over time is called__________. This process takes two forms. Which of these forms would start with bare rock? __________ Which would occur after a forest fire? __________ A relatively stable community that is the end product of this process is called a __________ community. A mowed lawn in suburbia is an example of a __________ community.

Review Questions

1. Define an ecological community, and list four important types of community interactions.
2. Describe four very different ways in which specific plants and animals protect themselves from being eaten. In each, describe an adaptation that might evolve in predators of these species that would overcome their defenses.
3. Define parasitism and mutualism and provide an example of each.
4. Define succession. Which type of succession would occur on a forest clear-cut (a region in which all the trees have been removed by logging) and why?
5. List two climax and two subclimax communities. How do they differ?

6. What is an invasive species? Why are they destructive? What general adaptations do invasive species possess?

Applying the Concepts

1. Herbivorous animals that eat seeds are considered by some ecologists to be predators of plants, and herbivorous animals that eat leaves are considered to be parasites of plants. Discuss the validity of this classification scheme.
2. An ecologist visiting an island finds two very closely related species of birds, one of which has a slightly larger bill than the other. Interpret this finding with respect to the competitive exclusion principle and the ecological niche, and explain both concepts.
3. Think about the case of the camouflaged frogfish and its prey. As the frogfish sits camouflaged on the ocean floor, wiggling its lure, a small fish approaches the lure and is eaten, while a very large predatory fish fails to notice the frogfish. Describe all the possible types of community interactions and adaptations that these organisms exhibit. Remember that predators can also be prey and that community interactions are complex!
4. Design an experiment to determine whether the kangaroo is a keystone species in the Australian outback.
5. Why is it difficult to study succession? Suggest some ways you would approach this challenge for a few different ecosystems.

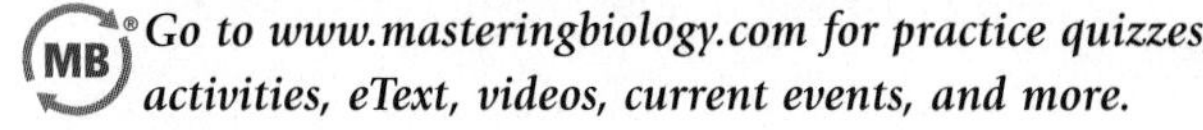

Go to www.masteringbiology.com for practice quizzes, activities, eText, videos, current events, and more.

How Do Ecosystems Work?

Chapter 28

▲ A grizzly bear intercepts a salmon on its spawning journey. Struggling up a waterfall, the salmon is attempting to reach the same streambed where it hatched years earlier.

Case Study

Dying Fish Feed an Ecosystem

SOCKEYE SALMON OF THE PACIFIC NORTHWEST have a remarkable life cycle. Hatching in shallow depressions in the gravel bed of a swiftly flowing stream, they follow the stream's path into ever-larger rivers that eventually enter the ocean. Emerging into estuaries—wetlands where fresh water and salt water mix—the salmon's remarkable physiology allows the fish to adapt to the change to salt water before they reach the sea. The small percentage of young salmon that evade predators grows to adulthood, feeding on crustaceans and smaller fish.

Years later, their bodies undergo another transformation. As they reach sexual maturity, a compelling instinctive drive—still poorly understood despite decades of research—lures them back to fresh water, but not just any stream or river will do. The salmon swim along the coast (probably navigating by sensing Earth's magnetic field) until the unique scent of their home stream entices them to swim inland. Battling swift currents, leaping up small waterfalls, undulating over shallow sandbars, and evading human traps, they carry their precious cargo of sperm and eggs back home to renew the cycle of life, dying after their mission has been accomplished.

The fishes' journey back to their birthplace is remarkable in another way. Nutrients almost always flow downstream, carried from the land into the ocean. But the salmon, filled with muscle and fat acquired from feeding in the ocean, not only battle against the flow of the current in their upstream journey, but also reverse the usual movement of nutrients. What awaits the salmon at their journey's end? How does their journey affect the web of life upstream?

At a Glance

28.1 HOW DO ENERGY AND NUTRIENTS MOVE THROUGH ECOSYSTEMS?

Two basic laws underlie ecosystem function: Nutrients constantly cycle and recycle within and among ecosystems, while energy moves through ecological communities (the various populations of interacting organisms that inhabit ecosystems) in a continuous one-way flow.

Nutrients are atoms and molecules that organisms obtain from their living or nonliving environment and that are required for survival. The same atoms have been sustaining life on Earth for about 3.5 billion years. The molecules that make up your body undoubtedly include oxygen atoms once trapped in primordial water and liberated for the first time by ancient photosynthetic bacteria. Some of your carbon atoms were exhaled by a dinosaur, and some of your nitrogen atoms were probably liberated by decomposer bacteria consuming the remnants of a salmon. Although nutrients may be transported, redistributed, or converted to different molecular forms, they do not leave Earth and are continually recycled through the nutrient cycles described in later sections.

Energy, in contrast, is continuously replenished. The activities of life—from the migration of salmon to the active transport of molecules through a cell membrane—are ultimately powered by sunlight. The solar energy that continuously bombards Earth is captured by photosynthetic organisms, then transformed through the myriad chemical reactions that energize life, and is finally converted to heat that radiates back into space.

28.2 HOW DOES ENERGY FLOW THROUGH ECOSYSTEMS?

Ninety-three million miles away from Earth, thermonuclear reactions in the sun convert hydrogen into helium, transforming a relatively small amount of matter into enormous quantities of energy. A tiny fraction of this energy reaches Earth in the form of electromagnetic waves, including heat, light, and ultraviolet energy. Of the energy that reaches Earth, much of it is reflected by the atmosphere, clouds, and Earth's surface. Still more is absorbed as heat by Earth and its atmosphere, leaving only about 1% to sustain life. Because only a fraction of this remaining solar energy is captured by photosynthetic organisms, the teeming life on this planet is supported by less than 0.03% of the energy reaching Earth from the sun.

Energy Enters Communities Primarily Through Photosynthesis

During photosynthesis, pigment molecules such as chlorophyll absorb specific wavelengths of sunlight (see Fig. 7-4). This solar energy is then used in reactions that store energy in chemical bonds. To forge new molecules, plants and other photosynthetic organisms acquire nutrients from the abiotic (nonliving) portions of ecosystems. They take up nitrogen and phosphorus from the soil and water, absorb carbon from the CO_2 in air, and derive oxygen from the air and from water molecules. Using energy from the sun and these inorganic nutrients, plants synthesize sugars, starches, proteins, nucleic acids, and all the other biological molecules they need to

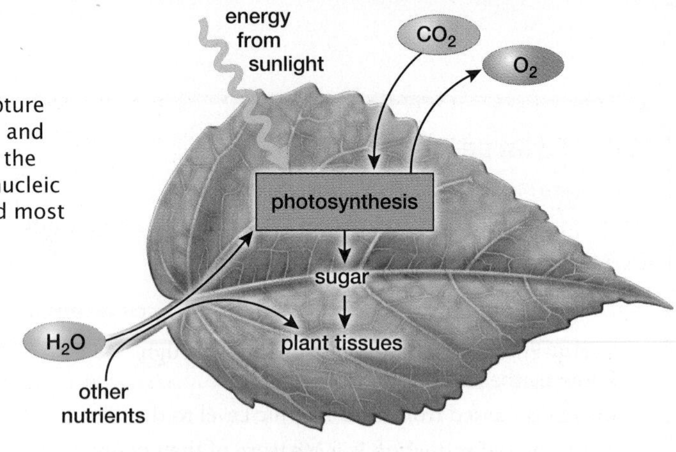

▶ **FIGURE 28-1 Producers** During photosynthesis, producers capture solar energy and release oxygen as a by-product. Using this energy and inorganic nutrients from the environment, producers synthesize all the molecules they need, including carbohydrates, fats, proteins, and nucleic acids. These molecules, in turn, provide nearly all of the energy and most of the nutrients for the rest of life on Earth.

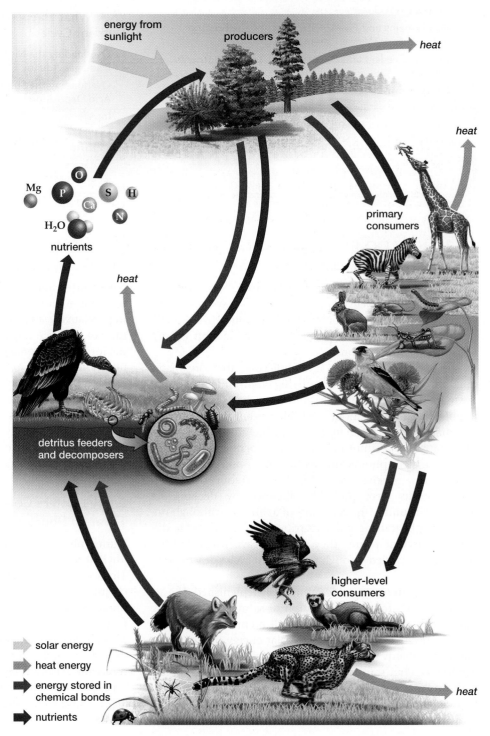

▶ **FIGURE 28-2 Energy flow, nutrient cycling, and feeding relationships in ecosystems** Energy steadily enters the producers as sunlight, which is captured by the producers during photosynthesis and stored in chemical bonds. The chemical energy is then passed through various trophic levels. At each level, some energy is lost as heat. Nutrients are continuously recycled.

QUESTION Why is some energy always lost as heat? Relate this loss to the second law of thermodynamics, as discussed on p. 99.

sustain life (**Fig. 28-1**). During photosynthesis, plants release the oxygen that most organisms require for the reactions that generate ATP. Thus, photosynthesizers serve as a conduit for both energy and nutrients into ecological communities.

Energy Is Passed from One Trophic Level to the Next

The complex interactions within biological communities, and between communities and their abiotic environments, determine the pathways followed by energy and nutrients as they move through ecosystems. **Figure 28-2** provides a preview of our study of how ecosystems function.

Energy flows through communities from photosynthetic producers through several levels of consumers. Each category of organisms is called a **trophic level** (literally, "feeding level"). Photosynthetic organisms, from oak trees in a forest to single-celled diatoms in the ocean, form the first trophic level, and are called **producers,** or **autotrophs** (Greek, "self-feeders"), because they produce food for themselves using inorganic nutrients and solar energy. In doing so, they directly or indirectly produce food for nearly all other forms of life as well. Organisms that cannot photosynthesize, called **consumers,** or **heterotrophs** (Greek, "other-feeders"), must acquire energy and most of their nutrients prepackaged in the molecules that comprise the bodies of other organisms.

Consumers occupy several trophic levels. Some consumers, called **primary consumers,** feed directly and exclusively on producers, the most abundant living energy source in any ecosystem. These **herbivores** (literally, "plant eaters"), which include animals such as grasshoppers, mice, and zebras, form the second trophic level. **Carnivores** (literally, "meat eaters"), such as spiders, hawks, cheetahs, and salmon, comprise the higher-level consumers. Carnivores act as **secondary consumers** when they prey on herbivores. Some carnivores at least occasionally eat other carnivores; when doing so, they occupy the fourth trophic level, and are called **tertiary consumers.**

Net Primary Production Is a Measure of the Energy Stored in Producers

The amount of life that a particular ecosystem can support is determined by the energy captured by the producers in that ecosystem. The energy that photosynthetic organisms store and make available to other members of the community over a given period is called **net primary production** (**Fig. 28-3**).

The net primary production of an ecosystem is influenced by many environmental variables, including the amount of nutrients available to the producers, the amount of sunlight reaching them, the availability of water, and the temperature. In the desert, for example, lack of water limits production. In the open ocean, light is a limiting factor in deep water, and lack of nutrients limits productivity in most surface water. In ecosystems where all resources are abundant, productivity is high; for example, in tropical rain forests and in **estuaries** (coastal areas where rivers meet the ocean, carrying nutrients washed from the land), as shown in Figure 28-3.

An ecosystem's contribution to Earth's overall productivity is determined both by the ecosystem's productivity and by the portion of Earth that it covers. Because open oceans, whose productivity is low, cover about 65% of Earth's surface, they contribute about 25% to Earth's total productivity. This is about the same overall contribution as tropical rain forests, whose productivity is very high but which cover only about 4% of Earth's surface.

Food Chains and Food Webs Describe the Feeding Relationships Within Communities

To illustrate who feeds on whom in a community, it is common to identify a representative organism within each trophic level

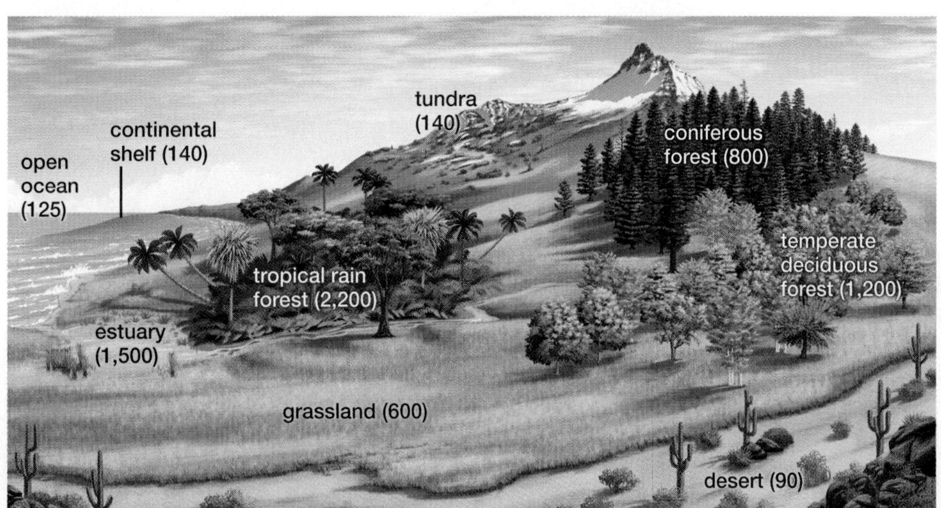

▲ **FIGURE 28-3 Net primary productivities of ecosystems** The average net primary production of some terrestrial and aquatic ecosystems is illustrated here, measured in grams of biological material produced per square meter per year. Notice the enormous differences.

QUESTION What factors contribute to these differences in productivity?

that is eaten by a representative of the level above it. This linear feeding relationship is called a **food chain.** Different ecosystems support radically different food chains. The dominant producers in land-based (terrestrial) ecosystems are plants (**Fig. 28-4a**). These support the familiar plant-eating insects, birds, and mammals, each of which may be preyed on by other insects, birds, or mammals. In contrast, the dominant producers in most aquatic food chains (such as in lakes and oceans; **Fig. 28-4b**) are microscopic protists and bacteria collectively called **phytoplankton** (literally, "plant drifters"). These producers support a diverse group of consumers called **zooplankton** (literally, "animal drifters"), which consist mainly of protists and small shrimp-like crustaceans. These are eaten primarily by fish, who are eaten by larger fish.

Animals in natural communities, however, often do not fit neatly into categories of primary, secondary, and tertiary consumers. A **food web** shows many interconnecting food chains, and more accurately describes the actual feeding relationships within a given community (**Fig. 28-5**). Some animals, such as raccoons, bears, rats, and humans, are **omnivores** (literally, "all eating"), and they act as primary, secondary, and occasionally tertiary consumers. A hawk, for instance, is a secondary consumer when it eats a mouse (an herbivore), and a tertiary consumer when it eats a robin that feeds on insects and earth-

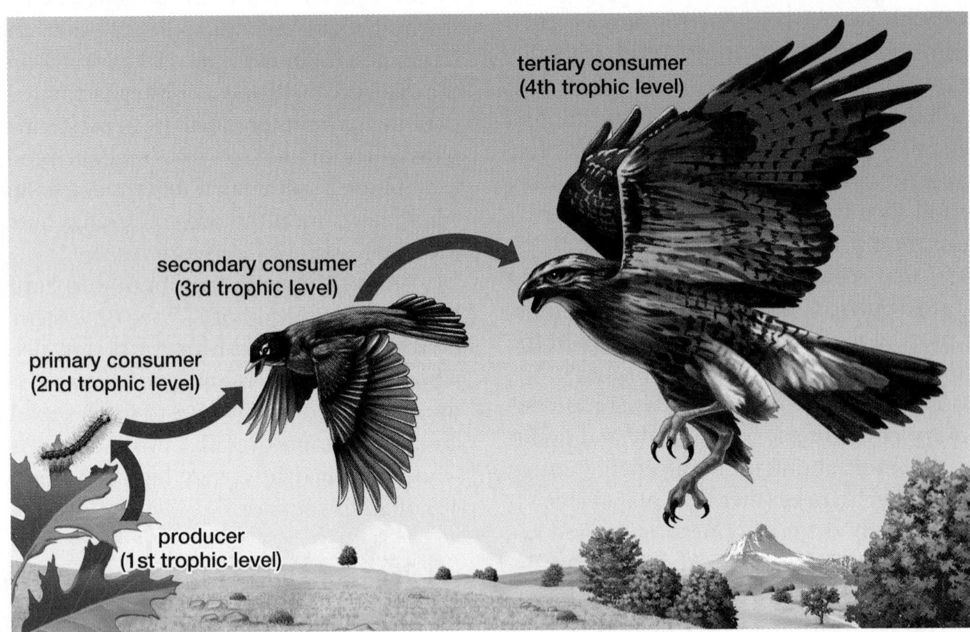

(a) A simple terrestrial food chain

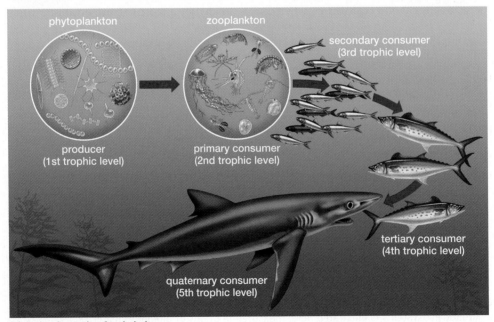

(b) A simple marine food chain

▲ **FIGURE 28-4 Food chains on land and sea**

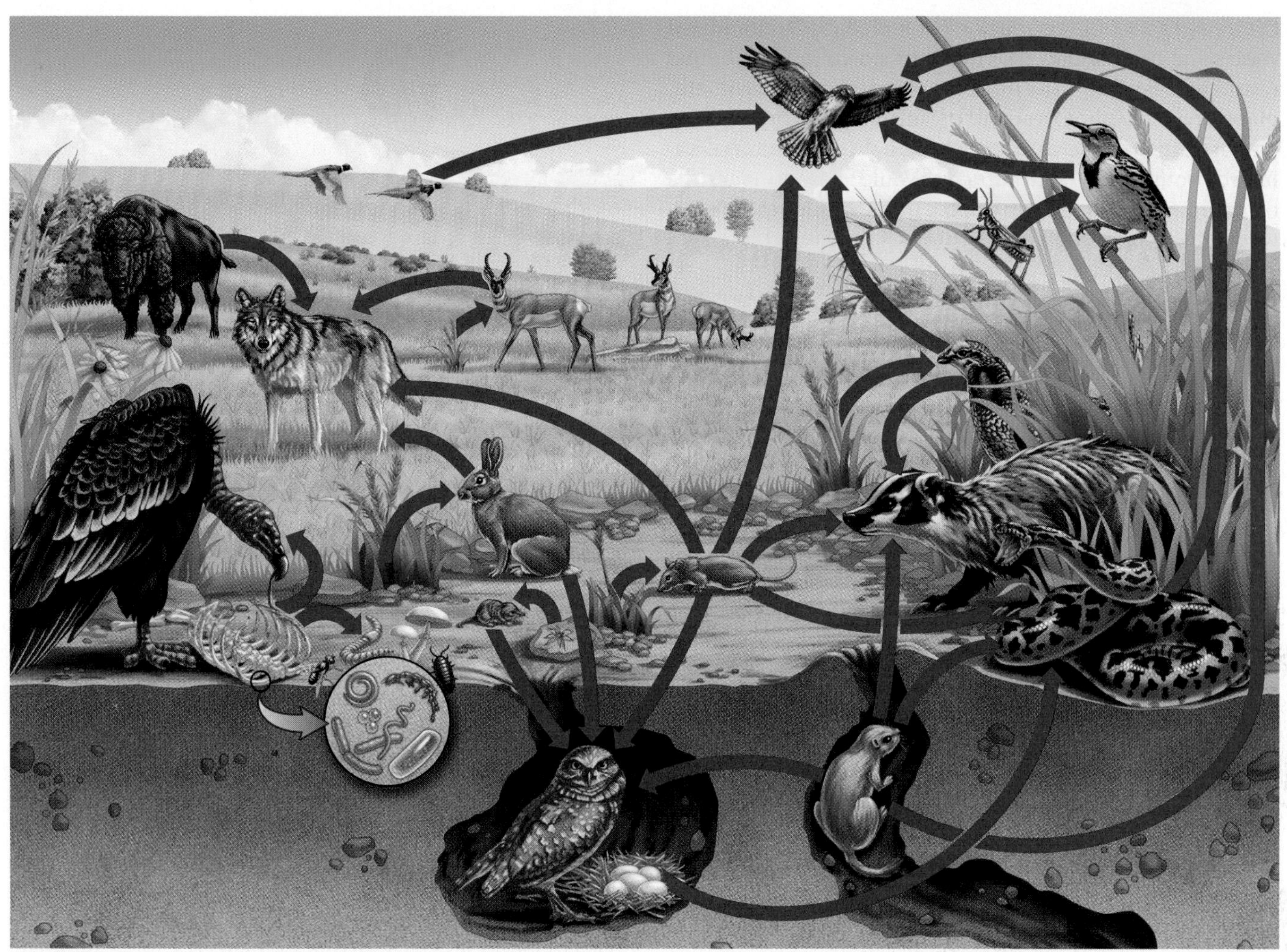

▲ **FIGURE 28-5 A simplified grassland food web** The animals pictured in the foreground include a vulture (a detritus-feeder), a bull snake, a ground squirrel, a burrowing owl, a badger, a mouse, a shrew (which looks like a small mouse but is carnivorous), and a wolf spider. In the middle distance you'll see a grouse, a meadowlark, a grasshopper, and a jackrabbit. In the far distance, look for pronghorn antelope, a hawk, a wolf, and bison.

EXERCISE Using this figure, generate a food chain with four trophic levels, and identify each level.

worms. When digesting a spider, a carnivorous plant such as the venus flytrap can tangle the food web further by acting as both a producer and a tertiary consumer.

Detritus Feeders and Decomposers Release Nutrients for Reuse

Among the most important strands in the food web are detritus feeders and decomposers. Both of these groups live on the refuse of life, such as fallen leaves and fruit, as well as the wastes and dead bodies of other organisms. **Detritus feeders** ("detritus" means "debris") are an army of mostly small and often unnoticed organisms, including certain mites and protists, nematode worms, earthworms, centipedes, and some insects and snails. A few large vertebrates such as vultures also fit this category (see Fig. 28-5). As they feed on organic debris, detritus feeders extract some of the energy stored there and excrete the rest in smaller pieces that provide food for other detritus feeders or for decomposers.

Decomposers are primarily fungi and bacteria. They differ from detritus feeders in that they secrete digestive enzymes outside of their bodies. These enzymes break down nearby organic matter, allowing the decomposers to absorb the nutrients they require, with the remaining elements and small molecules being released into the environment. Mushrooms in a lawn or the blue-gray fuzz you may notice on old bread are fungal decomposers hard at work, and the smelly slime on spoiled meat signals the presence of bacterial decomposers.

Through the activities of detritus feeders and decomposers, the bodies and wastes of living organisms are reduced to simple molecules—such as carbon dioxide, water, and minerals—that return to the atmosphere, soil, and water. By liberating nutrients for reuse, detritus feeders and decomposers perform the final stages of nutrient recycling and form a vital link in the nutrient cycles of ecosystems.

What would happen if detritus feeders and decomposers disappeared? This portion of the food web, although inconspicuous, is absolutely essential to life on Earth. Without it, communities would gradually be buried by accumulated wastes and dead bodies. The nutrients stored in these bodies would be unavailable to enrich the soil and water. Eventually, plants and other photosynthetic organisms would be unable to obtain enough nutrients to survive. With these producers eliminated, both energy and nutrients would cease to enter the community, and the higher trophic levels, including humans, would disappear as well.

Energy Transfer Through Trophic Levels Is Inefficient

As discussed in Chapter 6, a basic law of thermodynamics is that energy use is never completely efficient. For example, as your car burns gasoline, about 75% of the energy released is lost as heat. This is also true in living systems: Waste heat is produced by all the biochemical reactions that keep cells alive. For example, splitting the chemical bonds of adenosine triphosphate (ATP) to cause muscular contraction releases heat energy; this is why walking briskly on a cold day will warm you.

Energy transfer from one trophic level to the next is also quite inefficient. When a grasshopper (a primary consumer) eats the leaves of a plant (a producer), only some of the solar energy trapped by the plant is available to the insect. Some was used by the plant for cellular metabolic activities, and more was lost as heat during these processes. Some energy was converted into the chemical bonds of cellulose, which the grasshopper cannot digest. Therefore, only a fraction of the energy captured by the producers of the first trophic level can be used by organisms in the second trophic level. A robin that eats the grasshopper (the third trophic level) will not obtain all the energy that the insect acquired from the plant—some will have been used up to power hopping, flying, and the gnashing of mouthparts. Some energy will have been expended to construct the grasshopper's indigestible exoskeleton, and much of it will have been lost as heat. Likewise, the energy from the robin's body will not all be available to the hawk that eventually consumes it.

Energy Pyramids Illustrate Energy Transfer Between Trophic Levels

Although energy transfer among trophic levels within different communities varies significantly, studies of a variety of communities indicate that the net transfer of energy between trophic levels is roughly 10% efficient. This means that, in general, the energy stored in primary consumers is only about 10% of the energy stored in the bodies of producers. In turn, the bodies of secondary consumers possess roughly 10% of the energy stored in primary consumers. This inefficient energy transfer between trophic levels is called the "10% law." An **energy pyramid,** which shows maximum energy available at the base and steadily diminishing amounts at higher levels, illustrates the general energy relationships between trophic levels (**Fig. 28-6**). Ecologists sometimes measure **biomass,** which is the weight of living material (usually measured as dry weight within a given area), at each trophic level. Because the dry weight of organisms' bodies at each trophic level is roughly proportional to the amount of energy stored at that level, a biomass pyramid for a given community usually has the same general shape as its energy pyramid.

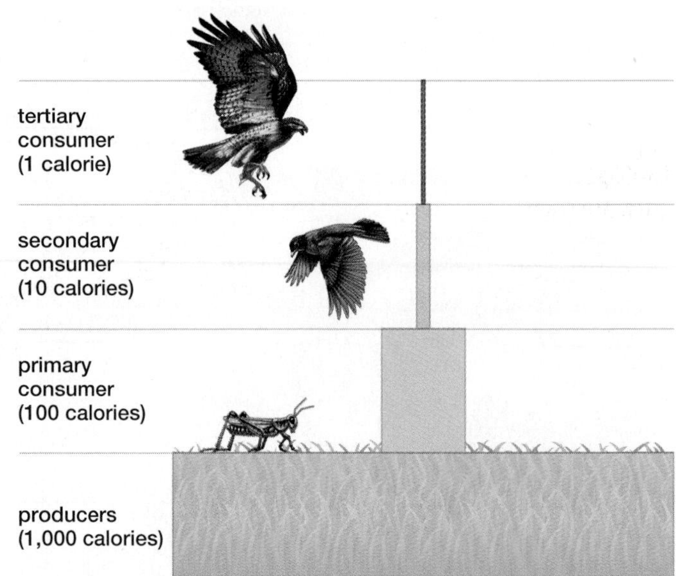

▲ **FIGURE 28-6 An energy pyramid for a grassland ecosystem** The width of each rectangle is proportional to the energy stored at that trophic level. Representative organisms for the first three trophic levels in a U.S. grassland ecosystem illustrated here are a grasshopper, a robin, and a red-tailed hawk.

What does this mean for community structure? If you wander through an undisturbed ecosystem, you will notice that the predominant organisms are plants, because they have the most energy available to them. The most abundant animals will be herbivores, and carnivores will be relatively rare, because there is far less energy available to support them. The inefficiency of energy transfer also has important implications for human food production. The lower the trophic level we utilize, the more food energy available to us; hence, far more people can be fed on grain than on meat.

Case Study continued
Dying Fish Feed an Ecosystem

Salmon are carnivores, eating small fish and large zooplankton. Overfishing and environmental disruption have led to a drastic reduction in their wild populations, which has in turn led to an increase in salmon farming operations. But farmed salmon are often fed fish meal and fish oil made from smaller, wild-caught fish. Because of the inefficiency of energy transfer, it takes 3 pounds of the wild-caught fish to raise 1 pound of farmed salmon. There is concern, therefore, that farming salmon is shifting the fishing pressure away from wild salmon onto the smaller ocean fish that are harvested to feed the salmon. These smaller fish are an important link in ocean food chains, providing food for larger ocean carnivores such as mackerel, tuna, whales, seals, and seabirds.

An unfortunate side effect of the inefficiency of energy transfer, coupled with human production and release of toxic chemicals, is that certain persistent toxic chemicals become increasingly concentrated in the bodies of animals that occupy increasingly high trophic levels, a process called **biological magnification.** Persistent poisons, such as mercury, are often taken up by producers, which are then passed on to primary consumers (who eat large numbers of producers) and then on to carnivores, who accumulate the poison from consumers below them. Long-lived carnivorous animals, such as swordfish, may accumulate enough mercury to pose a potential health hazard to humans, who act as high-level consumers when they eat these top predators (see "Health Watch: Food Chains Magnify Toxic Substances").

Health Watch

Food Chains Magnify Toxic Substances

In the United States, wildlife biologists during the 1950s and 1960s witnessed an alarming decline in populations of several predatory birds, especially fish-eaters such as bald eagles, cormorants, ospreys, and brown pelicans. The decline pushed some, including the brown pelican and the bald eagle, close to extinction. What caused this? The aquatic ecosystems supporting these birds had been sprayed with relatively low amounts of the pesticide DDT to control insects. In the tissues of these top predators, scientists found concentrations of DDT up to 1 million times greater than in the water where their fish prey lived; this impaired their ability to reproduce.

The birds were victims of biological magnification, the process by which toxic substances accumulate in increasingly high concentrations in animals occupying higher trophic levels. Fortunately, populations of predatory birds vulnerable to DDT have recovered significantly since the pesticide was banned in the United States in 1973.

Substances that undergo biological magnification (such as DDT) share two properties that make them dangerous. First, decomposer organisms cannot readily break them down into harmless substances—that is, they are not **biodegradable.** Second, they tend to be stored in living tissue (often in fat), accumulating over the years in the bodies of long-lived animals. Exposure to high levels of pesticides and other persistent pollutants has been linked to some types of cancer, infertility, heart disease, suppressed immune function, and neurological damage.

Mercury contamination is a particular cause for concern, because mercury is an extremely potent neurotoxin that accumulates in muscle as well as fat. Its high levels in a few types of long-lived ocean predators—such as swordfish, shark, and albacore tuna—have prompted the U.S. Food and Drug Administration to advise women of childbearing age and young children to avoid or limit their consumption of these types of fish, because the mercury levels may pose a potential health hazard. In the United States, coal-fired power plants are the largest single source of mercury contamination—atmospheric mercury can be wafted thousands of miles from these plants and be deposited in what should be pristine environments, such as in the Arctic. Inuit natives living north of the Arctic Circle have high levels of mercury and other pollutants from consuming large quantities of predatory marine mammals and fish.

A class of organic chemicals called endocrine disruptors—including some widely used pesticides, plasticizers (which make plastic flexible), and flame retardants—has become widespread in the environment. Like DDT, these chemicals accumulate in fat, and either mimic or interfere with the actions of animal hormones. There is compelling evidence that these chemicals are interfering with the reproduction and development of fish (including salmon), fish-eating birds such as cormorants, frogs (**Fig. E28-1**), salamanders, alligators, and many other animals. Endocrine disruptors are also suspected of causing reduced sperm counts in humans. Endocrine disruptors and other organic pollutants bioaccumulate in mussels that filter large quantities of water to obtain food. Zebra mussels (which have invaded the Great Lakes; see Chapter 27) are accumulating a variety of organic pollutants, which endanger the fish and birds that consume them.

Understanding the workings of food webs allows us to understand why bioaccumulation occurs, and why people, as well as wildlife, are susceptible. When we eat tuna or swordfish, for example, we act as tertiary or even quaternary consumers, and so are vulnerable to substances that bioaccumulate. In addition, the long human life span provides more time for substances to accumulate within the body. Threats to human health provide additional incentive to study what happens to chemicals released into the environment, and to restrict the manufacture and improper disposal of substances that bioaccumulate.

▲ **FIGURE E28-1 The price of pollution** Deformities such as the one exhibited by this frog found in Oregon have been linked to bioaccumulating chemicals. Abnormalities of the reproductive and immune systems are also common in many types of organisms exposed to these pollutants. Frogs, which have extremely thin skin and spend much of their lives in water, are particularly vulnerable to water-borne pollutants.

28.3 HOW DO NUTRIENTS CYCLE WITHIN AND AMONG ECOSYSTEMS?

In contrast to the energy of sunlight, nutrients do not flow down onto Earth in a steady stream from above. Nutrients are elements and small molecules that form the chemical building blocks of life. Some, called **macronutrients,** are required by organisms in large quantities. These include water, carbon, hydrogen, oxygen, nitrogen, phosphorus, sulfur, and calcium. **Micronutrients,** including zinc, molybdenum, iron, selenium, and iodine, are required only in trace quantities. **Nutrient cycles,** also called **biogeochemical cycles,** describe the pathways these substances follow as they move from the abiotic portions of ecosystems through communities and back to nonliving storage sites.

The major sources and storage sites of nutrients are called **reservoirs,** and they are almost always in the nonliving, or abiotic, environment. In the following sections, we describe the cycles of water, carbon, nitrogen, and phosphorus. You will see nutrient reservoirs indicated in white boxes in each nutrient cycle figure. Events that drive the movement of nutrients are in purple boxes, and the trophic levels through which nutrients move are in yellow boxes.

The Hydrologic Cycle Has Its Major Reservoir in the Oceans

The water cycle, or **hydrologic cycle** (Fig. 28-7), describes the pathway that water takes as it travels from its major reservoir—the oceans—through the atmosphere, to reservoirs in freshwater lakes, rivers, and groundwater, and then back again to the oceans. The hydrologic cycle differs from most other nutrient cycles in that the biotic portion of ecosystems plays only a small role—the fundamental process of the hydrologic cycle would continue even if life on Earth were to disappear.

The hydrologic cycle is driven by solar heat energy, which evaporates water and drives the winds that carry it as water vapor in the atmosphere. Gravity draws water back to Earth in the form of precipitation (mostly rain and snow), pulls it into the ground, and causes it to flow in rivers that empty into the oceans. Oceans cover about three-quarters of Earth's surface and contain more than 97% of Earth's water, with another 2% of the total water trapped in ice, leaving only 1% as liquid fresh water. Most evaporation occurs from the oceans, and most precipitation falls back onto them. Of the water that falls on land, some evaporates from the soil, plants, lakes, and streams; a portion runs back to the oceans; and a small amount enters

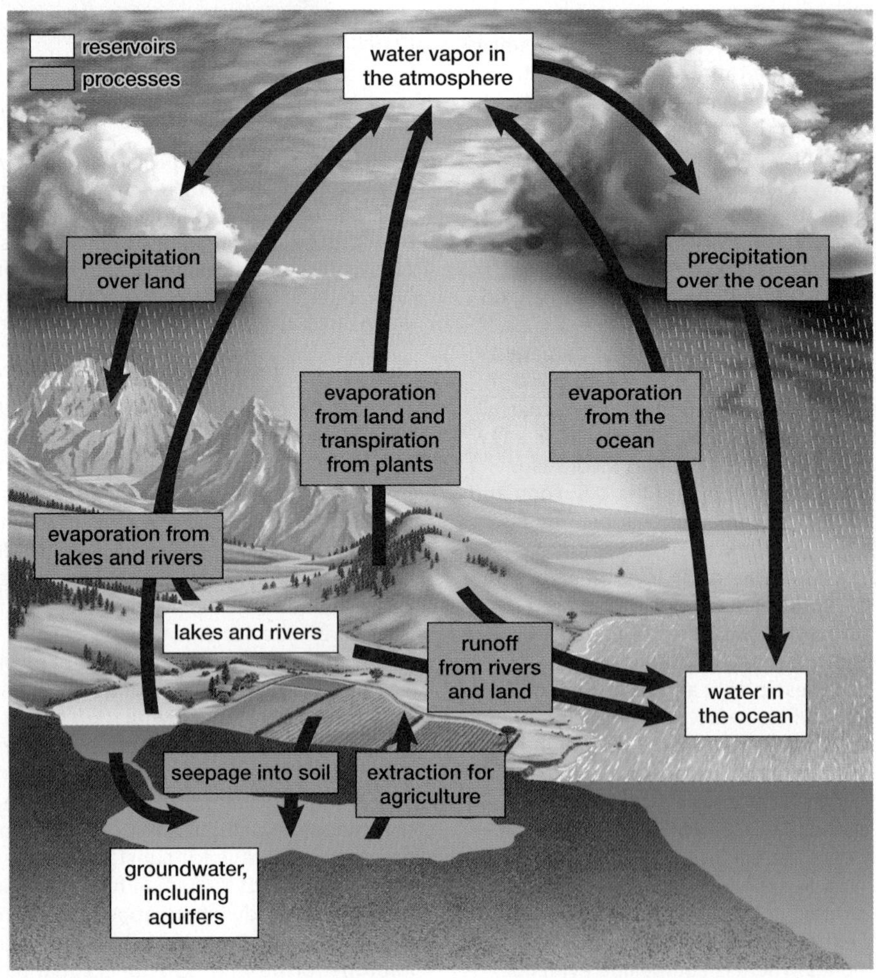

▲ **FIGURE 28-7 The hydrologic cycle**

natural underground reservoirs called **aquifers.** Aquifers are composed of water-permeable sediments such as silt, sand, or gravel, which are saturated with water. They are often tapped to supply water for irrigating crops.

Unfortunately, in many areas of the world—including China, India, Northern Africa, and the Midwestern United States—groundwater aquifers are being "mined" for agriculture; that is, water is being pumped out faster than it is being replenished. Underground water supplies are dropping rapidly in India, which relies heavily on groundwater for irrigation. India's population, which now exceeds 1 billion people, could suffer if its crop yields fall as groundwater becomes unavailable.

Because the bodies of living things are roughly 70% water, a tiny portion of the total water involved in the overall hydrologic cycle enters the living communities of terrestrial ecosystems. Some is absorbed by the roots of plants, and much of this is evaporated back to the atmosphere from their leaves through a process called **transpiration.** A relatively minuscule amount of water participates in the chemical reactions of photosynthesis, and is resynthesized and released during cellular respiration.

The hydrologic cycle is crucial for terrestrial communities because it continually restores the fresh water needed for land-based life. Water is a solvent for all of the other nutrients, and no nutrient can enter or leave the cells of an organism unless it is dissolved in water. As you study the nutrient cycles that follow, keep in mind that nutrients in soil must be dissolved in soil water to be taken up by the roots of plants or to be absorbed by bacteria. Plant leaves can only take up carbon dioxide gas after it has dissolved in a thin layer of water coating the cells inside the leaf. The hydrologic cycle doesn't depend on terrestrial organisms, but they would rapidly disappear without it.

The Carbon Cycle Has Major Reservoirs in the Atmosphere and Oceans

Chains of carbon atoms form the framework of all organic molecules, the building blocks of life. The **carbon cycle** (Fig. 28-8) describes the movement of carbon from its major short-term reservoirs in the atmosphere and oceans, through producers and into the bodies of consumers and detritus feeders, and then back again to its reservoirs. Carbon enters the living community when producers capture carbon dioxide (CO_2) during photosynthesis. On land, photosynthetic organisms acquire CO_2 from the atmosphere, where it represents 0.038% of all atmospheric gases. Aquatic producers such as phytoplankton obtain the CO_2 they need for photosynthesis from the water, where it is dissolved.

Producers return some CO_2 to the atmosphere or water during cellular respiration, and incorporate the rest of it into their bodies. Burning forests return carbon dioxide from these producers back to the atmosphere. When primary consumers eat producers, they acquire the carbon stored in the producers' tissues. As with producers, these herbivores and the organisms in higher trophic levels who consume them release CO_2 during

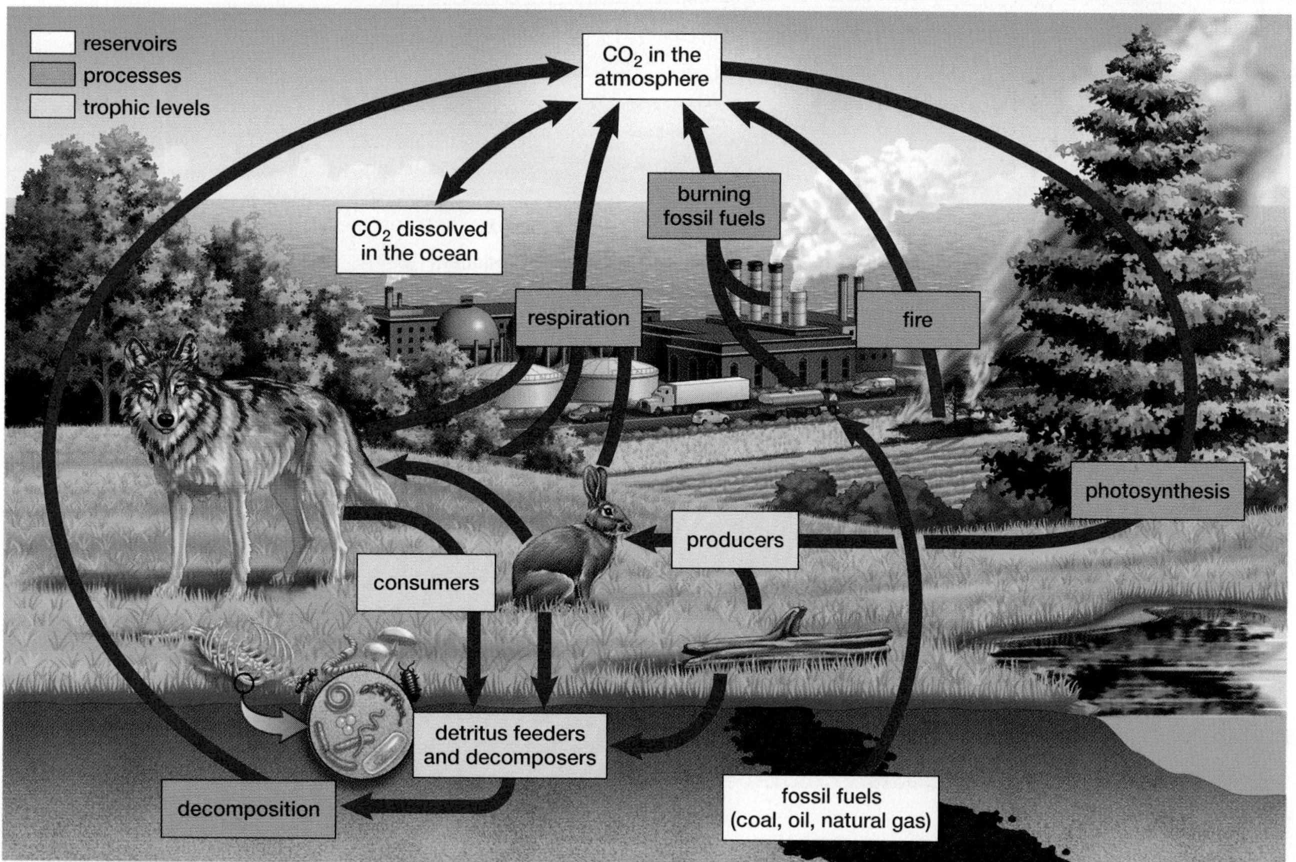

▲ **FIGURE 28-8 The carbon cycle**

respiration, excrete carbon compounds in their feces, and store the rest in their tissues. All living things eventually die, and their bodies are broken down by detritus feeders and decomposers. Cellular respiration by these organisms returns CO_2 to the atmosphere and oceans. CO_2 passes freely between these two great reservoirs (see Fig. 28-8). The complementary processes of uptake by photosynthesis and release by cellular respiration continually transfer carbon from the abiotic to the biotic portions of an ecosystem and back again.

Some carbon, however, cycles much more slowly. Much of Earth's carbon is bound up in limestone rock, formed from calcium carbonate ($CaCO_3$) deposited on the ocean floor in the shells of prehistoric phytoplankton. But because the movement of carbon from this source to the atmosphere and back again requires millions of years, this extremely long-term process makes very little contribution to the carbon cycling that supports ecosystems. Another long-term reservoir for carbon is in **fossil fuels,** which include coal, oil, and natural gas. These substances were produced over millions of years from the remains of prehistoric organisms buried deep underground and subjected to high temperature and pressure. In addition to carbon, the energy of prehistoric sunlight (originally captured by photosynthetic organisms) is trapped in these deposits. When humans burn fossil fuels to use this stored energy, CO_2 is released into the atmosphere, with potentially severe consequences, as described later in this chapter.

BioFlix™ The Carbon Cycle

The Nitrogen Cycle Has Its Major Reservoir in the Atmosphere

Nitrogen is a crucial component of amino acids, proteins, many vitamins, nucleotides (such as ATP), and nucleic acids (such as DNA). The **nitrogen cycle** (Fig. 28-9) describes the

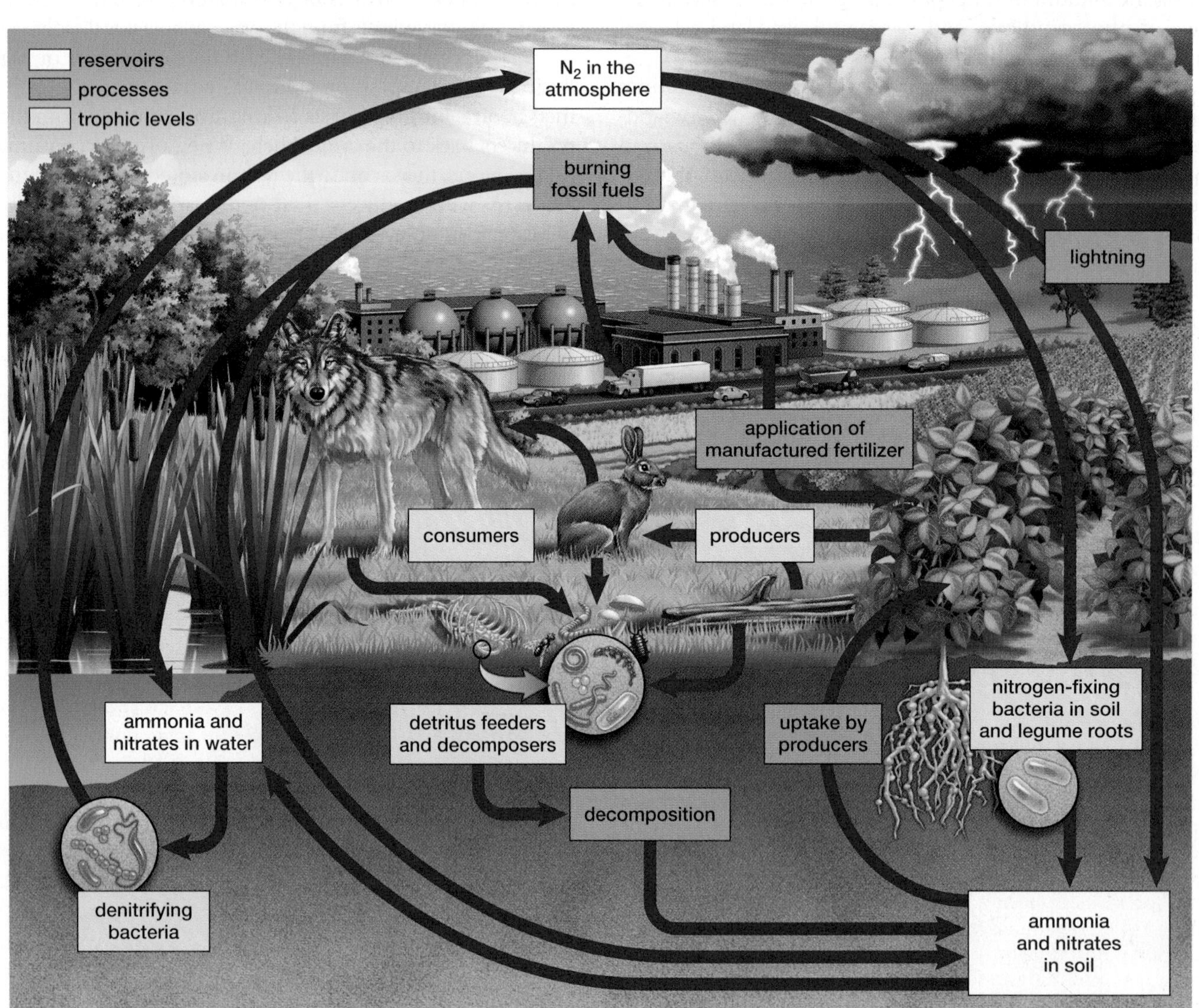

▲ **FIGURE 28-9 The nitrogen cycle**

QUESTION What incentives cause humans to capture nitrogen from the air and pump it into the nitrogen cycle? What are some consequences of human augmentation of the nitrogen cycle?

process by which nitrogen moves from its primary reservoir—nitrogen gas in the atmosphere—to reservoirs of ammonia and nitrate in soil and water, through producers and into consumers and detritus feeders, and then back again to its reservoirs.

The atmosphere contains about 78% nitrogen gas (N_2), but among all forms of life, only a few types of bacteria are able to convert N_2 into a form usable by plants and other producers. These microorganisms provide the major natural conduit between the atmospheric reservoir and ecological communities. In a process called **nitrogen fixation,** nitrogen-fixing bacteria in soil and water break the bonds in N_2 and combine it with hydrogen atoms to form ammonia (NH_3). Some nitrogen-fixing bacteria have entered a symbiotic association with plants in which the bacteria live in special swellings on roots (see Fig. 19-9). These plants, called **legumes** (including alfalfa, soybeans, clover, and peas), are extensively planted on farms, in part because they release excess ammonia produced by the bacteria, fertilizing the soil. Other bacteria in soil and water convert this ammonia to nitrate (NO_3^-), which producers can also use. Nitrates are also produced during electrical storms, when the energy of lightning combines nitrogen and oxygen gases to form nitrogen oxide compounds, which dissolve in rain. When the rain falls, it enriches the soil and water with this important nutrient.

Detritus feeders and decomposer bacteria also play a role in the nitrogen cycle, producing ammonia from the nitrogen-containing compounds in dead bodies and wastes. Ammonia and nitrate are absorbed by producers and are incorporated into a variety of biological molecules. These are passed through successively higher trophic levels as primary consumers eat the producers and are themselves eaten. At each trophic level, bodies and wastes are broken down by decomposers, which liberate ammonia back into the reservoir in soil and water. The nitrogen cycle is completed by **denitrifying bacteria.** These residents of wet soil, swamps, and estuaries break down nitrate, releasing nitrogen gas back into the atmosphere (see Fig. 28-9).

Fertilizer factories use energy from fossil fuels and N_2 from the atmosphere to synthesize ammonia, nitrate, and urea (an organic nitrogen compound also found in urine). The burning of fossil fuels combines atmospheric N_2 and O_2, generating nitrogen oxides that form nitrates. Nitrogen compounds introduced into ecosystems by fertilizing farm fields and burning fossil fuels now dominate the nitrogen cycle, creating serious environmental concerns, as discussed later in this chapter.

Case Study continued
Dying Fish Feed an Ecosystem

Salmon bodies contain high levels of nitrogen, and ecologists are interested in how much nitrogen their bodies contribute to the communities that surround their spawning streams. Nearly all nitrogen has a molecular weight of 14 (^{14}N), but a small amount is a heavier isotope (^{15}N). The ratio of these two isotopes is different in the ocean than it is in freshwater and terrestrial environments. Because salmon put on about 95% of their weight in the ocean, their bodies have the nitrogen isotope ratio of marine animals. Ecologists can measure the nitrogen in the vegetation around streams where salmon come to spawn and die. The percentage of nitrogen exhibiting the ocean ratio is assumed to come from salmon. Using this type of analysis, ecologists have found that the contribution of dying salmon to the plant nitrogen near their spawning streams can be as high as 70%.

The Phosphorus Cycle Has Its Major Reservoir in Rock

The **phosphorus cycle** (Fig. 28-10) describes the process by which phosphorus moves from its primary reservoir—phosphate-rich rock—to reservoirs of phosphate in soil and water, through producers and into consumers and detritus feeders, and then back to its reservoirs. Phosphorus is found in biological molecules including nucleotides (such as ATP), nucleic acids (such as DNA), and the phospholipids of cell membranes. It also forms a major component of vertebrate teeth and bones.

In contrast to carbon and nitrogen, phosphorus does not have an atmospheric reservoir. Throughout its cycle, phosphorus remains bound to oxygen in the form of phosphate (PO_4^{3-}). As phosphate-rich rocks are exposed by geological processes, some of the phosphate is dissolved by rain and flowing water, which carries it into soil, lakes, and the ocean, forming the reservoirs of phosphorus that are directly available to ecological communities. Dissolved phosphate is readily absorbed by consumers, which incorporate it into phosphate-containing biological molecules. From these producers, phosphorus is passed through food webs; at each level, excess phosphate is excreted. Ultimately, detritus feeders and decomposers return the phosphate to the soil and water, where it may then be reabsorbed by producers or may become bound to ocean sediment and eventually reformed into rock.

Some of the phosphate dissolved in fresh water is carried to the oceans. Although much of this phosphate ends up in marine sediments, some is absorbed by marine producers and is eventually incorporated into the bodies of invertebrates and fish. Some of these, in turn, are consumed by seabirds, which excrete large quantities of phosphorus back onto the land. At one time, seabird excrement (called "guano") deposited along the western coast of South America was collected and provided a major source of the world's phosphorus. As this resource was depleted, phosphorus-rich rock was mined, primarily to produce fertilizer. Soil that erodes from fertilized fields carries large quantities of phosphates into lakes, streams, and the ocean, where it stimulates the growth of producers. In lakes, phosphorus-rich runoff from land can stimulate such an excessive growth of

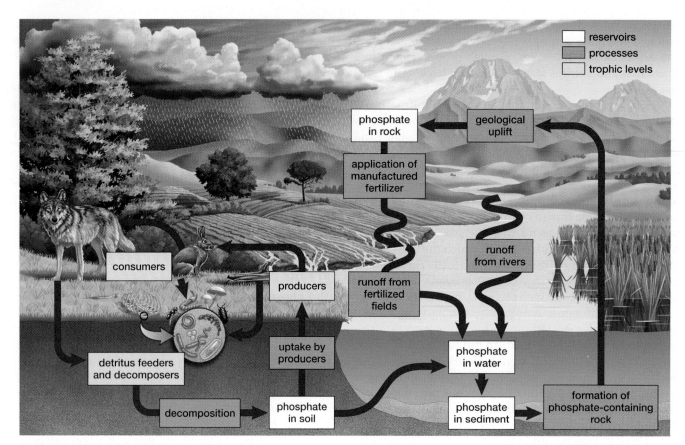

▲ **FIGURE 28-10 The phosphorus cycle**

algae and bacteria that its natural community interactions are disrupted.

28.4 WHAT HAPPENS WHEN HUMANS DISRUPT NUTRIENT CYCLES?

Many of the environmental problems that plague modern society have resulted from human interference in ecosystem function. Primitive peoples were sustained solely by the energy flowing from the sun, and they produced wastes that were readily taken back into the nutrient cycles. Through the mid-1800s, horses, oxen, and mules carried people, transported their goods, and pulled farmer's plows. The energy expended by these "beasts of burden" came from sunlight trapped by the plants that these animals consumed, and their wastes became fertilizer.

But as the population grew and technology increased, humans began to act more and more independently of these natural processes. The Industrial Revolution, which began in earnest in the mid-nineteenth century, resulted in a tremendous increase in our reliance on energy stored in fossil fuels for heat, light, transportation, industry, and agriculture. In mining and transporting these fuels, we have exposed ecosystems to a variety of substances that are foreign and often toxic to them, such as oil spilled into bodies of water (**Fig. 28-11**). Our reliance on fossil fuels has also disrupted the global nutrient cycles of nitrogen, phosphorus, sulfur, and carbon.

Overloading the Nitrogen and Phosphorus Cycles Damages Aquatic Ecosystems

In human-dominated ecosystems—such as farm fields, gardens, and suburban lawns—ammonia, nitrate, and phosphate are supplied by chemical fertilizers to stimulate plant growth. Each year, roughly 150 million tons of phosphate fertilizer are produced from mined phosphate rock. A similar amount of nitrogen-based fertilizer is manufactured using atmospheric nitrogen. These are applied to farm fields to help satisfy the agricultural demands of a growing population.

Water, driven by the hydrologic cycle, washes over the land, dissolving and carrying away enormous quantities of

▲ **FIGURE 28-11 A natural substance out of place** An oil-soaked bird on the shores of the Black Sea after an oil spill.

phosphate and nitrogen-based fertilizer. As the water drains into lakes, rivers, and ultimately the oceans, these fertilizers disrupt the delicate balance of food webs by overstimulating the growth of phytoplankton. This causes phytoplankton "blooms" that can turn reasonably clear water into an opaque green soup. As the phytoplankton die, their bodies sink into deeper water, where they provide a feast for decomposer bacteria. Cellular respiration by decomposer bacteria uses up most of the available dissolved oxygen. Deprived of oxygen, aquatic invertebrates and fish either die (contributing to the problem) or leave the area. A dramatic example occurs each year in the Gulf of Mexico off the coast of Louisiana, where the Mississippi river dumps enormous quantities of nitrates that have washed off fertilized farm fields in the Midwest. Each summer, this creates a "dead zone" covering 7,000 to 8,000 square miles, where oxygen levels are so low that few types of organisms can survive. Worldwide, dead zones are increasing rapidly in both size and number as agricultural activities intensify.

Overloading the Sulfur and Nitrogen Cycles Causes Acid Deposition

Although natural processes—such as the activity of nitrogen-fixing bacteria and decomposer organisms, fires, and lightning—produce nitrogen oxides and ammonia, about 60% of the nitrogen that is available to Earth's ecosystems now results from human activities. Burning fossil fuels combines atmospheric nitrogen with oxygen, producing most of the emissions of nitrogen oxides. Although sulfur is released as sulfur dioxide (SO_2) by volcanoes, hot springs, and decomposer organisms, human industrial activities—primarily burning sulfur-containing fossil fuels—account for about 75% of all the sulfur dioxide emissions worldwide. Excess production of nitrogen oxides and sulfur dioxide was identified in the late 1960s as the cause of a growing environmental threat—"acid rain," more accurately called **acid deposition.**

When combined with water vapor in the atmosphere, nitrogen oxides and sulfur dioxide are converted to nitric acid and sulfuric acid, respectively. Days later and often hundreds of miles from the source, these acids fall to Earth dissolved in rainwater. Acid deposition damages forests, can render lakes lifeless, and eats away at buildings and statues (**Fig. 28-12**). In the United States, the Northeast, Mid-Atlantic, upper Midwest, and Western regions and the state of Florida are the most vulnerable, because the rocks and soils that predominate in those areas have little ability to neutralize acids.

▲ **FIGURE 28-12 Acid deposition is corrosive** This lion statue was unveiled in 1867 outside the Leeds town hall in Yorkshire, England. It is carved from limestone, and has been severely eroded by acid deposition.

Acid Deposition Damages Life in Lakes and Forests

In the Adirondack Mountains of New York State, downwind from coal-burning power plants and industry in the Midwest, acid rain has rendered many lakes and ponds too acidic to support fish and the food webs that sustain them. Acid deposition increases the exposure of organisms to toxic metals—such as aluminum, mercury, lead, and cadmium—all of which are far more soluble in acidified water than in water of neutral pH. Aluminum dissolved from rock may inhibit plant growth and kill fish.

On land, acid rain leeches away essential plant nutrients such as calcium, magnesium, and potassium. It may also kill decomposer microorganisms, preventing the return of nutrients to the soil. Plants in acidified soil become weak and more vulnerable to infection and insect attack. Since 1965, scientists have witnessed the death of about half of the red spruce and beech trees and one-third of the sugar maples in the Green Mountains of Vermont. The snow, rain, and heavy fog that commonly cloak these mountaintops are highly acidic. For example, at a monitoring station atop Mount Mitchell in North Carolina, the pH of fog has been recorded at 2.9—more acidic than vinegar (**Fig. 28-13**).

Since 1990, government regulations have resulted in substantial reductions in emissions of both sulfur dioxide and nitrogen oxides from U.S. power plants. Air quality has improved, and rain has become less acidic. However, the total nitrogen oxide release remains high because more gasoline is being burned by automobiles, releasing more nitrogen oxides from this source. Damaged ecosystems recover slowly. Adirondack lakes are gradually becoming less acidic, although full recovery is still decades away. In the Southeast, some freshwater acid levels are still increasing. Many ecologists believe that far stricter controls on nitrogen emissions will be required to prevent further deterioration and to allow damaged ecosystems to recover.

Interfering with the Carbon Cycle Is Warming Earth's Climate

Between 345 million and 280 million years ago, huge quantities of carbon were diverted from the carbon cycle. This occurred under the warm, wet conditions of the Carboniferous

▲ **FIGURE 28-13 Acid deposition can destroy forests** Acid rain and acidic fog have destroyed this forest atop Mount Mitchell in North Carolina.

period, when bodies of prehistoric organisms were buried in sediments, escaping decomposition. Over time, heat and pressure converted their bodies (containing stored energy from sunlight) into fossil fuels such as coal, oil, and natural gas. Since the mid-1800s, however, we have increasingly relied on energy released by burning these fuels. One researcher estimates that a typical automobile gas tank holds the transformed remains of 100 tons of prehistoric life, largely microscopic phytoplankton. As we burn fossil fuels in our power plants, factories, and cars, we harvest the energy of prehistoric sunlight and release CO_2 into the atmosphere. Burning fossil fuels accounts for 80% to 85% of the CO_2 added to the atmosphere each year.

A second source of added atmospheric CO_2 is **deforestation,** which destroys tens of millions of forested acres annually and accounts for about 15% of CO_2 emissions. Deforestation is occurring principally in the Tropics, where rain forests are rapidly being converted to agricultural land to feed growing populations and supply the world's demand for biofuels, such as ethanol and biodiesel. The carbon stored in the massive trees in these forests returns to the atmosphere when they are cut down and burned.

Collectively, human activities are estimated to release more than 10 billion tons of carbon (in the form of CO_2) into the atmosphere each year, an amount that is increasing as the human population grows and as less-developed countries increase their living standards. About half of this carbon is absorbed into the oceans, plants, and soil, with the rest remaining in the atmosphere. As a result, since 1850 (when people began burning large quantities of fossil fuels during the Industrial Revolution) the CO_2 content of the atmosphere has increased by almost 37%—from 280 parts per million (ppm) to 383 ppm—and is growing by about 2 ppm annually (see Fig. 28-15b). Based on an analysis of gas bubbles trapped in ancient Antarctic ice, the atmospheric CO_2 content is now about 30% higher than at any time during the past 650,000 years.

Carbon Dioxide and Other Greenhouse Gases Trap Heat in the Atmosphere

The fate of sunlight entering Earth's atmosphere is shown in **Figure 28-14**. Some of the energy from sunlight is reflected back into space, bouncing off water vapor in clouds and other particles in the air (**Fig. 28-14** ❷). Most sunlight, however, reaches Earth, and is converted into heat (**Fig. 28-14** ❸) that is then radiated back toward space (**Fig. 28-14** ❹). Although most heat is released into space (**Fig. 28-14** ❺), carbon dioxide and several other **greenhouse gases** trap some of this heat in the atmosphere (**Fig. 28-14** ❻). This is a natural process called the **greenhouse effect,** which keeps our atmosphere relatively warm and allows life on Earth as we know it.

However, a large and growing body of evidence indicates that human activities have amplified the natural greenhouse effect, producing a phenomenon called **global warming.** For Earth's temperature to remain constant, the total energy entering and leaving Earth's atmosphere must be equal. As greenhouse gas levels rise, more heat is retained than is radiated back into space, causing Earth to warm. Although CO_2 accounts for most human emissions of greenhouse gases, other important greenhouse gases include methane (CH_4), released by agricultural activities, landfills, and coal mining; and nitrous oxide (N_2O), released by agricultural activities and burning fossil fuels.

Global Temperature Increases Have Paralleled Increases in Atmospheric Carbon Dioxide

The decade from 1998 to 2008 was the warmest ever recorded. Global surface temperature data are recorded from thousands of weather stations around the world and from satellites that measure ocean surface temperatures. Atmospheric CO_2 data are collected from air that has typically traveled for hundreds of miles before reaching the sampling site near the barren peak of

Have you ever wondered

What Is Your Carbon Footprint?

Each of us impacts Earth through our personal decisions and by the choices we make. A "carbon footprint" is a measure of the impact that human activities have on climate, based on the quantity of greenhouse gases they emit.

Our carbon footprint gives us a sense of our individual impacts. For example, each gallon of gasoline burned releases 19.6 pounds (8.9 kg) of CO_2 into the air. So, if your car gets 20 miles to the gallon, then each mile that you drive will add about a pound of CO_2 to the atmosphere.

The Web sites of the U.S. Environmental Protection Agency as well as for several environmental organizations provide household emissions calculators that allow you to estimate your carbon footprint, and provide advice on how you can reduce it. To get started, just type "carbon footprint" into your Internet browser.

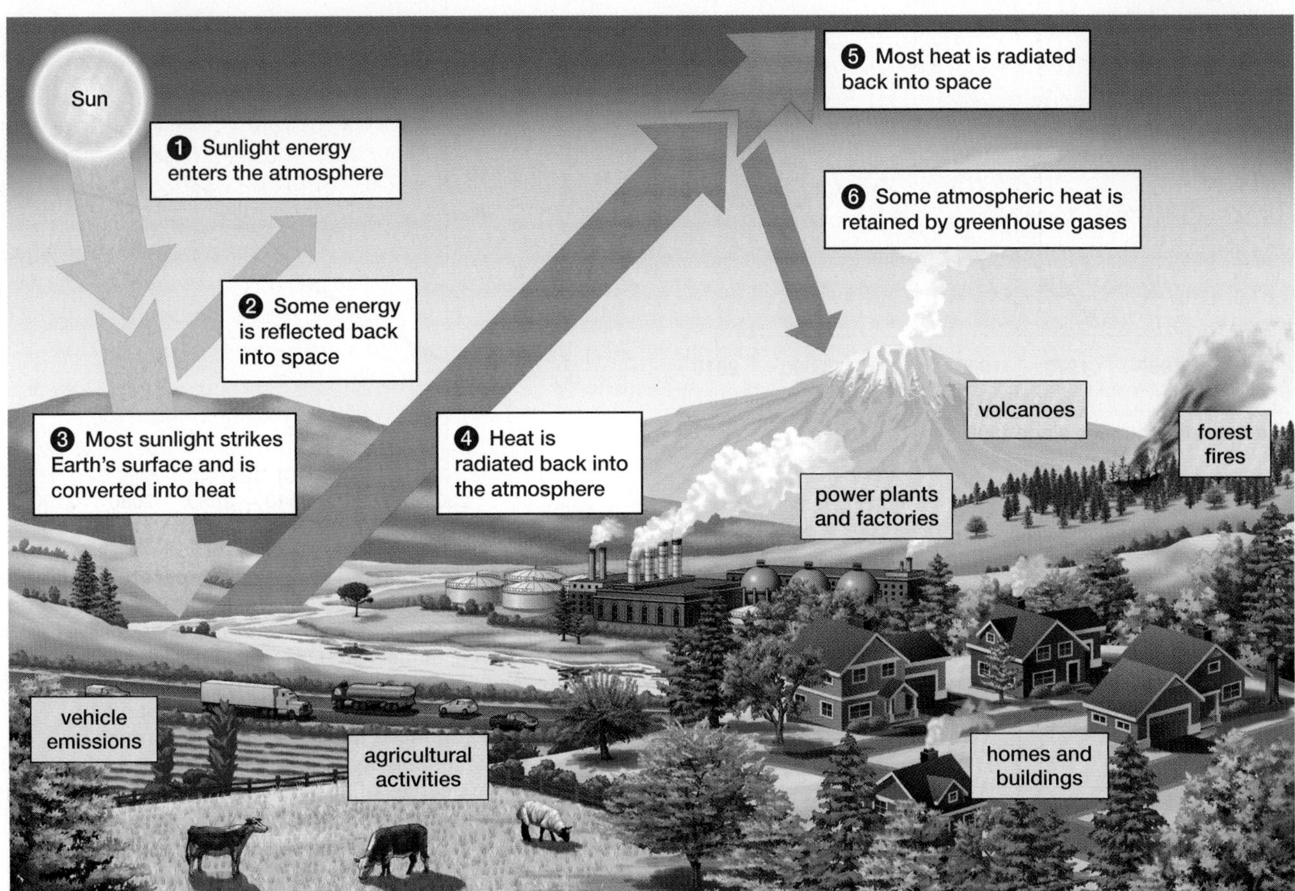

▲ **FIGURE 28-14 Greenhouse gases and global warming** Incoming sunlight warms Earth's surface and is radiated back to the atmosphere. Greenhouse gases, released by natural processes and substantially augmented by human activities (yellow rectangles), are absorbing increasing amounts of this heat, raising global temperatures.

Mauna Loa, a mountain in Hawaii. Here, numerous measurements are taken daily. Elaborate statistical procedures are used to eliminate readings that show high variability over a short time span, indicating that they have been influenced by nearby CO_2 sources. Historical and recent temperature records have revealed a global temperature increase (**Fig. 28-15a**) that parallels the measured rise in atmospheric CO_2 (**Fig. 28-15b**).

Any scientist will tell you that correlations between events do not prove that one event caused the other. However, the evidence that human activities are driving global warming is so compelling that the overwhelming majority of scientists accept this hypothesis and are convinced that urgent action is needed to slow the process. The Intergovernmental Panel on Climate Change (IPCC) is a consortium involving hundreds of scientists as well as other experts from 130 different nations. In their 2007 report, the IPCC predicted that even under the best-case scenario in which a concerted worldwide effort is made to reduce greenhouse gas emissions, the average global temperature will rise by at least 3.2 °F (1.8 °C) by the year 2100; a high-level emissions scenario projects an increase of 7.2 °F (4.0 °C) (**Fig. 28-16**). Small global temperature changes can have enormous impacts. For example, average temperatures during the peak of the last Ice Age (20,000 years ago) were only about 5 °C lower than at present.

Global Warming Will Have Many Consequences

As geochemist James White at the University of Colorado quipped, "If the Earth had an operating manual, the chapter on climate might begin with the caveat that the system has been adjusted at the factory for optimum comfort, so don't touch the dials." Perhaps the most visible effects of global warming are the melting of glaciers throughout the world (**Fig. 28-17**). For example, each year about 1.6 million people flock to Glacier National Park, Montana, named for its spectacular abundance of glaciers. Of the approximately 150 glaciers that graced its mountainsides in 1850, fewer than 30 remain.

More Extremes in Weather Are Predicted and May Already Be Occurring Many leading climatologists see evidence that global warming is already affecting our weather. Scientists at the National Center for Atmospheric Research have concluded that global warming has increased ocean surface temperatures beyond the range of normal fluctuations, contributing, since the mid-1960s, to an increase in severe hurricanes that generate higher wind speeds and greater rainfall. As the world warms, experts predict an increase in heat waves. Some world regions will experience more flooding, while in other areas, droughts are likely to last longer and to be more severe. Since the 1970s, the area of Earth impacted by severe drought has doubled as a

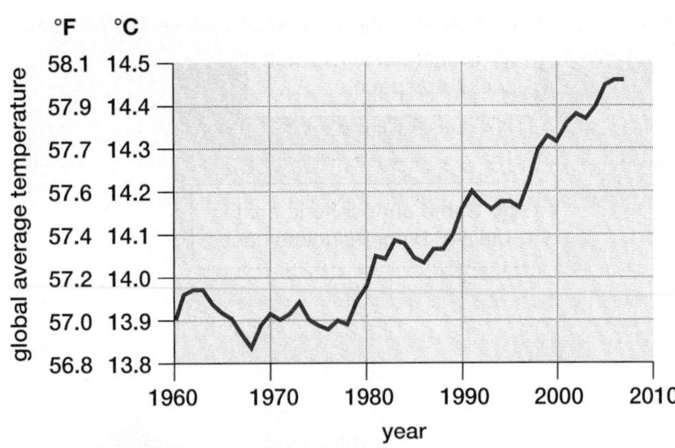

(a) Temperature change since 1960

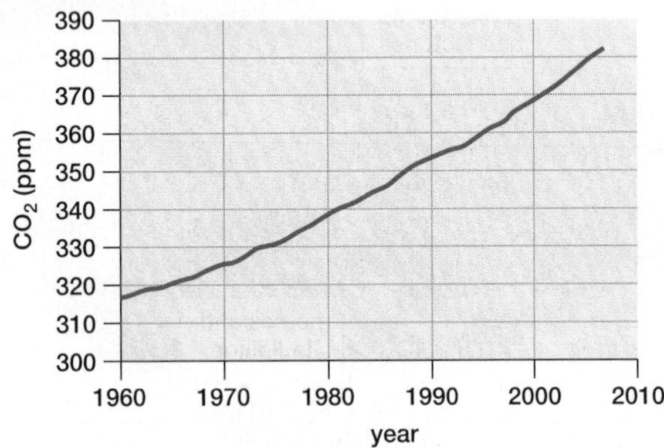

(b) Atmospheric CO_2 change since 1960

▲ **FIGURE 28-15 Global warming parallels atmospheric CO_2 increases (a)** Global surface temperatures. Because global temperature varies considerably from year to year, this temperature curve shows trends by averaging each successive year with the four years preceding it. **(b)** Yearly average CO_2 concentrations in parts per million These measurements were recorded at 11,155 feet (3,400 meters) above sea level, near the summit of Mauna Loa, Hawaii. Data from the National Oceanographic and Atmospheric Administration (NOAA).

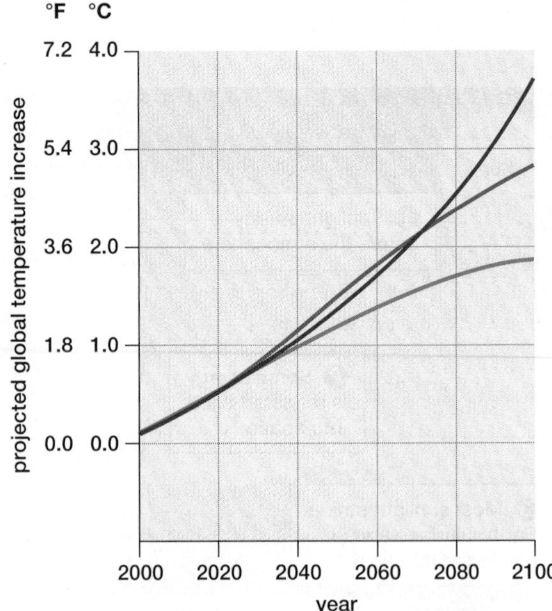

▲ **FIGURE 28-16 Projected range of temperature increases** The IPCC projections for the twenty-first century are based on three different scenarios of greenhouse gas emissions. The red, blue, and green data lines are projections based on high, moderate, and substantially reduced growth of greenhouse gas emissions, respectively. Even under the most optimistic assumptions, a continued increase in global warming is predicted. Global temperature change is expressed relative to a baseline of zero degrees of change. The baseline is the average temperature from 1980 to 1990. Data from IPCC. 2007. *Fourth assessment report: Summary for policymakers.*

result of increased temperatures and local decreases in rainfall. Agricultural disruption resulting from such extremes in weather could be disastrous for nations that are already struggling to grow enough food.

Community Interactions May Be Disrupted The predicted rapid rise in global temperature will exceed the rate at which natural selection can allow most organisms to adapt. The impact of global warming on forests could be profound. As the world warms, tree distributions will change based on their tolerance for heat. For example, sugar maples may disappear from northeastern U.S. forests, while some southeastern forests could be replaced by grasslands. Warming is contributing to massive destruction of evergreen trees in the western United

(a) Carroll Glacier, 1904

(b) Carroll Glacier, 2004

► **FIGURE 28-17 Glaciers are melting** Photos taken from the same vantage point in **(a)** 1904 and **(b)** 2004 document the retreat of the Carroll Glacier in Glacier Bay, Alaska.

States and Canada by the pine bark beetle, which is benefiting from warmer winters and longer reproductive seasons (see pp. 591–592). Coral reefs, already stressed by human activities, are likely to suffer further damage from warmer waters, which drive out the symbiotic algae that provide them with energy from photosynthesis. Further, as the oceans absorb more CO_2, their waters are becoming more acidic. Because acid dissolves limestone, this makes it more difficult for corals to form their limestone skeletons.

Spring is arriving earlier in Europe and in the northeastern United States. The growing season in Europe has increased by more than 10 days during the past three decades. In the northern hemisphere, animals as diverse as butterflies, birds, and whales are extending their ranges northward. The cumulative weight of data from diverse sources worldwide provides strong evidence that warming-related biological changes have begun. Global warming is also predicted to increase the range of tropical disease-carrying organisms, such as malaria-transmitting mosquitoes, with negative consequences for human health. Arctic and Antarctic wildlife from polar bears to penguins are among the hardest hit, as described in "Earth Watch: Poles in Peril."

Earth Watch

Poles in Peril

Nowhere are the impacts of global warming more visible than at the far ends of Earth, in our planet's unique ice-dominated ecosystems. The Antarctic Peninsula in the Southern Hemisphere is particularly vulnerable to global warming because its average year-round temperature hovers close to the freezing point. As global temperatures have risen, temperatures on the peninsula have increased at a far faster rate than the global average, and the flow of glacial ice into the sea has accelerated. Huge expanses of floating ice shelves have disintegrated, with far-reaching consequences. Sea ice creates conditions that favor abundant growth of phytoplankton and algae, which provide food for larval krill—shrimp-like zooplankton that are a keystone species in the Antarctic community. Krill comprise a major portion of the diet of seals, penguins, and several species of whales. Researchers are concerned that the impact of krill loss may reverberate throughout the Antarctic food web, harming all the other organisms who feed on them.

In the Northern Hemisphere, Greenland's ice cover is melting at record rates, and Arctic temperatures have increased by twice the global average during the past 50 years. During the twenty-first century, loss of sea ice during Arctic summers has repeatedly shattered records, and total summer ice cover has decreased by roughly 30% during the past 30 years. The loss of summer sea ice is bad news for polar bears and other marine mammals that rely on ice floes as nurseries for their young and as staging platforms for hunting fish or seals. As summer ice has diminished, both walrus and polar bear populations are moving onto land in record numbers to give birth, and the loss of ice platforms is reducing the expanse of ocean in which these top predators can hunt. In 2008, in response to its declining numbers and the projected continued loss of its habitat, the polar bear was added to the list of threatened species—the first to be listed primarily as a result of global warming. Complete loss of sea ice, which climate models predict will occur within the next century, could cause the extinction of polar bears in the wild (**Fig. E28-2**).

The thawing of Earth's northern regions and the melting of sea ice are cause for particular concern because both of these situations generate positive feedback loops that further accelerate global warming. Extensive areas of deep frozen soil called permafrost underlie Siberia and Alaska. These soils store huge deposits of organic matter from prehistoric plants and animals, which have not broken down because decomposer organisms are inhibited by the cold. Ominously, some of these regions are now thawing for the first time since the last ice age ended more than 10,000 years ago. In Siberia, for example, a region of frozen peat soil larger than Texas and Oklahoma combined is thawing, creating giant bogs that could release enormous quantities of CO_2 and methane gases as conditions become favorable for decomposers. Because methane is about 20 times as effective as carbon dioxide in trapping atmospheric heat, this is of particular concern. In a positive feedback loop, the greenhouse gases released by thawing permafrost will further warm the atmosphere, causing still more permafrost to thaw.

Sea ice reflects back most of the solar energy that hits it, but the ocean water that is exposed when ice disappears absorbs most solar energy, converting it to heat. This situation is a second example of positive feedback, because the heat absorbed by the exposed water will further warm Earth's surface, causing more ice to melt. Melting ice has also opened access to areas of the Arctic Ocean and its seabed that have been inaccessible for centuries. This has fueled a rush by several countries to lay claim to oil fields that are predicted to underlie the Arctic Ocean floor. Ironically, this suggests yet another type of positive feedback: Burning this oil will further accelerate the climate change that has now made it accessible.

▲ **FIGURE E28-2 Polar bears are on thin ice** The polar bear is now a threatened species as a result of global warming.

Case Study revisited
Dying Fish Feed an Ecosystem

Researchers investigating the sockeye salmon's return to an Alaskan stream witness an awesome sight. Hundreds of brilliant red bodies writhe in water so shallow that it barely covers them. A female beats her tail, excavating a shallow depression in the gravel where she releases her coral-colored eggs; meanwhile, a male showers them with sperm. But after their long and strenuous migration, these adult salmon are dying. Their flesh is tattered, their muscles wasted, and the final act of reproduction saps the last of their energy. Soon, the stream is clogged with dying, dead, and decomposing bodies—an abundance of nutrients unavailable at any other time of the year. Eagles, grizzly bears, and gulls gather to gorge themselves on the fleeting bounty. Flies breed in the carcasses, feeding spiders, birds, and trout. The breeding cycles of local mink populations have evolved around the event; females lactate at the time when the salmon provide them with abundant food. Bears drag the carcasses into the woods, often consuming only a part of each fish. The remnants become food for detritus feeders and decomposers, who liberate its nutrients to the soil where they will be taken up by plants and reenter nutrient cycles.

Because 95% of the mass of the salmon's body was accumulated while it lived in the ocean, the upstream migration of salmon represents an enormous transfer of nutrients in an unusual direction: upstream. Historically, researchers estimate that 500 million pounds of salmon migrated upstream in the U.S. Pacific Northwest each year, contributing hundreds of thousands of pounds of nitrogen and phosphorus to the region surrounding the Columbia River. Now, due to factors that include overfishing, river damming, diversion of water for irrigation, runoff from agriculture, and pollution of the estuaries (where several salmon species spend a significant part of their life cycle), migratory salmon populations in the region have declined by more than 90% in the past century. The web of life that relied on the mighty annual upstream flow of nutrients has been disrupted.

Consider This

Some salmon populations have been so thoroughly depleted that they are listed as endangered or threatened under the Endangered Species Act. Some people argue that because these salmon are also raised commercially in hatcheries and artificial ponds, they should not be afforded this protection. Meanwhile, researchers studying chinook salmon raised in hatcheries noted a 25% decline in the average size of eggs of hatchery-reared fish over just four generations. These eggs produce smaller juvenile fish. Based on this information, why is there a good case for protecting populations of wild salmon?

CHAPTER REVIEW

Summary of Key Concepts

28.1 How Do Energy and Nutrients Move Through Ecosystems?

Ecosystems are sustained by a continuous input of energy from sunlight and the constant recycling of nutrients. Energy enters the biotic portion of ecosystems when it is harnessed by photosynthetic organisms. Nutrients are obtained by organisms from their living and nonliving environment and are recycled within and among ecosystems.

28.2 How Does Energy Flow Through Ecosystems?

Energy enters ecosystems through photosynthesis. Photosynthetic organisms act as conduits of both energy and nutrients into biological communities. Energy is passed upward through trophic (feeding) levels. Autotrophs are photosynthetic and are called producers, the lowest trophic level. Among heterotrophs, herbivores form the second level: primary consumers. Carnivores act as secondary consumers when they prey on herbivores and as tertiary or higher-level consumers when they eat other carnivores. Omnivores, which consume both plants and other animals, occupy multiple trophic levels. Feeding relationships in which each trophic level is represented by one organism are called food chains. In natural ecosystems, feeding relationships are complex and are described as food webs. Detritus feeders and decomposers (which digest dead bodies and wastes) use and release the energy stored in these substances and liberate their nutrients, which then reenter nutrient cycles.

The higher the trophic level, the less energy available to sustain it. In general, only about 10% of the energy captured by organisms at one trophic level is converted to the bodies of organisms in the next higher level. As a result, plants are more abundant than herbivores, and herbivores are more common than carnivores. The storage of energy at each trophic level is illustrated graphically as an energy pyramid. This inefficiency of energy transfer through trophic levels leads to biological magnification, the process by which toxic substances accumulate in increasingly high concentrations in organisms occupying progressively higher trophic levels.

28.3 How Do Nutrients Cycle Within and Among Ecosystems?

A nutrient cycle depicts the movement of a particular nutrient from its reservoir, usually in the abiotic, or nonliving, portion of the ecosystem; through the biotic, or living, portion of the ecosystem; and back to its reservoir.

The major reservoir of water is the oceans. During the hydrologic cycle, solar energy evaporates water, which returns to Earth as precipitation. Water flows into lakes and underground aquifers, and is carried by rivers to the oceans. Relatively small amounts of water are passed through food webs.

Major short-term carbon reservoirs are the oceans and the atmosphere. Carbon enters producers through photosynthesis. From autotrophs, it is passed through the food web and released to the atmosphere as CO_2 during cellular respiration. Fossil fuels represent a long-term carbon reservoir; burning them is changing Earth's climate.

The major reservoir of nitrogen is the atmosphere. Plants get their nitrogen from nitrates and ammonia. Nitrogen gas is captured by nitrogen-fixing bacteria, which release ammonia. Other bacteria convert ammonia to nitrate, which can also be produced by lightning. Industrial processes manufacture fertilizer containing ammonia and nitrate. Nitrogen passes from producers to consumers and is returned to the environment through excretion and the activities of detritus feeders and decomposers. Nitrogen gas is returned to the air by denitrifying bacteria.

The reservoir for phosphorus is in rocks, as phosphate, which dissolves in water. Phosphate is absorbed by photosynthetic organisms and is passed through food webs. Some phosphate is excreted, and the rest is returned to the soil and water by decomposers. Some is carried to the oceans, where it is deposited in marine sediments. Humans mine phosphate-rich rock to produce fertilizer.

***BioFlix*™ The Carbon Cycle**

28.4 What Happens When Humans Disrupt Nutrient Cycles?

Environmental disruption occurs when human activities interfere with the natural functioning of ecosystems. Human activities release toxic substances and produce more nutrients than nutrient cycles can efficiently process. The use of enormous quantities of fertilizer by agricultural activities has disrupted many aquatic ecosystems. Nutrient-rich water washing into lakes, rivers, and the oceans causes excessive growth of aquatic plants and phytoplankton; their subsequent death and decomposition deplete oxygen, killing many aquatic organisms and producing "dead zones" in nearshore coastal waters. By burning fossil fuels, we have overloaded the natural cycles for sulfur, nitrogen, and carbon. Fossil fuel combustion releases sulfur dioxide and nitrogen oxides. In the atmosphere, these substances are converted to sulfuric acid and nitric acid, which fall to Earth as acid deposition. Acidification of many freshwater lakes in the eastern United States has substantially reduced their ability to support life. At high elevations, acid deposition has significantly damaged many eastern forests.

Burning fossil fuels has substantially increased atmospheric carbon dioxide, a greenhouse gas. This increase is correlated with increased global temperatures, leading nearly all scientists to conclude that global warming is due to human activities. Global warming is causing glaciers and ice sheets to melt, and is influencing the distribution and seasonal activities of wildlife. Scientists believe global warming is beginning to have a major impact on precipitation and weather patterns, with unpredictable results.

Key Terms

acid deposition *545*
aquifer *541*
autotroph *535*
biodegradable *539*
biogeochemical cycle *540*
biological magnification *539*
biomass *538*
carbon cycle *541*
carnivore *535*
consumer *535*
decomposer *537*
deforestation *546*
denitrifying bacteria *543*
detritus feeder *537*
energy pyramid *538*
estuary *535*
food chain *536*
food web *536*
fossil fuel *542*
global warming *546*
greenhouse effect *546*
greenhouse gas *546*
herbivore *535*
heterotroph *535*
hydrologic cycle *540*
legume *543*
macronutrient *540*
micronutrient *540*
net primary production *535*
nitrogen cycle *542*
nitrogen fixation *543*
nutrient *533*
nutrient cycle *540*
omnivore *536*
phosphorus cycle *543*
phytoplankton *536*
primary consumer *535*
producer *535*
reservoir *540*
secondary consumer *535*
tertiary consumer *535*
transpiration *541*
trophic level *535*
zooplankton *536*

Thinking Through the Concepts

Fill-in-the-Blank

1. Nearly all life gets its energy from __________, which is captured by the process of __________. In contrast, __________ are constantly recycled during processes called __________.
2. Photosynthetic organisms are called either __________ or __________. The energy that these store and make available to other organisms is called __________.
3. Feeding levels within ecosystems are also called __________. An illustration of these levels with only one organism at each level is called a(n) __________. Feeding relationships are most accurately depicted as __________.
4. In general, only about __________ percent of the energy available in one trophic level is captured by the level above it. This concept is based on what general physical principle? __________
5. Photosynthetic organisms are consumed by other organisms collectively called __________, or __________. Organisms in higher trophic levels are collectively called __________, or __________. Animals and protists that feed on wastes and dead bodies are called __________. The major types of decomposer organisms are __________ and __________.
6. During the nitrogen cycle, nitrogen gas is captured from its atmospheric reservoir by __________ in the soil, and is then returned to this reservoir by __________. The two forms of nitrogen that are used by plants are __________ and __________.
7. Two relatively short-term reservoirs for carbon are the __________ and __________. Carbon in these short-term reservoirs is in the form of __________. Two long-term reservoirs for carbon are __________ and __________.

Review Questions

1. What makes the flow of energy through ecosystems fundamentally different from the flow of nutrients?
2. What is an autotroph? What trophic level does it occupy, and what is its importance in ecosystems?
3. Define *net primary production*. Would you predict higher productivity in a farm pond or an alpine lake? Explain your answer.
4. List the first three trophic levels. Among the consumers, which are most abundant? Why would you predict that there will be a greater biomass of plants than herbivores in any ecosystem? Relate your answer to the "10% law."

5. How do food chains and food webs differ? Which is the more accurate representation of actual feeding relationships in ecosystems?
6. Define *detritus feeders* and *decomposers*, and explain their importance in ecosystems.
7. Trace the movement of carbon from one of its reservoirs through the biotic community and back to the reservoir. How have human activities altered the carbon cycle, and what are the implications for future climate?
8. Explain how nitrogen gets from the atmosphere into a plant's body.
9. Trace a phosphorus molecule from a phosphate-rich rock into a carnivore. What makes the phosphorus cycle fundamentally different from the carbon and nitrogen cycles?
10. Trace the movement of a water molecule from the ocean, through a plant body, and back to the ocean, describing all the intermediate stages and processes.

Applying the Concepts

1. **BioEthics** What could your college or university do to reduce its contribution to global warming? Be specific, and, if possible, offer practical alternatives to current practices.
2. Define and give an example of biological magnification. What qualities are present in materials that undergo biological magnification? In which trophic level are the problems worst, and why?
3. **BioEthics** Discuss the contribution of human population growth to (a) acid rain and (b) global warming.
4. Describe what would happen to a population of deer if all predators were removed and hunting banned. Include effects on vegetation as well as on the deer population itself. Relate your answer to carrying capacity as discussed in Chapter 26.

MB ***Go to www.masteringbiology.com for practice quizzes, activities, eText, videos, current events, and more.***

Earth's Diverse Ecosystems

Chapter 29

Case Study

The Birds and the Beans

WHAT DO CHOCOLATE AND COFFEE have in common? Some would say they are both necessities of life. People today continue a long-established tradition of enjoying these delicacies. Cocoa, the base product for chocolate, was consumed as early as 1000 B.C., and coffee, at around 800 A.D. People in the United States consume one-third of the world's annual coffee production, nearly 140 billion cups, and about one-quarter of its cocoa, or 800,000 tons.

Cocoa and coffee both come from seeds (beans) of rain-forest plants. Cocoa is produced from the beans of the cacao tree (*Theobroma cacao*) native to the lowland rain forests of South and Central America ("theobroma" means "food of the gods"). The coffee plant (*Coffea arabica*) originated in Ethiopia, Africa ("arabica" means "from Arabia"). Both cacao and coffee plants are now widely cultivated in the Tropics, including South and Central America, Africa, and Southeast Asia.

Until the 1960s, most of the world's cacao and coffee plants were cultivated in the shade of a variety of tree species, which provided a multilevel, diverse habitat that supported at least 180 species of birds, many of them migratory songbirds. The forest vegetation absorbed water and protected the soil from erosion. Complex community interactions kept the pests of cacao trees and coffee plants in check. Forest shade discouraged weed growth, and the rich community of decomposers and detritus feeders maintained nutrient cycles that made fertilizers unnecessary.

In the 1960s and 1970s, however, new varieties of cacao and coffee plants were developed that thrived in full sun and yielded more beans. As demand for chocolate and coffee increased, traditional shade farms and virgin rain forests were destroyed to make way for cacao and coffee monocultures. In the past 30 years, the switch to full-sun cacao and coffee plants has led to the destruction of significant areas of rain forest, as well as a substantial increase in the use of fertilizers, herbicides, and pesticides to compensate for the services that the shade farms once provided.

In the 1970s, biologists began to notice significant declines in populations of North American songbirds, including some thrushes, orioles, flycatchers, tanagers, vireos, and warblers. Why are the birds disappearing? How are North American songbirds and South and Central American cacao and coffee farms connected?

▲ A cacao pod is harvested. South American farmers are increasingly switching back to traditional sustainable methods of cacao farming. (Inset) An opened ripe cacao pod reveals the beans embedded in a sweet white pulp.

At a Glance

29.1 WHAT FACTORS INFLUENCE EARTH'S CLIMATE?

The distribution of life, particularly on land, is dramatically affected by both weather and climate. **Weather** refers to short-term fluctuations in temperature, humidity, cloud cover, wind, and precipitation in a region over periods of hours or days. **Climate,** in contrast, refers to patterns of weather that prevail over years or centuries in a particular region. The amount of sunlight and water and the range of temperatures determine the climate of a given region.

Both Climate and Weather Are Driven by the Sun

Both climate and weather are driven by a great thermonuclear engine: the sun. Solar energy reaches Earth in a range of wavelengths—from short, high-energy ultraviolet (UV) rays, through visible light, to the infrared wavelengths that we experience as heat (see Fig. 7-4). This solar energy drives the wind, ocean currents, and the global water cycle. Before it reaches Earth's surface, however, sunlight is modified by the atmosphere. A layer relatively rich in ozone (O_3) is found in the middle atmosphere, or stratosphere. This **ozone layer** absorbs much of the sun's high-energy UV radiation, which can damage biological molecules (see "Earth Watch: The Ozone Hole—A Puncture in Our Protective Shield").

Earth's Physical Features Also Influence Climate

Many physical factors influence climate. Among the most important are Earth's curvature and its tilted axis as it orbits around the sun. These factors cause uneven heating of the surface and seasonal changes in the directness of sunlight north and south of the equator. Uneven heating, in conjunction with Earth's rotation, generates air and ocean currents, which in turn are modified by irregularly shaped landmasses.

Earth's Curvature and Tilt Influence the Angle at Which Sunlight Strikes

Average yearly temperatures are determined by the amount of sunlight that reaches a given region, which in turn depends on latitude. Latitude is a measure of the distance north or south from the equator, expressed in degrees. The equator is defined as 0° latitude, and the poles are at 90° north and south latitudes. Sunlight hits the equator relatively directly throughout the year. With increasing distances north or south, however, it strikes Earth's surface less directly, spreading the same amount of sunlight over a larger area than at the equator, which decreases its intensity. In addition, sunlight

Have you ever wondered 

What Would Happen If Earth Were Not Tilted on Its Axis?

To visualize Earth as it rotates on its axis while orbiting the sun, imagine sticking a skewer straight through a tennis ball. The skewer creates an axis, and you can spin the tennis ball around this axis, which, to mimic Earth, would be tilted 23.5° from vertical.

If Earth were *not* tilted on its axis, however, the axis would remain perpendicular (at a 90° angle) to the plane of its orbit around the sun. In this situation, days and nights would each last 12 hours everywhere and all year round. Because the angle of sunlight and day length would remain unchanged as Earth orbited the sun, there would be no temperature changes over the course of the year, and therefore, no seasons. There would, however, be temperature differences in different locations; latitudes near the equator would remain uniformly warm, while temperatures would become increasingly cooler as one traveled away from the equator toward either the North or the South Pole.

Earth Watch

The Ozone Hole—A Puncture in Our Protective Shield

A fraction of the radiant energy produced by the sun—ultraviolet (UV) radiation—is so energetic that it can damage biological molecules. Excess UV radiation causes sunburn, premature aging of the skin, and skin cancer.

Fortunately, more than 97% of UV radiation is filtered out by an ozone-enriched layer of the stratosphere, called the ozone layer, which extends from about 10 to 15 miles (16 to 24 kilometers) above Earth. Ultraviolet light striking ozone and oxygen gas causes reactions that break down as well as regenerate ozone. In the process, the UV radiation is converted to heat, and the overall level of ozone remains reasonably constant—or at least did, until humans intervened.

In 1985, British atmospheric scientists published the startling news that springtime levels of stratospheric ozone over Antarctica had declined by more than 40% since 1977. In the **ozone hole** over Antarctica (**Fig. E29-1**), ozone now dips to about one-third of its pre-depletion levels. Photosynthesis by phytoplankton—the producers for marine ecosystems, and the basis for food webs that support penguins, seals, and whales—is reduced under the ozone hole above Antarctica. Although ozone layer depletion is most severe over Antarctica, the ozone layer is somewhat reduced over most of the world, including nearly all of the continental United States.

The ozone hole is caused primarily by human production and release of chlorofluorocarbons (CFCs). Developed in 1928, these gases were widely used in the production of foam plastic, as coolants in refrigerators and air conditioners, as aerosol spray propellants, and as cleansers for electronic parts. CFCs are very stable, and were considered safe. Their stability, however, has proved to be a major problem because they remain chemically unchanged as they slowly rise into the stratosphere. There, UV light causes them to break down and release chlorine atoms, which catalyze the breakdown of stratospheric ozone. Ice particles in clouds over the Arctic and Antarctic regions provide a surface on which the ozone-depleting reactions can occur.

Fortunately, major steps have been taken toward "plugging" the ozone hole. The 1987 Montreal Protocol and its subsequent amendments set limits and established phase-out periods for several ozone-depleting substances. In a remarkable worldwide effort, 191 countries have signed the treaty. The United States has now reduced its production of CFCs by nearly 98%. Ozone loss is no longer increasing, and some areas of the stratosphere are showing small improvements. But because CFCs persist for 50 to 100 years and take a decade or more to ascend into the stratosphere, significant recovery of the ozone layer could take several decades.

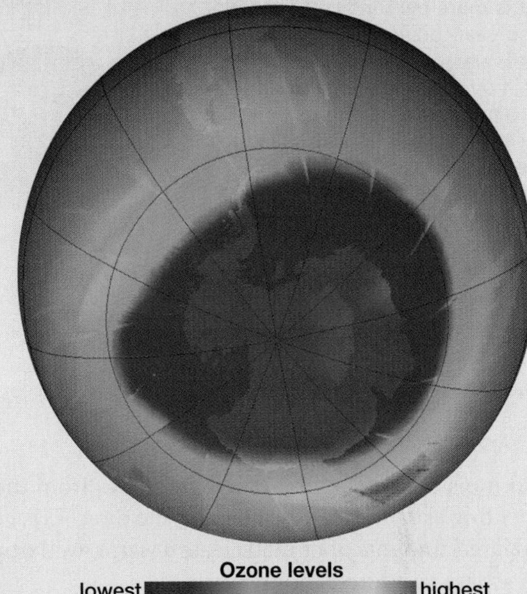

▲ **FIGURE E29-1 Satellite image of the Antarctic ozone hole** The Antarctic ozone hole, recorded in September 2009, is shown in blue and purple on this NASA satellite image. The 2009 ozone hole reached a maximum size of 6.3 million square miles (24 million square kilometers). The years 2002 and 2006 are tied for the largest hole, which covered 9.3 million square miles (29.5 million square kilometers). Image courtesy of NASA.

that reaches Earth at higher latitudes must travel through more of Earth's atmosphere, which reflects some sunlight back into space, further reducing its intensity.

Earth is also tilted on its axis, so as it makes its year-long journey around the sun, latitudes north and south of the equator experience regular, large changes in the angle of sunlight, resulting in pronounced seasons. When Earth's position in its orbit causes its Northern Hemisphere to be tilted toward the sun, this hemisphere receives the most direct sunlight and experiences summer. Simultaneously, the Southern Hemisphere is tilted away from the sun, and experiences winter (**Fig. 29-1**). Throughout the year, sunlight hits the equator with relatively little seasonal variation, so this region remains uniformly warm.

Air Currents Produce Broad Climatic Regions

Air currents are generated by Earth's rotation and by differences in temperature between different air masses (**Fig. 29-2**). Warm air is less dense than cold air, so as the sun's direct rays fall on the equator, heated air rises there. The warm air near the equator is also laden with water evaporated by solar heat. As the water-saturated air rises, it cools. Cool air cannot retain

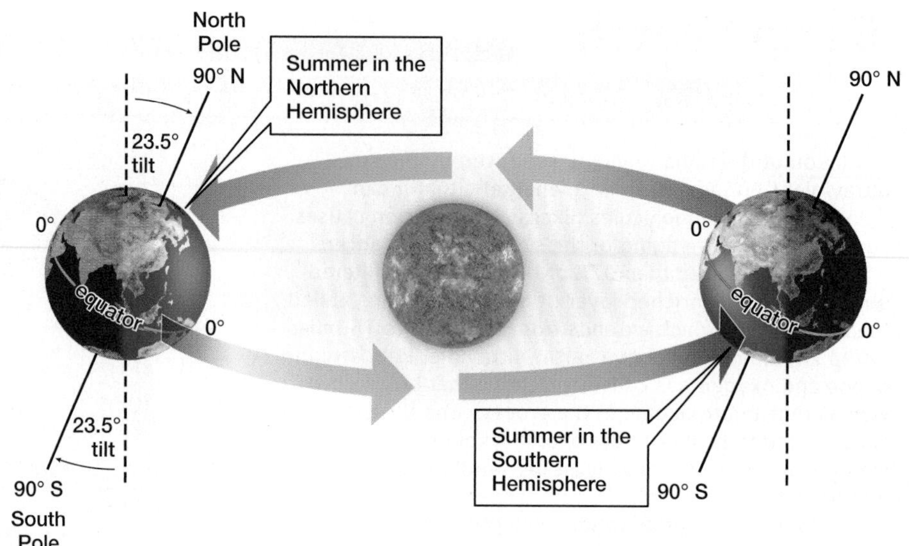

▶ FIGURE 29-1 Earth's curvature and tilt produce seasons and climate Temperatures are highest and most uniform at the equator and lowest and most variable at the poles. Equatorial sunlight is more perpendicular and hits this band year-round, while polar sunlight is seasonal, and its angle spreads it over a much larger land surface. The tilt of Earth on its axis causes seasonal variations in the range and directness of sunlight.

as much moisture, and so the water condenses from the rising air, falling as rain. Near the equator, the direct rays of the sun and large amounts of rainfall create a warm, wet, tropical climate, where rain forests flourish.

After the moisture has fallen from the rising equatorial air, cooler, drier air remains. The continuing upward flow of air from the equatorial region drives this cooler, drier air north and south. As the air approaches 30° N and 30° S latitudes, it cools enough to sink. As it sinks, the air is warmed by heat radiated from Earth. By the time it reaches the surface, it is both warm and very dry. Not surprisingly, the major deserts of the world are found at these latitudes (see Fig. 29-2). After reaching the desert surface, this warm, dry air flows back toward the equator, absorbing moisture as it travels. Farther north and south, this general circulation pattern is repeated, dropping moisture at around 60° N and 60° S, and creating extremely dry conditions at the North and South Poles.

Ocean Currents Moderate Coastal Climates

Ocean currents are driven by Earth's rotation, by winds, and by direct heating of water by the sun. Continents interrupt the currents, breaking them into roughly circular patterns

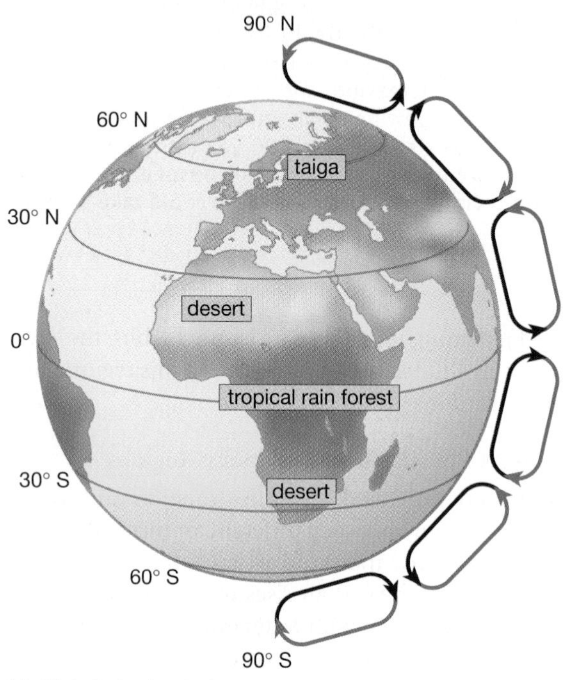

(a) Global air circulation patterns

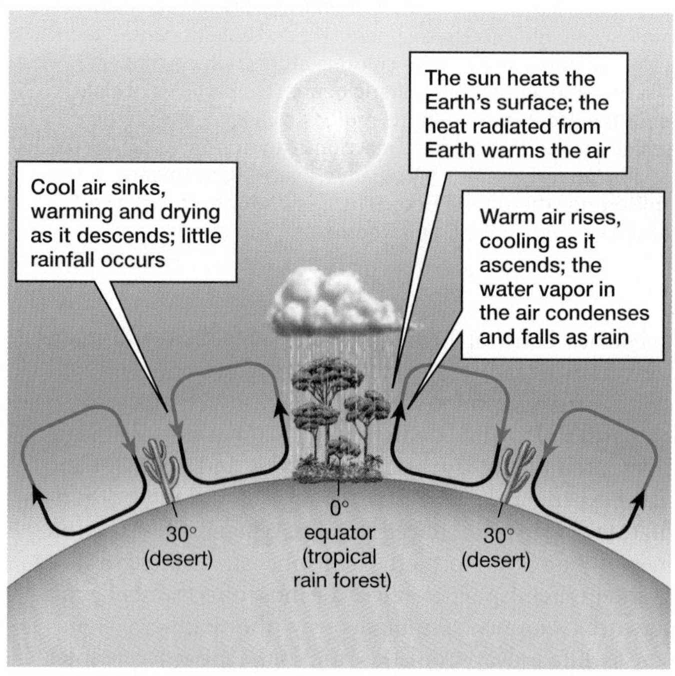

(b) Air circulation affects climate

▲ FIGURE 29-2 Air currents and climatic regions (a) Air currents rise and fall somewhat predictably with latitude. **(b)** Air circulation patterns produce broad climatic regions.

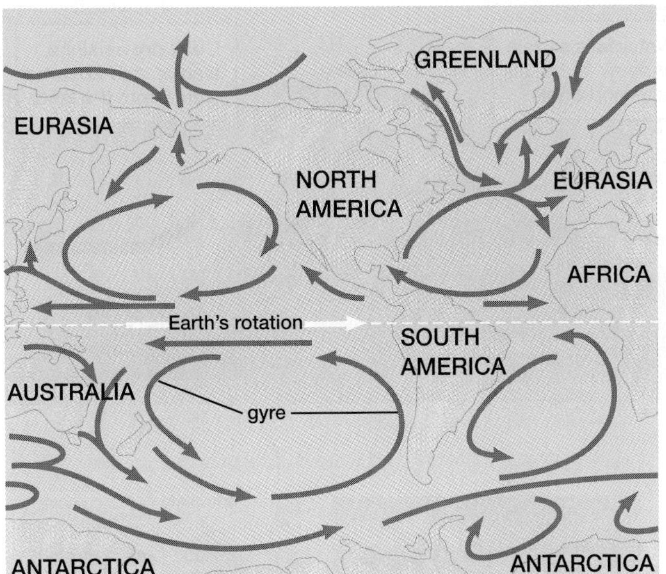

▲ **FIGURE 29-3 Ocean circulation patterns are called gyres** Gyres flow in clockwise circular patterns in the Northern Hemisphere, and counterclockwise in the Southern Hemisphere. These currents tend to distribute warmth from the equator to northern and southern coastal areas.

called **gyres,** which circulate clockwise in the Northern Hemisphere and counterclockwise in the Southern Hemisphere (**Fig. 29-3**). Because water both heats and cools more slowly than does land or air, oceans tend to moderate temperature extremes. Coastal areas, therefore, generally have less-variable climates than do areas near the center of continents. Ocean gyres further modify coastal climates by carrying warm water from equatorial regions to the coastlines of continents both north and south of the equator. This creates warmer, moister coastal climates than are found farther inland.

Continents and Mountains Complicate Weather and Climate

If Earth's surface were uniform, climate zones would occur in bands corresponding to latitude (see Fig. 29-2a). The presence of irregularly shaped continents (which heat and cool relatively quickly) amidst oceans (which heat and cool more slowly than landmasses) alters the flow of wind and water, and contributes to the irregular distribution of ecosystems.

Variations in elevation within continents further complicate the situation. As elevation increases, air becomes thinner and cools. The temperature drops approximately 3.5°F (2°C) for every 1,000 feet (305 meters) gained in elevation, which explains why snowcapped mountains are found even in the Tropics. **Figure 29-4** shows that increasing elevation and increasing latitude have similar effects on terrestrial ecosystems.

Mountains also modify rainfall patterns. When water-laden air is forced to rise as it meets a mountain, it cools. Cooling reduces the air's ability to retain water, and the water condenses as rain or snow on the windward (near) side of the mountain. The cool, dry air is warmed again as it travels down the far side of the mountain and it absorbs water from the land, creating a local dry area called a **rain shadow** (**Fig. 29-5**). For example, the Sierra Nevada range of the western United States wrings moisture from westerly winds blowing off the Pacific Ocean, producing the Mojave Desert in the mountains' rain shadow to the east.

29.2 WHAT CONDITIONS DOES LIFE REQUIRE?

From the lichens on Arctic rocks, to the thermophilic (literally "heat-loving") algae in the hot springs of Yellowstone National Park, to the bacteria thriving under the

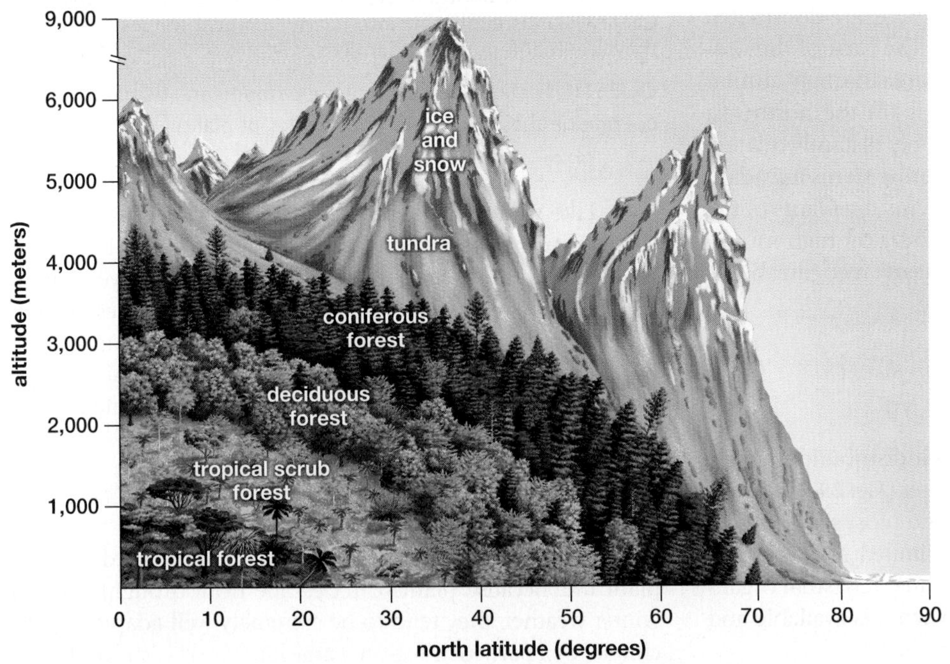

◀ **FIGURE 29-4 Effects of elevation on temperature** Climbing a mountain in the Northern Hemisphere is like heading north; in both cases, increasingly cool temperatures produce a similar series of biomes. Data from Körner, C. and Spehn, E.M. 2002. *Mountain Biodiversity: A Global Assessment.* New York, London. Parthenon Publishers.

Water vapor is carried from the ocean by the prevailing winds
Water falls as rain or snow as the air rises and cools
Cool, dry air sinks, warms, and absorbs water from the land
moist climate
dry climate in the rain shadow

▶ **FIGURE 29-5 Mountains create rain shadows**

pressure-cooker conditions of a deep-sea vent, to the tremendous diversity of ocean coral reefs and tropical rain forests, Earth teems with life. These habitats all share the ability to provide, to varying degrees, four fundamental requirements for life:

- Nutrients from which to construct living tissue.
- Energy to power metabolic activities.
- Liquid water to serve as a medium in which metabolic activities occur.
- Appropriate temperatures at which to carry out these processes.

As we will see in the following sections, these necessities are unevenly distributed over Earth's surface, limiting the types of organisms that can exist within the various terrestrial and aquatic ecosystems on Earth.

Communities within the various ecosystems are extraordinarily diverse, yet clear patterns emerge. Variations in temperature and in the availability of light, water, and nutrients shape the adaptations of organisms that inhabit each ecosystem. Desert communities, for example, are dominated by plants adapted to heat and drought. The cacti of the Mojave Desert in the American Southwest are strikingly similar to the euphorbias of the Canary Islands off the northwest African coast, although these plants are only distantly related. Their reduced leaves and thick, water-storing stems are adaptations for dry climates (**Fig. 29-6**). Likewise, the plants of the arctic tundra and those of the alpine tundra on high mountaintops show growth patterns that are clearly recognizable as adaptations to a cold, dry, windy climate.

(a) Cactus

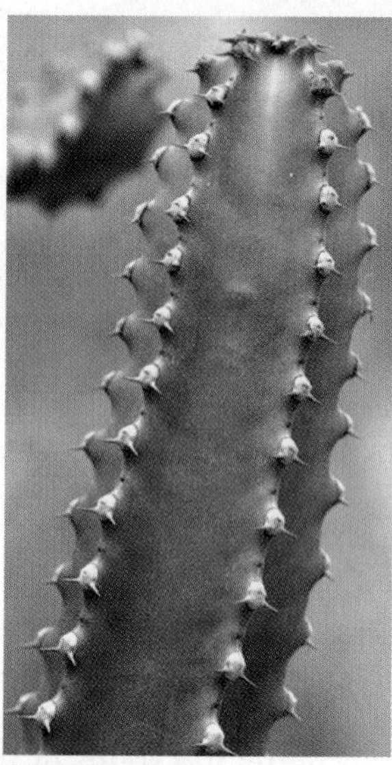
(b) Euphorbia

▲ **FIGURE 29-6 Environmental demands mold physical characteristics** Evolution in response to similar environments has molded the bodies of **(a)** this American cactus, and **(b)** this Canary Islands euphorbia into nearly identical shapes, although they are in different plant families.

QUESTION Describe the similar environmental demands that operate on these two different families of plants.

29.3 HOW IS LIFE ON LAND DISTRIBUTED?

Terrestrial organisms are restricted in their distribution largely by temperature and the availability of water (**Fig. 29-7**). Terrestrial ecosystems receive plenty of light, and most soils provide adequate nutrients. Water, however, is limited and very unevenly distributed, both in place and in time. Terrestrial organisms must be adapted to obtain water when it is available and to conserve it when it is scarce.

Like water, temperatures favorable to life are also unevenly distributed in place and time. At the South Pole, even in summer, the average temperature is usually well below freezing; not surprisingly, life is scarce there. Places such as central Alaska have favorable temperatures for plant growth only during their brief summer, whereas the Tropics have a uniformly warm, moist climate in which life abounds.

Terrestrial Biomes Support Characteristic Plant Communities

Terrestrial communities are dominated and defined by their plant life. Because plants can't escape from drought, sun, or winter weather, they tend to be extremely well adapted to the climate of a particular region. Large land areas with similar en-

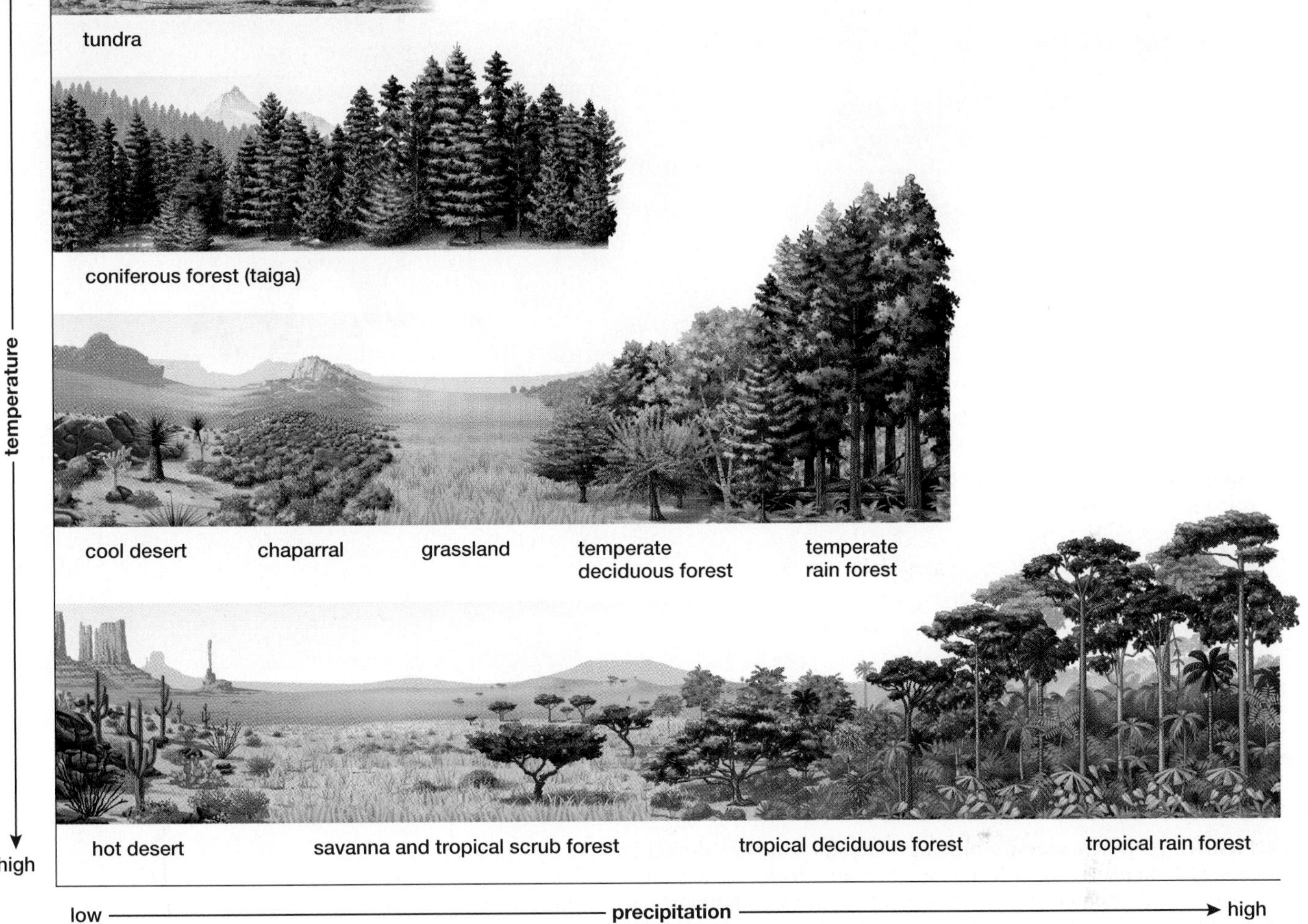

▲ **FIGURE 29-7 Rainfall and temperature influence the distribution of terrestrial ecosystems** Together, rainfall and temperature determine the soil moisture available to support plants. Look for ecosystems adapted to extreme conditions at the three corners of the diagram.

vironmental conditions and characteristic plant communities are called **biomes** (**Fig. 29-8**), which are generally named after the major type of vegetation found there.

The predominant vegetation of each biome is determined by the complex interplay of rainfall and temperature (see Fig. 29-7). These factors determine the soil moisture available to plants for metabolic activities and to replace water that evaporates from their leaves. In addition to the overall average rainfall and temperature, the way that these factors vary seasonally determines which plants can grow in a given region. Arctic tundra plants, for example, must be adapted to marshy conditions in the early summer, when the snow melts, but also to cold and extremely dry conditions for much of the rest of the year, when water is frozen solid and unavailable.

In the following sections, we discuss the major biomes, beginning at the equator and working our way poleward. We also discuss some of the impacts of human activities on these biomes. You will learn more about human impacts on Earth in Chapter 30.

Tropical Rain Forests

Near the equator, the temperature averages between 77 °F and 86 °F (25 °C and 30 °C) with little variation, and rainfall ranges from 100 to 160 inches (250 to 400 centimeters) annually. These evenly warm, evenly moist conditions combine to create the most productive biome on Earth, the **tropical rain forest**, dominated by huge broadleaf evergreen trees (**Fig. 29-9**). Extensive rain forests are found in Central and South America, Africa, and Southeast Asia.

Biodiversity refers to the total number of species within a given region (see pp. 582–583). Rain forests have the highest biodiversity of any ecosystem on Earth. Although rain forests cover only 6% of Earth's total land area, ecologists estimate that they are home to 5 to 8 million species, representing half to two-thirds of the world's biodiversity. For example, in a 3-square-mile (about 5-square-kilometer) tract of rain forest in Peru, scientists counted more than 1,300 butterfly species and 600 bird species. For comparison, the entire United States is home to only 400 butterfly species and 700 bird species.

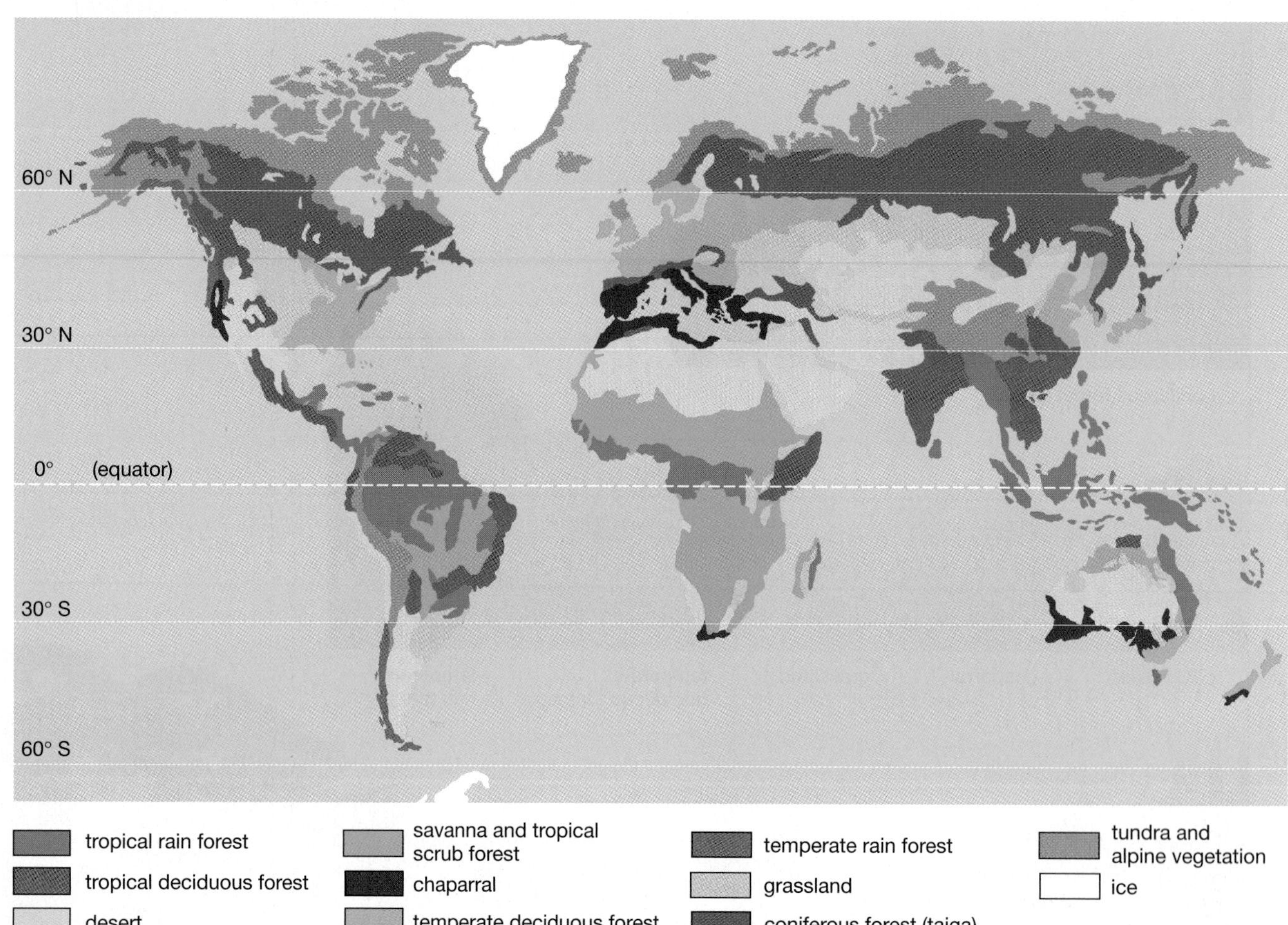

▲ **FIGURE 29-8 The world's biomes** Although mountain ranges and the immense size of continents complicate the distribution of biomes, look carefully and you will see some overall patterns. Tundras and coniferous forests are in the northernmost parts of the Northern Hemisphere, whereas the deserts of Mexico, the Sahara, Saudi Arabia, South Africa, and Australia are located around 30° N and 30° S latitudes. Tropical rain forests are found near the equator.

Tropical rain forests typically have several layers of vegetation. The tallest trees reach 150 feet (50 meters), towering above the rest of the forest. Below is a fairly continuous canopy of treetops at about 90 to 120 feet (30 to 40 meters). Another layer of shorter trees typically stands below the canopy. Huge woody vines grow up the trees. Collectively, these plants capture most of the sunlight. Many shorter plants have enormous leaves to trap the dim light that filters through to the forest floor.

Because edible plant material close to the ground in tropical rain forests is relatively scarce, much of the animal life—including a huge variety and number of birds, monkeys, and insects—inhabit the trees. Competition for the nutrients that do reach the ground is intense among both plants and animals. For example, researchers collecting monkey droppings found that hundreds of dung beetles converged on the droppings within minutes after they hit the ground. Almost as soon as soil bacterial and fungal decomposers release any nutrients from dead plants or animals, the living plants absorb them. Almost all of the nutrients in a rain forest are stored in the vegetation, leaving the soil relatively infertile and thin.

Human Impact Because of infertile soil and heavy rains, agriculture is risky and destructive in rain forests. If the trees are cut and carried away for lumber, few nutrients remain to support crops. If the trees are burned, thus releasing nutrients into the soil, the heavy year-round rainfall quickly dissolves the nutrients and carries them away, leaving the soil depleted after only a few seasons of cultivation.

Nevertheless, rain forests are being felled for lumber or burned down for ranching or farming at an alarming rate (**Fig. 29-10**). The demand for biofuels (fuels produced from biomass, including palm and soybean oil) is driving rapid destruction of rain forests to grow these crops. Estimates place annual rain-forest destruction at about 50,000 square miles—10 times the area of Connecticut, or a region the size of a football field felled *every second*. About half of the world's total rain forests are now gone. This destruction is cause for concern because rain forests harbor a substantial portion of Earth's remaining biodiversity. In addition, like all forests, rain forests absorb carbon dioxide and release oxygen. Researchers estimate that roughly 15% of the carbon dioxide released into the atmosphere comes from cutting and burning tropical rain forests, exacerbating global warming.

▲ **FIGURE 29-9 The tropical rain-forest biome** Towering trees draped with vines reach for the light in the dense tropical rain forest. Amid their branches dwells the most diverse assortment of life on Earth, including (clockwise from upper left) a tree-dwelling orchid, a leaf frog, and a fruit-eating toucan.

QUESTION How does a biome with such poor soil support the highest plant productivity and the greatest animal diversity on Earth?

▲ **FIGURE 29-10 Amazon rain forest cleared by burning** The burned area will be converted to ranching or agriculture, but both are doomed to fail eventually due to poor soil quality. Spreading fires and the smoke they produce also threaten adjacent forests and their myriad inhabitants.

Fortunately, some areas have been set aside as protected preserves, and some reforestation efforts are under way, with local residents becoming more involved in conservation efforts. Look for more information about loss of tropical rain forests and their biodiversity in Chapter 30.

Case Study continued
The Birds and the Beans

In 2008, a group representing farmers, scientists, cocoa industry leaders, and 14 African nations announced the first international sustainable cocoa farming plan for Africa, whose 2 million cocoa farmers produce 70% of the world's cocoa. The 30-year plan outlines specific steps to provide modern technology to cocoa farmers, helping increase their annual production and income. Some of the money obtained through the sale of the cocoa will go toward improving social services and toward rehabilitating the environment. The coalition plans for a widespread system of cocoa cropping that integrates the prevalent monoculture farms with mosaics of natural rain forests, and with rain forests that have been modified to sustainably support agriculture.

Tropical Deciduous Forests

Slightly farther from the equator, the rainfall is not nearly as constant, and there are pronounced wet and dry seasons. In these areas, which include much of India as well as parts of Southeast Asia, South America, and Central America, **tropical deciduous forests** grow. During the dry season, the trees cannot get enough water from the soil to compensate for evaporation from their leaves. As a result, the plants have adapted to the dry season by shedding their leaves, thereby minimizing water loss. If the rains fail to return on schedule, the trees delay the formation of new leaves until the drought ends.

Tropical Scrub Forests and Savannas

Along the edges of the tropical deciduous forest, reduced rainfall produces the **tropical scrub forest** biome, dominated by deciduous trees that are shorter and more widely spaced than in tropical deciduous forests. These trees often bear thorns, and grasses grow beneath them. Still farther from the equator, the climate grows drier, and grasses become the dominant vegetation, with only scattered trees; this biome is the **savanna** (**Fig. 29-11**).

Savanna grasslands typically have a rainy season during which virtually all of the year's scant precipitation falls—12 inches (30 centimeters) or less. When the dry season arrives, rain might not fall for months, and the soil becomes hard, dry, and dusty. Grasses are well adapted to this type of climate, growing very rapidly during the rainy season and dying back to drought-resistant roots during the arid times. Only a few specialized trees, such as the thorny acacia or the water-storing baobab, can survive the devastating savanna dry seasons. The African savanna has the most diverse and impressive array of large mammals on Earth. These mammals include numerous herbivores—such as antelope, wildebeest, water buffalo, elephants, and giraffes—and carnivores such as lions, leopards, hyenas, and wild dogs.

Human Impact Africa's rapidly expanding human population threatens the wildlife of the savanna. Poaching has driven the black rhinoceros to the brink of extinction (**Fig. 29-12**) and endangers the African elephant, a keystone species in this ecosystem. The abundant grasses that make the savanna a suitable habitat for so much wildlife also make it suitable for grazing domestic cattle. Fences erected to contain the cattle increasingly disrupt the migration of the great herds of wild herbivores as they search for food and water.

Deserts

Even drought-resistant grasses need at least 10 to 20 inches (25 to 50 centimeters) of rain a year, depending on the precipitation's seasonal distribution and the average temperature. Biomes where annual rainfall is 10 inches or less are called **deserts.**

Although we tend to think of them as hot, deserts are defined by their lack of rain rather than by their temperatures. In

▲ **FIGURE 29-11 The African savanna** Giraffes feed on savanna trees and share this biome with (clockwise from left) predatory lions, herds of zebra, and rare black rhinos.

▲ **FIGURE 29-12 Poaching threatens African wildlife** Rhinoceros horns, used for the handles of daggers and in traditional Asian medicines, fetch enormous prices and encourage poaching, even though the horns are composed primarily of keratin, a protein similar to that in fingernails. The black rhino is now nearly extinct.

the Gobi Desert of Asia, for example, temperatures average below freezing for half the year, while summer temperatures average 105 °F to 110 °F (41 °C to 43 °C). Desert biomes are found on every continent, typically at around 30° N and 30° S latitudes, and also in the rain shadows of major mountain ranges.

Deserts include a variety of environments. At one extreme are the Atacama Desert in Chile and parts of the Sahara Desert in Africa, where it almost never rains and no vegetation grows (**Fig. 29-13a**). More commonly, deserts are characterized by widely spaced vegetation and large areas of bare ground (**Fig. 29-13b**).

Cacti exhibit several adaptations for the dry desert environment. They have shallow spreading roots that absorb rainwater before it can evaporate from the soil surface. The thick stems of cacti and other succulents (plants with thick, fleshy stems or leaves) store water when it is available, for use during droughts. The spines of cacti are leaves modified to protect the cactus from predators, and to conserve water, presenting almost no surface area for evaporation. A waterproof, waxy coating on cactus stems (and on the leaves of many other desert plants) further reduces water loss. Deserts may get their annual rainfall in just a few storms, and specialized annual wildflowers take advantage of the brief period of moisture to sprout from seed, grow, flower, and produce seeds of their own in a month or less (**Fig. 29-14**).

Desert animals are also adapted to survive heat and drought. Few animals are seen during hot summer days. Many desert denizens take refuge from the heat of summer days in underground burrows that are relatively cool and moist. In North American deserts, nocturnal (night-active) animals include jackrabbits, bats, burrowing owls, and kangaroo rats (**Fig. 29-13c**). Reptiles such as snakes, turtles, and lizards adjust their activity cycles depending on the temperature. In summer, they may be active only around dawn and dusk. Kangaroo rats and many other small desert animals survive without ever drinking. They acquire water from their food and from water produced as a by-product of cellular respiration in their tissues. Larger animals, such as desert bighorn sheep, are dependent on permanent water holes during the driest times of the year.

Human Impact Desert ecosystems are fragile. Ecologists studying the soil of the Mojave Desert in southern California found tread marks left by tanks during army exercises in 1940. The desert soil is stabilized and enriched by microscopic cyanobacteria whose filaments intertwine among sand grains. The tanks, and now off-road vehicles that careen about the desert for recreation, destroy this crucial network. This damage allows the soil to erode and reduces nutrients available to the desert's slow-growing plants. Ecologists estimate that desert soil may require hundreds of years to fully recover from heavy vehicle use.

Human activities are also contributing to **desertification**, the process by which relatively dry, drought-prone regions are converted to desert as a result of drought coupled with overuse of the land. This overuse includes overharvesting of bushes and trees for firewood, overgrazing of grasses by livestock, and depletion of surface water and groundwater to grow crops. Loss of vegetation allows soil to erode and droughts to intensify,

(a) Sahara dunes

(b) Utah desert

(c) A kangaroo rat

▲ **FIGURE 29-13 The desert biome (a)** Under the most extreme conditions of heat and drought, deserts can be almost devoid of life, such as these sand dunes of the Sahara Desert in Africa. **(b)** Throughout much of Utah and Nevada, the Great Basin Desert presents a landscape of widely spaced shrubs, such as sagebrush and greasewood. **(c)** The kangaroo rat is an elusive inhabitant of North American deserts.

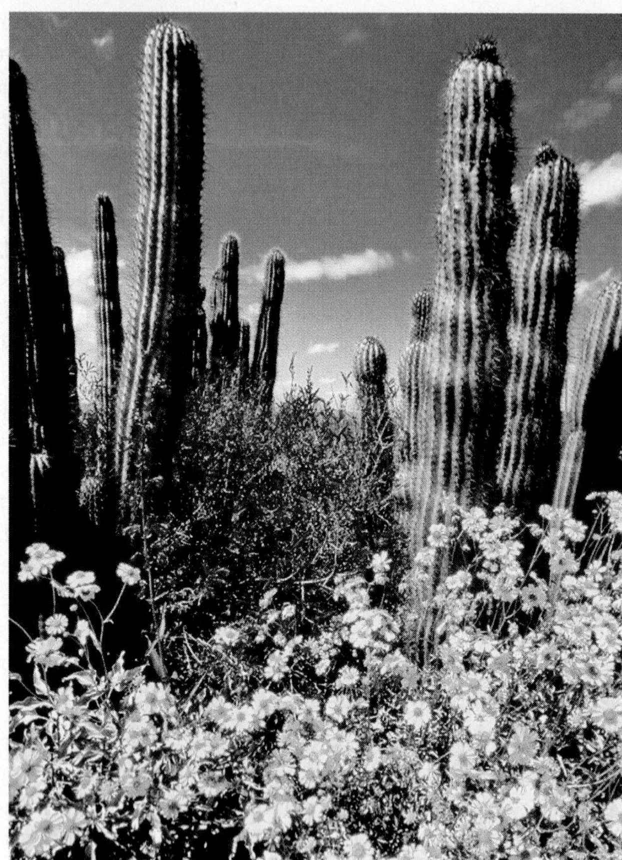

▲ **FIGURE 29-14 The Sonoran Desert** After a relatively wet spring, this Arizona desert is carpeted with wildflowers. Through much of the year—and sometimes for several years—annual wildflower seeds lie dormant, waiting for adequate spring rains to fall.

further decreasing the land's productivity. An example of desertification is found in the Sahel region just south of the Sahara Desert in Africa, which has been significantly overgrazed and degraded as its rapidly growing human population attempts to produce adequate food (**Fig. 29-15**).

▲ **FIGURE 29-15 Desertification** A rapidly growing human population coupled with drought and poor land use has reduced the ability of many dry regions, such as the Sahel in Africa, to support life.

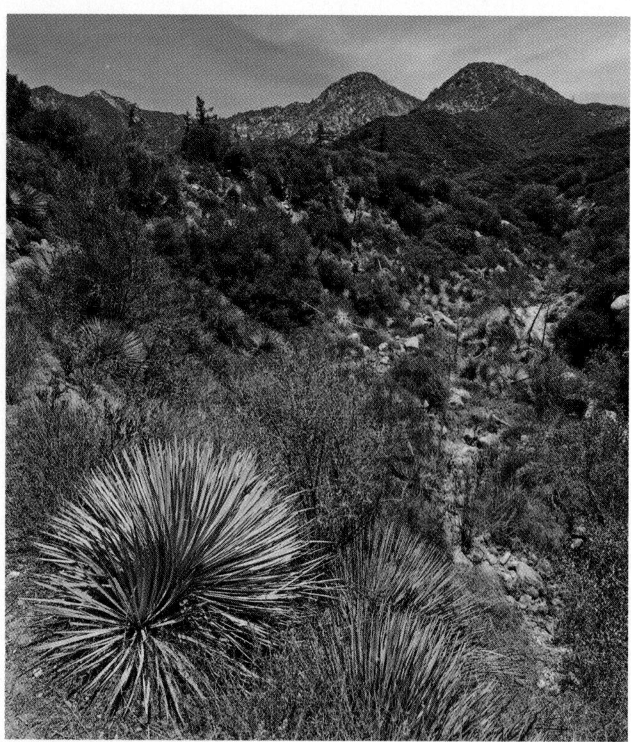

▲ **FIGURE 29-16 The chaparral biome** Limited to coastal regions and maintained by fires set by lightning, this hardy biome is characterized by drought-resistant shrubs and small trees, such as these seen in the western foothills of the San Gabriel mountains in Southern California.

Chaparral

Many coastal regions that border on deserts, such as in southern California and much of the Mediterranean, support a biome called **chaparral.** The annual rainfall in this biome is up to 30 inches (76 centimeters), nearly all of which falls during cool, wet winters. Summers are hot and dry. Chaparral plants consist mainly of drought-resistant shrubs and small trees (**Fig. 29-16**). Their leaves are usually small, and are often coated with tiny hairs or protective layers that reduce evaporation during the dry summer months. Chaparral is adapted to fire. Some of its hardy shrubs regrow from their roots after summer fires started by lightning. Others have seeds that are stimulated to germinate by the compounds in smoke.

Grasslands

In the United States, deserts occur in the rain shadows east of the Sierra Nevada and Rocky Mountains. Eastward, as the rainfall gradually increases, the land supports more and more grasses, giving rise to the prairies of the Midwest. Most **grassland,** or **prairie,** biomes are located in the centers of continents, such as North America and Eurasia, and receive 10 to 30 inches (25 to 75 centimeters) of rain annually. In general, they have a continuous cover of grass and virtually no trees, except along rivers.

In tallgrass prairie—found in Iowa, Missouri, and Illinois—grasses often reach up to 6 feet in height, with an acre of natural tallgrass prairie in the United States supporting 200 to 400 different species of native plants (**Fig. 29-17**). Shortgrass prairies, which receive less rainfall than tallgrass prairies, occur

▲ **FIGURE 29-17 Tallgrass prairie** In the central United States, moisture-bearing winds out of the Gulf of Mexico produce summer rains, allowing a lush growth of tall grasses and abundant wildflowers. Periodic fires, now carefully managed, prevent the encroachment of forest and maintain this subclimax biome.

QUESTION Why is tallgrass prairie one of the most endangered biomes in the world?

▲ **FIGURE 29-18 Shortgrass prairie** Shortgrass prairie is characterized by low-growing grasses. In addition to many wildflowers, life on the prairie includes (clockwise from left) bison (in preserves), pronghorns, and prairie dogs.

further west, extending from mid-Texas north into Canada (**Fig. 29-18**). Prairie dogs aerate the soil with their burrows, and provide food for eagles, foxes, coyotes, and bobcats. Pronghorns are seen in western U.S. grasslands, and bison survive in preserves. (A grassland food web is illustrated in Fig. 28-5.)

Why do grasslands lack trees? Water and fire are the crucial factors in the competition between grasses and trees. The hot, dry summers and frequent droughts of the shortgrass prairies can be tolerated by grass but are fatal to trees. In the more eastern tallgrass prairies, forests are the climax ecosystems. Historically, however, the trees were destroyed by frequent fires caused by lightning or deliberately set by Native Americans to maintain grazing land for bison. Although fire kills trees, the root systems of grasses survive.

Human Impact Grasses growing and decomposing for thousands of years produced the most fertile soil in the world. In the early nineteenth century, North American grasslands supported an estimated 60 million bison. Today, the midwestern U.S. prairies have been largely converted to farm and range land, and cattle have replaced the bison, which were driven to near extinction by overhunting.

Prairie dog colonies and the eagles and ferrets that hunted them have become a rare sight as their habitat has been displaced by ranching, and, more recently, by suburban sprawl. Wolves, once prevalent, have been eliminated from the prairies. In some regions, overgrazing has destroyed the native grasses, allowing woody sagebrush to flourish (**Fig. 29-19**). Undisturbed prairies are now largely confined to protected areas, and tallgrass prairie is now one of the most endangered ecosystems in the world. Only about 1% remains, in tiny remnants restored using native plant species and maintained by controlled burning.

Temperate Deciduous Forests

At their eastern edge, the North American grasslands merge into the **temperate deciduous forest** biome, also found in much of Europe and eastern Asia. More precipitation occurs in temperate deciduous forests than in grasslands (30 to 60 inches, or 75 to 150 centimeters). The soil retains enough moisture for trees to grow, shading out most grasses (**Fig. 29-20**).

Winters in the temperate deciduous forests often have long periods of below-freezing weather, when liquid water is not available. To reduce evaporation when water is in short supply,

▲ **FIGURE 29-19 Sagebrush desert or shortgrass prairie?** Biomes are influenced by human activities as well as by temperature, rainfall, and soil. The shortgrass prairie field on the right has been overgrazed by cattle, causing the grasses to be replaced by sagebrush.

▲ **FIGURE 29-20 The temperate deciduous forest biome** Temperate deciduous forests of the eastern United States are inhabited by (clockwise from left) white-tailed deer (this biome's largest herbivore) and birds such as this blue jay. In spring, a profusion of woodland wildflowers (such as these hepaticas) blooms briefly before the trees produce shading leaves.

the trees drop their leaves in the fall and remain dormant through the winter. During the brief time in spring when the ground has thawed but the emerging leaves have not yet blocked off all sunlight, abundant wildflowers grace the forest floor.

Insects and other arthropods are numerous and conspicuous in deciduous forests. The decaying leaf litter on the forest floor also provides food and habitat for bacteria, earthworms, fungi, and small plants. A variety of vertebrates—including mice, shrews, squirrels, raccoons, deer, bears, and many species of birds—dwell in the deciduous forests.

Human Impact Large predatory mammals such as black bears, wolves, bobcats, and mountain lions were formerly abundant in the eastern United States, but hunting and habitat loss have severely reduced their numbers; wolves have been completely eliminated. Deer thrive due to a lack of natural predators. Clearing for lumber, agriculture, and housing has dramatically reduced deciduous forests in the United States from their original extent, and virgin deciduous forests are almost nonexistent.

Temperate Rain Forests

On the U.S. Pacific Coast, from the lowlands of the Olympic Peninsula in Washington State to southeast Alaska, lies a **temperate rain forest** biome (**Fig. 29-21**), which supports populations of black bear, elk, and spotted owls. Temperate rain forests, which are relatively rare, are also located along the southeastern coast of Australia, the southwestern coast of New Zealand, and parts of Chile and Argentina. As in the tropical rain forest, temperate rain forests experience a tremendous amount of rain. In North America, these biomes typically receive more than 55 inches (140 centimeters) of rain annually, and the nearby ocean keeps the temperature moderate.

The abundance of water means that the trees have no need to shed their leaves in the fall, and almost all the trees are evergreens. In contrast to the broadleaf evergreen trees of the tropical rain forests, temperate rain forests are dominated by coniferous (literally "cone-bearing") trees. The forest floor and tree trunks are typically covered with mosses and ferns. Fungi thrive in the moisture and enrich the soil. As in tropical rain forests, so little light reaches the forest floor that tree seedlings usually cannot become established. Whenever one of the forest giants falls, however, it opens up a patch of light, and new seedlings quickly sprout, often right atop the fallen log.

Taiga

North of the grasslands and temperate forests stretches the **taiga** (also called the **northern coniferous forest; Fig. 29-22**). The taiga, which is the largest terrestrial biome on Earth, stretches across North America, Scandinavia, and Siberia, nearly encircling the globe. It includes parts of Alaska and the northern United States, and much of southern Canada (see Fig. 29-8).

Conditions in the taiga are much harsher than in temperate deciduous forests, with long, cold winters and short growing seasons. About 16 to 40 inches (40 to 100 centimeters) of precipitation occurs annually here, much of it as snow. Trees in the taiga are mostly conifers, such as spruce and fir. Their conical shape and narrow, stiff needles allow them to

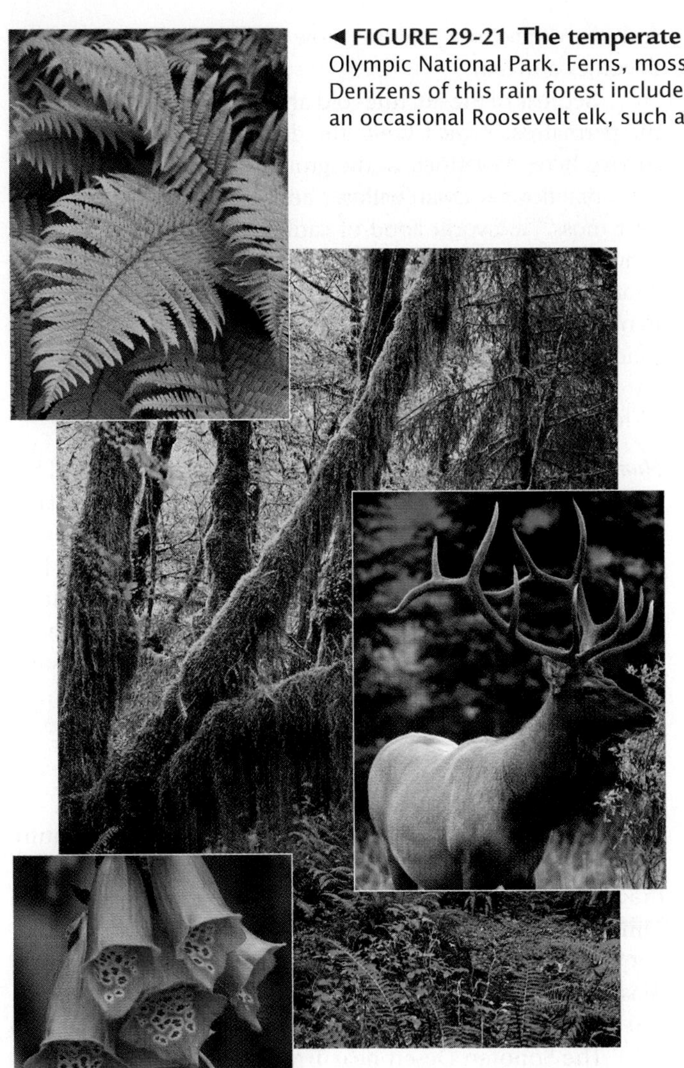

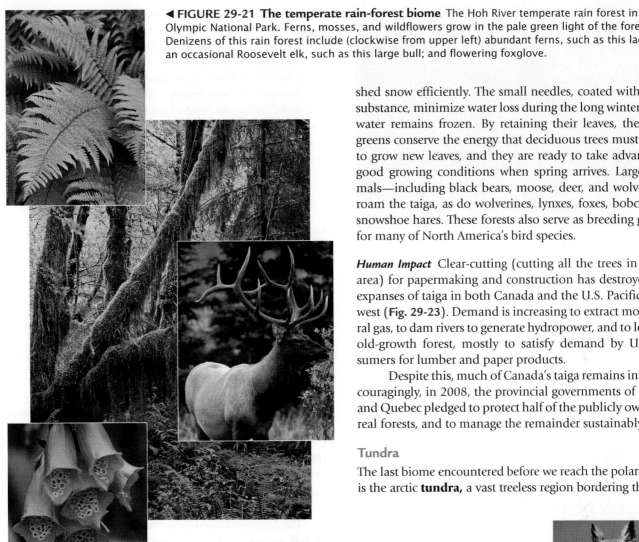

◀ **FIGURE 29-21 The temperate rain-forest biome** The Hoh River temperate rain forest in Olympic National Park. Ferns, mosses, and wildflowers grow in the pale green light of the forest floor. Denizens of this rain forest include (clockwise from upper left) abundant ferns, such as this lady fern; an occasional Roosevelt elk, such as this large bull; and flowering foxglove.

shed snow efficiently. The small needles, coated with a waxy substance, minimize water loss during the long winters, when water remains frozen. By retaining their leaves, these evergreens conserve the energy that deciduous trees must expend to grow new leaves, and they are ready to take advantage of good growing conditions when spring arrives. Large mammals—including black bears, moose, deer, and wolves—still roam the taiga, as do wolverines, lynxes, foxes, bobcats, and snowshoe hares. These forests also serve as breeding grounds for many of North America's bird species.

Human Impact Clear-cutting (cutting all the trees in a given area) for papermaking and construction has destroyed huge expanses of taiga in both Canada and the U.S. Pacific Northwest (**Fig. 29-23**). Demand is increasing to extract more natural gas, to dam rivers to generate hydropower, and to log more old-growth forest, mostly to satisfy demand by U.S. consumers for lumber and paper products.

Despite this, much of Canada's taiga remains intact. Encouragingly, in 2008, the provincial governments of Ontario and Quebec pledged to protect half of the publicly owned boreal forests, and to manage the remainder sustainably.

Tundra

The last biome encountered before we reach the polar ice caps is the arctic **tundra,** a vast treeless region bordering the Arctic

▶ **FIGURE 29-22 The taiga (northern coniferous forest) biome** The small needles and pyramidal shape of conifers allow them to shed heavy snows. (lower left) A Canada lynx catches a snowshoe hare. (upper right) A great horned owl watches for prey.

▲ **FIGURE 29-23 Clear-cutting** The taiga is vulnerable to clear-cutting, as seen in this forest in British Columbia. This is a relatively simple and inexpensive means of logging compared to selective harvesting of trees, but its environmental costs are high. Erosion diminishes the fertility of the soil, slowing new growth. Further, the dense stands of similarly aged trees that typically regrow are more vulnerable to attack by parasites than a natural stand of trees of various ages would be.

Ocean (**Fig. 29-24**). Conditions in the tundra are severe. Winter temperatures often reach −40°F (−55°C) or below, with howling winds. Precipitation averages 10 inches (25 centimeters) or less each year, making this region a "freezing desert." Even during the summer, the temperature can drop to freezing, and the growing season may last only a few weeks.

The cold climate of the arctic tundra results in **permafrost,** a permanently frozen layer of soil. Soil above the permafrost thaws each summer, usually to a depth of 2 feet (60 centimeters) or more. When the summer thaws arrive, the underlying permafrost limits the ability of soil to absorb the water from melting snow and ice, and so the tundra becomes a marsh.

Because of the extreme cold, the brief growing season, and the permafrost, which limits the depth of roots, trees cannot survive here. Nevertheless, the ground is carpeted with small perennial flowers, dwarf willows, and large lichens called "reindeer moss," a favorite food of caribou. The summer marshes provide superb mosquito habitat. These and other insects feed about 100 different species of birds, most of which migrate here to nest and raise their young during the brief summer feast. The tundra vegetation also supports lemmings, small rodents that are eaten by other tundra residents, including wolves, owls, arctic foxes, and grizzly bears.

Human Impact The tundra is among the most fragile of all the biomes because of its short growing season. A willow 4 inches (10 centimeters) high may be 50 years old. Alpine tundra is readily damaged by off-road vehicles and even hikers. Human activities in the arctic tundra can leave scars that persist for centuries. Fortunately for the arctic tundra inhabitants, the impact of civilization is localized around oil-drilling sites, pipelines, mines, and scattered military bases.

Rainfall and Temperature Limit the Plant Life of a Biome

Terrestrial biomes are greatly influenced by both temperature and rainfall (see Fig. 29-7). Because temperature influences the availability of water, areas that receive almost exactly the same rainfall can have remarkably different vegetation. For example, let's travel from southern Arizona to central Alaska, visiting ecosystems that each receive about 12 inches (30 centimeters) of rain annually.

The Sonoran Desert near Tucson, Arizona (see Fig. 29-14), has an average annual temperature of 68°F (20°C). The

► **FIGURE 29-24 The tundra biome** Life on the tundra is seen here in Denali National Park, Alaska, turning rich colors in the fall. (clockwise from left) Tundra animals, such as this caribou and arctic fox, can regulate blood flow in their legs, keeping them just warm enough to prevent frostbite while preserving precious body heat for the brain and other vital organs. Perennial plants such as this frost-covered bearberry grow low to the ground, avoiding the chilling tundra wind.

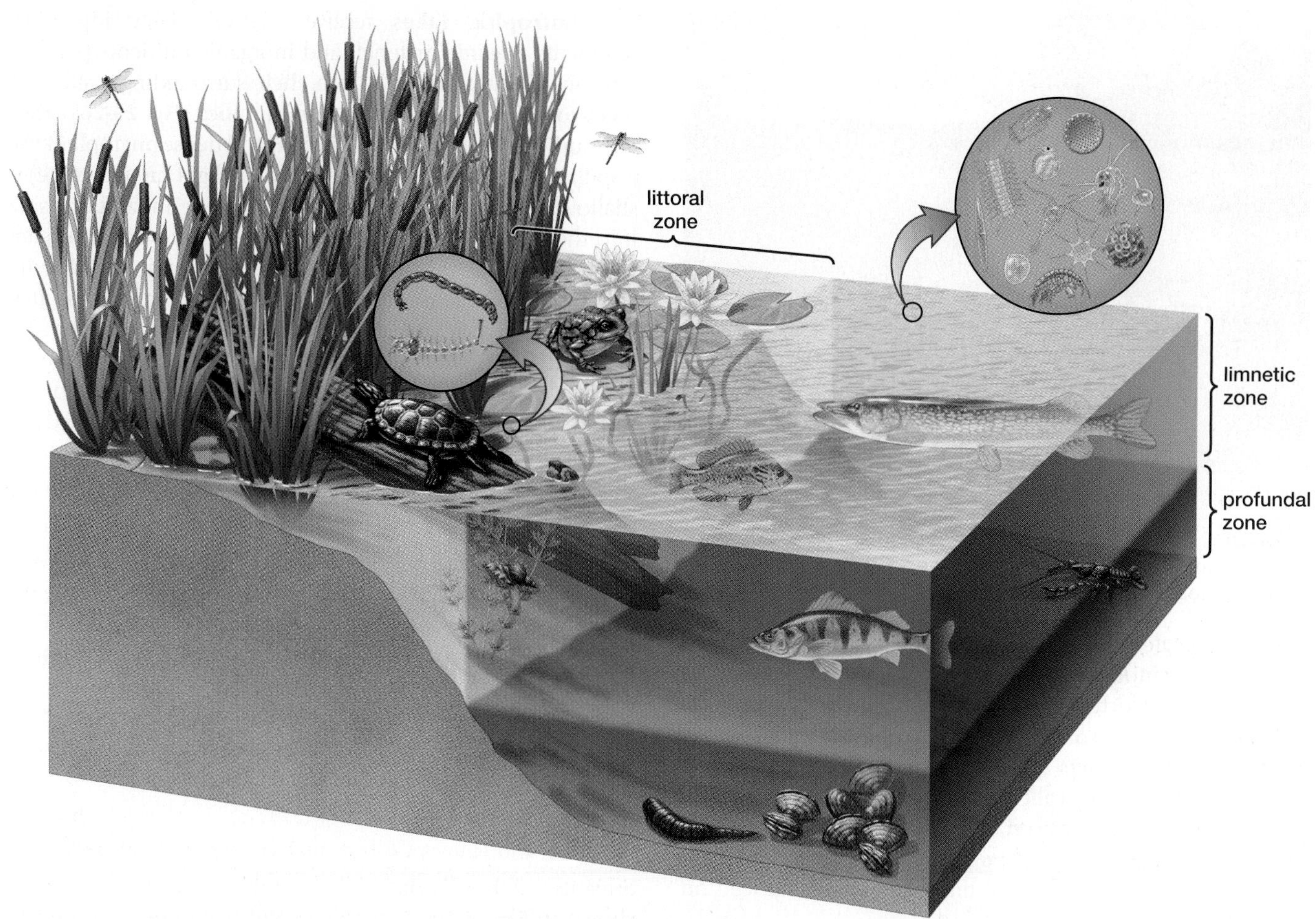

▲ **FIGURE 29-25 Lake life zones** A typical large lake has three life zones: a nearshore littoral zone with rooted plants, an open-water limnetic zone, and a deep, dark profundal zone.

landscape is dominated by giant saguaro cactus and low-growing, drought-resistant bushes. About 900 miles (1,500 kilometers) north, in eastern Montana, you will find short-grass prairie (see Fig. 29-18), largely because the average temperature is much lower—about 45°F (7°C). In the far north, central Alaska receives about the same annual rainfall but is covered with coniferous forest (see Fig. 29-22). As a result of the low average annual temperature (about 25°F, or −4°C), permafrost underlies some of the ground there. During the summer thaw, the taiga earns its Russian name "swamp forest," although its rainfall is about the same as in the Sonoran Desert.

29.4 HOW IS LIFE IN WATER DISTRIBUTED?

Of the four requirements for life, aquatic ecosystems provide abundant water and appropriate temperatures. Because water is slower to heat and cool than air, the temperatures of aquatic ecosystems are more moderate and vary less than those of terrestrial biomes. Light in aquatic ecosystems decreases with depth as it is absorbed by water and blocked by suspended particles. Nutrients in aquatic ecosystems tend to be concentrated in sediments near the bottom, so where nutrients are highest, the light levels are lowest.

Freshwater Ecosystems Include Lakes, Rivers, and Wetlands

The distribution, quantity, and type of life in a lake, river, or wetland largely depend on access to two limiting resources: light and nutrients. Lakes and rivers vary tremendously in depth, which affects both light and nutrient distribution. In contrast, shallow, nutrient-rich wetlands support a large diversity of life.

Large Lakes Have Distinct Life Zones

Freshwater lakes form when natural depressions fill with water from sources including groundwater seepage, streams, or runoff from rain or melting snow. Some lake beds were excavated by glaciers millennia ago, and others formed when landslides or debris deposited by slow-flowing rivers dammed up water behind them.

Large lakes in temperate climates have distinct zones of life (**Fig. 29-25**). Near the lake shore is a shallow **littoral zone,** where plants find both abundant light and nutrients from bottom sediments. Littoral zone communities are the most diverse parts of lakes. They support plants such as cattails, bullrushes, and water lilies, which anchor themselves to the bottom near the shore, and submerged plants and algae that flourish in deeper littoral waters.

▲ **FIGURE 29-26 A eutrophic lake** Rich in dissolved nutrients carried from the land, eutrophic lakes support dense growths of algae, phytoplankton, and both floating and rooted plants.

The greatest diversity of animal life is also found in the littoral zone. Littoral vertebrates include frogs, aquatic snakes, turtles, and fish (such as pike, bluegill, and perch); invertebrates include insect larvae, snails, flatworms, and crustaceans (such as crayfish). Littoral waters are also home to small, unattached organisms collectively called **plankton.** The two forms of plankton are **phytoplankton,** which includes photosynthetic protists and bacteria, and **zooplankton,** which includes nonphotosynthetic protists and tiny crustaceans that feed on phytoplankton.

As the water increases in depth, plants are unable to anchor to the bottom and still photosynthesize. This open-water region is divided into an upper limnetic zone and a lower profundal zone (see Fig. 29-25). In the **limnetic zone,** enough light penetrates to support photosynthesis. Here, plankton and fish predominate. Below this layer lies the **profundal zone,** where light is too weak to support photosynthesis. Organisms that inhabit the profundal zone are nourished by organic matter that drifts down from the littoral and limnetic zones, and by sediment washed in from the land. These inhabitants include a few fish that swim freely among the life zones, and detritus feeders and decomposers, such as crayfish, aquatic worms, clams, leeches, and bacteria,

Freshwater Lakes Are Classified According to Their Nutrient Content

Freshwater lakes are sometimes classified according to their nutrient levels as oligotrophic (Greek, "poorly fed") or eutrophic ("well fed"). Many fall in between and can be described as mesotrophic ("middle fed"). Here, we describe the two extremes.

Oligotrophic lakes are very low in nutrients and support relatively little life. Many were formed by glaciers that scraped depressions in bare rock and are fed by mountain streams. Because there is little sediment or microscopic life to cloud the water, oligotrophic lakes are clear, and light penetrates deeply. Fish such as trout, which require well-oxygenated water, thrive here.

Eutrophic lakes receive relatively large inputs of sediments, organic material, and inorganic nutrients (such as phosphates and nitrates) from their surroundings, allowing them to support dense plant communities (**Fig. 29-26**). They are murky from suspended sediment and dense phytoplankton populations, so the limnetic zone (where light can penetrate) is shallower. The dead bodies of limnetic zone inhabitants sink into the profundal zone, where they feed decomposer organisms. The metabolic activities of these decomposers use up oxygen, so that the profundal zone of eutrophic lakes is often very low in oxygen and supports little life.

Although large lakes may persist for millions of years, gradually, as nutrient-rich sediment accumulates, oligotrophic lakes tend to become eutrophic, a process called eutrophication. This same process—operating over geologic time for large lakes—may eventually cause lakes to undergo succession to dry land (see Fig. 27-18).

Human Impact Nutrients carried into lakes from farms, feedlots, sewage, and even fertilized suburban lawns accelerate eutrophication and disrupt normal community interactions. Lake Erie once suffered severe eutrophication due to phosphate-based detergents and runoff from fertilized farm fields, but agreements between the United States and Canada to reduce these pollutants have greatly improved Lake Erie's water quality and have helped protect other Great Lakes as well.

Streams and Rivers Collect and Transport Surface Water

Streams often originate in mountains, the source region shown in **Figure 29-27,** where runoff from rain and melting snow cascades over impervious rock. Little sediment reaches the streams here, phytoplankton is sparse, and the water is clear and cold. Algae adhere to rocks in the streambed, where insect larvae find food and shelter. Turbulence keeps mountain streams well oxygenated, providing a home for trout that feed on insect larvae.

At lower elevations, in the transition region, small side streams or tributaries merge, forming wider, slower-moving streams and small rivers. The water warms slightly, and more sediment is carried in, providing nutrients that allow aquatic plants, algae, and phytoplankton to proliferate. Fish such as black bass, bluegill, and yellow perch (which require less oxygen than trout do) are found here.

As the land becomes lower and flatter, the river warms, widens, and slows, meandering back and forth (see Fig. 29-27). When precipitation or snowmelt is high, the river may flood the surrounding flat land, called a floodplain, depositing a rich layer of sediment over the adjoining terrestrial ecosystem. Side streams carry in nutrient-rich sediment and deposit it in the riverbed. The water becomes murky with sediment and dense populations of phytoplankton. Decomposer bacteria deplete the oxygen in deeper water, but carp and catfish can still thrive where oxygen is relatively low.

Rivers drain into lakes or into other rivers that usually lead ultimately to oceans. Closer to sea level, the land often flattens and the river moves more slowly, depositing its sediment. The sediment interrupts the river's flow, breaking it

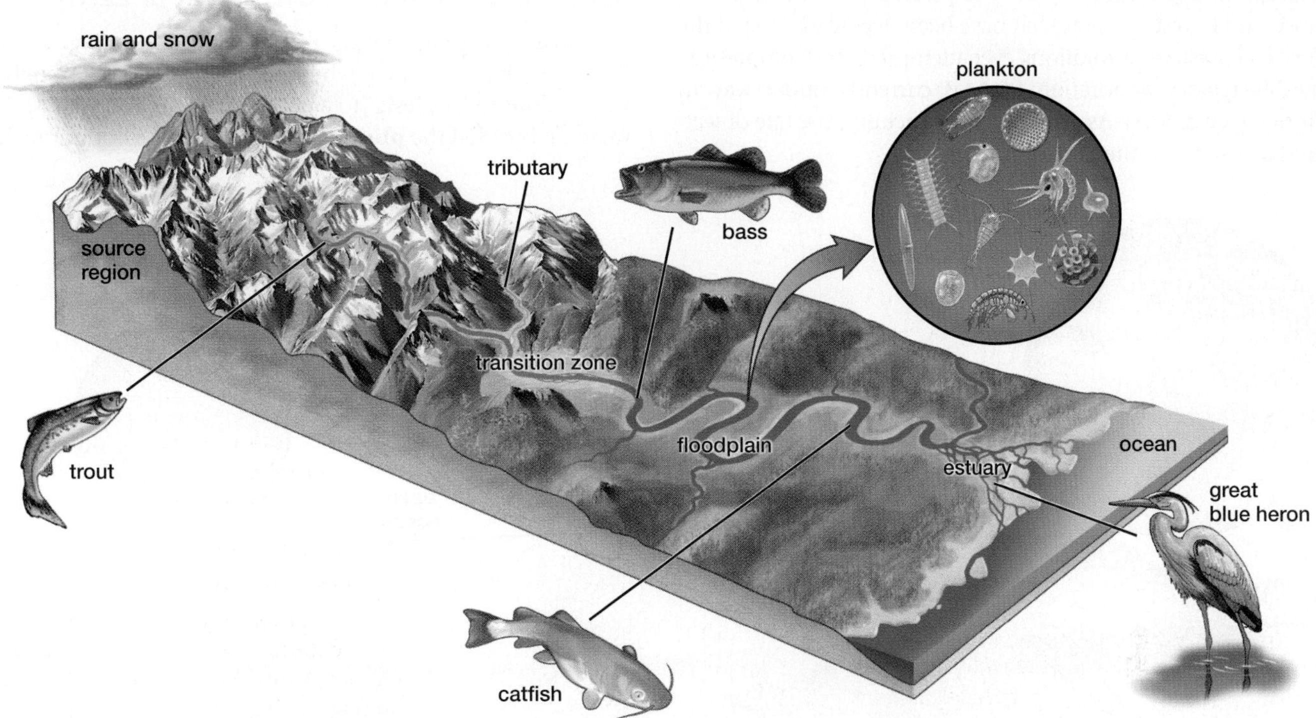

▲ **FIGURE 29-27 From stream to river to sea** At high elevations, precipitation feeds fast-flowing, clear streams, which grow and slow as they are joined by tributaries at lower elevations. Many become slow-moving rivers that weave through a floodplain, dropping rich sediment and creating wetlands that become estuaries where rivers meet the ocean. The communities of freshwater organisms change as the water flows from mountains to the ocean.

into small winding channels that form an estuary (described in greater detail later) as the river empties into the ocean.

Human Impact Rivers have been channelized (deepened and straightened; see Fig. E30-1 on p. 585) to facilitate boat traffic, to prevent flooding, and to allow farming along their banks. This has increased erosion, because water flows more rapidly in channelized rivers. Where natural flooding has been prevented, floodplain soil no longer receives the nutrients that floods formerly deposited there.

In the United States, northwestern Pacific and northeastern Atlantic salmon populations have been greatly reduced by hydroelectric dams, water diversion for agriculture, erosion from logging operations, and overfishing. On both coasts of the United States, state, federal, and local groups are working to restore the clean, free-flowing rivers and streams that support salmon and rich wildlife communities.

Freshwater Wetlands Are Called Marshes, Swamps, or Bogs

Freshwater **wetlands,** also called marshes, swamps, or bogs, are regions where soil is covered or saturated with water. Many aquatic and some terrestrial plant species thrive in wetlands, which support dense growths of algae and phytoplankton, as well as both floating and rooted plants, including water-tolerant grasses.

Freshwater wetlands are among the most productive ecosystems in North America. Many occur around the margins of lakes or in the floodplains of rivers. Wetlands act as giant sponges, absorbing water and then gradually releasing it into rivers or aquifers (underground reservoirs); this makes them important safeguards against flooding and erosion.

Wetlands also serve as giant natural water filters and purifiers. As water flows slowly through them, suspended particles fall to the bottom. Wetland plants and phytoplankton absorb nutrients such as nitrates and phosphates that have washed from the land. Toxic substances, including pesticides and heavy metals (such as lead and mercury), may be absorbed by wetland plants and sediments. Soil-dwelling bacteria break down some pesticides, rendering them harmless. Wetlands provide breeding grounds, food, and shelter for a great variety of birds (cranes, grebes, herons, kingfishers, ducks), mammals (beavers, otters), freshwater fish, and invertebrates such as crayfish and dragonflies.

Human Impact The extent of freshwater wetlands in the United States has been decreased by about half as a result of being drained and filled for agriculture, housing, and commercial uses. Destruction of wetlands makes nearby water more susceptible to pollutants, reduces wildlife habitat, and may increase the severity of floods.

Fortunately, many small communities in the United States have recognized the water-purifying capability of wetlands, and have constructed small wetlands to help clean wastewater. Additionally, local, state, and federal agencies have enacted laws and forged partnerships to protect existing wetlands and to restore some that have been degraded. One of the largest ecosystem restorations ever attempted, the Comprehensive Everglades Restoration Plan, is currently under way in Florida (see p. 585). As a result of these actions, the rate of wetland loss in the United States has declined.

Saltwater Ecosystems Cover Much of Earth

The oceans, like large lakes, also have life zones (**Fig. 29-28**). The upper layer of water, where the light is strong enough to support photosynthesis (to a depth of about 650 feet, or 200 meters), is called the **photic zone.** Below the photic zone lies

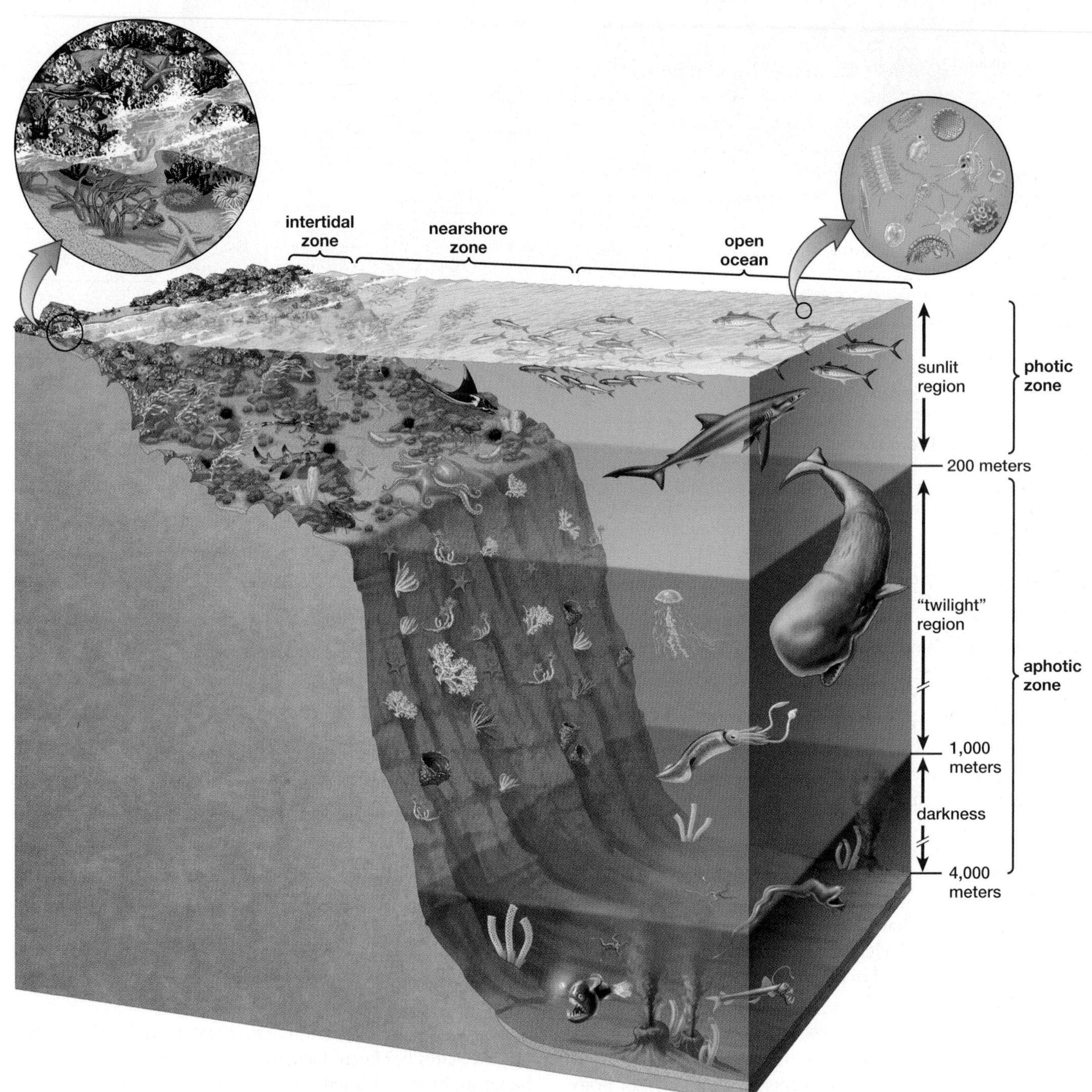

▲ **FIGURE 29-28 Ocean life zones** Photosynthesis can occur only in the sunlit photic zone, which includes the nearshore zone and the upper waters of the open ocean. Rough depths of various regions are shown, although these vary considerably depending on the clarity of the water; note that the depths are not drawn to scale. Nearly all the organisms that spend their lives in the twilight or midnight darkness of the aphotic zone rely on energy-rich material that drifts down from the photic zone. The average depth of the ocean floor is about 4,000 meters, but it reaches about 11,000 meters at its greatest depth, in the Mariana trench, located off Japan in the Pacific Ocean.

the **aphotic zone,** which extends to the ocean floor, reaching its deepest depth at about 36,000 feet (11,000 meters). Light in the aphotic zone is inadequate for photosynthesis; in its upper region, a murky twilight prevails, with constant midnight darkness in deeper waters. Nearly all the energy needed to support life must be extracted from the excrement and bodies of organisms that sink into the aphotic zone from the photic zone above.

As in lakes, most of the nutrients in the oceans are near the bottom, where there is not enough light for photosynthesis. Photic zone nutrients are constantly incorporated into the bodies of organisms and are then carried into the ocean depths when the creatures die. These nutrients are replenished from two major sources: runoff from the land and upwelling from the ocean depths. **Upwelling** brings cold, nutrient-laden water from the ocean depths to the surface. This happens when prevailing winds displace surface water, which is then replaced from below. Upwelling occurs around Antarctica and along western coastlines, including those of California, Peru, and West Africa.

Temperate Nearshore Ecosystems Include Estuaries, Intertidal Zones, and Kelp Forests

As in freshwater ecosystems, the major concentrations of life in the oceans are found where nutrients and light are both abundant, as occurs in the relatively shallow waters of the **nearshore zone,** where the land that forms continents slopes gradually downward. Such locations include **estuaries,** which are wetlands that form where rivers meet the oceans (**Fig. 29-29**). The waters of estuaries vary in salinity; high tides, for example, bring an influx of seawater. Estuaries support enormous biological productivity and diversity. Many commercially important species including shrimp, oysters, clams, crabs, and a variety of fish spend part of their lives in estuaries. The productivity of freshwater and saltwater wetlands is second only to that of rain forests.

The **intertidal zone** is a region alternately covered and uncovered by the rising and falling tides (see Fig. 29-28). Intertidal organisms on rocky shores must be adapted to withstand exposure to air as well as pounding waves. Barnacles (shelled crustaceans) and mussels (mollusks) filter phytoplankton from the water at high tide, and close their shells at low tide to resist drying. Further down, sea stars can be found prying open mussels to eat, sea urchins feast on algae coating the rocks, and anemones spread flower-like tentacles to catch passing shrimp and small fish.

Kelp forests—dense stands of kelp (an enormous brown alga)—are found throughout the world in cool waters of the nearshore zone enriched by nutrient upwelling (see Fig. 20-9b). Under ideal conditions, these remarkable protists can grow nearly 2 feet (about 50 centimeters) in a single day, and reach heights of over 60 feet (about 18 meters). Dense stands of kelp provide food and shelter for animals from nearly every phylum, including annelid worms, sea anemones, sea jellies, sea urchins, sea stars, snails, lobsters, crabs, fish, seals, and otters.

▲ **FIGURE 29-29 An estuary** Life flourishes in the shallow region where fresh river water mixes with seawater. Salt marsh grasses thrive here, providing shelter for fish and invertebrates that are eaten by these egrets and many other birds.

QUESTION Nearshore ecosystems have the highest productivity in the ocean. What factors explain this?

Human Impact Human population growth is increasing the conflict between preservation of coastal ecosystems as wildlife habitats and the development of these areas for energy extraction, housing, harbors, and marinas. Estuaries are threatened by runoff from farming operations, which can provide a glut of nutrients from fertilizer and livestock excrement. This fosters excessive growth of producers; as decomposers break down the producers' bodies when they die and sink to the bottom, the decomposers deplete the water of oxygen, killing both fish and invertebrates.

Fortunately, some efforts are being made to protect these ecosystems. In some areas, including off the coasts of California and New Zealand, marine reserves have been established, where fishing is prohibited, and kelp forests and their associated marine life are protected.

Coral Reefs

Coral reefs are complex formations that have accumulated over thousands of years from the calcium carbonate skeletons of corals (relatives of anemones). Coral reefs are most abundant in tropical waters of the Pacific and Indian Oceans, the Caribbean, and the Gulf of Mexico as far north as southern Florida, where the maximum water temperatures range between 72°F and 82°F (22°C and 28°C).

The tissues of reef-building corals harbor unicellular photosynthetic protists, called dinoflagellates, in a mutually beneficial relationship. The protists give many corals their diverse, brilliant colors (**Fig. 29-30**). Reef-building corals thrive within the photic zone, usually at depths of less than 130 feet (40 meters), where light penetrates the clear water and provides energy for photosynthesis. The dinoflagellates benefit from high nutrient and carbon dioxide levels in the coral tissues. In return, these protists provide food derived from photosynthesis to the corals.

▲ **FIGURE 29-30 Coral reefs** Coral reefs, composed of the bodies of corals, provide habitat for an extremely diverse community of fish and invertebrates. Many fish feed on the corals that form the multicolored background of this photo. An enormous variety of invertebrates inhabit the reef, including (clockwise from upper left) the spotted anemone shrimp, shown here on an anemone; a vase-like purple sponge; and the tiny blue-ringed octopus (6 inches long, fully extended)—one of the world's most venomous creatures.

QUESTION Why does "bleaching" threaten the life of a coral reef? What causes bleaching?

The coral skeletons provide anchorage, shelter, and food for an extremely diverse community of algae, fish, and invertebrates (such as shrimp, sponges, and octopuses; see Fig. 29-30). Coral reefs might be considered the "ocean's rain forests," since they are home to more than 90,000 known species, with probably 10 times that number yet to be identified.

Human Impact Anything that diminishes the water's clarity harms the coral's photosynthetic protist partners and hinders coral growth. Runoff from farming, agriculture, logging, and construction carries silt and excess nutrients that promote eutrophication, reducing sunlight and oxygen. Rain-forest logging dramatically increases erosion, sometimes destroying Earth's two most diverse ecosystems simultaneously.

In many tropical countries, mollusks, turtles, fish, crustaceans, and the corals themselves are harvested from reefs faster than they can reproduce. These are used for food or are sold to tourists, shell enthusiasts, and aquarium owners in developed countries. Removing predatory fish and invertebrates from reefs may disrupt the ecological balance of the community, allowing an explosion in populations of smothering algae, coral-eating sea urchins, or sea stars.

Complex interactions involving both human disturbances and global warming are hastening the demise of corals. Coral bleaching occurs when waters become too warm, causing the corals to expel their colorful photosynthetic protists and appear to be bleached (see Fig. 30-10c). The protists will return if the water cools, but if water temperatures remain too high, reef-building corals will gradually starve.

There is some good news, however. A fishing ban in reserves in the Florida Keys has begun to restore several important species. Additionally, many coral reefs are now protected within the world's two largest marine sanctuaries: Australia's Great

Barrier Reef Marine Park and the Northwestern Hawaiian Islands Marine National Monument. Collectively, about 20,000 species thrive in these two hotspots of biodiversity.

The Open Ocean

Beyond the coastal regions lie vast areas of the ocean in which the bottom is too deep to allow plants to anchor and still receive enough light to grow. Most life in the open ocean (**Fig. 29-31**; see also Fig. 29-28) is limited to the upper photic zone, where life-forms are **pelagic**—that is, free swimming or floating—for their entire lives. The food web of the open ocean is dependent on phytoplankton, which consists of microscopic photosynthetic bacteria and protists, primarily diatoms and dinoflagellates. These organisms are consumed by zooplankton, such as tiny shrimp-like crustaceans. Zooplankton, in turn, serve as food for larger invertebrates, small fish, and even marine mammals, such as the humpback whale (see Fig. 29-31).

The amount of pelagic life varies tremendously from place to place. The blue clarity of tropical waters is due to a lack of nutrients, which limits the concentration of phytoplankton in the water. Nutrient-rich waters that support a large phytoplankton community are greenish and relatively murky.

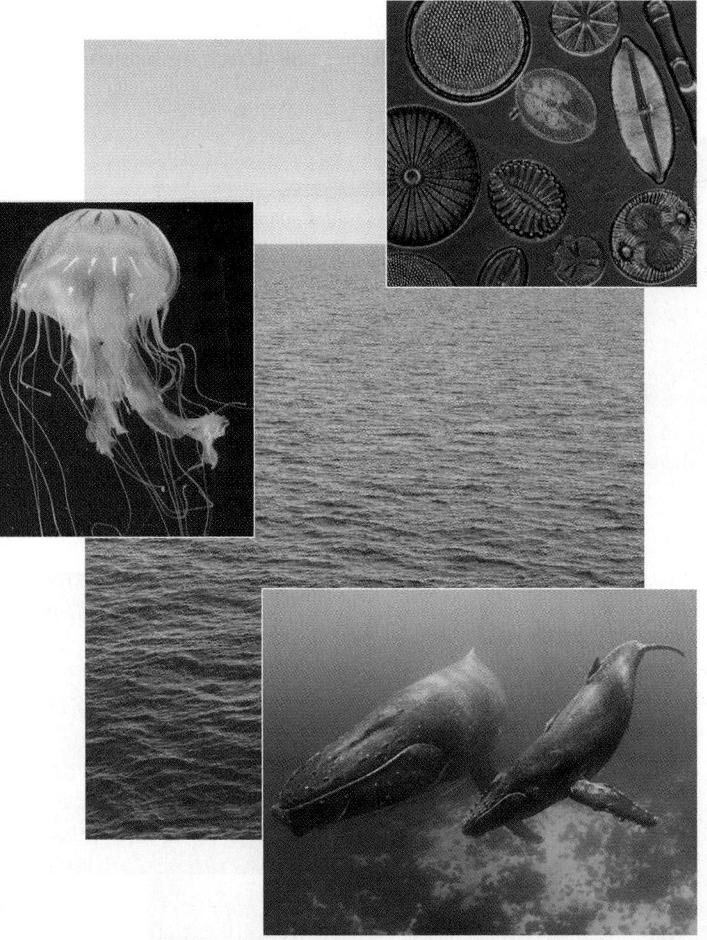

▲ **FIGURE 29-31 The open ocean** The open ocean supports abundant life in the photic zone, including (clockwise from left) a small sea jelly, which is a member of the zooplankton; phytoplankton, the ocean's photosynthetic producers; and marine mammals such as this humpback whale and its calf.

Human Impact Two major threats to the open ocean are pollution and overfishing. Oceangoing vessels dump millions of plastic containers overboard daily, and plastic refuse blows off the land. The plastic looks like food to unsuspecting sea turtles, gulls, porpoises, seals, and whales, many of which die after trying to consume it. Oil contaminates the open ocean from oil-tanker spills, runoff from improper disposal on land and leakage from offshore oil wells.

Increased demand for fish, fueled by growing human populations using more efficient fishing techniques, has caused fish to be harvested in unsustainable numbers (see p. 590). The once abundant cod populations of eastern Canada have nearly vanished despite more than a decade of severe fishing restrictions; the New England cod fishery may be destined to follow. Populations of haddock, swordfish, tuna, and many other types of seafood have also declined dramatically because of overfishing. Dredging for fish, scallops, and crabs has not only depleted many of these populations, but can seriously damage entire seafloor ecosystems. Many populations of sharks, which are keystone predators in ocean food webs, are now endangered due to overfishing. Because many types of sharks do not breed until they are more than a decade old, and because they produce small numbers of offspring, their populations are slow to recover.

Efforts are now being made to prevent overfishing. Many countries have established quotas on fish whose populations are threatened. Marine reserves, where fishing is prohibited, are increasingly being established throughout the world, causing substantial improvements in the diversity, number, and size of marine animals within these areas. Nearby areas also benefit because the reserves act as nurseries, helping to restore populations outside the reserves.

Case Study continued
The Birds and the Beans

Farming cacao and coffee plants while protecting rain forests helps to generate food sustainably. There is an urgent need to apply this concept to fisheries as well. The Marine Stewardship Council (MSC) is a nonprofit international organization that promotes and rewards sustainable fisheries. The MSC works with fisheries, scientists, conservation groups, and seafood companies, certifying individual fishery operations and placing its eco-label on certified seafood products. To be certified, a fishery must be operating at a level that sustains the population of fish or other seafood. It must also harvest its catch in a way that maintains the health of the ecosystem from which the food is gathered. The first fishery in Mexico to be certified harvests red rock lobster using traps that allow undersize lobster and other small animals that enter the traps to escape.

Unique Communities Cover the Ocean Floor

Because the amount of light in the aphotic zone is inadequate for photosynthesis, most energy in the aphotic zone comes from the excrement and dead bodies that drift down from above. Nevertheless, life is found here in amazing quantity and variety, including worms, sea cucumbers, sea stars, mollusks, squid, and fish of bizarre shapes (**Fig. 29-32**). Here, some animals generate their own light, a phenomenon known as bioluminescence. Some fish maintain colonies of bioluminescent bacteria in special visible chambers on their bodies. Bioluminescence may help bottom dwellers to attract prey, to see (many have large eyes), or to attract mates. Little is known of the behavior and ecology of these amazing and exotic creatures, which almost never survive being brought to the surface.

Recently, entire communities, including species new to science, have been found feeding on the dead bodies of whales, each of which brings an average of 40 tons of food down to the ocean floor (see Fig. 29-32). First, fish, crabs, worms, and snails swarm over the carcass, extracting nutrients from its flesh and bones. The bodies of bone-eating "zombie worms," first described in 2005, consist mostly of rootlike structures that tunnel into the bone and absorb nutrients, a feeding strategy never before observed (see Fig. 29-32). Anaerobic bacteria then continue bone breakdown, and a diverse community of clams, worms, mussels, and crustaceans moves in to feed on the bacteria during this stage of decomposition, which can last for decades.

Hydrothermal Vent Communities In 1977, geologists exploring the Galápagos Rift (an area of the Pacific floor where the plates that form Earth's crust are separating) found vents they called "black smokers" emitting superheated water blackened with sulfides and other minerals. Surrounding these vents was a rich **hydrothermal vent community** of pink fish, blind white crabs, enormous mussels, white clams, sea anemones, giant tube worms, and a snail sporting iron-laden armor plates (**Fig. 29-33**). Hundreds of new species have been found in these specialized habitats, which have now been discovered in many deep-sea regions where tectonic plates are separating, allowing material from Earth's interior to spew out.

In this unique ecosystem, sulfur bacteria serve as the producers. They harvest energy from a source that is deadly to most other forms of life—hydrogen sulfide discharged from cracks in Earth's crust. This process, called **chemosynthesis,** replaces photosynthesis in the vent communities, which flourish more than a mile below the ocean surface. Many vent animals consume the microorganisms directly, whereas others, such as the giant tube worm (which lacks a digestive tract), harbor the bacteria within their bodies and derive all their energy

▶ **FIGURE 29-32 Denizens of the deep** The skeleton of a whale provides an undersea nutrient bonanza. A "zombie worm" (bottom left) can insert its rootlike lower body deep into the bones of the decomposing whale carcass. Other denizens of the deep include (upper left) a nearly transparent deep ocean squid with short tentacles below bulging eyes, and a viperfish, whose huge jaws and sharp teeth allow it to grasp and swallow its prey whole.

◀ **FIGURE 29-33 Hydrothermal vent communities** "Black smokers" spew superheated water rich in minerals that provide both energy and nutrients to the vent community. Giant red tube worms may reach 9 feet (nearly 3 meters) in length and live up to 250 years. Parts of these worms are colored red by a hemoglobin-like molecule that traps hydrogen sulfide. (left) The foot of this vent community snail is protected by scales coated with iron sulfide.

from them. The worm, which can reach a length of 9 feet (2.7 meters), derives its red color from a unique form of hemoglobin that transports hydrogen sulfide to the symbiotic bacteria. These tube worms hold the record for invertebrate longevity—an estimated 250 years.

The bacteria and archaea that inhabit the vent communities can survive at remarkably high temperatures; some can live at 248°F (106°C; water at this depth can reach temperatures higher than the surface temperature boiling point, because of the tremendous pressure). Scientists are investigating how the enzymes and other proteins of these heat-loving microbes can continue to function at temperatures that would destroy the proteins in our bodies, and are looking for ways to put these amazing proteins to commercial use.

Case Study revisited
The Birds and the Beans

North American migratory songbirds are declining due to the loss of their habitat in three critical regions: along their migration routes, in their U.S. breeding territories, and in their winter rain-forest homes in Central and South America. Can the sustainable cultivation of coffee and cocoa help to preserve these birds? Traditional cultivation techniques—such as coffee grown in "rustic plantations" and cacao plants grown using a method called "cabruca farming"—provide hope.

Most coffee from Mexico is grown in shady plantations ("shade-grown coffee"), often under only a few types of trees that do not produce a sufficiently diverse habitat for rain-forest birds and other species. Certified "rustic" coffee plantations, on the other hand, must include at least 10 different tree species, and many plantations include far more (**Fig. 29-34**). In many cases, the canopy trees serve as an additional source of food or income to the farmers, providing citrus fruits, bananas, guavas, and lumber. Some rustic plantations provide a home for more than 150 different species of birds, which feast on the fruit or insects that thrive in the trees and moist soil.

As you delight in a chocolate bar, consider that deforestation has now claimed 93% of the roughly 300 million acres of rain forest that once towered over Brazil's

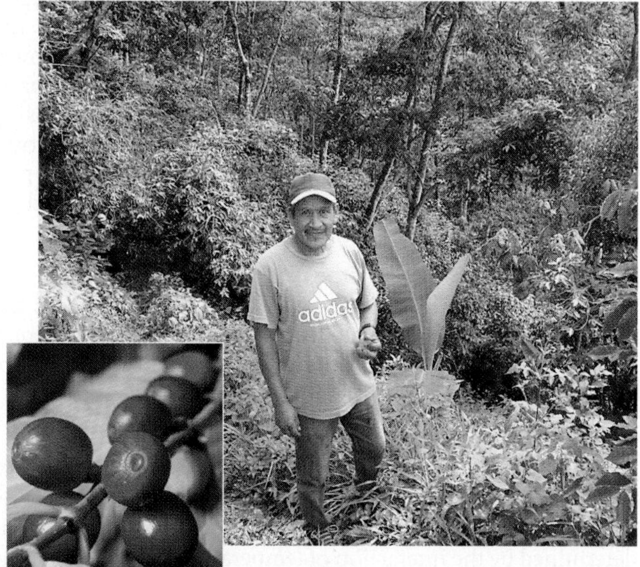

▲ **FIGURE 29-34 Rustic coffee plantations preserve biodiversity** A farmer stands in his rustic coffee plantation in Oaxaca, Mexico. Although coffee yields are lower in these diverse settings, rustic farmers can harvest a variety of fruits from the same forests in which they grow their coffee. (Inset) Coffee beans ripen on a coffee bush.

(continued)

Atlantic coast. Here, conservation programs are helping farmers reintroduce the traditional cabruca method of growing cacao trees, which involves thinning the rain forest and planting amidst its natural diversity. Joao Tavares, whose family has been using this technique for four generations, has planted many native trees on his farm, where orchids abound and the forest resounds with birdsong. "We understand that we have to preserve the cabruca," he explains, "even if you have less production." He adds that on farms where the forest has been cleared to plant dense stands of cacao trees, "you have more production, but you have lots of problems. You have more disease, more insects, so we decide to preserve."

Coffee and cacao importers and consumers are gradually awakening to the importance of these plantations in maintaining biodiversity, particularly that of migratory birds. The Smithsonian Migratory Bird Center certifies coffee plantations as Bird Friendly™ if they meet the high standards for forest diversity and canopy cover. The Rainforest Alliance certifies both cocoa and coffee grown sustainably. Several large corporations are working with cocoa- and coffee-producing countries, and with conservation organizations, to help make traditional farming methods profitable to farmers and to provide consumers with Earth-friendly choices.

The higher price of sustainably grown coffee and cocoa-based products reflects the value of the ecosystem services that rain forests provide, but discriminating consumers may find the flavors and environmental benefits well worth the cost. If you love to sit outdoors listening to birds as you drink coffee or cocoa, think about the relationship between them—is your drink helping to preserve rain forests?

Consider This

Most conservationists agree that wildlife habitat in less-developed countries will not be protected unless local residents actively support and participate in its preservation. Look up and describe other projects that fit this model of the sustainable use of rain forests or other endangered ecosystems, such as coral reefs.

CHAPTER REVIEW

Summary of Key Concepts

29.1 What Factors Influence Earth's Climate?

Weather refers to short-term fluctuations in temperature, humidity, cloud cover, wind, and precipitation in a region over periods of hours or days. Climate, in contrast, refers to patterns of weather that prevail over years or centuries in a particular region. Solar energy is distributed over a larger region north and south of the equator than at the equator. Thus, the equator is uniformly warm, while higher latitudes have lower overall temperatures. Earth's tilt on its axis causes dramatic seasonal variations at northern and southern latitudes. Rising warm air and sinking cool air in regular patterns from north to south produce areas of low and high moisture. These patterns are modified by the topography of continents and by proximity to oceans.

29.2 What Conditions Does Life Require?

The requirements for life on Earth include nutrients, energy, liquid water, and an appropriate temperature range. The interplay of these four factors shapes the characteristics and abundance of organisms in the different regions of Earth.

29.3 How Is Life on Land Distributed?

Of the four factors required for life, the two crucial limiting factors on land are temperature and water. On continents, large regions with similar climates will have similar vegetation, as determined by the interaction of temperature and rainfall or the availability of water. These regions are called biomes.

Tropical rain-forest biomes, located near the equator, are warm and wet, dominated by large broadleaf evergreen trees. Most nutrients are found in vegetation. Most animal life is arboreal. Rain forests, home to at least 50% of all species, are rapidly being cut for agriculture, although the soil is extremely poor. Slightly farther from the equator, wet seasons alternate with dry seasons during which trees shed their leaves, producing tropical deciduous forests.

Scrub forests and savannas receive less rain than tropical deciduous forests and have extended dry seasons. The African savanna is home to the world's most diverse and extensive herds of large mammals.

Most deserts, which receive less than 10 inches of rain annually, are located around 30° N and 30° S latitudes, or in the rain shadows of mountain ranges. In deserts, plants are widely spaced and have adaptations to conserve water. Animals have both behavioral and physiological mechanisms to avoid excessive heat and conserve water.

Chaparral exists in desert-like conditions that are moderated by their proximity to a coastline, allowing small trees and bushes to thrive.

Grasslands, also called prairies, are concentrated in the centers of continents. These biomes have a continuous grass cover and few trees. They produce the world's richest soils and have largely been converted to agriculture.

Temperate deciduous forests, whose broadleaf trees drop their leaves in winter to conserve moisture, dominate the eastern half of the United States, and are also found throughout most of Europe and in Eastern Asia. Higher precipitation occurs there than in the grasslands.

Wet temperate rain forests, dominated by evergreens, are found along the northern Pacific coast of the United States, the southeastern coast of Australia, and the southwestern coast of Chile.

The taiga or northern coniferous forest nearly encircles Earth below the arctic region. It is dominated by evergreen conifers whose small, waxy needles are adapted to conserve water and take advantage of the short growing season.

The tundra is a frozen desert where permafrost prevents the growth of trees and the bushes remain stunted. Nonetheless, diverse animal and perennial plant life flourishes in this fragile biome, which is found on mountain peaks and in the Arctic.

29.4 How Is Life in Water Distributed?

Of the four factors required for life, energy and nutrients are the two major limiting factors in the distribution and abundance of life in aquatic ecosystems. Nutrients are found in bottom sediments, washed in from surrounding land, or provided by upwelling in nearshore ocean waters.

In freshwater lakes, the nearshore littoral zone receives both sunlight and nutrients, and supports the most diverse lake community. The limnetic zone is the well-lit region of open water where photosynthetic protists can thrive. In the deep profundal zone of large lakes, light is inadequate for photosynthesis, and most energy is provided by detritus. Oligotrophic lakes are clear, low in nutrients, and support sparse communities. Eutrophic lakes are rich in nutrients and support dense communities. During succession, lakes tend to go from an oligotrophic to a eutrophic condition.

Streams begin at a source region, often in mountains, where water is provided by rain and snow. Source water is generally clear, high in oxygen, and low in nutrients. Streams join at lower elevations, carrying sediment from land and supporting a larger community in this transition region, where rivers form. On their way to lakes or oceans, rivers enter relatively flat floodplains, where they deposit nutrients, take a meandering path, and spill over the land when precipitation is high.

Most life in the oceans is found in shallow water, where sunlight can penetrate, and is concentrated near the continents, particularly in areas of upwelling, where nutrients are most plentiful. Temperate coastal waters include estuaries, highly productive areas where rivers meet the ocean. These sustain many bird species and marine organisms that form the basis of several important fisheries. The intertidal zone, alternately covered and exposed by tides, harbors organisms that can withstand waves and exposure to air. Kelp forests grow in cool, nutrient-rich coastal areas, and provide food and shelter for many fish and invertebrates, as well as seals and otters. Coral reefs, formed by the skeletons of corals, are primarily found in tropical, shallow seas. This complex habitat supports an extremely diverse undersea ecosystem.

In the open ocean, most life is found in the photic zone, where light supports photosynthesis by phytoplankton. In the lower aphotic zone, life is supported by nutrients that drift down from the photic zone.

The deep ocean floor lies within the aphotic zone. Here, many species are bioluminescent, and all are adapted to the tremendous water pressure. Whale carcasses provide a nutrient bonanza that supports a succession of unique communities over a span of many decades. Specialized vent communities, supported by chemosynthetic bacteria, thrive at great depths in superheated waters where Earth's crustal plates are separating.

Key Terms

aphotic zone *573*
biodiversity *559*
biome *559*
chaparral *564*
chemosynthesis *576*
climate *554*
coral reef *573*
desert *562*
desertification *563*
estuary *573*
eutrophic lake *570*
grassland *564*
gyre *557*
hydrothermal vent community *576*
kelp forest *573*
intertidal zone *573*
limnetic zone *570*
littoral zone *569*
nearshore zone *573*
northern coniferous forest *566*
oligotrophic lake *570*
ozone hole *555*
ozone layer *554*
pelagic *575*
permafrost *568*
photic zone *572*
phytoplankton *570*
plankton *570*
prairie *564*
profundal zone *570*
rain shadow *557*
savanna *562*
taiga *566*
temperate deciduous forest *565*
temperate rain forest *566*
tropical deciduous forest *562*
tropical rain forest *559*
tropical scrub forest *562*
tundra *567*
upwelling *573*
weather *554*
wetlands *571*
zooplankton *570*

Thinking Through the Concepts

Fill-in-the-Blank

1. The tilt of Earth on its axis produces __________. The zone around the equator is called the __________. Coastal climates are more moderate because of __________. A dry region on the side of a mountain range that faces away from the direction of prevailing winds is called a(n) __________.
2. Of the four major requirement of life, which two are most limited on land? __________, __________ Which two are most limited in aquatic ecosystems? __________, __________ The most biologically diverse terrestrial ecosystems are __________. The most biologically diverse aquatic ecosystems are __________.
3. In the __________ biome, most of the nutrients are found in the bodies of plants, rather than in the __________. In the __________ biome, there are pronounced seasons, and trees drop their leaves in winter. In the __________ biome, there is too little rain to support trees, but the soil is so rich that most of the biome has been converted to farmland. The __________ biome is characterized by less than 10 inches of rainfall annually. A biome in which grasses are the dominant vegetation and trees are widely spaced, with a huge diversity of large mammals, is the __________. Stunted vegetation grows in the "freezing desert" biome known as the __________.
4. The shallow portion of a large freshwater lake is called the __________. Photosynthetic plankton is called __________; nonphotosynthetic plankton is called __________. The open-water portion of a lake is divided into two zones, the upper __________ and the lower __________. Lakes that are low in nutrients are described as __________. Lakes high in nutrients are described as __________. The most diverse freshwater ecosystems are __________.
5. The primary producers of the open ocean are mainly __________. In the deep ocean, many fish produce light through __________. Hydrothermal vent communities are supported by bacteria that obtain energy from the process of __________, using the compound __________ as an energy source. Water may be at temperatures above the surface boiling point near hydrothermal vents, but it does not boil because of the __________.

Review Questions

1. Explain how air currents contribute to the formation of the Tropics and large deserts.
2. What are large, roughly circular ocean currents called? What effect do they have on climate, and where is that effect strongest?
3. What are the four major requirements for life? Which two are most often limiting in terrestrial ecosystems? In ocean ecosystems?
4. Explain why traveling up a mountain in the northern hemisphere takes you through biomes similar to those you would encounter by traveling north for a long distance.
5. Where are the nutrients of the tropical rain forest biome concentrated? Why is life in the tropical rain forest concentrated high above the ground?
6. Explain three undesirable effects of agriculture in the tropical rain-forest biome.
7. List some adaptations of desert cactus plants and desert animals to heat and drought.
8. What human activities damage deserts? What is desertification?
9. How are trees of the taiga adapted to a lack of water and a short growing season?
10. How do deciduous and coniferous biomes differ?
11. What single environmental factor best explains why the natural biome is shortgrass prairie in eastern Colorado, tallgrass prairie in Illinois, and deciduous forest in Ohio?
12. Where is life in the oceans most abundant, and why?
13. Distinguish among the limnetic, littoral, and profundal zones of lakes in terms of their location and the communities they support.
14. Distinguish between oligotrophic and eutrophic lakes. Describe a natural scenario and a human-created scenario under which an oligotrophic lake might be converted to a eutrophic lake.
15. Compare the source, transition, and floodplain zones of streams and rivers.
16. Distinguish between the photic and aphotic zones. How do organisms in the photic zone obtain nutrients? How are nutrients obtained in the aphotic zone?

Applying the Concepts

1. In which terrestrial biome is your college or university located? Discuss similarities and differences between your location and the general description of that biome in the text. In the city or town where your campus is located, how has human domination modified community interactions?
2. During the 1960s and 1970s, many parts of the United States and Canada banned the use of detergents containing phosphates. Until that time, almost all laundry detergents and many soaps and shampoos had high concentrations of phosphates. What environmental concern do you think prompted these bans, and what ecosystem has benefited most from the bans?
3. Global warming is expected to make most areas warmer, but it is also expected to change rainfall in much less predictable ways. Why is it especially important to be able to predict rainfall changes in tropical areas?
4. More northerly forests are far better able to regenerate after logging than are tropical rain forests. Explain why this is true.

MB® *Go to www.masteringbiology.com for practice quizzes, activities, eText, videos, current events, and more.*

Conserving Earth's Biodiversity

Chapter 30

Case Study

The Migration of the Monarchs

It may well be the greatest show on Earth. Each fall, the monarch butterflies of eastern North America—hundreds of millions of them—migrate south to spend the winter in just a handful of places in the mountains of central Mexico. For populations of monarchs in southeastern Canada, that means flying as far as 3,000 miles! Imagine insects weighing about half a gram each, spread out over millions of square miles during the summer months, flying for several weeks, and ending up in just a few clusters of trees, perhaps a dozen acres altogether. To make this story even more amazing, *none* of the monarchs that migrate to Mexico has ever been there before! Only their great-great grandparents, now long dead, had ever seen Mexico. How they accomplish this feat is a continuing puzzle for researchers.

But the alarming fact is that without these wintering sites in Mexico, the entire monarch population east of the Rocky Mountains would vanish. Conditions in the groves of fir and pine trees in central Mexico are just right for overwintering monarchs—a thick canopy of needles protects them from snow and rain, and the forests are cool enough to slow down their metabolism so that they don't starve to death but are not so cold that they freeze. These sites are so essential that Mexico and the United Nations have included most of the groves in a Monarch Biosphere Reserve, a legally protected area similar to a wildlife refuge or national park.

However, the monarch reserves are owned not by the Mexican government, but by the local people, most of whom are poor farmers called *campesinos*. The trees that are so essential to monarch survival are also an important economic resource for the *campesinos*, providing firewood and lumber. To complicate matters even further, rogue loggers, sometimes armed with automatic rifles, also covet the trees.

Monarch butterflies, *campesinos*, and loggers are symbolic of the dilemmas facing people worldwide—can the needs of the human population be met without destroying the natural environment? Should nature be preserved for its own sake, or, as some would say, for the enjoyment of a relatively wealthy few? Or does the environment provide services for all of us; services that can continue only if we behave as good stewards of the Earth? In this chapter, we will explore the discipline of conservation biology, which tries to answer these questions and to work out viable solutions for both the preservation and sustainable development of our planet.

▲ So many monarch butterflies cluster on their roosts in the mountains of central Mexico that their weight bends the branches of the trees.

At a Glance

30.1 WHAT IS CONSERVATION BIOLOGY?

Conservation biology is the branch of biology dedicated to understanding and preserving Earth's biological diversity. Biological diversity, or **biodiversity,** is simply the amazing variety of living organisms that inhabit Earth. Conservation biologists use the principles of biology to improve the well-being of life on Earth and maintain its diversity—both for Earth's own sake and for the benefits that biological diversity provides for people.

Conservation biologists study and seek to conserve biodiversity at different levels:

- **Genetic diversity** The success and survival of a species depend on the variety and relative frequencies of different alleles in its gene pool. Genetic diversity may be crucial for a species to adapt to changing environments.
- **Species diversity** The variety and relative abundance of the different species that comprise a community are important for the integrity and sometimes even the survival of the community.
- **Ecosystem diversity** Ecosystem diversity includes the variety of both communities and the nonliving environment on which the communities depend. Diverse communities protect ecosystems by providing services such as providing shade, degrading wastes, and generating oxygen.

Genetic and species diversity, and the resulting diversity of community interactions, are often required to maintain ecosystem function.

30.2 WHY IS BIODIVERSITY IMPORTANT?

Most people live in cities or suburbs. Our food comes mostly in packages from a supermarket. We may spend weeks without glimpsing an ecosystem in its natural state. So why should we care about preserving biodiversity? Many people would say that species and ecosystems are worth preserving for their own sake. Even if you disagree with that statement, a very practical reason for preserving biodiversity is simple self-interest: Ecosystems, both directly and indirectly, support us.

Ecosystem Services: Practical Uses for Biodiversity

In recent decades, scientists, economists, and policymakers have come to recognize that nature provides free but usually unrecognized benefits. These **ecosystem services** are the processes through which natural ecosystems sustain and enhance human life (**Fig. 30-1**). Ecosystem services include

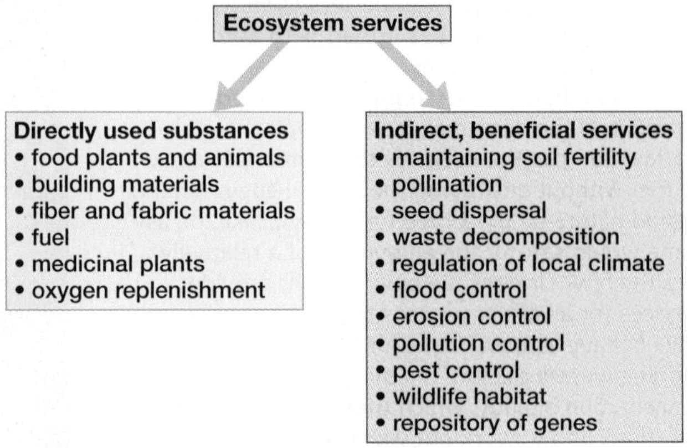

▲ **FIGURE 30-1 Ecosystem services**

purifying air and water, replenishing oxygen, pollinating plants and dispersing their seeds, generating soil and improving its fertility, providing wildlife habitat, detoxifying and decomposing wastes, controlling erosion and flooding, controlling pests, and providing recreational opportunities. These services are literally priceless because they sustain humanity. But because we don't pay for them, and their economic value is difficult to measure, ecosystem services are almost always ignored. When land is converted to housing, for example, there is usually no incentive for the developer to preserve ecosystems and their services, but there is considerable economic incentive to destroy them. Historically, people have seldom even attempted to weigh the true costs of the loss of ecosystem services against the economic benefits of altering the environment.

During the past two decades, however, some scientists have tried to calculate the value of ecosystem services. In 1997, an international team of ecologists, economists, and geographers calculated that ecosystem services provide about \$33 *trillion* in benefits to humanity every year, almost twice the world's gross national product. In 2005, the *Millennium Ecosystem Assessment*—a report resulting from 4 years of effort by over 1,300 scientists in 95 countries—concluded that 60% of all Earth's ecosystem services were being degraded or used in an unsustainable manner. These are only rough estimates, but they point to a fundamental problem: We rely on Earth's ecosystems for services of enormous value, but we aren't using these services in a way that can be sustained.

People Use Some Ecosystem Goods Directly

Healthy ecosystems provide a variety of resources directly to people. Nearly everyone can purchase wild-caught fish and other sea life that can thrive only in a healthy marine environment. Hunting for food and sport is important to the economy of many rural areas. In parts of Africa, Asia, and South America, many types of wild animals are harvested for food, and they provide an important source of protein for a growing and often poorly nourished population. In many less-developed countries, rural residents rely on wood from local forests for heat and cooking. Rain forests provide valuable hardwoods such as teak for consumers worldwide. Traditional medicines, used by about 80% of the world's people, are derived primarily from wild plants. More than three-quarters of the medicines most commonly prescribed in the United States contain active ingredients that are now—or were originally—extracted from natural sources, mostly plants.

Ecosystem Services Benefit People Indirectly

The indirect services provided by healthy, diverse ecosystems make a much greater contribution to human welfare than do goods harvested directly from nature. Here, we describe just a few important examples.

Soil Formation It can take hundreds of years to build up a single inch of soil. The rich soils of the Midwestern United States accumulated under natural grasslands over thousands of years. Farming has converted these grasslands into one of the most productive agricultural regions in the world.

▲ **FIGURE 30-2 Loss of flood control services** Conversion of natural ecosystems to agriculture contributed to flooding of the Missouri River after unusually heavy rains in 1993.

Soil—with its diverse community of decomposer and detritivore organisms (bacteria, fungi, worms, many insects, and others)—plays a major role in breaking down wastes and recycling nutrients. People rely on soils to decompose waste products from industry, sewage, agriculture, and forestry. Thus, soil serves some of the same functions as a water purification plant. Soil communities are also crucial to nearly every nutrient cycle. For example, nitrogen-fixing bacteria in soil convert atmospheric nitrogen into a form that plants can use.

Erosion and Flood Control Plants block wind that blows away loose soil. Their roots stabilize the soil and increase its ability to hold water, reducing both soil erosion and flooding. The massive flooding along the Missouri River in 1993, which resulted in \$12 billion in damage, was caused partly by conversion of the natural riverside forests, marshes, and grasslands to farmland. This greatly increased the runoff and accompanying soil erosion in the wake of heavy rains (**Fig. 30-2**).

Climate Regulation By providing shade, reducing temperatures, and serving as windbreaks, plant communities have a major impact on local climates. Forests dramatically influence the water cycle, returning water to the atmosphere through transpiration (evaporation from leaves). In the Amazon rain forest, one-third to one-half of the rain consists of water transpired by leaves. Extensive clear-cutting of rain forests can cause the local climate to become hotter and drier, making it harder for the ecosystem to regenerate and damaging nearby intact forests as well.

Forests also affect global climate. They absorb carbon dioxide from the atmosphere, storing the carbon in their trunks, roots, and branches. About 15% of the carbon dioxide produced by human activities results from deforestation; as the trees decompose or are burned, they release CO_2, which contributes to global warming.

Genetic Resources Crop plants, such as corn, wheat, and apples, have wild ancestors that humans have selectively bred

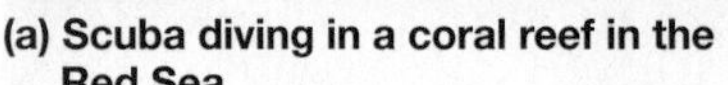

(a) Scuba diving in a coral reef in the Red Sea

(b) Viewing wildlife in Africa

(c) Spotting penguins in Antarctica

▲ **FIGURE 30-3 Ecotourism** Carefully managed ecotourism represents a sustainable use of natural ecosystems, generating revenue and providing an incentive to preserve wildlife habitat.

for centuries to produce modern domestic crops. According to the United Nations Food and Agriculture Organization, most of our food is supplied by only 12 crops. Many more wild plants might be developed into food sources that are more nutritious or better suited to local growing conditions. Researchers have identified genes in wild plants that might be transferred into crops to increase their productivity and to provide greater resistance to disease, drought, and salt accumulation in irrigated soil. For example, some wild relatives of wheat have considerable salt tolerance, and researchers are working to transfer the genes that confer the ability to thrive in salty water from these wild plants into domestic wheat. Because scientists have just begun to explore the genetic treasure house of biodiversity, it promises to become an increasingly important resource in the future—if it is preserved.

Recreation Many, perhaps most, people experience great pleasure in "returning to nature." In the United States, more than 450 million visitors flock to national parks and national forests each year. Hundreds of millions more go to wildlife refuges and state parks. In many rural areas, the local economy depends on money spent by visitors who come to hike, camp, hunt, fish, or photograph nature. Worldwide, the economic value of outdoor recreation has been estimated to be $3 trillion annually.

Ecotourism, in which people travel to observe unique biological communities, is a rapidly growing industry. Examples of ecotourism destinations include tropical coral reefs and rain forests, the Galápagos Islands, the African savanna, and even Antarctica (**Fig. 30-3**). More than 100,000 people visit the Monarch Biosphere Reserves in Mexico each year.

Ecological Economics Recognizes the Monetary Value of Ecosystem Services

The relatively new discipline of ecological economics attempts to determine the monetary value of ecosystem services and to assess the trade-offs that occur when natural ecosystems are damaged to make way for human activities. For example, a farmer planning to divert water from a wetland to irrigate a crop would traditionally weigh the monetary value of increased crop production against the cost of the project's labor and materials. If the loss of ecosystem services from the wetland (neutralizing pollutants, controlling floods, and providing breeding grounds for fish, birds, and many other animals) was factored into the decision, the wetland might well be more valuable than the crop. However, the benefits from development projects that damage ecosystems often go to individuals, whereas the costs are borne by society as a whole. Thus, in a market economy, it is difficult to apply the principles of ecological economics except for projects designed and funded by government agencies. In "Earth Watch: Restoring the Everglades," we describe a massive and costly project to undo human manipulation of the largest wetland ecosystem in the United States.

An excellent example of government planning to preserve ecosystem services comes from New York City, which obtains most of its water from the Catskill Mountains, 120 miles away in upstate New York (**Fig. 30-4**). In 1997, realizing that its water was being polluted by sewage and agricultural runoff as the Catskills were developed, city officials calculated that it would cost $6 billion to $8 billion to build a water filtration plant, plus an additional $300 million annually to

▲ **FIGURE 30-4 Ashokan Reservoir** Reservoirs in the Catskill Mountains supply New York City with extraordinarily clean water.

Earth Watch

Restoring the Everglades

In 1948, the U.S. Congress authorized the Central and Southern Florida Project, creating an extensive series of canals, levees, and other structures to control flooding, irrigate farms, and provide drinking water for new developments surrounding the Everglades, a massive wetland in central and southern Florida. In the 1960s, just north of the Everglades, the meandering, 103-mile-long Kissimmee River was straightened into a 56-mile-long canal, eliminating most of its surrounding wetlands (**Fig. E30-1**). As the Everglades and other wetlands diminished, so did the wildlife that depended on them. As native plants and animals declined, invasive species flourished. The natural water-purifying functions of the wetlands were also lost, compounding the pollution problem as new farms and cities sprang up. During the next 50 years, people became aware that a serious mistake had been made.

With half of the Everglades' original area converted to agriculture, housing, and other development, Florida and the U.S. government launched the Comprehensive Everglades Restoration Plan. Approved in 2000, the 30-year plan is intended to restore 18,000 square miles (over 11,500,000 acres) of wetlands. One of the largest ecosystem restorations ever attempted, the plan will remove 240 miles of canals and levees, reestablish natural river flow, restore wetlands, and recycle some waste water (**Fig. E30-2**). Over 400,000 acres of land will be purchased and restored. As a result of these efforts, Florida currently has over 41,000 acres of reconstructed wetlands, many of which act as giant water treatment areas. Eventually 47 miles of the Kissimmee River will be restored. Bird populations have already rebounded along restored portions of the river, and water quality has improved.

This 30-year program to undo human destruction of an ecosystem provides evidence that we are beginning to realize the economic and intrinsic values of natural communities.

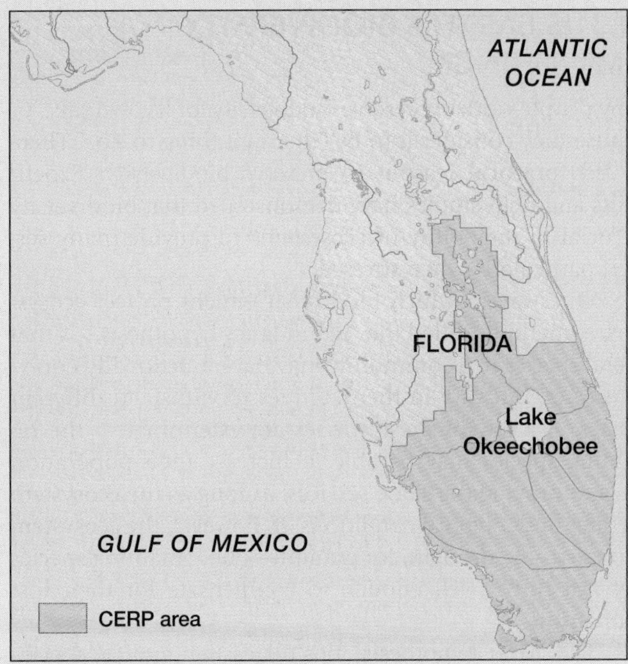

▲ **FIGURE E30-2 Restoring the Everglades** The area of Florida impacted by the Comprehensive Everglades Restoration Plan (CERP).

(a) Natural Kissimmee River before channelization

(b) Channelized Kissimmee River

▲ **FIGURE E30-1 Florida's Kissimmee River**

run it. Recognizing that the same water purification service was provided by the Catskill Mountains, city officials decided to invest in protecting them, purchasing large tracts of land and keeping them in a reasonably natural state. In 2007, the U.S. Environmental Protection Agency certified that the city would not need a filtration plant for at least another 10 years if the city continues to purchase land or acquire conservation easements in the Catskills watershed.

30.3 IS EARTH'S BIODIVERSITY DIMINISHING?

Many people work to preserve biodiversity for its own sake, or because they consider it to be "the right thing to do." There are also practical reasons to preserve biodiversity: Experiments and field studies have demonstrated that biodiversity is crucial to the ability of ecosystems to provide many services, particularly when stressed.

One way in which biodiversity might protect ecosystems, sometimes called the "redundancy hypothesis," is that several species in a community may have functionally equivalent roles, but vary in their abilities to withstand different stresses. If a few of these species are exterminated, the remaining species may be able to increase their population size and provide the same services, as long as the ecosystem operates under typical conditions. If, however, the ecosystem is stressed—by drought, for example—the remaining species may not thrive well enough to compensate for their lost compatriots.

The "rivet hypothesis" postulates that similar species may actually have somewhat different positions in the web of ecosystem stability. In an airplane wing, the loss of a couple of rivets may not be catastrophic, but lose rivets in strategic places and the entire wing falls apart. Similarly, in an ecosystem, the loss of a few critical species may cause collapse.

In a real ecosystem, some species, called *keystone species*, are neither redundant nor one of many rivets, but are fundamentally essential to the function of the ecosystem. (We will discuss keystone species in "Earth Watch: Restoring a Keystone Predator" on p. 594.) The bottom line is that species diversity is important to ecosystem function, whether the individual species are redundant, rivets, or keystones. Further, we often don't understand ecosystem function well enough to tell which species play which roles. Therefore, conservation biologists try to determine if biodiversity is diminishing, and if so, how to combat the loss.

Extinction Is a Natural Process, But Rates Have Risen Dramatically

The fossil record indicates that, in the absence of cataclysmic events, extinctions occur naturally at a very low rate, called the background extinction rate. However, the fossil record also provides evidence of five major **mass extinctions,** during which many species were eradicated in a relatively short time (see pp. 331–333 and Fig. 17-11). The most recent major extinction happened roughly 65 million years ago, abruptly ending the age of dinosaurs. The causes of mass extinctions are uncertain, but sudden changes in the environment, such as might be caused by enormous meteor impacts or rapid climate change, are the most likely explanations.

Most biologists have concluded that human activities are now causing a sixth mass extinction. Not all biologists, however, agree that current extinction rates are high enough to substantially reduce overall biodiversity. The lack of complete consensus reflects the difficulty of measuring extinction rates. Because biologists have identified only a fraction of Earth's species, it is difficult to establish the proportion of species that have already or may soon become extinct.

Extinctions of birds and mammals are best documented, although these represent only about 0.1% of the world's total species. Since the 1500s, we have lost about 2% of all mammal species and 1.3% of all bird species. The background extinction rate for birds is thought to be about one species every 400 years. However, in the past 500 years, the extinction rate has been about one species per year, and accelerating—almost entirely from human activities. The International Union for Conservation of Nature (IUCN)—the world's largest conservation network, consisting of member organizations in 140 countries, including 200 government agencies, over 800 nongovernmental conservation-related organizations, and about 11,000 scientists and other experts in over 160 countries—estimates that the current extinction rate for all species is 100 to 1,000 times the background rate.

The IUCN publishes an annual "Red List" that classifies at-risk species. Species may be described as **critically endangered, endangered,** or **vulnerable,** depending on how likely they are to become extinct in the near future. Species that fall into any of these categories are described as **threatened.** In 2008, the Red List contained 16,928 threatened species, including 12% of all birds, 21% of mammals, 5% of reptiles, and 31% of amphibians. **Figure 30-5** shows the 2008 Red List assessment for mammals. In the United

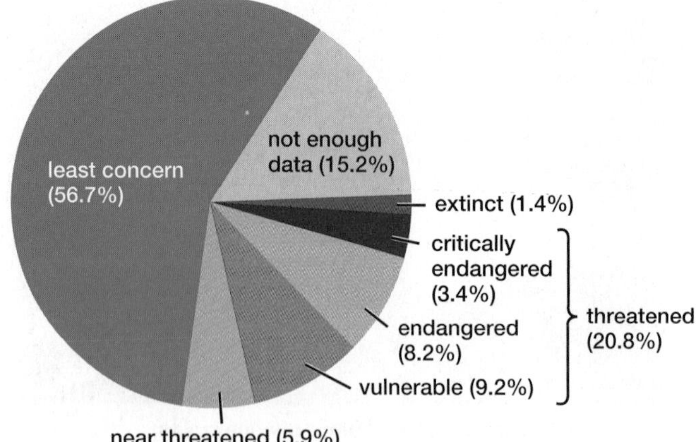

▲ **FIGURE 30-5 IUCN classification of mammals** Of the 5,487 species of mammals assessed by the IUCN in 2008, 20.8% are threatened; an additional 5.9% are "near threatened," which means that they are close to being vulnerable. Slightly more than half of the world's mammal species are considered safe ("least concern") for the time being. Data from the International Union for Conservation of Nature. 2008. *The IUCN red list of threatened species.*

States alone, there were almost 1,400 species on the U.S. Fish and Wildlife Service's 2008 list of threatened and endangered species. Many scientists fear that a large number of these endangered species are on their way to extinction. Why is this happening?

30.4 WHAT ARE THE MAJOR THREATS TO BIODIVERSITY?

Two major interrelated factors underlie the worldwide decline in biodiversity: (1) the increasingly large fraction of the Earth's resources used to support human beings, and (2) the direct impacts of human activities, such as habitat destruction, overexploitation of wild populations, invasive species, pollution, and global warming.

Humanity Is Depleting Earth's Ecological Capital

The human **ecological footprint** (see p. 504) estimates the area of Earth's surface required to produce the resources that we use and to absorb the wastes that we generate, expressed in acres of average productivity. A complementary concept, **biocapacity,** estimates the sustainable resources and waste-absorbing capacity actually available on Earth. While related to the concept of carrying capacity explained on pp. 494–495, both the footprint and biocapacity calculations are subject to change as new technologies influence the way people use resources. Scientists use the best estimates available, based primarily on statistics provided by international agencies such as the United Nations. The calculations are intended to be conservative and to avoid overstating human impacts. They also assume that humans can use the entire planet, without reserving any of it for the rest of life on Earth.

In 2005, the biocapacity available for each of Earth's 6.5 billion people was 5.2 acres (2.1 hectares), but the average human footprint was 6.7 acres (2.7 hectares). Therefore, we exceeded biocapacity by almost 30%; that is, in the long run, we would need about 1.3 Earths to support humanity at 2005 consumption and population levels (**Fig. 30-6**). Countries vary enormously in their ecological footprints, from about 12 to 24 acres for wealthy countries such as most of Europe, Canada, Australia, New Zealand, and the United States, to as little as 1 to 2 acres for poor countries such as most of those in Africa. Since these estimates were made, the human population has grown by over 250 million, while Earth's total biocapacity has not significantly increased.

Running such an "ecological deficit" is possible only on a temporary basis. Imagine a bank account that must support you for the rest of your life. If you preserve the capital and live on the interest, the account will sustain you indefinitely. But if you withdraw the capital to support an extravagant lifestyle or a growing family, you will soon run out of money. By degrading Earth's ecosystems, humanity is drawing down Earth's ecological capital. As the population grows and less-developed countries such as India and China (each with a population of over 1 billion) raise their living standards, the strain on Earth's resources will increase.

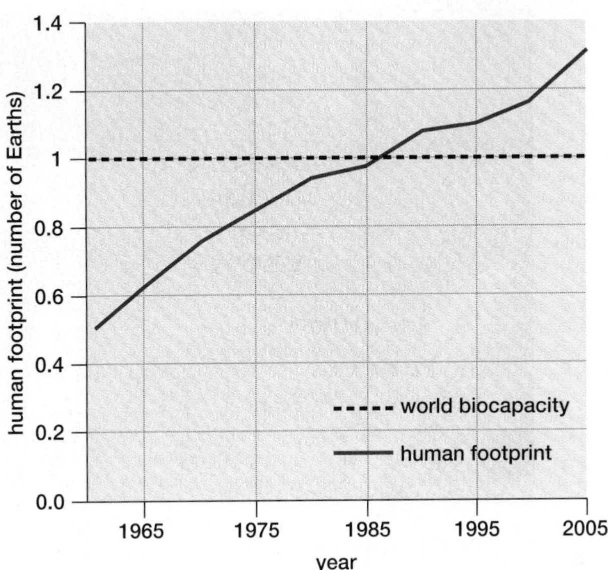

▲ **FIGURE 30-6 Human demand exceeds Earth's estimated biocapacity** Humanity's ecological footprint from 1961 to 2005, expressed as a fraction of Earth's total sustainable biocapacity (dashed line at 1.0). In 1961, we were using about half of Earth's biocapacity. By 2005, it would have required about 1.3 Earths to support us, at current rates of consumption, in a sustainable manner. Data from the World Wildlife Fund, the Zoological Society of London, and the Global Footprint Network. 2008. *The living planet report.*

Human Activities Directly Threaten Biodiversity

Habitat destruction, overexploitation, invasive species, pollution, and global warming pose the greatest hazards to natural populations. Imperiled species often face several of these threats simultaneously. For example, the major decline in frog populations worldwide has resulted from a combination of habitat destruction, invasive species, pollution, and a virulent fungal infection that many experts believe is linked to global warming (see "Earth Watch: Frogs in Peril" on pp. 456–457). Coral reefs, home to about one-third of marine fish species (over 4,000 species), suffer from a combination of overharvesting, pollution, and global warming. The IUCN estimates that as many as one-third of all reef-building corals are threatened with extinction.

Habitat Destruction Is the Most Serious Threat to Biodiversity

Since people began to farm around 11,000 years ago, Earth has lost about half of its total forest cover. Today, although many temperate zone forests are rebounding, tropical rain forests are being cut down at the rate of over 50,000 square miles each year—about the area of a football field every second. In addition to providing wood for export, rain-forest land is being extensively converted to agriculture to supply world demand for beef, coffee, soybeans, palm oil, sugarcane, and biofuels (**Fig. 30-7**; see also "Earth Watch: Biofuels—Are Their Benefits Bogus?" on p. 124). Because tropical rain forests are home to about 50% of Earth's plants and animals, their destruction is a primary cause of vanishing biodiversity.

(a) Cutting down forest

(b) Plantations seen from space

▲ **FIGURE 30-7 Habitat destruction** The loss of habitat due to human activities is the greatest single threat to biodiversity worldwide. **(a)** Clear-cutting a rain forest. **(b)** This image of soybean plantations created within a Bolivian rain forest was photographed by astronauts on the International Space Station in 2001.

Case Study continued

The Migration of the Monarchs

Monarch summer habitat is shrinking in the United States and Canada. Monarch caterpillars eat only milkweeds, most of which grow in disturbed areas such as roadsides and pastures. Herbicides and mowing have greatly reduced the wild milkweed population. If you have a garden, you can help monarch populations by planting milkweeds (*Asclepias* species). Most have beautiful pink or orange flowers, with lots of nectar for other butterfly species and hummingbirds as well.

The IUCN has identified habitat destruction as the leading threat to biodiversity worldwide as rivers are dammed, wetlands are drained, and grasslands and forests are converted to agriculture, roads, housing, and industry. Loss of habitat impacts over 85% of all threatened mammals, birds, and amphibians. Reptiles such as sea turtles also suffer. In Florida, seawalls built to protect coastal developments contribute to beach erosion and block female sea turtles as they seek nesting sites. In "Earth Watch: Saving Sea Turtles," you will learn about a successful and innovative turtle conservation program in Brazil.

Even when a natural ecosystem is not completely destroyed, it may become split into small pieces surrounded by regions devoted to human activities that are incompatible with the survival of many species (**Fig. 30-8**). This **habitat fragmentation** is a serious threat to wildlife. Some species of U.S. songbirds, such as the ovenbird and Acadian flycatcher, may need up to 600 acres of continuous forest to find food, mates, and breeding sites. In fragmented forest patches, their reproductive success plummets. Big cats are also threatened by habitat fragmentation. Mountain lions in Florida and near Los Angeles, California, are often killed while attempting to cross highways that cut through their ranges. In the 1970s, India established a series of forest reserves intended to protect the endangered Bengal tiger. Originally interconnected by forests, the reserves have now become islands in a sea of development, forcing the estimated 1,400 remaining tigers into isolated patches of woodland.

▲ **FIGURE 30-8 Habitat fragmentation** Fields isolate forest patches in Paraguay.

QUESTION Which types of species do you think are most likely to disappear from small patches of forest?

Earth Watch

Saving Sea Turtles

Six of the seven species of sea turtles are threatened with extinction. Why? Most sea turtles don't begin to breed until they are 20 to 50 years old. Then, when they reach reproductive age, the females swim hundreds, even thousands, of miles to reach their nesting grounds, often the same beaches on which they hatched. Dragging themselves ashore, they excavate a hole in the sand, deposit their eggs, and return to the sea (**Fig. E30-3a**). Baby turtles emerge after about 2 months and begin the difficult journey to adulthood. Seabirds and crabs attack them as they crawl back to the ocean (**Fig. E30-3b**). Once there, the hatchlings are a tasty morsel for fish. Although relatively few reach breeding age, under natural conditions, enough survive to maintain the turtle population.

Unfortunately, the turtles' nesting beaches attract poachers who find females and their eggs easy prey. Turtle meat and eggs are a delicacy for many people, their shells make beautiful jewelry, and their skin makes fancy leather. Turtles are also caught, both deliberately and accidentally, in fish lines and nets. Tourists are attracted to the same beaches as turtles are, and may frighten nesting females. Bright lights from beachfront developments may disorient hatchlings as they attempt to navigate toward the sea.

Since 1980, the conservation organization TAMAR (from the Portuguese ***ta***rtarugas ***mar***inhas, or sea turtle) has reduced these threats for the five species of sea turtles that nest along the Brazilian coast, becoming a model for integrated conservation efforts everywhere. TAMAR founders realized that fishermen and local villagers had to participate in the project or it would fail. Now, most of their employees are fishermen. Instead of hunting sea turtles, they now free turtles caught in nets and patrol the beaches during nesting season. TAMAR biologists tag females and trace their travels. The fishermen fend off the (now rare) turtle poachers, identify nests in risky locations, and relocate the eggs to better beach sites or to a nearby hatchery. Each year, TAMAR helps hundreds of thousands of hatchlings to reach the sea.

TAMAR has been successful because the project organizers have engaged local communities as partners in turtle protection. Money flows into the local economies as ecotourists flock to see baby turtles, visit turtle museums, buy souvenirs made by local residents, and learn about the program. TAMAR sponsors communal gardens, day-care centers, and environmental education activities. Recognizing that the economic benefits derived from preserving turtles far outweigh the money that can be made from hunting them, local residents eagerly participate in turtle conservation. The success of TAMAR not only underscores the need for community support for the sustainable use of any natural resource, but highlights how successful such efforts can be.

(a) A sea turtle excavating a nest

(b) A turtle hatchling heads for the sea

▲ **FIGURE E30-3 Endangered sea turtles (a)** A female green turtle scoops sand with powerful flippers, creating a cavity where she will bury about 100 eggs. **(b)** After incubating in the sand for about 2 months, the eggs hatch. Here a hatchling heads for the sea, where (if it survives) it will spend 20 to 50 years before reaching sexual maturity. Photo courtesy of Jim Watt/PacificStock.com

Habitat fragmentation may result in populations that are too small to survive. To be functional, a preserve must support a **minimum viable population (MVP),** the smallest isolated population that can persist in spite of natural events, including inbreeding, disease, fires, and floods. The MVP for any species is influenced by many factors, including the quality of the environment, the species' average life span, its fertility, and the number of young that typically reach maturity. Some wildlife experts believe that a minimum viable population of Bengal tigers must include at least 50 females—more than are found in most of India's tiger reserves.

Overexploitation Threatens Many Species

Overexploitation refers to hunting or harvesting natural populations at a rate that exceeds their ability to replenish their numbers. Overexploitation has increased as a growing demand for wild animals and plants has been coupled with technological advances that have greatly increased our efficiency at harvesting them. The IUCN estimates that overexploitation impacts about 30% of threatened mammals and birds.

Overfishing is the single greatest threat to marine life, causing dramatic declines of many species, including cod, sharks, red snapper, groupers, and swordfish. The UN Food and Agriculture Organization (FAO) estimates that about 25% of global fish populations are overexploited, and another 50% are being fished to their maximum sustainable yield. Other experts conclude that as much as 70% of the world's fisheries are overexploited. To complicate matters further, in 2008, the global fishing fleet was probably twice as large as was needed to land the fish that the ocean could sustainably produce. As a result, the FAO estimates that global fisheries actually lose about $50 billion a year, being heavily subsidized by governments (and therefore taxpayers). Rapidly growing populations in less-developed countries increase the demand for animal products, as hunger and poverty drive people to harvest all that can be sold or eaten, legally or illegally, without regard to its rarity. As Callum Rankine of the World Wildlife Fund explains, "It's extremely difficult to get people to live sustainably. Often they are just concerned with trying to live."

Rich consumers fuel the exploitation of some endangered animals by paying high prices for illegal products such as elephant-tusk ivory, rhinoceros horn, and exotic birds. The greatest threats to Bengal tigers in India are poachers who sell their skins and bones (which are used in some traditional Chinese medicines) for thousands of dollars. The demand for exotic wood by consumers in developed countries encourages unsustainable logging and tree poaching in tropical rain forests.

Invasive Species Displace Native Wildlife and Disrupt Community Interactions

Humans have transported a multitude of species around the world—everything from thistles to Japanese beetles to rats. In many cases, the introduced species cause no great harm. Sometimes, however, non-native species become invasive: They increase in number at the expense of native species, competing with them for food or habitat, or preying on them directly (see Chapter 27). Invasive species often make native species more vulnerable to extinction from other causes as well, such as disease or habitat destruction. Although scientists differ on exactly what constitutes an "invasive" species (for example, how large a population must exist, and how seriously it must impact a native species), the National Institute of Invasive Species Science lists almost 3,000 invasive species in the United States, mostly plants and insects. About half of all threatened U.S. species suffer from competition with or predation by invasive species.

Islands are particularly vulnerable to invasive species. Island populations of plants and animals are small, are often unique, and they have nowhere to go if conditions change. For example, the Hawaiian Islands have lost about 1,000 species of plants and animals since their settlement by humans, mostly caused by either overexploitation or competition and predation by invasive species. Most of the native wildlife of Hawaii remains in danger: As of 2008, the U.S. Fish and Wildlife Service reported that Hawaii had the largest number of threatened species of any state, with 394.

Many invasive species are transported unintentionally, but some are deliberately introduced. In Hawaii and other Pacific Islands, pigs and goats, released by early Polynesian settlers to provide food, have devastated native plants. Mongooses, small, cat-sized carnivores native to Asia and Africa, were deliberately imported in the 1800s to control accidentally introduced rats. Now both mongooses and rats pose major threats to Hawaii's native ground-nesting birds.

Lakes are also particularly vulnerable. The Great Lakes of the United States and Canada host dozens of invasive species, including the zebra mussel and lamprey. Lake Victoria in Africa was once home to about 400 to 500 different species of cichlid fish (**Fig. 30-9a**). Enormous predatory Nile perch (**Fig. 30-9b**) and much smaller plankton-feeding tilapia were introduced into Lake Victoria in the mid-1900s. Although researchers differ about the relative magnitudes of the threats, the combination of predation by Nile perch, competition from tilapia, pollution, and algal blooms (brought on by nutrients from surrounding farms draining into the lake) have caused the extinction of about 200 species of cichlids.

Pollution Is a Multifaceted Threat to Biodiversity

Pollution takes many forms, including synthetic chemicals such as plasticizers, flame retardants, and pesticides; naturally occurring substances such as mercury, lead, and arsenic; and high levels of nutrients, usually from sewage or agricultural runoff.

Synthetic chemicals are often lipid soluble. Even small amounts in the environment may accumulate to toxic levels in the fatty tissue of animals (see p. 539). In the mid-twentieth century, for example, the insecticide DDT accumulated in many predatory bird species, ultimately causing them to lay eggs with thin shells that could not withstand the pressures of the parents sitting on them during incubation. Now, an enormous controversy has arisen over a chemical called bisphenol A, which is commonly used in many plastics. Bisphenol A appears to mimic the actions of estrogen, and may cause reproductive effects in animals and people, although researchers disagree about whether current human exposures are high enough to cause harm.

Many heavy metals are naturally bound up in insoluble rocks, and thus rendered harmless. However, mining, industrial processes, and burning fossil fuels may release heavy metals into the environment. Even extremely low levels of several heavy metals, such as mercury and lead, are toxic to virtually all organisms.

Finally, nutrients in excessive amounts also become pollutants. For example, burning fossil fuels releases oxidized nitrogen and sulfur, disrupting their natural biogeochemical cycles and causing acid rain that threatens forests and lakes (see p. 545). As we noted earlier, agricultural runoff has probably contributed to the extinction of cichlids in Lake Victoria.

(a) Zebra obliquiden cichlid

(b) Nile perch

▲ **FIGURE 30-9 Invasive species endanger native wildlife (a)** Lake Victoria was home to hundreds of species of stunningly colored cichlid fish, such as the Zebra obliquiden pictured here. **(b)** The Nile perch, introduced into Lake Victoria for fishermen, has proven to be a disaster for native fish.

Global Warming Is an Emerging Threat to Biodiversity

The use of fossil fuels, coupled with deforestation, has substantially increased atmospheric carbon dioxide levels. As predicted by climatologists, this increase has been accompanied by an overall increase in global temperatures. In response to global warming, some species are shifting their ranges further toward the poles and many plants and animals are beginning springtime activities earlier in the year (see pp. 546–548).

The rapid pace of human-induced climate change challenges the ability of species to adapt. In May 2007, the United Nations International Day for Biological Diversity focused on the impacts of climate change on biodiversity, and concluded that global warming has already contributed to some extinctions and is likely to cause many more. Although it is difficult to predict the response of many species to global warming, likely impacts include:

- Deserts may become hotter and drier, making survival more difficult for their inhabitants.
- Warmer conditions will force species to retreat toward the poles or up mountains to stay within the climate zones in which they can survive and reproduce. Relatively immobile species, especially plants, may be unable to retreat fast enough to stay within a suitable temperature range, because they can only "move" during reproduction, when, for example, the wind or animals disperse their seeds.
- Cool habitat may disappear completely from mountaintops. Pikas, small relatives of rabbits that live above the treeline in the Rocky Mountains (**Fig. 30-10a**), face shrinking habitat as the mountains warm. Some local populations on isolated mountains have already vanished.
- Insect pests that were previously killed by frost or sustained freezes may spread and thrive. In the northern and central Rocky Mountains, pine bark beetles were formerly controlled partly by sustained, extremely cold weather in the winter. In the past 20 years, these beetles have reached epidemic levels, so that most of the mature lodgepole pines in the Rockies are expected to die within the next decade (**Fig. 30-10b**).
- Coral reefs require warm water, but too much warming causes bleaching and coral death (**Fig. 30-10c**). Coral reefs worldwide appear to be stressed already.

30.5 HOW CAN CONSERVATION BIOLOGY HELP TO PRESERVE BIODIVERSITY?

Research in conservation biology can help to devise strategies for conserving biodiversity. Four important goals of conservation biology are:

- To understand the impact of human activities on species, populations, communities, and ecosystems.
- To preserve and restore natural communities.
- To reverse the loss of Earth's biodiversity caused by human activities.
- To foster sustainable use of Earth's resources.

Within the life sciences, conservation biology enlists the help of ecologists, wildlife managers, geneticists, botanists, and zoologists. But effective conservation depends on expertise and support from people outside of biology as well. These include government leaders at all levels, who establish environmental policy and laws; environmental lawyers, who help enforce laws protecting species and their habitats; and ecological economists, who help place a value on ecosystem services. Social scientists provide insight into the ways that people in different cultural groups use their environments. Educators help students understand how ecosystems function, how they support human life, and how people can either disrupt or preserve them. Conservation organizations identify areas of concern, provide educational materials, and organize grassroots support by individuals. Finally, individual choices and actions ultimately determine whether conservation efforts succeed.

Conserving Wild Ecosystems

Because habitat destruction and fragmentation are key factors threatening biodiversity, habitat preservation is essential. Protected reserves, connected by wildlife corridors, are vital to conserving natural ecosystems.

Core Reserves Preserve All Levels of Biodiversity

Core reserves are natural areas protected from most human uses except low-impact recreation. Ideally, a core reserve encompasses enough space to preserve ecosystems with all their biodiversity. Core reserves should also be large enough to withstand storms, fires, and floods without losing species. "Earth Watch: Restoring a Keystone Predator" on p. 594 explains how reintroducing wolves to Yellowstone National Park, a core reserve in Wyoming, is restoring many community interactions and probably saving species.

To establish effective core reserves, conservationists must estimate the smallest areas required to sustain minimum viable populations of the species that require the most space. The sizes of minimum critical areas vary significantly among species, and also depend on the availability of food, water, and shelter. Thus, in general, large predators in arid environments will need a larger minimum critical area than small herbivores in lush environments. However, it is difficult to make precise estimates of minimum critical areas for individual species.

Corridors Connect Critical Animal Habitats

One fact stands out in estimating minimum critical areas, especially for reserves that include large predators: In today's crowded world, an individual core reserve is seldom large enough to maintain biodiversity and complex community interactions by itself. **Wildlife corridors,** which are strips of protected land linking core reserves, allow animals to move relatively freely and safely between habitats that would otherwise be isolated (**Fig. 30-11**). Corridors effectively increase the size of smaller reserves by connecting them. Both core reserves and corridors, ideally, are surrounded by buffer zones supporting human activities that are compatible with wildlife. Buffer zones prevent high-impact uses such as clearcutting, mining, freeways, and housing from impacting wildlife in the core region.

Although usually much wider, an effective wildlife corridor can sometimes be as narrow as an underpass beneath a highway. For example, in densely populated Southern California, plans for a development of over 1,000 new houses near San Diego were abandoned and freeway exits were closed after

(a) A pika gathers plants for the winter

(b) Pine bark beetles have killed these lodgepole pines

(c) Bleached corals (white) are usually dead or dying

▲ **FIGURE 30-10 Global warming threatens biodiversity**
(a) Pikas live near the tops of mountains in the Rockies; as the climate warms, suitable pika habitat may disappear right off the top of a mountain. **(b)** Pine bark beetles have killed many of the lodgepole pines on this hillside. Reddish-brown trees bear dead needles; in a year or two, the needles will fall off the lifeless trees. **(c)** Living corals usually contain photosynthetic algae that provide nourishment for the coral. When the water warms too much, corals lose their algae and become strikingly white; without the algae to help feed them, they often die.

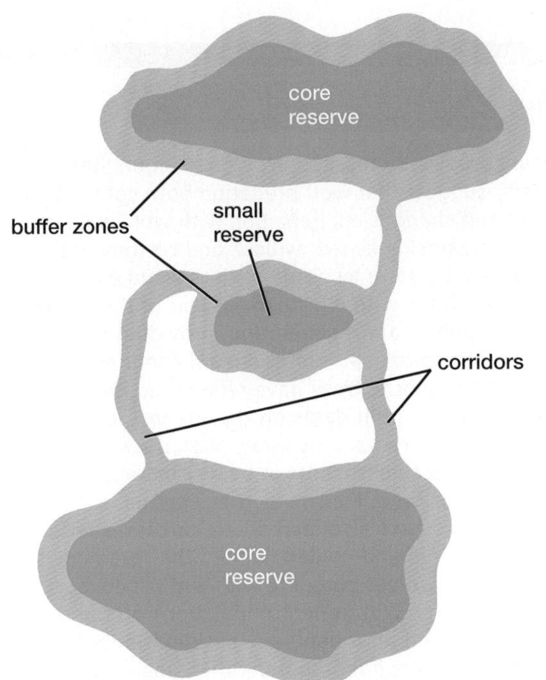

▲ **FIGURE 30-11 Corridors connect reserves**

QUESTION How do wildlife corridors reduce species extinction in small reserves?

wildlife biologists tracked a mountain lion that was using the Coal Canyon underpass to move between suitable habitat in the Chino Hills north of the freeway and the Santa Ana Mountains to the south. The underpass and its surroundings have been restored to a more natural state, encouraging mountain lions and other wild animals to cross safely beneath the freeway (**Fig. 30-12**). People, too, use the corridor—in fact, the wildlife corridor is now featured in local hiking guidebooks!

In the northern Rocky Mountains, a coalition of conservation groups and scientists has proposed a series of wildlife corridors linking existing core reserves, such as Yellowstone, Grand Teton, and Glacier National Parks, with nearby ecosystems. These interconnected habitats would sustain populations of grizzly bears, elk, and mountain lions.

Have you ever wondered?

What You Can Do to Prevent Extinctions?

You may feel helpless to prevent extinction, but in fact, you can help a lot. You can join organizations, such as The Nature Conservancy and the World Wildlife Fund, that work to protect endangered species. Another organization you might consider joining is Saving Species, founded by ecologist Stuart Pimm. Saving Species helped to purchase a pasture in Brazil to provide a wildlife corridor between two patches of habitat for the golden lion tamarin, an endangered monkey. In another South American country, they even paid off a local crime "godfather" to stop illegal logging. Now *that's* an innovative approach to protecting biodiversity!

(a) Radio-tagging a mountain lion

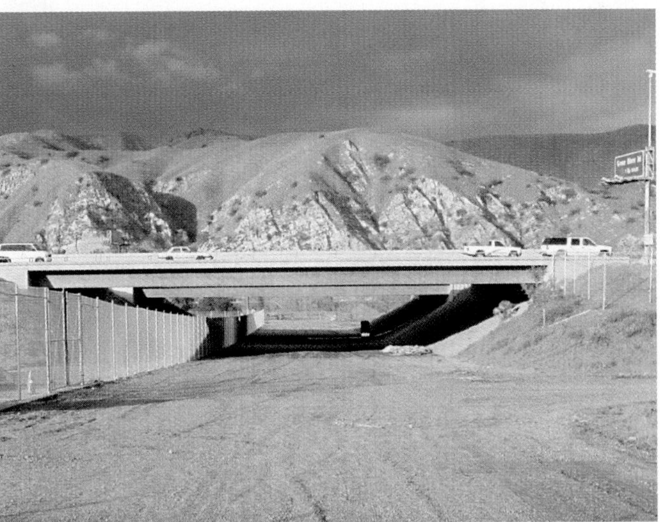

(b) Wildlife corridor beneath a freeway

▲ **FIGURE 30-12 Wildlife corridors connect habitats (a)** National Park Service wildlife biologists track mountain lions using radio collars, such as the one worn by this tranquilized animal. **(b)** Asphalt has been removed and traffic barred from the underpass at Coal Canyon beneath the Riverside Freeway near San Diego to allow mountain lions to move safely between habitats on either side.

Earth Watch

Restoring a Keystone Predator

The keystone sits at the top of a stone arch and holds all of the other pieces in place—remove the keystone and the entire arch collapses. Similarly, in a biological community, a **keystone species** is one that plays an essential role, usually more than might be predicted by the size of its population. Take away the keystone species, and the whole community changes drastically. Often, the keystone species in a community is a predator, whose hunting activities shape the structure of the community.

In Yellowstone National Park, recent research indicates that a predator long missing from the Park—the gray wolf—is a keystone species that regulates major aspects of the community, including its trees. How can wolves affect trees?

Considered a threat to elk and bison, wolves were deliberately exterminated from Yellowstone by 1928. Based on tree-ring data and aerial photographs, it appears that this event marked the beginning of the end for regeneration of aspen trees (**Fig. E30-4**). Aspen groves, which shelter a diverse community of plants and birds, have declined by more than 95% since the park was established in 1872. Elk, formerly the major prey of wolves in Yellowstone, ate nearly all the young aspen, as well as young willow and cottonwood trees.

In 1995 and 1996, after years of planning, study, and public comment, the U.S. Fish and Wildlife Service captured 31 wolves in Canada and released them into Yellowstone (**Fig. E30-4**, inset). Now, although the wolf population fluctuates from year to year, typically, there are 200 to 300 wolves in or near the park.

Recent studies by researchers from Oregon State University suggest that wolf predation both controls elk numbers and changes elk behavior. With wolves nearby, elk avoid streamside aspen, willow, and cottonwood groves where they are less able to detect and escape from wolves. Now, with elk steering clear of them, these plant communities are regenerating, providing more habitat for songbirds and better stream conditions for trout. As their favorite trees have rebounded, beaver have returned and built dams on the streams, creating wetland habitat for mink, muskrat, otter, ducks, and rare boreal toads. Succulent plants in beaver marshes are favored food for grizzly bears as they emerge from hibernation. Grizzlies also feed on elk carcasses left by wolves, as do bald and golden eagles. The researchers also found that wolves compete with and kill coyotes, which eat rodents. With rodent populations on the rise, red foxes that feed on rodents are proliferating, and biologists anticipate a rebound in other small predators such as weasels and wolverines.

Other keystone predators that may have similar effects include mountain lions in Zion National Park in Utah and Yosemite National Park in California; wolves in Banff and Jasper National Parks in Alberta, Canada; sea otters off the California coast; and starfish in Pacific tide pools. Sometimes, restoring just one or two species can have far-ranging impacts on the health of an entire ecosystem.

▲ **FIGURE E30-4 Impact of a keystone predator** Remnants of a once-thriving aspen grove in Yellowstone National Park attest to the lack of aspen regeneration since the early 1900s. (Photo courtesy of Dr. William Ripple.) (Inset) Wolves now roam Yellowstone, delighting visitors and exerting far-reaching effects on the community.

30.6 WHY IS SUSTAINABILITY THE KEY TO CONSERVATION?

Natural ecosystems share certain features that allow them to persist and flourish. Among the most important characteristics of sustainable ecosystems are:

- Diverse communities with complex community interactions.
- Relatively stable populations that remain within the carrying capacity of the environment.
- Recycling and efficient use of raw materials.
- Reliance on renewable sources of energy.

Environments that have been modified by human development often do not possess these qualities. As a result, many human-modified ecosystems may not be sustainable in the long run, and current modes of development and land use could lead to loss of biodiversity and ecosystem services. We humans must learn to meet our needs in ways that sustain the ecosystems on which we depend.

Sustainable Development Promotes Long-Term Ecological and Human Well-Being

Respect for nature's operating principles is central to sustainability. In the landmark document *Caring for the Earth*, the IUCN states that **sustainable development** "meets the needs of the present without compromising the ability of future generations to meet their own needs." It explains that "Humanity must take no more from nature than nature can replenish. This in turn means adopting lifestyles and development paths that respect and work within nature's limits. It can be done without rejecting the many benefits that modern technology has brought, provided that technology also works within those limits."

Unfortunately, in modern human society, "sustainable development" is almost an oxymoron, because "development" so often means replacing natural ecosystems with human infrastructure such as housing and retail developments. Traditionally, many economists and businesspeople have insisted that without continued growth, humanity cannot prosper. People in developed countries have, indeed, achieved economic growth and a high quality of life. But they have done so by exploiting, in an unsustainable manner, the direct and indirect services provided by ecosystems and by using large quantities of nonrenewable energy.

Now, however, evidence from all parts of the world shows that such activities are unraveling the tapestry of natural communities and undermining Earth's ability to support life. As individuals and governments recognize the need to change, an increasing number of projects are being developed that are intended to meet human needs sustainably. We describe a few of these in the following sections.

Biosphere Reserves Provide Models for Conservation and Sustainable Development

A world network of **Biosphere Reserves** has been designated by the United Nations. The goal of Biosphere Reserves is to maintain biodiversity and evaluate techniques for sustainable human development while preserving local cultural values. Biosphere Reserves consist of three regions. A central core reserve allows only research and sometimes tourism and some traditional sustainable cultural uses. A surrounding buffer zone permits low-impact human activities, such as recreation, research, and environmental education, and some development, such as carefully regulated forestry and grazing. Outside the buffer zone is a transition area that supports settlements, tourism, fishing, and agriculture, all (ideally) operated sustainably (**Fig. 30-13**). The first Biosphere Reserve was designated in the late 1970s, and there are now over 530 sites in 105 countries.

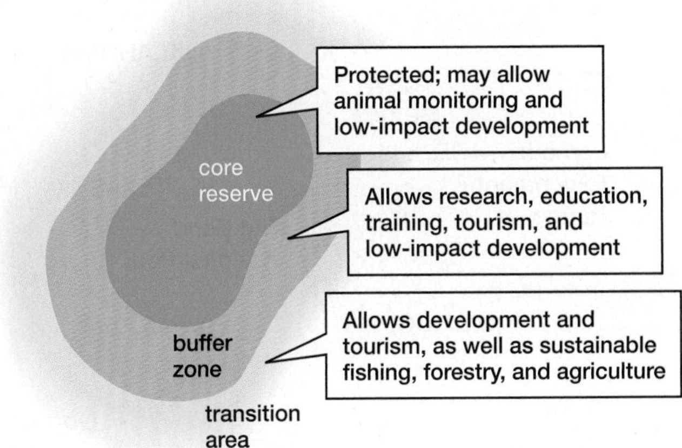

▲ **FIGURE 30-13 The design of an ideal Biosphere Reserve**

National governments nominate sites in their countries for Reserve designation and continue to own and manage them. This has greatly reduced opposition, but as a result of their voluntary nature, few completely adhere to the ideal Biosphere Reserve model. In the United States, most of the 47 biosphere core reserves are national parks and national forests. Much of the land in buffer and transition zones is privately owned, and the landowners may be unaware of its designation. Often, funding is inadequate to compensate them for restricting development or for promoting and coordinating sustainable development.

The Chihuahuan Desert Biosphere Reserve is an innovative regional reserve established in 1977 that actually consists of three separate reserves (**Fig. 30-14**). Big Bend National Park in Texas serves as the core reserve, supporting research and tourism

Case Study continued

The Migration of the Monarchs

"Monarch Biosphere Reserve" is a nice name, but often a designation like that isn't accompanied by any mechanism to make it a reality. However, the World Wildlife Fund has established a $5,000,000 trust fund that is used to help farmers find alternate sources of income rather than to log their land, thus helping to keep an intact buffer zone around the core monarch groves.

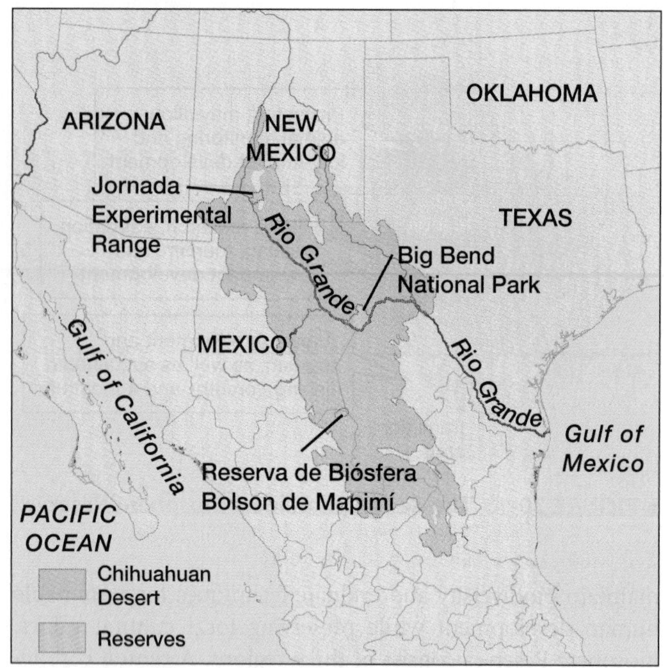

▲ **FIGURE 30-14 A unique biosphere reserve** This regional reserve within the Chihuahuan Desert (shown in brown) consists of three smaller reserves (green) in the United States and Mexico.

but allowing no private development. The Jornada portion of the reserve, located in New Mexico, is considered the buffer zone. Here, researchers investigate sustainable rangeland management in dry ecosystems. In Mexico, the Mapimi reserve serves as the transition area. More than 70,000 people live in this reserve, and scientists are working with them to shift to more sustainable farming practices, and to conserve desert species such as the endangered Bolson tortoise, the largest land reptile in North America. Scientists now hope to reintroduce this species to Big Bend National Park and to establish wildlife corridors linking these widely separated reserves.

Sustainable Agriculture Preserves Productivity with Reduced Impact on Natural Communities

The greatest loss of natural habitat occurs when people convert natural ecosystems to farms. For example, in the midwestern United States, tens of millions of acres of grasslands have been converted to farmland, principally for corn, wheat, and soybeans. Because farms typically grow only one or a few crops, and because most of the plant growth is harvested for human consumption, plant and animal diversity declines, compared to the natural habitat before the farms were established.

Farming is necessary to feed humanity. Further, to make a reasonable living, farmers must produce large amounts of food at low cost. This often leads to unsustainable practices that interfere with ecosystem services. For example, allowing fields to remain bare after harvesting often increases soil erosion, as wind and rain remove the exposed soil. Applying insecticides often indiscriminately kills pest insects, their natural predators, and pollinators. In many regions throughout the world, irrigation is depleting underground water supplies faster than they can be replenished by rain and snow. Further, because both underground and surface water supplies contain varying amounts of salt, evaporation of irrigation water often leaves enough salt behind to reduce soil fertility.

Fortunately, farmers are increasingly recognizing that sustainable agriculture ultimately saves money while preserving the land (**Table 30-1**). The **no-till** cropping technique, which leaves the residue of harvested crops in the fields to form

Table 30-1 Agricultural Practices Affect Sustainability

	Unsustainable Agriculture	Sustainable Agriculture
Soil erosion	Allows soil to erode far faster than it can be replenished because the remains of crops are plowed under, leaving the soil exposed until new crops grow.	Erosion is greatly reduced by no-till agriculture. Wind erosion is reduced by planting strips of trees as windbreaks around fields.
Pest control	Uses large amounts of pesticides to control crop pests.	Trees and shrubs near fields provide habitat for insect-eating birds and predatory insects. Reducing insecticide use helps to protect birds and insect predators.
Fertilizer use	Uses large amounts of synthetic fertilizer.	No-till agriculture retains nutrient-rich soil. Animal wastes are used as fertilizer. Legumes that replenish soil nitrogen (such as soybeans and alfalfa) are alternated with crops that deplete soil nitrogen (such as corn and wheat).
Water quality	Runoff from bare soil contaminates water with pesticides and fertilizers. Excessive amounts of animal wastes drain from feedlots.	Animal wastes are used to fertilize fields. Plant cover left by no-till agriculture reduces nutrient runoff.
Irrigation	May excessively irrigate crops, using groundwater pumped from natural underground storage at a rate faster than the water is replenished by rain or snow.	Modern irrigation technology reduces evaporation and delivers water only when and where it is needed. No-till agriculture reduces evaporation.
Crop diversity	Relies on a small number of high-profit crops, which encourages outbreaks of insects or plant diseases and leads to reliance on large quantities of pesticides.	Alternating crops and planting a wider variety of crops reduces the likelihood of major outbreaks of insects and diseases.
Fossil fuel use	Uses large amounts of nonrenewable fossil fuels to run farm equipment, produce fertilizer, and apply fertilizers and pesticides.	No-till agriculture reduces the need for plowing and fertilizing.

(a) Cotton seedlings emerge in a no-till field in North Carolina

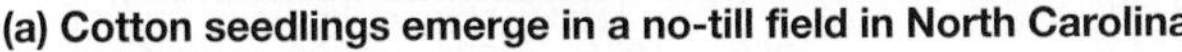

(b) The same field one month later

▲ **FIGURE 30-15 No-till agriculture (a)** A cover crop of wheat is killed with an herbicide. Cotton seedlings thrive amid the dead wheat, which anchors soil and reduces evaporation. **(b)** Later in the season, the same field shows a healthy cotton crop mulched by the dead wheat. Photos courtesy of Dr. George Naderman, Former Extension Soil Specialist (retired), College of Agriculture and Life Sciences, NC State University, Raleigh, NC

mulch for the next year's crops, represents one possible component of sustainable agriculture (**Fig. 30-15**). In 2007, no-till methods were used on about 65 million acres in the United States (about 24% of all croplands). Various reduced-tillage methods were used on another 50 million acres. No-till and reduced-till farming requires less plowing and harrowing, with an estimated savings of 3 to 14 gallons of diesel fuel per acre.

On the other hand, most no-till farmers spray herbicides to kill both weeds and the remains of last year's crop. Some of the herbicide inevitably blows off the fields and may damage nearby natural habitats. In addition, some herbicides may harm animals; although controversial, some studies indicate that atrazine, an herbicide commonly used in no-till agriculture, damages the reproductive systems of amphibians and perhaps other animals. The residues of last year's crops also may contain pathogens such as fungi, which would be reduced by plowing in conventional agriculture but may require the use of pesticides in a no-till field.

Organic farmers typically do not use synthetic herbicides, insecticides, or fertilizers. Some organic farmers use no-till methods, but most plow their fields at least every other year to help kill weeds. Organic farming relies on natural predators to control pests and on soil microorganisms to degrade animal and crop wastes, thereby recycling their nutrients. Diverse crops reduce outbreaks of pests and diseases that attack a single type of plant. There is an ongoing debate about the relative productivity of organic versus conventional farming, and whether organic farming with plowing or no-till farming with herbicides is better for the soil and the natural environment.

As you can see, how to implement sustainable agriculture is uncertain and controversial. In the best-case scenario, farmers would grow a variety of crops, using practices that retain soil fertility, with as little input of energy and potentially toxic chemicals as possible. Insect pests would be controlled by predators such as birds and predatory insects, and by crop rotation, so that pests that specialize on particular crop plants would not find a feast laid out for them every year. Fields would be relatively small, separated by strips of natural habitat for native plants and animals. In practice, there is substantial disagreement among both farmers and agricultural experts about whether all of these goals can be met while still yielding large harvests and keeping costs low.

Because the loss of ecosystem services is not factored into the costs of unsustainable farming practices, food produced unsustainably tends to be cheaper, at least in the short term. In the long term, of course, if typical commercial farming results in salty soils, outbreaks of crop diseases and pests, or loss of topsoil, then sustainable agriculture will be less expensive. Many projects, such as the University of California's Sustainable Agriculture Research and Education Program based in Davis, California, support research and educate farmers about the advantages of sustainable agriculture, how to practice it, and how to support it.

The Future Is in Your Hands

How should we manage our planet so that it provides a healthy, satisfying life for the current generation of people, while simultaneously retaining biodiversity and the resources needed for future generations? No one can give a single, simple answer to that question. However, two interacting issues must be considered: (1) What should human lifestyles look like and what technologies are appropriate to produce those lifestyles in a sustainable way, and (2) how many people can Earth support, at what lifestyle?

Changes in Lifestyle and Use of Appropriate Technologies Are Essential

The billions of people on Earth will never all agree on exactly what is needed for a happy, fulfilling life. Nearly everyone would agree, however, that a minimal lifestyle must include adequate food and clothing, clean air and water, good health care and working conditions, educational and career opportunities, and access to natural environments. Most of Earth's people live in less-developed countries and lack at least some of these necessities.

Without a sustainable approach to development, there can be no long-term improvement in the quality of human life. We must make choices about what technologies are sustainable in the future, and how to make the transition from the realities of today to a hoped-for tomorrow. For example, in the long run, unless energy sources such as nuclear fusion become a reality, sustainable living must rely on renewable energy sources—solar, wind, geothermal, and wave energy—that don't produce highly toxic wastes or more carbon dioxide than the planet can recycle. We must emulate natural ecosystems by recycling nonrenewable resources. Our choices as consumers can provide markets for foods and durable goods that are produced sustainably. Humans clearly have the ability to destroy nature, but we also have the ability and the need to protect it.

Human Population Growth Is Unsustainable

The root causes of environmental degradation are simple: too many people using too many resources and generating too much waste. As the IUCN eloquently stated in *Who Will Care for the Earth?* ". . . the central issue [is] how to bring human populations into balance with the natural ecosystems that sustain them."

In the long run, that balance cannot be achieved if the human population continues to grow. Given the lifestyle to which the vast majority of people on Earth aspire, many are convinced that the balance cannot be maintained even with our current population, and yet every year, we add 75 million to 80 million people. No matter how simple our diets, how efficient our housing, how low impact our farming techniques, or how much we reuse and recycle, population growth will eventually overwhelm our best efforts.

Let's return to our comparison of Earth's biocapacity and the human ecological footprint (**Fig. 30-16**). As you can see, the rapid increase in humanity's ecological footprint between 1961 and 2005 (red line; see also Fig. 30-6) is roughly paralleled by our rapid population increase (blue line). The ecological footprint *per person* (green line) has been almost constant for the past 35 years—in other words, the average person was using about the same amount of Earth's biocapacity in 2005 as in 1970. If the human population had not increased, the total human ecological footprint would still be well below Earth's biocapacity, but because there are so many more of us, the total human footprint has climbed far above Earth's biocapacity. Eliminating, and probably reversing, population growth is essential if we wish to improve the quality of life for the 6.8 billion of us already here, provide at least the potential for a similar quality of life for our descendants, and save what is left of Earth's biodiversity for future generations.

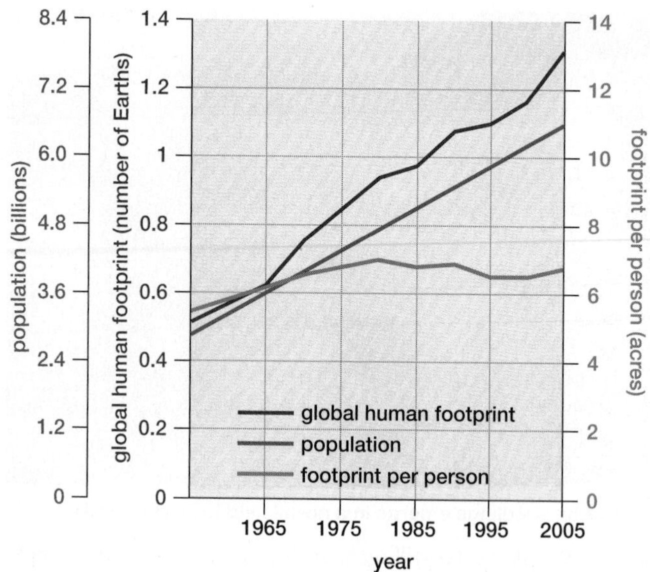

▲ **FIGURE 30-16 Human population growth threatens sustainability** Between 1961 and 2005, human population growth (blue line) increased at approximately the same rate as the global human ecological footprint (red line). The footprint per person (green line) has remained almost the same since 1970, meaning that the increase in our global footprint has resulted almost entirely from population growth. Data from the World Wildlife Fund, the Zoological Society of London, and the Global Footprint Network. 2008. *The living planet report*.

The Choices Are Yours

This chapter has provided some examples of human activities moving in the right direction. Look around your campus and community—what is being done sustainably? What isn't? What would it take to make the necessary changes? In "Links to Everyday Life: What Can Individuals Do?" we suggest some ways that individuals can live more sustainably and help protect life on Earth.

Links to *Everyday Life*

What Can Individuals Do?

"There are no passengers on spaceship earth. We are all crew."

—Marshall McLuhan

Sustainable living is, ultimately, an ethic that must permeate all levels of human society, beginning with individuals. The adage "Reduce, Reuse, and Recycle" provides excellent advice to minimize your impact on Earth's life-support systems. Here are some ways to make a difference:

Conserve Energy

- *Heating and cooling.* Don't heat your house to over 68°F in winter or air condition it below 78°F in summer. Turn down the heat or cooling while you're away. When you purchase or remodel a home, consider energy-efficient features such as passive solar heating, good insulation, an attic fan, double-glazed windows (with "low-E" coating to reduce heat transfer), and good weather stripping. Plant deciduous trees on the south side of your home for shade in summer (when the trees are covered with leaves) and sun in winter (after the leaves have dropped off). If possible, purchase renewable energy from your energy provider.
- *Hot water.* Take shorter showers and switch to low-flow shower heads. Wash only full loads in your washing machine and dishwasher; use cold water to wash clothes; don't prewash your dishes. Insulate and turn down the temperature on your water heater.
- *Appliances.* Compare Energy Star ratings when you choose a major appliance. Don't use your dryer in the summer—put up a clothesline. Turn off unused lights and appliances. Replace incandescent light bulbs with fluorescent or LED bulbs wherever possible.
- *Transportation.* Choose the most fuel-efficient car that meets your needs, and use it efficiently by combining errands. Use public transit, carpool, walk, bicycle, or telecommute when possible.

Conserve Materials

- *Recycle.* Look into recycling options in your community, and recycle everything that is accepted. Explore composting (there are excellent sites on the Internet). Support and encourage campus and community recycling efforts.
- *Buy recycled material.* Purchase recycled paper products. Decking and carpet are now made from recycled plastic bottles.
- *Reuse.* Reuse anything possible, such as manila envelopes, file folders, and both sides of paper. Refill your water bottle, and use refill cups when possible. Reuse your grocery bags. Give away—rather than throwing away—serviceable clothing, toys, and furniture. Make rags out of old clothes and use them instead of disposable cleaning materials.
- *Conserve water.* If you live in a dry area, plant drought-resistant vegetation around your home to reduce water usage.

Support Sustainable Practices

- *Food choices.* Buy locally and organically grown produce that does not require long-distance shipping. Look for shade-grown coffee with the "Bird-Friendly™" or "Rainforest Alliance Certified" seal of approval, and request it at your local Starbucks. Reduce meat, particularly beef, consumption. Search the Internet to find out which fish at your local supermarket have been harvested sustainably.
- *Limit or avoid the use of harmful chemicals.* Harsh cleaners, insecticides, and herbicides contaminate water and soil.

Magnify Your Efforts

- *Support organized conservation efforts.* Join conservation groups and donate money for conservation efforts. Find these on the Internet, and sign up for e-mail alerts that educate you about environmental legislation and make it easy for you to contact your government representatives and express your views. Join the Campus Climate Challenge and reduce energy use on your campus.
- *Volunteer.* Join grassroots efforts to change the world—this is where it all begins. Volunteer for local campus and community projects that improve the environment.
- *Make your vote count.* Investigate candidates' stands and voting records on conservation issues, and consider this information when making your candidate choices.
- *Educate.* Through your words and actions, share your concern for sustainability with your family, friends, and community. Write letters to the editor of your school or local newspaper, to local businesses, and to elected officials. Look for ways your campus could conserve energy, recruit other concerned students, and lobby for change.
- *Reduce population growth.* Consider the consequences of the enormous and expanding human population when you plan your family. Adoption, for example, allows people to have large families while simultaneously contributing to the welfare of humanity and the environment.

Case Study revisited

The Migration of the Monarchs

Sure, a monarch butterfly is beautiful, and a hundred million of them are spectacular. But can they help people to build a house or feed their families? Indeed they can, because it is not a case of monarchs and trees versus the local people. The reality is that the monarchs and trees can help the *campesinos* to prosper.

Aerial photographs show continued deforestation in the Monarch Biosphere Reserves, mostly by illegal loggers. But do these loggers replant the forest? No. In return for large profits for a few years, the loggers would leave the *campesinos* with bare slopes that cannot provide any of the dozens of ecosystem services typical of a healthy forest, such as preventing soil erosion, retaining and gradually releasing clean water, and providing wildlife habitat.

Fortunately, over 5 million trees have been planted in the past decade by a coalition of organizations, in collaboration with the Mexican government. These organizations include World Wildlife Fund Mexico, La Cruz Habitat Preservation Project, and the Michoacan Reforestation Fund, with the assistance of grants from American Forests, the National Fish and Wildlife Foundation, and the U.S. Fish and Wildlife Service. In an ideal world, helping the *campesinos* would be enough motive to replant the forests; in the real world, most of these organizations would not be involved were it not for the monarchs.

But how can the *campesinos* make a living in a Monarch Biosphere Reserve? The World Wildlife Fund, Alternare, and other groups are helping Mexican agricultural experts to train the *campesinos* in sustainable agriculture. One of the most profitable "crops" in the reserve is, ironically, trees. The soil and climate provide ideal conditions for rapid tree growth. Some conifers mature in less than 20 years. Planting seedlings today allows some harvesting in only 5 years, for firewood and Christmas trees. In 15 years, the trees are large enough for commercial lumber. If, meanwhile, the trees are continually replanted, the cycle can continue indefinitely, and the old-growth groves needed by the monarchs can be left alone.

Another source of income for the *campesinos* is ecotourism. More than 100,000 people come to the Monarch Biosphere Reserves each year to see the butterflies. Although ecotourism is not without its own problems, if properly regulated, it can both preserve the forest and provide significant income opportunities for the local people, who serve as guides to the monarch groves, and offer food, accommodations, and souvenirs for the tourists.

BioEthics Consider This

How do you think that society should deal with environmental problems that lack a "poster child" like the monarch butterfly? Are the ecosystem services provided by forests in, say, Guatemala or Borneo less important because they don't happen to be home to a few hundred million spectacular butterflies?

CHAPTER REVIEW

Summary of Key Concepts

30.1 What Is Conservation Biology?

Conservation biology is the branch of science that seeks to understand and conserve biodiversity, including diversity at the genetic, species, and ecosystem levels.

30.2 Why Is Biodiversity Important?

Biodiversity is a source of goods, such as food, fuel, building materials, and medicines. Biodiversity provides ecosystem services such as forming soil, purifying water, controlling floods, moderating climate, and providing genetic reserves and recreational opportunities. The emerging discipline of ecological economics attempts to measure the contribution of ecosystem goods and services to the economy, and estimates the costs of losing them to unsustainable development.

30.3 Is Earth's Biodiversity Diminishing?

Natural communities have a low background extinction rate. Many biologists believe that human activities are currently causing a mass extinction, increasing extinction rates by a factor of 100 to 1,000. Threatened species are designated as critically endangered, endangered, or vulnerable, depending on their likelihood of extinction in the near future. According to the IUCN Red List, in 2008, about 17,000 plants and animals were considered to be threatened with extinction.

30.4 What Are the Major Threats to Biodiversity?

The ecological footprint estimates the area of Earth required to support the human population at any given level of consumption and waste production. Biocapacity estimates the resources and waste-absorbing capacity actually available. The human footprint is exceeding Earth's biocapacity, leaving less and less to support other forms of life. Major threats to biodiversity include habitat destruction and fragmentation as ecosystems are converted to human uses; overexploitation as populations of wild animals and plants are harvested beyond their ability to regenerate; invasive species; pollution; and global warming.

30.5 How Can Conservation Biology Help to Preserve Biodiversity?

Conservation biology seeks to identify the diversity of life, explore the impact of human activities on natural ecosystems, and apply this knowledge to conserve species and foster the survival of healthy, self-sustaining communities. Conservation biology integrates knowledge from many areas of science and requires the efforts of government leaders, environmental lawyers, conservation organizations, and, most importantly, individuals. Conservation efforts include conserving wild ecosystems by establishing wildlife reserves connected by wildlife corridors, with the goal of preserving functional communities and self-sustaining populations.

30.6 Why Is Sustainability the Key to Conservation?

Sustainable development meets present needs without compromising the future. It requires that people maintain biodiversity, recycle raw materials, and rely on renewable resources. Biosphere Reserves promote conservation and sustainable development. A shift to sustainable farming is crucial for conserving soil and water, reducing pollution and energy use, and preserving biodiversity.

Human population growth is unsustainable and is driving consumption of resources beyond nature's ability to replenish them. We must bring our population into line with Earth's ability to support us, leaving room and resources for all forms of life. Individuals must make responsible reproductive choices and reduce resource consumption so that the demands of the human population do not exceed what Earth can sustain.

Key Terms

biocapacity *587*
biodiversity *582*
Biosphere Reserve *595*
conservation biology *582*
core reserve *592*
critically endangered species *586*
ecological footprint *587*
ecosystem services *582*
endangered species *586*
habitat fragmentation *588*
keystone species *594*
mass extinction *586*
minimum viable population (MVP) *589*
no-till *596*
overexploitation *590*
sustainable development *595*
threatened species *586*
vulnerable species *586*
wildlife corridor *592*

Thinking Through the Concepts

Fill-in-the-Blank

1. Three levels of biodiversity are __________, __________, and __________. If the population of a species becomes too small, it is likely to have lost much of its __________ diversity.
2. Products or processes by which functioning ecosystems benefit humans are collectively called __________. Four important examples of these benefits include __________, __________, __________, and __________.
3. Many of the benefits that humans derive from functioning ecosystems, such as purifying water, have traditionally been considered to be free. The discipline of __________ tries to quantify the monetary value of these benefits.
4. The major threats to biodiversity include __________, __________, __________, __________, and __________. For most endangered species, __________ is probably the major threat.
5. The smallest population of a species that is likely to be able to survive in the long term is called the __________. When suitable habitat for a given species is split up into areas that are too small to support a large enough population, this is called __________. One way in which conservation biologists seek to maintain large enough populations is to set up core reserves of suitable habitat, connected by __________.
6. A Native American saying tells us that "We do not inherit the Earth from our ancestors, we borrow it from our children." If this principle guided our activities, we would practice __________ development.

Review Questions

1. Define *conservation biology*. What are some of the disciplines it draws on, and how does each discipline contribute to it?
2. What are the three different levels of biodiversity, and why is each one important?
3. What is ecological economics? Why is it important?
4. List the types of goods and services that natural ecosystems provide.
5. What five specific threats to biodiversity are described in this chapter? Provide an example of each.
6. Why is the TAMAR turtle project a good model for conservation and sustainable development?
7. What types of evidence support the hypothesis that the wolf is a keystone species in Yellowstone National Park?

Applying the Concepts

1. **BioEthics** What are the ethical foundations of conservation biology? Do you agree with them? Why or why not?
2. List some reasons that the ecological footprints of U.S. residents are by far the largest in the world. Looking at your own life, how could you reduce the size of your footprint? How does the ecological footprint of U.S. residents extend into the Tropics?
3. Search for and describe some examples of habitat destruction, pollution, and invasive species in the region around your home or campus. Predict how each of these might affect specific local populations of native animals and plants.
4. Identify a dense suburban development near your home or school. Redesign it to make it into a sustainable development. (This would make a good group project.)
5. What economic arguments would conventional farmers be likely to raise against switching to organic farming techniques and other sustainable agricultural methods? What would be the advantages to farmers? How does this affect consumers?
6. In December 2007, the U.S. government passed legislation mandating the production of at least 36 billion gallons of biofuels by 2022. Discuss the use of biofuels from as many angles as possible.

MB ***Go to www.masteringbiology.com for practice quizzes, activities, eText, videos, current events, and more.***

Animal Anatomy and Physiology

The animal body is an exquisite expression of the elegance with which evolution has linked form to function.

Homeostasis and the Organization of the Animal Body

Chapter 31

▲ Three-hundred pound defensive ends couldn't knock Korey Stringer down, but strenuous exercising during hot weather did.

Case Study

Overheated

WHAT DID KOREY STRINGER, All-Pro offensive lineman for the Minnesota Vikings, Josh Fant, a freshman pitcher on his college baseball team, and Luke Roach, a marathon runner, all have in common? They died of heat stroke while exercising. How can that happen to young, healthy men?

Although "normal" human body temperature is often given as 98.6°F (37°C), real body temperatures in healthy people range between about 97° and 99°F. This narrow range of body temperatures is an example of homeostasis at work. Homeostasis is the ability to maintain a fairly constant internal condition in the face of varying external conditions, such as temperatures from below freezing to well over 100°F, and whether you're asleep or running a marathon.

Body temperatures much above 99°F indicate fever or hyperthermia. Fever is one of the body's defenses against infection; the body's "thermostat" in the brain turns up the heat in an attempt to combat invasion, usually by bacteria (see pp. 696–697). Hyperthermia (from Greek words meaning "excessive heat") arises when the body generates or absorbs more heat than it can get rid of, resulting in a dangerously elevated body temperature. Heat stroke is the often-deadly extreme of hyperthermia.

When the body temperature rises beyond about 104°F, a person becomes confused and disoriented, and may faint. Brain cells die, and the body often fails even to attempt to restore normal body temperature; in fact, one frequent symptom of heat stroke is failure to sweat! As a result, the body temperature continues to rise, sometimes soaring as high as 108°F, and death usually follows soon thereafter.

As you read this chapter, consider the many aspects of your physiology, in addition to temperature, that maintain homeostasis. How does the body recognize the "right" body temperature, or the correct concentration of sugar, salts, and oxygen in the blood? What kinds of actions does the body take to correct imbalances? What happens if imbalances cannot be corrected?

At a Glance

31.1 HOMEOSTASIS: HOW DO ANIMALS REGULATE THEIR INTERNAL ENVIRONMENT?

Whether you are sitting in your room, hiking through the desert, or shivering in a blizzard, most of the cells of your body—for instance, in your heart, brain, muscles, or bone marrow—maintain an almost constant temperature. Further, whether you are out in the air or swimming in a pool, the ocean, or the Great Salt Lake, those cells are bathed in extracellular fluid of an almost constant composition despite enormous differences in the water and salt content of the outside environment.

Humans, of course, are not the only animals to maintain benign, relatively stable internal conditions in the face of harsh, often rapidly changing external conditions. Kangaroo rats, for example, have the same salts in their bodily fluids as humans do, and at about the same concentrations, despite living in deserts so hot and dry that they may not find water to drink for weeks at a time. Salmon live most of their lives in salt water, but mate and spawn in fresh water without exploding from water seeping into their bodies by osmosis. A complete list of such accomplishments would be almost endless.

This "internal constancy" was first recognized by French physiologist Claude Bernard in the mid-nineteenth century. In the 1920s, Walter Cannon coined the term **homeostasis,** to describe the processes by which an organism maintains its internal environment within the narrow range of conditions necessary for optimal cell functioning in the face of a changing external environment. Although the word "homeostasis" (meaning "to stay the same") implies a static, unchanging state, the internal environment actually seethes with activity as the body continuously adjusts to internal and external changes.

The Internal Environment Is Maintained in a State of Dynamic Constancy

The internal state of an animal body can be described as a dynamic constancy. Many physical and chemical changes do occur (the dynamic aspect), but the net result of all this activity is that physical and chemical parameters are kept within the range that cells require to function (the constant aspect). Examples of conditions within an animal's body that are regulated by homeostatic mechanisms include the following:

- Temperature
- Water and salt concentrations
- Glucose concentrations
- pH (acid-base balance)
- Oxygen and carbon dioxide concentrations

Why are cells so particular about their surroundings? Under normal conditions, animal cells are constantly generating and using large quantities of ATP to sustain their life processes (see pp. 101–102). Continuous supplies of high-energy molecules (primarily glucose) and oxygen are required to carry out the series of reactions that generate most of this ATP. Thus, energy production helps explain the importance of both glucose and oxygen levels.

Each of these reactions is catalyzed by a specific protein whose ability to function depends on its three-dimensional structure, maintained, in part, by hydrogen bonds. These crucial but vulnerable bonds can be disrupted by an environment that is too hot, too salty, or too acidic or basic (see p. 51). The need to maintain these bonds and the protein function that depends on them helps explain the requirement for a narrow range of temperature, salt, and pH. And, as we will see in Chapter 33, reactions between carbon dioxide and water help to control the pH of the blood and extracellular fluid, so it is also crucial to regulate CO_2 concentrations.

Animals Vary in Their Homeostatic Abilities

In the external conditions that they normally encounter, many animals, such as most birds and mammals, are extremely proficient at maintaining homeostasis for all of the internal conditions listed above. Some animals, however, have reduced or absent homeostasis for one or more aspects of their internal environment. For example, many marine invertebrates, including a variety of snails, crabs, and worms, cannot regulate the overall concentration of their bodily fluids. When hot, dry air evaporates some of the water in a tide pool—increasing the salt concentration in the remaining water—or when heavy rains dilute the seawater in the pool, the body fluids of these animals become more or less salty, respectively, as their bodies lose or gain water. Marine worms living in mudflats experience low oxygen and high carbon dioxide levels twice a day during low tides. Typically, such animals move (if they can) or else partly shut down—for example, they can reduce their activity, which slows their metabolism and reduces their oxygen requirements—until better conditions return with the incoming tide.

Probably the most common and best studied variability in homeostasis among animals is a dramatic difference in their ability to regulate body temperature.

Animals Vary in How—and How Well—They Regulate Body Temperature

You are probably familiar with descriptions of mammals and birds as "warm-blooded," and reptiles, amphibians, fish, and invertebrates as "cold-blooded." In fact, the bodies of desert pupfish (**Fig. 31-1a**) may reach over 100°F (37.8°C) as their desert pools heat in the summer sun, so fish may not be cold-blooded at all. What about classifying animals by how constant their body temperatures are? Well, hummingbirds have body temperatures of about 104°F (40°C) while foraging during the day (**Fig. 31-1b**), but may cool down to 55°F (13°C) during the night to save energy.

To avoid such confusion, scientists often classify animals according to their major source of body warmth. Animals are **endotherms** (Greek for "inside heat") if they produce most of their heat by metabolic reactions; birds and mammals are endotherms. A few fish—tuna and some large sharks—as well as some butterflies and bees can also warm their bodies considerably by using metabolic heat. Animals are **ectotherms** (Greek for "outside heat") if they derive most of their heat from the environment. For example, insects and lizards often warm up by basking in the sun (**Fig. 31-1c**). Reptiles, amphibians, and most fish and invertebrates are ectotherms. In general, endotherms have higher metabolic rates than ectotherms do, allowing them to keep their bodies at consistently warm temperatures.

Ectotherms generally have lower and more variable body temperatures than endotherms do, because they are more dependent on environmental heat. However, through behavior or by occupying a very constant environment, ectotherm body temperatures may also remain quite stable. For example, the pupfish mentioned earlier can tolerate water temperatures ranging from 36° to 113°F (2.2° to 45°C), but can breed only within a narrow range of temperatures. During breeding season, a pupfish can regulate its temperature quite precisely by swimming to different areas of its pond or hot spring as the temperature changes. In the deep ocean, the temperature is so constant (typically from about 32° to 37.5°F, or 0° to 3°C) that ectothermic deep-sea fish experience little variation in body temperature throughout their lives.

Because warmer temperatures increase the rate of metabolic reactions, there are both benefits and costs to keeping warm. A major benefit is that a warm body usually can sense its environment better, respond more quickly, and move faster than a cold body. Hummingbirds, for example, need high body temperatures during the day for their amazing aerial acrobatics. The major cost, of course, is the energy required to maintain a high body temperature. If hummingbirds were to try to maintain a high body temperature throughout a very cool night, they would exhaust their energy reserves and starve. Many animals cool down at night and become relatively inactive, thereby conserving energy, but warm up and become active during the day. At night, butterflies and bees cool down and cannot fly; lizards are often too sluggish to hunt or escape predators. At daybreak, bees shiver and butterflies beat their wings to generate metabolic heat, while lizards seek a warm, sunlit stone, providing the heat they need to resume their active lifestyles.

How does an animal "know" the conditions within its body—its temperature or other variables—and how to adjust them when necessary? The internal environment is maintained by mechanisms collectively known as feedback systems.

Feedback Systems Regulate Internal Conditions

There are two types of feedback systems: (1) negative feedback systems, which counteract the effects of changes in the internal environment and are principally responsible for

(a) Desert pupfish

(b) Ruby-throated hummingbird

(c) Iguana

▲ **FIGURE 31-1 Warm-blooded or cold-blooded? (a)** Because "cold-blooded" fish such as this desert pupfish may be quite warm, and **(b)** "warm-blooded" animals like this hummingbird can become quite cool, scientists prefer to classify animals as endothermic or ectothermic, depending on the source of body warmth. **(c)** This iguana basking in the sun illustrates a behavioral mechanism that reptiles, which are ectotherms, use to regulate body temperature.

maintaining homeostasis; and (2) positive feedback systems, which drive rapid, self-limiting changes, such as those that occur when a mother gives birth.

Negative Feedback Reverses the Effects of Changes

The most important mechanism governing homeostasis is **negative feedback,** in which a change in the environment causes responses that "feed back" and counteract the change. The overall result of negative feedback is to return the system to its original condition. All negative feedback systems contain three principal components: a sensor, a control center, and an effector. The sensor detects the current condition, the control center compares that condition to a desired state called the set point, and the effector produces an output that restores the desired condition. Let's see how negative feedback systems work, first by using an everyday example (heating your home) and then by using a biological example (the control of body temperature).

In the negative feedback system that controls the temperature of your home on a cold day, the sensor is a thermometer, the control center is a thermostat, and the effector is a heater (**Fig. 31-2a**). The thermometer detects the room temperature, and sends that information to the thermostat, where the actual temperature is compared to the set point of the desired temperature (most thermostats, of course, have a thermometer built in). If the actual temperature is below the set point, then the thermostat signals the heater to turn on and generate heat. The heater warms the room, restoring the temperature to the set point, which causes the thermostat to turn off the heater.

Because none of these components is perfect or responds instantaneously, the room temperature actually gets a bit cooler than the set point before the thermostat turns on the heater, and a bit warmer than the set point before the heat stops. The temperature oscillating above and below the set point is an example of the dynamic constancy characteristic of homeostasis: The temperature actually does change over time (it is dynamic), but it always stays within limits (it is relatively constant).

Negative Feedback Maintains Body Temperature Although the systems are made of cells and fluids instead of electronics, wires, and heaters, people and other endothermic animals use conceptually similar negative feedback systems to maintain their internal temperature despite fluctuations in the temperature around them (**Fig. 31-2b**). The temperature control center is located in the hypothalamus, a part of the brain that controls many homeostatic responses. In most people, the temperature set point is between 97° and 99°F. Nerve endings in the hypothalamus, abdomen, skin, and large veins act as temperature sensors and transmit this information to

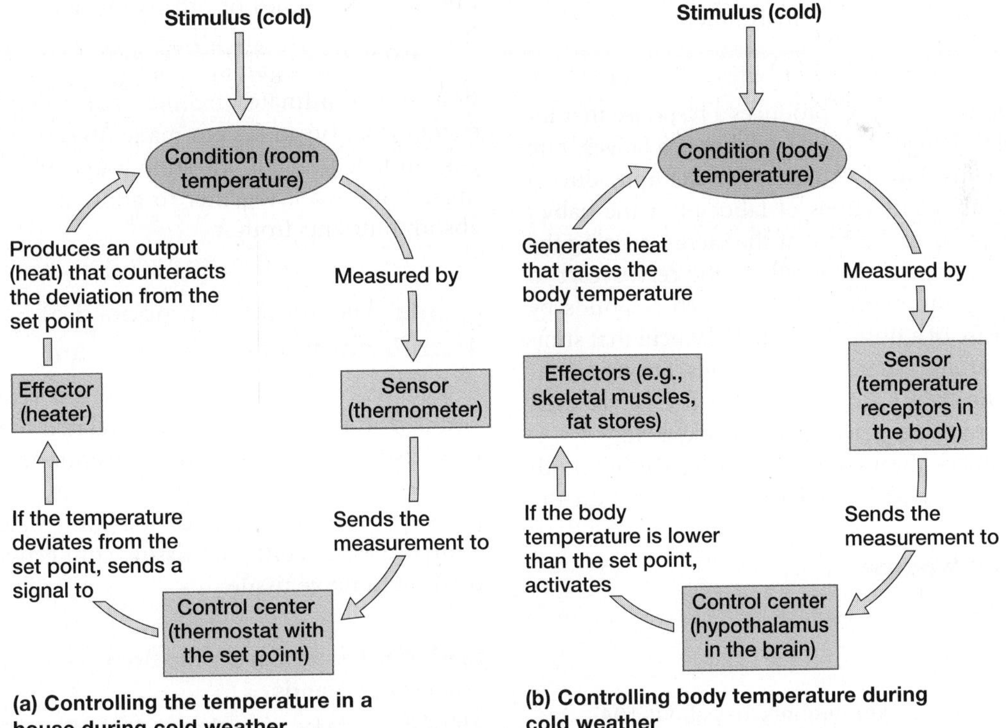

(a) Controlling the temperature in a house during cold weather

(b) Controlling body temperature during cold weather

▲ **FIGURE 31-2 Negative feedback maintains homeostasis** In negative feedback, responses to a stimulus (at the top of each diagram) counteract the effects of the stimulus. **(a)** Negative feedback regulates a specific condition (here, the temperature in your house) within narrow limits. **(b)** Most aspects of animal physiology, such as the control of body temperature, are regulated by negative feedback.

QUESTION What would happen if a cold, shivering mammal ingested a poison that destroyed all of its body's nerve endings that detect heat?

the hypothalamus. If the body temperature falls below the set point, the hypothalamus activates effector mechanisms that raise body temperature, such as increasing cellular metabolism by burning fats, shivering (small, rapid muscular contractions that produce heat), or moving to a warmer place. The blood vessels supplying nonvital areas of the body (such as the face, hands, feet, and skin) are constricted, reducing heat loss and diverting warm blood to inner regions (including the brain, heart, and other internal organs). When normal body temperature is restored, the sensors signal the hypothalamus to switch off these actions that generate and conserve heat.

The body's temperature control system can also act to reduce body temperature. If body temperature rises over the set point, the hypothalamus sends out signals that cause the blood vessels leading to the skin to dilate, allowing warm blood to flow to the skin, where the heat can be radiated out to the surrounding air. Sweat glands secrete fluid, further cooling the body by evaporating water from the skin. The fatigue and discomfort caused by elevated body temperature usually cause people to stop exercising, or at least to slow down, so the body generates less heat.

Negative feedback mechanisms abound in physiological systems. In the chapters that follow, you will find many examples of homeostatic control that operate by negative feedback, including the systems regulating blood oxygen content, water balance, blood-sugar levels, hormone levels, and other components of the internal environment.

Positive Feedback Enhances the Effects of Changes

In **positive feedback,** a change produces a response that intensifies the initial change. Positive feedback is relatively rare in biological systems, but does occur, for example, during childbirth. The early contractions of labor push the baby's head against the cervix at the base of the uterus, causing the cervix to stretch and open. Nerve cells in the cervix react to stretch by signaling the hypothalamus, which responds by triggering the release of a hormone called oxytocin that stimulates more and stronger uterine contractions. Stronger contractions cause the baby's head to stretch the cervix even more, which in turn causes release of more oxytocin.

Positive feedback always stops, eventually. During childbirth, the delivery of the baby relieves the pressure on the cervix, eliminating the stimulus triggering the positive feedback cycle.

Case Study continued

Overheated

When a person's body temperature reaches about 104° to 106°F, the body's ability to regulate its temperature malfunctions: Blood flow to the skin slows down, sweating is often reduced or stops altogether, and the individual's metabolic rate increases, generating more heat. Thus, body temperature increases even further. All too often, these runaway positive feedback loops are terminated only by death.

31.2 HOW IS THE ANIMAL BODY ORGANIZED?

Animals can maintain homeostasis only if all of their parts work together. Further, the animal body must perform dozens, perhaps hundreds, of functions simultaneously, to coordinate the activity of thousands to trillions of cells. For example, we have seen that the hypothalamus in the brain, temperature sensors in several parts of the body, and sweat glands and blood vessels in the skin are all crucial to controlling human body temperature—and this list by no means includes all of the parts of the body that are involved. This coordination of complex bodily systems is based on a simple organizational hierarchy of structures:

cells → tissues → organs → organ systems

An example of this hierarchy is illustrated in **Figure 31-3.** You learned in Chapter 1 (see pp. 2–3) that cells are the building blocks of all life. The animal body incorporates cells into **tissues;** each tissue is composed of dozens to billions of structurally similar cells that act in concert to perform a particular function. Tissues are the building blocks of **organs,** discrete structures that perform complex functions. Examples of organs include the stomach, small intestine, kidneys, and urinary bladder. Organs, in turn, are organized into **organ systems,** groups of organs that function in a coordinated manner. For example, the digestive system is an organ system made up of the mouth, esophagus, stomach, small intestine, large intestine, and other organs that work together to allow us to digest food and absorb nutrients from it.

Animal Tissues Are Composed of Similar Cells That Perform a Specific Function

A tissue is composed of cells that are similar in structure and perform a small number of related functions. A tissue may also include extracellular components produced by these cells, as in the case of cartilage and bone. Here we present a brief overview of the four major categories of animal tissues: epithelial tissue, connective tissue, muscle tissue, and nerve tissue.

Epithelial Tissue Covers the Body, Lines Its Cavities, and Forms Glands

All of the surfaces of the body—including the skin; the digestive, respiratory, and urinary tracts; and the circulatory system—are covered by epithelial tissue (**Fig. 31-4**). **Epithelial tissue** consists of sheets of cells firmly attached to one another by connections such as desmosomes and tight junctions (see pp. 92–93). Sheets of epithelial tissue are also attached to an underlying noncellular layer of fibrous proteins, called the basement membrane. Some of these protein fibers are secreted by

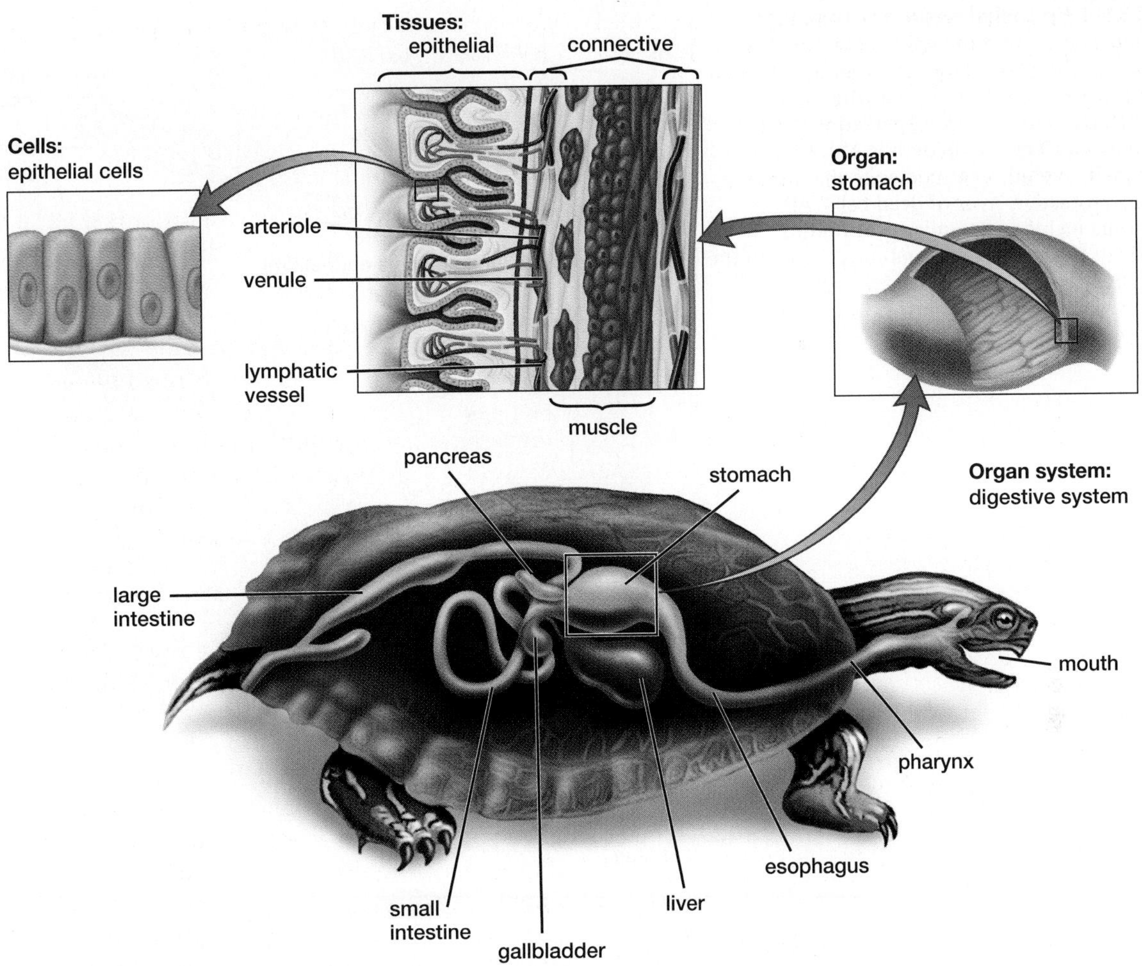

▲ **FIGURE 31-3 Cells, tissues, organs, and organ systems** The animal body is composed of cells, which make up tissues, which combine to form organs that work in harmony as organ systems.

the epithelial cells, whereas others are secreted by connective tissue cells (see the next section). The basement membrane provides support, flexibility, and strength to the epithelial sheets, which is obviously important in organs that are subject to stretch and strain, such as the skin. In the gallbladder and urinary bladder, tight junctions seal off the extracellular spaces between epithelial cells, which is crucial to prevent bile or urine, respectively, from leaking into the body.

Simple epithelium is only one cell thick. Simple epithelium lines most or all of the respiratory, digestive, urinary, reproductive, and circulatory systems. Simple epithelia vary in structure, depending on the organ and its function. For example, the simple epithelium that lines the lungs consists of a single layer of thin, flattened cells whose shape makes them ideal for allowing rapid diffusion of gases between the lungs and the bloodstream (**Fig. 31-4a**). The epithelium that lines the trachea (windpipe) is more complex, consisting of both short and elongated cells bearing cilia, interspersed with glandular cells that secrete mucus (**Fig. 31-4b**). The mucus traps inhaled debris, and the cilia sweep it out of the trachea.

Stratified epithelium, which is several cells thick and can usually withstand considerable wear and tear, is mostly found in the skin and just inside body openings that are continuous with the skin, such as the mouth and anus. Cells at the base of stratified epithelium divide rapidly, giving rise to daughter cells that form the surface layers. In the skin, the daughter cells become filled with strong proteins (mostly keratin) as they differentiate. Eventually these cells die and form a tough, flexible, waterproof layer on the surface of the skin (**Fig. 31-4c**).

An important property of both simple and stratified epithelia is that their cells are continuously lost and replaced. For example, consider the abuse suffered by the epithelia of your mouth and skin. Scalded by coffee and scraped by corn chips, the epithelium of the mouth would be destroyed within a few days if it did not replace itself continuously. The dead surface cells in the epithelium of the skin are worn away by rubbing on clothing, bath towels, tools, and furniture, and are replaced by new cells constantly produced in the dividing deeper layers, resulting in a complete replacement about every 2 weeks.

Glands are cells or groups of cells specialized to secrete (release) large quantities of substances outside the cell. Most, but not all, glands are composed of specialized epithelial cells. (Two glands, the posterior pituitary and adrenal medulla, are actually nervous, not epithelial, tissue.) Some glands are single cells, such as the mucus-secreting cells of the tracheal

▶ **FIGURE 31-4 Epithelial tissue (a)** Thin, flattened cells in a single layer form the epithelial tissue that lines the lungs, where exchange of gases by diffusion is important. **(b)** The trachea is lined with elongated epithelial cells bearing cilia, interspersed with cells that secrete mucus. **(c)** The epidermis of the skin consists of multilayered, stratified epithelial tissue covered by a protective layer of dead cells. All epithelial tissue includes a thin layer of fibrous proteins, called the basement membrane, beneath the epithelial cells.

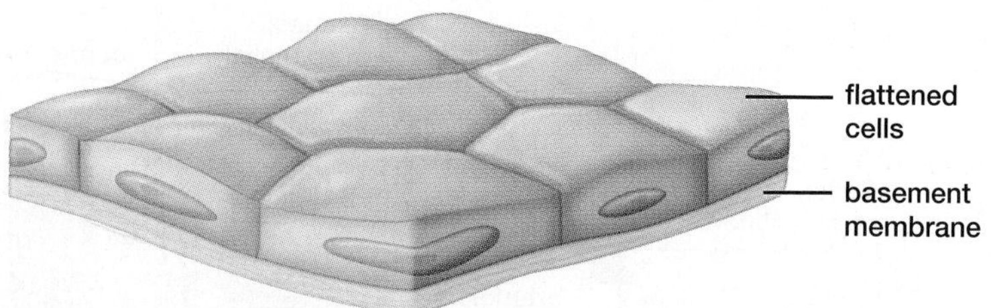

(a) Lining of the lungs (simple epithelium)

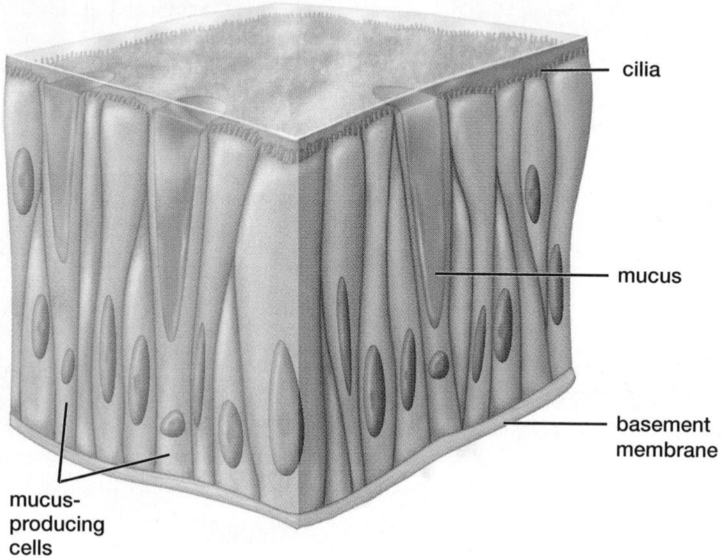

(b) Lining of the trachea (simple epithelium)

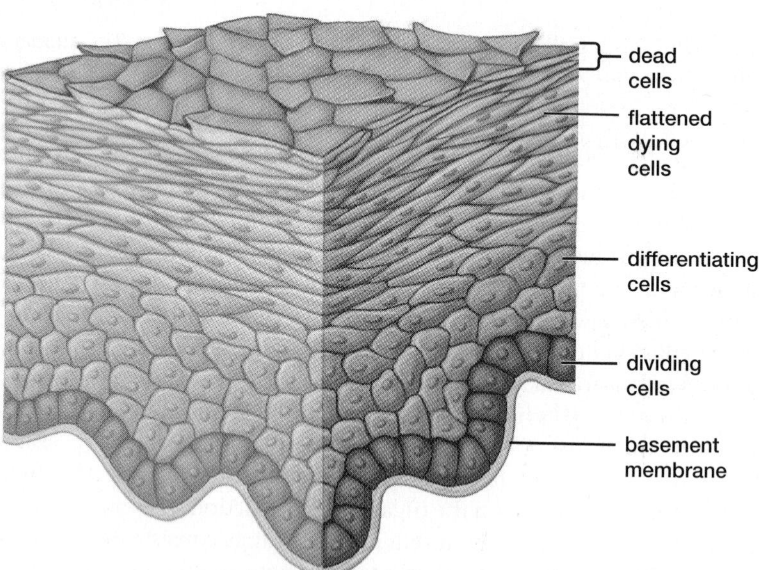

(c) Skin epidermis (stratified epithelium)

epithelium (see Fig. 31-4b). Most glands are multicellular, including those that secrete saliva, sweat, milk, and most hormones.

Glands are classified into two categories: exocrine glands and endocrine glands. **Exocrine glands** secrete substances into a body cavity or onto the body surface, usually through a narrow tube or duct. Examples of exocrine glands are sweat glands (see Fig. 31-11 later in this chapter) and mammary glands (which are actually modified sweat glands), salivary glands that release saliva into the mouth, and several glands that release digestive enzymes into the stomach or small intestine. **Endocrine glands** lack ducts. They secrete hormones into the extracellular fluid, from which the hormones diffuse into nearby capillaries. **Hormones** are chemicals that are produced in small quantities and are transported in the bloodstream to regulate the activity of other cells, often in distant parts of the body. Endocrine glands and their hormones are covered in detail in Chapter 37.

Connective Tissues Have Diverse Structures and Functions

Most types of **connective tissue** support and strengthen other tissues, and help to bind the cells of these other tissues together in coherent structures, such as skin or muscles. However, the characteristic that unifies all connective tissues is that a large proportion of connective tissue consists of a fluid—which may be almost as thin as water or as thick as molasses. This fluid contains a high concentration of proteins, the most abundant of which is usually collagen. Connective tissue cells are embedded in this fluid matrix, and produce the proteins contained in the fluid. Connective tissues fall into three main categories: loose connective tissue, dense connective tissue, and specialized connective tissue (**Fig. 31-5**).

Loose Connective Tissue This is the most abundant form of connective tissue, consisting of a thick fluid containing scattered cells that secrete protein fibers (**Fig. 31-5a**). This flexible tissue connects, supports, and surrounds other tissue types, and it forms an internal framework for organs such as the liver. Loose connective tissue combines with epithelial tissue to form membranes. (In biology, the term "membrane" means a sheet-like structure. This term is used for [1] the phospholipid bilayer that forms the plasma membrane, endoplasmic reticulum, and the outer layers of other organelles in a cell; and [2] thin multicellular structures in the body composed of epithelial and connective tissues, such as the skin, the mucous membranes that line the digestive and respiratory tracts, and membranes that attach many internal organs, including the intestines and stomach, to the abdominal wall.)

Dense Connective Tissue Dense connective tissue is packed with collagen fibers (**Fig. 31-5b**). In some, such as **tendons** (which connect muscles to bones) and **ligaments** (which connect bones to bones), the collagen fibers are arranged parallel to one another. In other types of dense connective tissue, such as the dermis of the skin and the capsules that surround many internal organs and muscles, the collagen fibers form an irregular meshwork. Both designs provide flexibility and strength. Tendons and ligaments have tremendous strength, but only in the direction in which the collagen fibers are oriented (which is why twisting a knee in the "wrong" direction can rupture a ligament). The irregular mesh of collagen in skin and muscle capsules resists stress coming from any direction, but of course isn't as strong as the parallel arrangement in tendons and ligaments.

Specialized Connective Tissues This diverse group includes cartilage, bone, fat, blood, and lymph. **Cartilage** consists of widely spaced cells surrounded by a thick, nonliving matrix composed of collagen secreted by the cartilage cells (**Fig. 31-5c**). Cartilage covers the ends of bones at joints, provides the supporting framework for the respiratory passages, supports the ear and

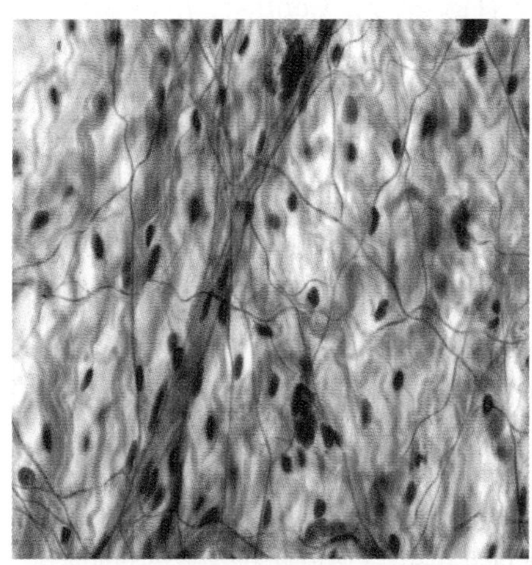

(a) Loose connective tissue

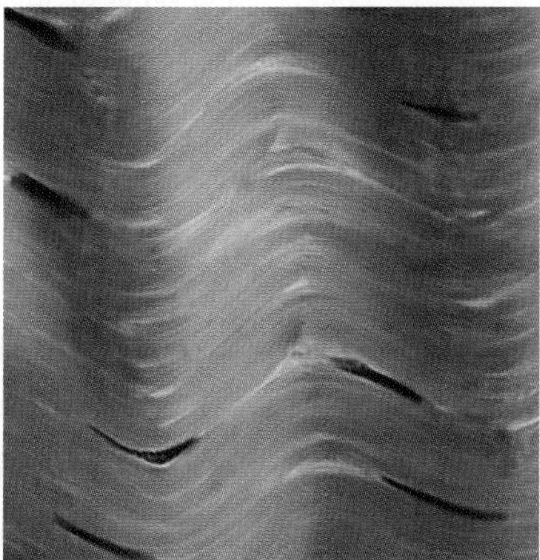

(b) Dense connective tissue

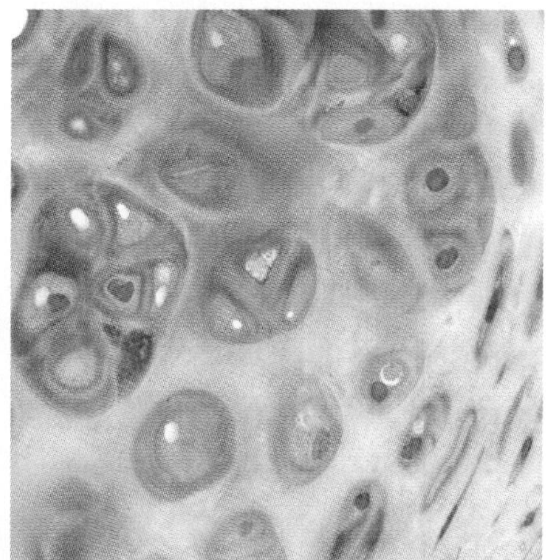

(c) Specialized connective tissue

▲ **FIGURE 31-5 Connective tissue (a)** Loose connective tissue underlies epithelial tissue and supports almost every organ in the body. **(b)** The dense connective tissue in tendons and ligaments is tremendously strong in the direction in which its protein fibers are aligned, but can be easily disrupted by forces acting perpendicular to the aligned fibers. **(c)** Cartilage is a specialized connective tissue consisting of scattered cells surrounded by the collagen and glycoproteins that they secrete.

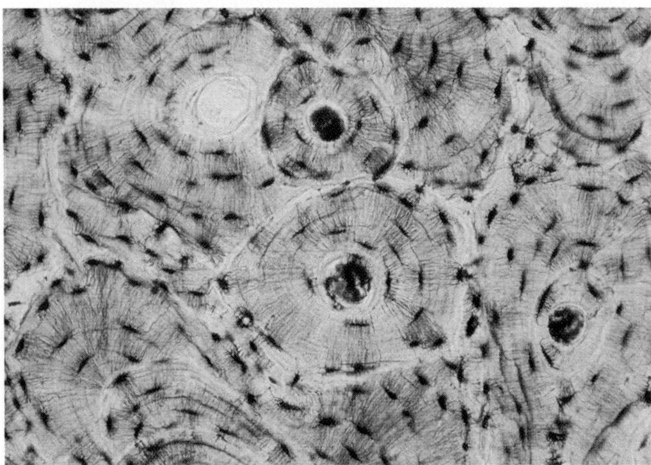

▲ **FIGURE 31-6 Bone** Concentric circles of bone (a specialized connective tissue), deposited around a central canal that contains blood vessels, are clearly visible in this micrograph. Individual bone cells appear as dark spots trapped in small chambers within the hard matrix that the cells themselves deposit.

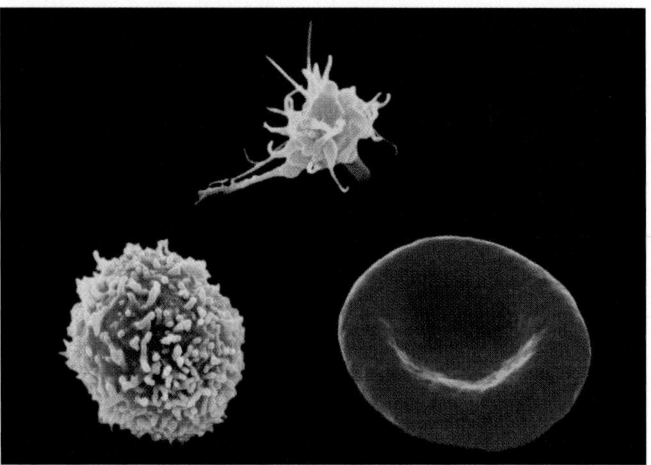

▲ **FIGURE 31-8 Blood cells** Blood contains white and red blood cells (bottom) and cell fragments called platelets (top), as shown in this colorized scanning electron micrograph. The cells are suspended in plasma, the fluid portion of this specialized connective tissue.

QUESTION Why is blood considered a form of connective tissue?

nose, and forms shock-absorbing pads between the vertebrae. Cartilage is fairly flexible, but can snap if bent too far.

Bone resembles cartilage, but its matrix is hardened by deposits of calcium phosphate. Bone forms in concentric circles around a central canal, which contains blood vessels (**Fig. 31-6**). (We will discuss cartilage and bone in Chapter 40.) Fat cells, collectively called **adipose tissue,** are modified for long-term energy storage (**Fig. 31-7**). Adipose tissue is especially important in the physiology of animals adapted to cold environments because it not only stores energy but also serves as insulation.

Although they are liquids, **blood** and **lymph** are considered specialized forms of connective tissues because they are composed largely of extracellular fluids in which proteins and individual cells are suspended. The cellular portion of blood consists of red blood cells (which transport oxygen), white blood cells (which fight infection), and cell fragments called platelets (which aid in blood clotting) (**Fig. 31-8**). These are all suspended in extracellular fluid called plasma. Lymph consists largely of fluid that has leaked out of blood capillaries (the smallest of the blood vessels) and is carried back to the circulatory system within lymph vessels. You will learn more about blood and lymph in Chapter 32.

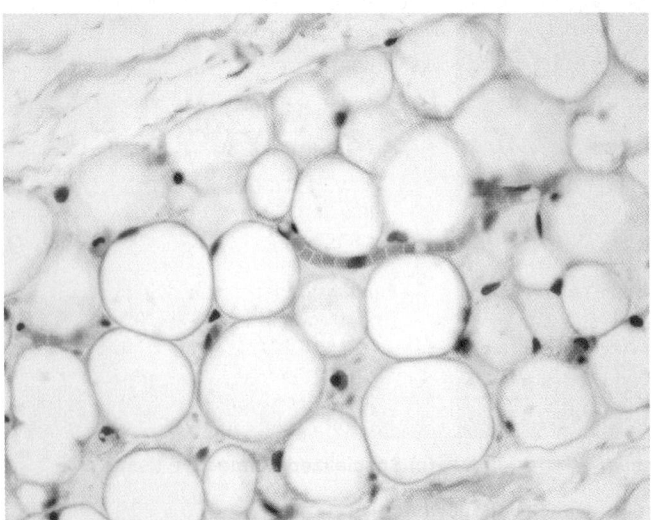

▲ **FIGURE 31-7 Adipose tissue** Adipose tissue is specialized connective tissue made up almost exclusively of fat cells. A droplet of oil occupies most of the volume of the cell.

Muscle Tissue Has the Ability to Contract

The long, thin cells of muscle tissue contract (shorten) when stimulated and then relax passively when the stimulation stops. There are three types of muscle tissue: skeletal, cardiac, and smooth muscle (**Fig. 31-9**). Contraction of **skeletal muscle** (**Fig. 31-9a**) is stimulated by the nervous system and is generally under voluntary, or conscious, control. As its name implies, its main function is to move the skeleton, as occurs when you walk or turn the pages of this text.

Cardiac muscle is located only in the heart (**Fig. 31-9b**). Unlike skeletal muscle, it is spontaneously active and involuntary (not under conscious control). Cardiac muscle cells are in-

Case Study continued
Overheated

During strenuous exercise, your skeletal muscles generate 10 to 15 times more heat than they do when you're simply resting—enough to raise your body temperature about 2°F every 5 minutes or so, if you can't get rid of the extra heat. If it is humid and your sweat doesn't evaporate fast enough, or if you're wearing a lot of clothing (let alone football pads), it won't take long to enter the first stages of heat stroke.

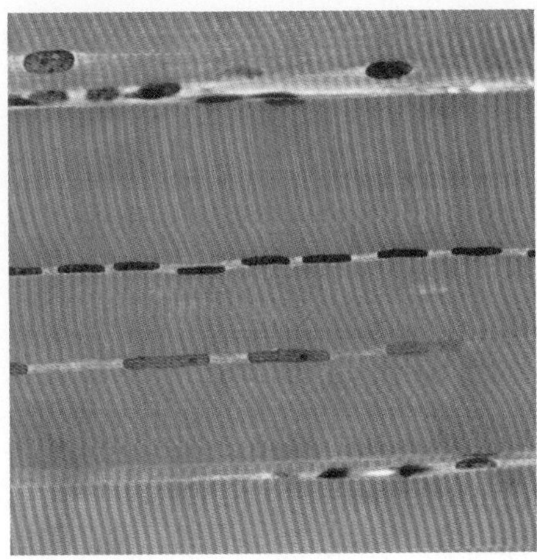
(a) Skeletal muscle

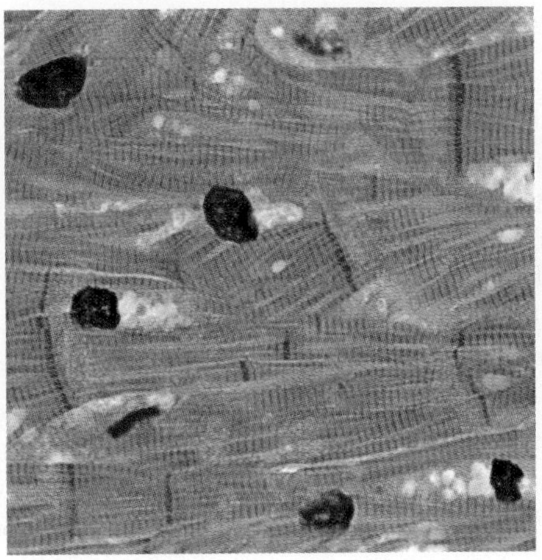
(b) Cardiac muscle

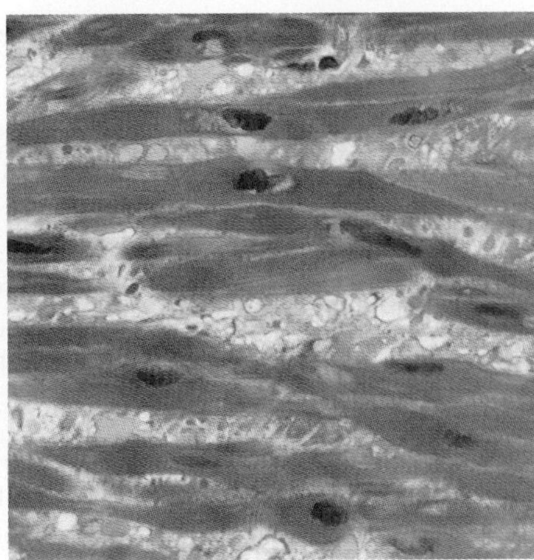
(c) Smooth muscle

▲ **FIGURE 31-9 Muscle tissue** Fibrous proteins inside muscle cells cause them to shorten, or contract, when they are stimulated. **(a)** In skeletal muscle, the regular arrangement of these proteins makes this type of muscle tissue appear striped, or "striated," when viewed under a microscope. **(b)** In cardiac muscle, the contractile proteins are also aligned in stripes. Clusters of gap junctions (dark bands between cells) interconnect the cytoplasm of adjacent cardiac muscle cells. **(c)** Smooth muscle fibers are spindle-shaped cells in which the contractile proteins are not consistently aligned; hence, they appear "smooth."

terconnected by gap junctions, through which electrical signals spread rapidly throughout the heart, stimulating the cardiac muscle cells to contract in a coordinated fashion.

Smooth muscle (Fig. 31-9c) is found throughout the body, embedded in the walls of the digestive and respiratory tracts, uterus, bladder, larger blood vessels, skin, and even in the iris of the eye. Smooth muscle produces slow, sustained contractions that may be stimulated by the nervous system, by stretch, or by hormones and a variety of other chemicals. Smooth muscle contraction is typically involuntary. You will learn more about muscles in Chapter 40.

Nerve Tissue Is Specialized to Produce and Conduct Electrical Signals

You owe your ability to sense and respond to the world to **nerve tissue,** which makes up the brain, the spinal cord, and the nerves that travel from them to all parts of the body. Nerve tissue is composed of two types of cells: nerve cells, also called neurons, and glial cells (Fig. 31-10). **Neurons** (Fig. 31-10a) are specialized to generate electrical signals and to conduct these signals to other neurons, muscles, or glands. A typical nerve cell consists of dendrites, which are specialized to receive signals from other neurons or from the environment; a cell body, which contains the nucleus and carries out most of the neuron's metabolism; an axon that carries the neuron's electrical signal to a muscle, gland, or other neuron; and the synaptic terminals that transmit information to the other cells. **Glial cells** (Fig. 31-10b) surround, support, electrically insulate, and protect neurons. Glial cells also regulate the composition of the extracellular fluid in the nervous system, allowing neurons to function optimally. We discuss nerve tissue in Chapter 38.

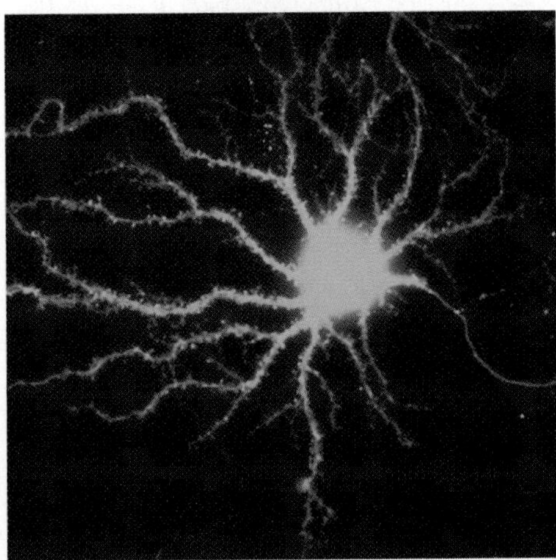
(a) Neuron

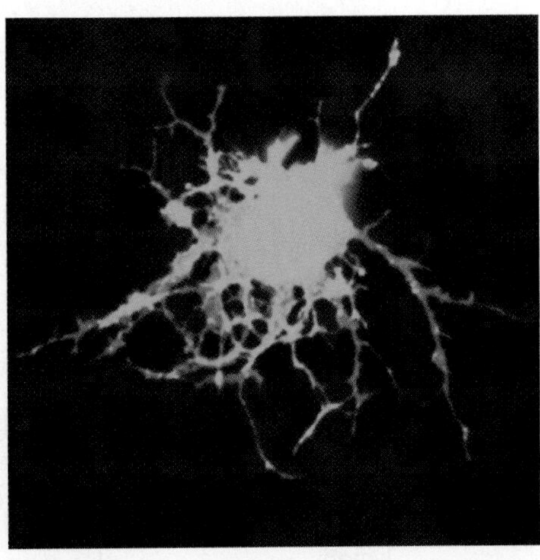
(b) Glial cell

◀ **FIGURE 31-10 Nerve tissue** Nerve tissue consists of neurons and glial cells. **(a)** Neurons are specialized to receive and transmit signals. This micrograph shows a human neuron that stimulates muscle cells to contract. **(b)** There are several different types of glial cells in the brain; this one, called an astrocyte, helps to nourish neurons and protect them from damage.

Organs Include Two or More Interacting Tissue Types

Organs are formed from at least two tissue types that function together. Most organs, however, consist of all four tissue types, with varying types and proportions of epithelial, connective, muscular, and nervous tissues. Most organs function as part of organ systems, which are discussed in other chapters of this unit. In the following section, we describe the skin as a representative organ that includes all four of the tissue types.

The Skin Illustrates the Properties of Organs

The skin consists of an outer layer of epithelial tissue underlain by connective tissue that contains a blood supply, a nerve supply, muscle, and glandular structures derived from the epithelium (**Fig. 31-11**). Far more than a simple covering for the body, the skin is essential for survival. For example, it is such an important barrier against the entry of disease-causing microorganisms and the evaporation of precious body water that the large-scale destruction of skin—as occurs with extensive burns—can prove fatal.

The **epidermis,** or outer layer of the skin, is a specialized stratified epithelial tissue (see Fig. 31-11). It is covered by a protective layer of dead cells produced by the underlying living epidermal cells. These dead cells are packed with the protein keratin, which helps make the skin elastic, tough, and relatively waterproof.

Immediately beneath the epidermis lies a layer of loose connective tissue, the **dermis.** Arterioles (small arteries) snake throughout the dermis, and carry blood pumped from the heart into a dense meshwork of capillaries that nourish both the dermal and epidermal tissue. The capillaries empty into venules (small veins) in the dermis. Loss of heat through the skin is regulated by neurons controlling the diameter of the arterioles. To conserve body heat, smooth muscles in the arterioles supplying the skin capillaries contract, reducing the diameter of the arterioles. To cool the body, these smooth muscles relax, allowing the arterioles to dilate and flood the capillary beds with blood, thus releasing excess heat. Lymph vessels collect and carry off extracellular fluid within the dermis. Various sensory nerve endings that are responsive to temperature, touch, pressure, vibration, and pain are scattered throughout the dermis and epidermis and provide information to the nervous system.

Specialized epithelial cells dip down from the epidermis into the dermis, forming **hair follicles.** Cells in the base of the follicles divide rapidly, and their daughter cells fill with

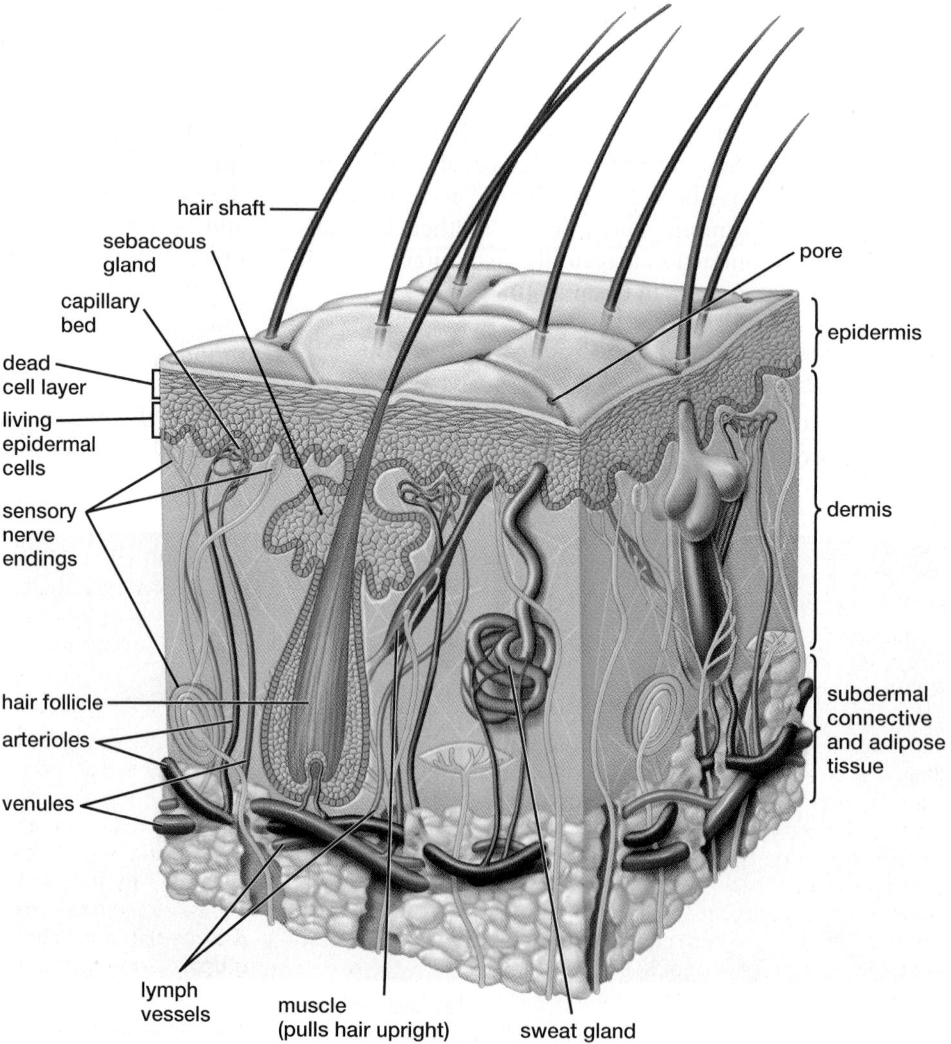

▶ **FIGURE 31-11 Skin is an organ** Mammalian skin, a representative organ, shown in cross-section. The skin contains epithelial, connective, muscle, and nervous tissue. A basement membrane of fibrous proteins holds the epidermis to the dermis, but is too thin to be seen at this scale.

Have you ever wondered 

Why You Get Goosebumps When You Are Frightened?

No matter what the temperature, you may get goosebumps when you're scared. This is a relic of our hairier evolutionary past. In frightening or threatening situations, many mammals erect their fur, which makes them appear larger. With any luck, this may prevent opponents from starting a fight with such an apparently large animal—after all, even the winner may come away with serious injuries.

keratin. As new cells form in the base of the follicle, older, dying cells are pushed upward to the surface of the skin, emerging as hairs.

The dermis is also packed with glands derived from epithelial tissue. Sweat glands produce watery secretions that cool the skin and excrete substances such as salts and urea. Sebaceous glands secrete an oily substance (sebum) that lubricates the epithelium.

Skin also contains muscle tissue. All skin contains muscle cells within the walls of the arterioles that supply blood to the skin. In addition, hairy skin has tiny muscles attached to the hair follicles that can cause the hairs of the skin to "stand on end" in response to signals from motor neurons. Most mammals are able to increase the height of their insulating fur in cold weather by erecting individual hairs, but this reaction is useless for retaining heat in people, who merely experience "goosebumps" when these tiny muscles contract.

Finally, below the dermis is a layer of adipose tissue interspersed with fibrous proteins produced by connective tissue cells. The fat helps to insulate the body, preserving heat and making it easier to maintain a constant body temperature in cool weather. The fat also serves as an energy storage depot.

Organ Systems Consist of Two or More Interacting Organs

Organ systems consist of two or more individual organs (in some cases, located in different regions of the body) that work together, performing a common function. An example is the digestive system, in which the mouth, esophagus, stomach, intestines, and other organs that supply digestive secretions, such as the liver and pancreas, all function together to convert food into nutrient molecules that can be absorbed into the bloodstream (see Fig. 31-3). The major organ systems of the vertebrate body and their representative organs and functions are described in **Table 31-1.**

Case Study revisited
Overheated

Exercise produces heat—in fact, about 70% to 80% of the energy used by your muscles is converted to heat, which is why shivering helps to warm you up on a cold day. Nevertheless, to function properly, the human body needs to stay within the range of 97° to 99°F, no matter how much you exercise. How does it do that? There are usually two heat-releasing responses during exercise. First, blood vessels in the skin relax, increasing blood flow to the surface of the body. If the air is cool enough and you aren't exercising too strenuously, you can probably give off the excess heat to the air. Second, if your temperature is still increasing, you begin to sweat. If the humidity is low enough, the water in the sweat evaporates from your skin, producing a dramatic cooling effect. But what if you're exercising really hard, or if you're really big, or it's hot and humid, or all of these circumstances occur simultaneously? Under these conditions, your body temperature may continue to increase, with fatal consequences.

Vikings offensive lineman Korey Stringer was fit, but huge—6 feet, 4 inches tall, weighing over 330 pounds. In August 2001, the Vikings held intense workouts, in full pads, under conditions of high heat and humidity. Stringer's strenuous exercise generated enormous amounts of heat. High temperatures meant that he couldn't radiate much of that heat to the air; high humidity meant that his sweat didn't evaporate very much, either. The pads insulated his body from whatever cooling the weather conditions might have allowed. Finally, as you learned in Chapter 5 (on pp. 91–92), as an object gets bigger, its internal volume and usually its mass increase more rapidly than its surface area. Stringer had huge muscles, generating lots of heat, but his body surface—his only pathway for heat loss—wasn't anywhere close to twice the area of an average man half his weight, with half of his heat-generating muscle mass. By the time Stringer was taken to a hospital, his body temperature was 108.8°F (over 42.5°C). Soon, he suffered multiple organ failure, as system after system in his body shut down. He died early the next morning.

Smaller athletes, like pitcher Josh Fant or marathoner Luke Roach, are usually at less risk of heat stroke, but can still succumb, either because of high temperatures while exercising (fairly common while playing baseball in summer), or extremely strenuous, prolonged exercise (such as running a marathon).

Heat stroke, of course, is not confined to athletes. In fact, when the air temperature becomes extremely high, especially coupled with high humidity, people can die of heat stroke sitting in their living rooms. Although athletes typically receive more coverage in the news, death from heat stroke is far more common among the elderly during summer heat waves, when the air temperature rises far above healthy body temperature and high humidity limits cooling through sweat. Farm workers, toiling under the hot sun, are also at considerable risk of heat stroke. According to the National Institute on Aging, hundreds of ordinary people, especially the elderly, die from heat stroke every summer.

BioEthics Consider This

Many climate models that forecast global warming also predict increased heat waves, with peak temperatures far exceeding the average warming. During the devastating European heat wave in the summer of 2003, between 35,000 and 52,000 "excess deaths" occurred, probably the vast majority from the heat. When economists and politicians try to calculate the costs of reducing the buildup of carbon dioxide that most climate scientists are convinced causes global warming, should excess deaths be calculated as a cost of inaction? If so, what value should we place on a life?

Table 31-1 Major Vertebrate Organ Systems

Organ System	Major Structures	Physiological Role
Integumentary system 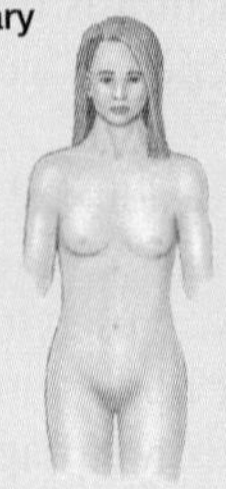	Skin, hair, nails, sensory receptors, various glands	Protects the underlying structures from damage; regulates body temperature; senses many features of the external environment
Circulatory system 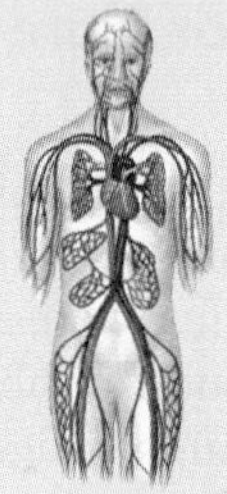	Heart, blood vessels, blood	Transports nutrients, gases, hormones, metabolic wastes; also assists in temperature control
Digestive system 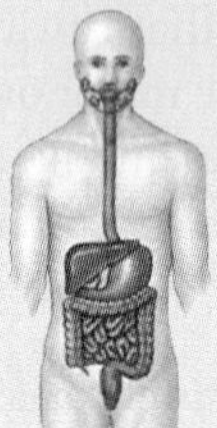	Mouth, esophagus, stomach, small and large intestines, glands producing digestive secretions	Supplies the body with nutrients that provide energy and materials for growth and maintenance
Nervous system 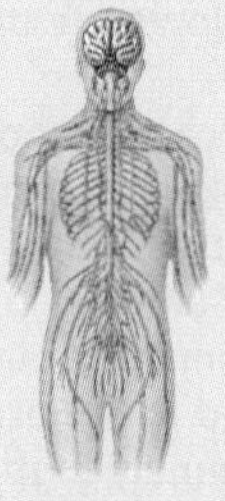	Brain, spinal cord, peripheral nerves	Controls physiological processes in conjunction with the endocrine system; senses the environment, directs behavior
Skeletal system 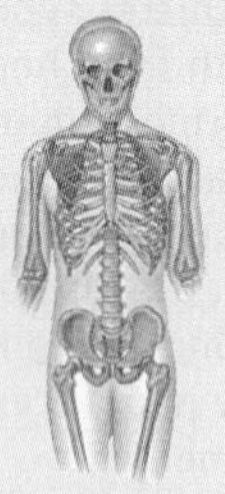	Bones, cartilage, tendons, ligaments	Provides support for the body, attachment sites for muscles, and protection for internal organs
Male reproductive system 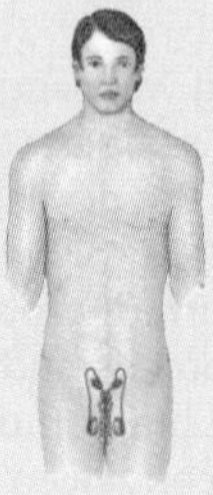	Testes, seminal vesicles, prostate gland, penis	Produces sperm and sex hormones, inseminates female
Respiratory system 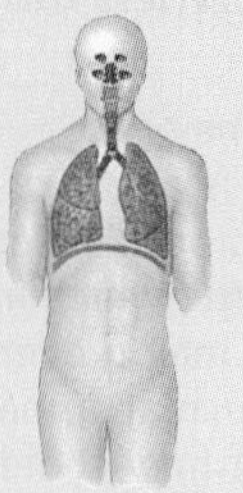	Nose, pharynx, trachea, lungs (mammals, birds, reptiles, amphibians), gills (fish and some amphibians)	Provides a large area for gas exchange between the blood and the environment; allows oxygen acquisition and carbon dioxide elimination
Lymphatic/ immune system 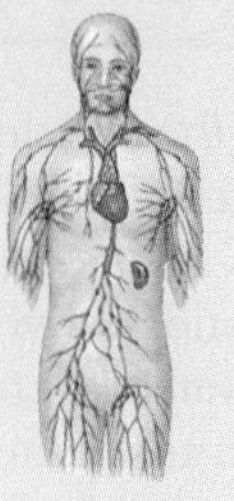	Lymph, lymph nodes and vessels, white blood cells	Carries fat and excess fluids to the blood; destroys invading microbes
Urinary system 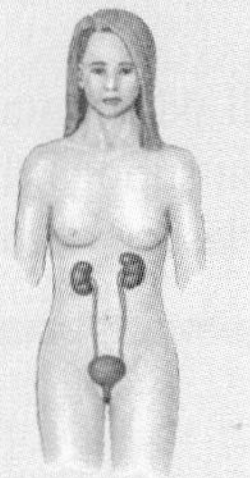	Kidneys, ureters, bladder, urethra	Maintains homeostatic conditions within the bloodstream; filters out cellular wastes, certain toxins, and excess water and nutrients
Endocrine system 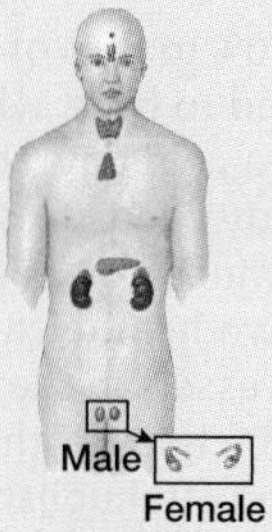	A variety of hormone-secreting glands and organs, including the hypothalamus, pituitary, thyroid, pancreas, adrenals, ovaries, and testes	Controls physiological processes, typically in conjunction with the nervous system
Muscular system 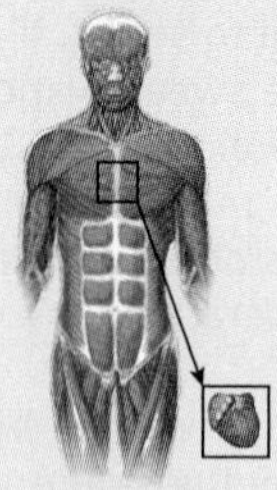	Skeletal muscle	Moves the skeleton
	Smooth muscle	Controls movement of substances through hollow organs (digestive tract, large blood vessels)
	Cardiac muscle	Initiates and implements heart contractions
Female reproductive system 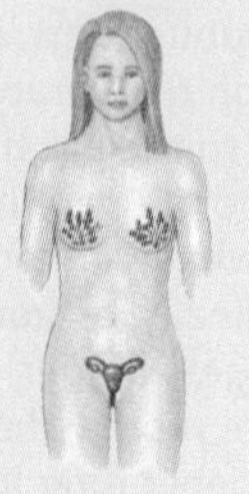	Ovaries, oviducts, uterus, vagina, mammary glands	Produces egg cells and sex hormones, nurtures developing offspring

CHAPTER REVIEW

Summary of Key Concepts

31.1 Homeostasis: How Do Animals Regulate Their Internal Environment?

Homeostasis refers to the dynamic equilibrium within the animal body in which physiological conditions including temperature, salt, oxygen, glucose, pH, and water levels are maintained within a range in which proteins can function and energy can be made available. Animals differ in temperature regulation. Ectotherms derive most of their body warmth from the environment and tend to tolerate larger extremes of body temperature. Endotherms derive most of their heat from metabolic activities and tend to regulate their body temperatures within a narrower range.

Homeostatic conditions are maintained through negative feedback, in which a change triggers a response that counteracts the change and restores conditions to a set point. There are a few instances of positive feedback, in which a change initiates events that intensify the change (such as uterine contractions leading to childbirth), but these situations are all self-limiting.

31.2 How Is the Animal Body Organized?

The animal body is composed of organ systems, each consisting of two or more organs. Organs, in turn, are made up of tissues. A tissue is a group of cells and extracellular material that form a structural and functional unit; it is specialized for a specific task. Animal tissues include epithelial, connective, muscle, and nerve tissue.

Epithelial tissue forms coverings over internal and external body surfaces and also gives rise to glands. Connective tissue usually contains considerable extracellular material in which cells and proteins are embedded, and includes the dermis of the skin, bone, cartilage, tendons, ligaments, fat, and blood. Muscle tissue is specialized to produce movement by contraction. There are three types of muscle tissue: skeletal, cardiac, and smooth. Nerve tissue, including neurons and glial cells, is specialized to generate and conduct electrical signals.

Organs include at least two tissue types that function together. Mammalian skin is a representative organ. The epidermis, an epithelial tissue, covers and protects the dermis beneath it. The dermis contains blood and lymph vessels, sweat and sebaceous glands, hair follicles, tiny muscles that erect the hairs, and a variety of sensory nerve endings. Animal organ systems include the integumentary, respiratory, circulatory, lymphatic/immune, digestive, urinary, nervous, endocrine, skeletal, muscular, and reproductive systems, summarized in Table 31-1.

Key Terms

adipose tissue *612*
blood *612*
bone *612*
cardiac muscle *612*
cartilage *611*
connective tissue *611*
dermis *614*
ectotherm *606*
endocrine gland *611*
endotherm *606*
epidermis *614*
epithelial tissue *608*
exocrine gland *611*
gland *609*
glial cell *613*
hair follicle *614*
homeostasis *605*
hormone *611*
ligament *611*
lymph *612*
negative feedback *607*
nerve tissue *613*
neuron *613*
organ *608*
organ system *608*
positive feedback *608*
simple epithelium *609*
skeletal muscle *612*
smooth muscle *613*
stratified epithelium *609*
tendon *611*
tissue *608*

Thinking Through the Concepts

Fill-in-the-Blank

1. The process by which the body maintains internal conditions within the narrow range required by cells to function is called __________. The most important feedback mechanism governing this process is __________.
2. The four levels of organization of the animal body, from the smallest to the most inclusive, are __________, __________, __________, and __________.
3. Fill in the appropriate tissue type: Supports and strengthens other tissues: __________; forms glands: __________; includes the blood: __________; includes the dermis of the skin: __________; covers the body and lines its cavities: __________; can shorten when stimulated: __________; includes glial cells: __________; includes adipose tissue: __________.
4. Glands with ducts connecting them to the epithelium are called __________ glands. Glands with no ducts are called __________ glands; most of these secrete __________ (a general term for a type of messenger molecule).
5. Fill in the appropriate type(s) of muscle (there may be more than one answer for some blanks): Contracts rhythmically and spontaneously: __________; is controlled voluntarily: __________; contains orderly arrangements of fibrous proteins, giving a striped appearance: __________; is not under voluntary control: __________; is found in the walls of the digestive tract: __________; moves the skeleton: __________.

Review Questions

1. Define *homeostasis*, and explain how negative feedback helps to maintain it. Describe one example of homeostasis in the human body.
2. Define and compare *ectotherms* and *endotherms*. Provide an example of each. Are "cold-blooded" and "warm-blooded" accurate ways to describe them? Explain.
3. Explain positive feedback, and provide one physiological example. Explain why this type of feedback is relatively rare in physiological processes.
4. Explain what goes on in your body to restore temperature homeostasis when you become overheated by exercising on a hot, humid day.
5. Describe the structure and functions of epithelial tissue.
6. What property distinguishes connective tissue from all other tissue types? List three general types of connective tissue, and briefly describe the function of each type.

7. Describe the skin as an organ. Include the various tissues that compose it, and briefly describe the role of each tissue.

Applying the Concepts

1. Why does life on land present particular difficulties in maintaining homeostasis?
2. Advertisements for a sports drink say that "water is not enough" for people who are exercising, and that it is important to consume a liquid that contains flavoring, carbohydrates, potassium, calcium, and sodium. Do you think that consuming such a sports drink during exercise can help maintain homeostasis? Explain your answer.
3. Third-degree burns are usually painless. Skin regenerates only from the edges of these wounds. Second-degree burns are often very painful. Skin regenerates from cells located at the burn edges, in hair follicles, and in sweat glands. First-degree burns are painful but heal rapidly from undamaged epidermal cells. Using this information, draw the depth of first-, second-, and third-degree burns on Figure 31-11.
4. Imagine you are a health care professional teaching a prenatal class for fathers. Create a design for a machine with sensors, electrical currents, motors, and so forth that would illustrate the feedback relationships involved in labor and childbirth that a layperson could understand.

MB® *Go to www.masteringbiology.com for practics quizzes, activities, eText, videos, current events, and more.*

Circulation

Chapter 32

Case Study

Running with Heart

BY JANUARY 2009, 64-year-old retired Harlem bookkeeper Donald Arthur had completed marathons in 26 states and was more than halfway to his goal of completing one in every state. But 13 years earlier, he found even chewing food exhausting, and walking a block could take him an hour. Arthur's heart was failing from *dilated cardiomyopathy*.

Dilated cardiomyopathy (based on Greek, meaning "expanded cardiac muscle disease") causes the heart muscle to become stretched and weakened, and renders it unable to adequately pump blood through the body, a condition called *heart failure*. Heart failure causes fluid to accumulate in the lungs, which, along with the inadequate pumping of the heart, deprives the body of much needed oxygen.

Although dilated cardiomyopathy strikes both sexes and people of all ages, its most common victims are middle-aged men. Some people inherit a tendency to develop this disorder, which is also associated with excessive alcohol consumption, the use of drugs such as cocaine and methamphetamine, and infections that attack heart valves or heart muscle. In serious cases such as Arthur's, lack of oxygen renders the victim short of breath even at rest, and makes any sort of exertion nearly impossible. Without dramatic intervention, the situation gradually worsens and is often fatal.

Donald Arthur's doctors had given him about 3 months to live when a fateful phone call came on a midsummer evening in 1996. Twenty-five-year-old Fitzgerald "Poochie" Gittens, a father of two, had been shot to death in a case of mistaken identity. His heart was a match for Arthur, and Gittens' grieving mother made the decision to donate it, so that her son's untimely death could preserve the life of a stranger.

Just 15 months after his transplant, Donald Arthur (a former self-proclaimed "confirmed couch potato") completed the first of his 10 New York marathons and began a new life. He became a founding member of Transplant Speakers International, an organization that helps to spread the word about organ and tissue donation. Arthur, who does his marathons at a rapid walk, says, "It's not the race, but the opportunity to spread the word and inspire people about organ transplants. You can't imagine the rewards."

How does heart failure differ from a heart attack? How do healthy circulatory systems supply the entire body with the oxygen and nutrients it needs to stay alive? Read on to find out.

▲ Another man's heart keeps Donald Arthur on the move and spreading the word about organ donation.

At a Glance

32.1 WHAT ARE THE MAJOR FEATURES AND FUNCTIONS OF CIRCULATORY SYSTEMS?

Billions of years ago, the first cells were nurtured by the primordial sea in which they evolved. The sea brought them nutrients, which diffused into the cells and washed away the wastes that diffused out. Today, microorganisms and some simple multicellular animals still rely almost exclusively on diffusion to exchange gases with the environment. Sponges, for example, circulate seawater through pores in their bodies, bringing the environment close to each cell (see Fig. 23-5). In more complex animals, individual cells are farther from the outside world, but they still require short diffusion distances, so that adequate nutrients reach them and they aren't poisoned by their own wastes. With the evolution of the circulatory system, a sort of "internal sea" was created that serves the same purpose as the sea did for the first cells. This internal sea transports food and oxygen close to each cell and carries away wastes produced by the cells.

All circulatory systems have three major parts:

- A pump, the **heart,** that keeps the blood circulating.
- A liquid, **blood,** that serves as a medium of transport.
- A system of tubes, **blood vessels,** that conducts the blood throughout the body.

Two Types of Circulatory Systems Are Found in Animals

The circulatory systems of animals take two different forms: open and closed. **Open circulatory systems** are present in many invertebrates, including arthropods (such as crustaceans, spiders, and insects) and mollusks (such as snails and clams). An animal with an open circulatory system has one or more simple hearts, a network of vessels, and a series of interconnected spaces within the body called a **hemocoel** (**Fig. 32-1a**). Within the hemocoel (which may occupy 20% to 40% of the body volume), tissues and internal organs are directly bathed in **hemolymph,** a fluid that functions both as blood and as the extracellular fluid that directly bathes all cells in multicellular organisms, as described later. In insects, the heart is a modified vessel with a series of contracting chambers (see Fig. 32-1a, right). When the chambers contract, valves in the heart are pressed shut, forcing the hemolymph out through vessels and into hemocoel spaces throughout the body. When the heart chambers relax, hemolymph is drawn back into them from the hemocoel.

Closed circulatory systems are present in some invertebrates, including the earthworm (**Fig. 32-1b**) and very active mollusks (such as squid and octopuses). Closed circulatory systems are also characteristic of all vertebrates, including

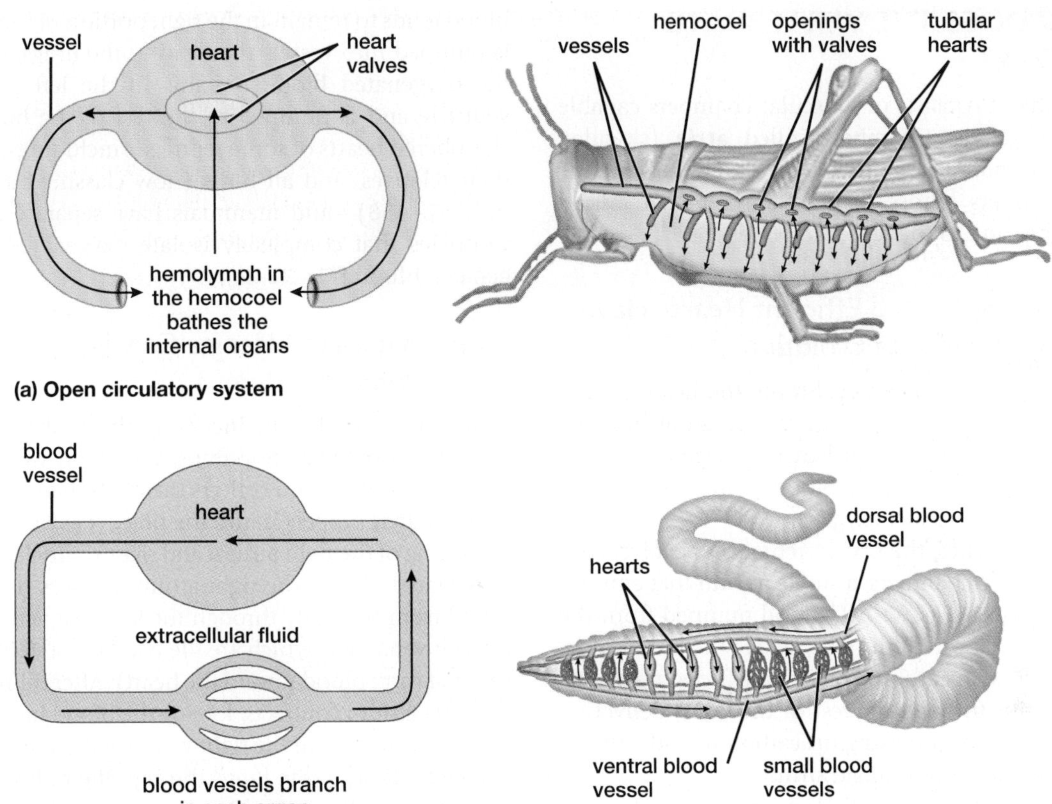

▲ **FIGURE 32-1 Open and closed circulatory systems** Arrows indicate the direction of blood or hemolymph flow. **(a)** (left) In the open circulatory systems of most invertebrates, a heart pumps hemolymph through vessels into the hemocoel, where blood directly bathes the other organs. (right) The grasshopper provides a good example of an open circulatory system (insect hemolymph lacks hemoglobin and is almost clear or very pale green). **(b)** (left) In a closed circulatory system, blood remains confined within the heart(s) and the blood vessels. (right) In the earthworm, five contractile vessels serve as hearts that pump blood through major ventral and dorsal vessels, from which smaller interconnecting vessels branch. Earthworm blood, like ours, contains red hemoglobin.

humans. In closed circulatory systems, the blood is confined to the heart and blood vessels, which branch elaborately throughout the tissues and organs of the body to allow the exchange of nutrients and wastes. Closed circulatory systems allow more rapid blood flow, more efficient transport of dissolved substances, and higher blood pressure than is possible in open systems. In the earthworm, five contractile vessels serve as hearts, pumping blood through major vessels from which smaller vessels branch (see Fig. 32-1b, right).

The Vertebrate Circulatory System Has Diverse Functions

The circulatory system supports all the other organ systems in the body. The circulatory system of vertebrates performs the following functions:

- Transports oxygen from the lungs or gills to the tissues, and transports carbon dioxide from the tissues to the lungs or gills.
- Distributes nutrients from the digestive system to all body cells.
- Transports waste products and toxic substances to the liver, where many of them are detoxified, and to the kidneys for excretion.
- Distributes hormones from the glands and organs that produce them to the tissues on which they act.
- Helps to regulate body temperature by adjustments in blood flow.
- Helps to heal wounds and prevent bleeding by creating blood clots.
- Protects the body from diseases by circulating white blood cells and antibodies.

In the following sections we examine the three parts of the circulatory system: the heart, the blood, and the blood vessels, with an emphasis on the human system. Finally, we describe the lymphatic system, which works closely with the circulatory system.

32.2 HOW DOES THE VERTEBRATE HEART WORK?

The vertebrate heart consists of muscular chambers capable of strong contractions. Chambers called **atria** (singular, atrium) collect blood. Atrial contractions send blood into the **ventricles,** chambers whose contractions circulate blood through the lungs and to the rest of the body.

Increasingly Complex and Efficient Hearts Have Arisen During Vertebrate Evolution

During the course of vertebrate evolution, the heart has become increasingly complex, with more separation between oxygenated blood (which has picked up oxygen from the lungs or gills) and deoxygenated blood (which, in passing through body tissues, has lost oxygen).

The hearts of fishes, the first vertebrates to evolve, consist of two contractile chambers: a single atrium that empties into a single ventricle (**Fig. 32-2a**). Blood pumped from the ventricle passes first through the gills, where the blood picks up oxygen and gives off carbon dioxide. The blood travels directly from the gills through the rest of the body, delivering oxygen to the tissues and picking up carbon dioxide. Blood from the body returns to the single atrium.

Over evolutionary time, as fish gave rise to amphibians and amphibians to reptiles, a three-chambered heart evolved, which consists of two atria and one ventricle (**Fig. 32-2b**). In the three-chambered hearts of amphibians and reptiles such as snakes, lizards, and turtles, deoxygenated blood from the body is delivered into the right atrium, while blood from the lungs enters the left atrium. Both atria empty into the single ventricle. Although some mixing occurs, the deoxygenated blood tends to remain in the right portion of the ventricle and is pumped into vessels that lead to the lungs, while most of the oxygenated blood remains in the left portion of the ventricle and is pumped to the rest of the body. The four-chambered hearts of some reptiles—including crocodiles and their relatives, and all birds (now classified as reptiles; see pp. 457–458)—and mammals have separate right and left ventricles that completely isolate oxygenated from deoxygenated blood (**Fig. 32-2c**).

Four-Chambered Hearts Consist of Two Separate Pumps

Four-chambered hearts, including those of humans, can be viewed as two separate pumps, each with two chambers. In each pump, an atrium collects the blood before passing it to a ventricle that propels it into the body (**Fig. 32-3**). One pump, consisting of the right atrium and right ventricle, deals with deoxygenated blood. The right atrium receives oxygen-depleted blood from the body through the superior vena cava and the inferior vena cava, which are the two largest **veins** (blood vessels that carry blood toward the heart). After filling with blood, the right atrium contracts, forcing the blood into the right ventricle. Contraction of the right ventricle then sends the oxygen-depleted blood to the lungs through the pulmonary **arteries** (blood vessels that carry blood away from the heart).

The other pump, consisting of the left atrium and left ventricle, deals with oxygenated blood. Oxygen-rich blood from the lungs enters the left atrium through pulmonary veins and is then squeezed into the left ventricle. A strong contraction of the left ventricle, the heart's most muscular chamber, sends the oxygenated blood coursing out through a major artery, the aorta, to the rest of the body.

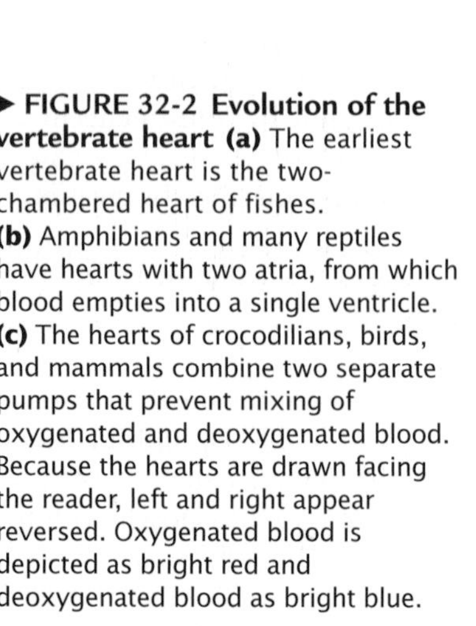

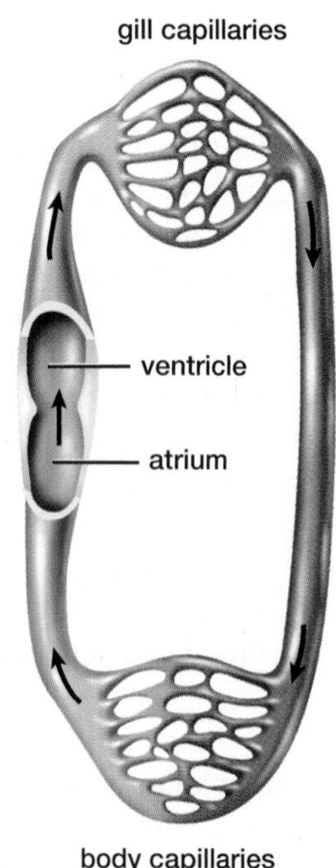

(a) Fish

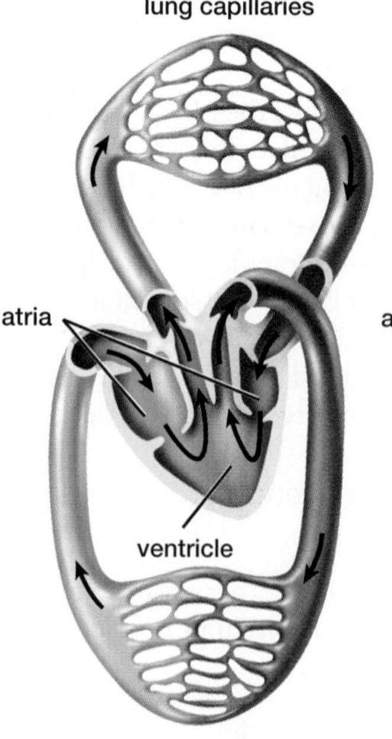

(b) Amphibians and some reptiles

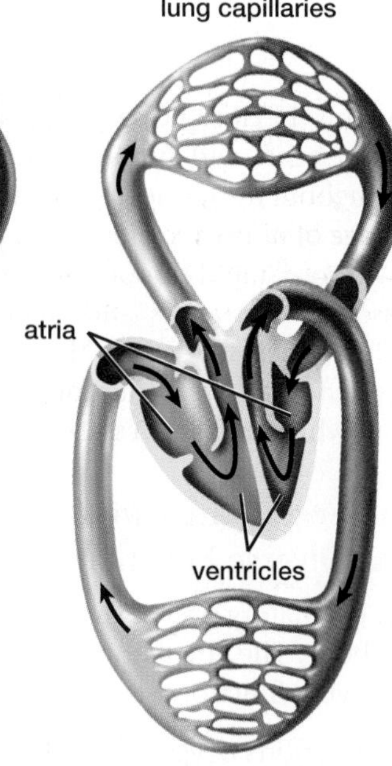

(c) Mammals, crocodiles, and birds

▶ **FIGURE 32-2 Evolution of the vertebrate heart (a)** The earliest vertebrate heart is the two-chambered heart of fishes. **(b)** Amphibians and many reptiles have hearts with two atria, from which blood empties into a single ventricle. **(c)** The hearts of crocodilians, birds, and mammals combine two separate pumps that prevent mixing of oxygenated and deoxygenated blood. Because the hearts are drawn facing the reader, left and right appear reversed. Oxygenated blood is depicted as bright red and deoxygenated blood as bright blue.

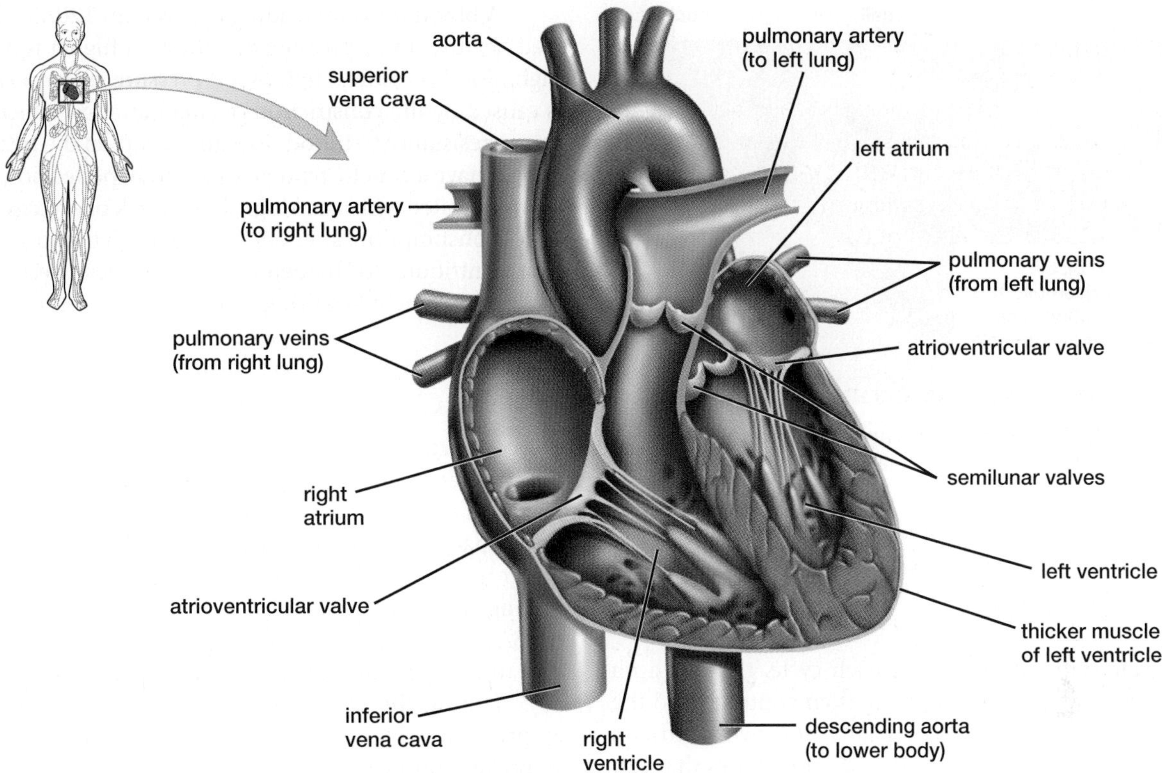

▲ **FIGURE 32-3 The human heart and its valves and vessels** The muscular wall of the left ventricle is thicker than the right because it must pump blood throughout the body. One-way semilunar valves separate the aorta from the left ventricle, and the pulmonary artery from the right ventricle. Atrioventricular valves separate each atrium from its corresponding ventricle.

Case Study continued
Running with Heart

Heart failure is a frequent and potentially fatal result of dilated cardiomyopathy. It occurs when the heart fails to circulate enough blood to meet the body's needs. Often, the stretched and weakened left ventricle can't pump out all the oxygenated blood that it receives from the lungs. As a result, blood backs up into the pulmonary veins, increasing the blood pressure in the lungs. This pressure forces fluid from the blood out through the walls of the lung capillaries (tiny, thin-walled vessels) and into the air spaces of the lungs. As in drowning, fluid in the lungs prevents adequate oxygen from reaching the blood that circulates through the lungs.

Valves Maintain the Direction of Blood Flow

When the ventricles contract, blood must be directed out through the arteries and must be prevented from flowing back up into the atria. Blood that has entered the arteries must also be prevented from flowing back into the ventricles as the heart relaxes. The directionality of blood flow is maintained by one-way valves (see Figs. 32-3 and 32-5). Pressure in one direction opens them easily, but reverse pressure forces them closed. **Atrioventricular valves** allow blood to flow from the atria into the ventricles, but not the reverse. **Semilunar valves** (from Latin, meaning "half-moon" valves, referring to their shape) allow blood to enter the pulmonary artery and the aorta when the ventricles contract, but prevent blood from returning as the ventricles relax.

Cardiac Muscle Is Present Only in the Heart

Most of the heart consists of a specialized type of muscle, **cardiac muscle,** found only in the heart. Each cardiac muscle cell is small, branched, and packed with an orderly array of protein strands that give it a striped appearance (**Fig. 32-4**; see also Fig. 31-9b). Cardiac muscle cells are linked to one another by **intercalated discs,** which appear as bands between the cells. Here, adjacent cell membranes are attached to one another by junctions called desmosomes (see Fig. 5-17a), which prevent the strong heart contractions from pulling the muscle cells apart. Intercalated discs also contain gap junctions that allow the electrical signals that trigger contractions to spread directly and rapidly from one muscle cell to adjacent ones. This causes the interconnected regions of heart muscle to contract almost synchronously.

The Coordinated Contractions of Atria and Ventricles Produce the Cardiac Cycle

The human heart beats about 100,000 times each day. Each heartbeat is actually a series of coordinated events, called the

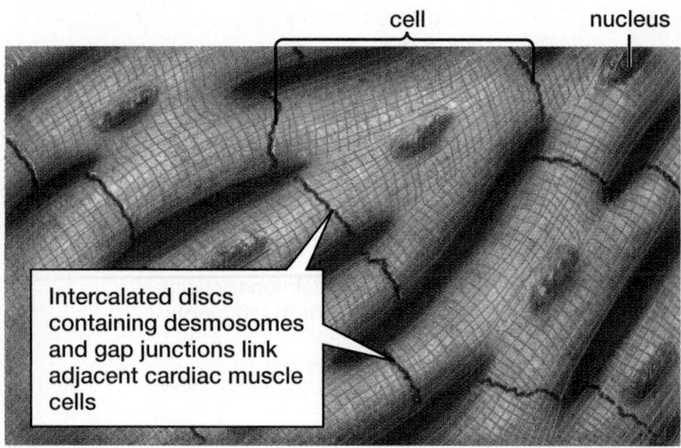
cell
nucleus
Intercalated discs containing desmosomes and gap junctions link adjacent cardiac muscle cells

▲ **FIGURE 32-4 The structure of cardiac muscle** Cardiac muscle cells are branched, and intercalated discs link adjacent cells.

QUESTION If a muscle is exercised regularly, it increases in size. Why is the resting heart rate of a well-conditioned athlete slower than the heart rate of a less active person?

cardiac cycle (**Fig. 32-5**). During each cycle, the two atria first contract in synchrony, emptying their contents into the ventricles. A fraction of a second later, the two ventricles contract simultaneously, forcing blood into arteries that exit the heart. Both atria and ventricles then relax briefly before the cardiac cycle repeats. In a typical resting person, the cycle occurs in just under 1 second, or about 70 times per minute. The cardiac cycle generates the forces that are measured when blood pressure is taken (**Fig. 32-6**). **Systolic pressure** (the higher of the two readings) is measured during ventricular contractions, and **diastolic pressure** is the minimum pressure in the arteries as the heart rests between contractions.

A blood pressure reading of less than 120/80 is considered healthy, and a pressure of 140/90 or higher is defined as high blood pressure. High blood pressure, or **hypertension,** is caused by the constriction of small arteries, which in turn causes resistance to blood flow and strain on the heart. Some people have a genetic tendency toward hypertension, but it is also associated with smoking, obesity, lack of exercise, high alcohol consumption, stress, and aging. High blood pressure, in turn, contributes to "hardening of the arteries," described later in "Health Watch: Repairing Broken Hearts" on pp. 632–633.

Electrical Impulses Coordinate the Sequence of Heart Chamber Contractions

The contraction of the heart is initiated and coordinated by a **pacemaker,** a cluster of specialized heart muscle cells that produces spontaneous electrical signals at a regular rate. The heart's pacemaker is the **sinoatrial (SA) node,** located in the upper wall of the right atrium (**Fig. 32-7**). The gap junctions linking adjacent cardiac cells allow the electrical signals from the SA node to pass freely and rapidly into the connecting cardiac muscle cells and then throughout the atria.

During the cardiac cycle, the atria contract first and empty their contents into the ventricles, then refill while the ventricles contract. This requires a slight delay between the atrial and the ventricular contractions. How is this accomplished? First, the SA node initiates a wave of contraction (**Fig. 32-7** ❶) that sweeps through the right and left atria, which contract in synchrony (**Fig. 32-7** ❷). The signal then reaches a barrier of electrically inexcitable tissue between the atria and the ventricles. Here, the excitation is channeled through the **atrioventricular (AV) node,** a small mass of specialized muscle cells located in the floor of the right atrium (**Fig. 32-7** ❸). The impulse is conducted slowly at the

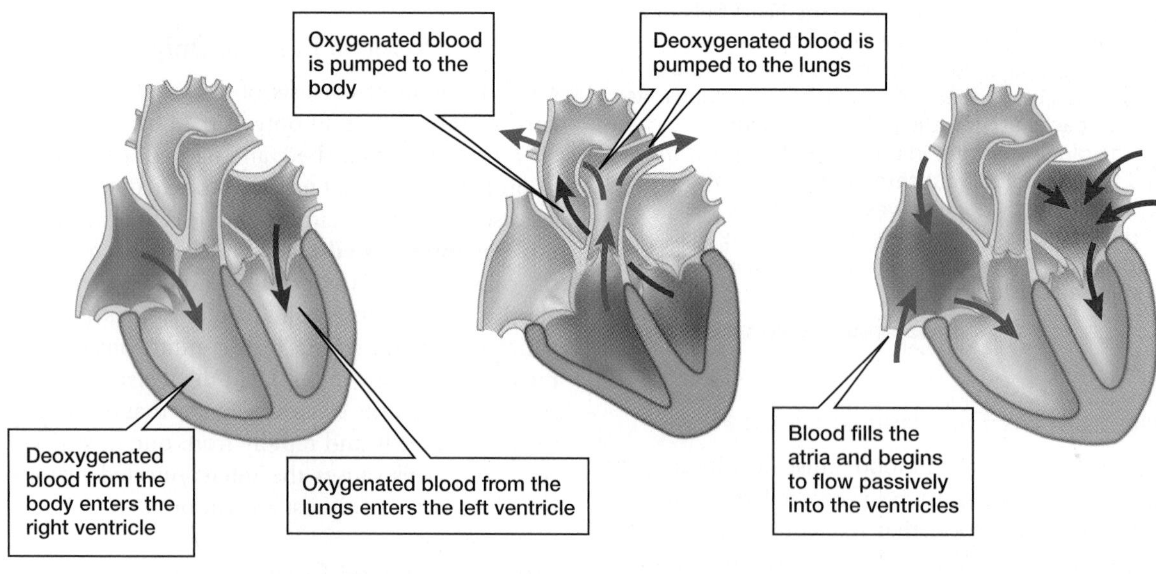
Oxygenated blood is pumped to the body
Deoxygenated blood is pumped to the lungs
Deoxygenated blood from the body enters the right ventricle
Oxygenated blood from the lungs enters the left ventricle
Blood fills the atria and begins to flow passively into the ventricles
❶ Atria contract, forcing blood into the ventricles
❷ Then the ventricles contract, forcing blood through the arteries to the lungs and the rest of the body
❸ The cycle ends as the heart relaxes

▲ **FIGURE 32-5 The cardiac cycle**

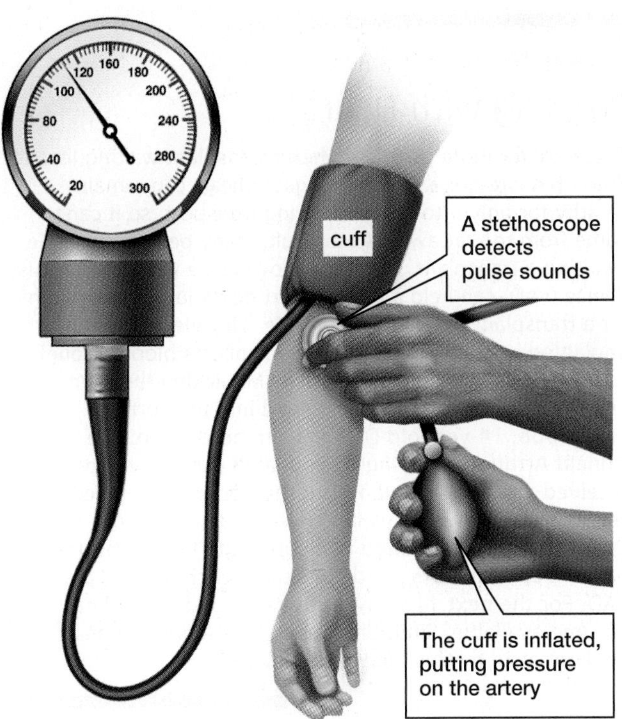

◀ **FIGURE 32-6 Measuring blood pressure** First, the cuff is inflated until its pressure closes off the arm's main artery, cutting off the flow of blood; this pressure is then gradually reduced. When rhythmic blood sounds are first heard through the stethoscope, the contractions of the left ventricle have overcome the pressure in the cuff, and the blood has resumed its flow. This pressure is recorded as the upper (and higher) reading: systolic pressure. Next, cuff pressure is further reduced until no pulse sounds are audible, indicating that blood is flowing continuously through the artery. In other words, the pressure between ventricular contractions is adequate to overcome the cuff pressure. This is the lower reading: diastolic pressure. The numbers are in millimeters of mercury, a standard measure of pressure, also used in barometers.

AV node, briefly postponing the ventricular contraction. This delay gives the atria time to complete the transfer of blood into the ventricles before ventricular contraction begins.

From the AV node, the signal to contract spreads along specialized tracts of rapidly conducting muscle fibers—starting with the thick cluster of fibers called the *atrioventricular bundle* (*AV bundle*), which sends branches to the lower portion of both ventricles (**Fig. 32-7** ❹). Here, the bundle of fibers branches further, forming *Purkinje fibers* that transmit

SA node
❶ An electrical signal from the sinoatrial (SA) node starts atrial contraction
❷ The electrical signal spreads through the atria, causing them to contract
❸ The signal enters the atrioventricular (AV) node, which transmits it to the AV bundle with a slight delay
❹ The signal travels through the AV bundle branches to the base of the ventricles
❺ Purkinje fibers transmit the signal to ventricular cardiac muscle cells, causing contraction from the base upwards
AV node
AV bundle
AV bundle branches
Inexcitable tissue separates the atria and ventricles
Purkinje fibers

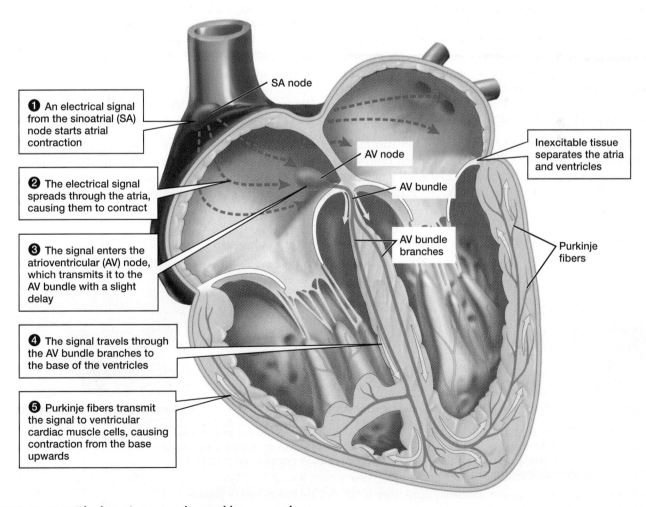

▲ **FIGURE 32-7 The heart's pacemaker and its connections**

the electrical signal to the surrounding cardiac muscle cells, sending a wave of contraction from the base of the ventricles upward within the ventricular walls (**Fig. 32-7** ⑤). This forces blood up into the pulmonary artery and the aorta.

A variety of disorders can interfere with the complex series of events that produce the normal cardiac cycle. When the pacemaker fails—or if other areas of the heart become more excitable and usurp the pacemaker's role—rapid, uncoordinated, weak contractions called **fibrillation** may occur. Fibrillation of the ventricles is soon fatal, because blood is not pumped by the quivering muscle. Fibrillation may be treated with a defibrillating machine, which applies a jolt of electricity to the heart, synchronizing the contraction of the ventricular muscle cells. If effective, this treatment allows the pacemaker to resume its normal coordinating function.

The Nervous System and Hormones Influence Heart Rate

The heart rate is finely tuned to the body's activity level, whether you are running to class or basking in the sun. On its own, the SA node pacemaker would maintain a steady rhythm of about 100 beats per minute. However, nerve impulses and hormones significantly alter the heart rate. In a resting individual, activity of the parasympathetic nervous system, which regulates body systems during periods of rest (see pp. 743–745), slows the heart rate to about 70 beats per minute (this resting rate is typically lower in athletes). When exercise or stress creates a demand for greater blood flow to the muscles, the sympathetic nervous system, which prepares the body for emergency action, accelerates the heart rate.

32.3 WHAT IS BLOOD?

Blood, sometimes called the "river of life," has two major components: (1) a liquid, called **plasma,** which comprises about 55% of the blood volume, and (2) the cell-based portion—red blood cells, white blood cells, and platelets—that are suspended in the plasma and constitute about 45% of the blood volume in men, and about 40% in women (**Fig. 32-8**). The average person has roughly 5.3 quarts (about 5 liters) of blood, so if you donate a pint of blood, you are giving only about 10% of your blood. The components of blood are summarized in **Table 32-1.**

Plasma Is Primarily Water in Which Proteins, Salts, Nutrients, and Wastes Are Dissolved

Although plasma is about 90% water, this clear, pale-yellow fluid contains more than 100 different types of molecules. The plasma transports hormones, nutrients, and cellular wastes, such as carbon dioxide. It also contains a variety of

Case Study continued

Running with Heart

There are far more people on heart transplant waiting lists than there are donors, and a donated heart can remain healthy for only 5 to 6 hours during transport, so it can't come from too far away. As a result, many people die while awaiting a transplant. For some, however, a ventricular assist device (VAD) can help a failing heart do its job long enough for a transplant to become available. This electric pump, implanted in the patient's abdomen, collects blood through a tube inserted into the left ventricle. Mimicking the normal heart rhythm, the VAD pumps blood into the aorta.

In 2008, 14-year-old D'Zhana Simmons (who, like Donald Arthur, was dying from dilated cardiomyopathy) received a heart transplant, but the donor heart failed and was removed just 2 days later. Her cardiac surgeon, Dr. Marco Ricci, fashioned two substitute heart "ventricles" out of a special fabric, and connected each to a separate VAD. For the next 118 days, until a second donor heart became available, Simmons survived without any heart at all, kept alive by this ingenious device.

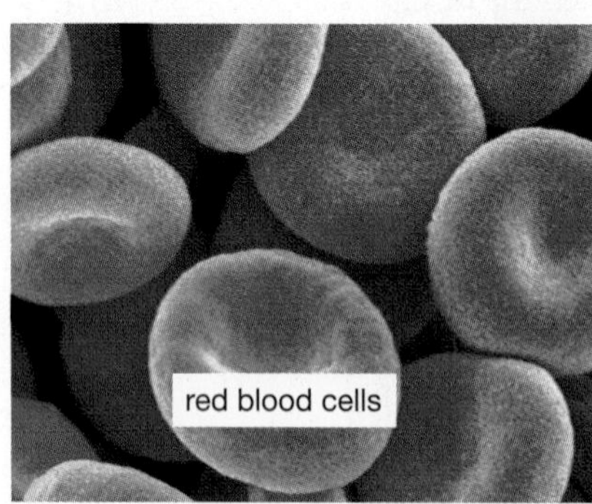

(a) Erythrocytes (red blood cells)

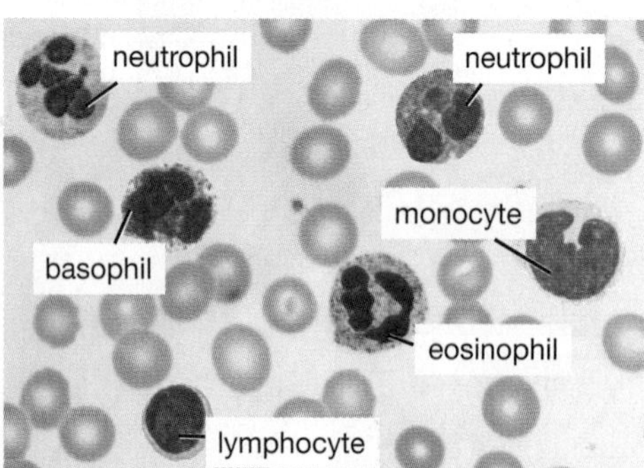

(b) Leukocytes (white blood cells)

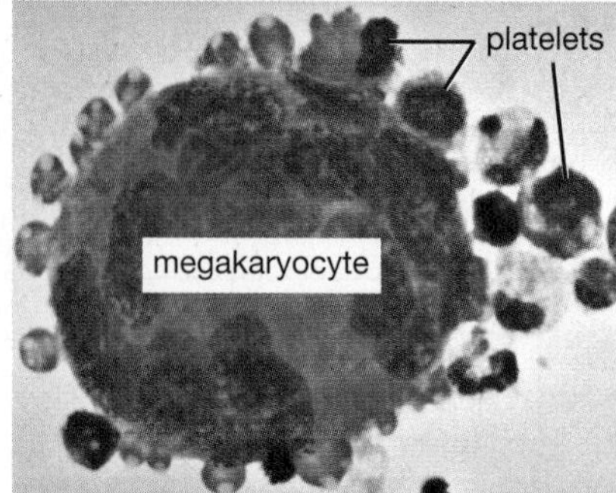

(c) Megakaryocyte forming platelets

▲ **FIGURE 32-8 Types of blood cells (a)** This false-color scanning electron micrograph clearly shows the biconcave disc shape of red blood cells. **(b)** This composite photo of stained white blood cells shows the five different types. **(c)** Platelets are membrane-enclosed pieces of cytoplasm, here shown budding off of a single megakaryocyte.

QUESTION Why does dietary iron deficiency cause anemia (too few red blood cells to provide adequate oxygen)?

Table 32-1 **Blood Components and Their Functions**

Plasma Components (about 55% of blood volume)	**Functions**
Water	Dissolves other components; gives blood its fluidity
Major Proteins	
Albumin	Maintains blood osmotic strength; binds and transports some hormones and fatty acids
Globulins	Serve as antibodies that fight infection; bind and transport some hormones, ions, and other molecules
Fibrinogen	Gives rise to fibrin, which promotes clotting
Ions (sodium, potassium, calcium, magnesium, chloride, bicarbonate, hydrogen)	Maintain pH; allow neuronal activity; allow muscle contraction; facilitate enzyme activity
Nutrients (simple sugars, amino acids, lipids, vitamins, oxygen)	Provide materials for cellular metabolism
Wastes (urea, carbon dioxide, ammonia)	By-products of cellular metabolism that are transported in blood to sites of elimination
Hormones	Signaling molecules that are transported in blood to their target cells
Cell-related Components (about 45% of blood volume)	**Functions**
Erythrocytes (5,000,000 per mm^3)	Transport oxygen
Leukocytes (5,000 – 10,000 per mm^3)	All fight infection and disease
Neutrophils	Engulf and destroy bacteria
Eosinophils	Kill parasites
Basophils	Produce inflammation
Lymphocytes	Mount an immune response
Monocytes	Mature into macrophages, which engulf debris, foreign cells, and foreign molecules
Platelets (250,000 per mm^3)	Essential for blood clotting

ions; some maintain blood pH, while others are crucial for the functioning of nerve and muscle cells or enzymes.

Proteins are the most abundant dissolved molecules by weight. The three most common plasma proteins are albumin, globulins, and fibrinogen. Albumin helps to maintain the blood's osmotic strength, thus preventing too much fluid from diffusing out of the plasma through capillary walls. Some globulins are antibodies that play an important role in the immune response (see pp. 698–699). Other globulins contribute to blood clotting, and still others bind and transport important substances that won't dissolve in the watery plasma, including certain hormones, vitamins, and fatty acids. Fibrinogen is important in blood clotting, described later in this chapter.

The Cell-Based Components of Blood Are Formed in Bone Marrow

Blood contains three cell-based components: red blood cells, white blood cells, and platelets. Only the white blood cells are complete, functional cells. Mature red blood cells are technically not cells because they lack a nucleus, which is lost during development. Platelets are actually small fragments of cells. All three components originate from blood stem cells that reside in bone marrow. **Stem cells** are unspecialized cells that can divide to produce offspring capable of maturing into one or more types of specialized cells.

Red Blood Cells Carry Oxygen from the Lungs to the Tissues

About 99% of all blood cells and nearly 45% of the total blood volume consists of oxygen-carrying red blood cells, also called **erythrocytes.** A red blood cell is shaped like a ball of clay squeezed between a thumb and forefinger (**Fig. 32-8a**). This shape, which results when the cell loses its nucleus during development, provides a larger surface area than would a spherical cell of the same volume, increasing the erythrocyte's ability to absorb and release oxygen through its plasma membrane.

The red color of erythrocytes is caused by the large, iron-containing protein **hemoglobin** (**Fig. 32-9**; see also Fig. 3-20), which transports almost all of the oxygen carried by blood. Each hemoglobin molecule can bind and carry four molecules of oxygen, one on each heme group (the red discs in Fig. 32-9). Hemoglobin takes on a bright cherry-red color when it binds oxygen, and it becomes a deeper red when oxygen is released, giving deoxygenated blood a bluish appearance when seen in the veins beneath the skin. For this reason, diagrams typically depict vessels carrying oxygenated blood in red (all arteries except the pulmonary arteries) and vessels carrying deoxygenated blood in blue (all veins except the pulmonary veins). Hemoglobin binds loosely to oxygen, picking it up in the capillaries of the lungs, where the oxygen concentration is high, and releasing it in other tissues of the body where the

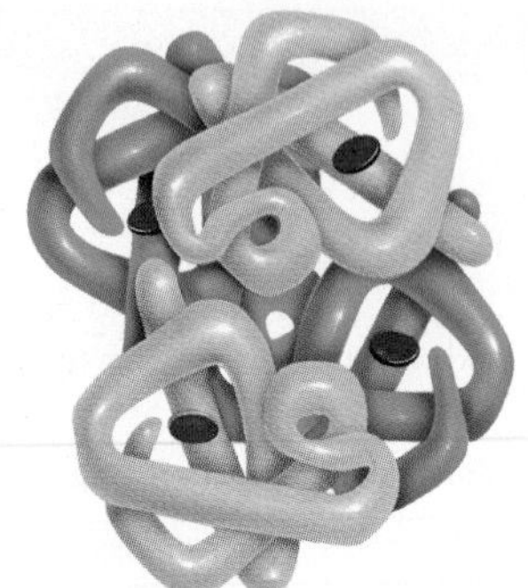

▲ **FIGURE 32-9 Hemoglobin**

Case Study continued
Running with Heart

Researchers have demonstrated that victims of heart failure often have elevated levels of erythropoietin, with the degree of increase of the hormone correlated with the severity of the heart failure, as their bodies attempt to compensate for lack of oxygen reaching their tissues. Administration of additional erythropoietin to heart failure patients may improve their condition.

oxygen concentration is lower. The role of blood in gas exchange is discussed further in Chapter 33.

Erythrocytes have an average life span of about 4 months. Every second, more than 2 million red blood cells (100 billion daily) die and are replaced by new ones from the bone marrow. Macrophages (literally, "big eaters") are white blood cells in the spleen and liver that engulf and break down the dead red blood cells. Iron from the erythrocytes is returned to the bone marrow where it is used in the synthesis of hemoglobin for new red blood cells. Although this recycling process is efficient, some iron is lost during bleeding from injury or from menstruation (in women), and small amounts are excreted daily, in feces. As a consequence, iron must be provided in the diet.

Negative Feedback Regulates Red Blood Cell Numbers

The number of red blood cells in the blood determines how much oxygen the blood can carry, and the red blood count is maintained by a negative feedback system that involves the hormone **erythropoietin.** Erythropoietin is produced by the kidneys and released into the blood in response to oxygen deficiency (**Fig. 32-10**). This lack of oxygen may be caused by a loss of blood, insufficient production of hemoglobin, high altitude (where less oxygen is available), or lung disease and heart failure, which interfere with gas exchange in the lungs. Erythropoietin stimulates rapid production of new red blood cells by the bone marrow. When the oxygen level is restored, erythropoietin production declines, and the rate of red blood cell production returns to normal.

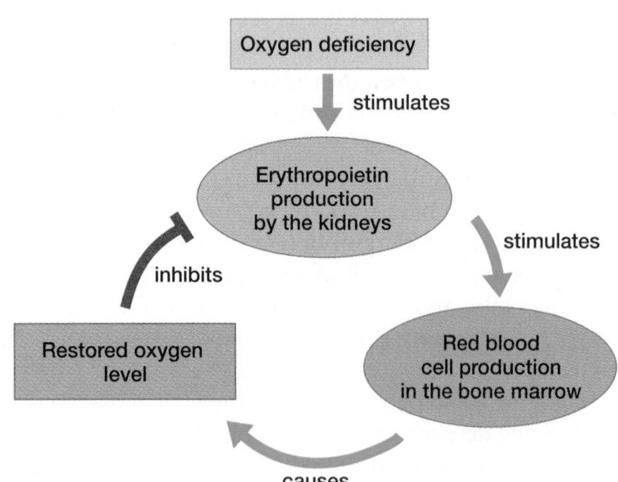

▲ **FIGURE 32-10 Red blood cell regulation by negative feedback**

QUESTION Some endurance athletes cheat by injecting large doses of erythropoietin. How does this provide a competitive advantage?

White Blood Cells Defend the Body Against Disease

There are five types of white blood cells, also called **leukocytes:** neutrophils, eosinophils, basophils, lymphocytes, and monocytes (**Fig. 32-8b**; see also Table 32-1). Their life spans range from hours to years, and together, they make up less than 1% of the cellular portion of the blood. All white blood cells help protect the body against disease. Monocytes, for example, enter tissues and transform into macrophages. At wounds, they engulf bacteria (**Fig. 32-11**)

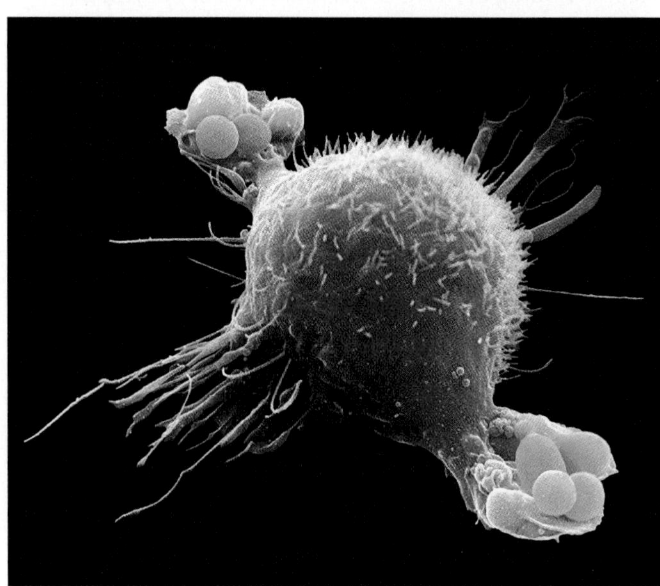

▲ **FIGURE 32-11 A white blood cell attacks bacteria** In this false-colored SEM, the large blue cell (a macrophage) has formed extensions of its cytoplasm that are engulfing the intestinal bacteria *Escherichia coli* (small green structures). Some forms of these intestinal bacteria can cause disease if they enter the bloodstream.

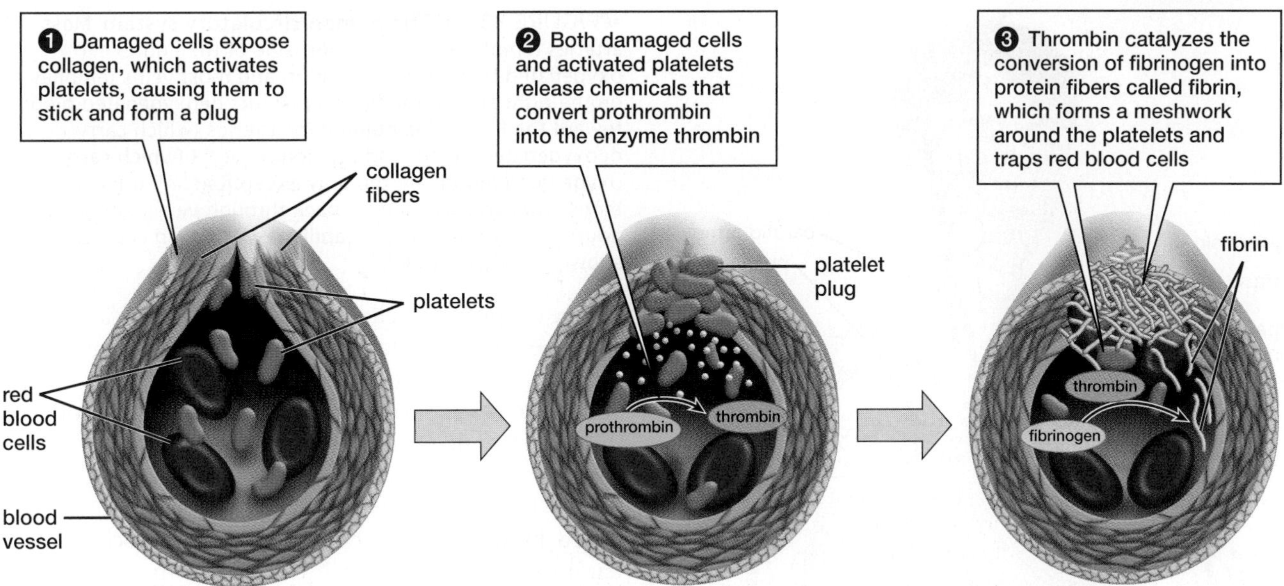
❶ Damaged cells expose collagen, which activates platelets, causing them to stick and form a plug
❷ Both damaged cells and activated platelets release chemicals that convert prothrombin into the enzyme thrombin
❸ Thrombin catalyzes the conversion of fibrinogen into protein fibers called fibrin, which forms a meshwork around the platelets and traps red blood cells
collagen fibers
platelets
red blood cells
blood vessel
platelet plug
prothrombin
thrombin
fibrin
thrombin
fibrinogen

▲ **FIGURE 32-12 Blood clotting** Injured tissue and adhering platelets cause a series of biochemical reactions among blood proteins that lead to clot formation. A simplified sequence is shown here.

and cellular debris. As previously noted, macrophages in the spleen and liver consume dead red blood cells.

Platelets Are Cell Fragments That Aid in Blood Clotting

Platelets are pieces of large cells called **megakaryocytes.** Megakaryocytes remain in the bone marrow, where they pinch off membrane-enclosed chunks of cytoplasm to form platelets (**Fig. 32-8c**). The platelets, which survive for about 10 days, enter the blood and play a central role in blood clotting.

Blood Clotting Plugs Damaged Blood Vessels

Blood clotting is a complex process that protects animals from loss of excessive amounts of blood, not only from trauma, but also from the minor wear and tear that occurs with normal activities. Clotting begins when blood comes in contact with injured tissue; for example, a break in a blood vessel wall. The rupture exposes collagen protein that causes platelets to adhere, forming a *platelet plug* that partially blocks the opening (**Fig. 32-12** ❶). Both the adhering platelets and the ruptured cells release a variety of substances, initiating complex sequences of reactions among circulating plasma proteins. One important result of these reactions is the production of the enzyme **thrombin** from its inactive form, prothrombin (**Fig. 32-12** ❷). Thrombin catalyzes the conversion of the soluble plasma protein **fibrinogen** into insoluble strands of a protein called **fibrin.** Fibrin strands adhere to one another, forming a fibrous network around the aggregated platelets (**Fig. 32-12** ❸).

This protein web traps blood cells (primarily erythrocytes) and more platelets (**Fig. 32-13**), increasing the density of the clot. Platelets adhering to the fibrous mass send out sticky projections that grip one another. The platelets then contract, pulling the fibrin web tighter, which forces fluid out.

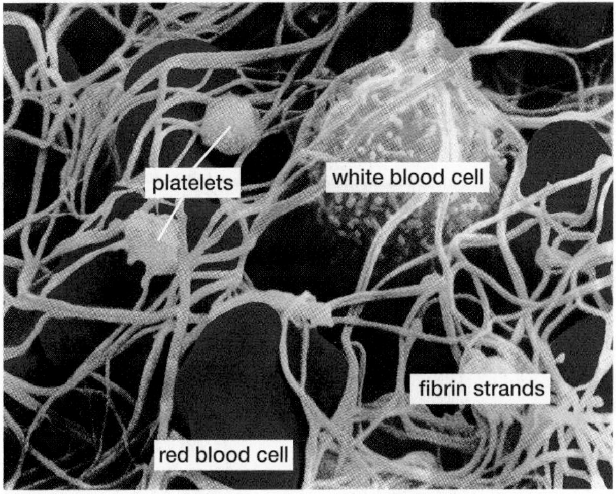
platelets
white blood cell
fibrin strands
red blood cell

▲ **FIGURE 32-13 A blood clot** In this false-colored SEM, threadlike fibrin proteins produce a tangled, sticky mass that traps blood cells and eventually forms a clot.

This creates a denser, stronger clot (on the skin it is called a scab), and also constricts the wound, pulling the damaged surfaces closer together, which promotes healing.

32.4 WHAT ARE THE TYPES AND FUNCTIONS OF BLOOD VESSELS?

Blood circulates throughout the body within a network of tubes, or blood vessels; some of the major blood vessels of the human circulatory system are illustrated in **Figure 32-14.** Blood leaving the heart travels from arteries to arterioles to capillaries, then into venules, and finally, to veins, which return it to the heart. With the exception of capillaries, blood vessels have a fundamentally similar structure, with three cellular layers. All blood vessels are lined with endothelial cells

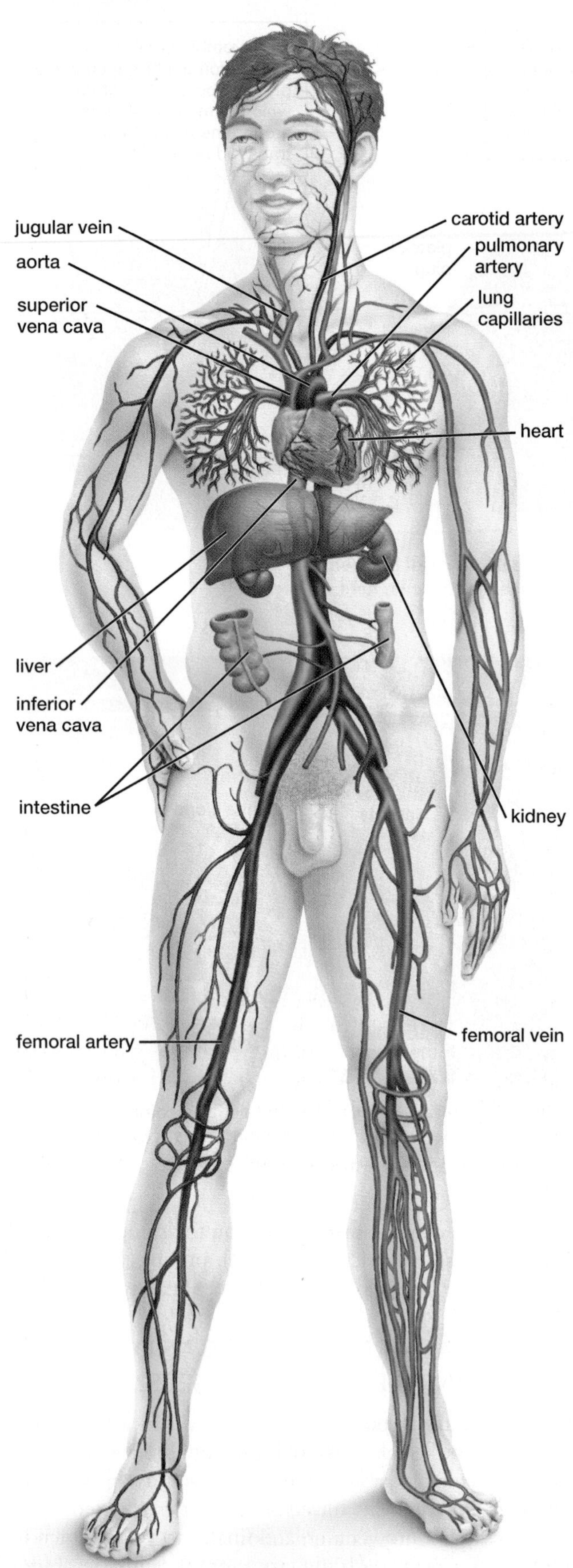

◀ **FIGURE 32-14 The human circulatory system** Most arteries (emphasized on the left side of the figure) carry oxygenated blood from the heart, and most veins (emphasized on the right side of the figure) conduct deoxygenated blood toward the heart. The pulmonary arteries (which carry deoxygenated blood) and pulmonary veins (which carry oxygenated blood) are the only exceptions. All organs receive blood from arteries, send it back through veins, and are nourished by microscopic capillaries. The lung capillaries are shown greatly enlarged.

(a type of epithelial cell; see pp. 608–610), but capillary walls consist only of a single endothelial cell layer. Larger vessels—including arteries, arterioles, veins, and venules—have two additional layers of cells; the second layer consists of smooth muscle cells, and the outermost layer is connective tissue. The structure of these vessels is shown in **Figure 32-15.**

What happens when blood vessels rupture, are narrowed by deposits of cholesterol, or are blocked by clots, damming the river of life? We explore these questions in "Health Watch: Repairing Broken Hearts" on pp. 632–633.

Arteries and Arterioles Carry Blood Away from the Heart

Arteries carry blood away from the heart. The walls of these vessels are thicker and far more elastic than are those of veins (see Fig. 32-15). With each surge of blood from the ventricles, the arteries expand slightly, like thick-walled balloons. As their elastic walls recoil between heartbeats, the arteries actually help pump the blood and keep it flowing steadily into the smaller vessels. Arteries branch into smaller diameter vessels called **arterioles,** which play a major role in determining how blood is distributed within the body, as described later.

Capillaries Allow Exchange of Nutrients and Wastes

Arterioles conduct blood into elaborate networks of tiny **capillaries** (see Fig. 32-15, middle), which are microscopically thin vessels. An important function of the circulatory system is to allow individual body cells to exchange nutrients and wastes with the blood by diffusion, and capillaries are the only vessels in which this exchange can occur. They are so narrow that red blood cells must pass through them single file (**Fig. 32-16**). Capillaries are also so numerous that most body cells are no more than 100 micrometers (slightly less than the thickness of four pages of this book) from a capillary. This is close enough for diffusion to work effectively. To supply all of the body's trillions of cells requires roughly 50,000 miles (about 80,500 kilometers) of capillaries—enough to encircle the globe twice. Both blood pressure and flow rate drop as blood moves through this narrow, lengthy capillary network, allowing more time for diffusion to occur.

With walls only one endothelial cell thick, capillaries are well adapted to their role of exchanging materials between the blood and the fluid bathing body cells. The relatively high pressure within the capillaries that branch directly from arterioles causes fluid to leak continuously from the blood plasma into the spaces surrounding the capillaries. The resulting

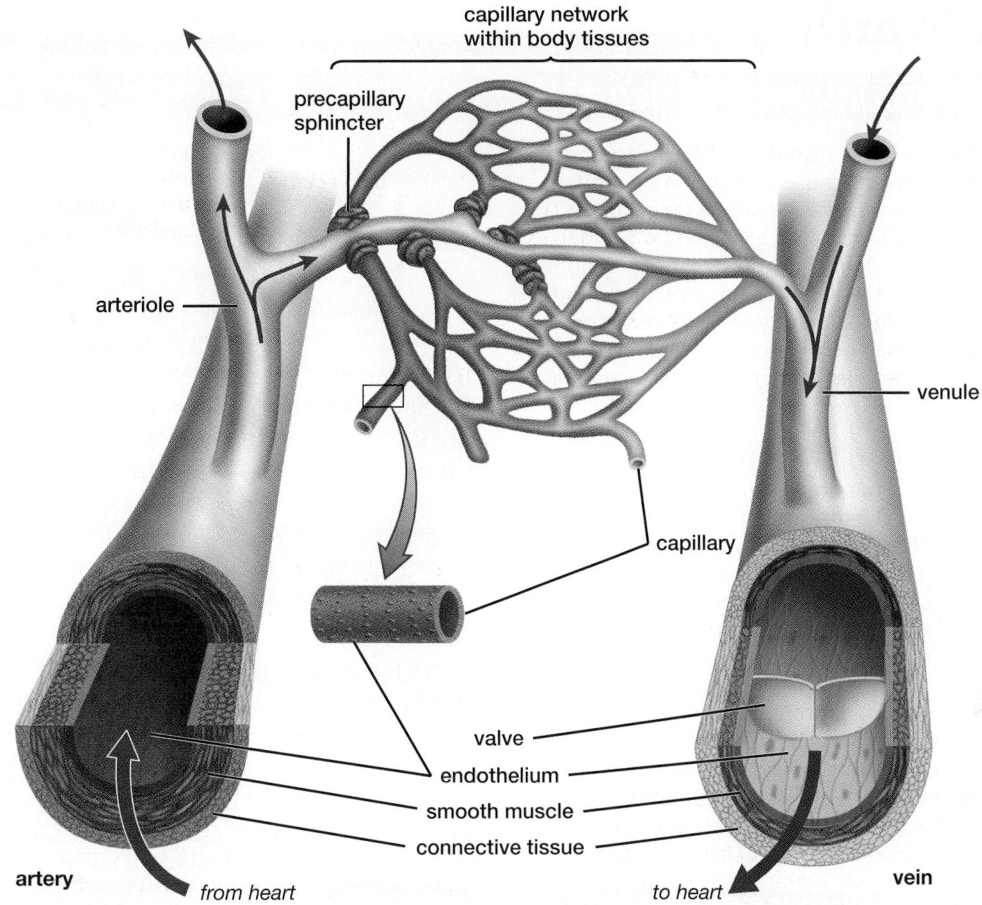

▲ **FIGURE 32-15 Structures and interconnections of blood vessels** Arteries and arterioles are more muscular than veins and venules. Oxygenated blood moves from arteries to arterioles to capillaries. Capillaries, which have walls only one cell thick, empty deoxygenated blood into venules, which empty into veins. Precapillary sphincters regulate the movement of blood from arterioles into capillaries.

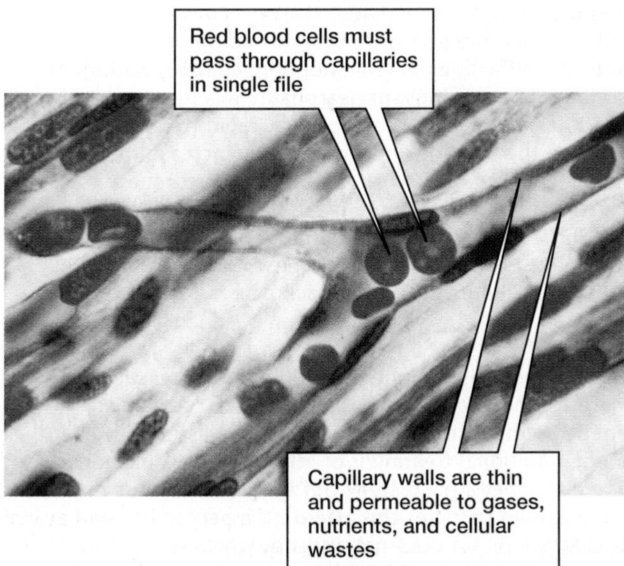

▲ **FIGURE 32-16 Red blood cells travel single file through a capillary**

QUESTION Why does oxygen move out of capillaries in body tissues, while carbon dioxide moves in (instead of the other way around)?

substance, called **extracellular fluid,** resembles plasma without the large plasma proteins. It consists primarily of water containing dissolved nutrients, hormones, gases, cellular wastes, and white blood cells. This medium acts as an intermediary between body cells and capillary blood, providing the cells with nutrients and accepting their wastes and other secretions.

Substances take various routes across the thin capillary walls. Gases, water, lipid-soluble hormones, and fatty acids can diffuse directly through the endothelial cell membranes. Small, water-soluble nutrients, such as salts, glucose, and amino acids, enter the extracellular fluid through narrow spaces between adjacent capillary cells. White blood cells can also ooze through these crevices. Some proteins may be ferried across the endothelial cell membranes in vesicles. As a result of the filtering action of the capillary walls, the composition of extracellular fluid differs from that of blood. Although the concentrations of ions and glucose are very similar, extracellular fluid lacks red blood cells and platelets and has far less protein than blood plasma.

Pressure within the capillaries drops as blood travels toward the venules, and the high osmotic pressure of the blood that remains inside the capillaries (due largely to the presence of albumin) draws water back into the vessels by osmosis as

Health Watch

Repairing Broken Hearts

Cardiovascular disease (CVD; disorders of the heart and blood vessels) is the leading cause of death in the United States. According to the American Heart Association, CVD claims roughly 900,000 lives each year—and no wonder. The heart must contract vigorously more than 2.5 billion times during an average lifetime without once stopping to rest, forcing blood through a lengthy network of vessels. Because these vessels may become constricted, weakened, or clogged, the cardiovascular system is a prime candidate for malfunction.

Atherosclerosis Obstructs Arteries

Atherosclerosis (from the Greek "athero," meaning "gruel" or "paste"; and "scleros," meaning "hard") is caused by deposits, called **plaques,** within the artery walls. These deposits cause the walls of the arteries to thicken and lose their elasticity. They tend to form in people with high levels of "bad" cholesterol (LDL) and low levels of "good" cholesterol (HDL). What is the difference between the two?

Cholesterol is transported through the bloodstream in packets called *lipoproteins,* which consist of cholesterol surrounded by a shell of proteins and phospholipids that make it soluble in the watery plasma. Although the cholesterol molecules are the same in both of these packets, low-density lipoproteins (LDL) have less total protein in their shells than do high-density lipoproteins (HDL), and the predominant types of protein differ between LDL and HDL. LDL carries cholesterol from the liver to the body cells, including the cells of the artery walls, where it may contribute to plaques. HDL transports cholesterol to the liver, thus removing it from the bloodstream and reducing overall blood cholesterol levels.

Plaque formation is likely to be initiated by minor damage to the endothelium lining the artery. The damaged endothelium attracts white blood cells, which burrow beneath it and ingest large quantities of LDL cholesterol and other lipids (**Fig. E32-1**). The bloated bodies of these macrophages contribute to a growing fatty core of plaque. Meanwhile, smooth muscle cells from below the endothelium migrate into the core, absorb more fat and cholesterol, and add to the plaque. They also produce proteins that form a fibrous cap that covers the fatty core.

As the plaque grows, its fibrous cap may rupture, exposing clot-promoting factors within the plaque. A blood clot then forms, which further obstructs the artery. The clot may completely block the artery (see Fig. E32-1) or it may dislodge and be carried in the blood until it blocks a narrower part of the artery. Arterial clots are responsible for the most serious consequences of atherosclerosis: heart attacks and strokes.

A **heart attack,** which can occur if an artery supplying the heart muscle is blocked, is the death of heart muscle cells from lack of oxygen when their blood supply is interrupted. Although heart attacks are the major cause of death from atherosclerosis, this disease causes plaques and clots to form in arteries throughout the body.

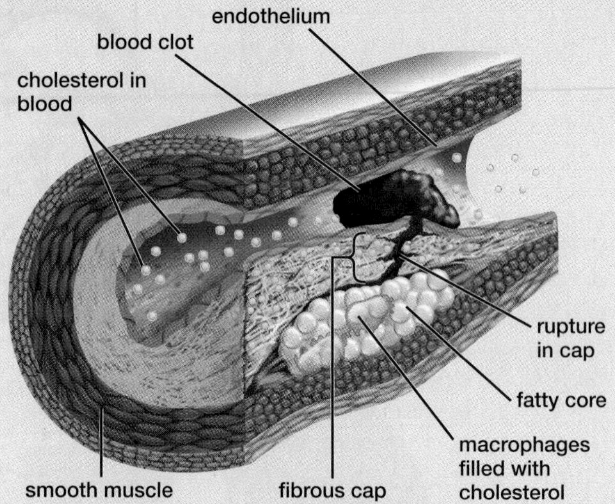

▲ **FIGURE E32-1 Plaque clogs an artery** Compare this artery to that shown in Figure 32-15. The yellow deposits inside this artery form a plaque. If the fibrous cap ruptures (as illustrated here), a blood clot forms, obstructing the artery.

A **stroke,** sometimes called a "brain attack," is the death of brain cells from lack of oxygen when their blood supply is interrupted. Depending on the extent and location of brain damage, strokes can cause a variety of neurological problems, including partial paralysis; difficulties in remembering, communicating, or learning; mood swings; or personality changes.

If a heart attack or stroke occurs, rapid treatment can minimize the damage and significantly increase the victim's chances of survival. Blood clots in coronary (heart) arteries or in the brain can be dissolved by injecting a "clot-busting" protein that breaks down the blood clot; this treatment works best if administered within a few hours after the attack occurs.

Treatment of Atherosclerosis

Atherosclerosis is promoted by high blood pressure, cigarette smoking, obesity, diabetes, lack of exercise, genetic predisposition, and high LDL-cholesterol levels in blood. Traditional treatment of atherosclerosis includes changes in diet and lifestyle; but if this fails, drugs may be prescribed to lower cholesterol. If a person has had a heart attack or suffers from **angina,** which is chest pain caused by insufficient blood flow to the heart, he or she

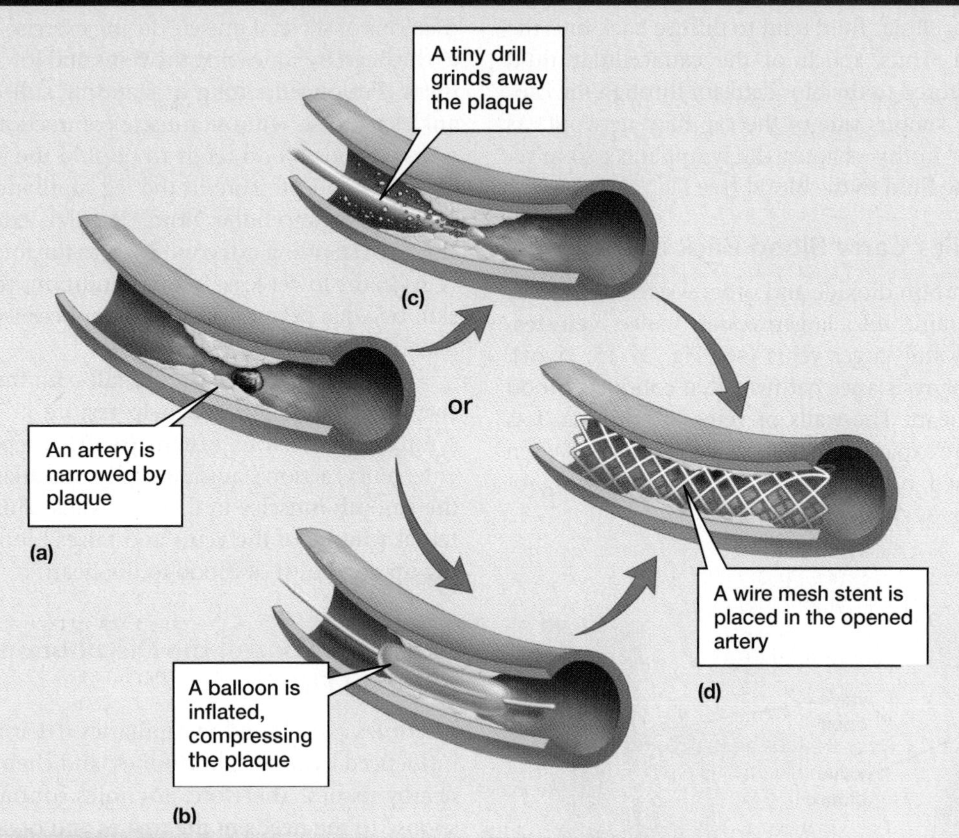

▲ **FIGURE E32-2 Angioplasty unclogs arteries (a)** A narrowed artery. **(b)** The artery may be opened by inflating a tiny balloon inside or **(c)** by drilling out the plaque. **(d)** Following angioplasty, a metal mesh stent is often inserted to maintain the opening.

may be a candidate for surgery to widen or bypass the obstructed artery.

Angioplasty refers to techniques that widen obstructed coronary arteries (**Fig. E32-2a**). To perform an angioplasty, a physician threads a flexible tube through an artery in the upper leg or arm and guides it into the clogged artery. The tube may have a small balloon at its tip, which is then inflated, compressing the plaque (**Fig. E32-2b**); or it may be equipped with tiny, rotating blades or a high-speed, diamond-coated drill bit (**Fig. E32-2c**), which shears off the plaque in microscopic pieces that are carried away in the blood. After doctors remove the plaque, they often insert a wire mesh tube, called a stent, into the artery to help keep it open (**Fig. E32-2d**). In more severe cases, coronary bypass surgery may be performed. This operation bypasses obstructed coronary arteries with segments of vein, usually taken from the patient's leg (**Fig. E32-3**), or segments of artery from the patient's forearm.

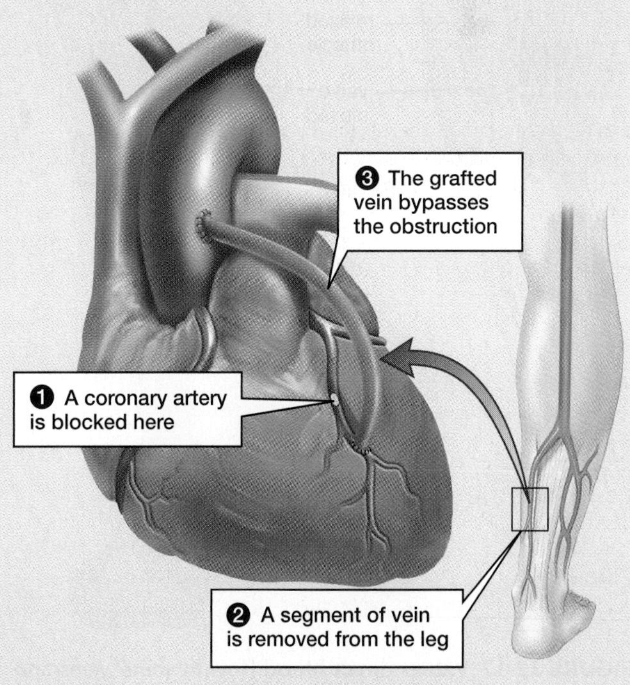

▲ **FIGURE E32-3 Coronary bypass surgery**

blood approaches the venous end of the capillaries. As water moves into the capillaries, diluting the blood, dissolved substances in the extracellular fluid tend to diffuse back into the capillaries as well. Thus, much of the extracellular fluid (about 85%) is restored to the bloodstream through the capillary walls on the venous side of the capillary network. As you will learn later in this chapter, the lymphatic system returns the remaining fluid to the blood (see Fig. 32-18).

Veins and Venules Carry Blood Back to the Heart

After picking up carbon dioxide and other wastes from cells, capillary blood drains into larger vessels called **venules,** which empty into still larger veins (see Fig. 32-15, right). Veins provide a low-resistance pathway that conducts blood back toward the heart. The walls of veins are thinner, less muscular, and more expandable than those of arteries. When veins are compressed, one-way valves keep blood flowing toward the heart (**Fig. 32-17**).

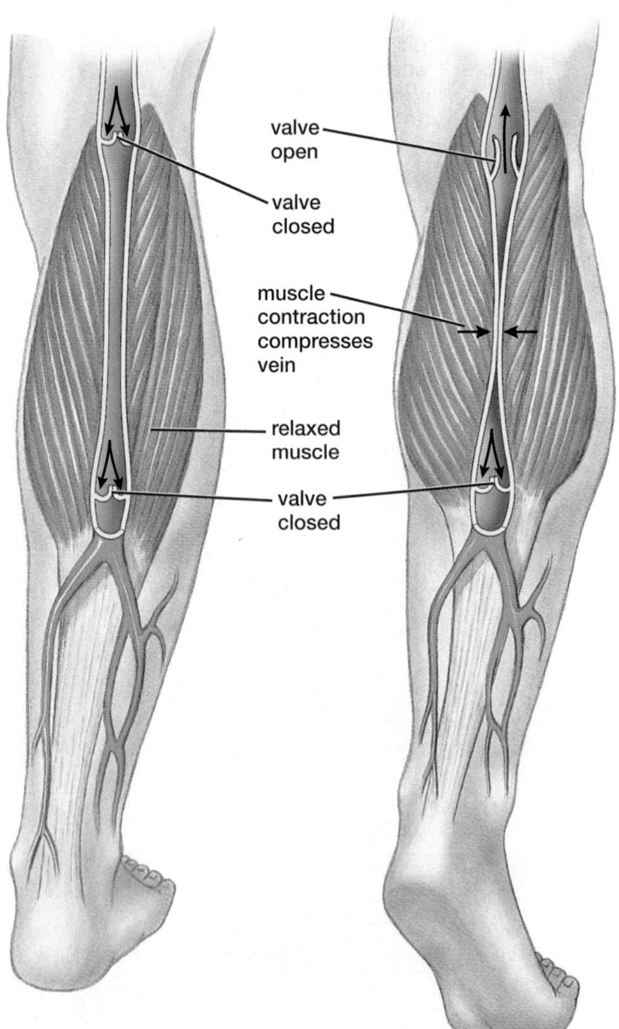

▲ **FIGURE 32-17 Valves direct blood flow in veins** Veins and venules have one-way valves that keep blood flowing in the proper direction. When the vein is compressed by nearby muscles, these valves allow blood to flow toward the heart, but clamp shut to prevent backflow.

Although blood pressure in the veins is low, pressure changes caused by breathing in and out, as well as the contractions of skeletal muscle during exercise, help return blood to the heart by squeezing the veins and forcing blood through them. Prolonged sitting or standing still can cause swollen ankles because, without muscle contractions to compress the veins, venous blood tends to pool in the lower legs. This increases blood pressure in the leg capillaries, which then absorb less extracellular fluid. Regular extended periods of sitting or standing can contribute to the formation of varicose veins in the lower legs. In this condition, veins just below the skin become permanently swollen because their valves have been stretched and weakened.

If blood pressure should fall—for instance, after extensive bleeding—veins can help restore it. In such cases, the sympathetic nervous system (which prepares the body for emergency action) automatically stimulates contraction of the smooth muscles in the vein walls. This decreases the internal volume of the veins and raises blood pressure, speeding up the return of blood to the heart.

Arterioles Control the Distribution of Blood Flow

Arterioles carry blood to capillaries. Their muscular walls are influenced by nerves, hormones, and chemicals produced by nearby tissues. Therefore, arterioles contract and relax in response to the needs of the tissues and organs they supply. In a suspense novel, you might read: "Her face paled as she gazed at the bloodstained floor." Skin becomes pale when the arterioles supplying the skin capillaries constrict because the nervous system has stimulated the smooth muscle in the arteriole walls to contract. This contraction raises blood pressure overall, but selective constriction also redirects blood to the heart and muscles, where it may be needed for vigorous action, and away from the skin, where it is less essential.

In extremely cold weather, fingers and toes can become frostbitten (damaged by freezing) because the arterioles that supply blood to the extremities constrict. Blood is shunted to vital organs, such as the heart and brain, which cannot function properly if their temperature drops. By minimizing blood flow to heat-radiating extremities, the body conserves heat. On a hot summer day, in contrast, you become flushed as arterioles in the skin expand and bring more blood to the skin capillaries. This enables the body to dissipate excess heat to the air outside, helping to maintain a relatively constant internal temperature.

The flow of blood in capillaries is further regulated by tiny rings of smooth muscle called **precapillary sphincters,** which surround the junctions between arterioles and capillaries (see Fig. 32-15). These open and close in response to local chemical changes that signal the needs of nearby tissues. For example, the accumulation of carbon dioxide, lactic acid, or other cellular wastes signals the need for increased blood flow to the tissues. These signals cause the precapillary sphincters and the muscles in the walls of nearby arterioles to relax, allowing more blood to flow through the capillaries.

Have you ever wondered?

How a Giraffe's Heart Can Pump Blood Up to Its Brain?

The long legs and 8-foot (2.5-meter) neck of a giraffe allow this amazing animal to browse for food high in trees, but these adaptations put enormous demands on its circulatory system. A giraffe's heart can meet these demands because it weighs about 22 pounds and is about 2 feet long, from top to bottom. If a giraffe weighed the same as an average person, its heart would be twice as large as the person's heart. The giraffe's heart rate (about 170 beats per minute) and blood pressure (about 280/140 mm Hg) are also roughly double those of a person. These adaptations help the blood make the long uphill journey to the giraffe's brain.

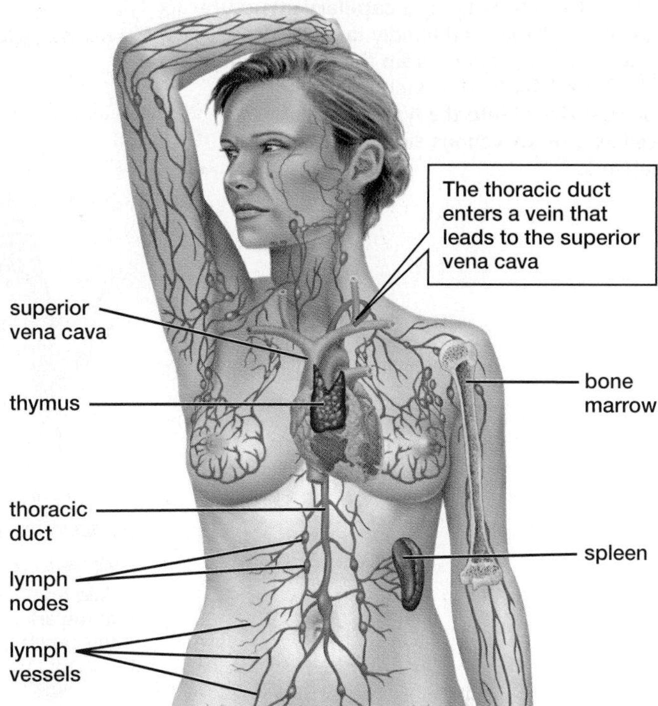

▲ **FIGURE 32-18 The human lymphatic system** Lymphatic vessels, lymph nodes, and two lymph organs, the thymus and spleen, are illustrated. Lymph is returned to the circulatory system through the thoracic duct.

32.5 HOW DOES THE LYMPHATIC SYSTEM WORK WITH THE CIRCULATORY SYSTEM?

The **lymphatic system** includes some organs, as well as an extensive system of lymphatic vessels, that eventually feed into the circulatory system (**Fig. 32-18**). This organ system serves to:

- Return excess extracellular fluid to the bloodstream.
- Transport fats from the small intestine to the bloodstream.
- Filter aged blood cells and other debris from the blood.
- Defend the body by exposing bacteria and viruses to white blood cells.

In the following sections, we emphasize the first three functions, in which the lymphatic system works intimately with the circulatory system. The role of the lymphatic system in defense of the body will be covered in Chapter 36.

Lymphatic Vessels Resemble the Capillaries and Veins of the Circulatory System

The smallest lymphatic vessels, called *lymphatic capillaries*, resemble blood capillaries in that they branch extensively throughout the body and their walls are only one cell thick. Lymphatic capillaries are far more permeable than blood capillaries, however, and are absent from bone and the central nervous system.

Unlike blood capillaries, which form a continuous interconnected network, lymphatic capillaries "dead-end" in the extracellular fluid surrounding body cells (**Fig. 32-19**). From the lymphatic capillaries, lymph is channeled into increasingly large lymphatic vessels, which resemble the veins of the circulatory system in that both have similar walls, and both possess one-way valves that control the direction of fluid movement (**Fig. 32-20**). As the larger lymphatic vessels fill, stretch stimulates contractions of smooth muscles in their walls, pumping the lymph toward larger vessels. As with veins, further impetus for lymph flow through lymphatic vessels comes from internal pressure changes caused by breathing and the contraction of nearby skeletal muscles during exercise.

The Lymphatic System Returns Extracellular Fluid to the Blood

As described earlier, dissolved substances are exchanged between the capillaries and body cells via extracellular fluid. This fluid is pressure filtered out of blood plasma through capillary walls. In an average person, each day, blood capillaries leak out about 3 or 4 more quarts (roughly 3 to 4 liters) of extracellular fluid than they reabsorb. One function of the lymphatic system is to return this excess fluid and its dissolved molecules to the blood.

As extracellular fluid accumulates around cells, increasing pressure forces it through flap-like openings between the cells of the lymphatic capillary walls. Acting like one-way doors, these valves allow substances to enter, but not leave. The lymphatic system transports this fluid—called **lymph** after it has entered lymphatic vessels—back to the circulatory system. The largest lymphatic vessels (such as the thoracic duct) empty into veins near the base of the neck, which merge into the superior vena cava that enters the heart (see Fig. 32-18). The importance of the lymphatic system in returning fluid to the bloodstream is illustrated by elephantiasis (**Fig. 32-21**). This disfiguring condition is caused by a parasitic roundworm that infects, scars, and blocks lymphatic vessels, preventing them from transporting the extracellular fluid back to the bloodstream.

► **FIGURE 32-19 Lymph capillary structure** Lymph capillaries end blindly in the body tissues. Here, pressure from the accumulation of extracellular fluid leaking from capillaries forces the fluid into the lymph capillaries as well as into the venous side of the capillary network.

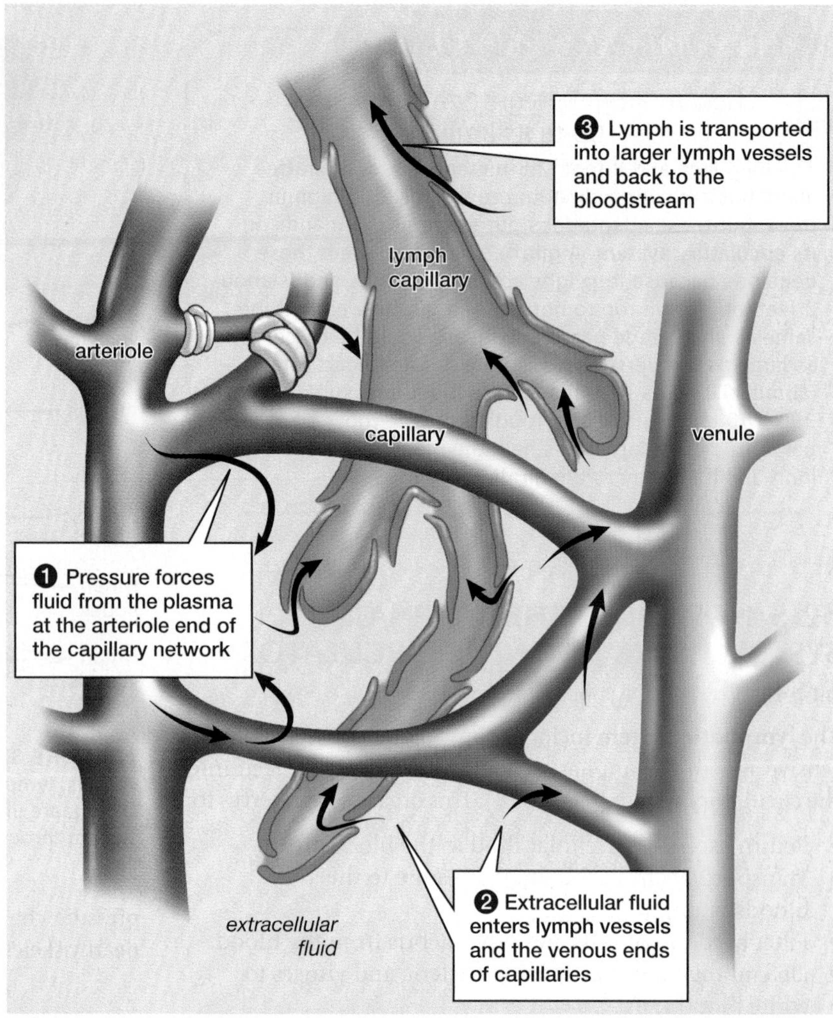
❸ Lymph is transported into larger lymph vessels and back to the bloodstream
lymph capillary
arteriole
capillary
venule
❶ Pressure forces fluid from the plasma at the arteriole end of the capillary network
extracellular fluid
❷ Extracellular fluid enters lymph vessels and the venous ends of capillaries

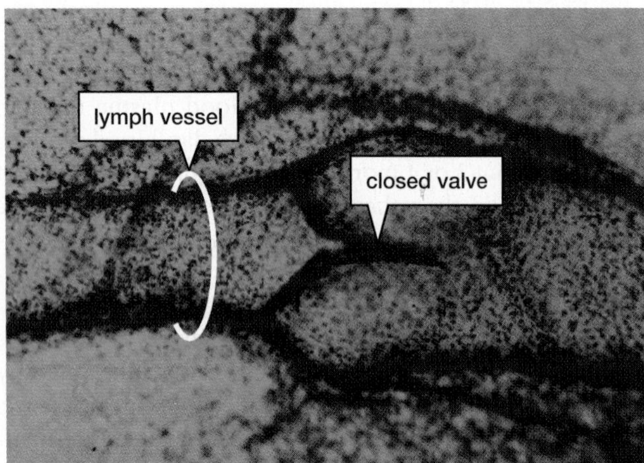
lymph vessel
closed valve

▲ **FIGURE 32-20 A valve in a lymph vessel** Like blood-carrying veins, lymphatic vessels have internal one-way valves that direct the flow of lymph toward the large veins into which they empty.

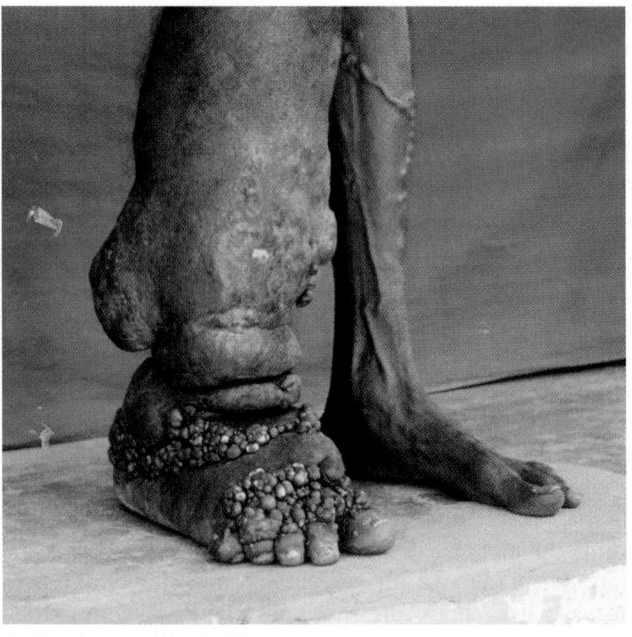

▲ **FIGURE 32-21 Elephantiasis results from blocked lymphatic vessels** When a parasitic worm scars lymphatic vessels, preventing fluid from returning to the bloodstream, the affected area can become massively swollen.

The Lymphatic System Transports Fats from the Small Intestine to the Blood

After a fatty meal, fat-transporting particles may make up 1% of the lymphatic fluid, giving it a milky white color. How does this occur? As you will learn in Chapter 34, the small intestine is richly supplied with lymph capillaries called *lacteals*. After absorbing digested fats, intestinal cells release fat-transporting particles into the extracellular fluid. These particles are too large to diffuse into blood capillaries but can easily move through the openings between lymphatic capillary cells. They are eventually released into the venous blood along with the lymph, as described earlier.

Lymphatic Organs Defend the Body and Filter the Blood

Organs of the lymphatic system are the tonsils, thymus, spleen, and hundreds of lymph nodes located along lymphatic vessels (see Fig. 32-18). All are important in the immune response (described in Chapter 36). The **spleen**—a fist-sized organ between the stomach and diaphragm that is supplied by vessels of both the lymphatic and circulatory systems—also plays an important role in filtering blood. The spleen's porous interior is lined with white blood cells, including macrophages, which engulf aged red blood cells and platelets, fragments of dead cells, and foreign matter, removing them from the blood.

Case Study revisited
Running with Heart

In October 2008, Donald Arthur crossed the finish line of the Atlantic City Marathon in New Jersey after 7 hours and 43 minutes; it was his 33rd marathon in his 23rd state. His finisher's medal around his neck, Arthur reached for his cell phone and called his heart donor's mother, Margaret, whom he has now adopted as if she were his own. "Hey, Mom. Finished," he said. "Yeah, Poochie and I did another one."

When Arthur was diagnosed with cardiomyopathy in 1989, he was far from unique. Each year, about 400,000 people in the United States are diagnosed with cardiomyopathy, which contributes to nearly 250,000 deaths. Roughly 2,200 heart transplants are performed annually in the United States, although thousands more victims of severe heart disease could benefit from this procedure if more donor hearts were available. About 70% of heart recipients survive for 5 years, and about half remain alive at the 10-year mark. Donald Arthur has beaten the odds.

Arthur's most memorable race was in 1999, when Poochie Gittens' brother, Mack, accompanied him in the New York Marathon. As they crossed the Verrazano Narrows Bridge, Donald grabbed Mack's hand and placed it on his chest, so that Mack could feel the rapidly beating heart. "This is your brother," he said. Looking back, Arthur recalls, "Maybe you could only see two people, but there were three of us running that day."

Consider This

Gittens' mother found it extremely difficult to allow her son's organs to be donated, but she has never regretted the decision. If more organs were donated, more people like Donald Arthur would be alive today. What arguments do potential donors use *against* organ donations? Would you want to donate your own, or have your loved one's organs donated, after death? For more information on organ donation, go to www.organdonor.gov.

CHAPTER REVIEW

Summary of Key Concepts

32.1 What Are the Major Features and Functions of Circulatory Systems?

Circulatory systems transport blood rich in dissolved nutrients and oxygen close to each cell, where nutrients can be released, and wastes are absorbed. All circulatory systems have three major parts: one or more hearts that pump blood, the blood itself, and a system of blood vessels. Invertebrates have open or closed circulatory systems. In open systems, found in most invertebrates, hemolymph is pumped by the heart into a hemocoel, where the hemolymph directly bathes the internal organs. A few invertebrates and all vertebrates have closed systems, in which blood is confined to the heart and blood vessels. Vertebrate circulatory systems transport gases, hormones, nutrients, and wastes; help regulate body temperature; and help to defend the body against disease.

32.2 How Does the Vertebrate Heart Work?

The vertebrate heart evolved from two chambers in fishes, to three in amphibians and some reptiles, to four in birds, crocodiles, and mammals. In the four-chambered heart, blood is pumped separately to the lungs and through the body, maintaining complete separation of oxygenated and deoxygenated blood. Deoxygenated blood is collected from the body in the right atrium and then passed to the right ventricle, which pumps it to the lungs. Oxygenated blood from the lungs enters the left atrium, passes to the left ventricle, and is pumped to the rest of the body.

The cardiac cycle consists of two stages: atrial contraction, followed by ventricular contraction. The direction of blood flow is maintained by valves within the heart. The contractions of the heart are initiated and coordinated by the sinoatrial node, the heart's pacemaker. The heart rate can be modified by the nervous system and by hormones such as epinephrine.

32.3 What Is Blood?

Blood is composed of both fluid plasma and cell-derived components. The fluid plasma consists of water that contains proteins, hormones, nutrients, gases, and wastes. Red blood cells, or erythrocytes, are packed with hemoglobin, which carries oxygen.

Their numbers are regulated by the hormone erythropoietin. There are five types of white blood cells, also called leukocytes, which fight infection. Platelets, which are fragments of megakaryocytes, are important for blood clotting.

32.4 What Are the Types and Functions of Blood Vessels?

Blood leaving the heart travels (in sequence) through arteries, arterioles, capillaries, venules, veins, and then back to the heart. Each vessel is specialized for its role. Elastic, muscular arteries conduct blood from the heart, leading to smaller arterioles that empty into capillaries. The microscopically narrow, thin-walled capillaries allow exchange of materials between the body cells and the blood. Venules and veins provide a path of low resistance back to the heart, with one-way valves that maintain the direction of blood flow. The distribution of blood is regulated by the constriction and dilation of arterioles under the influence of the sympathetic nervous system, and by local factors, such as the amount of carbon dioxide in the tissues. Local factors also regulate precapillary sphincters, which control blood flow to the capillaries.

32.5 How Does the Lymphatic System Work with the Circulatory System?

The human lymphatic system consists of the lymphatic vessels, tonsils, lymph nodes, thymus, and spleen. The lymphatic system removes excess extracellular fluid that leaks through blood capillary walls and delivers it back to the circulatory system. It transports fats to the bloodstream from the small intestine, and fights infection by contributing to the immune response. In the spleen, blood filters past macrophages, which remove debris and aging blood cells.

Key Terms

Thinking Through the Concepts

Fill-in-the-Blank

1. Fill in the following with the appropriate heart chamber, including the side it is on: Receives blood from the body: __________; has the thickest wall:__________; contains the SA node: __________; sends blood to the lungs: __________; pumps blood into the aorta: __________.
2. The heart's pacemaker is called the (complete term)__________. The pacemaker is composed of specialized __________. The pacemaker first sends impulses that stimulate contraction through both__________. The (complete term) __________ node introduces a delay in the transmission of pacemaker signals into the ventricles. Signals are conducted from this node directly into fibers called the__________. If the pacemaker loses control of heart contractions, ineffective quivering of the heart muscle, called__________, may occur.
3. Fill in the following with the appropriate type of blood vessel: Allow exchange of wastes and nutrients between blood and body cells:__________; have the thickest walls:__________; transport blood toward the heart: __________; receive blood from the capillaries:__________; deliver blood into capillaries:__________; carry blood away from the heart:__________.
4. During clot formation, damaged cells activate cell fragments called__________. These cell fragments release chemicals that cause the enzyme__________ to be produced. This enzyme catalyzes the conversion of a blood protein called__________ into strands of__________, which create a framework for the clot.
5. Fill in the following with the appropriate cell-based blood component, using the scientific term: Formed by megakaryocytes:__________; there are five types of these:__________; carry oxygen: __________; include macrophages:__________; make up about 45% of blood volume in males:__________; help defend against disease:__________; contain hemoglobin:__________.
6. Lymph is derived from __________ that has leaked out of blood __________. The lymphatic vessels with flap-like valves between their cells are called__________. Lymphatic vessels resembling veins have__________ that keep lymph flowing in the proper direction. The lymphatic organ that filters blood is the__________.

Review Questions

1. Trace the flow of blood through the mammalian circulatory system, starting and ending with the right atrium.
2. List three types of blood cells and describe their principal functions.
3. What are five functions of the vertebrate circulatory system?

4. In what way do veins and lymphatic vessels resemble one another? Describe how fluid is transported in each of these vessels.
5. Describe three important functions of the lymphatic system.
6. Distinguish among plasma, extracellular fluid, and lymph.
7. Describe veins, capillaries, and arteries, noting their similarities and differences.
8. Trace the evolution of the vertebrate heart from two chambers to four chambers.
9. Explain in detail the sequence of events that causes the mammalian heart to beat.
10. Describe the cardiac cycle, and relate the contractions of the atria and ventricles to the two readings taken during the measurement of blood pressure.
11. Describe how the number of red blood cells is regulated by a negative feedback system.
12. Describe the formation of an atherosclerotic plaque. What are the risks associated with atherosclerosis?

Applying the Concepts

1. Discuss the steps you can take now and in the future to reduce your risks of developing heart disease.
2. Considering the prevalence of cardiovascular disease and the high, increasing costs of treating it, certain treatments may not be available to all who might need them. What factors would you take into account if you had to ration cardiovascular procedures, such as heart transplants or bypass surgery?
3. Joe, an overweight 45-year-old executive of a major corporation who works 60-hour weeks, has been feeling chest pain when he tries to play basketball with his son on weekends. What treatments or lifestyle changes might Joe's physician recommend? If Joe's angina does not respond and becomes more severe, what treatment options might Joe's physician use? Explain how each option works.

MB *Go to www.masteringbiology.com for practics quizzes, activities, eText, videos, current events, and more.*

Respiration

Chapter 33

▲ Most students say they will quit smoking—later.

Case Study

Lives Up in Smoke

ON ALMOST ANY COLLEGE CAMPUS, you will find students clustered outside the building doors, sometimes shivering in the cold, lighting up between classes. Coughs punctuate their conversation. Like most adult smokers, many of these students picked up the habit in high school. "I started as a social smoker in high school," says a junior at the University of Illinois, quoted in the university's student newspaper. "The next thing you know, I'm addicted."

At 21, James was already a veteran smoker. "I had low self-esteem and despite knowing the dangers, I started smoking to fit in." By age 15, James often smoked 15 cigarettes a day; by age 19, the number had doubled. He had not yet reached 20 when his doctor told him that he had the lungs of a 40-year-old smoker.

Every day in the United States, several thousand teens under age 18 light up their first cigarette. For many, this is the beginning of a lifelong struggle with addiction. According to the Centers for Disease Control and Prevention, in 2007, about 20% of U.S. high school students smoked, down significantly from the peak rate of 36% in 1997. Still, about 6% of students graduate from high school already smoking a pack or more a day. On average, these young smokers will suffer from persistent coughs, more respiratory illnesses, and will be less able to exercise than nonsmokers. Their lungs may never achieve full normal development. Most of these students are aware of the dangers but say they will stop "when the time comes."

Are these new smokers likely to quit? If not, what are their chances of dying from smoking? Why is nicotine so addictive? What does the inside of a normal lung look like, and how is it changed by smoking? Why aren't our lungs outside our bodies, where they would be directly exposed to the air? In this chapter, we explore the specialized structures of respiratory systems.

At a Glance

33.1 WHY EXCHANGE GASES?

Late again! Sprinting up two flights of stairs to your classroom, your calves are burning. Remembering Chapter 8, you think, "Ah ha! That's lactic acid building up—my muscle cells are fermenting glucose because they can't get enough oxygen for cellular respiration." As you slip into your seat, quietly panting and feeling your heart pounding, the discomfort eases. Less exertion coupled with rapid breathing ensures that adequate oxygen (O_2) is now available. The lactic acid is being reconverted to pyruvate, which will be broken down into carbon dioxide (CO_2) and water, while providing additional energy.

You are experiencing firsthand the relationship between cellular respiration and the act of breathing, also called **respiration.** Each cell in your body (and in all organisms) must continuously expend energy to maintain itself. When you call on your muscles to carry you upstairs quickly, the demands are extreme. As cellular respiration converts the energy in nutrients (such as sugar) into ATP that can be used by cells, the process requires a steady supply of O_2 and generates CO_2 as a waste product. The rapid beating of your heart as you relax after your sprint upstairs reminds you that the circulatory system works in close harmony with the respiratory system. The circulatory system extracts O_2 from the air in your lungs, carries it within diffusing distance of each cell, and picks up CO_2 for release into the lungs, where it is then exhaled back to the atmosphere.

33.2 WHAT ARE SOME EVOLUTIONARY ADAPTATIONS FOR GAS EXCHANGE?

Gas exchange in all organisms ultimately relies on diffusion. Cellular respiration depletes O_2 and increases CO_2 levels, creating a concentration gradient that favors the diffusion of CO_2 out of cells and the diffusion of O_2 into them. Although animal respiratory systems are amazingly diverse, they all meet three requirements that facilitate diffusion:

- Respiratory surfaces remain moist, because living cell membranes are moist and so gases must be dissolved in water when they diffuse into or out of cells through cell membranes.
- Respiratory surfaces are very thin, to facilitate diffusion of gases through them.
- Respiratory systems have a sufficiently large surface area in contact with the environment to allow adequate gas exchange.

In the following sections, we examine diverse respiratory systems in animals, each shaped by the environment in which it evolved.

Some Animals in Moist Environments Lack Specialized Respiratory Structures

For some animals that live in moist environments, the outside of the body, covered by a thin, gas-permeable skin, provides an adequate surface area for the diffusion of gases. If the body is extremely small and elongated, as in microscopic roundworms, gases need to diffuse only a short distance to reach all the cells. Alternatively, an animal's body may be thin and flattened, as in flatworms, so that most cells are close to the moist skin through which gases diffuse (**Fig. 33-1a**). The relatively slow rate of gas exchange by diffusion may suffice even for a larger, thicker bodied organism if energy demands are sufficiently low. For example, sea jellies can be quite large, but cells that are far from the surface are relatively inert and require little O_2 (**Fig. 33-1b**). Another adaptation for gas exchange involves bringing the watery environment close to each cell. Sponges, for example, circulate seawater through channels within their bodies (**Fig. 33-1c**; see also Fig. 23-5).

Some animals combine a large skin surface—through which diffusion occurs—with a well-developed circulatory system. For example, in the earthworm, gases diffuse through

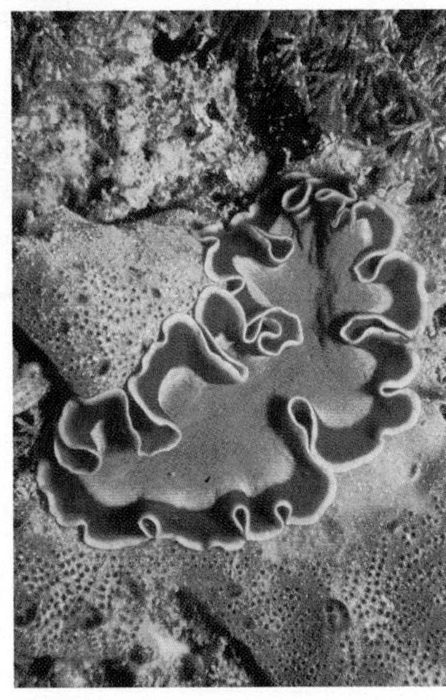

(a) Flatworm

(b) Sea jelly

(c) Sponge

▲ **FIGURE 33-1 Some animals lack specialized respiratory structures** Most animals that lack a respiratory system have low metabolic demands and large, moist body surfaces. **(a)** The flattened body of this marine flatworm exchanges gases with the water. **(b)** Cells in the bell-shaped body of a sea jelly have a low metabolic rate, and seawater flowing in and out of the bell during swimming allows adequate gas exchange. **(c)** Flagellated cells draw currents of oxygenated water through numerous pores in the body of the sponge, and expel it through one or more larger openings.

the moist skin and are distributed throughout the body by an efficient circulatory system (see Fig. 32-1b). Blood in skin capillaries rapidly carries off O_2 that has diffused through the skin, maintaining a concentration gradient that favors the inward diffusion of O_2. The worm's elongated shape ensures a large skin surface relative to its internal volume. To remain effective as a gas-exchange organ, the skin must stay moist; a dry earthworm will suffocate.

Respiratory Systems Facilitate Gas Exchange by Diffusion

Most animals have evolved specialized respiratory systems that interface closely with their circulatory systems to exchange gases between their cells and the environment. The transfer of gases between the environment and the body cells usually occurs in stages that alternate between bulk flow and diffusion. During **bulk flow,** liquids or gases move in bulk through relatively large spaces, from areas of higher pressure to areas of lower pressure, much as water moves from a faucet (high pressure) through an open hose (low pressure). Bulk flow contrasts with diffusion, in which molecules move individually from areas of higher concentration to areas of lower concentration (see pp. 84–87). For animals with well-developed respiratory systems—from insects to humans—gas exchange occurs in the following stages, illustrated for mammals in **Figure 33-2:**

1. Air or water, relatively high in O_2 and low in CO_2 is moved past a respiratory surface by bulk flow; this is usually facilitated by muscular movements, such as breathing.
2. O_2 and CO_2 are exchanged through the respiratory surface by diffusion; O_2 diffuses into the capillaries of the circulatory system, and CO_2 diffuses out.
3. Gases are transported between the respiratory system and the tissues by the bulk flow of blood as it is pumped throughout the body by the heart.
4. Gases are exchanged between the tissues and the circulatory system by diffusion. At the tissues, O_2 moves out of the capillaries and into nearby tissues, and CO_2 moves into capillaries from tissues. Both gases move along their diffusion gradients.

Gills Facilitate Gas Exchange in Aquatic Environments

Gills are the respiratory structures of many aquatic animals. The simplest type of gill, found in some amphibians (see Fig. 33-5a) and in nudibranch (literally, "naked gill") mollusks (**Fig. 33-3**), consists of many thin projections of the body surface that protrude into the surrounding water.

In general, gills are elaborately branched or folded to increase their surface area. Gills have a dense profusion of capillaries just beneath their delicate outer membranes, carrying blood close to the surface, where gas exchange occurs. The fish's body protects its delicate gill membranes beneath a bony flap, the *operculum*. Fish create a continuous current over their gills by pumping water into their mouths and ejecting it out through the operculum (see Fig. E33-2). Some fast swimmers, such as mackerel, tuna, and some species of sharks, rely heavily on swimming with their mouths open to create a water current over their gills.

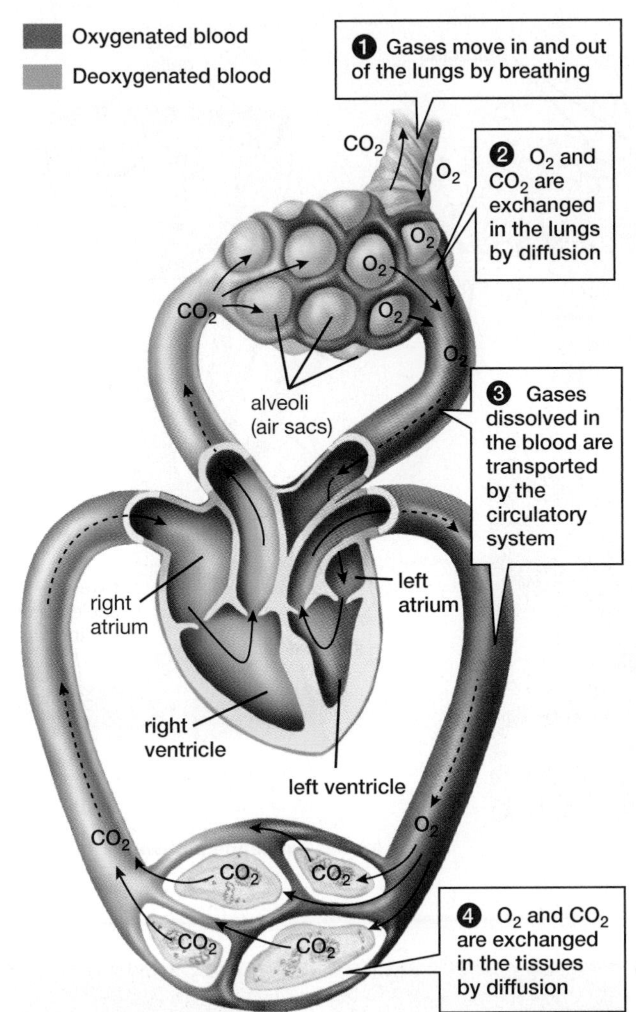

Oxygenated blood
Deoxygenated blood
1 Gases move in and out of the lungs by breathing
2 O_2 and CO_2 are exchanged in the lungs by diffusion
3 Gases dissolved in the blood are transported by the circulatory system
4 O_2 and CO_2 are exchanged in the tissues by diffusion
alveoli (air sacs)
right atrium
left atrium
right ventricle
left ventricle
CO_2
O_2

▲ **FIGURE 33-2 An overview of gas exchange in mammals**

Fish face a challenge in extracting O_2 from water. There are only about 3% as many oxygen molecules in a given volume of fresh water as there are in the same volume of air (ocean water contains even less). Because water is about 800 times as dense as air, pumping sufficient water over gills to obtain adequate oxygen uses up far more energy than does breathing air. In response to these challenges, fish have evolved a very efficient method, called **countercurrent exchange,** for exchanging gases with water. Within the gill, water and blood flow in opposite directions, maintaining a relatively constant concentration gradient, as described in "A Closer Look at Gills and Gases—Countercurrent Exchange" on pp. 646–647.

Terrestrial Animals Have Internal Respiratory Structures

Gills are useless in air because they collapse and dry out. Therefore, as animals made the transition from water to dry land over evolutionary time, natural selection favored respiratory structures whose thin surface membranes were protected, supported, and covered with a film of watery fluid to protect the living cell membranes through which gas exchange must

▲ **FIGURE 33-3 External gills on a mollusk** The feathery projections from the back of this nudibranch mollusk are used for gas exchange.

occur. Natural selection has produced a variety of these structures, including tracheae in insects and lungs in vertebrates.

Insects Respire Using Tracheae

Air enters and leaves the insect respiratory system through a series of openings called **spiracles,** located along each side of the body. Some large insects use pumping movements of their abdomens to enhance air movement in and out through the spiracles. The spiracles open into **tracheae** (singular, trachea), which are elaborately branching air tubes (**Figs. 33-4a,b**). Reinforced with chitin (a principal component of the insect's external skeleton), tracheae penetrate the body tissues and branch into microscopic channels called *tracheoles* (**Fig. 33-4c**). Tracheoles deliver air close to each body cell, minimizing diffusion distances for O_2 and CO_2.

Terrestrial Vertebrates Respire Using Lungs

Lungs are chambers containing moist respiratory surfaces that are protected within the body, where water loss is minimized and the body wall provides support. The first vertebrate lung probably appeared in a freshwater fish and consisted of an outpocketing of the digestive tract. This simple lung supplemented the gills, helping the fish survive in stagnant water, where O_2 is scarce.

Amphibians, which evolved from fish, straddle the boundary between aquatic and terrestrial life. Amphibians use gills during their aquatic larval (tadpole) stage, but most lose their gills and develop simple, sac-like lungs as they metamorphose into a more terrestrial adult form (**Figs. 33-5a,b**). Most amphibians also rely to a considerable extent on diffusion of gases through their thin, moist skin, which is rich in capillaries.

In reptiles (snakes, lizards, turtles, and birds) and in mammals, relatively waterproof skin covered with scales (**Fig. 33-5c**), feathers, or fur reduces water loss. This helps these animals survive in dry environments, but eliminates the skin as a respiratory organ. To compensate for the loss of gas-permeable skin,

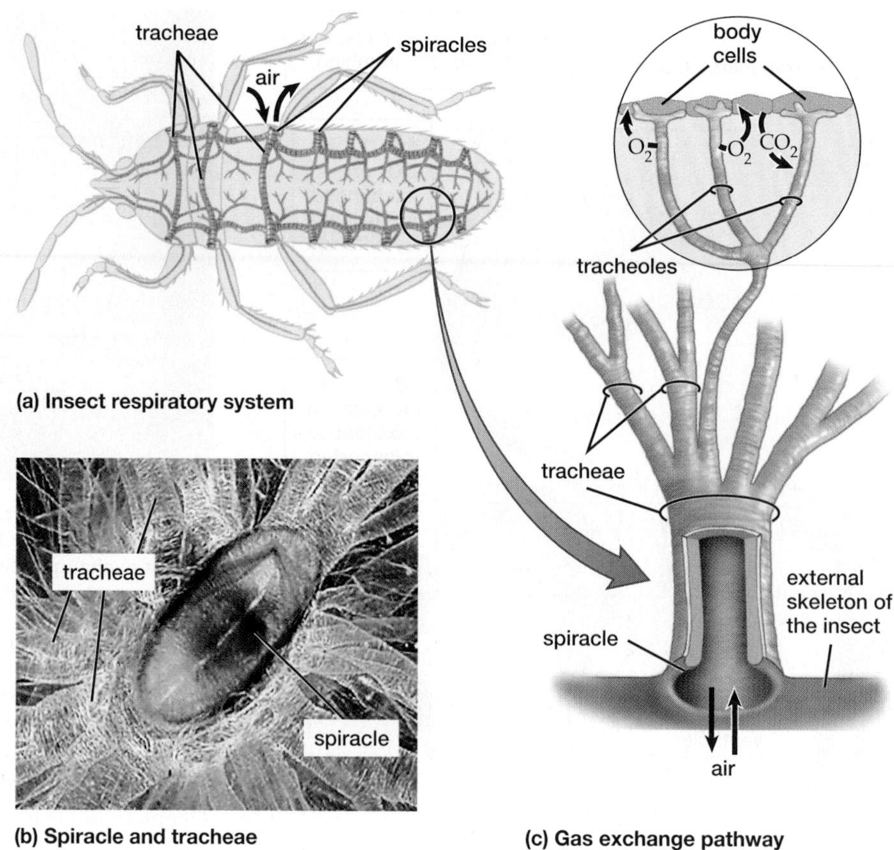

(a) Insect respiratory system

(b) Spiracle and tracheae

(c) Gas exchange pathway

▶ **FIGURE 33-4 Insects breathe using tracheae (a)** The tracheae of insects, such as this beetle, branch intricately throughout the body; air moves in and out through spiracles in the body wall. **(b)** This light microscope image shows the tracheae (blue-green) branching outward from the spiracle (brown). **(c)** An enlargement shows tracheae branching into microscopic tracheoles that conduct air to body cells for gas exchange.

the lungs of reptiles and mammals have a far larger surface area for gas exchange than do those of amphibians.

The bird lung has adaptations that allow exceptionally efficient gas exchange, providing adequate O_2 to support the enormous energy demands of flight. Birds differ from other vertebrates by using seven to nine inflatable air sacs, which do not exchange gases, but rather, serve as reservoirs for air. In **Figure 33-6a,** these have been diagrammatically grouped into the anterior air sacs (near the front of the lungs) and the posterior air sacs (near the rear of the lungs).

In contrast to mammalian lungs—which are highly flexible, and where gas exchange occurs in tiny, dead-end chambers—bird lungs are rigid and filled with barely visible,

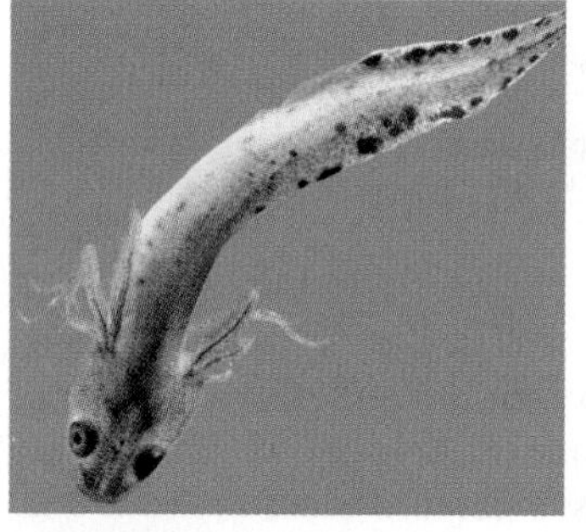

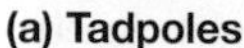

(a) Tadpoles

(b) Adult bullfrog

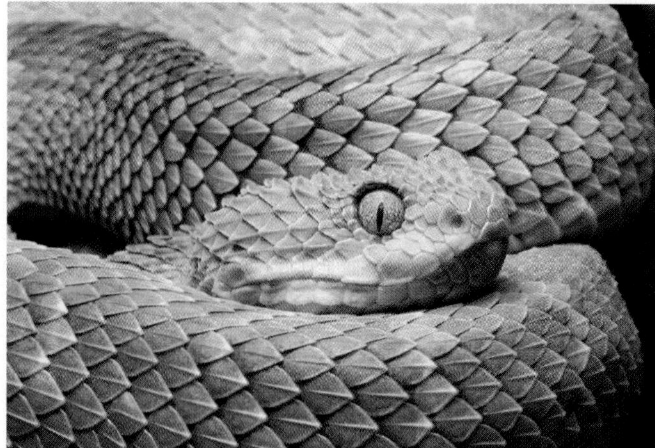

(c) Snake

▶ **FIGURE 33-5 Amphibians and reptiles have different respiratory adaptations (a)** The bullfrog, an amphibian, begins life as a fully aquatic tadpole with feathery external gills that will later become enclosed in a protective chamber. **(b)** During metamorphosis into an air-breathing adult, the frog's gills are lost and replaced by simple saclike lungs. In both tadpole and adult, gas exchange also occurs by diffusion through the skin, which must be kept moist to function as a respiratory surface. **(c)** Scaled terrestrial reptiles, such as this snake, cannot respire through their skin, so their lungs compensate with a larger surface area for gas exchange.

QUESTION How do the respiratory adaptations of amphibians influence the range of habitats in which amphibians are found?

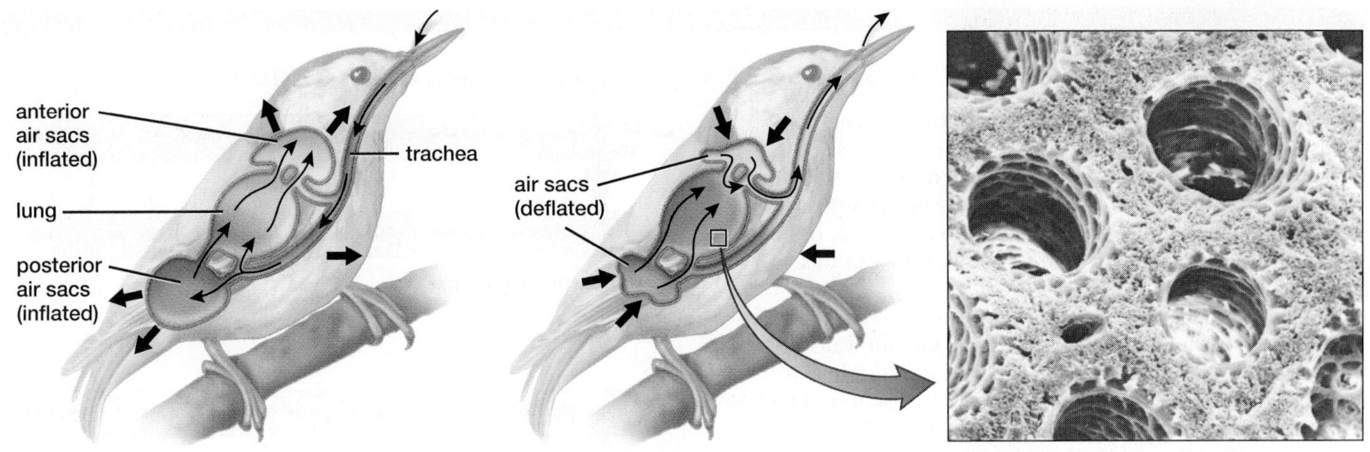

(a) Breathing in (b) Breathing out (c) SEM of parabronchi in lung tissue

▲ **FIGURE 33-6 The bird respiratory system is extremely efficient** In addition to their rather rigid lungs, birds have anterior and posterior groups of flexible air sacs that allow efficient gas exchange. In the figure, bold arrows indicate expansion and contraction of the chest and air sacs; thin arrows denote the pathway of air flow. The precise structure of the airways, which is not visible in this diagram, promotes unidirectional flow of air through the lungs without the use of valves. **(a)** Breathing in expands the air sacs, drawing fresh, oxygenated air (red) past the lungs and into the posterior air sacs, sending some of this air into the lungs, and moving "used" deoxygenated air (blue) out of the lungs and into the anterior air sacs. **(b)** Breathing out deflates the chest, compressing the air sacs and forcing used, deoxygenated air from the anterior sacs out through the nostrils, and fresh air out of the posterior sacs to fill the lungs. **(c)** Open-ended, tubular parabronchi conduct air through the lungs as it passes from posterior to anterior air sacs. The porous regions visible between the parabronchi are filled with capillaries and air spaces where gas exchange occurs.

perforated tubes called *parabronchi* (**Fig. 33-6c**). Parabronchi are open at both ends, allowing air to flow completely through the lungs, and are surrounded by tissue that is riddled with interconnected microscopic air spaces surrounded by a dense capillary network that allows gas exchange.

The unique organization of air sacs and lungs allows a one-way flow of fresh, oxygenated air through the lungs, from posterior to anterior, both as the bird inhales (breathes in) and exhales (breathes out). Inhalation causes all of the air sacs to inflate, drawing fresh air into the posterior sacs via a route that largely bypasses the lungs, filling the posterior air sacs and sending some of this fresh air forward into the lungs, where O_2 is extracted. Simultaneously, inhalation fills the anterior sacs with "used" air (low in O_2 and high in CO_2), which is pulled forward out of the lungs. As the bird exhales, all of the air sacs compress and deflate, forcing the used air from the anterior sacs out through the bird's nostrils, and fresh air from the posterior sacs into the lungs. Thus, bird lungs receive fresh air both as the bird inhales and as it exhales.

33.3 HOW DOES THE HUMAN RESPIRATORY SYSTEM WORK?

The respiratory system in humans and other lung-breathing vertebrates can be divided into two parts: the **conducting portion** and the **gas-exchange portion.** The conducting portion consists of a series of passageways that carry air into and out of the gas-exchange portion, where gases are exchanged with the blood in tiny sacs within the lungs.

The Conducting Portion of the Respiratory System Carries Air to the Lungs

The conducting portion carries air to the lungs; it also contains the apparatus that makes speaking possible. Air enters through the nose or the mouth, passes through the nasal cavity or oral cavity into a chamber, the **pharynx** (which is shared with the digestive tract), and then travels through the **larynx,** or "voice box," where sounds are produced (**Fig. 33-7**). The opening to the larynx is guarded by the **epiglottis,** a flap of tissue supported by cartilage. During normal breathing, the epiglottis is tilted upward (see Fig. 33-7), allowing air to flow freely into the larynx. During swallowing, the epiglottis folds downward and covers the larynx, directing substances into the esophagus (see Fig. 34-13). If an individual attempts to inhale and swallow at the same time, this reflex may fail and food can become lodged in the larynx, blocking air from entering the lungs. What should you do if you see this happen? Use the **Heimlich maneuver,** described in **Figure 33-8,** which is easy to perform and has saved countless lives.

Within the larynx are the **vocal cords,** bands of elastic tissue controlled by muscles. Muscular contractions can cause the vocal cords to partially obstruct the air passage through the larynx. Exhaled air causes the vocal cords to vibrate, producing the tones of speech or song. Stretching the cords changes the pitch of the tones, which can be articulated into words by movements of the tongue and lips.

Inhaled air travels past the larynx into the **trachea,** a flexible tube whose walls are reinforced with semicircular

A Closer Look At *Gills and Gases—Countercurrent Exchange*

An animal frequently needs to exchange heat energy or dissolved substances (solutes) between two parts of its body (for example, nutrients from the digestive tract into the blood) or between its body and the environment (for example, O_2 from the atmosphere into the blood, or heat from a warm rock into a lizard's body). How can this exchange occur?

Consider two fluids that differ in the concentration of a solute. If the fluids are separated only by thin walls that are permeable to the solute, then the solute will diffuse from the fluid with the higher solute concentration into the fluid with the lower solute concentration (see pp. 82–84). The rate of diffusion, and how much solute will pass across the walls, depends on several factors, such as the permeability of the walls, the concentration difference (gradient) between the two fluids, and the speed at which the molecules in each fluid are moving. The efficiency of solute transfer will also depend on movement of the two fluids relative to one another. Should they move in the same direction or in opposite directions?

To maximize transfer between them, the fluids should move in opposite directions, a process called countercurrent exchange. To help understand countercurrent exchange, let's first look at a contrasting situation, *concurrent exchange*, during which two fluids with large differences in solute concentration flow side by side *in the same direction* (**Fig. E33-1a**). In our diagram, the red color indicates high O_2 content (with a maximum of 100; bottom tube, left) and the blue color indicates low O_2 content (with a minimum of 0; top tube, left). Initially, a very steep gradient causes a large transfer of O_2 from the lower tube with a high O_2 content to the upper tube with a low O_2 content (thick red arrow, left). But over distance, the two fluids soon equalize (at 50; both tubes, right), at which point there is no further net change in either.

Figure E33-1b illustrates countercurrent exchange, in which fluids with different O_2 concentrations flow side by side *in opposite directions*, which maintains a constant gradient between them. For this situation to persist, the solution with the higher O_2 content must be continually replenished with O_2 (picked up from air or water, for example), while the solution with the lower O_2 content must be continuously depleted (taken up by body cells).

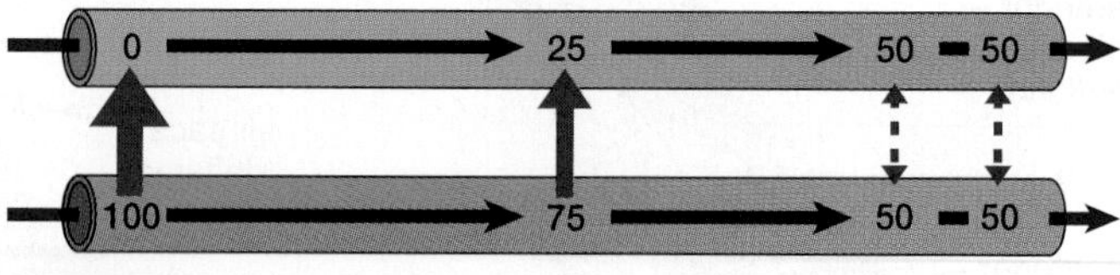

(a) Concurrent exchange

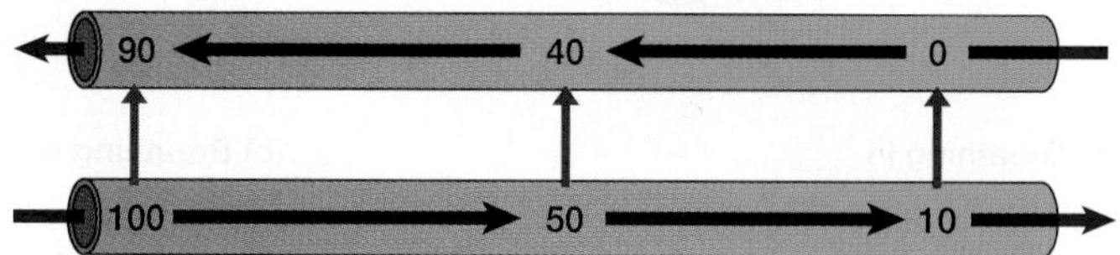

(b) Countercurrent exchange

▲ **FIGURE E33-1 Concurrent versus countercurrent exchange** The colors and numbers in this figure illustrate differences in the O_2 content (red = oxygenated; blue = deoxygenated) of the solutions flowing past one another. In each case, the lower tube has a higher concentration, and the O_2 is transferred to the upper tube. Red arrows indicate the direction of transfer, and the widths of the arrows reflect the rates of transfer. **(a)** During concurrent exchange, the concentrations of the two fluids equalize (50, purple color, right), which limits the net transfer of oxygen. **(b)** During countercurrent exchange, small but continuing concentration gradients are maintained along the length of the tubes. As the fluids flow past one another, the fluid in the upper tube that started with no oxygen (0, blue color, right) acquires nearly all the oxygen (90, red color, left) from the lower tube, whose oxygen content started high (100, red color, left) but ended very low (10, blue color, right).

Because the two fluids do not equilibrate, they are continuously transferring O_2 (or other solutes, or heat) as they travel past one another.

The arrows and numbers in Figure E33-1b show that the fluid in the lower tube becomes less concentrated due to outward diffusion as it flows from left to right past fluid in the upper tube, which is less concentrated still. In this example, by maintaining a concentration gradient of

bands of stiff cartilage. Within the chest, the trachea splits into two large branches called **bronchi** (singular, bronchus), one leading to each lung. Inside the lung, each bronchus branches repeatedly into ever smaller tubes. Finally, these divide into **bronchioles** that are only about 1 millimeter in diameter. The walls of bronchi and bronchioles are encased in smooth muscle, which regulates their diameter. During activities that require extra oxygen, such as exercise, the smooth muscle relaxes, allowing more air to enter. Bronchioles lead to microscopic **alveoli** (singular, alveolus), the tiny air sacs where gas exchange occurs (**Fig. 33-7b**).

During its passage through the conducting system, air is warmed and moistened. Much of the dust and bacteria it carries is trapped in mucus secreted by cells that line the respiratory passages. The mucus, with its trapped debris, is continuously swept upward toward the pharynx by cilia that line the bronchioles, bronchi, and trachea. Upon reaching the pharynx, the mucus is coughed up or swallowed. Smoking interferes

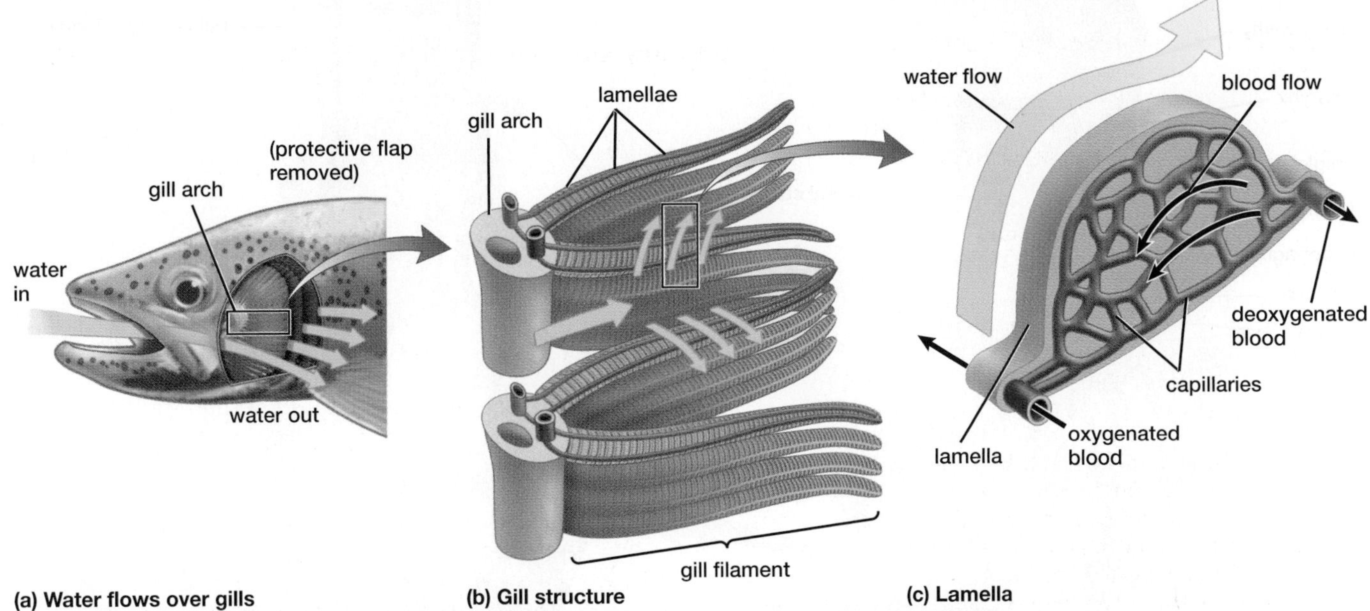

▲ **FIGURE E33-2 Gills exchange gases with water (a)** Fish pump water in through their mouths and out over their gills. Here, the protective operculum has been removed to reveal the gills. **(b)** Water flows past a dense array of paired filaments. **(c)** Lamellae protrude from each gill filament. Water flows over the lamellae in the opposite direction from the blood flowing through the lamellar capillary beds.

10 between the values in the lower and upper tubes, countercurrent flow allows a continuous transfer of fluids between the tubes (Fig. E33-1b, red arrows), and thus a greater overall flow of solutes compared to that of concurrent exchange. You can see that, in moving right to left, the fluid in the upper tube started with no O_2 (0, right), but ended up with nearly the same O_2 content (90) that the fluid in the lower tube began with (100, left).

Fish gills (**Fig. E33-2a**) use countercurrent exchange to promote diffusion of O_2 from the water into their gill capillaries, and CO_2 from their gill capillaries into the water. Fish gills consist of a series of filaments attached to bony gill arches (**Fig. E33-2b**). Each filament is served by a blood vessel that carries deoxygenated blood from the body, and another that carries oxygenated blood to the body. A series of lamellae (thin flaps of tissue) are arrayed across each filament between the incoming and outgoing vessels. A capillary bed in each lamella carries blood from the incoming vessel (low in O_2 and high in CO_2) to the outgoing vessel. Water (low in CO_2 and high in O_2) flows past the capillary beds in the direction opposite to the flow of blood (**Fig. E33-2c**).

This arrangement promotes a countercurrent exchange of gases between the blood and the water, with incoming blood from the body losing CO_2 and picking up O_2 from the passing water. As illustrated in Figure E33-1b, water and blood maintain a gradient of gases as they pass one another. This gradient favors continuous diffusion of O_2 from the water into the blood, and CO_2 from the blood into the water, all the way across each lamella. Countercurrent exchange is so efficient that some fish can extract 85% of the O_2 from the water flowing over their gills.

with this cleansing process by paralyzing the cilia (see "Health Watch: Smoking—A Life and Breath Decision" on p. 649).

Gas Exchange Occurs in the Alveoli

The lung provides an enormous moist surface for gas exchange. The dense, tree-like, branching system of bronchioles conducts air to the alveoli, which cluster around the end of each bronchiole like a bunch of grapes. In an average adult, the two lungs combined have approximately 300 million alveoli. These microscopic chambers (0.2 millimeter in diameter) give magnified lung tissue the appearance of a pink sponge.

Alveoli provide a huge surface area for diffusion, totaling about 1,500 square feet (roughly 145 square meters; about 80 times the skin surface area of a human adult). A network of capillaries covers most of the alveolar surface (see Fig. 33-7b). The walls of the alveoli consist of a single layer of epithelial cells. The **respiratory membrane,** through which

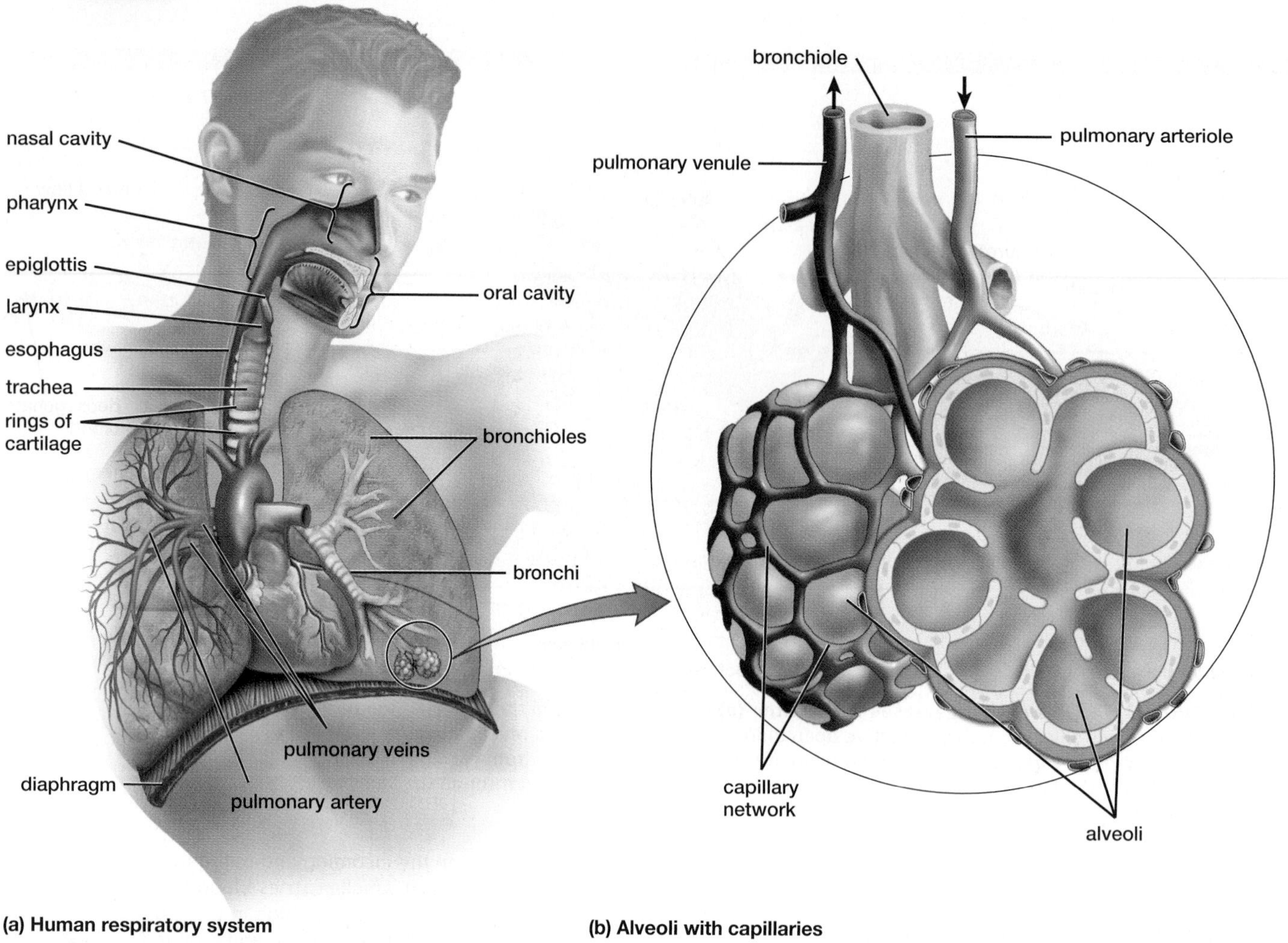

(a) Human respiratory system

(b) Alveoli with capillaries

▲ **FIGURE 33-7 The human respiratory system (a)** The major structures of the human respiratory system are illustrated here. The pulmonary artery carries deoxygenated blood (blue) to the lungs; the pulmonary vein carries oxygenated blood (red) back to the heart. **(b)** Close-up of alveoli (interiors shown in cutaway section) and their surrounding capillaries.

gases diffuse, consists of the epithelial cells of the alveolus and the endothelial cells that form the wall of each capillary, held together by protein fibers that these cells secrete. Because the alveolar walls and the adjacent capillary walls are each only one cell thick, gases must diffuse only a short distance to move between the air and the blood (**Fig. 33-9**). The alveoli are coated with a thin layer of watery fluid containing *surfactant* (a detergent-like substance composed of proteins and lipids), which prevents the alveolar surfaces from sticking together and collapsing when air is exhaled. Gases dissolve in this fluid as they pass in or out of the alveolar air.

Oxygen and Carbon Dioxide Are Transported in Blood Using Different Mechanisms

The respiratory and circulatory systems work in harmony to support cellular respiration. Blood picks up O_2 from the air in

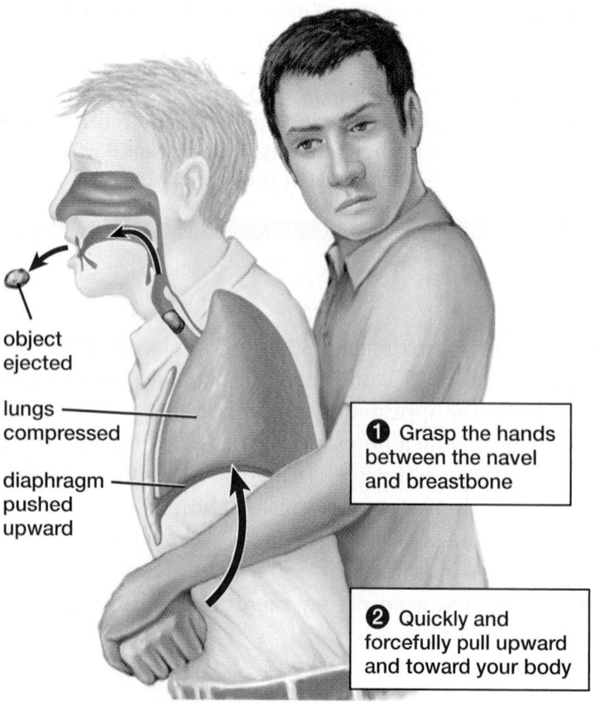

object ejected
lungs compressed
diaphragm pushed upward
❶ Grasp the hands between the navel and breastbone
❷ Quickly and forcefully pull upward and toward your body

◀ **FIGURE 33-8 The Heimlich maneuver can save lives** If a person is choking and unable to breathe, a rescuer can use the Heimlich maneuver to dislodge the object, by pushing rapidly upward and inward on the victim's diaphragm and forcing the air out of the victim's lungs. This action can be repeated if necessary.

Health Watch

Smoking—A Life and Breath Decision

About 440,000 people in the United States die of smoking-related diseases each year, including emphysema, chronic bronchitis, heart disease, stroke, lung cancer, and several other forms of cancer.

As smoke is inhaled, toxic substances such as nicotine and sulfur dioxide paralyze the cilia that line the respiratory tract; a single cigarette can inactivate them for a full hour. Because cilia remove inhaled particles, smoking inhibits them just when they are needed most. The visible portion of cigarette smoke consists of billions of microscopic carbon particles. Adhering to these particles are about 200 different toxic substances, of which more than a dozen are known or probable carcinogens (cancer-causing substances). With the cilia incapacitated, the particles stick to the walls of the respiratory tract and enter the lungs.

Cigarette smoke also impairs the white blood cells that defend the respiratory system by engulfing foreign particles and bacteria. Consequently, still more bacteria, dust, and smoke particles enter the lungs. In response to the irritation of cigarette smoke, the respiratory system produces more mucus, another method of trapping foreign particles. But without the cilia sweeping it along, the mucus builds up and can obstruct the airways, producing the familiar "smoker's cough." Microscopic smoke particles accumulate in the alveoli over the years until the lungs of a heavy smoker are literally blackened—compare the normal lung in **Figure E33-3a** to the diseased lung in **Figure E33-3b**. The longer the delicate tissues of the lungs are exposed to the carcinogens on the trapped particles, the greater the chance that cancer will develop (**Fig. E33-3c**).

Some smokers will develop chronic bronchitis, a persistent lung infection characterized by coughing, swelling of the lining of the respiratory tract, an increase in the production of mucus, and a reduction in the number and activity of cilia. The result is decreased air flow to the alveoli. Carbon monoxide, present in high levels in cigarette smoke, binds tenaciously to red blood cells in place of oxygen, reducing the blood's oxygen-carrying capacity and increasing the workload on the heart. Chronic bronchitis and emphysema compound this problem.

Smoking also promotes *atherosclerosis*, or thickening of the arterial walls by fatty deposits that can lead to heart attacks (see pp. 632–633). As a result, smokers are 70% more likely than nonsmokers to die of heart disease. The carbon monoxide in cigarette smoke may contribute to the reproductive problems experienced by women who smoke, because it deprives the developing fetus of oxygen. These complications include an increased incidence of infertility and miscarriages, lower birth weight babies, and, later, more learning and behavioral problems in children. In spite of this, at least 10% of pregnant women in the United States continue to smoke.

Passive smoking, or breathing secondhand smoke, poses real health hazards to both children and adults. Researchers have concluded that children whose parents smoke have decreased lung capacity and are more likely to contract bronchitis, pneumonia, ear infections, coughs, and colds. Children who grow up with smokers are more likely to develop asthma and allergies, and secondhand smoke increases the number and severity of asthma attacks. Nonsmoking spouses of smokers face a 30% higher risk of both heart attack and lung cancer than do spouses of nonsmokers. The Centers for Disease Control and Prevention estimates that secondhand smoke is responsible for about 49,000 tobacco-related deaths annually in the United States.

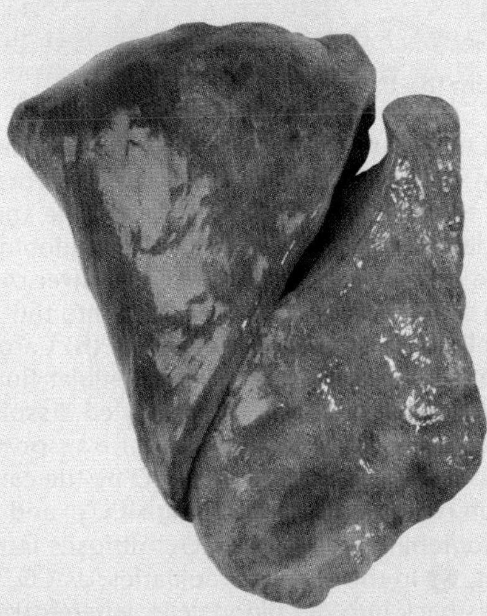

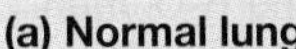

(a) Normal lung

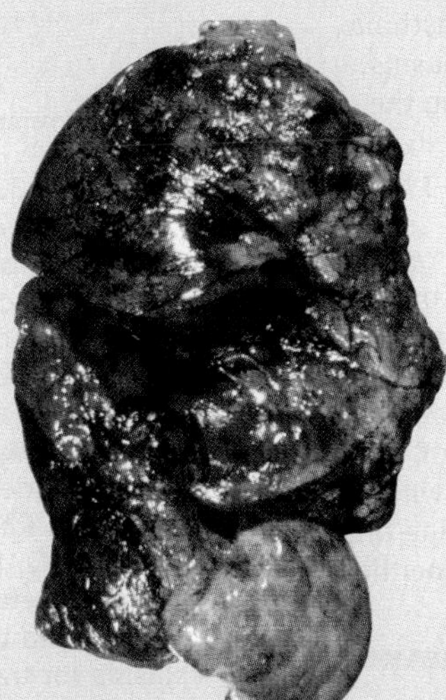

(b) A smoker's lung

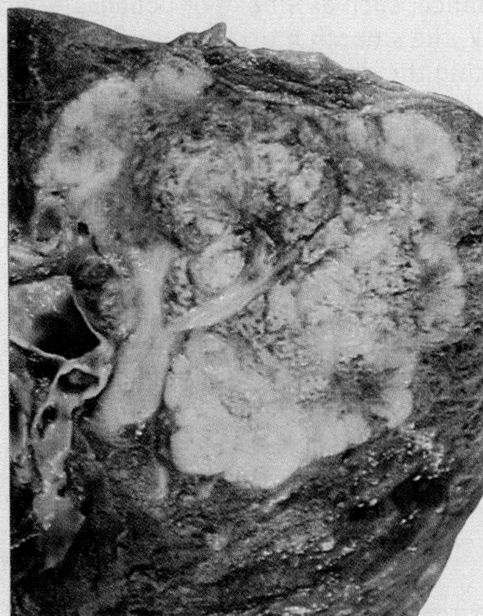

(c) A lung cancer

▲ **FIGURE E33-3 Smoking damages the lungs** **(a)** A normal lung. **(b)** A lung of a smoker who died of emphysema is both blackened and collapsed because its alveoli have ruptured. **(c)** A lung cancer is visible as a pale mass; the lung tissue surrounding it is blackened by trapped smoke particles.

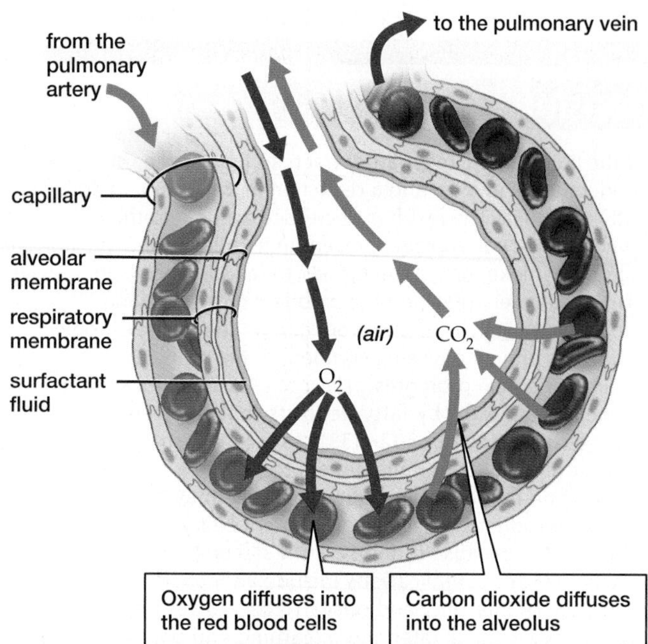

▲ **FIGURE 33-9 Gas exchange between alveoli and capillaries** The respiratory membrane consists of alveoli and capillary walls (each only one cell thick). Gases diffuse through this membrane between the lungs and the circulatory system. The inner alveolus is coated in a fluid containing a slippery surfactant that prevents the alveolar membranes from adhering to one another.

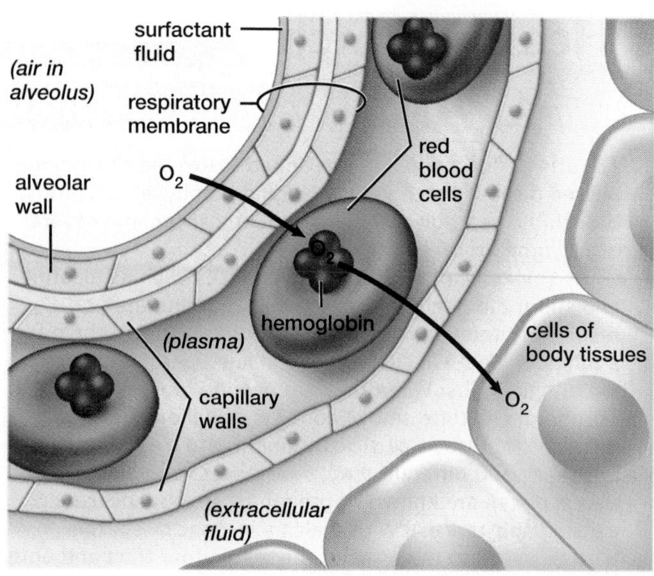

(a) O_2 transport from the lungs to the tissues

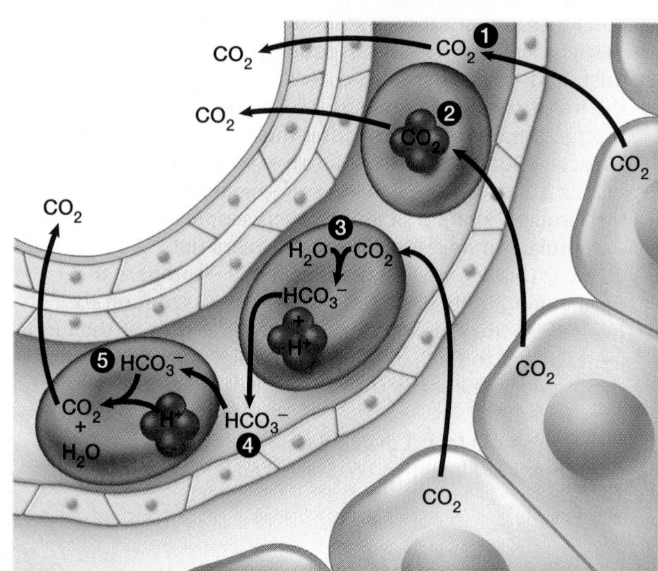

(b) CO_2 transport from the tissues to the lungs

▲ **FIGURE 33-10 Oxygen and carbon dioxide transport** **(a)** The high oxygen content of the air in the alveoli favors the diffusion of oxygen through the respiratory membrane and into the alveolar capillaries. Here, oxygen binds to hemoglobin and is transported to the cells of body tissues, whose lower oxygen content causes it to diffuse out of the capillaries, into the surrounding extracellular fluid, and into the cells. **(b)** Carbon dioxide diffuses from tissue cells through extracellular fluid and into capillaries. ❶ A small amount of CO_2 is carried dissolved in plasma. ❷ Some CO_2 binds to hemoglobin for transport. ❸ Most CO_2 from tissues is combined with H_2O by the carbonic anhydrase enzyme in red blood cells, forming HCO_3^- and H^+. The H^+ bind to hemoglobin and ❹ the HCO_3^- diffuses into the plasma for transport. ❺ In the alveolar capillaries, as CO_2 levels drop, HCO_3^- diffuses back into red blood cells, where carbonic anhydrase recombines them with H^+, forming H_2O and CO_2 ❺. The CO_2 diffuses into the plasma and then into the alveolus.

QUESTION Why is it important for the hydrogen ions (generated when bicarbonate is formed) to remain bound to hemoglobin?

Case Study continued
Lives Up in Smoke

Asthma occurs when smooth muscle tissue in the bronchioles becomes hyperexcitable and the production of mucus increases, often because of an allergy to an inhaled substance, such as smoke or pollen. During an asthma attack, the smooth muscle of the bronchioles spasms, reducing the diameter of the airways and causing the victim to struggle for breath. Asthma inhalers deliver medication that relaxes these muscles, reopening the airways.

By irritating the bronchioles, cigarette smoke greatly increases the likelihood of an asthma attack. Nicotine is so addictive, however, that some students carry both cigarettes and asthma inhalers. Recent research has shown that teenagers such as James, who smoke at least 300 cigarettes a year, have a fourfold increased risk of developing asthma. Teenagers who smoke, and whose mothers smoked during pregnancy, are nearly nine times as likely to develop asthma as are children with neither of these risk factors.

the lungs and supplies it to the body tissues, simultaneously absorbing CO_2 from the tissues and releasing it into the lungs. These exchanges occur because diffusion gradients favor them. In the lungs, O_2 is high and CO_2 is low (**Fig. 33-10**), whereas in body cells, CO_2 is high and O_2 is low (see Fig. 33-11).

Case Study continued
Lives Up in Smoke

The lungs of a smoker become clogged with mucus laden with toxic substances from cigarette smoke. The mucus causes irritation and frequent infections, which stimulate white blood cells to gather in the lungs. Here, they release enzymes that attack the walls of the alveoli, causing them to become brittle and rupture, causing emphysema. As emphysema progresses, the lungs of a smoker are transformed from resembling a healthy pink sponge to looking more like blackened swiss cheese. Loss of the alveoli, where gas exchange occurs, leads to oxygen deprivation throughout the body, and sometimes death. Nine out of 10 cases of emphysema are caused by smoking, and, although the resulting damage is permanent, people who stop smoking can greatly slow the progression of the disease.

Nearly all (about 98%) of the O_2 carried by blood is bound to **hemoglobin,** a large protein that gives red blood cells their color (see Fig. 33-10). Each hemoglobin molecule can carry up to four O_2 molecules, one bound to each of four iron-containing heme groups embedded in the protein (see Fig. 3-20). By removing oxygen from solution in the plasma, hemoglobin maintains the diffusion gradient that moves oxygen from the lungs into the plasma (where it is picked up by hemoglobin), allowing blood to carry far more oxygen than if the oxygen were simply dissolved in the plasma. As oxygen binds hemoglobin, the protein changes its shape, which alters its color; oxygenated blood is a bright cherry-red, and deoxygenated blood is maroon-red, which appears bluish through the skin. For this reason, oxygenated blood is depicted as red, and deoxygenated blood as blue.

Carbon dioxide from cellular respiration in body cells diffuses into nearby capillaries, and then is carried in the bloodstream to the respiratory membranes of the alveoli. Blood transports CO_2 in three different ways: roughly 10% is dissolved in the plasma (**Fig. 33-10b ❶**), about 20% is bound to hemoglobin (**Fig. 33-10b ❷**), and most of it (about 70%) combines with water to form bicarbonate ions (HCO_3^-) in the following reaction:

$$CO_2 + H_2O \longrightarrow H^+ + HCO_3^-$$

This reaction occurs in the red blood cells, where the necessary enzyme (carbonic anhydrase) is located (**Fig. 33-10b ❸**). Most HCO_3^- then diffuses into the plasma, while most of the H^+ remains in the red blood cells, bound to hemoglobin, which helps prevent the blood plasma from becoming too acidic (a serious condition that can lead to coma and death). The production of bicarbonate ions and the binding of some CO_2 to hemoglobin reduce the dissolved CO_2 in the blood plasma, thereby increasing the gradient for CO_2 to diffuse from body cells into the plasma.

The reaction producing bicarbonate ions is reversed as the blood flows through capillaries surrounding the alveoli, where CO_2 is low:

$$H^+ + HCO_3^- \longrightarrow CO_2 + H_2O$$

As CO_2 leaves the blood and diffuses into the alveoli, HCO_3^- diffuses back into the red blood cells (**Fig. 33-10b ❹**) where it recombines with H^+, regenerating CO_2 and H_2O (**Fig. 33-10b ❺**). The CO_2 then diffuses into the air in the alveoli, which is exhaled from the lungs, while the H_2O remains in the blood.

Air Is Inhaled Actively and Exhaled Passively

Breathing occurs in two stages: (1) **inhalation,** when air is drawn into the lungs, and (2) **exhalation,** when it is expelled from the lungs. Inhalation occurs when the chest cavity is enlarged. The lower boundary of the chest cavity is formed by a sheet of muscle, the **diaphragm,** which domes upward when relaxed. During inhalation, the diaphragm is contracted, which pulls it downward. The rib muscles also contract during inhalation, lifting the ribs up and outward. Both of these muscular movements enlarge the chest cavity (**Fig. 33-11a**). When the chest cavity is expanded, the lungs inflate within it, because an airless space with a layer of fluid seals them tightly against the inner wall of the chest. As the lungs expand, their increased volume draws in air. A puncture wound to the chest is dangerous in part because it can allow air to penetrate between the chest wall and the lungs, breaking the seal and preventing the lungs from inflating when the chest cavity expands.

Although air can be forcibly exhaled, exhalation usually occurs spontaneously when the muscles that cause inhalation are relaxed. As the diaphragm relaxes, it domes upward; at the same time, the ribs fall down and inward. Both of these movements decrease the size of the chest cavity and force air out of the lungs (**Fig. 33-11b**). Additional air can be forced out by

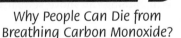

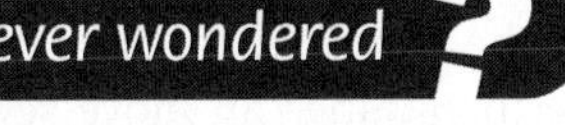

Why People Can Die from Breathing Carbon Monoxide?

Carbon monoxide (CO) is a toxic gas produced by combustion (as occurs in engines, furnaces, and the burning of charcoal and cigarettes) when the fuel is not completely burned to form carbon dioxide. Carbon monoxide can be deadly because it binds to the same sites on hemoglobin that bind O_2, but adheres more than 200 times more tightly, preventing the hemoglobin from transporting O_2. People can die of asphyxiation from breathing air with as little as 0.1% CO.

Hemoglobin bound to CO is bright red, like oxygenated hemoglobin, and because it binds so tightly, CO occupies far more binding sites on hemoglobin than does oxygen. As a result, although most victims of asphyxiation have bluish lips and nail beds because their hemoglobin is deoxygenated, the lips and nail beds of victims of CO poisoning are actually brighter red than normal.

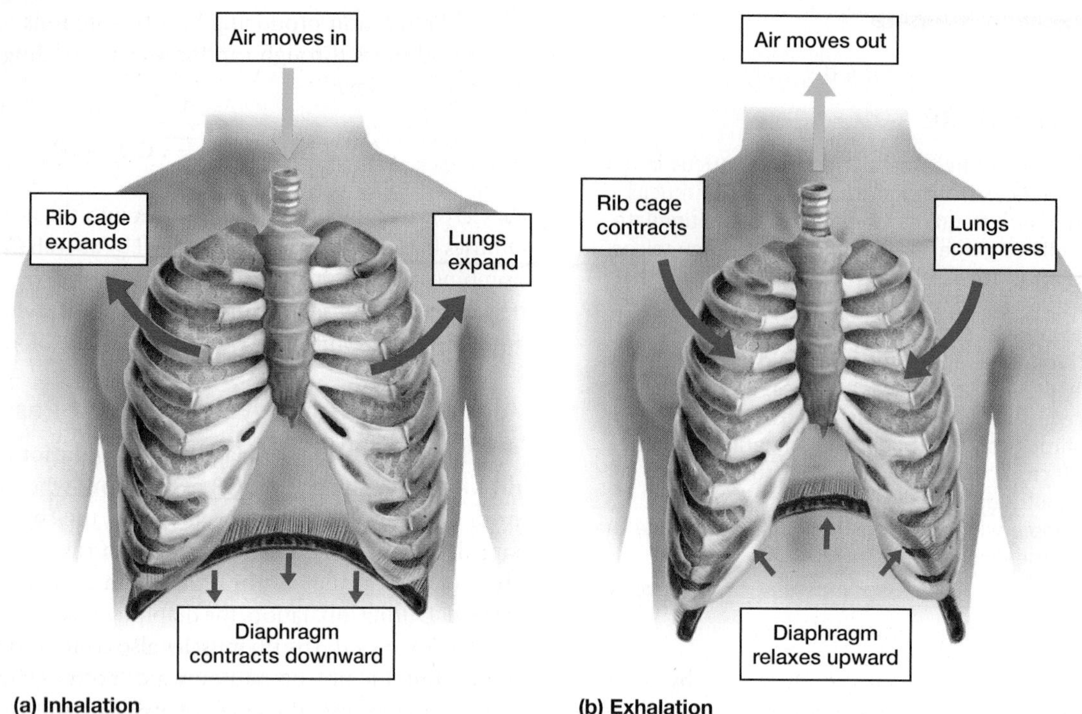

▲ **FIGURE 33-11 The mechanics of breathing (a)** During inhalation, rhythmic nerve impulses from the brain stimulate the diaphragm to contract (pulling it downward) and the muscles surrounding the ribs to contract (moving them up and outward). The result is an increase in the size of the chest cavity, which causes air to rush in. **(b)** During exhalation, these muscles relax, allowing the diaphragm to dome upward and the rib cage to collapse, forcing air out of the lungs.

QUESTION Why does contracting the diaphragm muscle (which decreases its size) increase the volume of the chest cavity?

contracting the abdominal muscles. After exhalation, the lungs still contain some air, which helps prevent the thin alveoli from collapsing and fills the spaces within the conducting portion of the respiratory system. A typical breath moves about a pint (500 milliliters) of "new" air into the respiratory system. Of this, only about three-quarters of a pint (350 milliliters) reaches the alveoli where gas exchange occurs. During exercise, deeper breathing can move several times more air than this.

BioFlix™ Gas Exchange

Breathing Rate Is Controlled by the Respiratory Center of the Brain

Imagine having to think about every breath. Fortunately, breathing occurs rhythmically and automatically without conscious thought. But, unlike the heart muscle, the muscles used in breathing are not self-activating; each contraction is stimulated by impulses from nerve cells. These impulses originate in the **respiratory center,** which is located in the medulla, a portion of the brain just above the spinal cord (see Fig. 38-12). Nerve cells in the respiratory center generate cyclic bursts of impulses (action potentials) that cause contraction (followed by passive relaxation) of the respiratory muscles.

The respiratory center receives input from several sources and adjusts the breathing rate and volume to meet the body's changing needs. The respiratory rate is primarily regulated by CO_2 receptors, also located in the medulla. These adjust the breathing rate to maintain a relatively constant, low level of CO_2 in the blood, while also ensuring that O_2 levels remain adequate. For example, if you run to class, your muscles need more ATP. This requires an increase in cellular respiration, which generates more CO_2. Receptors in the medulla detect the excess CO_2 and cause you to breathe faster and more deeply, eliminating more CO_2 and taking in more O_2. These receptors are extremely sensitive; an increase in CO_2 of only 0.3% can double the breathing rate. As a kind of backup system, there are also O_2 receptors in the aorta and carotid arteries that stimulate the respiratory center to increase the rate and depth of breathing if O_2 levels in the blood drop.

Case Study revisited

Lives Up in Smoke

"For some reason, a rolled-up paper with tobacco and nicotine inside is supposed to make you popular, or confident, or cool. It doesn't," says James, who finally gave up smoking on his third attempt, just before his 21st birthday. "I was tired of the need, the cravings, the smell, the taste. I was tired of the looks others gave me, the shame I felt, the pain inflicted on those I loved . . . the list is endless."

James' success is not the norm. Nearly all high school smokers say they will quit within 5 years, but more than half are still smoking 7 years later. The nicotine in tobacco is a powerfully addictive drug—as addictive as cocaine or heroin. Like cocaine and heroin, nicotine activates the brain's reward center, a cluster of cells deep within the brain. The reward center is normally activated by natural stimuli that help us to survive and reproduce—such as eating calorie-rich foods and engaging in sexual activity. But all addictive drugs also activate the reward center, often more intensely than natural stimuli. The brain adapts by becoming less sensitive to these drugs, requiring larger quantities to experience the same rewarding effect. These changes cause the reward center to feel understimulated in the absence of nicotine, producing withdrawal symptoms such as nicotine craving, depression, anxiety, irritability, difficulty concentrating, headaches, and disturbed sleep.

Unfortunately, adolescence is a time when the still-developing brain is particularly susceptible to nicotine addiction. Recent evidence suggests that teenagers who smoke only a few cigarettes a week can become dependent on the habit, experiencing cravings and other withdrawal symptoms when they try to quit. Although at least 70% of smokers would like to stop, only about half succeed. Of those who continue to smoke, about half will die from smoking-related illnesses.

If you are a smoker who kicks the habit, the rewards begin immediately. After 20 minutes, your blood pressure and pulse rate drop. After 8 hours, the level of carbon monoxide in your blood drops, allowing your blood O_2 to increase to normal levels. After a day, your chance of having a heart attack begins to decrease. After two days, you'll start getting more pleasure from your food as your senses of taste and smell increase. During the first few months, you will find it increasingly easy to exercise as your circulation and respiratory function improve. Your smoker's cough will gradually disappear, your sinus congestion will clear up, and you'll have more energy. After 5 years, you will have about half the risk of cardiovascular disease as does a continuing smoker. After 10 years, your risk of lung cancer will be about half that of a smoker, and your risk of cancers of the pancreas, kidney, bladder, esophagus, throat, and mouth also decline. After 20 to 30 years, your risk of death from all smoking-related causes will be nearly as low as that of people who have never smoked.

Consider This

A University of Illinois student says of smoking, "People are going to do it no matter what. . . . We have the right to choose how we live our lives." The National Institutes of Health reports that direct medical costs for treating illnesses caused by smoking are $75 billion annually in the United States. This cost is borne by society, whether through taxes that support Medicare and Medicaid, or through high insurance premiums paid by individuals and employers. Rising insurance rates have caused an increase in the percentage of uninsured families, who often get inadequate medical care as a result. Additionally, each year tens of thousands of U.S. nonsmokers die of lung cancer and cardiovascular disease that is attributable to secondhand smoke. With these facts in mind, discuss the assertion that people have the right to choose to smoke.

CHAPTER REVIEW

Summary of Key Concepts

33.1 Why Exchange Gases?

The respiratory system supports cellular respiration. Oxygen-rich air is inhaled and supplies O_2 to the blood, which carries it to cells throughout the body. Blood also picks up CO_2 (a product of cellular respiration) from body cells and transports it to the lungs, where it is released into the atmosphere.

33.2 What Are Some Evolutionary Adaptations for Gas Exchange?

The exchange of O_2 and CO_2 between the body and the environment occurs by diffusion across a moist surface. In watery environments, animals with very small or flattened bodies may lack specialized respiratory structures and instead rely exclusively on diffusion through the body surface. Animals with low metabolic demands or well-developed circulatory systems may also lack specialized respiratory structures. Larger, more active animals have evolved specialized respiratory systems. Animals in aquatic environments often possess gills, such as those of fish and amphibians. On land, respiratory surfaces must be protected and kept moist internally. This has selected for the evolution of tracheae in insects and lungs in terrestrial vertebrates.

The transfer of gases between respiratory systems and tissues requires both bulk flow and diffusion. Air or water moves by bulk flow past the respiratory surface; gases are also carried in blood by bulk flow. Gases move by diffusion across membranes between the respiratory system and the capillaries and between the capillaries and the tissues.

33.3 How Does the Human Respiratory System Work?

The human respiratory system includes a conducting portion that consists of the nose and mouth, pharynx, larynx, trachea, bronchi, and bronchioles, and a gas-exchange portion that is composed of alveoli. Blood, within a dense capillary network surrounding the alveoli, releases CO_2 to alveolar air and absorbs O_2 from it.

Most of the O_2 carried in the blood is bound to hemoglobin within red blood cells. By removing O_2 from solution in plasma, hemoglobin maintains a gradient that favors the diffusion of O_2 from the air in the alveoli into the blood. Hemoglobin then transports the O_2 to the body tissues, where it diffuses out. Carbon dioxide diffuses from the tissues into the blood. Most is transported as bicarbonate ions, some is bound to hemoglobin, and a little is carried dissolved in blood plasma.

Breathing involves inhalation, which actively draws air into the lungs by contracting the diaphragm and the rib muscles, expanding the chest cavity. Exhalation at rest is passive and occurs when these muscles are allowed to relax, reducing the volume of the chest cavity and expelling the air. Respiration is controlled by the medulla's respiratory center. The respiratory rate is controlled by receptors that monitor CO_2 and O_2 levels, primarily those in the medulla that monitor CO_2 levels.

BioFlix™ Gas Exchange

Key Terms

alveolus (plural, **alveoli**) *646*
bronchiole *646*
bronchus (plural, **bronchi**) *646*
bulk flow *642*
conducting portion *645*
countercurrent exchange *643*
diaphragm *651*
epiglottis *645*
exhalation *651*
gas-exchange portion *645*
gill *642*
Heimlich maneuver *645*
hemoglobin *651*
inhalation *651*
larynx *645*
lung *643*
pharynx *645*
respiration *641*
respiratory center *652*
respiratory membrane *647*
spiracle *643*
trachea *645*
tracheae *643*
vocal cords *645*

Thinking Through the Concepts

Fill-in-the-Blank

1. Which part of the conducting portion of the respiratory system is shared with the digestive tract? __________ What structure normally keeps food from entering the larynx? __________ After passing through the larynx, air travels in sequence through the __________, __________, and __________, after which it enters the gas-exchange portion of the respiratory system, which consists of the __________.
2. In the lungs, oxygen diffuses into blood vessels called __________. Most oxygen in the blood is carried by a protein called __________. Most carbon dioxide in the blood is transported in the form of __________ ions. Cells require oxygen and release carbon dioxide because they are generating energy using the process of __________.
3. The respiratory membrane is found in the __________. This membrane consists of the wall of a(n)__________and the wall of a(n) __________ joined by connective tissue. The respiratory membrane is (how many?) __________cells thick. The inside of the respiratory membrane is coated with a watery fluid containing __________.
4. Air is inhaled when the __________ and the rib muscles contract, making the chest cavity __________. In contrast, exhalation is a(n) __________ process, caused by allowing these muscles to __________. Breathing is driven by neurons of the __________ located in the __________ of the brain. The respiratory rate is increased when receptors detect an excess of __________ in the blood.
5. List three diseases that are far more common in smokers than in nonsmokers: __________, __________, __________. List three ailments (worsened by passive smoking) that are more common in children of smokers than nonsmokers: __________, __________, __________.

Review Questions

1. Describe countercurrent exchange. What are the benefits of this process? How does it work in fish gills?
2. Trace the route taken by air in the mammalian respiratory system, listing the structures through which it flows and the point at which gas exchange occurs.
3. Explain some characteristics of animals in moist environments that make specialized respiratory systems unnecessary.
4. How is breathing initiated? How is it modified, and why are these controls adaptive?
5. What events occur during human inhalation? Exhalation? Which of these is always an active process?
6. Describe the effects of smoking on the human respiratory system.
7. Compare bulk flow and diffusion. Explain how bulk flow and diffusion interact to promote gas exchange between air and blood and between blood and tissues.
8. Compare CO_2 and O_2 transport in the blood. Include the source and destination of each.
9. Explain how the structure and arrangement of alveoli make them well suited for their role in gas exchange.

Applying the Concepts

1. **BioEthics** Heart–lung transplants are performed in some cases in which both organs have been damaged by cigarette smoking. Based on your knowledge of the respiratory and circulatory systems, and the lifestyle choices that might damage these organ systems, what criteria would you use in selecting a recipient for such a transplant? Consider the scarcity of donor organs as you make your decision.
2. Nicotine is a drug in tobacco that is responsible for several of the effects that smokers crave. Discuss the advantages and disadvantages of low-nicotine cigarettes.
3. Discuss why a brief exposure to carbon monoxide is much more dangerous than a brief exposure to carbon dioxide.
4. Review insect and mammalian respiratory systems and circulatory systems (see Chapter 32). In what way does the respiratory system of an insect serve some of the functions of the circulatory system of mammals?
5. Mary, a strong-willed 3-year-old, threatens to hold her breath until she dies if she doesn't get her way. Can she carry out her threat? Explain.

MB *Go to www.masteringbiology.com for practice quizzes, activities, eText, videos, current events, and more.*

Nutrition and Digestion

Chapter 34

Case Study

Dieting to Death?

REFLECTING ON HER CAREER, former supermodel Carré Otis says, "The sacrifices I made were life-threatening. I had entered a world that seemed to support a 'whatever it takes' mentality to maintain abnormal thinness." For many models, performers, and others in the public eye, meeting expectations for thinness is a continuing battle that can lead to disaster. Having conquered anorexia, Carré Otis is now a spokesperson for the National Eating Disorders Association. She hopes to help others avoid the damage her body suffered. "It was common for the young girls I worked with to have a heart attack; if an eating disorder is not treated, it can be a fatal disease."

Eating disorders, which occur most frequently in women, include two particularly debilitating conditions: anorexia and bulimia. Celebrities who have battled these disorders include Jane Fonda, the late Princess Diana, Paula Abdul, Elton John, and the late model Ana Carolina Reston, one of several models who have died from complications of anorexia. People with anorexia experience an intense fear of gaining weight; although their bodies become skeletal, they see themselves as fat. In response, they eat very little food and often exercise compulsively. After burning off essentially all their body fat, the starving bodies of anorexics break down their muscle tissue to supply energy. The disorder disrupts digestive, reproductive, endocrine, and cardiac functions. The consequences can be disastrous.

People with bulimia (some of whom are anorexic, while others maintain a normal weight) engage in binge eating, consuming enormous amounts of food in a short period of time. To purge the food from their bodies, bulimics then induce vomiting or overdose with laxatives; they may also engage in excessive exercise.

How do eating disorders interfere with the body's ability to obtain nutrients, and what are the possible consequences? What are some treatments for under- and overeating? What nutrients do our bodies require, and how do our bodies extract nutrients from the food we eat? Read on to find out.

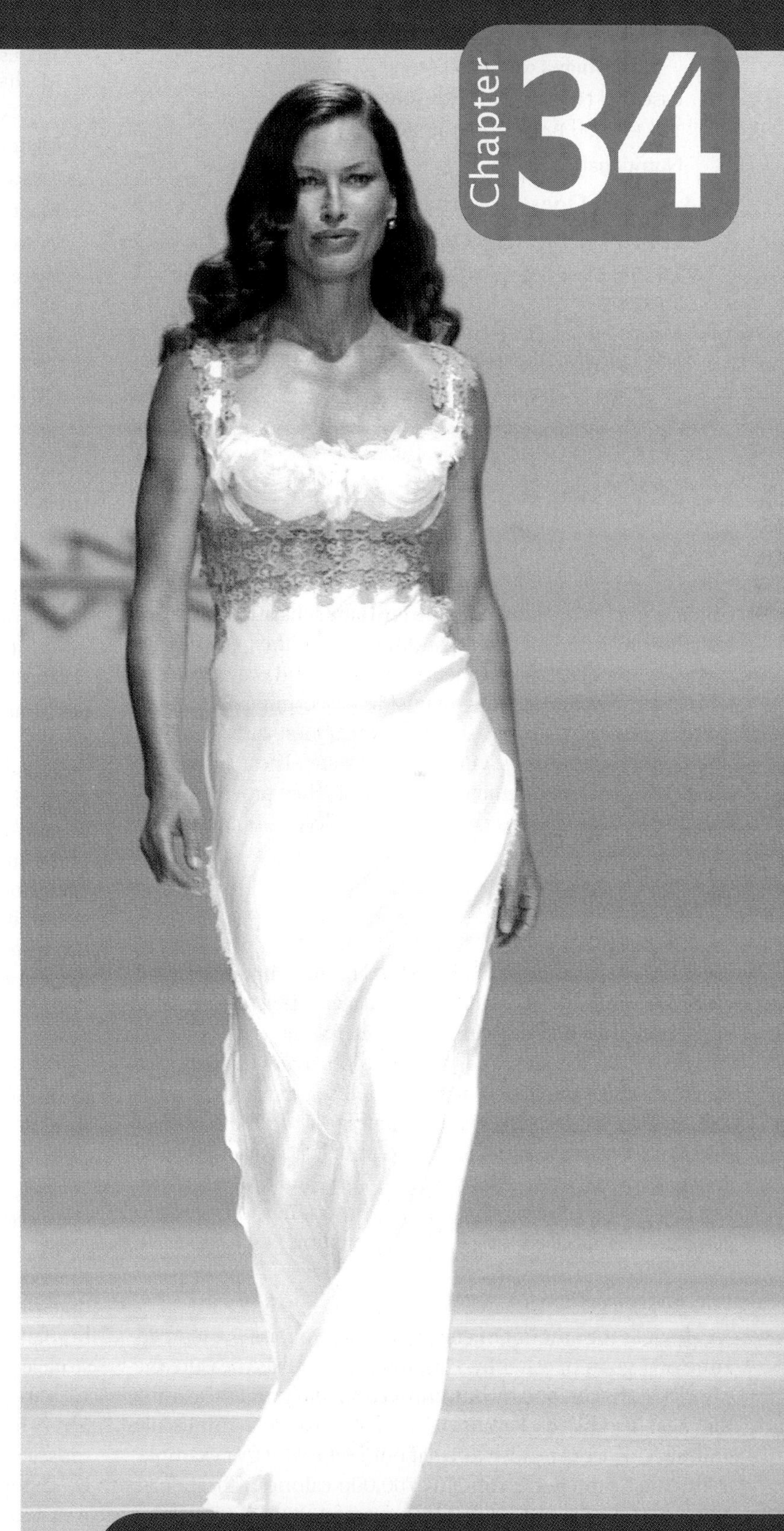

▲ Recovered anorexic Carré Otis is now the model of health.

At a Glance

34.1 WHAT NUTRIENTS DO ANIMALS NEED?

Whether you are eating cauliflower or a candy bar, your food contains important nutrients. **Nutrients** are substances obtained from the environment that organisms need for their growth and survival. Animals obtain nutrients from food and water; water itself is essential, and often contains dissolved minerals as well. Animal nutrients fall into six major categories: carbohydrates, lipids, proteins, minerals, vitamins, and water. These substances satisfy two basic requirements for survival: They provide energy and the raw materials to synthesize the molecules of life.

Most Energy Is Provided by Carbohydrates and Fats

Cells rely on a continuous supply of energy to maintain their incredible complexity and wide range of activities. Deprived of this energy, cells begin to die within minutes. Nutrients that provide energy are lipids (fats and oils), carbohydrates (sugars and starches), and proteins. Although people's diets vary considerably, in a typical U.S. diet, carbohydrates provide about 50% of the total energy; fats, about 35%; and proteins, about 15% (well above the world average for protein). These molecules are broken down by digestion, which frees their subunits to be used to generate ATP in cells during cellular respiration (see Chapter 8).

Energy from Nutrients Is Measured in Calories

A **calorie** is the amount of energy required to raise the temperature of 1 gram of water by 1 °C. In fact, scientists measure the calories stored in food by burning it completely and measuring the heat produced. Because a calorie is such a minuscule amount of energy relative to what our bodies use (for example, a Big Mac with cheese contains 700,000 calories), the calorie content of foods is measured in units of 1,000 calories (kilocalories), or **Calories** with a capital "C" (so the Big Mac contains 700 Calories). About 60% of the caloric energy in the food we eat is released as heat. The remaining 40% is available to generate ATP, which is continuously utilized during metabolic reactions including those that produce brain activity, muscle contraction, active transport across membranes, and the synthesis of new biological molecules.

The average human body at rest burns roughly 70 Calories per hour, but this value is influenced by several factors. For example, muscle tissue requires more Calories to maintain itself than does the same weight of fat tissue, so a muscular individual will burn more Calories just sitting than will a person of the same weight with a large amount of fat. People also differ in **metabolic rate,** the speed at which cellular reactions that release energy occur. Exercise significantly boosts Caloric requirements; well-trained athletes can temporarily burn nearly 20 Calories per minute during vigorous exercise. The Calorie content of some foods and the time required to burn off these Calories during various types of exercise is provided in **Table 34-1.**

Carbohydrates Are a Source of Quick Energy

Carbohydrates (see pp. 38–43) include sugars such as glucose, from which cells derive most of their energy; sucrose, which is table sugar; and polysaccharides, long chains of sugar molecules. Cellulose, starch, and glycogen are all polysaccharides composed of chains of glucose. Cellulose, the major structural component of plant cell walls, is the most abundant carbohydrate on the planet, but only a few types of animals are able to digest it, as described later. Starch is a major energy source for humans and many other animals, and the principal energy-storage material of plants.

Glycogen is used by animals for short-term energy storage. Animals, including humans, store glycogen in the liver and muscles. Athletes sometimes "carbo-load" before competing by eating meals rich in carbohydrates, such as potatoes and pasta, to store as much glycogen as possible. Although humans can accumulate hundreds of pounds of fat, most can store less

Table 34-1 Approximate Energy Expended by a 150-Pound Person for Different Activities

		Time to "Work Off"			
Activity	**Calories/Hour**	**500 Calories (Cheeseburger)**	**340 Calories (Ice cream cone)**	**70 Calories (Apple)**	**40 Calories (Broccoli, 1 cup)**
Running (6 mph)	700	43 min	26 min	6 min	3 min
Cross-country skiing (moderate)	560	54 min	32 min	7.5 min	4 min
Roller skating	490	1 hr 1 min	37 min	8.6 min	5 min
Bicycling (11 mph)	420	1 hr 11 min	43 min	10 min	6 min
Walking (3 mph)	250	2 hr	1 hr 12 min	17 min	10 min
Frisbee® playing	210	2 hr 23 min	1 hr 26 min	20 min	11 min
Studying	100	5 hr	3 hr	42 min	24 min

than a pound of glycogen. During exercise such as running, the body uses glycogen as a source of quick energy.

Fats and Oils Are the Most Concentrated Energy Source

Fats and oils are the most concentrated sources of energy, containing over twice as many Calories per unit weight as do carbohydrates or proteins (about 9 Calories per gram for fats compared to about 4 Calories per gram for proteins and carbohydrates).

When an animal's diet provides more energy than it expends, most of the excess carbohydrates and fats are stored as body fat. In addition to its high caloric content, fat is hydrophobic, so it neither attracts water nor dissolves in water, as carbohydrates and proteins do. For this reason, fat deposits do not cause extra water to accumulate in the body, allowing these deposits to store more Calories with less weight than do other molecules. Minimizing weight allows an animal to move faster (important for escaping predators and hunting prey) and to use less energy when it moves (important when food supplies are limited).

In addition to storing energy, fat deposits may provide insulation. Fat, which conducts heat at only one-third the rate of other body tissues, is often stored in a layer directly beneath the skin. Birds and mammals that live in polar climates or in cold ocean waters—such as penguins, seals, whales, and walruses—are particularly dependent on this insulating layer, which reduces the amount of energy they must expend to keep warm (**Fig. 34-1**).

Because humans evolved under the same food constraints as other animals, we have a strong tendency to eat when food is available—often eating more than we need, because we may require the energy later, when food is scarce. Some modern societies now have access to almost unlimited high-calorie food. In this environment, our natural tendency to overeat can become a liability, and some of us need to exert considerable willpower to avoid storing excessive amounts of fat and becoming overweight. The **body mass index (BMI)** is a common tool for estimating a healthy weight—a BMI between 18.5 and 24.9 is considered healthy. The U.S. Centers for Disease Control and Prevention estimate that about 33% of all U.S. adults are overweight (BMI between 25 and 29.9) and an additional 33% are obese (BMI of 30 or more); see "Health Watch: Eating to Death?" later in this chapter.

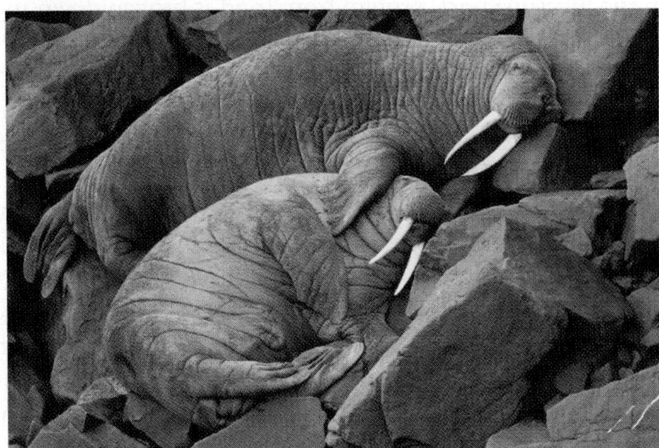

▲ **FIGURE 34-1 Fat provides insulation** These walruses can withstand the icy waters of the arctic seas because a thick layer of fat beneath the skin insulates them from the cold.

Case Study continued

Dieting to Death?

A BMI below 18.5 is considered underweight. Carrè Otis, who once carried less than 100 pounds on her 5' 10" frame, then had a BMI of 14.2. Recognizing that several models have died from complications of anorexia, the fashion trade show "Madrid Fashion Week" in Spain has banned any model with a BMI under 18 from its shows. Some fashion companies have followed suit, banning underweight models in their advertising.

Essential Nutrients Provide Raw Materials

To maintain health, our diets must provide our bodies with a variety of raw materials. Our cells can synthesize most of the molecules our bodies require, but they cannot synthesize certain raw materials, called **essential nutrients,** which must be supplied in the diet. For example, although carbohydrates are an important source of energy and part of a balanced diet, they are not considered essential nutrients because our bodies can synthesize those we need. Essential nutrients differ for different animals. Most mammals other than humans can synthesize vitamin C (ascorbic acid), and so for them, ascorbic acid is not an essential nutrient. Essential nutrients for humans include certain fatty acids and amino acids, a variety of minerals and vitamins, and water.

Certain Fatty Acids Are Essential in the Human Diet

Fats and oils are more than just a source of energy—some provide **essential fatty acids.** Essential fatty acids serve as raw materials used to synthesize molecules involved in a wide range of physiological activities: They help us to absorb fat-soluble vitamins (described later) and are important in cell division, fetal development, and the immune response. Important sources of essential fatty acids include fish oils, canola oil, soybean oil, flaxseed, and walnuts.

Amino Acids Form the Building Blocks of Protein

In the digestive tract, protein from food is broken down into its amino acid subunits, which can be used to synthesize new proteins. These perform many different roles in the body, acting as enzymes, receptors on cell membranes, oxygen transport molecules (hemoglobin), structural proteins (hair and nails), antibodies, and muscle proteins.

Humans are unable to synthesize 9 (adults) or 10 (infants) of the 20 different amino acids used in proteins. These **essential amino acids** must be obtained from protein-rich foods such as meat, milk, eggs, corn, beans, and soybeans. Because many plant proteins are deficient in some of the essential amino acids, vegetarians must consume a variety of plants (for example, legumes, grains, and corn) whose proteins collectively provide all of them. Protein deficiency can cause a debilitating condition called kwashiorkor (**Fig. 34-2**), which occurs most frequently in poverty-stricken countries.

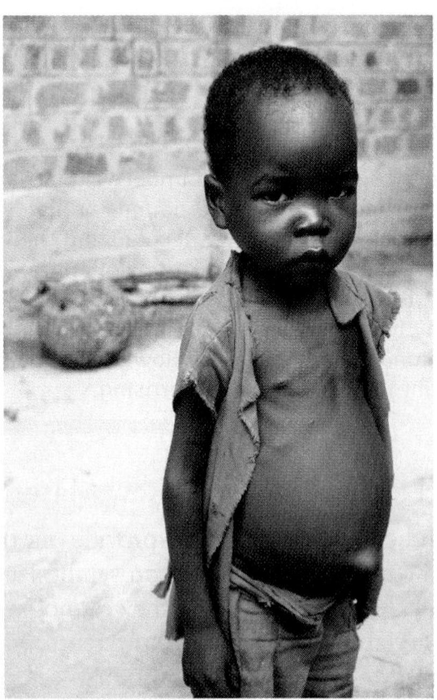

▲ **FIGURE 34-2 Kwashiorkor is caused by protein deficiency** In children with kwashiorkor, low levels of blood albumin protein decrease the blood's osmotic strength and allow fluids to leak from blood capillaries into the abdominal area, causing it to swell. Muscles are also wasted due to lack of protein.

Minerals Are Elements Required by the Body

Minerals are elements that play many crucial roles in animal nutrition. Because no organism can manufacture elements, all required minerals are essential nutrients that must be obtained from food or dissolved in drinking water. Minerals such as calcium, magnesium, and phosphorus are major constituents of bones and teeth. Sodium, calcium, and potassium are essential for muscle contraction and the conduction of nerve impulses. Iron is a central component of each hemoglobin molecule in blood, and iodine is found in hormones produced by the thyroid gland. We also require trace amounts of several other minerals, including zinc and magnesium (both are required for the function of some enzymes), copper (needed for hemoglobin synthesis), and chromium (used in the metabolism of sugar). The minerals required by humans, their sources, important functions in the body, and symptoms of severe deficiencies are provided in **Table 34-2.**

Vitamins Play Many Roles in Metabolism

"Take your vitamins!" is a familiar refrain in many households with children. But why are vitamins so important? **Vitamins** are a diverse group of organic molecules that animals require in small amounts for normal cell function, growth, and development. Many vitamins are required for the proper functioning of enzymes that control metabolic reactions throughout the body. Vitamins are essential nutrients that the body cannot synthesize (or cannot synthesize in adequate quantities to maintain health), so they must be obtained from the diet. The vitamins essential in human nutrition are listed in **Table 34-3.** Human vitamins are grouped into two categories: water soluble and fat soluble.

Water-Soluble Vitamins Water-soluble vitamins include vitamin C and the nine compounds that make up the B-vitamin complex. Because these substances dissolve in the watery blood plasma and are filtered out by the kidneys, they are not stored in appreciable amounts. Therefore, the body's supply of these vitamins must be constantly replenished by diet.

Most water-soluble vitamins act as *coenzymes;* that is, they work in conjunction with enzymes to promote chemical reactions that supply energy or synthesize biological molecules. Because each vitamin participates in several metabolic processes, a deficiency of a single vitamin can have wide-ranging effects (see Table 34-3). For example, deficiency of the B vitamin niacin causes the swollen tongue and skin lesions of pellagra (**Fig. 34-3**), as well as some nervous disorders. Folic acid, another B vitamin, is required to synthesize thymine, a component of DNA; folic acid deficiency impairs cell division throughout the body. As you might predict, it is particularly important for pregnant women to get enough folic acid to supply the rapidly growing fetus. Folic acid deficiency can also lead to a reduction in red blood cells, resulting in anemia. For folic acid to function properly, trace amounts of vitamin B_{12} are required. In the human diet, vitamin B_{12} is obtained only from eating animal protein (meat and dairy products) or supplemented foods.

Fat-Soluble Vitamins Fat-soluble vitamins can be stored in body fat and may accumulate in the body over time. The fat-soluble

Table 34-2 **Important Minerals for Humans**

Mineral	Dietary Sources	Important Roles in the Body	Deficiency Symptoms
Calcium	Milk, cheese, leafy vegetables	Helps in bone and tooth formation and maintenance; aids in blood clotting; contributes to nerve impulse transmission and muscle contraction	Stunted growth, rickets, osteoporosis
Phosphorus	Milk, cheese, meat, poultry, grains	Helps maintain pH of body fluids; contributes to bone and tooth formation; component of ATP and of phospholipids in cell membranes	Muscular weakness, weakening of bone
Potassium	Meats, milk, fruits	Helps maintain pH and osmotic strength of body fluids; important in nervous system activity	Nausea, muscular weakness, paralysis
Chlorine	Table salt	Helps maintain pH and osmotic strength of body fluids; component of HCl produced by gastric glands; important in nervous system activity	Muscle cramps, apathy, reduced appetite
Sodium	Table salt	Helps maintain pH and osmotic strength of body fluids; important in nervous system activity	Muscle cramps, nausea
Magnesium	Whole grains, leafy vegetables, dairy products, legumes, nuts	Helps to activate many enzymes	Tremors, muscle spasms and weakness, irregular heartbeat, hypertension
Iron	Meats, legumes, nuts, whole grains, leafy vegetables	Component of hemoglobin and many enzymes	Iron-deficiency anemia (weakness, reduced resistance to infection)
Fluorine	Fluoridated water, seafood	Component of teeth and bones	Increased tooth decay; may increase risk of osteoporosis
Zinc	Seafood, meat, cereals, nuts, legumes	Constituent of several enzymes; component of proteins required for normal growth, smell, and taste	Retarded growth, learning impairment, depressed immunity
Iodine	Iodized salt, seafood, dairy products, many vegetables	Component of thyroid hormones	Goiter (enlarged thyroid gland)
Chromium	Meats, whole grains	Helps maintain normal blood glucose levels	Elevated insulin in blood; increased risk of adult-onset diabetes

vitamins A, D, E, and K have a variety of functions (see Table 34-3). Vitamin A is used to synthesize the light-capturing molecule in the retina of the eye; a deficiency can lead to impaired night vision. Vitamin D is required for normal bone formation; a deficiency can lead to bone deformities, such as occur with the disorder rickets (**Fig. 34-4**). Vitamin D is produced naturally in skin exposed to sunlight, but people who get little exposure to sunlight may not synthesize enough. Researchers have discovered that many adult women, particularly those with dark skin (which reduces the penetration

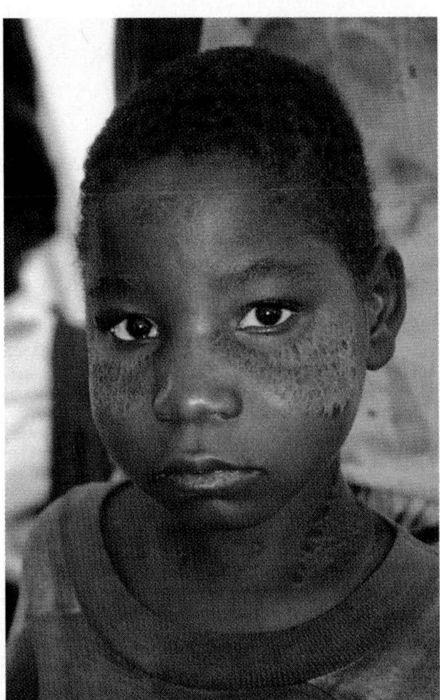

▲ FIGURE 34-3 **Pellagra is caused by niacin deficiency** Symptoms of pellagra include the scaly, reddish-brown skin lesions visible on this child's face and neck.

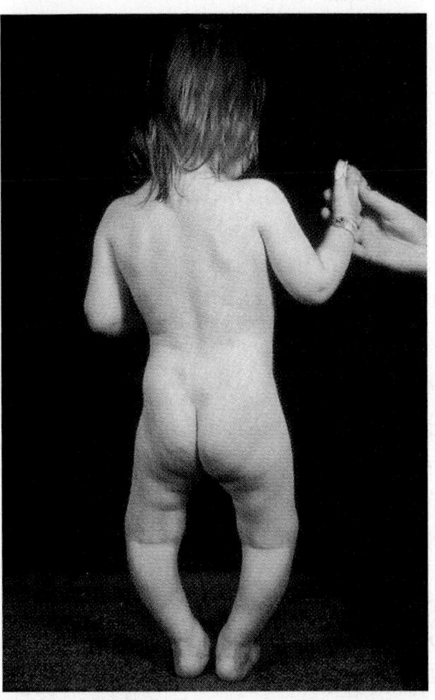

▲ FIGURE 34-4 **Rickets is caused by vitamin D deficiency** Lack of vitamin D in early childhood prevents bones from absorbing sufficient calcium, making them soft and unable to bear weight without bending. This can result in permanent bone deformity.

Table 34-3 **Important Vitamins for Humans**

Vitamin	Dietary Sources	Functions in Body	Deficiency Symptoms
Water Soluble			
B complex			
Vitamin B_1 (thiamin)	Meat, cereals, peas and soybeans, fish	Coenzyme in glucose metabolism	Beriberi (muscle weakness, peripheral nerve changes, heart failure)
Vitamin B_2 (riboflavin)	Shellfish, dairy products, eggs	Component of coenzymes involved in energy metabolism	Reddened lips, cracks at the corners of the mouth, light sensitivity, blurred vision
Niacin	Meats, fish, leafy vegetables	Component of coenzymes involved in energy metabolism	Pellagra (skin and gastrointestinal lesions, and nervous and mental disorders)
Vitamin B_5 (pantothenic acid)	Meats, whole grains, legumes, eggs; product of intestinal bacteria	Component of coenzyme A, with a role in cellular respiration	Fatigue, sleep disturbances, impaired coordination
Vitamin B_6 (pyridoxine)	Meats, whole grains, tomatoes, potatoes	Coenzyme in amino acid metabolism	Irritability, convulsions, skin disorders, increased risk of heart disease
Folic acid	Meats, green and leafy vegetables, whole grains, eggs; product of intestinal bacteria	Coenzyme in nucleic and amino acid metabolism	Anemia, gastrointestinal disturbances, diarrhea, retarded growth, birth defects in infants born to biotin-deficient mothers
Vitamin B_{12}	Meats, eggs, dairy products	Coenzyme in nucleic acid metabolism and other metabolic pathways	Anemia, neurological disturbances
Biotin	Legumes, vegetables, meats; product of intestinal bacteria	Coenzyme in amino acid metabolism and cellular respiration	Fatigue, depression, nausea, skin disorders, muscular pain
Vitamin C (ascorbic acid)	Citrus and other fruits, tomatoes, leafy vegetables	Helps maintain cartilage, bone, and dentin (hard tissue of teeth); helps in collagen synthesis	Scurvy (degeneration of skin, teeth, gums, and blood vessels, and epithelial hemorrhages)
Fat Soluble			
Vitamin A (retinol)	Green, yellow, and red vegetables; liver, fortified dairy products	Component of visual pigment; helps maintain skin and other epithelial cells; promotes normal development of teeth and bones	Night blindness, permanent blindness, increased susceptibility to infections
Vitamin D	Cod liver oil, eggs, fortified dairy products	Promotes bone growth and mineralization; increases calcium absorption	Rickets (bone deformities) in children, skeletal deterioration
Vitamin E (tocopherol)	Nuts, whole grains, leafy vegetables, vegetable oils	Antioxidant; may reduce cellular damage from free radicals	Neurological damage
Vitamin K	Leafy vegetables; product of intestinal bacteria	Important in blood clotting	Bleeding, internal hemorrhages

of sunlight), have inadequate levels of this vitamin. Breast-fed children born to vitamin D–deficient mothers are at particular risk for rickets. Vitamin E is an antioxidant, neutralizing free radicals that form in the body (see pp. 24–26), and vitamin K helps to regulate blood clotting.

The Human Body Is About Sixty Percent Water

A person can survive far longer without food than without water; this essential nutrient makes up roughly 60% of total body weight. All metabolic reactions occur in a watery solution, and water participates directly in hydrolysis reactions (see p. 38) that break down proteins, carbohydrates, and fats into simpler molecules. Water is the principal component of saliva, blood, lymph, extracellular fluid, and the cytoplasmic fluid within each cell. By sweating, people use the evaporation of water to keep from overheating. Urine, which is mostly water, is necessary to eliminate cellular waste products from the body (see Chapter 35).

Nutritional Guidelines Help People Obtain a Balanced Diet

Most people in the United States are fortunate to live amid an abundance of food. But the overwhelming variety of foods available in a typical U.S. supermarket, and the easy availability of fast food, can contribute to obesity and poor nutrition. To help people make informed choices, the U.S. government has placed nutritional guidelines, called "MyPyramid," on an interactive Web site. The Web site, which provides 12 individualized sets of nutritional recommendations, is a major departure from the familiar food pyramid. Check it out by entering "mypyramid" into an Internet search engine.

Still more nutritional information can be found on the labels of commercially packaged foods, which provide complete information about Calorie, fiber, fat, sugar, and vitamin content (**Fig. 34-5**). Fast-food restaurants often provide fliers and information on their Web sites that list nutritional information for their products.

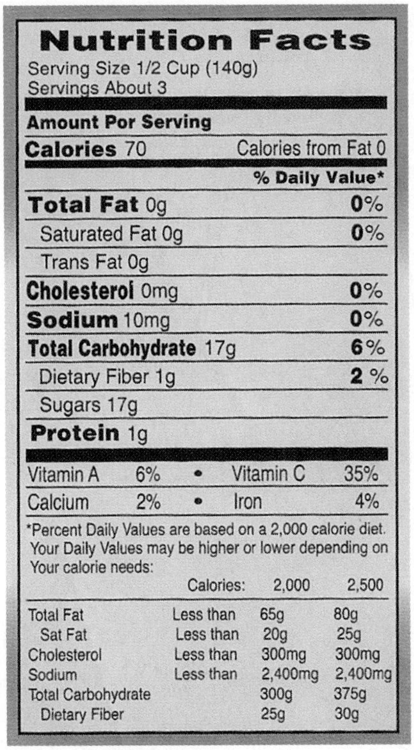

Nutrition Facts
Serving Size 1/2 Cup (140g)
Servings About 3

Amount Per Serving	
Calories 70	Calories from Fat 0
	% Daily Value*
Total Fat 0g	0%
Saturated Fat 0g	0%
Trans Fat 0g	
Cholesterol 0mg	0%
Sodium 10mg	0%
Total Carbohydrate 17g	6%
Dietary Fiber 1g	2 %
Sugars 17g	
Protein 1g	

Vitamin A	6%	•	Vitamin C	35%
Calcium	2%	•	Iron	4%

*Percent Daily Values are based on a 2,000 calorie diet. Your Daily Values may be higher or lower depending on Your calorie needs:

	Calories:	2,000	2,500
Total Fat	Less than	65g	80g
Sat Fat	Less than	20g	25g
Cholesterol	Less than	300mg	300mg
Sodium	Less than	2,400mg	2,400mg
Total Carbohydrate		300g	375g
Dietary Fiber		25g	30g

▲ **FIGURE 34-5 Food labeling** The U.S. government requires complete nutritional labeling of foods, as illustrated by this example. The weight (in grams) of various nutrients—such as fat, cholesterol, and sodium—is shown as a percentage of the recommended daily intake (% Daily Value), assuming a 2,000-Calorie diet.

QUESTION Does this item look like a healthy food choice? Explain.

34.2 HOW DOES DIGESTION OCCUR?

Digestion is the process that physically grinds up food and then chemically breaks it down. The animal **digestive system** takes in food and then digests its complex molecules into simpler molecules that it absorbs. The digestive system then expels leftover waste material from the body.

Animals eat the bodies of other organisms, but parts of these bodies resist becoming food. Each plant cell, for example, is supported by a wall of indigestible cellulose. Animal bodies may be covered with indigestible fur, scales, or feathers. In addition, the complex lipids, proteins, and carbohydrates in food do not occur in a form that can be used directly. These nutrients must be broken down before they can be absorbed and distributed to the cells of the animal that has consumed them, where they are recombined in unique ways. The digestive tracts of animals are diverse, having evolved to meet the challenges posed by a variety of diets. Amid this diversity, however, digestive systems must accomplish five specific tasks:

1. **Ingestion** Food is brought into the digestive tract through an opening, usually called a **mouth.**
2. **Mechanical digestion** The food is physically broken down into smaller pieces. The resulting particles have a greater surface area, allowing digestive enzymes to attack them more effectively.
3. **Chemical digestion** Particles of food are exposed to enzymes and other digestive secretions that break down large molecules into smaller subunits.
4. **Absorption** The small subunits are transported out of the digestive tract and into the body through cells lining the digestive tract. Although the digestive tract surrounds and acts on food, nutrients do not actually enter the body until they are absorbed.
5. **Elimination** Indigestible materials are expelled from the body.

In the following sections, we explore a few of the diverse mechanisms by which animal digestive systems accomplish these functions.

In Sponges, Digestion Occurs Within Single Cells

Sponges are the only animals that lack a digestive chamber, and rely exclusively on **intracellular digestion,** in which all digestion occurs within individual cells. As you might suspect, this limits their food to microscopic particles. Sponges attach permanently to rocks, circulating seawater bearing food particles through pores in their bodies (**Fig. 34-6** ❶). Specialized collar cells inside the sponge filter microscopic organisms from the water (**Fig. 34-6** ❷) and ingest them using the process of phagocytosis ("cell eating"; see p. 90). Once ingested by a cell, the food is enclosed in a **food vacuole** (**Fig. 34-6** ❸), a temporary space surrounded by a membrane. The vacuole fuses with a **lysosome,** a membrane-enclosed packet of digestive enzymes within the cell (**Fig. 34-6** ❹). Food is broken down within the vacuole into smaller molecules that can be absorbed into the cell cytoplasm. Undigested remnants are expelled by the cell using exocytosis (**Fig. 34-6** ❺) and are released through a large opening in the body wall (**Fig. 34-6** ❻).

The Simplest Digestive System Is a Chamber with One Opening

All other types of animals have evolved a chamber within the body where chunks of food are broken down by enzymes that act outside the cells, a process called **extracellular digestion.** One of the simplest of these chambers is found in cnidarians, such as sea anemones, *Hydra,* and sea jellies. The digestive chamber is called a **gastrovascular cavity,** and has a single opening through which food is ingested and wastes are ejected. The animal's stinging tentacles capture smaller animals and usher their prey through the mouth into the gastrovascular cavity (**Fig. 34-7** ❶). Gland cells lining the cavity secrete enzymes that begin digesting the prey (**Fig. 34-7** ❷). Nutritive cells lining the cavity then absorb the nutrients and also engulf partly digested food particles by phagocytosis. Further digestion is intracellular, within food vacuoles in the

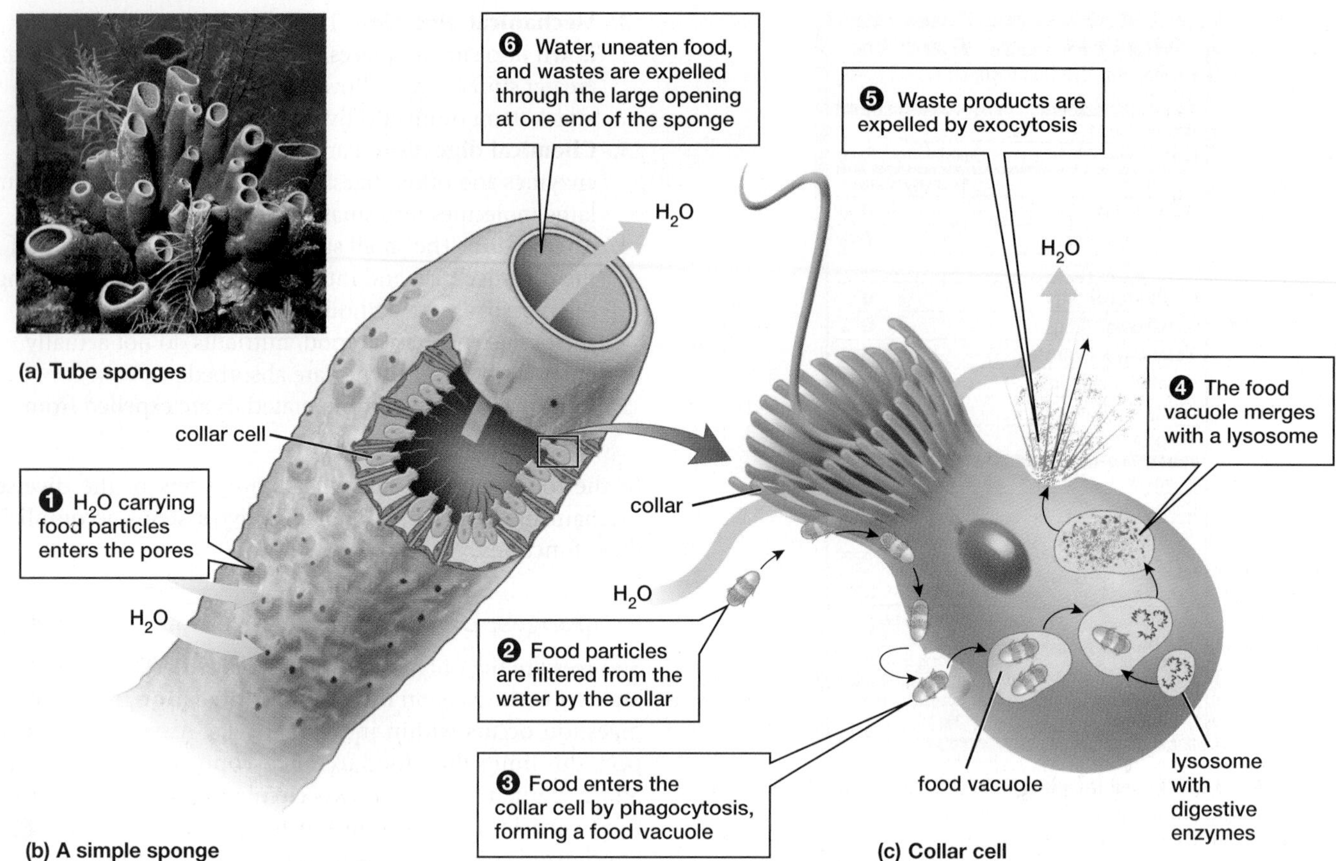

(a) Tube sponges
6 Water, uneaten food, and wastes are expelled through the large opening at one end of the sponge
5 Waste products are expelled by exocytosis
H_2O
H_2O
4 The food vacuole merges with a lysosome
collar cell
1 H_2O carrying food particles enters the pores
collar
H_2O
H_2O
2 Food particles are filtered from the water by the collar
3 Food enters the collar cell by phagocytosis, forming a food vacuole
food vacuole
lysosome with digestive enzymes
(b) A simple sponge
(c) Collar cell

▲ **FIGURE 34-6 Intracellular digestion in a sponge (a)** Tube sponges photographed in the Virgin Islands. **(b)** The anatomy of a simple sponge, showing the direction of water flow and the location of the collar cells. **(c)** Here, a single collar cell is enlarged to show the intracellular digestion of single-celled organisms, which are filtered from the water, trapped on the outside of the collar, engulfed, and digested.

nutritive cells (**Fig. 34-7** ❸). Undigested wastes are expelled back through the mouth, and so only one meal can be processed at a time.

Most Animals Have Tubular Digestive Systems with Specialized Compartments

A saclike digestive system is unsuitable for animals that must eat frequently. Most animals, including invertebrates such as mollusks, arthropods, echinoderms, and earthworms, as well as all vertebrates, have digestive systems that are basically one-way tubes that begin with a mouth and end with an anus. Specialized regions within the tube process the food in an orderly sequence, taking it in through the mouth, physically grinding it up, enzymatically breaking it down, absorbing the nutrients, and finally, expelling the wastes through the anus.

The earthworm provides a good example (**Fig. 34-8**). As it burrows, the worm ingests soil and bits of plant material that pass through the **esophagus,** a muscular tube leading from the mouth to the *crop*, an expandable sac where food is stored. The material is then gradually released into the *gizzard*, where ingested sand grains and muscular contractions grind it into smaller particles. In the intestine, enzymes attack the food particles, and the resulting small molecules are absorbed into the worm's body. The remaining soil containing undigested organic material is expelled through the anus.

Vertebrate Digestive Specializations

Different types of animals have radically different diets. **Carnivores,** such as wolves, cats, seals, and predatory birds, eat other animals. **Herbivores,** which eat only plants, include seed-eating birds; grazing animals, such as deer, camels, and cows; and many rodents, such as mice. **Omnivores,** such as humans, bears, and raccoons, consume and are adapted to digest both animal and plant sources of food. Specialized digestive tracts allow animals with different diets to extract the maximum amount of nutrients from their foods.

Teeth Accommodate Different Diets

Teeth are adapted to diet. The varied, omnivorous diet of humans has selected for our unique set of teeth. We have thin, flat incisors for shearing off food. Because we don't catch prey with our mouths, our canines are small and (in

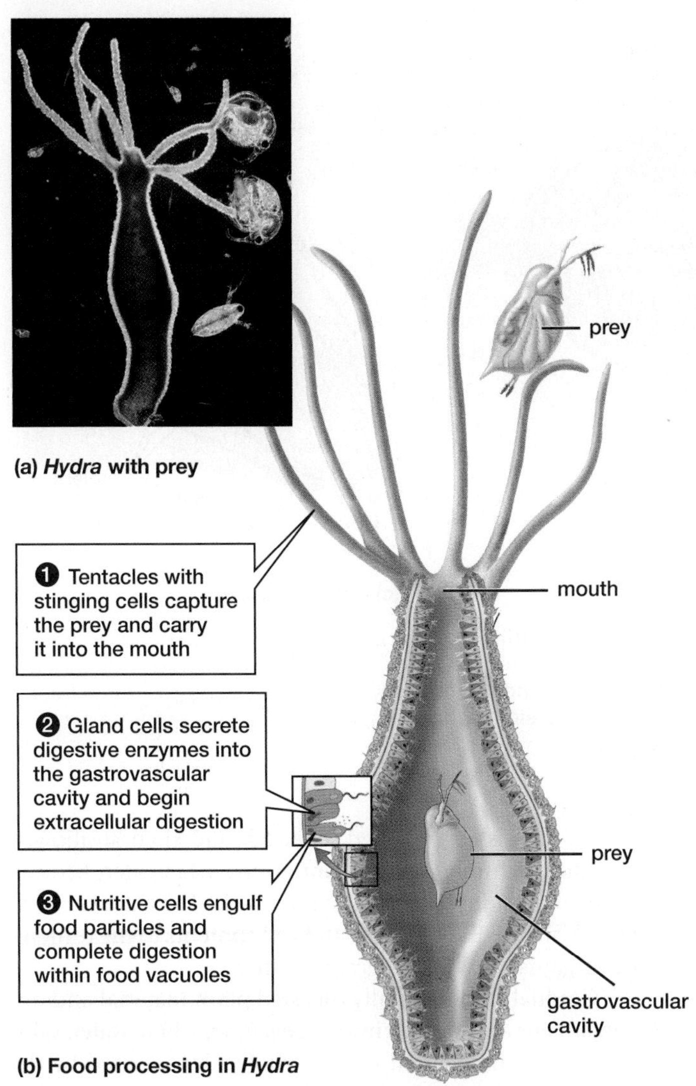
(a) Hydra with prey
prey
mouth
① Tentacles with stinging cells capture the prey and carry it into the mouth
② Gland cells secrete digestive enzymes into the gastrovascular cavity and begin extracellular digestion
③ Nutritive cells engulf food particles and complete digestion within food vacuoles
prey
gastrovascular cavity
(b) Food processing in Hydra

▲ **FIGURE 34-7 Digestion in a sac (a)** A *Hydra* has captured and ingested a waterflea (*Daphnia*, a tiny crustacean) into its gastrovascular cavity. **(b)** After nutrients are absorbed, undigested waste will be expelled out through the mouth.

some people) flat, like incisors. Our premolars and molars have relatively large, irregular surfaces for crushing and grinding (**Fig. 34-9a**).

If you have a dog, look carefully in its mouth (your cat probably won't allow this). Carnivores have very small incisors, but greatly enlarged canines for stabbing and tearing flesh. Their molars and premolars have sharp edges for shearing through tendon and bone. None of their teeth are adapted for grinding up plant material or chewing their prey, which they tend to swallow in chunks (**Fig. 34-9b**).

Herbivores, such as cows and horses, have reduced canines, but they have large, sharp incisors adapted for snipping plants. Their premolars and molars are big and flattened for grinding up tough plant material (**Fig. 34-9c**). The teeth of many grazing herbivores, including cattle and horses, grow continuously throughout their lives, compensating for the wear caused by a diet of abrasive plants, particularly grass.

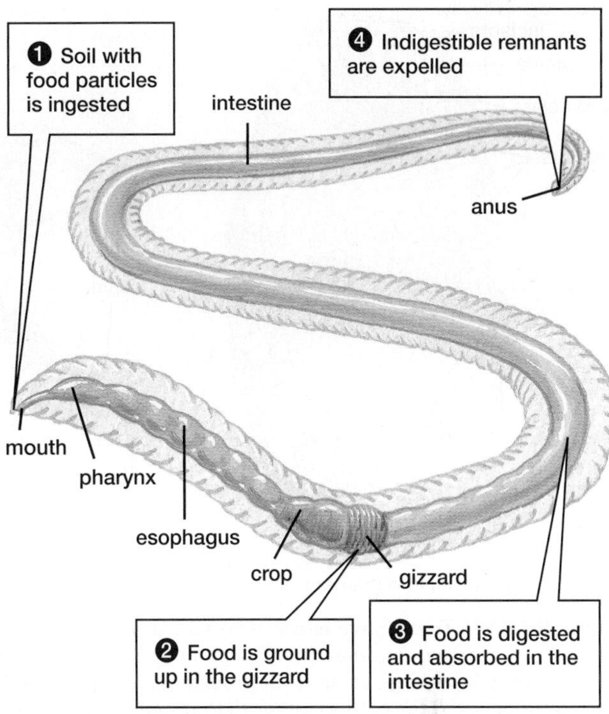
① Soil with food particles is ingested
④ Indigestible remnants are expelled
intestine
anus
mouth
pharynx
esophagus
crop
gizzard
② Food is ground up in the gizzard
③ Food is digested and absorbed in the intestine

▲ **FIGURE 34-8 A tubular digestive system** The earthworm has a one-way digestive system that passes food through a series of compartments, each specialized to play a specific role in breaking down food and absorbing it.

Birds' Stomachs Grind Food

Birds lack teeth and swallow their food whole, after which it passes through the muscular esophagus (**Fig. 34-10**). In seed-eating birds, the food is then stored and softened by water in a large, expandable crop. The food then passes gradually into two stomach chambers. The tube-like first chamber secretes protein-digesting enzymes that begin protein breakdown, while the second, the gizzard, is a thick-walled, muscular, grinding chamber lined with ridges or plates made of the protein keratin (which also forms the bird's beak). The gizzard crushes and grinds food using muscular contractions. Many birds swallow gritty sand or small stones that lodge in the gizzard and aid in the grinding process. From the gizzard, pulverized food particles are released into the small intestine, where they are further digested and their nutrients are absorbed.

Specialized Stomachs Allow Ruminants to Digest Cellulose

The cellulose surrounding each plant cell is potentially one of the most abundant food energy sources on Earth; nevertheless, if humans were restricted to a cow's diet of grass, we would soon starve. Although cellulose, like starch, consists of long chains of glucose molecules, because of the way the bonds link the glucose molecules together (see pp. 41–42), it resists the attack of animal digestive enzymes.

Ruminant animals can obtain energy from cellulose because their stomachs harbor symbiotic microorganisms that are able to break down this polysaccharide. Ruminants, which

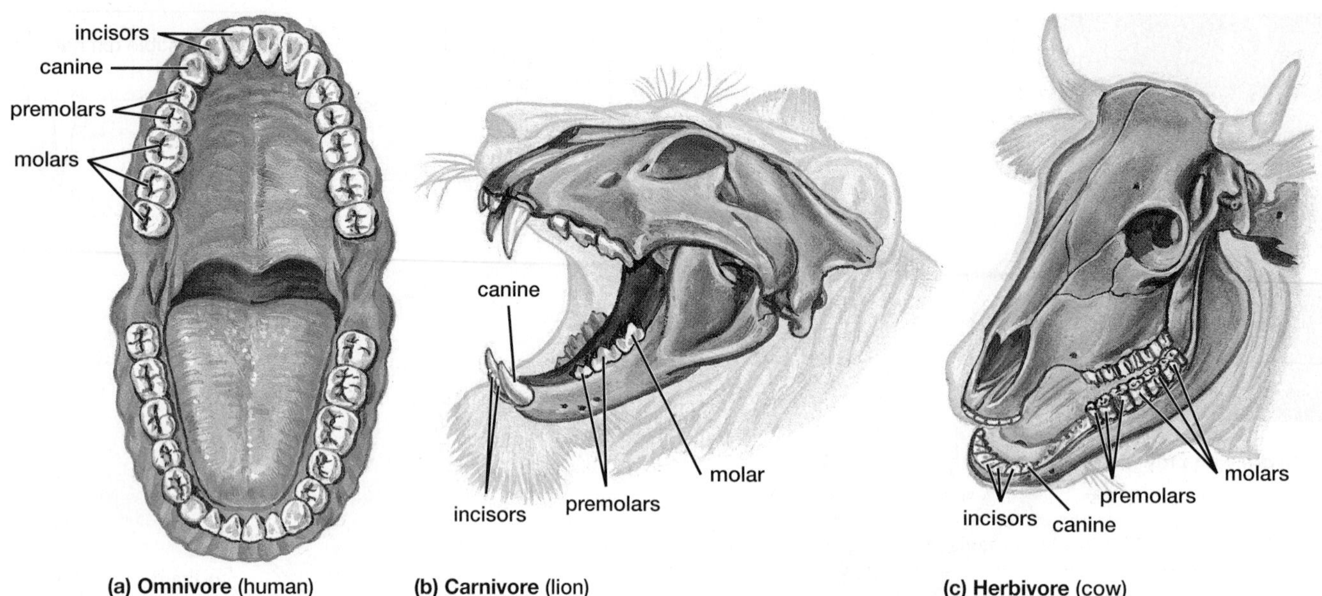

(a) **Omnivore** (human) (b) **Carnivore** (lion) (c) **Herbivore** (cow)

▲ FIGURE 34-9 **Teeth have evolved to suit different diets (a)** Humans have cutting incisors, reduced canines, and flattened premolars and molars for grinding up plant and animal food. **(b)** Carnivores have large canines for grasping and killing prey, reduced incisors, and premolars and molars adapted for cutting rather than grinding. **(c)** Herbivores have well-developed incisors for cutting plants, and enlarged premolars and molars for grinding them up.

include cows, sheep, goats, camels, and hippos, have multiple stomach chambers (**Fig. 34-11**). The first chamber is the *rumen*; a cow rumen can hold nearly 40 gallons (about 150 liters). This chamber is home to a variety of microorganisms that produce enzymes that break down and then ferment cellulose and other carbohydrates. During this process, they release small organic molecules that supply at least half of the cow's energy needs; most of these are absorbed through the rumen wall.

After partial digestion in the rumen, the plant material enters the reticulum, where it is formed into masses called cud. The cud is regurgitated, chewed, and then swallowed back into the rumen. (Ruminant animals can often be seen placidly *ruminating*, or chewing their cud.) The extra chewing exposes more of the cellulose and cell contents to the rumen's microorganisms, which digest it further.

Gradually, the partially digested plant material and microorganisms are released into the *omasum*, where water, salts, and the remaining small organic molecules released by the microorganisms are absorbed. From there, they enter the *abomasum*, where acid and protein-digesting enzymes are secreted and protein digestion begins. Here, the cow digests not only plant proteins, but also the microorganisms that accompany the partially digested food from the rumen. The cow then absorbs most of the products of digestion through the walls of its small intestine.

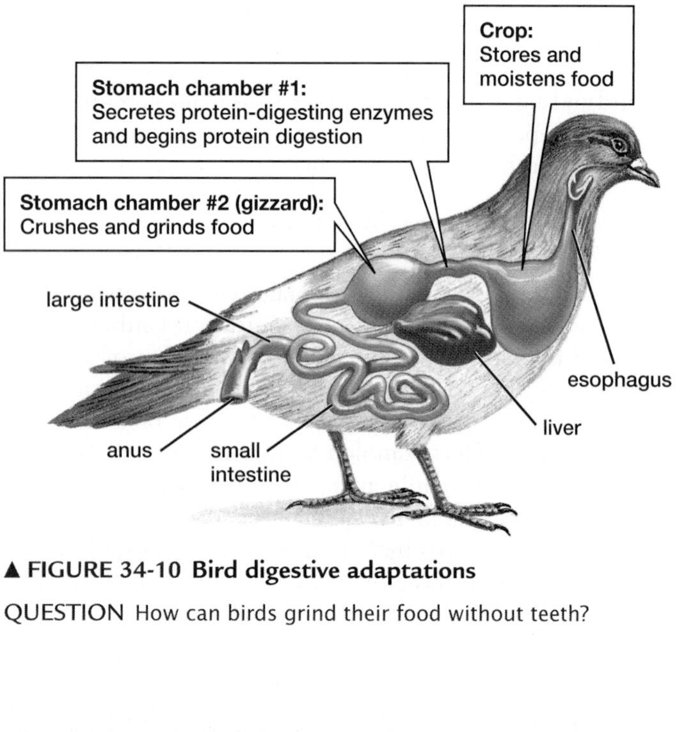

Crop: Stores and moistens food

Stomach chamber #1: Secretes protein-digesting enzymes and begins protein digestion

Stomach chamber #2 (gizzard): Crushes and grinds food

large intestine

esophagus

liver

anus

small intestine

▲ FIGURE 34-10 **Bird digestive adaptations**

QUESTION How can birds grind their food without teeth?

Small Intestine Length Is Correlated with Diet

Most digestion and absorption of nutrients occurs in the small intestine. Herbivores require a relatively long small intestine, which provides more opportunity to extract nutrients from their bulky diet of plant food. Carnivores, which consume a high-protein diet, have shorter small intestines than herbivores because proteins are relatively easy to digest, and protein digestion begins in the stomach, as described in more detail later.

The difference in small intestine length is strikingly illustrated during frog development. The juvenile tadpole is an algae-eating herbivore with an elongated small intestine. When it metamorphoses into a carnivorous (usually insect-eating) adult frog, the small intestine shortens to about one-third of its original length.

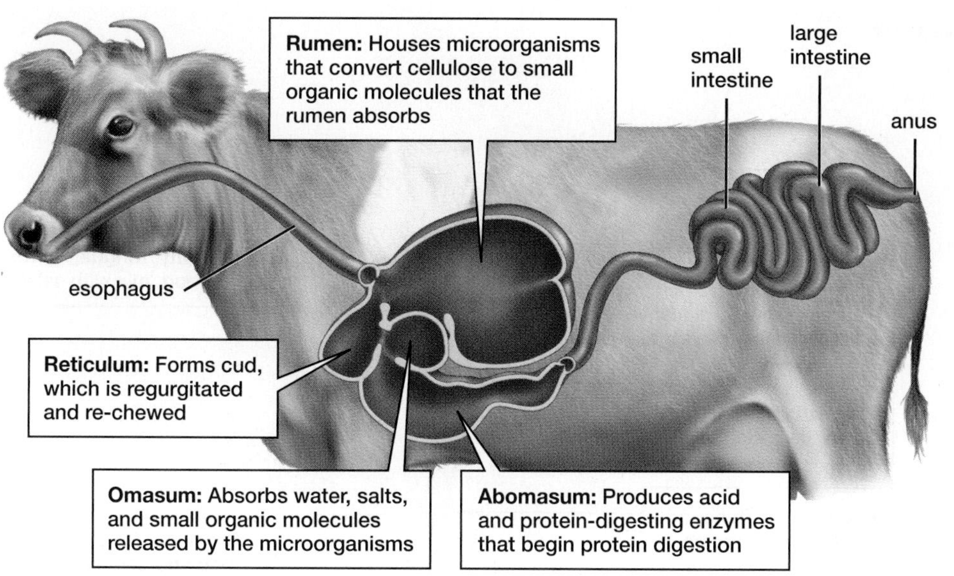
Rumen: Houses microorganisms that convert cellulose to small organic molecules that the rumen absorbs
small intestine
large intestine
anus
esophagus
Reticulum: Forms cud, which is regurgitated and re-chewed
Omasum: Absorbs water, salts, and small organic molecules released by the microorganisms
Abomasum: Produces acid and protein-digesting enzymes that begin protein digestion

◀ **FIGURE 34-11 Ruminants have a multichambered stomach**

QUESTION In addition to the ability to digest cellulose, what other nutritional benefits might ruminants gain by having microorganisms in their guts?

34.3 HOW DO HUMANS DIGEST FOOD?

The human digestive system (**Fig. 34-12**), which is adapted for processing the wide variety of foods in our omnivorous diet, provides a good example of the mammalian digestive system.

Mechanical and Chemical Digestion Begin in the Mouth

As you take a bite of food, your mouth waters and you begin chewing. This begins both the mechanical and the chemical digestion of food. While the teeth pulverize the food, the first phase of chemical digestion occurs as three pairs of salivary glands pour out saliva in response to the smell, feel, taste, and, if you're hungry, even the thought of food. Altogether, human salivary glands produce about 1 to 1.5 quarts (about 1.0 to 1.5 liters) of saliva daily.

Saliva has many functions. Water and mucus in saliva lubricate food to ease swallowing. Antibacterial agents in saliva help to guard against infection. Saliva also contains the digestive enzyme **amylase,** which begins breaking starches into sugars (**Table 34-4**). The water in saliva dissolves some molecules, such as acids and sugars, exposing them to clusters of taste receptor cells, called *taste buds,* on the tongue. These help to identify the type and the quality of the food.

The muscular tongue manipulates chewed food into a mass and presses it back into the **pharynx,** a cavity between the mouth and the esophagus (**Fig. 34-13a**). In addition to providing a passageway for food entering the digestive system, the pharynx also connects the nose and mouth with the larynx, which leads to the trachea, a tube that conducts air to the lungs. This arrangement occasionally causes problems, as anyone who has ever choked on a piece of food well knows. Normally, however, the swallowing reflex elevates the larynx so it meets the

Table 34-4 Digestive Secretions in Humans

Site of Digestion	Secretion	Source of Secretion	Role in Digestion
Mouth	Salivary amylase	Salivary glands	Breaks down starch into disaccharides
	Mucus, water	Salivary glands	Lubricates and dissolves food
Stomach	Hydrochloric acid	Cells lining the stomach	Allows pepsin to work; kills some bacteria; aids in mineral absorption
	Pepsin	Cells lining the stomach	Breaks down proteins into large peptides
	Mucus	Cells lining the stomach	Protects the stomach from digesting itself
Small intestine	Sodium bicarbonate	Pancreas	Neutralizes acidic chyme from the stomach
	Pancreatic amylase	Pancreas	Breaks down starch into disaccharides
	Protease	Pancreas	Breaks down proteins into large peptides
	Lipase	Pancreas	Breaks down lipids into fatty acids and glycerol
	Bile	Liver	Emulsifies lipids
	Peptidases	Small intestine	Split small peptides into amino acids
	Disaccharidases	Small intestine	Split disaccharides into monosaccharides
	Mucus	Small intestine	Protects the intestine from digestive secretions

▶ **FIGURE 34-12 The human digestive tract**

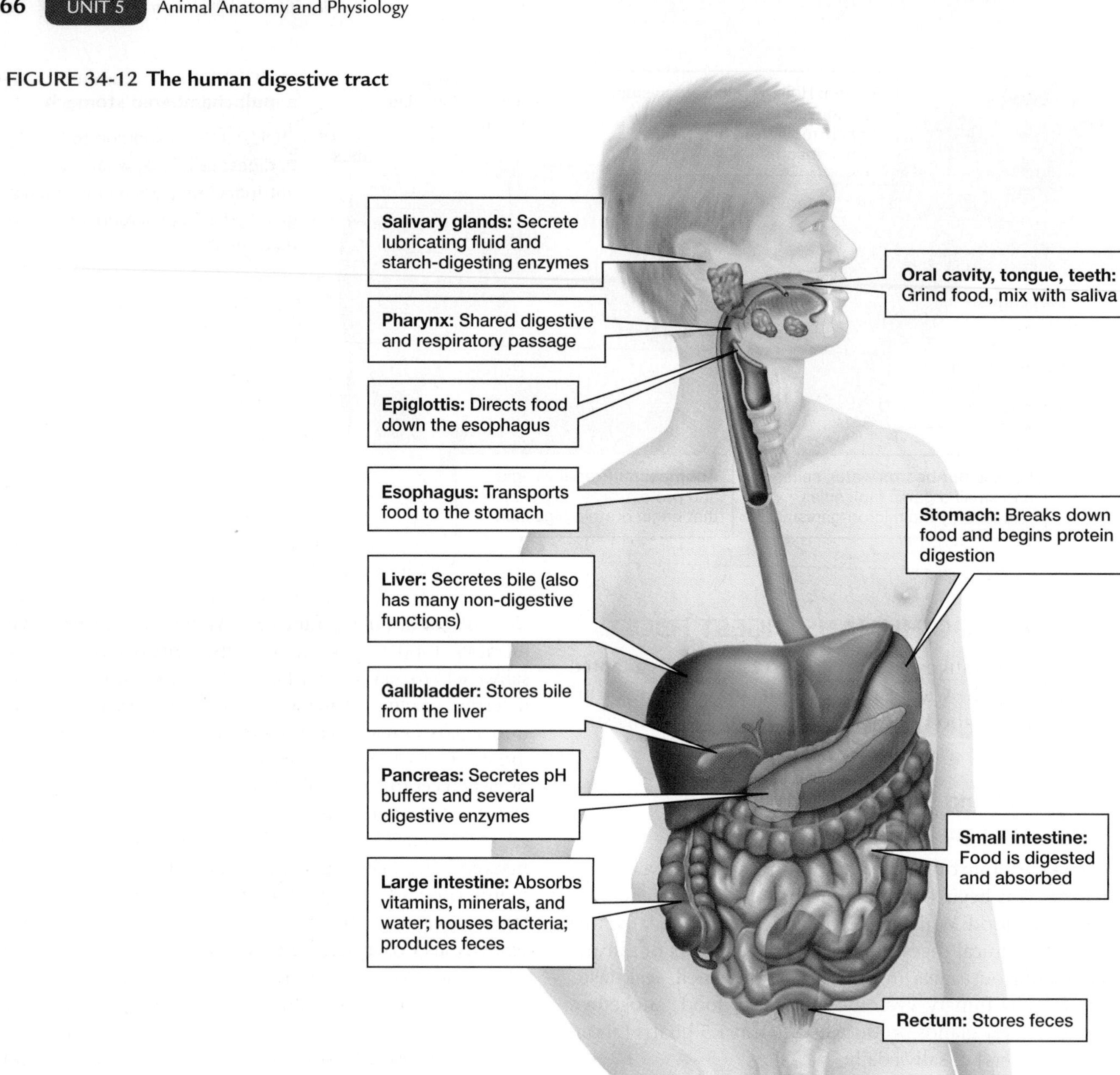
Salivary glands: Secrete lubricating fluid and starch-digesting enzymes
Oral cavity, tongue, teeth: Grind food, mix with saliva
Pharynx: Shared digestive and respiratory passage
Epiglottis: Directs food down the esophagus
Esophagus: Transports food to the stomach
Stomach: Breaks down food and begins protein digestion
Liver: Secretes bile (also has many non-digestive functions)
Gallbladder: Stores bile from the liver
Pancreas: Secretes pH buffers and several digestive enzymes
Small intestine: Food is digested and absorbed
Large intestine: Absorbs vitamins, minerals, and water; houses bacteria; produces feces
Rectum: Stores feces

epiglottis, a flap of tissue that blocks off the respiratory passages, directing food into the esophagus (**Fig. 34-13b**).

The Esophagus Conducts Food to the Stomach, Where Mechanical and Chemical Digestion Continue

Swallowing forces food into the esophagus, a muscular tube that propels food from the mouth to the stomach. Mucus secreted by cells that line the esophagus helps protect it from abrasion and also lubricates the food during its passage. Muscles surrounding the esophagus produce a wave of contraction that begins just above the swallowed mass and progresses down the esophagus, forcing the food toward the stomach. This muscular action, called **peristalsis,** occurs throughout the digestive tract, pushing food products through the esophagus, stomach, intestines, and finally, out through the anus. Peristalsis is so effective that a person can actually swallow when upside down.

The **stomach** in humans is a muscular sac with a folded inner lining that allows it to expand so that we can eat rather large, infrequent meals (**Fig. 34-14**). Some carnivores take this ability to an extreme. A lion, for instance, may consume 40 pounds (18 kilograms) of meat at one meal, and then spend the next few days digesting it. In adult humans, the stomach can comfortably hold about a quart (1 liter), although this

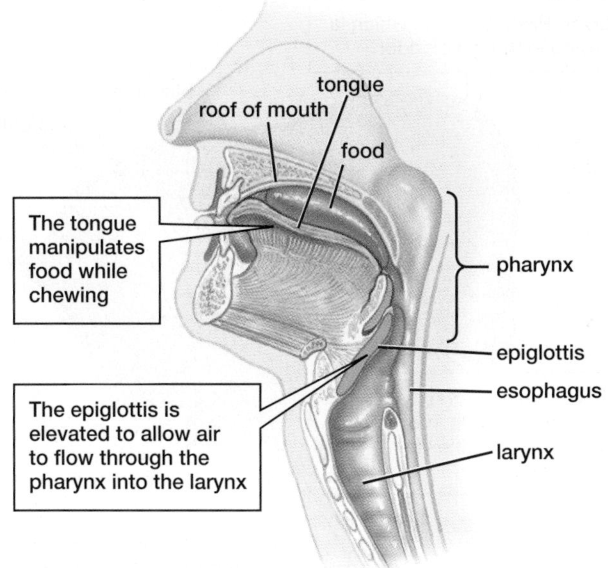
tongue
roof of mouth
food
The tongue manipulates food while chewing
pharynx
epiglottis
esophagus
The epiglottis is elevated to allow air to flow through the pharynx into the larynx
larynx
(a) Before swallowing

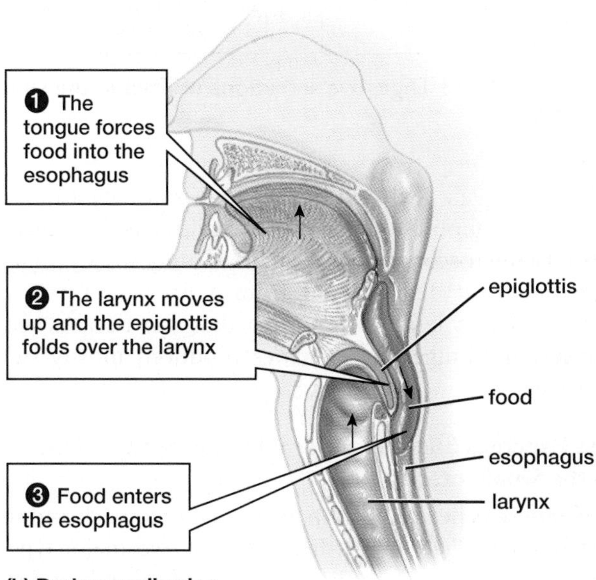
❶ The tongue forces food into the esophagus
❷ The larynx moves up and the epiglottis folds over the larynx
epiglottis
food
esophagus
❸ Food enters the esophagus
larynx
(b) During swallowing

▲ **FIGURE 34-13 The challenge of swallowing (a)** Swallowing is complicated by the fact that the esophagus (part of the digestive system) and the larynx (part of the respiratory system) both open into the pharynx. **(b)** During swallowing, the larynx moves upward beneath the epiglottis. The epiglottis folds down over the larynx, sealing off the opening to the respiratory system and directing food down the esophagus instead.

varies with body size. Food is retained in the stomach by two rings of circular muscles, called **sphincter muscles.** The sphincter at the top, called the *lower esophageal sphincter*, keeps food and stomach acid from sloshing up into the esophagus while the stomach churns. It opens briefly just after swallowing, allowing food to enter the stomach. A second sphincter, the *pyloric sphincter*, separates the lower portion of the stomach from the upper small intestine. This muscle regulates the passage of food into the small intestine.

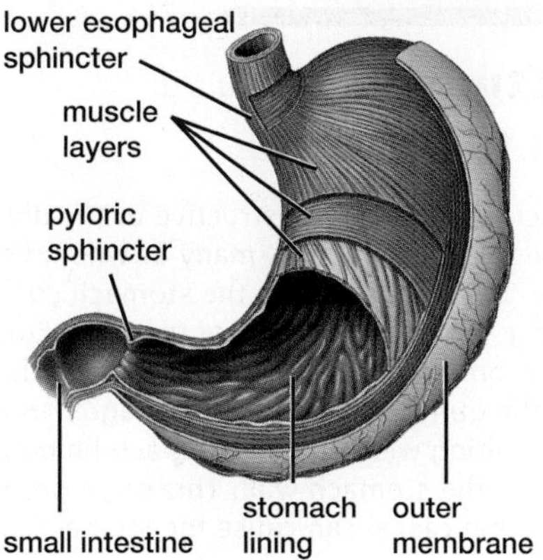
lower esophageal sphincter
muscle layers
pyloric sphincter
small intestine
stomach lining
outer membrane

▲ **FIGURE 34-14 The stomach**

The stomach has four distinct functions. First, it stores food and releases it gradually into the small intestine at a rate suitable to allow the small intestine to completely digest the food and absorb its nutrients. Second, the muscular stomach walls produce a variety of churning contractions that disrupt large chunks of food, breaking them up into much smaller pieces that are more readily attacked by digestive enzymes.

Third, the stomach begins protein breakdown using secretions from the gastric glands. The **gastric glands** are clusters of specialized epithelial cells that line millions of microscopic pits within the epithelial cell layer of the stomach lining. Gastric gland secretions include mucus, hydrochloric acid (HCl), and the protein pepsinogen. The hydrochloric acid gives the stomach fluid a very acidic pH of 1 to 3 (about the same as lemon juice). This destroys many microbes (such as bacteria and viruses) that are inevitably swallowed along with food. Pepsinogen is the inactive form of *pepsin*, a type of **protease**—a protein-digesting enzyme that breaks proteins into shorter chains of amino acids called peptides. The stomach's acidity converts pepsinogen into pepsin (which works best in this acidic environment). Pepsin then begins digesting the proteins in food. Why not secrete pepsin in the first place? Gastric glands secrete inactive pepsinogen because pepsin would digest the very cells that synthesize it. Mucus, secreted by most stomach epithelial cells, coats the stomach lining and serves as a barrier to self-digestion. The protection, however, is not perfect, and so cells of the stomach epithelium must be replaced every few days.

The gastric glands are also responsible for the fourth function of the stomach: secretion of the digestion-regulating hormone gastrin, which we'll describe in greater detail later.

As you may have noticed, the stomach produces all of the ingredients necessary to digest itself if its protective mucous barriers are breached. Indeed, this is what happens when a person develops ulcers, as described in "Scientific Inquiry: The Bacteria–Ulcer Link" on p. 669.

Food in the stomach is gradually converted to a thick, acidic liquid called **chyme,** which consists of digestive secretions and partially digested food. Peristaltic waves (about three per minute) then propel the chyme toward the small intestine,

Case Study continued
Dieting to Death?

Stomach acid can be very destructive to the digestive tracts of people with bulimia, many of whom vomit several times daily. The strong acid in the stomach contents dissolves the protective enamel of teeth, making them extremely prone to decay. Stomach acid also damages tissues of the gums, throat, and esophagus. In addition, frequent vomiting weakens the stomach lining, allowing acid to attack the stomach wall. This often produces ulcers, and, in extreme cases, can cause the stomach to rupture.

forcing about a teaspoon of chyme through the pyloric sphincter with each wave. Depending on the size of the meal and the type of food eaten, it takes roughly 4 hours to empty the stomach after a meal. Churning movements of an empty stomach are felt as hunger pangs.

Only a few substances, including alcohol and certain drugs, can enter the bloodstream through the stomach wall. Because food in the stomach slows alcohol absorption, the advice "never drink on an empty stomach" is based on sound physiological principles.

Most Chemical Digestion Occurs in the Small Intestine

The **small intestine** is a long, muscular tube that receives food from the stomach. The main functions of the small intestine are to chemically digest food into small molecules and to absorb these molecules into the body. After the stomach releases chyme into the small intestine, chemical digestion is accomplished with the aid of enzymes and other digestive secretions from three sources: the liver, the pancreas, and the cells lining the small intestine itself (**Fig. 34-15**). Most fat and carbohydrate digestion occurs in the small intestine, and the protein digestion that began in the stomach is completed here. Like the stomach, the small intestine is protected from digesting itself by mucus secreted by specialized cells in its lining.

The Liver and Gallbladder Provide Bile, Which Helps Break Down Fats

The **liver** is perhaps the most versatile organ in the body. The liver stores fats and carbohydrates for energy, regulates blood glucose levels, synthesizes blood proteins, stores iron and certain vitamins, converts toxic ammonia (released when amino acids are broken down) into urea (see p. 679), and detoxifies harmful substances such as nicotine and alcohol.

The role of the liver in digestion is to produce **bile,** a greenish liquid that contains substances synthesized from cholesterol, called bile salts. Bile is stored and concentrated in the **gallbladder,** and released through a tube called the *bile duct*. The bile duct empties into the first segment of the small intestine, a region called the **duodenum** (see Fig. 34-15), which is about 10 inches (25 centimeters) long.

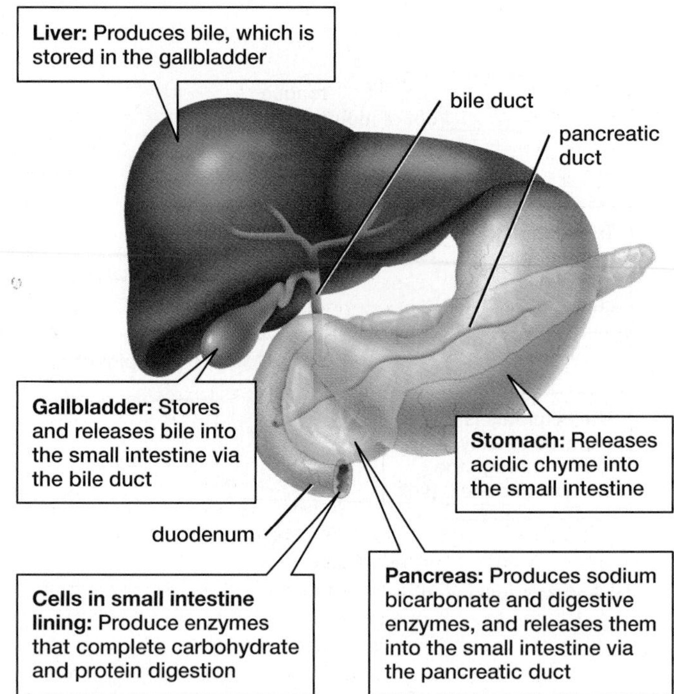

Liver: Produces bile, which is stored in the gallbladder
bile duct
pancreatic duct
Gallbladder: Stores and releases bile into the small intestine via the bile duct
Stomach: Releases acidic chyme into the small intestine
duodenum
Cells in small intestine lining: Produce enzymes that complete carbohydrate and protein digestion
Pancreas: Produces sodium bicarbonate and digestive enzymes, and releases them into the small intestine via the pancreatic duct

▲ **FIGURE 34-15 Digestive secretions utilized in the small intestine**

Bile salts have a hydrophilic end that is attracted to water, and a hydrophobic end that interacts with fats. They disperse fats into microscopic particles in the watery chyme, much as dish detergent disperses fat from a fry pan. Fat in tiny particles has a large surface area that can be attacked by **lipases,** lipid-digesting enzymes produced mostly by the pancreas.

The Pancreas Supplies Several Digestive Secretions to the Small Intestine

The **pancreas** lies in the loop between the stomach and small intestine (see Fig. 34-15). It consists of two major types of cells. One type produces hormones involved in blood sugar regulation (see Chapter 37), and the other produces a digestive secretion called **pancreatic juice.** About 1 quart (1 liter) of pancreatic juice enters the duodenum of the small intestine each day through the *pancreatic duct*. This secretion contains water, sodium bicarbonate (which neutralizes the acidic chyme), and several digestive enzymes, including pancreatic amylase, lipase, and proteases. Pancreatic digestive enzymes work best in the slightly alkaline (basic) environment created by the sodium bicarbonate in the pancreatic juice. Pancreatic amylase breaks down carbohydrates, lipase attacks lipids, and proteases break down proteins and peptides (see Table 34-4).

The Digestive Process Is Completed by Cells of the Intestinal Wall

The epithelium of the small intestine consists primarily of cells whose plasma membranes form fringes of microscopic projections called **microvilli** ("tiny hairs"), which produce

Scientific Inquiry

The Bacteria–Ulcer Link

Ulcers occur when localized areas of the tissue layers that line the stomach or duodenum become eroded (**Fig. E34-1**). Ulcer victims can experience burning pain, nausea, and, in severe cases, bleeding. Before the 1990s, doctors believed that most ulcers were caused mainly by overproduction of stomach acid, and treated their patients with antacids, a bland diet, and stress-reduction programs. The ulcers commonly recurred, however, when the treatment stopped. Now, bacteria-killing antibiotics are the standard treatment for most ulcers. How did researchers make the link between bacteria and ulcers?

In the 1980s, J. Robin Warren, a pathologist at the Royal Perth Hospital in Australia, noticed that samples of inflamed stomach tissue were consistently infected with a spiral-shaped bacterium. He discussed his work with Barry Marshall, then a trainee in internal medicine at the same hospital, and they collaborated to test the hypothesis that the bacterium, later named *Helicobacter pylori*, caused stomach inflammation and ulcers.

The medical community displayed a healthy skepticism about their ideas, since these bacteria are found in the stomachs of many people without ulcers. To demonstrate conclusively that a particular bacterium caused the disease, Warren and Marshall followed a protocol (developed by Robert Koch, a German microbiologist, in the 1880s) that researchers often use to find disease-causing microbes: First, confirm the presence of bacteria in all animals infected with the disease; second, grow the bacteria in culture; third, infect experimental animals with the cultured bacteria and demonstrate that they develop the disease; and fourth, re-isolate and culture the identical type of bacteria from the diseased animals.

Marshall and his team isolated bacteria from stomach samples, but were repeatedly frustrated in their attempts to culture *H. pylori*. Unbeknownst to Marshall, however, a technician was discarding the cultures after 2 days if no growth was visible. As is common in research, a chance mistake provided an opportunity for scientific insight when the technician left the cultures in an incubator over a holiday break, and the resulting 5-day-old cultures showed colonies of the slow-growing bacteria that the researchers were looking for.

Next, Marshall and his team attempted to infect piglets (experimental animals used instead of human subjects), but they proved resistant to *H. pylori*. To speed things along, Marshall took an unusual and dangerous approach to the problem: He experimented on himself. After undergoing an exam with an endoscope (a tiny camera threaded down his esophagus) that showed his own stomach to be free of inflammation, Marshall swallowed a culture of about a billion *H. pylori* grown from an ulcer patient. During the following week, he began feeling ill, and samples of his stomach tissue showed that the mucus-producing stomach lining was damaged, thinned, and heavily infected with the bacteria. This led to the hypothesis that taking antibiotics to kill the bacteria would relieve his symptoms, which is exactly what happened. Although this experiment was dangerous, had a sample size of one, and was not going to be repeated by others, it bolstered Marshall's hypothesis and set the stage for further research. After Marshall and independent researchers performed different types of studies using large sample sizes and adequate controls, the hypothesis was accepted.

Scientists now know that *H. pylori* colonize the protective mucus that coats the lining of the stomach and duodenum. In the process, these bacteria weaken the mucous layer and increase stomach acid production, making the stomach and duodenum more susceptible to attack by stomach acid and protein-digesting enzymes. The body's immune response to the infection further contributes to the tissue destruction. The U.S. Centers for Disease Control and Prevention report that the bacterium *H. pylori* causes about 90% of ulcers, and that most will be cured by a 2-week treatment with antibiotics.

For their findings, based on careful observations, chance, and the scientific method, Warren and Marshall were awarded the Nobel Prize for Physiology or Medicine in 2005.

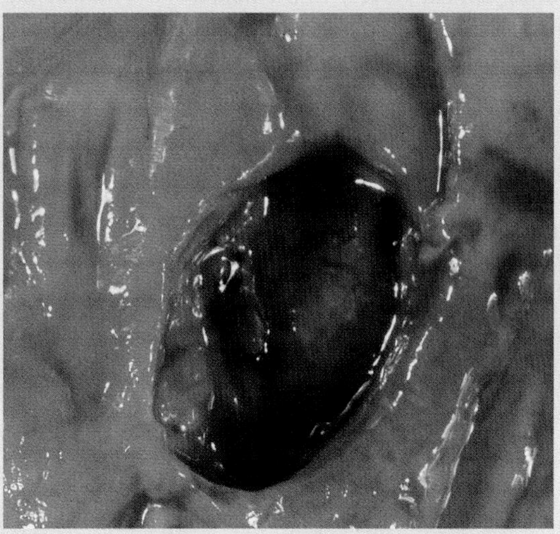

▲ **FIGURE E34-1 An ulcer** This ulcer was photographed through an endoscope.

a large surface area for absorption. These microvilli also bear enzymes that break down peptides into amino acids, and disaccharides into monosaccharides. An example is lactase, which splits lactose (milk sugar) into glucose and galactose.

Most Absorption Occurs in the Small Intestine

The small intestine is not only the principal site of chemical digestion, but is also the major site of nutrient absorption into the body.

The Intestinal Lining Provides a Huge Surface Area for Absorption

In a living human adult, the small intestine is about 1 inch (2.5 centimeters) in diameter and about 8 to 10 feet (2.5 to 3 meters) long. (Reports of 20 feet or more are based on measurements from cadavers, in which all muscle tone is lost.) In addition to being quite long, the small intestine has numerous folds and projections, giving it an internal surface area that is roughly 600 times that of a smooth tube of the same length (**Fig. 34-16a**). Tiny, finger-like protrusions called **villi** (singular,

Have you ever wondered

Why Some People Can't Digest Milk?

You or a friend may have **lactose intolerance,** a condition that causes bloating, gas pains, and diarrhea when milk or milk products are consumed. Lactose intolerance is caused by inadequate amounts of lactase, the enzyme that breaks down lactose, a sugar found in milk. Most mammals synthesize plenty of lactase as infants, when milk is the primary food source, but lose this ability after they are weaned, when milk is no longer available.

Ancient humans, like other mammals, lacked access to milk after weaning, and did not continue to secrete lactase. Today, most of the world's human population remains lactose intolerant. The ability to continue to secrete lactase in adulthood is a genetic property that was selected for in human populations (such as those in northern Europe) that domesticated cows and consumed their milk. People whose ancestors came from such populations may continue to enjoy milk products throughout their lives.

villus; from Latin, meaning "hair") cover the entire folded surface of the intestinal wall (**Fig. 34-16b,c**). Villi, which are about 1/25th of an inch (about 1 millimeter) long, make the intestinal lining appear velvety to the naked eye. Villi move gently back and forth in the chyme as it passes through the intestine, increasing their exposure to the molecules to be digested and absorbed. The plasma membranes of the epithelial cells covering the villi are folded into microvilli (**Fig. 34-16d**). Taken together, the specializations of the lining of the adult small intestine give it a surface area of about 2,700 square feet (about 250 square meters), almost the size of a tennis court.

Unsynchronized contractions of the circular muscles of the small intestine, called *segmentation movements,* slosh the chyme back and forth, bringing nutrients into contact with the enormous absorptive surface of the small intestine. When absorption is complete, coordinated peristaltic waves conduct the leftovers into the large intestine.

Nutrients Are Transported Through the Intestinal Wall in Several Ways

Each villus of the small intestine is provided with a rich supply of blood capillaries and a single lymph capillary, called a **lacteal,** that dead ends in the villus (see Fig. 34-16c). Nutrients absorbed by the small intestine include water, monosaccharides, amino acids and short peptides, fatty acids, vitamins, and minerals. Some nutrients enter the cells lining the small intestine by diffusion, and others by active transport. Water follows by osmosis. The water and most other nutrients then enter the blood capillaries of the villi.

Fatty acids released by the digestion of fats and oils take a distinctive pathway. Clustered together with cholesterol and fat-soluble vitamins, they diffuse directly through intestinal epithelial cell membranes. Inside the cells, these substances are assembled and coated with proteins to form particles called **chylomicrons,** which are released inside the villus. The chylomicrons, which are too large to enter blood capillaries, diffuse through the porous wall of the lacteal. They are transported in lymph by the lymphatic system, which eventually empties into a large vein near the heart (see Fig. 32-18).

In the body, excess fatty acids are reassembled into fat, which can accumulate to health-threatening levels, causing some obese people to resort to surgery to lose weight, as described in "Health Watch: Eating to Death?"

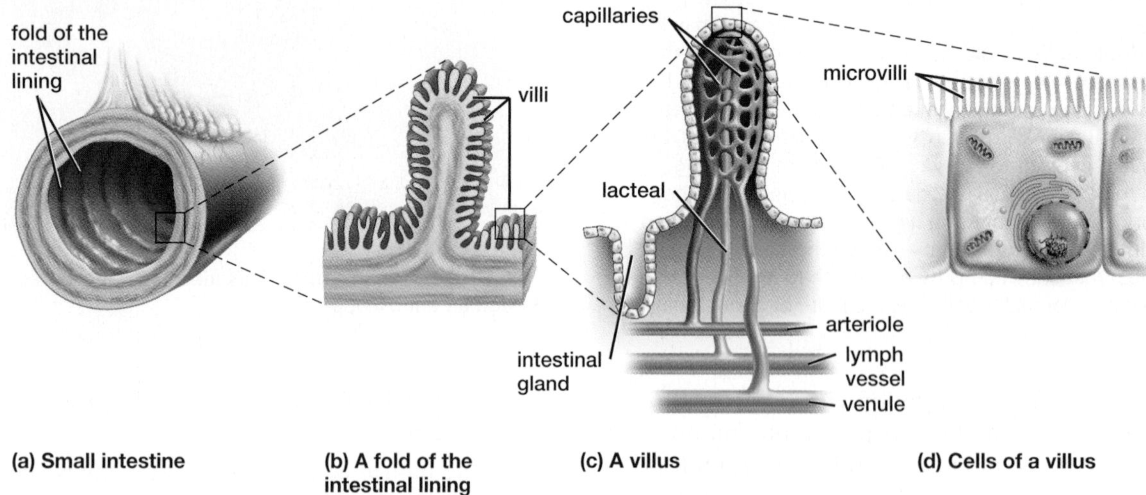

▲ **FIGURE 34-16 The structure of the small intestine (a)** Visible folds in the intestinal lining are carpeted with **(b)** tiny projections called villi, which extend from the folded intestinal lining. **(c)** Each villus contains a network of capillaries and a central lymph capillary called a lacteal. Most digested nutrients enter the capillaries, but fats enter the lacteal. **(d)** The plasma membranes of the epithelial cells covering each villus bear microvilli.

QUESTION What might the anatomy of the digestive system be like if the internal folds, villi, and microvilli of the small intestine had not evolved?

Health Watch

Eating to Death?

Patrick Deuel's doctor gave him an ultimatum: be hospitalized and lose weight, or die. At 1,072 pounds, he was imprisoned by his very size; in fact, a wall had to be removed to get him out of his bedroom for medical treatment. Although Deuel's case is extreme, obesity is a growing epidemic in many countries. In the United States, the percentage of overweight adults has more than doubled since 1980, and among children and adolescents, it has more than tripled.

Are some people simply born to be fat? If so, Deuel is one of them; his doctor described him as obese when he was only three months old. Researchers have found scores of different genes that influence weight, and it is clear that some people have far more difficulty remaining slim than others. But our genes haven't changed appreciably in the past 30 years, so the current epidemic of obesity is almost certainly caused by changes in our environment—which now offers an abundance of convenient, delicious, high-calorie foods—and our behavior, which has become more sedentary.

Obese individuals have a higher risk of liver disease, gallstones, sleep apnea (interruptions of breathing during sleep), diabetes, some cancers, arthritis, hypertension, heart disease, and stroke. People are described as *morbidly obese* when their weight poses a severe risk to their health and longevity. Nearly 6% of the U.S. population is morbidly obese, defined as having a BMI above 40, or being 100 pounds or more over ideal weight. Although the healthiest way to lose weight is to reduce caloric intake and increase exercise, many people who are morbidly obese have repeatedly failed in their attempts. This failure has spurred the development of surgical fixes that act directly on the digestive tract.

Although surgical approaches to weight loss are somewhat risky, doctors may recommend surgery for those whose obesity seems to pose an even more serious health risk than the surgery. After losing over 400 pounds on a supervised diet and exercise program, Deuel underwent gastric bypass surgery (**Fig. E34-2a**). This procedure blocks off all but a small pouch at the top of the stomach, and connects this pouch directly to the region of the small intestine below the duodenum. As a result, only very small amounts of solid food can be consumed at a sitting, and fewer calories are absorbed from food because the absorptive area of the small intestine is reduced. Although this surgery has proven effective in reducing weight, it restricts absorption of calcium, iron, and vitamins D and B_{12}, so vitamin and mineral supplementation is recommended to compensate for these losses.

The second most common form of weight-loss surgery is gastric band surgery. As shown in **Figure E34-2b,** this operation limits food intake by placing an inflatable band around the upper portion of the stomach, converting the food-collecting portion of the stomach into a pouch with a small opening through which digested food can pass. The band can be inflated or deflated to adjust the size of the opening from the pouch into the lower portion of the stomach. While some studies show that the gastric band produces less dramatic results than gastric bypass, it is simpler, reversible, and less likely to lead to nutrient deficiencies because it does not bypass any of the small intestine.

How is Deuel doing now? Although he reached a low of 370 pounds two years after gastric bypass surgery, five years after the surgery, he was back up to 560 pounds; not the man he used to be, but still morbidly obese. Although there are many weight-loss success stories, there are, unfortunately, no fail-safe cures.

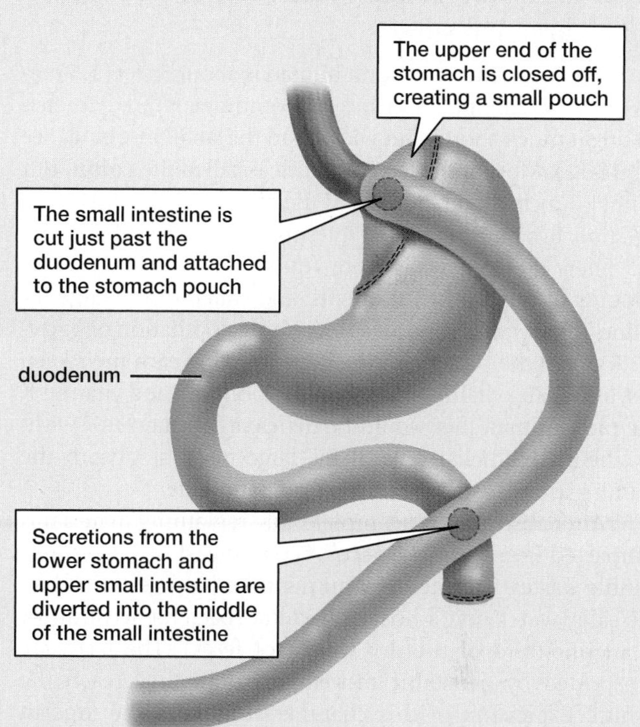
The upper end of the stomach is closed off, creating a small pouch
The small intestine is cut just past the duodenum and attached to the stomach pouch
duodenum
Secretions from the lower stomach and upper small intestine are diverted into the middle of the small intestine

(a) Gastric bypass surgery

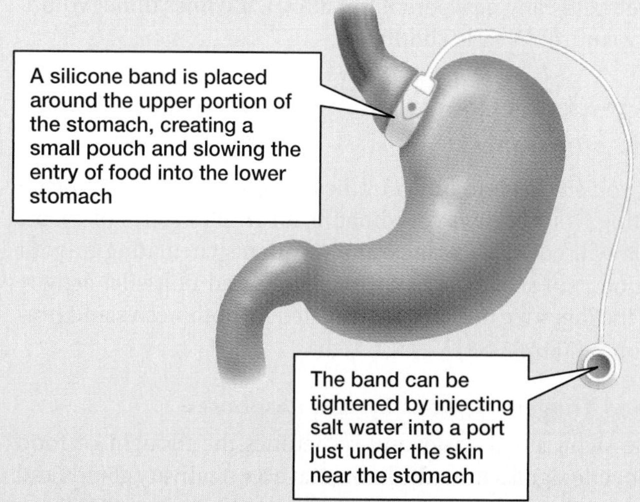
A silicone band is placed around the upper portion of the stomach, creating a small pouch and slowing the entry of food into the lower stomach
The band can be tightened by injecting salt water into a port just under the skin near the stomach

(b) Gastric band surgery

▲ **FIGURE E34-2 Weight-loss surgeries**

Table 34-5 Some Important Digestive Hormones in Humans

Hormone	Site of Production	Stimulus for Production	Effect
Gastrin	Stomach	Peptides and amino acids in the stomach	Stimulates acid secretion by cells in the stomach
Secretin	Small intestine	Acid in the small intestine	Stimulates bicarbonate production by the pancreas; increases bile output by the liver
Cholecystokinin	Small intestine	Amino acids and fatty acids in the small intestine	Stimulates the secretion of pancreatic enzymes and the release of bile from the gallbladder

Water Is Absorbed and Feces Are Formed in the Large Intestine

The **large intestine** in an adult human is about 5 feet (1.5 meters) long and about 2.5 inches (6.5 centimeters) in diameter, making it much shorter but wider than the small intestine (see Fig. 34-12). Most of the large intestine is called the **colon,** but its final 6-inch chamber is called the **rectum.** The leftovers of digestion flow into the large intestine: the indigestible cellulose fibers and cell walls from fruits and vegetables, small amounts of unabsorbed nutrients, and water. In the large intestine, these wastes support a flourishing population of bacteria. Some of the bacteria in the large intestine earn their keep by synthesizing vitamin B_{12}, thiamin, riboflavin, and vitamin K (a typical human diet would be deficient in vitamin K without them). Epithelial cells of the large intestine absorb the vitamins as well as most leftover water and salts.

After absorption is complete, any remaining material is compacted into semisolid **feces.** Feces consist of water, indigestible wastes, some leftover nutrients and indigestible fiber, bile salts, water, and a host of bacteria (bacteria account for about one-third of the dry weight of feces). The feces are transported by peristaltic movements until they reach the rectum. Expansion of this chamber stimulates the urge to defecate. The anal opening is controlled by two sphincter muscles—an inner one that is involuntary, and an outer muscle that can be consciously controlled. Although defecation is a reflex (as any new parent can attest), it comes under voluntary control in early childhood.

Digestion Is Controlled by the Nervous System and Hormones

As you sit down to lunch in the cafeteria and hungrily begin eating, your body is coordinating a complex series of events that will convert the meal into nutrients circulating in your blood. Not surprisingly, the secretions and muscular activity of the digestive tract are coordinated by both nerves and hormones (**Table 34-5**).

Food Triggers Nervous System Responses

The sight, smell, taste, and sometimes the thought of food generate signals from the brain that act on salivary glands and many other parts of the digestive tract, preparing it to digest and absorb food. For example, these nerve impulses cause the stomach to begin secreting acid and protective mucus. As food enters and moves through the digestive system, its bulk stimulates local nervous reflexes that cause peristalsis and segmentation movements.

Hormones Help Regulate Digestive Activity

Hormones secreted by the digestive system enter the bloodstream and circulate through the body, acting on specific receptors within the digestive tract. Like most hormones, they are regulated by negative feedback. For example, nutrients in chyme, particularly amino acids and peptides from protein digestion, stimulate cells in the stomach lining to release the hormone **gastrin** into the bloodstream (**Fig. 34-17**). Gastrin travels back to the stomach cells and stimulates further acid secretion, which promotes protein digestion. When the pH of the stomach reaches a low level (high acidity), this inhibits gastrin secretion, which in turn inhibits further acid production. Gastrin also stimulates muscular activity of the stomach, which helps it mechanically digest food and send chyme into the small intestine.

Two additional hormones, **secretin** and **cholecystokinin,** are released by cells of the duodenum in response to the acidity and nutrients of chyme, particularly peptides and fats. These hormones help regulate the chemical environment within the small intestine, and the rate at which the chyme enters, promoting optimal digestion and absorption of nutrients. Together, they increase bile production by the liver and bile release from the gallbladder. These hormones also increase production and release of pancreatic juice. In addition, secretin slows acid production by the stomach and inhibits its peristaltic contractions. This slows the rate at which chyme is forced into the small intestine, allowing more time for digestion and absorption to occur.

Scientists are discovering an ever-increasing array of hormones related to digestion and fat storage, some of which act on the brain, and many of which contribute in complex ways to a person's weight. One of these, leptin, which is secreted by body fat, is described in Chapter 37.

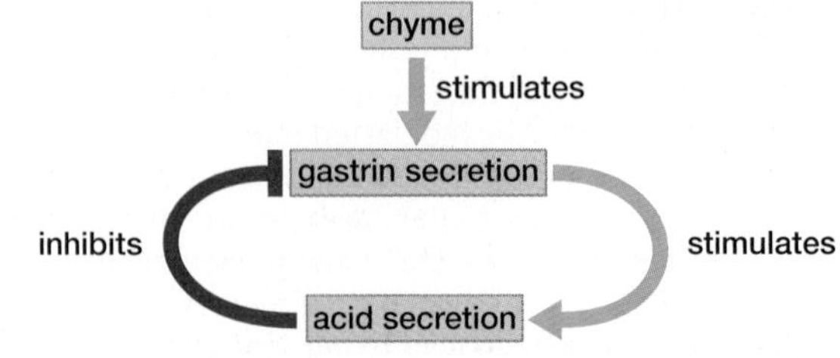

▲ **FIGURE 34-17 Negative feedback controls stomach acidity**

Case Study revisited
Dieting to Death?

Eating disorders, as you might predict, can cause severe malnutrition, and (in the most extreme cases) death from starvation. They also reduce the content of certain crucial ions in the blood, interfering with heart muscle contractions, which can cause death from cardiac arrest. Eating disorders also play havoc with the digestive system; most damage occurs because of frequent vomiting that exposes the upper digestive tract to stomach acids. The explosive pressure of repeated vomiting can even tear or rupture the esophagus—a true medical emergency.

The causes of eating disorders are numerous and poorly understood. Genes apparently play a role; people with a close relative who has an eating disorder are about five times as likely to develop such a disorder themselves. Mental problems (such as anxiety and depression) and personality traits (such as perfectionism, low self-esteem, and a high need for acceptance and achievement) seem to predispose people to eating disorders. Cases often begin during adolescence, when bodies and brains are undergoing rapid changes. Magazines, TV, and Internet ads bombard susceptible young people with the message that being skinny is a route to acceptance, beauty, and riches.

Jayne Gardham (**Fig. 34-18**), an aspiring British singer whose career has been derailed by anorexia, began dieting at age 15. Already slim, she entered a contest with three friends to see who could lose the most weight. "We'd seen the beautiful pop singers and models that are always in glossy magazines. And we wanted to be like them," she explains. As her friends dropped out, Jayne's dieting became a compulsion that she has been unable to conquer, even though she knows that her life is in danger. She describes anorexia's effects on her body: "My heart hammers in my chest and it's difficult for me even to get up the stairs. I haven't had a period in years, and its affecting my memory too. I can't remember the smallest things. I have downy hair all over my body now as I'm permanently cold. Even on a really hot day I have to have the heating on to keep me warm. I know I can't go on like this. And I've got to get better somehow, but I don't know how."

▲ **FIGURE 34-18 Anorexia** Jayne Gardham was a popular British singer before anorexia stopped her performances. Weighing about 70 pounds at 5′ 4″, she has a BMI of 12 in this photo.

Unfortunately, eating disorders are difficult to treat. Victims are given nutritional therapy to help them recover from malnutrition. Psychotherapy is usually necessary, and antidepressant drugs are helpful in some cases. Because many victims hide or deny their problems, and because treatment is expensive, the majority of sufferers are inadequately treated; fewer than half experience complete recovery. For Carré Otis, the wake-up call came when, at age 30, she required surgery to repair her heart, damaged from years of malnutrition. "I needed to make a change, or clearly my body would not hold up. At that moment, I finally realized how out of control I was, and knew that I was not ready to die. I was ready to embark on the road to recovery."

Consider This

Media that glamorize slimness have been blamed for the prevalence of eating disorders. Why do you think extreme thinness is made so appealing? Are there appropriate measures that a free society can take to reverse or limit this message? Do you support the idea of a minimum BMI for models?

CHAPTER REVIEW

Summary of Key Concepts

34.1 What Nutrients Do Animals Need?

All animals require nutrients that provide energy; for humans, these are primarily fats and carbohydrates, with a small percent derived from protein. Food energy is measured in Calories. Excess energy from food is stored in body fat, the most concentrated energy source. Each type of animal has specific requirements for essential nutrients. These cannot be synthesized by the animal's cells (either at all or in adequate quantities) but are required to synthesize molecules essential to cellular function. Essential nutrients for humans include essential fatty acids, essential amino acids, minerals, and vitamins that facilitate the diverse chemical reactions of metabolism. Water is essential in that it comprises about 60% of human body weight and is the medium in which all metabolic reactions occur.

34.2 How Does Digestion Occur?

Digestion is the mechanical and chemical breakdown of food, converting complex molecules into simpler molecules that can be used by the organism. Digestion at its simplest is entirely intracellular, as occurs within individual cells of sponges, which lack digestive systems. The simplest digestive system is the saclike gastrovascular cavity in organisms such as *Hydra*. Most digestive systems consist of a one-way tube along which specialized compartments process food in an orderly sequence. Specialized digestive systems allow different animals to utilize a wide variety of foods. Digestive systems must accomplish five tasks: ingestion of food, mechanical digestion, chemical digestion, absorption of nutrients, and elimination of wastes.

34.3 How Do Humans Digest Food?

In humans, digestion begins in the mouth, where food is physically broken down by chewing, and chemical digestion of starches is initiated by saliva. Food is then conducted to the stomach by peristaltic waves of the esophagus. In the acidic environment of the stomach, food is churned into smaller particles, and protein digestion begins. Gradually, acidic chyme is released into the small intestine, where it is neutralized by sodium bicarbonate from the pancreas. In the small intestine, secretions from the pancreas and liver, and from the cells of the intestinal epithelium, complete the chemical digestion of proteins, fats, and carbohydrates. In the small intestine, the simple molecular products of digestion are absorbed into the bloodstream, either directly via capillaries or indirectly through the lymphatic system, for distribution to the body cells. The colon portion of the large intestine absorbs most of the remaining water, salts, and vitamins manufactured by intestinal bacteria. The remaining wastes become feces that are passed to the rectum, whose distention triggers the urge to defecate through the anus.

Digestion is regulated by the nervous system and hormones. The smell and taste of food and the action of chewing trigger the secretion of saliva in the mouth and the production of gastrin by the stomach. Gastrin stimulates production of acid by the stomach. As chyme enters the small intestine, it stimulates the release of secretin and cholecystokinin by cells of the duodenum. These hormones increase the release of digestive secretions into the small intestine, and slow the emptying of the stomach.

Key Terms

absorption *661*
amylase *665*
bile *668*
body mass index (BMI) *657*
calorie *656*
Calorie *656*
carnivore *662*
chemical digestion *661*
cholecystokinin *672*
chylomicron *670*
chyme *667*
colon *672*
digestion *661*
digestive system *661*
duodenum *668*
elimination *661*
epiglottis *666*
esophagus *662*
essential amino acid *658*
essential fatty acid *658*
essential nutrient *657*
extracellular digestion *661*
feces *672*
food vacuole *661*
gallbladder *668*
gastric gland *667*
gastrin *672*
gastrovascular cavity *661*
herbivore *662*
ingestion *661*
intracellular digestion *661*
lacteal *670*
lactose intolerance *670*
large intestine *672*
lipase *668*
liver *668*
lysosome *661*
mechanical digestion *661*
metabolic rate *656*
microvillus (plural, microvilli) *668*
mineral *658*
mouth *661*
nutrient *656*
omnivore *662*
pancreas *668*
pancreatic juice *668*
peristalsis *666*
pharynx *665*
protease *667*
rectum *672*
ruminant *663*
secretin *672*
small intestine *668*
sphincter muscle *667*
stomach *666*
villus (plural, villi) *669*
vitamin *658*

Thinking Through the Concepts

Fill-in-the-Blank

1. Two general roles for nutrients are to provide ________ and ________. Nutrients required in small amounts that often serve as coenzymes are called ________. Nutrients that are elements are called ________. Nutrients that cannot be synthesized by the body are called ________ nutrients.
2. Sponges rely exclusively on ________ digestion. Cnidarians such as *Hydra* digest food in a(n) ________. Earthworms have a(n) ________ digestive system. Ruminants can break down ________ only because of microorganisms in their stomachs.
3. The enzyme called ________ is present in saliva and begins breaking down ________. Digestion of ________ begins in the stomach. Stomach acid converts the inactive substance ________ into the active enzyme ________. Nearly all fat digestion occurs in the ________.
4. The five major processes carried out by the digestive systems are ________, ________, ________, ________, and ________.
5. In humans, a cavity called the ________ is shared by the ________ and ________ systems. A flap called the ________ prevents food from entering the trachea during the act of ________. Food is pushed through the digestive system by muscular contractions called ________. Circular muscles that control movement into and out of organs such as the stomach are called ________.
6. Pancreatic juice is released into the ________ of the small intestine. Pancreatic juice contains ________ to neutralize the acidic ________ from the stomach, and enzymes that digest ________, ________, and ________.
7. Fats are dispersed by a secretion called ________. This secretion is produced by the ________ and is stored in the ________. Products of fat digestion are formed into chylomicrons inside intestinal ________ cells. The chylomicrons are released inside the villus and diffuse through the wall of the ________, which is part of the ________ system.

Review Questions

1. List the six general types of nutrients. Which two provide the most energy for humans? Which of these stores the most energy in the body, and why?
2. Give an example of a vertebrate and an invertebrate that both use gizzards. Describe the general structure and function of the gizzard and how it works.
3. Vertebrates can be grouped into three categories based on their diets; list and briefly define these categories, giving one example of each. Which group has the shortest small intestine, and why?
4. Why is the stomach both muscular and expandable?
5. List and describe the function of the three principal secretions of the stomach.
6. List the digestive substances secreted into the small intestine, and describe the origin and function of each.

7. Name and describe the muscular movements of the digestive system, where they occur, and their functions.
8. Vitamin C is an essential vitamin for humans but not for dogs. Certain amino acids are essential for humans but not for plants. Explain.
9. Name four structural or functional adaptations of the human small intestine that contribute to effective digestion and absorption.
10. Describe protein digestion in the stomach and small intestine.
11. Explain the role of the large intestine.

Applying the Concepts

1. The food label on a soup can shows that the product contains 10 grams of protein, 4 grams of carbohydrate, and 3 grams of fat. How many Calories are in this soup?
2. Small birds have high metabolic rates, efficient digestive tracts, and high-calorie diets. Some birds consume an amount of food equivalent to 34% of their body weight every day. Why do you think birds rarely consume leaves or grass?
3. Control of the human digestive tract involves several feedback loops and messages that coordinate activity in one chamber with those taking place in subsequent chambers. List the coordinating events described in this chapter in order, beginning with tasting, chewing, and swallowing a piece of meat, and ending with the undigested waste that enters the large intestine.
4. Symbiotic protozoa in the digestive tracts of termites produce cellulose-digesting enzymes that allow the termites to utilize wood as a food source. In return, termites provide protozoa with food and shelter. Imagine that the human species is gradually invaded, over many generations, by symbiotic protozoa capable of digesting cellulose. What adaptive changes in human body structure and function might be selected for as a result?

MB *Go to www.masteringbiology.com for practice quizzes, activities, eText, videos, current events, and more.*

The Urinary System

Chapter 35

Case Study

Paying It Forward

ANTHONY DEGIULIO HAD A DREAM—he wanted to save someone's life. At first, it was a nebulous ambition, but it started to take shape as he watched a segment of the TV show *60 Minutes* that highlighted living kidney donation. DeGiulio, who hadn't realized that a live person could donate a kidney, immediately saw this as a way to achieve his ambition. "I want to do this," he told his wife.

DeGiulio called New York-Presbyterian Hospital, and set in motion a series of events described as an "altruistic chain" that gave new life not to one person, but to four. The chain of events was possible because three people in the area needed kidneys, and each had a family member eager to donate, but were prevented from doing so because of incompatibility between their blood types. Barbara Asofsky, a nursery school teacher, had known for 5 years that she would need a kidney transplant. When Anthony DeGiulio's kidney was found to be a good match for Barbara, her husband, Douglas, was happy to donate the kidney he wanted to give to Barbara—but couldn't because of incompatibility—to a stranger instead. The fortunate stranger was Alina Binder, a student at Brooklyn College. Alina's father, Michael, was a good match for Andrew Novak, a telecommunications technician. Finally, Andrew's sister, Laura Nicholson, donated her kidney to Luther Johnson, a hotel kitchen steward.

Why are kidneys so important? Do kidney transplant recipients lead completely normal lives? How is a donor kidney removed, and what are the risks to the donor? For people whose kidneys have both failed, what are the alternatives to a kidney transplant?

▲ Barbara Asofsky hugs good Samaritan Anthony DeGiulio, whose kidney restored her health and started a chain of kidney donations.

At a Glance

35.1 WHAT ARE THE BASIC FUNCTIONS OF URINARY SYSTEMS?

Urinary systems help maintain **homeostasis**—the relatively constant internal environment required to preserve health and, ultimately, life (see p. 605). Urinary systems accomplish this by regulating the composition of the blood and **extracellular fluid** (the watery substance that bathes all cells) within the narrow bounds required for cellular metabolism. An essential element of homeostasis is water balance, which is crucial to maintaining the proper concentration, or **osmolarity,** of dissolved substances in cells and in their extracellular environment.

The second major function of urinary systems is **excretion,** a general term for eliminating wastes and excess substances from the body. Excretion also occurs through the respiratory system, where carbon dioxide is released to the environment, and through the digestive system, where undigested material is excreted in feces. Urinary systems produce **urine,** which contains the waste products of cellular metabolism. Whether we are looking at earthworms, fishes, or humans, urinary systems (often called "excretory systems," particularly in invertebrates) all perform similar functions using the same basic sequence of processes:

1. Either blood or extracellular fluid is filtered, removing water and small dissolved molecules.
2. Nutrients are selectively reabsorbed back into the filtered fluid.
3. Excess water, excess nutrients, and dissolved wastes are excreted from the body in urine.

To remove wastes, urinary systems must be in intimate association with the extracellular fluid. In some invertebrates, the excretory system filters this fluid directly. In most animals with a circulatory system, the excretory system filters blood, as explained in later sections.

35.2 WHAT ARE SOME EXAMPLES OF INVERTEBRATE EXCRETORY SYSTEMS?

The first animals to evolve lacked excretory systems, relying instead on diffusion and active transport through cell membranes to maintain homeostasis within individual cells, as sponges do today. The earliest excretory systems probably only served to maintain water balance, which is the primary function of the simple excretory system of flatworms.

Protonephridia Filter Extracellular Fluid in Flatworms

The excretory system of freshwater flatworms consists of **protonephridia** (Greek, "before the kidneys"; singular, protonephridium). Protonephridia are tubules that branch throughout the extracellular fluid that surrounds the flatworm's tissues. This simple excretory system serves primarily to collect excess water (which continually enters the body by osmosis) from the extracellular fluid. Ciliated cells (called "flame cells" because their beating cilia resemble a flickering flame) are spaced along the tubules (**Fig. 35-1a**). Flame cells produce a current that forces the watery urine out through excretory pores. The large body surface of flatworms also serves as an excretory structure through which most cellular wastes diffuse out.

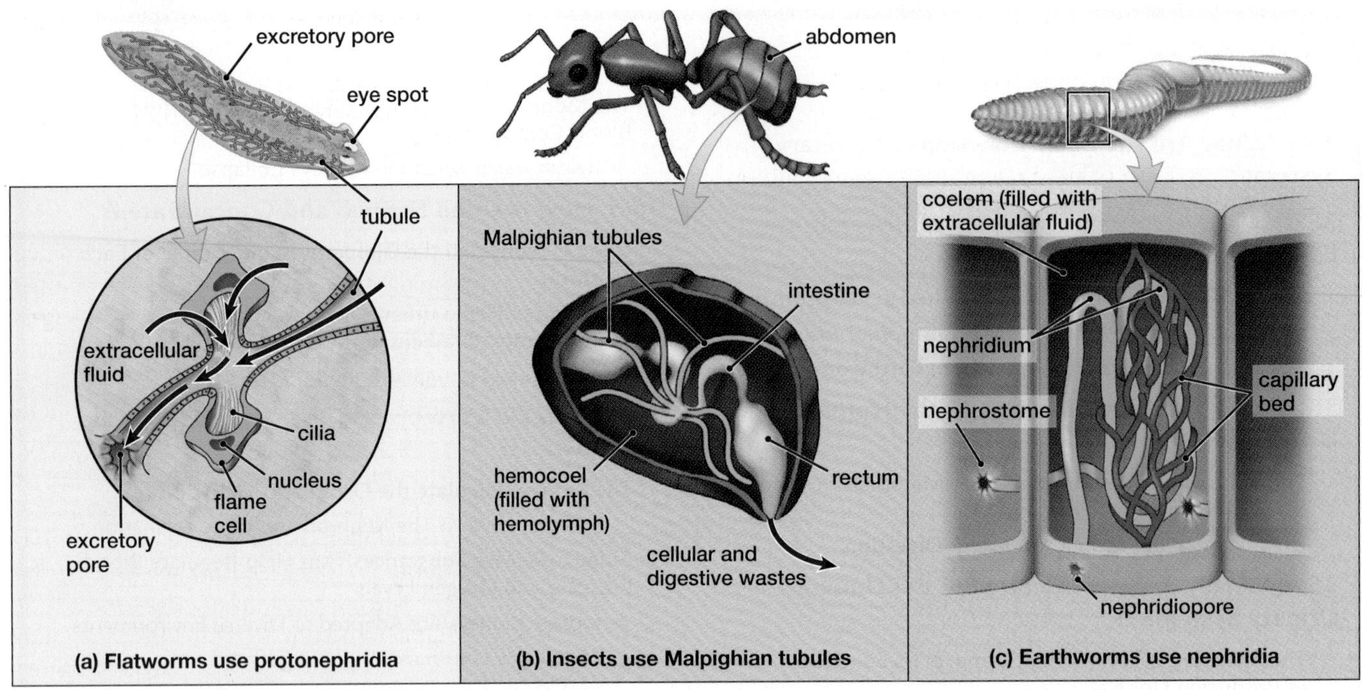

▲ **FIGURE 35-1 Some invertebrate excretory systems (a)** The protonephridia of freshwater flatworms consist of tubules that collect and excrete urine (primarily water) from the extracellular fluid. The water is propelled along the tubules to the excretory pores by ciliated flame cells. **(b)** Malpighian tubules of insects extend from the intestine and filter the hemolymph that fills the hemocoel. They produce concentrated urine, which is excreted with the feces. **(c)** Paired nephridia in each segment of the earthworm filter extracellular fluid from the coelom. Fluid enters the nephrostome, passes through the tubule, and exits the nephridiopore as urine.

Malpighian Tubules Filter the Hemolymph of Insects

Insects have an open circulatory system, in which hemolymph (a fluid that serves as both blood and extracellular fluid) fills the hemocoel (the body cavity) and bathes the internal tissues and organs directly. Insect excretory systems consist of **Malpighian tubules,** small tubes that extend outward from the intestine and end blindly within the hemolymph (**Fig. 35-1b**). Wastes and nutrients move from the hemolymph into the tubules by diffusion and active transport, and water follows by osmosis. The urine is conducted into the intestine, where important solutes are secreted into the hemolymph by active transport. Insects produce very concentrated urine, which is excreted along with feces.

Nephridia Filter Extracellular Fluid in Earthworms

In earthworms, mollusks, and several other invertebrates, excretion is performed by tubular structures called **nephridia** (singular, nephridium). In the earthworm, the body cavity (the coelom) is filled with extracellular fluid into which wastes and nutrients from the blood diffuse. Each nephridium begins with a funnel-like opening, the *nephrostome*, ringed with cilia that direct extracellular fluid into a narrow, twisted tubule surrounded by capillaries (**Fig. 35-1c**). As the fluid traverses the tubule, salts and other nutrients are reabsorbed back into the capillary blood, leaving water and wastes behind. The resulting urine is then excreted through an opening in the body wall called the *nephridiopore*. Each of the earthworm's many segments contains a pair of nephridia. As you study the kidney tubules of vertebrates, notice their similarities to nephridia.

35.3 WHAT ARE THE FUNCTIONS OF THE HUMAN URINARY SYSTEM?

Kidneys are the organs of the vertebrate urinary system where the blood is filtered and urine is produced. Because vertebrates live in such a wide variety of habitats, from salty oceans to searing deserts to mountain streams, they face radically different challenges in maintaining constant conditions within their bodies. As a result, vertebrate kidneys differ considerably in their homeostatic functions. In the sections that follow, we emphasize mammalian kidneys, with humans as an example.

The Kidneys of Humans and Other Mammals Perform Many Homeostatic Functions

The mammalian urinary system consists of the kidneys, ureters, bladder, and urethra. These organs filter the blood, collecting and then excreting the dissolved waste products in urine. During the filtration process, water and its dissolved molecules (except for large proteins) are forced out of the blood. The kidneys then return to the bloodstream nearly all of the water as well as those nutrients required by the body. The urine retains wastes, including urea produced by amino acid breakdown; excesses of water, salts, hormones, and some vitamins; and foreign substances, such as drugs and molecules produced when drugs are metabolized. The rest of the urinary system channels and stores the urine and eventually expels it from the body.

The mammalian urinary system helps maintain homeostasis in several ways; for example, by:

- Regulating blood levels of ions such as sodium, potassium, chloride, and calcium.
- Maintaining the proper pH of the blood by regulating hydrogen and bicarbonate ion concentrations.
- Regulating the water content of the blood.
- Retaining important nutrients such as glucose and amino acids in the blood.
- Eliminating cellular waste products such as urea.
- Secreting substances that help regulate blood pressure and blood oxygen levels.

Urea Is a Waste Product of Protein Digestion

An important function of most urinary systems is to eliminate nitrogenous (nitrogen-containing) wastes that are formed when cells break down amino acids (**Fig. 35-2**). In the digestive tract, proteins are broken down into their component amino acids, which are then transported in the blood and taken up by cells. Within cells, some amino acids are used to synthesize new proteins. Those that aren't required for protein synthesis have their amino groups removed, allowing the rest of the molecule to be used in synthesizing other molecules, or as a source of cellular energy. The liberated amino groups ($-NH_2$; see Table 3-1 on p. 38) are released by the cells and enter the bloodstream as **ammonia (NH_3),** which is toxic. The livers of humans and other mammals convert ammonia into **urea,** a far less toxic substance. Urea is filtered from the blood by the kidneys and then excreted in urine.

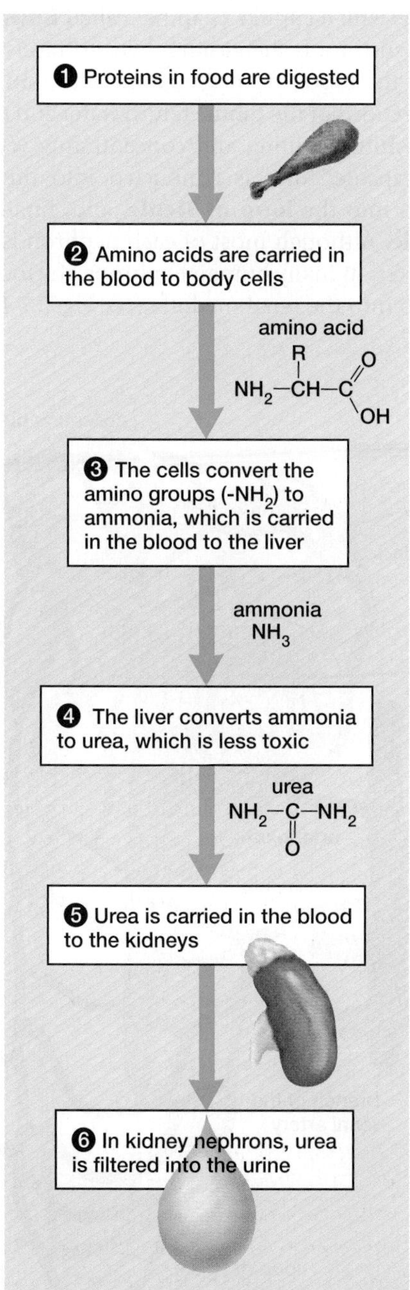

▲ **FIGURE 35-2 Urea formation and excretion**

35.4 WHAT ARE THE STRUCTURES OF THE HUMAN URINARY SYSTEM?

Human kidneys are paired organs located at about waist level on either side of the spinal column (**Fig. 35-3**). Each kidney is approximately 5 inches (12.5 centimeters) long, 3 inches (7.5 centimeters) wide, and 1 inch (2.5 centimeters) thick, and resembles a kidney bean in shape and color. Blood enters

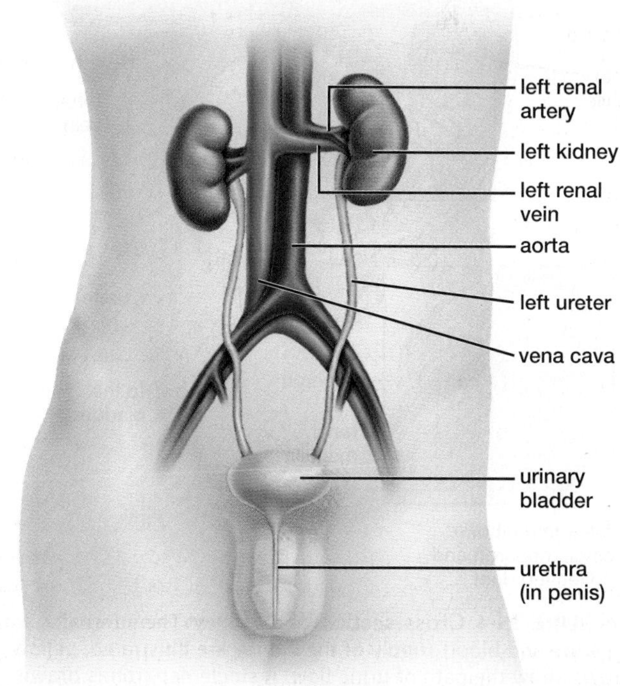

▲ **FIGURE 35-3 The human urinary system** A simplified diagram of the human urinary system and its blood supply

each kidney through a **renal artery.** After the blood has been filtered, it exits through the **renal vein** (see Fig. 35-3). Urine leaves each kidney through a narrow, muscular tube called the **ureter.** Using rhythmic contractions, the ureters transport urine to the urinary bladder (or simply, **bladder**), a hollow, muscular chamber that collects and stores urine.

The wall of the bladder, which contains smooth muscle, is capable of considerable expansion. When the bladder wall is stretched sufficiently by stored urine, receptors are activated that trigger reflexive smooth muscle contractions, which expel the urine. Urine is held in the bladder by two sphincter muscles. The internal sphincter, located where the bladder joins the urethra, opens automatically during these reflexive contractions. The external sphincter (located slightly below the internal sphincter) is under voluntary control, allowing the brain to suppress urination unless the bladder becomes overly full. The average adult bladder can hold, if necessary, about a pint (500 milliliters) of urine, but the desire to urinate is triggered by considerably smaller amounts. Urine exits the body via the **urethra,** a single narrow tube about 1.5 inches long in the female and about 8 inches long in the male (because it extends the length of the penis).

The Structure of the Kidney Supports Its Function of Producing Urine

Each kidney contains a solid outer layer consisting of the **renal cortex,** which overlies an inner layer called the **renal medulla.** The renal medulla surrounds a branched, funnel-like chamber called the **renal pelvis,** which collects urine and funnels it into the ureter (**Fig. 35-4**).

The renal cortex of each human kidney is packed with roughly 1 million microscopic filters, or **nephrons** (**Fig. 35-5**; see also Fig. 35-4, inset). A nephron has two major parts: the **glomerulus,** a dense knot of capillaries where fluid is filtered out of the blood through the porous capillary walls; and a long, twisted **tubule** (meaning "little tube"). Urine formation occurs in the tubule, which has four major sections. The tubule begins with a cuplike chamber called **Bowman's capsule,** which surrounds the glomerulus and receives fluid filtered out of the blood from the glomerular capillaries. The remaining sections of the tubule return water and nutrients to the blood, while retaining and concentrating wastes. From Bowman's capsule, fluid is conducted into the **proximal tubule,** then into the **loop of Henle,** and, finally, into the **distal tubule.** Although most of each nephron is located in the renal cortex, in many human nephrons, the loop of Henle extends deep into the renal medulla (see Fig. 35-4, inset).

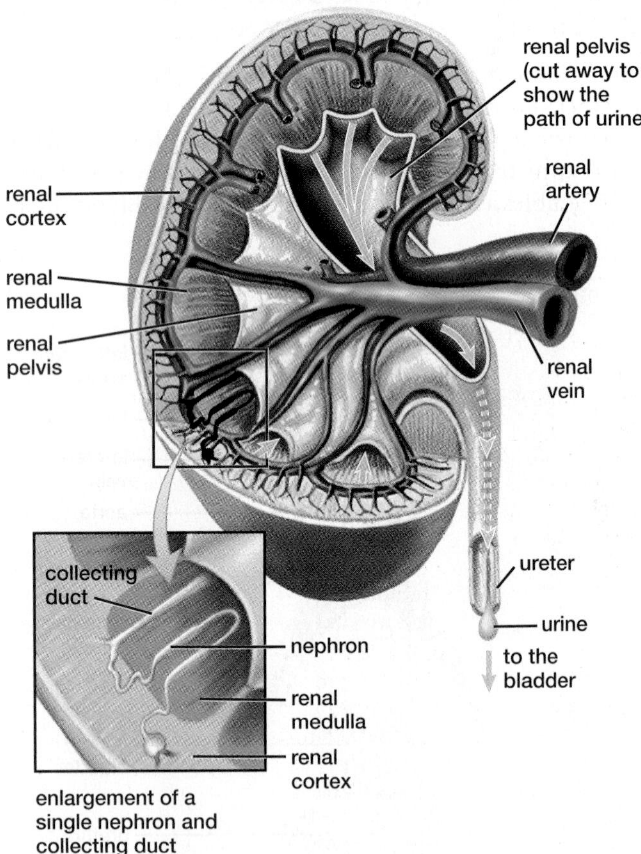

▲ **FIGURE 35-4 Cross-section of a kidney** The internal structure and blood supply of the kidney are illustrated; yellow arrows show the path of urine flow. A single nephron is drawn considerably larger than normal (and further enlarged in the inset) to show its location in the kidney and its relationship to the collecting duct. Collecting ducts empty the urine into the renal pelvis, which funnels urine into the ureter.

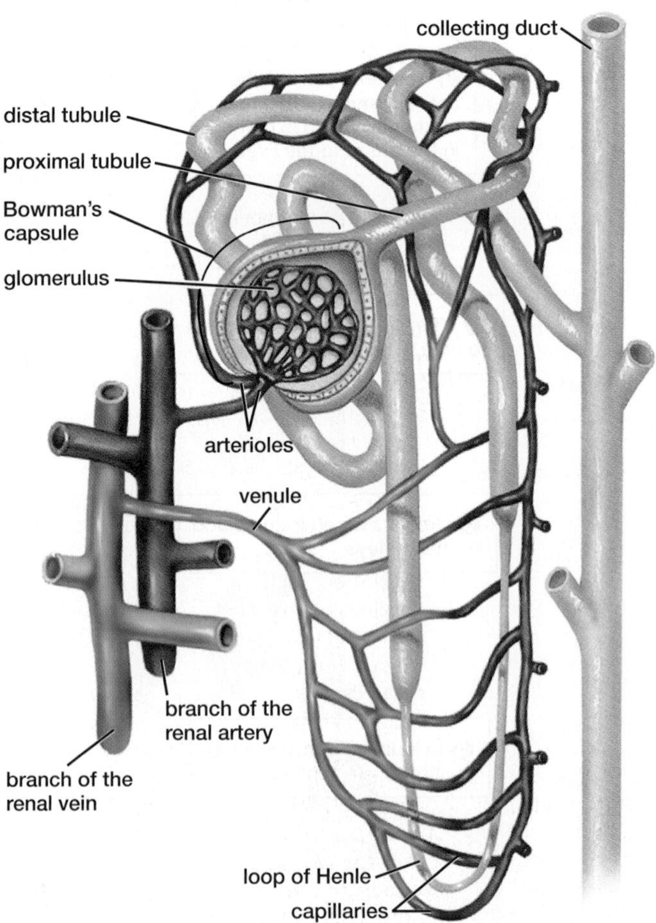

▲ **FIGURE 35-5 An individual nephron and its blood supply**

The distal tubule empties urine into a **collecting duct,** a larger tube that is not part of the nephron. Each of the thousands of collecting ducts within the kidney receives fluid from many nephrons. Collecting ducts gather urine from nephrons in the renal cortex, conduct it through the renal medulla, and empty it into the renal pelvis (see Fig. 35-4).

The Kidney's Blood Supply Allows It to Fine-Tune Blood Composition

To support their role of maintaining homeostasis, the kidneys have an enormous blood supply. Nearly one-quarter of the volume of blood pumped by each heartbeat travels through the kidneys, which receive more than 1 quart of blood (about 1 liter) every minute. This rapid flow of blood through the kidneys allows them to continuously maintain the composition of the blood within narrow limits. The immense filtering capacity of the kidneys explains why kidney donors and kidney recipients can each survive with only one kidney; half of the usual capacity is fully adequate to maintain homeostasis under most conditions. Should both of these versatile organs fail, however, death ensues quickly without medical intervention, as described in "Health Watch: When the Kidneys Collapse."

Health Watch

When the Kidneys Collapse

Each year in the United States, about 85,000 people die of kidney failure, also called end-stage renal disease (ESRD). The most common causes are diabetes and high blood pressure, both of which damage the glomerular capillaries, but the kidneys can also be damaged by infection or overdoses of some painkilling medicines. Kidney failure is typically treated by **hemodialysis**—a medical treatment during which wastes are removed from the blood by machine—or by kidney transplantation from a living or newly deceased donor. Although there are well over 300,000 patients in the United States being treated by hemodialysis, fewer than 20,000 kidney transplants are performed annually.

First used in 1945, hemodialysis operates on a simple principle: Substances will diffuse from areas of higher concentration to areas of lower concentration across a selectively permeable membrane. This process is called dialysis; therefore, the filtration of blood according to this principle is called hemodialysis ("hemo" means "blood").

During hemodialysis (often shortened to "dialysis"), the patient's blood is diverted from the body and pumped through narrow tubes made of a special cellophane membrane suspended in dialyzing fluid. Like glomerular capillaries, the membrane has pores too small to permit the passage of blood cells and large proteins, but large enough to pass small molecules such as water, sugar, salts, amino acids, and urea. Dialyzing fluid contains no waste products, but has normal blood levels of salts and nutrients; therefore, molecules whose concentrations are higher than normal in the patient's blood (such as the waste product urea) diffuse into the dialyzing fluid, which is continuously replenished to maintain the gradient. The patient usually remains attached to the dialysis machine for 4 to 6 hours, three times a week (**Fig. E35-1**). People relying on dialysis can survive for many years, but these patients' blood composition fluctuates, and toxic substances reach higher-than-normal levels between sessions.

Peritoneal dialysis is a less common but effective form of dialysis that can be done at home. Dialyzing fluid is pumped through a tube implanted directly into the abdominal cavity. The abdominal cavity is lined with a natural membrane called the peritoneum. Waste products from blood circulating in capillaries within the peritoneum gradually diffuse into the dialysis fluid, which is then drained out through the tube. Patients either replace the dialysis fluid about four times daily, or link the implanted tube to a machine that circulates the fluid through the abdominal cavity during the night.

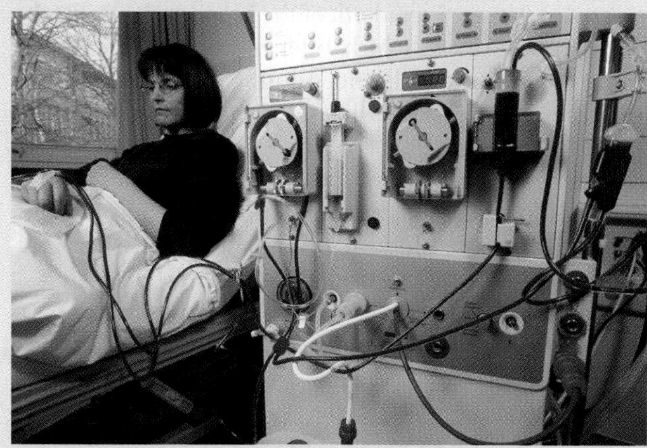

▲ **FIGURE E35-1 A patient on hemodialysis**

What might the future hold for victims of kidney failure who are unable to receive a transplant? At the University of Michigan, Dr. David Humes has developed a "renal assist device." Humes' device circulates blood through a cartridge filled with tiny tubes lined with billions of living kidney tubule cells, which are cultured from donated human kidneys that are unsuitable for transplantation. Clinical trials using the device on intensive-care patients whose kidneys were damaged demonstrated that the renal assist device—used in conjunction with dialysis—hastened the recovery of kidney function and improved survival compared to dialysis alone. This is because the living kidney cells perform a far greater range of functions than does dialysis, more closely mimicking a normal kidney. In the future, Humes hopes to develop a renal assist device that can be worn outside the body by people suffering from chronic kidney failure.

Meanwhile, researchers are also working on *xenotransplantation*, a process that would allow people to receive kidneys from animals, such as pigs, whose cells have been genetically modified to prevent the recipient's immune system from attacking and rejecting them.

Blood carried to the kidney by the renal artery enters each nephron through an arteriole associated with that nephron. Within Bowman's capsule, the arteriole (which is about one-tenth the diameter of an eyelash) branches further into a microscopic meshwork of capillaries that forms the glomerulus (see Fig. 35-5). The capillaries then empty into an outgoing arteriole (in contrast to most capillaries, which empty into venules; see pp. 629–631). Beyond the glomerulus, this arteriole branches into more capillaries that surround the tubule. These capillaries carry blood into a venule from which blood is delivered to the renal vein and then to the inferior vena cava (see Fig. 35-5).

35.5 HOW IS URINE FORMED AND CONCENTRATED?

Urine is produced in the nephrons of the kidneys in three stages: **filtration,** during which water and most small dissolved molecules are filtered out of the blood; **tubular reabsorption,** the process by which water and necessary nutrients are restored to the blood; and **tubular secretion,** during which wastes and excess ions that still remain in the blood are secreted into the urine.

As urine is formed, essentially all small organic nutrients—such as amino acids and glucose—are filtered out of, and then returned to, the blood. Large quantities of water and many ions—such as sodium (Na^+), chloride (Cl^-), potassium (K^+), calcium (Ca^{2+}), hydrogen (H^+), and bicarbonate (HCO_3^-)—are also filtered out, but their return rate is continuously adjusted to meet the body's changing needs. For example, blood pH must be tightly controlled by regulating the concentrations of H^+ (an acid) and HCO_3^- (a base) because the enzymes that regulate all the body's biochemical reactions function only within a narrow pH range.

The mechanisms of urine formation are described below, and are explained in greater detail in "A Closer Look at the Nephron and Urine Formation" later in this chapter.

Urine Is Formed in the Glomerulus and Tubule of Each Nephron

Filtration, the first step in urine formation, occurs when water carrying small dissolved molecules and ions is forced through the walls of the capillaries that form the glomerulus (**Fig. 35-6** ❶). During filtration, roughly 20% of the blood's fluid is filtered out because the blood pressure within the glomerular capillaries is far higher than in most body capillaries, and the glomerular capillary walls are far more porous than most capillary walls. Blood cells and large proteins are too large to leave the capillaries, however, and remain behind in the blood. The fluid filtered out of the glomerular capillaries, called **filtrate,** is collected in Bowman's capsule and then continues through the tubule.

Tubular reabsorption occurs primarily in the proximal tubule. Tubular reabsorption returns nearly all the organic nutrients (such as glucose, amino acids, and needed vitamins) and most of the ions (Na^+, Cl^-, K^+, Ca^{2+}, and HCO_3^-) to the blood. All of these substances move out of the filtrate through the walls of the tubule and into the extracellular fluid (**Fig. 35-6** ❷).

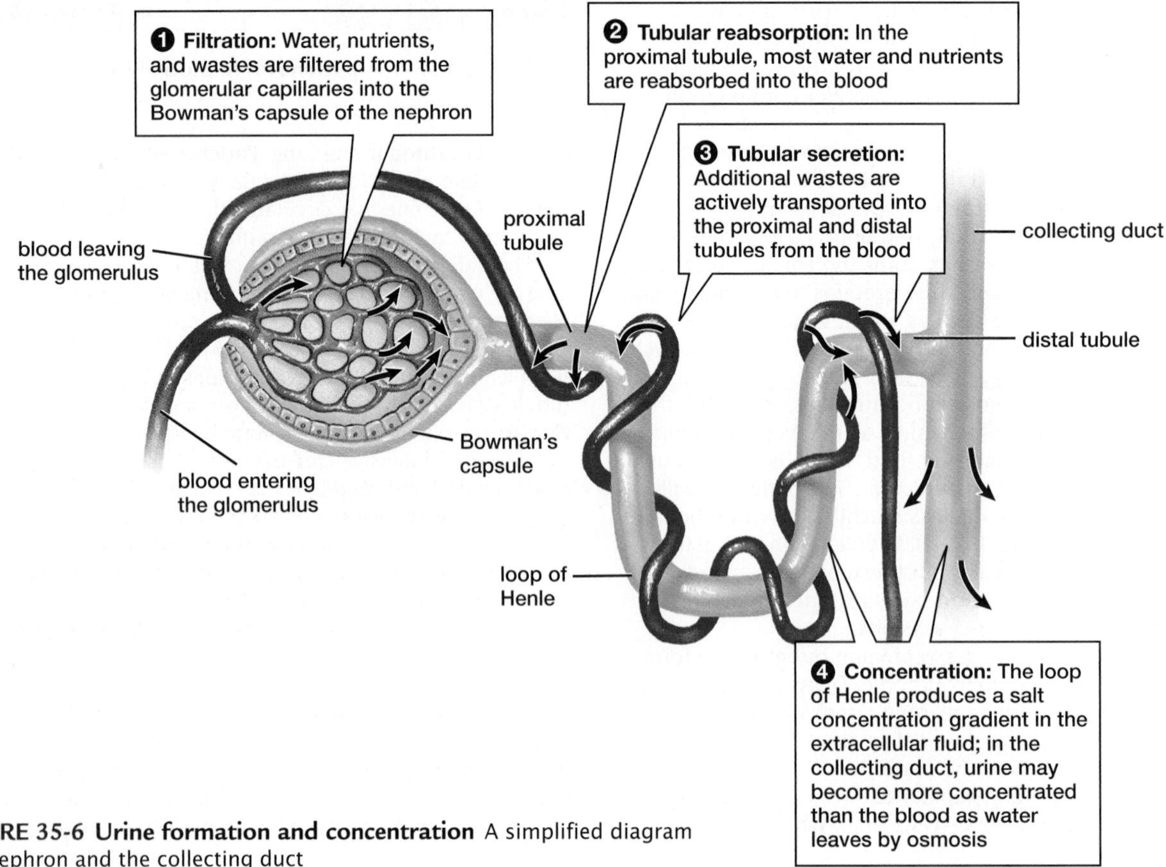
❶ Filtration: Water, nutrients, and wastes are filtered from the glomerular capillaries into the Bowman's capsule of the nephron
❷ Tubular reabsorption: In the proximal tubule, most water and nutrients are reabsorbed into the blood
❸ Tubular secretion: Additional wastes are actively transported into the proximal and distal tubules from the blood
blood leaving the glomerulus
proximal tubule
collecting duct
distal tubule
Bowman's capsule
blood entering the glomerulus
loop of Henle
❹ Concentration: The loop of Henle produces a salt concentration gradient in the extracellular fluid; in the collecting duct, urine may become more concentrated than the blood as water leaves by osmosis

▲ **FIGURE 35-6 Urine formation and concentration** A simplified diagram of the nephron and the collecting duct

Tubular reabsorption also restores most of the water that was filtered from the blood. The water follows the nutrients and ions by osmosis through **aquaporins** (proteins that form water pores; see pp. 84–86). From the extracellular fluid, the nutrients, ions, and water pass into the capillaries surrounding the tubule and are returned to the bloodstream.

During tubular secretion, remaining wastes and excess ions move from the blood into the proximal and distal tubules (**Fig. 35-6** ❸). Wastes secreted into the tubule for excretion include excess K^+ and H^+ (an excess of H^+ makes the blood too acidic), small quantities of ammonia, many drugs (such as penicillin and ibuprofen), food additives, pesticides, and toxic substances, such as nicotine from cigarette smoke. Tubular secretion, which occurs primarily by active transport, takes place in both the proximal and distal tubules. When the filtrate leaves the distal tubule, it has become urine.

The Loop of Henle Creates an Extracellular Concentration Gradient in the Renal Medulla

The function of the loop of Henle, the portion of the tubule that extends deep into the renal medulla, is twofold. First, some water and salt is reabsorbed from the filtrate as it passes through the loop (see Fig. E35-2). The most important function of the loop of Henle, however, is to create a high salt concentration in the extracellular fluid within the medulla (**Fig. 35-6** ❹).

To understand why this is important, we must start with an important role of the human kidney: water regulation. The kidneys help to maintain appropriate water content in body tissues (as described later) by producing dilute, watery urine when fluid intake is high, and producing concentrated urine, containing far less water, when fluid intake is low. The urine that exits the distal tubule and enters the collecting duct is extremely dilute (about one-third the osmolarity of blood plasma). If the collecting duct remains impermeable to water, a great deal of excess water will be excreted in this dilute urine.

Water can be conserved, however, by allowing it to move out of the collecting duct by osmosis down its concentration gradient; that is, from a high water concentration (low solute concentration) inside the duct to a low water concentration (high solute concentration) in the extracellular fluid surrounding the duct. The more concentrated the extracellular fluid, the more water that can leave the urine as it moves through the collecting duct. The water leaving the collecting duct is immediately carried away by the nearby capillaries, and so does not dilute the high solute concentration within the extracellular fluid of the medulla.

The loop of Henle produces and maintains a high salt concentration gradient in the extracellular fluid in the medulla by actively transporting salt out of the filtrate (described in greater detail in "A Closer Look at the Nephron and Urine Formation" on pp. 684–685). The collecting duct passes through this gradient as it conducts urine from the distal tubule (in the renal cortex) into the renal pelvis (below the renal medulla; see Fig. 35-4, inset). The permeability of the collecting duct to water is controlled by feedback mechanisms that regulate water balance, as we will describe in the next section.

Case Study continued

Paying It Forward

When a person's kidneys fail, wastes and water accumulate in the blood, and imbalances in ion concentrations occur. Patients awaiting kidney transplants are kept alive using hemodialysis, a treatment in which wastes are filtered from the blood by a machine (see "Health Watch: When the Kidneys Collapse" on p. 681).

Although people may remain on hemodialysis for many years, the treatment is far from ideal. Whereas healthy kidneys work nonstop, hemodialysis treatments are performed only about three times a week. Consequently, patients must carefully monitor their diets between hemodialysis sessions, because excess Na^+ can increase blood pressure, and an excessive buildup of K^+ may cause sudden cardiac arrest. Hemodialysis patients must also limit their fluid intake to compensate for the reduced ability of their kidneys to produce urine. If they fail to do so, their blood pressure will increase, and their hearts may become strained and weakened.

35.6 HOW DO VERTEBRATE KIDNEYS HELP MAINTAIN HOMEOSTASIS?

The entire plasma content of your blood is filtered through nephrons about 60 times daily; as a result, the kidney is able to fine-tune the composition of the blood to help maintain homeostasis throughout the body.

The Kidneys Regulate the Osmolarity of the Blood

One important function of the kidney is to regulate the water content of the blood. Human kidneys filter out about half a cup of fluid from the blood each minute. If the kidneys were unable to return this water to the blood, the rate of filtration would require that we drink nearly 50 gallons of water a day to replace the urine we would produce!

Consequently, the urinary system needs to restore nearly all (about 99%) of the water that is initially filtered out of the glomeruli. As described earlier, water reabsorption back into the blood occurs passively by osmosis as the filtrate travels through the tubule and collecting duct. When the filtrate reaches the distal tubule, about 80% of its water has already been reabsorbed. After this point, the amount of reabsorption is precisely regulated, helping to maintain blood osmolarity within narrow limits.

Osmosis along the tubule and in the collecting duct depends on the presence of aquaporins. Although aquaporins are abundant and permanent components of the membranes of the proximal tubule and the descending portion of the loop of Henle, their numbers in the distal tubule and collecting duct are regulated by **antidiuretic hormone (ADH;** "diuretic" means "increasing urine production," so an "antidiuretic" decreases urine production). ADH is secreted by the posterior pituitary gland and carried in the bloodstream. It

A Closer Look At *The Nephron and Urine Formation*

The complex structure of the nephron is finely adapted to its function. **Figure E35-2** illustrates the processes occurring in each portion. The osmolarity (solute concentration) of the blood and extracellular fluid (measured in milliosmols) is shown along the left vertical axis. The normal osmolarity of blood and extracellular fluid is regulated at about 300 milliosmols. Within the renal medulla, the osmolarity of the extracellular fluid reaches about 1,200 milliosmols. The following descriptions refer to circled numbers in the illustration.

1. During filtration, water and small dissolved substances—both wastes and nutrients—are forced out of the glomerular capillaries and into Bowman's capsule, where they are then funneled into the proximal tubule. This process produces filtrate, which resembles blood plasma without its large proteins, and which is the same osmolarity as the blood and extracellular fluid (about 300 milliosmols).
2. In the proximal tubule, the process of tubular reabsorption moves water and most dissolved nutrients from the filtrate back through the walls of the tubule and into the extracellular fluid, where they are reabsorbed into the capillary blood. Dissolved nutrients include amino acids, glucose, and various ions, such as sodium (Na^+), chloride (Cl^-), potassium (K^+), calcium (Ca^{2+}), and bicarbonate (HCO_3^-). Some of these are actively transported from the filtrate and others diffuse out; water (H_2O) follows by osmosis. Because both H_2O and its solutes have been reabsorbed, the osmolarity of the filtrate in the tubule remains about the same as the blood (about 300 milliosmols). Some wastes, including ammonia (NH_3), some drugs, and excess hydrogen ions (H^+), are actively transported by tubular secretion from the capillary blood into the proximal tubule.
3. The loop of Henle produces and maintains a salt (NaCl) concentration gradient in the extracellular fluid that surrounds it, with the highest concentration at the bottom of the loop. The descending portion of the loop of Henle is very permeable to H_2O, but not to NaCl or other dissolved substances. As the filtrate passes through the descending portion, H_2O leaves the tubule by osmosis as the osmolarity of the extracellular fluid increases. Loss of H_2O increases the concentration of the filtrate within the tubule. At the bottom of the loop of Henle, the filtrate reaches the same osmolarity as the surrounding extracellular fluid (about 1,200 milliosmols).
4. The thin portion of the ascending loop of Henle is impermeable to H_2O but permeable to NaCl. As the filtrate flows up the loop through the decreasing salt concentration in the extracellular fluid, NaCl diffuses out because NaCl is more concentrated in the filtrate than in the surrounding extracellular fluid (due to H_2O loss from the filtrate in the descending loop). Because H_2O cannot follow the NaCl, this process contributes to the high salt concentration of the extracellular fluid. The filtrate gradually becomes less concentrated than the surrounding extracellular fluid because it has lost NaCl but retained H_2O.
5. The thick portion of the ascending loop of Henle is also impermeable to H_2O. Here, NaCl is actively pumped out of the filtrate, contributing to the high osmolarity of the extracellular fluid while drastically reducing the osmolarity of the filtrate.
6. Tubular secretion (which occurs in both the proximal and distal tubules) actively transports excess H^+, K^+, and some drugs from the blood and into the distal tubule for excretion. Tubular reabsorption also continues here, with active pumping of Ca^{2+} and NaCl out of the filtrate. By the time it reaches the distal tubule, so much NaCl has been pumped out that the filtrate is about one-third the osmolarity of the blood (about 100 milliosmols, compared to normal blood osmolarity of 300 milliosmols).

   The water permeability of the distal tubule is controlled by ADH. If no ADH is present (which occurs after excess fluid intake), the filtrate remains at 100 milliosmols. At baseline ADH levels, some H_2O will leave the distal tubule by osmosis because the osmolarity of the extracellular fluid surrounding the distal tubule is higher than that of the filtrate. When ADH is elevated (which occurs when the body is dehydrated), water moves more freely out of the distal tubule, allowing the filtrate to equilibrate with the extracellular fluid at about 300 milliosmols.
7. Once the filtrate leaves the distal tubule, it is called urine. The collecting duct carries the urine through the increasing extracellular concentration gradient created by the loop of Henle within the renal medulla. At baseline ADH levels, some water can move out of the collecting duct by osmosis as the concentration of the surrounding extracellular fluid increases. When ADH is elevated, as occurs when the body is dehydrated, water moves freely out of the collecting duct and is restored to the blood via nearby capillaries.
8. The lower portion of the collecting duct is permeable to urea, allowing some urea to diffuse out along its concentration gradient, which further increases the osmolarity of the surrounding extracellular fluid. If ADH is present, H_2O continues to move out as well. In the most extreme cases of dehydration, the osmolarity of urine in the collecting duct can reach equilibrium with the high osmolarity of the extracellular fluid; in humans, this is about 1,200 milliosmols. If ADH is absent, as will occur if the blood contains excess water that needs to be excreted, the collecting duct remains impermeable to water, and the urine remains dilute.

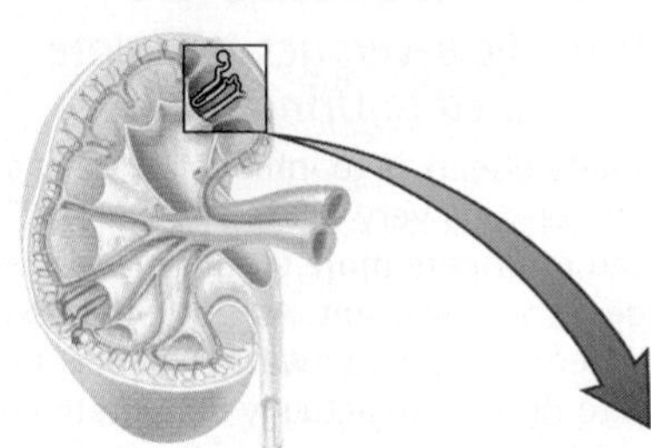

▼ **FIGURE E35-2 Details of urine formation** A single nephron, showing the movement of materials through different regions. The capillaries that surround the nephron (see Fig. 35-5) have been omitted for simplicity. The concentration of dissolved substances in the filtrate within the nephron is indicated by the intensity of yellow color; black arrows within the tubule indicate the direction of filtrate flow. Outside the nephron, darker shades of beige represent higher concentrations of salt and urea in the surrounding extracellular fluid. The dashed line marks the boundary between the renal cortex and the renal medulla. Asterisks indicate that water permeability in these regions is controlled by ADH.

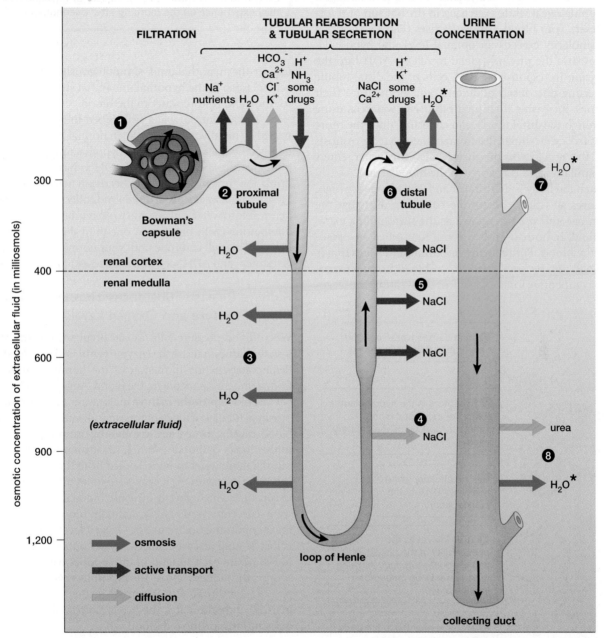

stimulates cells of the distal tubule and collecting duct to insert aquaporin proteins into their membranes; the abundance of aquaporin proteins determines the permeability of the membranes to water. This allows the distal tubule and collecting duct to fine-tune the water content of the urine.

Under normal conditions, some ADH is always present in the blood, so the distal tubule and collecting duct remain somewhat water permeable. Within the hypothalamus, receptors monitor blood osmolarity (the solute content of the blood), which increases when water is lost. For example, when your body becomes dehydrated, as might occur if you were dripping sweat while exercising in the hot sun or if lost in the desert, your blood osmolarity rises (**Fig. 35-7**). When blood osmolarity exceeds an optimal level, the hypothalamus stimulates the pituitary gland to release ADH into the bloodstream. In response to ADH, cells of the distal tubule and collecting duct insert additional aquaporins into their membranes, increasing their permeability to water. As urine flows through the distal tubule and collecting duct, the more concentrated extracellular fluid draws water out by osmosis through the aquaporins in their walls. The water enters nearby capillaries and is restored to the bloodstream.

When the body is dehydrated and ADH levels are high, water moves so readily out of the collecting duct that the urine can become as concentrated as the surrounding extracellular fluid; in humans, this is about four times the osmolarity of the blood. But reducing the amount of water that is excreted will not, by itself, restore blood to its normal osmolarity. To compensate for the water lost by sweating, the receptors in the hypothalamus simultaneously activate a "thirst center" (also in the hypothalamus) that stimulates the desire to drink and restore water to the blood.

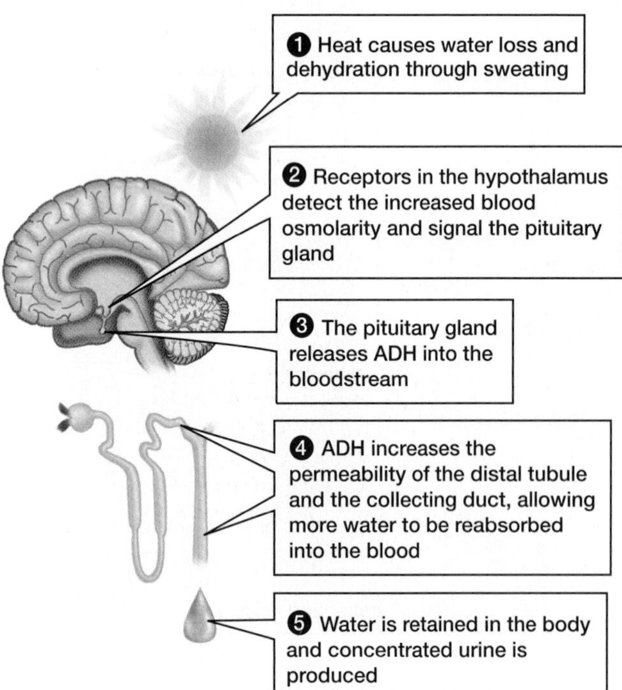

1 Heat causes water loss and dehydration through sweating
2 Receptors in the hypothalamus detect the increased blood osmolarity and signal the pituitary gland
3 The pituitary gland releases ADH into the bloodstream
4 ADH increases the permeability of the distal tubule and the collecting duct, allowing more water to be reabsorbed into the blood
5 Water is retained in the body and concentrated urine is produced

▲ **FIGURE 35-7 Dehydration stimulates ADH release and water retention**

QUESTION Describe the feedback process that would occur if you drank far more water than your body needed.

Have you ever wondered

Why Alcoholic Beverages Stimulate the Need to Urinate?

One of alcohol's many effects is to inhibit ADH release, and without ADH, urine remains very dilute and watery. This stimulates the need to urinate more than the water in the alcoholic beverages would warrant. As a result of excreting more water than needed to restore water balance, ironically, "having too much to drink" can actually dehydrate you. Dehydration contributes to the misery of the hangover you may experience on the morning after overindulging.

The amount of water reabsorbed into the blood is controlled by negative feedback. When normal blood osmolarity is detected by the receptors in the hypothalamus, they signal the pituitary to restore its release of ADH to baseline levels. If you drink too much water, ADH secretion will be temporarily blocked, making the distal tubule and collecting duct nearly impermeable to water, causing you to excrete large amounts of very dilute urine (with only about one-third the osmolarity of the blood). This will continue until your normal blood osmolarity is restored, and ADH secretion is increased to baseline levels.

Kidneys Release Substances That Help Regulate Blood Pressure and Oxygen Levels

When blood pressure falls, as can occur with excessive blood loss, the kidneys release the enzyme **renin** into the bloodstream. Renin catalyzes the formation of the hormone **angiotensin** from a protein circulating in the blood. Angiotensin helps combat low blood pressure in three major ways: (1) It stimulates the proximal tubules of the nephrons to reabsorb more Na^+ into the blood, causing more water to follow by osmosis (which reduces further losses of blood volume); (2) it stimulates ADH release, causing more water to be reclaimed from the distal tubule and collecting duct; and (3) it causes arterioles throughout the body to constrict, which directly increases blood pressure.

The kidneys also help maintain blood oxygen at levels that support the body's needs. Oxygen levels can fall if blood is lost, if lung disease reduces oxygen uptake, or at high altitudes, where each lungful of air supplies less oxygen. In response, the kidneys release the hormone **erythropoietin** (see p. 628). Erythropoietin is carried in the blood to the bone marrow and stimulates the bone marrow to produce more oxygen-transporting red blood cells.

Vertebrate Kidneys Are Adapted to Diverse Environments

Although we have described mammalian nephrons as having long loops of Henle that extend into the renal medulla, some mammalian nephrons are confined to the renal cortex—their

Case Study continued
Paying It Forward

Kidney failure almost always causes anemia, because the kidneys can no longer produce enough erythropoietin to stimulate adequate red blood cell production. Fortunately, human erythropoietin (manufactured by genetically engineered cells grown in culture) can be administered to patients who have anemia caused by kidney failure. However, a functioning kidney will better respond to the changing needs of the body, producing the proper amounts of this hormone at the proper times—a definite advantage for transplant recipients.

loops of Henle are very short and do not reach the renal medulla at all. The type of nephron that predominates in the kidneys of a given mammalian species depends on the availability of water in the natural habitat where the species evolved.

Mammalian Nephrons Are Adapted to the Availability of Water

Mammals adapted to dry climates generally have long loops of Henle. Longer loops of Henle allow a higher concentration of salt to be produced in the extracellular fluid of the kidney medulla, thus allowing more water to be reclaimed from the urine as it travels through the collecting ducts. The masters of urine concentration are desert rodents such as kangaroo rats, which can produce urine with 14 times the osmolarity of their blood (**Fig. 35-8**). Not surprisingly, kangaroo rats have only very long-looped nephrons. With their extraordinary ability to conserve water, kangaroo rats do not need to drink; they rely entirely on water contained in their food and from metabolic reactions that produce water.

In contrast, mammals adapted to habitats with an abundance of fresh water typically have short loops of Henle. For example, beavers, which live along streams, have only short-looped nephrons and can only concentrate their urine to about twice their blood osmolarity. Human kidneys have a mixture of long- and short-looped nephrons, and can concentrate urine to about four times the osmolarity of blood.

▲ **FIGURE 35-8 A well-adapted desert dweller** The desert kangaroo rat of the southwestern United States does not need to drink, partly because its long loops of Henle allow it to produce very concentrated urine.

Freshwater and Saltwater Environments Pose Special Challenges for Water Regulation

Animals that are constantly immersed in a solution that has either a lower (hypotonic) or a higher (hypertonic) osmolarity than their body fluids have evolved special mechanisms to maintain a homeostatic balance of water and salt within their bodies, a process called **osmoregulation.**

Freshwater fish, which live in a hypotonic environment, maintain plasma concentrations of dissolved substances (primarily salt) about four to six times the osmolarity of their freshwater surroundings. As these fish circulate water over their gills to exchange gases, some water continuously leaks into their bodies by osmosis, and salts diffuse out (**Fig. 35-9a**). Freshwater fish acquire salt from their food, and also through their gills, which use active transport to pump salt into their

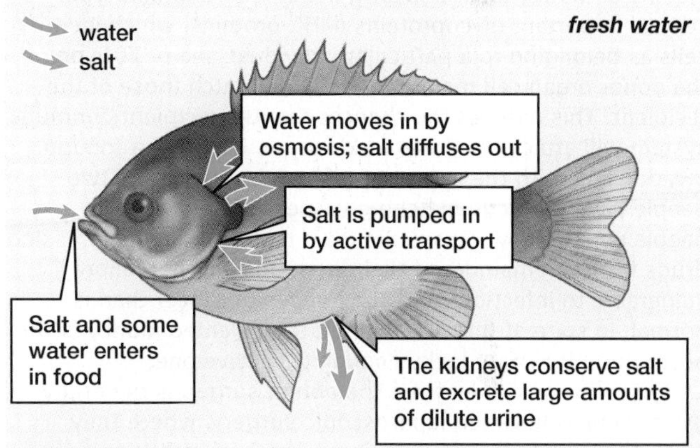

(a) Freshwater fish

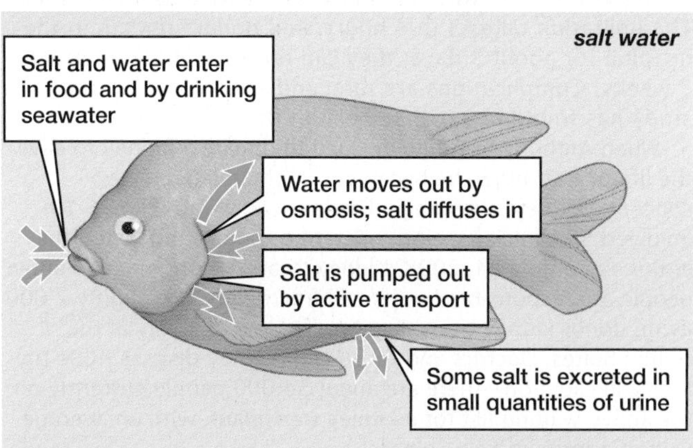

(b) Saltwater fish

▲ **FIGURE 35-9 Osmoregulation in fish** **(a)** Excretory systems of freshwater fish, such as this bluegill, must contend with large amounts of water entering their bodies by osmosis. **(b)** Excretory systems of saltwater fish, such as this Garibaldi, must conserve water, which constantly diffuses into the more salty surrounding seawater.

bodies from the surrounding water. To compensate for entry of excess water, freshwater fish never drink (although they take in some fresh water as they feed), and their kidneys retain salt and excrete large quantities of extremely dilute urine.

Saltwater fish live in a hypertonic environment; seawater has a solute concentration two to three times that of their body fluids. As a result, water is constantly leaving their tissues by osmosis, and salt is constantly diffusing in and being taken in with food. Most saltwater fish drink to restore their lost water, and the excess salt they take in is excreted by active transport through their gills (**Fig. 35-9b**).

Fish nephrons completely lack loops of Henle, and so fish cannot produce urine that is more concentrated than their blood. To conserve water, the kidneys of most saltwater fish excrete very small quantities of urine containing some salts that their gills cannot eliminate. The class of marine fishes that includes sharks and rays has evolved a different solution. These fishes store large quantities of urea in their tissues—enough to kill most other vertebrates. The stored urea gives their tissues approximately the same osmolarity as the surrounding seawater, so they avoid losing water by osmosis.

Case Study revisited

Paying It Forward

Since the 1950s, when living kidney donation was first recognized as a viable alternative to cadaver organ donors, family and friends have come forward to offer a kidney to victims of kidney failure. Ideally, in addition to blood type, several important glycoproteins (MHC proteins, which identify cells as belonging to a particular individual; see p. 701) on the donor organ cell membranes should match those of the recipient. This reduces the chances that the recipient's immune system will attack the donated kidney as if it were a foreign invader. But, with the exception of identical twins, no two people have perfectly matching tissues. This means that people with kidney transplants must take immune-suppressing drugs for the remainder of their lives, making them more vulnerable to infections and some types of cancer than is normal. In spite of this drawback, a transplanted kidney is the best option for those lucky enough to receive one.

To remove a kidney from the donor, surgeons generally use a technique called laparoscopic surgery, where they make three or four incisions about one-half inch long to insert surgical tools, including a tiny video camera. The kidney is removed through an incision about 2½ inches long. The operation takes 3 to 4 hours, and donors remain in the hospital for about 3 days; they can return to work in about 3 weeks. Complications are rare, and a large, long-term study has found no adverse health effects for the donor.

When Anthony DeGiulio decided to donate a kidney to save the life of a stranger, he became part of a relatively new concept: "kidney swapping." The four-way swap (**Fig. 35-10**) involved 50 clinicians, eight operating rooms, and lasted the entire day of July 24, 2008. Kidney swaps among unacquainted people are responsible for a small fraction of the roughly 7,000 living donor kidney transplants performed annually in the United States. Doctors and those with kidney disease hope the idea will catch on—there are about 54,000 people currently on the active waiting list for a kidney transplant, with an average wait of approximately 3 years.

When DeGiulio shared his plans to donate a kidney with his friends and family members, some of them told him he was "nuts." Anthony disagrees. "It's easily the greatest thing I've ever done in my life, and also the easiest decision I ever made," said the 32-year-old donor. "I sacrificed three days of my life, and this woman gets her life back."

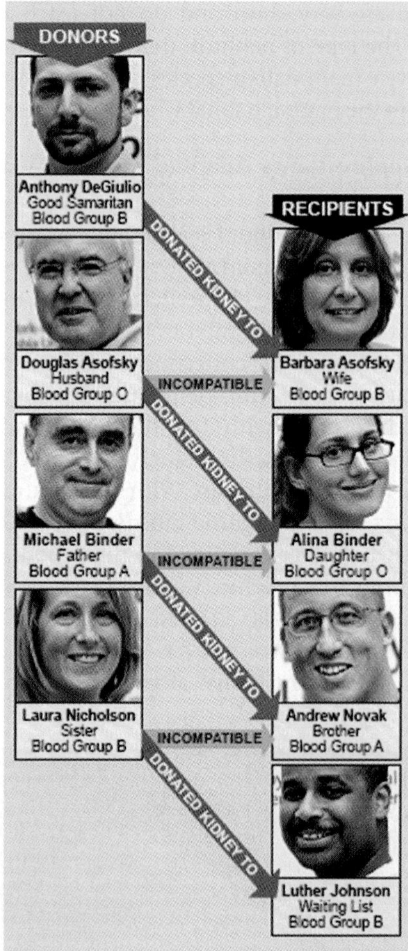

▲ **FIGURE 35-10 The chain of donations**

Consider This

Would you donate a kidney to a friend or family member whose kidneys were failing? Would you consider starting an "altruistic chain" by donating a kidney to a stranger?

CHAPTER REVIEW

Summary of Key Concepts

35.1 What Are the Basic Functions of Urinary Systems?

Urinary systems are essential to maintain homeostasis. All urinary (excretory) systems perform similar functions. First, the blood or extracellular fluid is filtered, removing water and small dissolved molecules; second, nutrients are reabsorbed from the filtrate; and third, any remaining water and dissolved wastes are excreted from the body.

35.2 What Are Some Examples of Invertebrate Excretory Systems?

The flatworm's simple excretory system consists of a network of tubules, called protonephridia, that branch through the body, collecting wastes and excess water. Flame cells create a current that forces the urine out of the body through excretory pores. Insects use Malpighian tubules that filter hemolymph within the hemocoel of their open circulatory systems. Malpighian tubules release concentrated urine into the intestine for elimination. Earthworms use tubules called nephridia to filter the extracellular fluid that fills the body cavity. Nutrients are reabsorbed into the extracellular fluid, and the urine is released through nephridiopores.

35.3 What Are the Functions of the Human Urinary System?

The urinary system eliminates foreign substances and cellular wastes—such as urea, which forms as a result of protein breakdown—while retaining vital nutrients in the blood. The urinary system plays a crucial role in homeostasis, regulating the water and ion content of the blood, as well as pH levels. The kidneys secrete substances that regulate blood pressure and blood oxygen levels.

35.4 What Are the Structures of the Human Urinary System?

The human urinary system consists of paired kidneys and ureters, a bladder, and a urethra. Kidneys produce urine, which is conducted by the ureters to the bladder, a storage organ. Distention of the muscular bladder wall triggers urination, during which urine passes out of the body through the urethra.

Each kidney consists of more than 1 million individual nephrons embedded in the outer renal cortex. Each nephron is composed of glomerular capillaries and a tubule that consists of Bowman's capsule, the proximal tubule, the loop of Henle, and the distal tubule. The loop of Henle often extends into an inner layer of the kidney, the renal cortex. Many nephrons empty urine into each collecting duct. Collecting ducts conduct urine through the medulla and empty into the renal pelvis, which funnels the urine into the ureter.

35.5 How Is Urine Formed and Concentrated?

Each nephron is served by an arteriole from the renal artery. The arteriole branches into the porous-walled capillary network of the glomerulus. There, water and small dissolved substances (both nutrients and wastes) are filtered from the blood, leaving proteins and blood cells behind. Past the glomerulus, filtered blood travels through capillaries that surround each tubule.

The filtrate is collected in Bowman's capsule and conducted through the tubule of the nephron. In the tubule (primarily the proximal tubule), water and other nutrients enter the surrounding capillaries and are restored to the blood through tubular reabsorption. Wastes remaining in the blood are pumped into the tubule by tubular secretion, which occurs in both the proximal and distal tubules. The function of the loop of Henle is primarily to generate a salt concentration gradient within the extracellular fluid of the renal medulla. After completing its passage through the tubule, urine enters the collecting duct, which passes through the concentration gradient in the renal medulla.

35.6 How Do Vertebrate Kidneys Help Maintain Homeostasis?

The water content of the blood is regulated by antidiuretic hormone (ADH), which is released into the blood by the posterior pituitary gland. ADH controls the density of aquaporins in the distal tubule and collecting duct membranes and, thus, regulates these membranes' permeability to water. Dehydration increases blood osmolarity and stimulates the release of ADH, which increases aquaporin density and, in turn, the amount of water absorption into the blood through the distal tubule and the collecting duct.

The kidneys help regulate blood pressure by secreting the enzyme renin when blood pressure falls. Renin catalyzes the formation of angiotensin from a blood protein. Angiotensin increases Na^+ reabsorption, stimulates ADH release, and constricts arterioles, elevating blood pressure. Erythropoietin, released from the kidneys when the oxygen content of the blood is low, stimulates the bone marrow to produce red blood cells.

Mammalian kidneys are adapted to the animal's environment. Animals that live where water is abundant tend to have short loops of Henle and produce dilute urine, whereas animals that live where water is scarce have very long loops of Henle and can produce very concentrated urine. Freshwater fish generate large quantities of dilute urine and actively transport salt in through their gill tissues. Saltwater fish drink seawater, actively transport salt out of their gill tissues, and produce very little urine.

Key Terms

ammonia *679*
angiotensin *686*
antidiuretic hormone (ADH) *683*
aquaporin *683*
bladder *680*
Bowman's capsule *680*
collecting duct *681*
distal tubule *680*
erythropoietin *686*
excretion *677*
extracellular fluid *677*
filtrate *682*
filtration *682*
glomerulus *680*
hemodialysis *681*
homeostasis *677*
kidney *678*
loop of Henle *680*
Malpighian tubule *678*
nephridium (plural, **nephridia**) *678*
nephron *680*
osmolarity *677*
osmoregulation *687*
protonephridium (plural, **protonephridia**) *677*
proximal tubule *680*
renal artery *680*
renal cortex *680*
renal medulla *680*
renal pelvis *680*
renal vein *680*
renin *686*
tubular reabsorption *682*
tubular secretion *682*
tubule *680*
urea *679*
ureter *680*
urethra **680**
urinary system *677*
urine *677*

Thinking Through the Concepts

Fill-in-the-Blank

1. Fill in the excretory organ of the following animals: earthworm: __________; insects: __________; flatworms: __________. Which of these organs most closely resembles the nephrons of vertebrates?__________
2. The vertebrate kidney consists of an outer layer called the __________, an underlying layer called the __________, and a central chamber, the __________. Urine is funneled from the central chamber into the __________, which leads to a storage organ, the __________, which empties through the __________.
3. The five parts of the nephron in the order that blood and its filtrate passes through them are __________, __________, __________, __________, __________.
4. Fill in with the appropriate term: Most tubular reabsorption occurs here: __________; final concentration of urine occurs here: __________; creates a concentration gradient in the extracellular fluid: __________; location of the concentration gradient within the kidney: renal __________.
5. Fill in the following substances: produced in the liver from ammonia and excreted in urine: __________; secreted into the tubule when blood pH is too low: __________; actively transported out of the ascending loop of Henle: __________; leaves the tubule by osmosis:__________.
6. The kidney helps maintain an internal balance called __________. To help regulate blood pressure, the kidney secretes the enzyme __________, which converts a substance circulating in the bloodstream into the hormone __________. To control oxygen levels, the kidneys secrete the hormone __________, which acts on the bone marrow to increase production of __________.
7. If blood osmolarity increases, receptors in the __________ detect this and signal the __________ gland to increase release of __________ hormone. This hormone acts on the walls of the __________ and the __________, causing them to insert more water pores called __________ into their membranes and, thus, cause the urine to become more __________.

Review Questions

1. Explain the two major functions of urinary systems.
2. Trace a urea molecule through the mammalian body, starting with an ammonia molecule in the bloodstream and ending outside the body.
3. What is the function of the loop of Henle? The collecting duct? Antidiuretic hormone?
4. Discuss the differences in function of the two major capillary beds in the kidneys: the glomerular capillaries and those surrounding the tubules.
5. Describe and compare the processes of filtration, tubular reabsorption, and tubular secretion.
6. Describe the role of the kidneys as organs of homeostasis.
7. Compare and contrast the excretory systems of humans, earthworms, and flatworms. In what general ways are they similar? How do they differ?

Applying the Concepts

1. Would you expect the restrooms in a restaurant with a bar to be used more than those in restaurants that didn't serve liquor? Explain.
2. Some "quick weight loss" diets are high in protein and very low in carbohydrates, and they require a person to drink more water than usual. Explain why the extra water is important.
3. Imagine yourself in a desert in the searing noonday sun. You've drained the last drops from your water bottle, perspiration drips from your face, and the dry air sucks water from your lungs with every breath. How does your urinary system respond?
4. In his poem "The Rime of the Ancient Mariner," Samuel Taylor Coleridge wrote: "Water, water, every where, nor any drop to drink." Seawater has more than four times the osmolarity of blood. Why can't a person keep from dying of thirst by drinking seawater?

MB *Go to www.masteringbiology.com for practice quizzes, activities, eText, videos, current events, and more.*

Defenses Against Disease

Chapter 36

Case Study

Flesh-Eating Bacteria

ON MAY 4, 1990, JIM HENSON, creator of the Muppets and the voice of Kermit the Frog, felt unusually tired and had a sore throat. By May 13, he felt considerably worse, and consulted a physician, who didn't find any signs of pneumonia or other serious illness. However, by May 15, he began coughing up blood. Only 3 hours later, at New York-Presbyterian Hospital, he couldn't breathe at all, and was placed on a respirator. Despite the best care and multiple antibiotics, Jim Henson died just 20 hours later.

Henson was killed by *Streptococcus pyogenes*. Chances are, you've been infected by this type of bacterium—it causes "strep throat." However, some strains of *S. pyogenes* are nastier than others, and some people are more susceptible than others. In these instances, *S. pyogenes* may cause one of two deadly streptococcal diseases. Henson died of streptococcal toxic shock syndrome when *S. pyogenes* invaded his lungs, causing massive inflammation and multiple organ failure.

In the other invasive streptococcal disease, *S. pyogenes* bacteria enter a cut or open sore, and rapidly destroy huge amounts of tissue, leading to the name "flesh-eating bacteria." Although this sounds overly dramatic, it isn't far off. In August 2007, Monica Jorge gave birth to a healthy baby girl by planned cesarean section. No one knows how the bacteria entered Ms. Jorge's body, but within hours, *S. pyogenes* had destroyed most of her abdominal wall. Before it was over, surgeons had to remove her uterus, ovaries, part of her large intestine, and large parts of both arms and both legs.

How can bacteria be so deadly, so fast? Why couldn't Henson's and Jorge's immune systems fight them off? Conversely, how does the human immune system usually eliminate the hordes of bacteria and other microbes that invade our bodies every day?

In this chapter, we will explore the intricate and usually highly effective defenses that the human body deploys to combat infection. We will see that the immune system recognizes bacteria and other microbes as "foreign," and mobilizes an army of cells and molecules to eliminate them from the body. As you learn about the immune response, consider a few questions: What is it about bacteria and other microbes that labels them as foreign? How does the immune system kill foreign invaders? How do you become immune to a disease? And how can microbes such as *S. pyogenes* fight back and evade the immune system?

▲Jim Henson, creator of the Muppets, was killed by a fast-acting, massive infection with *Streptococcus pyogenes*.

At a Glance

36.1 WHAT ARE THE MECHANISMS OF DEFENSE AGAINST DISEASE?

The environment teems with **microbes,** which include microscopic living organisms such as bacteria, protists, and many fungi, and also viruses, which are usually not considered to be alive. Most microbes live in water or in the soil. Most of the ones that live in animal bodies do no harm and may even be beneficial. For example, without cellulose-digesting bacteria in their digestive tracts, cattle would starve in the midst of a pasture full of grass. Some microbes, however, are **pathogens,** a term derived from Greek words meaning "to produce disease." Pathogens, of course, aren't trying to make us miserable. They're merely reproducing and seeking new hosts. In some cases, this process is mostly an inconvenience—when you have a cold, sneezing broadcasts viruses to everyone around you. Other diseases are far more dangerous. Cholera, for example, spreads by entering the water supply via diarrhea, and is often so devastating that the victim dies of dehydration.

Most microbial diseases, such as cholera, measles, plague, tetanus, and chicken pox, have been with us for hundreds or even thousands of years. In addition, new pathogens, or new, more deadly strains of familiar pathogens, may strike. These are called **emerging infectious diseases,** and are a high priority for research, prevention, and treatment.

Since the early 1980s, several viruses have emerged as serious threats to human health: HIV, Ebola virus, West Nile virus, SARS (severe acute respiratory syndrome), swine (or H1N1) flu, and bird (avian, or H5N1) flu. We are also threatened by deadly strains of bacteria. For example, the common intestinal bacterium *E. coli* is normally harmless—indeed beneficial, because it produces vitamin K in our large intestines. However, one strain, called O157:H7, which is usually spread by eating undercooked ground beef, can cause food poisoning, sometimes with fatal consequences. *Staphylococcus aureus* bacteria ("staph") occur frequently (and usually harmlessly) on the skin and in the nasal passages, but when they penetrate through the skin or mucous membranes, some strains cause fatal toxic shock syndrome or prolonged infections. Some strains are resistant to many commonly used antibiotics, making them difficult to treat. Some strains of *Mycobacterium tuberculosis* (which causes TB) are also resistant to most antibiotics. And, as discussed in the opening case study, the flesh-eating bacterium *Streptococcus pyogenes* can destroy much of a person's body in just a few hours.

Given the number and diversity of disease-causing organisms, you might wonder, "Why don't we get sick more often?" Over evolutionary time, animals and their pathogens have engaged in a constantly escalating battle. As animals evolve more sophisticated defense systems, pathogens in turn evolve more effective tactics for penetrating those defenses. This coevolutionary "arms race" has honed our defenses into a stunningly complex system that resists most attacks by microbes.

Vertebrate Animals Have Three Major Lines of Defense

Vertebrates (animals with backbones) have evolved three major forms of protection against disease (**Fig. 36-1**):

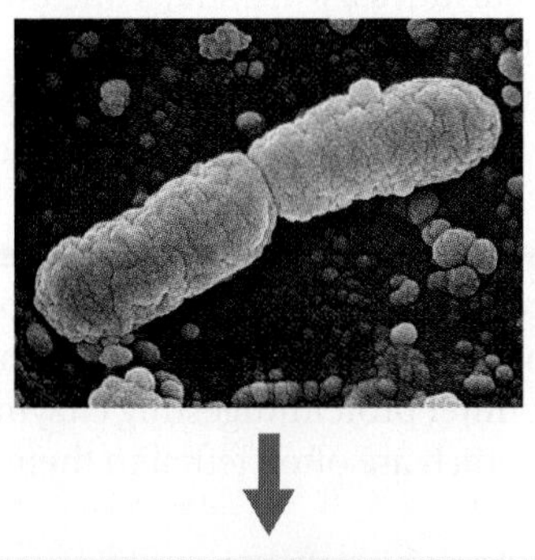

Nonspecific External Barriers
skin, mucous membranes

If these barriers are penetrated, the body responds with

Innate Immune Response
phagocytic and natural killer cells, inflammation, fever

If the innate immune response is insufficient, the body responds with

Adaptive Immune Response
cell-mediated immunity, humoral immunity

▲ **FIGURE 36-1 Levels of defense against infection**

- **Nonspecific external barriers** These barriers prevent most disease-causing microbes from entering the body. They are primarily anatomical structures, such as skin and cilia, and secretions, such as tears, saliva, and mucus. These barriers cover the external surfaces of the body and line the body cavities that are continuous with and therefore come in contact with the external environment, such as the surfaces of the respiratory, digestive, and urogenital tracts.
- **Nonspecific internal defenses** If the external barriers are breached, a variety of nonspecific internal defenses, collectively called the **innate immune response,** swing into action. Some white blood cells engulf foreign particles or destroy infected cells. Chemicals released by damaged body cells and proteins released by white blood cells trigger inflammation and fever (in animals that regulate their body temperature). Both the nonspecific external barriers and the nonspecific internal defenses operate regardless of the exact nature of the invader, repelling, killing, or neutralizing the threat.
- **Specific internal defenses** The final line of defense is the **adaptive immune response,** in which immune cells selectively destroy the specific invading toxin or microbe and then "remember" the invader, allowing a faster response if it reappears in the future. The major similarities and differences between innate immunity and adaptive immunity are described in **Table 36-1.**

The internal defenses involve a large number of different cell types, as briefly described in **Table 36-2.** Most of these cells and their roles in defending the body will be unfamiliar to you now, but the table will be a useful guide as you read the rest of the chapter.

Invertebrate Animals Possess the First Two Lines of Defense

Invertebrates possess the two nonspecific lines of defense just described. The various invertebrates exhibit an enormous range of physical barriers, ranging from external skeletons to

Table 36-1 Innate and Adaptive Immune Responses to Invasion

Characteristic	Innate Immune Response	Adaptive Immune Response
Specificity of response to invasion	Nonspecific	Specific for particular microbes
Speed of response to the first invasion by a given microbe	Immediate	Delay of several days to 2 weeks
Magnitude of the response to a second invasion by the same microbe	Identical to the first response	Enormously enhanced response; usually no disease symptoms occur (the animal has become immune to that specific microbe)
Distribution in the animal kingdom	Invertebrates and vertebrates	Vertebrates only

Table 36-2 The Body's Cellular Arsenal Against Disease

Type of Cell	Function
Neutrophils	White blood cells that engulf invading microbes
Dendritic cells	White blood cells that engulf invading microbes and present antigens to lymphocytes
Macrophages	White blood cells that engulf invading microbes and present antigens to lymphocytes
Natural killer cells	White blood cells that destroy infected or cancerous cells
Mast cells	Connective tissue cells that release histamine; important in the inflammatory response
B cells	White blood cells that produce antibodies
Memory B cells	Offspring of B cells that provide future immunity against invasion by the same antigen
Plasma cells	Offspring of B cells that secrete antibodies into the bloodstream
T cells	White blood cells that regulate the immune response or kill infected cells or cancerous cells
Cytotoxic T cells	T cells that destroy infected body cells or cancerous cells
Helper T cells	T cells that stimulate immune responses by both B cells and cytotoxic T cells
Memory T cells	Offspring of cytotoxic or helper T cells that provide future immunity against invasion by the same antigen
Regulatory T cells	T cells that suppress immune attack against the body's own cells and molecules; important in preventing autoimmune diseases

slimy secretions. Internally, invertebrates have white blood cells that attack pathogens and secrete proteins to neutralize these invaders or the toxins they release. Interestingly, the amino acid sequences of some defensive proteins, such as lysozyme (see below), are similar in vertebrates (including humans, birds, and fish) and invertebrates (including mollusks, worms, and insects). This suggests that the genes coding for these molecules have been passed down for hundreds of millions of years, from a common ancestor to most of today's diverse animal species.

36.2 HOW DO NONSPECIFIC DEFENSES FUNCTION?

The ideal defenses are barriers that prevent invaders from entering the body in the first place. However, if these barriers are breached, the body has several nonspecific methods of killing any of a wide variety of invading microbes.

The Skin and Mucous Membranes Form External Barriers to Invasion

In animal bodies, the first line of defense consists of the two surfaces with direct exposure to the environment: the skin and the mucous membranes of the digestive, respiratory, and urogenital tracts.

The Skin and Its Secretions Block Entry and Provide an Inhospitable Environment for Microbial Growth

The outer surface of human skin consists of dry, dead cells filled with tough proteins similar to those in hair and nails. Many microbes that land on this outer surface cannot obtain the water and nutrients they need to survive. The skin is also protected by secretions from sweat glands and sebaceous (oil) glands. These secretions contain natural antibiotics, such as lactic acid, that inhibit the growth of many bacteria and fungi. Despite these defenses, the skin is a veritable zoo of microbes, including both exotic types that are specialized for growth in the relative desert of the skin and the more familiar, and more dangerous, *Streptococcus* and *Staphylococcus* bacteria. However, with reasonable personal hygiene, their population is usually fairly low, and they seldom penetrate through the unbroken skin into the tissues below.

Antimicrobial Secretions, Mucus, and Ciliary Action Defend the Mucous Membranes Against Microbes

The warm, moist mucous membranes are far more hospitable to microbes than is the dry, oily skin, but they nonetheless deploy effective defenses. First, as their name implies, mucous membranes secrete mucus, which traps microbes that enter the body through the nose or mouth (**Fig. 36-2**). Further, mucus contains antibacterial proteins, including lysozyme, which kills bacteria by digesting their cell walls, and defensin, which makes holes in bacterial plasma membranes. Finally, cilia on the membranes sweep up the mucus, microbes and all, until it is either coughed or sneezed out of the body, or is swallowed.

If microbes are swallowed, they enter the stomach, where they encounter protein-digesting enzymes and extreme acidity, both of which are often lethal to them. Further along,

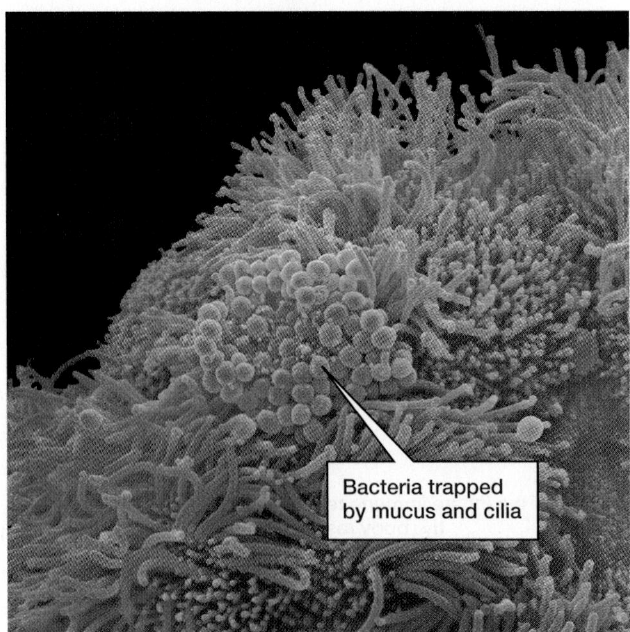

▲ **FIGURE 36-2 The protective function of mucus** Mucus traps microbes and debris in the respiratory tract. In this colored micrograph, bacteria are caught in mucus atop the cilia. The cilia then sweep both the mucus and bacteria out of the body.

the intestines contain bacteria that are normally harmless to people but that secrete substances that destroy invading bacteria or fungi. In the urinary tract, the slight acidity of urine inhibits bacterial growth. In females, acidic secretions and mucus help protect the vagina.

Finally, tears, urination, diarrhea, and vomiting all help to expel invaders. With diarrhea and vomiting, this is a double-edged sword: Although these unpleasant events leave fewer invaders inside the body, they also deplete the body's water supply—water loss through diarrhea kills hundreds of thousands of children each year in less-developed countries—and help to spread pathogens from one person to another.

Despite these defenses, many disease-causing microbes enter the body through the mucous membranes or through cuts in the skin.

The Innate Immune Response Combats Invading Microbes

Invading microbes that penetrate the skin or mucous membranes encounter the array of internal defenses collectively called innate immunity. Innate immune responses are nonspecific—that is, they attack many different types of microbes rather than targeting particular invaders. Innate immune responses fall into three major categories:

- **White blood cells** The body has a standing army of white blood cells, or **leukocytes,** many of which are specialized to attack and destroy invading cells or the body's own cells if they have been infected by viruses.
- **The inflammatory response** A wound, with its combination of tissue damage and often a massive invasion of microbes, provokes an inflammatory response, which recruits leukocytes to the site of injury and walls off the injured area, isolating the infected tissue from the rest of the body.
- **Fever** If a population of microbes succeeds in establishing a major infection, the body may produce a fever, which both slows down microbial reproduction and enhances the body's own fighting abilities.

Phagocytic Leukocytes and Natural Killer Cells Destroy Invading Microbes

The body contains several types of leukocytes, collectively known as **phagocytes,** which ingest foreign invaders and cellular debris by phagocytosis. Three important types of phagocytes are **macrophages** (literally, "big eaters"), **neutrophils,** and **dendritic cells.** These cells travel within the bloodstream, ooze through capillary walls, and patrol the body's tissues (**Fig. 36-3a**), where they consume bacteria and foreign substances that have penetrated the skin or mucous membranes (**Fig. 36-3b**). As described later, dendritic cells and macrophages also play a critical role in the adaptive immune response.

Natural killer cells are another type of leukocyte, which strike primarily at the body's own cells that have become cancerous or have been invaded by viruses. The surfaces of normal body cells display proteins of the major histocompatibility complex (MHC; see section 36.4). MHC proteins identify the body's cells as "self." Natural killer cells patrol the body, killing any "non-self" cells that they encounter, while sparing self cells. Both virus-infected cells and cancerous cells often lack certain MHC proteins, so they look like non-self to natural killer cells. Natural killer cells are not phagocytic; instead, they release proteins that bore holes in the membranes of infected or cancerous cells and then secrete enzymes through the holes. Shot full of holes and chewed on by enzymes, the attacked cells soon die.

Case Study continued
Flesh-Eating Bacteria

The intact skin is an effective barrier against flesh-eating bacteria. However, if the skin is breached, for example, by surgery or a wound, *S. pyogenes* may enter the body. Each *S. pyogenes* bacterium is surrounded by a capsule made of hyaluronic acid, which is also a major component of mammalian connective tissue. Hence, phagocytic cells often do not recognize the capsule as a foreign substance and, consequently, do not attack *S. pyogenes*.

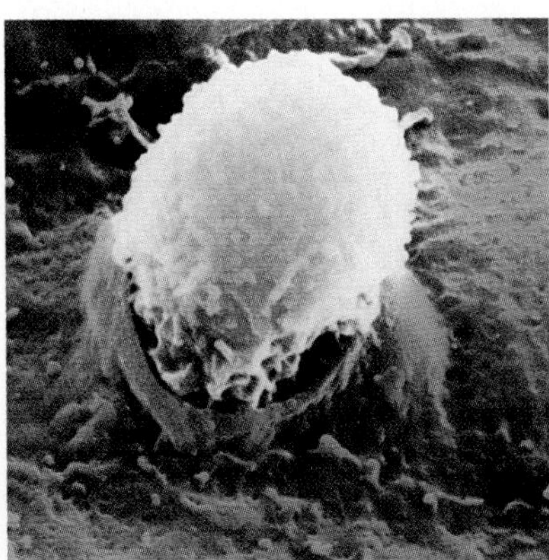

(a) A macrophage leaves a capillary and enters a wound

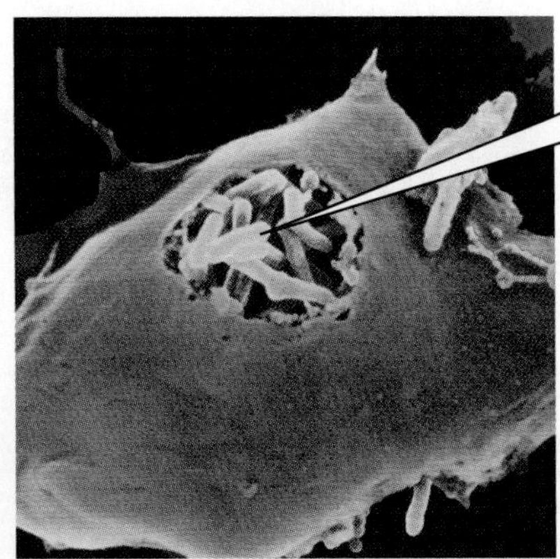

(b) A macrophage stuffed with bacteria that it has ingested

▲ FIGURE 36-3 The attack of the macrophages

How does the body benefit from killing virus-infected cells? Viruses enter a cell and force the cell's own metabolism to manufacture more viruses (see Fig. 13-2 and "A Closer Look at Virus Replication" on p. 366). Therefore, viral infections can be stopped by killing the infected cells before the viruses have had enough time to multiply and spread.

The Inflammatory Response Attracts Phagocytes to Injured or Infected Tissue

Whether you catch the flu or get a splinter in your finger, you will experience inflammation. The **inflammatory response** causes injured tissues to become warm, red, swollen, and painful (inflame literally means "to set on fire"). This defense mechanism has several functions: It attracts phagocytes to the area, promotes blood clotting, and causes pain, which stimulates protective behaviors.

The inflammatory response begins when damaged cells release chemicals that cause certain cells in the connective tissue, called **mast cells,** to release **histamine** and other substances (**Fig. 36-4**). Histamine relaxes the smooth muscle surrounding arterioles, thereby increasing blood flow, and makes capillary walls leaky. Extra blood flowing through leaky capillaries drives fluid from the blood into the wounded area. Thus, histamine accounts for the redness, warmth, and swelling of the inflammatory response. Other chemicals released by wounded cells and mast cells, and some substances produced by the microbes themselves, attract macrophages, neutrophils, and dendritic cells. These cells squeeze out through the leaky capillary walls and ingest bacteria, dirt, and cellular debris caused by the injury (see Figs. 36-3 and 36-4). In some cases, pus, a thick whitish mixture of dead bacteria, tissue debris, and living and dead white blood cells, may accumulate. (Not surprisingly, flesh-eating bacteria cause large amounts of pus to form in a wound; in fact, the species name *pyogenes* means "pus-producer.")

Meanwhile, other chemicals released by injured cells initiate blood clotting (see p. 629). Clots plug up damaged blood vessels, reducing blood loss and preventing microbes from entering the bloodstream. At the same time, clots seal off the wound from the outside world, limiting the entry of more microbes. Finally, pain is caused by swelling and some of the chemicals released by the injured tissue.

Fever Combats Large-Scale Infections

If invaders breach these defenses and mount a full-blown infection, they may trigger a **fever.** Although extremely high fevers can be dangerous, and even moderate ones are unpleasant, fever is an important part of the body's defense against infection. The onset of fever is controlled by the hypothalamus, the part of the brain housing the temperature-sensing nerve cells that are the body's thermostat. In humans, the thermostat is set at about 97° to 99°F (about 36° to 37°C). When they respond to an infection, certain macrophages release a protein called endogenous pyrogen (literally, "self-produced fire-maker"). Pyrogen travels in the bloodstream to the hypothalamus and raises the thermostat's set point. The body responds with increased fat metabolism (which generates more heat), constriction of blood vessels supplying the skin (which reduces heat loss through the skin), and behaviors such

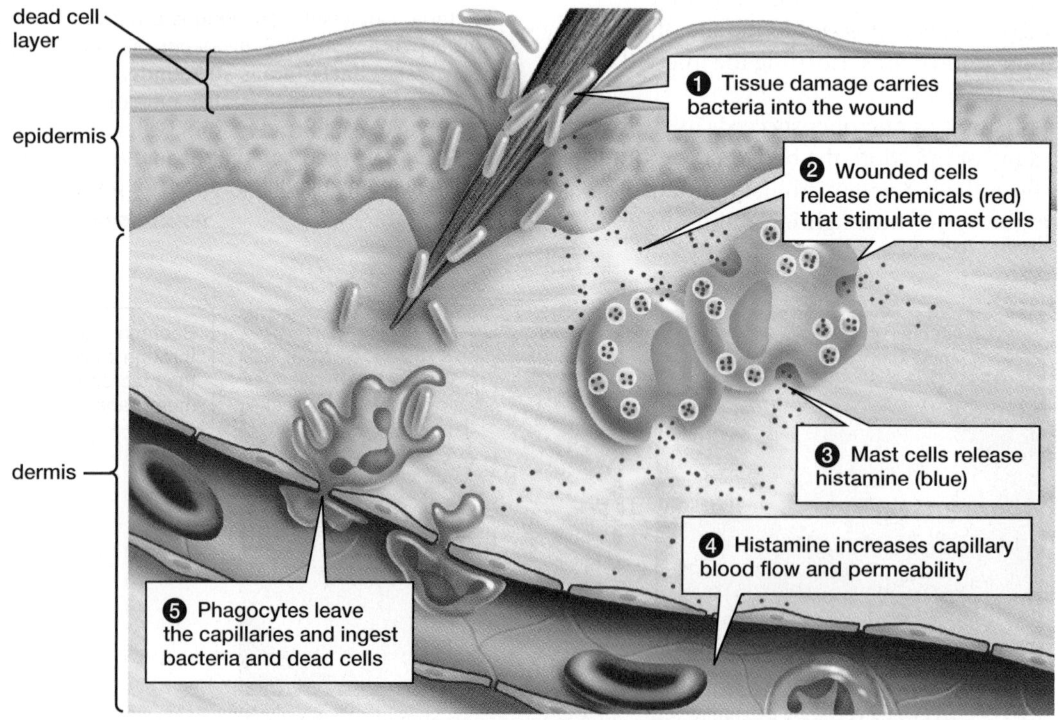
dead cell layer
epidermis
dermis
1 Tissue damage carries bacteria into the wound
2 Wounded cells release chemicals (red) that stimulate mast cells
3 Mast cells release histamine (blue)
4 Histamine increases capillary blood flow and permeability
5 Phagocytes leave the capillaries and ingest bacteria and dead cells

▲ **FIGURE 36-4 The inflammatory response**

QUESTION How does the inflammatory response help white blood cells to leave the capillaries?

as shivering and bundling under blankets. Pyrogens also cause other cells to reduce the concentration of iron in the blood.

An elevated body temperature increases the activity of phagocytic white blood cells while simultaneously slowing bacterial reproduction. The iron deficiency that accompanies a fever also hampers bacterial multiplication. Meanwhile, cells of the adaptive immune system multiply more rapidly, hastening the onset of an effective adaptive immune response. Fever also stimulates cells infected by viruses to produce a protein called interferon, which travels to other cells and increases their resistance to viral attack. Interferon also stimulates the natural killer cells that destroy virus-infected body cells.

Fever is an effective defense mechanism, as demonstrated by experiments in which volunteers were infected with cold viruses and then treated either with aspirin (which reduces fever) or with a placebo (an inactive substance that looks like the real drug and serves as a control). The people given aspirin had far more viruses in their noses, and they sneezed and coughed out more viruses than did those given the placebo. This finding supports the hypothesis that fever helps to control viral infections. Because very high fevers may be dangerous, whether you should take aspirin, acetaminophen, or ibuprofen when you have a fever is a decision that you should make in consultation with your physician.

36.3 WHAT ARE THE KEY COMPONENTS OF THE ADAPTIVE IMMUNE SYSTEM?

External barriers, phagocytic cells, natural killer cells, the inflammatory response, and fever are all nonspecific defenses. Their role is to block or overcome *any* microbes that invade the body. Unfortunately, the nonspecific defenses are not impregnable. Some microbes, such as *S. pyogenes*, have evolved adaptations that help to penetrate these defenses. When they are breached, the body mounts a highly specific and coordinated adaptive immune response directed against the *particular* organism that has successfully colonized the body.

The essential features of the adaptive immune response were recognized more than 2,000 years ago by the Greek historian Thucydides. He observed that, occasionally, a person would contract a disease, recover, and never catch that particular disease again—the person had become immune. With rare exceptions, immunity to one disease confers no protection against other diseases. Thus, the adaptive immune response attacks one specific type of microbe, overcomes it, and provides future protection against that microbe but no others.

The Adaptive Immune System Consists of Cells and Molecules Dispersed Throughout the Body

As befits a system that must patrol the entire body for invading microbes, the **adaptive immune system** (often called simply the **immune system**) is distributed throughout the body, with concentrations of cells in certain locations (**Fig. 36-5**). The immune system consists of three major components:

- **Immune cells** The adaptive immune response is produced by interactions among several types of white blood cells, including macrophages, dendritic cells, and lymphocytes. Some, such as macrophages and dendritic cells, play a role in both the innate and adaptive immune responses (see Table 36-2). The key cellular players in the adaptive immune response are the **lymphocytes.** There are two main types of lymphocytes: **B cells** and **T cells.** Both B cells and T cells arise from stem cells in the bone marrow. Some of these stem cells produce lymphocyte precursor cells. Some precursor cells complete their development in the bone marrow, becoming B (for *bone*) cells. Other precursor cells migrate from the marrow,

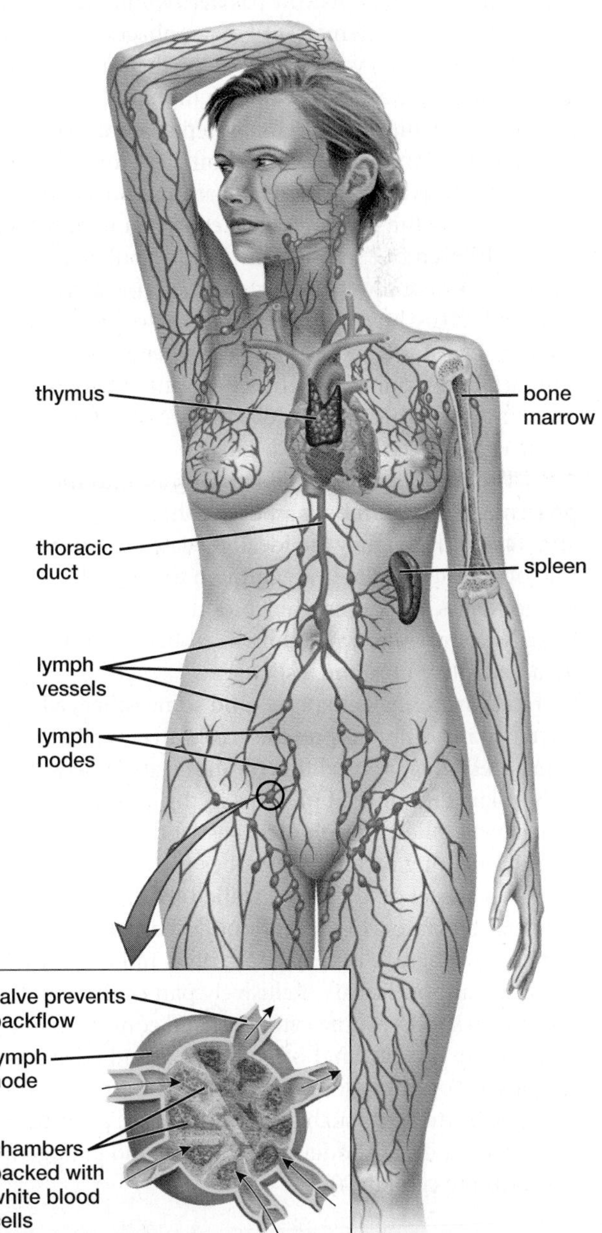

▲ **FIGURE 36-5 The lymphatic system contains much of the immune system** The cells of the immune system are formed in the bone marrow and mature either there or in the thymus. Most mature immune cells reside in the lymph nodes or spleen. As lymph travels through the lymph nodes, most microbial invaders are attacked and killed.

through the circulatory system, to the thymus (see below), where they develop into T (for *thymus*) cells.

- **Tissues and organs** The cells of the immune system are produced and reside in a variety of locations in the body, including the vessels of the lymphatic system, the lymph nodes, the thymus, the spleen, and patches of specialized connective tissue, such as the tonsils. Although the circulatory system is not usually considered part of the immune system, it carries immune cells and proteins throughout the body.

  The human body contains approximately 500 **lymph nodes** scattered along the lymph vessels (see Fig. 36-5). Lymph flows through narrow passages within the lymph nodes, which contain masses of macrophages and lymphocytes. When you have a disease that causes "swollen glands," these are actually swollen lymph nodes that have accumulated white blood cells, bacteria, debris from dead cells, and fluid in the process of combating the infection.

  The **thymus** is located beneath the breastbone slightly above the heart. It is quite large in infants and young children but starts to shrink after puberty. The thymus is essential for the development of some immune cells. The **spleen** is a fist-sized organ located on the left side of the abdominal cavity, between the stomach and diaphragm. The spleen filters blood, exposing it to white blood cells that destroy foreign particles and aged red blood cells.

  The **tonsils** are located in a ring around the pharynx (the uppermost part of the throat; see Figs. 33-7 and 34-12). Here, they are ideally located to sample microbes entering the body through the mouth. Macrophages and other white blood cells in the tonsils directly destroy many invading microbes and often start up an adaptive immune response.

- **Secreted proteins** Leukocytes and some other cells secrete many different proteins, collectively called **cytokines,** that are used for communication between cells. A large number of proteins in the blood, collectively called **complement,** assist the immune system in killing invading microbes (see section 36.5). Finally, a subset of leukocytes, called B cells, produces antibodies. As we will see, antibodies help the immune system to recognize invading microbes and destroy them. Although antibodies are exclusively part of the adaptive immune response, some cytokines and complement proteins are involved in both the innate and adaptive responses. For example, some cytokines stimulate nonspecific defenses such as inflammation, activation of natural killer cells, and fever (interferon and endogenous pyrogens are cytokines).

All adaptive immune responses include the same three steps. First, lymphocytes recognize an invading microbe and distinguish invader from self; second, they launch an attack; and third, they retain a memory of the invader that allows them to ward off future infections by the same type of microbe.

36.4 HOW DOES THE ADAPTIVE IMMUNE SYSTEM RECOGNIZE INVADERS?

To understand how the immune system recognizes invaders and initiates a response, we must answer three related questions: How do lymphocytes recognize foreign cells and molecules? How can lymphocytes produce specific responses to so many different types of cells and molecules? How do they avoid mistaking the body's own cells and molecules for invaders?

The Adaptive Immune System Recognizes Invaders' Complex Molecules

From the perspective of the immune system, bacteria and humans differ from one another because each contains specific, complex molecules that the other doesn't. Your immune cells distinguish your body's cells and molecules from those of all the other organisms on Earth, including other people, because some of your complex molecules are unique to you (unless you have an identical twin), and some of the complex molecules of all other organisms are unique to them. These large, complex molecules—usually proteins, polysaccharides, or glycoproteins—are called **antigens,** because they are "*anti*body *gen*erating" molecules; that is, they can provoke an immune response, including the production of antibodies.

Antigens are often located on the surfaces of invading microbes. Many viral antigens become incorporated into the plasma membranes of infected body cells. Viral or bacterial antigens are also "displayed" on the plasma membranes of dendritic cells and macrophages that engulf them. Other antigens, such as toxins released by bacteria, may be dissolved in the blood plasma, lymph, or other extracellular fluids.

Antibodies and T-Cell Receptors Recognize and Bind to Foreign Antigens

Lymphocytes generate two types of proteins that recognize, bind, and help to destroy specific antigens. **Antibody** proteins are produced by B cells and their offspring; **T-cell receptor** proteins are produced by T cells.

Antibodies Recognize and Help to Destroy Invaders

Antibodies are Y-shaped proteins composed of two pairs of peptide chains: one pair of identical large (heavy) chains and one pair of identical small (light) chains (**Fig. 36-6**). Both heavy and light chains consist of a **constant region,** which is similar in all antibodies of the same type, and a **variable region,** which differs among individual antibodies. The combination of light and heavy chains results in antibodies with two functional parts: the "arms" of the Y and the "stem" of the Y. The variable regions (located at the arm tips) form sites that bind antigens (see Fig. 36-6). Each binding site has a particular size, shape, and electrical charge, so only certain molecules can fit in and bind to the antibody. The sites are so specific that each antibody can bind only a few, very similar, types of antigen molecules.

Antibodies may function as receptors, binding to specific antigens and eliciting a response to them, or as effectors,

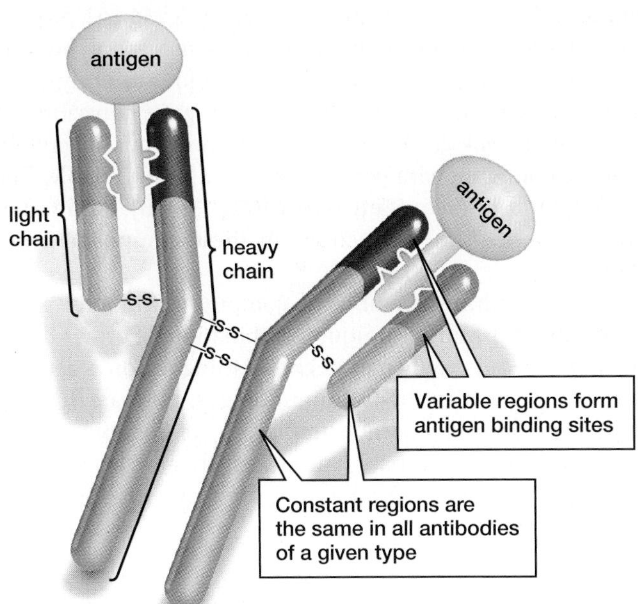

▲ **FIGURE 36-6 Antibody structure** Antibodies are Y-shaped proteins composed of two pairs of peptide chains (light chains and heavy chains). Constant regions on both chains form the stem of the Y. The variable regions on the two chains form a specific binding site at the end of each arm of the Y. Different antibodies have different variable regions, forming unique binding sites.

QUESTION Why do antibody molecules have both constant and variable regions?

helping to destroy the cells or molecules that bear the antigens. As a receptor, the stem of an antibody anchors the antibody in the plasma membrane of the B cell that produced it, while its two arms stick out from the B cell, sampling the blood and lymph for antigen molecules (**Fig. 36-7a**). When an arm of the antibody encounters an antigen with a compatible chemical structure, it binds to it. This binding triggers a response in the B cell, as we will describe shortly. As effectors, many antibodies are secreted into the bloodstream. There, they neutralize poisonous antigens (such as bacterial toxins), destroy microbes that bear antigens, or attract macrophages that engulf the antigen-bearing microbes (**Fig. 36-7b**). These functions of antibodies are described in section 36.5.

T-Cell Receptors Recognize Invaders and Help to Trigger an Immune Response

T-cell receptors are found only on the surfaces of T cells. Like antibodies, they consist of peptide chains that form highly specific binding sites for antigens. Unlike antibodies, T-cell receptors are never released into the bloodstream, and they do not directly contribute to the destruction of invading microbes or toxic molecules. As we describe later, a T-cell receptor triggers a response in its T cell when the receptor binds an antigen on the surface of a cancerous or infected cell, or on the surface of a dendritic cell or macrophage that has ingested an invading microbe.

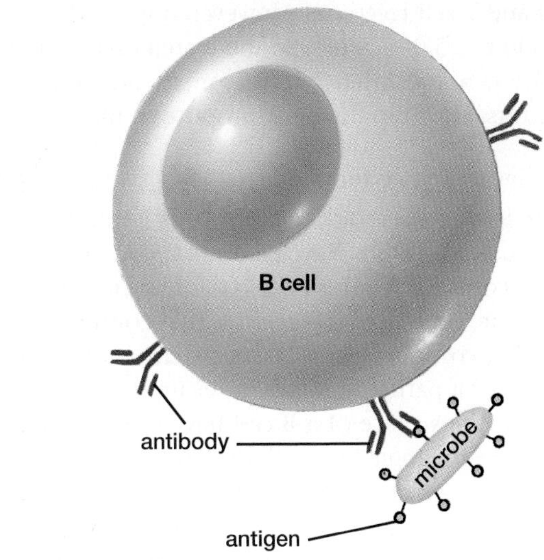

(a) Antibody receptor function

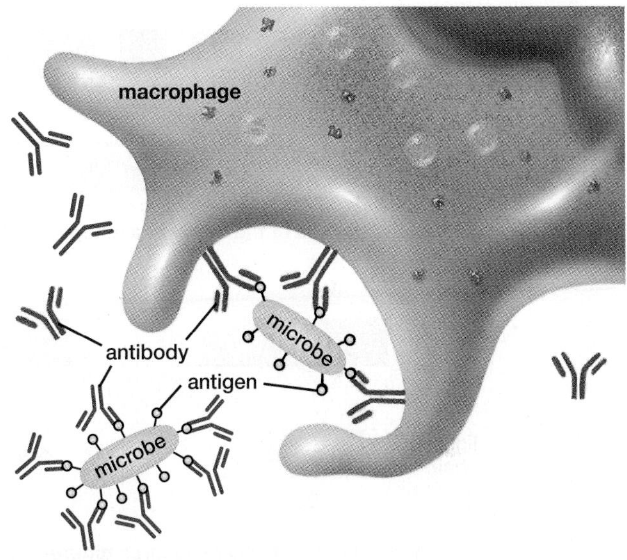

(b) Antibody effector function

▲ **FIGURE 36-7 Antibodies can serve as receptors or effectors (a)** As receptors, antibodies attached to the surface of a B cell recognize and bind to foreign antigens, triggering responses in the cell. **(b)** As effectors, circulating antibodies bind to foreign antigens on a microbe, helping the immune system to identify and destroy it. These antibodies may promote phagocytosis of the microbe by a macrophage (see also Fig. 36-11).

The Immune System Can Recognize Millions of Different Antigens

During your lifetime, your body will be challenged by a multitude of invaders. Your classmates may sneeze cold and flu viruses into the air you breathe. Your food may contain bacteria or molds, or the toxins they have released. A mosquito carrying West Nile virus may bite you. Fortunately, the adaptive immune system recognizes and responds to virtually all of the millions of potentially harmful antigens that you might encounter, because B and T cells produce millions of different antibodies and T-cell receptors. However, humans have only about 20,000 to 25,000 genes, so there aren't enough genes in the entire genome to encode this many antibodies and T-cell receptors. How can the body produce so many?

Antibody Genes Are Assembled from Segments of DNA

The answer is that there are no genes for entire antibody molecules. Instead, B cells have genes that code for *parts* of antibodies—constant regions (C), variable regions (V), and "joining" (J) or "diversity" (D) regions that connect the two (**Fig. 36-8**). The constant region in each chain is the same for any antibody of a particular type; for example, all antibodies located on the surface of a B cell have the same type of constant region for their heavy chain. Humans have about 200 genes for the variable region of the heavy chain, and 50 and 6 genes, respectively, for the diversity and joining regions. There are about 150 genes for the variable region of the light chain, and 5 genes for the joining region (**Fig. 36-8a**). As each B cell develops, it randomly cuts out and discards all but one gene of each type, and assembles two unique antibody genes from the genes it keeps—a heavy-chain gene consisting of one variable, one diversity, one joining, and one constant region; and a light-chain gene consisting of one variable, one joining, and one constant region (**Fig. 36-8b**). Antibodies are produced from instructions in these composite genes (**Fig. 36-8c**).

The random assembly of composite antibody genes yields about 3 million unique combinations. Further diversity arises because only part of each joining region is actually used in any given antibody. Immunologists estimate that perhaps 15 to 20 *billion* unique antibodies are possible. The result: Each B cell probably produces an antibody that is different from the one produced by every other B cell (except its own daughter cells).

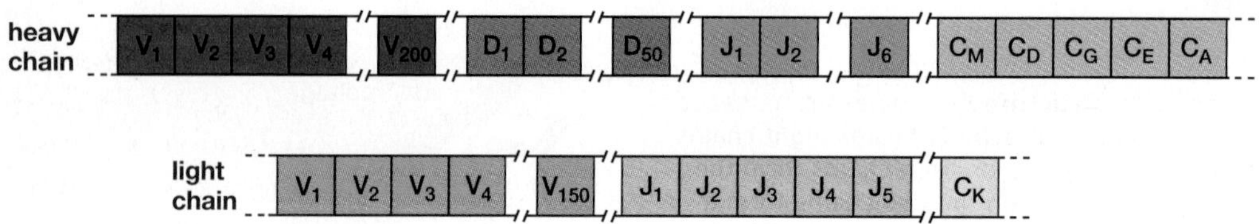

(a) Genes for parts of the heavy chain (top) and light chain (bottom) of antibodies

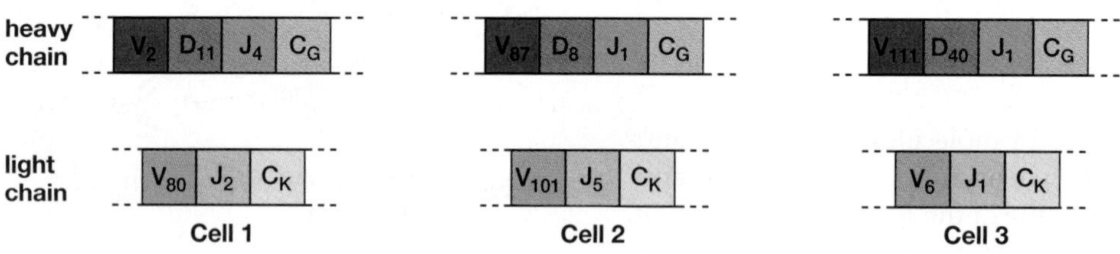

(b) Complete antibody genes in three different B cells

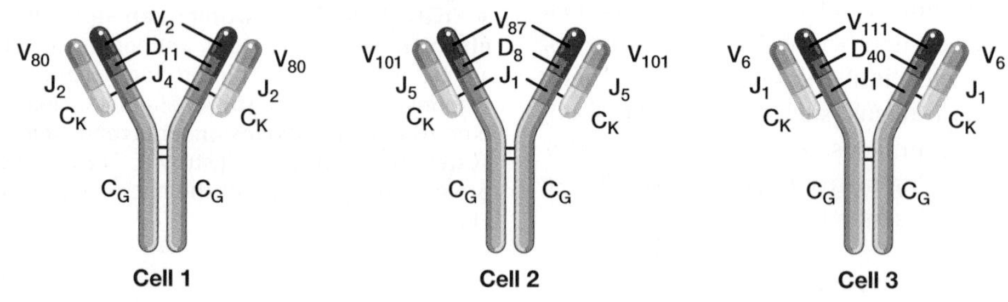

(c) Antibodies synthesized by these three B cells

▲ **FIGURE 36-8 Recombination produces antibody genes (a)** The precursor cells that give rise to B cells have many copies of genes that code for parts of antibodies, illustrated here as variable (V), diversity (D), joining (J), and constant (C) region genes. Each colored band represents an individual gene for an antibody part. There are five genes for the constant regions of the heavy chain and one gene for the constant region of the light chain. Only a few of the many possible variable, diversity, and joining genes are shown here. **(b)** Genes for antibody parts are spliced together to form genes for complete antibodies. For the heavy chain, there is one gene each for the variable, diversity, joining, and constant regions; for the light chain, one each for the variable, joining, and constant regions. In this illustration, the complete antibody genes assembled in cells 1, 2, and 3 are composed of different genes for the antibody parts, as is signified by the different numbers assigned to each gene type. **(c)** As a result, the variable regions for the antibodies produced by cells 1, 2, and 3 differ from one another.

It may be helpful to think of antibody production in terms of dealing cards for 5-card poker. Although a single deck contains only 52 cards, they can be dealt out in 2,598,960 unique 5-card hands. Similarly, each mature B cell is "dealt" an "antibody hand" consisting of a few parts randomly selected from pools of a few hundred parts, resulting in billions of possible unique antibodies.

T-cell receptors are made from different genes, but the process is similar. There are more parts available for constructing T-cell receptor genes, so there may be as many as a quadrillion different possible T-cell receptors!

Antibodies and T-Cell Receptors Are Not Tailor Made for Antigens

B and T cells do not design antibodies and T-cell receptors to fit invading antigens. Instead, the immune system randomly synthesizes millions of different antibodies and T-cell receptors. At any given time, the human body contains a standing army of perhaps 100 million different antibodies and even more T-cell receptors. This army is simply there, waiting, much like clothes lined up on racks in a department store. Given enough clothing from which to choose, most of us will find something suitable that fits. Similarly, antigens almost always encounter antibodies and T-cell receptors that will bind them.

The Immune System Distinguishes Self from Non-Self

The surfaces of the body's own cells bear large proteins and polysaccharides, just as those of microbes do. Some of these proteins, collectively called the **major histocompatibility complex (MHC),** are unique to each person (except identical twins, who have the same genes and, hence, the same MHC proteins). So why don't these self-antigens arouse an individual's own immune system? The key seems to be the continuous presence of the body's own antigens while the immune cells mature. Some newly formed immune cells do indeed produce antibodies or T-cell receptors that can bind the body's own proteins and polysaccharides, treating them as antigens. However, if these *immature* immune cells contact antigens that bind to their antibodies or T-cell receptors, they undergo apoptosis, or "programmed cell death," in which they essentially commit cellular suicide. Thus, one way that the immune system distinguishes self from non-self is by retaining only those immune cells that do not respond to the body's own molecules.

However, not all self-reactive B and T cells are eliminated in this way. Although no one understands the mechanism, **regulatory T cells** somehow prevent these remaining self-reactive lymphocytes from attacking the body and causing an autoimmune disease (see section 36.8).

Because an individual's MHC proteins are unique, they act as foreign antigens in other people's bodies. This is why organ transplants are sometimes rejected. Physicians must find a donor whose MHC proteins are as similar as possible to those of the recipient, but even with tissue matching, transplant patients must take drugs that suppress their immune systems. Without these drugs, the recipient's immune system attacks the foreign MHC proteins on the donor's cells, destroying the transplanted organ.

36.5 HOW DOES THE ADAPTIVE IMMUNE SYSTEM LAUNCH AN ATTACK?

The adaptive immune system simultaneously launches two types of attack against microbial invaders: humoral immunity and cell-mediated immunity. **Humoral immunity** is provided by B cells and the antibodies that they secrete into the bloodstream, which attack pathogens that are outside the body's cells. **Cell-mediated immunity** is produced by a type of T cell called the cytotoxic T cell, which attacks infected body cells, killing both the cell and any pathogens inside it. Although humoral and cell-mediated immunity are not completely independent, we consider them separately to make them easier to understand.

An Immune Response Takes Time to Develop

Although you have millions of different antibodies and T-cell receptors in your body, you have only one or a few cells bearing each type of antibody or T-cell receptor. The benefit to having millions of different types of cells, even with only a few cells of each type, is that almost any invader will provoke an immune response. The drawback is that the immune response takes some time to become effective, because cells recognizing the invader must multiply and differentiate. In fact, it usually takes 1 or 2 weeks to mount a really good immune response to the first exposure to an invading microbe (**Fig. 36-9**). In the meantime, you become ill, and may even die, if the development of the immune response loses the race to the multiplication of the microbe and the damage it causes to your body.

Humoral Immunity Is Produced by Antibodies Dissolved in the Blood

Each B cell bears its own unique antibodies on its surface. When an infection occurs, the antibodies borne by a few B cells can bind to antigens on the invader. Antigen–antibody binding causes these B cells, but no others, to divide rapidly. This process is called **clonal selection** because the resulting population of cells is composed of "clones" (genetically identical to the parent B cells) that have been "selected" to multiply by the presence of particular invading antigens (**Fig. 36-10**).

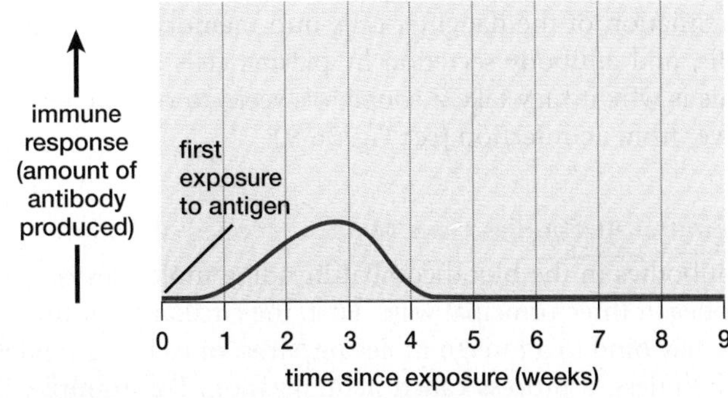

▲ **FIGURE 36-9 An effective immune response takes time to develop**

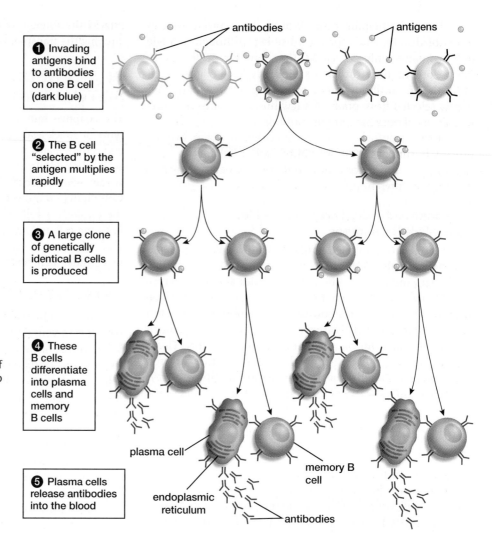

1 Invading antigens bind to antibodies on one B cell (dark blue)

2 The B cell "selected" by the antigen multiplies rapidly

3 A large clone of genetically identical B cells is produced

4 These B cells differentiate into plasma cells and memory B cells

5 Plasma cells release antibodies into the blood

antibodies

antigens

plasma cell

endoplasmic reticulum

memory B cell

antibodies

▶ **FIGURE 36-10 Clonal selection among B cells by invading antigens** The immune system contains millions of B cells, each with a unique antibody (top row). An invading antigen binds to the antibody on only one B cell, stimulating it to divide. Its daughter cells differentiate into plasma cells and memory B cells. Ordinary B cells are small, with little endoplasmic reticulum (top). As they differentiate into plasma cells, they become much larger (bottom), almost entirely filled with rough endoplasmic reticulum that synthesizes antibodies.

The daughter cells differentiate into two cell types: **memory B cells** and **plasma cells.** Memory B cells do not release antibodies, but they play an important role in future immunity to the invader that stimulated their production, as we will soon see. Plasma cells become enlarged and packed with rough endoplasmic reticulum, in which huge quantities of specific antibody proteins are synthesized (see Fig. 36-10). These antibodies are released into the bloodstream (hence the name "humoral" immunity; to the ancient Greeks, blood was one of the four "humors," or body fluids).

Clonal selection, multiplication of activated B cells, differentiation of the daughter cells into memory and plasma cells, and antibody secretion by plasma cells all take time. This is why it may take a couple of weeks to completely recover from an infection (see Fig. 36-9).

Humoral Antibodies Have Multiple Modes of Action

Antibodies in the blood combat invading molecules or microbes in three principal ways. First, the circulating antibodies may bind to a foreign molecule, virus, or cell, and render it harmless, a process called neutralization. For example, if the active site of a toxic enzyme in snake venom is covered with antibodies, it cannot harm your body (**Fig. 36-11**). Many viruses, such as those that cause rabies, gain entry into your body's cells when a protein on the virus binds to a specific receptor site on the surface of a cell. If antibodies cover up this viral protein, the neutralized virus cannot enter the cell.

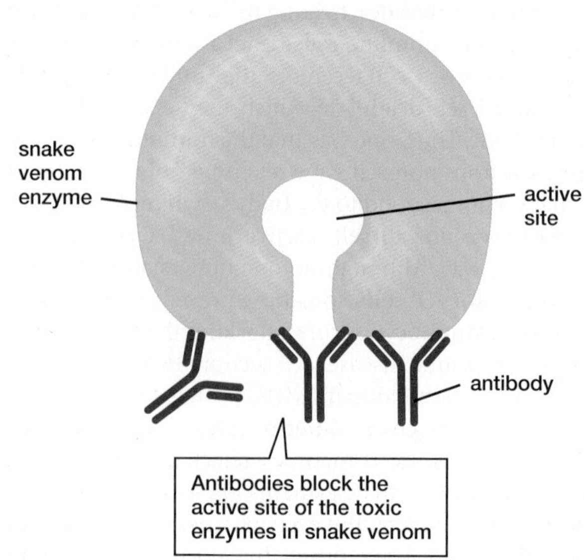

▲ **FIGURE 36-11 Antibodies neutralize toxic molecules**

Second, antibodies may coat the surface of invading molecules, viruses, or cells, and make it easier for phagocytic cells to destroy them (see Fig. 36-7b). Remember, the variable regions on the "arms" of an antibody bind to antigens on invaders, so the constant regions that make up the "stems" stick out into the blood or extracellular fluid. Macrophages recognize the antibody stems, engulf the antibody-coated invaders, and digest them.

Third, when antibodies bind to antigens on the surface of a microbe, the antibodies interact with complement proteins that are always present in the blood. Some complement proteins punch holes in the plasma membrane of the microbe, killing it. Other complement proteins promote phagocytosis of the invaders.

Case Study continued

Flesh-Eating Bacteria

Complement enhances phagocytosis of invading bacteria. However, *S. pyogenes* produces a protein on its surface that inhibits the binding of certain components of the complement system, helping to protect it from complement-stimulated phagocytosis.

Humoral Immunity Fights Invaders That Are Outside Cells

Because antibodies are large proteins that do not readily enter a cell through its plasma membrane, the humoral response is effective only against microbes or their toxins when they are outside of cells, generally in the blood or extracellular fluid. Bacteria, bacterial toxins, and some fungi and protists are usually vulnerable to the humoral immune response. Viruses are vulnerable when they are outside of the body's cells—for example, when they are spreading from cell to cell in the extracellular fluid—but are safe from antibody attack as long as they are inside a cell. Fighting viral infections, therefore, requires the help of the cell-mediated immune response.

Cell-Mediated Immunity Is Produced by Cytotoxic T Cells

Cell-mediated immunity, produced by **cytotoxic T cells,** is the body's primary defense against cells that are cancerous or have been infected by viruses. Although the process is complex, in essence, it works like this: When a cell is infected by a virus, some pieces of viral proteins are brought to the surface of the infected cell and "displayed" on the outside of the plasma membrane. Cytotoxic T cells drift about, occasionally bumping into the displayed viral proteins. If a cytotoxic T cell has a T-cell receptor that binds to the viral protein, the cytotoxic T cell squirts proteins onto the surface of the infected cell, punching holes in its plasma membrane and killing it. If the infected cell is killed before the virus has had enough time to finish multiplying, then no new viruses are produced, and the rest of the body's cells are spared new infection. Cancerous cells, too, often display unusual proteins on their surfaces that cytotoxic T cells recognize as foreign, and are killed as a result (**Fig. 36-12**).

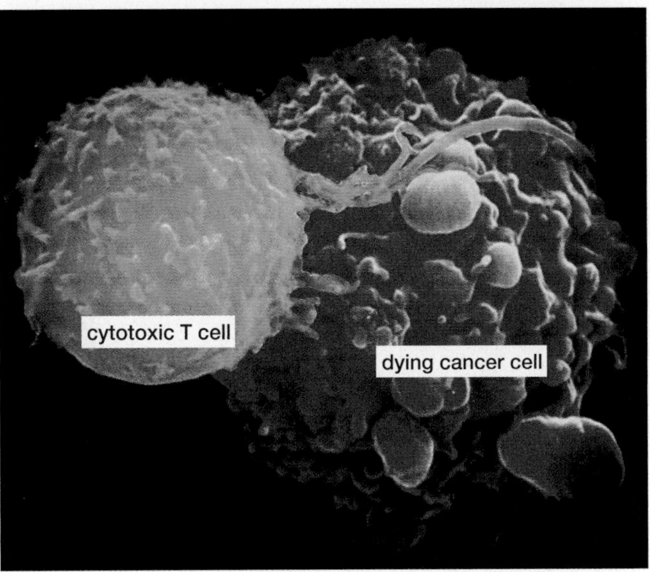

▲ **FIGURE 36-12 Cell-mediated immunity in action** A cytotoxic T cell attacks a cancer cell (red) in this scanning electron micrograph.

QUESTION Cancer consists of the body's own cells, so why does the immune system attack cancer cells?

Helper T Cells Enhance Both Humoral and Cell-Mediated Immune Responses

Although our earlier discussion implied that B cells and cytotoxic T cells fight microbial invasions by themselves, they are often quite ineffective without assistance from **helper T cells.** Helper T cells bear receptors that bind to antigens displayed on the surfaces of dendritic cells or macrophages that have engulfed and digested invading microbes. Only helper T cells bearing the correct T-cell receptor can bind to any particular antigen. When its receptor binds an antigen, a helper T cell multiplies rapidly. Its daughter cells differentiate and release cytokines that stimulate cell division and differentiation in both B cells and cytotoxic T cells. In fact, both B cells and cytotoxic T cells make a significant contribution to defense against disease only if they simultaneously bind antigen *and* receive stimulation by cytokines from helper T cells. As you may know, the human immunodeficiency virus (HIV), which causes AIDS, kills off helper T cells. Without helper T cells, the immune system cannot fight off diseases that would otherwise be trivial.

Figure 36-13 compares the humoral immune response with the cell-mediated immune response, and shows the role of helper T cells in both.

36.6 HOW DOES THE ADAPTIVE IMMUNE SYSTEM REMEMBER ITS PAST VICTORIES?

After recovering from a disease, you remain immune to that particular microbe for many years, perhaps a lifetime. The plasma cells and cytotoxic T cells that conquered the disease

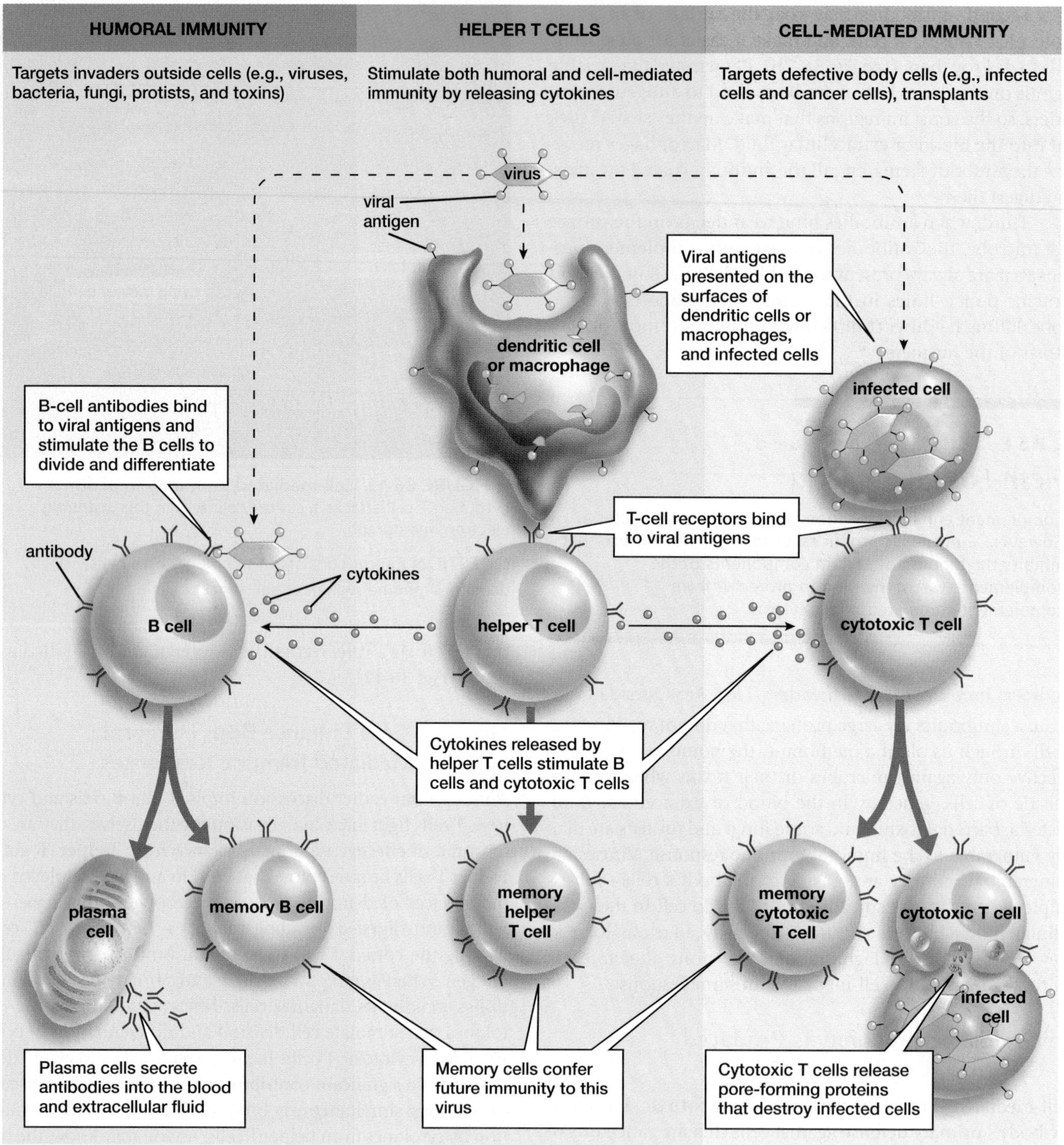

▲ **FIGURE 36-13 A summary of humoral and cell-mediated immune responses**

generally live only a few days. However, some of the daughter cells of the original B cells, cytotoxic T cells, and helper T cells that responded to the infection differentiate into memory B cells and **memory T cells** that survive for many years. Then, if the body is reinvaded by this same type of microbe, the memory cells recognize the invader and mount an immune response. Memory B cells rapidly produce a clone of plasma cells, secreting antibodies that combat this second invasion. Memory T cells produce clones of either helper T cells or cytotoxic T cells, also specific for the "remembered" invader. There are far more memory cells than original B, cytotoxic T, or helper T cells that responded to the first infection. Further, each memory cell responds faster than its original parent cells could. Therefore, upon a second infection, memory cells usually produce a second immune response that is so fast and so large that the body fends off the attack before you suffer any disease symptoms: You have become immune (**Fig. 36-14**).

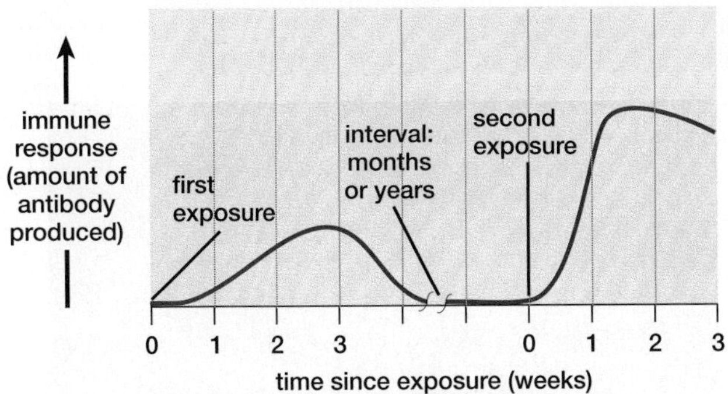

▲ **FIGURE 36-14 Acquired immunity** The immune system responds sluggishly to the first exposure to a disease-causing organism as B and T cells are selected and multiply. A second exposure activates memory cells formed during the first response, making the second response both faster and larger.

Acquired immunity confers long-lasting, perhaps lifelong, protection against many diseases—smallpox, measles, mumps, and chicken pox, for example. Acquired immunity may fail, however, if disease organisms mutate rapidly and produce new antigens to which the memory cells cannot respond (see "Health Watch: Exotic Flu Viruses" on p. 706).

36.7 HOW DOES MEDICAL CARE ASSIST THE IMMUNE RESPONSE?

For most of human history, the battle against disease was fought by the immune system alone. Now, however, the immune system has a powerful assistant: medical treatment. Here we describe two of the most important medical tools: antibiotics and vaccinations.

Antibiotics Slow Down Microbial Reproduction

Antibiotics are chemicals that help to combat infection by slowing down the multiplication of bacteria, fungi, or protists. Although antibiotics usually do not destroy every single disease-causing microbe in the body, they may kill enough of them to give the immune system time to finish the job. One problem with antibiotics, however, is that they are potent agents of natural selection, favoring the survival and reproduction of microbes that can withstand their effects (see the case study "Evolution of a Menace" in Chapter 15). The occasional mutant microbe that is resistant to an antibiotic will pass on the genes for resistance to its offspring. The result: Resistant microbes thrive while susceptible microbes die off. Eventually, many antibiotics become ineffective in treating diseases.

Antibiotics are not effective against viruses. Until recently, little could be done to help people suffering from viral infections except to treat the symptoms and hope that the immune system would triumph. Now, drugs are available that target different stages of the viral cycle of infection, which includes attachment to a host cell, replication of viral parts, assembly of the virus within the host cell, and the release of the virus into the extracellular fluid to infect new cells. These antiviral drugs are not prescribed for most viruses, but they are used to treat HIV, severe herpes virus infections, and in some cases, the flu virus.

Vaccinations Stimulate the Development of Memory Cells and Future Immunity Against Disease

A **vaccine** stimulates an immune response by exposing a person to antigens produced by a pathogen. Vaccines often consist of weakened or killed microbes (which, therefore, cannot cause disease) or some of the pathogen's antigens, which are synthesized using genetic engineering techniques. When the body is exposed to these antigens, it produces an army of memory cells that confer immunity against living, dangerous microbes of the same type. Just like immunity acquired by recovering from a real illness, immunity stimulated by an effective vaccine produces such a rapid and large immune response to a subsequent invasion by the living pathogen that you never experience any symptoms at all (see Fig. 36-14). Learn more about vaccines in "Scientific Inquiry: The Discovery of Vaccines" on p. 707.

36.8 WHAT HAPPENS WHEN THE IMMUNE SYSTEM MALFUNCTIONS?

Occasionally, the immune system launches inappropriate attacks, undermining health instead of promoting it. In addition, the immune system, like other body systems, is itself subject to disorders that decrease its effectiveness.

Allergies Are Misdirected Immune Responses

More than 35 million people in the United States suffer from **allergies,** which are immune reactions to harmless substances that are treated as if they were pathogens. Common allergies include those to pollen, mold spores, bee or wasp venoms, and some foods, such as milk, eggs, fish, wheat, tree nuts, and peanuts.

An allergic reaction begins when allergy-causing antigens, often called allergens, enter the body and bind to "allergy antibodies" on a special type of B cell (these are different from the antibodies that typically combat bacterial or viral infections). This B cell proliferates, producing plasma

Have you ever wondered 

Why You Get Colds So Often?

If the immune system is so good at remembering past assaults and developing long-lasting immunity, why do many people get colds so frequently? The problem is that there are at least 200 different viruses that cause cold symptoms (and possibly an equal number yet to be discovered), and immunity to one usually doesn't confer immunity to the others. Although older people typically develop immunity to a large number of cold viruses, and get fewer colds than young children, probably no one has lived long enough to catch, and become immune to, all of the viruses that cause colds.

Health Watch

Exotic Flu Viruses

Almost everyone has heard about avian flu, also called "bird flu," or H5N1 (**Fig. E36-1**). A few years ago, news reports suggested that we faced the real possibility of a worldwide epidemic that might kill millions of people. In 2009, "swine flu," or H1N1, reared its head, infecting tens of millions of people by the end of the year. Are avian flu and swine flu especially infectious or dangerous? How do they differ from regular human flu?

Every winter, a wave of "ordinary" human flu sweeps across the world. Hundreds of thousands of the elderly, newborn, or infirm die. Most healthy adults who catch the flu—millions of them each year—do not succumb, but they do suffer respiratory distress, fever, and muscle aches. They survive because their adaptive immune systems produce antibodies and cytotoxic T cells that eventually overcome the flu virus. Their immune systems also produce memory cells that lie in wait for the next flu season.

So why can you get the flu again and again? The reason is that flu viruses mutate very rapidly, so that this year's flu antigens aren't exactly the same as last year's. Memory cells from the previous year do respond, but not at maximum speed. Therefore, you get sick, but your partial immunity reduces the severity and duration of your illness.

In some years, however, the flu virus antigens are fundamentally different—not simply a couple of mutations in last year's flu. These new strains arise when mutations in flu viruses that normally infect other animals allow these viruses to infect people. This is where birds and pigs come in. Avian and swine flu viruses are so different from human flu viruses that people do not have partial immunity retained from past years' flu. With no useful memory from previous years, the adaptive immune system takes a week or two to do anything really effective against the virus. By then, depending on how deadly the virus is, it might be too late. In 1918, an avian flu virus mutated and caused the Spanish flu pandemic, killing more than 50 million people worldwide.

Is today's avian flu likely to cause a similar disaster? The first alarm occurred in 1997, when avian flu appeared in Hong Kong, infecting 18 people, 6 of whom died. Health authorities concluded that the flu came from infected chickens. To prevent further human disease, the government ordered every chicken in Hong Kong—about 1.5 million birds—to be killed. Nevertheless, avian flu continues to spread. By October 2008, infected migratory birds, including ducks and swans, had been found across Europe, including Russia, Greece, Italy, France, Denmark, Sweden, and Germany. And avian flu has killed people not only in southeast Asia, but as far away as Egypt, Iraq, and Turkey. As of December 2009, about 60% of the 467 confirmed human cases of avian flu had been fatal.

So far, today's avian flu isn't very efficient at passing from birds to people, and seems almost totally incapable of being transmitted from person to person. But a few more mutations might very well allow the avian flu virus to become efficient at moving between people. To make matters worse, today's avian flu virus seems deadlier than human flu viruses, attacking not only the respiratory tract, but the digestive tract, and even the brain. In the worst case, many millions of people might die.

To prevent a massive epidemic, governments and pharmaceutical companies are racing to develop vaccines against today's avian flu. Two were available as of 2008. Of course, these are the "wrong" vaccines, because they cannot target mutations that haven't yet happened. The vaccines would, however, provide partial immunity and thus might save millions of lives.

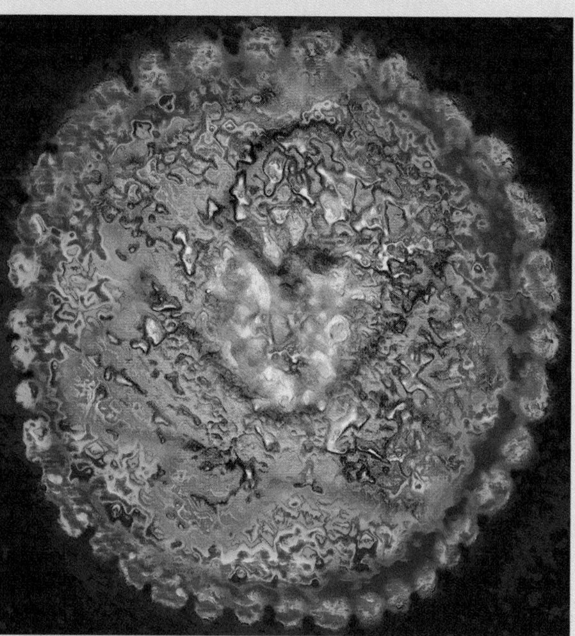

▲ **FIGURE E36-1 The avian flu virus—beautiful, but deadly**

The H1N1 swine flu is a combination of genes from human, avian, and pig viruses, and thus is also unfamiliar to the immune systems of most people. H1N1 flu is much more infectious to people than is avian flu, but is also much less deadly—at least so far. As of December 2009, the U.S. Centers for Disease Control and Prevention estimated that H1N1 flu had infected 35 to 70 million Americans (no one knows for sure), causing over 10,000 deaths, a fatality rate well under 0.1%. At least half of the people who died had other risk factors; apparently, obesity, diabetes, and pregnancy make people more susceptible to complications of H1N1 flu.

Public health officials have two principal concerns about the H1N1 flu. First, because very few people have even partial immunity against it, H1N1 seems likely to infect a much greater number of people than regular seasonal flu ever does. Therefore, even if H1N1 remains relatively mild, a very large number of deaths might occur. Second, what if H1N1 becomes more virulent? Combine hundreds of millions of cases worldwide with even a moderately virulent virus, and the death toll could be staggering. Fortunately, extremely rapid response by governments and the pharmaceutical industry has already resulted in vaccines against H1N1, which may well provide complete immunity. During the 2009–2010 flu season, there will not be enough for everyone to be vaccinated, particularly in poorer countries, but where they are available, these vaccines should greatly reduce both disease and death from the H1N1 flu.

Scientific Inquiry

The Discovery of Vaccines

Smallpox has been feared since ancient times. This highly contagious disease formerly killed about 30% of its victims and left most of the survivors disfigured with pitted skin. The pits are caused by blisters filled with pus. Writings from India dating back to 1100 B.C. describe protecting against smallpox by exposing healthy people to the fluid from the blisters of people with mild cases. Most recipients of these smallpox inoculations developed only mild symptoms and could resist later exposure to the disease.

In the early 1700s, Lady Mary Wortley Montague, wife of the English ambassador to Turkey, observed this procedure firsthand. Upon returning to England, she convinced some aristocrats, including some members of the British royal family, to inoculate their children. Deliberately exposing people to smallpox, however mild, was a frightening prospect to many. Patients would be sick for a few days to a couple of weeks, and 1% to 2% died—one of the sons of King George III of England died from the procedure. Nevertheless, smallpox inoculation became increasingly popular. For example, in 1777, George Washington ordered soldiers of the Continental Army to be inoculated.

Meanwhile, people noticed that milkmaids, who often contracted cowpox, seldom caught smallpox. In 1796, the English surgeon and experimental biologist Edward Jenner (who himself had suffered considerably from smallpox inoculation as a small boy) obtained fluid from cowpox blisters on milkmaid Sarah Nelmes and inoculated an 8-year-old boy, James Phipps, with this material (**Fig. E36-2**). A few months later, Jenner inoculated Phipps with smallpox, and the boy remained healthy. After repeating these results, Jenner published his findings in 1798. This procedure was rapidly adopted in Europe and eventually worldwide.

▲ **FIGURE E36-2 Jenner exposes a child to cowpox as a vaccine against smallpox**

Surprisingly, nearly a century passed before vaccination was applied to other infectious diseases. The French microbiologist Louis Pasteur, among the first to recognize the role of microbes in causing disease, experimented with chicken cholera in the late 1800s. He grew cholera bacteria in culture and found that they caused the fatal disease when injected into chickens. The story goes that when Pasteur injected chickens with bacteria from an old culture that had been left over the summer holidays, the chickens became ill but didn't die. Assuming something had gone wrong, he grew a fresh culture. As luck would have it, he inoculated some of the chickens that had survived the earlier injection with the spoiled culture. To his surprise, these chickens remained healthy, while "naïve" chickens died. Pasteur hypothesized that weakened bacteria in the spoiled culture protected against later infection by healthy bacteria. He coined the term "vaccine" (from the Latin "vacca," meaning "cow"), in recognition of Jenner's pioneering work with cowpox. Pasteur later applied the technique to anthrax in sheep and then to rabies in humans, saving a young boy who had been bitten by a rabid dog.

Pasteur's findings with chicken cholera were both a stroke of luck and a flash of genius. As Pasteur himself remarked, "Chance favors the prepared mind." This insight is just as relevant to scientific inquiry now as it was over a century ago.

cells that pour out allergy antibodies into the plasma (**Fig. 36-15**). The antibodies attach to mast cells, mostly in the respiratory and digestive tracts. If allergens later bind to these attached antibodies, they trigger the release of histamine, which causes leaky capillaries and other symptoms of inflammation (see Fig. 36-4). In the respiratory tract, histamine also increases the secretion of mucus. Thus, airborne substances such as pollen grains, which typically enter the nose and throat, often trigger allergic reactions that include the runny nose, sneezing, and congestion typical of "hay fever." Food allergies usually cause intestinal cramps and diarrhea. In severe cases, such as peanut allergies, the inflammatory response in the airways is so strong that the airways completely close, causing death by suffocation. More than 100 people in the United States die each year from peanut allergies.

An Autoimmune Disease Is an Immune Response Against the Body's Own Molecules

Our immune systems rarely mistake our own cells for invaders. Occasionally, however, something goes awry, and "anti-self" antibodies are produced. The result is an

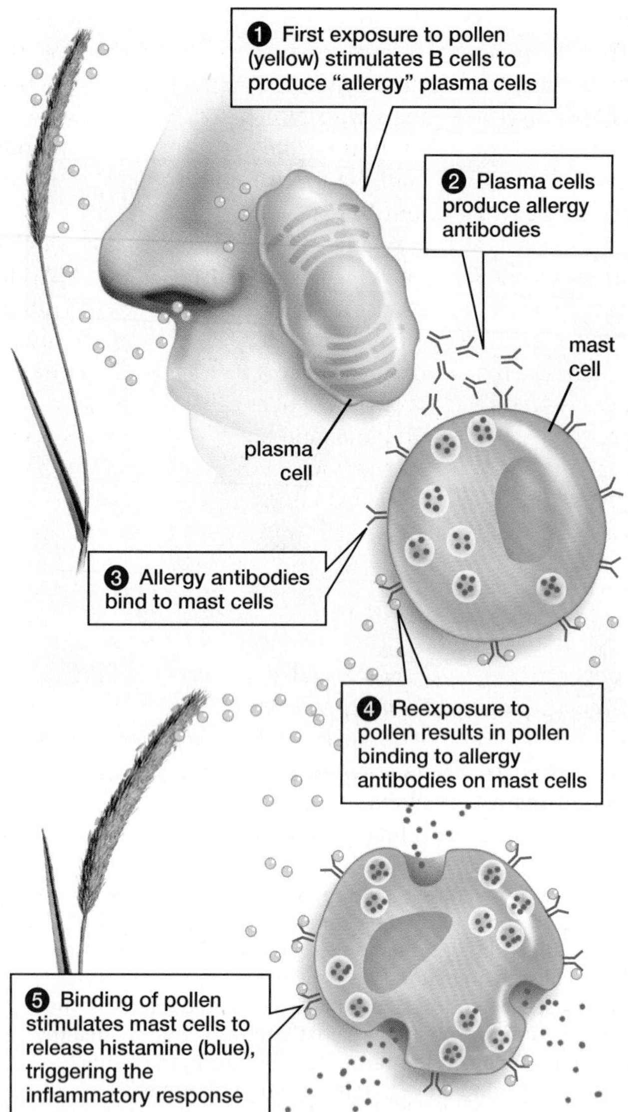
1 First exposure to pollen (yellow) stimulates B cells to produce "allergy" plasma cells
2 Plasma cells produce allergy antibodies
mast cell
plasma cell
3 Allergy antibodies bind to mast cells
4 Reexposure to pollen results in pollen binding to allergy antibodies on mast cells
5 Binding of pollen stimulates mast cells to release histamine (blue), triggering the inflammatory response

▲ **FIGURE 36-15 An allergic reaction to pollen**

autoimmune disease, in which the immune system attacks a component of one's own body. Some types of anemia, for example, are caused by antibodies that destroy a person's red blood cells. Many cases of type 1 diabetes begin when the immune system attacks the insulin-secreting cells of the pancreas. Other autoimmune diseases include rheumatoid arthritis (which affects the joints), myasthenia gravis (skeletal muscle), multiple sclerosis (central nervous system), and systemic lupus (almost any part of the body).

Unfortunately, at present, there are no known cures for autoimmune diseases. For some diseases, replacement therapy can alleviate the symptoms—for instance, by giving insulin to diabetics or blood transfusions to anemia victims. Alternatively, the autoimmune response can be reduced with drugs that suppress the immune response. Immune suppression, however, also reduces responses to the everyday assaults of disease microbes, so this therapy has major drawbacks.

Immune Deficiency Diseases Occur When the Body Cannot Mount an Effective Immune Response

There are two very different disorders in which the immune system cannot combat routine infections. About 1 in 80,000 children is born with severe combined immune deficiency, a group of genetic defects in which few or no immune cells are formed. Much more frequent is acquired immune deficiency syndrome (AIDS), in which a viral infection destroys a formerly functional immune system.

Severe Combined Immune Deficiency Is an Inherited Disorder

A child with **severe combined immune deficiency (SCID)** may survive the first few months of postnatal life, protected by antibodies acquired from the mother during pregnancy or in her milk. Once these antibodies are lost, however, common infections can prove fatal, because the child, lacking an immune system, cannot generate an effective immune response. One form of therapy is to transplant bone marrow (from which immune cells arise) from a healthy donor into the child. In some children, marrow transplants have resulted in enough immune cell production to confer normal immune responses. As we described in Chapter 13, genetic engineering has been used to create a functioning immune system in a handful of children with SCID (see p. 255).

AIDS Is an Acquired Immune Deficiency Disease

The most common immune deficiency disease is **acquired immune deficiency syndrome,** or **AIDS.** The World Health Organization estimated that in 2008, about 2 million people died of AIDS and nearly 2.7 million more became infected, bringing the total infected population to 33 million. AIDS is caused by **human immunodeficiency viruses (HIV).** These viruses undermine the immune system by infecting and destroying helper T cells, which stimulate both the cell-mediated and humoral immune responses (see Fig. 36-13). AIDS does not kill people directly, but AIDS victims become increasingly susceptible to other diseases as their helper T-cell populations decline. Genetic studies show that HIV is almost certainly derived from viruses that infect chimpanzees, but without giving them AIDS. The most recent research suggests that this ancestral virus mutated and acquired the ability to infect people in the early 1900s, although AIDS was not actually identified as a disease until the 1980s.

Because HIV cannot survive for long outside the body, it can be transmitted only by the direct contact of broken skin or mucous membranes with virus-laden body fluids, including blood, semen, vaginal secretions, and breast milk. HIV infection can be spread by sexual activity, by sharing needles among intravenous drug users, or through blood transfusions (this is rare in developed countries because all donated blood is now screened for anti-HIV antibodies). A woman infected with HIV can transmit the virus to her child during pregnancy, childbirth, or breast-feeding.

HIV enters a helper T cell and hijacks the cell's metabolic machinery, forcing it to make more viruses, which then emerge, taking an outer coating of T-cell membrane with them

(Fig. 36-16). Early in the infection, as the immune system fights the virus, the victim may develop a fever, rash, muscle aches, headaches, and enlarged lymph nodes. After several months, the rate of viral replication slows. Enough helper T cells remain that infected individuals are able to resist disease, and they generally feel quite well. In some cases, this condition persists for several years. If untreated, however, helper T cell levels continue to decline, severely weakening the immune response, and at this point the person is considered to have AIDS. As HIV levels skyrocket, they kill more helper T cells, and the person becomes easy prey for other infections. The life expectancy for untreated AIDS victims is about 1 to 2 years.

Several drugs can slow down the replication of HIV and thereby slow the progress of AIDS. Combinations of drugs targeting different stages of viral replication have been particularly effective, and a complete AIDS treatment regimen has now been combined into a once-a-day pill. Unfortunately, HIV can mutate into forms resistant to the drugs, and the drugs cause severe side effects in some patients. Nevertheless, HIV-positive individuals who receive the best medical care might now live out a normal life span, although nobody knows for sure, because the most effective drugs have been available only for about 10 years.

Clearly, the best solution would be to develop a vaccine against HIV. This is a major challenge, partly because HIV disables the immune response that a vaccine depends on. Further, HIV has an incredibly high mutation rate, perhaps a thousand times faster than that of flu viruses, which themselves have a high mutation rate. Single infected individuals may harbor different strains of HIV in blood and in semen because of mutations that occurred within their bodies after they were first infected. As of 2009, despite billions of dollars invested in research and clinical trials, no HIV vaccine has proven effective.

(a) HIV (red) on the surface of a helper T cell

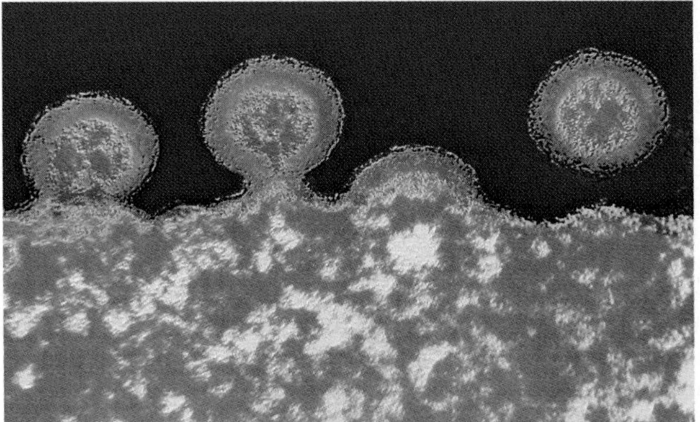

(b) HIV emerging from a helper T cell

▲ **FIGURE 36-16 HIV causes AIDS (a)** The red specks in this scanning electron micrograph are HIVs that have just emerged from the large, green helper T cell that was infected earlier. **(b)** In this higher-magnification transmission electron micrograph, HIVs are seen emerging from the helper T cell and acquiring an outer envelope of its plasma membrane (green) in the process. This will help them infect new cells.

36.9 HOW DOES THE IMMUNE SYSTEM COMBAT CANCER?

Cancer is one of the most dreaded words in the English language, and with good reason. More than 500,000 people in the United States will die of cancer this year, a fatality rate second only to heart disease. A sobering 40% of U.S. citizens will eventually contract some form of cancer. Cancers may be triggered by many causes, including environmental factors (for example, UV radiation or smoking), faulty genes, mistakes during cell division, and viruses. All of these triggers produce cancer by sabotaging the mechanisms that normally control the growth of the body's own cells.

The Immune System Recognizes Most Cancerous Cells as Foreign

Cancer cells form in our bodies every day. Fortunately, the immune system destroys nearly all of them before they have a chance to proliferate and spread. How are these cancer cells weeded out? Cancer cells are, of course, self, and the immune response usually does not respond to self. However, the very processes that cause cells to become cancerous often cause new and slightly different proteins to appear on their surfaces. Natural killer cells and cytotoxic T cells encounter these new proteins, recognize them as non-self antigens, and destroy the cancer cells (see Fig. 36-12). However, some cancer cells may evade detection because they do not bear antigens that allow the immune system to recognize them as foreign. Other cancers, such as leukemia, suppress the immune system. Still others simply grow so fast that the immune response can't keep up.

Vaccination Can Prevent Some Cancers

Some cancers are caused by viruses, including some cancers of the liver, mouth, throat, and penis; some types of leukemia; and probably all cases of cervical cancer. In the United States, two vaccines are available that help to prevent certain cancers: a vaccine against hepatitis B, which reduces the risk of liver cancer, and a vaccine against two human papilloma viruses, which together cause most cases of cervical cancer.

Vaccines May Someday Help to Cure Cancer

Researchers at the U.S. National Cancer Institute, universities, and pharmaceutical companies are developing "treatment vaccines" that may cure certain cancers. Some of these vaccines provide a patient with antigens commonly found on cells of the type of cancer that the patient has, often enhanced in various ways to try to boost the patient's immune response against the cancer. Clinical trials of this type of vaccine against prostate cancer and melanoma (a type of skin cancer) are in progress. Other treatment vaccines, also in clinical trials, consist of antigens from a patient's own tumor cells, often also enhanced to stimulate a stronger immune response. Still another approach is to take antigen-presenting dendritic cells from a patient, expose them to antigens from cancer cells, and force them to multiply rapidly in culture. The resulting daughter cells are then injected back into the patient. In principle, this large number of activated dendritic cells should stimulate the patient's own anticancer immune response.

Most Medical Treatments for Cancer Depend on Selectively Killing Cancerous Cells

In general, however, medicine does not have specific anticancer vaccines or other treatments. In a few cases, such as some types of breast cancer, the cancer cells are stimulated to divide by hormones, so drugs that block hormone action slow down the cancerous growth.

Attempts to eliminate cancer mostly focus on surgery, radiation, and chemotherapy. Surgically removing the tumor is the first step in treating many cancers, but it can be difficult to remove every bit of cancerous tissue. Tumors can be bombarded with radiation, which can destroy even microscopic clusters of cancer cells by disrupting their DNA, thus preventing their cell division and growth. Unfortunately, neither surgery nor radiation is effective against cancer that has spread throughout the body.

Chemotherapy is commonly used to supplement surgery or radiation, or to combat cancers that cannot be treated any other way. Chemotherapy drugs attack the machinery of cell division, and so they are somewhat selective for cancer cells, which divide more frequently than normal cells do. Unfortunately, chemotherapy inevitably also kills some healthy, dividing cells. Damage to dividing cells in patients' hair follicles and intestinal lining by chemotherapy produces its well-known side effects of hair loss, nausea, and vomiting.

Case Study revisited

Flesh-Eating Bacteria

Film-maker Woody Allen once said, "Just because you're paranoid doesn't mean they aren't out to get you." This certainly seems to apply to *S. pyogenes*, which avoids the immune response and damages the body in many different ways. In addition to the tactics described earlier, *S. pyogenes* releases several toxins, some that kill ordinary body cells and some that specifically kill phagocytic cells. It produces enzymes that trigger a cascade of reactions that dissolve blood clots, thereby making it easier for the bacteria to enter the bloodstream and spread throughout the body.

S. pyogenes also produces proteins that are "superantigens," which nonspecifically and massively activate the immune system—causing 10,000 to 100,000 times as much activation as a typical infection by other kinds of bacteria. This enormous immune response releases huge quantities of cytokines, causing the overwhelming inflammation of streptococcal toxic shock syndrome. Remember, inflammation causes leaky capillaries with increased blood flow, so affected tissues swell up with the extra fluid. If massive inflammation occurs in, for example, the lungs, then the victim cannot breathe, and may die. This is probably what happened to Jim Henson. The enormous loss of fluid from the circulatory system can also be fatal all by itself.

If *S. pyogenes* is so good at evading the immune response and killing cells, why aren't we all dead? After all, everyone gets cuts and scrapes, probably several each week. Fortunately, flesh-eating bacteria are pretty rare, so only a few, if any, usually enter any given wound. Despite the bacteria's best efforts to avoid detection and phagocytosis, the innate immune response almost always kills the handful of invading *S. pyogenes*, and the adaptive immune response mops up any that get past the initial defenses. But if a large number enter a wound, or if the immune system is weakened for some reason, then the dangers are very real. If such an infection isn't treated rapidly, it will probably be fatal.

BioEthics Consider This

Flesh-eating bacterial infections are quite rare—perhaps 500 to 1,000 cases occur each year in the United States. Although only a very tiny fraction of wounds develop severe infections with *S. pyogenes*, the death rate is very high—various studies estimate that 25% to 70% of the victims die. Many others, like Monica Jorge, survive only because the infected parts of their bodies are removed. As of 2009, *S. pyogenes* remains very susceptible to penicillin and related antibiotics. Any physician who suspects a flesh-eating infection begins antibiotic treatment immediately. By this time, of course, there is usually significant tissue damage, and sometimes the patient dies anyway.

Antibiotic treatment for anyone with an infected cut, even if there is no initial evidence of *S. pyogenes*, would certainly reduce suffering, disfigurement, and mortality. On the other hand, frequent use of antibiotics is a potent agent of natural selection for antibiotic-resistant bacteria. How do you think that the medical community should balance the benefits of antibiotic treatment for individuals against the overall risks to society?

CHAPTER REVIEW

Summary of Key Concepts

36.1 What Are the Mechanisms of Defense Against Disease?

First, nonspecific external barriers, including the skin and mucous membranes, prevent disease-causing organisms from easily entering the body. Second, nonspecific internal defenses, collectively called the called the innate immune response—consisting of white blood cells, inflammation, and fever—destroy microbes, toxins, and cancerous and infected body cells. Finally, the adaptive immune response selectively destroys the particular toxin or microbe and "remembers" the invader, allowing a faster response if the invader reappears in the future.

36.2 How Do Nonspecific Defenses Function?

The skin and its secretions physically block the entry of microbes into the body and inhibit their growth. The mucous membranes of the respiratory and digestive tracts secrete antibiotic substances and mucus that traps microbes. If microbes do enter the body, they are engulfed by phagocytic white blood cells. Natural killer cells secrete proteins that kill infected or cancerous cells. Injuries stimulate the inflammatory response, in which chemicals are released that attract phagocytic white blood cells, increase blood flow, and make capillaries leaky. Later, blood clots wall off the injury site. Fever is caused by endogenous pyrogens, chemicals that are released by white blood cells in response to infection. High temperatures inhibit bacterial growth and accelerate the immune response.

36.3 What Are the Key Components of the Adaptive Immune System?

The cells of the adaptive immune response are produced and located in bone marrow, thymus, spleen, and lymphatic vessels and nodes. These cells include macrophages, dendritic cells, and two types of lymphocytes: B cells and T cells. The cells secrete cytokine proteins that allow communication between cells. B cells also secrete antibodies that combat infection.

36.4 How Does the Adaptive Immune System Recognize Invaders?

Cells of the adaptive immune system recognize large, complex molecules, called antigens, produced by invading microbes and cancerous cells. Antibodies (on B cells) and T-cell receptors (on T cells) bind to specific antigens and trigger responses in their respective cell types. Antibodies are Y-shaped proteins composed of a constant region and a variable region. Each antibody has specific sites that bind only one or a few types of antigen. Each B cell synthesizes only one type of antibody, unique to that particular cell and its progeny. The millions of different antibodies arise from gene shuffling during B-cell development. T-cell receptors have a different structure but similar antigen binding and diversity.

Both foreign invaders and the body's own cells have antigens that can potentially bind antibodies and T-cell receptors. However, immature immune cells die if they bind antigen. Because the body's own proteins are continuously present during this time, self-reactive immune cells are usually destroyed. Therefore, only foreign antigens normally provoke an immune response.

36.5 How Does the Adaptive Immune System Launch an Attack?

Only those B and T cells that are activated by antigen binding multiply and produce a specific immune response to an invading microbe, a process called clonal selection. B cells give rise to plasma cells, which secrete antibodies into the bloodstream, causing humoral immunity. Antibodies destroy microbes or their toxins while they are outside of body cells. Cytotoxic T cells destroy some microbes, cancer cells, and virus-infected cells on contact, causing cell-mediated immunity. Helper T cells stimulate both the humoral and cell-mediated immune responses.

36.6 How Does the Adaptive Immune System Remember Its Past Victories?

Some progeny cells of both B and T cells are long-lived memory cells. If the same antigen reappears in the bloodstream, these memory cells are immediately activated, dividing rapidly and causing an immune response that is much faster and more effective than the original response.

36.7 How Does Medical Care Assist the Immune Response?

Antibiotics kill microbes or slow down their reproduction, allowing the body's defenses more time to respond and exterminate the invaders. Vaccinations are injections of antigens from disease organisms, in some cases the weakened or dead microbes themselves. An immune response is evoked by the antigens, providing memory and a rapid response should a real infection occur later.

36.8 What Happens When the Immune System Malfunctions?

Allergies are immune responses to normally harmless foreign substances. B cells treat these as antigens and produce "allergy antibodies" that bind to mast cells. When exposed to the antigen, the mast cells release histamine, causing a local inflammatory response. Autoimmune diseases arise when the immune system mistakes the body's own cells for foreign invaders and destroys them. Immune deficiency diseases occur when the immune system cannot respond strongly enough to ward off usually minor diseases. Immune deficiency diseases may be innate, such as severe combined immune deficiency (SCID), or acquired through viral infection, such as acquired immune deficiency syndrome (AIDS).

36.9 How Does the Immune System Combat Cancer?

Cancer is a population of the body's cells that multiplies without control. Cancerous cells may be recognized as "different" by the immune system and destroyed by natural killer cells and cytotoxic T cells. Cancer may be triggered by genetic factors, environmental factors, mistakes during cell division, or viruses. Vaccines can help to prevent certain virus-caused cancers. Other vaccines are under development to try to cure cancer once it occurs.

Key Terms

acquired immune deficiency syndrome (AIDS) *708*
adaptive immune response *693*
adaptive immune system *697*
allergy *705*
antibiotic *705*
antibody *698*
antigen *698*
autoimmune disease *708*
B cell *697*
cancer *709*
cell-mediated immunity *701*
clonal selection *701*

Thinking Through the Concepts

Fill-in-the-Blank

1. External defenses against microbial invasion include the __________, and the mucous membranes that line the __________, __________, and __________ tracts.
2. Nonspecific internal defenses against disease include __________, which engulf and digest microbes; __________, which destroy cells that have been infected by viruses; the __________, provoked by injury; and __________, an elevation of body temperature that slows microbial reproduction and enhances the body's defenses.
3. The specific immune response is stimulated when the body is invaded by complex proteins or polysaccharides collectively called __________. These molecules bind to one of two types of protein receptors of the immune system: __________ or __________.
4. An antibody consists of four protein chains, two __________ chains and two __________ chains. Each is composed of a(n) __________ region and a(n) __________ region. The __________ regions form the binding site for antigen.
5. __________ immunity is provided by B cells and their daughter cells, called __________, which secrete antibodies into the blood plasma. __________ immunity is provided by T cells. __________ T cells kill body cells that have been infected by viruses. __________ T cells produce cytokines that stimulate immune responses in both B cells and T cells. Protection against future invasions by microbes bearing the same antigens is provided by __________ cells of both B and T types.
6. In medical practice, a(n) __________ provides antigens that stimulate an immune response, to protect against infection without actually causing disease.
7. A(n) __________ occurs when the immune system produces a response to a harmless substance such as pollen. In a(n) __________, the immune system cannot mount an effective response, even to dangerous infections; these may be inborn or acquired. When the immune system attacks a person's own body, this is called a(n) __________.

Review Questions

1. List the human body's three lines of defense against invading microbes. Which are nonspecific (that is, act against all types of invaders) and which are specific (act only against a particular type of invader)?
2. How do natural killer cells and cytotoxic T cells destroy their targets?
3. Describe humoral immunity and cell-mediated immunity. Include in your answer the types of immune cells involved in each, the location of antibodies and receptors that bind foreign antigens, and the mechanisms by which invading cells are destroyed.
4. Diagram the structure of an antibody. What parts bind to antigens? Why does each antibody bind only to a specific antigen?
5. How does the immune system construct so many different antibodies?
6. How does the body distinguish "self" from "non-self"?
7. What are memory cells? How do they contribute to long-lasting immunity to specific diseases?
8. What is a vaccination? How does it confer immunity to a disease?
9. How does the inflammatory response help the body resist disease? What allergy symptoms does it cause?
10. Distinguish between autoimmune diseases and immune deficiency diseases, and give one example of each.
11. Describe the causes and eventual outcome of AIDS. How do AIDS treatments work? How is HIV spread?

Applying the Concepts

1. Why is it essential that antibodies and T-cell receptors bind only relatively large molecules (such as proteins) and not relatively small molecules (such as amino acids)?
2. The essay "Health Watch: Exotic Flu Viruses" states that the flu virus is different each year. If that is true, what good does it do to get a "flu shot" each winter?
3. Organ transplant patients often receive the drug cyclosporine. This drug inhibits the production of a cytokine that stimulates helper T cells to proliferate. How does cyclosporine prevent rejection of transplanted organs? Some patients who received successful transplants many years ago are now developing various kinds of cancers. Propose a hypothesis to explain this phenomenon.

MB *Go to www.masteringbiology.com for practice quizzes, activities, eText, videos, current events, and more.*

Chemical Control of the Animal Body: The Endocrine System

Chapter 37

Case Study

Anabolic Steroids—Fool's Gold?

ON JUNE 5, 2003, an anonymous whistle-blower tipped off the United States Anti-Doping Agency (USADA) that top track and field athletes were using a new, undetectable steroid nicknamed "The Clear." As proof, he sent the USADA a used syringe that he claimed contained The Clear, designed by chemist Patrick Arnold and manufactured by Victor Conte, owner of the Bay Area Laboratory Cooperative (BALCO). The whistle-blower turned out to be Trevor Graham, former coach of world record sprinters Marion Jones and Tim Montgomery—an ironic twist, as we will see shortly.

The USADA sent the syringe to Dr. Don Catlin and his team at the UCLA Olympic Analytical Laboratory. Catlin and his colleagues analyzed the tiny amount of the drug that could be rinsed from the syringe. Based on its chemical structure, Catlin named it tetrahydrogestrinone, or THG. THG belongs to a family of chemicals called anabolic steroids, performance-enhancing drugs whose chemical composition resembles that of the male sex hormone testosterone. Usually, anabolic steroids can be detected in urine months after a person stops using them. THG, however, could not—until Catlin's team developed a test for it.

By the end of 2003, Catlin's test found THG in samples from several champion track and field athletes. Hammer thrower Melissa Price, shot put champion Kevin Toth, and runners Regina Jacobs and Michelle Collins received 2- to 4-year bans from competition. The British Olympic Association permanently banned sprinter Dwain Chambers from its Olympic teams, although he was allowed to return to other competitions after 2 years.

Meanwhile, the BALCO scandal continued to spread. Trevor Graham and Marion Jones were convicted in 2008 of lying to federal investigators about BALCO. Jones and another U.S. sprinter, Antonio Pettigrew, confessed to drug use during the 2000 Olympics, and were stripped of their medals. Jones and Pettigrew also brought down the innocent with them—the International Olympic Committee has asked their teammates on gold-medal relay teams to return their medals. In Major League Baseball, home run king Barry Bonds was likewise caught in the BALCO scandal. Bonds was to be tried in March 2009 for perjury and obstruction of justice. However, the trial was postponed because of disputes about the admissibility of some of the prosecution's evidence.

Why do athletes take anabolic steroids and other hormones? Are their effects solely beneficial, from the athlete's point of view? As you read this chapter, notice how many different effects the same hormone can have throughout the body. Why might the multiple effects of anabolic steroids produce health risks?

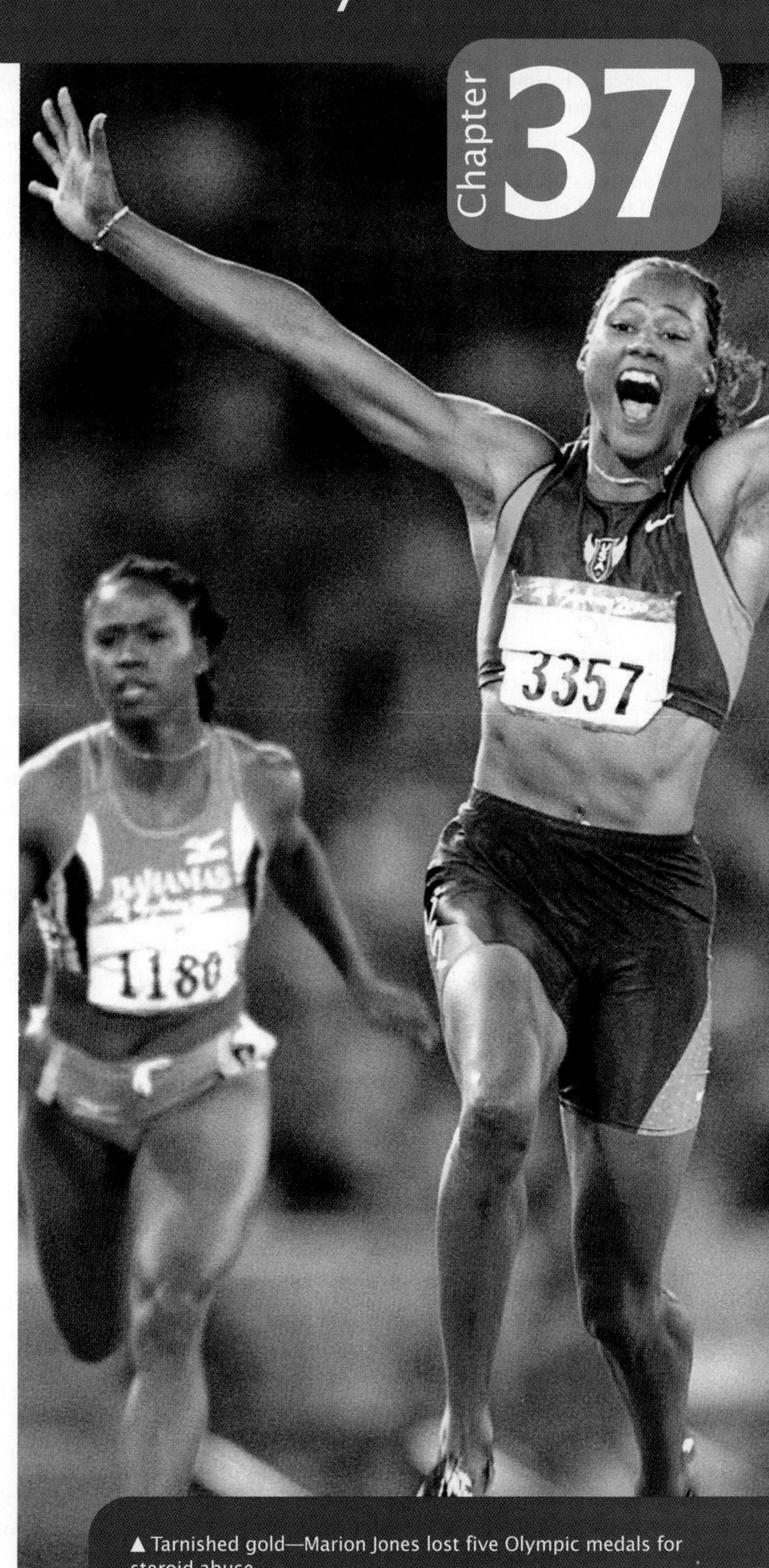

▲ Tarnished gold—Marion Jones lost five Olympic medals for steroid abuse.

At a Glance

37.1 HOW DO ANIMAL CELLS COMMUNICATE?

Like the players on a football team, the individual cells of an animal's body must communicate with one another, to ensure the proper functioning of the whole. Should leg muscles just hold up a standing body, or should they make the legs run? Should blood flow be directed to the intestines for digestion, or to the muscles for movement? These and hundreds of other decisions must be made and communicated throughout the body every minute. Methods of communication between cells fall into four broad categories: direct, synaptic, paracrine, and endocrine (**Table 37-1**).

In direct communication, some tissues, such as heart muscle, have gap junctions that directly link the insides of adjacent cells, allowing ions and electrical signals to flow between them (see Fig. 5-18a and pp. 623–626). This type of communication is very fast, but also has a very short range.

In the other three types of communication, "sending" cells release messenger chemicals through their plasma membranes. The chemicals then move to "receiving" cells and alter their physiology by binding to **receptors,** specialized proteins located either on the surface of, or inside, the receiving cells. When a messenger binds to a receptor, the recipient cell responds in a way that is determined by the messenger, the receptor, and the type of cell. These responses can be as varied as muscle contraction, secretion of milk in lactating women, and active transport of salt by cells in the kidney.

Every cell has dozens of receptors, each capable of binding a specific messenger and stimulating a particular response. Cells with receptors that bind a messenger molecule and respond to it are **target cells** for that messenger. Cells without the correct receptors cannot respond to the messenger and are not target cells. Therefore, a given cell can be a target cell for some messenger molecules but not others, depending on which receptors it has.

Synaptic, paracrine, and endocrine communication differ in speed and distance. Synaptic communication is used in the nervous system. Electrical signals within individual nerve cells can send information to the farthest reaches of the body in a fraction of a second. Then, the nerve cell communicates with a small number of other cells at junctions called synapses. At a synapse, a nerve cell elicits responses from a target cell by releasing chemicals called neurotransmitters across a tiny space between the end of a nerve cell and its target. These responses may be very brief, such as reflexes, or very long lasting, such as learning. We will explore the nervous system in detail in Chapter 38.

This chapter will focus on paracrine and endocrine communication. In paracrine communication, cells release chemicals that diffuse through the extracellular fluid to other cells in the immediate vicinity ("para" means "alongside" in Greek). They therefore influence only a small group of cells, but they do so quite rapidly, because the distances involved are very short.

Endocrine hormones, on the other hand, are released into the bloodstream, where they potentially can elicit a response in nearly every cell of the body. Endocrine hormones can move throughout the body in the circulatory system in only a few seconds—not as fast as the nervous system can send messages, and not fast enough to trigger escape from a predator, but quite rapidly nonetheless. Responses to endocrine hormones may last from a few seconds to a lifetime.

Local Hormones Diffuse to Nearby Target Cells

Many cells engage in paracrine communication, secreting **local hormones** into the extracellular fluid. Local hormones

Table 37-1 **How Cells Communicate**

Communication	Chemical Messengers	Mechanism of Transmission	Examples
Direct	Ions, small molecules	Direct movement through gap junctions linking the cytoplasm of adjacent cells	Ions flowing between cardiac muscle cells
Synaptic	Neurotransmitters	Diffusion from a neuron across a narrow space (synaptic cleft) to a cell bearing the appropriate receptors	Acetylcholine
Paracrine	Local hormones	Diffusion through extracellular fluid to nearby cells bearing the appropriate receptors	Prostaglandins
Endocrine	Hormones	Carried in the bloodstream to near or distant cells bearing the appropriate receptors	Insulin

include histamine, which is released as part of the allergic and inflammatory responses, and the cytokines by which cells of the immune system communicate with one another (see Chapter 36). Generally, local hormones have only short-range actions either because they are degraded soon after release, or because nearby cells take them up so fast that they cannot get very far from the cells that secrete them.

Prostaglandins are modified fatty acids that are important local hormones secreted by cells throughout the body. Prostaglandins have diverse roles, depending on both the type of prostaglandin and the target cell. For example, during childbirth, prostaglandins cause the cervix to dilate and help stimulate the muscles of the uterus to contract. Prostaglandins contribute to inflammation (such as occurs in arthritic joints) and pain sensations. Drugs such as aspirin, acetaminophen (Tylenol), and ibuprofen provide relief from these symptoms by blocking enzymes that synthesize prostaglandins.

Endocrine Hormones Are Transported to Target Cells Throughout the Body by the Circulatory System

Endocrine hormones are messenger molecules produced by the **endocrine glands.** An endocrine gland may be a well-defined cluster of cells whose principal function is hormone secretion, as is the case with the thyroid and pituitary glands. Alternatively, an endocrine gland may consist of clusters of cells, or even scattered individual cells, embedded in organs that have multiple functions, such as in the pancreas, stomach, ovary, or testes. In all cases, the secretory cells of an endocrine gland are embedded within a network of capillaries. The cells secrete their hormones into the extracellular fluid surrounding the capillaries. The hormones diffuse into the capillaries and are carried throughout the body by the bloodstream (**Fig. 37-1**).

A hormone in the bloodstream will reach nearly all of the body's cells, but only cells with receptors capable of binding that specific hormone can respond. The hormone **oxytocin,** for example, stimulates the contraction of uterine muscles during childbirth, because these muscle cells have receptors that bind oxytocin. However, oxytocin does not cause most of the other muscles of the body to contract, because their cells do not have the necessary receptors. A woman's uterine muscles, therefore, contain target cells for oxytocin, while her biceps do not.

The changes induced by hormones may be prolonged and irreversible, as in the onset of puberty or the transformation of a caterpillar into a butterfly. More typically, the changes are temporary and reversible, and help to regulate the physiological systems of the animal body with a time course of seconds to hours.

37.2 HOW DO ANIMAL HORMONES WORK?

The molecules used for vertebrate endocrine hormones are often evolutionarily ancient. Insulin, for example, is found not only in vertebrates but also in protists, fungi, and bacteria, although the function of insulin in most of these organisms is not known. Thyroid hormones have been found in invertebrates such as worms, insects, and mollusks, which do not have thyroid glands. In the few cases in which we understand how hormones work in the cells of invertebrates, the mechanisms are often similar to those in humans and other vertebrates.

▶ **FIGURE 37-1 Hormone release, distribution, and reception**

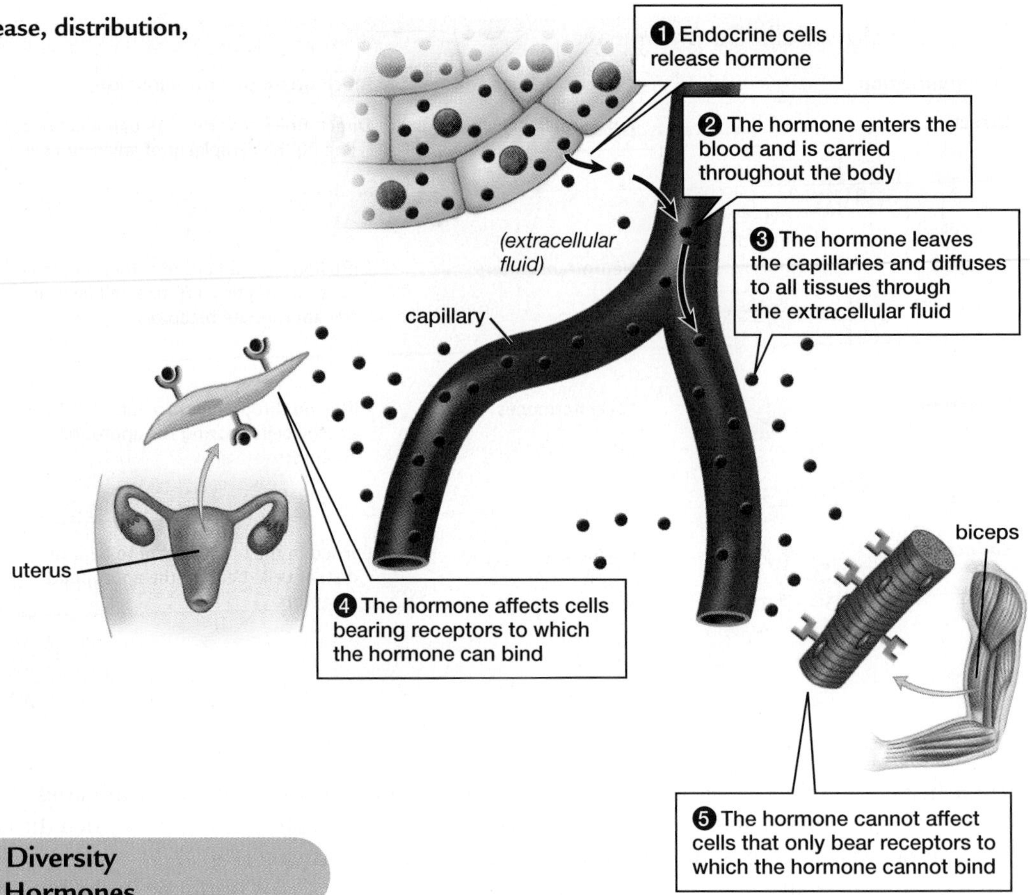

❶ Endocrine cells release hormone

❷ The hormone enters the blood and is carried throughout the body

❸ The hormone leaves the capillaries and diffuses to all tissues through the extracellular fluid

❹ The hormone affects cells bearing receptors to which the hormone can bind

❺ The hormone cannot affect cells that only bear receptors to which the hormone cannot bind

(extracellular fluid)

capillary

uterus

biceps

Table 37-2 The Chemical Diversity of Vertebrate Hormones

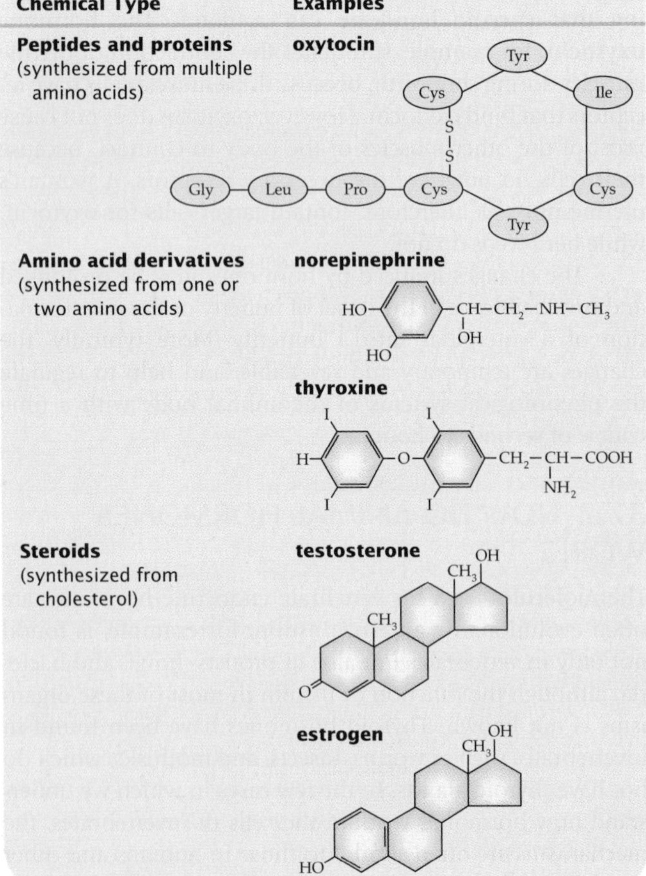

Chemical Type	Examples
Peptides and proteins (synthesized from multiple amino acids)	**oxytocin**
Amino acid derivatives (synthesized from one or two amino acids)	**norepinephrine** **thyroxine**
Steroids (synthesized from cholesterol)	**testosterone** **estrogen**

There are three classes of vertebrate endocrine hormones (**Table 37-2**): **peptide hormones,** which are chains of amino acids; **amino acid-derived hormones,** which are composed of one or two modified amino acids; and **steroid hormones,** which are synthesized from cholesterol.

Hormones Act by Binding to Receptors on or in Target Cells

Receptors for hormones are found in two general locations on target cells: on the plasma membrane, or inside the cell within the cytoplasm or the nucleus.

Peptide and Amino Acid Hormones Usually Bind to Receptors on the Surfaces of Target Cells

Peptide and amino acid-derived hormones are soluble in water but not in lipids. Therefore, these hormones cannot diffuse through the phospholipid bilayer of the plasma membrane. Most of these hormones bind to receptors on the surface of the target cell's plasma membrane (**Fig. 37-2** ❶). Hormone–receptor binding activates an enzyme that synthesizes a molecule, called a **second messenger,** inside the cell (**Fig. 37-2** ❷). A common second messenger is **cyclic adenosine monophosphate (cyclic AMP),** a nucleotide that regulates many cellular activities. The second messenger transfers the signal from the first messenger—the hormone—to other molecules within the cell, often activating specific intracellular enzymes (**Fig. 37-2** ❸), which then initiate a chain of biochemical reactions (**Fig. 37-2** ❹).

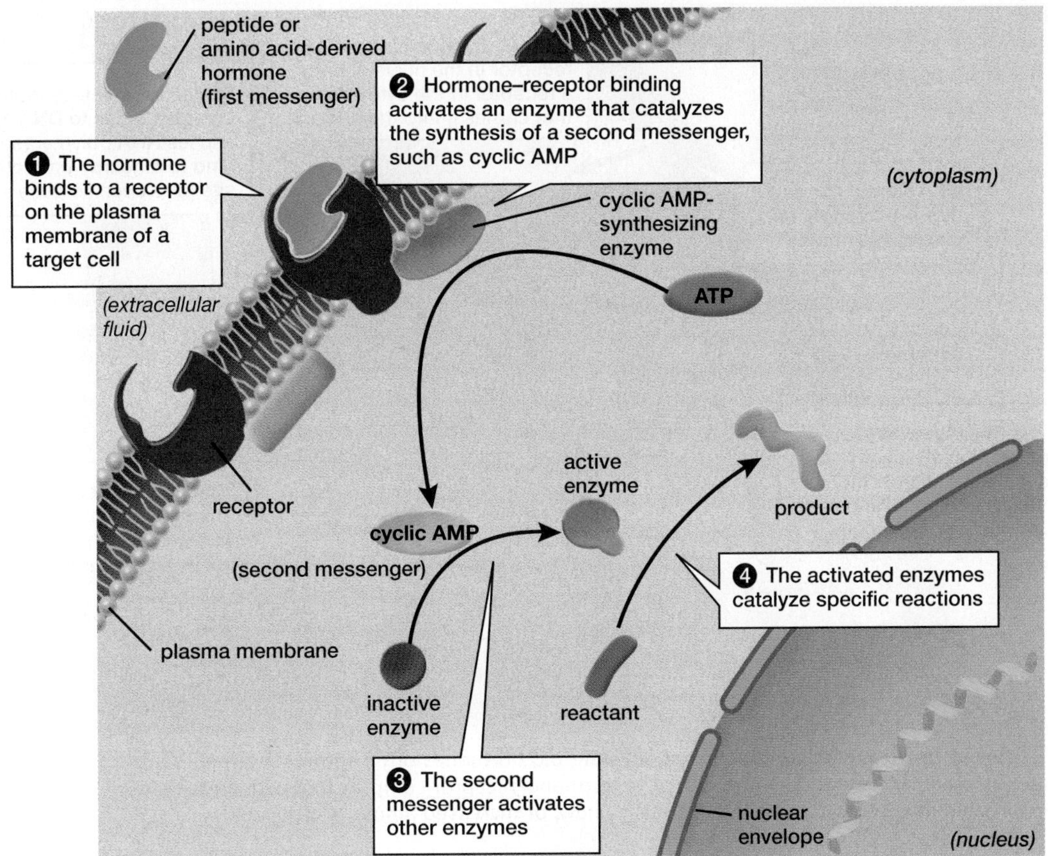

▲ **FIGURE 37-2 Actions of peptide and amino acid-derived hormones on target cells** Peptide and amino acid-derived hormones usually stimulate target cells by binding to receptors on the plasma membrane, which causes the cell to synthesize a second messenger molecule that sets off a cascade of intracellular biochemical reactions.

The reactions vary depending on the hormone, the second messenger, and the target cell. For example, epinephrine (also called adrenaline) stimulates the synthesis of cyclic AMP in both heart muscle cells and liver cells. The increased cyclic AMP acts differently, however, in the two cell types. Cyclic AMP causes heart muscle cells to contract more strongly, while in liver cells, it activates enzymes that break down glycogen (a starch-like polysaccharide) to glucose. Liver cells also respond to insulin, but through different receptors and intracellular reactions. Insulin both activates enzymes and stimulates gene transcription, leading to the removal of glucose from the blood and the conversion of glucose to glycogen.

Hormones that bind to receptors on the target cell's surface may exert rapid but short-lived effects, usually when they activate enzymes, or slower but more prolonged effects, usually when they affect gene transcription.

Steroid Hormones Usually Bind to Receptors Inside Target Cells

Steroid hormones are lipid soluble. Although they sometimes bind to receptors on the surface of a target cell, most steroid hormone action begins with the hormone diffusing through the plasma membrane (**Fig. 37-3** ❶). Once inside, these hormones attach to receptors inside target cells (**Fig. 37-3** ❷). The receptors are either in the nucleus or move into the nucleus after hormone binding. The hormone–receptor complex then binds to the DNA of the promoter region of specific genes (**Fig. 37-3** ❸), and stimulates the transcription of messenger RNA (**Fig. 37-3** ❹). The mRNA travels to the cytoplasm and directs protein synthesis (**Fig. 37-3** ❺). In hens, for example, the steroid hormone estrogen stimulates transcription of the albumin gene, causing the synthesis of albumin (egg white protein), which is packaged in the egg as a food supply for the developing chick.

Case Study continued
Anabolic Steroids—Fool's Gold?

Anabolic steroids and testosterone bind to the same intracellular receptors. However, many anabolic steroids activate these receptors at lower concentrations than testosterone can. Further, if you inject testosterone into your body, it will be completely metabolized within a few hours, whereas some anabolic steroids remain in your system for weeks. Therefore, the strength and duration of hormone–receptor action in target cells may be much greater for anabolic steroids than for natural testosterone.

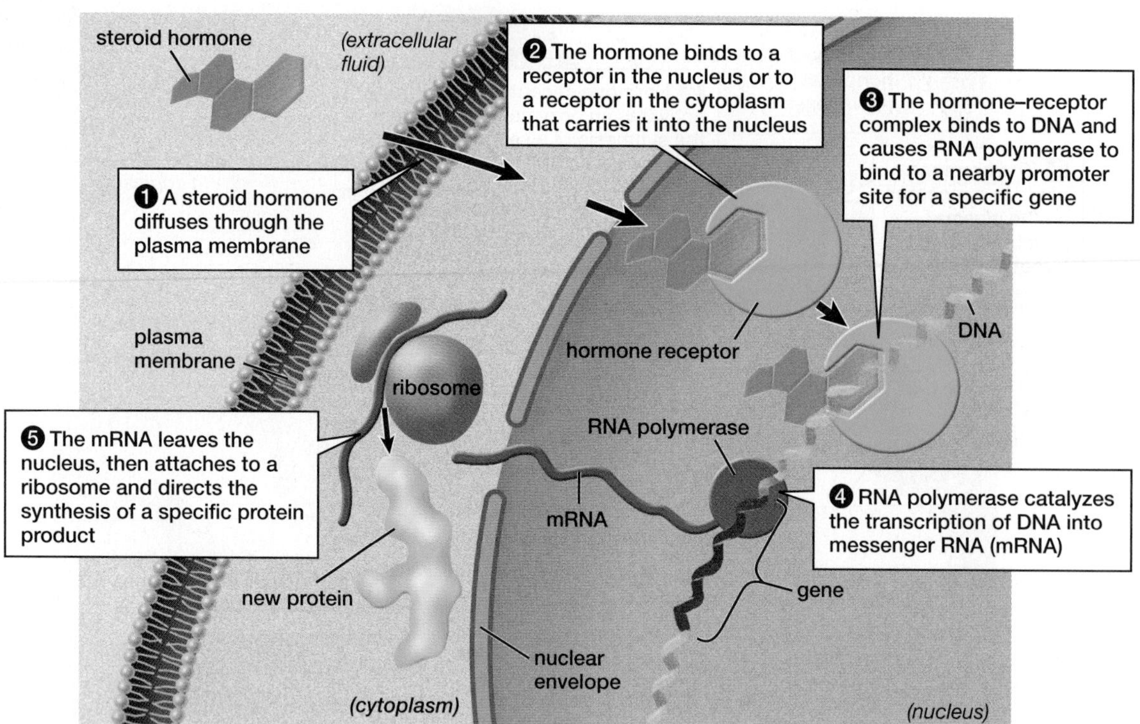

▲ **FIGURE 37-3 Steroid hormone action on target cells** Steroid hormones often stimulate target cells by binding to intracellular receptors, which creates a hormone–receptor complex that activates gene transcription and ultimately results in the synthesis of new, or increased amounts of, specific proteins.

Although it is not a steroid, thyroid hormone also acts intracellularly. Rather than simply diffusing across the plasma membrane, thyroid hormone is actively transported into many cell types. Once inside the cell, thyroid hormone binds to intracellular receptors and activates transcription of specific genes.

Hormones that bind to intracellular receptors may take several minutes or even days to exert their full effects.

Hormone Release Is Regulated by Feedback Mechanisms

The release of most hormones is controlled by negative feedback. As you may recall from Chapter 31, **negative feedback** is a response to a change that tends to counteract the change and restore the system to its original condition. For example, suppose you have jogged a few miles on a hot, sunny day and have lost a quart of water through perspiration. In response to the loss of water from your bloodstream, your pituitary gland releases antidiuretic hormone (ADH), which causes your kidneys to reabsorb water and to produce very concentrated urine (see pp. 683–686, and section 37.3). If, however, you arrive home and drink two quarts of water, you would more than replace the water you lost in sweat. If your body retained this extra water, it could potentially raise your blood pressure and damage your heart. Negative feedback, however, acts to restore the original condition by ensuring that ADH secretion is turned off when the water content of your blood returns to normal, as your kidneys eliminate the excess water. Look for other examples of negative feedback throughout this chapter.

In a few cases, hormone release is temporarily controlled by **positive feedback,** in which the response to a change enhances the change. For example, contractions of the uterus early in childbirth push the baby's head against the cervix (the ring of connective tissue between the uterus and the vagina), which causes the cervix to stretch. Stretching the cervix sends nervous signals to the mother's brain, which in turn causes the release of oxytocin. Oxytocin stimulates continued contractions of the uterine muscles, pushing the baby harder against the cervix, which stretches further, causing still more oxytocin to be released. However, positive feedback cannot continue indefinitely. In the case of childbirth, the positive feedback between oxytocin release and uterine contractions ends when the infant is born. After delivery, the cervix is no longer stretched, so oxytocin release stops.

Vertebrate and Invertebrate Endocrine Hormones Often Have Similar Mechanisms of Action

Invertebrates have peptide and steroid hormones, as well as some hormones belonging to completely different chemical classes. In many cases, invertebrate hormones trigger signaling mechanisms that are similar to those used by vertebrates.

Insect Molting Is Controlled by a Steroid Hormone

Insects are supported by an external skeleton composed of a rigid cuticle that they must molt periodically in order to grow.

Molting is controlled by the steroid hormone **ecdysone,** or molting hormone. As the old cuticle becomes tight, sensory cells stimulate release of another hormone that causes ecdysone secretion. Like many vertebrate steroid hormones, ecdysone acts on receptors located within the nucleus and affects gene transcription, initiating a complex process in which the epithelial cells detach from the old cuticle and secrete a soft new cuticle beneath it. The insect then expands its body by pumping itself full of air. This splits open the old cuticle and stretches out the new one to accommodate some future growth. As the insect emerges, it leaves an insect-shaped cuticle behind (**Fig. 37-4**). Researchers have exploited their understanding of this process to devise pesticides that bind to ecdysone receptors, stimulating them and causing larval insects to molt prematurely and die.

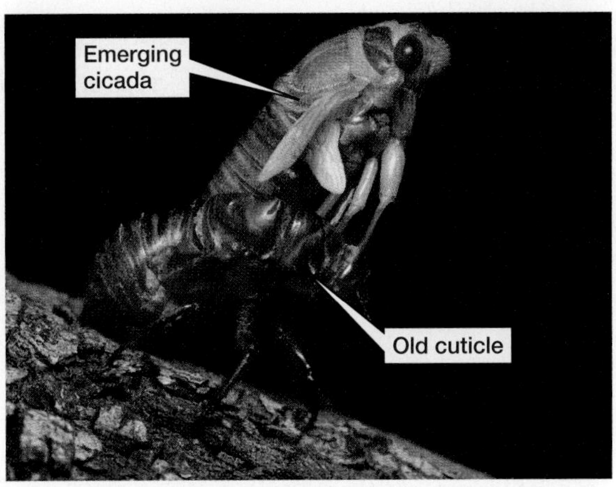

▲ **FIGURE 37-4 Insect molting** A cicada emerges from its shed cuticle.

37.3 WHAT ARE THE STRUCTURES AND FUNCTIONS OF THE MAMMALIAN ENDOCRINE SYSTEM?

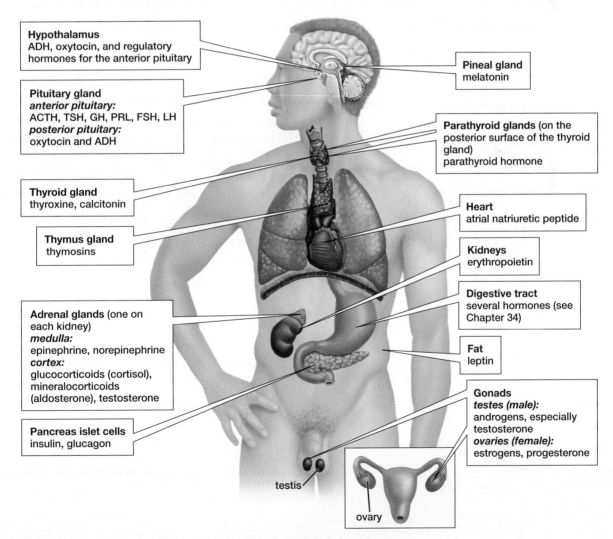

The mammalian **endocrine system** consists of the endocrine hormones and the glands that produce them. In the following sections, we will focus on the functions of the major endocrine glands and organs: the hypothalamus–pituitary complex, the thyroid gland, the pancreas, the sex organs, and the adrenal glands. These and other hormone-secreting organs are illustrated in **Figure 37-5** and are described in **Table 37-3.**

▲ **FIGURE 37-5 The major mammalian endocrine glands and their hormones**

Table 37-3 Major Mammalian Endocrine Glands and Hormones

Endocrine Gland	Hormone	Type of Chemical	Principal Function
Hypothalamus (to anterior pituitary)	Releasing and inhibiting hormones (at least seven)	Peptides	Releasing hormones stimulate the release of hormones from the anterior pituitary; inhibiting hormones inhibit the release of hormones from the anterior pituitary
Anterior pituitary	Follicle-stimulating hormone (FSH)	Peptide	Females: stimulates the growth of follicles in the ovary, secretion of estrogen, and perhaps ovulation Males: stimulates the development of sperm
	Luteinizing hormone (LH)	Peptide	Females: stimulates ovulation, the growth of the corpus luteum, and the secretion of estrogen and progesterone Males: stimulates the secretion of testosterone
	Thyroid-stimulating hormone (TSH)	Peptide	Stimulates the thyroid to release thyroxine
	Adrenocorticotropic hormone (ACTH)	Peptide	Stimulates the adrenal cortex to release hormones, especially glucocorticoids such as cortisol
	Prolactin (PRL)	Peptide	Stimulates milk synthesis in and secretion from the mammary glands
	Growth hormone (GH)	Peptide	Stimulates growth, protein synthesis, and fat metabolism; inhibits sugar metabolism
Hypothalamus (via posterior pituitary)	Antidiuretic hormone (ADH)	Peptide	Promotes reabsorption of water from the kidneys; constricts arterioles
	Oxytocin	Peptide	Females: stimulates contraction of uterine muscles during childbirth, milk ejection, and maternal behaviors Males: may facilitate ejaculation of sperm
Thyroid	Thyroxine	Amino acid derivative	Increases the metabolic rate of most body cells; increases body temperature; regulates growth and development
	Calcitonin	Peptide	Inhibits the release of calcium from bones; decreases the blood calcium concentration
Parathyroid	Parathyroid hormone	Peptide	Increases blood calcium by stimulating calcium release from bones, absorption by the intestines, and reabsorption by the kidneys
Pancreas	Insulin	Peptide	Decreases blood glucose levels by increasing uptake of glucose into cells and converting glucose to glycogen, especially in the liver; regulates fat metabolism
	Glucagon	Peptide	Converts glycogen to glucose, raising blood glucose levels
Testes[1]	Testosterone	Steroid	Stimulates the development of genitalia and male secondary sexual characteristics; stimulates the development of sperm
Ovaries[1]	Estrogen	Steroid	Causes the development of female secondary sexual characteristics and the maturation of eggs; promotes the development of the uterine lining
	Progesterone	Steroid	Stimulates the development of the uterine lining and the formation of the placenta
Adrenal cortex	Glucocorticoids (cortisol)	Steroid	Increase blood sugar; regulate sugar, lipid, and fat metabolism; have anti-inflammatory effects
Adrenal medulla	Epinephrine (adrenaline) and norepinephrine (noradrenaline)	Amino acid derivatives	Increase levels of sugar and fatty acids in the blood; increase metabolic rate; increase the rate and force of contractions of the heart; constrict some blood vessels
	Mineralocorticoids (aldosterone)	Steroid	Increase reabsorption of salt in the kidney
	Testosterone	Steroid	Causes masculinization of body features, growth
		Other Sources of Hormones	
Pineal gland	Melatonin	Amino acid derivative	Regulates seasonal reproductive cycles and sleep–wake cycles; may regulate onset of puberty
Thymus	Thymosin	Peptide	Stimulates maturation of T cells of the immune system
Kidney	Erythropoietin	Peptide	Stimulates red blood cell synthesis in the bone marrow
	Renin	Peptide	Acts on blood proteins to produce a hormone (angiotensin) that regulates blood pressure
Digestive tract[2]	Secretin, gastrin, cholecystokinin, and others	Peptides	Control secretion of mucus, enzymes, and salts in the digestive tract; regulate peristalsis
Heart	Atrial natriuretic peptide (ANP)	Peptide	Increases salt and water excretion by the kidneys; lowers blood pressure
Fat cells	Leptin	Peptide	Regulates appetite; stimulates immune function; promotes blood vessel growth; required for the onset of puberty

[1]See Chapters 41 and 42.

[2]See Chapter 34.

Hormones of the Hypothalamus and Pituitary Gland Regulate Many Functions Throughout the Body

The hypothalamus and pituitary gland coordinate the action of many key hormonal systems. The **hypothalamus** is a part of the brain that contains clusters of specialized nerve cells called neurosecretory cells. **Neurosecretory cells** synthesize peptide hormones, store them, and release them when stimulated. The **pituitary gland** is a pea-sized gland connected to the hypothalamus by a stalk. The pituitary consists of two distinct parts: the **anterior pituitary** and the **posterior pituitary** (**Fig. 37-6**). The anterior pituitary is a true endocrine gland, composed of several types of hormone-secreting cells enmeshed in a network of capillaries. The posterior pituitary, however, consists mainly of a capillary bed and the endings of neurosecretory cells whose cell bodies are in the hypothalamus. The hypothalamus controls the release of hormones from both parts of the pituitary.

Hypothalamic Hormones Control Hormone Release in the Anterior Pituitary

Neurosecretory cells of the hypothalamus produce at least seven hormones that regulate the release of hormones from the anterior pituitary (**Fig. 37-6** green step ❶). These hypothalamic hormones are called **releasing hormones** or **inhibiting hormones,** depending on whether they stimulate or inhibit, respectively, the release of a particular pituitary hormone. Releasing and inhibiting hormones are secreted into a capillary bed in the stalk connecting the hypothalamus to the pituitary, and travel a short distance through blood vessels to a second capillary bed that surrounds the endocrine cells of the anterior pituitary (**Fig. 37-6** green step ❷). There, they diffuse out of the capillaries and bind to receptors on the surfaces of the pituitary endocrine cells. Some of these hypothalamic hormones, such as growth hormone-releasing hormone, stimulate the release of pituitary hormones (**Fig. 37-6** green step ❸), while others inhibit the release of pituitary hormones.

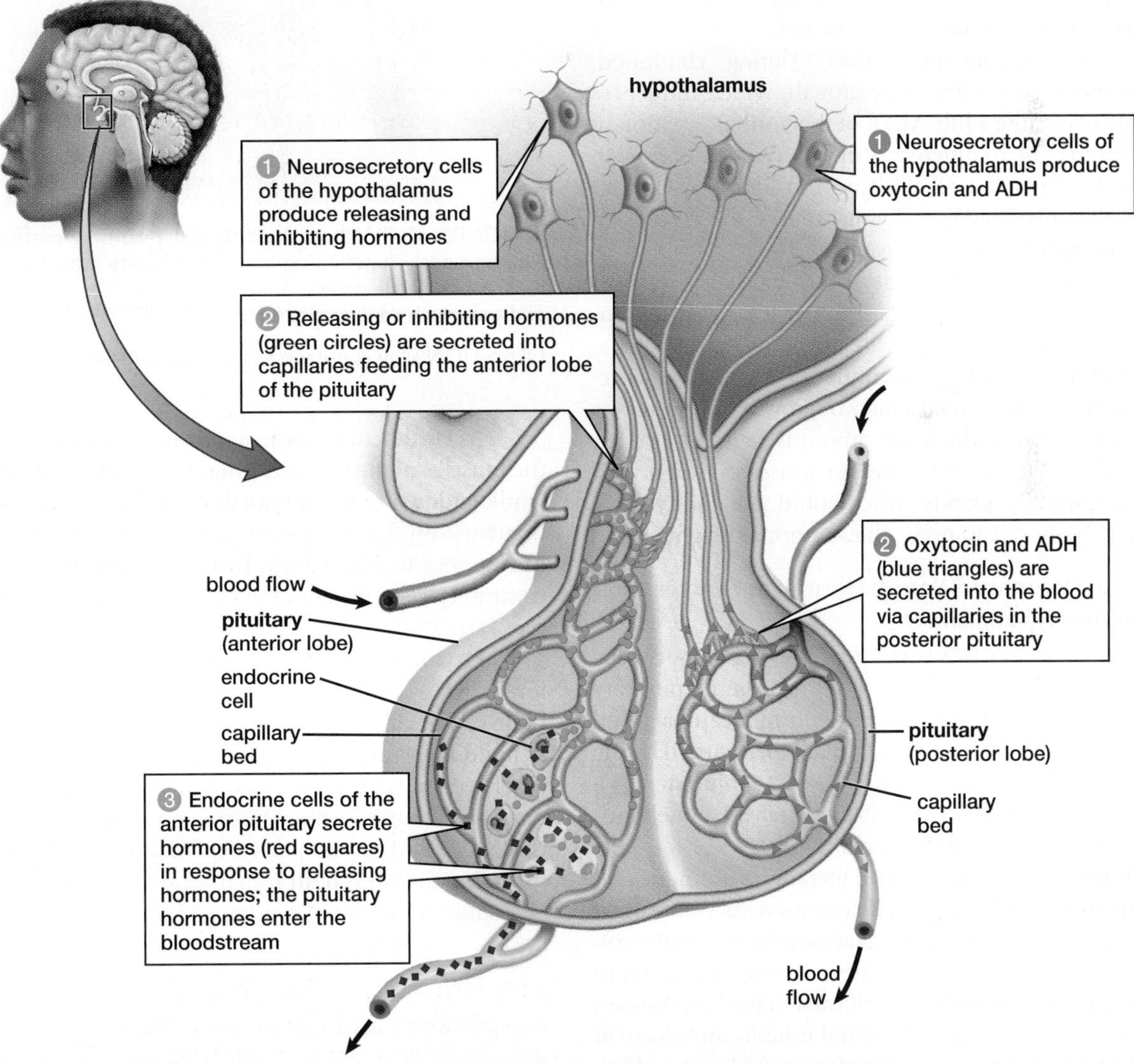
hypothalamus
1 Neurosecretory cells of the hypothalamus produce releasing and inhibiting hormones
1 Neurosecretory cells of the hypothalamus produce oxytocin and ADH
2 Releasing or inhibiting hormones (green circles) are secreted into capillaries feeding the anterior lobe of the pituitary
2 Oxytocin and ADH (blue triangles) are secreted into the blood via capillaries in the posterior pituitary
blood flow
pituitary (anterior lobe)
endocrine cell
capillary bed
pituitary (posterior lobe)
capillary bed
3 Endocrine cells of the anterior pituitary secrete hormones (red squares) in response to releasing hormones; the pituitary hormones enter the bloodstream
blood flow

▲ **FIGURE 37-6 The hypothalamus–pituitary system** The left side of the diagram (green circled numbers) shows the relationship between the hypothalamus and the anterior pituitary, and the right side (blue circled numbers) shows the relationship between the hypothalamus and the posterior pituitary. Releasing hormones are shown as green circles, hormones from the anterior pituitary as red squares, and hormones from the hypothalamus/posterior pituitary as blue triangles.

The Anterior Pituitary Produces and Releases Several Hormones

The anterior pituitary produces several peptide hormones. Four of these regulate hormone production in other endocrine glands. **Follicle-stimulating hormone (FSH)** and **luteinizing hormone (LH)** stimulate the production of sperm and testosterone in males and the production of eggs, estrogen, and progesterone in females. We will discuss the roles of FSH and LH in Chapter 41. **Thyroid-stimulating hormone (TSH)** stimulates the thyroid gland to release its hormones, and **adrenocorticotropic hormone (ACTH;** "hormone that stimulates the adrenal cortex") causes the release of the hormone cortisol from the adrenal cortex. We will examine the effects of thyroid and adrenal cortical hormones later in this chapter.

The remaining hormones of the anterior pituitary do not act on other endocrine glands. **Prolactin,** in conjunction with other hormones, stimulates the development of milk-producing mammary glands in the breasts during pregnancy. **Growth hormone** acts on nearly all the body's cells by increasing protein synthesis, promoting the use of fats for energy, and regulating carbohydrate metabolism. During childhood, growth hormone stimulates bone growth, which influences the ultimate size of the adult. Much of the normal variation in human height is due to differences in the secretion of growth hormone from the anterior pituitary. Too little growth hormone—or defective receptors for it—causes some cases of dwarfism; too much can cause gigantism (Fig. 37-7).

A major advance in the treatment of pituitary dwarfism occurred in 1981 when molecular biologists successfully inserted the gene for human growth hormone into bacteria, which then churned out large quantities of the hormone. Previously, the main commercial source of growth hormone was human cadavers, from which tiny amounts were extracted at great cost. Thanks to the new, cheaper source, children with underactive pituitary glands, who would previously have been extremely short, can now achieve normal height.

The Posterior Pituitary Releases Hormones Synthesized by Cells in the Hypothalamus

The hypothalamus contains two types of neurosecretory cells that grow thin fibers, called axons, into the posterior pituitary (Fig. 37-6 blue step ❶). These axons end in a capillary bed into which they release hormones that are then carried by the bloodstream to the rest of the body (Fig. 37-6 blue step ❷). These neurosecretory cells synthesize and release either antidiuretic hormone (ADH) or oxytocin.

Antidiuretic hormone (ADH; literally, "a hormone that reduces urination") helps prevent dehydration. As you learned in Chapter 35, ADH increases the water permeability of the collecting ducts of nephrons in the kidney, causing more water to be reabsorbed from the urine and returned to the bloodstream (see pp. 683–686). Interestingly, alcohol inhibits the release of ADH and increases urination. Therefore, people who drink strong alcoholic beverages may lose more water than they consume, and become dehydrated. In fact, dehydration contributes to the headache and generally miserable sensations known as a hangover.

▲ **FIGURE 37-7 When the anterior pituitary malfunctions** An improperly functioning anterior pituitary may produce too much or too little growth hormone. Too little causes dwarfism, and too much causes gigantism.

QUESTION Why is it easier to treat dwarfism than gigantism?

As we described earlier, oxytocin causes contractions of the muscles of the uterus during childbirth. It also triggers the "milk letdown reflex" in nursing mothers by causing muscle tissue within the mammary glands of the breasts to contract in response to stimulation by the suckling infant. This contraction ejects milk from the saclike milk glands into the nipples (Fig. 37-8).

Oxytocin also acts directly in the brain, causing behavioral effects. In rats, for example, injecting oxytocin into the brain causes virgin females to exhibit maternal behaviors, such as building a nest, licking other rats' pups, and retrieving pups that have strayed. Injecting oxytocin into the brains of virgin female sheep causes comparable maternal behaviors. In humans (both men and women), oxytocin may play a role in emotions, including trust and both romantic and maternal love (see the case study "How Do I Love Thee?" in Chapter 38).

The Thyroid and Parathyroid Glands Influence Metabolism and Calcium Levels

Lying at the front of the neck, nestled just below the larynx, the **thyroid gland** (Fig. 37-9) produces two hormones: thyroxine and calcitonin. The **parathyroid gland** consists of two pairs of small disks of endocrine cells, one pair on each side of the back of the thyroid gland. These cells release parathyroid hormone.

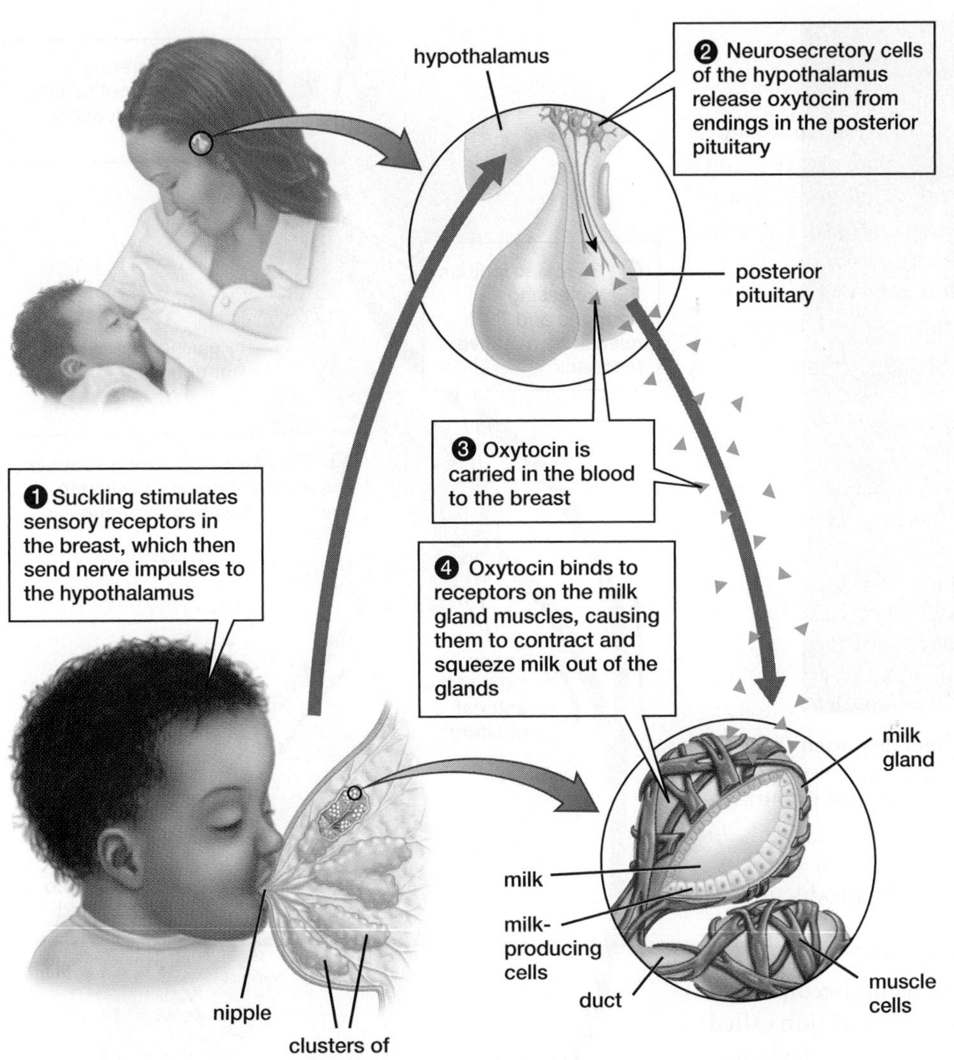
hypothalamus
❷ Neurosecretory cells of the hypothalamus release oxytocin from endings in the posterior pituitary
posterior pituitary
❸ Oxytocin is carried in the blood to the breast
❶ Suckling stimulates sensory receptors in the breast, which then send nerve impulses to the hypothalamus
❹ Oxytocin binds to receptors on the milk gland muscles, causing them to contract and squeeze milk out of the glands
milk gland
milk
milk-producing cells
duct
muscle cells
nipple
clusters of milk glands

◀ FIGURE 37-8 **Hormones and breast-feeding** Feedback between an infant and its mother regulates the control of milk letdown by oxytocin during breast-feeding. The cycle begins with the infant's suckling and continues until the infant is full and stops suckling. When the nipple is no longer stimulated, oxytocin release stops, the muscles relax, and milk flow ceases.

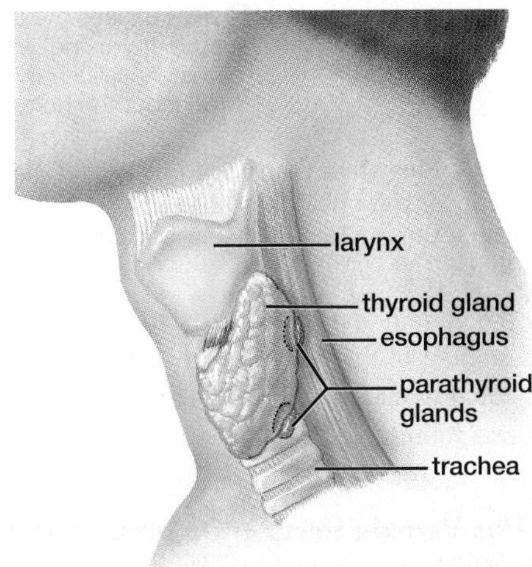
larynx
thyroid gland
esophagus
parathyroid glands
trachea

▲ FIGURE 37-9 **The thyroid and parathyroid glands** The thyroid gland wraps around the front of the larynx in the neck. The tiny parathyroid glands sit in the back of the thyroid.

Thyroxine Influences Energy Metabolism

Thyroxine, or thyroid hormone, is an iodine-containing amino acid derivative. Thyroxine works by binding to intracellular receptors that regulate gene activity. By stimulating the synthesis of enzymes that break down glucose and provide energy, and perhaps by direct actions on mitochondria, thyroid hormone elevates the metabolic rate of many body cells. In adults, levels of thyroxine determine the resting rate of cellular metabolism. Normal levels of thyroxine are required for mental alertness. Low thyroxine makes people feel mentally and physically sluggish. They may lose appetite but still gain weight, and they become less tolerant of cold (the body generates less of its own heat when its metabolic rate is low). Excess thyroxine leads to restlessness and irritability, increased appetite, and intolerance to heat.

In juvenile animals, including humans, thyroxine helps regulate growth by stimulating both metabolic rate and nervous system development. Undersecretion of thyroid hormone early in life can cause cretinism, a condition characterized by retardation of both mental and physical development. Fortunately, early diagnosis and thyroxine supplementation can reverse this condition.

▲ **FIGURE 37-10 Goiter** An iodine-deficient diet often causes enlargement of the thyroid gland. Seldom seen in developed countries, goiter is all too common in less-developed countries with populations that lack iodized salt in the diet.

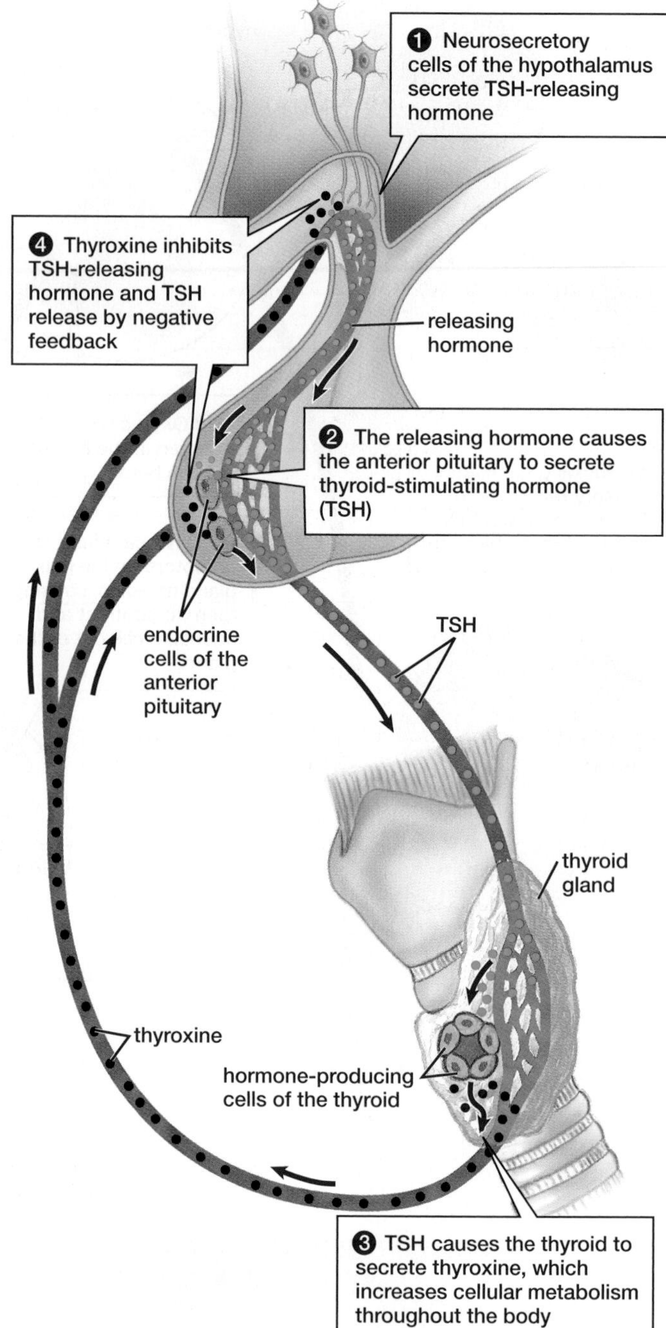
1 Neurosecretory cells of the hypothalamus secrete TSH-releasing hormone
4 Thyroxine inhibits TSH-releasing hormone and TSH release by negative feedback
releasing hormone
2 The releasing hormone causes the anterior pituitary to secrete thyroid-stimulating hormone (TSH)
endocrine cells of the anterior pituitary
TSH
thyroid gland
thyroxine
hormone-producing cells of the thyroid
3 TSH causes the thyroid to secrete thyroxine, which increases cellular metabolism throughout the body

▲ **FIGURE 37-11 Negative feedback in thyroid gland function** The concentration of thyroxine in the bloodstream regulates the secretion of TSH-releasing hormone and TSH by negative feedback.

QUESTION A common test of thyroid gland function is to measure the amount of thyroid-stimulating hormone circulating in the blood. What would you hypothesize if an abnormally high level of TSH was found?

An iodine-deficient diet can reduce the production of thyroxine and trigger a feedback that attempts to restore normal hormone levels by increasing the number of thyroxine-producing cells. The thyroid gland becomes enlarged and may form a bulge in the neck, a condition called **goiter** (**Fig. 37-10**). Even minor iodine deficiency, far below the deficit required to produce goiter, can cause inadequate synthesis of thyroxine. Worldwide, hundreds of millions of people, almost entirely in less-developed countries, have iodine-deficient diets. Iodine deficiency in pregnant women and young children is the leading preventable cause of mental retardation. Iodized salt is a simple, and cheap, solution to iodine deficiency, at a cost of less than $2 per ton of salt.

Thyroxine Release Is Controlled by the Hypothalamus and Anterior Pituitary

Levels of thyroxine in the bloodstream are fine-tuned by negative feedback loops. Thyroid stimulating hormone-releasing hormone (TSH-releasing hormone) produced by neurosecretory cells in the hypothalamus (**Fig. 37-11 ❶**) travels to the anterior pituitary and causes the release of TSH (**Fig. 37-11 ❷**). TSH travels in the bloodstream to the thyroid and stimulates the release of thyroxine (**Fig. 37-11 ❸**). Secretion of TSH-releasing hormone and TSH are regulated by negative feedback (**Fig. 37-11 ❹**)—adequate levels of thyroxine circulating in the bloodstream inhibit the secretion of both TSH-releasing hormone from the hypothalamus and TSH from the anterior pituitary, thus inhibiting further release of thyroxine from the thyroid gland.

Thyroxine Has Varied Effects in Different Vertebrates

Thyroxine governs many additional functions in non-human vertebrates. In amphibians, for example, thyroxine has the dramatic effect of triggering metamorphosis. In 1912, in one of the first demonstrations of the action of any hormone, tadpoles were fed minced horse thyroid. As a result, the tadpoles metamorphosed prematurely into miniature adult frogs. Thyroxine also regulates the seasonal molting of most vertebrates. From snakes to birds to your family dog, surges of thyroxine stimulate the shedding of skin, feathers, or hair.

Parathyroid Hormone and Calcitonin Regulate Calcium Metabolism

The proper concentration of calcium is essential to nerve and muscle function. **Parathyroid hormone** from the parathyroid gland and **calcitonin** from the thyroid work together to maintain nearly constant calcium levels in the blood and body fluids. The skeleton serves as a "bank" into which calcium can be deposited or withdrawn as necessary. If blood calcium levels drop, parathyroid hormone causes the bones to release calcium. It also causes the kidneys to reabsorb more calcium during urine production, and to return the calcium to the blood. The increased blood calcium then inhibits further release of parathyroid hormone in a negative feedback loop.

If blood calcium gets too high, the thyroid releases calcitonin, which inhibits the release of calcium from bone. In most vertebrates, calcitonin is important in regulating calcium concentrations in the blood, and may even promote bone growth. In humans, however, the actions of calcitonin appear to be minor compared to those of parathyroid hormone.

The Pancreas Has Both Digestive and Endocrine Functions

The **pancreas** produces both digestive secretions and hormones. As we described in Chapter 34, the pancreas produces bicarbonate and several enzymes that are released into the small intestine, promoting the digestion of food (see p. 668). The endocrine portion of the pancreas consists of clusters of **islet cells.** Each islet cell produces one of two peptide hormones: **insulin** or **glucagon.**

Insulin and Glucagon Control Glucose Levels in the Blood

Insulin and glucagon work in opposition to regulate carbohydrate and fat metabolism: Insulin reduces the blood glucose level, whereas glucagon increases it (**Fig. 37-12**). Together, the two hormones help keep the blood glucose level nearly constant. When blood glucose rises (for example, after you have eaten; **Fig. 37-12** ❶), the pancreas releases insulin (**Fig. 37-12** ❷). Insulin causes body cells to take up glucose (**Fig. 37-12** ❸) and either metabolize it for energy or convert it to fat or glycogen (a polymer of glucose that is stored in the liver and skeletal muscles). When blood glucose levels drop (for example, after you've skipped breakfast or run a 10-kilometer race; **Fig. 37-12** ❹ and ❺), insulin secretion is inhibited and glucagon secretion is stimulated (**Fig. 37-12** ❻). Glucagon activates an

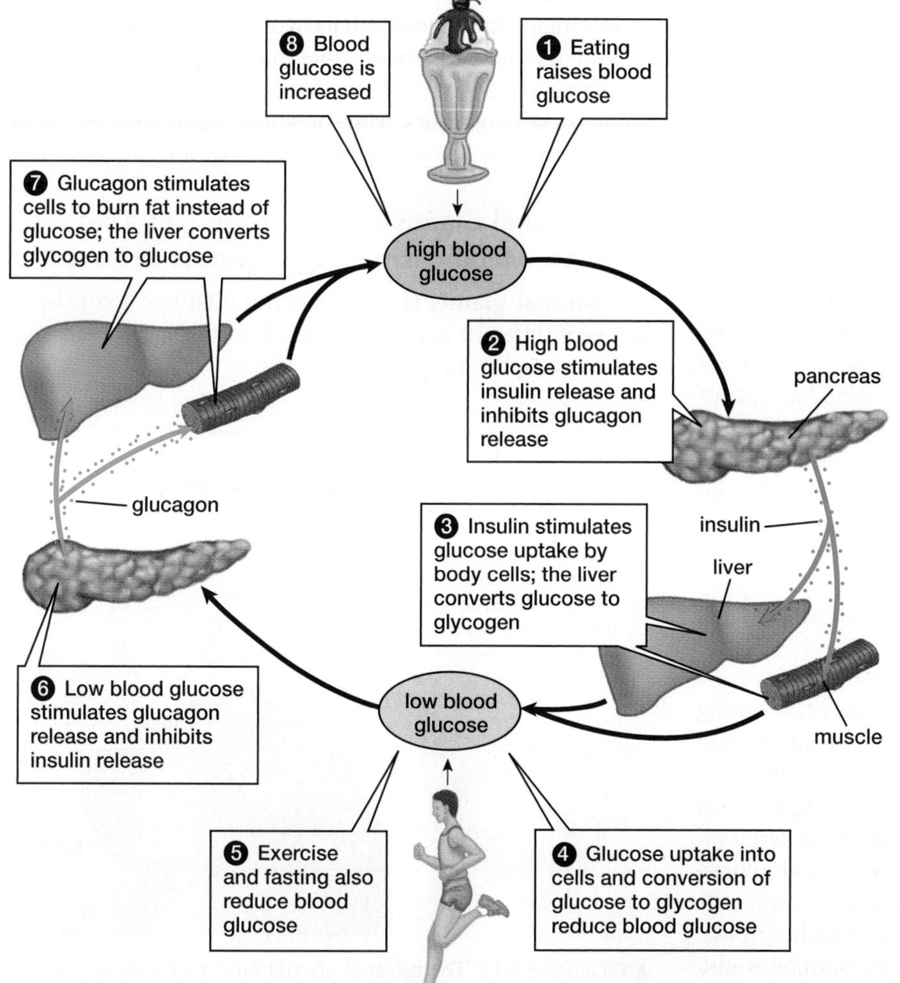

❽ Blood glucose is increased
❶ Eating raises blood glucose
❼ Glucagon stimulates cells to burn fat instead of glucose; the liver converts glycogen to glucose
high blood glucose
❷ High blood glucose stimulates insulin release and inhibits glucagon release
pancreas
glucagon
❸ Insulin stimulates glucose uptake by body cells; the liver converts glucose to glycogen
insulin
liver
❻ Low blood glucose stimulates glucagon release and inhibits insulin release
low blood glucose
muscle
❺ Exercise and fasting also reduce blood glucose
❹ Glucose uptake into cells and conversion of glucose to glycogen reduce blood glucose

◀ **FIGURE 37-12 The pancreas controls blood glucose levels** Pancreatic islets contain two populations of hormone-producing cells: One produces insulin (green arrows); the other produces glucagon (blue arrows). These two hormones cooperate in a two-part negative feedback loop to control blood glucose concentrations (black arrows).

QUESTION How would blood glucose be affected in a person who was born with a mutation that prevented glucagon receptors from binding glucagon?

enzyme in the liver that breaks down glycogen, releasing glucose into the blood (**Fig. 37-12** ❼). Glucagon also promotes fat breakdown, which releases fatty acids that can be metabolized for energy. These actions increase blood glucose levels (**Fig. 37-12** ❽), which inhibits glucagon secretion. If blood glucose levels rise too far, insulin is secreted once again.

BioFlix ™ Homeostasis: Regulating Blood Sugar

Diabetes Results from a Malfunctioning Insulin Control System

Lack of insulin production or the failure of target cells to respond to insulin results in **diabetes mellitus.** In either case, blood glucose levels are high and fluctuate with food intake. Because many body cells can't take up glucose unless they are stimulated by insulin, they rely heavily on fats as an energy source instead. This leads to high levels of lipids—including cholesterol—circulating in the blood. Severe diabetes causes fat deposits to accumulate in the blood vessels, predisposing individuals to high blood pressure and heart disease; more than 75% of diabetics die of cardiovascular disease. Fat deposits in small blood vessels in the kidneys and the retina of the eye may cause kidney failure and blindness, respectively. Insulin replacement therapy profoundly improves the health of diabetics, but requires daily blood testing and insulin injections, and can't fully mimic natural control of energy metabolism. Advances in the treatment of diabetes that may eliminate the daily inconvenience of insulin testing and administration and improve an individual's overall health are described in "Health Watch: Closer to a Cure for Diabetes."

The Sex Organs Produce Both Gametes and Sex Hormones

The sex organs do far more than just produce sperm or eggs. The **testes** in males and the **ovaries** in females are also important endocrine organs (see Fig. 37-5). The testes secrete several steroid hormones, collectively called **androgens.** The most important of these is **testosterone.** The ovaries secrete two types of steroid hormones: **estrogen** and **progesterone.** Although there is a surge of sex hormone production at puberty, sex hormones are present from the fetal stage onward. They influence development in both sexes, and they continue to affect both behavior and brain function throughout life. The roles of the sex hormones in sperm and egg production, the menstrual cycle, pregnancy, and development are discussed in Chapters 41 and 42.

Sex Hormone Levels Increase During Puberty

Sex hormones play a key role in puberty, the phase of life during which the reproductive systems of both sexes become mature and functional. Puberty begins when, for reasons not fully understood, the hypothalamus starts to secrete increasing amounts of releasing hormones, which in turn stimulate the anterior pituitary to secrete more luteinizing hormone (LH) and follicle-stimulating hormone (FSH). LH and FSH stimulate target cells in the testes or ovaries to produce higher levels of sex hormones.

The resulting increase in circulating sex hormones ultimately affects tissues throughout the body that contain the appropriate receptors. Both sexes develop pubic and underarm hair. Testosterone, secreted by the testes of males, stimulates the development of male secondary sexual characteristics, including body and facial hair, broad shoulders, and muscle growth. Testosterone also promotes sperm cell production. Estrogen from the ovaries in females stimulates breast development and the maturation of the female reproductive system, including egg production. Progesterone prepares the reproductive tract to receive and nourish the fertilized egg.

Studies in a wide variety of animals, as well as a few studies in humans, have revealed that environmental pollutants from agricultural and industrial activities can disrupt hormone signaling. These "endocrine disruptors" frequently mimic or block the actions of sex hormones, with sometimes devastating effects, as described in "Earth Watch: Endocrine Deception" on p. 728.

Case Study continued

Anabolic Steroids—Fool's Gold?

For athletes, the point of anabolic steroids is to mimic the muscle-building actions of testosterone, only more so. However, in both men and women, testosterone—especially in large amounts—also affects reproductive physiology, cholesterol metabolism, bone and hair growth, mood, and behavior. Anabolic steroids illustrate a problem faced by medical practitioners dating back to Hippocrates: It is extremely difficult to find a drug that does only one thing. In "Case Study Revisited: Anabolic Steroids—Fool's Gold?" at the end of the chapter, we describe a few of the common side effects of anabolic steroids.

The Adrenal Glands Secrete Hormones That Regulate Metabolism and Responses to Stress

The **adrenal glands** (Latin for "on the kidney") consist of two very different parts: the adrenal cortex and the adrenal medulla (**Fig. 37-13**).

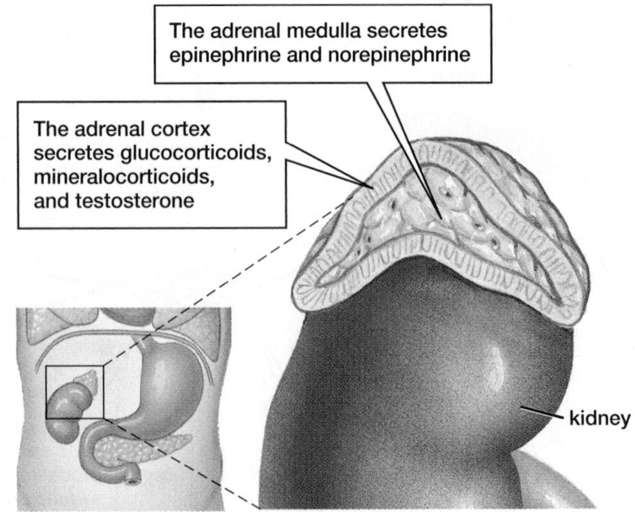
The adrenal medulla secretes epinephrine and norepinephrine
The adrenal cortex secretes glucocorticoids, mineralocorticoids, and testosterone
kidney

▲ **FIGURE 37-13 The adrenal glands** Atop each kidney sits an adrenal gland, composed of an outer cortex and an inner medulla.

Health Watch

Closer to a Cure for Diabetes

The most severe kind of diabetes, type 1, occurs when a person's immune system attacks and kills the insulin-producing islet cells of the pancreas. This form of diabetes often strikes early in life, triggering a lifelong daily regimen of blood tests and insulin injections. For the millions of people who suffer from this form of diabetes, recent research offers real hope.

In 1999, researchers led by James Shapiro of the University of Alberta, Canada, isolated islet cells from brain-dead donors, and injected them into veins leading to the livers of seven patients with type 1 diabetes. Some of the islet cells colonized the livers and started secreting insulin. Soon, all seven patients became completely independent of insulin injections for at least a year. Unfortunately, the recipients must take immunosuppressant drugs to keep their immune systems from rejecting the donated islet cells.

Immune suppression is a major threat to the recipients' health. Further, islet cell donors are in very short supply compared to the millions of type 1 diabetics. Researchers at Novocell, a biotech firm in San Diego, California, have made important strides toward bypassing both of these difficulties.

First, they encapsulated islet cells from deceased donors in polyethyleneglycol (PEG), an inert polymer that has been shown to be safe to use in humans. They injected the cells just beneath the skin of human volunteers and found that fairly small molecules, such as sugar and insulin, could diffuse into and out of the PEG capsules, but immune cells and antibodies couldn't enter and attack the transplanted islet cells. Therefore, immunosuppressants were not needed.

This technology still doesn't solve the problem of a shortage of transplantable islet cells. However, the Novocell team found a way to cause human embryonic stem (ES) cells (**Fig. E37-1**) to differentiate into insulin-secreting cells. Stem cells have the potential to continuously generate more cells, so the right stem-cell line could someday produce a virtually unlimited supply of transplantable, islet-like cells. Unfortunately, Novocell's ES-derived cells didn't completely mimic functional islet cells. Normally, islet cells form part of a negative feedback loop—too much glucose in the blood causes the islet cells to secrete lots of insulin, which stimulates other cells of the body to take up glucose. This, in turn, reduces the glucose concentration in the blood, so the islet cells slow down their insulin secretion (see Fig. 37-12). In contrast, Novocell's differentiated stem cells behaved like islet cells in fetuses—they produced insulin, but didn't release it in response to glucose.

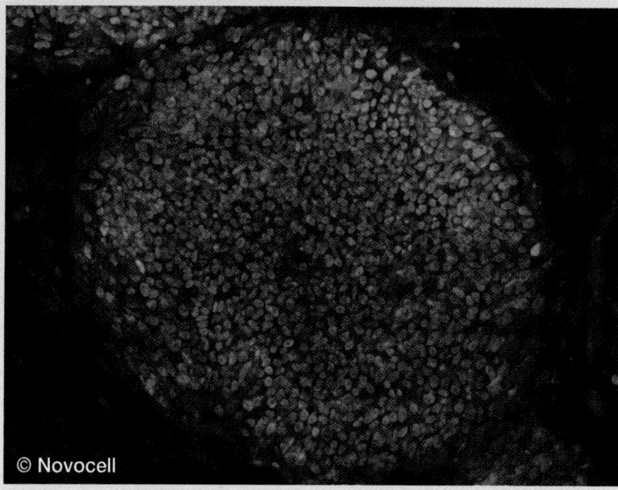

▲ **FIGURE E37-1 A potential cure for type 1 diabetes?** Soon, it may be possible to transform human embryonic stem cells such as these into fully functioning pancreatic islet cells. Photo © Novocell.

Hoping that differentiation in intact animals would work better than differentiation in a culture dish, the Novocell team tried implanting not-quite-fully differentiated human ES-derived cells into mice. Sure enough, the mice transformed the cells into functional islet cells, which release insulin in response to glucose.

There are still many obstacles to overcome, and researchers and physicians rightly caution that these treatments are many years away from clinical practice. Nevertheless, encapsulated ES-derived islet cells might one day prove to be the magic bullet against type 1 diabetes.

The Adrenal Cortex Produces Steroid Hormones

The outer layer of the adrenal gland forms the **adrenal cortex** ("cortex" is Latin for "bark"). The cortex secretes three types of steroid hormones: glucocorticoids, mineralocorticoids, and small amounts of testosterone. As their names imply, **glucocorticoids** help control glucose metabolism, while **mineralocorticoids** regulate salt metabolism.

Glucocorticoid release is stimulated by adrenocorticotropic hormone (ACTH) from the anterior pituitary, which in turn is stimulated by releasing hormones from the hypothalamus. Glucocorticoids are released in response to stimuli such as stress, trauma, or exposure to temperature extremes. **Cortisol** is by far the most abundant glucocorticoid. Cortisol increases blood glucose levels by stimulating glucose production, inhibiting the uptake of glucose by muscle cells, and promoting the use of fats for energy.

You may have noticed that many different hormones are involved in glucose metabolism: thyroxine, insulin, glucagon, epinephrine, and the glucocorticoids. Why? The reason can probably be traced to a metabolic requirement of the brain. Although most body cells can produce energy from fats and proteins as well as from carbohydrates, brain cells can burn only glucose. Thus, blood glucose levels cannot be allowed to fall too low, or brain cells rapidly starve, leading to unconsciousness and possibly death.

Earth Watch

Endocrine Deception

In recent decades, researchers have found that some synthetic organic compounds that enter the environment, especially aquatic ecosystems such as streams and lakes, can mimic or block the actions of certain hormones, most commonly estrogen, testosterone, or thyroxine. These **endocrine disruptors** include pesticides (DDT), herbicides (atrazine), plastics (bisphenol A, phthalates), flame retardants (polybrominated diphenyl ethers, or PBDEs), detergents (alkylphenol ethoxylates), sunscreens (oxybenzone), and polychlorinated biphenyls (PCBs, used in sealants, paints, and as insulating fluids in power transformers). Probably the most potent endocrine disruptor in the environment is ethinylestradiol, a synthetic estrogen commonly found in birth control pills.

In the most common form of endocrine disruption, a synthetic chemical enters cells and binds to estrogen receptors. Overactivation of estrogen receptors, inappropriate timing of receptor activation, or activation in males disrupts development. In a wide variety of animals, endocrine disruptors exert harmful effects, including feminization in males, masculinization in females, reproductive cancers, malformed sex organs, altered blood hormone levels, and reduced fertility. Not surprisingly, feminization of males—such as the presence of abnormal testes, and sometimes even egg production—is the most common effect of estrogenic endocrine disruptors.

In one of the best-known cases, agricultural runoff and a pesticide spill near Lake Apopka in Florida polluted the lake's water with large quantities of several endocrine disruptors, including DDT and its major breakdown products. Wildlife biologists noted an alarming decline in the alligator population of the lake. Many eggs were not hatching. Males had high estrogen, low testosterone, smaller penises, and abnormal testes. Females typically had exceptionally high estrogen levels and abnormal ovaries.

Although alarming, the Lake Apopka results were the result of massive exposure to endocrine disruptors. Do smaller doses, likely to be found in fairly clean water, also have harmful effects? Indeed they may. Scientists sampled a common minnow in Boulder Creek, which runs through Boulder, Colorado (**Fig. E37-2**). They found that downstream of the city's sewage outfall, more than 80% of the fish were females, whereas upstream of the outfall, half were males and half females. In lab studies, male fish exposed to effluent from the treatment plant were rapidly feminized, further implicating the wastewater. In 2007, the city of Boulder upgraded its sewage treatment plant, with the hope that most of the endocrine disruptors will be removed from the effluent.

Are endocrine disruptors harmful to people? Some, such as PCBs, certainly are. Others probably are harmful at high enough concentrations. The debates among toxicologists, reproductive biologists, industry, and government regulators focus on several questions: What concentrations are needed for human effects? Do those concentrations occur in humans? Do mixtures of several endocrine disruptors, all at very low concentrations, add up to harmful effects? How could we definitively prove either safety or harm, given that human experiments cannot be performed?

▲ **FIGURE E37-2 Boulder Creek** Although Boulder Creek hardly looks polluted, it contains chemicals that disrupt the reproductive systems of fish.

Some known endocrine disruptors, such as DDT and PCBs, have been banned in many countries. For suspected endocrine disruptors, the situation is more unsettled. In 2008, Canada banned the use of bisphenol A in plastic baby bottles. In 2010, the U.S. Food and Drug Administration, although not banning BPA, recommended that parents should limit their use of baby products that contain BPA. Meanwhile, plastic bottle manufacturer Nalgene will stop using bisphenol A, and Walmart will stop selling baby bottles containing bisphenol A. People disagree about whether the actions by Canada, Nalgene, and Walmart were based on sound scientific evidence, public opinion, or the "precautionary principle" that when harm is suspected, the burden of proof falls on those who claim that a product is safe.

In modern society, these debates play out in the popular press, so you can follow the arguments yourself, in newspapers, online, or in magazines such as *Scientific American*, *Science News*, or *New Scientist*.

Have you ever wondered?

Why You Often Get Sick When You're Stressed?

In addition to its effects on glucose metabolism, cortisol also inhibits the immune response. If you typically get sick during final exam week, stress-induced cortisol release may be responsible. Although finals are not usually life threatening, many other stresses, such as finding enough food or encountering predators, can be. Because activation of the immune system requires a lot of energy and makes people and other mammals feel sluggish, suppressing the immune response in favor of dealing with immediate stresses may save an animal's life.

Mineralocorticoid hormones regulate the mineral (salt) content of the blood. The most important mineralocorticoid is **aldosterone,** which helps to control sodium concentrations. Sodium ions are the most abundant positive ions in blood and extracellular fluid. The sodium ion gradient across plasma membranes (high in the extracellular fluid, low in the cytoplasm) is crucial to many cellular events, including the production of electrical signals by nerve cells. If blood sodium falls, the adrenal cortex releases aldosterone, which causes the kidneys and sweat glands to retain sodium. When salt and other sources of dietary sodium, combined with aldosterone-induced sodium conservation, return blood sodium concentrations back to normal levels, aldosterone secretion is shut off—another example of negative feedback.

In both women and men, the adrenal cortex also produces the male sex hormone testosterone, although in much smaller amounts than are produced by the testes. Tumors of the adrenal cortex can lead to excessive testosterone release, causing masculinization of women. Many of the "bearded ladies" who once appeared in circus sideshows probably had this condition.

The Adrenal Medulla Produces Amino Acid-Derived Hormones

The **adrenal medulla** is located in the center of each adrenal gland. It produces two hormones in response to stress or exercise—**epinephrine,** and a smaller quantity of **norepinephrine** (also called adrenaline and noradrenaline, respectively). These hormones prepare the body for emergency action. They increase heart and respiratory rates, increase blood pressure, cause blood glucose levels to rise, and direct blood flow away from the digestive tract and toward the brain and muscles. They also cause the air passages to the lungs to expand, allowing larger volumes of air to enter and leave the lungs. For this reason, epinephrine is often administered to asthmatics, whose airways become constricted during an asthma attack.

Embryologically, the adrenal medulla is actually part of the nervous system. It is activated by the sympathetic nervous system, which prepares the body for "fight or flight." We will discuss the sympathetic nervous system in Chapter 38.

Hormones Are Also Produced by the Pineal Gland, Thymus, Kidneys, Heart, Digestive Tract, and Fat Cells

The **pineal gland** is located between the two hemispheres of the brain (see Fig. 37-5). The pineal gland produces the hormone **melatonin,** an amino acid derivative. Melatonin is secreted in a daily rhythm, which in mammals is regulated by light entering the eyes. In some vertebrates, such as the frog, the pineal itself contains photoreceptive cells. The skull above it is thin, so the pineal can detect sunlight and thus day length. By responding to day lengths characteristic of different seasons, the pineal appears to regulate the seasonal reproductive cycles of many animals.

Despite years of research, the function of melatonin and the pineal gland in humans is still not well understood. In children, melatonin from the pineal may suppress the onset of puberty, but the mechanism remains unknown. Melatonin secretion from the pineal gland may influence sleep–wake cycles. Darkness increases melatonin production and bright light inhibits it, and there is some evidence that secretion of melatonin at night promotes sleep. Consequently, properly timed melatonin is sometimes used as a sleeping aid and to overcome jet lag, although not all experts agree on this approach.

The **thymus** is located in the chest cavity behind the breastbone (see Fig. 37-5). The thymus produces the hormone **thymosin,** which stimulates the development of specialized white blood cells (T cells) that play crucial roles in the immune response (see Chapter 36). The thymus is large in infants but, under the influence of sex hormones, decreases in size after puberty. As a result, the elderly produce fewer new T cells than adolescents do and, hence, are more susceptible to new diseases.

The kidneys produce **erythropoietin,** a peptide hormone that is released when the oxygen content of the blood is low. Erythropoietin stimulates the bone marrow to increase red blood cell production (see p. 628). The kidneys also produce an enzyme called **renin** in response to low blood pressure, for example, after profuse bleeding from a wound. Renin catalyzes the production of the hormone **angiotensin** from proteins in the blood. Angiotensin raises blood pressure by constricting arterioles. It also stimulates the release of aldosterone by the adrenal cortex, causing the kidneys to return sodium to the blood. A higher salt concentration attracts and retains water, increasing blood volume and pressure.

The stomach and small intestine produce a bewildering variety of peptide hormones that help regulate digestion. These hormones include gastrin, ghrelin, secretin, cholecystokinin, and several others, some of which we discussed in chapter 34 (see p. 672).

Even the heart releases a hormone. If the blood volume becomes too high—for example, if you drink too much water—the atria are overfilled, which stretches their walls and stimulates the release of **atrial natriuretic peptide (ANP).** ANP inhibits the release of ADH and aldosterone and increases the excretion of sodium. By reducing reabsorption of water and salt by the kidneys, ANP helps to lower blood volume.

Finally, who would have thought that fat could be an endocrine organ? In 1995 Jeffrey Friedman and colleagues at

Rockefeller University discovered the peptide hormone **leptin,** which is released by fat cells. Mice genetically engineered to lack the gene for leptin became obese (**Fig. 37-14**), and leptin injections caused them to lose weight. The researchers hypothesized that by releasing leptin, fat tissue "tells" the body how much fat it has stored and therefore how much to eat: If the mice already have lots of stored fat, then high leptin levels would cause them to eat less. Unfortunately, trials of leptin as a human weight-loss aid have not been encouraging. Many obese people have high levels of leptin but seem to be relatively insensitive to it. However, researchers have discovered other functions for leptin, which appears to stimulate the growth of new capillaries and to speed wound healing. It also stimulates the immune system and is required for the onset of puberty and the development of secondary sexual characteristics.

Many other cells in the body also produce hormones, and more hormones surely remain to be discovered.

▲ **FIGURE 37-14 Leptin helps regulate body fat** The mouse on the left has been genetically engineered to lack the gene for the hormone leptin.

Case Study revisited
Anabolic Steroids—Fool's Gold?

Like natural testosterone, anabolic steroids help to increase muscle mass. However, naturally occurring hormones are usually present in minuscule amounts. Taking relatively large doses, especially of synthetic hormones that are often more potent and longer lasting than the natural hormones they mimic, can mean trouble. Even if they're never caught and don't lose their medals, steroid abusers risk losing their health. Obviously, no one does clinical trials on secret, illegal drugs, so the side effects of anabolic steroids are often uncertain. However, the National Institute on Drug Abuse and the U.S. Anti-Doping Agency list many harmful effects.

In both sexes, anabolic steroids cause acne and may depress the immune system. Mood swings and sudden aggressiveness are sufficiently common to have produced the slang term "roid rage." Anabolic steroids have been linked to increases in blood pressure and decreases in the "good" form of cholesterol (HDL)—both are risk factors for heart attacks and strokes. Anabolic steroids can cause bone growth to shut down prematurely, so young steroid abusers may never reach their full potential height.

In males, anabolic steroids create a negative feedback effect that reduces the natural production of testosterone. Tricked by these testosterone mimics, the anterior pituitary releases smaller amounts of hormones that are required for testes development and sperm production, so the testes often shrink and sperm count drops. Finally, men produce enzymes that convert some anabolic steroids into estrogen, which may cause partial breast development.

In females, anabolic steroids promote male-like bodily changes, including deepening of the voice, increased facial hair, and even pattern baldness. Testosterone-like hormones also interfere with egg development and ovulation, often causing irregularities in the menstrual cycle.

Consider This

Some athletes say that they would willingly risk long-term damage to their bodies to win Olympic gold. Some commentators believe that there will always be successful, undetected, drug-abusing athletes. Therefore, they suggest that professional sports should let everyone use any drug they want, to level the playing field—if the athletes want to ruin their health, that's their problem. But what about high school athletes? Based on a poll taken in 2007, the National Institute on Drug Abuse concluded that about 2.3% of U.S. male high school seniors used anabolic steroids in the previous year. A poll of Southern California high school athletes found that only 1% admit to steroid use themselves, but 15% think that their teammates are using. Some of these students are as young as 14 or 15 years old, with bodies that are still developing rapidly. Should routine drug testing be conducted in high school sports to protect the athletes' health?

CHAPTER REVIEW

Summary of Key Concepts

37.1 How Do Animal Cells Communicate?

Within multicellular organisms, communication among cells occurs through gap junctions directly linking cells, by diffusion of chemicals to nearby cells (local hormones such as prostaglandins and neurotransmitters secreted by nerve cells), and by transport of chemicals within the bloodstream (endocrine hormones). Extracellular chemical messengers act selectively on target cells that bear specific receptors for the chemical. The vertebrate endocrine hormones are produced by glands embedded in capillary beds. The hormones are secreted into the extracellular fluid, diffuse into the capillaries, and are then transported in the bloodstream to other parts of the body.

37.2 How Do Animal Hormones Work?

Vertebrate hormones fall into one of three classes: peptides, amino acid derivatives, and steroids. Most hormones act on their target cells in one of two ways: (1) Peptide hormones and amino acid-derived hormones bind to receptors on the surfaces of target cells, and activate intracellular second messengers, such as cyclic AMP, which then alter the cell's metabolism. (2) Steroid hormones usually diffuse through the plasma membranes of their target cells and bind with receptors in the cytoplasm or nucleus. The hormone–receptor complex promotes the transcription of specific genes within the nucleus. Thyroid hormones are transported across the plasma membrane into cells, where they bind to receptors and influence gene transcription. Hormones in invertebrates include peptides and steroids, as well as some molecules not found in vertebrates. Their mechanisms of action are similar to those of vertebrate hormones.

Hormone action is commonly regulated through negative feedback, a process in which a hormone causes changes that inhibit further secretion of that hormone. In rare instances, such as childbirth, hormone release may be temporarily controlled by positive feedback, but in all known cases, hormone release eventually stops.

37.3 What Are the Structures and Functions of the Mammalian Endocrine System?

The major endocrine glands of the human body are the hypothalamus–pituitary complex, the thyroid and parathyroid glands, the pancreas, the sex organs, and the adrenal glands. The hormones released by these glands and their actions are summarized in Table 37-3. Other structures that produce hormones include the pineal gland, thymus, kidneys, digestive tract, heart, and fat cells.

BioFlix™ Homeostasis: Regulating Blood Sugar

Key Terms

adrenal cortex *727*
adrenal gland *726*
adrenal medulla *729*
adrenocorticotropic hormone (ACTH) *722*
aldosterone *729*
amino acid-derived hormone *716*
androgen *726*
angiotensin *729*
anterior pituitary *721*
antidiuretic hormone (ADH) *722*
atrial natriuretic peptide (ANP) *729*
calcitonin *725*
cortisol *727*
cyclic adenosine monophosphate (cyclic AMP) *716*
diabetes mellitus *726*
ecdysone *719*
endocrine disrupter *728*
endocrine gland *715*
endocrine hormone *715*
endocrine system *719*
epinephrine *729*
erythropoietin *729*
estrogen *726*
follicle-stimulating hormone (FSH) *722*
glucagon *725*
glucocorticoid *727*
goiter *724*
growth hormone *722*
hypothalamus *721*
inhibiting hormone *721*
insulin *725*
islet cell *725*
leptin *730*
local hormone *714*
luteinizing hormone (LH) *722*
melatonin *729*
mineralocorticoids *727*
negative feedback *718*
neurosecretory cell *721*
norepinephrine *729*
ovary *726*
oxytocin *715*
pancreas *725*
parathyroid gland *722*
parathyroid hormone *725*
peptide hormone *716*
pineal gland *729*
pituitary gland *721*
positive feedback *718*
posterior pituitary *721*
progesterone *726*
prolactin *722*
prostaglandin *715*
receptor *714*
releasing hormone *721*
renin *729*
second messenger *716*
steroid hormone *716*
target cell *714*
testis (plural, testes) *726*
testosterone *726*
thymosin *729*
thymus *729*
thyroid gland *722*
thyroid-stimulating hormone (TSH) *722*
thyroxine *723*

Thinking Through the Concepts

Fill-in-the-Blank

1. Hormones are molecules released by cells that are parts of the __________. These cells are embedded in capillary beds, so the hormones enter the bloodstream and move throughout the body. Only specific cells of the body, called __________, can respond to any given hormone, because only these cells bear proteins, called __________, that can bind the hormone.
2. Most hormones fall into three chemical classes: __________, __________, and __________. __________ and __________ are mostly water soluble and bind to receptors on the surfaces of cells. These typically stimulate the synthesis of intracellular molecules called __________, which activate enzymes and change the cell's metabolism. __________ are lipid soluble and bind to receptors in the cytoplasm or nucleus. The hormone–receptor complex typically binds to DNA and causes __________.
3. A part of the brain called the __________ controls the activity of the pituitary gland. Specialized nerve cells in this brain area, called __________, release the hormones __________ or __________ from the endings of their axons in the posterior lobe of the pituitary.
4. The major hormones produced by the anterior pituitary gland are (in any order): __________, __________, __________, __________, __________, and __________.
5. The pancreas releases the hormone __________ when blood glucose levels become too high; it causes many cells of the body to take up glucose. When the pancreas produces too little of this hormone, or body cells cannot respond to it, a disorder called __________ results. __________ is released when blood glucose levels become too low; it causes the liver to break down the starch-like storage molecule, __________, and release glucose into the blood.
6. The male sex organs, called the __________, release the sex hormone __________. The female sex organs, called the __________, release two hormones, __________ and __________.
7. The adrenal cortex releases three major types of steroid hormones: __________, __________, and __________. The adrenal medulla releases the amino acid-derived hormones __________ and __________.

Review Questions

1. What are the three types of molecules used as endocrine hormones in vertebrates? Give an example of each.
2. Which chemical class of hormones usually attaches to membrane receptors on target cells? What cellular events usually follow?
3. Which chemical class of hormones usually binds to receptors inside of target cells? What cellular events usually follow?
4. Diagram the process of negative feedback, and give an example of negative feedback in the control of hormone action.
5. What are the major endocrine glands in the human body, and where are they located?
6. Describe the structure and function of the hypothalamus–pituitary complex. Describe how releasing hormones regulate the secretion of hormones by cells of the anterior pituitary. Name the hormones of the anterior pituitary, and give one function of each.
7. Describe how the hormones of the pancreas act together to regulate the concentration of glucose in the blood.
8. Compare the adrenal cortex and adrenal medulla by answering the following questions: Where are they located within the adrenal gland? Which hormones do they produce? Which organs do their hormones target?

Applying the Concepts

1. A student decides to do a science project on the effect of the thyroid gland on frog metamorphosis. She sets up three aquaria with tadpoles. She adds thyroxine to the water of one, the drug thiouracil to a second, and nothing to the third. Thiouracil destroys thyroxine. Assuming that the student uses appropriate physiological concentrations, predict what will happen.
2. Suggest a hypothesis about the endocrine system to explain why many birds lay their eggs in the spring and why poultry farmers who produce eggs keep lights on at night.
3. **BioEthics** Some parents who are interested in college sports scholarships for their children are asking physicians to prescribe growth hormone treatments, even though their children are in the normal height range. What biological and ethical dilemmas does this raise for the parents, children, physicians, coaches, and college scholarship boards?
4. Why do you think that leptin might be required for the onset of puberty in girls?

MB *Go to www.masteringbiology.com for practice quizzes, activities, eText, videos, current events, and more.*

The Nervous System

Chapter 38

Case Study

How Do I Love Thee?

> "But soft! What light through yonder window breaks?
> It is the east, and Juliet is the sun."
> —*Romeo and Juliet*, Act II, scene II

IN SHAKESPEARE'S *Romeo and Juliet*, two teenagers fall in love the first time they meet. A few hours later, as Romeo watches Juliet gazing out of her window, he sees her as the sun that lights up his life. Shakespeare's *Romeo and Juliet* dramatically portrays the power of romantic love, for which the lovers defy their families, risk their fortunes and their futures, and finally, give up their lives.

Romance is, of course, not the only manifestation of love. A mother's love for her child is just as strong. People have also defined their lives and willingly died for love of God and love of country. Just what *is* love? Are all these types of love distinct, or are they related? What happens in the brain when two lovers meet, or a mother cradles her infant?

No one knows for sure—not in people, anyway. Perhaps surprisingly, neuroscientists know a lot about love—or at least pair bonding and sex—in a small rodent called the prairie vole. If Juliet had been a prairie vole, her first encounter with Romeo would have released a flood of oxytocin, the same hormone that causes uterine contractions during childbirth. The oxytocin would have bound to receptors in a few small areas of her brain, causing nerve cells to release dopamine, often called the reward chemical of the brain. She would have felt wonderful. What's more, she would have linked that euphoric feeling to Romeo. In a Romeo vole, some of the molecules and brain regions would have differed, but the end result would have been similar: a torrent of dopamine, giving him the ultimate high, and he would have believed that he could attain that feeling again only with Juliet. So, the two voles would mate and bond for life. They would build a nest, live together, and raise their young.

How do humans and other animals perceive their world? How do they evaluate what they perceive, and become calm or excited, fearful or smitten? How do they respond with appropriate behaviors such as resting, eating, or mating? Although most perceptions and behaviors are not completely understood, the answers to these questions lie in the nervous system.

▲ Love: "A fire sparkling in lovers' eyes . . . a madness most discreet," or just the right mixture of chemicals deep within lovers' brains? (Inset) Prairie voles provide insights into the neurochemical basis of emotional love.

At a Glance

38.1 WHAT ARE THE STRUCTURES AND FUNCTIONS OF NERVE CELLS?

The nervous system has two principal cell types: **neurons,** often called nerve cells, and **glia.** As we will see, neurons receive, process, and transmit information. Glia assist neuronal function in several ways, such as by providing nutrients, regulating the composition of the extracellular fluid in the brain and spinal cord, modulating communication between neurons, and speeding up the movement of electrical signals within neurons. Although glia are very important—the nervous system could not function without them—we will focus on the structure and function of neurons in this chapter.

The Functions of a Neuron Are Localized in Separate Parts of the Cell

A neuron must perform four functions:

1. Receive information from the internal or external environment, or from other neurons.
2. Process this information, often along with information from other sources, and produce an electrical signal.
3. Conduct the electrical signal, sometimes for a considerable distance, to a junction where it meets another cell.
4. Transmit information to other cells, such as other neurons or the cells of muscles or glands.

Although neurons vary enormously in structure, most vertebrate neurons have four distinct parts that carry out these four functions: dendrites, a cell body, an axon, and synaptic terminals (**Fig. 38-1**).

Dendrites Respond to Stimuli

Dendrites, which are branched tendrils protruding from the cell body, perform the "receive information" function (**Fig. 38-1** ❷ and ❼). Their branches provide a large surface area for receiving signals, either from the environment or from other neurons. Dendrites of sensory neurons have membrane adaptations that allow them to produce electrical signals in response to specific stimuli from the external environment—such as pressure, odor, or light—or the internal environment, such as body temperature, blood pH, or the position of a joint. Dendrites of neurons in the brain and spinal cord usually respond to chemicals, called **neurotransmitters,** that are released by other neurons.

The Cell Body Processes Signals from the Dendrites

Electrical signals travel down the dendrites and converge on the neuron's **cell body,** which performs the "process information" function (**Fig. 38-1** ❸). The cell body "adds up," or integrates, electrical signals that it receives from the dendrites. As we will see shortly, some of these signals are positive and some are negative. If their sum is sufficiently positive, the neuron will produce a large, rapid electrical signal called an **action potential** (**Fig. 38-1** ❹). The cell body also contains the organelles found in most cells, such as the nucleus, endoplasmic reticulum, and Golgi apparatus, and performs typical

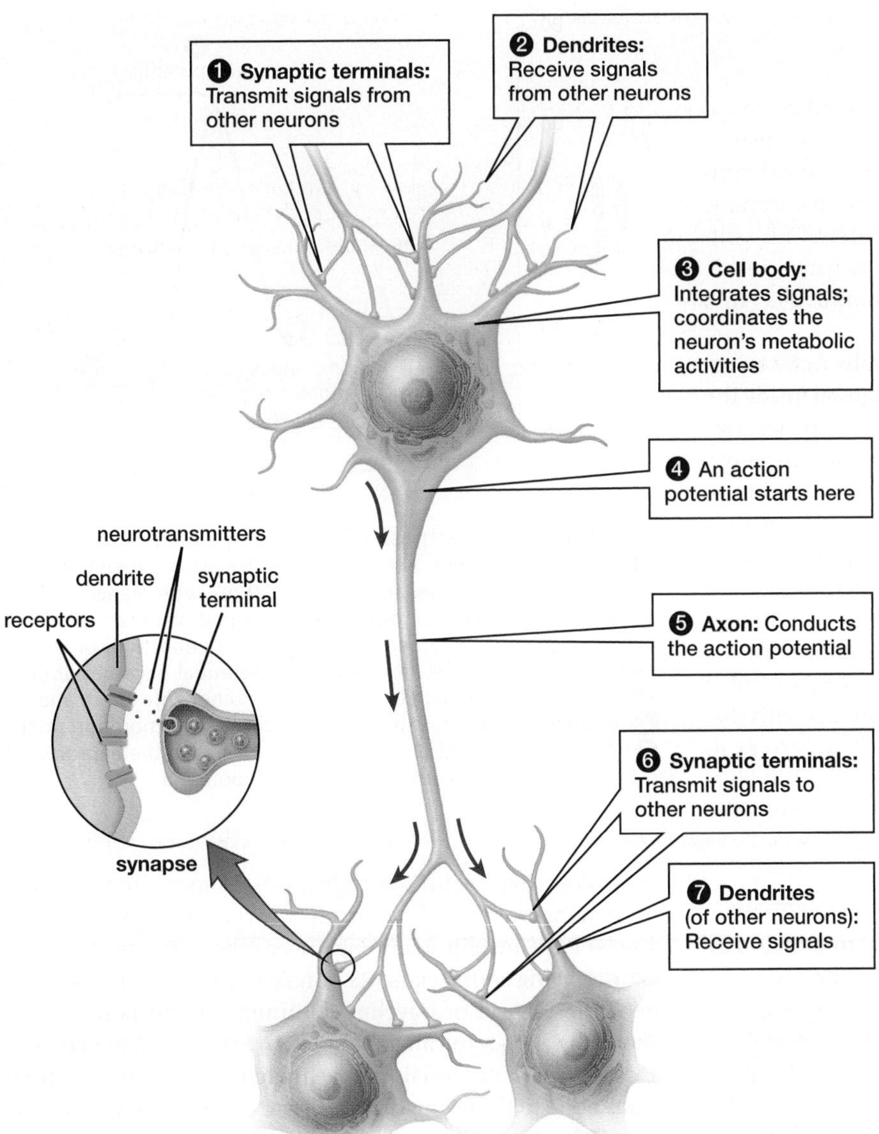
❶ Synaptic terminals: Transmit signals from other neurons
❷ Dendrites: Receive signals from other neurons
❸ Cell body: Integrates signals; coordinates the neuron's metabolic activities
❹ An action potential starts here
❺ Axon: Conducts the action potential
❻ Synaptic terminals: Transmit signals to other neurons
❼ Dendrites (of other neurons): Receive signals
neurotransmitters
dendrite
synaptic terminal
receptors
synapse

◀ **FIGURE 38-1 A neuron, showing its specialized parts and their functions** The red arrows indicate action potentials moving from the cell body down the axon to the synaptic terminals.

cellular activities such as synthesizing complex molecules and coordinating the cell's metabolism.

The Axon Conducts Action Potentials Long Distances

In a typical neuron, a long, thin strand called an **axon** extends outward from the cell body. The axon conducts action potentials (**Fig. 38-1** ❺) from the cell body to synaptic terminals at the axon's end, where it contacts other cells (**Fig. 38-1** ❶ and ❻). Single axons may stretch from your spinal cord to your toes, a distance of about 3 feet (approximately a meter), making neurons the longest cells in the body. Axons are typically bundled together into **nerves,** much like wires are bundled in an electrical cable. In vertebrates, axons bundled into nerves emerge from the brain and spinal cord and extend to all regions of the body.

At Synapses, Signals Are Transmitted from One Cell to Another

The site where a neuron communicates with another cell is called a **synapse.** A typical synapse consists of (1) the **synaptic terminal,** which is a swelling at the end of an axon of the "sending" neuron; (2) a dendrite or cell body of a "receiving" neuron, muscle, or gland cell; and (3) a small gap separating the two cells (**Fig. 38-1** ❻; see also Fig. 38-4 later in this chapter). Most synaptic terminals contain neurotransmitters that are released in response to an action potential reaching the terminal. The plasma membrane of the receiving neuron bears receptors that bind the neurotransmitters and stimulate a response in this cell. Therefore, at a synapse, the output of the first cell becomes the input to the second cell.

38.2 HOW DO NEURONS PRODUCE AND TRANSMIT INFORMATION?

Although there are many exceptions, as a general rule, information is carried *within* a neuron by electrical signals, and information is transmitted *between* neurons by neurotransmitters released from one neuron and received by a second neuron.

Electrical Signals Carry Information Within a Single Neuron

In the 1930s, biologists developed ways to record electrical events inside individual neurons. They found that an unstimulated, inactive neuron maintains a constant electrical voltage difference, or potential, across its plasma membrane, similar to the voltage across the poles of a battery. This voltage, called the **resting potential,** is always negative inside the cell and ranges from about −40 to −90 millivolts (mV; thousandths of a volt).

If the neuron is stimulated, either naturally or by an experimenter using an electrical current, the potential inside the neuron can be made either more negative or less negative (**Fig. 38-2**). If the potential becomes sufficiently less negative, it reaches a level called **threshold,** and triggers an action potential. During an action potential, the neuron's voltage rises rapidly to about +50 mV inside the cell. Action potentials last a few milliseconds (thousandths of a second) before the cell's negative resting potential is restored. The plasma membranes of axons are specialized to conduct action potentials from a neuron's cell body to the axon's synaptic terminals. Unlike electrical voltages in metal wires, which get smaller with distance, action potentials are conducted from cell body to axon terminal with no change in voltage. In "A Closer Look at Electrical Signaling in Neurons" on pp. 738–739, we examine the cellular mechanisms of resting and action potentials.

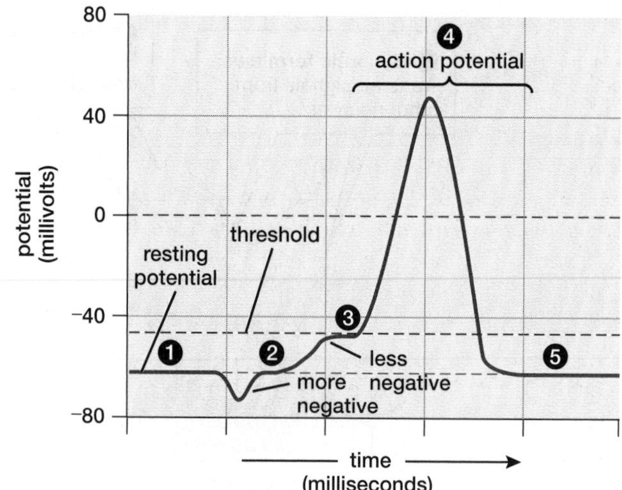

▲ **FIGURE 38-2 Electrical events during an action potential** ❶ A neuron maintains a voltage across its plasma membrane, called the resting potential, of about −60 mV with respect to the outside. ❷ Stimulation from the environment or other cells may make the neuron more negative (downward deflection) or less negative (upward deflection). ❸ If the potential becomes about 10 to 20 mV less negative, the neuron reaches threshold, and ❹ produces a brief positive potential called an action potential. ❺ After a millisecond or two, the voltage across the neuron's plasma membrane returns to the resting potential.

Myelin Speeds Up the Conduction of Action Potentials

How fast an action potential travels varies tremendously among axons. In general, the thicker the axon, the faster the action potential moves. A much more effective way of speeding conduction is to cover the axon with a fatty insulation called **myelin** (**Fig. 38-3**). Myelin is formed by glial cells—oligodendrocytes in the brain and spinal cord, and Schwann cells in the rest of the body—that wrap themselves around the axon, covering it with multiple layers of specially insulating plasma membrane, with scarcely any cytoplasm between the wraps. Each myelin wrapping covers about 0.2 to 2 mm of axon, leaving short segments of naked axon, called nodes, in between. Instead of traveling continuously but fairly slowly down the axon, typically about 3 to 6 feet (1 to 2 meters) per second, action potentials in myelinated axons "jump" rapidly from node to node, traveling at a rate of 10 to 330 feet (about 3 to 100 meters) per second (see Fig. 38-3).

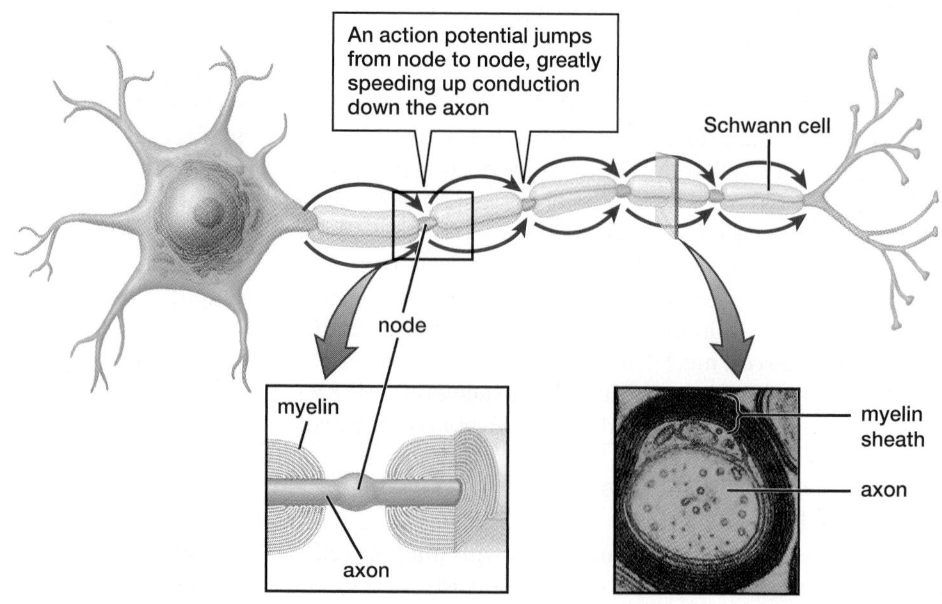

► **FIGURE 38-3 A myelinated axon** Many vertebrate axons are covered in myelin, which consists of insulating wrappings of membrane of specialized glial cells. Action potentials occur only at the nodes between each myelin wrapping, jumping from node to node (red arrows), with almost no time spent traveling beneath the myelin.

Neurons Use Chemicals to Communicate with One Another at Synapses

Think of an action potential as a packet of information moving down an axon. Once it reaches the synaptic terminal, this information must be transmitted to another cell, either another neuron or a cell in a muscle or gland. At what are called *electrical synapses*, electrical activity can pass directly from neuron to neuron through gap junctions connecting the insides of the cells (see p. 93). In the heart, electrical synapses interconnect cardiac muscle cells, helping to produce coordinated electrical activity, and hence contraction, of the heart. Although electrical synapses occur in many places in the mammalian brain, their significance in brain function remains poorly understood. Neurons far more frequently use chemicals to communicate with one another or with cells of muscles or glands. Here, we will confine our discussion to these chemical synapses.

In ordinary English, the word "transmit" means "to send something," and that is exactly what happens at a synapse, where the synaptic terminal of one neuron meets the dendrite of another (**Fig. 38-4**). The two neurons do not actually touch at a synapse: a tiny gap, the **synaptic cleft,** separates the first, or **presynaptic neuron,** from the second, or **postsynaptic neuron.** The presynaptic neuron sends neurotransmitter molecules across the gap to the postsynaptic neuron. Different neurons synthesize and respond to a vast array of neurotransmitters. **Table 38-1** lists a few well-known neurotransmitters and some of their functions.

A synaptic terminal contains scores of vesicles, each full of neurotransmitter molecules. When an action potential is initiated (**Fig. 38-4** ❶), it travels down an axon until it reaches its synaptic terminal (**Fig. 38-4** ❷). The inside of the terminal becomes positively charged, which triggers a cascade of changes that cause some of the vesicles to release neurotransmitters into the synaptic cleft (**Fig. 38-4** ❸). (Neurotransmitter release is actually a specialized case of exocytosis; see pp. 90–91.) The outer surface of the plasma membrane of the postsynaptic neuron, just across the synaptic cleft, is packed with receptor proteins that are specialized to bind the neurotransmitter released by the presynaptic neuron. The neurotransmitter molecules diffuse across the gap and bind to these receptors (**Fig. 38-4** ❹).

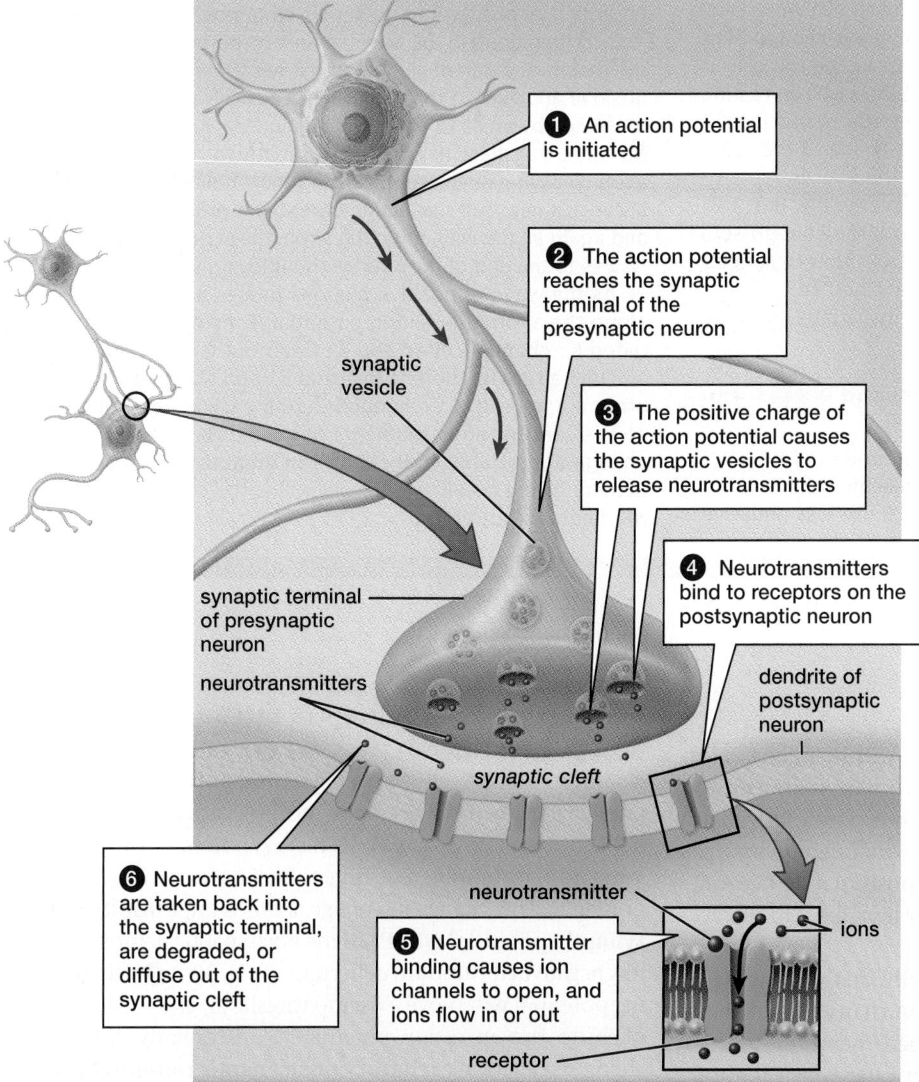
❶ An action potential is initiated
❷ The action potential reaches the synaptic terminal of the presynaptic neuron
synaptic vesicle
❸ The positive charge of the action potential causes the synaptic vesicles to release neurotransmitters
❹ Neurotransmitters bind to receptors on the postsynaptic neuron
synaptic terminal of presynaptic neuron
neurotransmitters
dendrite of postsynaptic neuron
synaptic cleft
❻ Neurotransmitters are taken back into the synaptic terminal, are degraded, or diffuse out of the synaptic cleft
neurotransmitter
ions
❺ Neurotransmitter binding causes ion channels to open, and ions flow in or out
receptor

◀ **FIGURE 38-4 The structure and function of the synapse**

QUESTION Imagine an experiment in which the neurons pictured here are bathed in a solution containing a nerve poison. The presynaptic neuron is stimulated and produces an action potential, but this does not result in a PSP in the postsynaptic neuron. When the experimenter adds some neurotransmitter to the synapse, the postsynaptic neuron still produces no PSP. How does the poison act to disrupt nerve function?

A Closer Look At *Electrical Signaling in Neurons*

Potassium Permeability Produces the Resting Potential

The resting potential is based on a balance between chemical and electrical gradients and is maintained by active transport and a membrane that is selectively permeable to specific ions. The ions of the cytoplasm consist mainly of positively charged potassium ions (K^+) and large, negatively charged organic molecules such as ATP and proteins, which cannot leave the cell (**Fig. E38-1a,** bottom). Outside the cell, the extracellular fluid contains mostly positively charged sodium ions (Na^+) and negatively charged chloride ions (Cl^-). The concentration gradients of Na^+ and K^+ are maintained by an active transport protein in the plasma membrane called the **sodium-potassium (Na^+-K^+) pump,** which simultaneously pumps K^+ into and Na^+ out of the cell.

In an unstimulated neuron, only K^+ can cross the plasma membrane, by traveling through specific membrane proteins called "resting" K^+ channels (see Fig. E38-1a, bottom). There are also "voltage-gated" Na^+ and K^+ channels in the membrane; as their name suggests, they have "gates" on their pores that are opened or closed by the voltage across the plasma membrane. In an unstimulated neuron, these voltage-gated Na^+ and K^+ channels are closed. We will describe the function of voltage-gated channels in a moment.

Because the K^+ concentration is higher inside the cell than outside, K^+ diffuses out of the cell through the resting K^+ channels, leaving the negatively charged organic ions behind (see Fig. E38-1a, bottom). As the inside of the cell becomes more and more negatively charged, K^+ is electrically attracted back into the cell. Eventually, the negative voltage inside the cell becomes large enough so that the rate of K^+ diffusing out is exactly counterbalanced by the rate of K^+ being pulled back in by electrical attraction (not shown in Fig. E38-1a). This negative voltage is the neuron's resting potential.

Changes in Permeability to Sodium and Potassium Produce the Action Potential

Action potentials occur when the resting potential is changed, becoming less negative and reaching the threshold voltage (usually about 10 to 20 mV less negative than the resting potential). At threshold, voltage-gated Na^+ channels open, allowing a rapid influx of Na^+, making the inside of the neuron positive (**Fig. E38-1b**).

Voltage-gated Na^+ channels stay open a very short time, and then spontaneously close. Meanwhile, voltage-gated K^+ channels open, allowing K^+ to flow out of the cell, restoring the negative resting potential (**Fig. E38-1c**).

Action Potentials Are Conducted Down Axons Without Changing Amplitude

Action potentials are all-or-none: if the neuron does not reach threshold, there will be no action potential; if threshold is reached, a full-size action potential will occur and travel the entire length of the axon.

An action potential usually starts where the axon emerges from a neuron's cell body. As Na^+ enters the axon here, its positive charge repels other positively charged ions in the axon's cytoplasm. Think of a pool table, with a dozen balls lined up in a row, touching each other. If you hit a ball on one of the ends with the cue ball, the ball on the opposite end instantly shoots off, and the balls in the middle remain where they were. Similarly, when Na^+ enters during an action potential, its positive charge repels other positively charged ions farther along in the axon, almost instantaneously causing the membrane potential of nearby areas to become more positive and exceed threshold. This causes Na^+ channels in these nearby areas to open, beginning a new action potential. Na^+ enters at these new locations slightly farther along the axon (see Fig. E38-1c, top), starting the whole process over. Because a new, full-size action potential is produced again and again all the way down the axon, the action potential travels to the end of the axon without losing voltage.

As the wave of positive charges passes a given point along the axon, the resting potential is restored as voltage-gated K^+ channels open and K^+ flows out (see Fig. E38-1c).

Only a tiny fraction of the total K^+ and Na^+ in and around a neuron is exchanged during each action potential, so the concentration gradients of K^+ and Na^+ do not change appreciably. In the long run (minutes to hours), the activity of the Na^+-K^+ pump maintains the concentration gradients of both ions.

Synapses Produce Excitatory or Inhibitory Postsynaptic Potentials

Recall that information is usually carried within a neuron by electrical signals. Fast, long-distance electrical signaling down an axon from a neuron's cell body to its presynaptic terminals is carried by action potentials. Therefore, synaptic activity typically alters the electrical activity of the postsynaptic neuron, making the neuron either more likely or less likely to fire action potentials.

At most synapses, the receptor proteins on the postsynaptic neuron are physically connected to ion channels that span the neuron's plasma membrane. When neurotransmitter molecules bind to these receptor proteins, they open the ion channels. Depending on which channels are associated with a specific receptor, ions such as Na^+, K^+, Ca^{2+}, or Cl^- may move through these channels (**Fig. 38-4** ❺), causing a small, brief change in voltage, called a **postsynaptic potential (PSP).** If the postsynaptic neuron becomes more negative (the downward deflection in Fig. 38-2 ❷), its resting potential moves farther away from threshold, reducing its likelihood of firing an action potential. This change in voltage is called an **inhibitory postsynaptic potential (IPSP).** If the postsynaptic neuron becomes less negative (the upward deflection in Fig. 38-2 ❷), then its resting potential will move closer to threshold, and it will be more likely to fire an action potential. Consequently, this voltage change is called an **excitatory postsynaptic potential (EPSP).**

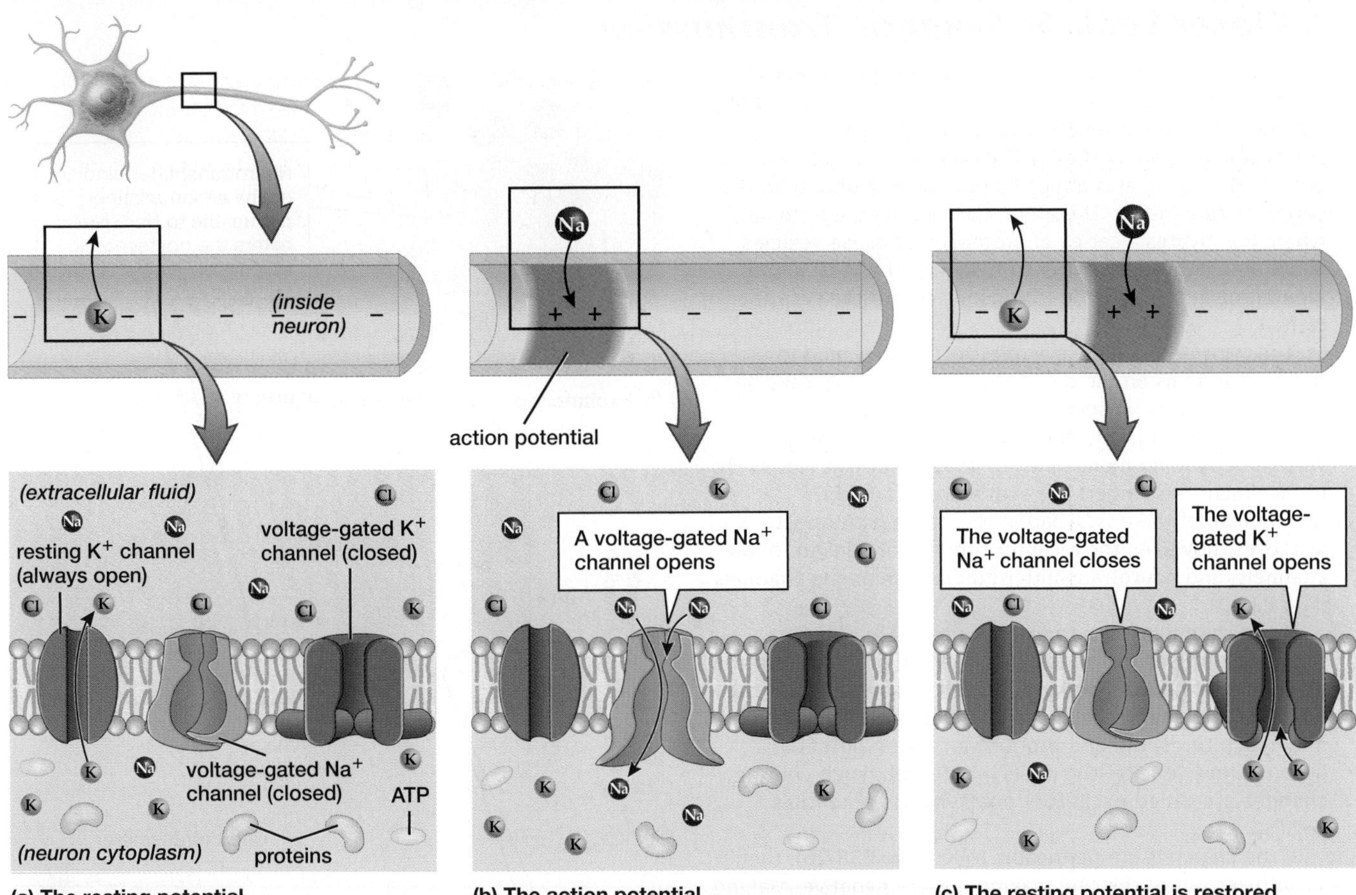

▲ **FIGURE E38-1 The ionic mechanisms underlying the resting and action potentials** The upper illustrations in each part of the figure show a section of an axon, with the important ion movements across the plasma membrane during **(a)** the resting potential, **(b)** the rising phase of an action potential, and **(c)** the falling phase of an action potential occurring in the boxed portion of the axon. In the upper illustration of part **(c),** a new action potential has begun, farther along in the axon (Na^+ enters the axon). The lower illustrations show the distribution of ions inside and outside the axon, the important ion channels controlling the resting and action potentials, and the movements of ions through the channels in the boxed portion of the axon.

BioFlix™ How Neurons Work

Table 38-1 Some Important Neurotransmitters

Neurotransmitter	Location in the Nervous System	Some Major Functions
Acetylcholine	Motor neuron-to-muscle synapses; autonomic nervous system, many areas of the brain	Activates skeletal muscles; activates target organs of the parasympathetic nervous system
Dopamine	Midbrain	Important in emotion, rewards, and the control of movement
Norepinephrine (noradrenaline)	Sympathetic nervous system	Activates target organs of the sympathetic nervous system
Serotonin	Midbrain, pons, and medulla	Influences mood and sleep
Glutamate	Many areas of the brain and spinal cord	Major excitatory neurotransmitter in the CNS
Glycine	Spinal cord	Major inhibitory neurotransmitter in the spinal cord
GABA (gamma amino butyric acid)	Many areas of the brain and spinal cord	Major inhibitory neurotransmitter in the brain
Endorphins	Many areas of the brain and spinal cord	Influence mood, reduce pain sensations
Nitric oxide	Many areas of the brain	Important in forming memories

A Closer Look At *Synaptic Transmission*

When an action potential reaches a presynaptic terminal, the positive charge inside the terminal opens a new set of voltage-gated ion channels, which are selectively permeable to calcium (Ca^{2+}). The concentration of Ca^{2+} outside the terminal is about 10,000 times higher than the concentration inside. Therefore, Ca^{2+} enters the terminal, where it activates a series of proteins that cause vesicles containing neurotransmitter to fuse with the presynaptic membrane and release their transmitter into the synaptic cleft.

The transmitters diffuse across the cleft and bind to receptor proteins on the postsynaptic cell. This typically has one of two effects. In some synapses, the result of transmitter–receptor binding is similar to what happens when a peptide hormone binds to its receptor (see Fig. 37-2): Intracellular messengers are synthesized, and the metabolism of the postsynaptic cell changes. At most synapses, however, the receptor proteins are linked to ion channels, and neurotransmitter binding opens the channels (**Fig. E38-2**).

If the channels are permeable to Na^+ (**Fig. E38-2a**), then Na^+ diffuses down its concentration gradient into the postsynaptic neuron, making the cell less negative. If the postsynaptic neuron becomes sufficiently less negative, it may reach threshold and produce an action potential. Because they "excite" the postsynaptic cell, such voltage changes are called excitatory postsynaptic potentials (EPSPs).

If the channels are permeable to K^+ (**Fig. E38-2b**), then K^+ diffuses out of the cell, making it more negative. Making the cell more negative inhibits the production of action potentials in the postsynaptic cell, so this voltage change is called an inhibitory postsynaptic potential (IPSP).

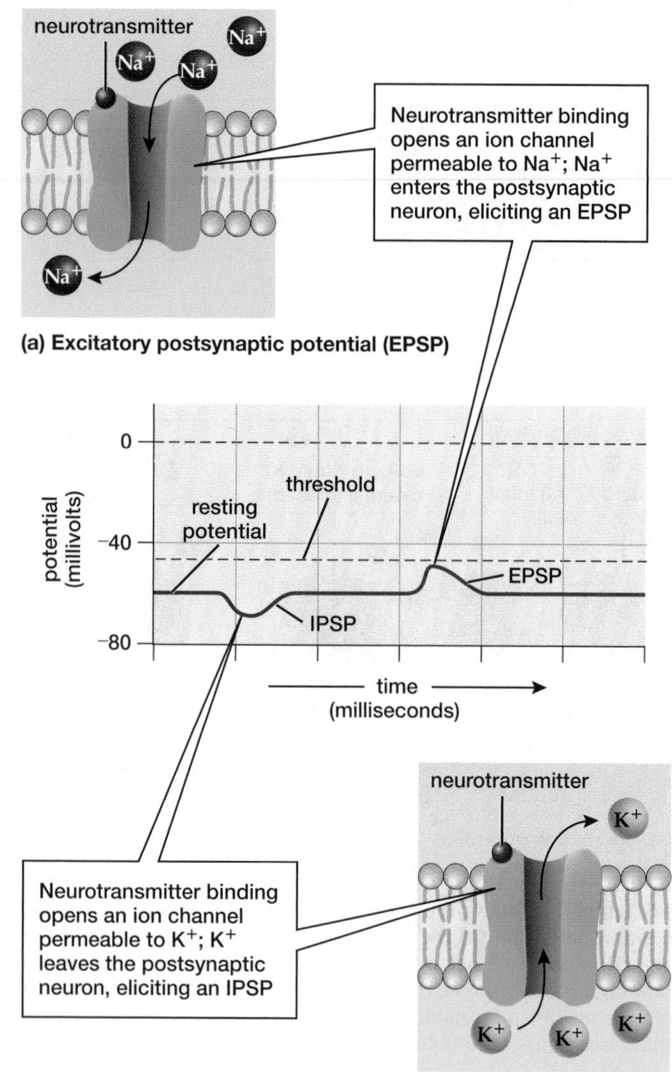
neurotransmitter
Na+
Neurotransmitter binding opens an ion channel permeable to Na+; Na+ enters the postsynaptic neuron, eliciting an EPSP
(a) Excitatory postsynaptic potential (EPSP)
0
-40
-80
potential (millivolts)
threshold
resting potential
EPSP
IPSP
time (milliseconds)
neurotransmitter
K+
Neurotransmitter binding opens an ion channel permeable to K+; K+ leaves the postsynaptic neuron, eliciting an IPSP
(b) Inhibitory postsynaptic potential (IPSP)

► **FIGURE E38-2 Neurotransmitter binding to receptor proteins opens ion channels (a)** The ionic mechanism of an EPSP. **(b)** The ionic mechanism of an IPSP.

The mechanisms by which neurotransmitters binding to receptors cause PSPs are explained in "A Closer Look at Synaptic Transmission."

Neurotransmitter Action Is Usually Brief

Consider what would happen if a presynaptic neuron started stimulating a postsynaptic cell and never stopped. You might, for example, contract your biceps muscle, flex your arm, and have it stay flexed forever! Not surprisingly, the nervous system has several ways of terminating neurotransmitter action. Some neurotransmitters—notably acetylcholine, the transmitter that stimulates skeletal muscle cells—are rapidly broken down by enzymes in the synaptic cleft. Many other neurotransmitters are transported back into the presynaptic neuron (**Fig. 38-4 ❻**). For all neurotransmitters, these mechanisms are supplemented by diffusion out of the synaptic cleft.

Summation of Postsynaptic Potentials Determines the Activity of a Neuron

Most postsynaptic potentials are small, rapidly fading signals, but they travel far enough to reach the cell body. There, they determine whether an action potential will be produced. How? The dendrites and cell body of a single neuron often receive EPSPs and IPSPs from the synaptic terminals of thousands of presynaptic neurons. The voltages of all the

Case Study continued

How Do I Love Thee?

Different animal species, and even different individuals within a species, have different numbers and types of receptors for neurotransmitters. Monogamous, pair-bonding prairie voles, for example, have a very high density of receptors for dopamine, the reward neurotransmitter, in certain parts of their brains. In contrast, another species of vole, the promiscuous montane vole, has much lower concentrations of dopamine receptors in these brain regions. Among humans, people with social phobias, who avoid many social situations for fear of embarrassment, have fewer dopamine receptors in reward areas of the brain than people who are good at social interactions.

PSPs that reach the postsynaptic cell body at about the same time are added up, a process called **integration.** If the excitatory and inhibitory postsynaptic potentials, when added together, raise the electrical potential inside the neuron above threshold, the postsynaptic cell will produce an action potential.

BioFlix™ How Synapses Work

38.3 HOW DO NERVOUS SYSTEMS PROCESS INFORMATION?

The individual neuron uses a simple language of action potentials, yet this basic language allows animals to perform complex behaviors. One key to the versatility of the nervous system is the presence of networks of neurons that range from dozens to millions of cells. As in computers, simple elements can perform amazing feats when connected properly.

Information Flow in the Nervous System Requires Four Basic Operations

At a minimum, a nervous system must be able to perform four operations:

1. Determine the type of stimulus.
2. Determine and signal the intensity of a stimulus.
3. Integrate information from many sources.
4. Initiate and direct appropriate responses.

The Nature of a Stimulus Is Determined by Connections Between the Senses and the Brain

If action potentials are the units of information of almost all neurons, and if all action potentials are basically the same, then how can the brain determine *what* a stimulus is—light, sound, hunger—or *how strong* a stimulus is?

All nervous systems interpret what a stimulus is by monitoring which neurons are firing action potentials. For example, your brain interprets action potentials that occur in the axons of your optic nerves (originating in the eye and traveling to the visual areas of the brain) as the sensation of light. Supposedly, a German physiologist once sat in a dark room and poked himself in the eye, slightly damaging his retina and producing action potentials that traveled to his brain. (As they say in TV ads showing cars speeding on racetracks, do *not* try this yourself!) The result? He "saw stars" because his brain interpreted action potentials in his optic nerve as light. Thus, you distinguish the sound of music from the taste of coffee, or the bitterness of coffee from the sweetness of sugar, because these different stimuli result in action potentials in different axons that connect to different areas of your brain.

The Intensity of a Stimulus Is Coded by the Frequency of Action Potentials

Because all action potentials are roughly the same size and duration, no information about the strength, or **intensity,** of a stimulus (for example, the loudness of a sound) can be encoded in a single action potential. Instead, intensity is coded in two other ways (**Fig. 38-5**). First, intensity can be signaled by the frequency of action potentials in a single neuron. The more intense the stimulus, the faster the neuron fires action potentials. Second, most nervous systems have many neurons that can respond to the same input. Stronger stimuli excite more of these neurons, whereas weaker stimuli excite fewer. Thus, intensity can also be signaled by the number of similar neurons that fire at the same time. A gentle touch may cause a single touch receptor in your skin to fire action potentials very slowly (**Fig. 38-5a**); a hard poke may cause several touch receptors to fire, some very rapidly (**Fig. 38-5b**).

The Nervous System Processes Information from Many Sources

Your brain is continuously bombarded by sensory stimuli from both inside and outside the body. The brain must evaluate these inputs, determine which ones are important, and decide how to respond. Nervous systems integrate information from many sources: a large number of neurons may funnel their signals to fewer neurons. For example, many sensory neurons may converge onto a small number of brain cells. Some of these brain cells act as "decision-making cells," adding up the postsynaptic potentials that result from the synaptic activity of the sensory neurons. Depending on their relative strengths (and other internal factors such as hormones or metabolic activity), they produce appropriate outputs.

The Nervous System Produces Outputs to Muscles and Glands

Action potentials from the decision-making neurons may travel to other parts of the brain, to the spinal cord, or to the sympathetic and parasympathetic nervous systems (described later). Ultimately, the output of the nervous system will stimulate activity in the muscles or glands that produce behaviors.

The same principles of connectivity and intensity coding that we just examined for sensory inputs are used for the brain's outputs. Which muscles or glands are activated is determined

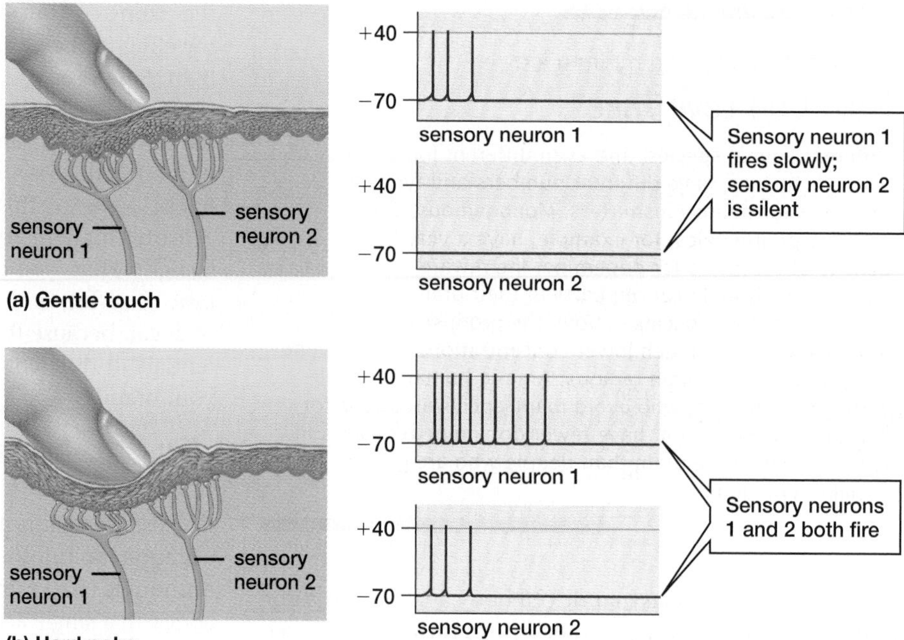

▶ **FIGURE 38-5 Signaling stimulus intensity** The intensity of a stimulus is signaled by the rate at which individual sensory neurons produce action potentials and by the number of sensory neurons activated. **(a)** In this example, a gentle touch activates only the closest sensory neuron, which fires action potentials at a low rate. **(b)** A hard poke activates nearby sensory neurons as well, causing the closest one to fire rapidly and more distant ones to fire more slowly.

QUESTION On the basis of this information, how do you think skin areas that are especially sensitive to touch differ from less sensitive areas?

by their connections to the brain or spinal cord. For example, the neurons that activate your biceps muscles are different than those that activate muscles in your face. How hard a muscle contracts is determined by how many neurons connect to it, and how fast those neurons fire action potentials.

38.4 HOW ARE NERVOUS SYSTEMS ORGANIZED?

Most behaviors are controlled by pathways composed of four elements:

1. **Sensory neurons** respond to a stimulus, either internal or external to the body.
2. **Interneurons** receive signals from sensory neurons, hormones, neurons that store memories, and many other sources. Based on this input, interneurons often activate motor neurons.
3. **Motor neurons** receive instructions from sensory neurons or interneurons and activate muscles or glands.
4. **Effectors,** usually muscles or glands, perform the response directed by the nervous system.

Simple behaviors, such as reflexes (see section 38.5), may be controlled by activity in as few as two or three neurons—a sensory neuron, a motor neuron, and perhaps an interneuron in between—ultimately stimulating a single muscle. Many reflexes do not use the brain at all. In humans, simple reflexes, such as the familiar knee-jerk or pain-withdrawal reflexes, are produced by neurons in the spinal cord.

Complex behaviors are organized by interconnected neural pathways, in which several types of sensory input (along with memories, hormones, and other factors) converge on a set of interneurons. By integrating the postsynaptic potentials from multiple sources, the interneurons "decide" what to do and stimulate motor neurons to direct the appropriate activity in muscles and glands. Hundreds, or even millions, of neurons, mostly in the brain, may be required to perform complex actions such as playing the piano, but the principles remain the same.

Complex Nervous Systems Are Centralized

In the animal kingdom, there are really only two nervous system designs: a diffuse nervous system, such as that of cnidarians (*Hydra*, jellyfish, and their relatives; **Fig. 38-6a**), and a centralized nervous system, found to varying degrees in more complex organisms.

Not surprisingly, nervous system design is highly correlated with an animal's lifestyle. Radially symmetrical cnidarians (see Fig. 23-2) have no "front end," so there has been no evolutionary pressure to concentrate the senses in one place. For example, a *Hydra* sits anchored to a rock at the bottom of a pond, so prey or predators are equally likely to come from any direction. Cnidarian nervous systems are composed of a network of neurons, often called a **nerve net,** woven through the animal's tissues. Here and there we find a cluster of neurons, called a **ganglion** (plural, ganglia), but nothing resembling a real brain.

Almost all other animals are bilaterally symmetrical, with definite head and tail ends. Because the head is usually the first part of the body to encounter food, danger, and potential mates, it is advantageous to have sense organs concentrated there. Sizable ganglia evolved that integrate the information gathered by the senses and direct appropriate actions. Over evolutionary time, the major sense organs of animals with complex nervous systems became localized in the head, and the ganglia became centralized into a brain. This trend, called cephalization, is clearly seen in the invertebrates (**Figs. 38-6b,c**). Cephalization reaches its peak in the vertebrates, in which nearly all the cell bodies of the nervous system are localized in the brain and spinal cord.

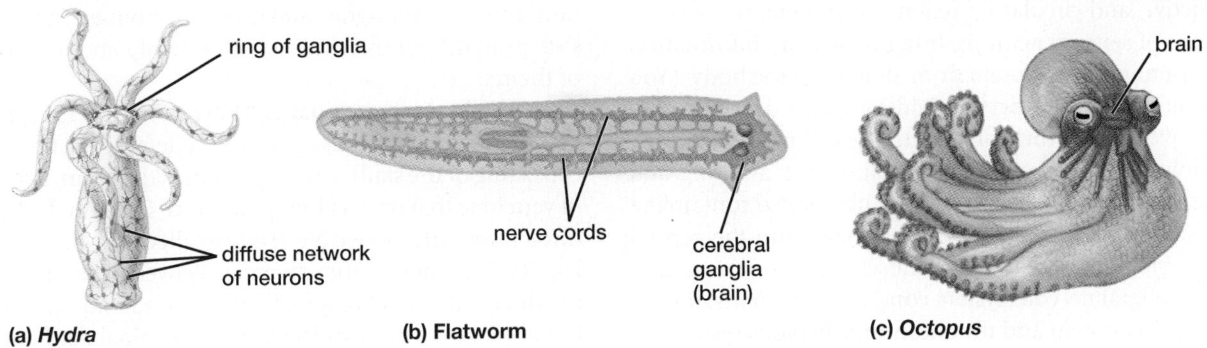

▲ **FIGURE 38-6 Nervous system organization (a)** The diffuse nervous system of *Hydra* contains a few concentrations of neurons at the bases of tentacles, but no brain. Neural signals are conducted in all directions throughout the body. **(b)** The flatworm has a nervous system that is less diffuse, with a cluster of ganglia in the head. **(c)** The genus *Octopus* has a large, complex brain and learning capabilities rivaling those of some mammals.

38.5 WHAT ARE THE STRUCTURES AND FUNCTIONS OF THE HUMAN NERVOUS SYSTEM?

The nervous system of all mammals, including humans, can be divided into two parts: central and peripheral. Each of these has further subdivisions (**Fig. 38-7**). The **central nervous system (CNS)** consists of the **brain** and **spinal cord.** The **peripheral nervous system (PNS)** consists of neurons that lie outside of the CNS and the axons that connect these neurons with the CNS. The cell bodies of neurons in the PNS are often located in ganglia alongside the spinal cord (see Figs. 38-9 and 38-10 later in the chapter) or in ganglia near target organs, such as ganglia in the head and neck that control the salivary glands (see Fig. 38-8).

The Peripheral Nervous System Links the Central Nervous System with the Rest of the Body

The nerves of the peripheral nervous system connect the brain and spinal cord with the rest of the body, including muscles, glands, sensory organs, and the digestive, respiratory, urinary,

The Nervous System

- **Central Nervous System (CNS)** — receives and processes information; initiates action
  - **Brain** — receives and processes sensory information; initiates responses; stores memories; generates thoughts and emotions
  - **Spinal Cord** — conducts signals to and from the brain; controls reflex activities
- **Peripheral Nervous System (PNS)** — transmits signals between the CNS and the rest of the body
  - **Motor Neurons** — carry signals from the CNS that control the activities of muscles and glands
    - **Somatic Nervous System** — controls voluntary movements by activating skeletal muscles
    - **Autonomic Nervous System** — controls involuntary responses by influencing organs, glands, and smooth muscle
      - **Sympathetic Division** — prepares the body for stressful or energetic activity; "fight or flight"
      - **Parasympathetic Division** — dominates during times of "rest and digest"; directs maintenance activities
  - **Sensory Neurons** — carry signals to the CNS from sensory organs

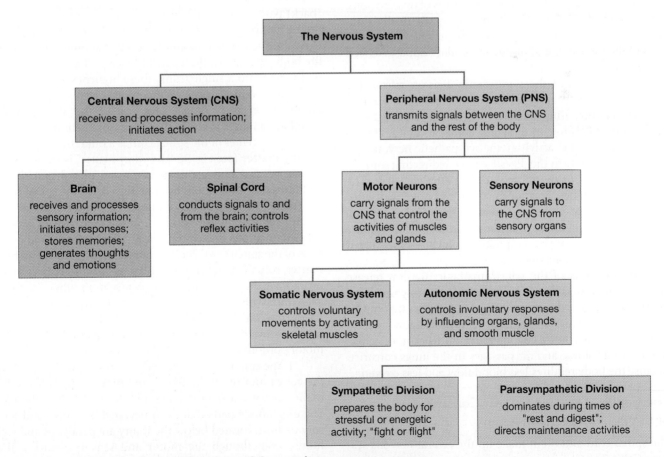

▲ **FIGURE 38-7 Organization and functions of the vertebrate nervous system**

reproductive, and circulatory systems. Peripheral nerves contain axons of sensory neurons, bringing sensory information to the central nervous system from all parts of the body. (You may think that these fibers should be called dendrites, because they carry information toward the cell body. However, neurobiologists call them axons because they are long and conduct action potentials.) Peripheral nerves also contain the axons of motor neurons that carry signals from the central nervous system to the glands and muscles. The motor portion of the peripheral nervous system consists of two parts: the somatic nervous system and the autonomic nervous system.

The Somatic Nervous System Controls Voluntary Movement

Motor neurons of the **somatic nervous system** form synapses with skeletal muscles and control voluntary movement. As you take notes, lift a coffee cup, or adjust your iPod, your somatic nervous system is in charge. The cell bodies of somatic motor neurons are located in the spinal cord (see Fig. 38-10 later). Their axons go directly to the muscles they control. We will examine muscles in Chapter 40.

The Autonomic Nervous System Controls Involuntary Actions

Motor neurons of the **autonomic nervous system** innervate the heart, smooth muscles, and glands, and produce mostly involuntary actions. The autonomic nervous system is controlled primarily by the hypothalamus, medulla, and pons, parts of the brain described later in this chapter. It consists of two divisions: the **sympathetic division** and the **parasympathetic division.** The two divisions of the autonomic nervous system innervate most of the same organs, but usually produce opposite effects (**Fig. 38-8**).

The neurons of the sympathetic division release the neurotransmitter norepinephrine (noradrenaline) onto their target organs, preparing the body for stressful or energetic activity, such as fighting, escaping, or taking an exam. During such "fight-or-flight" activities, the sympathetic nervous system reduces activity in the digestive tract, redirecting some of its blood supply to the muscles of the arms and legs. The heart rate accelerates. The pupils of the eyes open wider, admitting more light, and the air passages in the lungs expand, letting in more air. These things may also happen if you are suddenly asked to answer a question in class—especially if you don't know the answer!

The neurons of the parasympathetic division release acetylcholine onto their target organs. The parasympathetic division dominates during maintenance activities that can be carried on at leisure, often called "rest and digest." Under parasympathetic control, the digestive tract becomes active, the heart rate slows, and air passages in the lungs constrict, because the body requires less blood flow and less oxygen.

The Central Nervous System Consists of the Spinal Cord and Brain

The spinal cord and brain make up the central nervous system (CNS). The CNS receives and processes sensory information, generates thoughts, and directs responses. The CNS consists primarily of interneurons—probably about 100 billion of them!

The brain and spinal cord are protected from physical damage in three ways. The first line of defense is a bony armor, consisting of the skull, which surrounds the brain, and a chain of vertebrae that protect the spinal cord. Beneath the bones lie three layers of connective tissues called the meninges (see Fig. 38-12a later in the chapter). Between the layers of the meninges, the cerebrospinal fluid, a clear liquid similar to blood plasma, cushions the brain and spinal cord and nourishes the cells of the CNS.

The brain is also protected from potentially damaging chemicals in the bloodstream because the walls of brain capillaries are far less permeable than are the capillaries in the rest of the body. This **blood–brain barrier** selectively transports needed materials into the brain, while keeping many dangerous substances out. Generally, the blood–brain barrier keeps water-soluble substances from diffusing from the blood into the brain, but many lipid-soluble substances can still diffuse across the capillary walls.

The Spinal Cord Controls Many Reflexes and Conducts Information to and from the Brain

About as thick as your little finger, the spinal cord (**Fig. 38-9**) extends from the base of the brain to the lower back. Nerves carrying axons of sensory neurons emerge from the dorsal (back) part of the spinal cord, and nerves carrying axons of motor neurons emerge from the ventral (front) part. These nerves merge to form the spinal nerves that innervate most of the body. Because of their resemblance to the roots of a tree that merge into a single trunk, these branches are called the dorsal and ventral roots of the spinal nerves, respectively. Swellings on each dorsal root, called the **dorsal root ganglia,** contain the cell bodies of sensory neurons.

In the center of the spinal cord is a butterfly-shaped area of **gray matter** (see Fig. 38-9). ("Gray matter" is really a misnomer, because nervous tissue is mostly colored pinkish-tan and turns gray when it has been preserved.) In the spinal cord, gray matter consists of the cell bodies of motor neurons that control voluntary muscles and the autonomic nervous system, plus interneurons that communicate with the brain and other parts of the spinal cord. The gray matter is surrounded by **white matter,** containing myelin-coated axons of neurons that extend up or down the spinal cord (the fatty myelin sheaths color these axon tracts white). These axons carry sensory signals from internal organs, muscles, and the skin up to the brain. Axons also extend downward from the brain, carrying signals that direct the motor portions of the peripheral nervous system.

If the spinal cord is severed, sensory input from below the cut cannot reach the brain, and motor output from the brain cannot reach motor neurons located below the cut. Therefore, body parts that are innervated by motor and sensory neurons located below the injury are paralyzed and feel numb, even though the motor and sensory neurons, the spinal nerves, and the muscles remain intact.

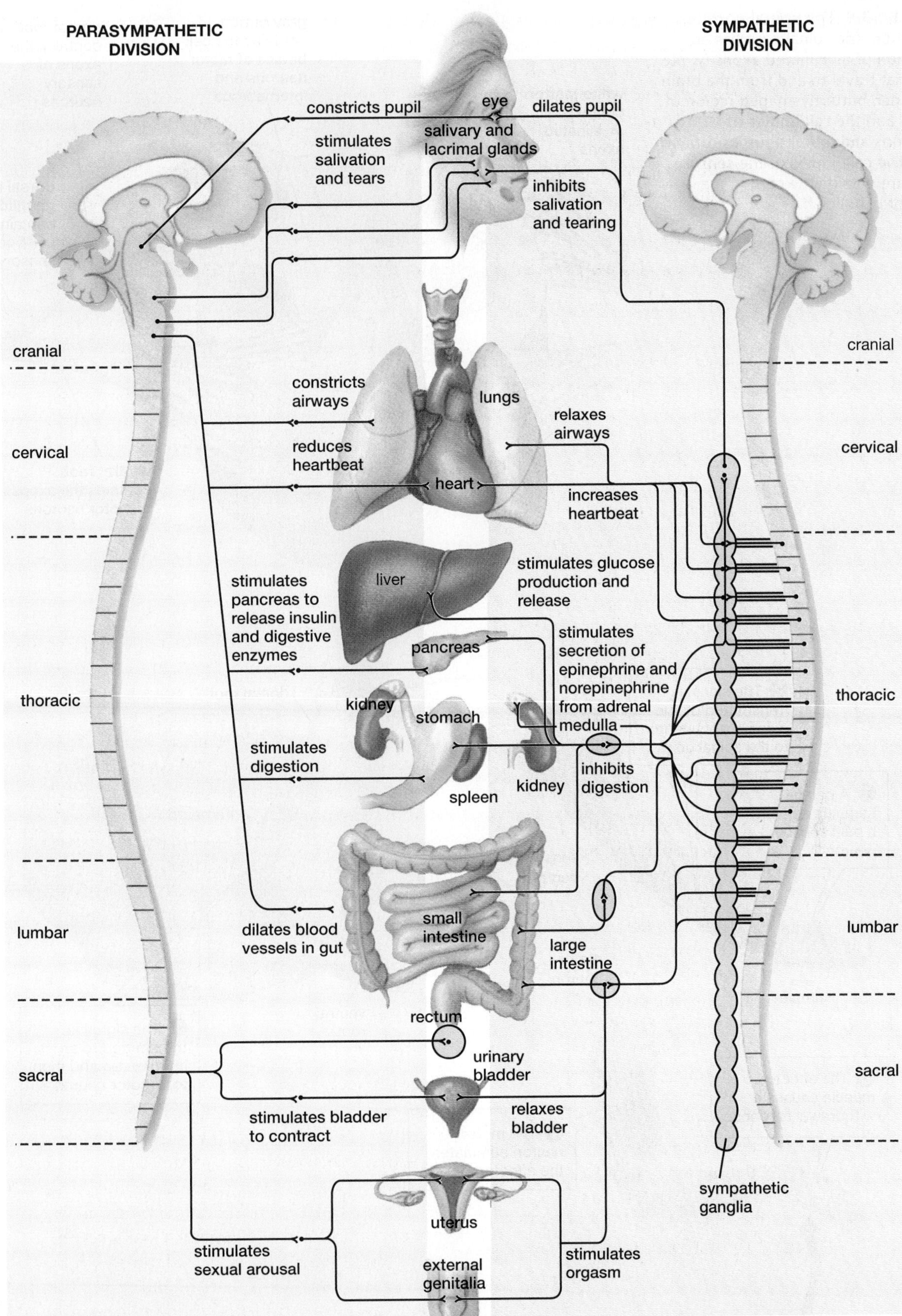

▲ **FIGURE 38-8 The autonomic nervous system** The autonomic nervous system has two divisions, sympathetic and parasympathetic, which supply nerves to many of the same organs but generally produce opposite effects. Activation of the autonomic nervous system is involuntarily commanded by signals from the hypothalamus.

▶ **FIGURE 38-9 The spinal cord** In cross-section, the spinal cord has an outer region of myelinated axons (white matter) that travel to and from the brain and an inner, butterfly-shaped region of dendrites and the cell bodies of interneurons and motor neurons (gray matter). The cell bodies of the sensory neurons are outside the cord in the dorsal root ganglion.

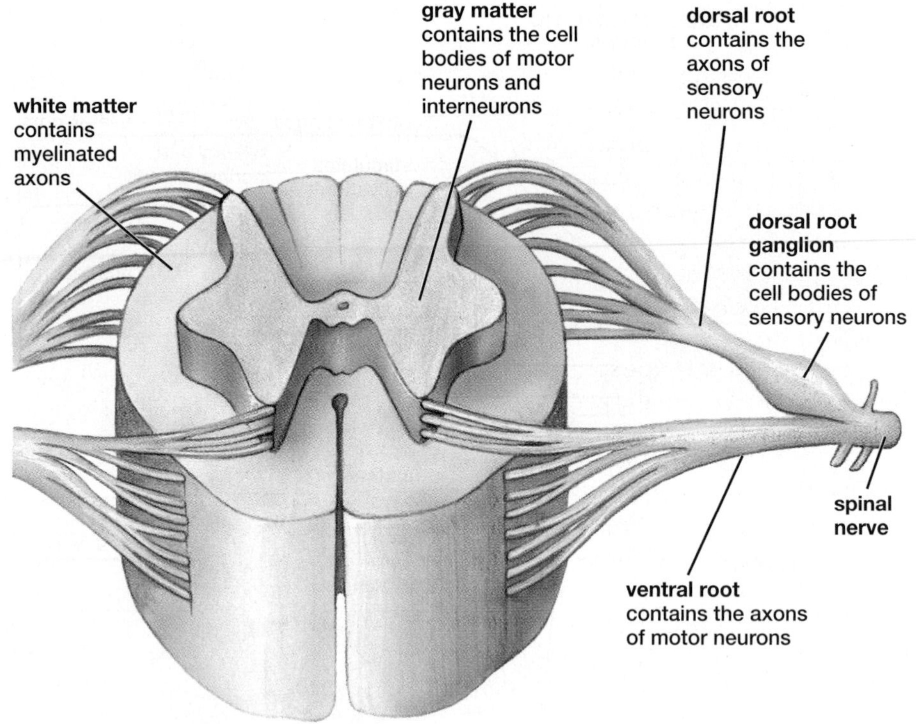
gray matter
contains the cell bodies of motor neurons and interneurons
dorsal root
contains the axons of sensory neurons
white matter
contains myelinated axons
dorsal root ganglion
contains the cell bodies of sensory neurons
spinal nerve
ventral root
contains the axons of motor neurons

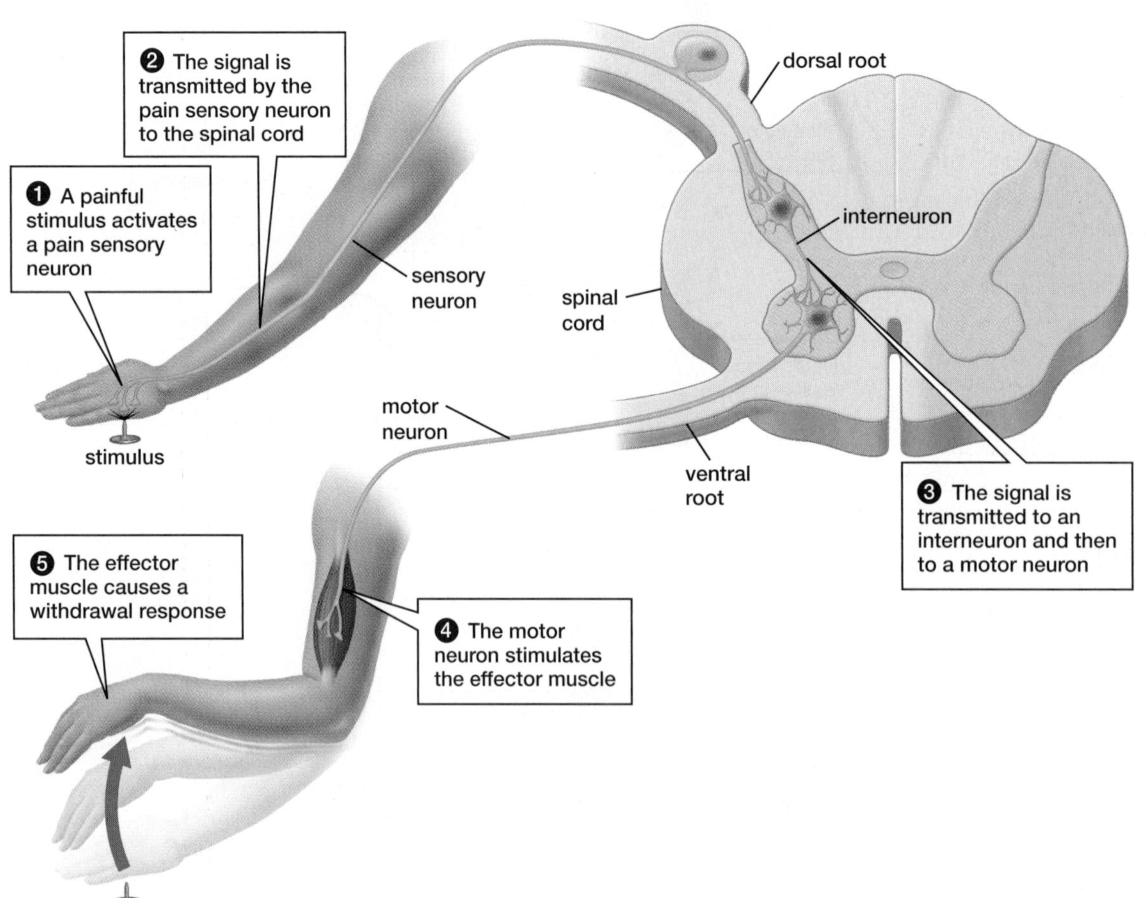
❷ The signal is transmitted by the pain sensory neuron to the spinal cord
dorsal root
❶ A painful stimulus activates a pain sensory neuron
interneuron
sensory neuron
spinal cord
stimulus
motor neuron
ventral root
❸ The signal is transmitted to an interneuron and then to a motor neuron
❺ The effector muscle causes a withdrawal response
❹ The motor neuron stimulates the effector muscle

▲ **FIGURE 38-10 The pain-withdrawal reflex**

QUESTION Why does a paralyzed victim of a spinal cord injury, when pricked with a pin on a paralyzed part of his or her body, often have a normal pain-withdrawal reflex but feel no pain?

The Neuronal Circuits for Many Reflexes Reside in the Spinal Cord

The simplest type of behavior is the **reflex,** a largely involuntary movement of a body part in response to a stimulus. In vertebrates, many reflexes are produced by the spinal cord and peripheral neurons, and do not use the brain.

Let's examine the pain-withdrawal reflex, which involves neurons of both the central and the peripheral nervous systems (**Fig. 38-10**). If you lean your hand on a tack, the tissue damage activates pain sensory neurons (**Fig. 38-10 ❶**). Action potentials in the axons of these pain sensory neurons travel up a spinal nerve and enter the spinal cord through a dorsal root (**Fig. 38-10 ❷**). Within the gray matter of the cord, the pain sensory neuron stimulates an interneuron, which stimulates a motor neuron (**Fig. 38-10 ❸**); you can see that interneurons are literally "between other neurons," in this case between a sensory neuron and a motor neuron. Action potentials in the axon of the motor neuron leave the spinal cord through a ventral root and travel in a spinal nerve to a skeletal muscle. The action potential stimulates the muscle (**Fig. 38-10 ❹**), which contracts, withdrawing your hand away from the tack (**Fig. 38-10 ❺**).

Many spinal cord interneurons also have axons that extend up to the brain. Action potentials in these axons inform the brain about pricked hands and may trigger more complex behaviors, such as shrieks and learning about the dangers of thumbtacks. The brain in turn sends action potentials down axons in the spinal cord white matter to interneurons and motor neurons in the gray matter. These signals from the brain can modify spinal reflexes. With enough training or motivation, you can suppress the pain-withdrawal reflex. To rescue a child from a burning crib, for example, you could reach into the flames.

Some Complex Actions Are Coordinated Within the Spinal Cord

The wiring for some fairly complex activities also resides within the spinal cord. All the neurons and interconnections needed for the basic movements of walking and running, for example, are contained in the spinal cord. The advantage of this semi-independent arrangement between brain and spinal cord is probably an increase in speed and coordination, because messages do not have to travel all the way up to the brain and back down again merely to swing a leg forward while walking. The brain's role in these semi-automatic behaviors is to initiate, guide, and modify the activity of spinal motor neurons, based on conscious decisions (where are you going; how fast should you walk?). To maintain balance, the brain also uses sensory input from the muscles to command motor neurons to adjust the way the muscles move.

The Brain Consists of Many Parts That Perform Specific Functions

All vertebrate brains consist of three major parts: the **hindbrain, midbrain,** and **forebrain** (**Fig. 38-11a**). Scientists believe that in the earliest vertebrates, these three anatomical divisions were also functional divisions: the hindbrain governed automatic behaviors such as breathing

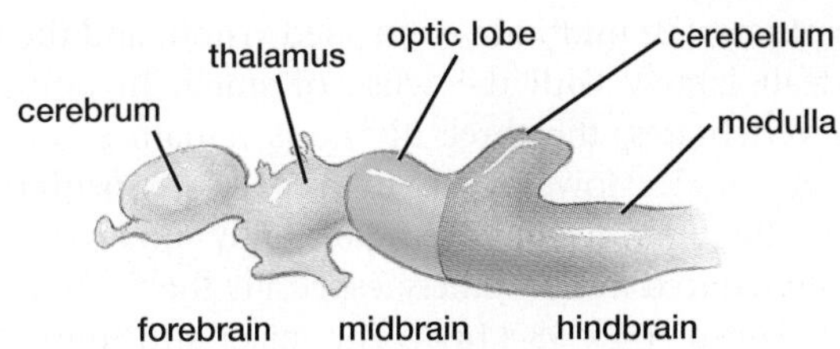

(a) Embryonic vertebrate brain

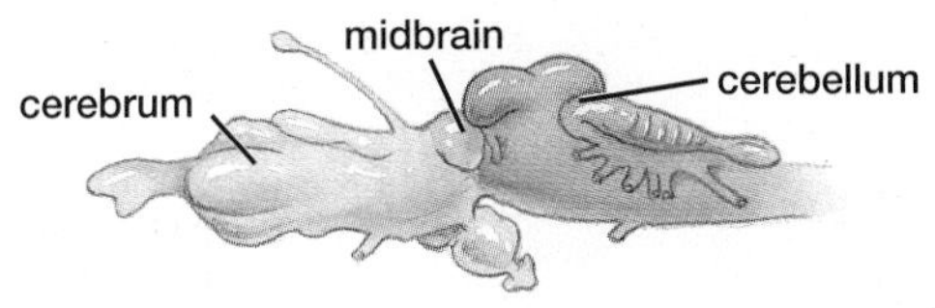

(b) Shark brain

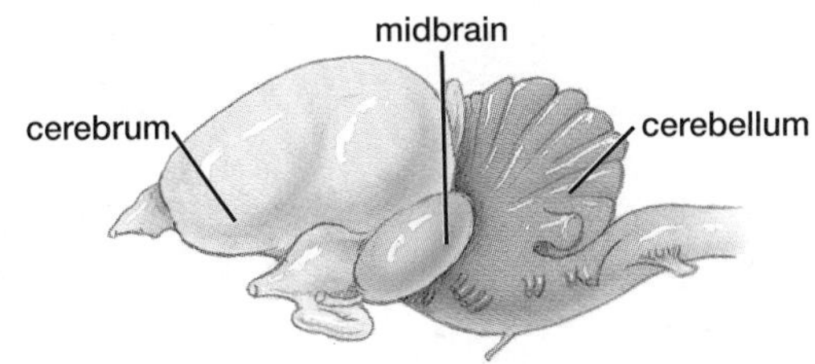

(c) Goose brain

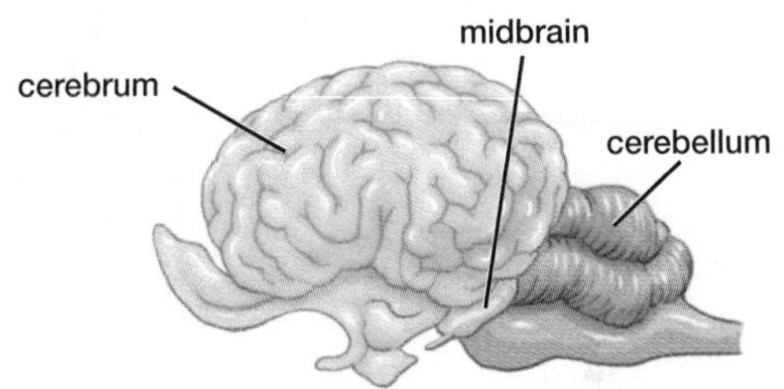

(d) Horse brain

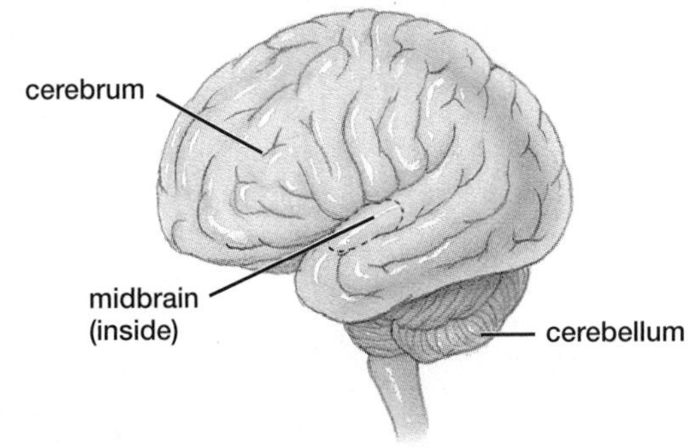

(e) Human brain

▲ **FIGURE 38-11 A comparison of vertebrate brains (a)** The brains of modern vertebrate embryos, thought to be similar to the brains of the distant ancestors of today's vertebrates, consist of three distinct regions: the forebrain, midbrain, and hindbrain. **(b)** The brain of an adult shark maintains this basic organization. **(c)** In the goose, the midbrain is reduced, and the cerebrum and cerebellum are larger. **(d, e)** In mammals, especially humans, the cerebrum is very large compared to the other brain regions.

and heart rate, the midbrain controlled vision, and the forebrain dealt largely with the sense of smell. In nonmammalian vertebrates, the three divisions remain prominent (**Figs. 38-11b,c**). However, in mammals—particularly in humans—the brain regions are significantly modified. Some have been reduced in size; others, especially the forebrain, are greatly enlarged (**Figs. 38-11d,e**). The major structures of the human brain are shown in **Figure 38-12**.

The Hindbrain Consists of the Medulla, Pons, and Cerebellum

In humans and other mammals, the hindbrain is comprised of the medulla, the pons, and the cerebellum (**Fig. 38-12a**). In both structure and function, the **medulla** is very much like an enlarged extension of the spinal cord. Like the spinal cord, the medulla has neuron cell bodies at its center, surrounded by a layer of myelin-covered axons. It controls sev-

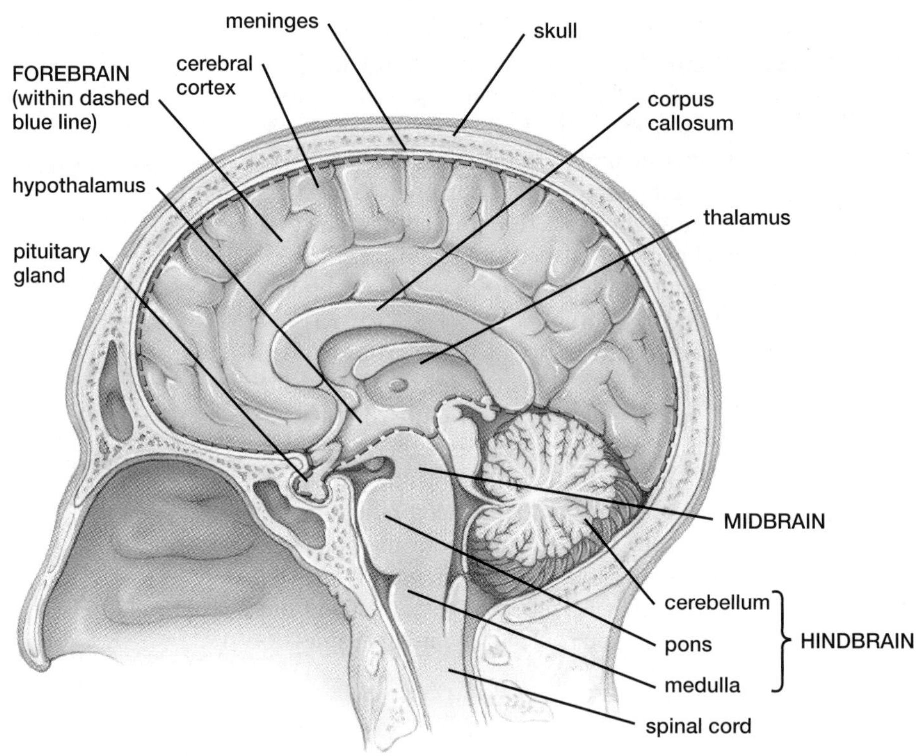

(a) A lateral section of the human brain

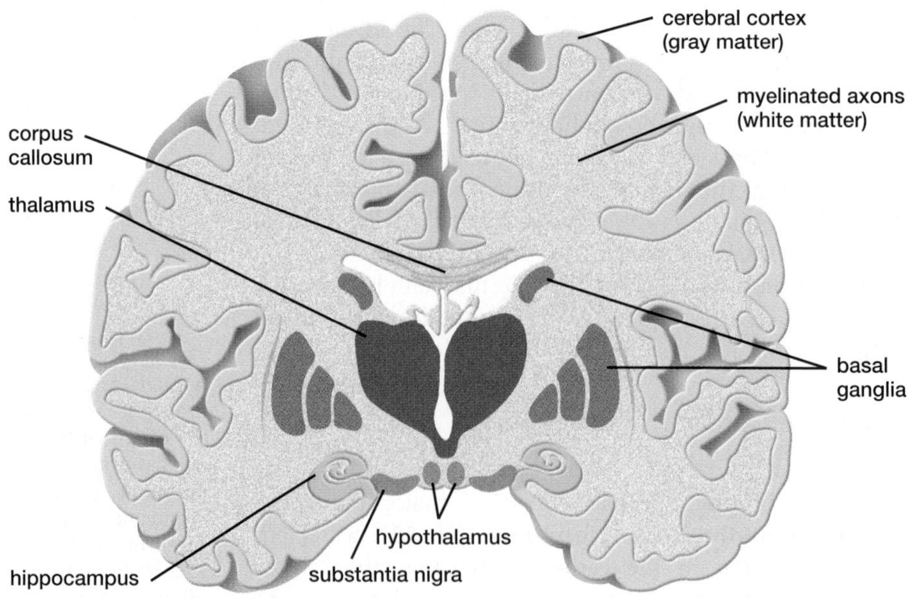

(b) A cross-section of the brain

▶ FIGURE 38-12 **The human brain**
Sections of the human brain **(a)** through the midline between the cerebral hemispheres, showing the hindbrain, midbrain, and forebrain, and **(b)** from ear to ear, showing mostly the cerebrum. Not all brain structures are visible in these sections.

eral automatic functions, such as breathing, heart rate, blood pressure, and swallowing. Certain neurons in the **pons,** located just above the medulla, appear to influence transitions between sleep and wakefulness and between stages of sleep. Other neurons in the pons influence the rate and pattern of breathing.

The **cerebellum** is crucial for coordinating movements of the body. It receives information both from command centers in the forebrain that control movement and from position sensors in muscles and joints. By comparing information from these two sources, the cerebellum guides smooth, accurate motions and body position. The cerebellum is also involved in *motor learning.* As you learn to write, throw a ball, or play the guitar, your forebrain directs your movements, often somewhat clumsily. After you have become proficient, your forebrain still "decides" what to do (for example, where to throw the ball), but your cerebellum becomes largely responsible for ensuring that the actions are carried out appropriately. Not surprisingly, the cerebellum is especially large in animals whose activities require coordinating fine movements or aerial maneuvers, such as bats and birds (see Fig. 38-11c).

The Midbrain Contains Clusters of Neurons That Contribute to Movement, Arousal, and Emotion

The midbrain is quite small in humans (see Figs. 38-11e and 38-12a). It contains an auditory relay center and clusters of neurons that control reflex movements of the eyes. For example, if you're sitting in class and someone races through the door, centers in your midbrain are alerted and direct your gaze to the new, and potentially interesting or threatening, visual stimulus. The midbrain also contains neurons that produce the transmitter dopamine. One of these clusters of neurons, called the substantia nigra (**Fig. 38-12b**), helps to control movement (see the description of the basal ganglia, below). Another cluster is an essential part of the "reward circuit" that is responsible for pleasurable sensations and, unfortunately, addiction, as we explore in "Health Watch: Drugs, Neurotransmitters, and Addiction" on p. 750.

Finally, the midbrain contains a portion of the **reticular formation.** The reticular formation consists of dozens of interconnected clusters of neurons in the medulla, pons, and midbrain, which send axons to the forebrain. These neurons receive input from virtually every sense, from every part of the body, and from many areas of the brain as well. The reticular formation plays a role in sleep and wakefulness, emotion, muscle tone, and some movements and reflexes. It filters sensory inputs before they reach the conscious regions of the brain, although the selectivity of the filtering seems to be set by higher brain centers, such as those that control conscious thought. Activities of the reticular formation allow you to read and concentrate in the presence of a variety of distracting stimuli, such as the music from your stereo and the smell of coffee. The fact that a mother wakens upon hearing the faint cry of her infant but sleeps through loud traffic noise outside her window testifies to the effectiveness of the reticular formation in screening inputs to the brain.

The Forebrain Includes the Thalamus, Hypothalamus, and Cerebrum

The **thalamus** is a complex relay station that channels sensory information from all parts of the body to the cerebral cortex (see Figs. 38-12a,b). In fact, input from all of the senses, except olfaction, passes through the thalamus on its way to the cerebral cortex. Signals traveling from the spinal cord, cerebellum, medulla, pons, and reticular formation also pass through the thalamus.

The **hypothalamus** (literally, "under the thalamus") contains many clusters of neurons. Some are neurosecretory cells that release hormones into the blood or control the release of hormones from the pituitary gland (see pp. 721–722). Other regions of the hypothalamus direct the activities of the autonomic nervous system. The hypothalamus, through its hormone production and neural connections, maintains homeostasis by influencing body temperature, food intake, water balance, heart rate, blood pressure, the menstrual cycle, and circadian rhythms.

The **cerebrum** consists of two **cerebral hemispheres.** Each hemisphere is composed of an outer **cerebral cortex,** several clusters of neurons beneath the cortex near the thalamus, and bundles of axons that both interconnect the two hemispheres and connect the hemispheres with the midbrain and hindbrain (see Figs. 38-12a,b).

Structures in the Interior of the Cerebrum Clusters of neurons in the **amygdala** (see Fig. 38-13) produce sensations of pleasure, fear, or sexual arousal when stimulated. Conscious humans whose amygdalas are electrically stimulated report feelings of rage or fear. Damage to the amygdala early in life eliminates the ability both to feel fear and to recognize fearful facial expressions in other people.

The **hippocampus** (see Fig. 38-12b) plays an important role in the formation of long-term memory, particularly of places, and is thus required for learning, as discussed in more detail later in this chapter. The hippocampus is evolutionarily very ancient—all vertebrates have a homologous part of the brain. What's more, the hippocampus-like region appears to be involved in "place learning" in most, or perhaps all, vertebrates. For example, some birds, such as jays and nutcrackers, store seeds for the winter and must remember where their caches are. These birds have a larger hippocampus than most other birds do. In humans, London taxi drivers, who must remember that city's seemingly random maze of streets, have larger-than-average hippocampi, showing that use can enlarge specific parts of the brain.

The **basal ganglia** consist of structures deep within the cerebrum (see Fig. 38-12b), as well as the substantia nigra in the midbrain. These structures are important in the overall control of movement. The motor part of the cerebral cortex directs specific movements, such as which motor neurons to fire to activate just the right muscles to pick up a pen. The basal ganglia appear to be essential to the decision to initiate a particular movement and suppress other movements. This is most clearly illustrated by two major disorders of the basal ganglia. In Parkinson's disease, the substantia nigra degenerates, and

Health Watch

Drugs, Neurotransmitters, and Addiction

Chances are, you know someone who is addicted. How can substances such as cocaine, alcohol, and nicotine so profoundly alter people's lives? The answer lies in the effects of these drugs on neurotransmitters and in how the nervous system adapts to those effects.

Many addictive drugs, including cocaine, methamphetamine (meth), and ecstasy (MDMA), target synapses in the brain's reward circuitry that use the transmitters dopamine or serotonin. Normally, after releasing transmitter, the presynaptic neurons in these synapses immediately pump the transmitter back in, thus limiting its effects. Cocaine, meth, and ecstasy block the pumps, or even make them work in reverse, causing spontaneous transmitter release. In either case, the concentration of transmitter increases around postsynaptic receptors, enhancing synaptic transmission. Because serotonin and especially dopamine make people feel good, cocaine and meth are highly rewarding and users want to repeat the experience. Meanwhile, the brain, recognizing that it is getting too much dopamine stimulation, reduces the number of dopamine receptors in the synapses. Therefore, the user feels "down" when he has only his body's own dopamine available. Over time, he needs more and more cocaine or meth just to feel OK, let alone to get high: He has become an addict (**Fig. E38-3**).

By blocking serotonin pumps, ecstasy causes a temporary massive increase in serotonin. Users report feelings of pleasure, increased energy, heightened sensory awareness, and improved rapport with other people, perhaps because the high serotonin levels increase oxytocin release in the brain. Evidence from both animal and human research suggests that ecstasy users may incur long-term damage to serotonin-producing neurons, and they may suffer deficits in learning and memory. Ecstasy may also damage dopamine-producing neurons.

Alcohol stimulates receptors for the neurotransmitter GABA (gamma amino butyric acid), enhancing inhibitory neuronal signals, and blocks receptors for glutamate, reducing excitatory signals. Together, these changes produce alcohol's well-known relaxing effects. However, when a person drinks frequently, the brain compensates by decreasing GABA receptors and increasing glutamate receptors. Without alcohol, an alcoholic feels jittery and nervous—in short, overstimulated. Alcohol also initially increases dopamine neurotransmission, so the drinker feels good, but then suppresses dopamine release. All of these effects combine to produce addiction—an alcoholic needs alcohol just to feel normal, and more and more of it to feel good.

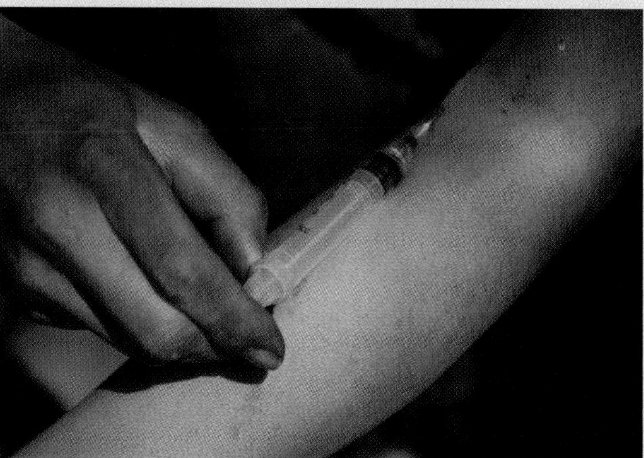

▲ **FIGURE E38-3 Addiction** An addict experiences extreme physical and emotional distress without his drug, because his nervous system has adapted to its presence.

Nicotine in cigarette smoke stimulates receptors that normally respond to acetylcholine. Overstimulation of these receptors activates other neurons that increase dopamine release, thereby contributing to smoking's pleasurable and addictive properties.

To overcome addiction, drug users must undergo the misery caused by a nervous system that is deprived of a drug to which it has adapted. Although transmitter release and receptor concentrations eventually return to normal, drug cravings often recur periodically, for unknown reasons. These drug cravings suggest that the brains of addicts have been permanently altered, in poorly understood, but obviously important, ways.

In some respects, love strongly resembles addiction—producing feelings of pleasure, trust, and euphoria. But is love really like addiction? Or is it the other way around? It is extremely unlikely that people have evolved reward centers and dopamine neurons so that we can become cocaine addicts. Rather, these circuits probably evolved to promote sex, pair bonding, and infant care. By activating these same circuits, cocaine, alcohol, and nicotine hijack essential brain functions, with destructive results.

affected people have a hard time starting a movement—the familiar "freezing" phenomenon. In Huntington's disease, basal ganglia in the cerebrum degenerate, and affected people make involuntary, undirected, flailing movements.

What is often called the **limbic system** is a diverse group of structures, including the hypothalamus, the amygdala, and the hippocampus, as well as nearby regions of the cerebral cortex, located in a ring between the thalamus and cerebral cortex (**Fig. 38-13**). These structures help to produce emotions and emotional behaviors, including fear, rage, calm, hunger, thirst, pleasure, and sexual responses. However, other brain regions are also involved in emotions, including other parts of the cerebral cortex, the midbrain, the hindbrain, and possibly even the spinal cord. Consequently, neuroscientists do not agree on which structures should be included in an "extended" limbic system, or even if the concept is useful, when so many brain regions are involved.

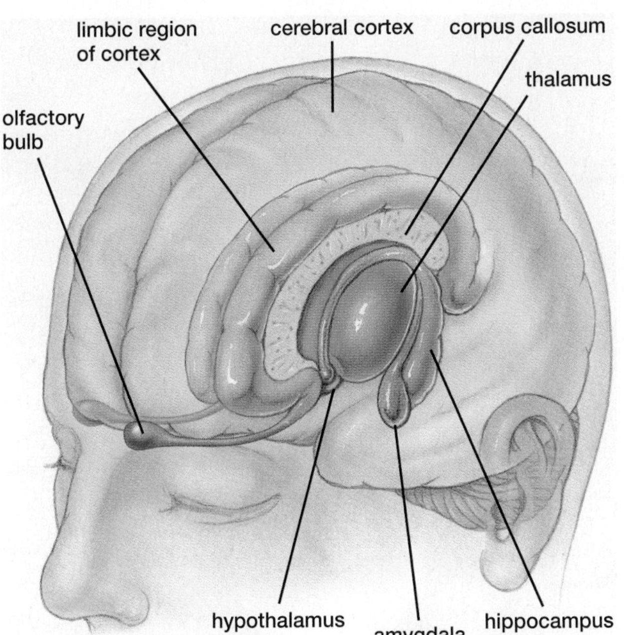

▲ **FIGURE 38-13 The limbic system** The limbic system includes several structures in the forebrain involved in producing emotions and responding to emotions perceived in others.

The Cerebral Cortex The cerebral cortex is the thin outer layer of each cerebral hemisphere, with billions of neurons packed in a highly organized way into a sheet just a few millimeters thick. The cortex is folded into **convolutions,** which are raised, wrinkled ridges that increase its surface area to over two square yards—about the area of the top of a single bed! Neurons in the cortex receive sensory information, process it, direct voluntary movements, create memories, and allow us to be creative and even envision the future. The cortexes in each hemisphere communicate with each other through a large band of axons, the **corpus callosum** (see Fig. 38-12).

The cerebral cortex is divided into four anatomical regions: the frontal, parietal, occipital, and temporal lobes (**Fig. 38-14**). The cortex can also be divided into functional areas. Primary sensory areas are regions where signals originating in sensory organs such as the eyes and ears are received and converted into subjective impressions—for instance, light and sound. Nearby association areas interpret the sounds as speech or music, and the visual stimuli as recognizable objects or words on this page. Association areas also link stimuli with memories stored in the cortex and generate commands to produce speech.

Primary sensory areas in the parietal lobe interpret sensations of touch that originate in all parts of the body; these body parts are "mapped" in an orderly sequence on the parietal cortex (see Fig. 38-14). In an adjacent region of the frontal lobe,

Have you ever wondered

How Con Men Fool Their Victims?

The basis of human society is probably trust—the belief that most other people will usually help you, rather than harm you. Oxytocin helps us to trust others. It quiets the amygdala, allowing us to trust rather than fear. It also helps us to feel good when we do nice things for other people. Unfortunately, con men take advantage of this. They seem kind, sometimes needy themselves, and pretend to offer their victim a reward in exchange for trust, thereby activating the victim's oxytocin trust system.

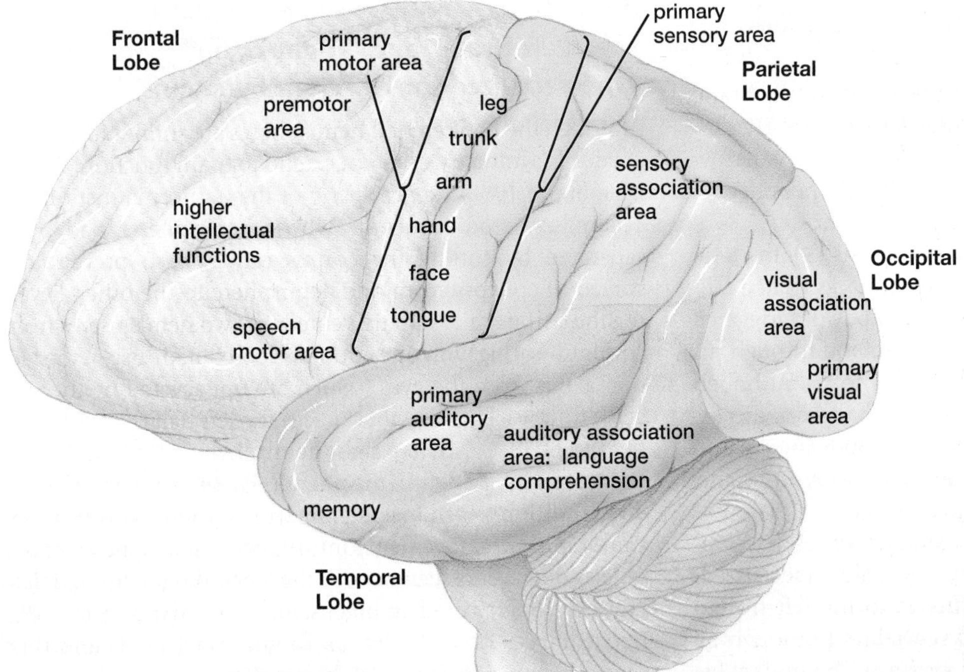

◀ **FIGURE 38-14 The cerebral cortex** A map of the human left cerebral cortex. A map of the right cerebral cortex would be similar, except that speech and language would not be as well developed.

Case Study continued
How Do I Love Thee?

Dopamine-containing neurons in the midbrain send axons to several parts of the forebrain. When people fall in love, oxytocin stimulates dopamine release in these areas, contributing to emotional attachment, lack of fear and critical judgement of the loved one, and fond memories of times spent together.

primary motor areas command movements in corresponding areas of the body by stimulating the motor neurons in the spinal cord that innervate muscles, allowing you to walk to class, throw a Frisbee®, or type a term paper. Like the primary sensory area, the primary motor area has adjacent association areas, including the motor association area (also called the premotor area), which directs the motor area to produce movements. Behind the bones of the forehead lie the association areas of the frontal lobe, which are important in complex reasoning functions such as short-term memory, decision making, predicting the consequences of actions, controlling aggression, planning for the future, and working for delayed rewards.

Damage to the cortex from trauma, stroke, or a tumor results in specific deficits, such as problems with speech, difficulty reading, or the inability to sense or move specific parts of the body, depending on where the damage occurs. Most brain cells of adults cannot be replaced, so if a brain region is destroyed, these deficits may be permanent. Fortunately, however, training can sometimes allow undamaged regions of the cortex to take control over and restore some of the lost functions.

How Do Neuroscientists Learn About the Functions of Brain Regions?

As we described the functions of different parts of the brain, you probably wondered, "How do they know *that*?" Historically, the functions of different parts of the brain were discovered by examining the behaviors and abilities of people who suffered brain injuries, often in accidents or wars.

Consider the case of Phineas Gage (**Fig. 38-15**). In 1848, Gage was setting an explosive charge to clear rocks from a railroad line under construction, when the gunpowder triggered prematurely. The blast blew a 13-pound steel rod through his skull, severely damaging both of his frontal lobes. Although Gage survived his wounds, his personality changed radically. Before the accident, Gage was conscientious, industrious, and well liked. After his recovery, he became impetuous, profane, and incapable of working toward a goal.

Studies of other people with brain injuries have revealed that many parts of the brain are highly specialized. One patient with very localized damage to the left frontal lobe was unable to name fruits and vegetables (although he could name everything else). Other victims of brain damage are unable to recognize faces, suggesting that the brain has regions specialized to recognize specific categories of things.

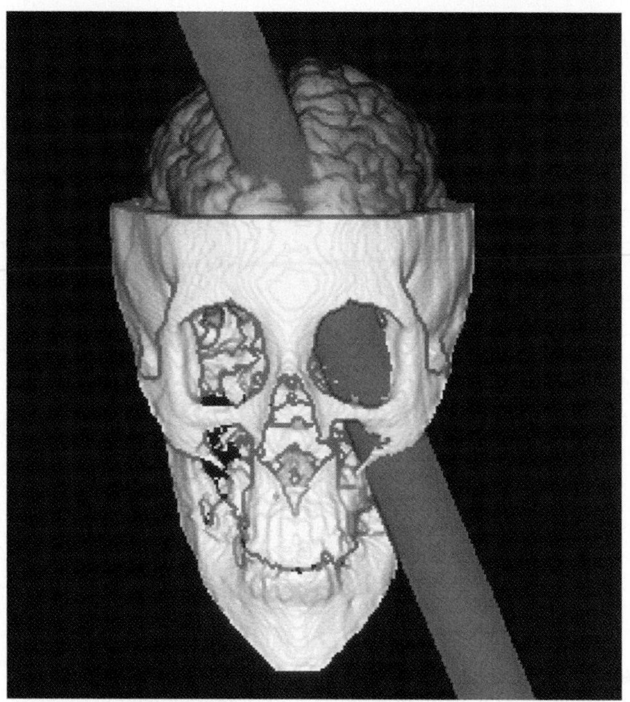

▲ **FIGURE 38-15 A revealing accident** Studies of the skull of Phineas Gage have enabled scientists to create this reconstruction of the path taken by the steel rod that was blown through his head by an explosion.

Rather than being restricted to examining people with brain injuries, modern neuroscience has powerful techniques for visualizing brain structure and activity, offering insights into the functioning of the human brain, intact or damaged. We explore one of these techniques in the "Scientific Inquiry: Neuroimaging—Observing the Brain in Action" on p. 754.

The "Left Brain" and "Right Brain" Are Specialized for Different Functions

Although the two cerebral hemispheres are similar in appearance, this symmetry does not extend to brain function. Beginning in the 1950s, Roger Sperry of the California Institute of Technology studied people whose hemispheres had been separated by cutting the corpus callosum to prevent the spread of epilepsy from one hemisphere to the other. Severing the corpus callosum prevented the two hemispheres from communicating with one another. Sperry made use of the fact that axons from the eyes (which are not severed by the surgery) follow a pathway that causes the left half of each visual field to be "seen" by the right hemisphere and the right half to be seen by the left hemisphere (**Fig. 38-16**). Under everyday conditions, even in someone with a severed corpus callosum, rapid eye movements inform both hemispheres about objects and events seen in both the left and right visual fields. However, Sperry used an ingenious device that projected different images onto the left and right visual fields and thus sent different signals to each hemisphere.

When Sperry projected an image of a nude figure onto only the left visual field, the subjects would blush and smile

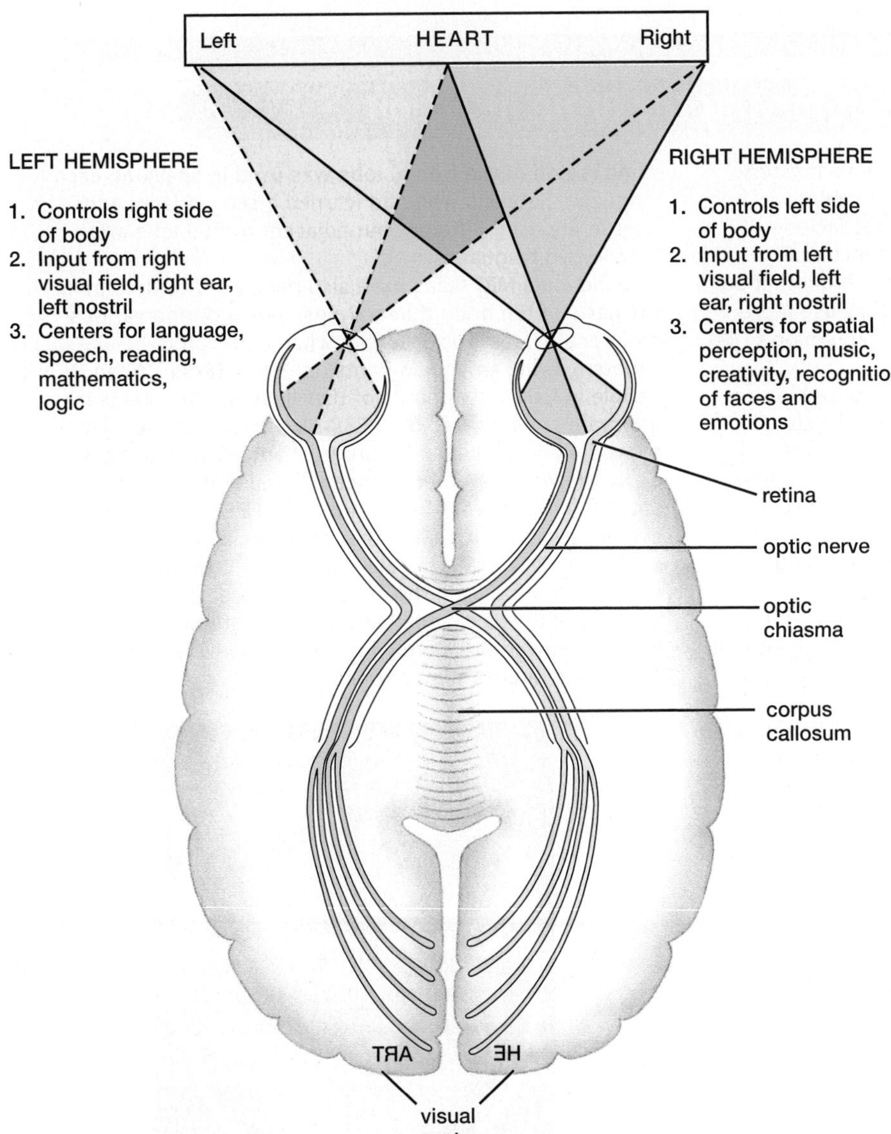

◀ **FIGURE 38-16 Specialization of the cerebral hemispheres** Each half of the retina of each eye "sees" the opposite visual field. The axons from the half-retinas that see the left visual field send information to the right hemisphere, and vice versa. Therefore, with a quick glance at the word "heart" (before you had a chance to move your eyes), the right hemisphere would perceive "he" and the left hemisphere would perceive "art." In addition to "seeing" different parts of the visual field, the two hemispheres typically control the opposite sides of the body and are specialized for a variety of functions.

but would claim to have seen nothing. The same figure projected onto the right visual field was readily described verbally. These and later experiments revealed that—in right-handed people—the left hemisphere is usually dominant in speech, reading, writing, language comprehension, mathematical ability, and logical problem solving. The right side of the brain is superior to the left in musical skills, artistic ability, recognizing faces, spatial visualization, and the ability to recognize and express emotions.

Recent experiments, however, indicate that the left–right dichotomy is not as rigid as was once believed. An individual who has suffered a stroke disrupting the blood supply to the left hemisphere typically shows symptoms such as loss of speaking ability. However, these deficits can often be partially overcome through training, even though the left hemisphere itself has not recovered. This observation suggests that the right hemisphere has some latent language capabilities.

Interestingly, women have a slightly larger corpus callosum than men do, suggesting a gender difference in the extent of interconnections between the two hemispheres. Imaging neural activity in the brains of normal subjects as they perform various mental tasks has provided further evidence of this difference. When subjects were asked to compare two word lists to find rhyming words, a specific region of the left cortex of male subjects became active, but in women, similar areas in both the left and right hemispheres were activated.

Learning and Memory Involve Biochemical and Structural Changes in Specific Parts of the Brain

Neurobiologists have made great strides in understanding the physical basis of learning and memory in recent decades. Learning has two phases: **working memory** and **long-term memory.** For example, if you look up a number in the phone book, you will probably remember the number long enough to dial it but will then promptly forget it. This is working memory. But if you call the number frequently, eventually you will remember the number more or less permanently—it has been stored in long-term memory.

Scientific Inquiry

Neuroimaging—Observing the Brain in Action

For most of human history, the brain has been a mystery. Now, however, imaging techniques provide exciting insights into brain function. These techniques include *PET* (positron emission tomography) and *fMRI* (functional magnetic resonance imaging). Regions of the brain that are most active have higher energy demands; they use more glucose and attract a greater flow of oxygenated blood than do less-active areas.

In typical PET scans, scientists inject the subject with a radioactive form of glucose, and then monitor levels of radioactivity that reflect differences in metabolic rate. These are translated by a computer into colors on images of the brain. By monitoring radioactivity while a specific task is performed, scientists can identify which parts of the brain are most active during that task. Functional MRI detects differences in the way oxygenated and deoxygenated blood respond to a powerful magnetic field. Active brain regions can be distinguished with fMRI without using radioactivity and over much shorter time spans than those required by PET.

Using fMRI or PET, researchers can observe changes as the brain performs a specific reasoning task or responds to an odor or a visual or auditory stimulus. For example, scientists have used brain scans to confirm that different aspects of the processing of language occur in distinct areas of the cerebral cortex (**Fig. E38-4**). Using fMRI, other researchers analyzed the frontal lobe areas used in generating words in individuals who spoke two languages. In subjects who had grown up speaking two languages, the same region of the frontal lobe was used in speaking each language. Subjects who had learned a second language later in life used different but adjacent frontal lobe areas for the two languages.

Functional MRI scans have also been used to determine the parts of the brain that are most active during various emotional states. For example, when a person is frightened, the amygdala lights up (**Fig. E38-5a**). When people in love view photos of their lovers, other areas in the brain are activated (**Fig. E38-5b**). Most of these same areas are also activated by drugs of abuse, such as cocaine.

By the way, never let anyone tell you that you use only a small fraction of your brain. Although PET and fMRI images sometimes make it appear that only a small area of the brain is active, this is because activity of other regions is subtracted out during the imaging process, to show where brain activity has changed as a result of stimulation. The images shown here merely highlight areas where activity is even more intense than normal.

Hearing Words

Seeing Words

Reading Words

Generating Verbs

0 max

▲ **FIGURE E38-4 Localization of language tasks** PET scans reveal the different cortical regions involved in language-related tasks, based on research by Marcus Raichle of the Washington University School of Medicine in St. Louis. The scale ranges from white (lowest brain activity) to red (highest).

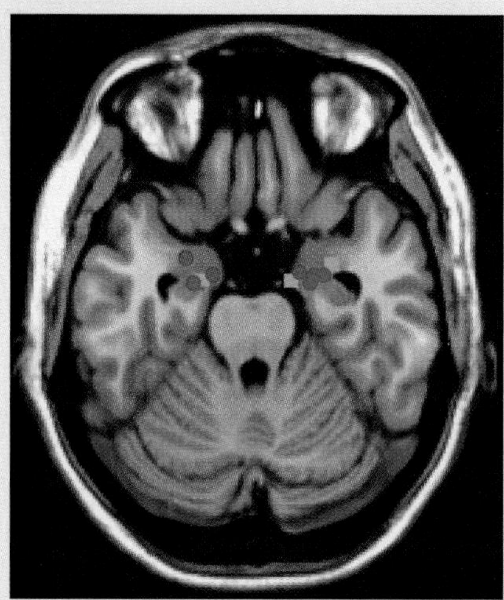

(a) Activation of the amygdala (red) by a frightening stimulus

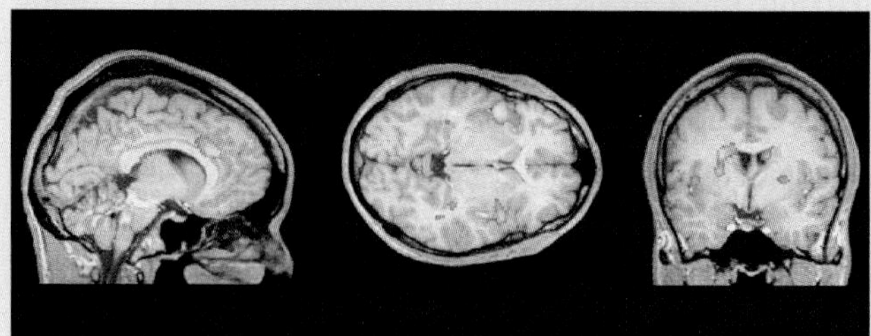

(b) Activation of the forebrain when a person views a photo of a loved one

▲ **FIGURE E38-5 Localization of emotions (a)** A frightening experience activates the amygdala, a part of the forebrain that apparently produces emotions such as fear and rage. **(b)** Looking at photos of lovers activates multiple parts of the brain, including areas in the cerebral cortex (left) and the basal ganglia (right).

The frontal and parietal lobes of the cerebral cortex, and some of the basal ganglia deep in the cerebrum, are the primary sites of working memory. Most working memory probably requires the repeated activity of a particular neural circuit in the brain. As long as the circuit is active, the memory stays. In other cases, working memory may rely on short-lived biochemical changes within neurons that temporarily strengthen synapses.

In contrast, long-term memory seems to be structural—the result, perhaps, of persistent changes in the expression of certain genes. It may require the formation of new, long-lasting synaptic connections between specific neurons or the long-term strengthening of existing but weak synapses (for example, increasing neurotransmitter release or increasing the number of receptors for the neurotransmitter). For many memories, including memories of places, facts, and specific events, converting working memory into long-term memory seems to involve the hippocampus, which is believed to process new memories and transfer them to the cerebral cortex for permanent storage. Although long-term memory probably resides in many areas of the cerebrum, some research suggests that the temporal and frontal lobes are particularly important. The cerebellum and basal ganglia are crucial for learning and storing habits and physical skills (motor learning).

Case Study revisited
How Do I Love Thee?

". . . heaven is here,
Where Juliet lives."
—*Romeo and Juliet*, Act III, scene III

What happens in the human brain when we fall in love? Although people aren't just big prairie voles, you may be surprised—and perhaps disappointed—to find out that people and prairie voles show striking similarities in both brain function and hormones during emotional encounters. For example, areas of the human brain that contain oxytocin and dopamine respond to pictures of faces of lovers and one's own children, but not to the faces of equally attractive and familiar people to whom the observer feels no emotional bond. Some of these areas are the same ones that are activated in prairie voles and that appear to be important in motivation and reward. In humans, as in voles, oxytocin probably plays an important role in attraction and commitment. Oxytocin also reduces stress and inhibits the amygdala, part of the brain involved in feeling fear. Oxytocin enhances trust, even between complete strangers. Finally, oxytocin levels increase in women and men during sexual encounters.

What about the different kinds of love? Brain scans reveal similarities as well as differences between romantic and parental love. Some of the same brain areas are activated by photos of lovers and children. Other areas are activated by one or the other, but not both. Still other brain areas, particularly those involved in critical decision making and social judgments, are turned off, at least while seeing the lover or child, who almost always seem to be better than they really are. This is probably especially important for parental love of newborns. Although often very cute, newborn infants can hardly reciprocate love the way adults can.

Will neurobiological explanations ruin the magic of love? Anthropologist Helen Fisher, who has extensively studied the neurobiology of love, doesn't think so. "You can know every ingredient in a piece of chocolate cake, and . . . it's [still] wonderful. In the same way, you can know all the ingredients of romantic love and still feel that passion."

Consider This

The lyrics for "Hooked on a Feeling" by B. J. Thomas include: " . . . I'm hooked on a feelin', high on believin' that you're in love with me." Is this a reasonable interpretation of romantic love?

CHAPTER REVIEW

Summary of Key Concepts

38.1 What Are the Structures and Functions of Nerve Cells?

Nervous systems are composed of cells called neurons. A neuron has four major functions, which are reflected in its structure: (1) dendrites receive information from the environment or from other neurons; (2) the cell body adds together electrical signals from the dendrites and from synapses on the cell body itself, and "decides" whether to produce an action potential (the cell body also coordinates the cell's metabolic activities); (3) the axon conducts the action potential to its output terminal—the synapse; and (4) synaptic terminals transmit the signal to other nerve cells, glands, or muscles.

38.2 How Do Neurons Produce and Transmit Information?

An unstimulated neuron maintains a negative resting potential inside the cell. Signals received from other neurons are small, rapidly fading changes in potential called postsynaptic potentials. Inhibitory and excitatory postsynaptic potentials (IPSPs and EPSPs) make the neuron less likely or more likely, respectively, to produce an action potential. If postsynaptic potentials, added together within the cell body, bring the neuron to threshold, an action potential will be triggered. The action potential is a wave of positive charge that travels, undiminished in magnitude, along the axon to the synaptic terminals.

A synapse consists of the synaptic terminal of the presynaptic neuron, a specialized region of the postsynaptic neuron, and the gap between them, called the synaptic cleft. Neurotransmitters from the presynaptic neuron, released in response to an action potential, diffuse across the synaptic cleft, bind to receptors in the postsynaptic cell's plasma membrane, and produce either an EPSP or an IPSP.

BioFlix™ How Neurons Work

BioFlix™ How Synapses Work

38.3 How Do Nervous Systems Process Information?

Information processing in the nervous system requires four operations. The nervous system must (1) determine the type of stimulus, (2) determine and signal the intensity of the stimulus, (3) integrate information from many sources, and (4) initiate and direct an appropriate response.

38.4 How Are Nervous Systems Organized?

Neural pathways usually have four elements: (1) sensory neurons, (2) interneurons, (3) motor neurons, and (4) effectors. Overall, nervous systems consist of numerous interconnected neural pathways, which may be diffuse (distributed throughout the body) or centralized, with the majority of the senses and nervous system in the head.

38.5 What Are the Structures and Functions of the Human Nervous System?

The nervous system of humans and other mammals consists of the central nervous system and the peripheral nervous system. The peripheral nervous system is divided into sensory and motor portions. The motor portion consists of the somatic nervous system (which controls voluntary movement) and the autonomic nervous system (which directs involuntary responses). The autonomic nervous system is further subdivided into the sympathetic and parasympathetic divisions.

The central nervous system consists of the brain and spinal cord. The spinal cord contains neurons that control voluntary muscles and the autonomic nervous system; neurons that communicate with the brain and other parts of the spinal cord; axons leading to and from the brain; and neural pathways for reflexes and certain simple behaviors.

The brain consists of three parts: the hindbrain, midbrain, and forebrain. Each is further subdivided into distinct regions. The hindbrain in humans consists of the medulla and pons, which control involuntary functions (such as breathing), and the cerebellum, which coordinates complex motor activities (such as typing). In humans, the small midbrain contains clusters of neurons that help to control movement, arousal, and emotion. The midbrain and hindbrain contain the reticular formation, which is a filter and relay for sensory stimuli. The forebrain includes the thalamus, a sensory relay station that shuttles information to and from conscious centers in the forebrain; the hypothalamus, which is responsible for maintaining homeostasis and directs much of the activity of the autonomic nervous system; and the cerebrum, the center for information processing, memory, and initiation of voluntary actions. The cerebral cortex includes primary sensory and motor areas, together with association areas that analyze sensory information and plan movements. A group of forebrain structures called the limbic system, together with many other parts of the brain, controls the perception and expression of emotions.

The cerebral hemispheres are specialized. In general, the left hemisphere dominates speech, reading, writing, language comprehension, mathematical ability, and logical problem solving. The right hemisphere specializes in recognizing faces and spatial relationships, producing artistic and musical abilities, and recognizing and expressing emotions.

Memory takes two forms: Short-term memory is electrical or chemical, whereas long-term memory probably involves structural changes that increase the effectiveness or number of synapses. The hippocampus is an important site for the transfer of information from short-term into long-term memory, often in the temporal and frontal lobes.

Key Terms

action potential *734*
amygdala *749*
autonomic nervous system *744*
axon *735*
basal ganglia *749*
blood–brain barrier *744*
brain *743*
cell body *734*
central nervous system (CNS) *743*
cerebellum *749*
cerebral cortex *749*
cerebral hemisphere *749*
cerebrum *749*
convolution *751*
corpus callosum *751*
dendrite *734*
dorsal root ganglion *744*
effector *742*
excitatory postsynaptic potential (EPSP) *738*
forebrain *747*
ganglion (plural, **ganglia**) *742*
glia *734*
gray matter *744*
hindbrain *747*
hippocampus *749*
hypothalamus *749*
inhibitory postsynaptic potential (IPSP) *738*
integration *741*
intensity *741*
interneuron *742*
limbic system *750*
long-term memory *753*
medulla *748*
midbrain *747*
motor neuron *742*
myelin *736*
nerve *735*
nerve net *742*
neuron *734*
neurotransmitter *734*
parasympathetic division *744*
peripheral nervous system (PNS) *743*
pons *749*
postsynaptic neuron *737*
postsynaptic potential (PSP) *738*
presynaptic neuron *737*
reflex *747*
resting potential *736*
reticular formation *749*
sensory neuron *742*
sodium-potassium (Na^+-K^+) pump *738*
somatic nervous system *744*
spinal cord *743*
sympathetic division *744*
synapse *735*
synaptic cleft *737*
synaptic terminal *735*
thalamus *749*
threshold *736*
white matter *744*
working memory *753*

Thinking Through the Concepts

Fill-in-the-Blank

1. The parts of an individual nerve cell, also called a(n) __________, are specialized to perform different functions. The "input" end of a nerve cell, called a(n) __________, receives information from the environment or from other nerve cells. The __________ contains the nucleus and other typical organelles of a eukaryotic cell. Electrical signals are sent down the __________, a long, thin strand that leads to the __________, where the nerve cell sends out its signal to other cells.
2. When they are not being stimulated, neurons have an electrical charge across their membranes called the resting potential. This potential is __________ (what charge?)

inside. When a neuron receives a sufficiently large stimulus, and reaches a potential called the __________, it produces an action potential. This causes the neuron to become __________ (what charge?) inside.

3. When an action potential reaches a synaptic terminal, it causes the release of a chemical called a(n)__________. This chemical binds to protein __________ on the postsynaptic cell, causing a change in potential. If the postsynaptic cell becomes less negative, this change in potential is called a(n) __________. If the postsynaptic cell becomes more negative, it is called a(n) __________.
4. The __________ is part of the peripheral nervous system that innervates skeletal muscles. Smooth muscles and glands are innervated by the __________, which has two divisions: the __________, active during "fight or flight," and the __________, active during "rest and digest."
5. The human hindbrain consists of three parts: the __________, the __________, and the __________. One of these, the __________, is important is coordinating complex movements such as typing.
6. The cerebral cortex consists of four lobes: the __________, __________, __________, and __________. The visual centers are located in the __________ lobe. The primary motor areas are in the __________ lobe.

Review Questions

1. List the four major parts of a neuron, and explain the specialized function of each part.
2. Diagram a synapse. How are signals transmitted from one neuron to another at a synapse?
3. How does the brain perceive the intensity of a stimulus? The type of stimulus?
4. What are the four elements of a neuron-to-muscle pathway? Describe how these elements function in the human pain-withdrawal reflex.
5. Draw a cross-section of the spinal cord. What types of neurons are located in the spinal cord? Explain why severing the cord paralyzes the body below the level where it is severed.
6. Describe the functions of the following parts of the human brain: medulla, cerebellum, reticular formation, thalamus, amygdala, and cerebral cortex.
7. What structure connects the two cerebral hemispheres? Describe the intellectual functions of the two hemispheres in most right-handed people.
8. Distinguish between long-term memory and working memory.

Applying the Concepts

1. **BioEthics** Parkinson's disease is caused by the degeneration of dopamine-producing cells in the substantia nigra, a small part of the midbrain that is important in controlling movement. Some physicians have reported improvement after injecting cells taken from the same general brain region of an aborted fetus into appropriate parts of the brain of a Parkinson's patient. Discuss this type of surgery from as many viewpoints as possible: ethical, financial, practical, and so on. Based on your responses, is fetal transplant surgery the answer to curing Parkinson's disease?
2. If the axons of human spinal cord neurons were unmyelinated, would you expect the spinal cord to be larger or smaller? Would you move faster or slower? Explain your answer.
3. What is the adaptive value of reflexes? Why couldn't all behaviors be controlled by reflexes?
4. You can buy oxytocin online, in the form of a cologne or perfume. The claim is that it will make other people trust you more, thus aiding both business and sexual conquests. Scientific research indeed shows that oxytocin enhances trust, although the researchers sprayed oxytocin into the nose, which provides direct access to the olfactory neurons and the brain. Assuming that wearing oxytocin cologne produces a high enough concentration of oxytocin in the air to affect anyone's brain (which it probably doesn't), do you think that it would be a worthwhile product? Why or why not?

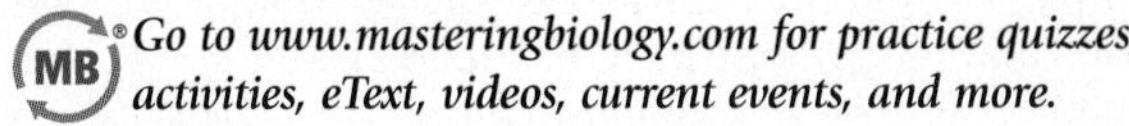

Go to www.masteringbiology.com for practice quizzes, activities, eText, videos, current events, and more.

The Senses

Chapter 39

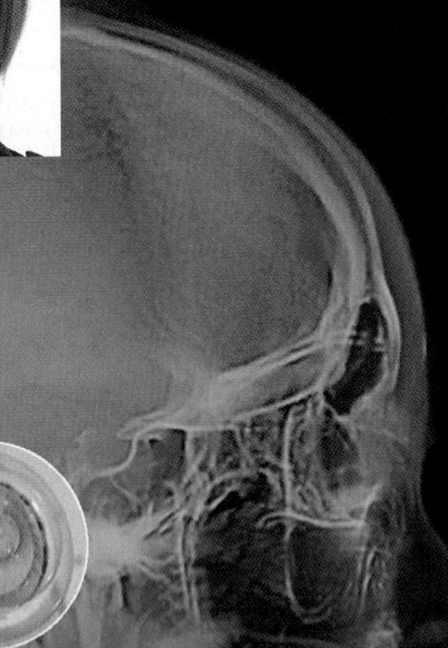

▲ For many deaf people, cochlear implants open the door to a new world of sound.

Case Study

Bionic Ears

JENNIFER THORPE (top right) lost her hearing when she was 4 years old. No obvious disease, no slow deterioration over time—one moment she could hear just fine, and a few minutes later, she was nearly deaf. Within a few months, she was totally deaf in one ear and had only minimal hearing in the other. With the help of a hearing aid in the still-functioning ear, and lessons in lip-reading and sign language, Jennifer went off to school, grew up, married, and became a mother. Then, suddenly, she lost most of the hearing in her "good" ear.

Samantha Brilling (bottom left) was born hard of hearing, although not totally deaf. She wore hearing aids for many years, but they couldn't provide normal hearing. Sounds that most of us take for granted were missing for Sam, including a lot of casual conversation and music.

Today, both Jennifer and Sam can hear, thanks to cochlear implants. As you will learn in this chapter, the cochlea is the part of the ear that converts sound vibrations to electrical signals, and triggers action potentials in the axons of the auditory nerve. These axons then carry information about sound to the brain. In a marvel of technology, a cochlear implant partially replaces cochlear function in a deaf person's ear by converting sound to electrical pulses that stimulate the axons of the auditory nerve.

Although neither has perfect hearing, Jennifer and Sam can now listen to music, carry on a conversation in a crowded room, and know when someone out of sight is calling to them. Sam, a college student, hears well enough that she doesn't need a note-taker or special amplification system to hear her instructors. In fact, Sam discovered that she can hear some fairly subtle sounds. As she puts it, "Dogs. . . . They are quite annoying. I can hear them when they lick themselves."

How do your ears produce the sensation of sound? How does a "bionic ear" replace a biological cochlea? How can electrical impulses, whether generated biologically or bionically, convey speech, music, and, yes, even the sound of a dog licking itself? And what about your other senses—how do electrical signals convey the odor of pine trees, the colors of blooming flowers, the silky feel of a baby's skin, or the "burn" of hot peppers?

At a Glance

39.1 HOW DOES THE NERVOUS SYSTEM SENSE THE ENVIRONMENT?

A **receptor** is a structure that changes when it is acted on by a stimulus and then produces some sort of response. Every cell has many types of receptor molecules, such as membrane proteins that bind hormones or neurotransmitters and T-cell receptors that bind foreign molecules (antigens) as part of the immune response. A **sensory receptor** is an entire specialized cell (often, but not always, a neuron) that produces an electrical signal in response to specific stimuli—that is, it translates environmental stimuli into the language of the nervous system. These stimuli may arise from the world around us (for example, light or sound) or from within our own bodies (for example, the position of a joint or the oxygen concentration in the blood). Sensory receptors are grouped into categories, according to the stimulus to which they respond (**Table 39-1**).

The Senses Inform the Brain About the Nature and Intensity of Environmental Stimuli

Encoding the nature of a stimulus—light, sound, odor, or touch, for example—begins with sensory receptor cells, each of which contains receptor molecules that respond to certain stimuli and not others. Each type of sensory receptor is linked to a specific set of axons that connect to particular places in the brain or spinal cord. In mammals, for example, neurons that detect odors send axons to a part of the brain called the olfactory bulb.

The linkage of stimulus type to sensory receptor to axons that lead to a particular brain region produces the first principle of sensory perception: The *type of stimulus* is encoded by the specific set of neurons in the brain that are activated.

Case Study continued
Bionic Ears

Cochlear implants work because the brain interprets action potentials in axons of the auditory nerve as sound, regardless of the actual stimulus triggering them. Alessandro Volta, who invented the battery in 1800, inadvertently discovered this principle. He stuck a metal rod in his ear and connected it to a battery. He felt a jolt and heard a sound like boiling water. Like Volta's rod, a modern cochlear implant stimulates action potentials in axons of the auditory nerve, although with much more sensitivity and precision.

Table 39-1 Some Vertebrate Receptor Types

Type of Receptor	Sensory Cell Type	Stimulus	Location
Thermoreceptor	Free nerve ending	Heat, cold	Skin
Mechanoreceptor	Hair cell	Vibration, motion, gravity	Inner ear
	Free nerve endings and endings surrounded by accessory structures (Pacinian corpuscle, Meissner's corpuscle, Ruffini corpuscle)	Vibration, pressure, touch	Skin
	Specialized nerve endings in muscles or joints (muscle spindle, Golgi tendon organ)	Stretch	Muscles, tendons
Photoreceptor	Rod, cone	Light	Retina of the eye
Chemoreceptor	Olfactory receptor	Odor (airborne molecules)	Nasal cavity
	Taste receptor	Taste (waterborne molecules)	Tongue and oral cavity
Pain receptor (nociceptor)	Free nerve ending	Chemicals released by tissue injury; excessive heat or cold; excessive stretch; acid; some environmental chemicals	Widespread in the body

Encoding the **intensity** of a stimulus—loud versus soft sounds, for instance—also begins with sensory receptor cells. When a sensory receptor is stimulated, it produces an electrical signal called a **receptor potential** (**Fig. 39-1**). In many sensory receptor neurons, the receptor potential may bring the neuron above threshold and trigger action potentials. The axons of these sensory receptor neurons often connect directly with the central nervous system (CNS).

Unlike action potentials (see Fig. 38-2), which are always the same size, receptor potentials vary in size with the intensity of a stimulus—the stronger the stimulus, the larger the receptor potential (**Fig. 39-2**). A small receptor potential may barely reach threshold and, hence, will produce only a few action potentials; a large receptor potential will go far above threshold, causing a high frequency of action potentials.

Some sensory receptors, such as those in the inner ear, do not have axons. Most of these receptors form synapses with neurons whose axons connect with the CNS (see Fig. 38-4 for a description of the anatomy and function of a synapse). Receptor potentials cause neurotransmitters to be released from the sensory receptors. The neurotransmitters produce excitatory postsynaptic potentials in the axon endings, where they stimulate action potentials that travel down the axon to the

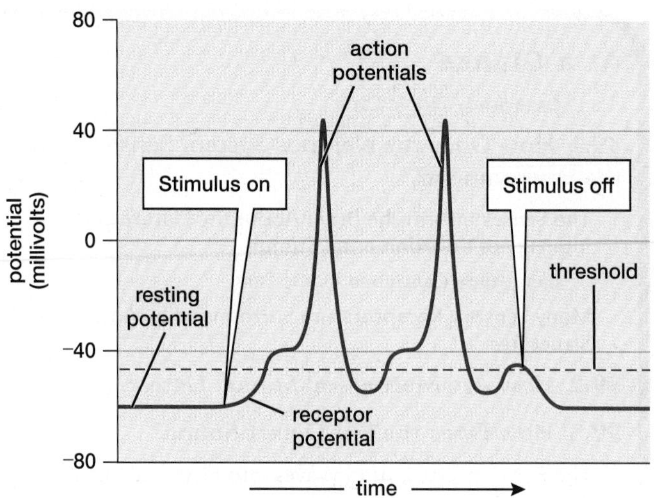

▲ **FIGURE 39-1 Converting an environmental stimulus to action potentials** In most sensory receptor neurons, an environmental stimulus generates a receptor potential that makes the resting potential less negative. If the receptor potential is large enough, it reaches threshold and triggers action potentials.

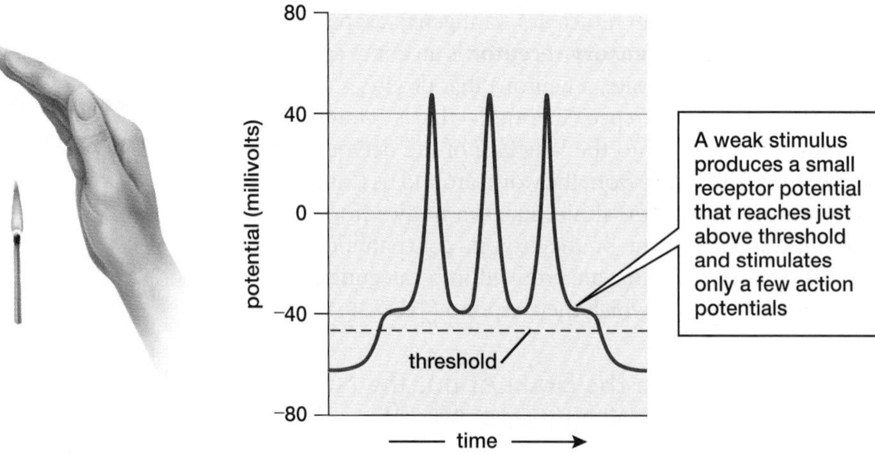

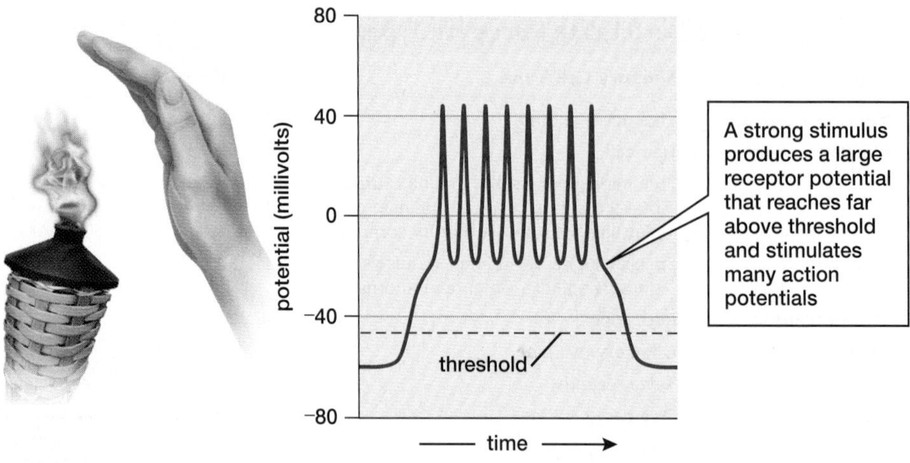

► **FIGURE 39-2 Stimulus intensity is encoded by the frequency of action potentials (a)** Weak stimuli cause small receptor potentials that barely reach threshold (dashed line) and produce only a few action potentials. **(b)** Strong stimuli cause large receptor potentials that reach far above threshold, producing many action potentials.

CNS. A strong stimulus causes a large receptor potential that releases a large amount of transmitter onto the neurons, producing a high frequency of action potentials.

The linkage of stimulus intensity to receptor potential size to action potential frequency produces the second principle of sensory perception: The *intensity of a stimulus* is encoded by the frequency of action potentials reaching the brain.

Many Sensory Receptors Are Surrounded by Accessory Structures

Some sensory receptors, called free nerve endings, consist of branching dendrites of sensory neurons. These may respond directly to touch, heat, cold, or pain. Many other sensory receptors are contained within structures that help them respond to a specific stimulus. These include receptors for vibration, pressure, sound, light, olfaction, and taste.

39.2 HOW ARE MECHANICAL STIMULI DETECTED?

Mechanoreceptors are found throughout the human body. They include receptors in the skin that respond to touch, vibration, or pressure; stretch receptors in many internal organs; position sensors in the joints; and receptors in the inner ear that allow us to detect sound (see section 39.3).

The skin of humans and most other vertebrates is exquisitely sensitive to touch. Embedded in the skin are several types of mechanoreceptor neurons, each with a dendrite that produces a receptor potential when its membrane is stretched or dented (**Fig. 39-3**). The dendrites of some touch receptors are free nerve endings that can produce sensations of touch, itching, or tickling. The endings of other receptors are enclosed in layers of connective tissue—for example, Pacinian corpuscles, which respond to changes in pressure such as rapid vibrations or a sharp poke; Meissner's corpuscles, which respond to light touch or slow vibrations; and Ruffini corpuscles, which respond to steady pressure. The density of mechanoreceptors in the skin varies tremendously over the surface of the body. Each square inch of fingertip has hundreds of touch receptors, but on the back, there may be less than one per square inch.

Mechanoreceptors in many hollow organs, such as the stomach, rectum, and urinary bladder, signal fullness by responding to stretch. Mechanoreceptors in the joints, also responding to stretch, let us know whether the joints are straight or bent, and by how much. Receptors in the muscles,

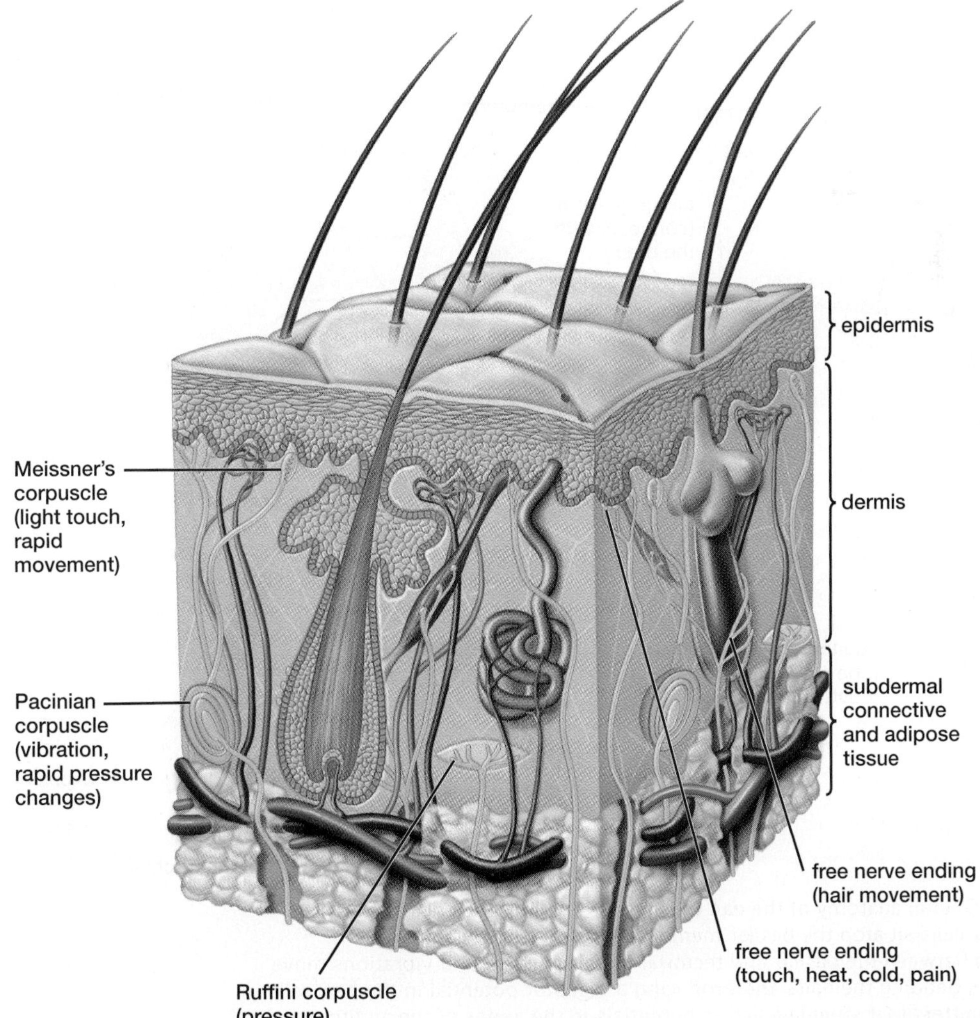

◀ FIGURE 39-3 **Receptors in the human skin** The diversity of receptors in the skin allows us to perceive mechanical stimuli such as touch, pressure, vibration, and tickling, as well as other sensations such as pain, heat, and cold.

called muscle spindles, tell the brain if the muscle is contracted or stretched, and by how much (muscle spindles are activated in the knee-jerk reflex so beloved by physicians and small boys). Another type of receptor, the Golgi tendon organ, located where a muscle attaches to a tendon, signals how much force the muscle is exerting. Joint and muscle receptors combine to inform the brain of the position, angle, and force of movement of arms, legs, hands, and feet—you don't have to look at them or consciously think about what they're doing. Imagine how annoying it would be if you had to watch your fork on its way to your mouth to avoid stabbing yourself in the face!

Mechanoreceptors are important to all animals. Spiders, for example, use mechanoreceptors in their legs to detect vibrations of their webs. They can tell fairly reliably whether the vibrations are from small fluttering objects that might be good to eat, like flies or moths; larger, possibly predatory, animals; or even potential mates. Cockroaches and many other insects can detect tiny puffs of air that move ahead of predators—much like the bow wave in front of a boat—and race off in the opposite direction. Fish have mechanoreceptors in organs called the lateral lines, which detect movement and vibration of the water around them. For some, such as blind cave fish that spend their lives in total darkness, the lateral lines are crucial to avoid bumping into their surroundings and to locate prey.

39.3 HOW DOES THE EAR DETECT SOUND?

Sound is produced by vibrating objects—drums, vocal cords, or the speakers of your CD player. The resulting vibrations, or sound waves, typically are transmitted through the air and are intercepted by our ears, which convert them to signals that our brains interpret as the direction, pitch, and loudness of sound. The mammalian ear consists of structures that transmit vibrations to specialized mechanoreceptor cells deep in the inner ear.

The Ear Converts Sound Waves into Electrical Signals

The ear of humans and other mammals consists of three parts: the outer, middle, and inner ear (**Fig. 39-4a**). The **outer ear** consists of the **pinna** and the **auditory canal.** The pinna, a flap of skin-covered cartilage attached to the surface of the head, collects sound waves. Humans and other large animals determine sound direction by differences in when sound arrives at the two ears, and in how loud it is in each ear. The shape of the pinna and, in many animals, the ability to swivel it around, further help to localize sounds.

The air-filled auditory canal conducts the sound waves to the **middle ear,** consisting of the **tympanic membrane,** or eardrum; three tiny bones called the **hammer** (malleus),

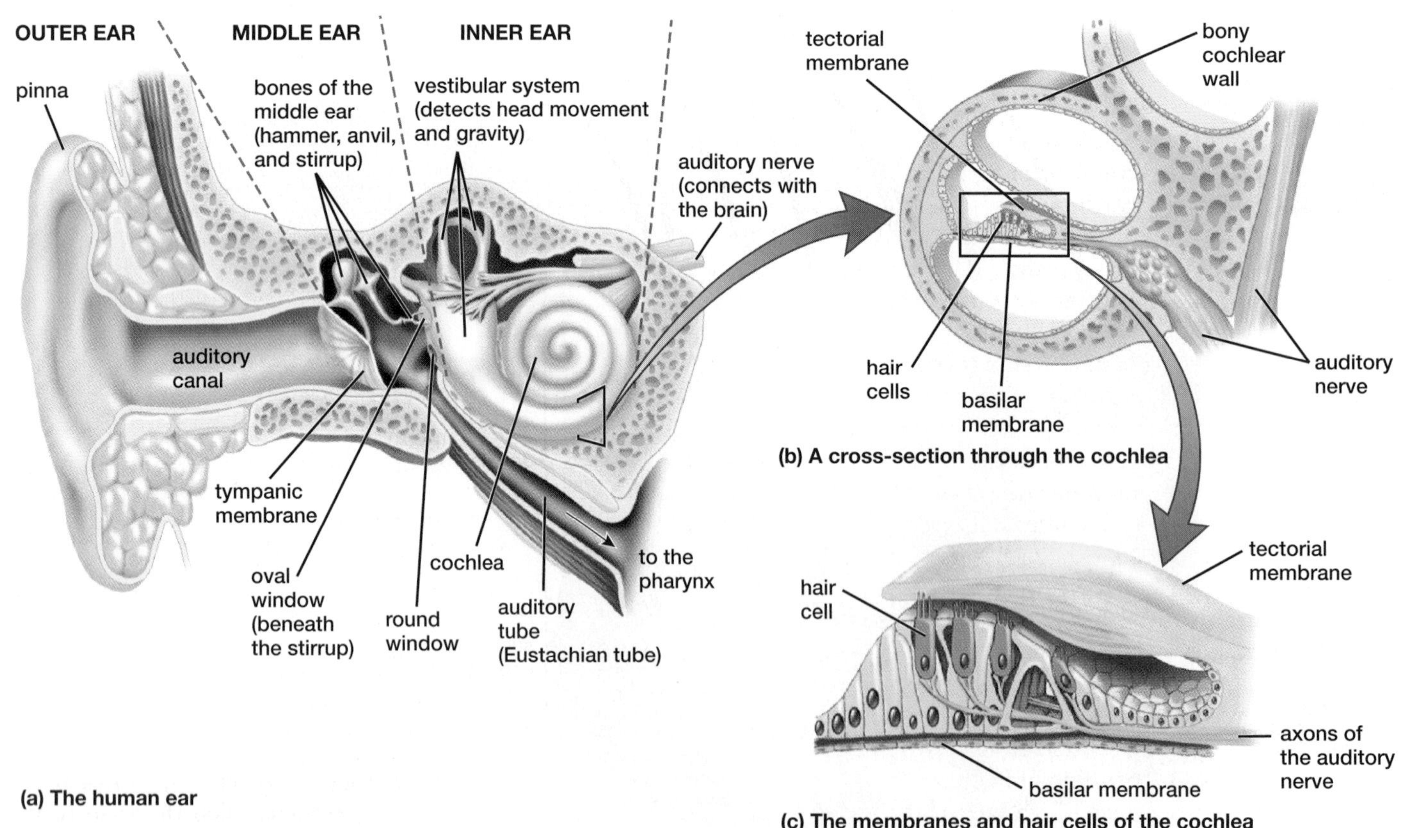

▲ **FIGURE 39-4 The human ear (a)** Overall anatomy of the ear. **(b)** In cross-section, the cochlea consists of three fluid-filled compartments; hair cells sit atop the basilar membrane in the central compartment. **(c)** The hairs of hair cells span the gap between the basilar and tectorial membranes. Sound vibrations move the membranes relative to one another, bending the hairs and producing a receptor potential in the hair cells. The hair cells then release neurotransmitters that stimulate action potentials in the axons of the auditory nerve.

anvil (incus), and **stirrup** (stapes); and the **auditory tube** (also called the Eustachian tube). The auditory tube connects the middle ear to the pharynx and equalizes the air pressure between the middle ear and the atmosphere. This tube may become swollen shut if you have a cold; if this happens, air pressure changes (such as those experienced during aircraft takeoff and landing) can be painful.

Sound waves traveling down the auditory canal vibrate the tympanic membrane, which in turn moves the hammer, the anvil, and the stirrup. These bones transmit vibrations to the **inner ear.** The fluid-filled hollow bones of the inner ear form the spiral-shaped **cochlea** ("snail" in Latin). The stirrup bone transmits sound waves to the fluid within the cochlea by vibrating a membrane in the cochlea called the **oval window.** The **round window** is a second membrane below the oval window; its movement allows the fluid within the cochlea to shift back and forth as the stirrup bone vibrates the oval window.

Vibrations Are Converted into Electrical Signals in the Cochlea

The cochlea, in cross-section, consists of three fluid-filled compartments. The central compartment houses the receptors and the supporting structures that activate them in response to sound vibrations. The floor of this central chamber is the **basilar membrane,** on top of which sit mechanoreceptors called **hair cells.** Hair cells have cell bodies topped by hairlike projections that resemble stiff cilia. Some of these hairs are embedded in a gelatinous structure called the **tectorial membrane,** which protrudes into the central canal (**Figs. 39-4b,c**).

The oval window passes vibrations from the bones of the middle ear to the fluid in the cochlea, which in turn vibrates the basilar membrane relative to the tectorial membrane. This movement bends the hairs of the hair cells, producing receptor potentials. The receptor potentials cause the hair cells to release neurotransmitters onto neurons whose axons form the **auditory nerve.** These axons produce action potentials that travel to auditory centers in the brain.

How do the structures of the inner ear allow us to perceive loudness (the magnitude of sound vibrations) and pitch (the musical note, or the frequency of sound vibrations)? Remember that the general principles of sensory perception are that the type of stimulus is encoded by which nerve cells fire action potentials, and the intensity of the stimulus is encoded by the frequency of action potentials. Soft sounds cause small vibrations of the tympanic membrane, the bones of the middle ear, the oval window, and the basilar membrane. Therefore, the hairs bend only a little, so the hair cells produce small receptor potentials that cause the release of a tiny bit of neurotransmitter and result in a low frequency of action potentials in axons of the auditory nerve. Loud sounds cause large vibrations, which cause greater bending of the hairs and a larger receptor potential, producing a high rate of action potentials in the auditory nerve. Very loud sounds can damage the hair cells (**Fig. 39-5a**), resulting in hearing loss, a fate suffered by some rock musicians and their fans. In fact, many sounds in our everyday environment have the potential to damage hearing, especially if they are prolonged (**Fig. 39-5b**).

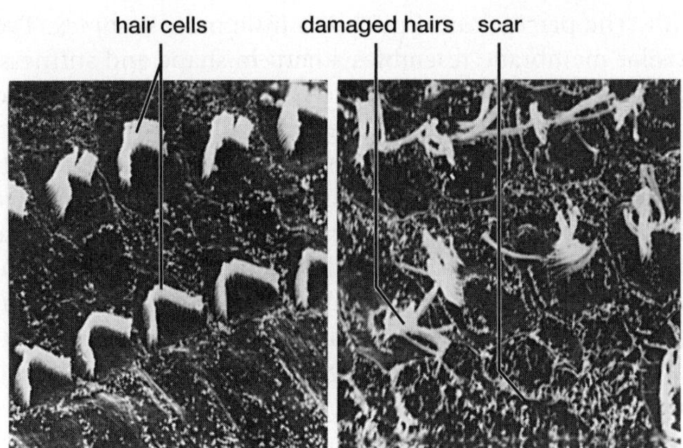

(a) Normal hair cells (left) and hair cells damaged by loud sound (right)

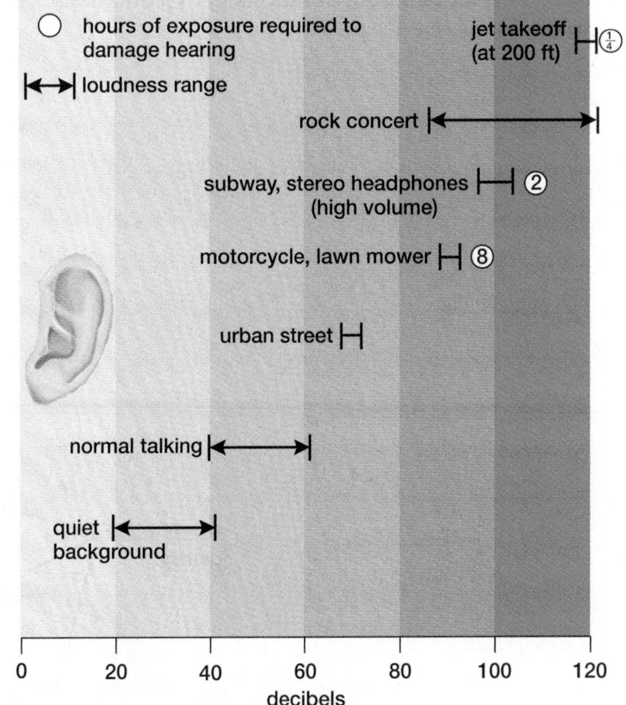

(b) The loudness of everyday sounds

▲ **FIGURE 39-5 Loud sounds can damage hair cells**
(a) Scanning electron micrographs (SEMs) show the effect of intense sound on the hair cells of the inner ear. (left) In a normal guinea pig, hairs emerge from each hair cell in a V-shaped pattern. (right) After 24-hour exposure to a sound level approached by loud rock music (2,000 vibrations per second at 120 decibels), many of the hairs are damaged or missing, leaving "scars." Hair cells in humans do not regenerate, so such hearing loss would be permanent. Source: SEMs by Robert S. Preston, courtesy of Professor J. E. Hawkins, Kresge Hearing Research Institute, University of Michigan Medical School.
(b) Sound levels of everyday noises and their potential to damage hearing. Sound intensity is measured in decibels on a logarithmic scale; a 20-decibel sound is 10 times as loud as a 10-decibel sound, a 30-decibel sound is 100 times as loud as a 10-decibel sound, and so on. You feel pain at sound intensities above 120 decibels. Source: Deafness Research Foundation, National Institute on Deafness and Other Communication Disorders.

The perception of pitch is a little more complex. The basilar membrane resembles a harp in shape and stiffness: narrow and stiff at the end near the oval window, wider and more flexible near the tip of the cochlea. In a harp, the short, tight strings produce high notes, and the long, looser strings produce low notes. In the basilar membrane, the progressive change in width and stiffness causes each portion to vibrate most strongly when stimulated by a particular frequency of sound: high notes near the oval window and low notes near the tip of the cochlea. The brain interprets signals originating in hair cells near the oval window as high-pitched sound, whereas signals from hair cells located progressively closer to the tip of the cochlea are interpreted as increasingly lower in pitch. Young people with undamaged cochleas can hear sounds from about 20 vibrations per second (very low bass) to about 20,000 vibrations per second (very, very high treble).

The Vestibular Apparatus Detects Gravity and Movement

The mammalian inner ear is also the site of the **vestibular apparatus,** consisting of the vestibule (a small chamber at the entrance to the apparatus) and the semicircular canals (**Fig. 39-6;** see also Fig. 39-4a).

The vestibule contains the **utricle** and **saccule,** which detect the direction of gravity and the degree of tilting of the head. Each consists of a cluster of hair cells. Their hairs are embedded in a gelatinous matrix containing tiny stones of calcium carbonate. The hairs in the utricle are vertical, whereas those in the saccule are horizontal. Gravity pulls the stones downward, causing the hairs to bend in various directions depending on the tilt of the head. People can detect about a half-degree tilt.

Beyond the vestibule are three **semicircular canals,** which detect head movement. Each semicircular canal consists

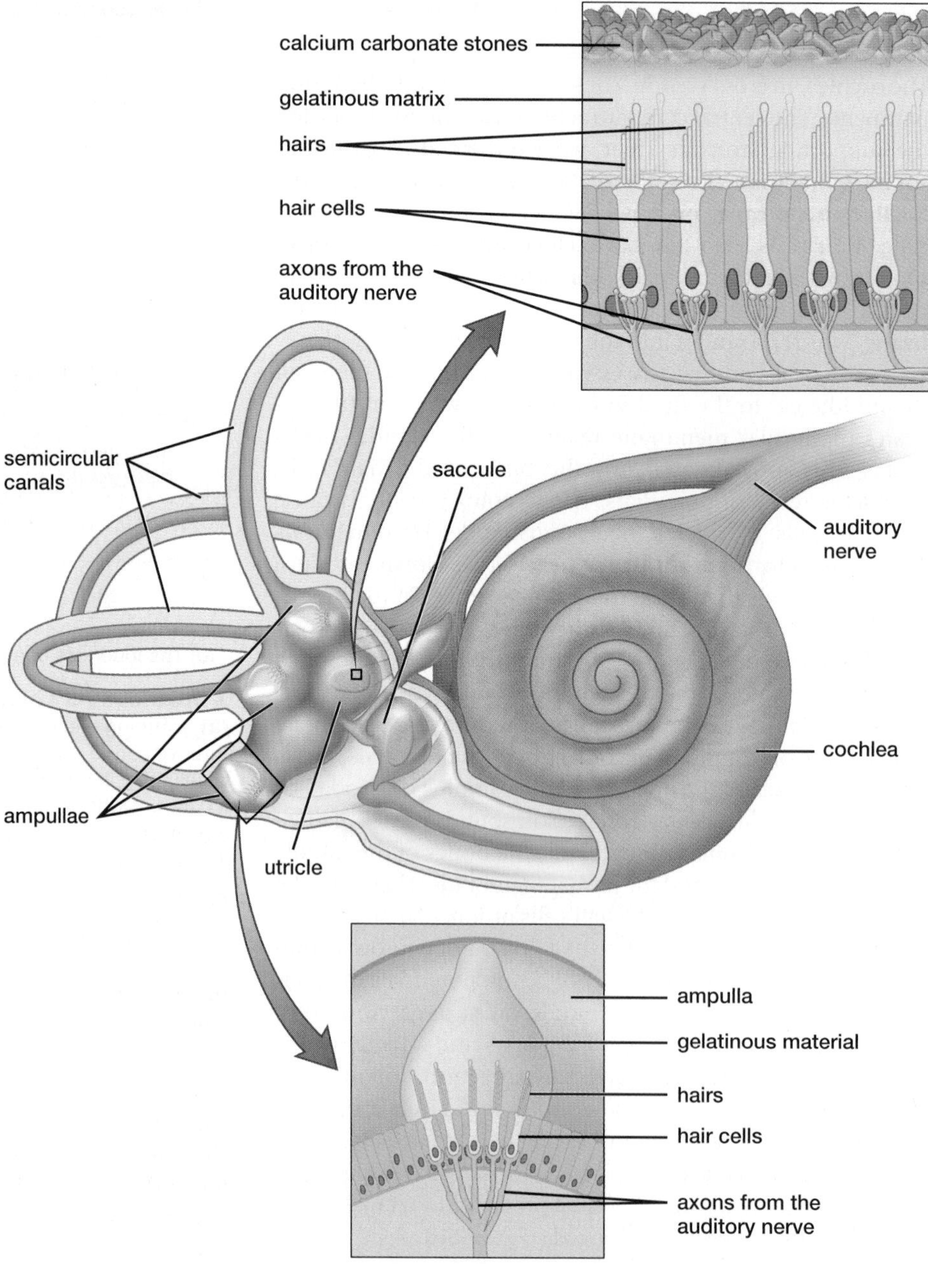

▶ **FIGURE 39-6 The vestibular apparatus** A cross-section of the vestibular apparatus. (top) The hairs of hair cells in the utricle and saccule bend under the weight of calcium carbonate stones, providing information about the direction of gravity and the tilt of the head. (bottom) The hairs of hair cells in the ampullae of the semicircular canals bend when head movement causes the fluid in the canals to slosh around.

of a fluid-filled tube with a bulge at one end, called an ampulla (plural, ampullae). Hair cells sit inside each ampulla, with their hairs embedded in a gelatinous capsule (but without the stones found in the utricle and saccule). Acceleration of the head—for example, if you shake your head "no" or lurch sideways in a roller coaster—pushes the fluid against the capsule, bending the hairs. The three semicircular canals are arranged perpendicularly to each other, similar to the X, Y, and Z axes of a 3-D graph, or the two walls and the floor in the corner of a room, allowing you to detect head movement in any direction.

As in the cochlea, axons from the auditory nerve innervate the hair cells of the vestibular apparatus (for this reason, the auditory nerve is technically named the "vestibulocochlear" nerve). When the hairs bend, the hair cells release neurotransmitter onto the endings of these axons, triggering action potentials in the axons that travel to balance and equilibrium centers in the brain.

39.4 HOW DOES THE EYE DETECT LIGHT?

Different animal species vary tremendously in how well they can see. All animal vision, however, uses receptor cells called **photoreceptors.** These cells contain receptor molecules called **photopigments** (because they are colored) that change shape when they absorb light. This shape change sets off chemical reactions inside the photoreceptors that ultimately produce receptor potentials.

The Compound Eyes of Arthropods Produce a Mosaic Image

Many arthropods, including insects and crustaceans, have **compound eyes,** which consist of a mosaic of many individual light-sensitive subunits called **ommatidia** (singular, ommatidium; **Fig. 39-7**). Each ommatidium functions as an individual light detector, like the pixels in a digital camera. Depending on the number of ommatidia, most arthropods probably see a reasonably accurate image of the world. However, the best arthropod eyes—in dragonflies, for example—contain only about 30,000 ommatidia. A human eye, in contrast, contains well over 100 million receptors. By human standards, therefore, the image formed by a compound eye is very grainy (**Fig. 39-8**). Compound eyes are excellent at detecting movement, as light and shadow flicker across adjacent ommatidia, which is an advantage in avoiding predators and in hunting. Many arthropods, such as bees and butterflies, also have good color perception. Some even see ultraviolet light, and can tell if light is polarized or not.

The Mammalian Eye Collects and Focuses Light, Converting It into Electrical Signals

Mammalian eyes consist of two major parts: (1) a variety of structures that hold the eye in a fairly fixed shape, control the amount of light that enters, and focus the light rays; and (2) the retina, which contains the photoreceptors that respond to the incoming light (**Fig. 39-9**).

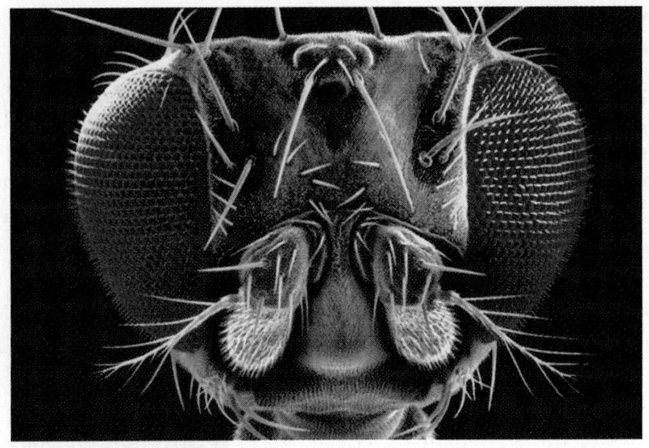

(a) Compound eyes

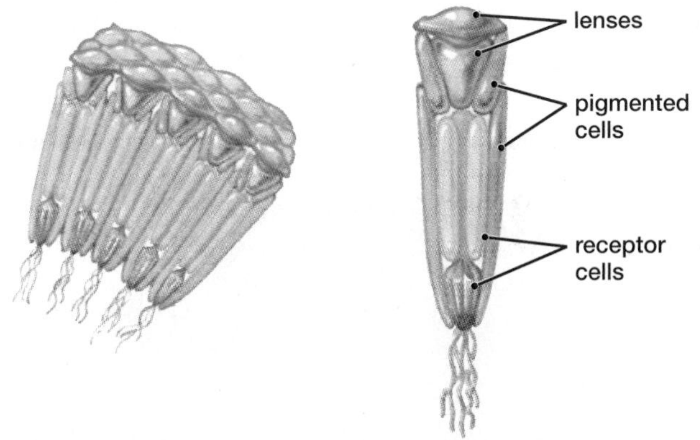

(b) Ommatidia

Single ommatidium

▲ **FIGURE 39-7 Compound eyes (a)** A scanning electron micrograph of the head of a fruit fly, showing a compound eye on each side of the head. **(b)** Each eye is made up of numerous ommatidia. Within each ommatidium are several receptor cells, capped by a lens. Pigmented cells surrounding each ommatidium prevent the passage of light to adjacent receptors.

Light entering the eye first encounters the **cornea,** a transparent covering over the front of the eyeball that collects light waves and begins to focus them. Behind the cornea, light passes through a chamber filled with a watery fluid called **aqueous humor,** which provides nourishment for both the lens and cornea. The amount of light entering the eye is adjusted by the **iris,** a pigmented muscular tissue. The iris regulates the size of the **pupil,** the opening in its center. Light passing through the pupil strikes the **lens,** a structure that resembles a flattened sphere, composed of transparent proteins. The lens is suspended behind the pupil by muscles that regulate its shape and allow fine focusing of the image. Behind the lens is a large chamber filled with **vitreous humor,** a clear jelly-like substance that helps to maintain the shape of the eyeball.

After passing through the vitreous humor, light hits the **retina.** Here, the light energy is converted into action potentials that are conducted to the brain. Behind the retina is the **choroid,** a darkly pigmented tissue. The choroid's rich blood supply helps nourish the cells of the retina. Its dark pigment absorbs stray light whose reflection inside the eyeball would interfere with clear vision. In nocturnal animals, the choroid

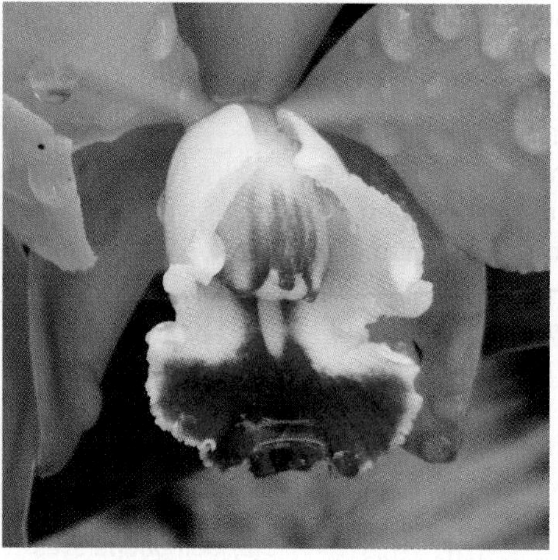
(a) A tropical orchid

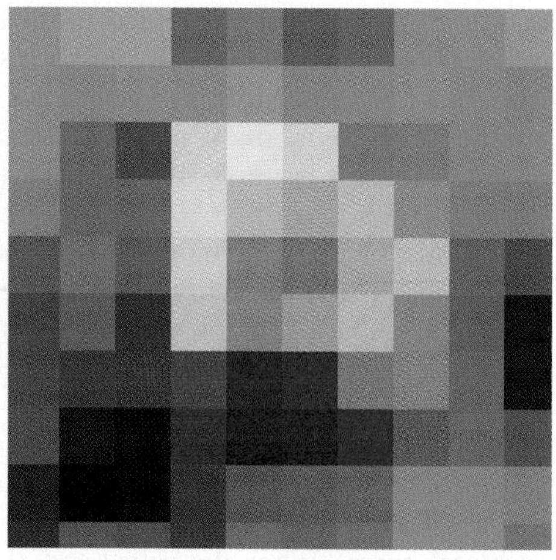
(b) The same orchid as seen by a bee

▶ **FIGURE 39-8 An insect's view of the world** At a distance of 2 or 3 feet, an insect—even one with relatively good vision, such as a bee or dragonfly—would see **(a)** an orchid as **(b)** a mosaic of individual blobs of color.

is often reflective rather than black. By reflecting light back through the retina, a mirror-like choroid gives the photoreceptors a second chance to capture scarce photons of light that they may have missed the first time through. At night, it's better to have vision that's a bit blurry than no vision at all!

Surrounding the outer portion of the eyeball is the **sclera,** a tough connective tissue layer that is visible as the white of the eye and is continuous with the cornea.

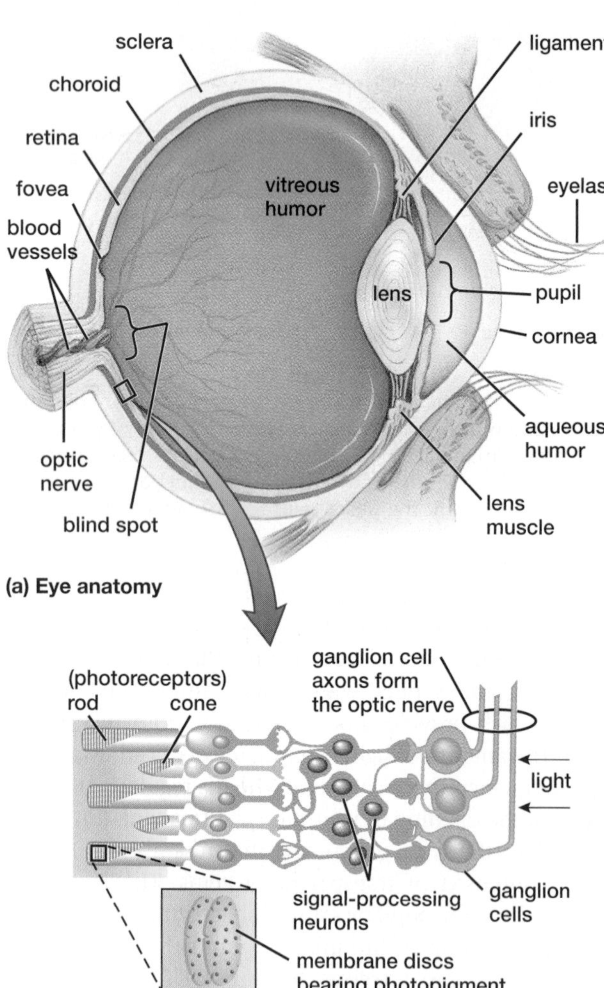

▲ **FIGURE 39-9 The human eye (a)** The anatomy of the human eye. **(b)** The human retina has photoreceptors (rods and cones), signal-processing neurons, and ganglion cells. Each rod and cone bears an extension packed with membranes in which light-sensitive molecules (photopigments) are embedded.

The Lens Focuses Light from Distant and Nearby Objects on the Retina

The visual image is focused most sharply on a small area of the retina called the **fovea.** Although focusing begins at the cornea, whose rounded contour bends light rays, the lens is responsible for final, sharp focusing. The shape of the lens is adjusted by its encircling muscle. When viewed from the side, the lens is either rounded, to focus on nearby objects, or flattened, to focus on distant objects (**Fig. 39-10a**).

If your eyeball is too long or your cornea is too rounded, you will be **nearsighted**—the light from distant objects will focus in front of the retina. The eyes of **farsighted** people, with eyeballs that are too short or corneas that are too flat, focus light from nearby objects behind the retina. These conditions can be corrected by either contact lenses or glasses of the appropriate shape (**Figs. 39-10b,c**). Both nearsightedness and farsightedness can also be corrected with laser surgery that reshapes the cornea. As people age, the lens within the eye stiffens, causing farsightedness because the lens can no longer curve enough to focus on nearby objects. By their mid-forties, most people require reading glasses for close work.

The Retina Detects Light and Produces Action Potentials in the Optic Nerve

The photoreceptors, called **rods** and **cones** because of their shapes, gather light at the rear of the retina (see Fig. 39-9b). Photoreception in both rods and cones begins with the absorption of light by photopigment molecules that are embedded in membranes of the photoreceptors, triggering chemical reactions that produce a receptor potential.

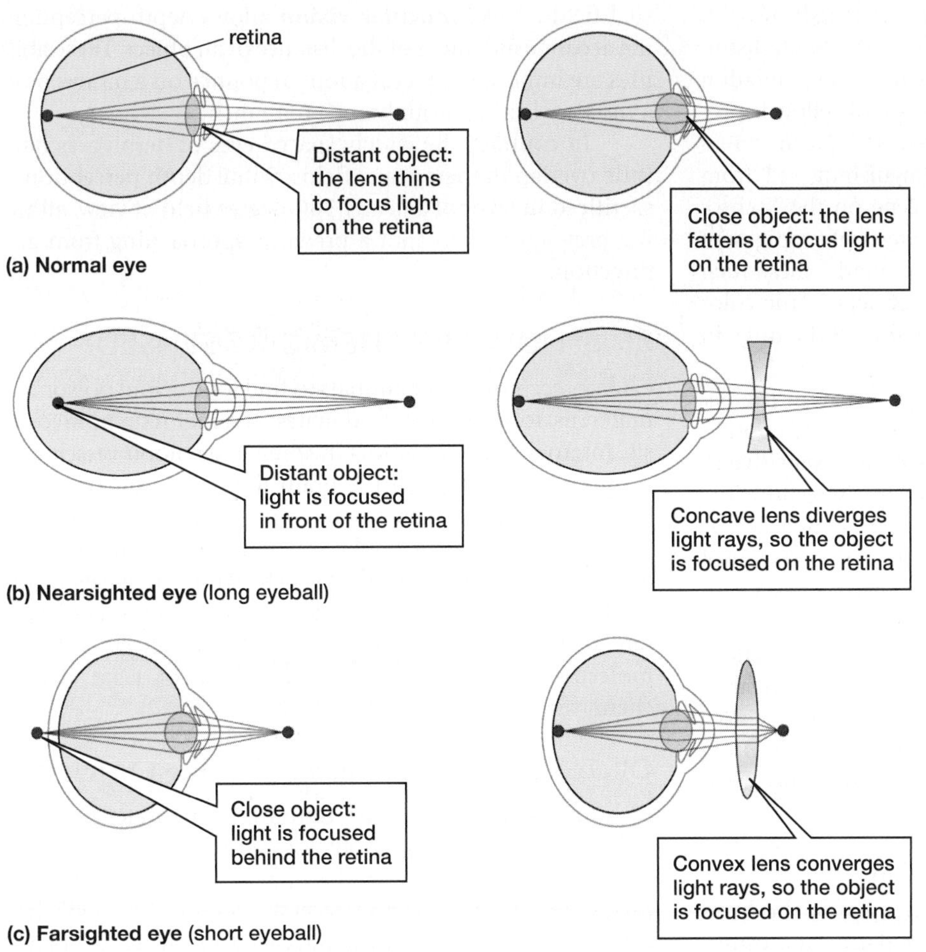
retina
Distant object: the lens thins to focus light on the retina
Close object: the lens fattens to focus light on the retina
(a) Normal eye
Distant object: light is focused in front of the retina
Concave lens diverges light rays, so the object is focused on the retina
(b) Nearsighted eye (long eyeball)
Close object: light is focused behind the retina
Convex lens converges light rays, so the object is focused on the retina
(c) Farsighted eye (short eyeball)

◀ **FIGURE 39-10 Focusing in the human eye (a)** The lens changes shape to focus on objects at different distances. **(b)** Nearsightedness is corrected by eyeglasses with concave lenses. **(c)** Farsightedness is corrected by eyeglasses with convex lenses.

QUESTION Today, many nearsighted and farsighted people choose to correct their vision problems with laser surgery on their corneas rather than with corrective lenses. For nearsightedness and farsightedness, respectively, how should corneas be reshaped to correct the problem?

Between the receptors and incoming light are several layers of neurons that process the signals from the photoreceptors. These neurons enhance our ability to detect edges, movement, and changes in light intensity. The retinal layer nearest the vitreous humor consists of **ganglion cells,** whose axons make up the **optic nerve.** Electrical signals from the photoreceptors and intervening neurons are converted to action potentials in the ganglion cells. To reach the brain, ganglion cell axons pass through the retina at a location called the **blind spot** (**Fig. 39-11**; see also Fig. 39-9a). This area lacks receptors, so objects focused there cannot be seen.

Rods and Cones Differ in Distribution and Light Sensitivity

Although cones are located throughout the retina, they are concentrated in the fovea, where the lens focuses images most sharply (see Figs. 39-9a and 39-11). The fovea looks like a dent near the center of the retina, because the layers of signal-processing neurons are spread apart, while still retaining their synaptic connections. This arrangement allows light to reach the cones of the fovea without having to pass through so many other cells.

Human eyes have three varieties of cones, each containing a slightly different photopigment. Each type of photopigment is

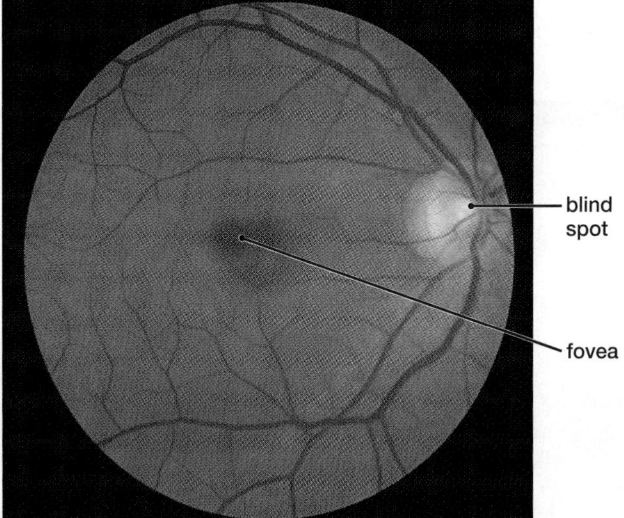
blind spot
fovea

▲ **FIGURE 39-11 The human retina** A portion of the human retina, photographed through the cornea and lens of a living person. The blind spot and fovea are visible. Blood vessels supply oxygen and nutrients; notice that they are dense over the blind spot (where they won't interfere with vision) and scarce near the fovea.

QUESTION Despite the presence of the blind spot, you do not consciously experience a "hole" in your vision. Why not?

most strongly stimulated by a particular wavelength of light, corresponding roughly to red, green, or blue. The brain distinguishes color according to the relative intensity of stimulation of different cones. For example, the sensation of yellow is produced by fairly equal stimulation of red and green cones. About 7% of men have difficulty distinguishing red from green, because they possess a defective gene on the X chromosome that encodes either the red or green photopigment (see p. 185). Although often called "color-blind," these men are more accurately described as "color-deficient." True color blindness, in which a person perceives the world only in shades of gray, is extremely rare.

Rods are most abundant outside the fovea, in the periphery of the retina. Rods are longer than cones and contain far more photopigment, so they are much more sensitive to light than cones are. Therefore, rods are largely responsible for our vision in dim light, such as moonlight. Unlike cones, all rods contain identical photopigments, so rods cannot provide color vision. In dim light, the world appears in shades of gray.

Not all vertebrates have both rods and cones. Those that are active almost entirely during the day (certain lizards, for example) may have all-cone retinas, whereas many nocturnal animals (such as rats and ferrets) and those dwelling in dimly lit habitats (such as deep-sea fishes) have mostly or entirely rods.

Binocular Vision Allows Depth Perception

Among mammals, the placement of the eyes on the head varies with the lifestyle of the animal. Predators and omnivores usually have both eyes facing forward (**Fig. 39-12a**), but most herbivores have one eye on each side of the head (**Fig. 39-12b**). The forward-facing eyes of predators and omnivores have slightly different but extensively overlapping visual fields. This **binocular vision** allows depth perception, the accurate judgment of the distance of an object. These abilities are important to a cat about to pounce on a mouse or to a monkey leaping from branch to branch.

In contrast, the widely spaced eyes of herbivores have little overlap in their visual fields. Some depth perception is sacrificed in favor of a nearly 360-degree field of view, allowing prey animals to spot a predator approaching from any direction.

39.5 HOW ARE CHEMICALS SENSED?

Chemoreceptors allow animals to find food, avoid poisonous materials, locate homes, find mates, and maintain homeostasis. Internal chemoreceptors in some large blood vessels and in the hypothalamus of the brain monitor levels of crucial molecules such as sugar, water, oxygen, and carbon dioxide in the blood. They also stimulate activities that maintain these levels within narrow limits. For perceiving chemicals in the external environment, terrestrial vertebrates have two separate senses: olfaction, the sense of smell, detects airborne molecules, whereas gustation, the sense of taste, detects chemicals dissolved in water or saliva.

Olfactory Receptors Detect Airborne Chemicals

In humans and most other terrestrial vertebrates, receptor cells for olfaction are neurons located in a patch of mucus-covered epithelial tissue in the upper portion of each nasal cavity (**Fig. 39-13**). Olfactory receptor neurons bear long dendrites that protrude into the nasal cavity and are embedded in the mucus. Odorous molecules in the air, such as those generated by coffee, diffuse into the layer of mucus and bind to receptor proteins on the dendrites. The olfactory receptor neurons send axons directly to the olfactory bulb in the brain.

(a) Binocular vision

(b) Almost 360° vision

▲ **FIGURE 39-12 Eye position differs in predators and prey (a)** Primates and most predators, such as this owl, have eyes in front of the head; both eyes can be focused on a target, providing binocular vision. **(b)** Most prey animals, such as rabbits, have eyes at the sides, which allow them to scan for predators.

QUESTION Why do some herbivorous or fruit-eating animals, such as monkeys and fruit bats, have both eyes in front?

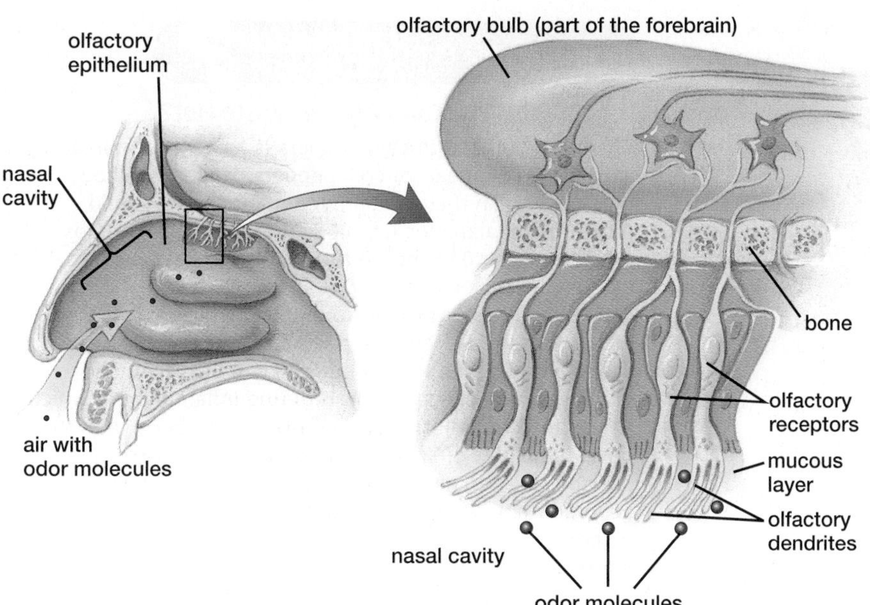

◀ **FIGURE 39-13 Human olfactory receptors** The receptors for olfaction in humans are neurons bearing microscopic hairlike projections that protrude into the nasal cavity. The projections are embedded in a layer of mucus in which odor molecules dissolve before contacting the receptors.

Humans produce about 350 to 400 different olfactory receptor proteins, each encoded by a separate gene, but each olfactory receptor neuron expresses only one kind of receptor protein. Each receptor protein is specialized to bind a particular type of molecule and cause the olfactory neuron to produce a receptor potential. If the receptor potential is large enough, it exceeds threshold, producing action potentials that travel along the neuron's axon to the brain. Many odors are complex mixtures of molecules that stimulate several receptor proteins, so our perception of odors arises as a result of the brain interpreting signals from many different olfactory receptor neurons.

As you know, many other animals detect odors better than we do. Dogs, for example, have 20 to 40 times more olfactory receptor neurons, and twice as many different types of receptor proteins. Not surprisingly, a dog's sense of smell is much more sensitive than ours.

Taste Receptors Detect Chemicals Dissolved in Liquids

The human tongue bears about 5,000 **taste buds,** embedded in small bumps, called papillae (**Fig. 39-14a**). A few taste buds are also found in the back of the mouth and in the pharynx. Each taste bud is a small pit in the epithelium, opening into the oral cavity through a taste pore. A taste bud contains a cluster of 50 to 150 cells of several types: supporting cells, stem cells, and taste receptor cells (**Fig. 39-14b**). Supporting cells act much like glial cells in the nervous system, regulating the composition of the extracellular fluid and helping the receptor cells to function properly. The stem cells divide to produce new receptor and support cells, following normal wear and tear or a close encounter with scalding-hot coffee. The taste receptor cells bear microvilli (thin projections of the plasma membrane) that protrude into the taste pore. Dissolved chemicals enter the pore and contact these microvilli. Although it was formerly thought that taste buds for specific tastes are concentrated on specific areas of the tongue, they are actually fairly evenly distributed.

There are five known tastes: the familiar sour, salty, sweet, and bitter, and a sensation known as *umami*, which is a Japanese word loosely translated as "delicious." (The umami receptor cell responds to the amino acid glutamate. Monosodium glutamate [MSG] has long been used as a seasoning, especially in Asian foods, because it enhances flavor.) Sour and salty sensations are caused by hydrogen ions or sodium ions, respectively, entering certain taste receptor cells through ion channels in the plasma membranes of the microvilli. Sweet, bitter, and umami sensations are caused by specific organic molecules binding to receptor proteins on the surface of the microvilli of other taste receptor cells, setting off a chain of intracellular reactions. In all cases, the taste receptor cells respond by producing a receptor potential.

We perceive a great variety of tastes through two mechanisms. First, a particular substance may stimulate two or more receptor types to different degrees, making the substance taste "sweet and sour," for example. Second, and more important, a substance being tasted usually releases molecules into the air inside the mouth. These odorous molecules diffuse to the olfactory receptors, which contribute an odor component to the basic flavor. (Recall from Chapter 34 that the mouth and nasal passages are connected; see pp. 665–667.)

To prove that what we call taste is really mostly smell, try holding your nose (and closing your eyes) while you eat different flavors of jelly beans. The flavors—from grape, cherry, and pear to buttered popcorn—will be indistinguishably sweet. Likewise, when you have a bad cold, usually tasty foods seem bland.

39.6 HOW IS PAIN PERCEIVED?

If you burn, cut, crush, or spill strong acid on your hand, you will experience the sensation of pain. Pain is a subjective feeling arising in the brain, produced by the stimulation of

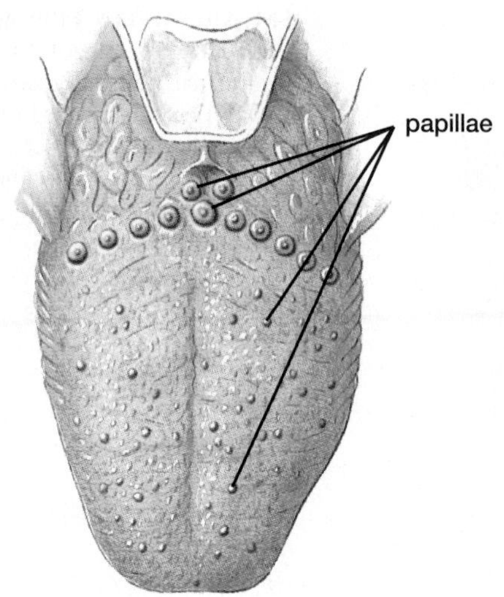

(a) The human tongue

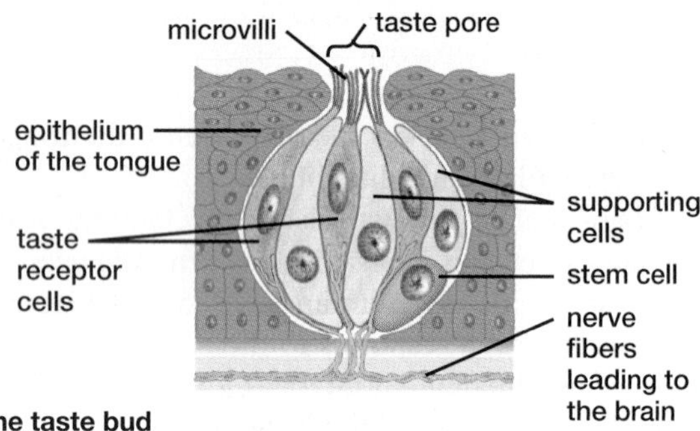

(b) One taste bud

▲ FIGURE 39-14 **Human taste receptors (a)** The human tongue is covered with papillae, bumps in which taste buds are embedded. **(b)** Each taste bud consists of taste receptor cells, supporting cells, and stem cells. Microvilli of taste receptor cells protrude into the taste pore. The microvilli bear protein receptors that bind tasty molecules, producing a receptor potential.

pain receptors (also called nociceptors). The pain-receptive parts of pain receptors are free nerve endings (much like dendrites). Pain receptors occur in most parts of the body.

Pain perception is crucially important to well-being and survival, teaching humans and other animals to avoid behaviors and objects that may cause bodily damage. A handful of people have been born who lack pain perception, and they typically suffer frequent injuries and even early death, because no one can consciously avoid all potentially damaging situations.

Have you ever wondered 

Why Chili Peppers Taste Hot?

The hot ingredient in chili peppers is a chemical called capsaicin. The "heat" of chili peppers depends on how much capsaicin they contain, ranging from very mild pimentos to jalapenos (about 20× hotter) to habeneros (50× hotter still) to the excruciatingly painful Naga (almost another 10× hotter). Capsaicin binds to, and activates, receptor proteins on many pain receptors, including some in the mouth. These receptors are also activated by heat above about 109°F (43°C), acid, and a variety of molecules produced during inflammation (which is why infected injuries often hurt). Remember one of the basic principles of sensory perception—*what type of stimulus* we perceive depends on *which sensory cells* fire action potentials. Because individual pain receptors respond to both damaging heat and capsaicin, the brain interprets both of these stimuli as burning pain.

Many Types of Damaging Stimuli Are Perceived as Painful

There are just about as many types of pain receptors as there are ways to harm your body. Some thermal nociceptors respond to high temperatures, with a typical threshold of about 109°F (43°C). Others respond to low temperatures, below about 59°F (15°C); this may not seem very cold, but remember, this is not air temperature, but rather, the temperature of the skin where the cold nociceptors are located. Still other pain receptors are mechanoreceptors that respond to excessive stretching, as might happen to your skin if you close a door on your finger. If you are cut or bruised, damaged cells release their contents, including potassium ions (K^+) and various enzymes. Potassium ions directly activate many nociceptors. Some of the enzymes convert proteins in the blood to bradykinin, a chemical that activates chemoreceptive pain receptors. Some single nociceptors respond to several damaging stimuli, including excessive heat, acid, or certain chemicals.

Case Study revisited

Bionic Ears

As you now know, sound waves stimulate vibration of the eardrum, the bones of the middle ear, the oval window, the fluid and membranes in the cochlea, and finally the hairs of the hair cells. Bending of the hairs causes the hair cells to release neurotransmitters onto the endings of auditory nerve axons, stimulating action potentials that travel to auditory centers of the brain. The frequency of action potentials informs the brain of the loudness of the sound, and which axons fire informs the brain about musical pitch.

A cochlear implant threads a set of 16 to 22 platinum wire electrodes through the cochlea (**Figure 39-15**). These electrodes are carefully arranged, from the beginning of the cochlea (near the oval window) to its tip. Each electrode can be activated independently of the others. In the sound-receiving unit, worn on the outside of the head, a tiny microphone connects to a microprocessor. When it receives input from the microphone, the microprocessor activates the

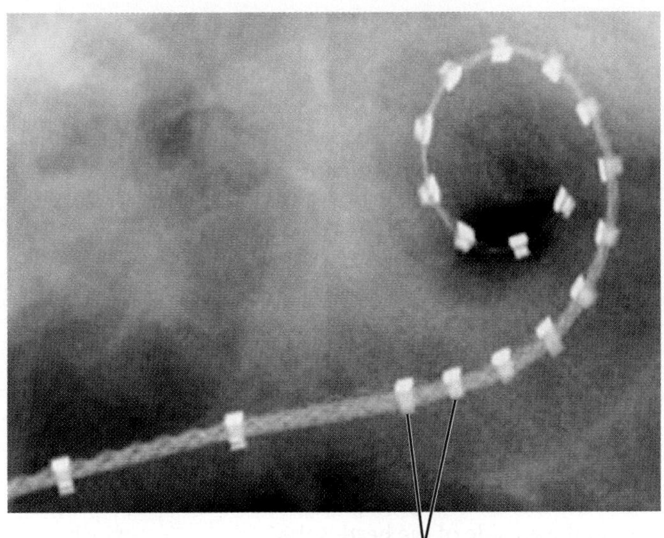

▲ **FIGURE 39-15 A cochlear implant** In this X-ray, each bright rectangle is an electrode. Each can be activated by electric current independently of the others, thereby stimulating only a small portion of the auditory nerve axons that innervate the cochlea. Uncoiled, the implant would be about an inch (25 millimeters) long.

implant, sending electric current to the electrodes in the cochlea, where they stimulate action potentials in the axons of the auditory nerve.

The microprocessor sorts sounds according to loudness and pitch. It sends electric currents to the electrodes with a strength that is proportional to the loudness of the sound (although the actual magnitude of the currents is always very tiny): the stronger the current, the faster the auditory nerve axons fire. For low notes, the microprocessor sends electrical current to the electrodes positioned near the tip of the cochlea. For high notes, the microprocessor sends current to the electrodes located near the oval window. These electrical currents stimulate action potentials in roughly the same axons that would have been stimulated by hairs cells in these locations in a functioning cochlea.

As wonderful as cochlear implants are for their users, they are still pretty unsophisticated compared to a functioning biological cochlea. The cochlea contains about 18,000 to 24,000 hair cells, about 3,500 of which are primarily involved in detecting sound and stimulating axons in the auditory nerve. Sixteen to 22 electrodes in an implant might seem hopelessly crude for listening to music, but the brain is a marvelous processor that can do wondrous things with very little information. When Jennifer first received her cochlear implant, she "felt vibrations inside my head, jackhammer-style." When Sam's implant was first turned on, "All I could hear were beeps for every syllable in a word." Over time, however, the vibrations and beeps resolved into useful sounds. As Sam puts it, "It takes time to 'train' your brain."

There may always be limitations to what cochlear implants can do. For example, the physics of electricity dictates that the electrodes in the cochlea must be relatively far apart, which sets an upper limit on how many there can be, which in turn restricts how well a user can hear music and perhaps detect emotional overtones in speech. But in Sam's words, her cochlear implant let her "into a New World."

BioEthics Consider This

Sam was 16 years old, Jennifer over 30, when they received their implants. They thought long and hard about the decision, and neither has any regrets. However, some very young children are now being given cochlear implants, long before they are old enough to understand their condition, the benefits of the implants, the risks of surgery, the possibility that better technologies may be developed in the future, and what a deaf life might be like. On the other hand, young children would probably adapt better to a cochlear implant than older people would. Is this a decision that parents should make for their little children, like whether to have vaccinations? Or should the use of cochlear implants wait until people have grown up and can make their own choices?

CHAPTER REVIEW

Summary of Key Concepts

39.1 How Does the Nervous System Sense the Environment?

Sensory receptors convert a stimulus from the internal or external environment to an electrical signal called a receptor potential. Either directly or indirectly, receptor potentials ultimately result in action potentials in specific axons that connect to appropriate brain regions. The type of stimulus is encoded by which neurons in the brain are activated. The intensity of a stimulus is encoded by the frequency of action potentials reaching the brain. Sensory receptors are categorized according to the stimulus to which they respond. Many sensory receptors are contained within accessory structures that help them to respond to a specific stimulus.

39.2 How Are Mechanical Stimuli Detected?

A variety of different mechanoreceptors detect stimuli such as touch, vibration, pressure, stretch, or sound. Some mechanoreceptors, including those for touch and the sensation of itching, are free nerve endings. Other mechanoreceptors are included in accessory structures that regulate which stimulus, such as pressure or sound, is detected.

39.3 How Does the Ear Detect Sound?

In the vertebrate ear, air vibrates the tympanic membrane, which transmits vibrations to the bones of the middle ear and then to the oval window of the fluid-filled cochlea. Within the cochlea, vibrations bend the hairs of hair cells, which are receptors located between the basilar and tectorial membranes. This bending produces receptor potentials in the hair cells that release neurotransmitters onto the endings of axons of the auditory nerve, causing action potentials that travel to auditory centers in the brain. The pitch (musical note) of sound is encoded by which hair cells are stimulated by a given frequency of sound vibrations. The intensity (loudness) of sound is encoded by the frequency of action potentials in the axons of the auditory nerve.

The vestibular apparatus consists of the utricle and saccule in the vestibule, which detect the direction of gravity and the orientation of the head, and the semicircular canals, which detect movement of the head.

39.4 How Does the Eye Detect Light?

In the vertebrate eye, light enters the cornea and passes through the pupil to the lens, which focuses an image on the fovea of the retina. Two types of photoreceptors, rods and cones, are located deep in the retina. They produce receptor potentials in response to light. These signals are processed through several layers of neurons in the retina and are translated into action potentials in the ganglion cells, whose axons make up the optic nerve leading to the brain. Rods are more abundant and more light sensitive than cones, providing black-and-white vision in dim light. Cones, which are concentrated in the fovea, provide color vision.

39.5 How Are Chemicals Sensed?

Terrestrial vertebrates detect chemicals in the external environment either by smell (olfaction) or by taste (gustation). Each olfactory or taste receptor cell type responds to only one or a few specific types of molecules, allowing discrimination among tastes and odors. The olfactory neurons of vertebrates are located in a tissue that lines the nasal cavity. Taste receptors are located within the taste buds on the tongue.

39.6 How Is Pain Perceived?

Nociceptors (pain receptors) respond to damaging stimuli such as cuts, burns, acid, or extremely high or low temperatures. Most nociceptors are free nerve endings with receptors that respond to chemicals from the environment (e.g., acid) or that are produced by the body during tissue damage. Some of these same nociceptors respond to temperature extremes.

Key Terms

anvil *763*
aqueous humor *765*
auditory canal *762*
auditory nerve *763*
auditory tube *763*
basilar membrane *763*
binocular vision *768*
blind spot *767*
choroid *765*
cochlea *763*
compound eye *765*
cone *766*
cornea *765*
farsighted *766*
fovea *766*
ganglion cell *767*
hair cell *763*
hammer *762*
inner ear *763*
intensity *760*
iris *765*
lens *765*
middle ear *762*
nearsighted *766*
ommatidium (plural, **ommatidia**) *765*
optic nerve *767*
outer ear *762*
oval window *763*
pain receptor *770*
photopigment *765*
photoreceptor *765*
pinna *762*
pupil *765*
receptor *759*
receptor potential *760*
retina *765*
rod *766*
round window *763*
saccule *764*
sclera *766*
semicircular canal *764*
sensory receptor *759*
stirrup *763*
taste bud *769*
tectorial membrane *763*
tympanic membrane *762*
utricle *764*
vestibular apparatus *764*
vitreous humor *765*

Thinking Through the Concepts

Fill-in-the-Blank

1. Sensory receptors respond to an appropriate stimulus with an electrical signal called a(n) __________. Larger stimuli cause these signals to be larger than the signals from small stimuli. Ultimately, the intensity of a stimulus is conveyed to the brain encoded by the __________ of action potentials in axons connected to specific sensory areas of the brain.
2. Sensory receptors are finely tuned to respond to specific stimuli. __________ respond to physical deformation, such as touch or pressure. __________ respond to light. The receptors in the taste buds and olfactory epithelium are examples of __________.
3. The mammalian ear consists of three parts: the outer, middle, and inner ear. The flap of skin-covered cartilage on the outside of the head is the __________, which collects sound waves and funnels them down the auditory canal to a flexible membrane called the __________. This connects to three small bones, the __________, __________, and __________. The final bone vibrates the __________, the beginning of the cochlea. Within the cochlea, the vibrations finally move the cilia of specialized mechanoreceptors called __________.
4. Light enters the human eye through the __________ and __________, and then passes through the pupil, which is a hole in the __________. Light then passes through the __________ and __________ before finally striking the retina. Both the __________ and __________ are involved in focusing light.
5. The retina of the human eye contains two types of photoreceptors, called __________ and __________. __________ are larger, are more sensitive to light, and provide black-and-white vision. __________ are mostly located in the central region of the retina, called the __________. These photoreceptors are smaller and less sensitive to light, but provide color vision.
6. In humans, the five principal kinds of taste sensations are __________, __________, __________, __________, and __________. Molecules that leave the food and enter the air inside the mouth are detected by the sense of __________, which actually plays a major role in the brain's perception of taste.

Review Questions

1. How do the senses encode the intensity of a stimulus? The type of stimulus?
2. What are the names of the specific receptors used for taste, vision, hearing, smell, and touch?
3. Why are we apparently able to distinguish hundreds of different flavors if we have only five types of taste receptors? How are we able to distinguish so many different odors?
4. Describe the structure and function of the various parts of the human ear by tracing the route of a sound wave from the air outside the ear to action potentials in the auditory nerve.
5. How does the structure of the inner ear allow for the perception of pitch? Of loudness?

6. Diagram the overall structure of the human eye. Label the cornea, iris, lens, sclera, retina, and choroid. Describe the function of each labeled structure.
7. How does the eye's lens change shape to allow focusing of distant objects? What defects make focusing on distant objects impossible, and what is this condition called? What type of lens can be used to correct it, and how does it do so?
8. List the similarities and differences between rods and cones.
9. Distinguish between taste and olfaction.

Applying the Concepts

1. It has been said that your sensory perceptions are purely a creation of your brain. Discuss this statement. Do you agree or disagree? What are the implications for communicating with other humans, with other animals, and with intelligent life elsewhere in the universe?
2. We don't merely identify odors. We also label them good or bad, fragrant or disgusting. What do you think might be the evolutionary advantage of emotional responses to odors? Do you think that all animals have the same emotional responses to odors that humans do?
3. Many people like spicy foods, but most other mammals don't. Historically, the general trend was that the hotter the climate, the spicier the foods a society traditionally favored. Using these facts as a starting point, what might be the selective advantage for plants that can manufacture spicy chemicals? Why might it be useful for humans to tolerate, and even enjoy, spicy food?
4. Recall the concepts underlying resting potentials, threshold, and action potentials discussed in Chapter 38. Why do you think that potassium ions, released from damaged cells, might activate pain receptors?

MB® *Go to www.masteringbiology.com for practice quizzes, activities, eText, videos, current events, and more.*

Action and Support: The Muscles and Skeleton

Chapter 40

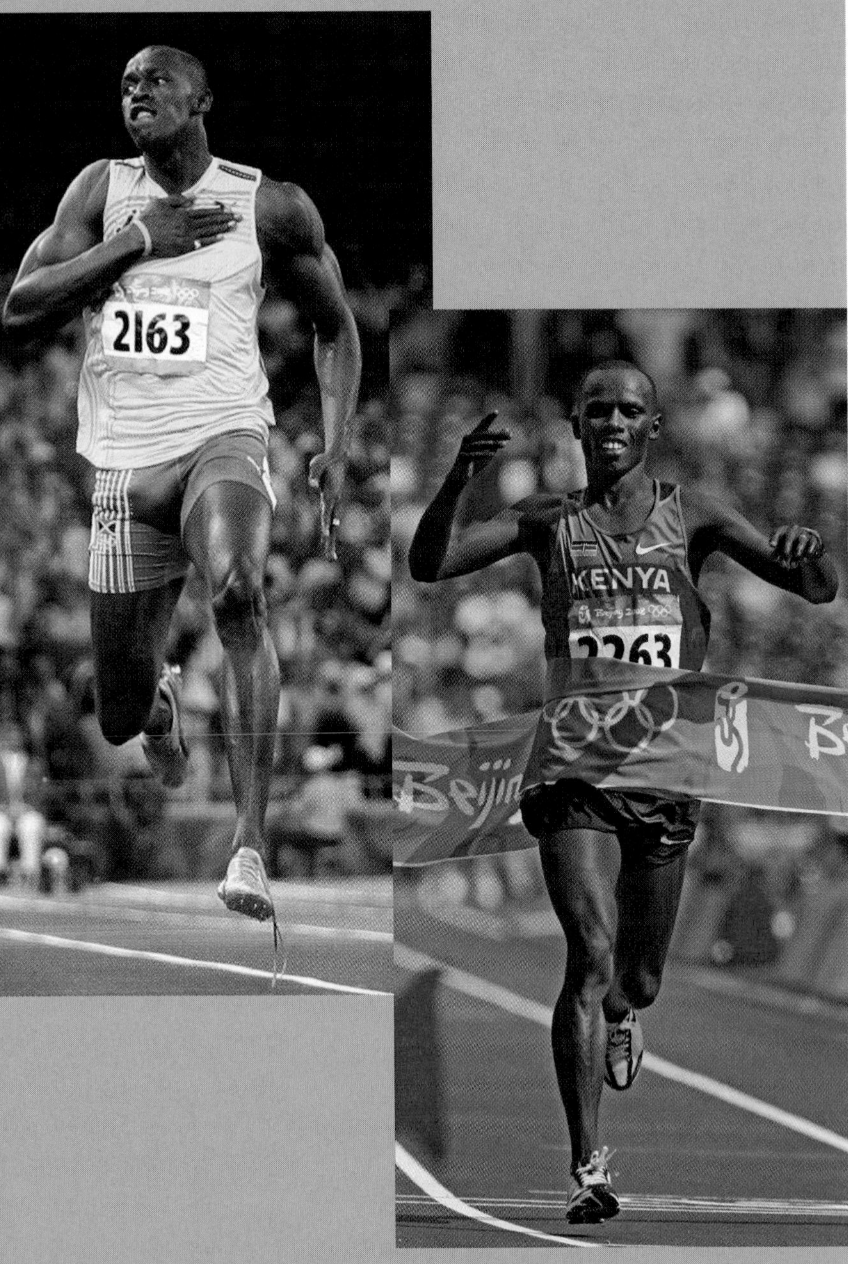

▲ Perfect coordination of muscles and skeletons: Usain Bolt (left) and Samuel Wansiru (right) both qualify as the world's fastest human, each in his own way.

Case Study

Muscles of Gold

WHEN USAIN BOLT OF JAMAICA (left) blasted out of the starting blocks at the 2008 Beijing Olympics, his leg muscles propelled his body faster than any human had ever gone before—100 meters in only 9.69 seconds. He could have run faster, but was so far ahead that he just cruised the last 20 meters to the finish line: Not only the gold medalist, but the world's fastest human.

Or is he? A few days later, Kenyan Samuel Wansiru (right) set a new Olympic record in the marathon—more than 26 miles in just 2 hours, 6 minutes, and 32 seconds. True, Wansiru's average speed was only about 18 seconds per 100 meters, but he kept this pace up for over 42,000 meters.

Could Bolt beat Wansiru in a marathon? No way. Nor could Wansiru even qualify for the Olympics in the 100-meter dash. Why not? For the obvious answer, compare their body types. Bolt weighs about 190 pounds, and Wansiru, just over 110. Bolt has both a larger skeleton and bigger muscles than Wansiru. Bolt's weight would be a serious problem during a marathon, but his large leg muscles are essential to powering off the blocks and driving to the finish of the 100 meters.

Bolt and Wansiru are both incredible athletes, but how did their bodies get to be so different? And what about Olympic weightlifters, like gold medalist Matthias Steiner (see Fig. E10-1)? Were these athletes "born" to their events? Or could Bolt have become a marathoner? Even more extreme, in another life, could Wansiru bulk up and compete with Steiner in weightlifting? As you read this chapter, consider the muscles of Usain Bolt, Samuel Wansiru, and your own legs. How do muscles contract? What supplies them with the energy they need? And why don't you have muscles of gold, like Bolt or Wansiru?

At a Glance

40.1 HOW DO MUSCLES AND SKELETONS WORK TOGETHER TO PROVIDE MOVEMENT?

Most animals have several types of muscles, specialized to perform functions such as moving the body, contracting the heart, or propelling food through the digestive tract. Here, we will focus on how muscles interact with skeletons to move an animal's body.

Despite enormous differences in body form and structure, nearly all animals—sea jellies, earthworms, crabs, horses, and people—move using the same fundamental mechanism: Contracting muscles exert forces on a structure that supports the body, called a **skeleton,** and cause the body to change shape. Imagine, for a moment, your body with its full complement of muscles, but no skeleton. Muscle contraction could cause your boneless body to twitch, and perhaps even flop around a bit, but there would be no coordinated movement—you couldn't walk, write, throw a ball, or even get up off the floor. And, of course, with a skeleton but no muscles, your body would remain in one position unless someone else moved it.

As we will see in section 40.3, muscles produce force by contracting. A muscle can only contract or not contract (relax)—it cannot use cellular energy to make itself longer. Muscle lengthening is passive, occurring when muscles relax and are stretched by other forces, such as contractions of other muscles, the weight of a limb, or pressure from food (in the digestive tract) or blood (in the heart).

Coordinated movement of an animal's body is produced by alternating contractions of muscles with opposing actions, called **antagonistic muscles.** As we will see shortly, antagonistic muscles typically make a tubular structure (such as the digestive tract or an entire tube-like animal body) alternately thinner and fatter, or move appendages such as arms, legs, or wings back and forth.

The Animal Kingdom Has Three Types of Coordinated Systems of Skeletons and Muscles

Animal skeletons come in three different forms: hydrostatic skeletons, exoskeletons, and endoskeletons. Antagonistic muscles act on each type of skeleton to provide movement (**Fig. 40-1**).

Worms, cnidarians (sea jellies, anemones, and their relatives), and many mollusks (snails, octopuses, and their relatives) have a **hydrostatic skeleton,** which is basically a sac or tube filled with liquid (**Fig. 40-1a**). "Hydrostatic" means roughly "to stand with water," and that is just what hydrostatic skeletons do. Think of a water-filled balloon: The balloon "stands up" because it contains water. Puncture the balloon and it collapses. Further, the volume of the balloon is fixed (because liquids cannot be compressed), but you can change its shape by squeezing it in various places.

An animal with a hydrostatic skeleton controls the overall shape of its body by using two sets of antagonistic muscles in its body wall—one circular, the other longitudinal. Consider a tubular animal such as an earthworm (see Fig. 40-1a). To move forward through its burrow, a worm uses wavelike, alternating contractions of longitudinal and circular muscles. To see how this works, let's start when the worm contracts longitudinal muscles in its front end. This end becomes shorter and fatter, pushing bristles in the earthworm's skin into the wall of the burrow. Other longitudinal muscles then contract in the middle and tail end of the worm, making the animal shorter, and finally pushing bristles in its tail into the burrow wall. The worm then contracts circular muscles in its front half, making that half of the worm longer and thinner, and simultaneously

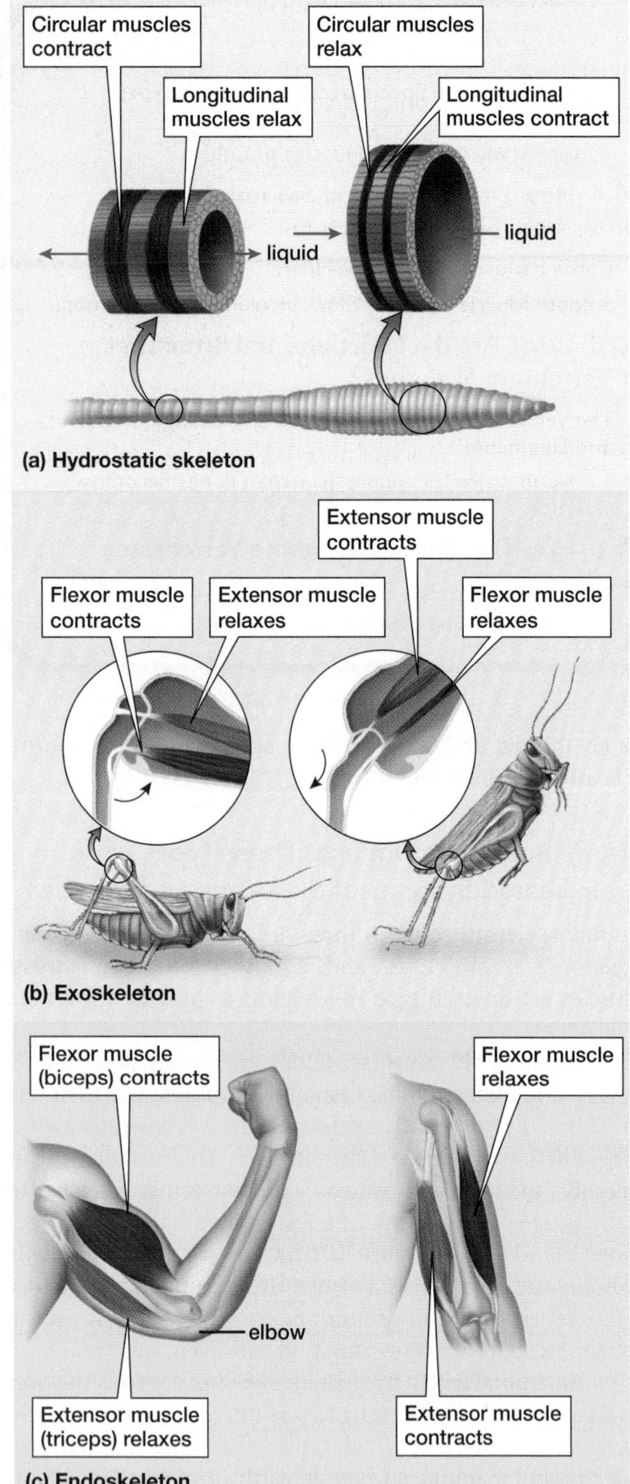

▲ **FIGURE 40-1 Antagonistic muscles move hydrostatic skeletons, exoskeletons, and endoskeletons (a)** A hydrostatic skeleton, as found in earthworms, is essentially a liquid-filled tube enclosed within walls containing antagonistic circular and longitudinal muscles. Contraction of the circular muscles makes the tube long and thin (left); contraction of the longitudinal muscles make the tube short and thick (right). **(b)** Antagonistic flexor and extensor muscles attach to the inner surfaces of an exoskeleton on opposite sides of a flexible joint. Alternating contractions of the antagonistic muscles bend and straighten the joint. **(c)** Antagonistic flexor and extensor muscles attach to an endoskeleton on opposite sides of the outer surfaces of joints.

releasing the front bristles from the wall. Because bristles anchor the tail, the head moves forward. When the worm is fully extended, longitudinal muscles in the head contract again, fattening and anchoring the head. Meanwhile, as the wave of circular muscle contraction moves down the worm, the tail gets thin and its bristles are pulled out of the wall. Longitudinal muscle contractions in the back half of the worm pull the tail up toward the head, until the tail gets fat enough to once again anchor the bristles in the wall. This cycle is repeated over and over as the worm crawls through the soil.

The bodies of arthropods (such as spiders, crustaceans, and insects) are encased by rigid **exoskeletons** (literally, "outside skeletons"; **Fig. 40-1b**). Because an exoskeleton cannot expand, an arthropod must periodically molt its exoskeleton so that it can grow (**Fig. 40-2**). Movement of an exoskeleton typically occurs only at **joints** in the legs, mouthparts, antennae, bases of the wings, and body segments, where thin, flexible tissue joins stiff sections of exoskeleton. Antagonistic muscles attach to opposite sides of the inside of a joint (**Fig. 40-1b**). Contraction of a **flexor** muscle bends a joint; contraction of an **extensor** muscle straightens a joint. Alternating contraction of the antagonistic muscles moves joints back and forth, allowing the animal to walk, fly, or spin a web.

Endoskeletons ("internal skeletons") are rigid structures found inside the bodies of echinoderms (sea stars and their relatives) and chordates (animals with a notochord,

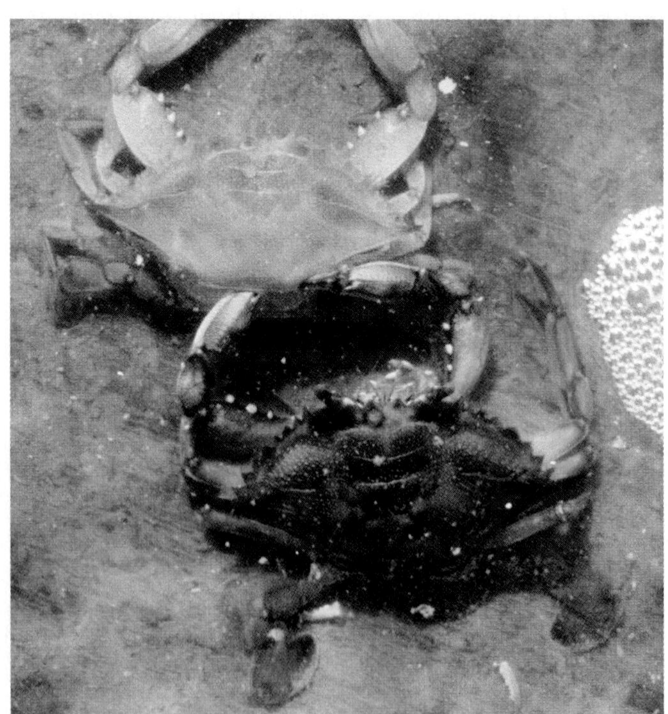

▲ **FIGURE 40-2 A crab molts its exoskeleton** Arthropods, such as this blue crab, must molt their exoskeletons in order to grow. Here a dark, newly molted crab has just crawled out of its old exoskeleton.

QUESTION Why are thick, armor-like exoskeletons found mostly in water-dwelling animals, whereas land-dwelling insects and spiders tend to have thinner exoskeletons?

most of which are vertebrates). We will examine vertebrate endoskeletons more closely in section 40.5. In animals with endoskeletons, movement occurs primarily at joints, where two parts of the skeleton are attached to one another, firmly but flexibly. Antagonistic muscles, such as the biceps (a flexor) and the triceps (an extensor), attach on opposite sides of the outside of a joint (in this case, the elbow; **Fig. 40-1c**). Antagonistic muscles move joints back and forth, or rotate them in one direction or the other.

40.2 WHAT ARE THE STRUCTURES OF VERTEBRATE MUSCLES?

The muscles of all animals have striking similarities in both the cellular components that produce contractions and in the structural arrangement of these components. The details of muscle structure and function, however, show a tremendous range of adaptations. For example, clams possess a special type of muscle that holds their shells tightly closed for hours using very little energy. Some flies have flight muscles that can contract 1,000 times per second. In this chapter, we will discuss the structure and function only of vertebrate muscles, using the human muscular system as an example.

Vertebrates have three types of muscle: skeletal, cardiac, and smooth. All work on the same basic principles but differ in function, appearance, and control (**Table 40-1**). **Skeletal muscle,** so named because it moves the skeleton, appears striped when viewed through a microscope, and is often referred to as striated (meaning "striped"; see Fig. 31-9a). Nearly all skeletal muscle is under voluntary, or conscious, control. Skeletal muscles can produce contractions ranging from quick twitches (as in blinking) to powerful, sustained tension (as in carrying an armload of textbooks). **Cardiac muscle,** which is also striated (see Fig. 31-9b), is located only in the heart. It is spontaneously active (that is, it initiates its own contractions), but it is also influenced by the nervous system and by

Table 40-1 Properties of the Three Muscle Types

	Type of Muscle		
Property	**Smooth**	**Cardiac**	**Skeletal**
Muscle appearance	Nonstriated	Striated	Striated
Cell shape	Tapered at both ends	Branched	Tapered at both ends
Number of nuclei	One per cell	One per cell	Many per cell
Speed of contraction	Slow	Intermediate	Slow to rapid
Contraction stimuli	Spontaneous, stretch, nervous system, hormones	Spontaneous	Nervous system
Function	Controls movement of substances through hollow organs and tubes	Pumps blood	Moves the skeleton
Under voluntary control?	No	No	Yes

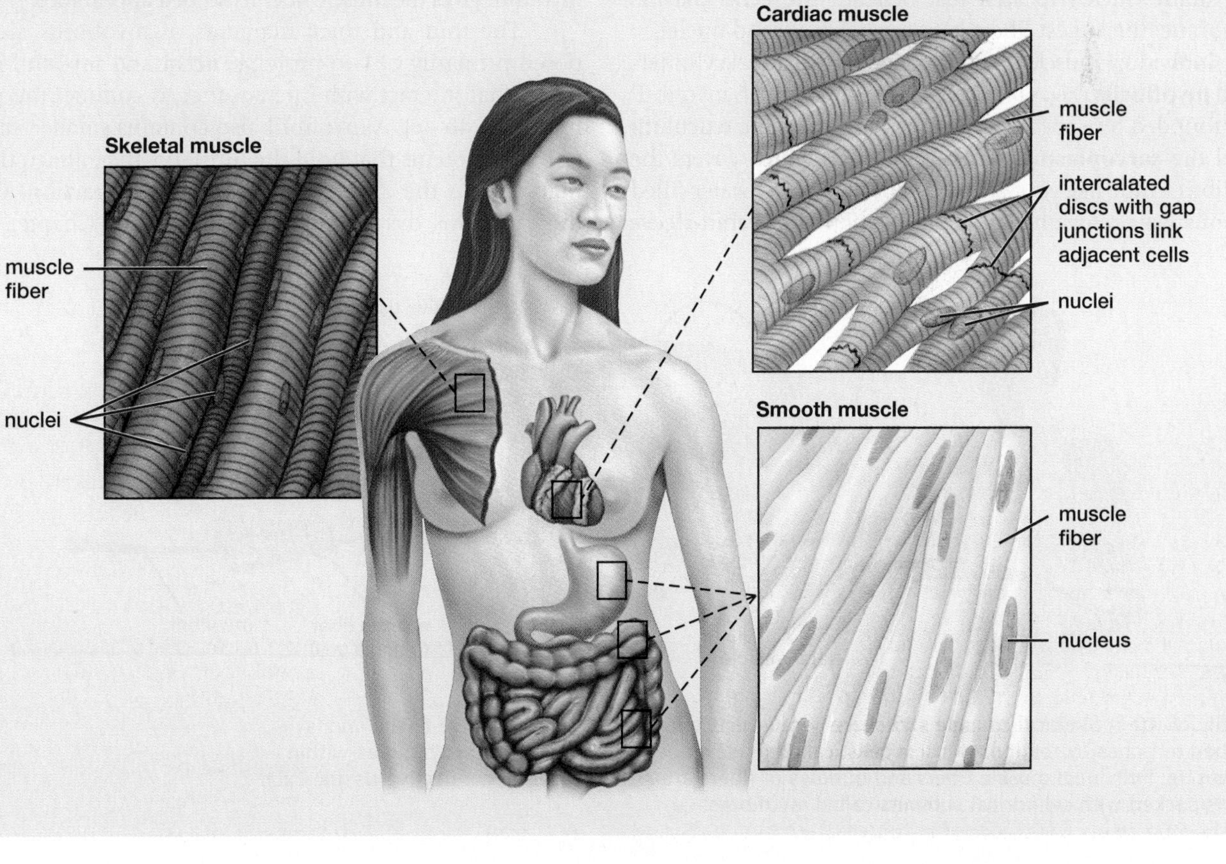

Skeletal muscle
muscle fiber
nuclei
Cardiac muscle
muscle fiber
intercalated discs with gap junctions link adjacent cells
nuclei
Smooth muscle
muscle fiber
nucleus

hormones. Cardiac muscle is usually not under voluntary control, although biofeedback training allows many people to regulate their heartbeat to a limited extent. **Smooth muscle** is not striated (see Fig. 31-9c). It surrounds large blood vessels and most hollow organs, producing slow, sustained contractions that usually cannot be controlled voluntarily.

The discussion that follows emphasizes skeletal muscle, although cardiac and smooth muscle have many similarities in structure and physiology.

Case Study continued

Muscles of Gold

Muscles can be incredibly strong—stronger than bones, tendons, and ligaments. Occasionally, weightlifters will dislocate a joint, tear a tendon clear off its attachment to a bone, or even break a bone. Anabolic steroids (see Chapter 37) appear to increase muscle strength more than skeletal strength, and as a consequence, make such injuries more likely.

Skeletal Muscles Have Highly Organized, Repeating Structures

Skeletal muscles consist of a series of nested, repeating parts, a little like Russian nesting matryoshka dolls (**Fig. 40-3**). Let's start at the outside and work our way in.

Skeletal muscles are encased in connective tissue sheaths and attached to the skeleton by tough, fibrous **tendons** (also a type of connective tissue). Within the muscle's outer sheath, individual muscle cells, called **muscle fibers,** are grouped into bundles by further coverings of connective tissue. Blood vessels and nerves pass through the muscle in the spaces between the bundles. Each individual muscle fiber also has its own thin connective tissue wrapping. These multiple connective tissue coverings, each connected to the others, provide the strength needed to keep the muscle from bursting apart during contraction.

Muscle fibers are among the largest cells in the human body. They range from 10 up to 100 micrometers in diameter (a bit smaller than the period at the end of this sentence), and some run the entire length of a muscle, so they can be over a foot (30 centimeters) long. Each skeletal muscle fiber contains many nuclei, located just beneath the cell's plasma membrane; the largest fibers have several thousand nuclei.

Individual muscle fibers contain many parallel cylinders called **myofibrils** (**Fig. 40-4;** see also Fig. 40-3). Each myofibril is surrounded by a specialized type of endoplasmic reticulum called the **sarcoplasmic reticulum (SR),** which covers the myofibril a little like your arm could be covered by water-filled balloons woven together to make a very peculiar shirt-sleeve (**Fig. 40-4a**). The sarcoplasmic reticulum consists of flattened, membrane-enclosed compartments (the balloons), filled with fluid containing a high concentration of calcium ions (the water inside the balloons). As we will see in section 40.3, calcium ions play a crucial role in muscle contraction. The plasma membrane that surrounds each muscle fiber tunnels deep into the inside of the cell at regular intervals, producing tubes called **T tubules.** T tubules encircle the myofibrils, running between, and closely attached to, segments of the SR (see Fig. 40-4a).

Each myofibril, in turn, consists of repeating subunits called **sarcomeres,** which are aligned end to end along the length of the myofibril (**Fig. 40-4b**), connected to one another by protein discs called **Z lines**—as if you had taken thousands of miniature soup cans and glued them together end to end. Within each sarcomere lies a precise arrangement of thin and thick protein filaments. Each **thin filament** is anchored to a Z line at one end. Suspended between the thin filaments are **thick filaments.** The regular arrangement of thin and thick filaments within each myofibril gives the muscle fiber its striped appearance.

The thin and thick filaments of myofibrils are composed primarily of two proteins, **actin** and **myosin,** respectively, that interact with one another to contract the muscle fiber (**Fig. 40-4c**). A myofibril also contains smaller amounts of other proteins that hold the fibril together, attach the thin filaments to the Z lines, and regulate contraction. One of these proteins, dystrophin, was introduced in Chapter 10 (see

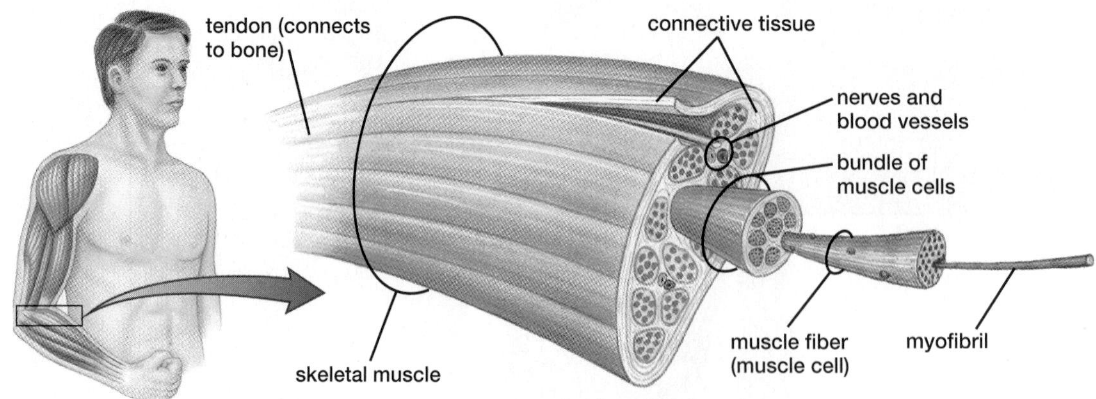

▲ FIGURE 40-3 **Skeletal muscle structure** A muscle is surrounded by connective tissue and is attached to bones by tendons. Muscle cells, called muscle fibers, are packaged into bundles within the muscle. Individual muscle fibers and bundles of fibers are also encased in connective tissue. Each fiber is packed with cylindrical subunits called myofibrils.

pp. 194–195). It binds thin filaments to proteins in the plasma membrane, which in turn are attached to extracellular proteins that surround the muscle fiber. Dystrophin helps to distribute the forces generated during muscle contraction, so that the fiber doesn't tear itself apart.

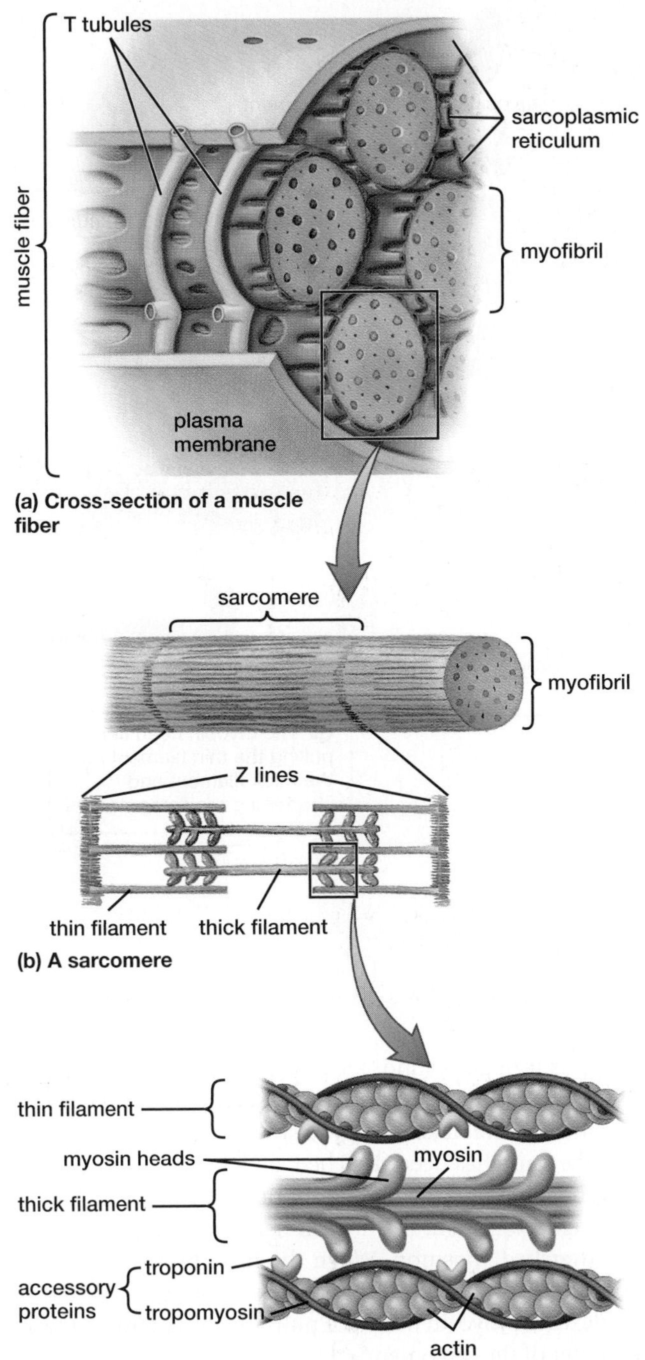

▲ **FIGURE 40-4 A skeletal muscle fiber (a)** Each muscle fiber is surrounded by plasma membrane that tunnels inside the fiber, forming T tubules. Sarcoplasmic reticulum surrounds each myofibril within the muscle cell. **(b)** Each myofibril consists of a series of subunits called sarcomeres, attached end to end by protein bands called Z lines. **(c)** Within each sarcomere are alternating thin filaments, composed of actin, troponin, and tropomyosin; and thick filaments, composed of myosin.

Individual actin proteins are nearly spherical (see Fig. 40-4c). A thin filament consists of two strands of actin proteins, wound about each other like two pearl necklaces twisted together. Accessory proteins called troponin and tropomyosin, which regulate contraction, lie atop the actin.

An individual myosin protein is shaped like a hockey stick—a head attached at an angle to a long shaft (see Fig. 40-4c). Unlike the blade of a hockey stick, however, the **myosin head** is hinged to the shaft, and can move back and forth. A thick filament consists of a bundle of myosin proteins, with the shafts in the middle of the bundle and the heads protruding out. The heads on the two ends of a thick filament are oriented in opposite directions (see Fig. 40-4b).

40.3 HOW DO SKELETAL MUSCLES CONTRACT?

Our description of muscle contraction will begin with the movements of the thin and thick filaments that cause an individual muscle fiber to shorten, continue with the control of individual muscle fibers by the nervous system, and conclude with a description of how the nervous system controls the strength and duration of contraction of entire muscles.

Muscle Fibers Contract Through Interactions Between Thin and Thick Filaments

The molecular architecture of thin and thick filaments allows them both to grip and to slide past one another, shortening the sarcomeres and producing muscle contraction by what is called the sliding filament mechanism (**Fig. 40-5**).

Each spherical actin protein has a binding site for a myosin head. In a relaxed muscle cell, however, these binding sites on actin are covered by tropomyosin, which prevents the myosin heads from attaching (**Fig. 40-5 ➊**). When a muscle contracts, tropomyosin moves aside, exposing the binding sites on the actin proteins. Myosin heads then bind to these sites, temporarily linking the thick and thin filaments (**Fig. 40-5 ➋**). The myosin heads flex, pulling on the thin filaments and causing them to slide a tiny distance along the thick filament (**Fig. 40-5 ➌**). The myosin heads on the two ends of each thick filament pull the thin filaments toward the middle of the sarcomere. Because the thin filaments are attached to the Z lines at the ends of the sarcomere, this movement shortens the sarcomere (**Fig. 40-6**). All of the sarcomeres of the entire muscle fiber shorten simultaneously, so the whole muscle fiber contracts a little. The myosin heads then release the thin filament, extend, reattach farther along the thin filament (**Fig. 40-5 ➍**), and flex again, shortening the muscle fiber a little more, much like a sailor hauling in a long anchor line a little at a time, hand over hand. The cycle repeats as long as the muscle is contracting.

Muscle Contraction Requires ATP

Contracting muscles need a lot of energy. As usual, the energy comes from ATP. You might think that the energy is used to flex the myosin head and pull the thin filament along. However, the energy of ATP is used not to flex the myosin head (see Fig. 40-5 ➌), but to extend it (see Fig. 40-5 ➍). This isn't as

► **FIGURE 40-5 The sliding filament mechanism of muscle contraction**

actin
thin filament
troponin
tropomyosin
binding sites
myosin head
myosin (part of a thick filament)

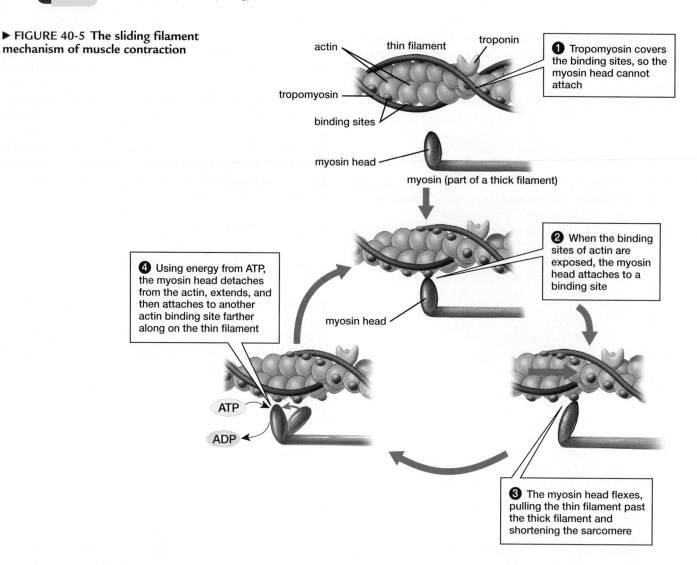

❶ Tropomyosin covers the binding sites, so the myosin head cannot attach

❷ When the binding sites of actin are exposed, the myosin head attaches to a binding site

myosin head

❸ The myosin head flexes, pulling the thin filament past the thick filament and shortening the sarcomere

❹ Using energy from ATP, the myosin head detaches from the actin, extends, and then attaches to another actin binding site farther along on the thin filament

ATP
ADP

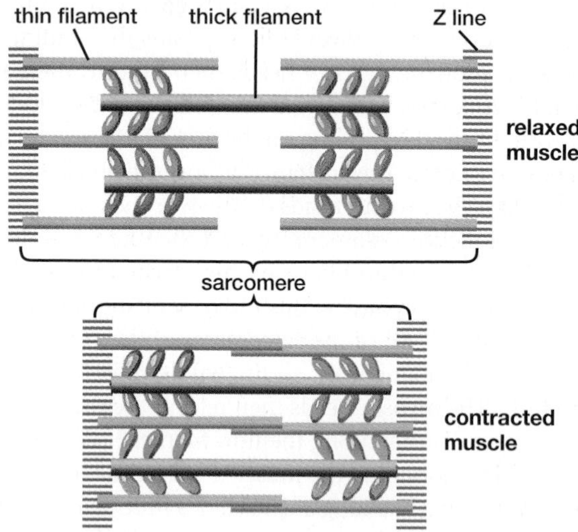

▲ **FIGURE 40-6 Filament sliding shortens sarcomeres**
Flexing of the myosin heads pulls the thin filaments toward the center of each sarcomere, shortening the sarcomere.

QUESTION Muscles typically generate maximum force when they start from a relaxed state (neither contracted nor stretched). Why would a highly stretched muscle generate less force when it starts contracting?

strange as it might sound. Picture shooting a pebble with an old-fashioned slingshot made of a Y-shaped stick and an elastic rubber band. It takes energy to stretch out the rubber band. Once the band has been stretched, all you need to do is to let go of the band and the pebble shoots out. Similarly, ATP energy is used to extend the myosin head, storing the energy in this "stretched" position. When the head binds to actin, it's like letting go of the rubber band in a slingshot: The stored energy flexes the myosin head and pulls the thin filament toward the center of the sarcomere.

There is another crucial role for ATP in muscle contraction. Once again, picture a sailor hauling in an anchor line. When he has pulled the line in as far as he can with one arm, he must release the rope before he can move this arm farther down and grasp the rope again for another pull. Similarly, when a myosin head has flexed and pulled on the thin filament, the head must release the actin before the head can extend and bind again at a second location a little farther along on the thin filament. When ATP binds to a myosin head, it causes the head to release actin. Only then can the energy of ATP be used to extend the head, storing that energy to use during the next pull on the thin filament.

A skeletal muscle's reserves of ATP are used up after only a few seconds of high-intensity exercise. Skeletal muscles also stock a supply of creatine phosphate, an energy-storage molecule that can donate a high-energy phosphate to ADP, thus regenerating ATP. However, creatine phosphate is also depleted rapidly. During brief, high-intensity exertion, muscle cells generate a bit more ATP using glycolysis, which does not require oxygen, but is also not very efficient (see pp. 129–130 and Fig. 8-2). For prolonged or low-intensity exercise, muscle cells produce ATP from glucose and fatty acids using cellular respiration, which requires a continuous supply of oxygen, delivered to the muscles by the cardiovascular system.

Have you ever wondered?

Why Rigor Mortis Occurs?

You've probably heard of rigor mortis, in which muscles become rigid after death. After death, there is no breathing and no heartbeat, so no oxygen reaches the muscles and very little ATP is synthesized. Rigor mortis occurs for two reasons, both related to the lack of ATP. First, ATP powers the Ca^{2+} pumps of the sarcoplasmic reticulum. Without ATP, Ca^{2+} leaking out of the SR cannot be pumped back in, so the Ca^{2+} concentration around the filaments remains high, and as a result, the myosin heads bind to actin. Second, ATP is required for the myosin heads to detach from the actin binding sites. Without ATP, all of the thin and thick filaments remain locked together, making the muscles stiff. Rigor mortis gradually fades, many hours later, as the muscle cells begin to decompose.

The Nervous System Controls Contraction of Skeletal Muscles

Skeletal muscle contraction is voluntary, controlled by the nervous system. We have already seen that moving the accessory proteins away from the binding sites on actin begins the cycle of myosin head movements that cause skeletal muscle fibers to contract. What links activity in the nervous system and the position of the accessory proteins?

Muscle fibers can fire action potentials, much like neurons can (see Fig. 38-2). As we shall see shortly, action potentials in muscle fibers cause the fibers to contract. The role of the nervous system is to trigger action potentials in muscle fibers.

Motor neurons, mostly in the spinal cord, send axons out to the skeletal muscles. These axons stimulate muscle fibers at specialized synapses called **neuromuscular junctions** (**Fig. 40-7**; see also Fig. 38-4, which shows a synapse between two nerve cells). All vertebrate neuromuscular junctions use the neurotransmitter acetylcholine (see Table 38-1). Each action potential in a motor neuron releases enough acetylcholine to produce a huge excitatory postsynaptic potential in the muscle fiber, bringing its membrane potential above threshold and triggering an action potential (**Fig. 40-7 ❶**).

Recall that the plasma membrane of a muscle fiber sends T tubules deep into the fiber and alongside the sarcoplasmic reticulum surrounding each myofibril. The muscle fiber's action potential moves down the T tubules to the SR (**Fig. 40-7 ❷**), where it causes calcium ions (Ca^{2+}) to be released from the SR into the cytoplasmic fluid surrounding the thin and thick filaments (**Fig. 40-7 ❸**). Ca^{2+} binds to the smaller accessory protein, troponin, causing it to pull the larger accessory protein, tropomyosin, off the actin binding sites (**Fig. 40-7 ❹**). With tropomyosin out of the way, myosin heads can bind to actin (**Fig. 40-7 ❺**). The myosin heads repeatedly attach, flex, release, extend, and reattach to actin, pulling the thin filaments toward the center of each sarcomere. A single action potential in a muscle fiber causes all of its sarcomeres to shorten simultaneously, slightly shortening the fiber.

What makes the fiber stop contracting? When the action potential in the muscle fiber is over (in just a few thousandths of a second), the SR stops releasing Ca^{2+}. Active transport proteins in the membrane of the sarcoplasmic reticulum pump Ca^{2+} back into the SR. Ca^{2+} leaves the accessory proteins, which move back over the actin binding sites. Therefore, the myosin heads can no longer attach to actin. Contraction stops within a few hundredths of a second.

BioFlix™ Muscle Contraction

Regulating the Intensity of Muscle Contraction

This description of muscle contraction—a motor neuron causes a single muscle fiber to contract a tiny bit, for a few hundredths of a second—probably doesn't seem to fit your everyday experiences, in which muscles often contract for several inches, and can stay contracted for several seconds. To control the force, distance, and duration of muscle contraction, you must be able to control how many muscle fibers in a single muscle contract, how much they contract, and how long they contract. How does this work?

First, a single motor neuron typically synapses with several muscle fibers in a single muscle. A motor neuron and all of the muscle fibers that it stimulates are called a **motor unit.** Motor units vary in size. In muscles used for fine control, such as those that move the eyes or fingers, motor units are small: A single motor neuron may synapse on just a few muscle fibers. In muscles used for large-scale movements, such as those of the thigh and buttocks, motor units are large: A single motor neuron may synapse on dozens or even hundreds of muscle fibers. To see the difference in control, try tying a piece of chalk to your knee and writing your name on a blackboard!

Second, the nervous system controls the strength of muscle contraction by varying both the number of muscle fibers stimulated and the frequency of action potentials in each fiber. Because motor neurons synapse on multiple muscle fibers in a given muscle, and because the muscle

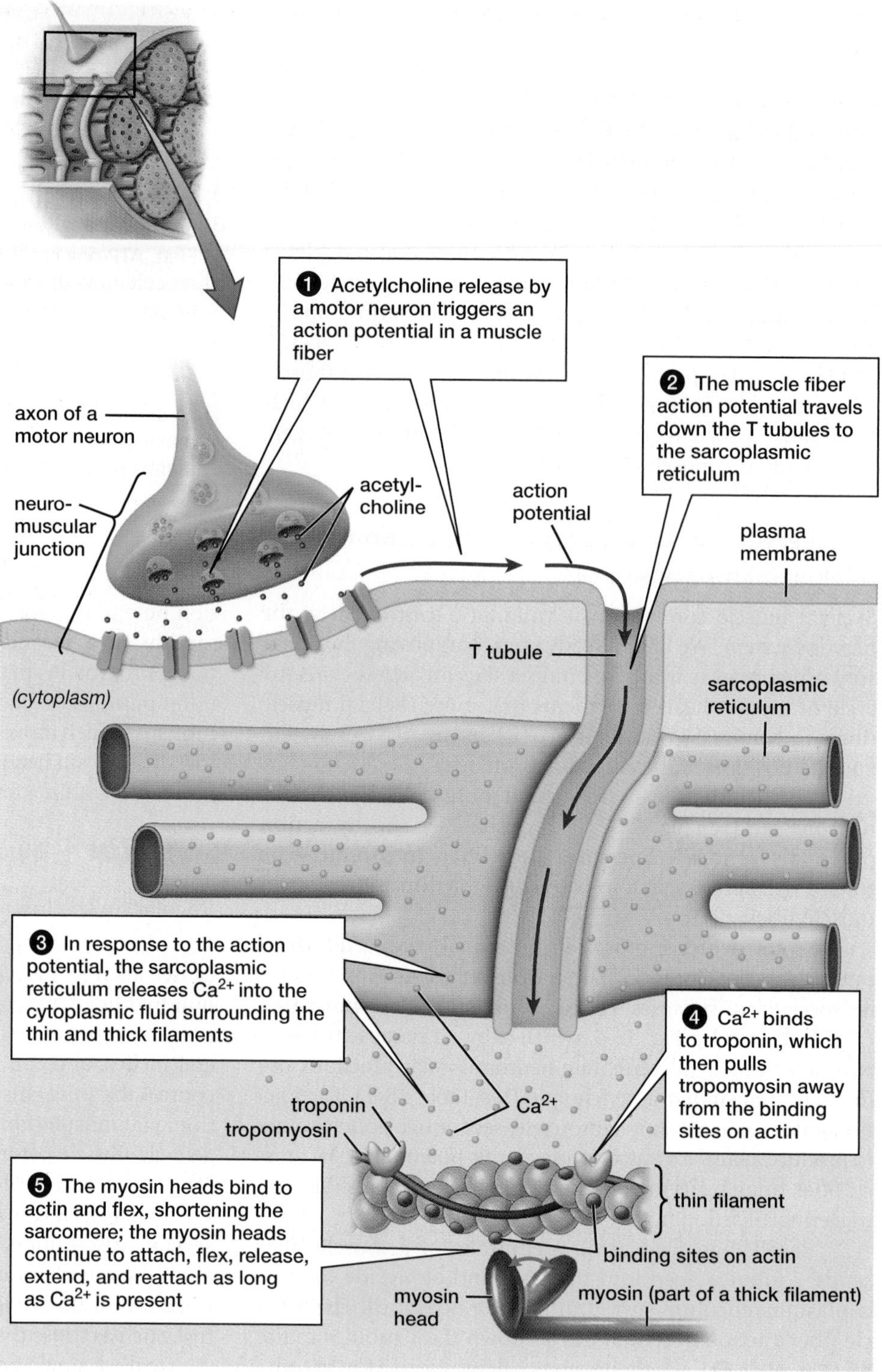

1 Acetylcholine release by a motor neuron triggers an action potential in a muscle fiber
2 The muscle fiber action potential travels down the T tubules to the sarcoplasmic reticulum
axon of a motor neuron
neuro-muscular junction
acetyl-choline
action potential
plasma membrane
T tubule
(cytoplasm)
sarcoplasmic reticulum
3 In response to the action potential, the sarcoplasmic reticulum releases Ca^{2+} into the cytoplasmic fluid surrounding the thin and thick filaments
4 Ca^{2+} binds to troponin, which then pulls tropomyosin away from the binding sites on actin
troponin
tropomyosin
Ca^{2+}
thin filament
5 The myosin heads bind to actin and flex, shortening the sarcomere; the myosin heads continue to attach, flex, release, extend, and reattach as long as Ca^{2+} is present
binding sites on actin
myosin head
myosin (part of a thick filament)

▶ **FIGURE 40-7 Activity in a motor neuron stimulates contraction of a skeletal muscle fiber**

fibers are attached to one another and to the muscle's tendons, a single action potential in a single motor neuron will cause some contraction of the entire muscle. The contractions caused by a single motor neuron firing multiple action potentials in rapid succession add up to a larger contraction. Simultaneously firing several motor neurons that stimulate multiple fibers in the same muscle will also cause a larger contraction of the muscle. Finally, rapid firing of all of the motor neurons that innervate all of the fibers in the muscle will cause a maximal contraction.

This general scheme should be familiar to you from Chapter 39. *Which muscles* contract is determined by which motor neurons fire action potentials. *How strongly* the muscles contract is determined by the number of motor neurons firing, how many muscle fibers each motor neuron innervates, and how fast the motor neurons fire.

Muscle Fibers Are Specialized for Different Types of Activity

Skeletal muscle fibers come in two basic types, slow twitch and fast twitch. Slow-twitch and fast-twitch fibers have different forms of myosin, causing them to contract slowly and more rapidly, respectively. However, there are many other differences as well.

Slow-twitch fibers contract with less power than fast-twitch muscles, but they can keep on contracting for a very long time. How? Slow-twitch muscles have lots of mitochondria and a plentiful blood supply that provides oxygen for cellular respiration in the mitochondria. Slow-twitch fibers are also thin. Thin fibers packed with mitochondria have fewer myofibrils, but they trade the resulting decreased power for rapid diffusion of oxygen in and wastes out. Thus, slow-twitch fibers produce abundant ATP and have fewer filaments to use it up, so they resist fatigue.

Fast-twitch fibers, on the other hand, contract more powerfully. They have a smaller blood supply, fewer mitochondria, and a larger diameter. Thick fibers with relatively few mitochondria have more myofibrils and are therefore more powerful. The extreme versions of fast-twitch fibers use mostly glycolysis for energy production, which does not require oxygen but supplies a lot less ATP than cellular respiration does. Fast-twitch fibers, therefore, fatigue more rapidly than do slow-twitch fibers.

Case Study continued

Muscles of Gold

The legs of champion sprinters like Usain Bolt have about 80% fast-twitch fibers, capable of the rapid, explosive contractions that are so essential to coming off the blocks. World-class marathoners like Samuel Wansiru, on the other hand, have about 80% slow-twitch fibers, which are less powerful but are capable of contracting again and again, each leg stepping over 10,000 times to complete a fast marathon. Both of these athletes probably have about the same number of muscle fibers in their legs, but Bolt has bigger muscles than Wansiru, because Bolt's fast-twitch fibers are much thicker, and Wansiru's slow-twitch fibers are very thin. Weightlifters like Matthias Steiner also have lots of fast-twitch muscles, and they are even bulkier than Bolt's.

40.4 HOW DO CARDIAC AND SMOOTH MUSCLES DIFFER FROM SKELETAL MUSCLE?

Although all muscle cells are built on the same general principles—filaments of actin and myosin attaching and sliding past one another—cardiac and smooth muscles differ significantly from skeletal muscles.

Cardiac Muscle Powers the Heart

Cardiac muscle, like skeletal muscle, is striated due to its regular arrangement of sarcomeres with their alternating thick and thin filaments (see Table 40-1). The fibers of cardiac muscle are branched, smaller than most skeletal muscle cells, and possess only a single nucleus. Because cardiac muscles must contract around 70 times each minute, and sometimes much faster, for your whole life, cardiac muscle fibers have enormous numbers of mitochondria, which occupy as much as 25% of the volume of the fibers. Unlike skeletal muscle fibers, cardiac muscle fibers can initiate their own contractions. This ability is particularly well developed in the specialized cardiac muscle fibers of the heart's pacemaker (see p. 624). Action potentials from the pacemaker spread rapidly through gap junctions in the intercalated discs that interconnect cardiac muscle fibers. Strong cell-to-cell attachments in the intercalated discs, called desmosomes, hold cardiac muscle fibers firmly to one another, preventing the forces of contraction from pulling them apart.

Smooth Muscle Produces Slow, Involuntary Contractions

Smooth muscle surrounds blood vessels and most hollow organs, including the uterus, bladder, and digestive tract. Smooth muscle cells are not striated, because the thin and thick filaments are scattered throughout the cells (see Table 40-1). Like cardiac muscle fibers, smooth muscle fibers each contain a single nucleus. Smooth muscle fibers are directly connected to one another by gap junctions, allowing the cells to contract in synchrony. Smooth muscle contraction is either slow and sustained (such as the constriction of arteries that elevates blood pressure during times of stress; see p. 634) or slow and wavelike (such as the waves that move food through the digestive tract; see p. 666). Smooth muscle stretches easily, as can be observed in the bladder, the stomach, and the uterus. Smooth muscle contraction is involuntary and can be initiated by stretching, by hormones, by signals from the autonomic nervous system (see pp. 744–745), or by a combination of these stimuli.

40.5 WHAT ARE THE FUNCTIONS AND STRUCTURES OF VERTEBRATE SKELETONS?

The bony endoskeleton of humans and other vertebrates serves a wide range of functions:

- The skeleton provides a rigid framework that supports the body and protects its internal organs. The brain and spinal cord are almost completely enclosed within the skull and vertebral column; the rib cage protects the lungs and the heart, while the pelvic girdle supports and partially protects abdominal organs.
- The skeleton allows locomotion. Different types of vertebrates have skeletons adapted for walking, running, jumping, swimming, flying, or combinations of these. Homologous bones may assume different forms, positions, and functions in the bodies of different vertebrate species (see Fig. 14-8).
- The skeleton participates in sensory function. Bones of the middle ear transmit sound vibrations between the eardrum and the cochlea (see pp. 762–763).
- Bones produce red blood cells, white blood cells, and platelets (see pp. 627–629) in red bone marrow,

located in the sternum (breastbone), ribs, upper arms and legs, and hips.

- Bones store calcium and phosphorus. They absorb and release these minerals as needed, maintaining a constant concentration in the blood (see p. 725).

The bones of terrestrial vertebrates can be grouped in two categories (**Fig. 40-8**). The **axial skeleton** includes the bones of the head, vertebral column, and rib cage. The **appendicular skeleton** includes the pectoral and pelvic girdles, and the appendages attached to them: the forelimbs

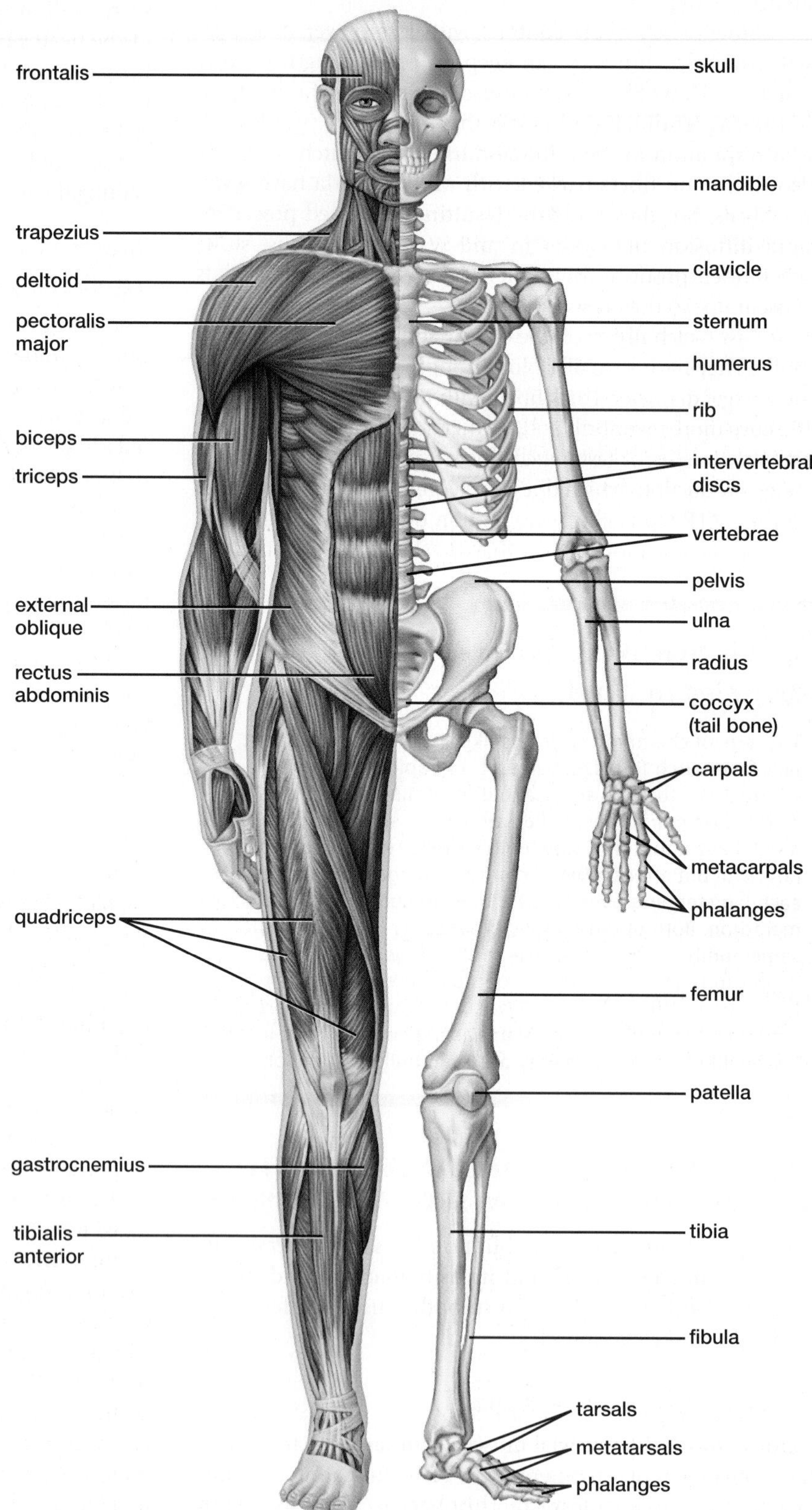

▶ **FIGURE 40-8 The human muscular and skeletal systems** The muscular and skeletal systems work together to move the body. Here, some of the major muscles and bones in the human body are illustrated. The human skeleton consists of 206 bones grouped into the axial skeleton (blue) and the appendicular skeleton (beige).

(in humans, arms and hands) and hind limbs (in humans, legs and feet). The pectoral girdle, which consists of the clavicle and scapula in humans, links the arms to the axial skeleton and provides attachment sites for muscles of the trunk and arms. Hip bones form the pelvic girdle, which links the legs to the axial skeleton, helps protect the abdominal organs, and forms attachment sites for muscles of the trunk and legs.

The Vertebrate Skeleton Consists of Cartilage, Bone, and Ligaments

Three types of connective tissue—cartilage, bone, and ligaments—make up the skeleton (see Chapter 31 for more information about connective tissue). All three consist of living cells embedded in a matrix of collagen protein, with various other substances included in the matrix. **Bone** includes large amounts of minerals composed mostly of calcium and phosphate and is, therefore, hard and rigid. **Cartilage** contains large amounts of glycoproteins (glucosamine, a popular dietary supplement, is a precursor for these glycoproteins) and often includes elastic fibers, which make some cartilage quite flexible (for example, in the pinna of the outer ear). **Ligaments,** which hold bones together at joints (see Fig. 40-11 later in the chapter), have small amounts of elastic fibers, but usually not very much.

Cartilage Provides Flexible Support and Connections

Cartilage plays many roles in the vertebrate skeleton. In some fishes, such as sharks and rays, the entire skeleton is composed of cartilage. During the embryonic development of other vertebrates, the skeleton, except for the skull and collarbone, is first formed from cartilage, which is later replaced by bone (**Fig. 40-9**). Cartilage also covers the ends of bones at joints (**Fig. 40-10**), supports the flexible portions of the nose and external ears, and provides the framework for the larynx, trachea, and bronchi of the respiratory system. In addition, cartilage forms the tough, shock-absorbing intervertebral discs between the vertebrae of the backbone (see Fig. 40-8).

The living cells of cartilage are called **chondrocytes.** These cells secrete the glycoproteins and collagen that make up most of the matrix of cartilage (see Fig. 40-10, left). No blood vessels penetrate cartilage. To exchange wastes and nutrients, chondrocytes rely on diffusion of materials through the collagen matrix. As you might predict, cartilage cells have a very low metabolic rate, so damaged cartilage repairs itself very slowly, if at all.

Bone Provides a Strong, Rigid Framework for the Body

A bone consists of a hard outer shell of **compact bone** that encloses **spongy bone** in its interior (see Fig. 40-10, middle). Compact bone is dense and strong and provides an attachment site for muscle. Compact bone develops as small tubes called osteons, with collagen and calcium phosphate surrounding a central canal containing blood vessels (see Fig. 40-10, right). Spongy bone consists of an open network of bony fibers. It is porous, lightweight, and rich in blood vessels. Bone marrow, where blood cells form, is found in the cavities of spongy bone. In contrast to cartilage, bone is well supplied with blood vessels.

There are three types of bone cells: **osteoblasts** (bone-forming cells), **osteocytes** (mature bone cells), and **osteoclasts** (bone-dissolving cells). Early in development, when bone replaces cartilage in the skeleton, osteoclasts invade and dissolve the cartilage. Osteoblasts then secrete a hardened matrix of bone and gradually become entrapped within it.

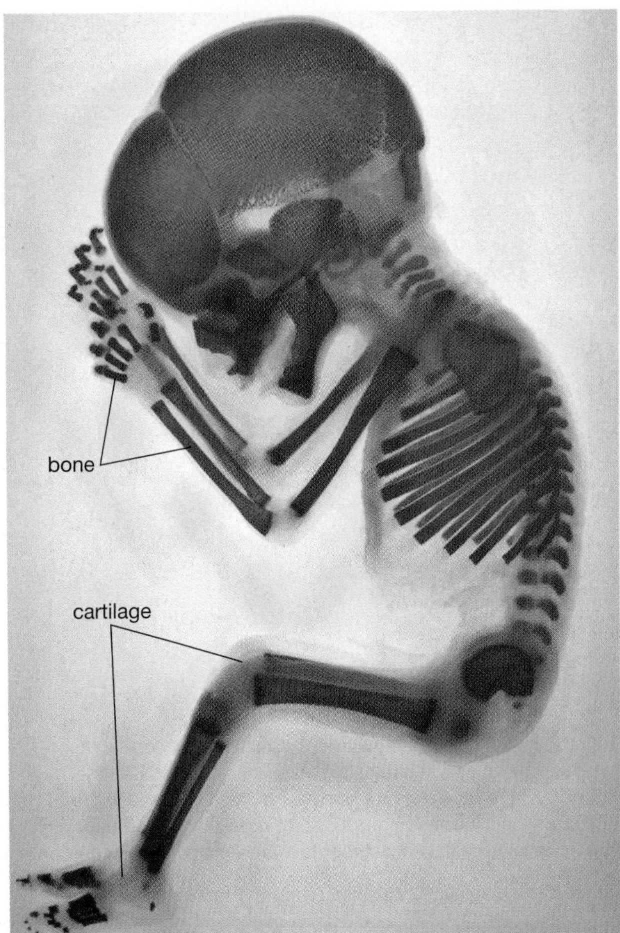

▲ **FIGURE 40-9 Bone replaces cartilage during development** In this 16-week-old human fetus, bone is stained magenta. The clear areas at the wrists, knees, ankles, elbows, and breastbone show cartilage that will later be replaced by bone.

As bones mature, the trapped osteoblasts mature into osteocytes. Although they are generally not capable of enlarging a bone, osteocytes are essential to bone health, because they constantly rework the calcium phosphate deposits, thereby preventing excessive crystallization that would make the bone brittle and likely to break.

Bone Remodeling Allows Skeletal Repair and Adaptation to Stresses

Each year, 5% to 10% of all the bone in your body dissolves and is replaced by the coordinated activity of osteoclasts, which secrete acid that dissolves small amounts of bone, and osteoblasts, which secrete new bone. This process, called bone remodeling, allows the skeleton to alter its shape in response to the demands placed on it. Bones that carry heavy loads or are subjected to extra stress become thicker, providing more strength and support. For example, a professional tennis player may have 30% more bone mass in his or her racket arm. Everyday stresses, such as walking, also help to maintain bone strength.

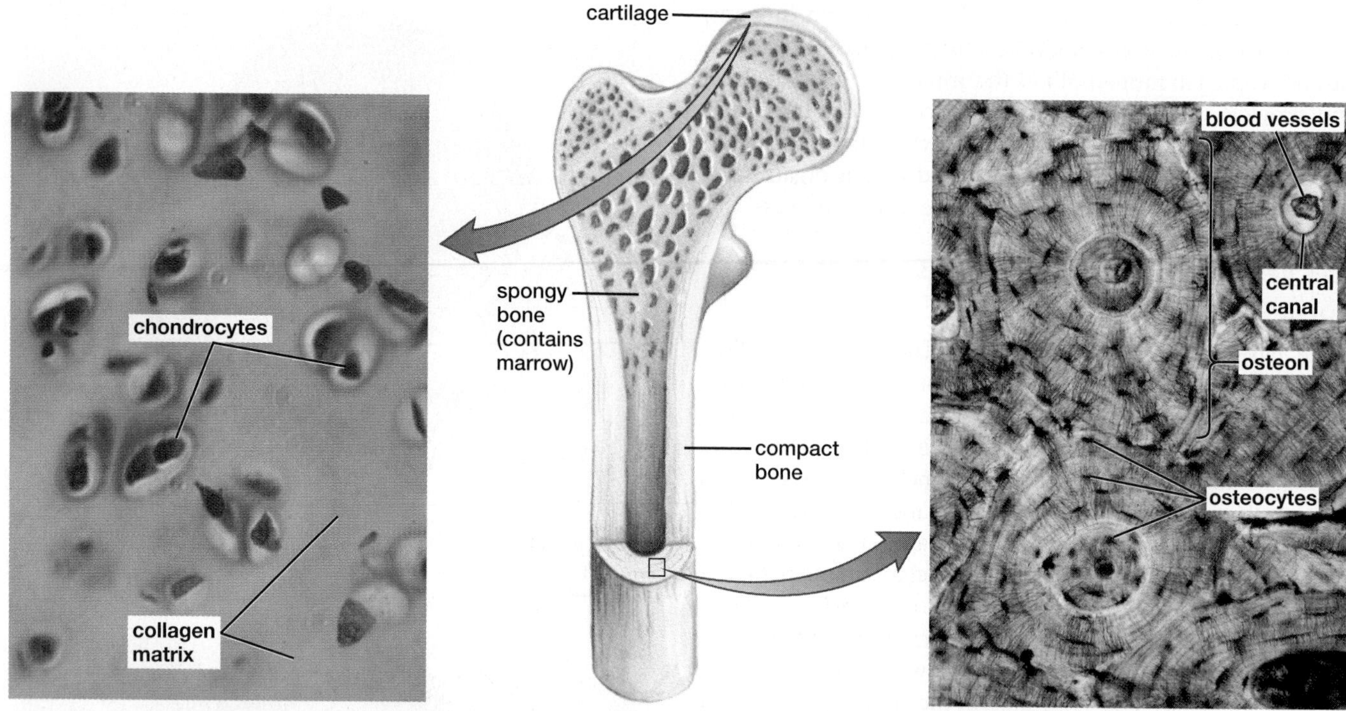
cartilage
chondrocytes
collagen matrix
spongy bone (contains marrow)
compact bone
blood vessels
central canal
osteon
osteocytes

▲ **FIGURE 40-10 Cartilage and bone** (Left) In cartilage, the chondrocytes, or cartilage cells, are embedded in an extracellular matrix of the protein collagen, which they secrete. (Middle) A bone consists of an outer layer of compact bone with spongy bone inside. For simplicity, blood vessels are not shown. (Right) Compact bone is composed of osteons, each including a central canal containing blood vessels. Mature bone cells, called osteocytes, are embedded in the concentric rings of bone material.

QUESTION Compact bone is stronger than spongy bone. Why not make the entire skeleton from compact bone?

Bone remodeling varies with age. Early in life, the activity of osteoblasts outpaces that of osteoclasts, allowing bones to become larger and thicker as a child grows. In the aging body, however, the balance of power shifts to favor osteoclasts, and bones become more fragile as a result. Although both sexes lose bone mass with age, this is typically more pronounced in women, as we explore in "Health Watch: Osteoporosis—When Bones Become Brittle."

The ultimate bone remodeling occurs after a fracture. Typically, a physician moves the ends of a broken bone back into their proper alignment and immobilizes the break with a cast or splint. The rest of the healing process is up to the body's own repair mechanisms, and usually takes about 6 weeks. A fracture ruptures the thin layer of connective tissue, rich in capillaries and osteoblasts, that surrounds the bone. Healing begins when a large blood clot surrounds the break (**Fig. 40-11 ❶**). Phago-

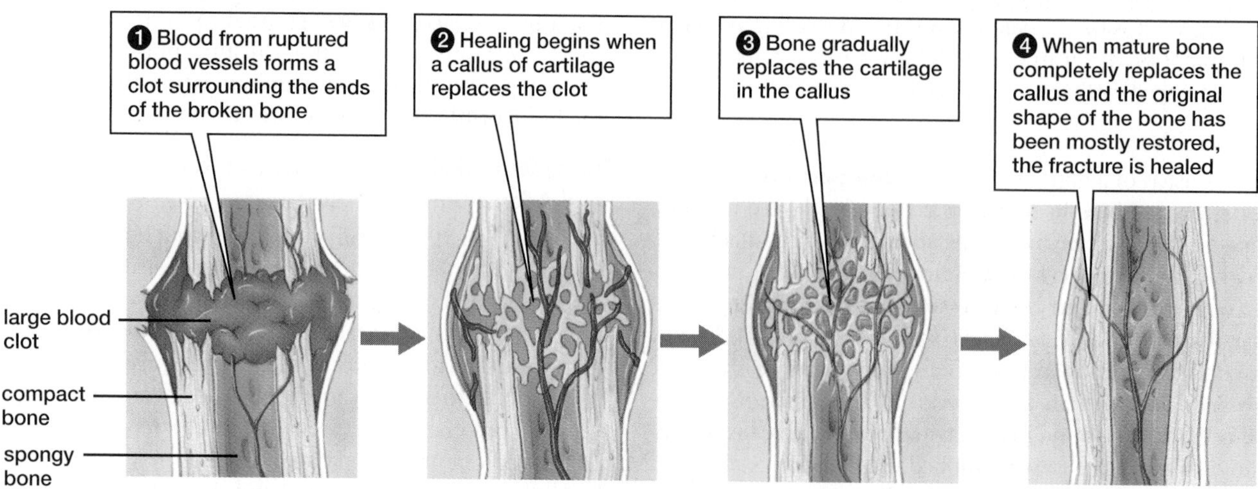
❶ Blood from ruptured blood vessels forms a clot surrounding the ends of the broken bone
❷ Healing begins when a callus of cartilage replaces the clot
❸ Bone gradually replaces the cartilage in the callus
❹ When mature bone completely replaces the callus and the original shape of the bone has been mostly restored, the fracture is healed
large blood clot
compact bone
spongy bone

▲ **FIGURE 40-11 Bone repair**

Health Watch

Osteoporosis—When Bones Become Brittle

Bone density in humans peaks between ages 25 and 35. In middle age, as the activity of osteoclasts starts to exceed that of osteoblasts, bone density begins a slow decline. In the 10 million Americans with **osteoporosis** (literally, "porous bones"), the loss is severe enough to weaken bones, making them vulnerable to fractures and deformities (**Fig. E40-1a**). In many cases, the vertebrae of individuals with osteoporosis compress, causing a hunchbacked appearance (**Fig. E40-1b**). In extreme cases, just lifting a shopping bag, opening a window, or even sneezing can break a bone. People with osteoporosis are very likely to break a bone during a fall. Hip fractures are particularly debilitating, especially in the elderly.

Women are about four times more likely than men to suffer from osteoporosis. Typically their bones are less massive than those of men to begin with. Further, the high estrogen levels of premenopausal women stimulate osteoblasts and help maintain bone density, but after menopause, estrogen production drops dramatically, and bone density decreases rapidly. In men, testosterone stimulates osteoblasts. Although testosterone levels decline with age, they usually do not drop as fast or as far as estrogen levels in women do after menopause. In both sexes, alcoholism and smoking also contribute to bone loss.

People with osteoporosis, in consultation with their physicians, may choose medical therapy to help maintain bone density. Usually, drugs that inhibit osteoclast activity, including Fosamax®, Actonel®, and Boniva®, are the frontline medications in treating osteoporosis. The hormone calcitonin also inhibits osteoclast activity, although some studies suggest that it is not as effective as these drugs for treating osteoporosis. Although not without other risks, estrogen replacement therapy (usually with very low amounts of estrogen) can slow bone loss in postmenopausal women, as can a drug, raloxifene, that mimics estrogen's effects on bone.

Can osteoporosis be prevented? Calcium, of course, is a major component of bone. In addition, bones thrive on moderate stress. Therefore, regular exercise throughout life, combined with a diet containing adequate calcium and vitamin D (which is required for calcium absorption from the intestine), helps to ensure that bone mass is as high as possible before age-related losses begin, and slows the rate of loss with aging. Older people tend to be less active, resulting in loss of bone minerals, but low-impact, weight-bearing exercise such as walking or dancing can reduce bone loss and even sometimes increase bone mass.

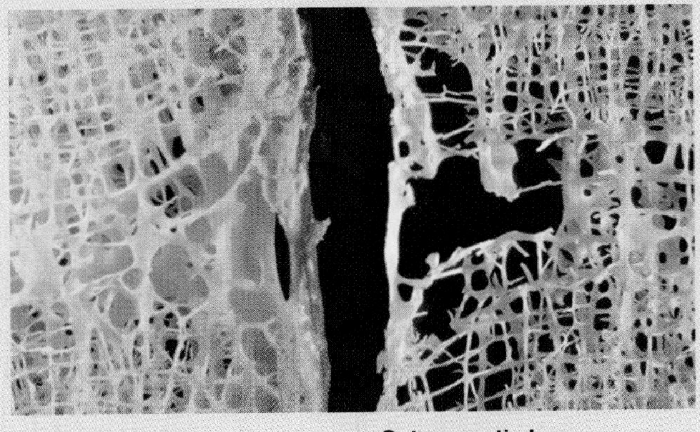

(a) Normal bone Osteoporotic bone

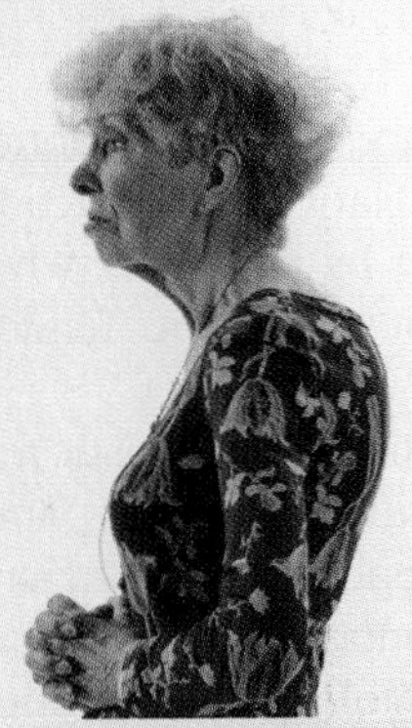

(b) A victim of osteoporosis

▲ **FIGURE E40-1 Osteoporosis (a)** Cross-section of (left) a normal bone compared with (right) a bone from a woman with osteoporosis. **(b)** The devastating effects of osteoporosis extend beyond a hunchbacked appearance; its victims are also at high risk for bone fractures.

cytic cells from the blood and osteoclasts from the damaged bone ingest cellular debris and dissolve bone fragments. Osteoblasts, in conjunction with cartilage-forming cells, secrete a callus, a porous mass of bone and cartilage that surrounds the break (**Fig. 40-11 ❷**). The callus replaces the original blood clot and temporarily holds the ends of the break together. Osteoclasts, osteoblasts, and capillaries invade the callus. Nourished by the capillaries, osteoclasts break down cartilage while osteoblasts add new bone (**Fig. 40-11 ❸**). Finally, osteoclasts remove excess bone, restoring the bone's original shape, but often leaving a slight thickening (**Fig. 40-11 ❹**).

40.6 HOW DO MUSCLES MOVE THE VERTEBRATE SKELETON?

As we noted in section 40.1, almost all animals move by the action of pairs of antagonistic muscles working on a skeleton. Here, we will focus on the arrangement and movement of muscles around the joints of the human skeleton, where two bones meet (**Fig. 40-12**). Not all joints are movable; for example, immobile joints called sutures join the bones of the skull. In movable joints, however, the portion of each bone that forms the joint is coated with a layer of cartilage, whose

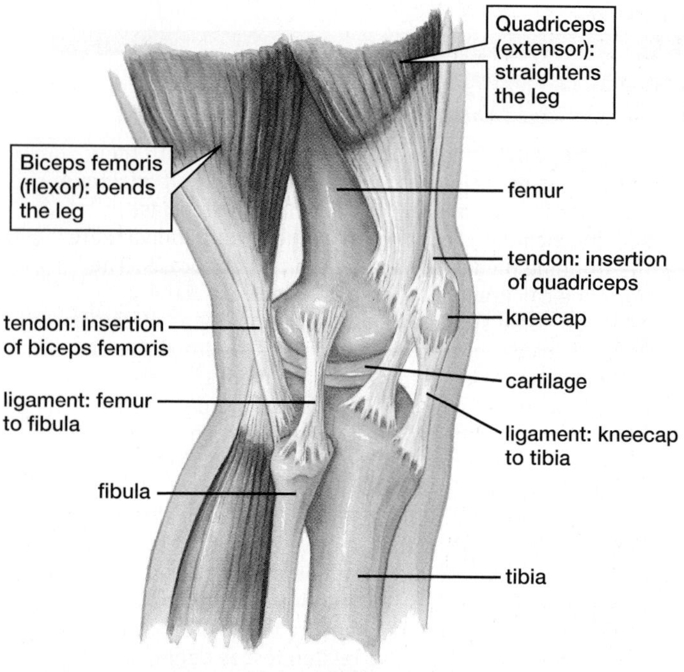

Quadriceps (extensor): straightens the leg
Biceps femoris (flexor): bends the leg
femur
tendon: insertion of quadriceps
tendon: insertion of biceps femoris
kneecap
cartilage
ligament: femur to fibula
ligament: kneecap to tibia
fibula
tibia

▲ **FIGURE 40-12 The human knee** The human knee, showing antagonistic muscles (the biceps femoris and the quadriceps of the thigh), tendons, and ligaments. The complexity of this joint, coupled with the extreme stresses placed on it during activities such as jumping, running, or skiing, makes it very susceptible to injury.

smooth, resilient surface allows the bone surfaces to slide past one another with relatively little friction (see Fig. 40-12). Joints are held together by ligaments, which are strong and flexible, but usually not very elastic. Tendons attach muscles to the bones.

When one of a pair of antagonistic muscles contracts, it moves the bone around its joint, and simultaneously stretches the relaxed opposing muscle (remember that muscles can only actively contract, not extend). Antagonistic muscles can cause a remarkable range of motions, depending on the configuration of a joint, including moving bones back and forth, moving them side to side, or rotating them.

Hinge joints are located in the elbows, knees, and fingers. Like a hinged door, these joints move in only two dimensions (**Fig. 40-13a**). In this case, the antagonistic muscle pair, the flexor and extensor muscles, lies in roughly the same plane as the joint (see Fig. 40-12). The tendon at one end of each muscle, called the **origin,** is fixed to a bone that remains stationary, while the other end, the **insertion,** is attached to the bone on the far side of the joint, which is moved by the

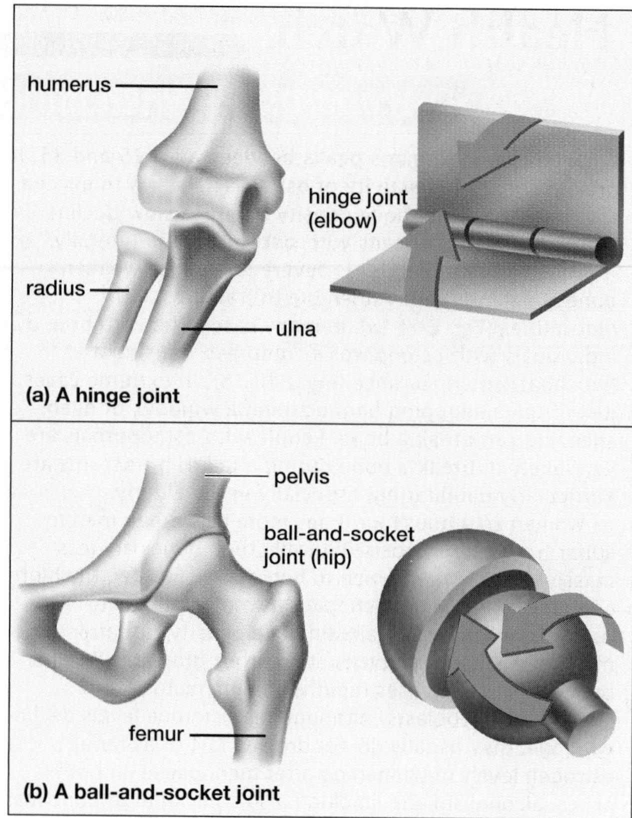

▲ **FIGURE 40-13 Hinge and ball-and-socket joints**
(a) The human elbow is a good example of a hinge joint. **(b)** The human hip can rotate because it has a ball-and-socket joint. This consists of the rounded end of the femur (the ball) that fits into a cuplike depression (the socket) in the pelvic bone.

muscle. When the flexor muscle contracts, it bends the joint; when the extensor muscle contracts, it straightens the joint. In Figure 40-12, for example, contraction of the biceps femoris (the flexor) bends the leg at the knee, while contraction of the quadriceps (the extensor) straightens it. Alternating contractions of flexor and extensor muscles cause the lower leg bones to swing back and forth at the knee joint.

Some other joints, such as those of the hip and shoulder, are **ball-and-socket joints,** in which the round end of one bone fits into a hollow depression in another (**Fig. 40-13b**). Ball-and-socket joints allow movement in several directions—compare the range of movement of your upper leg around its joint with your hip with the limited bending of your knee. The range of motion in ball-and-socket joints is made possible by at least two pairs of antagonistic muscles, oriented at angles to each other, which move the joint in three dimensions.

Case Study revisited
Muscles of Gold

We have seen that muscle fibers differ in their composition and capabilities. Fast-twitch fibers tend to be thick and contract rapidly, but fatigue readily. Slow-twitch fibers are thin and contract slowly, but resist fatigue. These facts raise a few interesting questions.

First, are the differences in fiber types what make Usain Bolt the Olympic champion in the 100- and 200-meter sprints, and Samuel Wansiru the gold medalist in the marathon? Yes and no. Without an overwhelming majority of fast-twitch muscle fibers, Bolt certainly would not have won gold in the sprints. However, all of the competitors in the sprint finals probably have just as high a proportion of fast-twitch fibers as Bolt does. All top marathoners are superficially very similar to Wansiru—extremely thin, with mostly slow-twitch muscle fibers in their legs. Therefore, the "right" types of muscle fibers are essential for Olympic glory, but aren't enough, on their own, to win the gold.

Second, are sprinters and marathoners born or made? Probably both. The proportions of fast-twitch and slow-twitch fibers vary greatly within the human population. Research published in the 1990s concluded that about half of the difference in fiber type is genetic and half is environmental. People with a high proportion of slow-twitch fibers are likely to excel in, and therefore favor, endurance sports, while those with more fast-twitch fibers will probably prefer sports requiring bursts of speed or strength. Intensive training generally does not change the number of muscle fibers a person has, but can convert some slow-twitch fibers into fast-twitch fibers, and vice versa. Bolt and Wansiru train very hard, and in very specialized ways, for their events. Therefore, although it would be difficult or impossible to do the appropriate experiments in humans, a world-class marathoner was probably both born with a tendency to develop slow-twitch muscles, and converted many fast-twitch fibers to slow-twitch through prolonged distance training. The opposite would be true for sprinters.

In addition, in 2009, geneticists specializing in sports and fitness concluded that at least 221 chromosomal genes and 18 mitochondrial genes contribute to athletic performance, so it's not just a matter of slow-twitch versus fast-twitch fibers, training regimes, or running through the pain that makes someone an elite athlete. No one knows which alleles of all of these genes are carried by Bolt, Wansiru, or any other elite athlete, but Bolt and Wansiru both undoubtedly have a superb set of fitness genes.

Finally, how do genetics, muscle fiber types, and training apply to regular people? Well, no one can reach the Olympics—let alone bring home the gold—without winning the genetic lottery, so you're probably out of luck. Further, being an Olympic athlete is a career, not a hobby. It takes many years of training both for fitness and technique just to have a chance. The days of the "born athlete" who wakes up one day and decides to win the Olympics are long gone.

What's particularly frustrating is that people differ dramatically in their response to training. On the same exercise regime, some people rapidly improve cardiovascular health, running endurance, or strength, whereas others show much less gain. Only a few of the genes responsible for these differences are known. One such gene, which produces a protein with the daunting name of peroxisome proliferator-activated receptor delta (PPARD), strongly influences the response to exercise in both mice and people. Mice engineered to produce abnormally large amounts of PPARD have a much higher proportion of slow-twitch fibers than normal mice, and can run twice as far. Mice genetically engineered to produce lots of the enzyme phosphoenolpyruvate carboxykinase run 10 to 20 times farther than normal mice can. Other genes, such as insulin-like growth factor-1, also influence muscle type and strength. When, or if, treatments based on such "fitness genes" will make it to clinical practice to cure disease, or to the track, giving gene-doping athletes an unfair advantage, is anyone's guess.

BioEthics Consider This

Many parents try to give their children a boost in life—a very understandable urge. A company called Atlas Sports Genetics will test little kids to see which allele they carry for a gene that encodes an important protein in muscle cells. One allele has been associated with fast-twitch muscles and ability in sports that require sprinting or strength. Presumably, kids homozygous for this allele should be steered into sprinting or football rather than distance running or soccer. However, about 18% of the population is homozygous for this allele, and hardly any of these people are superathletes. Do you think that such genetic testing is helpful, particularly for the kids? What if you could test for, say, 20 fitness genes?

CHAPTER REVIEW

Summary of Key Concepts

40.1 How Do Muscles and Skeletons Work Together to Provide Movement?

Muscles can only contract. Skeletal muscles work with the skeleton, which provides a framework against which contraction can move the body. Animals may have hydrostatic skeletons, exoskeletons, or endoskeletons. In all cases, pairs of antagonistic muscles act on the skeleton to move the body.

40.2 What Are the Structures of Vertebrate Muscles?

Vertebrates have three types of muscle: skeletal, cardiac, and smooth. Skeletal muscle is attached to bone and moves the skeleton when it contracts. Cardiac muscle, found only in the heart, pumps blood. Smooth muscle moves food through the digestive tract, adjusts blood vessel diameter, and contracts hollow organs such as the bladder and uterus.

Skeletal muscle fibers consist of myofibril subunits surrounded by sarcoplasmic reticulum. Myofibrils are composed of repeating subunits called sarcomeres, attached end to end. Each sarcomere consists of alternating thick filaments made of myosin and thin filaments made of actin and two accessory proteins (troponin and tropomyosin). Thin filaments are attached to proteins called Z lines that form the boundaries of sarcomeres.

40.3 How Do Skeletal Muscles Contract?

A motor neuron innervates each muscle fiber at a synapse called a neuromuscular junction. An action potential in the motor neuron causes it to release acetylcholine onto the muscle fiber, stimulating an excitatory postsynaptic potential that triggers an action potential in the fiber. This action potential stimulates the release of Ca^{2+} from the sarcoplasmic reticulum. Ca^{2+} causes the accessory proteins of the thin filament to move, revealing binding sites on actin to which myosin heads attach. The myosin heads flex repeatedly, causing the thin and thick filaments to slide past each other, shortening the sarcomere. Shortening sarcomeres shorten the entire muscle fiber. When the action potential is over, calcium is actively transported back into the sarcoplasmic reticulum, ending contraction. The energy for muscle contraction comes from ATP, which may be derived from cellular respiration, or, for bursts of strenuous activity, from glycolysis, which does not use oxygen.

The degree of muscle contraction is determined by the number of muscle fibers stimulated by a motor neuron and the frequency of action potentials in each fiber. Rapid firing in all of the fibers of a muscle causes maximum contraction.

BioFlix™ Muscle Contraction

40.4 How Do Cardiac and Smooth Muscles Differ from Skeletal Muscle?

Cardiac muscle is striated due to its regular arrangement of thick and thin filaments. Its cells contract rhythmically and spontaneously; these contractions are synchronized by electrical signals produced by specialized muscle fibers in the heart's pacemaker. Cardiac muscle fibers are interconnected by intercalated discs that contain many gap junctions. These conduct electrical signals, producing coordinated contraction.

Smooth muscle lacks organized sarcomeres. Like cardiac muscle, smooth muscle cells are electrically coupled by gap junctions. Smooth muscle surrounds hollow organs such as the uterus, digestive tract, bladder, arteries, and veins, producing involuntary, slow, and sustained or rhythmic contractions.

40.5 What Are the Functions and Structures of Vertebrate Skeletons?

The vertebrate skeleton provides support for the body, attachment sites for muscles, and protection for internal organs. Red blood cells, white blood cells, and platelets are produced in bone marrow. Bone acts as a storage site for calcium and phosphorus. The axial skeleton includes the skull, vertebral column, and rib cage. The appendicular skeleton consists of the pectoral and pelvic girdles and the bones of the arms, legs, hands, and feet.

The vertebrate skeleton consists of cartilage, bone, and ligaments. All consist of cells that secrete collagen and various other substances into an extracellular matrix. Cartilage is located at the ends of bones, and forms pads in many joints and the intervertebral discs. It also supports the nose, ears, and respiratory passages. During embryological development, cartilage is the precursor of bone. Ligaments connect bones at movable joints. Bone is formed by osteoblasts, which secrete a collagen matrix that becomes hardened by calcium phosphate. A typical bone consists of an outer shell of compact, hard bone, to which muscles are attached, and inner spongy bone, which may contain bone marrow. Bone remodeling occurs through the coordinated action of osteoclasts, which dissolve bone, and osteoblasts, which create new bone.

40.6 How Do Muscles Move the Vertebrate Skeleton?

In the vertebrate skeleton, movement occurs around joints. Tendons attach pairs of antagonistic muscles to the bones on either side of a joint. The contraction of one muscle moves the joint and stretches out its antagonistic muscle. At hinge joints, muscles are attached to a stationary bone at their origins and to a moving bone at their insertions. Contraction of the flexor muscle bends the joint; contraction of its antagonistic extensor straightens it.

Key Terms

actin *778*
antagonistic muscles *775*
appendicular skeleton *784*
axial skeleton *784*
ball-and-socket joint *788*
bone *785*
cardiac muscle *777*
cartilage *785*
chondrocyte *785*
compact bone *785*
endoskeleton *776*
exoskeleton *776*
extensor *776*
flexor *776*
hinge joint *788*
hydrostatic skeleton *775*
insertion *788*
joint *776*
ligament *785*
motor unit *781*
muscle fiber *778*
myofibril *778*
myosin *778*
myosin head *779*
neuromuscular junction *781*
origin *788*
osteoblast *785*
osteoclast *785*
osteocyte *785*
osteoporosis *787*
sarcomere *778*
sarcoplasmic reticulum (SR) *778*
skeletal muscle *777*
skeleton *775*
smooth muscle *778*
spongy bone *785*
T tubule *778*
tendon *778*
thick filament *778*
thin filament *778*
Z line *778*

Thinking Through the Concepts

Fill-in-the-Blank

1. The three types of skeletal systems found in animals are __________, __________, and __________. All three move the skeleton using pairs of __________ muscles.
2. __________ and __________ muscles are striped because they have a regular alignment of sarcomeres. __________ and __________ muscles are usually not under conscious control. __________ and __________ muscles are usually interconnected by gap junctions.
3. A skeletal muscle cell is called a(n) __________. It consists of bundles of cylindrical structures called __________, which in turn consist of much smaller cylinders, the __________, attached end to end at their __________.
4. Muscle contraction results from interactions between thin filaments, composed mainly of the protein __________, and thick filaments, composed mainly of the protein __________. The "heads" of the thick filament protein grab on to the thin filament protein and flex. The energy of __________ is used to extend the head, which stores that energy to power the flexing movement of the head, pulling the thin filament past the thick filament.
5. Motor neurons innervate muscle fibers with specialized synapses called __________. The release of acetylcholine

at these synapses ultimately triggers an action potential in the muscle fiber, which invades the interior of the fiber through __________, and stimulates the release of calcium ions from the __________.

6. Skeletons consist of three types of connective tissue: __________, __________, and __________. All consist of cells embedded in a matrix of __________ protein and other extracellular components, such as glycoproteins or calcium phosphate.
7. A(n) __________ joint moves in two dimensions, while a(n) __________ joint has significant motion in three dimensions. At a two-dimensional joint, the __________ muscle bends the joint and the __________ muscle straightens the joint.

Review Questions

1. Sketch a relaxed muscle fiber containing a myofibril, sarcomeres, and thick and thin filaments. How would a contracted muscle fiber look by comparison?
2. Describe the process of skeletal muscle contraction, beginning with an action potential in a motor neuron and ending with the relaxation of the muscle. Your answer should include the following words: *neuromuscular junction, T tubule, sarcoplasmic reticulum, calcium, thin filaments, binding sites, thick filaments, sarcomere, Z line,* and *active transport.*
3. Explain the following two statements: Muscles can only actively contract; muscle fibers lengthen passively.
4. What are the three types of skeletons found in animals? For one of them, describe how the muscles are arranged around the skeleton and how contractions of the muscles result in movement of the skeleton.
5. Compare the structures of the following pairs: spongy and compact bone, smooth and skeletal muscle, and cartilage and bone.
6. Explain the functions of osteoblasts and osteoclasts.
7. Describe a hinge joint and how it is moved by antagonistic muscles.

Applying the Concepts

1. Discuss some of the problems that would result if the human heart were made of skeletal muscle instead of cardiac muscle.
2. Myasthenia gravis is caused by the abnormal production of antibodies that bind to acetylcholine receptors on muscle cells and eventually destroy the receptors. The disease causes muscles to become flaccid, weak, or paralyzed. Drugs, such as neostigmine, that inhibit the action of acetylcholinesterase (the enzyme that breaks down acetylcholine) are used to treat myasthenia gravis. How does neostigmine restore muscle activity? Do you think that neostigmine could restore normal muscle function for an entire lifetime?
3. Would taking erythropoietin, a hormone that increases the number of red blood cells, primarily benefit sprinters or endurance athletes? Why? Explain your answer in terms of the characteristics of fast-twitch and slow-twitch muscle fibers.

MB *Go to www.masteringbiology.com for practice quizzes, activities, eText, videos, current events, and more.*

Animal Reproduction

Chapter 41

▲ Thanks to the expertise and hard work of reproductive biologists, Sumatran rhino Emi shows off her third calf, Harapan, at the Cincinnati Zoo.

Case Study

To Breed a Rhino

RHINOS ARE MASSIVE, often aggressive, and seemingly indestructible. Nevertheless, three of the five species are critically endangered. How can such animals be in danger of extinction? Their name—rhinoceros, or "nose horn"—gives you a clue.

Some cultures place enormous value on rhino horn. In Yemen, young men are traditionally given a ceremonial dagger, the jambiyya, with a handle made of rhino horn. In China and some other East Asian cultures, powdered rhino horn is believed to reduce fevers and cure rheumatism, gout, and many other disorders. (Contrary to popular belief, few people use rhino horn as an aphrodisiac.) Because rhinos are rare—the population of all five species combined is probably fewer than 20,000—rhino horn is expensive. The black market price is about $2,000 per horn from African rhinos and $14,000 per horn from Asian rhinos. In both sub-Saharan Africa and southeast Asia, a rhino poacher can make far more money selling a single horn than most people make in a year at a regular job.

Sumatran rhinos are particularly endangered, with a population below 300. Therefore, a concerted effort has been made to breed Sumatran rhinos in captivity, both to preserve them from extinction and to reintroduce them back into the wild if conditions permit.

Sounds simple, doesn't it? Put a male and female into the same pen, and let nature take its course. Unfortunately, this often doesn't work, particularly with large, aggressive loners like rhinos. It took years of effort by reproductive physiologist Terri Roth at the Cincinnati Zoo's Center for Conservation and Research of Endangered Wildlife to succeed in breeding Sumatran rhinos.

The 2008 Red List put out by the International Union for Conservation of Nature includes 187 other critically endangered mammals, including gazelles, tigers, bats, porpoises, and monkeys. Some can probably be saved only through captive breeding, artificial insemination, or *in vitro* fertilization. So far, captive breeding programs have staved off extinction for Pere David's deer, Przewalski's horse (which has been reintroduced into the wild, in Mongolia), and black-footed ferrets (also released into the wild, in Wyoming).

Think about Sumatran rhinos and other endangered mammals as you study reproduction. How do males and females identify each other as the same species, different genders, and ready to mate? Why do terrestrial animals have internal fertilization? How are sperm and eggs produced, and how does fertilization occur?

At a Glance

41.1 HOW DO ANIMALS REPRODUCE?

Animals reproduce either sexually or asexually. In **asexual reproduction,** a single animal produces offspring, usually through repeated mitotic cell divisions in some part of its body. The offspring are therefore genetically identical to the parent. In **sexual reproduction,** organs called **gonads** produce haploid sperm or eggs through meiotic cell division. A sperm and egg—usually from separate parents—fuse to produce a diploid fertilized egg, which then undergoes repeated mitotic cell divisions to produce an offspring. Because the offspring receives genes from both of its parents, it is not genetically identical to either.

In Asexual Reproduction, a Single Organism Reproduces by Itself

Asexual reproduction is efficient in effort (no need to search for mates, court members of the opposite sex, or battle rivals), materials (no wasted gametes), and genes (the offspring have all the genes of their parent). Efficiency has probably been the driving force behind the evolution of a variety of methods of asexual reproduction.

Budding Produces a Miniature Version of the Adult

Many sponges and cnidarians reproduce by **budding.** A miniature version of the adult—a bud—grows directly on the body of the adult (**Fig. 41-1**). When it has grown large enough, the bud breaks off and becomes independent.

▲ **FIGURE 41-1 Budding** The offspring of some cnidarians, such as these anemones, grow as buds sprouting from the body of the parent. When sufficiently developed, the buds break off and assume an independent existence.

QUESTION Which type of cell division gives rise to the cells of the bud's body?

Fission Followed by Regeneration Can Produce a New Individual

Many animals are capable of **regeneration,** the ability to regrow lost body parts. Regeneration is part of reproduction in species that reproduce by **fission.** Some annelids, flatworms, corals, sea jellies, and brittle stars reproduce by dividing into two or more pieces, each of which regenerates the missing parts of a complete body.

During Parthenogenesis, Eggs Develop Without Fertilization

The females of some animal species can reproduce by a process known as **parthenogenesis,** in which egg cells develop into viable offspring without being fertilized. In some species, parthenogenetically produced offspring are haploid. For example, male honeybees develop from unfertilized haploid eggs, and remain haploid; their diploid sisters develop from fertilized eggs. Some parthenogenetic fish, amphibians, and reptiles produce diploid offspring by doubling the number of chromosomes in the eggs, either before or after meiosis. In most species, all the resulting diploid offspring are female. In fact, some species of fish and some lizards, such as

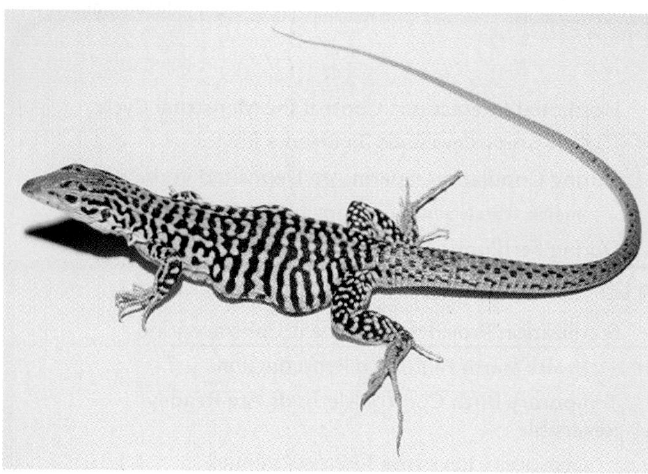

▲ **FIGURE 41-2 The whiptail lizard** This entire species consists of parthenogenetically reproducing females—and no males at all.

the whiptail of the southwestern United States and Mexico (**Fig. 41-2**), have populations consisting entirely of parthenogenetically reproducing females. Still other animals, such as some aphids, can reproduce either sexually or parthenogenetically, depending on environmental factors such as the season of the year and the availability of food (**Fig. 41-3**).

Sexual Reproduction Requires the Union of Sperm and Egg

No one is certain why sexual reproduction evolved in the first place, or why natural selection has made it by far the most common form of reproduction in animals. However, sex does have an important outcome: The genetic recombination that results from sexual reproduction creates novel genotypes—and therefore novel phenotypes—that are an important source of variation upon which natural selection may act.

In most animal species, an individual is either male or female, defined by the type of gamete that each produces. The female gonad, called the **ovary,** produces **eggs,** which are large, nonmotile haploid cells containing food reserves that provide nourishment for the **embryo,** an offspring in its early stages of development before birth or hatching. The male gonad, called the **testis** (plural, testes), produces small, motile haploid **sperm,** which have almost no cytoplasm and hence no food reserves. **Fertilization,** the union of sperm and egg, forms a diploid **zygote.**

In some animals, such as earthworms and many snails, single individuals produce both sperm and eggs. Such individuals are called **hermaphrodites.** Most hermaphrodites still engage in sex, with two individuals exchanging sperm (**Fig. 41-4**). Some hermaphrodites, however, can fertilize their own eggs. These animals, including tapeworms and some types of snails, are often relatively immobile and may find themselves isolated from other members of their species. Self-fertilization is advantageous under these circumstances.

For species with two separate sexes and hermaphrodites that cannot self-fertilize, successful reproduction requires that sperm and eggs from different animals be brought together for fertilization.

External Fertilization Occurs Outside the Parents' Bodies

In **external fertilization,** sperm and egg unite outside the bodies of the parents. Sperm and eggs are typically released into water, a process called **spawning,** and the sperm swim to reach the eggs. Because sperm and eggs are relatively short lived, spawning animals must synchronize their reproductive behaviors, both temporally (male and female spawn at the same time) and spatially (male and female spawn in the same place). Synchronization may be achieved through environmental cues, chemical signals, courtship behaviors, or some combination of these factors.

Most spawning animals rely on environmental cues to some extent. Seasonal changes in day length often stimulate the physiological changes required for breeding, which restricts

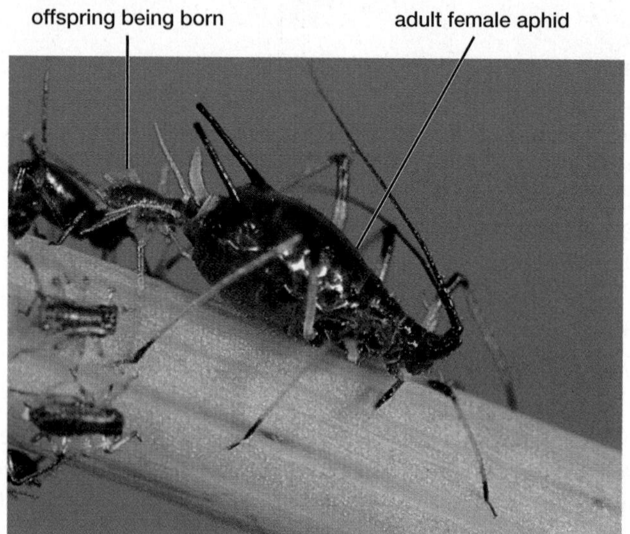

▲ **FIGURE 41-3 A female aphid gives birth** In spring and early summer, when food is abundant, aphid females reproduce parthenogenetically; in fact, the females are born pregnant! In fall, they reproduce sexually, as the females mate with males.

QUESTION Why do aphids switch to sexual reproduction in the fall?

▲ **FIGURE 41-4 Hermaphroditic earthworms exchange sperm**

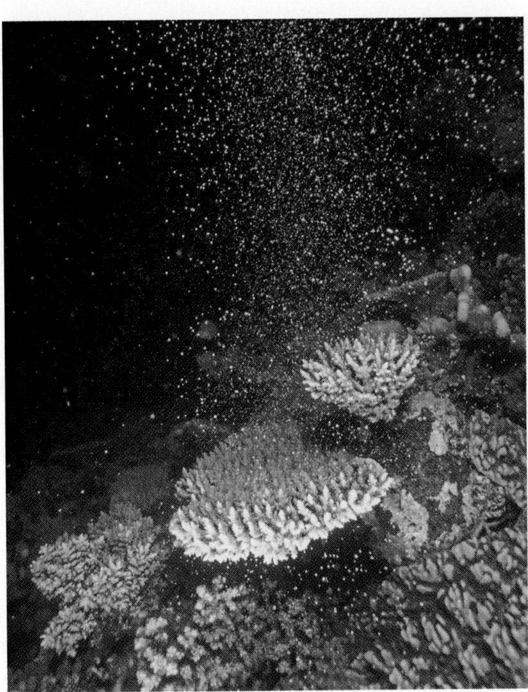

▲ **FIGURE 41-5 Environmental cues may synchronize spawning** In the Great Barrier Reef of Australia, thousands of corals spawn simultaneously, creating this "blizzard" effect. Spawning in these corals is linked to the phase of the moon.

▲ **FIGURE 41-6 Courtship rituals synchronize spawning** During spawning, male and female Siamese fighting fish perform a courtship dance that culminates with both partners releasing gametes simultaneously. The male (red with long fins) will collect the fertilized eggs in his mouth and spit them into the nest of bubbles visible above the courting fish. The male guards the eggs and the newly hatched babies in the nest until they are able to swim and feed for themselves.

QUESTION In addition to ensuring synchronized release of gametes, what other advantages do courtship rituals provide?

breeding to certain times of year. More precise synchrony, however, is required to coordinate the actual release of sperm and egg. For example, many coral species of Australia's Great Barrier Reef synchronize spawning by the phase of the moon, simultaneously releasing a blizzard of sperm and eggs into the water (**Fig. 41-5**).

Some animals communicate their sexual readiness to one another by sending visual, acoustic, or chemical signals. Often, a female simultaneously releases eggs and a chemical called a sex pheromone into the water. Nearby males, detecting the sex pheromone, immediately release their sperm. In many animals, the sperm are also lured by a chemical attractant produced by the eggs.

Synchronized timing alone does not guarantee efficient reproduction. In mobile animals, both temporal and spatial synchrony can be ensured by mating behaviors. Most fish, for example, have some form of courtship ritual in which the male and female come close together and release their gametes in the same place and at the same time (**Fig. 41-6**). Frogs and toads assume a characteristic mating pose called amplexus (**Fig. 41-7**). In shallow water near the edges of ponds and lakes, the male mounts the female and prods the sides of her abdomen. This stimulates her to extrude her eggs, which he fertilizes by releasing sperm above them.

Internal Fertilization Occurs Within the Female's Body

During **internal fertilization,** sperm are placed within the female's moist reproductive tract, where the egg is fertilized. Internal fertilization is an important adaptation to terrestrial life because sperm quickly die if they dry out. Even in aquatic environments, internal fertilization may increase the likelihood of success, because the sperm and eggs are confined in a small space rather than dispersed in a large volume of water.

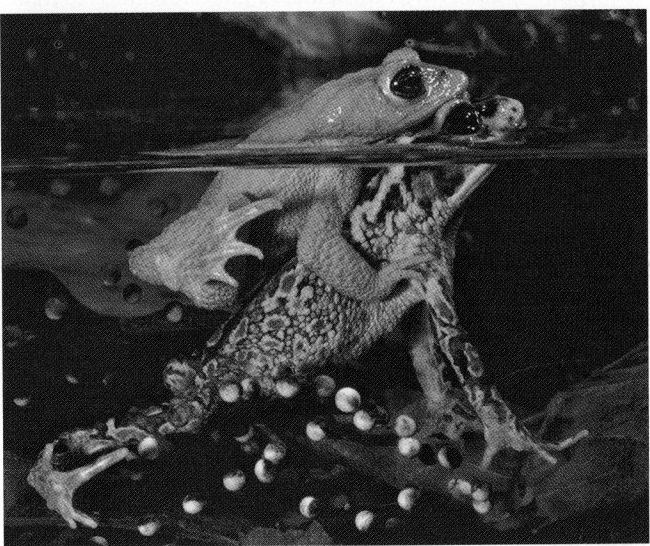

▲ **FIGURE 41-7 Golden toads mating** The smaller male clutches the female and stimulates her to release eggs. Golden toads lived in the cloud forests of Costa Rica, but have not been seen since 1989. They are probably extinct.

Internal fertilization usually occurs by **copulation,** in which the male deposits sperm directly into the female's reproductive tract (**Fig. 41-8**). In a variation of internal fertilization, males of some species package their sperm in a container called a **spermatophore** (Greek for "sperm carrier"). For example, some scorpions, grasshoppers, and salamanders do not copulate. The male simply drops a spermatophore on the ground,

▲ **FIGURE 41-8 Internal fertilization allows reproduction on land** Ladybugs mate on a blade of grass.

and if a female finds it, she fertilizes herself by inserting it into her reproductive cavity, where the sperm are released.

In most animals, neither sperm nor eggs live very long. Therefore, successful internal fertilization requires that **ovulation,** the release of a mature egg cell from the ovary of the female, occur shortly before or soon after sperm are deposited in the female's reproductive tract. Most mammals, for example, copulate when the female signals readiness to mate, which usually occurs about the same time as ovulation. In some animals, such as rabbits, courtship and mating stimulate ovulation, so new, healthy sperm and eggs are almost guaranteed to meet. In contrast, some female snails and insects have fairly short-lived eggs, but the males produce sperm that remain functional for days, weeks, or even months. After copulation, the female stores the sperm in a special sac inside her body and releases sperm whenever she produces eggs.

41.2 WHAT ARE THE STRUCTURES AND FUNCTIONS OF HUMAN REPRODUCTIVE SYSTEMS?

Humans and other mammals have separate sexes, copulate, and fertilize their eggs internally. Although most mammals reproduce only during certain seasons of the year, human reproduction is not restricted by season. Men produce sperm more or less continuously, and women ovulate about once a month.

The Ability to Reproduce Begins at Puberty

Sexual maturation in humans occurs at **puberty,** a stage of development characterized by rapid growth and the appearance of secondary sexual characteristics in both sexes. Although puberty generally begins in the early teens, it may occasionally start as early as age 8 or as late as age 15. In both sexes, brain maturation causes the hypothalamus to release **gonadotropin-releasing hormone (GnRH),** which stimulates the anterior pituitary to produce **luteinizing hormone (LH)** and **follicle-stimulating hormone (FSH).** Although LH and FSH derive their names from their functions in females, they are equally essential in males.

Case Study continued
To Breed a Rhino

Mating behaviors vary tremendously among animals. Sumatran rhinos urinate, raise their tails, vocalize, and bump each other on the head and genitals with their snouts. Sometimes this courtship gets violent enough to injure or even kill one or both animals, particularly if the female resists the male's advances and can't escape because the animals are confined in a pen.

These hormones stimulate the testes to produce the male sex hormone **testosterone** and the ovaries to produce the female sex hormone **estrogen.** In response to increases in testosterone, males develop secondary sexual characteristics: The **penis** (which deposits sperm in the female vagina) and testes enlarge; pubic, underarm, and facial hair appears; the larynx enlarges (deepening the voice); and muscular development increases. In response to increased estrogen (and other hormones that surge at puberty), females develop enlarged breasts, wider hips, pubic and underarm hair, and begin menstruating. Courtship behaviors also start to appear in both sexes.

The Male Reproductive System Includes the Testes and Accessory Structures

The male reproductive system includes the testes, which produce testosterone and sperm, glands that secrete substances that activate and nourish sperm, and tubes that store sperm and conduct them to the female reproductive system. (**Fig. 41-9** and **Table 41-1**).

Sperm Are Produced in the Testes

The testes are located in the **scrotum,** a pouch that hangs outside the main body cavity. This location keeps the testes about 1° to 6°F (about 0.5° to 3°C) cooler than the core of the body, depending on whether the man is standing (cooler) or sitting (warmer) and what type of clothing he wears. Cooler tempera-

Table 41-1 The Human Male Reproductive Tract

Structure	Function
Testis (male gonad)	Produces sperm and testosterone
Epididymis and vas deferens (ducts)	Store sperm; conduct sperm from the testes to the urethra
Urethra (duct)	Conducts semen from the vas deferens and urine from the urinary bladder to the tip of the penis
Penis	Deposits sperm in the female reproductive tract
Seminal vesicles (glands)	Secrete fluid into the semen
Prostate gland	Secretes fluids into the semen
Bulbourethral glands	Secrete fluid into the semen

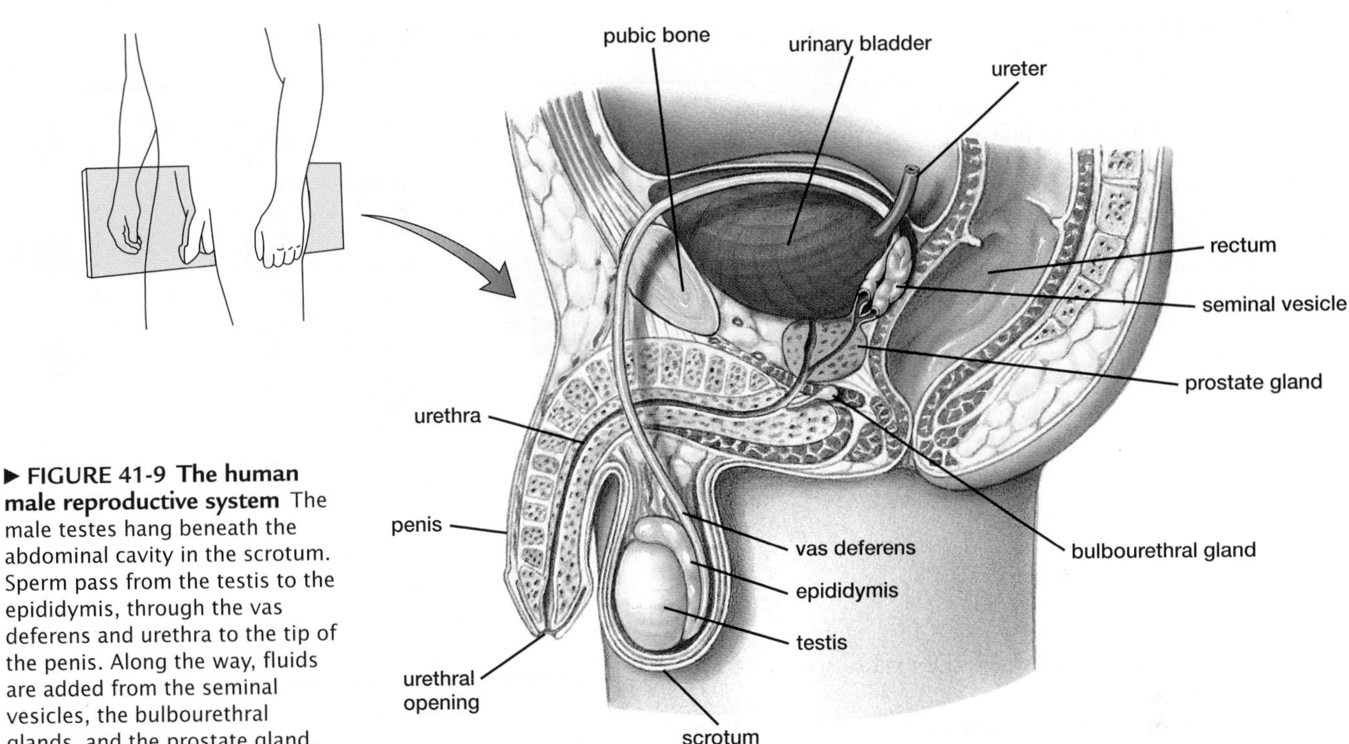

▶ **FIGURE 41-9 The human male reproductive system** The male testes hang beneath the abdominal cavity in the scrotum. Sperm pass from the testis to the epididymis, through the vas deferens and urethra to the tip of the penis. Along the way, fluids are added from the seminal vesicles, the bulbourethral glands, and the prostate gland.

tures promote sperm development. Coiled, hollow **seminiferous tubules,** in which sperm are produced, nearly fill each testis (**Fig. 41-10a**). **Interstitial cells,** which synthesize testosterone, are located in the spaces between the tubules (**Fig. 41-10b**).

Just inside the wall of each seminiferous tubule lie cells called **spermatogonia** (singular, spermatogonium), which give rise to sperm, and much larger **Sertoli cells,** which nourish the developing sperm and regulate their growth (see Fig. 41-10b). Spermatogonia are diploid cells that undergo mitotic cell division, forming two types of daughter cells (**Fig. 41-11**). One daughter cell remains a spermatogonium, ensuring a steady supply throughout a male's life. The other daughter cell becomes committed to **spermatogenesis,** the developmental processes that produce haploid sperm.

Spermatogenesis begins as the committed daughter cell differentiates into a **primary spermatocyte,** a large diploid cell that will undergo meiotic cell division (described on pp. 162–166). At the end of meiosis I, each primary spermatocyte gives rise to two haploid **secondary spermatocytes.** Each secondary spermatocyte undergoes meiosis II, producing two **spermatids.** Thus, each primary spermatocyte generates a total of four spermatids. Spermatids differentiate into sperm without further cell division. Spermatogonia, spermatocytes, and spermatids are enfolded in the Sertoli cells. As spermatogenesis proceeds, the developing sperm migrate to the central cavity of

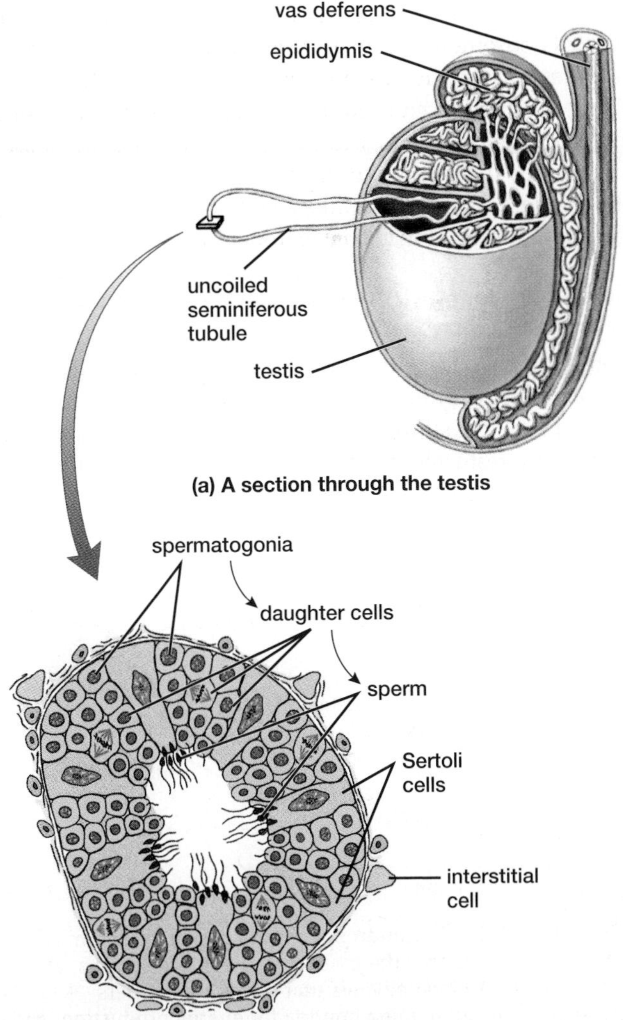

▶ **FIGURE 41-10 The anatomy of the testes (a)** A section of the testis, showing the seminiferous tubules, epididymis, and vas deferens. **(b)** The walls of the seminiferous tubules are lined with Sertoli cells and spermatogonia. As the spermatogonia divide, the daughter cells move inward. Mature sperm are freed into the central cavity, which is continuous with the inside of the epididymis. Testosterone is produced by the interstitial cells.

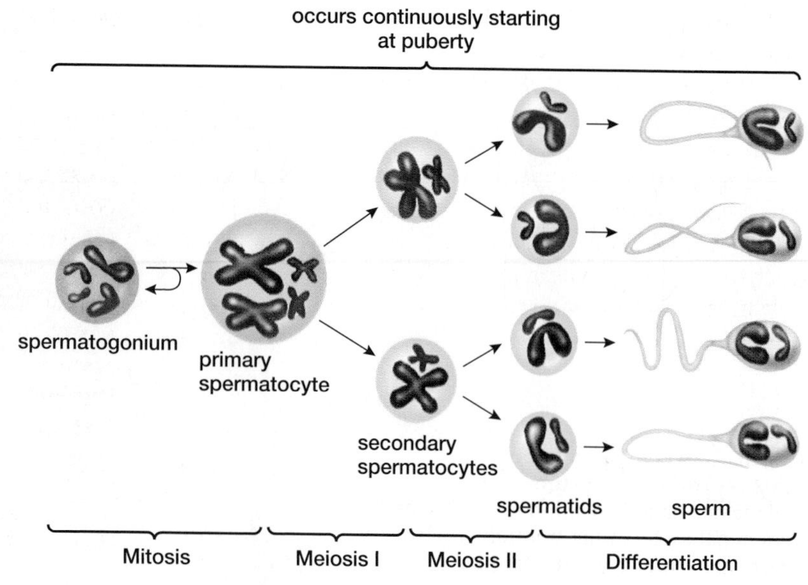

▶ **FIGURE 41-11 Spermatogenesis** A spermatogonium divides by mitotic cell division; one of its two daughter cells remains a spermatogonium (curved arrow), while the other enlarges and becomes a primary spermatocyte (straight arrow), which is committed to meiotic cell division followed by differentiation, producing haploid sperm. Although for clarity, only four chromosomes are shown, in humans, the diploid number is 46 and the haploid number is 23.

the seminiferous tubule into which the mature sperm are released (see Fig. 41-10b).

A human sperm (**Figure 41-12**) is unlike any other cell of the body. Most of the cytoplasm disappears, so the haploid nucleus nearly fills the sperm's head. Atop the nucleus lies a specialized lysosome called an **acrosome.** The acrosome contains enzymes that will dissolve protective layers around the egg, enabling the sperm to enter and fertilize it. Behind the head is the midpiece, which is packed with mitochondria. The mitochondria provide the energy needed to move the tail, which is actually a long flagellum. Whiplike movements of the tail propel the sperm through the female reproductive tract.

Hormones from the Anterior Pituitary and Testes Regulate Spermatogenesis

Spermatogenesis begins at puberty, when GnRH from the hypothalamus stimulates the anterior pituitary to produce LH and FSH. LH stimulates the interstitial cells of the testes to produce testosterone (**Fig. 41-13**). In combination with

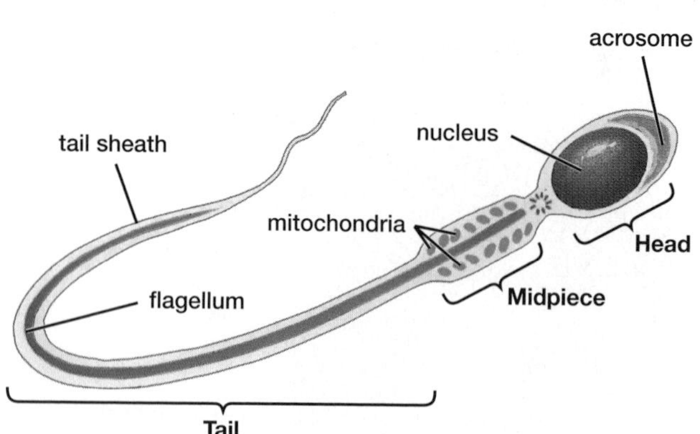

▲ **FIGURE 41-12 A human sperm cell** A mature sperm is a cell equipped with only the essentials: a haploid nucleus, an acrosome (containing enzymes that digest the barriers surrounding the egg), mitochondria for energy production, and a tail (a long flagellum) for locomotion.

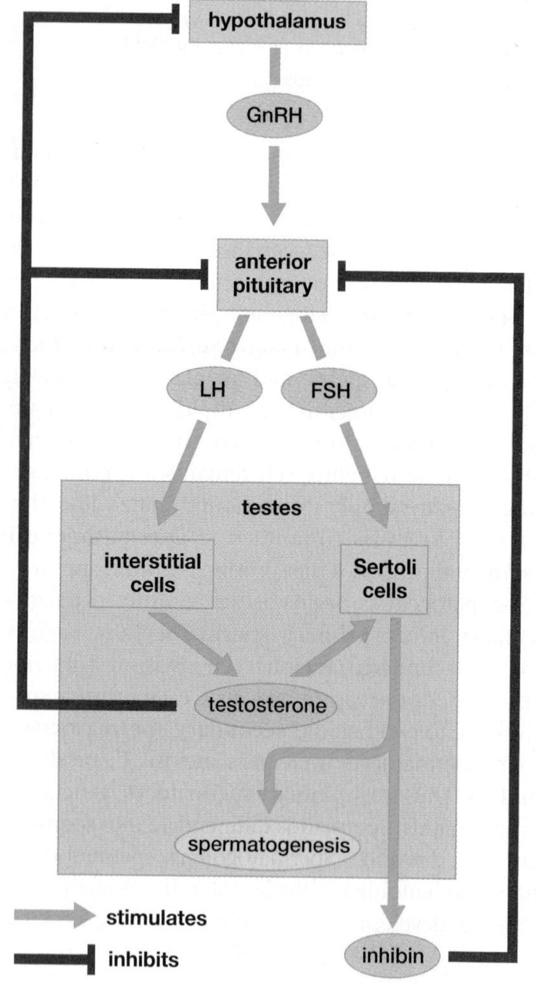

▲ **FIGURE 41-13 Hormonal control of spermatogenesis** Negative feedback interactions between the hypothalamus, anterior pituitary, and testes keep the concentration of testosterone and the production of sperm at nearly constant levels.

QUESTION Why would injections of testosterone suppress sperm production?

FSH, testosterone stimulates the Sertoli cells and promotes spermatogenesis.

Testicular function is regulated by negative feedback. Testosterone inhibits the release of GnRH, LH, and FSH, which limits further testosterone production and sperm development. The Sertoli cells, when stimulated by FSH and testosterone, secrete a hormone called inhibin, which inhibits FSH production by the anterior pituitary. This complex feedback process maintains relatively constant levels of testosterone and sperm production.

Accessory Structures Contribute to Semen and Conduct the Sperm Outside the Body

The seminiferous tubules merge to form the **epididymis,** a long, continuous, folded tube (see Figs. 41-9 and 41-10a). The epididymis leads into the **vas deferens,** a tube that carries sperm out of the scrotum. Most of the roughly hundred million sperm produced by a human male each day are stored in the vas deferens and epididymis. The vas deferens joins the **urethra,** which conducts urine out of the body during urination and sperm out of the body during ejaculation.

The fluid, called **semen,** is ejaculated from the penis, and consists of about 5% sperm, mixed with secretions from three types of glands that empty into the vas deferens or the urethra: the seminal vesicles, the prostate gland, and the bulbourethral glands (see Fig. 41-9 and Table 41-1). Fluid produced by the paired **seminal vesicles** comprises about 60% of the semen. This fluid is rich in fructose, a sugar that provides energy for the sperm. The fluid's slightly alkaline pH protects the sperm from the acidity of urine remaining in the man's urethra and from acidic secretions in the woman's vagina. It also contains prostaglandins (see p. 715), which stimulate uterine contractions that help to transport the sperm up the female reproductive tract. The **prostate gland** produces an alkaline, nutrient-rich secretion that comprises about 30% of the semen volume. The prostate fluid includes enzymes that increase the fluidity of the semen after it is released into the vagina, allowing the sperm to swim more freely. Paired **bulbourethral glands** secrete a small amount of alkaline mucus into the urethra, neutralizing remaining traces of acidic urine.

The Female Reproductive System Includes the Ovaries and Accessory Structures

The female reproductive system is almost entirely contained within the abdominal cavity (**Fig. 41-14** and **Table 41-2**). It consists of the ovaries and structures that accept sperm, conduct the sperm to the egg, and nourish the developing embryo.

Egg Production in the Ovaries Begins Before Birth

Oogenesis, the formation of egg cells, begins in the developing ovaries of a female embryo (**Fig. 41-15**). Oogenesis starts

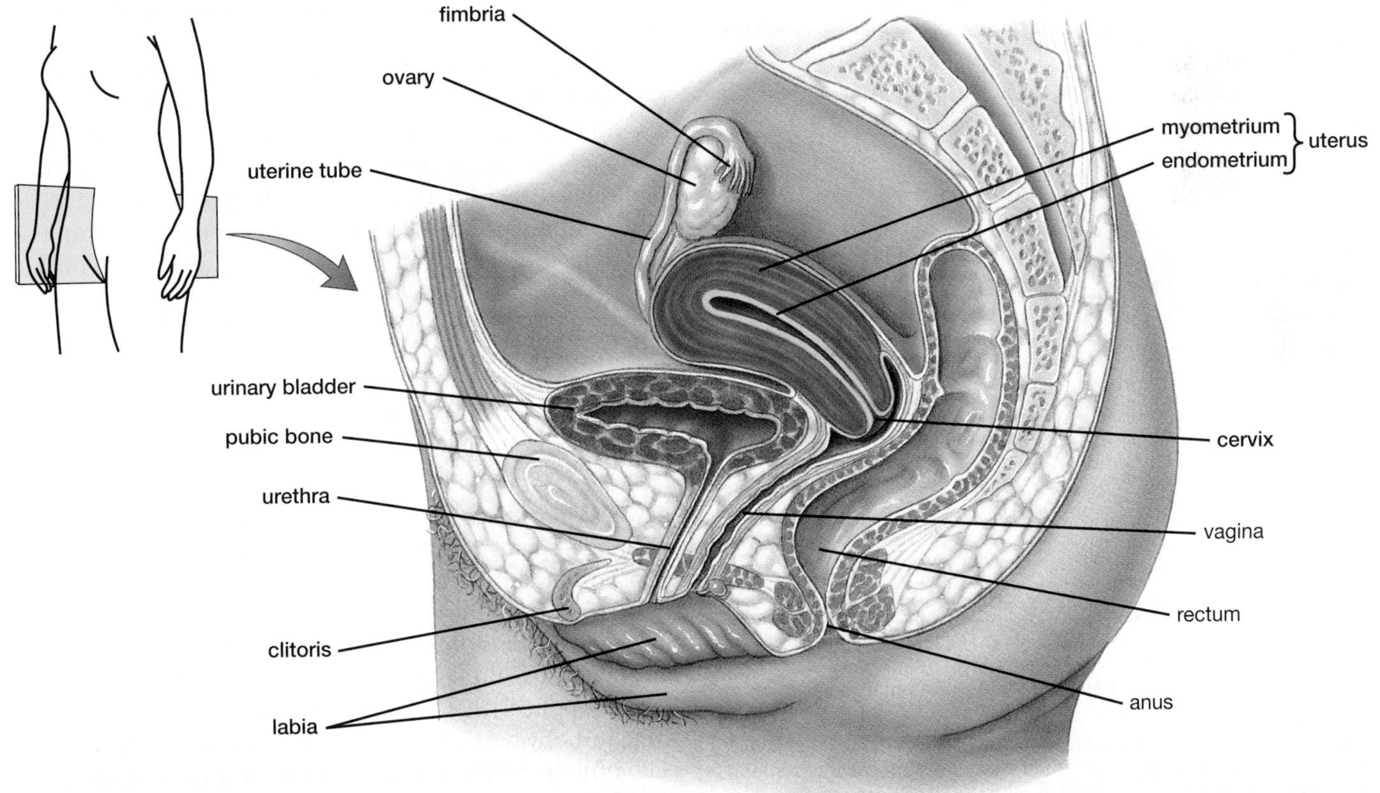

▲ **FIGURE 41-14 The human female reproductive system** Eggs are produced in the ovaries and enter the uterine tube. Sperm and egg usually meet in the uterine tube, where fertilization and very early development occur. The early embryo embeds in the lining of the uterus, where development continues. The vagina receives sperm and serves as the birth canal.

Table 41-2 The Human Female Reproductive Tract

Structure	Function
Ovary (female gonad)	Produces eggs, estrogen, and progesterone
Fimbriae (at the opening of the uterine tube)	Bear cilia that sweep the egg into the oviduct
Uterine tube	Conducts the egg to the uterus; site of fertilization
Uterus	Muscular chamber where the embryo develops
Cervix	Nearly closes off the outer end of the uterus
Vagina	Receptacle for semen; birth canal

with the formation of diploid cells called **oogonia** (singular, oogonium) as early as the 6th week of embryonic development. From about the 9th through the 20th weeks, the oogonia enlarge and differentiate, becoming **primary oocytes.** By about the 20th week, all of the primary oocytes have begun meiotic cell division, but this stops during prophase of meiosis I. None of the primary oocytes will resume meiotic cell division until puberty, probably 11 to 14 years later.

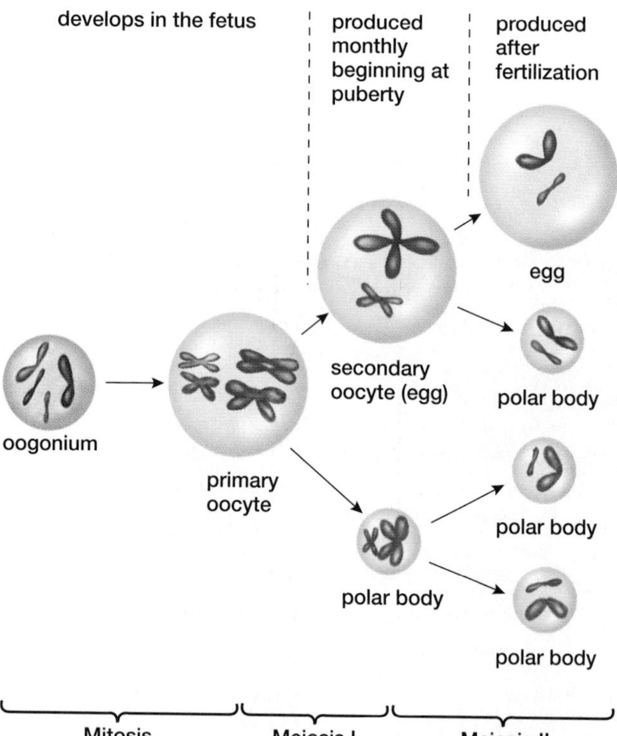

▲ **FIGURE 41-15 Oogenesis** An oogonium enlarges to form a primary oocyte. At meiosis I, almost all the cytoplasm is included in one daughter cell, the secondary oocyte. The other daughter cell is a small polar body that contains chromosomes but little cytoplasm. At meiosis II, almost all the cytoplasm of the secondary oocyte is included in the egg, and a second small polar body discards the remaining "extra" chromosomes. The first polar body may also undergo the second meiotic division. The polar bodies eventually degenerate. In humans, meiosis II does not occur unless a sperm penetrates the egg.

Therefore, a woman is born with her lifetime's supply of primary oocytes—about 1 to 2 million—and no new ones are generated later in life. Many of these die each day, but about 400,000 still remain at puberty. This is plenty, because only a few oocytes resume meiotic cell division during each month of a woman's reproductive span, from puberty to menopause at about age 50.

Surrounding each oocyte is a layer of smaller cells. Together, the oocyte and these accessory cells make up a **follicle** (**Fig. 41-16 ❶**). Roughly every month after puberty, the hormonal changes of the menstrual cycle stimulate the development of about a dozen follicles (**Fig. 41-16 ❷** and **❸**). The small follicle cells multiply, providing nourishment for the developing oocyte. In response to hormones secreted by the anterior pituitary, they also release estrogen into the bloodstream. Usually, only one follicle completely matures during each menstrual cycle. We will discuss the menstrual cycle in greater detail later in the chapter.

As a follicle develops, its primary oocyte completes meiosis I, dividing into a single **secondary oocyte** and a **polar body,** a small cell that is little more than a discarded set of chromosomes (see Fig. 41-15). Meiosis II will not occur unless the egg is fertilized. As the follicle matures, it becomes larger and fills with fluid. Ovulation occurs when the follicle erupts through the surface of the ovary, releasing its secondary oocyte (**Fig. 41-16 ❹**). For convenience, we will refer to the ovulated secondary oocyte as the egg.

Some of the follicle cells accompany the egg, but most of them remain in the ovary (**Fig. 41-16 ❺**), where they enlarge, forming a temporary gland called the **corpus luteum** (**Fig. 41-16 ❻**). The corpus luteum secretes both estrogen and a second hormone called **progesterone.** These hormones stimulate the development of the uterine lining. Both hormones also play crucial roles in controlling the menstrual cycle, as we will discuss shortly. If fertilization does not occur, the corpus luteum degenerates within a few days (**Fig. 41-16 ❼**).

Accessory Structures Include the Uterine Tubes, Uterus, and Vagina

Each ovary nestles within the open end of a **uterine tube** (also called the oviduct or Fallopian tube), which is fringed with ciliated "fingers" called fimbriae that nearly surround the ovary (see Fig. 41-14). The cilia create a current that sweeps the newly ovulated egg into the uterine tube. There, the egg may encounter sperm and be fertilized. Cilia lining the uterine tube transport the fertilized egg down the tube and into the **uterus.**

The wall of the uterus has two layers that correspond to its dual functions of nourishing the developing embryo and delivering a child. The inner lining, or **endometrium,** is richly supplied with blood vessels and with glands that secrete carbohydrates, lipids, and proteins. The endometrium will form the mother's contribution to the **placenta,** the structure that transfers oxygen, carbon dioxide, nutrients, and wastes between mother and embryo, as we will see in Chapter 42. The outer muscular wall of the uterus, called the **myometrium,** contracts during childbirth, expelling the infant out of the uterus.

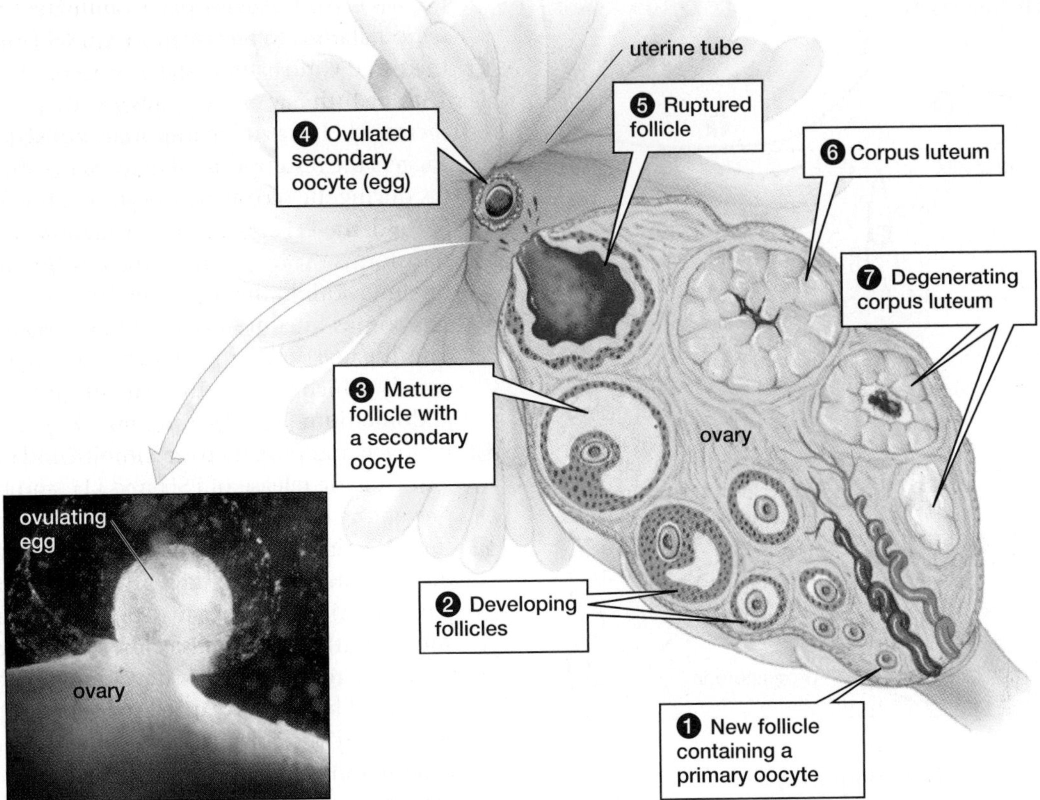

▲ **FIGURE 41-16 Follicle development** For convenience, this diagram shows all of the stages of follicle development throughout a complete menstrual cycle, going clockwise from lower right. In a real ovary, all the stages would never be present at the same time, and the follicle would not move around the ovary as it develops.

The outer end of the uterus is nearly closed off by the **cervix,** a ring of connective tissue that encircles a tiny opening. The cervix holds the developing baby in the uterus and then expands during labor, permitting passage of the child. Beyond the cervix is the **vagina,** which opens to the outside. The vaginal lining is acidic, which reduces the likelihood of infections. The vagina serves both as the receptacle for the penis and sperm during intercourse, and as the birth canal (see Fig. 41-14).

Developing follicles secrete estrogen, which stimulates the endometrium to become thicker and grow an extensive network of blood vessels and glands. After ovulation, estrogen and progesterone released by the corpus luteum further stimulate the endometrium. Thus, if an egg is fertilized, it encounters a rich environment for growth. If the egg is not fertilized, however, the corpus luteum disintegrates, estrogen and progesterone levels fall, and the overgrown endometrium disintegrates as well. The uterus then contracts (sometimes causing menstrual cramps) and squeezes out the excess endometrial tissue. This causes a flow of tissue and blood called **menstruation** (from the Latin "mensis," meaning "month").

Hormonal Interactions Control the Menstrual Cycle

The **menstrual cycle** (**Fig. 41-17**) actually consists of two cycles: (1) the ovarian cycle (**Fig. 41-17a**), in which interactions of hormones produced by the hypothalamus, the anterior pituitary gland, and the ovaries drive the development of follicles, maturation of oocytes, and conversion of the follicle cells after ovulation into the corpus luteum; and (2) the uterine cycle (**Fig. 41-17b**), in which estrogen and progesterone produced by the ovaries drive the development of the endometrium of the uterus. In the "typical" 28-day menstrual cycle, the beginning of menstruation is designated as day 1, because this is easily observed, even though the hormonal events that drive the cycle actually begin a day or two earlier. The numbers in the descriptions that follow correspond to the numbers in the illustration of the ovarian cycle in Figure 41-17a. The same number may appear in several panels when the description applies to several events.

1. The hypothalamus spontaneously releases GnRH (top panel). GnRH stimulates the anterior pituitary (second panel) to release FSH (blue line) and LH (red line). At this time the endometrium of the uterus is still being shed (step 1 in Fig. 41-17b).
2. FSH stimulates the development of several follicles within each ovary (third panel). The small follicle cells surrounding the oocyte secrete a small amount of estrogen (purple line, fourth panel). Under the combined influences of FSH and LH from the anterior pituitary and estrogen from the follicles, the follicles

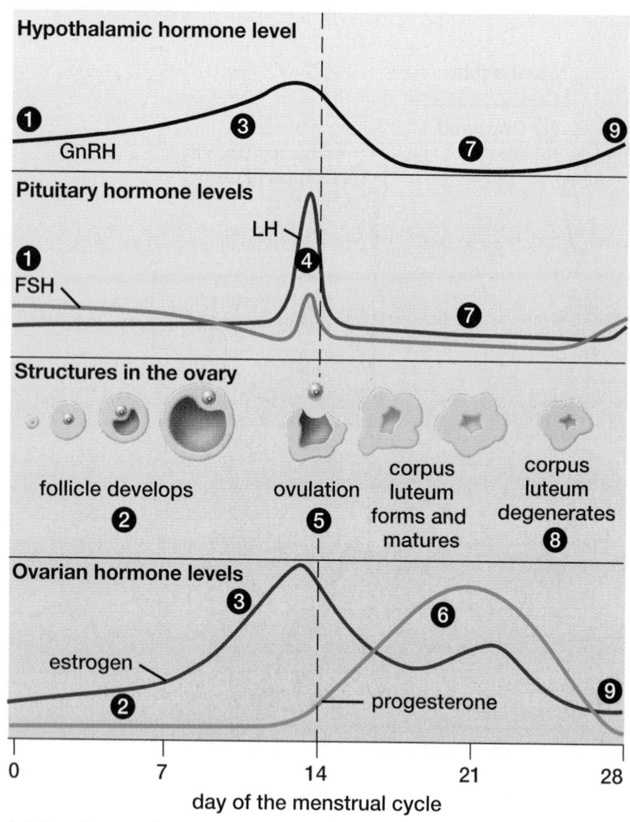

(a) Ovarian cycle

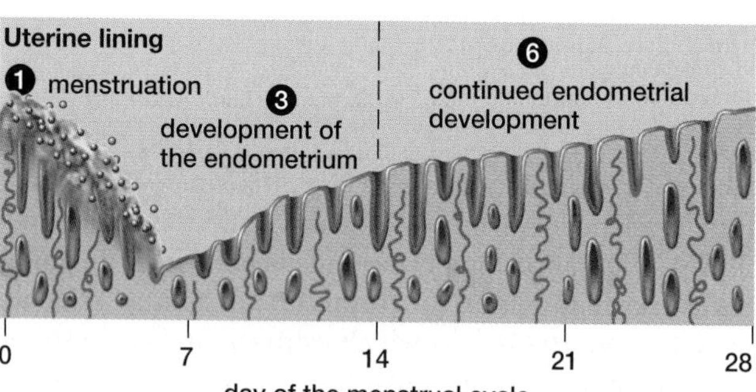

(b) Uterine cycle

▲ **FIGURE 41-17 Hormonal control of the menstrual cycle** The menstrual cycle consists of two related cycles: **(a)** the ovarian cycle and **(b)** the uterine cycle. These cycles are generated by interactions among the hormones of the hypothalamus, the anterior pituitary, and the ovaries. The circled numbers refer to the interactions described in the text.

QUESTION Most birth control pills contain estrogen and progesterone. Why would these hormones prevent pregnancy?

grow. The primary oocyte within each follicle enlarges, storing food and other substances that will be used by the fertilized egg during early development. Usually, only one follicle completes development each month.

3 The maturing follicle secretes increasing amounts of estrogen. Estrogen has three effects. First, it promotes the continued development of the follicle and of the primary oocyte within it (third panel). Second, it stimulates the growth of the endometrium (step 3 in Fig. 41-17b). Third, estrogen stimulates the hypothalamus to release more GnRH (top panel).

4 Increased GnRH stimulates a surge of LH at about the 13th or 14th day of the cycle (second panel). The increased LH has three important consequences. First, it triggers the resumption of meiosis I in the oocyte, producing the secondary oocyte and the first polar body.

5 Second, the LH surge causes ovulation, and third, it transforms the remnants of the follicle into the corpus luteum (both in the third panel).

6 The corpus luteum secretes both estrogen (purple line, fourth panel) and progesterone (green line, fourth panel), which further stimulate the growth of the endometrium (step 6 in Fig. 41-17b).

7 Estrogen and progesterone inhibit GnRH production, reducing the release of FSH and LH, and thereby preventing the development of more follicles.

8 If the egg is not fertilized, the corpus luteum starts to disintegrate about 12 days after ovulation (third panel). This occurs because the corpus luteum cannot survive without stimulation by LH (or a similar hormone released by the developing embryo, described below). Because estrogen and progesterone secreted by the corpus luteum shut down LH production, the corpus luteum actually causes its own destruction.

9 With the corpus luteum gone, estrogen and progesterone levels plummet, and most of the endometrium of the uterus disintegrates. Its blood and tissue are shed, forming the menstrual flow. The reduced levels of estrogen and progesterone no longer inhibit the hypothalamus, so the spontaneous release of GnRH resumes. GnRH stimulates the release of FSH and LH, initiating the development of a new set of follicles and restarting the cycle (back to step 1 in both the ovarian and uterine cycles).

The Menstrual Cycle Includes Both Positive and Negative Feedback

The hormones of the menstrual cycle are regulated by both positive and negative feedback. During the first half of the cycle, FSH and LH stimulate estrogen production by the follicles. High levels of estrogen then stimulate the midcycle surge of FSH and LH release. This early positive feedback causes hormone concentrations to reach high levels. During the sec-

Case Study continued

To Breed a Rhino

Female Sumatran rhinos are induced ovulators. That is, their eggs develop as in most other mammals, but the eggs are not released from the ovary unless the vagina is stimulated by copulation. This causes the female's pituitary gland to release a burst of LH, which triggers ovulation.

ond half of the cycle, negative feedback dominates. Estrogen and progesterone from the corpus luteum inhibit the release of GnRH, FSH, and LH. Without LH to keep it alive, the corpus luteum dies, shutting down progesterone production and greatly reducing estrogen production.

The Embryo Sustains Its Own Pregnancy

If fertilization occurs, the embryo prevents the negative feedback that would otherwise end the menstrual cycle. Shortly after the ball of cells formed by the dividing fertilized egg implants in the endometrium, it secretes an LH-like hormone called **chorionic gonadotropin (CG).** This hormone travels in the bloodstream to the ovary, where it keeps the corpus luteum alive. The corpus luteum continues to secrete estrogen and progesterone for a few months. These hormones continue to stimulate the development of the endometrium, nourishing the embryo and sustaining the pregnancy. Some CG is excreted in the mother's urine, where it can be detected to confirm pregnancy.

During Copulation, Sperm Are Deposited in the Vagina

The male role in copulation begins with erection of the penis. Before erection, the penis is relaxed (flaccid), because the smooth muscles surrounding the arterioles that supply it are chronically contracted, allowing little blood flow into the penis (**Fig. 41-18a**). Under psychological and physical stimulation, the nervous system releases a gas, called nitric oxide, onto the muscles of the arterioles. Nitric oxide activates an enzyme that synthesizes cyclic guanosine monophosphate (cyclic GMP), a chemical relative of cyclic adenosine monophosphate (see p. 51 and pp. 716–717). Cyclic GMP is an intracellular second messenger that causes smooth muscles to relax. Therefore, the arterioles dilate and blood flows into tissue spaces within the penis. As these tissues swell, they squeeze off the veins that drain the penis (**Fig. 41-18b**). Blood pressure increases, causing an erection. After the penis is inserted into the vagina, movements further stimulate touch receptors on the penis, triggering ejaculation. Muscles encircling the epididymis, vas deferens, and urethra contract, forcing semen out through the penis and into the vagina. A typical ejaculation consists of about 2 to 5 milliliters of semen containing about 100 million to 400 million sperm.

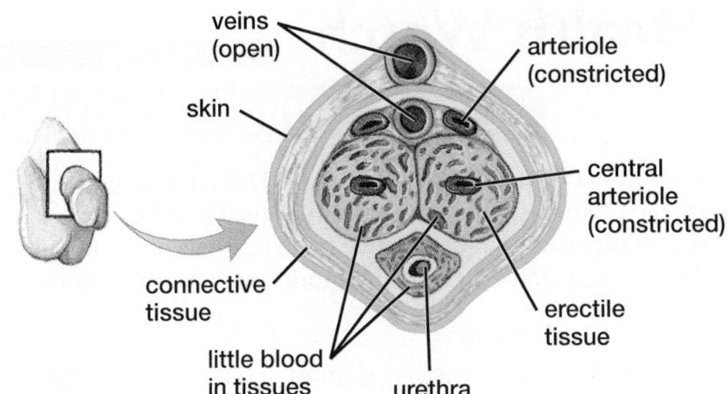

(a) Relaxed

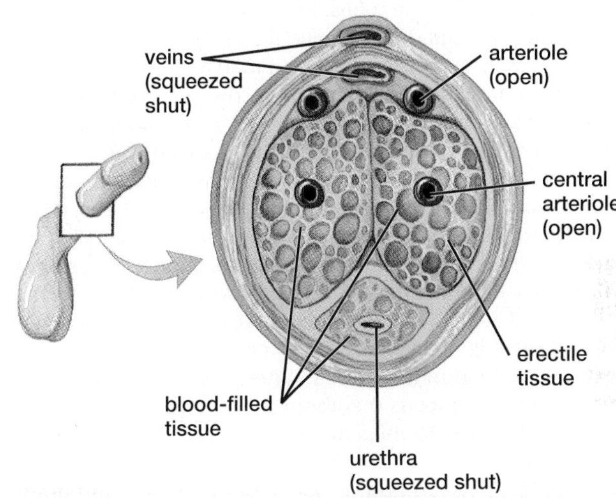

(b) Erect

▲ **FIGURE 41-18 Changes in blood flow in the penis cause erection (a)** Smooth muscles encircling the arterioles leading into the penis are usually contracted, limiting blood flow. **(b)** During sexual excitement, these muscles relax, and blood flows into spaces within the penis. The swelling penis squeezes off the veins through which blood would leave the penis, increasing its blood pressure and causing it to become elongated and firm.

In the female, sexual arousal causes increased blood flow to the vagina, to the paired folds of tissue called the **labia** (singular, labium), and to the **clitoris,** a small structure just in front of the vagina (see Fig. 41-14). Stimulation by the penis may result in orgasm, a series of rhythmic contractions of the vagina and uterus accompanied by intensely pleasurable sensations. Female orgasm is not necessary for fertilization, but the contractions of the vagina and uterus probably help to move sperm up toward the uterine tubes.

The intimate contact involved in copulation creates an ideal situation for transmitting disease organisms, as we describe in "Health Watch: Sexually Transmitted Diseases."

Have you ever wondered?

How Viagra® Works?

The concentration of cyclic GMP in smooth muscle cells is a function of how fast it is produced under nitric oxide stimulation, and how fast it is destroyed by an enzyme inside the cells. In men with erectile dysfunction, nitric oxide is still released in the penis during sexual excitement. However, not enough cyclic GMP is produced to fully relax the arteriole muscles. Therefore, an erection may not occur at all or may not be maintained for very long. Viagra® blocks the cyclic GMP-destroying enzyme, thereby allowing cyclic GMP concentrations to become higher for a longer period of time. Sustained, high levels of cyclic GMP cause the muscles of the arterioles in the penis to relax more thoroughly, so the penis becomes, and stays, erect long enough for successful copulation.

Health Watch

Sexually Transmitted Diseases

Sexually transmitted diseases (STDs) are caused by bacteria, viruses, protists, or arthropods that infect the sexual organs and reproductive tract. As their name implies, they are transmitted primarily through sexual contact.

Bacterial Infections

The U.S. Centers for Disease Control and Prevention (CDC) estimate that there are about 700,000 new cases of **gonorrhea** each year in the United States. Gonorrhea bacteria penetrate membranes lining the urethra, anus, cervix, uterus, uterine tubes, and throat. Males may experience painful urination and discharge of pus from the penis. Female symptoms are often mild and include a vaginal discharge or painful urination. Gonorrhea can be treated with antibiotics. However, because many infected individuals experience few or no symptoms, they may not seek treatment and, therefore, continue to infect their sexual partners. Gonorrhea can lead to infertility by blocking the uterine tubes with scar tissue. The bacteria also attack the eyes of newborns of infected mothers. Gonorrhea was once a major cause of blindness in infants. Today, most newborns are immediately given antibiotic eyedrops to kill the bacteria.

Syphilis is relatively rare; fewer than 50,000 cases occur each year in the United States. Syphilis bacteria penetrate the mucous membranes of the genitals, lips, anus, or breasts. Syphilis begins with a sore at the site of infection. If untreated, syphilis bacteria spread, damaging many organs including the skin, kidneys, heart, and brain, sometimes with fatal results. Syphilis is easily cured with antibiotics, but many people in the early stages of syphilis infection have mild symptoms and do not seek treatment. Syphilis can be transmitted to the embryo during pregnancy. Some infected infants are stillborn or die shortly after birth; others suffer damage to the skin, teeth, bones, liver, and central nervous system.

Chlamydia is the most common bacterial STD, with about 2.8 million new cases reported in the United States annually. Chlamydia causes inflammation of the urethra in males and of the urethra and cervix in females. In many cases, there are no obvious symptoms, so the infection goes untreated and spreads. The chlamydia bacterium can infect and block the uterine tubes, resulting in sterility. Chlamydial infection can cause eye inflammation in infants born to infected mothers and is a major cause of blindness in developing countries.

Viral Infections

Acquired immune deficiency syndrome, or **AIDS,** is caused by the human immunodeficiency virus (HIV; see pp. 708–709). It is spread primarily by sexual activity, contaminated blood and needles, and from mother to newborn. Syphilis increases the chance of HIV infection by two- to fivefold, because syphilis usually causes sores through which the HIV virus can enter the genitals. There is no cure for AIDS, but drug combinations can prolong life considerably.

Genital herpes is extremely common; the CDC estimates that 20% of American adults have had genital herpes infections. Genital herpes causes painful blisters on the genitals and surrounding skin and is transmitted primarily when blisters are present. Herpes virus remains in the body, emerging unpredictably, possibly in response to stress. Antiviral drugs can reduce the severity of outbreaks. A pregnant woman with an active case of genital herpes can transmit the virus to her developing embryo, in very rare cases causing mental or physical disability or stillbirth. Herpes can also be transmitted to babies during childbirth.

Human papillomavirus (HPV) (Fig. E41-1a) infects an estimated 50% of sexually active individuals, including 80% of women, at some time in their lives. Most experience no symptoms and never know that they have been infected. The virus may cause warts on the labia, vagina, cervix, or anus in females, and on the

During Fertilization, the Sperm and Egg Nuclei Unite

Neither sperm nor egg lives very long on its own. Sperm, under ideal conditions, may live for 2 to (rarely) 4 days inside the female reproductive tract, and an unfertilized egg remains viable for a day or so. During intercourse, the penis releases sperm into the vagina. The sperm move through the cervix, into the uterus, and finally, they enter the uterine tubes. If copulation occurs within a day or two of ovulation, the sperm may meet an egg in one of the uterine tubes.

When it leaves the ovary, the egg is surrounded by follicle cells. These cells, now called the **corona radiata,** form a barrier between the sperm and the egg (**Fig. 41-19a**). A second barrier, the jelly-like **zona pellucida** ("clear area"), lies between the corona radiata and the egg. In the uterine tube, hundreds of sperm reach the egg and encircle the corona radiata (**Fig. 41-19b**). Each sperm releases enzymes from its acrosome. These enzymes weaken both the corona radiata and the zona pellucida, allowing the sperm to wriggle through to the egg. If there aren't enough sperm, not enough enzyme is released, and none of the sperm will reach the egg. This may be the reason that natural selection has led to the ejaculation of so many sperm. Perhaps 1 in 100,000 reaches the uterine tube, and 1 in 20 of those encounters the egg, so only a few hundred of the several hundred million sperm that were ejaculated join in attacking the barriers around the egg.

When the first sperm finally contacts the egg's surface, the plasma membranes of egg and sperm fuse, and the sperm's head is drawn into the egg cytoplasm. As the sperm enters, it triggers two critical changes: First, vesicles near the surface of the egg release chemicals into the zona pellucida that reinforce it and prevent additional sperm from entering. Second, the egg

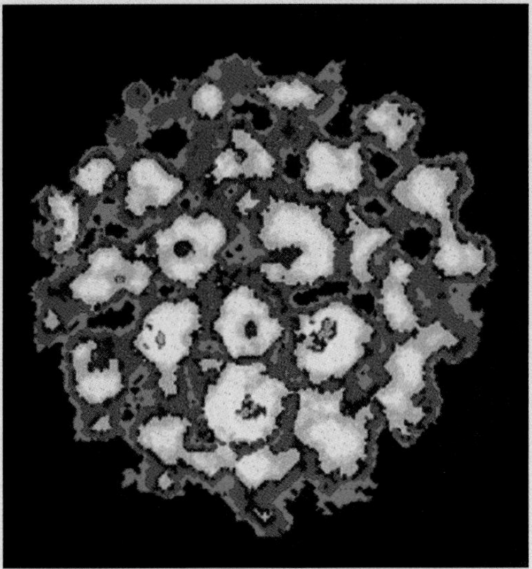

(a) The human papillomavirus

(b) Pubic lice

▲ **FIGURE E41-1 Agents of sexually transmitted diseases (a)** The DNA of some strains of the human papillomavirus becomes inserted into chromosomes of cells of the cervix. Proteins synthesized from the instructions in the viral DNA may promote uncontrolled cell division, causing cancer. **(b)** Pubic lice are called "crabs" because of their shape and the tiny claws on the ends of their legs, which they use to cling to pubic hair.

penis, scrotum, groin, or thighs in males. Certain strains of HPV cause most, and possibly all, cases of cervical cancer, which kills nearly 4,000 women each year in the United States. In 2006, the U.S. Food and Drug Administration approved a vaccine against the most common cancer-causing forms of HPV. HPV vaccines can prevent HPV infection, but cannot cure existing infections.

Protist and Arthropod Infections

Trichomoniasis is caused by a protist that colonizes the mucous membranes lining the urinary tract and genitals of both males and females. About 7.5 million Americans contract trichomoniasis each year. Symptoms may include itching or burning sensations, and, sometimes, a discharge from the penis or vagina. The protist is usually spread by intercourse, but can also be acquired from contaminated clothing and toilet articles. Trichomoniasis can be cured with the drug metronidazole. Prolonged infections can cause sterility.

Crab lice, also called pubic lice (Fig. E41-1b), are tiny insects that live and lay their eggs in pubic hair. Their mouthparts are adapted for penetrating skin and sucking blood and body fluids, a process that causes severe itching. "Crabs" can also spread infectious diseases. Crab lice can be killed by insecticides such as permethrin or pyrethrin.

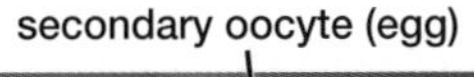

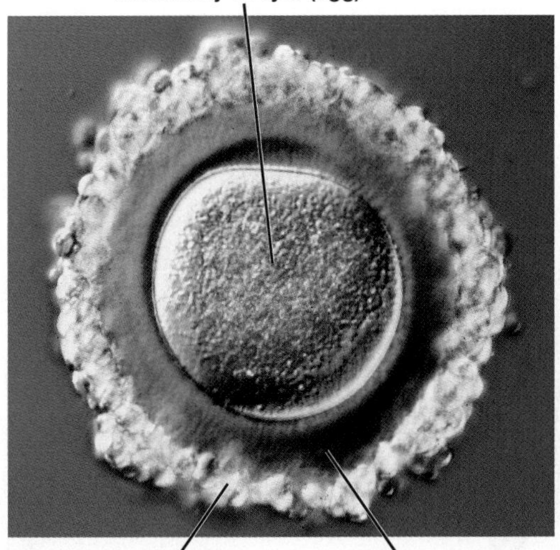

(a) An ovulated secondary oocyte

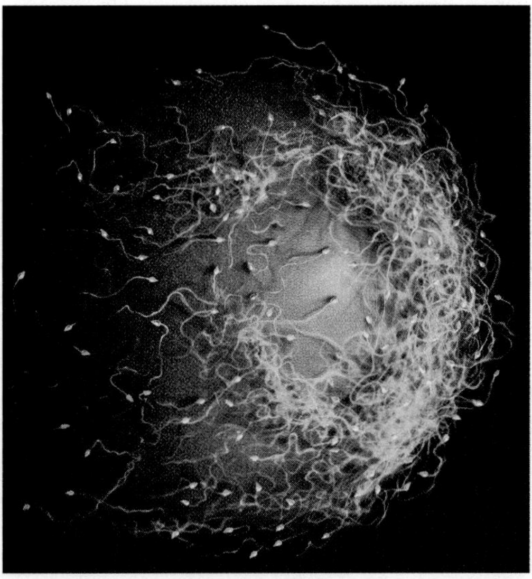

(b) Sperm surrounding an oocyte

◄ **FIGURE 41-19 The secondary oocyte and fertilization (a)** A human secondary oocyte (egg) shortly after ovulation. Sperm must digest their way through the corona radiata and the zona pellucida to reach the oocyte. **(b)** Sperm surround the oocyte, attacking the corona radiata and zona pellucida.

undergoes meiosis II. Fertilization occurs as the haploid nuclei of sperm and egg fuse, forming a diploid nucleus.

Defects in the male or female reproductive system can prevent fertilization. For example, a blocked uterine tube can prevent sperm from reaching the egg. A man with a low sperm count (fewer than 20 million sperm per milliliter of semen) may be unable to impregnate a woman through sexual intercourse because too few sperm reach the egg. Today, many couples who have difficulty conceiving a child seek help through artificial insemination or *in vitro* fertilization (see "Health Watch: High-Tech Reproduction").

41.3 HOW CAN PEOPLE LIMIT FERTILITY?

During most of human evolution, child mortality was high, and natural selection favored people who produced enough children to offset this high mortality rate. Today, however, most people do not need to have many children to ensure that a few will survive to adulthood. We nevertheless retain reproductive drives that are perhaps better suited to a more precarious existence. As a result, each week sees nearly 1.5 million new people added to our increasingly crowded planet. Controlling birth rates has become an environmental necessity. On the individual level, birth control allows people to plan their families and provide the best opportunities for themselves and their children.

Historically, limiting fertility has not been easy. In the past, women in some cultures have tried such inventive techniques as swallowing froth from the mouth of a camel or placing crocodile dung in the vagina. Since the 1970s, several effective techniques have been developed for **contraception** (the prevention of pregnancy). All forms of birth control have possible drawbacks. The choice of contraceptive method should always be made in consultation with a health professional.

Sterilization Provides Permanent Contraception

In the long run, the most foolproof and effortless method of contraception is **sterilization,** in which the pathways through which sperm or eggs must travel are interrupted (**Fig. 41-20**). In men, the vas deferens leading from each testis can be severed and the ends tied, clamped, or sealed in an operation called a vasectomy. Sperm are still produced, but they cannot reach the penis during ejaculation. The surgery is performed under a local anesthetic, in some cases requires no stitches, and has no known effects on health or sexual performance. Sperm are still produced, but they cannot leave the epididymis, where they die. Phagocytic white blood cells and the cells that make up the lining of the epididymis remove the debris.

The somewhat more complex operation of tubal ligation renders a woman infertile by clamping or cutting her uterine tubes, and tying or sealing off the cut ends. Ovulation still occurs, but sperm cannot travel to the egg, nor can the egg reach the uterus. An alternative consists of inserting tiny springlike structures into each uterine tube through the vagina and uterus. This procedure requires no incisions and only local anesthesia. The coil causes the uterine tube to form scar tissue that blocks passage of both sperm and eggs.

If a sterilized woman or man wishes to reverse the operation, a surgeon can attempt to reconnect the uterine tubes or vas deferens. About 70% to 90% of young women are able to become pregnant after their uterine tubes have been reconnected by a skilled, experienced surgeon. Pregnancy rates, however, vary dramatically according to the age of the woman (older women have a lower success rate) and the method of the original tubal ligation (higher success rates occur if the uterine tubes were clamped rather than cut or cauterized).

Reversing a vasectomy is somewhat more difficult. The vas deferens can usually be reconnected, and sperm reappear in the ejaculate in 70% to 98% of cases, depending mostly on the skill of the surgeon. However, the pregnancy rate is much lower, ranging from 30% to 75%. The longer the interval between vasectomy and reconnection, the lower the pregnancy rate, probably for multiple reasons, including the gradual development of an immune response against the man's sperm, with the result that he ejaculates damaged sperm.

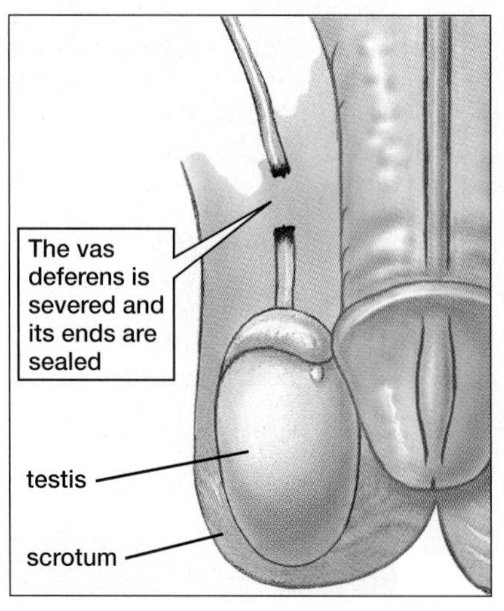
The vas deferens is severed and its ends are sealed
testis
scrotum

(a) Vasectomy

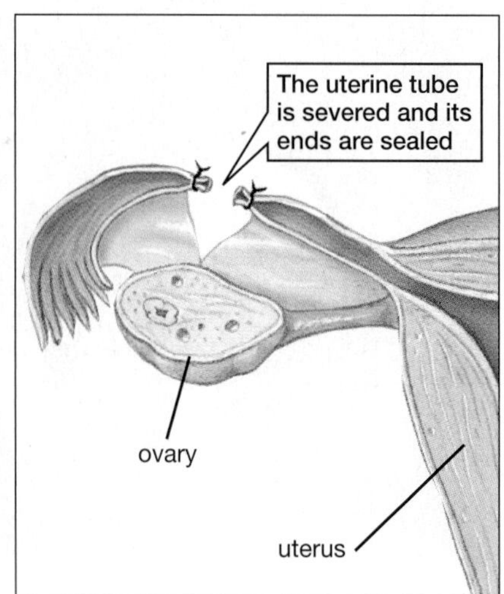
The uterine tube is severed and its ends are sealed
ovary
uterus

(b) Tubal ligation

▶ **FIGURE 41-20 Sterilization (a)** In a vasectomy, a surgeon cuts the vas deferens and ties, clamps, or seals the cut ends, preventing the passage of sperm. **(b)** In a tubal ligation, a surgeon clamps or cuts the uterine tube and ties or seals the cut ends, preventing sperm from reaching the oocyte and the oocyte from reaching the uterus.

Health Watch

High-Tech Reproduction

A surprisingly large number of couples have difficulty conceiving a child, but the ability to manipulate biological processes has led to major advances in assisted reproductive technologies. For women who do not ovulate regularly, injection of fertility drugs (which cause the release of extra FSH and LH) stimulate ovulation. However, fertility drugs often cause several eggs to be released simultaneously, with the result that the rate of multiple births in the United States has quadrupled since 1971. This is a significant drawback, because multiple births are much riskier than single births, for both the mother and her children.

Worldwide, well over 3 million people are now alive who were conceived by *in vitro* fertilization (IVF; literally, "fertilization in glass"). IVF usually starts with giving the woman drugs that stimulate follicle development. When the follicles are about to ovulate, a surgeon inserts a long needle into each ripe follicle and sucks out the oocyte. The oocytes are placed in a dish with freshly collected sperm. The fertilized eggs divide. At the two- to eight-cell stage (**Fig. E41-2**), usually two (in young women) to four embryos (in women over age 40) are gently sucked into a tube and expelled into the uterus. Transplanting more than one embryo increases the success rate, but also increases the probability of multiple births. Embryos that are not immediately implanted can be frozen for later use. (Nadya Suleman, the famous "Octomom," had six frozen embryos implanted at once; two of these presumably split after implantation, producing a total of eight embryos.) IVF costs about $10,000 to $15,000 per attempt. With an average success rate of about 30%, a typical conception via IVF costs about $30,000 to $50,000.

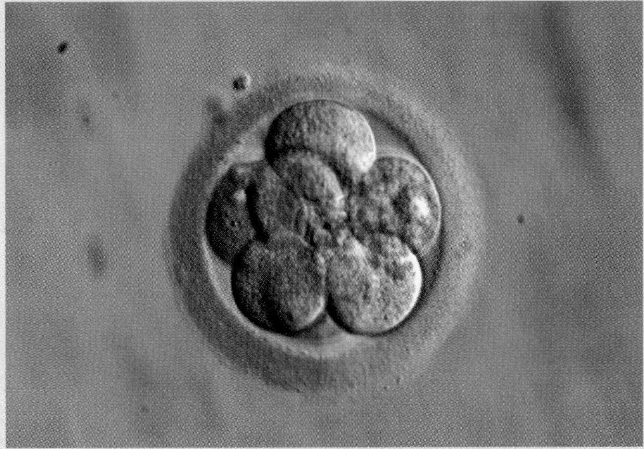

▲ **FIGURE E41-2 A human embryo ready for implantation**

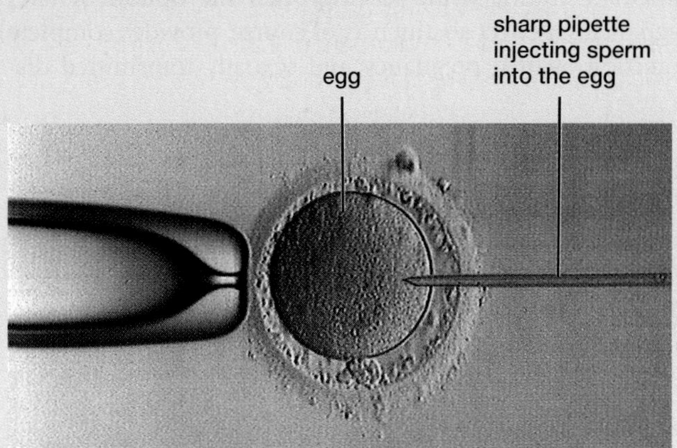

▲ **FIGURE E41-3 Intracytoplasmic sperm injection** An egg is held in place on the tip of a smooth glass pipette. A much smaller, sharp pipette injects a single sperm cell directly into the egg's cytoplasm.

Through intracytoplasmic sperm injection (ICSI), even men whose sperm are incapable of swimming or normal fertilization may be able to father children. In ICSI, immature sperm cells are extracted from the testes and injected through a tiny, sharp pipette directly into an egg's cytoplasm (**Fig. E41-3**).

Using sperm-sorting technology, parents can even change the odds of having a male or female child. This may be medically important if the parents are carriers of sex-linked disorders, but some parents are just seeking to balance their families. Because the Y chromosome is so small, sperm carrying a Y chromosome have 2.8% less DNA than X-bearing sperm, and this difference can be used to sort the sperm. So far, sperm sorting has been most successful at increasing the percentage of X sperm and thereby increasing the probability that a couple will have a girl. More recently, it has become possible to determine the gender of an IVF embryo before implantation. Although not yet routine, this method would allow a couple to select the gender of their child with absolute certainty.

As these examples show, biomedical technology has the potential to alter how humans reproduce (wealthy humans, at least). An article in the popular science magazine *New Scientist* half-jokingly suggested that in the not-too-distant future, sex might be for fun, but for reproduction, people would use IVF combined with preimplantation diagnosis of hereditary diseases, using genetic engineering to fix any disorders that are found (see p. 260). Assuming that such a future becomes possible, do you think that it's a good idea?

Temporary Birth Control Methods Are Readily Reversible

Temporary methods of birth control prevent pregnancy in the immediate future, while leaving open the option of later pregnancies. Perfect abstinence, of course, provides complete protection against pregnancy and sexually transmitted diseases (STDs). Other temporary birth control methods fall into three categories: (1) preventing ovulation, (2) preventing sperm and egg from meeting, and (3) preventing implantation in the uterus (**Table 41-3**).

Note that birth control methods do not protect against STDs, unless they prevent physical contact of the penis and vagina, as abstinence and condoms do.

Table 41-3 Methods of Temporary Contraception

Method	Description and Mechanism	Pregnancy Rate per Year[1]	Protection Against STDs
None	Regular intercourse with no contraception	About 95% for women under 25 years old, declining to about 45% by age 40	None
Hormonal Methods That Prevent Ovulation			
Birth control pill	Pill containing either synthetic estrogen and synthetic progesterone (combination pill) or progesterone only (minipill); taken daily	0.1% to 3%	None
Vaginal ring	Flexible plastic ring impregnated with synthetic estrogen and progesterone, inserted into the vagina around the cervix; replaced every 4 weeks	0.3% to 8%	None
Contraceptive patch	Skin patch containing synthetic estrogen and progesterone; replaced weekly	0.1% to 1%[2]	None
Birth control injection	Injection of synthetic progesterone that blocks ovulation; repeated at 3-month intervals	0.1% to 1%	None
Contraceptive implant	Small plastic rod impregnated with synthetic progesterone that blocks ovulation; replaced every 3 years	<0.1%	None
Methods That Prevent Sperm and Egg from Meeting			
Abstinence	No sexual activity	0%	Excellent
Condom (male)	Thin latex or polyurethane sheath placed over the penis just before intercourse, preventing sperm from entering the vagina; more effective when used in conjunction with spermicide	2% to 18%	Good
Condom (female)	Lubricated polyurethane pouch inserted into the vagina just before intercourse, preventing sperm from entering the cervix; more effective when used in conjunction with spermicide	5% to 21%	Good (probably about the same as a male condom)
Sponge	Domed disposable sponge, impregnated with spermicide, inserted in the vagina up to 24 hours before intercourse	9% to 20% (failure rates double after giving birth)	Poor
Diaphragm or cervical cap	Reusable, flexible, domed rubber-like barriers; spermicide is placed within the dome, and the diaphragm (larger) or cap (smaller) is fitted over the cervix just before intercourse	6% to 20%	Poor
Spermicide	Sperm-killing foam is placed in the vagina just before intercourse, forming a chemical barrier to sperm	6% to 26%	Probably none
Rhythm	Measuring the time since the last menstruation, or changes in body temperature and cervical mucus, to estimate the time of ovulation so that intercourse can be avoided during the fertile period	4% to 25% (rarely performed correctly)	None
Methods with Multiple Actions			
IUD (intrauterine device)	Small plastic device treated with hormones or copper and inserted through the cervix into the uterus	0.2% to 1%	None
"Morning after" pill (emergency contraception)	Concentrated dose of the hormones in birth control pills, taken within 72 hours after intercourse	5% to 40% (less effective the later the pill is taken after intercourse)	None

[1]The percentage of women becoming pregnant per year. The low numbers represent the pregnancy rate with consistent, correct contraceptive use; the higher numbers represent the pregnancy rate with more typical use that is not always consistent or correct.

[2]The patch is as effective as the pill, and more likely to be used properly; however, it may be less effective in women weighing more than 200 pounds.

Synthetic Hormones Prevent Ovulation

Birth control pills typically contain synthetic versions of estrogen and progesterone; minipills contain only progesterone. As you learned earlier in this chapter, follicle development is stimulated by FSH, and ovulation is triggered by a midcycle surge of LH. The estrogen in birth control pills prevents FSH release, so follicles do not develop. Even if one were to develop, the progesterone in the pill would suppress the surge of LH needed for ovulation. Progesterone also thickens the cervical mucus, making it more difficult for sperm to move from the vagina into the uterus, and slows down the movement of sperm and egg in the uterine tubes. Contraceptive patches, rings, injections, or implants of estrogen and progesterone are also available, as described in Table 41-3. These devices usually last from a few weeks to a few years.

Fertilization Will Not Occur If Sperm Cannot Contact the Egg

There are several ways to prevent the encounter of sperm and egg, including what are called barrier methods. Some barrier devices cover the cervix, preventing the entry of sperm into the uterus. A female condom completely lines the vagina. A male condom prevents sperm from being deposited in the vagina. Both types of condoms offer some protection against the spread of sexually transmitted diseases. Barrier devices are more effective when combined with a spermicide (sperm-killing substance), in case the barriers are breached and allow a few sperm to enter the vagina or uterus.

Less reliable techniques include the use of spermicides alone and the rhythm method (abstaining from intercourse during the ovulatory period of the menstrual cycle). The rhythm method has a high failure rate because the menstrual cycle often varies somewhat from month to month, making ovulation difficult to predict exactly. In addition, sperm can survive for several days in the female reproductive tract, so intercourse must be avoided for several days before ovulation.

Highly unreliable methods include withdrawal (removing the penis from the vagina before ejaculation) and douching (attempting to wash sperm out of the vagina before they have entered the uterus).

Some Birth Control Methods Work Through Multiple Mechanisms

Some birth control methods simultaneously prevent ovulation, hinder survival or motility of sperm, or prevent implantation of the embryo in the uterus. The intrauterine device (IUD)—a small copper or plastic loop, squiggle, or shield inserted through the cervix and into the uterus—interferes with sperm motility or survival and alters the uterine lining, lessening the likelihood of implantation. The "morning after" pill, which contains high doses of the same hormones as in birth control pills, acts primarily by delaying or preventing ovulation, but it may also interfere with the development of the corpus luteum and prevent implantation. Emergency contraceptive pills are usually effective if taken within 72 hours after intercourse, but ideally should be taken within 24 hours.

Male Birth Control Methods Are Under Development

You may have noticed that most birth control techniques are designed for use by women. There are probably three major reasons for this. First, the woman, not the man, becomes pregnant, bears the associated health risks of the pregnancy and childbirth, and usually plays a larger role in child care than men do. Second, it was relatively simple to design "use it and forget about it" birth control methods for women, including birth control pills and intrauterine devices, that do not seriously impede sexual desire or performance. This has proven to be much more difficult for men. Third, opinion surveys indicate that a significant number of men, often more than 25%, claim that they would never use hormonal birth control methods, even if they are assured that the hormones would not interfere with their sexual performance or other gender-related characteristics.

Nevertheless, researchers in several countries are working on male contraception. Among the options currently under investigation include reversible plugs injected into the vas deferens; substances, usually administered by injection, that block the action of GnRH or other pituitary hormones, thus preventing sperm from being produced (these often need to be supplemented with testosterone injections, because testosterone production would be inhibited as well; see Fig. 41-13); and vaccines that cause an immune response against sperm or other essential components of male reproduction. Note that all of these require injections or minor surgery. In addition, human trials of the hormonal methods generally give only 80% to 90% effectiveness, and often include side effects such as weight gain or adverse changes in blood cholesterol levels. Vaccines are unlikely to be 100% effective and may not be rapidly reversible.

Thus, the development and marketing of male contraceptives are hindered by difficulties of male acceptance, lower effectiveness than many female birth control methods, side effects, reversibility issues, and the fact that many require injection or minor surgery. It appears that the widespread availability of safe, effective, and convenient male contraception—except for condoms—is years away.

Abortion Removes the Embryo from the Uterus

Abortion is not a form of contraception, because it terminates rather than prevents pregnancy. It is usually performed by dilating the cervix and removing the embryo and placenta by suction. Most abortions are performed during the first 3 months of pregnancy. Alternatively, abortion can be induced during the first 7 weeks of pregnancy by the drug RU-486 (mifepristone). RU-486 blocks the actions of progesterone, which is essential to maintaining the uterine lining during pregnancy.

Abortion is a controversial procedure. Science can describe the events underlying pregnancy and embryonic development, but it cannot resolve ethical dilemmas, such as when an embryo becomes a "person" or about the relative rights of the embryo and the mother.

Case Study revisited

To Breed a Rhino

The Cincinnati Zoo's Sumatran rhinos, the cow Emi and the bull Ipuh, have now produced three calves. That means that about 1% of the world's Sumatran rhinos are Emi's offspring, thanks to Terri Roth and her staff in Cincinnati—not enough to stave off extinction, but a start. Sadly, Emi's role in rhino recovery has come to an end: In September 2009, she died in her sleep, at the age of 21. Her calves, however, may continue her legacy. Her eldest is back in Sumatra, in an international breeding program, and her second calf, Suci, still in Cincinnati, should be mature and able to breed in 2010.

Reproductive biology, particularly the discovery that Emi was an induced ovulator, was essential to breeding Emi. In hindsight, this makes sense. Sumatran rhinos are solitary wanderers, probably never very abundant. If a female were to develop a ripe follicle and immediately ovulate, she might not find a male in time, while the egg was still viable. Storing a ripe follicle in the ovary for a few days and ovulating soon after copulation increases the likelihood of fertilization.

With seriously endangered species like the Sumatran, Javan, and black rhinos, preserving genetic diversity is extremely important, so every male and every female need to have the chance to pass on their genes. To help achieve this goal, assisted reproductive technologies have become standard procedure with many endangered species. Probably the most common is artificial insemination, often using frozen sperm. In 2008, a white rhino calf was born in the Budapest Zoo. Its mother was artificially inseminated with sperm from a male at a zoo in Great Britain, whose sperm had been stored for 3 years in liquid nitrogen. In 2008, at the Smithsonian National Zoo, black-footed ferrets were born whose fathers died almost a decade ago. Several facilities, including the Audubon Nature Institute and the San Diego Zoo, maintain Frozen Zoos—liquid nitrogen storage tanks containing frozen sperm, tissue samples, and even embryos from endangered species.

Interspecies embryo transfer, in which an embryo from one species is implanted into the uterus of a related species, has proven successful with a domestic cat serving as the surrogate mother for an African wildcat, and an eland (a large, common antelope) serving as the surrogate mother for the endangered bongo antelope (**Fig. 41-21**).

Cloning endangered, or even extinct, animals is the most difficult technique, but several groups are working on it. As you may recall from Chapter 9, cloning involves replacing the nucleus of an egg with the nucleus of a body cell of a different animal. To be useful for endangered species, the egg would probably need to come from a related, common animal. African wildcat, gaur (a relative of the cow), and the recently extinct Pyrenean ibex (a relative of the goat) have all been cloned, but only the wildcat clones have lived and reproduced in their turn. Cloning rhinos probably wouldn't work, because they have no really close relatives to serve as egg donors or surrogate mothers—although perhaps the more abundant white rhino could be used as a donor and surrogate for critically endangered Sumatran, Javan, and black rhinos.

▲ **FIGURE 41-21 Interspecies embryo transfer** This eland antelope was the surrogate mother to the striped baby bongo.

BioEthics Consider This

Are these exotic methods of preserving endangered species worth it? Wouldn't it be better to preserve habitat and eliminate poaching? In Terri Roth's opinion, "We do not ever promote technology as a way to save species instead of saving habitat, because that's ridiculous." But Roth and her colleagues feel strongly that any technique is worth it to preserve species until "the wild is safe again and we can put them back out there." Other people believe that endangered species may need to be triaged, and some declared to be hopeless, so that the funds spent to keep a tiny population alive in zoos or liquid nitrogen could be freed to buy habitat or enforce hunting laws and protect species that have a better chance of avoiding extinction. What do you think?

CHAPTER REVIEW

Summary of Key Concepts

41.1 How Do Animals Reproduce?

Animals reproduce either sexually or asexually. Asexual reproduction—by budding, fission, or parthenogenesis—produces offspring that are usually genetically identical to the parent. In sexual reproduction, haploid gametes, usually from two separate parents, unite and produce an offspring that is genetically different from either parent.

During sexual reproduction, a male gamete (a small, motile sperm) fertilizes a female gamete (a large, nonmotile egg). Some species are hermaphroditic, producing both sperm and eggs, but most have separate sexes. Fertilization can occur outside the bodies of the animals (external fertilization) or inside the body of the female (internal fertilization). External fertilization must occur in water so that the sperm can swim to meet the egg. Internal fertilization generally occurs by copulation, in which the male deposits sperm directly into the female's reproductive tract.

41.2 What Are the Structures and Functions of Human Reproductive Systems?

The human male reproductive system consists of paired testes, which produce sperm and testosterone, accessory structures that conduct sperm to the female's reproductive system, and three sets of glands. The seminal vesicles, prostate gland, and bulbourethral glands secrete fluids that provide energy for the sperm, activate the sperm to swim, and provide the proper pH for sperm survival. Spermatogenesis and testosterone production are stimulated by FSH and LH, secreted by the anterior pituitary. Spermatogenesis and testosterone production are nearly continuous, beginning at puberty and lasting until death.

The human female reproductive tract consists of paired ovaries, which produce eggs as well as the hormones estrogen and progesterone, and accessory structures that conduct sperm to the egg and receive and nourish the embryo during prenatal development. Oogenesis, hormone production, and development of the lining of the uterus repeat in a monthly menstrual cycle. The cycle is controlled by hormones from the hypothalamus (GnRH), anterior pituitary (FSH and LH), and ovaries (estrogen and progesterone).

During copulation, the male ejaculates semen into the female's vagina. The sperm swim through the vagina and uterus into the uterine tube, where fertilization usually takes place. The unfertilized egg is surrounded by two barriers, the corona radiata and the zona pellucida. Enzymes released from the acrosomes in the heads of sperm digest these layers, permitting sperm to reach the egg. Only one sperm enters the egg and fertilizes it.

The ability to reproduce begins in puberty, when hypothalamic GnRH causes release of FSH and LH from the anterior pituitary. These, in turn, stimulate the gonads to produce testosterone (male) and estrogen (female). These induce secondary sexual characteristics and the development of sperm and eggs.

41.3 How Can People Limit Fertility?

Permanent contraception can be achieved by sterilization, usually by severing the vas deferens in males (vasectomy) or the uterine tubes in females (tubal ligation). Temporary contraception techniques include those that prevent ovulation by delivering estrogen and progesterone—for example, birth control pills, vaginal rings, contraceptive patches and implants, and hormone injections. Barrier methods, which prevent sperm and egg from meeting, include the diaphragm, the cervical cap, the sponge, and the condom, accompanied by spermicide. Spermicide alone is less effective. Withdrawal and douching are unreliable. The rhythm method, which has a high failure rate, requires abstinence around the time of ovulation. Intrauterine devices may block sperm movement and prevent implantation of the early embryo. Emergency contraception (the "morning after" pill) may prevent ovulation, the development of the corpus luteum, or implantation. Abortion is the removal of a developing embryo from the uterus.

Key Terms

acquired immune deficiency syndrome (AIDS) *804*
acrosome *798*
asexual reproduction *793*
budding *793*
bulbourethral gland *799*
cervix *801*
chlamydia *804*
chorionic gonadotropin (CG) *803*
clitoris *803*
cloning *810*
contraception *806*
copulation *795*
corona radiata *804*
corpus luteum *800*
crab lice *805*
egg *794*
embryo *794*
endometrium *800*
epididymis *799*
estrogen *796*
external fertilization *794*
fertilization *794*
fission *793*
follicle *800*
follicle-stimulating hormone (FSH) *796*
genital herpes *804*
gonad *793*
gonadotropin-releasing hormone (GnRH) *796*
gonorrhea *804*
hermaphrodite *794*
human papillomavirus (HPV) *804*
internal fertilization *795*
interstitial cell *797*
labium (plural, **labia**) *803*
luteinizing hormone (LH) *796*
menstrual cycle *801*
menstruation *801*
myometrium *800*
oogenesis *799*
oogonium (plural, **oogonia**) *800*
ovary *794*
ovulation *796*
parthenogenesis *793*
penis *796*
placenta *800*
polar body *800*
primary oocyte *800*
primary spermatocyte *797*
progesterone *800*
prostate gland *799*
puberty *796*
regeneration *793*
scrotum *796*
secondary oocyte *800*
secondary spermatocyte *797*
semen *799*
seminal vesicle *799*
seminiferous tubule *797*
Sertoli cell *797*
sexual reproduction *793*
sexually transmitted disease (STD) *804*
spawning *794*
sperm *794*
spermatid *797*
spermatogenesis *797*
spermatogonium (plural, **spermatogonia**) *797*
spermatophore *795*
sterilization *806*
syphilis *804*
testis (plural, **testes**) *794*
testosterone *796*
trichomoniasis *805*
urethra *799*
uterine tube *800*
uterus *800*
vagina *801*
vas deferens *799*
zona pellucida *804*
zygote *794*

Thinking Through the Concepts

Fill-in-the-Blank

1. Reproduction by a single animal, without the need for sperm fertilizing an egg, is called __________ reproduction. A(n) __________ is a new individual that grows on the body of the adult and eventually breaks off to become independent. During __________, an adult animal splits in two, with the two halves regenerating the rest of a complete organism.
2. In mammals, the male gonad is called the __________. It produces both sperm and the sex hormone __________. Within the male gonad, spermatogenesis occurs within the hollow, coiled structures called __________.
3. A sperm consists of three regions, the head, midpiece, and tail. The head contains very little cytoplasm, and consists mostly of the __________, in which the chromosomes are found, and a specialized lysosome, the __________. Organelles in the midpiece, the __________, provide energy for movement of the tail.

4. Sperm are stored in the __________ and __________ until ejaculation, when the sperm, mixed with fluids from three glands, the __________, __________, and __________, flow through the __________ to the tip of the penis.
5. In mammals, the female gonad is the __________. It produces eggs and two hormones, __________ and __________. Although usually referred to as an "egg," female mammals actually ovulate a cell called a __________, which has completed only meiosis I. Meiosis I also produces a much smaller cell, the __________, that serves mainly as a way to discard chromosomes. The "egg" is swept up by ciliated structures, called fimbriae, that form the entrance to the __________. Fertilization usually occurs in this structure. The fertilized egg then implants in the __________, where it will develop until birth.
6. Oocytes develop in a multicellular structure, the follicle. After ovulation, most of the follicle cells remain in the ovary, forming a temporary endocrine "gland," the __________. This "gland" disintegrates a few days after ovulation unless stimulated by a hormone, __________, secreted by the developing embryo.

Review Questions

1. List the advantages and disadvantages of asexual reproduction and sexual reproduction, including an example of an animal that uses each type.
2. Compare the structures of the egg and sperm. What structural modifications do sperm have that facilitate movement, energy use, and gaining access to the egg?
3. What is the role of the corpus luteum in a menstrual cycle? In early pregnancy? What determines its survival after ovulation?
4. Construct a chart of common sexually transmitted diseases. List the disease's name, cause (organism or virus), symptoms, and treatment.
5. List the structures, in order, through which a sperm passes, starting with the seminiferous tubules of the testis and ending in the uterine tube of the female.
6. Name the three accessory glands of the male reproductive tract. What are the functions of the secretions they produce?
7. Describe the interactions among hormones secreted by the pituitary gland and ovaries that produce the menstrual cycle.

Applying the Concepts

1. Discuss the most effective or appropriate method of birth control for each of the following couples: Couple A, who have intercourse three times a week but never want to have children; Couple B, who have intercourse once a month and may want to have children someday; and Couple C, who have intercourse three times a week and want to have children someday.
2. Would a hypothetical contraceptive drug that blocked receptors for FSH and LH be useful in males? How would it work? What side effects might it have?
3. **BioEthics** Think of all the ways a couple can produce a child, including *in vitro* fertilization using the couple's eggs and sperm, *in vitro* fertilization using a donor's sperm or egg, and insemination of a surrogate mother with sperm from the couple's husband. Imagine some more. Do these options present ethical issues for you? What legal issues might possibly arise? What medical issues?

MB® *Go to www.masteringbiology.com for practice quizzes, activities, eText, videos, current events, and more.*

Animal Development

Chapter 42

Case Study

The Faces of Fetal Alcohol Syndrome

"THE GUILT IS TREMENDOUS . . . I did it again and again. . . . I don't know how to tell them. It was something I could have prevented." Debbie, the young mother pictured here with her youngest child, Sabrina (upper photo), has had seven children. When she drank during pregnancy, so did her unborn children. Her daughter, Cory, has been diagnosed with **fetal alcohol syndrome (FAS)**, the most serious type of alcohol damage. At age 3, Cory was hyperactive and talked like a 1-year-old. Doctors believe that Cory's sister, Sabrina, is almost certainly a victim as well. She bears the features characteristic of FAS, including small eyes, a short nose, and a small head. At 7 months, she was weak, began having seizures, and could not eat solid food because she was unable to close her upper lip around a spoon.

John (lower photo) is another victim of FAS: His mother drank while she was pregnant with him, and she was drunk when she delivered him. John was fortunate, however, to have been adopted by a truly remarkable woman, Teresa Kellerman, who has become a leading parent advocate for children who have been damaged by alcohol. "Without intervention they end up homeless, jobless, addicted, arrested, pregnant or getting someone pregnant, living on the streets or dead, so many of them," says Kellerman, who founded the Fetal Alcohol Syndrome Community Resource Center in Tucson, Arizona.

Debbie entered rehabilitation and has resolved to stay sober and be a good parent to Sabrina and Cory. But even with the best parenting, John, Cory, Sabrina, and the thousands of other children born each year with FAS are unlikely ever to live without supervision, because the damage to their brains is irreparable. Tens of thousands of others with milder alcohol damage may become functional but will never reach their full potential.

As you study human development, think about how alcohol reaches a developing child when a pregnant woman drinks. How is the embryo nourished by its mother? To what extent is the embryo protected from environmental insults, whether alcohol and other drugs, toxins in cigarette smoke, or industrial pollutants? Are some stages of development more sensitive than others?

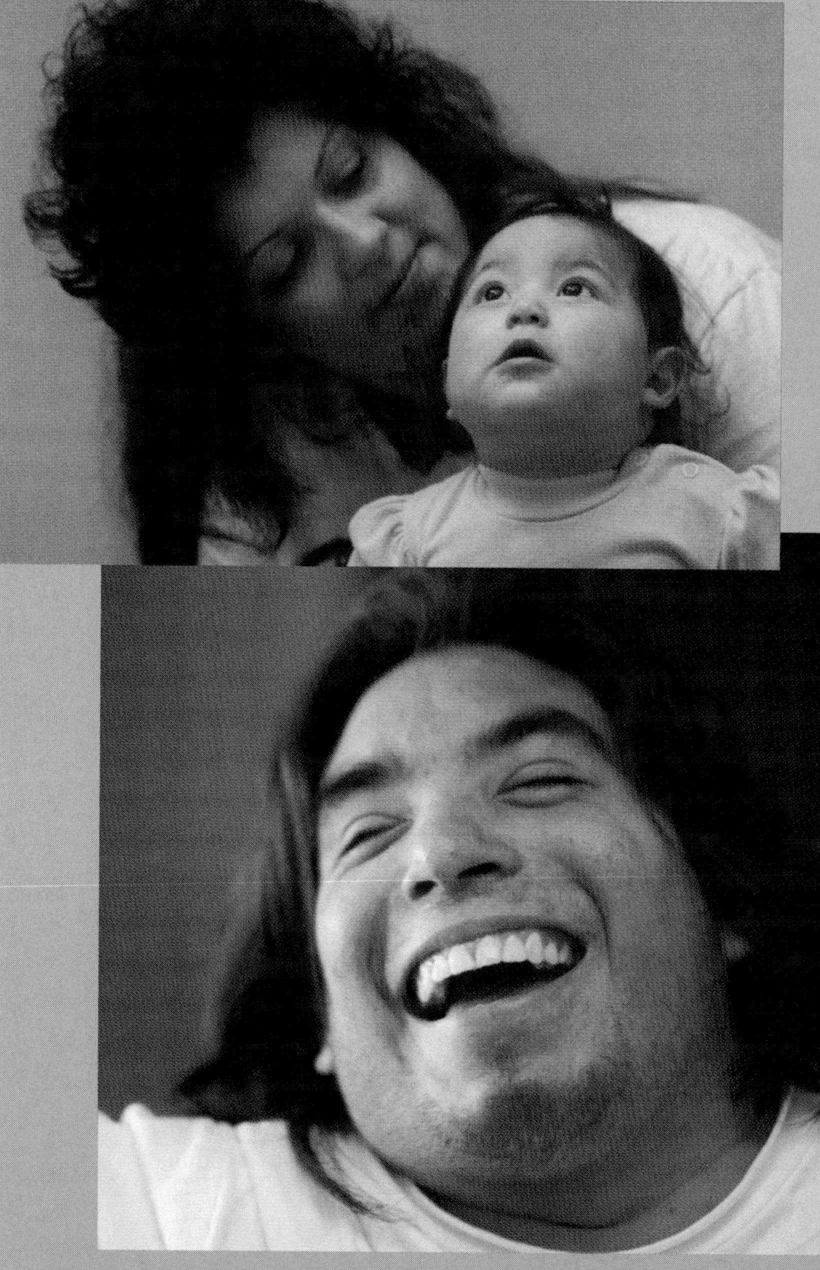

▲John, who struggles with the lasting effects of fetal alcohol syndrome, helps his adoptive mother Teresa Kellerman educate people about the dangers of drinking during pregnancy. (Upper photo) Debbie regrets drinking while she was pregnant with her daughter, Sabrina.

42.1 WHAT ARE THE PRINCIPLES OF ANIMAL DEVELOPMENT?

Development is the process by which a multicellular organism grows and increases in organization and complexity. Development is usually considered to begin with a fertilized egg and end with a sexually mature adult. Three principal mechanisms contribute to development. First, individual cells multiply. Second, some of their daughter cells **differentiate,** or specialize in both structure and function, for example, as nerve cells or muscle fibers. Third, as they differentiate, groups of cells move about and become organized into multicellular structures, such as a brain or a biceps muscle.

All of the cells of an individual animal's body are genetically identical. How, then, can they differentiate into remarkably different structures with distinct functions? As we will see, the solution is to use different genes in different places in an animal's body, and at different times during an animal's life.

In this chapter, we will focus on the early developmental stages, from fertilized egg through birth. We will describe the major structures that appear during early vertebrate development and the processes that control gene expression at this time. We begin with a brief survey of the variety of ways in which animals develop from a fertilized egg to an adult.

42.2 HOW DO INDIRECT AND DIRECT DEVELOPMENT DIFFER?

Human babies, puppies, and kittens obviously develop as they grow up. However, in most respects, baby mammals and reptiles (including birds) are miniature versions of the adults of their species, undergoing a process called **direct development.** The majority of animal species, however, undergo **indirect development,** in which the newborn has a very different body structure than the adult.

During Indirect Development, Animals Undergo a Radical Change in Body Form

Indirect development occurs in amphibians, such as frogs and toads, and in most invertebrates. The females of animals that undergo indirect development typically produce huge numbers of eggs, each containing a small amount of food reserve called **yolk.** The yolk nourishes the developing embryo until it hatches into a small, sexually immature, feeding stage called a **larva** (plural, larvae; **Fig. 42-1**). Because the parents usually provide these vulnerable offspring with neither food nor protection from predators, most die in their larval stage. After feeding for a few weeks to several years, the handful of survivors undergo a revolution in body form, or **metamorphosis,** and become sexually mature adults.

Most larvae not only look very different from the adults, but also play different roles in their ecosystems. For instance, most adult butterflies sip nectar from flowers and unintentionally pollinate the flowers in return. Their caterpillar larvae munch on leaves, often of specific host plants (see Fig. 42-1).

Although we tend to regard the adult form as the "real animal" and the larval stage as "preparatory," most of the life span of some animals, especially insects, is spent in the larval form. An extreme example is provided by the North American

(a) Caterpillar (larva)

(b) Butterfly (adult)

◀ **FIGURE 42-1 Indirect development** The larvae of animals with indirect development are very different from the adult form, in structure, behaviors, and ecological niches. **(a)** Caterpillars of butterflies, such as the tiger swallowtail caterpillar shown here, feed on leaves, usually of a limited number of host plant species. **(b)** The principal food of most adult butterflies, including tiger swallowtails, is flower nectar.

periodical cicadas. Depending on the species, periodical cicadas spend 12 or 16 years as underground larvae, sucking juices from plant roots, and only 4 to 6 weeks as adults, mostly mating and laying eggs.

Newborn Animals That Undergo Direct Development Resemble Miniature Adults

Other animals, including some snails and fish, and all mammals and reptiles (including birds), undergo direct development, in which the newborn animal closely resembles the adult (**Fig. 42-2**). As the young animal matures, it may grow much bigger, but it does not fundamentally change its body form.

Juveniles of directly developing species are typically much larger than larvae, so they need much more nourishment before emerging into the world. Two strategies have evolved that meet the embryo's food requirement. Birds and most other reptiles, and many fish, produce eggs that contain relatively large amounts of yolk. Mammals, some snakes, and a few fish have relatively little yolk in their eggs. Instead, the embryos are nourished within the mother's body. Providing food for directly developing embryos places great demands on the mother. Many of these offspring, such as those of birds and mammals, require additional care and feeding after birth, placing additional demands on one, and often both, parents. Relatively few offspring are produced, but a higher proportion reach adulthood, because the parents devote more resources to each individual.

42.3 HOW DOES ANIMAL DEVELOPMENT PROCEED?

Most of the mechanisms of development—the control of gene expression to allow differentiation of individual cells and entire body parts—are fundamentally similar in vertebrates and invertebrates, and in animals with indirect or direct development. Here, we will focus on the structures formed during vertebrate development.

Amphibians such as frogs, newts, and salamanders have long been favorite subjects for the study of development, because they can be induced to breed at any time of year, they have large, easily manipulated eggs and embryos, and the embryos develop in water, not inside their mothers. Most aspects of amphibian development are quite comparable to the development of other vertebrates, making most findings in amphibians applicable to other vertebrates, including humans. The descriptions and illustrations that follow will focus on amphibian development.

Cleavage of the Zygote Begins Development

The formation of an embryo begins with **cleavage,** a series of mitotic cell divisions of the fertilized egg, or **zygote** (**Fig. 42-3a**). The zygote is a very large cell. During cleavage, there is little or no cell growth between cell divisions, so as cleavage progresses, the available cytoplasm is split up into ever smaller cells, whose size gradually approaches that of cells in the adult. Eventually, a solid ball of small cells, the **morula** (**Fig. 42-3b**), is formed. As cleavage continues, a cavity opens within the morula, and the cells become the outer covering of a hollow structure called the **blastula** (**Fig. 42-3c**).

The details of cleavage differ by species and are partly determined by the amount of yolk, which hinders cytokinesis (cytoplasmic division). Eggs with extremely large yolks, such as a hen's egg, don't divide all the way through. Nevertheless, a hollow blastula (or its equivalent) is always produced. In birds and other reptiles, the blastula is flattened on top of the yolk.

Gastrulation Forms Three Tissue Layers

In amphibians and many other animals, the location of cells on the surface of the blastula forecasts their ultimate developmental fate in the adult. We have colored the cells of the blastula in

(a) Seahorses

(b) Snails

(c) Polar bears

◀ **FIGURE 42-2 Direct development** The offspring of animals with direct development closely resemble their parents from the moment of birth. **(a)** A male seahorse gives birth. The female seahorse deposits her eggs in his pouch, where he releases sperm to fertilize them. The fertilized eggs develop in the pouch for a few weeks, when vigorous muscular contractions of the pouch squirt out as many as 200 young. **(b)** Many land and freshwater snails hatch from small, yolk-rich eggs. **(c)** Mammalian mothers nourish their developing young within their bodies before birth, and with milk from their mammary glands after birth.

Figure 42-3c blue, yellow, and pink. These colors indicate the parts of the adult body that these cells are destined to produce during development (see below). In the blastula, all of these cells are on the surface (the lump of yellow cells on the bottom of the blastula is mostly yolk, and will gradually disappear during development). However, most of the structures that they will form are on the inside of the animal. These cells move to their proper destinations during the next stage of development, called **gastrulation** (literally, "producing the stomach").

Gastrulation begins when a dimple called the **blastopore** forms on one side of the blastula (see Fig. 42-3c). Surface cells then migrate through the blastopore in a continuous sheet, much as if you punched in an underinflated basketball. The resulting indentation enlarges to form a cavity that will eventually become the digestive tract (**Fig. 42-3d**).

The migrating cells form three tissue layers in the embryo, which is now called a **gastrula** (**Fig. 42-3e**). The cells that move through the blastopore to line the future digestive tract (yellow) are called **endoderm** (meaning "inner skin"); in addition to lining the digestive tract, endoderm also forms the lining of the respiratory tract and the liver and pancreas. The cells remaining on the outside (blue) are called **ectoderm** ("outer skin"); these mostly form surface structures such as the skin, hair, and nails, and the nervous system. Cells that migrate between the endoderm and ectoderm form the third layer (pink), the **mesoderm** ("middle skin"), which forms structures between these two, including muscles, the skeleton, and the circulatory system. **Table 42-1** lists the major structures produced from each layer.

Adult Structures Develop During Organogenesis

Organogenesis, the development of adult structures from the three embryonic tissue layers, proceeds by two major processes. First, a series of "master switch" genes turn on and

Table 42-1 Derivation of Adult Tissues from Embryonic Cell Layers

Embryonic Layer	Adult Tissue
Ectoderm	Epidermis of the skin; hair; lining of the mouth and nose; glands of the skin; nervous system
Mesoderm	Dermis of the skin; muscle, skeleton; circulatory system; gonads; kidneys; outer layers of the digestive and respiratory tracts
Endoderm	Lining of the digestive and respiratory tracts; liver; pancreas

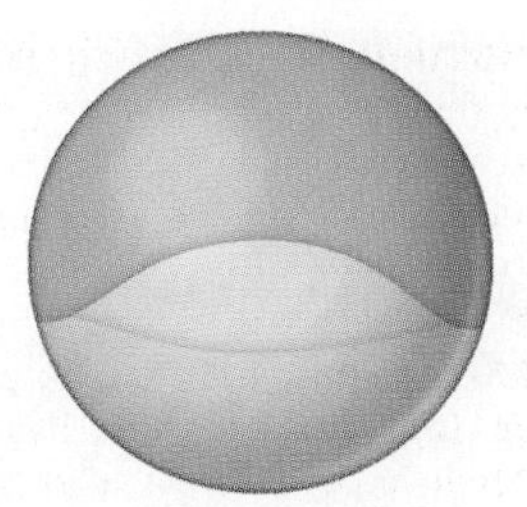
(a) Zygote

◀ **FIGURE 42-3 Gastrulation in the frog (a)** A frog zygote, or fertilized egg, is a huge cell, as large as 0.1 inch (3 millimeters) in diameter in some species. **(b)** The zygote divides to form a solid ball of cells, the morula. The morula is about the same size as the zygote, although its individual cells are smaller. **(c)** The frog blastula is a hollow ball of cells with a lump of yolk-containing cells at one end. The blastula is still about the same size as the zygote. During normal development, the positions of the cells in the frog blastula foretell their future developmental fate (see the color-code legend). **(d)** During gastrulation, surface cells migrate into the blastula to form the endoderm and mesoderm layers of the gastrula. The cells remaining on the surface will form ectoderm. **(e)** The resulting three-layered embryo is now termed a gastrula.

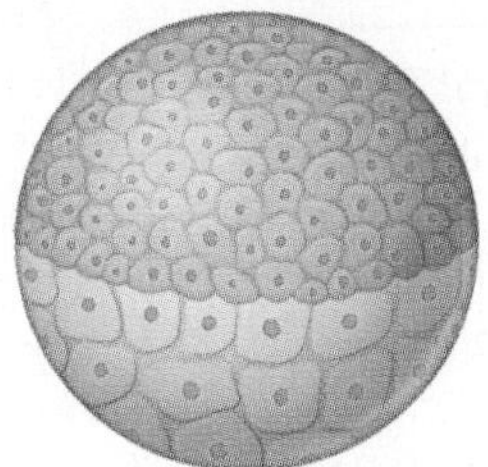
(b) Cleavage of the zygote forms a morula

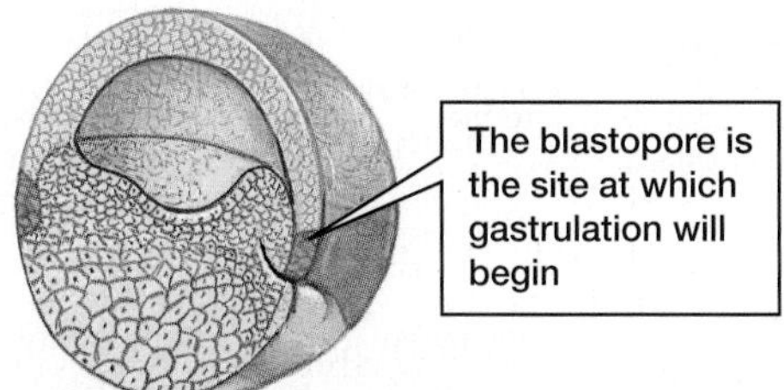
The blastopore is the site at which gastrulation will begin
(c) The blastula just before gastrulation

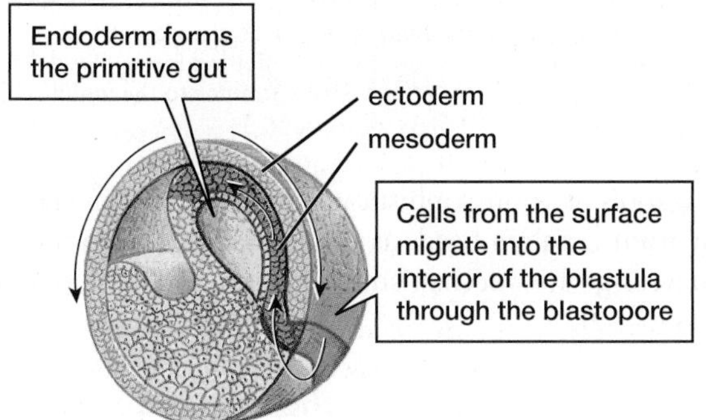
Endoderm forms the primitive gut
ectoderm
mesoderm
Cells from the surface migrate into the interior of the blastula through the blastopore
(d) Cells migrate at the start of gastrulation

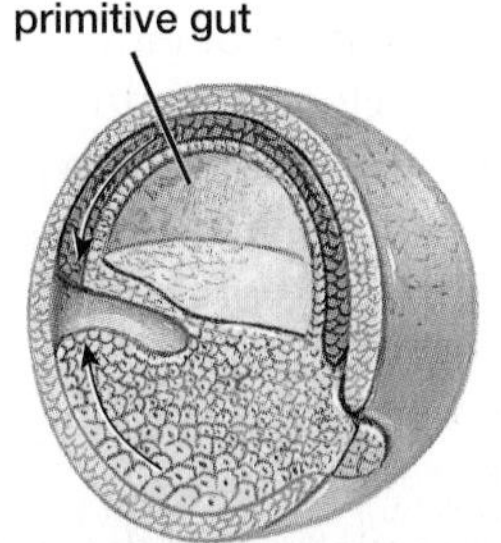
primitive gut
(e) A three-layered gastrula has formed

ectoderm mesoderm endoderm

off, each controlling the transcription and translation of the many genes involved in producing, say, an arm or a backbone. We will return to these master genes in section 42.4.

Second, organogenesis prunes away superfluous cells, much as Michelangelo chiseled away "extra" marble, leaving behind his statue of David. In development, this sculpting requires the death of excess cells. Embryonic vertebrates, for example, have far more motor neurons than adults do. However, embryonic motor neurons are programmed to die unless they form a synapse with a skeletal muscle, which releases a chemical that prevents the death of the neuron. As another example, all amphibians, reptiles, and mammals pass through embryonic stages in which they have tails and webbed fingers and toes. In humans, these stages appear during the fourth to seventh weeks of development. A few weeks later, the cells of the tail and webbing have died: The tail disappears, and the hands and feet have separate fingers and toes.

Development in Reptiles and Mammals Depends on Extraembryonic Membranes

Fish live and reproduce in water, usually by spawning. Although many amphibians spend their adult lives on land, they too lay their eggs in water. In both cases, the embryo obtains nutrients from the yolk of the egg and oxygen from the water, and releases its wastes, including carbon dioxide, into the water.

Fully terrestrial vertebrate life was not possible until the evolution of the **amniotic egg.** This innovation arose first in reptiles, and persists today in that group (including birds) and its descendants, the mammals. It allows members of these groups to complete their development into the adult form in their own "private pond." The amniotic egg is characterized by four **extraembryonic membranes:** the chorion, amnion, allantois, and yolk sac (**Table 42-2**). In reptiles, the **chorion** lines the shell and exchanges oxygen and carbon dioxide between the embryo and the air. The **amnion** encloses the embryo in a watery environment. The **allantois** surrounds and isolates wastes. The **yolk sac** contains the yolk.

In placental mammals (all mammals except marsupials, such as kangaroos, and monotremes, such as platypuses), the embryo develops within the mother's body until birth. Nevertheless, all four extraembryonic membranes still persist—remnants of the reptilian genetic program for development—and in fact are essential for development. Table 42-2 compares the structures and functions of these extraembryonic membranes in reptiles and mammals.

Table 42-2 **Vertebrate Embryonic Membranes**

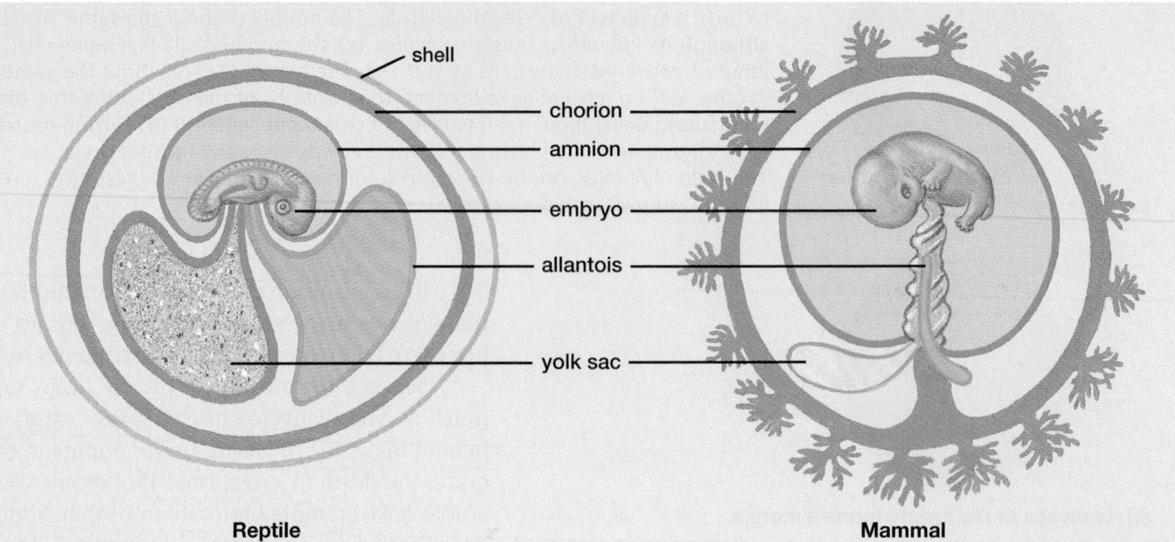

Membrane	Reptile Structure	Reptile Function	Mammal Structure	Mammal Function
Chorion	Membrane lining the inside of the shell	Acts as a respiratory surface; regulates the exchange of gases and water between the embryo and the air	Fetal contribution to the placenta	Provides for the exchange of gases, nutrients, and wastes between the embryo and the mother
Amnion	Sac surrounding the embryo	Encloses the embryo in fluid	Sac surrounding the embryo	Encloses the embryo in fluid
Allantois	Sac connected to the embryonic urinary tract; a capillary-rich membrane lining the inside of the chorion	Stores wastes (especially urine); acts as a respiratory surface	Membranous sac arising from the gut; varies in size	May store metabolic wastes; contributes to the umbilical cord blood vessels
Yolk sac	Membrane surrounding the yolk	Contains yolk as food; digests yolk and transfers its nutrients to the embryo	Small, membranous, fluid-filled sac	Helps absorb nutrients from the mother; forms blood cells; contributes to the umbilical cord

42.4 HOW IS DEVELOPMENT CONTROLLED?

A zygote contains all of the genes needed to produce an entire animal. Nearly every cell of the body also contains all of these genes—that's what makes cloning possible (see Chapter 9). In any given cell, however, some genes are used, or expressed, while others are not. The differentiation of cells during development arises because of these differences in gene expression.

You may recall from Chapter 12 that cells have several ways of controlling gene expression, including regulating which genes are transcribed into messenger RNA (mRNA). Proteins called transcription factors bind to DNA near the promoter regions where gene transcription begins. Different transcription factors bind to different genes and turn their transcription on or off. Which genes are transcribed determines the structure and function of the cell. This leads to one of the central questions about development: What causes different cells to transcribe different genes?

During the past decade, the techniques of molecular biology have revealed many of the genes and proteins involved in development, and some of the interactions among them. However, the fundamental principles were discovered about a century ago, using tools no more sophisticated than forceps, tiny knives, and hair from a baby's head, as we explore in "Scientific Inquiry: Unraveling the Mechanisms of Animal Development" on pp. 820–821.

Molecules Positioned in the Egg and Produced by Nearby Cells Control Gene Expression During Embryonic Development

In animal embryos, the differentiation of individual cells and the development of entire structures are driven by one or both of two processes: (1) the actions of gene-regulating substances inherited from the mother in her egg, and (2) chemical communication between the cells of the embryo.

Maternal Molecules in the Egg May Direct Early Embryonic Differentiation

Recall from Chapter 41 that essentially all of the cytoplasm in a zygote (a fertilized egg) was already in the egg before fertilization; the sperm contributed little more than a nucleus. In most invertebrates and some vertebrates, specific mRNA and protein molecules become concentrated in different places in the egg's cytoplasm during oogenesis. Some of these proteins are transcription factors that regulate which genes are turned on and off.

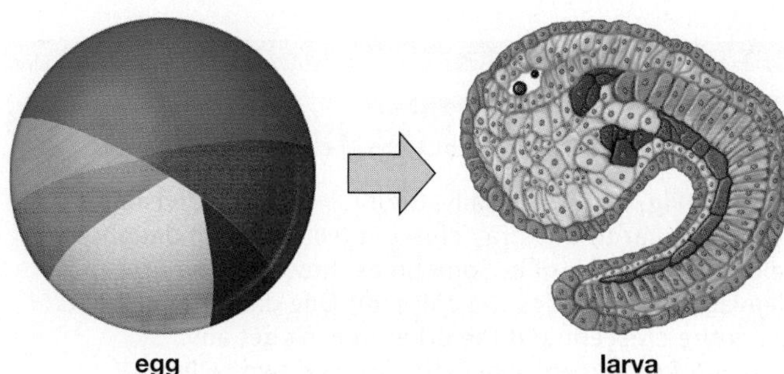

▲ **FIGURE 42-4 A "fate map" of the sea squirt egg** Gene-regulating substances in the cytoplasm of the egg of a sea squirt control early development. The coloring of the egg and larva shows which parts of the egg will give rise to which parts of the larva.

QUESTION If human development were as thoroughly determined as this, do you think that twins would be more or less common?

During the first few cleavage divisions after fertilization, the zygote and its daughter cells divide at specific places and in specific orientations. As a result, these early embryonic cells receive different maternal mRNAs and transcription factors. Therefore, different cells transcribe different genes, start differentiating into distinct cell types, and in many cases, ultimately give rise to specific adult structures. The positioning of maternal molecules in some eggs so strongly controls development that the egg can be mapped according to the major structures that will be produced by daughter cells inheriting each section of cytoplasm. For example, in the sea squirt egg diagrammed in **Figure 42-4,** cells that receive cytoplasm colored blue will form the skin, cells that receive cytoplasm colored green will form the nervous system, and so on.

In mammalian embryos, the current evidence indicates that all of the cells formed during cleavage are functionally equivalent. Which cells give rise to which parts of the later embryo appears to be a matter of chance, depending on where the cells happen to be located during the transition from the morula to the blastocyst (the mammalian equivalent of a blastula; see section 42.5).

Chemical Communication Between Cells Regulates Most Embryonic Development

At some point in all animal embryos, the developmental fate of each cell is determined by chemical interactions between cells, in a process called **induction.** During induction, certain cells release chemical messengers that alter the development of other, usually nearby, cells. Specific groups of genes are selectively activated in the recipient cells, causing them to differentiate in specific ways.

In amphibian embryos, it has been known since the early 1900s that a cluster of cells located near the blastopore (see Fig. 42-3c), called the *organizer*, determines whether nearby cells will become ectoderm or mesoderm, and even where the head and nervous system will form (see "Scientific Inquiry: Unraveling the Mechanisms of Animal Development"). How does the organizer accomplish this? The cells of the organizer release proteins given fanciful names such as Noggin, Cerberus (in Greek mythology, the three-headed dog that guarded the entrance to the underworld), and Dickkopf (meaning "thick-head" in German). These proteins interact with a number of other messengers to stimulate or repress the expression of genes in nearby cells, often genes that encode transcription factors and that therefore exert widespread effects on gene expression. Which constellations of genes are expressed then determines the structures and functions of the cells. As these cells differentiate, they in turn release other chemicals that alter the fate of still other cells, in a cascade that ultimately culminates in the development of the tissues and organs of the adult body.

Homeobox Genes Regulate the Development of Entire Segments of the Body

Homeobox genes are found in animals as diverse as fruit flies, frogs, and humans. Although their functions differ somewhat in different animals, homeobox genes code for transcription factors that regulate the transcription of many other genes. Each homeobox gene has major responsibility for the development of a particular region of the body. Homeobox genes were discovered in fruit flies, where specific mutations cause entire parts of the body to be duplicated, replaced, or omitted. For example, one mutant homeobox gene causes the development of an extra body segment, complete with an extra set of wings.

In all animals studied to date, the homeobox genes are arranged on the chromosomes in a head-to-tail order. Early in development, homeobox genes are transcribed in a specific sequence within the animal body. For example, "head" homeobox genes are transcribed in the anterior part of the embryo, and "tail" homeobox genes are transcribed in the tail, although multiple homeobox genes are usually expressed in any given region (**Fig. 42-5**).

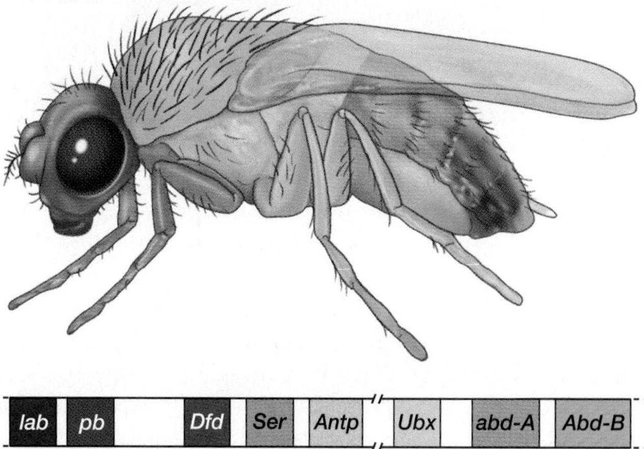

▲ **FIGURE 42-5 The sequence of homeobox genes on the chromosome corresponds to their role in the development of different body segments** In the fruit fly *Drosophila*, each homeobox gene is active in the body segment shown in the same color, in a head-to-tail order.

QUESTION Snakes have ribs all the way from just behind the head almost to the tip of the tail. Snakes also lack legs. Propose a possible genetic mechanism for this body structure, based on homeobox genes.

Scientific Inquiry

Unraveling the Mechanisms of Animal Development

Embryologists have long used amphibian eggs to study development. These eggs are very large (often over a millimeter in diameter) and many are pigmented in patterns that serve as natural markers for specific locations in the egg. They can also tolerate a lot of experimental manipulation and still develop fairly normally. In the 1920s, the German embryologist Hans Spemann used newt eggs to show that, at least in early embryos, all of the nuclei contain all of the genetic information needed to produce an entire embryo.

Molecules in the Egg Cytoplasm Regulate Early Development

Spemann wondered how genetically identical cells can produce structures as different as a brain, a leg, or a stomach. Might there be something in the egg cytoplasm that determines how genetic information is used?

Some fertilized amphibian eggs have three distinctively pigmented regions: dark on top, light on the bottom, and a crescent-shaped band of gray in between, on one side of the egg (**Fig. E42-1**). Usually, the first cleavage division is perpendicular to this gray crescent, so that both daughter cells receive some of it. Sometimes, however, the first cleavage plane misses the crescent: One daughter cell gets the entire crescent and the other doesn't get any.

Using fine forceps, Spemann lassoed two-celled salamander embryos with strands of hair from his baby daughter, Margrette, and tightened the nooses to separate the two cells. If both of the detached cells contained crescent material, both developed into normal embryos (**Fig. E42-1a**). But if only one cell contained crescent material, that cell would develop normally, whereas the other cell without crescent material would form a blob that Spemann called a "belly piece" (**Fig. E42-1b**). Spemann concluded that something, in a specific location of the cytoplasm, regulated the use of the genetic information that the two daughter cells received. In modern terms, gene-regulating molecules in specific locations in the egg cytoplasm altered gene expression and determined the developmental fates of the daughter cells.

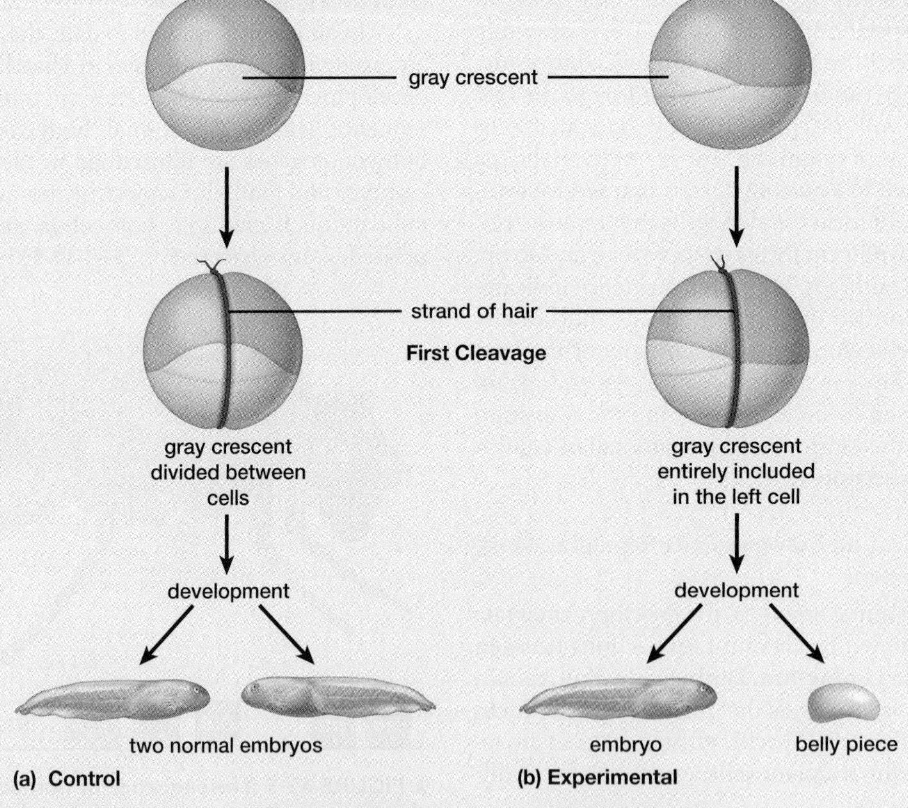

▲ **FIGURE E42-1 The egg cytoplasm controls early development in amphibians (a)** In salamanders and other amphibians, the first cleavage division usually bisects the gray crescent, so both daughter cells receive some of the crescent material. Separating the two cells produces two normal embryos. **(b)** Sometimes, the first division encloses the entire crescent in the cytoplasm of one daughter cell. If the two cells are separated, only the cell with the crescent material can form an embryo.

Certain Cells Determine the Development of Their Neighbors

What about later development? Surely the egg can't have infinitely precise positioning of myriads of molecules, dished out to daughter cells with such finesse that one particle of cytoplasm generates a leg, another an arm, and another an eye? Indeed not. Spemann and other embryologists had already done experiments in which bits of embryos, usually at the blastula stage, were transplanted to various locations in other embryos of the same species. They found that the fate of the transplanted cells was, in general, not predetermined: The transplants assumed the developmental fate of the area into which they were transplanted (**Fig. E42-2a**).

However, Spemann found that one small patch of cells, located near the blastopore (this patch is called the "dorsal lip"), followed its own developmental path, no matter where it was transplanted into the recipient embryo. In fact, transplanted dorsal lip cells caused the formation of a second head in the recipient embryo, and could actually produce a two-headed tadpole.

In the 1920s, Spemann and a graduate student, Hilde Mangold, performed a brilliant experiment showing that the transplanted dorsal lip cells actually changed the developmental fate of cells of the recipient embryo. Mangold cut out pieces of embryos from pale-colored newts and transplanted them into embryos of dark-colored newts, and vice versa. Transplants from the dorsal lip of the blastopore caused the surrounding tissue to form parts of a second head and, occasionally, to become almost complete secondary embryos (**Fig. E42-2b**). Spemann and Mangold concluded that the dorsal lip—now called Spemann's organizer—"organized" surrounding cells to form the head-to-tail axis of the embryo. We now know that Noggin, Cerberus, Dickkopf, and other molecules released by the organizer induce specific developmental pathways in nearby cells.

In 1935, Spemann was awarded the Nobel Prize for Physiology or Medicine. Hilde Mangold died tragically in a fire in 1924, and so she could not share the prize. (Nobel Prizes are not awarded posthumously.) In acknowledgment of Mangold's contributions, Spemann referred to "the experiments of Hilde Mangold" when describing these results.

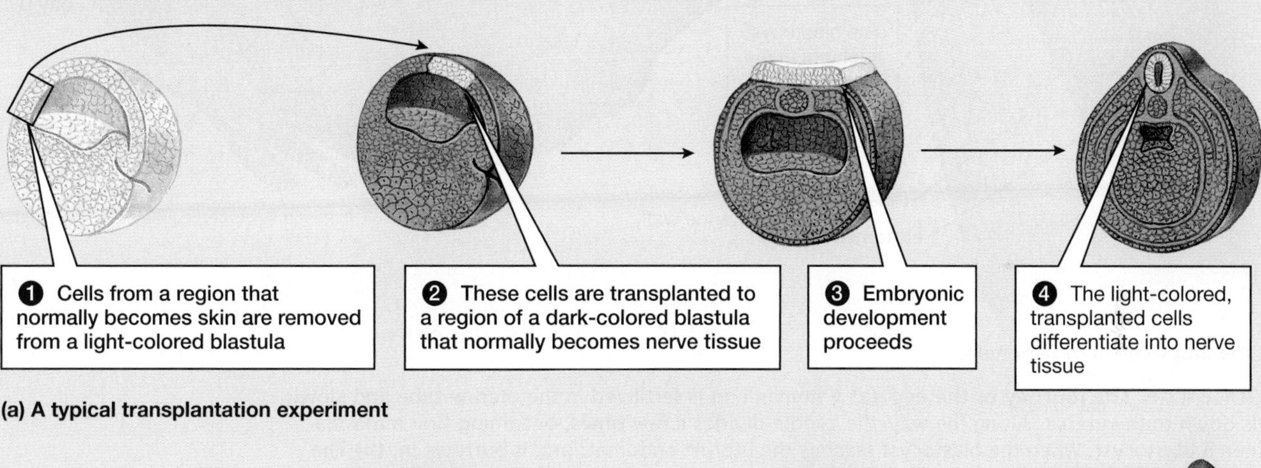

❶ Cells from a region that normally becomes skin are removed from a light-colored blastula
❷ These cells are transplanted to a region of a dark-colored blastula that normally becomes nerve tissue
❸ Embryonic development proceeds
❹ The light-colored, transplanted cells differentiate into nerve tissue

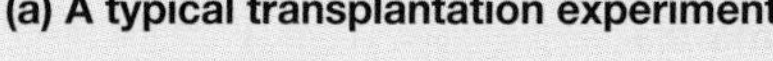
(a) A typical transplantation experiment

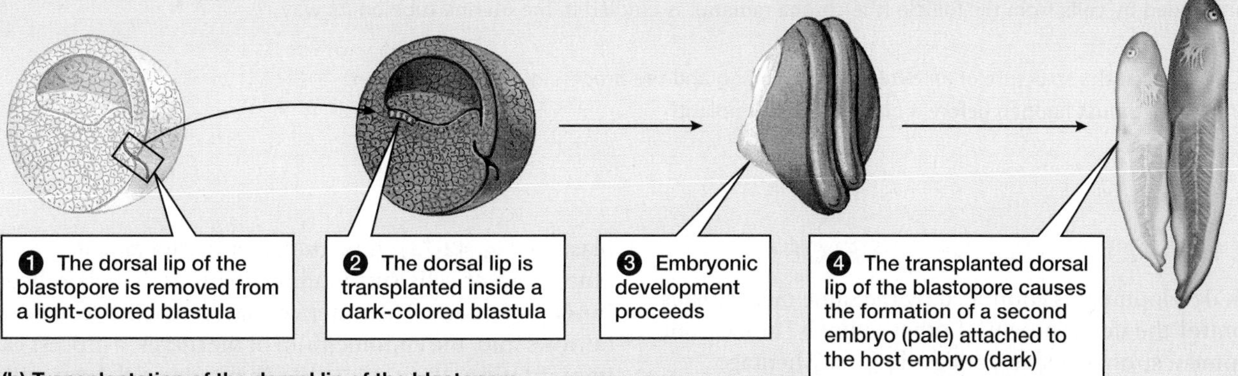

(b) Transplantation of the dorsal lip of the blastopore

▲ **FIGURE E42-2 Induction** During most of development, the fate of any given cell is strongly influenced by the cells that surround it. **(a)** When part of a blastula is transplanted into a second blastula, the surrounding cells usually induce the transplant to assume the characteristics of the region into which the transplant was placed. **(b)** A transplanted dorsal lip of the blastopore, however, induces the surrounding cells to change their fates, and develop into most of the structures of a secondary embryo.

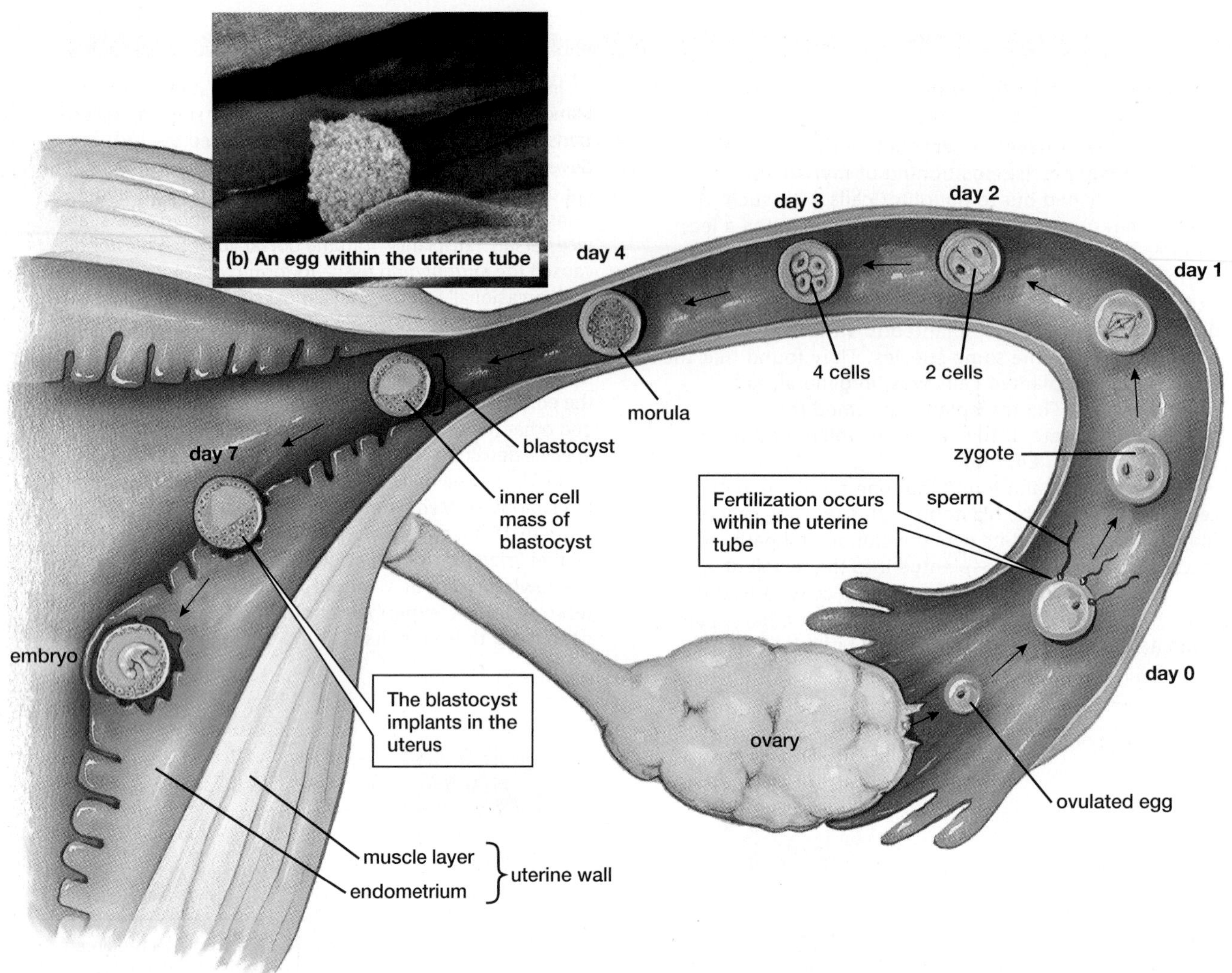

▲ FIGURE 42-6 **The journey of the egg** **(a)** A human egg is fertilized in the uterine tube and slowly travels down to the uterus. Along the way, the zygote divides a few times, becoming first a morula and then a blastocyst. When the blastocyst reaches the uterine endometrium, it burrows in. **(b)** The egg, surrounded by cells from the follicle (the corona radiata), is cradled in the uterine tube on its way to the uterus.

QUESTION Recall the structure of an ovulated human egg and the process of fertilization from Chapter 41. What must happen before a blastocyst can implant?

42.5 HOW DO HUMANS DEVELOP?

Human development is controlled by the same mechanisms that control the development of other animals. In fact, our development strongly reflects our evolutionary heritage.

Differentiation and Growth Are Rapid During the First Two Months

A human egg is usually fertilized in the uterine tube and undergoes a few cleavage divisions there, becoming a morula on its way to the uterus (Fig. 42-6). By about the fifth day after fertilization, the zygote has developed into a hollow ball of cells, known as a **blastocyst** (the mammalian version of a blastula; Fig. 42-7a). A mammalian blastocyst consists of an outer layer of cells surrounding a cluster of cells called the **inner cell mass.** The outer cell layer attaches to, and then burrows into, the endometrium of the uterus, a process called **implantation** (Figs. 42-7a,b). This outer cell layer will become the chorion. The complex intermingling of the chorion and the endometrium forms the placenta, which we will describe later in the chapter. The chorion also secretes chorionic gonadotropin (CG), which prevents the death of the corpus luteum (see pp. 802–803). The corpus luteum sustains the pregnancy by secreting progesterone and estrogen for the first couple of months, until the placenta takes over the secretion of these hormones.

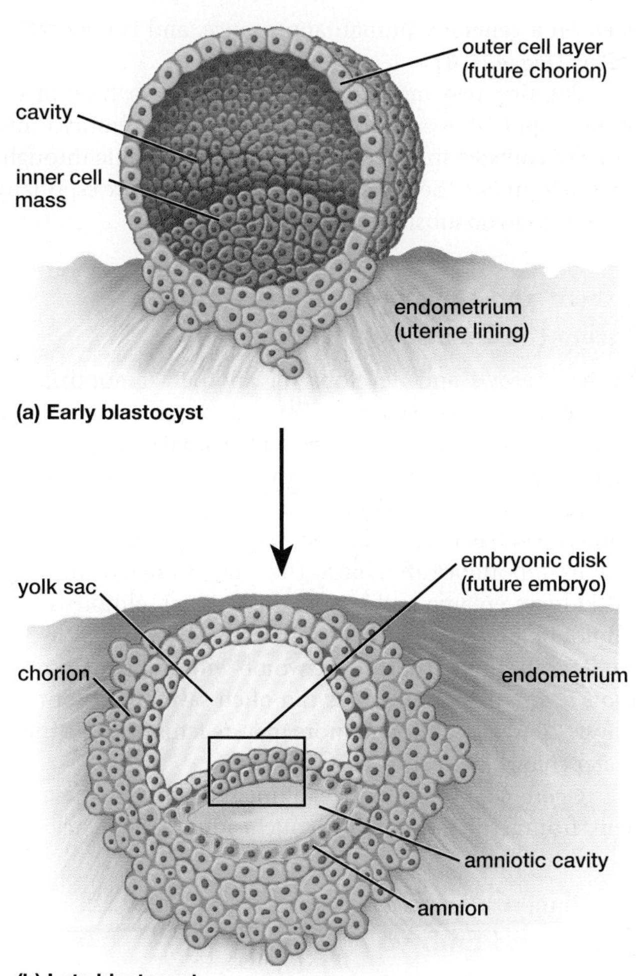

▲ **FIGURE 42-7 A blastocyst implants (a)** As it burrows into the uterine lining, the outer cell layer of the blastocyst forms the chorion, the embryonic contribution to the future placenta. **(b)** A few days later, the blastocyst has completely submerged beneath the uterine lining. The inner cell mass begins to develop into the yolk sac, the amnion, and the embryonic disk.

All of the cells of the inner cell mass have the potential to develop into any type of tissue. This remarkable flexibility allows the inner cell mass to produce both the entire embryo and the three remaining extraembryonic membranes. The inner cell mass is also the source of human embryonic stem cells, which researchers hope may one day be used to replace damaged adult tissues, as described in "Health Watch: The Promise of Stem Cells" on pp. 826–827.

After Implantation, Two Cavities Form and Gastrulation Occurs

During the second week, the inner cell mass grows and splits, forming two fluid-filled sacs that are separated by a double layer of cells called the **embryonic disk** (see Fig. 42-7b). One layer of cells is continuous with the yolk sac, although in placental mammals it contains no yolk. The second layer of cells is continuous with the amnion.

Gastrulation begins near the end of the second week after fertilization, and is usually complete by the end of the third week. The two layers of the embryonic disk separate slightly, and a slit forms in the cell layer on the "amnion side" of the disk. (This slit is the blastocyst's equivalent of the amphibian blastopore.) Cells migrate through the slit into the interior of the disk, forming mesoderm, endoderm, and the fourth extraembryonic membrane, the allantois. The cells remaining on the surface become ectoderm.

Body Structures Begin to Form During Weeks Three to Eight

During the third week of development, the embryo begins to form the spinal cord and brain. The heart starts to beat about the beginning of the fourth week. At this time, the embryo bulges into the uterine cavity, bathed in fluid contained within the amnion (**Fig. 42-8**). Meanwhile, the

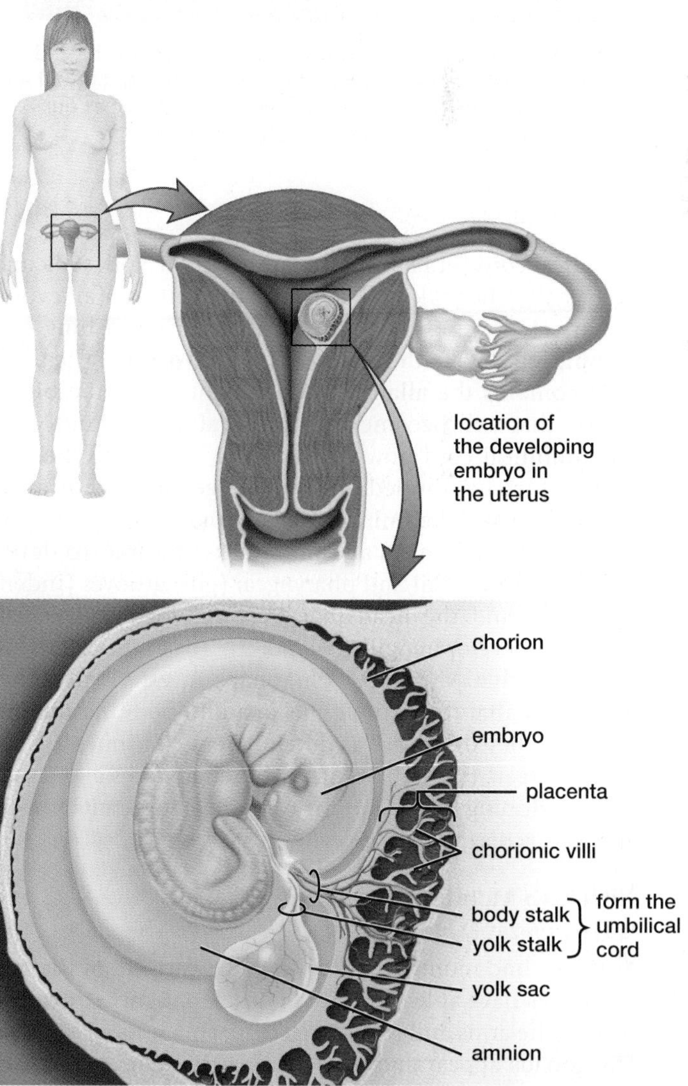

▲ **FIGURE 42-8 Human development during the fourth week** The embryo bulges into the uterus, and the placenta is restricted to one side. The body stalk and yolk stalk combine to form the umbilical cord, which exchanges wastes and nutrients between embryo and mother.

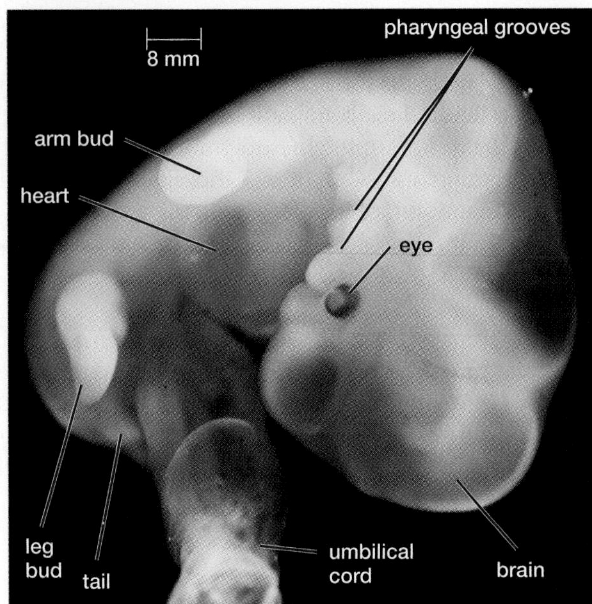

▲ **FIGURE 42-9 A 5-week-old human embryo** At the end of the fifth week, the human embryo is almost half head. A tail and pharyngeal (gill) grooves are clearly visible, evidence of our evolutionary relationship to other vertebrates.

umbilical cord forms from the fusion of the yolk stalk and body stalk. The yolk stalk connects the yolk sac (an evolutionary "leftover" from the yolk's role of nourishing embryonic reptiles) to the embryonic digestive tract. The body stalk contains the allantois, which contributes the blood vessels that will become the umbilical arteries and vein. The umbilical cord now connects the embryo to the placenta, which has formed from the merger of the chorion of the embryo and the lining of the uterus.

During the fourth and fifth weeks, the embryo develops a prominent tail and pharyngeal (gill) grooves (indentations behind the head that are homologous to a fish embryo's developing gills; **Fig. 42-9**). These structures are reminders that we share evolutionary ancestry with other vertebrates that retain their gills in adulthood. In humans, however, they disappear as development continues. By the seventh week, the embryo has rudimentary eyes and a rapidly developing brain, and the webbing between its fingers and toes is disappearing.

After Two Months, the Embryo Is Recognizably Human

As the second month draws to an end, nearly all of the major organs have at least begun to develop. Many structures, such as the arms and legs, are recognizably human in form. The gonads appear and develop into testes or ovaries. Sex hormones are secreted—either testosterone from the testes or estrogen from the ovaries. These hormones will affect the future development of the embryo; not only the reproductive organs, but also the brain and other parts of the body. After the second month of development, the embryo has taken on a generally human appearance, and is now called a **fetus** (**Fig. 42-10**).

The first two months of pregnancy are a time of extremely rapid differentiation and growth for the embryo, and a time of considerable danger. Although vulnerable throughout development, the rapidly developing organs are especially sensitive to toxic substances during this time.

Growth and Development Continue During the Last Seven Months

The fetus grows and develops for another 7 months. The brain continues to develop rapidly and the head remains disproportionately large. As the brain and spinal cord grow, they begin to generate behaviors. As early as the third month of pregnancy, the fetus can move and respond to stimuli. Some instinctive behaviors appear, such as sucking, which will have obvious importance after birth. The lungs, stomach, intestine, and kidneys enlarge and become functional, although they will not be used until after birth. Nearly all fetuses 32 weeks or older can survive outside the womb with some medical assistance, and heroic measures can often save infants born as early as 26 weeks, although more mature fetuses have a much greater chance of healthy survival.

Figure 42-11 summarizes human embryonic development, from fertilization to the conclusion of pregnancy.

The Placenta Exchanges Materials Between Mother and Embryo

During the first few days after implantation, the embryo obtains nutrients directly from the endometrium. During the following week or so, the **placenta** begins to develop from interlocking structures produced by the embryo and the

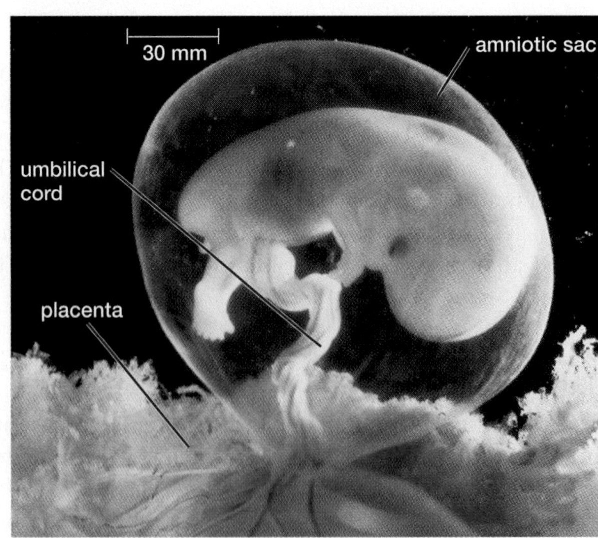

▲ **FIGURE 42-10 An 8-week-old human embryo** At the end of the eighth week, the embryo is clearly human in appearance and is now called a fetus. Most of the major organs of the adult body have begun to develop.

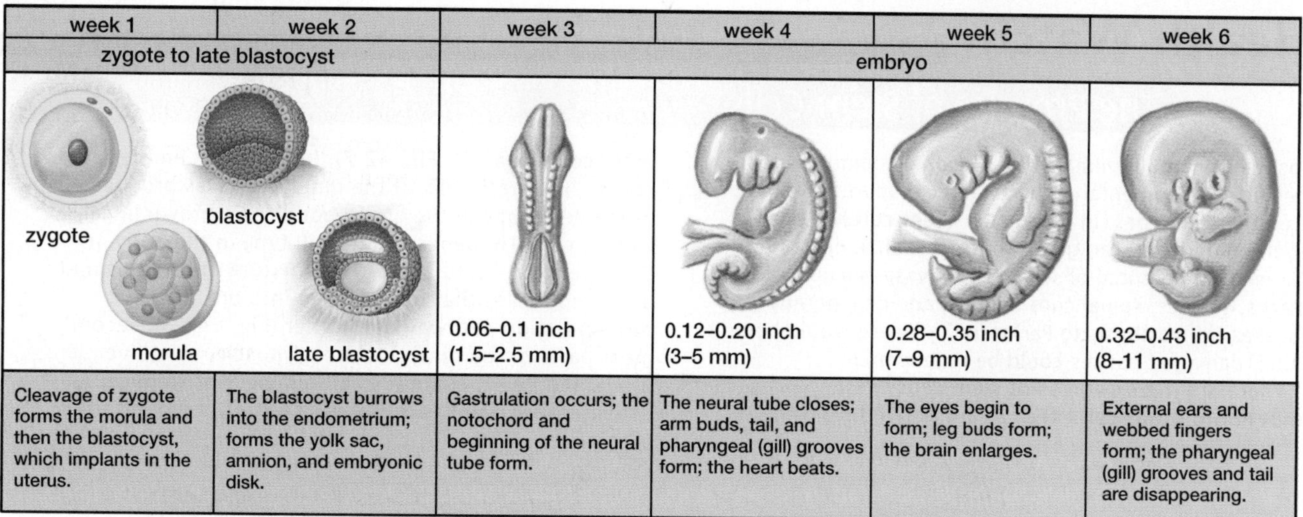

week 1	week 2	week 3	week 4	week 5	week 6
zygote to late blastocyst		embryo			
zygote, morula, blastocyst	late blastocyst	0.06–0.1 inch (1.5–2.5 mm)	0.12–0.20 inch (3–5 mm)	0.28–0.35 inch (7–9 mm)	0.32–0.43 inch (8–11 mm)
Cleavage of zygote forms the morula and then the blastocyst, which implants in the uterus.	The blastocyst burrows into the endometrium; forms the yolk sac, amnion, and embryonic disk.	Gastrulation occurs; the notochord and beginning of the neural tube form.	The neural tube closes; arm buds, tail, and pharyngeal (gill) grooves form; the heart beats.	The eyes begin to form; leg buds form; the brain enlarges.	External ears and webbed fingers form; the pharyngeal (gill) grooves and tail are disappearing.

week 7	week 8	week 10	week 12	week 16
embryo		fetus		
0.67–0.79 inch (1.7–2.0 cm)	0.90–1.10 inches (2.3–2.8 cm)	1.25–1.75 inches (3.2–4.4 cm)	2–3 inches (5–7.6 cm)	4–5 inches (10.2–12.7 cm)
Webbed toes form; bones begin to stiffen; the back straightens; the eyelids begin to form.	All the major organs begin to form; the arms can bend; fingers are distinct. Facial features and outer ears take shape.	After 8 weeks, the embryo is called a fetus. Red blood cells form; toes separate; eyelids have developed; major brain parts are present; the hands can form fists.	The neck is well defined; all organs are present; male or female genitals are present; arms and legs move; teeth begin to form; a heartbeat can be detected electronically.	Sucking and swallowing movements occur; the liver and pancreas begin functioning. The body has grown relative to the head; major organs continue developing. The mother may feel movement; weight is about 5 oz.

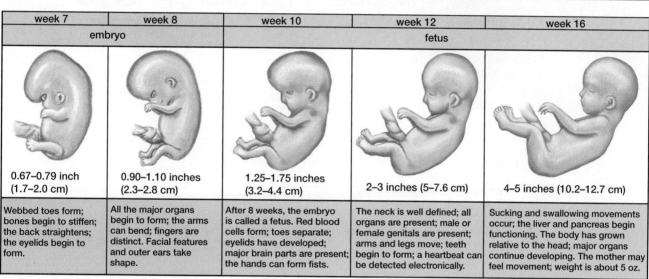

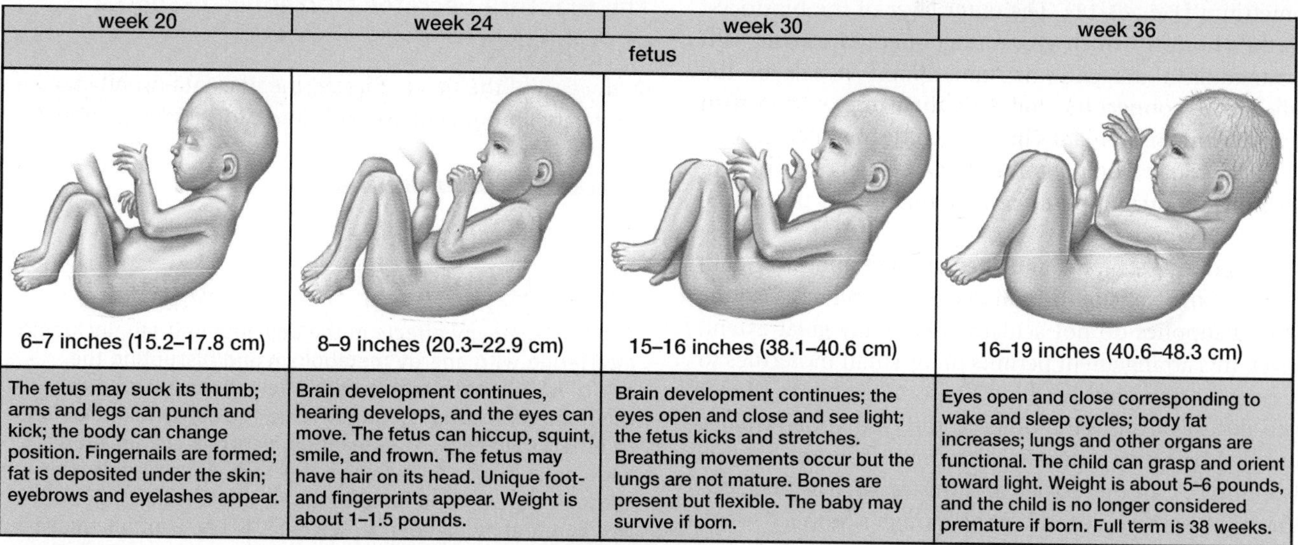

week 20	week 24	week 30	week 36
fetus			
6–7 inches (15.2–17.8 cm)	8–9 inches (20.3–22.9 cm)	15–16 inches (38.1–40.6 cm)	16–19 inches (40.6–48.3 cm)
The fetus may suck its thumb; arms and legs can punch and kick; the body can change position. Fingernails are formed; fat is deposited under the skin; eyebrows and eyelashes appear.	Brain development continues, hearing develops, and the eyes can move. The fetus can hiccup, squint, smile, and frown. The fetus may have hair on its head. Unique foot- and fingerprints appear. Weight is about 1–1.5 pounds.	Brain development continues; the eyes open and close and see light; the fetus kicks and stretches. Breathing movements occur but the lungs are not mature. Bones are present but flexible. The baby may survive if born.	Eyes open and close corresponding to wake and sleep cycles; body fat increases; lungs and other organs are functional. The child can grasp and orient toward light. Weight is about 5-6 pounds, and the child is no longer considered premature if born. Full term is 38 weeks.

▲ **FIGURE 42-11 A calendar of development from zygote to birth**

Health Watch

The Promise of Stem Cells

Most cells in adult animals have become committed to a specific developmental pathway and cannot differentiate into other cell types. In contrast, a **stem cell** has not differentiated, and can give rise to more than one cell type. The medical potential of stem cells is vast—victims of heart attacks, strokes, spinal cord injuries, and degenerative diseases from arthritis to Parkinson's disease would benefit if their damaged tissues could be regenerated.

There are three sources of stem cells. First, **embryonic stem cells (ESCs)** are derived from the inner cell mass (see Fig. 42-7). In the intact embryo, of course, the stem cells of the inner cell mass produce all of the cell types of the entire body. ESCs grown in cell culture can also generate any cell type in the body, if they are exposed to the correct mixture of proteins and other molecules that nudge them into one differentiation pathway or another (**Fig. E42-3**). Second, most parts of the body, including muscle, skin, liver, brain, and heart, contain small numbers of stem cells, usually called **adult stem cells (ASCs),** although they

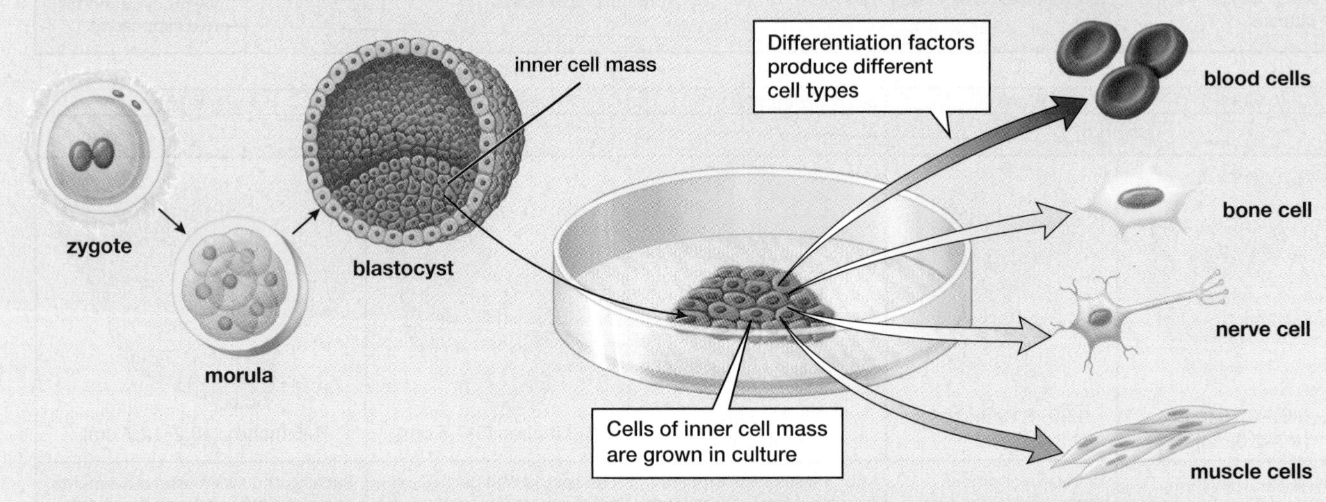

▲ **FIGURE E42-3 Culturing embryonic stem cells from the inner cell mass of a blastocyst**

endometrium (**Fig. 42-12**). The outer layer of the blastocyst forms the chorion, which grows finger-like **chorionic villi** that extend into the endometrium. Blood vessels of the umbilical cord connect the embryo's circulatory system with a dense network of capillaries in the villi. Meanwhile, implantation has eroded some of the blood vessels of the endometrium, producing pools of maternal blood that bathe the chorionic villi.

The embryo's blood and the mother's blood remain separated by the walls of the villi and their capillaries, so the two blood supplies do not actually mix to any great extent. However, this arrangement permits many small molecules to move between the mother's blood and the embryo's blood. Oxygen diffuses from the mother to the embryo. Nutrients, many aided by active transport, also travel from the mother to the embryo. Carbon dioxide and wastes, such as urea, diffuse from the embryo's blood to the mother's blood.

The membranes of the capillaries and chorionic villi act as barriers to some substances, including large proteins and cells. However, some disease-causing organisms and many harmful chemicals, including alcohol, can penetrate the placenta, as described in "Health Watch: The Placenta—Barrier or Open Door?" on p. 829.

The Placenta Secretes Hormones Essential to Pregnancy

By the end of the first trimester, the placenta usually secretes significant amounts of estrogen and progesterone, enough to

Case Study continued

The Faces of Fetal Alcohol Syndrome

Alcohol has many effects in the developing brain, including interfering with energy metabolism and disrupting the ability of cells to recognize their neighbors, migrate to their correct locations, and develop the correct connections to other brain cells. The resulting incorrect cell locations and connections can never be repaired. For example, the corpus callosum, the band of axons interconnecting the two cerebral hemispheres (see Fig. 38-12), is often smaller, and sometimes missing, in children with FAS, so the left and right sides of the brain cannot communicate properly with one another. Many brain cells die if correct connections are never made, resulting in a reduction in the size of many brain structures.

are also present in children. ASCs can differentiate into only a few cell types. Stem cells from bone marrow, which normally produce both red and white blood cells, have been used for decades in transplants to treat diseases such as leukemia.

Third, by inserting a handful of genes into them, adult (nonstem) cells can be transformed into **induced pluripotent stem cells (iPSCs),** which appear to be able to differentiate into all cell types. Up until early 2009, generating iPSCs required either at least one cancer-causing gene or the use of a potentially dangerous virus that inserts its DNA into the chromosomes of the host cell. Both of these techniques would probably make the resulting iPSCs unsafe for human applications. A major goal of iPSC researchers, therefore, is to produce iPSCs with methods that use neither viruses nor cancer-causing genes. In September 2009, a group at the Salk Institute and the University of California at San Diego reported that they had transformed fetal human neural stem cells into iPSCs using a single, noncarcinogenic gene. Although there are some differences between these iPSCs and ESCs, they offer the promise that it may someday be possible to make iPSCs safe for clinical use.

How might stem cells be used in medicine? In one experiment, researchers induced heart attacks in mice, and then injected ASCs into the damaged heart. The ASCs multiplied and partially repaired the heart muscle. Researchers also have been able to coax human ESCs to differentiate into insulin-secreting cells, and hope that this might be the first step to a cure for type 1 diabetes (see "Health Watch: Closer to a Cure for Diabetes" on p. 727).

However, there are significant obstacles to surmount before stem cells can become effective therapies. First, embryos must be destroyed to obtain ESCs (at least with today's technology). Many people object to this procedure, even when the embryos are "extras" that were created for *in vitro* fertilization and would be destroyed anyway.

Second, the immune system would reject ASCs or ESCs that are not genetically identical, or at least a very close match, to the recipient. This has led to proposals for **therapeutic cloning,** which would involve inserting the nucleus from one of a patient's own cells into an egg whose nucleus had been removed, to create ESCs that would not be rejected. Once again, an embryo would be destroyed to obtain ESCs, so this procedure is also controversial.

Third, are ESCs and iPSCs safe enough to use in people? In early 2009, the media reported that an Israeli boy had been taken to Russia and injected with ESCs three times since 2001 in an attempt to cure a rare brain disease. However, some of the ESCs grew into a tumor (benign, fortunately). Stem cell researchers and physicians stress that safety studies of stem cell therapies have not yet been carried out, so this procedure should never have been done. Some experimental stem cell therapies in mice do not induce tumors, so there's hope that stem cells can be made safe.

Despite these difficulties, researchers are enthusiastic about the future of stem cells for therapies. In the best-case scenario, if iPSCs can be made safe and still differentiate into any desired cell type, then the use of ESCs, the problem of rejection by the immune system, and the ethical issue of therapeutic cloning would all be solved.

sustain its own growth and development. About this time, the corpus luteum degenerates, so the future of the pregnancy now depends on the placental hormones. These hormones also stimulate development of the mammary glands. Finally, progesterone also inhibits premature contractions of the uterine muscles.

Pregnancy Culminates in Labor and Delivery

During the last months of pregnancy, the fetus usually becomes positioned head downward in the uterus, with the crown of the skull resting against (and being held up by) the cervix. Childbirth (**Fig. 42-13**) generally begins around the end of the ninth month. Birth results from a complex interplay between uterine stretching caused by the growing fetus, and maternal and fetal hormones that finally trigger **labor,** the contractions of the uterus that result in delivery of the child.

Unlike skeletal muscles, the smooth muscles of the uterus can contract spontaneously, and stretching enhances their tendency to contract. As the baby grows, it stretches the uterine muscles, which occasionally contract weeks before delivery. No one really knows what triggers labor in humans, but chemical signals from both the placenta and maturing fetus may be involved. Whatever the initial stimulus, the placenta releases prostaglandins, which make the uterine muscles more likely to contract. As the uterus contracts, it pushes the fetus' head against the cervix, stretching it. This has two effects. First, the cervix expands so that the fetus' head can fit through. Second, stretching the cervix sends nervous signals

Have you ever wondered 

Why Childbirth Is So Difficult?

Compared to the birth of other animals, childbirth is a lengthy ordeal. Why? Probably for two reasons. First, we walk upright, requiring a different shape to the pelvis as compared to four-footed animals. Thus, a woman's pelvic opening is both small (for her body size) and asymmetrical. Second, human infants have enormous heads, which actually must be squashed a bit to fit through the pelvic opening. Monkeys, with relatively small infant heads and large pelvic openings, give birth quickly and almost effortlessly. In contrast, human deliveries usually last several hours, cause considerable pain to the mother and probably the infant, and often require assistance. They don't call it "labor" for nothing!

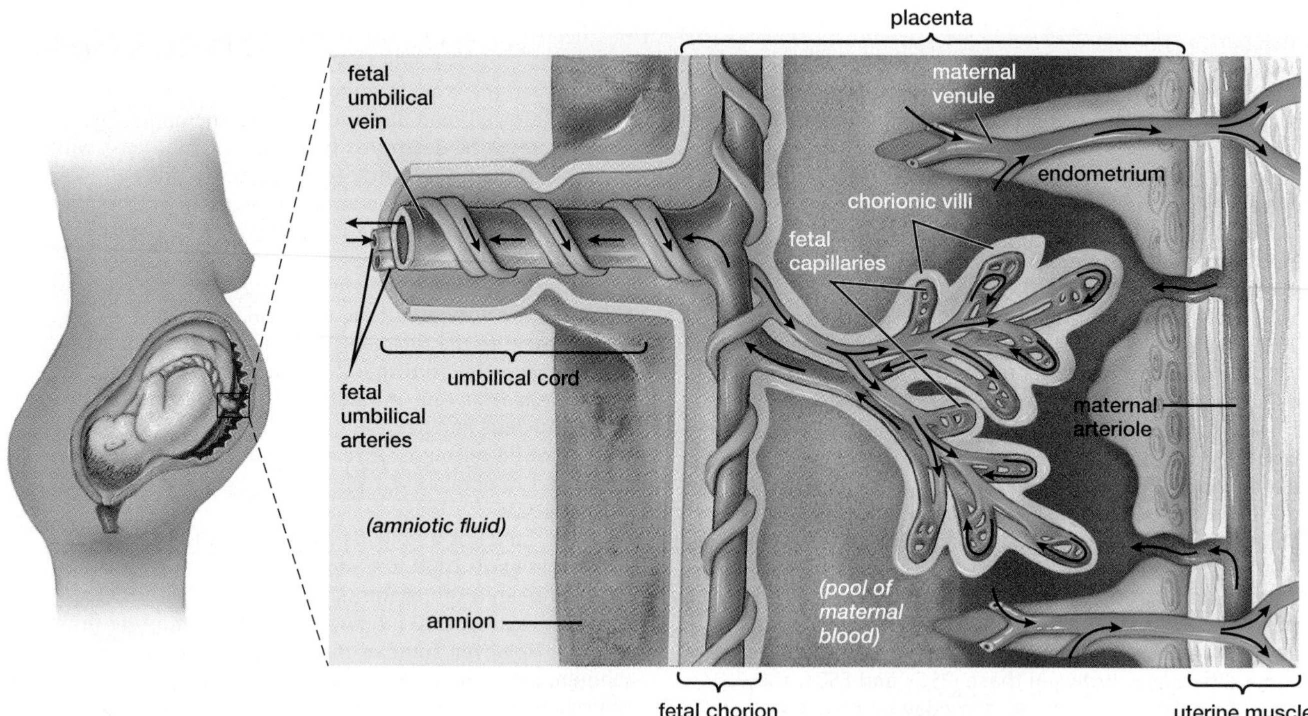

▲ **FIGURE 42-12 The placenta** The placenta allows exchange of wastes and nutrients between the fetal capillaries and the maternal blood pools, while keeping the fetal and maternal blood supplies separate. The umbilical arteries carry deoxygenated blood from the fetus to the placenta, and the umbilical vein carries oxygenated blood back to the fetus.

QUESTION A few types of mammals, such as kangaroos and platypuses, lack a placenta. What would you predict about the nature of development in nonplacental mammals?

to the mother's brain, causing the release of the hormone oxytocin (see Chapter 37). Oxytocin stimulates contractions of the uterine muscles, pushing the baby harder against the cervix, which stretches further, causing still more oxytocin to be released. This positive feedback cycle continues until the baby emerges from the vagina. The infant is in for a rude awakening. The uterus was soft, cushioned with fluid, and warm. Suddenly, the baby must breathe on its own, regulate its own body temperature, and suckle to obtain food.

After a brief rest following childbirth, the uterus resumes its contractions and shrinks remarkably. During these contractions, the placenta is sheared from the uterine wall and expelled through the vagina. The umbilical cord now releases prostaglandins that cause the muscles surrounding

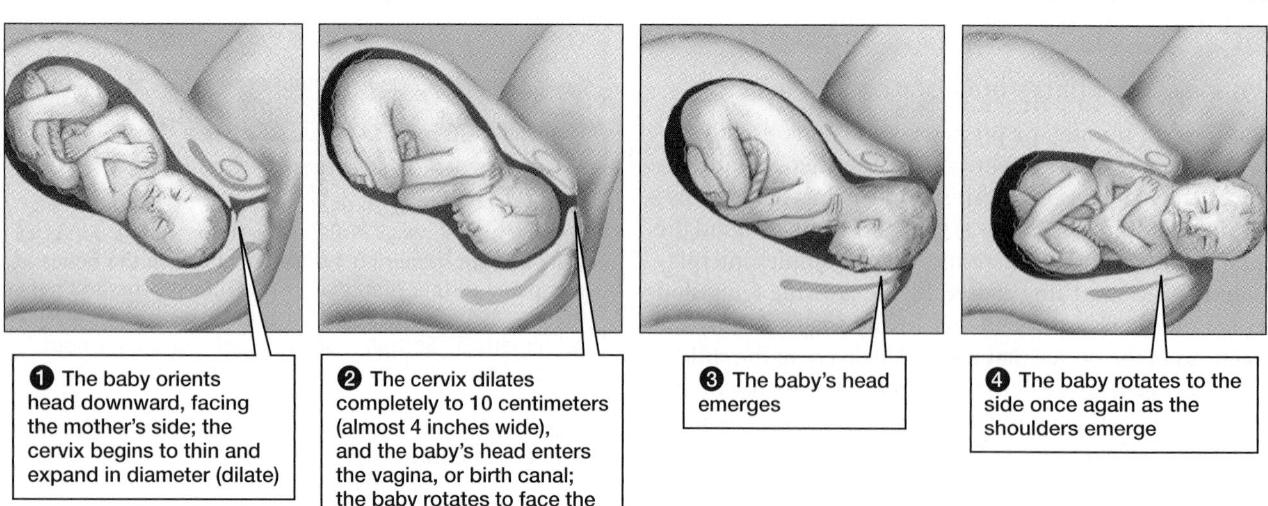

❶ The baby orients head downward, facing the mother's side; the cervix begins to thin and expand in diameter (dilate)
❷ The cervix dilates completely to 10 centimeters (almost 4 inches wide), and the baby's head enters the vagina, or birth canal; the baby rotates to face the mother's back
❸ The baby's head emerges
❹ The baby rotates to the side once again as the shoulders emerge

▲ **FIGURE 42-13 Human childbirth**

Health Watch

The Placenta—Barrier or Open Door?

The fetal and maternal blood supplies are separated by several membranes. However, many molecules, and even some infectious microbes, can pass through these thin membranes, moving from mother to fetus.

Microbes

Bacterial invaders include syphilis, some types of *Streptococcus* that cause blood infections, and *Listeria*, which in adults can cause fatal illness, usually transmitted in contaminated food. Viral invaders include those that cause genital herpes, German measles (rubella), hepatitis B, and HIV. Many of these microbes cause severe disorders in the fetus. For example, about 30% of fetuses infected with syphilis are stillborn, and about 40% suffer from mental retardation. *Listeria* can cause reduced fetal growth, premature delivery, and meningitis (a sometimes fatal infection of the membranes surrounding the brain). Infections with various types of viruses may cause premature birth, mental retardation, eye defects, and many other disorders.

Drugs

Many medicinal and recreational drugs are lipid soluble and can easily move across plasma membranes (see p. 84), including those of the placenta. The question is whether the drugs harm the fetus. Unfortunately, many people learn the answer the hard way. Thalidomide, for example, was commonly prescribed in Europe in the late 1950s and early 1960s as a sedative and antinausea medication to combat morning sickness. Its devastating effects on embryos were discovered only when many babies were born with missing or extremely abnormal limbs. In the late 1980s, the antiacne drug Accutane® was found to cause gross deformities in babies born to women using it.

Toxins in Cigarette Smoke

Many women smokers continue the habit during pregnancy, thus exposing their developing children to nicotine, carbon monoxide, and a host of carcinogens. As a result, these mothers have a higher incidence of miscarriages and tend to give birth to smaller infants, who have a higher risk of death shortly after birth. Children born to smokers also often have behavioral and intellectual deficits. See the "Health Watch: Smoking—A Life and Breath Decision" on p. 649 for more information about the health effects of smoking in children and adults.

Alcohol

Alcohol is soluble both in water and lipids. It dissolves in the blood and then easily passes through the placenta. As one FAS support group notes, "What you drink, baby drinks too." In fact, when a pregnant woman drinks, the blood of her unborn child has as much alcohol as her own. Worse, the fetus cannot metabolize alcohol as rapidly as an adult, so alcohol's effects in the fetus are greater and longer lasting.

More than 12% of U.S. women drink during pregnancy. Each year, they give birth to almost 40,000 infants with fetal alcohol spectrum disorders (FASD), which are a range of disabilities caused by prenatal alcohol exposure. Women who drink even lightly during pregnancy tend to have smaller babies. These children are also more likely to be anxious, depressed, or aggressive. If a pregnant woman drinks heavily on a regular basis (four or more drinks per day), or goes on alcoholic binges (five or more drinks at a time), her child has a significant risk of developing full-blown fetal alcohol syndrome, including below-average intelligence, hyperactivity, irritability, and poor impulse control. FAS children may also have reduced growth and defects of the heart and other organs. In extreme cases, children afflicted with FAS may have small, improperly developed brains (**Fig. E42-4**). The damage is irreversible and can be fatal.

FAS is the single most common cause of mental retardation in the United States. The U.S. Surgeon General advises pregnant women and those who are likely to become pregnant to avoid alcohol completely.

The Placenta Cannot Be Trusted to Protect the Embryo

A pregnant woman should assume that any drugs she takes will cross the placenta to her developing infant. Crucial stages of embryonic development can occur before a woman even realizes that she is pregnant, so any woman trying to become pregnant, or having unprotected sex, should take the same precautions as if she were already pregnant.

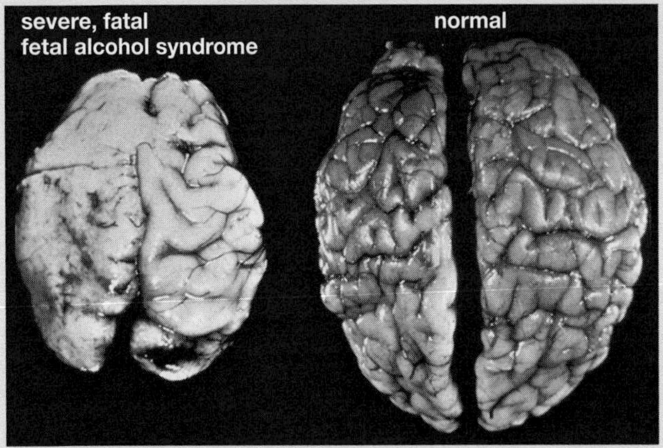

▲ **FIGURE E42-4 Alcohol impairs brain development** Comparing the brain of a 6-week-old child with severe fetal alcohol syndrome (left) and the brain of a normal child of the same age (right) shows the devastating effects of alcohol on the developing brain. (Photo courtesy of Dr. Sterling Clarren, University of Washington, Seattle, WA.)

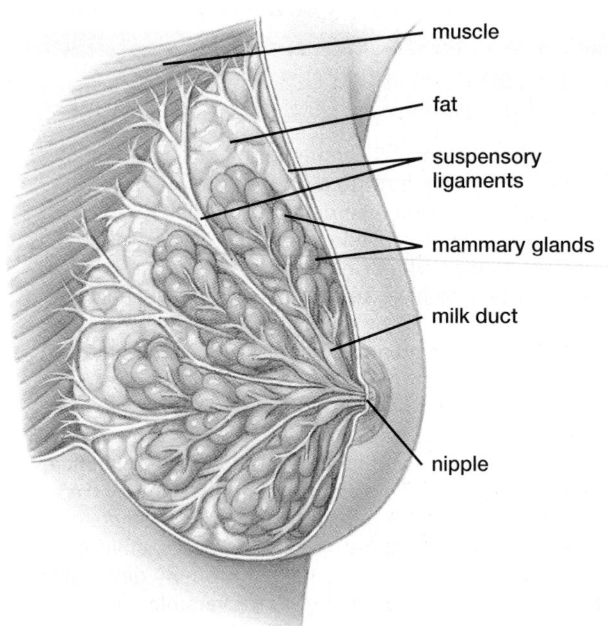

▲ **FIGURE 42-14 The structure of the mammary glands**
During pregnancy, fatty tissue, the milk-secreting glands, and the milk ducts all increase in size.

fetal blood vessels in the umbilical cord to contract and shut off blood flow. Although tying off the umbilical cord is standard practice, it is not usually necessary; if it were, other mammals would not survive birth.

Milk Secretion Is Stimulated by the Hormones of Pregnancy

During pregnancy, large quantities of estrogen and progesterone are secreted by the placenta. Estrogen and progesterone, acting together with several other hormones, stimulate the milk-producing **mammary glands** in the breasts to grow, branch, and develop the capacity to secrete milk. The mammary glands are arranged in a circle around the nipple, each with a milk duct leading to the nipple (Fig. 42-14).

Prolactin, a hormone secreted by the anterior pituitary gland, promotes both mammary gland development and the actual secretion of milk, a process called **lactation.** Prolactin release is stimulated by the high levels of estrogen produced by the placenta, so you might think that milk secretion would begin even before the child is born. However, lactation is inhibited by progesterone, which is also secreted by the placenta. During childbirth, the placenta is ejected from the uterus. Progesterone levels plummet, allowing prolactin to cause lactation.

Milk is released when the infant's suckling stimulates nerve endings in the nipples, which signal the hypothalamus to cause the pituitary gland to release an extra surge of prolactin and oxytocin. Oxytocin causes muscles surrounding the mammary glands to contract, ejecting the milk into the ducts that lead to the nipples (see Fig. 37-8). The prolactin surge stimulates rapid milk production for the next feeding.

During the first few days after birth, the mammary glands secrete a yellowish fluid called **colostrum.** Colostrum is high in protein and contains antibodies from the mother that help protect the infant against some diseases as its immune system is developing. Colostrum is gradually replaced by mature milk, which is higher in fat and milk sugar (lactose) and lower in protein.

42.6 IS AGING THE FINAL STAGE OF HUMAN DEVELOPMENT?

Aging (Fig. 42-15) is the gradual accumulation of damage to essential biological molecules, particularly DNA in both the nucleus and mitochondria, resulting in defects in cell functioning, declining health, and ultimately death. This damage—from natural errors in DNA replication, radiation from the sun or the rocks beneath our feet, and chemicals in food, cigarettes, and industrial products—begins at fertilization. Many biological molecules are harmed by free radicals (see pp. 24–25), some produced by environmental pollutants, but most by energy-generating reactions in our cells, especially in mitochondria.

Some molecular injury can be tolerated. Young bodies may be able to repair damage or compensate for it. As an animal ages, however, its repair abilities diminish; eventually, the body's tolerance for damage is exceeded. At the organismal level, aging is manifested in many ways: Muscle and bone mass are lost, skin elasticity decreases, reaction time slows, and senses such as vision and hearing become less acute. A less-robust immune response renders the aging individual more vulnerable to disease. Eventually, the individual can no longer fight off natural assaults, and death occurs.

For thousands of years, people have attempted to delay aging and extend the human life span. Modern medical care can prevent or cure many diseases, and can fix or replace some damaged organs (think bypass surgery and transplants). Some dietary changes, particularly not eating very much food (euphemistically called "caloric restriction"), can prolong life, at least in animal experiments. However, what seems to be the maximum human life span, about 130 years, has not changed.

▲ **FIGURE 42-15 Youth meets age**

Several evolutionary hypotheses suggest that aging is unavoidable. For example, natural selection favors organisms that leave the largest number of healthy, successful offspring. Even a hypothetically immortal animal that wouldn't die from "inside," so to speak, will eventually succumb to predation, accident, or disease. Therefore, perhaps natural selection favors devoting more of the body's resources to reproduction than to the continuous bodily repair required for immortality. The fact that humans can live so long after they stop reproducing is probably evidence of the selective advantage conferred by the care and teaching given to the young by their elders.

Case Study revisited

The Faces of Fetal Alcohol Syndrome

When John Kellerman was born, the delivery room reeked of the alcohol polluting his amniotic fluid: John was born drunk. Even today, without medication, his behavior resembles that of a drunk person: silly, volatile, and lacking impulse control.

The writings of people with FAS or FASD (fetal alcohol spectrum disorders) poignantly convey the day-to-day difficulties that these innocent people face over a lifetime. In his poem "Help," John wrote:

"When my brain is not working right,
I need to let someone I trust
Help me
To stay safe and get calm again.
When my brain is not working right,
I feel like I am on a FASD Train
Going downhill,
And the engineer is asleep
And I can't wake him up,
And I can't put on the brakes."

CJ, an FAS victim from Canada, explained:

"I am small, I have a different face . . . and Lots of Learning problems.
Moms do not do this on purpose
Do not be mean or mad or blame them
I am 17 years old and have had FAS all my life
I will have it forever
It will never go away
Don't be mean or mad or blame me. . . ."

BioEthics Consider This

In many states, women who use illegal drugs or abuse alcohol during pregnancy may be subject to legal action to protect fetal rights. How do you think society should approach the dilemma of pregnant women who often unwittingly damage their unborn children through alcohol or other drugs? Is this child abuse? Based on what you know about development, how would you deal with a friend who continued to smoke, drink, or take other drugs during pregnancy?

CHAPTER REVIEW

Summary of Key Concepts

42.1 What Are the Principles of Animal Development?

Development is the process by which an organism grows and increases in organization and complexity, using three main mechanisms: (1) Individual cells multiply; (2) daughter cells differentiate; and (3) groups of cells move about and become organized into multicellular structures.

42.2 How Do Indirect and Direct Development Differ?

Animals undergo either indirect or direct development. In indirect development, eggs (usually with relatively little yolk) hatch into larvae, which undergo metamorphosis to become adults with notably different body forms. In direct development, the newborn animal is sexually immature but otherwise resembles a small adult. Animals with direct development tend to produce either large, yolk-filled eggs or nourish the developing embryo within the mother's body.

42.3 How Does Animal Development Proceed?

Animal development occurs in several stages. *Cleavage and blastula formation:* A fertilized egg undergoes cell divisions with little intervening growth, so the daughter cells become smaller. Cleavage results in the formation of the morula, a solid ball of cells. A cavity then opens up within the morula, forming a hollow ball of cells called a blastula. *Gastrulation:* A dimple forms in the blastula, and cells migrate from the surface into the interior of the ball, eventually forming a three-layered gastrula. These three cell layers—ectoderm, mesoderm, and endoderm—give rise to all the adult tissues. *Organogenesis:* The cell layers of the gastrula form organs characteristic of the animal species (see Table 42-1).

In mammals and reptiles (including birds), extraembryonic membranes (the chorion, amnion, allantois, and yolk sac) encase the embryo in a fluid-filled space and regulate the exchange of nutrients and wastes between the embryo and its environment.

42.4 How Is Development Controlled?

All the cells of an animal body contain a full set of genetic information, yet cells are specialized for particular functions. Cells differentiate by stimulating and repressing the transcription of specific genes, which is controlled by (1) differences in gene-regulating substances inherited from the mother in her egg and/or (2) chemical communication between the cells of the embryo, a process called induction. A particularly important group of genes, called homeobox genes, regulates the differentiation of body segments and their associated structures, such as wings or legs.

42.5 How Do Humans Develop?

A human egg is fertilized in the uterine tube. The resulting zygote develops into a blastocyst and implants in the endometrium. The outer wall of the blastocyst will become the chorion and form the embryonic contribution to the placenta; the inner cell mass develops into the embryo and the three other extraembryonic membranes. During gastrulation, cells migrate and differentiate into ectoderm, mesoderm, and endoderm. During the third and fourth weeks, the endoderm forms a tube that will become the digestive tract, the heart begins to beat, and the rudiments of a nervous system appear. By the end of the second month, the major organs have begun to form, and the embryo—now called a fetus—appears human. In the next 7 months before birth, the fetus continues to grow; the lungs, stomach, intestine, kidneys, and nervous system enlarge, develop, and become functional. See Figure 42-11.

During pregnancy, mammary glands in the mother's breasts grow under the influence of estrogen, progesterone, and other hormones, such as prolactin. After about 9 months, uterine contractions are triggered by a complex interplay of uterine stretch and the release of prostaglandin and oxytocin. As a result, the uterus expels the baby and then the placenta. After its birth, the infant begins suckling and activates the release of prolactin, which stimulates milk production, and oxytocin, which triggers milk secretion.

42.6 Is Aging the Final Stage of Human Development?

Aging is the gradual accumulation of cellular damage (particularly to genetic material) over time that leads to loss of functionality of an organism and, eventually, to death. Some biologists hypothesize that aging results from natural selection favoring a sexually mature animal putting more resources into reproduction than into bodily repair.

Key Terms

adult stem cell (ASC) *826*
aging *830*
allantois *817*
amnion *817*
amniotic egg *817*
blastocyst *822*
blastopore *816*
blastula *815*
chorion *817*
chorionic villus (plural, **chorionic villi**) *826*
cleavage *815*
colostrum *830*
development *814*
differentiate *814*
direct development *814*
ectoderm *816*
embryonic disk *823*
embryonic stem cell (ESC) *826*
endoderm *816*
extraembryonic membrane *817*
fetal alcohol syndrome (FAS) *813*
fetus *824*
gastrula *816*
gastrulation *816*
homeobox gene *819*
implantation *822*
indirect development *814*
induced pluripotent stem cell (iPSC) *827*
induction *819*
inner cell mass *822*
labor *827*
lactation *830*
larva (plural, larvae) *814*
mammary gland *830*
mesoderm *816*
metamorphosis *814*
morula *815*
organogenesis *816*
placenta *824*
stem cell *826*
therapeutic cloning *827*
yolk *814*
yolk sac *817*
zygote *815*

Thinking Through the Concepts

Fill-in-the-Blank

1. A fertilized egg is a very large cell. One of the first events of animal development is the division of this cell, a process called __________, to produce a solid ball of much smaller cells, the __________. Soon, a cavity opens in this ball of cells, producing a hollow ball called the __________. Movements of cells in this hollow ball during gastrulation produce the three embryonic tissue layers, the __________, __________, and __________.
2. The evolution of the amniotic egg allows development to occur in fully terrestrial animals that do not lay eggs in water. The amniotic egg contains four extraembryonic membranes. In both reptiles and mammals, the __________ encloses the embryo in a watery environment. In mammals, the __________ forms the embryo's part of the placenta. The allantois contributes the blood vessels of the __________, which connects the embryo to the placenta. Finally, the __________ contains stored food in reptile eggs, but in mammals, this is "empty" and principally forms the lining of the digestive tract.
3. Undifferentiated cells that can multiply and produce daughter cells of many different types are called __________. Embryonic cells of this type can differentiate into any cell type of the body; they are derived from the __________.
4. Genes that control the development of entire body segments and their accompanying limbs (if any) are called __________. They usually code for proteins called __________ that regulate the transcription of many other genes.
5. Milk production in the mammary glands is stimulated by a hormone from the pituitary gland called __________. Another hormone, __________, stimulates contraction of both the uterine muscles during childbirth and the muscles around the mammary glands that cause milk to be ejected from the glands.

Review Questions

1. Distinguish between indirect and direct development, and give examples of each.
2. Describe the structure and function of the four extraembryonic membranes found in reptiles, including birds. Are these four present in placental mammals? In what ways are their roles similar in reptiles as compared to mammals? How do they differ?
3. What is gastrulation? Describe gastrulation in amphibians.
4. Name two structures derived from each of the three embryonic tissue layers—endoderm, ectoderm, and mesoderm.
5. Describe induction.
6. In humans, where does fertilization occur, and what stages of development occur before the fertilized egg reaches the uterus?
7. Describe how the human blastocyst gives rise to the embryo and its extraembryonic membranes.

8. Explain how the structure of the placenta prevents mixing of fetal and maternal blood while allowing for the exchange of substances between the mother and the fetus.
9. How do changes in the breast prepare a mother to nurse her newborn? How do hormones influence these changes and stimulate milk production?
10. Describe the events leading to the expulsion of the baby and the placenta from the uterus. Explain why this is an example of positive feedback.

Applying the Concepts

1. A researcher obtains two frog embryos at the gastrula stage of development. She carefully removes a cluster of cells from a location that she knows would normally become neural tube tissue and transplants it into the second gastrula in a location that would normally become skin. Does the second gastrula develop two neural tubes? Explain your answer.
2. In almost all animals that produce eggs with very little yolk, the blastula stage is roughly spherical. Mammalian eggs, however, contain virtually no yolk but have flat embryonic disks, which undergo gastrulation much like spherical blastulas do. Propose an explanation for this difference in shape.
3. **BioEthics** Some people feel that the production of induced pluripotent stem cells means that research on embryonic stem cells is unnecessary. Take a position on this issue and defend it.

MB *Go to www.masteringbiology.com for practice quizzes, activities, eText, videos, current events, and more.*

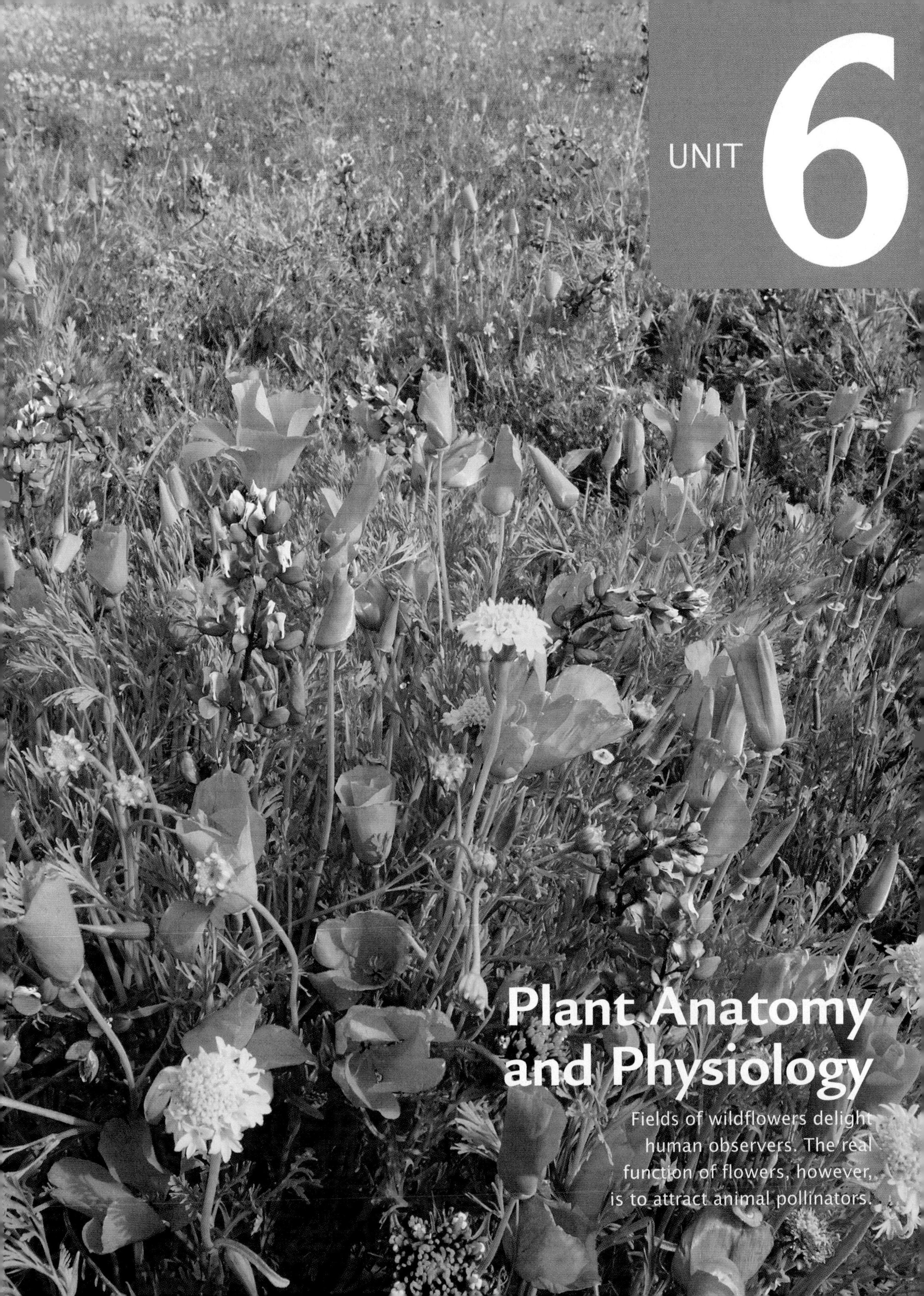

UNIT 6

Plant Anatomy and Physiology

Fields of wildflowers delight human observers. The real function of flowers, however, is to attract animal pollinators.

Plant Anatomy and Nutrient Transport

Chapter 43

▲ The flaming red colors of maple leaves in autumn delight human observers, but how do they benefit the trees?

Case Study

Why Do Leaves Turn Color in the Fall?

IT IS AUTUMN IN VERMONT, and undergraduate Will Young is cheerfully scaling ladders, collecting samples of brightly colored leaves. He is working with researchers from the University of Vermont and the U.S. Forest Service to help discover what environmental factors trigger the production of the red pigment, called anthocyanin, in autumn leaves. "This is Vermont," says Young. "Everyone cares about red leaves."

Tourists from around the world flock to the northeastern United States to enjoy the brilliant red, yellow, and orange autumn colors, and the mild sunny days and crisp chilly nights that bring on this display. Although more than 3 million "leaf peepers" pump at least $300 million into this small state's economy each year, the colors did not evolve to attract tourists. They presumably help the trees in some way—but how? The yellow and orange colors are no mystery. They have been there all along, helping with photosynthesis (see pp. 116–117). But anthocyanin is not involved in photosynthesis; it is newly synthesized by the aging leaves. Why would natural selection favor plants that spend energy synthesizing anthocyanin in leaves that are about to fall off?

In addition to questions of basic science, a practical concern has also arisen in states that rely on fall tourism to help support their economies: What impact will global warming have on fall colors? To investigate the environmental changes that trigger color change and leaf drop in the fall, researchers from the University of Vermont pump cooled antifreeze into tubes that coil around selected maple branches, chilling them to replicate extra-cold conditions. Cooling a branch causes its leaves to turn red and fall off, leaving the chilled branches bare while the uncooled neighboring branches remain festooned with yellow.

What are the functions of colorful leaf pigments? Why do leaves synthesize anthocyanin in the fall? How might global warming affect this process?

At a Glance

43.1 WHAT CHALLENGES ARE FACED BY ALL LIFE ON EARTH?

Most people admire fields of wildflowers and groves of giant sequoias, but seldom stop to think about the adaptations that allow plants to thrive. Plants cannot move to seek food or water, to escape predators, to avoid winter, or to find a mate. Yet, black spruce survive the harsh winters of central Alaska and the Yukon, mangroves perch in salt water along coastlines throughout the Tropics, cacti endure the searing heat of the Mojave Desert, and bristlecone pines in the White Mountains of California may live more than 4,000 years.

One of the best ways to appreciate plants is to consider how effectively they overcome the challenges encountered by all of life on Earth. To survive, organisms must:

- Obtain energy
- Obtain water and other nutrients
- Distribute water and nutrients throughout the body
- Exchange gases
- Support the body
- Grow and develop
- Reproduce

Evolution has produced a wide variety of distinctly different types of plants, as we described in Chapter 21. In this chapter, we will focus on the flowering plants, or angiosperms, which are the most widespread and diverse group of plants.

43.2 HOW ARE PLANT BODIES ORGANIZED?

The bodies of flowering plants consist of two major parts: the root system and the shoot system (**Fig. 43-1**). The **root system** consists of all the roots of a plant. **Roots** are branched portions of the plant body, usually embedded in the soil, that carry out six major functions. They:

- Anchor the plant in the ground
- Absorb water and minerals (plant nutrients) from the soil

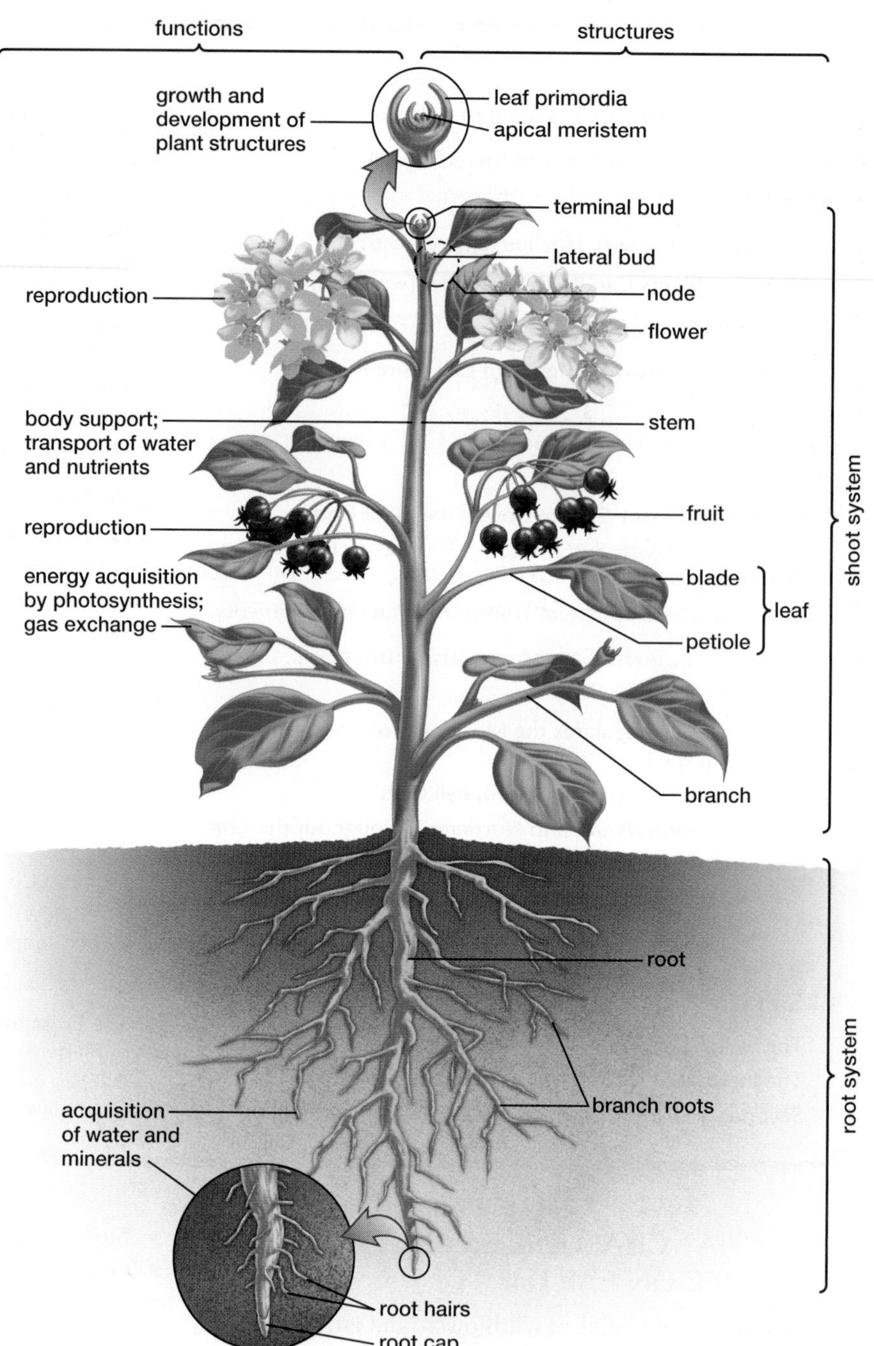

▶ **FIGURE 43-1 The structures and functions of a typical flowering plant** It will be helpful to refer back to this figure as we discuss each of these structures in the rest of the chapter.

- Store surplus food, principally carbohydrates, that were manufactured in the shoot during photosynthesis
- Transport water, minerals, sugars, and hormones to and from the shoot
- Produce some hormones
- Interact with soil fungi and bacteria that help provide nutrients to the plant

The **shoot system** is usually located aboveground. In flowering plants, the shoot system consists of buds, leaves, flowers, and fruits, all borne on various parts of stems (see Fig. 43-1). Buds give rise to leaves or flowers. Leaves are the principal sites of photosynthesis in most plants. Flowers are the plant's reproductive organs, producing male and female gametes, and helping them to reach one another. Flowers later produce seeds enclosed within fruits, which protect the seeds during their development and aid in seed dispersal. Stems, which are typically branched, usually elevate the leaves, flowers, and fruit above the ground. This helps the leaves to capture sunlight, and helps the flowers to attract animal pollinators or to release their pollen to the winds. Elevating the fruit helps to disperse the seeds. Some parts of the shoot system are specialized to transport water, minerals, and food molecules. Other parts produce hormones, which we will discuss in Chapter 45.

Flowering Plants Can Be Divided into Two Groups

The two broad groups of flowering plants are called monocots and dicots (**Fig. 43-2**). **Monocots** include lilies, daffodils, tulips, palm trees, and a wide variety of grasses (not only the familiar lawn grasses, but also wheat, rice, corn, oats, and bamboo). **Dicots** include virtually all "broad-leafed" plants, including deciduous trees and bushes, most vegetables, and many flowers in fields and gardens. Although there are differences between monocots and dicots in both root and shoot systems, the characteristic that gives the groups their names is the number of cotyledons. A cotyledon is the part of a plant embryo that absorbs and often stores food reserves in the seed and then transfers the food to the rest of the embryo when the seed sprouts, as we will describe in Chapter 44. When you eat a peanut, each half of the "nut" consists almost entirely of a cotyledon. As their names imply, monocots have a single cotyledon ("mono" means "one"), and dicots have two ("di" means "two").

43.3 HOW DO PLANTS GROW?

Animals and plants develop in dramatically different ways. One difference is the timing and distribution of growth. In most animals, the proportions of a newborn may differ from those of an adult (think of the large heads of babies), but all parts of a newborn's body become larger until they reach their adult size and structure, and then they usually stop growing. In contrast, flowering plants grow throughout their lives, never reaching a stable adult body form. Moreover, most plants grow longer or taller only at the tips of their branches and roots. For example, a swing tied to a tree branch or initials carved in tree bark do not move farther up from the ground as the tree grows.

During Plant Growth, Meristem Cells Give Rise to Differentiated Cells

Plants are composed of two fundamentally different types of cells: meristem cells and differentiated cells. **Meristem cells,** like the stem cells of animals, are unspecialized and are capable of mitotic cell division (see Chapter 9). Some of their daughter cells lose the ability to divide, and become **differentiated cells,** with specialized structures and functions. Continued divisions of meristem cells keep the plant growing throughout its life, whereas the differentiated daughter cells form the nongrowing parts of the plant, such as mature leaves.

Plants grow as a result of cell division and differentiation of meristem cells found in two general locations in the plant body. **Apical meristems** ("tip meristems") are located at the tips of roots and shoots (see Figs. 43-8 and 43-13). Growth produced by apical meristem cells is called **primary growth:** an increase in the height or length of a shoot or root and the development of specialized parts of the plant, such as leaves and buds. In both roots and shoots, primary growth

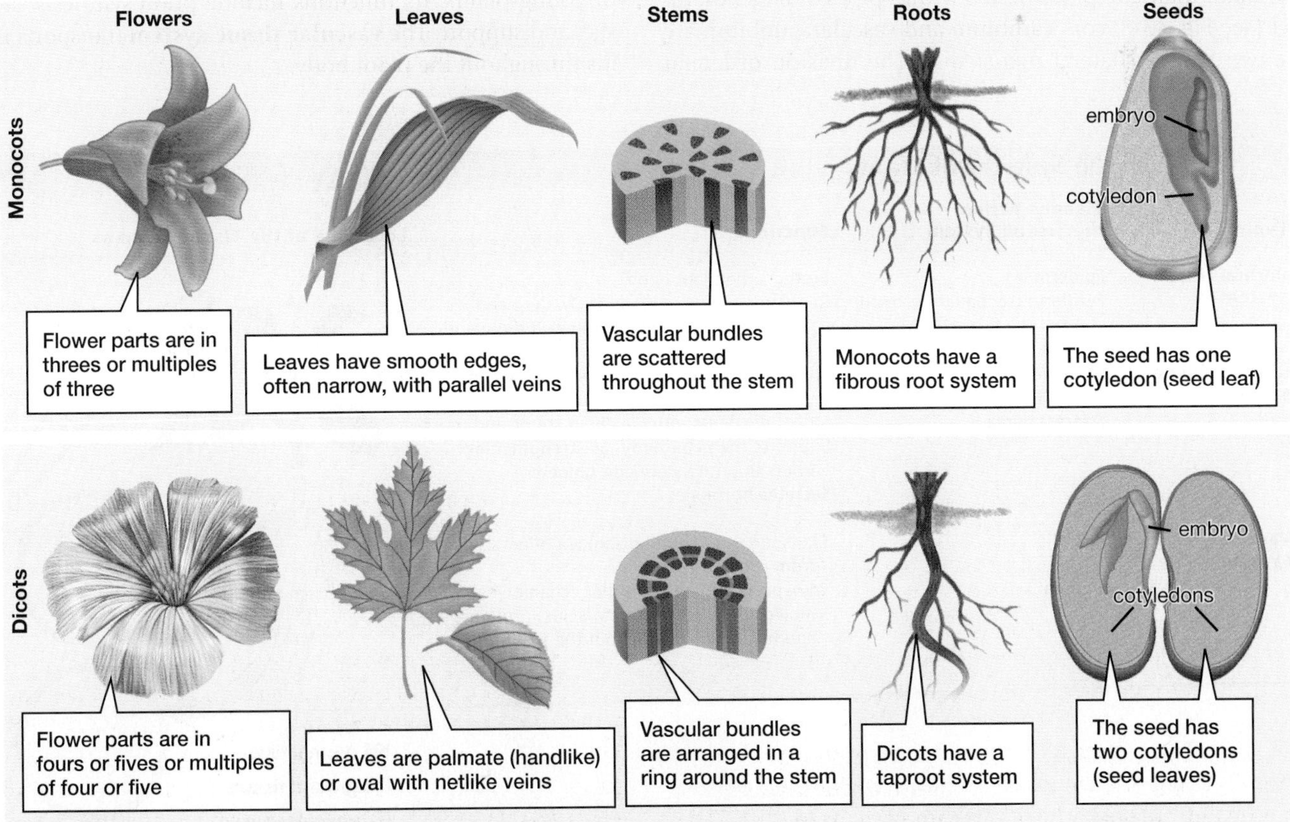

▲ **FIGURE 43-2 Characteristics of monocots and dicots**

Have you ever wondered

How Trees Can Live So Long?

Some trees, such as giant sequoias, redwoods, and bristlecone pines, can live more than 1,000 years; the oldest known bristlecone pine is about 4,800 years old! Most truly ancient trees have at least three of the following four characteristics. First, even relatively short-lived trees don't have localized, absolutely essential parts, like a brain or a heart. If a branch or part of a root dies, the rest of the tree lives on. Second, many trees don't really grow old. Because their meristems never stop producing new cells, they always have some young roots and branches. Third, many ancient trees produce resins or other substances that prevent attack from bacteria or fungi. Fourth, they often live in cold or dry environments, which limit the growth of bacteria and fungi and also limit undergrowth that might catch fire and burn down the trees. Barring accidents such as a lightning strike or major climate change, a bristlecone pine, which has all of these advantages, might live for tens of thousands of years.

also produces a complex arrangement of internal structures that we will explore in sections 43.5, 43.6, and 43.7. Primary growth from apical meristems explains why your childhood swing never got any higher off the ground—the tree grew taller only at the tips of its branches.

Lateral meristems ("side meristems"; also called **cambium**) are concentric cylinders of meristem cells, much like small-diameter pipes nested within pipes of larger diameter (see Fig. 43-8; cork cambium and vascular cambium are the two types of lateral meristems). The division of lateral meristem cells and differentiation of their daughter cells produce further concentric cylinders of **secondary growth,** typically an increase in the diameter and strength of roots and shoots. (In Fig. 43-8, cork, secondary xylem, and secondary phloem are the tissues formed by daughter cells of the lateral meristems.) Secondary growth occurs in woody plants, including deciduous trees, many shrubs, and most conifers such as pines and spruces. Some woody plants become very tall and thick and may live for hundreds, or even thousands, of years. Secondary growth is why the branch holding up your swing, and the trunk of the tree bearing the branch, both became thicker and stronger over the years.

Many plants do not undergo secondary growth. As you might predict, most plants that lack secondary growth are soft bodied, with flexible, usually fairly short, stems. These herbaceous, typically short-lived plants include lettuce, beans, lilies, and grasses.

43.4 WHAT ARE THE TISSUES AND CELL TYPES OF PLANTS?

As meristem cells differentiate, they produce a wide variety of cell types. When one or more specialized types of cells work together to perform a specific function, such as conducting water and minerals, they form a **tissue.** Functional groups of more than one tissue are called **tissue systems.** The plant body is composed of three tissue systems (**Table 43-1**). The **dermal tissue system** covers the outer surface of the plant body. The **ground tissue system** makes up most of the body of young plants. Its functions include photosynthesis, storage, and support. The **vascular tissue system** transports fluids throughout the plant body.

Table 43-1 Tissue Systems of Plants

Type	Tissues Within the Tissue System	Functions	Locations of the Tissue Systems
Dermal tissue system	Epidermis Periderm (secondary growth)	Protects the plant body Regulates the movement of O_2, CO_2, and water vapor between the air and the plant	
Ground tissue system	Parenchyma Collenchyma Sclerenchyma	Photosynthesizes; principally in leaves and young stems Stores nutrients; principally in stems and roots Supports the plant body, as strengthening fibers in both xylem and phloem Secretes hormones	
Vascular tissue system	Xylem Phloem	Transports water and dissolved minerals from root to shoot Transports sugars and other organic molecules, such as amino acids, proteins, and hormones throughout the plant body	

The Dermal Tissue System Covers the Plant Body

Plants have two types of dermal tissue: epidermal tissue and periderm. **Epidermal tissue** forms the **epidermis,** the outermost cell layer covering the leaves, stems, and roots of all young plants (see Figs. 43-6, 43-8, and 43-13). Epidermal tissue also covers flowers, seeds, and fruit. In herbaceous plants, the epidermis forms the outer covering of the entire plant body throughout its life. The epidermal tissue of the above-ground parts of a plant is generally composed of tightly packed, thin-walled cells, covered with a waterproof, waxy **cuticle** secreted by the epidermal cells (see Fig. 43-6). The cuticle reduces the evaporation of water from the plant, and helps protect it from the invasion of disease microorganisms. As we will see in sections 43.5, 43.6, and 43.9, adjustable pores regulate the movement of water vapor, O_2, and CO_2 across the epidermis of leaves and young stems. In contrast, the epidermal cells of roots are not covered with cuticle, which would prevent them from absorbing water and minerals.

Periderm replaces epidermal tissue on the roots and stems of woody plants as they age. Periderm is composed primarily of multiple layers of cork cells on the outside of the root or stem, and a layer of lateral meristem tissue called the cork cambium, which generates them (see pp. 847–848, and Figs. 43-8 and 43-9). Cork cells produce thick, waterproof cell walls as they grow, and then die when they reach maturity. They prevent water loss and protect the plant from damage. Because of the multiple layers of waterproof cork cells on their surfaces, root segments that are covered with periderm help to anchor the plant in the soil, but can no longer absorb water and minerals.

The Ground Tissue System Comprises Most of the Young Plant Body

The ground tissue system consists of all of the tissues of the plant body except dermal and vascular tissues. The three types of ground tissues are parenchyma, collenchyma, and sclerenchyma.

Parenchyma, the most abundant ground tissue, makes up most of the body of a young plant. The cells of parenchyma tissues, logically called parenchyma cells, have thin cell walls (see pp. 61–63) and are alive at maturity (**Fig. 43-3a**). They typically carry out most of the plant's metabolic activities, including photosynthesis (most of the cells of a leaf are parenchyma cells), secretion of hormones, and food storage. Potatoes, seeds, fruits, and storage roots such as carrots are packed with parenchyma cells that store various types of sugars and starches. Parenchyma cells also help to support the bodies of many plants, especially herbaceous plants. Some parenchyma cells can divide. Finally, in addition to making up much of the

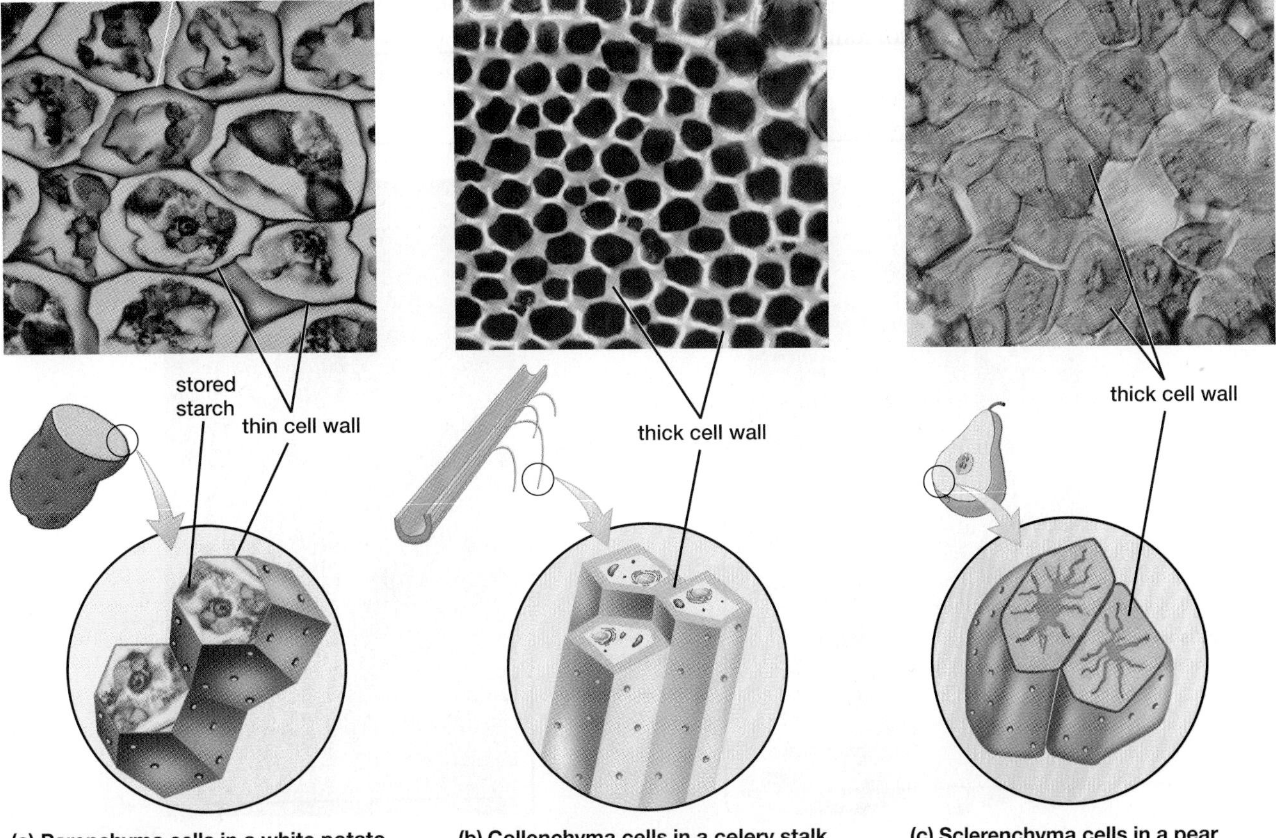

▲ **FIGURE 43-3 The structure of ground tissue (a)** Parenchyma cells are living, with thin cell walls. They perform many functions, including photosynthesis, hormone secretions, and storage. **(b)** Collenchyma cells are living and have thickened, but somewhat flexible, cell walls. They help support the plant body. **(c)** Sclerenchyma cells have thick, hard cell walls and die after they differentiate.

ground tissue system, parenchyma cells are found in periderm and vascular tissues.

Collenchyma tissue consists of cells that are typically elongated, with thickened, but still flexible, cell walls (**Fig. 43-3b**). Collenchyma cells are alive at maturity, but generally cannot divide. Collenchyma tissue provides support for the entire body of young and non-woody plants and for the leaf stalks, or petioles, of all plants. Celery stalks, which are actually extremely thick petioles, are supported by "strings" composed mostly of collenchyma cells.

Sclerenchyma tissue is composed of cells with thick, hardened cell walls (**Fig. 43-3c**). Like collenchyma, sclerenchyma cells support and strengthen the plant body; however, unlike collenchyma, they die after they differentiate. Their thick cell walls then remain as a source of support. Sclerenchyma cells form nut shells and the outer covering of peach pits. Scattered throughout the parenchyma cells in the flesh of a pear, sclerenchyma cells give pears their gritty texture. Sclerenchyma cells also support vascular tissues and form an important component of wood.

The Vascular Tissue System Transports Water and Nutrients

The vascular tissue system of plants serves a function somewhat like that of the blood vessels in animals—it conducts water and dissolved substances throughout the body. The vascular tissue system consists of two conducting tissues: xylem and phloem.

Xylem Transports Water and Dissolved Minerals from the Roots to the Rest of the Plant

Xylem transports water and dissolved minerals. As we will see in section 43.9, water and minerals travel only in one direction in xylem: from the roots up to all parts of the shoot system. In angiosperms, xylem contains supporting sclerenchyma fibers and two specialized conducting cell types: tracheids and vessel elements (**Fig. 43-4**). Both tracheids and vessel elements develop thick cell walls and then die as their final step of differentiation, leaving behind hollow tubes of nonliving cell wall.

Tracheids are thin, elongated cells, stacked atop one another. Their tapered, overlapping ends somewhat resemble the tips of hypodermic needles. The ends and sides of tracheids both contain **pits,** which are porous dimples in the walls that separate adjacent cells. Because the cell wall in a pit is both thin and porous, water and minerals can pass freely from one tracheid to another, or from a tracheid to an adjacent vessel element.

Vessel elements, which are larger in diameter than tracheids, form pipelines called **vessels.** Vessel elements are stacked end to end. Their adjoining end walls may be connected by fairly large holes, or the walls may disintegrate, leaving an open tube.

Phloem Transports Sugars and Other Organic Molecules Throughout the Plant Body

Phloem transports a solution containing a variety of organic molecules, including sugars, amino acids, and hor-

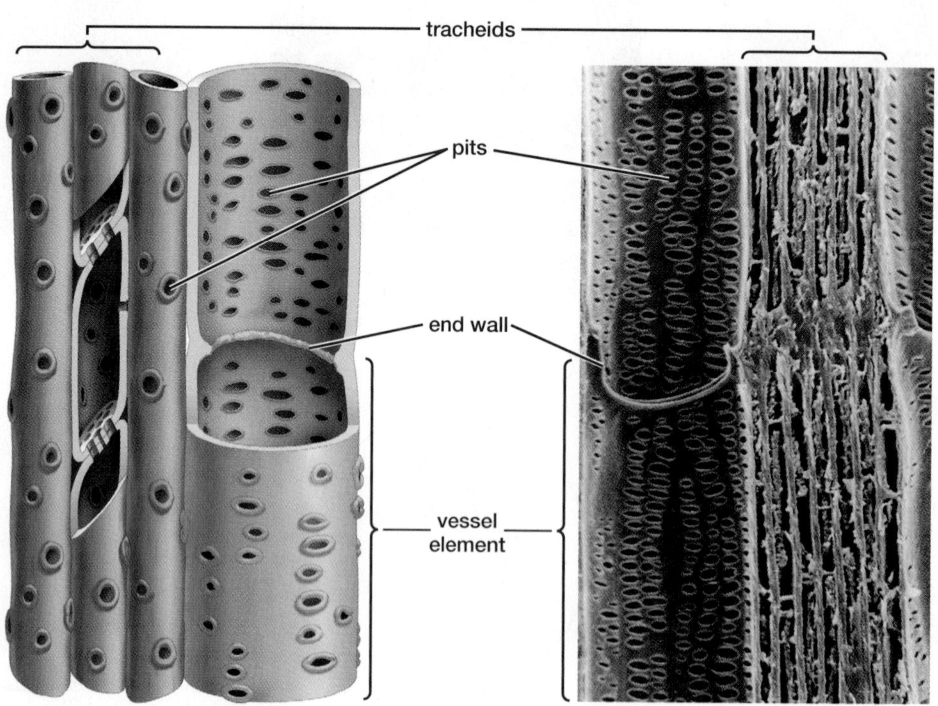

▲ **FIGURE 43-4 Xylem** Xylem contains two types of conducting cells: tracheids and vessel elements, as seen in a scanning electron micrograph on the right and the illustration on the left. Tracheids are thin with tapered ends; vessel elements are much larger in diameter and usually have blunt ends. Both the ends and sides of adjacent tracheids are connected by pits. Pits in their side walls also interconnect tracheids and vessel elements.

mones, from the structures that synthesize them to other structures that need them. Unlike xylem, in which fluids take a one-way trip from root to shoot, phloem transports fluids up or down the plant, to or from the leaves and roots, depending on the metabolic state of various parts of the plant at any given time.

Phloem consists mainly of two cell types: sieve-tube elements and companion cells (**Fig. 43-5**). **Sieve-tube elements** are joined end to end to form pipes called sieve tubes. As sieve-tube elements mature, they lose their nuclei and most other organelles, leaving behind only a thin layer of cytoplasm lining the plasma membrane. The junction between two sieve-tube elements is called a **sieve plate.** Here, membrane-lined pores connect the insides of the two sieve-tube elements, allowing fluid to move from one cell to the next.

As we will see in section 43.10, sieve-tube function requires an intact plasma membrane. How, then, can sieve-tube elements maintain and repair their plasma membranes when they lack nuclei and most other organelles? Life support for sieve-tube elements is provided by smaller, adjacent **companion cells,** which are connected to sieve-tube elements by pores called **plasmodesmata** (singular, plasmodesma; see p. 93). Companion cells help maintain the integrity of the sieve-tube elements by providing them with proteins and high-energy compounds such as ATP.

Like xylem, phloem also contains supporting sclerenchyma fibers.

43.5 WHAT ARE THE STRUCTURES AND FUNCTIONS OF LEAVES?

Leaves are the major photosynthetic structures of most plants. Their green color arises from light-absorbing chlorophyll molecules. The shapes and structures of leaves have evolved in response to the environmental challenges that plants face in obtaining the essentials for photosynthesis: sunlight, carbon dioxide (CO_2), and water. Water is absorbed from the soil by the roots, and is then transported to the leaves in the xylem. Assuming an adequate water supply, maximum photosynthesis would occur in a porous leaf (which would allow CO_2 to diffuse easily from the air into the leaf) with a large surface area (which would intercept the most sunlight). However, land plants cannot always obtain enough water from the soil. Consequently, on a hot, sunny day, a large, porous leaf would lose more water through evaporation than the plant could replace. The leaves of most flowering plants are an elegant compromise among these conflicting demands (**Fig. 43-6**): They have a large, mostly waterproof surface, with adjustable pores that can open and close to admit CO_2 or restrict water evaporation, as needed.

A typical angiosperm leaf consists of a broad, flat portion, the **blade,** connected to the stem by a stalk called the **petiole** (see Fig. 43-6). The petiole positions the blade, usually orienting the leaf for maximum exposure to the sun. Inside the petiole and blade, vascular tissues provide a conducting system between the leaf and the rest of the plant body.

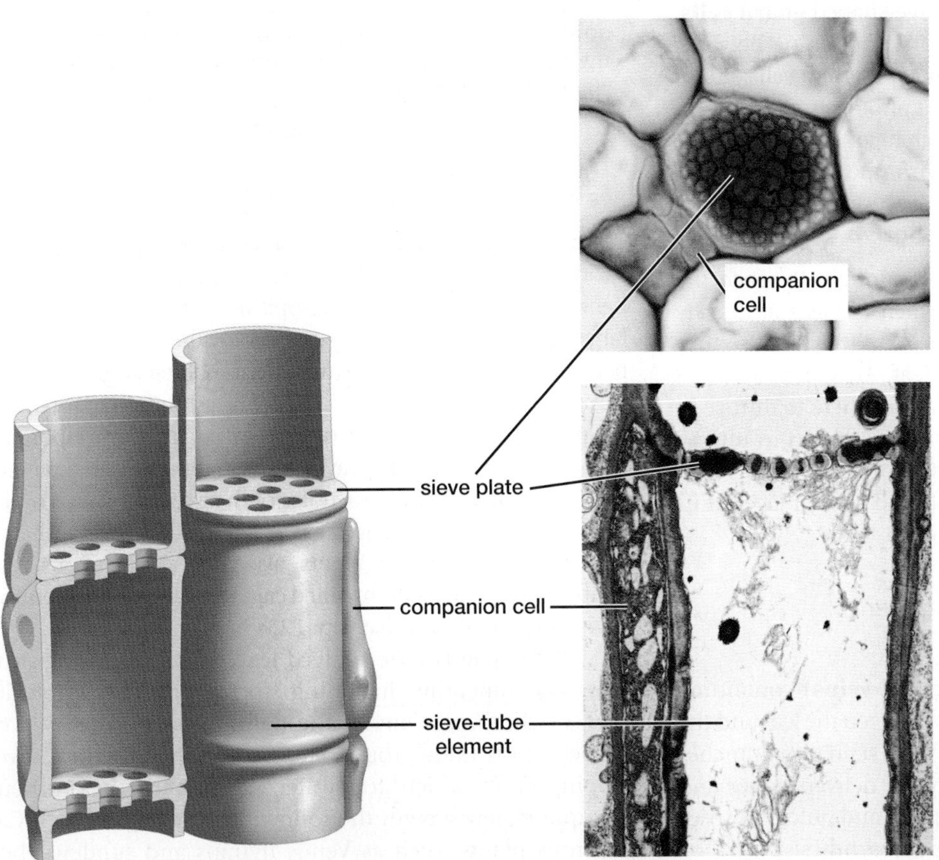

◀ **FIGURE 43-5 Phloem** Phloem includes sieve-tube elements and companion cells. Sieve-tube elements, stacked end to end, form the conducting system of phloem. Where they join at sieve plates, large membrane-lined pores allow fluid to move between them. Each sieve-tube element has a companion cell that nourishes it and regulates its function. The light micrographs on the right show a sieve plate (top) and a sieve-tube element and companion cell (bottom). Note the cytoplasm and organelles in the companion cell, compared to the almost-empty sieve-tube element.

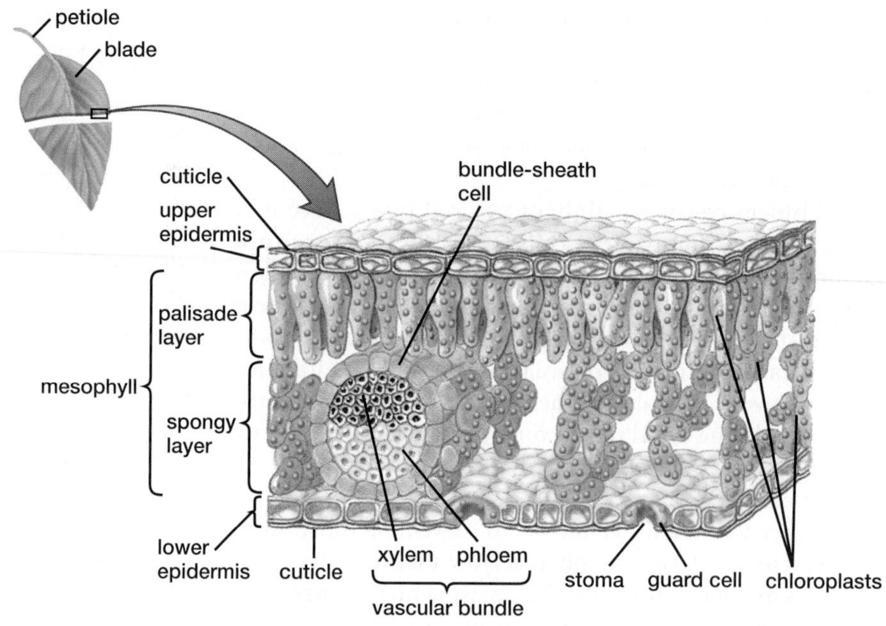

▶ **FIGURE 43-6 A typical dicot leaf** The cells of the epidermis lack chloroplasts and are transparent, allowing sunlight to penetrate to the chloroplast-containing mesophyll cells beneath. A waterproof cuticle covers the outside surfaces of the epidermal cells. The stomata that pierce the epidermis and the loose, open arrangement of the mesophyll cells ensure that CO_2 can diffuse into the leaf from the air and reach all of the photosynthetic cells.

The Epidermis Regulates the Movement of Gases into and out of the Leaf

The leaf epidermis consists of a layer of nonphotosynthetic, transparent cells that secrete a waxy cuticle on their outer surfaces. The cuticle is nearly waterproof and reduces the evaporation of water. The epidermis and cuticle are pierced by adjustable pores, the **stomata** (singular, stoma), which regulate the diffusion of CO_2, O_2, and water vapor into and out of the leaf. A stoma consists of two sausage-shaped **guard cells,** which enclose, and adjust the size of, an opening between them (see Fig. 43-6). Unlike the other epidermal cells, guard cells contain chloroplasts and carry out photosynthesis. In section 43.9, we will see how guard cells change shape to alter the size of the central opening to suit environmental conditions.

Photosynthesis Occurs in Mesophyll Cells

The transparent epidermal cells allow sunlight to reach the **mesophyll** ("middle of the leaf"), which consists of loosely packed cells containing chloroplasts. Mesophyll cells carry out most of the photosynthesis of a leaf. Air spaces between mesophyll cells allow CO_2 from the atmosphere to diffuse to each cell and O_2 produced during photosynthesis to diffuse away. Many leaves possess two types of mesophyll cells—an upper layer of columnar palisade cells and a lower layer of irregularly shaped spongy cells (see Fig. 43-6).

Veins Transport Water and Nutrients Throughout the Leaf

Vascular bundles (in leaves, also called **veins**) containing xylem and phloem conduct materials between the leaf and the rest of the plant body (see Fig. 43-6). Veins send thin branches close to each photosynthetic cell. Xylem delivers water and minerals to the mesophyll cells of the leaf, and phloem carries away the sugar they produce during photosynthesis.

Case Study continued
Why Do Leaves Turn Color in the Fall?

Photosynthesis relies on light energy trapped by pigments in the chloroplasts of the mesophyll cells—not only chlorophyll, but also orange and yellow carotenoid pigments (see p. 116), which absorb different wavelengths of light than chlorophyll does. As leaves age in the fall, chlorophyll is the first of the pigment molecules to break down. In leaves that turn yellow or orange, you are seeing the carotenoids, revealed when the more abundant green chlorophyll disappears.

Many Plants Produce Specialized Leaves

Temperature and the availability of water and light have exerted strong selection pressure on leaves. Plants growing in the dim light that reaches the floor of a tropical rain forest often have very large leaves, an adaptation demanded by the low light level and permitted by the abundant water in this habitat (**Fig. 43-7a**). The leaves of desert-dwelling cacti, in contrast, have been reduced to spines that provide almost no surface area for evaporation (**Fig. 43-7b**). The plump leaves of drought-resistant succulents store water in the central vacuoles of their cells, and are covered with a thick cuticle that greatly reduces water evaporation (**Fig. 43-7c**). Many succulents also use a water-conserving method of carbon fixation, called crassulacean acid metabolism, during photosynthesis (see pp. 123–124).

Some plants have evolved leaves with surprising structures and functions, including storing nutrients, capturing prey, or climbing. Onions and daffodil bulbs consist of extremely short underground stems enveloped by thick, overlapping leaves, which store water, sugars, and other nutrients to support new growth the following year (**Fig. 43-7d**). Carnivorous plants, such as Venus flytraps and sundews, bear

(a) Elephant ear leaves

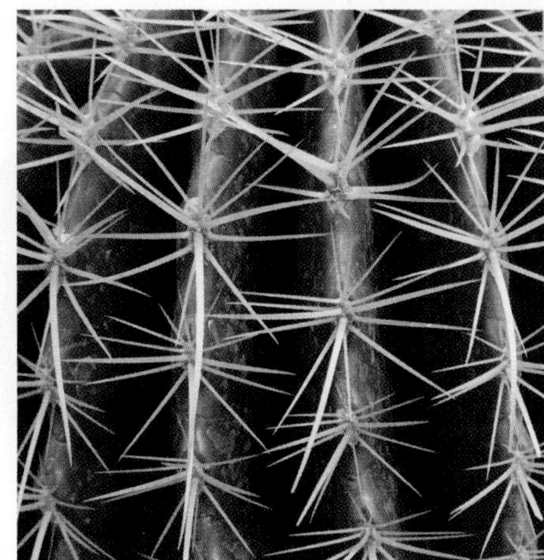

(b) Cactus spines

(c) Succulent leaves

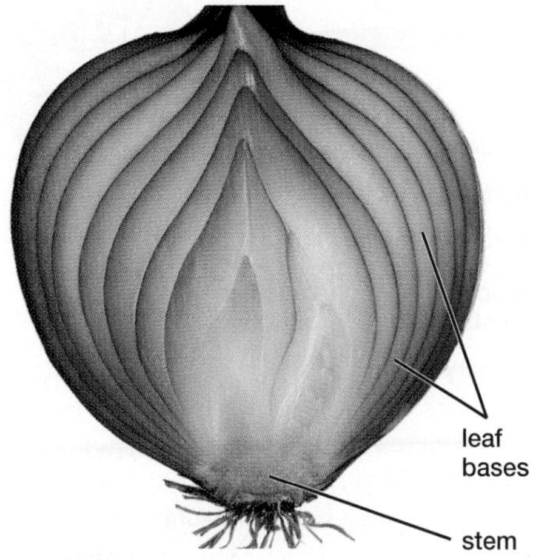

(d) Leaves of an onion

◀ **FIGURE 43-7 Specialized leaves** **(a)** Some plants that live on the rain-forest floor (such as this elephant ear) have enormous leaves to capture the limited light that filters through the taller trees. **(b)** Spines of desert cacti are nonphotosynthetic leaves whose surface area has been minimized, reducing evaporation and protecting the plant from browsing animals. **(c)** Succulent desert plants have fleshy leaves that store water from the infrequent rains. **(d)** An onion consists of a short central stem surrounded by thickened leaves that store water and food.

QUESTION As a gardener, should you harvest onions (which are biennials) at the end of their second year so they would be as large as possible?

leaves that are modified into snares that can trap and digest unwary insects (see the opening photo for Chapter 45). Pea plants produce thin, modified leaves called tendrils that coil around other plants, or the trellis in your garden, helping the plant to reach the sunlight.

43.6 WHAT ARE THE STRUCTURES AND FUNCTIONS OF STEMS?

The **stems** of the plant shoot system support and separate the leaves, lifting them into the sunlight and air. Stems also transport water and dissolved minerals from the roots up to the leaves, and transport sugars produced in the photosynthetic parts of the shoot to the roots and other parts of the shoot, such as buds, flowers, and fruits. In most dicots, stems undergo primary growth during their first year and at the tips of their branches throughout life. In perennial dicots, older stems and branches undergo secondary growth. The structures produced by primary and secondary growth are illustrated in Figure 43-8.

At the tip of a newly developing shoot sits a **terminal bud** consisting of apical meristem cells surrounded by a wrapping of **leaf primordia** (singular, primordium), or developing leaves, that were produced by the meristem. During primary growth, most of the daughter cells of the apical meristem differentiate into the specialized cell types of leaves, buds, and the structures of the lengthening stem.

The Surface Structures of the Stem

The petiole of a leaf is attached to the stem at a location called a **node** (see Fig. 43-8). When nodes first form from the apical meristem, they are extremely close together, but the stem between the nodes elongates to form naked **internodes** (meaning literally "between nodes"). As most dicot shoots grow, small clusters of meristem cells, the **lateral buds,** are left behind at the nodes, usually in the upper angle between the leaf petiole and the surface of the stem. Under appropriate hormonal conditions, lateral buds sprout and grow into branches (see Chapter 45 for more about the control of bud sprouting). As the branch grows, it duplicates the development of the stem, including producing new leaves and lateral buds. The vascular tissues of the stem connect with the developing vascular tissues of the branch.

During the reproductive season, usually spring or summer, meristem cells form **flower buds,** generally in the same locations as a terminal or lateral bud would otherwise develop.

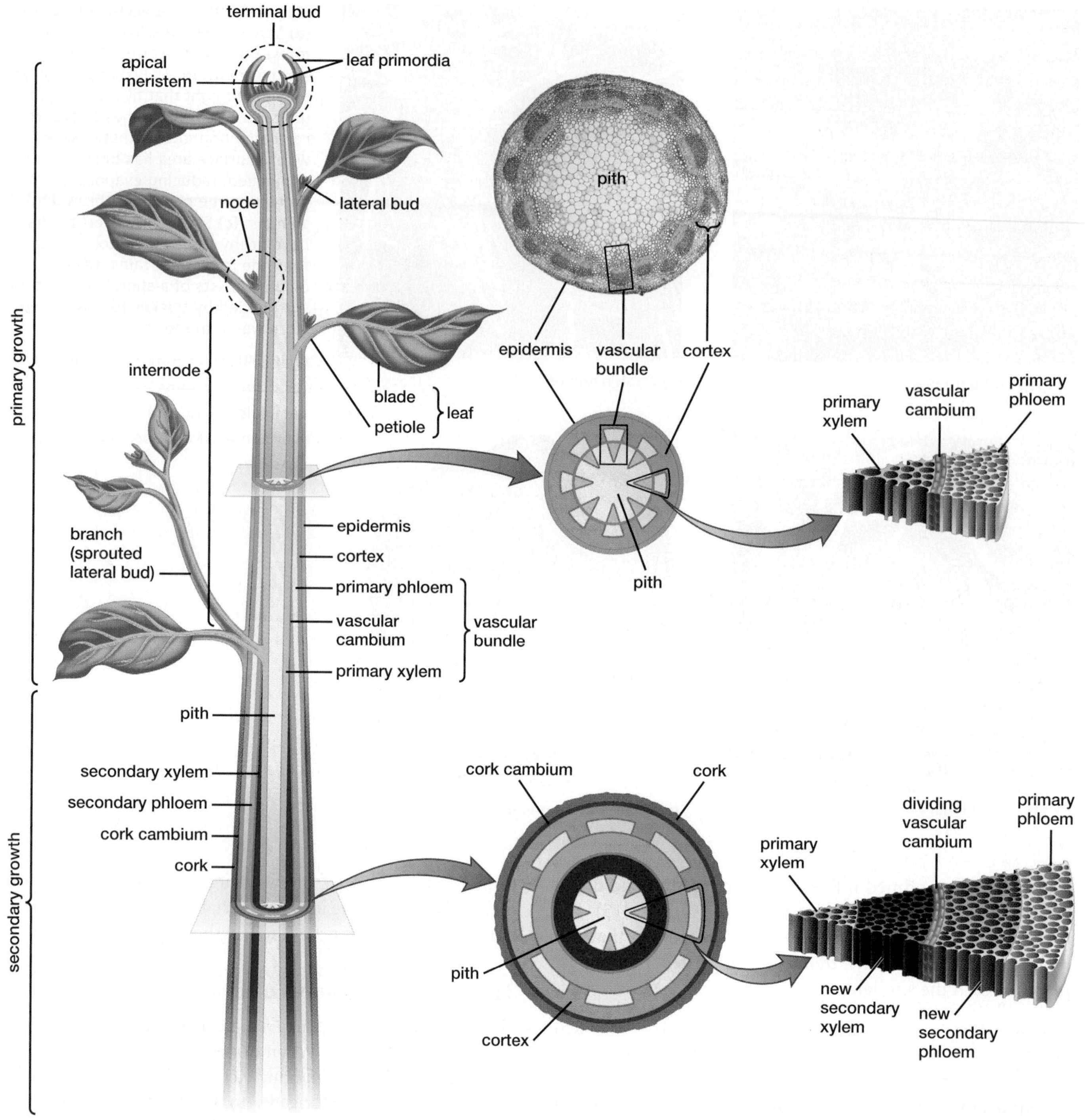

▲ **FIGURE 43-8 Primary and secondary growth in a dicot shoot** A dicot shoot after 2 years of growth, showing primary growth in the upper part of the shoot and secondary growth farther down. **(a)** A vertical slice of the stem, showing its internal structures. Cross-sections of **(b)** the entire stem and **(c)** an individual vascular bundle after primary growth (top) and secondary growth (bottom). During secondary growth, the vascular cambium forms a complete ring around the stem, so the resulting secondary xylem and phloem also form complete rings, rather than the discrete vascular bundles of primary xylem and phloem that form during primary growth.

The Internal Organization of the Stem

The apical meristem of a shoot also produces the internal structures of the stem, typically grouped into four tissues: epidermis, cortex, pith, and vascular tissues (see Fig. 43-8). Monocots and dicots differ somewhat in their arrangement of vascular tissues; we will discuss only dicot stems.

The Epidermis of the Stem Reduces Water Loss While Allowing Carbon Dioxide to Enter

The epidermis of a young stem, like that of leaves, secretes a waxy cuticle that retards water loss. As in leaves, the stem epidermis is perforated by stomata that regulate the movement of carbon dioxide, oxygen, and water vapor.

The Cortex and Pith Support the Stem, Store Food, and May Photosynthesize

Cortex and pith consist of parenchyma cells (ground tissue) that fill most of the young stem. In dicots, the **cortex** lies between the epidermis and the vascular tissues, and the **pith** fills the central part of the stem, surrounded by the vascular tissues. Cells of the cortex and pith perform three major functions: support, food storage, and (in some plants) photosynthesis:

- **Support** In very young stems, water filling the central vacuoles of cortex and pith cells causes turgor pressure (see Fig. 5-10). Turgor pressure stiffens the cells, much as air inflates a tire, and helps to hold the stem erect.
- **Storage** Parenchyma cells in both cortex and pith convert sugar into starch and store the starch as a food reserve.
- **Photosynthesis** In many young dicot stems, the outer cortex cells contain chloroplasts and carry out photosynthesis. In plants such as cacti, in which the leaves are reduced to spines, the cortex of the stem may be the only green, photosynthetic part of the plant.

Vascular Tissues in the Stem Transport Water, Dissolved Nutrients, and Hormones

Most young dicot stems contain bundles of vascular tissue, produced by primary growth from the apical meristem. Each bundle can be thought of as a rod of vascular tissue, often triangular in cross-section, extending the full height of the stem (**Fig. 43-8a**). A dozen or more bundles are arranged in a ring within the stem (**Fig. 43-8b**, top). The pointed inner edge of each bundle consists of primary xylem, whereas the rounded outside edge consists of primary phloem (**Fig. 43-8c**, top). (These are called primary xylem and primary phloem because they are produced from an apical meristem during primary growth.) Xylem carries water and dissolved nutrients up through the stem, supplying both stem and leaf cells. Phloem carries sugar from photosynthetic cells in the leaves and outer stem cortex to the roots, and to the pith and inner cortex, where some is metabolized for energy and the some is stored as starch. Between the primary xylem and primary phloem lies a thin strip of lateral meristem cells called the vascular cambium.

Secondary Growth Produces Thicker, Stronger Stems

In some conifers and perennial dicots, stems may survive for years, decades, or even centuries, becoming thicker and stronger each year. This secondary growth in stems results from cell division in the lateral meristems of the vascular cambium and cork cambium (see Fig. 43-8).

Vascular Cambium Produces Secondary Xylem and Secondary Phloem

The **vascular cambium** forms a cylinder of meristem cells that lies between the primary xylem and primary phloem (see Fig. 43-8a, bottom). Daughter cells of the vascular cambium produced toward the inside of the stem differentiate into secondary xylem; those produced toward the outside of the stem differentiate into secondary phloem (see Fig. 43-8c, bottom).

The cells of secondary xylem, with their thick cell walls, form the wood in perennial woody plants, including trees and woody shrubs. Young secondary xylem, called **sapwood,** transports water and minerals and is located just inside the vascular cambium (**Fig. 43-9**). Older secondary xylem, the **heartwood,** fills the central portion of older stems. Heartwood no longer carries water and solutes, but continues to provide support and strength to the stem.

Phloem cells are much weaker than xylem cells. Over time, the sieve-tube elements and companion cells are crushed between the hard xylem on the inside of the trunk and the tough cork on the outside. Only a thin strip of recently formed phloem remains alive and functioning.

In trees adapted to regions where there are pronounced seasons, cell division in the vascular cambium ceases during the cold of winter. In spring and summer, the cambium cells divide, forming new secondary xylem and phloem. In many trees, xylem cells formed in spring and early summer are large, with thin cell walls; xylem cells formed in late summer and early fall are smaller in diameter, with thicker cell walls. As a result, tree trunks in cross-section often show a pattern of **annual rings** of alternating pale regions (large cells with thin walls) and dark regions (small cells with thick walls), as shown in Figure 43-9. In many tree species, the approximate age can be determined by counting the annual rings. The widths of the rings also provide information about past climate, because wet years produce more growth and wider rings.

Cork Cambium Replaces the Epidermis with Cork

Epidermal cells are mature, differentiated cells that usually no longer divide. As new secondary xylem and phloem are added each year, increasing the diameter of a stem, the epidermis splits off. Apparently stimulated by hormones, some cells in the epidermis, cortex, or phloem (often all three at different times in the plant's life) become rejuvenated and form a layer of lateral meristem cells, the **cork cambium** (see Figs. 43-8b, bottom, and 43-9). These cells divide, forming daughter cells called **cork cells** or simply cork, with tough, waterproof cell walls that protect the trunk both from drying out and from physical damage. Cork cells die as they mature, and they may

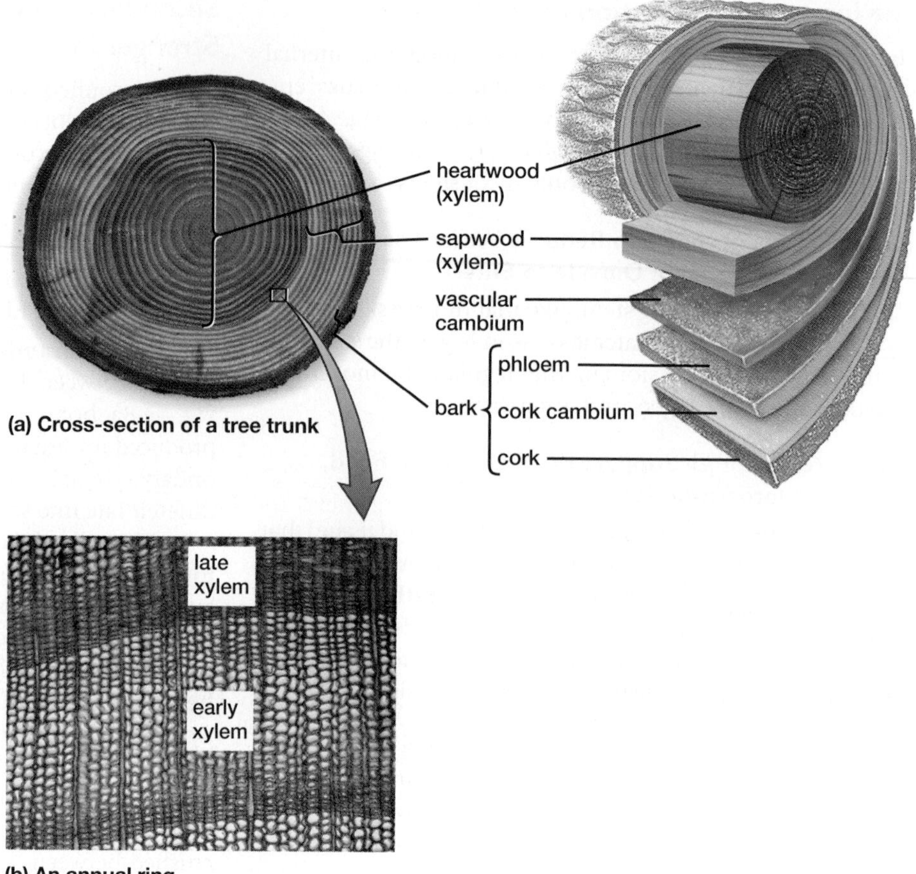

(a) Cross-section of a tree trunk

(b) An annual ring

▶ **FIGURE 43-9 Annual rings**
(a) Most trees in temperate climates form annual rings of secondary xylem. Early wood, formed during spring and early summer, and consisting of large cells with thin cell walls, is pale. Late wood, usually formed in summer and early autumn, and consisting of smaller cells with thicker walls, is dark. **(b)** The junction between spring and summer wood in a single annual ring.

QUESTION Would you expect a tree from a climate that was uniformly warm and wet all year to have growth rings? Explain your answer.

form a protective layer a couple of feet thick in some large tree species, such as the sequoia (**Fig. 43-10a**). As the trunk expands from year to year, the outermost layers of cork split apart or peel off, accommodating the growth. Corks used to plug wine bottles are made from the outermost layer of cork from cork oaks, carefully peeled off by harvesters so as not to harm the underlying cork cambium (**Fig. 43-10b**).

The common term **bark** includes all of the tissues outside the vascular cambium: phloem, cork cambium, and cork cells (see Fig. 43-9). Removing a strip of bark all the way around a tree, called girdling, kills the tree because it severs the phloem. Without phloem, sugars synthesized in the leaves cannot reach the roots. Deprived of energy, the roots can no longer take up minerals, and the tree dies.

Many Plants Produce Specialized Stems or Branches

Flowering plants have evolved a variety of stem specializations. For example, many stems are adapted for storage. The fat, bulging trunks of the baobab tree contain water-storing parenchyma cells, which allow it to thrive in climates with sporadic rainfall (**Fig. 43-11a**). Cactus stems both photosynthesize and store water. The common white potato is actually an underground stem, packed with parenchyma specialized to store starch. Each "eye" of the potato is a lateral bud, ready to sprout a stem branch when conditions become favorable. Strawberries send out horizontal stems, called runners, that spread along the soil, sprouting new strawberry plants (**Fig. 43-11b**).

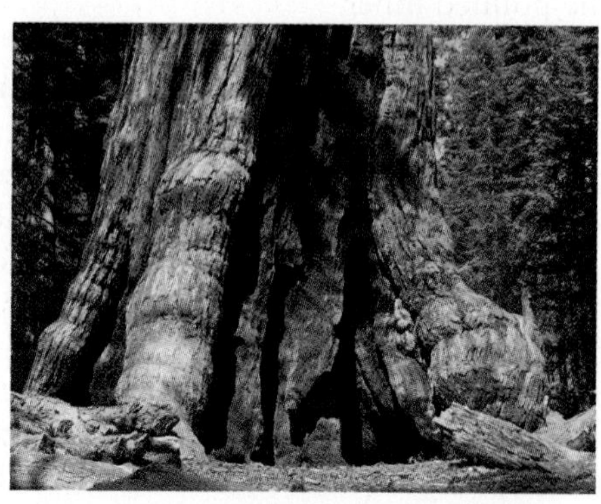

(a) Cork protects this giant sequoia tree

(b) Harvesting cork from a cork oak

▶ **FIGURE 43-10 Cork (a)** An ancient sequoia in the Sierra Nevada of California. The cork of a sequoia may be up to 2 feet thick (over half a meter). This massive, fire-resistant cork layer contributes to the sequoia's great longevity. Blackened areas are from past fires. **(b)** A layer of cork is stripped from a cork oak. The cork will regrow and can be harvested again in about a decade.

(c) Grape vine

(a) Baobab tree

(b) Strawberry plants

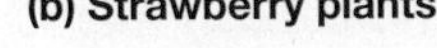

(d) Honeylocust tree

▲ **FIGURE 43-11 Specialized stems and branches (a)** The enormously expanded water-storing trunk of the baobab tree allows it to thrive in a dry climate. **(b)** Strawberry plants can reproduce using runners, which are horizontal stems. Where the node of a runner touches the soil, it may grow roots and develop into a complete plant. **(c)** Tendrils are specialized branches that allow grape vines to cling to trees or trellises. **(d)** Honeylocusts protect themselves with branches modified into strong, sharp, often forked thorns.

Certain branches of grapes and Boston ivy form grasping tendrils that coil around trees and trellises or adhere to buildings, giving the plant better access to sunlight (**Fig. 43-11c**). Thorns are branch adaptations that protect certain plants, such as hawthorns and honeylocust trees, from browsing by large herbivores (**Fig. 43-11d**).

43.7 WHAT ARE THE STRUCTURES AND FUNCTIONS OF ROOTS?

Roots anchor the plant in the soil, absorb water and minerals, and store water and food (primarily starches). Most dicots, such as carrots and dandelions, develop a **taproot system,** consisting of a central root with many smaller roots branching out from its sides (**Fig. 43-12a**). Most monocots, such as grasses and daffodils, produce a **fibrous root system** in which many roots of roughly equal size emerge from the base of the stem (**Fig. 43-12b**).

In young roots of both taproot and fibrous root systems, divisions of the apical meristem and differentiation of the resulting daughter cells give rise to four distinct regions (**Fig. 43-13**). Daughter cells produced on the lower side of the apical meristem differentiate into the root cap. Daughter cells produced on the upper side of the apical meristem differentiate into cells of the epidermis, cortex, and vascular cylinder.

(a) A taproot system

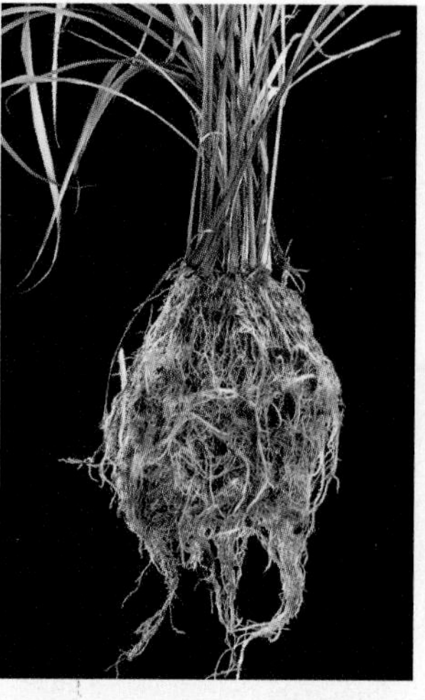

(b) A fibrous root system

▲ **FIGURE 43-12 Taproots and fibrous roots (a)** Dicots typically have a taproot system, consisting of a long central root with many smaller, secondary roots branching from it. **(b)** Monocots usually have a fibrous root system, with many roots of equal size.

The Root Cap Shields the Apical Meristem

The apical meristem is the source of primary growth in a root. The **root cap,** located at the very tip of the root, protects the apical meristem from being scraped off as the root pushes down between the rocky particles of the soil. Root-cap cells have thick cell walls and secrete a slimy lubricant that helps ease the root between soil particles. Nevertheless, root-cap cells wear away and must be continuously replaced by new cells from the apical meristem.

The Epidermis of the Root Is Permeable to Water and Minerals

A crucial function of most young roots is to absorb water and minerals from the soil. The root's outermost covering of cells is the epidermis, which is in contact with the soil and the water trapped among the soil particles. Unlike stem epidermis, which is covered in a waxy cuticle to reduce evaporation, root epidermis lacks a cuticle. Consequently, the walls of root epidermal cells are highly permeable to water and minerals. Further, many epidermal cells grow **root hairs** into the

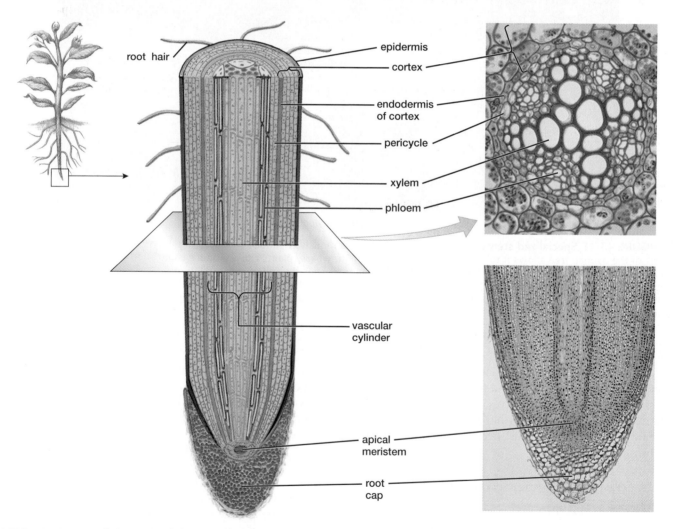

▲ **FIGURE 43-13 Primary growth in roots** Primary growth in roots results from mitotic cell division in the apical meristem near the tip. The root is composed of the root cap, epidermis, cortex, and vascular cylinder.

▲ FIGURE 43-14 **Root hairs** Root hairs, shown here in a sprouting radish, greatly increase a root's surface area, enhancing the absorption of water and minerals from the soil.

surrounding soil (**Fig. 43-14**; see also Fig. 43-13). By expanding the root's surface area, root hairs increase its ability to absorb water and minerals. Root hairs may add dozens of square meters of surface area to the roots of even small plants.

The Cortex Stores Food and Controls the Absorption of Water and Minerals into the Root

Cortex occupies most of the inside of a young root, between the epidermis and the vascular cylinder (see Fig. 43-13). Most of the cortex consists of large, loosely packed parenchyma cells with porous cell walls. Sugars produced in the shoot by photosynthesis are transported down to these cells, where they are converted to starch and stored. Roots of perennial plants store starches through the cold winter months, releasing them to power new growth of the shoot system in the spring. The cortex is especially large in roots specialized for carbohydrate storage, such as those of sweet potatoes, beets, carrots, and radishes (**Fig. 43-15**).

The innermost layer of cortex consists of a ring of small, closely packed cells called the **endodermis** that encircles the vascular cylinder (see Fig. 43-13). The cell wall of each endodermal cell contains a band of fatty, waterproof material called the **Casparian strip.** The Casparian strip is found between the cells (top, bottom, and both sides), but not on the inner or outer faces. Water and dissolved minerals can flow freely around both epidermal and cortex cells by moving through their porous cell walls. However, the Casparian strip blocks water and minerals from traveling between endodermal cells. We will explore the importance of the Casparian strip in section 43.8.

The Vascular Cylinder Contains Conducting Tissues and Forms Branch Roots

The **vascular cylinder** contains the conducting tissues of xylem and phloem. The outermost layer of the root vascular cylinder is the **pericycle,** located just inside the endodermis of the cortex and outside the xylem and phloem (see Fig. 43-13).

▲ FIGURE 43-15 **Specialized roots** Dicot taproots modified for nutrient storage include (left to right) beets, carrots, and radishes.

Pericycle cells work together with endodermal cells to regulate the movement of minerals and water into the xylem of the vascular cylinder.

The pericycle is also the source of branching in roots. Under the influence of plant hormones, pericycle cells start dividing, forming the apical meristem of a **branch root** (**Fig. 43-16**). Branch root development is similar to that of primary roots except that the branch must break out through the cortex and epidermis of the primary root. It does so both by crushing the cells that lie in its path and by secreting enzymes that digest them. The vascular tissues of the branch root connect with the vascular tissues of the primary root.

Roots May Undergo Secondary Growth

The roots of woody plants, including conifers and deciduous trees and shrubs, become thicker and stronger through secondary growth. Although there are some differences between

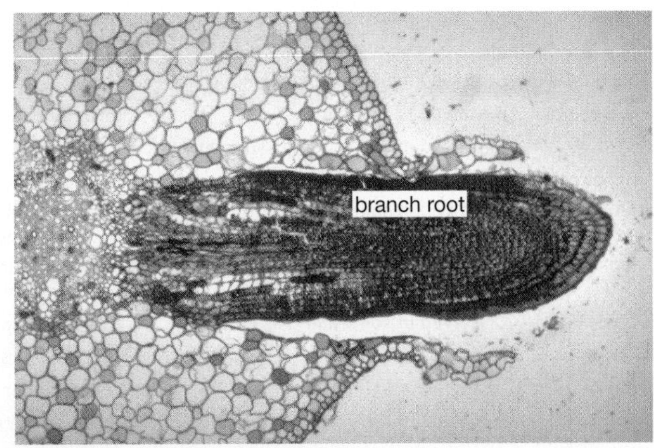

▲ FIGURE 43-16 **Branch roots** Branch roots emerge from the pericycle of a root. The center of this branch root is already differentiating into vascular tissue.

secondary growth in stems and roots, the essentials are similar: Vascular cambium produces secondary xylem and phloem in the interior of the root, and cork cambium produces a thick protective layer of cork cells on the outside.

43.8 HOW DO PLANTS ACQUIRE NUTRIENTS?

Nutrients are substances obtained from the environment that are required for the growth and survival of an organism (**Table 43-2**). Plants need only inorganic nutrients, because, unlike animals, plants can synthesize all of their own organic molecules. Some nutrients, called macronutrients, are required in large quantities; collectively these make up more than 99% of the dry weight of the plant body. Others, called micronutrients, are needed only in trace amounts.

Plants obtain carbon from carbon dioxide in the air, oxygen from the air or dissolved in water, and hydrogen from water. These three elements make up more than 95% of the mass of most plants. The other nutrients needed by plants are obtained by taking up **minerals** from the soil, either elements such as potassium (K^+) or calcium (Ca^{2+}), or small ionic compounds such as nitrate (NO_3^-) or phosphate (PO_4^{3-}).

Finally, a large part of the mass of a living plant is water. Water is also used to transport minerals, sugars, hormones, and other organic molecules throughout the plant body. Therefore, plants require large amounts of water *as water*, not only as a source of hydrogen and oxygen. For most plants, the primary source of water is the soil.

Roots Transport Minerals from the Soil into the Xylem of the Vascular Cylinder

Roots absorb minerals from the soil and transport them to the shoot. Soil consists of rock particles, air, water, and organic matter. Although the rock particles and the organic matter contain minerals, only minerals dissolved in the soil water can be taken up by roots. Because minerals are transported from root to shoot in the tracheids and vessel elements of xylem, a root must move minerals from the soil water to the xylem in the root's vascular cylinder. To understand how this occurs, let's begin by examining the structure of a root more closely (**Fig. 43-17**).

A young root is made up of (1) living cells; (2) extracellular space, mostly filled with the walls of these cells; and (3) the tracheids and vessel elements of xylem, which are dead and consist solely of cell walls. Most plant cell walls, including the walls of both the living cells and the tracheids and vessel elements, are very porous. Therefore, water and minerals can easily move through the extracellular space between the living cells, and from the extracellular space in the vascular cylinder into the insides of the tracheids and vessel elements. In roots, there is one exception to this general rule of porous cell walls: the Casparian strip. Like waterproof mortar in a brick wall, the Casparian strip is a waxy waterproofing on the top, bottom, and sides of the cells of the endodermis, but not on their inner or outer faces (**Fig. 43-17b**).

The Casparian strip divides the extracellular space of a root into two compartments, one outside the Casparian strip

Table 43-2 Essential Nutrients Required by Plants

Element*	Major Source	Function
Macronutrients		
Carbon	CO_2 in air	Component of all organic molecules
Oxygen	O_2 in air and dissolved in soil water	Component of all organic molecules
Hydrogen	Water in soil	Component of all organic molecules
Nitrogen	Dissolved in soil water (as nitrate and ammonia)	Component of proteins, nucleotides, and chlorophyll
Potassium	Dissolved in soil water	Helps control osmotic pressure, and regulates stomata opening and closing
Calcium	Dissolved in soil water	Component of cell walls; is involved in enzyme activation and the control of responses to environmental stimuli
Phosphorus	Dissolved in soil water (as phosphate)	Component of ATP, nucleic acids, and phospholipids
Magnesium	Dissolved in soil water	Component of chlorophyll; activates many enzymes
Sulfur	Dissolved in soil water (as sulfate)	Component of some amino acids and proteins; component of coenzyme A
Micronutrients		
Iron	Dissolved in soil water	Component of some enzymes; activates some enzymes; is required for chlorophyll synthesis
Chlorine	Dissolved in soil water	Helps maintain ionic balance across membranes; participates in splitting water during photosynthesis
Copper	Dissolved in soil water	Component of some enzymes; activates some enzymes
Manganese	Dissolved in soil water	Activates some enzymes; participates in splitting water during photosynthesis
Zinc	Dissolved in soil water	Component of some enzymes; activates some enzymes
Boron	Dissolved in soil water	Found in cell walls
Molybdenum	Dissolved in soil water	Component of some enzymes involved in nitrogen utilization

*Listed in approximate order of abundance in the plant body.

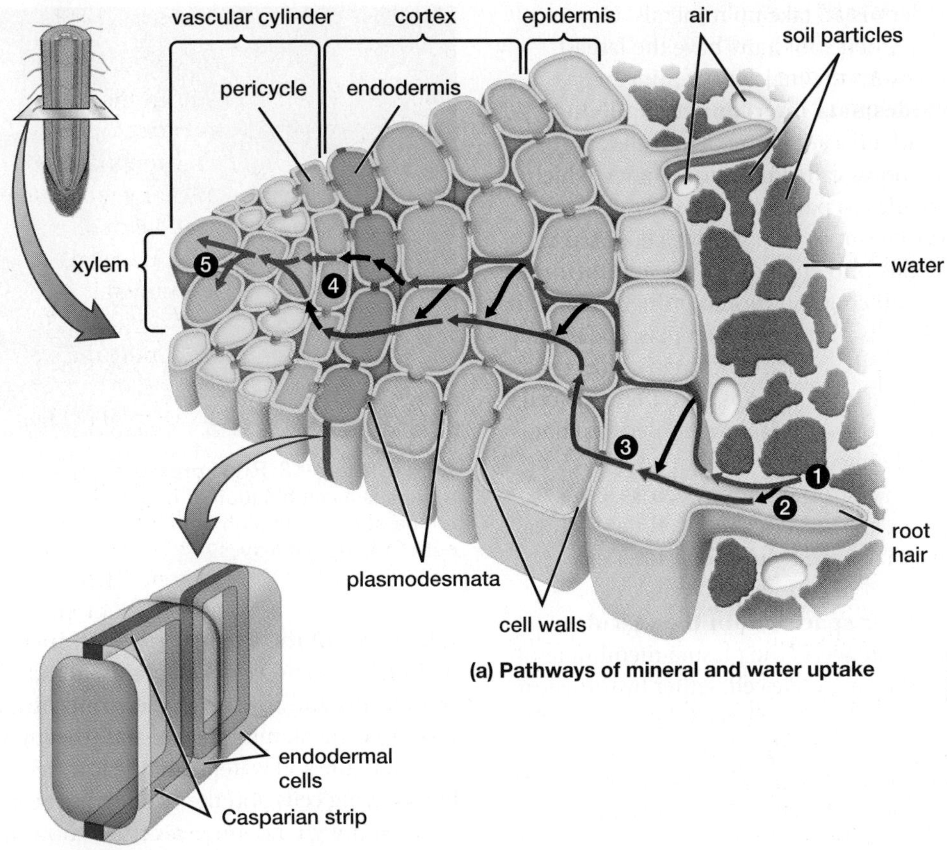

(a) Pathways of mineral and water uptake

(b) Endodermal cells, showing the Casparian strip

▲ FIGURE 43-17 **Mineral and water uptake by roots (a)** Mineral and water movement through a young root, from the soil water to the conducting cells of xylem. Minerals and water flowing through porous cell walls and extracellular space are shown as blue lines and arrows. Minerals and water crossing a plasma membrane are shown as black arrows. Minerals and water flowing from cell to cell through plasmodesmata are shown as red lines and arrows.

❶ Minerals dissolved in soil water fill the porous cell walls between the cells of the epidermis and cortex (blue pathway). The Casparian strip blocks water and minerals from moving between cells of the endodermis into the vascular cylinder.

❷ Minerals are actively transported across the plasma membranes of the root hairs, epidermal cells, cortex cells, and the outside face of the endodermal cells (black arrows). Water follows by osmosis.

❸ Minerals diffuse from cell to cell through plasmodesmata connecting the cytoplasm of epidermal, cortex, endodermal, and pericycle cells (red pathway).

❹ Minerals either diffuse or are actively transported across the plasma membranes of pericycle cells and the inside face of endodermal cells into the extracellular space inside the vascular cylinder (black arrows). Water follows by osmosis.

❺ Minerals and water enter the tracheids and vessel elements of xylem, passing freely from the extracellular space of the vascular cylinder through the pits in the walls of the tracheids and vessel elements (blue arrows).

(b) The location of the Casparian strip in the cell walls of endodermal cells.

and one inside the strip, in the vascular cylinder. Much like water in a bucket fills a sponge placed in the water, soil water fills the extracellular space of the outer compartment, carrying dissolved minerals with it (**Fig. 43-17a** ❶, blue pathway). The extracellular space of the inner compartment is continuous with the insides of the tracheids and vessel elements. To get from the soil water to the xylem, then, minerals must move from the outer compartment to the inner compartment, bypassing the Casparian strip.

To accomplish this, minerals absorbed by a root take a three-step journey (see Fig. 43-17a):

1. **From soil water into a living cell outside the Casparian strip.** Minerals in the soil water, including the water filling the outer compartment of extracellular space, are in contact with root hairs, epidermal cells, cortex cells, and the outer faces of endodermal cells. Minerals are taken up across the plasma membranes into the cytoplasm of these cells (**Fig. 43-17a** ❷, black arrows). The concentration of minerals in soil water is usually far lower than the concentration within plant cells, so most minerals are moved into the cells by active transport (see pp. 88–89). Cells of the epidermis,

cortex, and endodermis all take up minerals. However, epidermal cells and their root hairs have the largest surface area and absorb the majority of minerals.

2. **Through plasmodesmata interconnecting living cells.** The insides of adjacent living plant cells are connected by tiny pores called plasmodesmata, which allow small molecules to pass from the cytoplasm of one cell to the cytoplasm of neighboring cells. Active transport of minerals into epidermal cells and their root hairs produces a high concentration of minerals in their cytoplasm, so minerals diffuse through plasmodesmata from the epidermal cells, into and through cortex cells, into endodermal cells, and, perhaps, into pericycle cells (**Fig. 43-17a** ❸, red pathway). Mineral molecules that are absorbed directly into a cortex cell or an endodermal cell take a shorter journey across fewer layers of cells, but in all cases, this intracellular pathway bypasses the Casparian strip by moving minerals through the cytoplasm of endodermal cells.
3. **Into the extracellular space within the vascular cylinder.** Minerals exit across the plasma membrane of an endodermal cell or pericycle cell, either by diffusion or by active transport, into the extracellular space of the vascular cylinder (**Fig. 43-17a** ❹, black arrows). Remember that the tracheids and vessel elements consist only of porous cell walls, with no plasma membranes and no cytoplasm (see section 43.4). Therefore, minerals diffuse freely from the extracellular space into the tracheids and vessel elements of the xylem (**Fig. 43-17a** ❺, blue arrows).

The Casparian Strip Prevents the Loss of Minerals from the Vascular Cylinder

The Casparian strip may appear to be a major obstacle to mineral absorption, but in reality, the Casparian strip plays the essential role of leak-proofing the vascular cylinder. The soil water in the outer root compartment has a low concentration of minerals, and the inner compartment in the vascular cylinder acquires a much higher concentration of minerals, so a diffusion gradient exists that tends to drive the minerals back out of the vascular cylinder into the soil water. However, the Casparian strip blocks minerals from moving between the endodermal cells. Therefore, once inside the vascular cylinder, minerals can't leak back out again.

Roots Take Up Water from the Soil by Osmosis

As you know from Chapter 5, water moves passively across plasma membranes by osmosis, diffusing from areas of high free water concentration (with a low concentration of solutes) to areas of low free water concentration (caused by a high concentration of solutes; see p. 85). Soil water, with a low mineral concentration, has a high free water concentration. Minerals are actively transported into the outer cell layers of the root, move into the extracellular space of the vascular cylinder, and then into the tracheids and vessel elements of the xylem. The cytoplasm of the living cells and the extracellular space of the vascular cylinder, including the insides of the tracheids and vessel elements, all have high mineral concentrations and low free water concentrations. Thus, one or more plasma membranes separate the high free water concentration in the soil water from the low free water concentration in the living cells and the conducting cells of the xylem.

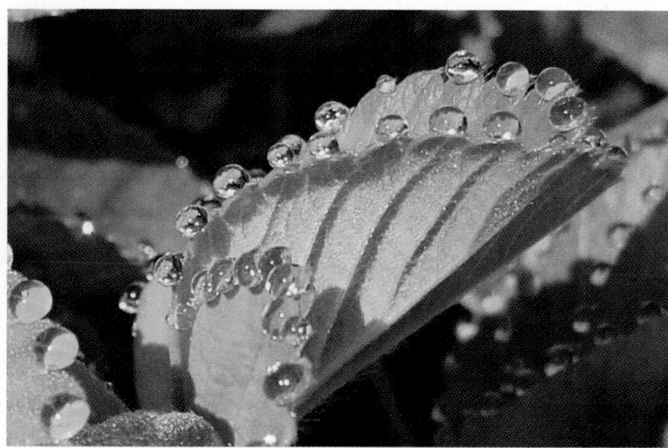

▲ **FIGURE 43-18 Root pressure** Sometimes, in early morning, if there is a combination of high soil water, relatively low temperatures, and high humidity, root pressure may force water out of the tips of leaves.

Although no one is certain how much water moves across which plasma membranes, the general principle is clear: Water moves by osmosis from the soil water, across plasma membranes (see Fig. 43-17a ❷, ❹), and into the tracheids and vessel elements (see Fig. 43-17a ❺). In some plants, this osmotic entry of water following mineral uptake is so powerful that it creates **root pressure:** Water entering the vascular cylinder actually pushes the solution of minerals up the root into the shoot. Sometimes, the effect of root pressure is visible as droplets of water, forced out of the tips of leaves (**Fig. 43-18**).

In most plants, under most conditions, osmosis following mineral uptake is not the major force for water entry into roots. Instead, water moving up the xylem, powered by the evaporation of water from the leaves, rapidly removes water from the tracheids and vessel elements in the roots. This provides a constant force pulling water molecules from the soil water, across the plasma membranes of root cells, and into the vascular cylinder. In section 43.9, we will see how water moves up the xylem, sometimes several hundred feet from the roots to the top of a tall tree.

Symbiotic Relationships Help Plants Acquire Nutrients

Many minerals, although abundant in rock particles, are too scarce in soil water to support plant growth. However, most plants have evolved mutually beneficial relationships with fungi that help them acquire these minerals. In addition, nitrogen is almost always in short supply in the soil. Plants called **legumes,** such as peas, soybeans, alfalfa, and clover, house bacteria in their roots that can convert nitrogen in the air into useful ammonia or nitrate.

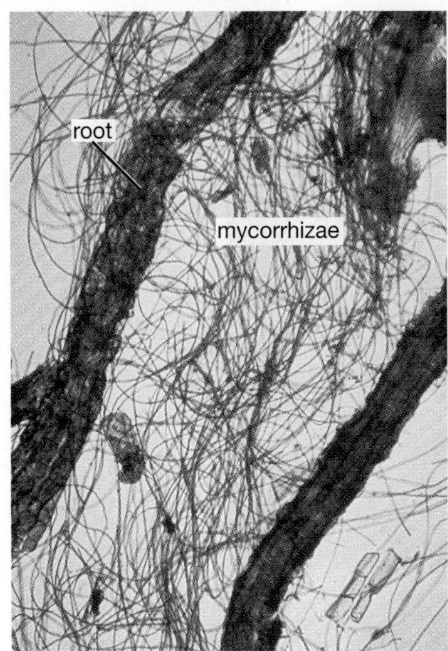

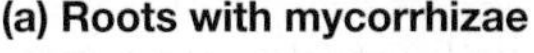
(a) Roots with mycorrhizae

(b) Mycorrhizae affect growth

◀ **FIGURE 43-19 Mycorrhizae: a root–fungus symbiosis** **(a)** A mesh of fungal strands surrounds and penetrates into a root. **(b)** Seedlings growing under identical conditions with (right) and without (left) mycorrhizal fungi illustrate the importance of mycorrhizae in plant nutrition.

QUESTION Based on what you've learned about root function, which part of the root system might you expect to be infected by mycorrhizal fungi?

Fungal Mycorrhizae Help Most Plants to Acquire Minerals

The roots of most land plants form symbiotic relationships with fungi. The resulting complexes, called **mycorrhizae** (literally "fungus root" in Greek), help the plant obtain scarce minerals from the soil. Microscopic fungal strands intertwine between the root cells and extend out into the soil (**Fig. 43-19**). The web of fungal filaments greatly increases the volume of soil from which minerals can be absorbed, compared to the volume in contact with the plant root alone. In addition, the fungus can extract some minerals, particularly phosphate, that are often bound in the rock particles of the soil and cannot be taken up by roots. Minerals absorbed by the fungus can then be transferred to the root. The fungus, in return, receives sugars, amino acids, and vitamins from the plant. In this way, both the fungus and the plant can grow in places where neither could survive alone, including deserts and high-altitude, rocky soils.

In some forests, mycorrhizae form an immense underground network that interlinks trees—even trees of different species. The fungi transfer sugars among the trees, causing those with access to abundant sunlight to subsidize their shaded neighbors. Thus, like an underground Robin Hood, the mycorrhizae transfer photosynthetic products from the rich to the poor.

Nitrogen-Fixing Bacteria Help Legumes to Acquire Nitrogen

Amino acids, nucleic acids, and chlorophyll all contain nitrogen, so plants need large amounts of it. Unfortunately, although nitrogen gas (N_2) makes up about 78% of the atmosphere and readily diffuses into the air spaces in the soil, plants can use nitrogen only in the form of ammonia (NH_3) or nitrate (NO_3^-). Some soil bacteria can perform **nitrogen fixation,** the conversion of N_2 into NH_3, which they then use to synthesize their own amino acids and nucleic acids. However, nitrogen fixation requires a lot of energy, using at least 12 ATPs to make only one molecule of NH_3. Consequently, these **nitrogen-fixing bacteria** don't routinely manufacture a lot of extra NH_3 to release into the soil.

Legumes and a few other plants enter into a mutually beneficial relationship with certain species of nitrogen-fixing bacteria. The bacteria enter the legume's root hairs and make their way into cortex cells. Both the bacteria and cortex cells multiply, forming a swelling, or **nodule,** composed of cortex cells full of bacteria (**Fig. 43-20**). The bacteria live off the root's food reserves. In fact, the bacteria obtain so much food that they produce more NH_3 than they need. The surplus NH_3 diffuses into the host cortex cells, providing the plant with usable nitrogen.

43.9 HOW DO PLANTS MOVE WATER AND MINERALS FROM ROOTS TO LEAVES?

In most plants, at least 90% of the water absorbed by the roots evaporates through the stomata of leaves. This evaporation, called **transpiration,** drives the movement of water upward through the plant body.

Water Movement in Xylem Is Explained by the Cohesion–Tension Theory

After entering the root xylem, water and minerals must be transported to the topmost parts of a plant. In redwood trees, this distance may be more than 350 feet (about 100 meters). Water flows up through the xylem from root to shoot. Because minerals are dissolved in the water, they are passively carried along as the water moves upward. But how do plants overcome the force of gravity and make water flow upward?

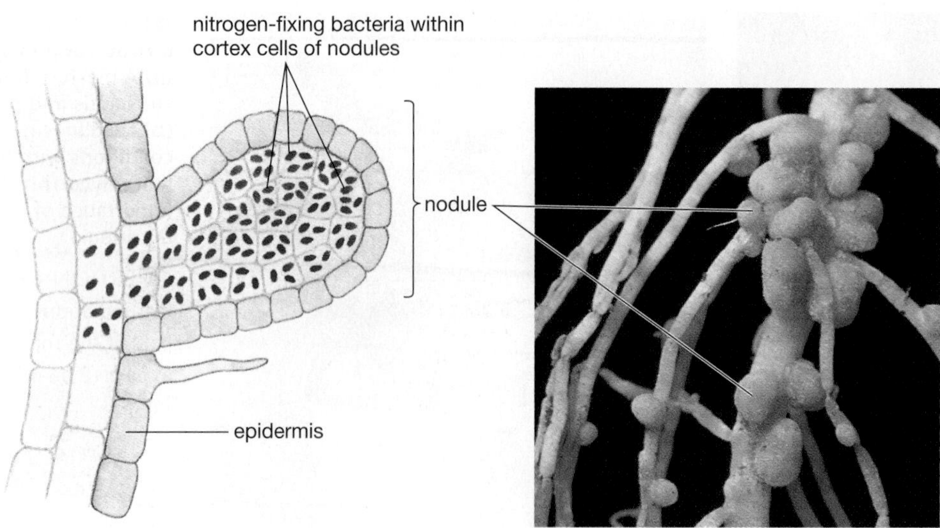

▲ **FIGURE 43-20 Root nodules house nitrogen-fixing bacteria**

According to the **cohesion–tension theory,** water is pulled up the xylem by transpiration from the leaves (**Fig. 43-21**). As its name suggests, this theory has two essential parts:

- **Cohesion** Attraction among water molecules holds water together, forming a chain-like column within the xylem tubes.
- **Tension** The chain of water is pulled up the xylem by the tension produced by water evaporating from the leaves.

Let's briefly examine both parts.

Hydrogen Bonds Between Water Molecules Produce Cohesion

Water is a polar molecule, with the oxygen end of the molecule bearing a slightly negative charge and the hydrogens bearing slightly positive charges (see Fig. 2-6). As a result, nearby water molecules attract one another electrically, forming hydrogen bonds (see Fig. 2-7). Just as individually weak cotton threads together make the strong fabric of your jeans, the network of individually weak hydrogen bonds in water produces a strong **cohesion,** the ability of a substance to resist being pulled apart. The column of water within the xylem is at least as strong—and as unbreakable—as a steel wire of the same diameter. This is the "cohesion" part of the theory: Hydrogen bonds among water molecules create a chain of water within the xylem, extending the entire height of the plant. Supplementing the cohesion between water molecules is adhesion resulting from hydrogen bonds between water molecules and the walls of xylem, helping to hold up the water chain.

Transpiration Produces the Tension That Pulls Water Upward

Transpiration provides the force for water movement—the "tension" part of the theory. Water evaporates from mesophyll cells into the air spaces within the leaf and through the stomata into the outside air (**Fig. 43-21** ❶). As water leaves the dehydrating mesophyll cells, their free water concentration decreases. Therefore, water moves by osmosis from nearby xylem, across the plasma membranes of the mesophyll cells, and into their cytoplasm.

Cohesion and Tension Work Together to Move Water Up the Xylem

Water molecules leaving the xylem are linked to other water molecules in the same xylem tube by hydrogen bonds—cohesion. Therefore, as water molecules leave the xylem to replace the water evaporating from mesophyll cells, they pull more water up the xylem—tension (**Fig. 43-21** ❷). The resulting pulling force in xylem is strong enough to lift water up more than 500 feet (over 150 meters), much taller than any living tree. This process continues all the way down to the roots, where water in the extracellular space of the vascular cylinder is pulled in through the porous pits in the walls of the vessel elements and tracheids of the xylem (**Fig. 43-21** ❸). Only the aboveground parts of a plant, usually the shoot, can transpire, so tension always pulls water up the xylem, never down. Therefore, water flow in xylem is unidirectional, from root to shoot.

A large maple tree may transpire about 250 gallons of water a day—almost a ton of water lifted up more than 50 feet, every day. Where does the energy come from? The sun. Sunlight warms both the leaves and the air, powering the evaporation of water from the leaves. Now imagine a whole forest, with each tree releasing hundreds of gallons of water into the air each day. This massive transpiration can have a major effect on the local climate, as we describe in "Earth Watch: The Surprising Impacts of Forests on Climate and Their Own Growth" on p. 858.

BioFlix™ Water Transport

Minerals Move Up the Xylem Dissolved in Water

The minerals absorbed by plant roots are charged ions, such as K^+, Ca^{2+}, PO_4^{3-}, or Cl^-. Ions dissolve in water because either

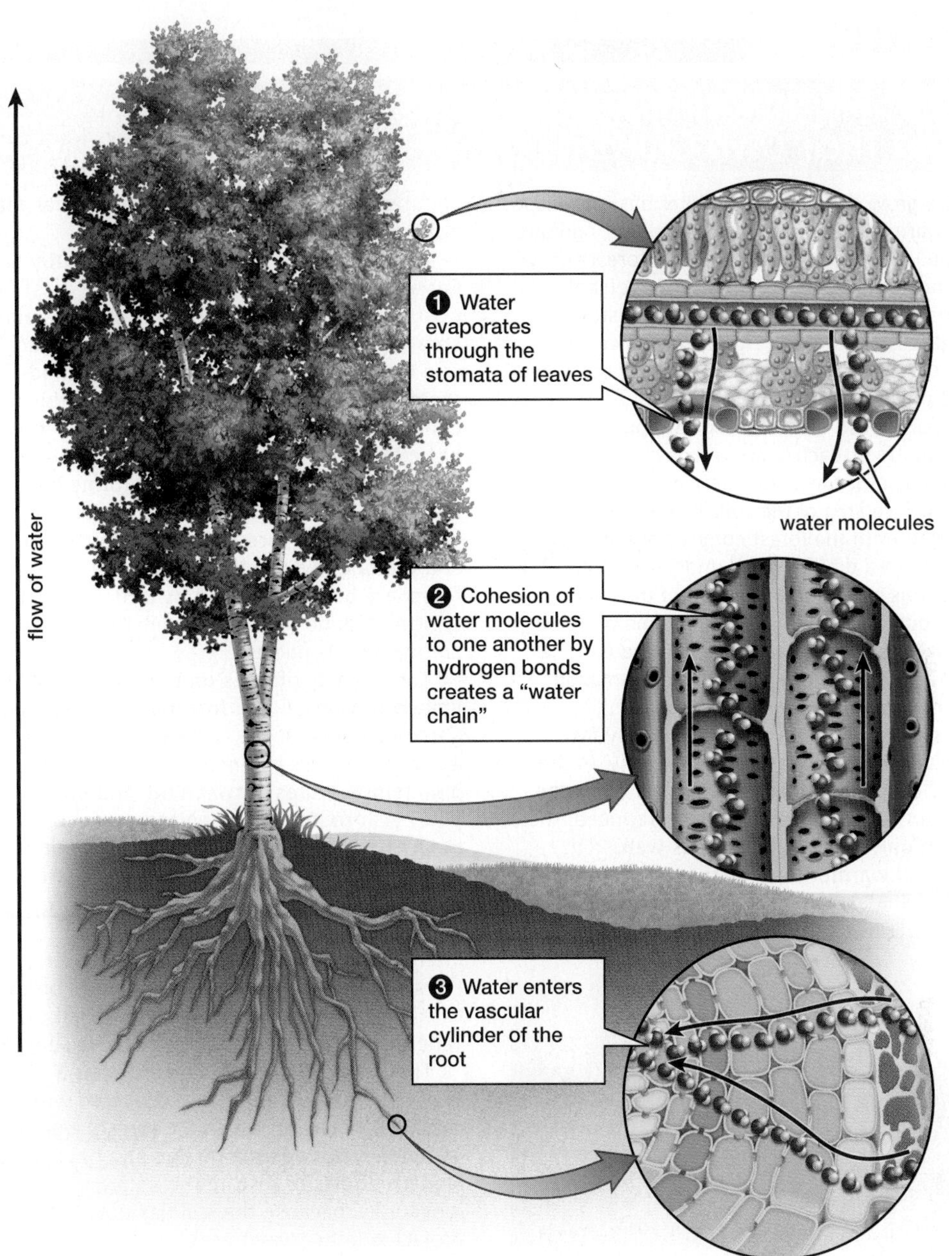

▲ **FIGURE 43-21 The cohesion–tension theory of water flow from root to leaf in xylem** ❶ As water molecules evaporate out of the leaves through transpiration, other water molecules replace them from the xylem of the leaf veins. ❷ As the top of the "water chain" is pulled up by evaporation, the rest of the chain, all the way down to the roots, comes along as well. ❸ As the molecules of the water chain travel up the xylem in the roots, the loss of water from the root xylem and the surrounding extracellular space causes water to enter from the soil, thus steadily replenishing the bottom of the chain.

the negative or the positive parts of water molecules are electrically attracted to the ions (see Fig. 2-9). In xylem, minerals are connected to the water molecules by electrical attraction and are carried along with the water from root to shoot.

Stomata Control the Rate of Transpiration

Transpiration is essential to plant life, because it powers the transport of water and minerals. Transpiration is also a great threat to plants, because it is by far the largest source of water loss—a loss that can be fatal in hot, dry weather. Water loss, of course, can be greatly reduced simply by closing the stomata through which the water evaporates. Closing the stomata, however, also prevents CO_2 from entering the leaf—and no CO_2 means no photosynthesis. Therefore, a plant must regulate its stomata to achieve a balance between acquiring carbon dioxide and losing water. In most plants, stomata open

Earth Watch

The Surprising Impacts of Forests on Climate and Their Own Growth

Tropical rain forests grow in the Amazon River basin because of the year-round warm temperatures and abundant rainfall (**Fig. E43-1**). As you learned in Chapter 29, rain forests grow where the prevailing air movements loft warm, moist air high into the atmosphere; as the air cools, the moisture condenses out and falls as rain (see Fig. 29-2). However, tropical rain forests are also a significant source of their own high humidity and plentiful rainfall.

In the Amazon rain forest, an acre of soil supports hundreds of towering trees, each bearing tens or even hundreds of thousands of leaves. The surface area of the leaves dwarfs the surface area of the soil. Therefore, much of the water evaporating from the forest comes from the trees—as little as 2 or 3 gallons a day for small understory trees, but as much as 200 gallons a day for large trees of the canopy. This evaporation produces perpetually high humidity. Some of the transpired water falls again as rain, right there in the rain forest. In fact, up to half of the rainfall in the Amazon region is water transpired by the trees themselves.

When rain forests are clear-cut, most frequently for agriculture, the local climate changes dramatically. In the intact forest, water evaporating from the leaves cools the air. As trees are removed and transpiration is reduced, the region becomes not only drier, but hotter as well. Decreased rainfall and warmer temperatures harm uncut adjacent forests, gradually changing their composition to vegetation that doesn't need as much moisture. If the clear-cut areas are abandoned, rain-forest trees may not be able to recolonize the land.

▲ **FIGURE E43-1 The Amazon rain forest** The rain-forest community helps produce and maintain its own environment.

Rain-forest transpiration affects nearby areas as well. For example, the Monteverde Cloud Forest Reserve in Costa Rica depends for its survival on an almost constant shroud of fog. Transpiration from Costa Rica's lowland rain forests pumps moisture into the winds flowing up the mountain slopes. As the air cools, this moisture condenses and forms fog. However, a century of logging has eliminated about 80% of Costa Rica's nearby lowland rain forests. As a result, the fog forms higher up the mountains. Mist-free conditions, which pose a threat to the fragile cloud forests, have become more common and prolonged since the 1970s. The lower parts of the forest are being colonized by bird species that are not typical of cloud forests. A similar trend has been seen among lizard species.

The impacts of trees on their own growth are not limited to tropical rain forests. On Tenerife, one of the Canary Islands off the northwest coast of Africa, rainfall averages less than 8 inches (20 cm) a year, with average high temperatures between 68° and 88°F (about 20° to 31°C) year-round. Nevertheless, some of Tenerife's mountain slopes support pine forests. How is this possible? Of course, it's a little cooler, with a little more rain, on the mountainsides than in the coastal lowlands. The major factor, however, is the forest itself. The trees intercept fogs drifting across the island from the Atlantic Ocean. Water from the fogs condenses on the trees and trickles down to the soil. More than half of the water in Tenerife's streams comes from condensing fog.

The nearby island of Lanzarote has a similar climate, and its mountains used to be covered with a similar fog-catching forest. However, the forests were almost completely cut down long ago and have not regrown. Modern efforts to restore the native trees have failed repeatedly, because the soil has now become too dry. A project is under way to capture the fogs with plastic mesh nets and drain the water into storage containers. The trapped water will be used to grow tree seedlings. When the trees become large enough, they should be able to capture their own moisture and become self-sustaining. Researchers are hopeful that a restored forest may one day capture enough water to replenish the aquifers (underground water supplies) that have been depleted since the forests were cut.

during the day, when sunlight can power photosynthesis, and close during the night, conserving water. However, if excessive evaporation threatens the plant with dehydration, the stomata will close regardless of the time of day.

Plants Regulate Their Stomata

A stoma consists of a central opening, or pore, surrounded by two guard cells (**Fig. 43-22**; see also Fig. 43-6). How do plants open and close their stomata? This is actually a two-part question: (1) Mechanically, how can the size of the opening be changed; and (2) physiologically, how do guard cells respond to stimuli, such as sunlight or dehydration, and adjust the size of the opening?

Plants don't have muscle cells, but some plant cells can change their volume and shape by taking up or losing water—this is the means by which guard cells adjust the size

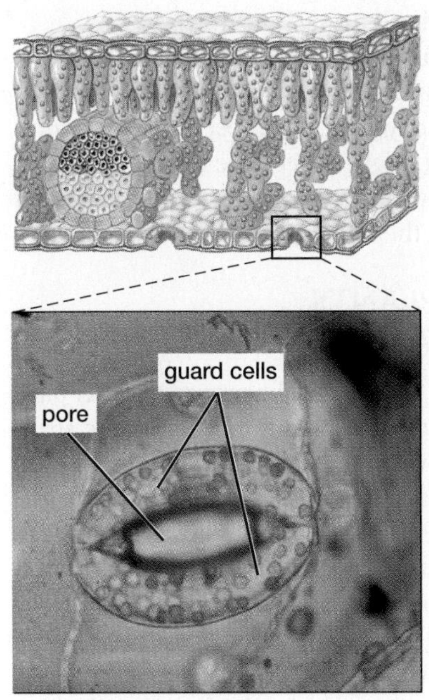

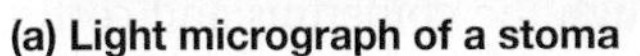

(a) Light micrograph of a stoma

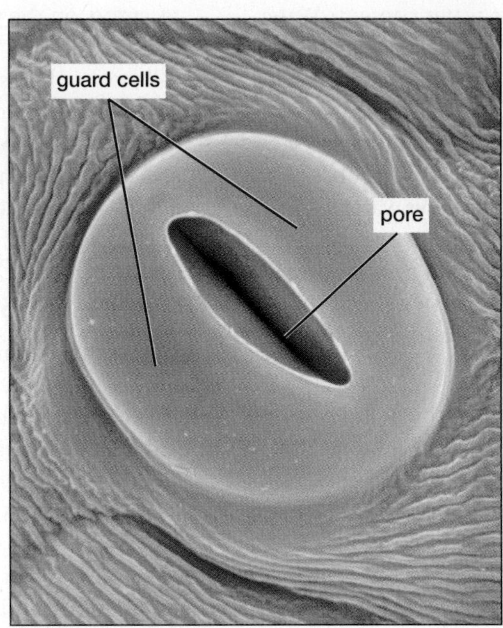

(b) SEM of a stoma

◀ **FIGURE 43-22 Stomata** Stomata as seen through **(a)** a light microscope and **(b)** a transmission electron microscope. The light micrograph shows that guard cells contain chloroplasts (green ovals inside the cells), unlike the other cells of the leaf epidermis, which do not.

of the opening of a stoma. Guard cells have both an unusual shape and a specific arrangement of cellulose fibers in their cell walls (**Fig. 43-23a**). The two guard cells of a stoma are curved slightly outward, like two sausage links tied together at both ends. Cellulose fibers in the guard cell walls encircle the cells, as if dozens of tiny belts were buckled firmly around each sausage. A stoma opens when its guard cells take up water, increasing the volume of the cells. The cellulose belts that surround the guard cells prevent the cells from getting fatter, and so they must get longer instead. The curved shape of the guard cells and their attachment at the ends means that they can get longer only by curving outward even more, which thus opens the central pore. A stoma closes when its guard cells lose water and decrease in volume. The cells become shorter, reducing their curvature and closing the pore.

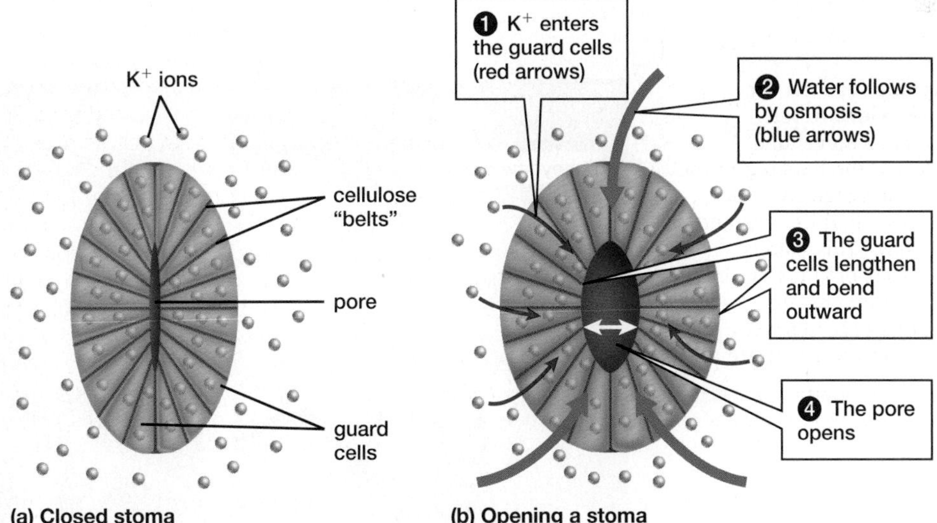

▲ **FIGURE 43-23 How guard cells open a stoma** Guard cells open and close the pore of a stoma as a result of changes in K^+ concentration, water content, and, hence, cell volume. **(a)** The structure of a stoma, showing the central pore closed. The guard cells have a relatively low K^+ concentration (about the same as other cells in the epidermis), a low water content, and a low volume. **(b)** When K^+ is transported into the guard cells, the resulting high K^+ concentration causes water to enter the guard cells and increase their volume, which opens the pore.

QUESTION When the stomata close, how is photosynthesis affected? How is the movement of water into the roots affected?

How does a guard cell change its water content? Water always moves across plasma membranes by osmosis, following a concentration gradient of dissolved solutes. Guard cells create osmotic gradients across their plasma membranes by adjusting the potassium (K^+) concentration in their cytoplasm in response to stimuli such as light and CO_2 concentrations (see below). Moving K^+ into the cell causes water to enter by osmosis, increasing the volume of the cells and opening the pore (**Fig. 43-23b**). Moving K^+ back out again causes water to leave by osmosis, shrinking the cells and closing the pore.

Physiologically, three important stimuli control K^+ movement into and out of guard cells:

- **Light** Light is absorbed by pigments in the guard cells, triggering a series of reactions that cause K^+ to enter the cells. Water follows by osmosis, and the guard cells swell, opening the pore. In the dark, the reactions promoting K^+ entry stop, and K^+ diffuses back out of the guard cells. Water follows out by osmosis, closing the pore.
- **Carbon dioxide** Carbon fixation during photosynthesis uses up CO_2, reducing its concentration within the leaf. Low CO_2 concentrations stimulate K^+ movement into the guard cells. Water enters by osmosis, opening the stoma. At night, when photosynthesis stops but cellular respiration continues, CO_2 levels rise. K^+ transport into the guard cells stops. K^+ diffuses out, water follows out by osmosis, and the guard cells close the pore.
- **Water** A stress hormone called abscisic acid (described in Chapter 45) is synthesized in many parts of a plant and moves throughout the plant body in the vascular tissues. When a plant begins to dry out, it synthesizes much more abscisic acid than usual, which binds to receptors on the guard cells and shuts down K^+ accumulation. K^+ diffuses out of the guard cells, water follows out by osmosis, and the guard cells close the stomata. Dangerous dehydration is most likely to occur on sunny days. As you might expect, the effects of abscisic acid are powerful, overcoming the effects of light and CO_2, and causing the stomata to close.

43.10 HOW DO PLANTS TRANSPORT SUGARS?

Sugars synthesized in the leaves must be moved to other parts of the plant, where they nourish nonphotosynthetic structures such as roots or flowers, or are stored in cortex cells of roots and stems. Sugar transport is the function of phloem.

Botanists studying phloem employ most unlikely lab assistants: aphids. Aphids are insects that feed by sucking fluids through a sharp hollow tube called a stylet, similar to the more familiar stylet used by mosquitoes to suck blood from their victims. Aphids, however, feed on the fluid in phloem. An aphid inserts its stylet through the epidermis and cortex of a young stem into a sieve tube (**Fig. 43-24**). As we will see, phloem fluid is under pressure, so the plant actually drives the fluid into the aphid's digestive tract; sometimes, the aphid inflates like a balloon. After an aphid has penetrated a sieve tube, botanists can anesthetize the aphid, remove most of its body, and leave the stylet in place. Phloem fluid flows out of the stylet for several hours. Chemical analysis shows that the fluid consists of water containing about 10% to 20% dissolved sugar (mostly sucrose), with lower amounts of other substances such as amino acids, proteins, and hormones. What drives the movement of this sugary solution in the phloem?

(a) An aphid sucks sap

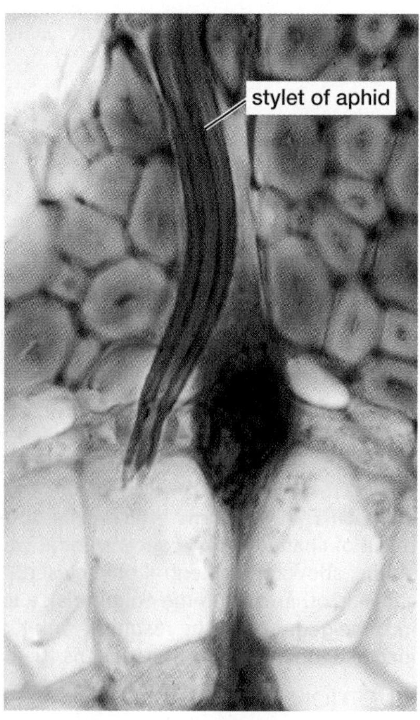

(b) A stylet penetrates into phloem

▶ **FIGURE 43-24 Aphids feed on the sugary fluid in phloem sieve tubes**
(a) When an aphid pierces a sieve tube, pressure in the tube forces the fluid out of the phloem and into the aphid's digestive tract. The aphid excretes excess fluid from its anus, as a sugar-rich "honeydew." This fluid is collected by certain species of ants that, in turn, defend the aphids from predators.
(b) The flexible stylet of an aphid passes through many layers of cells to penetrate a sieve-tube element.

The Pressure-Flow Theory Explains Sugar Movement in Phloem

The most widely accepted explanation for fluid transport in phloem is the **pressure-flow theory,** which states that differences in water pressure drive the flow of fluid through the sieve tubes. These water pressure differences are created indirectly by the production and use of sugar in different parts of the plant. Any part of a plant that synthesizes more sugar than it uses, such as a photosynthesizing leaf, is a sugar **source.** Any structure that uses more sugar than it produces is a sugar **sink.** Sinks include developing flowers and fruits, or cortex cells in a root converting sugar to starch. Sources and sinks can change with the seasons. For example, the roots of deciduous trees are sugar sinks during the summer, when they convert sucrose to starch for storage. The following spring, the roots become sugar sources, as they convert the starch back to sucrose, supplying energy for the development of new leaves.

According to the pressure-flow theory, phloem sieve tubes carry fluid away from sugar sources and toward sugar sinks, as illustrated by the numbered steps in **Figure 43-25**.

❶ Sugar produced by a source cell (for example, in a photosynthesizing leaf) is actively transported into a phloem sieve tube. This raises the sugar concentration in the phloem fluid in that portion of the sieve tube. ❷ Water from nearby xylem follows the sugar into the sieve tube by osmosis. Because their rigid cell walls prevent sieve-tube cells from expanding, water entering the sieve tube increases the pressure of the fluid. ❸ Water pressure drives the fluid through the phloem sieve tubes to regions of lower pressure. How is lower pressure created? ❹ Cells of a sugar sink (such as a fruit) actively transport sugar out of the phloem. Water follows by osmosis, producing lower water pressure in this part of the sieve tube. The phloem fluid thus moves from a source, where water pressure is high, to a sink, where water pressure is lower, carrying the dissolved sugar with it.

Note that sinks and sources of sucrose may occur in either roots or shoots, and that even the same structure may be a sink or a source depending on the season of the year (for example, roots in summer versus spring). The pressure-flow theory explains how phloem fluid can move either up or down the plant, but always from sugar sources to sugar sinks.

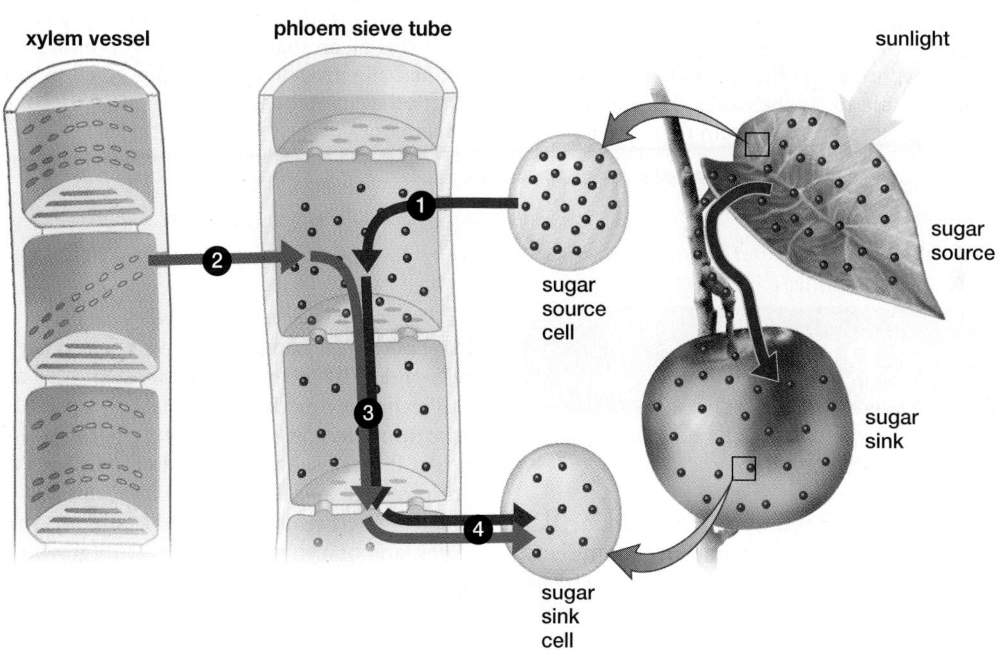

▲ **FIGURE 43-25 The pressure-flow theory of sugar transport in phloem** A photosynthesizing leaf is a sugar source, whereas a developing fruit is a sugar sink. Differences in water pressure drive phloem fluid from the leaf to the fruit. Numbers correspond to the explanations in the text. Red arrows show the movement of sugar; blue arrows show the movement of water. The gradient of blue color in the phloem sieve tube represents water pressure, which is higher in the source end of the tube and lower in the sink end of the same tube.

Case Study revisited

Why Do Leaves Turn Color in the Fall?

Of all the fall colors, the striking reds are the most intriguing. The role of anthocyanin pigments differs among species; in fact, some plants don't make them at all. However, plant scientists are learning why the leaves of many plants synthesize these red pigments just before they die: Anthocyanins act as sunscreen for leaves.

At the onset of autumn, as temperatures cool down but days remain fairly long and the sun is still very bright, a leaf's photosynthetic pathways become less efficient. The leaf is no longer able to use all of the light it absorbs. The excess light energy can harm the chloroplasts and reduce photosynthesis even more. Red, anthocyanin-containing leaves are better protected against intense light than are those without anthocyanin, and can carry out photosynthesis more effectively. In some plants, red leaves are predominant on the outer parts of the plant that receive the most sunlight, whereas shaded inner leaves become orange and yellow.

Why protect leaves that are just about to fall off? Leaves are rich storehouses of nutrients that a plant cannot afford to lose. In the autumn, complex organic molecules in leaves are broken down, and the resulting simpler molecules are transported to storage cells in stems and roots, where they remain throughout the winter. The plant must continue to photosynthesize in order to provide the energy needed for this breakdown and transport. Recently, researchers studied mutant plants that are unable to synthesize anthocyanin. These mutants retrieve far less nitrogen from their leaves under conditions of intense light and low temperatures than normal, anthocyanin-producing plants can. This finding indicates that anthocyanin indeed helps the plant to retrieve nutrients from its leaves.

This result dovetails nicely with research performed by Emily Habinck as an undergraduate at the University of North Carolina. Habinck measured the anthocyanin content and color intensity of red maples and sweetgum trees. She also measured the nutrient content of the soils in which they grew. Leaves with the most anthocyanin were growing in soils with the least nitrogen and other nutrients, suggesting that anthocyanin production is increased in plants that most need its protective properties as they store scarce nutrients.

Because the most intense hues of autumn seem to be stimulated by sunny days followed by cold nights, there is concern that global warming may dull the annual display. Some observers in Vermont, where fall temperatures have mostly been above average in recent years, believe the colors have dimmed with the warmer temperatures. Others think that fall colors appear later and less predictably than they did decades ago.

Consider This

Not all trees turn red in the fall. The Rocky Mountains, for example, are famous for their autumn display of golden yellow aspens, despite bright sunny days and cold nights. Does this mean that anthocyanins produced by maple trees do not actually provide any useful protection to their leaves? What other hypotheses might explain these observations? How might you test your hypotheses?

CHAPTER REVIEW

Summary of Key Concepts

43.1 What Challenges Are Faced by All Life on Earth?

All organisms must obtain energy, water, and other nutrients; distribute water and nutrients throughout the body; exchange gases with the environment; support their bodies; grow and develop; and reproduce.

43.2 How Are Plant Bodies Organized?

The body of a land plant consists of root and shoot systems. Roots are usually underground, and their functions include anchoring the plant in the soil; absorbing water and minerals; storing surplus photosynthetic products (sugar and starch); transporting water, minerals, sugars, and hormones; producing some hormones; and interacting with soil fungi and microorganisms that provide nutrients. Shoots are generally located aboveground and consist of stems, buds, leaves, and (in season) flowers and fruit. Shoot functions include photosynthesis, transport of materials, reproduction, and hormone synthesis.

The two principal groups of flowering plants are the monocots and dicots. Their major characteristics are shown in Fig. 43-2.

43.3 How Do Plants Grow?

Plant bodies are composed of two main classes of cells: meristem cells and differentiated cells. Meristem cells are undifferentiated and retain the capacity for mitotic cell division. Differentiated cells arise from divisions of meristem cells, become specialized for particular functions, and usually do not divide. Most meristem cells are located in apical meristems at the tips of roots and shoots and in lateral meristems (also called cambium) in the shafts of roots, stems, and branches. Primary growth (growth in length or height and the differentiation of parts) results from the division and differentiation of cells from apical meristems. Secondary growth (growth in diameter and strength) results from the division and differentiation of cells from lateral meristems.

43.4 What Are the Tissues and Cell Types of Plants?

Plant bodies consist of three tissue systems: the dermal, ground, and vascular systems (see Table 43-1). The dermal tissue system forms the outer covering of the plant body. The dermal tissue system of leaves and of primary roots and stems is usually a single cell layer of epidermis. Dermal tissue after secondary growth, called periderm, consists mostly of a covering of cork cells and the lateral meristem that produces them, called the cork cambium.

The ground tissue system consists of various cell types including parenchyma, collenchyma, and sclerenchyma cells. Most are involved in photosynthesis, support, or storage. Ground tissue makes up most of the body of a young plant during primary growth.

The vascular tissue system consists of xylem, which transports water and minerals from the roots to the shoots, and phloem, which transports water, sugars, and other organic molecules throughout the plant body.

43.5 What Are the Structures and Functions of Leaves?

Leaves, the major photosynthetic organs of plants, consist of a petiole and a blade. The petiole attaches the blade to a stem or branch. The blade consists of an outer epidermis covered by a waterproof cuticle; mesophyll cells, which have chloroplasts and carry out photosynthesis; and vascular bundles of xylem and phloem, which carry water, minerals, and photosynthetic products to and from the leaf. The epidermis is perforated by adjustable pores called stomata that regulate the exchange of gases and water.

43.6 What Are the Structures and Functions of Stems?

Primary growth in dicot stems results in a structure consisting of an outer epidermis, usually covered by a waterproof cuticle; supporting, storage, and sometimes photosynthetic cells of cortex beneath the epidermis; vascular tissues of xylem and phloem; and supporting and storage cells of pith at the center. Leaves and lateral buds are found at nodes along the surface of the stem. Under the proper hormonal conditions, lateral buds sprout into a branch. Secondary growth in stems results from cell divisions in the vascular cambium and cork cambium. Vascular cambium produces secondary xylem and secondary phloem, increasing the stem's diameter. Cork cambium produces waterproof cork cells that cover the outside of the stem.

43.7 What Are the Structures and Functions of Roots?

Primary growth in roots results in a structure consisting of an outer epidermis and an inner vascular cylinder of xylem and phloem, with cortex between the two. The apical meristem near the tip of the root is protected by the root cap. Cells of the root epidermis absorb water and minerals from the soil. Root hairs are projections of epidermal cells that increase the surface area for absorption. Most cortex cells store surplus sugars (usually in the form of starch) produced by photosynthesis. The innermost layer of cortex cells is the endodermis, which controls the movement of water and minerals from the soil into the vascular cylinder. The outer layer of the vascular cylinder is the pericycle, which helps to regulate mineral transport into the vascular cylinder and also produces branch roots.

43.8 How Do Plants Acquire Nutrients?

Most minerals are taken up from the soil water by active transport into the root hairs. These minerals diffuse into the root through plasmodesmata to the endodermis and pericycle. There, they are actively transported into the extracellular space of the vascular cylinder. The minerals diffuse from this extracellular space into the tracheids and vessel elements of xylem.

Because mineral uptake results in a higher concentration of minerals (and a lower concentration of free water) in the vascular cylinder than in the soil water, water follows the high mineral concentration into the vascular cylinder, across plasma membranes of epidermal, cortex, or endodermal cells, by osmosis. Water is rapidly transported up the xylem of the vascular cylinder, which also reduces the concentration of free water in the vascular cylinder and further promotes water influx into the vascular cylinder by osmosis. Water transport in xylem is usually the most important factor driving water uptake in roots.

Many plants have fungi associated with their roots, forming root–fungus complexes called mycorrhizae that help absorb soil nutrients. Nitrogen can be absorbed only as ammonium or nitrate, both of which are scarce in most soils. Legumes have evolved a cooperative relationship with nitrogen-fixing bacteria that invade legume roots. The plant provides the bacteria with sugars, and the bacteria use some of the energy in those sugars to convert atmospheric nitrogen to ammonia, which the plant then absorbs.

43.9 How Do Plants Move Water and Minerals from Roots to Leaves?

The cohesion–tension theory explains xylem function: Water molecules bind to one another by means of hydrogen bonds. The resulting cohesion holds together the water within xylem tubes almost as if it were a solid chain. As water molecules evaporate from the leaves during transpiration, the hydrogen bonds pull other water molecules up the xylem to replace them. This movement is transmitted down the xylem to the root, where water loss from the vascular cylinder promotes water movement across the endodermis from the soil water by osmosis. Minerals move up the xylem dissolved in the water.

Stomata in leaves and young epidermis control the evaporation of water (transpiration). The opening of a stoma is regulated by the shape and volume of the guard cells that form the pore. Open stomata allow more rapid transpiration. In most plants, stomata open during the day (admitting carbon dioxide needed for photosynthesis); the stimuli that trigger opening of the stomata are light and a low carbon dioxide concentration in the guard cells. If the plant is losing too much water, a hormone called abscisic acid is released, causing the guard cells to close the stomata.

BioFlix ™ Water Transport

43.10 How Do Plants Transport Sugars?

The pressure-flow theory explains sugar transport in phloem. Parts of the plant that synthesize sugar (for example, leaves) export sugar into the sieve tube. High sugar concentrations attract water to enter the sieve tube by osmosis, increasing the local water pressure in the phloem. Other parts of the plant (for example, fruits) consume sugar, reducing the sugar concentration in the sieve tube and causing water to leave the tube by osmosis, which reduces pressure. Water and dissolved sugar move in the sieve tubes from high pressure to low pressure.

Key Terms

annual ring *847*
apical meristem *839*
bark *848*
blade *843*
branch root *851*
cambium *840*
Casparian strip *851*
cohesion *856*
cohesion–tension theory *856*
collenchyma *842*
companion cell *843*
cork cambium *847*
cork cell *847*
cortex *847*
cuticle *841*
dermal tissue system *840*
dicot *839*
differentiated cell *839*
endodermis *851*
epidermal tissue *841*
epidermis *841*
fibrous root system *849*
flower bud *845*
ground tissue system *840*
guard cell *844*
heartwood *847*
internode *845*
lateral bud *845*
lateral meristem *840*
leaf *843*
leaf primordium (plural, **primordia**) *845*
legume *854*
meristem cell *839*
mesophyll *844*

mineral *852*
monocot *839*
mycorrhizae *855*
nitrogen fixation *855*
nitrogen-fixing bacterium *855*
node *845*
nodule *855*
nutrient *852*
parenchyma *841*
pericycle *851*
periderm *841*
petiole *843*
phloem *842*
pit *842*
pith *847*
plasmodesma (plural, **plasmodesmata**) *843*
pressure-flow theory *861*
primary growth *839*
root *837*
root cap *850*
root hair *850*
root pressure *854*
root system *837*
sapwood *847*
sclerenchyma *842*
secondary growth *840*
shoot system *838*
sieve plate *843*
sieve-tube element *843*
sink *861*
source *861*
stem *845*
stoma (plural, **stomata**) *844*
taproot system *849*
terminal bud *845*
tissue *840*
tissue system *840*
tracheid *842*
transpiration *855*
vascular bundle *844*
vascular cambium *847*
vascular cylinder *851*
vascular tissue system *840*
vein *844*
vessel *842*
vessel element *842*
xylem *842*

Thinking Through the Concepts

Fill-in-the-Blank

1. Plants grow through division of __________ cells and differentiation of the resulting daughter cells. These dividing cells reside in two locations: at the tip of a shoot or root, called the __________, and in cylinders along the sides of roots and stems, called __________. Which is responsible for primary growth? __________ Which is responsible for secondary growth? __________
2. The three tissue systems of a plant body are __________, __________, and __________. Which covers the outside of the plant body? __________ Which conducts water, minerals, and sugars within the plant body? __________ Which stores starches? __________
3. Water travels upward through plant roots and shoots within hollow tubes of __________, which contains two types of conducting cells, __________ and __________. Water molecules within these tubes are interconnected by forces called __________, which allow a chain of water molecules to be pulled up the plant, driven by the evaporation of water from the leaves, a process called __________. The __________ in the epidermis of a leaf control the rate of evaporation from the leaf.
4. __________ transports sugars and other organic molecules from sources of sugar to sinks of sugar. The conducting cell type in this tissue is the __________, which is supported by adjacent cells called __________.
5. The very tip of a young root is protected by the cells of the __________. The surface area of a young root is increased by projections from epidermal cells called __________, which move minerals from the soil water into their cytoplasm by the process of __________. Many young roots have a mutually beneficial relationship with fungi to help absorb some minerals; these root–fungus complexes are called __________.

Review Questions

1. Describe the locations and functions of the three tissue systems in land plants.
2. Distinguish between primary growth and secondary growth, and describe the cell types involved in each.
3. Distinguish between meristem cells and differentiated cells.
4. Diagram the internal structure of a dicot root after primary growth, labeling and describing the function of epidermis, cortex, endodermis, pericycle, xylem, and phloem. What tissues are located in the vascular cylinder?
5. Diagram the internal structure of a dicot stem after primary growth, labeling and describing the function of the epidermis, cortex, pith, xylem, and phloem.
6. How do xylem and phloem differ?
7. What are the main functions of roots, stems, and leaves?
8. What types of cells form root hairs? What is the function of root hairs?
9. Diagram the structure of a leaf. What structures regulate water loss and CO_2 absorption by a leaf?
10. Describe the daily cycle of the opening and closing of stomata.
11. Describe how water and minerals are absorbed by a root.
12. Describe how water and minerals move in xylem. Why is water and mineral movement in xylem unidirectional?
13. Describe how sugars are moved in phloem, and why phloem fluid may move up or down the plant.

Applying the Concepts

1. An important goal of molecular botanists is to insert the genes for nitrogen fixation into crop plants such as corn or wheat (see pp. 248–250). Why would the insertion of such genes be useful? What changes in farming practices would this technique allow?
2. Chapter 2 describes the unusual characteristics of water. Discuss several ways in which the evolution of vascular plants has been influenced by water's special characteristics.
3. A major environmental problem is desertification, in which overgrazing by cattle or other animals removes most of the vegetation in an area, and the region becomes drier and less able to support plants as a result. Explain this phenomenon based on your understanding of transpiration and how water moves through plants.
4. Grasses (monocots) form their primary meristem near the ground surface rather than at the tips of branches the way dicots do. How does this feature allow you to grow a lawn and mow it every week in the summer? What would happen if you had a dicot lawn and tried to mow it?

MB *Go to www.masteringbiology.com for practice quizzes, activities, eText, videos, current events, and more.*

Plant Reproduction and Development

Chapter 44

Case Study

Some Like It Hot—and Stinky!

FOR THE MOST PART, flowers are pretty, delicate, fragrant, and fairly small. And the bodies of virtually all plants remain at the same temperature as their environment. But not all flowering plants obey these rules: Behold *Amorphophallus titanum*, known in its native Sumatra as bunga bangkai, the corpse flower. (People seldom translate its scientific name in print, but you can probably figure it out.)

Five to 10 feet tall, the corpse flower gets its common name from its odor, variously described as resembling rotting fish, rotting pumpkin, or just generic rotting carrion. As if that isn't enough, the corpse flower also gets hot. In fact, the flower produces pulses of warm water vapor, at temperatures as high as 97°F (36°C). In the Sumatran forest, the huge flower acts like a chimney, blasting its smelly steam high into the air and dispersing it throughout the vicinity.

A variety of insects, including flies and carrion beetles, are attracted by the smell of rotting flesh. They swarm around a carcass and lay their eggs in it. When the eggs hatch, the larvae eat the decaying meat, then eventually pupate and hatch into another generation of flies and beetles. When a corpse flower emits "eau de carcass," it attracts these scavengers.

Several other flowers are warm, or stinky, or both, although none is as impressive as the corpse flower. The related dead-horse arum smells up Corsica and other islands in the northern Mediterranean. The stinking corpse lily (see the opening photo for Chapter 21) shares Sumatran forests with the corpse flower. In South Africa, the starfish flower attracts flies with its five-armed, putrid flowers—and believe it or not, some people raise these as houseplants! Several relatives of the corpse flower, including the philodendron of tropical rain forests and the skunk cabbage of deciduous forests in eastern North America and Asia, bear heat-generating flowers that are not particularly offensive (although the leaves of skunk cabbages really do smell bad if they are damaged).

Why do plants produce flowers? Specifically, why would a plant produce an enormous, hot, putrid flower? How does a plant benefit by deceiving beetles and flies into mistaking its flower for a mass of rotting flesh? Read on.

▲ *Amorphophallus titanum* is truly breathtaking—in more ways than one! Often more than 6 feet tall, its flower emits odors reminiscent of a rotting carcass.

At a Glance

44.1 HOW DO PLANTS REPRODUCE?

Many plants can reproduce either sexually or asexually. During asexual reproduction, part of an existing plant uses mitotic cell division to produce a new plant. Asexually produced offspring, therefore, are genetically identical to the parent. Examples of asexually reproducing plants include aspen trees, which sprout new trunks from their roots (see Fig. 9-2). The arching branches of blackberries and the spreading horizontal runners of strawberries (see Fig. 43-11b) can root where they touch the ground and produce new plants. The bulbs of tulips, daffodils, and amaryllis can reproduce by growing new, smaller bulbs.

Evolutionary biologists disagree on what factors, or combination of factors, might have favored the evolution of sexual reproduction. One hypothesis is that the genetic variation caused by crossing over during meiosis and the fusion of gametes from two different parents (see pp. 168–169) produces offspring with new traits that allow them to adapt to changing environments or to new, slightly different habitats. Another hypothesis proposes that the genetic diversity produced by sexual reproduction helps to reduce parasitism, because the parasites must continually scramble to overcome ever-changing host defenses. In addition, some DNA repair occurs only while chromosomes are crossing over, which takes place mostly during meiosis (see pp. 189–191). Whether these or other advantages prove to be most important, the fact is that most eukaryotes, including all of the plants mentioned above, reproduce sexually at least some of the time.

The Plant Sexual Life Cycle Alternates Between Diploid and Haploid Stages

You may recall from Chapter 9 that the sexual life cycles of organisms vary in the timing of when mitotic and meiotic cell division occur, and whether the adult body forms (which produce reproductive cells) are haploid or diploid (see Fig. 9-17). The plant sexual life cycle is called **alternation of generations,** because it alternates between two distinct, multicellular reproductive stages—one diploid and one haploid—that give rise to each other.

Figure 44-1 provides an overview of the plant sexual life cycle, using angiosperms (flowering plants) as an example. The multicellular diploid form is called the **sporophyte.** In angiosperms, sporophytes are the plants of gardens, orchards, forests, and fields that produce flowers. The diploid body form is named the sporophyte, or "spore-bearing plant," because it produces specialized reproductive cells, often called *mother cells* (**Fig. 44-1** ❶), that undergo meiotic cell division to form haploid cells called **spores** (**Fig. 44-1** ❷). In angiosperms, spores are formed in the male and female reproductive structures of the flower. Why are these cells spores, and not gametes? Gametes do not divide; rather, they fuse to form a diploid cell, the **zygote** (see below). Spores do not fuse to form a diploid cell; instead, spores undergo mitotic cell division to produce a multicellular haploid body form called the **gametophyte** (the "gamete-bearing plant"; **Fig. 44-1** ❸).

Angiosperms and gymnosperms (conifers and their relatives) produce separate male and female gametophyte stages. In some other types of plants, a single gametophyte

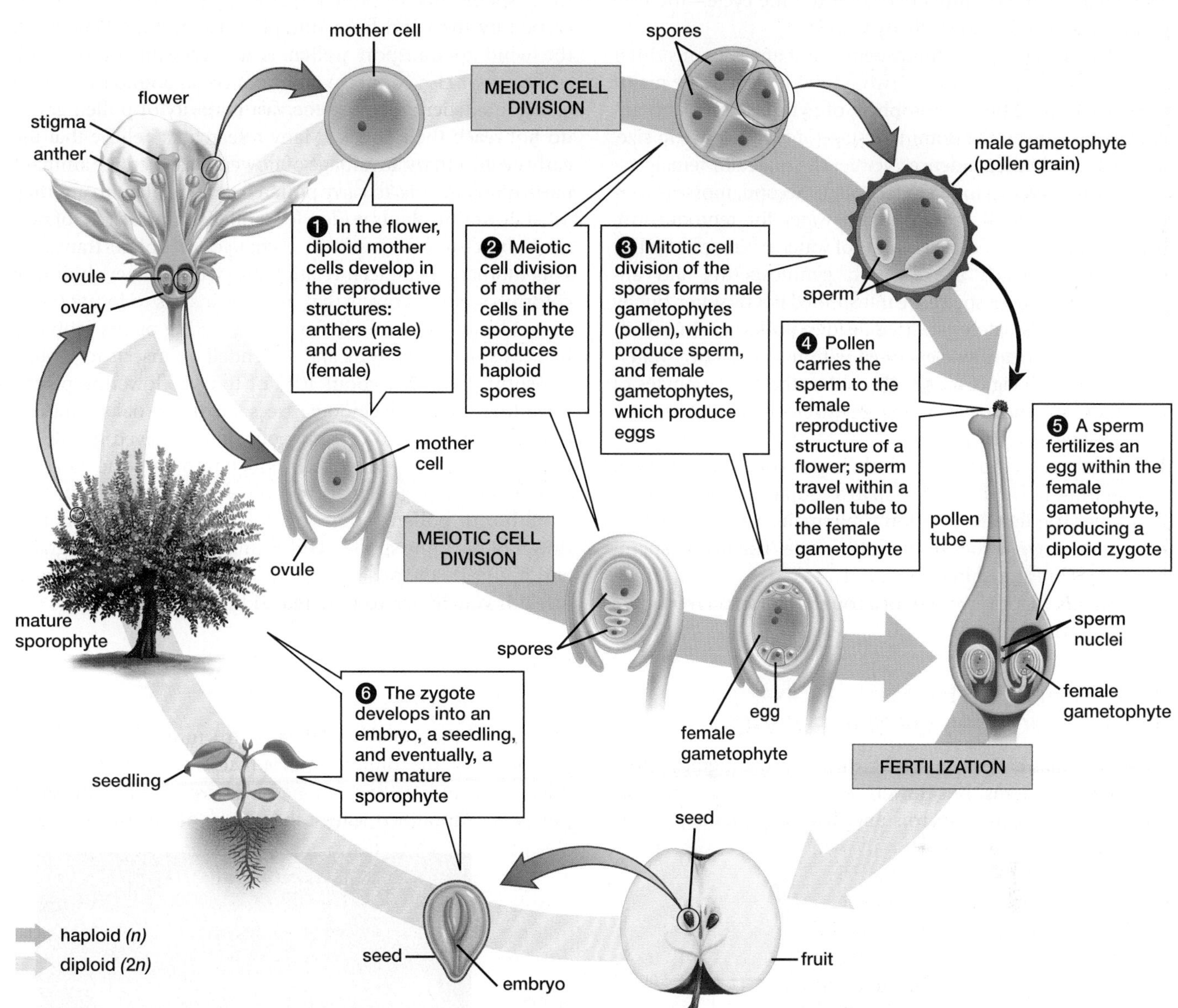

▲ **FIGURE 44-1 Alternation of generations is the sexual life cycle of a flowering plant**

may produce both sperm and eggs, although in separate reproductive structures (see Fig. 21-8). In either case, some of the cells of the gametophyte differentiate into sperm or eggs; because they are already haploid, the cells of the gametophyte can produce gametes without going through meiotic cell division. Eggs are retained within the female reproductive structures, so sperm must travel to the egg, usually either by swimming through a film of water (mosses and ferns) or carried within a pollen grain (gymnosperms and angiosperms; **Fig. 44-1 ❹**). A sperm fertilizes an egg, producing a diploid zygote (**Fig. 44-1 ❺**). The zygote undergoes repeated mitotic cell divisions followed by differentiation of the resulting daughter cells to form an embryo and eventually a new adult sporophyte plant (**Fig. 44-1 ❻**).

Although alternation of generations is the sexual life cycle of all plants, the relative size, complexity, and life span of the sporophyte and gametophyte stages vary considerably among different types of plants. In mosses and liverworts, the gametophyte stage is an independent plant that dominates the life cycle (see Fig. 21-6). Sperm fertilize eggs that are retained in the gametophyte. The resulting zygote develops into a sporophyte that grows directly on the gametophyte, and relies on the gametophyte for nourishment. The sporophyte is never an independent plant.

In ferns, as in mosses and liverworts, sperm fertilize eggs retained in an independent gametophyte, and the zygote begins growing on the gametophyte (see Fig. 21-8). However, eventually, the sporophyte develops its own roots and leaves, and becomes the dominant stage of the life cycle—the fern commonly seen in moist, shady woods.

The life cycles of gymnosperms and angiosperms differ from the cycles of mosses, liverworts, and ferns in three major ways. First, the diploid sporophyte of gymnosperms and angiosperms is by far the dominant stage of the life cycle in size, longevity, and independence. Neither the male nor female gametophyte is ever an independent plant. Second, mosses, liverworts, and ferns all require liquid water for reproduction, because sperm swim through films of water, or are splashed by raindrops, to reach the eggs. In gymnosperms and angiosperms, however, sperm are transported not by water, but by wind or animals, safely packaged inside a **pollen grain,** which is the male gametophyte enclosed within a protective, watertight jacket (see Fig. 44-1 ❷, ❸). Third, the gametophytes of gymnosperms and angiosperms are extremely small. The male gametophyte (pollen grain) consists of only three to six cells. The female gametophyte is a little larger, consisting of seven cells in most flowering plants, up to a couple of thousand cells in a pine. The female gametophyte remains sheltered within the reproductive tissues of the sporophyte. The gymnosperm life cycle is illustrated in Figure 21-11.

In this chapter, we will focus our attention on reproduction in flowering plants.

44.2 WHAT IS THE FUNCTION AND STRUCTURE OF THE FLOWER?

Flowering plants have many features that distinguish them from gymnosperms, but flowers are the most conspicuous. Although both gymnosperms and flowering plants package their sperm inside pollen grains, gymnosperm pollen is carried by the wind from one plant to another. Relying on the wind to transport pollen is a successful reproductive strategy—or else conifers would not be so abundant—but it is also inefficient, because the vast majority of pollen grains do not reach their targets. Many researchers believe that the early evolutionary advantage of flowers was to entice animals, particularly insects, to carry pollen from one plant to another. Most flowering plants trade a few pollen grains or a sip of nectar as lunch for an insect in exchange for pollen transport. This mutually beneficial relationship led to the evolution of colorful, scented flowers that help the insects to locate them.

The ancestors of all modern flowering plants probably had conspicuous flowers and depended on insects for pollen transport. However, about 10% of today's flowering plants, including many deciduous trees such as oaks, maples, birches, poplars, and cottonwoods, have evolved greatly reduced flowers and release their pollen into the wind. Grasses, sagebrush, and ragweed also produce prodigious amounts of wind-borne pollen. Some of the pollen reaches another flower of the same species. Unfortunately, other pollen grains end up inside the noses of allergy sufferers, as we explore in "Health Watch: Are You Allergic to Pollen?".

Flowers Are the Reproductive Structures of Angiosperms

Flowers are the sexual reproductive structures of angiosperms, produced by the diploid sporophyte. A **complete flower** (**Fig. 44-2**), such as that of a petunia, rose, or lily, consists of four sets of modified leaves: the sepals, petals, stamens, and

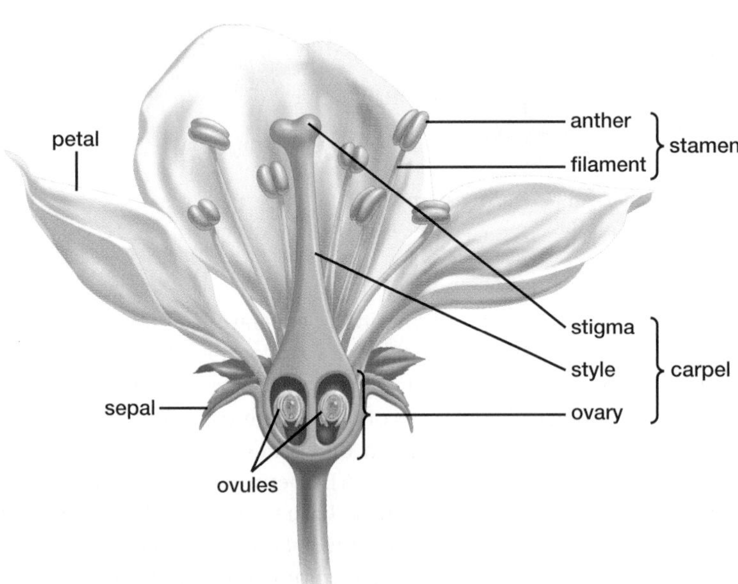

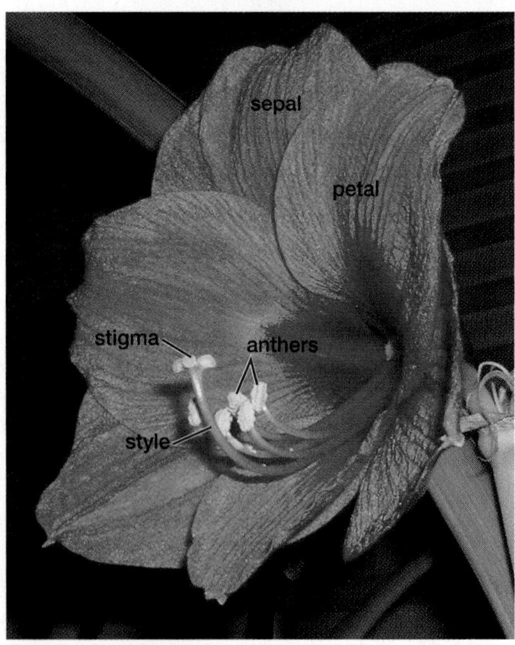

▲ **FIGURE 44-2 A complete flower (a)** A complete flower has four parts: sepals, petals, stamens (the male reproductive structures), and at least one carpel (the female reproductive structure). This illustration shows a complete dicot flower. **(b)** The amaryllis is a complete monocot flower, with three sepals (virtually identical to the petals), three petals, six stamens, and three carpels (fused into a single structure). The anthers are well below the stigma, making self-pollination unlikely.

Health Watch

Are You Allergic to Pollen?

Wind pollination can succeed only if plants release huge quantities of pollen into the air. Unfortunately for allergy sufferers, people often inhale these microscopic male gametophytes. Proteins in pollen coats activate the immune systems of sensitive individuals, causing itchy eyes, runny noses, burning throats, coughing, and sneezing. If you are among these unfortunate people, your own immune system is creating all these symptoms in an attempt to rid you of harmless pollen, which it mistakes for dangerous parasites (see Fig. 36-15).

People who suffer from "hay fever" are usually sensitive only to specific types of pollen. In temperate climates, springtime sufferers may be allergic to tree pollen, whereas summertime allergies may be caused by grasses. In the United States, however, the most common cause of hay fever is not hay (grass), but ragweed, which pollinates during late summer and fall (**Fig. E44-1**). The inconspicuous flowers of a single ragweed plant can release a million pollen grains daily; collectively, ragweed is estimated to release 100 million tons of pollen in the United States each year. Ragweed pollen has been collected 400 miles out to sea and 2 miles up in the atmosphere, so it is nearly impossible to avoid ragweed pollen completely.

Plants pollinated by bees and other animals are rarely a cause of allergies, because their pollen is sticky and produced in small amounts. Goldenrod, for example, which blooms during the ragweed season and is conspicuously yellow, is often blamed for allergies that are actually caused by ragweed. In fact, goldenrod's yellow blooms attract bee and butterfly pollinators, and most people can enjoy them in perfect comfort (**Fig. E44-2**).

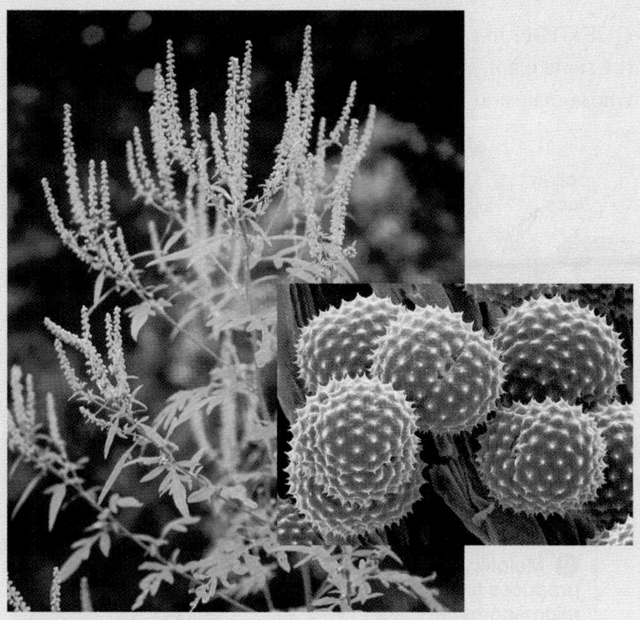

▲ **FIGURE E44-1 Inconspicuous ragweed flowers and their pollen**

▲ FIGURE E44-2 **Goldenrod** Like most colorful, insect-pollinated flowers, goldenrod seldom causes allergies in people.

carpels. The **sepals** are located at the base of the flower. As you learned in Chapter 43, there are two major groups of flowering plants, the dicots and monocots (see Fig. 43-2). In dicots, sepals are usually green and leaflike (**Fig. 44-2a**); in monocots, sepals typically resemble the petals (**Fig. 44-2b**). In either case, sepals surround and protect the flower bud as the remaining three structures develop. Just above the sepals are the **petals,** which are often brightly colored and fragrant, advertising the location of the flower to potential pollinators.

The male reproductive structures, the **stamens,** are attached just above the petals. Each stamen usually consists of a slender **filament** bearing an **anther** that produces pollen. In the center of the flower are one or more female reproductive structures, called **carpels.** A typical carpel is somewhat vase shaped, with a sticky **stigma** mounted atop an elongated **style. Pollination** occurs when pollen from an anther of a stamen lands on the stigma of a carpel. The style connects the stigma with the bulbous **ovary** at the base of the carpel (see Fig. 44-2a). Inside the ovary are one or more **ovules;** a female gametophyte develops inside each ovule. After fertilization, each ovule will become a **seed,** consisting of a small, embryonic plant and stored food for the embryo. The ovary will develop into a **fruit** enclosing the seeds.

Incomplete flowers lack one or more of the four floral parts. For example, grass flowers (see Fig. 44-6 later in the chapter) lack both petals and sepals. Other incomplete flowers lack either the male stamens or the female carpels. In such cases, the flowers are described as **imperfect,** as well as incomplete. Plant species with imperfect flowers produce separate male and female flowers, sometimes on a single plant, as occurs in zucchini (**Fig. 44-3**) and the corpse flower. Other species with imperfect flowers produce male and female flowers on separate plants. Only female American holly trees, for example, produce the decorative red fruit, so female trees are often favored as ornamentals (as long as a few males are available nearby for pollination).

▲ **FIGURE 44-3 Male and female imperfect flowers** Plants of the squash family, such as zucchinis, bear separate female (left) and male (right) flowers. Note the small zucchini (actually a fruit) forming from the ovary at the base of the female flower.

QUESTION In species with separate male and female flowers on the same plant, why would natural selection favor individuals whose male and female flowers bloom at different times?

The Pollen Grain Is the Male Gametophyte

The male gametophytes, or pollen grains, develop within the anthers of a flower of the sporophyte plant (**Fig. 44-4**). Each anther consists of four chambers called pollen sacs. Within each pollen sac, hundreds to thousands of diploid **microspore mother cells** develop (**Fig. 44-4 ①**). Each microspore mother cell undergoes meiotic cell division (see pp. 162–165) to produce four haploid **microspores** (**Fig. 44-4 ②**). Each microspore then undergoes one mitotic cell division to produce

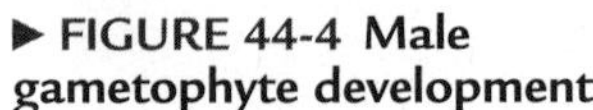

▶ **FIGURE 44-4 Male gametophyte development**

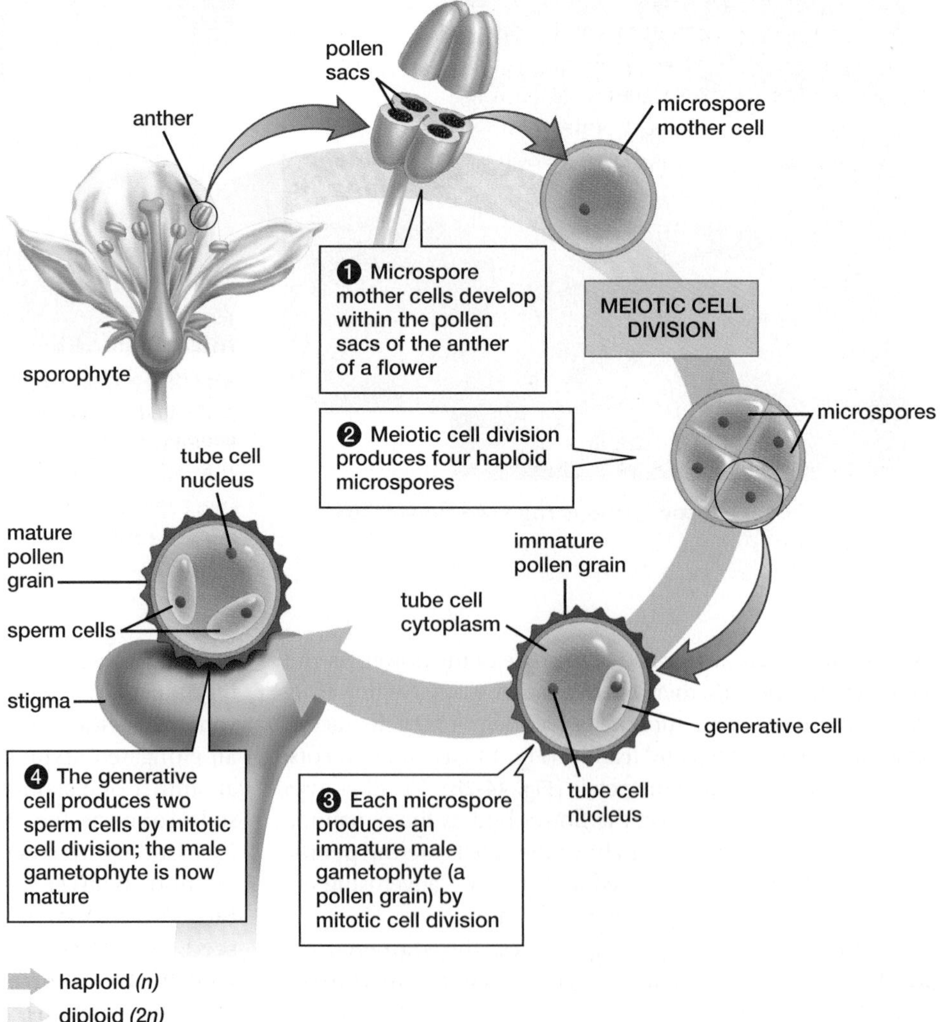
pollen sacs
anther
microspore mother cell
① Microspore mother cells develop within the pollen sacs of the anther of a flower
MEIOTIC CELL DIVISION
sporophyte
microspores
② Meiotic cell division produces four haploid microspores
tube cell nucleus
mature pollen grain
sperm cells
immature pollen grain
tube cell cytoplasm
stigma
generative cell
④ The generative cell produces two sperm cells by mitotic cell division; the male gametophyte is now mature
③ Each microspore produces an immature male gametophyte (a pollen grain) by mitotic cell division
tube cell nucleus
haploid (n)
diploid (2n)

Case Study continued
Some Like It Hot—and Stinky!

The corpse flower is actually a mass of separate male and female flowers. In principle, sperm from the male flowers could fertilize eggs in the female flowers of the same plant. However, as you may recall from Chapters 10 and 15, mating among close relatives can lead to genetic defects, as offspring become homozygous for dysfunctional recessive alleles of important genes. Self-fertilization, of course, is the ultimate in mating with a relative. The corpse lily avoids self-fertilization by opening its female flowers a day earlier than its male flowers. The female flowers usually wilt before the male flowers mature. Therefore, pollinators seldom transfer pollen from an anther to a stigma on the same plant.

an immature pollen grain—the male gametophyte—consisting of two cells: a large **tube cell** occupying most of the volume of the pollen grain and a smaller **generative cell** that resides within the cytoplasm of the tube cell (**Fig. 44-4** ❸). As the gametophyte matures, mitotic cell division of the generative cell produces two haploid sperm cells (**Fig. 44-4** ❹). The pollen grain is surrounded by a tough, waterproof coat, often sculpted with an elaborate pattern of pits and protrusions characteristic of the plant species (**Fig. 44-5**). The coat protects the sperm during their journey to the sometimes-distant female carpel.

When the pollen has matured, the pollen sacs of the anther split open. In wind-pollinated flowers, such as those of grasses (**Fig. 44-6**) and oaks, the anthers usually protrude from small, often inconspicuous flowers. The slightest breeze carries off the pollen grains; most are lost, but a few reach and pollinate other flowers of the same species. In animal-pollinated flowers, the pollen adheres weakly to the anther until the pollinator comes along and brushes or picks it off.

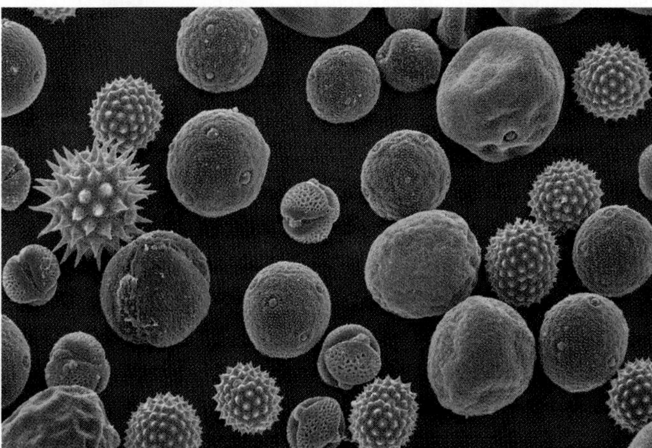

▲ **FIGURE 44-5 Pollen grains** The tough outer coats of many pollen grains are elaborately sculptured in species-specific shapes and patterns, as shown in this colorized scanning electron micrograph.

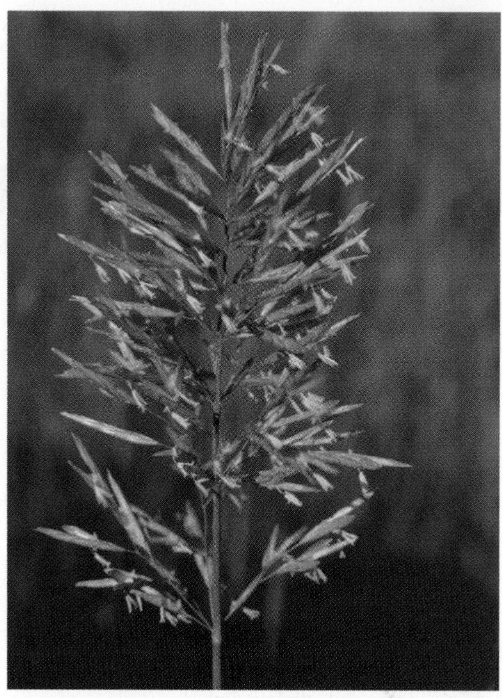

▲ **FIGURE 44-6 Wind-pollinated flowers** The flowers of grasses and many deciduous trees are wind pollinated, with anthers (yellow structures) exposed to the wind.

QUESTION Would you expect wind-pollinated flowers to have colorful petals and sweet scents? Why or why not?

The Female Gametophyte Forms Within the Ovule

Within the ovary of a carpel, masses of cells differentiate into ovules (**Fig. 44-7**). Depending on the plant species, a carpel may have as few as one ovule or as many as several dozen. Each young ovule consists of protective outer layers called **integuments,** which surround a single, diploid **megaspore mother cell** (**Fig. 44-7** ❶). The megaspore mother cell undergoes meiotic cell division, producing four haploid **megaspores** (**Fig. 44-7** ❷). Only one megaspore survives; the other three degenerate. The nucleus of the surviving megaspore undergoes three rounds of mitosis, producing eight haploid nuclei (**Fig. 44-7** ❸). Plasma membranes and cell walls then form, dividing the cytoplasm into the seven (not eight) cells that make up the female gametophyte (**Fig. 44-7** ❹). There are three small cells at each end, each with one nucleus, and one large central cell with two nuclei. The **egg** is one of the three cells at the lower end, located near an opening in the integuments of the ovule.

Pollination of the Flower Leads to Fertilization

Pollination is necessary for fertilization, but these are two distinct events, just as copulation is a separate event from fertilization in mammals. Pollination occurs when a pollen grain lands on the stigma of a flower of the same plant species, beginning a remarkable series of events (**Fig. 44-8** ❶). The pollen grain absorbs water from the stigma. The tube cell, which

ovule
megaspore mother cell
1 A megaspore mother cell develops within each ovule of the ovaries of a flower
ovary
integuments
MEIOTIC CELL DIVISION
4 Cytoplasmic division produces the seven cells of the mature female gametophyte
central cell with two nuclei
female gametophyte
egg cell
megaspores
2 Meiotic cell division produces four haploid megaspores; three degenerate
3 The single remaining megaspore forms eight nuclei by mitosis
haploid (n)
diploid (2n)

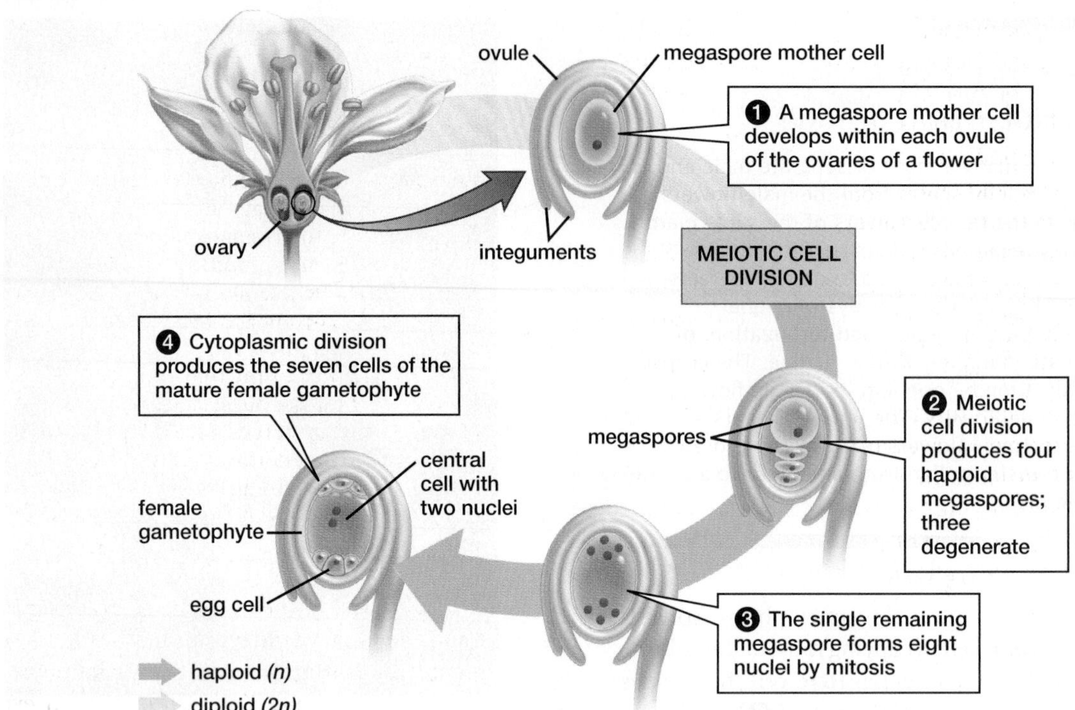

▲ **FIGURE 44-7 Female gametophyte development**

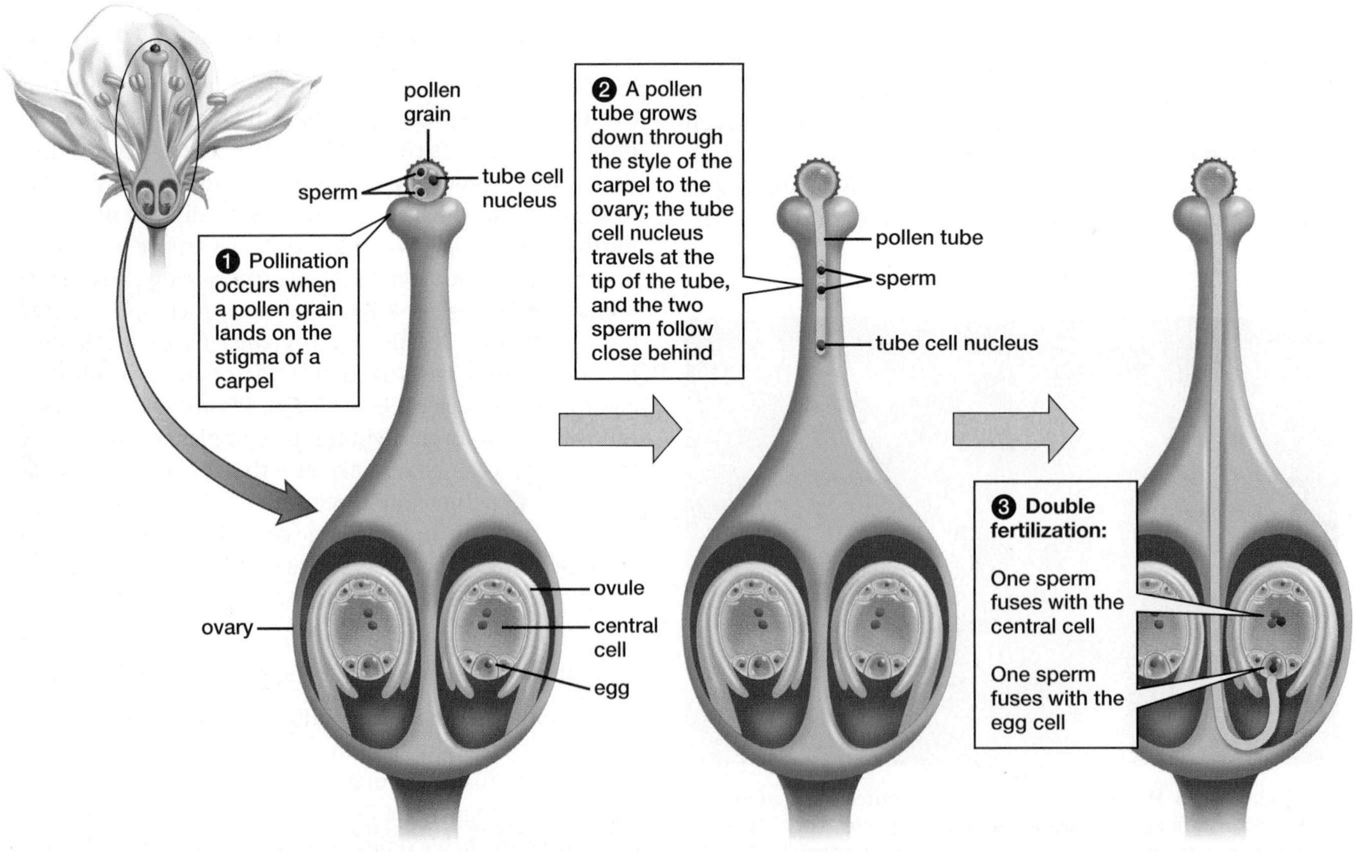

pollen grain
sperm
tube cell nucleus
1 Pollination occurs when a pollen grain lands on the stigma of a carpel
2 A pollen tube grows down through the style of the carpel to the ovary; the tube cell nucleus travels at the tip of the tube, and the two sperm follow close behind
pollen tube
sperm
tube cell nucleus
ovule
ovary
central cell
egg
3 Double fertilization:
One sperm fuses with the central cell
One sperm fuses with the egg cell

▲ **FIGURE 44-8 Pollination and fertilization of a flower**

makes up most of the pollen grain (see Fig. 44-4), then breaks out of the pollen coat and elongates, burrowing through the style and producing a tube that will conduct sperm down the style and into an ovule within the ovary (**Fig. 44-8** ❷).

If all goes well, the pollen tube reaches an opening in the integuments of an ovule and enters the female gametophyte. The tip of the tube ruptures, releasing the two sperm. In a process unique to flowering plants, called **double fertilization,** both sperm fuse with cells of the female gametophyte (**Fig. 44-8** ❸). One sperm fertilizes the egg, producing a diploid zygote that will develop into an embryo and eventually into a new sporophyte. The second sperm enters the large central cell, where its nucleus fuses with the two nuclei already present, forming a triploid nucleus, containing three sets of chromosomes. Through repeated mitotic cell divisions, the central cell will develop into the triploid **endosperm,** a food-storage tissue within the seed. The other five cells of the female gametophyte degenerate soon after fertilization.

Have you ever wondered?

When Is a Fruit a Vegetable?

To a biologist, a fruit is the ripened, seed-containing ovary of an angiosperm, often including some other parts of the flower. A vegetable is a nonreproductive part of a plant, such as a leaf or root. The culinary and legal worlds, however, beg to differ. Many foods that a biologist would call fruits are deemed vegetables by chefs and cookbooks: tomatoes, zucchinis, cucumbers, bell and chili peppers, and eggplants, to name just a few. The U.S. Supreme Court ruled on the status of the tomato in 1893, calling it a vegetable, at least for tax purposes, reasoning that vegetables are served as part of a main course, and fruits are served for dessert. Arkansas decisively straddles the fence: In 1987, the South Arkansas Vine Ripe Pink Tomato was named both the state fruit and the state vegetable!

44.3 HOW DO FRUITS AND SEEDS DEVELOP?

After double fertilization, the female gametophyte and the surrounding integuments of the ovule develop into a seed. The seed is surrounded by the ovary, which will form a fruit. Having already served their functions of attracting pollinators and producing pollen, petals and stamens shrivel and fall away as the fruit enlarges.

The Fruit Develops from the Ovary

When you eat a fruit, you are consuming the plant's ripened ovary, sometimes accompanied by other flower parts (**Fig. 44-9**). In a bell pepper, for example, the edible flesh develops from the wall of the ovary. Each of the seeds develops from an individual ovule within the ovary. As you will learn in section 44.6, fruits are not always edible; some are hard, feathery, winged, spiked, adhesive, or even explosive. These various shapes, colors, and textures all serve the same function: They help disperse seeds away from the parent plant, in many cases taking advantage of the mobility of animals (see "Earth Watch: Pollinators, Seed Dispersers, and Ecosystem Tinkering" on p. 874).

The Seed Develops from the Ovule

Three distinct developmental processes transform an ovule into a seed (**Fig. 44-10a**). First, the integuments of the ovule thicken, harden, and become the **seed coat** that surrounds and protects the seed. Second, the triploid central cell divides

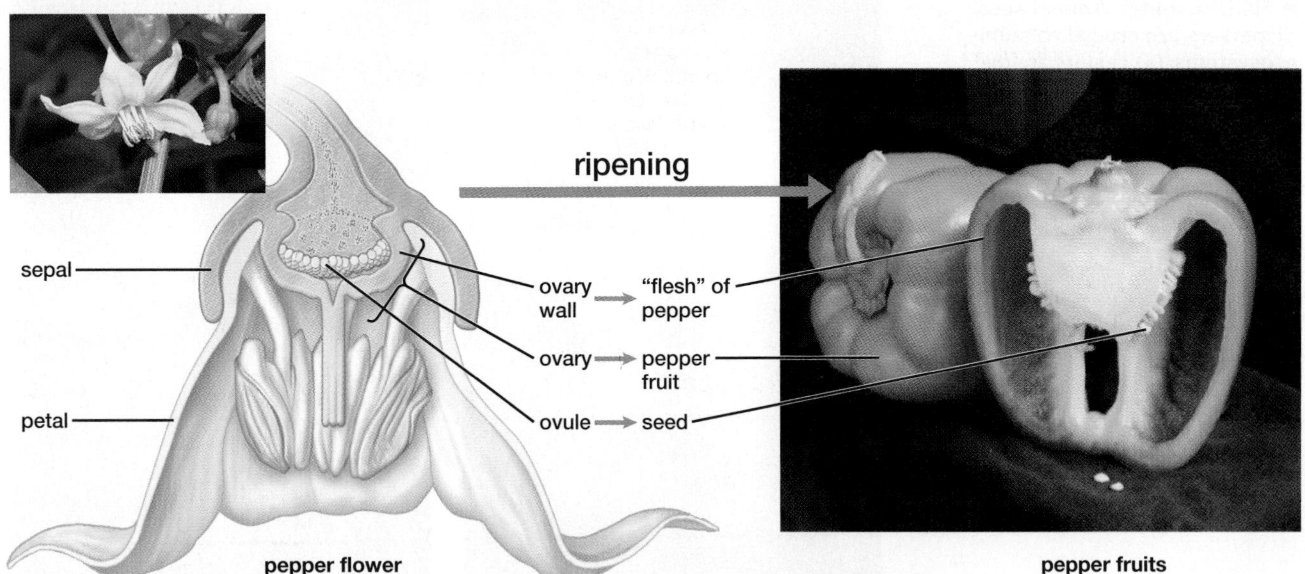

▲ **FIGURE 44-9 Development of fruit and seeds in a bell pepper** Fruits and seeds develop from flower parts. The ovary wall ripens into the fruit flesh. Each bell pepper ovary houses many ovules, which develop into seeds.

Earth Watch

Pollinators, Seed Dispersers, and Ecosystem Tinkering

Flowering plants dominate most terrestrial ecosystems largely because of mutually beneficial relationships with animals that pollinate their flowers and disperse their seeds. If pollinators or seed dispersers are eliminated, intentionally or accidentally, entire ecosystems may be endangered.

Sound alarmist? Consider Madagascar. On this island off the coast of Africa, researchers have identified more than 20 tree species that depend primarily on primates called lemurs for seed dispersal (**Fig. E44-3a**). But a burgeoning human population is destroying lemur habitat, and many species are endangered. When lemurs disappear, so do these trees.

Monkeys and fruit-eating bats are important agents of seed dispersal in some tropical forests (**Fig. E44-3b**). For example, biologist Donald Thomas discovered that after passing through a bat's digestive tract, nearly all the seeds of certain trees germinate; in contrast, seeds planted directly from fruit have only a 10% germination rate. Unfortunately, many fruit-eating and seed-dispersing animals such as monkeys, deer, and tapir have been overhunted. Fruit-eating bats are threatened by habitat destruction as land is cleared for agriculture. As a result, many tropical fruits rot on the forest floor or send up doomed sprouts under the shade of their parents, as their dispersal has been dramatically reduced. As Alejandro Estrada of the University of Mexico put it, "The continued existence of tropical forests whose primates and . . . birds and bats have been shot is just as precarious as if their trees had been chain-sawed and bulldozed."

Although domesticated "ecosystems"—large-scale farms—rely on humans for seed dispersal, they can likewise be vulnerable to the loss of pollinators. Some crops, such as corn and wheat, are pollinated by the wind, but most fruits and nuts, and many vegetables, depend on pollination by introduced European honeybees. The U.S. Department of Agriculture estimates that bee pollination provides about $15 billion in crop value, with honeybees responsible for most of that.

Honeybees have been in trouble for some time, plagued with mites (tiny relatives of spiders that feed on the body fluids of honeybees), viruses, fungi, and pesticides. However, in the fall of 2006, whole colonies began to disappear, leaving only the queen and larvae behind, a phenomenon named colony collapse disorder. Some beekeepers lost as many as 90% of their colonies that winter. No one really knows why colony collapse disorder occurs. Hypotheses include particularly virulent strains of viruses or bacteria, exploding populations of mites, or combinations of disease and pesticide exposure that have only recently reached critical levels.

There are many native bee species throughout the world—about 2,000 in California alone—many of which pollinate flowers. If colony collapse disorder continues, can these native bees take up the slack? Probably not; at least not right away. Populations of many native bees have been severely reduced by competition from honeybees and by habitat loss as enormous, intensively managed farms have replaced the hedgerows, meadows, and forest edges that formerly supported the wildflowers on which native bees depend for food.

Flowering plants, their pollinators, and their seed dispersers often form an intricate, interconnected web, with each supporting the others. As ecologist Aldo Leopold wrote in *A Sand County Almanac*, "To keep every cog and wheel is the first precaution of intelligent tinkering." In many of our ecosystems, including domesticated ones, we may lose some critical cogs and wheels, not realizing how important they are.

▶ **FIGURE E44-3 Animal seed dispersers are crucial to some ecosystems (a)** Lemurs are the principal pollinators for many tree species on the island of Madagascar. **(b)** A bat eats a ripe fig in Kenya. Without bats and other seed-dispersing animals, some types of tropical forest communities might not survive.

(a) Brown lemur

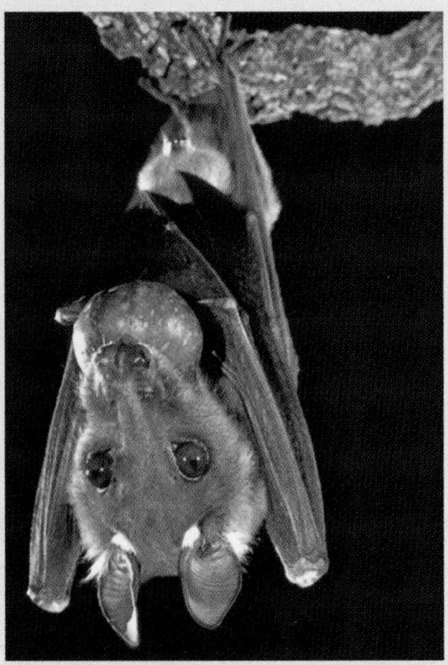

(b) Fruit-eating bat

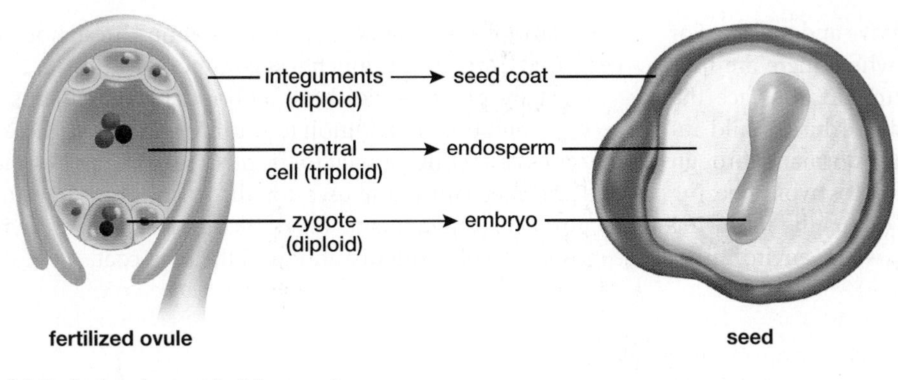

(a) Early development of the seed

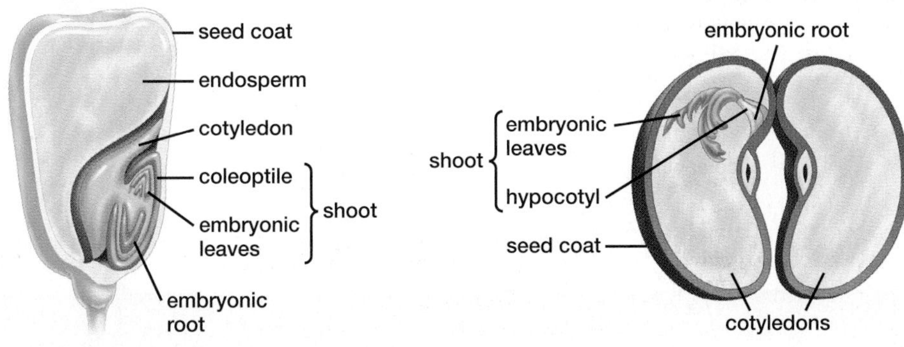

(b) Corn seed (monocot)

(c) Bean seed (dicot)

◀ FIGURE 44-10 **Seed development** **(a)** Seed formation begins after one sperm fuses with the egg, forming a diploid zygote, and the second sperm fuses with the two nuclei of the central cell. The endosperm develops from the triploid central cell, which undergoes many mitotic cell divisions as it absorbs nutrients from the parent plant. The embryo develops from the zygote. The integuments of the ovule develop into the seed coat. **(b)** Monocot seeds, such as corn, have a single cotyledon and usually retain most of their endosperm until germination. **(c)** Dicot seeds, such as beans, have two cotyledons that usually absorb most of the endosperm before germination. Dicot seeds, therefore, are typically made up mostly of the two large cotyledons.

rapidly. The resulting daughter cells absorb nutrients from the parent plant, forming a food-filled endosperm. Third, the zygote develops into the embryo.

As the seed matures, the embryo begins to differentiate into shoot and root (**Fig. 44-10b,c**). The shoot portion includes one or two **cotyledons** that absorb food molecules from the endosperm and transfer them to other parts of the embryo. Monocot seeds usually differ from dicot seeds in several respects. The seeds of monocots, as their name implies, have a single cotyledon ("mono" means "one"; see Fig. 44-10b). In most monocots, including grasses, rice, corn, and wheat, the cotyledon absorbs some endosperm during development, but most of the endosperm remains in the mature seed until the seed sprouts. Flour made from wheat or rice is ground-up endosperm. We sometimes consume the wheat embryo separately as "wheat germ." The monocot embryo is enclosed in a pair of sheaths, one surrounding the developing root and a second, called the **coleoptile,** surrounding the developing shoot tip.

Dicot seeds have two cotyledons ("di" means "two"; see Fig. 44-10c). In the seeds of most dicots, including peas, beans, peanuts, walnuts, and squash, the cotyledons absorb most of the endosperm during seed development, so the mature seed is virtually filled with embryo, particularly its cotyledons. If you strip the thin seed coat from a bean or peanut, you will find that the inside splits easily into two halves; each is a cotyledon. The tiny white nub adhering to one of the cotyledons is the rest of the embryo.

In both monocots and dicots, the embryonic shoot consists of two regions. Below the attachment point of the cotyledons, but above the root, is the **hypocotyl** ("hypo" is Greek, meaning "below"); above the cotyledons, the shoot is called the **epicotyl** ("epi" means "above"). At the tip of the epicotyl lies the apical meristem of the shoot; its daughter cells will later differentiate into the specialized cell types of stem, leaves, and flowers (see Chapter 43). In some embryos, one or two developing leaves may already be growing from the epicotyl.

44.4 HOW DO SEEDS GERMINATE AND GROW?

Germination, often called sprouting, occurs when the embryonic plant within a seed grows, breaks out of the seed, and forms a seedling. Seeds need warmth and moisture to germinate. But even under ideal conditions, many newly matured seeds do not germinate immediately. Instead, they enter a period of **dormancy** during which they will not sprout. Dormant seeds are typically able to resist adverse environmental conditions such as freezing and drying.

Seed Dormancy Helps Ensure Germination at an Appropriate Time

Seed dormancy solves two problems. First, it prevents seeds from germinating within a moist fruit, where the emerging plant might be consumed by a fruit-eating animal or attacked by mold growing on decomposing fruit. Even if they survived, multiple seedlings germinating within a fruit would grow in a dense cluster, competing for nutrients and light. Second, environmental conditions that are suitable for seedling germination (such as warmth and moisture) may not coincide with conditions that allow the seedling to survive and mature. For example, seeds in temperate climates (where there are four distinct seasons) mature in the late summer and face the harsh winter to come. Most do not germinate during mild autumn weather. Instead, the seeds remain dormant through both autumn and winter, so that the embryos avoid freezing as tender sprouts. Germination usually occurs the following spring. In warm, moist, tropical regions, where environmental conditions are suitable for germination throughout the year, seed dormancy is much less common.

Many plant species have additional requirements for seed germination, in addition to warmth and water. These requirements are finely tuned to the plant's native environment and the mechanisms it uses for dispersal. The three most common requirements to break seed dormancy are drying, exposure to cold, and disruption of the seed coat.

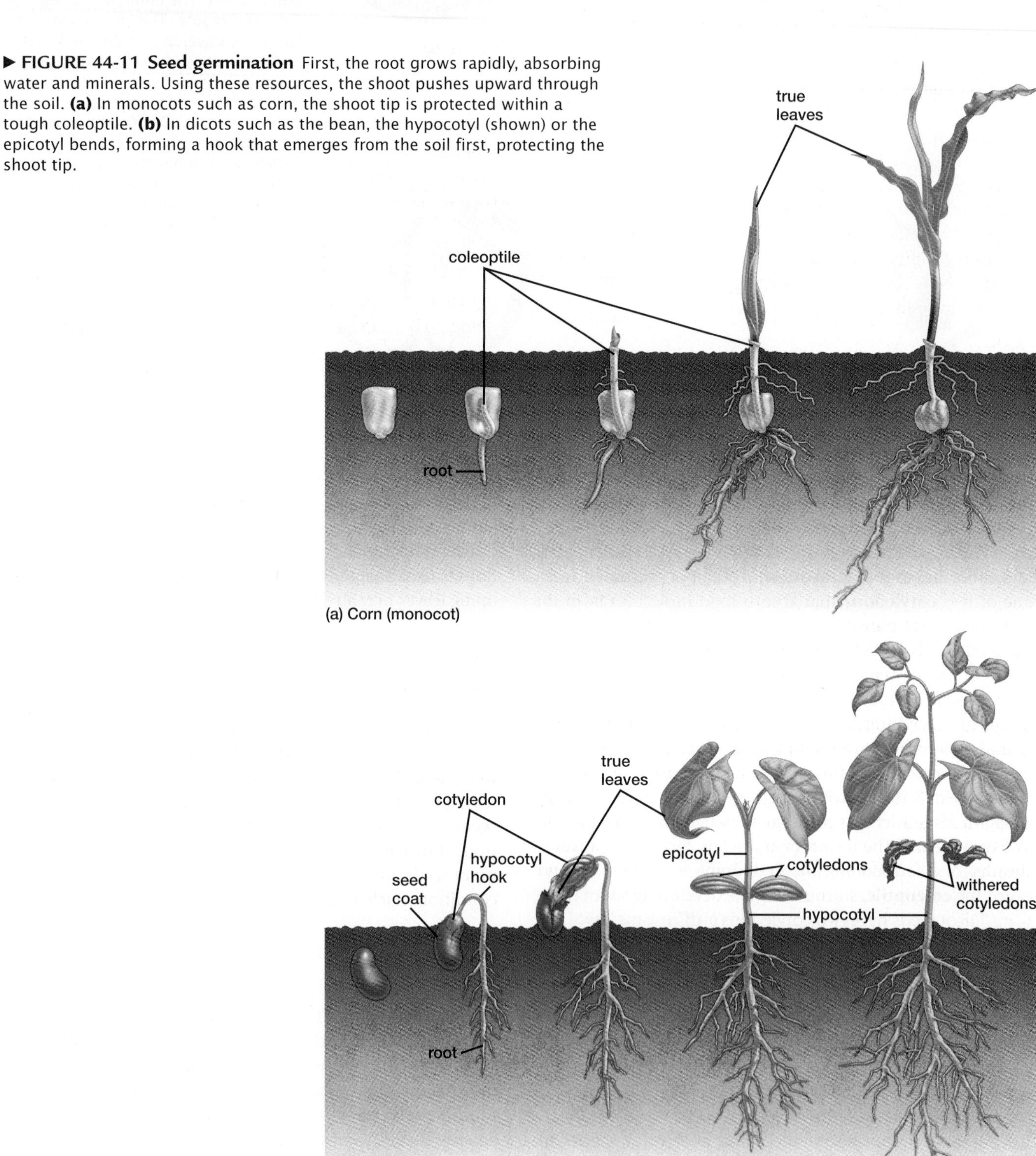

▶ **FIGURE 44-11 Seed germination** First, the root grows rapidly, absorbing water and minerals. Using these resources, the shoot pushes upward through the soil. **(a)** In monocots such as corn, the shoot tip is protected within a tough coleoptile. **(b)** In dicots such as the bean, the hypocotyl (shown) or the epicotyl bends, forming a hook that emerges from the soil first, protecting the shoot tip.

- **Drying** Seeds that require drying often are dispersed by fruit-eating animals that cannot digest the seeds. The seeds are excreted and exposed to air, where they dry out. Later, when temperature and moisture levels are favorable, they germinate.
- **Cold** Seeds of many temperate and arctic plants will not germinate unless they are exposed to prolonged subfreezing temperatures, followed by sufficient warmth and moisture. This ensures that seeds released in mild autumn weather do not immediately germinate. Requiring a substantial cold spell keeps them from sprouting until the following spring.
- **Seed coat disruption** The seed coat may need to be weathered or partially digested before germination can occur. Some coats contain chemicals that inhibit germination. In deserts, for example, years may go by without enough water for a plant to complete its life cycle. The seed coats of many desert plants have water-soluble chemicals that inhibit germination, and only a hard rainfall can wash away enough of the inhibitors to allow sprouting.

During Germination, the Root Emerges First, Followed by the Shoot

During germination, the embryo absorbs water, which makes it swell and burst its seed coat. The root usually emerges first and grows rapidly, absorbing water and minerals from the soil (**Fig. 44-11**). Much of the water is transported to the shoot, where the cells elongate and push upward through the soil toward the light.

The energy for germination ultimately comes from the endosperm of the seed. Recall that monocot seeds retain most of their endosperm until germination. During germination, the cotyledon digests the endosperm, absorbing its nutrients and transferring them to the growing embryo. In dicot seeds, the cotyledons have already absorbed most of the endosperm long before germination, so the cotyledons merely transfer these nutrients to the embryo as germination occurs.

Most seeds are buried in soil to some extent, so the embryo, especially its apical meristems, must be protected from being damaged by sharp soil particles during germination. The apical meristem of a root tip is protected by a root cap throughout the life of the plant (see Fig. 43-13). Shoots, however, are underground for only a brief time during germination, and so they need only temporary protection. In monocots, the coleoptile encloses the shoot tip like a glove around a finger (**Fig. 44-11a**), and pushes aside soil particles as the tip grows. Once out in the air, the coleoptile degenerates, allowing the shoot to emerge. The cotyledon stays belowground in the remnants of the seed.

In dicots, which lack coleoptiles, the shoot forms a hook in the epicotyl or hypocotyl (**Fig. 44-11b**). The bend of the hook, encased in thick-walled epidermal cells, forces its way up through the soil, clearing the path for the downward-pointing apical meristem and its delicate new leaves. In dicots with hypocotyl hooks, such as beans and squash, the elongating shoot carries the cotyledons out of the soil into the air. These aboveground cotyledons typically become green and photosynthetic, and transfer both previously stored food and newly synthesized sugars to the shoot (**Fig. 44-12**). Eventually, the cotyledons wither away. In dicots with epicotyl hooks, the cotyledons remain below the ground, shriveling up as the embryo absorbs their stored food. In all dicots, the shoot straightens after it emerges, orienting its true leaves toward the sunlight and undergoing primary growth, as discussed in Chapter 43.

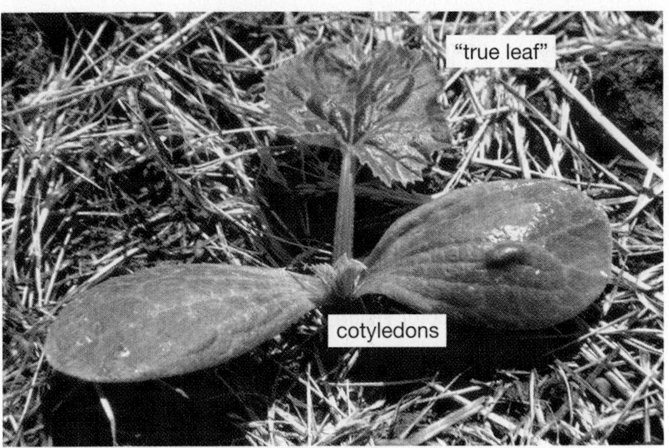

▲ **FIGURE 44-12 Cotyledons nourish the developing plant** In some dicots, such as the zucchini shown here, the cotyledons emerge from the soil, expand, and photosynthesize. The first true leaf (the crinkled central leaf) develops a bit later. Eventually, the cotyledons shrivel up.

44.5 HOW DO PLANTS AND THEIR POLLINATORS INTERACT?

Plants and their pollinators have coevolved; that is, each has acted as an agent of natural selection on the other. Animal-pollinated flowers evolved traits that attract useful pollinators and frustrate undesirable visitors that might eat nectar or pollen without fertilizing the flower. The pollinators evolved senses and behaviors that help them to locate and identify nutritious flowers and extract the nectar or pollen. Animal-pollinated flowers can be loosely grouped into three categories, depending on the benefits, real or perceived, that they offer to potential pollinators: food, sex, or a nursery.

Some Flowers Provide Food for Pollinators

Many flowers provide food for foraging animals such as beetles, bees, moths, butterflies, or hummingbirds. In return, the animals unwittingly distribute pollen from flower to flower. Most flying pollinators locate flowers from a distance because the flower colors contrast with the mass of green leaves surrounding them. Therefore, flower colors have often coevolved to match the color vision of their pollinators.

For example, bees have good color vision, but do not see the same range of colors that humans do (**Fig. 44-13**). Typically, bees do not perceive red as a distinct color, but

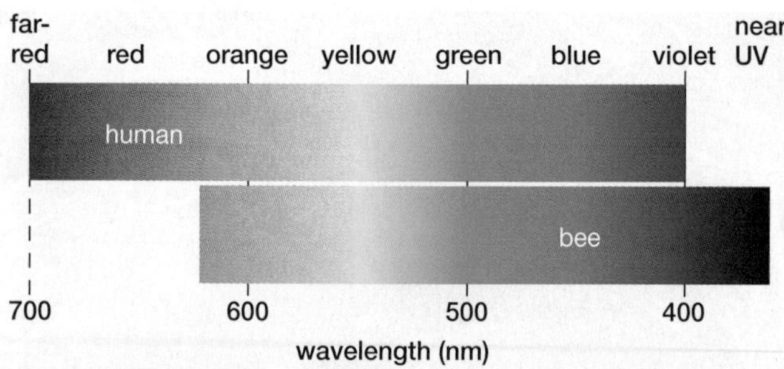

(a) A comparison of color vision in humans and bees

human vision

bee vision

(b) Flower color patterns seen by humans and bees

▲ **FIGURE 44-13 Ultraviolet patterns guide bees to nectar** **(a)** The spectra of color vision for humans and bees overlap considerably, but are not identical. Humans (top) are sensitive to red, which bees (bottom) do not perceive; bees can see near-UV light, which is invisible to the human eye. **(b)** Many flowers photographed (left) under ordinary daylight and (right) under UV light show striking differences in color patterns. Bees can see the UV patterns that presumably direct them to the nectar- and pollen-containing centers of the flowers.

they can see ultraviolet light. Bee-pollinated flowers must look brightly colored *to a bee,* so these flowers are usually white, blue, yellow, or orange. Many have markings that reflect UV light, including central spots or lines pointing toward the center. We can also thank the bees for most of the sweet-smelling flowers because "flowery" odors attract these pollinators.

Bee-pollinated flowers have structural adaptations that help ensure pollen transfer. In the Scotch broom flower, for example, nectar forms in a crevice between enclosing petals. In newly opened flowers, pollen-laden stamens lurk within the crevice. When a bee visits a young flower, the stamens emerge, brushing pollen onto its back as the bee's weight deflects the petals downward (**Fig. 44-14**). In older flowers, the carpel's style elongates, pushing the sticky stigma out through the crevice; therefore, when a pollen-coated bee delves for nectar, she leaves pollen behind on the stigma.

Case Study continued
Some Like It Hot—and Stinky!

How do corpse flowers get hot? They have special mechanisms for disconnecting cellular respiration from ATP synthesis. In most cells, cellular respiration uses about 40% of the energy in glucose to synthesize ATP, with the rest given off as heat (see p. 128). Hot flowers, on the other hand, synthesize very little ATP; instead, almost all of the energy in glucose is released as heat, causing the flower to warm up.

Many flowers adapted for moth and butterfly pollinators have nectar-containing tubes that accommodate the long tongues of these insects. Flowers pollinated by night-flying moths open only in the evening. Most are white, which makes them more visible in the dark. Some also give off strong, musky odors that attract moths. Bat-pollinated flowers are usually also white and open at night. Beetles and flies often feed on animal wastes or carrion, so flowers pollinated by beetles and flies frequently smell like dung or rotting flesh. These flowers usually deceive their pollinators by smelling like a nutritious meal but offering no food at all. Some, including the corpse flower, also heat up. The warmth attracts pollinators and helps broadcast the flower's foul scents.

Hummingbirds are one of the few important vertebrate pollinators (**Fig. 44-15a**), although some mammals also pollinate flowers (**Fig. 44-15b**). Because hummingbirds have a poor sense of smell, hummingbird-pollinated flowers seldom synthesize fragrant chemicals. However, they often produce more

▶ **FIGURE 44-14 "Pollinating" a pollinator (a)** In Scotch broom flowers, the bee finds nectar near the junction of the top and bottom petals. **(b)** The bee's weight deflects the bottom petals downward, causing pollen-laden stamens to pop up and cover the bee's hairy back with pollen. The bee will carry the pollen to other Scotch broom flowers, leaving some on their stigmas.

(a) A bee on a Scotch broom flower

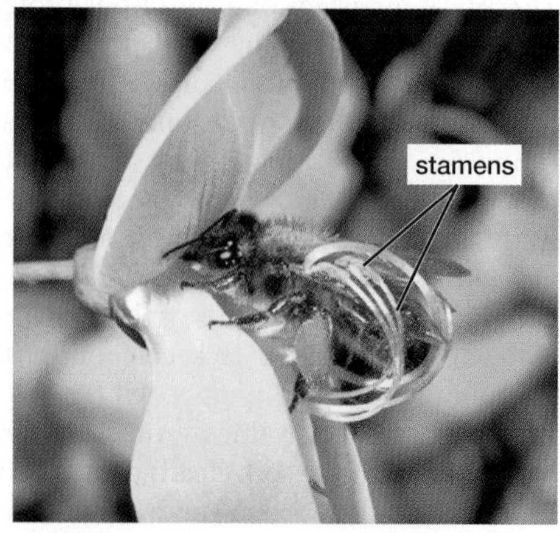

(b) The flower deposits pollen on the bee

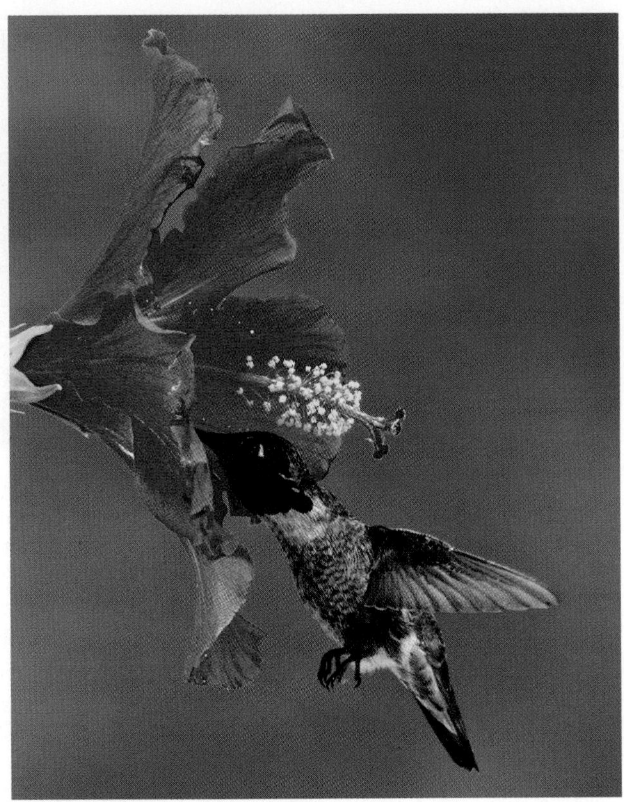
(a) Hummingbird

(b) Honey possum

◀ **FIGURE 44-15 Vertebrate pollinators (a)** A hummingbird feeds at a hibiscus flower. Notice how the anthers are positioned to deposit pollen on its head. **(b)** As the honey possum of Australia stuffs its face into this flower, pollen adheres to its muzzle and whiskers. A visit to another flower will transfer the pollen.

QUESTION Why have many plants that are pollinated by hummingbirds evolved flowers shaped like long, narrow tubes?

nectar than other flowers, because hummingbirds need more energy than insects do and will favor flowers that provide it. Hummingbird-pollinated flowers may have a deep, tubular shape that accommodates the birds' long bills and tongues. In addition, they are often red, which is attractive to hummingbirds but is not seen well by bees (see Fig. 44-13).

Some Flowers Are Mating Decoys

A few plants, most notably some orchids, take advantage of the mating drive and stereotyped behaviors of male wasps, bees, and flies. These orchid flowers mimic female wasps, bees, or flies both in scent (the orchids release a sexual attractant similar to that produced by the female insect) and shape (**Fig. 44-16**). The males land atop these "females" and attempt to copulate, but get only a packet of pollen for their efforts. As they repeat their attempts on other orchids of the same species, the pollen packet is transferred.

▲ **FIGURE 44-16 Sexual deception promotes pollination** This male wasp is trying to copulate with an orchid flower. The result is successful reproduction—by the orchid, not the wasp!

Some Flowers Provide Nurseries for Pollinators

Perhaps the most elaborate relationships between plants and pollinators occur in a few cases in which insects fertilize a flower and then lay their eggs in the flower's ovary. This arrangement occurs between milkweeds and milkweed bugs, figs and fig wasps, and yuccas and yucca moths (**Fig. 44-17**). For example, the yucca moth's remarkable behaviors result in the pollination of yuccas and a well-stocked pantry for its own offspring. A female moth visits a yucca flower, collects pollen, and rolls it into a compact ball. She carries the pollen ball to another yucca flower, drills a hole in the ovary wall, and lays her eggs inside the ovary. Then she smears the pollen ball all over the stigma of the flower. By pollinating the yucca, the moth ensures that the plant will provide a supply of developing seeds for its caterpillar offspring. Because the caterpillars eat only a fraction of the seeds, the yucca also reproduces successfully. The mutual adaptation of yucca and yucca moth is so complete that neither can reproduce without the other.

▲ **FIGURE 44-17 A mutually dependent relationship** **(a)** Yuccas bloom on the dry plains of eastern Colorado in early summer. **(b)** A yucca moth places pollen on the stigma of a yucca flower.

(a) Dandelion fruits

(b) Maple fruits

▲ **FIGURE 44-18 Wind-dispersed fruits** **(a)** Dandelion fruits have filamentous tufts that catch the breezes. **(b)** Maple fruits resemble miniature glider–helicopters, whirling away from the tree as they fall.

EXERCISE To see how the wings aid in seed dispersal, take two maple fruits and pluck the wings off one. Hold both fruits over your head and drop them. Compare where they land. Try this on both a calm and a windy day.

44.6 HOW DO FRUITS HELP TO DISPERSE SEEDS?

A plant benefits if its seeds are dispersed far enough away so that its offspring don't compete with it for light and nutrients. In addition, plant predators, from insects to mammals, are often abundant nearby, or are attracted to, the parent plants. While the parent may be able to withstand some browsing, a seedling is likely to be killed. Finally, plant species will be more successful and widespread if they at least occasionally disperse their seeds to distant habitats. In flowering plants, fruits use a fascinating variety of mechanisms to disperse seeds.

Explosive Fruits Shoot Out Seeds

A few plants develop explosive fruits that eject their seeds far away from the parent plant. Dwarf mistletoes, common parasites of trees, produce fruits that shoot sticky seeds more than 60 feet (about 20 meters). If a seed strikes a nearby tree, it sticks to the bark and germinates, sending rootlike fibers into the vascular tissues of its host, from which it draws its nourishment. Because the proper germination site for a mistletoe seed is not the ground but a tree limb, shooting the seeds, not dropping them, is clearly useful. Other plants with explosive fruits include the touch-me-not, witch hazel, and squirting cucumber.

Lightweight Fruits May Be Carried by the Wind

Dandelions, milkweeds, elms, and maples produce lightweight fruits with large wind-catching surfaces. Each hairy tuft on a dandelion puff is a separate fruit bearing a single small seed that it can carry for miles if the winds cooperate (**Fig. 44-18a**). The single wing of the maple fruit, in contrast, causes its seed to spin like a propeller as it falls, usually taking it only a few yards from its parent tree (**Fig. 44-18b**).

Floating Fruits Allow Water Dispersal

Many fruits can float on water for a time and may be carried in streams or rivers, although this is usually not their principal method of dispersal. The coconut fruit, however, is a champion floater. Round, buoyant, and waterproof, the coconut drops from its parent palm, often near a sandy shore. It may germinate in place, or it may be washed out to sea and float for weeks or months until it washes ashore on some distant isle (**Fig. 44-19**). There, it may germinate, possibly establishing a new coconut colony where none previously existed.

Clingy or Edible Fruits Are Dispersed by Animals

Clingy fruits, such as burdocks, burr clover, foxtails, and sticktights, grasp animal fur (or human clothing) with prongs, hooks, spines, or adhesive hairs (**Fig. 44-20**). The parent plant holds its ripe fruit very loosely, so that even slight contact with fur pulls the fruit off the plant, leaving it stuck to the animal. Some of these fruits later fall off as the animal rolls on the ground, brushes against objects, grooms itself, or sheds its fur.

▲ **FIGURE 44-19 Water-dispersed fruit** This coconut may have been washed ashore after a long journey at sea. Coconut "meat" and coconut "milk" are two different types of endosperm. The large size and massive food reserves of coconuts are probably adaptations for successful germination and growth on barren, sandy beaches.

QUESTION Although many fruits can float, why is water seldom their primary method of dispersal?

Unlike these hitchhiker fruits, edible fruits benefit both the plant and the animal disperser. The plant stores sugars, starches, and appealing flavors in a fleshy fruit that surrounds the seeds, enticing hungry animals (**Fig. 44-21**). Some edible fruits, including peaches, plums, and avocados, contain large, hard seeds that animals usually do not eat. Other fruits, such as blackberries, raspberries, strawberries, tomatoes, and peppers, have small seeds that animals swallow. These seeds are eventually excreted unharmed (see "Earth Watch: Pollinators, Seed Dispersers, and Ecosystem Tinkering" on page **874**). Some have seed coats that must be scraped or weakened by passing through an animal's digestive tract before they will germinate. Besides being transported away from its parent plant, a seed that is swallowed and excreted benefits in another way: It ends up with its own supply of fertilizer!

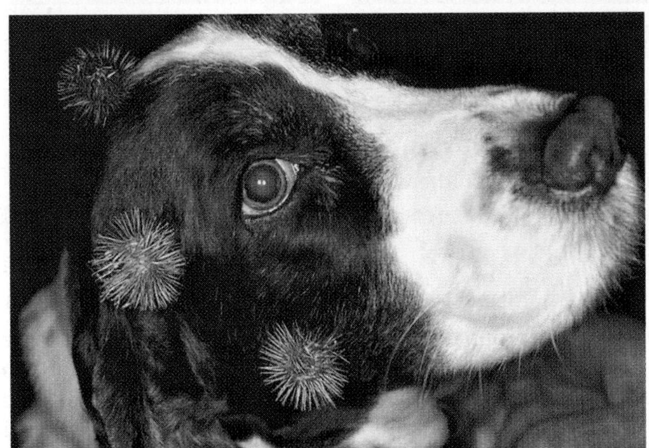

▲ **FIGURE 44-20 The cocklebur fruit uses hooked spines to hitch a ride on furry animals** This dog—or its owner—will eventually dislodge the cockleburs, often far away from the parent plant.

▲ **FIGURE 44-21 The colors of ripe fruits attract animals** A bright red raspberry fruit has attracted a resplendent quetzal in Costa Rica. Only ripe fruits containing mature seeds are sweet and brightly colored, appealing to animals that feed on them and disperse their seeds.

It is often important for the right type of animal to eat the fruit and seeds. For example, a graduate student studying the spicy seeds of chili peppers found that the burning sensation discourages local mammals from eating the fruit but does not deter birds, who are insensitive to it. He further discovered that the digestive tracts of mammals destroy chili seeds, but seeds that have passed through a bird's digestive tract germinate at three times the rate of those that just fall to the ground.

Case Study revisited

Some Like It Hot—and Stinky!

The corpse flower, starfish flower, stinking corpse lily, and dead-horse arum are pollinated by carrion-loving beetles and flies. Chemical analysis shows that the dead-horse arum produces some of the same putrid chemicals as decaying carcasses do. Flies and beetles can't tell the difference. What's more, decaying carcasses warm up, a by-product of the metabolism of bacteria decomposing the flesh. Heat production is probably another aspect of corpse mimicry; experiments have shown that flies prefer warm, smelly flowers to cool, equally smelly ones.

The dead-horse arum has a particularly sophisticated bloom that ensures not just pollination, but cross-pollination. Its bloom consists of numerous, separate, female and male flowers contained in a chamber. The female flowers blossom on the first day that the plant blooms. This is also when the plant emits its putrid odor and warms up. Lured by the smell and the warmth, flies enter the chamber. Spines at the entrance keep them from leaving again. As they blunder about in the chamber, the flies shower the female flowers with any pollen that they may have picked up from an earlier visit to another arum. By the next morning, the female flowers have wilted, the spines have collapsed, the bloom no longer smells, and the male flowers have matured. As the flies leave the chamber, the male flowers dust them with pollen. Later, some of these same flies will be deceived by, and pollinate, other dead-horse arums.

The corpse flower and dead-horse arum seem to offer no reward in exchange for pollination. This is not always the case with warm flowers, however. Many species of philodendron, plants found in tropical rain forests and many people's houses, produce hot flowers (up to 114°F for one species) with mildly pleasant aromas. The flowers of some species serve as warm, cozy orgy rooms for beetles, who crawl into the bloom in the early evening and spend the night mating (**Fig. 44-22**). The beetles use only half as much energy in a philodendron bloom as they would in the cool night air outside. The beetles also feed on pollen and other parts of the flower. Philodendrons, therefore, give the beetles real rewards in exchange for their services as pollinators.

Consider This

Heat-producing flowers are rare, and many are members of evolutionarily ancient groups. Some botanists hypothesize that warmth was an early innovation to attract beetle pollinators. Today, most plants bear "fast-food" flowers: They supply their pollinators with a sip of nectar, and then send them on their way with a dusting of pollen. Compare the pollination strategies used by the dead-horse arum and the philodendron with the more common fast-food flowers. Suggest evolutionary reasons why fast-food flowers predominate today.

(a) A newly opened philodendron flower

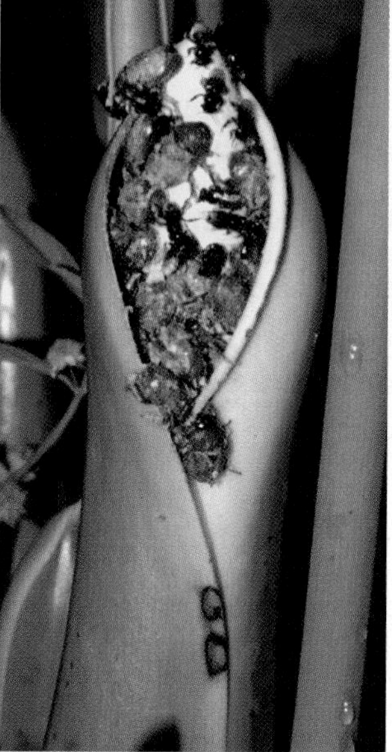

(b) Beetles swarm on a philodendron flower

▲ FIGURE 44-22 **A beetle orgy (a)** This philodendron flower produces scent on the first day of blooming. **(b)** Then it heats up, attracting beetles that congregate, feed, mate, and conserve energy on the warm blossom.

CHAPTER REVIEW

Summary of Key Concepts

44.1 How Do Plants Reproduce?

The sexual life cycle of plants, called alternation of generations, includes both a multicellular diploid form (the sporophyte generation) and a multicellular haploid form (the gametophyte generation). Meiotic cell division in cells of the diploid sporophyte produces haploid spores. The spores undergo mitotic cell division to produce the haploid gametophyte generation. Reproductive cells of the gametophyte differentiate into sperm and eggs, which fuse to produce a diploid zygote. Mitotic cell division of the zygote gives rise to another sporophyte generation. In mosses and liverworts, the gametophyte is the dominant stage; the sporophyte develops on the gametophyte and never lives independently. In ferns, the sporophyte is the dominant stage; both the gametophyte and sporophyte are independent structures. In gymnosperms and angiosperms, the sporophyte is the dominant stage; the gametophytes are very small and never live independently.

44.2 What Is the Function and Structure of the Flower?

Complete flowers consist of four parts: sepals, petals, stamens (male reproductive structures), and carpels (female reproductive structures). The sepals form the outer covering of the flower bud. Most petals (and in some cases, the sepals) are brightly colored and attract pollinators to the flower. The stamen consists of a filament that bears an anther, in which pollen develops. The carpel consists of the ovary, in which one or more female gametophytes develop, and a style that bears a sticky stigma to which pollen adheres during pollination. Most flowers are pollinated by animals and have colorful, often scented, petals. Some flowers are wind pollinated; these are usually small, with prominent anthers but often lacking petals.

The male gametophyte of flowering plants is the pollen grain. The diploid microspore mother cell undergoes meiotic cell division to produce four haploid microspores. Each undergoes mitotic cell division to form the pollen grain. The immature pollen grain consists of a tube cell and a generative cell that will later divide to produce two sperm cells.

The female gametophyte develops within the ovules of the ovary. A diploid megaspore mother cell undergoes meiotic cell division to form four haploid megaspores. Three degenerate; the fourth undergoes mitotic divisions to produce the eight nuclei of the female gametophyte. One of these becomes the egg; another becomes a large central cell with two nuclei, and the rest degenerate.

When a pollen grain lands on a stigma, its tube cell grows through the style to the female gametophyte. The two sperm cells travel down the style within the tube, eventually entering the female gametophyte. One sperm fuses with the egg to form a diploid zygote, which will give rise to the embryo. The other sperm fuses with the two nuclei of the central cell, producing a triploid cell that will give rise to the endosperm, a food-storage tissue within the seed.

44.3 How Do Fruits and Seeds Develop?

The function of the fruit is to disperse seeds. A fruit is a ripened ovary, often with contributions from other parts of the flower. Seeds develop from the ovules. The integuments of the ovules form a protective seed coat. Within the seed coat, a seed contains an embryo, consisting of an embryonic root and embryonic shoot, including the cotyledon (one in monocots, two in dicots), and the food-storing endosperm.

44.4 How Do Seeds Germinate and Grow?

Seed germination requires warmth and moisture. Energy for germination comes from food stored in the endosperm, which is transferred to the embryo by the cotyledons. Seeds may remain dormant for some time after fruit ripening, particularly in temperate climates. To break dormancy and germinate, some seeds require drying, exposure to cold, or disruption of the seed coat. The root emerges first from the germinating seed, absorbing water and nutrients that are transported to the shoot. Monocot shoots are protected by a coleoptile that covers the shoot tip during germination. Dicot shoots form epicotyl or hypocotyl hooks that break through the soil during germination, loosening the soil and protecting the delicate shoot tip.

44.5 How Do Plants and Their Pollinators Interact?

Plants and their animal pollinators have coevolved, acting as agents of natural selection on one another. Flowers attract animals with scent, food such as nectar, and appropriate colors and shapes that make them visible and accessible to their pollinators. Some flowers deceive pollinators, attracting insects with food scents or the shape of a mate. Some plants and their pollinators, such as the yucca plant and yucca moth, are completely dependent on one another.

44.6 How Do Fruits Help to Disperse Seeds?

Fruits disperse seeds in many ways:

- Explosive fruits shoot seeds away from the parent plant.
- Lightweight fruits are carried by the wind.
- Floating fruits are dispersed by water.
- Clinging fruits adhere to animals.
- Edible fruits are eaten by animals with little or no harm to the seeds.

Key Terms

alternation of generations *869*
anther *869*
carpel *869*
coleoptile *875*
complete flower *868*
cotyledon *875*
dormancy *875*
double fertilization *873*
egg *871*
endosperm *873*
epicotyl *875*
filament *869*
flower *868*
fruit *869*
gametophyte *866*
generative cell *871*
germination *875*
hypocotyl *875*
imperfect flower *870*
incomplete flower *870*
integument *871*
megaspore *871*
megaspore mother cell *871*
microspore *870*
microspore mother cell *870*
ovary *869*
ovule *869*
petal *869*
pollen grain *868*
pollination *869*
seed *869*
seed coat *873*
sepal *869*
spore *866*
sporophyte *866*
stamen *869*
stigma *869*
style *869*
tube cell *871*
zygote *866*

Thinking Through the Concepts

Fill-in-the-Blank

1. The plant sexual life cycle is called __________. The diploid generation is called the __________. Reproductive cells in this stage of the cycle produce spores through __________ (type of cell division). The spores germinate to produce the haploid generation, called the __________.

2. In a flowering plant, the male gametophyte is the __________. It is formed in the __________ of a flower. Pollination occurs when pollen lands on the __________ of a flower of the same plant species. The pollen grain grows a tube through the __________ of the carpel to the ovary at the base of the carpel. The tube enters an ovule through an opening in the __________ of the ovule.

3. Double fertilization in flowering plants occurs when one sperm cell fuses with an egg to form a diploid cell, the __________. The other sperm fuses with the two nuclei of the central cell of the female gametophyte. This triploid cell will divide mitotically to form a food storage organ called the __________. This food is transferred to the developing plant embryo by structures called __________.

4. The fruit of a flowering plant is formed from the __________ of a flower, possibly with additional contributions from other flower parts. The seed forms from the __________. The seed coat develops from the outer covering, or __________, of this structure.

5. __________ is the emergence of the embryo from a seed. Usually, the __________ (structure of the embryo) emerges first. In monocots, the shoot is protected by a sheath called the __________. In dicots, a(n) __________ in the hypocotyl or epicotyl protects the apical meristem and developing leaves.

Review Questions

1. Diagram the plant sexual life cycle. Which stages are haploid and which are diploid? At which stage are gametes formed?
2. Diagram a complete flower. Where are the male and female gametophytes formed?
3. How does an egg develop within the female gametophyte? How does double fertilization occur?
4. What is a pollen grain, and how is it formed?
5. What are the parts of a seed, and how does each part contribute to the development of a seedling?
6. Describe the characteristics you would expect to find in flowers that are pollinated by the wind, beetles, bees, and hummingbirds, respectively.
7. What is the endosperm? From which cell of the female gametophyte is it derived? Is endosperm usually more abundant in the mature seed of a dicot or of a monocot?
8. Describe three mechanisms whereby seed dormancy is broken in different types of seeds. How are these mechanisms related to the typical environment of the plant?
9. How do monocot and dicot seedlings protect the delicate shoot tip during seed germination?
10. Describe three types of fruits and the mechanisms whereby these fruit structures help disperse their seeds.

Applying the Concepts

1. A friend gives you some seeds to grow in your yard. When you plant some, nothing happens. What might you try to get the seeds to germinate?
2. Charles Darwin once described a flower that produced nectar at the bottom of a tube nearly a foot deep (30 centimeters). He predicted that there must be a moth or other animal with a 30-centimeter-long "tongue" to match. He was right; it's a moth. Such specialization almost certainly means that this particular flower could be pollinated only by that specific moth. What are the advantages and disadvantages of such specialization?
3. Many plants that we call weeds were brought from another continent either accidentally or purposefully. In their new environment, they have few competitors or animal predators, so they tend to grow in such large numbers that they displace native plants. Think of several ways in which humans become involved in plant dispersal. To what degree do you think humans have changed the distributions of plants? In what ways is this change helpful to humans? In what ways is it a disadvantage?

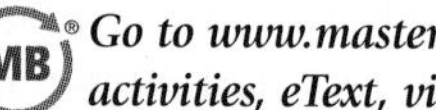

Go to www.masteringbiology.com for practice quizzes, activities, eText, videos, current events, and more.

Plant Responses to the Environment

Chapter 45

Case Study

Predatory Plants

IN A MARSH, plants are hungry—not for sunlight, but for nitrogen. Marshes tend to be acidic, and acidic conditions discourage the growth of nitrogen-fixing bacteria, which capture atmospheric nitrogen and transform it into a form that plants can use (see p. 855). Nitrogen is abundant in the proteins of animal bodies, however, and some bog-dwelling plants have evolved carnivorous lifestyles to satisfy their nitrogen needs.

In a South Carolina marsh, a fly lands on the clamshell-like leaves of a seemingly harmless Venus flytrap plant. Suddenly, the leaves snap together and their spiked edges mesh, trapping the hapless insect. During the next week or so, enzymes digest the protein from the fly's body, and the leaf absorbs the nitrogen-containing molecules before the trap reopens to attract its next victim.

Nearby, a lacewing insect lands on stalks bearing sweet droplets that resemble dew glistening in the sunlight—only to find itself struggling helplessly in a sticky mass secreted by a sundew leaf (see Fig. 45-15). Its struggle stimulates the sundew to secrete a cocktail of digestive enzymes into the goo. These enzymes rapidly break down the insect's body, and the leaves absorb the liberated nitrogen compounds.

Beneath the surface of the marsh, still another drama unfolds. A bladderwort plant dangles hundreds of pear-shaped chambers into the water (see Fig. 45-16). Each is sealed by a watertight trapdoor whose lower edge is fringed with bristles. A tiny water flea (related to shrimp, but barely visible to us) swims by, brushing the hairs. Within one-sixtieth of a second, the animal is sucked into the chamber.

How do these predatory plants detect their prey and then move quickly enough to trap them? How do more ordinary plants sense environmental stimuli and respond to them? In this chapter, you will find some answers.

▲ A fly enters a Venus flytrap.

45.1 WHAT ARE SOME MAJOR PLANT HORMONES?

Although plants seem relatively inert, they respond to their environment in sophisticated ways. Plants sense and react to stimuli that include touch, gravity, moisture, light, and day length. Like animals, plants use **hormones**—chemicals that are secreted by cells and transported to other cells, where they exert specific effects—to convey messages within their bodies. Hormones exert their effects by binding to specific receptors on target cells and initiating a series of biochemical reactions. Some hormones move to nearby target cells; others are transported in vessels, either blood vessels in animals, or xylem and phloem in plants.

Often produced in response to environmental stimuli, **plant hormones** stimulate growth, development, and even aging in their target cells, influencing every aspect of plant life cycles. Each plant hormone can elicit a variety of responses depending on the type of cell it stimulates, the stage of the plant life cycle, the concentration of the hormone, the presence and concentrations of other hormones, and the species of plant.

In this chapter, we focus on six important types of plant hormones that are active during various stages of flowering plant life cycles: auxins, gibberellins, cytokinins, ethylene, abscisic acid, and florigens. Auxins, gibberellins, and cytokinins are families of molecules that share some chemical properties, whereas florigens are a group of very similar proteins. In contrast, ethylene and abscisic acid do not vary among plants. Here, we briefly review some of the roles of hormones in plant physiology (**Table 45-1**).

Table 45-1 Some Major Plant Hormones and Their Functions

Hormone	Some Major Effects	Major Sites of Synthesis
Auxins	Promote cell elongation in shoots Inhibit growth of lateral buds (apical dominance) Promote root branching Control phototropism and gravitropism in shoots and roots Stimulate vascular tissue development Stimulate fruit development Delay senescence of leaves and fruit	Shoot apical meristem
Gibberellins	Stimulate stem elongation by promoting cell division and cell elongation Stimulate bud sprouting, flowering, fruit development, and seed germination	Shoot apical meristem Young leaves Plant embryos
Cytokinins	Stimulate cell division throughout the plant Stimulate lateral bud sprouting Inhibit formation of branch roots Delay senescence of leaves and flowers	Root apical meristem
Ethylene	Promotes growth of shorter, thicker stems in response to mechanical disturbance Stimulates ripening in some fruits Promotes senescence in leaves Promotes leaf and fruit drop	Throughout the plant, particularly during stress and aging
Abscisic acid	Causes stomata to close Inhibits stem growth and stimulates root growth in response to drought Maintains dormancy in buds and seeds	Throughout the plant
Florigens	Stimulate flowering in response to day length	Mature leaves

Auxins promote or inhibit elongation in different target cells. In the shoot, high concentrations of auxin cause cells to elongate. In roots, low auxin levels stimulate elongation, whereas slightly higher concentrations inhibit it. By regulating cell elongation, auxin causes the shoot to bend toward light, and the root to bend toward the pull of gravity. Auxin inhibits the sprouting of lateral buds to form stem branches, but stimulates root branching by increasing cell division in lateral roots. Auxin also stimulates the differentiation of vascular tissues (xylem and phloem) within newly formed plant parts. In the mature plant, auxin stimulates fruit development, and it prevents both fruits and leaves from falling prematurely. The primary site of auxin synthesis is the shoot apical meristem (the growing shoot tip); from here, auxin is carried throughout the plant body. Some auxin is also synthesized in young leaves, as well as in developing seeds and plant embryos where it influences development and germination.

Synthetic auxin (2,4-D) is widely used to kill dicot plants by disrupting the normal balance among auxin and other plant hormones. It is also used commercially to promote root formation in plant cuttings, to stimulate fruit development, and to delay fruit fall. Auxin was the first plant hormone to be recognized, as described in "Scientific Inquiry: How Were Plant Hormones Discovered?" on pp. 890–891.

Gibberellins are primarily active in plant shoots. They promote stem elongation by increasing both cell division and cell elongation. Gibberellins also stimulate bud sprouting, flowering, fruit production and development, and seed germination. Gibberellins are produced in the shoot apical meristem, in young leaves, and in plant embryos within seeds.

This type of hormone is named after the fungus *Gibberella fujikuroi,* which causes "foolish seedling" disease in rice. This disease stimulates plants to grow exceptionally tall. In the 1930s, scientists isolated substances produced by the fungus that stimulated plant growth. These and similar molecules are now called gibberellins.

Cytokinins promote cell division, which is required for the growth of all plant tissues. The major site of cytokinin synthesis is in root apical meristems, although some is also generated in the shoot. In roots, cytokinins inhibit the formation of branches; in shoots, however, they promote formation of branches by stimulating cell division in lateral bud meristems. Cytokinins cause nutrients to be transported into plant leaves, stimulating chlorophyll production and delaying aging. Commercially, cytokinin is sprayed on cut flowers to keep them fresh.

Ethylene is an unusual plant hormone because it is a gas. Produced in most plant tissues, it is released in response to a wide range of environmental stimuli and has diverse effects on its target cells. It is also recognized as a plant "stress hormone"; that is, it is synthesized by plant tissues to stimulate adaptive changes in response to wounding, flooding, drought, and temperature extremes. Ethylene also stimulates the formation of weak-celled abscission layers, which allow leaves, flower petals, and fruit to drop off at the appropriate times of year. Ethylene is best known, and most commercially valuable, for its ability to cause certain fruits to ripen.

Ethylene's role in leaf drop led to its discovery. In the 1800s, "coal gas" street lamps came into wide use (coal gas is a mixture of volatile flammable chemicals captured from coal baked at high temperatures). People observed that plants growing near leaky coal gas pipes lost their leaves prematurely. In 1901, the Russian botanist Dimitri Neljubov, then a graduate student, tested the components of coal gas on plants, and discovered that ethylene was the active ingredient, responsible for a variety of changes in plant growth.

Abscisic acid, synthesized in tissues throughout the plant body, helps plants to withstand unfavorable environmental conditions. As you learned in Chapter 43, abscisic acid causes stomata to close when water is scarce (see pp. 858–860). It also promotes root growth and inhibits stem growth under dry conditions. Abscisic acid helps to maintain dormancy in seeds at times when germination would lead to death. The name "abscisic acid" was based on the mistaken hypothesis that it caused leaf abscission, now known to be a function of ethylene.

Florigens (literally "flower producers") are hormones synthesized in leaves that control the timing of flowering in response to environmental cues. Florigens are a family of very similar proteins that travel within the phloem from leaves to the apical meristem, where they induce flowering. Florigens, which plant physiologists had been trying to isolate for 70 years, were discovered in 2007.

45.2 HOW DO HORMONES REGULATE PLANT LIFE CYCLES?

The life cycle of each plant species results from a complex interplay between its genetic information and its environment. Hormones provide an important link between these factors because they are often produced in response to environmental stimuli, and they influence the activity of genes. In the sections that follow, we summarize the effects of the best-studied hormones on plant life cycles.

Each Plant Life Cycle Begins with a Seed

The timing of seed germination is crucial to ensure that new seedlings will have adequate water and appropriate temperatures to thrive. Two hormones play major roles in seed germination: abscisic acid and gibberellin.

Abscisic Acid Maintains Seed Dormancy

Although a warm autumn day provides ideal conditions for germination, the seeds of temperate plants remain dormant until the following spring. Abscisic acid enforces seed dormancy by slowing the metabolism of the embryo. Hence, before germination can occur, the amount of abscisic acid within the seed must first be reduced. For example, in the seeds of many plants that must survive a winter, a period of cold temperatures is required for germination. In such plants, prolonged cold causes a reduction in the amount of abscisic acid within the seed, preparing the plant to germinate during the warming days of spring.

In deserts, where lack of water is the greatest threat to seedling survival, some seeds have high levels of abscisic acid in their seed coats. A hard rain will wash the hormone away, allowing germination at a time when enough moisture is available for growth, briefly carpeting the desert with wildflowers (**Fig. 45-1**). In contrast, some plants—particularly those of grasslands, chaparral, and certain forests—require fire to germinate, as described in "Earth Watch: Where There's Smoke, There's Germination."

Gibberellin Stimulates Germination

Falling levels of abscisic acid and rising levels of gibberellin allow germination. Gibberellin is produced by the embryo and activates genes that code for enzymes. These enzymes break down the starch reserves of the endosperm, releasing sugar that is used by the developing embryo as a source of both energy and carbon atoms to synthesize biological molecules.

Auxin Controls the Orientation of the Sprouting Seedling

Nourished by the endosperm, the growing embryo breaks out of its seed coat, and the new seedling immediately faces a crucial dilemma: Which way is up? Light and gravity provide important directional cues. Auxin controls positive **phototropism** (growth toward light) in shoots and

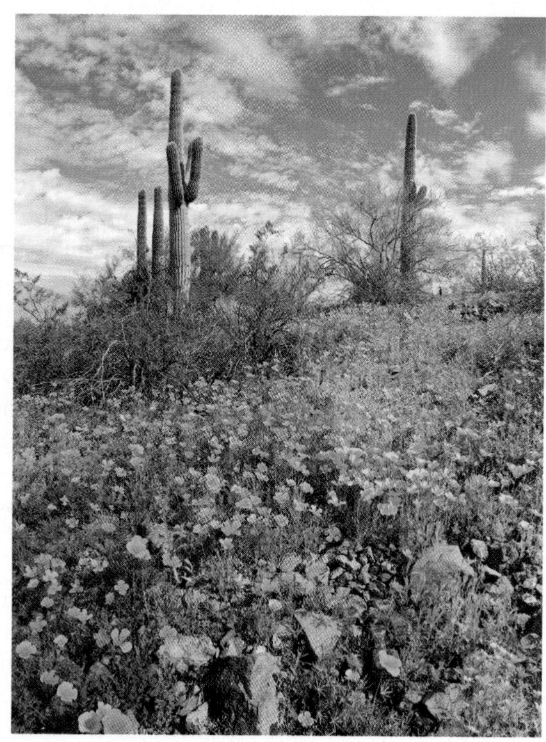

▲ **FIGURE 45-1 A desert in bloom** After abundant rainfall, flowering desert plants often germinate in large numbers, grow rapidly, and carpet the desert floor with blossoms, as shown here in Picacho Peak State Park, Arizona.

Earth Watch

Where There's Smoke, There's Germination

Wildfires are necessary to maintain ecosystems such as tallgrass prairie and coastal chaparral (see pp. 564–565), and to regenerate aging climax forests. Without human intervention, lightning-set fires naturally sweep through temperate forests every few decades, keeping them relatively open and providing a diversity of habitats. But people suppressed fires in the United States for most of the 20th century. As a result, many forests are now overly dense, with less healthy trees competing for space, water, nutrients, and light. When fire comes, it burns with an unnatural intensity. Far from being restored, the forest is devastated.

Land plants have been evolving in the presence of fire throughout their evolutionary history. As a result, many plants are adapted to take advantage of the open space, light, and nutrients that become newly available in the aftermath of a fire. Many species of fire-adapted plants are stimulated to germinate by chemicals in smoke. Nitrogen dioxide gas in smoke triggers germination in a variety of chaparral species (**Fig. E45-1**). Australian researchers recently discovered a group of simple, related compounds (called "karrikins," from the Aboriginal word for smoke) that are released by burning cellulose. These molecules trigger germination in several smoke-responsive species. Plant germination in response to chemicals in smoke ensures that fire-blackened landscapes will speedily regain their carpet of green. Recognizing the value of fire, forest managers now often allow forest fires to continue burning if they don't endanger human lives and property.

▲ **FIGURE E45-1 Fires help maintain chaparral ecosystems**

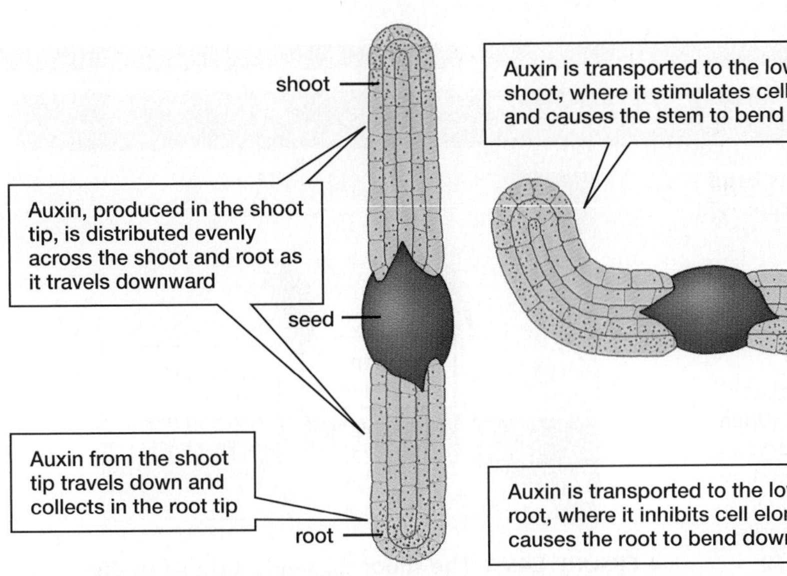

(a) The shoot and root are oriented vertically

(b) The shoot and root are oriented horizontally

(c) A tomato shoot displays negative gravitropism in darkness

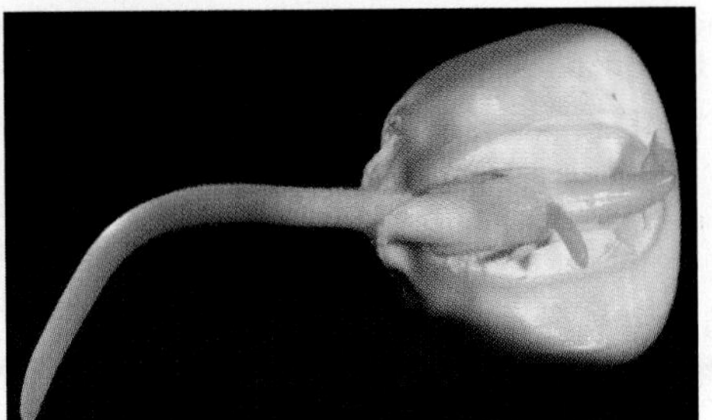

(d) A corn root displays positive gravitropism

◀ FIGURE 45-2 **Auxin causes positive gravitropism**

gravitropism (growth toward or away from gravity) in shoots and roots.

Auxin Stimulates Shoot Elongation Away from Gravity and Toward Light

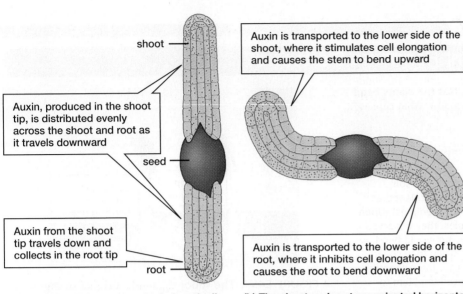

As seeds germinate, gravitropism and phototropism work together to cause the emerging shoot to grow upward. Auxin is synthesized in the shoot apical meristem, and stimulates cells of the stem to elongate. If the stem is vertical, auxin is distributed evenly among cells below the meristem (**Fig. 45-2a**). If the stem is horizontal, its cells detect gravity and transport auxin to the lower side of the stem (**Fig. 45-2b**). In response, the lower cells elongate rapidly, bending the stem upward and away from the pull of gravity *(negative gravitropism)*. When the shoot tip becomes vertical, auxin becomes evenly distributed and the stem grows straight up out of the soil. Mature plants temporarily placed in darkness in a horizontal position will exhibit negative gravitropism (**Fig. 45-2c**). Roots use positive gravitropism to orient downward into the soil (**Fig. 45-2d**).

Auxin also mediates phototropism, a directional response to light. Auxin accumulates in the side of the shoot away from the light, causing this side to elongate and bending the shoot toward the light. When the shoot points directly toward the light, auxin becomes evenly distributed, and bending ceases. Seedlings on a windowsill clearly demonstrate *positive phototropism*, which dominates negative gravitropism in the presence of abundant sunlight (**Fig. 45-3**).

Auxin Stimulates Root Elongation Toward Gravity

Auxin is transported from the shoot tip down into the root tip. If the root is horizontal, cells in the root tip sense the direction of gravity and cause auxin transport to the lower side

▲ FIGURE 45-3 **Radish seedlings show positive phototropism**

Scientific Inquiry

How Were Plant Hormones Discovered?

Anyone who keeps houseplants knows that the plants bend toward sunlight streaming through a window. What causes this phototropism?

Charles and Francis Darwin Determined the Source of the Signal

In England, around 1880, Charles Darwin and his son Francis investigated phototropism using grass coleoptiles (protective sheaths that surround monocot seedlings). Upon illuminating the sprout from various angles, they observed that a region a few millimeters below the tip would bend, pointing the tip toward the light source. When they covered the tip of the coleoptile with an opaque cap, the coleoptile didn't bend (**Fig. E45-2**). A clear cap allowed the stem to bend, and this occurred even when the bending region was covered by an opaque sleeve (**Fig. E45-3**). The Darwins concluded that the tip of the coleoptile perceives the direction of light, although bending occurs farther down. This suggested that the tip transmits information about light direction down to the bending region.

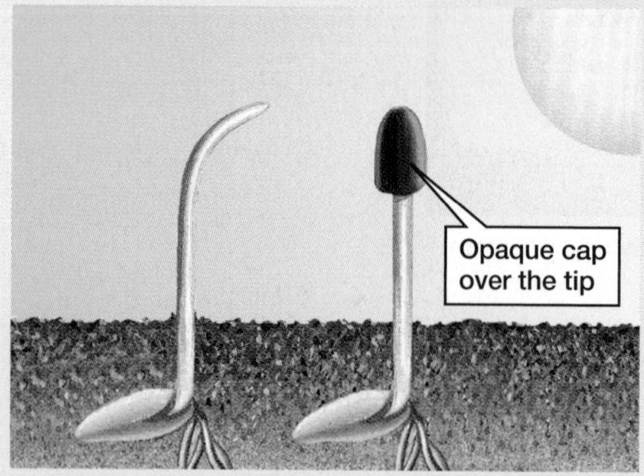

▲ FIGURE E45-2 **The shoot tip senses light**

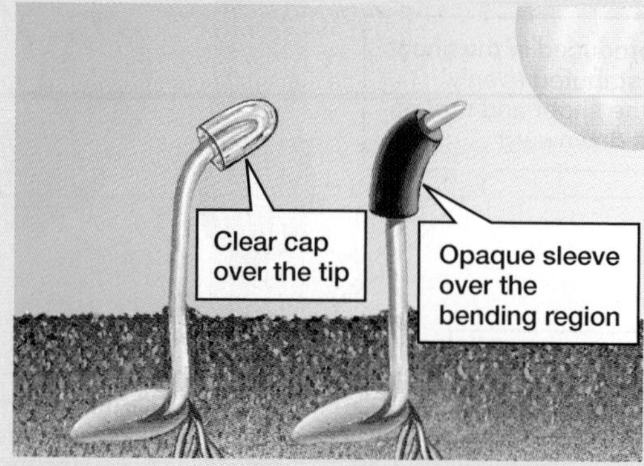

▲ FIGURE E45-3 **The shoot tip sends a signal to the bending region**

Peter Boysen-Jensen Demonstrated That the Signal Is a Chemical

In 1913, Peter Boysen-Jensen of Denmark cut the tips off coleoptiles and found that the remaining stump neither elongated nor bent toward the light. If he replaced the tip and placed the patched-together coleoptile in the dark, it elongated straight up. In the light, it showed normal phototropism. When he inserted a thin layer of gelatin that prevented direct contact but permitted substances to diffuse between the severed tip and the stump, he observed the same elongation and bending. In contrast, an impenetrable barrier eliminated these responses (**Fig. E45-4**). Boysen-Jensen concluded that the factor is a diffusing chemical.

Frits Went Collected the Chemical and Named It Auxin

In 1926, Frits Went, working in the Netherlands, placed the tips of oat coleoptiles on agar, allowing the

of the root. There, auxin inhibits root cell elongation, causing the root to exhibit positive gravitropism, bending downward in the direction of gravity (see Fig. 45-2d). When the root tip points down, the auxin becomes equally distributed, causing the root to grow straight downward.

How Do Plants Sense Gravity?

Although scientists have not yet definitively answered this question, some evidence suggests that specialized starch-filled plastids called statoliths contribute to the response. Statoliths, found in certain cells of both the root and shoot, always settle into the lower part of the cell, shifting position if the plant is tilted (**Fig. 45-4**). The statoliths may then cause auxin to accumulate on the same side of the cell. However, some mutant plants that lack statoliths still exhibit gravitropism, suggesting that other mechanisms may contribute to this response.

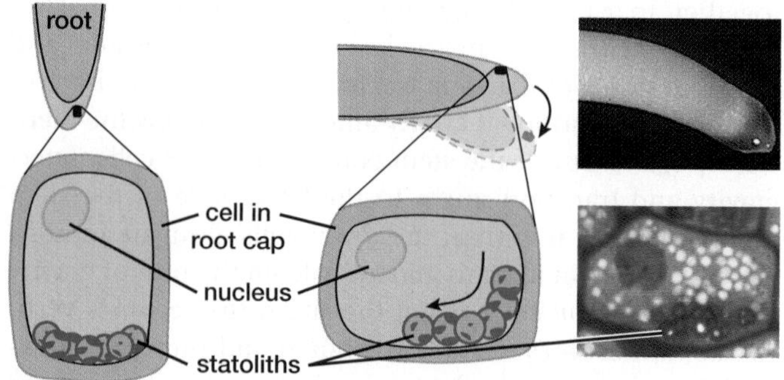

▲ FIGURE 45-4 **Statoliths may be gravity detectors**
Statoliths in root caps fall to the lowest part of the root cells. In a vertically oriented root tip (left), this corresponds to elongation straight down into the soil. Orienting the root horizontally (middle) causes the statoliths to fall downward and the root tip to bend down. These events are shown in light micrographs (right) of the bending root tip (top) and root cap cells (bottom), stained to show the statoliths (dark granules) that have accumulated on the cells' downward sides.

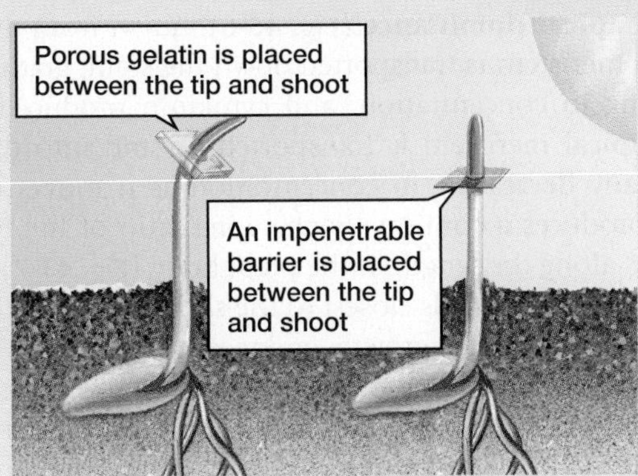

▲ **FIGURE E45-4 A chemical diffuses down the coleoptile**

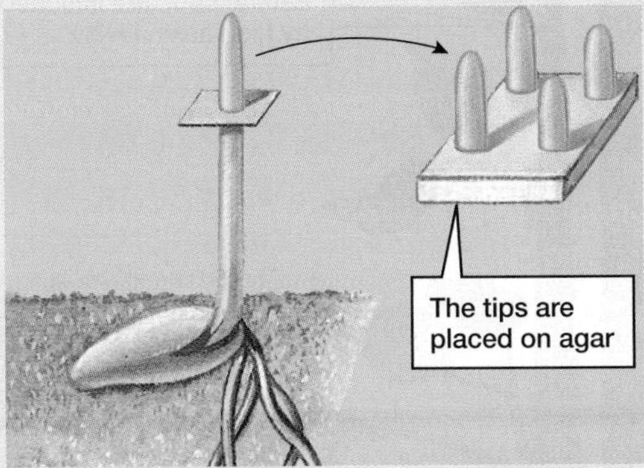

▲ **FIGURE E45-5 The chemical accumulates in agar**

unidentified chemical to migrate into this gelatinous material (**Fig. E45-5**). He then cut up the agar and placed small pieces on the tops of coleoptile stumps in darkness. A piece of agar placed squarely atop the stump caused it to elongate straight up; all of the stump cells received equal amounts of the chemical and elongated at the same

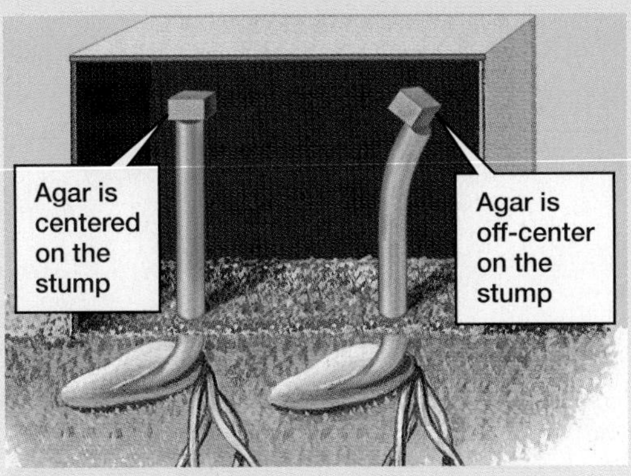

▲ **FIGURE E45-6 The chemical stimulates growth**

rate. When Went placed agar on only one side of the cut stump, the stump bent away from the side with agar (**Fig. E45-6**). Confirming the hypothesis that a diffusing chemical stimulated elongation, he named it "auxin," from a Greek word meaning "to increase."

Kenneth Thimann Determined the Chemical Structure of Auxin

Working at Caltech in the early 1930s, Kenneth Thimann purified auxin and determined its chemical structure. He and other researchers then determined which features of the molecule were important for its action in plants. This knowledge led to the production of synthetic auxins, such as the herbicide 2,4-D, developed in 1946.

The story of auxin, from its discovery to its commercial application, spanned two continents over a 60-year period. It relied on clear communication among scientists, and illustrates how each new scientific discovery leads to more questions. Research on auxin continues throughout the world today, steadily opening new lines of inquiry.

The Growing Plant Responds to Environmental Pressures

As a seed germinates, both root and shoot push against the surrounding soil, and the pressure induces ethylene production. Ethylene causes both the root and shoot to slow their elongation and become thicker, stronger, and better able to force their way through the soil. In dicots, ethylene also causes the emerging shoot to form a hook, so that the tender new leaves point downward and the bend of the hook protects them as it forces its way up through the soil (see Fig. 44-11b).

Some plants display **thigmotropism,** a directional movement or change in growth in response to touch. Most thigmotropic plants wrap tendrils (modified leaves or stems) around supporting structures (**Fig. 45-5**). This occurs when cell elongation is inhibited on the side of the tendril in contact with the object. The hormonal control of this form of grasping thigmotropism is poorly understood, but a reasonable hypothesis is that ethylene is produced by cells touching an object, inhibiting cell elongation and causing the tendril to curl around it.

Auxin and Cytokinin Control Stem and Root Branching

The size of the root and shoot systems of plants must be balanced so that as the shoot grows, the roots also grow enough to provide adequate anchorage, water, and nutrients. Both auxin and cytokinin promote this balance by controlling the branching of stems and roots. In stems, auxin inhibits the growth of lateral buds to form branches, whereas cytokinin promotes this process. In contrast, cytokinin stimulates root branching, whereas auxin inhibits it.

Auxin Inhibits and Cytokinin Stimulates Stem Branching

Gardeners know that pinching back the tip of a growing plant causes it to become bushier. This happens because the apical

▲ **FIGURE 45-5 A twining stem illustrates thigmotropism** Cells on the side of the stem that contact a narrow object (such as a stem or branch of a nearby plant) are inhibited from elongating, as illustrated by this morning glory. Greater elongation of cells away from the object causes the stem to continuously bend, encircling the object.

▲ **FIGURE 45-6 Apical dominance** The bean plant on the left has produced a single stem that is exhibiting apical dominance by suppressing side branches. The bean plant on the right has had its apical meristem cut off, allowing lateral buds to sprout into side branches.

meristem in the growing tip releases auxin, which suppresses the sprouting of lateral buds into branches, a phenomenon called **apical dominance** (**Fig. 45-6**). Auxin from the shoot apical meristem is transported down the stem, gradually decreasing in concentration, and cytokinin produced by the root apical meristem is transported up and into the stem, gradually decreasing in concentration as it moves upward. This produces a continually changing ratio of the two hormones along the length of the plant body (**Fig. 45-7**).

The lateral buds closest to the shoot tip receive enough auxin to inhibit their growth, and receive very little cytokinin; as a consequence, they remain dormant. Lower buds, however, are exposed to less auxin and more cytokinin and, thus, are stimulated to grow into branches. In some plants, this

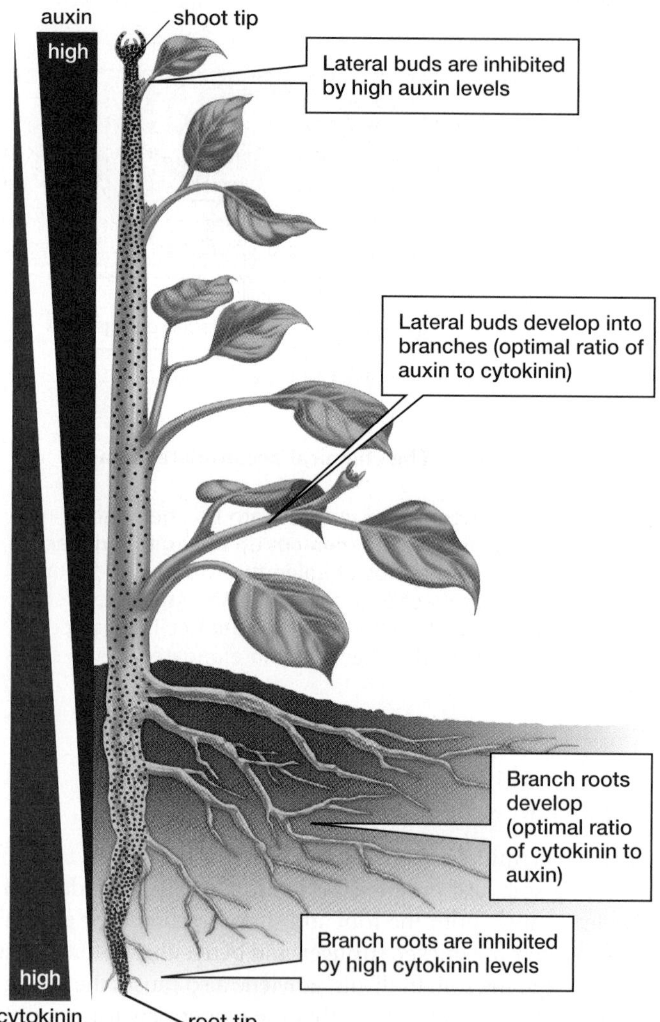

▲ **FIGURE 45-7 The role of auxin and cytokinin in lateral bud sprouting** Auxin (blue) and cytokinin (red) control the sprouting of lateral buds and also the development and branching of lateral roots. Auxin, produced by the shoot apical meristem, moves downward; cytokinin, produced by the apical meristem in the root tip, moves upward.

QUESTION If you removed a plant's shoot apical meristem and applied auxin to the cut surface, how would you expect the lateral buds to respond?

gradient of auxin to cytokinin ratios produces an orderly progression of branch sprouting from the bottom to the top of the shoot (see Fig. 45-7)

Auxin Stimulates and Cytokinin Inhibits Root Branching

Auxin, transported down from the shoot apical meristem, stimulates root pericycle cells to divide and form branch roots (see Fig. 43-16). Commercial auxin powder allows gardeners to produce a new plant by dipping the cut end of a stem into the auxin and placing it in soil or water, where it will develop roots. In contrast, cytokinin produced in the root apical meristem inhibits root branching. Lower roots receive more cytokinin and less auxin, while roots closer to the shoot receive more auxin and less cytokinin. The gradients and counteracting effects of auxin and cytokinin help to keep the size of the root and shoot systems in balance with one another (see Fig. 45-7).

Plants Sense and Respond to Light and Darkness

The timing of flowering and seed production must be finely tuned to the environment. In temperate climates, plants must flower early enough so that their seeds mature before the killing frosts of late autumn. Depending on how quickly the seeds develop, flowering may occur in spring, as it does in oaks; in summer, as in roses; or even in autumn, as in asters.

What environmental cues do plants use to determine the seasons? Most cues, such as temperature or water availability, vary quite unpredictably in temperate climates. The only truly consistent and reliable cue is day length. Days that progressively increase in length mean that spring and summer are approaching; shortening days signal the onset of fall and winter.

Botanists have long recognized that some plants, called **day-neutral plants** (such as roses, tomatoes, cucumbers, and corn), flower independently of day length, as long as they are mature and environmental conditions are favorable (**Fig. 45-8**). Other plants were categorized as long-day plants or short-day plants according to the day length during their flowering period. But further investigation revealed that flowering in such plants is controlled not by the duration of uninterrupted daylight, but by the duration of uninterrupted darkness. **Long-day plants** (short-night plants, including iris, lettuce, spinach, and hollyhocks) flower only when uninterrupted darkness is shorter than a species-specific duration (see Fig. 45-8). **Short-day plants** (long-night plants, such as cockleburs, chrysanthemums, asters, potatoes, and goldenrod) flower when uninterrupted darkness exceeds a species-specific duration (see Fig. 45-8). For example, a cocklebur (a long-night plant) flowers only if uninterrupted darkness lasts more than 8.5 hours, whereas spinach, a long-day plant, flowers only if uninterrupted darkness lasts 10 hours or less; both will flower with 9-hour nights. How do these plants sense the duration of uninterrupted darkness?

The Photopigment Phytochrome Detects Light

Organisms detect light using molecules called *photopigments*, which absorb specific wavelengths of light (we perceive light wavelengths as colors; see Fig. 7-4). As they absorb light

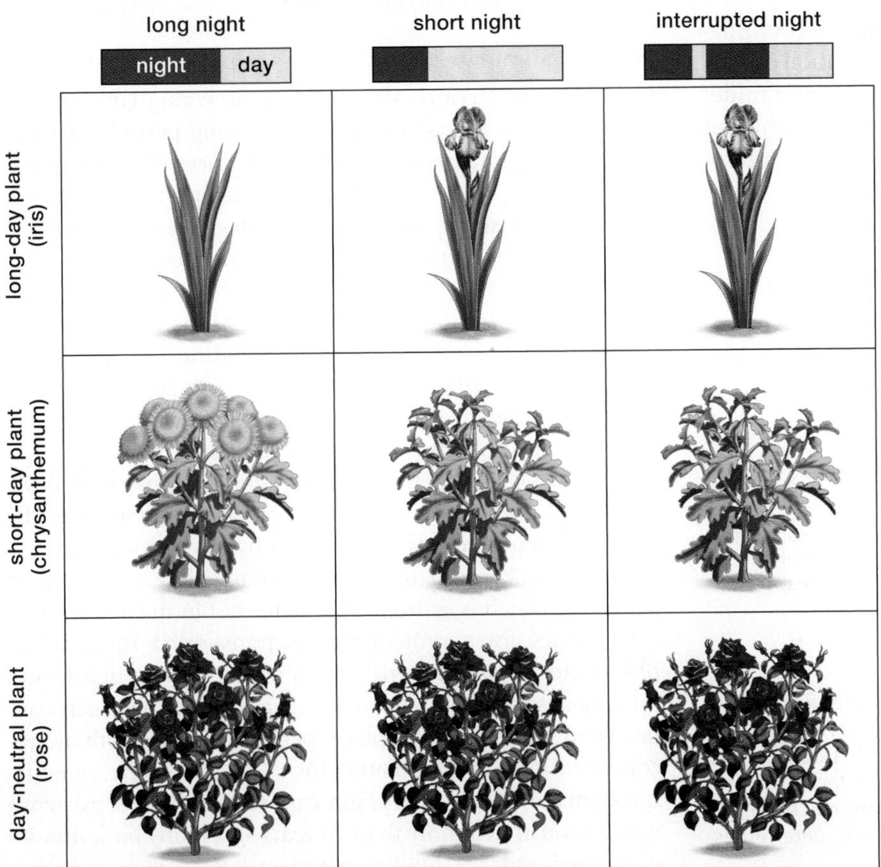

◄ **FIGURE 45-8 The effects of night length on flowering** Examples of long-day, short-day, and day-neutral plants are illustrated here, with and without flowers, depending on the length of their exposure to continuous darkness.

energy, photopigments alter their chemical configurations, initiating a series of biochemical reactions that cause a response in the organism. The eyes of animals contain photopigments that allow vision, whereas the bodies of plants contain a photopigment called **phytochrome** (literally, "plant color") that detects light and darkness, allowing plants to respond appropriately.

Plants measure the duration of darkness using an internal **biological clock** that relies on complex biochemical reactions still under investigation. If a plant in darkness is exposed to light of a certain wavelength, its phytochrome molecules will reset the clock, something like stopping and resetting a stopwatch. Thus, if an 8-hour night is interrupted after 4 hours with a few minutes of white light, the plant will perceive only the uninterrupted 4 hours of darkness after the interval of light.

How does phytochrome respond to light? Phytochrome occurs in two similar forms that can be converted back and forth from one to the other. The most stable form, called P_r, is inactive and absorbs red light (r), and it appears turquoise blue in concentrated solutions. As it absorbs red light, P_r is rapidly converted into the active form, called P_{fr}, and takes on a more greenish hue. This active form influences (stimulates or inhibits) a response to light such as flowering, as described shortly. The active P_{fr} absorbs far-red light (fr), which converts it back to inactive P_r (**Fig. 45-9**). Because sunlight consists of all wavelengths of light, including both red and far red, in full sun, both forms will be present; about 60% of the phytochrome will be in its active P_{fr} form, causing the appropriate responses. Over a period of hours in darkness, P_{fr} spontaneously reverts back to the inactive and stable P_r.

Plants use the two forms of phytochrome in combination with their biological clocks to detect the duration of continuous darkness. A cocklebur, for example, will flower under a schedule of 16 hours of darkness and 8 hours of light, but interrupting the 16-hour dark period with just a minute or two of white light (causing production of some P_{fr}) prevents flowering by causing the plant to perceive a shorter night. In contrast, a nighttime exposure to far-red light—which is not absorbed by P_r—has no effect on cocklebur flowering. However, a brief exposure to pure red light—which is absorbed by P_r, converting it into active P_{fr}—inhibits flowering. This means that, in some way, P_{fr} resets the plant's biological clock.

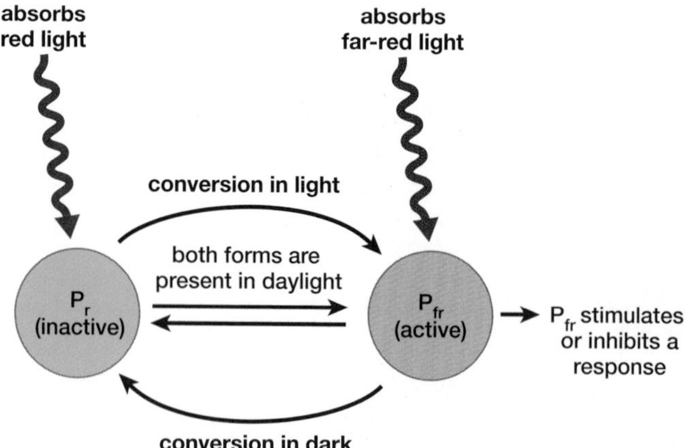

▲ **FIGURE 45-9 The light-sensitive phytochrome pigment**
Phytochrome exists in inactive (P_r) and active (P_{fr}) forms. In darkness, P_{fr} spontaneously converts to inactive P_r. Upon absorbing red light, P_r is converted into active P_{fr}. Upon absorbing far-red light, P_{fr} is transformed back into inactive P_r. In daylight, both forms are present and the active P_{fr} determines the plant response.

Phytochrome Influences Other Responses of Plants to Their Environment

Because plants cannot survive without adequate light, the light-responsive P_{fr} influences several plant responses. For example, P_{fr} inhibits shoot elongation. Because P_{fr} spontaneously reverts to P_r in the dark, a shoot emerging in the darkness of the soil will contain no inhibitory P_{fr} and, consequently, will elongate very rapidly. Once it reaches the sunlight, however, P_{fr} is formed in the shoot and slows elongation, preventing the seedling from growing too rapidly and becoming weak and spindly. P_{fr} also causes the hook in the stem of dicot seedlings to straighten, which orients the newly formed leaves upward toward the light. This active phytochrome also stimulates the new leaves to grow and produce chlorophyll.

Seedlings growing beneath other plants will be exposed primarily to far-red light, because the green chlorophyll in the leaves above them absorbs most of the red light but allows far-red light to pass through. Far-red light converts the inhibitory P_{fr} into inactive P_r, and so the shaded seedlings elongate rapidly, which tends to bring them out of the shade and into sunlight. There, the conversion of P_r back to P_{fr} slows their growth, allowing their stems to grow in diameter and strengthen.

Florigen Stimulates Flowering in Response to Light Cues

In the 1930s, research revealed that flowering could be induced by exposing spinach leaves to long periods of light, even if the apical meristem (which produces the flower) was covered. Later, scientists induced flowering in the cocklebur by exposing a single leaf to long nights. These and other experiments demonstrated that, for flowering, the leaves are the light-responsive portion of the plant. They produce a substance that is transmitted through the phloem to the apical meristem, where it induces flower formation.

For about 70 years, plant physiologists sought but were unable to isolate this elusive signaling molecule, which they called florigen. During the past decade, however, molecular biological techniques have revealed specific genes that play a central role in flowering. These genes are influenced by the plant's biological clock (which is set by phytochrome).

In 2007, researchers demonstrated that one of these genes, *FT*, which is active in leaves but not in the apical meristem, codes for a protein that is transported through the phloem to the apical meristem. Here, the FT protein activates other genes that cause flowering. The pathway has been confirmed in several different plant species, including both monocots and dicots. It appears that the FT protein is the signaling molecule, produced in the light-exposed leaf, whose existence was demonstrated more than 70 years ago. Many botanists are now convinced that the elusive florigen has finally been found.

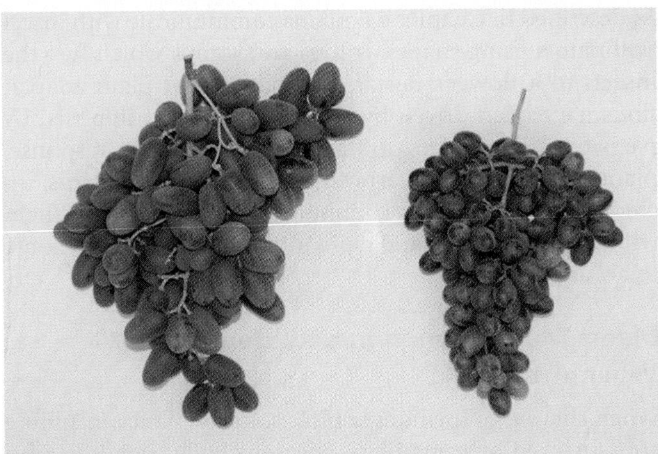

▲ **FIGURE 45-10 Commercial uses for plant hormones** The grapes on the left were sprayed with gibberellin, producing looser clusters of larger grapes. Those on the right have developed naturally, forming smaller grapes in tighter clusters.

QUESTION Agricultural biotechnologists have developed genetically modified tomato plants in which ethylene production is blocked. Why might such a plant be valuable to tomato growers?

Hormones Coordinate the Development of Seeds and Fruit

Seed and fruit development are influenced by auxin and gibberellin. Auxin promotes the growth of the ovary, causing it to store food materials and to develop into the mature fruit. Synthetic auxin may be sprayed on fruits to cause them to grow larger and sweeter than they would under natural conditions, or to prevent the fruit from dropping prematurely. Gibberellin likewise contributes to fruit growth. It is commercially applied to grapes, which grow larger and form looser clusters as a result (**Fig. 45-10**).

In nature, not surprisingly, seed maturation and fruit ripening are closely coordinated, because most plants use their fruits to disperse seeds. Animals seldom eat unripe fruit because, in general, unripe fruits are inconspicuously colored (often green, like the rest of the plant), hard, bitter, and in some cases, poisonous. As the seeds mature, however, the fruit ripens and becomes softer, as enzymes weaken its cell walls; sweeter, as starches are converted to sugar; and more brightly and conspicuously colored, as green chlorophyll is broken down, and yellow, orange, and red pigments are revealed or synthesized. These features attract animals that will disperse the seeds (**Fig. 45-11**), just as they attract human consumers to the produce section of the supermarket.

In fruits such as bananas, apples, pears, tomatoes, and avocados, the changes that accompany ripening are stimulated by ethylene. As these fruits ripen, they themselves release ethylene, which hastens the ripening of other fruits nearby. The discovery of the role of ethylene in ripening has revolutionized modern fruit marketing. Bananas grown in Central America can be picked green and tough, then shipped to North American markets, where they are ripened with ethylene. Green tomatoes, which withstand transport better than ripe ones, are often ripened using ethylene (but never seem as flavorful as those grown in home gardens). Not all fruits ripen when treated with ethylene. Those that don't include strawberries, grapes, and cherries, and so shipping these ripe, soft fruits to markets without damage remains a challenge.

Senescence and Dormancy Prepare the Plant for Winter

In autumn, under the influence of shortening days and falling temperatures, plants undergo **senescence,** a genetically programmed series of events that prepares the plant for winter. Ethylene production increases, and the production of auxin and cytokinin (which delay senescence) declines. During senescence, starches and chlorophyll in the leaf are broken down into simpler molecules that are transported to the stem and roots for winter storage.

Although many hormones interact to control senescence, ethylene seems particularly important. For example, ethylene promotes the breakdown of chlorophyll and triggers the production of enzymes that weaken a layer of cells, called the **abscission layer,** located where fruit or leaf stalks join the plant stem (**Fig. 45-12**). This allows the aging leaf or ripe fruit to drop at the appropriate time of year. Environmental stresses may also cause a rapid increase in ethylene production. This is what causes leaves to undergo premature senescence and drop off when they are infected by microorganisms or exposed to temperature extremes, or when they are subjected to drought—a phenomenon you may observe if you forget to water your houseplants.

In the autumn, new buds become tightly wrapped and dormant rather than developing into leaves or branches, as they would during spring and summer. Dormancy in buds, as in seeds, is enforced by abscisic acid. The plant's metabolism slows, and it enters its long winter sleep, awaiting signals of warmth, moisture, and longer spring days before awakening once again.

▲ **FIGURE 45-11 Ripe fruit becomes attractive to animal seed dispersers** A desert tortoise is attracted to a prickly pear cactus fruit.

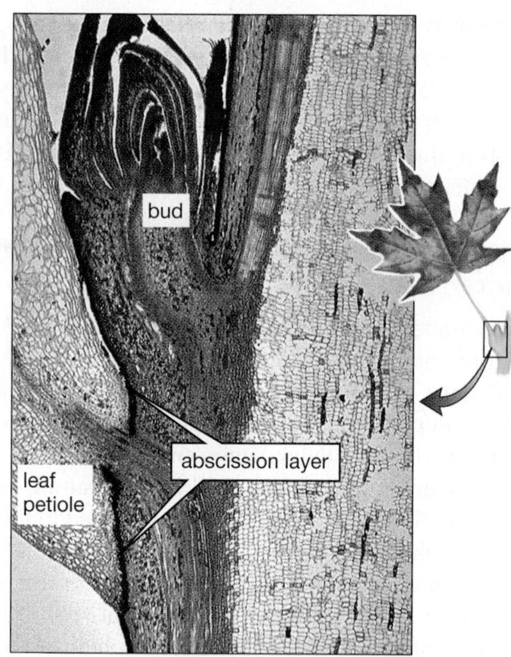

▲ **FIGURE 45-12 The abscission layer** This light micrograph of the base of a maple leaf clearly shows the abscission layer. A lateral bud is visible above the senescing leaf petiole.

45.3 HOW DO PLANTS COMMUNICATE AND CAPTURE PREY?

Flowering plants have been coevolving with animals and microorganisms for well over 400 million years, and have adapted to these organisms in numerous and complex ways. As described in Chapter 44, plants communicate with insect pollinators using shapes, colors, and scents, which lure the insects to a flower's nectar. But many other plant adaptations have been driven by selection pressures imposed by parasitism, predation, or limited nutrients. In response, plants have evolved an array of sophisticated behaviors, including plant-to-plant communication, that helps them withstand predation, and rapid movements that allow some to act as predators.

Plants May Summon Insect "Bodyguards" When Attacked

When chewed on by hungry insects, many plants, including corn attacked by caterpillars, release volatile chemicals into the air, producing what could be called a chemical "cry for help" (**Fig. 45-13**). The corn alarm signal is stimulated by a compound called volicitin in the saliva of the caterpillar; damage from other causes (such as hail) will not stimulate this response. Parasitic female wasps are attracted to the corn's alarm call and lay their eggs in the caterpillar's body, which becomes food for the emerging wasp larvae.

Similarly, lima bean plants that are attacked by spider mites release a chemical that attracts a carnivorous mite that preys on the spider mite. Wild tobacco plants munched on by hornworms (hawk moth caterpillars) release different chemicals at different times. During the day, when parasitic wasps are active and seeking hornworm caterpillars, the plants produce wasp-attracting chemicals. At night, when adult hawk moths are active, the tobacco plants release chemicals that deter the moths from laying eggs on them.

► **FIGURE 45-13 A chemical cry for help**

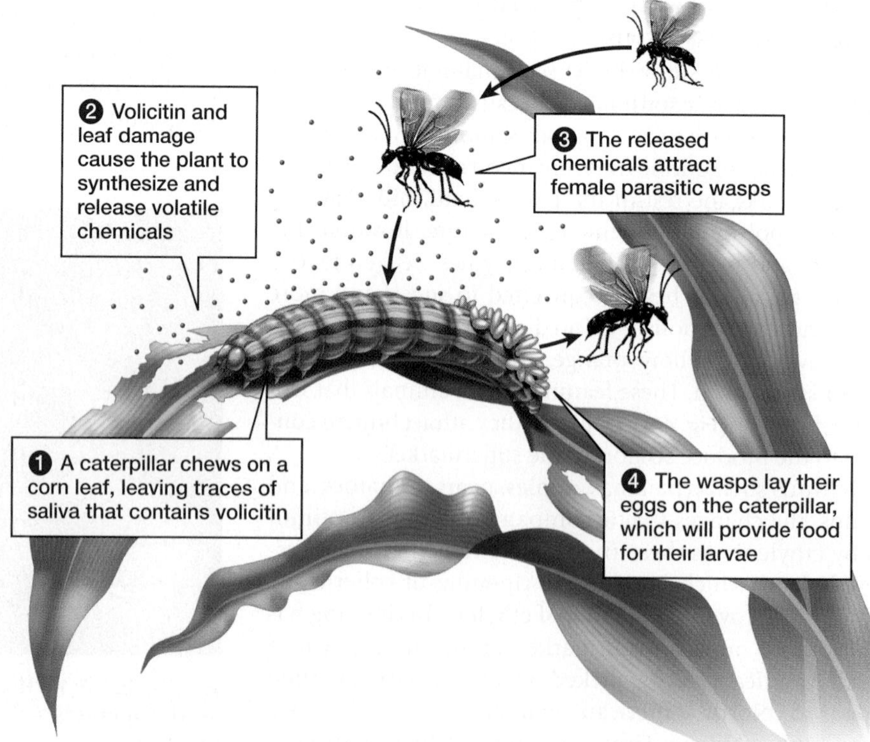

2 Volicitin and leaf damage cause the plant to synthesize and release volatile chemicals
3 The released chemicals attract female parasitic wasps
1 A caterpillar chews on a corn leaf, leaving traces of saliva that contains volicitin
4 The wasps lay their eggs on the caterpillar, which will provide food for their larvae

Case Study continued
Predatory Plants

Whereas some plants summon insects for help, others summon them for food, luring them with sugary secretions and bright colors. The inner edges of Venus flytrap leaves are studded with nectar glands that produce a sweet-smelling sugar lure, and their inner surfaces are often red, giving them a flower-like appearance. Enzymes secreted by the leaves of the flytrap digest the trapped insect bodies, and the plants feast on the nitrogen-rich nutrients. Sundews attract their insect prey using an appealing scent, red hairs, and sticky droplets that resemble nectar (see Fig. 45-15).

Have you ever wondered ?
What People Took for Pain Before Aspirin Was Invented?

Hippocrates, the Greek physician born in 460 B.C., wrote of a powder made from the bark and leaves of willow trees that relieved headache, pain, and fever. Native Americans independently discovered the pain-relieving power of willow bark, which they chewed or made into a tea. In the early 1800s, chemists isolated the active compound, salicin, from willow bark. Although this chemical is synthesized by the willow to deter herbivores, in the human body it is converted to salicylic acid, which reduces pain. Unfortunately, salicylic acid administered orally can cause extreme stomach upset. In 1897, the German chemist Felix Hoffmann, hoping to relieve his father's arthritis pain, modified salicylic acid to acetylsalicylic acid, which is far gentler on the stomach. In 1899, the Bayer company (Hoffmann's employer) marketed the drug, calling it "aspirin."

Attacked Plants Defend Themselves

Leaf damage from insects causes many plants to produce a signaling molecule, which then travels through the plant body, where it stimulates responses that make the plant more distasteful, more difficult to eat, or more toxic. For example, when tobacco leaves are damaged, they produce more nicotine (a poison used commercially as an insecticide). Radish plants attacked by caterpillars produce a bitter-tasting chemical and grow more spiny hairs on their leaves. Researchers found that the seedlings of these damaged plants were less attractive to predators than seedlings from uninjured parent plants. Thus, plants may not only defend themselves, but apparently also pass on a chemical signal within their seeds that causes specific genes to be activated, triggering the development of more defenses in their offspring.

Plants also have a type of immune response that helps defend them against infectious microorganisms. In response to infection, many plants produce salicylic acid, which stimulates production of a variety of proteins that help them fight their current infection and future infections as well.

Wounded Plants Warn Their Neighbors

If plants under siege can summon help, might neighboring plants also get the message? Evidence is accumulating that many healthy plants (including willow, alder, and birch trees) sense chemicals released by members of their species that are wounded by insects. The salicylic acid produced by the infected plants is converted to methyl salicylate (wintergreen), a highly volatile compound, some of which may be absorbed by nearby plants, generating an immune response that makes them better able to resist the infection. These neighbors boost their defenses and, thus, reduce their vulnerability to the predator. This communication may also cross species lines. In one study, wild tobacco plants that were downwind from wounded sagebrush (which releases volatile chemicals in response to insect attack) became more resistant to insects than were tobacco plants downwind from healthy sagebrush.

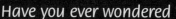

As one researcher observed, "Plants can't run away and they can't make noises. But they are wonderful chemists." Perhaps human chemists, learning from plants, will design chemicals that will enable farmers of the future to protect their crops from predators and diseases using natural signaling substances instead of toxic pesticides.

Sensitive Plants React to Touch

If you touch a sensitive plant *(Mimosa)*, the rows of leaflets along each side of the petiole immediately fold together, and the petiole droops (**Fig. 45-14**). This rapid and dramatic

(a) Before the leaves are touched

(b) After the leaves are touched

◀ **FIGURE 45-14 A rapid response to touch (a)** A leaf of the sensitive plant (*Mimosa*) consists of an array of leaflets emerging from a central stalk, which is attached to the main stem by a short petiole. **(b)** Touching causes the leaflets to fold together.

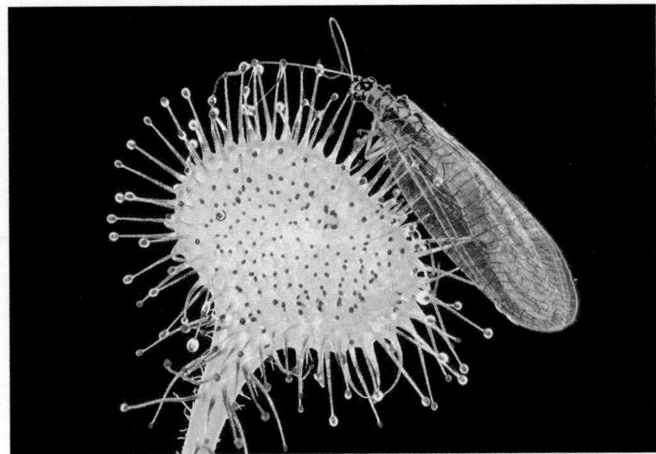

▲ **FIGURE 45-15 A sundew with its insect prey** The leaf of this sundew (genus *Drosera*) uses sticky hairs to capture a lacewing insect.

movement is an example of thigmotropism (see p. 891), which may surprise and discourage leaf-eating insects. It is stimulated by electrical signals conducted through specialized "motor cells" located at the base of each leaflet and where the petiole joins the stem. The signal causes the motor cells to increase their permeability to potassium, which they normally maintain at a high internal concentration. As potassium ions diffuse out, water follows by osmosis. This causes the motor cells to shrink, pulling the leaflets together and causing the petioles to droop rapidly.

Carnivorous Sundews and Bladderworts Respond Rapidly to Prey

Plants are not only excellent chemists; some have also evolved ingenious prey-catching behavior. Sundew leaves attract and then trap insects with red hairs whose tips secrete sweet glistening gluey globules (**Fig. 45-15**). The movements of a trapped insect trigger thigmotropism in the hairs, which bend toward the insect over the next few minutes. This response is caused by vibrations from the insect's struggle to escape, which open ion channels in the hairs, producing an electric current. The current causes the stimulated hairs to curl toward the source of the vibration, smothering the insect with their secretions and digesting it with enzymes.

▲ **FIGURE 45-16 The bladderwort snares tiny aquatic organisms** The bladders on this bladderwort (genus *Utricularia*) are clearly visible. (Inset) A water flea has triggered this bladder to expand rapidly. If the animal were smaller, it would have been sucked inside and digested. This particular bladder, however, has bitten off more than it can chew, and the water flea has become wedged in its trapdoor.

The world's speediest plant is the predatory aquatic bladderwort (**Fig. 45-16**). The trapdoor covering each hollow bladder is hinged at the top, opens inward, and is sealed shut by sticky secretions. Cells lining the bladder actively transport ions out of the water in the bladder, causing water to leave the bladder by osmosis. The reduced volume of water pulls the bladder walls inward under tension, setting the trap. If a water flea or other tiny aquatic organism bumps into the bristles that surround the trapdoor, the bristles push the door inward and break the seal around it. In a split second, the bladder walls spring outward to their resting position, suddenly increasing the volume of the bladder and sucking in the water flea to its death and eventual digestion.

Case Study revisited
Predatory Plants

How do the leaves of a Venus flytrap close quickly enough to trap a fly? This plant engineering problem recently caught the attention of a Harvard researcher, Lakshminarayanan Mahadevan. His team painted dots of fluorescent paint on the flytrap leaves, and then tracked them with a video camera as the leaves closed. Using these data in a computer simulation, the researchers concluded that a pair of flytrap leaves resembles a tennis ball that has been cut nearly in half, leaving only a hinge section. If the two halves of the ball were turned inside-out, they would be placed under tension, so that it would take only the slightest pressure to cause the halves to snap back together as they resumed their original shapes. The researchers hypothesized that open flytrap leaves are under a similar type of tension, probably due to the compression of cells in the central mesophyll layer. When an insect bumps into the hairs, it causes these cells to rapidly absorb water and swell, effectively "popping" each leaf from a slightly convex to a slightly concave shape that closes around the insect (**Fig. 45-17**).

▲ FIGURE 45-17 Success!

This hypothesis suggests that the open leaves store potential energy that is released as the leaves snap shut. If nothing is trapped, however, reopening the leaves takes a few days, and uses large quantities of ATP. To help prevent this waste of energy, the plant has a "fail-safe" mechanism that helps prevent it from closing on an inanimate object. For the leaves to clamp shut, either two hairs must be touched simultaneously, or one hair must be touched twice within about 20 seconds. But how do the hairs transform the touch stimulus into an electrical signal? Hairs bumped by a nectar-seeking fly open channels that allow ions to flow in, generating an electrical signal. This sets off a chain of events that causes the trap to close in less than half a second.

But the flytrap still holds mysteries. How do the sensory hairs "know" which ones have been touched and at what interval? How does the electrical signal cause cells to absorb water? Science is a never-ending quest for deeper understanding, because the answer to one question immediately raises more.

Consider This

Many wetlands in the United States are threatened by runoff water from nearby farms, which may be heavily fertilized or rich in animal wastes. Carnivorous plants thrive in nitrogen-poor marshes partly because other species, which can't trap nitrogen-rich food, cannot compete with them. Explain why runoff from farms poses a threat to carnivorous plants in nearby wetlands.

CHAPTER REVIEW

Summary of Key Concepts

45.1 What Are Some Major Plant Hormones?

Plants have evolved the ability to sense and react to environmental stimuli including touch, gravity, moisture, light, and day length. Plant hormones, secreted in response to changes in the environment, influence every stage of plant life cycles. Six major types of plant hormones that are important in the flowering plant life cycle are auxins, gibberellins, cytokinins, ethylene, abscisic acid, and florigens. Some major effects and sites of synthesis of these hormones are summarized in Table 45-1.

45.2 How Do Hormones Regulate Plant Life Cycles?

Hormones, which are usually secreted in response to environmental factors, generally exert their effects by influencing gene activity; this, in turn, regulates plant growth and development. Dormancy in seeds is enforced by abscisic acid. Falling levels of abscisic acid and rising levels of gibberellin trigger germination. As the seedling grows, auxin causes the shoot to grow toward light and away from the pull of gravity, and it causes the roots to grow toward gravity. Plant roots may detect gravity, in part, by using statoliths, which may influence auxin distribution. Ethylene slows elongation and promotes thickening of both roots and shoots in the seedling. Ethylene may also stimulate plants to wrap tendrils around supporting objects.

Branching in stems and roots results from changes in the ratio of auxin to cytokinin and in differing responses of roots and shoots to these hormones. Auxin is produced in the shoot apical meristem and transported downward, and cytokinin is synthesized in the root apical meristem and transported upward. Auxin inhibits stem branching (producing apical dominance), but stimulates root branching. Cytokinin inhibits root branching, but stimulates stem branching.

Plants detect light and darkness using the photopigment phytochrome. The stable, inactive form (P_r) absorbs red light and is converted into the unstable active form (P_{fr}), which absorbs far-red light and converts back to (P_r). Phytochrome causes the hook of the emerging shoot to straighten and the seedling to elongate. It also triggers the expansion of new leaves and the synthesis of chlorophyll. In conjunction with the plant's biological clock, leaf phytochrome measures uninterrupted darkness and helps control flowering in long-day and short-day plants. In response to day length, the flowering hormone florigen is produced in plant leaves and travels to the flower bud in the apical meristem where it initiates flowering.

Seed and fruit growth and development are influenced by auxin, cytokinin, and gibberellin. In some fruits, ethylene triggers ripening. During senescence, fruit and leaves fall in response to ethylene, which weakens their abscission layers. Buds become dormant as a result of high concentrations of abscisic acid.

45.3 How Do Plants Communicate and Capture Prey?

Some plants under attack by insects release volatile chemicals that attract other insects that prey on or parasitize the plant predators. Chemicals released by injured or infected plants may also stimulate neighboring plants to increase their defenses. Damaged plants use internal chemical signals to strengthen their own defenses and, sometimes, to produce better-defended offspring.

The aquatic bladderwort generates tension within its bladders, released when a bladder is touched by prey, expanding the bladder and sucking in water and prey. Sensitive plant leaves fold together when touch sensors in the leaves generate electrical signals that cause specialized cells to rapidly lose water.

Key Terms

abscisic acid *887*
abscission layer *895*
apical dominance *892*
auxin *887*
biological clock *894*
cytokinin *887*
day-neutral plant *893*
ethylene *887*
florigen *887*
gibberellin *887*
gravitropism *889*
hormone *886*
long-day plant *893*
phototropism *888*
phytochrome *894*
plant hormone *886*
senescence *895*
short-day plant *893*
thigmotropism *891*

Thinking Through the Concepts

Fill-in-the-Blank

1. Auxin produced by the apical meristem of the shoot __________ sprouting of lateral branches; this phenomenon is called __________. In roots, auxin __________ development of side branches. The discovery of auxin by the Darwins was based on its ability to stimulate __________.
2. The hormone __________ was discovered by researchers investigating "foolish seedling" disease in rice. During stem elongation, this hormone stimulates both cell __________ and cell __________.
3. The hormone __________ is a gas. During senescence, this hormone stimulates the formation of the __________, which allows dead leaves and ripe fruits to fall. It hastens the __________ of fruits such as __________ and __________.
4. The major site of cytokinin synthesis is the __________. This hormone__________ the sprouting of lateral roots, but __________ the sprouting of lateral __________.
5. The hormone __________ causes stomata to close when water is scarce. It maintains __________ in seeds during unfavorable environmental conditions.

Review Questions

1. How does auxin cause positive phototropism and negative gravitropism in shoots?
2. What is apical dominance? How do auxin and cytokinin interact in determining the growth of lateral buds?
3. Why are short-day plants most accurately called "long-night" plants? At what season of the year would you expect them to flower? Explain why.
4. What is phytochrome? How do the two forms of phytochrome help control the plant life cycle?
5. Which hormones cause fruit development? Which hormone causes ripening in bananas?
6. What is senescence? Describe some changes that accompany leaf senescence in the fall. Which hormone causes abscission?
7. What is a major commercial use of gibberellin? Of ethylene?
8. Describe one example of a chemical defense mechanism in plants.
9. Describe how a sensitive plant closes its leaves. Why might this behavior have evolved?
10. What is the advantage of predatory behavior in plants? What habitats favor this behavior and why?

Applying the Concepts

1. Suppose you got a job in a greenhouse in which the owner was trying to start the flowering of chrysanthemums (a long-night plant) for Mother's Day (in mid-May). You accidentally turned on the light in the middle of the night. Would you be likely to lose your job? Why or why not? What would happen if you turned on the lights in the day?
2. A student reporting on a project said that one of her seeds did not grow properly because it was planted upside down, so that it got confused and tried to grow down. Should the teacher accept this explanation? Why or why not?
3. Bean sprouts, such as those you might eat in a salad, need to be grown in the dark to form their long stems. If they are grown in the light, they will be short and green. Why do seedlings grow long and spindly in the dark? What advantages would growing this way in darkness have in nature?
4. Suppose that on August 4, you discover that both a long-night plant and a short-night plant have bloomed in your garden. Discuss how it is possible for both plants to bloom.
5. What did the Darwins, Boysen-Jensen, and Went each contribute to our understanding of phototropism? Do their experiments truly prove that auxin is the hormone that controls phototropism? What other experiments using more modern techniques would help investigate the role of auxins in phototropism?

MB® *Go to www.masteringbiology.com for practice quizzes, activities, eText, videos, current events, and more.*

Appendix I

Biological Vocabulary: Common Roots, Prefixes, and Suffixes

Biology has an extensive vocabulary often based on Greek or Latin rather than English words. Rather than memorizing every word as if it were part of a new, foreign language, you can figure out the meaning of many new terms if you learn a much smaller number of word roots, prefixes, and suffixes. We have provided common meanings in biology rather than literal translations from Greek or Latin. For each item in the list, the following information is given: meaning; part of word (prefix, suffix, or root); example from biology.

a–, an–: without, lack of (prefix); *abiotic,* without life

acro–: top, highest (prefix); *acrosome,* vesicle of enzymes at the tip of a sperm

ad–: to (prefix); *adhesion,* property of sticking to something else

allo–: other (prefix); *allopatric* (literally, "different fatherland"), restricted to different regions

amphi–: both, double, two (prefix); *amphibian,* a class of vertebrates that usually has two life stages (aquatic and terrestrial; e.g., a tadpole and an adult frog)

andro: man, male (root); *androgen,* a male hormone such as testosterone

ant–, anti–: against (prefix); *antibiotic* (literally "against life"), a substance that kills bacteria

antero–: front (prefix or root); *anterior,* toward the front of

aqu–, aqua–: water (prefix or root); *aquifer,* an underground water source, usually rock saturated with water

apic–: top, highest (prefix); *apical meristem,* the cluster of dividing cells at the tip of a plant shoot or root

arthr–: joint (prefix); *arthropod,* animals such as spiders, crabs, and insects, with exoskeletons that include jointed legs

–ase: enzyme (suffix); *protease,* an enzyme that digests protein

auto–: self (prefix); *autotrophic,* self-feeder (e.g., photo-synthetic)

bi–: two (prefix); *bipedal,* having two legs

bio–: life (prefix or root); *biology,* the study of life

blast: bud, precursor (root); *blastula,* embryonic stage of development, a hollow ball of cells

bronch: windpipe (root); *bronchus,* a branch of the trachea (windpipe) leading to a lung

carcin, –o: cancer (root); *carcinogenesis,* the process of producing a cancer

cardi, –a–, –o–: heart (root); *cardiac,* referring to the heart

carn–, –i–, –o–: flesh (prefix or root); *carnivore,* an animal that eats other animals

centi–: one hundredth (prefix); *centimeter,* a unit of length, 1 one-hundredth of a meter

cephal–, –i–, –o–: head (prefix or root); *cephalization,* the tendency for the nervous system to be located principally in the head

chloro–: green (prefix or root); *chlorophyll,* the green, light-absorbing pigment in plants

chondr–: cartilage (prefix); *Chondrichthyes,* class of vertebrates including sharks and rays, with a skeleton made of cartilage

chrom–: color (prefix or root); *chromosome,* a threadlike strand of DNA and protein in the nucleus of a cell (*Chromosome* literally means "colored body," because chromosomes absorb some of the colored dyes commonly used in microscopy.)

–cide: killer (suffix); *pesticide,* a chemical that kills "pests" (usually insects)

–clast: break down, broken (root or suffix); *osteoclast,* a cell that breaks down bone

co–: with or together with (prefix); *cohesion,* property of sticking together

coel–: hollow (prefix or root); *coelom,* the body cavity that separates the internal organs from the body wall

contra–: against (prefix); *contraception,* acting to prevent conception (pregnancy)

cortex: bark, outer layer (root); *cortex,* outer layer of kidney

crani–: skull (prefix or root); *cranium,* the skull

cuti: skin (root); *cuticle,* the outermost covering of a leaf

–cyte, cyto–: cell (root or prefix); *cytokinin,* a plant hormone that promotes cell division

de–: from, out of, remove (prefix); *decomposer,* an organism that breaks down organic matter

dendr: treelike, branching (root); *dendrite,* highly branched input structures of nerve cells

derm: skin, layer (root); *ectoderm,* the outer embryonic germ layer of cells

deutero–: second (prefix); *deuterostome* (literally, "second opening"), an animal in which the coelom is derived from the gut

di–: two (prefix); *dicot,* an angiosperm with two cotyledons in the seed

diplo–: both, double, two (prefix or root); *diploid,* having paired homologous chromosomes

dys–: difficult, painful (prefix); *dysfunction,* an inability to function properly

eco–: house, household (prefix); *ecology,* the study of the relationships between organisms and their environment

ecto–: outside (prefix); *ectoderm,* the outermost tissue layer of animal embryos

–elle: little, small (suffix); *organelle* (literally, "little organ"), a subcellular structure that performs a specific function

end–, endo–, ento–: inside, inner (prefix); *endocrine,* pertaining to a gland that secretes hormones inside the body

epi–: outside, outer (prefix); *epidermis,* outermost layer of skin

equi–: equal (prefix); *equidistant,* the same distance

erythro–: red (prefix); *erythrocyte,* red blood cell

eu–: true, good (prefix); *eukaryotic,* pertaining to a cell with a true nucleus

ex–, exo–: out of (prefix); *exocrine,* pertaining to a gland that secretes a substance (e.g., sweat) outside of the body

extra–: outside of (prefix); *extracellular,* outside of a cell

–fer: to bear, to carry (suffix); *conifer,* a tree that bears cones

gastr–: stomach (prefix or root); *gastric,* pertaining to the stomach

–gen–: to produce (prefix, root, or suffix); *antigen,* a substance that causes the body to produce antibodies

glyc–, glyco–: sweet (prefix); *glycogen,* a starch-like molecule composed of many glucose molecules bonded together

gyn, –o: female (prefix or root); *gynecology,* the study of the female reproductive tract

haplo–: single (prefix); *haploid,* having a single copy of each type of chromosome

hem–, hemato–: blood (prefix or root); *hemoglobin,* the molecule in red blood cells that carries oxygen

hemi–: half (prefix); *hemisphere,* one of the halves of the cerebrum

herb–, herbi–: grass (prefix or root); *herbivore,* an animal that eats plants

hetero–: other (prefix); *heterotrophic,* an organism that feeds on other organisms

hom–, homo–, homeo–: same (prefix); *homeostasis,* to maintain constant internal conditions in the face of changing external conditions

hydro–: water (usually prefix); *hydrophilic,* being attracted to water

hyper–: above, greater than (prefix); *hyperosmotic,* having a greater osmotic strength (usually higher solute concentration)

hypo–: below, less than (prefix); *hypodermic,* below the skin

inter–: between (prefix); *interneuron,* a neuron that receives input from one (or more) neuron(s) and sends output to another neuron (or many neurons)

intra–: within (prefix); *intracellular,* pertaining to an event or substance that occurs within a cell

iso–: equal (prefix); *isotonic,* pertaining to a solution that has the same osmotic strength as another solution

–itis: inflammation (suffix); *hepatitis,* an inflammation (or infection) of the liver

kin–, kinet–: moving (prefix or root); *cytokinesis,* the movements of a cell that divide the cell in half during cell division

lac–, lact–: milk (prefix or root); *lactose,* the principal sugar in mammalian milk

leuc–, leuco–, leuk–, leuko–: white (prefix); *leukocyte,* a white blood cell

lip–: fat (prefix or root); *lipid,* the chemical category to which fats, oils, and steroids belong

–logy: study of (suffix); *biology,* the study of life

lyso–, –lysis: loosening, split apart (prefix, root, or suffix); *lysis,* to break open a cell

macro–: large (prefix); *macrophage,* a large white blood cell that destroys invading foreign cells

medulla: marrow, middle substance (root); *medulla,* inner layer of kidney

mega–: large (prefix); *megaspore,* a large, haploid (female) spore formed by meiotic cell division in plants

–mere: segment, body section (suffix); *sarcomere,* the functional unit of a vertebrate skeletal muscle cell

meso–: middle (prefix); *mesophyll,* middle layers of cells in a leaf

meta–: change, after (prefix); *metamorphosis,* to change body form (e.g., developing from a larva to an adult)

micro–: small (prefix); *microscope,* a device that allows one to see small objects

milli–: one-thousandth (prefix); *millimeter,* a unit of measurement of length; 1 one-thousandth of a meter

mito–: thread (prefix); *mitosis,* cell division (in which chromosomes appear as threadlike bodies)

mono–: single (prefix); *monocot,* a type of angiosperm with one cotyledon in the seed

morph–: shape, form (prefix or root); *polymorphic,* having multiple forms

multi–: many (prefix); *multicellular,* pertaining to a body composed of more than one cell

myo–: muscle (prefix); *myofibril,* protein strands in muscle cells

neo–: new (prefix); *neonatal,* relating to or affecting a newborn child

neph–: kidney (prefix or root); *nephron,* functional unit of mammalian kidney

neur–, neuro–: nerve (prefix or root); *neuron,* a nerve cell

neutr–: of neither gender or type (usually root); *neutron,* an uncharged subatomic particle found in the nucleus of an atom

non–: not (prefix); *nondisjunction,* the failure of chromosomes to distribute themselves properly during cell division

oligo–: few (prefix); *oligomer,* a molecule made up of a few subunits (see also *poly–*)

omni–: all (prefix); *omnivore,* an animal that eats both plants and animals

oo–, ov–, ovo–: egg (prefix); *oocyte,* one of the stages of egg development

opsi–: sight (prefix or root); *opsin,* protein part of light-absorbing pigment in eye

opso–: tasty food (prefix or root); *opsonization,* process whereby antibodies and/or complement render bacteria easier for white blood cells to engulf

–osis: a condition, disease (suffix); *atherosclerosis,* a disease in which the artery walls become thickened and hardened

oss–, osteo–: bone (prefix or root); *osteoporosis;* a disease in which the bones become spongy and weak

para–: alongside (prefix); *parathyroid,* referring to a gland located next to the thyroid gland

pater, patr–: father (usually root); *paternal,* from or relating to a father

path–, –i–, –o–: disease (prefix or root); *pathology,* the study of disease and diseased tissue

–pathy: disease (suffix); *neuropathy,* a disease of the nervous system

peri–: around (prefix); *pericycle,* the outermost layer of cells in the vascular cylinder of a plant root

phago–: eat (prefix or root); *phagocyte,* a cell (e.g., some types of white blood cell) that eats other cells

–phil, philo–: to love (prefix or suffix); *hydrophilic* (literally, "water loving"), pertaining to a water-soluble molecule

–phob, phobo–: to fear (prefix or suffix); *hydrophobic* (literally, "water fearing"), pertaining to a water-insoluble molecule

photo–: light (prefix); *photosynthesis,* the manufacture of organic molecules using the energy of sunlight

–phyll: leaf (root or suffix); *chlorophyll,* the green, light-absorbing pigment in a leaf

–phyte: plant (root or suffix); *gametophyte* (literally, "gamete plant"), the gamete-producing stage of a plant life cycle

plasmo, –plasm: formed substance (prefix, root, or suffix); *cytoplasm,* the material inside a cell

ploid: chromosomes (root); *diploid,* having paired chromosomes

pneumo–: lung (root); *pneumonia,* a disease of the lungs

–pod: foot (root or suffix); *gastropod* (literally, "stomach-foot"), a class of mollusks, principally snails, that crawl on their ventral surfaces

poly–: many (prefix); *polysaccharide,* a carbohydrate polymer composed of many sugar subunits (see also *oligo–*)

post–, postero–: behind (prefix); *posterior,* pertaining to the hind part

pre–, pro–: before, in front of (prefix); *premating isolating mechanism,* a mechanism that prevents gene flow between species, acting to prevent mating (e.g., having different courtship rituals or different mating seasons)

prim–: first (prefix); *primary cell wall,* the first cell wall laid down between plant cells during cell division

pro–: before (prefix); *prokaryotic,* pertaining to a cell without (that evolved before the evolution of) a nucleus

proto–: first (prefix); *protocell,* a hypothetical evolutionary ancestor to the first cell

pseudo–: false (prefix); *pseudopod* (literally, "false foot"), the extension of the plasma membrane by which some cells, such as *Amoeba,* move and capture prey

quad–, quat–: four (prefix); *quaternary structure,* the "fourth level" of protein structure, in which multiple peptide chains form a complex three-dimensional structure

ren: kidney (root); *adrenal,* gland attached to the mammalian kidney

retro–: backward (prefix); *retrovirus,* a virus that uses RNA as its genetic material; this RNA must be copied "backward" to DNA during infection of a cell by the virus

sarco–: muscle (prefix); *sarcoplasmic reticulum,* a calcium-storing, modified endoplasmic reticulum found in muscle cells

scler–: hard, tough (prefix); *sclerenchyma,* a type of plant cell with a very thick, hard cell wall

semi–: one-half (prefix); *semiconservative replication,* the mechanism of DNA replication, in which one strand of the original DNA double helix becomes incorporated into the new DNA double helix

–some, soma–, somato–: body (prefix or suffix); *somatic nervous system,* part of the peripheral nervous system that controls the skeletal muscles that move the body

sperm, sperma–, spermato–: seed (usually root); *gymnosperm,* a type of plant producing a seed not enclosed within a fruit

stasis, stat–: stationary, standing still (suffix or prefix); *homeostasis,* the physiological process of maintaining constant internal conditions despite a changing external environment

stoma, –to–: mouth, opening (prefix or root); *stoma,* the adjustable pore in the surface of a leaf that allows carbon dioxide to enter the leaf

sub–: under, below (prefix); *subcutaneous,* beneath the skin

sym–: same (prefix); *sympatric* (literally "same father"), found in the same region

tel–, telo–: end (prefix); *telophase,* the last stage of mitosis and meiosis

test–: witness (prefix or root); *testis,* male reproductive organ (derived from the custom in ancient Rome that only males had standing in the eyes of the law; *testimony* has the same derivation)

therm–: heat (prefix or root); *thermoregulation,* the process of regulating body temperature

trans–: across (prefix); *transgenic,* having genes from another organism (usually another species); the genes have been moved "across" species

tri–: three (prefix); *triploid,* having three copies of each homologous chromosome

–trop–, tropic: change, turn, move (suffix); *phototropism,* the process by which plants orient toward the light

troph: food, nourishment (root); *autotrophic,* self-feeder (e.g., photosynthetic)

ultra–: beyond (prefix); *ultraviolet,* light of wavelengths beyond the violet

uni–: one (prefix); *unicellular,* referring to an organism composed of a single cell

vita: life (root); *vitamin,* a molecule required in the diet to sustain life

–vor: eat (usually root); *herbivore,* an animal that eats plants

zoo–, zoa–: animal (usually root); *zoology,* the study of animals

Appendix II

Periodic Table of the Elements

atomic number (number of protons)

element (chemical symbol)

atomic mass (total mass of protons + neutrons + electrons)

1 **H** 1.008																	2 **He** 4.003
3 **Li** 6.941	4 **Be** 9.012											5 **B** 10.81	6 **C** 12.01	7 **N** 14.01	8 **O** 16.00	9 **F** 19.00	10 **Ne** 20.18
11 **Na** 22.99	12 **Mg** 24.31											13 **Al** 26.98	14 **Si** 28.09	15 **P** 30.97	16 **S** 32.07	17 **Cl** 35.45	18 **Ar** 39.95
19 **K** 39.10	20 **Ca** 40.08	21 **Sc** 44.96	22 **Ti** 47.87	23 **V** 50.94	24 **Cr** 52.00	25 **Mn** 54.94	26 **Fe** 55.85	27 **Co** 58.93	28 **Ni** 58.69	29 **Cu** 63.55	30 **Zn** 65.39	31 **Ga** 69.72	32 **Ge** 72.61	33 **As** 74.92	34 **Se** 78.96	35 **Br** 79.90	36 **Kr** 83.80
37 **Rb** 85.47	38 **Sr** 87.62	39 **Y** 88.91	40 **Zr** 91.22	41 **Nb** 92.91	42 **Mo** 95.94	43 **Tc** (98)	44 **Ru** 101.1	45 **Rh** 102.9	46 **Pd** 106.4	47 **Ag** 107.9	48 **Cd** 112.4	49 **In** 114.8	50 **Sn** 118.7	51 **Sb** 121.8	52 **Te** 127.6	53 **I** 126.9	54 **Xe** 131.3
55 **Cs** 132.9	56 **Ba** 137.3	57 ***La** 138.9	72 **Hf** 178.5	73 **Ta** 180.9	74 **W** 183.8	75 **Re** 186.2	76 **Os** 190.2	77 **Ir** 192.2	78 **Pt** 195.1	79 **Au** 197.0	80 **Hg** 200.6	81 **Tl** 204.4	82 **Pb** 207.2	83 **Bi** 209.0	84 **Po** (209)	85 **At** (210)	86 **Rn** (222)
87 **Fr** (223)	88 **Ra** (226)	89 **†Ac** (227)	104 **Rf** (261)	105 **Db** (262)	106 **Sg** (263)	107 **Bh** (264)	108 **Hs** (265)	109 **Mt** (268)	110 **Ds** (281)	111 **Rg** (280)	112 ** (277)	113 **	114 ** (285)	115 **			

***Lanthanide series**	58 **Ce** 140.1	59 **Pr** 140.9	60 **Nd** 144.2	61 **Pm** (145)	62 **Sm** 150.4	63 **Eu** 152.0	64 **Gd** 157.3	65 **Tb** 158.9	66 **Dy** 162.5	67 **Ho** 164.9	68 **Er** 167.3	69 **Tm** 168.9	70 **Yb** 173.0	71 **Lu** 175.0
†Actinide series	90 **Th** 232.0	91 **Pa** 231	92 **U** 238.0	93 **Np** (237)	94 **Pu** (244)	95 **Am** (243)	96 **Cm** (247)	97 **Bk** (247)	98 **Cf** (251)	99 **Es** (252)	100 **Fm** (257)	101 **Md** (258)	102 **No** (259)	103 **Lr** (262)

The periodic table of the elements was first devised by Russian chemist Dmitri Mendeleev. The atomic numbers of the elements (the numbers of protons in the nucleus) increase in normal reading order: left to right, top to bottom. The table is "periodic" because all the elements in a column possess similar chemical properties, and such similar elements therefore recur "periodically" in each row. For example, elements that usually form ions with a single positive charge—including H, Li, Na, K, and so on—occur as the first element in each row.

The gaps in the table are a consequence of the maximum numbers of electrons in the most reactive, usually outermost, electron shells of the atoms. It takes only two electrons to completely fill the first shell, so the first row of the table contains only two elements, H and He. It takes eight electrons to fill the second and third shells, so there are eight elements in the second and third rows. It takes 18 electrons to fill the fourth and fifth shells, so there are 18 elements in these rows.

The lanthanide series and actinide series of elements are usually placed below the main body of the table for convenience—it takes 32 electrons to completely fill the sixth and seventh shells, so the table would become extremely wide if all of these elements were included in the sixth and seventh rows.

The important elements found in living things are highlighted in color. The elements in pale red are the six most abundant elements in living things. The elements that form the five most abundant ions in living things are in purple. The trace elements important to life are shown as dark blue (more common) and lighter blue (less common). For radioactive elements, the atomic masses are given in parentheses, and represent the most common or the most stable isotope. Elements indicated as double asterisks have not yet been named.

Appendix III

Metric System Conversions

To Convert Metric Units:	Multiply by:	To Get English Equivalent:
Length		
Centimeters (cm)	0.3937	Inches (in)
Meters (m)	3.2808	Feet (ft)
Meters (m)	1.0936	Yards (yd)
Kilometers (km)	0.6214	Miles (mi)
Area		
Square centimeters (cm^2)	0.155	Square inches (in^2)
Square meters (m^2)	10.7639	Square feet (ft^2)
Square meters (m^2)	1.1960	Square yards (yd^2)
Square kilometers (km^2)	0.3831	Square miles (mi^2)
Hectare (ha) (10,000 m^2)	2.4710	Acres (a)
Volume		
Cubic centimeters (cm^3)	0.06	Cubic inches (in^3)
Cubic meters (m^3)	35.30	Cubic feet (ft^3)
Cubic meters (m^3)	1.3079	Cubic yards (yd^3)
Cubic kilometers (km^3)	0.24	Cubic miles (mi^3)
Liters (L)	1.0567	Quarts (qt), U.S.
Liters (L)	0.26	Gallons (gal), U.S.
Mass		
Grams (g)	0.03527	Ounces (oz)
Kilograms (kg)	2.2046	Pounds (lb)
Metric ton (tonne) (t)	1.10	Ton (tn), U.S.
Speed		
Meters/second (mps)	2.24	Miles/hour (mph)
Kilometers/hour (kmph)	0.62	Miles/hour (mph)

To Convert English Units:	Multiply by:	To Get Metric Equivalent:
Length		
Inches (in)	2.54	Centimeters (cm)
Feet (ft)	0.3048	Meters (m)
Yards (yd)	0.9144	Meters (m)
Miles (mi)	1.6094	Kilometers (km)
Area		
Square inches (in^2)	6.45	Square centimeters (cm^2)
Square feet (ft^2)	0.0929	Square meters (m^2)
Square yards (yd^2)	0.8361	Square meters (m^2)
Square miles (mi^2)	2.5900	Square kilometers (km^2)
Acres (a)	0.4047	Hectare (ha) (10,000 m^2)
Volume		
Cubic inches (in^3)	16.39	Cubic centimeters (cm^3)
Cubic feet (ft^3)	0.028	Cubic meters (m^3)
Cubic yards (yd^3)	0.765	Cubic meters (m^3)
Cubic miles (mi^3)	4.17	Cubic kilometers (km^3)
Quarts (qt), U.S.	0.9463	Liters (L)
Gallons (gal), U.S.	3.8	Liters (L)
Mass		
Ounces (oz)	28.3495	Grams (g)
Pounds (lb)	0.4536	Kilograms (kg)
Ton (tn), U.S.	0.91	Metric ton (tonne) (t)
Speed		
Miles/hour (mph)	0.448	Meters/second (mps)
Miles/hour (mph)	1.6094	Kilometers/hour (kmph)

Metric Prefixes

Prefix			Meaning
giga-	G	10^9 =	1,000,000,000
mega-	M	10^6 =	1,000,000
kilo-	k	10^3 =	1,000
hecto-	h	10^2 =	100
deka-	da	10^1 =	10
		10^0 =	1
deci-	d	10^{-1} =	0.1
centi-	c	10^{-2} =	0.01
milli-	m	10^{-3} =	0.001
micro-	μ	10^{-6} =	0.000001

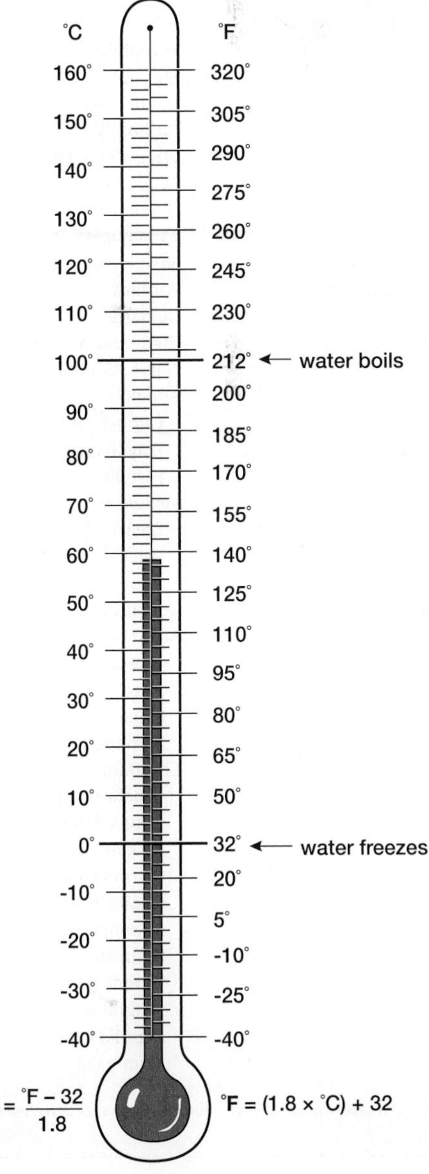

Appendix IV Classification of Major Groups of Eukaryotic Organisms*

Kingdom	Phylum or Class	Common Name
Excavata	Parabasalia	parabasalids
	Diplomonadida	diplomonads
Euglenozoa	Euglenida	euglenids
	Kinetoplastida	kinetoplastids
Stramenopila	Oomycota	water molds
	Phaeophyta	brown algae
	Bacillariophyta	diatoms
Alveolata	Apicomplexa	sporozoans
	Pyrrophyta	dinoflagellates
	Ciliophora	ciliates
Rhizaria	Foraminifera	foraminiferans
	Radiolaria	radiolarians
Amoebozoa	Tubulinea	amoebas
	Myxomycota	acellular slime molds
	Acrasiomycota	cellular slime molds
Rhodophyta		red algae
Chlorophyta		green algae
Plantae	Bryophyta	liverworts, mosses
	Pteridophyta	ferns
	Coniferophyta	evergreens
	Anthophyta	flowering plants
Fungi	Chytridiomycota	chytrids
	Zygomycota	zygote fungi
	Glomeromycota	glomeromycetes
	Ascomycota	sac fungi
	Basidiomycota	club fungi
Animalia	Porifera	sponges
	Cnidaria	hydras, sea anemones, sea jellies, corals
	Ctenophora	comb jellies
	Platyhelminthes	flatworms
	Annelida	segmented worms
	Oligochaeta	earthworms
	Polychaeta	tube worms
	Hirudinea	leeches
	Mollusca	mollusks
	Gastropoda	snails
	Pelecypoda	mussels, clams
	Cephalopoda	squid, octopuses
	Nematoda	roundworms
	Arthropoda	arthropods
	Insecta	insects
	Arachnida	spiders, ticks
	Crustacea	crabs, lobsters
	Myriapoda	millipedes, centipedes
	Chordata	chordates
	Urochordata	tunicates
	Cephalochordata	lancelets
	Myxini	hagfishes
	Petromyzontiformes	lampreys
	Chondrichthyes	sharks, rays
	Actinopterygii	ray-finned fishes
	Actinistia	coelacanths
	Dipnoi	lungfishes
	Amphibia	amphibians (frogs, salamanders)
	Reptilia	reptiles (turtles, crocodiles, birds, snakes, lizards)
	Mammalia	mammals

*This table lists only those taxonomic categories described in the textbook.

Glossary

abiotic (ā-bī-ah′-tik): nonliving; the abiotic portion of an ecosystem includes soil, rock, water, and the atmosphere.

abscisic acid (ab-sis′-ik): a plant hormone that generally inhibits the action of other hormones, enforcing dormancy in seeds and buds and causing the closing of stomata.

abscission layer: a layer of thin-walled cells, located at the base of the petiole of a leaf, that produces an enzyme that digests the cell walls holding the leaf to the stem, allowing the leaf to fall off.

absorption: the process by which nutrients enter the body through the cells lining the digestive tract.

accessory pigment: a colored molecule other than chlorophyll *a*, that absorbs light energy and passes it to chlorophyll *a*.

acellular slime mold: a type of organism that forms a multinucleate structure that crawls in amoeboid fashion and ingests decaying organic matter; also called *plasmodial slime mold*. Acellular slime molds are members of the protist clade Amoebozoa.

acid deposition: the deposition of nitric or sulfuric acid, either in rain (acid rain) or in the form of dry particles, as a result of the production of nitrogen oxides or sulfur dioxide through burning, primarily of fossil fuels.

acid: a substance that releases hydrogen ions (H+) into solution; a solution with a pH less than 7.

acidic: referring to a solution with an H^+ concentration exceeding that of OH-; referring to a substance that releases H^+.

acquired immune deficiency syndrome (AIDS): an infectious disease caused by the human immunodeficiency virus (HIV); attacks and destroys T cells, thus weakening the immune system.

acrosome (ak′-rō-sōm): a vesicle, located at the tip of the head of an animal sperm, that contains enzymes needed to dissolve protective layers around the egg.

actin (ak′-tin): a major muscle protein whose interactions with myosin produce contraction; found in the thin filaments of the muscle fiber; see also *myosin*.

action potential: a rapid change from a negative to a positive electrical potential in a nerve cell. An action potential travels along an axon without a change in amplitude.

activation energy: in a chemical reaction, the energy needed to force the electron shells of reactants together, prior to the formation of products.

active site: the region of an enzyme molecule that binds substrates and performs the catalytic function of the enzyme.

active transport: the movement of materials across a membrane through the use of cellular energy, normally against a concentration gradient.

adaptation: a trait that increases the ability of an individual to survive and reproduce compared to individuals without the trait.

adaptive immune response: a response to invading toxins or microbes in which immune cells are activated by a specific invader, selectively destroy that invader, and then "remember" the invader, allowing a faster response if that type of invader reappears in the future; see also *innate immune response*.

adaptive immune system: a widely distributed system of organs (including the thymus, bone marrow, and lymph nodes), cells (including macrophages, dendritic cells, B cells, and T cells), and molecules (including cytokines and antibodies) that work together to combat microbial invasion of the body; the adaptive immune system responds to and destroys specific invading toxins or microbes; see also *innate immune response*.

adaptive radiation: the rise of many new species in a relatively short time; may occur when a single species invades different habitats and evolves in response to different environmental conditions in those habitats.

adenine (A): a nitrogenous base found in both DNA and RNA; abbreviated as A.

adenosine diphosphate (a-den′-ō-sēn dī-fos′-fāt; ADP): a molecule composed of the sugar ribose, the base adenine, and two phosphate groups; a component of ATP.

adenosine triphosphate (a-den′-ō-sēn trī-fos′-fāt; ATP): a molecule composed of the sugar ribose, the base adenine, and three phosphate groups; the major energy carrier in cells. The last two phosphate groups are attached by "high-energy" bonds.

adhesion: the tendency of polar molecules (such as water) to adhere to polar surfaces (such as glass).

adipose tissue (a′-dī-pōs): tissue composed of fat cells.

adrenal cortex: the outer part of the adrenal gland, which secretes steroid hormones that regulate metabolism and salt balance.

adrenal gland: a mammalian endocrine gland, adjacent to the kidney; secretes hormones that function in water regulation and in the stress response.

adrenal medulla: the inner part of the adrenal gland, which secretes epinephrine (adrenaline) and norepinephrine (noradrenaline) in the stress response.

adrenocorticotropic hormone (a-drēn-ō-kor-tik-ō-trō′-pik; ACTH): a hormone, secreted by the anterior pituitary, that stimulates the release of hormones by the adrenal cortex, especially in response to stress.

adult stem cell (ASC): any stem cell not found in an early embryo; can divide and differentiate into any of several cell types, but usually not all of the cell types of the body.

aerobic: using oxygen.

age structure: the distribution of males and females in a population according to age groups.

aggression: antagonistic behavior, normally among members of the same species, often resulting from competition for resources.

aggressive mimicry (mim′ik-rē): the evolution of a predatory organism to resemble a harmless animal or a part of the environment, thus gaining access to prey.

aging: the gradual accumulation of damage to essential biological molecules, particularly DNA in both the nucleus and mitochondria, resulting in defects in cell functioning, declining health, and ultimately death.

albinism: a recessive hereditary condition caused by defective alleles of the genes that encode the enzymes required for the synthesis of melanin, the principal pigment in mammalian skin and hair; albinism results in white hair and pink skin.

alcoholic fermentation: a type of fermentation in which pyruvate is converted to ethanol (a type of alcohol) and carbon dioxide, using hydrogen ions and electrons from NADH; the primary function of alcoholic fermentation is to regenerate NAD+ so that glycolysis can continue under anaerobic conditions.

aldosterone: a hormone, secreted by the adrenal cortex, that helps regulate ion concentration in the blood by stimulating the reabsorption of sodium by the kidneys and sweat glands.

alga (al′-ga; pl., **algae, al′-jē):** any photosynthetic protist.

allantois (al-an-tō′-is): one of the embryonic membranes of reptiles (including birds) and mammals; in reptiles, serves as a waste-storage organ; in mammals, forms most of the umbilical cord.

allele (al-ēl′): one of several alternative forms of a particular gene.

allele frequency: for any given gene, the relative proportion of each allele of that gene in a population

allergy: an inflammatory response produced by the body in response to invasion by foreign materials, such as pollen, that are themselves harmless.

allopatric speciation (al-ō-pat′-rik): the process by which new species arise following physical separation of parts of a population (geographical isolation).

allosteric regulation: the process by which enzyme action is enhanced or inhibited by small organic molecules that act as regulators by binding to the enzyme at a regulatory site distinct from the active site, and altering the shape and/or function of the active site.

alternation of generations: a life cycle, typical of plants, in which a diploid sporophyte (spore-producing) generation alternates with a haploid gametophyte (gamete-producing) generation.

altruism: a behavior that benefits other individuals while reducing the fitness of the individual that performs the behavior.

alveolate (al-vē′-ō-lāt): a member of the Alveolata, a large protist clade. The alveolates, which are characterized by a system of sacs beneath the cell membrane, include ciliates, dinoflagellates, and apicomplexans.

alveolus (al-vē′-ō-lus; pl., **alveoli):** a tiny air sac within the lungs, surrounded by capillaries, where gas exchange with the blood occurs.

amino acid: the individual subunit of which proteins are made, composed of a central carbon atom bonded to an amino group ($—NH_2$), a carboxyl group ($—COOH$), a hydrogen atom, and a variable group of atoms denoted by the letter *R*.

amino acid-derived hormone: a class of hormone that is synthesized by the body from single amino acids. Examples include epinephrine and thyroxine.

ammonia: NH_3; a highly toxic nitrogen-containing waste product of amino acid breakdown. In the mammalian liver, it is converted to urea.

amniocentesis (am-nē-ō-sen-tē′-sis): a procedure for sampling the amniotic fluid surrounding a fetus: A sterile needle is inserted through the abdominal wall, uterus, and amniotic sac of a pregnant woman, and 10 to 20 milliliters of amniotic fluid are withdrawn. Various tests may be performed on the fluid and the fetal cells suspended in it to provide information on the developmental and genetic state of the fetus.

amnion (am′-nē-on): one of the embryonic membranes of reptiles (including birds) and mammals; encloses a fluid-filled cavity that envelops the embryo.

amniotic egg (am-nē-ōt′-ik): the egg of reptiles, including birds; contains a membrane, the amnion, that surrounds the embryo, enclosing it in a watery environment and allowing the egg to be laid on dry land.

amoeba: an amoebozoan protist that uses a characteristic streaming mode of locomotion by extending a cellular projection called a *pseudopod*. Also known as *lobose amoebas.*

amoeboid cell: a protist or animal cell that moves by extending a cellular projection called a pseudopod.

amoebozoan: a member of the Amoebozoa, a protist clade. The amoebozoans, which generally lack shells and move by extending pseudopods, include the lobose amoebas and the slime molds.

amphibian: a member of the chordate clade Amphibia, which includes the frogs, toads, and salamanders, as well as the limbless caecelians.

amygdala (am-ig′-da-la): part of the forebrain of vertebrates that is involved in the production of appropriate behavioral responses to environmental stimuli.

amylase (am′-i-lās): an enzyme, found in saliva and pancreatic secretions, that catalyzes the breakdown of starch.

anaerobe (an–ə–rōb): an organism that can live and grow in the absence of oxygen.

anaerobic: not using oxygen.

analogous structure: structures that have similar functions and superficially similar appearance but very different anatomies, such as the wings of insects and birds. The similarities are the result of similar environmental pressures rather than a common ancestry.

anaphase (an′-a-fāz): in mitosis, the stage in which the sister chromatids of each chromosome separate from one another and are moved to opposite poles of the cell; in meiosis I, the stage in which homologous chromosomes, consisting of two sister chromatids, are separated; in meiosis II, the stage in which the sister chromatids of each chromosome separate from one another and are moved to opposite poles of the cell.

androgen: a male sex hormone.

angina (an-jī′-nuh): chest pain associated with reduced blood flow to the heart muscle; caused by an obstruction of the coronary arteries.

angiosperm (an′-jē-ō-sperm): a flowering vascular plant.

angiotensin (an-jē-ō-ten′-sun): a hormone that functions in water regulation in mammals by stimulating physiological changes that increase blood volume and blood pressure.

annual ring: a pattern of alternating light (early) and dark (late) xylem in woody stems and roots; formed as a result of the unequal availability of water in different seasons of the year, normally spring and summer.

antagonistic muscles: a pair of muscles, one of which contracts and in so doing extends the other, relaxed muscle; this arrangement makes possible movement of the skeleton at joints.

anterior pituitary: a lobe of the pituitary gland that produces prolactin and growth hormone as well as hormones that regulate hormone production in other glands.

anther (an′-ther): the uppermost part of the stamen, in which pollen develops.

antheridium (an-ther-id′-ē-um; pl., **antheridia):** a structure in which male sex cells are produced; found in nonvascular plants and certain seedless vascular plants.

antibiotic: chemicals that help to combat infection by destroying or slowing down the multiplication of bacteria, fungi, or protists.

antibody: a protein, produced by cells of the immune system, that combines with a specific antigen and normally facilitates the destruction of the antigen.

anticodon: a sequence of three bases in transfer RNA that is complementary to the three bases of a codon of messenger RNA.

antidiuretic hormone (an-tē-dī-ūr-et′-ik; ADH): a hormone produced by the hypothalamus and released into the bloodstream by the posterior pituitary when blood volume is low; increases the permeability of the distal tubule and the collecting duct to water, allowing more water to be reabsorbed into the bloodstream.

antigen: a complex molecule, normally a protein or polysaccharide, that stimulates the production of a specific antibody.

antioxidant: any molecule that reacts with free radicals, neutralizing their ability to damage biological molecules. Vitamins C and E are examples of dietary antioxidants.

anvil: the second of the small bones of the middle ear, linking the tympanic membrane (eardrum) to the oval window of the cochlea; also called the *incus*.

aphotic zone: the region of the ocean below 200 m, where sunlight does not penetrate.

apical dominance: the phenomenon whereby a growing shoot tip inhibits the sprouting of lateral buds.

apical meristem (āp′-i-kul mer′-i-stem): the cluster of meristem cells at the tip of a shoot or root (or one of their branches).

apicomplexan (ā-pē-kom-pleks′-an): a member of the protist clade Apicomplexa, which includes mostly parasitic, single-celled eukaryotes such as *Plasmodium*, which causes malaria in humans. Apicomplexans are part of a larger group known as the alveolates.

appendicular skeleton (ap-pen-dik′-ū-lur): the portion of the skeleton consisting of the bones of the extremities and their attachments to the axial skeleton; the appendicular skeleton therefore consists of the pectoral and pelvic girdles and the arms, legs, hands, and feet.

aquaporin: a channel protein in the plasma membrane of a cell that is selectively permeable to water.

aqueous humor (ā′-kwē-us): the clear, watery fluid between the cornea and lens of the eye; nourishes the cornea and lens.

aquifer (ok′–wi–fer): an underground deposit of fresh water, often used as a source for irrigation.

archaea: prokaryotes that are members of the domain Archaea, one of the three domains of living organisms; only distantly related to members of the domain Bacteria.

Archaea: one of life's three domains; consists of prokaryotes that are only distantly related to members of the domain Bacteria.

archegonium (ar-ke-gō′-nē-um; pl., **archegonia):** a structure in which female sex cells are produced; found in nonvascular plants and certain seedless vascular plants.

arteriole (ar-tēr′-ē-ōl): a small artery that empties into capillaries; constriction of arterioles regulates blood flow to various parts of the body.

artery (ar′-tuh-rē): a vessel with muscular, elastic walls that conducts blood away from the heart.

arthropod: a member of the animal phylum Arthropoda, which includes the insects, spiders, ticks, mites, scorpions, crustaceans, millipedes, and centipedes.

artificial selection: a selective breeding procedure in which only those individuals with particular traits are chosen as breeders; used mainly to enhance desirable traits in domesticated plants and animals; may also be used in evolutionary biology experiments.

ascomycete: a member of the fungus phylum Ascomycota, whose members form sexual spores in a saclike case known as an ascus.

ascus (as′-kus; pl., **asci):** a saclike case in which sexual spores are formed by members of the fungus phylum Ascomycota.

asexual reproduction: reproduction that does not involve the fusion of haploid gametes.

atherosclerosis (ath′-er-ō-skler-ō′-sis): a disease characterized by the obstruction of arteries by cholesterol deposits and thickening of the arterial walls.

atom: the smallest particle of an element that retains the properties of the element.

atomic mass: the total mass of all the protons, neutrons, and electrons within an atom

atomic nucleus: the central part of an atom that contains protons and neutrons.

atomic number: the number of protons in the nuclei of all atoms of a particular element.

ATP synthase: a channel protein in the thylakoid membranes of chloroplasts and the inner membrane of mitochondria that uses the energy of H+ ions moving through the channel down their concentration gradient to produce ATP from ADP and inorganic phosphate.

atrial natriuretic peptide (ANP) (ā′-trē-ul nā-trē-ū-ret′-ik; ANP): a hormone, secreted by cells in the mammalian heart, that reduces blood volume by inhibiting the release of ADH and aldosterone.

atrioventricular (AV) node (ā′-trē-ō-ven-trik′-ū-lar nōd): a specialized mass of muscle at the base of the right atrium through which the electrical activity initiated in the sinoatrial node is transmitted to the ventricles.

atrioventricular valve: a heart valve that separates each atrium from each ventricle, preventing the backflow of blood into the atria during ventricular contraction.

atrium (ā′-trē-um; pl., **atria):** a chamber of the heart that receives venous blood and passes it to a ventricle.

attachment protein: a protein in the plasma membrane of a cell that attaches either to the cytoskeleton inside the cell, to other cells, or to the extracellular matrix.

auditory canal (aw′-di-tor-ē): a relatively large-diameter tube within the outer ear that conducts sound from the pinna to the tympanic membrane.

auditory nerve: the nerve leading from the mammalian cochlea to the brain; it carries information about sound.

auditory tube: a thin tube connecting the middle ear with the pharynx, which allows pressure to equilibrate between the middle ear and the outside air; also called the *Eustachian tube.*

autoimmune disease: a disorder in which the immune system attacks the body's own cells or molecules.

autonomic nervous system: the part of the peripheral nervous system of vertebrates that synapses on glands, internal organs, and smooth muscle and produces largely involuntary responses.

autosome (aw′-tō-sōm): a chromosome that occurs in homologous pairs in both males and females and that does not bear the genes determining sex.

autotroph (aw′-tō-trōf): literally, "self-feeder"; normally, a photosynthetic organism; a producer.

auxin (awk′-sin): a plant hormone that influences many plant functions, including phototropism, gravitropism, apical dominance, and root branching.

axial skeleton: the skeleton forming the body axis, including the skull, vertebral column, and rib cage.

axon: a long extension of a nerve cell, extending from the cell body to synaptic endings on other nerve cells or on muscles.

B cell: a type of lymphocyte that matures in the bone marrow, and that participates in humoral immunity; gives rise to plasma cells, which secrete antibodies into the circulatory system, and to memory cells.

bacteria (sing., bacterium): prokaryotes that are members of the domain Bacteria, one of the three domains of living organisms; only distantly related to members of the domain Archaea.

Bacteria: one of life's three domains; consists of prokaryotes that are only distantly related to members of the domain Archaea.

bacteriophage (bak-tir′-ē-ō-fāj): a virus that specifically infects bacteria.

ball-and-socket joint: a joint in which the rounded end of one bone fits into a hollow depression in another, as in the hip; allows movement in several directions.

bark: the outer layer of a woody stem, consisting of phloem, cork cambium, and cork cells.

Barr body: a condensed, inactivated X chromosome in the cells of female mammals, which have two X chromosomes.

basal body: a structure derived from a centriole that produces a cilium or flagellum and anchors this structure within the plasma membrane.

basal ganglia: several clusters of neurons in the interior of the cerebrum, plus the substantial nigra in the midbrain, that function in the control of movement. Damage to or degeneration of one or more basal ganglia causes disorders such as Parkinson's disease and Huntington's disease.

base: (1) a substance capable of combining with and neutralizing H+ ions in a solution; a solution with a pH greater than 7; (2) one of the nitrogen-containing, single- or double-ringed structures that distinguishes one nucleotide from another. In DNA, the bases are adenine, guanine, cytosine, and thymine.

basic: referring to a solution with an H+ concentration less than that of OH-; referring to a substance that combines with H+.

basidiomycete: a member of the fungus phylum Basidiomycota, which includes species that produce sexual spores in club-shaped cells known as basidia.

basidiospore (ba-sid′-ē-ō-spor): a sexual spore formed by members of the fungus phylum Basidiomycota.

basidium (pl., **basidia):** a diploid cell, typically club-shaped, formed by members of the fungus phylum Basidiomycota; produces basidiospores by meiosis.

basilar membrane (bas′-eh-lar): a membrane in the cochlea that bears hair cells that respond to the vibrations produced by sound.

behavior: any observable activity of a living animal.

bilateral symmetry: a body plan in which only a single plane through the central axis will divide the body into mirror-image halves.

bile (bīl): a liquid secretion, produced by the liver, that is stored in the gallbladder and released into the small intestine during digestion; consists of a complex mixture of bile salts, water, other salts, and cholesterol.

binary fission: the process by which a single bacterium divides in half, producing two identical offspring.

binocular vision: the ability to see objects simultaneously through both eyes, providing greater depth perception and more accurate judgment of the size and distance of an object than can be achieved by vision with one eye alone.

binomial system: the method of naming organisms by genus and species, often called the scientific name, usually using Latin words or words derived from Latin.

biocapacity: an estimate of the sustainable resources and waste-absorbing capacity actually available on Earth. Biocapacity calculations are subject to change as new technologies change the way people use resources.

biodegradable: able to be broken down into harmless substances by decomposers.

biodiversity: the diversity of living organisms; measured as the variety of different species, the variety of different alleles in species' gene pools, or the variety of different communities and nonliving environments in an ecosystem or in the entire biosphere.

biofilm: a community of prokaryotes of one or more species, in which the prokaryotes secrete and are embedded in slime that adheres to a surface.

biogeochemical cycle: the pathways of a specific nutrient (such as carbon, nitrogen, phosphorus, or water) through the living and nonliving portions of an ecosystem; also called a *nutrient cycle.*

biological clock: a metabolic timekeeping mechanism found in most organisms, whereby the organism measures the approximate length of a day (24 hours) even without external environmental cues such as light and darkness.

biological magnification: the increasing accumulation of a toxic substance in progressively higher trophic levels.

biomass: the total weight of all living material within a defined area.

biome (bī′-ōm): a terrestrial ecosystem that occupies an extensive geographical area and is characterized by a specific type of plant community; for example, deserts.

bioremediation: the use of organisms to remove or detoxify toxic substances in the environment.

biosphere (bī′-ō-sfēr): that part of Earth inhabited by living organisms; includes both living and nonliving components.

Biosphere Reserve: designated by the United Nations, a Biosphere Reserve is a region intended to maintain biodiversity and evaluate techniques for sustainable human development while maintaining local cultural values.

biotechnology: any industrial or commercial use or alteration of organisms, cells, or biological molecules to achieve specific practical goals.

biotic (bī-ah′-tik): living.

biotic potential: the maximum rate at which a population is able to increase, assuming ideal conditions that allow a maximum birth rate and minimum death rate.

birth rate: the number of births per individual in a specified unit of time, such as a year.

bladder: a hollow muscular organ that stores urine.

blade: the flat part of a leaf.

blastocyst (blas′-tō-sist): an early stage of human embryonic development, consisting of a hollow ball of cells, enclosing a mass of cells attached to its inner surface, which becomes the embryo.

blastopore: the site at which a blastula indents to form a gastrula.

blastula (blas′-tū-luh): in animals, the embryonic stage attained at the end of cleavage, in which the embryo usually consists of a hollow ball with a wall that is one or several cell layers thick.

blind spot: the area of the retina at which the axons of the ganglion cells merge to form the optic nerve; because there are no photoreceptors in the blind spot, objects focused at the blind spot cannot be seen.

blood: a specialized connective tissue, consisting of a fluid (plasma) in which blood cells are suspended; carried within the circulatory system.

blood clotting: a complex process by which platelets, the protein fibrin, and red blood cells block an irregular surface in or on the body, such as a damaged blood vessel, sealing the wound.

blood vessel: any of several types of tubes that carry blood throughout the body.

blood–brain barrier: relatively impermeable capillaries of the brain that protect the cells of the brain from potentially damaging chemicals that reach the bloodstream.

body mass index (BMI): a number derived from an individual's weight and height that is used to estimate body fat. The formula is weight (in kg) /height^2 (in meters^2).

bone: a hard, mineralized connective tissue that is a major component of the vertebrate endoskeleton; provides support and sites for muscle attachment.

boom-and-bust cycle: a population cycle characterized by rapid exponential growth followed by a sudden massive die-off; seen in seasonal species, such as many insects living in temperate climates, and in some populations of small rodents, such as lemmings.

Bowman's capsule: the cup-shaped portion of the nephron in which blood filtrate is collected from the glomerulus.

brain: the part of the central nervous system of vertebrates that is enclosed within the skull.

branch root: a root that arises as a branch of a preexisting root; occurs through divisions of pericycle cells and subsequent differentiation of the daughter cells.

bronchiole (bron′-kē-ōl): a narrow tube, formed by repeated branching of the bronchi, that conducts air into the alveoli.

bronchus (bron′-kus; pl., **bronchi):** a tube that conducts air from the trachea to each lung.

budding: asexual reproduction by the growth of a miniature copy, or bud, of the adult animal on the body of the parent. The bud breaks off to begin independent existence.

buffer: a compound that minimizes changes in pH by reversibly taking up or releasing $H+$ ions.

bulbourethral gland (bul-bō-ū-rē′-thrul): in male mammals, a gland that secretes a basic, mucous-containing fluid that forms part of the semen.

bulk flow: the movement of many molecules of a gas or liquid in unison (in bulk, hence the name) from an area of higher pressure to an area of lower pressure.

bundle sheath cells: cells that surround the veins of plants; in C_4 (but not in C_3) plants, bundle sheath cells contain chloroplasts.

C_3 pathway: in photosynthesis, the cyclic series of reactions whereby carbon from carbon dioxide is fixed as phosphoglyceric acid, the simple sugar glyceraldehyde-3-phosphate is generated, and the carbon-capture molecule, RuBP, is regenerated.

C_3 plant: a plant that relies on the C_3 pathway to fix carbon.

C_4 pathway: the series of reactions in certain plants that fixes carbon dioxide into a four-carbon molecule, which is later broken down for use in the Calvin cycle of photosynthesis. This reduces wasteful photorespiration in hot, dry environments.

C_4 plant: a plant that relies on the C_4 pathway to fix carbon.

calcitonin (kal-si-tōn′-in): a hormone, secreted by the thyroid gland, that inhibits the release of calcium from bone.

calorie (kal′-ō-rē): the amount of energy required to raise the temperature of 1 gram of water by 1 degree Celsius.

Calorie: a unit used to measure the energy content of foods; it is the amount of energy required to raise the temperature of 1 liter of water 1 degree Celsius; also called a *kilocalorie*, equal to 1,000 calories.

Calvin cycle: in photosynthesis, the cyclic series of reactions whereby carbon from carbon dioxide is fixed as phosphoglyceric acid, the simple sugar glyceraldehyde-3-phosphate is generated, and the carbon-capture molecule, RuBP, is regenerated.

cambium (kam′-bē-um; pl., **cambia):** a lateral meristem, parallel to the long axis of roots and stems, that causes secondary growth of woody plant stems and roots. See *cork cambium; vascular cambium.*

camouflage (cam′-a-flaj): coloration and/or shape that renders an organism inconspicuous in its environment.

cancer: a disease in which some of the body's cells escape from normal regulatory processes and divide without control.

capillary: the smallest type of blood vessel, connecting arterioles with venules; capillary walls, through which the exchange of nutrients and wastes occurs, are only one cell thick.

carbohydrate: a compound composed of carbon, hydrogen, and oxygen, with the approximate chemical formula $(CH_2O)_n$; includes sugars, starches, and cellulose.

carbon cycle: the biogeochemical cycle by which carbon moves from its reservoirs in the atmosphere and oceans through producers and into higher trophic levels, and then back to its reservoirs.

carbon fixation: the process by which carbon derived from carbon dioxide is captured in organic molecules during photosynthesis.

cardiac cycle (kar′-dē-ak): the alternation of contraction and relaxation of the heart chambers.

cardiac muscle (kar′-dē-ak): the specialized muscle of the heart; able to initiate its own contraction, independent of the nervous system.

carnivore (kar′-neh-vor): literally, "meat eater"; a predatory organism that feeds on herbivores or on other carnivores; a secondary (or higher) consumer.

carotenoid (ka-rot′-en-oid): a red, orange, or yellow pigment, found in chloroplasts, that serves as an accessory light-gathering pigment in thylakoid photosystems.

carpel (kar′pel): the female reproductive structure of a flower, composed of stigma, style, and ovary.

carrier: an individual who is heterozygous for a recessive condition; a carrier displays the dominant phenotype but can pass on the recessive allele to offspring.

carrier protein: a membrane protein that facilitates the diffusion of specific substances across the membrane. The molecule to be transported binds to the outer surface of the carrier protein; the protein then changes shape, allowing the molecule to move across the membrane.

carrying capacity *(K)*: the maximum population size that an ecosystem can support for a long period of time without damaging the ecosystem; determined primarily by the availability of space, nutrients, water, and light.

cartilage (kar′-teh-lij): a form of connective tissue that forms portions of the skeleton; consists of chondrocytes and their extracellular secretion of collagen.

Casparian strip (kas-par′-ē-un): a waxy, waterproof band, located in the cell walls between endodermal cells in a root, that prevents the movement of water and minerals into and out of the vascular cylinder through the extracellular space.

catalyst (kat′-uh-list): a substance that speeds up a chemical reaction without itself being permanently changed in the process; a catalyst lowers the activation energy of a reaction.

cell: the smallest unit of life, consisting, at a minimum, of an outer membrane that encloses a watery medium containing organic molecules, including genetic material composed of DNA.

cell body: the part of a nerve cell in which most of the common cellular organelles are located; typically a site of integration of inputs to the nerve cell.

cell cycle: the sequence of events in the life of a cell, from one cell division to the next.

cell division: splitting of one cell into two; the process of cellular reproduction.

cell plate: in plant cell division, a series of vesicles that fuse to form the new plasma membranes and cell wall separating the daughter cells.

cell theory: the scientific theory stating that every living organism is made up of one or more cells; cells are the functional units of all organisms; and all cells arise from preexisting cells.

cell-mediated immunity: an adaptive immune response in which foreign cells or substances are destroyed by contact with T cells.

cellular respiration: the oxygen-requiring reactions, occurring in mitochondria, that break down the end products of glycolysis into carbon dioxide and water while capturing large amounts of energy as ATP.

cellular slime mold: a type of organism consisting of individual amoeboid cells that can aggregate to form a sluglike mass, which in turn forms a fruiting body. Cellular slime molds are members of the protist clade Amoebozoa.

cellulose: an insoluble carbohydrate composed of glucose subunits; forms the cell wall of plants.

central nervous system (CNS): in vertebrates, the brain and spinal cord.

central vacuole (vak′-ū-ōl): a large, fluid-filled vacuole occupying most of the volume of many plant cells; performs several functions, including maintaining turgor pressure.

centriole (sen′-trē-ōl): in animal cells, a short, barrel-shaped ring consisting of nine microtubule triplets; a pair of centrioles is found near the nucleus and may play a role in the organization of the spindle; centrioles also give rise to the basal bodies at the base of each cilium and flagellum that give rise to the microtubules of cilia and flagella.

centromere (sen′-trō-mēr): the region of a replicated chromosome at which the sister chromatids are held together until they separate during cell division.

cephalization (sef-ul-ī-zā′-shun): concentration of sensory organs and nervous tissue in the anterior (head) portion of the body.

cerebellum (ser-uh-bel′-um): the part of the hindbrain of vertebrates that is concerned with coordinating movements of the body.

cerebral cortex (ser-ē′-brul kor′-tex): a thin layer of neurons on the surface of the vertebrate cerebrum, in which most neural processing and coordination of activity occurs.

cerebral hemisphere: one of two nearly symmetrical halves of the cerebrum, connected by a broad band of axons, the corpus callosum.

cerebrum (ser-ē′-brum): the part of the forebrain of vertebrates that is concerned with sensory processing, the direction of motor output, and the coordination of most of the body's activities; consists of two nearly symmetrical halves (the hemispheres) connected by a broad band of axons, the corpus callosum.

cervix (ser′-viks): a ring of connective tissue at the outer end of the uterus that leads into the vagina.

channel protein: a membrane protein that forms a channel or pore completely through the membrane and that is usually permeable to one or to a few water-soluble molecules, especially ions.

chaparral: a biome located in coastal regions, with very low annual rainfall; is characterized by shrubs and small trees.

checkpoint: a mechanism in the eukaryotic cell cycle by which protein complexes in the cell determine whether the cell has successfully completed a specific process that is essential to successful cell division, such as the accurate replication of chromosomes.

chemical bond: an attraction between two atoms or molecules that tends to hold them together. Types of bonds include covalent, ionic, and hydrogen.

chemical digestion: the process by which particles of food within the digestive tract are exposed to enzymes and other digestive fluids that break down large molecules into smaller subunits.

chemical energy: a form of potential energy that is stored in molecules and may be released during chemical reactions.

chemical reaction: the process that forms and breaks chemical bonds that hold atoms together in molecules.

chemiosmosis (ke-mē-oz-mō′-sis): a process of ATP generation in chloroplasts and mitochondria. The movement of electrons down an electron transport system is used to pump hydrogen ions across a membrane, thereby building up a concentration gradient of hydrogen ions; the hydrogen ions diffuse back across the membrane through the pores of ATP-synthesizing enzymes; the energy of their movement down their concentration gradient drives ATP synthesis.

chemosynthesis (kē-mō-sin-the-sis): the process of oxidizing inorganic molecules, such as hydrogen sulfide, to obtain energy. Producers in hydrothermal vent communities, where light is absent, use chemosynthesis instead of photosynthesis.

chemosynthetic (kēm′-ō-sin-the-tik): capable of oxidizing inorganic molecules to obtain energy.

chiasma (kī-as′-muh; pl., chiasmata): a point at which a chromatid of one chromosome crosses with a chromatid of the homologous chromosome during prophase I of meiosis; the site of exchange of chromosomal material between chromosomes.

chitin (kī′-tin): a compound found in the cell walls of fungi and the exoskeletons of insects and some other arthropods; composed of chains of nitrogen-containing, modified glucose molecules.

chlamydia (kla-mid′-ē-uh): a sexually transmitted disease, caused by a bacterium, that causes inflammation of the urethra in males and of the urethra and cervix in females.

chlorophyll (klor′-ō-fil): a pigment found in chloroplasts that captures light energy during photosynthesis; chlorophyll absorbs violet, blue, and red light but reflects green light.

chlorophyll *a* (klor′-ō-fil): the most abundant type of chlorophyll molecule in photosynthetic eukaryotic organisms and in cyanobacteria; chlorophyll *a* is found in the reaction centers of the photosystems.

chloroplast (klor′-ō-plast): the organelle in plants and plantlike protists that is the site of photosynthesis; surrounded by a double membrane and containing an extensive internal membrane system that bears chlorophyll.

cholecystokinin (kō′-lē-sis-tō-ki′-nin): a digestive hormone produced by the small intestine that stimulates the release of pancreatic enzymes.

chondrocyte (kon′-drō-sīt): a living cell of cartilage. Together with their extracellular secretions of collagen, chondrocytes form cartilage.

chorion (kor′-ē-on): the outermost embryonic membrane in reptiles (including birds) and mammals. In reptiles, the chorion functions mostly in gas exchange; in mammals, it forms most of the embryonic part of the placenta.

chorionic gonadotropin (kor–ē–on′–ik gō–na–dō–trō′–pin; CG): a hormone secreted by the chorion (one of the fetal membranes) that maintains the integrity of the corpus luteum during early pregnancy.

chorionic villus (kor-ē-on-ik; pl., **chorionic villi):** in mammalian embryos, a finger-like projection of the chorion that penetrates the uterine lining and forms the embryonic portion of the placenta.

chorionic villus sampling (kōr–ē–on′–ik; CVS): a procedure for sampling cells from the chorionic villi produced by a fetus: A tube is inserted into the uterus of a pregnant woman, and a small sample of villi is suctioned off for genetic and biochemical analyses.

choroid (kor′-oid): a darkly pigmented layer of tissue behind the retina; contains blood vessels and also pigment that absorbs stray light.

chromatid (krō′-ma-tid): one of the two identical strands of DNA and protein that forms a duplicated chromosome. The two sister chromatids of a duplicated chromosome are joined at the centromere.

chromatin (krō′-ma-tin): the complex of DNA and proteins that makes up eukaryotic chromosomes.

chromosome (krō′-mō-sōm): a DNA double helix together with proteins that help to organize and regulate the use of the DNA.

chylomicron: a particle produced by cells of the small intestine, consisting of proteins, triglycerides, and cholesterol; transports the products of lipid digestion into the lymphatic system and ultimately into the circulatory system.

chyme (kīm): an acidic, souplike mixture of partially digested food, water, and digestive secretions that is released from the stomach into the small intestine.

chytrid: a member of the fungus phylum Chytridomycota, which includes species with flagellated swimming spores.

ciliate (sil′-ē-et): a member of a protist group characterized by cilia and a complex unicellular structure. Ciliates are part of a larger group known as the alveolates.

cilium (sil′-ē-um; pl., **cilia):** a short, hair-like, motile projection from the surface of certain eukaryotic cells that contains microtubules in a 9 + 2 arrangement. The movement of cilia may propel cells through a fluid medium or move fluids over a stationary surface layer of cells.

citric acid cycle: a cyclic series of reactions, occurring in the matrix of mitochondria, in which the acetyl groups from the pyruvic acids produced by glycolysis are broken down to CO_2, accompanied by the formation of ATP and electron carriers; also called the Krebs cycle.

clade: a group that includes all the organisms descended from a common ancestor, but no other organisms; a monophyletic group.

class: in Linnaean classification, the taxonomic rank composed of related orders. Closely related classes form a phylum.

cleavage: the early cell divisions of embryos, in which little or no growth occurs between divisions; reduces the cell size and distributes gene-regulating substances to the newly formed cell.

climate: patterns of weather that prevail for long periods of time (from years to centuries) in a given region.

climax community: a diverse and relatively stable community that forms the endpoint of succession.

clitoris: an external structure of the female reproductive system that is composed of erectile tissue; a sensitive point of stimulation during sexual response.

clonal selection: the mechanism by which the adaptive immune response gains specificity; an invading antigen elicits a response from only a few lymphocytes, which proliferate to form a clone of cells that attack only the specific antigen that stimulated their production.

clone: offspring that are produced by mitosis and are, therefore, genetically identical to each other.

cloning: the process of producing many identical copies of a gene; also the production of many genetically identical copies of an organism.

closed circulatory system: a type of circulatory system, found in certain worms and vertebrates, in which the blood is always confined within the heart and vessels.

closed system: a hypothetical space where neither energy nor matter can enter or leave.

club fungus: a fungus of the phylum Basidiomycota, whose members (which include mushrooms, puffballs, and shelf fungi) reproduce by means of basidiospores.

clumped distribution: the distribution characteristic of populations in which individuals are clustered into groups; the groups may be social or based on the need for a localized resource.

cochlea (kahk′-lē-uh): a coiled, bony, fluid-filled tube found in the mammalian inner ear; contains mechanoreceptors (hair cells) that respond to the vibration of sound.

codominance: the relation between two alleles of a gene, such that both alleles are phenotypically expressed in heterozygous individuals.

codon: a sequence of three bases of messenger RNA that specifies a particular amino acid to be incorporated into a protein; certain codons also signal the beginning or end of protein synthesis.

coelom (sē′-lōm): in animals, a space or cavity, lined with tissue derived from mesoderm, that separates the body wall from the inner organs.

coenzyme: an organic molecule that is bound to certain enzymes and is required for the enzymes' proper functioning; typically, a nucleotide bound to a water-soluble vitamin.

coevolution: the evolution of adaptations in two species due to their extensive interactions with one another, such that each species acts as a major force of natural selection on the other.

cohesion: the tendency of the molecules of a substance to stick together.

cohesion–tension theory: a model for the transport of water in xylem; water is pulled up the xylem tubes, powered by the force of evaporation of water from the leaves (producing tension) and held together by hydrogen bonds between nearby water molecules (cohesion).

coleoptile (kō-lē-op′-tīl): a sheath surrounding the shoot in monocot seedlings that protects the shoot from abrasion by soil particles during germination.

collecting duct: a tube within the kidney that collects urine from many nephrons and conducts it through the renal medulla into the renal pelvis. Urine may become concentrated in the collecting ducts if antidiuretic hormone (ADH) is present.

collenchyma (kōl-en′-ki-muh): an elongated plant cell type, with thickened, flexible cell walls, that is alive at maturity and that supports the plant body.

colon: the longest part of the large intestine; does not include the rectum.

colostrum (kō-los′-trum): a yellowish fluid, high in protein and containing antibodies, that is produced by the mammary glands before milk secretion begins.

communication: the act of producing a signal that causes a receiver, normally another animal of the same species, to change its behavior in a way that is, on average, beneficial to both signaler and receiver

community: all the interacting populations within an ecosystem.

compact bone: the hard and strong outer bone; composed of osteons. Compare with *spongy bone*.

companion cell: a cell adjacent to a sieve-tube element in phloem; involved in the control and nutrition of the sieve-tube element.

competition: interaction among individuals who attempt to utilize a resource (for example, food or space) that is limited relative to the demand for that resource.

competitive exclusion principle: the concept that no two species can simultaneously and continuously occupy the same ecological niche.

competitive inhibition: the process by which two or more molecules that are somewhat similar in structure compete for the active site of an enzyme.

complement: a group of blood-borne proteins that participate in the destruction of foreign cells, especially those to which antibodies have bound.

complementary base pair: in nucleic acids, bases that pair by hydrogen bonding. In DNA, adenine is complementary to thymine and guanine is complementary to cytosine; in RNA, adenine is complementary to uracil, and guanine to cytosine.

complete flower: a flower that has all four floral parts (sepals, petals, stamens, and carpels).

compound: a substance whose molecules are formed by different types of atoms; can be broken into its constituent elements by chemical means.

compound eye: a type of eye, found in many arthropods, that is composed of numerous independent subunits called *ommatidia*. Each ommatidium contributes a piece of a mosaic-like image perceived by the animal.

concentration: the number of particles of a dissolved substance in a given unit of volume.

concentration gradient: the difference in concentration of a substance between two parts of a fluid or across a barrier such as a membrane.

conclusion: in the scientific method, a decision about the validity of a hypothesis, made on the basis of experiments or observations.

conducting portion: the portion of the respiratory system in lung-breathing vertebrates that carries air to the alveoli.

cone: a cone-shaped photoreceptor cell in the vertebrate retina; not as sensitive to light as are the rods. The three types of cones are most sensitive to different colors of light and provide color vision; see also *rod*.

conifer (kon′-eh-fer): a member of a group of nonflowering vascular plants whose members reproduce by means of seeds formed inside cones; retains its leaves throughout the year.

conjugation: in prokaryotes, the transfer of DNA from one cell to another via a temporary connection; in single-celled eukaryotes, the mutual exchange of genetic material between two temporarily joined cells.

connective tissue: a tissue type consisting of diverse tissues, including bone, cartilage, fat, and blood, that generally contain large amounts of extracellular material.

conservation biology: the application of knowledge from ecology and other areas of biology to understand and conserve biodiversity.

constant region: the part of an antibody molecule that is similar in all antibodies of a given class.

constant-loss population: A population characterized by a relatively constant death rate; constant-loss populations have a roughly linear survivorship curve.

consumer: an organism that eats other organisms; a heterotroph.

contest competition: a mechanism for resolving intraspecific competition by using social or chemical interactions to limit access of some individuals to important limited resources, such as food, mates, or territories.

contraception: the prevention of pregnancy.

contractile vacuole: a fluid-filled vacuole in certain protists that takes up water from the cytoplasm, contracts, and expels the water outside the cell through a pore in the plasma membrane.

control: that portion of an experiment in which all possible variables are held constant; in contrast to the "experimental" portion, in which a particular variable is altered.

convergent evolution: the independent evolution of similar structures among unrelated organisms as a result of similar environmental pressures; see *analogous structures*.

convolution: a folding of the cerebral cortex of the vertebrate brain.

copulation: reproductive behavior in which the penis of the male is inserted into the body of the female, where it releases sperm.

coral reef: an ecosystem created by animals (reef-building corals) and plants in warm tropical waters.

core reserve: a natural area protected from most human uses that encompasses enough space to preserve all of the biodiversity of the ecosystems in that area.

cork cambium: a lateral meristem in woody roots and stems that gives rise to cork cells.

cork cell: a protective cell of the bark of woody stems and roots; at maturity, cork cells are dead, with thick, waterproof cell walls.

cornea (kor′-nē-uh): the clear outer covering of the eye, located in front of the pupil and iris; begins the focusing of light on the retina.

corona radiata (kuh-rō′-nuh rā-dē-a′-tuh): the layer of cells surrounding an egg after ovulation.

corpus callosum (kor′pus kal-ō′-sum): the band of axons that connects the two cerebral hemispheres of vertebrates.

corpus luteum (kor′-pus loo′-tē-um): in the mammalian ovary, a structure that is derived from the follicle after ovulation and that secretes the hormones estrogen and progesterone.

cortex: the part of a primary root or stem located between the epidermis and the vascular cylinder.

cortisol (kor′-ti-sol): a steroid hormone released into the bloodstream by the adrenal cortex in response to stress. Cortisol helps the body cope with short-term stressors by raising blood glucose levels; it also inhibits the immune response.

cotyledon (kot-ul-ē′don): a leaflike structure within a seed that absorbs food molecules from the endosperm and transfers them to the growing embryo; also called *seed leaf*.

countercurrent exchange: a mechanism for the transfer of some property, such as heat or a dissolved substance, from one fluid to another, generally without the two fluids actually mixing; in countercurrent exchange, the two fluids flow past one another in opposite directions, and they transfer heat or solute from the fluid with the higher temperature or higher solute concentration to the fluid with the lower temperature or lower solute concentration.

coupled reaction: a pair of reactions, one exergonic and one endergonic, that are linked together such that the energy produced by the exergonic reaction provides the energy needed to drive the endergonic reaction.

covalent bond (kō-vā′-lent): a chemical bond between atoms in which electrons are shared.

crab lice: an arthropod parasite that can infest humans; can be transmitted by sexual contact.

craniate: an animal that has a skull.

critically endangered species: a species that faces an extreme risk of extinction in the wild in the immediate future.

cross-fertilization: the union of sperm and egg from two individuals of the same species.

crossing over: the exchange of corresponding segments of the chromatids of two homologous chromosomes during meiosis I; occurs at chiasmata.

cuticle (kū′-ti-kul): a waxy or fatty coating on the surfaces of aboveground epidermal cells of many land plants; aids in the retention of water.

cyclic adenosine monophosphate (cyclic AMP): a cyclic nucleotide, formed within many target cells as a result of the reception of amino acid derivatives or peptide hormones, that causes metabolic changes in the cell.

cytokine (sī′-tō-kīn): any of several chemical messenger molecules released by cells that facilitate communication with other cells and transfer signals within and between the various systems of the body. Cytokines are important in cellular differentiation and the adaptive immune response.

cytokinesis (sī-tō-ki-nē′-sis): the division of the cytoplasm and organelles into two daughter cells during cell division; normally occurs during telophase of mitosis.

cytokinin (sī-tō-ki′-nin): a group of plant hormones that promotes cell division, fruit development, and the sprouting of lateral buds; also delays the senescence of plant parts, especially leaves.

cytoplasm (sī′-tō-plaz-um): all of the material contained within the plasma membrane of a cell, exclusive of the nucleus.

cytoplasmic fluid: the fluid portion of the cytoplasm; the substance within the plasma membrane exclusive of the nucleus and organelles; also called cytosol.

cytosine (C): a nitrogenous base found in both DNA and RNA; abbreviated as C.

cytoskeleton: a network of protein fibers in the cytoplasm that gives shape to a cell, holds and moves organelles, and is typically involved in cell movement.

cytotoxic T cell: a type of T cell that, upon contacting foreign cells, directly destroys them.

daughter cell: one of the two cells formed by cell division.

day-neutral plant: a plant in which flowering occurs as soon as the plant has undergone sufficient growth and development, regardless of day length.

death rate: the number of deaths per individual in a specified unit of time, such as a year.

decomposer: an organism, usually a fungus or bacterium, that digests organic material by secreting digestive enzymes into the environment, in the process liberating nutrients into the environment.

deductive reasoning: the process of generating hypotheses about the results of a specific experiment or the nature of a specific observation.

deforestation: the excessive cutting of forests. In recent years, deforestation has occurred primarily in rain forests in the Tropics, to clear space for agriculture.

dehydration synthesis: a chemical reaction in which two molecules are joined by a covalent bond with the simultaneous removal of a hydrogen from one molecule and a hydroxyl group from the other, forming water; the reverse of hydrolysis.

deletion mutation: a mutation in which one or more pairs of nucleotides are removed from a gene.

demographic transition: a change in population dynamic in which a fairly stable population with both high birth rates and high death rates experiences rapid growth as death rates decline, and then returns to a stable (although much larger) population as birth rates decline.

demography: the study of the changes in human numbers over time, grouped by world regions, age, sex, educational levels, and other variables.

denature: to disrupt the secondary and/or tertiary structure of a protein while leaving its amino acid sequence intact. Denatured proteins can no longer perform their biological functions.

denatured: having the secondary and/or tertiary structure of a protein disrupted, while leaving the amino acid sequence unchanged. Denatured proteins can no longer perform their biological functions.

dendrite (den′-drīt): a branched tendril that extends outward from the cell body of a neuron; specialized to respond to signals from the external environment or from other neurons.

dendritic cell (den-drit′-ick): a type of phagocytic leukocyte that presents antigen to T and B cells, thereby stimulating an adaptive immune response to an invading microbe.

denitrifying bacteria (dē-nī′-treh-fī-ing): bacteria that break down nitrates, releasing nitrogen gas to the atmosphere.

density-dependent: referring to any factor, such as predation, that limits population size to an increasing extent as the population density increases.

density-independent: referring to any factor, such floods or fires, that limits a population's size regardless of its density.

deoxyribonucleic acid (dē-ox-ē-rī-bō-noo-klā′-ik; DNA): a molecule composed of deoxyribose nucleotides; contains the genetic information of all living cells.

dermal tissue system: a plant tissue system that makes up the outer covering of the plant body.

dermis (dur′-mis): the layer of skin beneath the epidermis; composed of connective tissue and containing blood vessels, muscles, nerve endings, and glands.

desert: a biome in which less than 10 inches (25 centimeters) of rain fall each year; characterized by cacti, succulents, and widely spaced, drought-resistant bushes.

desertification: the process by which relatively dry, drought-prone regions are converted to desert as a result of drought and overuse of the land, for example, by overgrazing or cutting of trees.

desmosome (dez′-mō-sōm): a strong cell-to-cell junction that attaches adjacent cells to one another.

detritus feeder (de-trī′-tus): one of a diverse group of organisms, ranging from worms to vultures, that live off the wastes and dead remains of other organisms.

deuterostome (doo′-ter-ō-stōm): an animal with a mode of embryonic development in which the coelom is derived from outpocketings of the gut; characteristic of echinoderms and chordates.

developing country: a country (including most in Central and South America, Asia, and Africa) that has a relatively low average living standard and limited access to modern technology and medical care.

development: the process by which an organism proceeds from fertilized egg through adulthood to eventual death.

diabetes mellitus (di-uh-bē′-tēs mel-ī′-tus): a disease characterized by defects in the production, release, or reception of insulin; characterized by high blood glucose levels that fluctuate with sugar intake.

diaphragm (dī′-uh-fram): in the respiratory system, a dome-shaped muscle forming the floor of the chest cavity; when the diaphragm contracts, it flattens, enlarging the chest cavity and causing air to be drawn into the lungs.

diastolic pressure (dī′-uh-stal-ik): the blood pressure measured during relaxation of the ventricles; the lower of the two blood pressure readings.

diatom (dī′-uh-tom): a member of a protist group that includes photosynthetic forms with two-part glassy outer coverings; important photosynthetic organisms in fresh water and salt water. Diatoms are part of a larger group known as the stramenopiles.

dicot (dī′-kaht): short for dicotyledon; a type of flowering plant characterized by embryos with two cotyledons, or seed leaves, that are usually modified for food storage.

differentiate: the process whereby a cell becomes specialized in structure and function.

differentiated cell: a mature cell specialized for a specific function; in plants, differentiated cells normally do not divide.

diffusion: the net movement of particles from a region of high concentration of that particle to a region of low concentration, driven by the concentration gradient; may occur entirely within a fluid or across a barrier such as a membrane.

digestion: the process by which food is physically and chemically broken down into molecules that can be absorbed by cells.

digestive system: a group of organs responsible for ingesting food, digesting food into simple molecules that can be absorbed into the circulatory system, and expelling undigested wastes from the body.

dinoflagellate (dī-nō-fla′-jel-et): a member of a protist group that includes photosynthetic forms in which two flagella project through armor-like plates; abundant in oceans; can reproduce rapidly, causing "red tides." Dinoflagellates are part of a larger group known as the alveolates.

diploid (dip′-loid): referring to a cell with pairs of homologous chromosomes.

diplomonad: a member of a protist group characterized by two nuclei and multiple flagella. Diplomonads, which include disease-causing parasites such as *Giardia*, are part of a larger group known as the excavates.

direct development: a developmental pathway in which the offspring is born as a miniature version of the adult and does not radically change in body form as it grows and matures.

directional selection: a type of natural selection that favors one extreme of a range of phenotypes.

disaccharide (dī-sak′-uh-rīd): a carbohydrate formed by the covalent bonding of two monosaccharides.

disruptive selection: a type of natural selection that favors both extremes of a range of phenotypes.

distal tubule: the last section of a mammalian nephron, following the loop of Henle and emptying urine into a collecting duct; most tubular secretion and a small amount of tubular reabsorption occur in the distal tubule.

disturbance: any event that disrupts an ecosystem by altering its community, its abiotic structure, or both; disturbance precedes succession.

disulfide bond: the covalent bond formed between the sulfur atoms of two cysteines in a protein; typically causes the protein to fold by bringing otherwise distant parts of the protein close together.

DNA helicase: an enzyme that helps unwind the DNA double helix during DNA replication.

DNA ligase: an enzyme that bonds the terminal sugar in one DNA strand to the terminal phosphate in a second DNA strand, creating a single strand with a continuous sugar-phosphate backbone.

DNA polymerase: an enzyme that bonds DNA nucleotides together into a continuous strand, using a preexisting DNA strand as a template.

DNA probe: a sequence of nucleotides that is complementary to the nucleotide sequence of a gene or other segment of DNA under study; used to locate the gene or DNA segment during gel electrophoresis or other methods of DNA analysis.

DNA profile: the pattern of short tandem repeats of specific DNA segments; using a standardized set of 13 short tandem repeats, DNA profiles identify individual people with great accuracy.

DNA replication: the copying of the double-stranded DNA molecule, producing two identical DNA double helices.

DNA sequencing: the process of determining the order of nucleotides in a DNA molecule.

domain: the broadest category for classifying organisms; organisms are classified into three domains: Bacteria, Archaea, and Eukarya.

dominance hierarchy: a social structure that arises when the animals in a social group establish individual ranks that determine access to resources; ranks are usually established through aggressive interactions.

dominant: an allele that can determine the phenotype of heterozygotes completely, such that they are indistinguishable from individuals homozygous for the allele; in the heterozygotes, the expression of the other (recessive) allele is completely masked.

dormancy: a state in which an organism does not grow or develop; usually marked by lowered metabolic activity and resistance to adverse environmental conditions.

dorsal root ganglion: a ganglion, located on the dorsal (sensory) branch of each spinal nerve, that contains the cell bodies of sensory neurons.

double fertilization: in flowering plants, the fusion of two sperm nuclei with the nuclei of two cells of the female gametophyte. One sperm nucleus fuses with the egg to form the zygote; the second sperm nucleus fuses with the two haploid nuclei of the central cell to form a triploid endosperm cell.

double helix (hē′-liks): the shape of the two-stranded DNA molecule; similar to a ladder twisted lengthwise into a corkscrew shape.

Down syndrome: a genetic disorder caused by the presence of three copies of chromosome 21; common characteristics include mental retardation, distinctively shaped eyelids, a small mouth with protruding tongue, heart defects, and low resistance to infectious diseases; also called *trisomy 21*.

duodenum: the first section of the small intestine, in which most food digestion occurs; receives chyme from the stomach, buffers and digestive enzymes from the pancreas, and bile from the liver and gallbladder.

duplicated chromosome: a eukaryotic chromosome following DNA replication; consists of two sister chromatids joined at the centromeres.

early-loss population: a population characterized by a high birth rate, a high death rate among juveniles, and lower death rates among adults; early-loss populations have a concave survivorship curve.

ecdysone (ek-dī′-sōn): a steroid hormone that triggers molting in insects and other arthropods.

ecological footprint: the area of productive land needed to produce the resources used and absorb the wastes (including carbon dioxide) generated by an individual person, or by an average person of a specific part of the world (for example, an individual country), or of the entire world, using current technologies.

ecological niche (nitch): the role of a particular species within an ecosystem, including all aspects of its interaction with the living and nonliving environments.

ecology (ē-kol′-uh-jē): the study of the interrelationships of organisms with each other and with their nonliving environment.

ecosystem (ē′-kō-sis-tem): all the organisms and their nonliving environment within a defined area.

ecosystem services: the processes through which natural ecosystems and their living communities sustain and fulfill human life. Ecosystem services include purifying air and water, replenishing oxygen, pollinating plants, reducing flooding, providing wildlife habitat, and many more.

ectoderm (ek′-tō-derm): the outermost embryonic tissue layer, which gives rise to structures such as hair, the epidermis of the skin, and the nervous system.

ectotherm: an animal that obtains most of its body warmth from its environment; body temperatures of ectotherms vary with the temperature of their surroundings.

effector (ē-fek′-tor): a part of the body (normally a muscle or gland) that carries out responses as directed by the nervous system.

egg: the haploid female gamete, usually large and nonmotile; contains food reserves for the developing embryo.

electromagnetic spectrum: the range of all possible wavelengths of electromagnetic radiation, from wavelengths longer than radio waves, to microwaves, infrared, visible light, ultraviolet, x-rays, and gamma rays.

electron carrier: a molecule that can reversibly gain or lose electrons. Electron carriers generally accept high-energy electrons produced during an exergonic reaction and donate the electrons to acceptor molecules that use the energy to drive endergonic reactions.

electron shell: a region in an atom within which electrons orbit; each shell corresponds to a fixed energy level at a given distance from the nucleus.

electron transport chain (ETC): a series of electron carrier molecules, found in the thylakoid membranes of chloroplasts and the inner membrane of mitochondria, that extract energy from electrons and generate ATP or other energetic molecules.

electron: a subatomic particle, found in an electron shell outside the nucleus of an atom, that bears a unit of negative charge and very little mass.

element: a substance that cannot be broken down, or converted, to a simpler substance by ordinary chemical means.

elimination: the expulsion of indigestible materials from the digestive tract, through the anus, and outside the body.

embryo (em′-brē-ō): in animals, the stages of development that begin with the fertilization of the egg cell and end with hatching or birth; in mammals, the early stages in which the developing animal does not yet resemble the adult of the species.

embryonic disk: in human embryonic development, the flat, two-layered group of cells that separates the amniotic cavity from the yolk sac; the cells of the embryonic disk produce most of the developing embryo.

embryonic stem cell (ESC): a cell derived from an early embryo that is capable of differentiating into any of the adult cell types.

emerging infectious disease: a previously unknown infectious disease (one caused by a microbe), or a previously known infectious disease whose frequency or severity has significantly increased in the past two decades.

emigration (em-uh-grā′-shun): migration of individuals out of an area.

endangered species: a species that faces a high risk of extinction in the wild in the near future.

endergonic (en-der-gon′-ik): pertaining to a chemical reaction that requires an input of energy to proceed; an "uphill" reaction.

endocrine disrupter: an environmental pollutant that interferes with endocrine function, often by disrupting the action of sex hormones.

endocrine gland: a ductless, hormone-producing gland consisting of cells that release their secretions into the extracellular fluid from which the secretions diffuse into nearby capillaries; most endocrine glands are composed of epithelial cells.

endocrine hormone: a molecule produced by cells of endocrine glands and released into the circulatory system. An endocrine hormone causes changes in target cells that bear specific receptors for the hormone.

endocrine system: an animal's organ system for cell-to-cell communication; composed of hormones and the cells that secrete them.

endocytosis (en-dō-sī-tō′-sis): the process in which the plasma membrane engulfs extracellular material, forming membrane-bound sacs that enter the cytoplasm and thereby move material into the cell.

endoderm (en′-dō-derm): the innermost embryonic tissue layer, which gives rise to structures such as the lining of the digestive and respiratory tracts.

endodermis (en-dō-der′-mis): the innermost layer of small, close-fitting cells of the cortex of a root that form a ring around the vascular cylinder; see also *Casparian strip*.

endometrium (en-dō-mē′-trē-um): the nutritive inner lining of the uterus.

endoplasmic reticulum (en-dō-plaz′-mik re-tik′-ū-lum; ER): a system of membranous tubes and channels in eukaryotic cells; the site of most protein and lipid synthesis.

endoskeleton (en′-dō-skel′-uh-tun): a rigid internal skeleton with flexible joints that allow for movement.

endosperm: a triploid food storage tissue in the seeds of flowering plants that nourishes the developing plant embryo.

endospore: a protective resting structure of some rod-shaped bacteria that withstands unfavorable environmental conditions.

endosymbiont hypothesis: the hypothesis that certain organelles, especially chloroplasts and mitochondria, arose as mutually beneficial associations between the ancestors of eukaryotic cells and captured bacteria that lived within the cytoplasm of the pre-eukaryotic cell.

endotherm: an animal that obtains most of its body heat from metabolic activities; body temperatures of endotherms usually remain relatively constant within a fairly wide range of environmental temperatures.

energy pyramid: a graphical representation of the energy contained in succeeding trophic levels, with maximum energy at the base (primary producers) and steadily diminishing amounts at higher levels.

energy-requiring transport: the transfer of substances across a cell membrane using cellular energy; includes active transport, endocytosis, and exocytosis.

energy: the capacity to do work.

entropy (en′-trō-pē): a measure of the amount of randomness and disorder in a system.

environmental resistance: any factor that tends to counteract biotic potential, limiting population growth and the resulting population size.

enzyme (en′zīm): a biological catalyst, usually a protein, that speeds up the rate of specific biological reactions.

epicotyl (ep′-ē-kot-ul): the part of the embryonic shoot located above the attachment point of the cotyledons but below the tip of the shoot.

epidermal tissue: dermal tissue in plants that forms the epidermis, the outermost cell layer that covers leaves, young stems, and young roots.

epidermis (ep-uh-der′-mis): in animals, specialized stratified epithelial tissue that forms the outer layer of the skin; in plants, the outermost layer of cells of a leaf, young root, or young stem.

epididymis (e-pi-di′-di-mus): a series of tubes that connect with and receive sperm from the seminiferous tubules of the testis, and empty into the vas deferens.

epiglottis (ep-eh-glah′-tis): a flap of cartilage in the lower pharynx that covers the opening to the larynx during swallowing; directs food into the esophagus.

epinephrine (ep-i-nef′-rin): a hormone, secreted by the adrenal medulla, that is released in response to stress and that stimulates a variety of responses, including the release of glucose from the liver and an increase in heart rate.

epithelial tissue (eh-puh-thē′-lē-ul): a tissue type that forms membranes that cover the body surface and line body cavities, and that also gives rise to glands.

equilibrium population: a population in which allele frequencies and the distribution of genotypes do not change from generation to generation.

erythrocyte (eh-rith′-rō-sīt): a red blood cell, which contains the oxygen-binding protein hemoglobin and thus transports oxygen in the circulatory system.

erythropoietin (eh-rith′-rō-pō-ē′-tin): a hormone produced by the kidneys in response to oxygen deficiency; stimulates the production of red blood cells by the bone marrow.

esophagus: a muscular, tubular portion of the mammalian digestive tract located between the pharynx and the stomach; no digestion occurs in the esophagus.

essential amino acid: an amino acid that is a required nutrient; the body is unable to manufacture essential amino acids, so they must be supplied in the diet.

essential fatty acid: a fatty acid that is a required nutrient; the body is unable to manufacture essential fatty acids, so they must be supplied in the diet.

essential nutrient: any nutrient that cannot be synthesized by the body, including certain fatty acids and amino acids, vitamins, minerals, and water.

estrogen: a female sex hormone, produced by follicle cells of the ovary, that stimulates follicle development, oogenesis, the development of secondary sex characteristics, and growth of the uterine lining.

estuary (es′–choō–ār–ē): a wetland formed where a river meets the ocean; the salinity is quite variable, but lower than in seawater and higher than in fresh water.

ethology (ē-thol′-ō-jē): the study of animal behavior.

ethylene: a plant hormone that promotes the ripening of some fruits and the dropping of leaves and fruit; promotes senescence of leaves.

euglenid (ū′-gle-nid): a member of a protist group characterized by one or more whiplike flagella, which are used for locomotion, and by a photoreceptor, which detects light; are photosynthetic. Euglenids are part of a larger group known as euglenozoans.

euglenozoan: a member of the Euglenozoa, a protist clade. The euglenozoans, which are characterized by mitochondrial membranes that appear under the microscope to be shaped like a stack of disks, include the euglenids and the kinetoplastids.

Eukarya (ū-kar′-ē-a): one of life's three domains; consists of all eukaryotes (plants, animals, fungi, and protists).

eukaryote (ū-kar′-ē-ōt): an organism whose cells are eukaryotic; plants, animals, fungi, and protists are eukaryotes.

eukaryotic (ū-kar-ē-ot′-ik): referring to cells of organisms of the domain Eukarya (plants, animals, fungi, and protists). Eukaryotic cells have genetic material enclosed within a membrane-bound nucleus and contain other organelles.

eutrophic lake: a lake that receives sufficiently large inputs of sediments, organic material, and inorganic nutrients from its surroundings to support dense communities, especially of plants and phytoplankton; contains murky water with poor light penetration.

evolution: (1) the descent of modern organisms, with modification, from preexisting life-forms; (2) the theory that all organisms are related by common ancestry and have changed over time; (3) any change in the genetic makeup (the proportions of different genotypes) in a population from one generation to the next.

excavate: a member of the Excavata, a protist clade. The excavates, which generally lack mitochondria, include the diplomonads and the parabasalids.

excitatory postsynaptic potential (EPSP): an electrical signal produced in a postsynaptic cell that makes the resting potential of the postsynaptic neuron less negative and, hence, makes the neuron more likely to produce an action potential.

excretion: the elimination of waste substances from the body; can occur from the digestive system, skin glands, urinary system, or lungs.

exergonic (ex-er-gon′-ik): pertaining to a chemical reaction that releases energy (either as heat or in the form of increased entropy); a "downhill" reaction.

exhalation: the act of releasing air from the lungs, which results from a relaxation of the respiratory muscles.

exocrine gland: a gland that releases its secretions into ducts that lead to the outside of the body or into a body cavity, such as the digestive or reproductive system; most exocrine glands are composed of epithelial cells.

exocytosis (ex-ō-sī-tō′-sis): the process in which intracellular material is enclosed within a membrane-bound sac that moves to the plasma membrane and fuses with it, releasing the material outside the cell.

exon: a segment of DNA in a eukaryotic gene that codes for amino acids in a protein (see also intron).

exoskeleton (ex′-ō-skel′-uh-tun): a rigid external skeleton that supports the body, protects the internal organs, and has flexible joints that allow for movement.

experiment: in the scientific method, the use of carefully controlled observations or manipulations to test the predictions generated by a hypothesis.

exponential growth: a continuously accelerating increase in population size; this type of growth generates a curve shaped like the letter "J."

extensor: a muscle that straightens (increases the angle of) a joint.

external fertilization: the union of sperm and egg outside the body of either parent.

extinction: the death of all members of a species.

extracellular digestion: the physical and chemical breakdown of food that occurs outside a cell, normally in a digestive cavity.

extracellular fluid: fluid that bathes the cells of the body; in mammals, extracellular fluid leaks from capillaries and is similar in composition to blood plasma, but lacking the large proteins found in plasma.

extraembryonic membrane: in the embryonic development of reptiles (including birds) and mammals, one of the chorion (functions in gas exchange), amnion (provision of the watery environment needed for development), allantois (waste storage), or yolk sac (storage of the yolk).

facilitated diffusion: the diffusion of molecules across a membrane, assisted by protein pores or carriers embedded in the membrane.

family: in Linnaean classification, the taxonomic rank composed of related genera. Closely related families make up an order.

farsighted: the inability to focus on nearby objects, caused by the eyeball being slightly too short or the cornea too flat.

fat (molecular): a lipid composed of three saturated fatty acids covalently bonded to glycerol; fats are solid at room temperature.

fatty acid: an organic molecule composed of a long chain of carbon atoms, with a carboxylic acid (COOH) group at one end; may be saturated (all single bonds between the carbon atoms) or unsaturated (one or more double bonds between the carbon atoms).

feces: semisolid waste material that remains in the intestine after absorption is complete and is voided through the anus. Feces consist principally of indigestible wastes and bacteria.

feedback inhibition: in enzyme-mediated chemical reactions, the condition in which the product of a reaction inhibits one or more of the enzymes involved in synthesizing the product.

fermentation: anaerobic reactions that convert the pyruvic acid produced by glycolysis into lactic acid or alcohol and CO_2, using hydrogen ions and electrons from NADH; the primary function of fermentation is to regenerate NAD+ so that glycolysis can continue under anaerobic conditions.

fertilization: the fusion of male and female haploid gametes, forming a zygote.

fetal alcohol syndrome (FAS): a cluster of symptoms, including mental retardation and physical abnormalities, that occur in infants born to mothers who consumed large amounts of alcoholic beverages during pregnancy.

fetus: the later stages of mammalian embryonic development (after the second month for humans), when the developing animal has come to resemble the adult of the species.

fever: an elevation in body temperature caused by chemicals (pyrogens) that are released by white blood cells in response to infection.

fibrillation (fi-bri-lā′-shun): the rapid, uncoordinated, and ineffective contraction of heart muscle cells.

fibrin (fī′-brin): a clotting protein formed in the blood in response to a wound; binds with other fibrin molecules and provides a matrix around which a blood clot forms.

fibrinogen (fī-brin′-ō-jen): the inactive form of the clotting protein fibrin. Fibrinogen is converted into fibrin by the enzyme thrombin, which is produced in response to injury.

fibrous root system: a root system, commonly found in monocots, that is characterized by many roots of approximately the same size arising from the base of the stem.

filament: in flowers, the stalk of a stamen, which bears an anther at its tip.

filtrate: the fluid produced by filtration; in the kidneys, the fluid produced by the filtration of blood through the glomerular capillaries.

filtration: within Bowman's capsule in each nephron of a kidney, the process by which blood is pumped under pressure through permeable capillaries of the glomerulus, forcing out water and small solutes, including wastes and nutrients.

first law of thermodynamics: the principle of physics that states that within any closed system, energy can be neither created nor destroyed, but can be converted from one form to another; also called the law of conservation of energy.

fission: asexual reproduction by dividing the body into two smaller, complete organisms.

fitness: the reproductive success of an organism, relative to the average reproductive success in the population.

flagellum (fla-jel′-um; pl., **flagella):** a long, hair-like, motile extension of the plasma membrane; in eukaryotic cells, it contains microtubules arranged in a 9 + 2 pattern. The movement of flagella propel some cells through fluids.

flavin adenine dinucleotide (FAD or $FADH_2$): an electron carrier molecule produced in the mitochondrial matrix by the Krebs cycle; subsequently donates electrons to the electron transport chain.

flexor: a muscle that flexes (decreases the angle of) a joint.

florigen: one of a group of plant hormones that may stimulate or inhibit flowering in response to day length.

flower: the reproductive structure of an angiosperm plant.

flower bud: a cluster of meristem cells (a bud) that forms a flower.

fluid mosaic model: a model of cell membrane structure; according to this model, membranes are composed of a double layer of phospholipids in which various proteins are embedded. The phospholipid bilayer is a somewhat fluid matrix that allows the movement of proteins within it.

fluid: a liquid or gas.

follicle: in the ovary of female mammals, the oocyte and its surrounding accessory cells.

follicle-stimulating hormone (FSH): a hormone, produced by the anterior pituitary, that stimulates spermatogenesis in males and the development of the follicle in females.

food chain: a linear feeding relationship in a community, using a single representative from each of the trophic levels.

food vacuole: a membranous sac, within a single cell, in which food is enclosed. Digestive enzymes are released into the vacuole, where intracellular digestion occurs.

food web: a representation of the complex feeding relationships within a community, including many organisms at various trophic levels, with many of the consumers occupying more than one level simultaneously.

foraminiferan (for-am-i-nif′-er-un): a member of a protist group characterized by pseudopods and elaborate calcium carbonate shells. Foraminiferans are generally aquatic (largely marine) and are part of a larger group known as rhizarians.

forebrain: during development, the anterior portion of the brain. In mammals, the forebrain differentiates into the thalamus, the limbic system, and the cerebrum. In humans, the cerebrum contains about half of all the neurons in the brain.

fossil: the remains of a dead organism, normally preserved in rock; may be petrified bones or wood; shells; impressions of body forms, such as feathers, skin, or leaves; or markings made by organisms, such as footprints.

fossil fuel: a fuel such as coal, oil, and natural gas, derived from the remains of ancient organisms.

founder effect: the result of an event in which an isolated population is founded by a small number of individuals; may result in genetic drift if allele frequencies in the founder population are by chance different from those of the parent population.

fovea (fō′-vē-uh): in the vertebrate retina, the central region on which images are focused; contains closely packed cones.

free nucleotides: nucleotides that have not been joined together to form a DNA or RNA strand.

free radical: a molecule containing an atom with an unpaired electron, which makes it highly unstable and reactive with nearby molecules. By removing an electron from the molecule it attacks, it creates a new free radical and begins a chain reaction that can lead to the destruction of biological molecules crucial to life.

fruit: in flowering plants, the ripened ovary (plus, in some cases, other parts of the flower), which contains the seeds.

functional group: one of several groups of atoms commonly found in an organic molecule, including hydrogen, hydroxyl, amino, carboxyl, and phosphate groups, that determine the characteristics and chemical reactivity of the molecule.

gallbladder: a small sac located next to the liver that stores and concentrates the bile secreted by the liver. Bile is released from the gallbladder to the small intestine through the bile duct.

gamete (gam′-ēt): a haploid sex cell, usually a sperm or an egg, formed in sexually reproducing organisms.

gametophyte (ga-mēt′-ō-fīt): the multicellular haploid stage in the life cycle of plants.

ganglion (gang′-lē-un; pl., **ganglia):** a cluster of neurons.

ganglion cell: in the vertebrate retina, the nerve cells whose axons form the optic nerve.

gap junction: a type of cell-to-cell junction in animals in which channels connect the cytoplasm of adjacent cells.

gas-exchange portion: the portion of the respiratory system in lung-breathing vertebrates where gas is exchanged in the alveoli of the lungs.

gastric gland: one of numerous small glands in the stomach lining; contains cells that secrete mucus, hydrochloric acid, or pepsinogen (the inactive form of the protease pepsin).

gastrin: a hormone produced by the stomach that stimulates acid secretion in response to the presence of food.

gastrovascular cavity: a saclike chamber with digestive functions, found in some invertebrates such as cnidarians (sea jellies, anemones, and related animals); a single opening serves as both mouth and anus.

gastrula (gas′-troo-luh): in animal development, a three-layered embryo with ectoderm, mesoderm, and endoderm cell layers. The endoderm layer usually encloses the primitive gut.

gastrulation (gas-troo-la′-shun): the process whereby a blastula develops into a gastrula, including the formation of endoderm, ectoderm, and mesoderm.

gel electrophoresis: a technique in which molecules (such as DNA fragments) are placed in wells in a thin sheet of gelatinous material and exposed to an electric field; the molecules migrate through the gel at a rate determined by certain characteristics, most commonly size.

gene flow: the movement of alleles from one population to another owing to the migration of individual organisms.

gene pool: the total of all alleles of all genes in a population; for a single gene, the total of all the alleles of that gene that occur in a population.

gene: the unit of heredity; a segment of DNA located at a particular place on a chromosome that encodes the information for the amino acid sequence of a protein and, hence, particular traits.

generative cell: in flowering plants, one of the haploid cells of a pollen grain; undergoes mitotic cell division to form two sperm cells.

genetic code: the collection of codons of mRNA, each of which directs the incorporation of a particular amino acid into a protein during protein synthesis or causes protein synthesis to start or stop.

genetic drift: a change in the allele frequencies of a small population purely by chance.

genetic engineering: the modification of the genetic material of an organism, usually using recombinant DNA techniques.

genetic recombination: the generation of new combinations of alleles on homologous chromosomes due to the exchange of DNA during crossing over.

genital herpes: a sexually transmitted disease, caused by a virus, that can cause painful blisters on the genitals and surrounding skin.

genotype (jēn′-ō-tip): the genetic composition of an organism; the actual alleles of each gene carried by the organism.

genus (jē-nus): in Linnaean classification, the taxonomic rank composed of related species. Closely related genera make up a family.

germination: the growth and development of a seed, spore, or pollen grain.

gibberellin (jib-er-el′-in): one of a group of plant hormones that stimulates seed germination, fruit development, flowering, and cell division and elongation in stems.

gill: in aquatic animals, a branched tissue richly supplied with capillaries around which water is circulated for gas exchange.

gland: a cluster of cells that are specialized to secrete substances such as sweat, mucus, enzymes, or hormones.

glia: cells of the nervous system that provide nutrients for neurons, regulate the composition of the extracellular fluid in the brain and spinal cord, modulate communication between neurons, and insulate axons, thereby speeding up the conduction of action potentials. Also called *glial cells.*

glial cell: a cell of the nervous system that provides support and insulation for neurons, and regulates the composition of the extracellular fluid in the nervous system.

global warming: a gradual rise in global atmospheric temperature as a result of an amplification of the natural greenhouse effect due to human activities.

glomeromycete: a member of the fungus phylum Glomeromycota, which includes species that form mycorrhizal associations with plant roots and that form bush-shaped branching structures inside plant cells.

glomerulus (glō-mer′-ū-lus): a dense network of thin-walled capillaries, located within the Bowman's capsule of each nephron of the kidney, where blood pressure forces water and small solutes, including wastes and nutrients, through the capillary walls into the nephron.

glucagon (gloo′-ka-gon): a hormone, secreted by the pancreas, that increases blood sugar by stimulating the breakdown of glycogen (to glucose) in the liver.

glucocorticoid (gloo-kō-kor′-tik-oid): a class of hormones, released by the adrenal cortex in response to the presence of ACTH, that makes additional energy available to the body by stimulating the synthesis of glucose.

glucose: the most common monosaccharide, with the molecular formula $C_6H_{12}O_6$; most polysaccharides, including cellulose, starch, and glycogen, are made of glucose subunits covalently bonded together.

glycerol (glis′-er-ol): a three-carbon alcohol to which fatty acids are covalently bonded to make fats and oils.

glycogen (glī′-kō-jen): a highly branched polymer of glucose that is stored by animals in the muscles and liver and metabolized as a source of energy.

glycolysis (glī-kol′-i-sis): reactions, carried out in the cytoplasm, that break down glucose into two molecules of pyruvic acid, producing two ATP molecules; does not require oxygen but can proceed when oxygen is present.

glycoprotein: a protein to which a carbohydrate is attached.

goiter: a swelling of the thyroid gland caused by iodine deficiency, which affects the functioning of the thyroid gland and its hormones.

Golgi apparatus (gōl′-jē): a stack of membranous sacs, found in most eukaryotic cells, that is the site of processing and separation of membrane components and secretory materials.

gonad: an organ where reproductive cells are formed; in males, the testes, and in females, the ovaries.

gonadotropin-releasing hormone (gō-na-dō-trō′-pin; GnRH): a hormone produced by the neurosecretory cells of the hypothalamus, which stimulates cells in the anterior pituitary to release follicle-stimulating hormone (FSH) and luteinizing hormone (LH). GnRH is involved in the menstrual cycle and in spermatogenesis.

gonorrhea (gon-uh-rē′-uh): a sexually transmitted bacterial infection of the reproductive organs; if untreated, can result in sterility.

gradient: a difference in concentration, pressure, or electrical charge between two regions.

grassland: a biome, located in the centers of continents, that primarily supports grasses; also called a *prairie.*

gravitropism: growth with respect to the direction of gravity.

gray matter: the outer portion of the brain and inner region of the spinal cord; composed largely of neuron cell bodies, which give this area a gray color.

greenhouse effect: the process in which certain gases such as carbon dioxide and methane trap sunlight energy in a planet's atmosphere as heat; the glass in a greenhouse does the same. The result, global warming, is being enhanced by the production of these gases by humans.

greenhouse gas: a gas, such as carbon dioxide or methane, that traps sunlight energy in a planet's atmosphere as heat; a gas that participates in the greenhouse effect.

ground tissue system: a plant tissue system, consisting of parenchyma, collenchyma, and sclerenchyma cells, that makes up the bulk of a leaf or young stem, excluding vascular or dermal tissues. Most ground tissue cells function in photosynthesis, support, or carbohydrate storage.

growth hormone: a hormone, released by the anterior pituitary, that stimulates growth, especially of the skeleton.

growth rate: a measure of the change in population size per individual per unit of time.

guanine (G): a nitrogenous base found in both DNA and RNA; abbreviated as G.

guard cell: one of a pair of specialized epidermal cells that surrounds the central opening of a stoma in the epidermis of a leaf or young stem, and regulates the size of the opening.

gymnosperm (jim′-nō-sperm): a nonflowering seed plant, such as a conifer, gnetophyte, cycad, or gingko.

gyre (jīr): a roughly circular pattern of ocean currents, formed because continents interrupt the flow of the current; rotates clockwise in the Northern Hemisphere and counterclockwise in the Southern Hemisphere.

habitat fragmentation: the process by which human development and activities produce patches of wildlife habitat that may not be large enough to sustain minimum viable populations.

habituation (heh-bich-oo-ā′-shun): a type of simple learning characterized by a decline in response to a repeated stimulus.

hair cell: a type of mechanoreceptor cell found in the inner ear that produces an electrical signal when stiff hairlike cilia projecting from the surface of the cell are bent. Hair cells in the cochlea respond to sound vibrations; those in the vestibular system respond to motion and gravity.

hair follicle (fol′-i-kul): a cluster of specialized epithelial cells located in the dermis of mammalian skin, which produces a hair.

hammer: the second of the small bones of the middle ear, linking the tympanic membrane (eardrum) with the oval window of the cochlea; also called the *malleus.*

haploid hap′-loid: referring to a cell that has only one member of each pair of homologous chromosomes.

Hardy–Weinberg principle: a mathematical model proposing that, under certain conditions, the allele frequencies and genotype frequencies in a sexually reproducing population will remain constant over generations.

heart: a muscular organ responsible for pumping blood within the circulatory system throughout the body.

heart attack: a severe reduction or blockage of blood flow through a coronary artery, depriving some of the heart muscle of its blood supply.

heartwood: older secondary xylem that usually no longer conducts water or minerals, but that contributes to the strength of a tree trunk.

heat of fusion: the energy that must be removed from a compound to transform it from a liquid into a solid at its freezing temperature.

heat of vaporization: the energy that must be supplied to a compound to transform it from a liquid into a gas at its boiling temperature.

Heimlich maneuver: a method of dislodging food or other obstructions that have entered the airway.

helix (hē′-liks): a coiled, springlike secondary structure of a protein.

helper T cell: a type of T cell that helps other immune cells recognize and act against antigens.

hemocoel (hē′-mō-sēl): a cavity within the bodies of certain invertebrates in which a fluid, called hemolymph, bathes tissues directly; part of an open circulatory system.

hemodialysis (hē-mō-dī-al′-luh-sis): a procedure that simulates kidney function in individuals with damaged or ineffective kidneys; blood is diverted from the body, artificially filtered, and returned to the body.

hemoglobin (hē′-mō-glō-bin): the iron-containing protein that gives red blood cells their color; binds to oxygen in the lungs and releases it in the tissues.

hemolymph: in animals with an open circulatory system, the fluid that is located within the hemocoel and that bathes all the body cells, therefore serving as both blood and extracellular fluid.

hemophilia: a recessive, sex-linked disease in which the blood fails to clot normally.

herbivore (erb′-i-vor): literally, "plant-eater"; an organism that feeds directly and exclusively on producers; a primary consumer.

hermaphrodite (her-maf′-ruh-dit′): an organism that possesses both male and female sexual organs. Some hermaphroditic animals can fertilize themselves; others must exchange sex cells with a mate.

hermaphroditic (her-maf′-ruh-dit′-ik): possessing both male and female sexual organs. Some hermaphroditic animals can fertilize themselves; others must exchange sex cells with a mate.

heterotroph (het′-er-ō-trōf′): literally, "other-feeder"; an organism that eats other organisms; a consumer.

heterozygous (het′-er-ō-zī′-gus): carrying two different alleles of a given gene; also called *hybrid.*

hindbrain: the posterior portion of the brain, containing the medulla, pons, and cerebellum.

hinge joint: a joint at which the bones fit together in a way that allows movement in only two dimensions, as at the elbow or knee.

hippocampus (hip-ō-kam′-pus): the part of the forebrain of vertebrates that is important in emotion and especially learning.

histamine (his′-ta-mēn): a substance released by certain cells in response to tissue damage and invasion of the body by foreign substances; promotes the dilation of arterioles and the leakiness of capillaries and triggers some of the events of the inflammatory response.

homeobox gene (hō′-mē-ō-boks): a sequence of DNA coding for a transcription factor protein that activates or inactivates many other genes that control the development of specific, major parts of the body.

homeostasis (hōm-ē-ō-stā′-sis): the maintenance of the relatively constant internal environment that is required for the optimal functioning of cells.

hominin: a human or a prehistoric relative of humans; the oldest known hominin is *Sahelanthropus*, whose fossils are more than 6 million years old.

homologous chromosome: a chromosome that is similar in appearance and genetic information to another chromosome with which it pairs during meiosis; also called *homologue*.

homologous structure: structures that may differ in function but that have similar anatomy, presumably because the organisms that possess them have descended from common ancestors.

homologue (hō-′mō-log): a chromosome that is similar in appearance and genetic information to another chromosome with which it pairs during meiosis; also called *homologous chromosome*.

homozygous (hō-mō-zī′-gus): carrying two copies of the same allele of a given gene; also called *true-breeding*.

hormone: a chemical that is secreted by one group of cells and transported to other cells, whose activity is influenced by reception of the hormone.

host: the prey organism on or in which a parasite lives; the host is harmed by the relationship.

human immunodeficiency virus (HIV): a pathogenic virus that causes acquired immune deficiency syndrome (AIDS) by attacking and destroying the immune system's helper T cells.

human papillomavirus (pap-il-lo′-ma; HPV): a virus that infects the reproductive organs, often causing genital warts; causes most, if not all, cases of cervical cancer.

humoral immunity: an immune response in which foreign substances are inactivated or destroyed by antibodies that circulate in the blood.

Huntington disease: an incurable genetic disorder, caused by a dominant allele, that produces progressive brain deterioration, resulting in the loss of motor coordination, flailing movements, personality disturbances, and eventual death.

hybrid: an organism that is the offspring of parents differing in at least one genetically determined characteristic; also used to refer to the offspring of parents of different species.

hydrogen bond: the weak attraction between a hydrogen atom that bears a partial positive charge (due to polar covalent bonding with another atom) and another atom (oxygen, nitrogen, or fluorine) that bears a partial negative charge; hydrogen bonds may form between atoms of a single molecule or of different molecules.

hydrologic cycle (hī–drō–loj′–ik): the biogeochemical cycle by which water travels from its major reservoir, the oceans, through the atmosphere to reservoirs in freshwater lakes, rivers, and groundwater, and back into the oceans. The hydrologic cycle is driven by solar energy. Nearly all water remains as water throughout the cycle (rather than being used in the synthesis of new molecules).

hydrolysis (hī-drol′-i-sis): the chemical reaction that breaks a covalent bond by means of the addition of hydrogen to the atom on one side of the original bond and a hydroxyl group to the atom on the other side; the reverse of dehydration synthesis.

hydrophilic (hī-drō-fil′-ik): pertaining to molecules that dissolve readily in water, or to molecules that form hydrogen bonds with water; polar.

hydrophobic (hī-drō-fō′-bik): pertaining to molecules that do not dissolve in water or form hydrogen bonds with water; nonpolar.

hydrophobic interaction: the tendency for hydrophobic molecules to cluster together when immersed in water.

hydrostatic skeleton (hī-drō-stat′-ik): in invertebrate animals, a body structure in which fluid-filled compartments provide support for the body and change shape when acted on by muscles, which alters the animal's body shape and position, or causes the animal to move.

hydrothermal vent community: a community of unusual organisms, living in the deep ocean near hydrothermal vents, that depends on the chemosynthetic activities of sulfur bacteria.

hypertension: arterial blood pressure that is chronically elevated above the normal level.

hypertonic (hī-per-ton′-ik): referring to a solution that has a higher concentration of dissolved particles (and therefore a lower concentration of free water) than has the cytoplasm of a cell.

hypha (hī′-fuh; pl., **hyphae):** in fungi, a threadlike structure that consists of elongated cells, typically with many haploid nuclei; many hyphae make up the fungal body.

hypocotyl (hī′-pō-kot-ul): the part of the embryonic shoot located below the attachment point of the cotyledons but above the root.

hypothalamus: (hī-pō-thal′-a-mus): a region of the brain that controls the secretory activity of the pituitary gland; synthesizes, stores, and releases certain peptide hormones; directs autonomic nervous system responses.

hypothesis (hī-poth′-eh-sis): in the scientific method, a supposition based on previous observations that is offered as an explanation for an observed phenomenon and is used as the basis for further observations or experiments.

hypotonic (hī-pō-ton′-ik): referring to a solution that has a lower concentration of dissolved particles (and therefore a higher concentration of free water) than has the cytoplasm of a cell.

immigration (im-uh-grā′-shun): migration of individuals into an area.

immune system: a system of cells, including macrophages, B cells, and T cells, and molecules, such as antibodies and cytokines, that work together to combat microbial invasion of the body.

imperfect flower: a flower that is missing either stamens or carpels.

implantation: the process whereby the early embryo embeds itself within the lining of the uterus.

imprinting: a type of learning in which an animal acquires a particular type of information during a specific sensitive phase of development.

incomplete dominance: a pattern of inheritance in which the heterozygous phenotype is intermediate between the two homozygous phenotypes.

incomplete flower: a flower that is missing one of the four floral parts (sepals, petals, stamens, or carpels).

indirect development: a developmental pathway in which an offspring goes through radical changes in body form as it matures.

induced pluripotent stem cell (ploo-rē-pō′-tent; iPSC): a type of stem cell produced from nonstem cells by the insertion of a specific set of genes that cause the cells to become capable of unlimited cell division and to be able to be differentiated into many different cell types, possibly any cell type of the body.

induction: the process by which a group of cells causes other cells to differentiate into a specific tissue type.

inductive reasoning: the process of creating a generalization based on many specific observations that support the generalization, coupled with an absence of observations that contradict it.

inflammatory response: a nonspecific, local response to injury to the body, characterized by the phagocytosis of foreign substances and tissue debris by white blood cells and by the walling off of the injury site by the clotting of fluids that escape from nearby blood vessels.

ingestion: the movement of food into the digestive tract, usually through the mouth.

inhalation: the act of drawing air into the lungs by enlarging the chest cavity.

inheritance: the genetic transmission of characteristics from parent to offspring.

inhibiting hormone: a hormone, secreted by the neurosecretory cells of the hypothalamus, that inhibits the release of specific hormones from the anterior pituitary.

inhibitory postsynaptic potential (IPSP): an electrical signal produced in a postsynaptic cell that makes the resting potential more negative and, hence, makes the neuron less likely to fire an action potential.

innate (in-āt′): inborn; instinctive; an innate behavior is performed correctly the first time it is attempted.

innate immune response: nonspecific defenses against many different invading microbes, including phagocytic white blood cells, natural killer cells, the inflammatory response, and fever.

inner cell mass: in human embryonic development, the cluster of cells, on the inside of the blastocyst, that will develop into the embryo.

inner ear: the innermost part of the mammalian ear; composed of the bony, fluid-filled tubes of the cochlea and the vestibular apparatus.

inorganic: describing any molecule that does not contain both carbon and hydrogen.

insertion: the site of attachment of a muscle to the relatively movable bone on one side of a joint.

insertion mutation: a mutation in which one or more pairs of nucleotides are inserted into a gene.

insight learning: a type of learning in which a problem is solved by understanding the relationships among the components of the problem rather than through trial and error.

insulin: a hormone, secreted by the pancreas, that lowers blood sugar by stimulating many cells to take up glucose and by stimulating the liver to convert glucose to glycogen.

integration: the process of adding up all of the electrical signals in a neuron, including sensory inputs and postsynaptic potentials, to determine the output of the neuron (action potentials and/or synaptic transmission).

integument (in-teg′-ū-ment): in plants, the outer layers of cells of the ovule that surround the female gametophyte; develops into the seed coat.

intensity: the strength of stimulation or response.

intercalated disc: junctions connecting individual cardiac muscle cells that serve both to attach adjacent cells to one another and to allow electrical signals to pass between cells.

intermediate filament: part of the cytoskeleton of eukaryotic cells that is composed of several types of proteins and probably functions mainly for support.

intermembrane space: the fluid-filled space between the inner and outer membranes of a mitochondrion.

internal fertilization: the union of sperm and egg inside the body of the female.

interneuron: in a neural network, a nerve cell that is postsynaptic to a sensory neuron and presynaptic to a motor neuron. In actual circuits, there may be many interneurons between individual sensory and motor neurons.

internode: the part of a stem between two nodes.

interphase: the stage of the cell cycle between cell divisions in which chromosomes are duplicated and other cell functions occur, such as growth, movement, and acquisition of nutrients.

interspecific competition: competition among individuals of different species.

interstitial cell (in-ter-sti′-shul): in the vertebrate testis, a testosterone-producing cell located between the seminiferous tubules.

intertidal zone: an area of the ocean shore that is alternately covered by water during high tides and exposed to the air during low tides.

intracellular digestion: the chemical breakdown of food within single cells.

intraspecific competition: competition among individuals of the same species.

intron: a segment of DNA in a eukaryotic gene that does not code for amino acids in a protein (see also exon).

invasive species: organisms with a high biotic potential that are introduced (deliberately or accidentally) into ecosystems where they did not evolve and where they encounter little environmental resistance and tend to displace native species.

inversion: a mutation that occurs when a piece of DNA is cut out of a chromosome, turned around, and reinserted into the gap.

invertebrate (in-vert′-uh-bret): an animal that lacks a vertebral column.

ion (ī-on): a charged atom or molecule; an atom or molecule that has either an excess of electrons (and, hence, is negatively charged) or has lost electrons (and is positively charged).

ionic bond: a chemical bond formed by the electrical attraction between positively and negatively charged ions.

iris: the pigmented muscular tissue of the vertebrate eye that surrounds and controls the size of the pupil, through which light enters the eye.

islet cell: a cell in the endocrine portion of the pancreas that produces either insulin or glucagon.

isolating mechanism: a morphological, physiological, behavioral, or ecological difference that prevents members of two species from interbreeding.

isotonic (ī-sō-ton′-ik): referring to a solution that has the same concentration of dissolved particles (and therefore the same concentration of free water) as has the cytoplasm of a cell.

isotope (ī′-suh-tōp): one of several forms of a single element, the nuclei of which contain the same number of protons but different numbers of neutrons.

J-curve: the J-shaped growth curve of an exponentially growing population in which increasing numbers of individuals join the population during each succeeding time period.

Jacob syndrome: a set of characteristics typical of human males possessing one X and two Y chromosomes (XYY); most XYY males are phenotypically normal, but XYY males have a higher than average incidence of high testosterone levels, severe acne, and above-average height.

joint: a flexible region between two rigid units of an exoskeleton or endoskeleton, allowing for movement between the units.

karyotype: a preparation showing the number, sizes, and shapes of all of the chromosomes within a cell.

kelp forest: a diverse ecosystem consisting of stands of tall brown algae and associated marine life. Kelp forests occur in oceans worldwide in nutrient-rich cool coastal waters.

keratin (ker′-uh-tin): a fibrous protein in hair, nails, and the epidermis of skin.

keystone species: a species whose influence on community structure is greater than its abundance would suggest.

kidney: one of a pair of organs of the excretory system that is located on either side of the spinal column and filters blood, removing wastes and regulating the composition and water content of the blood.

kin selection: a type of natural selection that favors traits that enhance the survival or reproduction of an individual's relatives, even if the traits reduce the fitness of the individuals bearing them.

kinetic energy: the energy of movement; includes light, heat, mechanical movement, and electricity.

kinetochore (ki-net′-ō-kor): a protein structure that forms at the centromere regions of chromosomes; attaches the chromosomes to the spindle.

kinetoplastid: a member of a protist group characterized by distinctively structured mitochondria. Kinetoplastids are mostly flagellated and include parasitic forms such as *Trypanosoma*, which causes sleeping sickness. Kinetoplastids are part of a larger group known as euglenozoans.

kingdom: the second broadest taxonomic category, consisting of related phyla. Related kingdoms make up a domain.

Klinefelter syndrome: a set of characteristics typically found in individuals who have two X chromosomes and one Y chromosome; these individuals are phenotypically males but are sterile and have several female-like traits, including broad hips and partial breast development.

Krebs cycle: a cyclic series of reactions, occurring in the matrix of mitochondria, in which the acetyl groups from the pyruvic acids produced by glycolysis are broken down to CO_2, accompanied by the formation of ATP and electron carriers; also called the citric acid cycle.

labium (pl., labia): one of a pair of folds of skin of the external structures of the mammalian female reproductive system.

labor: a series of contractions of the uterus that result in birth.

lactation: the secretion of milk from the mammary glands.

lacteal (lak-tēl′): a lymph capillary; found in each villus of the small intestine.

lactic acid fermentation: anaerobic reactions that convert the pyruvic acid produced by glycolysis into lactic acid, using hydrogen ions and electrons from NADH; the primary function of lactic acid fermentation is to regenerate NAD+ so that glycolysis can continue under anaerobic conditions.

lactose (lak′-tōs): a disaccharide composed of glucose and galactose; found in mammalian milk.

lactose intolerance: the inability to digest lactose (milk sugar) because lactase, the enzyme that digests lactose, is not produced in sufficient amounts; symptoms include bloating, gas pains, and diarrhea.

lactose operon: in prokaryotes, the set of genes that encodes the proteins needed for lactose metabolism, including both the structural genes and a common promoter and operator that control transcription of the structural genes.

large intestine: the final section of the digestive tract; consists of the colon and the rectum, where feces are formed and stored.

larva (lar′-vuh; pl., **larvae):** an immature form of an animal that subsequently undergoes metamorphosis into its adult form; includes the caterpillars of moths and butterflies, the maggots of flies, and the tadpoles of frogs and toads.

larynx (lar′-inks): the portion of the air passage between the pharynx and the trachea; contains the vocal cords.

late-loss population: a population in which most individuals survive into adulthood; late-loss populations have a convex survivorship curve.

lateral bud: a cluster of meristem cells at the node of a stem; under appropriate conditions, it grows into a branch.

lateral meristem: a meristem tissue that forms cylinders parallel to the long axis of roots and stems; normally located between the primary xylem and primary phloem (vascular cambium) and just outside the phloem (cork cambium); also called *cambium*.

law of conservation of energy: the principle of physics that states that within any closed system, energy can be neither created nor destroyed, but can be converted from one form to another; also called the first law of thermodynamics.

law of independent assortment: the independent inheritance of two or more traits, assuming that each trait is controlled by a single gene with no influence from gene(s) controlling the other trait; states that the alleles of each gene are distributed to the gametes independently of the alleles for other genes; this law is true only for genes located on different chromosomes or very far apart on a single chromosome.

law of segregation: the principle that each gamete receives only one of each parent's pair of alleles of each gene.

laws of thermodynamics: the physical laws that define the basic properties and behavior of energy.

leaf: an outgrowth of a stem, normally flattened and photosynthetic.

leaf primordium (pri-mor′-dē-um; pl., **pri-mor-dia):** a cluster of dividing cells, surrounding a terminal or lateral bud, that develops into a leaf.

learning: the process by which behavior is modified in response to experience.

legume (leg′-ūm): a member of a family of plants characterized by root swellings in which nitrogen-fixing bacteria are housed; includes peas, soybeans, lupines, alfalfa, and clover.

lens: a clear object that bends light rays; in eyes, a flexible or movable structure used to focus light on the photoreceptor cells of the retina.

leptin: a peptide hormone released by fat cells that helps the body monitor its fat stores and regulate weight.

leukocyte (loo′-kō-sīt): any of the white blood cells circulating in the blood.

lichen (lī′-ken): a symbiotic association between an alga or cyanobacterium and a fungus, resulting in a composite organism.

life table: a data table that groups organisms born at the same time and tracks them throughout their life span, recording how many continue to survive in each succeeding year (or other unit of time). Various parameters such as sex may be used in the groupings. Human life tables may include many other parameters (such as socioeconomic status) used by demographers.

ligament: a tough connective tissue band connecting two bones.

light reactions: the first stage of photosynthesis, in which the energy of light is captured as ATP and NADPH; occurs in thylakoids of chloroplasts.

lignin: a hard material that is embedded in the cell walls of vascular plants and that provides support in terrestrial species; an early and important adaptation to terrestrial life.

limbic system: a diverse group of brain structures, mostly in the lower forebrain, that includes the thalamus, hypothalamus, amygdala, hippocampus, and parts of the cerebrum and is involved in basic emotions, drives, behaviors, and learning.

limnetic zone: the part of a lake in which enough light penetrates to support photosynthesis.

linkage: the inheritance of certain genes as a group because they are parts of the same chromosome. Linked genes do not show independent assortment.

lipase (lī′-pās): an enzyme that catalyzes the breakdown of lipids such as fats.

lipid (li′-pid): one of a number of organic molecules containing large nonpolar regions composed solely of carbon and hydrogen, which make lipids hydrophobic and insoluble in water; includes oils, fats, waxes, phospholipids, and steroids.

littoral zone: the part of a lake, usually close to the shore, in which the water is shallow and plants find abundant light, anchorage, and adequate nutrients.

liver: an organ with varied functions, including bile production, glycogen storage, and the detoxification of poisons.

lobefin: fish with fleshy fins that have well-developed bones and muscles. Lobefins include two living clades: the coelacanths and the lungfishes.

local hormone: a general term for messenger molecules produced by most cells and released into the cells' immediate vicinity. Local hormones, which include prostaglandins and cytokines, influence nearby cells bearing appropriate receptors.

locus (plural, loci): the physical location of a gene on a chromosome.

logistic population growth: population growth characterized by an early exponential growth phase, followed by slower growth as the population approaches its carrying capacity, and finally reaching a stable population at the carrying capacity of the environment; this type of growth generates a curve shaped like a stretched-out letter "S."

long-day plant: a plant that will flower only if the length of uninterrupted darkness is shorter than a species-specific critical period; also called a *short-night plant*.

long-term memory: the second phase of learning; a more-or-less permanent memory formed by a structural change in the brain, brought on by repetition.

loop of Henle (hen′-lē): a specialized portion of the tubule of some nephrons in birds and mammals that creates an osmotic concentration gradient in the fluid immediately surrounding it. This gradient in turn makes possible the production of urine that is more osmotically concentrated than blood plasma.

lung: in terrestrial vertebrates, one of the pair of respiratory organs in which gas exchange occurs; consists of inflatable chambers within the chest cavity.

luteinizing hormone (loo′-tin-īz-ing; LH): a hormone produced by the anterior pituitary that stimulates testosterone production in males and the development of the follicle, ovulation, and the production of the corpus luteum in females.

lymph (limf): a pale fluid found within the lymphatic system; composed primarily of interstitial fluid and white blood cells.

lymph node: a small structure located on a lymph vessel, containing macrophages and lymphocytes (B and T cells). Macrophages filter the lymph by removing microbes; lymphocytes are the principal components of the adaptive immune response to infection.

lymphatic system: a system consisting of lymph vessels, lymph capillaries, lymph nodes, and the thymus and spleen; helps protect the body against infection, absorbs fats, and returns excess fluid and small proteins to the blood circulatory system.

lymphocyte (lim′-fō-sit): a type of white blood cell (natural killer cell, B cell, or T cell) that is important in either the innate or adaptive immune response.

lysosome (lī′-sō-sōm): a membrane-bound organelle containing intracellular digestive enzymes.

macronutrient: a nutrient required by an organism in relatively large quantities.

macrophage (mak′-rō-fāj): a type of white blood cell that engulfs microbes and destroys them by phagocytosis; also presents microbial antigens to T cells, helping stimulate the immune response.

major histocompatibility complex (MHC): a group of proteins, normally located on the surfaces of body cells, that identify the cell as "self"; also important in stimulating and regulating the immune response.

Malpighian tubule (mal-pig′-ē-un): the functional unit of the excretory system of insects, consisting of a small tube protruding from the intestine; wastes and nutrients move from the surrounding blood into the tubule, which drains into the intestine, where nutrients are reabsorbed back into the blood, while the wastes are excreted along with feces.

maltose (mal′-tōs): a disaccharide composed of two glucose molecules.

mammal: a member of the chordate clade Mammalia, which includes vertebrates with hair and mammary glands.

mammary gland (mam′-uh-rē): a milk-producing gland used by female mammals to nourish their young.

marsupial (mar-soo′-pē-ul): a member of the clade Marsupialia, which includes mammals whose young are born at an extremely immature stage and undergo further development in a pouch, where they remain attached to a mammary gland; kangaroos, opossums, and koalas are marsupials.

mass extinction: a relatively sudden extinction of many species, belonging to multiple major taxonomic groups, as a result of environmental change. The fossil record reveals five mass extinctions over geologic time.

mast cell: a cell of the immune system that releases histamine and other molecules used in the body's response to trauma and that are a factor in allergic reactions.

matrix: The fluid contained within the inner membrane of the mitochrondrion.

mechanical digestion: the process by which food in the digestive tract is physically broken down into smaller pieces.

medulla (med-ū′-luh): the part of the hindbrain of vertebrates that controls automatic activities such as breathing, swallowing, heart rate, and blood pressure.

megakaryocyte (meg-a-kar′-ē-ō-sīt): a large cell type in the bone marrow, which pinches off pieces of itself that enter the circulation as platelets.

megaspore: a haploid cell formed by meiotic cell division from a diploid megaspore mother cell; through mitotic cell division and differentiation, it develops into the female gametophyte.

megaspore mother cell: a diploid cell, within the ovule of a flowering plant, that undergoes meiotic cell division to produce four haploid megaspores.

meiosis (mī-ō′-sis): in eukaryotic organisms, a type of nuclear division in which a diploid nucleus divides twice to form four haploid nuclei.

meiotic cell division: meiosis followed by cytokinesis.

melatonin (mel-uh-tōn′-in): a hormone, secreted by the pineal gland, that is involved in the regulation of circadian cycles.

memory B cell: a type of white blood cell that is produced by clonal selection as a result of the binding of an antibody on a B cell to an antigen on an invading microorganism. Memory B cells persist in the bloodstream and provide future immunity to invaders bearing that antigen.

memory T cell: a type of white blood cell that is produced by clonal selection as a result of the binding of a receptor on a T cell to an antigen on an invading microorganism. Memory T cells persist in the bloodstream and provide future immunity to invaders bearing that antigen.

menstrual cycle: in human females, a roughly 28-day cycle during which hormonal interactions among the hypothalamus, pituitary gland, and ovary coordinate ovulation and the preparation of the uterus to receive and nourish a fertilized egg. If pregnancy does not occur, the uterine lining is shed during menstruation.

menstruation: in human females, the monthly discharge of uterine tissue and blood from the uterus.

meristem cell (mer′-i-stem): an undifferentiated cell that remains capable of cell division throughout the life of a plant.

mesoderm (mēz′-ō-derm): the middle embryonic tissue layer, lying between the endoderm and ectoderm, and normally the last to develop; gives rise to structures such as muscles, the skeleton, the circulatory system, and the kidneys.

mesophyll (mez′-ō-fil): loosely packed, usually photosynthetic cells located beneath the epidermis of a leaf.

messenger RNA (mRNA): a strand of RNA, complementary to the DNA of a gene, that conveys the genetic information in DNA to the ribosomes to be used during protein synthesis; sequences of three bases (codons) in mRNA that specify particular amino acids to be incorporated into a protein.

metabolic pathway: a sequence of chemical reactions within a cell, in which the products of one reaction are the reactants for the next reaction.

metabolic rate: the speed at which cellular reactions that release energy occur.

metabolism: the sum of all chemical reactions that occur within a single cell or within all the cells of a multicellular organism.

metamorphosis (met-a-mor′-fō-sis): in animals with indirect development, a radical change in body form from larva to sexually mature adult, as seen in amphibians (e.g., tadpole to frog) and insects (e.g., caterpillar to butterfly).

metaphase (met′-a-fāz): in mitosis, the stage in which the chromosomes, attached to spindle fibers at kinetochores, are lined up along the equator of the cell; also the approximately comparable stages in meiosis I and meiosis II.

microbe: a microorganism.

microfilament: part of the cytoskeleton of eukaryotic cells that is composed of the proteins actin and (in some cases) myosin; functions in the movement of cell organelles, locomotion by extension of the plasma membrane, and sometimes contraction of entire cells.

micronutrient: a nutrient required by an organism in relatively small quantities.

microspore: a haploid cell formed by meiotic cell division from a microspore mother cell; through mitotic cell division and differentiation, it develops into the male gametophyte.

microspore mother cell: a diploid cell contained within an anther of a flowering plant; undergoes meiotic cell division to produce four haploid microspores.

microtubule: a hollow, cylindrical strand, found in eukaryotic cells, that is composed of the protein tubulin; part of the cytoskeleton used in the movement of organelles, cell growth, and the construction of cilia and flagella.

microvillus (mī-krō-vi′-lus; pl., **microvilli):** a microscopic projection of the plasma membrane, which increases the surface area of a cell.

midbrain: during development, the central portion of the brain; contains an important relay center, the reticular formation.

middle ear: the part of the mammalian ear composed of the tympanic membrane, the Eustachian tube, and three bones (hammer, anvil, and stirrup) that transmit vibrations from the auditory canal to the oval window.

mimicry (mim′-ik-rē): the situation in which a species has evolved to resemble something else, typically another type of organism.

mineral: an inorganic substance, especially one in rocks or soil, or dissolved in water. In nutrition, minerals such as sodium, calcium, and potassium are essential nutrients that must be obtained in the diet.

mineralocorticoids: steroid hormones produced by the adrenal cortex that regulate salt retention in the kidney, thereby regulating the salt concentration in the blood and other extracellular fluids.

minimum viable population (MVP): the smallest isolated population that can persist indefinitely and survive likely natural events such as fires and floods.

mitochondrion (mī-tō-kon′-drē-un; pl., **mitochondria):** an organelle, bounded by two membranes, that is the site of the reactions of aerobic metabolism.

mitosis (mī-tō′-sis): a type of nuclear division, used by eukaryotic cells, in which one copy of each chromosome (already duplicated during interphase before mitosis) moves into each of two daughter nuclei; the daughter nuclei are therefore genetically identical to each other.

mitotic cell division: mitosis followed by cytokinesis.

molecule (mol′-e-kūl): a particle composed of one or more atoms held together by chemical bonds; the smallest particle of a compound that displays all the properties of that compound.

molt: to shed an external body covering, such as an exoskeleton, skin, feathers, or fur.

monocot: short for monocotyledon; a type of flowering plant characterized by embryos with one seed leaf, or cotyledon.

monomer (mo′-nō-mer): a small organic molecule, several of which may be bonded together to form a chain called a polymer.

monosaccharide (mo-nō-sak′-uh-rīd): the basic molecular unit of all carbohydrates, normally composed of a chain of carbon atoms bonded to hydrogen and hydroxyl groups.

monotreme: a member of the clade Monotremata, which includes mammals that lay eggs; platypuses and spiny anteaters are monotremes.

morula (mor′-ū-luh): in animals, an embryonic stage during cleavage, when the embryo consists of a solid ball of cells.

motor neuron: a neuron that receives instructions from sensory neurons or interneurons and activates effector organs, such as muscles or glands.

motor unit: a single motor neuron and all the muscle fibers on which it forms synapses.

mouth: the opening through which food enters a tubular digestive system.

multicellular: many-celled; most members of the kingdoms Fungi, Plantae, and Animalia are multicellular, with intimate cooperation among cells.

multiple alleles: many alleles of a single gene, perhaps dozens or hundreds, as a result of mutations.

muscle fiber: an individual muscle cell.

muscle tissue: tissue composed of one of three types of contractile cells (smooth, skeletal, or cardiac).

mutation: a change in the base sequence of DNA in a gene; often used to refer to a genetic change that is significant enough to alter the appearance or function of the organism.

mutualism (mū′-choo-ul-iz-um): a symbiotic relationship in which both participating species benefit.

mycelium (mī-sēl′-ē-um; pl., **mycelia):** the body of a fungus, consisting of a mass of hyphae.

mycorrhiza (mī-kō-rī′-zuh; pl., **mycorrhizae):** a symbiotic association between a fungus and the roots of a land plant that facilitates mineral extraction and absorption.

myelin (mī′-uh-lin): a wrapping of insulating membranes of specialized nonneural cells around the axon of a vertebrate nerve cell; increases the speed of conduction of action potentials.

myofibril (mū-ō-fī′-bril): a cylindrical subunit of a muscle cell, consisting of a series of sarcomeres, surrounded by sarcoplasmic reticulum.

myometrium (mī-ō-mē′-trē-um): the muscular outer layer of the uterus.

myosin (mī′-ō-sin): one of the major proteins of muscle, the interaction of which with the protein actin produces muscle contraction; found in the thick filaments of the muscle fiber; see also *actin*.

myosin head: the part of a myosin protein that binds to the actin subunits of a thick filament; flexion of the myosin head moves the thin filament toward the center of the sarcomere, causing muscle fiber contraction.

Na^+-K^+ pump: an active transport protein that uses the energy of ATP to transport Na^+ out of a cell and K^+ into a cell; produces and maintains the concentration gradients of these ions across the plasma membrane, such that the concentration of Na^+ is higher outside a cell than inside, and the concentration of K^+ is higher inside a cell than outside.

natural causality: the scientific principle that natural events occur as a result of preceding natural causes.

natural killer cell: a type of white blood cell that destroys some virus-infected cells and cancerous cells on contact; part of the innate immune system's nonspecific internal defense against disease.

natural selection: the unequal survival and reproduction of organisms with different phenotypes, caused by environmental forces. Natural selection refers specifically to cases in which the differing phenotypes are heritable; that is, they are caused at least partly by genetic differences, with the result that better adapted phenotypes become more common in the population.

nearshore zone: the region of coastal water that is relatively shallow but constantly submerged, and that can support large plants or seaweeds; includes bays and coastal wetlands.

nearsighted: the inability to focus on distant objects caused by an eyeball that is slightly too long or a cornea that is too curved.

negative feedback: a physiological process in which a change causes responses that tend to counteract the change and restore the original state. Negative feedback in physiological systems maintains homeostasis.

nephridium (nef-rid′-ē-um; pl., **nephridia):** an excretory organ found in earthworms, mollusks, and certain other invertebrates; somewhat resembles a single vertebrate nephron.

nephron (nef′-ron): the functional unit of the kidney; where blood is filtered and urine is formed.

nerve: a bundle of axons of nerve cells, bound together in a sheath.

nerve cord: a major nervous pathway consisting of a cord of nervous tissue extending lengthwise through the body, paired in many invertebrates and unpaired in chordates.

nerve net: a simple form of nervous system, consisting of a network of neurons that extends throughout the tissues of an organism such as a cnidarian.

nerve tissue: the tissue that makes up the brain, spinal cord, and nerves; consists of neurons and glial cells.

net primary productivity: the energy stored in the autotrophs of an ecosystem over a given time period.

neuromuscular junction: the synapse formed between a motor neuron and a muscle fiber.

neuron (noor′-on): a single nerve cell.

neurosecretory cell: a specialized nerve cell that synthesizes and releases hormones.

neurotransmitter: a chemical that is released by a nerve cell close to a second nerve cell, a muscle, or a gland cell and that influences the activity of the second cell.

neutral mutation: a mutation that does not detectably change the function of the encoded protein.

neutron: a subatomic particle that is found in the nuclei of atoms, bears no charge, and has a mass approximately equal to that of a proton.

neutrophil (nū′-trō-fil): a type of white blood cell that engulfs invading microbes and contributes to the nonspecific defenses of the body against disease.

nicotinamide adenine dinucleotide (NAD^+ or NADH): an electron carrier molecule produced in the cytoplasmic fluid by glycolysis and in the mitochondrial matrix by the Krebs cycle; subsequently donates electrons to the electron transport chain.

nitrogen cycle: the biogeochemical cycle by which nitrogen moves from its primary reservoir of nitrogen gas in the atmosphere via nitrogen-fixing bacteria to reservoirs in soil and water, through producers and into higher trophic levels, and then back to its reservoirs.

nitrogen fixation: the process that combines atmospheric nitrogen with hydrogen to form ammonium (NH_4^+).

nitrogen-fixing bacterium: a bacterium that possesses the ability to remove nitrogen (N_2) from the atmosphere and combine it with hydrogen to produce ammonium (NH_4^+).

no-till: a method of growing crops that leaves the remains of harvested crops in place, with the next year's crops being planted directly in the remains of last year's crops without significant disturbance of the soil.

node: in plants, a region of a stem at which the petiole of a leaf is attached; usually, a lateral bud is also found at a node.

nodule: a swelling on the root of a legume or other plant that consists of cortex cells inhabited by nitrogen-fixing bacteria.

noncompetitive inhibition: the process by which an inhibitory molecule binds to a site on an enzyme that is distinct from the active site. As a result, the enzyme's active site is distorted, making it less able to catalyze the reaction involving its normal substrate.

nondisjunction: an error in meiosis in which chromosomes fail to segregate properly into the daughter cells.

nonpolar covalent bond: a covalent bond with equal sharing of electrons.

nonvascular plant: a plant that lacks lignin and well-developed conducting vessels. Nonvascular plants include mosses, hornworts, and liverworts.

norepinephrine (nor-ep-i-nef-rin′): a neurotransmitter, released by neurons of the parasympathetic nervous system, that prepares the body to respond to stressful situations; also called *noradrenaline*.

northern coniferous forest: a biome with long, cold winters and only a few months of warm weather; dominated by evergreen coniferous trees; also called *taiga*.

notochord (nōt′-ō-kord): a stiff but somewhat flexible, supportive rod that extends along the head-to-tail axis and is found in all members of the phylum Chordata at some stage of development.

nuclear envelope: the double-membrane system surrounding the nucleus of eukaryotic cells; the outer membrane is typically continuous with the endoplasmic reticulum.

nuclear pore complex: an array of proteins that line pores in the nuclear membrane and control which substances enter and leave the nucleus.

nucleic acid (noo-klā′-ik): an organic molecule composed of nucleotide subunits; the two common types of nucleic acids are ribonucleic acid (RNA) and deoxyribonucleic acid (DNA).

nucleoid (noo-klē-oid): the location of the genetic material in prokaryotic cells; not membrane enclosed.

nucleolus (noo-klē′-ō-lus; pl., **nucleoli):** the region of the eukaryotic nucleus that is engaged in ribosome synthesis; consists of the genes encoding ribosomal RNA, newly synthesized ribosomal RNA, and ribosomal proteins.

nucleotide substitution: a mutation in which a single base pair in DNA has been changed.

nucleotide: a subunit of which nucleic acids are composed; a phosphate group bonded to a sugar (deoxyribose in DNA), which is in turn bonded to a nitrogen-containing base (adenine, guanine, cytosine, or thymine in DNA). Nucleotides are linked together, forming a strand of nucleic acid, by bonds between the phosphate of one nucleotide and the sugar of the next nucleotide.

nucleus (atomic): the central region of an atom, consisting of protons and neutrons.

nucleus (cellular): the membrane-bound organelle of eukaryotic cells that contains the cell's genetic material.

nutrient: a substance acquired from the environment and needed for the survival, growth, and development of an organism.

nutrient cycle: the pathways of a specific nutrient (such as carbon, nitrogen, phosphorus, or water) through the living and nonliving portions of an ecosystem; also called a *biogeochemical cycle*.

observation: in the scientific method, the recognition of and a statement about a specific phenomenon, usually leading to the formulation of a question about the phenomenon.

oil: a lipid composed of three fatty acids, some of which are unsaturated, covalently bonded to a molecule of glycerol; oils are liquid at room temperature.

oligotrophic lake: a lake that is very low in nutrients and hence supports little phytoplankton, plant, and algal life; contains clear water with deep light penetration.

ommatidium (ōm-ma-tid′-ē-um; pl., **ommatidia):** an individual light-sensitive subunit of a compound eye; consists of a lens and several receptor cells.

omnivore: an organism that consumes both plants and animals.

oogenesis (ō-ō-jen′-i-sis): the process by which egg cells are formed.

oogonium (ō-ō-gō′-nē-um; pl., **oogonia):** in female animals, a diploid cell that gives rise to a primary oocyte.

open circulatory system: a type of circulatory system found in some invertebrates, such as arthropods and most mollusks, that includes an open space (the hemocoel) in which blood directly bathes body tissues.

operant conditioning: a laboratory training procedure in which an animal learns to perform a response (such as pressing a lever) through reward or punishment.

operator: a sequence of DNA nucleotides in a prokaryotic operon that binds regulatory proteins that control the ability of RNA polymerase to transcribe the structural genes of the operon.

operon (op′–er–on): in prokaryotes, a set of genes, often encoding the proteins needed for a complete metabolic pathway, including both the structural genes and a common promoter and operator that control transcription of the structural genes.

optic nerve: the nerve leading from the eye to the brain; it carries visual information.

order: in Linnaean classification, the taxonomic rank composed of related families. Related orders make up a class.

organ: a structure (such as the liver, kidney, or skin) composed of two or more distinct tissue types that function together.

organ system: two or more organs that work together to perform a specific function; for example, the digestive system.

organelle (or-guh-nel′): a membrane-enclosed structure found inside a eukaryotic cell that performs a specific function.

organic molecule: a molecule that contains both carbon and hydrogen.

organic: describing a molecule that contains both carbon and hydrogen.

organism (or′-guh-niz-um): an individual living thing.

organogenesis (or-gan-ō-jen′-uh-sis): the process by which the layers of the gastrula (endoderm, ectoderm, mesoderm) rearrange to form organs.

origin: the site of attachment of a muscle to the relatively stationary bone on one side of a joint.

osmolarity: a measure of the total number of dissolved solute particles in a solution.

osmoregulation: homeostatic maintenance of the water and salt content of the body within a limited range.

osmosis (oz-mō′-sis): the diffusion of water across a differentially permeable membrane, normally down a concentration gradient of free water molecules. Water moves into the solution that has a lower concentration of free water from a solution that has a higher concentration of free water.

osteoblast (os′-tē-ō-blast): a cell type that produces bone.

osteoclast (os′-tē-ō-klast): a cell type that dissolves bone.

osteocyte (os′-tē-ō-sī t): a mature bone cell.

osteoporosis (os′-tē-ō-por-ō′-sis): a condition in which bones become porous, weak, and easily fractured; most common in elderly women.

outer ear: the outermost part of the mammalian ear, including the external ear and auditory canal leading to the tympanic membrane.

oval window: the membrane-covered entrance to the cochlea.

ovary: (1) in animals, the gonad of females. (2) in flowering plants, a structure at the base of the carpel that contains one or more ovules and develops into the fruit.

ovary: in animals, the gonad of females.

overexploitation: hunting or harvesting natural populations at a rate that exceeds those populations' ability to replenish their numbers.

ovulation: the release of a secondary oocyte, ready to be fertilized, from the ovary.

ovule: a structure within the ovary of a flower, inside which the female gametophyte develops; after fertilization, develops into the seed.

oxytocin (oks-ē-tō′-sin): a hormone, released by the posterior pituitary, that stimulates the contraction of uterine and mammary gland muscles.

ozone hole: a region of severe ozone loss in the stratosphere caused by ozone-depleting chemicals; maximum ozone loss occurs from September to early October over Antarctica.

ozone layer: the ozone-enriched layer of the upper atmosphere (stratosphere) that filters out much of the sun's ultraviolet radiation.

pacemaker: a cluster of specialized muscle cells in the upper right atrium of the heart that produce spontaneous electrical signals at a regular rate; the sinoatrial node.

pain receptor: a receptor cell that stimulates activity in the brain that is perceived as the sensation of pain; responds to very high or very low temperatures, mechanical damage (such as extreme stretching of tissue), and/or certain chemicals, such as potassium ions or bradykinin, that are produced as a result of tissue damage; also called *nociceptor*.

pancreas (pan′-krē-us): a combined exocrine and endocrine gland located in the abdominal cavity next to the stomach. The endocrine portion secretes the hormones insulin and glucagon, which regulate glucose concentrations in the blood. The exocrine portion secretes pancreatic juice (a mixture of water, enzymes, and sodium bicarbonate) into the small intestine; the enzymes digest fat, carbohydrate, and protein; the bicarbonate neutralizes acidic chyme entering the intestine from the stomach.

pancreatic juice: a mixture of water, sodium bicarbonate, and enzymes released by the pancreas into the small intestine.

parabasalid: a member of a protist group characterized by mutualistic or parasitic relationships with the animal species inside which they live. Parabasalids are part of a larger group known as the excavates.

parasite (par′-uh-sīt): an organism that lives in or on a larger organism (its host), harming the host but usually not killing it immediately.

parasympathetic division: the division of the autonomic nervous system that produces largely involuntary responses related to the maintenance of normal body functions, such as digestion; often called the *parasympathetic nervous system*.

parathyroid gland: one of four small endocrine glands, embedded in the surface of the thyroid gland, that produces parathyroid hormone, which (with calcitonin from the thyroid gland) regulates calcium ion concentration in the blood.

parathyroid hormone: a hormone released by the parathyroid gland that stimulates the release of calcium from bone.

parenchyma (par-en′-ki-muh): a plant cell type that is alive at maturity, normally with thin cell walls, that carries out most of the metabolism of a plant. Most dividing meristem cells in a plant are parenchyma.

parthenogenesis (par-the-nō-jen′-uh-sis): an asexual specialization of sexual reproduction, in which a haploid egg undergoes development without fertilization.

passive transport: the movement of materials across a membrane down a gradient of concentration, pressure, or electrical charge without using cellular energy.

pathogen: an organism (or a toxin) capable of producing disease.

pathogenic (path′-ō-jen-ik): capable of producing disease; referring to an organism with such a capability (a pathogen).

pedigree: a diagram showing genetic relationships among a set of individuals, normally with respect to a specific genetic trait.

pelagic (puh-la′-jik): free-swimming or floating.

penis: an external structure of the male reproductive and urinary systems; serves to deposit sperm into the female reproductive system and deliver urine to the outside of the body.

peptide (pep′-tīd): a chain composed of two or more amino acids linked together by peptide bonds.

peptide bond: the covalent bond between the nitrogen of the amino group of one amino acid and the carbon of the carboxyl group of a second amino acid, joining the two amino acids together in a peptide or protein.

peptide hormone: a hormone consisting of a chain of amino acids; includes small proteins that function as hormones.

pericycle (per′-i-sī-kul): the outermost layer of cells of the vascular cylinder of a root.

periderm: the outer cell layers of roots and stems that have undergone secondary growth; consists primarily of cork cambium and cork cells.

peripheral nervous system (PNS): in vertebrates, the part of the nervous system that connects the central nervous system to the rest of the body.

peristalsis: rhythmic coordinated contractions of the smooth muscles of the digestive tract that move substances through the digestive tract.

permafrost: a permanently frozen layer of soil, usually found in tundra of the arctic or high mountains.

petal: part of a flower, typically brightly colored and fragrant, that attracts potential animal pollinators.

petiole (pet′-ē-ōl): the stalk that connects the blade of a leaf to the stem.

pH scale: a scale, with values from 0 to 14, used for measuring the relative acidity of a solution; at pH 7 a solution is neutral, pH 0 to 7 is acidic, and pH 7 to 14 is basic; each unit on the scale represents a tenfold change in $H+$ concentration.

phagocyte (fā′-gō-sīt): a type of immune system cell that destroys invading microbes by using phagocytosis to engulf and digest the microbes. Also called a *phagocytic cell*.

phagocytosis (fa-gō-sī-tō′-sis): a type of endocytosis in which extensions of a plasma membrane engulf extracellular particles, enclose them in a membrane-bound sac, and transport them into the interior of the cell.

pharyngeal gill slit (far-in′-jē-ul): one of a series of openings, located just posterior to the mouth, that connects the throat to the outside environment; present (as some stage of life) in all chordates.

pharynx (far′-inks): in vertebrates, a chamber that is located at the back of the mouth, shared by the digestive and respiratory systems; in some invertebrates, the portion of the digestive tube just posterior to the mouth.

phenotype (fēn′-ō-tīp): the physical characteristics of an organism; can be defined as outward appearance (such as flower color), as behavior, or in molecular terms (such as glycoproteins on red blood cells).

pheromone (fer′-uh-mōn): a chemical produced by an organism that alters the behavior or physiological state of another member of the same species.

phloem (flō′-um): a conducting tissue of vascular plants that transports a concentrated solution of sugars (primarily sucrose) and other organic molecules up and down the plant.

phospholipid (fos-fō-li′-pid): a lipid consisting of glycerol bonded to two fatty acids and one phosphate group, which bears another group of atoms, typically charged and containing nitrogen. A double layer of phospholipids is a component of all cellular membranes.

phospholipid bilayer: a double layer of phospholipids that forms the basis of all cellular membranes. The phospholipid heads, which are hydrophilic, face the water of extracellular fluid or the cytoplasm; the tails, which are hydrophobic, are buried in the middle of the bilayer.

phosphorus cycle (fos′–for–us): the biogeochemical cycle by which phosphorus moves from its primary reservoir—phosphate-rich rock—to reservoirs of phosphate in soil and water, through producers and into higher trophic levels, and then back to its reservoirs.

photic zone: the region of an ocean where light is strong enough to support photosynthesis.

photon (fō′-ton): the smallest unit of light energy.

photopigment (fō′-tō-pig-ment): a chemical substance in a photoreceptor cell that, when struck by light, changes shape and produces a response in the cell.

photoreceptor: a receptor cell that responds to light; in vertebrates, rods and cones.

photorespiration: a series of reactions in plants in which O_2 replaces CO_2 during the Calvin cycle, preventing carbon fixation; this wasteful process dominates when C_3 plants are forced to close their stomata to prevent water loss.

photosynthesis: the complete series of chemical reactions in which the energy of light is used to synthesize high-energy organic molecules, usually carbohydrates, from low-energy inorganic molecules, usually carbon dioxide and water.

photosystem: in thylakoid membranes, a cluster of chlorophyll, accessory pigment molecules, proteins, and other molecules that collectively capture light energy, transfer some of the energy to electrons, and transfer the energetic electrons to an adjacent electron transport chain.

phototropism: growth with respect to the direction of light.

phylogeny (fī-lah′-jen-ē): the evolutionary history of a group of species.

phylum (fī-lum): in Linnaean classification, the taxonomic rank composed of related classes. Related phyla make up a kingdom.

phytochrome (fī′-tō-krōm): a light-sensitive plant pigment that mediates many plant responses to light, including flowering, stem elongation, and seed germination.

phytoplankton (fī′-tō-plank-ten): photosynthetic protists that are abundant in marine and freshwater environments.

pigment molecule: a light-absorbing, colored molecule, such as chlorophyll, carotenoid, or melanin molecules.

pineal gland (pī-nē′-al): a small gland within the brain that secretes melatonin; controls the seasonal reproductive cycles of some mammals.

pinna: a flap of skin-covered cartilage on the surface of the head that collects sound waves and funnels them to the auditory canal.

pinocytosis (pi-nō-sī-tō′-sis): the nonselective movement of extracellular fluid, enclosed within a vesicle formed from the plasma membrane, into a cell.

pioneer: an organism that is among the first to colonize an unoccupied habitat in the first stages of succession.

pit: an area in the cell walls between two plant cells where the two cells are separated only by a thin, porous cell wall.

pith: cells forming the center of a root or stem.

pituitary gland: an endocrine gland, located at the base of the brain, that produces several hormones, many of which influence the activity of other glands.

placenta (pluh-sen′-tuh): in mammals, a structure formed by a complex interweaving of the uterine lining and the embryonic membranes, especially the chorion; functions in gas, nutrient, and waste exchange between embryonic and maternal circulatory systems, and also secretes the hormones estrogen and progesterone, which are essential to maintaining pregnancy.

placental (pluh-sen′-tul): referring to a mammal possessing a complex placenta (that is, species that are not marsupials or monotremes).

plankton: microscopic organisms that live in marine or freshwater environments; includes phytoplankton and zooplankton.

plant hormone: a chemical produced by specific plants cells that influences the growth, development, or metabolic activity of other cells, typically some distance away in the plant body.

plaque (plak): a deposit of cholesterol and other fatty substances within the wall of an artery.

plasma: the fluid, noncellular portion of the blood.

plasma cell: an antibody-secreting descendant of a B cell.

plasma membrane: the outer membrane of a cell, composed of a bilayer of phospholipids in which proteins are embedded.

plasmid (plaz′-mid): a small, circular piece of DNA located in the cytoplasm of many bacteria; usually does not carry genes required for the normal functioning of the bacterium, but may carry genes, such as those for antibiotic resistance, that assist bacterial survival in certain environments.

plasmodesma (plaz-mō-dez′-muh; pl., **plasmodesmata):** a cell-to-cell junction in plants that connects the cytoplasm of adjacent cells.

plasmodium (plaz-mō′-dē-um): a sluglike mass of cytoplasm containing thousands of nuclei that are not confined within individual cells.

plastid (plas′-tid): in plant cells, an organelle bounded by two membranes that may be involved in photosynthesis (chloroplasts), pigment storage, or food storage.

plate tectonics: the theory that Earth's crust is divided into irregular plates that are converging, diverging, or slipping by one another; these motions cause continental drift, the movement of continents over Earth's surface.

platelet (plāt′-let): a cell fragment that is formed from megakaryocytes in bone marrow; platelets lack a nucleus; they circulate in the blood and play a role in blood clotting.

pleated sheet: a form of secondary structure exhibited by certain proteins, such as silk, in which many protein chains lie side by side, with hydrogen bonds holding adjacent chains together.

pleiotropy (ple′-ō-trō-pē): a situation in which a single gene influences more than one phenotypic characteristic.

point mutation: a mutation in which a single base pair in DNA has been changed.

polar body: in oogenesis, a small cell, containing a nucleus but virtually no cytoplasm, produced by both the first meiotic division (of the primary oocyte) and the second meiotic division (of the secondary oocyte).

polar covalent bond: a covalent bond with unequal sharing of electrons, such that one atom is relatively negative and the other is relatively positive.

pollen: the male gametophyte of a seed plant; also called a *pollen grain.*

pollen grain: the male gametophyte of a seed plant.

pollination: in flowering plants, when pollen grains land on the stigma of a flower of the same species; in conifers, when pollen grains land within the pollen chamber of a female cone of the same species.

polygenic inheritance: a pattern of inheritance in which the interactions of two or more functionally similar genes determine phenotype.

polymer (pah′-li-mer): a molecule composed of three or more (perhaps thousands) smaller subunits called monomers, which may be identical (for example, the glucose monomers of starch) or different (for example, the amino acids of a protein).

polymerase chain reaction (PCR): a method of producing virtually unlimited numbers of copies of a specific piece of DNA, starting with as little as one copy of the desired DNA.

polyploidy (pahl′-ē-ploid-ē): having more than two sets of homologous chromosomes.

polysaccharide (pahl-ē-sak′-uh-rīd): a large carbohydrate molecule composed of branched or unbranched chains of repeating monosaccharide subunits, normally glucose or modified glucose molecules; includes starches, cellulose, and glycogen.

pons: a portion of the hindbrain, just above the medulla, that contains neurons that influence sleep and the rate and pattern of breathing.

population bottleneck: the result of an event that causes a population to become extremely small; may cause genetic drift that results in changed allele frequencies and loss of genetic variability.

population cycle: regularly recurring, cyclic changes in population size.

population: all the members of a particular species within an ecosystem, found in the same time and place and actually or potentially interbreeding.

positive feedback: a physiological mechanism in which a change causes responses that tend to amplify the original change.

post-anal tail: a tail that extends beyond the anus, and contains muscle tissue and the most posterior part of the nerve cord; found in all chordates at some stage of development.

posterior pituitary: a lobe of the pituitary gland that is an outgrowth of the hypothalamus and that releases antidiuretic hormone and oxytocin.

postmating isolating mechanism: any structure, physiological function, or developmental abnormality that prevents organisms of two different species, once mating has occurred, from producing vigorous, fertile offspring.

postsynaptic neuron: at a synapse, the nerve cell that changes its electrical potential in response to a chemical (the neurotransmitter) released by another (presynaptic) cell.

postsynaptic potential (PSP): an electrical signal produced in a postsynaptic cell by transmission across the synapse; it may be excitatory (EPSP), making the cell more likely to produce an action potential, or inhibitory (IPSP), tending to inhibit an action potential.

potential energy: "stored" energy; normally chemical energy or energy of position within a gravitational field.

prairie: a biome, located in the centers of continents, that primarily supports grasses; also called *grassland*.

precapillary sphincter (sfink′-ter): a ring of smooth muscle between an arteriole and a capillary that regulates the flow of blood into the capillary bed.

predation (pre-dā′-shun): the act of eating another living organism.

predator: an organism that eats other organisms.

prediction: in the scientific method, a statement describing an expected observation or the expected outcome of an experiment, assuming that a specific hypothesis is true.

premating isolating mechanism: any structure, physiological function, or behavior that prevents organisms of two different species from exchanging gametes.

pressure-flow theory: a model for the transport of sugars in phloem; the movement of sugars into a phloem sieve tube causes water to enter the tube by osmosis, while the movement of sugars out of another part of the same sieve tube causes water to leave by osmosis. The resulting pressure gradient causes the bulk movement of water and dissolved sugars from the end of the sieve tube into which sugar is transported (a source) toward the end of the sieve tube from which sugar is removed (a sink).

presynaptic neuron: a nerve cell that releases a chemical (the neurotransmitter) at a synapse, causing changes in the electrical activity of another (postsynaptic) cell.

prey: organisms that are eaten, and often killed, by another organism (a predator).

primary consumer: an organism that feeds on producers; an herbivore

primary electron acceptor: A molecule in the reaction center of each photosystem that accepts an electron from one of the two reaction center chlorophyll *a* molecules and transfers the electron to an adjacent electron transport chain.

primary growth: growth in length and development of the initial structures of plant roots and shoots; results from cell division of apical meristems and differentiation of the daughter cells.

primary oocyte (ō′-ō-sīt): a diploid cell, derived from the oogonium by growth and differentiation, that undergoes meiotic cell division, producing the egg.

primary spermatocyte (sper-ma′-tō-sīt): a diploid cell, derived from the spermatogonium by growth and differentiation, that undergoes meiotic cell division, producing four sperm.

primary structure: the amino acid sequence of a protein.

primary succession: succession that occurs in an environment, such as bare rock, in which no trace of a previous community is present.

primate: a member of the mammalian clade Primates, characterized by the presence of an opposable thumb, forward-facing eyes, and a well-developed cerebral cortex; includes lemurs, monkeys, apes, and humans.

prion (prē′-on): a protein that, in mutated form, acts as an infectious agent that causes certain neurodegenerative diseases, including kuru and scrapie.

producer: a photosynthetic organism; an autotroph.

product: an atom or molecule that is formed from reactants in a chemical reaction.

profundal zone: the part of a lake in which light is insufficient to support photosynthesis.

progesterone (prō-ge′-ster-ōn): a hormone, produced by the corpus luteum in the ovary, that promotes the development of the uterine lining in females.

prokaryote (prō-kar′-ē-ōt): an organism whose cells are prokaryotic (their genetic material is not enclosed in a membrane-bound nucleus and they lack other membrane-bound organelles); bacteria and archaea are prokaryotes.

prokaryotic (prō-kar-ē-ot′-ik): referring to cells of the domains Bacteria or Archaea. Prokaryotic cells have genetic material that is not enclosed in a membrane-bound nucleus; they also lack other membrane-bound organelles.

prolactin: a hormone, released by the anterior pituitary, that stimulates milk production in human females.

promoter: a specific sequence of DNA at the beginning of a gene, to which RNA polymerase binds and starts gene transcription.

prophase (prō′-fāz): the first stage of mitosis, in which the chromosomes first become visible in the light microscope as thickened, condensed threads and the spindle begins to form; as the spindle is completed, the nuclear envelope breaks apart, and the spindle microtubules invade the nuclear region and attach to the kinetochores of the chromosomes. Also, the first stage of meiosis:In meiosis I, the homologous chromosomes pair up, exchange parts at chiasmata, and attach to spindle microtubules; in meiosis II, the spindle re-forms and chromosomes attach to the microtubules.

prostaglandin (pro-stuh-glan′-din): a family of modified fatty acid hormones manufactured by many cells of the body.

prostate gland (pros′-tāt): a gland that produces part of the fluid component of semen; prostatic fluid is basic and contains a chemical that activates sperm movement.

protease (prō′-tē-ās): an enzyme that digests proteins.

protein: a polymer composed of amino acids joined by peptide bonds.

protist: a eukaryotic organism that is not a plant, animal, or fungus. The term encompasses a diverse array of organisms and does not represent a monophyletic group.

protocell: the hypothetical evolutionary precursor of living cells, consisting of a mixture of organic molecules within a membrane.

proton: a subatomic particle that is found in the nuclei of atoms; it bears a unit of positive charge, and has a relatively large mass, roughly equal to the mass of the neutron.

protonephridium (prō-tō-nef-rid′-ē-um; pl., **protonephridia):** the functional unit of the excretory system of some invertebrates, such as flatworms; consists of a tubule that has an external opening to the outside of the body, but lacks an internal opening within the body. Fluid is filtered from the body cavity into the tubule by a hollow cell at the end of the tubule, such as a flame cell in flatworms, and released outside the body.

protostome (prō′-tō-stōm): an animal with a mode of embryonic development in which the coelom is derived from splits in the mesoderm; characteristic of arthropods, annelids, and mollusks.

protozoan (prō-tuh-zō′-an; pl., **protozoa):** a nonphotosynthetic, single-celled protist.

proximal tubule: the initial part of a mammalian nephron, between Bowman's capsule and the loop of Henle; most tubular reabsorption and a small amount of tubular secretion occur in the proximal tubule.

pseudocoelom (soo′-dō-sēl′-ōm): in animals, a "false coelom," that is, a space or cavity, partially but not fully lined with tissue derived from mesoderm, that separates the body wall from the inner organs; found in roundworms.

pseudoplasmodium (soo′-dō-plaz-mō′-dē-um): an aggregation of individual amoeboid cells that form a sluglike mass.

pseudopod (sood′-ō-pod): an extension of the plasma membrane by which certain cells, such as amoebas, locomote and engulf prey.

puberty: a stage of development characterized by sexual maturation, rapid growth, and the appearance of secondary sexual characteristics.

Punnett square method: a method of predicting the genotypes and phenotypes of offspring in genetic crosses.

pupa (pl., **pupae):** a developmental stage in some insect species in which the organism stops moving and feeding and may be encased in a cocoon; occurs between the larval and the adult phases.

pupil: the adjustable opening in the center of the iris, through which light enters the eye.

quaternary structure (kwat′-er-nuh-rē): the complex three-dimensional structure of a protein consisting of more than one peptide chain.

question: in the scientific method, a statement that identifies a particular aspect of an observation that a scientist wishes to explain.

radial symmetry: a body plan in which any plane along a central axis will divide the body into approximately mirror-image halves. Cnidarians and many adult echinoderms exhibit radial symmetry.

radioactive: pertaining to an atom with an unstable nucleus that spontaneously disintegrates, with the emission of radiation.

radiolarian (rā-dē-ō-lar′-ē-un): a member of a protist group characterized by pseudopods and typically elaborate silica shells. Radiolarians are largely aquatic (mostly marine) and are part of a larger group known as rhizarians.

rain shadow: a local dry area, usually located on the downwind side of a mountain range that blocks the prevailing moisture-bearing winds.

random distribution: the distribution characteristic of populations in which the probability of finding an individual is equal in all parts of an area.

reactant: an atom or molecule that is used up in a chemical reaction to form a product.

reaction center: two chlorophyll *a* molecules and a primary electron acceptor complexed with proteins and located near the center of each photosystem within the thylakoid membrane. Light energy is passed to one of the chlorophylls, which donates an energized electron to the primary electron acceptor, which then passes the electron to an adjacent electron transport chain.

receptor: (1) a protein, located in a membrane or the cytoplasm of a cell, that binds to specific molecules (for example, a hormone or neurotransmitter), triggering a response in the cell, such as endocytosis, changes in metabolic rate, cell division, or electrical changes; (2) a cell that responds to an environmental stimulus (chemicals, sound, light, pH, and so on) by changing its electrical potential.

receptor potential: an electrical potential change in a receptor cell, produced in response to the reception of an environmental stimulus (chemicals, sound, light, heat, and so on). The size of the receptor potential is proportional to the intensity of the stimulus.

receptor protein: a protein, located in a membrane or the cytoplasm of a cell, that binds to specific molecules (for example, a hormone or neurotransmitter), triggering a response in the cell, such as endocytosis, changes in metabolic rate, cell division, or electrical changes.

receptor-mediated endocytosis: the selective uptake of molecules from the extracellular fluid by binding to a receptor located at a coated pit on the plasma membrane and pinching off the coated pit into a vesicle that moves into the cytoplasm.

recessive: an allele that is expressed only in homozygotes and is completely masked in heterozygotes.

recognition protein: a protein or glycoprotein protruding from the outside surface of a plasma membrane that identifies a cell as belonging to a particular species, to a specific individual of that species, and in many cases to one specific organ within the individual.

recombinant DNA: DNA that has been altered by the addition of DNA from a different organism, typically from a different species.

recombination: the formation of new combinations of the different alleles of each gene on a chromosome; the result of crossing over.

rectum: the terminal portion of the vertebrate digestive tube where feces are stored until they are eliminated.

reflex: a simple, stereotyped movement of part of the body that occurs automatically in response to a stimulus.

regeneration: the regrowth of a body part after loss or damage; also, asexual reproduction by means of the regrowth of an entire body from a fragment.

regulatory gene: in prokaryotes, a gene encoding a protein that binds to the operator of one or more operons, controlling the ability of RNA polymerase to transcribe the structural genes of the operon.

regulatory T cell: a type of T cell that suppresses the adaptive immune response, especially by self-reactive lymphocytes, and appears to be important in the prevention of autoimmune disorders.

releasing hormone: a hormone, secreted by the hypothalamus, that causes the release of specific hormones by the anterior pituitary.

renal artery: the artery carrying blood to each kidney.

renal cortex: the outer layer of the kidney, in which the largest portion of each nephron is located, including Bowman's capsule and the distal and proximal tubules.

renal medulla: the layer of the kidney just inside the renal cortex, in which loops of Henle produce a highly concentrated extracellular fluid, allowing the production of concentrated urine.

renal pelvis: the inner chamber of the kidney, in which urine from the collecting ducts accumulates before it enters the ureter.

renal vein: the vein carrying blood away from each kidney.

renin: in mammals, an enzyme that is released by the kidneys when blood pressure falls. Renin catalyzes the formation of angiotensin, which causes arterioles to constrict, thereby elevating blood pressure.

replacement-level fertility (RLF): the average number of offspring per female that is required to maintain a stable population.

repressor protein: in prokaryotes, a protein encoded by a regulatory gene, which binds to the operator of an operon and prevents RNA polymerase from transcribing the structural genes.

reproductive isolation: the failure of organisms of one population to breed successfully with members of another; may be due to premating or postmating isolating mechanisms.

reptile: a member of the chordate group that includes the snakes, lizards, turtles, alligators, birds, and crocodiles.

reservoir: the major source and storage site of a nutrient in an ecosystem, normally in the abiotic portion.

resource partitioning: the coexistence of two species with similar requirements, each occupying a smaller niche than either would if it were by itself; a means of minimizing the species' competitive interactions.

respiration: in terrestrial vertebrates, the act of moving air into the lungs (inhalation) and out of the lungs (exhalation); during respiration, oxygen diffuses from the air in the lungs into the circulatory system and carbon dioxide diffuses from the circulatory system into the air in the lungs.

respiratory center: a cluster of neurons, located in the medulla of the brain, that sends rhythmic bursts of nerve impulses to the respiratory muscles, resulting in breathing.

respiratory membrane: within the lungs, the fusion of the epithelial cells of the alveoli and the endothelial cells that form the walls of surrounding capillaries.

resting potential: an electrical potential, or voltage, in unstimulated nerve cells; the inside of the cell is always negative with respect to the outside.

restriction enzyme: an enzyme, usually isolated from bacteria, that cuts double-stranded DNA at a specific nucleotide sequence; the nucleotide sequence that is cut differs for different restriction enzymes.

restriction fragment length polymorphism (RFLP): a difference in the length of DNA fragments that were produced by cutting samples of DNA from different individuals of the same species with the same set of restriction enzymes; fragment length differences occur because of differences in nucleotide sequences, and hence in the ability of restriction enzymes to cut the DNA, among individuals of the same species.

restriction fragment: a piece of DNA that has been isolated by cleaving a larger piece of DNA with restriction enzymes.

reticular formation (reh-tik′-ū-lar): a diffuse network of neurons extending from the hindbrain, through the midbrain, and into the lower reaches of the forebrain; involved in filtering sensory input and regulating what information is relayed to conscious brain centers for further attention.

retina (ret′-in-uh): a multilayered sheet of nerve tissue at the rear of camera-type eyes, composed of photoreceptor cells plus associated nerve cells that refine the photoreceptor information and transmit it to the optic nerve.

rhizarian: a member of Rhizaria, a protist clade. Rhizarians, which use thin pseudopods to move and capture prey and which often have hard shells, include the foraminiferans and the radiolarians.

ribonucleic acid (RNA) (rī-bō-noo-klā′-ik; RNA): a molecule composed of ribose nucleotides, each of which consists of a phosphate group, the sugar ribose, and one of the bases adenine, cytosine, guanine, or uracil; involved in converting the information in DNA into protein; also the genetic material of some viruses.

ribosomal RNA (rī–bō–sō′–mul; rRNA): a type of RNA that combines with proteins to form ribosomes.

ribosome (rī′–bō–sōm): a complex consisting of two subunits, each composed of ribosomal RNA and protein, found in the cytoplasm of cells or attached to the endoplasmic reticulum, that is the site of protein synthesis, during which the sequence of bases of messenger RNA is translated into the sequence of amino acids in a protein.

ribozyme: an RNA molecule that can catalyze certain chemical reactions, especially those involved in the synthesis and processing of RNA itself.

RNA polymerase: in RNA synthesis, an enzyme that catalyzes the bonding of free RNA nucleotides into a continuous strand, using RNA nucleotides that are complementary to those of the template strand of DNA.

rod: a rod-shaped photoreceptor cell in the vertebrate retina, sensitive to dim light but not involved in color vision; see also *cone*.

root: the part of the plant body, normally underground, that provides anchorage, absorbs water and dissolved nutrients and transports them to the stem, produces some hormones, and in some plants serves as a storage site for carbohydrates.

root cap: a cluster of cells at the tip of a growing root, derived from the apical meristem; protects the growing tip from damage as it burrows through the soil.

root hair: a fine projection from an epidermal cell of a young root that increases the absorptive surface area of the root.

root pressure: pressure within a root caused by the transport of minerals into the vascular cylinder, accompanied by the entry of water by osmosis.

root system: all of the roots of a plant.

round window: a flexible membrane at the end of the cochlea opposite the oval window that allows the fluid in the cochlea to move in response to movements of the oval window.

rubisco: in the carbon fixation step of the Calvin cycle, the enzyme that catalyzes the reaction between ribulose bisphosphate (RuBP) and carbon dioxide, thereby fixing the carbon of carbon dioxide in an organic molecule; short for ribulose bisphosphate carboxylase.

ruminant (roo′-min-ant): an herbivorous animal with a digestive tract that includes multiple stomach chambers, one of which contains cellulose-digesting bacteria, and that regurgitates the contents ("cud") of the first chamber for additional chewing ("ruminating").

S-curve: the S-shaped growth curve produced by logistic population growth, usually describing a population of organisms introduced into a new area; consists of an initial period of exponential growth, followed by a decreasing growth rate, and finally, relative stability around a growth rate of zero.

sac fungus: a member of the fungus phylum Ascomycota, whose members form spores in a saclike case called an ascus.

saccule: a patch of hair cells in the vestibule of the inner ear; bending of the hairs of the hair cells permits detection of the direction of gravity and the degree of tilt of the head.

sapwood: young secondary xylem that transports water and minerals in a tree trunk.

sarcomere (sark′-ō-mēr): the unit of contraction of a muscle fiber; a subunit of the myofibril, consisting of thick and thin filaments, bounded by Z lines.

sarcoplasmic reticulum (sark′-ō-plas′-mik re-tik′-ū-lum; SR): the specialized endoplasmic reticulum in muscle cells; forms interconnected hollow tubes. The sarcoplasmic reticulum stores calcium ions and releases them into the interior of the muscle cell, initiating contraction.

saturated: referring to a fatty acid with as many hydrogen atoms as possible bonded to the carbon backbone (therefore, a saturated fatty acid has no double bonds in its carbon backbone).

savanna: a biome that is dominated by grasses and supports scattered trees; typically has a rainy season during which most of the year's precipitation falls, followed by a dry season during which virtually no precipitation occurs.

scientific method: a rigorous procedure for making observations of specific phenomena and searching for the order underlying those phenomena.

scientific name: the two-part Latin name of a species; consists of the genus name followed by the species name.

scientific theory of evolution: the theory that modern organisms descended, with modification, from preexisting life-forms.

scientific theory: a general explanation of natural phenomena developed through extensive and reproducible observations; more general and reliable than a hypothesis.

sclera: a tough, white connective tissue layer that covers the outside of the eyeball and forms the white of the eye.

sclerenchyma (skler-en′-ki-muh): a plant cell type with thick, hardened cell walls that normally dies as the last stage of differentiation; may support or protect the plant body.

scramble competition: a free-for-all scramble for limited resources among individuals of the same species.

scrotum (skrō′-tum): in male mammals, the pouch of skin containing the testes.

second law of thermodynamics: the principle of physics that states that any change in a closed system causes the quantity of concentrated, useful energy to decrease and the amount of randomness and disorder (entropy) to increase.

second messenger: an intracellular chemical, such as cyclic AMP, that is synthesized or released within a cell in response to the binding of a hormone or neurotransmitter (the first messenger) to receptors on the cell surface; brings about specific changes in the metabolism of the cell.

secondary consumer: an organism that feeds on primary consumers; a carnivore.

secondary growth: growth in the diameter and strength of a stem or root due to cell division in lateral meristems and differentiation of their daughter cells.

secondary oocyte (ō′-ō-sīt): a large haploid cell derived from the diploid primary oocyte by meiosis I.

secondary spermatocyte (sper-ma′-tō-sīt): a large haploid cell derived from the diploid primary spermatocyte by meiosis I.

secondary structure: a repeated, regular structure assumed by a protein chain, held together by hydrogen bonds; for example, a helix.

secondary succession: succession that occurs after an existing community is disturbed—for example, after a forest fire; secondary succession is much more rapid than primary succession.

secretin: a hormone produced by the small intestine that stimulates the production and release of digestive secretions by the pancreas and liver.

seed: the reproductive structure of a seed plant, protected by a seed coat; contains an embryonic plant and a supply of food for it.

seed coat: the thin, tough, and waterproof outermost covering of a seed, formed from the integuments of the ovule.

segmentation (seg-men-tā′-shun): an animal body plan in which the body is divided into repeated, typically similar units.

selectively permeable: the quality of a membrane that allows certain molecules or ions to move through it more readily than others.

self-fertilization: the union of sperm and egg from the same individual.

semen: the sperm-containing fluid produced by the male reproductive tract.

semicircular canal: in the inner ear, one of three fluid-filled tubes, each with a bulge at one end containing a patch of hair cells; movement of the head moves fluid in the canal and consequently bends the hairs of the hair cells.

semiconservative replication: the process of replication of the DNA double helix; the two DNA strands separate, and each is used as a template for the synthesis of a complementary DNA strand. Consequently, each daughter double helix consists of one parental strand and one new strand.

semilunar valve: a valve located between the right ventricle of the heart and the pulmonary artery, or between the left ventricle and the aorta; prevents the backflow of blood into the ventricles when they relax.

seminal vesicle: in male mammals, a gland that produces a basic, fructose-containing fluid that forms part of the semen.

seminiferous tubule (sem-i-ni′-fer-us): in the vertebrate testis, a series of tubes in which sperm are produced.

senescence: in plants, a specific aging process, typically including deterioration and the dropping of leaves and flowers.

sensory neuron: a nerve cell that responds to a stimulus from the internal or external environment.

sensory receptor: a cell (typically, a neuron) specialized to respond to particular internal or external environmental stimuli by producing an electrical potential.

sepal (sē′-pul): one of the group of modified leaves that surrounds and protects a flower bud; in dicots, usually develops into a green, leaflike structure when the flower blooms; in monocots, usually similar to a petal.

septum (pl., **septa**)**:** a partition that separates the fungal hypha into individual cells; pores in septa allow the transfer of materials between cells.

Sertoli cell: in the seminiferous tubule, a large cell that regulates spermatogenesis and nourishes the developing sperm.

severe combined immune deficiency (SCID): a disorder in which no immune cells, or very few, are formed; the immune system is incapable of responding properly to invading disease organisms, and the individual is very vulnerable to common infections.

sex chromosome: either of the pair of chromosomes that usually determines the sex of an organism; for example, the X and Y chromosomes in mammals.

sex-linked: referring to a pattern of inheritance characteristic of genes located on one type of sex chromosome (for example, X) and not found on the other type (for example, Y); in mammals, in almost all cases, the gene controlling the trait is on the X chromosome, so this pattern is often called X-linked. In X-linked inheritance, females show the dominant trait unless they are homozygous recessive, whereas males express whichever allele, dominant or recessive, that is found on their single X chromosome.

sexual reproduction: a form of reproduction in which genetic material from two parent organisms is combined in the offspring; usually, two haploid gametes fuse to form a diploid zygote.

sexual selection: a type of natural selection that acts on traits involved in finding and acquiring mates.

sexually transmitted disease (STD): a disease that is passed from person to person by sexual contact; also known as *sexually transmitted infection (STI)*.

shoot system: all the parts of a vascular plant exclusive of the root; normally aboveground. Consists of stem, leaves, buds, and (in season) flowers and fruits; functions include photosynthesis, transport of materials, reproduction, and hormone synthesis.

short tandem repeat (STR): a DNA sequence consisting of a short sequence of nucleotides (usually 2 to 5 nucleotides in length) repeated multiple times, with all of the repetitions side by side on a chromosome; variations in the number of repeats of a standardized set of 13 STRs produce DNA profiles used to identify people by their DNA.

short-day plant: a plant that will flower only if the length of uninterrupted darkness exceeds a species-specific critical period; also called a *long-night plant*.

sickle-cell anemia: a recessive disease caused by a single amino acid substitution in the hemoglobin molecule. Sickle-cell hemoglobin molecules tend to cluster together, distorting the shape of red blood cells and causing them to break and clog capillaries.

sieve plate: in plants, a structure between two adjacent sieve-tube elements in phloem, where holes formed in the cell walls interconnect the cytoplasm of the sieve-tube elements.

sieve-tube element: in phloem, one of the cells of a sieve tube.

simple diffusion: the diffusion of water, dissolved gases, or lipid-soluble molecules through the phospholipid bilayer of a cellular membrane.

simple epithelium: a type of epithelial tissue, one cell layer thick, that lines many hollow organs such as those of the respiratory, digestive, urinary, reproductive, and circulatory systems.

sink: in plants, any structure that uses up sugars or converts sugars to starch, and toward which phloem fluids will flow.

sinoatrial (SA) node (sī′-nō-āt′-rē-ul nōd): a small mass of specialized muscle in the wall of the right atrium; generates electrical signals rhythmically and spontaneously and serves as the heart's pacemaker.

skeletal muscle: the type of muscle that is attached to and moves the skeleton and is under the direct, normally voluntary, control of the nervous system; also called *striated muscle*.

skeleton: a supporting structure for the body, on which muscles act to change the body configuration; may be external or internal.

small intestine: the portion of the digestive tract, located between the stomach and large intestine, in which most digestion and absorption of nutrients occur.

smooth muscle: the type of muscle that surrounds hollow organs, such as the digestive tract, bladder, and blood vessels; not striped in appearance (hence the name "smooth") and normally not under voluntary control.

solute: a substance dissolved in a solvent.

solution: a solvent containing one or more dissolved substances (solutes).

solvent: a liquid capable of dissolving (uniformly dispersing) other substances in itself.

somatic nervous system: that portion of the peripheral nervous system that controls voluntary movement by activating skeletal muscles.

source: in plants, any structure that actively synthesizes sugar, and away from which phloem fluid will be transported.

spawning: a method of external fertilization in which male and female parents shed gametes into water, and sperm must swim through the water to reach the eggs.

speciation: the process of species formation, in which a single species splits into two or more species.

species (spē′-sēs): the basic unit of taxonomic classification, consisting of a population or group of populations that evolves independently of other populations. In sexually reproducing organisms, a species can be defined as a population or group of populations of organisms that interbreed freely with one another under natural conditions but that do not interbreed with members of other populations.

specific heat: the amount of energy required to raise the temperature of 1 gram of a substance by 1 °C.

sperm: the haploid male gamete, normally small, motile, and containing little cytoplasm.

spermatid: a haploid cell derived from the secondary spermatocyte by meiosis II; differentiates into the mature sperm.

spermatogenesis: the process by which sperm cells form.

spermatogonium (pl., **spermatogonia**)**:** a diploid cell, lining the walls of the seminiferous tubules, that gives rise to a primary spermatocyte.

spermatophore: a package of sperm formed by the males of some invertebrate animals; the spermatophore can be inserted into the female reproductive tract, where it releases its sperm.

sphincter muscle: a circular ring of muscle surrounding a tubular structure, such as the esophagus, stomach, or intestine; contraction and relaxation of a sphincter muscle controls the movement of materials through the tube.

spinal cord: the part of the central nervous system of vertebrates that extends from the base of the brain to the hips and is protected by the bones of the vertebral column; contains the cell bodies of motor neurons that form synapses with skeletal muscles, the circuitry for some simple reflex behaviors, and axons that communicate with the brain.

spindle microtubule: microtubules organized in a spindle shape that separate chromosomes during mitosis or meiosis.

spiracle (spi′-ruh-kul): an opening in the body wall of insects through which air enters the tracheae.

spleen: the largest organ of the lymphatic system, located in the abdominal cavity; contains macrophages that filter the blood by removing microbes and aged red blood cells, and lymphocytes (B and T cells) that reproduce during times of infection.

spongy bone: porous, lightweight bone tissue in the interior of bones; the location of bone marrow. Compare with *compact bone.*

spontaneous generation: the proposal that living organisms can arise from nonliving matter.

sporangium (spor-an′-jē-um; pl., **sporangia):** a structure in which spores are produced.

spore: (1) in plants and fungi, a haploid cell capable of developing into an adult without fusing with another cell (without fertilization); (2) in bacteria and some other organisms, a stage of the life cycle that is resistant to extreme environmental conditions.

sporophyte (spor′-ō-fīt): the multicellular diploid stage in the life cycle of a plant; produces haploid, asexual spores through meiosis.

stabilizing selection: a type of natural selection that favors the average phenotype in a population.

stamen (stā′-men): the male reproductive structure of a flower, consisting of a filament and an anther, in which pollen grains develop.

starch: a polysaccharide that is composed of branched or unbranched chains of glucose molecules; used by plants as a carbohydrate-storage molecule.

start codon: the first AUG codon in a messenger RNA molecule.

startle coloration: a form of mimicry in which a color pattern (in many cases resembling large eyes) can be displayed suddenly by a prey organism when approached by a predator.

stem: the portion of the plant body, normally located aboveground, that bears leaves and reproductive structures such as flowers and fruit.

stem cell: an undifferentiated cell that is capable of dividing and giving rise to one or more distinct types of differentiated cell(s).

sterilization: a generally permanent method of contraception in which the pathways through which the sperm (vas deferens) or egg (oviducts) must travel are interrupted; the most effective form of contraception.

steroid: a lipid consisting of four fused carbon rings, with various functional groups attached.

steroid hormone: a class of hormone whose chemical structure (four fused carbon rings with various functional groups) resembles cholesterol; steroids, which are lipids, are secreted by the ovaries and placenta, the testes, and the adrenal cortex.

stigma (stig′-muh): the pollen-capturing tip of a carpel.

stirrup: the third of the small bones of the middle ear, linking the tympanic membrane with the oval window; the stirrup is directly connected to the oval window; also called the *stapes.*

stoma (stō′-muh; pl., **stomata):** an adjustable opening in the epidermis of a leaf or young stem, surrounded by a pair of guard cells, that regulates the diffusion of carbon dioxide and water into and out of the leaf or stem.

stomach: the muscular sac between the esophagus and small intestine where food is stored and mechanically broken down and in which protein digestion begins.

stop codon: a codon in messenger RNA that stops protein synthesis and causes the completed protein chain to be released from the ribosome.

stramenopile: a member of Stramenopila, a large protist clade. Stramenopiles, which are characterized by hair-like projections on their flagella, include the water molds, the diatoms, and the brown algae.

strand: a single polymer of nucleotides; DNA is composed of two strands wound about each other in a double helix; RNA is usually single stranded.

stratified epithelium: a type of epithelial tissue composed of several cell layers, usually strong and waterproof; mostly found on the surface of the skin.

stroke: an interruption of blood flow to part of the brain caused by the rupture of an artery or the blocking of an artery by a blood clot. Loss of blood supply leads to rapid death of the area of the brain affected.

stroma (strō′-muh): the semifluid material inside chloroplasts in which the thylakoids are located; the site of the reactions of the Calvin cycle.

structural gene: in the prokaryotic operon, the genes that encode enzymes or other cellular proteins.

style: a stalk connecting the stigma of a carpel with the ovary at its base.

subclimax: a community in which succession is stopped before the climax community is reached; it is maintained by regular disturbance—for example, a tallgrass prairie maintained by periodic fires.

substrate: the atoms or molecules that are the reactants for an enzyme-catalyzed chemical reaction.

succession (suk-seh′-shun): a structural change in a community and its nonliving environment over time. During succession, species replace one another in a somewhat predictable manner until a stable, self-sustaining climax community is reached.

sucrose: a disaccharide composed of glucose and fructose.

sugar-phosphate backbone: a chain of sugars and phosphates in DNA and RNA; the sugar of one nucleotide bonds to the phosphate of the next nucleotide in a DNA or RNA strand. The bases in DNA or RNA are attached to the sugars of the backbone.

sugar: a simple carbohydrate molecule, either a monosaccharide or a disaccharide.

surface tension: the property of a liquid to resist penetration by objects at its interface with the air, due to cohesion between molecules of the liquid.

survivorship curve: the curve that results when the number of individuals of each age in a population is graphed against their age, usually expressed as a percentage of their maximum life span.

survivorship table: a data table that groups organisms born at the same time and tracks them throughout their life span, recording how many continue to survive in each succeeding year (or other unit of time). Various parameters such as sex may be used in the groupings. Human life tables may include many other parameters (such as socioeconomic status) used by demographers.

sustainable development: human activities that meet present needs for a reasonable quality of life without exceeding nature's limits and without compromising the ability of future generations to meet their needs.

sympathetic division: the division of the autonomic nervous system that produces largely involuntary responses that prepare the body for stressful or highly energetic situations; often called the *sympathetic nervous system.*

sympatric speciation (sim-pat′-rik): the process by which new species arise in populations that are not physically divided; the genetic isolation required for sympatric speciation may be due to ecological isolation or chromosomal aberrations (such as polyploidy).

synapse (sin′-aps): the site of communication between nerve cells. At a synapse, one cell (presynaptic) normally releases a chemical (the neurotransmitter) that changes the electrical potential of the second (postsynaptic) cell.

synaptic cleft: in a synapse, a small gap between the presynaptic and postsynaptic neurons.

synaptic terminal: a swelling at the branched ending of an axon; where the axon forms a synapse.

syphilis (si′-ful-is): a sexually transmitted bacterial infection of the reproductive organs; if untreated, can damage the nervous and circulatory systems.

systematics: the branch of biology concerned with reconstructing phylogenies and with naming clades.

systolic pressure (sis′ -tal -ik): the blood pressure measured at the peak of contraction of the ventricles; the higher of the two blood pressure readings.

T cell: a type of lymphocyte that matures in the thymus, and that recognizes and destroys specific foreign cells or substances or that regulates other cells of the immune system.

T tubule: a deep infolding of the muscle plasma membrane; conducts the action potential inside a cell.

T-cell receptor: a protein receptor, located on the surface of a T cell, that binds a specific antigen and triggers the immune response of the T cell.

taiga (tī′-guh): a biome with long, cold winters and only a few months of warm weather; dominated by evergreen coniferous trees; also called *northern coniferous forest.*

taproot system: a root system, commonly found in dicots, that consists of a long, thick main root and many smaller lateral roots that grow from the main root.

target cell: a cell on which a particular hormone exerts its effect.

taste bud: a cluster of taste receptor cells and supporting cells that is located in a small pit beneath the surface of the tongue and that communicates with the mouth through a small pore.

taxonomy (tax-on′-uh-mē): the branch of biology concerned with naming and classifying organisms.

tectorial membrane (tek-tor′-ē-ul): one of the membranes of the cochlea in which the hairs of the hair cells are embedded. In sound reception, movement of the basilar membrane relative to the tectorial membrane bends the hairs.

telomere (tē-le-mēr): the nucleotides at the end of a chromosome that protect the chromosome from damage during condensation, and prevent the end of one chromosome from attaching to the end of another chromosome.

telophase (tēl-ō-fāz): in mitosis and both divisions of meiosis, the final stage, in which the spindle fibers usually disappear, nuclear envelopes re-form, and cytokinesis generally occurs. In mitosis and meiosis II, the chromosomes also relax from their condensed form.

temperate deciduous forest: a biome having cold winters and warm summers, with enough summer rainfall for trees to grow and shade out grasses; characterized by trees that drop their leaves in winter (deciduous trees), an adaptation that minimizes water loss when the soil is frozen.

temperate rain forest: a temperate biome with abundant liquid water year-round, dominated by conifers.

template strand: the strand of the DNA double helix from which RNA is transcribed.

tendon: a tough connective tissue band connecting a muscle to a bone.

terminal bud: meristem tissue and surrounding leaf primordia that are located at the tip of a plant shoot or a branch.

territoriality: the defense of an area in which important resources are located.

tertiary consumer (ter′-shē-er-ē): a carnivore that feeds on other carnivores (secondary consumers).

tertiary structure (ter′-shē-er-ē): the complex three-dimensional structure of a single peptide chain; held in place by disulfide bonds between cysteines.

test cross: a breeding experiment in which an individual showing the dominant phenotype is mated with an individual that is homozygous recessive for the same gene. The ratio of offspring with dominant versus recessive phenotypes can be used to determine the genotype of the phenotypically dominant individual.

testis (pl., **testes**)**:** the gonad of male animals.

testosterone: in vertebrates, a hormone produced by the interstitial cells of the testis; stimulates spermatogenesis and the development of male secondary sex characteristics.

thalamus: the part of the forebrain that relays sensory information to many parts of the brain.

therapeutic cloning: the production of a clone for medical purposes. Typically, the nucleus from one of a patient's own cells would be inserted into an egg whose nucleus had been removed; the resulting cell would divide and produce embryonic stem cells that would be compatible with the patient's tissues and therefore would not be rejected by the patient's immune system.

thick filament: in the sarcomere, a bundle of myosin that interacts with thin filaments, producing muscle contraction.

thigmotropism: growth in response to touch.

thin filament: in the sarcomere, a protein strand that interacts with thick filaments, producing muscle contraction; composed primarily of actin, plus the accessory proteins troponin and tropomyosin.

threatened species: all species classified as critically endangered, endangered, or vulnerable.

threshold: the electrical potential at which an action potential is triggered; the threshold is usually about 10 to 20 mV less negative than the resting potential.

thrombin: an enzyme produced in the blood as a result of injury to a blood vessel; catalyzes the production of fibrin, a protein that assists in blood clot formation.

thylakoid (thī′-luh-koid): a disk-shaped, membranous sac found in chloroplasts, the membranes of which contain the photosystems, electron transport chains, and ATP-synthesizing enzymes used in the light reactions of photosynthesis.

thymine (T): a nitrogenous base found only in DNA; abbreviated as T.

thymosin (thī-mō-sin): a hormone, secreted by the thymus, that stimulates the maturation of T lymphocytes of the immune system.

thymus (thī′-mus): an organ of the lymphatic system that is located in the upper chest in front of the heart and that secretes thymosin, which stimulates maturation of T lymphocytes of the immune system.

thyroid gland: an endocrine gland, located in front of the larynx in the neck, that secretes the hormones thyroxine (affecting metabolic rate) and calcitonin (regulating calcium ion concentration in the blood).

thyroid-stimulating hormone (TSH): a hormone, released by the anterior pituitary, that stimulates the thyroid gland to release hormones.

thyroxine (thī-rox′-in): a hormone, secreted by the thyroid gland, that stimulates and regulates metabolism.

tight junction: a type of cell-to-cell junction in animals that prevents the movement of materials through the spaces between cells.

tissue: a group of (normally similar) cells that together carry out a specific function; a tissue may also include extracellular material produced by its cells.

tissue system: a group of two or more tissues that together perform a specific function.

tonsil: a patch of lymphatic tissue, located at the entrance to the pharynx, that contains macrophages and lymphocytes; destroys many microbes entering the body through the mouth and stimulates an adaptive immune response to them.

trachea (trā′-kē-uh): in terrestrial vertebrates, a flexible tube, supported by rings of cartilage, that conducts air between the larynx and the bronchi.

tracheae (tra′-ke-ē): the respiratory organ of insects, consisting of a set of air-filled tubes leading from openings in the body called spiracles and branching extensively throughout the body.

tracheid (trā′-kē-id): an elongated cell type in xylem, with tapered ends that overlap the tapered ends of other tracheids, forming tubes that transport water and minerals. Pits in the cell walls of tracheids allow easy movement of water and minerals into and out of the cells, including from one tracheid to the next.

trans fat: a type of fat produced during the process of hydrogenating oils that may increase the risk of heart disease. The fatty acids of trans fats include an unusual configuration of double bonds that is not normally found in fats of biological origin.

transcription: the synthesis of an RNA molecule from a DNA template.

transfer RNA (tRNA): a type of RNA that binds to a specific amino acid, carries it to a ribosome, and positions it for incorporation into the growing protein chain during protein synthesis. A set of three bases in tRNA (the anticodon) is complementary to the set of three bases in mRNA (the codon) that codes for that specific amino acid in the genetic code.

transformation: a method of acquiring new genes, whereby DNA from one bacterium (normally released after the death of the bacterium) becomes incorporated into the DNA of another, living, bacterium.

transgenic: referring to an animal or a plant that contains DNA derived from another species, usually inserted into the organism through genetic engineering.

translation: the process whereby the sequence of bases of messenger RNA is converted into the sequence of amino acids of a protein.

translocation: a mutation that occurs when a piece of DNA is removed from one chromosome and attached to another chromosome.

transpiration (trans′-per-ā-shun): the evaporation of water through the stomata, chiefly in leaves.

transport protein: a protein that regulates the movement of water-soluble molecules through the plasma membrane.

trial-and-error learning: a type of learning in which behavior is modified in response to the positive or negative consequences of an action.

trichomoniasis (trik-ō-mō-nī′-uh-sis): a sexually transmitted disease, caused by the protist *Trichomonas*, that causes inflammation of the mucous membranes that line the urinary tract and genitals.

triglyceride (trī-glis′-er-īd): a lipid composed of three fatty-acid molecules bonded to a single glycerol molecule.

trisomy 21: see *Down syndrome*.

trisomy X: a condition of females who have three X chromosomes instead of the normal two; most such women are phenotypically normal and are fertile.

trophic level: literally, "feeding level"; the categories of organisms in a community, and the position of an organism in a food chain, defined by the organism's source of energy; includes producers, primary consumers, secondary consumers, and so on.

tropical deciduous forest: a biome, warm all year-round, with pronounced wet and dry seasons; characterized by trees that shed their leaves during the dry season (deciduous trees), an adaptation that minimizes water loss.

tropical rain forest: a biome with evenly warm, evenly moist conditions year-round, dominated by broadleaf evergreen trees; the most diverse biome.

tropical scrub forest: a biome, warm all year-round, with pronounced wet and dry seasons (drier conditions than in tropical deciduous forests); characterized by short, deciduous, often thorn-bearing trees with grasses growing beneath them.

true-breeding: pertaining to an individual all of whose offspring produced through self-fertilization are identical to the parental type. True-breeding individuals are homozygous for a given trait.

tube cell: the outermost cell of a pollen grain, containing the sperm. When the pollen grain germinates, the tube cell produces a tube penetrating through the tissues of the carpel, from the stigma, through the style, and to the opening of an ovule in the ovary.

tubular reabsorption: the process by which cells of the tubule of the nephron remove water and nutrients from the filtrate within the tubule and return those substances to the blood.

tubular secretion: the process by which cells of the tubule of the nephron remove wastes from the blood, actively secreting those wastes into the tubule.

tubule (toob′-ūl): the tubular portion of the nephron; includes a proximal portion, the loop of Henle, and a distal portion. Urine is formed from the blood filtrate as it passes through the tubule.

tundra: a biome with severe weather conditions (extreme cold and wind, and little rainfall) that cannot support trees.

turgor pressure: pressure developed within a cell (especially the central vacuole of plant cells) as a result of osmotic water entry.

Turner syndrome: a set of characteristics typical of a woman with only one X chromosome; women with Turner syndrome are sterile, with a tendency to be very short and to lack typical female secondary sexual characteristics.

tympanic membrane (tim-pan′-ik): the eardrum; a membrane that stretches across the opening of the middle ear and transmits vibrations to the bones of the middle ear.

unicellular: single-celled; most members of the domains Bacteria and Archaea and the kingdom Protista are unicellular.

uniform distribution: the distribution characteristic of a population with a relatively regular spacing of individuals, commonly as a result of territorial behavior.

unsaturated: referring to a fatty acid with fewer than the maximum number of hydrogen atoms bonded to its carbon backbone (therefore, an unsaturated fatty acid has one or more double bonds in its carbon backbone).

upwelling: an upward flow that brings cold, nutrient-laden water from the ocean depths to the surface.

urea (ū-rē′-uh): a water-soluble, nitrogen-containing waste product of amino acid breakdown; one of the principal components of mammalian urine.

ureter (ū′-re-ter): a tube that conducts urine from a kidney to the urinary bladder.

urethra (ū-rē′-thruh): the tube leading from the urinary bladder to the outside of the body; in males, the urethra also receives sperm from the vas deferens and conducts both sperm and urine (at different times) to the tip of the penis.

urinary system: the organ system that produces, stores, and eliminates urine, which contains cellular wastes, excess water and nutrients, and toxic or foreign substances. The urinary system is critical for maintaining homeostatic conditions within the bloodstream. In mammals, it includes the kidneys, ureters, bladder, and urethra.

urine: the fluid produced and excreted by the urinary system, containing water and dissolved wastes, such as urea.

uterine tube: the tube leading from the ovary to the uterus, into which the secondary oocyte (egg cell) is released; also called the *oviduct*, or, in humans, the *Fallopian tube*.

uterus: in female mammals, the part of the reproductive tract that houses the embryo during pregnancy.

utricle: a patch of hair cells in the vestibule of the inner ear; bending of the hairs of the hair cells permits detection of the direction of gravity and the degree of tilt of the head.

vaccine: an injection into the body that contains antigens characteristic of a particular disease organism and that stimulates an immune response appropriate to that disease organism.

vacuole (vak′-ū-ōl): a vesicle that is typically large and consists of a single membrane enclosing a fluid-filled space.

vagina: the passageway leading from the outside of a female mammal's body to the cervix of the uterus; serves as the receptacle for semen and as the birth canal.

variable: a factor in a scientific experiment that is deliberately manipulated in order to test a hypothesis.

variable region: the part of an antibody molecule that differs among antibodies; the ends of the variable regions of the light and heavy chains form the specific binding site for antigens.

vas deferens (vaz de′-fer-enz): the tube connecting the epididymis of the testis with the urethra.

vascular bundle: a strand of xylem and phloem in leaves and stems; in leaves, commonly called a *vein*.

vascular cambium: a lateral meristem that is located between the xylem and phloem of a woody root or stem and that gives rise to secondary xylem and phloem.

vascular cylinder: the centrally located conducting tissue of a young root; consists of primary xylem and phloem, surrounded by a layer of pericycle cells.

vascular plant (vas′-kū-lar): a plant that has conducting vessels for transporting liquids; also called a tracheophyte.

vascular tissue system: a plant tissue system consisting of xylem (which transports water and minerals from root to shoot) and phloem (which transports water and sugars throughout the plant).

vein: in vertebrates, a large-diameter, thin-walled vessel that carries blood from venules back to the heart; in plants, a vascular bundle in a leaf.

ventricle (ven′-tre-kul): the lower muscular chamber on each side of the heart that pumps blood out through the arteries. The right ventricle sends blood to the lungs; the left ventricle pumps blood to the rest of the body.

venule (ven′-ūl): a narrow vessel with thin walls that carries blood from capillaries to veins.

vertebral column (ver-tē′-brul): a column of serially arranged skeletal units (the vertebrae) that enclose the nerve cord in vertebrates; the backbone.

vertebrate: an animal that has a vertebral column.

vesicle (ves′-i-kul): a small, membrane-bound sac within the cytoplasm.

vessel: in plants, a tube of xylem composed of vertically stacked vessel elements with heavily perforated or missing end walls, leaving a continuous, uninterrupted hollow cylinder.

vessel element: one of the cells of a xylem vessel; elongated, dead at maturity, with thick lateral cell walls for support but with end walls that are either heavily perforated or missing.

vestibular apparatus: part of the inner ear, consisting of the vestibule and the semicircular canals, involved in the detection of gravity, tilt of the head, and movement of the head.

vestigial structure (ves-tij′-ē-ul): a structure that serves no apparent purpose but is homologous to functional structures in related organisms and provides evidence of evolution.

villus (vi′-lus; pl., **villi):** a finger-like projection of the wall of the small intestine that increases its absorptive surface area.

viroid (vī′-roid): a particle of RNA that is capable of infecting a cell and of directing the production of more viroids; responsible for certain plant diseases.

virus (vī′-rus): a noncellular parasitic particle that consists of a protein coat surrounding genetic material; multiplies only within a cell of a living organism (the host).

vitamin: one of a group of diverse chemicals that must be present in trace amounts in the diet to maintain health; used by the body in conjunction with enzymes in a variety of metabolic reactions.

vitreous humor (vit′-rē-us): a clear, jelly-like substance that fills the large chamber of the eye between the lens and the retina; helps to maintain the shape of the eyeball.

vocal cords: a pair of bands of elastic tissue that extend across the opening of the larynx and produce sound when air is forced between them. Muscles alter the tension on the vocal cords and control the size and shape of the opening, which in turn determines whether sound is produced and what its pitch will be.

vulnerable species: a species that is likely to become endangered unless conditions that threaten its survival improve.

waggle dance: a symbolic form of communication used by honeybee foragers to communicate the location of a food source to their hive mates.

warning coloration: bright coloration that warns predators that the potential prey is distasteful or even poisonous.

water mold: a member of a protist group that includes species with filamentous shapes that give them a superficially fungus-like appearance. Water molds, which include species that cause economically important plant diseases, are part of a larger group known as the stramenopiles.

wax: a lipid composed of fatty acids covalently bonded to long-chain alcohols.

weather: short-term fluctuations in temperature, humidity, cloud cover, wind, and precipitation in a region over periods of hours to days.

wetland: a region (sometimes called a marsh, swamp, or bog) in which the soil is covered by, or saturated with, water for a significant part of the year.

white matter: the portion of the brain and spinal cord that consists largely of myelin-covered axons and that gives these areas a white appearance.

wildlife corridor: a strip of protected land linking larger areas. Wildlife corridors allow animals to move freely and safely between habitats that would otherwise be isolated by human activities.

work: energy transferred to an object, usually causing the object to move.

working memory: the first phase of learning; short-term memory that is electrical or biochemical in nature.

X chromosome: the female sex chromosome in mammals and some insects.

xylem (zī-lum): a conducting tissue of vascular plants that transports water and minerals from root to shoot.

Y chromosome: the male sex chromosome in mammals and some insects.

yolk: protein-rich or lipid-rich substances contained in eggs that provide food for the developing embryo.

yolk sac: one of the embryonic membranes of reptiles (including birds) and mammals. In reptiles, the yolk sac is a membrane surrounding the yolk in the egg; in mammals, it forms part of the umbilical cord and the digestive tract but does not contain yolk.

Z line: a fibrous protein structure to which the thin filaments of skeletal muscle are attached; forms the boundary of a sarcomere.

zona pellucida (pel-oo′-si-duh): a clear, noncellular layer between the corona radiata and the egg.

zooplankton: nonphotosynthetic protists that are abundant in marine and freshwater environments.

zygomycete: a fungus of the phylum Zygomycota, which includes the species that cause fruit rot and bread mold.

zygosporangium (zī′-gō-spor-an-jee-um): a tough, resistant reproductive structure produced by fungi in the phylum Zygomycota; encloses diploid nuclei that undergo meiosis and give rise to haploid spores.

zygote (zī′-gōt): in sexual reproduction, a diploid cell (the fertilized egg) formed by the fusion of two haploid gametes.

Answers to Selected Questions

Chapter 1

Figure Caption Questions

Figure 1-1 Global warming will impact the entire biosphere. The growing human population—whose energy use is driving global warming and whose demand for space and resources is causing widespread habitat destruction—would also be an appropriate answer.

Figure 1-3 You might answer "no" at this point, but keep this art in mind as you read the section on endosymbiosis in Chapter 17.

Figure 1-5 The antibacterial chemicals produced by fungi probably evolved because they improved the fungi's ability to compete with bacteria for access to resources such as food and space (by excluding bacteria from areas where the fungi are present).

Figure E1-1 Redi's experiment demonstrated that the maggots were caused by something that was excluded by a gauze cover, but the possibility remained that some agent other than flies produced the maggots. An effective follow-up experiment might involve a series of closed, meat-containing systems that were identical in all respects other than the addition of a single possible causal element. Perhaps one container would have flies added, one roaches, one dust or soot, and so on. And, of course, a control would have nothing added.

Fill-in-the-Blank

1. atom; cell; carbon, organic molecules; tissues
2. scientific theory; hypothesis; scientific method
3. evolution; natural selection
4. deoxyribonucleic acid, DNA; genes
5. homeostasis; stimuli; energy, materials; organized, complex

Chapter 2

Figure Caption Questions

Figure 2-2 Atoms with outer shells that are not full are unstable. They become stable by filling (or emptying) their outer shells. Except for hydrogen, atoms in biological molecules have outer shells that are stable when filled with eight electrons.

Figure 2-4 Free radicals have atoms (often oxygen atoms) with one or more unpaired electrons in their outer shells, making them very unstable and causing them to capture electrons from nearby molecules to complete their outer shells. This can cause damage to biological molecules, including DNA, which are critical to proper cellular functioning.

Figure E2-2 Go to Table 2-1 and compare the atomic mass of H_2 and He (note that the mass of H_2 would be double the mass of H, and so would have an atomic mass of 2). Because H_2 is lighter than He, it would float more readily in air (which is composed primarily of N_2 and O_2).

Fill-in-the-Blank

1. protons, neutrons; atomic number; electrons, electron shells
2. ion; positive; negative; ionic
3. isotopes; energy; radioactive
4. inert; reactive; share
5. free radicals; DNA; antioxidants
6. polar; hydrogen; cohesion

Chapter 3

Figure Caption Questions

Figure 3-7 During hydrolysis of sucrose, water would be split; a hydrogen atom from water would be added to the oxygen from glucose (that formerly linked the two subunits) and the remaining OH from water would be added to the carbon (formerly bonded to oxygen) of the fructose subunit.

Figure 3-12 The enzyme would hydrolyze the triglyceride into a glycerol and three fatty acids, using up three water molecules.

Figure 3-15 As lipids, steroids are soluble in the lipid-based cell membrane and can cross it (and the nuclear membrane) to act inside the cell. Other types of hormones (mostly peptides) are not lipid soluble and cannot easily cross the cell membrane.

Figure 3-20 Heat energy can break chemical bonds, and the hydrogen bonds that account for secondary (and higher level) protein structure are especially susceptible to heat. Because a protein's functional ability usually depends on its shape, breaking shape-controlling hydrogen bonds disrupts function.

Fill-in-the-Blank

1. monomer, polymers; polysaccharides; hydrolysis; cellulose, starch, glycogen; sucrose, lactose, maltose
2. oil; wax; fat; steroids; cholesterol; phospholipid
3. dehydration, water; amino acids; primary; helix, pleated sheet; pleated sheet; denatured
4. double bonds (or double covalent bonds); hydrogen bonds; disulfide bonds; hydrogen bonds; peptide bonds
5. ribose or deoxyribose sugar, base, phosphate; adenosine triphosphate (ATP); adenine, guanine, cytosine, thymine; deoxyribonucleic acid (DNA), ribonucleic acid (RNA); phosphate
6. genetic material (DNA, RNA, or nucleic acids); vCJD; bovine spongiform encephalitis (BSE); scrapie

Chapter 4

Figure Caption Questions

Figure 4-4 Of the four structures listed, only the ribosome is found in all of the main branches of life (i.e., found in bacteria, archaea, and all eukaryotes). Thus, ribosomes must have been present in the common ancestor of all living cells, and nuclei, mitochondria, and chloroplasts must have arisen later.

Figure 4-7 Fluid would not move upward because the beating of the flagella would direct fluid straight out from the cell membranes. Mucus and trapped particles would accumulate in the trachea.

Figure 4-9 By forming chromosomes during cell division, the DNA can be parcelled evenly to the resulting daughter cells, with each receiving a complete copy of the genetic information.

Figure 4-13 Because membranes are fundamentally similar, they can merge with one another. This allows molecules in vesicles to be transferred from one membrane-enclosed structure (such as endoplasmic reticulum) to another (such as the Golgi apparatus). The material can be exported from the cell when the vesicle merges with the plasma membrane.

Fill-in-the-Blank

1. phospholipids, proteins; phospholipid; protein; protein
2. cytoskeleton; microfilaments, intermediate filaments, microtubules; microtubules
3. ribosomes; endoplasmic reticulum; nucleolus; Golgi apparatus; cell wall; messenger RNA
4. rough endoplasmic reticulum; vesicles, Golgi apparatus; carbohydrate; plasma
5. mitochondria; chloroplasts; cell wall; nucleoid; cilia; cytoplasm
6. mitochondria, chloroplasts; double, ATP, DNA; size

Chapter 5

Figure Caption Questions

Figure 5-9 The distilled water made the solution hypotonic to the blood cells, causing enough water to flow in to burst their fragile cell membranes.

Figure 5-10 The rigid cell wall of plant cells counteracts the pressure exerted by water entering by osmosis, so the cell does not burst. Animal cells lack a cell wall, and when placed in a highly hypotonic solution will absorb water by osmosis until the cell membrane bursts.

Figure 5-15 Exocytosis uses cellular energy, whereas diffusion is passive. In addition, while materials move through a membrane during diffusion out of a cell, in exocytosis, materials are expelled without passing directly through the plasma membrane. Exocytosis allows materials that are too large to pass through membranes to be eliminated from the cell.

Fill-in-the-Blank

1. phospholipids; receptor, recognition, enzymes, attachment, transport
2. selectively permeable; diffusion; osmosis; aquaporins; active transport
3. kinks (or bends or double bonds); pressure, temperature
4. channel, carrier; simple, lipids
5. active transport, endocytosis, exocytosis; active transport
6. simple diffusion; simple diffusion; facilitated diffusion (may also move by active transport); facilitated diffusion
7. endocytosis; yes; pinocytosis, phagocytosis; vesicles

Chapter 6

Figure Caption Questions

Figure 6-6 Other sources of energy that might overcome activation energy include mechanical energy (e.g., shaking), electricity, and radiation.

Figure 6-10 No, endergonic reactions do not occur spontaneously because they require an input of energy, whether or not a catalyst is present.

Fill-in-the-Blank

1. created, destroyed; kinetic, potential
2. more, less; organized; entropy
3. activation; electron shells; heat or movement of molecules
4. exergonic; endergonic; exergonic; endergonic; coupled
5. adenosine triphosphate; energy carrier; adenosine diphosphate, phosphate; energy
6. proteins; activation energy; active site; shape, electrical charges

Chapter 7

Figure Caption Questions

Figure 7-9 No. Because oxygen is generated during the light reactions, these bubbles would not form in darkness.

Figure 7-11 The C_4 pathway is less efficient than the C_3 pathway; C_4 uses one extra ATP per CO_2 molecule (to regenerate PEP from pyruvate). Thus, when CO_2 is abundant and photorespiration is not a problem, C_3 plants produce sugar at a lower energy cost, and they outcompete C_4 plants.

Fill-in-the-Blank

1. stomata, oxygen (O_2), carbon dioxide (CO_2); water loss (evaporation); chloroplasts, mesophyll; bundle sheath
2. red, blue, violet; green; carotenoids
3. photosystem II, chlorophyll *a*, primary electron acceptor; electron transport chain (ETC, or ETC II); H+ (hydrogen ions); chemiosmosis
4. water (H_2O), carbon dioxide (CO_2); carbon fixation; rubisco, oxygen
5. both; both; C_4 plants, CAM plants
6. ATP, NADPH, Calvin; RuBP; G3P, glucose

Chapter 8

Figure Caption Questions

Figure 8-6 Without oxygen, ATP production halts. Oxygen is the final acceptor in the electron transport chain. If it is not present, electrons cannot proceed along the chain, and production of ATP by chemiosmosis ceases.

Figure 8-10 In oxygen-rich environments, aerobic bacteria would prevail because their respiration produces far more ATP per glucose molecule. Some anaerobes would survive but would generate far less energy, while others would be poisoned by the oxygen. In oxygen-poor environments, aerobic bacteria are limited by the oxygen shortage (unless they can switch to fermentation), and anaerobes would prevail.

Fill-in-the-Blank

1. glycolysis, cellular respiration; cytoplasmic fluid, mitochondria; cellular respiration
2. anaerobic; glycolysis, two; fermentation, NAD
3. ethanol, carbon dioxide; lactic acid; lactic acid
4. erythropoietin (EPO), red blood, oxygen; cellular respiration
5. matrix, intermembrane space, concentration gradient; chemiosmosis; ATP synthase
6. Krebs, citric acid; acetyl CoA; two; NADH, $FADH_2$

Chapter 9

Figure Caption Questions

Figure 9-8 If the sister chromatids of one replicated chromosome failed to separate, then one daughter cell would not receive any copy of that chromosome, whereas the other daughter cell would receive both copies.

Figure 9-15 If one pair of homologues failed to separate at anaphase I, one of the resulting daughter cells (and the gametes produced from it) would have both homologues and the other daughter cell (and the gametes produced from it) would not have any copies of that homologue.

Fill-in-the-Blank

1. binary fission
2. mitotic, differentiation
3. homologues OR homologous chromosomes; autosomes; sex chromosomes
4. prophase, metaphase, anaphase, telophase; cytokinesis; telophase
5. kinetochore; polar
6. four; meiosis I; gametes OR sperm and eggs
7. prophase; chiasmata; crossing over
8. shuffling of homologues; crossing over; fusion of gametes

Chapter 10

Figure Caption Questions

Figure 10-8 Half of the gametes produced by a *Pp* plant will have the *P* allele, and half will have the *p* allele. All of the gametes produced by a *pp* plant will have the *p* allele. Therefore, half of the offspring of a *Pp* × *pp* cross will be *Pp* (purple) and half will be *pp* (white), whereas all of the offspring of a *PP* × *pp* cross will be *Pp* (purple). See Figure 10-9.

Figure 10-11 A plant with wrinkled green seeds has the genotype *ssyy*. A plant with smooth yellow seeds could be *SSYY*, *SsYY*, *SSYy*, or *SsYy*. Set up four Punnett squares to see if the smooth yellow plant's genotype can be revealed by a test cross.

Figure 10-12 Chromosomes, not individual genes, assort independently during meiosis. Therefore, if the genes for seed color and seed shape were on the same chromosome, they would tend to be inherited together, and would not assort independently.

Figure 10-24 One of Victoria and Albert's sons, Leopold, had hemophilia. To be male, Leopold must have inherited Albert's Y chromosome. The X chromosome, not the Y chromosome, bears the gene for blood clotting, so Leopold must have inherited the hemophilia allele from his mother, Victoria.

Fill-in-the-Blank

1. locus; alleles; mutations
2. genotype; phenotype; heterozygous
3. independently; as a group; linked
4. an X and a Y, two X; sperm
5. sex linked
6. incomplete dominance; codominance; polygenic inheritance

Chapter 11

Figure Caption Questions

Figure 11-5 It takes more energy to break apart a C–G base pair, because these are held together by three hydrogen bonds, compared with the two hydrogen bonds that bind A to T.

Figure E11-7 DNA polymerase always moves in the 3′ to 5′ direction on a parental strand. Because the two strands of a DNA double helix are oriented in opposite directions, the 5′ direction on one strand leads toward the replication fork and the 5′ direction on the other strand leads away from the fork. Therefore, DNA polymerase must move in opposite directions on the two strands.

Fill-in-the-Blank

1. nucleotides; sugar (deoxyribose), phosphate, base (order not important)
2. phosphate, sugar (order not important); double helix
3. thymine, cytosine; complementary
4. semiconservative
5. DNA helicase; DNA polymerase; DNA ligase
6. mutations; nucleotide substitution, point mutation

Chapter 12

Figure Caption Questions

Figure 12-3 RNA polymerase always travels in the 3′ to 5′ direction on the template strand. Because the two DNA strands run in opposite directions, if the other DNA strand were the template strand, then RNA polymerase must travel in the opposite direction (that is, right to left in this illustration).

Figure 12-4 Cells produce far larger amounts of some proteins than others. Obvious examples include cells that produce antibodies or protein hormones, which they secrete into the bloodstream in vast quantities, affecting functioning throughout the body. If a cell needs to produce more of certain proteins, it will probably synthesize more mRNA, which will then be translated into that protein.

Figure 12-7 Grouped in codons, the original mRNA sequence visible here is CGA AUC UAG UAA. Changing all G to U would produce the sequence CUA AUC UAU UAA. The two changes are in the first codon (CGA to CUA) and the third codon (UAG to UAU). Refer to the genetic code illustrated in Table 12-3. First, CGA encodes arginine, while CUA encodes leucine, so the first G → U change would substitute leucine for arginine in the protein. Second, UAG is a stop codon, but UAU encodes tyrosine. Therefore, the second G → U change would add tyrosine to the protein instead of stopping translation. The final codon in the illustration, UAA, is a stop codon, so the new protein would end with tyrosine.

Figure E12-1 The mold would be able to grow if any of ornithine, citrulline, or arginine were added to the medium.

Fill-in-the-Blank

1. transcription; translation; ribosome
2. messenger RNA; transfer RNA; ribosomal RNA (order not important); microRNA
3. three; codon; anticodon
4. RNA polymerase; template; promoter; termination signal
5. start, stop; transfer; peptide
6. point; insertion; deletion

Chapter 13

Figure Caption Questions

Figure 13-3 *Thermus aquaticus* lives in hot springs, and its DNA polymerase has evolved to be efficient at high temperatures. The DNA polymerase would probably work more slowly at both higher and lower temperatures than the springs in which *T. aquaticus* resides; 70 °C is close to optimal.

Figure 13-7 As they do for other genes, each person normally has two copies of each STR gene, one on each of a pair of homologous chromosomes. A person may be homozygous (two copies of the same allele) or heterozygous (one copy of each of two alleles) for each STR. The bands on the gel represent individual alleles of an STR gene. Therefore, a single person can have one band (if homozygous) or two bands (if heterozygous). If a person is homozygous for an STR allele, then he has two copies of the same allele. The DNA from both (identical) alleles will run in the same place on the gel and, therefore, that (single) band will have twice as much DNA as each of the two bands of DNA from a heterozygote. The more DNA, the brighter the band.

Figure 13-11 A heterozygote has one normal globin allele and one sickle-cell allele. Therefore, MstII would produce two DNA bands from the normal allele and one from the sickle-cell allele, for a total of three bands of different sizes. The amount of DNA in each of the bands from heterozygotes would be half as much as from the homozygotes, because DNA from heterozygotes contains only one copy of each type of allele, whereas homozygous normal DNA has two copies of the normal allele, and homozygous sickle-cell DNA has two copies of the sickle-cell allele. Therefore, bands from heterozygotes should be about half as bright as bands from homozygotes.

Fill-in-the-Blank

1. genetically modified organisms OR transgenic organisms
2. transformation; plasmids
3. polymerase chain reaction
4. short tandem repeats (STRs); length, or size, or number of repeats; DNA profile
5. electrophoresis OR gel electrophoresis; DNA probe, base pairing, OR hydrogen bonds

Chapter 14

Figure Caption Questions

Figure 14-6 No. Mutations, the ultimate source of the phenotypic variation on which natural selection acts, occur in all organisms, including those that reproduce asexually.

Figure 14-7 No. Evolution can include changes in traits that are not revealed in morphology, such as physiological systems and metabolic pathways. More generally, evolution in the sense of changes in a species' gene pool is inevitable in all lineages; genetic evolution is not necessarily reflected in morphological change.

Figure 14-9 Possibilities include coccyx (tailbone)—homologous with tail bones of a cat (or any tetrapod); goosebumps—homologous with erectile hairs of a chimp (or any mammal; used for aggressive displays and insulation); appendix—homologous with the cecum of a rabbit (and other herbivorous mammals; extension of large intestine used for storage); wisdom teeth—homologous with grinding rear molars of a leaf-eating monkey (or other mammals); ear-wiggling muscles (the muscles around the ear that some people can use to make their ears move)—homologous with muscles in a dog (and other mammals) that enable the external ear to orient toward sound.

Figure 14-10 Analogous. Bird tails consist of feathers; dog tails consist of bone/muscle/skin.

Fill-in-the-Blank

1. wing, arm; analogous, convergent; vestigial
2. common ancestor; amino acids, ATP
3. catastrophism; uniformitarianism; old
4. evolution; mutations, DNA
5. natural selection; artificial selection
6. many traits are inherited; Gregor Mendel

Chapter 15

Figure Caption Questions

Figure 15-3 The surviving colonies would be in different places on each treated plate, and there would be a different number of surviving colonies on each plate (because the antibiotic-caused mutations would arise unpredictably, depending on which bacteria happened to interact with the antibiotic such that mutations were caused). Another possibility would be for all of the colonies to survive (if the antibiotic always caused mutations in every colony).

Figure 15-5 For a locus with two alleles, one dominant and one recessive, there are two possible phenotypes. A mating between a heterozygote and a homozygote-recessive yields offspring with a 50:50 ratio of the two phenotypes.

	B	*b*
b	*Bb* (black)	*bb* (brown)
b	*Bb* (black)	*bb* (brown)

Figure 15-6 Allele *B* should behave roughly as it does in the size 4 population, its frequency drifting to fixation or loss in almost all cases. But, because the population is a bit larger, the allele should, on average, take a longer time (greater number of generations) to reach fixation or loss. The longer period of drift should also allow for more reversals of direction (e.g., frequency drifting down, then up, then down again, etc.) than in the size 4 population.

Figure 15-7 Mutations inevitably and continually add variability to a population and, after the population becomes larger, the counteracting, diversity-reducing effects of drift decrease. The net result is an increase in genetic diversity.

Figure 15-11 Greater for males. A female's reproductive success is limited by her maximum litter size, but a male's potential reproductive success is limited only by the number of available females. When, as in bighorn sheep, males battle for access to females, the most successful males can impregnate many females, whereas unsuccessful males may not fertilize any females at all. Thus, the difference between the most and least successful male can be very large. In contrast, even the most successful female can have only one litter of offspring per breeding season, which is not that many more offspring than a female who fails to reproduce.

Figure 15-13 There is always a limit to directional selection. As a trait becomes more extreme, eventually the cost of increasing it further outweighs the benefits (for example, the cost of obtaining extra food may outweigh the benefit of larger size), or it may be physically impossible for the trait to become more extreme (for example, a limb's length may be limited by the maximum length that a bone can attain without breaking under its own weight).

Fill-in-the-Blank

1. Hardy–Weinberg principle, equilibrium, allele; no
2. alleles; nucleotides; mutations; homozygous, heterozygous
3. genotype, phenotype; phenotype
4. genetic drift; small; founder effect, population bottleneck; founder effect
5. the same species; natural selection, coevolution; adaptations
6. reproduction; environment

Chapter 16

Figure Caption Questions

Figure 16-9 Possibilities include continental drift; climate changes (especially glacial advances) that cause habitat fragmentation; formation of islands by volcanic activity or rising sea level; movements of organisms to existing islands (including "islands" of isolated habitats such as lakes, mountaintops, deep-ocean vents); and formation of barriers to movement (e.g., new mountain ranges, deserts, rivers). These processes are indeed sufficiently common and widespread to account for a multitude of speciation events over the history of life.

Figure 16-10 The key question is whether the two populations inhabiting two species of trees (apple and hawthorn) interbreed. Tests might involve careful observation of flies under natural conditions, lab experiments in which captive flies of the two types are provided with opportunities to interbreed, or genetic comparisons to determine the degree of gene flow between the two types of flies.

Figure 16-14 Natural selection cannot look forward and ensure that the only traits that evolve are those that ensure survival of the species as a whole. Instead, natural selection ensures only the preservation of traits that help individuals survive and reproduce more successfully than individuals lacking the trait. So if, in a particular species, highly specialized individuals survive and reproduce better than less-specialized individuals, the specialized phenotype will eventually dominate, even if it ultimately puts the species at greater risk of extinction.

Fill-in-the-Blank

1. populations, independently; reproductive isolation; asexually
2. behavioral isolation; hybrid inviability; temporal isolation; gametic incompatibility; mechanical incompatibility
3. genetically isolated, diverge; allopatric speciation; genetic drift, natural selection
4. adaptive radiation; habitat (or environment)
5. small, specialized; habitat destruction

Chapter 17

Figure Caption Questions

Figure 17-2 The presence of oxygen would prevent the accumulation of organic compounds by quickly oxidizing them or their precursors. All of the successful abiotic synthesis experiments used oxygen-free "atmospheres."

Figure 17-5 The bacterial sequence would be most similar to that of the plant mitochondrion, because (as the descendant of the immediate ancestor of the mitochondrion) the bacterium shares with the mitochondrion a more recent common ancestor than with the chloroplast or the nucleus.

Figure 17-8 Most likely because of competition with seed plants, which had not yet arisen during the period when ferns and club mosses reached large sizes. After seed plants arose, competition from them eventually eliminated other types of plants from many ecological niches, presumably including those niches that favored evolution of large size.

Figure 17-9 No. The mudskipper merely demonstrates the plausibility of a hypothetical intermediate step in the proposed scenario for the origin of land-dwelling tetrapods. But the existence of a modern example similar in form to the hypothetical intermediate form does not provide information about the actual identity of that intermediate form.

Figure 17-11 On average, the rate at which new species arise has been larger than the rate at which species have gone extinct.

Figure 17-19 The African replacement hypothesis. These fossils are the oldest modern humans found so far, and their presence in Africa suggests that modern humans were present in Africa before they were present anywhere else, which, if true, would mean that they originated in Africa.

Figure E17-1 713 million years old. (1:1 ratio means that ½ of the original uranium-235 is left, so its half-life has passed.)

Fill-in-the-Blank

1. anaerobic; photosynthesize; poisonous (toxic, harmful), aerobic (cellular) respiration, energy
2. RNA (ribonucleic acid), enzymes (catalysts), ribozymes
3. eukaryotic; endosymbiotic; aerobic; DNA (deoxyribonucleic acid)
4. swimming, moist (wet); pollen
5. conifers; wind; cold, northern; flowers, insects; efficient
6. arthropods, exoskeletons, drying
7. reptiles, egg; skin; lungs

Chapter 18

Figure Caption Questions

Figure E18-4 Fungi and animals are monophyletic; protists, great apes, seedless plants, and prokaryotes are not. (Descendants of the most recent common ancestor of protists include all the other eukaryotes; for great apes, include humans; for prokaryotes, include eukaryotes; for seedless plants, include the seed plants).

Fill-in-the-Blank

1. taxonomy; systematics; clade
2. genus, species; Latin; capitalized, italic
3. domain, kingdom, phylum, class, order, family, genus, species; Archaea, Bacteria, Eukarya
4. anatomy, DNA sequence; reproduce asexually; phylogenetic species concept, phylogeny (evolutionary tree)
5. 1.5 million, 7 million, 100 million

Chapter 19

Figure Caption Questions

Figure 19-4 Protective structures like endospores are most likely to evolve in environments in which protection is especially advantageous. Compared to other environments inhabited by bacteria, soils are especially vulnerable to drying out, which can be fatal to unprotected bacteria. Bacteria that could resist long dry periods would gain an evolutionary advantage.

Figure 19-5 Enzymes from bacteria that live in hot environments are active at high temperatures (temperatures that usually denature enzymes in organisms from more temperate environments). This ability to function at high temperatures makes the enzymes useful in test tube reactions (such as PCR) that are run at high temperatures.

Figure 19-7 The main advantage is efficiency. In binary fission, every individual produces new individuals. In sexual reproduction, only some individuals (e.g., females) produce offspring. So the average individual produces twice as many offspring by fission as it would by sexual reproduction.

Figure 19-9 The concentration of nitrogen gas would increase, because the major process for removing atmospheric nitrogen would end, whereas the processes that add nitrogen gas to the atmosphere would continue.

Figure 19-12 Viruses lack ribosomes and the rest of the "machinery" required to manufacture proteins.

Figure 19-13 Many viruses replicate by integrating their genetic material into the host cell's genome. Thus, if biotechnologists can insert foreign genetic material into such a virus, the virus will naturally tend to transfer the foreign genes to the cells they infect.

Fill-in-the-Blank

1. bacteria, cell walls, archaea
2. smaller; rod-shaped, corkscrew-shaped, spherical
3. flagella; biofilms; endospores
4. anaerobic; photosynthetic
5. binary fission, conjugation
6. nitrogen-fixing; cellulose
7. tuberculosis, cholera, bubonic plague, syphilis, gonorrhea, pneumonia, Lyme disease, etc.; meat, produce
8. DNA, RNA, protein; host; bacteriophage

Chapter 20

Figure Caption Questions

Figure 20-2 Sex is the process that combines the genomes of two different individuals. In plants and animals, this mixing of genomes occurs only during reproduction. But in many protists (and prokaryotes), genome mixing may occur through conjugation and other processes that take place independently of reproduction (which in many cases occurs by simple mitotic cell division).

Figure 20-7 Convergent evolution. Water molds and fungi live in similar environments and acquire nutrients in similar ways. These ecological similarities have fostered the evolution of superficially similar structures, even though the two taxa are only distantly related.

Fill-in-the-Blank

1. pseudopods; decomposers, parasites
2. algae; protozoa
3. secondary endosymbiosis, photosynthetic protist
4. diplomonad (or excavate); kinetoplastid (or euglenozoan); apicomplexan (or alveolate)
5. water mold (or oomycete or stramenopile); amoebozoan
6. dinoflagellates, diatoms; green algae

Chapter 21

Figure Caption Questions

Figure 21-5 Bryophytes lack lignin (which provides stiffness and support) and conducting cells (which transport materials to distant parts of the body). Conducting cells and stiff stems seem to be required to achieve more than minimal height.

Figure 21-7 All of the pictured structures are sporophytes. In ferns, horsetails, and club mosses, the gametophyte is small and inconspicuous.

Figure 21-9 The most common adaptations are hard, protective shells and incorporation of toxic and/or distasteful chemicals.

Figure 21-12 Both types of pollination persist in angiosperms because the cost–benefit balance and, therefore, the most adaptive pollination system differ, depending on the ecological circumstances of a species.

Type of Pollination	Advantages	Disadvantages
Wind	Not dependent on the presence of animals; no investment in nectar or showy flowers; pollen can disperse over large distances	A larger investment in pollen because most fails to reach an egg; higher chance of failure to fertilize any egg
Animal	Each pollen grain has a much greater chance of reaching a suitable egg	Depends on the presence of animals; must invest in nectar and showy flowers

Fill-in-the-Blank

1. green algae; nonvascular plants, vascular plants; embryos, alternation of generations
2. cuticle, stomata, water; conducting cells, lignin; water, nutrients
3. swim to the egg; seeds, pollen; gymnosperms, angiosperms; attract pollinators; facilitate seed dispersal
4. hornworts, liverworts, mosses; club mosses, horsetails, ferns; angiosperms

Chapter 22

Figure Caption Questions

Figure 22-1 Its filamentous shape helps the fungal body to penetrate and extend into its food sources, and is also a shape that maximizes the ratio of surface area to interior volume (which maximizes the area available for absorbing nutrients). The extreme thinness of the filaments ensures that no cell is very far from the surface at which nutrients are absorbed.

Figure 22-8 No. The common source of the two hyphae means that they will both inherit the same mating type. Hyphae must be of different mating types in order to reproduce sexually.

Figure 22-9 Haploid (although there might be some diploid cells in the basidia concealed in the gills of these haploid mushrooms).

Figure 22-19 In nature, bacteria compete with fungi for access to food and living space. The antibiotic chemicals produced by fungi serve as a defense against competition from bacteria.

Fill-in-the-Blank

1. reproduction; spores
2. mycelium, hyphae, septa; chitin
3. mitotic, one; meiotic, zygote, one
4. glomeromycetes; chytrids; basidiomycetes
5. lichens; mycorrhizae; endophytes
6. wood (cellulose and lignin); yeasts; athlete's foot, jock itch, vaginal infections, ringworm, histoplasmosis, valley fever (many possible answers)

Chapter 23

Figure Caption Questions

Figure 23-4 Sponges are "primitive" only in the sense that their lineage arose early in the evolutionary history of animals and their body plan is comparatively simple. But early origin and simplicity do not determine effectiveness, and the sponge body plan and way of life are clearly suitable for excellent survival and reproduction in many habitats.

Figure 23-6 (a) polyp, (b) medusa, (c) polyp, (d) medusa

Figure 23-10 Parasitic tapeworms have no gut and absorb nutrients across their body surfaces. Their ribbon-like shape maximizes surface area for absorption and allows the worm body to extend through the greatest possible area of the host's body (to be in contact with as many nutrients as possible).

Figure 23-11 Two openings allow one-way travel of food through the gut. One-way movement allows more efficient digestion than two-way movement; digestive waste from which all nutrients have been extracted can be excreted quickly without the need for reverse travel back along the gut, and food can be processed more quickly.

Figure 23-12 Water travels easily through the moist epidermis of a leech. When a high concentration of salt is dissolved in the moisture on the outside of a leech's body, water moves rapidly out of the leech's body by osmosis, dehydrating and ultimately killing the animal.

Fill-in-the-Blank

1. three, endoderm, mesoderm, ectoderm; two, mesoderm
2. cephalized; sensing the environment, ingesting food; cavity, mesoderm
3. protostome; deuterostome, echinoderms, chordates
4. invertebrates, vertebrates; invertebrates; sponges, single-celled organisms; cnidarians; annelids
5. bivalves, gastropods, cephalopods; arthropods; insects, arachnids, crustaceans
6. annelids, arthropods (chordates also correct); closed circulatory; open circulatory, hemocoel
7. mollusks, cnidarians, echinoderms, arthropods (chordates also correct)

Chapter 24

Figure Caption Questions

Figure 24-7 A freshwater fish's body is immersed in a hypotonic solution, so water tends to continuously enter the body by osmosis. The physiological challenge is to get rid of all this excess water. For a saltwater fish, the challenge is reversed. The surrounding solution is hypertonic, so water tends to leave the body. The physiological challenge is to retain sufficient water.

Figure 24-9 One advantage is that adults and juveniles occupy different habitats and therefore do not compete with one another for resources (the niche occupied by an individual over its lifetime is broadened).

Figure 24-12 Flight is a very expensive trait (consumes a lot of energy, requires many special structures). In circumstances in which the benefits of flight are low, such as in habitats without predators or in species whose size is very large, natural selection may favor individuals that forgo an investment in flight, and flightlessness can arise.

Fill-in-the-Blank

1. hollow, dorsal; anus, notochord
2. tunicates, lancelets, hagfishes (order not important); skull; hagfishes
3. lungfishes; cartilage; ray-finned fishes; jaws
4. mammals, amphibians, reptiles, amphibians
5. monotremes; salamanders; bats

Chapter 25

Figure Caption Questions

Figure 25-1 One possibility is that the variation necessary for selection has never arisen. (If no members of the species by chance gain the ability to discriminate between their own chicks and cuckoo chicks, then selection has no opportunity to favor the novel behavior.) Another possibility is that the cost of the behavior is relatively low. (If parasitism by cuckoos is rare, a parent that feeds any begging chick in its nest is, on average, much more likely to benefit than suffer.)

Figure 25-7 Because the crossbreeding experiment shows that differences in orientation direction between the two populations stem from genetic differences, birds from the western population should orient in a southwesterly direction, regardless of the environment in which they are raised.

Figure 25-21 If females do indeed gain any benefits from their mates, those benefits would necessarily be genetic (because the male provides no material benefits). Male fitness may vary, and to the extent that this fitness can be passed to offspring, females would benefit by choosing the most fit males. If a male's fitness is reflected in his ability to build and decorate a bower, females would benefit by preferring to mate with males that build especially good bowers.

Figure 25-24 Canines forage mainly by smell; apes are mostly visual foragers. Modes of sexual signaling are affected not only by the nature of the information to be encoded, but also by the sensory biases and sensitivities of the species involved. Communication systems may evolve to take advantage of traits that originally evolved for other functions.

Fill-in-the-Blank

1. genes, environment; stimulus; innate
2. energy, predators, injury; practice, skills, adult; brains, learning
3. habituation, repeated; imprinting; sensitive period
4. aggressive; displays, injuring (wounding, damaging); weapons (such as fangs, claws), larger
5. territoriality (territorial behavior); mate, raise young, feed, store food; male; same species
6. For pheromones, any of the following advantages is correct: long lasting, use little energy to produce, species specific, don't attract predators, may convey messages over long distances. Any of the following disadvantages is correct: cannot convey changing information, conveys fewer different types of information. For visual displays, any of the following advantages is correct: instantaneous, can change rapidly, many different messages can be sent, is quiet. Any of the following disadvantages is correct: makes animals conspicuous to predators, generally ineffective in darkness or dense vegetation, requires that recipient be in visual range (close by).

Chapter 26

Figure Caption Questions

Figure 26-4 Many variables interact in complex ways to produce real population cycles. Weather, for example, affects the lemmings' food supply and thus their ability to survive and reproduce. Predation of lemmings is influenced by both the number of predators and the availability of other prey, which in turn is influenced by multiple environmental variables.

Figure 26-13 Emigration relieves population pressure in an overpopulated area, spreading the migrating animals into new habitats that may have more resources. Human emigration within and between countries is often driven by the desire or need for more resources, although social factors—such as wars and religious or racial persecution—also fuel human emigration. (This is a subjective question that can lead to discussion of the extent to which overpopulation drives human emigration.)

Figure 26-16 This is a subjective question. Many scenarios are possible based on different assumptions about increases in technology, changes in birth and death rates, and the resiliency of the ecosystems that sustain us.

Figure 26-20 When fertility exceeds RLF, there are more children than parents. As the additional children mature and become parents themselves, this more numerous generation produces still more additional children, and so on.

Figure 26-21 High birth rates in developing countries are sustained by cultural expectations, lack of health education, and lack of access to contraception. Lower birth rates in developed countries are encouraged by easy access to contraception, the relatively high cost of raising children, and more varied career opportunities for women. (Students should be able to expand on, or add to, these factors.)

Figure 26-22 U.S. population growth is in the rapidly rising "exponential" phase of the S-curve. Stabilization will require some combination of reduction in immigration rates and birth rates. An increase in death rates is less likely, but cannot be ruled out entirely in any future scenario. (Students should be encouraged to speculate about the timing and the reason for various time frames.)

Fill-in-the-Blank

1. Any two of the following: freezing, drought, flood, fire, storms, habitat destruction; predation and competition
2. survivorship curves; early-loss; late-loss
3. exponential; no; J-curve; no
4. carrying capacity; logistic, S-
5. clumped; uniform; random
6. pyramid (triangle); stable; pyramidal (triangular)
7. births, immigrants, deaths, emigrants; decreases, increases, increases

Chapter 27

Figure Caption Questions

Figure 27-1 Because they did not evolve in the habitat to which they were introduced, invasive species may occupy a niche that is nearly identical to that of a native species; for example, zebra mussels compete with other freshwater mussels or clams. A successful invasive species may have adaptations that allow it to outcompete the native species, such as a higher growth rate or reproductive rate. Further, all native species are likely to have local predators, whereas the introduced species may not. The absence of predators would also help the invasive species outcompete a native species if they occupy very similar niches.

Figure 27-3 Examples of coevolution between predators and prey include the keen eyesight of the hawk and the camouflage coloration of a mouse. Grasses have evolved tough silicon embedded in their leaves, and grazing herbivores have teeth that grow continuously throughout life so that they are not completely worn down by the abrasive grasses. A moth may detect a bat's echolocation signals and fly erratically, while the bat shifts its echolocation pitch to a pitch that the moth cannot detect. Milkweed plants have evolved a toxin to deter predators, and monarch butterfly caterpillars are not susceptible to the toxin.

Figure 27-5 Although scientists have not observed these examples of evolution as they occurred, a likely scenario is that chance mutation caused some animals to more closely resemble their surroundings than did other members of their species. These individuals were less likely to be seen and eaten by predators. As a result, more survived and reproduced, passing their genetically determined appearance on to their offspring. Over large spans of time, more chance mutations that enhanced their resemblance to the environment further reduced predation, again selecting for individuals bearing these genetic traits.

Figure 27-12 Although tasty prey have often evolved to blend in with their surroundings, poisonous prey species (such as the monarch caterpillar, which stores milkweed toxin in its body) frequently advertise their presence with bright "warning colors." These bright colors make it easy for predators to learn to actively avoid animals bearing them.

Figure 27-15 Forests in which fires are suppressed grow extremely dense stands of trees, which tend to be less healthy because they are competing with one another for light, nutrients, and water. A fire under these conditions will burn hotter and more extensively, leaving fewer living plants and seeds to recolonize the area, slowing succession. Smaller, more frequent fires clear spaces within a forest and allow patches of localized succession to occur, increasing the habitat for a variety of animals and increasing the species diversity of plants.

Fill-in-the-Blank

1. natural selection; coevolution; competition, predation, parasitism, mutualism
2. carnivores, herbivores; camouflage; larger, abundant (or numerous)
3. niches; interspecific competition; resource partitioning
4. warning coloration; startle coloration; Batesian mimicry; aggressive mimicry
5. mutualism; parasitism; predation; parasitism; mutualism; competition
6. succession; primary succession; secondary succession; climax; subclimax

Chapter 28

Figure Caption Questions

Figure 28-2 Whenever energy is utilized, the second law of thermodynamics applies: When energy is converted from one form to another, the amount of useful energy decreases. Much of this energy is lost as heat.

Figure 28-3 On land, high productivity is supported by optimal temperatures for plant growth, a long growing season, and plenty of moisture, such as is found in rain forests. Lack of water limits desert productivity. In aquatic ecosystems, high productivity is supported by an abundance of nutrients and adequate light, such as is found in estuaries. Lack of nutrients limits the productivity of the open ocean in the photic zone.

Figure 28-5 There are many possibilities, including many combinations that are not linked by arrows in the figure. One example is grass (producer) → grasshopper (primary consumer) → meadowlark (secondary consumer) → hawk (tertiary consumer). Another is grass (producer) → grasshopper (primary consumer) → spider (secondary consumer) → shrew (tertiary consumer).

Figure 28-9 Humanity's need to grow crops to feed our growing population has led to the trapping of nitrogen for fertilizer using industrial processes. Additionally, large-scale livestock feedlots generate enormous amounts of nitrogenous waste. Nitrogen oxides are also generated when fossil fuels are burned in power plants, vehicles, and factories, as well as when forests are burned. Consequences include the overfertilization of lakes and rivers, and the creation of dead zones in nearshore coastal waters that receive excessive nutrient runoff from land. Another important consequence is acid deposition, in which nitrogen oxides formed by combustion produce nitric acid in the atmosphere; this acid is then deposited on land or in water.

Fill-in-the-Blank

1. sunlight, photosynthesis; nutrients, nutrient cycles OR biogeochemical cycles
2. autotrophs, producers; net primary productivity
3. trophic levels; food chain; food webs
4. 10; the second law of thermodynamics
5. herbivores, primary consumers; heterotrophs, consumers; detritus feeders; fungi, bacteria
6. nitrogen-fixing bacteria, denitrifying bacteria; ammonia, nitrates
7. atmosphere, oceans; CO_2 (carbon dioxide); limestone, fossil fuels

Chapter 29

Figure Caption Questions

Figure 29-6 Both live in arid environments, so their evolution has selected for fleshy, water-storing bodies. In each case, leaf area has been reduced, limiting evaporation. Both of these plants would be attractive to desert animals because they store water; as a result of this predation pressure, both carry defensive spines.

Figure 29-9 Most nutrients in tropical rain forests are not stored in the soil. The optimal temperature and moisture of tropical climates allow plants to make such efficient use of nutrients that nearly all nutrients are stored in plant bodies and, to a lesser extent, in the bodies of the animals they support.

Figure 29-17 Tallgrass prairie is vulnerable to two major encroachments. First, if humans suppress natural wildfires, then adequate rainfall allows forests to take over. Second, these biomes provide the world's most fertile soils and have excellent growing conditions for farming, which has now displaced native vegetation.

Figure 29-29 Nearshore ecosystems have an abundance of the two limiting factors for life in water: nutrients and light to support photosynthetic organisms. Upwelling from ocean depths or runoff from the land (or both) provide nutrients. The shallow water in these areas allows adequate light to penetrate to support rooted plants and anchored algae, which in turn provide food and shelter for a wealth of other marine life.

Figure 29-30 Bleaching refers to the loss of symbiotic algae that normally inhabit the corals' tissues. The algae (dinoflagellates) provide the corals with energy-rich compounds that store energy from photosynthesis. Loss of these symbiotic algae can eventually kill the coral. Bleaching is a common response to water that is excessively warm and, thus, global warming may contribute to the demise of coral reefs.

Fill-in-the-Blank

1. seasons; Tropics; ocean currents OR the presence of the ocean; rain shadow
2. appropriate temperatures, liquid water; light, nutrients; tropical rain forests; coral reefs
3. tropical rain forest, soil; deciduous forest; shortgrass prairie; desert; savanna; tundra
4. littoral zone; phytoplankton, zooplankton; limnetic zone, profundal zone; oligotrophic; eutrophic; wetlands
5. phytoplankton; bioluminescence; chemosynthesis, hydrogen sulfide; high pressure

Chapter 30

Figure Caption Questions

Figure 30-8 Any rare organism faces a greater chance of extinction in a small reserve. For example, large carnivores need a large prey population to support them (see energy pyramids in Chapter 28), which in turn needs a large area of suitable habitat. Large herbivores are likely to be rare for similar reasons; for example, it takes a very large area of vegetation to support a population of elephants. A chance event (storm, fire, disease) may well kill all of the members of such a population in a small reserve.

Figure 30-11 If a species becomes extinct in one of the small reserves, members of the species in nearby reserves can move through the corridor and recolonize the now-vacant reserve.

Fill-in-the-Blank

1. genetic, species, ecosystem; genetic
2. ecosystem services; many possible answers, including food, wood, medicines, soil formation, erosion and flood control, climate regulation, genetic resources, recreation
3. ecological economics
4. habitat destruction, overexploitation, invasive species, pollution, global warming; habitat destruction
5. minimum viable population; habitat fragmentation; wildlife corridors
6. sustainable

Chapter 31

Figure Caption Questions

Figure 31-2 If the mammal's heat-sensing nerve endings were rendered nonfunctional, the nervous system would not send a signal to the hypothalamus when the body reached the set-point temperature. Consequently, the hypothalamus would send continuous "turn on" signals to the body's heat-generating and heat-retention mechanisms, which would continue to increase body temperature indefinitely (presumably until stored energy reserves are exhausted or the animal dies).

Figure 31-8 Blood is considered a form of connective tissue because it is composed largely of a matrix of extracellular fluid (plasma) in which cells and proteins are suspended.

Fill-in-the-Blank

1. homeostasis; negative feedback
2. cells, tissues, organs, organ systems
3. connective; epithelial; connective; connective; epithelial; muscle; nerve; connective
4. exocrine; endocrine, hormones
5. cardiac; skeletal; cardiac and skeletal; cardiac and smooth; smooth; skeletal

Chapter 32

Figure Caption Questions

Figure 32-4 Due to an exercise-induced increase in muscle size, the hearts of well-conditioned athletes are larger than those of sedentary people, so a greater volume of blood is pumped with each heartbeat. This allows the body's resting demand for oxygen to be met by a slower heart rate.

Figure 32-8 Iron is a key component of hemoglobin, which is necessary for building red blood cells capable of transporting oxygen. When red blood cells die, their iron is recycled, which allows new red blood cells to be made. However, iron recycling in the body is not 100% efficient, so dietary iron is necessary to replace that lost through excretion or bleeding. Otherwise, more red blood cells will die than new red blood cells can be produced, resulting in anemia.

Figure 32-10 The hormone erythropoietin stimulates production of additional red blood cells. These extra cells increase the blood's capacity to carry oxygen to muscles, thereby increasing the amount of time that the muscles can work without becoming fatigued, which occurs when their oxygen supply is depleted and they can no longer carry out cellular respiration to produce ATP.

Figure 32-16 The principles of diffusion (see pp. 82–83) tell us that molecules move down gradients from high to low concentration. Because the tissues through which capillaries pass generally consume oxygen, the concentration of oxygen is higher inside the capillaries than outside them, and oxygen diffuses along its concentration gradient from the inside of the capillary to the outside. Conversely, tissues generally produce carbon dioxide, so the concentration of carbon dioxide is higher outside the capillaries than inside them, and carbon dioxide diffuses along its concentration gradient from the outside of the capillary to the inside.

Fill-in-the-Blank

1. right atrium; left ventricle; right atrium; right ventricle; left ventricle
2. sinoatrial node; cardiac muscle cells; atria; atrioventricular; AV bundle (atrioventricular bundle); fibrillation
3. capillaries; arteries; veins; venules; arterioles; arteries
4. platelets; thrombin; fibrinogen, fibrin
5. platelets; leukocytes; erythrocytes; leukocytes; erythrocytes; leukocytes; erythrocytes
6. extracellular fluid, capillaries; lymphatic capillaries; one-way valves; spleen

Chapter 33

Figure Caption Questions

Figure 33-5 Because juveniles of many amphibian species breathe through gills, and because both juveniles and adults often rely on supplemental respiration through the skin, most amphibians live in moist habitats, such as marshes, swamps, ponds, and rain forests.

Figure 33-10 Body cells can only function within a narrow pH range, so blood pH must be tightly controlled. If hydrogen ions entered the plasma along with the bicarbonate ions, the pH (a function of the hydrogen ion concentration) of the plasma would become acidic.

Figure 33-11 Like any skeletal muscle, the diaphragm muscle is longer when relaxed than when contracted. At rest, the diaphragm is domed upward into the chest cavity, reducing the cavity's volume. Contracting the diaphragm muscle makes it shorter, which can occur only by flattening its domed shape, thereby increasing the volume of the chest cavity.

Fill-in-the-Blank

1. the pharynx; the epiglottis; trachea, bronchi, bronchioles, alveoli
2. capillaries; hemoglobin; bicarbonate; cellular respiration
3. alveoli; alveolus, capillary; two; surfactant
4. diaphragm, expand (larger also acceptable); passive, relax; respiratory center, medulla; carbon dioxide
5. lung cancer, chronic bronchitis, asthma, emphysema (any three); ear infections, asthma, coughs, colds, allergies, bronchitis (any three)

Chapter 34

Figure Caption Questions

Figure 34-5 This food contains relatively little total fat, and the fats that are present are primarily unsaturated; there are no trans fats (see p. 46 for more information on types of fat). The food has a substantial percentage of the recommended daily amounts of a variety of vitamins and iron. Yes, it would be a good contribution to a balanced diet. Note, however, that it lacks any significant protein and contains very little fiber.

Figure 34-10 Birds grind food using a muscular gizzard, often containing small sharp stones that substitute for teeth.

Figure 34-11 Because the microorganism populations live in the first chamber of the ruminant's digestive system, many of them are passed to the other chambers, where they are digested and become an additional source of nutrients, particularly organic nutrients that the microbes synthesize but that the cow cannot synthesize or obtain in the food it eats.

Figure 34-16 A likely outcome is that the needed absorptive surface area would have been achieved by some other sort of adaptation, perhaps a much longer intestine and a correspondingly larger abdominal cavity to hold it. One could also speculate about other features that might have been required to accommodate the extra long intestine (e.g., skeletal and muscular features to support it).

Fill-in-the-Blank

1. energy, raw materials; vitamins; minerals; essential
2. intracellular; gastrovascular cavity; tubular; cellulose
3. amylase, starch; protein; pepsinogen, pepsin; small intestine
4. ingestion, mechanical digestion, chemical digestion, absorption, elimination
5. pharynx, digestive, respiratory; epiglottis, swallowing; peristalsis; sphincter muscles
6. duodenum; sodium bicarbonate, chyme, carbohydrates, lipids (OR fats and oils), proteins
7. bile (OR bile salts); liver, gallbladder; epithelial; lacteal, lymphatic

Chapter 35

Figure Caption Questions

Figure 35-7 If you consumed excess water, a drop in blood osmolarity would occur that would be detected by receptors in the hypothalamus. Activation of these receptors would cause less ADH to be released from the posterior pituitary. In the kidneys, less ADH would result in fewer aquaporins in the membranes of the distal tubules and collecting ducts, making them almost impermeable to water. This would result in production of large amounts of watery urine until normal blood osmolarity was restored.

Fill-in-the-Blank

1. nephridia; Malpighian tubules; protonephridia; nephridia
2. renal cortex, renal medulla, renal pelvis; ureter, bladder, urethra
3. glomerulus, Bowman's capsule, proximal tubule, loop of Henle, distal tubule
4. proximal tubule; collecting duct; loop of Henle; medulla
5. urea; H+ (hydrogen ions); NaCl (salt); water
6. homeostasis; renin, angiotensin; erythropoietin, red blood cells
7. hypothalamus, posterior pituitary, antidiuretic (ADH); distal tubule, collecting duct, aquaporins, concentrated

Chapter 36

Figure Caption Questions

Figure 36-4 In most capillaries, there are small gaps between the cells that make up the capillary walls (see Chapter 32). The inflammatory response causes these gaps to become larger, which allows white blood cells to move more easily between cells of the capillary walls.

Figure 36-6 The combination of constant and variable components allows an antibody to accomplish not only functions that are unique to a particular antibody molecule (i.e., recognize and bind to a specific antigen), but also functions that are common to all antibodies in a given class (i.e., carry out the appropriate response to neutralize the invader, such as enhancing phagocytosis). Variable regions provide the unique properties; constant regions provide the universal properties.

Figure 36-12 Cytotoxic T cells are activated by antigens on the target cell's surface. The plasma membranes of cancer cells often bear distinctive, abnormal molecules that are not recognized as "self" by the immune system; thus, they act as antigens that stimulate the adaptive immune response to kill the cancerous cells.

Fill-in-the-Blank

1. skin; digestive, respiratory, urogenital (order not important)
2. phagocytes; natural killer cells; inflammatory response; fever
3. antigens; antibodies, T-cell receptors (order not important)
4. light, heavy (order not important); variable, constant (order not important); variable
5. humoral, plasma cells; cell-mediated; cytotoxic; helper; memory
6. vaccine
7. allergy; immune deficiency; autoimmune disease

Chapter 37

Figure Caption Questions

Figure 37-7 Dwarfism can be treated by administering growth hormone to affected individuals; it is much more difficult to halt the production of excess hormone (especially of a pituitary hormone such as growth hormone). Disruption of the pituitary is likely to disrupt the release of many other hormones as well. (Some cases of gigantism are caused by pituitary tumors, which can be treated by surgery.)

Figure 37-11 There are several possible hypotheses, including: (1) A tumor in the anterior pituitary caused multiplication of the TSH-secreting cells; (2) a tumor in the hypothalamus caused multiplication of the cells that secrete TSH-releasing hormone; and (3) an iodine-deficient diet caused the thyroid to produce too little thyroxine, so that the negative feedback system regulating TSH release is shut off.

Figure 37-12 If the receptors could not bind glucagon, then target cells, such as those in the liver, would not be able to respond to glucagon. When blood glucose levels are too low, liver cells and other cells cannot convert glycogen to glucose, so blood glucose levels would stay low.

Fill-in-the-Blank

1. endocrine system; target cells, receptors
2. peptide, amino acid derived, steroid (order not important); peptide hormones, amino acid-derived hormones (order not important); second messengers; steroid hormones; changes in gene transcription (OR synthesis of messenger RNA)
3. hypothalamus; neurosecretory cells, oxytocin, antidiuretic hormone (order of hormones not important)
4. adrenocorticotropic hormone, follicle-stimulating hormone, luteinizing hormone, growth hormone, prolactin, thyroid-stimulating hormone
5. insulin; diabetes mellitus; glucagon, glycogen
6. testes, testosterone; ovaries, estrogen, progesterone
7. glucocorticoids, mineralocorticoids, testosterone (order of hormones not important); norepinephrine, epinephrine (order of hormones not important)

Chapter 38

Figure Caption Questions

Figure 38-4 Based on the evidence, the toxin must act on the postsynaptic part of the synapse, by blocking neurotransmitter receptors. If the toxin acted by blocking release of neurotransmitters from the presynaptic neuron, then adding neurotransmitter to the synapse would have generated a postsynaptic action potential. If the toxin acted by preventing sodium channels from opening, then it would not have been possible to stimulate an action potential in the presynaptic neuron.

Figure 38-5 Sensitive areas have a higher density of touch-sensitive sensory neurons. This would allow two types of increased sensitivity: (1) The higher density of receptors would mean that touching any particular place on the sensitive skin area would be more likely to stimulate at least one touch receptor; and (2) a higher density of receptors would also mean that a touch of a particular force would stimulate a larger number of neurons. The touch thus would generate a larger number of action potentials in the sensitive area and would be perceived as a stronger stimulus.

Figure 38-10 A severed or damaged spinal cord prevents the sensation of pain from being relayed to the brain but does not disrupt the reflex circuit, which requires transmission within only a small portion of the spinal cord.

Fill-in-the-Blank

1. neuron; dendrite; cell body; axon, synaptic terminal
2. negative; threshold; positive
3. neurotransmitter; receptors; excitatory postsynaptic potential; inhibitory postsynaptic potential
4. somatic nervous system; autonomic nervous system, sympathetic, parasympathetic
5. cerebellum, pons, medulla (order not important); cerebellum
6. frontal, parietal, temporal, occipital (order not important); occipital; frontal

Chapter 39

Figure Caption Questions

Figure 39-10 For nearsightedness, the cornea is flattened (made less convex), which causes the light rays to converge less before they reach the lens. For farsightedness, the edges of the cornea are reduced, which leaves the cornea with a more rounded, convex shape that causes the light rays to converge more before they reach the lens.

Figure 39-11 You do not experience a hole in your vision because (1) the blind spots in your two eyes do not see the same place in your visual field, so one of the eyes always sees the "hole" in the visual field of the other eye; and (2) your eyes constantly move around, so no part of the visual field is in the blind spot of either eye for very long. The brain "fills in" the missing visual information with what the eye saw just a fraction of a second before.

Figure 39-12 Distance perception is useful for many behaviors, not just predation. Fruit-eating bats fly around during the daytime, and binocular vision is useful for avoiding obstacles such as branches. Monkeys often jump from branch to branch, and need to know how far it is to the next branch.

Fill-in-the-Blank

1. receptor potential; frequency
2. mechanoreceptors; photoreceptors; chemoreceptors
3. pinna, eardrum (tympanic membrane); hammer (malleus), anvil (incus), stirrup (stapes) (order *is* important); oval window; hair cells
4. cornea, aqueous humor, iris; lens, vitreous humor; cornea, lens (order not important)
5. rods, cones (order not important); rods; cones, fovea
6. sweet, salty, sour, bitter, umami (order not important); olfaction OR smell

Chapter 40

Figure Caption Questions

Figure 40-2 Water helps to support some of the weight of a thick exoskeleton. For land-dwelling arthropods, the weight of a thick exoskeleton would require much more energy expenditure just to stand, let alone move.

Figure 40-6 A highly stretched muscle would have the thin and thick filaments pulled so far out that not all of the myosin heads would overlap a thin filament and, hence, could not attach to actin when a contraction begins. Fewer myosin heads pulling on the thin filaments would generate less force of contraction than if all of the heads could pull on the thin filaments.

Figure 40-10 Two reasons: (1) Compact bone is much heavier than spongy bone, so it would require much more energy for an animal to stand or move; and (2) compact bone, being virtually solid, would leave no room for bone marrow, which is where blood cells are produced.

Fill-in-the-Blank

1. hydrostatic skeleton, endoskeleton, exoskeleton (order not important); antagonistic
2. skeletal, cardiac (order not important); cardiac, smooth (order not important); cardiac, smooth (order not important)
3. muscle fiber; myofibrils, sarcomeres, Z lines
4. actin, myosin; ATP
5. neuromuscular junctions; T tubules, sarcoplasmic reticulum
6. cartilage, bone, ligaments (order not important); collagen
7. hinge, ball-and-socket; flexor, extensor

Chapter 41

Figure Caption Questions

Figure 41-1 Mitotic cell division.

Figure 41-3 Asexual reproduction produces offspring that are genetically identical to the mother, all containing all of her genes. This is an efficient way for her to maximize the number of her genes in the population, but provides no significant opportunity for any of her offspring to have new genotypes that might enhance survival or reproduction. Because the weather is favorable and food is so abundant in the spring and summer, natural selection is not very strong, so this is probably not a major difficulty. When autumn arrives, weather and food become more problematic. Sexual reproduction produces genetically variable offspring, some of which may be better adapted to the conditions and be favored by natural selection.

Figure 41-6 Courtship rituals help ensure that animals mate with other individuals of their own species. Courtship rituals are reproductive isolating mechanisms that help prevent potentially disadvantageous hybrid mating. The courtship activities in some species stimulate ovulation.

Figure 41-13 Testosterone inhibits the release of GnRH and FSH. Without adequate FSH, the Sertoli cells are not stimulated and, hence, spermatogenesis slows down or stops.

Figure 41-17 Estrogen and progesterone inhibit the release of GnRH, FSH, and LH. Without enough FSH, follicles will not develop. Even if a follicle somehow does develop, without LH, the follicle will not mature, meiosis I will not be completed, and ovulation will not occur. Without an ovulated secondary oocyte, fertilization and pregnancy cannot occur.

Fill-in-the-Blank

1. asexual; bud; fission
2. testis; testosterone (androgen also acceptable); seminiferous tubules
3. nucleus, acrosome; mitochondria
4. epididymis, vas deferens, seminal vesicles, prostate gland, bulbourethral gland (order of glands not important), urethra
5. ovary; estrogen, progesterone (order not important); secondary oocyte; polar body; oviduct OR uterine tube OR fallopian tube; uterus
6. corpus luteum; chorionic gonadotropin

Chapter 42

Figure Caption Questions

Figure 42-4 Twins would probably never occur, because even the first cell division of cleavage would produce daughter cells with different developmental fates, with the result that each of the daughter cells would be able to give rise to some, but not all, of the structures of the human adult.

Figure 42-5 *Ribs:* Homeobox genes are usually active in a head-to-tail sequence, such that gene A might be active in the head region, gene B would be active in the torso, gene C would be active in the pelvic region, and so on. If, for example, gene B causes production of ribs, then perhaps in snakes, gene B is active all the way from the end of the (very short) neck to the tail segments.

Lack of legs: A specific homeobox gene would normally be responsible for triggering the production of the entire leg (or at least most of it). A severe mutation (for example, a premature stop codon) in this particular homeobox gene would prevent leg formation.

Figure 42-8 The ovulated egg is surrounded by the zona pellucida and the corona radiata (through which the sperm must penetrate to fertilize the egg). However, it is the outer layer of the blastocyst that attaches to, and implants in, the uterus. Therefore, the blastocyst must emerge out of the zona pellucida and corona radiata at some point before implantation.

Figure 42-13 Monotremes such as platypuses lay eggs; the embryos are not nourished inside the mother at all. In marsupials such as kangaroos, the embryo is enclosed within an amnion that segregates it from the mother's blood (and thus from attack by the mother's immune system), but nutrients cannot pass across the amnion. However, marsupials produce only a primitive placenta-like structure, much like a large yolk sac, rather than a true placenta that allows efficient exchange of nutrients, wastes, and gases between embryo and mother. Therefore, marsupial offspring are born after a short gestation, in a much less developed state than are the offspring of placental mammals. Development continues outside the uterus (typically inside an external pouch).

Fill-in-the-Blank

1. cleavage, morula; blastula; ectoderm, mesoderm, endoderm (order not important)
2. amnion; chorion; umbilical cord; yolk sac
3. stem cells; inner cell mass
4. homeobox genes; transcription factors
5. prolactin; oxytocin

Chapter 43

Figure Caption Questions

Figure 43-7 The onion stores nutrients for use by the shoot produced during the second year, after which the plant dies. An onion bulb at the end of the second year would be practically nonexistent.

Figure 43-9 No. In a climate that was truly uniformly warm and wet year-round, the vascular cambium would divide at the same rate year-round, and the daughter secondary xylem cells would grow the same way year-round. All of the secondary xylem cells would have the same size and cell wall thickness, and so they would all be the same color.

Figure 43-19 The fungus acquires sugars from the plant root. Cortex cells typically receive sugars and store them (possibly converting them to starch). Therefore, the fungus would likely benefit most from a close association with cortex cells. Cortex cells can also take up minerals, so any minerals made available by the fungus could be taken up by the cortex cells.

Figure 43-23 CO_2 is required for photosynthesis. When stomata close, photosynthesis in the leaf slows and eventually stops because closed stomata prevent CO_2 in the air from entering the leaf and reaching the mesophyll cells where photosynthesis occurs. Closed stomata also greatly reduce transpiration from the leaf. Because transpiration provides the driving force for the upward movement of water in xylem, which in turn provides the major driving force for water entry into the roots, closing the stomata would slow down water entry into the roots.

Fill-in-the-Blank

1. meristem; apical meristem, lateral meristem; apical meristem; lateral meristem
2. dermal, ground, vascular (order not important); dermal; vascular; ground
3. xylem, tracheids, vessel elements (order not important for the last two answers); hydrogen bonds, transpiration; stomata
4. phloem; sieve tube element, companion cells
5. root cap; root hairs, active transport; mycorrhizae

Chapter 44

Figure Caption Questions

Figure 44-3 Separate bloom times reduce the chances of self-pollination and, hence, self-fertilization, which would result in inbreeding. Plants that do not inbreed in many cases have more successful offspring. Inbred individuals are more likely to have homozygous recessive alleles that harm the plant's growth or reproduction.

Figure 44-6 Colorful petals and sweet scents have evolved because they attract pollinators; thus, they would not be useful to a wind-pollinated plant, which has no need to attract animal pollinators. Further, petals, scents, and nectar require energy to produce, so they would be selected against. Therefore, wind-pollinated plants would not be expected to have them.

Figure 44-15 When nectar is at the base of a long, tubular flower, it is accessible to hummingbirds (which have long bills and tongues), but not to most insects and other small animals that might like to eat it. A flower whose nectar is inaccessible to insects is more likely to have nectar available for a visiting hummingbird; it is thus more likely that hummingbirds will learn to repeatedly visit this type of flower, ensuring that the flowers will be pollinated and the plant will reproduce.

Figure 44-18 *Hints:* How do maple fruits "fly"? Do they fly further with or without their wings? Does wind make them fly further? If so, how do the wings affect their flight distance on a windy day?

Figure 44-19 Water flows only downhill. Therefore, floating fruits move downhill. A species using water as its primary means of dispersal will move downhill generation after generation, until eventually its most common habitat would be the ocean shore. This, of course, is precisely where coconut palms are found.

Fill-in-the-Blank

1. alternation of generations; sporophyte; meiotic cell division; gametophyte
2. pollen grain; anthers; stigma; style; integuments
3. zygote; endosperm; cotyledons
4. ovary; ovule; integuments
5. germination; root; coleoptile; bend OR hook

Chapter 45

Figure Caption Questions

Figure 45-7 The added auxin would replace the auxin normally produced by the apical meristem of the intact shoot tip, so apical dominance would persist. Lateral buds would be inhibited from sprouting into branches.

Figure 45-10 Ripe tomatoes are often damaged in shipping. Blocking ethylene production would prevent ripening. Green tomatoes could be ripened after they reach the market by exposing them to ethylene.

Fill-in-the-Blank

1. inhibits, apical dominance; stimulates; positive phototropism
2. gibberellin; divide, auxin
3. ethylene; abscission layer; ripening, (any two of the following are correct, but this listing from the text is not exclusive) bananas, apples, pears, tomatoes, OR avocados
4. root apical meristem; inhibits, stimulates, branches OR buds
5. abscisic acid; dormancy

Photo Credits

Detailed Contents Photos: p. v: Tischenko Irina/Shutterstock; p. vi: Victoria Alexandrova/Shutterstock; p. vii: Dr. Jeremy Burgess/Science Photo Library/Photo Researchers, Inc.; p. viii: Image courtesy Indigo® Instruments: http://www.indigo.com; p. ix: Photos.com; p. x: Photos.com; p. xi: Allen Blake Sheldon/Allen Blake Sheldon Nature Photography; p. xii: Roland Birke/Peter Arnold, Inc.; p. xiii: Meppu/iStockphoto.com; p. xiv: David Fleetham/Alamy Images; p. xv: Andy Sands/Nature Picture Library; p. xvi: Jan Martin Will/Shutterstock; p. xvii: Dr. David M. Phillips/Visuals Unlimited/Getty Images; p. xviii: Michael Durham/Minden Pictures; p. xix: James Cavallini/Photo Researchers, Inc.; p. xx: Dennis Kunkel/Phototake NYC; p. xxi: H. Stanley Johnson/Purestock/SuperStock p. xxii: John Cancalosi/Photolibrary.com

Unit Openers: Unit 1: Dennis Kunkel/Phototake NYC; Unit 2: GK Hart/Vicky Hart/Photonica/Getty Images; Unit 3: Frans Lanting/CORBIS; Unit 4: Georgette Douwma/Getty Images; Unit 5: Marty Sohl Photography; Unit 6: Purestock/Getty Images

Chapter 1 opener: WHO/Christopher Black; opener inset: Frederick A. Murphy/Centers for Disease Control; 1-2: Electron micrograph by Wm. P. Wergin, courtesy E. H. Newcomb, University of Wisconsin-Madison/Biological Photo Service; 1-3: Hunter Cole; 1-5: Christine Case/Visuals Unlimited; 1-6: Frank Cucinotta/NASA; 1-7: Francois Gohier/Photo Researchers, Inc.; 1-8: Wim van Egmond/Getty Images Inc. Visuals Unlimited; 1-9: AP Wide World Photos; E1-3: Frans Lanting/CORBIS; E1-3 inset: Unclesam/Fotolia; E1-4: Mark R/Shutterstock

Chapter 2 opener: Underwood & Underwood/CORBIS; 2-4: Mediacolor's/Alamy Images; 2-5c: Piotr Sosnowski/Wikimedia Commons/Wikemedia Commons; 2-8a: Stephen Dalton/Photo Researchers, Inc.; 2-8b: Teresa and Gerald Audesirk; 2-10: Renn Sminkey/CDV LLC, Creative Digital Visions; 2-13a: Comstock Images/Jupiter Unlimited; 2-13b: Hans Huber/Photoshot Archive Creative; E2-1c: Warren Grant/National Institute of Health; E2-2: UPI/CORBIS-NY; E2-3: Don Mason/Brand X Pictures/Jupiter Images - PictureArts Corporation/Brand X Pictures Royalty Free; E2-3 inset: iStockphoto.com

Chapter 3 opener: gordondix/iStockphoto.com; 3-8a: Dr. Jeremy Burgess/Science Photo Library/Photo Researchers, Inc.; 3-9a: Mike Norton/Shutterstock; 3-9b: Jeremy Burgess/Science Photo Library/Photo Researchers, Inc. 3-9c: Biophoto Associates/Photo Researchers, Inc.; 3-10: Mark Jones/Minden Pictures; 3-11a: Sylvie Bouchard/Fotolia; 3-11b: Tischenko Irina/Shutterstock; 3-16a: Tracy Morgan/Dorling Kindersley; 3-16b: gabriela schaufelberger/iStockphoto.com; 3-16c: Xiaodong Ye/iStockphoto.com; E3-1: Renn Sminkey/Creative Digital Visions, Pearson Education; E3-2: GJLP/Photo Researchers, Inc.; E3-3: Coka/Fotolia

Chapter 4 opener: © 2006 Advanced Tissue Sciences, a Division of Advanced BioHealing; opener inset: © 2006 Advanced Tissue Sciences, a Division of Advanced BioHealing; 4-5b: Molecular Probes, Inc.; 4-6: William L. Dentler/Biological Photo Service; 4-7a: Ellen R. Dirksen/Visuals Unlimited; 4-7b: Yorgos Nikas/Getty Images Inc.; 4-8b: Dr. Elena Kiseleva/SPL/Photo Researchers, Inc.; 4-9: Lester V. Bergman/Bettmann/CORBIS; 4-10: Oscar L. Miller, Jr., University of Virginia; 4-11b: R. Bolender and Donald Fawcett/Visuals Unlimited; 4-12: Lee W. Wilcox; 4-16: The Keith R. Porter Endowment for Cell Biology; 4-17: © E.H. Newcomb & W.P. Wergin/Biological Photo Service, Terraphotographics; 4-18: Biophoto Associates/Photo Researchers, Inc.; 4-19b: CNRI/Science Photo Library/Photo Researchers, Inc.; 4-19c: Visuals Unlimited/CORBIS; 4-19d: John Cardamone,Jr./Biological Photo Service; 4-19e: "TEM showing the thylakoid membranes of strain CCMEE 5410". Phycobilisomes were not observed (Bar, 100nm). From PNAS 01/18/05; Vol 102 (3) page 852, Fig 2A by Scott R. Miller, et al. Copyright © 2005 National Academy of Sciences, U.S.A.; 4-20: Sam Ogden/Photo Researchers, Inc.; E4-1aT: Plate XI, Micrographia by Robert Hooke 1665 Project Gutenberg.; E4-1aB: Cecil Fox/Science Source/Photo Researchers, Inc.; E4-1bT: Brian J. Ford; E4-1bB: Brian J. Ford; E4-1c: Courtesy of DOE's Pacific Northwest National L/Pacific Northwest National Laboratory; E4-2a: M.I. Walker/Photo Reserachers, Inc.; E4-2b: Manfred Kage/Peter Arnold, Inc.; E4-2c: David M. Phillips/Visual Unlimited; E4-2d: National Library of Medicine

Chapter 5 opener: Daniel J. Lyons/Photoshot Holdings Ltd/Bruce Coleman USA; opener inset: Larry F. Jemigan/Photolibrary.com; 5-9a: Joseph Kurantsin-Mills, The George Washington Univeristy Medical Center; 5-9b: Joseph Kurantsin-Mills, The George Washington Univeristy Medical Center; 5-9c: Joseph Kurantsin-Mills, The George Washington Univeristy Medical Center; 5-10a: Nigel Cattlin/Holt Studios International/Photo Researchers, Inc.; 5-10b: Nigel Cattlin/Holt Studios International/Photo Researchers, Inc.; 5-12b: BCC Microimaging, Corp. Reproduced with permission.; 5-13b1: M.M. Perry and A.B. Gilbert, Journal of Cell Science, 39:257-272 (1979). © 1979 The Company of Biologists Limited.; 5-13b2: M.M. Perry and A.B. Gilbert, Journal of Cell Science, 39:257-272 (1979). © 1979 The Company of Biologists Limited.; 5-13b3: M.M. Perry and A.B. Gilbert, Journal of Cell Science, 39:257-272 (1979). © 1979 The Company of Biologists Limited.; 5-13b4: M.M. Perry and A.B. Gilbert, Journal of Cell Science, 39:257-272 (1979). © 1979 The Company of Biologists Limited.; 5-14b: Visuals Unlimited/CORBIS; 5-14c: Dennis Kunkel/Phototake NYC; 5-15: L. A. Hufnagel, Ultrastructural Aspects of Chemoreception in Ciliated Protists (Ciliophora), Journal of Electron Microscopy Technique, 1991. Photomicrograph by Jurgen Bohmer and Linda Hufnagel, University of Rhode Island.; 5-19a: American College of Physicians; 5-19b: Photo Courtesy of Justin Schwarts, www.rattlesnakebite.org; E5-1: Victoria Alexandrova/Shutterstock; E5-2: Stephen J. Boitano/Getty Images; E5-3a: Reproduced by permission from: Water permeability of aquaporin-I (AQPI) expressed in Xenous oocytes. Fig 3, part D of Trends in Biochem Science 1994 Oct; 19(10):421-5 by Chrispeels MJ and Agre P. Copyright © 1994 by Elsevier Science Ltd. Photo courtesy of G.M. Preston

Chapter 6 opener: Jason DeCrow/AP Wide World Photos; 6-1: Tim Davis/CORBIS; E6-1: Baumgartner Olivia/CORBIS; UN6-1: Gail Shumway/Getty Images

Chapter 7 opener: Joe Tucciarone; 7-1a: Nikolay Okhitin/Shutterstock; 7-1c: J. Michael Eichelberger/Visuals Unlimited; 7-2a: Dr. Jeremy Burgess/Science Photo Library/ Photo Researchers, Inc.; 7-2b: Dr. Jeremy Burgess/Photo Researchers, Inc. 7-5: Teresa and Gerald Audesirk; 7-9: Colin Milkins/Oxford Scientific Films/Animals Animals/Earth Scenes; E7-1: Daniel Beltra/Getty Images

Chapter 8 opener: Newscom; 8-10: Julián Rovagnati/Shutterstock; E8-3: William Leaman /Alamy Images; E8-4: Gordon Swanson/Shutterstock

Chapter 9 opener: Young Ho/SIPA Press; 9-2a: David Scharf/Peter Arnold, Inc.; 9-2b: M. I. Walker/Photo Researchers, Inc.; 9-2c: Carolina Biological Supply Company/Phototake NYC; 9-2d: Teresa and Gerald Audesirk; 9-4: Andrew Syred/Photo Researchers, Inc.; 9-6: CNRI/Science Photo Library/Photo Researchers, Inc.; 9-8a: Michael W. Davidson, Florida State University/Molecular Expressions (http://www.microscopy.fsu.edu).; 9-8b: Michael W. Davidson, Florida State University/Molecular Expressions (http://www.microscopy.fsu.edu).; 9-8c: Michael W. Davidson, Florida State University/Molecular Expressions (http://www.microscopy.fsu.edu).; 9-8d: Michael W. Davidson, Florida State University/Molecular Expressions (http://www.microscopy.fsu.edu).; 9-8e: Michael W. Davidson, Florida State University/Molecular Expressions (http://www.microscopy.fsu.edu).; 9-8f: Michael W. Davidson, Florida State University/Molecular Expressions (http://www.microscopy.fsu.edu).; 9-8g: Michael W. Davidson, Florida State University/Molecular Expressions (http://www.microscopy.fsu.edu).; 9-8h: Michael W. Davidson, Florida State University/Molecular Expressions (http://www.microscopy.fsu.edu).; E9-1: Photograph courtesy of The Roslin Institute

Chapter 10 opener: Ronald Modra/Sports Illustrated; 10-2: Archiv/Photo Researchers, Inc.; 10-15: Image courtesy Indigo® Instruments: http://www.indigo.com/; 10-19b: Sarah Leen/National Geographic Image Collection; 10-20: David Hosking /Alamy Images; 10-22a: Rick Guidotti for Positive Exposure; 10-22b: Jeremy Voisey/iStockphoto.com; 10-23a: Dennis Kunkel/Phototake NYC; 10-23b: Wlater Reinhart/Phototake NYC; 10-24: Gunn and Stewart/Mary Evans/Photo Researchers, Inc.; 10-26a: CNRI/Science Photo Library/Photo Researchers, Inc.; 10-26b: philidor/Fotolia E10-1: VES HERMAN/Landov Media; E10-2a: Custom Medical Stock Photo; E10-2b: Custom Medical Stock Photo

Chapter 11 opener: Yann Arthus-Bertrand/CORBIS/Yann Arthus-Betrand/Bettmann/CORBIS; 11-4a: Photo by M.H.F. Wilkins. Courtesy of King's College London, England; 11-4b: The Nobel Foundation 2008; 11-4c: Jenifer Glynn/National Library of Medicine; 11-5c: Michael Freeman/Phototake NYC; 11-7: Dr. Gopal Murti/Science Photo Library/Photo Researchers, Inc.; 11-9: Stuart Isett/Polaris Images; E11-3: National Cancer Institute

Chapter 12 opener: Photograph by Pete Moss, Courtesy Sony BMG and Luke Martineau; 12-4: Oscar L. Miller, Jr., University of Virginia; 12-5b: From Hamkalo and Miller, Electronmicroscopy of Genetic Material, Figure 6a, p. 379. Reproduced with permission, from Annual Review of Biochemistry,

Volume 42, © 1973 by Annual Reviews, Inc.; 12-11: From Fig. 3A (top): Nature Genetics, Vol 30, pp. 167-174 of Xist RNA, Anton Wutz, Theodore P. Rasmussen & Rudolf Jaenisch Figure 3. "Analysis of expression and localization of mutated Xist RNAs.; 12-12: Photos.com; 12-13: Delly Carr/Sportshoot; E12-2: Richard Jorgensen; E12-3: Howard W. Jones, Jr., M.D., Eastern Virginia Medical School E12-4a: Courtesy International Registry of Werner Syndrome, http://www.wernersyndrome.org; E12-4b: Courtesy International Registry of Werner Syndrome, http://www.wernersyndrome.org

Chapter 13 opener: Lisa Poole/AP Wide World Photos; 13-1a: Stanley N. Cohen/Science Photo Library/Photo Researchers, Inc.; 13-7: Dr. Margaret Kline, National Institute of Standards and Technology.; 13-8: Monsanto Company; 13-13: Photo courtesy of Promega Corporation; 13-15: Uli Holz/The Innocence Project www.innocenceproject.org; E13-1: Janet Chapple, author of "Yellowstone Treasures: The Traveler's Companion to the National Park."; E13-2: Photos._com; E13-3: Courtesy Syngenta; E13-4: LADA/Photo Researchers, Inc.

Chapter 14 opener: Peter Lillie/Photolibrary.com; 14-4a: Scott Orr/iStockphoto.com; 14-4b: Woudloper/Wikimedia Commons. Collection of Universiteit Utrecht.; 14-4c: Chip Clark/Chip Clark; 14-6: James Carmichael/ NHPA/Photo Reasearchers, Inc.; 14-10a: Stephen Dalton/Photo Researchers, Inc.; 14-10b: Stephen Dalton/Photo Researchers, Inc.; 14-11a: Photo Lennart Nilsson/Albert Bonniers Forlag AB; 14-11b: Photo Lennart Nilsson/Albert Bonniers Forlag AB; 14-11c: Photo Lennart Nilsson/Albert Bonniers Forlag AB; 14-13a: JohnPitcher/iStockphoto.com; 14-13b: Courtesy of Timothy O'Keefe; 14-14: Courtesy of Sean Earnshaw & Anne Magurran/ University of St. Andrews, Scotland; 14-15: Patti Murray/Animals Animals/ Earth Scenes; 14-16: Vem/Photo Researchers, Inc.; E14-1: Bettmann/CORBIS; E14-2: Anthony Bannister/NHPA/Photo Researchers, Inc.; E14-3: Gilles Gonthier/Wikemedia Commons

Chapter 15 opener: David Scharf/Peter Arnold, Inc.; 15-4: Dr. Jeremy Burgess/Photo Researchers, Inc.; 15-7b: PureStockX; 15-8a: Alan Mason Chesney Medical Archives of the John Hopkins Institutions; 15-8b: SPL/Photo Researchers, Inc.; 15-9: Johann Schumacher/Peter Arnold, Inc.; 15-10a: Shem Compion/Getty Images; 15-10b: Gunnari/Fotolia; 15-11: Sumio Harada/Minden Pictures; 15-12: Sophie-Laure RAPHAEL/Fotolia; 15-14: Thomas B. Smith, Ph.D./Professor, Ecology and Evolutionary Biology and Director, Center for Tropical Research/UCLA Institute of the Environment; E15-1: Frans Lanting/CORBIS

Chapter 16 opener: Claro Cortes IV/Reuters Limited; 16-1a: Rick& Nora Bowers/VIREO; 16-1b: Bob Steele/VIREO; 16-2a: Wayne Lankinen/Valan Photos; 16-2b: Edgar T. Jones/Photoshot Holdings Ltd./Bruce Coleman; 16-3a: Pat & Tom Leeson/Photo Researchers, Inc.; 16-3b: Pat & Tom Leeson/Photo Researchers, Inc.; 16-4: Tim Laman/NGS Image Collection; 16-5a: Gary A. Monroe; 16-5b: Hey Paul/Wikemedia Commons; 16-6: Loic Degen; 16-7a: From: Nature 203 425:679, Fig 1a Single-gene speciation by left-right reversal. By Rei Ueshima and Takahiro Asami; 16-7b: From: Nature 2003 425:679, Fig 1b & c "Single-gene speciation by left-right reversal. By Rei Ueshima and Takahiro Asami." Mirror-image species and copulation in snails"; 16-7c: From: Nature 2003 425:679, Fig 1b & c "Single-gene speciation by left-right reversal. By Rei Ueshima and Takahiro Asami." Mirror-image species and copulation in snails"; 16-8: Gerard Lacz/Animals Animals/Earth Scenes; 16-12a: Dani/Jeske/Animals Animals/Earth Scenes; 16-12b: A.C. Medeiros/Dr. Gerald D. Carr, PhD; 16-12c: Jack Jeffrey/Photo Resource Hawaii/Alamy Images; 16-12d: Dr. Gerald D. Carr, PhD; 16-13: Tom McHugh/Steinhart Aquarium/Photo Researchers, Inc.; 16-14: Allen Blake Sheldon/Allen Blake Sheldon Nature Photography; E16-1: Mileniusz Spanowiez/WCS; E16-2: Martin Harvey/Peter Arnold, Inc.

Chapter 17 opener: National Geographic Image Collection; 17-4: Hybrid painting, 204, for Scientific Americn. © 2006 by Don Dixon/cosmographica .com; 17-6: Roland Birke/Peter Arnold, Inc.; 17-7a: Chase Studio; 17-7b: Peter Halasz/Wikemedia Commons; 17-7c: Yves Grau/iStockphoto .com; 17-7d: Douglas Faulkner/Photo Researchers, Inc.; 17-8: Illustration by Ludek Pesek/Science Photo Library/Photo Researchers, Inc.; 17-9: Terry Whittake/Photo Researchers, Inc.; 17-10: Illustration by Chris Butler/Science Photo Library/Photo Researchers, Inc.; 17-13a: Tom McHugh/Chicago Zoological Park/Photo Researchers, Inc.; 17-13b: Frans Lanting/CORBIS; 17-13c: Anup Shah/Minden Pictures; 17-14: © Michel Brunet/M.P.F.T.; 17-15: © Michel Brunet/M.P.F.T.; 17-17: David Frayer, Dept of Anthropology, University of Kansas; 17-18: Jerome Chatin/ZUMA Press

Chapter 18 opener: KRISTIN MOSHER/DanitaDelimont.com; 18-1a: Wayne Lankinen/Photoshot Holdings Ltd./Bruce Coleman; 18-1b: Nancy Nehring/iStockphoto.com; 18-1c: Maslowski/Photo Researchers, Inc.; 18-3a: C. Steven Murphree/Biological Photo Service; 18-3b: Dr. Greg Rouse, Department of Invertebrate Zoology, National Museum of Natural History, Smithsonian Institution; 18-3c: Dr. Jeremy Burgess/SPL/Photo Researchers, Inc.; 18-5a: Hans Gelderblom/Getty Images; 18-5b: Reprinted by permission of Springer-Verlag from W.J. Jones, J.A. Leigh, F. Mayer, C.R. Woese, and R.S. Wolfe, Methanococcus jannaschii sp. nov., an extremely thermophilic methanogen from a submarine hydrothermal vent. Archives of Microbiology 136:254-261 (1983). © 1983 by Springer-Verlag GmbH & Co KG. Image courtesy of W. Jack Jones.; 18-8: Zig Koch/Kino Fotoarqiovo

Chapter 19 opener: AP Wide World Photos; 19-1a: Visuals Unlimited/CORBIS; 19-1b: SPL/Photo Researchers, Inc.; 19-1c: Scott Camazine/Photo Researchers, Inc.; 19-2a: Karl O. Stetter, Univeristy of Regensburg, Germany; 19-3: Eye of Science/Photo Researchers, Inc.; 19-4: Visuals Unlimited/CORBIS; 19-5: Meppu/iStockphoto.com; 19-6: Visuals Unlimited/CORBIS; 19-7: David Scharf/Peter Arnold, Inc.; 19-8: Dennis Kunkel/Phototake NYC; 19-9a: Nigel Cattlin/Photo Researchers, Inc.; 19-9b: Steve Gschmeissner/Photo Researchers, Inc.; 19-11a: MANFRED KAGE/Peter Arnold, Inc.; 19-11b: Dept. of Microbiology, Biozentrum/Photo Researchers, Inc.; 19-11c: Dr. Linda Stannard UCT/Photo Researchers, Inc.; 19-11d: Dr. Linda Stannard UCT/Photo Researchers, Inc.; 19-13: Oliver Meckes/Ottawa/Photo Researchers, Inc.; 19-14: EM Unit. VLA/Photo Researchers, Inc.; E19-1: Jack Puccio/iStockphoto.com

Chapter 20 opener: Olivier Digoit/Alamy Images; 20-1: M.I. Walker/Photo Researchers, Inc.; 20-2a: Carolina Bilogical Supply Company/Phototake, NYC; 20-2b: Eric Grave/Science Source/Photo Researchers, Inc.; 20-3: P. M. Motta and F. M. Magliocca/Science Photo Library/Photo Researchers, Inc.; 20-4: David M. Phillips/The Population Control/Photo Researchers, Inc.; 20-6: Oliver Meckes/Photo Researchers, Inc.; 20-7: William Merrill, Penn State University; 20-8: Manfred Kage/Peter Arnold, Inc.; 20-9a: D.P. Wilson/Eric and David Hosking/Photo Researchers, Inc.; 20-9b: Flip Nicklin/Minden Pictures; 20-10: Visuals Unlimited/CORBIS; 20-11: Pete Atkinson/Photo Researchers, Inc.; 20-14: Oliver Meckes & Nicole Ottawa/Eye of Science/Photo Researchers, Inc.; 20-15a: Andrew Syred/Photo Researchers, Inc.; 20-15b: Manfred Kage/Peter Arnold, Inc.; 20-16: Dennis Kunkel/Phototake NYC; 20-17a: P.W. Grace/Science Source/Photo Researchers, Inc.; 20-17b: Ray Simons/Photo Researchers, Inc.; 20-18 inset: Carolina Biological/Visuals Unlimited/Getty Images; 20-19: Francis Abbott/Nature Picture Library; 20-20a: Ray Simons/Photo Researchers, Inc.; 20-20b: Ross Armstrong/AGE Fotostock America

Chapter 21 opener: Frans Lanting/CORBIS; 21-2: Steve Ringman/ Newscom; 21-3: Andre Seale/AGE Fotostock America; 21-5a: Lee W. Wilcox; 21-5b: Carolina Biological Supply Company/Phototake NYC; 21-5c: JC Schou/Biopix; 21-5d: Glen Allison/Photographer's Choice/Getty Images Inc.; 21-6: JC Schou/Biopix; 21-7a: Miika Silfverberg/Wikemedia Commons; 21-7b: Courtesy of Milton Rand; 21-7c: Mauritius-images/Photoshot Archive Creative; 21-7d: Konrad Wothe/Minden Pictures; 21-8: Ed Reschke/Peter Arnold, Inc.; 21-9b: Brian Jackson/iStockphoto.com; 21-9c: Khoroshunova Olga/Shutterstock; 21-10a: Maurice Nimmo/SPL/A-Z Botanical Collection, Ltd./Photo Reserachers, Inc.; 21-10b: Teresa and Gerald Audesirk; 21-10c: Peter Johnson/CORBIS; 21-10d: Image100/Alamy Images; 21-11T: Dr. William M. Harlow/Photo Researchers, Inc.; 21-11B: Gilbert S. Grant/Photo Researchers, Inc.; 21-12a: Tony Wharton/CORBIS; 21-12b: David Dare Parker/OnAsia Images Pte Ltd.; 21-12b inset: Matt Jones/Auscape International Pty. Ltd.; 21-12c: JC Schou/ Biopix; 21-12d: Larry West/Photo Researchers, Inc.

Chapter 22 opener: Philippe Clement/Nature Picture Library; 22-1a: Robert & Linda Mitchell Photography; 22-1b: Elmer Koneman/Visuals Unlimited; 22-2: Cabisco/Phototake NYC; 22-3a: Jeff Lepore/Photo Researchers, Inc.; 22-3b: G.L. Barron/Biological Photo Service; 22-5: Thomas J. Volk, TomVolkFungi._net; 22-6T: Gregory G. Dimijian/Photo Researchers; 22-6B: Carolina Biological Supply Company/Phototake NYC; 22-7: Mark Brundrett; 22-8: S Lowry/University of Ulster/Getty Images; 22-9a: Scott Camazine/Photo Researchers, Inc.; 22-9b: Lee Collins/Wikemedia Commons; 22-9c: Michael & Patricia Fogden/CORBIS; 22-10: Darrell Hensley, Ph.D., University of Tennessee, Entomology & Plant Pathology; 22-11: Andrew Syred/Photo Researchers, Inc.; 22-12a: W.K. Fletcher/Photo Researchers, Inc.; 22-12b: Bill Beatty/Animals Animals/Earth Scenes; 22-14a: Robert Lee/Photo Researchers, Inc.; 22-14b: Robert & Linda Mitchell Photography; 22-15: Stanley L. Flegler/Visuals Unlimited; 22-16: Visuals Unlimited/CORBIS; 22-17a: Emanuele Biggi/Photolibrary.com; 22-17b: Hugh Sturrock/University of Edingurgh; 22-18: David M. Phillips/Visuals Unlimited; 22-19: Teresa and Gerald Audesirk; 22-20: M. Viard/Jacana/Photo Researchers, Inc.; E22-1: Matt Meadows/Peter Arnold, Inc.

Chapter 23 opener: Dr. Tsunemi Kubodera/AP Wide World Photos; 23-4a: Masa Ushioda/Image Quest Marine; 23-4b: John Anderson/Alamy Images; 23-4c: Stephen Frink/CORBIS; 23-6a: Gregory Ochocki/Photo Researchers, Inc.; 23-6b: Mark Webster/Photolibrary.Com; 23-6c: Constantinos Petrinos/Nature Picture Library; 23-6d: DAVID DOUBILET/National Geographic Image Collection; 23-8a: Visuals Unlimited/CORBIS; 23-9a: Dr. Richard Kessel & Dr. Gene Shih/Visuals Unlimited/Getty Images, Inc.; 23-9b: M.I. (Spike) Walker/Alamy Images; 23-9c: Dr. Wolfgang Seifarth; 23-10L: Martin Rotker/Phototake NYC; 23-10R:

Dennis Kunkel Microscopy/Phototake NYC; 23-12a: Tim Rock/Peter Arnold, Inc.; 23-12b: Peter Batson/Image Quest Marine; 23-12c: J. H. Robinson/Photo Researchers, Inc.; 23-14a: Ray Coleman/Photo Researchers, Inc.; 23-14b: Juuyoh TANAKA/Wikemedia Commons; 23-15a: Fred Bavendam/Minden Pictures; 23-15b: Ed Reschke/Peter Arnold, Inc.; 23-16a: Fred Bavendam/Minden Pictures; 23-16b: Kjell B. Sandved/Photo Researchers, Inc.; 23-16c: David Fleetham/Alamy Images; 23-17: Reg Morrison/Auscape/Miden Pictures; 23-18: Brian1442/Wikemedia Commons; 23-20: David Scharf/Peter Arnold, Inc.; 23-21a: Carolina Biological Supply Company/Phototake NYC; 23-21b: Adrian Hepworth/Photoshot Archive Creative; 23-21c: Stephen Dalton/Photo Researchers, Inc.; 23-21d: Dean Evangelista/ Shutterstock; 23-22a: Audrey Snider-Bell/Shutterstock; 23-22b: Tim Flach/Getty Images Inc.; 23-22c: Teresa and Gerald Audesirk; 23-23a: Tom McHugh/Photo Researchers, Inc.; 23-23b: Tom Brakefield/CORBIS; 23-24a: Tom Branch/Photo Researchers, Inc.; 23-24b: Peter J. Bryant/Biological Photo Service; 23-24c: Carolina Biological Supply Company/Phototake NYC; 23-24d: Marevision/age fotostock/ SuperStock; 23-25: Tom E. Adams/Peter Arnold, Inc.; 23-26a: Carolina Biological Supply Company/Phototake NYC; 23-26b: Reproduced by permission from Howard Shiang, D.V.M., Journal of the American Veterinary Medical Association 163:981, Oct. 1973.; 23-27a: Teresa and Gerald Audesirk; 23-27b: Andrew J. Martinez/Photo Researchers, Inc.; 23-27c: Chris Newbert/Photoshot Holdings Ltd./Bruce Coleman; 23-28b: Michael Male/Photo Researchers, Inc.; 23-29: Dr. Tsunemi Kuboders/National Science Museum/AP Wide World Photos; E23-1: Volker Steger/Photo Researchers, Inc.

Chapter 24 opener: Peter Scoones - TCL/Getty Images; 24-2: John Giannicchi/Science Source/Photo Researchers, Inc.; 24-3a: Erling Svensen/UWPhoto ANS; 24-3b: Tom McHugh/Photo Researchers, Inc.; 24-4: Tom McHugh/Steinhart Aquarium/Photo Researchers, Inc.; 24-5: Tom Stack/Tom Stack & Associates; 24-5 inset: Dr. John D. Cunningham/Alamy Images; 24-6a: Jeffrey L. Rotman/CORBIS; 24-6b: David Hall/Photo Researchers, Inc.; 24-7a: Peter David/Getty Images Inc.; 24-7b: Mike Neumann/Photo Researchers, Inc.; 24-7c: Stephen Frink/Getty Images Inc.; 24-8a: Tom McHugh/Photo Researchers, Inc.; 24-8b: Alan Root/Survival Anglia/OSF/ Photolibrary.Com; 24-9a: Breck P. Kent/Animals Animals/Earth Scenes; 24-9b: Kenneth H. Thomas/Photo Researchers, Inc.; 24-9c: Cosmos Blank/National Audubon Society/Photo Researchers, Inc.; 24-9d: Michael & Patricia Fogden/CORBIS; 24-10a: Joseph T. & Suzanne L. Collins/Photo Researchers; 24-10b: Roger K. Burnard/Biological Photo Service; 24-10c: Frans Lanting/ CORBIS; 24-11: Mark Deeble & Victoria Stone/ Photolibrary.com; 24-12a: OBXbchcmbr/ iStockphoto.com; 24-12b: Carolina Biological Supply Company/ Phototake NYC; 24-12c: J & C Sohns/AGE Fotostock America; 24-13: Tom McHugh/Photo Researchers, Inc.; 24-14a: Dave Watts/Nature Picture Library; 24-14b: Craig Ingram/Almay Images; 24-15a: Mark Newman; 24-15b: Dave Watts/Alamy Images; 24-15c: Klein/Peter Arnold, Inc.; 24-16a: Flip Nicklin/ Minden Pictures; 24-16b: Jonathan Watts/Science Photo Library/Photo Researchers, Inc.; 24-16c: C. and M. Denis-Huot/Peter Arnold, Inc.; 24-16d: S. R. Maglione/Photo Researchers, Inc.; E24-1: STANLEY BREEDEN/National Geographic Image Collection

Chapter 25 opener: JEWEL SAMAD/Getty Images; opener inset: Andy Sands/Nature Picture Library; 25-1a: Eric and David Hosking/Frank Lane Picture Agency Limited; 25-1b: Eric and David Hosking/Frank Lane Picture Agency Limited; 25-3(1): Caroll Boltin; 25-3(2): Caroll Boltin; 25-3(3): Caroll Boltin; 25-3(4): Caroll Boltin; 25-3(5): Caroll Boltin; 25-4: Bijlsma & Ruiter/Foto Natura/Minden Pictures; 25-6: Thomas McAvoy/Getty Images/Time Life Pictures 25-8: Renee Lynn/CORBIS; 25-9: Photograph by Joanna M. Setchell, of a female mandrill housed at the Centre Internationale de Recherches Medicales, Franceville, Gabon.; 25-10: Michael & Patricia Fogden/Minden Pictures; 25-11: Ingo Arndt/Minden Pictures; 25-12: Stuart Wilson/Photo Researchers; 25-13a: Creatas/AGE Fotostock America; 25-13b: Hans Pfletschinger/Peter Arnold, Inc.; 25-14a: Ingo Arndt/Nature Picture Library; 25-14b: Richard Herrmann; 25-15: Australian National University, HO/AP Wide World Photos; 25-16: blickwinkel/Alamy Images; 25-17: Michael K. Nichols/National Geographic Image Collection; 25-18: William Leaman/Alamy Images; 25-20: John Anderson/Alamy Images; 25-21a: © Konrad Wothe/Minden Pictures; 25-21b: Tui De Roy/Minden Pictures; 25-22: Anne et Jacques Six; 25-24a: Rolf Nussbaumer/Alamy Images; 25-24b: Teresa and Gerald Audesirk; 25-25a: Frans Lanting/CORBIS; 25-25b: Matthias Breiter/Minden Pictures; 25-25c: Michio Hoshino/Minden Pictures; 25-26: Fred Bruemmer/Peter Arnold, Inc.; 25-28: © Raymond Mendeez/Animals Animals/Earth Scenes; 25-29: Photo Lennart Nilsson/Albert Bonniers Forlag; 25-30: William P. Fifer, New York State Psychiatric Insitute, Columbia University; 25-31: Reproduced by permission from Low-, normal-, high- and perfect-symmetry versions of a male face From Fig 1 page 235 in Animal Behaviour, 202, 64, 233-238 by Koehler N, Rhodes, G, & Simmons L.W. Copyright © 2006 by Elsevier Science Ltd. Image courtesy of Animal Behavior.; E25-1: CHAMUSSY/SIPA Press

Chapter 26 opener: Photolibrary._com; 26-10a: Tom McHugh/Photo Researchers, Inc.; 26-10b: Winfried Wisniewski/Photolibrary.com; 26-12a: Ed Reschke /Peter Arnold, Inc.; 26-12b: Scott Camazine/Photo Researchers, Inc.; 26-13: Jean Francois Hellio & Nicolas Van Ingen/Photo Researchers, Inc.; 26-14a: Jurgen Freund/Nature Picture Library; 26-14b: Mark Tomalty/Masterfile Corporation; 26-14c: raciro/iStockphoto.com; E26-1: Florida Fish and Wildlife Conservation Commission's Fish and Wildlife Research Institute; E26-2: David Keith Jones/Images of Africa Photobank/Alamy; E26-2 inset: FINBARR O'REILLY/Reuters /Landov Media

Chapter 27 opener: David Wong; 27-3a: James R. Page; 27-3b: Stephen Dalton/Photo Researchers, Inc.; 27-3c: Michiel Schaap/Minden Pictures; 27-4a: Art Wolfe/Art Wolfe, Inc.; 27-4b: Frans Lanting/CORBIS; 27-5a: Ingo Arndt/Nature Picture Library; 27-5b: Paul A. Zahl/Photo Researchers, Inc.; 27-5c: Ray Coleman/Photo Researchers, Inc.; 27-5d: Geoffrey Bryant /Photo Researchers; 27-6a: Jean-Paul Ferrero and Jean-Michel Labat/Auscape International Proprietary Ltd.; 27-6b: Charles V. Angelo/Photo Researchers, Inc.; 27-7: Thomas Marent/Minden Pictures; 27-8a: Millard Sharp/Photo Researchers; 27-8b: Mark Cassino/ SuperStock; 27-9aL: goran cakmazovic/Shutterstock; 27-9aR: David Jones—www.mybitoftheplanet.com 2001; 27-9bL: Carmela Leszcynski/Animals Animals/Earth Scenes; 27-9bR: Joe McDonald/CORBIS; 27-10a: Zig Leszczynski/ Animals Animals/Earth Scenes; 27-10b: Thomas Marent/Minden Pictures; 27-10c: Jeff Lepore/Photo Researchers, Inc.; 27-11a: Robert P. Carr/Photoshot Holdings Ltd./Bruce Coleman; 27-11b: T. Hoffmeister/B. Roitberg; 27-12a: Thomas Eisner and Daniel Aneshansley, Cornell University; 27-12b: Frans Lanting/CORBIS; 27-13a: Teresa Audesirk; 27-13b: W. Gregory Brown/Animals Animals/Earth Scenes; 27-14a: Lee Prince/Shutterstock; 27-14b: Tish1/Shutterstock; 27-15aL: E.R. Degginger/Animals Animals/Earth Scenes; 27-15aR: Steffen Foerster/Shutterstock; 27-15bL: Photo Juerg Alean, Glaciers Online, http://www.glaciers-online.net; 27-15bR: Photo Juerg Alean, Glaciers Online, http://www.glaciers-online.net; 27-15cL: Raymond German/Bettmann/ CORBIS; 27-15cR: Jim Zipp/Photo Researchers, Inc.; 27-18a: Robert & Linda Mitchell Photography; 27-18b: Robert & Linda Mitchell Photography; 27-19: Dave Brenner/University of Michigan Sea Grant Program; E27-1a: Ian Waldie/Getty Images; E27-1b: Chuck Pratt/Photoshot Holdings Ltd. Bruce Coleman; E27-1c: Frank Vetere/Acclaim Images; E27-2: Stephen P. Yanoviak

Chapter 28 opener: Shutterstock; 28-11: Alexander Nemenov/AFP/Newscom; 28-12: Adam Hart-Davis /Photo Researchers, Inc.; 28-13: Will McIntyre/Photo Researchers, Inc.; 28-17a: W.O. Field, 1941/U.S. Geological Survey, Denver; 28-17b: Bruce Molnia, 2004/U.S. Geological Survey, Denver; E28-1: Frans Lanting/CORBIS; E28-2: Jan Martin Will/Shutterstock

Chapter 29 opener: Florian Kopp/Imagebroker.net/Photolibrary.com; opener inset: Tomasz Zajaczkowski/iStockphoto; 29-6a: Teresa and Gerald Audesirk; 29-6b: Tom McHugh/Photo Researchers, Inc.; 29-9BL: Schafer and Hill/Peter Arnold, Inc.; 29-9TL: Shiell, Richard/Animals Animals/Earth Scenes; 29-9M: Martin Harvey/CORBIS; 29-9R: Michael Fogden/ Photolibrary.Com; 29-10: J. Maier, Jr./The Image Works; 29-11M: Erik Kolstad/ iStockphoto.com; 29-11L: EcoPrint/Shutterstock; 29-11BR: Federico Rizzato/Fotolia; 29-11TR: Dirk Freder/iStockphoto.com; 29-12: James Hancock/Photo Researchers, Inc.; 29-13a: Vladimir Wrangel/ Shutterstock; 29-13b: Louise A Heusinkveld/Alamy Images; 29-13c: Tom McHugh/Photo Researchers, Inc.; 29-14: William H. Mullins/Photo Researchers, Inc.; 29-15: Mike Goldwater/Getty Images; 29-16: Photoshot Archive Creative; 29-17: Bonnie Harper Lore/Lady Bird Johnson Wildflower Center; 29-18M: Bill Brooks/Alamy Images; 29-18L: Olivierl /Dreamstime; 29-18BR: Eric Gevaert/ Shutterstock; 29-18TR: W. Perry Conway/Aerie Nature Series; 29-19: Teresa and Gerald Audesirk; 29-20M: Aleksander Bolbot/Shutterstock; 29-20L: Paul Tessier/iStockphoto.com; 29-20BR: Greg Okimi/iStockphoto.com; 29-20TR: Francis Bosse/ iStockphoto.com; 29-21M: Ro-Ma Stock/Corbis RF/Alamy Images Royalty Free; 29-21BL: Photos._com; 29-21TL: Jon Meier/iStockphoto.com; 21R: Mark Wallace/Alamy Images; 29-22M: Jim Brandenburg/Minden Pictures; 29-22B: Marty Stouffer Productions/Animals Animals/Earth Scenes; 29-22T: Wm. Bacon III/National Audubon Society/Photo Researchers, Inc.; 29-23: Neil Rabinowitz/CORBIS; 29-24M: Lisa Dearing/Alamy Images; 29-24L: Michael Giannechini/Photo Researchers, Inc.; 29-24BR: Larry Trupp/Alamy Images; 29-24TR: Dmitry Deshevykh/iStockphoto.com; 29-26: Michael Gadomaski/Animals Animals/Earth Scenes; 29-29: Lee Feldstein/iStockphoto.com; 29-30M: Franklin J. Viola/Viola's Photo Visions; 29-30BL: David & Debi Henshaw/Alamy Images; 29-30TL: Dennis Sabo/iStockphoto.com; 29-30R: Photodisc/Alamy Images; 29-31M: Aldo Ottaviani/iStockphoto.com; 29-31L: Michael Durham/National Geographic Image Collection; 29-31BR: Michael Patrick O'Neill/Photo Researchers, Inc.; 29-31TR: Jan Hinsch/Photo Researchers, Inc.; 29-32M: Dr. Craig R. Smith; 29-32BL: Greg Rouse; 29-32TL: Peter Batson/Image Quest Marine; 29-32R: Kelvin Aitken/Image Quest Marine; 29-33M: Verena Tunnicliffe, University of Victoria; 29-33 inset: Reproduced with permission from Waren, A. Bengston, S., Goffredi, S. K. and Van Dover, C. L. 203. A hot-vent gastropod with iron sulfide dermal sclerites. Science, 32, p. 107. Photo courtesy of Dr. Michel Segonzac/Ifremer, France. Copyright 203 American Association for the Advancement of Science.; 29-34M: Reprinted with permission from The Rodale Institute's® NewFarm.org.

www.newfarm.org; 29-34 inset: hvoya/Shutterstock; E29-1: NASA/Goddard Space Flight Center

Chapter 30 opener: Gregory G. Dimijian/Photo Researchers; 30-2: Courtesy of Missouri Department of Transportation; 30-3a: Martin Strmko/ iStockphoto.com; 30-3b: Richard Nowitz/National Geographic Image Collection; 30-3c: Dream Pictures/Stewart Cohel/Photographer's Choice/Getty Images; 30-4: Judy Glasel/The Image Works; 30-7a: Mark Edwards/Peter Arnold, Inc.; 30-7b: Earth Sciences & Image Analysis Lab/NASA/Johnson Space Center; 30-8: © Bruno Locatelli; 30-9a: Jason Selong; 30-9b: Franz Pagot/Alamy Images; 30-10a: Ed Reschke/Peter Arnold, Inc.; 30-10b: Jen Chase/AP Wide World Photos; 30-10c: Jessie Kneeland/Woods Hole Oceanographic Institution; 30-12a: Eric York/Grand Canyon National Park; 30-12b: (2003) California State Parks; 30-15a: Dr. George Naderman, Former Extension Soil Specialist (retired), College of Agriculture and Life Sciences, NC State University, Raleigh, NC; 30-15b: Dr. George Naderman, Former Extension Soil Specialist (retired), College of Agriculture and Life Sciences, NC State University, Raleigh, NC; E30-1a: South Florida Water Management District; E30-1b: South Florida Water Management District; UN30-1: Tom & Pat Leeson/Photo Researchers, Inc. E30-3a: © Morales/Age Fotostock/SuperStock, Inc.; E30-3b: Jim Watt/PacificStock.com; E30-4M: Photo by Eric Larsen, courtesy Bill Ripple/OSU http://www.cof.orst.edu/wolves; E30-4 inset: Barry O'Neill/Yellowstone National Park

Chapter 31 opener: Al Messerschmidt/Getty Images; 31-1a: Tom McHugh/ Photo Researchers, Inc.; 31-1b: kwan tse/Shutterstock; 31-1c: Ivan Cholakov Gostock-dot-ne/ Shutterstock; 31-5a: Pearson Science; 31-5b: Ed Reschke/Peter Arnold; 31-5c: Pearson Science; 31-6: Ed Reschke/Peter Arnold; 31-7: Nina Zanetti/Pearson Education/Pearson Science; 31-8: Dr. Stanley Flegler/Visuals Unlimited/Getty Images; 31-9a: Innerspace Imaging/Photo Researchers, Inc.; 31-9b: Ed Reschke/Peter Arnold, Inc.; 31-9c: G W Willis /Photolibrary.com; 31-10a: Bernhard Götze and Michael Kiebler, MPI for Developmental Biology, Tübingen, Germany; 31-10b: Jurjen Broeke/Wikipedia, The Free Encyclopedia

Chapter 32 opener: Rudy Archuleta/Redux Pictures; 32-8a: Susumu Nishinaga/Photo Researchers, Inc.; 32-8b: Dr. Don Rubbelke; 32-8c: Kenneth Kaushansky, M.D., University of California, San Diego Department of Medicine; 32-11: Dr. David M. Phillips/Visuals Unlimited/Getty Images; 32-13: © CNRI/Photo Researchers, Inc.; 32-16: Photo Lennart Nilsson/Albert Bonniers Forlag; 32-20: John D. Cunningham/Visuals Unlimited/Getty Images; 32-21: World Health Organization/Image Library, Special Programme for Research and Training in Tropical Diseases/Photo by Chandran

Chapter 33 opener: Studio MPM/Stone/Getty Images; 33-1a: Leslie Newman & Andrew Flowers/Photo Researchers, Inc.; 33-1b: Frederic Pacorel/Getty Images; 33-1c: Andrew J. Martinez/Photo Researchers, Inc.; 33-3: Watkins, Bruce/Animals Animals/Earth Scenes; 33-4b: Michael Davidson at Molecular Expressions; 33-5a: Kim Taylor and Jane Burton © Dorling Kindersley; 33-5b: Design Pics/SuperStock; 33-5c: Mark Kostich/iStockphoto.com; 33-6c: Hans-Rainer Duncker, University of Giessen, Germany; E33-3a: Matt Meadows/Peter Arnold, Inc.; E33-3b: © Southern Illinois University/Photo Researchers, Inc.; E33-3c: Bill Travis, M.D./National Cancer Institute

Chapter 34 opener: Schito/Olympia/SIPA Press; 34-1: Charles Krebs/Getty Images; 34-2: Howard Brinton/General Board of Global Ministries; 34-3: Pascale Delchevalerie/Medecins Sans Frontiers; 34-4: Biophoto Associates/ Photo Researchers, Inc.; 34-5: Teresa Audesirk/Teresa and Gerald Audesirk; 34-6a: Dennis Sabo/Shutterstock; 34-7a: Roland Birke/Okapia/Photo Researchers, Inc.; 34-18: Gary Roberts/Rex Features USA Ltd.; E34-2: Dr. E. Walker/Science Photo Library/Photo Researchers, Inc.

Chapter 35 opener: Susan Watts/New York Daily News; 35-8: Michael Durham/Minden Pictures; 35-10,1: Susan Watts/New York Daily News; 35-10,2: Susan Watts/New York Daily News; 35-10,3: Susan Watts/New York Daily News; 35-10,4: Susan Watts/New York Daily News; 35-10,5: Susan Watts/New York Daily News; 35-10,6: Susan Watts/New York Daily News; 35-10,7: Susan Watts/New York Daily News; 35-10,8: Susan Watts/New York Daily News; E35-1: Ronald Frommann/laif/Redux Pictures

Chapter 36 opener: Bettmann/CORBIS; 36-1: David Scharf/Peter Arnold, Inc.; 36-2: Juergen Berger/Photo Researchers, Inc; 36-3a: NIBSC/Science Photo Library/Photo Researchers, Inc.; 36-3b: S.H.E. Kaufmann & J.R. Golecki/Science Photo Library/Photo Researchers, Inc.; 36-12: Dr. Andrejs Liepins/Photo Researchers, Inc.; 36-16a: National Institute for Biological Standards and Control (U.K.)/Science Photo LibraryPhoto Researchers, Inc.; 36-16b: National Institute for Biological Standards and Control (U.K.)/Science Photo LibraryPhoto Researchers, Inc.; E36-1: James Cavallini/Photo Researchers, Inc.; E36-2: Bettmann/CORBIS

Chapter 37 opener: UPI Photos/Newscom; 37-4: Kent Wood/Photo Researchers; 37-7: Image courtesy of RINGLING BROS. AND BARNUM & BAILEY® THE GREATEST SHOW ON EARTH®.; 37-10: Alison Wright/CORBIS; 37-14: Photo courtesy of The Jackson Laboratory, Bar Harbor, Maine; E37-1: © 2008 Novocell; E37-2: Sheila Murphy/Biological Resources Division, U.S. Geological Survey

Chapter 38 opener: Miramar Films, Universal Pictures/Laurie Sparham/ Picture Desk, Inc./Kobal Collection; opener inset: Michael R. Jeffords; 38-3: C. Raines/Visuals Unlimited; 38-15: From H. Damasio, T. Grabowski, R. Frank, A.M. Galaburda, A.R. Damasio, The Return of Phineas Gage: Clues about the brain from the skull of a famous patient. Science 264:1102-1105, 1994. Department of Neurology and Image Analysis Facility, University of Iowa.; E38-3: Xinhua/Landov Media; E38-5a: Reprinted by permission from Nature V2; May 2001, P357; Fig 3a (left), Andrew J. Calder, et al, Neuropsychology of Fear and Loathing. Copyright © 2001 Macmillan Magazines Limited; E38-5b: From NeuroReport Vol II, #17,27 November 2000 Fig 3a (middle and right). Andreas Bartels & Zemi Scki, The Neural Basis of Romantic Love.

Chapter 39 opener: Advanced Bionics LLC; opener inset a: Jennifer Thorpe; opener inset b: Samantha Brilling/Advanced Bionics LLC; 39-5aL: Scanning electron micrographs by Dr. Robert S. Preston and Prof. Joseph E. Hawkins, Kresge Hearing Research Institute, University of Michigan Medical School.; 39-5aR: Scanning electron micrographs by Dr. Robert S. Preston and Prof. Joseph E. Hawkins, Kresge Hearing Research Institute, University of Michigan Medical School.; 39-7a: Dennis Kunkel/Phototake NYC; 39-8a: Gerry Audesirk/Teresa and Gerald Audesirk; 39-8b: Gerry Audesirk/Teresa and Gerald Audesirk; 39-11: Susan Eichorst/Teresa and Gerald Audesirk; 39-12a: Francois Etienne du Plessis/Shutterstock; 39-12b: Sekernas/Dreamstime; 39-15: Advanced Bionics LLC

Chapter 40 openerL: Anja Niedringhaus/AP Wide World Photos; openerR: Terry Scmitt/UPI/LandovMedia; 40-2: Mary Hollinger, NODC Biologist/ National Oceanic and Atmospheric Administration; 40-9: Photo Researchers, Inc.; 40-10L: Wayne Schaefer, University of Washington; 40-10R: Michael Abbey/Photo Researchers, Inc.; E40-1a: Michael Klein/Peter Arnold, Inc.; E40-1b: Yoav Levy/Phototake NYC;

Chapter 41 opener: wAlanb/Flickr/Wikimedia Commons; 41-1: David Wrobel/Visuals Unlimited/Getty Images; 41-2: Photograph courtesy of Charles J. Cole; 41-3: Peter J. Bryant/Biological Photo Service; 41-4: © Robin Chittenden/Frank Lane Picture Agency/CORBIS All Rights Reserved; 41-5: Jurgen Freund/Nature Picture Library; 41-6: Anne et Jacques Six; 41-7: Michael Fogden/Oxford Scientific Films/Animals Animals/Earth Scenes; 41-8: isgaby/iStockphoto.com; 41-16L: C. Edelmann/La Villette/Photo Researchers, Inc.; 41-19a: Photo Lennart Nilsson/Albert Bonniers Forlag; 41-19b: Francis Leroy, Biocosmos/SPL/Custom Medical Stock Photo; 41-21: Dr. Betsy Dresser, Audubon Nature Institute, New Orleans; E41-1a: Dr. Linda Stannard, UCT/Photo Researchers, Inc.; E41-1b: Eye of Science/Photo Researchers, Inc.; E41-2: ekem/Courtesy, RWJMS IVF Program/Wikipedia, The Free Encyclopedia; E41-3: Haag and Kropp Mauritius, GMBH/Phototake NYC

Chapter 42 openerT: © 2006 Tucson Citizen; openerB: © 2006 Tucson Citizen; 42-1a: H. Stanley Johnson/Purestock/SuperStock; 42-1b: Doug Lemke/ Shutterstock; 42-2a: Shinji Kusano/Nature Production/Minden Pictures; 42-2b: Kim Taylor/npl/Minden Pictures; 42-2c: Outdoorsman/Dreamstime; 42-6b: Photo Lennart Nilsson/Albert Bonniers Forlag AB; 42-9: Photo Lennart Nilsson/Albert Bonniers Forlag AB; 42-10: Photo Lennart Nilsson/Albert Bonniers Forlag AB; 42-15: Jerry Gay/Getty Images Inc.-Stone Allstock; E42-4: Sterling K. Clarren, M.D., Children's Hospital & Regional Medical Center, University of Washington School of Medicine

Chapter 43 opener: Mark Hamblin/Photolibrary.com; 43-3a: George Wilder/Visuals Unlimited; 43-3b: George Wilder/Visuals Unlimited; 43-3c: George Wilder/Visuals Unlimited; 43-4R: Dr. Richard Kessel & Dr. Gene Shih/Visuals Unlimited/Getty Images, Inc.; 43-5T: Lee. W. Wilcox; 43-5B: Biophoto Associates/Photo Researchers; 43-7a: Emmanuel LATTES/Alamy Images; 43-7b: almagami/Shutterstock; 43-7c: Photos.com; 43-7d: Gusto Productions/Photo Researchers, Inc.; 43-8T: Ed Reschke/Peter Arnold; 43-9a: Kim Taylor/Minden Pictures; 43-9b: Herve Conge/Phototake NYC; 43-10a: Kenneth Sponsler/Shutterstock; 43-10b: M. Newton/Alamy Images; 43-11a: Bernard Gagnon/Wikipedia, The Free Encyclopedia; 43-11b: Kenneth W. FInk/ Photo Researchers, Inc.; 43-11c: Charles D. Winters/Photo Researchers, Inc.; 43-11d: KentZilla/Wikimedia Commons; 43-12a: Lynwood M. Chace/Photo Researchers, Inc.; 43-12b: Stefan Diller/Photo Researchers, Inc.; 43-13T: Ed Reschke/Peter Arnold, Inc.; 43-13B: M. I. Walker/Photo Researchers; 43-14: Teresa and Gerald Audesirk; 43-15: Teresa and Gerald Audesirk; 43-16: J. Robert Waaland, University of Washington/Biological Photo Service; 43-18: Noah Elhardt/Wikimedia Commons; 43-19a: Biophoto Associates/Science Source/Photo Researchers, Inc.; 43-19b: R. Roncadori/Visuals Unlimited; 43-20: Hugh Spencer/National Audubon Society/Photo Researchers, Inc.; 43-22a: Ray Simon/Photo Researchers, Inc.; 43-22b: Dr. Jeremy Burgess/Science Photo Library/Photo Researchers, Inc.; 43-24a: Martin H. Zimmermann, Harvard Forest, Harvard University; 43-24b: Martin H. Zimmermann, Harvard Forest, Harvard University; E43-1: Jordi Bas Casas/Photoshot Archive Creative

Chapter 44 opener: Lucy Nicholson/Agence France Presse/Getty Images; 44-2b: Heather Audesirk; 44-3: Teresa and Gerald Audesirk; 44-5: David

Scharf /Photo Researchers, Inc.; 44-6: Teresa and Gerald Audesirk; 44-9L: David Rhodes/Purdue University/Dept of Horticulture and Landscape Architecture; 44-9R: Terry Audesirk; 44-12: Teresa and Gerald Audesirk; 44-13bL: Leonard Lessin/Photo Researchers, Inc.; 44-13bR: Leonard Lessin/Photo Researchers, Inc.; 44-14a: Nuridsany et Perennou/The National Audubon Society Collection/Photo Researchers, Inc.; 44-14b: Nuridsany et Perennou/The National Audubon Society Collection/Photo Researchers, Inc.; 44-15a: Anthony Mercieca/Photo Researchers, Inc.; 44-15b: Oxford Scientific Films/Animals Animals/Earth Scenes; 44-16: Dr. J. A. L. Cooke/Oxford Scientific Films/Annimals Animals/Earth Scenes; 44-17a: Philippe Clement/Nature Picture Library; 44-17b: M. P. L. Fodgen/Photoshot Archive Creative; 44-18a: Teresa and Gerald Audesirk; 44-18b: Stephen Maka/CORBIS- NY; 44-19: Teresa and Gerald Audesirk; 44-20: Scott Camazine/Photo Researchers, Inc.; 44-21: MICHAEL AND PATRICIA FOGDEN/National Geographic Image Collection; 44-22a: Marc Gibernau (CNRS-France). From: Science News. Vol. 164, No. 24 December 13, 2003, p. 379; 44-22b: Marc Gibernau (CNRS-France). From: Science News. Vol. 164, No. 24 December 13, 2003, p. 379; E44-1: John Kaprielian/Photo Reserachers, Inc.; E44-1inset: © David Scharf/Peter Arnold, Inc.; E44-2: Mark Cassino/SuperStock; E44-3a: Pete Oxford/Minden Pictures; E44-3b: Merlin D. Tuttle/Bat Conservation International/Photo Researchers, Inc.

Chapter 45 opener: David M Dennis/Photolibrary.com; 45-1: Anton Foltin/Shutterstock; 45-2c: Lee W. Wilcox; 45-2d: Malcolm B. Wilkins; 45-3: Lee W. Wilcox; 45-4T: Michael L. Evans; 45-4B: Michael L. Evans; 45-5: Adam Jones/Visuals Unlimited/Getty Images; 45-6: Nigel Cattlin/Alamy Images; 45-10: Photo Stephen Vasquez and Matthew Fidelibus © UC Regents; 45-11: John Cancalosi/Photolibrary.com; 45-12: Biophoto Associates/Photo Researchers, Inc.; 45-14a: Ed Reschke/Peter Arnold; 45-14b: Ed Reschke/Peter Arnold; 45-15: Kim Taylor/Photoshot Holdings Ltd./Bruce Coleman; 45-16: John Clegg/Photo Researchers, Inc.; 45-16inset: Kim Taylor/Photoshot Holdings Ltd./Bruce Coleman; 45-17: David M Dennis/Photolibrary.com; E45-1: David McNew/Getty Images

Index

B

C

D

E

I

T